CENOZOIC MAMMALS OF AFRICA

Cenozoic Mammals of Africa

Edited by

LARS WERDELIN AND WILLIAM JOSEPH SANDERS

UNIVERSITY OF CALIFORNIA PRESS
Berkeley Los Angeles London

University of California Press, one of the most distinguished university presses in the United States, enriches lives around the world by advancing scholarship in the humanities, social sciences, and natural sciences. Its activities are supported by the UC Press Foundation and by philanthropic contributions from individuals and institutions. For more information, visit www.ucpress.edu.

For digital version, see the UC Press Website (www.ucpress.edu).

University of California Press
Berkeley and Los Angeles, California
University of California Press, Ltd.
London, England

Library of Congress Cataloging-in-Publication Data

Werdelin, Lars.
 Cenozoic mammals of Africa / edited by Lars Werdelin and William Joseph Sanders.
  p.   cm.
 Includes bibliographical references and index.
  ISBN 978-0-520-25721-4 (cloth : alk. paper) —
ISBN 978-0-520-94542-5 (e-book)
  1. Mammals, Fossil—Africa.   2. Paleontology—Cenozoic.
3. Mammals—Evolution—Africa—History.   I. Sanders, William Joseph.   II. Title.
QE881.W47   2010
569.096—dc22

16   15   14   13   12   11   10
10   9   8   7   6   5   4   3   2   1

Cover illustration: Scene from an early Oligocene ecosystem, Fayum, Egypt. *Arsinoitherium zitteli, Aegyptopithecus zeuxis* (in the tree) and Creodonta indet. (foreground). Artwork by Mauricio Antón.

Time, like an ever-rolling stream,
Bears all its sons away. . . .

—ISAAC WATTS, *The Psalms of David*, 1719

Dedicated to the memory of the late Francis Clark Howell (1925–2007), whose development of interdisciplinary and integrative approaches in the study of African mammalian paleontology led the way for us all,

and

to the many unheralded African field assistants whose numerous fossil discoveries have provided vital contributions to our understanding of mammalian evolution on the continent.

CONTENTS

CONTRIBUTORS

PETER ANDREWS, Natural History Museum, London, UK, pjandrews@uwclub.net

MIRANDA J. ARMOUR-CHELU, College of Medicine, Department of Anatomy, Laboratory of Evolutionary Biology, Howard University, Washington, DC, USA, machelu@comcast.net

ROBERT J. ASHER, Department of Zoology, University of Cambridge, Cambridge, UK, r.asher@zoo.cam.ac.uk

D. MARGARET AVERY, Cenozoic Studies, Iziko South African Museum, Cape Town, South Africa, mavery@iziko.org.za

RAYMOND L. BERNOR, College of Medicine, Department of Anatomy, Laboratory of Evolutionary Biology, Howard University, Washington, DC; National Science Foundation, Sedimentary Geology and Paleobiology Program, Arlington, Virginia, USA, rbernor@nsf.gov

LAURA C. BISHOP, Research Centre in Evolutionary Ecology and Palaeoecology, School of Natural Sciences and Psychology, Liverpool John Moores University, Liverpool, UK, L.C.Bishop@ljmu.ac.uk

JEAN-RENAUD BOISSERIE, UMR CNRS 6046, IPHEP, Bâtiment Sciences Naturelles, Faculté des Sciences Fonda-mentales et Appliquées, Université de Poitiers, Poitiers, France, jean.renaud.boisserie@univ-poitiers.fr

DAVID A. BURNEY, National Tropical Botanical Garden, Kalaheo, Hawaii, USA, dburney@ntbg.org

PERCY M. BUTLER, School of Biological Sciences, Royal Holloway University of London, London, UK, percy@butler92.freeserve.co.uk

THURE E. CERLING, Department of Geology and Geophysics and Department of Biology, University of Utah, Salt Lake City, Utah, USA, tcerling@mines.utah.edu

H. BASIL S. COOKE, 2133 154 Street, White Rock, British Columbia V4A 4S5, Canada

MARGERY C. COOMBS, Department of Biology and Graduate Program in Organismic and Evolutionary Biology, University of Massachusetts, Amherst, Massachusetts, USA, mccc@bio.umass.edu

SUSANNE M. COTE, Department of Anthropology and Peabody Museum, Harvard University, Cambridge, Massachusetts, USA, scote@post.harvard.edu

CYRILLE DELMER, UMR 5143 du CNRS "Paléobiodiversité et Paléoenvironnements"; Département Histoire de la Terre, Muséum National d'Histoire Naturelle, Paris, France, delmer@mnhn.fr

PETER B. DEMENOCAL, Department of Earth and Environmental Sciences, Lamont Doherty Earth Observatory, Palisades, New York, USA, peter@ldeo.columbia.edu

CHRISTIANE DENYS, Museum National d'Histoire Naturelle, Paris, France, denys@mnhn.fr

JEREMY DESILVA, Department of Anthropology, Boston University, Boston, Massachusetts, USA, jdesilva@bu.edu

DARYL P. DOMNING, Laboratory of Evolutionary Biology, Department of Anatomy, Howard University, Washington, DC, USA, ddomning@fac.howard.edu

SARAH J. FEAKINS, Department of Earth Sciences, University of Southern California, Los Angeles, California, USA, feakins@usc.edu

JOHN G. FLEAGLE, Department of Anatomical Sciences, Health Sciences Center, Stony Brook University, Stony Brook, New York, USA, jfleagle@notes.cc.sunysb.edu

STEPHEN FROST, Department of Anthropology, University of Oregon, Eugene, Oregon, USA, sfrost@uoregon.edu

TIMOTHY J. GAUDIN, Department of Biological & Envi-ronmental Sciences, University of Tennessee, Chattanooga, Tennessee, USA, Timothy-Gaudin@utc.edu

ALAN W. GENTRY, Department of Palaeontology, Natural History Museum, London, UK, Alantgentry@aol.com

DENIS GERAADS, UPR 2147 CNRS, Paris, France, denis.geraads@evolhum.cnrs.fr

EMMANUEL GHEERBRANT, UMR 5143 du CNRS "Paléobiodiversité et Paléoenvironnements"; Département Histoire de la Terre, Muséum National d'Histoire Naturelle, Paris, France, gheerbra@mnhn.fr

HENRY GILBERT, Department of Anthropology, California State University, Hayward, California; Human Evolution Research Center, University of California, Berkeley, California, USA, henry.gilbert@fossilized.org

PHILIP D. GINGERICH, Museum of Paleontology, University of Michigan, Ann Arbor, Michigan, USA, gingeric@umich.edu

LAURIE R. GODFREY, Department of Anthropology, University of Massachusetts, Amherst, Massachusetts, USA, lgodfrey@anthro.umass.edu

MARC GODINOT, École Pratique des Hautes Études, UMR 7207, Paléontologie, Département d'Histoire de la Terre, Muséum National d'Histoire Naturelle, Paris, France, godinot@mnhn.fr

GREGG F. GUNNELL, Museum of Paleontology, University of Michigan, Ann Arbor, Michigan, USA, ggunnell@umich.edu

MERCEDES GUTIÉRREZ, Department of Anthropology, Washington University in St. Louis, St. Louis, Missouri, USA, mgutierrez@wustl.edu

JOHN M. HARRIS, George C. Page Museum, Los Angeles, California, USA, jharris@usc.edu

TERRY HARRISON, Center for the Study of Human Origins, Department of Anthropology, New York University, New York, New York, USA, terry.harrison@nyu.edu

PATRICIA A. HOLROYD, Museum of Paleontology, University of California, Berkeley, California, USA, pholroyd@berkeley.edu

NINA G. JABLONSKI, Department of Anthropology, The Pennsylvania State University, University Park, Pennsylvania, USA, ngj2@psu.edu

BONNIE F. JACOBS, Huffington Department of Earth Sciences, Southern Methodist University, Dallas, Texas, USA, bjacobs@mail.smu.edu

WILLIAM L. JUNGERS, Department of Anatomical Sciences, School of Medicine, Stony Brook University, Stony Brook, New York, USA, William.Jungers@sunysb.edu

THOMAS M. KAISER, Biocenter Grindel and Zoological Museum, University of Hamburg, Hamburg, Germany, thomas.kaiser@uni-hamburg.de

JOHN KAPPELMAN, Department of Anthropology and Archeology, University of Texas, Austin, Texas, USA, jkappelman@mail.utexas.edu

MEAVE G. LEAKEY, National Museums of Kenya, Nairobi, Kenya; Department of Anthropology, Stony Brook University, Stony Brook, New York, USA, meave.leakey@stonybrook.edu

NAOMI E. LEVIN, Department of Earth and Planetary Sciences, Johns Hopkins University, Baltimore, Maryland, USA, nlevin3@jhu.edu

MARGARET E. LEWIS, Biology Program, School of Natural and Mathematical Sciences, The Richard Stockton College of New Jersey, Pomona, New Jersey, USA, Margaret.Lewis@stockton.edu

FABRICE LIHOREAU, Laboratoire de Phylogenie, Paleobiologie, et Paleontologie, Université de Montpellier II, Montpellier, France, lihoreau@isem.univ-montp2.fr

LAURA M. MACLATCHY, Department of Anthropology, University of Michigan, Ann Arbor, Michigan, USA, maclatch@umich.edu

ELLEN R. MILLER, Department of Anthropology, Wake Forest University, Winston-Salem, North Carolina, USA, millerer@wfu.edu

MICHAEL MORLO, Abt. Messelforschung, Forschungsinstitut Senckenberg, Frankfurt am Main, Germany, Michael.Morlo@senckenberg.de

EILEEN M. O'BRIEN, Department of Science, Engineering, and Technology, Gainesville State College, Oakwood, Georgia, USA, eobrien@gsc.edu

AARON D. PAN, Fort Worth Museum of Science and History, Fort Worth, Texas, USA, apan@fwmsh.org

TIMOTHY C. PARTRIDGE, 12 Cluny Road, Forest Town, Johannesburg, South Africa, tcp@iafrica.com

BENJAMIN H. PASSEY, Department of Geology and Geophysics, University of Utah, Salt Lake City, Utah; Department of Earth and Planetary Sciences, Johns Hopkins University, Baltimore, Maryland, USA, bhpassey@gmail.com

STÉPHANE PEIGNÉ, Museum National d'Histoire Naturelle, Département Histoire de la Terre, UMR 5143 du CNRS "Paléobiodiversité et Paléoenvironnements," Paris, France, peigne@mnhn.fr

D. TAB RASMUSSEN, Department of Anthropology, Washington University in St. Louis, St. Louis, Missouri, USA, dtrasmus@artsci.wustl.edu

HARUO SAEGUSA, Museum of Nature and Human Activities, Hyogo, Japan, saegusa@nat-museum.sanda.hyogo.jp

WILLIAM J. SANDERS, Museum of Paleontology, University of Michigan, Ann Arbor, Michigan, USA, wsanders@umich.edu

ELLEN SCHULZ, Biocenter Grindel and Zoological Museum, University of Hamburg, Hamburg, Germany, ellen.schulz@uni-hamburg.de

CHRISTOPHER R. SCOTESE, Department of Earth and Environmental Sciences, University of Texas, Arlington, Texas, USA, cscotese@uta.edu

ERIK R. SEIFFERT, Department of Anatomical Science, Stony Brook University, Stony Brook, New York, USA, erik.seiffert@stonybrook.edu

ELWYN L. SIMONS, Division of Fossil Primates, Duke Lemur Center, Durham, North Carolina, USA, esimons@duke.edu

NIKOS SOLOUNIAS, New York College of Osteopathic Medicine, Department of Anatomy, Old Westbury, New York, USA, nsolouni@iris.nyit.edu

LARS WERDELIN, Department of Palaeozoology, Swedish Museum of Natural History, Stockholm, Sweden, werdelin@nrm.se

ELEANOR WESTON, Department of Palaeontology, Natural History Museum, London, UK, e.weston@nhm.ac.uk

ALISA J. WINKLER, Roy M. Huffington Department of Earth Sciences, Southern Methodist University, Dallas, Texas; Department of Cell Biology, University of Texas Southwestern Medical Center, Dallas, Texas, USA, awinkler@smu.edu

BERNARD WOOD, Anthropology Department, George Washington University, Washington, D.C.; Human Origins Program, National Museum of Natural History, Smithsonian Institution, Washington, D.C., USA, bernardawood@gmail.com

IYAD S. ZALMOUT, Museum of Paleontology, University of Michigan, Ann Arbor, Michigan, USA, zalmouti@umich.edu

PREFACE

H. B. S. Cooke

The two decades following the end of World War II saw exciting new discoveries of hominids in South Africa and in East Africa. Those discoveries stimulated interest in the associated fauna of fossil mammals that could help in defining age relationships and environmental conditions. In addition, there followed a more systematic approach to the geological framework within each site and the intercorrelation between sites at distant locations. Radiometric dating at Olduvai in 1957 changed the then-prevailing concept of the antiquity of some fossil hominids and focused attention on the importance of the East African deposits in establishing a dated temporal sequence based on the volcanic deposits intercalated within the sedimentary strata. In the early 1960s, the introduction of paleomagnetic orientation determination led to development of a paleomagnetic time scale, controlled by radiometric dates. This proved to be a useful tool in correlation between sites and offered a method for the relative dating of sites that lacked volcanic rocks that could be dated more directly.

Armed with these advances, by the mid-1970s it was clearly necessary to review and consolidate the scattered and often fragmented taxonomic interpretations then available for many African mammal fossil groups so as to provide a firmer basis for evolutionary studies in the future. This led to production of

Evolution of African Mammals, published in 1978. Since that time many new sites have been discovered, filling in some of the former gaps in our knowledge. New fossils found at these and at established sites have changed or modified previous interpretations of taxonomic relationships and evolutionary trends. In addition, the application of geochemical trace element analyses of tuffs consolidated and refined the East African correlations and has opened the way for even more refined intersite comparisons. Most striking over the past years has been a shift in studying fossils from a largely descriptive taxonomy to a more analytical approach, including consideration of faunal associations, their distribution in time and space, and the environmental and climatic factors that prevailed and changed through time. Faunal turnover is being evaluated based on our more refined time scale and more stable taxonomy so that African prehistory has become more a study of paleobiology than mere paleontology.

This new holistic approach requires a firm basis on which geologists, paleobiologists, anthropologists, and others can build as they more forward into the 21st century. This new volume, *Cenozoic Mammals of Africa*, will help to further the science as a whole and lead to a unified picture of the total environment in which our ancestors and the precursors of the vast and magnificent African fauna existed.

ACKNOWLEDGMENTS

Conversations with numerous colleagues in field, university, and museum settings about the need for a contemporary summary of the African mammalian fossil record and related systematic, paleobiological, and chronostratigraphic developments sparked the conception of this book. The actual undertaking of the *Cenozoic Mammals of Africa* project was due to the encouragement and advocacy of our science publisher, Chuck Crumly, who urged us to move ahead without hesitation and tirelessly shepherded the project. It was to our great good fortune that Chuck suggested that we work together; our individual skills allowed us to focus on different aspects of the project, which has resulted in a better end product than either of us could have produced separately.

Along with our gratitude to Chuck Crumly, without whose cajoling and focus it would have been impossible to keep such a massive undertaking on track, we are especially grateful to others who have had vital roles in various aspects of the book production process: at University of California Press, Jenny Wapner, acquisitions; Kate Hoffman and Francisco Reinking, project editors for science; and Heather Vaughan, publicist; and at Michael Bass Associates, Aline Magee, production manager. Their professionalism, keen efforts, and guidance have been invaluable. We are also very thankful to the regents of the University of California, who gave their approval to the project and financially supported it. In part, they were informed by the comments of three anonymous reviewers of the book proposal and outline, and thus we are equally thankful to those reviewers for their confidence. Additionally, we are appreciative of Gregg Gunnell for his editorial advice and friendship, Bonnie Miljour for her many expert improvements of figure artwork, Kristine DeLeon for help in obtaining permissions for figure reproductions and editorial assistance, Basil Cooke and Jonathan Dutton (Vince Maglio) for their encouragement, and the many who generously provided their time and expertise to anonymously review chapter manuscripts.

No one who has visited Africa and encountered its sounds, scents, people, and sights could fail to be touched profoundly by the experience, and so the book is as much a celebration of Africa and its natural history as it is a synthesis of the mammalian fossil record of the continent. Those who have nurtured our intellectual and sentimental attachment to Africa and provided opportunities for field and museum research pertinent to the book are too many to completely enumerate, but notably they include, for W.J.S., Joan Eimer, the late Karl Sanders, Terry Harrison, Philip Gingerich, Yohannes Haile-Selassie, Laura MacLatchy, Clifford Jolly, the late Jean de Heinzelin, Noel Boaz, Erik Seiffert, Elwyn Simons, Ellen Miller, and John Kappelman; and for L.W., Meave and Louise Leakey, Alan Walker, Kaye Reed, the late Charles Lockwood, Brett Hendey, Jan Bergström, Stefan Bengtson, and Åsa Nilsonne. W.J.S. is especially grateful to his wife Mayra Rodríguez for indulging a continuing infatuation with the prehistory of Africa and for her loving encouragement. We are deeply thankful to Samson Tesgaye, Mamitu Yilma, Muluneh Marian, Ato Jara, Menkir Bitew, Alemu Admasu, Mulugeta Feseha, Yohannes Haile-Selassie, and Zelalem Assefa (Addis Ababa, Ethiopia), Charles Msuya, Amandus Kweka, Michael Mbago, and the late Christine Kiyembe (Dar es Salaam, Tanzania), Mary Muungu, Emma Mbua, Fredrick Kyalo Manthi, and Meave Leakey (Nairobi, Kenya), Robert Kityo, Abiti Nelson, Sara Musalizi, and Ezra Musiime (Kampala, Uganda), Ron Mininger (Kinshasa, Democratic Republic of Congo), Graham and Margaret Avery, Kerwin van Willingh, and Thalassa Matthews (Cape Town, South Africa), Phillip Tobias, Ron Clarke, Beverly Kramer, Christine Steininger, Bruce Rubidge, and Mike Raath (Johannesburg, South Africa), Heidi Fourie (Pretoria, South Africa), Ahmed el-Barkooky, Mohammed Korani, and Fathi Ibrahim Imbabi (Cairo, Egypt), and Jerry Hooker and Alan Gentry (London, United Kingdom) for their hospitality, help with local logistics, and invaluable assistance in accessing fossil collections.

Finally, we reserve our greatest appreciation and heartfelt thanks for our contributors, who enthusiastically embraced the project from the start and worked with dedication and thoroughness. We are indebted to them for sharing their knowledge and producing such extraordinary chapters. Some authors took on chapters very late in the process, and their grace under pressure and achievements should be noted, especially Laura MacLatchy ("Hominini") and Erik Seiffert

("Chronology of Paleogene Mammal Localities"). A number of contributors to the *Evolution of African Mammals* reprised their efforts in the current volume, including Percy Butler, Alan Gentry, Elwyn Simons, Peter Andrews, Daryl Domning, John Harris, and Basil Cooke, providing a sense of continuity in endeavor and several centuries of combined acumen. It was an inspiration to collaborate with them and their fellow authors—kudos to all.

Funding for this project was provided by Swedish Research Council grants to L.W., and by the College of Literature, Science, and the Arts, and the Museum of Paleontology, The University of Michigan, to W.J.S.

William J. Sanders and Lars Werdelin

INTRODUCTION

William J. Sanders and Lars Werdelin

Africa, including Madagascar and other surrounding islands, is the second largest continent, accounting for 20% of the world's landmass, and, spanning the equator, it is the only continent to occupy both northern and southern temperate zones (O'Brien and Peters, 1999; Saarinen et al., 2006; Schlüter, 2008). Today, Africa is home to over 1,100 mammalian species, or nearly a quarter of all living mammals (Kingdon, 1997; Medellín and Soberón, 1999). These species comprise an impressive variety of physical types, ranging from armored pangolins, to web-bodied anomalures and retrotusked warthogs, to long-snouted sengis. They vary in size from tiny mouse shrews, at less than 10 g, to bush elephants that exceed 6,000 kg in mass (Kingdon, 1997). Visitors to the continent are familiar with the vast herds of antelopes and other savanna and woodland denizens such as rhinos, zebra, giraffes, lions, cheetahs, and baboons but may be less knowledgeable about the great variety of many smaller mammals, such as guenons, duikers, genets, and mongooses, or more secretive lowland and gallery forest species like the okapi and giant forest hog. In a continent of Africa's magnitude, geographic placement, and ecological and geomorphological heterogeneity, however, such diversity is unsurprising (Medellin and Soberón, 1999).

Modern Africa can be divided into numerous physiographic regions whose contrasts in structure and elevation—over half the continent sits above 1,000 m—relate to vast differences in soil types, climate, and vegetation cover (O'Brien and Peters, 1999) that are conducive to local endemism and the formation of distinctively different local and regional faunas (Kingdon, 1989). The remarkably rich mammalian presence in modern Africa is due not only to ecological, geographic, and geomorphological factors, however, but also to an admixture of old endemics, filter-route and sweepstakes immigrants, episodes of faunal turnover facilitated by the opening of new corridors, and bursts of in situ evolution.

Factors of Past Diversity

As impressive as is the contemporary mammalian diversity of Africa, it is dwarfed by that of the Cenozoic and especially the Miocene. Africa began to separate from Gondwana in the mid-Cretaceous and from then on was largely an island continent, until the late Oligocene or early Miocene, about 75 million years later (Cox, 2000; Gheerbrant and Rage, 2006; Partridge, this volume, chapter 1). Several important trans-Tethyan dispersals from Laurasia occurred during the Paleogene. These included condylarths, strepsirrhine, and anthropoid primates, and probably precursors to afrotherians (sirenians, hyraxes, proboscideans, tubulidentates, macroscelideans, tenrecoids, chrysochlorids, as well as some extinct groups), initially by way of a Mediterranean Tethyan Sill and later by an "Iranian" route (Gheerbrant and Rage, 2006). These early Tertiary taxa would have found dense, close-canopied forests, a warm, equable climate with little latitudinal gradation, with many free niches and relatively weak competition, conducive to adaptive radiation and establishment as early endemics of Africa (Janis, 1993; Kennett, 1995; Denton, 1999; Gheerbrant and Rage, 2006; Feakins and deMenocal, this volume, chapter 4). Although Africa remained geographically secluded in the Eocene and its fauna strongly endemic (Cooke, 1968; Coryndon and Savage, 1973; Maglio, 1978; Holroyd and Maas, 1994; Gheerbrant, 1998), additional immigrants arriving during this epoch from Eurasia included ctenodactylid, phiomyid, and ischyromyid rodents; anthracotheres; pangolins; didelphids; and pantolestids (Janis, 1993). The flowering of new and old endemic African faunas is evident in the late Eocene–early Oligocene Egyptian Fayum, where a cohort of primitive anthropoid primates co-occurred with immense arsinoitheres and numerous species of proboscideans, hyraxes, and anthracotheres (Gagnon, 1997, and several chapters in this volume). These taxa continued to flourish and evolve on the continent until they were replaced or augmented by episodic invasion of Eurasian taxa such as giraffids, carnivorans, rhinos, bovids, suids, and tragulids between the terminal Oligocene and middle Miocene (Maglio, 1978; Pickford, 1981; Bernor et al., 1987; Guérin and Pickford, 2003; Sanders et al., 2004; Rasmussen and Gutiérrez, 2009; and several chapters in this volume). Much of what we today consider to be a distinctive African mammalian fauna ultimately derives from this early Neogene upheaval.

Subsequent origination and diversification of African mammals is often placed in the context of climate change and habitat fragmentation, and is seen as resulting from "climatic forcing" or as a response to environmental instability, such as increased seasonality and secular changes in climatic variability (e.g., deMenocal, 2004). The most significant mammalian faunal transformation in Africa after the early to middle Miocene Eurasian influx began in the late Miocene and accelerated through the Plio-Pleistocene. It is correlated with global cooling, greater seasonality, strong uplift of rift shoulders in eastern Africa, more widespread occurrence of arid conditions, ecological fragmentation, and decrease in atmospheric CO_2 and associated increase of C_4 plant biomass such as grasses and sedges (Cerling et al., 1993, 1997, 1998, and this volume, chapter 48; Partridge et al., 1995, Partridge, this volume, chapter 1). Successful lineages, including elephants, suids, several tribes of bovids, hippos, and some equids and rhinos, were able to selectively capitalize on the expansion of grazing resources (Cerling et al., 2003, this volume, chapter 48).

Thus, the modern composition of African mammalian faunas is an amalgam of old Gondwanan- and Laurasian-derived endemics and their descendants, taxa originating from repeated episodes of Eurasian incursions, and more recent evolutionary response to rapidly changing and oscillating ecological conditions during the Plio-Pleistocene. The buffering effects of relatively mild climates and a large, heterogeneous landmass distributed about the equator combined to promote the preservation of ancient lineages, species abundance, and great taxonomic diversity through the temporal vastness of the Cenozoic (see Andrews and O'Brien, this volume, chapter 47, for a discussion of factors contributing to species richness).

Progress and Motivation

Just over 30 years ago, the publication of *Evolution of African Mammals* (Maglio and Cooke, 1978) offered the first modern overview of the abundant fossil record of these African mammals and attempted to document their prehistoric diversity. Although a chapter was dedicated to Mesozoic mammals, the emphasis was on Cenozoic taxa, and the coverage of the volume predominantly featured Neogene species supplemented by information about Paleogene mammals from the Egyptian Fayum, reflecting the then-current scope of knowledge. The strength of the volume derived from the systematic organization and detailed review of taxa in chronologic, taxonomic, and geographic frameworks. As a result, *Evolution of African Mammals* became a vital guide for many involved in field- and specimen-based research on African mammalian fossils, and it remained a useful reference even after new classifications had begun to supplant those in the book. It is no exaggeration to say that several generations of students first learned to identify and interpret the fossils they were finding during the day by poring over its pages by lantern at night in camps spread from the Transvaal to the Semliki to the Fayum.

The present volume was undertaken with the goal of providing a successor to this classic book by constructing an updated, comprehensive review of the African mammalian fossil record, focused particularly on the evolution and systematics of taxa from the last 65 million years, since African Mesozoic mammals were recently reviewed by Kielan-Jaworowska et al. (2004). In the three decades that have passed since the previous summary of the African mammalian fossil record, drastic changes have taken place in our

understanding of the interrelationships of higher groups of mammals (e.g., Springer et al., 1997, 2003; Stanhope et al., 1998; Eizirik et al., 2001; Murphy et al., 2001; Scally et al., 2001). One of the most intriguing hypotheses to result from these technical advances has been the grouping of such anatomically and ecologically diverse mammals as aardvarks, sengis, elephants, hyraxes, sea cows, golden moles, and tenrecs into a higher superorder, the apparently endemic African clade Afrotheria. Morphological and genetic evidence in support of this clade is critically evaluated by Asher and Seiffert (this volume, chapter 46), who offer optimism that the increased pace of Paleogene African exploration and discovery will redress the unevenness of the fossil records of individual afrothere taxa and further improve our understanding of the interrelationships of afrotheres, their extinct relatives, and the events that contributed to their diversification. Another surprising implication from molecular studies—that cetaceans are not only close relatives of artiodactyls but likely systematically embedded within the Artiodactyla as a sister taxon to hippos—has been validated by recent paleontological discoveries (Gingerich, this volume, chapter 45). Other chapters in this volume also at least implicitly evoke the operational power that new, combined molecular and paleontological/anatomical approaches offer for phylogenetic resolution.

In addition, refinements in absolute dating techniques and paleomagnetic sequence stratigraphy (Feibel, 1999; Ludwig and Renne, 2000), as well as long-term efforts at site-specific and regional stratigraphic resolution, for example, in the Omo, Turkana Basin, Middle Awash, Lukeino, and early Miocene Kisingiri sequence (e.g., Brown, 1994; Drake et al., 1988; Feibel et al., 1989; WoldeGabriel et al., 1990, 2001; Haileab and Brown, 1992; Walter and Aronson, 1993; Renne et al., 1999; Deino et al., 2002; McDougall and Brown, 2006), to highlight a few, have produced far more accurate dating of volcanics, better correlation between rock units of sites and basins, and more secure chronostratigraphic contexts for mammalian faunas. The scope of these advances is well documented in separate chapters in this volume on Paleogene (Seiffert, chapter 2) and Neogene (Werdelin, chapter 3) African chronostratigraphy.

Intensified paleontological field survey, driven in large part by the search for human and ape predecessors, has vastly expanded the scope and quality of the African mammalian fossil record. This has led to much better understanding of faunas from previously poorly known areas such as the Saharan Atlas (Mahboubi et al., 1986), the Western Rift (Boaz, 1990, 1994), north-central Tanzania (Harrison, 1997), the Malawi Rift (Bromage et al., 1995), and Central Africa (Vignaud et al., 2002), as well as greater clarity about faunal succession from sites already well documented (e.g., Lothagam, Kenya [Leakey and Harris, 2003], Fayum, Egypt [Gagnon, 1997], and the Middle Awash, Ethiopia [Haile-Selassie et al., 2004]). For these reasons, the context, extent, and systematic organization of African fossil mammals all differ substantially in this volume from the synopses in *Evolution of African Mammals*.

Comparison of faunal chapters in the current volume with their precursor sections in *Evolution of African Mammals* underscores the tremendous improvements in the documentation and interpretation of the African mammalian fossil record that have occurred over a relatively very short time, mostly within the last 10 years. Most relevant to us is the group to which we belong, the Homonini. Commenting

54 years after the discovery of the Taung child in South Africa, Clark Howell (1978:154) wrote, "The ever-accelerating pace of intensified field researches suggests that . . . the African continent will afford the fullest documentation of hominid biological and cultural evolution in the world of the late Cenozoic." These were prescient words, because field and laboratory efforts of the last 30+ years have provided a much fuller and more informed account of the temporal depth of the Hominini and the complexities of the phylogenetic journey traveled by our evolutionary precursors, from ecologically marginal apes to dominant, culturally elaborate, and intellectually impressive humans.

At the time Howell was writing his summary, the earliest well-accepted hominin was the mid- to late Pliocene *Australopithecus africanus*, though he believed that late Miocene isolated dental remains from Lukeino, Kenya, also belonged to our lineage. Hominins were also suspected from older deposits, dating back to 9.8 Ma. Diversity of known species was modest, including *A. africanus*, *A. boisei*, *A. crassidens*, *A. robustus*, *H. habilis*, *H. erectus*, and three fossil subspecies of humans, *H. sapiens rhodensiensis/neanderthalensis/afer* (Howell, 1978). Additional hominins had been discovered from the mid- to late Pliocene Laetolil Beds, Tanzania and Hadar Fm., Ethiopia, but their taxonomy and systematic affinities had not yet been well worked out. Since that time, the pace and breadth of fossil hominin discoveries have extended the tribe's secure record back to the late Miocene, possibly as early as 7 Ma; confirmed Africa as the locus for all major anatomical (e.g., bipedality, hypermegadonty, increase in brain size and complexity) and cultural innovations (e.g., diversification and standardization of tool types, expression of ritual in art and burial practices, spatial organization of living sites) associated with the tribe; and shown Africa to be the initial source area for nearly all repeated geographic expansions of hominin species, such as the early Pleistocene cosmopolitan distribution of *Homo erectus* across the Old World and late-middle and late Pleistocene excursions from the continent to other parts of the globe of anatomically modern humans (MacLatchy et al., this volume, chapter 25). Beginning in 1978, new African species added to the hominin roster include *Australopithecus afarensis* (1978), *Homo rudolfensis* (1986), *Australopithecus anamensis* (1995), *Ardipithecus ramidus* (1995), *Australopithecus bahrelghazeli* (1996), *Australopithecus garhi* (1999), *Orrorin tugenensis* (2001), *Kenyanthropus platyops* (2001), *Sahelanthropus tchadensis* (2002), and *Ardipithecus kadabba* (2004). *Homo rudolfensis* now encompasses part of what had been *Homo habilis*, *Homo heidelbergensis* has (for many researchers) replaced *H. sapiens rhodensiensis* taxonomically, and the presence of Neanderthals in Africa is no longer credibly recognized (MacLatchy et al., this volume, chapter 25, and references therein). Moreover, at the time that the chapters of the present volume had gone into production, the long awaited description and interpretation of *Ardipithecus ramidus* were published in a series of papers in the journal *Science* (summarized by Lovejoy et al., 2009), suggesting a novel and controversial explication of our early evolutionary development that could not have been anticipated in 1978.

The effort to recover African nonhuman primates has led to achievements rivaling those for the hominins, reflected in the devotion of the present volume to individual suborders and superfamilies. There are separate chapters on Malagasy subfossil lemurs and Paleogene prosimians, which were either not covered or poorly known in 1978, as well as discrete chapters on cercopithecoids, later Tertiary lorisiformes, Paleogene anthropoids, and late Oligocene-Miocene apes. The fossil record of these groups has markedly expanded in the last 30 years, and they have now received much-needed systematic reorganization. The new documentation of Paleogene prosimians and accounting of the broad radiation and temporally deep presence of anthropoids on the continent show the particular importance of North Africa for primate evolution, including the roots of our own ancestry (e.g., Sigé et al., 1990; Simons, 1992; Gheerbrant et al., 1993; Simons et al., 1995; Simons and Seiffert, 1999; Adaci et al., 2007).

Progress in African mammalian paleontology over the last three decades is not only seen in accounts of relative newcomers to the African fossil record, such as hominins, but also exemplified by the current accounting of more deeply endemic taxa, such as proboscideans. These had their origin far back in the African Paleogene and have survived on the continent to the present. Since 1978, the proboscidean fossil record has been pushed back in time more than 20 million years, from the late Eocene to the late Paleocene (Gheerbrant, 2009), and by conservative count has added 14 new genera and 29 new species (Sanders et al., this volume, chapter 15). Among the taxa added since 1978 are phosphatheres, daouitheres, and numidotheres, which account for more than a third of the temporal range of fossil proboscideans. These groups have completely altered conceptions of the primitive morphological condition of the order. Detailed phylogenetic analyses of a richer fossil record (e.g., Tassy, 1996; Shoshani, 1996) have produced better resolution of proboscidean relationships, and more precise geographic documentation and temporal calibration of proboscidean fossil occurrences (Sanders et al., this volume, chapter 15) have provided a more secure framework for studying their evolution and a better appreciation of episodes of phylogenetic radiation. As a result, the importance of Africa as the center of proboscidean origins and most of the major evolutionary events of the order has been firmly established.

Each of the remaining taxonomic chapters, which comprise the central section of the book, reveals evidence of equally impressive novel systematic and paleobiological interpretations of their constituent taxa, based on new discoveries similar in scope to those documented for hominins and proboscideans. The taxonomic core of the book is subdivided into sections on metatherians (one chapter), afrotheres (10 chapters), euarchontoglires (nine chapters), and laurasiatheres (20 chapters), and it is set between an introductory section on the physical and temporal contexts of African Cenozoic mammalian evolution (five chapters) and several concluding chapters offering broader perspectives from which to examine aspects of African mammalian phylogeny, faunal composition, and adaptive transformation (three chapters). Taxonomic chapters are organized primarily taxonomically and systematically, and are accompanied by figures highlighting salient distinctions of particular taxa and tables summarizing the taxonomy, geological age, and site occurrences of those taxa (and which are annotated by key references). Attempts were made to be comprehensive in compiling this information, but, to paraphrase words attributed to Pliny the Elder, "something new is always coming out of Africa," so no such effort can possibly remain exhaustive for long. Nonetheless, the chapters represent what is currently known about the fossil record and systematic organization of Cenozoic African mammals, and they offer an abundance of resources.

Current Perspectives

Despite the massive influx of new data over the past 30 years, all is not well in the field of African mammal paleontology. Many groups are simply studied by too few people relative to the massive amount of material collected over the years. In some cases, there are only one or two specialists available to undertake the sort of reviews that this book contains. Viewed in this context, the search for human and primate origins is both a boon and a bane. It is a boon in the sense that the vast majority of the material reviewed in this volume was derived from field campaigns in search of fossils that might reveal human and primate origins. Without this focus, we would have much less material to study. It is a bane in that the extreme focus on one aspect of the fossil record has meant that many groups, viewed as relatively uninteresting to this field, have been left unstudied or have only, and increasingly, been studied on the basis of what they can tell us about human origins. There is nothing intrinsically wrong with this approach, of course, and it has generated a tremendous amount of information about the paleoecology and evolution of some selected groups of mammals. However, it has also meant that taxonomic and systematic studies of these and other groups have been relatively neglected. We hope to show with the data compiled in this volume that such information has a broad range of uses, not only for the study of human and primate evolution, but also for understanding biodiversity and migration patterns through time, and the relationship between climate and faunal evolution at the macroscale that has had broad success on other continents (e.g., Graham et al., 1996; Barnosky, 2001; Eronen et al., 2009). It is our hope that this volume will encourage those interested in specimen-based research on African mammals to take up this field of inquiry, and also that it will provide students of our discipline the same practical utility as its predecessor.

Literature Cited

Adaci, M., R. Tabuce, F. Mebrouk, M. Bensalah, P.-H. Fabre, L. Hautier, J.-J. Jaeger, V. Lazzari, M. Mahboubi, L. Marivaux, O. Otero, S. Peigné, and H. Tong. 2007. Nouveaux sites à vertébrés paléogènes dans la région des Gour Lazib (Sahara nord-occidental Algérie). *Comptes Rendus Palevol* 6:535–544.

Barnosky, A. D. 2001. Distinguishing the effects of the Red Queen and Court Jester on Miocene mammal evolution in the northern Rocky Mountains. *Journal of Vertebrate Paleontology* 21:172–185.

Bernor, R. L., M. Brunet, L. Ginsburg, P. Mein, M. Pickford, F. Rögl, S. Sen, F. Steininger, and H. Thomas. 1987. A consideration of some major topics concerning Old World Miocene mammalian chronology, migrations and paleogeography. *Geobios* 20:431–439.

Boaz, N. T. (ed.). 1990. *Evolution of Environments and Hominidae in the African Western Rift Valley.* Virginia Museum of Natural History Scientific Publication Series, Martinsville, 356 pp.

Boaz, N. T. 1994. Significance of the Western Rift for hominid evolution; pp. 321–344 in R. S. Corruccini and R. L. Ciochon (eds.), *Integrative Paths to the Past: Paleoanthropological Advances in Honor of F. Clark Howell.* Prentice Hall, Englewood Cliffs, N.J.

Bromage, T. G., F. Schrenk, and Y. M. Juwayeyi. 1995. Paleobiogeography of the Malawi Rift: Age and vertebrate paleontology of the Chiwondo Beds, northern Malawi. *Journal of Human Evolution* 28:37–57.

Brown, F. H. 1994. Development of Pliocene and Pleistocene chronology of the Turkana Basin, East Africa, and its relation to other sites; pp. 285–312 in R. S. Corruccini and R. L. Ciochon (eds.), *Integrative Paths to the Past: Paleoanthropological Advances in Honor of F. Clark Howell.* Prentice Hall, Englewood Cliffs, N.J.

Cerling, T. E., J. R. Ehleringer, and J. M. Harris. 1998. Carbon dioxide starvation, the development of C_4 ecosystems, and mammalian evolution. *Philosophical Transactions of the Royal Society, London* 353:159–171.

Cerling, T. E., J. M. Harris, and M. G. Leakey. 2003. Isotope paleoecology of the Nawata and Nachukui Formations at Lothagam, Turkana Basin, Kenya; pp. 605–624 in M. G. Leakey and J. M. Harris (eds.), *Lothagam: The Dawn of Humanity in Eastern Africa.* Columbia University Press, New York.

Cerling, T. E., J. M. Harris, B. J. MacFadden, M. G. Leakey, J. Quade, V. Eisenmann, and J. R. Ehleringer. 1997. Pattern and significance of global ecologic change in the Late Neogene. *Nature* 389:153–158.

Cerling, T. E., Y. Wang, and J. Quade. 1993. Expansion of C_4 ecosystems as an indicator of global ecological change in the late Miocene. *Nature* 361:344–345.

Cooke, H. B. S. 1968. The fossil mammal fauna of Africa. *Quarterly Review of Biology* 43:234–264.

Coryndon, S. C., and R. J. G. Savage. 1973. The origin and affinities of African mammal faunas. *Special Papers in Paleontology (The Paleontological Association, London)* 12:121–135.

Cox, C. B. 2000. Plate tectonics, seaways and climate in the historical biogeography of mammals. *Memórias do Instituto, Oswaldo Cruz, Rio de Janeiro* 95:509–516.

Deino, A. L., L. Tauxe, M. Monaghan, and A. Hill. 2002. $^{40}Ar/^{39}Ar$ geochronology and paleomagnetic stratigraphy of the Lukeino and lower Chemeron Formations at Tabarin and Kapcheberek, Tugen Hills, Kenya. *Journal of Human Evolution* 42:117–140.

deMenocal, P. B. 2004. African climate change and faunal evolution during the Pliocene-Pleistocene. *Earth and Planetary Letters* 220:3–24.

Denton, G. H. 1999. Cenozoic climate change; pp. 94–114 in T. G. Bromage and F. Schrenk (eds.), *African Biogeography, Climate Change, and Human Evolution.* Oxford University Press, New York.

Drake, R. E., J. A. Van Couvering, M. H. Pickford, G. H. Curtis, and J. A. Harris. 1988. New chronology for the early Miocene mammalian faunas of Kisingiri, western Kenya. *Journal of the Geological Society, London* 145:479–491.

Eizirik, E., W. J. Murphy, and S. J. O'Brien. 2001. Molecular dating and biogeography of the early placental mammal radiation. *Journal of Heredity* 92:212–219.

Eronen, J. T., M. M. Ataabadi, A. Micheels, A. Karme, R. L. Bernor, and M. Fortelius. 2009. Distribution history and climatic controls of the late Miocene Pikermian chronofauna. *Proceedings of the National Academy of Sciences* 106:11867–11871.

Feibel, C. S. 1999. Tephrostratigraphy and geological context in paleoanthropology. *Evolutionary Anthropology* 8:87–100.

Feibel, C. S., F. H. Brown, and I. McDougall. 1989. Stratigraphic context of fossil hominids from the Omo Group deposits: Northern Turkana Basin, Kenya and Ethiopia. *American Journal of Physical Anthropology* 78:595–622.

Gagnon, M. 1997. Ecological diversity and community ecology in the Fayum sequence (Egypt). *Journal of Human Evolution* 32:133–160.

Gheerbrant, E. 1998. The oldest known proboscidean and the role of Africa in the radiation of modern orders of placentals. *Bulletin of the Geological Society of Denmark* 44:181–185.

———. 2009. Paleocene emergence of elephant relatives and the rapid radiation of African ungulates. *Proceedings of the National Academy of Sciences, USA* 106:10717–10721.

Gheerbrant, E., and J.-C. Rage. 2006. Paleobiogeography of Africa: How distinct from Gondwana and Laurasia? *Palaeogeography, Palaeoclimatology, Palaeoecology* 241:224–246.

Gheerbrant, E., H. Thomas, J. Roger, S. Sen, and Z. Al-Sulaimani. 1993. Deux nouveaux primates dans l'Oligocène inférieur de Taqah (Sultanat d'Oman): Premiers adapiformes (?Anchomomyini) de la Péninsule Arabique. *Palaeovertebrata* 22:141–196.

Graham, R. W., E. L. J. Lundelius, M. A. Graham, E. K. Schoeder, R. S. I. Toomey, E. Anderson, A. Barnosky, J. A. Burns, C. S. Churcher, D. K. Grayson, R. D. Guthrie, C. R. Harington, G. T. Jefferson, L. D. Martin, H. G. McDonald, R. E. Morlan, H. A. J. Semken, S. D. Webb, L. Werdelin, and M. C. Wilson. 1996. Spatial response of mammals to Late Quaternary environmental fluctuations. *Science* 272:1601–1606.

Guérin, C., and M. Pickford. 2003. *Ougandatherium napakense* nov. gen. nov. sp., le plus ancien Rhinocerotidae Iranotheriinae d'Afrique. *Annales de Paléontologie* 89:1–35.

Haileab, B., and F. H. Brown. 1992. Turkana Basin–Middle Awash Valley correlations and the age of the Sagantole and Hadar Formations. *Journal of Human Evolution* 22:453–468.

Haile-Selassie, Y., G. WoldeGabriel, T. D. White, R. L. Bernor, D. Degusta, P. R. Renne, W. K. Hart, E. Vrba, S. Ambrose, and

F. C. Howell. 2004. Mio-Pliocene mammals from the Middle Awash, Ethiopia. *Geobios* 37:536–552.

Harrison, T. (ed.). 1997. *Neogene Paleontology of the Manonga Valley, Tanzania: A Window into the Evolutionary History of East Africa.* Plenum Press, New York, 418 pp.

Holroyd, P. A., and M. C. Maas. 1994. Paleogeography, paleobiogeography, and anthropoid origins; pp. 297–334 in J. G. Fleagle and R. F. Kay (eds.), *Anthropoid Origins.* Plenum Press, New York.

Howell, F. C. 1978. Hominidae; pp. 154–248 in V. J. Maglio and H. B. S. Cooke (eds.), *Evolution of African Mammals.* Harvard University Press, Cambridge.

Janis, C. M. 1993. Tertiary mammal evolution in the context of changing climates, vegetation, and tectonic events. *Annual Review of Ecological Systematics* 24:467–500.

Kennett, J. P. 1995. A review of polar climatic evolution during the Neogene, based on the marine sediment record; pp. 49–64 in E. S. Vrba, G. H. Denton, T. C. Partridge, and L. H. Burckle (eds.), *Paleoclimate and Evolution, with Emphasis on Human Origins.* Yale University Press, New Haven.

Kielan-Jaworowska, Z., R. L. Cifelli, and Z.-X. Luo. 2004. *Mammals from the Age of Dinosaurs. Origins, Evolution, and Structure.* Columbia University Press, New York, 630 pp.

Kingdon, J. 1989. *Island Africa: The Evolution of Africa's Rare Animals and Plants.* Princeton University Press, Princeton, N.J., 287 pp.

———. 1997. *The Kingdon Field Guide to African Mammals.* Academic Press, San Diego, 464 pp.

Leakey, M. G., and J. M. Harris (eds.). 2003. *Lothagam: The Dawn of Humanity in Eastern Africa.* Columbia University Press, New York, 678 pp.

Lovejoy, C. O., G. Suwa, S. W. Simpson, J. H. Matternes, and T. D. White. 2009. The great divide: *Ardipithecus ramidus* reveals the postcrania of our last common ancestors with African apes. *Science* 326:100–106.

Ludwig, K. R., and P. R. Renne. 2000. Geochronology on the paleoanthropological time scale. *Evolutionary Anthropology* 9:101–110.

Maglio, V. J. 1978. Patterns of faunal evolution; pp. 603–619 in V. J. Maglio and H. B. S. Cooke (eds.), *Evolution of African Mammals.* Harvard University Press, Cambridge.

Maglio, V. J. and H. B. S. Cooke. 1978. *Evolution of African Mammals.* Harvard University Press, Cambridge, 641 pp.

Mahboubi, M., R. Ameur, J. Y. Crochet, and J. J. Jaeger. 1986. El Kohol (Saharan Atlas, Algeria): A new Eocene mammal locality in northwestern Africa. *Palaeontographica Abt. A* 192:15–49.

McDougall, I., and F. H. Brown. 2006. Precise $^{40}Ar/^{39}Ar$ geochronology for the upper Koobi Fora Formation, Turkana Basin, northern Kenya. *Journal of the Geological Society, London* 163:205–220.

Medellín, R. A., and J. Soberón. 1999. Predictions of mammal diversity on four land masses. *Conservation Biology* 13:143–149.

Murphy, W. J., E. Eizirik, W. E. Johnson, Y. P. Zhang, O. A. Ryder, and S. J. O'Brien. 2001. Molecular phylogenetics and the origins of placental mammals. *Nature* 409:614–618.

O'Brien, E. M., and C. R. Peters. 1999. Landforms, climate, ecogeographic mosaics, and the potential for hominid diversity in Pliocene Africa; pp. 115–137 in T. G. Bromage and F. Schrenk (eds.), *African Biogeography, Climate Change, and Human Evolution.* Oxford University Press, New York.

Partridge, T. C., B. A. Wood, and P. B. deMenocal. 1995. The influence of global climatic change and regional uplift on large-mammalian evolution in East and southern Africa; pp. 331–355 in E. S. Vrba, G. H. Denton, T. C. Partridge, and L. H. Burckle (eds.), *Paleoclimate and Evolution, with Emphasis on Human Origins.* Yale University Press, New Haven.

Pickford, M. 1981. Preliminary Miocene mammalian biostratigraphy for Western Kenya. *Journal of Human Evolution* 10:73–97.

Rasmussen, D. T., and M. Gutiérrez. 2009. A mammalian fauna from the late Oligocene of northwestern Kenya. *Palaeontographica Abt. A* 288:1–52.

Renne, P. R., G. WoldeGabriel, W. K. Hart, G. Heiken, and T. D. White. 1999. Chronostratigraphy of the Miocene-Pliocene Sagantole Formation, Middle Awash Valley, Afar Rift, Ethiopia. *Geological Society of America Bulletin* 111:869–885.

Saarinen, T. F., M. Parton, and R. Billberg. 2006. Relative size of continents on world sketch maps. *Cartographica: The International Journal for Geographic Information and Geovisualization* 33:37–48.

Sanders, W. J., J. Kappelman, and D. T. Rasmussen. 2004. New large-bodied mammals from the late Oligocene site of Chilga, Ethiopia. *Acta Paleontologica Polonica* 49:365–392.

Scally, M., O. Madsen, C. J. Douady, W. W. de Jong, M. J. Stanhope, and M. S. Springer. 2001. Molecular evidence for the major clades of placental mammals. *Journal of Mammalian Evolution* 8:239–277.

Schlüter, T. 2008. *Geological Atlas of Africa: With Notes on Stratigraphy, Tectonics, Economic Geology, Geohazards, Geosites and Geoscientific Education of Each Country.* 2nd ed. Springer, Berlin, 307 pp.

Shoshani, J. 1996. Para-or monophyly of the gomphotheres and their position within Proboscidea; pp. 149–177 in J. Shoshani and P. Tassy (eds.), *The Proboscidea: Evolution and Palaeoecology of Elephants and Their Relatives.* Oxford University Press, Oxford.

Sigé, B., J.-J. Jaeger, J. Sudre, and M. Vianey-Liaud. 1990. *Altiatlasius koulchii* n. gen. n. sp, primate omomyidé du Paléocène du Maroc, et les origins des euprimates. *Palaeontographica Abt. A* 214:31–56.

Simons, E. L. 1992. Diversity in the early Tertiary anthropoidean radiation in Africa. *Proceedings of the National Academy of Sciences, USA* 89:10743–10747.

Simons, E. L., and E. R. Seiffert. 1999. A partial skeleton of *Proteopithecus sylviae* (Primates, Anthropoidea): First associated dental and postcranial remains of an Eocene anthropoidean. *Comptes Rendus de l'Académie des Sciences, Paris, Série IIA,* 329:921–927.

Simons, E. L., D. T. Rasmussen, and P. D. Gingerich. 1995. New cercamoniine adapid from Fayum, Egypt. *Journal of Human Evolution* 29:577–589.

Springer, M. S., G. C. Cleven, O Madsen, W. W. de Jong, V. G. Waddell, H. M. Amrine, and M. J. Stanhope. 1997. Endemic African mammals shake the phylogenetic tree. *Nature* 388:61–64.

Springer, M. S., W. J. Murphy, E. Eizirik, and S. J. O'Brien. 2003. Placental mammal diversification and the Cretaceous-Tertiary boundary. *Proceedings of the National Academy of Sciences, USA* 100:1056–1061.

Stanhope, M. J., V. G. Waddell, O. Madsen, W. de Jong, S. B. Hedges, G. C. Cleven, D. Kao, and M. S. Springer. 1998. Molecular evidence for multiple origins of Insectivora and for a new order of endemic African insectivore mammals. *Proceedings of the National Academy of Sciences, USA* 95:9967–9972.

Tassy, P. 1996. Who is who among the Proboscidea? pp. 39–48 in J. Shoshani and P. Tassy (eds.), *The Proboscidea: Evolution and Palaeoecology of Elephants and Their Relatives.* Oxford University Press, Oxford.

Vignaud, P., P. Duringer, H. T. Mackaye, A. Likius, C. Blondel, J-R. Boisserie, L. de Bonis, V. Eisenmann, M.-E. Etienne, D. Geraads, F. Guy, T. Lehmann, F. Lihoreau, N. Lopez-Martinez, C. Mourer-Chauviré, O. Otero, J.-C. Rage, M. Schuster, L. Viriot, A. Zazzo, and M. Brunet. 2002. Geology and palaeontology of the Upper Miocene Toros-Menalla hominid locality, Chad. *Nature* 418:152–155.

Walter, R. C., and J. L. Aronson. 1993. Age and source of the Sidi Hakoma Tuff, Hadar Formation, Ethiopia. *Journal of Human Evolution* 25:229–240.

WoldeGabriel, G., J. L. Aronson, and R. C. Walter. 1990. Geology, geochronology, and rift basin development in the central sector of the Main Ethiopia Rift. *Geological Society of America Bulletin* 102:439–458.

WoldeGabriel, G., Y. Haile-Selassie, P. R. Renne, W. K. Hart, S. H. Ambrose, B. Asfaw, G. Heiken, and T. White. 2001. Geology and paleontology of the late Miocene Middle Awash Valley, Afar Rift, Ethiopia. *Nature* 412:175–178.

PHYSICAL AND TEMPORAL SETTING

ONE

Tectonics and Geomorphology of Africa during the Phanerozoic

TIMOTHY C. PARTRIDGE

Plate Tectonic Setting

The Cenozoic evolution of Africa cannot be comprehended satisfactorily without reference to the early history of the Gondwana supercontinent and the events that occurred during and following its fragmentation. Gondwana first formed during the Neoproterozoic Pan-African–Brazilian orogeny (720–580 Ma [million years ago]; see Unrug, 1996; Caby, 2003). The closing of the Paleotethys gulf during the collision of Laurasia with Gondwana in the late Palaeozoic completed the growth phase (figure 1.1).

The northern margin of Africa was first created from late Permian times onward by the opening of the Neotethys seaway. At much the same time, compression along the southern margin of the supercontinent initiated the rise of the Cape Fold Mountains. However, complete isolation of the African continent occurred only during the Cretaceous. Along the east coast of southern Africa, rifting, driven by mantle-plume activity, began as early as 183 Ma and was followed by fissure volcanism. The separation of Africa from Antarctica followed in the interval between 157 and 153 Ma. Along the west coast of southern Africa, dykes dated to 132 Ma marked the start of the Etendeka volcanism of Namibia that preceded the first divergence at around 123 Ma in the Neocomian. Hot-spot activity continued in the area under the influence of the Tristan da Cunha mantle plume and generated the important oceanic highs of the Walvis Ridge and Rio Grande Rise. Detachment along the west African coast was, however, diachronous; thus, transcurrent shearing west of the Gulf of Guinea began only in Aptian times (~100 Ma).

The volcanism that preceded rifting along both coasts of southern Africa had one particularly important consequence: a combination of thermal effects and magmatic underplating created a high "rim bulge" that resulted in the presence of significant coastal escarpments when separation occurred. High postrifting coastal hinterlands coincided largely with the area occupied by the African Superswell that, as will be discussed later, gave rise to "High Africa" in the south and east of the continent (in contrast to the less prominent hypsography of "Low Africa" to the north and west). Although evidence now points to the development (or perhaps rejuvenation) of the superswell during the Neogene, it is significant that more than 100 million years of erosion was unable to reduce totally the high relief—and vestiges of the associated drainage—that came into being just before Africa became isolated from the remainder of Gondwanaland. In contrast to these areas of early volcanism, the margins formed by shearing did not experience uplift prior to continental separation (e.g., that of West Africa).

The tectonic history of Africa during Cretaceous and Cenozoic times is characterized by long periods of extensional stress as the other Gondwana continents drifted away from it. These periods were punctuated by relatively short intervals in which a compressive stress regime was (at least locally) dominant. It was during these periods of crustal shortening that the major structural features of the Late Cretaceous and Cenozoic were superimposed on the Gondwana mosaic of stable cratons and intervening Pan-African orogenic belts (discussed later). Figure 1.2 indicates the principal elements of this tectonic framework. During the early part of this period, prior to 85 Ma, the motion of Africa relative to the Earth's axis of rotation took the form of a slow clockwise rotation around a pole located in the mid-Atlantic (Burke, 1996) and an equally gradual northward drift; indeed, during the entire post-Cretaceous period, this northward movement amounted to no more than about 14° of latitude.

The short intervals of crustal compression afford good illustrations of the influence of global events on Africa's tectonic history. In North Africa, the period of mid-Cretaceous quiescence that saw major marine transgressions both from the Neotethys and the Atlantic came to an end at about 85 Ma in the Santonian (i.e., just before the time of marine magnetic anomaly 34). This anomaly marks a significant change in the poles of rotation and a slowing in the rate of seafloor spreading in the Atlantic, and it is also correlated with the opening of the North Atlantic and the reactivation of far-field push from oceanic ridges. At much the same time, India and Madagascar separated from the east coast of Africa, the Mascarene Basin of the Indian Ocean began to open, and the Alpine chain of Europe was first uplifted (Guiraud and Bosworth, 1997). Although not all of these events were directly related, the analogy of a tectonic domino effect may

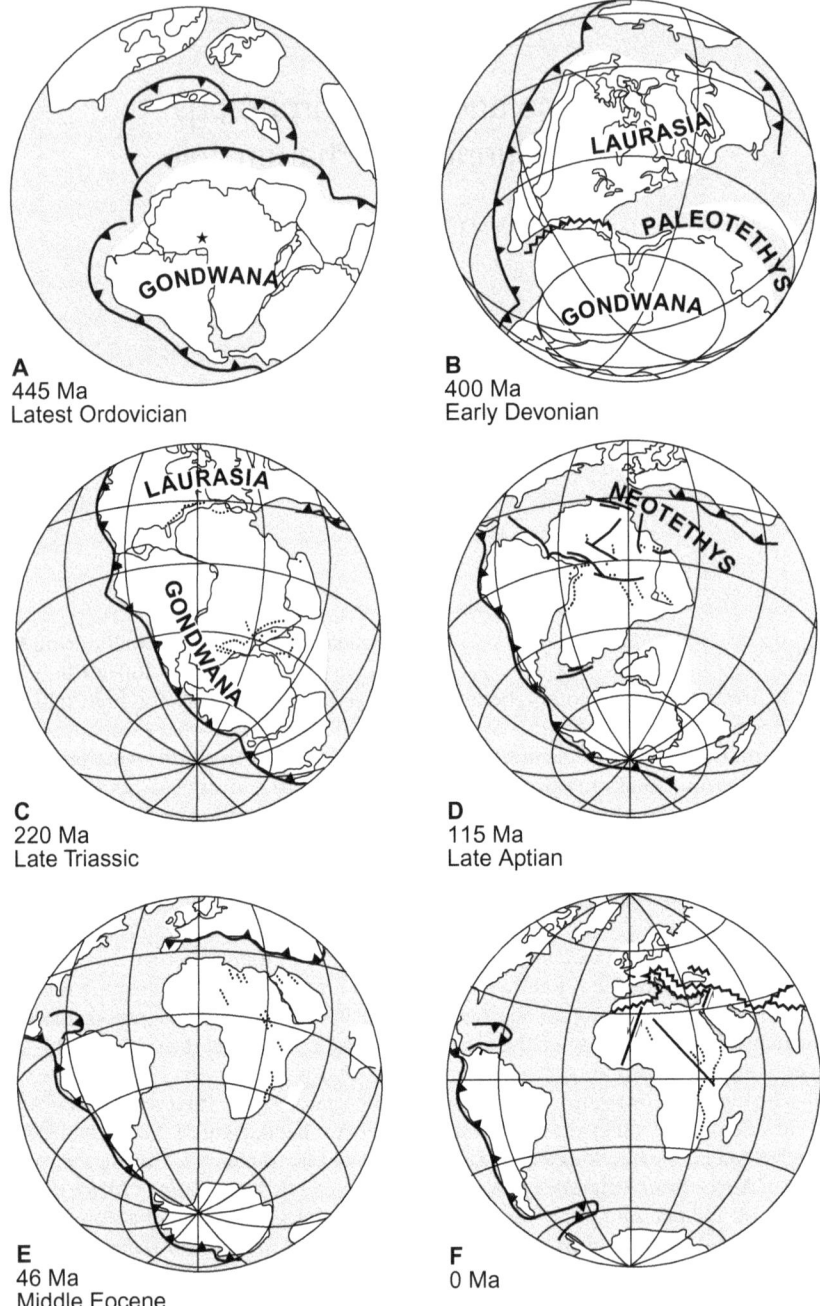

FIGURE 1.1 Palaeogeographic-palaeotectonic reconstructions showing Gondwana breakup. A) Reconstruction at ~445 Ma (latest Ordovician); black star shows the South Pole position; after Konate et al., 2003. B) Reconstruction at ~400 Ma (Early Devonian); wavy line is the Acadian collision. C) Reconstruction at ~220 Ma (Late Triassic); dotted lines correspond to major Karoo rifts in Africa-Arabia. D) reconstruction at ~115 Ma (late Aptian); dotted line represents major Late Jurassic–Early Cretaceous rifts and fault zones in Africa-Arabia. E) Reconstruction at ~46 Ma (Lutetian), with major late Senonian to early Eocene rifts in Africa-Arabia. F) Present day, with major Oligocene to Recent rifts and fault zones in Africa; wavy line is the Alpine fold-thrust belt. After Guiraud et al., 2005. A-F, copyright © 2010 Elsevier, B.V.

not be inappropriate. This global plate reorganization led to collision between the African-Arabian and Eurasian plates; its association with the Cretaceous Normal Magnetic Quiet zone indicates an origin in the Earth's lower mantle and core (Guiraud and Bosworth, 1997). The results in North Africa were crustal shortening, accompanied by dextral transpression, basin inversion, the creation of narrow fold-related relief, the uplift of extensive areas accompanied by renewed rifting, and volcanism in the intercratonic Pan-African belts.

One of these rift features was the northwest/southeast trenching series of basins extending from Sudan to the Lamu Embayment of Kenya, which is now crossed by the line of the later East African Rift System. Important is the fact that the Santonian rifting was passive in nature; that is, it was driven by plate boundary forces and not associated with local mantle plumes.

Between the Santonian event and the end of the Cretaceous, there was a return to an extensional stress regime, but

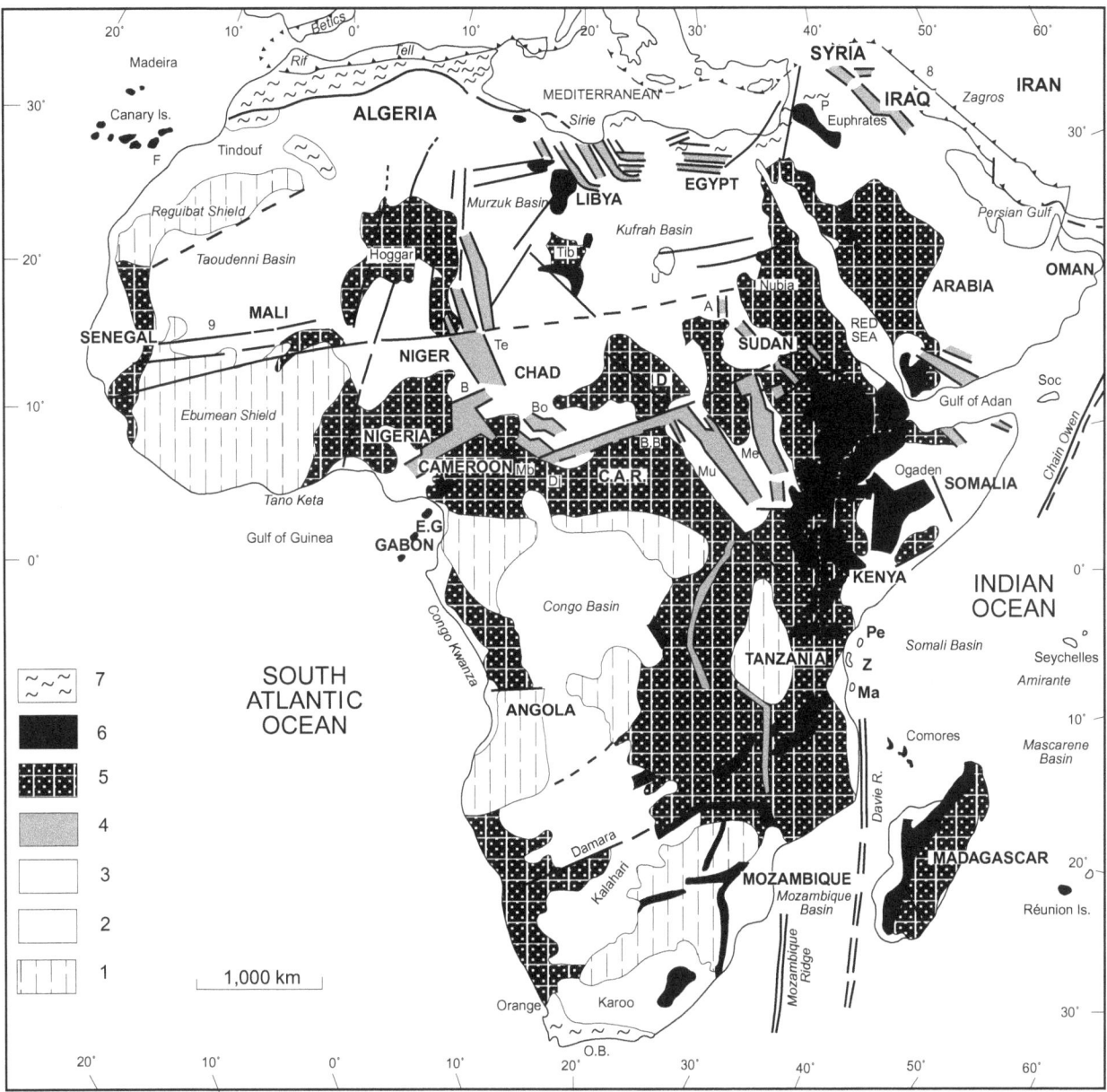

FIGURE 1.2 Major tectonic localities of Africa and Arabia.

ABBREVIATIONS 1 = Archean cratonic area; 2 = Proterozoic belt (mainly Pan African); 3 = Phanerozoic rift; 4 = Karoo rift; 5 = Mesozoic and/or Cenozoic rift; 6 = Mesozoic or Cenozoic magmatism; 7 = Phanerozoic fold belt; 8 = Alpine thrust front; 9 = major fault zone; A = Abyad; B = Bornu; B-B = Birao-Bagarra; Bo = Bongor; C.A.R. = Central African Republic; C.V.I. = Cape Verde Islands; D = Darfur; Dj = Djerem; E.G. = Equatorial Guinea; F = Fuerteventura; Iv.C. = Ivory Coast; M = Massirah Island; Ma = Mafia Island; Mb = Mbere; O.B. = Outeniqua Basin; P. = Palmyrides; Pe = Pemba Island; Soc. = Socotra Island; Tib. = Tibesti; U. = Uweinat; Z = Zanzibar. After Guiraud and Bosworth, 1997. A–F, copyright © 2010 Elsevier, B.V.

from that time onward, all blocks within the African Plate assumed a counterclockwise rotation. All of the major basins of northern Africa subsided, and the Kalahari Basin began to form. There was also a resurgence of magmatic activity and volcanism in northern Africa. This period of quiescence ended around the terminal Cretaceous when a brief compressional event accentuated a number of the Santonian structures and generated uplift of as much as 400 m in the Kenya Dome.

The second major compressional event to be felt across the face of Africa dates to the late Eocene (about 37 Ma) and corresponds with the onset of collision between the African/Arabian Plate and Eurasia. This event caused Africa to come almost completely to rest with respect to the underlying

mantle-plume system and other thermal anomalies in the mantle, and it initiated a wholly new stress regime across the continent. As is discussed later, this led to a new style of active rifting in North Africa that was directly associated with plume activity. These events in the north were accompanied by a continent-wide rejuvenation of the Pan-African basin-and-swell structure.

With the impending collision of North Africa with the Iberian margin of Europe, the African Plate became directly involved in the Alpine orogeny; this marked the beginning of the Atlas Event in which crustal shortening of about 25 km continued until the mid-Miocene and created the Atlas and Tell ranges of Morocco and Algeria.

It was at about this time that the African Plate began to suffer the first major disruption since its isolation in the course of Gondwana rifting. This was associated with the development of the East African Rift System (EARS), beginning around 45 Ma with major flood basalt volcanism in Ethiopia. Rifting began in the north and extended southward, and it has remained active to the present day; the EARS now extends over more than 3,200 km from the Afar triple junction, formed by the Red Sea and the Gulf of Aden, to the lower Zambezi Valley of southern Africa. South of Lake Turkana in Kenya, the rift bifurcates into eastern (Kenya) and western (Gregory) branches around the Nyanza Craton. Volcanism and rifting along the EARS were, for the most part, diachronous. Major faulting occurred in Ethiopia in the early Neogene; and by 6 Ma, a major set of eastward dipping faults defined a series of half-grabens in Kenya. Conjugate faulting between 5.5 and 3.7 Ma created a full graben morphology in this area. The EARS is of fundamental importance in the Cenozoic evolution of Africa. Its formation divided much of northeastern Africa into two new tectonic plates, the Nubian and the Somalian, as well as giving rise to the smaller Rovuma and Victoria microplates. These divisions were driven by regional asthenospheric upwelling and mantle flow (Calais et al., 2006), but while considerable lithospheric thinning is evident, crustal extension has probably amounted to no more than 20–30 km.

The final chapter in this brief chronicle of major plate-tectonic influence on Africa is the closing of the Neotethys to form the Mediterranean Sea when Arabia and Asia collided between 16.5 and 15 Ma.

Legacy of Early Gondwana Events

Africa is overwhelmingly a continent of old landscapes. It may surprise some readers that few of its important landscape features have their origin in the Quaternary—contra the beliefs of many Northern Hemisphere geomorphologists. Some important elements of its architecture in fact predate the Cenozoic. Among these is Africa's unique basin-and-swell structure, which had its origin in a great tectonothermal event in central Gondwana (the Pan-African–Brazilian orogeny) that dates to between 720 and 580 Ma. During this fundamentally important period, mountain building accompanied by strike-slip faulting occurred over wide areas in zones of lithospheric weakness (Guiraud et al., 2005). Subsequent erosion erased the Pan-African relief, but the underlying structural welts, separating rigid cratonic blocks, remained subject to renewed faulting and provided loci for magmatic activity (in North Africa) and widespread preferential uplift. The reactivation of these Pan-African zones throughout Phanerozoic times is manifested today in the broad basin-and-swell structure that characterizes the modern face of Africa (figure 1.3).

Another important legacy from prerifting times was the rise of the Cape Fold Mountains that later came to flank the southern margin of the African continent. The shelf sediments of the Cape Supergroup were deposited off the passive margin of the Kalahari Craton in the early Paleozoic. As shorelines migrated northward, an active margin developed in the south leading to a rise of parallel ranges of resistant sandstone and quartzite around the end of the Permian (de Wit and Ransome, 1992). Although subsequently unconformably overlain in places by softer sedimentary rocks of the Karoo Supergroup, these mountains have been reexposed in

FIGURE 1.3 The "basin-and-swell" structure of Africa. After Holmes, 1944.

subsequent cycles of erosion and form an important physiographic province (and a major center of biodiversity) at the southern extremity of Africa.

Major elements of the drainage of the continent also owe their origin to events that predated or accompanied the onset of rifting. In southern Africa, continental breakup was preceded by the rise of two mantle plumes. The associated surface doming exerted an important (although not exclusive) influence on the predominantly centripetal drainage that characterized this portion of Gondwana (Moore and Blenkinsop, 2002). The Karoo Plume, probably centered beneath the east coast at about present latitude 22°S, was the earlier, resulting in widespread outpourings of basalt with an age of around 183 Ma. While most drainage lines in the hinterland transgress the supposed zone of influence of the plume and follow preexisting structural features, the dome associated with the 40–50 Ma younger Parana Plume was located a little farther to the north and clearly exerted a more profound influence. Evidence of both radial and dome-flank components can be seen in the drainage net reconstructed for the Early Cretaceous around this feature—the former in the headwaters of the Congo (Zaire), upper Zambezi, Cuando, and Okavango rivers; and the latter in the inferred courses of the Kalahari and Karoo rivers (Moore and Blenkinsop, 2002; de Wit, 1999). The dominant influence of the doming associated with the Parana Plume provides a ready explanation for the preponderance of eastward-flowing rivers across a wide belt south of the Congo Basin.

In West Africa, structures created during continental rifting exerted a similarly important control over major elements of the drainage (Potter and Hamblin, 2006). The Benue River occupies a Cretaceous failed rift (aulacogen) that is the inland arm of a triple junction created during the separation of Africa from South America. The Niger, directed inland from its source by the rift-margin bulge of the Fouta Djallon Highlands (Sierra Leone), was subsequently captured by a coastal river occupying another aulacogen (the Bida Rift) to create the unusual, arc-shaped river system that now crosses several West African countries.

Early History of the Continent and Its Shelves

Several events that followed the separation of Africa from the remainder of Gondwana must be mentioned briefly because of their important influence on the structural and geomorphological history of the Cenozoic. There is little doubt that, resulting from the uplifts associated with the mantle plumes discussed in the foregoing section, elevations along the newly formed coasts of southern Africa must have been relatively high (Partridge and Maud [1987] have estimated hinterland elevations of 1,800–2,400 m). This helps to explain the ensuing massive influx of Cretaceous sediment onto the surrounding submarine shelves; this terrigenous input peaked in Albian to Turonian times (~110–90 Ma) and was replaced by slow carbonate deposition only after the end of the Cretaceous. As noted earlier, this early offshore loading would have generated a response in the form of further uplift of the hinterland, although recent models suggest that this would not have been of great magnitude (Séranne and Anka, 2005). This prolonged period of denudation removed up to 3 km of sediment from the coastal hinterland of southern Africa, in the process driving back the marginal, rift- generated escarpment by up to 200 km in the eastern hinterland and about 50 km inland of the west coast. In the interior, valley-flank recession reduced most areas susceptible to erosion to a gently undulating plain by the end of the Cretaceous (Partridge and Maud, 1987).

North of the Walvis Ridge off Namibia, and along the coasts of West and northeastern Africa, the offshore evidence paints a very different picture. Along the equatorial western margin, the early Cretaceous was characterized by evaporite and carbonate deposition on a shallow, slowly subsiding shelf, which was progressively replaced by terrigenous sediments from late Turonian times onward. But by the end of the Cretaceous, this input had dropped significantly and did not resume until the beginning of the Neogene, after which massive fans grew at an increasing rate off the mouths of all major rivers. An analogous sequence of events can be reconstructed for the coastal Lamu and Tana basins and the adjacent Kenya shelf, where a long period of postrifting carbonate sedimentation gave way to rapid terrigenous accumulation in response to the increase in local relief along the rising flanks of the EARS from Eocene times onward.

The early history of the coast of North Africa followed yet another pattern. Here large continental basins developed during and after rifting and became hosts to deep sequences of terrigenous sediment; during Cenomanian times, the sea invaded the northern African platform and led to the accumulation of evaporite and neritic carbonate sequences on these earlier deposits (Guiraud et al., 2005). Although uplift and basin inversion occurred during the Santonian tectonic event, large areas of North Africa continued to be covered by shallow seas. It was only after the uplifts caused by the late Eocene compressional event that the sea receded and fluviatile-lacustrine sedimentation resumed in a series of smaller inland basins. The rate of sedimentation in these depositories tended to increase from Miocene times onward in response to regional-scale uplifts associated with a resurgence in Neogene magmatic activity and uplift within the Pan-African swells.

This brief review of offshore deposition around the margin of Africa highlights major regional differences in the early geomorphic evolution of the continent. In southern Africa, where prerifting elevations were high as a result of uplifts generated by the Parana and Karoo plumes, rapid scarp recession occurred

during the Cretaceous, and concomitant loading of the offshore shelves resulted in further modest uplifts within the coastal hinterlands. In the equatorial and northern parts of the continent, by contrast, smaller rim bulges and generally lower inland relief led to terrigenous sedimentary outputs that were several orders of magnitude lower. The end result in both areas was, however, the same: the creation by the end of the Cretaceous of a low-relief landscape of coalescing plains occasionally punctuated by high-lying residual massifs. This surface was the product of prolonged erosion in the course of which fluvial activity was accentuated by several discrete periods of regional uplift (e.g., during periods of kimberlite emplacement and alkali volcanism around 120, 90, and 67 Ma, as well as the continentally important Santonian Event of 85 Ma). By the beginning of the Cenozoic, large-scale erosion had ceased in most areas, which then underwent deeply penetrative weathering under the influence of the torrid, late Mesozoic climate. The resulting thick kaolinitic regolith (with bauxite development in places) was invariably capped by duricrusts ranging from silcrete in southwestern Africa, the Sahara, and parts of East Africa to laterite in the equatorial belt and the hinterland of West Africa. The coupling of deep weathering and mature duricrusts high in the landscape is diagnostic of this Mesozoic–early Cenozoic land surface.

Cenozoic Tectonic Evolution

The Paleocene and early Eocene were, for the most part, periods of tectonic quiescence throughout Africa, except in the northeast of the continent. In the Sirt and El Gindi basins of northern Egypt, renewed rifting disrupted the shallow marine platforms that characterized the early Paleocene; in the process, several ridges were uplifted and appeared as islands (Guiraud and Bosworth, 1999). These events were accompanied by alkaline magmatic activity that persisted until the middle Eocene. At the same time, a major volcanic province developed in Ethiopia during the interval 45–33 Ma; this magmatic activity was a precursor to the first major faulting that began at the northern end of the EARS around 20 Ma (Ebinger et al., 2000).

In North Africa, a major compressive event (the Pyrenean-Atlas event) occurred in the late Eocene at about 37 Ma. An important result of the ongoing collision between the African and Eurasian plates, this event was driven by changes in the rates of opening of various sectors of the Atlantic Ocean. Among the consequences were the initiation of subduction of the Maghrebian Tethys beneath the Iberian Balearic margin (Frizon de Lamotte et al., 2000), major folding in the Saharan Atlas, and thrusting in the Tellian domain of the northwest African hinterland (Guiraud et al., 2005). In adjoining areas of the plate, fault zones were rejuvenated as strike-slip structures, and a new stress regime was imposed across the African plate.

From late Eocene times onward, the northeastern areas of Africa were dominated by the opening of the Red Sea and Gulf of Aden, as well as the development of the EARS. By the late Oligocene (~25 Ma), rifting had outlined the entire Red Sea and had spread to the Afar. Only by 12 Ma had faulting extended as far south as the Kenya Rift.

The most important consequence of Africa's collision with Eurasia from the Eocene onward was the slowing of movement of the African Plate with respect to the underlying mantle. Since that time, most perturbations have been from below (i.e., expressed in vertical uplift) because the lithosphere has remained all but stationary in relation to the various mantle plumes (Burke, 1996). This contrasts with the earlier evolution

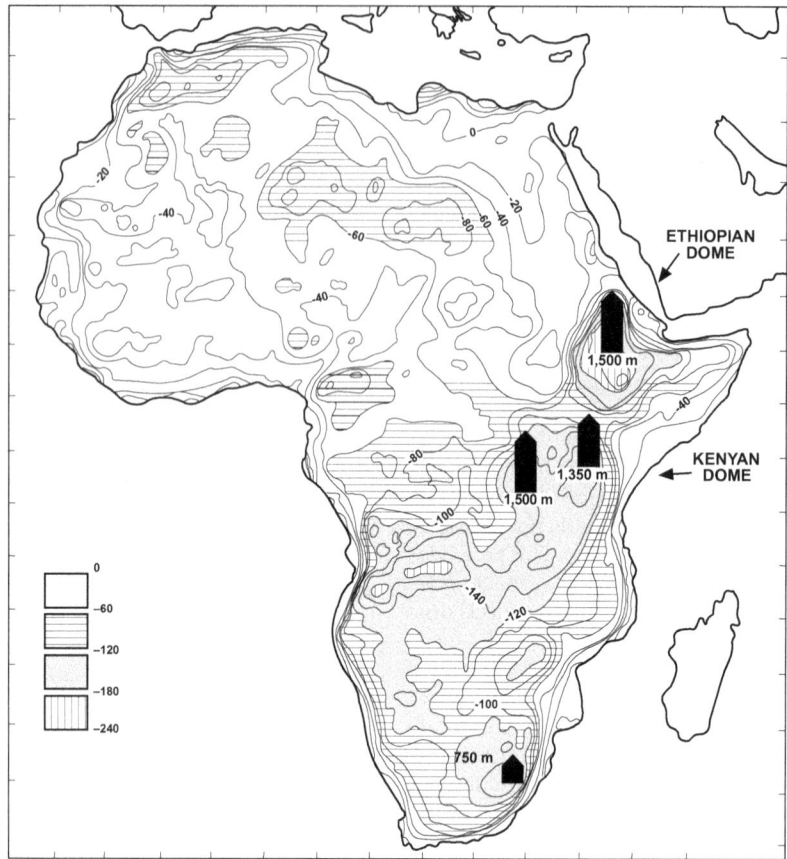

FIGURE 1.4 Bouguer gravity map of Africa showing typical late Neogene uplifts associated with principal negative gravity anomalies. After Sletlene et al., 1973, and Partridge et al., 1996. With permission of Yale University Press.

of the continent during which influences along the plate margins (e.g., ridge push) dominated tectonic events. The relatively large magmatic provinces that developed in response to this change were the foci of extensive domal uplifts that reduced the size of basins and caused several to become terrestrial rather than marine depositories. While magmatic and tectonic activity in several of these areas has spanned the entire Eocene–Recent interval, the greatest activity was in the early Miocene (Guiraud et al., 2005). Burke (1996) draws attention to the fact that most of the plumes have been associated with the Pan-African welts where the lithospheric mantle is sufficiently thin to be penetrated. Uplifts have involved partial melting of the mantle and are associated with notable density deficiencies (figure 1.4). Of significance is the fact that volcanic activity has been restricted to only a few of the low-density anomalies (e.g., in the environs of the EARS); the largest of these anomalies occur in the southern part of Africa where no Cenozoic magmatic provinces are located. The large South and East African density anomalies coincide closely with the African Superswell of Nyblade and Robinson (1994) (figure 1.5). As has been mentioned previously, the area of the superswell defines "High Africa," with its elevated plateaus and prominent escarpments, as distinct from "Low Africa" to the north, where the regional relief seldom exceeds 500 m.

The most significant feature of the Paleogene is, however, the presence of a widespread Oligocene offshore unconformity that persists around the entire African margin. This hiatus coincides with the first appearance of ice on Antarctica, which led to a 30- to 90-m drop in sea level and accelerated current erosion in some areas. Off equatorial West Africa,

up to 500 m of sediment was removed during this period (Séranne and Anka, 2005). Of importance is the fact that, contra Burke (1996), the Oligocene in central and southern Africa was not characterized by significant tectonism: recent evidence is consistent in referring widespread renewed uplift to the early Neogene.

The evidence for Miocene tectonism on a subcontinental to regional scale is persuasive. Potter and Hamblin (2006) have, indeed, argued for a worldwide tectonic interval during the Miocene. Lunde et al. (1992) and Hudec and Jackson (2004) document early Miocene uplift, accompanied by the erosion of 1,000–2,000 m from parts of the hinterland, from the marine succession in the Kwanza Basin off the coast of Angola; while Lavier et al. (2000, 2001) conclude that the northern Angola and Congo margins were uplifted in the Burdigalian (~18 Ma) and, to a somewhat lesser extent, in the Tortonian (11–7 Ma). Thermochronological studies (fluid inclusion and apatite fission track analyses) confirm the occurrence of about 500 m of Miocene uplift in the coastal basin of Gabon and Angola (Walgenwitz et al., 1990, 1992). Off the Limpopo and Tugela rivers of southeastern Africa, which today carry about one-third of the sediment load from that area, Neogene terrigenous sediments were redistributed under the influence of the vigorous Agulhas Current; there is nonetheless abundant evidence for increased sediment inputs during Mio-Pliocene times (Dingle et al., 1983). In the coastal hinterland of the Algoa Basin along the same coast late Miocene marine sediments have been elevated to 400 m AMSL along the flanks of a coastal monocline extending as far north as Swaziland (Partridge, 1998). Deformation of the

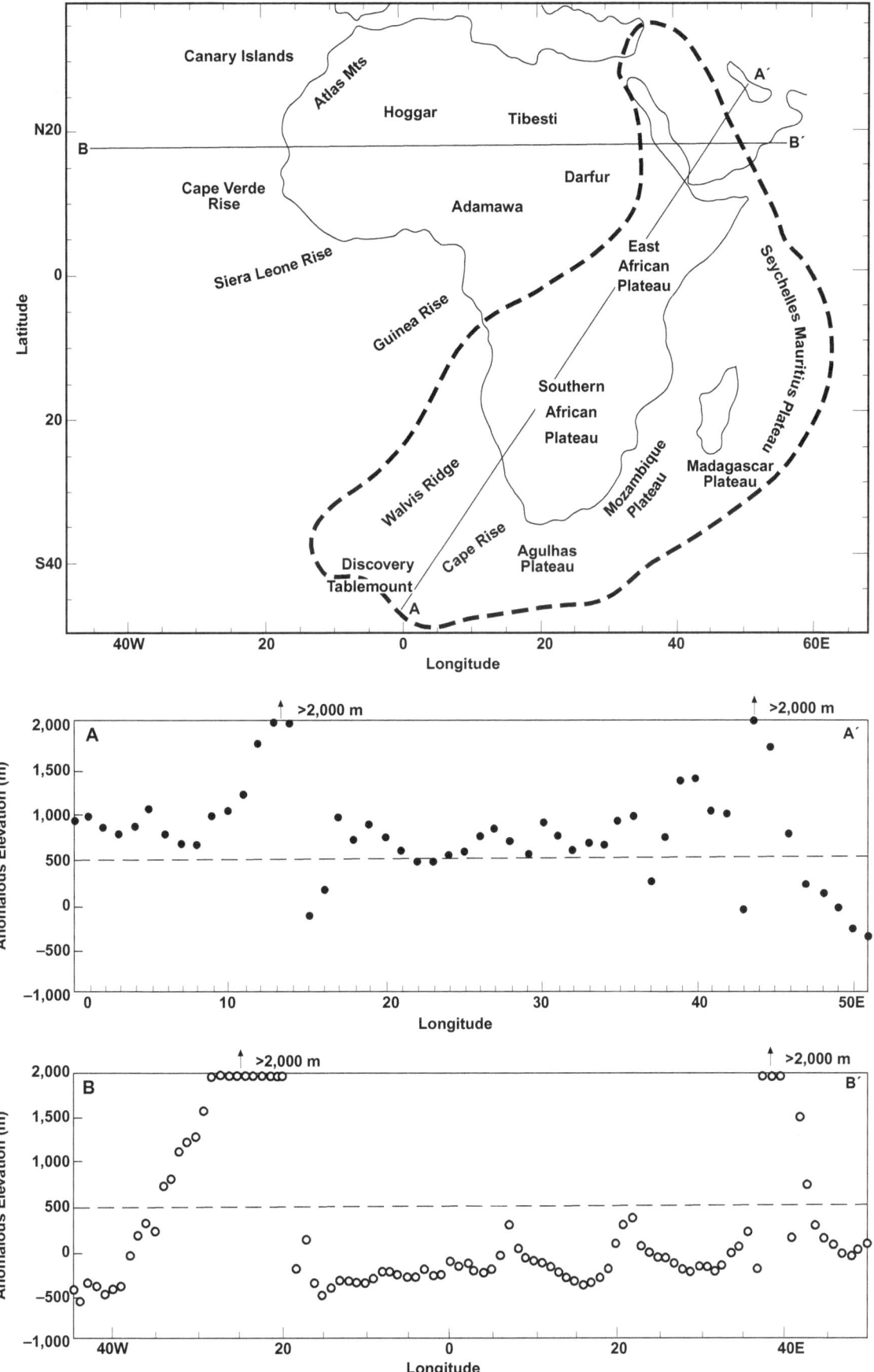

FIGURE 1.5 The African Superswell (upper diagram, outlined by bold dashed line). A-A′ and B-B′ (lower two diagrams) are cross sections along the two transects. After Nyblade and Robinson, 1994. Courtesy of American Geophysical Union.

late Mesozoic/early Cenozoic surface, with its deep kaolinization and diagnostic duricrusts, is another indicator: divergence between this datum and the mid-Miocene surface reaches about 250 m near the axis of maximum monoclinal uplift, indicating a moderate tectonic pulse in the early Miocene. The mid-Miocene surface, with its patches of marine sediment up to 20 km inland of the present coastline, has, in turn, been upwarped by between 600 and 900 m along the crest of much the same axis.

The extent of the Mio-Pliocene uplift within the southeastern hinterland of South Africa has been questioned by Burke (1996) on the grounds that such major deformations seldom occur within so short a time frame. It must be noted, however, that Pliocene upwarping along the flanks of the EARS in Kenya has been estimated at 1,500 m (Baker et al., 1972), and it has recently been shown, on the basis of dated markers, that 1,170 m of that movement occurred over an interval of 600 kyr between 3.21 and 2.65 Ma (Veldkamp et al., 2007).

In other areas of Africa, the Miocene disturbances were associated with the reactivation of old structures such as the Benue, Luangwa, and Zambezi rifts (Burke, 1996). In coastal areas of the Congo, uplift began in the mid-Miocene (~16 Ma) and increased until about 11 Ma; after a short interval, uplift was resumed in the late Miocene and reached a maximum by the start of the Pliocene (Lavier et al., 2001), mirroring the twofold movements that occurred along the southeastern coast of South Africa. A similar sequence of events along the Angolan margin further elevated marine shelves first exposed above sea level during the Oligocene-Miocene interval.

All of the foregoing evidence confirms that multiphase uplifts of considerable magnitude occurred in southern Africa, East Africa, and some magmatic provinces of North Africa during the Miocene, and continued into the early Pliocene in some areas. Those in South and East Africa gave rise to the African Superswell, which is higher than the normal elevation of Africa and the surrounding continental shelves by an average of 500 m and exceeds 1,000 m over large areas (Nyblade and Robinson, 1994). These elevated areas are associated with an anomalously hot (and thus lower-density) mantle at depth (Hartley et al., 1996; Lithgow-Bertelloni and Silver, 1998). As will be discussed later, the diachronous nature of these Miocene movements was expressed in the generation of discrete scarps separating areas of partial planation. These, together with yet higher-lying remnants of the end-Cretaceous surface, give to Africa the unique multistoried morphology and the anomalous hypsography that characterize large areas of the continent.

Before concluding our discussion of Africa's tectonic history, mention must be made of late-phase events along its northern coast. A change in plate motion during the Tortonian (~8.5 Ma) generated dextral transpression along the Africa-Arabian plate margin (Guiraud and Bellion, 1995), resulting in renewed thrusting along the southern front of the Maghrebian Alpine Belt and the rejuvenation of intraplate fracture zones in Hoggar, Egypt, and Sudan. The Tethys, which was closed off by collision between Arabia and Eurasia between 16.5 and 15 Ma to form the Mediterranean, narrowed further during Messinian times (5.6–4.8 Ma) converting the Mediterranean into an enclosed inland basin (Ruddiman et al., 1989). Ensuing evaporation lowered its surface level by some 2,500 m, creating huge evaporite deposits and initiating the cutting of major canyons in the North African hinterland.

The East African Rift System

The East African Rift System (EARS) extends for more than 3,200 km from the Afar triple junction (which marks its convergence with rift arms occupied by the Red Sea and Gulf of Aden) to the lower Zambezi Valley in southern Africa. In the north, the rift crosses the Ethiopian Plateau, whose high elevations reflect the combined effects of Eocene plume-driven volcanism and uplift. South of the Turkana depression in Kenya, the rift bifurcates into eastern (Kenya) and western (Gregory) branches around the Nyanza Craton, which coincides in part with the East African Plateau; the latter was uplifted by more than 400 m in the Late Cretaceous (Foster and Gleadow, 1996). This early broad doming was probably occasioned by the development of deep-seated thermal anomalies in the mantle and may represent an early stage in the formation of the African Superswell. Throughout most of their length, the rifts are fairly narrow (<100 km); but north of the point of bifurcation, the Kenya Rift widens to about 300 km in a complex zone of interaction between the separate Ethiopian and Kenya rift systems and older, cross-cutting Mesozoic rift structures (Ebinger et al., 2000).

The importance of the EARS for paleontology stems from the occurrence, along both branches, of lakes whose sediments contain important fossil and environmental archives. The Gregory Rift contains a number of large and deep lakes (Albert, Edward, Kivu, and Tanganyika) which preserve thick sedimentary sequences dating back to the Miocene, while the Kenya Rift is characterized by a series of small evaporative lakes, most of which formed during the Pliocene. Lake Malawi, beyond the point where the rifts reunite, occupies much of the southern sector of the EARS and, like the lakes of the Gregory Rift, contains a long sedimentary sequence. Lake Turkana, in the depressed area between the Ethiopian Plateau and the East African Plateau, is fed by the Omo River. Previously of much greater extent, it is flanked by dissected Plio-Pleistocene lacustrine sequences that have yielded abundant mammalian fossils, including a number of highly important hominid specimens. Similar fossiliferous deposits flank the northward-flowing Awash River, which crosses the floor of the Ethiopian sector of the rift.

Important in the evolution of the EARS has been the superimposition of additional relief along the rift flanks during the evolution of the system. This has occurred through both uplift of the rift shoulders under the influence of the underlying thermal anomalies and the construction of volcanic edifices. The latter account for 140,000 km^3 in the vicinity of the Kenya Dome alone. These late (and in some cases ongoing) additions to the local relief were of sufficient magnitude to affect local climates by increasing orographic precipitation or creating rain shadows.

The sequence of events that have created the EARS has been reviewed by Baker et al. (1988), Partridge et al. (1995), and Ebinger et al. (2000). Both structurally and magmatically controlled processes have been involved, and the evolution of the entire system has been markedly diachronous. In Ethiopia, large-scale volcanism preceded rifting in the interval 45–33 Ma. The first faulting may have occurred in Oligocene times, but it was only in the early to mid-Miocene (20–14 Ma) that uplift and rift faulting occurred on a large scale (Ebinger et al., 2000). These movements uplifted the lateritized African Surface on the 1,000-km-wide Afar Plateau by about 500 m. They were followed by fluvio-lacustrine deposition in the Afar Basin spanning the Middle and Upper Miocene. An interval of alkali

basalt and trachyte volcanism in the mid-Miocene was followed by renewed uplift of 1,000–1,500 m in the Pliocene
(Baker et al., 1972; Adamson and Williams, 1987; Denys et al.,
1986), during which the rift margins were heightened, the
Awash graben was lowered, and there was a major influx of
sediments into the Middle Awash Valley and the Hadar Basin.

In the broad zone of rifting to the south of the Ethiopian
Plateau, volcanism and faulting first occurred between 18 Ma
and 10 Ma; over time both the active fault and volcanic centers shifted eastward and brought the later Pliocene to Recent
rifts into closer alignment with those to the north and south
(Ebinger et al., 2000). Fluvio-lacustrine deposition in the
Omo-Turkana basin, which occupies the western part of this
sector, began in the mid-Miocene and has continued inter-
mittently until the present.

On the East African Plateau to the south, narrow belts of elevated topography occur along the shoulders of both the Kenya
and Gregory rifts. These rift-flank ridges are 100–200 km wide
and have been uplifted above the surrounding plateau (itself
elevated by Cretaceous movements) by an average of about
1,000 m. In the Kenya Rift, south of about 3°N, doming of
300–500 m preceded the formation of eastward dipping
faults that began at about 12 Ma (Baker et al., 1988; Ebinger
et al., 2000). Half-graben formation continued until around
6 Ma, after which conjugate faults developed between 5.5 and
3.7 Ma, creating a full graben morphology. Important is the
fact that a considerable proportion of the topography above
the upwarped rift shoulders was created by volcanic activity:
flood phonolites from about 15 Ma and trachytes and basalts
after about 10 Ma (Shackleton, 1978; Baker, 1986; Mohr, 1987;
Baker et al., 1988). Further volcanic centers were added
around 5 Ma ago; this phase of magmatic activity was accompanied by further uplift of the rift shoulders, which totaled
1,200–1,500 m in places (the abundance of well-dated lavas
and sedimentary sequences in proximity to the rift has
enabled these movements to be bracketed in time with reasonable accuracy on the basis of the deformation of planation
surfaces of known age; see, e.g., Baker et al., 1972; Saggerson
and Baker, 1965; Ebinger, 1989; Fairhead, 1986; and Pickford
et al., 1993) (figure 1.6).

By 2.6 Ma, the central sector of the Kenya Rift had become
further segmented by west-dipping faults, creating the 30-km-
wide intrarift Kinangop Plateau and the tectonically active
40-km-wide inner rift. This Pliocene interval of rift evolution
saw the development of a number of discrete fault basins with
local relief of up to 1,800 m between floor and rift shoulder.
This topography was accentuated in some areas by renewed
faulting as recently as 1.2 Ma. Sedimentation within the rift
basins began as early as the Middle Miocene in the Baringo-
Bogoria area; the oldest sequences in the Kenya and Tanzania
segments of the rift are, however, no older than early Pliocene
(~5 Ma).

The evolution of the Western (Gregory) Rift differed from
that of the Kenya Rift in that early Miocene doming around
the incipient rift was of smaller amplitude and volcanic activity began later (21–10 Ma) and was restricted mainly to fault-
bounded basins (Ebinger, 1989; Ebinger et al., 1989; Pickford
et al., 1993). Further uplift of the rift shoulders, mainly in the
3–2 Ma interval, increased their elevation by about 1,500 m
but reached 4,300 m in the Ruwenzori Mountains. In the process, the rift became segmented into discrete basins, the floors
of some of which extend below sea level. All of the deep lakes
that characterize the Western Rift have sedimentary sequences
that span a considerable proportion of the Neogene. The

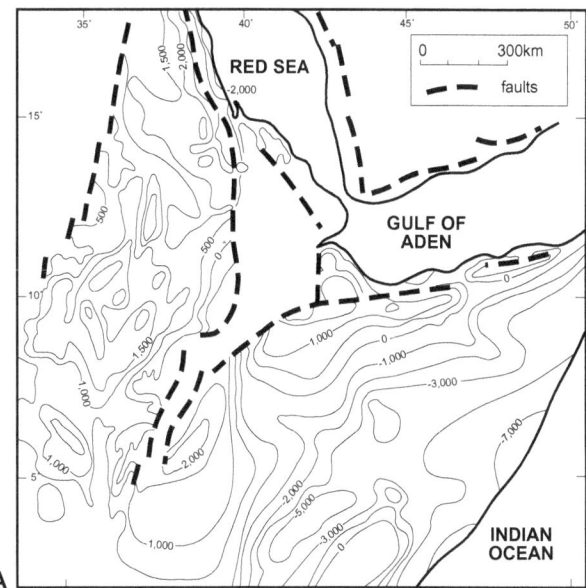

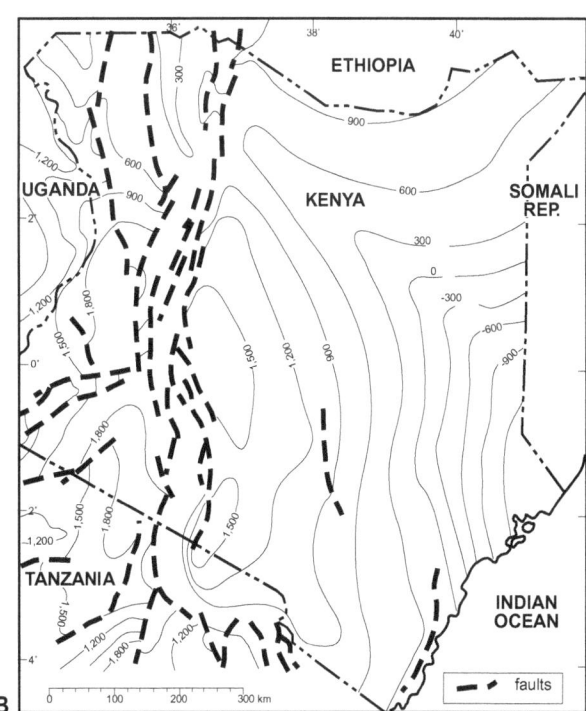

FIGURE 1.6 A) Isohypsals of the Precambrian basement adjacent to
the Ethiopian Rift. B) Isobases of the sub-Miocene erosion surface
in Kenya (equivalent to Post–African I surface in southern Africa).
Elevations in meters. After Baker et al., 1972. With permission of the
Geological Society of America.

main events in the evolution of both rifts and the intervening 1,300–km-wide East African Plateau are summarized in
figure 1.7.

Viewed as a whole, the EARS propagated southward as a
result of the replacement of lithospheric material by buoyant,
hot asthenosphere (Ebinger et al., 2000). These authors conclude, further, that the Paleogene flood basalts of Ethiopia
have a definite mantle-plume geochemical signature. As a
result of northward movement of the African Plate relative to
a hot-spot reference frame, the stem of the same plume now
probably lies beneath the East African Plateau. Despite the

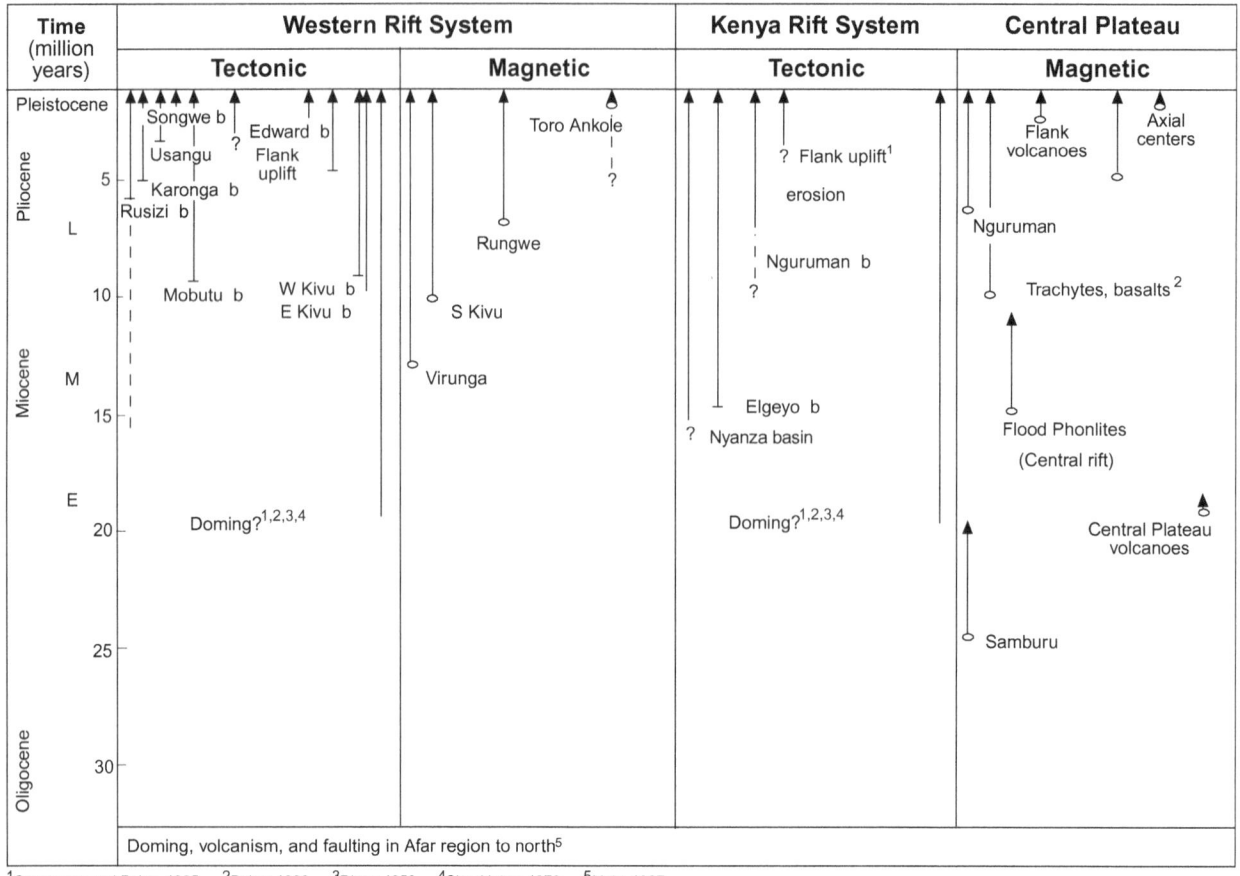

FIGURE 1.7 Chronological constraints on vertical movements, volcanic activity, and basinal subsidence within the East African Plateau region. Lines indicate approximate time span of activity; dashes and question marks are used where dating is uncertain. After Ebinger, 1989, using data from Saggerson and Baker (1965), Baker (1986), Dixey (1956), Shackleton (1978), and Mohr (1987). With permission of the Geological Society of America.

considerable lithospheric thinning that has occurred during these events, crustal extension across most areas has been modest, amounting to no more than about 20–30 km.

Macroscale Geomorphic Evolution during the Cenozoic

During the early Cenozoic significant areas of central North Africa were inundated in the course of a major marine transgression. The epicontinental sea eventually surrounded the Hoggar Massif, but with the onset of the Laramide Event in the late Paleocene, some east-west ridges were uplifted along the North African Tethyan margin (Guiraud et al., 2005). The uplifts that followed the 37 Ma Eocene compressive event caused the sea to regress, leaving a mosaic of shallow gulfs and enclosed terrestrial depositories. The early drainage of North and West Africa was, at that time, probably toward the Mediterranean, with the palaeo-Benue River constituting an important element in West Africa (Burke, 1996). The remainder of the African continent was characterized by a series of gently undulating plains of low relief forming subdued interfluves between major drainage lines. Isolated massifs, invariably formed of or capped by resistant rocks, stood above these plains; examples are the basalt- and sandstone-capped Lesotho Highlands of southern Africa, the resistant quartzite ridges of the Cape Fold Mountains in the southwestern extremity of the continent, the Guinea Highlands of West

Africa, and the gently upwarped Kenya Dome, which straddled the line of the more recent Kenya Rift of East Africa. A seaward-facing escarpment (the Great Escarpment) that forms a great arc within the hinterland of southern Africa (Angola, Namibia, South Africa, and Zimbabwe) separated an inland plateau area from the surrounding, low-lying coastal plain. This escarpment was the legacy of uplifts associated with the Parana and Karoo plumes that created areas of unusually high relief along these incipient continental margins of Africa prior to rifting. Two major drainages of southern Africa, the Kalahari and Karoo rivers, had cut (or maintained) courses through the high rim bulge to debouch into the newly formed Atlantic Ocean (de Wit, 1993). The Kalahari River probably included drainages that now constitute headwaters of the Zambezi, Okavango, and Limpopo rivers. An early capture by the upper Limpopo River is thought to have diverted these headwaters into the Indian Ocean; it has been remarked that the reduced size of the present Limpopo catchment is inconsistent with its large, partially elevated delta (Hartnady, 1990). In Central Africa, the Congo (Zaire) River had been actively contributing sediment to the offshore shelf since the Cretaceous from a basin adjoining those that fed into the Tethys.

Of note is the fact that, by the Late Cretaceous or early Paleocene, an enclosed basin had begun to form within the western hinterland of southern Africa. Known as the Kalahari Basin, this zone of regional subsidence now covers much of Botswana and parts of western Zambia and southern Angola.

Its topographic isolation may have been due, in part, to early uplift along the east-northeast/west-southwest trending Griqualand-Transvaal Axis (du Toit, 1933) and the parallel Central Angolan Axis, which now forms the southern watershed of the Congo Basin. Both axes are associated with clusters of kimberlite and alkali-volcanic pipes that were emplaced in phases spanning the period 120–67 Ma; it is conceivable that events associated with these volcanic pulses may have driven the local axial uplifts. The sedimentary infilling of the Kalahari Basin buries downwarped and silcrete-capped remnants of the African Surface; in the Congo (Zaire) Basin, in contrast, subsidence began earlier, and the African Surface cuts across an infill of Cretaceous sediments. Continuing sedimentation led to the accumulation of the Gres Polymorphes series on this surface (Giresse, 2005).

THE AFRICAN SURFACE

Of African geomorphology, Burke (1996) remarks, "Interest focuses naturally on the extensive erosional surfaces which, when compared with those of other continents, are surely distinct if not uniquely well-developed features of African geology. Linked to these surfaces are the scarps that separate them, especially those escarpments of regional extent" (p. 363). The gently undulating end-Cretaceous/early Cenozoic surface has been named by King (1949) the African Surface, a name perpetuated by Partridge and Maud (1987) and Burke (1996). Its Pan-African extent is remarkable. Now uplifted by Miocene and early Pliocene movements amounting to hundreds (and, locally, thousands) of meters, its dissected remnants appear as flat or gently inclined cappings on high-lying topography; their accordance over hundreds of kilometers is readily apparent in topographic sections. In susceptible lithologies, these cappings overlie deeply kaolinized saprolite, the thickness of which is usually tens and, more rarely, hundreds of meters. The composition of the cappings themselves varies in relation to climatic and pedogenetic conditions that prevailed regionally or locally: under humid tropical conditions (e.g., in the northeast, west, center, and southeast) laterite predominates, sometimes in association with bauxite; in the north, southwest, and semiarid to arid areas of the eastern hinterland, silcrete is widespread. In areas little affected by later tectonism (or where large areas have been vertically uplifted) surface gradients defined by these remnants are small to imperceptible, but where warping was more localized, the African Surface has sometimes been tilted to slopes of several tens of meters per kilometer. Figure 1.8 shows the deformations that have occurred in response to axial monoclinal warping in the southeastern hinterland of South Africa.

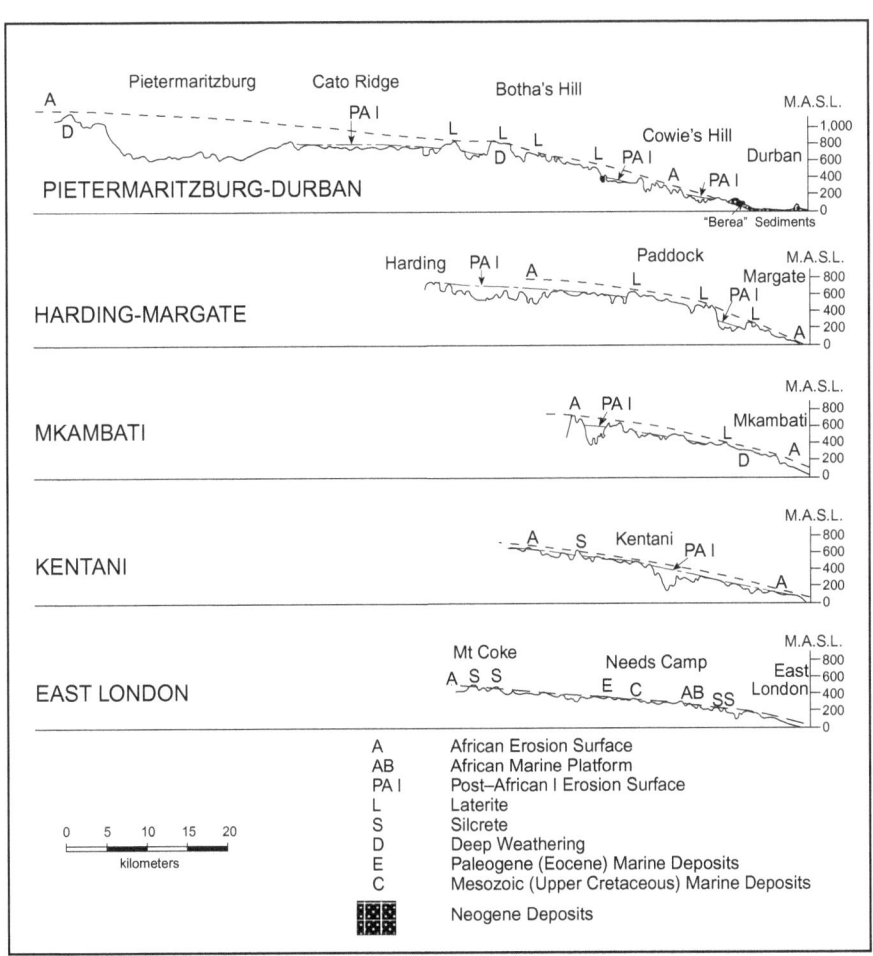

FIGURE 1.8 Sections drawn inland from the coast across the southeastern hinterland of southern Africa showing erosional remnants, duricrusts associated with deep weathering, and marine deposits from Mesozoic and early Cenozoic transgressions. After Partridge and Maud, 1987. Copyright © 2010 South African Journal of Geology.

The antiquity of the African Surface is now well established. Around the coastline of Africa, terrigenous lower and Middle Cretaceous sequences, referable to the period during which the postrifting relief of the continent was almost entirely consumed by erosion, are present. These are thickest and show the highest sedimentation rates around the margins that were driven to the highest prerifting elevations by the Karoo and Parana plumes. Direct dating evidence is available also from numerous localities in the continental interior. This includes dates from numerous penecontemporaneous diatremes that preserve crater facies of Late Cretaceous age (Partridge and Maud, 2000), diagnostic fossil wood in relict channels crossing remnants of the surface (Partridge and Maud, 2000), $^{40}Ar/^{39}Ar$ dates for manganiferous pisolites within its laterite cappings in West Africa that place them within the Paleocene (Colin et al., 2005), fission-track ages of ~60 Ma from below the laterized surface in Kenya (Foster and Gleadow, 1996), and the occurrence of laterite caps (often with associated deep weathering and bauxite development) beneath dated lavas in the Afar of northeast Africa (Drury et al., 1994) and other areas of North Africa such as the Jos Plateau, the Hoggar, and the Tibesti Mountains (Burke, 1996). Furthermore, on former coastal margins where the African Surface has been uplifted by Neogene movements, distal remnants of it are overlain by transgressive marine sediments of Late Cretaceous to Eocene age (Partridge and Maud, 2000; Burke, 1996). Thus, Africa, at the beginning of the Cenozoic, was a continent of subdued relief whose surface was heavily armored by duricrusts and crossed by a dense network of mature rivers—the legacy of the humid and temperate climates that characterized much of the Cretaceous. Extensive remnants of the African Surface, now uplifted to varying degrees, are preserved on the Jos Plateau of Nigeria, the Guinea Highlands, the Fouta Djallon Plateau of Sierra Leone, the southern watershed of the Congo Basin in southern Zaire and northern Zambia, the Angolan Highlands, the interior plateau of Namibia, and the Bushmanland Plain and the Highveld of the interior of South Africa. Significant areas of the pristine surface have also survived on the shoulders of the EARS in a belt extending from the Ethiopian Highlands to the southern end of Lake Malawi.

NEOGENE SCARP FORMATION AND PEDIMENTATION

The influence of the major compressional event of the Eocene was restricted largely to the North African seaboard and adjacent areas. Of much more widespread influence was the global drop in sea level of the early Oligocene that resulted in the exposure of large areas of the continental shelves around Africa and initiated the cutting of coastal gorges along major rivers. These valleys were later rejuvenated and extended in response to Miocene tectonic events. As has been discussed previously, renewed tectonism within the interior of the continent was restricted chiefly to pulses within the Miocene, some of which continued into the early Pliocene.

In North Africa, uplifts associated with regional magmatic provinces saw the recession of the sea and the development of intervening fluviatile and lacustrine basins. Both uplifted areas and the associated volcanic massifs were attacked by erosion into the newly created relief; this resulted in the formation of extensive escarpments, the crests of which were capped by "African" duricrusts (e.g., the Massak Mallat escarpment in Niger [Busche, 2001] and those fringing the Hoggar and Tibesti mountains). Toward the end of the Miocene a

major river, the "Eonile" was established to the west of the uplifted rift shoulders of the Red Sea–Gulf of Suez and north of the Jebel Uweinat and Darfur swells (Said, 1981). The resulting canyon exceeded in length and depth the Grand Canyon of the western United States.

In West Africa, the African Surface was upwarped to elevations of between 200 and 700 m. Subsequent scarp formation and recession within its lateritized and bauxite-rich weathering profile created a younger surface that became capped by quartz-rich ferricrete (Colin et al., 2005). The localization of axes of uplift within the coastal hinterland created discrete basins, each with an internal drainage system (e.g., the Niger and Chad basins, and possibly, for a short period, the Zaire Basin; see Burke, 1996). Capture through these coastal swells eventually reconnected most of these basins to the Atlantic Ocean, although the Chad Basin has remained isolated except during highstands of Lake Megachad. In the Zaire (Congo) Basin, Miocene uplift of the margins stimulated new tributary development through the Albertine Rift and captured the southern part of the Chad drainage (Giresse, 2005). In the process, the drainage net shifted into near-present alignments, and a younger surface was cut to the new base levels; this surface is overlain by the Neogene Sables Ocres series. Subsequent incision in response to late Miocene movements has adjusted channel gradients in the coastal hinterland to present sea level.

South of the margins of the Congo Basin, Miocene uplift of the African Superswell has created higher escarpments and led to greater divergence between planation surfaces. In South Africa, modest uplift at the beginning of the Miocene (200–300 m) was concentrated in the southeastern coastal hinterland. The resulting low escarpments separate the African surface from a partially planed "Post–African I" surface (Partridge and Maud, 1987, 2000). Potassium-bearing cryptomelane in local weathering profiles on this surface have given $^{40}Ar/^{39}Ar$ ages of 12–15 Ma (van Niekerk et al., 1999). In the western parts of South Africa, where the uplift was smaller, downcutting in major rivers such as the Orange was followed by the accumulation of deposits—now preserved as high terraces—containing faunal remains dating to 17–19 Ma; deposits of similar age are preserved in defunct drainages crossing the Post–African I surface to the south of the Orange River (Corvinus and Hendey, 1978; Pickford et al., 1996). The largest areas of Post–African I planation in South Africa occur in the Northern Cape and the Bushveld Basin. However, throughout Africa, this undulating surface, below scarps bounding African remnants, is widely represented. Particularly extensive areas flank the central cores of the Congo and Kalahari basins and parallel the coast of West Africa. Others occur in eastern Chad, northern Sudan, and Egypt.

A much larger uplift at the end of the Miocene or start of the Pliocene further elevated the southern African sector of the Superswell (figure 1.9). A similar dichotomy of movements has been documented in areas of Angola and Gabon, although the evidence does not everywhere support the occurrence of clearly identifiable pulses that are expressed in distinct topographic breaks (Séranne and Anka, 2005; Lavier et al., 2001). In South Africa, uplift was again concentrated in the southeastern hinterland along the Ciskei-Swaziland axis and totaled between 600 m and 900 m. The ubiquitous response was the cutting of major gorges along the eastward-flowing rivers; steepening of channel gradients in their west-flowing counterparts produced lesser incision and terrace formation. Only in areas of susceptible rock (the Lowveld

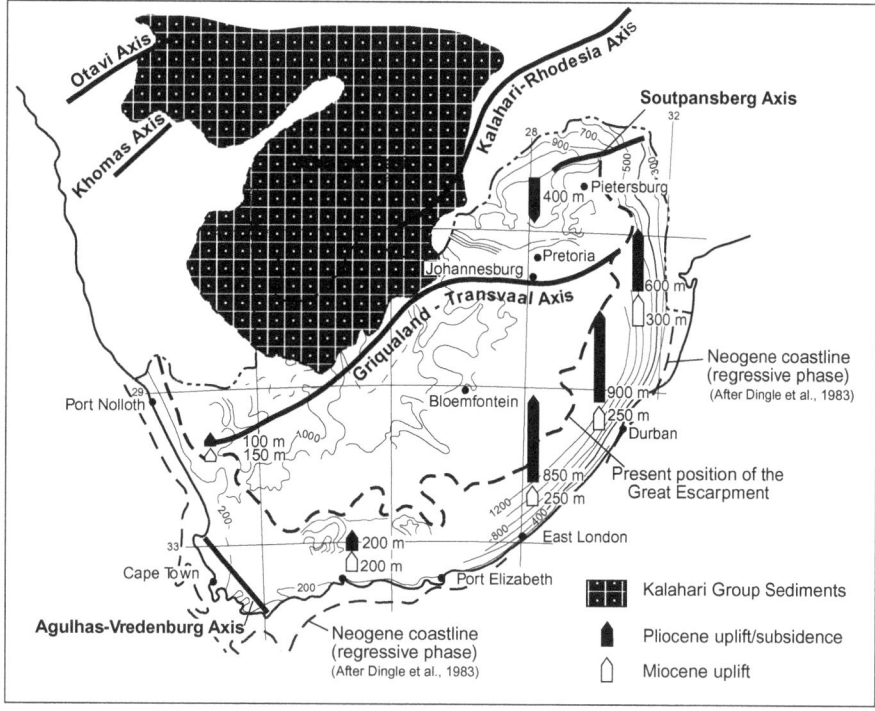

FIGURE 1.9 Generalized contours on Post–African I (Miocene) erosion surface in southern Africa. Open arrows indicate the amplitude of early Miocene uplift; solid arrows show Mio-Pliocene uplift or subsidence. The present position of the Great Escarpment is shown by the broken line. After Partridge and Maud, 1987; interior axes of uplift are after du Toit, 1933. Copyright © 2010 South African Journal of Geology.

of Swaziland and Mpumalanga, and the Algoa Basin) did any degree of planation occur (the "Post–African II Surface"). The generation of numerous falls and nick points along major rivers bears testimony to the young age of this movement; they represent an early stage in the formation of a new series of erosional escarpments and pediment benches around the flanks of the Superswell.

A PHILOSOPHICAL DIGRESSION

The African Surface is the earliest epicontinental land surface recognizable across the breadth of Africa. But in almost every area of the continent, one or more lower bevels—each with its characteristic surface manifestations—is preserved in the landscape. These are the product of regional uplifts that have generated new pulses of erosion. The occurrence within the same landscape of surfaces of different ages is, as King argued as early as 1949, possible only if the dominant mode of landscape evolution is the formation and lateral recession of escarpments, rather than surface downwearing. This model of African landscape evolution has been challenged in several recent publications, including Fleming et al. (1999), Brown et al. (2002) and van der Beck et al. (2002). Because their theoretical approach to landscape studies has a significant following, it is important that attention should be drawn to its incompatibility with the large body of evidence reviewed in this chapter. These authors base their opposition on data from apatite fission-track analysis and numerical modeling of landscape development. The limitations of both techniques are well-known. For example, Burke (1996) stresses that the low geothermal gradients that characterize the cratonic areas of Africa make apatite fission-track analysis a poor tool for distinguishing postrifting events, and the validity of results from numerical models that purport to predict the dominant mode of landscape evolution are overwhelmingly dependent on the "erodability constant" (L_f) that is assigned by the operator. The widespread occurrence of hard capping layers in the African landscape that ensure the long-term survival of escarpments makes this a particularly difficult challenge. But the problem runs much deeper than one of technical limitations and preferences. Moore and Blenkinsop (2006) point out that the results put forward by these authors (many of whom build on the earlier views of Gilchrist and Summerfield, 1990, 1994), run counter to a substantial body of data available from within the landscapes themselves, including continent-wide evidence for the former extent of resistant cap rocks, the coexistence of surfaces of demonstrably different age, and the uplift of areas associated with well-dated marine deposits. It is a pity that facts derived from informed field observation are being ignored by these workers in favor of highly subjective output from currently voguish models.

THE ARID LANDSCAPES
OF THE NORTH AND SOUTHWEST

Although the deserts of Africa are relatively young (the Sahara more so than the Namib), it seems appropriate to consider the evolution of their landscapes separately. As has already been noted, continental deposits first began to accumulate in the Sahara from the Eocene onward. These early sequences include bauxite and lateritic duricrusts as far north as 23° (in contrast to their present limit of 8°), reflecting both the subsequent northward drift of Africa and Cenozoic climate change (Le Houérou, 1997). By the Oligocene, the climate had dried sufficiently in some areas for the development of a savanna vegetation, but the first true aridity appeared in the mid- to Upper Pliocene. By the end of the Pliocene, major sand seas had begun to develop, and the formation of calcrete and gypsum duricrusts was widespread (Le Houérou, 1997).

Over large areas of the Sahara, the Upper Cretaceous Nubian Sandstone is sufficiently resistant to sustain prominent escarpments. Some of these escarpments flank silcrete-capped remnants of the African Surface. Around the Tibesti and Hoggar mountains, two pedimented surfaces are preserved below these escarpments: the upper is covered by penecontemporaneous basalts dated to 8.4–7.9 Ma, while the lower is covered by Plio-Pleistocene ignimbrites (Busche, 1976).

A unique feature of the Sahara is the occurrence, over an area of some 2.5 million km², of large-scale rotational landslides along scarp edges (Busche, 2001). These landslide fringes are up to 3 km wide and are sometimes continuous over hundreds of kilometers. Busche considers these features to be a contemporaneous response (probably of late Pliocene age) to the wetting and weakening of soft strata beneath the caprock during a markedly more humid interval.

The Namib Desert, spanning the coasts of southern Angola, Namibia, and the northern part of the Atlantic seaboard of South Africa, mostly coincides with a subaerial erosional platform cut below the Great Escarpment in response to early Miocene uplift of the African Superswell. The persistence of hyperarid conditions, virtually from the time of formation of this surface, is confirmed by the occurrence of a thick, cemented aeolian sequence (the Tsondab Sandstone) in the central Namib. Eggshells of extinct giant avians, the oldest dating to 16 Ma, are preserved within these deposits. Uplift of the area in Mio-Pliocene times triggered the cutting of gorges across the Namib by rivers originating in better-watered areas to the east (e.g., the Kuiseb Canyon). As in parts of the Sahara, the central Namib is today occupied by a sand sea that includes some of the highest mobile dune ridges in the world.

Literature Cited

Adamson, D. A., and M. A. J. Williams. 1987. Geological setting of Pliocene rifting and deposition in the Afar depression of Ethiopia. *Journal of Human Evolution* 16:597–610.

Baker, B. H. 1986. Tectonics and volcanism of the southern Kenya Rift Valley and its influence on rift sedimentation; pp. 45–57 in L. E. Frostick, R. W. Renaut, I. Reid, and J. J. Tiercelin (eds.), *Sedimentation in the African Rifts*. Special Publication of the Geological Society of London, 25.

Baker, B. H., J. G. Mitchell, and L. A. J. Williams. 1988. Stratigraphy, geochronology and volcano-tectonic evolution of the Kdong-Naivasha-Kinangop region, Gregory Rift Valley, Kenya. *Journal of the Geological Society of London* 145:107–116.

Baker, B. H., P. A. Mohr, and L. A. J. Williams. 1972. *The Geology of the Eastern Rift System of Africa*. Geological Society of America Special Paper No. 136, 67 pp.

Brown, R. W., M. A. Summerfield, and A. J. W. Gleadow. 2002. Denudational history along a transect across the Drakensberg Escarpment of southern Africa derived from apatite fission-track thermochronology. *Journal of Geophysical Research* 107:(B12) 2350, DOI:10.1029/2001 JB 000745.

Burke, K. 1996. The African plate. *South African Journal of Geology* 99:341–409.

Busche, D. 1976. Pediments and climate. *Palaeoecology of Africa* 9:20–24.

Busche, D. 2001. Early Quaternary landslides and their significance for geomorphic and climatic history. *Journal of Arid Environments* 49:429–448.

Caby, R. 2003. Terrane assembly and geodynamic evolution of central-western Hoggar: A synthesis. *Journal of African Earth Sciences* 37:133–159.

Calais, E., C. Hartnady, C. Ebinger, and J. M. Nocquet. 2006. Kinematics of the East African Rift from GPS and earthquake slip vector data; pp. 9–22 in G. Yirgu, C. J. Ebinger, and P. K. H. Maguire (eds.), *Evolution of the Rift Systems with the Afar Volcanic Province, Northeast Africa*. Special Publication of the Geological Society of London, 259.

Colin, F., A. Beauvais, G. Ruffet, and O. Hénocque. 2005. First ⁴⁰Ar/³⁹Ar geochronology of lateritic manganiferous pisolites: Implications for the Paleogene history of a West African landscape. *Earth and Planetary Science Letters* 238:172–188.

Corvinus, G., and Q. B. Hendey. 1978. A new Miocene locality at Arrisdrift in Namibia (South West Africa). *Neues Jahrbuch für Geologie und Paläontologie, Monatshefte* 4:193–205.

de Wit, M. C. J. 1993. Cainozoic Evolution of Drainage Systems in the North-western Cape. Unpublished Ph.D. dissertation, University of Cape Town, 348 pp.

de Wit, M. C. J. 1999. Post-Gondwana drainage and the development of diamond placers in western South Africa. *Economic Geology* 94:721–740.

de Wit, M. J., and I. G. D. Ransome. 1992. Regional inversion tectonics along the southern margin of Gondwana; pp. 15–21 in M. J. de Wit and I. G. D. Ransome (eds.), *Inversion Tectonics of the Cape Fold Belt, Karoo and Cretaceous Basins of Southern Africa*. A. A. Balkema, Rotterdam, 269 pp.

Denys, C., J. Chorowitz, and J. J. Tiercelin. 1986. Tectonic and environmental control on rodent diversity in the Plio-Pleistocene sediments of the African Rift System; pp. 363–372 in L. E. Frostick, R. W. Renaut, I. Reid, and J. J. Tiercelin (eds.), *Sedimentation in the African Rifts*. Special Publication of the Geological Society of London, 25.

Dingle, R. V., W. G. Siesser, and A. R. Newton. 1983. *Mesozoic and Tertiary Geology of Southern Africa*. A. A. Balkema, Rotterdam, 375 pp.

Dixey, F. 1956. *The East African Rift System*. Supplementary Bulletin No. 1, Overseas Geological Mineral Resources, H.M. Stationery Office, London, 71 pp.

Drury, S. A., S. P. Kelley, S. M. Berhe, R. E. L. Collier, and M. Abraha. 1994. Structures related to the Red Sea evolution in northern Eritrea. *Tectonics* 13:1371–1380.

du Toit, A. L. 1933. Crustal movement as a factor in the geographical evolution of South Africa. *South African Geographical Journal* 16:3–20.

Ebinger, C. J. 1989. Tectonic development of the western branch of the East African Rift System. *Bulletin of the Geological Society of America* 101:885–903.

Ebinger, C. J., A. L. Deino, R. E. Drake, and A. L. Tesha. 1989. Chronology of volcanism and rift propagation: Rungwe volcanic province, East Africa. *Journal of Geophysical Research* 94:15785–15803.

Ebinger, C. J., T. Yemane, D. J. Harding, S. Tesfaye, S. Kelley, and D. C. Rex. 2000. Rift deflection, migration and propagation: Linkage of the Ethiopian and Eastern rifts, Africa. *Bulletin of the Geological Society of America* 112:163–176.

Fairhead, J. D. 1986. Geophysical controls on sedimentation within the African Rift Systems; pp. 19–27 in L. E. Frostick, R. W. Renaut, I. Reid, and J. J. Tiercelin (eds.), *Sedimentation in the African Rifts*. Special Publication of the Geological Society of London, 25.

Fleming, A., M. A. Summerfield, J. O. Stone, L. K. Fifield, and R. G. Creswell. 1999. Denudation rates for the southern Drakensberg escarpment, S.E. Africa, derived from in-situ produced cosmogenic ³⁶Cl: Initial results. *Journal of the Geological Society of London* 156:209–212.

Foster, D. A., and J. W. Gleadow. 1996. Structural framework and denudation history of the flanks of the Kenya and Anza rifts, East Africa. *Tectonics* 15:258–271.

Frizon de Lamotte, D., B. Saint Bezar, R. Bracène, and E. Mercier. 2000. The two main steps of Atlas building and geodynamics of the western Mediterranean. *Tectonics* 19:740–761.

Gilchrist, A. R., and M. A. Summerfield. 1990. Differential denudation and flexural isostasy in formation of rifted margins. *Nature* 346:739–742.

Gilchrist, A. R., and M. A. Summerfield. 1994. Tectonic models of passive margin evolution and their implications for theories of long-term landscape development; pp. 58–84 in M. J. Kirkby (ed.), *Process Models and Theoretical Geomorphology*. John Wiley, New York.

Giresse, P. 2005. Mesozoic-Cenozoic history of the Congo Basin. *Journal of African Earth Sciences* 43:301–315.

Guiraud, R., and Y. Bellion. 1995. Late Carboniferous to Recent geodynamic evolution of the West Gondwanian cratonic Tethyan margins; pp. 101–124 in A. Nairn, J. Dercourt, and B. Vrielynck (eds.), *The Ocean Basins and Margins, Volume 8: The Tethys Ocean.* Plenum Press, New York.

Guiraud, R., and W. Bosworth. 1997. Senonian basin inversion and rejuvenation of rifting in Africa and Arabia: Synthesis and implications to plate-scale tectonics. *Tectonophysics* 282:39–82.

Guiraud, R., and W. Bosworth. 1999. Phanerozoic geodynamic evolution of northeastern Africa and the northwestern Arabian platform. *Tectonophysics* 315:73–108.

Guiraud, R., W. Bosworth, J. Thierry, and A. Delplanque. 2005. Phanerozoic geological evolution of Northern and Central Africa: An overview. *Journal of African Earth Sciences* 43:83–143.

Hartley, R., A. B. Watts, and J. D. Fairhead. 1996. Isostasy of Africa. *Earth and Planetary Science Letters* 137:1–18.

Hartnady. C. J. H. 1990. Seismicity and plate boundary evolution in southeastern Africa. *South African Journal of Geology* 93:473–484.

Holmes, A. 1944. *Principles of Physical Geology.* Thomas Nelson and Sons, Edinburgh, 532 pp.

Hudec, M. R., and M. P. A. Jackson. 2002. Structural segmentation, inversion and salt tectonics on a passive margin: Evolution of the Inner Kwanza Basin, Angola. *Bulletin of the Geological Society of America* 114:1222–1244.

King, L. C. 1949. On the ages of African land-surfaces. *Quarterly Journal of the Geological Society of London* 104:439–453.

Konate, M., M. Guiraud, M. Lang, and M. Yahaya. 2003. Sedimentation in the Kandi extensional basin (Benin and Niger): Fluvial and marine deposits related to the Late Ordovician deglaciation in West Africa. *Journal of African Earth Sciences* 36:185–206.

Lavier, L. L., M. S. Steckler, and F. Brigaud. 2000. An improved method for reconstruction of the stratigraphy and bathymetry of continental margins: Application to the Cenozoic tectonic and sedimentary history of the Congo margin. *Bulletin of the American Association of Petroleum Geologists* 84:923–939.

Lavier, L. L., M. S. Steckler, and F. Brigaud. 2001. Climatic and tectonic control on the Cenozoic evolution of the West African margin. *Marine Geology* 178:63–80.

Le Houérou, H. N. 1997. Climate, flora and fauna changes in the Sahara over the past 500 million years. *Journal of Arid Environments* 37:619–647.

Lithgow-Bertelloni, C., and P. G. Silver, 1998. Dynamic topography, plate driving forces and the African Superswell. *Nature* 395:269–272.

Lunde, G., K. Aubert, O. Lauritzen, and E. Lorange. 1992. Tertiary uplift of the Kwanza basin in Angola; pp. 99–117 in R. Curnell (ed.), *Géologie africaine.* Elf Aquitaine, Boussens, France.

Mohr, P. A. 1987. Structural style of continental rifting in Ethiopia: Reverse décollements. *Eos (American Geophysical Union Transactions)* 68:721–729.

Moore, A., and T. Blenkinsop. 2002. The role of mantle plumes in the development of continental-scale drainage patterns: The southern African example revisited. *South African Journal of Geology* 105:353–360.

Moore, A., and T. Blenkinsop. 2006. Scarp retreat vs. pinned drainage in the formation of the Drakensberg escarpment, southern Africa. *South African Journal of Geology* 109:599–610.

Nyblade, A. A., and S. W. Robinson. 1994. The African Superswell. *Geophysical Research Letters* 21:765–768.

Partridge, T. C. 1998. Of diamonds, dinosaurs and diastrophism: 150 million years of landscape evolution in southern Africa. *South African Journal of Geology* 101:167–184.

Partridge, T. C., and R. R. Maud. 1987. Geomorphic evolution of southern Africa since the Mesozoic. *South African Journal of Geology* 90:179–208.

Partridge, T. C., and R. R. Maud. 2000. Macro-scale geomorphic evolution of southern Africa; pp. 3–18 in T. C. Partridge and R. R. Maud (eds.), *The Cenozoic of Southern Africa.* Oxford University Press, New York.

Partridge, T. C., B. A. Wood, and P. B. deMenocal. 1995. The influence of global climatic change and regional uplift on large-mammalian evolution in East and southern Africa; pp. 331–355 in E. S. Vrba, G. H. Denton, T. C. Partridge, and L. H. Burckle (eds.), *Paleoclimate and Evolution with Special Emphasis on Human Origins.* Yale University Press, New Haven, CT.

Pickford, M., B. Senut, and D. Hadoto. 1993. *Geology and Palaeobiology of the Albertine Rift Valley, Uganda-Zaire. Volume 1: Geology.* International Centre for Training and Exchanges in the Geosciences (CIFEG), Occasional Paper 1993/24. Orleans, France, 190 pp.

Pickford, M., D. Gommery, J. Morales, D. Soria, M. Nieto, and J. Ward. 1996. Preliminary results of new excavations at Arrisdrift, Middle Miocene of southern Namibia. *Comptes Rendus de l'Académie des Sciences, Paris* 322:991–996.

Potter, P. E., and W. K. Hamblin. 2006. *Big Rivers Worldwide.* Brigham Young University Geology Studies 48, Provo, Utah, 78 pp.

Ruddiman, W. F., M. E. Raymo, D. G. Martinson, B. M. Clement, and J. Backman. 1989. Pleistocene evolution: Northern Hemisphere ice sheets and North Atlantic Ocean. *Palaeoceanography* 4:353–412.

Saggerson, E. P., and B. H. Baker. 1965. Post-Jurassic erosion surfaces in eastern Kenya and their deformation in relation to rift structure. *Quarterly Journal of the Geological Society of London* 121:51–72.

Said, R. 1981. *The Geological Evolution of the River Nile.* Springer-Verlag, Berlin, 151 pp.

Séranne, M., and Z. Anka. 2005. South Atlantic continental margins of Africa: A comparison of the tectonic vs. climate interplay on the evolution of equatorial west Africa and S. W. Africa margins. *Journal of African Earth Sciences* 43:283–300.

Shackleton, R. M. 1978. Structural development of the East African rift system; pp. 20–28 in W. W. Bishop (ed.), *Geological Background to Fossil Man.* Scottish Academic Press, Edinburgh.

Sletlene, L., L. E. Wilcox, R. S. Blouse, and J. R. Sanders. 1973. A Bouguer gravity map of Africa. DMAAC Technical Paper 73-003, Defense Mapping Agency, St. Louis, MO.

Unrug, R. 1996. The assembly of Gondwanaland: Scientific results of IGCP 288, Gondwanaland sutures and mobile belts. *Episodes* 19:11–20.

van der Beek, P., M. A. Summerfield, J. Braun, R. W. Brown, and A. Fleming. 2002. Modelling post-breakup landscape development and denudational history across the southeastern African (Drakensberg Escarpment) margin. *Journal of Geophysical Research* 107:(B12) 2350, DOI: 10.1029/2001 JB 000744.

van Niekerk, H. S., N. J. Beukes, and J. Gutzmer. 1999. Post-Gondwana pedogenic ferromanganese deposits, ancient soil profiles, African land surfaces and palaeoclimatic change on the Highveld of South Africa. *Journal of African Earth Sciences* 29:761–781.

Veldkamp, A., E. Buis, J. R. Wijbrans, D. O. Olago, E. H. Boshoven, M. Maree, and R. M. van den Berg van Saparoea. 2007. Late Cenozoic fluvial dynamics of Tana River, Kenya: An uplift dominated record. *Quaternary Science Reviews* 26:2897–2912.

Walgenwitz, F., J. P. Richert, and P. Charpentier. 1992. Southwest border of African plate–thermal history and geodynamical implications; pp. 234–254 in W. Poag and P. de Graciansky (eds.), *Geological Evolution of the Atlantic Continental Rises.* Van Nostrand Reinhold, New York.

Walgenwitz, F., M. Pagel, A. Mayer, H. Maluski, and P. Monie. 1990. Thermo-chronological approach to reservoir diagenesis in the offshore Angola Basin. *Bulletin of the American Association of Petroleum Geologists* 74:547–569.

TWO

Chronology of Paleogene Mammal Localities

ERIK R. SEIFFERT

The Afro-Arabian fossil record of Paleogene mammalian evolution is poor when compared with those from northern continents, but it is steadily improving. Age constraints on early African mammal localities have historically been very poor as well, and this problem has allowed for remarkably different chronological interpretations (e.g., Rasmussen et al., 1992; Godinot, 1994; Gheerbrant et al., 1998). Fortunately, significant biochronological improvements have been made over the course of the last half-decade, but correlation of Paleogene records to the global timescale remains inadequate when compared with the relatively solid biochronostratigraphic framework that has been developed for the African Neogene. The earliest part of the African record is particularly difficult to interpret, as it consists largely of isolated occurrences from poorly dated horizons; taxonomy is often highly speculative due to inadequate material, and practically nothing is known about species ranges. The quality of the Paleogene record is best in its later third, thanks to decades of work in the late Eocene (Priabonian) and early Oligocene (Rupelian) deposits of northern Egypt (Simons et al., 2008). Recent paleontological exploration in other parts of Africa (Kappelman et al., 2003; Stevens et al., 2008; Rasmussen and Gutiérrez, 2009), combined with reanalysis of chronological data from Egypt and Oman (Seiffert, 2006; Seiffert et al., 2008), has helped to make the Oligocene the most thoroughly sampled Paleogene epoch on the continent. Mammalian collections from the earliest Eocene, and the early-middle Eocene boundary, are also being actively augmented by ongoing work (Gheerbrant et al., 2003; Adaci et al., 2007), but there remains a large gap in the Afro-Arabian record during the Lutetian-Bartonian (between ~45 and 37 Ma), and there are no placental-bearing localities on the landmass that are older than late Paleocene.

Latest Paleocene–Earliest Eocene Sites

The oldest placental mammal-bearing fossil localities in Afro-Arabia— the Sidi Chennane quarries of the Ouled Abdoun Basin in Morocco, and Adrar Mgorn 1, Adrar Mgorn 1bis, Ihadjamène, Ilimzi, and Talazit, in the Ouarzazate Basin in Morocco—are estimated to be late Paleocene (Thanetian) in age (figure 2.1) based on combined evidence from magnetostratigraphy and selachians (Gheerbrant et al., 1993b, 1998; Solé et al., 2009 see also table 2.1). The Sidi Chennane sites have yielded remains of relatively large mammals, such as the hyaenodontid *Lahimia* (Solé et al., 2009) and the basal proboscidean *Eritherium*

(Gheerbrant, 2009), while the Ouarzazate sites have produced a diverse assemblage of micromammals (Gheerbrant, 1992; Gheerbrant et al., 1998; Seiffert, this volume, chap. 28) and occasional remains of larger species, including the primate *Altiatlasius* (Sigé et al., 1990). The Ouarzazate localities fall within a zone of reversed polarity that has been correlated with Chron C24r, which spans the Paleocene-Eocene boundary; therefore the latest Paleocene estimate (i.e., 56.5–55.8 Ma) for the sites hinges on the biostratigraphic evidence provided by the selachian fauna (e.g., the absence of carcharhinids such as *Physogaleus*). Mammals from younger horizons in the Ouled Abdoun basin, are estimated to be earliest Eocene based on selachians (Gheerbrant et al., 2003), and they include proboscideans (Gheerbrant et al., 2002, 2005), a hyracoid similar to *Seggeurius amourensis* (Gheerbrant et al., 2003), a primitive hyaenodontid creodont (Gheerbrant et al., 2006), and three enigmatic mammals—*Abdounodus*, *Ocepeia* (Gheerbrant et al., 2001), and another alleged "condylarth" represented by a partial mandibular ramus bearing a single damaged tooth (Gheerbrant et al., 2003). The locality N'Tagourt 2 extends the micromammal record in the Ouarzazate Basin into the early Eocene (Ypresian) (Gheerbrant, 1993); the site has also produced a tethythere, *Khamsaconus* (Sudre et al., 1993). N'Tagourt 2 falls within a zone of reversed polarity that has been correlated with C23r (Gheerbrant et al., 1998), suggesting that it is about 52–52.5 Ma in age.

El Kohol, in northern Algeria, is likely to be early Eocene in age but is probably slightly younger than the Ouled Abdoun sites. The site is best known for having produced a large sample of the primitive proboscidean *Numidotherium* (Mahboubi et al., 1986), which is clearly quite derived when compared with earliest Eocene *Daouitherium* (Gheerbrant et al., 2002) and *Phosphatherium* (Gheerbrant et al., 2005) from Ouled Abdoun. El Kohol has also produced the primary record of the basal hyracoid *Seggeurius* (Court and Mahboubi, 1993), which, as noted earlier, may be present at Ouled Abdoun as well (Gheerbrant et al., 2003), and the "insectivore" *Garatherium*, which is also known from the late Paleocene and early Eocene of the Ouarzazate Basin (Gheerbrant et al., 1998; Seiffert, this volume, chap. 28). Rodents and primates are notably absent at El Kohol, although this may simply be due to inadequate sampling of the micromammal fauna. Charophytes from El Kohol include *Nitellopsis (Tectochara) dutempleii* and *Peckichara* aff. *disermas*, which appear to be consistent with an early to mid-Ypresian age (Mebrouk et al., 1997).

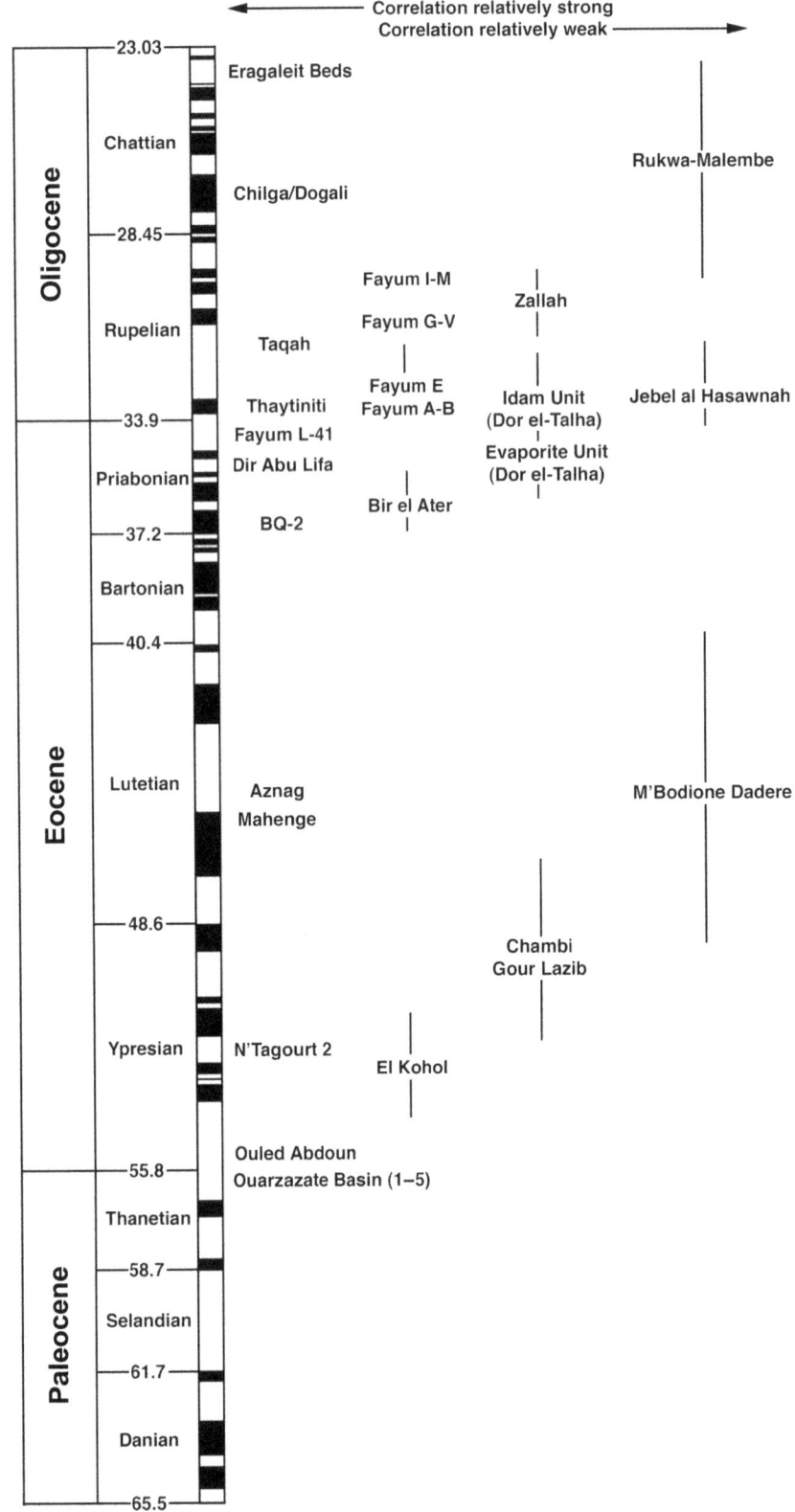

FIGURE 2.1 Approximate correlation of terrestrial mammal localities to the global timescale (timescale based on Gradstein et al., 2004). Localities in the first column are temporally well constrained, while the correlation of localities further to the right is progressively less clear. Vertical bars indicate approximate ranges of temporal uncertainty. "Ouarzazate Basin (1-5)" includes the latest Paleocene localities Adrar Mgorn 1, Adrar Mgorn 1bis, Ihadjamène, Ilimzi, and Talazit, in the Ouarzazate Basin of Morocco. The Malian localities In Tafidet and Tamaguilelt are not shown as they are of uncertain temporal position within the Eocene.

TABLE 2.1

Approximate chronological ordering (from youngest at top to oldest at bottom)
of terrestrial mammal localities and the lines of evidence that have been used for age estimates

Locality			Country	Radiometric	Magnetostratigraphy	Foraminifera	Charophytes	Selachians	Mammals
Lothidok/Nakwai			Kenya	X					X
Dogali			Eritrea	X					
Chilga			Ethiopia	X	X				X
Rukwa →			Tanzania						X
Malembe →			Angola	X					X
Jebel Qatrani I–M			Egypt		X				X
Jebel Qatrani G–V			Egypt		X				X
Zallah			Libya			X			X
Taqah			Oman		X	X			X
Jebel Qatrani E			Egypt		X				X
Jebel el Hasawnah			Libya						X
Thaytiniti			Oman		X	X			X
Jebel Qatrani A–B			Egypt		X				X
Idam Unit →			Libya						X
Jebel Qatrani L–41			Egypt		X				X
Evaporite Unit			Libya						X
Dir Abu Lifa			Egypt		X	X			X
Bir el Ater			Algeria						X
Birket Qarun			Egypt		X	X			X
M'Bodione Dadere			Senegal						X
In Tafidet			Mali						X
Tamaguilelt →			Mali						X
Aznag			Morocco			X		X	X
Mahenge			Tanzania	X					
Chambi			Tunisia				X		X
Gour Lazib			Algeria				X		X
El Kohol			Algeria				X		X
N'Tagourt 2			Morocco		X		X	X	X
Ouled Abdoun 1			Morocco					X	X
Ouarzazate LP			Morocco		X			X	X
Ouled Abdoun 2			Morocco					X	

NOTE: "Ouled Abdoun 1" includes the Grand Daoui quarries; "Ouled Abdoun 2" includes the Sidi Chennane quarries; "Ouarzazate LP" includes the latest Paleocene localities Adrar Mgorn 1, Adrar Mgorn 1bis, Ihadjamène, Ilimzi, and Talazit, in the Ouarzazate Basin of Morocco. Dashed lines in second column indicate which localities are of approximately the same age; arrows to the right of some localities point to column showing broader correlation.

Sites from Near the Early-Middle Eocene Boundary

The Gour Lazib complex in the Hammada du Dra region of western Algeria has recently produced a diverse mammalian fauna, most of which is undescribed at present. A preliminary faunal list has been made available (Adaci et al., 2007). These new collections should significantly advance our understanding of early Cenozoic mammalian evolution in Afro-Arabia. On the basis of the older and far less complete collections from the area (Sudre, 1979), it had previously been suggested that the Gour Lazib mammals might be contemporaneous with part of the late Eocene–early Oligocene Jebel Qatrani Formation in northern Egypt (Rasmussen et al., 1992), but it is now clear that the Fayum mammal faunas, and those from Gour Lazib, are significantly different. For instance, the Gour Lazib sites have produced first appearances of azibiid primates, zegdoumyid rodents, and a strange mammal called *Helioseus*, none of which have been found in either the Jebel Qatrani or older Birket Qarun deposits of Egypt. The Gour Lazib sites appear to document a time period prior to the arrival of the rodent clade Hystricognathi, which is first recorded in the earliest late Eocene of Egypt (Sallam et al. 2009). The late Eocene-to-Recent anomaluroid rodents of Afro-Arabia (Jaeger et al., 1985) might be descended from the more primitive zegdoumyids (Vianey-Liaud and Jaeger, 1996; Marivaux et al., 2005). Adaci et al. (2007) report three new primate genera from the Gour Lazib complex in addition to the already described *Algeripithecus*, *Azibius*, *Dralestes*, and *Tabelia* (the latter two genera are now considered invalid; see Tabuce et al., 2009), but they do not assign these taxa to a lower taxonomic rank, which leaves their biochronological significance uncertain.

An important independent source of chronological information is the charophyte assemblage from the Gour Lazib area, which has been interpreted as supporting an early or early-middle Eocene age for the mammals (Mebrouk et al., 1997; Mebrouk and Feist, 1999). With new collections an early Eocene estimate has gained new support through the combined presence of *Maedleriella lavocati*, *Maedleriella cristellata*, *Stephanocaris acris*, and *Nitellopsis* (*Tectochara*) *thaleri* (Adaci et al., 2007)—a species for which a late Ypresian charophyte zone exists (Gradstein et al., 2004). The biochronological significance of *Raskyella peckii meridionale*, which was previously interpreted as supporting a mid-Ypresian age (Mebrouk et al., 1997) has been weakened, however, by its presence in the Lutetian-early Bartonian Peguera Formation of southwest Mallorca (Martín-Closas and Ramos, 2005).

The Gour Lazib hyracoids have been the primary source for biochronological debate among specialists (Rasmussen et al., 1992; Godinot, 1994; Tabuce et al., 2001). The genera *Megalohyrax*, *Microhyrax*, *Titanohyrax*, possibly *Bunohyrax* and *Thyrohyrax*, and a new genus have been reported from the area (Adaci et al., 2007), but all of the described fossils aside from those of *Microhyrax* are isolated teeth, and their assignment to the genera *Megalohyrax* and *Titanohyrax* (otherwise known from the latest Eocene to the late Oligocene) is dubious at present. The only genus that is represented by sufficient material (partial mandibles, a maxilla, and postcranial elements, Tabuce et al., 2007) is *Microhyrax*, which clearly is a very primitive form that falls outside the clade containing genera from the Jebel Qatrani Formation

(Seiffert, 2007b). *Titanohyrax mongereaui*, which is now known from Gour Lazib and Chambi in Tunisia (Court and Hartenberger, 1992), does bear similarities to titanohyracids from the Jebel Qatrani Formation, but it is much smaller and clearly more primitive. If *T. mongereaui* is a titanohyracid, it documents a very early phase in the group's evolution. Adaci et al. (2007) report the discovery of cranial material and upper and lower jaws of "*Megalohyrax*" *gevini* from the Glib Zegdou Formation that should help to clarify that species' phylogenetic affinities.

The Chambi locality in Tunisia has produced the charophyte *Raskyella* cf. *sahariana*, apparently very similar to *Raskyella sahariana* from the middle member of the Glib Zegdou Formation (Mebrouk et al., 1997; Adaci et al., 2007) and, as noted, the hyracoid *Titanohyrax mongereaui*. Zegdoumyid rodents (Vianey-Liaud et al., 1994) and a primate similar to *Algeripithecus* (Godinot and Mahboubi, 1994; Seiffert et al., this volume, chap. 22) are also present, indicating that the Gour Lazib and Chambi localities are probably similar in age, and definitely older than the oldest Fayum mammals. The presence of adapiform (Court, 1992) primates at Chambi—which have not yet been documented in the Gour Lazib complex—could indicate a slightly younger age for the Tunisian site.

Problematic West African Localities

Two Eocene mammal localities in Mali, In Tafidet and Tamaguilelt, have proven very difficult to date. The In Tafidet site has produced remains of *Moeritherium*, which otherwise makes its first appearance in the transitional middle-late Eocene sediments of the Fayum Depression, but the genus likely has a long evolutionary history extending back well into the middle Eocene, and so it is of limited biochronological utility. The age of In Tafidet has been estimated as late Eocene (Arambourg et al., 1951) or middle Eocene (Gheerbrant et al., 1998). A condensed phosphate conglomerate at Tamaguilelt has produced a hyracoid incisor and a few edentulous mandibles that have been interpreted as proboscidean (O'Leary et al., 2006); the layer that produced the fossils has been identified as early Eocene (Patterson and Longbottom, 1989; Moody and Sutcliffe, 1993; O'Leary et al., 2006) or middle Eocene (Gheerbrant et al., 1998) in age. Well cuttings from near the village of M'Bodione Dadere in Senegal yielded half a molar of *Moeritherium* and two fragments of smaller "condylarth" mammal teeth (Gorodiski and Lavocat, 1953; Sudre, 1979). Based on associated bivalves, the mammals are estimated to be middle Eocene (Lutetian) in age (Gorodiski and Lavocat, 1953; Gheerbrant et al., 1998).

Aznag and Mahenge

The Aznag locality in the Ouarzazate Basin of Morocco, and the Mahenge locality in Tanzania, are the only definitively middle Eocene (Lutetian) land mammal sites in Africa, based on foraminifera (Tabuce et al., 2005) and a U-Pb date (Gunnell et al., 2002), respectively. It is unfortunate, given how poorly sampled the middle Eocene is in Afro-Arabia, that neither locality has been very productive. Mahenge has only produced a single bat, *Tanzanycteris* (Gunnell et al., 2002), which is unlike all other known African bats, while Aznag has produced 15 isolated mammal teeth, most of which are broken (Tabuce et al., 2005). Among the Aznag teeth is a bat molar, a rodent incisor, two primate molar fragments, an "insectivore" molar and premolar, and two molars that have been interpreted as "condylarthran." It is difficult to draw any meaningful conclusions from this sample.

Late Eocene Sites

After a long gap spanning almost 10 million years, through the upper Lutetian and the entire Bartonian, the detailed record of Afro-Arabian mammal evolution picks up again at the base of the Priabonian (~37 Ma), with the oldest (Birket Qarun) terrestrial mammal faunas in the Fayum succession (Seiffert et al., 2008). The Birket Qarun fauna includes the first records of hystricognathous rodents in Afro-Arabia, and the oldest ptolemaiids, lorisiform primates, and definitive anomaluroid rodents. Other members of the fauna document intermediate stages in the evolution of endemic clades: two derived proboscideans are present (*Moeritherium* and a small species of *Barytherium*), a hyracoid that is intermediate in morphology between *Microhyrax* and *Seggeurius* and the Jebel Qatrani genera, a primitive herodotiine, and hyaenodontid creodonts. Primates, including anthropoids (Seiffert et al., 2005), are the most common members of the fauna. The earliest Priabonian estimate for the Birket Qarun localities is based on magnetostratigraphy (Seiffert et al., 2005, 2008; Seiffert, 2006), sequence stratigraphy (Gingerich, 1992; Seiffert et al., 2005), and foraminifera from adjacent rocks (Strougo and Haggag, 1984; Haggag, 1985; Strougo and Boukhary, 1987; Gingerich, 1992). The zone of normal polarity recorded at the primary locality, BQ-2, has been correlated with Chron C17n.1n (Seiffert et al., 2005; Seiffert, 2006).

The mammals from the Bir el Ater (or Nementcha) locality (Coiffait et al., 1984) in northern Algeria are very similar to those from the Birket Qarun localities, but the Algerian locality is estimated to be slightly younger based on the first recorded appearance of Anthracotheriidae, an immigrant artiodactyl group from Asia. The records of the proboscidean *Moeritherium*, the anthropoid primate *Biretia*, and primitive anomaluroid and hystricognathous rodents are most similar to those from Birket Qarun Locality 2 (BQ-2) in the Fayum.

The few records of terrestrial or amphibious mammals from the mid-Priabonian Dir Abu Lifa Member of the Qasr el-Sagha Formation in the Fayum area (*Moeritherium*, *Barytherium grave*, anthracotheriids) resemble those from the Evaporite Unit in the Dor el-Talha area in Libya (Savage, 1969; Rasmussen et al., 2008). The two horizons are probably of about the same age, although it is conceivable that the Evaporite Unit correlates with part of the lower sequence of the Jebel Qatrani Formation. The Idam Unit exposed at Dor el-Talha very likely correlates with some part of the Jebel Qatrani lower sequence; in Egypt, the oldest definitive records of taxa that are also recovered from the Idam Unit (the proboscideans *Palaeomastodon* and *Phiomia*, and the embrithopod *Arsinoitherium*) occur just above an unconformity (above Quarry L-41) that has been attributed to earliest Oligocene nearshore erosion (Seiffert, 2006), and so it is possible that the Idam Unit is early Oligocene in age.

Quarry L-41 in the Jebel Qatrani Formation appears to be of terminal Eocene age (Seiffert, 2006). This locality marks the last appearance in northern Africa of galagid lorisiforms, plesiopithecids, djebelemurines, adapids, and herodotiines, and the first appearance of the rodent *Gaudeamus* phiomyid, and phiocricetomyine rodents, derived macroscelideans (*Metoldobotes*), and the first definitive records of the hyracoid genera *Antilohyrax*, *Bunohyrax*, *Megalohyrax*, *Saghatherium*, *Thyrohyrax*, and *Titanohyrax*. Quarry L-41 falls within a zone of reversed polarity (Kappelman et al., 1992) that has been correlated with Chron C13r (Seiffert, 2006).

Early Oligocene Sites

It has been argued that a major unconformity just above Quarry L-41 is due to erosion associated with the rapid drop in sea level that occurred in the earliest Oligocene (Seiffert, 2006). The oldest productive mammal sites above this unconformity are Quarries A and B, which mark the definitive first appearances of the proboscideans *Palaeomastodon* and *Phiomia*, the embrithopod *Arsinoitherium*, and the rodent genus *Phiomys*. Embrithopods might be early Eocene immigrants, given their first Afro-Arabian appearance in the middle Eocene and presence on northern continents earlier in the Eocene (Radulescu et al., 1976; Sen and Heintz, 1979; Maas et al., 1998; Pickford et al., 2008; Sanders et al., this volume chp. 12). The younger Jebel Qatrani Quarry E shares a number of mammalian taxa with Quarries A and B, but has produced more small mammals, including primates (Simons and Kay, 1983, 1988; Simons et al., 1987; Rasmussen and Simons, 1988; Holroyd, 1994; Seiffert et al., 2007).

A very similar mammal assemblage to those from Quarries A, B, and E has been recovered from the Thaytiniti locality in Oman in association with the foraminiferan *Nummulites fichteli*, an early Oligocene index fossil (Roger et al., 1993; Serra-Kiel et al., 1998; Thomas et al., 1999). The zone of normal polarity at the Thaytiniti locality has recently been correlated with Chron C13n (Seiffert, 2006). The presence of the hyracoid *Saghatherium antiquum* at the Jebel Hasawnah locality in Libya (Gheerbrant et al., 2007), a taxon that is also present at Quarries A, B, and E, suggests that that locality is early Oligocene (perhaps 31–33 Ma) in age.

Localities near the base of the Jebel Qatrani upper sequence, in particular Quarries G and V, record the first appearance of derived propliopithecid catarrhine and parapithecine parapithecid anthropoids and the rodent *Metaphiomys*, among other taxa, and the last appearance of the *Saghatherium*-*Selenohyrax* hyracoid lineage. The sediments at Quarry G and V are both of reversed polarity and have been correlated with the upper part of Chron C12r (Seiffert, 2006). The localities are probably about 31 million years old. The Taqah locality in Oman combines taxa typical of the Jebel Qatrani upper sequence (parapithecine parapithecids ["cf. *Apidium*"], a derived propliopithecid, and a derived rodent that was placed in the genus *Metaphiomys*) with taxa otherwise only known from the Jebel Qatrani lower sequence (djebelemurine strepsirrhines, various primitive phiomyid rodents, the hyracoid *Thyrohyrax meyeri*)(Thomas et al., 1991, 1992; Gheerbrant et al., 1993a, 1995; Holroyd, 1994; Pickford et al., 1994; Thomas et al., 1999). At present there is a major sampling gap in the Jebel Qatrani Formation between Quarry E and Quarry G, and it may be that the Taqah fauna is to some degree characteristic of this missing time slice; the later occurrence of some mammalian taxa at Taqah could also conceivably be due to latitudinal contraction of species ranges in association with early Oligocene cooling (Seiffert, 2007a).

The youngest diverse mammal faunas from the Jebel Qatrani Formation, from Quarries I and M, are in a zone of normal polarity that has been correlated with Chron C11n (Seiffert, 2006) and are estimated to be 29–30 million years old. The I and M faunas do not differ greatly from those at Quarries G and V, but they do mark the first and only appearances in the Jebel Qatrani Formation of the herpetotheriine marsupial *Peratherium* (Hooker et al., 2008), the primate *Afrotarsius* (Simons and Bown, 1985), and the distinctive rodents *"Paraphiomys" simonsi* and *Phiocricetomys* (Holroyd, 1994).

Malembe and Rukwa

Two areas in sub-Saharan Africa, Malembe in Angola (Pickford, 1986) and the Rukwa basin in Tanzania (Stevens et al., 2008) have produced mammal faunas of Oligocene age, but more informative mammalian remains or other chronological data will be needed before they can be assigned to either the early or late Oligocene. The Malembe fauna includes a possible propliopithecid catarrhine, embrithopods, and proboscideans (Pickford, 1986), while the Rukwa basin has produced phiomorph rodents, an anthropoid humerus, a hyracoid and undescribed macroscelideans (Stevens et al., 2008). The rodent *Metaphiomys* cf. *beadnelli* from the Rukwa locality TZ-01 is similar to *M. beadnelli* from the upper levels of the Jebel Qatrani Formation (Stevens et al., 2006). The anthropoid humerus from the same locality is similar to that of *Qatrania* (Fleagle and Simons, 1995), which in the Fayum area is also found in levels as young as Quarry M (Simons and Kay, 1988). Radiometric dating indicates that the mammals from the Rukwa basin are older than ~24.9 Ma (Roberts et al., in press).

Late Oligocene Sites

Mammalian evolution during the late Oligocene is now documented in at least three areas, all in east Africa. The oldest sites are in the highlands of Ethiopia (Chilga; Kappelman et al., 2003) and Eritrea (Dogali; Shoshani et al., 2006). The Chilga beds have produced a diverse proboscidean and hyracoid fauna, as well as a large species of *Arsinoitherium* (Kappelman et al., 2003; Sanders et al., 2004), while the only taxon known from Dogali is the proboscidean *Eritreum*. The Chilga localities are perhaps the best dated in the Paleogene of Africa, being between 27 and 28 Ma based on both radioisotopic and magnetostratigraphic data (Kappelman et al., 2003). A tuff with a ^{40}Ar/^{39}Ar date of 27.36 ± 0.11 Ma provides a maximum age for the youngest Chilga mammals, and all localities fall within normal polarity zone Chron C9n. A minimum age for the Dogali locality is provided by a ^{40}Ar/^{39}Ar date of 26.8 ± 1.5 Ma.

The youngest Paleogene localities in Africa outcrop west of Lake Turkana in northern Kenya (Leakey et al., 1995; Rasmussen and Gutiérrez, 2009). Until recently the only described mammal from this area was the catarrhine primate *Kamoyapithecus* (Leakey et al., 1995), but more recently a diverse mammal fauna has been described that includes proboscideans, hyracoids, *Arsinoitherium*, catarrhine primates, creodonts, anthracotheres, phiomorph rodents, and an immigrant carnivoran, *Mioprionodon* (Rasmussen and Gutiérrez, 2009; Rasmussen and Gutiérrez, this volume, chap. 13). These mammals have been found in two areas, near Lothidok Hill and farther south in the Nakwai region, which appear to be of about the same age based on shared species. K/Ar dates for over- and underlying basalts constrain the Lothidok mammals to be between 24 and 27 million years in age (Boschetto et al., 1992).

NOTE ADDED IN PROOF

Pickford et al. (2008) have recently reported a Paleogene mammal fauna from Namibia for which they suggested a Lutetian age. On the basis of the published mammalian fauna

I consider a Priabonian, or younger, age for the Black Crow Locality to be more likely; the other sites appear to be considerably younger.

Literature Cited

Adaci, M., R. Tabuce, F. Mebrouk, M. Bensalah, P.-H. Fabre, L. Hautier, J.-J. Jaeger, V. Lazzari, M. Mahboubi, L. Marivaux, O. Otero, S. Péigne, and H. Tong. 2007. Nouveaux sites à vertébrés paléogènes dans la région des Gour Lazib (Sahara nord-occidental, Algérie). *Comptes Rendus Palevol* 6:535–544.

Arambourg, C., J. Kikoine, and R. Lavocat. 1951. Découverte du genre *Moeritherium* Andrews dans le Tertiaire continental du Soudan. *Comptes Rendus de l'Académie des Sciences, Paris* 233:68–70.

Boschetto, H. B., F. H. Brown, and I. McDougall. 1992. Stratigraphy of the Lothidok range, northern Kenya, and K/Ar ages of its Miocene primates. *Journal of Human Evolution* 22:47–71.

Coiffait, P.-É., B. Coiffait, J.-J. Jaeger, and M. Mahboubi. 1984. Un nouveau gisement à Mammifères fossiles d'âge Éocène supérieur sur le versant sud des Nementcha (Algérie orientale): Découverte des plus anciens Rongeurs d'Afrique. *Comptes Rendus de l'Académie des Sciences, Paris,* Série II, 299:893–898.

Court, N. 1992. An enigmatic new mammal from the Eocene of North Africa. *Journal of Vertebrate Paleontology* 13:267–269.

Court, N., and J.-L. Hartenberger. 1992. A new species of the hyracoid mammal *Titanohyrax* from the Eocene of Tunisia. *Palaeontology* 35:309–317.

Court, N., and M. Mahboubi. 1993. Reassessment of Lower Eocene *Seggeurius amourensis*: Aspects of primitive dental morphology in the mammalian order Hyracoidea. *Journal of Paleontology* 67:889–893.

Fleagle, J. G., and E. L. Simons. 1995. Limb skeleton and locomotor adaptations of *Apidium phiomense*, an Oligocene anthropoid from Egypt. *American Journal of Physical Anthropology* 97:235–289.

Gheerbrant, E. 1992. Les mammifères paléocenes du Bassin d'Ouarzazate (Maroc). I. Introduction général et palaeoryctidae. *Palaeontographica, Abt. A* 224:67–132.

———. 1993. Premières données sur les mammifères "insectivores" de l'Yprésien du Bassin d'Ouarzazate (Maroc: site de N'Tagourt 2). *Neues Jahrbuch für Geologie und Paläontologie, Abhandlungen* 187:225–242.

———. 2009. Paleocene emergence of elephant relatives and the rapid radiation of African ungulates. Proceedings of the National Academy of Sciences, USA. 106:10717–10721.

Gheerbrant, E., H. Cappetta, M. Feist, J.-J. Jaeger, J. Sudre, M. Vianey-Liaud, and B. Sigé. 1993b. La succession des faunes de vertébrés d'âge paléocene supérieur et éocène inférieur dans le bassin d'Ouarzazate, Maroc: Contexte géologique, portée biostratigraphique et paléogéographique. *Newsletters on Stratigraphy* 28:33–58.

Gheerbrant, E., M. Iarochène, M. Amaghzaz, and B. Bouya. 2006. Early African hyaenodontid mammals and their bearing on the origin of the Creodonta. *Geological Magazine* 143:475–489.

Gheerbrant, E., S. Peigné, and H. Thomas. 2007. Première description du squelette d'un hyracoïde paléogène: *Saghatherium antiquum* de l'Oligocène inférieur de Jebel al Hasawnah, Libye. *Palaeontographica, Abt. A* 279:93–145.

Gheerbrant, E., J. Sudre, H. Capetta, M. Iarochène, M. Amaghzaz, and B. Bouya. 2002. A new large mammal from the Ypresian of Morocco: Evidence of surprising diversity of early proboscideans. *Acta Palaeontologica Polonica* 47:493–506.

Gheerbrant, E., J. Sudre, H. Capetta, C. Mourer-Chauviré, E. Bourdon, M. Iarochène, M. Amaghzaz, and B. Bouya. 2003. Les localités à mammifères des carrières de Grand Daoui, bassin des Ouled Abdoun, Maroc, Yprésien: Premier état des lieux. *Bulletin de la Société Géologique de France* 174:279–293.

Gheerbrant, E., J. Sudre, M. Iarochène, and A. Moumni. 2001. First ascertained African "condylarth" mammals (primitive ungulates: cf. Bulbulodentata and cf. Phenacodonta) from the earliest Ypresian of the Ouled Abdoun Basin, Morocco. *Journal of Vertebrate Paleontology* 21:107–118.

Gheerbrant, E., J. Sudre, S. Sen, C. Abrial, B. Marandat, B. Sigé, and M. Vianey-Liaud. 1998. Nouvelles données sur les mammifères du Thanétien et de l'Yprésien du Bassin d'Ouarzazate (Maroc) et leur contexte stratigraphique. *Palaeovertebrata* 27:155–202.

Gheerbrant, E., J. Sudre, P. Tassy, M. Amaghzaz, B. Bouya, and M. Iarochene. 2005. Nouvelles données sur *Phosphatherium escuilliei*

(Mammalia, Proboscidea) de l'Éocène inférieur du Maroc, apports à la phylogénie des Proboscidea et des ongulés lophodontes. *Geodiversitas* 27:239–333.

Gheerbrant, E., H. Thomas, J. Roger, S. Sen, and Z. Al-Sulaimani. 1993a. Deux nouveaux primates dans l'Oligocene inferieur de Taqah (Sultanat d'Oman): Premiers adapiformes (?Anchomomyini) de la peninsule arabique? *Palaeovertebrata* 22:141–196.

Gheerbrant, E., H. Thomas, S. Sen, and Z. Al-Sulaimani. 1995. Nouveau primate Oligopithecinae (Simiiformes) de l'Oligocène inférieur de Taqah, Sultanat d'Oman. *Comptes Rendus de l'Académie des Sciences Paris*, Série IIA, 321:425–432.

Gingerich, P. D. 1992. Marine mammals (Cetacea and Sirenia) from the Eocene of Gebel Mokattam and Fayum, Egypt: Stratigraphy, age, and paleoenvironments. *University of Michigan Papers in Paleontology* 30:1–84.

Godinot, M. 1994. Early North African primates and their significance for the origin of Simiiformes (= Anthropoidea); pp. 235–296 in J. G. Fleagle and R. F. Kay (eds.), *Anthropoid Origins*. Plenum Press, New York.

Godinot, M., and M. Mahboubi. 1994. Les petits primates simiiformes de Glib Zegdou (Éocène inférieur à moyen d'Algérie). *Comptes Rendus de l'Académie des Sciences, Paris*, Série II, 319:357–364.

Gorodiski, A., and R. Lavocat. 1953. Première découverte de mammifères dans le Tertiaire (Lutétien) du Sénégal. *Comptes Rendus Sommaires de la Société Géologique de France* 15:314–316.

Gradstein, F., J. Ogg, and A. Smith. 2004. *A Geological Time Scale 2004*. Cambridge University Press, Cambridge, 589 pp.

Gunnell, G. F., B. F. Jacobs, P. S. Herendeen, J. J. Head, E. Kowalski, C. P. Msuya, F. A. Mizambwa, T. Harrison, J. Habersetzer, and G. Storch. 2002. Oldest placental mammal from sub-Saharan Africa: Eocene microbat from Tanzania: Evidence for early evolution of sophisticated echolocation. *Palaeontologia Electronica* 5:1–10.

Haggag, M. A. Y. 1985. Middle Eocene planktonic foraminifera from Fayoum area, Egypt. *Revista España de Micropaleontologia* 17:27–40.

Hartenberger, J.-L., and B. Marandat. 1992. A new genus and species of an early Eocene primate from North Africa. *Human Evolution* 7:9–16.

Holroyd, P. A. 1994. An examination of dispersal origins for Fayum Mammalia. Unpublished PhD dissertation, Duke University, Durham, N.C., 328 pp.

Hooker, J. J., M. R. Sánchez-Villagra, F. J. Goin, E. L. Simons, Y. Attia, and E. R. Seiffert. 2008. The origin of Afro-Arabian "didelphimorph" marsupials. *Palaeontology* 51:635–648.

Jaeger, J.-J., C. Denys, and B. Coiffait. 1985. New Phiomorpha and Anomaluridae from the late Eocene of north-west Africa: Phylogenetic implications; pp. 567–588 in W. P. Luckett and J.-L. Hartenberger (eds.), *Evolutionary Relationships among Rodents: A Multidisciplinary Analysis*. Plenum Press, New York.

Kappelman, J., E. L. Simons, and C. C. Swisher, III. 1992. New age determinations for the Eocene-Oligocene boundary sediments in the Fayum Depression, northern Egypt. *Journal of Geology* 100:647–668.

Kappelman, J., D. T. Rasmussen, W. J. Sanders, M. Feseha, T. Bown, P. Copeland, J. Crabaugh, J. Fleagle, M. Glantz, A. Gordon, B. Jacobs, M. Maga, K. Muldoon, A. Pan, L. Pyne, B. Richmond, T. Ryan, E. R. Seiffert, S. Sen, L. Todd, M. C. Wiemann, and A. Winkler. 2003. Oligocene mammals from Ethiopia and faunal exchange between Afro-Arabia and Eurasia. *Nature* 426:549–552.

Leakey, M. G., P. S. Ungar, and A. Walker. 1995. A new genus of large primate from the Late Oligocene of Lothidok, Turkana District, Kenya. *Journal of Human Evolution* 28:519–531.

Maas, M. C., J. G. M. Thewissen, and J. Kappelman. 1998. *Hypsamasia seni* (Mammalia, Embrithopoda) and other mammals from the Eocene Kartal Formation of Turkey. *Bulletin of the Carnegie Museum of Natural History* 34:286–297.

Mahboubi, M., R. Ameur, J.-Y. Crochet, and J.-J. Jaeger. 1986. El Kohol (Saharan Atlas, Algeria): A new Eocene mammal locality in north-western Africa. *Palaeontographica*, Abt. A 192:15–49.

Marivaux, L., S. Ducrocq, J. J. Jaeger, B. Marandat, J. Sudre, Y. Chaimanee, S. T. Tun, W. Htoon, and A. N. Soe. 2005. New remains of *Pondaungimys anomaluropsis* (Rodentia, Anomaluroidea) from the latest middle Eocene Pondaung Formation of Central Myanmar. *Journal of Vertebrate Paleontology* 25:214–227.

Martín-Closas, C., and E. Ramos. 2005. Paleogene charophytes of the Balearic Islands (Spain). *Geologica Acta* 3:39–58.

Mebrouk, F., and M. Feist. 1999. Nouvelles charophytes de l'Eocène continental de l'Algérie. *Géologie Méditerranéenne* 26:29–45.

Mebrouk, F., M. Mahboubi, M. Bessedik, and M. Feist. 1997. L'apport des charophytes à la stratigraphie des formations continentales paléogènes de l'Algérie. *Geobios* 30:171–177.

Moody, R. T. J., and P. J. C. Sutcliffe. 1993. The sedimentology and palaeontology of the Upper Cretaceous–Tertiary deposits of central West Africa. *Modern Geology* 18:539–554.

O'Leary, M. A., E. M. Roberts, M. Bouare, F. Sissoko, and L. Tapanila. 2006. Malian Paenungulata (Mammalia: Placentalia): New African afrotheres from the early Eocene. *Journal of Vertebrate Paleontology* 26:981–988.

Patterson, C., and A. E. Longbottom. 1989. An Eocene amiid fish from Mali, West Africa. *Copeia* 4:827–836.

Pickford, M. 1986. Première découverte d'une faune mammalienne terrestre paléogène d'Afrique sub-saharienne. *Comptes Rendus de l'Académie des Sciences, Paris*, Série II, 302:1205–1210.

Pickford, M., B. Senut, J. Morales, P. Mein, and I. M. Sanchez. 2008. Mammalia from the Lutetian of Namibia. *Memoir of the Geological Survey of Namibia* 20:465–514.

Pickford, M., H. Thomas, S. Sen, J. Roger, E. Gheerbrant, and Z. Al-Sulaimani. 1994. Early Oligocene Hyracoidea (Mammalia) from Thaytiniti and Taqah, Dhofar Province, Sultanate of Oman. *Comptes Rendus de l'Académie des Sciences, Paris*, Série II, 318:1395–1400.

Radulescu, C., G. Iliesco, and M. Iliesco. 1976. Decouverte d'un Embrithopode nouveau (Mammalia) dans la Paléogène de la dépression de Hateg (Roumanie) et considération générales sur la géologie de la région. *Neues Jahrbuch für Geologie und Paläontologie, Monatshefte* 11:690–698.

Rasmussen, D. T., T. M. Bown, and E. L. Simons. 1992. The Eocene-Oligocene transition in continental Africa; pp. 548–566 in D. R. Prothero and W. A. Berggren (eds.), *Eocene-Oligocene Climatic and Biotic Evolution*. Princeton University Press, Princeton.

Rasmussen, D. T., and M. Gutiérrez. 2009. A mammalian fauna from the late Oligocene of northern Kenya. *Palaeontographica, Abt. A.* 288:1–52.

Rasmussen, D. T., and E. L. Simons. 1988. New specimens of *Oligopithecus savagei*, early Oligocene primate from the Fayum, Egypt. *Folia Primatologica* 51:182–208.

Rasmussen, D. T., S. O. Tshakreen, M. M. Abugares, and J. B. Smith. 2008. Return to Dor al-Talha: Paleontological reconnaissance of the early Tertiary of Libya; pp. 181–196 in J. G. Fleagle and C. C. Gilbert (eds.), *Elwyn L. Simons: A Search for Origins*. Springer, New York.

Roberts, E., P. M. O'Connor, N. J. Stevens, M. D. Gottfried, Z. A. Jinnah, S. Ngasala, A. M. Choh, and R. A. Armstrong. In press. Sedimentology and depositional environments of the Red Sandstone Group, Rukwa Rift Basin, southwestern Tanzania: New insight into Cretaceous and Paleogene terrestrial ecosystems and tectonics in sub-equatorial Africa. *Journal of African Earth Sciences*.

Roger, J., S. Sen, H. Thomas, C. Cavelier, and Z. Al Sulaimani. 1993. Stratigraphic, paleomagnetic and paleoenvironmental study of the early Oligocene vertebrate locality of Taqah (Dhofar, Sultanate of Oman). *Newsletters on Stratigraphy* 28:93–119.

Sallam, H.M., Seiffert, E.R., Steiper, M.E., and Simons, E.L. 2009. Fossil and molecular evidence constrains scenarios for the early evolutionary and biogeographic history of hystricognathous rodents. *Proceedings of the National Academy of Sciences, USA.* 106:16722–16727.

Sanders, W. J., D. T. Rasmussen, and J. Kappelman. 2004. New large-bodied mammals from the late Oligocene site of Chilga, Ethiopia. *Acta Palaeontologica Polonica* 49:365–392.

Savage, R. J. G. 1969. Early Tertiary mammal locality in southern Libya. *Proceedings of the Geological Society, London* 1657:167–171.

Seiffert, E. R. 2006. Revised age estimates for the later Paleogene mammal faunas of Egypt and Oman. *Proceedings of the National Academy of Sciences, USA* 103:5000–5005.

Seiffert, E. R. 2007a. Evolution and extinction of Afro-Arabian primates near the Eocene-Oligocene boundary. *Folia Primatologica* 78:314–327.

———. 2007b. A new estimate of afrotherian phylogeny based on simultaneous analysis of genomic, morphological, and fossil evidence. *BMC Evolutionary Biology* 7:224.

Seiffert, E. R., T. M. Bown, W. C. Clyde, and E. L. Simons. 2008. Geology, paleoenvironment, and age of Birket Qarun Locality 2 (BQ-2), Fayum Depression, Egypt; pp. 71–86 in J. G. Fleagle and C. C. Gilbert (eds.), *Elwyn L. Simons: A Search for Origins*. Springer, New York.

Seiffert, E. R., E. L. Simons, W. C. Clyde, J. B. Rossie, Y. Attia, T. M. Bown, P. Chatrath, and M. Mathison. 2005. Basal anthropoids from Egypt and the antiquity of Africa's higher primate radiation. *Science* 310:300–304.

Seiffert, E. R., E. L. Simons, T. M. Ryan, T. M. Bown, and Y. Attia. 2007. New remains of Eocene and Oligocene Afrosoricida (Afrotheria) from Egypt, and the origin(s) of afrosoricid zalambdodonty. *Journal of Vertebrate Paleontology* 27:963–972.

Sen, S., and E. Heintz. 1979. *Palaeomasia kansui* Ozansoy 1966, embrithopode (Mammalia) de l'Eocène d'Anatolie. *Annales de Paléontologie* 65:73–91.

Serra-Kiel, J., L. Hottinger, E. Caus, K. Drobne, C. Ferràndez, A. K. Jauhri, G. Less, R. Pavlovec, J. Pignatti, J. M. Samsó, H. Schaub, E. Sirel, A. Strougo, Y. Tambareau, J. Tosquella, and E. Zakrevskaya. 1998. Larger foraminiferal biostratigraphy of the Tethyan Paleocene and Eocene. *Bulletin de la Société Géologique de France* 169:281–299.

Shoshani, J., R. C. Walter, M. Abraha, S. Berhe, P. Tassy, W. J. Sanders, G. H. Marchant, Y. Libsekal, T. Ghirmai, and D. Zinner. 2006. A proboscidean from the late Oligocene of Eritrea, a "missing link" between early Elephantiformes and Elephantimorpha, and biogeographic implications. *Proceedings of the National Academy of Sciences, USA* 103:17296–17301.

Sigé, B., J.-J. Jaeger, J. Sudre, and M. Vianey-Liaud. 1990. *Altiatlasius koulchii* n. gen. n. sp., primate omomyidé du Paléocène du Maroc, et les origines des euprimates. *Palaeontographica, Abt. A* 214:31–56.

Simons, E. L., and T. M. Bown. 1985. *Afrotarsius chatrathi*, first tarsiiform primate (?Tarsiidae) from Africa. *Nature* 313:475–477.

Simons, E. L., T. M. Bown, and D. T. Rasmussen. 1987. Discovery of two additional prosimian primate families (Omomyidae, Lorisidae) in the African Oligocene. *Journal of Human Evolution* 15:431–437.

Simons, E. L., P. Chatrath, C. C. Gilbert, and J. G. Fleagle. 2008. Five decades in the Fayum; pp. 51–70 in J. G. Fleagle and C. C. Gilbert (eds.), *Elwyn L. Simons: A Search for Origins*. Springer, New York.

Simons, E. L., and R. F. Kay. 1983. *Qatrania*, new basal anthropoid primate from the Fayum, Oligocene of Egypt. *Nature* 304:624–626.

———. 1988. New material of *Qatrania* from Egypt with comments on the phylogenetic position of the Parapithecidae (Primates, Anthropoidea). *American Journal of Primatology* 15:337–347.

Solé, F., E. Gheerbrant, M. Amaghzaz, and B. Bouya. 2009. Further evidence of the African antiquity of hyaenodontid ('Creodonta', Mammalia) evolution. *Zoological Journal of the Linnean Society* 156:827–846.

Stevens, N. J., M. D. Gottfried, E. M. Roberts, S. Kapilima, S. Ngasala, and P. M. O'Connor. 2008. Paleontological exploration in Africa: A view from the Rukwa Rift Basin of Tanzania; pp. 159–180 in J. G. Fleagle and C. C. Gilbert (eds.), *Elwyn L. Simons: A Search for Origins*. Springer, New York.

Stevens, N. J., P. M. O'Connor, M. D. Gottfried, E. M. Roberts, S. Ngasala, and M. R. Dawson. 2006. *Metaphiomys* (Rodentia: Phiomyidae) from the Paleogene of southwestern Tanzania. *Journal of Paleontology* 80:407–410.

Strougo, A., and M. A. Boukhary. 1987. The Middle Eocene-Upper Eocene boundary in Egypt: Present state of the problem. *Revue de Micropaléontologie* 30:122–127.

Strougo, A., and M. A. Y. Haggag. 1984. Contribution to the age determination of the Gehannam Formation in the Fayum Province, Egypt. *Neues Jahrbuch für Geologie und Paläontologie, Monatshefte* 1984:46–52.

Sudre, J. 1979. Nouveaux mammifères éocènes du Sahara occidental. *Palaeovertebrata* 9:83–115.

Sudre, J., J.-J. Jaeger, B. Sigé, and M. Vianey-Liaud. 1993. Nouvelle données sur les condylarthres du Thanétien et de l'Yprésien du Bassin d'Ouarzazate (Maroc). *Geobios* 26: 609–615.

Tabuce, R., S. Adnet, H. Cappetta, A. Noubhani, and F. Quillevere. 2005. Aznag (bassin d'Ouarzazate, Maroc), nouvelle localité à sélaciens et mammifères de l'Eocène moyen (Lutétien) d'Afrique. *Bulletin de la Société Géologique de France* 176:381–400.

Tabuce, R., M. Mahboubi, and J. Sudre. 2001. Reassessment of the Algerian Eocene hyracoid *Microhyrax*. Consequences on the early diversity and basal phylogeny of the Order Hyracoidea (Mammalia). *Eclogae Geologicae Helvetiae* 94:537–545.

Tabuce, R., L. Marivaux, M. Adaci, M. Bensalah, J.-L. Hartenberger, M. Mahboubi, F. Mebrouk, P. Tafforeau, and J.-J. Jaeger. 2007. Early Tertiary mammals from North Africa reinforce the molecular Afrotheria clade. *Proceedings of the Royal Society B* 274:1159–1166.

Tabuce, R., L. Marivaux, R. Lebrun, M. Adaci, M. Bensalah, P.-H. Fabre, E. Fara, H. G. Rodrigues, L. Hautier, J.-J. Jaeger, V. Lazzari, F., Mebrouk, S. Peigné, J. Sudre, P. Tafforeau, X. Valentin, and M. Mahboubi. 2009. Anthropoid versus strepsirhine status of the African Eocene primates Algeripithecus and Azibius: craniodental evidence. Proceedings of the Royal Society B 276: 4087–4094.

Thomas, H., J. Roger, S. Sen, and Z. Al-Sulaimani. 1992. Early Oligocene vertebrates from Dhofar (Sultanate of Oman); pp. 283–293 in A. Sadek (ed.), *Geology of the Arab World*. Cairo University, Cairo.

Thomas, H., J. Roger, S. Sen, M. Pickford, E. Gheerbrant, Z. Al-Sulaimani, and S. Al-Busaidi. 1999. Oligocene and Miocene terrestrial vertebrates in the southern Arabian peninsula (Sultanate of Oman) and their geodynamic and palaeogeographic settings; pp. 430–442 in P. J. Whybrow and A. Hill (eds.), *Fossil Vertebrates of Arabia*. Yale University Press, New Haven.

Thomas, H., S. Sen, J. Roger, and Z. Al-Sulaimani. 1991. The discovery of *Moeripithecus markgrafi* Schlosser (Propliopithecidae, Anthropoidea, Primates), in the Ashawq Formation (Early Oligocene of Dhofar Province, Sultanate of Oman). *Journal of Human Evolution* 20:33–49.

Vianey-Liaud, M., and J.-J. Jaeger. 1996. A new hypothesis for the origin of African Anomaluridae and Graphiuridae (Rodentia). *Palaeovertebrata* 25:349–358.

Vianey-Liaud, M., J.-J. Jaeger, J.-L. Hartenberger, and M. Mahboubi. 1994. Les rongeurs de l'Eocène d'Afrique nord-occidentale (Glib Zegdou [Algérie] et Chambi [Tunisie]) et l'origine des Anomaluridae. *Palaeovertebrata* 23:93–118.

THREE

Chronology of Neogene Mammal Localities

LARS WERDELIN

While the Neogene record of fossil mammals in Africa is considerably more substantial than that of the Paleogene (see Seiffert, this volume, chap. 2), it is still far from the relative completeness of the record in, for example, Europe and North America. Though in some regions of Africa this degree of completeness has in fact been matched, in other regions the record is virtually nonexistent. Thus, many West and Central African countries are completely devoid of a Neogene mammal record, while other countries have only a handful of localities that at best serve as a modest window into what might exist there. Even in regions or countries where there is a record, such as South Africa, that record is very uneven, with some time intervals well represented and others devoid of fossil mammals.

In addition to the unevenness of the record, the quality of dating is also uneven. In some regions, such as the Turkana Basin of northern Kenya and southern Ethiopia, the dating is unsurpassed, as this region has more or less been a testing ground for radiometric and tephrochronologic dating methods (see, e.g., Brown 1994). In other regions, such as the Miocene of southern and northern Africa, dating is poor and controversial, dependent mainly on biochronologic correlations with faunas outside Africa. This approach naturally becomes increasingly problematic as the distance to other continents increases. Finally, the dating of karstic cave sites such as the australopithecine sites in Gauteng and Limpopo provinces of South Africa is notoriously difficult and rife with controversy.

The brief overview that follows can only discuss a selection of sites and provide what amounts to a consensus view of their chronology. For additional localities and alternative chronologies, I refer to the published literature, as well as to other chapters in this volume. This chapter is organized geographically into the major regions of northern, eastern, and southern Africa, and within each of these by age from oldest to youngest.

Much to do with the dating of the sites discussed herein is controversial and subject to considerable debate. This chapter is an attempt to provide a single, coherent whole, but it is not intended to provide answers to ongoing debates. Therefore, the editors have made no attempt to force conformity between the various faunal chapters of the book

and this chapter. However, in many cases it will be possible to move from a citation of a site in a faunal chapter to this chapter to see if any controversy surrounds the dating of that site.

North African Localities (Figure 3.1)

MIOCENE

The North African Miocene is represented by numerous sites ranging from early Miocene sites such as Gebel (or Jebel) Zelten (Savage and Hamilton, 1973) to sites near the Miocene-Pliocene boundary such as As-Sahabi (Boaz 1987). The absence of radiometrically datable material means that the vast majority of these sites are dated on the basis of biostratigraphic correlations with European faunas. Because of the proximity to Europe and the often close phylogenetic relationships between North African and European mammals, this is less of a problem than it is for sub-Saharan localities. In addition, many North African sites have substantial rodent faunas, making biostratigraphic correlations with Europe, where local stratigraphies are often entirely based on rodents, more secure.

The oldest North African Miocene sites are Gebel Zelten in Libya (Savage and Hamilton, 1973) and Siwa (Hamilton, 1973) and Wadi Moghara (Miller, 1999), both in Egypt. Wadi Moghara is here taken to include the nearby Wadi Faregh. None is older than the Burdigalian, leaving a gap from the late Rupelian to the early Burdigalian (>10 million years) with no North African fossil mammal sites (cf. Seiffert, this volume, chap. 2). The three sites mentioned are of similar age and share some large mammal species, such as *Brachypotherium snowi*, and *Afromeryx africanus*. Gebel Zelten also has a substantial micromammal fauna (Wessels et al., 2003). The site was originally thought to extend over only a relatively short span of time, but recent studies on the microfauna (Wessels et al., 2003) suggest that at least three time periods are represented at the locality, the oldest dating to 19–18 Ma and the youngest to 15–14 Ma—that is, extending into the Langhian (lower middle Miocene). The implications of this for the dating of Siwa and Wadi Moghara are not entirely clear, though the latter is still likely

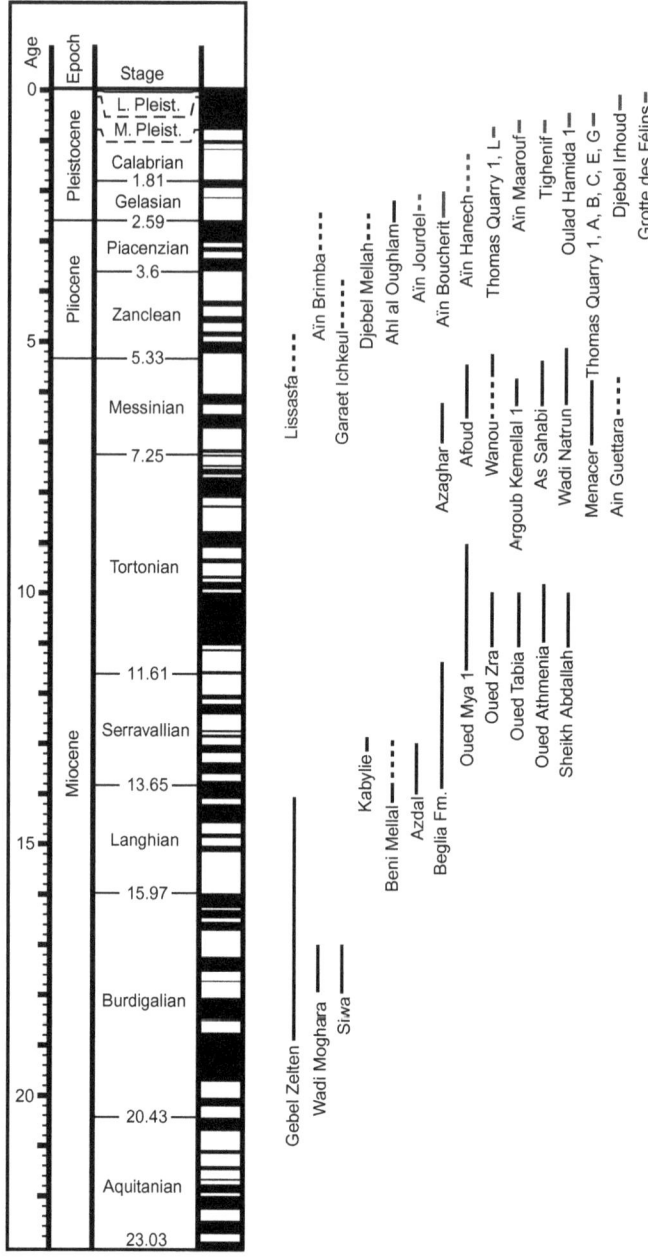

FIGURE 3.1 Approximate correlation of North African terrestrial mammal localities to the global timescale (timescale based on Gradstein et al., 2004). For information about dating methods, see the text.

to be correlated with the oldest time period of Gebel Zelten. The age of Siwa is more problematic as it is a small fauna composed entirely of very large mammals, but an early Miocene date seems implied.

The middle Miocene of North Africa is also poor in mammal faunas. The first discovered of these may have been the poorly known site of Kabylie, Algeria, where *Gomphotherium pygmaeus* was first found (Depéret, 1897). The site is correlated with the middle Miocene (ca. 13 Ma) site of Ngenyin, Kenya (Pickford, 2004). The best-known middle Miocene locality, and the one with the largest fauna, however, is Beni Mellal, Morocco (Lavocat, 1961). Lavocat dated Beni Mellal on the basis of rodent biostratigraphy to ca. 14 Ma (i.e., latest Langhian), but the carnivorans described by Ginsburg (1977) suggest an age no greater than MN 8 (i.e., Serravallian). This

discrepancy will surely be resolved through renewed analysis of this fauna, not least because some of the carnivorans suggest unique connections with Europe. For example, if the identification is correct, Beni Mellal has the only African record of *Mustela* or at least a musteline mustelid. A third middle Miocene fauna is the micromammal locality of Azdal, Morocco, in the Aït Kandoula Basin. Benammi et al. (1996) date the locality on the basis of rodent biostratigraphy and magnetostratigraphy to 14–13 Ma—that is, probably lowermost Serravallian, and broadly coeval with Beni Mellal, a date that is underscored by the presence at both of *Mellalomys punicus*.

The late Miocene of North Africa is richer in mammal faunas, including both Vallesian (~lower Tortonian) and Turolian (~upper Tortonian–Messinian) localities. The oldest and perhaps richest of these is the Beglia Formation, Tunisia, which includes both middle Miocene (Serravallian) and late Miocene levels. This formation is poorly constrained temporally, and parts are likely to be as old as upper Langhian (Mahjoub and Khessibi, 1988), though the bulk of the vertebrate faunas are close to the middle-late Miocene boundary (e.g., Pickford, 1990, 2000). The Beglia Formation includes a broad range of vertebrates, including fish (Greenwood, 1972), crocodiles (Llinás-Agrasar, 2003), birds (Vickers-Rich, 1972), and mammals (e.g., Black, 1972; Robinson, 1972; Kurtén, 1976). Another poorly time-constrained large mammal fauna from North Africa is Oued Mya 1, Algeria (Sudre and Hartenberger, 1992). This site is dominated by *Hipparion* and *Aceratherium* and includes the aberrant amphicyonid *Myacyon dojambir* (Werdelin and Peigné, this volume, chap. 32). Another rich Vallesian site is Bou Hanifia (Oued el Hamman, Algeria), which includes substantial faunas of both large and small mammals (Arambourg, 1968; Ameur, 1984). Other Vallesian localities (e.g., Oued Zra, Oued Tabia, both in Morocco, and Oued Athmenia in Algeria) mainly include rodents. The most interesting of these micromammal faunas may be Sheikh Abdallah, Egypt, which includes specimens referred to a galagid primate (Pickford et al., 2006; see also Harrison, this volume, chap. 20).

The Turolian of North Africa is relatively rich in mammal localities, though few of these include large mammals. The best known are probably the youngest, As-Sahabi, Libya (Heinzelin and El-Arnauti, 1987) and Wadi Natrun, Egypt (Stromer, 1911). Both of these are late Messinian in age and may include early Pliocene elements. Sahabi is notable for including a number of typically Eurasian "Pikermian" faunal elements such as the hyenid *Adcrocuta eximia* (Howell, 1987; Lihoreau et al. 2006). Another Turolian large mammal site is Menacer (formerly Marceau), Algeria (Thomas and Petter, 1986), which has been dated by correlation with nearby marine sediments to foraminifer stage N17 (upper Tortonian/lower Messinian). In addition to these sites, there are a number of micromammal sites in the North African Turolian. These are mostly dated on the basis of rodent biostratigraphy and correlation with Europe, and they include sites such as Afoud 1, 2, and 5, Wanou, Azaghar, and Argoub Kemellal 2, in the Aït Kandoula Basin, Morocco (Benammi et al. 1996), and Aïn Guettara, also in Morocco (Brandy and Jaeger, 1980). All of these sites are Messinian in age and include elements that can be correlated with the European rodent stratigraphy.

PLIO-PLEISTOCENE

The North African Pliocene, especially the early part, is relatively poorly represented compared to the late Miocene and Pleistocene. Site dating is complicated by the referral of many sites to the "Villafranchian" (e.g., Fournet, 1971; Arambourg, 1979), although the correlation with the European Villafranchian is far from clear. Only a few sites with good mammal faunas are known, and the majority of these are poorly dated. A possible early Pliocene micromammal site is Lissasfa, Morocco, though the dating is uncertain (Geraads, 1998), and the site may be latest Miocene. Another site of similar or slightly younger age is Garaet (Lac) Ichkeul, Morocco (Arambourg, 1979; Benammi et al., 1996). A classic "Villafranchian" locality in North Africa is Aïn Brimba, Tunisia, which has yielded a substantial fauna of large mammals, including both ungulates and carnivores (Arambourg, 1979). The locality is of somewhat uncertain absolute age but is probably somewhere around 3 Ma—that is, older than Ahl al Oughlam (discussed later) and thus well below the Plio-Pleistocene boundary. A site of similar though slightly younger age is Djebel Mellah, Tunisia (Fournet, 1971). However, the richest and most important Pliocene site in North Africa is Ahl al Oughlam, which has yielded the largest Neogene fauna in North Africa, with about 55 species of mammals, as well as birds and reptiles (Geraads, 2006). The site includes representatives of nearly every major African mammal group and is especially rich in carnivorans (including the only African record of a walrus; see Werdelin and Peigné, this volume, chap. 32) and murid rodents (Geraads, 1995, 1997). The locality is biostratigraphically dated to the late Pliocene (i.e., 2.5 Ma). Interestingly, despite the extensive fauna, there is no record of a hominin from Ahl al Oughlam, providing strong evidence that our ancestors had not reached northern (or at least northwestern) Africa at this time.

North Africa has far too many Pleistocene localities to mention here, so only a handful of especially interesting ones will be highlighted. Several of these are of particular importance as recording the first appearance of various hominins in North Africa. Two of the oldest Pleistocene sites are Aïn Jourdel and Aïn Boucherit, Algeria, which on biochronological grounds must be younger than Ahl al Oughlam (Geraads et al., 2004c), though by how much is debatable. An age of between 2.3 and 2 Ma seems reasonable. The most important early Pleistocene site in North Africa may, however, be Aïn Hanech, Algeria (near Aïn Boucherit). This site includes stone tools, making it possibly the oldest archeological site in North Africa. Unfortunately, there is considerable disagreement over the age of Aïn Hanech (as well as Aïn Boucherit), which is variously dated to the Olduvai subchron (1.97–1.78 Ma) or to ca. 1.2 Ma (for discussion, see Sahnouni and de Heinzelin, 1998; Geraads 2002; Geraads et al., 2002; Sahnouni et al., 2002, 2004). Another site that may be only slightly younger at ca. 1 Ma is Thomas Quarry 1, level L, Morocco (Geraads, 2002), which includes some large mammals and an important though small micromammal fauna.

Important middle Pleistocene sites in North Africa include the hominin sites of Aïn Maarouf and Tighenif (= Ternifine = Palikao), both in Morocco, dating to 0.8–0.6 Ma, with Aïn Maarouf being slightly the younger. Other sites of similar or somewhat younger age include Thomas Quarry 1, Levels A, B, C, E, and G and Oulad Hamida 1 (*Homo erectus* cave [= Thomas III] and rhino cave), Morocco. Several of these sites include records of the giant gelada baboon, *Theropithecus oswaldi*. A somewhat younger site, probably datable to the boundary between the middle and upper Pleistocene, is the Jebel Irhoud hominin site, Morocco (Amani, 1991; Amani and Geraads, 1993; Geraads and Amani, 1998).

Finally, late Pleistocene sites in North Africa are numerous. Many include hominin remains and/or tools, and some also include immigrants from Europe, such as brown bear, *Ursus arctos* (Hamdine et al., 1998). A recently described site of particular interest is Grotte des Félins, near Dar Bouazza, Morocco (Raynal et al., 2008).

Southern African Localities (Figure 3.2)

MIOCENE

With the exception of the rich site of Arrisdrift, Namibia (Hendey, 1978; Pickford et al., 1996), the Miocene is poorly represented in southern Africa. Most of what we know comes from the pioneering surveys of the Namib Desert and coastal Namaqualand by Pickford, Senut, and colleagues (e.g., Pickford

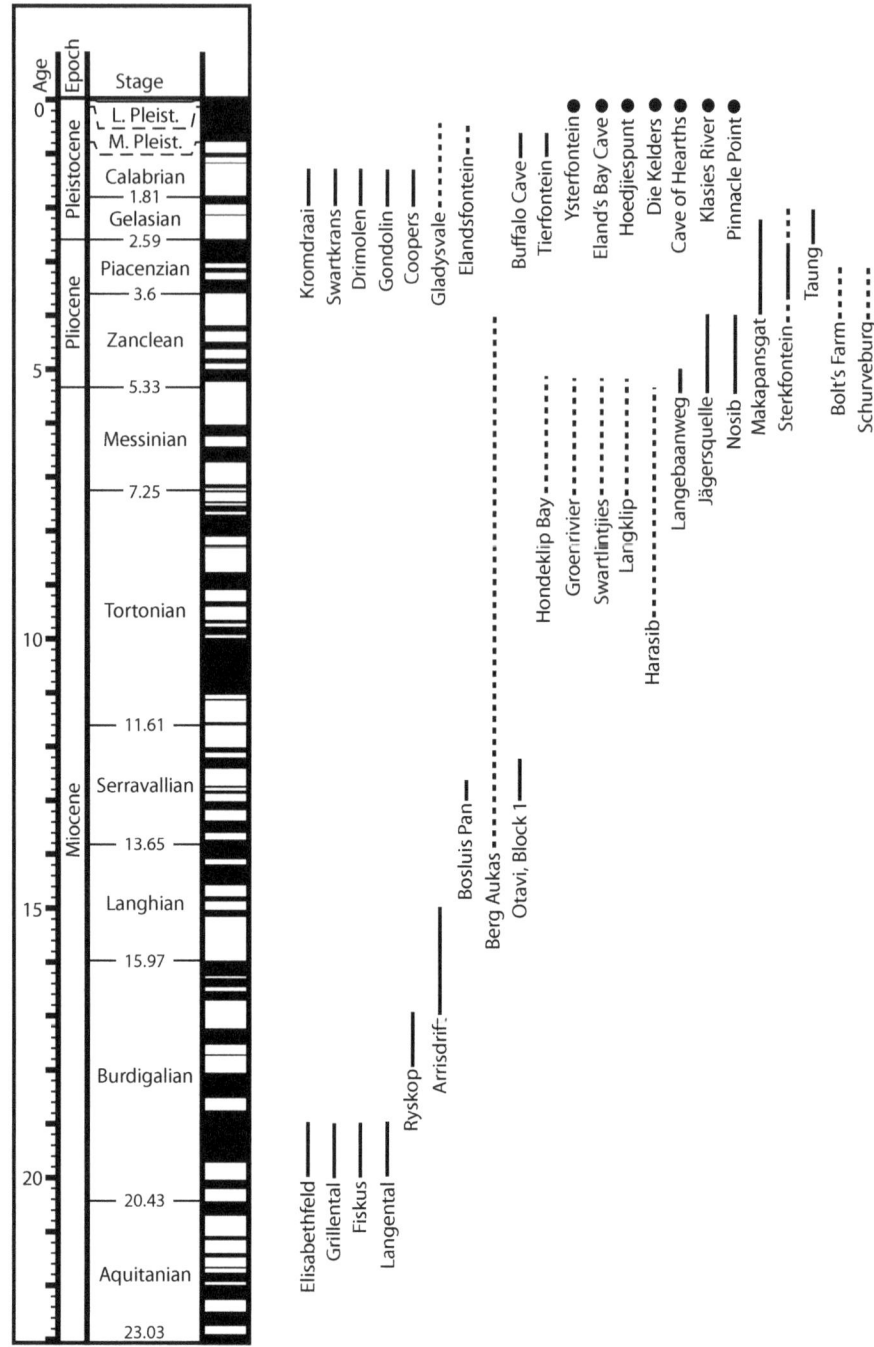

FIGURE 3.2 Approximate correlation of southern African terrestrial mammal localities to the global timescale (timescale based on Gradstein et al., 2004). For information about dating methods, see the text.

and Senut, 1997). Most of these faunas are relatively limited in extent and difficult to date, but they provide a series of windows into mammal evolution in southern Africa that has the potential to change our view of the pattern of mammalian evolution on the continent of Africa as a whole (Pickford, 2004), although it would be desirable to test these hypotheses against Miocene faunas from sites in the eastern part of southern Africa (e.g., Mozambique, Zimbabwe).

The oldest Miocene sites in southern Africa are probably a series of sites in the Sperrgebiet, Namibia, including Elisabethfeld, Grillental, Fiskus, and Langental, all dated biostratigraphically to Pickford's (1981) Faunal Set I, currently approximately

dated 20–19 Ma (i.e., middle Burdigalian). These, along with Ryskop, which has yielded the earliest record of a pinniped in Africa (Pickford and Senut, 2000), are the only localities in southern Africa that with some certainty can be dated to the lower Miocene.

The middle Miocene is better represented in terms of fauna, if not in terms of the number of localities, due to the dominating influence of Arrisdrift. This site can be biostratigraphically dated to ca. 17.5–17 Ma (i.e., the upper Burdigalian; Pickford and Senut, 2003). The fauna from this site is extensive, including pedetids (Senut, 1997), tragulids (Morales et al., 2003), and pikas (Mein and Pickford, 2003), as well as numerous carnivores

(Morales et al. 1998, 2003). Another middle Miocene site is Bosluis Pan, South Africa, which can be dated to ca. 13 Ma based on the presence of *Gomphotherium pygmaeus* (Pickford, 2005). Other sites that include middle Miocene faunal elements are parts of longer sequences, or series of faunas. Such is the case with Berg Aukas (Pickford et al. 1994), which includes middle Miocene to (probably) Pliocene faunas. Perhaps the most important and interesting such locality, however, is Otavi, Namibia, which includes a series of faunal blocks with dates from the lower to middle Serravallian to the Holocene. The hominoid *Otavipithecus namibiensis* was recovered from Block 1, dated to the later Serravallian (ca. 13–12 Ma); Conroy et al., 1992.

PLIO-PLEISTOCENE

In the southern African Plio-Pleistocene, we encounter some of the most intractable dating problems in all of African paleontology, the South African hominin cave sites including Makapansgat, Limpopo Province, and the many sites in the Sterkfontein Valley, "Cradle of Humankind" World Heritage Site, Gauteng Province. In the following, I shall not attempt to recapitulate the convoluted history of the dating of these sites or, for the most part, the current controversies surrounding this topic (and especially Sterkfontein). This would be the topic of a separate chapter. I shall simply try to provide some reasonable estimates of the dating of the major faunas and members of these sites, based on the latest analyses, noting at the same time that the dates given here are likely to change as new data are gathered and new dating techniques developed.

Before discussing the hominin-bearing localities, there are a few early Pliocene ones that, while not necessarily better dated, have not generated the same degree of controversy, presumably because they do not include any hominin fossils. These include Jägersquelle and Nosib in the Otavi Mountains of Namibia (Senut et al. 1992) but, more importantly, the earliest Pliocene (ca 5.2–5.0 Ma on the basis of marine-terrestrial correlation) locality of Langebaanweg in the Cape Province of South Africa (cf. various contributions in *African Natural History* 2:173–202). This is probably the richest fossil vertebrate site in Africa and one of the richest in the world, with hundreds of thousands of fossils recovered, representing more than 200 species including over 80 species of mammals.

The hominin-bearing localities can conveniently be separated into two groups, those that on present data are mainly Pliocene in age (Makapansgat, Sterkfontein, Schurveburg, Bolt's Farm) and those that are wholly or to a large extent Pleistocene in age (Taung, Swartkrans, Kromdraai, Coopers, Gladysvale, Drimolen, Gondolin). Among the former group, Schurveberg and Bolt's Farm are very poorly dated, and their placement in this group is mainly by biostratigraphic comparison with the other two localities. Many attempts have been made to date both Makapansgat and Sterkfontein. Of the two, Makapansgat seems more securely dated by magnetostratigraphy, which indicates dates of >3.5 Ma for Mb. 1, 3.5–3.2 Ma for Mb. 2, 3.2–3.1 Ma for Mb. 3, and <2.5 Ma for Mb. 4 (Partridge et al. 2000).

Of all the hominin-bearing localities, that subject to the most debate is probably Sterkfontein, not least because of the find (as yet undescribed) of the "Littlefoot" hominin skeleton. Partridge et al. (2000) carried out magnetostratigraphic analyses of sediments from Mb. 2 (from which "Littlefoot" comes) at Sterkfontein, concluding that the Gauss/Gilbert boundary (3.58 Ma; Cande and Kent, 1995) is located within or near Mb. 2B, and the

Mammoth (3.33–3.22 Ma) and Kaena (3.11–3.04 Ma) reversed subchrons are near Member 2C. This would then provide an age of ca. 4–3 Ma for Mb. 2, with an age for Mb. 3 of >2.5 Ma. However, Berger et al. (2002) challenged this interpretation and suggested that what Partridge et al. took to be the Gauss/Gilbert boundary is, in fact, the Gauss/Kaena boundary (about 0.5 Ma younger). This changes the remainder of the interpretation as well, such that what was taken to be the Mammoth and Kaena subchrons are instead two reversed intervals of the Matuyama chron separated by the Réunion normal subchron. Under this interpretation, Mb. 2 would be dated <3.1 Ma to <2 Ma. Partridge et al. (2003) in turn challenged this on the basis of cosmogenic burial dates for ^{26}Al and ^{10}Be, which suggest that what was originally taken to be the Gauss/Gilbert boundary is instead the boundary between the Gilbert reversed chron and the Cochiti normal subchron (4.18 Ma), making Mb. 2 0.5 Ma older than in their previous interpretation. Other, more recent, and as yet unpublished studies suggest that Sterkfontein might in fact be much younger than this, with an age similar to that of the other Cradle of Humankind localities. Thus, the dating of Sterkfontein is at present more uncertain than ever.

Of the remaining hominin localities, Taung may be the oldest. This locality includes two distinct fossil-bearing units, called the Hrdlicka and Dart deposits. Of these, the Hrdlicka deposits are dated biostratigraphically to ca. 2.5–2.3 Ma, while the Dart deposits are slightly older by perhaps 200,000 years (McKee, 1993).

None of the other sites is well constrained biostratigraphically. The best is probably Kromdraai B, dated to ca. 1.9–1.6 Ma (with Kromdraai A about the same age; Thackeray et al. 2001). The other sites mentioned also seem for the most part to fall within the time interval 2–1.5 Ma (e.g., Drimolen: Keyser et al., 2000; Gondolin: Watson, 1993). However, any or all of them may (and probably do) include deposits that are older and/or younger than this, and the difficult dating of these sites remains a serious impediment to understanding interregional faunal evolution in Africa.

There are, of course, numerous middle and late Pleistocene localities in southern Africa. As in the case of North Africa, these are many too many to review here, so only a modest selection of interesting sites can be mentioned. A crucial locality that suffers the usual southern African problem of uncertain dating is Elandsfontein in the Cape Province, South Africa. This has been dated biostratigraphically to ca. 1–0.6 Ma (e.g., Klein et al., 2007). However, the fauna includes the sabercat *Megantereon*, which is otherwise extinct in Africa by ca. 1.4 Ma. Thus, either there is some admixture of older elements in the Elandsfontein fauna, or southernmost Africa was a last refuge for this genus in Africa, and fortuitous sampling has revealed its presence at Elandsfontein. Either of these options seems possible. Another middle Pleistocene locality, dated on bio- and magnetostratigraphy, is Buffalo Cave, Limpopo, South Africa (Herries et al., 2006). This locality includes sediments dating from the Olduvai event (1.95–1.78 Ma) to the Brunhes/Matuyama boundary (780,000 yrs), with the faunal-bearing parts dated between the Jaramillo event (1.07–0.99 Ma) and the Brunhes/Matuyama. A similar age may possibly apply to Tierfontein in the Orange Free State, South Africa (Cooke, 1974).

Late Pleistocene localities include Ysterfontein, Elands Bay Cave, Hoedjiespunt, Die Kelders, Cave of Hearths, Klasies River, and Pinnacle Point, all in South Africa and all important to the study of the emerging behavioral repertoire of *Homo sapiens* (Klein et al., 2004, 2007; Marean et al., 2007).

Eastern African Localities

WESTERN KENYAN EARLY AND MIDDLE MIOCENE (FIGURE 3.3)

The Kenyan early and middle Miocene can be said to be the foundation for understanding this part of the stratigraphic column across all of eastern Africa. However, it is a shaky foundation, as only a few of the many localities can be said to be firmly dated, and many are subject to alternative interpretations in which their ages are older or younger by several million years. Western Kenya includes a large number of lower and middle Miocene localities, and in order to make this brief presentation tractable, I shall subdivide the region into nine areas, more or less according to the scheme of Pickford (1986a), and discuss the localities and stratigraphy within each of these separately. There is much current work

on the faunas and dating of these localities, and it is to be expected that dates will change, though hopefully at least relative positions within each area will not. Most of the areas include sediments that are of lower middle Burdigalian in age, but some have been dated to the Langhian.

Pickford's (1986a) first area is the Songhor-Koru-Muhoroni area. The oldest locality in this area, and the oldest in the west Kenyan Miocene, is Meswa Bridge in the Muhoroni Agglomerate, bracketed in age between 23.5 and 19.6 Ma, with a most likely age of 23–22 Ma (Pickford and Tassy, 1980; Pickford and Andrews, 1981). The site includes the proboscidean *Eozygodon morotoensis*, which it shares with Moroto in Uganda, recently dated by [40]Ar/[39]Ar incremental heating to >20 Ma (Gebo et al., 1997), but also the ptolemaiid *Kelba quadeemae*, which it shares with the younger west Kenyan localities of Songhor, Legetet, Chamtwara, Rusinga Island,

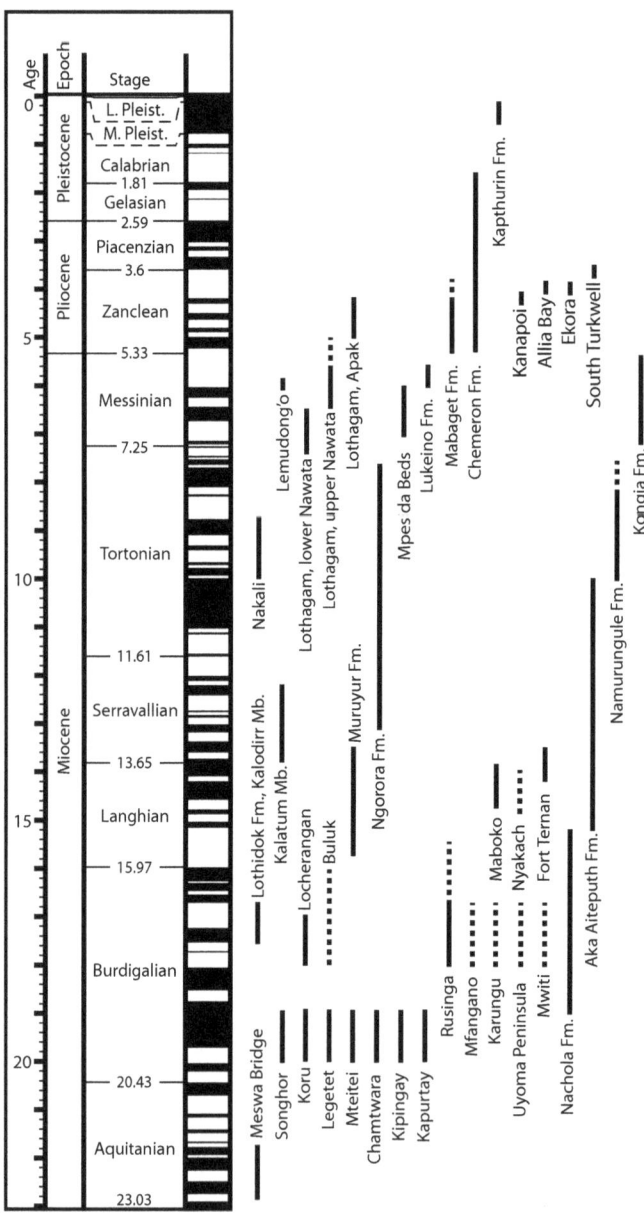

FIGURE 3.3 Approximate correlation of Kenyan terrestrial mammal localities, exclusive of the Koobi Fora and Nachukui Formations, to the global timescale (timescale based on Gradstein et al., 2004). For information about dating methods, see the text.

and Mfangano Island (Cote et al. 2007; Gunnell et al., this volume, chap. 7), tying Meswa Bridge firmly into the west Kenyan Miocene sequence. Most of the sites in the Songhor-Koru-Muhoroni area are similar in age to one another and date between 20 and 19 Ma (Bishop et al., 1969), tending toward the younger end of this range. These include localities in the Koru and Legetet formations, as well as Songhor, Mteitei, Chamtwara, Kipingay, and Kapurtay. They are all placed in Pickford's (1981) Faunal Set I.

The second area is Rusinga Island. The complex lower Miocene succession at Rusinga Island was subdivided by Van Couvering (1972) into two groups: the Rusinga Group and the Kisingiri Group. Although the sediments of the Kisingiri Group are fossiliferous and have been collected on Rusinga Island as well as the Uyoma Peninsula (discussed later), the vast majority of fossils, including two species of *Proconsul* as well as other primates, at Rusinga come from Rusinga Group sediments. The Rusinga Group in turn was subdivided into six formations (Van Couvering, 1972; Pickford, 1986a), of which most are fossiliferous, though the majority of fossils have come from the Hiwegi and Kulu Formations. Despite this complicated lithology, most of the lower Miocene sequence at Rusinga Island is currently thought to have been deposited over about 0.5 Ma around 17.8 Ma (Drake et al., 1988), with deposition in the Kulu Formation possibly continuing until about 16 Ma.

The third area is Mfangano Island. Sediments on this island are broadly the same as those of Rusinga (see Drake et al., 1988, for correlation between Rusinga Island sediments and those of Karungu, Uyoma, and Mfangano). Thus, the age of the fossils from Mfangano is thought to be the same as on Rusinga—around 17.9 Ma (Drake et al., 1988). The same can be said for the majority of sediments on the Uyoma Peninsula (Pickford's area 4), Karungu (Pickford's area 5), and Gwasi-Homa Bay (Pickford's area 6), as well as Mwiti (= Kajong).

Pickford's area 7 is the Maboko-Ombo area. The sediments on Maboko Island, another important primate locality, are younger than those of the preceding areas. Feibel and Brown (1991) discuss the age of these sediments. They sampled two levels, a tuff from bed 8 and a phonolite flow that caps the sedimentary sequence. Taking into account possible contamination from older sediments, these authors suggest ca. 14.7 Ma for the maximum age of bed 8 and a best estimate of 13.8 Ma for the phonolite flow. The majority of Maboko fossils come from beds 3 and 5, which are slightly younger than bed 8, but not by much. Thus, the Maboko sequence, which also includes Ombo (Bishop, 1967), Kaloma, and Majiwa (Andrews et al., 1981), is probably constrained to within the middle and upper Langhian.

The eighth area is the Sondu-Kerichiro-Muhoroni area, including the Nyakach Formation. The Nyakach Formation is not well dated, though it does have a date of 13.4 ± 1.3 Ma from a tuff in the middle of the sequence (Pickford, 1986a). However, the Nyakach fauna is thought by Pickford (e.g., 1986a) to be older than Fort Ternan, which is dated ca. 14 Ma (see later discussion) and an age similar to that of Maboko (i.e., in the Langhian) seems reasonable.

The ninth, and for our purposes the last, area (area 10, the Homa Peninsula, does not include any Miocene fossils) is Fort Ternan. This much discussed locality, which includes the pivotal hominoid *Kenyapithecus wickeri* (see Harrison, this volume, chap. 24), has been generally thought to be ca. 14 Ma or slightly older. However, recent dates on anorthoclases and biotites from a phonolite at the locality using both whole-rock K/Ar and single-crystal ^{40}Ar/^{39}Ar methods gave a mean date for layers over- and underlying the bone beds of

13.7 ± 0.3 and 13.8 ± 0.3 Ma (Pickford et al., 2006). This would place the locality in the uppermost Langhian, possibly extending into the lowermost Serravallian.

TUGEN HILLS SEQUENCE (FIGURE 3.3)

The extensive fossiliferous sediments in the Tugen Hills, Baringo District, Kenya, include localities dated from the Burdigalian to the late Pleistocene. It is subdivided into a series of formations and beds, the most important of which are (from oldest to youngest): Muruyur Formation, Ngorora Formation, Mpesida Beds, Lukeino Formation, Mabaget Formation (= Chemeron Formation, northern extension), Chemeron Formation, and Kapthurin Formation (not all of these have been formally defined).

The Muruyur Beds interfinger with phonolite flows of the early middle Miocene. Perhaps the most important locality in these beds is Kipsaraman (Behrensmeyer et al., 2002; Pickford and Kunimatsu, 2005). The fossiliferous levels of this locality belong to Members 1–3 of the lower part of the Muruyur Formation These levels are dated to the early Langhian, between ca. 15.8 and 15.4 Ma (Behrensmeyer et al., 2002). The Muruyur Beds otherwise extend upward in the sequence to ca. 13.5 Ma.

The Ngorora Formation is underlain by the Tiim Phonolites, which have a date of 13.15 Ma, and the capping Ewalel Phonolite is associated with dates of 7.6 and 7.2 Ma (Deino et al., 1990); that is, the Ngorora Formation extends from the mid-Serravallian to the latest Tortonian. The Ngorora Formation includes a number of important localities, especially in its lower part where, for example, locality BPRP #38 in the Kabarsero type section of the Formation has yielded specimens of cercopithecine monkeys and hominoids (Hill et al., 2002).

A trachyte flow just below the base of the Mpesida Beds has been dated 7.0 ± 0.4 Ma (Chapman and Brook, 1978; Kingston et al., 2002), while trachytes above the sedimentary sequence have been variously dated between 6.2 and 6.36 Ma. This suggests that the Mpesida Beds are certainly bracketed between 7 and 6 Ma and that the fossil material is older than 6.37 Ma (early Messinian).

The Mpesida Beds are followed by the Lukeino Formation. Indeed, some sedimentary exposures previously thought to belong in the latter may turn out to be part of the former (Kingston et al., 2002). The base of the Lukeino Formation (Kapgiyma Mb.) is set by Sawada et al. (2002) to the boundary between chrons C3An.1r and C3An.1n, dated to 6.14 Ma (Cande and Kent, 1995). ^{40}Ar/^{39}Ar dates at the top of the Kapcheberek Mb. (top of the Lukeino Formation) have a mean of 5.73 ± 0.05 Ma, with the capping Kaparaina Basalt slightly younger (Sawada et al., 2002), leaving the Lukeino Formation entirely within the Messinian. The oldest specimens of the hominin *Orrorin tugenensis* are from the Lukeino Formation.

The Mabaget Formation overlies the Lukeino Formation. Only its base is reasonably well dated, since it is underlain by the Kaparaina Basalt, the top of which is dated to ca. 5.3 Ma. The age of top of the Formation is not well established, but it must be younger than 4.3 Ma on the basis of dates on volcanic tuffs within the Formation (Deino et al., 2002).

The better-defined Chemeron Formation is partly laterally equivalent to the Mabaget Formation. The base of the Chemeron Formation is just over 5.3 Ma (Deino et al., 2002), while the top of the formation is dated to 1.60 ± 0.05 Ma on the basis of ^{40}Ar/^{39}Ar dates on a tuff at the top of the formation overlying the Ndau Trachymugerite. This tuff is in turn unconformably overlain by the Kapthurin Formation (Deino and Hill, 2002).

The hiatus between the Chemeron and Kapthurin Formations is approximately 1 Ma, on the basis of dates on the Kaseurin Basalt at the base of the latter formation that have a mean of 0.61 ± 0.04 Ma (Deino and McBrearty, 2002). The minimum age of the Kapthurin Formation is 0.235 ± 0.002 Ma based on dates on a tephra from the top part of the formation (Deino and McBrearty, 2002).

NORTHERN KENYA MIOCENE (FIGURE 3.3)

Among the most important Miocene sequences in Africa is that from the Lothidok Range, west of Lake Turkana in northern Kenya, which extends from the late Oligocene (see Seiffert this volume, chap. 2) to the late Serravallian. This fossiliferous area was first described by Arambourg (1943), while the stratigraphy and dating of the sequence is discussed by Boschetto et al. (1992). The Kalakol Basalts, in which the late Oligocene Eragaleit Beds sit, cross the Oligocene-Miocene boundary. The base of the overlying Lothidok Formation is dated to slightly less than 17.9 Ma based on dates from a Kalakol Basalt flow just below the base of the formation. The Lothidok Formation is subsequently divided into (from oldest to youngest) the Moruorot, Kalodirr, Naserte, Lokipenata, and Kalatum Mbs. The Moruorot Mb. has not been dated, but the Kalodirr tuffs near the base of the Kalodirr Mb. have mean ages of around 17.5 Ma (Boschetto et al., 1992). The boundary between the Kalodirr and Naserte Mbs is marked by the Naserte Tuffs, which have been dated to a mean age of 16.8 ± 0.2 Ma. The Kalodirr Mb. includes the important mammal sites of Moruorot, Kalodirr, and Kanukurinya (see, e.g., Leakey and Leakey, 1986a, 1986b, 1987). Near the top of the formation are the fossil sites of Esha and Atirr, which lie within the Kalatum Mb. The base of the latter is dated to ca. 13.8 Ma based on ages of the Kalatum phonolite. The Kamurunyang, ca. 50 m below the top of the formation is dated to a maximum of 13.2 ± 0.2 Ma, while the minimum age for the Kalatum Mb. (and Lothidok Formation as a whole) is close to 12.2 Ma, based on dates for the overlying Loperi Basalts (Boschetto et al., 1992).

A Miocene site west of Lake Turkana outside the Lothidok Formation is Locherangan, which has been dated biostratigraphically to between 18 and 17 Ma (Anyonge, 1991).

Another Miocene locality in northern Kenya is Buluk, to the east of Lake Turkana. A series of dates obtained from ignimbrites, basalts, and tuffs in various members of the Gum Dura Formation overlying the fossiliferous layers at Buluk gave dates between 15.8 and 17.2 Ma, suggesting that the fossils are slightly older than this, but they are unlikely to be older than 18 Ma (McDougall and Watkins, 1985).

SAMBURU HILLS SEQUENCE (FIGURE 3.3)

Another important Mocene sedimentary succession in Kenya is the Samburu Hills sequence south of Lake Turkana. This succession includes a series of formations (from oldest to youngest): Nachola, Aka Aiteputh, Namurungule, Kongia, and Tirr Tirr (Sawada et al., 1998). The last two are separated from the older formations and from each other by hiatuses. The Nachola Formation has been bracketed through a sequence of dates to between ca. 19 and 15.5 Ma (Sawada et al., 1998, 2006). The recently described hominoid Nacholapithecus was found at a level close to the top of this formation.

The Aka Aiteputh Formation is less well dated but can be bracketed between ca. 15.5 and 10 Ma, while the Namurungule Formation is dated near its base to ca. 10 Ma, but the top of the formation has only been dated as older than 7.5 Ma,

based on the age of the base of the Kongia Formation and the hiatus between them. Nevertheless, the Namurungule Formation lies entirely within the Tortonian. The Samburupithecus specimen found there has been dated to close to 9.5 Ma (Sawada et al., 2006).

The base of the Kongia Formation is dated to ca. 7.2 Ma and the top to ca. 5.3 Ma, which means that it covers nearly all of the Messinian, while the Tirr Tirr Formation, which unconformably overlies the Kongia, is dated 4.1 Ma at its base and extends into the late Pleistocene.

OTHER KENYAN SITES (FIGURE 3.3)

Another important Kenyan Miocene locality is Nakali near Maralal, from whence Nakalipithecus was recently reported (Kunimatsu et al., 2007). This locality includes three members, and the fossils have been recovered from the Upper Mb. Dates on anorthoclase in the uppermost part of the Lower Mb. average 9.86 ± 0.09 Ma, while a date from the Middle Mb. is 10.10 ± 0.12 Ma. These dates, together with magnetostratigraphy, place the Upper Mb. (which exhibits reversed magnetic polarity) in Chron C5n.1r (9.88–9.92 Ma) and the hominoid bed within chronozone C5n.1n (9.74–9.88 Ma), both therefore in the Tortonian.

A recently discovered and studied locality from the Kenyan late Miocene is Lemudong'o, in Narok District, west of Nairobi. Four tuffs bracket the fossiliferous deposits at the locality. These have been dated by $^{40}Ar/^{39}Ar$ to between 6.04 ± 0.019 Ma and 6.108 ± 0.018 Ma (Deino and Ambrose, 2007).

Finally, the important locality of Lothagam, west of Lake Turkana, includes members dated to both the Miocene and the Pliocene (McDougall and Feibel, 1999). The lowermost beds at Lothagam are the Nabwal Arangan Beds, which are mostly middle Miocene, between ca. 14 and 12 Ma, with an uppermost basalt dated to 9.1 Ma. The main fossiliferous deposits consist of the lower and upper members of the Nawata Formation, and the Apak Mb. of the Nachukui Formation. The lower Nawata is dated on the basis of tuffaceous horizons to between 7.4 ± 0.1 Ma and 6.5 ± 0.1 Ma (the latter also the age of the lower Nawata–upper Nawata boundary). The upper boundary of the Nawata Formation is poorly constrained but is thought to be approximately 5 Ma. A tuff in the Apak Mb. of the Nachukui Formation is dated 4.22 ± 0.22 Ma, while the overlying Lothagam Basalt is dated to 4.20 ± 0.22 Ma (McDougall and Feibel, 1999).

Kenya, of course, has many fossil localities apart from those that are mainly Miocene and the Turkana Basin sequences discussed elsewhere herein. One western Kenyan area that includes numerous, mainly Pleistocene sites is the Homa Peninsula, which includes such localities as Kanam, Kanjera, and Rawi (Ditchfield et al., 1999). The Rawi Formation is dated to the uppermost Gauss Chron (i.e., latest Piacenzian). Kanjera south is suggested to include the base of the Olduvai Subchron (1.95 Ma) as well as the Réunion Subchron (2.150–2.140 Ma), while Kanjera north is thought to include the Jaramillo Subchron (1.070–0.990 Ma) and the Brunhes-Matuyama boundary at 0.78 Ma. Kanam includes several levels, but they are as yet not well dated. Other formations in the area include the Kasibos Formation, which is placed in the Jaramillo subchron; the Apoko Formation, which is thought to be late middle Pleistocene; and the Abundu Formation, which is early Pleistocene in age.

The archeological site of Chesowanja has a minimum date provided by the Chesowanja Basalt (1.42 ± 0.07 Ma) and a maximum biostratigraphic age of 1.98, but the latter figure is probably much too high, and the maximum age of Chesowanja may be closer to 1.6 Ma (Gowlett et al., 1981).

Olorgesailie includes a series of important archeological and paleontological sites that have been excavated for over 50 years. The principal exposures at this locality approximately span the time period 1–0.5 Ma (Deino and Potts, 1990).

UGANDA (FIGURE 3.4)

Localities in Uganda are with a few exceptions not as well studied as those in Kenya, and controversy surrounds the chronostratigraphy of many of them. Among the most important are the lower Miocene (?upper Aquitanian–lower Burdigalian) localities of Napak and Moroto. The former includes a number of fossiliferous deposits that are intercalated with tuffs. Dates on these tuffs are 19.5 ± 2 Ma and 18.3 ± 0.4, though evidence suggests that the latter date is anomalously young. On the basis of current evidence, the fossiliferous horizons are judged to be between 20 and 19 Ma (Bishop et al., 1969; Senut et al., 2000; MacLatchy et al., 2006). The dating of Moroto I and II is currently controversial. Early attempts at biostratigraphic correlation suggested an age as young as 16.5–14.5 Ma. This was later revised to ca. 17.5 Ma (Pickford et al., 2003). However, $^{40}Ar/^{39}Ar$ incremental heating dating of the Moroto I capping lava gave an isochron age of 20.61 ± 0.05 Ma, while the Moroto II lava gave an age of >20 Ma, though demonstrating isotopic

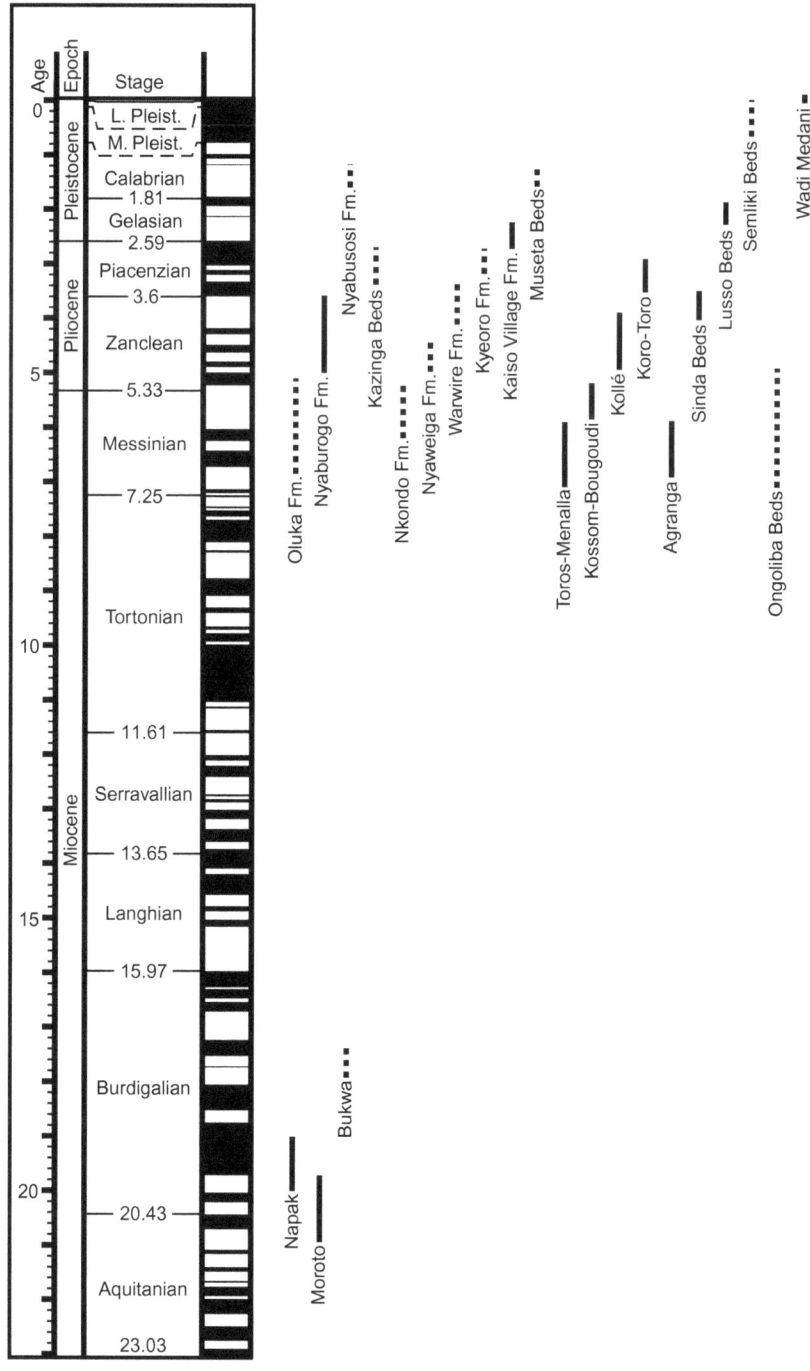

FIGURE 3.4 Approximate correlation of Ugandan and central African terrestrial mammal localities to the global timescale (timescale based on Gradstein et al., 2004). For information about dating methods, see the text.

disturbance (Gebo et al., 1997). Thus, on present evidence, Moroto could be older than the localities in the Songhor area placed within Pickford's (1981) Faunal Set I.

Another early Miocene site in Uganda is Bukwa. This locality has been thought to be as much as 22 Ma, but more recent biostratigraphic correlation suggests it to be similar in age to Rusinga or slightly younger, at ca. 17.5 Ma (Pickford, 2002).

In addition to these localities, there are numerous fossiliferous localities in the Ugandan part of the Albertine Rift Valley, as summarized by Pickford et al. (1993). The stratigraphy and dating of these localities is discussed in Pickford et al. (1993) but is still poorly known and at times inconsistent. The following is a brief summary of the current state of knowledge.

Pickford et al. (1993) distinguish three fossiliferous areas. The first of these is the Kisegi-Nyabusosi area in the southern Albertine Rift. The lowermost formation in this area is the probably middle Miocene Kisegi Formation, from which few fossils have been found. The likewise poorly collected Kakara Formation is next. This formation is probably of Tortonian age, though more exact dates are not available. The overlying Oluka Formation is richer in fossils. This formation is likely to be Messinian in age. The oldest Pliocene formation in the area is the Nyaburogo Formation, with a date of about 5–3.5 Ma. Overlying this is the Nyakabingo Formation, which has yielded too few fossils for dating. However, the Nyabusosi Formation overlying it has been assigned an age of ca. 1.5 Ma. The Rwebishengo Beds, finally, are of late Pleistocene age.

The second area is that surrounding Lake Edward. In this area there is only one currently datable unit, the Kazinga Beds. This unit has yielded a small fauna, including *Ugandax* and *Hippopotamus*, and may be of lower Pliocene (middle Zanclean) age.

The third and final area in the Ugandan part of the Albertine Rift is the Nkondo-Kaiso area. This area has yielded the richest fossil assemblages in the western rift. The lowermost beds in this area belong to the Nkondo Formation, divided into the Nkondo and Nyaweiga Mbs. The former has yielded a rich mammal fauna that may be correlated with the Lukeino Formation in Kenya—to the middle Messinian, ca. 6 Ma. The overlying Nyaweiga Mb. is similar but clearly younger and may be dated to the lower Zanclean, ca. 5–4.5 Ma.

The Warwire Formation is relatively poorly dated, but most of the fossils come from beds underlying the Warwire Tuff, thought by correlation with the Lomogol Tuff to be about 3.6 Ma old, so the fossils may be up to 4 Ma in age. The next formation, the Kyeoro, includes few mammals but has been suggested to be ca. 3 Ma. The overlying Kaiso Village Formation, on the other hand, is rich in mammal fossils. It has been suggested (Gentry and Gentry, 1978) to be similar in age to the Shungura Formation, Mb. F of the Omo Group—ca. 2.35 Ma. The Museta Beds, finally, have an age similar to that of the Nyabusosi Formation—ca. 1.5 Ma.

TANZANIA (FIGURE 3.5)

There are fewer well dated sites with mammal faunas in Tanzania than in Kenya and Uganda, but the country does include two of the most important Plio-Pleistocene localities in all of Africa: Laetoli and Olduvai.

Laetoli consists of a series of beds, none of which are precisely dated. The three main fossiliferous beds are the Lower and Upper Laetolil Beds, and the Upper Ndolanya Beds. The Lower Laetolil Beds are of uncertain age, but a maximum age of

ca. 4.3 Ma and a minimum age of 3.8 Ma (base of the Upper Laetolil Beds) have been suggested (Drake and Curtis, 1987). The Upper Laetolil Beds are somewhat better constrained, with a minimum age of ca. 3.4 Ma to go with the maximum age of 3.8 Ma. The Upper Ndolanya Beds are dated to ca. 2.7–2.6 Ma, though these dates are not tightly constrained (Drake and Curtis, 1987).

Olduvai has been extensively studied for its archeological contexts and less so for its paleontology. The locality is subdivided into Beds I–IV and the Masek Beds (Hay, 1976; Potts, 1988). The beds are overall not particularly well dated, but the following dates are currently more or less accepted: Bed I is between 1.87 and 1.7 Ma; Bed II, 1.7–1.2 Ma; Bed III, 1.2–ca. 0.8 Ma; Bed IV, ca. 0.8 Ma–ca 0.6 Ma; and the Masek Beds, ca. 0.6 Ma–0.4 Ma.

A third important sedimentary sequence in Tanzania is the deposits of the Manonga Valley (Harrison, 1997). These comprise the Wembere-Manonga Formation, with three members, none of which is firmly dated. The Ibole Mb. is ca. 5.5–5 Ma in age, the Tinde Mb. 5–4.5 Ma, and the Kiloleli Mb. ca. 4.5–4 Ma (Harrison and Baker, 1997).

There are, of course many other localities, especially archeological ones, in Tanzania. I will mention only two here. Peninj is well known for the presence of *Paranthropus boisei*. This was found in outcrops of the Humbu Formation, dated to ca. 2–1.3 Ma on the basis of magneto- and biostratigraphy. The Basal Sandy Clay, from which the hominin remains derive, are no younger than 1.6 Ma (Domínguez-Rodrigo, 1996). The much younger archeological site of Isimila has been dated through uranium series dating of bone to ca. 250,000 years (Howell et al., 1972).

MALAWI (FIGURE 3.5)

Malawi has not been as well studied as other Rift Valley countries. However, fossiliferous exposures have been found at Chiwondo. These consist of several levels and are subdivided into the northern Karonga and southern Uraha localities. The units within them are dated on biostratigraphy, where unit 2 (lower) is dated ca. 4 Ma, while unit 3A has faunas dated to ca. 3.8–2 Ma and younger and to 1.6 Ma and younger (Bromage et al., 1995).

CENTRAL AFRICA (FIGURE 3.4)

Central Africa (broadly speaking) has a small but growing number of fossil mammal localities, some of which are among the most important in their time periods. The localities that have garnered the most attention in recent years are those from Chad. The oldest of these is Toros-Menalla, a collection of sites from the late Miocene that include, among a very diverse mammal fauna, specimens of *Sahelanthropus tchadensis*. Dating of these sites has been controversial and mainly based on biostratigraphic correlation with East African sites (Vignaud et al., 2002). Recently, an attempt to date these sites using cosmigenic nuclides of $^{10}Be/^9Be$ resulted in dates of 7.2–6.8 Ma (Lebatard et al., 2008), which is roughly in line with the biostratigraphic data.

A somewhat younger locality is Kossom-Bougoudi, which is dated biostratigraphically to the uppermost Messinian. The site is older than Kollé (discussed later) and younger than Sahabi (Brunet et al., 2000; Brunet, 2001). The Kollé site, on the other hand, is biostratigraphically dated to the lower Zanclean, ca. 5–4 Ma (Brunet et al., 1998). Finally, the youngest site in the Chad sequence is Koro-Toro, source of

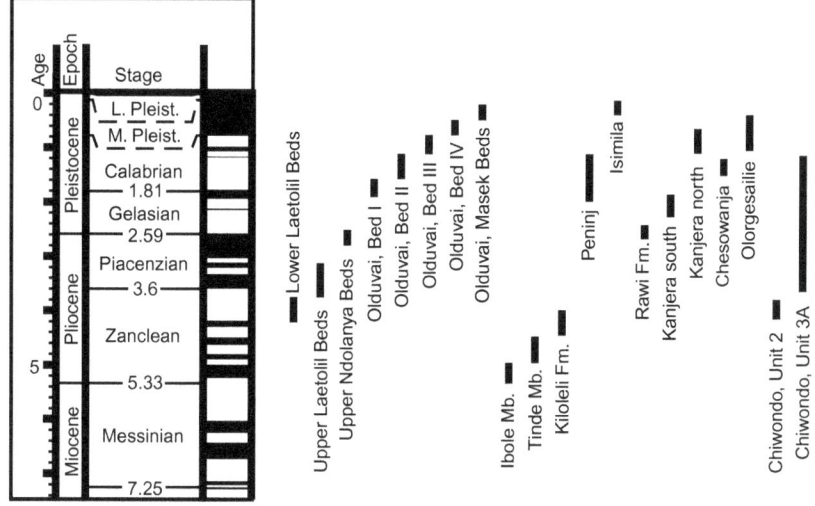

FIGURE 3.5 Approximate correlation of Tanzanian and Malawian terrestrial mammal localities to the global timescale (timescale based on Gradstein et al., 2004). For information about dating methods, see the text.

Australopithecus bahrelghazali. This locality has been biostratigraphically dated to 3.5–3 Ma and recently by $^{10}Be/^{9}Be$ to 3.58 ± 0.27 Ma (Brunet et al., 1997; Lebatard et al., 2008). A final Chadian site is Agranga, which has a date similar to Toros-Menalla at 7–6 Ma (Lihoreau, 2003).

A second important sequence of localities in Central Africa is the Western Rift of the Democratic Republic of Congo. Like its counterpart in Uganda, this sequence is poorly dated. Boaz (1994) divides it into two main regions, the lower Semliki near Lake Albert and the upper Semliki near Lake Edward. The lower Semliki sediments include a sequence of beds, from the Edo Beds (oldest), over the Mohari Beds, Kabuga Beds, Sinda Beds, Ndirra Beds, and Katomba Beds (youngest except for overlying, unnamed late Pleistocene/Holocene sediments). Of these only the Sinda Beds have a date: ca. 4.1 Ma, based on a tuff that can be correlated with the Moiti Tuff at Koobi Fora (Boaz, 1994). Boaz (1994) considers the majority of the Sinda Beds to overlie this tuff. On the other hand, Makinouchi et al. (1992) place the largest part of the Sinda Beds in the Miocene, but it appears that their Sinda Beds lower member is the equivalent of the Edo Beds of Boaz (1994) and their middle member equivalent to the Mohari Beds, so the discrepancy may be minor.

The upper Semliki sediments are divided into the Lusso Beds, the Semliki Beds, late Pleistocene/Holocene terraces, and Katwe Ash (Boaz, 1994). The Lusso Beds are suggested on biostratigraphy to be 2.3–2 Ma (i.e., lower Pleistocene), while the Semliki Beds are suggested to possibly be of middle Pleistocene age.

A few other central African localities may be mentioned. The Ongoliba Beds in the Democratic Republic of Congo are suggested by Pickford to be Messinian in age. Wadi Medani in the Sudan is a late Pleistocene locality including a skull of *Colobus* (Simons, 1967; Jablonski and Frost, this volume, chap. 23). Finally, Malembe, in the Cabinda Province, Angola, is a very interesting Oligocene/early Miocene site with a diverse fauna (Hooijer, 1963; Pickford, 1986b; Seiffert, this volume, chap. 2).

HORN OF AFRICA (FIGURE 3.6)

This region includes many important mammal localities, mainly from the late Miocene onward. Much of the work is ongoing, and dates and stratigraphy are not always published.

The following discussion represents a more or less current state of affairs and is very far from complete or definitive.

Hadar is one of the best-known localities in the region. The Hadar Formation includes three major units. The Sidi Hakoma Mb. lies between the Sidi Hakoma Tuff (3.4 Ma) and Triple Tuff 4 (3.22 Ma), the Denen Dora Mb. lies between Triple Tuff 4 and the Kada Hadar Tuff (3.18 Ma), and the Kada Hadar Mb. lies between the Kada Hadar Tuff and tuff BKT-3 (ca 2.9 Ma) (Walter, 1994).

Fejej is one of the few lower Miocene sites in the region and has a date from a capping lava of 16.18 ± 0.05 Ma (Richmond et al., 1998).

Gona, a study area south of Hadar, includes sediments that extend from the late Miocene to the late Pleistocene. Best known are horizons with the earliest stone tools, dated to ca. 2.6 Ma. The chronology of the late Miocene Adu-Asa Formation at Gona has recently been published (Kleinsasser et al., 2008).

The archeological site of Melka Kunturé includes several fossiliferous levels, of which the richest is Garba-IV, dated litho- and biostratigraphically to between 1.0 and 0.8 Ma (Geraads et al., 2004b; Raynal et al., 2004). Somewhat younger is Asbole, a site in the lower Awash Valley with an extensive mammal fauna. The fossil levels lie between a conglomerate and the Bironita Tuff, which is probably correlated with a Middle Awash tuff dated between 0.74 and 0.55 Ma, while the conglomerate is unlikely to be older than 1 Ma. Geraads et al. (2004a) suggest an age of between 0.8 and 0.6 Ma for the fauna.

Dikika, where recently a juvenile hominin was found, spans the Hadar (ca 3.8–2.9 Ma) and Busidima Formations (2.7–<0.6 Ma). The DIK-1 locality where the hominin was found has an age of 3.4–3.22 Ma; that is, it is in the Sidi Hakoma Mb. (Wynn et al., 2006).

Whole rock K/Ar dating on lava flows from Chorora in the Awash Basin of Ethiopia has given dates in the Tortonian, at 11–10 Ma (Geraads et al., 2002), making this one of the few pre-Messinian Neogene localities in the region.

The Middle Awash series of sedimentary deposits is one of the most significant Plio-Pleistocene sequences in Africa, where a large number of hominin taxa and specimens have been recovered (WoldeGabriel et al., 2001; White et al., 2006). The oldest formation in the Middle Awash study area is the Adu-Asa Formation. This formation has been dated by $^{40}Ar/^{39}Ar$

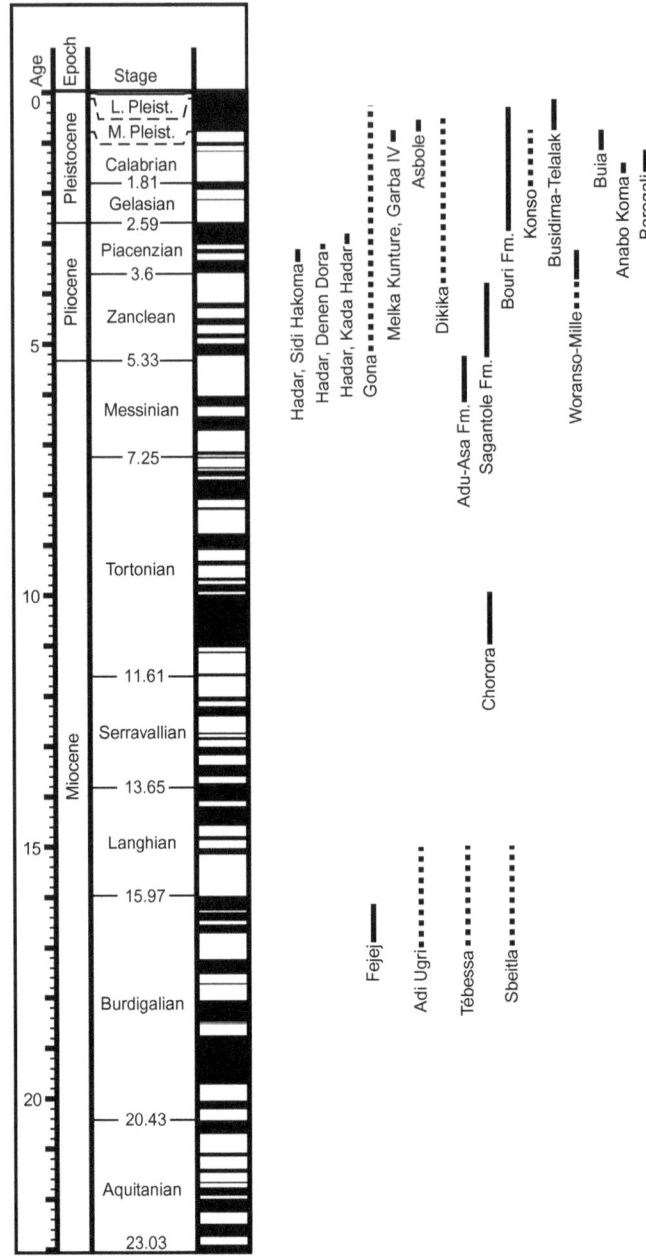

FIGURE 3.6 Approximate correlation of terrestrial mammal localities from the Horn of Africa, exclusive of the Shungura Formation, to the global timescale (timescale based on Gradstein et al., 2004). For information about dating methods, see the text.

using incremental heating techniques on tephras and lavas. Basaltic lavas dated 6.33 ± 0.07 and 6.16 ± 0.06 Ma underlie the lowermost Saraitu Mb. and provide a maximum age for the formation. The minimum age of the fossil-bearing horizons is given by a basalt at Saitune Dora dated 5.54 ± 0.17 Ma.

The Adu-Asa Formation is overlain by the Sagantole Formation, which includes a series of members dated between ca. 5.55 (lower Kuseralee Mb.) to ca. 3.8 Ma (upper Belohdehlie Mb.). On the eastern side of the Awash River, the poorly dated Pliocene Matabaietu Formation is overlain by the Wehaietu Formation, which extends into the middle Pleistocene. On the western side of the Awash River, the Bouri Formation encompasses sediments from the latest Pliocene/earliest Pleistocene (ca. 2.5 Ma) Hata Mb. over the early/middle Pleistocene Daka Mb., to the late Pleistocene Herto Mb. (Gilbert, 2003).

Another interesting locality with an extensive fauna is Konso, Ethiopia. At this locality fossils come from a series of dated intervals ranging in age from ca. 1.9 Ma to ca. 0.8 Ma, though the main fossil-bearing horizons are around 1.5–1.4 Ma in age (Katoh et al., 2002, Suwa et al. 2003). A younger locality is Busidima-Telalak (Alemseged and Geraads, 2000), which includes sediments dated biostratigraphically to the middle Pleistocene, ca. 800,000–200,000 yrs.

Finally, the recently opened studies of the Woranso-Mille area are providing another valuable sequence of fossiliferous sediments. Preliminary results indicate that the main part of the fossil-bearing sediments are ca. 3.7–3.5 Ma in age (Haile-Selassie et al., 2007), though older sediments are also present in the area.

Eritrea also has a number of important paleontological localities. The oldest of these (excluding Oligocene localities;

see Seiffert, this volume, chap. 2; Sanders et al., this volume, chap. 12) are Adi Ugri, Tébessa, and Sbeitla, which on the basis of biostratigraphy are more or less similar in age to Gebel Zelten at ca. 17–15 Ma (Vialli, 1966). Another important locality in Eritrea is Buia, which includes several units dated between ca. 1.3 and 0.9 Ma. A hominin fossil was found at the top of the Alat Formation in Buia, dated ca. 1 Ma (Abbate et al., 2004).

Finally, Djibouti, though less studied, also has fossil mammal sites. Two of the more important are Anabo Koma, biostratigraphically dated to ca. 1.6 Ma (De Bonis et al., 1988), and Barogali, an elephant butchery site, dated to ca. 1.6–1.3, also on biostratigraphy (Chavaillon et al., 1987).

TURKANA BASIN (FIGURES 3.3 AND 3.7)

The Turkana Basin in northern Kenya and southern Ethiopia includes perhaps the best known of all African sedimentary sequences and certainly those that have the most exact and detailed dating. The major localities in the basin are in the Koobi Fora, West Turkana, and Omo sedimentary sequences of the Omo group, but there are also a number of other Plio-Pleistocene localities in the basin that must be mentioned.

One such is Kanapoi, in the southwestern part of the basin. Fossils at Kanapoi, including *Australopithecus anamensis*, are found in a constrained interval of the exposures. All specimens come from above a pumiceous tuff dated 4.17 ± 0.03 Ma and nearly all from below the Kanapoi Tuff, dated 4.07 ± 0.02 Ma (Leakey et al., 1995, 1998; Feibel, 2003). A few specimens come from above this tuff but are likely to be close to it in age. Adjacent to Kanapoi is Ekora, which is slightly younger (Behrensmeyer, 1976). Allia Bay, to the southeast, has yielded fossils that come from in or near the Moiti Tuff, dated to 3.92 ± 0.04 Ma (Leakey et al., 1995).

A number of isolated sites to the west of Lake Turkana have yielded mammal fossils. These include South Turkwel, with sediments dated to between 3.58 and 3.2 Ma (Ward et al., 1999), and Eshoa Kakurongori and Nakoret, dated to ca. 3 and ca. 2 Ma, respectively (M. G. Leakey, pers. comm.). Rawe, also from the west side of Lake Turkana, has a similar date to Nakoret.

Finally, the Koobi Fora, West Turkana, and Omo sequences of the Omo Group have been so extensively studied that it is pointless to discuss them in detail here (see, e.g., Brown, 1994; McDougall and Brown, 2006, 2008). Their dating and correlation of the Shungura, Koobi Fora, and Nachukui formations to each, as well as to Hadar, are shown in figure 3.7.

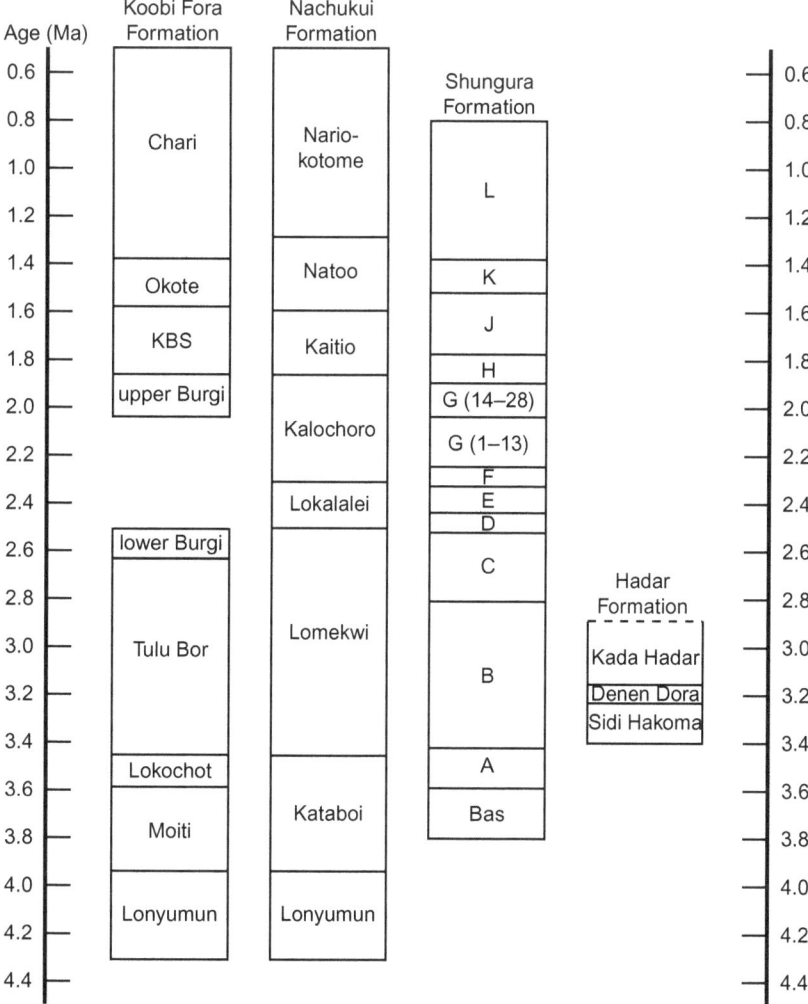

FIGURE 3.7 Correlation of the Koobi Fora, Nachukui, and Shungura Formation sedimentary sequences to each other, with the Hadar Formation at Hadar shown for comparison. From numerous sources, including Feibel et al. (1989), Brown (1994), McDougall and Brown (2006, 2008), Walter (1994).

The Usno Formation is approximately equivalent to the lower part of the Shungura Formation and dated to ca. 3.3–3.0 Ma, while the Mursi Formation, which is associated with the Mursi Basalt, is dated to ca. 4.3 Ma or slightly less.

ACKNOWLEDGMENTS

I would like to thank Bill Sanders for his herculean efforts on behalf of this volume. The many people who did the work on which this chapter is based are the ones who deserve the credit; I can only at best reiterate and at worst misinterpret what they have done. Special thanks to Martin Pickford for compiling the BACCM bibliography at www.cifeg.org, which made finding correct references a tractable job. This chapter was written while I was on sabbatical at the School of Human Evolution and Social Change, Arizona State University. My work is funded by the Swedish Research Council.

Literature Cited

Abbate, E., B. Woldehaimanot, P. Bruni, P. Falorni, M. Papini, M. Sagri, S. Girmay, and T. M. Tecle. 2004. Geology of the *Homo*-bearing Pleistocene Dandiero Basin (Buia Region, Eritrean Danakil Depression). *Rivista Italiana di Paleontologia e Stratigrafia* 110, suppl.:5–34.

Alemseged, Z., and D. Geraads. 2000. A new middle Pleistocene fauna from the Busidima-Telalak region of the Afar, Ethiopia. *Comptes Rendus de l'Académie des Sciences, Paris, Sciences de la Terre et des Planètes* 331:549–556.

Amani, F. 1991. La faune de la grotte à Hominidé du Jebel Irhoud (Maroc). Unpublished PhD dissertation, Université de Rabat, Morocco, 229 pp.

Amani, F., and D. Geraads. 1993. Le gisement moustérien du Djebel Irhoud, Maroc: Précisions sur la faune et la biochronologie, et description d'un nouveau reste humain. *Comptes Rendus de l'Académie des Sciences, Paris*, Série II, 316:847–852.

Ameur, R. 1984. Découverte de nouveaux rongeurs dans la formation miocène de Bou Hanifia (Algérie occidentale). *Geobios* 7:167–175.

Andrews, P., G. E. Meyer, D. R. Pilbeam, J. A. Van Couvering, and J. A. H. Van Couvering. 1981. The Miocene fossil beds of Maboko Island, Kenya: Geology, age, taphonomy and palaeontology. *Journal of Human Evolution* 10:35–48.

Anyonge, W. 1991. Fauna from a new lower Miocene locality west of Lake Turkana, Kenya. *Journal of Vertebrate Paleontology* 11:378–390.

Arambourg, C. 1943. Contribution à l'étude géologique du bassin du lac Rodolphe et de la basse vallée de l'Omo : Mission Scientifique Omo 1932–1933. I: Géologie et Paléontologie. *Editions Museum National d'Histoire Naturelle, Paris* 2:157–230.

———. 1968. Un suidé fossile nouveau du Miocène supérieur de l'Afrique du Nord. *Bulletin de la Société Géologique de France* 10:110–115.

———. 1979. Vertébrés villafranchiens d'Afrique du Nord. *Édition de la Fondation Singer-Polignac* 10:1–126.

Behrensmeyer, A. K. 1976. Lothagam Hill, Kanapoi and Ekora: a general summary of stratigraphy and fauna; pp. 163–170 in Y. Coppens, F. C. Howell, G. L. Isaac, and R. E. F. Leakey (eds.), *Earliest Man and Environments in the Lake Rudolf Basin*. University of Chicago Press, Chicago.

Behrensmeyer, A. K., A. L. Deino, A. Hill, J. D. Kingston and J. J. Saunders. 2002. Geology and geochronology of the Middle Miocene Kipsaramon site complex, Muruyur Beds, Tugen Hills, Kenya. *Journal of Human Evolution* 42:11–38.

Benammi, M., M. Calvo, M. Prévot, and J.-J. Jaeger. 1996. Magnetostratigraphy and paleontology of Aït Kandoula Basin (High Atlas, Morocco) and the African-European late Miocene terrestrial fauna exchanges. *Earth and Planetary Science Letters* 145:15–29.

Berger, L., R. Lacruz, and D. De Ruiter. 2002. Brief communication: Revised age estimates of *Australopithecus*-bearing deposits at Sterkfontein, South Africa. *American Journal of Physical Anthropology* 119:192–197.

Bishop, W. W. 1967. The later Tertiary in East Africa: Volcanics, sediments, and faunal inventory; pp. 31–54 in W. W. Bishop and J. D. Clark (eds.), *Background to Evolution in Africa*. University of Chicago Press, Chicago.

Bishop, W. W., F. J. Miller, and J. A. Fitch. 1969. New potassium-argon age determinations relevant to the Miocene fossil mammal sequence in East Africa. *American Journal of Science* 267:669–699.

Black, C. 1972 A new species of *Merycopotamus* (Artiodactyla: Anthracotheriidae) from the late Miocene of Tunisia. *Notes du Service Géologique, Tunis 37, Travaux Géologiques* 6:5–39.

Boaz, D. D. 1987. Taphonomy and paleoecology at the Pliocene site of Sahabi, Libya; pp. 337–348 in N. T. Boaz, A. El-Arnauti, A. W. Gaziry, J. De Heinzelin, and D. D. Boaz (eds.), *Neogene Paleontology and Geology of Sahabi*. Liss, New York.

———. 1994. Significance of the Western Rift for hominid origins; pp. 321–343 in R. S. Corruccini and R. L. Ciochon (eds.), *Integrative Paths to the Past: Paleoanthropological Advances in Honor of F.C. Howell*. Prentice Hall, Englewood Cliffs, N.J.

Boschetto, H. B., F. H. Brown and I. McDougall 1992. Stratigraphy of the Lothidok Range, northern Kenya, and K/Ar ages of its Miocene primates. *Journal of Human Evolution* 22:47–71.

Brandy, L. D., and J.-J. Jaeger. 1980. Les échanges de faunes terrestres entre l'Europe et l'Afrique nord-occidentale au Messinien. *Comptes Rendus de l'Académie des Sciences, Paris* 291:465–468.

Bromage, T. G., F. Schrenk, and Y. M. Juwayeyi. 1995. Paleobiogeography of the Malawi Rift: Age and vertebrate paleontology of the Chiwondo Beds, northern Malawi. *Journal of Human Evolution* 28:37–57.

Brown, F. H. 1994. Development of Pliocene and Pleistocene chronology of the Turkana Basin, East Africa, and its relation to other sites; pp. 285–312 in R. S. Corruccini and R. L. Ciochon (eds.), *Integrative Paths to the Past. Paleoanthropological Advances in Honor of F. Clark Howell*. Prentice Hall, Englewood Cliffs, N.J.

Brunet, M. 2001. Chadian australopithecines: Biochronology and environmental context; pp. 103–106 in P. V. Tobias, M. A. Raath, J. Moggi-Cecchi, and G. A. Doyle (eds.), *Humanity from African Naissance to Coming Millennia*. Florence University Press, Florence.

Brunet, M., A. Beauvilain, D. Geraads, F. Guy, M. Kasser, H. T. Mackaye, L. MacLatchy, G. Mouchelin, J. Sudre, and P. Vignaud. 1997. Tchad: Un nouveau site à hominidés Pliocène. *Comptes Rendus de l'Académie des Sciences* 324:341–345.

———. 1998. Tchad: Découverte d'une faune de mammifères du Pliocène inférieur. *Comptes-Rendus Hebdomadaires des Séances de l'Académie des Sciences, Paris* 326:153–158.

Brunet, M., and MPFT. 2000. Chad: Discovery of a vertebrate fauna close to the Mio-Pliocene boundary. *Journal of Vertebrate Paleontology* 20:205–209.

Cande, S. C., and D. V. Kent. 1995. Revised calibration of the geomagnetic polarity timescale for the Late Cretaceous and Cenozoic. *Journal of Geophysical Research* 100(B4):6093–6095.

Chapman, G. R., and M. Brook. 1978. Chronostratigraphy of the Baringo Basin, Kenya Rift Valley; pp. 207–223 in W. W. Bishop (ed.), *Geological Background to Fossil Man*. Scottish Academic Press, Edinburgh.

Chavaillon, J., J.-L. Boisaubert, M. Faure, C. Guérin, C. Ma, B. Nickel, G. Poupeau, P. Rey, and S. A. Warsama. 1987. Le site de dépeage pléistocène *Elephas recki* de Barogali, République de Djibouti, nouveaux résultats et datations. *Comptes Rendus de l'Académie des Sciences, Paris* 305:1259–1266.

Conroy, G. C., M. Pickford, B. Senut, J. Van Couvering, and P. Mein. 1992. *Otavipithecus namibiensis*, first Miocene hominoid from southern Africa. *Nature* 356:144–148.

Cooke, H. B. S. 1974. The fossil mammals of Cornelia, O.F.S., South Africa. *Memoirs of the National Museum, Bloemfontein* 9:63–84.

Cote, S., L. Werdelin, E. R. Seiffert, and J. C. Barry. 2007. Additional material of the enigmatic early Miocene mammal *Kelba* and its relationship to the order Ptolemaiida. *Proceedings of the National Academy of Sciences, USA* 104:5510–5515.

De Bonis, L., D. Geraads, J.-J. Jaeger, and S. Sen. 1988. Vertébrés Pleistocènes de Djibouti. *Bulletin de la Société Géologique de France* 4 (série 8):323–334.

Deino, A. L., and S. H. Ambrose. 2007. ^{40}Ar/^{39}Ar dating of the Lemudong'o late Miocene fossil assemblages, southern Kenya Rift. *Kirtlandia* 56:65–71.

Deino, A., and A. Hill. 2002. ^{40}Ar/^{39}Ar dating of Chemeron Formation strata encompassing the site of hominid KNM-BC 1, Tugen Hills, Kenya. *Journal of Human Evolution* 42:141–151.

Deino, A. L., and S. McBrearty. 2002. ^{40}Ar/^{39}Ar dating of the Kapthurin Formation, Baringo, Kenya. *Journal of Human Evolution* 42:185–210.

Deino, A., and R. Potts 1990. Single-crystal ^{40}Ar/^{39}Ar dating of the Olorgesailie formation, southern Kenya Rift. *Journal of Geophysical Research* 95(B6):8453–8470.

Deino, A., L. Tauxe, M. Monaghan, and R. Drake. 1990. Single crystal $^{40}Ar/^{39}Ar$ ages and the litho- and paleomagnetic stratigraphies of the Ngorora Formation, Kenya. *Journal of Geology* 98:567–587.

Deino, A. L., L. Tauxe, M. Monaghan, and A. Hill. 2002. $^{40}Ar/^{39}Ar$ geochronology and paleomagnetic stratigraphy of the Lukeino and lower Chemeron Formations at Tabarin and Kapcheberek, Tugen Hills, Kenya. *Journal of Human Evolution* 42:117–140.

Depéret, C. 1897. Découverte du *Mastodon angustidens* dans l'étage cartennien de Kabylie. *Bulletin de la Société Géologique de France* 25(série 3):518–521.

Ditchfield, P., J. Hicks, T. Plummer, L. C. Bishop, and R. Potts. 1999. Current research on the Late Pliocene and Pleistocene deposits north of Homa Mountain, southwestern Kenya. *Journal of Human Evolution* 36:123–150.

Domínguez-Rodrigo, M. 1996. La cronología del Grupo Peninj, al oeste del Lago Natrón (Tanzania): Revisión de las discordancias bioestratigrafícas. *Complutum* 7:7–15.

Drake, R., and G. H. Curtis. 1987. K-Ar geochronology of the Laetoli fossil localities; pp. 48–52 in M. D. Leakey and J. M. Harris (eds.), *Laetoli: A Pliocene Site in Northern Tanzania*. Clarendon Press, Oxford.

Drake, R. E., J. A. Van Couvering, M. H. Pickford, G. H. Curtis, and J. A. Harris. 1988. New chronology for the early Miocene mammalian faunas from Kisingiri, western Kenya. *Journal of the Geological Society, London* 145:479–491.

Feibel, C. S. 2003. Stratigraphy and depositional setting of the Pliocene Kanapoi Formation, lower Kerio Valley, Kenya. *Contributions in Science* 498:9–20.

Feibel, C. S., and F. H. Brown. 1991. Age of the primate-bearing deposits on Maboko Island, Kenya. *Journal of Human Evolution* 21:221–225.

Feibel, C. S., F. H. Brown, and I. McDougall. 1989. Stratigraphic context of fossil hominids from the Omo Group deposits: Northern Turkana Basin, Kenya and Ethiopia. *American Journal of Physical Anthropology* 78:595–622.

Fournet, A. 1971. Les Gisements à faune Villafranchienne de Tunisie. *Notes du Service Géologique de Tunisie* 34:53–69.

Gebo, D. L., L. MacLatchy, R. Kityo, A. Deino, J. Kingston, and D. Pilbeam. 1997. A hominoid genus from the early Miocene of Uganda. *Science* 276:401–404.

Gentry, A. W., and A. Gentry, A. 1978. Fossil Bovidae (Mammalia) of Olduvai Gorge, Tanzania. *Bulletin of the British Museum (Natural History) Geology* 29:289–446; 30:1–83.

Geraads, D. 1995. Rongeurs et insectivores (Mammalia) du Pliocène final de Ahl al Oughlam (Casablanca, Maroc). *Geobios* 28:99–115.

———. 1997. Carnivores du Pliocène terminal de Ahl al Oughlam (Casablanca, Maroc). *Geobios* 30:127–164.

———. 1998. Rongeurs du Mio-Pliocène de Lissasfa (Casablanca, Maroc). *Geobios* 31:229–245.

———. 2002. Plio-Pleistocene mammalian biostratigraphy of Atlantic Morocco. *Quaternaire* 13:43–53.

———. 2006. The late Pliocene locality of Ahl al Oughlam, Morocco: Vertebrate fauna and interpretation. *Transactions of the Royal Society of South Africa* 61:97–101.

Geraads, D., Z. Alemseged, and H. Bellon. 2002. The Late Miocene mammalian fauna of Chorora, Awash Basin, Ethiopia: Systematics, biochronology and the $^{40}K–^{40}Ar$ ages of associated volcanics. *Tertiary Research* 21:113–122.

Geraads, D., Z. Alemseged, D. H. Reed, J. G. Wynn, and D. C. Roman. 2004a. The Pleistocene fauna (other than Primates) from Asbole, lower Awash Valley, Ethiopia, and its environmental and biochronological implications. *Geobios* 37:697–718.

Geraads, D., and F. Amani. 1998. Bovidae (Mammalia) du Pliocène final d'Ahl al Oughlam, Casablanca, Maroc. *Paläontologische Zeitschrift* 72:191–205.

Geraads, D., V. Eisenmann, and G. Petter. 2004b. The large mammal fauna of the Oldowan sites of Melka Kunturé; pp. 169–192 in J. Chavaillon and M. Piperno (eds.), *Studies on the Early Paleolithic Site of Melka Kunturé, Ethiopia*. Istituto Italiano di Preistoria e Protostoria, Florence.

Geraads, D., J.-P. Raynal, and V. Eisenmann. 2004c. The earliest human occupation of North Africa: A reply to Sahnouni et al. (2002). *Journal of Human Evolution* 46:751–761.

Gilbert, W. H. 2003. The Daka Member of the Bouri Formation, Middle Awash Valley, Ethiopia: Fauna, Stratigraphy, and Environment. Unpublished PhD dissertation, University of California, Berkeley, 374 pp.

Ginsburg, L. 1977. Les carnivores du Miocène de Beni Mellal (Maroc). *Géologie Méditerranéenne* 4:225–240.

Gowlett, J., J. Harris, D. Walton, and B. Wood. 1981 Early archaeological sites, hominid remains and traces of fire from Chesowanja, Kenya. *Nature* 294:125–129.

Gradstein, F., J. Ogg, and A. Smith. 2004. *A Geologic Time Scale 2004*. Cambridge University Press, Cambridge, 589 pp.

Greenwood, P. H. 1972. Fish fossils from the late Miocene of Tunisia. *Notes du Service Géologique de Tunisie* 37:47–72.

Haile-Selassie, Y., A. Deino, B. Saylor, M. Umer, and B. Latimer. 2007. Preliminary geology and paleontology of new hominid-bearing Pliocene localities in the central Afar region of Ethiopia. *Anthropological Science* 115:215–222.

Hamdine, W., M. Thévenot, and J. Michaux. 1998. Histoire récente de l'ours brun au Maghreb. *Comptes Rendus de l'Academie des Sciences, Paris, Sciences de la Vie* 321:565–570.

Hamilton, W. R. 1973. A Lower Miocene mammalian fauna from Siwa, Egypt. *Palaeontology* 16:275–281.

Harrison, T. (ed.) 1997. *Neogene Paleontology of the Manonga Valley, Tanzania: A Window into the Evolutionary History of East Africa*. Plenum Press, New York, 418 pp.

Harrison, T., and E. Baker. 1997. Paleontology and biochronology of fossil localities in the Manonga Valley, Tanzania; pp. 361–393 in T. Harrison (ed.), *Neogene Paleontology of the Manonga Valley, Tanzania: A Window into the Evolutionary History of East Africa*. Plenum Press, New York.

Hay, R. L. 1976. *Geology of the Olduvai Gorge*. University of California Press, Berkeley, 203 pp.

Heinzelin, J. de, and A. El-Arnauti. 1987. The Sahabi Formation and related deposits; pp. 1–22 in N. T. Boaz, A. El-Arnauti, A. W. Gaziry, J. de Heinzelin, and D. D. Boaz (eds.), *Neogene Paleontology and Geology of Sahabi*. Liss, New York.

Hendey, Q. B. 1978 Preliminary report on the Miocene vertebrates from Arrisdrift, South West Africa. *Annals of the South African Museum* 76:1–1.

Herries, A. I. R., K. E. Reed, K. L. Kuykendall, and A. G. Latham. 2006. Speleology and magnetobiostratigraphic chronology of the Buffalo Cave fossil site, Makapansgat, South Africa. *Quaternary Research* 66:233–45.

Hill, A., M. G. Leakey, J. D. Kingston, and S. Ward. 2002. New cercopithecoids and a hominoid from 12.5 Ma in the Tugen Hills succession, Kenya. *Journal of Human Evolution* 42:75–93.

Hooijer, D. A. 1963. Miocene Mammalia of Congo. *Annales, Musée Royal de l'Afrique Centrale*, Séries in 8° 46:1–77.

Howell, F. C. 1987. Preliminary observations on Carnivora from the Sahabi Formation (Libya); pp. 153-181 in N. T. Boaz, A. El-Arnauti, A. W. Gaziry, J. de Heinzelin, and D. D. Boaz (eds.), *Neogene Paleontology and Geology of Sahabi*. Liss, New York.

Howell, F. C., G. H. Cole, M. R. Kleindienst, B. J. Szabo, and K. P. Oakley. 1972. Uranium-series dating of bone from the Isimila prehistoric site, Tanzania. *Nature* 237:51–52.

Katoh, S. 2002. Tephrostratigraphic correlation of the Konso and Turkana Basins; pp. 223–225 in G. Suwa (ed.), *Morphological and Behavioral Evolution of Early Hominids*. Report of the Japan Society for Promotion of Science, Tokyo.

Keyser A. W., C. G. Menter, J. Moggi-Cecchi, T. R. Pickering, and L. R. Berger. 2000. Drimolen: A new hominid- bearing site in Gauteng, South Africa. *South African Journal of Science* 96:193–197.

Kingston, J., B. F. Jacobs, A. Hill, and A. Deino. 2002. Stratigraphy, age and environments of the late Miocene Mpesida Beds, Tugen Hills, Kenya. *Journal of Human Evolution* 42:95–116.

Klein, R. G., G. Avery, K. Cruz-Uribe, D. Halkett, J. E. Parkington, T. Steele, T. P. Volman, and R. J. Yates. 2004. The Ysterfontein 1 Middle Stone Age site, South Africa, and early human exploitation of coastal resources. *Proceedings of the National Academy of Sciences, USA* 101:5708–5715.

Klein, R. G., G. Avery, K. Cruz-Uribe, and T. E. Steele. 2007. The mammalian fauna associated with an archaic hominin skullcap and later Acheulian artifacts at Elandsfontein, Western Cape Province, South Africa. *Journal of Human Evolution* 52:164–186.

Kleinsasser, L. L., J. Quade, W. C. McIntosh, N. E. Levin, S. W. Simpson, and S. Semaw. 2008. Stratigraphy and geochronology of the late Miocene Adu-Asa Formation at Gona, Ethiopia; pp. 33–65 in J. Quade and J. G. Wynn (eds.), *The Geology of Early Humans in the Horn of Africa*. Geological Society of America Special Paper 446.

Kunimatsu, Y., M. Nakatsukasa, Y. Sawada, T. Sakai, M. Hyodo, H. Hyodo, T. Itaya, H. Nakaya, H. Saegusa, A. Mazurier, M. Saneyoshi, H. Tsujikawa, A. Yamamoto, and E. Mbua. 2007. A new Late Miocene great ape from Kenya and its implications for the origins of African great apes and humans. *Proceedings of the National Academy of Sciences, USA* 104:19220–19225.

Kurtén, B. 1976. Fossil Carnivora from the Late Tertiary of Bled Douarah and Cherichira, Tunisia. *Notes du Service Géologique de Tunisie* 42:177–214.

Lavocat, R. 1961. Le gisement de Vertébrés miocènes de Beni Mellal (Maroc). Etude systematique de la faune de Mammifères. *Notes et Mémoires du Service Géologique du Maroc, Rabat* 155:29–94.

Leakey, R. E., and M. G. Leakey. 1986a. A new Miocene hominoid from Kenya. *Nature* 324:143–146.

———. 1986b. A second new Miocene hominoid from Kenya. *Nature* 324:146–148.

———. 1987. A new Miocene small-bodied ape from Kenya. *Journal of Human Evolution* 16:369–387.

Leakey, M. G., C. S. Feibel, I. McDougall, and A. Walker. 1995. New four million-year-old hominid species from Kanapoi and Allia Bay, Kenya. *Nature* 376:565–571.

Leakey, M. G., C. S. Feibel, I. McDougall, C. Ward, and A. Walker. 1998. New specimens and confirmation of an early age for *Australopithecus anamensis. Nature* 393:62–66.

Lebatard, A.-E., D. L. Bourlès, P. Duringer, M. Jolivet, R. Braucher, J. Carcaillet, M. Schuster, N. Arnaud, P. Monié, F. Lihoreau, A. Likius, H. T. Mackaye, P. Vignaud, and M. Brunet. 2008. Cosmigenic nuclide dating of *Sahelanthropus tchadensis* and *Australopithecus bahrelghazali*: Mio-Pliocene hominids from Chad. *Proceedings of the National Academy of Sciences, USA* 105:3226–3231.

Lihoreau, F. 2003. Systématique et paléoécologie des Anthracotheriidae (Artiodactyla; Suiformes) du Mio-Pliocène de l'Ancien Monde: Implications paléobiogéographiques. Unpublished PhD dissertation, University of Poitiers, France.

Lihoreau, F., J.-R. Boisserie, L. Viriot, Y. Coppens, A. Likius, H. T. Mackaye, P. Tafforeau, P. Vignaud, and M. Brunet. 2006. Anthracothere dental anatomy reveals a late Miocene Chado-Libyan bioprovince. *Proceedings of the National Academy of Sciences, USA* 103:8763–8767.

Llinás-Agrasar E. 2003. New fossil crocodilians from the middle/upper Miocene of Tunisia. *Annales de Paléontologie* 89:103–110.

MacLatchy L., A. Deino, and J. Kingston. 2006. An updated chronology for the early Miocene of NE Uganda. *Journal of Vertebrate Paleontology* 26(suppl. to 3).93A.

Mahjoub, M. N., and M. Khessibi. 1988: Sedimentological study of sandy and shaly deposits (Beglia Formation) in Cap Bon area. *AAPG Bulletin* 72:8.

Makinouchi, T., S. Ishida, Y. Sawada, N. Kuga, N. Kimura, Y. Orihashi, B. Bajope, M. wa Yemba, and H. Ishida. 1992. Geology of the Sinda-Mohari Region, haut-Zaire Province, eastern Zaire. *African Study Monographs*, suppl. vol. 17:3–18.

Marean, C. W., M. Bar-Matthews, J. Bernatchez, E. Fisher, P. Goldberg, A. I. R. Herries, Z. Jacobs, A. Jerardino, P. Karkanas, N. Mercier, T. Minichillo, P. J. Nilssen, E. Thompson, C. Tribolo, H. Valladas, I. Watts, and H. M. Williams. 2007. Early human use of marine resources and pigment in South Africa during the middle Pleistocene. *Nature* 449:905–908.

McDougall, I. and F. H. Brown. 2006. Precise ^{40}Ar/^{39}Ar geochronology for the upper Koobi Fora Formation, Turkana Basin, northern Kenya. *Journal of the Geological Society, London* 163:205–220.

———. 2008. Geochronology of the pre-KBS Tuff sequence, Omo Group, Turkana Basin. *Journal of the Geological Society, London* 165:549–562.

McDougall, I., and C. S. Feibel. 1999. Numerical age control for the Miocene-Pliocene succession at Lothagam, a hominoid-bearing sequence in the northern Kenya Rift. *Journal of the Geological Society, London* 156:731–745.

McDougall, I., and R. T. Watkins. 1985. Age of hominoid-bearing sequence at Buluk, northern Kenya. *Nature* 318: 175–178.

McKee, J. K. 1993 The faunal age of the Taung hominid deposit. *Journal of Human Evolution* 25:363–376

Mein, P., and M. Pickford. 2003. Fossil picas (Ochotonidae, Lagomorpha, Mammalia) from the basal Middle Miocene of Arrisdrift, Namibia. *Memoirs of the Geological Survey of Namibia* 19:171–176.

Miller, E. R. 1999. Faunal correlation of Wadi Moghara, Egypt: Implications for the age of *Prohylobates tandyi. Journal of Human Evolution* 36:519–533.

Morales, J., M. Pickford, S. Fraile, M. Salesa, and D. Soria. 2003. Creodonta and Carnivora from Arrisdrift, early Middle Miocene of southern Namibia. *Memoirs of the Geological Survey of Namibia* 19:177–194.

Morales, J., M. Pickford, D. Soria, and S. Fraile. 1998. New carnivores from the basal Middle Miocene of Arrisdrift, Namibia. *Eclogae Geologicae Helvetiae* 91:27–40.

Morales, J., D. Soria, I. M. Sánchez, V. Quiralte, and M. Pickford. 2003. Tragulidae from Arrisdrift, basal Middle Miocene, southern Namibia. *Memoirs of the Geological Survey of Namibia* 19:359–369.

Partridge, T. C., D. E. Granger, M. W. Caffee, and R. J. Clarke. 2003. Lower Pliocene hominid remains from Sterkfontein. *Science* 300:607–612.

Partridge, T. C., A. G. Latham, and D. Heslop. 2000. Appendix on magnetostratigraphy of Makapansgat, Sterkfontein, Taung and Swartkrans; pp. 126–129 in T. C. Partridge and R. R. Maud (eds.), *The Cenozoic of Southern Africa*. Oxford University Press, Oxford.

Partridge, T. C., J. Shaw, and D. Heslop. 2000. Note on recent magnetostratigraphic analyses in Member 2 of the Sterkfontein Formation; pp. 129–130 in T. C. Partridge and R. R. Maud (eds.), *The Cenozoic of Southern Africa*. Oxford University Press, Oxford.

Pickford, M. 1981. Preliminary Miocene mammalian biostratigraphy for western Kenya. *Journal of Human Evolution* 10:73 97.

———. 1986a. Cainozoic palaeontological sites of western Kenya. *Münchner Geowissenschaftliche Abhandlungen* 8:1–151.

———. 1986b. Première découverte d'une faune mammalienne terrestre paléogène d'Afrique sub-saharienne. *Comptes Rendus de l'Académie des Sciences, Paris* 302:1205–1210

———. 1990. Révision des Suidés de la Formation de Beglia (Tunisie). *Annales de Paléontologie* 76:133–141.

———. 2000. Crocodiles from the Beglia Formation, middle/late Miocene boundary, Tunisia, and their significance for Saharan palaeoclimatology. *Annales de Paléontologie* 86:59–67.

———. 2002. Early Miocene grassland ecosystem at Bukwa, Mount Elgon, Uganda. *Comptes Rendus Palevol* 1:213–219.

——— 2004. Southern Africa: A cradle of evolution. *South African Journal of Science* 100:205–214.

———. 2005. *Choerolophodon pygmaeus* (Proboscidea: Mammalia) from the Middle Miocene of Southern Africa. *South African Journal of Science* 101:175–177.

Pickford, M., and P. Andrews. 1981. The Tinderet Miocene sequence in Kenya. *Journal of Human Evolution* 10:11–33.

Pickford, M., and Y. Kunimatsu. 2005. Catarrhines from the Middle Miocene (ca. 14.5 Ma) of Kipsaraman, Tugen Hills, Kenya. *Anthropological Science* 113:189–224.

Pickford, M., P. Mein, and B. Senut. 1994. Fossiliferous Neogene karst fillings in Angola, Botswana and Namibia. *South African Journal of Science* 90:227–230.

Pickford, M., Y. Sawada, R. Tayama, Y.-K. Matsuda, T. Itaya, H. Hyodo, and B. Senut. 2006. Refinement of the age of the Middle Miocene Fort Ternan Beds, western Kenya, and its implications for Old World biochronology. *Comptes Rendus Geoscience* 338:545–555.

Pickford, M, and B. Senut. 1997. Cainozoic mammals from coastal Namaqualand, South Africa. *Palaeontologia Africana* 34:199–217.

———. 2000. Geology and paleobiology of the Namib Desert, southwestern Africa. *Memoirs of the Geological Survey of Namibia* 18:1–155.

———. 2003. Miocene paleobiology of the Orange River Valley, Namibia. *Memoirs of the Geological Survey of Namibia* 19:1–22.

Pickford, M., B. Senut, D. Gommery, and E. Musiime. 2003. New Catarrhine fossils from Moroto II early Middle Miocene (ca. 17.5 Ma) Uganda. *Comptes Rendus Palevol* 2:649–662.

Pickford, M., B. Senut, and D. Hadoto. 1993. Geology and palaeobiology of the Albertine Rift Valley, Uganda-Zaire. *CIFEG Occasional Publications* 24:1–190.

Pickford, M., B. Senut, P. Mein, D. Gommery, J. Morales, D. Soria, M. Nieto, and J. Ward. 1996. Preliminary results of new excavations at Arrisdrift, middle Miocene of southern Namibia. *Comptes-Rendus Hebdomadaires des Séances de l'Académie des Sciences, Paris* 322:991–996.

Pickford, M., and P. Tassy. 1980. A new species of *Zygolophodon* (Mammalia, Proboscidea) from the Miocene hominoid localities of Meswa Bridge and Moroto (East Africa). *Neues Jahrbuch für Geologie und Paläontologie, Abhandlungen* 1980:235–251.

Pickford, M., H. Wanas, and H. Soliman. 2006. Indications for a humid climate in the Western Desert 11–10 Myr ago: Evidence from Galagidae (Primates, Mammalia). *Comptes Rendus Palevol* 5:935–943.

Potts, R. 1988. *Early Hominid Activities at Olduvai*. Aldine de Gruyter, New York, 396 pp.

Raynal, J.-P., F. Amani, D. Geraads, M. El Graoui, L. Magoga, J.-P. Texier, and F.-Z. Sbihi-Alaoui. 2008. La grotte des félins, site paléolithique du Pléistocène supérieur à Dar Bouazza (Maroc). *L'Anthropologie* 112:182–200.

Raynal, J.-P., G. Kieffer, and G. Bardin. 2004. Garba IV and the Melka Kunture Formation. A preliminary lithostratigraphic approach; pp. 137–166 in J. Chavaillon and M. Piperno (eds.), *Studies on the Early Paleolithic Site of Melka Kunture, Ethiopia*. Istituto Italiano di Preistoria e Protostoria, Florence.

Richmond, B. G., J. G. Fleagle, J. Kappelman, and C. C. Swisher III. 1998. First hominoid from the Miocene of Ethiopia and the evolution of the catarrhine elbow. *American Journal of Physical Anthropology* 105:257–277.

Robinson, P. 1972. *Pachytragus solignaci*, a new species of caprine bovid from the late Miocene Beglia Formation of Tunisia. *Notes du Service Géologique de Tunisie* 37:73–94.

Sahnouni, M., and J. De Heinzelin. 1998. The site of Aïn Hanech revisited: New investigations at this lower Pleistocene site in northern Algeria. *Journal of Archaeological Science* 25:1083–1101.

Sahnouni, M., D. Hadjouis, J. van der Made, A. Derradji, A. Canals, M. Medig, H. Belahrech, Z. Harichane, and M. Rabhi. 2002. Further research at the Oldowan site of Aïn Hanech, north-eastern Algeria. *Journal of Human Evolution* 43:925–937.

———. 2004. On the earliest human occupation in North Africa: a response to Geraads et al. *Journal of Human Evolution* 46:763–775.

Savage, R. J. G., and W. R. Hamilton. 1973. Introduction to the Miocene mammal faunas of Gebel Zelten, Libya. *Bulletin of the British Museum (Natural History), Geology* 22:513–527.

Sawada, Y., M. Pickford, T. Itaya, T. Makinouchi, M. Tateishi, K. Kabeto, S. Ishida, and H. Ishida. 1998. K-Ar ages of Miocene Hominoidea (*Kenyapithecus* and *Samburupithecus*) from Samburu Hills, Northern Kenya. *Comptes Rendus de l'Académie des Sciences, Paris* 326:445–451.

Sawada, Y., M. Pickford, B. Senut, T. Itaya, M. Hyodo, T. Miura, C. Kashine, T. Chujo, and H. Fujii. 2002. The age of *Orrorin tugenensis*, an early hominid from the Tugen Hills, Kenya. *Comptes Rendus Palevol* 1:293–303.

Sawada, Y., M. Saneyoshi, K. Nakayama, T. Sakai, T. Itaya, M. Hyodo, Y. Mukokya, M. Pickford, B. Senut, S. Tanaka, T. Chujo, and H. Ishida. 2006. The ages and geological backgrounds of Miocene hominoids *Nacholapithecus*, *Samburupithecus* and *Orrorin* from Kenya; pp. 71–96 in H. Ishida, R. Tuttle, M. Pickford, N. Ogihara, and M. Nakatsukasa (eds.), *Human Origins and Environmental Backgrounds*. Springer, New York.

Senut, B. 1997. Postcranial morphology and springing adaptations in Pedetidae from Arrisdrift, Middle Miocene (Namibia). *Palaeontologia Africana* 34:101–109.

Senut, B., M. Pickford, D. Gommery, and Y. Kunimatsu. 2000. Un nouveau genre d'hominoïde du Miocène inférieur d'Afrique orientale: *Ugandapithecus major* (Le Gros Clark & Leakey, 1950). *Comptes Rendus de l'Académie des Sciences, Paris* 331:227–233.

Senut, B., M. Pickford, P. Mein, G. Conroy, and J. Van Couvering. 1992. Discovery of twelve new Late Cainozoic fossiliferous sites in palaeokarsts of the Otavi Mountains, Namibia. *Comptes Rendus de l'Academie des Sciences, Paris* 314:727–733

Simons, E. L. 1967. A fossil *Colobus* skull from the Sudan (Primates, Cercopithecidae). *Postilla* 111:1–12.

Stromer, E. 1911. Fossile Wirbeltier-Reste aus dem Uadi Fâregh und Uadi Natrûn in Ägypten. *Abhandlungen der Senckenbergischen Naturforschenden Gesellschaft* 29:99–132.

Sudre, J., and J.-L. Hartenberger. 1992. Oued Mya 1, nouveau gisement de mammifères du Miocène supérieur dans le sud Algérien. *Geobios* 25:553–565.

Suwa, G., H. Nakaya, B. Asfaw, H. Saegusa, A. Amzaye, R. T. Kono, Y. Beyene, and S. Katoh. 2003. Plio-Pleistocene terrestrial mammal assemblage from Konso, southern Ethiopia. *Journal of Vertebrate Paleontology* 23:901–916.

Thackeray, J. F., D. J. de Ruiter, L. R. Berger, and N. J. van der Merve 2001. Hominid fossils from Kromdraai: A revised list of specimens discovered since 1938. *Annals of the Transvaal Museum* 38:43–56.

Thomas, H., and G. Petter. 1986. Révision de la faune de mammifères du Miocène supérieur de Menacer (ex-Marceau), Algérie: Discussion sur l'âge du gisement. *Geobios* 19:357–373.

Van Couvering, J. A. 1972. Geology of Rusinga Island and correlation of the Kenyan mid-Tertiary fauna. Unpublished PhD dissertation, Cambridge University, Cambridge.

Vialli, V. 1966. Sul rinvenimento di Dinoterio (*Deinotherium* cf. *hobleyi* Andrews) nelle ligniti di Ath Ugri (Eritrea). *Giornale di Geologia (Bologna)* 33:447–458.

Vickers-Rich, P. 1972. A fossil avifauna from the upper Miocene Beglia Formation of Tunisia. *Notes du Service Géologique de Tunisie* 35:29–66

Vignaud, P., P. Duringer, H. T. Mackaye, A. Likius, C. Blondel, J.-R. Boisserie, L. de Bonis, V. Eisenmann, M.-E. Etienne, D. Geraads, F. Guy, T. Lehmann, F. Lihoreau, N. Lopez-Martinez, C. Mourer-Chauviré, O. Otero, J.-C. Rage, M. Schuster, L. Viriot, A. Zazzo, and M. Brunet. 2002. Geology and paleontology of the Upper Miocene Toros-Menalla hominid locality, Chad. *Nature* 418:152–155.

Walter, R. C. 1994. The age of Lucy and the First Family: Single crystal $^{40}Ar/^{39}Ar$ dating of the Denen Dora and lower Kada Hadar Members of the Hadar Formation, Ethiopia. *Geology* 22:6–10.

Ward, C. V., M. G. Leakey, B. Brown, F. Brown, J. Harris, and A. Walker. 1999. South Turkwel: A new Pliocene hominid site in Kenya. *Journal of Human Evolution* 36:69–95.

Watson, V. 1993. Glimpses from Gondolin: A faunal analysis of a fossil site near Broederstroom, Transvaal, South Africa. *Palaeontologia Africana* 30:35–42.

Wessels, W., O. Fejfar, P. Pélaez-Campomanes, A. J. Van der Meulen, and H. De Bruijn. 2003. Miocene small mammals from Jebel Zelten, Libya. *Coloquios de Paleontología*, suppl. vol. 1:699–715.

White, T. D., G. WoldeGabriel, B. Asfaw, S. Ambrose, Y. Beyene, R. L. Bernor, J.-R. Boisserie, B. Currie, H. Gilbert, Y. Haile-Selassie, W. K. Hart, L. J. Hlusko, F. C. Howell, R. T. Kono, T. Lehmann, A. Louchart, C. O. Lovejoy, P. R. Renne, H. Saegusa, E. S. Vrba, H. Wesselman, and G. Suwa. 2006. Asa Issie, Aramis and the origin of *Ardipithecus*. *Nature* 440:883–889.

WoldeGabriel, G., Y. Haile-Selassie, P. R. Renne, W. K. Hart, S. H. Ambrose, B. Asfaw, G. Heiken, and T. D. White. 2001. Geology and paleontology of the late Miocene Middle Awash Valley, Afar rift, Ethiopia. *Nature* 412:175–178.

Wynn, J. G., Z. Alemseged, R. Bobé, D. Geraads, D. Reed and D. Roman. 2006. Geological and palaeontological context of a Pliocene juvenile hominin at Dikika, Ethiopia. *Nature* 443:332–336.

FOUR

Global and African Regional Climate during the Cenozoic

SARAH J. FEAKINS AND PETER B. DEMENOCAL

The last 65 Ma of Earth's history, the Cenozoic, has been a time characterized by significant climate change. Major global changes included massive tectonic reorganization, a reduction in atmospheric pCO_2 (Pagani et al., 1999; Pearson and Palmer, 2000), and a dramatic cooling of global climate, plunging the world from generally warm conditions into the repeated glacial-interglacial cycles of the ice age (Zachos et al., 2001). Deep-sea oxygen isotope records record global cooling of up to 8°C in the early Cenozoic, heralding the development of major ice sheets on Antarctica from 35 Ma, which further intensified the global cooling trend and culminated in cyclical Northern Hemisphere glaciation during the past 3 Ma (figure 4.1). Many events in global tectonics and high latitude climate had significant effects on Cenozoic climate evolution. These are well described elsewhere (e.g., Kennett, 1995; Denton, 1999; Zachos et al., 2001) and are summarized in figure 4.1.

In this chapter, we focus on three revolutions in climate research that have dramatically altered our perception of global and African climate. First, the discovery that large magnitude climate events occurred abruptly, sometimes in as little as decades, has prompted high-resolution paleoclimate reconstructions and new conceptions of climate dynamics, revealing significant climate variability at times that were previously thought to be quiescent (e.g., the Holocene). On longer timescales, high-resolution oxygen isotope stratigraphies have also revealed transient events in the early Cenozoic (Zachos et al., 2001). These discoveries have revolutionized theories of climate change and demonstrated the need for high-resolution reconstruction of climate variability on 10^0- to 10^5-year timescales.

Second, recent climate studies have revealed significant tropical climate variability. Modern observational climate data have indicated that the largest mode of global interannual climate variability is the El Niño Southern Oscillation (ENSO) in the tropical Pacific (Ropelewski and Halpert, 1987; Trenberth et al., 1998). Large amplitude tropical environmental variability has also been reconstructed in the paleoclimate record. In particular, revised estimates of tropical sea surface temperatures (SSTs) during global cool and warm events have revealed significant tropical sensitivity to global climate change (e.g., Pearson et al., 2001; Lea et al., 2003).

Revised tropical SST reconstructions have implications both for local climate interpretations and for global dynamical predictions, leading to new perspectives on the nature of Cenozoic climate change.

Third, the role of the tropics in global climate change has been reconceptualized. Rather than being a passive responder to changes in the high-latitude cryosphere, tropical climate variability may be at least partially decoupled from high-latitude climate. For example, there is considerable evidence that precessional variations in insolation may directly influence the intensity of African precipitation, independent of high-latitude climate variability (Rossignol-Strick, 1983; Partridge et al., 1997; Denison et al., 2005). The tropics may even have driven global climate change. For example, ENSO generates global teleconnections that have been observed in the instrumental record (Cane and Zebiak, 1985; Cane and Clement, 1999), and evidence for tropical initiation of past global climate changes comes from both paleoclimate and modeling analyses (Linsley et al., 2000; Clement et al., 2001; Hoerling et al., 2001; Yin and Battisti, 2001). This chapter provides a synthesis of climate data from a tropical perspective that offers new insights into aspects of Cenozoic African environmental change.

Modern African Climate

Precipitation is the critical interannual variable in African climate. Seasonal variations in the position of the Intertropical Convergence Zone (ITCZ) exert a significant control on the seasonal pattern of precipitation maxima across much of Africa. Figure 4.2 shows the major atmospheric circulation regimes for average conditions in July/August and January that illustrate climatic zones and provide a basis for understanding climatic variability (Nicholson, 2000). Distinct atmospheric circulation systems affect North Africa, West and Central Africa, East Africa, and southern Africa; they are separated in large part by the ITCZ. These regions experience characteristic patterns of interannual variability, teleconnections, and surface characteristics (Janowiak, 1988; Semazzi et al., 1988).

The northern coast of Africa has a Mediterranean climate receiving winter precipitation supplied by the mid-latitude

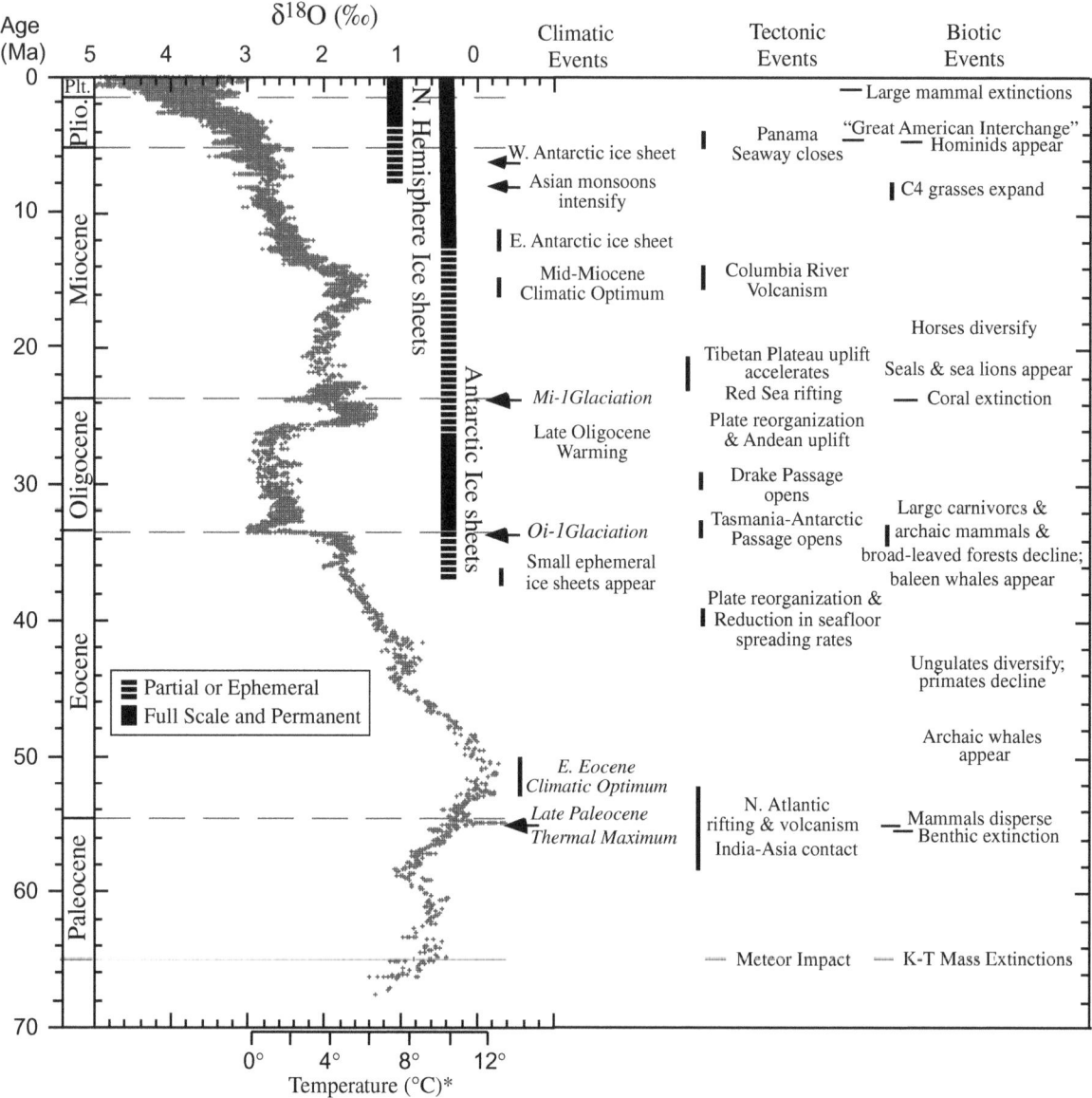

FIGURE 4.1 Global deep-sea oxygen isotope records for the last 65 Ma based on data compiled from more than 40 DSDP/ODP marine sites, together with some key tectonic and biotic events. The temperature scale refers to an ice-free ocean and thus only applies prior to the onset of large-scale glaciations (~35Ma). From the early Oligocene to the present, much of the oxygen isotope variability reflects changes in ice volume. Reprinted in part with permission from Zachos et al. (2001). © 2001 AAAS.

westerlies (Nicholson, 2000). Interannual to interdecadal variability is influenced by regional and global atmospheric teleconnections associated with the North Atlantic Oscillation (Hurrell, 1995) and to a variable extent by ENSO (Knippertz et al., 2003). Greenland ice core data indicate that these patterns of variability have persisted for several hundred years (Dansgaard et al., 1993).

West Africa is dominated by summer monsoonal precipitation associated with the northward migration of the ITCZ (figure 4.2). Tropical Atlantic SSTs have been shown to exert primary control on the strength of West African monsoon and Sahelian precipitation at interannual to interdecadal timescales (Rowell et al., 1995; Giannini et al., 2003). Warm SST anomalies in the Gulf of Guinea reduce the land-sea temperature contrast and weaken the monsoon. Convection cells remain over the ocean, increasing rainfall to coastal regions and decreasing rainfall to the Sahel.

Variations in continental heating also influence the land-sea temperature contrast and strength of the monsoon on millennial timescales (deMenocal et al., 2000; Liu et al., 2003). Central African rainfall is also negatively correlated with equatorial Atlantic SSTs but positively correlated with subtropical Atlantic SSTs (Nicholson and Entekhabi, 1987; Camberlin et al., 2001), with seasonal rainfall maxima associated with the passage of the ITCZ (figure 4.2). The hydrogen isotopic composition of plant leaf wax biomarkers in the Congo Fan indicate that this relationship has persisted for the past 20 ka (Schefuss et al., 2005).

Southern Africa has a strong precipitation gradient from >1,000 mm per year in the east to <20 mm per year in the west (Nicholson, 2000). Most precipitation falls in the austral summer as convective rainfall, particularly in the southeast. In the southwest, the precipitation pattern is more complex, with rainfall maxima associated with the seasonal peak in

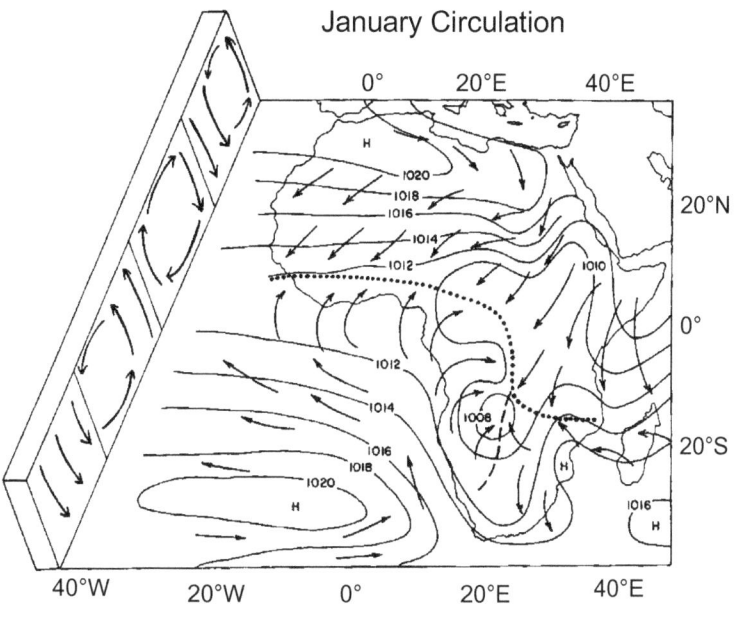

January Circulation

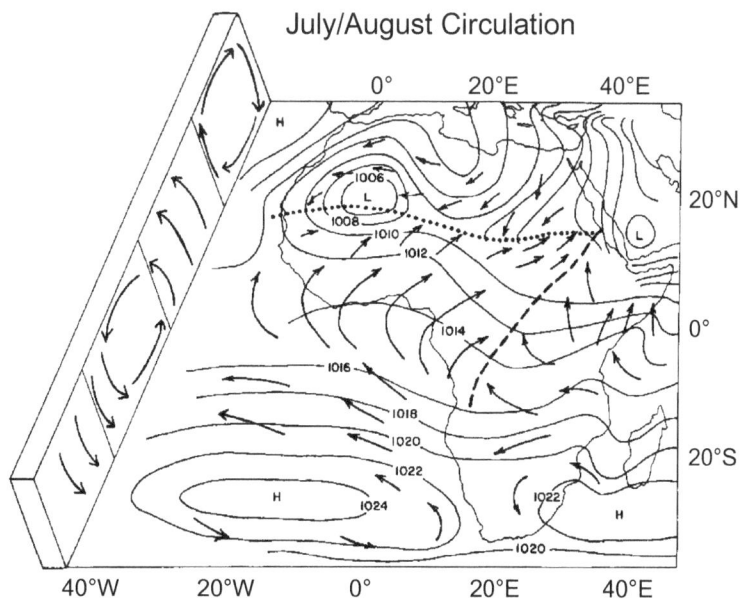

July/August Circulation

FIGURE 4.2 Schematic of the general patterns of winds, pressure, and convergence over Africa. Dotted lines indicate the ITCZ; dashed lines indicate other convergence zones. Reprinted with permission from Nicholson (2000). © 2000 Elsevier.

SSTs in Benguela coastal regions. Summer heating creates low pressure over central Africa that pulls in moisture from the western Indian Ocean. Significant interannual variability in precipitation is dominated by western Indian Ocean SSTs, with variability that is strongly related to ENSO in the Pacific (Cane et al., 1994; Goddard and Graham, 1999; figure 4.3). El Niño years are anomalously dry in southern Africa between December and February. These interannual observations suggest that past variations in Indian and Pacific SSTs are likely to have had a significant influence on southern African rainfall, particularly in the southeast, whereas variations in the strength of upwelling and the Benguela current may have been more important for southwest African climate.

East Africa receives most of its precipitation in April to March and September to November, associated with the biannual passage of the ITCZ and seasonally reversing winds

(figure 4.2). The primary moisture source is the central Indian Ocean via the southeasterly trade winds. Although the relative humidity is moderately high, precipitation in East Africa is low because of regional atmospheric circulation (Rodwell and Hoskins, 1996). Subsidence associated with the Indian Monsoon system inhibits convection, and the Ethiopian highlands constrict the southeasterly flow, resulting in southwesterly moisture divergence feeding the Indian Monsoon. On interannual timescales, Indian and Pacific ocean SSTs determine precipitation variability (figure 4.3; Goddard and Graham, 1999). These anomalous features of East African circulation suggest that uplift of the Himalayas and Ethiopian Highlands would have driven a gradual aridification of equatorial East Africa during the Cenozoic with superimposed climate variability resulting from SST variability in the Indian and Pacific oceans.

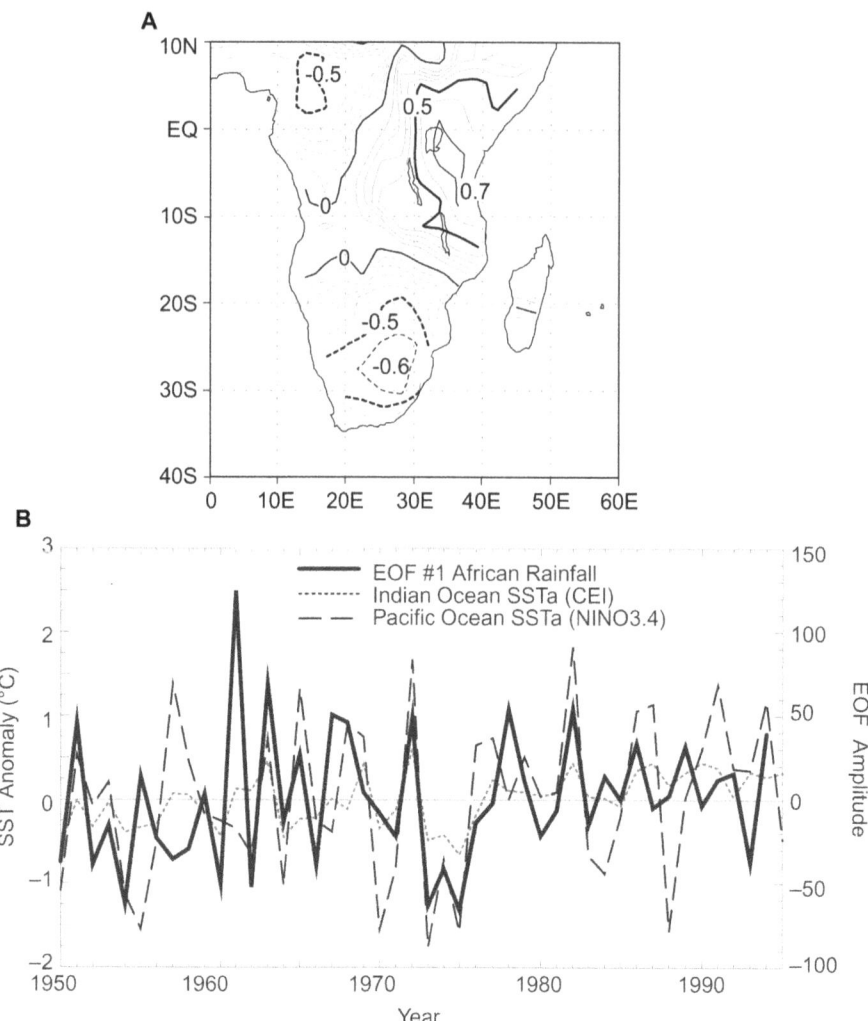

FIGURE 4.3 SSTs and southern African climate (from Goddard and Graham, 1999). A) First empirical orthogonal function (EOF) of observed rainfall anomalies for November to December–January 1950–1995, plotted as correlations between the rainfall anomalies and the EOF temporal function. B) Time series for amplitude of first EOF of African rainfall (thick solid line), amplitude on right axis. Also plotted are time series of sea surface temperature (SST) indices for the Indian Ocean, central equatorial Indian index (CEI; 15°S–0; 50°E–80°E, dotted line) and the Pacific Ocean, NINO3.4 (5°S–5°N; 170°W–110°W, dashed line).

ENSO SST variability in the tropical Pacific is the dominant mode of global interannual (2–7 yr) climate variability, with well-documented regional and global climate effects. El Niño events most strongly influence East and South Africa with strengthened upper westerly winds that lead to decreased rainfall in South Africa (December–March) and increased rainfall in East Africa (October–December—see figure 4.3; Hastenrath et al., 1993). North and West Africa are partially influenced by ENSO variability. In the Sahel, El Niño events tend to correlate with dry conditions (July–September). Sustained El Niño or La Niña–like conditions may also explain global climate patterns during paleoclimate events such as the warm, wet mid-Pliocene or be a possible trigger mechanism for abrupt climate events such as the Younger Dryas (Clement et al., 2001; Molnar and Cane, 2002; Wara et al., 2005; Ravelo et al., 2006). Observational and modeled studies of modern seasonal to interdecadal variability have provided important insights into the regional and global climate parameters that influence African climate variability. These fresh perspectives may

provide new answers about the global signature of Cenozoic paleoclimate events, particularly in Africa.

Cenozoic Climate Change

ABRUPT EVENTS IN THE PALEOCENE

One of the most dramatic events of the entire Cenozoic occurred in the late Paleocene (ca. 55 Ma). Antarctic and deep ocean temperatures rose by more than 6°C in less than 10 ka, creating a dramatic warming event, the Paleocene-Eocene Thermal Maximum (PETM) that lasted for 50 ka (figure 4.1; Stott and Kennett, 1990). This abrupt warming took place in an already warm era when global carbon dioxide levels were extremely high (>2,000 ppm), the poles were ice free (Zachos et al., 2001), sea levels were high, and a large marine transgression covered most of northern Africa (65–50 Ma; Le Houérou, 1997). This large-amplitude, abrupt event has been documented in a Tunisian record and would likely have had a significant impact on nearby continental climate (Bolle et al., 1999). These dramatic climate events have been linked to the Cretaceous-Paleogene extinction event and subsequent stabilization and rapid speciation, particularly within 10–100 ka after the PETM, when most of the modern orders of mammals appeared (Gingerich, 2004).

REVISED TROPICAL TEMPERATURES IN THE EOCENE

The Eocene is thought to be the warmest epoch of the Cenozoic, although climate data for this period are sparse, especially in Africa. Warm deep ocean temperatures (>10°C) are recorded in benthic foraminiferal oxygen isotopes during the early Eocene climatic optimum (54–50 Ma; figure 4.1). Early estimates of tropical temperatures based on foraminiferal oxygen isotopes suggested unexpectedly cool temperatures (15°–23°C), resulting in meridional temperature gradients that could not be reconciled with known dynamical mechanisms (Zachos et al., 1994). Cool tropical temperature estimates may have been biased by diagenesis or winter foraminifera growth, and recently revised tropical SST estimates (>28°C) are consistent with dynamic predictions based on high-latitude warming (Kobashi et al., 2001; Pearson et al., 2001). African continental environments appear to have been warm and wet during the Eocene. Bauxite, iron, and lateritic deposits at paleolatitude 5°–15°N indicate a humid Eocene climate (Guiraud, 1978). Paleobotanical remains from a middle Eocene crater lake in Tanzania (12°S) suggest high rainfall (640–780 mm per year) and woodland vegetation (Jacobs and Herendeen, 2004). A generally warm and wet African climate during the Eocene is consistent with a strong moisture source to the atmosphere provided by warm SSTs.

OLIGOCENE ANTARCTIC GLACIATION AND SOUTHERN AFRICAN CLIMATE

Significant southern African climate change at the Eocene-Oligocene boundary is indicated by seismic evidence from the Zaire (Congo) deep-sea fan. Marine sediments indicate a shift from pelagic sedimentation during Eocene greenhouse conditions to dramatically increased continental erosion associated with uplift in southern Africa and Oligocene global cooling (Anka and Séranne, 2004). The growth of the first permanent Antarctic ice sheet (35–26 Ma; figure 4.1) led to the development of the cold Benguela Current and associated increase in southern African aridity. Productivity proxies indicate that Benguela coastal upwelling intensified in the mid-Miocene with the second phase of Antarctic ice sheet growth (Diester-Haass et al., 1990; Robert et al., 2005). Dust records indicate that southwest African aridity increased after 9.6 Ma, and between 8.9 and 6.9 Ma, with significant variability after 6.5 Ma closely associated with the intensity of Benguela upwelling and ultimately the history of Antarctic glaciation (Robert et al., 2005). These marine records indicate that southwest African aridity developed in the Oligocene and Miocene, closely related to the intensity of the Benguela upwelling, which strengthened at times of increased equator-to-pole temperature gradients.

MID-MIOCENE CLIMATE CHANGE IN EAST AFRICA

The significant climate events that affected East Africa in the mid-Miocene are relatively well documented compared to earlier ones. Forested conditions in the early Miocene gave way to mixed grassland and forest in the mid-Miocene. Fossil plants indicate the earliest C3 grasslands in Africa at Fort Ternan, Kenya, ca. 14 Ma (Retallack, 1992). Isotopic studies

confirm that mid-Miocene grasslands were C3 (Cerling et al., 1997a; Feakins et al., 2005; see figure 4.4). Wet rain forest survived in many areas, including in northwestern Ethiopia (8 Ma; Yemane et al., 1987) and in the Tugen Hills, Kenya (8–6 Ma; Kingston et al., 2002) indicating mixed savanna and forest habitats across East Africa.

C4 grasses appear in East Africa in the mid-late Miocene (Cerling et al., 1997b). C4 grasses replace C3 plants as the most significant dietary component of northeast African grazing mammals between 8 and 6 Ma (Cerling et al., 1997b). Faunal assemblages at Lothagam, in the western Turkana Basin, also support a transition from C3 forest to a mixed C3 and C4 savanna mosaic between 8 and 4 Ma (Leakey et al., 1996). However, soil carbonate and leaf wax biomarker isotopic data indicate that C3 vegetation remained a significant component of regional vegetation during the Pliocene (Wynn, 2004; Feakins et al., 2005; figure 4.4). These vegetation reconstructions indicate that C4 plants may only have expanded to become a dominant component of the landscape in the late Pliocene and Pleistocene, much later than they appeared as a significant component of the diets of certain grazing mammals.

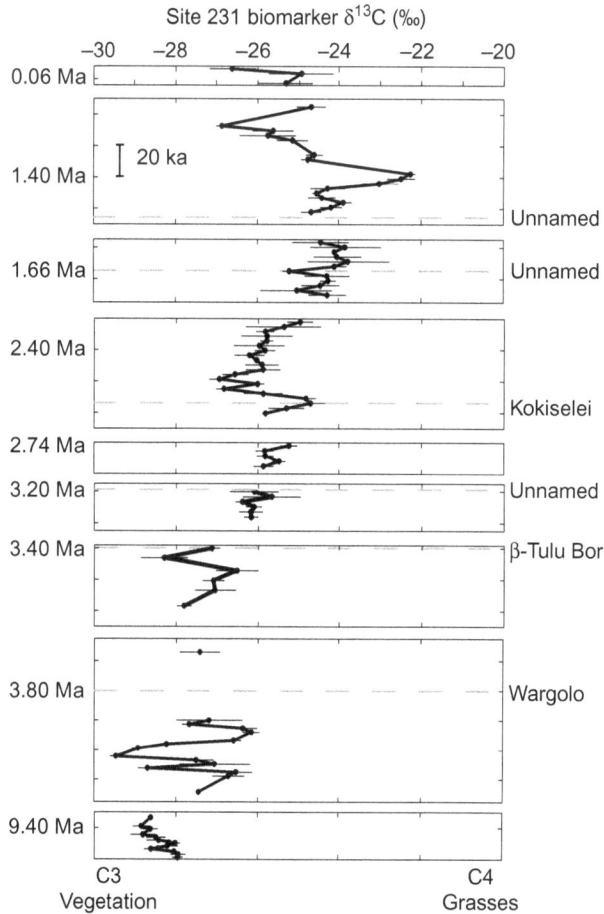

FIGURE 4.4   Carbon isotopic values of C_{30} n-alkanoic acids from DSDP Site 231 for nine 20- to 100-kyr time intervals spanning the late Neogene (near 9.4, 3.8, 3.4, 3.2, 2.7, 2.4, 1.7, 1.4, and 0.1 Ma; from Feakins et al., 2005). Mean $\delta^{13}C$ and 1σ analytical errors are shown. Vegetation changes are inferred based on $\delta^{13}C$ values of these C_{30} n-alkanoic acids, which have been identified as having a terrestrial vegetation source. Tephra age constraints are shown with dashed lines (Feakins et al., 2007). Foraminiferal $\delta^{18}O$ suggests interglacial timing of the upper interval rather than the glacial age (0.06 Ma) indicated by the interpolated age model. Reprinted from Nicholson, 2000, © 2000 with permission from Elsevier.

It has been suggested that the mid-Miocene appearance of the C4 photosynthetic pathway may be linked to declining pCO_2 levels (Cerling et al., 1997b). However, alkenone-based pCO_2 reconstructions do not support this explanation and instead indicate that pCO_2 levels rose from a low of 180 ppm at 15 Ma, to 260–300 ppm between 8 and 6 Ma (Pagani et al., 1999). Alternatively, uplift of the Himalayas and resultant intensification of the Indian Monsoon (9–6 Ma; Molnar et al., 1993) may have altered the seasonality of precipitation in the region with increased aridity driving a shift to almost exclusively C4 vegetation in the summer precipitation regime in Pakistan (Quade and Cerling, 1995), and a mixed C3 and C4 vegetation in East Africa (Cerling et al., 1997b; Feakins et al., 2005; figure 4.4). Elsewhere, at Langebaanweg, South Africa, C3 vegetation remained dominant in the diet of grazing mammals at 5 Ma (Franz-Odendaal et al., 2002). Changing precipitation regimes (linked to regional circulation patterns; figure 4.2) would explain why C4 vegetation did not uniformly expand across Africa between 8 and 6 Ma and may also explain the late Pliocene and Pleistocene increase in C4 vegetation in East Africa (figure 4.4).

African environments also experienced cyclical precipitation variability during the late Neogene. Organic rich sapropel deposits in the Mediterranean indicate times of high runoff from the Nile catchment (Rossignol-Strick, 1985; Sachs and Repeta, 1999). These sapropel deposits occur at precessional minima indicating northeast African climate sensitivity to orbital variations in the seasonal distribution of insolation (Rossignol-Strick, 1985). Precessional frequency sapropel deposits are reported from at least 10 Ma onward (Hilgen, 1991; Hilgen et al., 1995; Krijgsman et al., 1995). Similarly, precessional cyclicity in terrigenous dust flux to marine sediments off West Africa is reported in the late Miocene indicating dramatic variability in dust availability in the source region or transport efficiency (Tiedemann et al., 1994; deMenocal, 1995; deMenocal and Bloemendal, 1995). These records suggest that precession provided the fundamental pacing of African humid-arid cycles during the Miocene.

A WARM AND WET MID-PLIOCENE

Global SST reconstructions indicate that the Pliocene included extended periods both warmer and cooler than today, with low-amplitude orbital frequency variability (Pliocene Research, Interpretation and Synoptic Mapping Project [PRISM]; Dowsett et al., 1996). Humid conditions leading up to the Pliocene are recorded in central and eastern North Africa during the Zeit Wet Phase (7.5–5.5 Ma) with an expanded Lake Chad and increased Nile runoff (Griffin, 2002). Even during the Messinian salinity crisis (6.7–5.33 Ma), when sea levels in the Mediterranean were minimal or completely dry, conditions in North Africa were wet (deMenocal and Bloemendal, 1995; Hilgen et al., 1995; Griffin, 1999).

The mid-Pliocene (4.5–3 Ma) was characterized by warmer conditions (+3°C) on average globally, higher sea levels (+10–20 m), reduced Antarctic ice cover, and percentage higher pCO_2 (Ravelo et al., 2004). The mid-Pliocene appears to have been broadly wetter throughout much of Africa consistent with PRISM model predictions in scenarios with increased meridional circulation and higher pCO_2 (Haywood and Valdes, 2004). For example, pollen records indicate that North Africa was wetter in the mid-Pliocene (Dupont and Leroy,

1995), and PRISM models predict that North Africa was warmer (by 5°C) and wetter (by 400–1,000 mm per year) relative to today (Haywood et al., 2000).

In the equatorial Pacific, the east to west SST gradient resembled a permanent El Niño, in contrast to the mean La Niña state in the modern ocean (Cannariato and Ravelo, 1997; Chaisson and Ravelo, 2000; Wara et al., 2005; Ravelo et al., 2006). Although the atmospheric teleconnection mechanisms associated with a modern El Niño event cannot simply be extrapolated to longer timescales, in many regions Pliocene climate appears to be roughly analogous to that observed in a modern El Niño (Molnar and Cane, 2002). In Africa, modern El Niño conditions are associated with anomalously wet conditions in East Africa and dry conditions in southeast Africa. We do find evidence for wet conditions in East Africa during the Pliocene, although it is hard to separate out the effects of warmer SSTs around Africa and teleconnections from the Pacific. Flora characteristic of the modern West African rain forest are found in East Africa around 3.4 Ma (Bonnefille and Letouzey, 1976; Bonnefille, 1987); soil carbonate records of vegetation in East and Central Africa indicate C3 vegetation and humid conditions (Cerling et al., 1977); lakes freshened in the Afar region of northeast Africa (Gasse, 1990), and Lake Tanganyika in southeast Africa expanded at about 3.6 Ma (Cohen et al., 1997). In southern Africa, there are fewer records, although vegetation appears to be relatively close to modern (Scott, 1995).

Most terrestrial records are at too low resolution to identify variability in the Pliocene. Marine records indicate that Pliocene African climate variability was dominated by precessional frequency (19–21 ka) variations in the strength of monsoonal precipitation. Precessional cycles in dust concentration (varying by a factor of 2–5) are seen in marine sediments off West and East Africa between 5 Ma and 2.8 Ma (figure 4.5), indicating dramatic variability in dust availability in the source region or transport efficiency (Tiedemann et al., 1994; deMenocal, 1995; deMenocal and Bloemendal, 1995). Evidence for precessionally driven precipitation changes in the Nile catchment in northeast Africa are seen in the organic-rich sapropel layers of the eastern Mediterranean throughout the Pliocene (Rossignol-Strick, 1983; Hilgen, 1991). Pollen from a terrestrial site at Hadar, Ethiopia, indicates an abrupt change in forest cover ca. 3.3 Ma (Bonnefille et al., 2004) that is consistent with environmental change during part of a precessional cycle identified in sapropel and dust records. Precessional variations in C3/C4 vegetation type are also seen in leaf wax biomarker records from marine sediments off northeast Africa ca. 3.8–3.7 Ma (figure 4.4; Feakins et al., 2005) and off southwest Africa ca. 2.56–2.51 Ma (Denison et al., 2005). These marine records clearly indicate that precession dominated the pacing of precipitation variations across Africa during the Pliocene.

PLIO-PLEISTOCENE ENVIRONMENTAL CHANGE

Major global climate events at the end of the Pliocene warm phase were not synchronous and instead occurred in a series of regional events (Ravelo et al., 2004). Tectonic processes caused significant reorganization of tropical ocean circulation during the late Pliocene. The restriction of the Panamanian seaway (4.5–4 Ma) caused changes in Atlantic circulation and an increase in meridional overturning (Haug and Tiedemann, 1998; Haug et al., 2001). The northward migration of New Guinea led to restriction of the Indonesian Seaway (4–3 Ma); models predict that this would likely have

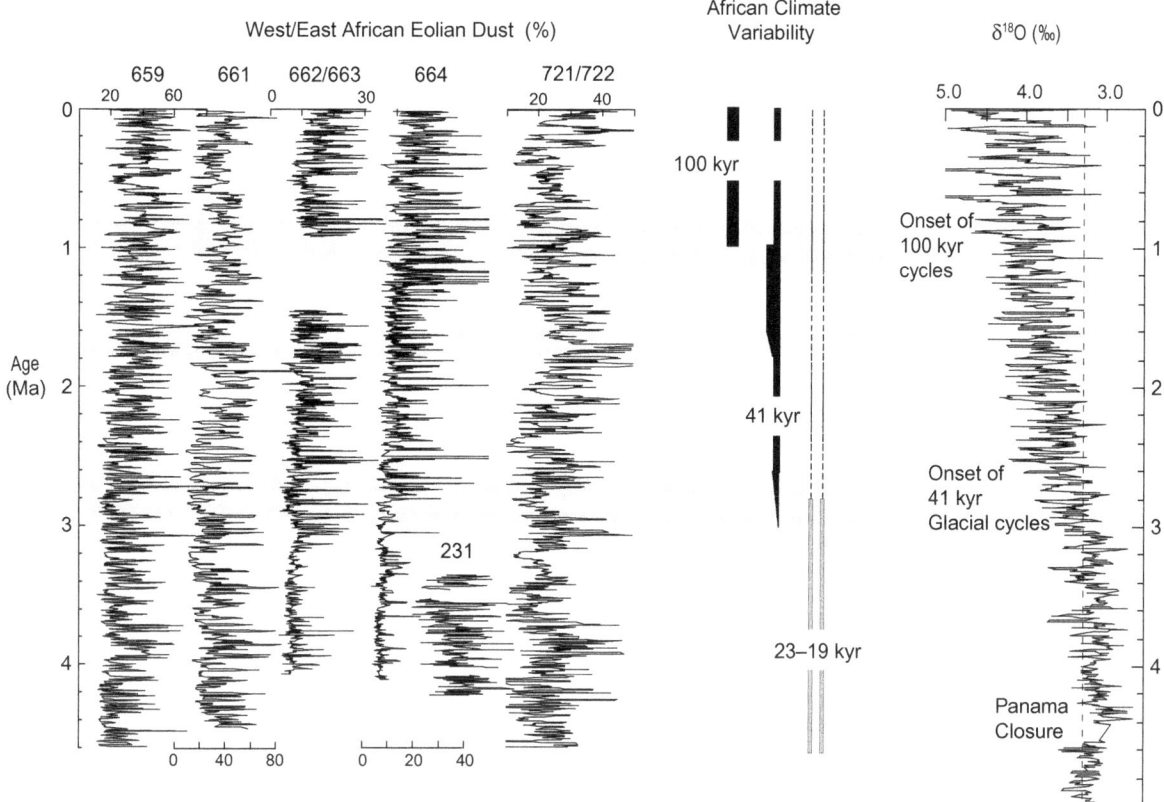

West/East African Eolian Dust (%)

African Climate Variability

δ¹⁸O (‰)

FIGURE 4.5   Pliocene-Pleistocene records of eolian dust deposition at seven DSDP/ODP sites off western and eastern subtropical Africa, together with summary indication of spectral analysis of those same dust records, and a stacked benthic oxygen isotope record. Reprinted with permission from deMenocal (2004). © 2004 Elsevier.

caused a cooling of Indian Ocean SSTs and a reduction of East African precipitation (Cane and Molnar, 2001). This predicted change in Indian Ocean SSTs is one possible explanation for the C4 vegetation expansion in East Africa after ca. 3.4 Ma seen in the leaf wax biomarker record (figure 4.4).

Significant Northern Hemisphere Glaciation began and intensified between 3.2 and 2.6 Ma (figure 4.1; Shackleton, 1995; Lisiecki and Raymo, 2005). Ocean temperatures cooled, and obliquity (41 ka) paced northern hemisphere glacial cycles commenced and intensified between 3.2 and 2.6 Ma (Shackleton, 1995; Lisiecki and Raymo, 2005). As the high latitudes cooled, most records indicate that Hadley circulation strengthened, trade winds intensified, and subtropical regions became more arid and more variable. Aridity and wind strength increased in North Africa at 2.8 ± 0.2 Ma as indicated by pollen (Dupont and Leroy, 1995) and dust records of West and East Africa (figure 4.5; Tiedemann et al., 1994; deMenocal, 1995; deMenocal and Bloemendal, 1995). In southern Africa, SSTs cooled with intensified upwelling, leading to greater aridity (Marlow et al., 2000). In contrast, lake levels in the Baringo-Bogoria Basin, Kenya, and Gadeb, Ethiopia, apparently record a wet interval from 2.7 to 2.5 Ma (Trauth et al., 2005, and references therein) indicating that perhaps not all of Africa experienced increased aridity at this time.

The onset of Northern Hemisphere glaciation signaled a change in the periodicity of some features of African climate variability. Dust records off West and East Africa document a shift from precession paced humid-arid cycles before 2.8 Ma, to obliquity frequency after ca. 2.8 Ma, suggesting a glacial control on either transport strength or source aridity (figure 4.5;

deMenocal, 1995, 2004). Similarly, leaf wax biomarker records of southwest African vegetation document C3/C4 cycles in tune with obliquity paced Atlantic SST variations in the mid-Pleistocene (Schefuss et al., 2003), suggesting that glacial-interglacial cycles influenced African climate in both the Southern and Northern hemispheres.

Not all aspects of African climate were dominated by changes in the high latitudes, however. A second biomarker record from southwest Africa indicates that vegetation changes continued to be dominated by precessional timing shortly after the onset of Northern Hemisphere glaciation (2.56–2.51 Ma; Denison et al., 2005). Precessional frequency precipitation variations are also recorded in the last 200 ka of the Pleistocene in Tswaing Impact Crater in South Africa (Partridge et al., 1997) and in various East African lakes (Trauth et al., 2001). Finally, sapropel stratigraphy indicates dominantly precessional timing of precipitation variations in northeast Africa throughout the Pliocene and Pleistocene (Rossignol-Strick, 1983; Tuenter et al., 2003). Therefore, despite evidence that Northern Hemisphere glacial cycles led African aridity, there are many counterindications of independent precessional pacing of African climate, particularly in those proxies that directly relate to precipitation.

The tropical Pacific was also partially decoupled from high-latitude climate. Despite significant high-latitude changes ca. 2.8 Ma, tropical Pacific SST gradients appeared to have remained largely stable with El Niño–like conditions until ca. 2 Ma (Chaisson and Ravelo, 2000; Wara et al., 2005; Ravelo et al., 2006). The reorganization of the tropical Pacific ca. 2 Ma occurred at a time when high-latitude climate was relatively invariant (Wara et al., 2005). A cooling of eastern Pacific SSTs ca. 2 Ma relative to warm SSTs in the western Pacific (La Niña–like conditions) indicates the initiation of Walker circulation and the likely beginning of ENSO variability.

This reorganization of the tropical Pacific, from an El Niño–like to a La Niña–like mean state, may have produced climate repercussions with a global signature, since the same mechanisms that generate interannual ENSO variability in modern climates may also produce variability on longer time-scales (Cane and Zebiak, 1985; Clement et al., 2001; Molnar and Cane, 2002). Around this time, biostratigraphic events mark the Plio-Pleistocene boundary and local cooling in the Mediterranean (1.77 Ma; Raffi et al., 1993), and an increase in dust flux off Africa records increased aridity (deMenocal, 1995). A C4 expansion is seen in the Turkana Basin, East Africa (2–1.7 Ma; Cerling et al., 1977), and organic carbon concentrations dramatically increase in marine sediments off West Africa (2.45–1.7 Ma; Wagner, 2002). These records appear to indicate an arid shift in North African climate ca. 2 Ma coincident with a major reorganization of the tropical ocean-atmosphere system to a mean La Niña–like mean state.

COOL AND DRY CONDITIONS DURING THE LAST GLACIAL MAXIMUM

Most terrestrial paleoclimate research for Africa has focused on the Last Glacial Maximum (LGM) to the present, for which the geomorphological evidence is typically best preserved and reconstructions are within the range of radiocarbon and optical dating. Tropical paleoclimate records for the LGM have been reconsidered since CLIMAP concluded that there was minimal cooling (<2°C) or even a slight warming in the tropics (CLIMAP-project, 1976). Tropical SST reconstructions from the Atlantic (Guilderson et al., 1994; Emiliani, 1995; Lea et al., 2003), Indian (Sonzogni et al., 1998; Visser et al., 2003), and Pacific oceans (Emiliani, 1995; Prahl et al., 1995) have now demonstrated a 2°–6°C glacial cooling relative to modern. In addition, many modeling studies support the evidence for cooler tropical temperatures during glacials (Rind and Peteet, 1985; Pinot et al., 1999). Terrestrial temperature reconstructions also indicate significant cooling at the LGM. In East Africa cooling estimates include 5°–6°C from tropical snowline depressions (Rind and Peteet, 1985), 2°–8°C from pollen assemblages (Coetzee, 1967; Chalie, 1995), and 3°–5°C in Lake Malawi from the new TEX$_{86}$ molecular paleothermometer (Powers et al., 2005). In South Africa, 5°C glacial cooling is estimated from speleothems (Talma and Vogel, 1992; Holmgren et al., 2003) and groundwater (Kulongoski et al., 2004).

However, aridity changes may be more critical than temperature changes for flora and fauna in African environments. Modern climate patterns would predict reduced precipitation in most regions as a result of cooler glacial SSTs. Various terrestrial records indicate broad arid phases between 23 and 17 ka during the LGM. Fossil sand dunes indicate that the Sahara expanded southward during the LGM (Grove and Warren, 1968; Mauz and Felix-Henningsen, 2005). Terrigenous dust (Tiedemann et al., 1994; deMenocal, 1995), pollen (Dupont and Leroy, 1995; Hooghiemstra et al., 1998), and leaf wax biomarkers (Zhao et al., 2003) transported to marine sediments off northwest Africa indicate increased aridity and trade wind strength during the last glacial. Other terrigenous material reaching marine sediments include *Melosira,* fresh-water diatoms, deflated from dry West African lake beds

during the LGM (Pokras and Mix, 1987), and grass phytoliths that indicate increased aridity in cold stages (Abrantes, 2003). Lake levels were almost uniformly low across Africa (Street and Grove, 1979). Leaf wax biomarker hydrogen isotope reconstructions from the Congo fan record arid conditions in Central Africa (Schefuss et al., 2005). In southern Africa, fossil dunes in the Mega Kalahari (Stokes et al., 1997) and sedimentation in the Tswaing Impact Crater (Partridge et al., 1997) indicate arid conditions 26–10 ka. In East Africa, Lakes Victoria (Talbot and Laerdal, 2000) and Albert (Beuning et al., 1997) record two prolonged droughts between 18 and 12.5 ^{14}C ka that led to complete desiccation of both lakes. Lake levels in Lake Tanganyika dropped by ~350 m (Scholz et al., 2003). A diatom record from Lake Massoko, Tanzania, shares the pattern of climate variability seen in the northeast African lakes with dry conditions up to 15 ka (Barker et al., 2003). Pollen-based precipitation reconstructions from Burundi also indicate an arid LGM (Bonnefille and Chalie, 2000). These precipitation records correlate well with alkenone-derived SST reconstructions in the southwest Indian Ocean (Sonzogni et al., 1998), suggesting that cool SSTs resulted in reduced East African precipitation during the LGM. These many records indicate arid conditions throughout much of Africa at the LGM.

THE HOLOCENE

The early Holocene was anomalously wet in most of Africa. Lake levels across Africa are almost uniformly higher and indicate a 150- to 400-mm annual increase in precipitation in the Sahara (Street and Grove, 1979; Street-Perrott and Harrison, 1984). Fluvial sediments, lacustrine carbonates, freshwater algae, faunal records, paleosols, and stabilized fossil dunes all indicate that the early to mid-Holocene was more humid than the present (e.g., Haynes and Mead, 1987; Kropelin and Soulie-Marsche, 1991). Marine records off West Africa indicate that the "African Humid Period" extended from 15 to 5 ka in North Africa (Pokras and Mix, 1987; deMenocal et al., 2000). While records from South Africa are comparatively rare, stalagmites in Makapansgat Valley and pollen records from Wonderkrater have revealed generally warm conditions between 10 and 6 ka (Scott, 1999; Holmgren et al., 2003).

High-resolution paleoclimate records are revealing new details of Holocene climate change. A tropical ice core, recovered from Mount Kilimanjaro, Tanzania, has provided a continuous and detailed record of East African climate variability throughout the entire Holocene (Thompson et al., 2002). The ice core oxygen isotope and aerosol records indicate an abrupt cooling and drying event ca. 8.3 ka that corresponds to a cooling event in the North Atlantic Ocean (Alley et al., 1997). Further arid shifts are observed between 6.4 and 5.2 ka and at 4 ka (Thompson et al., 2002).

The late Holocene has been comparatively arid in much of Africa. An abrupt transition has been identified in many tropical records ca. 4 ka (Marchant and Hooghiemstra, 2004). Late Holocene arid conditions in East Africa are recorded in Arabian Sea dust records (Davies et al., 2002) and lake-level low stands (Halfman and Johnson, 1988; Talbot and Laerdal, 2000). In West Africa, there is evidence for a southerly shift of arid vegetation zones (Kutzbach and Street-Perrot, 1985; Lezine, 1989) and low lake levels since mid-Holocene times (Street and Grove, 1979; Street-Perrott and Perrott, 1990). This mid-Holocene transition to arid conditions in Africa is largely independent of high-latitude climate change (Marchant and Hooghiemstra, 2004). Modeling studies suggest that this humid-arid transition is strongly dependent on a nonlinear climate response to precessional insolation forcing (Claussen et al., 1999). Thus in the Holocene, like much of the late Neogene, we find evidence for precessional forcing of African climate change.

Literature Cited

Abrantes, F. 2003. A 340,000 year continental climate record from tropical Africa: News from opal phytoliths from the equatorial Atlantic. *Earth and Planetary Science Letters* 209:165–179.

Alley, R. B., P. A. Mayewski, T. Sowers, M. Stuiver, K. C. Taylor, and P. U. Clark. 1997. Holocene climatic instability: A prominent, widespread event 8200 yr ago. *Geology* 25:483–486.

Anka, Z., and M. Séranne. 2004. Reconnaissance study of the ancient Zaire (Congo) deep-sea fan (ZaiAngo Project). *Marine Geology* 209:223–244.

Barker, P., D. Williamson, F. Gasse, and E. Gibert. 2003. Climatic and volcanic forcing revealed in a 50,000-year diatom record from Lake Massoko, Tanzania. *Quaternary Research* 60:368–376.

Beuning, K. R. M., M. R. Talbot, and K. Kelts. 1997. A revised 30,000-year paleoclimatic and paleohydrologic history of Lake Albert, East Africa. *Palaeogeography, Palaeoclimatology, Palaeoecology* 136:259–279.

Bolle, M. P., T. Adatte, G. Keller, K. Von Salis, and S. Burns. 1999. The Paleocene-Eocene transition in the southern Tethys (Tunisia): Climatic and environmental fluctuations. *Bulletin de la Société Géologique de France* 170:661–680.

Bonnefille, R. 1987. Forest and climatic history during the last 40,000 yrs in Burundi. *Comptes Rendus de l'Académie des Sciences, Série II,* 305:1021–1026.

Bonnefille, R., and F. Chalie. 2000. Pollen-inferred precipitation time-series from equatorial mountains, Africa, the last 40 kyr BP. *Global and Planetary Change* 26:25–50.

Bonnefille, R., and R. Letouzey. 1976. Fruits fossiles d'*Antrocaryon* dans la vallée de l'Omo (Ethiopie). *Adansonia* 16:65–82.

Bonnefille, R., R. Potts, F. Chalie, D. Jolly, and O. Peyron. 2004. High-resolution vegetation and climate change associated with Pliocene *Australopithecus afarensis*. *Proceedings of the National Academy of Sciences, USA* 101:12125–12129.

Camberlin, P., S. Janicot, and I. Poccard. 2001. Seasonality and atmospheric dynamics of the teleconnection between African rainfall and tropical sea-surface temperature: Atlantic vs. ENSO. *International Journal of Climatology* 21:973–1005.

Cane, M., and A. C. Clement. 1999. A role for the tropical Pacific coupled ocean-atmosphere system on Milankovitch and millennial timescales. Part II: Global impacts; pp. 373–383 in P. Clark, R. S. Webb, and L. D. Keigwin (eds.), *Mechanisms of Global Climate Change at Millennial Time Scales.* American Geophysical Union, Washington, D.C.

Cane, M. A., G. Eshel, and R. Buckland. 1994. Forecasting Zimbabwean maize yield using eastern equatorial Pacific sea surface temperature. *Nature* 370:204–206.

Cane, M. A., and P. Molnar. 2001. Closing of the Indonesian Seaway as the missing link between Pliocene East African aridification and the Pacific. *Nature* 411:157–162.

Cane, M. A., and S. Zebiak. 1985. A theory for El Niño and the Southern Oscillation. *Science* 228:1085–1087.

Cannariato, K. G., and A. C. Ravelo. 1997. Pliocene-Pleistocene evolution of eastern tropical Pacific surface water circulation and thermocline depth. *Paleoceanography* 12:805–820.

Cerling, T. E., J. M. Harris, S. H. Ambrose, M. G. Leakey, and N. Solounias. 1997a. Dietary and environmental reconstruction with stable isotope analyses of herbivore tooth enamel from the Miocene locality of Fort Ternan, Kenya. *Journal of Human Evolution* 33:635–650.

Cerling, T. E., J. M. Harris, B. J. MacFadden, M. G. Leakey, J. Quade, V. Eisenmann, and J. R. Ehleringer. 1997b. Global vegetation change through the Miocene/Pliocene boundary. *Nature* 389:153–158.

Cerling, T. E., R. L. Hay, and J. R. O'Neil. 1977. Isotopic evidence for dramatic climatic changes in East-Africa during Pleistocene. *Nature* 267:137–138.

Chaisson, W. P., and A. C. Ravelo. 2000. Pliocene development of the east-west hydrographic gradient in the equatorial Pacific. *Paleoceanography* 15:497–505.

Chalie, F. 1995. Paleoclimates of the southern Tanganyika Basin over the last 25-thousand years: Quantitative reconstruction from the statistical treatment of pollen data. *Comptes Rendus de l'Académie des Sciences,* Série II, 320:205–210.

Claussen, M., C. Kubatzki, V. Brovkin, A. Ganapolski, P. Hoelzmann, and H.-J. Pachur. 1999. Simulation of an abrupt change in Saharan vegetation in the mid-Holocene. *Geophysical Research Letters* 26:2037–2040.

Clement, A. C., M. A. Cane, and R. Seager. 2001. An orbitally driven tropical source for abrupt climate change. *Journal of Climate* 14:2369–2375.

CLIMAP-project. 1976. The surface of the ice-age Earth. *Science* 191:1131–1136.

Coetzee, J. 1967. Pollen analytical studies in East and southern Africa. *Palaeoecology of Africa* 3:1–46.

Cohen, A., K. E. Lezzar, J. J. Tiercelin, and M. Soreghan. 1997. New palaeogeographic and lake-level reconstructions of Lake Tanganyika: Implications for tectonic, climatic and biological evolution in a rift lake. *Basin Research* 9:107–132.

Dansgaard, W. S., S. Johnsen, H. B. Glausen, D. Dahl-Jensen, N. S. Gundestrup, C. U. Hammer, C. S. Hvidberg, J. Steffesen, A. E. Sveinsbjörnsdottir, J. Jouzel, and G. Bond. 1993. Evidence for general instability of past climate from a 250-kyr ice core record. *Nature* 364:218–220.

Davies, G. R., S. J. A. Jung, and G. M. Ganssen. 2002. Abrupt climate change in N-Africa: The asynchronous termination of the African Humid Period. *Geochimica et Cosmochimica Acta* 66(15A):A170–A170.

deMenocal, P. B. 1995. Plio-Pleistocene African climate. *Science* 270:53–59.

———. 2004. African climate change and faunal evolution during the Pliocene-Pleistocene. *Earth and Planetary Science Letters* 220:3–24.

deMenocal, P., and J. Bloemendal. 1995. Plio-Pleistocene climatic variability in subtropical Africa and the paleoenvironment of hominid evolution: A combined data-model approach; pp. 262–288 in E. Vrba, G. Denton, T. Partridge, and L. Burckle (eds.), *Paleoclimate and Evolution: With Emphasis on Human Origins.* Yale University Press, New Haven.

deMenocal, P. B., J. Ortiz, T. Guilderson, J. Adkins, M. Sarnthein, L. Baker, and M. Yarusinsky. 2000. Abrupt onset and termination of the African Humid Period: Rapid climate responses to gradual insolation forcing. *Quaternary Science Reviews* 19:347–361.

Denison, S., M. A. Maslin, C. Boot, R. Pancost, and V. Ettwein. 2005. Precession-forced changes in South West African vegetation during Marine Isotope Stages 101–100 (~2.56–2.51 Ma). *Palaeogeography Palaeoclimatology Palaeoecology* 220:375–386.

Denton, G. H. 1999. Cenozoic climate change; pp. 94–114 in T. Bromage and F. Schrenk (eds.), *African Biogeography, Climate Change and Human Evolution.* Oxford University Press, Oxford.

Diester-Haass, L., P. Meyers, and P. Rothe. 1990. Miocene history of the Benguela current and Antarctic ice volumes: evidence from rhythmic sedimentation and current growth across the Walvis Ridge (DSDP Sites 362 and 532). *Paleoceanography* 5:685–707.

Dowsett, H., J. Barron, and R. Z. Poore. 1996. Middle Pliocene sea-surface temperatures: A global reconstruction. *Marine Micropaleontology* 27:13–25.

Dupont, L., and S. Leroy. 1995. Steps towards drier climatic conditions in Northwestern Africa during the Upper Pliocene; pp. 289–297 in E. Vrba, G. Denton, T. Partridge, and L. Burckle (eds.), *Paleoclimate and Evolution: With Emphasis on Human Origins.* Yale University Press, New Haven.

Emiliani, C. 1995. Tropical paleotemperatures. *Science* 268:1264.

Feakins, S. J., F. H. Brown, and P. B. deMenocal. 2007. Plio-Pleistocene Microtephra in DSDP Site 231, Gulf of Aden. *Journal of African Earth Sciences* 48:341–352

Feakins, S. J., P. B. deMenocal, and T. I. Eglinton. 2005. Biomarker records of Late Neogene changes in East African vegetation. *Geology* 33:977–980.

Franz-Odendaal, T. A., J. A. Lee-Thorp, and A. Chinsamy. 2002. New evidence for the lack of C-4 grassland expansions during the early Pliocene at Langebaanweg, South Africa. *Paleobiology* 28:378–388.

Gasse, F. 1990. Tectonic and climatic controls on lake distribution and environments in Afar from Miocene to Present; pp. 19–41 in B. Katz (ed.), *Lacustrine Basin Exploration: Case Studies and Modern Analogs.* Memoirs of the American Association of Petroleum Geologists, 50.

Giannini, A., R. Saravan, and P. Chang. 2003. Oceanic forcing of Sahel rainfall on interannual to interdecadal time scales. *Science* 302:1027–1030.

Gingerich, P. D. 2004. Paleogene vertebrates and their response to environmental change. *Neues Jahrbuch für Geologie und Paläontologie, Abhandlungen* 234:1–23.

Goddard, L., and N. E. Graham. 1999. Importance of the Indian Ocean for simulating rainfall anomalies over eastern and southern Africa. *Journal of Geophysical Research* 104:19,099–19,116.

Griffin, D. L. 1999. The late Miocene climate of northeastern Africa: Unravelling the signals in the sedimentary succession. *Journal of the Geological Society, London* 156:817–826.

———. 2002. Aridity and humidity: Two aspects of the late Miocene climate of North Africa and the Mediterranean. *Palaeogeography, Palaeoclimatology, Palaeoecology* 182:65–91.

Grove, A. T., and A. Warren. 1968. Quaternary landforms and climate on the south side of the Sahara. *Geographical Journal* 134:194–208.

Guilderson, T. P., R. P. Fairbanks, and J. L. Rubenstone. 1994. Tropical temperature variations since 20,000 years ago: modulating interhemispheric climate change. *Science* 263:663–665.

Guiraud, R. 1978. Le Continental Intercalaire en Algérie. *Annales de la Faculte des Sciences de Dakar* 31:85–87.

Halfman, J, and T. Johnson. 1988. High-resolution record of cyclic climatic change during the past 4 ka from Lake Turkana, Kenya. *Geology* 16:496–500.

Hastenrath, S., A. Nicklis, and L. Greischar. 1993. Atmospheric-hydrospheric mechanisms of climate anomalies in the western equatorial Indian Ocean. *Journal of Geophysical Research* 98:219–235.

Haug, G. H., and R. Tiedemann. 1998. Effect of the formation of the Isthmus of Panama on Atlantic Ocean thermohaline circulation. *Nature* 393:673–676.

Haug, G. H., R. Tiedemann, R. Zahn, and A. C. Ravelo. 2001. Role of Panama uplift on oceanic freshwater balance. *Geology* 29:207–210.

Haynes, C. V., and A. R. Mead. 1987. Radiocarbon dating and paleoclimatic significance of sub-fossil *Limicolaria* in NW Sudan. *Quaternary Research* 28:86–89.

Haywood, A., B. Sellwood, and P. Valdes. 2000. Regional warming: Pliocene (3 Ma) paleoclimate of Europe and Mediterranean. *Geology* 38:1063–1066.

Haywood, A., and P. Valdes. 2004. Modelling Pliocene warmth: Contribution of atmosphere, oceans and cryosphere. *Earth and Planetary Science Letters* 218:363–377.

Hilgen, F. J. 1991. Astronomical calibration of Gauss to Matuyama sapropels in the Mediterranean and implication for the geomagnetic polarity time scale. *Earth and Planetary Science Letters* 104:226–244.

Hilgen, F., W. Krijgsman, C. Langereis, L. J. Lourens, A. Santarelli, and W. J. Zachariasse. 1995. Extending the astronomical (polarity) time scale into the Miocene. *Earth and Planetary Science Letters* 136:495–510.

Hoerling, M. P., J. W. Hurrell, and T. Xu. 2001. Tropical origins for recent North Atlantic climate change. *Science* 292:90–92.

Holmgren, K., J. Lee-Thorp, G. Cooper, K. Lundblad, T. C. Partridge, L. Scott, R. Sithaldeen, A. Talma, and P. Tyson. 2003. Persistent millennial-scale climatic variability over the past 25,000 years in Southern Africa. *Quaternary Science Reviews* 22:2311–2326.

Hooghiemstra, H., R. Bonnefille, and H. Behling. 1998. Intertropical last glacial and Holocene climatic change. *Review of Palaeobotany and Palynology* 99:75–76.

Hurrell, J. W. 1995. Decadal trends in the North Atlantic Oscillation: Regional temperatures and precipitation. *Science* 269:676–679.

Jacobs, B. F. and P. Herendeen. 2004. Eocene dry climate and woodland vegetation in tropical Africa reconstructed from fossil leaves from northern Tanzania. *Palaeogeography, Palaeoclimatology, Palaeoecology* 213:115–123.

Janowiak, J. 1988. An investigation of interannual rainfall variability in Africa. *Journal of Climate* 1:240–255.

Kennett, J. P. 1995. A review of polar climatic evolution during the Neogene, based on the marine sediment record; pp. 49–64 in E. Vrba, G. Denton, T. Partridge, and L. Burckle (eds.), *Paleoclimate and Evolution: With Emphasis on Human Origins.* Yale University Press, New Haven.

Kingston, J., B. F. Jacobs, A. Hill, and A. Deino. 2002. Stratigraphy, age and environments of the late Miocene Mpesida Beds, Tugen Hills, Kenya. *Journal of Human Evolution* 42:92–116.

Knippertz, P., U. Ulbrich, F. Marques, and J. Corte-Real. 2003. Decadal changes in the link between El Niño and springtime North Atlantic

oscillation and European–North African rainfall. *International Journal of Climatology* 23:1293–1311.

Kobashi, T., E. L. Grossman, T. E. Yancey, and D. T. Dockery III. 2001. Reevaluation of conflicting Eocene tropical temperature estimates: Molluscan oxygen isotope evidence for warm low latitudes. *Geology* 29:983–986.

Krijgsman, W., F. Hilgen, C. Langereis, A. Santarelli, and W. Zachariasse. 1995. Late Miocene magnetostratigraphy, biostratigraphy and cyclostratigraphy in the Mediterranean. *Earth and Planetary Science Letters* 136:475–494.

Kropelin, S., and I. Soulie-Marsche. 1991. Charophyte remains from Wadi Howar as evidence for deep mid-Holocene freshwater lakes in the eastern Sahara of northwest Sudan. *Quaternary Research* 36:210–223.

Kulongoski, J., D. Hilton, and E. Selaolo. 2004. Climate variability in the Kotswana Kalahari from the late Pleistocene to the present day. *Geophysical Research Letters* 31:L10204. doi:10.1029/2003GL019238.

Kutzbach, J. E., and F. A. Street-Perrot. 1985. Milankovitch forcing of fluctuations in the level of tropical lakes from 18 to 0 kyr BP. *Nature* 317:130–134.

Lea, D. W., D. Pak, L. Peterson, and K. Hughen. 2003. Synchroneity of tropical and high-latitude Atlantic temperatures over the last glacial termination. *Science* 301:1361–1364.

Leakey, M. G., C. S. Feibel, R. L. Bernor, J. M. Harris, T. E. Cerling, K. M. Stewart, G. W. Storrs, A. Walker, L. Werdelin, and A. J. Winkler. 1996. Lothagam: A record of faunal change in the late Miocene of East Africa. *Journal of Vertebrate Paleontology* 16:556–570.

Le Houérou, H. 1997. Climate, flora and fauna changes in the Sahara over the past 500 million years. *Journal of Arid Environments* 37:619–647.

Lezine, A. M. 1989. Late Quaternary vegetation and climate of the Sahel. *Quaternary Research* 32:317–334.

Linsley, B. K., G. Wellington, and D. P. Schrag. 2000. Decadal sea surface temperature variability in the subtropical south Pacific from 1726–1997 AD. *Science* 290:1145–1148.

Lisiecki, L., and M. Raymo. 2005. A Pliocene-Pleistocene stack of 57 globally distributed benthic $\delta^{18}O$ records. *Paleoceanography* 20:1–17.

Liu, Z., B. Otto-Bleisner, J. Kutzbach, L. Li, and C. Shields. 2003. Coupled climate simulation of the evolution of global monsoons in the Holocene. *Journal of Climate* 16:2472–2490.

Marchant, R., and H. Hooghiemstra. 2004. Rapid environmental change in African and South American tropics around 4000 years before present: A review. *Earth-Science Reviews* 66:217–260.

Marlow, J. R., C. B. Lange, G. Wefer, and A. Rosell-Melé. 2000. Upwelling intensification as a part of the Pliocene-Pleistocene climate transition. *Science* 290:2288–2291.

Mauz, B., and P. Felix-Henningsen. 2005. Palaeosols in Saharan and Sahelian dunes of Chad: Archives of Holocene North African climate changes. *The Holocene* 15:453–458.

Molnar, P., and M. A. Cane. 2002. El Niño's tropical climate and teleconnections as a blueprint for pre-Ice Age climates. *Paleoceanography* 17:1021. doi:10.1029/2001PA000663.

Molnar, P., P. England, and J. Martinod. 1993. Mantle dynamics, uplift of the Tibetan Plateau and the Indian Monsoon. *Reviews of Geophysics* 31:357–396.

Nicholson, S. 2000. The nature of rainfall variability over African on time scales of decades to millennia. *Global and Planetary Change* 26:137–158.

Nicholson, S., and D. Entekhabi. 1987. Rainfall variability in equatorial and southern Africa: Relationships with sea surface temperatures along the southwestern coast of Africa. *Journal of Applied Meteorology* 26:561–578.

Pagani, M., M. Arthur, and K. H. Freeman. 1999. Miocene evolution of atmospheric carbon dioxide. *Paleoceanography* 14:273–292.

Partridge, T., P. B. deMenocal, S. Lorentz, M. Paiker, and J. Vogel. 1997. Orbital forcing of climate over South Africa: A 200,000-year rainfall record from the Pretoria Saltpan. *Quaternary Science Reviews* 16:1125–1133.

Pearson, P., P. Ditchfield, J. Singano, K. Harcourt-Brown, C. Nicholas, R. Olsson, N. J. Shackleton, and M. Hall. 2001. Warm tropical sea surface temperatures in the Late Cretaceous and Eocene epochs. *Nature* 413:481–487.

Pearson, P., and M. Palmer. 2000. Atmospheric carbon dioxide concentrations over the past 60 million years. *Nature* 406:695–699.

Pinot, S., G. Ramstein, S. P. Harrison, I. C. Prentice, J. Guiot, M. Stute, and S. Joussaume. 1999. Tropical paleoclimates at the Last Glacial

Maximum: Comparison of Paleoclimate Modeling Intercomparison Project (PMIP) simulations and paleodata. *Climate Dynamics* 15:857–874.

Pokras, E., and A. C. Mix. 1987. Earth's precession cycle and Quaternary climatic change in tropical Africa. *Nature* 326:486–487.

Powers, L., T. Johnson, J. Werne, I. Castaneda, E. Hopmans, J. S. Damsté, and S. Schouten. 2005. Large temperature variability in the southern African tropics since the Last Glacial Maximum. *Geophysical Research Letters* 32 doi:10.1029/ 2004GL02214.

Prahl, F., N. Pisias, M. Sparrow, and A. Sabin. 1995. Assessment of sea surface temperature at 42°N in the California current over the last 30 000 years. *Paleoceanography* 4:763–773.

Quade, J. and T. E. Cerling. 1995. Expansion of C-4 grasses in the Late Miocene of Northern Pakistan: Evidence from Stable Isotopes in Paleosols. *Palaeogeography, Palaeoclimatology, Palaeoecology* 115:91–116.

Raffi, I., J. Backman, D. Rio, and N. J. Shackleton. 1993. Plio-Pleistocene nannofossil biostratigraphy and calibration to oxygen isotope stratigraphies from DSDP site 607 and ODP site 677. *Paleoceanography* 8:387–408.

Ravelo, A. C., P. S. Dekens, and M. McCarthy. 2006. Evidence for El Niño–like conditions during the Pliocene. *GSA Today* 16:4–11. doi: 10.1130/1052-5173.

Ravelo, A., D. Andreasen, M. Lyle, A. Olivarez-Lyle, and M. Wara. 2004. Regional climate shifts caused by gradual global cooling in the Pliocene epoch. *Nature* 429:263–267.

Retallack, G. J. 1992. Middle Miocene fossil plants from Fort Ternan (Kenya) and evolution of African grasslands. *Paleobiology* 18:383–400.

Rind, D. and D. Peteet. 1985. Terrestrial conditions at the last glacial maximum and CLIMAP sea-surface temperature estimates: Are they consistent? *Quaternary Research* 24:1–22.

Robert, C., L. Diester-Haass, and J. Paturel. 2005. Clay mineral assemblages, siliciclastic input and paleoproductivity at ODP Site 1085 off Southwest Africa: A late Miocene–early Pliocene history of Orange river discharges and Benguela current activity, and their relation to global sea level change. *Marine Geology* 216:221–238.

Rodwell, M., and B. Hoskins. 1996. Monsoons and the dynamics of deserts. *Quarterly Journal of the Royal Meteorological Society* 122:1385–1404.

Ropelewski, C., and M. Halpert. 1987. Global and regional scale precipitation patterns associated with the El Niño/Southern Oscillation. *Monthly Weather Review* 115:1606–1627.

Rossignol-Strick, M. 1983. African monsoons, an immediate climatic response to orbital insolation forcing. *Nature* 304:46–49.

———. 1985. Mediterranean Quaternary sapropels, an immediate response of the African Monsoon to variation of insolation. *Palaeogeography Palaeoclimatology Palaeoecology* 49:237–263.

Rowell, D., C. Folland, K. Maskell, and M. Ward. 1995. Variability of summer rainfall over tropical north Africa (1906–92): Observations and modelling. *Quarterly Journal of the Royal Meteorological Society* 121:669–704.

Sachs, J. P., and D. J. Repeta. 1999. Oligotrophy and nitrogen fixation during eastern Mediterranean sapropel events. *Science* 286:2485–2488.

Schefuss, E., S. Schouten, J. H. F. Jansen, and J. S. S. Damsté. 2003. African vegetation controlled by tropical sea surface temperatures in the mid-Pleistocene period. *Nature* 422:418–421.

Schefuss, E., S. Schouten, and R. Schneider. 2005. Climatic controls on central African hydrology during the past 20,000 years. *Nature* 437:1003–1006.

Scholz, C., J. King, G. Ellis, P. Swart, J. C. Stager, and S. Colman. 2003. Paleolimnology of Lake Tanganyika, East Africa over the past 100 kyr. *Journal of Paleolimnology* 30:139–150.

Scott, L. 1995. Pollen evidence for vegetational and climatic change in southern Africa during the Neogene and Quaternary; pp. 65–76 in E. Vrba, G. Denton, T. Partridge, and L. Burckle (eds.), *Paleoclimate and Evolution: With Emphasis on Human Origins.* Yale University Press, New Haven.

———. 1999. The vegetation history and climate in the Savanna Biome, South Africa since 190 ka: A comparison of pollen data from the Tswaing Crater (the Pretoria Saltpan) and Wonderkrater. *Quaternary International* 57:215–223.

Semazzi, F., V. Mehta, and Y. Sud. 1988. An investigation of the relationship between sub-Saharan rainfall and global sea surface temperatures. *Atmosphere-Ocean* 26:118–138.

Shackleton, N. 1995. New data on the evolution of Pliocene climatic variability; pp. 242–248 in E. Vrba (eds.), *Paleoclimate and Evolution: With Emphasis on Human Origins.* Yale University Press, New Haven.

Sonzogni, C., E. Bard, and F. Rostek. 1998. Tropical sea surface temperature during the last glacial period: A view based on alkenones in Indian ocean sediments. *Quaternary Science Reviews* 17:1185–1201.

Stokes, S., D. S. G. Thomas, and R. Washington. 1997. Multiple episodes of aridity in southern Africa since the last interglacial period. *Nature* 388:154–158.

Stott, L., and J. P. Kennett. 1990. Antarctic Paleogene planktonic foraminifer biostratigraphy, ODP Leg 113, Sites 689 and 690. *Proceedings of the Ocean Drilling Program, Scientific Results* 113:549–569.

Street, F. A., and A. T. Grove. 1979. Global maps of lake-level fluctuations since 30,000 years BP. *Quaternary Research* 12:83–118.

Street-Perrott, F. A., and S. A. Harrison. 1984. Temporal variations in lake levels since 30,000 yr BP: An index of the global hydrological cycle; pp. 118–129 in J. E. Hansen and T. Takahashi (eds.), *Climate Processes and Climate Sensitivity.* American Geophysical Union, Washington, D.C.

Street-Perrott, F., and R. A. Perrott. 1990. Abrupt climate fluctuations in the tropics: The influence of Atlantic Ocean circulation. *Nature* 343:607–612.

Talbot, M., and T. Laerdal. 2000. The Late Pleistocene–Holocene palaeolimnology of Lake Victoria, East Africa, based upon elemental and isotopic analyses of sedimentary organic matter. *Journal of Paleolimnology* 23:141–164.

Talma, A. S., and J. C. Vogel. 1992. Late Quaternary palaeotemperatures derived from a speleothem from Cango Caves, Cape Province, South Africa. *Quaternary Research* 37:203–213.

Thompson, L., E. Mosley-Thompson, M. Davis, K. Henderson, H. Brecher, V. Zagorodnov, T. Mashiotta, L. Ping-Nan, V. Mikhalenko, D. Hardy, and J. Beer. 2002. Kilimanjaro ice core records: Evidence of Holocene climate change in tropical Africa. *Science* 298:589–593.

Tiedemann, R., M. Sarnthein, and N. J. Shackleton. 1994. Astronomical timescale for the Pliocene Atlantic $\delta^{18}O$ and dust flux records of ODP Site 659. *Paleoceanography* 9:619–638.

Trauth, M., A. Deino, and M. Strecker. 2001. Response of the East African climate to orbital forcing during the last interglacial (130–117 ka) and the early last glacial (117–60 ka). *Geology* 29:499–502.

Trauth, M., M. A. Maslin, A. Deino, and M. Strecker. 2005. Late Cenozoic moisture history of East Africa. *Science* 5743:2052–2053.

Trenberth, K., W. Branstator, D. Karoly, A. Kumar, N. Lau, and C. Ropelewski. 1998. Progress during TOGA in understanding and modeling global teleconnections associated with tropical sea surface temperatures. *Journal of Geophysical Research* 103:14291–14324.

Tuenter, E., S. Weber, F. Hilgen, and L. Lourens. 2003. The response of the African summer monsoon to remote and local forcing due to precession and obliquity. *Global and Planetary Change* 36:219–235.

Visser, K., R. Thunell, and L. Stott. 2003. Magnitude and timing of temperature change in the Indo-Pacific warm pool during deglaciation. *Nature* 421:152–155.

Wagner, T. 2002. Late Cretaceous to early Quaternary organic sedimentation in the eastern Equatorial Atlantic. *Palaeogeography Palaeoclimatology Palaeoecology* 179:113–147.

Wara, M., A. C. Ravelo, and M. Delaney. 2005. Permanent El Niño–like conditions during the Pliocene Warm Period. *Science* 309:758–761.

Wynn, J. G. 2004. Influence of Plio-Pleistocene aridification on human evolution: Evidence from paleosols of the Turkana Basin, Kenya. *American Journal of Physical Anthropology* 123:106–118.

Yemane, K., C. Robert, and R. Bonnefille. 1987. Pollen and clay mineral assemblages of a late Miocene lacustrine sequence from the northwestern Ethiopian highlands. *Palaeogeography Palaeoclimatology Palaeoecology* 60:123–141.

Yin, J., and D. Battisti. 2001. The importance of tropical sea surface temperature patterns in simulations of last glacial maximum climate. *Journal of Climate* 14:565–581.

Zachos, J. C., L. Stott, and K. Lohmann. 1994. Evolution of early Cenozoic marine temperatures. *Paleoceanography* 9:353–387.

Zachos, J. C., M. Pagani, L. Sloan, E. Thomas, and K. Billups. 2001. Trends, rhythms and aberrations in global climate 65 Ma to Present. *Science* 292:686–692.

Zhao, M. X., L. Dupont, G. Eglinton, and M. Teece. 2003. n-alkane and pollen reconstruction of terrestrial climate and vegetation for NW Africa over the last 160 kyr. *Organic Geochemistry* 34:131–143.

FIVE

A Review of the Cenozoic Vegetation History of Africa

BONNIE F. JACOBS, AARON D. PAN, AND CHRISTOPHER
R. SCOTESE

The aim of this chapter is to review and interpret the Cenozoic paleobotanical record of Africa. Ideally, we want to present a dynamic view of plant community and ecosystem change through time, so that the evolutionary and biogeographic history of Cenozoic African mammals can be considered in the context of the communities to which they belonged. To facilitate this goal, we discuss environmental change in the context of major physiographic change such as graben formation associated with rifting in East Africa, and show paleobotanical sites in their correct position on paleogeographic maps (citations provided in table 5.1). However, spatial and temporal coverage are uneven, allowing detailed paleoenvironmental reconstruction for some localities and only the most general inferences for most time intervals.

A common challenge to providing a floral context for the mammalian fossil record stems from the fossilization process itself. The circumstances in which bones and plants become fossils are not often the same, and consequently they are rarely found together. Bones decay readily in acidic conditions, but acidity favors the preservation of organic matter. Movement of mineral-rich waters through the soil can favor the lithification of bone, but oxidizing conditions hasten the decay of both pollen and leaves. Occasionally, the chemical environment favors preservation of plant and vertebrate fossils in the same deposit, although the quality of each may be somewhat compromised; small or delicate bones are absent, and microscopic detail is lacking from the plants.

In addition to the universal problem of imperfections in the fossilization process, Africa is disappointingly undersampled. This vast continent, roughly three times the area of the United States, has so far been documented by only a handful of Paleogene plant and vertebrate localities, and it has a Neogene record heavily biased toward the depositional basins of the East African Rift (figures 5.1–5.5). However, continued exploration is filling some of these gaps. Moreover, we are fortunate that the sampling scale of most fossil localities is at the plant community level, and larger-scale changes took place one community at a time. Thus, as Africa becomes better sampled, the uneven record will ultimately become a more complete narrative of dynamic change at the community and ecosystem levels.

Paleobotany and Paleogeography

THE PALEOCENE AND EOCENE

Africa's physical setting at the beginning of the Cenozoic was characterized by a lack of topography (other than the massifs of Tibesti [Chad] and Hoggar [Algeria]), isolation of the continent from other landmasses, and connection with the Arabian plate. Africa was south of its current position by at least 10° and rotated slightly clockwise relative to today (Scotese, 2005; Lofty and van der Voo, 2007). Consequently, the broadest part of the continent was between about 0° and 15° N latitude, and consisted of lowlands without obstructions to moist air masses (figure 5.1). As a result, rain forest would be expected across that region. Although there is no unequivocal evidence for rain forest plant communities at the low latitudes, fossils of some plant families today associated with tropical forests do occur. For example, Paleocene pollen from Nigeria and Cameroon and wood from Niger and Mali document Fabaceae (legumes, pollen and wood), Arecaceae (palms, pollen), Euphorbiaceae (wood, pollen), Meliaceae (mahogany, wood), and pollen of the genus *Ctenolophon* in the Linaceae family (Koenigeur, 1971; Adegoke et al., 1978; Salard-Cheboldaeff 1979, 1981; Crawley, 1988). Although most of these taxa occur fairly commonly in West African pollen assemblages, they do not all co-occur, and these pollen assemblages include other taxa with unknown affinities such as *Retitricolpites clarensis*, *Saturna enigmaticus*, and *Scabratriporites annelus*. Thus, the argument for Paleocene tropical forest is currently weak, even in low-latitude, well-sampled West Africa.

Among data from the northern margins of the continent, Méon (1990) points to a gradual, but substantial, transition in Tunisia from Late Cretaceous pollen assemblages having biogeographic affinities with Africa, South America, and Eurasia, to assemblages dominated by Eurasian taxa by the Paleocene. These northern pollen taxa are also present in a Paleocene pollen sequence from Morocco (Ollivier-Pierre, 1982). They include several species in the Juglandaceae (walnut and hickory family) and the *Normapolles* group, a polyphyletic form genus characterizing a large Northern Hemisphere palynoprovince during the Late Cretaceous and early Paleogene, probably including members of or precursors to the birch, hickory, and

TABLE 5.1
African plant fossil localities
Locality numbers correspond to sites on the paleogeographic maps in figures 5.1–5.5.

Locality Number	Flora(s)	Age	Paleo-coordinates	References
1	Track of Hammam Mellegue, El Kef, Tunisia	Maastrichtian–Danian	N 28.5, E 6.0	Méon, 1990
2	Bir Abu Munqar horizon (part of the Dakhla Formation), Egypt	Danian	N 17.0, E 23	Gregor and Hagn, 1982
3	Egoli-1 borehole, Nigeria	Late Cretaceous–Paleocene	S 0.5, W 1.0	Van Hoeken–Klinkenberg, 1966
4	Lower Esna Shales (Kharga Oasis locality), Egypt	Danian	N 16.0, E 25.5	Chandler, 1954
5	Lower Esna shales (Kossier/Quseir locality), Egypt	Danian	N 16.5, 29.0	Chandler, 1954
6	Nubian sandstone (Wadi Zeraib), Egypt	Paleocene	N 16.5, E 29.0	Seward, 1935
7	Gbekebo-1 borehole, Nigeria	Late Cretaceous–Eocene	S 1.0, W 1.0	Van Hoeken–Klinkenberg, 1966
8	Kerri-Kerri Formation (Bh. 1138–Bh. 1350), Nigeria	Paleocene	N 3.0, E 5.0	Adegoke et al., 1978
9	Krebb de Sessao, Niger	Paleocene	N 10.0, W 0.5	Koeniguer, 1971
10	Owan-1 borehole, Nigeria	Late Cretaceous–Paleocene	S 0.5, W 1.0	Van Hoeken–Klinkenberg, 1966
11	Kwa-Kwa borehole, Cameroon	Maastrichtian–Early Miocene	S 4.0, E 3.0 (50 Ma); N 1.0, E 6.0(30 Ma); N 3.0, E 8.0 (20 Ma)	Salard-Cheboldaeff, 1979, 1981; Guinet and Salard-Cheboldaeff, 1975
12	Kharga borehole 1, Egypt	Late Cretaceous–Eocene	N 16.0, E 25.5	Kedves, 1971
13	Walalane borehole, Senegal	Paleocene	N 11.0, W 20.5	Caratini and Tissot, 1991
14	Tambacounda borehole, Senegal	Late Paleocene–Middle Eocene and Late Miocene	N 9.5, W 18.0	Medus, 1975
15	Mahenge flora, Tanzania	Middle Eocene (~46 Ma)	S 14.5, E 26.5	Herendeen and Jacobs, 2000; Kaiser et al., 2006
16	Songo borehole, Sierra Leone	Eocene	N 4.0, W 19.0	Chesters, 1955
17	Wadi el Rayan, Egypt	Lutetian	N 20.0, E 25.0	Chandler, 1954
18	Kaninah Formation, Yemen	Middle Eocene	N 1.5, E 39.0	As-Saruri et al., 1999
19	Graret el Gifa Formation, Libya	Late Lutetian–Late Eocene	N 22.0, E 13.0	Privé-Gill et al., 1993
20	Ogwashi-Asaba Formation (SM 10 borehole), Nigeria	Late Eocene	S 1.0, E 0.5	Jan du Chêne et al., 1978
21	Ogwashi-Asaba Formation (SM 5 borehole), Nigeria	Late Eocene	S 1.0, 0.5	Jan du Chêne et al., 1978
22	Ogwashi-Asaba Formation (SM 6 borehole), Nigeria	Late Eocene	S 1.0, E 0.5	Jan du Chêne et al., 1978
23	Jebel Qatrani Formation, Egypt	Early Oligocene	N 26.5, E 27.5	Kräusel, 1939; Bown et al., 1982; Tiffney, 1991

Locality Number	Flora(s)	Age	Paleo-coordinates	References
24	Taqah locality, Oman	Early Oligocene	N 10.5, E 48.0	Privé-Gill et al., 1993; Thomas et al., 1999
25	Djebel Coquin Ouest, Libya	Oligocene	N 23.0, E 15.0	Louvet and Mouton, 1970
26	Dor el Abd, Libya	Oligocene	N 27.0, E 14.5	Koeniguer, 1967
27	Lower and Upper Gezira Formations, Sudan	Oligocene or Early Miocene	N 12.0, E 30.0	Awad, 1994; Awad and Breir, 1993
28	Zella el Ghetia, Libya	Oligocene or Oligo-Miocene	N 25.5, E 15.0	Dupéron-Laudoueneix and Dupéron, 1994; Louvet, 1973
29	Chilga Basin (macrofossil assemblages), Ethiopia	Late Oligocene (28–27 Ma)	N 9.0, E 33.5	Jacobs et al., 2005; García Massini et al., 2006; Pan et al., 2006; Pan, 2007
30	Chilga Basin (Hauga Section), Ethiopia	Late Oligocene (28–27 Ma)	N 9.0, E 33.5	Yemane et al., 1987
31	Loperot-I drill well pollen assemblage, Kenya	Late Oligocene (~25 Ma)	S 1.0, E 31.0	Vincens et al., 2006
32	Ekokor no. 1 borehole, Nigeria	Neogene	N 4.5, E 5.0	Legoux, 1978
33	Napak, Uganda	Early Miocene (20–19 Ma)	N 1.0, E 32.0	This publication
34	Raghama or Dhaylan Formation, Saudi Arabia	Miocene (most likely) or Oligocene (possibly)	N 23.0, E 34.5	Privé-Gill et al., 1999
35	Bugishu flora, Uganda	Early Miocene	S 0.5, E 32.0	Chaney, 1933
36	Bukwa locality, Uganda	Early Miocene (22 or 17.5 Ma)	S 0.0, E 32.5	Hamilton, 1968; Pickford, 2002; Winkler et al., 2005
37	Gebel El-Khashab Southern Petrified Forest, Egypt	Early Miocene	N 29.0, E 29.5	El-Saadawi, 2004
38	Gebel Ruzza, Egypt	Early Miocene	N 29.0, E 29.0	Kamal El-Din and El-Saadawi, 2004
39	Mt. Elgon (near Butandiga), Uganda	Early Miocene	S 0.0, E 32.0	Bancroft, 1932a
40	Mt. Elgon (near Buyobo), Uganda	Early Miocene	S 0.0, E 32.0	Bancroft, 1935
41	Mt. Elgon (Walasi Hill), Uganda	Early Miocene	S 0.0, E 32.0	Bancroft, 1935
42	Noordhoek sediments, South Africa	Early Miocene	S 35.0, E 16.0	Coetzee and Muller, 1984
43	Omo Basin (Omo bridge), Ethiopia	Early Miocene	N 7.0, E 36.0	Lemoigne, 1978
44	Rusinga flora, Kenya	Early Miocene (17.8 Ma)	S 2.0, E 32.0	Chesters, 1957
45	Fejej (site FJ-18), Ethiopia	Early to middle Miocene (16.1 Ma)	N 3.0, E 34.0	Tiffney et al., 1994
46	Djebel Zelten, Libya	Early middle Miocene	N 27.5, E 19.0	Boureau et al., 1983; Privé-Gill et al., 1993
47	Blue Nile Valley, Ethiopia	Miocene or Pliocene	N 9.0, E 36.5	Prakash et al., 1982
48	Kikongo, Uganda	Miocene (?)	S 2.0, E 32.0	Bancroft, 1932b
49	Mero camp fossils, Democratic Republic of Congo	Middle Tertiary (?Miocene-underlying basalt)	S 4.0, E 26.0	Lakhanpal, 1966
50	Molale Deposits, Ethiopia	Probably Miocene	N 9.0, E 38.0	Lemoigne et al., 1974
51	Mush Valley (Site A), Ethiopia	Probably Miocene	N 8.5, E 38.0	Lemoigne et al., 1974

TABLE 5.1 (CONTINUED)

Locality Number	Flora(s)	Age	Paleo-coordinates	References
52	Mush Valley (Site B), Ethiopia	Miocene	N 8.5, E 38.0	Lemoigne et al., 1974
53	Bulbulla, Ethiopia	Miocene	N 6.5, E 35.0	Lemoigne, 1978
54	Fort Ternan, Kenya	Middle Miocene (14 Ma)	S 1.0, E 34.0	Retallack, 1992
55	Welkite, Ethiopia	Miocene	N 8.0, E 37.0	Lemoigne, 1978
56	Ngorora Formation, Kenya	Middle Miocene (12.2 Ma)	N 0.5, E 35.0	Jacobs and Kabuye, 1987
57	Ediagbor borehole, Nigeria	Neogene	N 6.0, E 5.0	Legoux, 1978
58	Erema borehole, Nigeria	Neogene	N 5.0, E 6.0	Legoux, 1978
59	Jatumi borehole, Nigeria	Neogene	N 5.5, E 4.5	Legoux, 1978
60	Obagi borehole, Nigeria	Neogene	N 5.0, E 6.0	Legoux, 1978
61	Obodo borehole, Nigeria	Neogene	N 5.5, E 4.5	Legoux, 1978
62	Sombreiro River borehole, Nigeria	Neogene	N 5.0, E 6.0	Legoux, 1978
63	Elandsfontyn Formation (S1 borehole), South Africa	Late Miocene	S 34.0, E 17.0	Coetzee and Rogers, 1982
64	Karugamania beds, Democratic Republic of Congo	Miocene	N 1.0, E 30.0	Lakhanpal and Prakash, 1970
65	Niger Delta drill samples, Nigeria	Middle or late Miocene–Pleistocene	N 7.0, E 5.0 (20 Ma); N 4.5, E 7.0 (10 Ma); N 5.0, E 8.0 (0 Ma)	Morley and Richards, 1993
66	Anloua Basin (Neogene leaf localities), Cameroon	Miocene or Pliocene?	N 7.0, E 13.0	Brunet et al., 1986
67	Bignona borehole, Senegal	Late Miocene (~10–5 Ma)	N 13.0, W 17.0	Medus, 1975
68	Nkondo Formation, Uganda	Late Miocene–Pliocene (6.5–5 Ma)	N 1.0, E 30.0	Dechamps et al., 1992; Pickford, 2002
69	Upper Laetolil Beds, Tanzania	Pliocene	S 3.0, E 35.0	Bonnefille and Riollet, 1987
70	Mount Cameroon fossil plants, Cameroon	Late Miocene or Pliocene	N 4.0, E 8.5 (10 Ma); N 4.0, E 9.0 (0 Ma)	Menzel, 1920
71	Rusizi Valley (Ru. 231 borehole), Burundi	Pliocene-Pleistocene	S 3.0, E 29.0	Sah, 1967
72	Sahabi Formation, Libya	Pliocene	N 30.0, E 21.0	Boaz et al., 1987
73	Ziguinchor borehole, Senegal	Late Miocene, possibly middle to late Eocene–early Miocene	N 9.0, W 21.0 (50 Ma); N 12.5, W 17.0 (10 Ma)	Medus, 1975,
74	Omo Basin ('Brown Sands' locality), Ethiopia	Pliocene (~3.42 Ma)	N 5.5, E 36.0	Bonnefille and Letouzey, 1976; Williamson, 1985
75	Omo Basin (Shungura Formation), Ethiopia	Pliocene (~3.42 Ma)	N 5.0, E 36.0	Bonnefille and Letouzey, 1976; Williamson, 1985
76	Omo Basin (Jimma), Ethiopia	Pliocene–Pleistocene	N 8.5, E 37.0	Lemoigne, 1978

Locality Number	Flora(s)	Age	Paleo-coordinates	References
77	Kaiso Village Formation (Hohwa Member), Uganda	Pliocene-Pleistocene (2.6 Ma)	N 1.5, E 31.0	Dechamps et al., 1992; Pickford, 2002
78	Lusso Beds (Upper Semliki area), Democratic Republic of Congo	Pleistocene (2.3–2.0 Ma)	N 0.0, E 30.0	Dechamps and Maes, 1990
79	Olduvai Beds I and II, Tanzania	Pleistocene	S 3.0, E 35.5	Bonnefille and Riollet, 1980; Bonnefille et al., 1982
80	Koobi Fora Formation (KFFP 1, Ileret area), Kenya	Pleistocene (1.6–1.5 Ma)	N 4.0, E 36.0	Bonnefille, 1976
81	Region V (off the coast), Namibia	Lutetian (46–42 Ma)	S 36.5, E 6.0	Scott et al., 2006
82	Koingnaas locality, South Africa	Paleocene or Eocene	S 39.0, E 7.0	Zavada and de Villiers, 2000

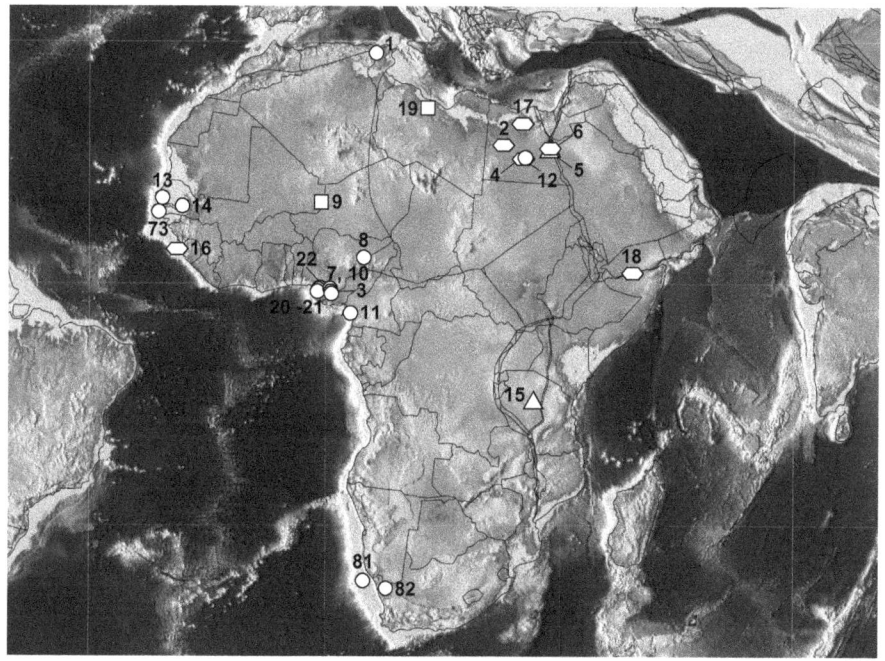

FIGURE 5.1 Paleogeographic map showing Paleocene and Eocene paleobotanical sites. The map represents Africa at 50 Ma. Note the isolation of Africa (the Afro-Arabian continent) from the other continents and few highlands. Numbers and symbols indicate fossil plant localities and their geographic position at the time of deposition. Shapes correspond to particular kinds of plant fossils: circles, pollen; triangles, leaves; squares, wood; hexagons, fruits/seeds; and stars, assemblages with a combination of two or more kinds. Locality numbers on the maps correspond to those in table 5.1, which includes citations.

witch-hazel families (Betulaceae, Hamamelidaceae, and Juglandaceae). A few northern elements are documented in Cretaceous–early Eocene strata of Egypt, including typically Eurasian families and genera such as Fagaceae fruits (oaks and relatives; Gregor and Hagn, 1982), Carya-like pollen (hickory/pecans; Kedves, 1971), Normapolles pollen (Kedves, 1971), and Betulaceae-like pollen (birches; Kedves, 1971). Tropical elements present among these Egyptian paleofloras include the seeds or fruits of a fan palm (subfamily Coryphoideae; Gregor and Hagn, 1982; Pan et el., 2006), Icacinaceae (Chandler, 1954;

Gregor and Hagn, 1982), Erythropalum (Olacaceae; Gregor and Hagn, 1982), Anonaspermum (custard apple family; Chandler, 1954), and Euphorbiaceae (Chandler, 1954). Interestingly, these taxa are also represented among plant fossils from Paleogene floras of southern England, where their tropical and Indo-Malaysian affinities were recognized (Chandler, 1964). A leaf flora reported from near Quseir, Egypt, in deposits referred to as the Nubian Sandstone, consists of large leaves thought to represent trees of tropical Asian affinity (e.g., Dipterocarpaceae; Seward, 1935). Their precise age and taxonomic placement are uncertain, but on the basis of their physiognomy, these assemblages likely represent a tropical to subtropical mesophytic plant community.

There is evidence for mangrove communities at essentially all Paleocene and Eocene coastal sites. Mangrove species are salt-tolerant trees and shrubs represented by a variety of plant families and, depending on the species, are tolerant of fully marine to estuarine conditions. They are currently limited to tropical latitudes where they trap sediment and prevent coastal erosion, thereby creating their own complex ecosystems. The extinct pollen species *Spinizonocolpites* and *Proxapertites*, related to the now endemic Indian mangrove palm subfamily Nypoideae, dominate Paleogene mangrove communities (e.g., Van Hoeken–Klinkenberg, 1966; Medus, 1975; Jan du Chêne et al., 1978; Adegoke et al., 1978; Salard-Cheboldaeff, 1979, 1981; Kaska, 1989: Morley, 2000). In addition, *Nypa* fruits are known from Paleocene and Eocene localities in Egypt and Senegal (Fritel, 1921; Kräusel, 1939; Chandler 1954; Gregor and Hagn, 1982). Mangrove palms in general are quite common in Paleocene and Eocene pollen samples, co-occurring, especially during the middle Eocene, with grass pollen among other taxa, including cycads (Adegoke et al., 1978; Salard-Cheboldaeff, 1979, 1981; Kedves, 1971; Morley, 2000). Morley (2000) suggests these assemblages could represent freshwater palm swamp environments, if not open landscapes.

The Eocene physical setting was similar to that of the Paleocene, although the African plate continued its northward movement to approximately 8° south of its modern position. Volcanism in southern Ethiopia, thought to be associated with the Kenyan plume ultimately responsible for doming and rifting along the East African Rift, began as early as 45 Ma, propagating to the south-southwest as the African plate moved north-northeastwardly (George et al., 1998). This early Kenyan plume activity is not thought to have altered the landscape as much as the later Oligocene massive flood basalts of the Ethiopian-Yemeni plateau associated with the Afar plume (Ebinger et al., 1993).

Eocene paleobotanical sites are known from the far southwestern edge of the continent to the northeastern coast of Egypt, but they are very sparsely distributed and rarely provide information about the continental interior. In addition to the pollen of mangrove and other palms along the coasts, West African pollen records from coastal Cameroon, Nigeria, and Gabon (Boltenhagen, 1965; Van Hoeken–Klinkenberg, 1966; Jan du Chêne et al., 1978; Salard-Cheboldaeff, 1981; Salard-Cheboldaeff and Dejax, 1991) include a diversity of angiosperm pollen without modern botanical affinities, especially in the early Eocene. One exception includes the earliest Eocene, late Eocene, and Oligocene occurrences of *Anacolosidites* pollen in the Maastrichtian–early Miocene Kwa Kwa core from Cameroon (Jan du Chêne et al., 1978; Salard-Cheboldaeff, 1979). A recent analysis of the taxonomic position of numerous fossils attributed to *Anacolosidites* confirms the attribution of Eocene West African pollen to this fossil genus, which is related to the extant tribe Anacolosae (Olacaceae; Malecot and Lobreau-Callen, 2005). Today, all species of the three living Anacolosae genera (*Cathedra*, *Anacolosa*, and *Phanerodiscus*) are found in tropical wet forests of Africa, South America, and Asia (Malecot and Lobreau-Callen, 2005).

By the late Eocene, pollen floras from West Africa include many genera that can be found today in lowland tropical forests, and at the same time grass pollen diminishes (Salard-Cheboldaeff, 1990; Morley, 2000). Seeds of *Anonaspermum* from Sierra Leone (Chesters, 1955) and Yemen may also be representative of a forest environment, but living members of the Annonaceae family can occupy a variety of habitats.

The Eocene and Oligocene of North Africa are represented largely by fossil woods, which are relatively common there. However, a specimen's age may be poorly constrained if found on the surface as many are, and its botanical affinity cannot be assumed based on its name (e.g., *Detarioxylon aegyptiacum* from Libya is estimated to be Oligocene and is not necessarily the genus *Detarium*; see Boureau et al., 1983; Privé Gill et al., 1999). Nevertheless, fossil wood can provide valuable information about vegetation structure, and some anatomical characters can be used to infer paleoclimate (Wheeler et al., 2007). Boureau et al. (1983) interpreted the Paleogene woods of Africa north of the equator as representing a band of forest along the northern (Tethyan) coast, and drier woodland or more open area inland on the basis of nearest living relatives such as *Combretum*, thought to be related to the wood taxon *Combretoxylon*.

Paleogene data from the eastern part of the continent were obtained in part by petroleum exploration in central Sudan. A brief palynological summary of more than 4,000 well samples spanning the Late Cretaceous to (minimally) the late Oligocene (Kaska, 1989) indicates that the Paleogene is generally characterized by an abundance of palm species, legumes, and ferns. The late Eocene/Oligocene is represented by *Peregrinipollis nigericus* (syn. *Brachystegia*; Morley, 2000) and *Striatricolpites catatumbus* (syn. *Crudia*; Morley, 2000), legumes that serve as biochronological markers (Kaska, 1989). The legume family, which is diverse and ecologically important today, and was first documented unequivocally in Africa in the Paleocene of Mali (Crawley, 1988), was also important at most Paleogene paleobotanical sites.

Southern Africa is represented by two interesting occurrences of Paleocene-Eocene Asteraceae (sunflower family) pollen, tribe Mutisieae, from off the coasts of northwestern South Africa and southern Namibia (Zavada et al., 2000; Scott et al., 2006). These are the earliest definitive records for this currently large cosmopolitan family, and they have important implications for the evolution of South Africa's unique Cape flora. Scott et al. (2006) suggest that the presence of Mutisieae pollen in southwestern Africa may signify the establishment of arid climate in that region by the early Eocene, but not all genera in this tribe live in arid climates today. For example, the genus *Stenopadus* currently occurs in the highland and lowland forests of Guyana (Rull, 2005). Sedimentological data are consistent with an arid climate reconstruction; however, Scott et al. (2006) point out the contradictory co-occurrence of water-dependent fern spores.

The middle Eocene Mahenge site, north-central Tanzania (approximately 12°S paleolatitude), represents a more definitive early example of Cenozoic aridity and woodland vegetation structurally similar to the miombo woodlands found in that area today (Herendeen and Jacobs, 2000; Jacobs and Herendeen, 2004; Kaiser et al., 2006). In addition to plant fossils, this 46-Ma kimberlite crater-lake has produced abundant fish (Murray, 2000, 2001, 2003; Murray and Budney, 2002; Murray and Wilson, 2005), a frog with soft parts preserved (Baez 2000), and a bat (Gunnell et al., 2003). At the family level, the dominance of Fabaceae (*Acacia*, *Aphanocalyx*, *Bauhinia*, cf. *Cynometra*, and other unknowns) differs little from modern woodlands. Mean annual paleoprecipitation estimates range from about 640–800 mm, approximately equivalent to modern levels (660 mm/yr). These were calculated from leaf fossils using regression equations derived from the relationship between modern leaves and climate in tropical regions (Jacobs and Herendeen 2004). Using the same approach, the total wet months precipitation (all months averaging ≥ 50 mm) can be calculated, yielding a range from about 630 to 700 mm, indicating that a slightly larger proportion of annual precipitation occurred in the dry months compared with today (Jacobs and Herenedeen, 2004). The lake sediments are comprised largely of dolomite, providing an independent indication of aridity, and support the notion that tropical forest did not extend southward (at least not continuously) through the tropical zone.

To summarize the Eocene, palynological information from tropical western Africa seems to indicate forested conditions, at least during the early and late Eocene, with the possibility of more open environments (perhaps wetlands) dominated by palms and grasses in the middle Eocene. Similarly, middle Eocene pollen samples from Egypt are dominated by monocots such as palms and grasses, but cycads are also present (Kedves, 1971). At approximately the same time, macrofossils from Tanzania document woodlands representing a seasonally dry climate. Throughout the Eocene, mangrove palms grew at tropical latitudes along the coasts, but middle Eocene pollen from coastal southern Namibia and northwestern South Africa may represent shrub-dominated arid lands.

THE OLIGOCENE

African Oligocene paleogeography differs little from the Eocene with respect to continental position. However, regional physiography was significantly altered by the eruption of massive basalt traps in Ethiopia and Yemen (prior to rifting of the Arabian plate away from the African plate) caused by activity of the Afar plume between about 33 and 29 Ma (Hofmann et al., 1997; Ukstins et al., 2002). Much more massive and widespread than the basalts associated with earlier Kenyan plume activity in southern Ethiopia (45 Ma; Davidson and Rex, 1980; Ebinger et al., 1993), these eruptions resulted in a stack of volcanic sediments as much as 4 km thick, which today forms the ~2-km-thick Ethiopian and Yemeni plateaus (Ukstins et al., 2002). Faulting and Ethiopian rift graben formation did not take place until the late Oligocene and early Miocene, and it was associated with the separation of the Arabian Peninsula from Africa at the Afar triple junction. Thus, one may envision an Oligocene landscape across most of Ethiopia and Yemen as a vast plateau comprised of basalts and other volcanogenic sediments. These regional, tectonically driven, physiographic changes undoubtedly had consequences for floral and faunal ecology and evolution. The effects of contemporaneous and subsequent global climate changes associated with Oligocene and later polar glaciations further complicated regional environmental change in the East African highlands and Rift Valley.

The scarcity of Oligocene paleobotanical sites in Africa is illustrated in figure 5.2. In West Africa, the Cameroon (Kwa Kwa) core indicates a significant increase in diversity of angiosperm pollen from the late Eocene into the Oligocene, among which are many genera and families typical of extant West African lowland forests (Salard-Cheboldaeff, 1978, 1979, 1981). This change was interpreted as indicating an expansion and diversification of lowland (Guineo-Congolian) forest, due in

part to the migration of this locality into the equatorial zone with northward movement of the African plate (Salard-Cheboldaeff, 1981; Salard-Cheboldaeff and Dejax, 1991). However, despite the apparent success of angiosperm forest species across the Eocene-Oligocene boundary in West Africa, several palm pollen taxa disappear at this time (Salard-Cheboldaeff, 1981; Salard-Cheboldaeff and Dejax, 1991; Morley, 2000). This observation is noteworthy considering the very limited role played by palms in African forests today compared with other wet tropical regions of the world (Moore, 1973; Richards, 1996; Morley, 2000; Pan et al., 2006). The record of grass pollen declines significantly for the middle Eocene and is absent in middle Oligocene sites, supporting the notion that regional vegetation was predominantly forest. The band of forest along the coastal regions described for the Eocene of northern Africa, as documented by fossil woods, is thought to have narrowed in the Oligocene as the continent drifted northward (Koeniguer, 1967; Louvet and Mouton, 1970; Louvet, 1973; Boureau et al., 1983; Dupéron-Laudoueneix and Dupéron, 1994). Examples of taxa interpreted as indicating an interior zone of dry forest, woodland, or wooded savanna communities (although evidence for grass is absent) include *Combretoxylon (Anogeissuxylon) bussoni* and the leaflets of legume species including *Detariophyllum* (similar to living *Detarium microcarpum*), *Pterocarpus* (similar to living *Pterocarpus erinaceus*), and *Caesalpinities* (Louvet and Mouton, 1970).

Approximately contemporaneous earliest Oligocene plant fossils from the Fayum Depression, Egypt (Jebel Qatrani Formation), and Oman (Taqah)(figure 5.2), document coastal mangrove vegetation and gallery forests (Bown et al., 1982; Tiffney, 1991; Privé-Gill et al., 1993; Wing et al., 1995). The typically halophytic fern *Acrostichum* occurs at both localities, as do woods of the tropical forest family Monimiaceae, represented in the Fayum by *Atherospermoxylon* (Kräusel, 1939) and in Oman by *Xymaloxylon zeltenense* Louvet (thought by Privé-Gill et al. [1993] to be synonymous with the Libyan Eocene *Flacourtioxylon [Monimiaxylon] gifaense* Louvet). Fayum plant localities also produce wood of *Gynotrochoxylon* (Tamaricaceae; Dupéron-Laudoueneix and Dupéron, 1995), providing additional evidence of brackish or saline water, the fruit of *Epipremnum*, a vine interpreted as indicating tall trees, and seeds of *Canarium*, Annonaceae, and Icacinaceae, which, with ichnofossils of flank-buttressed trees, were taken to indicate forest (at least gallery forest) vegetation (Bown et al., 1982; Tiffney, 1991; Wing et al., 1995). The nature of vegetation in the interfluves is not clear.

The middle and late Oligocene of eastern Africa are represented by fossiliferous intertrap sediments from the Chilga area of northwestern Ethiopia, and palynomorph assemblages from subsurface cores of Sudan (El Gezira) and Kenya (Lokichar Basin). The Ethiopian Chilga strata are well dated at between 28 and 27 Ma based on ^{40}Ar/^{39}Ar ages and paleomagnetic reversal stratigraphy (Kappelman et al., 2003). These deposits have produced leaf, flower, and fruit compressions with cuticle preserved, silicified in situ assemblages of trees, leaf assemblages encased in tuffaceous sediments, and fruit and seed assemblages that co-occur with vertebrate fauna (Kappelman et al., 2003; Jacobs et al., 2005; Pan et al., 2006; Garcia Massini et al., 2006). Palynological assemblages from outcrops reported by Yemane et al. (1985, 1987) as late Miocene are actually Oligocene and stratigraphically equivalent to the vertebrate and macrofloral beds (Kappelman et al., 2003). Together, these plant assemblages document rich forest vegetation in a physical setting characterized by relatively

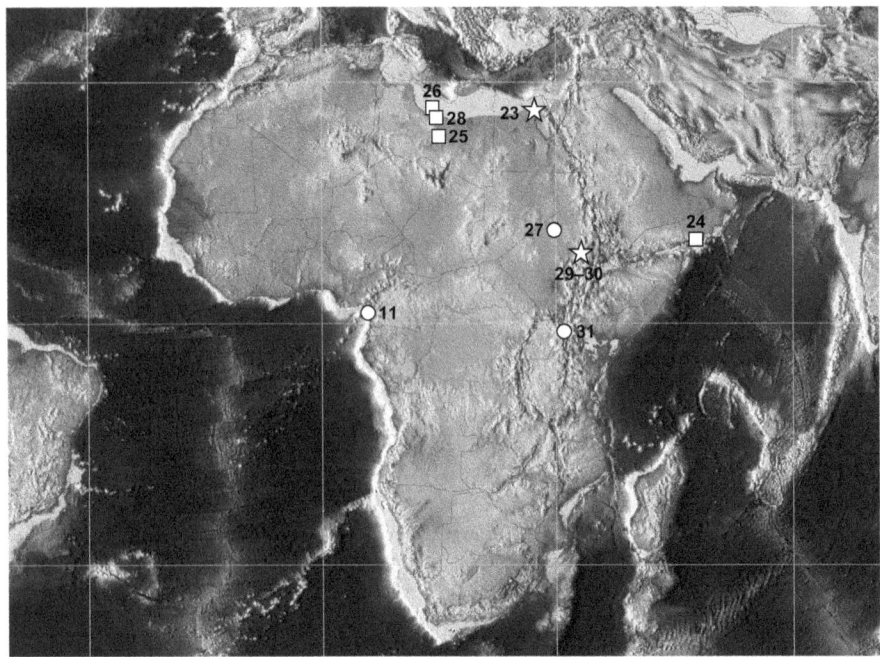

FIGURE 5.2 Paleogeographic map showing Oligocene paleobotanical sites. The map represents Africa at 30 Ma. Note the Ethiopian flood basalt plateau. For full explanation, see caption for figure 5.1. For numbered sites, see table 5.1.

low relief and a high water table, and they provide a rare and increasingly detailed view of contemporaneous floral and faunal assemblages (Jacobs et al., 2005). Detailed stratigraphic mapping and characterization of paleosols document temporal and spatial variations in water table in the context of episodic input of fine volcanic ash. Dynamic changes in plant community composition and structure during the one million years represented by the sedimentary strata were in part or wholly determined by these local to regional physical processes (Jacobs et al., 2005; Garcia Massini, in press). Botanical affinities of the flora lay with Central and West Africa, on the one hand, and the currently disjunct East African coastal and Eastern Arc forests, on the other, representing a historical biogeographic link between the two regions (Pan, 2007). For example, fossils of *Afzelia* (Fabaceae) document a possible ancestor for apparently vicariant species of this genus in today's western and eastern forests (Pan, 2007). Other taxa with living relatives in lowland or submontane forests of tropical Africa (and Madagascar) include *Cola*, the endemic palm genus *Eremospatha*, *Sorindeia* (Anacardiaceae), *Vepris* (Rutaceae), *Ocotea* (Lauraceae), *Dioscorea* section *Lasiophyton* (Dioscoreaceae), *Tetracera* (Dilleniaceae), and *Scadoxus* (Amaryllidaceae) (Pan et al., 2006; Pan, 2007; Pan and Jacobs, 2009; Pan, in press). Several fern species are also known from both macrofossil remains and spores (García Massini et al., 2006, García Massini and Jacobs, in press). Trees measured from in situ silicified forests were approximately 20–35 m in height, based on an allometric relationship between stem diameter and tree height (Rich et al., 1986; Niklas, 1994), and an average of 3 m apart from one another (Jacobs et al., 2005). Paleoprecipitation estimates based on a leaf assemblage from along the Guang River are between about 1,200 and 1,500 mm per year, derived from the same regression equations used for the Eocene Mahenge site (mentioned earlier) and the overlapping distributions of living African relatives (Pan, 2007).

Core sediments of "Oligo-Miocene" age from Sudan (Awad and Breir, 1993) document angiosperm pollen interpreted as representing lowland forest families (Meliaceae, Malvaceae *sensu lato*, Clusiaceae, Fabaceae, and Pedaliaceae), and a relatively rich fern spore assemblage including the families Cyatheaceae (tree ferns), Polypodiaceae, Adiantaceae, Lindsayaceae, and Parkeriaceae. Similarly, pollen core samples from Lokichar Basin in northern Kenya, cited as Oligocene–early Miocene, are dominated by angiosperms having modern botanical affinities with both Guineo-Congolian and Zambesian communities, and representing primarily (semideciduous) forest structure or, less likely, humid woodlands (Vincens et al., 2006). Ferns are also common in these assemblages, but grass pollen counts are low, lending support to the notion that the environment was wet and vegetated by forest communities. The age constraints on these Oligo-Miocene pollen assemblages from Sudan and northern Kenya are imprecise, but they are placed near the Paleogene-Neogene boundary. The reconstructed vegetation from currently semiarid regions of northern Kenya and central Sudan includes a seasonally dry forest that was wet enough to support a diverse fern flora including tree ferns.

The occurrence of palms in nearly all of the fossiliferous deposits at Chilga (about 100 km^2), Lokichar, and palynological samples from West Africa, is significant, because while they occur commonly today in tropical forests elsewhere in the world, they have yet to be found in equal abundance or diversity at African Neogene localities, and they are infrequent, absent, or in monotypic stands in African forests today (Pan et al., 2006). Thus, palms from Chilga and other wet-tropical Paleogene localities across the continent document their greater ecological role relative to the Neogene or today (Pan et al., 2006). In this regard, Paleogene forest community ecology differed significantly from modern, and this has implications for understanding trophic relationships among Paleogene Afro-Arabian endemic mammals, many of which are absent by the Neogene.

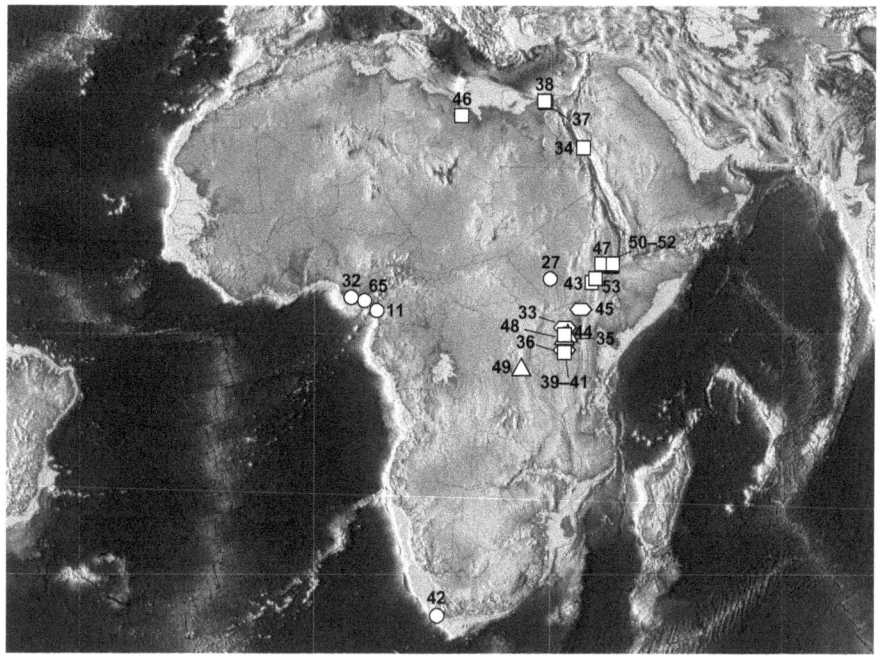

FIGURE 5.3 Paleogeographic map showing early Miocene paleobotanical sites. The map represents Africa at 20 Ma. Note the large number of plant localities associated with the East African Rift System. For full explanation, see caption for figure 5.1. For numbered sites, see table 5.1.

THE MIOCENE

The transition to the Neogene is marked by continued tectonically driven physiographic evolution in eastern Africa, which had profound effects on the landscape, climate, and vegetation. Wolfenden et al. (2004) document a long separation between the time of initial rifting in the southern Red Sea and Gulf of Aden, about 28 Ma, and extension along the main Ethiopian Rift (forming the third arm of a triple junction) about 11 Ma. The main East African Rift axis propagated southward during the middle and late Miocene from Ethiopia to Kenya, and Tanzania (Logatchev et al., 1972; Woldegabriel et al., 1990; Ebinger et al., 1993; Wolfenden et al., 2004; Chorowicz, 2005). The Western Rift Valley basins of Lakes Kivu, Albert, and Malawi were established between about 10 and 5 Ma as reflected by the age of their sediments (Chorowicz, 2005). The maps depicting Neogene paleobotanical sites in figures 5.3, 5.4, and 5.5 clearly show sample bias toward East Africa as valley formation provided accommodation space for rift sediments and opportunities for fossilization.

The development of complex physiography contributed to the formation of arid regions in East Africa, as can be understood from features of modern climate. Moisture brought inland from the Atlantic or Indian Ocean is largely intercepted by the highlands, leaving the rift valleys in local rain shadows. In addition, the yearly distribution of rainfall across the latitudinal span of the East African rifts (approximately 12°N–15°S) is further complicated by the effects of the highlands, which hinder north-south movement of air masses associated with the Intertropical Convergence Zone (ITCZ; Griffiths, 1992, see also Feakins and deMenocal, this volume, chap. 4). This factor is relevant to understanding the regional effects of Neogene global climate change, because the position of the ITCZ is also influenced by the extent of polar ice caps. Furthermore, regional climate would likely have been

affected by progressive closure of the Paratethys from the Oligocene through the Miocene. This would have diminished moisture available to northeasterly air masses during Northern Hemisphere winters, presumably causing a decrease in winter rains to the Horn of Africa (Meulenkamp and Sissingh, 2003; Pan, 2007). Thus, understanding the causes of regional or continental-scale Neogene vegetation (and faunal) changes, largely documented by sediments from the East African Rift, is complicated by the impact of global climate change on regional and local vegetation in the context of the evolution of complex topography and tectonic history. Add to this mix the potential ecological and evolutionary impact of Eurasian faunal immigration, and one should expect Neogene plant community evolution in Africa, especially in eastern Africa, to be complex as well. As intra- and intersite sampling increases, paleoecology will be understood at a finer scale, allowing spatial and temporal variation to be documented more thoroughly than currently. Our review of the Neogene attempts to document what is known, but ironically, because of improvement of the fossil plant record as one moves toward the Recent, summarizing each and every locality becomes unreasonable. Those records that are of particular importance for understanding paleoecological change or that document associations between plants and vertebrates will be discussed, and patterns of change will be reviewed.

Miocene vegetation is documented in West Africa by the youngest segment of the Kwa Kwa pollen core from coastal Cameroon (Salard-Cheboldaeff, 1979, 1981) and pollen assemblages from the Niger Delta (Legoux, 1978; Morley and Richards, 1993). Salard-Cheboldaeff (1979, 1981), and Legoux (1978) document palms, mangrove pollen (of the modern mangrove family, Rhizophoraceae) for the first time since the disappearance of mangrove palms in the early Oligocene, and a host of other modern families present today in West African forest vegetation. Among these taxa is pollen of the legume

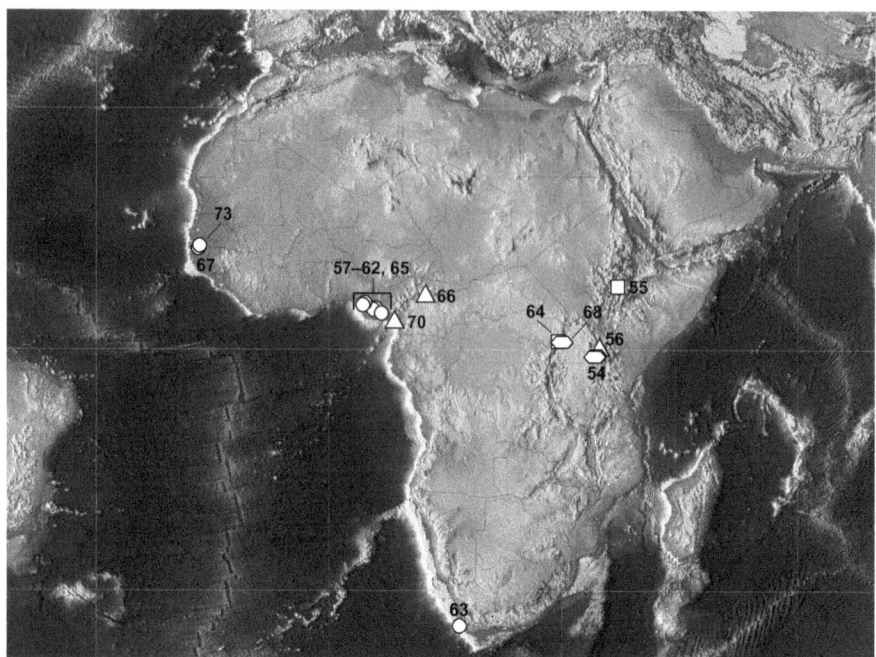

FIGURE 5.4 Paleogeographic map of late Miocene paleobotanical sites. The map represents Africa at 10 Ma. For full explanation, see caption for figure 5.1. For numbered sites, see table 5.1.

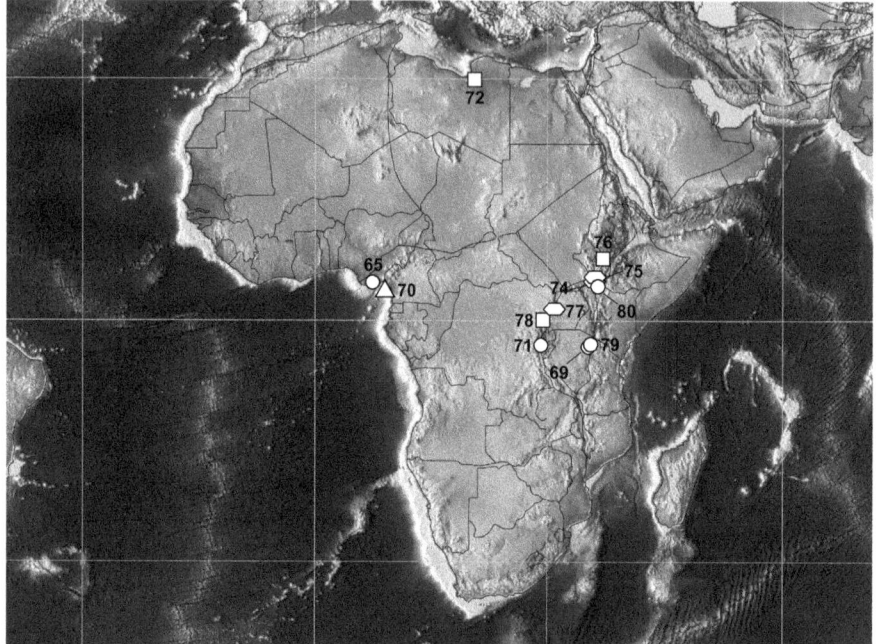

FIGURE 5.5 Paleogeographic map showing Pliocene and Pleistocene paleobotanical sites. The map represents Africa's present geographic position. For full explanation, see caption for figure 5.1. For numbered sites, see table 5.1.

genus *Sindora*, of which one species remains today near the coast of Gabon and Congo. Interestingly, *Sindora* wood was documented from Saudi Arabia (late Oligocene or Miocene) by Privé-Gill et al. (1999), providing a biogeographic link between the single African species and several species found today in Asia and the American tropics. This forest genus no longer occurs on the arid Arabian Peninsula.

Morley and Richards (1993) document an early Miocene to Holocene record of change in grass pollen and charred grass cuticle abundances in core samples from the Niger Delta as proxies

for changing terrestrial grassland dominance (and presumed aridity). The early Miocene, lacking charred cuticle and with very low grass pollen percentages, is interpreted as having a relatively wet climate and little if any grass-dominated communities of the sort found today. By the middle Miocene, grass pollen and charred cuticle increase somewhat, and by the late Miocene they are both abundant (Morley and Richards, 1993). This is interpreted as signaling the onset of pronounced seasonality of rainfall and the emergence of large areas having savanna or other grass-dominated communities (Morley and Richards, 1993).

Among the oldest records of abundant grass pollen are those from the early or middle Miocene of Saudi Arabia, where at least one of five core samples contains 68% grass pollen and 26% nonforest taxa (Whybrow and McClure, 1981). While the grasses could have been aquatic species, such high relative percentages usually indicate open vegetation with grass as an important component (e.g., Vincens et al., 1997, 2000).

Wheeler et al. (2007) report a middle Miocene assemblage of silicified woods from Fejej, in southernmost Ethiopia. The relatively high specific gravities and narrow vessels of the 12 wood types indicate an environment with a significant dry season. The plant community was reconstructed as a deciduous forest or woodland. Interestingly, fruits and seeds identified from Fejej (Tiffney et al., 1994) are not the same as the identified woods but are consistent with the environmental interpretation.

A recurring aspect of many East African early and middle Miocene paleofloras is the presence of taxa related to plants found today in dry or wet forests of Central and West Africa, or the disjunct forest communities of coastal eastern Africa and the Eastern Arc Mountains. For example, fossil woods from the eastern and western Ethiopian Plateau, thought to be early Miocene, were interpreted as representing dry forest having taxa found in West Africa today (Lemoigne et al., 1974; Lemoigne, 1978). A small assemblage of flowers, fruit, and leaves from Bukwa, Uganda, that underlie a dated (22 Ma) volcanic deposit including *Cola* (now interpreted to be *Pterygota* [Pan and Jacobs, 2009]) and *Bersama* was interpreted as similar to current Ugandan or West African forest (Walker, 1968, 1969; Brock and McDonald, 1969). A separate unit at Bukwa preserves an assemblage of monocot leaves (mainly grasses) and rhizomes identified to *Juncellus laevigatus*, an aquatic plant (Hamilton, 1968). Pickford (2002) later interpreted a snail fauna from Bukwa as indicative of an in situ grassland at this locality, but this is not consistent with the plant fossils, which support an interpretation of forest and aquatic habitats. Another early Miocene locality from Mount Elgon was reported by Bancroft (1933), who identified silicified wood specimens as related to the living Dipterocarpaceae, a diverse family typical of lowland forests in Southeast Asia.

Other early Miocene assemblages include a small collection of leaves, fruit, and wood from Bugishu, Uganda (associated with early Miocene volcanics of Mount Elgon), that can be found today in dry forest or woodland communities. The flora was interpreted by Chaney (1933) as representing woodland savanna or savanna based on the size of the leaves, although there was no evidence of grass fossils.

Lakhanpal (1966) and Lakhanpal and Prakash (1970) reported leaves and woods, respectively, from the Republic of Congo. Lakhanpal (1966) documented *Sclerosperma*, a palm today usually found in monotypic stands in swampy areas of Central and West Africa. This is among the relatively rare citations for Neogene palm fossils, which are much more common in the late Cretaceous and Paleogene. The woods reported by Lakhanpal and Prakash (1970) were associated with early Miocene mammal fossils and were interpreted as representing forest vegetation. New, rare fruits and seeds from the early Miocene (19 Ma) primate locality, Napak, Uganda (e.g., Bishop, 1958), include *Antrocaryon* (Anacardiaceae, cashew family), another Central and West African forest taxon, reported here for the first time.

Fruits and seeds from Rusinga Island in Lake Victoria (17.8 Ma; Drake et al., 1988) (Chesters, 1957; Andrews and Van Couvering, 1975) include several taxa found today in moist or dry forests of Central and West Africa, or their eastern outliers. The Rusinga paleoflora is of particular interest as it represents a large collection and is directly associated with the primate genus

Proconsul. The assemblage has been interpreted both as a gallery forest based on the abundance of vines along with arboreal taxa (Chesters, 1957) and as a wet lowland forest based on the floral and faunal assemblages (Andrews and Van Couvering, 1975). More recent analyses of plant fossil assemblages from specific stratigraphic horizons at Rusinga are interpreted as representing litter from a closed woodland (Collinson et al., 2009). Taphonomic analysis indicates variation in environment across the landscape, and the likelihood that the fruit and seed collections published by Chesters (1957) represent a mixture of forest and woodland communities (Collinson et al., 2009).

The middle Miocene is represented in East Africa by sites in Kenya, including the widely studied and variously interpreted locality, Fort Ternan (14 Ma; Shipman et al., 1981). Much of the attention stems from the occurrence of the primate, *Kenyapithecus*, originally thought to be an early representative of the human lineage (Leakey, 1968). A large fauna, grass fossils, and one pollen assemblage have been interpreted variously as representing grassland (Retallack et al., 1990; Dugas and Retallack, 1993) and open or closed woodland (Shipman et al., 1981; Shipman, 1986; Bonnefille, 1984; Cerling et al., 1991, 1997). Grass macrofossils are most likely from an area associated with high water table rather than vast grassland or wooded grassland as originally envisioned by Retallack et al. (1990). The pollen sample indicates the presence of montane forest in the highlands, while local vegetation was interpreted as open woodland (Bonnefille, 1984). This appears to be consistent with paleoenvironmental interpretations based on faunal assemblages (Andrews and Nesbit Evans, 1979; Shipman, 1986; Kappelman, 1991) and carbon isotopic analyses of mammalian tooth enamel (Cerling et al., 1991, 1997).

Kenya's Tugen Hills provide a sedimentary sequence that has produced both plant and vertebrate fossils, including hominins. Paleobotanical sites include Kabarsero, a diverse autochthonous assemblage of leaves, twigs, and fruit preserved in an airfall or minimally reworked tuff, dated at 12.6 Ma (Hill et al., 1985; Jacobs and Kabuye, 1987, 1989; Jacobs and Winkler 1992; Jacobs and Deino, 1996). The plant fossils represent a forest akin to those found in Central Africa today, including typical genera such as *Cola* and *Pollia*, a forest floor herb (Jacobs and Kabuye, 1989). Another Tugen Hills locality, Waril, dated at approximately 10 Ma based on relative stratigraphic position, represents a plant community dominated by a species of small-leafleted legume indicative of a seasonally dry climate (Jacobs, 1999). Kapturo, dated at 6.8 Ma, represents woodland to dry forest vegetation (Jacobs, 1999). These localities and others from the Tugen Hills, including fossil woods preserved in situ and representing upland forest vegetation (Kingston et al., 2002), indicate that a variety of environments were present in the evolving rift between about 13 and 6 Ma (see also Kingston et al., 1994). The sequence of lowland or submontane forest at 12.6 Ma, seasonally arid woodland or wooded savanna at about 10 Ma, and dry forest to woodland and upland forests at 6.8 Ma among localities that are within 10 km of each other demonstrates that paleoenvironmental change was not unidirectional from forest to savanna but reflected the combined effects of global climate change and regional physiographic development.

THE PLIOCENE

The Pliocene, a critical time interval for human evolution, is unfortunately not a good time for the preservation of plant fossils, which could potentially provide paleoanthropologists with a fine-scale view of landscape variation and ecosystem primary producers. Oxidizing conditions, brought about by alternating wet and dry intervals, breaks down plant organic matter, including relatively robust pollen grains. As relief in the rift became more pronounced, increasing opportunities for fluvial and lacustrine sedimentation, organic preservation became less likely due to periodic desiccation and oxidation as rain shadows developed and glaciations at both poles increased episodic aridity at low latitudes. Despite these challenges, many terrestrial pollen samples have been studied, and new methods are yielding more detailed results. A summary of the East African Plio-Pleistocene terrestrial pollen record (Bonnefille, 1995) documented variation on the theme of grass dominance in pollen samples from the Omo, Hadar (Ethiopia), Turkana (Kenya), and Laetoli (Tanzania) Basins—all areas with important hominin records. Open woodland, wooded grassland, and grasslands surrounded these basins to varying degrees, and forest communities varied in their distance from sample localities (Bonnefille, 1995). A notable stratigraphic sequence from the Ethiopian Highlands at Gadeb dated at approximately 2.5 Ma contained pollen of Afro-alpine communities, documenting much cooler conditions, 5°–6° C lower than modern mean annual temperature. These sediments were correlated with lowland assemblages from Lake Turkana that indicated drying conditions, typical for glacial intervals (Bonnefille, 1995). In these earlier studies, the Hadar pollen record (3.7–3.2 Ma) documented the presence of species found today in both humid evergreen upland and lowland forests, associated with those from seasonally dry bushland and woodland communities. A fruit of the forest plant, *Antrocaryon*, was reported from contemporaneous sediments of the Omo Valley, Ethiopia, suggesting a possible outlier of Guineo-Congolian lowland forest in the East African Rift (Bonnefille and Letouzey, 1976). More recently, fine-scale analysis of pollen assemblages from exposed lake sediments at Hadar, representing about 20,000 years during the middle Pliocene (about 3.3 Ma), documented variation in the biomes present near the ancient lake (Bonnefille et al., 2004). The pollen taxa, assigned to plant functional types, were linked to biomes and these to climate, using multivariate statistical methods. The authors conclude that Hadar was near the "limit of the WAMF [warm mixed forest] biome adjacent to xerophytic cool steppe along an escarpment slope" (Bonnefille et al., 2004, p. 12126). Rainfall in this region would have been nearly twice the modern, and mean annual temperature was cooler by about 8°–11°C, in part due to higher elevation of the area during the Pliocene.

A new approach to understanding paleoecology at Laetoli, Tanzania, an important Pliocene (3.5–3.76 Ma) hominin site, was devised by Andrews and Bamford (2008), who established an association between the variation in plant communities across the modern landscape in the Laetoli area and specific soil, topographic, and geologic characteristics. The geology and geomorphology among Laetoli sediments provided information about past conditions, and these were used to reconstruct vegetation across the paleolandscape. First, assuming no change in climate, they found that Laetoli would have been a mosaic of low and tall deciduous woodland with woodlands or forests in riparian settings. New scenarios for vegetation distribution and variation on the landscape are hypothesized using the same technique, but assuming a change in climate parameters, which can come from independent proxies. Evidence that the climate was wetter at Laetoli during the early Pliocene yields hypothesized recon-

structed vegetation consisting of less distance to the montane forest in the eastern highlands, more extensive woodland vegetation supported away from stream beds, and grass-dominated wetland areas. This more detailed reconstruction is further developed by sparse fossil woods (including potential species composition data; Bamford, pers. comm., June 2008) and is consistent with previous palynological work (Bonnefille and Riollet, 1987) and more recent evidence based on carbon isotopes from fossil tooth enamel (Kingston and Harrison, 2007).

THE PLEISTOCENE

The Pleistocene record is dominated by palynological samples from modern lakes with sediments that date back thousands, or even hundreds of thousands, of years, and samples from offshore drill cores. These records, which can be continuous, can also be analyzed for oxygen and carbon isotopic variations providing paleoclimate data from independent lines of evidence (e.g., Leroy and Dupont, 1997; Dupont et al., 2000). These data are abundant, and any detailed discussion is beyond the scope of this chapter (for African pollen database information, see Vincens et al., 2007).

Marine palynological studies document broad-scale changes, particularly off the coast of West Africa, where interpretations are not hindered by the kind of complex topography present in East Africa. The long-term record indicates that overall cooling of the deep Atlantic Ocean is correlated with aridification of northwestern Africa (Leroy and Dupont, 1997). At a finer scale, coastal cores from West Africa indicate expansion of dry forest and savanna at the expense of rain forest around the Gulf of Guinea during the last glacial stage, but Guinean and Congolian rain forests were probably not completely separated from each other, an important finding for biogeographers and plant evolutionary ecologists (Dupont et al., 2000). In eastern Africa, the general circulation effects of global cooling were imposed on rift lakes and their surroundings as recorded by pollen records (e.g., Vincens, 1993; Mohammed and Bonnefille, 1996), which document significant cooling at high elevations and drying at low altitudes during glacial stages. In southern Africa, late Pleistocene and Holocene records have been widely sampled and reported by Scott (e.g., Scott, 1996, 1999, 2002) and Coetzee before him (e.g., Coetzee, 1978), where glacial cycles had a significant impact on vegetation at these higher latitudes.

Discussion and Conclusions

During the course of the Cenozoic, significant changes in Africa's physical setting contributed to ecosystem evolution, as did the local and regional effects of global climate change and biotic interactions. A generally low-elevation landscape across Africa's tropical latitudes during the Paleocene and Eocene may imply that lowland tropical wet forest would be present. However, the ecological significance of paleobotanical data is equivocal because macrofossil evidence is scarce, and pollen data include many taxa without known modern relatives. Certainly, mangrove vegetation related to the modern tropical Asian palm, *Nypa*, was present along Africa's coasts at low latitudes in the Paleocene and Eocene, and pollen evidence of other palms is common. Interestingly, during the middle Eocene, a time of global relative warmth, there is evidence for seasonally dry climate from central Tanzania and open habitat in West Africa and Egypt, where grass and cycad pollen are noted. Interpreting these data as evidence of heat or water stress is tantalizing, but the environmental significance of grass and cycad pollen from the latter sites is poorly understood (Kedves, 1971; Salard-Cheboldaeff, 1979, 1981; Jacobs, 2004). By the late Eocene, pollen evidence from West Africa documents dramatically increasing diversity of forest taxa continuing into the Oligocene (Salard-Cheboldaeff, 1981), but North Africa is thought to have become increasingly latitudinally stratified, with coastal forests (biogeographically related to Eurasian paleofloras) and inland communities of woodland (e.g., Boureau et al., 1983).

The African plate was isolated in the Paleogene but moved northward and rotated counterclockwise slightly, eventually making contact with the Eurasian plate during the Oligocene. Consequently, a land connection was established in the earliest Miocene by closure of the Paratethys Seaway. This had biogeographic implications and may have diminished winter rainfall to the Horn of Africa, affecting plant community composition (Pan, 2007). In addition, Antarctic ice buildup at the Eocene-Oligocene boundary, demonstrated elsewhere by a number of proxies (e.g., Prothero and Berggren, 1992; Zachos et al., 2001), may have had a significant impact on Africa's biota, but evidence of this is limited. An important decline in palm richness in West African pollen cores is evident at and near the boundary, and it may help explain the relatively minor role of palms in Neogene and modern African forests (Moore, 1973; Morley, 2000; Pan et al., 2006).

The Neogene is characterized by the spread of grass-dominated environments and by development of the East African Rift Valley, which had a profound impact on local and regional plant community composition and distribution. Most early Miocene tropical African paleobotanical sites record forest vegetation, but by the middle Miocene grass-dominated ecosystems begin to spread, as evidenced by grass cuticles and pollen in core sediments of the Niger Delta (Morley and Richards, 1993). East Africa's complex topographic changes associated with rift development resulted in a range of plant community types during the middle and late Miocene (Kingston et al., 1994; Jacobs et al., 1999; Kingston et al., 2002). These included forests with Central and West African botanical affinities, seasonally dry woodlands, and montane or submontane forests. Thus, within the context of long-term increasing seasonality of rainfall across the tropical region of the continent beginning perhaps as early as the late Oligocene, plant communities in East Africa also reflect local and regional topographic change.

Limited macrofossil evidence from the Pliocene documents the rare occurrence of some relictual lowland forest taxa in the East African Rift. Pliocene and Pleistocene pollen records from across Africa show variations on a more modern theme of plant community composition, and variations that are relatively minor compared with the larger scale of change discussed for the previous 60 million years. Pollen records reflect the influence of grass dominance across much of the continent and document change in upland or lowland forest distribution and composition in response to global climate change.

Documentation of the Cenozoic vegetation history of Africa continues to improve, with new localities being discovered each year and new approaches applied to understanding the paleoecology of ancient ecosystems and their evolution. The improved record over the last decade has allowed for the documentation of paleoecology and paleoclimate from multiple proxies and more informed hypotheses of faunal change in an environmental context (e.g., Potts, 2007). A more complete record for the entire Cenozoic will allow for more sophisticated analyses (e.g., Garcin et al., 2006) and ultimately an accurate understanding of Africa's biotic evolution and paleoclimatic history.

Literature Cited

Adegoke, O. S., R. E. Jan Du Chêne, A. E. Agumanu, and P. O. Ajayi. 1978. Palynology and age of the Kerri-Kerri Formation, Nigeria. *Revista Española de Micropaleontologia* 10:267–283.

Andrews, P., and M. Bamford. 2008. Past and present vegetation ecology of Laetoli, Tanzania. *Journal of Human Evolution* 54:78–98.

Andrews, P., and E. Nesbit Evans. 1979. The environment of *Ramapithecus* in Africa. *Paleobiology* 5:22–30.

Andrews, P., and J. A. H. Van Couvering. 1975. Palaeoenvironments in the East African Miocene. *Contributions to Primatology* 5:62–103.

As-Saruri, M. L., P. J. Whybrow, and M. E. Collinson. 1999. Geology, fruits, seeds, and vertebrates (?Sirenia) from the Kaninah Formation (middle Eocene), Republic of Yemen; pp. 443–453 in P. J. Whybrow and A. Hill (eds.), *Fossil Vertebrates of Arabia*. Yale University Press, New Haven.

Awad, M. Z. 1994. Stratigraphic, palynological and paleoecological studies in the East-Central Sudan (Khartoum and Kosti Basins), Late Jurassic to Mid-Tertiary. *Berliner Geowissenschaftliche Abhandlungen* 161:1–163.

Awad, M. Z., and F. E. R. Breir. 1993. Oligo-Miocene to Quaternary palaeoenvironment in Gezira area, central Sudan; pp. 465–470 in U. Thorweihe and H. Schandelmeier (eds.), *Geoscientific Research in Northeast Africa*. Balkema, Rotterdam.

Baez, A. M. 2000. An Eocene pipine frog from North-central Tanzania. *Journal of Vertebrate Paleontology* 20(suppl. to no. 3):28A.

Bancroft, H. 1932a. A fossil cyatheoid stem from Mount Elgon, East Africa. *New Phytologist* 31:241–253.

———. 1932b. Some fossil dicotyledonous woods from the Miocene (?) beds of East Africa. *Annals of Botany* 46:745–767.

———. 1933. A contribution to the geological history of the Dipterocarpaceae. *Geologiska Föreningens i Stockholm Förhandlingar* 55:59–100.

———. 1935. Some fossil dicotyledonous woods from Mount Elgon, East Africa, 1. *American Journal of Botany* 22:164–183.

Bishop, W. W. 1958. Miocene Mammalia from the Napak volcanics, Karamoja, Uganda. *Nature* 182:1480–1482.

Boaz, N. T., A. El-Arnauti, A. W. Gaziry, J. de Heinzelin, and D. D. Boaz (eds.). 1987. *Neogene Paleontology and Geology of Sahabi*. Liss, New York.

Boltenhagen, E. 1965. Introduction à la palynologie stratigraphique du bassin sedimentaire de L'Afrique equatoriale. Extr. 1° Coll. Afr. Micropaleontol. *Mémoires du Bureau de Recherches Géologiques et Minières* 32:305–326.

———. 1984. Cenozoic vegetation and environments of early hominids in East Africa; pp. 579–612 in R. O. Whyte (ed.), *The Evolution of the East Asian Environment: Volume II. Paleobotany, Paleozoology, and Paleoanthropology*. University of Hong Kong, Hong Kong.

———. 1995. A reassessment of the Plio-Pleistocene pollen record of East Africa; pp. 299–310 in E. S. Vrba, G. H. Denton, T. C. Partridge and L. H. Burckle (eds.), *Paleoclimate and Evolution: With Emphasis on Human Origins*. Yale University Press, New Haven.

Bonnefille, R., and R. Letouzey. 1976. Fruits fossiles d'*Antrocaryon* dans la vallée de L'Omo (Ethiopie). *Adansonia* 16:65–82.

Bonnefille, R., D. Lobreau, and G. Riollet. 1982. Pollen fossile de *Ximenia* (Olacaceae) dans le Pléistocène Inférieur d'Olduvai en Tanzanie: Implications paleoécologiques. *Journal of Biogeography* 9:469–486.

Bonnefille, R., R. Potts, F. Chalie, D. Jolly, and O. Peyron. 2004. High-resolution vegetation and climate change associated with Pliocene *Australopithecus afarensis*. *Proceedings of the National Academy of Sciences, USA* 101:12125–12129.

Bonnefille, R., and G. Riollet. 1980. Palynologie, végétation et climats de Bed I et Bed II à Olduvai, Tanzanie. *Actes du 8° Congrès Panafricaine de Préhistoire, Études Quaternaires*, 123–127.

———. 1987. Palynological spectra from the Upper Laetolil Beds; pp. 52–61 in M. D. Leakey and J. M. Harris, (eds.), *Laetoli, a Pliocene Site in Northern Tanzania*. Clarendon Press, Oxford.

Boureau, E., M. Cheboldaeff-Salard, J.-C. Koeniguer, and P. Louvet. 1983. Évolution des flores et de la vegetation Tertiaires en Afrique, au nord de l'Équateur. *Bothalia* 14:355–367.

Bown, T. M., M. J. Kraus, S. L. Wing, J. G. Fleagle, B. H. Tiffney, E. L. Simons, and C. F. Vondra. 1982. The Fayum primate forest revisited. *Journal of Human Evolution* 11:603–632.

Brock, P. W. G., and R. Macdonald. 1969. Geological environment of the Bukwa mammalian fossil locality, Eastern Uganda. *Nature* 223:593–596.

Brunet, M., Y. Coppens, D. Pilbeam, S. Djallo, K. Behrensmeyer, A. Brillanceau, W. Downs, M. Dupéron, G. Ekodeck, L. Flynn, E. Heintz, J. Hell, Y. Jehenne, L. Martin, C. Mosser, M. Salard-Cheboldaeff, S. Wenz, and S. Wing. 1986. Les formations sédimentaires continentales du Crétacé et du Cénozoïque camerounais: Premiers résultats d'une prospection paléontologique. *Comptes Rendus de l'Académie des Sciences,* Série II, 303:425–428.

Caratini, C., and C. Tissot. 1991. Paleocene palynoflora from Walalane borehole, Senegal. *Palaeoecology of Africa* 22:123–133.

Cerling, T. E., J. M. Harris, S. H. Ambrose, M. G. Leakey, and N. Solounias. 1997. Dietary and environmental reconstruction with stable isotope analyses of herbivore tooth enamel from the Miocene locality of Fort Ternan, Kenya. *Journal of Human Evolution* 33:635–650.

Cerling, T. E., J. Quade, S. H. Ambrose, and N. E. Sikes. 1991. Fossil soils, grasses, and carbon isotopes from Fort Ternan, Kenya: Grassland or woodland? *Journal of Human Evolution* 21:295–306.

Chandler, M. E. J. 1954. Some Upper Cretaceous and Eocene fruits from Egypt. *Bulletin of the British Museum (Natural History) Geology* 2:147–187.

———. 1964. *The Lower Tertiary Floras of Southern England, Vol. IV, A Summary and Survey of Findings in the Light of Recent Botanical Observations.* British Museum (Natural History), London, 162 pp.

Chaney, R. W. 1933. A Tertiary flora from Uganda. *Journal of Geology* 41:702–709.

Chesters, K. I. M. 1955. Some plant remains from the Upper Cretaceous and Tertiary of West Africa. *Annals and Magazine of Natural History* 8:498–503.

———. 1957. The Miocene flora of Rusinga Island, Lake Victoria, Kenya. *Palaeontographica, Abt. B* 101:30–71.

Chorowicz, J. 2005. The East African rift system. *Journal of African Earth Sciences* 43:379–410.

Coetzee, J. A. 1978. Late Cainozoic palaeoenvironments of southern Africa; pp. 115–127 in E. M. Van Zinderen Bakker (ed.), *Antarctic Glacial History and World Palaeoenvironments.* Balkema, Rotterdam.

Coetzee, J. A., and J. Muller. 1984. The phytogeographic significance of some extinct Gondwana pollen types from the Tertiary of the southwestern Cape (South Africa). *Annals of the Missouri Botanical Garden* 71:1088–1099.

Coetzee, J. A., and J. Rogers. 1982. Palynological and lithological evidence for the Miocene palaeoenvironment in the Saldanha region (South Africa). *Palaeogeography, Palaeoclimatology, Palaeoecology* 39:71–85.

Collinson, M. E., P. Andrews, and M. Bamford. 2009. Taphonomy of the early Miocene Hiwegi Formation, Rusinga Island, Kenya. *Journal of Human Evolution* 57:149–162.

Crawley, M. 1988. Paleocene wood from the Republic of Mali. *Bulletin of the British Museum (Natural History) Geology* 44:3–14.

Davidson, A., and D. Rex. 1980. Age of volcanism and rifting in southwestern Ethiopia. *Nature* 283:657–658.

Dechamps, R., and F. Maes. 1990. Woody plant communities and climate in the Pliocene of the Semliki Valley, Zaire; pp. 71–94 in N. T. Boaz, (ed.), *Evolution of Environments and Hominidae in the African Western Rift Valley.* Virginia Museum of Natural History Press, Martinsville.

Dechamps, R., B. Senut, and M. Pickford. 1992. Fruits fossiles pliocènes et pléistocènes du Rift Occidental Ougandais. Signification paléoenvironnementale. *Comptes Rendus de l'Académie des Sciences,* Série II, 314:325–331.

Drake, R. E., J. A. Van Couvering, M. H. Pickford, G. H. Curtis, and J. A. Harris. 1988. New chronology for the early Miocene mammalian faunas of Kisingiri, western Kenya. *Journal of the Geological Society, London* 145:479–491.

Dugas, D. P., and G. J. Retallack. 1993. Middle Miocene fossil grasses from Fort Ternan, Kenya. *Journal of Paleontology* 67:113–128.

Dupéron-Laudoueneix, M., and J. Dupéron. 1994. Inventory of Mesozoic and Cenozoic woods from equatorial and north equatorial Africa. *Review of Palaeobotany and Palynology* 84:439–480.

Dupont, L. M., S. Jahns, F. Marret, and S. Ning. 2000. Vegetation change in equatorial West Africa: Time-slices for the last 150 ka. *Palaeogeography, Palaeoclimatology, Palaeoecology* 155:95–122.

Ebinger, C. J., T. Yemane, G. Woldegabriel, J. L. Aronson, and R. C. Walter. 1993. Late Eocene: Recent volcanism and faulting in the southern main Ethiopian rift. *Journal of the Geological Society, London* 150:99–108.

El-Saadawi, W., S. G. Youssef, M. M. Kamal-El-Din. 2004. Fossil palm woods of Egypt: II. Seven Tertiary *Palmoxylon* species new to the country. *Review of Palaeobotany and Palynology* 129:199–211.

Fritel, P. H. 1921. Sur deux fruites fossiles trouvés au Sénégal, dans l'Eocène moyen. *Bulletin du Comité d'Études Historiques et Scientifiques de l'Afrique Occidentale Française* 4:549–552.

García Massini, J. L., and B. F. Jacobs. 2009. Cretaceous-Cenozoic record of ferns in Africa; pp. 201–220 in X. van der Burgh, J. van der Maeson and J.-M. Onana (eds.) *Systematics and Conservation of African Plants.* Royal Botanic Gardens, Kew.

García Massini, J. L., B. F. Jacobs, A. Pan, N. Tabor, and J. Kappelman. 2006. The occurrence of the fern *Acrostichum* in Oligocene volcanic strata of the northwestern Ethiopian plateau. *International Journal of Plant Sciences* 167:909–918.

García Massini, J. L., B. Jacobs and N. Tabor. in press. Paleoenvironmental reconstruction of Late Oligocene terrestrial volcaniclastic deposits from the northwestern Ethiopian Plateau. *Palaeontologia Electronica.*

Garcin, Y., A. Vincens, D. Williamson, J. Guiot, and G. Buchet. 2006. Wet phases in tropical southern Africa during the last glacial period. *Geophysical Research Letters* 33:L07703.

George, R., N. Rogers, and S. Kelley. 1998. Earliest magmatism in Ethiopia: Evidence for two mantle plumes in one flood basalt province. *Geology* 26:923–926.

Gregor, H.-J., and H. Hagn. 1982. Fossil fructifications from the Cretaceous-Paleocene Boundary of SW Egypt (Danian, Bir Abu Munqar). *Tertiary Research* 4:121–147.

Griffiths, J. F. 1972. *Climates of Africa.* Elsevier, Amsterdam, 604 pp.

Guinet, P., and M. Salard-Cheboldaeff. 1975. Grains de pollen du Tertiaire du Cameroun pouvant être rapportés aux Mimosacées. *Boissiera* 24:21–28.

Gunnell, G., B. F. Jacobs, P. S. Herendeen, J. J. Head, E. Kowalski, C. P. Msuya, F. A. Mizambwa, T. Harrison, J. Habersetzer, and G. Storch. 2003. Oldest placental mammal from sub-Saharan Africa: Eocene microbat from Tanzania: Evidence for early evolution of sophisticated echolocation. *Palaeontologia Electronica* 5(3):10 pp.; http://palaeo-electronica.org/paleo/2002_2/africa/issue2_02.htm.

Hamilton, A. 1968. Some plant fossils from Bukwa. *Uganda Journal* 32:157–164.

Herendeen, P. S., and B. F. Jacobs. 2000. Fossil legumes from the middle Eocene (46.0 Ma) Mahenge flora of Singida, Tanzania. *American Journal of Botany* 87:1358–1366.

Hill, A., R. Drake, L. Tauxe, M. Monaghan, J. C. Barry, A. K. Behrensmeyer, G. Curtis, B. F. Jacobs, L. Jacobs, N. Johnson, and D. Pilbeam. 1985. Neogene palaeontology and geochronology of the Baringo Basin, Kenya. *Journal of Human Evolution* 14:759–773.

Hofmann, C., V. Courtillot, G. Féraud, P. Rochette, G. Yirgu, E. Ketefo, and R. Pik 1997. Timing of the Ethiopian flood basalt event and implications for plume birth and global change. *Nature* 389:838–841.

Jacobs, B. F. 1999. Estimation of rainfall variables from leaf characters in tropical Africa. *Palaeogeography, Palaeoclimatology, Palaeoecology* 145:231–250.

———. 2004. Palaeobotanical studies from tropical Africa: Relevance to the evolution of forest, woodland and savannah biomes. *Philosophical Transactions of the Royal Society of London B* 359:1573–1583.

Jacobs B. F., and A. L. Deino. 1996. Test of climate-leaf physiognomy regression models, their application to two Miocene floras from Kenya, and $^{40}Ar/^{39}Ar$ dating of the Late Miocene Kapturo site. *Palaeogeography, Palaeoclimatology, Palaeoecology* 123:259–271.

Jacobs, B. F., and P. S. Herendeen. 2004. Eocene dry climate and woodland vegetation in tropical Africa reconstructed from fossil leaves from northern Tanzania. *Palaeogeography, Palaeoclimatology, Palaeoecology* 213:115–123.

Jacobs, B. F., and C. H. S. Kabuye. 1987. A middle Miocene (12.2 my old) forest in the East African Rift Valley, Kenya. *Journal of Human Evolution* 16:147–155.

———. 1989. An extinct species of *Pollia* Thunberg (Commelinaceae) from the Miocene Ngorora Formation, Kenya. *Review of Palaeobotany and Palynology* 59:67–76.

Jacobs, B. F., J. D. Kingston, and L. L. Jacobs. 1999. The origin of grass-dominated ecosystems. *Annals of the Missouri Botanical Garden* 86:590–643.

Jacobs, B. J., N. Tabor, M. Feseha, A. Pan, J. Kappelman, T. Rasmussen, W. Sanders, M. Wiemann, J. Crabaugh, and J. L. García Massini. 2005. Oligocene terrestrial strata of northwestern Ethiopia: A preliminary report on paleoenvironments and paleontology. *Palaeontologica Electronica* 8(1):19 pp.; http://palaeo-electronica.org/paleo/2005_1/jacobs25/issue1_05.htm.

Jacobs, B. F., and D. A. Winkler. 1992. Taphonomy of a middle Miocene autochthonous forest assemblage, Ngorora Formation, central Kenya. *Palaeogeography, Palaeoclimatology, Palaeoecology* 99:31–40.

Jan du Chêne, R. E., M. S. Onyike, and M.A. Sowunmi. 1978. Some new Eocene pollen of the Ogwashi-Asaba Formation, South-Eastern Nigeria. *Revista Española de Micropaleontologia* 10:285–322.

Kaiser, T. M., J. Ansorge, G. Arratia, V. Bullwinkel, G. F. Gunnell, P. S. Herendeen, B. Jacobs, J. Mingram, C. Msuya, A. Musolff, R. Naumann, E. Schulz, and V. Wilde. 2006. The maar lake of Mahenge (Tanzania): Unique evidence of Eocene terrestrial environments in sub-Sahara Africa. *Zeitschrift der Deutschen Gesellschaft für Geowissenschaften* 157:411–431.

Kamal El-Din, M. M., and W. S. El-Saadawi. 2004. Two leguminosae woods from the Miocene of Gebel Ruzza, Egypt. *IAWA Journal* 25:471–483.

Kappelman, J. 1991. The paleoenvironment of *Kenyapithecus* at Fort Ternan. *Journal of Human Evolution* 20:95–110.

Kappelman, J., D. T. Rasmussen, W. J. Sanders, M. Feseha, T. Bown, P. Copeland, J. Crabaugh, J. Fleagle, M. Glantz, A. Gordon, B. Jacobs, M. Maga, K. Muldoon, A. Pan, L. Pyne, B. Richmond, T. Ryan, E. R. Seiffert, S. Sen, L. Todd, M. C. Wiemann, and A. Winkler. 2003. New Oligocene mammals from Ethiopia and the pattern and timing of faunal exchange between Afro-Arabia and Eurasia. *Nature* 426:549–552.

Kaska, H. V. 1989. A spore and pollen zonation of early Cretaceous to Tertiary nonmarine sediments of Central Sudan. *Palynology* 13:79–90.

Kedves, M. 1971. Présence de types sporomorphes importants dans les sediments pre-quaternaires Egyptiens. *Acta Botanica Academiae Scientiarum Hungaricae* 17:371–378.

Kingston, J. D., and T. Harrison. 2007. Isotopic dietary reconstructions of Pliocene herbivores at Laetoli: Implications for early hominin paleoecology. *Palaeogeography, Palaeoclimatology, Palaeoecology* 243:272–306.

Kingston, J. D., B. F. Jacobs, A. Hill, and A. Deino. 2002. Stratigraphy, age and environments of the late Miocene Mpesida Beds, Tugen Hills, Kenya. *Journal of Human Evolution* 42:95–116.

Kingston, J. D., B. D. Marino, and A. Hill. 1994. Isotopic evidence for Neogene hominid paleoenvironments in the Kenya Rift Valley. *Science* 264:955–959.

Koeniguer, J. C. 1967. Étude paléoxylologique de la Libye. I. Sur un bois fossile de l'Oligocène de Dar el Abd (Syrte): *Bridelioxylon arnouldi* n. sp. II. Sur la présence de *Dombeyoxylon oweni* (Carr.) Kräusel, 1939, dans le Tertiaire de la Syrte. III. Sur la présence de *Sapindoxylon* sp. dans le Tertiaire du nord du Tibesti. *91e Congrès National des Sociétés Savants, Rennes, 1966, Sciences* 3:153–172.

———. 1971. Sur les bois fossiles du Paleocène de Sessao (Niger). *Review of Palaeobotany and Palynology* 12:303–323.

Kräusel, R. 1939. Ergebnisse der Forschungsreisen Prof. E. Stromers in den Wüsten Ägyptens. *Abhandlungen der Bayerischen Akademie der Wissenschaften, Neue Folge* 47:1–140.

Lakhanpal, R.N. 1966. Some middle Tertiary plant remains from South Kivu, Congo. *Annales du Musée Royal de l'Afrique Central*, Série in 8°, 52:21–30.

Lakhanpal, R. N., and U. Prakash. 1970. Cenozoic plants from Congo, I. Fossil woods from the Miocene of Lake Albert. *Annales du Musée Royal de l'Afrique Central*, Série in 8°, 64:1–20.

Leakey, L. S. B. 1968. Bone smashing by late Miocene Hominidae. *Nature* 218:528–530.

Legoux, O. 1978. Quelques espèces de pollen caractéristiques du Néogène du Nigeria. *Bulletin du Centre Recherche Exploration-Production Elf-Aquitaine* 2:265–317.

Lemoigne, Y. 1978. Flores tertiaires de la haute vallée de l'Omo (Ethiopie). *Palaeontographica, Abt. B* 165:89–157.

Lemoigne, Y., J. Beauchamp, and E. Samuel. 1974. Etude paléobotanique des dépôts volcaniques d'âge tertiaire des bordures est et ouest du système des rifts éthiopiens. *Geobios* 7:267–288.

Leroy, S. A. G., and L. M. Dupont. 1997. Marine palynology of the ODP Site 658 (NW Africa) and its contribution to the stratigraphy of Late Pliocene. *Geobios* 30:351–359.

Lofty, H., and R. Van der Voo. 2007. Tropical northeast Africa in the middle-Late Eocene: Paleomagnetism of the marine-mammals sites and basalts in the Fayum province, Egypt. *Journal of African Earth Sciences* 47:135–152.

Logatchev, N. A., V. V. Beloussov, and E. E. Milanovsky. 1972. East African Rift development. *Tectonophysics* 15:71–81.

Louvet, P. 1973. Sur les affinités des flores tropicales ligneuses africaines tertiaire et actuelle. *Bulletin de la Société Botanique de France* 120:385–395.

Louvet, P., and J. Mouton. 1970. La flore oligocène du Djebel Coquin (Libye). *95e Congrès National des Sociétés Savants, Reims, 1970, Sciences* 3:79–96.

Malécot, V. and D. Lobreau-Callen. 2005. A survey of species assigned to the fossil pollen genus *Anacolosidites*. *Grana* 44:314–336.

Medus, J. 1975. Palynologie de sediments tertiaires du Sénégal meridional. *Pollen et Spores* 17:545–601.

Menzel, P. 1920. Über Pflanzenreste aus Basalttuffen des Kamerungebietes. *Beiträge zur geologischen Erforschung der deutschen Schützgebiete* 18:2–32.

Méon, H. 1990. Palynologic studies of the Cretaceous-Tertiary boundary interval at El Kef outcrop, northwestern Tunisia: Paleogeographic implications. *Review of Palaeobotany and Palynology* 65:85–94.

Meulenkamp, J. E., and W. Sissingh. 2003. Tertiary palaeogeography and tectonostratigraphic evolution of the Northern and Southern Peri-Tethys platforms and the intermediate domains of the African-Eurasian convergent plate boundary zone. *Palaeogeography, Palaeoclimatology, Palaeoecology* 196:209–228.

Mohammed, M. U., and R. Bonnefille. 1996. A late Glacial/late Holocene pollen record from a highland peat at Tamsaa, Bale Mountains, south Ethiopia. *Global and Planetary Science Change* 16:17:121–129.

Moore, H. E. J. 1973. Palms in the tropical forest ecosystems of Africa and South America; pp. 63–88 in B. J. Meggers, E. S. Ayesu, and W. D. Duckworth, (eds.), *Tropical Forest Ecosystems in Africa and South America: A Comparative Review.* Smithsonian Institution Press, Washington, D.C.

Morley, R. J. 2000. *Origin and Evolution of Tropical Rain Forests.* Wiley, Chichester, 362 pp.

Morley, R. J., and K. Richards. 1993. Gramineae cuticle: A key indicator of Late Cenozoic climatic change in the Niger Delta. *Review of Palaeobotany and Palynology* 77:119–127.

Murray, A. M. 2000. Eocene Cichlid fishes from Tanzania, East Africa. *Journal of Vertebrate Paleontology* 20:651–664.

———. 2001. The oldest fossil cichlids (Teleostei: Perciformes): Indication of a 45 million-year-old species flock. *Proceedings of the Royal Society of London B* 268:679–684.

———. 2003. A new characiform fish (Teleostei: Ostariophysi) from the Eocene of Tanzania. *Canadian Journal of Earth Science* 40:473–481.

Murray, A. M., and L. A. Budney. 2002. An Eocene catfish (Claroteidae: Chrysichthys) from an East African crater lake. *Journal of Vertebrate Paleontology* 22(suppl. to no. 3):91A.

Murray, A., and M. V. H. Wilson. 2005. Description of a new Eocene osteoglossid fish and additional information on *Singida jacksonoides* Greenwood and Patterson, 1967 (Osteoglossomorpha), with an assessment of their phylogenetic relationships. *Zoological Journal of the Linnean Society* 144:213–228.

Niklas, K. J. 1994. Predicting the height of fossil plant remains: An allometric approach to an old problem. *American Journal of Botany* 81:1235–1242.

Ollivier-Pierre, M.-F. 1982. La microflore du Paleocène et de l'Eocène des series phosphates des Ganntour (Maroc). *Bulletin des Sciences Géologiques* 35:117–127.

Pan, A. D. 2007. The Late Oligocene (28–27 Ma) Guang River flora from the northwestern plateau of Ethiopia. Unpublished PhD dissertation, Southern Methodist University, Dallas, 219 pp.

Pan, A.D. in press. Rutaceae leaf fossils from the late Oligocene (27–23 Ma) Guang River flora of northwestern Ethiopia. *Review of Palaeobotany and Palynology.*

Pan, A. D. and B. F. Jacobs. 2009. The earliest record of the genus *Cola* (Malvaceae *sensu lato*: Stercolioideae) from the late Oligocene (28–27 Ma) of Ethiopia and leaf characteristics within the genus. *Plant Systematics and Evolution* 283:247–262.

Pan, A., B. F. Jacobs, J. Dransfield, and W. J. Baker. 2006. The fossil history of palms (Arecaceae) in Africa and new records from the late Oligocene (28–27 Mya) of north-western Ethiopia. *Botanical Journal of the Linnean Society* 151:69–81.

Pickford, M. 2002. Early Miocene grassland ecosystem at Bukwa, Mount Elgon, Uganda. *Comptes Rendus Palevol* 1:1–6.

Potts, R. 2007. Environmental hypotheses of Pliocene human evolution; pp. 25–49 in Bobe, R., Z. Alemseged, and A. K. Behrensmeyer (eds.), *Hominin Environments in the East African Pliocene: An Assessment of the Faunal Evidence*. Springer, Dordrecht.

Prakash, U., N. Awasthi, and Y. Lemoigne. 1982. Fossil dicotyledonous woods from the Tertiary of Blue Nile Valley, Ethiopia. *The Palaeobotanist* 30:43–59.

Privé-Gill, C., G. A. Gill, H. Thomas, J. Roger, S. Sen, and E. G. Z. Al-Sulaimani. 1993. Premier bois fossile associé aux primates oligocènes du Dhofar (Taqah, Sultanat d'Oman). *Comptes Rendus de l'Académie des Sciences*, Série II, 316:553–559.

Privé-Gill, C., H. Thomas, and P. Lebret. 1999. Fossil wood of Sindora (Leguminosae, Caesalpiniaceae) from the Oligo-Miocene of Saudi Arabia: Paleobiogeographical considerations. *Review of Palaeobotany and Palynology* 107:191–1999

Prothero, D. R., and W. A. Berggren. 1992. *Eocene-Oligocene Climatic and Biotic Evolution*. Princeton University Press, Princeton, 568 pp.

Retallack, G. J. 1992. Middle Miocene fossil plants from Fort Ternan (Kenya) and evolution of African grasslands. *Paleobiology* 18:383–400.

Retallack, G. J., D. P. Dugas, and E. A. Bestland. 1990. Fossil soils and grasses of a middle Miocene East African grassland. *Science* 247:1325–1328.

Rich, P. M., K. Helenurm, D. Kearns, S. R. Morse, M. W. Palmer, and L. Short. 1986. Height and stem diameter relationships for dicotyledonous trees and arborescent palms of Costa Rican tropical wet forest. *Bulletin of the Torrey Botanical Club* 115:241–246.

Richards, P. W. 1996. *The Tropical Rain Forest*. Cambridge University Press, Cambridge, 575 pp.

Rull, V. 2005. Vegetation and environmental constancy in the Neotropical Guayana Highlands during the last 6000 years? *Review of Palaeobotany and Palynology* 135:205–222.

Sah, S. C. D. 1967. Palynology of an Upper Neogene profile from Rusizi Valley (Burundi). *Musée Royal de l'Afrique Centrale* 57:1–173.

Salard-Cheboldaeff, M. 1978. Sur la palynoflore Maestrichtienne et Tertiaire du bassin sedimentaire littoral du Cameroun. *Pollen et Spores* 20:215–260.

———. 1979. Palynologie Maestrichtienne et Tertiaire du Cameroun. Étude Qualitative et Repartition Verticale des Principales Espèces. *Review of Paleobotany and Palynology* 28:365–388.

———. 1981. Palynologie Maestrichtienne et Tertiaire du Cameroun: Resultats botaniques. *Review of Palaeobotany and Palynology* 32:401–439.

———. 1990. Intertropical African palynostratigraphy from Cretaceous to Late Quaternary times. *Journal of African Earth Sciences* 11:1–24.

Salard-Cheboldaeff, M., and J. Dejax. 1991. Evidence of Cretaceous to Recent West African intertropical vegetation from continental sediment spore-pollen analysis. *Journal of African Earth Sciences* 12:353–361.

Scotese, C. R. 2005. Point Tracker for Windows. Arlington, Tex., Paleomap Project (www.scotese.com/software.htm).

Scott, L. 1996. Palynology of hyrax middens: 2000 years of palaeoenvironmental history in Namibia. *Quaternary International* 33:73–79.

———. 1999. The vegetation history and climate in the Savanna Biome, South Africa, since 190,000 Ka: A comparison of pollen data from the Tswaing Crater (the Pretoria Saltpan) and Wonderkrater. *Quaternary International* 57–58:215–223.

———. 2002. Grassland development under glacial and interglacial conditions in southern Africa: Review of pollen, phytolith and isotope evidence. *Paleogeography, Paleoclimatology, Paleoecology* 177:47–57.

Scott, L., A. Cadman, and I. McMillan. 2006. Early history of Cainozoic Asteraceae along the southern African west coast. *Review of Palaeobotany and Palynology* 142:47–52.

Seward, A. C. 1935. Leaves of dicotyledons from the Nubian Sandstone of Egypt. *Geological Survey of Egypt*, 21 pp.

Shipman, P. 1986. Paleoecology of Fort Ternan reconsidered. *Journal of Human Evolution* 15:193–204.

Shipman, P., A. Walker, J. A. Van Couvering, and P. Hooker. 1981. The Fort Ternan hominoid site, Kenya: Geology, age, taphonomy and paleoecology. *Journal of Human Evolution* 10:49–72.

Thomas, H., J. Roger, S. Sen, M. Pickford, E. Gheerbrant, Z. Al-Sulaimani, and S. Al-Busaidi. 1999. Oligocene and Miocene terrestrial vertebrates in the southern Arabian Peninsula (Sultanate of Oman) and their geodynamic and palaeogeographic settings; pp. 430–442 in P. J. Whybrow, and A. Hill, (eds.), *Fossil Vertebrates of Arabia*. Yale University Press, New Haven.

Tiffney, B. H. 1991. Paleoenvironment of the Oligocene Jebel Qatrani Formation, Fayum Depression, northern Egypt, based on floral remains. *Geological Society of America, Abstracts with Programs* 23:456.

Tiffney, B. H., J. G. Fleagle, and T. M. Bown. 1994. Early to middle Miocene angiosperm fruits and seeds from Fejej, Ethiopia. *Tertiary Research* 15:25–42.

Ukstins, I. A., P. R. Renne, E. Wolfenden, J. Baker, D. Ayalew, and M. Menzies. 2002. Matching conjugate volcanic rifted margins: ^{40}Ar/^{39}Ar chrono-stratigraphy of pre- and syn-rift bimodal flood volcanism in Ethiopia and Yemen. *Earth and Planetary Science Letters* 198:289–306.

Van Hoeken-Klinkenberg, P. M. J. 1966. Maastrichtian Paleocene and Eocene pollen and spores from Nigeria. *Leidse Geologische Mededelingen* 38:37–44.

Vincens, A. 1993. Nouvelle séquence pollinique du Lac Tanganyika: 30,000 ans d'histoire botanique et climatique du Bassin Nord. *Review of Palaeobotany and Palynology* 78:381–394.

Vincens, A., M. A. Dubois, B. Guillet, G. Achoundong, G. Buchet, V. Kamgang Kabeyene Beyala, C. de Namur, and B. Riera. 2000. Pollen-rain-vegetation relationships along a forest–savanna transect in southeastern Cameroon. *Review of Palaeobotany and Palynology* 110:191–208.

Vincens, A., A.-M. Lezine, G. Buchet, D. Lewden, A. Le Thomas, and contributors. 2007. African pollen database inventory of tree and shrub pollen types. *Review of Palaeobotany and Palynology* 145:135–141.

Vincens, A., I. Ssemmanda, M. Roux, and D. Jolly. 1997. Study of the modern pollen rain in Western Uganda with a numerical approach. *Review of Palaeobotany and Palynology* 96:145–168.

Vincens, A., J.-J. Tiercelin, and G. Buchet. 2006. New Oligocene–early Miocene microflora from the southwestern Turkana Basin: Palaeoenvironmental implications in the northern Kenya Rift. *Palaeogeography, Palaeoclimatology, Palaeoecology* 239:470–486.

Walker, A. 1968. The lower Miocene fossil site of Bukwa, Sebei. *Uganda Journal* 32:149–156.

———. 1969. Fossil mammal locality on Mount Elgon, eastern Uganda. *Nature* 223:591–593.

Wheeler, E. A., M. C. Wiemann, and J. G. Fleagle. 2007. Woods from the Miocene Bakate Formation, Ethiopia. Anatomical characteristics, estimates of original specific gravity and ecological inferences. *Review of Palaeobotany and Palynology* 146:193–207.

Whybrow, P. J. and H. A. McClure. 1981. Fossil mangrove roots and palaeoenvironments of the Miocene of the eastern Arabian peninsula. *Palaeogeography, Palaeoclimatology, Palaeoecology* 32:213–225.

Williamson, P. G. 1985. Evidence for an early Plio-Pleistocene rainforest expansion in East Africa. *Nature* 315:487–489.

Wing, S. L., S. T. Hasiotis, and T. M. Bown. 1995. First ichnofossils of flank-buttressed trees (late Eocene), Fayum Depression, Egypt. *Ichnos* 3:281–286.

Winkler, A. J., L. MacLatchy, and M. Mafabi. 2005. Small rodents and a lagomorph from the early Miocene Bukwa Locality, Eastern Uganda. *Palaeontologica Electronica* 8(1):12 pp.; http://palaeo-electronica.org/paleo/2005_1/winkler24/issue1_05.htm.

Woldegabriel, G., J. L. Aronson, and R. C. Walter. 1990. Geology, geochronology, and rift basin development in the central sector of the Main Ethiopia Rift. *Geological Society of America Bulletin* 102:439–458.

Wolfenden, E., C. J. Ebinger, G. Yirgu, A. L. Deino, and D. Ayalew. 2004. Evolution of the northern Main Ethiopian rift: Birth of a triple junction. *Earth and Planetary Science Letters* 224:213–228.

Yemane, K., R. Bonnefille, and H. Faure. 1985. Palaeoclimatic and tectonic implications of Neogene microflora from the northwestern Ethiopian highlands. *Nature* 318:653–656.

Yemane, K., C. Robert, and R. Bonnefille. 1987. Pollen and clay mineral assemblages of a Late Miocene lacustrine sequence from the northwestern Ethiopian highlands. *Palaeogeography, Palaeoclimatology, Palaeoecology* 60:123–141.

Zachos, J. C., M. Pagani, L. C. Sloan, E. Thomas, and K. Billups. 2001. Trends, rhythms, and aberrations in global climate 65 Ma to present. *Science* 292:686–693.

Zavada, M. S., and S. E. de Villiers. 2000. Pollen of the Asteraceae from the Paleocene-Eocene of South Africa. *Grana* 39:39–45.

THE FAUNA: TAXONOMIC ACCOUNTS

PART ONE

METATHERIA

SIX

Marsupialia

GREGG F. GUNNELL

The fossil record of pouched mammals (cohort Marsupialia) in Africa is extremely limited, although it could be relatively ancient if a tooth described as a marsupial from the Cretaceous of Madagascar truly represents the cohort (Krause, 2001; but see Averianov et al., 2003). Crochet (1984) suggested that the African radiation of marsupials must have extended into the Turonian (late Cretaceous), but there is no fossil evidence outside of the Madagascar tooth to lend support to this contention (Jaeger and Martin, 1984; Bown and Simons, 1984a, 1984b; Jacobs et al., 1988; Flynn et al., 1999).

The Cenozoic record of African marsupials is nearly as limited as that of the Mesozoic. Proposed records include samples from the North African early Eocene (Tunisia and Algeria), late Eocene (Egypt), early Oligocene (Egypt), and from the Arabian peninsula, early Oligocene (Oman).

Systematic Paleontology

Cohort MARSUPIALIA Illiger, 1811
Order DIDELPHIMORPHIA Gill, 1872
Family HERPETOTHERIIDAE Trouessart, 1879
Subfamily HERPETOTHERIINAE Trouessart, 1879
Genus *PERATHERIUM* Aymard, 1850
PERATHERIUM AFRICANUM Simons and Bown, 1984
Figure 6.1A

Synonyms Peratherium africanus, Simons and Bown, 1984; Didelphidae, Bown and Simons, 1984a; *Qatranitherium africanum*, Crochet et al., 1992; *Qatranitherium* aff. *africanum*, Crochet et al., 1992; *Quatranitherium*, Szalay, 1994; *Peratherium africanum*, Hooker et al., 2008.

Diagnosis Modified from Simons and Bown, 1984, Crochet et al., 1992, and Hooker et al., 2008. Moderate-sized species of *Peratherium*; M1—3 transversely broad with deep ectoflexus; upper molar metacone larger than paracone; stylar cusp B large and confluent with cusp A; m1—2 with entoconid close to trigonid and connected to it by a high entocristid; m1—3 talonids short; m1–3 hypoconulids vertical, as large as and taller than entoconids; molars increase in size posteriorly; p1—2 diastema small; p2 with distinct paraconid.

Description The holotype has m1—3 in place and clearly shows an alveolus for m4. One of the referred specimens from Quarry M (DPC 3120) preserves a medially inflected mandibular angular process, another marsupial character state. An upper dentition described by Hooker et al. (2008) clearly shows that *Peratherium africanum* was dilambdodont and that it possessed enlarged upper molar metaconids, both herpetotheriid characters.

Age Early Oligocene, Rupelian (31.4–29.8 Ma; Seiffert, 2006).

African Occurrence Fayum Depression, Egypt, and Oman, Arabian Peninsula.

Remarks The type (CGM 40236) and referred specimens of *Peratherium africanum* from Fayum Quarry M are the only certain Cenozoic marsupials known from Africa. *Peratherium africanum* also is represented by more complete specimens (including both upper and lower dentitions) than any of the other potential African marsupials (isolated teeth and a single jaw fragment with two teeth). Simons and Bown (1984) placed these specimens in the genus *Peratherium*, therefore implying that they belonged in the subfamily Herpetotheriinae. Crochet et al. (1992) proposed a new generic name, *Qatranitherium*, for the Quarry M specimens along with a single left dp3 from Taqah in Oman, and placed this taxon in the family Peradectidae. Their new genus was distinguished from *Peratherium* (and thus from herpetotheriines) by having lower molars that increase in length posteriorly, relatively short molar talonids and labially placed cristid obliquae, by the presence of a p1–2 diastema, and by the presence of high and subequal entoconid and hypoconulid. Hooker et al. (2008) described additional marsupial specimens from Quarry M including upper dentitions that indicate a close relationship between the Fayum taxon and *Peratherium*, in particular with *P. lavergnense* known from the late Bartonian (MP16) to early Rupelian (MP21) in France. These authors place *Qatranitherium* within *Peratherium* as a junior synonym and make a convincing case that these Fayum specimens represent a herpetotheriine rather than a peradectid.

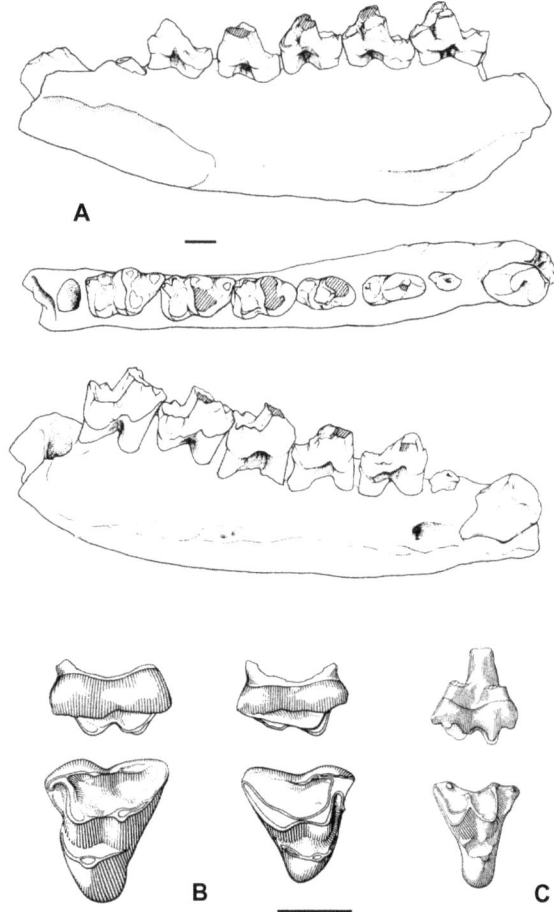

FIGURE 6.1 A) *Peratherium africanum*, right dentary c1-m3 (holotype, CGM 40236), in medial, occlusal, and lateral views; adapted from Simons and Bown, 1984; permission for use granted by the American Society of Mammalogists and Allen Press, Inc. B) *Kasserinotherium tunisiense* left M3 (holotype, EY 10, on left) and right M1 (EY 12, on right) in buccal and occlusal views; adapted from Crochet, 1986. C) *Garatherium mahboubii*, right M3 (holotype) in buccal and occlusal views; adapted from Mahboubi et al., 1983. Scale bars = 1 mm. B, C permission courtesy of Comptes Rendus de l'Academie des Sciences, Paris.

Cohort ?MARSUPIALIA Illiger, 1811
Genus *GARATHERIUM* Crochet, 1984
GARATHERIUM MAHBOUBII Crochet, 1984
Figure 6.1C

Diagnosis Crochet (1984). Stylar cusp B slightly lower than paracone but taller than cusp A; stylar cusp C poorly defined; preparacrista concave with the lowest portion placed approximately equally distant from stylar cusp B and the paracone; conules distinct, extending from the walls of the paracone and metacone; protocone high; protofossa deep and narrow.

Description The holotype (and only specimen) of *Garatherium mahboubii* is a single upper molar from the early Eocene of Algeria. The lingual part of the tooth is very restricted anteroposteriorly, and there is no trace of a hypocone. The conules are low and small, and the stylar cusps small and not basally inflated.

Age Late early Eocene, Ypresian (50 Ma).

African Occurrence El Kohol, Algeria, and Bassin d'Ouarzazate, Morocco.

Remarks *Garatherium* was originally described as a peradectine didelphid (Mahboubi et al., 1983; Crochet, 1984). Gheerbrant (1995) described an additional tooth of a probable new species of *Garatherium* from Paleocene deposits in Morocco and argued that both the original and new species of *Garatherium* were not marsupials but adapisoriculid lipotyphlans instead. It is doubtful that *Garatherium* represents a marsupial whatever its true affinities are. With nothing more than two isolated teeth to go on, its affinities remain vague.

Genus *KASSERINOTHERIUM* Crochet, 1986
KASSERINOTHERIUM TUNISIENSE Crochet, 1986
Figure 6.1B

Diagnosis Small size; upper molars lacking distinct stylar cusps; not dilambdodont; conules absent.

Description The two known upper molars have moderate stylar shelves but lack development of any distinct stylar cusps. The paracone and metacone are very basally inflated, and the protocone is set close to the bases of the buccal cusps, leaving a very small protofossa.

Age Early Eocene, Ypresian (52 Ma).

African Occurrence Chambi, Tunisia.

Remarks *Kasserinotherium tunisiense* is known by only two upper teeth (Crochet, 1986). McKenna and Bell (1997) were unable to place this taxon among marsupials, instead assigning it to the Supercohort Theria along with several other enigmatic taxa. Given the primitive and simple nature of these teeth, it is difficult to know where *Kasserinotherium* belongs, and there is no compelling reason to believe it represents a marsupial. Like *Garatherium*, the lower teeth of *Kasserinotherium* remain unknown.

Genus *GHAMIDTHERIUM* Sánchez-Villagra et al., 2007
GHAMIDTHERIUM DIMAIENSIS Sánchez-Villagra et al., 2007

Diagnosis Lower molars with trigonids open and uncompressed; metaconids low relative to protoconids but slightly taller than paraconids; hypoconulid lingual and relatively small, connected to hypoconid; buccal cingulids weak and noncontinuous; precingulids well developed.

Description The holotype specimen of *Ghamidtherium* is a right lower dentary that includes two molars and alveoli for double-rooted teeth anterior and posterior to these molars. Sánchez-Villagra et al. (2007) interpreted the preserved teeth as m2–3 with the alveoli being for m1 and m4, thus suggesting that *Ghamidtherium* was a marsupial. An additional lower molar from the same locality as the holotype was interpreted as a possible m1 of *Ghamidtherium* by Sánchez-Villagra et al. (2007). It is similar to the two molars in the holotype, but the trigonid is more widely open and is narrower than the talonid. However, it differs from the molars of *Ghamidtherium* in having a postcristid that extends to the base of the entoconid and not to the hypoconulid. This could simply represent variation within the molar series, or it could suggest that the isolated tooth is from another, similar taxon.

Age Late Eocene, Priabonian (37 Ma; Seiffert, 2006).

African Occurrence Fayum Depression, Locality BQ-2, Egypt.

Remarks In addition to the holotype and referred lower molar of *Ghamidtherium dimaiensis*, Sánchez-Villagra et al. (2007) also described two upper molars from BQ-2. The upper teeth do not belong to *G. dimaiensis* but conceivably could belong to a taxon in the same clade. In some features, the upper molars resemble those of bats (as do the lower molars), and it is possible that *Ghamidtherium* and its potential relatives should all be assigned to Chiroptera instead of being included within Marsupialia. Examination of enamel micro-structure in the referred possible m1 was equivocal as well (Sánchez-Villagra et al., 2007), with some features being more commonly found within marsupials and others being more typical of lipotyphlans and chiropterans. More complete specimens of all taxa are required to be more certain of the affinities of any of these animals from BQ-2.

General Discussion and Summary

The only confirmed occurrence of a marsupial in the African Cenozoic is that of *Peratherium africanum* from Quarry M in the Jebel Qatrani Formation, Fayum Depression, Egypt. *Peratherium* also may have been present on the Arabian Peninsula at Taqah in Oman. The fact that *P. africanum* closely resembles some European *Peratherium* species indicates that some faunal interchange was occurring between Europe and North Africa in the early Oligocene (Simons and Bown, 1984; Hooker et al., 2008). Other, earlier records from Egypt, Tunisia and Algeria either are not likely to represent marsupials or cannot be confidently assigned to that group. If there was a long record of marsupial evolution in Africa as proposed by Crochet (1984), it remains hidden at this time.

ACKNOWLEDGMENTS

I thank Lars Werdelin and Bill Sanders for the invitation to participate in this volume. I have benefited from discussions with Erik Seiffert, Bernard Sigé, Jerry Hooker, and Thierry Smith. An anonymous reviewer improved the manuscript.

Literature Cited

Averianov, A. O., J. D. Archibald, and T. Martin. 2003. Placental nature of the alleged marsupial from the Cretaceous of Madagascar. *Acta Palaeontologica Polonica* 48:149–151.

Bown, T. M., and E. L. Simons. 1984a. First record of marsupials (Metatheria: Polyprotodonta) from the Oligocene in Africa. *Nature* 308:447–449.

———. 1984b. African marsupials—vicariance or dispersion? Reply. *Nature* 312:379–380.

Crochet, J.-Y. 1984. *Garatherium mahboubii* nov. gen., nov. sp., marsupial de l'Éocène inférieur d'El Kohol (Sud-Oranais, Algérie). *Annales de Paléontologie* 70:275–294.

———. 1986. *Kasserinotherium tunisiense* nov. gen., nov. sp., troisième marsupial découvert en Afrique (Eocène inférieur de Tunisie). *Comptes Rendus de l'Académie des Sciences, Paris* 302:923–926.

Crochet, J.-Y., H. Thomas, S. Sen, J. Roger, E. Gheerbrant, and Z. Al-Sulaimani. 1992. Découverte d'un Péradectidé (Marsupialia) dans l'Oligocène inférieur du Sultanat d'Oman: Nouvelles données sur la paléobiogéographie des Marsupiaux de la plaque arabo-africaine. *Comptes Rendus de l'Académie des Sciences, Paris* 314:539–544.

Flynn, J. J., J. M. Parrish, B. Rakotosamimanana, W. F. Simpson, and A. R. Wyss. 1999. A middle Jurassic mammal from Madagascar. *Nature* 401:57–60.

Gheerbrant, E. 1995. Les mammifères Paléocènes du Bassin D'Ouarzazate (Maroc). III. Adapisoriculidae et autres mammifères (Carnivora, ?Creodonta, Condylarthra, ?Ungulata et incertae sedis). *Palaeontographica, Abt. A* 237:39–132.

Hooker, J. J., M. R. Sánchez-Villagra, F. G. Goin, E. L. Simons, Y. Attia, and E. R. Seiffert. 2008. The origin of Afro-Arabian "didelphimorph" marsupials. *Palaeontology* 51:635–648.

Jacobs, L. L., J. D. Congleton, M. Brunet, J. Dejax, L. J. Flynn, J. V. Hell, and G. Mouchelin. 1988. Mammal teeth from the Cretaceous of Africa. *Nature* 336:158–160.

Jaeger, J.-J., and M. Martin. 1984. African marsupials: Vicariance or dispersion? *Nature* 312:379.

Krause, D. W. 2001. Fossil molar from a Madagascan marsupial. *Nature* 412:497–498.

Mahboubi, M., R. Ameur, J.-Y. Crochet, and J.-J. Jaeger. 1983. Première découverte d'un marsupial en Afrique. *Comptes Rendus de l'Académie des Sciences, Paris* 297:691–694.

McKenna, M. C., and S. Bell. 1997. *Classification of Mammals above the Species Level*. Columbia University Press, New York, 631 pp.

Sánchez-Villagra, M. R., E. R. Seiffert, T. Martin, E. L. Simons, G. F. Gunnell, and Y. Attia. 2007. Enigmatic new mammals from the late Eocene of Egypt. *Paläontologische Zeitschrift* 81:406–415.

Seiffert, E. R. 2006. Revised age estimates for the later Paleogene mammal faunas of Egypt and Oman. *Proceedings of the National Academy of Sciences, USA* 103:5000–5005.

Simons, E. L., and T. M. Bown. 1984. A new species of *Peratherium* (Didelphidae; Polyprotodonta): the first African marsupial. *Journal of Mammalogy* 65:539–548.

Szalay, F. S. 1994. *Evolutionary History of the Marsupials and an Analysis of Osteological Characters*. Cambridge University Press, Cambridge, 481 pp.

PART TWO

AFROTHERIA

SEVEN

Ptolemaiida

GREGG F. GUNNELL, PHILIP D. GINGERICH,
AND PATRICIA A. HOLROYD

Ptolemaiidae is an enigmatic family of fossil mammals that has proven difficult to classify. Osborn (1908) proposed the family based on *Ptolemaia lyonsi*, represented by a single lower jaw from Oligocene deposits in the Fayum Depression, Egypt. Osborn declined to place this family in a higher taxon, noting instead that it likely represented a new mammalian order. Schlosser (1910, 1911) suggested that ptolemaiids might be best placed in the order Creodonta, a notion unenthusiastically supported by Matthew (1918). Schlosser (1922) later moved *Ptolemaia* to the Pantolestidae, an idea initially endorsed by Van Valen (1966) but later questioned (Van Valen, 1967). Butler (1969) noted that the deciduous teeth described by Schlosser (1911) as *Ptolemaia* (now *Qarunavus*) shared several features in common with Miocene elephant shrews (Macroscelidea). Simons and Gingerich (1974) named an additional ptolemaiid genus, *Qarunavus*, and noted similarities between ptolemaiids, pantolestids, and tubulidentates. Bown and Simons (1987) added a third ptolemaiid genus, *Cleopatrodon*, and placed the family provisionally within Pantolesta (*sensu* McKenna, 1975). Russell and Godinot (1988) suggested that ptolemaiids might be related to paroxyclaenids (specifically the subfamily Merialinae), a European family of Pantolesta. Simons and Bown (1995) erected a new mammalian order, Ptolemaiida, for the family Ptolemaiidae and stated that it may trace its ancestry to Pantolesta but was sufficiently distinct from any other group of mammals to warrant separate ordinal status. Gagnon (1997) classed ptolemaiids as insectivores without discussion, while McKenna and Bell (1997) placed Ptolemaiidae in the order Cimolesta and suborder Pantolesta. Recently, Nishihara et al. (2005) have suggested that ptolemaiids, given their possible relationships with macroscelideans and tubulidentates, may ultimately be included in Afrotheria. Cote et al. (2007) also supported inclusion of Ptolemaiida within Afrotheria, perhaps as sister group to an Afroinsectiphillia clade (including aardvarks, elephant shrews, golden moles, and tenrecs). Seiffert (2007), in a broader analysis of afrotherians, found *Kelba* to be best placed as a stem member of Tubulidentata, clearly nesting within the afrotherian radiation. Ptolemaiida are known from Eocene and Oligocene deposits in Egypt and early Miocene localities in Kenya and Uganda (see table 7.1).

Systematic Paleontology

Order PTOLEMAIIDA Simons and Bown, 1995
Family PTOLEMAIIDAE Osborn, 1908

Included Genera Ptolemaia, Cleopatrodon.

Genus *PTOLEMAIA* Osborn, 1908
Figure 7.1B

Diagnosis Family Ptolemaiidae differs from Kelbidae in having a P4 metaconule present, short upper molars lacking mesial and distal cingula, p4 lacking a cristid obliqua but having a large hypoconid, and lower molars with constricted trigonids and short talonids. *Ptolemaia* differs from other ptolemaiids and *Qarunavus* in having molars decreasing in size progressively from anterior to posterior; further differs from *Qarunavus* in having undifferentiated paraconids and metaconids on lower molars and lacking buccal and lingual basal cingulids; differs from *Cleopatrodon* in having mandibular ascending ramus angled posteriorly, in lacking p1 (often p2 as well), in having a diastema between c1 and anterior-most premolar, relatively larger p3–4, and less well-defined molar cusps; differs from *Kelba* in possessing antero-posteriorly compressed upper premolars and molars, in having P3 lacking protocone, more bunodont, and relatively larger premolars.

Description *Ptolemaia* was a moderate sized mammal with enlarged P4/p4 and M1/m1 (m1 ranges from 7.75 to 9.6 mm in length). The premolars increase in size posteriorly, while molars decrease in size posteriorly. The single known skull (Simons and Bown, 1995) is long (approximately 178 mm) and low, with an elongate snout and a relatively small braincase. There are no postcranial remains of ptolemaiids known.

Age Early Oligocene, Rupelian (33.7–31.3 Ma).

African Occurrence Fayum Depression, Egypt only.

Remarks *Ptolemaia* includes two species, the genotype species *P. lyonsi* and *P. grangeri*. *Ptolemaia lyonsi* differs from *P. grangeri* in being somewhat smaller, in lacking p2, and in having a relatively smaller m3 that retains two roots and has a lower and weaker talonid. *P. lyonsi* is known for certain only by the holotype specimen. A second specimen (YPM 18117) was

TABLE 7.1

Occurrences and ages of Ptolemaiida

Taxon	Occurrence (Site, Location)	Stratigraphic Unit	Age	Primary Reference
		PTOLEMAIIDAE		
Ptolemaia lyonsi	Quarry A, Egypt	Jebel Qatrani Fm., Lower Sequence	Early Oligocene (33.7 Ma)	Osborn, 1908
Ptolemaia grangeri	Quarry V, Egypt	Jebel Qatrani Fm., Middle Sequence	Early Oligocene (31.3 Ma)	Bown & Simons, 1987
Cleopatrodon ayeshae	Quarry V, Egypt	Jebel Qatrani Fm., Middle Sequence	Early Oligocene (31.3 Ma)	Bown & Simons, 1987
Cleopatrodon robusta	Quarry I, Egypt	Jebel Qatrani Fm., Upper Sequence	Early Oligocene (29.8 Ma)	Bown & Simons, 1987
		KELBIDAE		
Kelba quadeemae	Meswa Bridge, Kenya	Muhoroni Agglomerate	Early Miocene (22 Ma)	Cote et al., 2007
	Songhor, Kenya		Early Miocene (19.5 Ma)	Cote et al., 2007
	Legetet, Kenya	Legetet Fm.	Early Miocene (19.5 Ma)	Cote et al., 2007
	Chamtwara, Kenya		Early Miocene (19.5 Ma)	Cote et al., 2007
	Napak, Uganda		Early Miocene (19.5 Ma)	Savage, 1965
	Rusinga Island, Kenya	Hiwegi Fm.	Early Miocene (18.3 Ma)	Savage, 1965
	Mfwanganu Island, Kenya	Hiwegi Fm.	Early Miocene (18.3 Ma)	Savage, 1965
	BPRP Site 38, Kenya		Middle Miocene (12.5 Ma)	Cote et al., 2007
		FAMILY UNCERTAIN		
Qarunavus meyeri	Quarry A, Egypt	Jebel Qatrani Fm., Lower Sequence	Early Oligocene (33.7 Ma)	Simons & Gingerich, 1974
Ptolemaiid indet.	Fayum Quarry L-41, Egypt	Jebel Qatrani Fm., Lower Sequence	Late Eocene (34 Ma)	Gagnon, 1997
Ptolemaiid indet.	Fayum Quarry BQ-2, Egypt	Birket Qarun Fm.	Late Eocene (37 Ma)	Cote et al., 2007

referred to this species by Simons and Gingerich (1974), but it differs from *P. lyonsi* in retaining p2, like in *P. grangeri*, and might be better assigned to this latter species.

Genus *CLEOPATRODON* Bown and Simons, 1987
Figure 7.1C

Diagnosis Differs from *Ptolemaia* in having a relatively longer jaw lacking anterior diastema, more vertical ascending ramus, retaining p1–2, p3–4 relatively smaller, p3 with only two cusps, m1–2 equivalent in size, m3 larger; differs from *Qarunavus* in having closely appressed molar paraconids and metaconids, shallower trigonid fovea and talonid basin, lacking large lingual talonid notches on m1–2, lacking molar precingulids, and having relatively wider molars.

Description Cleopatrodon ranges in size from that of *P. lyonsi* to somewhat larger, but both species, *C. ayeshae* and

C. robusta, possess more robust posterior premolars and molars than those of *Ptolemaia*.

Age Early Oligocene, Rupelian (31.3–29.8 Ma).

African Occurrence Fayum Depression, Egypt only.

Remarks Cleopatrodon ayeshae is one of the better-known ptolemaiids, being represented by 18 specimens from Quarry V. In fact, of the 47 known specimens representing the family Ptolemaiidae, 38 come from Quarry V (the other 20 representing *P. grangeri*). Quarry V is unique among Fayum localities in preserving fossils in a hardened sandstone (Bown and Simons, 1987). This unusual depositional environment suggests that the relatively high abundance of ptolemaiids at Quarry V may be the result of sampling derived from relatively rare paleohabitats. In turn, this indicates that ptolemaiids may have preferred habitats different from those typically sampled in the Fayum sequence.

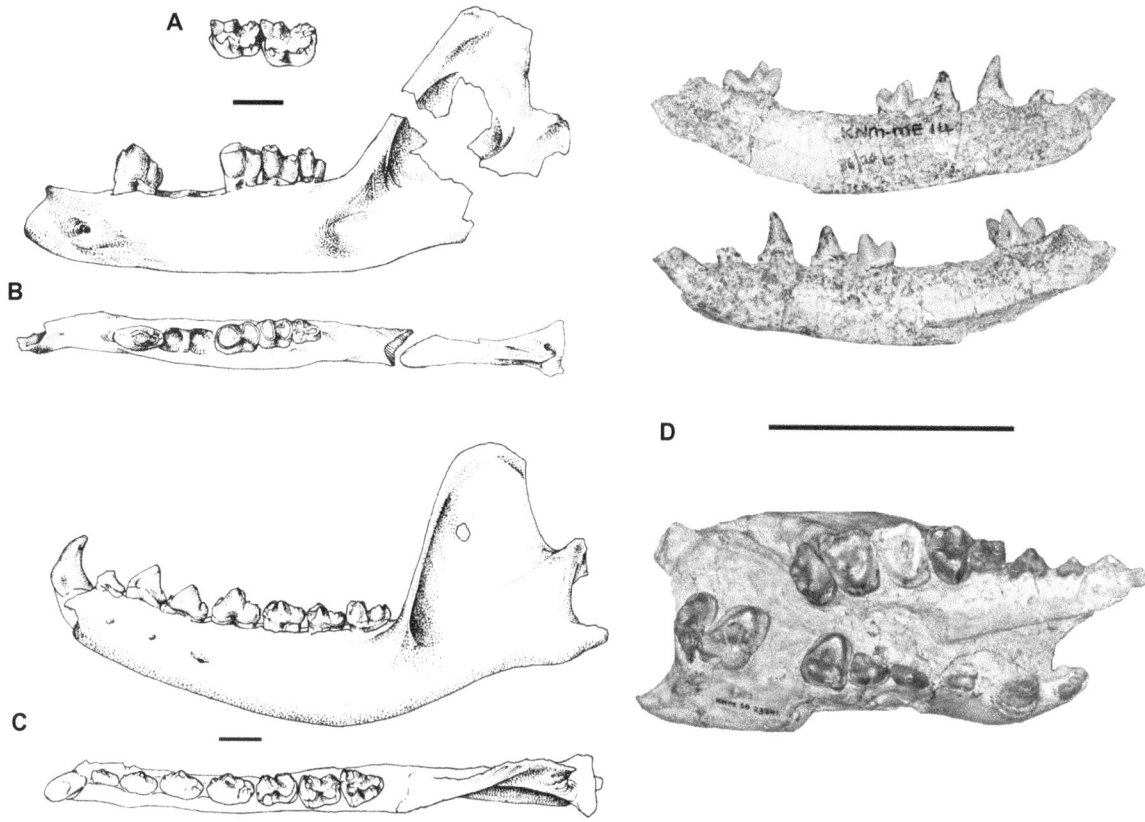

FIGURE 7.1 Ptolemaiida. A) *Qarunavus meyeri*, left m1–2 (part of holotype, BMNH M-10189), in occlusal view. B) *Ptolemaia lyonsi*, left dentary p3, m1–3 (holotype, AMNH 13269) in lateral and occlusal views. C) *Cleopatrodon ayeshae*, left dentary c1–m3 (holotype, CGM 40242) in lateral and occlusal views. D) *Kelba quadeemae*, left dentary c1–p4, m3 (KNM ME 14) in medial and lateral views and palate with right P1–M3 and left I1–P4, M2–3 (KNM SO 23296) in occlusal view. Scale bars in A–C = 10 mm; scale bar in D = 5 cm. Figures 7.1A–7.1C adapted from Bown and Simons (1987), © Copyright 2008 The Society of Vertebrate Paleontology. Reprinted and distributed with permission of the Society of Vertebrate Paleontology. Figure 7.1D photographs courtesy of L. Werdelin.

Family KELBIDAE Cote et al., 2007

Included Genera Kelba.

Genus *KELBA* Savage, 1965
Figure 7.1D

Diagnosis Modified from Cote et al. (2007). Family Kelbidae differs from Ptolemaiidae in having upper molars with mesial and distal cingula well developed, paracone and metacone widely separated, premetacrista oriented labially, postprotocrista extending to lingual flank of metacone, and mesostyles present on M1–2; p4 cristid obliqua sharply defined and talonid basin relatively large; cuspules developed on lower molar entocristids (also present in *Qarunavus*). Further differs from other ptolemaiidans in having less bunodont premolars (except for p4/P4) and lower molars, in having p2 anteriorly inclined with low anterior shelf, in having posterior premolars smaller relative to molars, in having diastemata between canine and anterior lower premolars, and in lacking extreme lower molar trigonid constriction and vespiform molar shape. Further differs from *Ptolemaia* in lacking extreme premolar hypertrophy and progressive reduction of molar size posteriorly, in having lingually longer upper premolars and molars (not compressed anteroposteriorly like *Ptolemaia*, especially P4 and M2), in having P3 with a small, posterolingually placed protocone that is appressed to the base of the paracone, in having P4 with a well-developed

metacone, and in having upper molars with a pericone. Is similar to *Cleopatrodon* in having molariform p4 but differs in having a more developed p4 talonid basin; similar to *Qarunavus* in having well-developed lower molar trigonids with distinct paraconid and metaconid, but differs in having trigonid distinctly separated from talonid basin by high posterior trigonid wall (postvallid), unlike *Qarunavus*.

Description When originally described by Savage (1965) *Kelba quadeemae* was based on three isolated upper molars from three separate early Miocene localities (Rusinga and Mfwanganu Islands in Kenya, and Napak in Uganda). Savage described the holotype as representing a right M2, interpreting the tooth to have an expanded parastyle and a distinct hypocone. New specimens of *Kelba* described by Cote et al. (2007) include a palate and a nearly complete lower jaw. The palate of *Kelba* from Songhor shows that the holotype is in fact a left M2 (possibly M1) with an enlarged metastyle and a distinct pericone instead of hypocone.

Age Early to middle Miocene (22–12.5 Ma).

African Occurrence Meswa Bridge, Songhor, Legetet, Chamtwara, Rusinga Island, Mfwanganu Island, and BPRP Site 38, all in Kenya; Napak, Uganda.

Remarks Savage (1965) placed *Kelba* in the condylarthran family Arctocyonidae, which he believed belonged in Creodonta. Van Valen (1967) suggested that *Kelba* should be placed in Ptolemaiidae but also thought it could be a pantolestid. Ultimately in his classification (Van Valen, 1967:260), he included *Kelba* within ptolemaiids. Savage

(1978), noting that a complete upper dentition was now known (although undescribed), concurred that *Kelba* might be closely related to ptolemaiids but did not rule out possible pantolestid or "aberrant condylarth" affinities (Savage, 1978:251). In describing upper teeth of Paleogene ptolemaiids for the first time, Bown and Simons (1987) made comparisons with *Kelba* but chose not to place it definitively within Ptolemaiidae. Recent work by Cote et al. (2007) provides convincing evidence that *Kelba* represents a distinct family within Ptolemaiida, Kelbidae. Their phylogenetic analysis suggests that *Kelba* along with an, as yet undescribed, ptolemaiidan from Fayum Quarry BQ-2 (37 Ma) form a clade that is the sister group to a clade (family Ptolemaiidae) formed by *Ptolemaia* and *Cleopatrodon*. The phylogenetic position of *Qarunavus* was unresolved in their analysis due to the subadult nature of the known specimens.

Family Indet.
Genus *QARUNAVUS* Simons and Gingerich, 1974
Figure 7.1A

Diagnosis Differs from *Ptolemaia* in having a reduced number of incisors, well-developed lower molar trigonids with distinct paraconids and metaconids, and distinct buccal and lingual basal cingulids; differs from *Cleopatrodon* in having narrower molars relative to length, distinct, large, lingual talonid notches, relatively longer talonids, and well-developed precingulids on lower molars.

Description It is difficult to make direct comparisons between *Qarunavus meyeri* and other ptolemaiids because the only known specimens were relatively young individuals with none of the permanent dentition anterior to m1 erupted (e.g., *Kelba* and *Qarunavus* have little basis for comparison because known specimens do not preserve any teeth in common). The first two lower molars are similar in size and have distinct paraconids, relatively large and lingual hypoconulids, and distinct anterior and buccal cingulids.

Age Early Oligocene, Rupelian (33.7 Ma).

African Occurrence Fayum Depression, Egypt only.

Remarks The holotype (BMNH M-10189) of *Q. meyeri* is a left dentary with dp2–4, m1–2 that was collected from Fayum Quarry A. Schlosser (1911) figured a specimen from Quarry A that he assigned to *P. lyonsi*. This specimen consists of a right dentary preserving the same teeth as the holotype of *Q. meyeri* and resembling it in nearly every detail. There is a good chance that these specimens may be from the same individual, but they have not been directly compared to see if this determination could be made. Schlosser (1911) noted the presence of other specimens from Quarry A that he attributed to *P. lyonsi*. These may also represent *Q. meyeri* but await additional study.

ADDITIONAL PTOLEMAIIDA SPECIMENS

Sudre et al. (1993) described an isolated lower premolar from the latest Paleocene locality of Talazit in Morocco as an indeterminate arctocyonid condylarthran. Although relatively small (5 mm in length) and heavily worn, the tooth is not unlike that of p3–4 in known ptolemaiids and could potentially represent the earliest occurrence of the group. If ptolemaiids are afrotheres as has been suggested (Nishihara et al., 2005; Seiffert, 2007; Cote et al., 2007), then a late Paleocene presence in North Africa is certainly plausible.

Gagnon (1997) noted the presence of a ptolemaiid from the late Eocene Quarry L-41 (34 Ma), near the base of the lower sequence of the Jebel Qatrani Formation. Isolated teeth of ptolemaiids have also been recovered from locality BQ-2 (37 Ma) in the Birket Qarun Formation at a horizon slightly lower than Quarry L-41 (Seiffert et al., 2008). The BQ-2 specimens have been included within the family Kelbidae by Cote et al. (2007); however, none of the material has been described as yet, so we prefer to place it in family uncertain until descriptions have been provided.

General Discussion

Ptolemaiids have a restricted distribution both temporally and geographically. They are best known from an approximately four-million-year time interval in the early Oligocene from three stratigraphically successive quarries in the Jebel Qatrani Formation (Seiffert, 2006), Fayum Depression, Egypt. Quarry A (33.7 Ma) contains the oldest taxonomically assignable records with two species (*Ptolemaia lyonsi* and *Qarunavus meyeri*) represented by four (possibly only three) specimens. Quarry V (31.3 Ma) has the largest number of ptolemaiid specimens ($n = 38$) from any of the quarries and also contains two taxa, *Ptolemaia grangeri* and *Cleopatrodon ayeshae*. The latest Egyptian occurrence of ptolemaiids is at Fayum Quarry I (29.8 Ma), represented by two specimens of *Cleopatrodon robusta* (Bown and Simons, 1987). Ptolemaiids are extremely rare in the African Miocene, represented by 10 specimens from Kenya and 1 specimen from Uganda. All of these Miocene specimens were assigned to *Kelba quadeemae* by Cote et al. (2007), but these authors did note that the sample was quite variable and might represent more than one species.

Bown (1982) described a trace fossil (mammalian burrow) that he speculated might have been made by a ptolemaiid. The burrow intersects a fossil termite nest and suggests that whoever dug the burrow did so in search of termites. This evidence suggests either that ptolemaiidans were partially myrmecophagous or that the burrow was produced by another mammal (perhaps a tubulidentate, if, in fact, ptolemaiids turn out to not be stem tubulidentates; see Holroyd, this volume, chap. 10).

Ptolemaiids remain difficult to classify. They are very poorly represented in general; even where they are relatively abundant (Quarry V), they are still only known by fragmentary teeth and jaws for the most part. A definitive understanding of their phylogenetic relationships and functional morphology awaits more complete specimens.

ACKNOWLEDGMENTS

We thank Lars Werdelin and Bill Sanders for the invitation to participate in this volume. We also thank an anonymous reviewer for helpful comments.

Literature Cited

Bown, T. M. 1982. Ichnofosssils and rhizoliths of the nearshore fluvial Jebel Qatrani Formation (Oligocene), Fayum Province, Egypt. *Palaeogeography, Palaeoclimatology, Palaeoecology* 40:255–309.

Bown, T. M., and E. L. Simons. 1987. New Oligocene Ptolemaiidae (Mammalia: ?Pantolesta) from the Jebel Qatrani Formation, Fayum Depression, Egypt. *Journal of Vertebrate Paleontology* 7:311–324.

Butler, P. M. 1969. Insectivores and bats from the Miocene of East Africa: New material; pp. 1–37 in L. S. B. Leakey (ed.), *Fossil Vertebrates of Africa*, vol. 1. Academic Press, New York.

Cote, S., L. Werdelin, E. R. Seiffert, and J. C. Barry. 2007. Additional material of the enigmatic early Miocene mammal *Kelba* and its relationship to the order Ptolemaiida. *Proceedings of the National Academy of Sciences, USA* 104:5510–5515.

Gagnon, M. 1997. Ecological diversity and community ecology in the Fayum sequence (Egypt). *Journal of Human Evolution* 32:133–160.

Matthew, W. D. 1918. A revision of the Lower Eocene Wasatch and Wind River faunas. Part V. Insectivora (continued), Glires, Edentata. *Bulletin of the American Museum of Natural History* 38:565–657.

McKenna, M. C. 1975. Toward a phylogenetic classification of the Mammalia; pp. 21–46 in P. W. Luckett and F. S. Szalay (eds.), *Phylogeny of the Primates*. Plenum Press, New York.

McKenna, M. C., and S. K. Bell. 1997. *Classification of Mammals above the Species Level*. Columbia University Press, New York, 631 pp.

Nishihara, H., Y. Satta, M. Nikaido, J. G. M. Thewissen, M. J. Stanhope, and N. Okada. 2005. A retroposon analysis of afrotherian phylogeny. *Molecular Biology and Evolution* 22:1823–1833.

Osborn, H. F. 1908. New fossil mammals from the Fayum Oligocene, Egypt. *Bulletin of the American Museum of Natural History* 24:265–272.

Russell, D. E., and M. Godinot. 1988. The Paroxyclaenidae (Mammalia) and a new form from the early Eocene of Palette, France. *Paläontologische Zeitschrift* 62:319–331.

Savage, R. J. G. 1965. Fossil mammals of Africa: 19. The Miocene Carnivora of East Africa. *Bulletin of the British Museum (Natural History), Geology* 10:239–318.

———. 1978. Carnivora; pp. 249–267 in V. J. Maglio and H. B. S. Cooke (eds.), *Evolution of African Mammals*. Harvard University Press, Cambridge.

Schlosser, M. 1910. Über einige fossile Säugetiere aus dem Oligocän von Ägypten. *Zoologische Anzeiger* 35:500–508.

———. 1911. Beiträge zur Kenntnis der oligozänen Landsäugetiere aus dem Fayum, Ägypten. *Zeitschrift für Paläontologie und Geologie, Oesterreich-Ungarns und des Orients* 24:51–167.

———. 1922. Mammalia; pp. 402–689 in K. A. von Zittel (ed.), *Grundzüge der Paläontologie (Paläozoologie), II. Abteilung: Vertebrata*. 4th ed. Oldenbourg, Munich.

Seiffert, E. R. 2006. Revised age estimates for the later Paleogene mammal faunas of Egypt and Oman. *Proceedings of the National Academy of Sciences, USA* 103:5000–5005.

———. 2007. A new estimate of afrotherian phylogeny based on simultaneous analysis of genomic, morphological, and fossil evidence. *BMC Evolutionary Biology* 7:224; doi:10.1186/1471-2148-7-224.

Seiffert, E. R., T. M. Bown, W. C. Clyde, and E. L. Simons. 2008. Geology, paleoenvironment, and age of Birket Qarun Locality 2 (BQ-2), Fayum Depression, Egypt; pp. 71–86 in J. G. Fleagle and C. C. Gilbert (eds.), *Elwyn Simons: A Search for Origins, Developments in Primatology: Progress and Prospects*. Springer, New York.

Simons, E. L., and T. M. Bown. 1995. Ptolemaiida, a new order of Mammalia—with description of the cranium of *Ptolemaia grangeri*. *Proceedings of the National Academy of Sciences, USA* 92:3269–3273.

Simons, E. L., and P. D. Gingerich. 1974. New carnivorous mammals from the Oligocene of Egypt. *Annals of the Geological Survey of Egypt* 4:157–166.

Sudre, J., J.-J. Jaeger, B. Sigé, and M. Vianey-Liaud. 1993. Nouvelles données sur les Condylarthres du Thanétien et de l'Yprésien du Bassin D'Ouarzazate (Maroc). *Geobios* 26:609–615.

Van Valen, L. 1966. Deltatheridia, a new order of mammals. *Bulletin of the American Museum of Natural History* 132:1–126.

———. 1967. New Paleocene insectivores and insectivore classification. *Bulletin of the American Museum of Natural History* 135:217–284.

EIGHT

Macroscelidea

PATRICIA A. HOLROYD

The distribution and broader relationships of the Macroscelidea, or sengis, have been the subject of two recent reviews (Butler, 1995; Holroyd and Mussell, 2005) that discuss their interrelationships and the relationships of the order to other placentals.

This contribution focuses on the stratigraphic and geographic distribution of the individual genera within the order and summarizes the morphologic characteristics that distinguish the subfamilies and genera that constitute the Macroscelidea. All known sengis are classified in the order Macroscelidea and family Macroscelididae. Currently, the order is comprised exclusively of fossil and recent taxa that occur in Africa, although some Holarctic taxa may ultimately be recognized as stem-macroscelideans or as more closely related to sengis than to other extant taxa (see discussions in Simons et al., 1991; Tabuce et al., 2001, 2007; Asher et al., 2003; Zack et al., 2005). Noncrown forms are first recognized in the Eocene of North Africa. Taxa assignable to the extant subfamilies are subsequently known from the latest Paleogene to Pleistocene from East and southern Africa (table 8.1, figure 8.1) with scant undescribed Miocene-Pleistocene records in North Africa.

Systematic Paleontology

Order MACROSCELIDEA Butler, 1956
Family MACROSCELIDIDAE Bonaparte, 1838
Subfamily HERODOTIINAE Simons, Holroyd and
Bown, 1991

Diagnosis After Simons et al. (1991) and Tabuce et al. (2001). Differs from other macroscelideans in having brachyodont upper molars with a wide buccal cingulum, P4 incompletely molariform with an anteriorly projecting parastyle, hypocone smaller than protocone on P4–M2, M1–2 with short hypoloph and protoloph, m1–2 with a small paraconid and a low cristid obliqua that does not ascend the posterior trigonid wall; m3 present but very reduced, lower molars with low and labiolingually oriented paracristid, strong p3 entocristid and large p3 hypoconid, and p4 metaconid directly lingual to protoconid.

Included Genera Herodotius (type genus), *Chambius*, *Nementchatherium.*

Genus *NEMENTCHATHERIUM* Tabuce et al., 2001

Type and Only Known Species Nementchatherium senarhense Tabuce et al., 2001.

Diagnosis for Genus and Species From Tabuce et al. (2001). Differs from *Chambius* in having p4 more molarized with a more developed metaconid, P4 with a postprotocrista linked to the distal cingulum, and in lacking conules and postprotocrista on M1–2. Differs from *Herodotius* in smaller size, less molariform p4, P4 with crestiform metacone and lacking hypocone, M1–2 with larger hypocone and higher centrocrista.

Description Nementchatherium senarhense is currently only known from isolated teeth representing p4, m1 or m2, m3, P4, and M1 or M2.

Genus *CHAMBIUS* Hartenberger, 1986

Type and Only Known Species Chambius kasserinensis Hartenberger, 1986.

Diagnosis for Genus and Species Emended based on Simons et al. (1991) and Tabuce et al. (2001). Differs from both *Herodotius* and *Nementchatherium* in having less molariform P4 and in retaining conules and a postprotocrista on M1–2. Differs from *Herodotius* in retaining m1–2 hypoconulid, larger m1 paraconid and m3 talonid, retaining conules on P4–M2.

Description Chambius kasserinensis is currently known from a maxilla bearing P4–M3, dentaries with p4–m3, an astragalus and calcaneum, all from Chambi (Hartenberger, 1986; Tabuce et al., 2007). Dentally, *Chambius* is bunodont with semimolarifom P4/p4 and is phenetically similar to small "condylarths." The astragalus and calcaneum possess characteristics associated with limits on lateral motion and associated with cursorial or saltatorial mammals (Tabuce et al., 2007). Adaci et al. (2007) questionably assign to this genus an isolated m3 from HGL04 at Gour Lazib, Algeria. However, they note that it is morphologically distinct from and more primitive than *C. kasserinensis* and may represent either a new species in this genus or a distinct, new genus.

Genus *HERODOTIUS* Simons, Holroyd, and Bown, 1991
Figure 8.2A

Type and Only Known Species Herodotius pattersoni Simons, Holroyd, and Bown, 1991.

TABLE 8.1
Major occurrences and ages of African sengis

Taxon	Occurrence (Site, Locality)	Stratigraphic Unit	Age
HERODOTIINAE			
Nementchatherium senarhense	Bir El Ater, Algeria		Late middle to late Eocene
Chambius kasserinensis	Chambi, Tunisia		Early Eocene
?Chambius	Gour Lazib, Algeria	Glib Zegdou Fm.	Early–middle Eocene
Herodotius pattersoni	L-41, Fayum, Egypt	Lower sequence, Jebel Qatrani Fm	34.8–33.7 Ma (Chron 13r)
METOLDOBOTINAE			
Metoldobotes stromeri	Quarry M, Fayum, Egypt[1]	Upper sequence, Jebel Qatrani Fm.	30–29 Ma
MACROSCELIDINAE			
Elephantulus brachyrhynchus	Mumbwa Caves, Zambia		170–40 ka
E. fuscus	Makapansgat, South Africa		3.5–2.5 Ma
	Swartkrans, South Africa		2–1.5 Ma
	Sterkfontein (Mbr. 5), South Africa		2–1.5 Ma
	Olduvai (FLK NI, NNI, I), Tanzania		1.87–0.6 Ma
E. broomi	Olduvai (Layers 2, 3, 5), Tanzania		1.87–0.6 Ma
	Makapansgat, South Africa		3.5–2.5 Ma
E. antiquus	Makapansgat, South Africa		3.5–2.5 Ma
	Bolt's Farm (Camp #56), SA		Late Pliocene
	Rodent Cave, Buxton Limeworks & Dump, SA		Early Pleistocene
	Swartkrans		2–1.5 Ma
	cf. Olduvai FLK N1, Tanzania		1.87–0.6 Ma
E. brachyrhynchus	Kabwe, Zambia		Middle Pleistocene
	Twin Rivers, Zambia		Middle or late Pleistocene
E. intufi	Swartkrans (Mbrs. 1–3), South Africa		2–1.5 Ma
	Sterkfontein (Mbr. 5), South Africa		2–1.5 Ma
E. rupestris	Hoedjiespunt 1, South Africa		300–200 ka
Elephantulus sp.	Kromdraai, South Africa		1.9–1.6 Ma
	Buxton Limeworks, SA		Early Pleistocene
Macroscelides proboscideus	Makapansgat, South Africa		3.5–2.5 Ma
	Swartkrans, South Africa		2–1.5 Ma
	Kromdraai A & B, South Africa		1.87–0.6 Ma
	Olduvai, Tanzania		1.87–0.6 Ma
Paleothentoides africanus	Klein Zee, Namibia		Pliocene– early Pleistocene?
Hiwegicyon juvenalis	Loc. R3 (type) & Kaswanga, Kenya	Hiwegi Fm.	17.8 ± 0.5 Ma
Pronasilio ternanensis	Fort Ternan, Kenya	Hiwegi Fm.	14.2 Ma
Miosengi butleri	Kalodirr, Kenya	Lothidok Fm., Kalodirr Mbr.	17.5–16.8 Ma

Taxon	Occurrence (Site, Locality)	Stratigraphic Unit	Age
	MYOHYRACINAE		
Myohyrax oswaldi	Karungu (type), Kenya		17.8 ± 0.5 Ma
	Chamtwara & Songhor, Kenya	Kapurtay Agglomerates	20–19 Ma
	Elisabethfeld, Namibia		20–19 Ma
	Rusinga & Kaswanga, Kenya		17.8 ± 0.5 Ma
	Mfangano, Kenya	Kiahera Fm.	17.8 ± 0.5 Ma
	Fort Ternan, Kenya		13.8 ± 0.3 Ma
	Bosluis Pan, South Africa		ca. 13 Ma
	Arrisdrift, Namibia		17.5–17 Ma
	Auchas Mine and Bohrloch des Betriebes, Namibia		
Protypotheroides beetzi	Langental, Namibia		20–19 Ma
	MYLOMYGALINAE		
Mylomygale spiersi	"Cave near Taung & Sterkfontein," South Africa		Early Pleistocene?
	RHYNCHOCYONINAE		
Rhynchocyon pliocaenicus	Laetoli, Tanzania	Upper Laetolil Beds	3.6–3.5 Ma
Miorhynchocyon clarki	LG, Loc. 25, Kenya	Koru Fm.	20–19 Ma
	LG, Loc. 10, 21, 29, Kenya	Legetet Fm.	20–19 Ma
	Chamtwara & Songhor, Kenya	Kapurtay Agglomerates	20–19 Ma
	Rusinga, Kenya (multiple locs.)	Hiwegi Fm.	17.8 ± 0.5 Ma
	Rusinga, Kenya (Rs 903.50)	Kulu Fm.	17.8 ± 0.5 Ma
	Karungu, Kenya (? referred)		18.6–17.2 Ma
M. rusingae	Songhor	Kapurtay Agglomerates	20–19 Ma
	Mfangano, Kenya	Kiahera Fm.	17.8 ± 0.5 Ma
	Locs R1, R3, R3a, Kenya		17.8 ± 0.5 Ma
	Fort Ternan, Kenya	Hiwegi Fm.	14.2 Ma
M. meswae	Meswa Bridge, Kenya	Muhoroni Agglomerate	23–22 Ma
M. gariepensis	Arrisdrift, Namibia		17.5–17 Ma
Miorhynchocyon sp.	Moroto II, Uganda		20–17.5 Ma
	Napak, Uganda		20–17.5 Ma

SOURCE: Dates based on Seiffert (this volume, chap. 2) and Werdelin (this volume, chap. 3); age of Zambian sites from Avery, 2007.
[1]Type locality only known to be Jebel Qatrani Fm.; distribution based on referred material.

Diagnosis for Genus and Species Emended from Simons et al. (1991). Differs from *Chambius* in having m1–2 hypoconulid lacking, slightly narrower p3 talonid with poorer separation between entoconid and hypoconulid; more lingually placed p3 hypoconid, smaller m1 paraconid and m3 talonid, more weakly developed crests, well-developed cingulids on M1–2, conules on P4–M2 lacking, relatively larger M3. Differs from *Nementchatherium* in larger size, more molariform p4/P4 with P4 bearing small hypocone, M1–2 with smaller hypocones and lower centrocrista.

Description *Herodotius pattersoni* is currently only known from two lower jaws and a maxillary fragment preserving part of the facial region. Additional jaws and cranial material are under study and noted by Seiffert (2003); these fossils promise to considerably expand our understanding of herodotiines and their relationships to macroscelideans and other afrotheres.

Subfamily METOLDOBOTINAE Simons et al., 1991

Diagnosis for Subfamily After Simons et al. (1991). Differs from all other macroscelideans in its large size and in having single rooted p1–p2, enlarged procumbent i3 with lingual groove, and m3 lost and coupled with profound reduction of m2. Differs from Herotodiinae (and resembles Miocene to

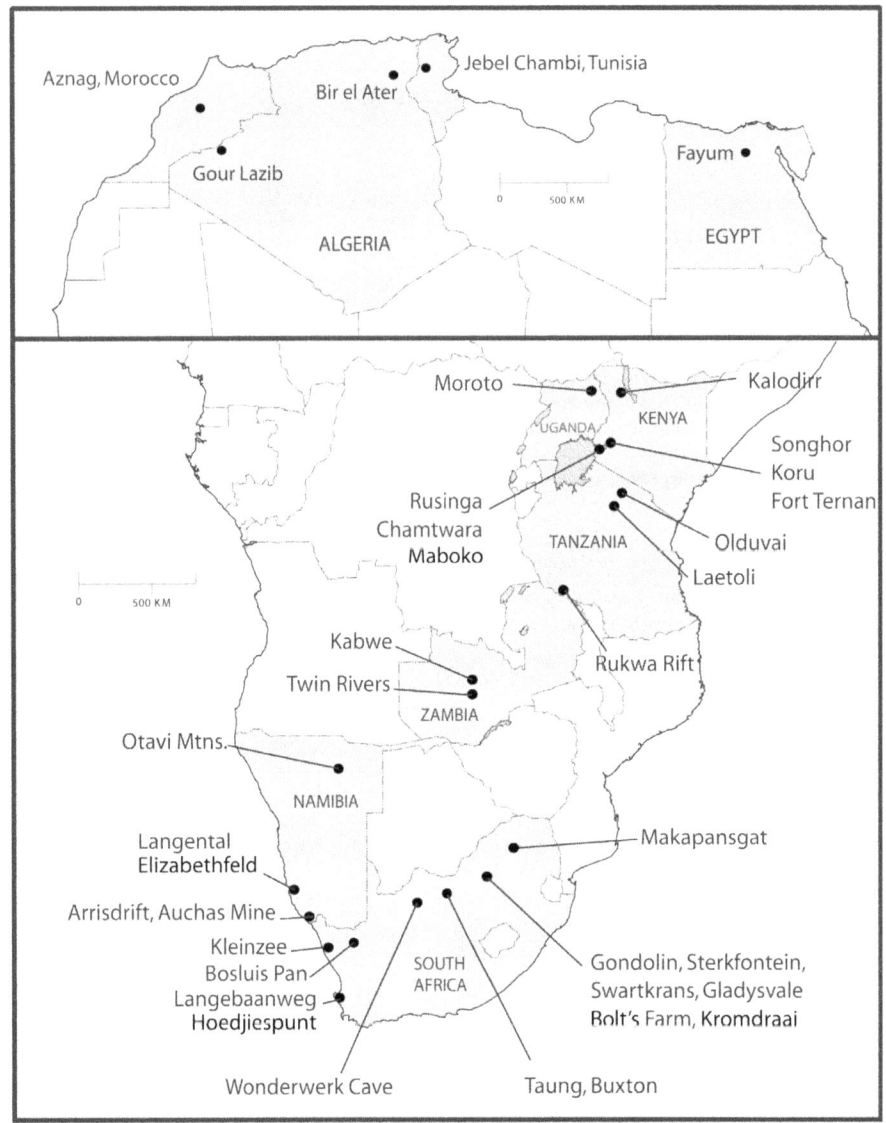

FIGURE 8.1 Maps showing distribution of fossil macroscelideans; see table 8.1 for taxonomic occurrences and ages of sites. A) Paleogene sites of North Africa; B) Late Paleogene to Pleistocene sites of sub-Saharan Africa.

Recent macroscelideans) in absence of upper molar paraconule and cingula; presence of upper molar anteroloph, p4 metaconid positioned distolingual to protoconid, M1–2 hypocone equal to protocone in size. Differs from Myohyracinae, Mylomygalinae, Rhynchocyoninae, and Macroscelidinae in retaining an upper molar metaconule and in lacking hypsodont molars with reentrant folds, an upper molar hypoloph, and prismatic cheek teeth.

Type and Only Included Genus Metoldobotes.

Genus *METOLDOBOTES* Schlosser, 1910
Figure 8.2B

Type and Only Known Species Metoldobotes stromeri Schlosser, 1910.

Diagnosis As for subfamily.

Discussion Currently the taxon is known from the type lower jaw from an unspecified level in the largely early Oligocene Jebel Qatrani Formation and from an isolated molar referred to this taxon by Simons et al. (1991) from

Quarry M in the upper sequence of the Jebel Qatrani Formation. In his analysis of afrotherian mammals, Seiffert (2003) scored characters for an undescribed late Eocene species of *Metoldobotes*.

Subfamily MACROSCELIDINAE Bonaparte, 1838

Diagnosis Differs from other macroscelidids, where known, in having palate with large vacuities, M3 absent. Differs from Mylomygalinae, Myohyracinae, and Rhynchocyoninae as discussed later under those taxa. Differs from Metoldobotinae and Herodotiinae as discussed earlier.

Included Genera Macroscelides, Petrodromus, Elephantulus, Palaeothentoides, Hiwegicyon, Pronasilio, Miosengi.

Genus *MACROSCELIDES* Smith, 1829

Diagnosis Modified after Corbet and Hanks (1968) and Butler and Greenwood (1976). Differs from other macroscelidines (where known) in having enlarged auditory bullae

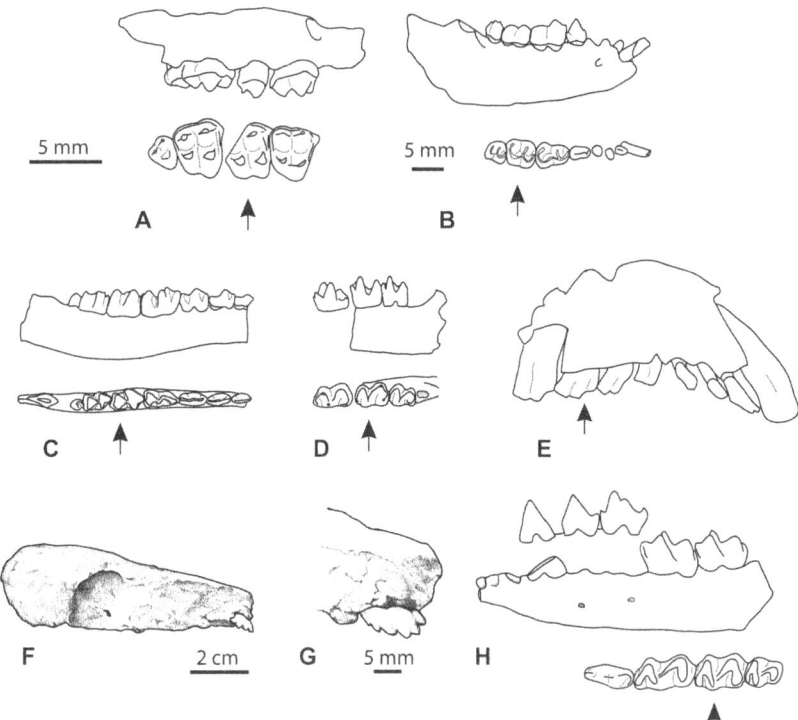

FIGURE 8.2 Representative macroscelidean taxa; for each taxon, the first molar is indicated by an arrow. A, C–E, and H to same scale. A) Herodotiinae: *Herodotius pattersoni* upper dentition in lateral and occlusal views. B) Metoldobotinae: *Metoldobotes stromeri* lower dentition in lateral and occlusal view after Schlosser, 1911. C) Macroscelidinae: *Palaeothentoides africanus* lower dentition in lingual and occlusal view, after Patterson, 1965: figure 2, plate 1. D) Macroscelidinae: *Pronasilio ternanensis* lower dentition in lingual and occlusal views, after Butler, 1984: figure 7. E) Myohyracinae: *Myohyrax oswaldi* maxilla in lateral view showing enlarged incisors, after Senut, 2003: plate 1, figure 2b. F–G) Rhynchocyoninae: cf. *Miorhynchocyon gariepensis*, lateral view of skull (F) after Senut, 2003: plate 4, figure 3; and enlarged anterolateral view (G), after Senut, 2003: plate 4, figure 5. H) Rhynchocyoninae: *Rhynchocyon pliocaenicus* lower dentition in lateral and occlusal views, after Butler, 1987: figure 4.1; composite dentitions based on holotype mandible and paratypes.

involving the occipital, squamosal, and parietal, more hypsodont molars; and differs from *Elephantulus* in greater height of ascending ramus relative to the length of the jaw.

Discussion Records of *Macroscelides* are much less common than those of *Elephantulus* in Plio-Pleistocene assemblages, and all are assigned to the living species. Based on the presence of *Macroscelides* at Makapansgat (Butler and Greenwood, 1976), this genus appears to have split from other macroscelidines by the late Pliocene.

Genus *PETRODROMUS* Peters, 1846

Diagnosis After Corbet and Hanks (1968). Differs from other macroscelidines (where known) in having hallux absent, larger body size; I1 prominent and twice as long as I2; I3 double-rooted. Additionally, differs in having simple, nonsemimolariform p2–p3.

Discussion To date, no fossils have been reported of *Petrodromus*.

Genus *ELEPHANTULUS* Thomas and Schwann, 1906

Selected Synonymy Elephantomys Broom, 1937.
Diagnosis Differs from *Macroscelides* in its lower-crowned cheek teeth and by its less crowded anterior dentition; differs

from *Pronasilio* in its higher-crowned and more anteriorly leaning cheek teeth; differs from *Palaeothentoides* and *Miosengi* as discussed below.

Included Extinct Species E. broomi (= *Elephantomys langi* Broom, 1937), *E. antiquus, E. fuscus leakeyi* (extinct subspecies of living species).

Living Species Also Reported in the Fossil Record E. brachyrhynchus, E. rupestris, E. intufi.

Discussion Elephantulus is the most common sengi recovered in Southern African Plio-Pleistocene assemblages (e.g., Butler and Greenwood, 1976; Avery, 2000, 2001, 2003, 2007). Only a small proportion of these fossils has been formally described (Olduvai and Makapansgat specimens by Butler and Greenwood, 1976), and some taxa reported in faunal lists are extralimital records of the living species (e.g., Kabwe, Avery, 2003; Hoedjiespunt 1, Matthews et al., 2005). The oldest well-dated record of the genus would appear to be from Makapansgat.

Elephantulus antiquus is the largest of the three fossil species, and *E. broomi* and *E. fuscus leakeyi* differ primarily in the presence or absence of m3 (see discussion in Avery, 1998, regarding difficulties in distinguishing the latter two species). *E. fuscus leakeyi* and *E. broomi* are considered to be closely related to a part of the complex of living species of South African *Elephantulus*.

The extant *Elephantulus rozeti* is the single North African species attributed to this genus. Butler and Greenwood (1976) noted that *E. rozeti* differs from *E. antiquus* in having relatively more reduced P2/p2–P3/p3 that are separated by diastemata. M2 is also more reduced and the anterior fossa of M1 is shallower than the posterior fossa, disappearing at an early stage of wear. More recently, Douady et al.'s (2003) molecular phylogeny demonstrated that *E. rozeti* from North Africa is the sister taxon of *Petrodromus*. As of this writing, this species has not been formally assigned to either *Petrodromus* or a new genus.

Genus *PALAEOTHENTOIDES* Stromer, 1932
Figure 8.2C

Type and Only Known Species Palaeothentoides africanus Stromer, 1932.

Diagnosis for Genus and Species Modified after Patterson (1965). p1 two-rooted and nonincisiform; p2–p3 with anterior cusps little separated from protoconids, p3 lacking metaconid and entoconid, p4 narrow with metaconid lingual and posterior to protoconid, p4 cristid obliqua swollen and nearly filling trigonid, m1–2 with shallow clefts between paraconid and metaconid; differs from other macroscelidines in having p4 with labial swelling between trigonid and talonid and posteriorly positioned metaconid (Novacek, 1984).

Description Palaeothentoides is known only from two dentaries and the tooth positions p1–m3. A third dentary with m2–m3 is now lost (Patterson, 1965).

Genus *HIWEGICYON* Butler, 1984

Type and Only Known Species Hiwegicyon juvenalis Butler, 1984.

Synonymy Macroscelididae of uncertain genus and species Butler, 1969.

Diagnosis for Genus and Species After Butler (1984). m1 moderately high crowned, cristid obliqua of dp4 runs to middle of protolophid, no metastylid on dp3–dp4, dp3 has metaconid but lacks protostylid, anterior lower teeth probably crowded and procumbent.

Discussion The type and only known specimen is a juvenile, and it shares some superficial resemblances to both *Pronasilio* and *Miorhynchocyon*. It is here placed in the Macroscelidinae based on its greater resemblance to *Pronasilio* (Butler, 1984).

Genus *PRONASILIO* Butler, 1984
Figure 8.2D

Type and Only Known Species Pronasilio ternanensis Butler, 1984.

Diagnosis for Genus and Species Modified after Butler (1984). p4–m2 moderately hypsodont; cristid obliqua directed toward metaconid, talonid lower than trigonid, m3 present, p4 trigonid only moderately extended and incompletely molariform with reduced entoconid. Differs from *Elephantulus* in having upright crowns that do not lean anteriorly.

Discussion Pronasilio is known from three dentary fragments. In those characters than can be evaluated, it parallels *Rhynchocyon* in its more lingually directed cristid obliqua but is a much higher-crowned form. However, it appears to be more primitive than other macroscelidines in the incomplete molarization of p4 (Butler, 1984).

Genus *MIOSENGI* Grossman and Holroyd, 2009

Type and Only Known Species Miosengi butleri, Grossman and Holroyd, 2009.

Diagnosis for Genus and Species After Grossman and Holroyd (2009). Differs from other macroscelidines in relatively lower crown height and posterior wall of m2 talonid anteriorly inclined and posteriorly rounded (vs. vertical). Differs from *Palaeothentoides* and is similar to other macroscelidines in smaller size; differs from *Macroscelides*, *Petrodromus*, and some *Elephantulus* species in retaining m3; differs from *Macroscelides* in having relatively wider molars relative to length and a lower crown. Differs from *Pronasilio* in having an m2 slightly mesiodistally longer but with a lower crown; m2 lacking anterobuccal cingulum; more gracile dentary. *Miosengi butleri* further differs from *Petrodromus*, *Paleothentoides*, and *Elephantulus* but is similar to *Macroscelides* and *Pronasilio*, in retaining a salient hypoconulid.

Discussion Miosengi is known only from a dentary with m2 and an m3 alveolus and a premolar that may represent p3 or p2. Based on comparisons with other macroscelidines, *Miosengi* is more primitive in its molar morphology than other macroscelidines, but the premolars are most similar to those of *Paleothentoides* and *Elephantulus*, suggesting a possible closer relationship to those taxa.

Subfamily MYOHYRACINAE Andrews, 1914

Diagnosis After Novacek (1984). Differs from other macroscelidids in having I1–2 enlarged and procumbent with enamel restricted to labial faces, dentary below posterior cheek teeth very deep and robust, and fossettes on crowns of upper cheek teeth.

Included Genera Myohyrax, Protypotheroides.

Genus *MYOHYRAX* Andrews, 1914
Figure 8.2E

Type and Only Known Species Myohyrax oswaldi Andrews, 1914 (= *M. doederleini* Stromer, 1926).

Diagnosis After Senut (2003). Presence of cement not constant in upper cheek teeth; P4 and M1 largest teeth; 2 mental foramina in dentary constant; upper incisors very curved and possess four digitations; parastyle of upper cheek teeth always lower than the paracone; lingual posterior fossette generally absent in M2; M3 variably present. Smaller than *Protypotheroides*.

Discussion Butler (1984) observed that the Namibian material ascribed to *M. doederleini* by Stromer (1926) and synonymized with *M. oswaldi* by Whitworth (1954) is actually smaller than the East African sample of *M. oswaldi* and thought the matter deserved further investigation, but also noted that one jaw from Napak, Uganda, might be specifically different from *M. oswaldi*. *Myohyrax* is now known from more than 100 specimens from Namibia and is the most common macroscelidean from the Miocene of Southwest Africa.

Genus *PROTYPOTHEROIDES* Stromer, 1922

Type and Only Known Species Protypotheroides beetzi Stromer, 1922 (= *Myohyrax osborni* Hopwood, 1929 (part), pp. 6–8, figs. 5–6; *Myohyrax beetzi* Whitworth, 1954).

Diagnosis After Butler (1984). Differs from *Myohyrax* in larger size, presence of fossettids lacking cementum on lower molars (one each in the trigonid and talonid).

Subfamily MYLOMYGALINAE Patterson, 1965

Type and Only Known Genus Mylomygale
Diagnosis In part after Novacek (1984). Differs from Myohyracinae in having anterior teeth small and crowded, m3 absent, p4–m2 tall, columnar and strongly compressed anteroposteriorly and bearing very deep lingual and labial reentrant folds, and alveolar border concave below p4–m2.

Discussion Corbet and Hanks (1968) noted similarities to *Macroscelides*, especially *M. proboscideus*, and Butler (1995) suggested Mylomygalinae might be synonymized with Macroscelidinae. Novacek's (1984) analysis suggests that *Mylomygale* is more closely related to Myohyracinae than to Macroscelidinae based on their shared hypsodonty and development of a deep, convex lower border on the dentary.

Genus *MYLOMYGALE* Broom, 1948

Diagnosis As for subfamily.
Included Species Mylomygale spiersi Broom, 1948.
Description Mylomygale is known solely from the type lower jaw with partial p3, p4–m2 and alveoli for the anterior teeth, plus an isolated p4. This limited material suggests it is a hypsodont form with a shortened face.

Discussion The provenance of the type and only specimen is simply "cave near Taung and Sterkfontein," and the taxon has typically been referred to as early Pleistocene in the literature. As many of the sites in this vast area of South Africa are now thought to be Pliocene as well as Pleistocene in age, no age can be assigned to this taxon with confidence.

Subfamily RHYNCHOCYONINAE Gill, 1872

Diagnosis Composite of characters as given by Novacek (1984) and Butler (1995). Differs from other Macroscelididae in having accessory cusp on posterior crest of protoconid in nonmolariform lower premolars and deciduous molars, anterior edge of coronoid process more gently sloping; differs from Macroscelidines in having broad face and palate lacking palatal vacuities and having long infraorbital canal that opens anteriorly above p3. m3/M3 absent, p4/P4 fully molariform and larger than m1/M1, and P3 semimolariform. Lower canine smaller than p1.
Included Genera Rhynchocyon, Miorhynchocyon.

Genus *RHYNCHOCYON* Peters, 1847
Figure 8.2H

Diagnosis Differs from *Miorhynchocyon* in having higher crowned teeth, cristid obliqua directed toward metaconid or metastylid; differs at least from *M. gariepensis* in having reduced upper incisors.
Included Fossil Species R. pliocaenicus Butler, 1987.
Description Rhynchocyon pliocaenicus is known from a total of six specimens from the Upper Laetolil Beds and is a very rare element in the fauna (Butler, 1987; Winkler, in press). It differs from the extant species in being approximately 20% smaller, having a p2 with distinct posterior heel, anterobuccal cingula present on p4–m2 (rare in extant spp.) and posterior cingula present on p4–m1, and having

the protoloph of P4 and M1 connect to the tip of the paracone versus connecting with the anterior cingulum (Winkler, in press). In most of these features, Winkler noted it is similar to *Miorhynchocyon* but is generally less brachyodont than the Miocene genus.

Genus *MIORHYNCHOCYON* Butler and Hopwood, 1957
Figures 8.2F, 8.2G

Type Species M. clarki (Butler and Hopwood, 1957).
Included Species M. rusingae (Butler, 1969), *M. meswae* (Butler, 1984), *M. gariepensis* Senut, 2003.
Diagnosis Emended after Butler (1984). Differs from *Rhynchocyon* in having upper incisors unreduced (at least in specimens provisionally assigned to *M. gariepensis*), lower crowned cheek teeth, cristid obliqua directed toward the middle of the protoconid-metaconid crest (vs. lingually and connected to the metaconid or metastylid), metastylid absent on dp4 and m1, paraconid of p4 and m1 higher and more lingually situated.

Description The species of *Miorhynchocyon* differ from one another primarily in size and relative sizes of cusps and stylids and are described in detail in Butler (1984) and Senut (2003). *M. clarki* and *M. gariepensis* are the best-known species, being represented by cranial, dental, and postcranial remains.

Miorhynchocyon clarki is generally considered the most primitive member of this genus due to its smaller size and having P2 parastyle present, M2 less reduced, flatter facial region, more divergent tooth rows, extra nasal processes of premaxilla well-developed; facial part of lachrymal less extensive; olfactory bulbs placed further back in relation to the orbit. However, *M. gariepensis* is similar to *M. clarki* in size and is notable in having p/4 talonid and trigonid equal in width and in which the talonid is distinctly lower than the trigonid, all primitive characters. The relationships among *Miorhynchocyon* species or of the genus to extant *Rhynchocyon* have not yet been explored in detail. *M. rusingae* has been suggested to be more closely related to *Rhynchocyon* based on its larger size and slightly more molariform premolars (Butler, 1978), suggesting that *Miorhynchocyon* is paraphyletic with respect to the extant genus and the lineage leading to the modern taxon might diverge in the early Miocene.

Discussion In addition to specimens attributed to *Miorhynchocyon* species, undescribed specimens attributed to *Miorhynchocyon* sp. from the early Miocene of Uganda have been noted from Moroto (Pickford et al., 2003) and Napak (Winkler, in press).

The status of *Miorhynchocyon* as a distinct genus has been questioned. Some authors (Novacek, 1984; McKenna and Bell, 1997; Holroyd and Mussell, 2005) have considered it a synonym of the extant genus *Rhynchocyon*, based on Novacek's (1984) observation that the living and fossil species comprised a monophyletic group or represent what we would now term a stem-based taxon. Butler (1978) avers that *Miorhynocyon* is sufficiently distinct from later occurring species of *Rhynchocyon* to merit a distinct genus. While not explicitly stated, the crux of this argument is whether the species attributed to *Miorhynchocyon* should be united with *Rhynchocyon* species within a stem-based concept of the genus or whether some or all of the Miocene species are distinct. While I prefer to unite all these species within a stem-based concept of the genus, I readily admit that we currently do not know enough about the relationships among the Miocene to recent rhynchocyonines to justify either taxonomic decision. Future work will hopefully address this issue through explicit phylogenetic analyses that may determine whether some species now placed in *Miorhynchocyon* are more closely related to the extant species.

Macroscelidea, Undetermined

In addition to the records assigned to genus and/or species already discussed here, there are a number of additional, undescribed records of Macroscelidea. In East Africa, Stevens et al. (2008) report the existence of new genera and species of macroscelideans from the late Oligocene of the Rukwa Rift Basin, Tanzania from Unit II of the Red Sandstone Group. In North Africa, Macroscelididae are reported to be present in karsts of late Miocene (11–10 Ma) age in the Western Desert of Egypt (Pickford et al., 2006) and to be present in approximately 2 Ma deposits at Irhoud-Ocre, Morocco (J.-J. Jaeger, pers. comm. in Butler, 1995). In southwestern Africa, Pickford and Senut (2002) note the presence of sengis from the Miocene of Berg Aukas, the Plio-Pleistocene Kaokoland karsts of Namibia, and the late Pleistocene sites Asis Ost & Kombat E 900, Otavi Mountains, Namibia, as well as Plio-Pleistocene caves of the Humpata Plateau in southern Angola (Pickford et al., 1992). In South Africa, Avery (2007) lists *Macroscelides* and/or *Elephantulus* from middle Pleistocene Wonderwerk Cave, Northern Cape Province, and Sénégas et al. (2005) note rare macroscelidids from the early Pleistocene mine dump at Gondolin in the Sterkfontein area.

Discussion

There has been no comprehensive phylogeny of fossil and recent sengis, but a number of contributions have phylogenetically tested (Tabuce et al., 2001; Douady et al. 2003) or posited (Corbet and Hanks, 1968; Butler and Greenwood, 1976; Novacek, 1984) relationships among selected taxa among the Macroscelidea. Figure 8.3 is an interpretive summary of these relationships. The distribution of characters among the subfamily-level clades and among genera is generally well supported. Major areas for future research are filling in the morphologic transition between the Herodotiinae and later occurring macroscelideans; developing a comprehensive morphologic data set for testing relationships among the named subfamilies; and establishing the relationships among the fossil and extant species placed in *Elephantulus*, particularly in light of the finding that *E. rozeti* appears to be more closely related to *Petrodromus* (Douady et al., 2003). Further study and resolution of these relationships should provide important insights into the historical biogeography of the late Paleogene through Pleistocene diversification of the macroscelidines.

A potentially more vexing problem is recognizing and understanding the timing and place of the early diversification of Macroscelidea. Several authors (Hartenberger, 1986; Simons et al., 1991; Tabuce et al., 2001; Zack et al., 2005; Tabuce et al., 2007; Penkrot et al., 2008) have noted the phenetic similarities and/or phylogenetic signal in the dental and

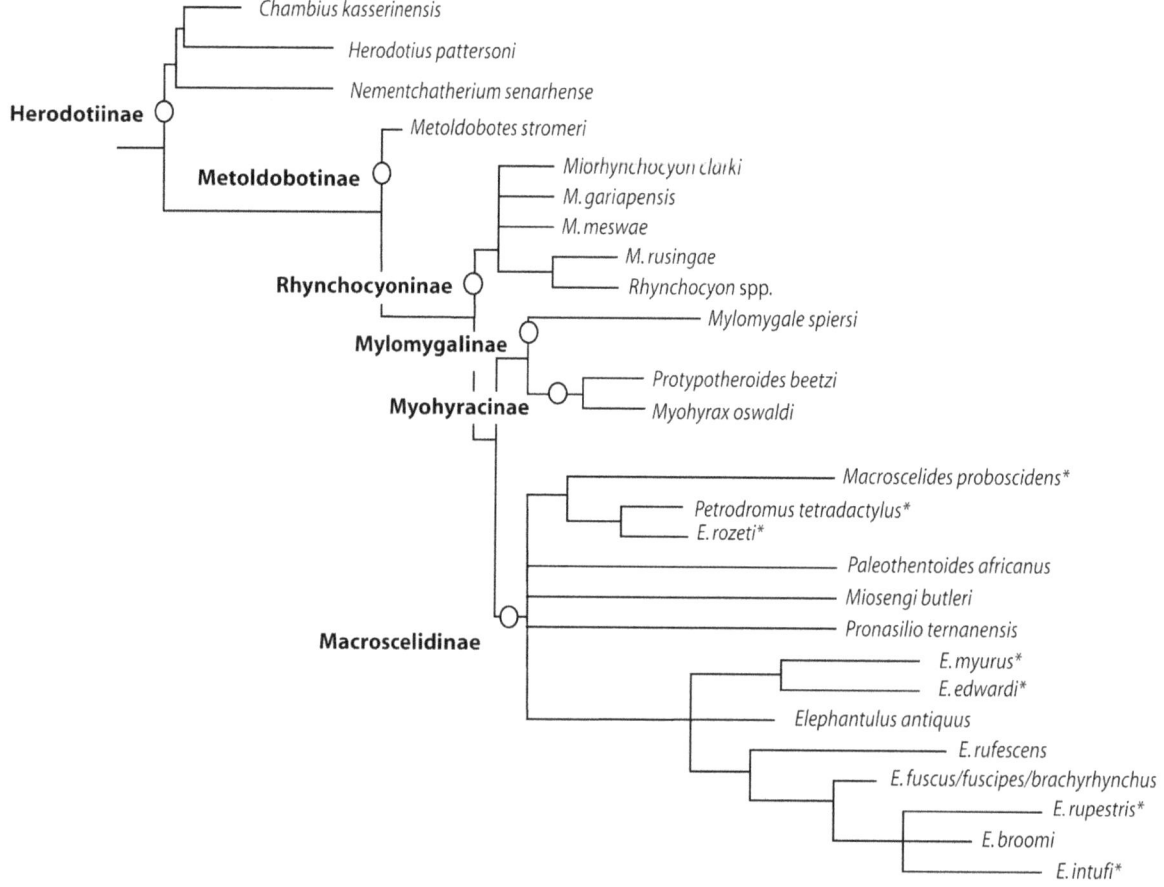

FIGURE 8.3 An interpretive summary of hypotheses of relationships among sengis and Paleogene splits as determined by Tabuce et al. (2001), Miocene diversifications after Novacek, 1984; relationships among extant species included in molecular analyses (indicated by asterisk) after Douady et al., 2003; and position of fossil species of extant genera after Butler and Greenwood, 1976.

postcranial morphology of macroscelideans and the paraphyletic or polyphyletic grouping Condylarthra, most notably European lousinine and North American apheliscine "hyopsodontids." These similarities have even led some workers to erroneously consider early macroscelideans to be condylarths (e.g., McKenna and Bell, 1997, for *Chambius*), and the likely close relationships of some Holarctic taxa to macroscelideans suggests complex biogeographic relationships and trans-Tethyan dispersal patterns (e.g., Zack et al., 2005; Tabuce et al., 2007; Penkrot et al., 2008) that are not recoverable from studies including only extant forms.

However, more problematic is recognizing taxa that may be macroscelideans when only represented by fragmentary remains. Tabuce et al. (2005) recognized this problem in describing isolated specimens from the early middle Eocene (approx. 45.8–43.6 Ma) of Aznag, Morocco. Among this assemblage, a single tooth assigned to Hyopsodontidae shares some similarities with *Chambius* but cannot be attributed to that taxon or to Herodotiinae given our current limited knowledge of the morphology of Paleogene macroscelideans. Accepting Macroscelidea as part of Afrotheria suggests a ghost lineage of several million years (see discussion in Holroyd and Mussell, 2005), complicating the process of recognizing fossil sengis. Work to better understand the placement of Macroscelidea among other Recent afrotherian orders and extinct placentals is an area of active research (e.g., Asher et al., 2003; Asher, 2007; Sánchez-Villagra et al., 2007; Seiffert, 2007; Asher and Lehmann, 2008; Penkrot et al., 2008). This ongoing work promises not only to better establish phylogenetic relationships among placental mammals but also to provide much new morphologic information that may ultimately reveal morphological characteristics that will allow us to more readily recognize sengis in the fossil record.

ACKNOWLEDGMENTS

I want to thank the editors of this volume, Lars Werdelin and William Sanders, for their kind invitation to participate, and Robert Asher, Ari Grossman, and Nancy Stevens for helpful discussions of sengis. J. Howard Hutchison assisted with figure 8.2. Elwyn Simons, Paul Murphey, Jaelyn Eberle, Peter Robinson, Toni Culver, Ivy Rutzky, and Eileen Westwig provided access to collections in their care, and Percy Butler highlighted the taxonomic problems associated with *Miorhynchocyon* in a review of another essay.

NOTE ADDED IN PROOF

After submission of this chapter, several important papers on sengis were published that add significantly to the fossil record of the group. From North Africa, Pickford et al. (2008b) and Wanas et al. (2009) describe in more detail the late Miocene (11–10 Ma) karsts in the Western Desert of Egypt that have yielded macroscelidids and illustrate specimens assigned to *Rhynchocyon*.

From Namibia, the first middle Eocene macroscelidean and a large sample of Miocene taxa are newly described. Pickford et al. (2008a) report a single upper molariform tooth that they ascribe the Macroscelididae. Based on the lack of buccal cingula seen in herodotiines, low metaconule, and general proportions, this unnamed taxon may represent the oldest member of the crown Macroscelididae. Senut (2008) has reported a diverse macroscelidid fauna from the lower Miocene of the Northern Sperrgebiet in Namibia. *Myohyrax oswaldi* is

represented by skulls and associated skeletons, and a second larger species of *Myohyrax*, *M. pickfordi*, is represented by a suite of maxillae and dentaries. *Protypotheroides beetzi* is present, and the new specimens allow the diagnosis to be emended. Two new genera of rhynchocyonine are named: *Brachyrhynchocyon* and *Hypsorhynchocyon*. *Brachyrhynchocyon* is, as the name suggests, a more brachyodont form with smooth (rather than sharp) premolars, large multipectinate upper incisors, and lacking m3. This genus is founded on the new species *B. jacobi*, and *Miorhynchocyon garapiensis* (Senut, 2003) is transferred to this genus. *Hypsorhynchocyon burrelli* is based on a single dentary with p4–m2 that is notable for its greater crown height and a more vertically ascending ramus than seen in other rhynchocyonines. The Sperrgebiet assemblages also include macroscelidid postcrania, many likely ascribable to *Protypotheroides beetzi*, and are under further study.

Literature Cited

Adaci, M., R. Tabuce, F. Mebrouk, M. Bensalah, P.-H. Fabre, L. Hautier, J.-J. Jaeger, V. Lazzari, M. Mahboubi, L. Marivaux, O. Otero, S. Peigné, and H. Tong. 2007. Nouveaux sites à vertébrés paléogènes dans la région des Gour Lazib (Sahara nord-occidental, Algérie). *Comptes Rendus Palevol* 6:535–544.

Asher, R. J. 2007. A web-database of mammalian morphology and a reanalysis of placental phylogeny. *BMC Evolutionary Biology* 7:106; doi:10.1186/1471-2148-7-108.

Asher, R. J., and T. Lehmann. 2008. Dental eruption in afrotherian mammals. *BMC Biology* 6:14; doi:10.1186/1741-7007-6-14.

Asher, R. J., M. J. Novacek, and J. H. Geisler. 2003. Relationships of endemic African mammals and their fossil relatives based on morphological and molecular evidence. *Journal of Mammalian Evolution* 10:131–194.

Avery, D. M. 1998. An assessment of the lower Pleistocene micromammalian fauna from Swartkrans Members 1–3, Gauteng, South Africa. *Geobios* 31:393–414.

———. 2000. Notes on the systematics of micromammals from Sterkfontein, Gauteng, South Africa. *Palaeontologia Africana* 36:83–90.

———. 2001. The Plio-Pleistocene vegetation and climate of Sterkfontein and Swartkrans, South Africa, based on micromammals. *Journal of Human Evolution* 41:113–132.

———. 2003. Early and Middle Pleistocene environments and hominid biogeography: Micromammalian evidence from Kabwe, Twin Rivers, and Mumbwa Caves in central Zambia. *Palaeogeography, Palaeoclimatology, Palaeoecology* 189:55–69.

———. 2007. Pleistocene micromammals from Wonderwerk Cave, South Africa: Practical issues. *Journal of Archaeological Science* 34:613–625.

Butler, P. M. 1978. Insectivora and Chiroptera; pp. 56–68 in V. J. Maglio and H. B. S. Cooke (eds.), *Evolution of African Mammals*. Harvard University Press, Cambridge.

———. 1984. Macroscelidea, Insectivora and Chiroptera from the Miocene of East Africa. *Palaeovertebrata* 14:117–200.

———. 1987. Fossil insectivores from Laetoli; pp. 85–87 in M. D. Leakey and J. M. Harris, eds. Laetoli, *A Pliocene Site in Northern Tanzania*. Oxford University Press, Oxford.

———. 1995. Fossil Macroscelidea. *Mammal Review* 25:3–14.

Butler, P. M., and M. Greenwood, M. 1976. Elephant-shrews (Macroscelididae) from Olduvai and Makapansgat; pp. 1–56 in R. J. G. Savage and S. C. Coryndon (eds.), *Fossil Vertebrates of Africa*, vol. 4. Academic Press, London.

Corbet, G. B., and J. Hanks. 1968. A revision of the elephant-shrews, family Macroscelididae. *Bulletin of the British Museum (Natural History), Zoology* 16:45–111.

Douady, C. J., F. Catzeflis, J. Raman, M. S. Springer, and M. J. Stanhope. 2003. The Sahara as a vicariant agent, and the role of Miocene climatic events, in the diversification of the mammalian order Macroscelidea (elephant shrews). *Proceedings of the National Academy of Sciences, USA* 100:8325–8330.

Grossman, A., and P. A. Holroyd, P. A. 2009. *Miosengi butleri*, gen. et sp. nov., (Macroscelidea) from the Kalodirr Member, Lothidok Formation, early Miocene of Kenya. *Journal of Vertebrate Paleontology* 29:957–960.

Hartenberger, J.-L. 1986. Hypothèse paléontologique sur l'origine des Macroscelidea (Mammalia). *Comptes Rendus de l'Académie des Sciences, Paris, Sciences de la Terre et des Planètes* 302:247–249.

Holroyd, P. A., and J. C. Mussell. 2005. Macroscelidea and Tubulidentata; pp. 71–84 in K. D. Rose and J. D. Archibald (eds.), *Origin and Relationships of the Major Extant Clades: The Rise of Placental Mammals*. Johns Hopkins University Press, Baltimore.

Matthews, T., C. Denys, and J. E. Parkington. 2005. Paleoecology of the micromammals from the late middle Pleistocene site of Hoedjiespunt 1 (Cape Province, South Africa). *Journal of Human Evolution* 49:432–451.

McKenna, M. C. and S. K. Bell. 1997. *Classification of Mammals above the Species Level*. Columbia University Press, New York. 631 pp.

Novacek, M. J. 1984. Evolutionary stasis in the elephant-shrew, *Rhynchocyon*; pp. 4–22 in N. Eldredge and S. M. Stanley (eds.), *Living Fossils*. Springer, New York.

Patterson, B. 1965. The fossil elephant shrews (Family Macroscelididae). *Bulletin of the Museum of Comparative Zoology* 133:297–336.

Penkrot, T. A., S. P. Zack, K. D. Rose, and J. I. Bloch. 2008. Postcranial morphology of Apheliscus and Haplomylus (Condylarthra, Apheliscidae): Evidence for a Paleocene Holarctic origin of Macroscelidea; pp. 73–106 in E. J. Sargis and M. Dagosto (eds.), *Mammalian Evolutionary Morphology: A Tribute to Frederick S. Szalay*. Springer, New York.

Pickford, M., and B. Senut. 2002. *The Fossil Record of Namibia*. Ministry of Mines and Energy, Geological Survey of Namibia, Windhoek, 39 pp.

Pickford, M., P. Mein, and B. Senut. 1992. Primate bearing Plio-Pleistocene cave deposits of Humpata, Southern Angola. *Human Evolution* 7:17–33.

Pickford, M. B. Senut, D. Gommery, and E. Musiime. 2003. New catarrhine fossils from Moroto II, early middle Miocene (ca 17.5 Ma) Uganda. *Comptes Rendus Palevol* 2:649–662.

Pickford, M., Senut, B., Morales, J., Mein, P., and Sanchez, I. M. 2008a. Mammalia from the Lutetian of Namibia. *Memoirs of the Geological Survey of Namibia* 20:465–514.

Pickford, M., Wanas, H., Mein, P., and Soliman, H. 2008b. Humid conditions in the Western Desert of Egypt during the Vallesian (Late Miocene). *Bulletin of the Tethys Geological Society* 3:63–79.

Pickford, M., H. Wanas, and H. Soliman. 2006. Indications for a humid climate in the Western Desert of Egypt 11–10 Myr ago: Evidence from Galagidae (Primates, Mammalia). *Comptes Rendus Palevol* 5:935 943.

Sánchez-Villagra, M. R., Y. Narita, and S. Kuratani. 2007. Thoracolumbar vertebral number: The first skeletal synapomorphy for afrotherian mammals. *Systematics and Biodiversity* 5:1–7.

Schlosser, M. 1911. Beiträge zur Kenntnis der oligozänen Landsäugetiere aus dem Fayum, Ägypten. *Beiträge zur Paläontologie und Geologie Österreich-Ungarns* 24:51–167.

Seiffert, E. R. 2003. A phylogenetic analysis of living and extinct afrotherian mammals. Unpublished PhD dissertation, Duke University.

———. 2007. A new estimate of afrotherian phylogeny based on simultaneous analysis of genomic, morphological, and fossil evidence. *BMC Evolutionary Biology* 7:224 doi:10.1186/1471-2148-7-224.

Sénégas, F., E. Paradis, and J. Michaux. 2005 Homogeneity of fossil assemblages extracted from mine dumps: an analysis of Plio-Pleistocene fauna from South African caves. *Lethaia* 38:315–322.

Senut, B. 2003. The Macroscelididae from the Miocene of the Orange River, Namibia. *Memoirs of the Geological Survey of Namibia* 19:119–141.

———. 2008. Macroscelididae from the lower Miocene of the Northern Sperrgebiet, Namibia. *Memoirs of the Geological Survey of Namibia* 20:185–225.

Simons, E. L., P. A. Holroyd, and T. M. Bown. 1991. Early Tertiary elephant-shrews from Egypt and the origin of the Macroscelidea. *Proceedings of the National Academy of Sciences, USA* 88:9734–9737.

Stevens, N. J., M. D. Gottfried, E. M. Roberts, S. Kapilima, S. Ngawala, and P. M. O'Connor. 2008. Paleontological Exploration in Africa: a view from the Rukwa Rift Basin of Tanzania; pp. 159–180, in J. G. Fleagle and C. C. Gilbert (eds.), *Elwyn Simons: A Search for Origins; Developments in Primatology: Progress and Prospects*. Springer, New York.

Stromer, E. 1926. Reste Land- und Süsswasser-bewohnender Wirbeltiere aus den Diamantfeldern Deutsch-Südwestafrikas; pp. 107–153 in E. Kaiser, ed. *Die Diamantenwurste Südwestafrikas*, Vol. 2. Dietrich, Riemer, Berlin.

Tabuce, R., B. Coiffait, P.-E. Coiffait, M. Mahboubi, and J.-J. Jaeger. 2001. A new genus of Macroscelidea (Mammalia) from the Eocene of Algeria: A possible origin for elephant-shrews. *Journal of Vertebrate Paleontology* 21:535–546.

Tabuce, R., L. Marivaux, M. Adaci, M. Bensalah, J.-L. Hartenberger, M. Mahboubi, F. Mebrouk, P. Tafforeau, and J.-J. Jaeger. 2007. Early Tertiary mammals from North Africa reinforce the molecular Afrotheria clade. *Proceedings of the Royal Society of London B* 274:1159–1166.

Wanas, H. A., Pickford, M., Mein, P., Soliman, H., and Segalen, L. 2009. Late Miocene karst system at Sheikh Abdalla, between Bahariya and Farafra, Western Desert, Egypt: Implications for palaeoclimate and geomorphology. *Geologica Acta* 7:475–487. doi:10.1344/105.000001450.

Winkler, A. J. In press. Fossil Macroscelididae from Laetoli, Tanzania; in T. Harrison, ed., *Paleontology and Geology of Laetoli, Tanzania: Human Evolution in Context. Volume 2: Fossil Hominins and the Associated Fauna*. Springer, Dordrecht.

Whitworth, T. 1954. The Miocene hyracoids of East Africa. *Fossil Mammals of Africa* 7:1–58.

Zack, S. P., T. A. Penkrot, L. I. Bloch, and K. D. Rose. 2005. Affinities of 'hyopsodontids' to elephant shrews and a Holarctic origin of Afrotheria. *Nature* 434:497–501.

NINE

Tenrecoidea

ROBERT J. ASHER

Tenrecs (Tenrecidae) and golden moles (Chrysochloridae), grouped together in the taxon Tenrecoidea by McDowell (1958) or alternatively "Afrosoricida" by Stanhope et al. (1998), both comprise diverse groups of small, insectivoran-grade placental mammals. As defined by Bronner and Jenkins (2005), there are 10 genera and 30 species of living tenrecs, 27 of which are from Madagascar and three from Central and West Africa. Bronner and Jenkins list nine genera of living golden moles containing 21 species, all but three of which are from continental Africa near to or south of the 20th parallel. Neither group has a good fossil record. The earliest, undisputed fossils are from the early Miocene of East Africa (Butler, 1984), although a tentative identification has recently been made of fossil tenrec and golden mole jaws from the Eo-Oligocene of Egypt (Seiffert et al., 2007). For most of the 20th century, tenrecs and golden moles were regarded as members of the Insectivora or Lipotyphla, along with hedgehogs, shrews, moles, and *Solenodon* (Butler, 1988). Currently, tenrecoids (as I will subsequently refer to tenrecs and golden moles collectively) are recognized as part of the Afrotheria (Stanhope et al., 1998). Because the fossil record of both groups is so poor, the following discussion conveys their history in the context of their extant diversity.

INSTITUTIONAL ABBREVIATIONS

BMNH, The Natural History Museum, London; KNM, National Museums of Kenya, Nairobi; MCZ, Museum of Comparative Zoology, Cambridge, USA; TM, Transvaal Museum, Pretoria; ZMB, Zoologisches Museum, Berlin.

Systematic Paleontology

Family CHRYSOCHLORIDAE
Figures 9.1–9.3

Content and Distribution Golden moles have a particularly complicated taxonomic history. Species- and generic-level treatments differ considerably in recent revisions of the group. For example, the species first named *Chrysochloris leuchorhina* by its author (Huet, 1885), now known by its corrected specific epithet *leucorhinus*, has over the years been assigned to the genera *Amblysomus* (Simonetta, 1968;

Petter, 1981), *Chlorotalpa* (Meester et al., 1986), *Calcochloris* (Bronner and Jenkins, 2005) and *Huetia* (Asher et al., in press). Previous authorities have used a distinct number of genera to categorize a roughly similar hypodigm of extant chrysochlorids: Petter (1981) used five, Hutterer (1993) seven, and Bronner and Jenkins (2005) nine. Recent advances in the understanding of golden moles concern their morphometric (Bronner, 1995a, 1995b), karyotypic (Bronner, 1995a), locomotor (Fielden et al., 1992), and cranial (Mason, 2003) diversity.

Until 2010 (Asher et al., in press), Bronner (1991, 1995a) had published the only character-based phylogenies of the family to date. These sample either hyoid (Bronner, 1991) or chromosome (Bronner, 1995a) morphology from less than a dozen of the currently recognized 21 species. Other hypotheses of chrysochlorid interrelationships exist (e.g., Simonetta, 1968; reproduced here in figure 9.1) but are not character based. In addition, Bronner (1995b: figures 9.10 and 9.11 therein) presented in his PhD dissertation a phylogeny based on 14 terminal taxa sampled for 15 morphometric and discrete characters of the cranium, dentition, and chromosomes (also reproduced here in figure 9.1).

Regardless of the preferred generic-level taxonomy, the distribution of chrysochlorid genera presents quite a biogeographic puzzle (figure 9.2). Members of the genus *Chrysochloris* are separated by over 3,000 km: *C. asiatica* in the western cape and *C. stuhlmanni* near Lakes Victoria and Malawi in East Africa. Patchy occurrences of *Calcochloris* are separated by at least 2,000 km: *C. obtusirostris* in southern Mozambique and "*C.*" *leucorhinus* in the Democratic Republic of the Congo (DRC) and Angola. The alternative classification of *leucorhinus* in the genus *Chlorotalpa* (Hutterer, 1993) does not improve the geographic cohesiveness of this taxon. In fact, neither *Chrysochloris* nor *Calcochloris* (as defined by Bronner and Jenkins, 2005) is supported as a monophyletic genus following Simonetta (1968) and Bronner (1995b: his figure 9.10; although the weighted analysis in his figure 9.11 does support a *C. leucorhinus-obtusirostris* clade).

There are four extinct chrysochlorid species named to date, with the recent addition of a possible nonzalambdodont, stem chrysochlorid genus from the Oligocene (Seiffert et al., 2007). Starting with the more complete material, *Prochrysochloris miocaenicus* is represented by cranial fragments from Kenyan

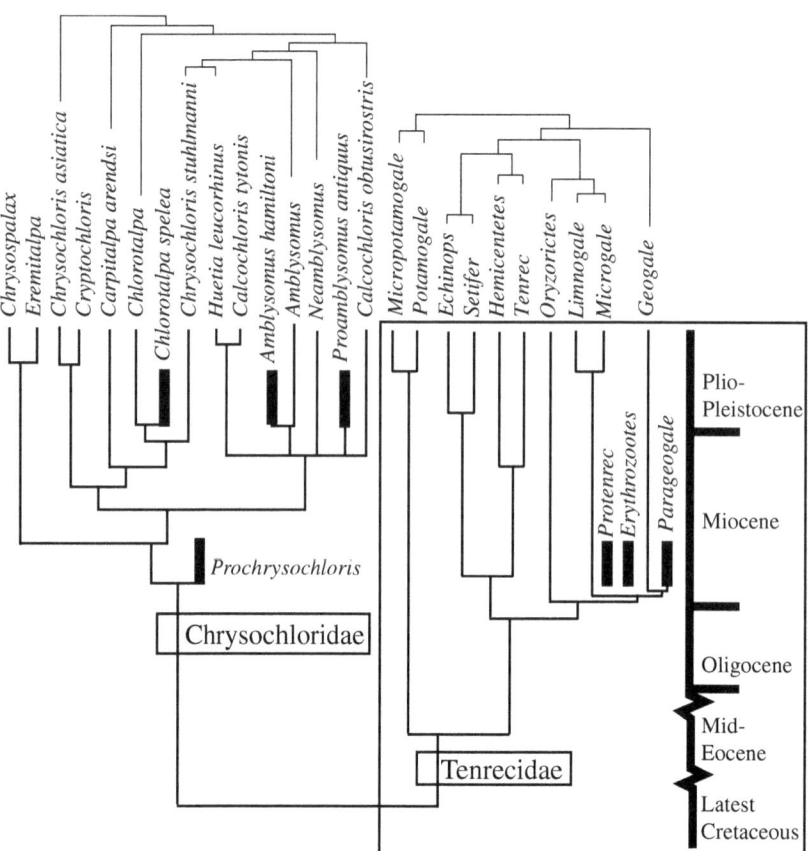

FIGURE 9.1 Phylogeny of tenrecoid (or "afrosoricid") afrotherians. Tenrecid interrelations based on Asher and Hofreiter (2006)(lower cladogram) and Olson and Goodman (2003)(upper cladogram). Olson and Goodman (2003) have a larger sample of *Microgale* than that depicted here and suggest furthermore that *Limnogale* nests within it. The affinities of *Protenrec* and *Erythrozootes* are not yet established. Chrysochlorid presented here is based on the intuitive, non-character-based approximation of Simonetta (1968: figure 2; lower tree; but see Asher et al., in press) with the fossils positioned as hypothesized by Broom (1941) and Butler (1978, 1984). The upper cladogram is based on the character-based phylogeny of Bronner (1995b: figure 9.10). Divergence estimates for the Tenrecidae are based on the molecular clock analysis of Poux et al. (2005). Except for the temporal ranges of fossils, those for chrysochlorids are arbitrary in this figure.

Miocene localities such as Songhor, Legetet, and Chamtwara (Butler, 1984). The only other record of *Prochrysochloris* is from the Miocene of southern Namibia (Mein and Pickford, 2003). Two fossil chrysochlorid species described by Broom (1941) are known from the Plio-Pleistocene of South Africa: *Chlorotalpa spelea* (figure 9.3E) from Sterkfontein and *Proamblysomus antiquus* (figure 9.3D) from "one of the small caves at Mr. Bolt's workings at Sterkfontein" (Broom, 1941:215). DeGraaff (1958) named "*Chrysotricha*" *hamiltoni* (placed by Butler [1978] in the genus *Amblysomus*) based on a skull from Makapansgat, "found amongst the sorted material from the dumps" (DeGraaff, 1958:21). Broom (1948:11) mentioned a golden mole skull from Kromdraai, with "a number of teeth well preserved" and which "possibly belongs to the species *Proamblysomus antiquus*." This specimen was also referred to by DeGraaff (1958) but, frustratingly, now appears to be lost. Several of the currently extant golden mole species also possess a fossil record in the Plio-Pleistocene of South Africa (Avery, 2000, 2001; Asher and Avery, in press). Most recently, Seiffert et al. (2007) named *Eochrysochloris tribosphenus* from the lower Oligocene of the Fayum, Egypt, and placed it in the Chrysochloridea of Broom (1916). The

published record of this taxon consists of two mandibular fragments, one preserving the m2 and another with p4 and a partial p3, both preserving several alveoli. As its specific epithet implies, this taxon retains a small talonid basin, based on which Seiffert et al. (2007; see also Seiffert, this volume, chap. 16) argue for the convergent evolution of zalambdodonty in extant tenrecs and golden moles.

Morphological Diversity Identification keys for golden moles include Roberts (1951), Simonetta (1968), and Meester et al. (1986) and focus on size, dental formula, skull shape, and the morphology of the malleus, molars, and manus to distinguish species. Easiest among the nine genera to identify are the two species of *Chrysospalax*, which are much larger than other species and show conspicuous lambdoid plates on the posterior skull for attachment of temporalis and nuchal musculature (figure 9.3A). *Chrysochloris* and *Cryptochloris* show an enlarged epitympanic recess that bulges anteriorly in the posterior compartment of the temporal fossa (figure 9.3B). This houses an elongate, club-shaped malleus. Although with less visible hypertrophy on the exterior of the skull, the mallei of *Eremitalpa* and *Chrysospalax* are also enlarged

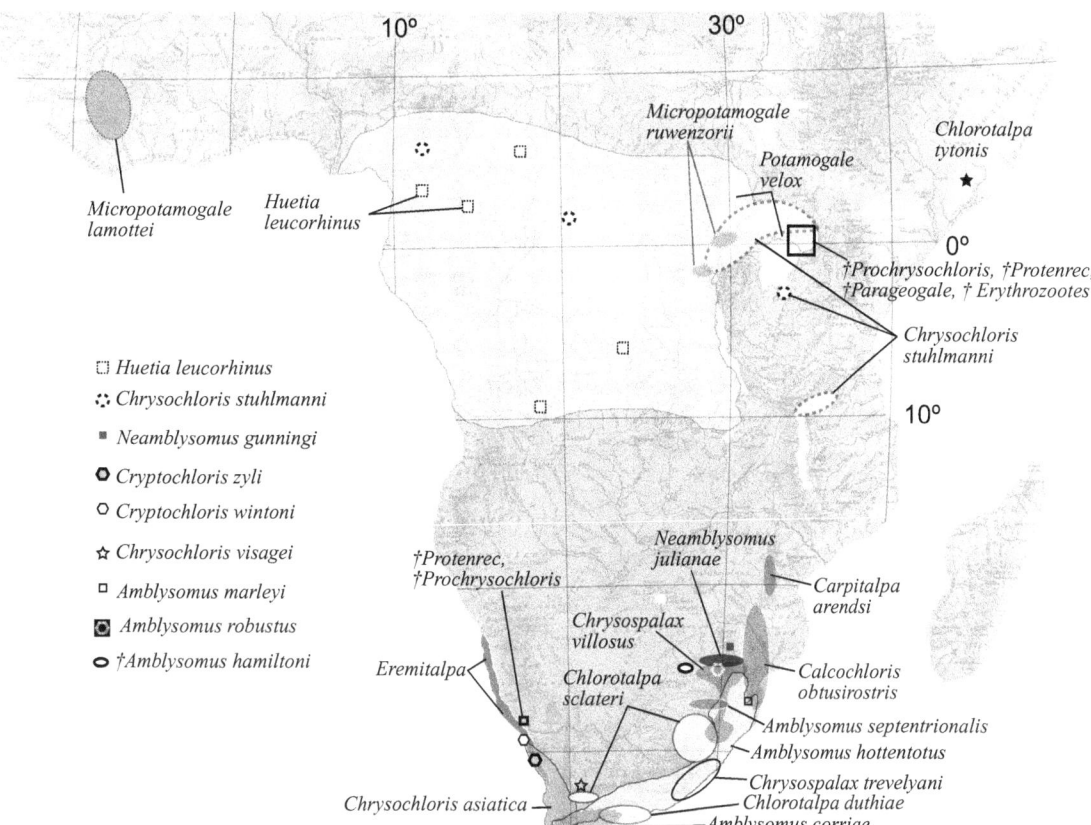

FIGURE 9.2 Distribution of golden moles and African tenrecs. The former is based on Bronner (1995b), Duncan and Wrangham (1971), Butler (1984), and the collections of the Zoologisches Museum Berlin, which houses specimens of *C. stuhlmanni* from "Ussagara" in northern Tanzania (ZMB 76775) and "Kissingi, Deutsch Ostafrika" (ZMB 76774), probably near Gisenye in north-western Rwanda, which would extend the historical range of *C. stuhlmanni* to include the area south of Lake Victoria. Tenrecid distribution is based on Kingdon (1974), Vogel (1983), Butler (1984), Nicoll and Rathbun (1990), Nowak (1999), and Mein and Pickford (2003).

but are pea shaped rather than elongate (Mason, 2003). *Amblysomus*, *Calcochloris*, and *Chlorotalpa* show a much smaller ossicular chain. Most species of *Amblysomus* have only two molars, yielding a total of 36 teeth in the upper and lower jaws, and show small, single-cusped lower molar talonids (figure 9.3F). *Chrysochloris asiatica*, *Cryptochloris*, *Eremitalpa*, and *Huetia leucorhinus* lack talonids altogether; the latter taxon lacks upper molar protocones as well.

Every species of golden mole is adept at burrowing. Among the small taxa, only *Eremitalpa granti* and *Cryptochloris* do not regularly dig durable burrows, inhabiting instead the sand dunes of the Atlantic coast (figure 9.2) where they forage at the surface (Gasc et al., 1986; Fielden et al., 1990). *Eremitalpa* shows the greatest anatomical specialization to this locomotor style—for example, in the morphology of its carpus (Kindahl, 1949). Rather than a single hypertrophied digit as in most other chrysochlorids, *Eremitalpa* shows three similarly developed terminal phalanges on the second, third, and fourth digits (Hickman, 1990). All golden moles show a reduced, buttonlike first phalanx and lack the fifth digit altogether.

The fossil taxon *Prochrysochloris* shows the characteristic teardrop shape of a golden mole skull, tapered anteriorly and rounded posteriorly. It is also zalambdodont, lacking upper molar metacones. *Prochrysochloris* has 10 teeth in each quadrant, showing prominent molar protocones and larger talonids than in extant golden moles (figures 9.3C, G). The temporal fossa does not show hypertrophied lambdoid plates as in

Chrysospalax (figure 9.3A); nor does it exhibit any swelling for a hypertrophied malleus, as in *Chrysochloris stuhlmanni* (figure 9.3B). *Prochrysochloris* also shows clear sutures between the nasal, maxilla, and frontals, demonstrating, for example, the posterior incursion of the maxilla into the orbital mosaic.

The types of both Sterkfontein fossils are largely edentulous, and Broom was himself unsure if *Proamblysomus* possessed nine or 10 teeth in each quadrant. He did, however, note that unlike other golden moles, *Proamblysomus* exhibited a laterally elongate auditory bulla (figure 9.3D). Regarding his *C. spelea*, Broom (1941) remarked only that "the fossil form is longer and narrower" than the living *Chlorotalpa sclateri*. The type of *C. spelea* (TM 1572) preserves only the right P2 but shows alveoli for 10 teeth in each quadrant. A third fossil golden mole skull kept with the *C. spelea* type in the TM collections (TM 1572-327) most likely has only nine alveoli; this difference may be sufficient to warrant recognition of another species from Sterkfontein. *Amblysomus hamiltoni* also shows nine teeth in each dental quadrant.

TENRECS (TENRECIDAE)
Figures 9.1, 9.2, and 9.4

Content and Distribution Tenrecs are better understood than golden moles in terms of their intergeneric relations. However, their fossil record is just as scanty, with only four published species in three genera (Butler, 1984; Mein and

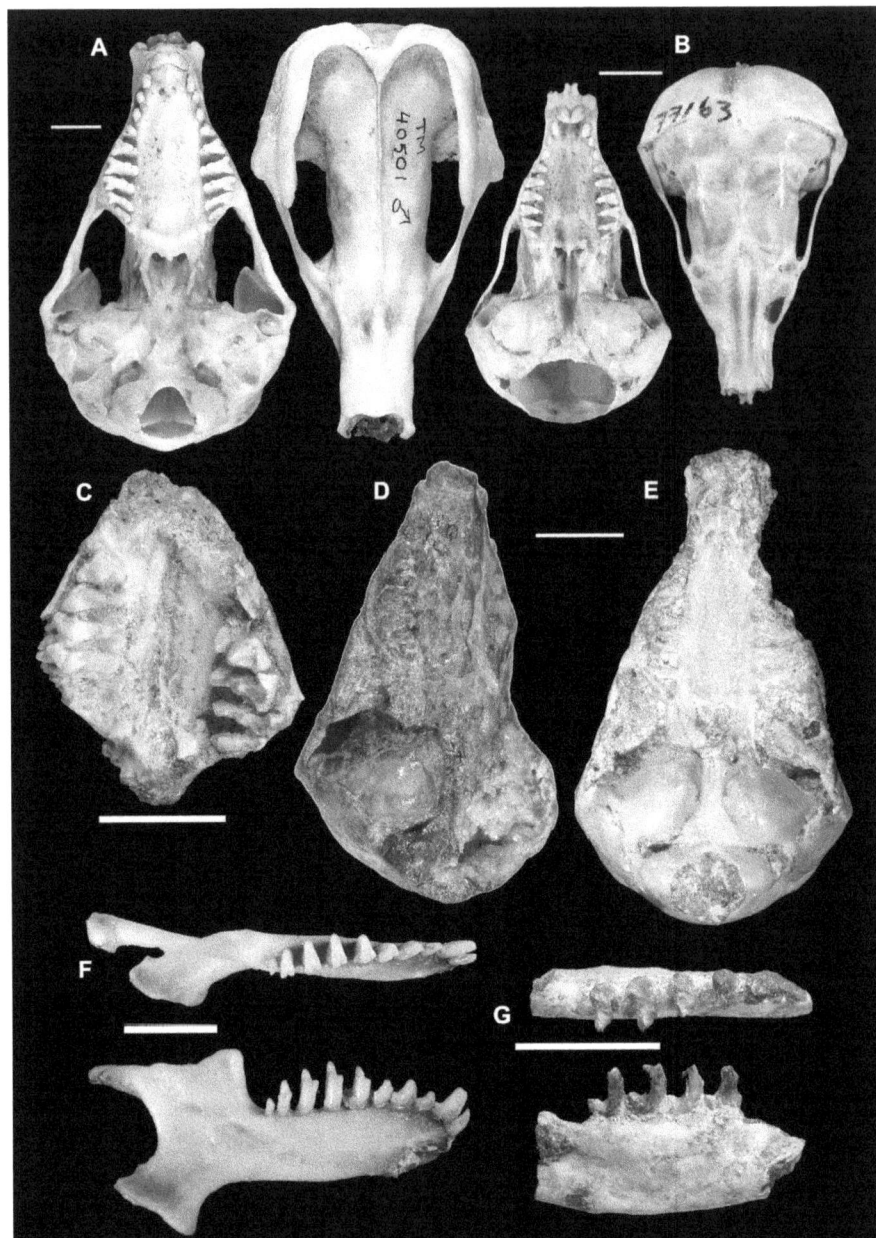

FIGURE 9.3   Selected fossil and extant golden moles: ventral and dorsal views of A) *Chrysospalax treve-lyani* (TM 40501) and B) *Chrysochloris stuhlmanni* (ZMB 77163); ventral views of C) †*Prochrysochloris miocaenicus* from Songhor, exposing right upper P3–M2 and left upper P4–M3 (KNM-SO 1412); D) †*Proamblysomus antiquus* (TM 1573); and E) †*Chlorotalpa spelea* (TM 1572); F) occlusal and internal views of *Amblysomus gunningi* (TM 40772); and G) †*Prochrysochloris miocaenicus* from Legetet, (KNM-LG 1531). Scale bars indicate 5 mm; note that D and E are photographed at the same scale.

Pickford, 2003), again with the recent, possible additions of two genera from the Eo-Oligocene boundary of the Fayum (Seiffert et al., 2007). Following Bronner and Jenkins (2005), six of the 10 extant genera (*Potamogale, Echinops, Setifer, Tenrec, Limnogale, Geogale*) are monotypic, and three (*Micropotamogale, Hemicentetes, Oryzorictes*) have two species each. The genus *Microgale* is by far the most taxonomically diverse. MacPhee (1987) recognized just 10 of 22 species, the validity of which have been complicated due primarily to variation in the appearance of the permanent dentition. Since his revision, some *Microgale* species have been reestablished and a few new ones described (cf. Jenkins, 2003; Olson et al., 2004). Bronner and Jenkins (2005) recognize 18 species of *Microgale*.

Extinct tenrecid species are similar to golden moles in their distribution: *Protenrec tricuspis* (figure 9.4F, G), *Erythrozootes chamerpes* (figure 9.4C, H; Protenrecinae), and *Parageogale aletris* (figure 9.4E; Geogalinae) are known from early Miocene localities in southwest Kenya and Uganda, including Chamtwara, Legetet, Songhor, Napak, and Rusinga Island (Butler, 1984). *Protenrec butleri* has recently been named from the Miocene of southern Namibia, from the same locality that produced *Prochrysochloris* sp. (Mein and Pickford, 2003). Seiffert et al. (2007) suggested that two new genera, *Widanelfarasia* and *Jawharia*, known from several jaws and teeth, may nest within the Tenrecidae (and at the base of the Oligocene would therefore comprise the oldest record of the group).

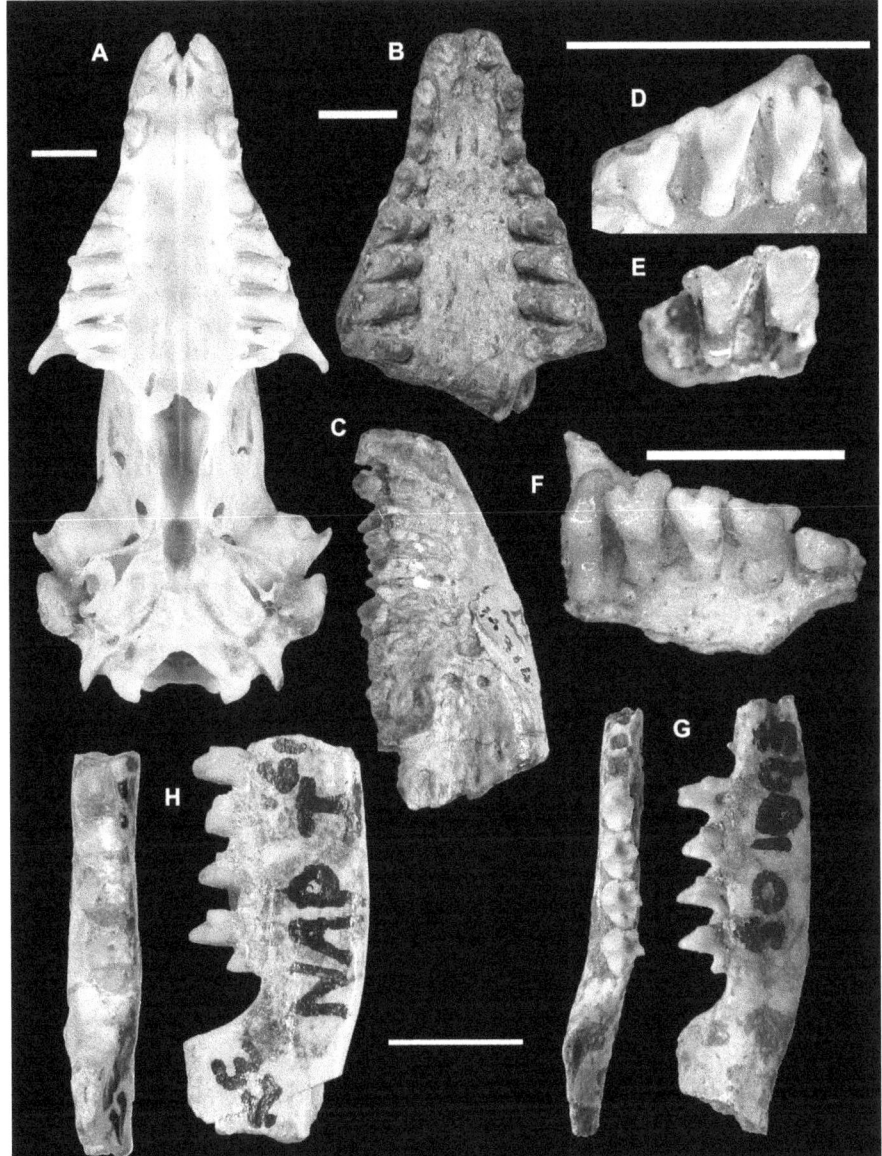

FIGURE 9.4   Selected fossil and extant tenrecs: A) ventral view of *Setifer setosus* (ZMB 44293); B) ventral and C) lateral views of †*Erythrozootes chamerpes* (BMNH M14314); D) *Geogale aurita* left upper P4–M3 (MCZ 45044); E) †*Parageogale aletris* from Chamtwara, left upper M1–2 (KNM-CA 1548; the P4 from this specimen illustrated by Butler [1984: figure 18E] is currently missing); F) †*Protenrec tricuspis* from Napak IV, right upper P3–M3 (BMNH M43552); G) †*Protenrec tricuspis* from Songhor, left lower p4–m3 (KNM-SO 1093); H) †*Erythrozootes chamerpes* from Napak I, left lower p4–m3 (BMNH M21831). Scale bars indicate 5 mm; note that B–C, D–E, and G–H are photographed at the same scale.

However, Seiffert et al. (2007) did not rule out the possibility that one or both comprised sister taxa to a tenrec–golden mole clade. In contrast to the golden moles, there are to date no fossil tenrecs known from South Africa. Nor are there distinct fossil tenrec species from Madagascar (Goodman et al., 2003). Grandidier (1928) described "*Cryptogale australis*" based on subfossil cranial remains from Andrahomana cave, near Fort Dauphin (Taolanaro) in the southeast. However, following Heim de Balsac (1972), this specimen is currently regarded as an individual of the extant *Geogale aurita*.

Only three living tenrec species are known from the African mainland: *Potamogale velox* from a large (and noncontiguous) region of tropical river systems spanning Nigeria, the DRC, western Kenya, Zambia, and Angola (figure 9.2). *Micropotamogale ruwenzorii* shows a very small, patchy distribution west of Lake Victoria in northeastern DRC and Uganda, possibly extending also into Burundi and Rwanda (Nicoll and Rathbun, 1990). Over 4,000 km to the west, *M. lamottei* is known near the intersection of Liberia, Guinea, and the Ivory Coast (Vogel, 1983). African otter shrews are placed in the Potamogalinae as a subfamily of the Tenrecidae (e.g., McDowell, 1958; Butler, 1984; Hutterer, 1993; Nowak 1999; Asher, 1999; Olson and Goodman 2003; Bronner and Jenkins 2005; Asher and Hofreiter, 2006). In contrast, Salton and Szalay (2004), Seiffert et al. (2007), and Salton and Sargis (2008) are the only recent authors to use the taxonomy of Simpson (1945) and Eisenberg and Gould (1970), according to which potamogalines are elevated to family level, with their "Tenrecoidea" equivalent to the Tenrecidae of other authors.

All but five of the 27 tenrec species on Madagascar are present in the humid forest along the island's east coast. Most of these are also present in Madagascar's central and northern highlands (see Goodman, 2003a, table 13.7 therein). Three species, *Geogale aurita* (figure 9.4D), *Echinops telfairi*, and *Microgale nasoloi* are known only from the relatively arid southwest. *Setifer setosus* (figure 9.4A), *Tenrec ecaudatus*, and three species of *Microgale* (*brevicaudata*, *longicaudata*, and *talazaci*) are found throughout the island in most or all of the habitat zones defined by Goodman (2003a). *Tenrec* is also present, presumably introduced by humans, in the nearby islands of Mauritius, Reunion, and the Comoros (Nicoll, 2003).

The eight Malagasy genera consist of "spiny tenrecs" (*Echinops*, *Setifer*, *Tenrec*, and *Hemicentetes*) and "soft tenrecs" (*Microgale*, *Limnogale*, *Oryzorictes*, *Geogale*). Some uncertainty persists regarding the affinities of *Geogale*. In the analysis of Asher and Hofreiter (2006) it is supported as sister taxon to the extinct *Parageogale*, to which the other East African fossils form a weakly supported clade, followed by *Microgale* and *Limnogale* to the exclusion of *Oryzorictes* (figure 9.1). Regarding the extant genera, Poux et al. (2008) support a similar topology, with *Geogale* as the sister taxon to Oryzorictines to the exclusion of spiny tenrecs. In contrast, Olson and Goodman (2003) placed *Geogale* as the sister taxon to all other Malagasy tenrecs (figure 9.1). The monophyly of a spiny tenrec clade is supported by analyses of both DNA sequences (Olson and Goodman, 2003; Poux et al., 2005, 2008; Asher and Hofreiter, 2006) and morphology (Asher, 1999). In contrast, morphology and DNA sequences do not produce the same signal regarding the semiaquatic *Limnogale*. I have previously argued for a *Limnogale*–African potamogaline clade based on shared possession of a number of cranial characters (Asher, 1999; see later discussion). However, these morphological similarities appear to be homoplastic based on the near perfect support indices for a *Microgale-Limnogale* clade using not only independent sequence data sets (Poux et al., 2005, 2008) but also combined analyses of DNA and morphology (Olson and Goodman, 2003; Asher and Hofreiter, 2006). In fact, two publications that sample multiple species of *Microgale* indicate that *Limnogale* actually falls within the former genus (Olson and Goodman, 2003; Poux et al., 2008).

Morphological Diversity The niche of semiaquatic carnivory has been independently occupied by the Malagasy *Limnogale* and African potamogalines (Olson and Goodman, 2003; Asher and Hofreiter, 2006). Several species are adept at digging and climbing; among these the "rice-tenrec" *Oryzorictes* is the most fossorial (Goodman, 2003b). Long-tailed shrew tenrecs (*Microgale longicaudata*) and the lesser Malagasy hedgehog (*Echinops telfairi*) are semiarboreal; the former species is aided in this endeavor by its prehensile tail (Jenkins, 2003: figure 13.28).

The African "otter-shrews" are dedicated semiaquatic faunivores. All possess syndactyl pedal digits II and III, which are used for grooming and evidently play a major role in maintaining the animal's health. Captivity has been associated with distress and inadequate grooming, leading to the inability to maintain water resistance, temperature regulation, and, ultimately, death (Nicoll, 1985). The largest of the three African species (*Potamogale velox*) swims like a teleost, using lateral movements of its back and enlarged, mediolaterally compressed tail (Kingdon, 1974). As it swims, its non-webbed feet are tucked ventral to the pelvis and are not used in locomotion (Dobson, 1883; Kingdon, 1974). Both species

of *Micropotamogale* have a much smaller tail than *P. velox*. Presumably this is accompanied by a distinct locomotor style involving (for example) the webbed hind feet of *M. ruwenzorii*. Interestingly, *M. lamottei* has neither digital webbing nor pronounced flattening of the tail but is nevertheless adept at locomotion in a riverine habitat.

The "web-footed" tenrec *Limnogale* (Benstead and Olson, 2003) resembles *Micropotamogale ruwenzorii* in possessing an ovoid tail and webbed feet. It shares with all potamogalines enlarged hypoglossal foramina, reduction of the lacrimal foramina, and a shortened frontal bone. The latter two characters can be interpreted as susceptible to homoplasy among semiaquatic mammals, possibly associated with reduction of the lacrimal canal and olfactory lobe of the brain, respectively (Voss, 1988). However, nontenrecid semiaquatic faunivores do not consistently resemble *Limnogale* and potamogalines regarding these characters (Sánchez-Villagra and Asher, 2002). In addition, some specimens of terrestrial *Microgale* and *Geogale* also show enlarged hypoglossal foramina (Olson and Goodman, 2003). Histological examinations of soft tissues show that the hypoglossal nerve is not correspondingly enlarged in potamogalines, *Microgale*, or *Geogale* (Asher, 2001).

The two monotypic genera of Malagasy "hedgehogs" (*Setifer* and *Echinops*) are externally quite similar to their European erinaceid analogs. Both taxa are smaller than European *Erinaceus* but show overlap with smaller individuals of the Afro-Asian erinaceids *Atelerix* and *Hemiechinus*. Malagasy "hedgehogs" will also roll up into a ball of spines when threatened. The other two tenrecines, *Hemicentetes* and *Tenrec*, also possess a spiny pelage but do not share the hedgehog-like appearance of *Echinops* and *Setifer*. Adult *Hemicentetes* resemble juvenile *Tenrec*, both showing longitudinal, dorsal coloration and the "stridulating organ" that consists of thickened quills on the middorsum that can rub together and produce high-pitched signals (Eisenberg and Gould, 1970). Both *Hemicentetes* and *Tenrec* have barbed, detachable spines, most densely distributed around the dorsum of the neck. *Tenrec* grows to a considerably larger size than *Hemicentetes*, with adults resembling *Didelphis* in overall appearance but with a reduced tail and a relatively larger head. Captive specimens tend to keep growing as long as the animal retains teeth with which to masticate. Body temperature in tenrecs fluctuates daily, although less so in females during pregnancy (Nicoll, 2003; Stephenson, 2003). Interestingly, breeding males appear to maintain a slightly lower body temperature than females or nonbreeding males, possibly to maintain sperm motility with the testicles located intraabdominally, adjacent to the kidneys (Nicoll, 2003).

At least one of the fossil tenrecs exhibits cranial similarities to an extant tenrecid lineage: *Parageogale aletris* (first described as *Geogale aletris* in Butler and Hopwood, 1957) resembles the living *Geogale aurita* in small size, molar morphology (figures 9.4D and 9.4E), possession of two premaxillary teeth, broad interincisal distance, rostrum shape, and in having a long infraorbital canal (Butler, 1984, 1985; Asher and Hofreiter, 2006). These similarities led Butler (e.g., 1984, 1985, and references cited therein) to support a sister-taxon relationship to the exclusion of other living and fossil tenrecs between the living *Geogale* and the East African fossil *Parageogale*. Poduschka and Poduschka (1985) disputed this interpretation, primarily because the type specimen of *Parageogale* (BMNH M33046, a rostral skull fragment) is poorly preserved and, in their opinion, not diagnostically a tenrecid. However, they did not offer an alternative interpretation, nor did they adequately consider other fossils mentioned by Butler (1969, 1984), such as a maxilla fragment identified as *Parageogale* (KNM-CA 1548) clearly showing a zalambdodont M1 and partial M2 (figure 9.4E), that closely resemble the M1–2 of modern *Geogale* in both size and morphology (figure 9.4D). Cladistic analyses testing the proposed *Geogale-Parageogale* clade (Asher and Hofreiter, 2006; Seiffert et al., 2007) support Butler's interpretation.

ACKNOWLEDGMENTS

I am grateful to Lars Werdelin and Bill Sanders for the opportunity to contribute to this volume and for constructive critiques of the manuscript. For assistance and discussion during museum visits in South Africa, I thank Theresa Kearny, Thomas Lehmann, Stephanie Potze, Denise Hamerton, Margaret Avery, Kerwin van Willingh, and Gary Bronner. I thank Emma Mbua, Susy Cote, and Eleanor Weston for assistance in Nairobi. Paula Jenkins, Ralf Britz, and Marcelo Sánchez also deserve thanks for their hospitality during my visits to London. I thank Rainer Hutterer for helpful comments on the manuscript. For financial support I thank the Deutsche Forschungsgemeinschaft (grant AS 245/2–1), the European Commission's Research Infrastructure Action via the SYNTHESYS Project (GB-TAF 218, SE-TAF 4069), and the National Science Foundation USA (DEB 9800908).

Literature Cited

Asher, R. J. 1999. A morphological basis for assessing the phylogeny of the "Tenrecoidea" (Mammalia, Lipotyphla). *Cladistics* 15:231–252.

———. 2001. Cranial anatomy in tenrecid insectivorans: Character evolution across competing phylogenies. *American Museum Novitates* 3352:1–54.

Asher, R. J., and D. M. Avery. 2010. New golden moles (Afrotheria, Chrysochloridae) from the Pliocene of South Africa. *Paleontologia Electronica* 13 (1), 3A.

Asher, R. J., and M. Hofreiter. 2006. Tenrec phylogeny and the noninvasive extraction of nuclear DNA. *Systematic Biology* 55:181–194.

Asher, R. J., S. Maree, G. Bonner, N. C. Bennett, P. Bloomer, P. Czechowski, M. Meyer, and M. Hofreiter. In press. A phylogenetic estimate for golden moles (Mammalia, Afrotheria, Chrysochloridae). BMC Evolutionary Biology.

Avery, D. M. 2000. Micromammals; pp. 305–338 in T. C. Partridge and R. R. Maud (eds.), *The Cenozoic of Southern Africa*. Oxford Monographs on Geology and Geophysics 40. Oxford University Press, New York.

———. 2001. The Plio-Pleistocene vegetation and climate of Sterkfontein and Swartkrans, South Africa, based on micromammals. *Journal of Human Evolution* 41:113–132.

Benstead, J. P., and L. E. Olson. 2003. *Limnogale mergulus*, web-footed tenrec or aquatic tenrec; pp. 1267–1273 in S. M. Goodman and J. P. Benstead (eds.), *The Natural History of Madagascar*. University of Chicago Press, Chicago.

Bronner, G. N. 1991. Comparative hyoid morphology of nine chrysochlorid species (Mammalia: Chrysochloridae). *Annals of the Transvaal Museum* 35:295–311.

———. 1995a. Cytogenetic properties of nine species of golden moles (Insectivora: Chrysochloridae). *Journal of Mammalogy* 76:957–971.

———. 1995b. Systematic revision of the golden mole genera *Amblysomus*, *Chlorotalpa*, and *Calcochloris* (Insectivora, Chrysochloromorpha, Chrysochloridae). Unpublished PhD dissertation, University of Natal, Durban.

Bronner, G. N., and P. Jenkins. 2005. Order Afrosoricida; pp. 71–81 in D. E. Wilson and D. M. Reeder (eds.), *Mammal Species of the World*. Johns Hopkins University Press, Baltimore.

Broom, R. 1916. On the structure of the skull in *Chrysochloris*. *Proceedings of the Zoological Society of London* 1916: 449–459.

———. 1941. On two Pleistocene golden moles. *Annals of the Transvaal Museum* 20:215–216.

———. 1948. Some South African Pliocene and Pleistocene mammals. *Annals of the Transvaal Museum* 21:1–38.

Butler, P. M. 1969. Insectivores and bats from the Miocene of East Africa: New material; pp. 1–37 in L. S. B. Leakey (ed.), *Fossil Vertebrates of Africa*, vol. 1. Academic Press, London.

———. 1978. Insectivora and Chiroptera; pp. 56–68 in V. J. Maglio and H. B. S. Cooke (eds.), *Evolution of African Mammals*. Harvard University Press, Cambridge.

———. 1984. Macroscelidea, Insectivora, and Chiroptera from the Miocene of East Africa. *Palaeovertebrata* 14:117–200.

———. 1985. The history of African insectivores. *Acta Zoologica Fennica* 173:215–217.

———. 1988. Phylogeny of the insectivores; pp. 117–141 in M. J. Benton (ed.), *The Phylogeny and Classification of the Tetrapods*, vol. 2. Clarendon Press, Oxford.

Butler, P. M., and A. T. Hopwood. 1957. Insectivora and Chiroptera from the Miocene rocks of Kenya colony. *Fossil Mammals of Africa, British Museum (Natural History)* 13:1–35.

DeGraaff, G. 1958. A new chrysochlorid from Makapansgat. *Palaeontologia Africana* 5:21–27.

Dobson, G. E. 1883. *A Monograph of the Insectivora, Systematic and Anatomical: Part II. Including the Families Potamogalidae, Chrysochloridae, and Talpidae.* Van Voorst, London, 172 pp.

Duncan, P., and R. W. Wrangham. 1971. On the ecology and distribution of subterranean insectivores in Kenya. *Journal of Zoology, London* 164:149–163.

Eisenberg, J. F., and E. Gould. 1970. The tenrecs: a study in mammalian behavior and evolution. *Smithsonian Contributions to Zoology* 27:1–137.

Fielden, L. J., G. C. Hickman, and M. R. Perrin. 1992. Locomotory activity in the Namib Desert golden mole *Eremitalpa granti namibensis* (Chrysochloridae). *Journal of Zoology* 226:329–344.

Fielden, L. J., M. R. Perrin, and G. C. Hickman. 1990. Feeding ecology and foraging behaviour of the Namib Desert golden mole, *Eremitalpa granti namibensis* (Chrysochloridae). *Journal of Zoology* 220:367–389.

Gasc, J. P., F. K. Jouffroy, S. Renous, and F. von Blottnitz. 1986. Morphofunctional study of the digging system of the Namib Desert golden mole (*Eremitalpa granti namibensis*): Cinefluorographical and anatomical analysis. *Journal of Zoology* 208:9–35.

Goodman, S. M. 2003a. Checklist to the extant land mammals of Madagascar; pp. 1187–1191 in S. M. Goodman and J. P. Benstead (eds.), *The Natural History of Madagascar*. University of Chicago Press, Chicago.

———. 2003b. *Oryzorictes*, mole tenrec or rice tenrec; pp. 1278–1281 in S. M. Goodman and J. P. Benstead (eds.), *The Natural History of Madagascar*. University of Chicago Press, Chicago.

Goodman, S. M., J. U. Ganzhorn, and D. Rakotodravony. 2003. Introduction to the mammals; pp. 1159–1186 in S. M. Goodman and J. P. Benstead (eds.), *The Natural History of Madagascar*. University of Chicago Press, Chicago.

Grandidier, G. 1928. Description de deux nouveaux Mammifères Insectivores de Madagascar. *Bulletin du Museum National d'Histoire Naturelle* 34:63–70.

Heim de Balsac, H. 1972. Insectivores; pp. 629–660 in R. Battistini and G. Richard-Vindard (eds.), *Biogeography and Ecology in Madagascar*. Junk, The Hague.

Hickman, G. C. 1990. The Chrysochloridae: studies toward a broader perspective of adaptation in subterranean mammals; pp. 23–48 in E. R. Nevo and O. A. Reig (eds.), *Evolution of Subterranean Mammals at the Organismal and Molecular Levels*. Liss, New York.

Huet, M. 1885. Note sur une espèce nouvelle de Chrysochlore de la côte du Golfe de Guinée et sur les insectivores du même genre faisant partie de la collection du Muséum d'Histoire Naturelle. *Nouvelles Archives du Museum d'Histoire Naturelle*, Série II, 8:1–15.

Hutterer, R. 1993. Order Insectivora; pp. 69–130 in D. E. Wilson and D. M. Reeder (eds.), *Mammal Species of the World*. Smithsonian Institution Press, Washington, D.C.

Jenkins, P. 2003. *Microgale*, shrew tenrecs; pp. 1273–1278 in S. M. Goodman and J. P. Benstead (eds.), *The Natural History of Madagascar*. University of Chicago Press, Chicago.

Kindahl, M. 1949. The embryonic development of the hand and foot of *Eremitalpa* (*Chrysochloris granti* [Broom]). *Acta Zoologica* 25:133–152.

Kingdon, J. 1974. *East African Mammals: An Atlas of Evolution in Africa*, Vol. IIA. University of Chicago Press, Chicago, 340 pp.

MacPhee, R. D. E. 1987. The shrew tenrecs of Madagascar: Systematic revision and Holocene distribution of *Microgale* (Tenrecidae, Insectivora). *American Museum Novitates* 2889:1–45.

Mason, M. 2003. Morphology of the middle ear of golden moles (Chrysochloridae). *Journal of Zoology* 260:391–403.

McDowell, S. B. 1958. The Greater Antillean insectivores. *Bulletin of the American Museum of Natural History* 115:115–213.

Meester, J. A. J., I. L. Rautenbach, N. J. Dippenaar, and L. M. Baker. 1986. Classification of southern African mammals. *Transvaal Museum Monograph* 5:1–359.

Mein, P., and M. Pickford. 2003. Insectivora from Arrisdrift, a basal Middle Miocene locality in southern Namibia. *Memoir of the Geological Survey of Namibia* 19:143–146.

Nicoll, M. E. 1985. The biology of the giant otter shrew *Potamogale velox*. *National Geographic Research* 21: 331–337.

———. 2003. *Tenrec ecaudatus*, Tenrec, *Tandraka, Trandraka*; pp. 1283–1287 in S. M. Goodman and J. P. Benstead (eds.), *The Natural History of Madagascar*. University of Chicago Press, Chicago.

Nicoll, M. E., and G. B. Rathbun. 1990. *African Insectivora and Elephant-shrews: An Action Plan for Their Conservation*. IUCN, Gland, Switzerland, 53 pp.

Nowak, R. M. 1999. *Walker's Mammals of the World*, 6th ed. Johns Hopkins University Press, Baltimore, 2015 pp.

Olson, L. E., and S. M. Goodman. 2003. Phylogeny and biogeography of tenrecs; pp. 1235–1242 in S. M. Goodman and J. P. Benstead (eds.), *The Natural History of Madagascar*. University of Chicago Press, Chicago.

Olson, L. E., S. M. Goodman, and A. D. Yoder. 2004. Illumination of cryptic species boundaries in long-tailed shrew tenrecs (Mammalia: Tenrecidae; *Microgale*): New insights into geographic variation and distributional constraints. *Biological Journal of the Linnean Society* 83:1–22.

Petter, F. 1981. Remarques sur le systématique des chrysochlorides. *Mammalia* 45:49–53.

Poduschka, W., and C. Poduschka. 1985. Zur Frage des Gattungsnamens von "*Geogale*" *aletris* Butler und Hopwood, 1957 (Mammalia, Insectivora) aus dem Miozän Ostafrikas. *Zeitschrift für Säugetierkunde* 50:129–140.

Poux C., O. Madsen, J. Glos, W. W. de Jong, and M. Vences. 2008. Molecular phylogeny and divergence times of Malagasy tenrecs: Influence of data partitioning and taxon sampling on dating analyses. *BMC Evolutionary Biology* 8:102.

Poux, C., O. Madsen, E. Marquard, D. R. Vieites, W. W. De Jong, and M. Vences. 2005. Asynchronous colonization of Madagascar by the four endemic clades of primates, tenrecs, carnivores, and rodents as inferred from nuclear genes. *Systematic Biology* 54:719–730.

Roberts, A. 1951. *Mammals of South Africa*. Trustees of the Mammals of South Africa Book Fund, Pretoria, 700 pp.

Salton J. A., and E. J. Sargis. 2008. Evolutionary morphology of the Tenrecoidea (Mammalia) carpal complex. *Biological Journal of the Linnean Society* 93:267–288.

Salton J. A., and Szalay F. S. 2004. The tarsal complex of Afro-Malagasy Tenrecoidea: A search for phylogenetically meaningful characters. *Journal of Mammalian Evolution* 11:73–104.

Sánchez-Villagra, M. R., and R. J. Asher. 2002. Cranio-sensory adaptations in small, faunivorous mammals, with special reference to olfaction and the trigeminal system. *Mammalia* 66:93–109.

Seiffert E. R., E. L. Simons, T. M. Ryan, T. M. Bown, and Y. Attia. 2007. New remains of Eocene and Oligocene Afrosoricida (Afrotheria) from Egypt, with implications for the origin(s) of afrosoricid zalambdodonty. *Journal of Vertebrate Paleontology* 27: 963–972.

Simonetta, A. M. 1968. A new golden mole from Somalia with an appendix on the taxonomy of the family Chrysochloridae (Mammalia: Insectivora). *Monitore Zoologico Italiano* 2 (suppl.):27–55.

Simpson, G. G. 1945. The principles of classification and a classification of mammals. *Bulletin of the American Museum of Natural History* 85:1–350.

Stanhope, M. J., V. G. Waddell, O. Madsen, W. W. de Jong, S. B. Hedges, G. C. Cleven, D. Kao, and M. S. Springer. 1998. Molecular evidence for multiple origins of the Insectivora and for a new order of endemic African mammals. *Proceedings of the National Academy of Sciences, USA* 95:9967–9972.

Stephenson, P. J. 2003. *Hemicentetes*, streaked tenrecs, *Sora, Tsora*; pp. 1281–1283 in S. M. Goodman and J. P. Benstead (eds.), *The Natural History of Madagascar*. University of Chicago Press, Chicago.

Vogel, P. 1983. Contribution a l'écologie et a la zoogéographie de *Micropotamogale lamottei* (Mammalia, Tenrecidae). *Revue d'Écologie (La Terre et la Vie)* 38:37–48.

Voss, R. S. 1988. Systematics and ecology of ichthyomyine rodents (Muroidea): Patterns of morphological evolution in a small adaptive radiation. *Bulletin of the American Museum of Natural History* 188:260–493.

TEN

Tubulidentata

PATRICIA A. HOLROYD

Aardvarks, or antbears, are the least diverse extant order and one of the most ancient eutherian lineages, although the fossil record offers us few clues to their early evolution. Holroyd and Mussell (2005) reviewed hypotheses for the phylogenetic position of tubulidentates, Lehmann et al. (2005) reviewed the taxonomic history of the order and presented the first species level analysis of tubulidentate morphological evolution; and Lehmann (2006) provides an overview of the aardvark record throughout the Old World. This contribution focuses on a review of their African distribution (figure 10.1)

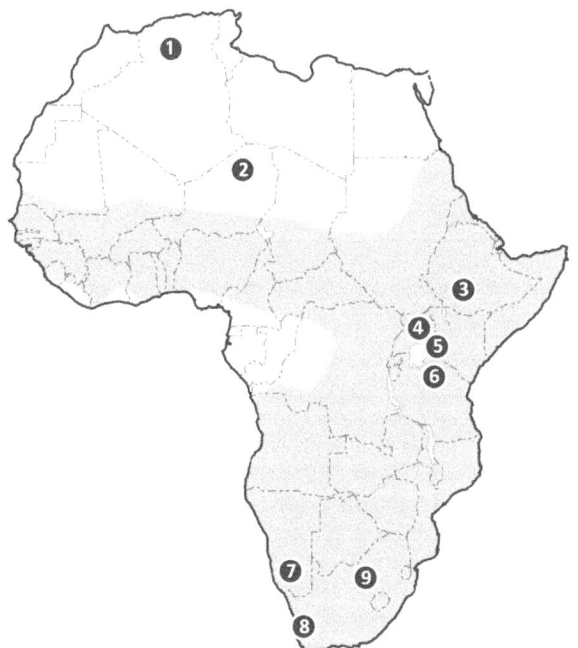

FIGURE 10.1 Map showing distribution of fossil tubulidentates detailed in table 10.1; area shown in gray is the distribution of modern *Orycteropus afer*. 1) *Orycteropus*, Pleistocene, Algeria; 2) *Orycteropus*, early Pliocene, Chad; 3) *Orycteropus*, Pliocene-Pleistocene, Ethiopia; 4) *Myorycteropus*, early Miocene, Uganda; 5) numerous sites of Miocene to Pleistocene age; (*Orycteropus*, *Leptorycteropus*, and *Myorycteropus*), Kenya; 6) *Orycteropus*, Pleistocene, Tanzania; 7) *Orycteropus* Namibia; 8) *Orycteropus*, early Pliocene, South Africa; 9) *Orycteropus*, Pliocene-Pleistocene, South Africa.

and clarifying taxonomy, based on a discussion of characters originally used to recognize taxa and the analysis of Lehmann et al. (2005).

Systematic Paleontology

Order TUBULIDENTATA Huxley, 1872
Family ORYCTEROPODIDAE Gray, 1821

Orycteropodids share a dentition consisting of teeth that are ever-growing, fused dentine columns surrounded by cement. Where known, all have elongated snouts with enlarged olfactory regions and postcranial features associated with digging, related to their inferred habits of burrowing for shelter and digging for ants and termites, as seen in extant *Orycteropus afer*.

Genus *MYORYCTEROPUS* MacInnes, 1956
Figures 10.2D and 10.2G

Selected Synonymy Orycteropus africanus (MacInnes, 1956), Pickford, 1975.
Only Known Species Myorycteropus africanus MacInnes, 1956.
Age and Occurrence Early Miocene, Kenya and Uganda; see table 10.1.
Emended Diagnosis After Pickford (1975) for species and MacInnes (1956) and Lehmann et al. (2005) for genus. Sixty percent as large as *O. afer*; ascending ramus of jaw lies at about a 45º angle with the alveolar border of horizontal ramus (figure 10.2); distal breadth of humerus about 48% of length; third trochanter occupies about 22% of femur length. Differs from similarly sized *O. chemeldoi* in having shorter and smaller teeth. M3 distinctly bilobed. Forelimbs generally more robust (e.g., relatively wider distal radius, more robust ulna), and digging features accentuated. Differs from *Orycteropus* and "*Leptorycteropus*" in having proximal humerus that is wider anteroposteriorly than mediolaterally, curved humeral shaft, distinct femoral neck, and possessing a cnemial tibial tuberosity, all presumably primitive retentions.
Remarks Pickford (1975) synonymized *Myorycteropus* with *Orycteropus* and referred to this species a number of new specimens, primarily postcranial. However, the phylogenetic

TABLE 10.1
Major occurrences and ages of African tubulidentates

Taxon	Occurrence	Stratigraphic Unit	Age
Myorycteropus africanus	Napak IV, Uganda		20–19 Ma
	Rusinga, Kenya	Hiwegi Fm.	17.8 ± 0.5 Ma
	Mfangano, Kenya		17.8 ± 0.5 Ma
Leptorycteropus guilielmi	Lothagam, Kenya	Nawata Fm., lower part	7.4–6.5 Ma
Orycteropus minutus	Songhor, Kenya		20–19 Ma
	Mfangano, Kenya		17.8 ± 0.5 Ma
	Rusinga, Kenya		17.8 ± 0.5 Ma
	Arrisdrift, Namibia		17.5–17 Ma
O. chemeldoi	Tugen Hills, Kenya	Ngorora Fm.	13.15–7.6 Ma
	Ft. Ternan, Kenya		13.8 ± 0.3 Ma
O. crassidens	Rusinga & Kanjera, Kenya		Early Pleistocene
O. mauritanicus	Bou Hanifa, Algeria		Late Miocene
O. djourabensis	Kollé, Chad		Early Pliocene (5–4 Ma)
	Koobi Fora , Kenya	Koobi Fora Fm.	Early Pleistocene 4.1–4.2 Ma
O. cf. *O. djourabensis*	Asa Issie, Ethiopia		
O. abundulafus	Kossom Bougoudi, KB03, Chad		~5 Ma (Mio-Plio boundary)
O. afer	Algeria		Late Pleistocene
	Lainyamok, Kenya		Mid- to late Pleistocene
	Equus Cave, Sea Harvest Site, South Africa		Late Pleistocene
O. cf. *O. afer*	Langebaanweg, South Africa	Varswater Fm.	5.2–5.0 Ma
Orycteropus sp. (small), undetermined	Lukeino, Kenya	Lukeino Fm.	6.1–5.7 Ma
Orycteropus sp. (large), undetermined	Meswa Bridge, Kenya	Muhoroni Agglomerate	23–22 Ma
		Nawata Fm.	7.4–5 Ma
	Lothagam, Kenya	Lukeino Fm.	6.1–5.7 Ma
	Lukeino, Kenya	Okote Mbr.,	4–3.4 Ma
	East Turkana, Area 250, Kenya	Koobi Fora Fm.	4–2.6 Ma
	Laetoli, Tanzania		4–2.6 Ma
	Omo, Ethiopia		3.5–2.5 Ma
	Makapansgat, South Africa		2–1.5 Ma
	Swartkrans, South Africa		1.54–1.38 Ma
	Ileret, Area 11, Kenya		Pleistocene
	Kanjera, Kenya		Early Pleistocene
	Konso, Ethiopia		

SOURCE: Based on Leakey (1931, 1951, 1987), Romer (1938), Clark (1942), Dietrich (1942), Peabody (1954), MacInnes (1956), Arambourg (1959), Kitching (1963), Butzer (1971), Hendey (1973), Howell and Coppens (1974), Patterson (1975), Pickford (1975, 1994, 1996), Pickford and Andrews (1981), Potts et al. (1988), Klein et al. (1991), Asfaw et al. (1992), Grine and Klein (1993), Pickford and Senut (2002), Milledge (2003), Lehmann (2004, 2008), and Lehmann et al. (2005, 2006). Numeric age estimates from Werdelin (this volume, chap. 3).

analysis of Lehmann et al. (2005) demonstrates that *M. africanus* lacks several key features that unite the other tubulidentate species. While it is only retention of presumably plesiomorphic characters that distinguishes it from *Orycteropus*, it indicates that at least two distinct lineages of orycteropodids existed by the early Miocene.

Genus *ORYCTEROPUS* C. Geoffroy Saint-Hilaire, 1796
Figure 10.2A–10.2C, 10.2E, 10.2H, and 10.2I

Selected Synonymy Leptorycteropus Patterson, 1975:186.
Included African Species Orycteropus afer (Pallas, 1766); *O. crassidens* MacInnes, 1956; *O. mauritanicus* Arambourg, 1959; *O. minutus* Pickford, 1975; *O. chemeldoi* Pickford, 1975; *O. djourabensis* Lehmann, Vignaud, Mackaye, Brunet, 2004; *O. abundulafus* Lehmann, Vignaud, Likius, and Brunet, 2005.

Age and Occurrence Early Miocene to recent; see table 10.1.
Diagnosis Differs from both *Leptorycteropus* and *Myorycteropus* in larger size (figure 10.2). Differs from *Myorycteropus* in having proximal humerus that is subequal in anteroposterior and mediolateral dimensions, straight humeral shaft, indistinct femoral neck that is confluent with femoral head, and cnemial tuberosity lacking. Differs from *Leptorycteropus* in more robust limbs and smaller tooth size relative to the size of the animal.

Description Orycteropus spp. are differentiated primarily based on size, the relative proportions of the dentition, and aspects of the lower dentition (figure 10.2). Postcranial differences are also key, but too few African species are known from reliably attributed postcrania to form a basis for ready differentiation among species.

The oldest records of *Orycteropus* are from the early Miocene of Kenya (Pickford, 1975) in the form of the poorly known

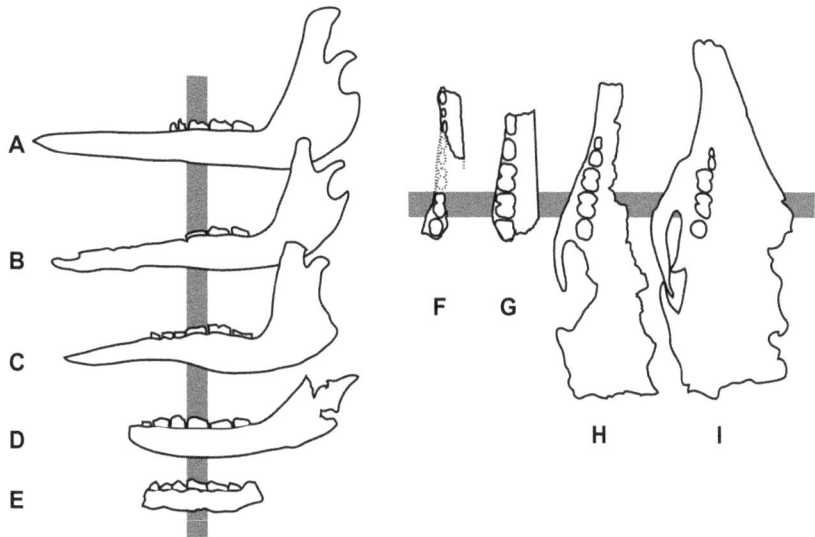

FIGURE 10.2 Representative tubulidentate taxa, scaled to equal m1 and M2 size to illustrate differences in the relative proportions of the dentition and orientation of the ascending ramus. A) *Orycteropus afer*; B, H) *O. crassidens*; C, I) *O. abundulafus*; D, G) *Myorycteropus africanus*; E) *O. chemeldoi*; F) *Leptorycteropus guilelmi*.

O. minutus. The type specimen comprises articulated right metacarpals II and III (of which only MCII has been illustrated), and the hypodigm consists of other podial elements, two isolated molars, and an acetabulum. These remains suggest a species approximately 50% the size of *O. afer,* based on the molars.

In the middle Miocene, the *Orycteropus* lineage is represented by *O. chemeldoi* in Kenya, which is larger than *O. minutus* and is distinguished from other *Orycteropus* spp. by having extremely narrow m1–m2 and is about 66% the size of *O. afer* (Pickford, 1975; Lehmann et al., 2004).

By the late Miocene and into the early Pliocene, several species and at least two distinct lineages appear to be present among African *Orycteropus.* In East Africa, both small and large species are known, but many of the fossils are too fragmentary to assign to species. During this time, the best-represented *Orycteropus* species are those from North Africa. In the upper Miocene of Algeria, *O. mauritanicus* is distinguished by having the longest teeth being m1/M1 (vs. m2/M1 in *Myorycteropus* and m2/M2 in other *Orycteropus*) and in retaining a long astragalar neck. *O. mauritanicus* appears to be more closely related to the later occurring *O. afer* and *O. crassidens* by the following shared derived features: anterior border of orbit above M3 (rather than M2), rectilinear posterior palatine rim, flat glenoid cavity lacking tubercle, proximal tibial epiphysis with four (rather than three) lobes and elongated tibial tuberosity (Lehmann et al., 2005). In Chad, nearly complete skeletons have been recovered of *O. abundulafus* (from near the Miocene-Pliocene boundary) and the slightly younger Pliocene *O. djourabensis. O. abundulafus* is approximately 75% the size of *O. afer* and appears to be most closely related to the European species *O. gaudryi,* based on a suite of shared derived cranial and postcranial features, including a bowed palatine border located closer to M3, associated with a less elongate and more slender facial region; an elongate depression developed in the pterygoid and alisphenoid anterior to the foramen ovale for attachment of the medial and lateral pterygoid muscles as well as a fused mandibular symphysis, indicating greater masticatory power than in other *Orycteropus* spp.; development of an oblique rim on the radius

and possessing a triangular, proximally unbounded olecranon fossa of the humerus, and more slender metacarpals (Lehmann et al., 2005).

The differentiation of the extant species lineage appears to have occurred by the early Pliocene with *Orycteropus afer* remains described from South Africa (Makapansgat, Langebaanweg, and Baard's Quarry: Kitching, 1963; Pickford, 2005; Swartkrans Member 1: Lehmann, 2004). However, these do differ from modern forms: the 1.8 Ma specimens from Swartkrans are smaller in size than extant *O. afer* (Lehmann, 2004), and Lehmann (2008) does not consider the specimens from Langebaanweg to be sufficiently diagnostic to be ascribed with confidence. These fossils led Pickford (2005) to suggest that South Africa may be the area of origin of the extant species, which then spread north during the Pliocene and/or Pleistocene into its current range as these areas became more arid through time.

Lehmann (2008) has also suggested that *Orycteropus* had a more complex biogeographic history than previously thought, based on recent finds of *O. djourabensis* in the Pleistocene of East Turkana and *O.* cf. *djourabensis* in the early Pliocene of Ethiopia. These finds suggest that the *O. djourabensis* lineage is the first aardvark with an extensive geographic range and that it also overlapped with other distinct (but specifically indeterminate) *Orycteropus* in East Africa, and it may have even shared this range with *O. crassidens* during the early Pleistocene.

Orycteropus crassidens has been the most taxonomically controversial species. Described as a distinct species by MacInnes (1956) based on having relatively larger teeth than *O. afer,* Pickford (1975) synonymized the two species. However, Lehmann et al. (2005) demonstrated that while *O. crassidens* appears to be the sister taxon of *O. afer,* it can be distinguished based on the angle of the zygomatic arch and the orientation of the maxillary alveoli as originally noted by MacInnes (1956). Recognizing *O. crassidens* as a distinct lineage is important not simply for taxonomic purposes, but because it illuminates complex biogeographic and phylogenetic patterns in the evolution of aardvarks.

The *Orycteropus afer* lineage is distinct from other *Orycteropus* spp. except *O. crassidens* in having a shortened astragalus in which breadth is subequal to width. It differs from other

Orycteropus spp. (where known) except *O. crassidens* and *O. mauritanicus* in having anterior border of orbit above M3 (rather than M2), rectilinear posterior palatine rim, flat glenoid cavity lacking tubercule, proximal tibial epiphysis with four (rather than three) lobes and an elongated tibial tuberosity. Furthermore, Lehmann (2008) has noted a shared trend toward large teeth in both *O. crassidens* and *O. djourabensis*, which is not observed in *O. afer*, which he avers may be useful in future work to elucidate the evolution of *O. afer*.

Remarks Providing a meaningful diagnosis of the genus *Orycteropus* is difficult; no differential diagnosis has been previously published. The characters provided here are based on the diagnoses of *Leptorycteropus* (Patterson, 1975) and *Myorycteropus* (MacInnes, 1956), with additional characters from the character matrix of Lehmann et al. (2005).

In addition to the single extant species and six African fossil species, six Eurasian species of the genus have also been described and range in age from middle Miocene to Pliocene in age (see Lehmann et al., 2005), although some of these likely represent a new genus (see Lehmann, 2008). In addition to the named species given earlier, many early Miocene to Pleistocene records of *Orycteropus* sp. have been reported (see table 10.1).

Genus *LEPTORYCTEROPUS* Patterson, 1975
LEPTORYCTEROPUS GUILIELMI (Patterson, 1975)
Figure 10.2F

Age and Occurrence Lothagam, lower Nawata, Kenya.

Diagnosis After Patterson (1975). Differs from *Orycteropus* in having a weakly developed deltoid crest on the humerus and tibia-fibula diaphysis straight, limbs generally more slender than both *Myorycteropus* and *Orycteropus*, especially much narrower across distal extremities than in *Myorycteropus*. Differs from *Orycteropus* except *O. pottieri* in retaining a canine, and teeth large relative to the size of the animal (figure 10.2).

Description Patterson (1975) provided a comprehensive description of the type and only known specimen, which comprises parts of the dentary and maxilla and a series of postcranial elements.

Remarks In their phylogenetic analysis, Lehmann et al. (2005) were unable to resolve a monophyletic *Orycteropus* to the exclusion of *Leptorycteropus*, suggesting that it is more closely related to that genus than to *Myorycteropus* and potentially is more closely related to some species among *Orycteropus*.

Discussion

Relationships among tubulidentate species are slowly becoming clearer, as are their broader relationships within Eutheria. Lehmann et al. (2005) have presented the most comprehensive phylogeny among tubulidentates, shown in figure 10.3. Relationships among the earliest members of the order are the least well resolved, reflecting how little of their anatomy is yet known. Holroyd and Mussell (2005) reviewed various hypotheses for the phylogenetic placement of Tubulidentata within eutherian mammals. The most highly supported are those that posit tubulidentates as part of Afrotheria and those that suggest that they arose from within the Condylarthra, a wastebasket group of Paleogene ungulate-grade mammals. These hypotheses are not mutually exclusive, and both suggest that there is a long ghost lineage for tubulidentates in Africa, perhaps extending back to the Cretaceous. One possible sighting in this long blank record is a trace fossil from the late Eocene of Egypt. Bown (1982: figure 14a) described and figured a mammalian burrow intersecting and exiting an underground termite nest. He suggested that this burrow might be attributable to a ptolemaiid, a group now generally regarded as related to the otterlike pantolestids (but see Gunnell et al., this volume, chap. 7). An alternative hypothesis is that this burrow may have been created by a late Eocene tubulidentate. If Paleogene tubulidentates were small in body size, fairly gracile but possessed digging adaptations, and had developed an appetite for termites (as their early Miocene relatives appear to have), then it is plausible that they may have preyed on subterranean termitaria. This hypothesis for the phylogenetic affinities of this trace maker is highly conjectural, but the fact that even the early Miocene forms are well adapted for digging (Pickford, 1975) and myrmecophagy suggests that their Paleogene relatives may have shared these features and exploited the diverse termite faunas that the trace fossil record indicates was present in the African Paleogene (Genise and Bown, 1994).

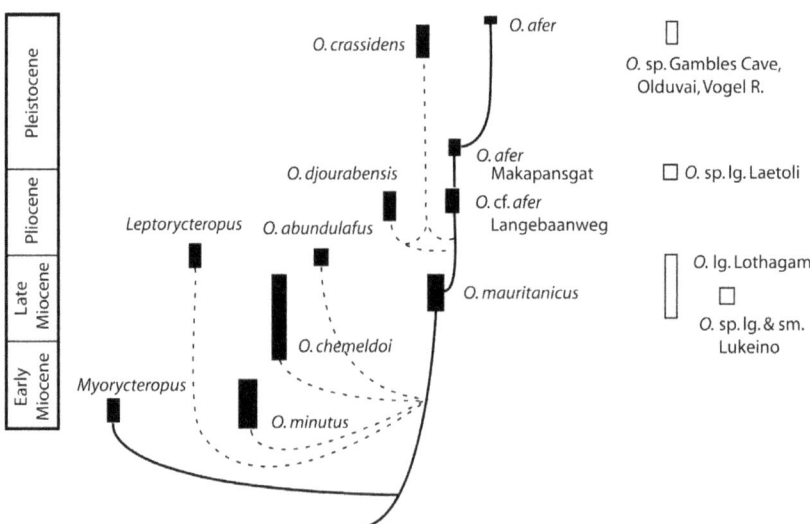

FIGURE 10.3 Hypothesis of relationships among Orycteropodidae, with ranges of taxa shown, modified after Lehmann et al., 2005.

It is ironic that, other than overall size, it is aspects of the masticatory apparatus (angle of ascending ramus, dental proportions) that form the basis for distinguishing species among fossil aardvarks, as we really have no idea how they use them, even in the notoriously shy living aardvark. Fortelius et al. (2003) suggested that the unusual tubulidentate structures may be related to a diet rich in formic acid, which affects enamel more than dentine, and Patterson (1975) discussed at length the possible coevolutionary relationship between aardvarks and the fruit *Cucumis*, despite the observation that they don't seem to chew them. An equally probable conjecture is that offered by Prinz et al. (2003), who observed that mastication of insects prior to digestion confers nutritional advantage through increased release of biomolecules and that the wear pattern of aardvark teeth to form a mortar and pestle shape suggests that they may masticate their prey prior to digestion.

ACKNOWLEDGMENTS

I thank the editors for the invitation to participate in this volume, an anonymous reviewer for helpful comments, and Thomas Lehmann and Thomas Stidham for recent literature and discussion.

NOTE ADDED IN PROOF

After submission of this chapter, Lehmann (2009) published a significant revision of the Tubulidentata based on a phylogenetic analysis of 39 cranial, postcranial, and dental characters of nine species. This analysis revalidates the distinctiveness of *Leptorycteropus* and *Myorycteropus* and recovers them as distinctive lineages basal to other aardvarks. He also transfers all non-African records of aardvarks to the new genus *Amphiorycteropus* as well as the African species *A. abundulafus* and *A. mauritanicus*. New or emended diagnoses are provided for all genera based on the results of this new analysis.

Literature Cited

Arambourg, C. 1959. Vertébrés continentaux du Miocène supérieur de l'Afrique du Nord. *Publication du Service de Carte Géologique, Algérie, N. S., Paléontologie* 4:42–53.

Asfaw, B., Y. Beyene, G. Suwa, R. Walter, T. White, G. WoldeGabriel, and T. Yemane. 1992. The earliest Acheulean from Konso-Gardula. *Nature* 360:732–735.

Bown, T. M. 1982. Ichnofossils and rhizoliths of the nearshore fluvial Jebel Qatrani Formation (Oligocene), Fayum Province, Egypt. *Palaeogeography, Palaeoclimatology, Palaeoecology* 40:255–309.

Butzer, K. W. 1971. Another look at the australopithecine cave breccias of the Transvaal. *American Anthropologist* 73:1197–1201.

Clark, J. D. 1942. Further excavations (1939) at the Mumbwa Caves, Northern Rhodesia. *Transactions of the Royal Society of South Africa* 29:133–201.

Dietrich, W. O. 1942. Ältestquartäre Säugetiere aus der südlichen Serengeti, Deutsch-Ostafrika. *Paleontographica, Abt. A* 94:43–133.

Fortelius, M., S. Numella, and S. Sen. 2003. Orycteropodidae (Tubulidentata); pp. 194–201 in M. Fortelius, J. Kappelman, S. Sen, and R. Bernor (eds.), *Geology and Paleontology of the Miocene Sinap Formation*. Columbia University Press, New York.

Genise, J. F., and T. M. Bown. 1994. New trace fossils of termites (Insecta: Isoptera) from the late Eocene–early Miocene of Egypt, and the reconstruction of ancient isopteran social behavior. *Ichnos* 3:155–183.

Grine, F. E., and R. G. Klein. 1993. Late Pleistocene humans remains from the Sea Harvest Site, Saldanha Bay, South Africa. *South African Journal of Science* 89:145–152.

Hendey, Q. B. 1973. Fossil occurrences at Langebaanweg, Cape Province. *Nature* 244:13–14.

Holroyd, P. A., and J. C. Mussell. 2005. Macroscelidea and Tubulidentata; pp. 71–83 in K. D. Rose and J. D. Archibald (eds.), *The Rise of Placental Mammals: Origins and Relationships of the Major Extant Clades*. Johns Hopkins University Press, Baltimore.

Howell, F. C., and Y. Coppens. 1974. Les faunes de mammifères fossiles des formations Plio-Pléistocenes de l'Omo en Ethiopie (Tubulidentata, Hyracoidea, Lagomorpha, Rodentia, Chiroptera, Insectivora, Carnivora, Primates). *Comptes Rendus Hebdomadaires des Séances de l'Académie des Sciences, Serie D, Sciences Naturelles* 278:2421–2424.

Kitching, J. W. 1963. A fossil *Orycteropus* from the limeworks quarry, Makapansgat, Potgietersrus. *Paleontologia Africana* 8:119–121.

Klein, R. G., K. Cruz-Uribe, and P. B. Beaumont. 1991. Environmental, ecological and paleoanthropological implications of the late Pleistocene mammalian fauna from Equus Cave, Northern Cape Province, South Africa. *Quaternary Research* 36:94–119.

Leakey, L. S. B. 1931. *The Stone Age Cultures of Kenya Colony*. Cambridge University Press, Cambridge, 287 pp.

———. 1951. *Olduvai Gorge: A Report on the Evolution of the Hand-axe Culture in Beds I–IV*. Cambridge University Press, Cambridge, 163 pp.

Leakey, M. G. 1987. Fossil aardvarks from the Laetolil Beds; pp. 297–300 in M. D. Leakey and J. M. Harris (eds.), *Laetoli: A Pliocene Site in Northern Tanzania*. Clarendon Press, Oxford.

Lehmann, T. 2004. Fossil aardvark (*Orycteropus*) from Swartkrans Cave, South Africa. *South African Journal of Science* 100:311–314.

———. 2006. Biodiversity of the Tubulidentata over geological time. *Afrotherian Conservation* 4:6–11.

———. 2008. Plio-Pleistocene aardvarks (Mammalia, Tubulidentata) from East Africa. *Fossil Record* 11:67–81.

———. 2009. Phylogeny and systematics of the Orycteropodidae (Mammalia, Tubulidentata). *Zoological Journal of the Linnean Society* 155:649–702.

Lehmann, T., P. Vignaud, A. Likius, and M. Brunet. 2005. A new species of Orycteropodidae (Mammalia, Tubulidentata) in the Mio-Pliocene of northern Chad. *Zoological Journal of the Linnean Society* 143:109–131.

Lehmann, T., P. Vignaud, A. Likius, H. T. Mackaye, and M. Brunet. 2006. A sub-complete fossil aardvark (Mammalia, Tubulidentata) from the Upper Miocene of Chad. *Comptes Rendus Palevol* 5:693–703.

Lehmann, T., P. Vignaud, H. T. Mackaye, and M. Brunet. 2004. A fossil aardvark (Mammalia, Tubulidentata) from the Lower Pliocene of Chad. *Journal of African Earth Sciences* 40:201–217.

MacInnes, D. G. 1956. Fossil Tubulidentata from east Africa. *Fossil Mammals of Africa* 10:1–38.

Milledge, S. A. H. 2003. Fossil aardvarks from the Lothagam Beds; pp. 363–368, in M. G. Leakey and J. M. Harris (eds.), *Lothagam: The Dawn of Humanity in Eastern Africa*. Columbia University Press, New York.

Patterson, B. 1975. The fossil aardvarks (Mammalia: Tubulidentata). *Bulletin of the Museum of Comparative Zoology* 147:185–237.

Peabody, F. E. 1954. Travertines and cave deposits of the Kaap Escarpment of South Africa, and the type locality of *Australopithecus africanus* Dart. *Bulletin of the Geological Society of America* 65:671–706.

Pickford, M. 1975. New fossil Orycteropodidae (Mammalia, Tubulidentata) from East Africa. *Netherlands Journal of Zoology* 25:57–88.

———. 1994. Tubulidentata of the Albertine rift valley, Uganda; pp. 261–262 in B. Senut and M. Pickford (eds.), *Geology and Palaeobiology of the Albertine Rift Valley, Uganda-Zaire: Volume II, Palaeobiology*. Publication Occasionnelle 29—Centre International pour la Formation et les Echanges Géologiques.

———. 1996. Tubulidentata (Mammalia) from the middle and upper Miocene of southern Namibia. *Comptes Rendus de l'Académie des Sciences, Serie II, Sciences de la Terre et des Planètes* 322:805–810.

———. 2005. *Orycteropus* (Tubulidentata, Mammalia) from Langebaanweg and Baard's Quarry, early Pliocene of South Africa. *Comptes Rendus Palevol* 4:715–726.

Pickford, M., and P. Andrews. 1981. The Tinderet Miocene sequence in Kenya. *Journal of Human Evolution* 10:11–33.

Pickford, M., and B. Senut. 2002. *The Fossil Record of Namibia*. Ministry of Mines and Energy, Geological Survey of Namibia, Windhoek, 39 pp.

Potts, R., P. Shipman, and E. Ingall. 1988. Taphonomy, paleoecology, and hominids of Lainyamok, Kenya. *Journal of Human Evolution* 17:596–614.

Prinz, J. F., C. J. L. Silwood, A. W. D. Claxson, and M. Grootveld. 2003. Simulated digestion status of intact and exoskeletally-punctured insects and insect larvae: A spectroscopic investigation. *Folia Primatologica* 74:126–140.

Romer, A. S. 1938. Mammalian remains from some Paleolithic stations in Algeria. *Logan Museum Bulletin* 5:165–184.

Bibymalagasia (Mammalia *Incertae Sedis*)

LARS WERDELIN

The order Bibymalagasia consists of the single genus *Plesiorycteropus*, with two species, *P. madagascariensis* and *P. germainepetterae* (MacPhee, 1994). As suggested by the generic name, *Plesiorycteropus* has traditionally been considered a member of the order Tubulidentata, family Orycteropodidae (aardvarks) ever since its initial description by Filhol (1895; Lamberton, 1946; Patterson, 1975; Thewissen, 1985). However, MacPhee (1994) carried out an exhaustive anatomical and phylogenetic study of the available material of *Plesiorycteropus* and found significant cause to question this traditional assignment. Instead, he found that the systematic position of *Plesiorycteropus* was unstable in his analyses. These gave no reason to believe that *Plesiorycteropus* was any closer to Tubulidentata than to several other ungulates (*sensu lato*) and that many of the characters allying *Plesiorycteropus* with tubulidentates may be functional convergences due to a fossorial lifestyle. As a consequence, MacPhee erected the new order Bibymalagasia for *Plesiorycteropus* to highlight its uncertain phylogenetic position. A number of subsequent authors who have studied tubulidentate relationships have accepted this reasoning, including Holroyd and Mussell (2005), Rose et al. (2005), and Lehmann (2009). In this volume, *Plesiorycteropus* (and Bibymalagasia) is therefore discussed here, instead of with the Tubulidentata (Holroyd, this volume, chap. 10).

ABBREVIATION

MNHN: Natural History Museum, Paris.

Systematic Paleontology

Order BIBYMALAGASIA MacPhee, 1994
Family PLESIORYCTEROPODIDAE Patterson, 1975
Genus *PLESIORYCTEROPUS* Filhol, 1895

Synonymy Majoria Thomas, 1915; *Hypogeomys* G. Grandidier, 1912 *(partim).*

Diagnosis MacPhee (1994). A eutherian distinguished from all other known eutherians (including all recognized tubulidentates) by the following combination of nonprimitive features: (1) mandibular fossa large and flat, restricted to facies articularis of squamosal (i.e., with no involvement of the zygomatic process of squamosal); (2) nasals markedly widened rostrally; (3) neural arches of posterior thoracic and all lumbars pierced by large longitudinal channels (transarcual canals); (4) ischial tuberosities highly modified, expanded, and caudally flattened; and (5) posteromedial process of astragalus present and very large (may have acted as a pulley for flexor tibialis).

Included Species P. madagascariensis Filhol, 1895; *P. germainepetterae* MacPhee, 1994.

Age and Occurrence Late Quaternary, Madagascar.

Remarks Ideas on the affinities of *Plesiorycteropus* have varied somewhat over the years. In his original description (unillustrated), Filhol (1895) suggested closest affinities with *Orycteropus*, and it was not until Lamberton (1946) that any serious study of the subject was made. Lamberton suggested several options regarding the affinities of *Plesiorycteropus*, of which the idea of edentate (in the old sense of a group encompassing xenarthrans, tubulidentates, and pholidotans) is of particular historic interest (see MacPhee, 1994, for a longer discussion). The second major study of *Plesiorycteropus* was published by Patterson (1975), who in his revision of the fossil Tubulidentata discussed the affinities of *Plesiorycteropus* to this group, concluding that it was an early offshoot of that order. Patterson's conclusions were iterated in his 1978 chapter in the precursor to this volume, and they were also accepted by Thewissen (1985), who nevertheless produced new anatomical data on *Plesiorycteropus*. Then, in 1994 MacPhee provided the first serious challenge to the tubulidentate status of *Plesiorycteropus* since Lamberton (1946).

PLESIORYCTEROPUS MADAGASCARIENSIS Filhol, 1895

Synonymy Majoria rapeto Thomas, 1915; *"Hypogeomys" boulei* G. Grandidier, 1912.

Holotype MNHN 328, a partial cranium of a probable subadult individual (see MacPhee, 1994: figure 2).

Age and Distribution Late Quaternary, Madagascar; several sites including Belo (west coast), Ambolisatra (southwestern coast), Antsirabe (central island), Ampasambazimba (central island), Anjohibe (northwest); see Godfrey et al., this volume, chap. 21: figure 21.1.

Diagnosis MacPhee (1994). Differs from smaller *P. germainepetterae* n. sp. (q.v.) in the following combination of traits: (1) braincase larger and less globose; (2) orbital constriction less pronounced; (3) temporal lines higher; (4) pseudoglenoid process less prominent; (5) small ?vascular foramen absent adjacent to foramen ovale; (6) rostral and caudal tympanic processes of petrosal more developed; (7) temporal tubercle faint; (8) dorsal profile of nuchal crest straight; and (9) third trochanter of femur larger.

PLESIORYCTEROPUS GERMAINEPETTERAE MacPhee, 1994

Holotype MNHN 327, adult partial cranium lacking facial region (MacPhee, 1994: figure 5).

Age and Distribution Late Quaternary, Madagascar; type locality "central island," probably present at a number of other localities.

Diagnosis MacPhee (1994). Differs from *P. madagascariensis* in being smaller overall (by 8%–19% for linear measurements) and in the following combination of traits: (1) braincase smaller and more globose; (2) orbital constriction more pronounced (extends across cranium); (3) temporal lines lower; (4) pseudoglenoid process more prominent; (5) small ?vascular foramen present adjacent to foramen ovale; (6) rostral and caudal tympanic processes of petrosal less developed; (7) temporal tubercle prominent; (8) dorsal profile of nuchal crest indented; and (9) third trochanter of femur smaller.

Remarks The separate status of the two species is arguable, as the differences could be accounted for by sexual dimorphism and individual variation.

Discussion

MacPhee's (1994) argument for erecting the order Bibymalagasia rested mainly on the shifting position of *Plesiorycteropus* in his various phylogenetic analyses, as well as the possibility that in a data set loaded with characters of potential functional significance, any and all characters could be due to functional convergence as well as phylogenetic affinity. MacPhee's analysis was carried out well before the advent of the Afrotheria concept (Stanhope et al., 1998), and it is interesting in this context that many of his analyses based on the data matrix he generated (but not that of Novacek [1989]) retrieve a monophyletic Afrotheria (albeit including only *Plesiorycteropus*, Tubulidentata, Hyracoidea, and in some topologies the "condylarthran" *Meniscotherium*).

Given this result, it is of obvious interest to consider the position of *Plesiorycteropus* in a broader sampling of characters and Afrotherian and other taxa. This was done by Asher et al. (2003). Their analyses, both morphological alone (Asher et al., 2003: figure 2) and combined morphological and molecular (Asher et al., 2003: figure 5) tend to unite *Plesiorycteropus* with *Orycteropus* (with some variation depending on the parameter set used; see Asher et al., 2003, for discussion). A similar result

was obtained by Zack et al. (2005) on the basis of slightly different character and taxon sampling.

The weight of the evidence at this point favors a position of *Plesiorycteropus* as (at least) the sister taxon to Tubulidentata. Further analyses are required to determine whether this position is stable. Seiffert (2007) undertook a major study of afrotherian interrelationships but unfortunately did not include *Plesiorycteropus* in his taxon sample. Doing so may be illuminating. Finally, the material of *Plesiorycteropus* is quite young and may possibly yield DNA that would further illuminate the position of *Plesiorycteropus* and the status of the order Bibymalagasia.

Literature Cited

Asher, R. J., M. J. Novacek, and J. H. Geisler. 2003. Relationships of endemic African mammals and their fossil relatives based on morphological and molecular evidence. *Journal of Mammalian Evolution* 10:131–194.

Filhol, H. 1895. Observations concernant les mammifères contemporains des *Aepyornis* à Madagascar. *Bulletin du Muséum d'Histoire Naturelle, Paris* 1:12–14.

Holroyd, P. A., and J. C. Mussell. 2005. Macroscelidea and Tubulidentata; pp. 71–83 in K. D. Rose and J. D. Archibald (eds.), *The Rise of Placental Mammals: Origins and Relationships of the Major Extant Clades.* Johns Hopkins University Press, Baltimore.

Lamberton, C. 1946. Contribution à la connaissance de la faune subfossile de Madagascar. Note XV: Le *Plesiorycteropus madagascariensis* Filhol. *Bulletin de l'Académie Malgache* 25:25–53 (1942–43).

Lehmann, T. 2009. Phylogeny and systematics of the Orycteropodidae (Mammalia, Tubulidentata). *Zoological Journal of the Linnean Society* 155:649–702.

MacPhee, R. D. E. 1994. Morphology, adaptations and relationships of *Plesiorycteropus*, and a diagnosis of a new order of eutherian mammals. *Bulletin of the American Museum of Natural History* 220:1–214.

Novacek, M. J. 1989. Higher mammal phylogeny: The morphological-molecular synthesis; pp. 421–435 in B. Fernholm, K. Bremer, and H. Jörnvall (eds.), *The Hierarchy of Life: Molecules and Morphology in Phylogenetic Analysis.* Elsevier, Amsterdam.

Patterson, B. 1975. The fossil aardvarks (Mammalia: Tubulidentata). *Bulletin of the Museum of Comparative Zoology* 147:185–237.

———. 1978. Pholidota and Tubulidentata; pp. 268–278 in V. J. Maglio and H. B. S. Cooke (eds.), *Evolution of African Mammals.* Harvard University Press, Cambridge.

Rose, K. D., R. J. Emry, T. J. Gaudin, and G. Storch. 2005. Xenarthra and Pholidota; pp. 106–126 in K. D. Rose and J. D. Archibald (eds.), *The Rise of Placental Mammals: Origins and Relationships of the Major Extant Clades.* Johns Hopkins University Press, Baltimore.

Seiffert, E. R. 2007. A new estimate of afrotherian phylogeny based on simultaneous analysis of genomic, morphological, and fossil evidence. *BMC Evolutionary Biology* 7:224.

Stanhope, M. J., V. G. Wadell, O. Madsen, W. W. DeJong, S. B. Hedges, G. C. Cleven, D. Kao, and M. S. Springer. 1998. Molecular evidence for multiple origins of the Insectivora and for a new order of endemic African mammals. *Proceedings of the National Academy of Sciences, USA* 95:9967–9972.

Thewissen, J. G. M. 1985. Cephalic evidence for the affinities of Tubulidentata. *Mammalia* 49:257–284.

Zack, S. P., T. A. Penkrot, J. I. Bloch, and K. D. Rose. 2005. Affinities of "hyopsodontids" to elephant shrews and a Holarctic origin of Afrotheria. *Nature* 434:497–501.

TWELVE

Embrithopoda

WILLIAM J. SANDERS, D. TAB RASMUSSEN,
AND JOHN KAPPELMAN

Embrithopoda is represented in the Afro-Arabian fossil record by *Arsinoitherium*, named after the Ptolemaic Egyptian queen Arsinöe (Beadnell, 1902), and *Namatherium*, named after the region in Namibia in which it occurs (Pickford et al., 2008). Physically impressive, *Arsinoitherium* superficially resembled extant rhinos in the ornamentation of its cranium by massive, protuberant horns, and by the great magnitude of its skeletal frame (figure 12.1; Cooke, 1968). Arsinoitheres were endemic to Afro-Arabia during the middle Eocene through late Oligocene and are best known from the Fayum, Egypt (table 12.1). Taxa in the older embrithopod subfamily Palaeoamasinae, however, have been recovered from early to late Eocene sites in Eurasia (Ozansoy, 1966; Radulesco et al., 1976; Sen and Heintz, 1979; Radulesco and Sudre, 1985; Kappelman et al., 1996; Maas et al., 1998). The last known occurrence of embrithopods was at Lothidok, Kenya, of latest Oligocene age (Boschetto et al., 1992; Gutiérrez and Rasmussen, 2007).

Conflicting ideas about embrithopod relationships (e.g., Andrews, 1906; Gregory, 1910; Simpson, 1945; McKenna, 1975) have been addressed by recent morphologic and phylogenetic analyses (Court, 1989, 1990, 1992a; Asher et al., 2003; Gheerbrant et al., 2005; Asher, 2007; Seiffert, 2007; Tabuce et al., 2007). These studies reaffirm Simpson's (1945) earlier classification of arsinoitheres (and, by extension, Embrithopoda) in Paenungulata, with proboscideans, sirenians, desmostylians, and hyraxes. Nonetheless, debate continues about whether arsinoitheres are more closely related to proboscideans (e.g., Asher, 2007), sirenians (e.g., Seiffert, 2007), or are a more distant sister taxon to tethytheres (e,g., Gheerbrant et al., 2005; Tabuce et al., 2007). Molecular analyses of extant taxa suggest that paenungulates belong with elephant shrews, African insectivorans, and aardvarks in the clade Afrotheria (Murphy et al., 2001; Springer et al., 2003), whose modern biogeography is largely African (Asher et al., 2003).

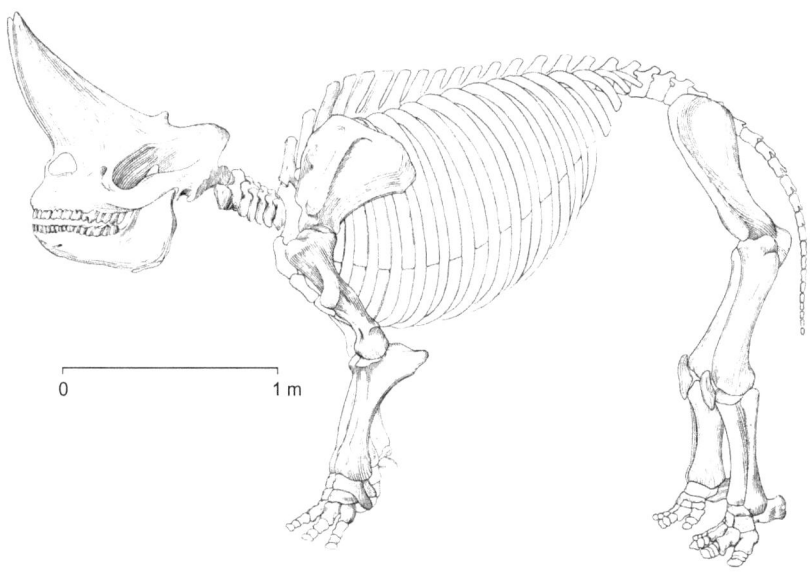

FIGURE 12.1 Skeletal reconstruction of *Arsinoitherium zitteli*, modified after Andrews, 1906.

Taxon	Occurrence (Site, Locality)	Stratigraphic Unit	Age	Key References
Arsinoitheriidae			Middle Eocene– late Oligocene	
Namatherium blackcrowense	Black Crow, Sperrgebiet, Namibia (type)	Black Crow Carbonate	Middle Eocene (Lutetian)	Pickford et al., 2008
Arsinoitherium zitteli	Fayum, Egypt (type)	Gebel el Qatrani Fm.	Late Eocene– early Oligocene, ca. 34.0–30.0 Ma	Beadnell, 1902; Lankester, 1903; Andrews, 1906; Simons, 1968; Coryndon and Savage, 1973; Tanner, 1978; Holroyd and Maas, 1994; Gagnon, 1997; Seiffert, 2006
Arsinoitherium sp.	Aydim area, Dhofar, Oman	Moosak Mb., Aydim Fm.	Late Eocene	Al-Sayigh et al., 2008
	Malembe, Angola		Early Oligocene	Pickford, 1986, 1987
	Taqah and Thaytiniti, Oman	Shizar Mb., Ashawq Fm.	Early Oligocene, 31.5–31.0 Ma and 33.7–33.3 Ma	Thomas et al., 1999; Seiffert, 2006
	Dor el Talha, Libya	Idam Unit	Early Oligocene	Wight, 1980
	Lothidok, Kenya	Eragaleit beds	Late Oligocene, 27.5–24.0 Ma	Boschetto et al., 1992; Gutiérrez and Rasmussen, 2007
Arsinoitherium giganteum	Chilga, Ethiopia (type)	Chilga Fm.	Late Oligocene, 28–27 Ma	Kappelman et al., 2003; Sanders et al., 2004

Competing ecomorphological hypotheses reconstruct *Arsinoitherium* as fully terrestrial, based on conjecture that their heavy dental wear was caused by an abrasive diet of forage in open country settings (Thenius, 1969), or as partially or primarily aquatic, based on occurrences in fluvial environments (Moustapha, 1955; Sen and Heintz, 1979) or postcranial functional morphology (Court, 1993). While known occurrences (figure 12.2) sample a range of well-watered paleoenvironments (Wight, 1980; Bown et al., 1982; Olson and Rasmussen, 1986; Pickford, 1987; Bown and Kraus, 1988; Gagnon, 1997; Jacobs et al., 2005), terrestrial taxa from the same localities do not support the latter hypotheses. In addition, results of stable isotope analyses on *Arsinoitherium* from the lower sequence of the Gebel el Qatrani Fm., Fayum, Egypt, indicate that it was a terrestrial C_3 feeder (Clementz et al., 2008).

ABBREVIATIONS

BM(NH), the Natural History Museum, London (formerly the British Museum [Natural History]); CH, Chilga specimens housed in the National Museum, Addis Ababa, Ethiopia; H, height; I/i, upper/lower incisor; L, length; Ma, mega annum (10^6 years); mm, millimeters; M/m, upper/lower molar; P/p, upper/lower premolar; W, width.

Systematic Paleontology

Order EMBRITHOPODA Andrews, 1906
Family ARSINOITHERIIDAE Andrews, 1904
Genus *NAMATHERIUM* Pickford et al., 2008
NAMATHERIUM BLACKCROWENSE Pickford et al., 2008
Figure 12.3

Age and Occurrence Middle Eocene, southern Africa (table 12.1).

Diagnosis Based on Pickford et al. (2008). Differs from Eurasian embrithopods by greater molar hypsodonty, absence of interloph crest in upper molars, and lingual offset of M3s, and from *Arsinoitherium* by lesser hypsodonty and smaller size of molars, monolophodont upper premolars with shallow posterior fossettes, strong lateral flare of zygomatic arches (figure 12.3), and more anterior position of the infraorbital foramen and anterior margin of the orbits.

Description The species is known from a partial skull (BC 13′08) found in the Black Crow carbonate at Black Crow, Sperrgebiet, Namibia (Pickford et al., 2008). Although very fragmentary, it preserves good dental details and much of the palate and zygomatic arches. The width of the palate has been exaggerated by separation of the maxilla at the midline suture (figure 12.3). Teeth preserved include left M1–M3 and right P3–M3; these are smaller than cheek teeth of *Arsinoitherium* (length of M1–M3 = 105.9–106.3 mm in *Namatherium* [Pickford et al., 2008] and 155.0–232.0 mm in *Arsinoitherium* [Andrews, 1906]). The root of the zygomatic arch is at P4, and the infraorbital foramen opens above P4. The most impressive feature of the specimen is the great lateral flare of the zygomatic arches. As in *Arsinoitherium*, the nasal aperture is retracted. Upper premolars are monolophodont and simpler than in later arsinoitheres. It is unknown whether they possessed frontal and nasal horns. Other craniodental distinctions between *Namatherium* and *Arsinoitherium* are detailed in Pickford et al. (2008); despite these, it is clear that they are more similar to one another than either is to Eurasian embrithopods.

Remarks This extraordinary find extends the fossil record of embrithopods in Africa back in time considerably. In

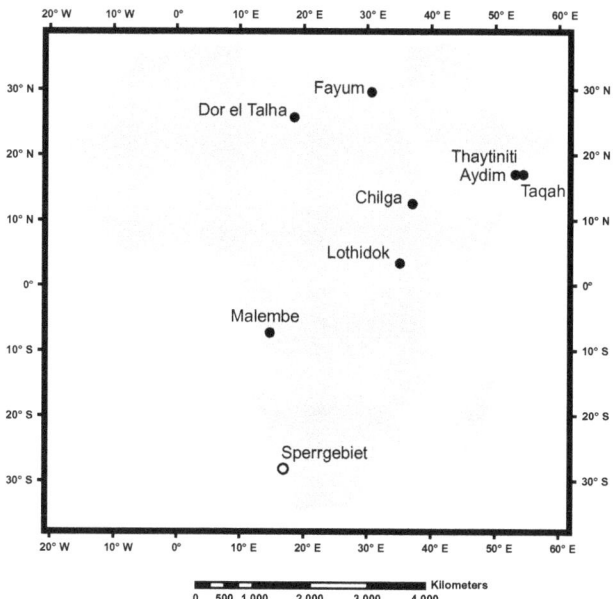

FIGURE 12.2 Map of Afro-Arabian occurrences of *Arsinoitherium* (closed circles) and *Namatherium* (open circle)(see table 12.1).

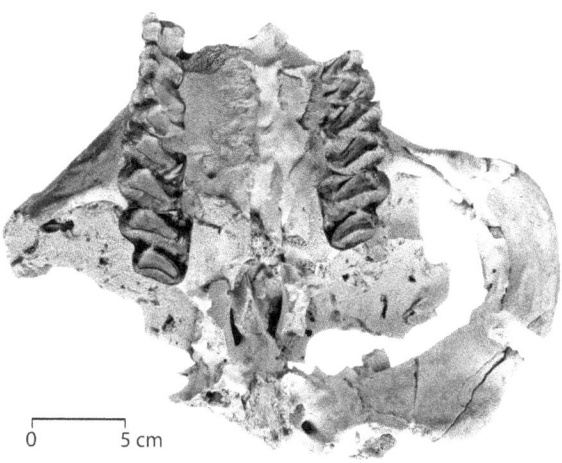

FIGURE 12.3 Holotype cranium BC 13'08 of *Namatherium black-crowense*, in palatal view. Photograph courtesy of Martin Pickford.

addition, it will be useful for further systematic assessment of arsinoitheres. For example, the morphology of the zygomatic arch and surrounding areas of the cranium is reminiscent of moeritheres, and details of the M3 root system recall the condition in *Phosphatherium*, reviving ideas of a close phylogenetic relationship between embrithopods and proboscideans (Pickford et al., 2008).

Genus *ARSINOITHERIUM* Beadnell, 1902
ARSINOTHERIUM ZITTELI Beadnell, 1902
Figure 12.4

Partial Synonymy Arsinoitherium andrewsi, Lankester, 1903; *A. andrewsi*, Andrews, 1904; *Arsinoitherium* sp. indet., Osborn, 1908; *A. andrewsi*, Simons, 1968; *A. andrewsi*, Tanner, 1978.

Age and Occurrence Late Eocene–early Oligocene, northern Africa, ?western Africa, and ?Arabia (table 12.1).

Diagnosis Based on Beadnell (1902); Andrews (1906); Radulesco and Sudre (1985); Court (1989, 1992a, 1992b, 1993). Very large, graviportal mammal with paired frontonasal horns (the anterior pair much larger and considerably more projecting), high-crowned bilophodont molars in continuous series with the rest of the dentition, extensive mandibular symphysis extending as far posterior as m2, and tarsus with a number of features shared with proboscideans, including presence of a large tuberculum mediale on the astragalus. Larger and more hypsodont than palaeoamasines, with relatively shorter premolars, and larger and more hypsodont than *Namatherium*, with zygomatic arches far less projecting laterally and more complex premolars.

Description Well-preserved crania representing an ontogenetic series, and the mandible, complete dentitions, and nearly all postcranial elements have been recovered for *A. zitteli* (Andrews, 1906; Court, 1989, 1992a, 1993). Detailed descriptions of these remains (Andrews, 1904, 1906; Court, 1989, 1992a, 1992b, 1993) are abridged here.

The cranium is relatively elongate and narrow (L = 740–770 mm; W = 316–335 mm; Andrews, 1906; Tanner, 1978), with an anteriorly slanted occipital planum, large, posteriorly projecting occipital condyles, zygomatic arches that rise steeply posterodorsally but do not flare widely laterally, an anteriorly highly arched palate, deeply excavated temporal fossae, large narial openings separated in adults by a prenasal bar, and paired frontonasal horns (figures 12.4A, 12.4B). The anterior horns dominate the face and are principally formed of the nasal bones. Internally, they are hollow and communicate with the frontal sinus system, as do the much smaller posterior horns, which are derived from the frontal bones. Externally, the horns are marked by fine vascular grooves, suggesting that in life they were covered with keratinous sheaths. Posterior retraction of the narial openings and the presence of a large, single incisive foramen may be indicative of a small proboscis or mobile upper lip (Court, 1992a) and the arching of the palate may have accommodated an extensible, prehensile tongue (Andrews, 1906), which would have been critical for securing browse, in light of the relatively small dimensions of the anterior dentition.

The dentaries of arsinoitheres are elongate (L = 525–730 mm; Andrews, 1906; Tanner, 1978) and relatively shallow and slender, considering crown height and the overall size of the animal. The modest height of the corpora is contrasted by the great elevation of the rami and coronoid processes, and expansive development of the masseteric fossae (figure 12.4C). The mandibular condyle is divided into two adjacent facets (Court, 1992b). The symphysis is extended as far posteriorly as m2 (figure 12.4D; L = 180–215 mm; Andrews, 1906; Tanner, 1978), and there is a single mandibular foramen in each dentary, usually below p3 in adults.

Arsinoitheres possess a full dentition, with a formula of 3-1-4-3/3-1-4-3 (figures 12.4E, 12.4F; Court, 1992b). The first incisor is larger than the peglike I2 and I3, but it is not particularly prominent. The canine resembles the incisors and projects to the same level. None of these teeth are procumbent or form tusks, and it is unlikely that they were employed for social display or as weapons. There is no diastema in either the upper or lower dental series, except for a narrow gap in the midline between the teeth of the right and left sides. Considering the Paleogene age of *Arsinoitherium*, the molars are remarkably hypsodont (Tanner, 1978), and there is some evidence of incomplete closure of cheek tooth roots (pers. obs.), suggestive of delayed tooth development.

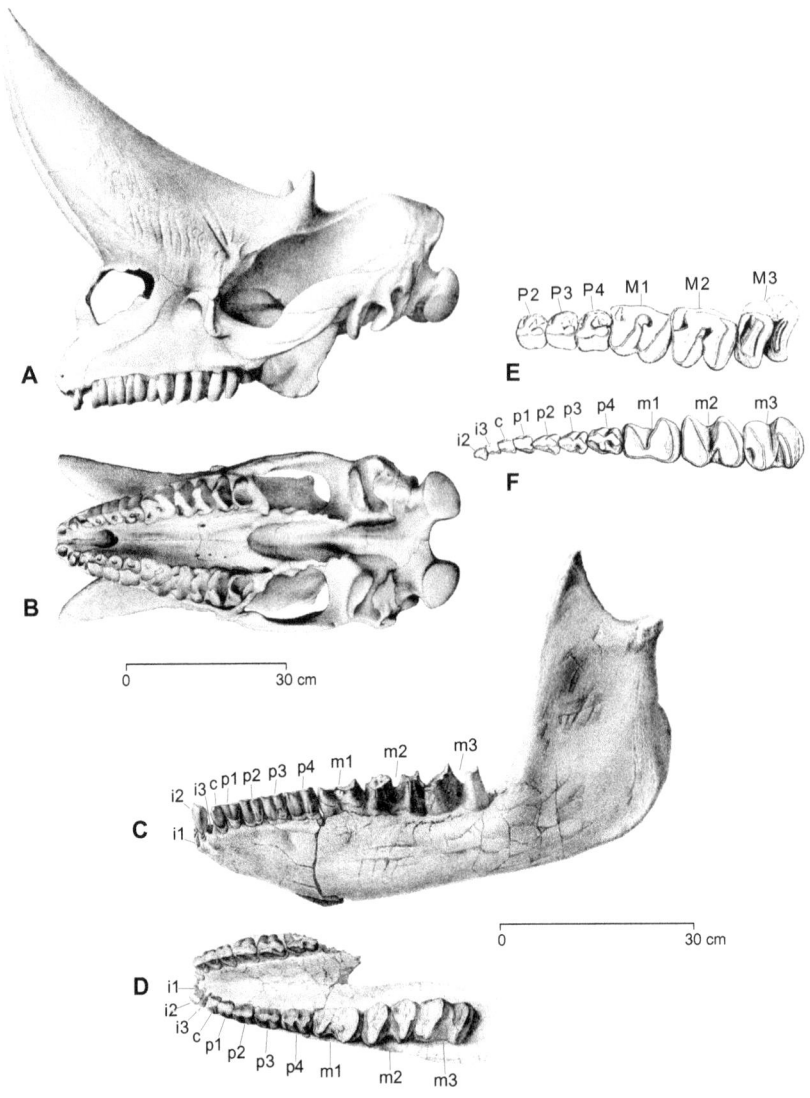

FIGURE 12.4 Craniodental elements of *Arsinoitherium zitteli*, from Andrews (1906). A) Cranium, specimen BM(NH) 8463, left lateral view. B) Cranium, specimen BM(NH) 8463, inferior view. C) Mandible, specimen BM(NH) 8461, left lateral view. D) Mandible, specimen BM(NH) 8461, superior view. E) Upper cheek teeth, left P2-M3, occlusal view (reversed). F) Lower dental series, left i2-m3, occlusal view (reversed).

The molars are comprised of two strong transverse crests or pillars separated by deep transverse valleys and, in the lowers, weakly interconnected by a cristid obliqua (Court, 1992b). With advanced wear, the transverse meta- and hypolophids of the lower molars formed double V-shaped enamel figures (Andrews, 1906). Molars usually have a prominent paracrista(id) partially separated from the anteriormost loph(id) by a deep fossa. The lower premolars are bilobed and simple; in the upper premolars, the paracone and metacone are connected by an ectoloph, and the protocone and hypocone are fused into an entoloph. In heavy wear, the hypocone and metacone merged, but they otherwise were separated by a deep valley (Court, 1992b). Together, the teeth compose a continuous dental battery that exhibits a strong transverse wear gradient.

The postcranium of arsinoitheres exhibits graviportal adaptations for stable support of a heavy, barrel-shaped body on pillarlike limbs and short, pentadactyl feet (figure 12.1). Configuration of manus and pes elements suggests that the feet were carried in plantigrade position (Court, 1993). These features are consistent with body mass estimations of 510–1,500 kg for *A. zitteli* and 1,760–1,960 kg for *"A. andrewsi,"* based on regression formulas for proboscidean long bone dimensions (see Christiansen, 2004). These estimates indicate animals in the size range of extant black rhinos to the upper limits of white rhinos (Kingdon, 1997). The pelvis has widely flared ilia but surprisingly delicate auricular articulations for the sacrum, an extremely short symphysis, and little angulation of the ischial tuberosities posterior to the acetabulae (Court, 1993). The calcaneum differs in form from that of elephants, having a relatively shorter, robust heel process, and a larger sustentacular than ectal articulation for the astragalus. There is greater similarity between the astragali of elephants and arsinoitheres, with both sharing prominent medial tuberosities, expansive, flattened tibial articulations, and absence of a neck; they are, however, distinguished by their calcaneal articular proportions and conformation of articulations for the fibula and cuboid (Andrews, 1906; Court,

1993). Reconstruction of the articulated configuration of carpal joints suggests that the forefoot of arsinoitheres was more laterally oriented and plantigrade than that of elephants, and that they also had more divergent toes that could probably be flexed to a greater degree than those of elephants (Court, 1993).

Joint and process configurations of the cervical vertebrae in arsinoitheres suggest that the head was habitually carried in a raised position and that the skull was capable of moving over a greater lateral than vertical range (Court, 1993). The cervical vertebrae and neck are short, pre- and postzygapophyses of the posterior thoracic and lumbar vertebrae were tightly embraced laterally and dorsally in articulation, and it is likely that the lumbar region was short and further stabilized by close approximation of the ribs to the iliac crests (Court, 1993). Although isolated sacral vertebrae have been found, no fused sacrum is known for arsinoitheres (Andrews, 1906; Court, 1993), which is inexplicable given the magnitude of forces that must have been transmitted between the hind limbs and trunk via the pelvic girdle.

Remarks A handful of cranial, dental, and postcranial specimens, among the largest of their respective elements found in the Fayum and contemporaneous with remains of *A. zitteli* (Tanner, 1978), were assigned to *Arsinoitherium andrewsi* by Lankester (1903) and Andrews (1906). It is more likely, however, that these remains represent outliers at the upper end of the size range of *A. zitteli*. The metric variation for the combined dental sample of *A. zitteli* and *"A. andrewsi"* is comfortably within a range expected for a sexually dimorphic species (Sanders et al., 2004) and is comparable with observed dental variation in modern rhino species, in which males may be more than 50% larger than females (Nowak and Paradiso, 1983). Large herbivores with cranial ornaments suggesting mating competition are almost always sexually dimorphic (Sanders et al., 2004), and this appears to be borne out in the Egyptian arsinoithere sample by the much larger, more sharply pointed horns in the adult crania of presumed males (Andrews, 1906).

The sizable *Arsinoitherium zitteli* sample extends nearly throughout the entire Gebel el-Qatrani Fm. sequence, stratigraphically from just above locality L-41 to Quarry M (Tanner, 1978; Gagnon, 1997; E. Seiffert, pers. comm.), dated from ca. 34.0–30.0 Ma (Seiffert, 2006). *Arsinoitherium* from other sites of similar age (table 12.1), not assigned to species because of their fragmentary nature, could also belong to *A. zitteli*.

ARSINOITHERIUM GIGANTEUM Sanders et al., 2004

Partial Synonymy *Arsinoitherium* sp. nov., Kappelman et al., 2003.

Age and Occurrence Late Oligocene, eastern Africa (table 12.1).

Diagnosis Based on Sanders et al. (2004). Of generally greater size (table 12.2) and with higher-crowned teeth than *Arsinoitherium zitteli* (including *"A. andrewsi"*) and all other embrithopods.

Description Based on Sanders et al. (2004). Ubiquitous, isolated bony remains, and a number of partial skeletons in association from localities at Chilga, Ethiopia, show that *Arsinoitherium giganteum* was anatomically similar to *A. zitteli*, but larger and with relatively longer distal leg segments. Body mass estimates for *A. giganteum*, calculated from long bone

regressions for proboscideans (see Christiansen, 2004), range from 1,970 to 2,370 kg, at or above the upper end of size range of white rhinos (Kingdon, 1997) and *A. zitteli*, and they are slightly heavier than estimates for *"A. andrewsi."*

The species is marked by its exceptional hypsodonty, far exceeding that of the largest teeth in *A. zitteli*. Lower premolars, especially p3 and p4, are more squared in occlusal dimensions than homologues in *A. zitteli*, with widths approximating lengths.

Composite skeletal proportions of *A. giganteum* suggest that tibial length comprises a higher percentage of femoral length in this species than in *A. zitteli* (ca. 61% vs. 53%–55%), due to relative increase in tibial length rather than femoral shortening (table 12.2).

Remarks *Arsinoitherium giganteum* co-occurs at Chilga with a fauna predominantly comprised of large-bodied paenungulates (Kappelman et al., 2003; Sanders et al., 2004). The Chilga mammals lived at an elevation of approximately 1,000 m (Kappelman et al., 2003), making it difficult to envision arsinoitheres as semiaquatic animals specialized for life in coastal swampland environments (see Court, 1993). The only geologically younger remains of arsinoitheres have been recovered from the Eragaleit Beds at Lothidok, Kenya (Gutiérrez and Rasmussen, 2007) and dated to the interval 27.5–24.0 Ma (Boschetto et al., 1992).

Discussion

Perhaps as remarkable as the horned crania of *Arsinoitherium* is its apparently bifunctional dental battery. Analysis of occlusal dynamics indicates that these animals had a masticatory system in which molars sliced through bulky, pliant food items, while premolars more finely reduced these items through crushing and grinding (Court, 1992b). This masticatory system incorporates conspicuous attachment sites for the temporalis and masseter muscles, dual-faceted mandibular condyles, continuous tooth series, and unusual bilophodonty of the molars. These features have been interpreted as specializations for feeding "very selectively on specific plants or parts of plants, . . . perhaps on bulky fruits" (Court, 1992b:109). Fossil remains of such food items occur abundantly in the arsinoithere sites of Fayum, Egypt (Bown et al., 1982), and Chilga, Ethiopia (Jacobs et al., 2005).

Arsinoitheres were among the largest land mammals of their time in Afro-Arabia, rivaled in size only by barytheres and palaeomastodonts, yet surprisingly seem to have been poorly structured for prolonged walking or running, particularly because of weak sacral design, short stride length, and inferred feeble development of pectoralis and subscapularis muscles, which act as dorsal stablizers of the scapula against ground reaction forces (Court, 1993). The skeletal imprint of graviportal features, however, is consistent with terrestrial quadrupedalism. Results of stable isotope analyses suggest that *Arsinoitherium* was a terrestrial browser (Clementz et al., 2008).

The higher-order systematic relationships of *Arsinoitherium*, particularly within Paenungulata, remain unresolved, due to the highly specialized nature of their skeletons. As the earliest known proboscideans, sirenians, hyraxes, and embrithopods were already quite distinct morphologically, their common ancestry certainly must have predated the Eocene. Since palaeoamasines are only "dental" taxa, however, it is difficult to reconstruct or

TABLE 12.2
Dimensions (in mm) of selected postcranial elements in *Arsinoitherium*
Bold numerals are the largest dimension in each category.

Fragment	*Arsinoitherium zitteli*	*"Arsinoitherium andrewsi"*	*Arsinoitherium giganteum*
Femur			
Length	533–758	800	**830**
Head diameter	90–134	135	140–**180**
Proximal width	155–233	**280**	240
Midshaft width	87–137	**163**	140
Distal width	131–188	**203**	160
Condyles width	106–151	**157**	130–140
Tibia			
Length	286–415	—	**510**
Proximal width	115–180	—	—
Midshaft width	55–85	—	**90**
Distal width	92–135	—	**160**
Fibula			
Length	377–410	—	—
Distal width	75–93	—	—
Acetabulum (Innominate)			
Height	114–124	—	**160**
Width	96–99	—	**100**
Midcervical Vertebra (C3 or C4)			
Ventral length	—	—	65
Middle length	33–47	48	—
Dorsal length	—	—	**66**
Width	136–145	**185**	165
Height	110	**152**	135
Humerus			
Length	500–615	—	—
Head diameter	114–149	—	—
Midshaft width	99–133	**180**	—
Distal articulation width	136–169	**230**	—
Ulna			
Length	370–505	**600**	—
Proximal articulation width	123–180	—	—
Midshaft width	72–103	—	—
Distal articulation width	95–121	—	—
Radius			
Length	300–410	—	—
Proximal articulation width	73–100	—	—
Distal articulation width	79–116	—	—

SOURCE: Data for *Arsinoitherium zitteli* and *"A. andrewsi"* are from Andrews (1906).

recognize a primitive ancestral morphotype for the clade, though the recent find of *Namatherium* may help in this regard (Pickford et al., 2008).

ACKNOWLEDGMENTS

Funding for field and museum research was generously provided by grants from the National Science Foundation, the National Geographic Society, the Jacob and Frances Sanger Mossiker Chair in the Humanities, and the University Research Institute of the University of Texas at Austin (to J.K.), and by a Scott Turner Award in Earth Science from the Department of Geological Sciences, University of Michigan (to W.J.S.). We are grateful to Bonnie Miljour for preparing figures 12.1 and 12.4, to Iyad Zalmout for assistance with figure 12.2, to Martin Pickford for figure 12.3, and to Pat Holroyd for insightful review of the manuscript. We extend our thanks to Lars Werdelin for his editorial guidance and expertise.

Literature Cited

Al-Sayigh, A. R., S. Nasir, A. S. Schulp, and N. J. Stevens. 2008. The first described *Arsinoitherium* from the Upper Eocene Aydim Formation of Oman: Biogeographic implications. *Palaeoworld* 17:41–46.

Andrews, C. W. 1904. Further notes on the mammals of the Eocene of Egypt:Part II. *Geological Magazine Series* 5(1):157–162.

———. 1906. *A Descriptive Catalogue of the Tertiary Vertebrata of the Fayum, Egypt.* British Museum of Natural History, London, 324 pp.

Asher, R. J. 2007. A web-database of mammalian morphology and a reanalysis of placental phylogeny. *BMC Evolutionary Biology* 7:10 pp.

Asher, R. J., M. J. Novacek, and J. H. Geisler. 2003. Relationships of endemic African mammals and their fossil relatives based on morphological and molecular evidence. *Journal of Mammalian Evolution* 10:131–194.

Beadnell, H. J. C. 1902. *A Preliminary Note on Arsinoitherium zitteli Beadnell, from the Upper Eocene Strata of Egypt.* Egyptian Survey Department, Public Works Ministry, Cairo, 4 pp.

Boschetto, H. B., F. H. Brown, and I. McDougall. 1992. Stratigraphy of the Lothidok Range, northern Kenya, and K/Ar ages of its Miocene primates. *Journal of Human Evolution* 22:47–71.

Bown, T. B., and M. J. Kraus. 1988. Geology and paleoenvironment of the Oligocene Jebel Qatrani Formation and adjacent rocks, Fayum Depression, Egypt. *U.S. Geological Survey, Professional Paper* 1452:1–60.

Bown, T. B., M. J. Kraus, S. L. Wing, J. G. Fleagle, B. H. Tiffney, E. L. Simons, and C. F. Vondra. 1982. The Fayum forest revisited. *Journal of Human Evolution* 11:603–632.

Christiansen, P. 2004. Body size in proboscideans, with notes on elephant metabolism. *Zoological Journal of the Linnean Society* 140:523–549.

Clementz, M. T., P. A. Holroyd, and P. L. Koch. 2008. Identifying aquatic habits of herbivorous mammals through stable isotope analysis. *Palaios* 23:574–585.

Cooke, H. B. S. 1968. The fossil mammal fauna of Africa. *Quarterly Review of Biology* 43:234–264.

Coryndon, S. C., and R. J. G. Savage. 1973. The origins and affinities of African mammal faunas. *Special Papers in Palaeontology* 12:121–135.

Court, N. 1989. Morphology, functional morphology and phylogeny of *Arsinoitherium* (Mammalia, Embrithopoda). Unpublished PhD dissertation, University of Bristol, 372 pp.

———. 1990. Periotic anatomy of *Arsinoitherium* (Mammalia, Embrithopoda) and its phylogenetic implications. *Journal of Vertebrate Paleontology* 10:170–182.

———. 1992a. The skull of *Arsinoitherium* (Mammalia, Embrithopoda) and the higher order interrelationships of ungulates. *Palaeovertebrata* 22:1–43.

———. 1992b. A unique form of dental bilophodonty and a functional interpretation of peculiarities in the masticatory system of *Arsinoitherium* (Mammalia, Embrithopoda). *Historical Biology* 6:91–111.

———. 1993. Morphology and functional anatomy of the postcranial skeleton in *Arsinoitherium* (Mammalia, Embrithopoda). *Palaeontographica, Abt. A* 226:125–169.

Gagnon, M. 1997. Ecological diversity and community ecology in the Fayum sequence (Egypt). *Journal of Human Evolution* 32:133–160.

Gheerbrant, E., J. Sudre, P. Tassy, M. Amaghzaz, B. Bouya, and M. Iarochène. 2005. Nouvelles données sur *Phosphatherium escuillei* (Mammalia, Proboscidea) de l'Éocène inférieur du Maroc, apports à la phylogénie des Proboscidea et des ongulés lophodontes. *Geodiversitas* 27:239–333.

Gregory, W. K. 1910. The orders of mammals. *Bulletin of the American Museum of Natural History* 27:1–524.

Gutiérrez, M., and D. Rasmussen. 2007. Late Oligocene mammals from northern Kenya. *Journal of Vertebrate Paleontology* 27 (suppl. to no. 3):85A.

Holroyd, P. A., and M. C. Maas. 1994. Paleogeography, paleobiogeography, and anthropoid origins; pp. 297–334 in J. G. Fleagle and R. F. Kay (eds.), *Anthropoid Origins.* Plenum Press, New York.

Jacobs, B., N. Tabor, M. Feseha, A. Pan, J. Kappelman, T. Rasmussen, W. Sanders, M. Wiemann, J. Crabaugh, and J. L. G. Massini. 2005.

Oligocene terrestrial strata of northwestern Ethiopia: A preliminary report on paleoenvironments and paleontology. *Palaeontologia Electronica* 8(1), 25A:19 pp.

Kappelman, J., M. C. Maas, S. Sen, B. Alpagut, M. Fortelius, and J.-P. Lunkka. 1996. A new early Tertiary mammalian fauna from Turkey and its paleogeographic significance. *Journal of Vertebrate Paleontology* 16:592–595.

Kappelman, J., D. T. Rasmussen, W. J. Sanders, M. Feseha, T. Bown, P. Copeland, J. Crabaugh, J. Fleagle, M. Glantz, A. Gordon, B. Jacobs, M. Maga, K. Muldoon, A. Pan, L. Pyne, B. Richmond, T. Ryan, E. Seiffert, S. Sen, L. Todd, M. C. Wiemann, and A. Winkler. 2003. Oligocene mammals from Ethiopia and faunal exchange between Afro-Arabia and Eurasia. *Nature* 426:549–552.

Kingdon, J. 1997. *The Kingdon Field Guide to African Mammals.* Academic Press, San Diego, 464 pp.

Lankester, E. R. 1903. A new extinct monster. *Sphere (London)* 1903:238.

Maas, M. C., J. G. M. Thewissen, and J. Kappelman. 1998. *Hypsamasia seni* (Mammalia; Embrithopoda) and other mammals from the Eocene Kartal Formation of Turkey. *Bulletin of the Carnegie Museum of Natural History* 34:286–297.

McKenna, M. C. 1975. Towards a phylogenetic classification of the Mammalia; pp. 21–46 in W. P. Luckett and F. S. Szalay (eds.), *Phylogeny of the Primates: A Multidisciplinary Approach.* Plenum Press, New York.

Moustapha, W. 1955. An interpretation of *Arsinoitherium. Bulletin, Institut d'Egypte* 36:111–118.

Murphy, W. J., E. Eizirik, S. J. O'Brien, O. Madsen, M. Scally, C. J. Douady, E. Teeling, O. A. Ryder, M. J. Stanhope, W. W. DeJong, and M. S. Springer. 2001. Resolution of the early placental mammal radiation using Bayesian phylogenetics. *Science* 294:2348–2351.

Nowak, R. M., and J. L. Paradiso. 1983. *Walker's Mammals of the World,* 4th ed. Johns Hopkins University Press, Baltimore, 1,362 pp.

Olson, S. L., and D. T. Rasmussen. 1986. Paleoenvironments of the earliest hominoids: new evidence from the Oligocene avifauna of the Fayum Depression, Egypt. *Science* 233:1202–1204.

Osborn, H. F. 1908 New fossil mammals from the Fayûm Oligocene, Egypt. *Bulletin of the American Museum of Natural History* 24:265–272.

Ozansoy, F. 1966. Türkiye Senozoik çaglarinda fosil insan formu problemi ve biostratigrafik dayanaklari. Ankara University D.T.C.F. Yayinlari 172, 104 pp.

Pickford, M. 1986. Première découverte d'une faune mammalienne terrestre paléogène d'Afrique sub-saharienne. *Comptes Rendus de l'Academie des Sciences, Paris,* Série II, 19:1205–1210.

———. 1987. Recognition of an early Oligocene or late Eocene mammal fauna from Cabinda, Angola. *Musée Royal de l'Afrique Centrale (Belgique), Rapport Annuel du Département de Géologie et de Mineralogie 1985–1986:*89–92.

Pickford, M., B. Senut, J. Morales, P. Mein, and I. M. Sanchez. 2008. Mammalia from the Lutetian of Namibia. *Memoir of the Geological Survey of Namibia* 20:465–514.

Radulesco, C., G. Iliesco, and M. Iliesco. 1976. Un Embrithopode nouveau (Mammalia) dans le Paléogène de la dépression de Hateg (Roumanie) et la géologie de la région. *Neues Jahrbuch für Geologie und Paläontologie, Monatshefte* 11:690–698.

Radulesco, C., and J. Sudre. 1985. *Crivadiatherium iliescui* n. sp., nouvel Embrithopode (Mammalia) dans le Paléogène ancien de la Dépression de Hateg (Roumanie). *Palaeovertebrata* 15:139–157.

Rasmussen, D. T., and M. Gutiérrez. 2009. A mammalian fauna from the late Oligocene of northwestern Kenya. *Palaeontographica, Abt. A* 288:1–52.

Sanders, W. J., J. Kappelman, and D. T. Rasmussen. 2004. New large-bodied mammals from the late Oligocene site of Chilga, Ethiopia. *Acta Palaeontologica Polonica* 49:365–392.

Seiffert, E. R. 2006. Revised age estimates for the later Paleogene mammal faunas of Egypt and Oman. *Proceedings of the National Academy of Sciences, USA* 103:5000–5005.

———. 2007. A new estimate of afrotherian phylogeny based on simultaneous analysis of genomic, morphological, and fossil evidence. *BMC Evolutionary Biology* 7:13 pp.

Sen, S., and E. Heintz. 1979. *Palaeoamasia kansui* Ozansoy 1966, Embrithopode (Mammalia) de l'Éocène d'Anatolie. *Annales de Paléontologie (Vertébrés)* 65:73–91.

Simons, E. L. 1968. Early Cenozoic mammalian faunas. Fayum Province, Egypt. Part I, African Oligocene mammals: Introduction, history of study, and faunal succession. *Peabody Museum of Natural History, Yale University Bulletin* 28:1–21.

Simpson, G. G. 1945. The principles of classification and a classification of mammals. *Bulletin of the American Museum of Natural History* 85:1–350.

Springer, M. S., W. J. Murphy, E. Eizirik, and S. J. O'Brien. 2003. Placental mammal diversification and the Cretaceous Tertiary boundary. *Proceedings of the National Academy of Sciences, USA* 100:1056–1061.

Tabuce, R., C. Delmer, and E. Gheerbrant. 2007. Evolution of the tooth enamel microstructure in the earliest proboscideans (Mammalia). *Zoological Journal of the Linnean Society* 149:611–628.

Tanner, L. G. 1978. Embrithropoda *[sic]*; pp. 277-283 in V. J. Maglio and H. B. S. Cooke (eds.), *Evolution of African Mammals*. Harvard University Press, Cambridge.

Thenius, E. 1969. Stammesgeschichte der Säugetiere. *Handbuch der Zoologie* 8(48):369–722. Thomas, H., J. Roger, S. Sen, C. Bourdillon-de-Grissac, and Z. Al-Sulaimani. 1989. Découverte de vertébrés fossiles dans l'Oligocène inférieur du Dhofar (Sultanat d'Oman). *Geobios* 22:101–120.

Thomas, H., J. Roger, S. Sen, M. Pickford, E. Gheerbrant, Z. Al-Sulaimani, and S. Al-Busaidi. 1999. Oligocene and Miocene terrestrial vertebrates in the southern Arabian Peninsula (Sultanate of Oman) and their geodynamic and palaeogeographic settings; pp. 430–442 in P. J. Whybrow and A. Hill (eds.), *Fossil Vertebrates of Arabia*. Yale University Press, New Haven.

Wight, A. W. R. 1980. Paleogene vertebrate fauna and regressive sediments of Dur at Talhah, southern Sirt Basin, Libya; pp. 309–325 in M. J. Salem and M. T. Busrewil (eds.), *The Geology of Libya*, vol. 1. Academic Press, London.

THIRTEEN

Hyracoidea

D. TAB RASMUSSEN AND MERCEDES GUTIÉRREZ

In the first decade of the 20th century, British zoologist Charles Andrews described new fossil mammals from Egypt that demonstrated Africa had once harbored an archaic, endemic fauna very different from the continent's modern mammal communities (Andrews and Beadnell, 1902; Andrews, 1903, 1904a, 1904b, 1906, 1907). This fauna, we now know, characterized much or all of the continent during the early Tertiary, a time when Africa was isolated from Eurasia by the Tethys Sea (Kappelman et al., 2003; Jaeger, 2003; Stevens et al., 2004). The order Hyracoidea was a central component of this endemic fauna, a startling realization given the inauspicious nature of the living hyraxes, which today include only a few small species in three genera.

The African fossil record reveals dozens of hyracoids that ranged in size from that of small rabbits upward to that of modern Sumatran rhinos (Schwartz et al., 1995). Some fossil forms have teeth so bunodont and piglike in arrangement that they were mistaken by experts as the teeth of pigs (Suidae)(Andrews, 1906; Hooijer, 1963). Other hyracoids possess the smooth, crescentic dental blades of leaf-eating specialists. Hyracoid locomotor adaptations are remarkably diverse, including rhino-sized graviports, small scrambling forms, runners, jumpers, and if one includes extant forms, even the arboreal genus *Dendrohyrax* (Whitworth, 1954; Schwartz et al., 1995; Rasmussen and Simons, 2000; Thomas et al., 2004). The ecological diversity of early Tertiary hyracoids is attested to by the fact that more than half a dozen common species co-existed in a single ecological community at some fossil quarry assemblages (figure 13.1; Gagnon, 1997). Hyracoids were indeed the dominant "ungulates" of archaic Africa. They represent the ecological analogs of perissodactyls and artiodactyls evolving at the same time in the northern continents. They were, as Andrews (1906) noticed immediately, "a very important factor in the fauna." A century later, the full extent of hyracoid diversity is still poorly understood; paleontological discoveries continue to unveil new hyracoid taxa and novel adaptations at an ever-increasing rate.

Hyracoids have played an important role in studies of mammalian evolution for another reason: they are related to elephants. This was a provocative idea entertained even before Andrews's time and well before the advent of modern DNA analyses (Gill, 1870; Cope, 1882; Schlosser, 1923; Weitz, 1953). The phylogenetic link between hyracoids and proboscideans has often seemed nonintuitive to many observers because of the utter contrast between the massive size and specialized morphology of elephants in comparison to the simple, furry, generalized shape of the living hyracoids. As conveyed by the genus name *Procavia* (one of the living forms), modern hyraxes resemble nothing so much in outward appearance as primitive guinea pigs (*Cavia*). But evidence from molecular and morphological analyses firmly supports a close relationship between elephants and hyracoids (Shoshani, 1986; Shoshani and McKenna, 1998; Rasmussen et al., 1990; Novacek 1992, Springer et al., 1997). This has been reflected in a classification that places Hyracoidea and Proboscidea, along with Sirenia and the extinct order Embrithopoda, in the superorder Paenungulata (Simpson, 1945).

The primary intraordinal phylogenetic divisions within Hyracoidea, however, have received much less attention than have their ties to elephants. As a consequence, family-level taxonomy of the group has been particularly inconsistent. Some researchers have hesitated to develop a balanced and cogent classification until the phylogeny of all groups has been worked out reliably (Rasmussen, 1989; DeBlieux et al., 2006). As a result, taxonomic lumping has been used as a conservative measure at times to avoid errors, and in other cases, raising one or another particular group under taxonomic study into one well-defined family has merely left other hyracoids in taxonomic limbo (e.g., Pickford and Fischer, 1987). Clearly, the attempt of linking different radiations that are distributed through vast periods of time in a phylogenetically reliable way has been problem. For this chapter, we have compared a substantial number of fossil hyracoids aiming to get a better understanding of primitive dental structure in the order (Court and Mahboubi, 1993; Tabuce et al., 2000, 2001), and we have utilized new discoveries of early Tertiary diversity to build on the taxonomic revision of Pickford (2004) and to present a new, phylogenetically consistent, family-level classification of African hyracoids.

A few technical notes concerning our review include the following items. Age estimates for North African sites follow the recent revisions of Seiffert (2006). In our taxonomic diagnoses, we have utilized simple differential diagnoses that are designed to highlight just a few key morphological features that discriminate the taxon in question from relevant relatives; more complete description and illustration of known

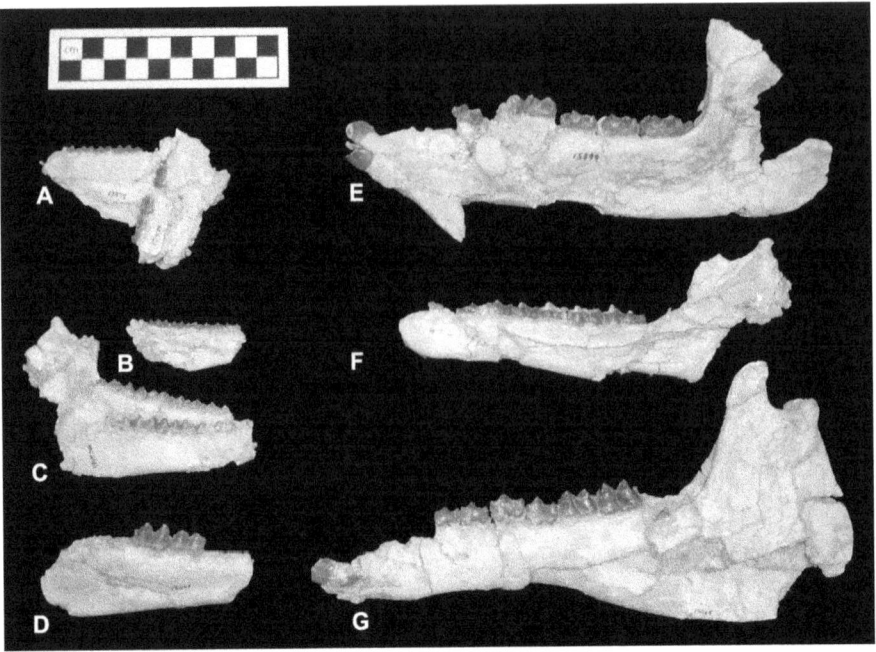

FIGURE 13.1   Hyracoid diversity represented in a single community of late Eocene mammals (Quarry L-41, Jebel Qatrani Formation, Fayum, Egypt). Mandibles of seven taxa all illustrated to the same scale: A) *Saghatherium bowni* (DPC 17979, right and left p4-m3), the most common small mammal at the quarry; B) *Thyrohyrax meyeri* with right p1-m3, the smallest Fayum hyracoid (DPC 20304); C) *Thyrohyrax litholagus* with right and left p1-m3 (DPC 16783); D) a small, unnamed species of *Bunohyrax* with right m1-2 (DPC 18667); E) an unnamed species of *Titanohyrax* with left i1, right i1-2, p2-m3 (note the spatulate i2 characteristic of the genus; DPC 15399); F) *Antilohyrax pectidens*, a genus unique to Quarry L-41 (left c-m3, DPC 17373); G) *Megalohyrax* aff. *M. eocaenus*, the most common large hyracoid at the quarry (juvenile with left i1, p1-m2, m3 unerupted; DPC 13065).

morphology for each taxon can be found in the cited literature. We refer to the large cusp on the distolingual corner of upper molars as the hypocone, even though fossil evidence suggests that in hyracoid history this cusp was derived from what was originally a metaconule. Hypocones have evolved independently in many mammalian lineages (Fortelius, 1985; Hunter and Jernvall, 1995; Jernvall, 1995) by several different means (from lingual cingulum, by splitting from the protocone, by elaboration of the metaconule, among others).

ABBREVIATIONS

GSNW, Geological Survey of Namibia, Windhoek; AMNH, American Museum of Natural History, New York; BMNH, British Museum (Natural History), London; CGM, Cairo Geological Museum, Cairo; KNM, Kenya National Museums, Nairobi; UO, University of Oran, Algeria; USTL, University of Sciences and Techniques of Languedoc, Montpellier, France.

Systematic Paleontology

We have chosen to rely on dental characteristics to make most of our family-level taxonomic judgments because tarsal bones of known taxonomic association, although useful, are rarely available in the fossil record. We suggest that there are two fundamentally different organizational plans for the lower molars of hyracoids. In one of these, the lower molars tend to be bunodont, and the lingual cusps stand alone, not connected by crests or lophs to the buccal cusps (Rasmussen and Simons, 1988). This pattern is similar to what is seen among certain lineages of early artiodactyls, in which the buccal cusps may develop crescents around the lingual cusps but never connect

directly to them by a crest. We call this pattern the "artiodactyl-type" of molar tooth. This arrangement stands in contrast to the teeth of Perissodactyla, in which the protoconid connects to the metaconid at the back of the trigonid, and the hypoconid connects to the entoconid at the back of the talonid. This is a common dental pattern within hyracoids, including some early Tertiary forms as well as modern species. We refer to this as the "perissodactyl-type" tooth. We believe these two patterns are fairly conservative evolutionarily, based on their distributions among ungulate-like mammals, and because within hyracoid lineages there does not seem to be much change from one to the other. Obviously, this is a hypothesis that will be tested as new fossils are discovered.

Most analyses of hyracoid evolution have relied on the view that the artiodactyl-type pattern is primitive among hyracoids, especially those that are bunodont in crown structure (*Seggeurius, Geniohyus, Bunohyrax*). This is supported by (1) the greater importance of the artiodactyl-type groups in the early radiation of hyracoids; (2) the similarity to the bunodont, nonlophed molars of hyracoid outgroups such as palaeomastodontids and sirenians (although teeth of arsinoitheres, numidotheres, and barytheres are lophed); and (3) the observation that across mammal taxa, the presence of cross lophs is derived. In this chapter, the taxa with artiodactyl-type teeth are all placed in the family Geniohyidae. All other hyracoids known from the fossil record are hypothesized to share a common ancestor. While it is possible that the perissodactyl-type tooth may have evolved more than once in Hyracoidea (e.g., independently in saghatheriids and titanohyracids), at present we believe there is no positive evidence to support that idea. Table 13.1 presents a new, formal taxonomy of African fossil taxa belonging to the mammalian order Hyracoidea.

Family . . . . Geniohyidae Andrews, 1906
    Genus . . . . *Seggeurius* Crochet, 1986
    Seggeurius amourensis
      Crochet, 1986
    Genus . . . . *Geniohyus* Andrews, 1904
    Geniohyus mirus Andrews,
      1904
    Geniohyus magnus (Andrews,
      1904)
    Geniohyus diphycus
      Matsumoto, 1926
    Genus . . . . *Bunohyrax* Schlosser, 1910
    Bunohyrax matsumotoi Tabuce
      et al., 2000
    Bunohyrax fajumensis Andrews,
      1904
    Bunohyrax major (Andrews,
      1904)
    Genus . . . . *Pachyhyrax* Schlosser, 1910
    Pachyhyrax crassidentatus
      Schlosser, 1910
    Genus . . . . *Brachyhyrax* Pickford, 2004
    Brachyhyrax aequatorialis
      Pickford, 2004
    Brachyhyrax oligocenus
      Rasmussen and Gutiérrez,
      2009
Family . . . . Saghatheriidae Andrews,
      1906
    Genus . . . . *Microhyrax* Sudre, 1979
    Microhyrax lavocati Sudre, 1979
    Genus . . . . *Saghatherium* Andrews and
      Beadnell, 1902
    Saghatherium bowni Rasmussen
      and Simons, 1991
    Saghatherium antiquum
      Andrews and Beadnell,
      1902
    Saghatherium humarum
      Rasmussen and Simons,
      1988
    Genus . . . . *Selenohyrax* Rasmussen and
      Simons, 1988
    Selenohyrax chatrathi
      Rasmussen and
      Simons, 1988
    Genus . . . . *Thyrohyrax* Meyer, 1973
    Thyrohyrax meyeri
      Rasmussen, 1991
    Thyrohyrax domorictus Meyer,
      1973
    Thyrohyrax litholagus
      Rasmussen, 1991
    Thyrohyrax pygmaeus
      (Matsumoto, 1922)
    Thyrohyrax kenyaensis
      Rasmussen and Gutiérrez,
      2009
    Thyrohyrax microdon
      Rasmussen and Gutiérrez,
      2009

    Genus . . . . *Megalohyrax* Andrews, 1903
    Megalohyrax eocaenus
      Andrews, 1903
    Megalohyrax gevini Sudre, 1979
Family . . . . Titanohyracidae
      Matsumoto, 1926
    Genus . . . . *Titanohyrax* Matsumoto,
      1922
    Titanohyrax andrewsi
      Matsumoto, 1922
    Titanohyrax mongereaui Sudre,
      1979
    Titanohyrax angustidens
      Rasmussen and Simons,
      1988
    Titanohyrax tantulus Court
      and Hartenberger, 1992
    Titanohyrax ultimus
      Matsumoto, 1922
    Genus . . . . *Antilohyrax* Rasmussen and
      Simons, 2000
    Antilohyrax pectidens
      Rasmussen and Simons,
      2000
    Genus . . . . *Afrohyrax* Pickford, 2004
    Afrohyrax championi
      (Arambourg, 1933)
    Afrohyrax sp. nov. Rasmussen
      and Gutiérrez, 2009
Family . . . . Pliohyracidae Osborn, 1899
    Genus . . . . *Meroehyrax* Whitworth,
      1954
    Meroehyrax bateae
      Whitworth, 1954
    Meroehyrax kyongoi
      Rasmussen and Gutiérrez,
      2009
    Genus . . . . *Prohyrax* Stromer, 1924
    Prohyrax hendeyi Pickford,
      1994
    Prohyrax tertiarius Gray, 1868
    Genus . . . . *Parapliohyrax* Lavocat, 1961
    Parapliohyrax mirabilis
      Lavocat, 1961
    Parapliohyrax ngororaensis
      Pickford and Fisher, 1987
Family . . . . Procaviidae Thomas, 1892
    Genus . . . . *Heterohyrax* Gray, 1868
    Heterohyrax auricampensis
      Rasmussen et al., 1996
    Genus . . . . *Dendrohyrax* Gray, 1868
      (contains extant taxa
      only)
    Genus . . . . *Procavia* Storr, 1780
    Procavia antiqua Broom,
      1934
    Procavia transvaalensis Shaw,
      1937
    Genus . . . . *Gigantohyrax* Kitching, 1965
    Gigantohyrax maguirei
      Kitching, 1965

Family GENIOHYIDAE Andrews, 1906

The few genera we place in Geniohyidae are divergent from each other in several important ways (Court and Mahboubi, 1993), but they all share the primitive artiodactyl-type molar pattern, tend to have bunodont molars and simple premolars, and defy placement in clades with more specialized relatives. We suggest that *Seggeurius* is primitive for the group. *Bunohyrax* is a typical generalized form, represented by several species that vary considerably in size. *Geniohyus* is an odd form with derived features of the teeth and jaws (Court and Mahboubi, 1993). *Brachyhyrax* is a latest surviving genus of geniohyid, occuring in the late Oligocene and early Miocene. Oligocene *Pachyhyrax* and some of its unnamed relatives represent a specialized radiation within the family in which the molars become crestier and the premolars are molarized, but done in a manner that is clearly in parallel to the cresty teeth found in other hyracoid families.

Type Genus Geniohyus Andrews, 1904.

Included Genera Seggeurius, Geniohyus, Bunohyrax, Pachyhyrax, Brachyhyrax.

Distribution Eocene to early Miocene of North and East Africa, with one record from the Bugti beds of Pakistan (Pickford, 1986b).

Diagnosis Differs from all other hyracoid families by the absence of typical connecting crests or lophs between the buccal and lingual cusps of the lower molars, instead, a distinct V-shaped furrow or gutter separates hypoconid and entoconid. The midline may be bridged by short, descending postprotocristids and postmetacristids, which meet in the midline at a V-shaped notch, or by accompanying spurs off the talonid cusps (usually mesiolingually directed spurs off the hypoconid and mesiobuccally directed ones off the entoconid). This family further differs from others in that the molar cusps are symmetrically arranged on the buccal and lingual halves of the tooth. Differs from all families except primitive species of Saghatheriidae in the simple, nonmolariform premolars. Dental formula 3.1.4.3/3.1.4.3.

Genus *SEGGEURIUS* Crochet, 1986

Type and Only Known Species Seggeurius amourensis Crochet, 1986.

Distribution Eocene of El Kohol, Algeria.

Diagnosis Modified from Crochet's diagnosis in Mahboubi et al. (1986) and from Court and Mahboubi (1993). Small brachydont geniohyid with shallow mandibular corpus, extremely simple premolars (uppers lacking clear differentiation of paracone and metacone, hypocone absent, lowers lacking entoconid and with straight hypocristids). Differs from *Bunohyrax* and *Pachyhyrax* but resembles *Geniohyus* in the simple premolars. Differs from *Geniohyus mirus* in having double-rooted upper and lower canines with no diastema between canines and premolars, and in lacking the unique gnathic features of that genus. Further differs from *Geniohyus* in having the lingual cusps of the lower molars taller and more compressed than the buccal ones. Is unique in having extreme compression of anterior trigonid (lack of paraconid or paracristid), the cristid obliqua connecting mesially to the protoconid apex, small upper molar metacone lacking postmetacrista, cuspate mesostyle with distally trending crest (continuous with distal cingulum in M1), hypocone wider than protocone bearing long prehypocrista that recurves to abut the base of protocone on M3.

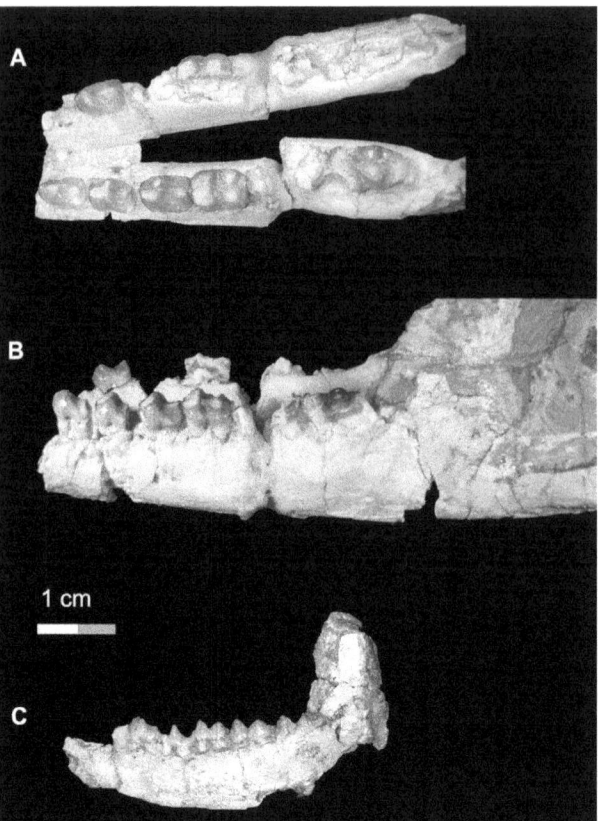

FIGURE 13.2 Small, primitive hyracoids from the Eocene of Algeria. A) Occlusal view and B) lateral view of *Seggeurius amourensis* (left p2-m1, m3, and damaged m2; right p3, m1, and other root fragments). Note the simple premolars with straight postcristids, which contrast with the quadrate, bunodont m1. C) Left mandible of *Microhyrax lavocati* with p3-m3. This is the smallest species of fossil hyracoid known; it is a primitive member of Saghatheriidae.

SEGGEURIUS AMOURENSIS Crochet, 1986
Figures 13.2A and 13.2B

Holotype UO-K206, left M3.

Distribution and Diagnosis Same as for genus.

Remarks There is a consistent agreement among researchers that *Seggeurius* possesses primitive dental structure for Hyracoidea. Mahboubi et al. (1986) and Court and Mahboubi (1993) pointed out that *Seggeurius* is more primitive than younger *Geniohyus* from the Fayum, Egypt. The premolars are less molarized than those of any other hyracoid except for *Geniohyus*. The upper M3 bears a postprotocrista that links to the hypocone. The arrangement of lower molar cusps—with the lingual ones tall and compressed, while the buccal ones are broader and slope up gradually from the buccal tooth border—was used by Court and Mahboubi (1993) to affine *Seggeurius* to saghatheriids. This arrangement is also found in the geniohyids *Bunohyrax* and *Pachyhyrax*.

The mandible of *Seggeurius amourensis* lacks an internal mandibular chamber, as well as the broad lingual fossa of the corpus seen in *Geniohyus mirus*. Notably, however, a small depression or fossa has been described low on the lingual face of the ascending ramus, which gradually becomes shallower anteriorly to a position under the m2. This has been interpreted as possibly being the primitive condition for the odd hyracoid trait of an internal mandibular chamber (Court and Mahboubi, 1993; DeBlieux et al., 2006), which in its most derived expression, in taxa such as *Thyrohyrax*, becomes an expanded internal cavity that opens lingually through a small window (DeBlieux et al., 2006). However, the distribution and expression of a lingual hollow or cavity is complex, and the fossa-like condition seen in *Seggeurius* and *Geniohyus* could be derived from an internal mandibular chamber or may be an independent development not related to the internal chamber.

Genus *GENIOHYUS* Andrews, 1904

Type Species Geniohyus mirus Andrews, 1904.
Included Species G. mirus, G. diphycus, G. magnus.
Distribution Early Oligocene, Jebel Qatrani Formation, Egypt.
Diagnosis Variably sized geniohyids with brachydont molars and very simple lower premolars (lacking a metaconid except for slight development on p4, slight development of talonid basin, a single small hypoconid positioned at the back of the tooth connected to the protoconid by a straight crest), and simple upper premolars (lacking hypocones and mesostyles). Lower molars with compressed, symmetrical trigonid (i.e., with premetacristid rather than an open flexid on lingual side).
Remarks The conception of *Geniohyus* should continue to rely heavily on what is known of *G. mirus*. There are problems with allocation of the other two smaller species (*G. diphycus* and *G. magnus*) to this genus, rather than to *Bunohyrax* or a new genus. *G. mirus* seems to be unique in ways that are not yet demonstrable for the other species. As Meyer (1978) noted, "many specimens placed in *Geniohyus* by Schlosser and by Matsumoto show no difference in tooth morphology from those placed in *Bunohyrax*"; these authors relied instead on the occurrence or shape of an internal mandibular chamber, a feature now known to be sexually dimorphic (DeBlieux et al., 2006).

GENIOHYUS MIRUS Andrews, 1904
Figures 13.3A, 13.3E, and 13.4B

Synonymy Saghatherium majus Andrews, 1906:91 *pars*, plate VI, figure 5; *non Geniohyus mirus* Matsumoto, 1926:269, figures 5–8.
Holotype CGM-8634, fragment of a right ramus with root of i2 and p1 to m3.
Distribution Early Oligocene, Jebel Qatrani Formation, Egypt.
Diagnosis Differs from all other fossil hyracoids in having its two-rooted lower canine separated from the premolar by a diastema. Further differs from other hyracoids in having an enormous lingual fossa on the lingual side of the mandibular corpus, a thickened torus of bone at the front of the fossa gives the mandible a distinctive shape. The symphyseal region is very narrow. The lower molar cusps of this species are more brachydont than those of *Seggeurius* (which has higher cusps), and they further differ in having extremely symmetrical lingual and buccal cusps on the lower molars, and small accessory cuspules located around the primary cusps. The cusps of this species are greatly inflated, and crown volume presents a full appearance.
Remarks This is the best-known species of the genus; indeed, if more material of other taxa becomes available, it may prove to be the only species properly attributed here. *G. mirus* is remarkable for its combination of apparently primitive dental features (diastema between canine and premolar, simple premolar structure, bunodont molar cusps) and odd, apparent

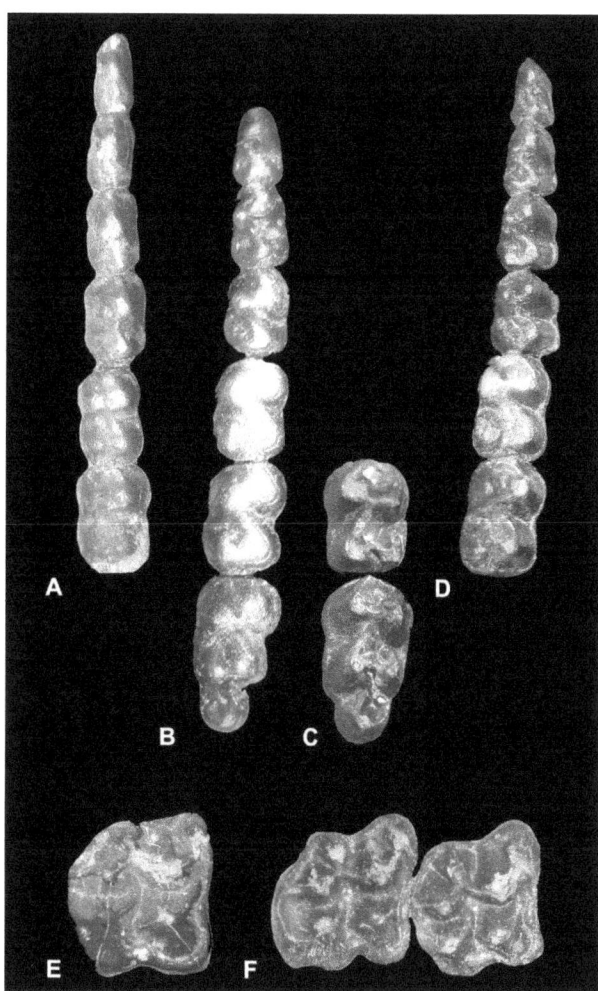

FIGURE 13.3 Representatives of the family Geniohyidae. A) Holotype of *Geniohyus mirus* in occlusal view showing right p1–m2; notice the similarity in the simple premolars and quadrate molars to *Seggeurius*, figure 13.2. B) *Bunohyrax fajumensis* of the Fayum (right p2–m3; notice the development of talonid basins on the premolars and the less inflated molar crowns in comparison to *Geniohyus*). C) *Bunohyrax* sp. from the Upper Oligocene of Chilga, Ethiopia. D) *Pachyhyrax crassidentatus* from Quarry I, Fayum (DPC 7514, right p1–m2; note the molarized premolars and the molar entoconids isolated from the buccal cusps) E) Maxillary M2 of *Geniohyus mirus* from the Fayum; note the smaller parastyle than in *Bunohyrax* and the hypocone that has not migrated far from its original position as a metaconule. F) Maxillary M2–3 of *Bunohyrax* sp. from Quarry L-41 (DPC 20719).

specializations (large mandibular fossa, very symmetrical buccal-lingual cusp pairs, molar cuspules). In their analysis of *Seggeurius*, Court and Mahboubi (1993) emphasized the specialized (or at least odd) nature of *G. mirus*. While this taxon has often been held up as the most primitive Fayum hyracoid, in the light of *Seggeurius* and *Bunohyrax*, it might be best to view *G. mirus* as a hyracoid that has become more specialized towards extremely inflated crowns and unique mandibular structures.

From an adaptive point of view, *Geniohyus* does share obvious similarities to the teeth of suids and tayassuids, suggesting a similar diet of tubers, fallen fruit, and other food items requiring crushing. It is a rare taxon in the Fayum collections, with the best-known specimens still being the two lower jaws described by Andrews (1906, 1907), suggesting that it may have been a unique component of the faunas at

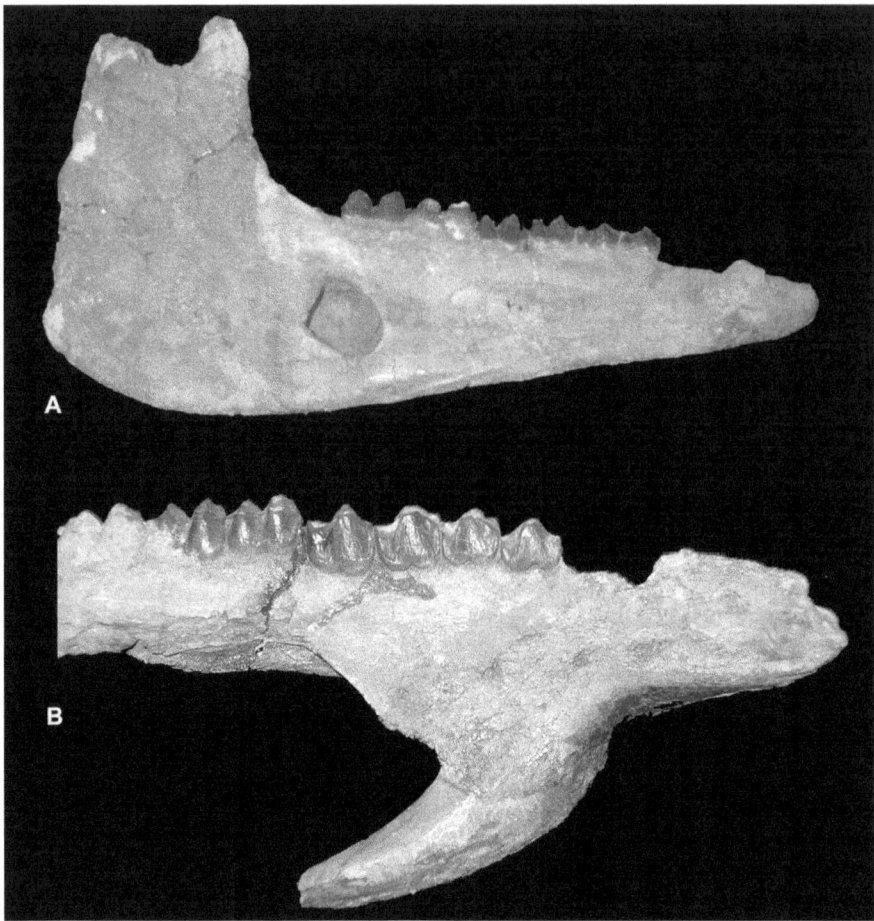

FIGURE 13.4 Mandibles of geniohyids. A) *Pachyhyrax crassidentatus* in lingual view, showing the round mandibular foramen that opens into a hollow chamber within the mandibular corpus (DPC 7514); B) *Geniohyus mirus*, showing the abrupt deepening of the corpus, which then forms a robust torus beneath a thin plate of bone covering the buccal side of a broad lingual fossa (holotype).

Quarries A and B (which occur in the Lower Sequence of the Jebel Qatrani Formation at levels that have not been worked much in recent decades; Bown and Kraus, 1988). The upper molar named *Saghatherium majus* by Andrews clearly belongs to this species and therefore represents our only certain view of the upper dentition of the genus. No postcranial elements have been allocated to this genus.

GENIOHYUS DIPHYCUS Matsumoto, 1926

Holotype AMNH 13349, left ramus with symphysis and p1 to m2.

Distribution Early Oligocene, Jebel Qatrani Formation, Egypt.

Diagnosis Small species of the genus. Differs from *G. mirus* in its smaller size, the lack of diastemata, the better development of paraconids on the lower premolars, the lack of mesostyles on the upper molars, and in mandible shape (Meyer, 1978).

Remarks Given the uncertainty about referred specimens, the diagnosis should rely exclusively on the holotype for now. See comment later under *G. magnus*.

GENIOHYUS MAGNUS (Andrews, 1904)

Synonymy Saghatherium antiquum Andrews, 1903:340, figure 2 *non* Andrews and Beadnell, 1902; *S. magnum* Andrews,

1904:214; *S. magnum* Andrews, 1906:89 *pars* plate VI, figure 3 *non* figure 4; *S. antiquum* Schlosser, 1911:110, 113 *pars.*; *S. majus* Schlosser, 1911:110, 114 *pars.*

Holotype BMNH M8398, right maxilla with I1, C to M3.

Distribution Exact locality of type unknown, but almost certainly from the Lower Sequence of the Jebel Qatrani Formation, Egypt. Referred specimens tentatively allocated to this species from Quarry V, Upper Sequence, Jebel Qatrani Formation.

Diagnosis Small species of *Geniohyus*.

Remarks The type specimen was first studied by Andrews, who recognized this upper dentition as hyracoid and placed it in *Saghatherium* (rather than with the lower teeth of *Geniohyus*). Matsumoto (1926) allocated the specimen to *Geniohyus*. Matsumoto (1926) and Meyer (1978) both made favorable comparisons between this type and the maxillary dentition of *Thyrohyrax pygmaeus*, which at one point or another in its history has been placed in *Geniohyus*, *Megalohyrax* and *Pachyhyrax* (Meyer, 1978; Rasmussen and Simons, 1988). This complete upper-right dentition has proven difficult to interpret, in part because of heavy occlusal wear on all teeth but P4 and M3. The holotypes of *G. mirus*, *G. diphycus*, *Bunohyrax fajumensis*, and *B. major* are all specimens of lower dentitions. This much can be said of the *G. magnus* type: the molars are bunodont and simple, the premolars have only slight development of a hypocone, the canine is double rooted and in contact with the P1, and the rostrum is short with only small diastemata between I1 and I2, and I2 and I3, the latter of which is in contact with C. This assemblage of features indicates a geniohyid, based on the bundont molars and the simple premolars.

Maxillary and mandibular specimens of a similar hyracoid with nearly identical upper tooth measurements were reported from Quarry V, at a much higher stratigraphic level than the origin of Andrews's specimens (Rasmussen and Simons, 1988). The lower dentition of this sample is like those of *Geniohyus* and *Bunohyrax* in a general sense: the premolars are simple, the molars bunodont. However, the Quarry V geniohyid has proportionally higher cusps and more acute crests than those of *G. mirus*. Until a better sample of stratigraphically controlled specimens that match well with Andrews's type are recovered from the Lower Sequence, his original odd, lone holotype will continue to be enigmatic.

Genus *BUNOHYRAX* Schlosser, 1910

Synonymy Geniohyus Andrews, 1906, *pars.*; *Geniohyus* Schlosser, 1910, 1911, *pars.*; *Geniohyus* Matsumoto, 1926 *pars.*; *non Bunohyrax* Whitworth, 1954.
Type Species Bunohyrax fajumensis (Andrews, 1904).
Included Species B. fajumensis, B. major, B. matsumotoi.
Distribution Eocene and Oligocene, North Africa.
Diagnosis Variably sized geniohyids with bunodont dentitions. Differs from saghatheriids in having a deep, V-shaped furrow separating the lingual cusps from the buccal cusps on the lower molars. Differs from *Geniohyus* in its less symmetrical lingual-buccal cusp pairs on the lower molars. The lingual ones (metaconid, entoconid) are more buccolingually compressed and rise steeply from the tooth base, while the buccal ones (protoconid, hypoconid) are broader and slope up more gradually from the base, which bears a moderate cingulum. The buccal cusps are distinctly more bunoselenodont than the very bunodont ones of *Geniohyus*. Further differs from *Geniohyus* and *Seggeurius* in having the cusps placed further apart, and in having more molariform premolars. Differs from *Pachyhyrax* in being less selenodont and having simpler premolars. Dental formula 3.1.4.3/3.1.4.3.

BUNOHYRAX MATSUMOTOI Tabuce et al., 2000

Holotype UO 84-372, right M3.
Distribution Middle to late Eocene of Bir El Ater, Algeria.
Diagnosis Small species of the genus. Differs from *B. major* by its less inflated cusps and by its smaller hypoconulid lobe on M3. Further differs from *B. fajumensis* in having a smaller spur on the mesial side of the protocone, a less W-shaped ectoloph on M3, and a weaker spurlike hypoconulid on m1–2.
Remarks This species is smaller and more primitive than the Fayum members of the genus. It joins *Seggeurius amourensis* and *Microhyrax lavocati* as one of three minute, primitive hyracoids that are valuable in assessing the primitive morphology for known Paleogene hyracoids (Tabuce et al., 2000).

BUNOHYRAX FAJUMENSIS Andrews, 1904
Figure 13.3B

Synonymy Geniohyus fajumensis Andrews, 1904:162; *G. fajumensis* Andrews, 1906: 195, plate XIX, figure 2; *Saghatherium majus* Andrews, 1906:91 *pars.*; *G. minutus* Schlosser, 1910:503, *nomen nudum*; *Geniohyus* aff. *G. mirus* Andrews,

1904:162; *Geniohyus micrognathus* Schlosser, 1911:123, plate X, figures 1, 2; *G. subgigas* Matsumoto, 1926:266, figure 4; *G. mirus* Matsumoto, 1926 (*non* Andrews):269, figures 5–8; *Bunohyrax affinis* Matsumoto, 1926:309, figures 18, 19; *Megalohyrax suillus* Matsumoto, 1926:319 (*non* Schlosser).

Holotype BMNH M8435, fragment of a right ramus with p1–p4.
Distribution Early Oligocene, Jebel Qatrani Formation, Egypt.
Diagnosis A medium-sized species of the genus, differing from *B. matsumotoi* in its larger size and from *B. major* in its smaller size. Further differs from *B. major* in its much less inflated cusps. Differs from all other known hyracoids in having a single-rooted lower canine that is separated from the p1 by a diastema (the canine of *G. mirus* is isolated but double rooted). As in other hyracoids, the upper canine is double rooted.
Remarks This is the best-known species of the genus. Matsumoto (1926) described a well-preserved, entire cranium, which shows a long, tubular snout. This species has an internal mandibular chamber and lingual fenestra that could parallel the case of *Thyrohyrax litholagus* and *Meroehyrax eocaenus* in being sexually dimorphic (DeBlieux et al., 2006). An undescribed species from Quarry L-41 (late Eocene) is closely related to but older than *B. fajumensis*. Like its odd relative, *G. mirus*, this species suggests dietary adaptations resembling those of generalized suiform artiodactyls. A specimen from Cabinda, Angola, may represent a sub-Saharan record of this species or a close relative (Pickford, 1986a).

BUNOHYRAX MAJOR (Andrews, 1904)

Synonymy Geniohyus major Andrews, 1904:212; *G. major* Andrews, 1906:196, figure 63; *G. gigas* Matsumoto, 1926:264, figures 1–3.
Holotype CGM 8980, left ramus with p1 to p3.
Distribution Early Oligocene, Jebel Qatrani Formation, Egypt; possible occurrence in Malembe, Angola (Pickford, 1986b).
Diagnosis Very large species of the genus that further differs from *B. fajumensis* in its much more inflated premolar and molar cusps.
Remarks This is another rare Fayum species. It represents the most extreme development of low-crowned bunodont tooth cusps among all known hyracoids to date. No crania or postcrania are known.

BUNOHYRAX, new species
Figures 13.1D and 13.3F

Distribution Late Eocene, Quarry L-41, Egypt.
Remarks This is a relatively uncommon species at Quarry L-41 and differs from *B. fajumensis* by its smaller size.

Genus *PACHYHYRAX* Schlosser, 1910

Synonymy Megalohyrax Matsumoto, 1922, 1926, *pars.*
Type species Pachyhyrax crassidentatus Schlosser, 1910.
Included species P. crassidentatus and a new species from the late Oligocene, Ethiopia.
Distribution Oligocene of Egypt and Ethiopia.
Diagnosis Medium to large geniohyids differing from more primitive members of the family in having a greater degree of molarization of premolars and development of bunoselendont molars. Postcristid of the lower premolars is

slightly V-shaped in occlusal view in p1, and very V-shaped in p2–4. There is a distinct metaconid separated from the protoconid on p2–4. On upper premolars, hypocone is slightly developed on P1, well developed on P2–4. Buccal cusps of lower molars with well-developed crescentic ridges; accessory ridges descend near the midline, and the gaps formed between them outline deep pits at the tooth midline. Upper molars with well-developed ectoloph ridges. Small, simple accessory ridges descend off lingual side of ectoloph distal to paracone and metacone. Protocone and hypocone have moderately developed precristae and postcristae, arranged diagonally to mesiodistal axis of tooth.

Remarks The more selenodont development of the lower molar buccal cusps, combined with the lack of cristids connecting to the lingual cusps, gives this genus a distinctive appearance. The buccal crescents curve around the lingual cusps and give the lower molars an anthracothere-like appearance, in contrast to the more bunodont, suidlike teeth of *Geniohyus* and *Bunohyrax*. *Pachyhyrax* occurs alongside anthracotheres in the early Oligocene of Egypt and the late Oligocene of Ethiopia. The premolars are more molarized than is seen in other geniohyids and are decidedly more so than those of the large saghatheriid *Megalohyrax*, with which it co-occurs. This is a unique and morphologically distinctive Oligocene genus that could have been derived from a form resembling *Bunohyrax fajumensis* of the Fayum's Lower Sequence. A specialization shared between the two is the distinctive lingual tilt of the lower buccal cusps as they rise from the basal cingulum. A hypothetical morphocline sequence of lower tooth characters (not necessarily a phylogenetic link) ranking the geniohyid dentition (the artiodactyl-type tooth) from primitive small size, bunodonty, and simple premolars to more derived, anthracothere-like bunoselenodonty with molarized premolars would be *Seggeurius*, *Bunohyrax*, *Pachyhyrax*.

PACHYHYRAX CRASSIDENTATUS Schlosser, 1910
Figures 13.3D and 13.4A

Holotype Uncataloged specimen Staatliches Museum für Naturkunde, Ludwigsburg, West Germany, associated left M1, M2, M3, right P3, M2, and p4.

Distribution Early Oligocene, Jebel Qatrani Formation, Egypt.

Diagnosis Differs from new species of *Pachyhyrax* from Chilga, Ethiopia, in being more robust.

Remarks The preceding comments about the genus refer primarily to this species. This is a common species restricted to the Upper Sequence of the Jebel Qatrani Formation, where it occurs with other distinctive mammalian taxa unique to this level, such as *Thyrohyrax domorictus*, and the primates *Aegyptopithecus zeuxis* and *Apidium phiomense*.

PACHYHYRAX, new species

Distribution Late Oligocene of Chilga, Ethiopia.

Remarks The new species from Chilga differs from *P. crassidentatus* of the Fayum in being slightly smaller and more gracile, with lighter, delicate crests (Kappelman et al., 2003).

New genus and species, aff. PACHYHYRAX

Distribution Late Oligocene of Chilga, Ethiopia.

Remarks This new genus is considerably larger than *P. crassidentatus*. It shows extreme elongation of the upper molars, which have greater mesiodistal length than buccolingual breadth, which is very unusual for hyracoids. The ectoloph and the pre- and postcristae of the protocones and metacones are well developed and elongated. The lower molars are large, robust, and heavily crested. This genus remains unnamed pending the recovery of a suitable holotype. To date, only incomplete, isolated teeth have been found. This taxon, which is apparently represented by a single species at Chilga (Kappelman et al., 2003), represents a dental adaptation that emphasizes the addition of shearing crests to an underlying bunodont pattern by elongating the crests of the upper molars in a buccolingual axis, rather than connecting buccal and lingual cusps together with lophs.

Genus BRACHYHYRAX Pickford, 2004

Synonymy Bunohyrax sp. Whitworth, 1954:25, plate 7, figure 3.

Type Species Brachyhyrax aequatorialis Pickford, 2004.

Included Species B. aequatorialis, B. oligocenus.

Distribution Early Miocene of Kenya (Songhor) and Uganda (Napak); late Oligocene of Kenya (Losodok, Benson's Site).

Diagnosis Differs from most nongeniohyid hyracoids in having brachydont upper molars and brachyselenodont lower molars. In particular, the upper molars have brachyodont buccal cusps and a weak metastyle, the longitudinal central valley of upper molars is almost straight and unbroken by cristae, and the upper premolars are simple with small metacones and hypocones. Differs from other geniohyids in having upper molars with square occlusal outlines, and with reduced parastyles. Differs from geniohyids besides *Pachyhyrax* in having well-developed pre- and postcristae on the protocone and hypocone that are lined up with each other. Differs from *Pachyhyrax* in having the ectoloph less developed and in lacking lingual spurs on paracone and metacone.

BRACHYHYRAX AEQUATORIALIS Pickford, 2004

Synonymy Bunohyrax sp. Whitworth, 1954:25, figure 3.

Holotype BMNH M21340, left upper molar, probably M2 (field number Sgr 311'49, not Sgr 311'48 as reported in Whitworth).

Distribution and Diagnosis Same as for genus.

Remarks Whitworth attributed two teeth to *Bunohyrax* sp. (M21340 and KNM RU 295; see Pickford, 2004). Later researchers were skeptical about the generic attribution of these specimens (Meyer, 1978; Rasmussen, 1989), and Pickford (2004) was persuasive in excluding the second of Whitworth's teeth from Hyracoidea. Pickford allocated additional teeth to the taxon represented by M21340 and demonstrated that these are hyracoid. Among the features of the lower teeth allocated to *Brachyhyrax*, Pickford emphasized (1) the mesiodistally shorter crescents associated with the protocone in contrast to the longer one of the hypoconid, and (2) the presence of a slight rib on the lingual side of the hypoconid (which he called a mesohypocristid), which extends to the base of the talonid. Both features differ from the condition observed in the common Miocene titanohyracid *Afrohyrax* but resemble features found in *Geniohyus* and *Pachyhyrax*.

BRACHYHYRAX OLIGOCERUS Rasmussen and Gutiérrez, 2009

Holotype KNM-NW 22593, right and left p1–m3.

Diagnosis Rasmussen and Gutiérrez, 2009. Differs from all hyracoids but geniohyids in having the entoconid separated from the hypoconid by a distinct v-shaped notch, and by having low-crowned, bunodont molar cusps. Differs from *Bunohyrax* and other geniohyids by the reduced parastyle and the extremely brachydont premolars and molars. Differs from *Seggeurius* and *Geniohyus* in its more molarized premolars, and from *Pachyhyrax* by its less molarized premolars. Further differs from *Geniohyus* in having the primary cusps plased less closely together and in the absence of small cuspules on the lower molars. Further differs from *Pachyhyrax* in its more bunodont cusps, and lack of development of high crests and extra ridges. Differs from *Brachyhyrax aequatorialis* in its less reduced parastyle, the more distally positioned entoconid of the lower molars, and in its smaller size.

Distribution Late Oligocene, Benson's Site, Kenya.

Remarks This species is smaller than *B. aequatorialis* and more primitive in dental morphology. It retains, for example, a relatively larger parastyle than that seen in the Miocene species. The new species is represented by all the lower cheek teeth and several upper molars. It confirms that *Brachyhyrax* is a geniohyid, showing the typical lower molars structure of the family, simple lower premolars, and square, bunodont upper molars with reduced styles.

Family SAGHATHERIIDAE Andrews, 1906

In some ways, the primitive members of this family resemble geniohyids, but in all cases, saghatheriids are characterized by the linking of lingual and buccal molar cusps by smooth cross lophs (this is the perissodactyl-type tooth we described earlier). The most primitive member of this family is probably *Microhyrax*, which despite its small size and relatively brachydont teeth, nevertheless has "the buccal and lingual cusps clearly connected and not separated by a deep furrow" (Tabuce et al., 2001:539). Saghatheriids range in size from small *Microhyrax* to large *Megalohyrax*, and they display a fair range of dental variation reflecting dietary adaptations. The few taxa known postcranially seem to have generalized skeletons, in contrast to titanohyracids.

Type Genus Saghatherium Andrews and Beadnell, 1902.

Included Genera Microhyrax, Saghatherium, Selenohyrax, Thyrohyrax, Megalohyrax.

Distribution Eocene and Oligocene of North Africa and Arabian Peninsula, late Oligocene of Ethiopia.

Diagnosis Differs from geniohyids in having connecting cristids linking protoconid to metaconid, and hypoconid to entoconid. Differs from titanohyracids in being less lophoselenodont, in having mesiodistally short, slightly pectinate lower incisors, and in lacking cursorial specializations of the postcranial skeleton. Differs from pliohyracids and procaviids in having relatively low-crowned premolars and molars, and in often having an internal mandibular chamber with lingual fenestra.

Remarks In general, this is a bunoselenodont to selenodont group of hyracoids with primitive incisors, moderately to highly molarized premolars, with a primitive dental formula of 3.1.4.3/3.1.4.3. Saghatheriids are the most common and diverse hyracoids at most sites of the African early Tertiary. (We remove *Meroehyrax* from this family and place it in Pliohyracidae, based on new finds in Kenya.)

Genus *MICROHYRAX* Sudre, 1979

Type and Only Known Species Microhyrax lavocati Sudre, 1979.

Distribution Middle Eocene of Gour Lazib and Glib Zegdou, Algeria.

Revised Diagnosis Very small hyracoid with slender mandible. The protocristid linking protoconid to metaconid, and the hypocristid linking hypoconid to entoconid, are low but complete, with the buccal and lingual cusps clearly connected and not separated by a furrow; this distinguishes *Microhyrax* from all geniohyids. Further differs from primitive geniohyids *Seggeurius, Geniohyus,* and *Bunohyrax,* in having slightly molarized premolars with individualized metaconid and simple talonid. Differs from other saghatheriids in being more brachydont, with a weakly W-shaped ectoloph on the upper molars and small styles.

MICROHYRAX LAVOCATI Sudre, 1979
Figure 13.2C

Synonymy Hyracoid indet. Gevin et al., 1975:967; Hyracoid indet. Sudre, 1975:355.

Holotype Right mandible with posterior part of p2 and p3–m3; GL2-2, temporarily deposited in the Paleontology collection of USTL.

Distribution and Diagnosis Same as for genus.

Remarks The recent discussions of primitive hyracoid dental morphology by researchers working on Eocene *Microhyrax, Seggeurius,* and *B. matsumotoi* have contributed significantly to our understanding of the basal radiation of known fossil hyracoids (Court and Mahboubi, 1993; Tabuce et al., 2000). While the differences among these small Eocene taxa may seem slight, this is to be expected near the base of what later becomes a significant evolutionary radiation.

Genus *SAGHATHERIUM* Andrews and Beadnell, 1902

Type Species Saghatherium antiquum Andrews and Beadnell, 1902.

Included Species S. bowni, S. antiquum, S. humarum.

Distribution Late Eocene and early Oligocene of Egypt; early Oligocene of Libya and Oman.

Diagnosis Small saghatheriids with very deep mandibles, short symphyseal region, heavily built molars, and distinctive upper cheek teeth. Differs from other saghatheriids in having the upper molars with a heavily folded buccal surface of the ectoloph (rounded ribs associated with each of the three styles and two major cusps separated by furrows), and in having lingual spurs off the ectoloph, robust lingual cusps, upper premolars with distinct, projecting parastyles and small hypocones. Further differs from *Thyrohyrax* in having lower molars with the primary cusps placed close together (rather than on the tooth margin), and with the base of the crown inflated. Two species (*S. bowni* and *S. antiquum*) have sexual dimorphism in body size and jaw depth, but lack the dimorphic mandibular chamber of most saghatheriids.

SAGHATHERIUM BOWNI Rasmussen and Simons, 1991
Figures 13.1A and 13.5E

Holotype CGM 41881, left maxilla with intact C-M2, and broken M3.

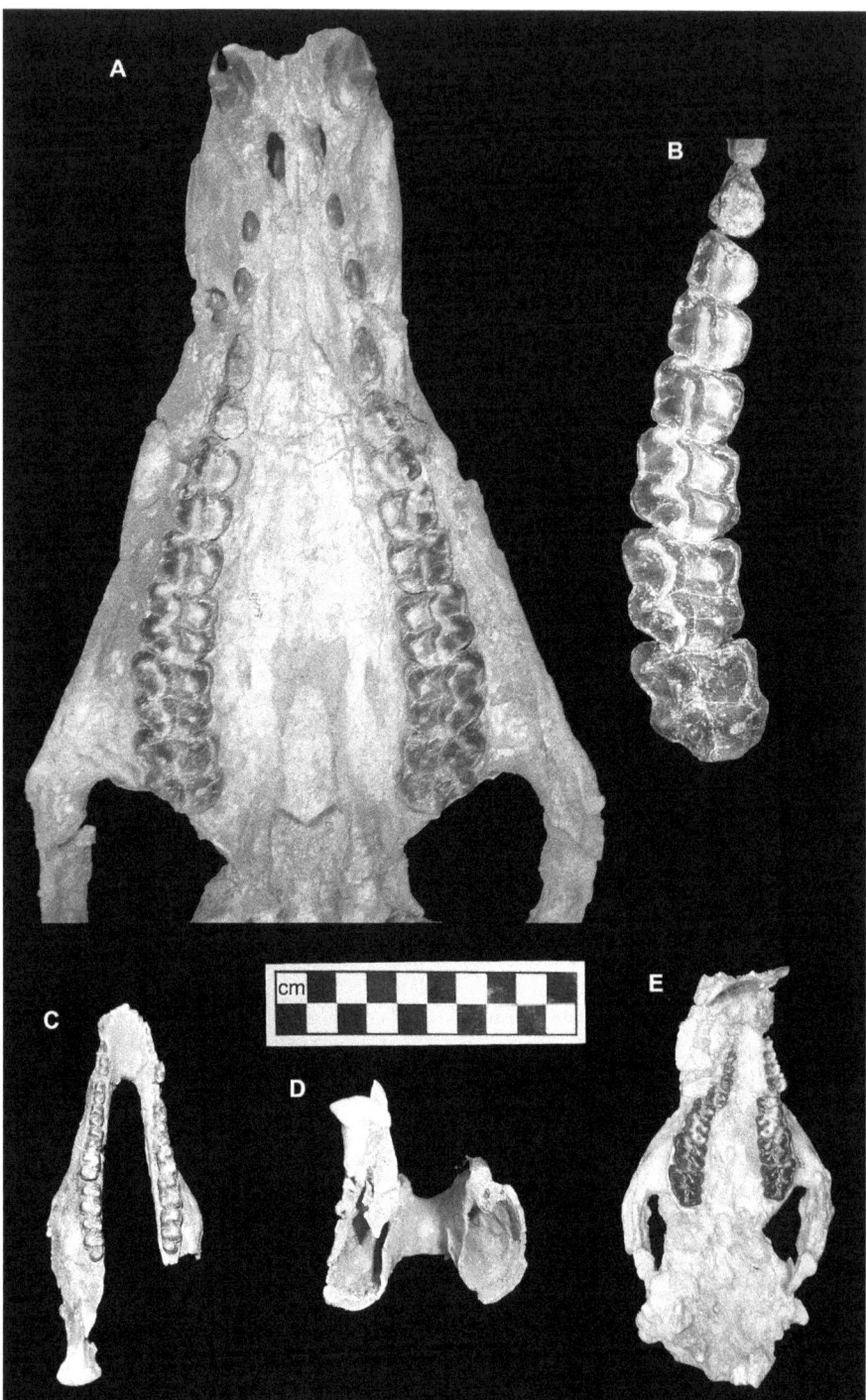

FIGURE 13.5 Representatives of the family Saghatheriidae. A and B) *Megalohyrax eocaenus* showing the entire maxillary dentition and a close-up view of the right cheek teeth (DPC 6640, Quarry I). Notice the large tusklike i1, the small buttons of i2 and i3 separated from other teeth by diastemata, the premolariform canine, and the progressively molarized premolars; the molars have moderately developed ectoloph, rounded styles, and protocone and hypocone of subequal size. C and D) Mandible of *Thyrohyrax domorictus* in occlusal and posterior views showing the inflated hollow chamber inside the corpus (DPC 5683, Quarry I). E) Cranium of *Saghatherium bowni* showing the tusks and right and left cheek tooth series from canine to m3 (DPC 13039, Quarry L-41).

Distribution Late Eocene, Quarry L-41, Jebel Qatrani Formation, Egypt; referred material from the early Oligocene, Ashawq Formation, Oman.

Diagnosis Smallest species of *Saghatherium* with the simplest premolars.

Remarks This is the oldest, smallest, and most primitive species of *Saghatherium* known. It is the most common

mammal at Quarry L-41, one of the most important paleontological sites in all of the African early Tertiary. *S. bowni* and its slightly younger relative, *S. antiquum*, are unusual among saghatheriids for lacking the internal chamber and lingual fenestra of the mandibular corpus. Instead, these two taxa show a pattern of sexual dimorphism in which one sex was larger with much deeper mandibles than the other (Rasmussen and Simons, 1991). These two taxa were presumably not ancestral to younger *S. humarum* because that Oligocene taxon has the chamber. Abundant cranial and postcranial specimens are known for *S. bowni* but these remain unstudied. Teeth of a small *Saghatherium* from Oman have been referred to this species (Pickford et al., 1994; Thomas et al., 1999).

SAGHATHERIUM ANTIQUUM Andrews and Beadnell, 1902

Synonymy Saghatherium minus Andrews and Beadnell, 1902:7; *S. magnum* Andrews, 1906:89 *pars*, plate VI, figure 4 (*non* Andrews, 1904); *S. minus* Andrews, 1906:89; *S. majus* Andrews, 1906:91 *pars* (*non* plate VI, figure 5); *S. magnum* Schlosser, 1910:503; *S. majus* Schlosser, 1910:503; *S. magnum* Schlosser, 1911:113 *pars; S. majus* Schlosser, 1911:114, plate X, figure 7; *S. macrodon* Matsumoto, 1926:333, figure 29 (*non* p. 334, figure 30); *S. euryodon* Matsumoto, 1926:337, figures 31, 32; *S. minus* Schlosser, 1911:112; *S. annectens* Matsumoto, 1926:342; *S. sobrina* Matsumoto, 1926:347.

Holotype CGM 8635, right maxilla with P1 to M3, left maxilla with M1 to M3, and portions of the skull.

Distribution Early Oligocene, Jebel Qatrani Formation, Egypt; a similar form, possibly conspecific, occurs at Jebel al Hasawnah, Libya.

Diagnosis Large species of *Saghatherium* with deep mandibular corpus lacking an internal chamber or lingual fenestra. Differs from older *S. bowni* in its larger size, differs from younger *S. humarum* in the absence of the chamber.

Remarks This was the best-known small-bodied hyracoid from the early Tertiary from the time of its discovery in the earliest 20th century until the description of *Thyrohyrax* (Meyer, 1978). The taxonomy of this species has been plagued by problems related to not understanding associations of upper and lower teeth, misinterpreting sexual dimorphism, misidentifying tooth positions on some specimens, and oversplitting by earlier taxonomists (see Rasmussen and Simons, 1991). At this point, it seems that *S. bowni* and *S. antiquum* represent two well-defined, cohesive species, possibly in an ancestor-descendent relationship, that occur in Egypt's latest Eocene and earliest Oligocene, respectively.

One of the most important recent contributions to knowledge of fossil hyracoids is the description of nearly complete skeletons of *Saghatherium* from Gebel al Hasawnah, Libya (Thomas et al., 2004; Gheerbrant et al., 2007). These skeletons are very similar to those of extant procaviids, but with a few interesting differences. The tarsus and manus are longer than those of procaviids, and their feet are digitigrade with clawlike toes. The first digit of the manus is retained. In procaviids, all toes but the second one are hooflike, and there is reduction of the first digit. These features suggest that digitigrade cursoriality was primitive for the order (Thomas et al., 2004; Gheerbrant et al., 2007), while the scrambling and climbing of the living hyraxes is a specialization.

The tarsus and manus of the Libyan *Saghatherium* possess the taxeopode or serial arrangement of bones, which contrasts with the interlocking, diplarthral arrangement of most ungulate mammals. This arrangement presumably allows for more flexibility of the midtarsal and midcarpal joints (Fischer, 1986). More important than its functional correlates, however, is that taxeopody in cursorial *Saghatherium* and in the even more specialized cursor *Antilohyrax* indicate that this bone arrangement is a primitive, conservative feature of the order. It is probably a reliable synapomorphy of paenungulates, as graviportal proboscideans share the trait (Rasmussen et al., 1990; Thomas

et al., 2004). Another distinctive hyracoid feature evident in the Libyan *Saghatherium* is in the ankle joint; the medial malleolus of the tibia is greatly enlarged and fits within a broad, concave fossa on the astragalus. This seems to be another specialized character linking primitive paenungulates (Court, 1993).

SAGHATHERIUM HUMARUM Rasmussen and Simons, 1988

Holotype CGM 42849, mandible with symphysis and right i2, c-m3.

Distribution Early Oligocene, Quarry V, Jebel Qatrani Formation, Egypt.

Diagnosis Species of *Sagatherium* with an internal chamber in the mandible. Further differs from *S. bowni* and *S. antiquum* in having a short diastema between incisors and cheek teeth. Differs from *Thyrohyrax* in relatively small size of premolars relative to molars, marked lingual tilt of the buccal surface of the premolars, relatively broad molars with marked buccal cingulum, less molariform premolars, weaker development of high, lophlike cristids on lower molars, presence of large spurs on lingual side of ectoloph on upper molars, and in having a mandible with a moderately inflated chamber rather than greatly inflated one.

Remarks This species and *Selenohyrax chatrathi* are two species unique to Quarry V, a cemented, coarse, pale sandstone quarry unlike the loose, heterogeneous colorful sands of other Upper Sequence quarries. Primate taxa from Quarry V are also unique, suggesting that Quarry V is of a significantly distinctive time horizon or paleoenvironment than the younger, better-known Quarries I and M. *Saghatherium humarum* retains the primitive saghatheriid characters of having an internal mandibular chamber and diastemata between incisors and cheek teeth, even though it is the youngest species of the genus known.

Genus *SELENOHYRAX* Rasmussen and Simons, 1988

Type and Only Known Species Selenohyrax chatrathi Rasmussen and Simons, 1988.

Distribution Early Oligocene, Quarry V, Jebel Qatrani Formation, Egypt.

Diagnosis Small hyracoid with selenodont dentition. Differs from all other hyracoid genera in extreme development of selenodonty (in the descriptive sense of looking like crescent moons, not in the sense of modern selenodont artiodactyls): cristid obliqua and hypocristid together form a smooth, curved, continuous crest with only a slight prominence for the hypoconid, and the paracristid and protocristid form a similar crest from paraconid to metaconid. Further differs from *Titanohyrax* in its much smaller size, lack of metastylids on lower molars and premolars, presence of prominent spurs on lingual side of hypoconids on molars and premolars and on lingual side of protoconid on molars, presence of a large selenodont hypoconulid on m3, and approximately equal size of protocone and hypocone, which are strongly crested. Differs from the slightly smaller *Sagatherium* in having longer, narrower, more selenodont lower molars.

SELENOHYRAX CHATRATHI Rasmussen and Simons, 1988

Holotype CGM 52850, partial mandible with right p2–m3 and a small portion of the horizontal ramus.

Distribution and Diagnosis Same as for genus.

Remarks This hyracoid shows a unique dental adaptation in which smooth, crescentic, sharp-edged ridges have developed across each of the primary buccal cusps of the lower molars. These do not form flat lophs but, rather, arched crescents. The two primary lingual cusps are reduced in size and are incorporated into the distolingual end of the crescents. Upper dentitions that have been recovered at Quarry V since the original description of the genus show mesiodistally long molars with very elongated, cresty ectolophs. The buccal face of the ectoloph is folded, as in *Saghatherium*. This unique genus seems to be a specialized browser derived from a form like *Saghatherium*. It also shows some interesting resemblances to primitive members of Pliohyracidae (see later discussion).

Genus *THYROHYRAX* Meyer, 1973
Figures 13.1B, 13.1C, 13.5C, and 13.5D

Type Species Thyrohyrax domorictus Meyer, 1973.
Included Species T. meyeri, T. litholagus, T. pygmaeus, T. domorictus, T. kenyaensis, T. microdon.
Distribution Late Eocene and Oligocene of Egypt, with additional material from the early Oligocene of Oman and the late Oligocene of Kenya.
Diagnosis Small saghatheriids with long narrow symphyses, relatively well-molarized premolars, and distinctive selenolophodont cheek teeth, characterized by high, flat zig-zag lophs. Differs from *Saghatherium* in having lower molar crowns less inflated basally, individual cusps less inflated and placed near the crown margin, cristids more lophlike (flatter). Further differs from *Saghatherium* in having simpler, more quadratic upper molars, ectolophs without elaborate buccal folding or lingual spurs, and premolars that resemble the molar pattern more closely. Differs from *Megalohyrax* in its more lophlike cristids, the proportionally narrow, higher molar crowns, the much more molarized premolars, and its much smaller size.
Remarks When Meyer (1973, 1978) diagnosed this genus and showed how it differed from *Saghatherium*, it seemed that *Thyrohyrax* might be the lone small saghatheriid of the Fayum's Upper Sequence, while *Saghatherium* was restricted to the Lower Sequence. Since then, however, abundant *Thyrohyrax* have been identified in the Lower Sequence, and *S. humarum* has been found at Quarry V. It is now clear that these are two parallel radiations of small-bodied saghatheriids, with *Thyrohyrax* showing the development of lophodont dental structure remarkably similar to that of early Eocene horses (e.g., *Hyracotherium*), while *Saghatherium* has more bunoselenodont teeth. Details of morphological divergence, parallelism in molarization of premolars, changes in body size, and abundance of taxa in the Fayum section are covered elsewhere (Rasmussen and Simons, 1991). One clade of *Thyrohyrax* (containing *T. meyeri* and *T. domorictus*) is notable for the extreme, balloon-like inflation of the mandibular chamber, and its expansion well up into the ascending ramus (Meyer, 1973; De Blieux et al., 2006).

THYROHYRAX MEYERI Rasmussen and Simons, 1991
Figure 13.1B

Holotype CGM 41882, partial mandible with p1 to m3.
Distribution Late Eocene, Quarry L-41, Jebel Qatrani Formation, Egypt; referred material from the early Oligocene, Ashawq Formation, Oman.
Diagnosis Small species of the genus with simple premolars. Further differs from *T. litholagus* and *T. pygmaeus* in its more inflated internal mandibular chamber.

Remarks This species has lower molars that are only slightly differentiated from those of the most primitive species of *Saghatherium* (*S. bowni*) with which it occurs at Quarry L-41. The premolars of these two species are less molarized than later members of their two lineages. This suggests an evolutionary divergence of these two taxa from each other at a point not too much earlier than the late Eocene. *Saghatherium* and *Thyrohyrax* resemble each other but differ from older *Microhyrax* in having higher, better-developed cristids linking buccal to lingual cusps in the perissodactyl-type tooth. A morphocline of lower molar structure among these three taxa would be, from primitive to derived, *M. lavocati*, *S. bowni*, and *T. meyeri*. Teeth of a small *Thyrhyrax* from Oman have been referred to this species (Pickford et al., 1994; Thomas et al., 1999).

THYROHYRAX DOMORICTUS Meyer, 1973
Figures 13.5C and 13.5D

Holotype CGM 40001, partial right ramus with p1 to p4, part of m2, and m3.
Distribution Early Oligocene, Jebel Qatrani Formation, Egypt.
Diagnosis Highly lophoselenodont species, with the flattest lophs, most molarized premolars and the greatest expansion of the internal mandibular chamber of any known member of the genus.
Remarks This is the most common hyracoid of the Upper Sequence quarries (I and M). The mammalian community at these quarries is very well-known and serves as our best view of the African early Oligocene (e.g., Gagnon, 1997). *Thyrohyrax domorictus* is absent from the late Oligocene of Chilga, Ethiopia, where other comparably sized mammals are also lacking (suggesting taphonomic bias), but a very similar species of *Thyrohyrax* (*T. kenyaensis*) is present in the late Oligocene of northwestern Kenya.

THYROHYRAX LITHOLAGUS Rasmussen and Simons, 1991
Figure 13.1C

Holotype CGM 41880, mandible with p3 to m3.
Distribution Late Eocene, Quarry L-41 Jebel Qatrani Formation, Egypt.
Diagnosis Medium-sized species of the genus with only a small internal chamber of the mandibular corpus. Differs from *T. meyeri* in its greater size, from *T. pygmaeus* in its smaller size, and from *T. meyeri* and *T. domorictus* in its small mandibular chamber.
Remarks This species, like *T. meyeri*, is known from a large sample of fossils from Quarry L-41. It shows the typical dental features of *Thyrohyrax*, but at a somewhat larger size and without the greatly inflated chamber.

THYROHYRAX PYGMAEUS (Matsumoto, 1922)

Synonymy Saghatherium magnum Andrews, 1907:99, figure 2 (*non* Andrews, 1904); *Megalohyrax pygmaeus* Matsumoto, 1922:840, figure 1; *M. pygmaeus* Matsumoto, 1926:321, figures 24, 25 ; *Pachyhyrax pygmaeus* Meyer, 1978:300, figure 14.3.
Holotype AMNH 14454, anterior portion of skull and right ramus.
Distribution Early Oligocene, Jebel Qatrani Formation, Egypt.

Diagnosis Differs from other species of *Thyrohyrax* by its notably larger size. Further differs from *T. meyeri* and *T. domorictus* by the small mandibular chamber. Further differs from *T. domorictus* in its more molarized premolars.

Remarks This taxon was first described long before Meyer (1978) diagnosed *Thyrohyrax*. Specimens were originally confused with *Saghatherium*, but later it was recognized as distinct from it and was named as a small species of *Megalohyrax* (hence, "pygmaeus"; Matsumoto, 1922, 1926). Later, it was included in the genus *Pachyhyrax* by Meyer (1978), who had misinterpreted the genus (Rasmussen and Simons, 1988). With the discovery of *T. litholagus*, the identity of *T. pygmaeus* became obvious—it is a slightly younger, larger species related to *T. litholagus*, possibly a direct descendant. This situation is directly analogous to the relationship between *S. bowni* of quarry L-41 and larger *S. antiquum* of the traditional collecting areas in the upper part of the Lower Sequence (see earlier discussion). *T. pygmaeus* is represented by a very well-preserved cranium, which shows a distinctive, corrugated texture of the bone on the frontal and zygomatic regions. The cranium shows the typical pattern seen in early Tertiary hyracoids of a long, tubular rostrum, small braincase, typical hyracoid basicranium and temporomandibular joint, and full eutherian dentition with continuously growing upper incisors.

THYROHYRAX KENYAENSIS Rasmussen and Gutiérrez, 2009

Holotype KNM-NW 22545, left maxilla with P3, dP4, M1, and M2.

Diagnosis Rasmussen and Gutiérrez, 2009. A medium-sized species of *Thyrohyrax*, similar in size to *T. domorictus* of the Early Oligocene. Differs from *T. domorictus* and other Fayum species of the genus in having the lingual cusps of the upper molars positioned more closely to the buccal pair, with a longer and more gradual slope off the lingual face of the lingual cusps. Further differs from *T. domorictus* in lacking a paraconule on the preprotocrista; in having higher crowns, especially at the ectoloph; in the absence of a cingulum on the buccal side of the paracone and metacone; and in having somewhat simpler premolars. Differs from *T. pygmaeus* in its smaller size and less molariform premolars. Differs from *T. meyeri* and *T. microdon* in its much greater size.

Distribution Late Oligocene, Benson's Site, Kenya

Remarks This species, known from upper and lower dentitions, is about the size of *T. domorictus* of the early Oligocene. It differs from that species in a few points of dental comparison. This species and the next show that *Thyrohyrax* survived until the very latest Oligocene, just before the Oligocene-Miocene transition. The presence of *Thyrohyrax*, *Megalohyrax*, *Pachyhyrax*, and their kin, in the late Oligocene of Ethiopia and Kenya, suggests that these three genera, the most common hyracoids of the Fayum's youngest stratigraphic levels, were characteristic of the African Oligocene.

THYROHYRAX MICRODON Rasmussen and Gutiérrez, 2009

Holotype KNM-NW 22526, paired maxillae with roots or tooth bases of left P1–M3, right P2–M3, crowns of left M3 and right P4, partial crowns of right P3 and M1.

Diagnosis Rasmussen and Gutiérrez, 2009. A tiny species of *Thyrohyrax*, differing from all others by its smaller size. Further differs from *T. domorictus* in its less pronounced buccal cingulum, less cuspate primary cusps on the lower molars, and reduced tooth size gradient from front to back (i.e. posterior

teeth are relatively smaller than anterior teeth). Further differs from *T. meyeri*, the next smallest species of the genus, in its more molarized premolars.

Distribution Late Oligocene, Benson's Site, Kenya

Remarks This is a tiny species of the genus, smaller even than *T. meyeri* of the late Eocene. It differs from that Eocene species in having a highly lophodont, zig-zag lower molar structure more similar to derived members of the genus, such as *T. domorictus*, than to *T. meyeri*.

Genus *MEGALOHYRAX* Andrews, 1903

Synonymy *Mixohyrax* Schlosser, 1910, *Mixohyrax* Schlosser, 1911 (*non Megalohyrax* Schlosser).

Type Species *Megalohyrax eocaenus* Andrews, 1903.

Included Species *M. eocaenus*, *M. gevini*, *Megalohyrax* new species, from Chilga, Ethiopia.

Distribution Eocene and Oligocene of North Africa, late Oligocene of Ethiopia.

Diagnosis Very large saghatheriid with bunoselenodont molars and relatively simple premolars. Differs from *Thyrohyrax* in its more bunodont cusps connected by looping crests (rather than more straight-sided teeth with high lophs). Differs from *Pachyhyrax* in having complete cristids connecting protoconid to metaconid, and hypoconid to entoconid, in its simpler premolars, and in its slightly larger size. Differs from titanohyracids in its much less selenodont molars, far simpler premolars, and the absence of the anterior dental specializations of that group (discussed later).

Remarks This is a common hyracoid that shows very little morphological change through time and space. By inference, it was very successful in its ecological role in early Tertiary Africa. *Megalohyrax* occurs in the late Eocene of Quarry L-41 and in the uppermost quarries of the early Oligocene. The new species from Chilga, a few million years younger and thousands of kilometers away, is diagnosed on only the slightest dental differences. Over the time interval represented by the relatively static *Megalohyrax*, there are wholesale changes in other mammals, including other saghatheriids and catarrhine primates.

MEGALOHYRAX EOCAENUS Andrews, 1903
Figures 13.1G, 13.5A, and 13.5B

Synonymy *Megalohyrax minor* Andrews, 1904:213; *Megalohyrax minor* Andrews, 1906:97 *pars*, plate VII, figure 1, *non* figures 2, 3; *Mixohyrax niloticus* Schlosser, 1910:503, *nomen nudum*; *Mixohyrax suillus* Schlosser, 1910:503, *nomen nudum*; *Mixohyrax andrewsi* Schlosser, 1910:503; *Mixohyrax andrewsi* Schlosser, 1911:115, plate X, figures 9–11; *Mixohyrax niloticus* Schlosser, 1911:116, plate XI, figure 9, plate XII, figures 3, 6, plate XV, figures 1, 4, 8; *Mixohyrax suillus* Schlosser, 1911:118, plate X, figure 6; *Megalohyrax minor* Matsumoto, 1922:842; *Megalohyrax niloticus* Matsumoto, 1922:843; *Megalohyrax suillus* Matsumoto, 1922:843; *Bunohyrax major* Matsumoto, 1926:300 *pars*; *Megalohyrax minor* Matsumoto, 1926:313, figures 20–22; *Megalohyrax niloticus* Matsumoto, 1926:316, figure 23.

Holotype BMNH M8502, right maxilla with C to M3.

Distribution Late Eocene and early Oligocene, Jebel Qatrani Formation, Egypt.

Diagnosis Differs from the one tooth on which *M. gevini* is based in having the hypocone and protocone share equal size and very similar structure.

Remarks This species shapes our understanding of large early Tertiary hyracoids. It is a common member of most

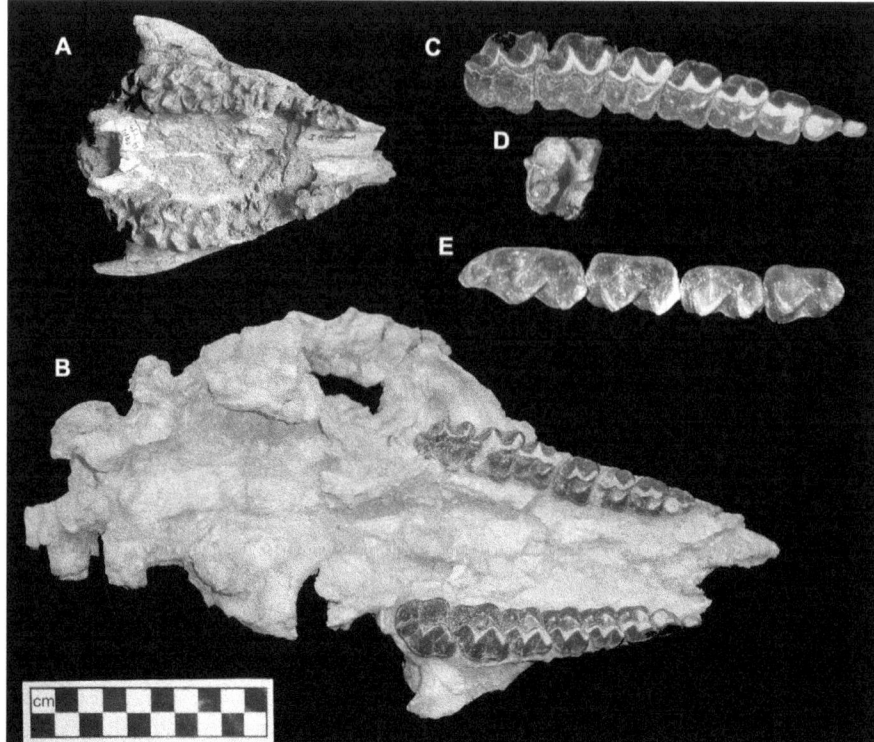

FIGURE 13.6 Representatives of the Titanohyracidae. A) partial cranium of *Afrohyrax championi* (KNM-MT-1) in occlusal view and a close-up of m2 from the same specimen (D). B) Cranium of *Titanohyrax* sp. from Quarry L-41, Fayum (right P1-M3, left C-M3), in ventral view, and with a close-up of the right cheek tooth series, canine to M3 (C). Notice the better development of the ectoloph, the sharper molar styles, and the more molarized premolars than in *Megalohyrax* (figure 13.5). E) Mandibular dentition of *Titanohyrax* sp. (DPC 12348, right p3-m3, Quarry L-41).

Fayum quarries. It has a relatively primitive dentition compared to those of other saghatheriids in the extremely simple premolars and the low-slung cross lophs, which contrast to the high, flat lophids of taxa such as *Thyrohyrax* and *Titanohyrax*. The cranium is very long, flat, and low, with a tubular snout (like *Bunohyrax*) (Thewissen and Simons, 2001). The anterior dentition contains several diastemata. The undescribed postcrania of *Megalohyrax* contrast with those of titanohyracids in lacking evidence of adaptation to cursoriality (Whitworth, 1954; Rasmussen and Simons, 2000). The relative stability of this species in time and space suggests that it was generalized enough to withstand variation in climate and local habitat conditions. *Megalohyrax* could be considered the typical large hyracoid of the African Paleogene.

MEGALOHYRAX GEVINI Sudre, 1979

Synonymy Hyracoid indet. (locus 2), Gevin et al., 1975:967; Hyracoid indet. (locus 2), Sudre, 1975:355.

Holotype Left M2 (or M1), GL2-3, temporarily deposited at USTL.

Distribution Eocene, Gour Lazib, Algeria.

Diagnosis Differs from *M. eocaenus* in that the hypocone is notably smaller and more mesiodistally compressed than the protocone, as well as possibly somewhat larger overall.

Remarks This is a taxon based on a single tooth. The small hypocone of this lone tooth is potentially a shared similarity to titanohyracids, which have a distinctly compressed hypocone, but most of the specimen does not look as structurally derived as those of later titanohyracids. Whether this tooth represents a species of *Megalohyrax* or a primitive member of

the titanohyracid clade probably cannot be determined without further material. Either way, it is notable as a chronologically early record of a large-bodied hyracoid.

Another record of a large *Megalohyrax*-like hyracoid comes from the late Eocene or early Oligocene of Libya (Arambourg and Magnier, 1961; see Rasmussen, 1989). This form may also be closer to *Titanohyrax* than to *Megalohyrax*.

Family TITANOHYRACIDAE Matsumoto, 1926

Type Genus Titanohyrax Matsumoto, 1922.

Included Genera Titanohyrax, Antilohyrax, Afrohyrax.

Distribution Eocene and Oligocene of North Africa, and early Miocene of Kenya.

Diagnosis Medium to large hyracoids that differ from geniohyids and saghatheriids in having highly lophoselenodont cheek teeth, specialized lower incisors, short, tapered rostra, and cursorial adaptations of the postcranium. The upper molars have well-developed W-shaped ectolophs, small compressed hypocones in contrast to large protocones, highly molarized premolars, lack of large diastemata in the anterior dentition, and spatulate or hyperpectinate lower incisors.

Genus TITANOHYRAX Matsumoto, 1922
Figures 13.1E, 13.6B, 13.6C, and 13.6E

Synonymy Megalohyrax Andrews, 1906, *pars* (*non* Andrews, 1903); *Megalohyrax* Schlosser, 1910; *Megalohyrax* Schlosser, 1911.

Type Species Titanohyrax andrewsi Matsumoto, 1922.

Included Species T. tantulus, T. andrewsi, T. mongereaui, T. ultimus, T. angustidens.

Distribution Eocene and Oligocene of North Africa.

Diagnosis Medium to very large titanohyracids. Differs from *Antilohyrax* in having spatulate (not hyperpectinate) lower incisors, well-developed metastylids, relatively higher-crowned cheek teeth. Differs from *Afrohyrax* in having mesiodistally broad, spatulate incisors.

TITANOHYRAX TANTULUS Court and Hartenberger, 1992

Synonymy Cf. *Pachyhyrax* and cf. *Saghatherium* Hartenberger et al., 1985:649.

Holotype CBI 42, right juvenile maxillary fragment with a fragment of DP3 and complete DP4, M1, M2.

Distribution Middle Eocene, Jebel Chambi, Tunisia.

Diagnosis Differs from other species of the genus in its much smaller size. Differs from all species but large *T. mongereaui* in the relatively low-crowned molars and basally rounded upper molar styles.

Remarks This species shows key, distinctive features of the titanohyracid clade at a very early date, roughly contemporary with small, primitive forms such as *Seggeurius, Microhyrax,* and *Bunohyrax matsumotoi,* and probably considerably older than *Thyrohyrax meyeri, Saghatherium bowni,* and all Fayum species of *Geniohyus.* This indicates that the basal radiation of geniohyids, saghatheriids, and titanohyracids occurred no later than the middle Eocene, and the degree of specialization already evident in *T. tantulus* suggests the divergence event may have been significantly older than that.

TITANOHYRAX ANDREWSI Matsumoto, 1922

Synonymy Megalohyrax minor Andrews, 1906:97, plate VII, figures 2, 3 (*non* Andrews, 1904), *M. minor* Schlosser, 1910:502 (*non* Andrews, 1904); *M. eocaenus* Schlosser, 1910:502 (*non* Andrews, 1903); *M. palaeotherioides* Schlosser, 1910:502 *nomen nudum*; *M. minor* Schlosser, 1911:105; *M. eocaenus* Schlosser, 1911:105, plate XI, figure 7; *M. palaeotherioides* Schlosser, 1911:106, plate XI, figure 1, plate XII, figure 1; *Tytanohyrax schlosseri* Matsumoto, 1922:847; *Titanohyrax palaeotherioides* Matsumoto, 1922:847; *T. schlosseri* Matsumoto, 1926:325; *T. palaeotherioides* Matsumoto, 1926:326, figures 26–28.

Holotype CGM 8822-3, partial mandible with left i1–i2, p3–m2, and right p3–m3.

Distribution Early Oligocene, Jebel Qatrani Formation, Egypt.

Diagnosis Medium-sized species of the genus. Differs from *T. tantulus* in its much larger size, and from *T. ultimus* in its much smaller size and much more distinct metastylids. Differs from *T. angustidens* in its relatively broader, more robust lower molars with heavy buccal cingulum, in having mesiodistally compressed trigonids, and in its lower-crowned upper molars.

Remarks This taxon, a fossil of which was first illustrated by Andrews as an upper dentition of *Megalohyrax,* has a remarkably complex taxonomic history, which is surprising given the obvious differences between *Megalohyrax* and *Titanohyrax.* Meyer (1978) finally resolved the last of the problems regarding nomenclatural priority and holotype allocations of the several species proposed for these two genera. *T. andrewsi* can perhaps be considered a "typical" member of the genus: it is fairly well-known and shows the characteristic dental attributes, including the highly selenodont teeth, the distinctive metastylids, the compressed hypocones, and the spatulate incisors. There is a new, undescribed species very similar to *T. andrewsi* known from Quarry L-41 (late Eocene) and therefore older than *T. andrewsi.* The Quarry L-41 form is

smaller and has lower-crowned molars than *T. andrewsi.* Both *T. andrewsi* and the L-41 species lack internal mandibular chambers in known specimens.

TITANOHYRAX MONGEREAUI Sudre, 1979

Synonymy Hyracoid indet. Gevin et al., 1975:967; Hyracoid indet. Sudre, 1975:355.

Holotype GZ-I, M2 (or M3), upper left accompanied by distal part of the preceding molar.

Distribution Middle Eocene, Glib Zegdou, Algeria.

Diagnosis Intermediate in size (based on the one tooth) between *T. andrewsi* and *T. ultimus.* Differs from other large species of the genus in being more brachydont.

Remarks Like *M. gevini,* this taxon is diagnosed on the basis of a single upper molar, this one poorly preserved. The tooth shows a well-developed W-shaped ectoloph and a hypocone reduced in size relative to the protocone. These characters may indicate that the tooth does represent a titanohyracid, while the brachydonty suggests it may be primitive with respect to other members of the clade. As with *M. gevini,* which *T. mongereaui* resembles, the only substantive conclusion is that very large hyracoids had evolved by the middle Eocene.

TITANOHYRAX ULTIMUS Matsumoto, 1922

Holotype BMNH M12057, right M2.

Distribution Early Oligocene, Jebel Qatrani Formation, Egypt.

Diagnosis Differs from other species of *Titanohyrax* and all other African hyracoids by its much larger size, which is approached by the Pliocene pliohyracids from Eurasia.

Remarks Another giant hyracoid based on a one-tooth holotype, this species is represented by a few other specimens of both upper and lower molars, but all heavily worn. The upper molars have the characteristic arrangement of protocone and hypocone, and W-shaped ectoloph as other species of the genus, the lowers show the compressed trigonid and well-developed metastylid. This animal was the size of small modern rhinoceroses (Schwartz et al., 1995). It was part of a terrestrial megafauna in the early Oligocene that also included *Arsinoitherium* and palaeomastodontids.

TITANOHYRAX ANGUSTIDENS Rasmussen and Simons, 1988

Synonymy Megalohyrax palaeotherioides Schlosser, 1911; *Titanohyrax palaeotherioides* (Matsumoto, 1926) in part; *T. andrewsi* (Meyer, 1978) in part.

Holotype CGM 42848, partial left mandible with p2–p4, m1–m2, unerupted m3, an internal mandibular fenestra, much of the horizontal and ascending rami, and intact coronoid process and articular condyle.

Distribution Early Oligocene, Jebel Qatrani Formation, Egypt.

Diagnosis Differs from *T. andrewsi* and other species of the genus in having relatively longer, narrower lower molars, trigonid not compressed mesiodistally, weak or absent buccal cingulum on lower molars, and presence of an internal chamber and lingual fenestra in the mandibular corpus. Upper molars differ from those of *T. andrewsi* in having more prominent metastyles on the molars and premolars, and in being higher crowned. Differs from *T. ultimus* and *T. mongereaui* in its much smaller size and from *T. tantulus* in being larger.

Remarks This species seems to be adaptively divergent from the pattern of developing larger size and more robust molars as seen in *T. andrewsi* and *T. ultimus*. In contrast, *T. angustidens* has relatively tall, narrow molars that look delicate. It occurred alongside a diverse fauna of large hyracoids, including *T. ultimus*, *Megalohyrax eocaenus*, and *Pachyhyrax crassidentatus*.

TITANOHYRAX, new species
Figures 13.6B, 13.6C, and 13.6E

Distribution Late Eocene, Quarry L-41, Jebel Qatrani Formation, Egypt.

Remarks This species of *Titanohyrax* closely resembles *T. andrewsi* but for its smaller size and less robust build. It is represented by nearly complete dentitions and numerous postcrania.

Genus *ANTILOHYRAX* Rasmussen and Simons, 2000

Type and Only Known Species Antilohyrax pectidens.

Distribution Late Eocene, Quarry L-41, Jebel Qatrani Formation, Egypt.

Diagnosis Differs from all known hyracoids in the mesiodistally elongated, hyperpectinate dental comb formed by the lower incisors, the double-condylar astragalus-navicular joint, and the tibia and fibula that are extensively fused together. Further differs from *Titanohyrax* in the absence of metastylids on the lower molars.

ANTILOHYRAX PECTIDENS Rasmussen and Simons, 2000
Figure 13.1F

Holotype CGM 42205, mandible with right i1, c-m3 and left i1, c-m3.

Distribution and Diagnosis As for the genus.

Remarks This is one of the few early Tertiary hyracoids to be described on the basis of dentitions, associated postcrania, and cranial remains (Rasmussen and Simons, 2000; DeBlieux and Simons, 2002). The skeletal elements that have been studied to date indicate a surprisingly specialized form, with limb proportions, size and function similar to that of the living springbok (Rasmussen and Simons, 2000). The condylar development at the front of the astragalus allowing midtarsal flexion and extension departs far from the primitive hyracoid condition, as do also the extensively fused tibia and fibula. Additional remains of *Antilohyrax* recovered since its original description support the interpretation of this unusual hyracoid as a cursorial leaper. The initial erroneous interpretation that the animal lacked upper incisors (Rasmussen and Simons, 2000) was corrected by DeBlieux and Simons (2002).

The postcranial and incisor specializations of *Antilohyrax*, along with the early age of *T. tantalus*, and the species level diversity of *Titanohyrax*, all suggest an ancient evolutionary history of this specialized hyracoid clade, and this is true whether *T. mongereaui* proves to be a titanohyracid or not.

Genus *AFROHYRAX* Pickford, 2004

Synonymy Pliohyrax championi Arambourg, 1933; *Megalohyrax championi* (Arambourg), Whitworth, 1954:6–22, plates 1–4; *Megalohyrax* sp. (cf. *M. pygmaeus* Matsumoto), Whitworth, 1954:23, plate 7, figure 2; *Pachyhyrax championi* (Arambourg), Meyer, 1978:301; *M. championi* (Arambourg), Rasmussen, 1988.

Type Species Afrohyrax championi (Arambourg, 1933).

Included Species A similar, smaller, unnamed species occurs in the late Oligocene, Kenya.

Distribution Early Miocene of Kenya and Uganda (Rusinga Island, Maboko, Songhor, Moruorot, Karungu, Mfwanganu Island, Bukwa), and late Oligocene of northwestern Kenya.

Diagnosis Differs from other titanohyracids in having typical hyracoid lower incisors that are tall and mesiodistally short, with slight pectinations at the apex. Differs from species of *Titanohyrax* except for *T. angustidens* in having an internal mandibular chamber and lingual fenestra, and from all species except for *T. tantalus* in having rounded, swollen bases of the molar styles.

Remarks This Miocene hyracoid resembles early Tertiary hyracoids in a number of key dental features: the ectoloph rises steeply from the buccal edge of the tooth and forms a W shape in occlusal view, the parastyles and mesostyles have sharp ridges buccally the metastyles are only slightly developed, buccal ribs or folds of the paracone and metacone are reduced or absent, the hypocone is much smaller than the protocone, and the prehypocrista is short. The posterior premolars are well molarized. In addition, the head of the astragalus of *Afrohyrax* is similar to that of *Antilohyrax* and an undescribed specimen of *Titanohyrax*, in being lined up with the trochlea, rather than offset significantly to the medial side.

Arambourg (1933) put this genus in *Pliohyrax* along with large-bodied, hypsodont forms from the Pliocene of Eurasia. Whitworth was conservative in lumping it with Fayum *Megalohyrax*. Meyer (1978) thought he had solved the problem by placing it in *Pachyhyrax*, but his conception of that genus was faulty. Rasmussen (1989) tentatively followed Whitworth (1954) in suggesting that titanohyracid characters may have been attained independently of that clade. Finally, Pickford (2004) formally provided the moniker *Afrohyrax* and placed it with titanohyracids.

AFROHYRAX CHAMPIONI (Arambourg, 1933)
Figures 13.6A and 13.6D

Synonymy Pliohyrax championi Arambourg, 1933; *Megalohyrax championi* (Arambourg), Whitworth, 1954:6–22, plates 1–4; *Megalohyrax* sp. (cf. *M. pygmaeus* Matsumoto), Whitworth, 1954:23, plate 7, figure 2; *Pachyhyrax championi* (Arambourg), Meyer, 1978:301; *M. championi* (Arambourg), Rasmussen, 1989.

Distribution and Diagnosis Same as for genus.

Remarks Whitworth (1954) worked out the paleobiology of this animal quite well. It is a cursorial animal of at least partially open country with dietary specializations indicating some degree of folivory. It has no obvious direct precursors in the Eocene or early Oligocene, as all titanohyracids of that epoch are specialized in the incisors *(T. andrewsi, Antilohyrax)*, the postcranium *(Antilohyrax)*, or molar hypsodonty *(T. angustidens)*, in ways that seem to exclude them from direct ancestry. The only known titanohyracid from which one could draw *Afrohyrax* without much evolutionary reversal would be primitive *T. tantalus*.

An isolated record of *"Pachyhyrax* aff. *championi* " from the early middle Miocene of As Sarrar, Saudi Arabia, may represent this species (Thomas et al., 1982; Whybrow and Clements, 1999). A few dental specimens from the late Oligocene of Kenya are very similar to *A. championi* except for their significantly smaller size.

AFROHYRAX new species Rasmussen and Gutiérrez, 2009

Distribution Late Oligocene of northwestern Kenya.

Remarks A slightly smaller species of *Afrohyrax* closely related to *A. championi* is known from dental and postcranial specimens from late Oligocene beds.

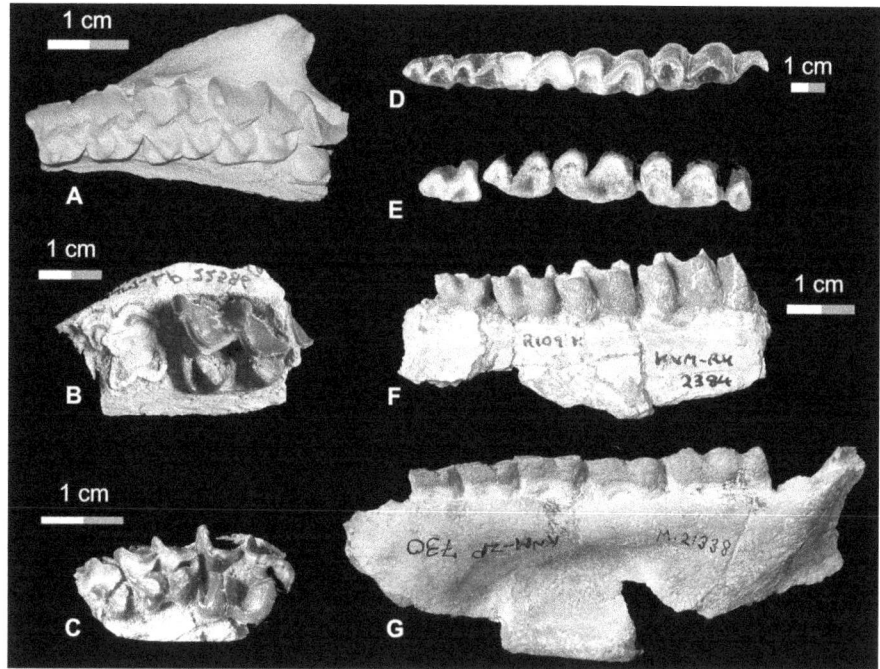

FIGURE 13.7   Representatives of the family Pliohyracidae. A) *Prohyrax tertiarius* from the middle Miocene of Namibia (holotype, left P3-M2, partial M3, photo of cast); B) *Meroehyrax* sp. from Benson's Site, late Oligocene of Kenya (right M2, image reversed for comparison); C) *Meroehyrax* sp. from Benson's Site (left dP3-4, M1); D) *Parapliohyrax ngororaensis* from the middle Miocene of Kenya (KNM-BN 1741, right p3-m3); E and F) *Meroehyrax bateae* from Rusinga Island, early Miocene of Kenya, left mandible in occlusal (reversed) and lateral views (KNM-RU 2384, p3-m2, fragment of m3); G) *Meroehyrax bateae* from Rusinga Island showing the broad lingual fossa of the mandibular corpus below the cheek teeth (holotype, right p3-m3, photo of cast).

Family *PLIOHYRACIDAE* Osborn, 1899

Type Genus Pliohyrax Osborn 1899.
Included Genera Meroehyrax, Prohyrax, Parapliohyrax (Africa); Pliohyrax, Kvabebihyrax, Postschizotherium, Sogdohyrax (Eurasia).
Distribution Late Oligocene to Pliocene, East and South Africa (also Pliocene to Pleistocene of Europe and Asia).
Diagnosis Compared to early Tertiary hyracoid families, the African members of Pliohyracidae have hypsodont teeth, especially buccally (but not developed to the extreme extent seen in the Eurasian Plio-Pleistocene forms). The cresting of the buccal side of both upper and lower molars is highly emphasized compared to the reduced cusp size and minimal cresting on the lingual side of the teeth. The ectolophs are elongated mesiodistally compared to the relatively shorter lingual edge of the tooth. The disparity in length is accommodated by having the large, extended parastyle overlap the lingually deflected metastyle on the buccal side of the tooth, to the extent that the mesial edge of the parastyle lies closer to the mesostyle of the preceding tooth than it does to the distal border of that tooth. Hypocones are about the same size as the protocones, but they are shifted buccally and distally very close to the metacones. The effect on tooth size is that upper molars are mesiodistally long, buccolingually narrow, with the paracone-protocone pair buccolingually broader than the metacone-hypocone pair. The ectoloph constitutes a continuous high crest, while the lingual cusps stand relatively isolated. The M3 of *Parapliohyrax* is bifurcated. The lower molars also have a high, crested buccal edge, with the lingual cusps relatively small and incorporated by short lophids into the buccal cristids.
Remarks The pliohyracids of Africa *(Meroehyrax, Prohyrax, Parapliohyrax)* have been linked confidently to the later

radiation of large-bodied, extremely hyposodont hyracoids from Asia (Pickford and Fischer, 1987; Pickford et al., 1997). Africa is clearly the root of this radiation. A small primitive species of *Meroehyrax* appears in the late Oligocene of Kenya, a larger species of the same genus in the early Miocene. The similar genus *Prohyrax* is known from the middle to late Miocene of southern Africa. A larger, more derived taxon, *Parapliohyrax*, appears in the middle Miocene and persists until the late Miocene. These African pliohyracids are unique in their dental adaptations: a single set of very hypsodont, elongated blades has evolved on both the upper and lower teeth (the ectoloph of the uppers, with reduced lingual cusps, and the continous, smooth, W-shaped cristids associated with protoconid and hypoconid on the lower teeth).

Genus *MEROEHYRAX* Whitworth, 1954
Figures 13.7B, 13.7C, and 13.7E–13.7G

Synonymy Prohyrax sp. Meyer, 1978:309, figure 14.10. Meyer (1978) correctly placed *Meroehyrax* in Pliohyracinae, but it was reallocated to Saghatheriidae by Pickford (2004); new specimens from the late Oligocene of Kenya confirm its pliohyracid affinities.
Type Species Meroehyrax bateae Whitworth 1954.
Included Species M. bateae, M. kyongoi.
Distribution Early Miocene of Kenya and Uganda; late Oligocene of Kenya.
Diagnosis Small pliohyracid with lingual cusps of the upper molars separated from the ectoloph by a furrow. Differs from saghatheriids in having very high-crowned upper and lower molars, buccolingually narrow upper and lower molars, reduction in size of the protocone and hypocone, and

incorporation of noncuspate lingual conids into high crescentic lophs of the lower molars. Further differs from *Prohyrax* in lacking faint buccal ribs for the paracone and metacone, and in having a better-developed metastyle. These features result in *Meroehyrax* having smoother, U-shaped curves on the buccal wall of the ectoloph.

MEROEHYRAX BATEAE Whitworth, 1954
Figures 13.7E–13.7G

Holotype BMNH M21338, right ramus with p3 to m3.
Distribution Miocene of Rusinga Island, Kenya, and Bukwa, Uganda.
Diagnosis Differs *M. kyongoi* by its larger size.
Remarks Pickford (2004) revised *Meroehyrax* based on new specimens from Uganda, and he reevaluated various specimens tentatively assigned to *Meroehyrax* over the years. An m3 from Uganda is very similar to that in the holotype described by Whitworth (1954), and upper dentitions of an appropriate size from the same deposits are therefore allocated to *Meroehyrax* as well. Pickford (2004) suggested that *Meroehyrax* may lie near the origin of the procaviid clade.

MEROEHYRAX KYONGOI Rasmussen and Gutiérrez, 2009

Holotype KNM-NW 22558A, right mandible fragment with P3 in crypt, dP4, M1.
Diagnosis A small species of *Meroehyrax* that differs from *M. bateae* in its narrower, straight-sided lower teeth and gracile upper molars. Differs from all hyracoids except other pliohyracids in its hypsodont, narrow lower cheek teeth; the lack of inflation of lower molar cusps; the molar trigonids and talonids that are formed by smoothly rounded lophs; the upper cheek teeth that are long and buccolingually narrow with high, bowed ectoloph and small lingual cusps. Differs from *Parapliohyrax* and *Prohyrax* in its smaller size, less hypsodont teeth, and the lack of a contact between the upper lingual cusps (protocone, hypocone) and the lingual side of the ectoloph. Further differs from *M. bateae* in having lower molars less inflated at the base, a longer talonid on m3 with more open crescents, and in having no lingual fossa of the mandible in the three specimens for which this feature can be assessed.
Distribution Late Oligocene of northwestern Kenya.
Remarks This species is known from more than a dozen specimens of upper and lower teeth. The lower teeth make it directly comparable to the type and referred specimens of *M. bateae*, and the uppers make it comparable to the type of *Prohyrax tertiarius*. The results suggest that the genera *Meroehyrax* and *Prohyrax* are very similar to each other, with the former being geologically older, smaller and dentally more primitive.

Genus *PROHYRAX* Stromer, 1926

Type Species Prohyrax tertiarius (Stromer,1924)
Included Species P. tertiarius, P. hendeyi.
Distribution Miocene of Africa.
Diagnosis Differs from other pliohyracids except *Meroehyrax* in being less hypsodont. Further differs from *Parapliohyrax* in lacking an arched palate and in the less extreme development of metastyle bifurcation and facial fossae. Differs from *Meroehyrax* in having the protocone and hypocone physically connected to the ectoloph, rather than standing freely, in having the lingual cones more elongated and bladelike (rather than short and pyramidal).

PROHYRAX TERTIARIUS Stromer, 1924
Figure 13.7A

Holotype 1926 X 10, Staatliche Sammlung für Paläontologie und historische Geologie, Munich, left maxilla with I1, P3, and part of M3.
Distribution Miocene of Elisabethfeld, Namibia; Jebel Zelten, Libya; Langental, Namibia.
Diagnosis Differs from *P. hendeyi* by its smaller size, less inflated styles on the upper molars, and shorter M3 lacking a talon (Pickford, 1994).
Remarks Prohyrax has long been enigmatic with regard to *Meroehyrax* because the former was described based on upper teeth and the latter by heavily worn lower teeth. New collections from Kenya allow comparison to both and confirm that *Prohyrax* is very similar to *Meroehyrax* (Meyer, 1978). The close relationship between *Prohyrax* and *Parapliohyrax* has been well documented (Pickford et al., 1997). Based on current evidence we can conclude that *Prohyrax* is a middle to late Miocene genus only slightly modified from an ancestor resembling early Miocene *Meroehyrax*.

PROHYRAX HENDEYI Pickford, 1994

Holotype GSNW AD363, skull lacking a few upper teeth.
Distribution Middle Miocene of Arrisdrift, Namibia.
Diagnosis Differs from *P. tertiarius* in its larger size, more inflated styles of the upper molars, and elongated M3 with an extended talon (Pickford, 1994).
Remarks This is a very well-known species of the genus and, indeed, one of the best-known fossil hyracoids. It is represented by nearly complete dentitions, crania, and numerous postcrania (including the scapula, humerus, radius, ulna, femur, tibia, and foot elements). The dentition shows a nearly closed tooth row but for a short diastema between I1 and I2, and between i2 and i3. The upper central incisors are tusklike, while those of I2, I3, and C are premolariform. The upper premolars are squarish in outline, with an extended parastyle. The molars differ notably in shape from the premolars, being mesiodistally long and buccolingually narrow, with an elaborate ectoloph-bearing large styles. The lower i1 and i2 are projecting teeth similar to those of procaviids. The i3 and c are premolariform. The skull is relatively long and narrow, and the rostrum is long, high, and gently tapering, compared to the reduced snout of procaviids. The orbit is closed behind by a postorbital bar. There is no internal chamber or lingual fossa of the mandible (unlike many early Tertiary forms), nor are there palatine pockets like those of *Parapliohyrax*.

The large dental sample of *P. hendeyi* from Arrisdrift (Hendey, 1978) allows population variation to be assessed. Pickford (1994) analyzed mandible depth and tooth size and concluded that a single species of sexually dimorphic hyracoid was represented. This pattern of sexual dimorphism (with one sex being larger and deeper jawed, but lacking sexually dimorphic mandibular chambers) is a similarity to some species of *Saghatherium* and *Selenohyrax chatrathi* (Rasmussen and Simons, 1991; DeBlieux et al., 2006).

The postcranium of *P. hendeyi* resembles that of modern procaviids and Eocene *Saghatherium* in many features. Pickford (1994) interpreted the limb bones as indicating a more cursorial animal than procaviids with, however, a notable degree of flexibility in the ankle and elbow. The long bones are straighter than those of procaviids, and the fibula is completely unfused to the tibia.

Genus *PARAPLIOHYRAX* Lavocat, 1961

Type Species *Parapliohyrax mirabilis* Lavocat, 1961.
Included Species *P. mirabilis*, *P. ngororaensis*.
Distribution Middle to late Miocene of Morocco, Tunisia, Libya, Kenya, Namibia, and South Africa.
Diagnosis Large hyracoids with complete eutherian dentition, closed tooth rows, deeply excavated, arched palates, and an M3 with a well-developed bifurcated distal extension of the ectoloph forming a third lobe. Further differs from *Prohyrax* in having a well-developed infralacrimal fossa, and mandibles with more or less well-developed lingual fossae like those of *Meroehyrax*.

PARAPLIOHYRAX MIRABILIS Lavocat, 1961

Holotype Mandible with cheek teeth, Beni Mellal, Morocco.
Distribution Middle to late Miocene of North Africa (Morocco, Libya).
Diagnosis Differs from *P. ngororaensis* in its smaller size.
Remarks This species is known from the type material and several referred jaws from Morocco and Libya (Lavocat, 1961; Ginsburg, 1977; Pickford, 1994). The dental structure of this taxon is fairly similar to the much larger species *P. ngororaensis* and to species of *Prohyrax*. The geographic range of *P. mirabilis* along the southern Mediterranean is of interest because of the later radiation of larger, more specialized pliohyracids in southern Europe and the Near East, along the northern Mediterranean coast. *P. mirabilis* is sufficiently primitive in structure to have given rise to the more derived pliohyracids of Eurasia.

PARAPLIOHYRAX NGORORAENSIS Pickford and Fischer 1987
Figures 13.7D and 13.8A–13.8D

Holotype KNM-BN 1741, fragmented skull and associated mandible comprising the following pieces: left maxilla with P1–M3, right maxilla fragment with I1, I3, and C, right maxilla fragment with P4–M3 (M3 broken); crushed mandible comprising both rami with all teeth except for the left I, and canine.
Distribution Middle to late Miocene of Kenya, Namibia and South Africa.
Diagnosis Differs from *P. mirabilis* in its larger size, in having the canine and i3 curved sharply downward (forming a notch in the occusal plane of the tooth row), in having premolariform upper I3 and lower i3 (modified from Pickford and Fischer, 1987).
Remarks This is a moderately large hyracoid with very unusual dental features. While the emphasis on buccal crests is typical of earlier members of this family, there are unique specializations of the anterior teeth not seen elsewhere. The i1 and i2 have high crowns that project considerably, while the i3 is small and premolariform. The jaw under i3 and c drops so that the crowns of these teeth are aligned below the occlusal plane of the other cheek teeth. A paleobiological interpretation of these features has not yet been presented. The species has also been recorded at Berg Aukas, Namibia (Pickford, 1996). A distal metapodial from the late Miocene of South Africa has been allocated to this species (Pickford, 2003). Teeth from the late Miocene Samburu Hills of northern Kenya also seem to be related to this species (Tsujikawa, 2005).

Family PROCAVIIDAE Thomas 1892

Type Genus *Procavia* Storr, 1780.
Included Genera *Heterohyrax*, *Dendrohyrax*, *Procavia*, *Gigantohyrax*.

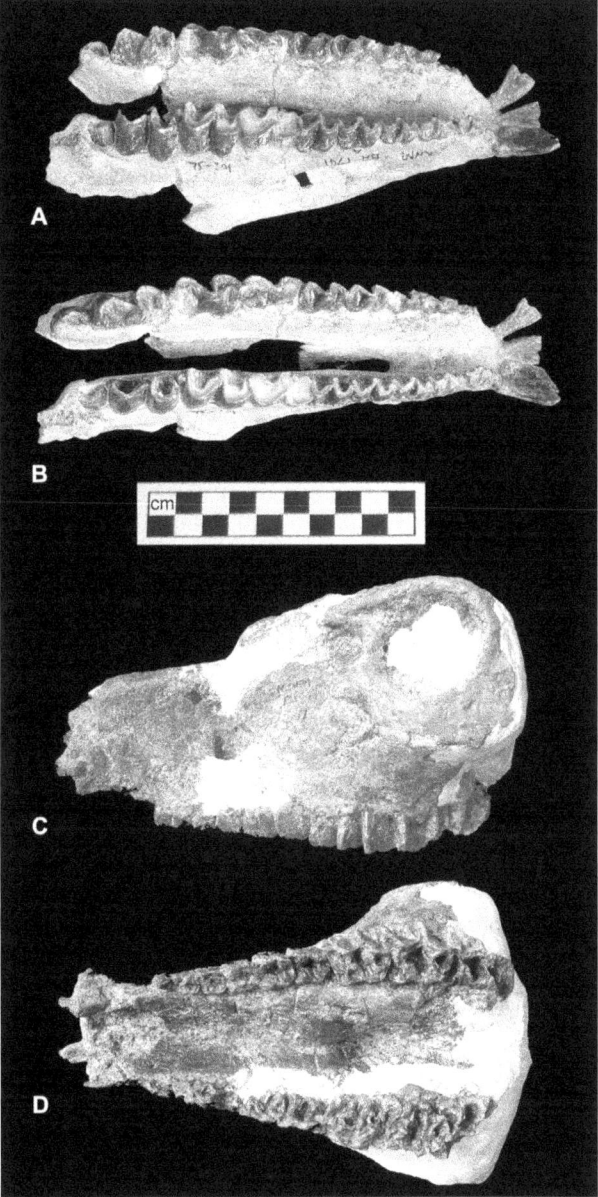

FIGURE 13.8 Specimens of *Parapliohyrax ngororaensis*. A and B) Mandible in oblique and occlusal views (KNM-BN 1741, right i1-m3, left i?-2, p1-m3); C and D) facial cranium in lateral and ventral views (KNM-BN 207, left I3-M3, partial right dentition). Notice the raised, round orbits with strong postorbital bar.

Distribution Miocene of Namibia, Pliocene and Pleistocene of South Africa, and recent of Africa and southwestern Asia.
Diagnosis Small to medium-sized hyraxes with molars ranging from brachydont to fairly hypsodont. Differs from other hyracoid families in their typically reduced (but variable) dental formula of 1.0.4.3/2.0.4.3. Differs from other families in lacking hypoconulids on m3. Differs from many early Tertiary forms in lacking internal mandibular chambers and lingual fenestra or fossae.
Remarks The family Procaviidae (Thomas, 1892; Barry and Shoshani, 2000; Pickford, 2005) first appears in the fossil record in the late Miocene. There is one convincing record of an early procaviid from the late Miocene of Nakali, Kenya (Fischer, 1986), and a much more complete record of associated teeth and postcrania of a primitive species attributable to

Heterohyrax in mine tailings of late Miocene age from northern Namibia (Rasmussen et al., 1996). In contrast to the ungulate-like nature of many earlier fossil hyracoids (Pickford, 1994; Rasmussen and Simons, 2000; Thomas et al., 2004), procaviids have short carpals and tarsals that function for scrambling and climbing on rocks and in trees (in the case of *Dendrohyrax*). With the arrival and diversification in Africa of artiodactyls and perissodactyls during the early Miocene, it appears that the procaviid radiation evolved as a set of novel adaptive solutions to habitat types not usually utilized by the other ungulates. The family consists of three extant genera containing about seven or so species (Starck, 1995). Two of the extant genera are also known from the fossil record, with an additional genus *(Gigantohyrax)* known exclusively by fossils.

Genus *HETEROHYRAX* Gray, 1868

Synonymy See expanded synonymy of living genera in Meyer (1978) and Barry and Shoshani (2000).
Type Species *Heterohyrax brucei* Gray, 1868.
Included Species *H. brucei* (extant), *H. auricampensis*.
Distribution Upper Miocene to Recent of Africa and the Sinai.
Diagnosis *Heterohyrax* has cheek teeth substantially less hypsodont than those of *Procavia* and *Gigantohyrax*; it is more hypsodont than *Dendrohyrax* and lacks features of flexibility in the forearm related to pronation and supination.

HETEROHYRAX AURICAMPENSIS Rasmussen, Pickford, Mein, Senut and Conroy, 1996

Holotype GSNW BER I 31′91a, partial right mandible with p3–m2, broken-off roots in the alveoli for p1–p2, and part of the alveolus for i2.
Distribution Late Miocene of Berg Aukas, Namibia.
Diagnosis Differs from all hyracoids except members of the family Procaviidae by the absence of i3, absence of upper and lower canines, and absence of a hypoconulid on the lower third molar. Differs from *Procavia*, *Dendrohyrax*, and *Heterohyrax brucei* in the less molarized upper and lower second premolars, and further differs from *H. brucei* in its basally more inflated molar cusps. Differs from *Procavia* in the shape of the lower molar crests, which form nearly perpendicular angles at the protoconid, where the paracristid and the protocristid meet, and also at the hypoconid, where the cristid obliqua and the hypocristid meet (unlike the less angled, crescentic, open curves formed by the same crests in *Procavia*). In addition, it differs from *Procavia* in its lower-crowned upper and lower molars, more inflated molar cusps, fewer foldings of the buccal wall of the upper molar ectolophs, and absence of lingual cingula. Differs from *Dendrohyrax* in its higher-crowned molars, the M3 hypocone that is relatively closely appressed to the metacone, the more dorsoventrally compressed radius, and the absence on the calcaneum of a medially extended shelf of the anterior astragalar facet.
Remarks Fortunately, the species is represented by much of the dentition along with several key postcranial elements. The teeth are primitive with respect to extant *Heterohyrax*, being slightly more brachdont. In this respect, they approach those of *Dendrohyrax*, but the Namibian hyrax has a flattened radius unsuitable for pronation and supination, unlike the proximally rounder radius of *Dendrohyrax*. This combination makes *H. auricampensis* somewhat of a mosaic of primitive morphological features when compared to the living species.

Genus *DENDROHYRAX* Gray, 1868

Synonymy See Meyer (1978).
Type Species *Dendrohyrax arboreus* Smith, 1827.
Included Species *D. arboreus*, *D. dorsalis*, *D. validus* (all extant).
Remarks Until recently, no fossil record of this genus had been recorded. However, Pickford (2007) described two late Miocene species of this genis, *D. samueli* and *D. validus*. We refer to Pickford (2005, 2007) for descriptions and details.

Genus *PROCAVIA* Storr, 1780

Synonymy See Meyer (1978).
Type Species *Procavia capensis* Storr, 1780.
Included species *P. antiqua*, *P. transvaalensis* (both fossil); *P. capensis*, *P. habessinica*, *P. johnstoni*, *P. ruficeps* (all extant).
Distribution Plio-Pleistocene of South Africa, recent of Africa and southwestern Asia.
Diagnosis From Schwartz (1997). Small to medium-sized procaviids with hypsodont dentition, dental arcade convergent anteriorly, right and left upper incisors close together, p1–4 shorter than m1–3, lower incisors subequal in size, i2 larger than i1, premolars molariform, no hypoconulids on m3.

PROCAVIA ANTIQUA Broom, 1934
Figure 13.9D–13.9F

Synonymy "Hyrax" sp., Wells, 1939, p. 365; *Procavia robertsi* Broom and Schepers, 1946:79; *P. robertsi* Broom, 1948:33.
Holotype Lost, but a sample collected with the type has been relocated and analyzed (Schwartz, 1997).
Lectotype Palatal specimen of Albany Museum Catalog No. 4194A, Albany Museum, Grahamstown, South Africa; posterior portion of the palate preserving left M1–M3 and right M2–M3.
Distribution Pliocene of South Africa.
Diagnosis Small *Procavia* with hypsodont dentition, dental formula 1.0.4.3/2.0.4.3, skull slightly smaller and lighter than in extant species, more rounded posterior border on I1, strong mesostyles on M2, well-developed hypocones on M3, bunodont protocone, longer ectoloph in comparison with the lingual border on all maxillary molars, and well-developed hypostyles on M3.
Remarks The taxonomic assessment of this species has always centered on the subtle size differences among it, the extant species, and fossil *P. transvaalensis* (Churcher, 1956; McMahon and Thackeray, 1994). The latest quantitative analysis of fossil material from several Plio-Pleistocene sites and extant samples showed that earlier material from the older members of Swartkrans and Taung were statistically separable from modern forms and from those from the youngest member of Swartkrans; temporally intermediate fossils from Kromdraai were metrically intermediate. The older material represents *P. antiqua* and is characterized by a shorter ectoloph on the upper molars, strong M2 mesostyles, a better-developed hypocone on M3, and more brachydont dentition (Schwartz, 1997).

PROCAVIA TRANSVAALENSIS Shaw, 1937
Figures 13.9A–13.9C

Synonymy *Procavia obermeyerae* Broom, 1937:766, figure 8B; *P. obermeyerae* Broom and Schepers, 1946:119.

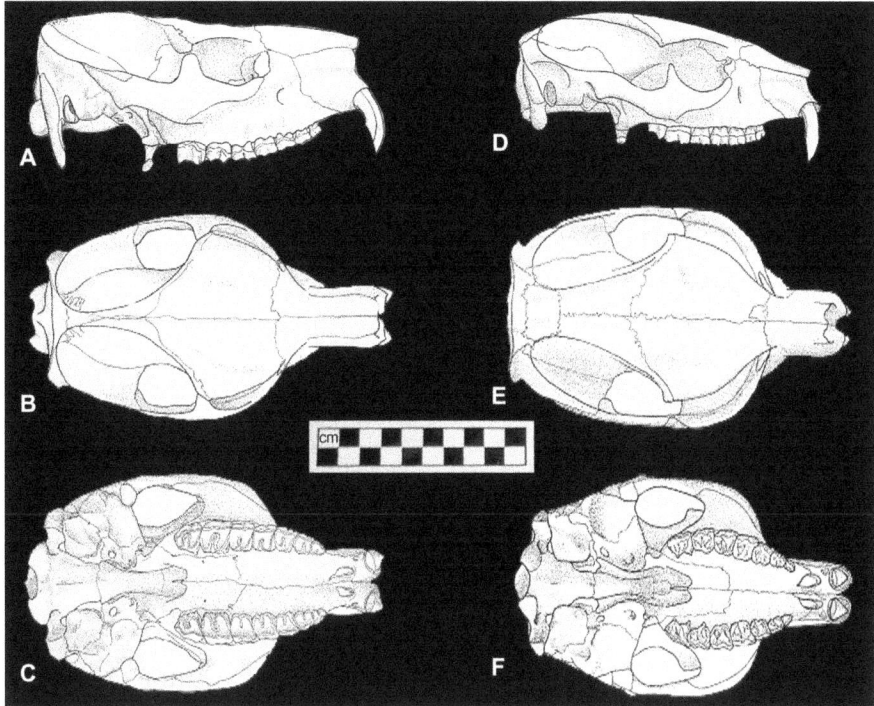

FIGURE 13.9 Crania of *Procavia transvaalensis* (A-C) and *P. antiqua* (D-F) in side view (top), dorsal view (middle), and ventral view (bottom). Notice the short, pinched rostra, and the reduced dental formula when compared to saghatheriids and pliohyracids. Reproduced from Churcher, 1956. Annals of the Transvaal Museum.

Holotype No. 20, Oral and Dental Hospital, University of the Witwatersrand, Johannesburg, Republic of South Africa; right ramus with complete dentition.

Distribution Plio-Pleistocene of Taung, Swartkrans, Sterkfontein, Cooper's, Kromdraai, Makapansgat, and Bolt's Farm.

Diagnosis Large species of *Procavia*, 50% greater in linear dimensions than any living species. The lower incisors are subequal in size, P1 and p1 are retained throughout life, interparietal free from parietals and supraoccipital, postorbital bar open, upper molars with heavier cingulum and more corrugated buccal surface of ectoloph than the extant forms.

Remarks *P. transvalensis* has been considered a specialized hyracoid of steppe habitats not as closely related to *P. capensis* as is *P. antiqua* (Churcher, 1956).

Genus *GIGANTOHYRAX* Kitching, 1965

Type and Only Known Species *Gigantohyrax maguirei* Kitching, 1965.

Distribution Pliocene of South Africa and Ethiopia.

Diagnosis Large procaviids with moderately hypsodont, lophoselenodont dentition. Upper dental formula 1.0.4.3, I2, I3, and upper canine lost, I1 massive and short, P1 long, with very small protocone and paracone, slight parastyle, large hypocone. The more distal premolars (P2–P4) are molariform but lack a mesostyle.

GIGANTOHYRAX MAGUIREI Kitching, 1965

Holotype BPI M8230, Bernard Price Institute for Paleontological Research, Johannesburg, South Africa, anterior two-thirds of a skull with full upper dentition.

Diagnosis As for genus.

Remarks Undescribed specimens attributable to this genus have been recovered at East African Pliocene sites (Howell and

Coppens, 1974, Brown et al., 1985). While the name of this genus suggests the evolution of gigantic hyracoids in the procaviid radiation of the Plio-Pleistocene, this is not true. In reality, *Gigantohyrax* is much smaller than early Tertiary forms like *Megalohyrax* and *Titanohyrax*, representing only a modest increase in size over *Procavia*. The relatively small size of African procaviids during the Plio-Pleistocene contrasts dramatically with the radiation of hyracoids in Asia occurring at the same time, which included truly gigantic, semiaquatic, hippolike pliohyracids with hypsodont teeth of a type that never appear in Africa (see references in Rasmussen, 1989; Fischer and Heizmann, 1992; Pickford et al., 1997).

Discussion

Hyracoids are one of the most diverse and abundant mammals of the early Tertiary. They were one of the essential "ungulate" components of archaic African communities. When Africa docked with the northern continents near the Oligocene-Miocene boundary, there was a notable decrease in diversity of hyracoids. At the broadest level, a view of hyracoid evolution may be considered in the framework of three evolutionary radiations.

The first of these was the radiation of archaic geniohyids, saghatheriids, and titanohyracids of the Eocene and Oligocene. By "archaic," we mean mammalian taxa present on island Africa before the Miocene. These taxa were extremely diverse in their ecomorphological adaptations and in their species-level diversity. They formed a key component of early Tertiary mammalian communities at the time. A corollary of this is that the early Tertiary hyracoids are critical for assessments of biostratigraphy, community structure, and evolutionary dynamics in archaic, isolated Africa, before the closing of the Tethys Sea.

The second radiation involved the emergence of more hypsodont forms, the early pliohyracids, that would eventually give rise to important lineages of large-bodied, extremely hypsodont hyracoids in Eurasia. The late Oligocene and Miocene members of this clade in Africa are clearly ancestral in a general sense to the later Plio-Pleistocene diversification of giant pliohyracids. Although these early pliohyracids were widespread in Africa and quite common at some sites, they were never particularly diverse. The family Pliohyracidae represents one of the few successful radiations of archaic African taxa into the broader ecological stage of Eurasia (along with Proboscidea).

The third hyracoid radiation was that of the procaviids in the late Miocene through the Plio-Pleistocene. This radiation represents a novel adaptation to marginal habitats and food resources eschewed by the immigrant artiodactyls and perissodactyls. This radiation is less diverse taxonomically and ecomorphologically than were the previous two major radiations. We are lucky to have the few remaining hyracoids of this family on Earth. An interesting thought experiment for paleontologists to consider is to evaluate whether we would have reconstructed the phylogeny and paleobiology of this order correctly if there were no living species to provide information on soft tissue, behavior, ecology, and DNA.

ACKNOWLEDGMENTS

We thank above all the field paleontologists who have recovered fossil hyracoids in Africa; you know who you are. We are especially grateful to Elwyn Simons, who has invited us to participate in the Fayum project over the years and has provided access to the Fayum collection at Duke University. We also thank Friderun Ankel-Simons for her scholarly collegiality and for help with data collection at Duke University. We thank Prithijit Chatrath, who through his management of the Fayum field project, as well as his collecting and preparatory skills, has been responsible for the addition of more data to the fossil record of hyracoids than any other person; his aid has been invaluable in studying the Fayum collections. Don DeBlieux has provided valuable discussion on hyracoid evolution over the years and has prepared many important specimens. We thank Erik Seiffert for his helpful discussions on late Oligocene hyracoids and his valuable role on the Fayum project. We thank Susy Cote for helpful discussion and information about early Miocene hyracoids. Many other international scholars have provided useful information and discussion of hyracoids over the years. We are grateful to Rodolphe Tabuce for providing the photographs of *Seggeurius* and *Microhyrax*, and Dr. Kurt Heißig of the Bayerische Staatssammlung für Paläontologie und Geologie, Munich, for sending a cast of the type of *Prohyrax tertiarius*. We also thank Director Mamitu Yilma and the staff of the Ethiopian National Museum in Addis Ababa, Mary Muugu and Drs. Emma Mbua and Meave Leakey of the National Museums of Kenya, and the late Dr. Yousry Attiya and his staff at the Cairo Geological Museum. We thank William Sanders and Lars Werdelin for inviting us to write this chapter.

Note

After submission of this chapter, Pickford (2009) described two species of fossil African hyraxes, *Regubahyrax selleyi* from Libya, and *Prohyrax bukwaensis* from Uganda.

Literature Cited

Andrews, C. W. 1903. Notes on an expedition to the Fayum, Egypt, with descriptions of some new mammals. *Geological Magazine* 4:337–343.

———. 1904a. Further notes on the mammals of the Eocene of Egypt, II. *Geological Magazine* 5:157–162.

———. 1904b. Further notes on the mammals of the Eocene of Egypt, III. *Geological Magazine* 5:211–215.

———. 1906. *A Descriptive Catalogue of the Tertiary Vertebrata of the Fayum, Egypt*. British Museum (Natural History), London, 324 pp.

———. 1907. Note on some vertebrate remains collected in the Fayum, Egypt. *Geological Magazine* 5:97–100.

Andrews, C. W., and H. J. L Beadnell. 1902. *A Preliminary Note on Some New Mammals from the Upper Eocene of Egypt*. Survey Department, Public Works Ministry, Cairo, 9 pp.

Arambourg, C. 1933. Mammifères miocènes du Turkana (Afrique orientale). *Annales de Paléontologie* 22:121–148.

Arambourg, C., and P. Magnier 1961. Gisements de vertébrés dans le bassin tertaire de Syrte (Libie). *Comptes Rendus Hebdomadaires des Séances de l'Académie des Sciences* 252:1181–1183.

Barry, R. E., and J. Shoshani. 2000. *Heterohyrax brucei. Mammalian Species* 645:1–7.

Bown, T. M., and M. J. Kraus. 1988. Geology and paleoenvironment of the Oligocene Jebel Qatrani Formation and Adjacent Rocks, Fayum Depression, Egypt. *U.S. Geological Survey Professional Paper* 1452:1–60.

Brown, F. H., I. McDougall, T. Davies, and R. Maier. 1985. An integrated Plio-Pleistocene chronology for the Turkana Basin; pp. 82–90 in E. Delson (ed.), *Ancestors: The Hard Evidence*. Liss, New York.

Churcher, C. S. 1956. The fossil Hyracoidea of the Transvaal and Taung deposits. *Annals of the Transvaal Museum* 22:477–501.

Cope, E. D. 1882. The classification of the ungulate Mammalia. *Proceedings of the American Philosophical Society* 20:438–461.

Court, N. 1993. Morphology and functional anatomy of the postcranial skeleton in *Arsinoitherium* (Mammalia, Embrithopoda). *Palaeontographica, Abt. A* 226:125–169.

Court, N., and M. Mahboubi. 1993. Reassessment of Lower Eocene *Seggeurius amourensis*: Aspects of primitive dental morphology in the mammalian order Hyracoidea. *Journal of Paleontology* 67:889–893.

DeBlieux, D. D., M. R. Baumrind, E. L. Simons, P. S. Chatrath, G. E. Meyer, and Y. S. Attia 2006. Sexual dimorphism of the internal mandibular chamber in Fayum Pliohyracidae (Mammalia). *Journal of Vertebrate Paleontology* 26:160–169.

DeBlieux, D. D., and E. L. Simons. 2002. Cranial and dental anatomy of *Antilohyrax pectidens*: A late Eocene hyracoid (Mammalia) from the Fayum, Egypt. *Journal of Vertebrate Paleontology* 22:122–136.

Fischer, M. S. 1986. Die Stellung der Schiefer (Hyracoidea) im phylogenetischen System der Eutheria: Zugleich ein Beitrag zur Anpassungsgeschichte der Procaviidae. *Courier Forschungsinstitut Senckenberg* 84:1–132.

Fischer, M. S., and E. P. J. Heizmann 1992. Über neogene Hyracoiden. Die Gattung *Pliohyrax*. *Neues Jahrbuch für Geologie und Paläontologie, Abhandlungen* 186:321–344.

Fortelius, M. 1985. Ungulate cheek teeth: Developmental, functional and evolutionary interrelations. *Acta Zoologica Fennica* 180:1–76.

Gagnon, M. 1997. Ecological diversity and community ecology in the Fayum sequence (Egypt). *Journal of Human Evolution* 32:133–160.

Gheerbrant, E., S. Peigné, and E. Thomas. 2007. Première description du squelette d'un mammifère hyracoïde paléogène: *Saghatherium antiquum* de l'Oligocène inférieur de Jebel al Hasawnah, Libye. *Palaeontographica, Abt. A* 279:93–145.

Gill, T. 1870. On the relations of the orders of mammals. *Proceedings of the American Association for the Advancement of Science* 19:267–270.

Ginsburg, L. 1977. L'Hyracoïde (Mammifère subongulé) du Miocène de Béni Mellal (Maroc). *Géologie Méditerranéenne* 4:241–254.

Hendey, Q. B. 1978. Preliminary report on the Miocene vertebrates from Arrisdrift, South West Africa. *Annals of the South African Museum* 76:1–41.

Hooijer, D. A. 1963. *Miocene Mammalia of Congo*. Museum Royal d'Afrique Centrale, Tervuren, Belgique, Annual Series in 8°, 46 pp.

Howell, F. C., and Y. Coppens. 1974. Les faunes de mammifères fossiles des formations Plio-Pléistocènes de l'Omo en Ethiopie (Tubulidentata, Hyracoidea, Lagomorpha, Rodentia, Chiroptera, Insectivora, Carnivora, Primates). *Comptes Rendus de l'Académie des Sciences, Paris* 278:2421–2424.

Hunter, J. P., and J. Jernvall. 1995. The hypocone as a key innovation in mammalian evolution. *Proceedings of the National Academy of Sciences, USA* 92:10718–10722.

Jaeger, J. J. 2003. Mammalian evolution—isolationist tendencies. *Nature* 426:509.

Jernvall, J. 1995. Mammalian molar cusp patterns: Developmental mechanisms of diversity. *Acta Zoologica Fennica* 198:1–61.

Kappelman, J., D. T. Rasmussen, W. Sanders, M. Feseha, T. Bown, P. Copeland, J. Crabaugh, J. Fleagle, M. Glantz, A. Gordon, B. Jacobs, M. Maga, K. Muldoon, A. Pan, L. Pyne, B. Richmond, T. Ryan, E. Seiffert, S. Sen, L. Todd, M. C. Wiemann, and A. Winkler. 2003. Oligocene mammals from Ethiopia and faunal exchange between Afro-Arabia and Eurasia. *Nature* 426:549–552.

Lavocat, R. 1961. Le gisement de vertébrés miocènes de Béni Mellal (Maroc) : Part 2. Etude systematique de la des de mammifères. *Notes et Mémoires du Service Géologique, Maroc* 155:29–92.

Mahboubi, M., R. Ameur, J.-Y. Crochet, and J. J. Jaeger. 1986. El Kohol (Saharan Atlas, Algeria), a new Eocene mammal locality in northwestern Africa: stratigraphic, phylogenetic and paleobiogeographical data. *Palaeontographica Abt. A* 192:15–49.

Matsumoto, H. 1922. *Megalohyrax* Andrews and *Titanohyrax* gen. nov., a revision of the genera of hyracoids from the Fayum, Egypt. *Proceedings of the Zoological Society of London* 1921:839–850.

———. 1926. Contribution to the knowledge of the fossil Hyracoidea of the Fayum, Egypt, with description of several new species. *Bulletin of the American Museum of Natural History* 56:253–350.

McMahon, C. R., and J. F. Thackeray. 1994. Plio-Pleistocene Hyracoidea from Swartkrans Cave, South Africa. *South African Journal of Zoology* 29:40–45.

Meyer, G. E. 1973. A new Oligocene hyrax from the Jebel Qatrani Formation, Fayum, Egypt. *Postilla* 163:1–11.

———. 1978. Hyracoidea; pp. 284–314 in V. J. Maglio, and H. B. S. Cooke (eds.), *Evolution of African Mammals.* Harvard University Press, Cambridge.

Novacek, M. J. 1992. Fossils, topologies, missing data, and the higher level phylogeny of eutherian mammals. *Systematic Biology* 41:58–73.

Pickford, M. 1986a. Première découverte d'une faune mammalienne terrestre paléogène d'Afrique sub-saharienne. *Comptes Rendus de l'Académie des Sciences, Paris,* Série II, 302:1205–1210.

———. 1986b. Première découverte d'une hyracoïde paléogène en Eurasie. *Comptes Rendus de l'Académie des Sciences, Paris,* Série II, 303:1251–1254.

———. 1994. A new species of *Prohyrax* (Mammalia, Hyracoidea) from the middle Miocene of Arrisdrift, Namibia. *Communications of the Geological Survey of Namibia* 9:43–62.

———. 1996. Pliohyracids (Mammalia, Hyracoidea) from the upper Middle Miocene at Berg Aukas, Namibia. *Comptes Rendus de l'Académie des Sciences, Paris* 322:501–505.

———. 2003. Giant dassie (Hyracoidea, Mammalia) from the middle Miocene of South Africa. *South African Journal of Science* 99:366–367.

———. 2004. Revision of the early Miocene Hyracoidea (Mammalia) of East Africa. *Comptes Rendus Palevol* 3:675–690.

———. 2005. Fossil hyraxes (Hyracoidea: Mammalia) from the late Miocene and Plio-Pleistocene of Africa, and the phylogeny of the Procaviidae. *Palaeontologia Africana* 41:141–169.

———. 2007. Late Miocene procaviid hyracoids (Hyracoidea: *Dendrohyrax*) from Lemudong'o, Kenya. *Kirtlandia* 56:106–111.

———. 2009. New Neogene hyracoid specimens from the Peri-Tethyan region and East Africa. *Paleontological Research* 13:265–278.

Pickford, M., and M. Fischer. 1987. *Parapliohyrax ngororaensis,* a new hyracoid from the Miocene of Kenya, with an outline of the classification of Neogene Hyracoidea. *Neues Jahrbuch für Geologie und Paläontologie, Abhandlungen* 175:207–234.

Pickford, M., S. Moyà Solà, and P. Mein. 1997. A revised phylogeny of Hyracoidea (Mammalia) based on new specimens of Pliohyracidae from Africa and Europe. *Neues Jahrbuch für Geologie und Paläontologie, Abhandlungen* 205:265–288.

Pickford, M., H. Thomas, S. Sen, J. Roger, E. Gheerbrant, and Z. Al-Sulaimai. 1994. Early Oligocene Hyracoidea (Mammalia) from Thaytiniti and Taqah, Dhofar Province, Sultanate of Oman. *Comptes Rendus de l'Académie des Sciences, Paris,* Série II, 318:1395–1400.

Rasmussen, D. T. 1989. The evolution of the Hyracoidea: a review of the fossil evidence; pp. 57–78 in Prothero, D. R. and R. M. Schoch (eds.), *The Evolution of Perissodactyls.* Oxford University Press, New York.

Rasmussen, D. T. and M. Gutiérrez, 2009. A mammalian fauna from the late Oligocene of northwestern Kenya. *Palaeontographica Abt. A* 288:1–52.

Rasmussen, D. T., M. Gagnon, and E. L. Simons. 1990. Taxeopody in the carpus and tarsus of Oligocene Pliohyracidae (Mammalia: Hyracoidea) and the phyletic position of hyraxes. *Proceedings of the National Academy of Sciences, USA* 87:4688–4691.

Rasmussen, D. T., M. Pickford, P. Mein, B. Senut, and G. C. Conroy. 1996. Earliest known procaviid hyracoid from the late Miocene of Namibia. *Journal of Mammalogy* 77:745–754.

Rasmussen, D. T., and E. L. Simons. 1988. New Oligocene hyracoids from Egypt. *Journal of Vertebrate Paleontology* 8:67–83.

———. 1991. The oldest Egyptian hyracoids (Mammalia: Pliohyracidae): New species of *Saghatherium* and *Thyrohyrax* from the Fayum. *Neues Jahrbuch für Geologie und Paläontologie, Abhandlungen* 182:187–209.

———. 2000. Ecomorphological diversity among Paleogene hyracoids (Mammalia): A new cursorial browser from the Fayum, Egypt. *Journal of Vertebrate Paleontology* 20:167–176.

Schlosser, M. 1923. *Grundzüge der Paläontologie (Paläozoologie) von Karl A. von Zittel, II Abteilung-Vertebrata.* Neuarbeitet von F. Broili und M. Schlosser. Oldenbourg, Munich, 706 pp.

Schwartz, G. T. 1997. Re-evaluation of the Plio-Pleistocene Hyraxes (Mammalia: Procaviidae) from South Africa. *Neues Jahrbuch für Geologie und Paläontologie, Abhandlungen* 206:365–383.

Schwartz, G. T., D. T. Rasmussen, and R. J. Smith. 1995. Body size diversity and community structure of fossil hyracoids. *Journal of Mammalogy* 76:1088–1099.

Seiffert, E. R. 2006. Revised age estimates for the later Paleogene mammal faunas of Egypt and Oman. *Proceedings of the National Academy of Sciences, USA* 103:5000–5005.

Shoshani, J. 1986. Mammalian phylogeny: Comparison of morphological and molecular results. *Molecular Biology and Evolution* 3:222–242.

Shoshani, J., and M. C. McKenna. 1998. Higher taxonomic relationships among extant mammals based on morphology, with selected comparisons of results from molecular data. *Molecular Phylogenetics and Evolution* 9:572–584.

Simpson, G. G. 1945. The principles of classification and a classification of mammals. *Bulletin of the American Museum of Natural History* 85:1–350.

Springer, M. S., G. C. Cleven, O. Madsen, W. W. de Jong, V. G. Waddell, H. M. Amrine, and M. J. Stanhope. 1997. Endemic African mammals shake the phylogenetic tree. *Nature* 388:61–64.

Starck, D. 1995. Ordo 27. Hyracoidea; pp. 930–947 in D. Starck (ed.), *Lehrbuch der Speziellen Zoologie: Band II. Wirbeltiere. Teil 5/2: Säugetiere.* Fischer, Jena.

Stevens, N. J., M. D. Gottfried, P. M. O'Connor, E. M. Roberts, and S. Kapilima. 2004. A new Paleogene fauna from the East African Rift, southwestern Tanzania. *Journal of Vertebrate Paleontology* 24 (suppl.):118A.

Tabuce, R., B. Coiffait, P. E. Coiffait, M. Mahboubi, and J.-J. Jaeger. 2000. A new species of *Bunohyrax* (Hyracoidea, Mammalia) from the Eocene of Bir El Ater (Algeria). *Comptes Rendus de l'Académie des Sciences, Paris, Sciences de la Terre et des Planètes* 331:61–66.

Tabuce, R., M. Mahboubi, and J. Sudre. 2001. Reassessment of the Algerian Eocene Hyracoid *Microhyrax.* Consequences on the early diversity and basal phylogeny of the Order Hyracoidea (Mammalia). *Eclogae Geologicae Helvetiae* 94:537–545.

Thewissen, J. G. M., and E. L. Simons. 2001. Skull of *Megalohyrax eocaenus* (Hyracoidea, Mammalia) from the Oligocene of Egypt. *Journal of Vertebrate Paleontology* 21:98–106.

Thomas, O. 1892. On the species of the Hyracoidea. *Proceedings of the Zoological Society of London* 1892:50–76.

Thomas, H., E. Gheerbrant, and J. M. Pacaud. 2004. Découverte de squelettes subcomplets de mammifères (Hyracoidea) dans le Paléogène d'Afrique (Libye). *Comptes Rendus Palevol* 3:209–217.

Thomas, H., J. Roger, S. Sen, M. Pickford, E. Gheerbrant, Z. Al-Sulaimani, and S. Al-Busaidi. 1999. Oligocene and Miocene terrestrial vertebrates in the southern Arabian Peninsula (Sultanate of Oman) and their geodynamic and palaeogeographic settings; pp. 430–443 in P. J. Whybrow and A. Hill (eds.), *Fossil Vertebrates of Arabia.* Yale University Press, New Haven.

Thomas, H., S. Sen, M. Khan, B. Battail, and G. Ligabue. 1982. The Lower Miocene fauna of Al-Sarrar (eastern province, Saudi Arabia). *ATLATL, Journal of Saudi Arabian Archaeology* 5:109–136.

Tsujikawa, H. 2005. The updated Late Miocene large mammal fauna from Samburu Hills, northern Kenya. *African Study Monographs* 32:1–50.

Weitz, B. 1953. Serological relationships of hyrax and elephant. *Nature* 171:261.

Whitworth, T. 1954. The Miocene hyracoids of East Africa with some observations on the order Hyracoidea. *Fossil Mammals of Africa, British Museum (Natural History)* 7:1–70.

Whybrow, P. J., and D. Clements. 1999. Arabian Tertiary fauna, flora, and localities; pp. 460–473 in P. J. Whybrow and A. Hill (eds.), *Fossil Vertebrates of Arabia.* Yale University Press, New Haven.

FOURTEEN

Sirenia

DARYL P. DOMNING, IYAD S. ZALMOUT,
AND PHILIP D. GINGERICH

The order Sirenia is the only extant group of mammals adapted to feed exclusively on aquatic plants. In view of the worldwide abundance of aquatic macrophytes and the few other large herbivores competing for this resource, it is noteworthy that Recent sirenians comprise only three genera and five species. One of these, Steller's sea cow (*Hydrodamalis gigas*) of the North Pacific, was exterminated by humans in the 18th century. Uniquely among sirenians, the Steller's sea cow was adapted to cold-temperate climates and a diet of kelp and other algae (Domning, 1978b). All the living sirenians are tropical forms that feed preferentially on angiosperms, and this appears to have been the primitive condition for the order.

The Indian Ocean and West Pacific tropics are today inhabited by a single species, *Dugong dugon*, distributed in nearshore marine waters from East Africa and the Red Sea to Japan, Micronesia, and Australia. The three species of manatees *(Trichechus)* occur on both sides of the tropical Atlantic: *T. manatus* in fresh and salt water from the southeastern United States through the Caribbean to beyond the eastern tip of Brazil, *T. senegalensis* in rivers and coastal waters of West Africa, and *T. inunguis* in the Amazon Basin of South America (Bertram and Bertram, 1973).

The fossil record of sirenians is extensive but uneven both geographically and taxonomically, and many of the described genera are monotypic—though none of the higher taxa, in welcome contrast to the situation 30 years ago (Domning, 1978a). The order seems to have reached its peak diversity in the Miocene with about a dozen known genera.

We provisionally follow Sickenberg (1934) and Simpson (1945) in recognizing four families of sirenians (figure 14.1): Prorastomidae, Protosirenidae, Trichechidae, and Dugongidae. The former two are known only from Eocene deposits; the trichechids (including the extinct Miosireninae) have a scanty Oligocene to Recent record; while the dugongids, comprising the majority of known forms, are increasingly well documented from the middle Eocene to the present and are divided into three subfamilies (Halitheriinae, Dugonginae, and Hydrodamalinae). Cladistic analyses (Domning, 1994; Sagne, 2001) indicate that at least the Halitheriinae, and some or all of the families, apart from the Trichechidae, are paraphyletic, but they do not agree on which.

The prorastomids (*Prorastomus* Owen, 1855, and *Pezosiren* Domning, 2001, from Jamaica) were amphibious quadrupeds with resemblances to condylarths, but their inflated rostrum, retracted nares, extensive pachyosteosclerosis, and other traits clearly mark them as sirenians.

Protosiren is more derived, with complete hindlimbs but a weak sacroiliac joint that probably precluded quadrupedal locomotion on land. Cranially, it appears to be a good structural ancestor for the other two families, though too late in time for actual ancestry of at least the Dugongidae. Its vertebrae and ribs, however, show some peculiarities that may represent an evolutionary dead end: the vertebrae have a large keyhole-like vertebral foramen, and the ribs lack synovial joints; instead, rib capitulum and tuberculum are connected to the vertebral body by massive cartilage.

The Miosireninae of Abel, 1919 (*Anomotherium* Siegfried, 1965 and *Miosiren* Dollo, 1889), now placed in the Trichechidae (Domning, 1994, 1996), are an aberrant late Oligocene to mid-Miocene lineage known only from rare finds in northwestern Europe. This distribution of trichechids is not surprising if both miosirenines and modern manatees are derived from Tethyan protosirenids, as some evidence indicates (Sagne, 2001). Although seagrasses were available in the North Sea area, the miosirenines have massively reinforced palates that suggest adaptation for crushing invertebrates—a deviation from the herbivorous sirenian norm that is nevertheless plausible, given their distribution around the present North Sea, since modern manatees and dugongs are known to supplement their diets with animal protein on the cooler borders of their range (Preen, 1995).

The more typical manatees (Trichechinae) are only a little more common as fossils, appearing in the middle Miocene and in South America, and evidently remaining confined to that continent until the Pliocene. Alone among sirenians, they evolved unlimited horizontal replacement of the molars (an adaptation to an abrasive diet of aquatic true grasses) and reduced their number of cervical vertebrae to six. In the late Pliocene or early Pleistocene, they reached West Africa by waif dispersal across the tropical Atlantic (Domning, 2005). No fossils of manatees have yet been found in Africa. Hatt (1934) provides a useful summary of the anatomy and distribution of the West African species *T. senegalensis* Link, 1795.

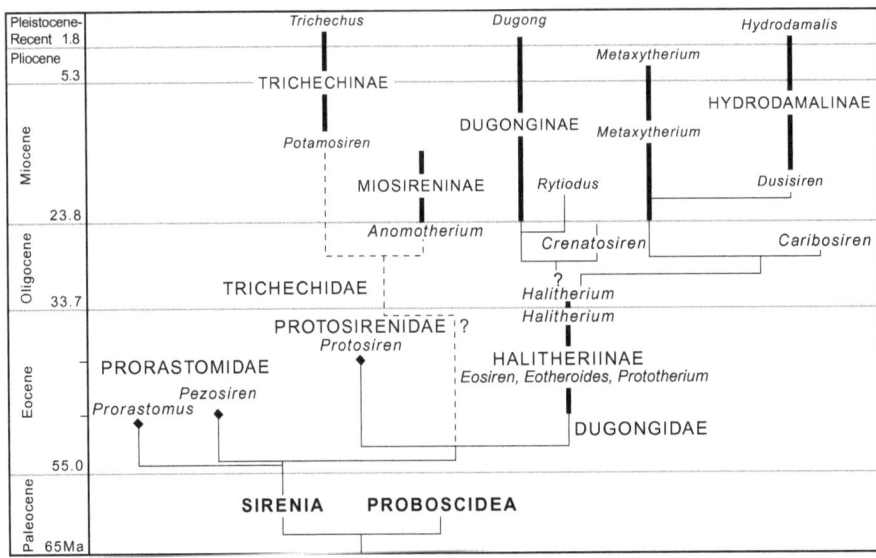

FIGURE 14.1 Phylogeny of the Sirenia. Known stratigraphic ranges indicated by bold lines; dashed lines indicate uncertain phylogenetic relationships; closed diamonds represent extinct families (modified from Gheerbrant et al., 2005).

The dugongids, the most prehistorically diverse and widely distributed sirenian family, maintained a pantropical distribution throughout the Tertiary. The basal forms are placed in the paraphyletic Halitheriinae (middle Eocene–Pliocene), and were generalized seagrass eaters that gave rise to two other subfamilies. The Dugonginae (late Oligocene–Recent) are pantropical animals, most with large tusks, that are thought to have characteristically specialized on rhizomes of the larger seagrasses. The Hydrodamalinae (early Miocene–Recent; extinct) are a North Pacific endemic lineage that increased in body size and lost their teeth as the climate grew colder and kelps replaced seagrasses in their habitat.

The dugongines flourished in the late Oligocene and early Miocene, with their apparent center of diversity in the West Atlantic–Caribbean region, although they had appeared in southern Europe, North Africa, and India by the early Miocene. The living *Dugong dugon* (Müller, 1776) of the Indopacific realm, with no Tertiary fossil record, was assumed to have evolved from these latter forms east of Suez until a well-preserved skull (still in a private collection and unfortunately undescribed) of its cladistically closest known relative was discovered in a deposit in Florida only two million years old. This strongly suggests dispersal (in one direction or the other) between Florida and the Indian Ocean in the geologically recent past, most likely by way of the Cape of Good Hope, which is near 35°S latitude, only about 9° south of the southernmost modern record of the dugong in Africa (Maputo or Delagoa Bay, Mozambique, 26°S; Dutton, 2004). There supposedly even exist unpublished records from the early 20th century of transient dugongs in South Africa (V. Cockcroft, pers. comm. to D.P.D., December 8, 1991). Hence a global climate only slightly warmer than the present would make it feasible for dugongs to round the Cape. The eventual discovery of their fossils in West Africa has been predicted (Domning, 1995).

Both dugongs and manatees have traditionally been hunted by humans (see, e.g., Petit, 1927), and archaeologists working at coastal or riverine sites in tropical Africa should expect to encounter sirenian bones (for illustrations see Kaiser, 1974).

Structure

The Sirenia, together with the Cetacea, are the only obligatorily aquatic mammals, and today they and cetaceans share fusiform, nearly hairless bodies, lack of external ear pinnae, short necks, paddlelike forelimbs, loss of hindlimbs (except for a greatly reduced pelvis with no bony connection to the vertebrae), stiffening of the body by a subdermal connective-tissue sheath of helically wound fibers, and horizontally expanded tail fins for main propulsion (Domning, 2000, 2001a). The caudal fin is triangular and cetacean-like in dugongids, but it retains a more primitive paddle shape in manatees. Unlike many cetaceans, however, sirenians lack dorsal fins and show no aptitude or adaptations for echolocation. All body hairs are sinus hairs and act as specialized tactile organs—especially the elaborately developed vibrissae, some of which are also prehensile, together with the greatly developed upper lip (Marshall et al., 1998).

As slow-swimming, shallow-diving herbivores with low metabolic rates and energy-poor diets, sirenians have refined the art of hydrostasis to a unique degree in order to minimize energy expenditure in locomotion. The massive skeleton displays both pachyostosis (increase in volume of individual elements, especially the ribs) and osteosclerosis (increased proportion of compact to cancellous bony tissue). These conditions, referred to as pachyosteosclerosis, provide ballast, and together with the horizontal diaphragm, elongate lungs, and monopodially branching bronchial trees, they help to maintain fore-and-aft trim, keeping the body close to a horizontal position (Domning and Buffrénil, 1991). In addition, the degree of downward flexion of the rostrum and anterior mandible is adjusted to the feeding habits of the species: *Dugong*, with a very downturned snout, is adapted to habitats (seagrass meadows) where the dominant growth habit of plants is short and bottom hugging, whereas sirenians with shallower snout deflections are found in habitats with abundant floating or near-surface vegetation (Domning, 1977).

All adequately known Eocene sirenians, including *Prorastomus* (Savage et al., 1994), *Protosiren*, *Eotheroides*, and *Eosiren*

(Sickenberg, 1934), had a dental formula of 3.1.5.3/3.1.5.3. This soon became reduced, first by loss of P5/p5 and retention of DP5/dp5 in the adult dentition, then by progressive loss of the anterior permanent and deciduous premolars and the canines and incisors (Domning et al., 1982). The first upper incisors are usually retained as a pair of tusks. They are sexually dimorphic in *Dugong dugon* and possibly in Pliocene *Metaxytherium*, but the sexes never seem to differ to the extent of presence versus complete absence of tusks. In all sirenians the fronts of the upper and lower jaws bear tough pads that serve (with the prehensile lips and vibrissae) to pull food into the mouth, with or without the help of the flippers.

While halitheriines always retain an adult cheek dentition of at least M2–3 (more commonly DP5–M3 in Neogene forms) and the teeth retain the primitive bunobilophodont-brachydont condition, the other subfamilies show more specialization: M3 of *Miosiren* is reduced to a simple, stout peg; adult *Dugong dugon* retain only M2–3, lacking roots, enamel, and (after initial wear) cusps; *Hydrodamalis* was completely toothless; and trichechines, after reaching a halitheriine-like degree of tooth reduction by the middle Miocene, evolved unlimited numbers of supernumerary molars (Domning, 1982; Domning and Hayek, 1984), in contrast to elephants, with which they are often misleadingly compared.

The most common sirenian fossils are fragments of the swollen, pachyosteosclerotic ribs, which are usually indeter-minable. Unlike most other fossil mammals, the teeth are of limited diagnostic value compared with portions of the skull, especially the skull roof and rostrum.

Fossil Sirenia in Africa

The first sirenian fossils reported from Africa (table 14.1, figure 14.2), like most of the subsequent ones, were Egyptian: indeterminate fragments from somewhere in the Nile Valley (Blainville, 1840, 1844), Ain Musa near Cairo (Fraas, 1867), and from the Isthmus of Suez (Gervais, 1872).

Later came the discovery of *Eotheroides aegyptiacum* (Owen, 1875) in the middle Eocene Nummulitic Beds of the Mokattam Hills near Cairo. Subsequent discoveries were predominantly from the late Eocene marine strata of the Fayum. Other sirenian records, however (mostly indeterminate), have gradually emerged from the Miocene of Madagascar (Collignon and Cottreau, 1927) and the Congo (Dartevelle, 1935); the Pliocene of Morocco (Ennouchi, 1954); the Eocene, Oligocene, and Miocene of Libya (Savage and White, 1965; Savage, 1967, 1969, 1971); the Eocene and Oligocene of Somalia (Savage, 1969); the Oligocene and Miocene of Tunisia (Robinson and Black, 1969; Savage, 1969); the Eocene of Togo (Gingerich et al., 1992), and the Eocene of Madagascar (Samonds et al., 2009 see table 14.1, figure 14.2). Good material of *Metaxytherium serresii* from Sahabi, Libya, formerly

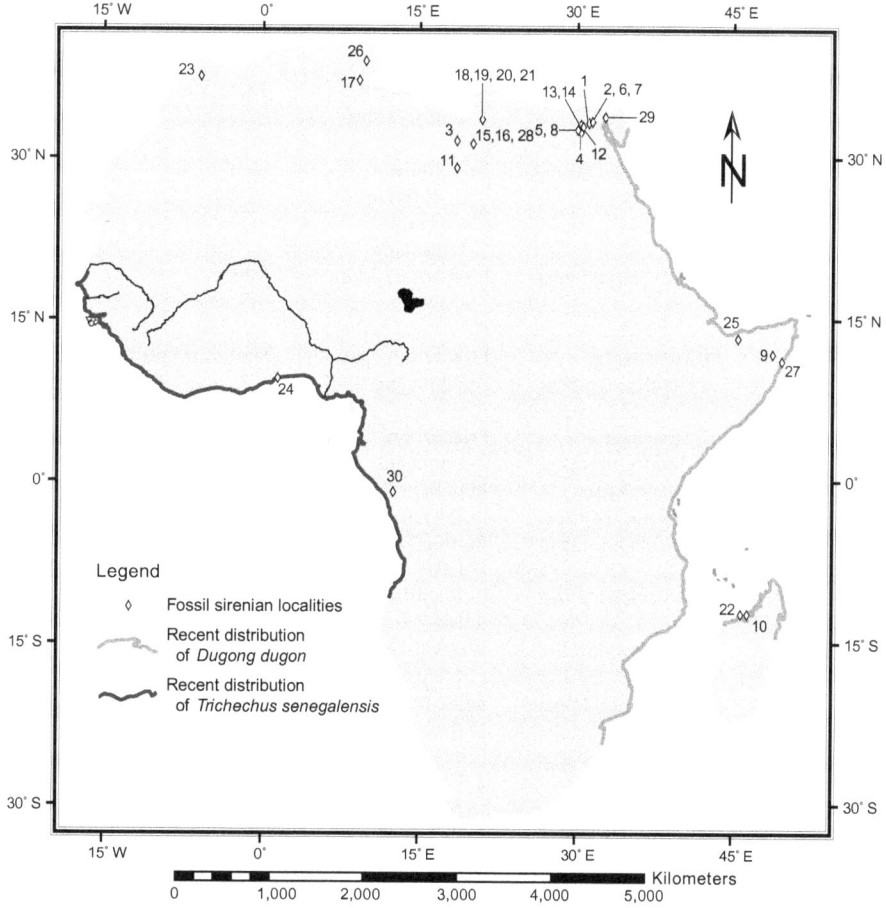

FIGURE 14.2 African distribution of fossil and living sirenians known to date. Fossil localities presented as open diamonds associated with numbers; Recent distribution of *Dugong dugon* (light color) and *Trichechus senegalensis* (dark color) highlighted as narrow strips along shores and rivers and in inland areas.

TABLE 14.1
African fossil sirenian taxa, localities, ages, and distributions published to date

Location	Taxon	Locality	Age	Material	Author
PROTOSIRENIDAE, MIDDLE–LATE EOCENE					
1	Protosiren?	Left side of the Lower Nile Valley, Egypt	Lutetian?, Mokattam Limestone?	Vertebra and rib fragments	Blainville, 1840, 1844
2	Protosiren fraasi	Jabal Mokattam, Egypt	Lutetian, Mokattam Limestone	Skulls and postcrania	Filhol, 1878; Abel, 1907
3	Protosirenidae, n.gen. n.sp.	Bu el Haderait, Libya	Lutetian	Partial skeletons	Savage and White, 1965; Savage, 1971, 1977; Heal, 1973
4	Protosiren sp.	Wadi Rayyan, Fayum, Egypt	Lutetian, Wadi Rayyan Series	Scapula	Zalmout and Gingerich, in prep.
5	Protosiren smithae	Wadi Hitan, Fayum, Egypt	Priabonian, Birket Qarun Fm.	Skulls and skeleton	Domning and Gingerich, 1994
DUGONGIDAE, MIDDLE EOCENE–RECENT					
6	Eosiren abeli	Jabal Mokattam, Cairo, Egypt	Lutetian, Mokattam Limestone	Skull, mandible, teeth, vertebrae (mostly destroyed)	Sickenberg, 1934
7	Eotheroides aegyptiacum	Jabal Mokattam, Egypt	Lutetian, Mokattam Limestone	Skulls and postcrania	Owen, 1875; Abel, 1913; Zdansky, 1938; Sickenberg, 1934
8	Eotheroides sp.	Wadi Hitan, Fayum, Egypt	Priabonian, Birket Qarun Fm.	Skeletons	Zalmout and Gingerich, 2005
9	Dugongidae	Callis, Somalia	Middle Eocene, Carcar Series	Teeth, rib	Savage, 1969, 1977; Savage and Tewari, 1977
10	Eotheroides lambondrano	Ampazony, Majunga, Madagascar	Bartonian-Priabonian, Nummulitic Limestone	Skull	Samonds et al., 2009
11	Dugongidae	Dor el Talha, Libya	Priabonian, Idam Unit	Ribs	Savage, 1969, 1971, 1977; Heal, 1973; Wight, 1980
12	Eosiren libyca	Qasr el Sagha, Fayum, Egypt	Priabonian, Qasr el Sagha Fm.	Skulls and postcrania	Andrews, 1902; Sickenberg, 1934; Siegfried, 1967
13	Eosiren stromeri	Qasr el Sagha, Fayum, Egypt	Priabonian, Qasr el Sagha Fm.	Skull and postcrania	Sickenberg, 1934
14	Eosiren imenti	Jabal Qatrani, Fayum, Egypt	Rupelian, Jabal Qatrani Fm.	Skull and some ribs	Domning et al., 1994
15	Rytiodus n.sp.	Jabal Zaltan, Libya	Burdigalian, Marada Fm.	Skulls, mandibles, etc.	Heal, 1973
16	Metaxytherium sp.	Jabal Zaltan, Libya	Burdigalian, Marada Fm.	Fragments of 2 skulls	Heal, 1973
17	Dugongidae	Bled ed Douarah, Tunisia	Serravallian	Teeth	Robinson and Black, 1969
18	Metaxytherium sp.?	Qasr Sahabi, Libya	Late Miocene [Fm. M]	Articulated skeletons	Domning and Thomas, 1987
19	Metaxytherium sp.?	Qasr Sahabi, Libya	Late Tortonian or early Messinian? [Fm. P]	Sirenian ribs	de Heinzelin and El Arnauti, 1987
20	Metaxytherium serresii	Qasr Sahabi, Libya	Early Messinian [Mbr. T]	Skull fragments, mandibles, postcrania	Domning and Thomas, 1987

Location	Taxon	Locality	Age	Material	Author
21	*Metaxytherium serresii*	Qasr Sahabi, Libya	Early Messinian? [Mbr. U-2; also U-1]	Partial skeleton	Domning and Thomas, 1987
22	Dugongidae	Ile Makamby (E side of the island), Madagascar	Oligocene/ Miocene, Lower Terrigenous Unit and Upper Nummulitic Marl Unit	Skullcap	Collignon and Cottreau, 1927; Samonds et al., 2009
23	*Metaxytherium* cf. *serresii*	Dar bel Hamri, Morocco	Pliocene	Skullcap, tooth	Ennouchi, 1954
		SIRENIA INDET.			
24	Sirenian remains	Kpogamé-Hahotoé, Togo	Lutetian, Kpogamé-Hahotoé Phosphate level, middle Eocene	Thoracic vertebrae, rib heads	Gingerich et al., 1992
25	Sirenia indet.	25 km SE of Berbera, Somalia	Middle Eocene [Daban Series]	Ribs	Macfadyen 1952; Savage and Tewari, 1977; Savage, 1969
26	Sirenia indet.	Djebel ech Cherichira, Tunisia	Oligocene	Skeleton	Savage, 1969
27	Sirenia?	Bedeil, Somalia	Oligocene	Tusk	Savage, 1969
28	Sirenia indet.	Jabal Zaltan, Libya	Burdigalian, Marada Fm.	Sirenian remains	Heal, 1973
29	Sirenia indet.	Isthmus of Suez, Egypt	Miocene	Fragmentary ribs	Gervais, 1872
30	Sirenia indet.	Malembe, Congo	Miocene	Ribs	Dartevelle, 1935

considered early Pliocene, is now assigned to the late Miocene (Domning and Thomas, 1987; Boaz et al., 2004; Carone and Domning, 2007).

Although the Egyptian sirenians (now including new middle Eocene and early Oligocene specimens) have been extensively described, important new Eocene and Miocene taxa from Libya (Heal, 1973) still await formal publication, and the sirenian history of the rest of the continent has barely begun to be investigated.

INSTITUTIONAL ABBREVIATIONS

BMNH (= NHML), British Museum of Natural History (= Natural History Museum), London (England); CGM, Cairo Geological Museum, Cairo (Egypt); SMNS, Staatliches Museum für Naturkunde, Stuttgart (Germany); UM, Museum of Paleontology, University of Michigan, Ann Arbor (USA).

Systematic Paleontology

For complete synonymies and bibliography, see Domning (1996).

Order SIRENIA Illiger, 1811
Family PROTOSIRENIDAE Sickenberg, 1934
Figures 14.3A, 14.3B, 14.5A, and 14.5B

Based on one genus and four species, this Tethyan Eocene group is known with assurance only from Africa and Asia (Abel, 1907; Domning and Gingerich, 1994; Gingerich et al.,

1995; Zalmout et al., 2003b). Reports of protosirenids from North America by Domning et al. (1982) require further study to clarify the identity of these taxa.

Members of this group are the largest of all known Eocene sirenians. The skull roof is robust and stout, and very broad across the supraorbital processes; the nasal bones are reduced and separated in some species; the alisphenoid canal is well defined; the rostrum is slightly deflected; the palatal region of the rostrum is wide; and the cheek teeth are large. Postcranial elements are osteosclerotic, middle thoracic vertebrae have a keyhole-shaped vertebral foramen, and the heads of the ribs display a cartilaginous rather than fully ossified articular surface. As shown by the well-preserved sacrum, pelvis, femur, patella, tibia, and fibula of *P. smithae*, *Protosiren* had well-developed hindlimbs and may have maintained some terrestrial mobility, perhaps in pinniped-like fashion, but could not fully support its body weight on its legs on land.

Genus *PROTOSIREN* Abel, 1907
PROTOSIREN FRAASI Abel, 1907
Figures 14.2, 14.3A, and 14.5A

Age and Occurrence Lutetian of the Mokattam Hills near Cairo (table 14.1, figure 14.2).

Diagnosis Tusk alveoli small. Nasals contact each other in midline. Orbital opening small and dorsoventrally compressed. Pelvis has a large obturator foramen and a deep acetabulum.

Description Rostrum is slightly deflected, premaxillary-maxillary suture is shifted posteriorly, and alisphenoid canal is present. Squamosal does not contribute to the back of the

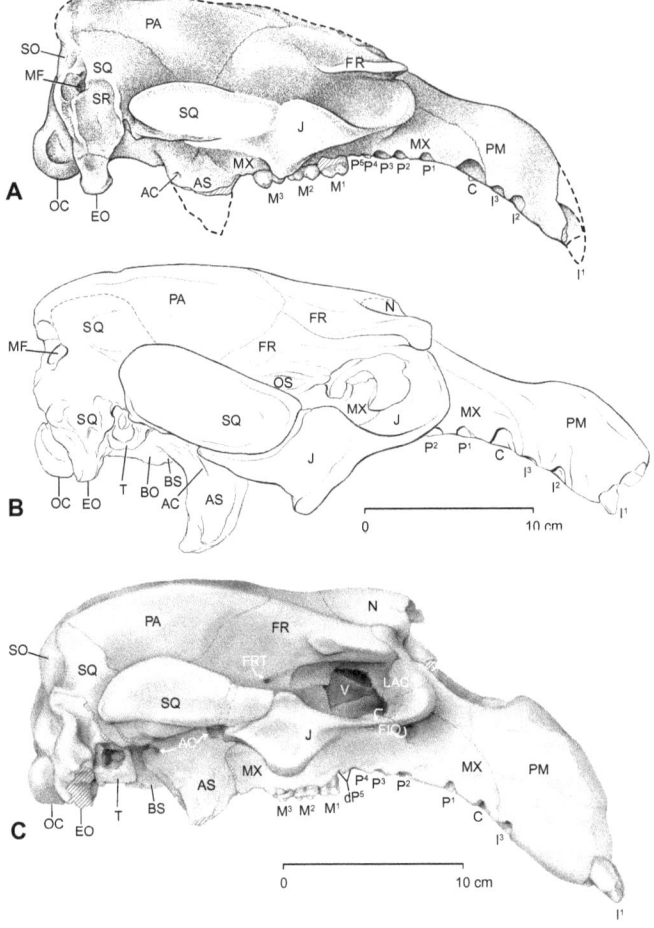

FIGURE 14.3 Cranial outlines of some African Paleogene sirenians (Protosirenidae and Halitheriinae) in lateral views. A) Lutetian *Protosiren fraasi* (reconstructed from CGM 10171 and SMNS 10576; reconstruction of Gingerich et. al. 1994); B) Priabonian *Protosiren smithae* (CGM 43392; Domning and Gingerich, 1994); C) Priabonian *Eotheroides* sp. (UM 101219; Zalmout and Gingerich, 2005).

ABBREVIATIONS: AC, alisphenoid canal; AS, alisphenoid; BO, basioccipital; BS, basisphenoid; C^1, upper canine; DP, deciduous premolar; EO, exoccipital; FIO, infraorbital foramen; FR, frontal; FRT, foramen rotundum; I^1, etc., upper incisor; J, jugal; LAC, lacrimal; M^1, etc., upper molar or alveoli; MES, mesethmoid; MF, mastoid foramen; MX, maxilla; N, nasal; OC, occipital condyle; OS, orbitosphenoid; P^1, etc., upper premolar alveoli; PA, parietal; PM, premaxilla; SO, supraoccipital; SQ, squamosal; SR, sigmoidal ridge; T, tympanic; V, vomer.

skull, squamosal and supraoccipital are completely separated by a parietal process, and posttympanic process of squamosal is absent. In contrast to nearly all other sirenians, dorsal surface of endocranium is smooth in all known specimens, with no bony falx cerebri, bony tentorium, or internal occipital protuberance (Gingerich et al., 1994). The pelvis has a rodlike ilium with a spindlelike end.

Remarks Protosiren fraasi is known from skulls, a dentary, thoracic vertebrae, ribs, and innominates from the Lutetian of the Mokattam Hills near Cairo. The holotype skull and mandible (CGM 10171 = 42297) were erroneously described as "*Eotherium aegyptiacum* (?)" by Andrews (1906: text figures 66 and 67). Innominates referred to "*Eotherium*" *aegyptiacum* (Abel, 1904: plate 7, figure 1) also pertain to *P. fraasi*.

PROTOSIREN SMITHAE Domning and Gingerich, 1994
Figures 14.2, 14.3B, and 14.5B

Age and Occurrence Early Priabonian of the Birket Qarun Formation, about 40 km west of Lake Qarun, 160 km SW of Cairo (table 14.1, figure 14.2).

Diagnosis Upper incisor tusks enlarged. Rostral masticating surface trapezoidal. Rostral deflection increased. Pterygoid process long dorsoventrally. Nasals separated by frontals. Supraorbital processes massive. Exoccipitals widely separated in the dorsal midline. Sternebrae short and blocklike, resembling those of terrestrial mammals.

Description The holotype (CGM 42292) is a mature adult represented by a skull and lower jaw with the upper and lower teeth, most of the vertebral column, some ribs, and front and hindlimbs. The premaxillae are short, rostral deflection is about 60°, and the nasals are separated by the frontals in dorsal view (best seen in UM 101224, Domning and Gingerich, 1994). The endocranial surface is smoothly concave with no bony falx cerebri; however, in UM 101224 the falx and tentorium are very faintly marked (Domning and Gingerich, 1994). An alisphenoid canal is present and well developed. Cervical vertebrae are short with flattened endplates, the neural spine is bifurcated on the posterior cervicals and the first thoracics, and the vertebral foramina are wide. Scapulae are sickle shaped with a narrow supraspinous fossa. Innominates have a rodlike ilium, broad and flattened ischium, well-developed pubis and pubic symphysis, reduced obturator foramen, and shallow acetabulum. The femur is 140 mm long with distinct greater and lesser trochanters and no third trochanter, the femoral head is oval, and the shaft is flattened anteroposteriorly. The patellar surface and distal condyles have a wide and shallow saddle shape allowing movement of the distal limb.

Remarks Protosiren smithae from the Priabonian of the Western Desert of Egypt is intermediate between amphibious and fully aquatic sirenians. Domning and Gingerich (1994) commented that the retention of functional hip and knee articulations and well-developed tibia and fibula indicate that this taxon retained a mobile foot.

PROTOSIRENIDAE, n. gen. n. sp. Heal, 1973
Figure 14.2

Age and Occurrence Middle Eocene (Lutetian) of Bu el Haderait, Libya (table 14.1, figure 14.2).

Diagnosis The largest protosirenid to be described to date (total length of the skull is 45 cm, sagittal length of skull roof approximately 25 cm); nasals large; sagittal length of parietals much greater than that of frontals; length of m2 = 26 mm. Five lower premolars were likely present.

Description The premaxillary symphysis forms one-third of the premaxillary length, the external nares are wide, the maxilla is almost straight, the lacrimal bone separates the premaxilla from the frontal, nasals and frontals are enlarged, frontals are heavy and thick, and their supraorbital processes extend anterolaterally to reach the front of the nasals and lacrimals; parietal and frontal meet along a transversely wide and shallow suture, the zygomatic process of the squamosal is short and laterally compressed, the pterygoid process is short, an alisphenoid canal is present and its openings are narrow, and the tentative dental formula is 3.1.4.3/3.1.4(5?).3.

Postcranial elements are not well known; however, a recovered atlas and an anterior thoracic vertebra are larger than those of any other Eocene sirenians. The thoracic vertebra shows characteristics of typical protosirenid postcrania with its large vertebral canal and high neural spine.

Remarks Described in an unpublished doctoral dissertation (Heal, 1973) and represented by good cranial material, this is the largest known Eocene sirenian. One specimen has an apparently supernumerary last lower molar (m4?) with a circular crown, reminiscent of the reduced M3 of *Miosiren kocki* (Sickenberg, 1934).

Family DUGONGIDAE Gray, 1821
Subfamily HALITHERIINAE Carus, 1868

The Dugongidae are the most diverse and successful group of sirenians. They made their first appearance in the middle and late Eocene of the Old World Tethys. Three major genera of Dugongidae dominate the Eocene record: *Eotheroides*, *Eosiren*, and *Prototherium*. These are fully aquatic with very reduced pelvic and femoral features, including: reduced length of the ischium and ilium, a diminutive obturator fora-

men, and reduction or loss of the pubic symphysis, presaging the complete loss of function in hindlimbs.

Genus *EOTHEROIDES* Palmer, 1899

The first diagnosis of *Eotheroides* ("*Eotherium*"), by Owen (1875), was based on a cranial endocast from the Lutetian of the Mokattam Hills. The endocast (NHML 46722) contains enough information to visualize the braincase. Later, more elements were recovered and assigned to this taxon. These include partial skulls and skeletons (Abel, 1913). So far the genus is known from the Lutetian of the Mokattam Hills near Cairo (Owen, 1875; Abel, 1913) and the Priabonian Birket Qarun Formation of the Fayum (Zalmout and Gingerich, 2005).

EOTHEROIDES AEGYPTIACUM (Owen, 1875)
Figure 14.2

Age and Occurrence Lutetian of the Mokattam Formation near Cairo (table 14.1, figure 14.2).

Diagnosis Falx, bony tentorium, and internal occipital protuberance present. Nasals are long and in contact along the midline; however, their posterior extremities are separated slightly by the frontals (though not as much as in *Protosiren* or *Eosiren*). The lacrimal foramen opens laterally (Abel, 1913: plate 2). The palate is broad and its posterior border, which is formed by the palatines, lies abaft the rear of the toothrow. Anterior ribs are dense, thickened and swollen (pachyosteosclerotic), having a banana-like shape (see Abel, 1919:836, figure 663).

Description The rostrum is deflected, the premaxillary-maxillary suture lies below the premaxillary symphysis, an alisphenoid canal is absent, the posttympanic process of the squamosal is present, the dental formula is 2-3.1.5.3/3.1.4-5.3, the fifth permanent premolar is lost, and the trirooted DP5 is retained and not replaced.

Remarks Eotheroides aegyptiacum is distinctive among Eocene sirenians in having extensively pachyosteosclerotic banana-like ribs. New records from Priabonian beds in the Fayum, including articulated and more complete skeletons of an undescribed *Eotheroides* species, will reveal more information about the evolutionary history of this group.

EOTHEROIDES sp. nov.
Figures 14.2, 14.3C, 14.5C, and 14.5D

Age and Occurrence Early Priabonian of the Birket Qarun Formation (table 14.1, figure 14.2).

Diagnosis Infraorbital foramen enlarged. Nasals highly arched upward, rising higher than parietals. Lacrimal faces posteriorly and is less exposed laterally, and lacrimal foramen absent. Pelvis has a short clublike ilium, flat and robust ischium, tiny obturator foramen, and extremely long and narrow pubic bone (longer than any pubic bone of other species).

Description I3 is absent, DP5 and dp5 are not replaced, and the dental formula is 2.1.5.3/3.1.5.3. Mandibles are straight and parallel to each other. Anterior and middle ribs are swollen; posterior ribs are slender.

Remarks This species is based on marvelous material including fairly complete postcranial skeletons. The most characteristic element that distinguishes this species from the associated sirenian taxa is the clublike ilium. Reduction in the number of upper incisors and the overall reduction in the pelvic girdle represent derived conditions for *Eotheroides*.

Genus *EOSIREN* Andrews, 1902
Figure 14.2

Age and Occurrence Middle Eocene to early Oligocene (table 14.1, figure 14.2), known from the Mokattam Hills near Cairo (Lutetian), Qasr el-Sagha in the Fayum (Priabonian), and Jabal Qatrani, north of Birket Qarun in the Fayum area (Rupelian).

Diagnosis Rostrum strong and enlarged, premaxillary symphysis extends more than one-third of the skull length, supracondylar fossa of the occipital is deep and extends across entire width of the occipital condyle, anterior border of coronoid process extends slightly anterior to base of the process, and cheek teeth are enlarged. Ribs are osteosclerotic but gracile compared to other sirenians. Scapulae are flat and broad. Innominates are very reduced; ilium is short and gracile, obturator foramen is either closed or small, and acetabulum (figures 14.5E, 14.5F) is shallow and oval. Femora are reduced as well, with rounded cross section at the middle of the shaft. Dental formula is usually 1-3.1.5.3/3.1.5.3.

EOSIREN ABELI Sickenberg, 1934
Figure 14.2

Age and Occurrence Lutetian of the Mokattam Hills near Cairo (table 14.1, figure 14. 2).

Diagnosis M2 differs in shape and size from *Eotheroides aegyptiacum* and *Protosiren fraasi*, and is slightly smaller than M2 of *Eosiren libyca*. The skull is short, and so is the braincase (the median length of the brain endocast is 52 mm, which is 10 mm shorter than *E. aegyptiacum*). The end of the olfactory bulb is notably steeper, and overall the brain morphology shows a great affinity with the braincase of *E. libyca* described (reconstructed) by Andrews (1906, text figure 65, page 202).

Description The most informative specimen (now destroyed) was Abel's (1913:309) individual VI of *"Eotherium" aegyptiacum*, which consisted of a skull (premaxilla, maxilla, skull roof, occipitals and basicranium, ear apparatus including tympanics and ossicles), lower jaw, and an atlas and some other vertebrae. A squamosal and some vertebrae representing this species are illustrated in Sickenberg (1934: plate 4, figures 3, 11). Only cranial elements seem to be characteristic in this species; vertebrae and ribs are similar to Qasr el-Sagha *Eosiren*.

Remarks The holotype was a right M2 destroyed during World War II along with a referred skull and other cranial and postcranial elements collected from the Mokattam Hills near Cairo. *E. abeli* is the third sirenian taxon in the Mokattam Hills. Sickenberg (1934) mentioned that finding *Eosiren* in the Mokattam Hills is not remarkable since the precursor of the Fayum *Eosiren* should have been present in older strata.

EOSIREN LIBYCA Andrews, 1902
Figures 14.2, 14.4A, 14.5E, and 14.4F

Age and Occurrence Late Priabonian of Qasr el-Sagha Formation, north of Lake Qarun, Fayum area, Egypt (table 14.1, figure 14.2).

Diagnosis Differs from *Eosiren stromeri* in having a narrow palate, M2 smaller than M3, larger and more prominent tusk alveoli, and smaller temporal fossa. Frontals are shorter or slightly longer than parietals. Innominates are vestigial like those in *Halitherium* and have their obturator foramen almost closed.

Description Nasals are shortened and are still joined in the midline with a slight posterior separation by the frontal. The nasal process of the premaxilla overlaps less than one-third of the anteroposterior length of the supraorbital process. A

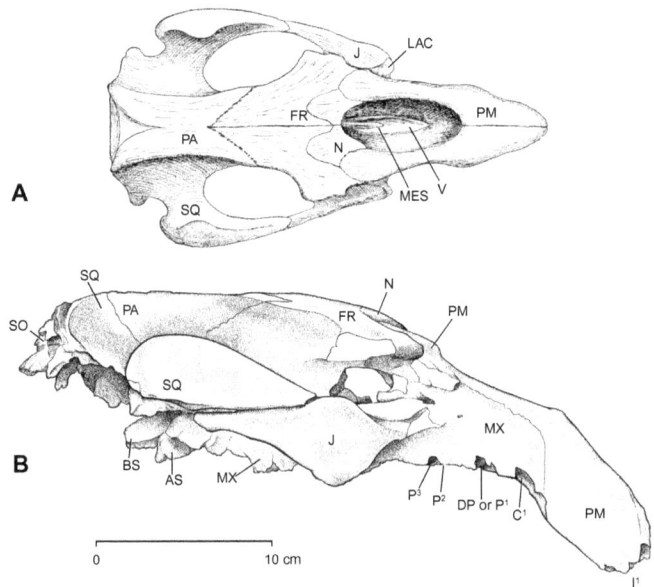

FIGURE 14.4 Eocene and Oligocene *Eosiren* from the Fayum area in Egypt. A) Dorsal view of Priabonian *Eosiren libyca* of Qasr el Sagha Formation (Andrews, 1902; illustration of Abel, 1928); B) lateral view of Rupelian *Eosiren imenti* (CGM 40210, Domning et al., 1994). Abbreviations as in figure 14.3. Scale bar refers only to B.

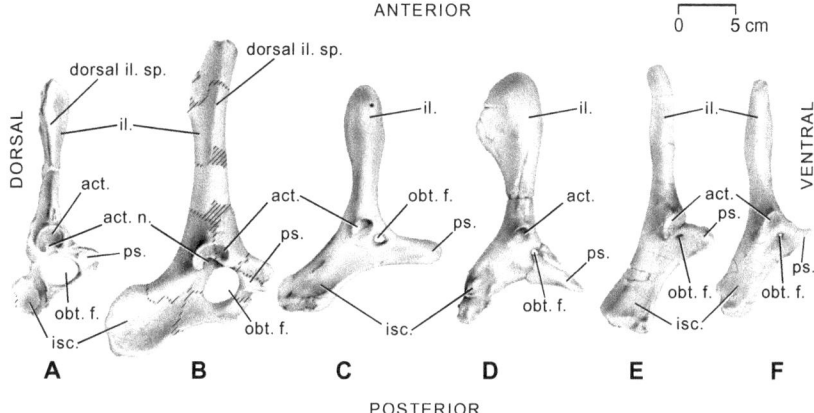

FIGURE 14.5   Right innominate bones of Paleogene African sirenians. A) *Protosiren fraasi* (SMNS 43976A, from Abel, 1904: plate 7, figure 1); B) *Protosiren smithae* (Domning and Gingerich, 1994, CGM 43392); C) *Eotheroides* sp. (Zalmout and Gingerich 2005 UM 97514); D) *Eotheroides* sp. (Zalmout and Gingerich 2005 UM 101219); E) *Eosiren libyca* (UM 101226); F) *Eosiren libyca* (CGM 29774).

ABBREVIATIONS: act., acetabulum; act. n., acetabular notch; dorsal il. sp. dorsal iliac spine; il., ilium; is., ischium; obt. f., obturator foramen; ps., pubis; ps. sym., pubic symphysis.

distinct crista intratemporalis is present, the parietal roof is usually bilaterally convex, and the ventral process of the jugal is positioned slightly abaft the dorsal process. Dental formula is 3.1.5.3/3.1.5.3.

Remarks This species is the best-known and best-documented dugongid from the late Priabonian Qasr el-Sagha Formation of the Fayum area.

EOSIREN STROMERI (Sickenberg, 1934)
Figure 14.2

Age and Occurrence Late Priabonian of Qasr El-Sagha Formation, north of Lake Qarun, Fayum area, Egypt (table 14.1, figure 14.2).

Diagnosis Frontals are much longer than parietals along midline. M2 is larger than M3. I1 is reduced although the premaxillary symphysis is heavy and enlarged. Anteroposterior length of the zygomatic process of the squamosal seems to be greater than in *E. libyca*.

Description According to Sickenberg (1934:130–140), *E. stromeri* is very similar in overall morphology to *E. libyca*. However, *E. stromeri* has a wider skull and a narrower palate than *E. libyca*, and an M2 that is larger than M1 and M3. Sickenberg (1934) also mentioned that the skull of this species is shorter than the skulls of *E. libyca*. Nevertheless, UM 100137, which has an enlarged M2 and narrow palate, has a skull length that may exceed 350 mm, compared to an observed maximum length of 305 mm in *E. libyca* (CGM-C.10054).

Remarks Illustrations and reconstructions in Sickenberg (1934) and Kretzoi (1941), showing the premaxillae separated from frontals in dorsal view, are erroneous. All examined specimens, including the holotypes of both *E. libyca* and *E. stromeri*, show that the premaxillae and frontals are always in contact. Minor differences between *E. libyca* and *E. stromeri* could reflect merely individual variation, but we provisionally consider the two species to be valid.

EOSIREN IMENTI Domning, Gingerich, Simons and
Ankel-Simons, 1994
Figures 14.2 and 14.4B

Age and Occurrence Rupelian red beds of the Jabal Qatrani Formation (Bown and Kraus, 1988), north of Lake Qarun, Fayum area, Egypt (table 14.1, figure 14.2).

Diagnosis Differs from *Eosiren libyca* in having more elongate skull and more overlap of premaxilla and frontal, narrower palate, separated nasals, no distinct crista intratemporalis, sharp and upraised temporal crests separated by broadly concave parietal roof, more anteriorly located ventral process of jugal, and loss of I2–3. Differs from *Halitherium taulannense* in having more overlap of premaxilla and frontal, separated nasals, no distinct crista intratemporalis, more anteriorly located ventral process of jugal, and loss of I2–3. Differs from *H. schinzii* in retaining canines and DP or P1.

Description *Eosiren imenti* is the most derived form of *Eosiren*. It has an elongate skull (400 mm in total length; see Domning el al., 1994, for complete skull measurements). The premaxillary symphysis is enlarged relative to the cranium, the masticating surface of the rostrum is trapezoidal, the nasal opening is expanded anteroposteriorly, the supraorbital process is well developed, and the cranial vault is square in cross section with smoothly concave roof and sharp edges. The preorbital process of the jugal is flattened against the maxilla, and the jugal is separated from the premaxilla by a small portion of the maxilla. The upper dental formula (based on the only specimen, CGM 40210) is 1.1.5?.3. Tentorium osseum, transverse sulcus, internal occipital protuberance, and bony falx cerebri are prominent. *Eosiren* and early *Halitherium* are closely similar, generalized halitheriines with moderately deflected rostra and with tusks that were persistently small in *Eosiren* but became medium sized in *Halitherium*.

Genus HALITHERIUM Kaup, 1838
Figure 14.2

Age All the supposed African records of *Halitherium* are said to be Miocene, but this genus is not validly recorded after the Oligocene (table 14.1, figure 14.2).

African Occurrence All supposed African specimens are here considered to be indeterminate sirenians, with the exception of a dugongid from Madagascar that is potentially determinable (see later discussion).

Description Halitherium spp. are generalized halitheriines with moderate rostral deflection and a subconical tusk with an alveolus roughly one-half the length of the premaxillary symphysis. Single-rooted permanent premolars 2–4/2–4 are retained.

Remarks Halitherium is the common Oligocene sirenian of Europe (Lepsius, 1882; Spillmann, 1959). As the senior available generic name proposed for a fossil sirenian, *Halitherium* has also served as a wastebasket name for a variety of Eocene to Pliocene specimens, many fragmentary and indeterminable. Today it nominally comprises three valid Old World species, all European: *H. taulannense* Sagne, 2001 (late Eocene); the type species *H. schinzii* (Kaup, 1838)(early Oligocene); and *H. christolii* Fitzinger, 1842 (late Oligocene). Even the last of these is conceivably assignable to *Metaxytherium*. Specimens referred to *Halitherium* have also been found in South Carolina (USA) and Puerto Rico.

Genus *METAXYTHERIUM* Christol, 1840 (= *Felsinotherium* Capellini, 1872)
METAXYTHERIUM sp.
Figure 14.2

Age Early Miocene (Burdigalian).

African Occurrence Jabal Zaltan, Libya (table 14.1, figure 14.2), yielded fragments of two skulls described in an unpublished doctoral dissertation (Heal, 1973:141). Its age would make this animal a contemporary of the European *M. krahuletzi* Depéret, 1895, but the taxonomic assignment requires corroboration (Domning and Pervesler, 2001:41).

Description This genus is characterized in part by absence of permanent premolars, a rather strongly deflected rostrum, long and thin nasal processes of the premaxillae, a flat or convex frontal roof, and (in species earlier than the latest Miocene) small subconical tusks.

Remarks Metaxytherium is the most common fossil sirenian in the Neogene of Europe, where it is represented by four chronospecies (Domning and Thomas, 1987). It also occurs in the New World and is probably derived from *Halitherium*.

METAXYTHERIUM SERRESII (Gervais, 1847)

Partial Synonymy Halitherium serresii Gervais, 1847; *Felsinotherium serresii* Zigno, 1878.

Age Late Miocene–early Pliocene.

African Range North Africa (Sahabi, Libya; ?Dar bel Hamri, Morocco)

Diagnosis Differs from both earlier and later European species of the genus in its smaller body size, and in having a tusk alveolus roughly one-half the length of the premaxillary symphysis (i.e., longer than in earlier species but smaller than in the later, middle Pliocene one). *M. serresii* is also intermediate in degree of reduction of the supracondylar fossa.

Description Remains of *Metaxytherium serresii* from Sahabi include: partial skulls, cranial elements, cheek teeth, mandibles, and some postcranial elements (Domning and Thomas, 1987). The rostrum of this species is strongly deflected downward (50°–75°), and incisor alveoli extend about half the length of the premaxillary symphysis. Nasals are reduced compared to *Halitherium*. Frontals are relatively flat, and supraorbital process has a distinct posterolateral corner. Palatine extends forward on the palate to about the level of the

front of M1 and rear edge of the zygomatic-orbital bridge. The mandible has a deep and strongly arched corpus.

The manubrium has a broad anterior tongue that is expanded slightly at front; the xiphisternum is wide anteriorly with a small demifacet. Innominates show apparent sexual dimorphism (Domning and Thomas, 1987); supposed females have long and narrow innominates while supposed male innominates have broad ischiac and pubic regions.

Remarks The reduced body size of this species, later reversed in the lineage, is interpreted as ecophenotypic dwarfing related to the late Tortonian-Messinian salinity crises in the Mediterranean Basin (Bianucci et al., 2008). The type locality of this species (Montpellier, France) is its latest-known occurrence (early Pliocene); the earliest occurrence is in Calabria, Italy (latest Tortonian; Carone and Domning, 2007). The relatively abundant Sahabi material is now considered early Messinian rather than post-Messinian. The Moroccan record is based only on a lower molar and a skullcap (Ennouchi, 1954); their age and identity are uncertain.

Subfamily DUGONGINAE Gray, 1821
(including Rytiodontinae Abel, 1914)
Genus *RYTIODUS* Lartet, 1866

Diagnosis According to Abel (1928), *Rytiodus* has a strongly deflected rostrum, a large lacrimal lacking any duct, large flattened incisor tusks, and an unreduced and complex M3.

RYTIODUS sp. nov. Heal, 1973
Figure 14.2

Age Early Miocene (Burdigalian).

African Occurrence Jabal Zaltan, Libya (table 14.1, figure 14.2).

Diagnosis The premaxillary rami abut against the nasals; the lacrimal is triangular.

Description Highly derived dugongine with short, strongly deflected rostrum and large, self-sharpening bladelike tusks with paper-thin enamel on the medial surface. Nasal processes of premaxillae short, thick, and abut against triangular nasal bones. Frontal roof concave. Temporal crests pronounced, concave laterad, meet in midline. Sagittal length of skull roof approximately 26 cm; length of m3 = 29.5 mm. Other character states listed by Bajpai and Domning (1997, table 3).

Remarks Described in an unpublished doctoral dissertation (Heal, 1973) and represented by good cranial material. The referral to *Rytiodus*, a poorly known dugongine from the Aquitanian of France, needs corroboration, but this is at least a closely allied and conceivably descendant form. Delfortrie's (1880: plate 5) restoration of the type species *R. capgrandi* with a straight, undeflected rostrum is incorrect, as shown by the strongly downturned anterior end of its maxilla.

Indeterminate Sirenia of Assorted Ages

At this point may be listed several African occurrences of Tertiary sirenians that, unfortunately, do not yet include diagnostic material (table 14.1, figure 14.2).

EOCENE

Isolated cheek teeth and other remains have been collected from the Carcar Series near Callis, Somalia (Savage, 1977; not near Mogadishu as stated; Dr. A. Azzaroli, *in litt.*, 24 September 1986), and further fragments from the middle Eocene Nautilus Beds southeast of Berbera, Somalia (Macfadyen, 1952). Upper Eocene beds at Dor el Talha, Libya, have also yielded rib fragments (Savage, 1971; Heal, 1973).

Recent investigations in the Cenozoic deposits of Madagascar (Samonds et al., 2009) yielded an Eocene sirenian, known from a fairly complete skull, that represents a new species, *Eotheroides lambondrano*.

OLIGOCENE

A tusk, possibly sirenian, has been reported from Bedeil, Somalia (Savage, 1969). A skeleton has been found at Djebel ech Cherichira, Tunisia (Savage, 1969).

MIOCENE

Dugongid teeth have been collected from the Serravallian Beglia Formation, Bled ed Douarah, Tunisia (Robinson and Black, 1969). Three other, supposedly Miocene records from the African region have been referred to *Halitherium*, but none is sufficiently diagnostic to justify assignment to this Eocene and Oligocene north-Tethyan genus:

1. Gervais (1872:341) records ribs of "*Halithérium*" from "dépôts à *Carcharodon megalodon*" at Chalouf (= El Shallûfa), Isthmus of Suez, received by the Paris Museum. He also (p. 352) alludes to other ribs from Lower Egypt cited by Blainville. In fact, Blainville (1840:43, 51) was originally inclined to refer the fragmentary vertebrae and ribs in question to a pinniped, but later (1844:119–120) concluded that they were sirenian. These remains, from the "left bank of the Nile Valley" and of uncertain age and unknown present location, appear to be the earliest-recorded sirenian remains from Africa.

2. A fragmentary skull and skeleton from Makamby Island on the northwest coast of Madagascar were recorded as *Halitherium* sp. by Collignon and Cottreau (1927), who considered the deposits Miocene (Burdigalian to Helvetian). They supposed that the latest European record of *Halitherium* was "*H.*" *bellunense* from the basal Burdigalian of Italy, but the latter species is probably a dugongine. The Madagascar skullcap is elongated, with temporal crests meeting in the midline, somewhat resembling that of the early Oligocene *H. schinzii*, but its apparently late date indicates a different generic referral. The published drawing appears to show a concave frontal roof, suggesting that this animal may in fact be a dugongine. It needs to be compared with the *Rytiodus* from Libya (see earlier discussion).

3. Ribs of probable Burdigalian age from Malembe, Angola, were referred to *Halitherium*(?) sp. (Dartevelle, 1935), but the only tenable identification is Sirenia indet.

General Discussion

Sirenians first appear in the Eocene, and the abundance of their remains in the middle and upper Eocene of North Africa led earlier writers to label that area (or even the Fayum in particular) as the center of origin of the group. This was corroborated by recognition of anatomical similarities between sirenians and other characteristically African groups, such as proboscideans and hyracoids. Now, however, it seems more prudent to be less geographically specific. The earliest and most primitive known sirenian fossils are from the late early or early middle Eocene of Jamaica (Savage et al., 1994; Domning, 2001a), with possible middle Eocene records from Israel (Goodwin et al., 1998) and Jordan (Zalmout et. al. 2003a); and the presently known distribution of Eocene sirenians from the Caribbean to the East Indies (Domning et al., 1982; Ivany et al., 1990; Domning 2001a) points to "the shores of the Tethyan Seaway, probably in the Old World," as the specification of choice for a "center" of origin. Significantly, the marine angiosperms (seagrasses) also show signs of an originally Tethyan distribution, although they entered the water during the Cretaceous, well before the appearance of sirenians (Hartog, 1970).

Today, both morphological and molecular evidence support the concept of an Afro-Asian Tethytheria (Sirenia + Proboscidea, plus the extinct Embrithopoda, Desmostylia, and Anthracobunidae). Indeed, Tethytheria is the best-supported supraordinal monophyletic grouping among the ungulates. Molecular (though not morphological) support is even stronger for the concept of Paenungulata (Tethytheria + Hyracoidea). The interordinal relationships within these groupings, however, are not well resolved, nor are the (presumably Paleocene) origins of the groupings themselves, due to paucity of primitive fossils for some of the orders (Gheerbrant et al., 2005). Surprisingly, molecular studies strongly support placement of the Paenungulata within the Afrotheria rather than the ungulates. If this proves true, we will have returned to thinking of an African origin for sirenians.

Sagne (2001) has pointed out a persistent zoogeographic division between sirenian faunas on the north shore of the Paleogene Tethys (*Sirenavus, Prototherium, Halitherium*) and the south shore (*Protosiren, Eotheroides, Eosiren*). Although these and other Eocene sirenians still require extensive taxonomic revision, a picture seems to be emerging of largely separate evolutionary histories on either side of Tethys during this period, with the northern lineage leading to *Halitherium* being distinct from the African *Eosiren* at least as early as the Priabonian.

Fossil Sirenia from Africa have been collected from shallow marine, lagoonal, estuarine, deltaic, and riverine environments. The African landmass was changing during the early and middle Cenozoic; the closure of the Tethyan Sea in the north, compression and formation of the Syrian Arc in the northeast, and opening of the Red Sea in the east, were undoubtedly responsible for the diversity of the ecological settings. The result was a notable diversification of the African (Tethyan) lineages. Climate and eustacy undoubtedly played a role in this as well. *Protosiren, Eotheroides*, and *Eosiren* are known from middle and late Eocene and in one case Oligocene beds. By the late Oligocene and early Miocene, the African continent had its final form, with a notable drop in sea level and domination of deltaic and riverine environments (Jabal Qatrani and Moghara formations in Egypt and Sahabi Formation in Libya). Generally speaking, these are freshwater deposits with minor pulses of marine incursion. Isotopic analysis of dental enamel from contemporaneous Eocene sirenians in Egypt (Clementz et al., 2006) has produced isotopic signatures of marine settings, suggesting that these marine herbivores were adapted to a predominantly marine seagrass-based diet. These authors concluded that the low variation in the $\delta^{13}C$ values between these Eocene sirenians suggests that their dietary preferences were highly focused.

The ecology and geography of Recent Sirenia partly reflect the Recent distribution of seagrasses: *Dugong* and some *Trichechus* both depend largely on this food source, and the majority of extinct sirenians evidently did so as well. Seagrass fossils were found in association with sirenian remains from the Priabonian of the Fayum area and comprised two genera: *Thalassodendron* and *Cymodocea* (Zalmout and Gingerich, 2004). This is the only record of fossil seagrass from Africa, and it surely does not give us an adequate picture of the Eocene diversity of these important marine plants in the Tethyan realm, where they seem to have originated.

Summary

Protosirenidae, Dugongidae, and Trichechidae are the only sirenian families that are known to have lived in the near-shore habitats of the African continent, although it is not unlikely that the Prorastomidae were once represented there as well. The Dugongidae account for more than 85% of Africa's sirenian fossil record. No trichechid fossils have been found in Africa at all, but the living West African manatee probably arrived there from the New World in the late Pliocene or Pleistocene.

The abundance of Eocene sirenians in northeastern Africa (now the Libyan Desert) gives us our best sample to date of the diversity of sirenians that once inhabited the Tethyan Seaway (including Southeast Asia, South Europe, and North America). For the remainder of the Cenozoic the sirenian fossil record in Africa is much less complete, but Miocene records of *Rytiodus* and *Metaxytherium* indicate continuing faunal connections between Europe and North Africa. Further collecting in marine Tertiary strata of Africa will undoubtedly expand the faunal list of African Sirenia.

African Sirenia provide some of the best documentation of reduction and loss of the hindlimbs in formerly terrestrial vertebrates as a progressive adaptation for aquatic life in coastal and off-shore environments. Preserved pelvic girdles and limb bones of Protosirenidae and Dugongidae show that there was a gradual reduction in their size with loss of some features associated with hindlimb functions. Unlike the most primitive sirenian from the middle Eocene of Jamaica (*Pezosiren portelli* Domning, 2001a, 2001b) with its quadrupedal body form, middle-late Eocene *Protosiren* from Egypt show intermediate characteristics between semiaquatic and fully aquatic forms. Among Eocene sirenians, *Eotheroides* and *Eosiren* have the most reduced pelvic and femoral features, including: reduced length of the ischium and ilium, a diminutive obturator foramen, and unfused and distinctly separated left and right pubic bones that must have connected to one another by ligaments or cartilage, presaging the complete loss of function in the hindlimbs. Later sirenians (Oligocene to Recent) show further reduction of the pelvis, as seen for example in the innominates of the Sahabi *Metaxytherium* (Domning and Thomas, 1987).

ACKNOWLEDGMENTS

We thank the Cairo Geological Museum staff for granting us access to their collection. The National Geographic Society and National Science Foundation have been generous in supporting fieldwork during which most of the material reported and illustrated here was collected. Most of the Fayum Eocene fossils at the University of Michigan were carefully prepared by William J. Sanders. We thank Bonnie Miljour for the fine drawings in figures 14.3, 14.4, and 14.5. We extend our thanks to Lars Werdelin and William J. Sanders for their editorial guidance; also thanks to anonymous reviewers for the constructive suggestions.

Literature Cited

Abel, O. 1904. Die Sirenen der mediterranen Tertiärbildungen Österreichs. *Abhandlungen der Kaiserlich-Königlichen Geologischen Reichsanstalt (Wien)* 19:1–223.
——. 1907. Die Stammesgeschichte der Meeressäugetiere. *Meereskunde* 1:1–36.
——. 1913. Die eozänen Sirenen der Mittelmeerregion. Erster Teil: Der Schädel von *Eotherium aegyptiacum*. *Palaeontographica* 59:289–360 (title page bears date of 1912).
——. 1919. *Die Stämme der Wirbeltiere: Vereinigung Wissenschaftlicher Verleger*. Walter de Gruyter, Berlin, 914 pp.
——. 1928. Vorgeschichte der Sirenia; pp. 475–504 in M. Weber (ed.), *Die Säugetiere. Einführung in die Anatomie und Systematik der recenten und fossilen Mammalia*. Fischer, Jena.
Andrews, C. W. 1902. Preliminary note on some recently discovered extinct vertebrates from Egypt (Part III). *Geological Magazine* 9:291–295.
——. 1906. *A Descriptive Catalogue of the Tertiary Vertebrata of the Fayum, Egypt*. British Museum (Natural History), London, 324 pp.
Bajpai, S., and D. P. Domning. 1997. A new dugongine sirenian from the early Miocene of India. *Journal of Vertebrate Paleontology* 17:219–228.
Bertram, G. C. L., and C. K. R. Bertram. 1973. The modern Sirenia: their distribution and status. *Biological Journal of the Linnean Society* 5:297–338.
Bianucci, G., G. Carone, D. P. Domning, W. Landini, L. Rook, and S. Sorbi. 2008. Peri-Messinian dwarfing in Mediterranean *Metaxytherium* (Mammalia: Sirenia): Evidence of habitat degradation related to the Messinian Salinity Crisis; pp. 145–157 in N. T. Boaz, A. El-Arnauti, P. Pavlakis, and M. J. Salem (eds.), *Circum-Mediterranean Geology and Biotic Evolution During the Neogene Period. The Perspective from Libya*. Garyounis Scientific Bulletin, Special Issue 5.
Blainville, H. M. D. de. 1840. *Ostéographie, Livre 7, Des Phoques (G. Phoca, L.)*. Arthus Bertrand, Paris.
——. 1844. *Ostéographie, Livre 15, Des Lamantins (Buffon), (Manatus, Scopoli), ou gravigrades aquatiques*. Arthus Bertrand, Paris.
Boaz, N. T., A. El-Arnauti, J. Agusti, R. L. Bernor, P. Pavlakis, and L. Rook. 2004. Temporal, lithostratigraphic, and biochronologic setting of As Sahabi Formation, north-central Libya. *Abstracts, Third Symposium on Geology of East Libya, Benghazi, Libya, Nov. 21–23, 2004*, p. 21.
Bown, T. B., and M. J. Kraus. 1988. Geology and paleoenvironment of the Oligocene Jebel Qatrani Formation and adjacent rocks, Fayum Depression, Egypt. *United States Geological Survey, Professional Paper* 1452:1–60.
Carone, G., and D. P. Domning. 2007. *Metaxytherium serresii* (Mammalia: Sirenia): First pre-Pliocene record, and implications for Mediterranean paleoecology before and after the Messinian salinity crisis. *Bollettino della Societa Paleontologica Italiana* 46:55–92.
Clementz, M. T., A. Goswami, P. D. Gingerich, and P. L. Koch. 2006. Isotopic records from early whales and sea cows: Contrasting patterns of ecological transition. *Journal of Vertebrate Paleontology* 26:355–370.
Collignon, M., and J. Cottreau. 1927. Paléontologie de Madagascar. XIV. Fossiles du Miocène marin. *Annales de Paléontologie* 16:135–171.
Dartevelle, E. 1935. Les premiers restes de mammifères du tertiaire du Congo: La faune Miocène de Malembe. *Comptes-Rendus du Congrès National des Sciences, Bruxelles* 1:715–720.
Delfortrie, E. 1880. Découverte d'un squelette entier de *Rytiodus* dans le falun Aquitanien. *Actes de la Societe Linnéenne de Bordeaux* 34:131–144.
Dollo, L. 1889. Première note sur les siréniens de Boom. *Bulletin de la Société Belge de Géologie, de Paléontologie et d'Hydrologie* 3:415–421.
Domning, D. P. 1977. An ecological model for Late Tertiary sirenian evolution in the North Pacific Ocean. *Systematic Zoology* 25:352–362.
——. 1978a. Sirenia; pp. 573–581 in V. J. Maglio and H. B. S. Cooke (eds.), *Evolution of African Mammals*. Harvard University Press, Cambridge.

———. 1978b. Sirenian evolution in the North Pacific Ocean. *University of California Publications in Geological Sciences* 118:1–176.

———. 1982. Evolution of manatees: A speculative history. *Journal of Paleontology* 56:599–619.

———. 1994. A phylogenetic analysis of the Sirenia. *Proceedings of the San Diego Society of Natural History* 29:177–189.

———. 1995. What do we know about the evolution of the dugong? Pp. 23–24 in *Mermaid Symposium: First International Symposium on Dugong and Manatees, November 15–17, 1995, Toba, Mie, Japan: Abstracts.* Toba, Toba Aquarium.

———. 1996. Bibliography and Index of the Sirenia and Desmostylia. *Smithsonian Contributions to Paleobiology* 80:1–611.

———. 2000. The readaptation of Eocene sirenians to life in water; in J.-M. Mazin, V. de Buffrénil, and P. Vignaud (eds.), *Secondary Adaptation of Tetrapods to Life in Water. Historical Biology (Special Issue)* 14:115–119.

———. 2001a. The earliest known fully quadrupedal sirenian. *Nature* 413:625–627.

———. 2001b. Sirenians, seagrasses, and Cenozoic ecological change in the Caribbean; in W. Miller III and S. E. Walker (eds.), *Cenozoic Paleobiology: The Last 65 Million Years of Biotic Stasis and Change. Palaeogeography, Palaeoclimatology, Palaeoecology* 166:27–50.

———. 2005. Fossil Sirenia of the West Atlantic and Caribbean region. VII. Pleistocene *Trichechus manatus* Linnaeus, 1758. *Journal of Vertebrate Paleontology* 25:685–701.

Domning, D. P., and V. de Buffrénil. 1991. Hydrostasis in the Sirenia: Quantitative data and functional interpretations. *Marine Mammal Science* 7:331–368.

Domning, D. P., and P. D. Gingerich. 1994. *Protosiren smithae*, new species (Mammalia, Sirenia), from the late middle Eocene of Wadi Hitan, Egypt. *Contributions from the Museum of Paleontology. University of Michigan* 29:69–87.

Domning, D. P., P. D. Gingerich, E. L. Simons, and F. A. Ankel-Simons. 1994. A new early Oligocene dugongid (Mammalia, Sirenia) from Fayum Province, Egypt. *Contributions from the Museum of Paleontology* (University of Michigan) 29:89–108.

Domning, D. P., and L. Hayek. 1984. Horizontal tooth replacement in the Amazonian manatee *(Trichechus inunguis). Mammalia* 48:105–127.

Domning, D. P., G. S. Morgan, and C. E. Ray. 1982. North American Eocene sea cows (Mammalia: Sirenia). *Smithsonian Contribution to Paleobiology* 52:1–69.

Domning, D. P., and P. Pervesler. 2001. The osteology and relationships of *Metaxytherium krahuletzi* Depéret, 1895 (Mammalia: Sirenia). *Abhandlungen der Senckenbergischen Naturforschenden Gesellschaft* 553:1–89.

Domning, D. P., and H. Thomas. 1987. *Metaxytherium serresii* (Mammalia: Sirenia) from the Lower Pliocene of Libya and France: A reevaluation of its morphology, phyletic position, and biostratigraphic and paleoecological significance; pp. 205–232 in N. T. Boaz, A. El-Arnauti, A.W. Gaziry, J. de Heinzelin, and D. D. Boaz (eds.), *Neogene Paleontology and Geology of Sahabi.* Liss, New York.

Dutton, T. P. 2004. Dugong (*Dugong dugon*) population trends in the Bazaruto Archipelago National Park, Mozambique, 1990–2003. *Sirenews* (IUCN/SSC Sirenia Specialist Group) 41:12–14.

Ennouchi, E. 1954. Un sirénien, *Felsinotherium* cf. *serresi*, à Dar bel Hamri. *Service Géologique du Maroc* 121:77–82.

Filhol, Henri. 1878. Note sur la découverte d'un nouveau mammifère marin (*Manatus coulombi*) en Afrique, dans les carrières de Mokattan près du Caire. *Bulletin de la Société philomatique, Paris.* II (7):124–125.

Fraas, O. 1867. *Aus dem Orient: Geologische Beobachtungen am Nil, auf der Sinai-Halbinsel und in Syrien.* Ebner and Seubert, Stuttgart, 222 pp.

Gervais, P. 1872. Travaux récents sur les sirénides vivants et fossiles (analyse des publications de Mm. Van Beneden, E. Lartet, Delfortrie, Capellini, etc.). *Journal de Zoologie* 1:332–353.

Gheerbrant, E., D. P. Domning, and P. Tassy. 2005. Paenungulata (Sirenia, Proboscidea, Hyracoidea, and relatives); pp. 84–105 in K. D. Rose and J. D. Archibald (eds.), *The Rise of Placental Mammals: Origin and Relationships of the Major Extant Clades.* Johns Hopkins University Press, Baltimore.

Gingerich, P. D., M. Arif, M. A. Bhatti, H. A. Raza, and S. M. Raza. 1995. *Protosiren* and *Babiacetus* (Mammalia, Sirenia and Cetacea) from the middle Eocene Drazinda Formation, Sulaiman Range, Punjab (Pakistan). *Contributions from the Museum of Paleontology* (University of Michigan) 29:331–357.

Gingerich, P. D., H. Cappetta, and M. Traverse. 1992. Marine mammals (Cetacea and Sirenia) from the middle Eocene of Kpogamé-Hahotoé in Togo (abstract). *Journal of Vertebrate Paleontology 12* (suppl. to no. 3):29A–30A.

Gingerich, P. D., D. P. Domning, C. E. Blane, and M. Uhen. 1994. Cranial morphology of *Protosiren fraasi* (Mammalia, Sirenia) from the Middle Eocene of Egypt: A new study using computed tomography. *Contributions from the Museum of Paleontology* (University of Michigan) 29:41–67.

Goodwin, M. B., D. P. Domning, J. H. Lipps, and C. Benjamini. 1998. The first record of an Eocene (Lutetian) marine mammal from Israel. *Journal of Vertebrate Paleontology* 18:813–815.

Hartog, C. den. 1970. The Sea-Grasses of the World. *Koninklijke Nederlandse Akademie van Wetenschappen Verhandelingen, Afdeling Natuurkunde* 59:1–275.

Hatt, R. T. 1934. A manatee collected by the American Museum Congo Expedition, with observations on the Recent manatees. *Bulletin of the American Museum of Natural History* 66:533–566.

Heal, 1973. Contributions to the study of sirenian evolution. Unpublished PhD dissertation, University of Bristol, England.

Heinzelin, J. de, and A. El-Arnauti. 1987. The Sahabi Formation and related deposits; pp. 1–21 in N. T. Boaz, A. El-Arnauti, A. W. Gaziry, J. de Heinzelin, and D. D. Boaz (eds.), *Neogene Paleontology and Geology of Sahabi.* Liss, New York.

Ivany, L., R. W. Portell, and D. S. Jones. 1990. Animal-plant relationships and palaeobiogeography of an Eocene seagrass community from Florida. *Palaios* 5:244–258.

Kaiser, H. E. 1974. *Morphology of the Sirenia: A Macroscopic and X-ray Atlas of the Osteology of Recent Species.* Karger, New York, 76 pp.

Kretzoi, M. 1941. *Sirenavus hungaricus* n.g. n. sp., ein neuer Prorastomide aus dem Mitteleozän (Lutetium) von Felsögalla in Ungarn. *Annales Musei Nationalis Hungarici, Pars Mineralogica, Geologica et Palaeontologica* 34:146–156.

Lepsius, G. R. 1882. *Halitherium Schinzi*, die fossile Sirene des Mainzer Beckens. Eine vergleichend-anatomische Studie. *Abhandlungen des Mittelrheinischen Geologischen Vereins* 1:1–200.

Macfadyen, W. A. 1952. Note on the geology of the Daban area and the localities of the described nautiloids. In O. Haas and A. K. Miller, Eocene Nautiloids of British Somaliland. *Bulletin of the American Museum of Natural History* 99:347–349.

Marshall, C. D., L. A. Clark, and R. L. Reep. 1998. The muscular hydrostat of the Florida manatee *(Trichechus manatus latirostris)*: A functional morphological model of perioral bristle use. *Marine Mammal Science* 14:290–303.

Owen, R. 1875. On fossil evidences of a sirenian mammal (*Eotherium aegyptiacum*, Owen) from the Nummulitic Eocene of the Mokattam Cliffs, near Cairo. *Quarterly Journal of the Geological Society of London* 31:100–105.

Petit, G. 1927. Nouvelles observations sur la pêche rituelle du dugong à Madagascar. *Bulletins et Mémoires de la Société d'Anthropologie de Paris* 7:246–250.

Preen, A. R. 1995. Diet of dugongs: Are they omnivores? *Journal of Mammalogy* 76:163–171.

Robinson, P., and C. C. Black. 1969. Note préliminaire sur les vertébrés fossiles du Vindobonien (formation Béglia), du Bled Douarah, Gouvernorat de Gafsa, Tunisie). *Notes du Service Géologique (Tunisie)* 31:67–70.

Sagne, C. 2001. La diversification des siréniens à l'Éocène (Sirenia, Mammalia): Étude morphologique et analyse phylogenetique du sirénien de Taulanne, *Halitherium taulannense.* Unpublished PhD dissertation, Muséum National d'Histoire Naturelle, Paris, 2 vols.

Samonds, K. E., I. S. Zalmout, M. T. Irwin, D. W. Krause, R. R. Rogers, and L. L. Raharivony. 2009. *Eotheroides lambondrano*, new middle Eocene seacow (Mammalia, Sirenia) from the Mahajanga Basin, northwestern Madagascar. *Journal of Vertebrate Paleontology* 29:1233–1243.

Savage, R. J. G. 1967. Early Miocene mammal faunas of the Tethyan region. *Systematics Association Publication* 7:247–282.

———. 1969. Early Tertiary mammal locality in southern Libya. *Proceedings of the Geological Society of London* 1657:167–171.

———. 1971. Review of the fossil mammals of Libya. In *Symposium on the Geology of Libya, University of Libya,* pp. 215–225.

———. 1977. Review of early Sirenia. *Systematic Zoology* 25:344–351.

Savage, R. J. G., D. P. Domning, and J. G. M. Thewissen. 1994. Fossil Sirenia of the West Atlantic and Caribbean region. V. The most primitive known sirenian, *Prorastomus sirenoides* Owen, 1855. *Journal of Vertebrate Paleontology* 14:427–449.

Savage, R. J. G., and B. S. Tewari. 1977. A new sirenian from Kutch India. *Journal of the Palaeontological Society of India* 20:216–218.

Savage, R. J. G., and M. E. White. 1965. Exhibit: Two mammal faunas from the early Tertiary of central Libya. *Proceedings of the Geological Society of London* 1623:89–91.

Sickenberg O. 1934. Beiträge zur Kenntnis Tertiärer Sirenen. I. Die Eozänen Sirenen des Mittelmeergebietes; II. Die Sirenen des Belgischen Tertiärs. *Mémoires du Museé Royal d'Histoire Naturelle de Belgique* 63:1–352.

Siegfried, P. 1965. *Anomotherium langewieschei* n. g. n. sp. (Sirenia) aus dem Ober-Oligozän des Dobergs bei Bunde/Westfalen. *Paläontographica, Abt. A* 124:116–150.

———. 1967. Das Femur von *Eotheroides libyca* (Owen)(Sirenia). *Paläontologisches Zeitschrift* 41:165–172.

Simpson, G. G. 1945. The principles of classification and a classification of mammals. *Bulletin of the American Museum of Natural History* 85:1–350.

Spillmann, F. 1959. Die Sirenen aus dem Oligozän des Linzer Beckens (Oberösterreich), mit Ausführungen über "Osteosklerose" und "Pachyostose." *Österreichische Akademie der Wissenschaften, Mathematisch-Naturwissenschaftliche Klasse, Denkschriften* 110:1–68.

Wight, A. W. R. 1980. Paleogene vertebrate fauna and regressive sediments of Dur at Talhah, southern Sirt Basin, Libya; pp. 309–325 in M. J. Salem and M. T. Busrewil (eds.), *The Geology of Libya*, vol. 1. Academic Press, London.

Zalmout, I. S., and P. D. Gingerich. 2004. Paleobiology of Tethyan seacows (Sirenia, Mammalia) from the marine Eocene Fayum Province, Egypt (abstract). p. B-31 in K. Bice, M.-P. Aubry, and K. Ouda (eds.), *Climate and Biota of the Early Paleogene: V. Abstract and Program Book.* Luxor.

———. 2005. Eocene Sirenia of Egyptian Tethys: Aquatic Adaptations. *Journal of Vertebrate Paleontology* 25 (suppl. to no. 3):133A.

Zalmout, I. S., P. D, Gingerich, H. Mustafa, A. Smadi, and A. Khammash. 2003a. Cetacea and Sirenia from the Eocene Wadi Esh-Shallala Formation of Jordan. *Journal of Vertebrate Paleontology* 23 (suppl. to no. 3):113A.

Zalmout, I. S., M. Ul-Haq, and P. D. Gingerich. 2003b. New species of *Protosiren* (Mammalia, Sirenia) from the early middle Eocene of Balochistan (Pakistan). *Contributions from the Museum of Paleontology* (University of Michigan) 31:79–87.

Zdansky, O. 1938. *Eotherium majus* sp.n., eine neue Sirene aus dem Mitteleozän von Aegypten. *Palaeobiologica* 6:429–434.

FIFTEEN

Proboscidea

WILLIAM J. SANDERS, EMMANUEL GHEERBRANT,
JOHN M. HARRIS, HARUO SAEGUSA, AND CYRILLE DELMER

Proboscideans are represented in modern Africa by the savanna and forest elephants *Loxodonta africana* and *L. cyclotis*, whose sub-Saharan distribution is increasingly fragmented and threatened (Kingdon, 1997). These species and the Asian elephant *Elephas maximus* constitute the last remnants of a once-flourishing order that enjoyed its maximum diversity in the Miocene (Shoshani and Tassy, 1996; Todd, 2006). More advanced species are easily recognized as proboscideans by their large, projecting tusks, enormous, pneumatized crania with retracted nasal apertures, massive bodies and graviportal postcranial adaptations, but archaic forms are identifiable to the order only by more subtle features (Mahboubi et al., 1986; Gheerbrant et al., 2002, 2005). Together with the extinct embrithopods and desmostylians, as well as sirenians and hyraxes, proboscideans belong to a larger grandorder, the Paenungulata (Simpson, 1945). In turn, paenungulates, elephant shrews, tenrecs, chrysochlorids, and aardvarks are placed in the superordinal clade Afrotheria (Asher et al., 2003; Tabuce et al., 2008; Asher and Seiffert, this volume, chap. 46), primarily based on the results of molecular analyses (e.g., Springer et al., 1997, 2003; Stanhope et al., 1998; Murphy et al., 2001; Scally et al., 2001) and more tenuously on morphological criteria (Asher et al., 2003; Sánchez-Villagra et al., 2007; Seiffert, 2007; Tabuce et al., 2007b; Asher and Lehmann, 2008).

The molecular data have been interpreted as suggesting Cretaceous origins for Afrotheria and an early Paleocene divergence of Proboscidea (e.g., Eizirik et al., 2001; Springer et al., 2003). Such ancient divergence estimates, however, seem overly reliant on assumptions of molecular rate homogeneity and are discordant with the paleontological evidence (Allard et al., 1999). In fact, the oldest unequivocal fossil evidence of proboscideans, from North Africa, dates only to the latest Paleocene (Gheerbrant et al., 2002, 2003; Gheerbrant, 2009). Subsequent to this first appearance, Proboscidea remained an endemic Afro-Arabian order until the late Oligocene (Antoine et al., 2003), and nearly all of its important phylogenetic events and adaptive radiations appear to have initiated in this part of the Old World.

Since the last major review of African Proboscidea (Coppens and Beden, 1978, Harris, 1978, and Coppens et al., 1978 in Maglio and Cooke, 1978), a wealth of novel comparative morphological and phylogenetic studies (e.g., Tassy, 1981, 1982, 1988, 1990; Court, 1994a, 1994b; Shoshani, 1996; Pickford, 2001; Sanders, 2004; Gheerbrant et al., 2005) have established a more reliable systematic context of proboscidean relationships, and intensified fieldwork has greatly expanded the number of proboscidean taxa known from the continent (table 15.1). In addition, improvements in radiometric dating and magnetostratigraphy (Feibel, 1999; Ludwig and Renne, 2000) have refined temporal calibration of African Cenozoic geological sequences. As a result, it is now possible to construct a far more informed and comprehensive classification of the order (table 15.1) and more precise synopses of African proboscidean chronostratigraphy and geographic distribution (tables 15.2–15.6). Along with similar progress in paleoecology, biogeography, climatology, and tectonics, these advances facilitate an increasingly accurate recounting of the interplay between changes in the physical landscape and environment, faunal composition, and African proboscidean evolution covering a span of 55 million years.

Proboscideans are one of the oldest surviving and most speciose mammalian groups to have inhabited Africa. The present account documents their evolutionary history, from small, condylarth-like phosphatheres through a series of adaptive diversifications and extinctions to the elephants, the largest and ecologically most dominant extant terrestrial megaherbivores (Eltringham, 1992; Shoshani and Tassy, 1996). Among the advances covered are the identification of a new, late Paleogene cohort of basal taxa (phosphatheres, daouitheres, numidotheres) that has led to a revised definition of Proboscidea and altered conceptions of its ancestral condition; unequivocal recognition of barytheres, deinotheres, and moeritheres as proboscideans; new information about the early fossil records of moeritheres, deinotheres, mammutids, and gomphotheres that phylogenetically connects them with more archaic proboscidean taxa; and isotopic support for adaptive hypotheses about the evolutionary transformation of the craniodental apparatus in elephants. Based on this account, it is clear that the broad sweep of proboscidean evolution is essentially an African story, and that proboscideans epitomize African Mammalia.

TABLE 15.1
Classification and temporal distribution of Afro-Arabian proboscideans

Starred taxa (*) are those identified since the last major review of African Proboscidea (in Maglio and Cooke, 1978).

Abbreviations: e, early; l, late; m, middle; E, Eocene; M, Miocene; O, Oligocene; P, Pliocene; Pal, Paleocene Pl, Pleistocene; R, Recent.

?Order Proboscidea Illiger, 1811
Family . . . .?Phosphatheriidae Gheerbrant,
Sudre, Tassy, Amaghzaz,
Bouya, and Iarochène, 2005
Genus . . . .*Khamsaconus* Jaeger, Sigé, and
Vianey-Liaud, 1993*
Khamsaconus bulbosus Jaeger
et al., 1993* eE

Order Proboscidea Illiger, 1811

Suborder. . . . ."Plesielephantiformes"
Shoshani et al.,
2001a*

Family .incertae sedis Gheerbrant,
2009
Genus . . . .*Eritherium* Gheerbrant, 2009*
E. azzouzorum Gheerbrant,
2009* lPal
Family .Phosphatheriidae
Gheerbrant, Sudre, Tassy,
Amaghzaz, Bouya, and
Iarochène, 2005*
Genus . . . .*Phosphatherium* Gheerbrant,
Sudre, and Cappetta, 1996*
P. escuilliei Gheerbrant et al.,
1996* eE
Family .incertae sedis
Gen. et sp. indet., probably
nov. O'Leary, Roberts,
Bouare, Sissoko, and
Tapanila, 2006* eE
Family .incertae sedis, probably nov.
Gheerbrant et al.,
2005*
Genus . . . .*Daouitherium* Gheerbrant and
Sudre in Gheerbrant et al.,
2002*
D. rebouli Gheerbrant and
Sudre in Gheerbrant et al.,
2002* eE
Superfamily Barytherioidea Andrews, 1906
Family .Numidotheriidae Shoshani
and Tassy, 1992*
Genus . . . .*Numidotherium* Jaeger in
Mahboubi et al., 1986*
N. koholense Jaeger in
Mahboubi et al., 1986* eE
N. savagei Court, 1995*
(= *Arcanotherium savagei*
(Delmer, 2009)* lE or eO
Family .Barytheriidae Andrews, 1906
Genus . . . .*Barytherium* Andrews, 1901b
Barytherium sp. indet. (Birket
Qarun, Fayum), probably
nov.* lE
B. grave (Andrews, 1901a) lE

Suborder. .incertae sedis

Superfamily Moeritherioidea Andrews, 1906
Family . . . . .Moeritheriidae Andrews,
1906
Genus . . . .*Moeritherium* Andrews, 1901d
Moeritherium sp. indet.
(Birket Qarun, Fayum),
probably nov.* lE
M. chehbeurameuri Delmer,
Mahboubi, Tabuce, and
Tassy, 2006* lE
M. lyonsi Andrews, 1901d lE
M. trigodon Andrews, 1904b eO

Suborder. .incertae sedis

Superfamily Deinotherioidea Bonaparte, 1845

Family . . . . .Deinotheriidae Bonaparte,
1845
Subfamily. . . . . .Chilgatheriinae Sanders,
Kappelman, and
Rasmussen, 2004*
Genus . . . .*Chilgatherium* Sanders et al.,
2004*
C. harrisi Sanders et al.,
2004* lO
Subfamily. . . . . .Deinotheriinae Bonaparte,
1845
Genus . . . .*Prodeinotherium* Éhik, 1930
Prodeinotherium sp. indet.
Rasmussen and Gutiérrez,
2009* lO
P. hobleyi (Andrews, 1911)
e–mM
Genus . . . .*Deinotherium* Kaup, 1829
D. bozasi Arambourg, 1934a
lM–ePl

Suborder. .Elephantiformes Tassy, 1988

Family . . . . .Palaeomastodontidae
Andrews, 1906
Genus . . . .*Palaeomastodon* Andrews,
1901d
P. beadnelli Andrews, 1901d eO
P. sp. nov. A Sanders et al.,
2004* lO
P. sp. nov. B Sanders et al.,
2004* lO
Genus . . . .*Phiomia* Andrews and
Beadnell, 1902
P. serridens Andrews and
Beadnell, 1902 eO
P. major Sanders et al.,
2004* lO
Superfamily Elephantoidea Gray, 1821
Family . . . . .Mammutidae Hay, 1922
Genus . . . .*Losodokodon* Rasmussen and
Gutiérrez, 2009*
L. lodosokius Rasmussen and
Gutiérrez, 2009* lO

INSTITUTIONAL ABBREVIATIONS

BMNH, The Natural History Museum, London (formerly the British Museum [Natural History]); CGM, Cairo Geological Museum; DOG, field number, Dogali, Eritrea; DPC, Duke University Primate Center; EP, Tanzanian National Museums (Eyasi Plateau); KBA, Kanam, Kenya; KI, Uganda Museum, Kampala (Kisegi-Nyabusosi); KK, field designation, Kakesio, Tanzania; KNM, National Museums of Kenya: -AT, Aterir; -BC, Chemeron; -KP, Kanapoi; -LT, Lothagam; -ME, Meswa Bridge; -MI, Mwiti; -MP, Mpesida Beds; -NK, Lemudong'o; -RU, Rusinga, Kenya; -TH, Tugen Hills; L followed by a number series (e.g., L 124-1), Ethiopian National Museums (Middle Awash); NAP, Napak, Uganda; NK, Uganda Museum, Kampala (Nkondo-Kaiso area, Nyawiega); PQ-L followed by a number series (e.g., SAM-PQ-L 2562), Langebaanweg; SAM, Iziko South African Museum; UM, University of Michigan Museum of Paleontology; WM, Tanzanian National Museums (Wembere-Manonga).

DENTAL ABBREVIATIONS

C or c, upper or lower canine; DI, upper deciduous incisor; di, lower deciduous incisor; DP or dp, deciduous premolar (e.g., DP3 is the upper third deciduous premolar and dp3 is the lower third deciduous premolar); ET, enamel thickness; H, height; HI, hypsodonty index, H × 100/W; I or i, incisor (e.g., I2 is the upper second incisor and i2 is the lower second incisor); L, length; l., left; LF, lamellar frequency, number of loph(id)s or plates per 100 mm; M or m, molar (e.g., M1 is the upper first molar, and m1 is the lower first molar); P or p, premolar (e.g., P3 is the upper third premolar, and p3 is lower third premolar); r., right; W, width; x, as in x3x, denotes a tooth comprised of three lophs with an anterior and posterior cingulum; +, indicates a missing portion of a tooth, and that the original dimension was greater.

DENTAL DEFINITIONS

ABAXIAL CONELET The outer, main cone in each half-loph(id) (Tassy, 1996a)

ACCESSORY CENTRAL CONULES Enamel-covered pillars situated at the anterior and/or posterior faces of the loph(id)s or plates, or in the transverse valleys, partially blocking them centrally (Tobien, 1973b)

ADAXIAL CONELET(S) The inner, or meso-, conelet(s) in each half-loph(id) (Tassy, 1996a)

ANANCOIDY Alternation of paired half-loph(id)s, in which lingual half-loph(id)s are anterior to buccal half-loph(id)s (Tobien, 1973b)

CHEVRONING The arrangement of half-loph(id)s to occlusally form an anteriorly pointing V, or chevron (Tobien, 1975)

CHOERODONTY Occurrence of accessory tubercles within transverse valleys (Osborn, 1942)

CRESCENTOIDS Enamel crests running from the apices of abaxial conelets of pretrite half-loph(id)s to the bottom of transverse valleys, and ending near the middle axis of the crown (Tobien, 1975)

INTERMEDIATE MOLARS DP4/dp4, M1/m1, and M2/m2

POSTTRITE Refers to the less worn half of each loph(id), which is lingual in lower and buccal in upper molars (Vacek, 1877)

PRETRITE Refers to the more worn half of each loph(id), which is buccal in lower and lingual in upper molars (Vacek, 1877)

PTYCHODONTY Plication or infolding of enamel borders with grooving of the sides of the molars (Osborn, 1942)

TREFOIL A tripartite enamel wear figure of a half-loph(id) formed by the conelets and associated anterior and posterior accessory central conules

ZYGODONT CRESTS Enamel crests running from the apices of the abaxial conelets of the posttrite half-loph(id)s to the bottom of the transverse valleys, and ending near the middle axis of the crown (Tobien, 1975).

Systematic Paleontology

?Order PROBOSCIDEA Illiger, 1811
Family ?PHOSPHATHERIIDAE Gheerbrant, Sudre, Tassy, Amaghzaz, Bouya, and Iarochène, 2005
Genus *KHAMSACONUS* Jaeger, Sigé, and Vianey-Liaud in Sudre et al., 1993
KHAMSACONUS BULBOSUS Jaeger, Sigé, and Vianey-Liaud in Sudre et al., 1993
Figures 15.1A and 15.1B

Age and Occurrence Early Eocene, northern Africa (table 15.2).

Diagnosis Gheerbrant et al. (1998). Smallest known tethythere and possible proboscidean, with DP4 especially similar to DP4 in *Phosphatherium escuilliei* in its incipient bilophodont structure, absence of conules, and presence of a large postentoconule, but much smaller and of more primitive construction, including a lower, more bunodont crown, poorly advanced lophodonty, larger postentoconule, and more differentiated preparacrista.

Description Sudre et al. (1993); Gheerbrant et al. (1998). Only a diminutive DP4 (L = 4.66 mm; W = 3.78 mm; figures 15.1A, 15.1B) is known for this species. This tooth is molariform with inflated, blunt cusps, and a distinct buccal cingulum. The protoloph and metaloph are incipient and separated by a distinct interloph, which is continuous with a deeply notched entoflexus. The protocone and hypocone are subequal in size. The postentoconule is large, bulbous, and located behind the hypocone and metacone. A well-developed distocrista links the postentoconule and postmetacrista, and a sizable preparacrista is connected to the parastyle.

Remarks There is a strong resemblance between DP4 in *Khamsaconus* and *Phosphatherium* (Gheerbrant et al., 1998), particularly in the absence of conules, occlusal outline, development of an incipient protoloph and metaloph, absence of a lingual cingulum, and reduced postprotocrista. The presence of a distocrista is a proboscidean feature. Although cladistic analysis suggests a close affinity between it and *Phosphatherium* (Gheerbrant et al., 2005), *Khamsaconus* is very primitive and too poorly known for definite referral to Proboscidea, and the spatial arrangement of its enamel types lacks the synapomorphic condition of the earliest definitive proboscideans (Tabuce et al., 2007a).

Order PROBOSCIDEA Illiger, 1811
Suborder "PLESIELEPHANTIFORMES" Shoshani, Shoshani, and Tassy, 2001a

"Plesielephantiform" proboscideans derive from the early Paleogene of a geographically isolated Afro-Arabia (Coryndon and Savage, 1973; Maglio, 1978; Holroyd and Maas, 1994; Rögl, 1998; Harzhauser et al., 2002). This proposed suborder includes primitive proboscideans that have bilophodont molars, such as phosphatheres and barytherioids (Shoshani et al., 2001a), and if valid should also incorporate daouitheres,

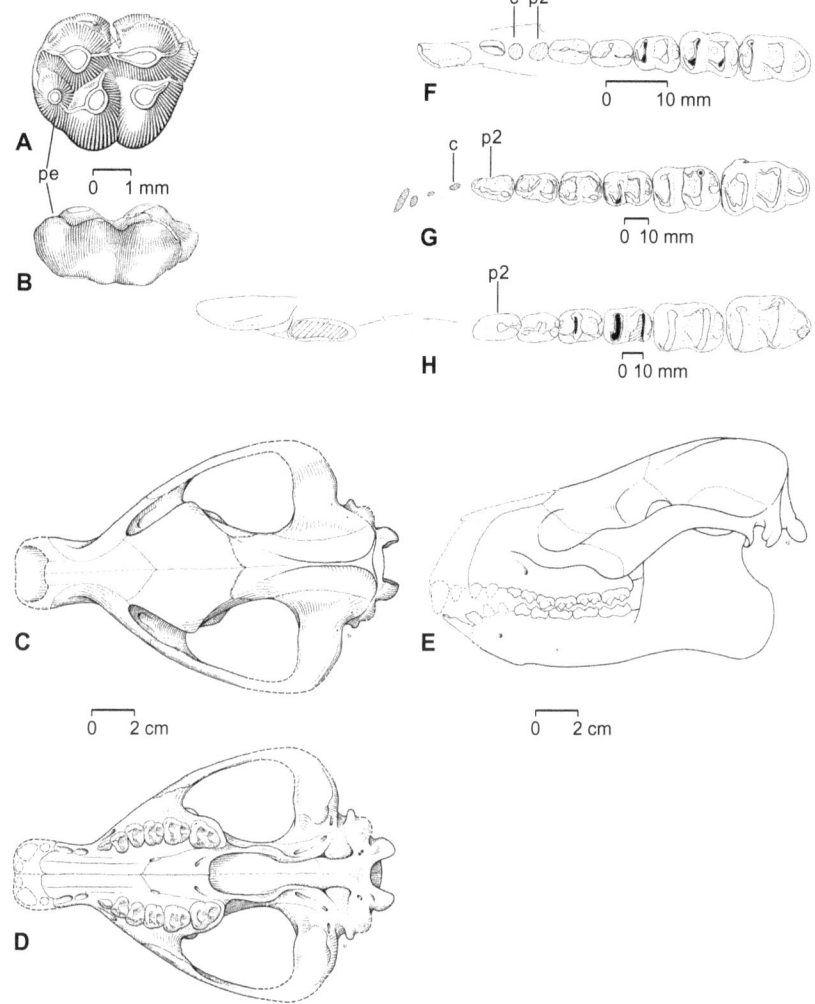

FIGURE 15.1 Aspects of ?proboscidean and "plesielephantiform" craniodental morphology. A) Right DP4, *Khamsaconus bulbosus*, occlusal view (Sudre et al., 1993: figure 2); anterior is to the left. B) Right DP4, *K. bulbosus*, lingual view. C) Cranial reconstruction, *Phosphatherium escuilliei*, dorsal view (Gheerbrant et al., 2005: figure 8A). D) Cranial reconstruction, *P. escuilliei*, ventral view (Gheerbrant et al., 2005: figure 8B). E) Skull reconstruction, *P. escuilliei*, lateral view (Gheerbrant et al., 2005: figure 8C). F) Reconstruction, left lower tooth series, *P. escuilliei*, occlusal view (Gheerbrant et al., 2005: figure 14). G) Reconstruction, right lower tooth series, *Daouitherium rebouli*, occlusal view (Gheerbrant et al., 2005: figure 27). H) Reconstruction, left lower tooth series, *Numidotherium koholense*, occlusal view (Gheerbrant et al., 2005: figure 27). A, B, copyright © Elsevier B. V. C-H, copyright permission, © Publications Scientifiques du Muséum national d'Histoire naturelle, Paris.

ABBREVIATIONS: c, lower canine; p2, second lower premolar; pe, postentoconule.

which share gnathic and dental features with these taxa (Gheerbrant et al., 2002). Deinotheres, however, should not be included in "Plesielephantiformes," as they appear to be more closely related to Elephantiformes (Gheerbrant et al., 2005).

The primitive lophodont taxa included in "Plesielephantiformes" do not constitute a monophyletic clade, but rather consist of a sequential suite of stem groups at the base of more advanced proboscidean groups (Gheerbrant et al., 2005). However, the bilophodont molars of "Plesielephantiformes" (Shoshani et al., 2001a) indicate a lophodont ancestry for the Proboscidea (Gheerbrant et al., 2005). These taxa are notably distinguished from Elephantiformes and deinotheres by inferred absence of a trunk and lesser development of tusks. Furthermore, despite remarkable similarity between barythere and deinotheriine molars (Harris, 1978), the dental morphology of the earliest deinotheres indicates descent from a bunolophodont ancestor (Sanders et al., 2004) and independent acquisition of lophodonty.

Family PHOSPHATHERIIDAE Gheerbrant et al., 2005

This early Eocene African family is comprised of some of the oldest and most primitive recognizable proboscideans. Comprised of only one or two species (depending on the status of *Khamsaconus*), Phosphatheriidae is nonetheless among the best-known early representatives of the modern orders of ungulates, documented by good skull and dental material. The discovery of phosphatheres and relegation of South Asian anthracobunids, once thought to be archaic proboscideans (Wells and Gingerich, 1983; West, 1983, 1984; Gingerich et al., 1990; Kumar, 1991; Shoshani et al., 1996), to the phylogenetic fringes of Tethytheria (Gheerbrant et al., 2005; Tabuce

TABLE 15.2
Major occurrences and ages of Afro-Arabian "Plesielephantiformes" and moeritheres

? = Attribution or occurrence uncertain; alt. Alternatively.

Taxon	Occurrence (Site, Locality)	Stratigraphic Unit	Age	Key References
?PROBOSCIDEA, EARLY EOCENE ?PHOSPHATHERIIDAE, EARLY EOCENE				
Khamsaconus bulbosus	N'Tagourt2, Ouarzazate Basin, Morocco (type)	Ait Ouarithane or Jbel Ta'louit Fm.	Early Eocene (early Ypresian, ca. 55 Ma)	Sudre et al., 1993
FAMILY INDET., LATE PALEOCENE				
Eritherium azzouzorum	NE Ouled Abdoun Basin, Morocco (type)	phosphate beds II	late Paleocene	Gheerbrant, 2009
PHOSPHATHERIIDAE, EARLY EOCENE				
Phosphatherium escuilliei	Grand Daoui Quarries, Ouled Abdoun Basin, Morocco (type)	"Intercalaire couches II-I"	Early Eocene (early Ypresian, ca. 55 Ma)	Gheerbrant et al., 1996, 1998, 2003, 2005; Gheerbrant, 1998
FAMILY INDET., EARLY EOCENE				
Gen. et sp. indet., probably nov.	Tamaguélelt, Mali	Phosphate beds, Tamaguélelt Formation	Early Eocene (Ypresian)	Patterson and Longbottom, 1989; Moody and Sutcliffe, 1993; O'Leary et al., 2006
FAMILY INDET., PROBABLY NOV., EARLY EOCENE				
Daouitherium rebouli	Grand Daoui Quarries, Ouled Abdoun Basin, Morocco (type)	"Intercalaire couches II-I"	Early Eocene (early Ypresian, ca. 55 Ma)	Gheerbrant et al., 2002
NUMIDOTHERIIDAE, EARLY EOCENE–LATE EOCENE OR EARLY OLIGOCENE				
Numidotherium koholense	El Kohol, Algeria (type)	El Kohol Fm.	Early Eocene (late Ypresian)	Mahboubi et al., 1984, 1986
Numidotherium savagei (= "*Barytherium* small species") (= "*Arcanotherium savagei*")	Dor el Talha (Dur at Talhah), Libya (type)	Evaporite Unit	Latest Eocene or early Oligocene	Arambourg and Magnier, 1961; Savage, 1969, 1971; Wight, 1980; Court, 1995; Shoshani et al., 1996; Le Blanc, 2000; Delmer, 2009
BARYTHERIIDAE, LATE EOCENE–EARLY OLIGOCENE				
Barytherium sp. indet.	Fayum, Egypt	Birket Qarun Fm.	Late Eocene, ~37 Ma, early Priabonian	Seiffert, 2006
Barytherium grave	Fayum, Egypt (type)	Qasr el Sagha Fm.	Late Eocene	Andrews, 1901a, 1901b, 1904b, 1906; Simons, 1968; Harris, 1978
	Dor el Talha (Dur at Talhah), Libya	Idam Unit	Early Oligocene	Arambourg and Magnier, 1961; Savage, 1969, 1971; Wight, 1980; Shoshani et al., 1996
?*Barytherium* sp. indet. ("cf. Barytherioidea")	Thaytiniti, Oman	Shizar Mb., Ashawq Fm.	Late Eocene, 33.7–33.3 Ma	Thomas et al., 1989, 1999; Seiffert, 2006

Taxon	Occurrence (Site, Locality)	Stratigraphic Unit	Age	Key References
MOERITHERIIDAE, LATE EOCENE–EARLY OLIGOCENE				
Moeritherium sp. indet.	Fayum, Egypt	Birket Qarun Fm.	Late Eocene, ~37 Ma, early Priabonian	Seiffert, 2006
Moeritherium chehbeurameuri	Bir el Ater, Algeria (type)	Sandy, fluvio-deltaic member	Late Eocene, ?Priabonian	Delmer et al., 2006
	?Kenchella, Algeria		Late Eocene, ?Priabonian	Pickford and Tassy, 1980; Delmer et al., 2006
Moeritherium lyonsi (including "M. gracile," "M. ancestrale," "M. pharaonensis," and "M. latidens")	Fayum, Egypt (type)	Qasr el Sagha Fm.	Late Eocene	Andrews, 1901a, 1901b, 1902, 1906; Matsumoto, 1923; Petronievics, 1923; Osborn, 1936; Deraniyagala, 1955; Coppens and Beden, 1978; Holroyd et al., 1996
	Dor el Talha (Dur at Talhah), Libya	Evaporite Unit	Latest Eocene or early Oligocene	Savage, 1971; Wight, 1980; Le Blanc, 2000
Moeritherium trigodon (including "M. andrewsi" and "M. trigonodon")	Fayum, Egypt (type)	Gebel el Qatrani Fm.	Early Oligocene, ca. 33–30 Ma	Andrews, 1904a, 1906; Schlosser, 1911; Matsumoto, 1923; Osborn, 1936; Coppens and Beden, 1978; Seiffert, 2006
	Dor el Talha (Dur at Talhah), Libya	Idam Unit	Early Oligocene	Savage, 1971; Wight, 1980; Le Blanc, 2000

et al., 2007b) have returned the focus on proboscidean origins to Africa.

Genus *PHOSPHATHERIUM* Gheerbrant, Sudre, and Cappetta, 1996
PHOSPHATHERIUM ESCUILLEI Gheerbrant, Sudre, and Cappetta, 1996
Figures 15.1C–15.1F, 15.2A, and 15.2B

Age and Occurrence Early Eocene, northern Africa (table 15.2).

Diagnosis Modified from Gheerbrant et al. (2005). Differs from *Khamsaconus* by larger size, smaller postentoconule in DP4, lower bunodonty, enamel type development (Tabuce et al., 2007), and more developed lophodonty. Distinct from other proboscideans in numerous primitive features of the skull and dentition, including: face twice as long as the braincase; long nasals located anteriorly on the cranium; very weakly pneumatized cranial bones; presence of hypoglossal foramen; strong postorbital constriction; low braincase; narrow mandibular corpus; short, unfused mandibular symphysis; dental formula; and small tooth size (figures 15.1C–15.1E, 15.2A, 15.2B). Substantially smaller than *Daouitherium*, with single-rooted p2, simpler lower premolars, lesser size difference between the molars, lower mandibular condyle, and narrower mandibular corpus.

Description Gheerbrant et al. (1998, 2005). The cranium is relatively long (L = 170 mm; W = 100–120 mm, H = 63 mm) and condylarth-like, with an elongated face, narrow snout, long, high maxillae (suggesting long nasals), and a large, unretracted nasal fossa (figures 15.1C–15.1E). The basicranium is moderately elevated above the level of the palate. Infraorbital foramina are large and slightly above the P2–P3 boundary. The development of the infraorbital fossa and foramen and high position of the maxilla suggest a mobile snout and upper lip. The orbit opens above P4, and exhibits a lacrimal foramen. The zygomatics extend broadly above P2–M2 and flare widely laterally. Strongly compressed bilaterally, the braincase has a small cerebral cavity, and is ornamented by salient sagittal and nuchal crests. The glenoid fossa is vast and supported by a very robust basis of the squamosal; these are probably proboscidean synapomorphies. There is a small but broad postglenoid process.

The dentary is brevirostrine, with the symphysis extending to p3, and coronoid process moderately high above a low condyle. Mandibular foramina are small, and positioned below c and p4–m1. The angular processes protrude behind the articular condyles as wide blades. Estimated length of the mandible is 110–115 mm.

The dental composition of *Phosphatherium* is very primitive, particularly in the inferred retention of P1 and c (or p1); the preferred dental formula is I3/2-C1/1-P4/3-M3/3 (figure 15.1F), and the most reasonable alternative is I3/2-C1/0-P4/4-M3/3 (Gheerbrant et al., 2005). The lower incisors are i1–2. Molars are bilophodont, with sharp, continuous loph(id)s. Molar size increases posteriorly through the series, most markedly between M1–M2 and m1–m2. In upper molars, the protoloph is wider than the

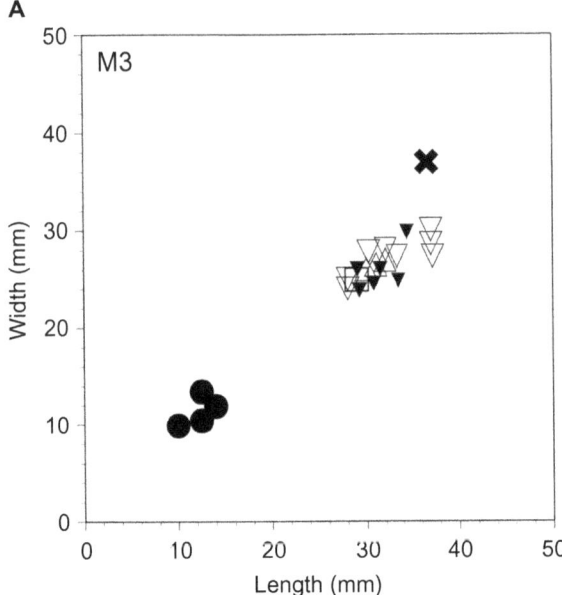

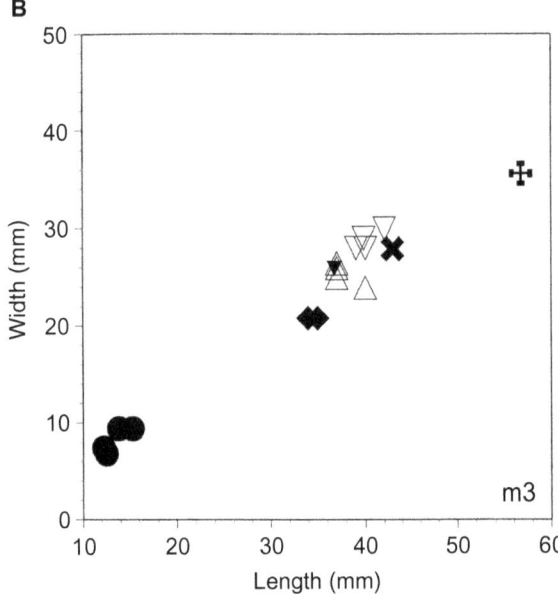

FIGURE 15.2 Bivariate plots of M3 (A) and m3 (B) crown length versus width in phosphatheres, daouitheres, numidotheres, and moeritheres. Comparative dimensions supplementing original measurements are from Andrews (1906), Matsumoto (1923), Tobien (1971), Mahboubi et al. (1986), Court (1995), Gheerbrant et al. (2002, 2005), and Delmer et al. (2006). Symbols: closed circle, *Phosphatherium escuilliei*; closed diamond, *Daouitherium rebouli*; X, *Numidotherium koholense*; cross, *Numidotherium savagei*; open square, *Moeritherium chehbeurameuri*; open inverted triangle, *M. lyonsi*; open triangle, *M. trigodon*.

metaloph. Lower molars have an elongate occlusal outline, with independent distocingulids. In m1–2, the distocingulid bears a hypoconulid and a variable entoconulid. In m3, the hypoconulid contributes to a pronounced third, distal lobe, but there is no distinct postentoconulid.

The lower central incisor is typically enlarged. Its crown is not very large, but the root is hypertrophied. The i2 is smaller than i1 and spatulate (figure 15.1F). Both are slightly procum-

bent. The p3 and p4 are similar: simple elongate teeth, each with a high, sharp, transversely compressed protoconid, crestiform talonid, and a long mesiodistal crest linking the trigonid to the talonid. The p4 is slightly larger and more molariform than p3. Length of p2–m3 is about 50 mm (Gheerbrant et al., 2005).

In the upper tooth row, P1 is separated from P2 by a short diastema. P2 is a small, simple, two-rooted tooth. It is elongated and sharp. Similar in size and morphology, P3 and P4 are molariform and three-rooted. Each has a well-developed protocone. Three buccal cusps are apparent: a large paracone, parastyle, and small metacone. A low, continuous protoloph links the protocone to the paracone. C and P1 were evidently medium- to large-sized teeth.

Remarks Except for *Eritherium* and possibly for *Khamsaconus*, *Phosphatherium* is the most primitive and smallest known proboscidean, with an estimated body mass of 10–15 kg (Gheerbrant et al., 1996; Gheerbrant, 1998). Despite its archaic morphology, *Phosphatherium* exhibits a number of proboscidean synapomorphies, including a well-developed zygomatic process of the maxillary contributing significantly to the lower border of the orbit and to the zygomatic arch, large pars mastoidea in the periotic, labial position of the hypoconulid, true lophodonty, enlarged i1, loss of i3 and p1 or c, and frontal contact with the squamosal (Gheerbrant et al., 1998, 2005). Coeval with *Khamsaconus*, *Daouitherium*, and an unnamed taxon from Mali, *Phosphatherium* belongs to the oldest radiation of the order. The existence of such diversity in North Africa at the base of the Eocene suggests that the origin of Proboscidea is even older and likely African (Gheerbrant et al., 1996). In fact, the recent discovery of the very diminutive *Eritherium azzouzorum* from the Ouled Abdoun basin, Morocco, dated to the late Paleocene (Gheerbrant, 2009), appears to confirm the deeper antiquity of proboscideans in Africa. The age and morphology of *Phosphatherium*, along with evidence from other early Eocene proboscideans, favor a true lophodont ancestral morphotype for Proboscidea (Gheerbrant et al., 1996, 1998).

Family INCERTAE SEDIS
GEN. ET SP. INDET., PROBABLY NOV. O'Leary, Roberts, Bouare, Sissoko, and Tapanila, 2006

Age and Occurrence Early Eocene, northern Africa (table 15.2).

Diagnosis O'Leary et al. (2006). Intermediate in size between *Phosphatherium* and *Daouitherium*.

Description O'Leary et al. (2006). Known only from a toothless dentary from Mali which preserves alveoli for p3–m3, the base of the ramus, and part of the symphysis. The ramus is rooted lateral to m3 and may have been posteriorly inclined. The corpus is robust with a thick symphysis that is strongly angled medially and extends posteriorly as far as the mesial edge of p4. Posteriorly, the corpus widens lateral to the alveoli for m2–3. Alveoli suggest that the teeth were double rooted. Three mandibular foramina are present, approximately along a horizontal line above midheight of the corpus.

Remarks The posterior expansion of the corpus is typical of basal proboscideans, and the position of the ramus resembles the condition in *Phosphatherium*, *Daouitherium*, and *Numidotherium koholense* (O'Leary et al., 2006). Although this specimen was originally placed in Numidotheriidae (O'Leary et al., 2006), its overall morphology more closely resembles that of phosphatheres and daouitheres.

Family INCERTAE SEDIS
Genus *DAOUITHERIUM* Gheerbrant and Sudre in
Gheerbrant et al., 2002
DAOUITHERIUM REBOULI Gheerbrant and Sudre in
Gheerbrant et al., 2002
Figures 15.1F, 15.1G, and 15.2B

Age and Occurrence Early Eocene, northern Africa
(table 15.2).

Diagnosis Modified from Gheerbrant et al. (2002). Primitive proboscidean with bilophodont molars similar to those in other "plesielephantiforms." Differs from *Phosphatherium* and more closely resembles barytherioids in its larger tooth size (figure 15.2B), more molarized premolars, deeper mandibular corpus, and higher mandibular ramus. Distinguished from barytherioids by retention of four teeth anterior to p2 and its more primitive anterior dentition lacking a diastema in the dentary. The large hypoconids in p2 and p3, and extension of enamel onto the buccal side of anterior roots in its premolars, are probably autapomorphic.

Description Gheerbrant et al. (2002). No cranium is known for this species. The dentary is large (L ≥ 240 mm) and robust with a tall ramus (H ≥ 120 mm), and the height of the condyle above the tooth row (at least 50 mm) suggests that the basicranium was elevated well above the level of the palate. Although the symphysis is missing, crowding of anterior teeth indicates that the mandible was brevirostrine.

The lower dental formula is not known with certainty but may be I3-C1-P3-M3 (alternatively, I3-C0-P4-M3 or I2-C1-P4-M3), possibly preserving one lower incisor more than *Phosphatherium* (figures 15.1F, 15.G). The alveoli for the anterior dentition show that the first two incisors were upright and that i1 was larger than i2. Alveoli for i3 and the canine are diminutive. The p2 is a simple, cutting tooth with a triangular occlusal outline and similar in size to p3–4. The p3 is submolariform with a well-developed, basined talonid, and like p4 has a paraconid and metaconid. The p4 is molariform and nearly bilophodont, though the hypolophid is small.

Molar size increases from m1–3, especially between m1 and m2 (figure 15.1G). The hypolophid is wider than the protolophid. The occlusal outline is elongated. The first and second molars have well-developed distal cingulids; in m2 it bears a small hypoconulid behind the hypoconid, and in m3 this is extended to form a third, distal lobe. Length of p2–m3 = 140 mm.

The only upper tooth preserved for *Daouitherium* is a P3 or P4 with a submolariform morphology similar to that in *Phosphatherium*, but with a less distinct metacone.

Remarks *Daouitherium rebouli* is the oldest known large-bodied (tapir-sized) mammal from Africa and appears to have been sympatric with *Phosphatherium escuilliei*. Although its dental formula obviates descent from *Phosphatherium*, *Daouitherium* is in most other morphological aspects more derived than that taxon in the direction of the barytherioids (Gheerbrant et al., 2002, 2005). Its stratigraphic association with *Phosphatherium* was unexpected, but nevertheless strengthens the case for an old African origin and basal radiation of Proboscidea (Gheerbrant et al., 2002).

Superfamily BARYTHERIOIDEA Andrews, 1906

Numidotherium and *Barytherium* constitute this superfamily, which is united by similarities in craniodental and postcranial morphology. Shared postcranial features such as limb

proportions and orientation of articular facets suggest similar postures, unaffected by significant differences in size and weight. *Barytherium* evolved to elephantine size, and even the more lightly built numidotheriids were larger than *Phosphatherium* and *Daouitherium*. Barytherioids contrast with phosphatheriids and daouitheres especially in the more derived configuration of their skulls and reduction in number of teeth, and occur later in geological time (table 15.2).

Family NUMIDOTHERIIDAE Shoshani and Tassy, 1992
Genus *NUMIDOTHERIUM* Jaeger in Mahboubi et al., 1986
NUMIDOTHERIUM KOHOLENSE Jaeger in
Mahboubi et al., 1986
Figures 15.1H, 15.2A, 15.2B, 15.3A–15.3C

Age and Occurrence Early Eocene, northern Africa
(table 15.2).

Diagnosis Based in part on Mahboubi et al. (1986). Smallest of the barytherioids (shoulder height about 1 m), with an elevated, moderately pneumatized cranium that broadens at the level of the frontal and has an anteriorly placed nasal opening. Resembles other "plesielephantiforms" in the bilophodonty of its molars, and *Barytherium* in particular in the relatively large size of its head. I1-3 and C are retained (the latter reduced), and the lower dentition anterior to p2 includes only two incisors (figure 15.1H). Tooth size greater than in *Phosphatherium* and *Daouitherium* (figures 15.2A, 15.2B).

Description Mahboubi et al. (1984, 1986); Court (1994a). The cranium is characterized by the great height of the glenoid and occipital condyles above the palatal plane, the vertical disposition of the basicranium (strongly angled relative to the palate), an extensive submaxillary fossa, projecting zygomatic apophyses, and upwardly sweeping, laterally flared zygomatic arches (figures 15.3A, 15.3B). Bounded posteriorly by postorbital apophyses of the frontal, the orbits are situated above P3–4 and are continuous with capacious temporal fossae. The nasal fossa is much higher than broad; the premaxillary is short and high, contributing to a narrow snout; and the infraorbital foramen is positioned low on the cranium, above P2. The nuchal planum is angled forward and bordered superiorly by a crest, but there is no sagittal crest.

The dentary has a massive, tall ramus that dominates the relatively slender corpus (figure 15.3C). The body of the dentary decreases markedly in height anterior to p2. There is a mandibular foramen low on the corpus, below p2–3, and a long diastema separates the cheek teeth from the incisors. The symphysis is extended posteriorly to p4.

The adult dental formula is I3/2-C1/0-P3/3-M3/3, with all teeth in occlusion simultaneously. The cheek teeth are preceded by three deciduous premolars in each jaw quadrant. In the upper dentition, I2 is the largest incisor and caninelike, with a high crown (about 45 mm) covered with thin enamel. A sizable diastema separates the incisors from the premolars. The P2 has a triangular outline, with a low protocone and single main outer cusp. The P3 is wider than long, with a feeble anterior transverse loph, and no hypocone; P4 is similar in morphology but larger. Upper molars are bilophodont, very tapirlike, and become progressively larger from M1 to M3. They are wider anteriorly than posteriorly. The two main buccal cusps in each possess a postparacrista and postmetacrista, respectively.

The most salient feature of the lower dentition is the tusklike, semiprocumbent central incisor (crown length about 35 mm; figures 15.1H, 15.3C). This and the smaller lateral

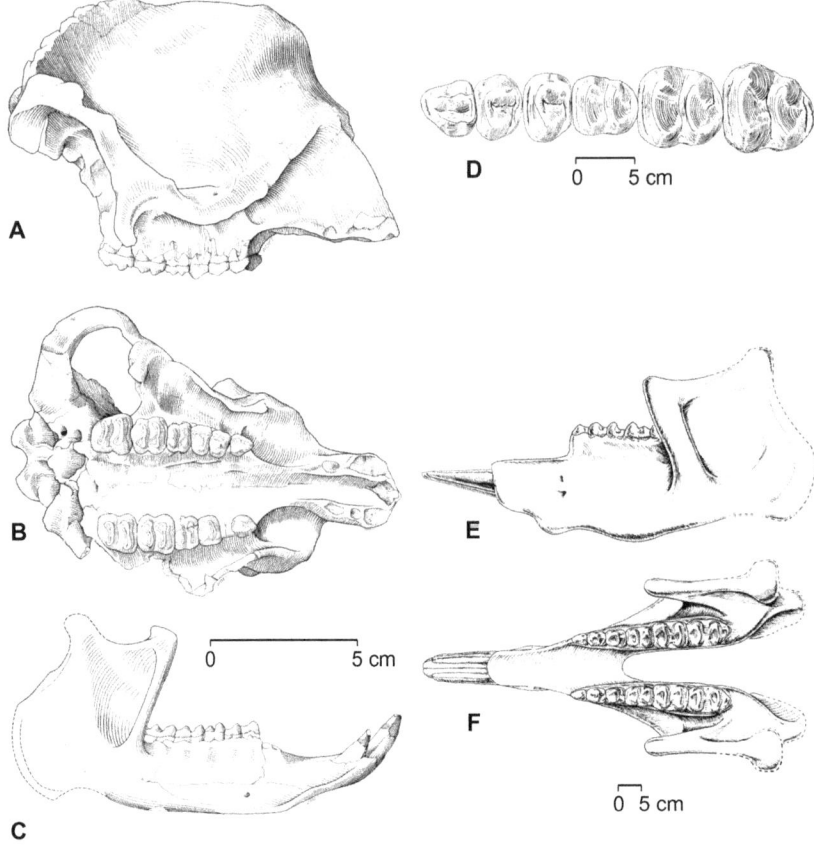

FIGURE 15.3 Aspects of barytherioid cranial, mandibular, and dental morphology. A) Cranium, *Numidotherium koholense*, lateral view (from Mahboubi et al., 1986:26, figure 4). B) Cranium, *N. koholense*, ventral view (Mahboubi et al., 1986:29, figure 7). C) Mandible, *Numidotherium koholense*, lateral view (Mahboubi et al., 1986:30, figure 8). D) Left P2–M3, *Barytherium grave*, occlusal view (Delmer, 2005). E) Mandibular reconstruction, *B. grave*, lateral view (Delmer, 2005). F) Mandibular reconstruction, *B. grave*, dorsal view (Delmer, 2005). A-C, with permission of E. Schweizerbart'sche Verlagsbuchhandlung OHG. D-F, courtesy of C. Delmer.

incisor are spatulate. The lateral incisor bears numerous distolabial serrations. The premolars are moerithere-like, longer than broad, with a prominent median cusp located anteriorly, and much lower hypoconids located posteriorly, on the buccal side of the talonids. Masticatory wear on the root of p2 suggests ventrolateral movement of the dentary, consistent with a behavior such as stripping leaves from slender branches (Court, 1993). The lower molars are rectangular in occlusal shape, with two main lophids and strong distocingulids closely appressed to the posterior lophid, except in m3, where there is a prominent posterior shelf or talonid. These molars increase in length from m1 to m3, especially between m1 and m2.

Postcranials are well represented and exhibit proportions and morphology typical of graviportal animals with plantigrade manus and pes. Nevertheless, to Court (1994a) the configuration of joint complexes suggested an ambulatory or semisprawling mode of locomotion, rather than a parasagittal recruitment of limbs. The humerus is equal in length (L = 300 mm) to the forearm bones and is robust in construction, with a strong deltoid ridge, deep olecranon fossa, broad epicondyles, and a superiorly oriented head. It primitively retains an entepicondylar foramen. The radius and ulna are distally fused in a pronated position. The ulnar olecranon process is very low and posteriorly projecting, permitting full extension of the forelimb. A fragmentary pelvic bone shows the acetabulum to be oriented downward. The ilium is broad. The femur is

straight and much longer (405 mm) than the tibia (260 mm). Its head is globular, directed upward, and has a deep fovea. The neck is short and flattened, the greater trochanter is low and massive, and there is a third trochanter. The astragalus and calcaneum anticipate the morphology of these elements in the palaeomastodonts: the astragalus has a prominent medial process, the ectal facet is larger than the sustentacular facet, and the neck is short and narrow; in the calcaneum, the tuber calcanei is massive and connected to the rest of the element by a flattened neck.

Remarks Numidotherium koholense encompasses the most extensive early proboscidean fossil collection (Mahboubi et al., 1986; Court, 1994a). It was one of the largest terrestrial mammals of its time, initiating an impressive size increase among proboscideans. Nonetheless, it is much smaller than *Barytherium grave*, from which it also differs in having only two roots on its lower cheek teeth, and a nonbilophodont p4 (Mahboubi et al., 1986). In addition, in *B. grave* the proportions of the ectal and sustentacular facets are reversed, there is no entepicondylar foramen or fovea capitis femoris, the distal femoral condyles are more symmetrical, and the scapular coracoid process is more prominent (Andrews, 1906; Mahboubi et al., 1986; Court, 1994a).

The proboscidean status of *N. koholense* is marked by the strong development of I2 and i1, the elevation of the cranial glenoid and external auditory meatus, the structure of the astragalus, the anterior position of the orbit, and the absence of a condylian foramen (Mahboubi et al., 1984). Its position within Proboscidea vis-à-vis *Moeritherium* has been debated (e.g., Tassy and Shoshani, 1988; Tassy, 1996c), but it appears to belong to a more primitive radiation (Court, 1992, 1994a, 1994b; Gheerbrant et al., 2005). The presence of a subdivided perilymphatic foramen in the periotics of *N. koholense* and *Prorastomus sirenoides* (Court, 1990; Court and Jaeger, 1991), the most primitive known sirenian, shows that its absence in later sirenians and proboscideans is a convergence, rather than a synapomorphy (Court, 1994b).

NUMIDOTHERIUM SAVAGEI Court, 1995
Figure 15.2B

Partial Synonymy *Barytherium* sp., Harris, 1978; *Barytherium* sp., Wight, 1980; *Barytherium* sp. nov., Shoshani et al., 1989; small *Barytherium*, Shoshani et al., 1996.

Age and Occurrence Late Eocene or early Oligocene, northern Africa (table 15.2).

Diagnosis Based on Court (1995); Delmer (2005). Resembles *Numidotherium koholense* but was larger (1.0–1.5 m shoulder height and approximately 200 kg; Shoshani et al., 1996), with bigger teeth (figure 15.2B). In addition, the ramus is rooted more anteriorly on the dentary, the metaconid is better developed in the premolars, the ulna is more robust with a broader distal epiphysis, and the carpus is wider anteriorly than in *N. koholense*. The scapular supraglenoid tubercle is smaller than in *Barytherium grave* and lacks the medially deflected coracoid process of *N. koholense*.

Description Based in part on Court (1995). The dentary is fragmentary but corresponds closely with the morphology in *N. koholense*, except for the position of the ramus. It is less robust overall than the dentary of *Barytherium*.

The lower dental formula is inferred to be I2-C0-P3-M3, all in occlusion at the same time in adults. The central incisors are large, spatulate teeth (H = 52 mm; mesiodistal W = 31 mm; labiolingual W = 27 mm) with mammillons ornamenting the distolabial margins of the crown. Except for the worn apex, the crown is covered in enamel. The p2 is triangular in occlusal outline, with a high protoconid and accompanying metaconid, and a much lower talonid hosting a buccally placed hypoconid. The p3 is more rectangular with a more equal-sized protoconid and metaconid. In the p4, there is greater separation between the protoconid and metaconid, which form a true lophid connected by a transverse ridge. The molars are bilophodont, with the protolophid higher than the hypolophid, and well-developed distocristids. In m3, the distocristid is transformed into a large hypoconulid lobe. A prominent cristid descends anteromedially into the transverse valley from the hypoconid, in each molar.

M1–3 are bilophodont, each with prominent mesial and lingual cingulae and a weak entoflexus. The M3 has a strong distocrista, with a distinct paraconule and metaconule. M1–2 exhibit well-defined postparacrista, -protocrista, and -hypocrista.

Postcranially, *N. savagei* shares traits characteristic of *N. koholense*, including an ulnar styloid process, probable possession of a free os centrale in the carpus, flangelike development of a femoral third trochanter, and asymmetry of the femoral distal condyles.

Remarks This species was originally placed in *Barytherium* (Savage, 1969; Harris, 1978; Shoshani et al., 1989; Shoshani et al., 1996), which seemed reasonable prior to the discovery and full appreciation of the distinctness of *Numidotherium*

koholense, and given the abundance of barytheres at the site (Savage, 1971). Morphological contrasts between *N. savagei* and *N. koholense* may be accounted for by an increase in body size but alternatively could mark intergeneric distinctions (Delmer, 2005, 2009). This is supported by the observation that *N. savagei* shares an arrangement of patterns of enamel types ("Schmelzmuster") with elephantoids but not with *N. koholense* (Tabuce et al., 2007b). Finds associated with *N. savagei* include aquatic plants, fish, turtles, crocodiles, sirenians, and cetaceans (Shoshani et al., 1996), indicating an amphibious habitus for the species.

Family BARYTHERIIDAE Andrews, 1906
Genus BARYTHERIUM Andrews, 1901
BARYTHERIUM GRAVE (Andrews, 1901)
Figures 15.2, 15.3D–15.3F

Partial Synonymy *Bradytherium grave*, Andrews, 1901; *Barytherium grave*, Andrews, 1901.

Age and Occurrence Late Eocene–early Oligocene, northern Africa (table 15.2).

Diagnosis Based in part on Harris (1978); Court (1995); Shoshani et al. (1996). Very large-bodied animal (shoulder height approximately 2–3 m; body mass about 3–4 metric tons) with graviportal postcranial adaptations. Characterized by a massive dentary bearing a large, projecting central incisor, and a cranium with anteriorly located orbits, weakly retracted nares, and extensive pneumatization. Differs from *Numidotherium* in body size, the absence of an entepicondylar foramen on the humerus, more massive supraglenoid tubercle on the scapula, and radius not fused distally with the ulna. Teeth larger than other "plesielephantiforms" (compare figures 15.2 and 15.6).

Description Based in part on Andrews (1906); Harris (1978); Shoshani et al. (1996). Cranial remains of this species are fragmentary, but preserve enough structure to show that the zygomatics are robust and flaring; the orbits are set anterior to the premolars, and that the external auditory meatus sits at a level slightly higher than that of the orbit. There is a long diastema between P2 and the incisors. The cranium is elevated in a manner similar to that in *Numidotherium*, with the basicranium strongly angled relative to the palate.

The mandible is deep, with a fused, spoutlike symphysis extending posteriorly to p4 or m1 (figures 15.3E, 15.3F). The ramus is high and is rooted on the corpus at the level of m2. An extensive diastema separates the cheek teeth from the incisors. There are multiple mandibular foramina, the largest of which is located below p2 and above a sizable ventral tuberosity.

The dental formula is I2/2-C0/0-P3/3-M3/3, with all teeth simultaneously in occlusion in adults. The molars are bilophodont with strong distal cingulae(ids), except for m3, which has a prominent posterior heel (figures 15.3D, 15.3F). Cheek tooth occlusal morphology is very similar to that in *Numidotherium*. Incisors project from the lower jaw in a hippolike manner (figures 15.3E, 15.3F), with the central incisor dominant and shearing against the upper lateral incisor. The upper incisors are vertically inserted into their alveoli; the lateral incisor is chisel edged and much larger than I1, and it has enamel restricted to its mesial cutting face. In contrast to deinotheres, the cheek teeth of barytheres were not bifunctional, and served primarily in a simple shearing action (Harris, 1978), probably of leafy matter in a manner similar to that of numidotheres (see Court, 1993).

The most salient features of the postcranial skeleton of *Barytherium* include an enlarged, hooklike scapular coracoid process; sustentacular facet larger than the ectal facet in the astragalus, which exhibits a well-developed medial tubercle; massive humerus greatly expanded distally, longer than the forearm, and with a deep olecranon fossa; radius with an extensive articulation for the humerus and fixed (but not fused) on the ulna in a pronated position; long, anteroposteriorly flattened femur with a third trochanter; and carpal-metacarpal configuration suggesting plantigrade disposition of the manus. An incomplete calcaneum from Fayum, Egypt (UM 13973), has a large fibular facet facing obliquely upward and laterally, a prominent ventral tuberosity below the fibular facet, a transversely compressed neck, and it differs from the calcaneum of *Numidotherium* by its considerably more expanded tuber calcis.

Remarks Barytheres have had a complicated taxonomic history, in part due to their enigmatic morphology that has been more fully comprehended systematically only with the relatively recent discoveries of *Numidotherium* and *Phosphatherium*. First discovered in the Qasr el Sagha Formation, Fayum, Egypt, they were initially placed in a subdivision of Amblypoda, the Barypoda (later Barytheria; Andrews, 1904b), but soon thereafter were transferred to the Proboscidea (Andrews, 1906). Although eventually barytheres were placed in their own suborder within Proboscidea (Simpson, 1945), they have occasionally been excluded from the order (Osborn, 1936, 1942; Deraniyagala, 1955; Harris, 1978). More rigorous reassessment of barytheres,

aided by an expanded fossil sample from Dor el Talha, Libya, has demonstrated unequivocally the proboscidean status of *Barytherium* (Tassy, 1982, 1985, 1996c). Features supporting this interpretation include pneumatization of the cranium, elevated nasal opening, hypertrophy of I2, and development of the medial astragalar tuberosity (Shoshani et al., 1996). Based on near identity of posterior molar morphology, barytheres and deinotheriines were regarded as closely affiliated (Harris, 1978), but new discoveries, including those of numidotheres (Mahboubi et al., 1984) and archaic deinotheres (Sanders et al., 2004), suggest a more distant relationship.

The size and slight retraction of the nares imply that *Barytherium* possessed an enlarged proboscis (but probably not a trunk). Recovery of fossil material from near-shore marine settings, and the forward position of the orbit are evidence of at least a semi-aquatic mode of life. The feet and anterior dentition of this odd proboscidean resemble those of hippos, further indicating that it may have occupied a similar, amphibious ecological niche. Stable isotopic analyses support this interpretation and indicate that they browsed on freshwater plants (Liu et al., 2008).

Suborder INCERTAE SEDIS
Superfamily MOERITHERIOIDEA Andrews, 1906
Family MOERITHERIIDAE Andrews, 1906
Figures 15.4A and 15.4B

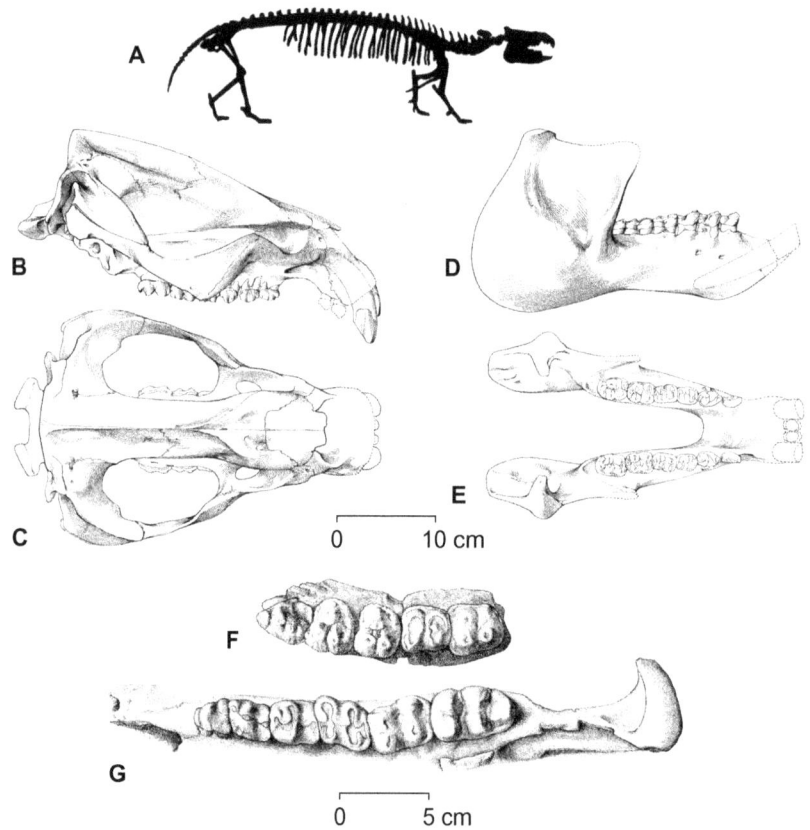

FIGURE 15.4 Aspects of moerithere skeletal and dental morphology. A) Silhouette, moerithere skeletal reconstruction, lateral view (Tobien, 1971: figure 7). B) Cranial reconstruction, *Moeritherium trigodon*, lateral view (Osborn, 1936: figure 42A). C) Cranial reconstruction, *M. trigodon*, dorsal view (Osborn, 1936: figure 42A1). D) Mandibular reconstruction, *M. trigodon*, lateral view (Osborn, 1936: figure 42B). E) Mandibular reconstruction, *M. trigodon*, dorsal view (Osborn, 1936: figure 42B1). F) Right P2–M2, occlusal view (Andrews, 1906: plate X, figure 2). G) Left dentary with p2–m3, occlusal view (Andrews, 1906: plate X, figure 1). A, © 2010 The Society of Vertebrate Paleontology. Reprinted and distributed with permission of the Society of Vertebrate Paleontology. B-E, courtesy of The American Museum of Natural History.

Moeritheres from late Eocene–early Oligocene localities in the Fayum, Egypt, were the first proboscideans with non-elephant-like morphology to be discovered (Andrews, 1901a, 1901b, 1901c, 1901d). Their unusual anatomy has hindered efforts to interpret their intraordinal relationships. Similar in size to a large pig (0.5–0.7 m tall at the shoulders), moeritheres were short-legged with an elongate body (figure 15.4A), with very short-faced, low crania (figure 15.4B; Coppens and Beden, 1978). To Andrews (1906), this unusual morphology indicated that moeritheres were familially distinct from the rest of Proboscidea, though he did note the structural antecedants of more advanced elephantiforms in their dentitions, particularly the enlarged second incisors (Andrews, 1901b).

Subsequently, morphological similarities to sirenians were emphasized, and the inclusion of moeritheres in the Proboscidea was debated (Osborn, 1909, 1921, 1936; Schlosser, 1911; Matsumoto, 1923; Deraniyagala, 1955, Tobien, 1971, Savage, 1976). The systematic uncertainties of moeritheres were acknowledged by Coppens and Beden (1978:333), who could not decide whether moeritheres should be attributed to "Proboscidea, Sirenia, or Desmostylia *incertae sedis*." It was only with the description of a new moerithere cranium from Dor el Talha, Libya, and cladistic treatment of the genus that these strange mammals were more firmly recognized as true proboscideans (Tassy, 1979b, 1981). Their proboscidean synapomorphies include hypertrophied I2s, raised external auditory openings, laterally flared zygomatic arches, an enlarged scapular coracoid process, pronated position of the radius, and a well-developed astragalar medial tubercle (Shoshani et al., 1996).

Found in marine and deltaic sediments along what was once the shoreline of the Tethys Sea (Shoshani et al., 1996; Delmer et al., 2006), moeritheres probably had an amphibious mode of life (Andrews, 1906; Matsumoto, 1923; Coppens and Beden, 1978), with feeding habits similar to those of sirenians or hippos (Osborn, 1909, 1936). Along with barytheres, isotopic analyses also indicate that moeritheres were at least semi-aquatic and consumed freshwater plants (Liu et al., 2008). They are represented by a lone genus, *Moeritherium*, exclusive to the late Paleogene of Africa (table 15.2; Shoshani et al., 1996). Although the genera *Anthracobune* and *Lamhidania* from the middle Eocene of South Asia were proposed to belong in Moeritheriidae (West, 1983, 1984), this hypothesis was overturned, and these taxa were reassigned to their own family, Anthracobunidae (Wells and Gingerich, 1983; Shoshani et al., 1996).

Moeritheres have long been considered to represent the ancestral stock of Proboscidea, due to their possession of primitive features such as elongate, low crania with anteriorly placed external nares and a sagittal crest (Matsumoto, 1923; Tassy and Shoshani, 1988; Shoshani et al., 1996; Tassy, 1996c). However, the discovery of phosphatheriids, daouitheres (Gheerbrant et al., 1996, 2002) and a new, primitive moerithere species with near-lophodont cheek teeth (Delmer et al., 2006), as well as further systematic study (Gheerbrant et al., 2005), reveal a more derived phylogenetic position for *Moeritherium*, with bunolophodonty of the cheek teeth of advanced species shared with elephantiforms.

Genus *MOERITHERIUM* Andrews, 1901
Figures 15.4B–15.4G

Moeritherium includes odd-shaped, semiaquatic species with anatomical convergences on sirenians, desmostylians (Tassy, 1981), and hippos (Osborn, 1936). Species from Fayum, Egypt, and Dor el Talhah, Libya, are especially close morphologically (Andrews, 1906; Matsumoto, 1923; Tassy, 1981). The cranium is remarkable, with its tubular, elongated cerebral cavity, long sagittal crest, and very forward position of the orbits (figures 15.4B, 15.4C). The latter are confluent with capacious temporal fossae and are not bounded posteriorly by postorbital processes. Pneumatization is weak. The auditory meatus is located higher on the cranium than the level of the orbit, but the basicranium is not very elevated or strongly angled relative to the palatal plane. Zygomatic arches are massive. The anterior, low placement of the nasal aperature indicates absence of a trunk (Osborn, 1936). Upper incisors and canines are vertically implanted in their alveoli.

The mandibles are brevirostrine and massively built (figure 15.4D), with lateral broadening of the corpus beginning at the level of p2 and increasing posteriorly. Symphyses are short, fused, and extended back to p3–p4 (figure 15.4E). The anterior border of the ramus is inclined anteriad, and rises at the level of m2–m3. The lower incisors are procumbent.

The moerithere dental formula is I3/2-C1/0-P3/3-M3/3. The permanent teeth were in occlusion simultaneously, and although the second upper and lower incisors are enlarged, they were not evergrowing. First incisors are reduced but evidently were functional, and I3 and the canine are also much reduced. Large diastemata are present between the anterior dentition and premolars. The P2 has an elongated, triangular occlusal outline and consists of an outer row of cusps with a narrow posterolingual shelf. P3 is more square in shape and has an anterior transverse loph and posterior buccal cusp; P4 is morphologically similar but slightly smaller. Upper and lower molars are tetrabunodont and bunolophodont or sublophodont (Osborn, 1936), with buccal and lingual half-loph(id)s separated by a median sulcus (figures 15.4F, 15.4G). There may be substantial postmetaloph ornamentation; M1–M2 metalophs exhibit an association of a convolute (*sensu* Gräf, 1957) and a distocrista, while only the latter is present in M3. Tooth size increases gradually from M1 to M3.

The p2 is small and narrow, with a raised anterior cusp and strong talonid exhibiting a median longitudinal crest. By comparison, the p3 and p4 are broader, with an anterior transverse lophid and a low, more extensive talonid. Lower premolars each have a protostylid (*sensu* Tassy, 1982) on the buccal side of the protolophid. The first and second lower molars have substantial distocingulids, and the m3 has a nascent third lophid (figure 15.4G).

In moeritheres, the postcranium is decidedly not graviportal. The femoral and humeral heads face medially and posteriorly, respectively, rather than upward. Limbs are short, relative to the length of the body (Coppens and Beden, 1978). There are at least 19 thoracic vertebrae, 4 lumbars, 3 fused sacrals, and a short tail (Andrews, 1906). The scapula bears a strong, hooked coracoid process, with supraspinous and infraspinous fossae of similar size; the humerus lacks an epicondylar crest; the radius appears to have been habitually pronated; the femur lacks a third trochanter and bears an enlarged lesser trochanter, which extends posteriorly; and the astragalus has a well-developed medial tubercle (Shoshani et al., 1996). The innominate resembles the pelvis of early sirenians, expecially in its expanded ischium and narrow ilium (Andrews, 1906). The olecranon process of the ulna is high and not reflected posteriorly, indicating a habitually flexed posture of the forearm on the humerus. Humeral length is documented as 240–260 mm, and femoral length as approximately 270 mm (Andrews, 1906).

MOERITHERIUM CHEHBEURAMEURI Delmer, Mahboubi, Tabuce and Tassy, 2006
Figure 15.2B

Partial Synonymy Mastodon turicensis, Gaudry, 1891; *Moeritherium trigodon*, Schlesinger, 1912.

Age and Occurrence Late Eocene, northern Africa (table 15.2).

Diagnosis Delmer et al. (2006). Small moerithere (figure 15.2B) with nearly lophodont dentition.

Description Delmer et al. (2006). This species is known from isolated teeth. I2 is diminutive, triangular in cross section, and has a complete covering of enamel and a linguodistally oriented wear facet. Molars are bilophodont and have thin, occasionally rugose enamel. P3 is very wide and lacks a buccal cingulum. The upper molars are nearly lophodont, with the main cusps only separated by a small groove in the center of each loph. They have a strong lingual cingulum, and incipient metaconule and postprotocrista. M1 is substantially smaller than M2. The p3 is poorly molarized and has a narrow protolophid, with the metaconid very small and not separable from the protoconid, and is slightly smaller than p4, which has a strongly developed metaconid. The m1 is rectangular in occlusal view, with no trace of a buccal cingulum. In occlusal view, p4 is longer than wide, with a small paraconid that is distinct from the protolophid.

Remarks The Bir el Ater, Algeria moerithere specimens assigned to *M. chehbeurameuri* are among the oldest yet recovered, with new, undescribed moerithere fossils from the Fayumian Birket Qarun Fm. probably slightly more ancient (table 15.2; Seiffert, 2006).

MOERITHERIUM LYONSI Andrews, 1901

Partial Synonymy Moeritherium lyonsi, Andrews, 1901a, 1901b, 1901c, 1901d; *M. gracile*, Andrews, 1902, 1906; *M. gracile*, Matsumoto, 1923; *M. ancestrale*, Petronievics, 1923; *M. pharoahensis*, Deraniyagala, 1955; *M. latidens*, Deraniyagala, 1955; *Moeritherium* cf. *lyonsi*, Tassy, 1981.

Age and Occurrence Late Eocene, northern Africa (table 15.2).

Diagnosis Moeritherium lyonsi (and *M. trigodon*) differ from *M. chehbeurameuri* in the more pronounced bunolophodonty of their cheek teeth. *Moeritherium gracile* is synonymized with *M. lyonsi* as a smaller morph of the same species, possibly due to sexual dimorphism (Coppens and Beden, 1978). Crania vary in length from 314 to 370 mm (Andrews, 1906; Tassy, 1981). *Moeritherium lyonsi* has been described as having relatively broader lower molars than *M. trigodon* (Osborn, 1936), but this may not be the case consistently. *Moeritherium lyonsi* is diagnosed by having simple upper molars with no accessory conules, and lower molars each with a low crescentoid blocking the transverse valley between the first and second lophids, and a talonid composed of a stout, centrally located cusp (Tobien, 1978).

MOERITHERIUM TRIGODON Andrews, 1904
Figures 15.2A and 15.2B

Partial Synonymy Moeritherium trigonodon, Andrews, 1906; *M. andrewsi*, Schlosser, 1911; *M. andrewsi*, Matsumoto, 1923.

Age and Occurrence Early Oligocene, northern Africa (table 15.2).

Diagnosis M. andrewsi is synonymized with *M. trigodon* as a larger morph of the same species, possibly due to sexual dimorphism (Coppens and Beden, 1978). The dental size variation within and between *M. lyonsi* and *M. trigodon* appears insubstantial (figures 15.2A, 15.2B) and could readily fit within the range of variation of other proboscidean species. In contrast, these species may be separable by differences in cranial size, as crania of the latter exceed 380 mm in length (Matsumoto, 1923), reaching an estimated measurement of 440 mm in the massively built skull of AMNH 13430. According to Tobien (1978), in *M. trigodon* M1 and M2 have a posterior fovea behind the second loph, there are zygodont crests on the posterior slope of the first loph in M2 and M3, and the posttrite half-lophs are more crestlike than in *M. lyonsi*. In addition, in *M. trigodon* the lower molars have more conelets in each half-lophid (Tobien, 1978).

Remarks Moeritheres are among the most unusual proboscideans, due to their specialized adaptations to a semi-aquatic life. Dental morphology in *M. chehbeurameuri* suggests derivation from a true lophodont taxon, though this remains to be confirmed in the moerithere tooth sample from the older Birket Qarun localities. Skull morphology in *M. lyonsi* and *M. trigodon* also points to an ancestry rooted in "Plesielephanti-formes"; their crania are elongated versions of *Phosphatherium* crania. While the skeletal anatomy of these moeritheres appears too apomorphic to provide direct ancestry to elephantiforms, the bunolophodonty and occlusal organization of their molars anticipate the development of palaeomastodont cheek teeth. It has also been argued that deinothere cheek teeth were evolutionarily transformed from a moerithere-like bunolophodonty (Sanders et al., 2004), but there are no skulls available of the earliest deinotheres to further assess the degree of their relationship to moeritheres.

Suborder INCERTAE SEDIS
Superfamily DEINOTHERIOIDEA Bonaparte, 1845
Family DEINOTHERIIDAE Bonaparte, 1845
Figure 15.5

Deinotheres were specialized browsing proboscideans characterized by low-crowned, bilophodont cheek teeth (but trilophodont deciduous fourth premolars and first molars) and retention of only lower tusks (figure 15.5). Although Mio-Pliocene deinotheres were first documented from Africa in the early 1900s (Andrews, 1911; Haug, 1911), more ancient confamilials eluded search until their recovery in a late Oligocene mammalian fossil assemblage from Chilga, Ethiopia nearly a century later (Kappelman et al., 2003).

The presence of the small, primitive *Chilgatherium* in this assemblage (Sanders et al., 2004) points to an African origin for deinotheres. The larger *Prodeinotherium hobleyi* is characteristic of early Miocene assemblages of eastern Africa (Harris, 1978), but representatives of the genus thereafter migrated to Eurasia, where it is represented in European MN4 sites by *P. bavaricum* (Tassy, 1989) and in the early Miocene of South Asia by *P. pentapotamiae* (Forster Cooper, 1922). During the mid- to late Miocene, *Prodeinotherium* species were replaced by larger, more progressive species of *Deinotherium* (Harris, 1978; Sanders, 2003). *Deinotherium indicum* became extinct in South Asia around 7 Ma (Barry and Flynn, 1989), and *D. giganteum* (as well as *D. gigantissimum*, if it is a separate species from *D. giganteum*) died out in Europe by the end

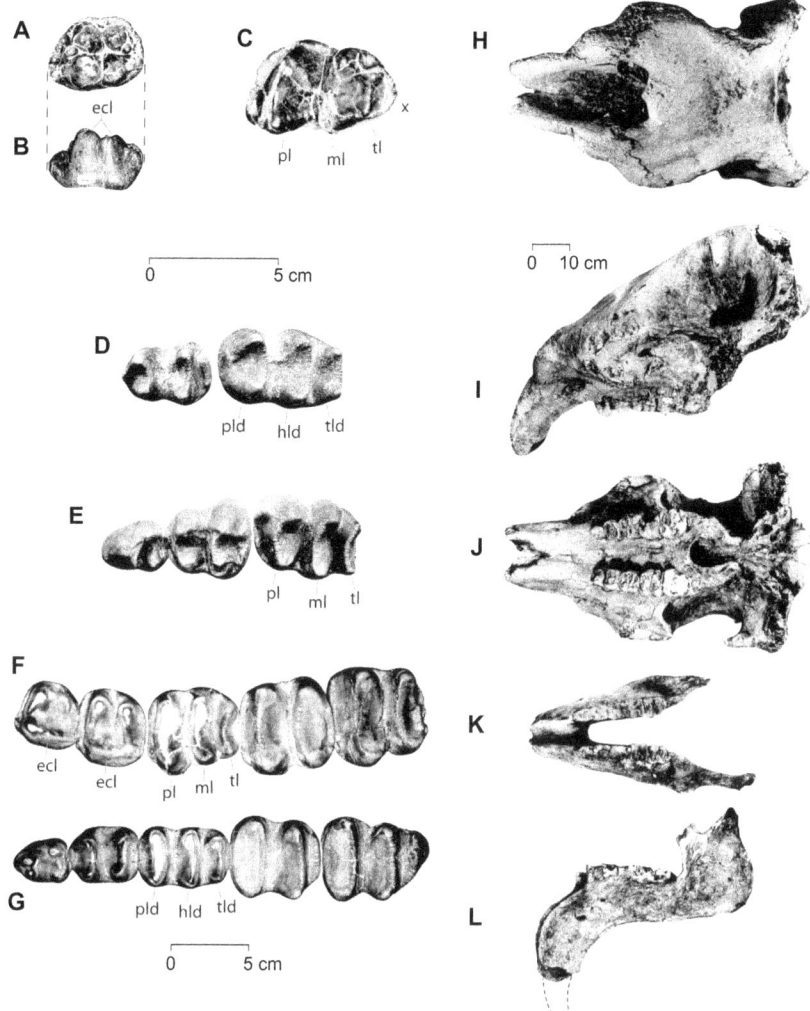

FIGURE 15.5 Aspects of deinotherioid cranial, mandibular, and dental morphology. Anterior is to the left in all specimens. A–E to same scale. A) Occlusal view, right P3, CH9-22, *Chilgatherium harrisi*. B) Buccal view, right P3, CH9-22, *C. harrisi*. C) Occlusal view, M3, CH35-1, *C. harrisi*. D) Occlusal view, dp3–4, NAP I 152'99, *Prodeinotherium hobleyi*. E) Occlusal view, DP2–4, NAP I 152'99, *P. hobleyi*. F) Occlusal view, P3–M3, M15713, *P. hobleyi*. G) Occlusal view, p3–m3, M15713, *P. hobleyi*. H) Dorsal view, cranium M26665, *P. hobleyi*. I) Right lateral view, cranium M26665, *P. hobleyi* (reversed). J) Ventral view, cranium M26665, *P. hobleyi*. K) Occlusal view, mandible with right and left p3–m3 6412:10, *P. hobleyi*. L) Right lateral view, mandible 6412:10, *P. hobleyi* (reversed), lower tusk reconstructed. H-L, courtesy of John Harris.

ABBREVIATIONS: ecl, ectoloph; hld, hypolophid; ml, metaloph; pl, protoloph; pld, protolophid; tl, tritoloph; tld, tritolophid; x, anterior or posterior cingulum(id).

of the Miocene (G. Markov, pers. comm.), but *D. bozasi* persisted in Africa until about 1 Ma (Beden, 1985; Behrensmeyer et al., 1995).

Despite having barythere-like posterior molars, deinotheres appear to be more closely related to elephantiforms than to "plesielephantiforms," based on features such as hypolophid higher than the protolophid, presence of a hypocone in P3 and P4, and a suite of postcranial synapomorphies, including an enlarged ectal facet and reduced fibular facet in the astragalus (Harris, 1976; Tassy, 1996c; Gheerbrant et al., 2005). Harris's (1978) hypothesis that advanced, lophodont deinotheres derived from a moerithere-like bunolophodont ancestor is supported by the new finds from Chilga, Ethiopia (Sanders et al., 2004). In this revised phylogenetic scheme, deinothere lophodonty is secondarily derived in comparison with the condition of cheek teeth in "Plesielephantiformes" (Gheerbrant et al., 2005).

Subfamily CHILGATHERIINAE Sanders, Kappelman and
Rasmussen, 2004
Figures 15.6A and 15.6B

These are diminutive deinotheres, with teeth smaller than in *Prodeinotherium* and *Deinotherium* (figures 15.6A, 15.6B). This monogeneric, monospecific subfamily also differs from deinotheriines in the following features: P3 with bunodont cusps that are more independent in occlusal distribution and that crowd the talonid basin and with a weakly formed ectoloph; m2 with poor expression of cristids; m2, m3, and M3 with at least incipient development of tritoloph(id)s. They are

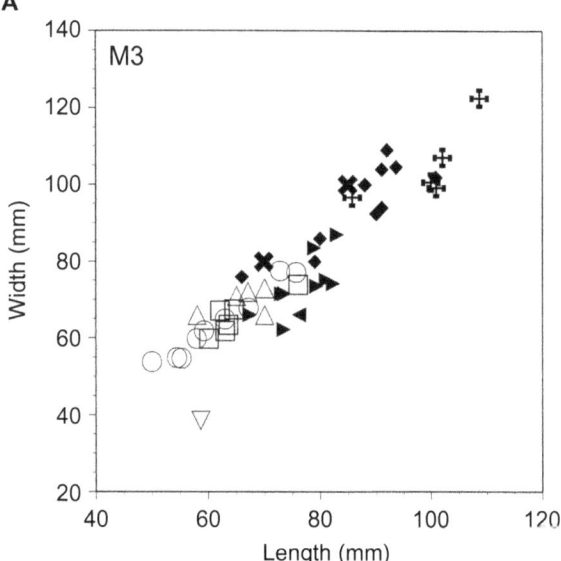

A

M3

(Bivariate plot: Length (mm) on x-axis from 40 to 120, Width (mm) on y-axis from 20 to 140)

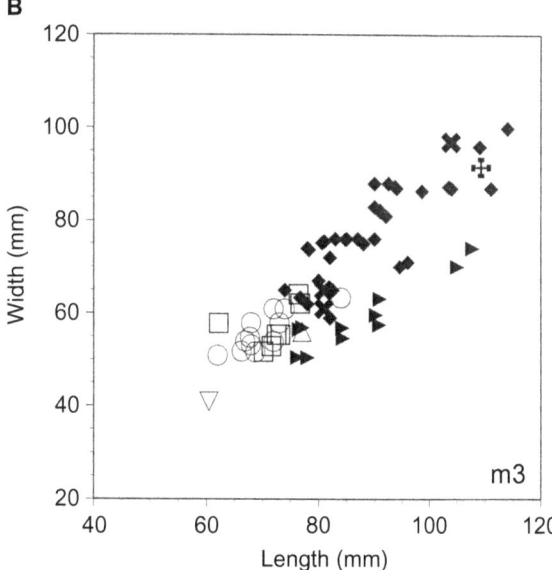

B

m3

(Bivariate plot: Length (mm) on x-axis from 40 to 120, Width (mm) on y-axis from 20 to 120)

FIGURE 15.6 Bivariate plots of M3 (A) and m3 (B) crown length versus width in deinotheres and barytheres. Comparative dimensions supplementing original measurements are from Bachmann (1875), Weinsheimer (1883), Roger (1886), Andrews (1911), Forster Cooper (1922), Palmer (1924), Éhik (1930), MacInnes (1942), Gräf (1957), Sahni and Tripathi (1957), Symeonidis (1970), Harris (1973, 1977a, 1983, 1987a), Gaziry (1976), Tobien (1988), Tsoukala and Melentis (1994), Huttunen (2000), Sach and Heizmann (2001), Sanders (2003), Sanders et al. (2004), and Delmer (2005).

SYMBOLS: inverted open triangle, *Chilgatherium harrisi*; open circle, *Prodeinotherium hobleyi*; open square, *P. bavaricum* (including "*P. hungaricum*"); open triangle, *P. pentapotamiae*; cross, *Deinotherium bozasi*; closed diamond, *D. giganteum* (including "*D. levius*"); X, *D. indicum*; right-facing closed triangle, *Barytherium grave*; Q, left-facing closed triangle, *Barytherium* sp. indet. (Birket Qarun Fm., Fayum, Egypt).

distinguished from *Phosphatherium*, *Daouitherium*, *Numidotherium*, and *Barytherium* (all with bilophodont molars) by development of the m2 distocristid into a third lophid; by development of a postentoconulid in m3; by greater expression of lingual cusps (tetradonty) in P3 (shared with other deinotheres); and by the bunodont (P3) and bunolophodont condition of cheek teeth (Sanders et al., 2004).

Genus *CHILGATHERIUM* Sanders, Kappelman, and Rasmussen, 2004
CHILGATHERIUM HARRISI Sanders, Kappelman, and Rasmussen, 2004
Figures 15.5A–15.5C

Partial Synonymy Deinotheriidae gen. et sp. nov., Kappelman et al., 2003.

Age and Occurrence Late Oligocene, eastern Africa (table 15.3).

Diagnosis As for the subfamily.

Description Sanders et al. (2004). Known only by a small dental sample. In occlusal view, the P3 resembles those in deinotheriines, with an anterior projection on its buccal side, but the cusps are singularly inflated and transversely separated (figure 15.5A). The paracone and metacone abut to form a weak ectoloph, without mammillons (figure 15.5B). The M3 has buccolingually continuous lophs but retains individual cusps at the corners of these lophs. Its tritoloph is low, formed of a transverse row of apical digitations or mammillons, and is closely appressed to the metaloph (figure 15.5C). The metaloph and tritoloph enclose a small basin, as in some posterior molars of *Moeritherium* (Tobien, 1978). A low, arced cingulum is located distal to the tritoloph (figure 15.5C).

The m2 has a more substantial tritolophid, with no distocingulid posterior to it. In m3, the tritolophid is weakly expressed, and its outer cusps are only tenuously connected transversely by several mammillons. The main lophids in both m2 and m3 form transversely continuous crests anchored by individual cusps. Cristids are weakly developed in these teeth.

Remarks Cheek tooth morphology of *Chilgatherium* provides support for Harris's (1969, 1975, 1978) prescient hypothesis that the tritoloph of m1 developed by hypertrophy of the distocingulid, and that the tritoloph of M1 evolved via exaggeration of postmetaloph ornamentation.

Recovery of *Chilgatherium* from Chilga, Ethiopia fills part of the substantial temporal gap between early Miocene deinotheres and early Oligocene moeritheres, and extends the geological age of deinotheres about five million years farther back in time than was previously known. The similarly dated Erageleit Beds at Lothidok, Kenya (Boschetto et al., 1992; Leakey et al., 1995) have not yielded chilgatheres among its mammalian assemblage. The Chilga specimens also clarify the phylogenetic position of deinotheres as part of a separate, later proboscidean radiation from the initial diversification of "plesielephantiforms."

Subfamily DEINOTHERIINAE Bonaparte, 1845
Figures 15.5D–15.5J and 15.5L

Large herbivorous, graviportal proboscideans. Dental formulae 0.0.3/1.0.3 for deciduous teeth and 0.0.2.3/1.0.2.3 for the permanent dentition; second deciduous premolars and third premolars with well-developed external crests; deciduous fourth premolars and first molars trilophodont (figures 15.5D–15.5G); remainder of the cheek teeth bilophodont. Horizontal tooth replacement not developed so that all permanent teeth may be erupted at the same time (figure 15.5J). Mandibular symphysis and lower tusks curved downward so that the tusk tips are vertically or near vertically aligned (figure 15.5L). Skull low with deep rostral trough, retracted external nares, low orbit, inclined occiput, high occipital condyles, elongate paroccipital processes, and diploë (figures 15.5H, 15.5I).

TABLE 15.3
Major occurrences and ages of Afro-Arabian deinotheres
? = Attribution or occurrence uncertain; alt. Alternatively

Taxon	Occurrence (Site, Locality)	Stratigraphic Unit	Age	Key References
	DEINOTHERIIDAE, LATE–OLIGOCENE–EARLY PLEISTOCENE CHILGATHERIINAE, LATE OLIGOCENE			
Chilgatherium harrisi	Chilga, Ethiopia (type)	Chilga Fm.	Late Oligocene, 28–27 Ma	Kappelman et al., 2003; Sanders et al., 2004
	DEINOTHERIINAE, EARLY MIOCENE–EARLY PLEISTOCENE			
Prodeinotherium hobleyi	Adi Ugri, Eritrea		Early Miocene	Vialli, 1966; Harris, 1978
	Sbeitla, Tunisia		Early Miocene	Vialli, 1966; Harris, 1978
	Tébessa, Algeria		Early Miocene	Brives, 1919; Vialli, 1966; Harris, 1978
	Moroto II, Uganda		>20.6 Ma (alt. ca. 17.5 Ma)	Bishop and Whyte, 1962; Bishop, 1967; Gebo et al., 1997; Pickford et al., 2003
	Arrisdrift, Namibia	Arrisdrift Gravel Fm.	ca. 20–19 Ma	Anonymous, 1976; Harris, 1977a
	Napak, Uganda		ca. 20.0 Ma	Bishop, 1964; Pickford, 2003; MacLatchy et al., 2006
	Koru, Kenya	Koru Fm.	19.5 Ma	Bishop, 1967; Harris, 1978; Pickford, 1986a
	Bukwa, Uganda		ca. 19.5–19.1 Ma (alt. 23.0 Ma)	Walker, 1969; Harris, 1978; Pickford, 1981; Drake et al., 1988; MacLatchy et al., 2006
	Samburu Hills, Kenya	Nachola Fm.	19.2–15.0 Ma	Pickford et al., 1987
	Wadi Moghara, Egypt	Moghara Fm.	18–17 Ma	Osborn, 1936; Harris, 1978; Miller, 1996, 1999
	Kalodirr, Kenya	Tiati Grits	18.0–16.0 Ma (alt. 17.7–16.6 Ma)	Drake et al., 1988
	Lothidok, Kenya	Moruorot Mb.	17.9–17.5 Ma	Bishop, 1967; Madden, 1972; Pickford, 1981; Tassy, 1986; Boschetto et al., 1992
	Karangu, Kenya (type)	Karangu Fm.	17.8 Ma (alt. 22.5 Ma)	Andrews, 1911; Bishop, 1967; Pickford, 1981; Tassy, 1986; Drake et al., 1988
	Rusinga, Kenya	Wayondo, Hiwegi, and Kulu Fms.	Slightly >17.8 Ma; 17.8 Ma; slightly <17.8 Ma	MacInnes 1942; Bishop, 1967; Harris, 1969, 1973; Pickford, 1981, 1986a; Tassy, 1986
	Mfwangano, Kenya	Hiwegi Fm.	17.8 Ma	Bishop, 1967; Pickford, 1986a; Drake et al., 1988
	Arongo Uyoma (Chianda), Kenya		Early Miocene, cf. Hiwegi and Kulu Fms.	Pickford, 1986a
	Buluk (West Stephanie), Kenya		>17.3 Ma (alt. 16–15 Ma)	Harris and Watkins, 1974; Pickford, 1981; McDougall and Watkins, 1985; Tassy, 1986

TABLE 15.3 (CONTINUED)

Taxon	Occurrence (Site, Locality)	Stratigraphic Unit	Age	Key References
P. hobleyi continued	Mwiti, Kenya		ca. 17 Ma	Savage and Williamson, 1978; Tassy, 1986; Drake et al., 1988
	Loperot, Kenya		Early Miocene, ca. 17 Ma	Bishop, 1967; Savage and Williamson, 1978; Pickford, 1981; Leakey and Leakey, 1986; Tassy, 1986
	As Sarrar, Saudi Arabia	Dam Fm.	Early middle Miocene, ca. 17–15 Ma (alt. ca. 19–17 Ma)	Thomas et al., 1982; Whybrow and Clements, 1999
	Gebel Zelten, Libya	Qaret Jahanneam Mb., Marada Fm.	ca. 16.5 Ma	Savage and White, 1965; Savage, 1967; Harris, 1973; Sanders, 2008a
	Maboko, Kenya	Maboko Fm.	ca. 16 Ma (alt. slightly >14.7 Ma)	MacInnes, 1942; Bishop, 1967; Tassy, 1986; Feibel and Brown, 1991
	Majiwa, Kenya	Maboko Fm.	ca. 16 Ma (alt. slightly >14.7 Ma)	Andrews et al., 1981
	Muruyur, Tugen Hills, Kenya	Mururyur Beds	15.5 Ma (alt. 14.3–13.2 Ma)	Bishop, 1972; Harris, 1978
	Beni Mellal, Morocco		Middle Miocene, ca. 14 Ma	Lavocat, 1961; Remy, 1976
	Ft. Ternan, Kenya	Ft. Ternan Beds	14.0–13.9 Ma	Shipman et al., 1981; Cerling et al., 1997
	Ombo, Kenya		Middle Miocene	Bishop, 1967; Tassy, 1986
	?Sinda Area, Lower Semliki, Democratic Republic of Congo	Upper Mb., Sinda Beds	?Middle Miocene	Hooijer, 1963; Makinouchi et al., 1992; Yasui et al., 1992
	Alengerr, Kenya	Alengerr Beds	ca. 14.0–12.5 Ma	Bishop et al., 1971; Bishop, 1972; Harris, 1978
	Gebel Cherichera, Tunisia	Beglia Fm.	ca. 13-11 Ma	Harris, 1978
	Ngorora, Tugen Hills, Kenya	Ngorora Fm.	Within 13.0–8.5 Ma interval	Bishop et al., 1971; Hill et al., 1985; 1986; Hill, 2002
Prodeinotherium sp. indet.	Lothidok, Kenya	Eragaleit Beds	Late Oligocene, 27.5–24.0 Ma	Boschetto et al., 1992; Rasmussen and Gutiérrez, 2009
Deinotherium bozasi	Ngorora, Tugen Hills, Kenya	Ngorora Fm.	Within 13.0–8.5 Ma interval	Harris, 1983; Hill, 2002
	Nakali, Kenya	Nakali Fm.	9.9–9.8 Ma	Harris, 1978; Nakatsukasa et al., 2007
	Samburu Hills, Kenya	Lower, Upper Mbs., Namurungule Fm.	ca. 9.5 Ma	Nakaya et al., 1984; Sawada et al., 1998; Tsujikawa, 2005b
	Lothagam, Kenya	Lower and Upper Mbs., Nawata Fm.; Apak and Kaiyumung Mbs., Nachukui Fm.	At least 7.4–5.0 Ma; 6.5–5.0; ca. 3.5–3.0 Ma	McDougall and Feibel, 2003; Harris, 2003
	Middle Awash, Ethiopia	Adu Asa Fm., Kuseralee Mb., Sagantole Fm.	ca. 6.3–5.6 Ma; ca. 5.6–5.2 Ma	Kalb et al., 1982; Renne et al., 1999; WoldeGabriel et al., 2001; Haile-Selassie et al., 2004

Taxon	Occurrence (Site, Locality)	Stratigraphic Unit	Age	Key References
D. *bozasi* continued	Tugen Hills, Kenya	Lukeino and Chemeron Fms.	6.2–5.6 Ma; 5.6–1.6 Ma	Bishop et al., 1971; Harris, 1977b; Hill et al., 1985, 1986; Deino et al., 2002; Hill, 2002
	Kanam East, West, Central, Kendu Bay, Kenya	Kanam Fm.	Late Miocene to early Pliocene	Kent, 1942; MacInnes, 1942; Pickford, 1986a; Ditchfield et al., 1999
	Bala, Homa Peninsula, Kenya	Kanam Fm.	Late Miocene to early Pliocene	Pickford, 1986a
	Wadi Natrun, Egypt		Late Miocene or early Pliocene	Bernor and Pavlakis, 1987
	Manonga Valley, Tanzania	Tinde Mb., Wembere-Manonga Fm.	ca. 5.0–4.5 Ma	Harrison and Baker, 1997; Sanders, 1997
	As Duma, Gona Western Margin, Ethiopia		4.5–4.3 Ma	Semaw et al., 2005
	Aramis, Middle Awash, Ethiopia	Aramis Mb., Sagantole Fm.	4.4 Ma	WoldeGabriel et al., 2001
	Kanapoi, Kenya	Kanapoi Fm.	4.2–4.1 Ma	Feibel 2003; Harris et al., 2003
	Omo, Ethiopia (type)	Mursi, Usno, and Shungura Fms.	>4.15 Ma; 3.6–2.7 Ma; 3.6–1.16 Ma	Arambourg, 1934a, 1934b; Arambourg et al., 1969; Beden, 1975, 1976; Brown et al., 1985; Feibel et al., 1989; Alemseged, 2003
	West Turkana, Kenya	Kataboi, Lomekwi, Lokalalei, Kalochoro, Kaitio, and Natoo Mbs., Nachukui Fm.	4.1–1.3 Ma	Harris et al., 1988a, 1988b; Brugal et al., 2003
	East Turkana (Ileret, Allia Bay, Koobi Fora), Kenya	Koobi Fora Fm.	4.1–1.3 Ma	Harris, 1983; Brown, 1994; Behrensmeyer et al., 1997
	Ekora, Kenya		ca. 4.0–3.75 Ma (slightly younger than 4.0 Ma)	Patterson et al., 1970; Behrensmeyer, 1976
	Karonga, Uraha, Malawi	Chiwondo Beds	ca. 4.0–3.0 Ma	Bromage et al., 1995
	Laetoli, Tanzania	Upper Unit, Laetolil Beds; ?Upper Ndolanya Beds	3.8–3.5 Ma; ?~2.7–2.6 Ma	Drake and Curtis, 1987; Harris, 1987a; Sanders, 2005
	Hadar, Ethiopia	Sidi Hakoma, Denen Dora, Kada Hadar Mbs., Hadar Fm.	3.4–2.9 Ma	Taieb et al., 1976; White et al., 1984; Bonnefille et al., 2004
	Praia de Morrungusu, Mozambique		Late Pliocene or early Pleistocene	Harris, 1977a
	Olduvai Gorge, Tanzania	Beds I and II	~1.87–1.6 Ma (alt. 2.10–1.75 Ma)	Osborn, 1936; Leakey, 1965; Hay, 1976; Tamrat et al., 1995
	Chesowanja, Kenya		>1.42 Ma	Bishop et al., 1971; Hooker and Miller, 1979; Beden, 1985
	Marsabit Road, Kenya		Early Pleistocene	Gentry and Gentry, 1969; Harris, 1978
	Kanjera, Kenya	Kanjera Fm.	1.1–1.0 Ma	Behrensmeyer et al., 1995

Genus *PRODEINOTHERIUM* Éhik, 1930
Figures 15.5I, 15.5L, 15.6A, and 15.6B

Small deinotheriines (figure 15.6A, 15.6B). Dental formulas as for the family; second–third molars with well-defined postmetaloph-posthypolophid ornamentation. Skull rostrum turned down parallel to the mandibular symphysis (figures 15.5I, 15.5L); rostral trough and external nares narrow; preorbital swelling close to top of orbit; external nares more anteriorly sited than in *Deinotherium* and nasal bones with anterior median projection; skull roof relatively longer and wider than in *Deinotherium*; occiput more vertically inclined; occipital condyles sited more ventrally than in *Deinotherium*; paroccipital processes shorter. Postcranial skeleton graviportally adapted; scapula with well-developed spine and stout acromion and metacromion; tarsals and carpals narrow but not dolichopodous.

PRODEINOTHERIUM HOBLEYI (Andrews, 1911)
Figures 15.5I, 15.5K, and 15.5L

Partial Synonymy Deinotherium hobleyi, Andrews, 1911; *D. cuvieri*, Brives, 1919; *D. bavaricum*, Gräf, 1957; *Deinotherium* sp., Vialli, 1966; *D. cuvieri*, Savage, 1967; *Prodeinotherium hobleyi*, Harris, 1973.

Age and Occurrence Early–middle Miocene, eastern and northern Africa (table 15.3).

Diagnosis Andrews (1911). Differs from *Prodeinotherium bavaricum* by having a more distinct anterointernal cusp in p3, a proportionately less elongate p4, and a more distinct outer tubercle in the talonid of m3.

Description Harris (1973, 1978). Crania are long and low, pneumatized, and have retracted nasal openings suggesting development of a proboscis. The nasals contact the frontal. Although the basicranium is not strongly angled upward, the occipital condyles and external auditory meatus on each side are much higher than the palatal plane (figure 15.5I). Skull length reaches 940 mm.

Mandibles have a broad, elongate, and strongly downcurved symphysis that extends posteriorly to p3 (figure 15.5K). The corpus appears relatively slender, and the ramus anteroposteriorly broad but not high, with a modestly sized coronoid process (figure 15.5L).

The i2s are oval in cross section and lack enamel; conversely, di2s are covered in enamel. These tusks are usually nearly vertical. Wear facets may be present near their tips, indicating their importance for food acquisition (but likely not for digging).

Abundant postcrania of *P. hobleyi* have been recovered from Gebel Zelten, Libya. These are readily distinguishable from those of elephantoids, but exhibit similar graviportal adaptations, such as the upward alignment of the femoral head and greater length of the femur than the tibia. Height of the spinous processes of the anterior thoracic vertebrae, projection of paroccipital processes, and elevation of the occipital condyles suggest that these deinotheres had greater capacity for rotation of the head about the condyles than did elephantoids. The astragalus is relatively narrower than in gomphotheres, but has larger, more vertically oriented articulations for the fibula and tibial malleolus, and an unusual, posteromedially projecting tubercle.

Remarks The oldest African deinotheriines are documented from late Oligocene- early Miocene sites in Kenya and Uganda (table 15.3). *Prodeinotherium hobleyi* is a common element of early Miocene assemblages of eastern Africa (table 15.3), but thereafter declines in relative abundance. The only known crania are from Gebel Zelten in Libya (Harris, 1973), but even partial deinotheriine molars are unmistakable by virtue of their size, smoothness, and characteristic loph(id) shape. *Prodeinotherium hobleyi* teeth of middle Miocene age (e.g., Maboko, Gebel Zelten) are slightly larger than those from the early Miocene (e.g., Rusinga). Most African deinotheriine teeth can be readily allocated to *P. hobleyi* or *D. bozasi* on an overall size basis, although some teeth or tooth fragments from late Miocene horizons appear to be of intermediate size, as might be expected in an ancestor-descendant relationship. Other recognized congeners include the type species from Europe, *Prodeinotherium bavaricum* (von Meyer, 1831), the South Asian species *P. pentapotamiae* (Falconer, 1868), and possibly *P. sinense* from China (Qiu et al., 2007).

Genus *DEINOTHERIUM* Kaup, 1829
Figures 15.6A and 15.6B

Large deinotheres (figures 15.6A, 15.6B). Dental formulae as for the family; tendency for the development of subsidiary styles on P3–4 and for simplification of the postmetaloph ornamentation of second and third molars, compared to *Prodeinotherium*. Rostral trough and external nares wide; preorbital swelling sited anteriorly on the rostrum; skull roof short and narrow at the temporal fossae; occiput slopes gently posteriorly; occipital condyles elevated above the level of the external auditory meatus; and paroccipital processes very elongate. Postcranial skeleton with supposed cursorial modifications to graviportal structure; scapular spine reduced with no acromion or metacromion; carpals and tarsals narrow with dolichopodous metapodials exhibiting functional tetradactyly (Harris, 1978).

DEINOTHERIUM BOZASI Arambourg, 1934

Partial Synonymy Deinotherium giganteum, Joleaud, 1928; *D. hopwoodi*, Osborn, 1936; *D. giganteum* var. *bozasi*, Dietrich, 1942.

Age and Occurrence Late Miocene–early Pleistocene, eastern and southern Africa (table 15.3).

Diagnosis Species of *Deinotherium* with teeth of similar size to *D. giganteum* but not found outside Africa. Skull rostrum turned steeply down comparable to that of *Prodeinotherium hobleyi* and in contrast to that of *D. giganteum*. External nares and rostral trough narrower than *D. giganteum*; preorbital swelling reduced and located just in advance of P3; occiput steeply inclined; nasal bones with slight anterior median projection. Mandibular symphysis flexed at right angles.

Description Harris (1976, 1978, 1983). The cranium of *D. bozasi* is relatively higher than in *Prodeinotherium*, with a shorter, narrower roof and more projecting paroccipital processes; cranial length is greater, exceeding 1,100 mm; and the external nares are more posteriorly retracted, suggesting the presence of a longer proboscis. Despite these differences, overall morphological affinity of the cranium of *D. bozasi* is closer to that of *P. hobleyi* than with Eurasian *Deinotherium*.

The mandibular symphysis is relatively shorter than in *P. hobleyi*, but more abruptly flexed downward, and lower tusks are slightly curved posteriorly under the symphysis.

In the astragalus, the posteromedial process is reduced and the tibial facet is flatter, relative to the condition in *Prodeinotherium*. Also, the scapular spine is reduced, and there is no metacromion or acromion. Reconstruction of manus configuration suggests that it was more digitigrade than in *Prodeinotherium*.

Remarks The first specimen of *D. bozasi* to be figured was an isolated upper molar from the Omo, Ethiopia, that Haug (1911) wrongly identified as a lower. Joleaud (1928) attributed the specimen to *Deinotherium giganteum*. Arambourg (1934a, 1934b) mentioned this tooth and described the lectotype mandible when creating the species *D. bozasi*. Subsequently, Haug's (1911) specimen was identified as the holotype of *D. bozasi* (Arambourg, 1947:249), but it was never accessioned into the collections of the Museum National d'Histoire Naturelle, and its current whereabouts are unknown (P. Tassy, pers. comm.).

Similar-aged species of *Deinotherium* are also known from Europe (type species *D. giganteum* Kaup, 1829, possibly including *D. gigantissimum* Stefanescu, 1892) and South Asia (*D. indicum* Falconer, 1845), but thus far only *Prodeinotherium* has been identified in Asia (Qiu et al., 2007). *Deinotherium bozasi* was the last surviving deinothere species and is only represented by small numbers of specimens. Most records are based on teeth or tooth fragments, but crania are known from the younger part of the Koobi Fora Formation (Harris, 1976) and from Hadar. The species has been recovered from as far south as Malawi (Bromage et al. 1995) and Mozambique (Harris, 1977a), but remains from North Africa have so far proved elusive (but see Remy, 1976). As with *Prodeinotherium hobleyi* teeth, there seems to be a temporal size gradient in *D. bozasi* teeth, those from Kanapoi being smaller than those from younger horizons in the Lake Turkana Basin (Harris et al., 2003). The last occurrences of *D. bozasi* appear to have been in Omo Shungura Member K (Beden, 1985) and at Kanjera (Behrensmeyer et al., 1995).

The low-crowned, lophodont teeth of deinotheres are superficially similar to those of tapirs. The lophs have beveled cutting edges that are maintained, though at different angles, throughout the life of the individual for processing the food prior to digestion (Harris 1975). There is an overall increase in size from the earliest representatives of *Prodeinotherium* to the latest representatives of *Deinotherium,* but deinotheriine teeth remain essentially unchanged in morphology from the early Miocene to the early Pleistocene. The conservative nature of their low-crowned lophodont teeth suggests that deinotheriines were well adapted to a browsing diet. Stable isotope analysis confirms that deinotheriines maintained a C_3 diet from their earliest record until their extinction in the early Pleistocene, thereby contrasting with the anancine gomphotheres and elephantids that transitioned to a C_4 diet during the late Miocene (Cerling et al. 2005).

Tobien (1962) noted that the manus and pes of *Deinotherium giganteum* were narrower than those of contemporary elephantoids, and suggested that this feature was a cursorial adaptation. However, this seems less likely in the face of Christiansen's (2004) interpretation that *Deinotherium* was appreciably larger than extant elephants, approaching the size of *Mammuthus trogontherii*, and that the largest individuals of *D. giganteum* would have reached 20,000 kg. Even though their conservative brachyodont teeth suggest that that deinotheriines had a more nutritious diet and greater digestive efficiency than extant elephants, they would have required a substantial amount of daily fodder. Craniodental features, including modest sized lower tusks and a probably short proboscis, suggest that deinotheres were acquiring this forage in jungle and densely vegetated gallery forest habitats (Harris, 1978). This may explain why *D. bozasi* is not found in the more arid northern and southern portions of the continent and why the species became extinct as equatorial habitats became more open and drier. It may also explain why only small numbers of deinotheres are encountered at fossiliferous localities after the early Miocene.

Suborder ELEPHANTIFORMES Tassy, 1988

Elephantiformes consists of palaeomastodonts and elephantoids (mammutids, gomphotheres, stegodonts, and elephants). These elephant-like taxa became the dominant proboscideans in Afro-Arabia during the late Paleogene-Neogene, supplanting numidotheres, barytheres, and moeritheres. United by elongation of the face, posterior shift of the orbits and retraction of the nasal opening (signaling the development of a trunk), and elongation of the mandibular symphysis (Tassy, 1994b), the group is also characterized by trends for increase in body size, expansion of cranial diploë, and impressive enlargement and projection of tusks. Despite their remarkable morphology, and in contrast to their striking diversity and nearly worldwide dispersal during the Miocene (Shoshani and Tassy, 1996; Todd, 2006), Elephantiformes survives today only as Asian and African elephants, the last of the proboscideans.

Family PALAEOMASTODONTIDAE Andrews, 1906

Palaeomastodonts of the Afro-Arabian Oligocene (table 15.4) represent the first comprehensive appearance of elephant-like morphology among proboscideans. Of considerable dimensions, ranging in size from slightly larger than moeritheres to the magnitude of modern Asian elephants (Andrews, 1906; Christiansen, 2004), their postcranial skeletons are correspondingly graviportal (Andrews, 1906).

Palaeomastodont skulls are primitive, with small, low braincases, sagittal crests, and modest pneumatization (Andrews, 1906, 1908). More advanced features shared with elephantoids include short nasals, retraction of the nasal aperature, and backward shift of the orbits to a position above the molars, consistent with the presence of at least a small trunk. Occipital condyles project markedly posteriorly and are strongly convex dorsoventrally, suggesting that the head was capable of moving over a wide vertical arc (Andrews, 1906).

The palaeomastodont dental formula is I1/1-C0/0-P3/2-M3/3, preceded in each jaw quadrant by three deciduous premolars and a "milk" tusk (Andrews, 1906, 1908). Adult teeth were in occlusion simultaneously. Their molars are bunolophodont and very brachyodont, and basically three-lophed.

Palaeomastodont dentaries are conjoined anteriorly by an elongated, nearly straight symphysis which projects beyond the rostrum. Their i2s form moderate-sized, procumbent tusks closely appressed to one another in the midline, and their I2s are oriented down and slightly laterally. The former have a flattened, pyriform shape in cross section; the latter have a lateral enamel band and are more rounded and wider dorsally than ventrally. There are extensive diastemas between the tusks and the anteriormost premolars. Their low-crowned teeth and existence in well-watered forested and woodland conditions (Wight, 1980; Bown et al., 1982; Olson and Rasmussen, 1986; Pickford, 1987b; Bown and Kraus, 1988; Gagnon, 1997; Jacobs et al., 2005) suggest that palaeomastodonts were browsers.

Palaeomastodont taxonomy has remained unstable since their initial discovery in the early 1900s (Moustafa, 1974a,

TABLE 15.4
Major occurrences and ages of Afro-Arabian palaeomastodonts
? = Attribution or occurrence uncertain; alt. Alternatively.

Taxon	Occurrence (Site, Locality)	Stratigraphic Unit	Age	Key References
	PALAEOMASTODONTIDAE, ?LATEST EOCENE–LATE OLIGOCENE			
Palaeomastodon beadnelli (including "P. parvus" and "P. intermedius")	Fayum, Egypt (type)	Gebel el Qatrani Fm.	Early Oligocene, ca. 33–30 Ma	Andrews, 1901b, 1905, 1906; Matsumoto, 1922, 1924; Simons, 1968; Seiffert, 2006
Palaeomastodon sp. nov. A	Chilga, Ethiopia	Chilga Fm.	Late Oligocene, 28–27 Ma	Kappelman et al., 2003; Sanders et al., 2004
Palaeomastodon sp. nov. B	Chilga, Ethiopia	Chilga Fm.	Late Oligocene, 28–27 Ma	Kappelman et al., 2003; Sanders et al., 2004
Palaeomastodon sp. indet.	Zella, Libya		?Early Oligocene	Arambourg and Magnier, 1961; Savage, 1971
Phiomia serridens (including "Ph. minor," "Ph. wintoni," and "Ph. osborni")	Fayum, Egypt (type)	Gebel el Qatrani Fm.	Early Oligocene, ca. 33–30 ma	Andrews and Beadnell, 1902; Andrews, 1904a, 1905, 1906; Matsumoto, 1922, 1924; Simons, 1968; Seiffert, 2006
	Zella, Libya		?Early Oligocene	Arambourg and Magnier, 1961; Savage, 1971
("Ph. minor" and "Ph. wintoni")	Dor el Talha (Dur at Talhah), Libya	Idam Unit	Early Oligocene	Savage, 1971; Wight, 1980; LeBlanc, 2000
("Ph. osborni")	Gebel Bon Gobrine, Tunisia		?Early Oligocene (alt. late Oligocene)	Arambourg and Burollet, 1962
Phiomia major	Chilga, Ethiopia (type)	Chilga Fm.	Late Oligocene, 28–27 Ma	Kappelman et al., 2003; Sanders et al., 2004
Phiomia sp. indet.	Dor el Talha (Dur at Talhah), Libya	Evaporite Unit	Latest Eocene or early Oligocene	Savage, 1971; Wight, 1980; LeBlanc, 2000
	?Malembe, Angola		Early Oligocene	Pickford, 1986a, 1987a
	Taqah, Oman	Shizar Mb., Ashawq Fm.	Early Oligocene, 31.5–31.0 Ma	Thomas et al., 1999; Whybrow and Clements, 1999; Seiffert, 2006

1974b; El-Khashab, 1979). Known predominantly from the Fayum, Egypt, palaeomastodonts have alternatively been placed largely or completely in a single genus, *Palaeomastodon* (e.g., Andrews, 1906; Lehmann, 1950; Coppens et al., 1978), separated among a number of species in two genera, *Palaeomastodon* and *Phiomia* (e.g., Schlosser, 1905, 1911; Matsumoto, 1922, 1924; Osborn, 1936; Tobien, 1971, 1978; Moustafa, 1974b; El-Khashab, 1979), and have been further divided at even higher taxonomic levels (Moustafa, 1974b; Kalandadze and Rautian, 1992; McKenna and Bell, 1997; Shoshani and Tassy, 2005).

A number of hypotheses have also been proposed about the phylogenetic relationships of palaeomastodonts. In the most prominent of these, *Palaeomastodon* was considered ancestral to mammutids, and *Phiomia* ancestral to gomphotheriid elephantoids (Matsumoto, 1924; Tobien, 1971, 1978), while oth-ers posited a special ancestor-descendant relationship between *Phiomia* and amebelodonts (Osborn, 1919, 1936; Borissiak, 1929; Tobien, 1973a). Most recent, parsimony-based treatments of proboscidean phylogeny have rejected these ideas and recognized *Phiomia* and elephantoids, including mammutids, as sister taxa, and ranked *Palaeomastodon* as their immediate outgroup (Tassy, 1994b, 1996c; Shoshani, 1996).

Genus *PALAEOMASTODON* Andrews, 1901
PALAEOMASTODON BEADNELLI Andrews, 1901
Figures 15.7A, 15.7B, 15.8A, and 15.8B

Partial Synonymy Palaeomastodon parvus, Andrews, 1905; *P. barroisi* (in part), Pontier, 1907; *P. intermedius*, Matsumoto, 1922; *Palaeomastodon* (in part), Lehmann, 1950; *Palaeomast-odon (Palaeomastodon) beadnelli*, Coppens et al., 1978.

Age and Occurrence Early Oligocene, northern Africa (table 15.4).

Diagnosis Based in part on Andrews (1901c, 1901d); Matsumoto (1922, 1924); Moustafa (1974b); El-Khashab (1979); Sanders et al. (2004). Distinguished from *Phiomia* by greater length of premolar series, relative to molar row; shorter symphysis (figures 15.8A, 15.8B); wider palate; less complete development of third loph(id)s; absence or less prominent appearance of accessory central conules; more lophodont molar crowns, and tendency for chisel-like wear on loph(id) faces. Molars smaller than those of *Palaeomastodon* sp. nov. A and B (figure 15.7A). Along with *Phiomia serridens*, discernible from moeritheres by anterior projection of i2s, larger size of cheek teeth, and loss of I1, I3, C, i1, and p2, and from nearly all elephantoids by smaller tooth size (figures 15.7A, 15.17B).

Description The skull of *Palaeomastodon beadnelli* is imperfectly known, but similar to that of *Phiomia serridens* (see figure 15.8), except for differences in palatal and symphyseal proportions (Andrews, 1906, 1908; Moustafa, 1974b; El-Khashab, 1979).

Molars wear like lophodont teeth, producing sharp ridges apically, and accessory conules are low and crescentoid-like where present (Matsumoto, 1924). Zygodont crests may be prominent on the posttrite outer conelets of half-loph(id)s. Third loph(id)s of molars may be poorly formed, with posttrite last half-loph(id)s composed of only a diminutive conelet that is considerably smaller than its pretrite counterpart, yielding loph(id) formulae of 2 1/2. Loph(id) formulae for P2-4 and p3-4 are 1, 1 1/2, and 1, 2, respectively.

Palaeomastodont postcranial remains are uncommon, and do not differ significantly in structure between *Palaeomastodon* and *Phiomia* (Andrews, 1906, 1908). Recovered elements are elephant-like, and appear to have been adapted for graviportal support of heavy bodies (Andrews, 1906). For example, the femur is pillar-like with an upwardly facing head; the head of the humerus also faces primarily upward; the humerus has a large and deep olecranon fossa to help stabilize the ulna in extension; and the greatly expanded olecranon process of the ulna is posteriorly reflected and low, permitting its vertical extension on the humerus.

PALAEOMASTODON SP. NOV. A Sanders, Kappelman and
Rasmussen, 2004
Figure 15.7A

Partial Synonymy aff. *Palaeomastodon* sp. nov. A, Kappelman et al., 2003.

Age and Occurrence Late Oligocene, eastern Africa (table 15.4).

Description Molars are morphologically identical to those of *P. beadnelli* (figure 15.8F), except for their larger size (figure 15.7A).

PALAEOMASTODON SP. NOV. B Sanders, Kappelman and
Rasmussen, 2004
Figures 15.7A and 15.8G

Partial Synonymy Mammutid, Sanders and Kappelman, 2001; aff. *Palaeomastodon* sp. nov. B, Kappelman et al., 2003.

Age and Occurrence Late Oligocene, eastern Africa (table 15.4).

Description Similar to *P. beadnelli* in the weak development of the posttrite side of the tritoloph in M3, yielding a loph

formula of 2 1/2, but much larger in crown dimensions (figure 15.7A). The M3 is unique among palaeomastodonts in its rectangular occlusal outline, the pronounced development of its posterior crescentoid, zygodont crests, and cingulae, and its strong enamel rugosity (figure 15.8G; Sanders et al., 2004).

Genus *PHIOMIA* Andrews and Beadnell, 1902
PHIOMIA SERRIDENS Andrews and Beadnell, 1902
Figures 15.8A–15.8E

Partial Synonymy Palaeomastodon minor, Andrews, 1904a; *Palaeomastodon wintoni*, Andrews, 1905; *Palaeomastodon minus*, Andrews, 1905; *Palaeomastodon barroisi* (in part), Pontier, 1907; *Phiomia osborni*, Matsumoto, 1922; *Phiomia (minus) minor*, Matsumoto, 1922; *Phiomia wintoni*, Matsumoto, 1922; *Palaeomastodon* (in part), Lehmann, 1950; *Palaeomastodon (Phiomia) serridens*, Coppens et al., 1978.

Age and Occurrence Early Oligocene, northern Africa (table 15.4).

Diagnosis Matsumoto (1922, 1924). Similar in size range to *P. beadnelli*, but with a greater tendency for complete molar trilophodonty, more bunolophodont wear, stronger development of central accessory conules, and inferred greater length of the mandibular symphysis (figures 15.8A, 15.8B).

Description The most common palaeomastodont craniodental remains from the Fayum are those of *Phiomia wintoni* (Andrews, 1906), synonymized here under *Phiomia serridens*. These include well preserved mandibles and the most complete cranium known of a palaeomastodont (figure 15.8C, 15.8E), with dimensions of L = 700 mm, and W = 447 mm, proportionally similar to dimensions of a large cranium of *P. beadnelli* (BMNH 8464), with L = 635 mm and W = 420 mm (Andrews, 1906, 1908). Zygomatics are stout and flare laterally widely; the basicranium is not raised much above the level of the palate; skull height is low; and pneumatization is modest. The palate is relatively long, due to persistence of complete premolar-molar series in adults. The associated mandible has a length of 470 mm, including a slightly downturned symphysis that extends for 140 mm (Andrews, 1908). Mandibles of palaeomastodont species differ primarily in symphyseal size, inferred to be longer in *Phiomia serridens* because of the closer approximation of the posterior rim of the symphysis to its anteriormost premolar (see figures 15.8A, 15.8B, 15.8D; Matsumoto, 1922, 1924).

PHIOMIA MAJOR Sanders, Kappelman and Rasmussen, 2004
Figures 15.7A, 15.7B, 15.9A, and 15.10A

Age and Occurrence Late Oligocene, eastern Africa (table 15.4).

Diagnosis Sanders et al. (2004). Largest species of *Phiomia*, with molar size range surpassing that of Fayum palaeomastodonts (figures 15.7A, 15.7B); symphysis and incisors much longer than in other palaeomastodonts; distinguished from *Palaeomastodon* by absence of posttrite cristae, stronger development of accessory central conules, presence of a central conelet in the posterior loph of P4, and full molar trilophodonty.

Description Cheek teeth morphologically similar to those of *Phiomia serridens*. Upper tusks curve downward and outward; of flattened pyriform shape proximally, higher than wide, becoming more rounded toward the tip, with enamel along their outer face, and long for a palaeomastodont (450 mm, compared with a range of 197–250 mm for Fayum *Phiomia serridens*; Sanders et al., 2004). Lower tusks straighter, with a flattened, pyriform shape in cross section, becoming rounder

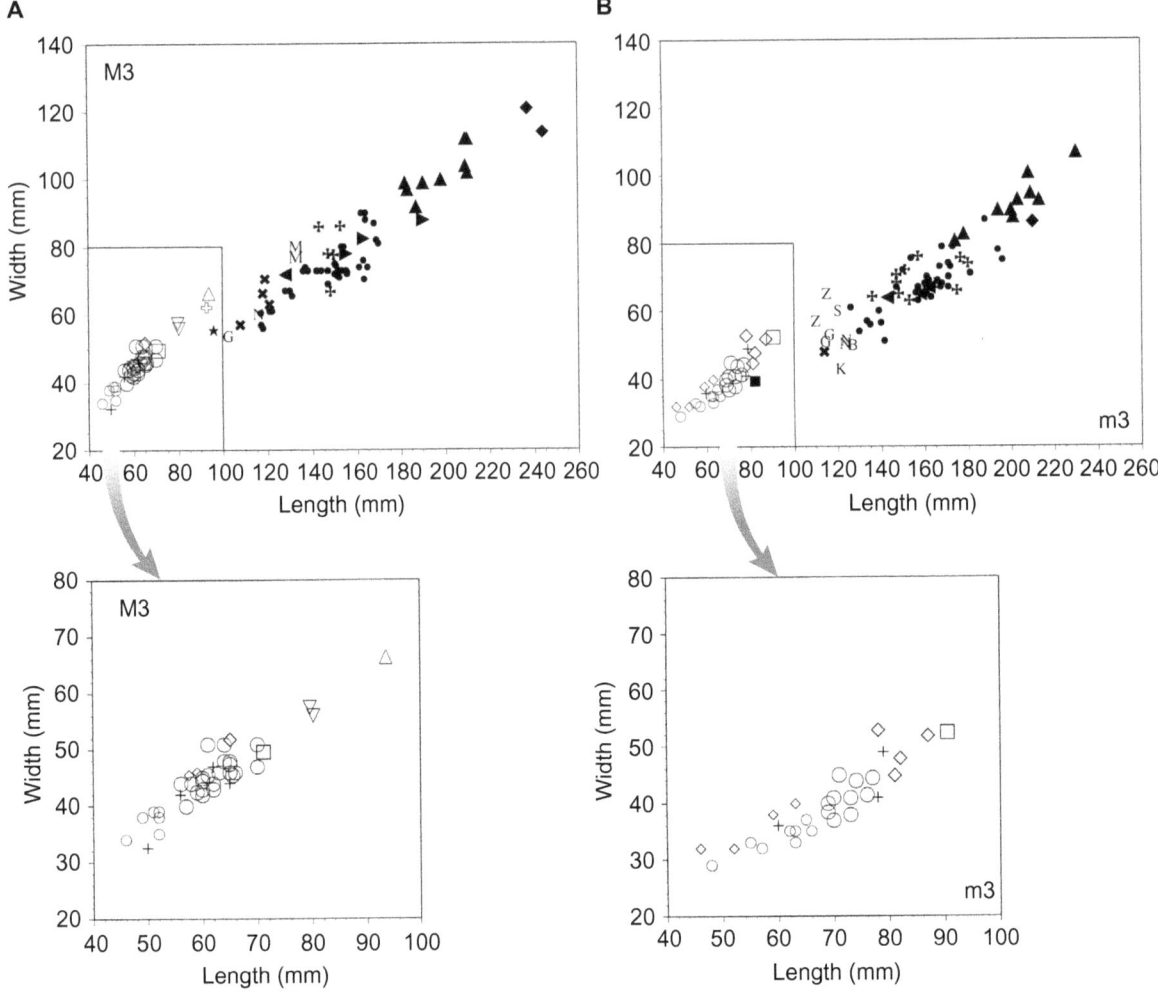

FIGURE 15.7 Bivariate plots of M3 and m3 crown length versus width in palaeomastodonts and Afro-Arabian and selected Eurasian gomphotheriids. A) M3 (palaeomastodont M3 data points detailed in lower plot); B) m3 (palaeomastodont m3 data points detailed in lower plot). Comparative dimensions supplementing original measurements are from Andrews (1906), Fourtau (1918), Forster Cooper (1922), Matsumoto (1924), Lehmann (1950), Bergounioux and Crouzel (1959), Arambourg (1961), Hamilton (1973), Gaziry (1976, 1987a), Tassy (1983b, 1985, 1986), Gentry (1987), Roger et al. (1994), Göhlich (1998), Sanders and Miller (2002), Pickford (2003, 2004, 2005a), Sanders (2003), and Sanders et al. (2004).

SYMBOLS: small open diamond, *Palaeomastodon wintoni* (female); large open diamond, *Palaeomastodon wintoni* (male); inverted open triangle, *Palaeomastodon* sp. nov. A (Chilga, Ethiopia); open triangle, *Palaeomastodon* sp. nov. B (Chilga, Ethiopia); small open circle, *Phiomia serridens* (female); large open circle, *Phiomia serridens* (male); open square, *Phiomia major*; +, Fayum palaeomastodont gen. et sp. indet. (Lehmann, 1950); open cross, *Hemimastodon*; star, cf. *Gomphotherium* sp. nov., Gebel Zelten, Libya; closed square, *Eritreum melakeghebrekristosi*; left-facing closed triangle, *Gomphotherium cooperi*; closed triangle, *G. sylvaticum*; M, *Gomphotherium* sp. (Mwititype); closed circle, *G. angustidens* (Europe, including *G. "subtapiroideum"*); cross, *G. angustidens libycum* (Wadi Moghara, Gebel Zelten, Ad Dabtiyah, Gebel Cherichera); right-facing closed triangle, *G. browni*; B, *Gomphotherium pygmaeus*, Bosluis Pan, South Africa; G, "pygmy" gomphothere, Ghaba, Oman; K, *Gomphotherium pygmaeus* (type), Kabylie, Algeria; N, *Gomphotherium pygmaeus*, Ngenyin, Tugen Hills, Kenya; S, "pygmy" gomphothere, Siwa, Egypt; Z, "pygmy" gomphothere, Gebel Zelten, Libya; closed diamond, *Tetralophodon* sp. nov.; X, ?*Tetralophodon* sp. nov. (Chorora).

toward the tip, longitudinally torqued, set close to one another about the midline, and also much longer than those in Fayum *Phiomia serridens* (460 mm compared to maximum of 250 mm; Sanders et al., 2004). Symphysis very long (382 mm, compared with a range of 137–275 mm for Fayum *Phiomia serridens*; Sanders et al., 2004), with a shallow midline channel running its length.

Remarks Differences in molar occlusal morphology and wear patterns in particular support division of palaeomastodonts into two genera. Further sorting of palaeomastodonts within each genus has been based largely on cheek tooth size (Matsumoto, 1922, 1924), resulting in recognition of as many as eight sympatric species in the Fayum (Simons, 1968), though

it is difficult to imagine how so many large, morphologically similar species could coexist ecologically. Morphometric differences in the Fayum palaeomastodont dental sample can be accommodated within two genera composed of one sexually dimorphic species each (see Schlosser, 1905, 1911); molar dimensions of *P. beadnelli* and *Ph. serridens* are bimodal and do not exceed the ranges of other, presumably dimorphic proboscidean species (figures 15.7A, 15.7B).

New fossil evidence appears to contradict parsimony-based phylogenies that place *Palaeomastodon* as the sister taxon to a clade of *Phiomia* + elephantoids. Rather, traits in the M3 of "*Palaeomastodon* sp. nov. B," from the late Oligocene of Chilga, Ethiopia, suggest a closer relationship between *Palaeomastodon*

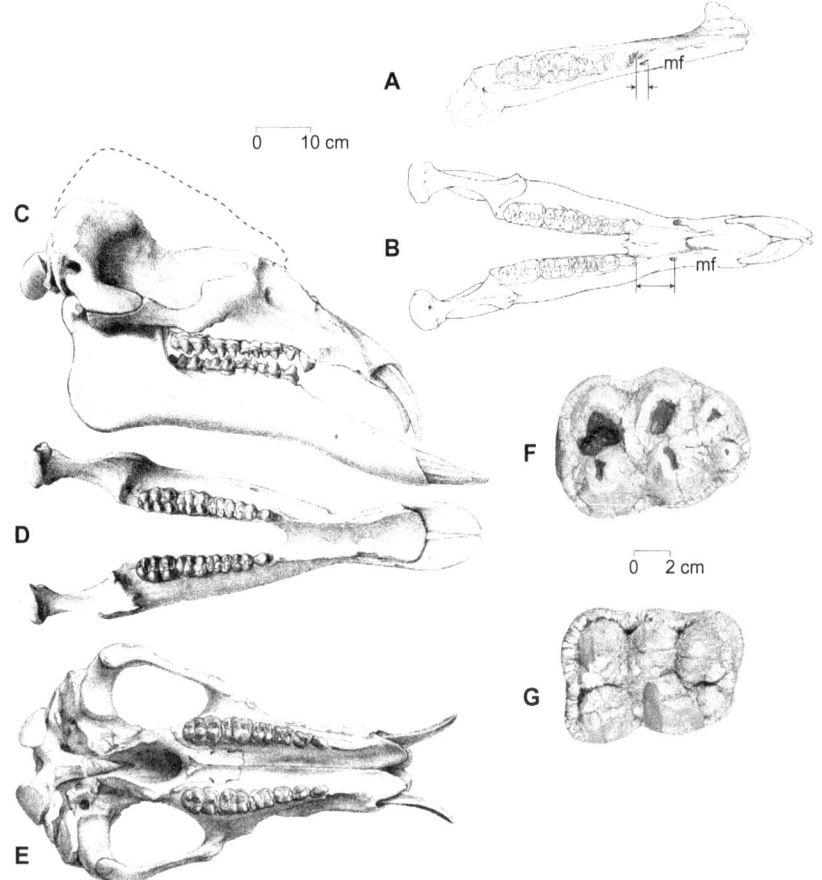

FIGURE 15.8   Aspects of palaeomastodont skull morphology. Anterior is to the right. A) Occlusal view, left dentary (reversed), C10014, *Palaeomastodon beadnelli* (type; Osborn, 1936: plate 90B). B) Occlusal view, mandible, AMNH 13468, *Phiomia serridens* ("*Ph. osborni*" [type]; Osborn, 1936: figure 90A). Arrows indicate greater distance between p3 alveolus and main mandibular foramen than in *P. beadnelli* (consistent with inferred longer symphysis). C) Right lateral view, reconstruction of skull of *Phiomia serridens* (Andrews, 1908: plate 31, figure 1). D) Occlusal view, reconstruction of mandible of *Phiomia serridens* (Andrews, 1908: plate 31, figure 2). E) Ventral view, reconstruction of cranium of *Phiomia serridens* (reversed; Andrews, 1908: plate 31, figure 3). F) Occlusal view, right M3, CH35-V-23, *Palaeomastodon* sp. nov. A. G) Occlusal view, reconstructed cast of right M3, CH14-11, *Palaeomastodon* sp. nov. B. A, B courtesy of the American Museum of Natural History.

and Mammutidae (Sanders et al., 2004). Even more compelling as a link between these taxa is the recent identification of diminutive molars (figures 15.9A, 15.10A) with distinct mammutid features, from the Eragaleit beds at Lothidok, Kenya, placed in the new species *Losodokodon losodokius* (Gutiérrez and Rasmussen, 2007; Rasmussen and Gutiérrez, 2009), dated to the latest Oligocene, ca. 27.5–24.0 Ma (Boschetto et al., 1992). This suggests that the phylogenetic split between mammutids and gomphotheres traces back at least to the beginning of the Oligocene. If so, Elephantiformes will require major systematic revision.

Superfamily ELEPHANTOIDEA Gray, 1821

This very diverse superfamily is composed of mammutids, gomphotheriids, stegodonts, and elephants (and possibly the controversial, primitive South Asian species *Hemimastodon crepusculi*) (Tassy, 1988). Elephantoidea has alternatively been used to refer to a more restricted grouping of tetralophodonts, stegodonts, and elephants (Shoshani and Tassy, 2005), but this usage destabilizes the communicative property of the superfamily as it was originally defined.

Elephantoidea has been circumscribed by a set of traits that includes lengthening of tusks and "conveyor belt" functional succession of cheek teeth (Tassy, 1994b). Some of these features likely related to increased emphases on the trunk for manipulation of objects, on the incisors for social display and interaction, and on delayed ontogenetic development of teeth, in association with increases in body size, life span, behavioral complexity, and dietary flexibility. It is possible, however, that horizontal succession of cheek teeth evolved in parallel among elephantoids, as this mechanism had apparently not yet appeared in the earliest mammutids or gomphotheriids.

Family MAMMUTIDAE Hay, 1922

This family is comprised of elephantoids whose fossil record extends to the terminal Oligocene. By that time, divergence in occlusal morphology between mammutids and gomphotheriids was already marked, suggesting a prior time of divergence. Mammutids first occurred in Africa, emigrated into Europe by the end of the early Miocene, and by the middle Miocene had spread throughout Eurasia and into the New World (Tassy, 1986;

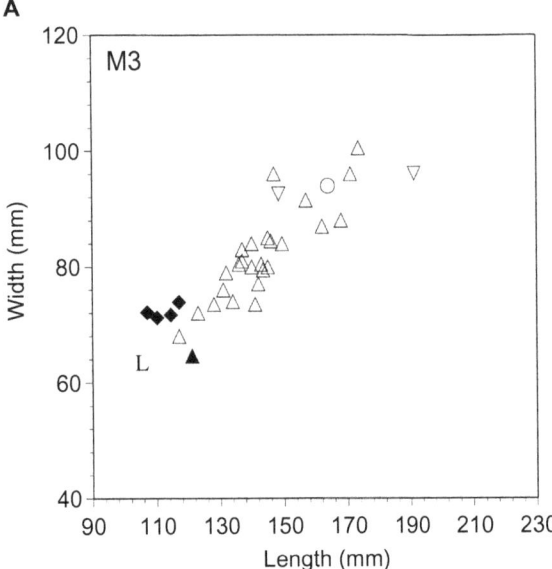

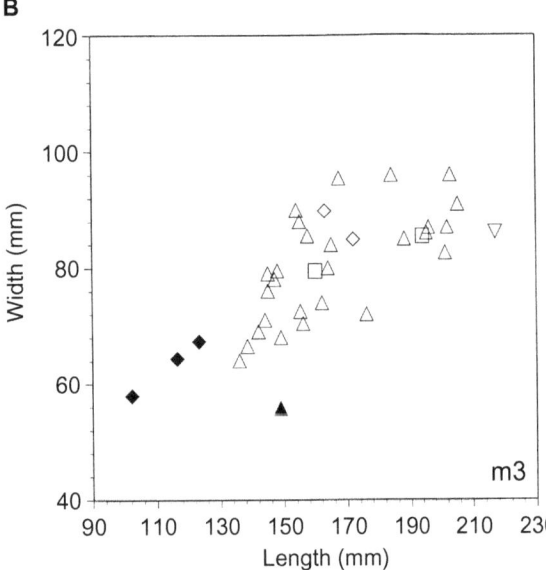

FIGURE 15.9 Bivariate plots of M3 (A) and m3 (B) crown length versus width in African and selected Eurasian mammutids. Comparative dimensions supplementing original measurements are from Osborn (1936), Tassy (1977a, 1983a, 1985), Göhlich (1998), Sanders and Miller (2002), and M. Gutiérrez (pers. comm.).

SYMBOLS: L, *Losodokodon losodokius*, Eragaleit Beds, Lothidok, Kenya; closed diamond, *Eozygodon morotoensis*; closed triangle, *Zygolophodon aegyptensis*; open triangle, *Z. turicensis*; open diamond, *Z. atavus*; inverted open triangle, *Z. metachinjiensis*; open square, *Z. gobiensis*; open circle, *Z. gromovae*.

Mazo, 1996; Saunders, 1996; Tobien, 1996). Nevertheless, despite their initial success, mammutids were progressively marginalized in African faunas throughout the Miocene in comparison with and perhaps as the result of the increasing diversity and importance of gomphotheriids (Coppens et al., 1978; Tobien, 1996).

Mammutid molars are nonbunodont and distinguished by mesoconelets markedly lower than principal, outer conelets (zygolophodonty), mesiodistal narrowing of half-loph(id) pairs apically to form relatively compressed transverse crests, lack of accessory conules in interloph(id)s, and presence of

posttrite zygodont crests. Pretrite crescentoids may also extend low into interloph(id)s. These features likely functioned in vertical shearing (Tobien, 1996), and would have been effective for mastication of leafy vegetation.

Genus *EOZYGODON* Tassy and Pickford, 1983
EOZYGODON MOROTOENSIS (Pickford and Tassy, 1980)
Figures 15.9A, 15.9B, and 15.10B–15.10G

Partial Synonymy Mastodont nov. gen. Bishop, 1967; *Zygolophodon* aff. *turicensis* Tassy, 1979a; *Zygolophodon morotoensis* Pickford and Tassy, 1980; *Eozygodon morotoensis* Tassy and Pickford, 1983.

Age and Occurrence Early Miocene, eastern and southern Africa (table 15.5).

Diagnosis Large-bodied mammutid with relatively diminutive teeth (figures 15.9A, 15.B); third molars and I2 smaller than those of most confamilials. Upper molars relatively wide (figures 15.9A, 15.10E, 15.10F). Tusks *Phiomia*-like, dissimilar to those in other mammutids. Loph(id) apices and zygodont crests of unworn molars finely crenulated. Pretrite crescentoids weak or absent.

Description Cranium short, wide, moderately high; basicranium set well above the level of the palate; occipital condyles high and project far posterior to the occiput; rostrum relatively short; palate deep; nasal aperture broad; orbit above M1–M2; infraorbital foramen above P4; large nuchal fossa; upper tusks oriented downward and slightly laterally; upper tooth rows most divergent at M2 (Pickford 2003). Upper tusks short (KNM-ME 7543, L = 510 mm), strongly curved (figure 15.10B) and ovoid to pyriform in cross section, higher than wide (-ME 7543, proximal H = 64.5 mm, W = 51.0 mm), with enamel band ventrolaterally. Lower tusks straighter than uppers (figures 15.10B–15.10D), pyriform in cross section, with distinct dorsomedial longitudinal sulcus, higher than wide and slightly flattened (KNM-ME 19, proximal H = 58.2 mm, W = 36.1 mm). Cheek teeth simultaneously in wear. P4 bilophodont, occlusally square in shape. Molars occlusally rectangular, trilophodont (except m3, which has a diminutive fourth lophid; figure 15.10G), strongly zygolophodont, with broad, open transverse valleys between loph(id)s, and without cementum; posttrite zygodont crests narrow but sharp; median sulci deep; upper cheek tooth crowns surrounded by moderately prominent beaded cingular ribbons (Pickford and Tassy, 1980; Tassy and Pickford, 1983; Pickford, 2003).

Postcrania: astragalus high, with prominent medial tubercle and extensive facet for the tibial malleolus; phalanges high and narrow. Femoral length >1,000 mm, suggesting an animal comparable to the American mastodon in body size (Tassy and Pickford, 1983; see Christiansen, 2004).

Remarks The best represented mammutid in the African fossil record is *Eozygodon morotoensis* (Pickford and Tassy, 1980; Tassy and Pickford, 1983; Tassy, 1986; Pickford, 2003). Documented from several early Miocene sites (table 15.5), *Eozygodon* is the oldest recorded elephantoid in that epoch (Tassy, 1986, 1996b). The reported presence of this species at Wadi Moghara, based on an edentulous dentary (Pickford, 2003), cannot be confirmed. Absence or poor development of fourth loph(id)s in M3/m3, small tooth size (figures 15.9A, 15.9B), strong upper molar cingulae, *Phiomia*-like tusks, and a suite of postcranial features identify this species as the most primitive of the family (Tassy and Pickford, 1983; Tassy, 1986). Molars from Meswa Bridge, Kenya are more primitive than those of Moroto, Uganda in having stronger cingulae, and weaker expression of zygodont

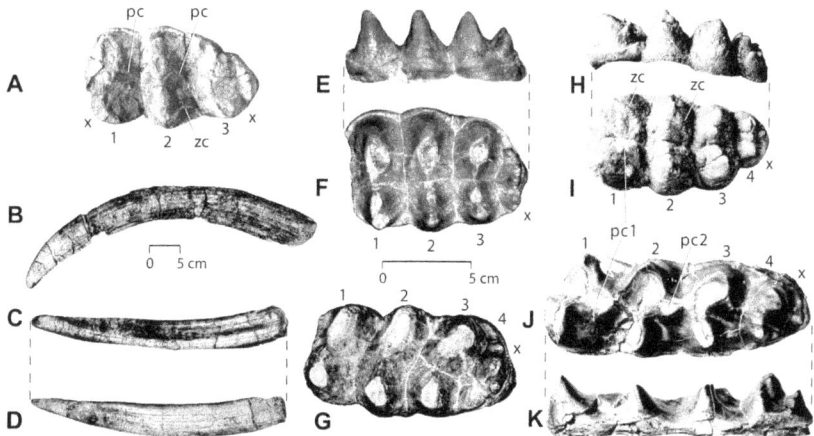

FIGURE 15.10 Aspects of mammutid dental morphology. Anterior is to the left in all specimens. A) Occlusal view, right M3, KNM-LS 18244, *Losodokodon losodokius* (courtesy of M. Gutiérrez). B) Lateral view, right I2, KNM-ME 7543, *Eozygodon morotoensis*. C) Medial view, right i2, KNM-ME 7543, *E. morotoensis*. D) Lateral view, right i2, KNM-ME 7543, *E. morotoensis*. E) Lateral view, right M3, KNM-ME 7545, *E. morotoensis*. F) Occlusal view, right M3, KNM-ME 7545, *E. morotoensis*. G) Occlusal view, right m3, KNM-ME 7547, *E. morotoensis*. H) Medial view, left M3, DPC 12598, *Zygolophodon aegyptensis*. I) Occlusal view, left M3, DPC 12598, *Zygolophodon aegyptensis*. J) Occlusal view, right m3, DPC 9009, *Z. aegyptensis* (type). K) Medial view, right m3, DPC 9009, *Z. aegyptensis* (type).

ABBREVIATIONS: pc, posterior crescentoid; x, anterior or posterior cingulum(id); zc, zygodont crest; 1, 2, 3, . . . , first, second, third, . . . loph(id).

crests and crescentoids. Paleoecological inference suggests that *E. morotoensis* was an inhabitant of dry, dense forests with open areas nearby (Tassy and Pickford, 1983).

Genus *ZYGOLOPHODON* Vacek, 1877
ZYGOLOPHODON AEGYPTENSIS Sanders and Miller, 2002
Figures 15.9A, 15.9B, and 15.10H–K

Age and Occurrence Early Miocene, northern Africa (table 15.5).

Diagnosis Anthony and Friant (1940); Tobien (1975, 1996); Tassy (1977a, 1985); Tassy and Pickford (1983); Sanders and Miller (2002). Mammutid with small molars; third molars narrower than those of confamilials (figures 15.9A, 15.9B). Distinguished from *Eozygodon* by stronger development of pretrite crescentoids, and greater expression of fourth loph(id)s in M3/m3 (figures 15.10H–15.10K), from other congeners by mesiodistally wider interlophids in m3 and anterior convexity of m3 lophids three and four, and from *Mammut* by wider median sulci, absence or trace only of cementum, stronger expression of pretrite crescentoids and cingulae(ids), smaller size, and less mediodistally attenuated loph(id) apices.

Description Only known from dentition. Unworn halfloph(id)s are formed of large outer conelets and smaller, lower mesoconelets; half-lophs are divided by deep, narrow median longitudinal sulci. Lower third molar with four lophids and low postcingulid, and prominent pretrite crescentoids. Upper third molars of putative males are substantially wider and have broader fourth lophs and blunter crests, crescentoids, and loph apices, than M3 of putative females. Loph(id)s are teat shaped in lateral outline (figure 15.10H).

ZYGOLOPHODON TURICENSIS (Schinz, 1824)

Diagnosis Tassy (1985). Molars larger, with more pronounced anterior and posterior pretrite crescentoids than in *Z. aegyptensis*. The lower third molar may exhibit a fifth lophid.

Description Pickford (2007). Two M2s from the Tugen Hills, Kenya are very large for *Z. turicensis* but otherwise resemble the molars of the robust morph of this species. They are trilophodont, with apically anteroposteriorly compressed lophs, and with salient pretrite anterior and posterior crescentoids. The posttrite cusps display zygodont crests. Transverse valleys are broad. A low cingulum runs along the lingual margin of the crown in each of these M2s.

Remarks The sparse remains of *Zygolophodon aegyptensis* unequivocally mark the presence of the genus in Africa. Until recently, the occurrence of *Zygolophodon* in Africa had been hinted at only by a paltry sample of broken teeth, including molar fragments from Gebel Cherichera, Tunisia (Tassy, 1985; Thomas and Petter, 1986), and Gebel Zelten. Of less reliable affinity is a d2 referred to ?*Zygolophodon* cf. *turicensis* from the late Miocene site of Menacer (ex-Marceau), Algeria (Thomas and Petter, 1986; see also Arambourg, 1959), alternatively considered a gomphothere tooth, as is a DP4 from Rusinga, which was originally attributed to *Zygolophodon* (Pickford and Tassy, 1980; Tassy, 1986). A presumed zygolophodont tooth from Khenchella, Tunisia (Gaudry, 1891), may be a moerithere molar (Pickford and Tassy, 1980). More confidence can be given a broken m3 from Daberas Mine, Namibia, which exhibits features typical of *Zygolophodon* and is the first evidence of the genus in southern Africa (Pickford, 2005b), and the M2s from the Tugen Hills, Kenya attributed to *Z. turicensis* (Pickford, 2007). In addition, a fragmentary mammutid P4 from the Moruorot Mb. at Lothidok, Kenya, recently placed in *Eozygodon* (Tassy, 1986; Tobien, 1996; Pickford, 2003), should be returned to its original assignment in *Zygolophodon* (Madden, 1980). Together, this modest sample indicates a pan-African distribution of the genus by the start of the middle Miocene.

Zygolophodon aegyptensis particularly resembles *Z. gromovae* from the middle Miocene of Ulan Tologoj, western Mongolian People's Republic (Dubrovo, 1974). Its presence in North Africa helps document the rich interconnections that existed between Afro-Arabia and Eurasia during the early Miocene. Geological

TABLE 15.5
Major occurrences and ages of Afro-Arabian mammutids and gomphotheriids
? = Attribution or occurrence uncertain; alt. alternatively.

Taxon	Occurrence (Site, Locality)	Stratigraphic Unit	Age	Key References
MAMMUTIDAE, LATE OLIGOCENE–?LATE MIOCENE				
Losodokodon losodokius	Lothidok, Kenya	Eragaleit Beds	Late Oligocene, 27.5–24.0 Ma	Boschetto et al., 1992; Gutiérrez and Rasmussen, 2007; Rasmussen and Gutiérrez, 2009
Eozygodon morotoensis	Meswa Bridge, Kenya	Muhoroni Agglomerate	23.5–19.6 Ma (probably 23.0–22.0 Ma)	Bishop et al., 1969; Pickford and Tassy, 1980; Pickford and Andrews, 1981; Tassy and Pickford, 1983; Tassy, 1986
	Elisabethfeld, Namibia		ca. 21 Ma	Pickford, 2003
	Moroto I (type) and II, Uganda		>20.6 Ma (alt. ca. 17.5 Ma)	Pickford and Tassy, 1980; Tassy and Pickford, 1983; Pickford et al., 1986; Gebo et al., 1997; Pickford, 2003, 2007; Pickford et al., 2003; Pickford and Mein, 2006
	Auchas, Namibia	Arrisdrift Gravel Fm.	ca. 20–19 Ma	Pickford, 2003
Zygolophodon aegyptensis	Wadi Moghara, Egypt (type)	Moghara Fm.	ca. 18–17 Ma	Miller, 1996, 1999; Sanders and Miller, 2002
Zygolophodon turicensis	Tugen Hills, Kenya	Mb. A, Ngorora Fm.	ca. 13 Ma	Pickford, 2007
Zygolophodon sp. indet.	Lothidok 4, Kenya	Moruorot Mb.	17.9–17.5 Ma	Tassy, 1986; Boschetto et al., 1992; Tobien, 1996; Pickford, 2003
	?Rusinga, Kenya	Hiwegi Fm.	17.8 Ma	Pickford and Tassy, 1980; Pickford, 1981, 1986b; Drake et al., 1988
	Daberas Mine, Namibia		early or middle Miocene, ca. 17–14 Ma	Pickford and Senut, 2000; Pickford, 2005a
	Gebel Zelten, Libya	Qaret Jahanneam Mb., Marada Fm.	ca. 16.5 Ma	Savage and Hamilton, 1973; Sanders, 2008a
	Gebel Cherichera, Tunisia	Beglia Fm.	ca. 13–11 Ma (alt. ?early Miocene)	Errington de la Croix, 1887; Robinson, 1974; Robinson and Black, 1974; Tassy, 1985; Thomas and Petter, 1986
	?Menacer (ex-Marceau), Algeria		?Late Miocene	Arambourg, 1959; Thomas and Petter, 1986; Pickford, 2007

Taxon	Occurrence (Site, Locality)	Stratigraphic Unit	Age	Key References
	FAMILY INCERTAE SEDIS, LATE OLIGOCENE			
cf. *Gomphotherium* sp.	Chilga, Ethiopia	Chilga Fm.	28–27 Ma	Kappelman et al., 2003; Sanders et al., 2004
	?Gebel Zelten, Libya	?Marada Fm.	?Late Oligocene or basal early Miocene	Pickford, 2003; Sanders, 2008a
Eritreum melakeghebrekristosi	Dogali, Eritrea	Dogali Fm.	26.8 Ma	Shoshani et al., 2001b, 2006
	GOMPHOTHERIIDAE, EARLY MIOCENE–LATE PLIOCENE			
	GOMPHOTHERIINAE, EARLY MIOCENE–MIDDLE MIOCENE			
	"*GOMPHOTHERIUM ANNECTENS* GROUP," EARLY MIOCENE			
Gomphotherium sp.	Songhor, Kenya		Early Miocene, 19.5 Ma	Bishop et al., 1969; Pickford and Andrews, 1981; Pickford, 1986b
	Mfwangano, Kenya	Hiwegi Fm.	17.8 Ma	Drake et al., 1988
	Mwiti, Kenya	Mwiti 5	ca. 17 Ma	Drake et al., 1988
	"*GOMPHOTHERIUM ANGUSTIDENS* GROUP," EARLY–LATE MIOCENE			
Gomphotherium angustidens libycum	Wadi Moghara, Egypt (type)	Moghara Fm.	ca. 18–17 Ma	Fourtau, 1918; Miller, 1996, 1999; Sanders and Miller, 2003
	Ad Dabtiyah, Saudi Arabia	Dam Fm.	Early middle Miocene (alt. ca. 19–17 Ma)	Gentry, 1987; Whybrow et al., 1987; Whybrow and Clements, 1999; Sanders and Miller, 2002
	?Fejej, Ethiopia	Bakate Fm.	>16.18 Ma	Tiffney et al., 1994; Richmond et al., 1998
	Gebel Zelten, Libya	Qaret Jahanneam Mb., Marada Fm.	ca. 16.5 Ma	Hormann, 1963; Savage and Hamilton, 1973; Coppens et al., 1978; Pickford, 2003; Sanders, 2008a
	?Al Jadidah, Saudi Arabia	Hofuf Fm.	Middle Miocene, ca. 14 Ma	Whybrow and Clements, 1999
	Gebel Cherichera, Tunisia	Beglia Fm.	ca. 13–11 Ma	Gaudry, 1891; Robinson, 1974; Robinson and Black, 1974; Pickford, 2003
	Gebel el Hendi, Testour, Tunisia		Late Miocene	Robinson and Black, 1973
Gomphotherium sp. indet.	As-Sarrar, Saudi Arabia	Dam Fm.	Early middle Miocene, 17–15 Ma (alt. ca. 19–17 Ma)	Thomas et al., 1982; Whybrow and Clements, 1999
	"'PYGMY' *GOMPHOTHERIUM* GROUP"			
Gomphotherium pygmaeus	Kabylie, Algeria (type)		?Middle Miocene	Depéret, 1897; Bergounioux and Crouzel, 1959; Coppens et al., 1978; Pickford, 2004

TABLE 15.5 (CONTINUED)

Taxon	Occurrence (Site, Locality)	Stratigraphic Unit	Age	Key References
G. pygmaeus continued	Bosluis Pan, South Africa		Middle Miocene, ca. 16 Ma	Senut et al., 1996; Pickford, 2005b
	Ngenyin, Tugen Hills, Kenya	Mb. A, Ngorora Fm.	ca. 13 Ma	Pickford, 2004
Gomphotherium sp. indet.	Ghaba, Oman	Dam Fm.	Latest early Miocene or early middle Miocene	Roger et al., 1994; Pickford, 2003
	Siwa, Egypt		Early Miocene	Hamilton, 1973; Coppens et al., 1978
	Gebel Zelten, Libya	Qaret Jahanneam Mb., Marada Fm.	ca. 16.5 Ma	Arambourg, 1961; Coppens et al., 1978; Gaziry, 1987a; Sanders, 2008a

AMEBELODONTINAE, EARLY–LATE MIOCENE

Taxon	Occurrence (Site, Locality)	Stratigraphic Unit	Age	Key References
Progomphotherium maraisi	Auchas, Namibia (type)	Arrisdrift Gravel Fm.	ca. 20–19 Ma	Pickford, 2003
	? Moroto II		>20.6 Ma (alt. ca. 17.5 Ma)	Gebo et al., 1997; Pickford, 2003; Pickford et al., 2003; Pickford and Mein, 2006
	?Karangu, Kenya	Karangu Fm.	17.8 Ma	Drake et al., 1988; Pickford, 2003
cf. Archaeobelodon	?Legetet, Kenya	Legetet Fm.	Early Miocene, ca. 20 Ma	Pickford, 1986b; Tassy, 1986
	?Songhor, Kenya		Early Miocene, 19.5 Ma	Bishop et al., 1969; Pickford and Andrews, 1981; Pickford, 1986b
	Wadi Moghara, Egypt	Moghara Fm.	ca. 18–17 Ma	Miller, 1996, 1999; Sanders and Miller, 2002
	Rusinga, Kenya	Hiwegi Fm.	17.8 Ma	MacInnes, 1942; Tassy, 1979b, 1984, 1986; Drake et al., 1988
		Kulu Fm.; immediately suprajacent to the Hiwegi Fm.	Slightly <17.8 Ma	Pickford, 1981; Tassy, 1986
	Arongo Uyoma (Chianda), Kenya		Early Miocene, cf. Hiwegi and Kulu Fms.	Pickford, 1986b
Archaeobelodon filholi	?Napak, Uganda		Early Miocene, ca. 20.0 Ma	Bishop, 1964; Bishop et al., 1969; Pickford and Andrews, 1981; Pickford, 2003; MacLatchy et al., 2006
	Buluk (West Stephanie), Kenya		18.0–17.2 Ma (alt. 16–15 Ma)	Harris and Watkins, 1974; Pickford, 1981; McDougall and Watkins, 1985; Tassy, 1986
	Mwiti, Kenya	Mwiti 1	ca. 17 Ma	Tassy, 1986; Drake et al., 1988
	?Nachola, Kenya	Aka Aiteputh Fm.	15.5 Ma	Pickford, 2003
Afromastodon coppensi	Arrisdrift, Namibia (type)	Arrisdrift Gravel Fm.	Early middle Miocene, ca. 16.0 Ma (alt. ca. 17.5–17.0 Ma)	Corvinus and Hendey, 1978; Hendey, 1978b; Pickford, 2003

Taxon	Occurrence (Site, Locality)	Stratigraphic Unit	Age	Key References
Amebelodontinae cf. *A. coppensi*	BUR1, Burji, Ethiopia		ca. 17–15 Ma	Suwa et al., 1991; Pickford, 2003
Protanancus macinnesi	Kalodirr, Kenya	Tiati Grits	ca. 18.0–16.0 Ma (alt. 17.7–16.6 Ma)	Drake et al., 1988
	Maboko, Kenya (type)	Maboko Fm.	ca. 16 Ma (alt. slightly >14.7 Ma)	MacInnes, 1942; Tassy, 1986; Feibel and Brown, 1991
	Kaloma, Kenya	Maboko Fm.	Middle Miocene	Pickford, 1982
	Majiwa, Kenya	Maboko Fm.	Middle Miocene	Pickford, 1981, 1982; Tassy, 1986
	Kipsaraman, Tugen Hills, Kenya	Muruyur Fm.	15.5 Ma	Hill, 1999; Pickford, 2003
	Nyakach, Kenya		ca. 15 Ma	Tassy, 1986; Pickford, 2003
	Ft. Ternan, Kenya	Ft. Ternan Beds	14.0–13.9 Ma	Shipman et al., 1981; Tassy, 1986
	Sinda River area, near Ongoliba (Semliki 531), Democratic Republic of Congo	?Sinda Beds	?Middle Miocene	Lepersonne, 1959; Hooijer, 1963; Boaz, 1994
	?Ombo, Kenya		Middle Miocene	Van Couvering and Van Couvering, 1976; Pickford, 1981; Tassy, 1986
	Alengerr, Kenya		<14.0 Ma, >12.4 Ma	Bishop, 1972; Tassy, 1986
Amebelodon cyrenaicus	Sahabi, Libya (type)	Sahabi Fm.	Late Miocene–early Pliocene, ca. 5.2 Ma	Gaziry, 1982, 1987b; Heinzelin and El-Arnauti, 1987; Tassy, 1999; Bernor and Scott, 2003; Sanders, 2008b
Platybelodon sp.	Loperot, Kenya		Early Miocene, ca. 17 Ma	Maglio, 1969a; Coppens et al., 1978; Tassy, 1986; Pickford, 2003
?Amebelodontinae (?*Platybelodon*)	Ad Dabtiyah, Saudi Arabia	Dam Fm.	Early middle Miocene (alt. ca. 19–17 Ma)	Hamilton et al., 1978; Whybrow and Clements, 1999
?Amebelodontinae gen. et sp. indet.	As Sarrar, Saudi Arabia	Dam Fm.	Early middle Miocene, ca. 17–15 Ma (alt. ca. 19–17 Ma)	Thomas et al., 1982; Whybrow and Clements, 1999
CHOEROLOPHODONTINAE, EARLY–LATE MIOCENE				
Afrochoerodon kisumuensis	Wadi Moghara, Egypt	Moghara Fm.	ca. 18–17 Ma	Miller, 1996, 1999; Sanders and Miller, 2002
	Maboko, Kenya (type)	Maboko Fm.	ca. 16 Ma (alt. slightly >14.7 Ma)	MacInnes, 1942; Tassy, 1977b, 1979b, 1986; Feibel and Brown, 1991
	Kaloma, Kenya	Maboko Fm.	Middle Miocene	Pickford, 1982
	Majiwa, Kenya	Maboko Fm.	Middle Miocene	Pickford, 1982
	Kipsaraman and Cheparawa, Tugen Hills, Kenya	Muruyur Fm.	15.5 Ma	Bishop, 1972; Tassy, 1986; Hill, 1999; Behrensmeyer et al., 2002; Pickford, 2001, 2003
	Bosluis Pan, South Africa		ca. 16–15 Ma	Senut et al., 1996; Pickford, 2005b

TABLE 15.5 (CONTINUED)

Taxon	Occurrence (Site, Locality)	Stratigraphic Unit	Age	Key References
Choerolophodon zaltaniensis	Gebel Zelten, Libya (type)	Qaret Jahanneam Mb., Marada Fm.	ca. 16.5 Ma	Savage and Hamilton, 1973; Gaziry, 1987a; Pickford, 1991b; Sanders, 2008a
Choerolophodon ngorora primitive morph.	Tugen Hills, Kenya	Mbs. A–D, Ngorora Fm.	ca. 13–11 Ma	Tassy, 1986
	Ft. Ternan, Kenya	Ft. Ternan Beds	14.0–13.9 Ma	Shipman et al., 1981; Tassy, 1986
Choerolophodon ngorora advanced morph	Tugen Hills, Kenya (type)	Mb. E , Ngorora Fm.	10.5 Ma	Maglio, 1974; Tassy, 1986
	?Mbagathi, Kenya	Kirimun Fm.	?Middle Miocene (alt. ?early Miocene)	Pickford, 1981; Ishida and Ishida, 1982; Tassy, 1986
	Samburu Hills, Kenya	Lower Mb., Namurungule Fm.	ca. 9.5 Ma	Tsujikawa, 2005b
	Nakali, Kenya	Nakali Fm.	9.9–9.8 Ma	Aguirre and Alberdi, 1974; Aguirre and Leakey, 1974; Tassy, 1986; Pickford, 2003; Nakatsukasa et al., 2007
Choerolophodon sp. indet.	Majiwa, Kenya		Middle Miocene	Pickford, 1981; Tassy, 1986
	Gebel Cherichera, Tunisia	Beglia Fm.	ca. 13–11 Ma	Robinson, 1974; Robinson and Black, 1974
	Henchir Beglia, Tunisia	Beglia Fm.	ca. 13–11 Ma	Robinson, 1974; Robinson and Black, 1974
	?Bled Douarah, Tunisia	Beglia Fm.	ca. 13–11 Ma	Robinson and Black, 1974; Geraads, 1989
	?Gebel Krechem el Artsouma, Tunisia	Segui Fm.	Late Miocene, ca. 11 Ma	Geraads, 1989

TETRALOPHODONTINAE

Taxon	Occurrence (Site, Locality)	Stratigraphic Unit	Age	Key References
Tetralophodon sp. nov.	Samburu Hills, Kenya	Lower, Upper Mbs., Namurungule Fm.	ca. 9.5 Ma	Nakaya et al., 1984, 1987; Sawada et al., 1998; Tsujikawa, 2005a, b
Tetralophodon sp. indet.	Gebel Cherichera, Tunisia	Beglia Fm.	ca. 13–11 Ma	Bergounioux and Crouzel, 1956
	?Bled Douarah, Tunisia	Beglia Fm.	ca. 13–11 Ma	Robinson and Black, 1974; Geraads, 1989
	Gebel Krechem el Artsouma, Tunisia	Segui Fm.	Late Miocene, ca. 11 Ma	Geraads, 1989
	?Gebel Sémène, Tunisia		?Late Miocene	Bergounioux and Crouzel, 1956
	?Zidania, Morocco		?Late Miocene	Coppens et al., 1978
	?Smendou, Algeria		?Late Miocene	Coppens et al., 1978
	?Chorora, Ethiopia	Chorora Fm.	Late Miocene, 11–10 Ma	Sickenberg and Schönfeld, 1975; Tiercelin et al., 1979; Geraads et al., 2002
	Kisegi-Nyabusosi area, Uganda	Kakara Fm.	Late Miocene, ca. ?9.0 Ma	Pickford et al., 1993; Tassy, 1999
	Sinda River, Democratic Republic of Congo	Sinda Beds	?Late Miocene	Hooijer, 1963; Madden, 1977, 1982
Gen. and sp. indet. (tetralophodont form)	Tugen Hills, Kenya	Mb. D, Ngorora Fm.	Ca. 11 Ma	Tassy, 1979b, 1986
	?Nakali, Kenya	=Kabarsero Fm.	Ca. 9.5 Ma	Pickford, 2003

Taxon	Occurrence (Site, Locality)	Stratigraphic Unit	Age	Key References
	ANANCINAE, LATE MIOCENE–LATE PLIOCENE			
Anancus kenyensis	Lothagam, Kenya	Lower and Upper Mbs., Nawata Fm.	7.4–5.0 Ma	McDougall and Feibel, 2003; Tassy, 2003
	Toros-Menalla, Chad		ca. 7.0–6.0 Ma	Vignaud et al., 2002
	Tugen Hills, Kenya	Mpesida Beds	ca. 7.0–6.0 Ma	Tassy, 1986; Kingston et al., 2002
		Lukeino Fm.	6.2–5.6 Ma	Hill et al., 1985, 1986; Tassy, 1986; Hill, 2002
	Middle Awash, Ethiopia	Adu Asa Fm.	ca. 6.3–5.6 Ma	Kalb and Mebrate, 1993; Haile-Selassie, 2001; WoldeGabriel et al., 2001; Haile-Selassie et al., 2004
		Kuseralee Mb., Sagantole Fm.	ca. 5.6–5.2 Ma	Kalb and Mebrate, 1993; Renne et al., 1999; Haile-Selassie, 2001; Haile-Selassie et al., 2004
		Haradaso Mb., Sagantole Fm.	ca. 5.0–4.4 Ma (probably about 4.8 Ma)	Kalb and Mebrate, 1993; Renne et al., 1999
	Lemudong'o, Kenya	Lemundong'o Fm.	6.1–6.0 Ma	Ambrose et al., 2003; Saegusa and Hlusko, 2007
	Manonga Valley, Tanzania	Ibole Mb., Wembere-Manonga Fm.	ca. 5.5–5.0 Ma	Harrison and Baker, 1997; Sanders, 1997
	Kossom Bougoudi, Chad		ca. 5.3 Ma	Brunet et al., 2000; Brunet, 2001
	Kanam East (type) and West, Kenya	Kanam Fm.	Late Miocene to earliest Pliocene	MacInnes, 1942; Tassy, 1986; Ditchfield et al., 1999
	Nkondo, Uganda	Nkondo Fm.	ca. 5.0–4.0 Ma	Tassy, 1995
	As Duma, Gona Western Margin, Ethiopia		4.5–4.3 Ma	Semaw et al., 2005
	Endolele, Tanzania	Lower Unit, Laetolil Beds	?ca. 4.3 Ma	Drake and Curtis, 1987; Sanders, in press
	Galili, Ethiopia	Dhidinley Mb., Mt. Galili Fm.	Early Pliocene (alt. ca. 4.17–4.07 Ma)	Kullmer et al., 2008
Anancus sp. nov.	Tugen Hills, Kenya	Chemeron Fm.	ca. 5.0–4.0 Ma	Hill et al., 1985, 1986
	Lothagam, Kenya	Apak Mb., Nachukui Fm., and unknown horizon(s)	5.0–4.2 Ma	McDougall and Feibel, 2003; Tassy, 2003
	Kollé, Chad		5.0–4.0 Ma	Brunet, 2001
	Nkondo-Kaiso Area, Nyaweiga, Uganda	Nyaweiga Mb., Nkondo Fm.	5.0–4.0 Ma	Cooke and Coryndon, 1970; Tassy, 1995
	Aterir, Kenya	Aterir beds	ca. 4.5 Ma (alt. slightly <4.0 Ma)	Bishop et al., 1971; Hill, 1994
	Kiloleli; Ngofila; Beredi South, Manonga Valley, Tanzania	Kiloleli Mb., Wembere-Manonga Fm.	ca. 4.5–4.0 Ma	Harrison and Baker, 1997; Sanders, 1997
	Middle Awash, Ethiopia	Aramis, Beidareem, and Adgantole Mbs., Sagantole Fm.	4.4–4.3 Ma	Kalb and Mebrate, 1993; Renne et al., 1999; White et al., 2006

TABLE 15.5 (CONTINUED)

Taxon	Occurrence (Site, Locality)	Stratigraphic Unit	Age	Key References
A. sp. nov. continued	Kakesio, Tanzania	Lower Unit, Laetolil Beds	ca. 4.3 Ma	Drake and Curtis, 1987; Harris, 1987b; Hay, 1987; Sanders, 2005, in press
	Kanapoi, Kenya	Kanapoi Fm.	4.2–4.1 Ma	Feibel, 2003; Harris et al., 2003
	Sinda River, Lower Semliki, Democratic Republic of Congo	Sinda Beds	ca. 4.1 Ma	Hooijer, 1963; Yasui et al., 1992; Boaz, 1994
	Ekora, Kenya		ca. 4.0–3.75 Ma (slightly younger than 4.0 Ma)	Behrensmeyer, 1976; Kalb and Mebrate, 1993
	Laetoli, Tanzania	Upper Unit, Laetolil Beds	3.8–3.5 Ma	Drake and Curtis, 1987; Harris, 1987b; Sanders, in press
Anancus capensis	Langebaanweg, South Africa (type)	Quarry E, Quartzose Sand Mb. and Pelletal Phosphorite Mb., Varswater Fm.	Early Pliocene, ca. 5.0 Ma	Hendey, 1981; Sanders, 2006, 2007
Anancus petrocchii	Sahabi, Libya (type)	Sahabi Fm.	Late Miocene–early Pliocene, ca. 5.2 Ma	Petrocchi, 1954; Heinzelin and El-Arnauti, 1987; Bernor and Scott, 2003; Sanders, 2008b
Anancus osiris	Wadi Natrun, Egypt		Late Miocene or early Pliocene	Coppens et al., 1978; Geraads, 1982; Thomas et al., 1982
	Giza, Egypt (type)		Late Pliocene	Arambourg, 1945, 1970
	Grombalia, Tunisia		Late Pliocene	Arambourg, 1970
	Aïn Boucherit, Algeria		Late Pliocene; 2.32 Ma (alt. ca. 2.0 Ma)	Arambourg, 1970; Geraads and Amani, 1998; Geraads, 2002; Sahnouni et al., 2002
	Lac Ichkeul, Tunisia		Late Pliocene	Arambourg, 1970
	Ferryville, Tunisia		Late Pliocene	Depéret et al., 1925; Coppens et al., 1978
	Hamada Damous, Tunisia		Late Pliocene	Fournet, 1971
	Gebel Melah		Late Pliocene	Fournet, 1971
	Ahl al Oughlam, Casablanca, Morocco		Late Pliocene, ca. 2.5 Ma	Geraads and Metz-Muller, 1999; Geraads, 2002
	Fouarat, Morocco		Late Pliocene	Arambourg, 1970
Anancus sp. indet.	Aramis, Middle Awash, Ethiopia	Haradaso Mb., Sagantole Fm.	ca. 4.4 Ma	WoldeGabriel et al., 1994
	Omo, Ethiopia	Mursi Fm.	>4.15 Ma	Beden, 1976; Feibel et al., 1989
	Karonga, Uraha, Malawi	Chiwondo Beds	>4.0 Ma	Mawby, 1970; Bromage et al., 1995
	Makapansgat, South Africa	Mb. 4	<3.1 Ma	Cooke, 1993; Partridge et al., 2000
	Baard's Quarry, South Africa	Lower Level	Late Pliocene	Hendey, 1978a, 1981

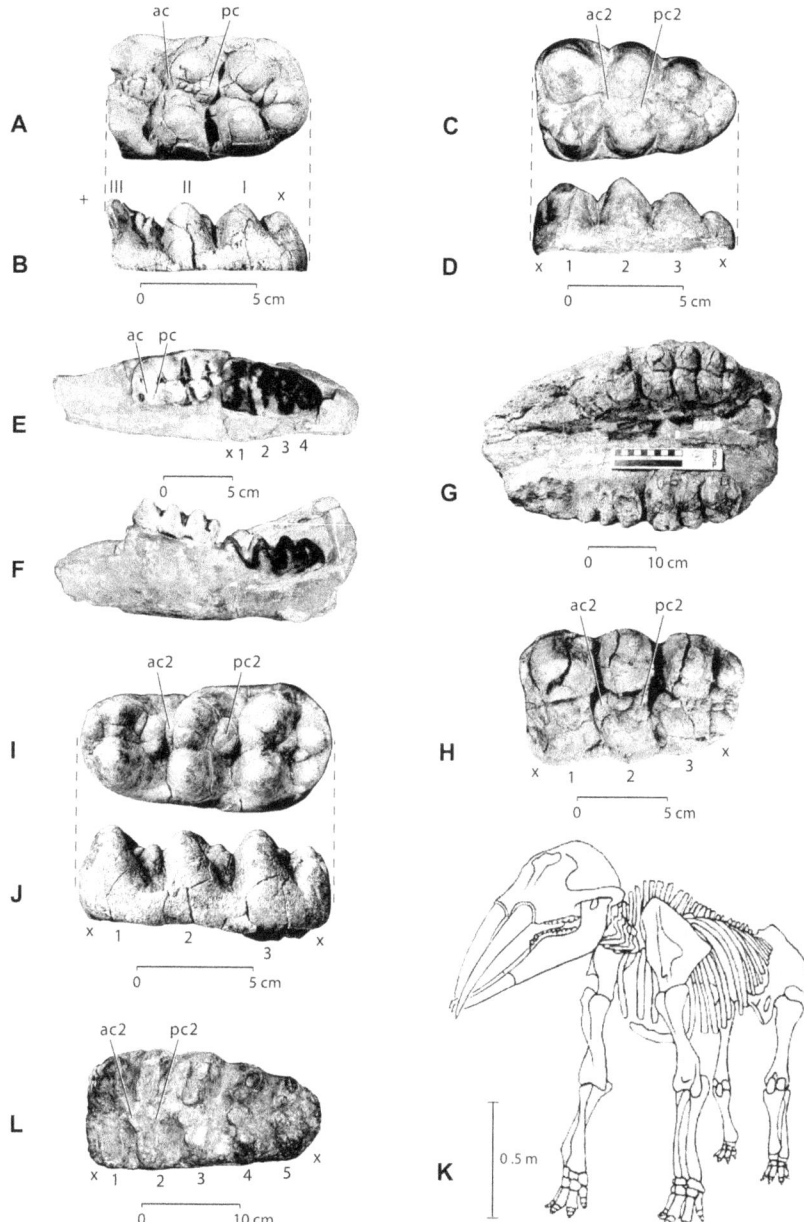

FIGURE 15.11 Aspects of gomphothere dental and skeletal morphology. Anterior is to the left in all specimens. A) Occlusal view, right m3 (cast), CH14-V14, cf. *Gomphotherium* sp. nov., Chilga, Ethiopia. B) Right lateral view, right m3 (cast), CH14-V14, cf. *Gomphotherium* sp. nov., Chilga, Ethiopia (reversed). C) Occlusal view, left M3, M21866, ?cf. *Gomphotherium* sp. nov. D) Right lateral view, left M3, M21866, ?cf. *Gomphotherium* sp. nov. E) Occlusal view, left dentary with m2-m3, DOG87.1, *Eritreum melakeghebrekristosi* (type specimen). F) Right lateral view (reversed), left dentary with m2–m3, DOG87.1, *Eritreum melakeghebrekristosi* (type specimen). G) Occlusal view, palate with right and left M2–M3, KNM-MI 1, *Gomphotherium* sp., Mwiti, Kenya. H) Occlusal view, left M3, KNM-MI 1, *Gomphotherium* sp., Mwiti, Kenya. I) Occlusal view, right m2, DPC 12926. *Gomphotherium angustidens libycum.* J) Left lateral view, right m2, DPC 12926. *Gomphotherium angustidens libycum.* K) Skeletal reconstruction, *Gomphotherium pygmaeus* (modified from Pickford, 2004: figure 3). L) Occlusal view, left M3, KI 6492, *Tetralophodon* sp. nov. K, courtesy of Martin Pickford.

ABBREVIATIONS: ac, pretrite anterior accessory central conule; pc, pretrite posterior accessory central conule; x, anterior or posterior cingulum(id); 1, 2, 3, . . . , first, second, third, . . . loph(id); I, II, III . . . , loph(id)s counted from the posterior end of the crown.

and paleoecological evidence suggest that *Z. aegyptensis* inhabited well-watered, forested landscapes (Said, 1962; Bown et al., 1982; Pickford, 1991b).

Family INCERTAE SEDIS
CF. *GOMPHOTHERIUM* SP. NOV. Sanders, Kappelman and
Rasmussen, 2004
Figures 15.11A–15.11D

Age and Occurrence Late Oligocene, eastern Africa, ?northern Africa (table 15.5).

Description Very small elephantoid, at the low end of dental size range for gomphotheres, but with larger molars than palaeomastodonts. Premolars and molars simultaneously in occlusion. Lower third molar with at least three lophids and a prominent postcingulid formed of two stout conelets (figures 15.11A, 15.11B). Molars with small anterior and posterior accessory central conules throughout the pretrite side, and a low cingular ribbon rimming the buccal margin of the crown. Half-lophids composed of a single, large abaxial conelet and a smaller adaxial conelet. No cementum. The upper fourth premolar resembles P4 in *Gomphotherium* in the presence of a central cusp in its posterior loph, and prominent enamel swelling on the anterior face of the last pretrite half-loph (Tassy, 1985; Sanders et al., 2004). An M3 from Gebel Zelten, Libya (M21866) is a good fit morphologically for the m3 from Chilga, Ethiopia, and may belong in this taxon. It is subtriangular in occlusal view, small, and has three lophs and a low, stout heel, lophs composed of massive main conelets superficially subdivided from diminutive mesoconelets, and pretrite accessory central conules accompanying lophs 1 and 2 anteriorly and posteriorly, and loph 3 anteriorly (figures 15.11C, 15.11D).

Remarks Oldest known possible gomphothere, documented with certainty only from the Chilga region, Ethiopia, from sites also containing palaeomastodonts (Sanders et al., 2004). Prior to its discovery, gomphothere-like proboscideans had only been known from Neogene horizons. This new taxon from Chilga, along with *Eritreum*, suggests an autochthonous African origination of elephantoids unrelated to the early Miocene faunal turnover in Afro-Arabia that was marked by an influx of Eurasian taxa (Kappelman et al., 2003).

Genus *ERITREUM* Shoshani et al., 2006
ERITREUM MELAKEGHEBREKRISTOSI Shoshani et al., 2006
Figures 15.7B, 15.11E, and 15.11F

Age and Occurrence Late Oligocene, eastern Africa (table 15.5).

Diagnosis Very diminutive elephantoid (body mass estimate, 600 kg), with molars in size range of palaeomastodonts and smaller than those of other gomphotheres and mammutids (figure 15.7B). Anterior cingulid reduced relative to condition in palaeomastodonts. Further distinguished from palaeomastodonts by presence of fourth lophid in m3, and trefoil wear patterns throughout the length of molar crowns.

Description Lower tusks pyriform in cross section, and higher than wide. Mandibular corpus wider than high; symphysis like that of *Phiomia* in proportions. Intermediate molars trilophodont with a thick postcingulid. Lower third molar with four lophids but no appreciable postcingulid (figure 15.11E). Pretrite anterior and posterior accessory central conules throughout molar crowns contribute to trefoil enamel patterns in wear. Adaxial conelets smaller than abaxial conelets, but more equal in height than in mammutids. No cementum. Emergence of m3 was delayed until intermediate molars were well in wear (figure 15.11F).

Remarks Known from an isolated discovery, *Eritreum* documents an intermediate stage in the evolution of palaeomastodonts to more derived gomphotheres, and further points to the Horn of Africa as a locus of pre-Miocene elephantoid origination.

Family GOMPHOTHERIIDAE Hay, 1922

Arguably the most successful of proboscidean families, of remarkable taxonomic diversity and geographic and temporal extent. Gomphotheriids may have first appeared in the Horn of Africa during the late Oligocene, depending on the affinities of new taxa from Eritrea and Ethiopia (Sanders and Kappelman, 2001; Shoshani et al., 2001b; Kappelman et al.,

2003; Sanders et al., 2004; Shoshani et al., 2006), underwent a series of adaptive radiations during the Miocene, and persisted in Africa until the late Pliocene (Kalb and Mebrate, 1993). Along with dramatic molar size increase over that of their palaeomastodont precursors, gomphothere-grade dentition was achieved by substantial enlargement and differentiation of tusk forms, and by elaboration of molars through the addition of loph(id)s, cementum, and accessory central conules. Several gomphothere lineages independently lost lower tusks, and further morphological differentiation among taxa was achieved by occlusal rearrangement of transverse half-loph(id) pairs into anancoid, pseudoanancoid, and chevron patterns. Elephants are thought to have evolved from a gomphotheriid in the late Miocene, a transformation largely achieved through radical reorganization of molar construction and chewing dynamics (Maglio, 1972).

Afro-Arabian constituents of Gomphotheriidae include gomphotheriines, anancines, tetralophodonts, choerolophodontines, and amebelodontines (but see Shoshani, 1996; Tassy, 1996c; Shoshani and Tassy, 2005). These taxa have been tenuously linked only by a small set of traits, such as enlarged central accessory conules, broad narial opening, pyriform cross section of lower tusks, and brachyodont, bunolophodont molars with a tendency for enamel wear figures to incorporate accessory conules into trefoil patterns (Coppens et al., 1978; Shoshani, 1996; Tassy, 1996c).

Subfamily GOMPHOTHERIINAE Hay, 1922

This subfamily, and particularly its core genus *Gomphotherium*, serve as wastebasket taxa, encompassing species that lack specializations (e.g., "shovel tusks") characteristic of more clearly definable gomphotheriids, such as amebelodonts, and which primitively retain features like trilophodonty and lower tusks with pyriform cross sections. *Gomphotherium* has been divided gradistically into a primitive "*G. annectens* group," including *G. annectens* (Asia), *G. cooperi* (South Asia), *G. sylvaticum* (Europe), and *Gomphotherium* sp. (Africa), and a more advanced "*G. angustidens* group," distinguished by differences in cranial vault height, width of the nasal opening, cross-sectional shape of lower tusks, degree of development of accessory conules, and angle of downturn of the symphysis (Tassy, 1985, 1986, 1994a, 1996c). Also included in the subfamily are the so-called pygmy gomphotheres.

Genus *GOMPHOTHERIUM* Burmeister, 1837
["*GOMPHOTHERIUM ANNECTENS* GROUP" Tassy, 1985]
GOMPHOTHERIUM SP. Tassy, 1979a
Figures 15.11G and 15.11H

Partial Synonymy Progomphotherium maraisi (in part), Pickford, 2003:228.

Age and Occurrence Early Miocene, eastern Africa (table 15.5).

Description Primitive trilophodont gomphothere with only nascent development of a fourth loph in M3, and little or no expression of independent accessory central conules (figures 15.11G, 15.11H). Loph(id)s formed of massive, low main conelets accompanied by more diminutive mesoconelets; enamel very thick; no molar cementum. M3s are relatively wide and larger than those of *P. maraisi*. Lower tusks modest in size, pyriform in cross section, and longitudinally torqued.

Remarks This taxon was contemporaneous with other species of the "*G. annectens* group," indicating that these gomphotheres dispersed widely throughout the Old World rapidly after the group first appeared in Africa (Sanders and Miller, 2002). It also overlapped temporally with more advanced congeners, but the early success of the group apparently did not continue into the middle Miocene (table 15.5).

["*GOMPHOTHERIUM ANGUSTIDENS* GROUP" Tassy, 1985]
GOMPHOTHERIUM ANGUSTIDENS LIBYCUM (Fourtau, 1918)
Figures 15.7, 15.11I, and 15.11J

Partial Synonymy Mastodon angustidens var. *libyca*, Fourtau, 1918:84; *Mastodon spenceri*, Fourtau, 1918:89; *Rhyncotherium spenceri*, Osborn, 1936:485; *Trilophodon angustidens* var. *libycus*, Osborn, 1936:260; *Serridentinus* sp., Hormann, 1963:92; *Mastodon angustidens*, Savage, 1971:221; *Gomphotherium angustidens*, Hamilton, 1973:276; *G. angustidens*, Tobien, 1973a:214–215; *G. angustidens*, Coppens et al., 1978:342–343; "*Gomphotherium*" *pygmaeus* (in part), Coppens et al., 1978:344, 346; *G. cooperi*, Gentry, 1987; *G. spenceri*, Savage, 1989:594; *G. cooperi*, Savage, 1989:596; *G. angustidens*, Savage, 1989:594; *Archaeobelodon* aff. *A. filholi*, Tiffney et al., 1994:27; *G. angustidens libycum*, Sanders and Miller, 2002:389; *Afromastodon lybicus*, Pickford, 2003:231, 233; ?*Eozygodon* sp., Pickford, 2003:230–231; *Gomphotherium* sp nov. (in part), Pickford, 2003:233; *Afrochoerodon zaltaniensis* (in part), Pickford, 2003:231, 233.

Age and Occurrence Early–?late Miocene, northern Africa and Arabia (table 15.5).

Diagnosis Sanders and Miller (2002). Differs from more advanced forms of *G. angustidens* by variable expression of fourth loph in M3 and occasional absence of fourth lophid in m3; from *G. steinheimense* and *G. browni* by smaller size (figure 15.7) and lesser subdivision of adaxial and abaxial conelets, and from "pygmy" gomphotheres by larger size and lesser development of cementum. Distinguished from *G. annectens* grade gomphotheres by more pronounced development of anterior and posterior accessory central conules, and less massive structure of loph(id)s (figures 15.11I, 15.11J).

Description Medium-sized gomphothere. The dentition is most like that of European *G. angustidens* in size (figure 15.7) and in the degree of expression of pretrite accessory conules, particularly posterior to loph(id)s, contributing to trefoil enamel patterns with wear. There are no posttrite accessory conules. Half-loph(id)s are separated by a median sulcus, and each is usually composed of a large, outer main conelet accompanied by a much more diminutive mesoconelet. Mesoconelets are aligned transversely side by side and may be slightly more anterior than their main conelets (figure 15.11I). Enamel is thick (as much as 6–7 mm in third molars), and cementum is slight to absent. Intermediate molars are trilophodont.

Remarks The type series and referred specimens from Wadi Moghara, Egypt, have had a varied taxonomic history and have been divided among several species, but there are no compelling reasons to separate these fossils or remove them from *Gomphotherium*. *Gomphotherium angustidens libycum* co-occurs at Moghara with *Archaeobelodon, Zygolophodon, Afrochoerodon,* and *Prodeinotherium* (Harris, 1978; Sanders and Miller, 2002), but these are distinct morphologically. Gomphothere specimens from a handful of other Afro-Arabian sites (table 15.5) identify closely with the Moghara sample, including molars from Ad Dabtiyah, Saudi Arabia. A recent proposal to refer *G. angustidens libycum* to a new species, *Afromastodon libycus* (Pickford, 2003), is baseless. *Afromastodon* clearly belongs in the Amebelodontinae (see later discussion), and unlike molars of *A. coppensi*, in the Moghara and allied material half-loph(id)s

are not transversely offset, and there are no posttrite accessory central conules.

An age-grade series of jaws with *Gomphotherium*-type molars from Gebel Zelten clarifies the mandibular morphology of *G. angustidens libycum* and strengthens the connection between the Moghara and Zelten samples (Sanders, 2008a). These jaws have wide symphyseal gutters, symphyseal angulation that increased substantially downward with age, large arterial foramina that communicate medially from the mandibular canal, just posterior to the i2 alveoli, and three mandibular foramina: a large opening below the anteriormost cheek tooth, a smaller foramen lateral to the symphysis, and a capacious ("torpedo-tube") opening anterior and ventral to that, marking the anteriormost extent of the mandibular canal (Sanders, 2008a). Tassy (1985) speculated that a dentary from Moghara with similar morphology might have choerolophodontine affinities, but the Zelten specimens show that broad guttering and downward angulation of the symphysis are also present in some forms of *Gomphotherium*.

["'PYGMY' *GOMPHOTHERIUM* GROUP"]
GOMPHOTHERIUM PYGMAEUS (Depéret, 1897)
Figures 15.7A, 15.7B, and 15.11K

Partial Synonymy Mastodon angustidens var. *pygmaeus*, Depéret, 1897; *Phiomia pygmaeus*, Osborn, 1936:246–247, figures 186, 187A, A1–A4; *Trilophodon olisiponensis* var. *pygmaeus*, Bergounioux and Crouzel, 1959: 102; "*Gomphotherium*" *pygmaeus*, Coppens et al., 1978:343, 346; *Choerolophodon pygmaeus*, Pickford, 2004, 2005a.

Holotype University of Lyon, Institute of Geology, N° 1678, partial r. m3.

Type Locality Kabylie, Algeria.

Age and Occurrence Middle Miocene, northern, eastern, southern Africa (table 15.5).

Referred Specimens (Bosluis Pan, South Africa) SAM-Q 2516, incomplete l. m3; (Ngenyin, Tugen Hills, Kenya) Bar 801'02, l. metacarpal; Bar 802'02, two I2 frags.; Bar 803'02, r. m2; Bar 804'02, l. m2; Bar 805'02, l. and r. m3; Bar 806'02, l. m1; Bar 807'02, l. M1; Bar 808'02, l. M2; Bar 809'02, l. and r. M3; Bar 812'02, r. calcaneum; Bar 814'02, prox. phalanx; Bar 816'02, partial r. femur; Bar 817'02, r. MT IV; Bar 818'02, l. magnum; Bar 821'02, l. trapezoid; Bar 822'02, l. lunate; Bar 823'02, r. lat. cuneiform; Bar 827'02, prox. phalanx frag.; Bar 828'02, prox. phalanx; Bar 829'02, prox. phalanx; Bar 830'02, r. trapezium; Bar 831'02, l. trapezium; Bar 832'02, r. MC IV; Bar 1995'02, Bar 1996'02, l. cuboid; l. navicular; Bar 1998'02, prox. phalanx frag.; Bar 2002'02 r. distal tibia; Bar 2003'02 l. distal fibula.

Diagnosis Gomphothere of small size; molars larger than those of palaeomastodonts but near lower extreme of range for *Gomphotherium* (figure 15.7B); intermediate molars trilophodont and third molars with four loph(id)s; crowns may be well invested with cementum and exhibit choerodonty and ptychodonty; crowding of half-loph(id)s causes main conelets to converge strongly toward the midline apically and produces weak chevroning of posterior loph(id)s (though unlike choerolophodont molars pre- and posttrite mesoconelets are transversely aligned); pretrite accessory central conules present throughout the crown, particularly posterior to loph(id)s.

Description The type specimen, a right m3 missing its anterior lophid, preserves three lophids and a low, substantial distal cingulid composed of several large conelets. Its transverse valleys are filled with cementum, but even so it is apparent that the pre- and posttrite half-lophids are each comprised of a large, outer conelet and more diminutive mesoconelet, and that they are transversely aligned. Also, it is possible to see the outline of pretrite anterior and posterior accessory central conules, which in wear would have contributed to trefoil enamel figures. Because of the narrowness of the crown, the main, outer conelets of each lophid are closely convergent apically.

Morphometrically quite similar to the Kabylie tooth are molars from Bosluis Pan, South Africa, and Ngenyin, Kenya, and particularly in the latter the trefoil arrangement of main conelets and accessory conules can be confirmed. Although crowding of loph(id)s produces weak chevroning in these specimens, they exhibit no sign of the typical choerolophodont occlusal pattern, marked by advancement of the pretrite mesoconelet anterior to the posttrite half-loph(id). An I2 fragment from Ngenyin attributed to this species (Pickford, 2004) is D shaped, typical of *Gomphotherium*, and is too abraded to know whether it originally possessed a band of enamel.

A wealth of postcrania from Ngenyin, Kenya are associated with the dental sample. These are similar to corresponding skeletal elements in *Gomphotherium*, and indicate an animal 1.5 m tall at the shoulder, with graviportal adaptations for support of the body and semicursorial locomotor capabilities (figure 15.11K; Pickford, 2004).

A lower molar from Arrisdrift, Namibia, attributed to this species (Pickford, 2005a) is a dp4 rather than an m1, and as such, its size is not remarkably small. It should be returned to its original assignment in *Afromastodon coppensi* (Pickford, 2003).

Remarks Diminutive gomphothere molars from a number of sites (figures 15.7A, 15.7B) have previously been included in *G. pygmaeus*, based primarily on size (e.g., Coppens et al., 1978). Inclusion of dentition of all small Afro-Arabian gomphotheres, however, would make this an unrealistically heterogeneous species (see later discussion). Roger et al. (1994) considered *G. pygmaeus* a *nomen dubium* due to what they considered the inadequacy of the type specimen, but the addition of the Ngenyin fossils and reconsideration of the peculiar combination of traits in the Kabylie molar demonstrate the validity of the type and taxon. Because of weak chevroning and thick cementum, it has also been suggested that *G. pygmaeus* is a choerolophodont (Tobien, 1973a; Pickford, 2004). This argument is unconvincing, as the former is an artefact of occlusal crowding and small size, and the latter is highly homoplasic among gomphotheriids.

GOMPHOTHERIUM SP. INDET.

Partial Synonymy Trilophodon pygmaeus, Arambourg, 1961:108; *Mastodon pygmaeus*, Savage, 1971:221; *Gomphotherium angustidens*, Hamilton, 1973:276; "*Gomphotherium*" *pygmaeus*, Coppens et al., 1978:344, 346; *Gomphotherium angustidens pasalarensis*, Gaziry, 1987a:76; Elephantoidea gen. et sp. indet., Roger et al., 1994:14, plate I, figures 2–4.

Age Early–middle Miocene, northern Africa and Arabia (table 15.5).

Description Gomphotherium-type molars from a few sites can be grouped together based on their small size and common differences from *G. pygmaeus*, including lesser degree of occlusal crowding and chevroning, and lack of thick cementum. This group has trilophodont intermediate molars and, at least incipiently, four loph(id)s in its third molars. Among this sample, cheek teeth from Ghaba, Oman, are more derived than those of Siwa, Egypt, or Gebel Zelten, Libya, by the greater subdivision of their half-loph(id)s into at least two conelets each, and by

a more balanced trefoil pattern of pretrite main conelets and accessory conelets throughout their crowns. Such distinctions may warrant species-level separation of these site samples.

Remarks The occurrences of these specimens and *G. pygmaeus* suggest that "pygmy" gomphotheres were pan-Afro-Arabian in distribution and taxonomically diverse, but not common. It is uncertain whether they represent true cases of dwarfing, or primitively retained an ancestral condition of small size. Among known gomphotheriids, their affinities are unquestionably closest to *Gomphotherium*.

Subfamily AMEBELODONTINAE Barbour, 1927

The origin of amebelodonts and the oldest records of a number of amebelodont genera can be traced to the early Miocene in Africa (MacInnes, 1942; Maglio, 1969a; Tassy, 1984, 1985, 1986; Sanders and Miller, 2002; Pickford, 2003). The group subsequently flourished in early–middle Miocene faunas of Europe (Tassy, 1985), South Asia (Tassy, 1983a), and Asia (Osborn, 1936; Guan, 1996), and it enjoyed particular success in North America during the late Miocene (Osborn, 1936; Fisher, 1996; Lambert and Shoshani, 1998).

The subfamily includes the "shovel tuskers," so called because of the great broadening and dorsoventral flattening of their i2s, accompanied in some instances by internal development of dentinal tubules and rods. Among these are the genera *Amebelodon* and most notably *Torynobelodon* and *Platybelodon* (Osborn and Granger, 1931, 1932; Osborn, 1936; Lambert, 1990, 1992). More plesiomorphic members of the subfamily do not possess these traits to any great extent, if at all.

Along with flattening of i2s, Amebelodontinae is also characterized by tendencies for strong pretrite wear patterns in molars; enlargement of pretrite accessory central conules; presence of posttrite accessory central conules and, in some taxa, of secondary, posttrite trefoils; pseudoanancoid offset of pre- and posttrite half-loph(id)s; relatively narrow molar crowns; and prominent splanchnocrania, especially in the peri- and prenasal regions (Tobien, 1973a; Tassy, 1984, 1985, 1986).

There is little doubt that the success of amebelodonts was largely due to the evolutionary transformation of their lower tusks into powerful tools for the efficient acquisition of plant materials, such as tough subsurface and aquatic browse (Lambert, 1992), that may have been difficult or impossible for other proboscideans to access. Such specialization likely fostered niche displacement between amebelodonts and choerolophodonts, permitting them to widely share early–middle Miocene African habitats (table 15.5). Nevertheless, despite these formidable dental adaptations, by the onset of the Pliocene amebelodonts had disappeared everywhere across their extensive geographic range. Compelling evidence has been presented for the effects of cooler, drier climatic regimes and consequent transition to steppe habitats as prevalent factors contributing to the extinction of the subfamily in North America at the end of the late Miocene (Fisher, 1996). Comparable changes across African landscapes at the close of the Miocene (Cerling et al., 1997; Pagani et al., 1999) may have played a similar role in the demise of amebelodonts on that continent.

Genus *PROGOMPHOTHERIUM* Pickford 2003
PROGOMPHOTHERIUM MARAISI Pickford 2003
Figures 15.12A, 15.12B, 15.13A, and 15.13B

Age and Occurrence Early Miocene, southern and ?eastern Africa (table 15.5).

Diagnosis Pickford (2003). External nares proportionally higher and slightly narrower than in gomphotheriines. Zygomatic process of maxilla projects from face at right angle. Marked angulation between dorsal profile of neurocranium and splanchnocranium. Basicranium nearly in same plane as the palate. Mandibular symphysis relatively brevirostrine, not spatulate anteriorly, and massively constructed for a gomphothere. Molars very small (figures 15.12A, 15.12B).

Description Pickford (2003). Neurocranium heavily pneumatized; rostrum elongated and rostral trough broad; upper tusks socketed parallel to each other; nasals overhang rostrum; external nares about the same width as the rostrum; cranium low and elongated in lateral profile (figure 15.13A). Upper tusks oval in cross section, with greatest diameter dorsoventral and an enamel band ventrolaterally. Lower tusks suboval in cross section, with greatest diameter vertical and mesial face flat or concave. Cheek teeth simultaneously in wear. Molar half-loph(id)s constructed of massive main conelets and poorly differentiated mesoconelets. Lower third molar with four lophids and no appreciable postcingulid; upper third molar variably has three lophs or an incipient fourth loph (figure 15.13B). Intermediate molars trilophodont. Anterior and especially posterior accessory central conules are weakly developed, and enamel trefoil figures may not be distinguishable with wear. Variable development of posttrite accessory central conules (figure 15.13B).

Remarks To Pickford (2003), the craniodental morphology of this diminutive species, barely more advanced than that of palaeomastodonts, suggested its inclusion with primitive early Miocene gomphotheres in the "*Gomphotherium annectens* group." However, features such as facial elongation, close spacing between temporal fossae, low alveolar height of the maxillae, "hourglass" dorsal profile of the cranium, and size and configuration of the nasal opening, as well as development of posttrite accessory conules, suggest that *Progomphotherium* may instead belong in Amebelodontinae. The presence of three lophs in M3, suboval cross sectional shape of the i2s, and inferred short length of the symphysis indicate that if this attribution is correct, *P. maraisi* is the most primitive amebelodont known.

Molars from Moroto II, Uganda, have been assigned to *P. maraisi* (Pickford and Mein, 2006). The pronounced wear asymmentry and angulation between pre- and posttrite molar sides, poor expression of pretrite accessory conules, and development of posttrite accessory conules in the Moroto specimens are features typical of primitive *Archaeobelodon*, supporting the identification of *P. maraisi* as an amebelodont. Molar fragments from Napak, Uganda assigned to *P. maraisi* (Pickford, 2003) have larger accessory conules, and are more similar to *Archaeobelodon filholi* teeth.

Genus *ARCHAEOBELODON* Tassy, 1984
Figure 15.12A

Diagnosis Archaic amebelodonts with relatively slender symphyses that are little angled on mandibular corpora. Lower tusks more flattened than those of *Afromastodon*, but not as wide as those in *Protanancus*, *Amebelodon*, or *Platybelodon*. Upper tusks are robust, without longitudinal torque, and have a band of enamel. P2 is retained. Molars narrow (figure 15.12A) and very brachyodont. Posttrite accessory central conules variably present and diminutive.

CF. *ARCHAEOBELODON* SP. Tassy, 1984

Partial Synonymy Trilophodon angustidens kisumuensis (in part), MacInnes, 1942: plate III, figures 14–16; plate V,

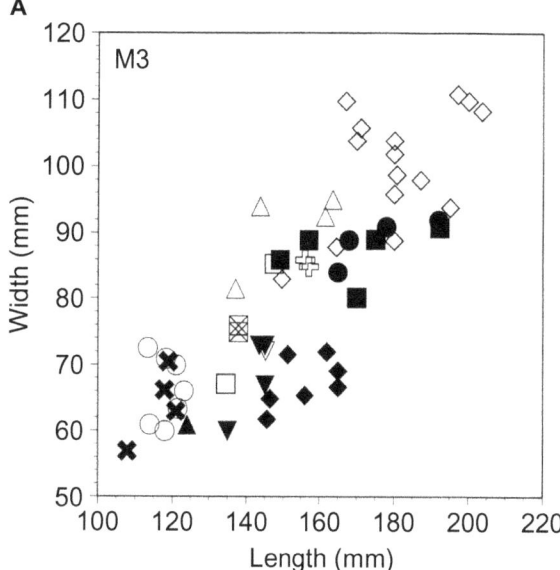

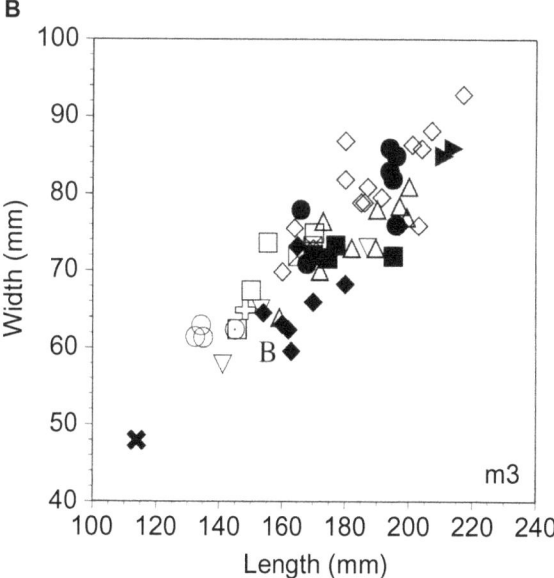

FIGURE 15.12 Bivariate plots of M3 (A) and m3 (B) crown length versus width in African and selected Eurasian amebelodonts and choerolophodonts. Comparative dimensions supplementing original measurements are from Forster Cooper (1922), Tiercelin et al. (1979), Tassy (1983a, 1983b, 1985, 1986), Gaziry (1976, 1987a, 1987b), Suwa et al. (1991), Pickford (2001, 2003), Sanders and Miller (2002), and Sanders (2003).

SYMBOLS: open circle, *Afrochoerodon kisumuensis*; open circle with dot, "*Choerolophodon*" *palaeindicus*; open square with x, *Choerolophodon zaltaniensis*; open square, *C. ngorora*; open cross, cf. *Choerolophodon* (S. Asia); inverted open triangle, *C. anatolicus*; open triangle, *C. pentelici*; open diamond, *C. corrugatus*; B, Burji gomphotheriid; X, *Progomphotherium maraisi*; inverted closed triangle, cf. *Archaeobelodon*; closed triangle, *A. filholi*; closed diamond, *Protanancus macinnesi*; closed square, *P. chinjiensis*; closed circle, *Afromastodon coppensi*; right-facing closed triangle, *Amebelodon cyrenaicus*.

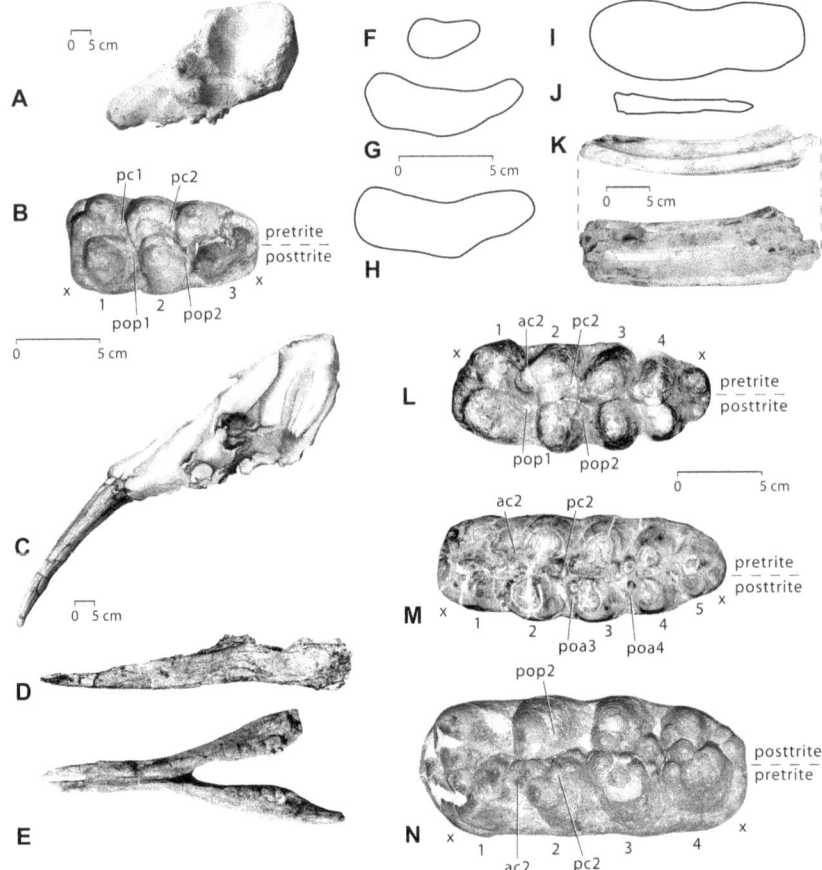

FIGURE 15.13 Aspects of amebelodontine craniodental morphology. A) Right lateral view (reversed), cranium, AM 1'95, holotype of *Progomphotherium maraisi* (see Pickford, 2003: plate 6). B) Right M3, AM 1'95, part of holotype of *Progomphotherium maraisi* (see Pickford, 2003: plate 7). C) Drawing, right lateral view (reversed), cranium, KNM-MI 7532, *Archaeobelodon filholi*. D) Left lateral view, mandible, KNM-MI 7532, *Archaeobelodon filholi*. E) Occlusal view, mandible, KNM-MI 7532, *Archaeobelodon filholi*. F–J, i2 cross sections, medial side to the left, all to same scale. F) KNM-MI 7532, *Archaeobelodon filholi*. G) M15532 (KBA 109), *Protanancus macinnesi*. H) Semliki n° 531A, *Protanancus macinnesi*. I) Sahabi 481P34A, *Amebelodon cyrenaicus* (reversed). J) MCZ 38-64K, *Platybelodon* sp. K) Left i2 fragment, lateral view (top) and dorsal view (bottom), M15533, *Protanancus macinnesi*. L) Occlusal view, right M3, M15525, part of holotype of *Protanancus macinnesi*. M) Occlusal view, right m3, M15438, *Protanancus macinnesi*. N) Occlusal view, left m3, PQAD 257, *Afromastodon coppensi* (see Pickford, 2003: plate 14).

ABBREVIATIONS: ac, pretrite anterior accessory central conule; pc, pretrite posterior accessory central conule; poa, posttrite anterior accessory central conule; pop, posttrite posterior accessory central conule; x, anterior or posterior cingulum(id); 1, 2, 3, . . . , first, second, third, . . . loph(id).

figures 3, 4, 5, 6, 9; plate VI, figure 8; ?Mammutid aff. *Zygolophodon*, Van Couvering and Van Couvering, 1976; cf. *Gomphotherium angustidens* (in part), Coppens et al., 1978.

Age and Occurrence Early Miocene, northern and eastern Africa (table 15.5).

Description Small amebelodont with trilophodont intermediate molars. M3 with four lophs. Molars have massive main conelets. Pretrite accessory conules may be absent (e.g., KNM-RU 4423; Tassy, 1986) or small and contribute to a trefoil arrangement (e.g., CGM 30892; Sanders and Miller, 2002). When present, posttrite accessory conules are diminutive. Worn molars exhibit a distinctive asymmetry between larger, more deeply excavated pretrite half-loph(id)s and posttrite half-loph(id)s. A p3 from Rusinga, Kenya is bilophodont and has tiny accessory conules posterior to its proto- and metaconids (Tassy, 1986).

ARCHAEOBELODON FILHOLI (Frick, 1933)
Figures 15.13C–15.13F

Partial Synonymy Platybelodon kisumuensis, Harris and Watkins, 1974; *Gomphotherium kisumuensis*, Savage and Williamson, 1978; *Platybelodon* sp., Pickford, 1981; *Archaeobelodon* aff. *filholi*, Tassy, 1984, 1985, 1986.

Age and Occurrence Early Miocene, eastern Africa (table 15.5).

Description The skull morphology of this taxon is documented in an associated cranium and mandible from Mwiti, Kenya (Tassy, 1986). Compared with the neurocranium, the face is relatively massive, with an elongate prenasal area and wide nasal region. The orbit is low and situated in a line posterior to the molar row; overall, the cranium is low and the basicranium is not very raised (figure 15.13C). The mandible is longirostral, with a transversely narrow symphysis, and is nearly straight along the length of its ventral margin (figures 15.13D, 15.13E).

Upper tusks have a lateral band of enamel and are relatively short and robust; the I2 of the skull has dimensions of L = 610 mm, and W = 110 mm, H = 102 mm at the alveolus (Tassy, 1986). Cross-sectional shape is ovoid. Lower tusks have a flattened pyriform shape in cross section, with a shallow longitudinal sulcus dorsally (figure 15.13F), and are torqued along their length, resembling i2s of *Phiomia major*.

Third and fourth premolars are bilophodont and typical for gomphotheres, with low, narrow cingulae(ids) and cusps much higher and larger anteriorly than posteriorly (Tassy, 1986).

Intermediate molars are trilophodont, and third molars have loph(id) formulas of x4x. Enamel is thick (4.0–6.4 mm), and cusps are low and massive (Tassy, 1986). Cementum is variably present but not extensively distributed on molar crowns. Molars are relatively narrow, with strong development of pretrite anterior and posterior accessory central conules, and much smaller posttrite accessory central conules. Pseudoanancoidy of half-loph(id)s varies but is generally weak.

Remarks The African *Progomphotherium* and *Archaeobelodon* samples mark the beginnings of the Amebelodontinae, and its primitive status is signaled by the retention of features such as low, massive conelets, symphyses projecting horizontally from mandibular corpora, modest flattening of lower tusks, poor or moderate development of accessory central conules, relatively small size, and low crania. *Archaeobelodon* migrated to Europe shortly after its origin in Africa (Tassy, 1985, 1986), and the morphological identity of East African specimens with European material indicates conspecificity.

Genus *AFROMASTODON* Pickford, 2003
AFROMASTODON COPPENSI Pickford, 2003
Figures 15.12A, 15.12B, and 15.13N

Partial Synonymy Gomphotherium cf. *angustidens*, Corvinus and Hendey, 1978; Gomphotheriidae, gen. et. sp. indet., Hendey, 1978b, *Protanancus macinnesi*, Tassy, 1985, 1986.

Age and Occurrence Middle Miocene, southern Africa (table 15.5).

Diagnosis M3/m3 differ from those of *Protanancus macinnesi* (Tassy, 1986; Pickford, 2003) by their lack of fifth loph(id)s, relatively greater width, and larger size (figures 15.12A, 15.12B), though are similar in occlusal organization. Molars are also larger than those of *Archaeobelodon* (figure 15.12A). Lower tusks lack enamel, are ovoid in cross section (Pickford, 2003, table 11), and have height/width ratios above the proportional range of variation for other amebelodonts (see later discussion). Upper tusks D shaped in cross section, with an enamel band along their lateral surfaces. Differs from South Asian *Protanancus chinjiensis* in having less pronounced pseudoanancoidy of half-loph(id)s, but similar in M3 size and proportions (figures 15.12A, 15.12B).

Description Afromastodon coppensi is represented by a modest dental sample, including upper and lower bilophodont deciduous third premolars, trilophodont intermediate molars, and third molars with loph(id) formulas of x4x (figure 15.13N).

Molars are sexually bimodal in size, bunolophodont and very brachyodont, with thick enamel and little or no cementum. Pre- and posttrite half-loph(id)s are separated by deep median longitudinal sulci (Pickford, 2003). In m2, the third lophid is widest; each half-lophid is composed of a main, outer conelet and a more diminutive mesoconelet; pretrite accessory conules are

well developed and may be multiplied posteriorly, forming trefoil enamel figures in wear; and small accessory central conules are present posterior to posttrite half-lophids. In M2, crown shape is more rectangular; pretrite half-lophs are located slightly more mesially than their posttrite counterparts; and posttrite posterior accessory central conules are also present. There is also a slight offset of half-loph(id)s in third molars, with the main conelet of the pretrite side anterior to the main conelet of the posttrite side in M3. The condition is reversed in m3. In m3, the main conelets are large and dominate much smaller mesoconelets (which are transversely adjacent to one another); pretrite accessory conules are prominent and may be multiplied posteriorly; and posttrite half-lophids may have small anterior and posterior accessory conules (figure 15.13N; Hendey, 1978b: figure 7). Although third molars develop strong trefoil wear patterns on their pretrite sides, occlusally worn specimens do not develop trefoil enamel figures along the opposing sides, despite the occurrence of accessory central conules in posttrite half-loph(id)s (Pickford, 2003: plate 15, figure 1).

Remarks The slight transverse dislocation of molar half-loph(id)s and occurrence of posttrite accessory central conules led to the initial identification of *A. coppensi* as *Protanancus macinnesi* (Tassy, 1985, 1986). Generic-level distinction of this species is justified only if unassociated lower tusks attributed to *A. coppensi* truly belong to the taxon. In cross section, these are rounded rather than flattened, an unexpected shape for an amebelodont. Otherwise, the morphological resemblance of the molars to those of *P. macinnesi* and proportional similarity with molars of *P. chinjiensis* supports affinity with *Protanancus*. Biochronological correlation of the Arrisdrift fauna with eastern African faunal sets indicates an early middle Miocene date near to that of Maboko, Kenya, ca. 16 Ma (Corvinus and Hendey, 1978; Hendey, 1978b), suggesting that on the continent, evolutionary trajectories of amebelodonts may have diverged regionally.

It is conceivable that an enigmatic elephantoid individual from Burji, Ethiopia, also dated to the early middle Miocene (table 15.5), has affinity with *A. coppensi*. Originally attributed to *Choerolophodon* (Suwa et al., 1991), this proboscidean has features more typical of amebelodonts, including a narrow m3 (figure 15.12B), posttrite accessory conules, and an upper tusk with an enamel band. The lower tusk of this individual is ovoid in cross section, proportionally similar to those of *A. coppensi* (see later discussion).

Genus *PROTANANCUS* Arambourg, 1945
PROTANANCUS MACINNESI Arambourg, 1945
Figures 15.12A, 15.13G, 15.13H, and 15.13K–15.13M

Partial Synonymy Trilophodon angustidens kisumuensis (in part), MacInnes, 1942; *Trilophodon angustidens kisumuensis* (in part) Arambourg, 1945; *Protanancus macinnesi*, Arambourg, 1945; *Trilophodon angustidens* cf. *kisumuensis* (in part), Hooijer, 1963; *Protanancus macinnesi*, Leakey, 1967; *Gomphotherium angustidens* (in part), Maglio, 1973: figure 22; *Platybelodon kisumuensis* (in part), Tobien, 1973a; *Platybelodon kisumuensis* (in part), Van Couvering and Van Couvering, 1976; cf. *Gomphotherium angustidens* (in part), Coppens et al., 1978; *Protanancus macinnesi*, Tassy, 1979a; *Platybelodon*, Shipman et al., 1981; *Trilophodon angustidens*, Boaz, 1994.

Age and Occurrence Middle Miocene, eastern Africa (table 15.5).

Diagnosis Tassy (1979a, 1984, 1985, 1986). Amebelodont with powerful upper incisors that curve strongly outwardly and have a band of enamel. Lower tusks flattened and dorsally hollowed, broader than in *Archaeobelodon*, and distinguished from i2s in *Platybelodon* by their lack of dentinal tubules and rods, and less severe compression. Differs from *P. chinjiensis* in having a shorter face, narrower, smaller molars, and less pseudoanancoid dislocation of third molar half-loph(id)s. Has smaller, narrower M3s (figure 15.12A), and wider, flatter i2s than *A. coppensi*. Posttrite accessory central conules are very small and irregularly distributed.

Description A fragmentary cranium from Fort Ternan, Kenya, preserves a small, narrow tympanic bulla, a large glenoid and postglenoid region lacking a postglenoid fossa, a wide palate, and a basicranium only slightly raised. Enough of the cranium is preserved to show the position of the infraorbital foramen above M2, and that the face was short (Tassy, 1986).

In cross section, lower tusks are very compressed dorsoventrally (at midtusk, specimen BMNH 15532 [KBA 109] has W = 83.5 mm and H = 30.0 mm) and are dorsally concave and ventrally convex (figures 15.13G, 15.13H, 15.13K). They are thicker medially, and exhibit polished wear facets at opposing surfaces of the distal end of the tusks, dorsomedially and ventrolaterally. In dorsal view, their tips form a V shape in the midline. Upper tusks are massive (KBA 110 measures L = 1055 mm and has basal diameters of 105 mm × 95 mm) and are pear shaped in cross section, each with an extensive enamel band along the lateral surface and torque about the longitudinal axis (MacInnes, 1942).

Protanancus macinnesi appears to have retained second–fourth deciduous premolars, and third and fourth premolars. The dp2 has two, narrowly transversely appressed anterior cusps, and much lower, paired posterior cusps. The p3 is bilophodont, and intermediate molars are trilophodont with postcingula(ids) formed of several prominent conelets. Lower third molars have lophid formulas of x5x, with very prominent distal cingulids, and may be longitudinally curved (figure 15.13M). In the sample from Maboko, Kenya, M3s have loph formulas of x4x, while M3s from the younger site of Fort Ternan, Kenya, have a nascent fifth loph. In M3, pretrite half-lophs may be slightly anterior to corresponding posttrite half-lophs, and in m3 the reverse occurs, but unlike the condition in choerolophodonts, the mesoconelets of half-loph(id)s are aligned transversely, chevroning is largely absent, and trefoil wear patterns of pretrite half-loph(id)s are generally well developed. On the posttrite side, small anterior accessory central conules may be present in m3s, and diminutive posterior accessory central conules are variably located in M3s (figure 15.13L). Cementum is present in the floor of transverse valleys and occasionally along the sides of loph(id)s; enamel is thick (4.5–6.0 mm) and unfolded; and molars are brachyodont and relatively narrow.

The few postcranials tentatively attributable to *P. macinnesi* include an axis vertebra, several femora distinguished from those of choerolophodonts by relatively small third trochanters, symmetrical condyles, and modest epicondyles, and possibly a radius and several ulnae, all from Fort Ternan, Kenya (Tassy, 1986).

Remarks *Protanancus macinnesi* is the oldest species of the genus, and with *P. chinjiensis* of South Asia constituted an intermediate evolutionary stage between *Archaeobelodon* and *Amebelodon* (Tassy, 1986). The genus extended into the late Miocene in South Asia, but evidently not in Africa (Tassy, 1983a, 1984, 1985, 1986). *Protanancus macinnesi* co-occurs widely with choerolophodonts in East African fossil sites (table 15.5) and is readily distinguishable from them by greater body size and possession of conspicuous lower tusks. However, the pseudo-anancoidy of its molars may be mistaken for the offset and chevroning of half-loph(id)s in choerolophodonts; this has led to minor confusion in the assignment of a few specimens from Maboko, Kenya (see later discussion; see also Pickford, 2001:4). Wear patterns and the V-shaped configuration of lower tusk tips in *P. macinnesi* indicates that these teeth were used primarily for slicing through tough vegetation (Lambert, 1992).

Genus AMEBELODON Barbour, 1927
AMEBELODON CYRENAICUS Gaziry, 1987
Figure 15.13I

Partial Synonymy *Amebelodon* sp., Gaziry, 1982; "*Amebelodon* sp.," Tassy, 1986; *Amebelodon (Konobelodon) cyrenaicus*, Lambert, 1990.

Age and Occurrence Late Miocene or early Pliocene, northern Africa (table 15.5).

Diagnosis Gaziry, 1987b) A large species of *Amebelodon*, with trilophodont m1, tetralophodont m2, five to six lophids in m3, extensive molar cementum, and lower incisors wider and thicker than those of other Old World congeners, with internal dentinal tubules.

Description The only reliable record of this genus in Africa, and youngest occurrence worldwide (Lambert, 1990) is from Sahabi, Libya (Gaziry, 1982, 1987b). Its presence is marked by i2 fragments that are flat and broad, with ventral and dorsal longitudinal sulci (Sanders, 2008b). Dimensions of the most complete specimen are L = +420 mm; W = 127 mm; H = 44 mm. Proportionally, these are consistent with an attribution to *Amebelodon*, and are less dorsoventally compressed than in *Platybelodon* (figure 15.13I). In cross section, centrally large dentinal tubules and rods are apparent in these tusks, surrounded by dentinal laminae. Ventral tusk surfaces are abraded and polished.

Molars attributed to the species have strongly developed pretrite trefoils, with anterior and posterior accessory central conules, and are relatively long and narrow but lack posttrite accessory conules, and do not exhibit pseudoanancoid offset of half-lophids (Gaziry, 1982, 1987b).

Remarks *Amebelodon* species, best known from North America and Asia, are characterized by elongated, moderately expanded mandibular symphyses, and elongate, flattened lower tusks with dorsal and ventral sulci (Barbour, 1930; Frick, 1933; Osborn, 1936; Tobien et al., 1986; Guan, 1996; Lambert and Shoshani, 1998). Body size estimates for North American *Amebelodon* indicate a weight of ca. 3,000–4,500 kg and height at the pelvis of about 2.3–2.7 m, approximately equivalent to that of the extant Asian elephant (Lambert, 1990; Christiansen, 2004). The cranium is not well-known, but at least in Asian *A. tobieni* it is laterally compressed and anteriorly elongated (Guan, 1996). Dentinal rods and tubules are variably present in i2 (Osborn and Granger, 1931; Lambert, 1990; Lambert and Shoshani, 1998; *contra* Tassy, 1985, 1986), possibly convergent on the condition of i2s in *Platybelodon* and *Torynobelodon* (Lambert, 1990, 1992).

The genus is subdivided into the subgenera *A. (Amebelodon)* and *A. (Konobelodon)*, the latter being distinguished by tetralophodonty of intermediate molars, third molars with six loph(id)s, and lower tusks with dentinal tubules and rods (Lambert, 1990). Based on these criteria, the Sahabi amebelodont belongs in *A. (Konobelodon)*. The lophid formulas of the Sahabi amebelodonts, and presence of dentinal rods and tubules in their i2s are derived features for *Amebelodon* (Lambert and Shoshani, 1998). These apomorphies are coupled with features that are curiously plesiomorphic for such a late-occurring amebelodont, including absence of posttrite accessory conules, and transversely straight alignment of half-lophids (Tassy, 1985; Lambert and Shoshani, 1998). Alternatively, it is possible that the molars belong to a different, unidentified elephantoid than do the tusks (Sanders, 2008b). Tassy (1999) has pointed out that the flattened Sahabi *A. cyrenaicus* tusks are very similar to those of European and South and western Asian *"Mastodon" grandincisivus*, which (unlike the Sahabi sample) also has molars typical for amebelodonts, with features such as pseudo-anancoid, offset half-loph(id)s and posttrite accessory central conules. If at least the Sahabi tusks of *A. cyrenaicus* are conspecific with *"Mastodon" grandincisivus*, the latter species nomen would have priority (Tassy, 1999).

Wear facets and polishing on the tips of lower tusks in *Amebelodon* have been correlated with a wide range of feeding behaviors, including digging and shoveling in abrasive substrates, vegetation stripping, and bark scraping (Barbour, 1930; Lambert, 1992). Although the tusk fragments of the Sahabi amebelodonts do not preserve their tips, the restriction of abrasion to the ventral side hints at a scooping or shoveling function.

Genus *PLATYBELODON* Borissiak, 1928
PLATYBELODON SP. Maglio, 1969
Figure 15.13J

Partial Synonymy *Platybelodon* sp. cf. *P. grangeri*, Van Couvering and Van Couvering, 1976; ?*Platybelodon*, Coppens et al., 1978; *Platybelodon* sp., Tassy, 1986.

Age and Occurrence Early Miocene, eastern Africa, ?Arabia (table 15.5).

Diagnosis Based on Osborn and Granger (1931, 1932); Osborn (1936). Amebelodonts with highly derived lower tusks and mandibular symphyses that together approximate the form of a shovel. Differ from other amebelodonts in having relatively short, widely flared symphyses that are deeply excavated dorsally, and which accommodate lower incisors uniquely characterized by their extreme degree of dorsoventral compression, breadth, and internal composition of numerous, small and closely compacted dentinal tubules.

Description A single fragment of a lower incisor marks the presence of *Platybelodon* at Loperot, Kenya (Maglio, 1969a). It is extremely flattened dorsoventrally, with dimensions of W = 78.5 mm, and H = 14.6 mm medially and 8.5 mm laterally. The i2 compression index (H × 100/W) for this specimen is 19, well below the range of 35–50 for *P. macinnesi*, 68–84 for *A. coppensi*, 85 for the Burji gomphotheriid, 48–52 for African *Archaeobelodon*, and 35 for *A. cyrenaicus*, but comparable to Asian and North American *Platybelodon* and *Torynobelodon* (Osborn and Granger, 1931, 1932; Osborn, 1936; Guan, 1996). In cross section, the tusk is shallowly concave dorsally, and nearly horizontally flat on its ventral surface (figure 15.13J). Internally, the tusk is composed of a complex system of fine dentinal tubules and rods, invested within and surrounded by laminar dentine (Maglio, 1969a; Coppens et al., 1978). These features, along with the degree of compression, are diagnostic of *Platybelodon*.

Remarks The Loperot *Platybelodon* is the oldest instance of the genus, and it may also be present at Ad Dabtiyah, Saudi Arabia, of similar antiquity (Hamilton et al., 1978; Whybrow and Clements, 1999), indicating an Afro-Arabian origin, though it soon dispersed to Asia (Guan, 1996). By the late Miocene, the genus had reached North America, along with a cohort of other amebelodonts (Frick, 1933; Osborn, 1936; Fisher, 1996; Lambert, 1996; Lambert and Shoshani, 1998). The antiquity of the Loperot *Platybelodon*, combined with penecontemporaneous occurrences of other amebelodont taxa (table 15.5), shows that subfamilial diversity was achieved early on in the phylogenetic history of the group, largely through a temporally condensed adaptive radiation rooted in Africa.

The lower incisors in the "shovel tusker" *Platybelodon* have been hypothesized as functioning to shovel, scoop, dig, and dredge (Osborn, 1936) during feeding on soft vegetation in swampy or lowland aquatic habitats (Coppens et al., 1978). This interpretation was supported by the observation of beveling along the ventral tip of i2 (Osborn and Granger, 1932; Tobien, 1986). The complex of dentinal tubules, surrounded by dentinal laminae, was thought to have resisted abrasion and to have provided strength to the tusk during these sorts of activities (Osborn and Granger, 1931; see Lambert, 1990). A more recent analysis of tusk wear surfaces, however, indicates that the i2s of *Platybelodon* were more likely used as scythes to cut tough vegetation, rather than for shoveling (Lambert, 1992).

Subfamily CHOEROLOPHODONTINAE Gaziry, 1976

Comprised of the genera *Afrochoerodon* and *Choerolophodon*, this subfamily evidently originated in Africa during the early Miocene. By the end of the early Miocene, choerolophodonts had dispersed into South Asia and the eastern Mediterranean region, and persisted throughout much of the Old World until the close of the epoch (Tassy, 1977b, 1983b, 1985; Pickford, 2001; Sanders, 2003). Paleoecological data suggest that African species inhabited primarily closed, wet forest and woodland habitats (Savage and Hamilton, 1973; Pickford, 1985; Jacobs and Kabuye, 1987; Sanders and Miller, 2002; Tsujikawa, 2005a), and presumably were browsers, though choerolophodonts from Fort Ternan, Kenya may have lived in more open conditions (Evans et al., 1981; Shipman et al., 1981; Shipman, 1986).

Subfamilial features include distally upcurved I2s without enamel bands; orbits in line with the posterior end of the last molars; nonflaring zygomatic arches; elongation of the splanchnocranium; deep, wide symphyseal gutters; and loss of i2s and permanent premolars (Tassy, 1985, 1986). Central accessory conules and pretrite mesoconelets are high and large. Molars are specially characterized by chevroning of half-loph(id) pairs and may be choerodont and ptychodont. Half-loph(id)s are transversely offset, with the mesoconelet of each pretrite half-loph(id) projecting anterior to the posttrite half-loph(id) and to its own paired abaxial conelet (Tassy, 1985, 1986). These dental features are less prominently expressed in earlier species. *Choerolophodon* is contrasted with the more primitive *Afrochoerodon* by smaller temporal fossae and less convergent temporal lines, a more elongated basicranium, orbits positioned higher on the cranium, a more horizontal face, more pronounced chevroning, and stronger expression of choero-, ptycho-, and cementodonty, warranting at least generic-level separation (Pickford, 2001; *contra* Shoshani and Tassy, 2005).

Genus *AFROCHOERODON* Pickford, 2001
AFROCHOERODON KISUMUENSIS (MacInnes, 1942)
Figures 15.12A, 15.12B, 15.13J, and 15.14A–15.14D

Partial Synonymy *Trilophodon angustidens kisumuensis* (in part), MacInnes, 1942:51; plate 3, figures 13, 17, 18; plate 4,

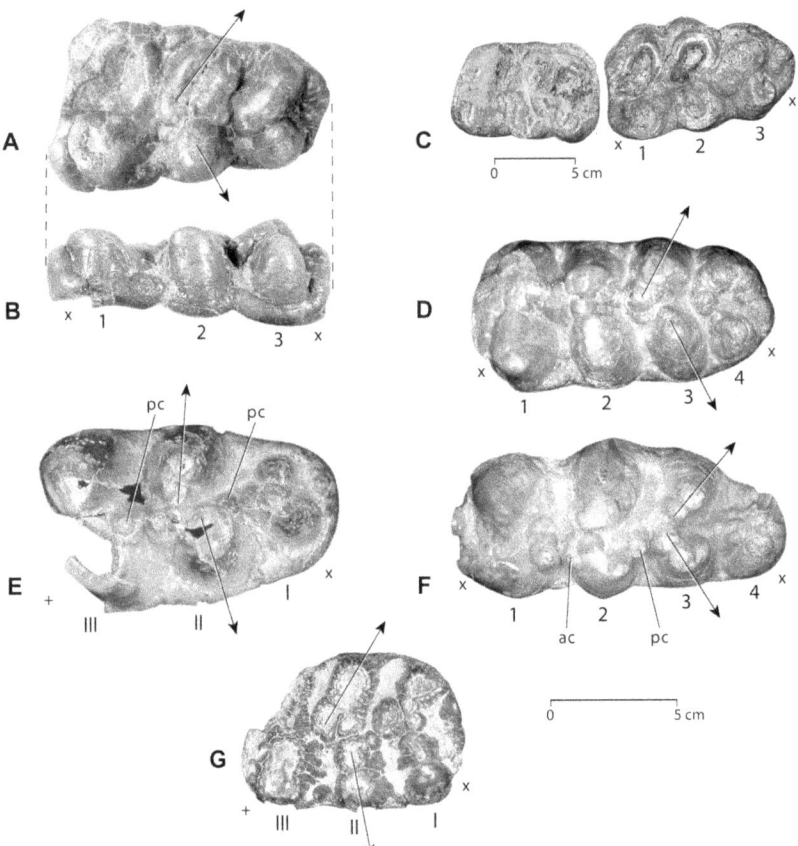

FIGURE 15.14 Aspects of choerolophodont molar morphology. Oblique lines indicate chevroning of pre- and posttrite half-loph(id)s. Anterior is to the left in all specimens. A) Occlusal view, right M3, DPC 14584, *Afrochoerodon kisumuensis*. B) Lateral view, right M3, DPC 14584, *A. kisumuensis*. C) Occlusal view, right M2–3, M15524, *A. kisumuensis* (type). D) Occlusal view, right m3, M15529, *A. kisumuensis*. E) Occlusal view, left ?M3, no number (Gebel Zelten, Libya), *Choerolophodon zaltaniensis*. F) Occlusal view, left m3, KNM-FT 2835, *C. ngorora*. G) Occlusal view, right M1, KNM-NA 4, *C. ngorora*.

ABBREVIATIONS: ac, pretrite anterior accessory central conule; pc, pretrite posterior accessory central conule; x, anterior or posterior cingulum(id); +, indicates a missing portion of a tooth; 1, 2, 3, ..., first, second, third, . . . loph(id).

figures 5, 6, 7, 8; plate 6, figure 7; *Protanancus* (in part), Arambourg, 1945:491; *Platybelodon kisumuensis* (in part), Tobien, 1973a:261; *Choerolophodon kisumuensis*, Tassy, 1977b:2488; *Afrochoerodon kisumuensis*, Pickford, 2001; *Afrochoerodon* aff. *kisumuensis*, Pickford, 2005a.

Age and Occurrence Early to middle Miocene, northern, eastern and southern Africa (table 15.5).

Diagnosis Tassy (1986); Pickford (2001); Sanders and Miller (2002). Small choerolophodont, with third molars below size range for other subfamilials (figures 15.12A, 15.12B; specimen KBA 202 = BMNH 15539, a large m3 from Maboko, belongs instead in *Protanancus macinnesi*—*contra* Tassy, 1986; see Pickford, 2001; specimen KBA 004 = BMNH 15529, from Maboko, is an m3 and not an M3—*contra* Tassy, 1986). Molar occlusal morphology simple, with little or no choerodonty, ptychodonty, or cementum. Chevroning of lophs modest throughout upper molar crowns. M3 with as few as three lophs. Pretrite half-loph(ids) with greatly enlarged anterior accessory central conules, which may be laterally shifted in m3, and irregular presence of smaller posterior accessory central conules.

Description Cranium high and short, with anteroposteriorly short but steeply inclined facial region; basicranium short; temporal lines converge closely behind the orbits and temporal fossae large; occipital surface nearly as high as wide and flaring laterally (occipital flanges almost platelike); orbits set low; voluminous juga for tusks dominate the facial region; palatines extend beyond distal end of M3s; zygomatic arches massive but not widely prominent; infraorbital foramen positioned above mesial end of M3 (Tassy, 1986; Pickford, 2001). Upper tusks large (diameter 90 mm) and almost circular in cross section, with no enamel band, initially flaring laterally and downward from widely separated alveoli (Tassy, 1986; Pickford, 2001). Nothing is known about symphyseal angulation or lower tusks in this species.

Intermediate molars are trilophodont, with greatest width at second loph(id)s. Cusps are massive, bulbous, and low crowned. Enamel is thick and unfolded. Upper third molars have three lophs or a fourth, smaller loph, and a diminutive posterior cingulum (figures 15.14A–15.14C). Mild chevroning occurs at the posterior end of crown, with pretrite mesoconelets more mesial than principal conelets. Posttrite half-lophs are constructed of a single large conelet or superficially subdivided into two conelets. Lower third molars have four lophids and a small posterior cingulid (figure 15.14D). In m3, pretrite mesoconids are anterior to main conelets and may be separated from them by enlarged, laterally set anterior accessory central conules (Tassy, 1986; Pickford, 2001; Sanders and Miller, 2002).

Proboscidean postcranials from Maboko, Kenya have not been definitively sorted among the taxa present there (table 15.5); however, an isolated elephantoid astragalus from the site has been tentatively attributed to *A. kisumuensis* based on its small size (MacInnes, 1942; Tassy, 1986). In contrast, limb bones and metacarpals from Maboko are slender but from an animal of considerable size, with a pelvic height of about 6–7 feet (MacInnes, 1942).

Remarks The antiquity of this species suggests an African origin for the subfamily. However, it is similar dentally and may be synonymous with *Choerolophodon palaeindicus* from Dera Bugti, Pakistan (Tassy, 1985; Sanders and Miller, 2002), which has nomenclatural priority if this is the case. *Choerolophodon palaeindicus*, whose exact provenance is unknown but is likely of comparable antiquity (Forster Cooper, 1922; Raza and Meyer, 1984; Tassy, 1985, 1986), should be transferred to *Afrochoerodon*. The primitiveness of *A. kisumuensis* among choerolophodonts is evident in the occlusal simplicity, small size, and mild chevroning of its molars (Tassy, 1979a).

Genus *CHOEROLOPHODON* Schlesinger, 1917
CHOEROLOPHODON ZALTANIENSIS Gaziry, 1987
Figures 15.12A and 15.14E

Partial Synonymy Afrochoerodon zaltaniensis, Pickford, 2001; *C. zaltaniensis*, Pickford, 2004.

Age and Occurrence Middle Miocene, northern Africa (table 15.5).

Diagnosis Differs from *A. kisumuensis* in variably greater expression of choero- and ptychodonty, slightly more cementum invested in transverse valleys, stronger chevroning of half-loph(id)s, invariant presence of a fourth loph in M3, and larger size of molars (figure 15.12A). In contrast to *Choerolophodon ngorora*, there are in *C. zaltaniensis* no m3s with five lophids, and molar choerodonty and cementum are generally less well developed.

Description Composed primarily of a small sample of teeth from Gebel Zelten, Libya. An adult dentary from the site is too incomplete to accurately ascertain symphyseal angulation or the absence of lower incisors. A juvenile choerolophodont dentary exhibits a wide symphyseal gutter, but is broken anteriorly (Gaziry, 1987a). No cranium has been recovered.

Molars are moderate sized, and in morphology and proportions more resemble teeth of *C. ngorora* than of *A. kisumuensis* (figure 15.12A). Intermediate molars are trilophodont. Upper third molars have a loph formula of x4x, though the fourth loph is narrow, and m3 has four lophids. Enamel is rugose but unfolded. In some molars, multiple accessory conules may be distributed within the transverse valleys (Gaziry, 1987a: figure 5a). In M3, small pretrite accessory conules are present posterior to lophs 1–3 (figure 15.14E). Particularly posterior to the first loph(id), pretrite anterior central accessory conules and mesoconelets are aligned oblique to the main axis of the crown, the main conelets are posterior to the mesoconelets, together contributing to the formation of loph(id)s into markedly anteriorly pointing V's (Gaziry, 1987a).

Remarks The original generic designation is retained for this species, rather than following Pickford's (2001) transfer of it to *Afrochoerodon* (but see Pickford, 2004), because it more closely resembles other species of *Choerolophodon* in chevroning, cementum development, ptychodonty, and upper molar choerodonty. In contrast, in these features and especially in cranial

morphology, *A. kisumuensis* stands apart from other choerolophodonts, except its near contemporaries *Afrochoerodon chioticus* from Chios, Greece (Tobien, 1980; Pickford, 2001), and *"C." palaeindicus* (Forster Cooper, 1922; Raza and Meyer, 1984; Tassy, 1985, 1986; Sanders and Miller, 2002).

CHOEROLOPHODON NGORORA (Maglio, 1974)
Figures 15.12A, 15.12B, 15.14F, and 15.14G

Partial Synonymy Gomphotherium ngorora, Maglio, 1974; *Gomphotherium*, Andrews and Walker, 1976:300; *Choerolophodon ngorora*, Tassy, 1977b; *"Gomphotherium" ngorora*, Coppens et al., 1978; *C. ngorora*, Tassy, 1986; *C. ngorora*, Nakaya, 1993; *Afrochoerodon ngorora*, Pickford, 2001; *Afrochoerodon* sp. nov. [specimens from Mbs. A–D, Ngorora Fm.], Pickford, 2004; *C. ngorora* [specimens from Mb. E, Ngorora Fm.], Pickford, 2004; *Choerolophodon* sp., Tsujikawa, 2005b.

Age and Occurrence Middle to early late Miocene, eastern Africa (table 15.5).

Diagnosis Molars larger and with more complex occlusal morphology than those of *A. kisumuensis* (figures 15.12A, 15.12B, 15.14F, 15.14G), and contrast with those of *C. zaltaniensis* in having folded enamel and (occasionally) greater expression of choerodonty. Variable development of a fifth lophid in m3. Tendency for upper molars to retain a pretrite posterior central accessory conule behind loph 2, and for pretrite anterior accessory central conules to be independent of their corresponding mesoconelets (Tassy, 1986).

Description Symphyseal region of mandible elongate and strongly downturned. Upper tusks curved upward, with no enamel band, but may have an enamel cap; they are ovoid and slightly flattened dorsoventrally in cross section (Tassy, 1986). No cranial remains have been recovered for this species, and there is no evidence of lower tusks.

DP2 is tiny, dominated by an anteriorly situated, mesiodistally compressed loph, and has a diminutive central conule. These features resemble DP2 in Eurasian choerolophodonts (Tassy, 1986). The dp3 is bilophodont and also typical of choerolophodonts: the crown narrows to a point mesially, the cusps of its lophids are offset anteroposteriorly to one another, and it sports large central accessory conules. Intermediate molars are trilophodont. The loph formula for M3 is x4x, and m3 has four-five lophids (figure 15.14F). Mesoconelets are small, but higher than the main conelets, as are the accessory central conules. Anterior accessory conules are massive, while posterior accessory conules are reduced in importance and may be limited to the anterior end of the crown or absent. Chevroning is marked (Tassy, 1986). Enamel is strongly rugose and may be moderately to finely folded (figure 15.14G); cementum may be thick; and in some molars (particularly uppers), multiple enamel tubercles may be present in transverse valleys (see Tassy, 1986: figure 27).

Proboscidean postcrania from Fort Ternan include long bones, carpals, and tarsals (Tassy, 1986). Among those tentatively assigned to *C. ngorora* are femora with salient third trochanters and medial epicondyles, and humeri with reduced trochlear separation between the medial and lateral condyles and with high, capacious epicondyles. The calcaneum is moderate in size for an elephantoid (L = 180.6 mm) but robustly built, with a massive tuber calcanei and a large fibular facet, closely resembling calcanei of Eurasian choerolophodonts (Tassy, 1986).

Remarks The molars of *C. ngorora* are separable into two subsamples (table 15.5): a late middle Miocene morph similar to teeth of *C. zaltaniensis*, and a more advanced, early late Miocene morph. The cheek teeth of the younger morph are larger, higher crowned, and have more finely folded enamel. These differences may signal speciation of the lineage around 11 Ma, but the more advanced subsample is too small and fragmentary for formal acknowledgment of such an event (but see Pickford, 2004). The relationship of similar-aged choerolophodonts from North Africa (table 15.5) to *C. ngorora* has yet to be established.

Despite evolving more complex, durable molars, by ≈9 Ma choerolophodonts had vanished from Africa (Tassy, 1986), but there is no obvious explanation for their demise. While there is evidence for increased aridity and seasonality in Africa from the middle to late Miocene (Pagani et al., 1999), the disappearance of choerolophodonts from the African continent apparently did not coincide with an abrupt shift in vegetation (Cerling et al., 1993; Kingston et al., 1994), or with the emergence of a cohort of new proboscidean rivals, and remains unexplained.

Subfamily TETRALOPHODONTINAE Van der Maarel, 1932
Genus *TETRALOPHODON* Falconer, 1857
TETRALOPHODON SP. NOV. Tsujikawa, 2005
Figures 15.7A, 15.7B, and 15.11L

Partial Synonymy Tetralophodon longirostris, Bergounioux and Crouzel, 1956; *Trilophodon angustidens* cf. *kisumuensis*, Hooijer, 1963:32, plate I, figure 2; *Tetralophodon* sp. indet., Madden, 1977; ?*Tetralophodon* cf. *longirostris*, Coppens et al., 1978:346; Proboscidea indet., Tiercelin et al., 1979:257; *Stegotetrabelodon grandincisivum*, Madden, 1982; *Tetralophodon* sp., Nakaya et al., 1984; *Tetralophodon* cf. *longirostris*, Geraads, 1989; *Stegotetrabelodon* n. sp.?, Geraads et al., 2002:114.

Age and Occurrence ?Late middle–early late Miocene, eastern and ?northern Africa (table 15.5).

Diagnosis Based in part on Nakaya et al. (1984, 1987); Geraads (1989); Nakaya (1993). Large gomphotheriid with tetralophodont intermediate molars and third molars with five-six loph(id)s (figure 15.11L). Enamel massively thick. Second-fourth permanent premolars are retained; upper and lower tooth formulae 1-0-3-3, 1-0-2-3, respectively. Occasional development of secondary trefoils on posttrite half-loph(id)s. Differs from anancine gomphotheres in straighter transverse alignment of half-loph(id)s, and from European tetralophodonts by greater angulation of the basicranium, relative to the palate, and absence of cementum.

Description Tetrabelodont, with rounded upper tusks lacking an enamel band and thick lower tusks rounded to pyriform in cross section (the cross sections of putative rostral incisor alveoli from Gebel Sémène are actually those of i2s—*contra* Bergounioux and Crouzel, 1956). The symphysis is long and strongly downturned (Nakaya et al., 1984).

Molar loph(id)s are brachyodont and formed of four to six conelets, with mesoconelets rivaling main conelets in size, and are accompanied by accessory central conules. Lamellar frequency is low (2.3 in m2), enamel thickness is extreme (10.9 mm in m2), and molars are very large (figure 15.7A, 15.7B; Nakaya et al., 1984, 1987). An isolated m3 from Chorora, Ethiopia may also belong in this taxon. It has six lophids, well-developed pretrite posterior accessory central conules, and diminutive posttrite accessory central conules (Coppens and Tassy in Tiercelin et al., 1979).

Permanent premolars P2–4 and p4 have been recovered for this taxon. These are bilophodont teeth (Nakaya et al., 1984;

Geraads, 1989), and resemble permanent premolars in *Stegotetrabelodon*.

Remarks Tetralophodon is well represented in Eurasia by several late middle–late Miocene (Astaracian-Turolian) species and seems likely to have emigrated from there into Africa (Coppens et al., 1978; Tassy, 1985; Tobien et al., 1988). Morphologically and chronostratigraphically, African *Tetralophodon* is appropriately situated to be ancestral to elephants.

Evidence from the Samburu Hills, Kenya suggests that *Tetralophodon* may have encountered a variety of habitats, including woodlands and savanna (Nakaya, 1993; Tsujikawa, 2005a), but dental isotopic analysis reflects a narrower preference for (possibly xeric) browse with only a small component of C_4 vegetation in its diet (Cerling et al., 2003).

An M3 (KI 64′92) from the Western Rift (table 15.5; figure 15.11L) has been referred to as a primitive elephant (Tassy, 1995). Nevertheless, it should remain in *Tetralophodon* (Pickford et al., 1993), because its overall morphology, including a low number of lophs (x5x), division of the crown by a marked median sulcus, enlarged mesoconelets, presence of anterior and posterior pretrite accessory central conules, wide spacing of lophs (LF = 2.6), brachyodont crown (HI = 53), and low number of conelets per loph (Tassy, 1995), closely resembles the condition of typical tetralophodont gomphotheriid molars.

GEN. ET SP. INDET. (tetralophodont form) Tassy, 1986

Age and Occurrence Late Miocene, eastern Africa (table 15.5).

Description Several specimens from the early late Miocene of eastern Africa appear to be of a different form of tetralophodont than that present at Samburu Hills, notably a partial cranium and incomplete mandible from the Tugen Hills, Kenya. The cranium is reconstructed as having been relatively high, with a short face and angled basicranium (Tassy, 1986). It retains DI2 and I2, as well as DP2–DP4; these latter teeth are bilophodont, trilophodont, and tetralophodont, respectively, and are characterized by a pseudoanancoid offset of pre- and posttrite half-lophs and weak expression of accessory central conules (Tassy, 1986). The mandible is reconstructed as having been brevirostrine with a shallow symphyseal gutter, and carries a p3 typical of tetralophodont elephantoids (Tassy, 1986).

Remarks These specimens superficially resemble anancine gomphothere cheek teeth in the transverse offset of their half-lophs, and the apparent brevirostrine condition of the mandible. They are sufficiently unique and isolated to refrain from assigning them to a particular taxon. Because of their potential phylogenetic role in the emergence of elephants and *Anancus*, it is unfortunate that nonanancine tetralophodonts are so poorly represented in the African fossil record.

Subfamily ANANCINAE Hay, 1922
Figures 15.15A–15.15F

Anancine gomphotheres constitute a monogeneric group (*Anancus* spp.) of Eurasian origin, possibly derived from *Tetralophodon,* that was widely dispersed throughout the Old World during the late Miocene–late Pliocene (Tobien, 1973a; Coppens et al., 1978; Mebrate and Kalb, 1985; Tassy, 1985, 1986, 1996c; Tobien et al., 1988; Metz-Muller, 1995; Kalb et al., 1996a; Shoshani, 1996; Göhlich, 1999; G. Markov, pers. comm.). Immigration of anancines into Africa toward the end of the Miocene coincided with the first appearance of

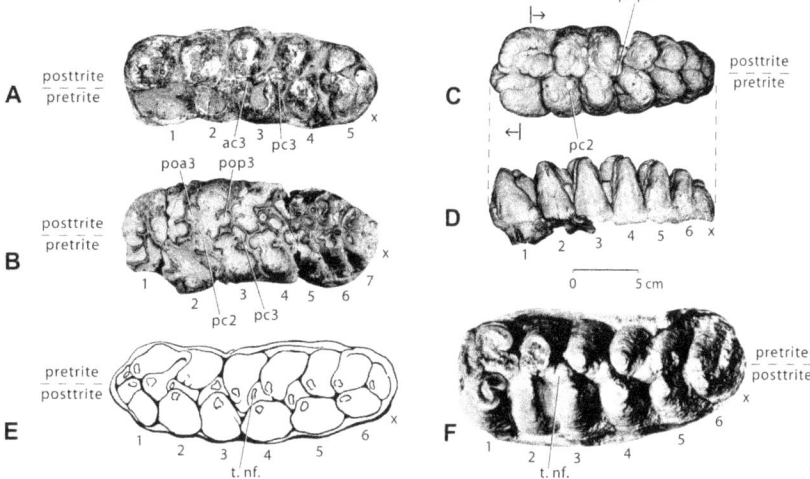

FIGURE 15.15 Aspects of anancine gomphothere molar morphology. Anterior is to the left in all specimens. A) Left m3, KNM-LU 57, *Anancus kenyensis*; B) left m3, EP 197/05, *A.* sp. nov. (type); C) left M3, SAM-PQ-L 41692, *A. capensis* (type), occlusal view (arrows denote relative position of pre- and posttrite half-lophs; in upper molars, pretrite half-lophs are offset anterior to posttrite half-lophs); D) left M3, SAM-PQ-L 41692, *A. capensis* (type), lingual view; E) right m3, Sahabi molar no. 8, *A. petrocchii*; F) right m3, 1956-4: A1, *A. osiris*.

ABBREVIATIONS: ac, pretrite anterior accessory central conule; pc, pretrite posterior accessory central conule; poa, posttrite anterior accessory central conule; pop, posttrite posterior accessory central conule; x, anterior or posterior cingulum(id); t. nf., "tubercle de néoformation" (Arambourg, 1945, 1970); 1, 2, 3, ..., first, second, third, . . . loph(id).

elephants, with whom they remained ubiquitously sympatric until their demise (Tassy, 1986). Although *Anancus* survived well into the late Pliocene in northern and southern Africa, it disappeared elsewhere on the continent during the mid-Pliocene (table 15.5; Tassy, 1986; Kalb et al., 1996b).

African anancines are characterized by short, wide crania with domed, elevated vaults, very raised bases, enlarged tympanic bullae, straight upper tusks lacking enamel, mandibles with brevirostrine symphyses and no lower tusks, and tetralophodont-pentalophodont intermediate molars (Petrocchi, 1954; Coppens, 1967; Coppens et al., 1978; Tassy, 1985, 1986; Kalb and Mebrate, 1993). Their bunolophodont cheek teeth are typified by anancoidy of pre- and posttrite half-loph(id)s: in upper molars, pretrite half-lophs are anterior to their paired posttrite half-lophs, while the reverse occurs in lower molars (figures 15.15A–15.15F; Coppens et al., 1978; Mebrate and Kalb, 1985; Tassy, 1985, 1986; Kalb and Mebrate, 1993). Molar occlusal complexity increased independently in several lineages of *Anancus* and included addition of loph(id)s and elaboration of accessory conules and conelets in transverse valleys and postcingulae(ids) (Coppens et al., 1978; Mebrate and Kalb, 1985; Tassy, 1985, 1986; Kalb and Mebrate, 1993; Sanders, 1997).

The forelimbs of anancine gomphotheres have been reconstructed as habitually flexed, suggesting that they were ground-level feeders (Ferretti and Croitor, 2001, 2005). This conforms with results from stable isotope analyses on tooth enamel from a number of African sites indicating that these proboscideans were grazers with a predominantly C_4-plant-based diet (Cerling et al., 1999, 2003; Zazzo et al., 2000; Harris et al., 2003; Semaw et al., 2005), except at Langebaanweg, where C_3 grasses are inferred to have been prevalent (Franz-Odendaal et al., 2002). Over the course of the Pliocene, this dietary emphasis may have contributed

to their eventual decline, placing them in increasingly crowded competition with an expanding cohort of hypsodont elephants and other ungulate grazers (see Cerling et al., 2003).

Genus *ANANCUS* Aymard, 1855
ANANCUS KENYENSIS (MacInnes, 1942)
Figures 15.15A, 15.16A, and 15.16B

Partial Synonymy Pentalophodon sivalensis kenyensis, MacInnes, 1942; *Anancus arvernensis* subsp., Dietrich, 1943; *Anancus kenyensis*, Arambourg, 1947; *Anancus osiris* (in part), Coppens et al., 1978; *Anancus* cf. *A. kenyensis*, Haile-Selassie et al., 2004; *Anancus* sp. indet., Haile-Selassie et al., 2004.

Age and Occurrence Late Miocene to early Pliocene, eastern and Central Africa (table 15.5).

Diagnosis MacInnes (1942); Tassy (1986). Tetralophodont intermediate molars. Anancoidy not pronounced. Enamel unfolded to coarsely folded, and very thick (5.0–7.0 mm in third molars). Crown morphology simple. In lateral view, loph(id)s are bulbous and massive. Posttrite posterior accessory conules restricted to mesial half of crown in m3, and usually in M3, as well. Permanent premolars retained.

Description Crania of *A. kenyensis* from Chad, Kenya, and Ethiopia have vertical basicrania nearly perpendicular to their palates, orbits anterior to the molars, robust, broad zygomatic arches, and lateral alisphenoid processes developed as pillars (Coppens, 1967; Tassy, 1986; Kalb and Mebrate, 1993). Adult upper tusks are rounded in cross section, and exhibit marked interindividual contrasts in size, attributable to sexual dimorphism (Tassy, 1986; Kalb and Mebrate, 1993). Mandibles of *A. kenyensis* have short molar alveolar segments, high rami, and anteroposteriorly narrow, transversely broad condyles (Tassy, 1986; Haile-Selassie, 2001). The forward rotation of

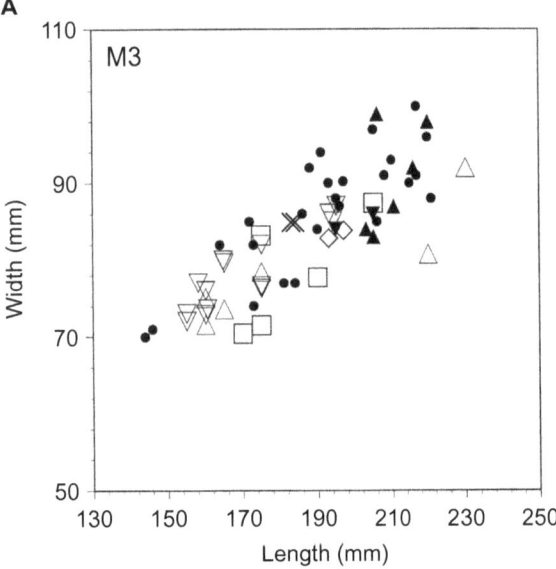

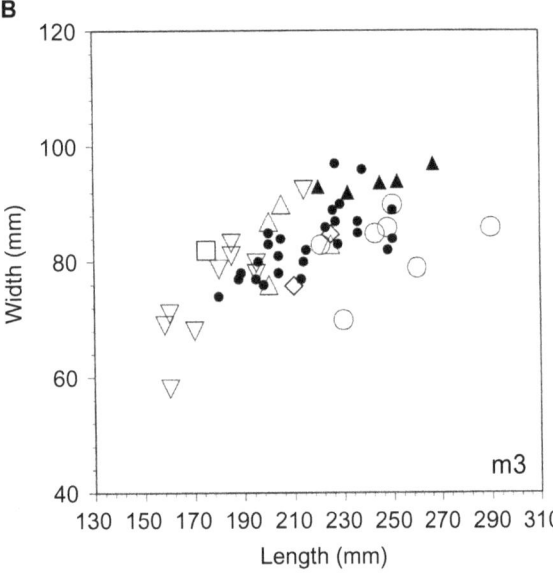

FIGURE 15.16 Bivariate plots of M3 (A) and m3 (B) crown length versus width in African and selected Eurasian anancine gomphotheres. Comparative dimensions supplementing original measurements are from Arambourg (1945, 1970), Petrocchi (1954), Tassy (1986, 1995), Tobien et al. (1988), Boeuf (1992), and Metz-Muller (1995).

SYMBOLS: inverted open triangle, *Anancus kenyensis*; open triangle, *A.* sp. nov. open square, *A. capensis*; open diamond, *A. osiris*; open circle, *A. petrocchii*; X, *A. perimensis*; closed circle, *A. arvernensis*; closed triangle, *A. sinensis*.

emerging cheek teeth precluded the simultaneous occlusion of the entire molar series.

Molar proportions and occlusal morphology vary widely within *A. kenyensis*. Third molar size extends to the low end of the range for the genus, but there is substantial overlap in dimensions among African congeners (figures 15.16A, 15.16B). Pre- and posttrite posterior accessory conules are often present at least in the first three loph(id)s, especially in intermediate molars, but their expression may be more limited in m3. Cementum ranges in thickness from trace amounts to filling transverse valleys. Third molars modally have five and

intermediate molars four loph(id)s, with a few exceptions (figure 15.15A; Tassy, 1986, 1995, 2003). Diminutive M3s from the Kuseralee Mb. of the Sagantole Fm., Middle Awash, Ethiopia have six lophs, and have been differentiated from *A. kenyensis* as "*Anancus* sp. indet." (Haile-Selassie, 2001; Haile-Selassie et al., 2004). However, their simple crown morphology and associated tetralophodont intermediate molars suggest instead that they are a variant of *A. kenyensis*.

Remarks Anancus kenyensis is the oldest representative of the subfamily in Africa (table 15.5), and constituted a progressively evolving lineage. Evidence for this is documented by directional increase in occlusal complexity, crown size, and loph(id) number in the abundant molar sample from the Middle Awash, Ethiopia, between 6.3 and 4.3 Ma (Mebrate and Kalb, 1985). Schema partitioning the lineage into pseudotaxonomic units include those of Mebrate and Kalb (1985), who divided the Middle Awash sequence into four time-successive stages, and Tassy (1986), who separated the species into an *A. kenyensis* "kenyensis-morph" and an *A. kenyensis* "*petrocchii*-morph." It is more appropriate to term these forms "primitive-morph" and "advanced-morph," respectively, to avoid confounding pentalophodont eastern African anancines with the pentalophodont but distinct species *A. petrocchii* from Sahabi, Libya. Differences between these morphs are profound and warrant more formal taxonomic division. Accordingly, they are here assigned to separate species, with the "primitive-morph" retained as *A. kenyensis*.

ANANCUS sp. nov. Sanders, in press
Figures 15.15B, 15.16A, and 15.16B

Partial Synonymy Trilophodon angustidens cf. *kisumuensis* (in part), Hooijer, 1963; *Anancus osiris*, Coppens, 1965; *Anancus osiris*, Servant-Vildary, 1973; *Anancus kenyensis* (in part), Coppens et al., 1978; *Anancus kenyensis* (in part; *A. kenyensis* "*petrocchii*-morph"), Tassy, 1986; *Anancus* sp. (Sagantole-type), Kalb and Mebrate, 1993; *Anancus* cf. *Anancus* sp. (Sagantole-type), Sanders, 1997; *Anancus kenyensis*, Harris and Leakey, 2003; *Anancus kenyensis* (in part), Tassy, 2003; *Anancus kenyensis*, Mackaye et al., 2005; *Anancus osiris*, Mackaye et al., 2005.

Age and Occurrence Early–mid-Pliocene, eastern and Central Africa (table 15.5).

Diagnosis Pentalophodont intermediate molars; third molars with six or seven loph(id)s. Anancoidy well expressed. Occlusal morphology usually complex; posttrite and pretrite accessory conules may extend to the back half of molar crowns (figure 15.15B). Accessory conules may be doubled. Talonids may be crowded with many conelets. Moderately worn half-loph(id)s with coarsely to finely folded enamel.

Description An m3 from Laetoli, Tanzania (EP 197/05) is a very advanced specimen, with seven lophids, strong anancoidy, and a complex occlusal pattern of anterior and posterior central accessory conules throughout nearly the entire extent of both the pre- and posttrite sides (figure 15.15B). An M3 from nearby Kakesio, Tanzania, is also strongly anancoid, with thick (4.8–5.5 mm), coarsely folded enamel, seven lophs, posterior accessory conules fused to worn pretrite half-lophs 1–5, and doubled posterior accessory conules associated with posttrite half-lophs 1–4. Cementum is thinly distributed on loph walls and in the transverse valleys. Molars from Kanapoi and Aterir, Kenya, exhibit similar morphological organization. The dental sample from Chemeron differs in having little or no folding of enamel wear figures, thicker enamel, and only six loph(id)s in third molars. Conversely, third molars from the Aramis and

Beidareem Mbs. of the Sagantole Fm., Middle Awash, Ethiopia, are more complex in occlusal morphology, with thinner enamel, greater enamel folding, thicker cementum, and pre- and posttrite accessory conules distributed throughout the length of the crown.

Remarks *Anancus* sp. nov. is largely composed of the "advanced morph" of the *A. kenyensis* lineage (see Tassy, 1986). Progression from tetralophodonty to pentalophodonty in East-Central African anancines occurred in the 5.0–4.5 Ma interval. Division of the lineage, however, is complicated by considerable morphometric overlap among temporally successive site samples (figures 15.16A, 15.16B), as well as by a large degree of intra-site variation, and further taxonomic refinements may be necessary. Nonetheless, striking differences in occlusal complexity leave little doubt about the specific-level separation of *A. kenyensis* and at least the more advanced representatives of *A.* sp. nov.

ANANCUS CAPENSIS Sanders, 2007
Figures 15.15C and 15.15D

Partial Synonymy Gomphotheriidae gen. et sp. indet., Hendey, 1976; *Anancus* sp., Coppens et al., 1978; *Anancus* sp., Hendey, 1981; *Anancus* sp. nov., Sanders, 2006.

Age and Occurrence Early Pliocene, southern Africa (table 15.5).

Diagnosis Sanders (2007). Contrasted with *A. petrocchii* and *A.* new species by its tetralophodont intermediate molars. More advanced than *A. kenyensis* and *A. osiris* in having more complex molars, with greater folding of enamel, six lophs in M3 (figures 15.15C, 15.15D), an incipient seventh lophid in m3, more pronounced anancoidy, and posttrite posterior accessory conules not restricted to the mesial half of crown in third molars (may extend throughout the crown in M3).

Description Sanders (2006, 2007). Cementum coats sides of molar loph(id)s and may fill transverse valley floors. Intermediate molars with four loph(id)s, half-loph(id)s each consisting of two conelets, and posttrite posterior accessory conules to the third or fourth loph(id). Pretrite posterior accessory conules variable, and may occur from loph(id)s 1–3. Enamel thick (about 4.5 mm in M1, 5.0–6.0 mm in M2) and lightly to moderately folded. Third molars low crowned (H = 57–58 mm), with moderately to well folded, thick enamel (4.5–6.0 mm), pre- and posttrite posterior accessories well distributed throughout the crown and occasionally accompanied by multiple additional accessory conules, exhibiting low lamellar frequencies (3.5–4.0), and with a postcingulum(id) formed of three to six conelets.

Remarks Molars of anancine gomphotheres from Langebaanweg exhibit a novel mix of primitive (low crowned, tetralophodonty, thick enamel) and advanced (enamel folding, complex distribution of accessory conules, six to seven third molar loph(id)s, pronounced anancoidy) features (Sanders, 2006). This mix suggests that the progressive evolutionary transformation to *A. capensis* from a primitive, *A. kenyensis*–like molar pattern occurred independently from that in eastern African *A.* sp. nov. in which increasing crown complexity was accompanied by pentalophodonty, and northern African *A. petrocchii*, in which the acquisition of pentalophodonty was not associated with more intricate occlusal morphology (but see below). Overall, the degree of morphological complexity of the Langebaanweg specimens is similar to that in cheek teeth from Kakesio, Tanzania and Aterir, Kenya, ca. 4.5–4.0 Ma. Only a single anancine molar is known from Baard's Quarry at Langebaanweg, and it differs from molars of *A. capensis* in having laterally elongated, anteroposteriorly compressed posttrite half-lophids. Unfortunately, it is too fragmentary to assign more precisely than to *Anancus* sp. (table 15.5). Further study is necessary to determine if specimens from Karonga and Uraha, Malawi attributed to *A. kenyensis* and *A.* sp. (table 15.5; Mawby, 1970; Bromage et al., 1995) might instead belong in *A. capensis*.

ANANCUS PETROCCHII Coppens, 1965
Figures 15.15E and 15.16B

Partial Synonymy *Pentalophodon sivalensis*, Petrocchi, 1943; *Anancus (Pentalophodon) petrocchii*, Coppens, 1965; *Anancus osiris* (in part), Arambourg, 1970.

Age and Occurrence Late Miocene, northern Africa (table 15.5).

Diagnosis Distinguished from *A. kenyensis*, *A. capensis*, and *A. osiris* by its pentalophodont intermediate molars, and from *A.* sp. nov. by the simplicity of its occlusal morphology (no accessory conules in m3) and massiveness of its pyramidal lophids. Lower third molars relatively narrow and range to substantially larger size than those of other African congeners (figure 15.16B).

Description Mandible brevirostrine, with no lower tusks. Apices of lower molar conelets converge strongly toward the midline of the crown. Third molars are composed of six lophids and a simple postcingulid formed of a large, single conelet (Petrocchi, 1954; Coppens, 1965; Coppens et al., 1978). Posttrite half-lophids have two conelets each; corresponding pretrite half-lophids have one. A low, narrow cingular ribbon surrounds the molar crowns (figure 15.15E). Molars are very brachyodont. Anancoidy is not marked.

Remarks Represented solely by lower teeth and jaws from the latest Miocene–earliest Pliocene site of Sahabi, Libya, as depicted by Petrocchi (1943, 1954), this species is advanced only in the pentalophodonty of its intermediate molars. Otherwise, the simplicity of its molar crowns clearly distinguishes it from *A.* sp. nov. (the "*A. kenyensis* 'petrocchii'-morph'" of Tassy, 1986) and mediates against any special relationship between the taxa. However, reexamination of Sahabi anancine molars shows that the occlusal morphology is more complex than previously described. Small pre- and posttrite accessory conules are present throughout upper and lower molar crowns, and may be accompanied by coarsely folded enamel in worn specimens (Sanders, 2008b). *Anancus petrocchii* has been posited as the terminal derivation of *A. osiris* (Coppens, 1965; Coppens et al., 1978), but this idea is untenable, as the latter species is primarily known from younger deposits (table 15.5).

ANANCUS OSIRIS Arambourg, 1945
Figure 15.15F

Partial Synonymy *Mastodon arvernensis*, Depéret et al., 1925; *Anancus arvernensis*, Dietrich, 1943; *Anancus (Mastodon) arvernensis*, Ennouchi, 1949.

Age and Occurrence Late Pliocene, northern Africa (table 15.5).

Diagnosis Modified from Coppens (1965); Coppens et al. (1978); Tassy (1986). Tetralophodont intermediate molars; molars with heavy, pyramidal loph(id)s and simple crowns (third molars with few pretrite and no posttrite accessory conules), but with conspicuous anancoidy of half-loph(id)s. Higher crowned than *A. petrocchii*.

Description The type M3, from Gizeh, Egypt, has five lophs and plain crown morphology, lacking posterior accessory conules. Each pretrite half-loph is formed of a large outer cone-let and diminutive mesoconelet, and has a large "tubercle de néoformation" (figure 15.15F; Arambourg, 1945, 1970), or anterior accessory conule, that projects anterior to the accompanying posttrite half-loph. This feature is occasionally seen in other species of *Anancus*, particularly *A. kenyensis*. Cementum is not apparent, and the postcingulum is uncomplicated, of only a few conelets. Cementum may be more heavily distributed in molars from other localities (Pickford, 2003). Other third molars, such as those from Aïn Boucherit, Algeria, have equally basic occlusal organization, though they possess five and a half or six loph(id)s (Arambourg, 1970). Second molars, in contrast, may have pretrite posterior accessory conules throughout the crown, and even posttrite accessories (Arambourg, 1970).

Remarks Anancus osiris is primarily a late Pliocene North African species (table 15.5; Tassy, 1986) that did not attain the occlusal complexity of sub-Saharan congeners, and whose relationship to other African members of the genus remains obscure. Although it has been hypothesized that *A. osiris* and *A. kenyensis* derived from different Eurasian ancestors (Tassy, 1986), the phylogenetic analysis on which this is based is less than satisfactory, as presumed synapomorphies linking particular African and Eurasian species are prone to homoplasic expression. Nonetheless, even though the fossil record is presently too incomplete to resolve these relationships, its emerging pattern is sufficient to infer independent evolution of African anancines within at least three major zones: North, East-Central, and South African.

Family STEGODONTIDAE Osborn, 1918

The family Stegodontidae is composed of the archaic genus *Stegolophodon* ("stegolophodonts"), and the more derived genus *Stegodon* ("stegodonts"). The oldest representative of the family is *Stegolophodon nasaiensis* from the early Miocene of Thailand (Tassy et al., 1992), and the youngest is *Stegodon orientalis*, which survived until the Holocene in China (Ma and Tang, 1992; Saegusa, 1996). It is predominantly an Asian family geographically and in terms of faunal importance (Coppens et al., 1978; Saegusa et al., 2005). *Stegodon* also occurred in the latest Miocene-Pliocene of Africa, where it was rare and limited in distribution (table 15.6). Although the oldest example of *Stegodon* was reported to be from ~7-Myr-old sediments in Kenya (Sanders, 1999), stegodont localities in Yunnan, China, are now correlated to 9 Ma (Saegusa et al., 2005), strengthening the position that these proboscideans are immigrants to Africa.

Stegodon is characterized by extremely brachyodont molars convergent on those of elephants in having platelike loph(id)s. These may be composed of numerous, bilaterally compressed conelets ("mammillae") and mesiodistally separated by transverse valleys that are Y shaped in cross section (Osborn, 1942; Coppens et al., 1978; Kalb et al., 1996a; Saegusa, 1996). In addition, *Stegodon* crania have high nasal openings, which separate short foreheads from long lower faces, reduced facial crests, and steeply vertical lateral walls of infraorbital canals (Osborn, 1942; Saegusa, 1987). Apart from the development of molar plates, numerous other homoplasies are shared by elephants and stegodonts, including anteroposterior compression of the cranium, elevation of the parietals and occipital, loss of lower tusks and concomitant shortening of the mandibular symphysis, lack of median molar sulci, inflation of mesoconelets, and absence of accessory central conules

(Osborn, 1942; Saegusa, 1987, 1996; Kalb et al., 1996a). Molar wear analysis indicates that, as in elephants, stegodont mastication was achieved via fore-aft shearing (Saegusa, 1996). Because of these similarities to elephants, stegodonts have played a significant role in the development of ideas about the origin of elephants since the time of Clift (1828), who first described these proboscideans as intermediates between "mastodonts" and elephants.

More recently, stegodontids were regarded as a group close to mammutids (Maglio, 1970a, 1973; Tobien, 1975; Coppens et al., 1978), but now are considered a sister taxon of tetralophodont gomphotheres and elephants (Tassy and Darlu, 1986; Tassy, 1990, 1996c; Kalb and Mebrate, 1993; Shoshani, 1996; Shoshani et al., 1998). Given their early Miocene divergence from the dental morphological pattern conserved in gomphotheriids, one of us (W.J.S.) feels that it is alternatively possible that stegodontids are the sister taxon to all gomphotheriids + elephants (figure 15.20, later). Monophyly of the family has been defended by Tassy (1990, 1996c), Saegusa (1987, 1996), Shoshani (1996), and Shoshani et al. (1998). By contrast, Kalb and Mebrate (1993) and Kalb et al. (1996a) allocate *Stegodon* to the Elephantinae. Both hypotheses are only weakly supported by parsimony analysis, because of heavy reliance on dental features. Chronostratigraphy and biogeography of stegodontids and early elephants, however, are concordant with monophyly of the Stegodontidae and Asian origin of stegodonts. Advanced stegolophodonts are stratigraphically succeeded by primitive, morphologically similar stegodonts in late Miocene formations in Thailand and Myanmar (Saegusa et al., 2005; Takai et al., 2006), while African proboscideans initially reported as stegolophodonts (Petrocchi, 1954; Singer and Hooijer, 1958; Hooijer, 1963) have been reidentified as early elephants (Maglio and Hendey, 1970; Maglio, 1973; Coppens et al., 1978).

Stegodonts have been thought of as forest-dwelling browsers (Osborn, 1921, 1942), an interpretation supported by their extreme brachyodonty and by dental isotopic analysis of late Miocene South Asian material (Cerling et al., 1999). Their relatively short, massive bodies and closely parallel, gently upcurved tusks have also been described as adaptations to dense forest (Osborn, 1921). However, carbon isotopic studies of fossil teeth reveal stegodonts from Central Africa to have been surprisingly eclectic feeders, ranging from browsers to grazers during the late Miocene, to early Pliocene mixed feeders and grazers, to grazers during the mid-Pliocene (Zazzo et al., 2000). This apparent shift in dietary preferences may signal increased ecological pressures over time to compete for C_4 resources, particularly with more hypsodont ungulates, including elephants, possibly the ultimate cause of their disappearance from the continent.

Genus *STEGODON* Falconer and Cautley, 1847
STEGODON KAISENSIS Hopwood, 1939
Figures 15.17A and 15.17B

Partial Synonymy Stegodon fuchsi, MacInnes, 1942; *S. kaisensis*, Cooke and Coryndon, 1970; *Primelephas gomphotheroides* (in part; Nyawiega, Uganda), Maglio, 1973:20; *Stegodon* sp., Beden, 1975; *S. kaisensis* "Nkondo stage" and "Warwire stage," Tassy, 1995.

Age and Occurrence Late Miocene–late Pliocene, eastern and Central Africa (table 15.6).

Diagnosis Emended diagnosis based on observations by H. Saegusa. While nearly all of the stegodont specimens from

TABLE 15.6
Major occurrences and ages of Afro-Arabian stegodonts and elephants
? = attribution or occurrence uncertain; alt. = alternatively.

Taxon	Occurrence (Site, Locality)	Stratigraphic Unit	Age	Key References
	STEGODONTIDAE, LATE MIOCENE–LATE PLIOCENE			
Stegodon cf. *S. kaisensis*	Tugen Hills, Kenya	Mpesida Beds	Ca. 7.0 Ma	Sanders, 1999; Kingston et al., 2002
Stegodon kaisensis "Nkondo stage"	Shoshomagai 2, Inolelo 3, Manonga Valley, Tanzania	Ibole Mb., Wembere-Manonga Fm.	Ca. 5.5–5.0 Ma	Harrison and Baker, 1997
	Kossom Bougoudi, Chad		Ca. 5.3 Ma (alt. 6.0–5.0 Ma)	Brunet et al., 2000; Zazzo et al., 2000; Brunet, 2001; Fara et al., 2005
	"Kaiso Village," south of the Howa River, probably Nkondo area, Uganda (type)	?Nkondo Fm.	Ca. 5.0 Ma	Hopwood, 1939; Cooke and Coryndon, 1970; Coppens et al., 1978; Sanders, 1990; Tassy, 1995
	Kaiso Central; Kisegi Wasa/N. Nyabrogo; Nyawiega site I, Uganda	Lower Kaiso Fm. (= Nkondo Fm.)	Ca. 5.0 Ma (alt. 6.0 Ma)	Cooke and Coryndon, 1970; Sanders, 1990, 1999; Pickford et al., 1993
	Kazinga-Kisenyi Area, Kazinga Channel, Uganda	Kazinga Beds	Ca. 5.0 Ma	MacInnes, 1942; Cooke and Coryndon, 1970; Sanders, 1990; Pickford et al., 1993; Tassy, 1995
	Nkondo-Kaiso Region, Uganda	Nkondo Fm.	Ca. 5.0 Ma	Pickford et al., 1993; Tassy, 1995
	Kollé, Chad		Ca. 5.0–4.0 Ma	Brunet et al., 1998
Stegodon kaisensis "Warwire stage"	Nkondo-Kaiso Region, Uganda	Warwire Fm.	Ca. 3.5–3.0 Ma	Pickford et al., 1993; Tassy, 1995
	Koro Toro, Chad		Ca. 3.5–3.0 Ma	Brunet, 2001; Fara et al., 2005
	Omo, Ethiopia	Upper Mb. B, Shungura Fm.	2.95–2.85 Ma	Beden, 1975, 1976; Alemseged, 2003
Stegodon sp. indet.	Sinda River, Lower Semliki, Democratic Republic of Congo	Sinda Beds	Ca. 4.1 Ma	Yasui et al., 1992
	Laetoli, Tanzania	Upper Unit, Laetolil Beds	3.8–3.5 Ma	Drake and Curtis, 1987; Sanders, 2005, in press
	Ishango 11 and Senga near Sn1, Upper Semliki, Democratic Republic of Congo	Lusso Beds	Late Pliocene	Sanders, 1990
	ELEPHANTIDAE, LATE MIOCENE–PRESENT			
	STEGOTETRABELODONTINAE, LATE MIOCENE–EARLY PLIOCENE			
Stegotetrabelodon orbus	Lothagam, Kenya (type)	Lower and Upper Mbs., Nawata Fm.; Apak Mb., Nachukui Fm.	At least 7.4–5.0 Ma; 6.5–5.0 Ma; 5.0–4.2 Ma	Maglio, 1970a, 1973; Maglio and Ricca, 1977; McDougall and Feibel, 2003; Tassy, 2003
	Tugen Hills, Kenya	Mpesida Beds	Ca. 7.0 Ma	Hill et al., 1985, 1986; Kingston et al., 2002

TABLE 15.6 (CONTINUED)

Taxon	Occurrence (Site, Locality)	Stratigraphic Unit	Age	Key References
S. orbus continued		Lukeino Fm.	6.2–5.6 Ma	Hill et al., 1985, 1986; Tassy, 1986; Hill, 2002
	Manonga Valley, Tanzania	Ibole Mb., Wembere-Manonga Fm.	Ca. 5.5–5.0 Ma	Harrison and Baker, 1997; Sanders, 1997
Stegotetrabelodon syrticus	Sahabi, Libya (type)	Sahabi Fm.	Late Miocene–early Pliocene, ca. 5.2 Ma	Petrocchi, 1943, 1954; Maglio, 1970; de Heinzelin and El-Arnauti, 1987; Bernor and Scott, 2003; Sanders, 2008b
Stegotetrabelodon sp. indet.	?Lemudong'o, Kenya	Lemundong'o Fm.	6.1–6.0 Ma	Ambrose et al., 2003; Saegusa and Hlusko, 2007
	Shuwaihat and Jebel Barakah, Abu Dhabi	Baynunah Fm.	Late Miocene, ca. 6.0 Ma	Glennie and Evamy, 1968; Madden et al., 1982; Hailwood and Whybrow, 1999; Tassy, 1999
	Kanam East and Central, Kenya	Kanam Fm.	Late Miocene to earliest Pliocene	MacInnes, 1942; Ditchfield et al., 1999
	Nyabusosi, Uganda	Kakara Fm., Lower and Upper Oluka Fm.	Late Miocene	Pickford et al., 1993; Tassy, 1995

ELEPHANTINAE, LATE MIOCENE–PRESENT

Taxon	Occurrence (Site, Locality)	Stratigraphic Unit	Age	Key References
Primelephas korotorensis	Lothagam, Kenya	Lower and Upper Mbs., Nawata Fm.; Apak Mb., Nachukui Fm.	At least 7.4–5.0 Ma; 6.5–5.0 Ma; 5.0–4.2 Ma	Maglio, 1970a, 1973; Maglio and Ricca, 1977; Leakey et al., 1996; McDougall and Feibel, 2003; Tassy, 2003
	Toros Menalla, Chad		7.0–6.0 Ma	Brunet et al., 2000; Mackaye et al., 2008
	Tugen Hills, Kenya	Lukeino Fm.	6.2–5.6 Ma	Hill et al., 1985, 1986; Tassy, 1986; Hill, 2002; Sanders, 2004
		Lower part, Chemeron Fm.	5.3–4.0 Ma	Hill et al., 1985, 1986; Deino et al., 2002
	Middle Awash, Ethiopia	Saraitu, Adu Dora, and Asa Koma Mbs., Adu-Asa Fm.	6.3–5.6 Ma	Mebrate, 1983; Kalb and Mebrate, 1993; Kalb, 1995; Haile-Selassie, 2001; WoldeGabriel et al., 2001; Haile-Selassie et al., 2004
		Kuseralee Mb., Sagantole Fm.	5.6–5.2 Ma	Mebrate, 1983; Kalb and Mebrate, 1993; Renne et al., 1999; Haile-Selassie, 2001; WoldeGabriel et al., 2001; Haile-Selassie et al., 2004
	"Galili Area," Mulu Basin, Ethiopia	Asa Koma Mb., Adu-Asa Fm.	5.8–5.6 Ma	Haile-Selassie, 2000; WoldeGabriel et al., 2001
	Manonga Valley, Tanzania	Ibole Mb., Wembere-Manonga Fm.	Ca. 5.5–5.0 Ma	Harrison and Baker, 1997; Sanders, 1997

Taxon	Occurrence (Site, Locality)	Stratigraphic Unit	Age	Key References
P. korotorensis continued	Nyabusosi, Uganda	Lower Oluka Fm.	Late Miocene	Pickford et al., 1993; Tassy, 1995
	North and South Nyabrogo, Uganda	Lower horizons, Kaiso Fm.	Late Miocene–early Pliocene	Cooke and Coryndon, 1970
	Kossom Bougoudi, Chad		Ca. 5.3 Ma (alt. 6.0–5.0 Ma)	Brunet et al., 2000; Zazzo et al., 2000; Brunet, 2001; Fara et al., 2005; Mackaye et al., 2008
	Kollé, Chad		Ca. 5.0–4.0 Ma	Brunet et al., 1998; Mackaye et al., 2008
	Koulà, Chad (type)		?Early Pliocene	Coppens, 1965; Maglio, 1970a, 1973; Coppens et al., 1978
	Kolinga, Chad		?Early Pliocene (alt. latest Miocene)	Coppens, 1967; Maglio, 1973; Coppens et al., 1978
Stegodibelodon schneideri	Menalla, Chad (type)		?Early Pliocene	Coppens, 1972; Coppens et al., 1978
	Kolinga I, Chad		?Early Pliocene (alt. latest Miocene)	Coppens, 1972; Coppens et al., 1978
	Kollé, Chad		5.0–4.0 Ma	Brunet, 2001; Fara et al., 2005; Mackaye et al., 2005

"LOXODONTA ADAURORA GROUP"

Taxon	Occurrence (Site, Locality)	Stratigraphic Unit	Age	Key References
Loxodonta adaurora adaurora	?Kanam East, Kenya	Kanam Fm.	Latest Miocene–earliest Pliocene	MacInnes, 1942; Ditchfield et al., 1999
	?Bugoma, Kaiso Central, Nyawiega, Uganda	Lower Kaiso Beds (= Nkondo Fm.)	Ca. 5.0 Ma (alt. 6.0 Ma)	Cooke and Coryndon, 1970; Sanders, 1990; Pickford et al., 1993; Boaz, 1994
	Tugen Hills, Kenya	Chemeron Fm.	Early Pliocene (within 5.3–4.0 Ma interval)	Maglio, 1973; Hill et al., 1985, 1986; Deino et al., 2002
	Middle Awash Valley, Ethiopia	Aramis and Beidareem Mbs., Sagantole Fm.	4.4–4.3 Ma	Mebrate, 1983; Kalb and Mebrate, 1993; Renne et al., 1999
	Kanapoi, Kenya (type)	Kanapoi Fm.	4.2–4.1 Ma	Maglio, 1970a, 1973; Feibel, 2003; Harris et al., 2003
	Omo, Ethiopia	Mursi Fm.	Early Pliocene, >4.15 Ma	Heinzelin, 1983; Beden, 1987a; Feibel et al., 1989
		Mb. B, Shungura Fm.	3.4–2.85 Ma	Heinzelin, 1983; Beden, 1987a; Alemseged, 2003
	West Turkana, Kenya	Kataboi and lower Lomekwi Mbs., Nachukui Fm.	Ca. 4.10–3.0 Ma	Harris et al., 1988a, 1988b
	Ekora, Kenya		Ca. 4.0–3.75 Ma (slightly younger than 4.0 Ma)	Maglio, 1970a, 1973; Behrensmeyer, 1976
	Allia Bay, Kenya	Moiti and Lokochot Mbs., Koobi Fora Fm.	Ca. 3.9–3.4 Ma	Beden, 1983; Harris, 1983; Brown, 1994
	Lothagam, Kenya	Kaiyumung Mb., Nachukui Fm. (=Unit 3)	Ca. 3.5–3.0 Ma	Coppens et al., 1978; Feibel, 2003; McDougall and Feibel, 2003

TABLE 15.6 (CONTINUED)

Taxon	Occurrence (Site, Locality)	Stratigraphic Unit	Age	Key References
L. adaurora adaurora; continued	Hadar, Ethiopia	Sidi Hakoma and Denen Dora Mbs., Hadar Fm.	3.40–3.18 Ma	White et al., 1984; Bonnefille et al., 2004
L. adaurora kararae	Koobi Fora, Kenya (subspecies type)	Upper Burgi Mb., Koobi Fora Fm.	2.0–1.88 Ma	Beden, 1983; Feibel et al., 1989
	Karonga, Malawi	Chiwondo Beds	Ca. 2.4–2.3 Ma	Bromage et al., 1995
	Omo, Ethiopia	Mb. E, Shungura Fm.	2.40–2.36 Ma	Heinzelin, 1983; Beden, 1987a; Alemseged, 2003
	?Middle Awash, Ethiopia	Matabaietu Fm.	Ca. 2.3–2.0 Ma	Kalb and Mebrate, 1993

"LOXODONTA EXOPTATA–L. AFRICANA GROUP"

Taxon	Occurrence (Site, Locality)	Stratigraphic Unit	Age	Key References
Loxodonta sp. indet.	Toros Menalla, Chad		7.0–6.0 Ma	Brunet et al., 2000
	Tugen Hills, Kenya	Lukeino Fm.	6.2–5.6 Ma	Hill et al., 1985, 1986; Tassy, 1986; Hill, 2002
	Lothagam, Kenya	Apak Mb., Nachukui Fm.	5.0–4.2 Ma	McDougall and Feibel, 2003; Tassy, 2003
	Kollé, Chad		5.0–4.0 Ma	Brunet et al., 1998; Brunet, 2001
	Konso, Ethiopia	Intervals 1 and 4, Konso Fm.	Ca. 1.91 Ma and 1.43 Ma	Suwa et al., 2003
	West Turkana, Kenya	Kaitio Mb., Nachukui Fm.	1.9–1.6 Ma	Brugal et al., 2003
Loxodonta cookei	Tugen Hills, Kenya	?Lukeino Fm.	6.2–5.6 Ma	Hill et al., 1985, 1986; Tassy, 1986; Hill, 2002
		Chemeron Fm.	5.3–4.0 Ma	Hill et al., 1985, 1986; Deino et al., 2002
	Nkondo-Kaiso Region, Uganda	Nkondo Fm.	Ca. 5.0 Ma (alt. ca. 6.0 Ma)	Pickford et al., 1993; Tassy, 1995
	Langebaanweg, South Africa (type)	Quarry E, Quartzose Sand Mb. and Pellatal Phosphate Mb., Varswater Fm.	Early Pliocene, ca. 5.0 Ma	Hendey, 1981; Sanders, 2006, 2007
	?Endolele, Tanzania	?Lower Unit, Laetolil Beds	?>4.3 Ma	Sanders, 2005, in press
Loxodonta exoptata	?Manonga Valley, Tanzania	Kilolei Mb., Wembere-Manonga Fm.	Ca. 4.5–4.0 Ma	Harrison and Baker, 1997; Sanders, 1997
	Kakesio (and Esere, Emboremony, and Noiti), Tanzania	Lower Unit, Laetolil Beds	Ca. 4.3–3.8 Ma	Drake and Curtis, 1987; Harris, 1987b; Hay, 1987; Sanders, 2005
	Kanapoi, Kenya	Kanapoi Fm.	4.2–4.1 Ma	Feibel, 2003; Harris et al., 2003
	Karonga, Uraha, Malawi	Chiwondo Beds	Ca. 4.0–3.0 Ma	Bromage et al., 1995
	Allia Bay, Kenya	Lokochot and Tulu Bor Mbs., Koobi Fora Fm.	~3.5–2.6 Ma	Beden, 1983; Harris, 1983; Brown, 1994
	Laetoli, Tanzania (type)	Upper Unit, Laetolil Beds; Upper Ndolanya Beds	3.8–3.5 Ma; ~2.7–2.6 Ma	Beden, 1987b; Drake and Curtis, 1987; Harris, 1987b; Harrison, 2002; Sanders, 2005, in press
	Koro Toro, Chad		3.5–3.0 Ma	Brunet et al., 1995; Brunet, 2001

Taxon	Occurrence (Site, Locality)	Stratigraphic Unit	Age	Key References
L. exoptata continued	West Turkana, Kenya	lower Lomekwi Mb., Nachukui Fm.	Ca. 3.36–3.0 Ma	Harris et al., 1988a, 1988b
	Omo, Ethiopia	Mb. A, Shungura Fm.	Ca. 3.5–3.4 Ma	Heinzelin, 1983; Beden, 1987a; Alemseged, 2003
	Hadar, Ethiopia	Denen Dora Mb., Hadar Fm.	3.22–3.18 Ma	White et al., 1984; Bonnefille et al., 2004
	?Middle Awash, Ethiopia	Matabaietu Fm.	Ca. 2.3–2.0 Ma	Kalb and Mebrate, 1993; Kalb, 1995
Loxodonta atlantica angammensis	Omo, Ethiopia	Mbs. D, F, Shungura Fm.	2.52–2.40 Ma and 2.36–2.33 Ma	Beden, 1987a; Feibel et al., 1989; Alemseged, 2003
	Angamma-Yayo, Chad (subspecies type)		?Early Pleistocene	Coppens, 1965; Beden, 1987a
L. atlantica atlantica	Ternifine, Algeria (type)		Middle Pleistocene	Pomel, 1879; Osborn, 1942; Maglio, 1973; Coppens et al., 1978
	Oued Constantine, Algeria	Maison Carrée Fm.	Middle Pleistocene	Maglio, 1973; Coppens et al., 1978
	Sablière de Palekao, Algeria		Middle Pleistocene	Maglio, 1973
	Sidi Abderrahmane, Algeria		Middle Pleistocene	Maglio, 1973; Coppens et al., 1978; Raynal et al., 2004a
L. atlantica zulu	Elandsfontein, South Africa		Ca. 1.0–0.6 Ma (alt. ca. 0.7–0.4 Ma)	Maglio, 1973; Klein and Cruz-Uribe, 1991; Klein et al., 2007
	Zululand, southeast coast of Africa, South Africa (subspecies type)		Middle Pleistocene	Scott, 1907; Osborn, 1942; Maglio, 1973
Loxodonta africana (type locality probably Cape Colony, South Africa [Osborn, 1942: p. 1197])	Middle Awash, Ethiopia	?Wehaietu Fm.	Between 0.8 and 0.2 Ma	Kalb and Mebrate, 1993
	Ounaianga Kebir I, Chad		?Middle Pleistocene	Joleaud and Lombard, 1933a, 1933b; Coppens, 1967
	Kanjera, Kenya	Apoko Fm.	Ca. 0.5 Ma	Plummer and Potts, 1989; Behrensmeyer et al., 1995
	Duinefontein 2, Elandsfontein, South Africa		270 Ka	Klein et al., 2007
	Omo, Ethiopia	Mb. III, Kibish Fm.	130–75 Ka	Assefa et al., 2008
	Elands Bay Cave, Elandsfontein, South Africa		13,600–7,900 y BP	Klein et al., 2007
	Bone Circle assemblage, Elandsfontein, South Africa		12,000–10,000 y BP	Klein and Cruz-Uribe, 1991
	Omo, Ethiopia	Mb. IV, Kibish Fm.	<10 Ka	Assefa et al., 2008
Mammuthus subplanifrons	Middle Awash, Ethiopia	?Kuseralee and Aramis Mbs., Sagantole Fm.	5.77–5.18 Ma; ca. 4.4 Ma	Mebrate, 1983; Kalb and Mebrate, 1993; Renne et al., 1999; Haile-Selassie, 2001
	Langebaanweg, South Africa	Quarry E, Quartzose Sand Mb., Varswater Fm.	Early Pliocene, ca. 5.0 Ma	Maglio and Hendey, 1970; Hendey, 1981; Sanders, 2006, 2007

TABLE 15.6 (CONTINUED)

Taxon	Occurrence (Site, Locality)	Stratigraphic Unit	Age	Key References
M. subplanifrons continued	?Nyawiega, Uganda	Lower Kaiso Beds (= Nkondo Fm.)	Ca. 5.0 Ma (alt. 6.0 Ma)	Cooke and Coryndon, 1970; Pickford et al., 1993
	?Nkondo-Kaiso Area, Uganda	Lower Kaiso Beds (= Nkondo Fm.)	Ca. 5.0 Ma (alt. 6.0 Ma)	Pickford et al., 1993; Tassy, 1995
	Vaal River, South Africa (type)	Middle Terrace	?Early Pliocene	Osborn, 1928, 1934
	Virginia, Orange Free State, South Africa		?Early Pliocene	Meiring, 1955; Maglio, 1973
	?Uraha, Malawi	Chiwondo Beds	Ca. ≥4.0 Ma	Mawby, 1970; Bromage et al., 1995
	?Ishasha River, Virunga National Park, Democratic Republic of Congo	Kaiso Group sediments	Early Pliocene	Vanoverstraeten et al., 1990
Mammuthus africanavus	Lac Ichkeul, Tunisia (type)		Mid–late Pliocene	Arambourg, 1970; Coppens et al., 1978
	Aïn Brimba, Tunisia		Mid–late Pliocene	Arambourg, 1970
	Kebili, Tunisia		Mid–late Pliocene	Maglio, 1973; Coppens et al., 1978
	Garet et Tir, Algeria		Mid–late Pliocene	Arambourg, 1970; Coppens et al., 1978
	Aïn Boucherit, Algeria	Oued Boucherit Fm.	Late Pliocene; 2.32 Ma (alt. ca. 2.0 Ma)	Arambourg, 1970; Coppens et al., 1978; Geraads and Amani, 1998; Sahnouni and Heinzelin, 1998; Geraads, 2002; Sahnouni et al., 2002
	Oued Akrech, Morocco		Mid–late Pliocene	Arambourg, 1970; Coppens et al., 1978
	Fouarat, Morocco		Mid–late Pliocene	Arambourg, 1970
	Goz-Kerki, Chad (70 km SE of Koro-Toro)		Ca. 3.5–3.0 Ma	Coppens, 1965; Brunet, 2001; Fara et al., 2005
	Koulà, Chad		Ca. 3.5–3.0 Ma	Coppens, 1965; Maglio, 1973; Coppens et al., 1978; Brunet, 2001; Fara et al., 2005
	Ouadi-Derdemy, Chad		Ca. 3.5–3.0 Ma	Coppens, 1965; Maglio, 1973; Coppens et al., 1978; Brunet, 2001; Fara et al., 2005
	Toungour, Chad (30–50 km W of Koro-Toro)		Ca. 3.5–3.0 Ma	Coppens, 1965; Maglio, 1973; Coppens et al., 1978; Brunet, 2001; Fara et al., 2005
Mammuthus meridionalis	Aïn Hanech, Algeria	Aïn Hanech Fm.	Between 1.95 and 1.78 Ma (alt. ca. 1.2 Ma)	Arambourg, 1970; Maglio, 1973; Sahnouni and Heinzelin, 1998; Geraads et al., 2004; Sahnouni et al., 2002, 2004
Mammuthus sp. indet.	Djebel Bel Hacel, Algeria		Early Pleistocene	Arambourg, 1970
	Hadar, Ethiopia	Sidi Hakoma and Denen Dora Mbs., Hadar Fm.	3.40–3.18 Ma	White et al., 1984; Bonnefille et al., 2004

Taxon	Occurrence (Site, Locality)	Stratigraphic Unit	Age	Key References
Elephas ekorensis	?Lothagam, Kenya	Apak Mb., Nachukui Fm.	5.0–4.2 Ma	Beden, 1985; McDougall and Feibel, 2003; Tassy, 2003
	Kanapoi, Kenya	Kanapoi Fm.	4.2–4.1 Ma	Maglio, 1970a, 1973; Feibel, 2003; Harris et al., 2003
	Ekora, Kenya (type)		Ca. 4.0–3.75 Ma (slightly younger than 4.0 Ma)	Maglio, 1970a, 1973; Behrensmeyer, 1976
	Dikika, Ethiopia (DIK-1)	Sidi Hakoma Mb., Hadar Fm.	3.40–3.22 Ma	Wynn et al., 2006
	Hadar, Ethiopia	Sidi Hakoma and Denen Dora Mbs., Hadar Fm.	3.40–3.18 Ma	White et al., 1984; Bonnefille et al., 2004
	Omo, Ethiopia	Usno Fm.	Within 3.6–2.7 Ma interval	Butzer, 1971; Heinzelin, 1983; Beden, 1987a
Elephas recki brumpti	Bolt's Farm, South Africa		Ca. 5.0–4.0 Ma (alt. 3.4–2.9 Ma)	Maglio, 1973; Cooke, 1993; Sénégas and Avery, 1998
	?Middle Awash, Ethiopia	?Aramis Mb., Sagantole Fm.	Ca. 4.4 Ma (alt. 4.1–3.8 Ma)	Kalb and Mebrate, 1993; Renne et al., 1999
	Koobi Fora and Allia Bay, Kenya	Lokochot Mb., Koobi Fora Fm. (= "Kubi Algi Fm., Zone B")	~3.50–3.36 Ma	Beden, 1983, 1985; Harris, 1983; Brown, 1994
	?Goz-Kerki, Chad (70 km SE of Koro-Toro)		Ca. 3.5–3.0 Ma	Coppens, 1965; Maglio, 1973; Brunet, 2001; Fara et al., 2005
	?Koulà, Chad		Ca. 3.5–3.0 Ma	Coppens, 1965; Maglio, 1973; Brunet, 2001; Fara et al., 2005
	?Ouadi-Derdemy, Chad		Ca. 3.5–3.0 Ma	Coppens, 1965; Maglio, 1973; Coppens et al., 1978; Brunet, 2001; Fara et al., 2005
	Omo, Ethiopia (subspecies type)	Upper Mb. A, Mb. B, Shungura Fm.	Ca. 3.5–2.85 Ma	Beden, 1980, 1987a; Feibel et al., 1989; Alemseged, 2003
	Dikika, Ethiopia (DIK-1)	Sidi Hakoma Mb., Hadar Fm.	3.40–3.22 Ma	Wynn et al., 2006
	West Turkana, Kenya	lower and upper Lomekwi Mb., Nachukui Fm.	3.36–2.5 Ma	Harris et al., 1988a, 1988b
	Hadar, Ethiopia	Denen Dora Mb., Hadar Fm.	3.22–3.18 Ma	White et al., 1984; Bonnefille et al., 2004
E. recki shungurensis	Koobi Fora and Allia Bay, Kenya	Tulu Bor and Upper Burgi Mbs., Koobi Fora Fm.	3.36–1.88 Ma	Beden, 1980, 1983; Harris, 1983; Feibel et al., 1989; Brown, 1994
	Hadar, Ethiopia	Kada Hadar Mb., Hadar Fm.	3.18–2.9 Ma	White et al., 1984; Bonnefille et al., 2004; Wynn et al., 2006
	Tugen Hills, Kenya	Chemeron Fm. (loc. 91)	Ca. 2.85–2.40 Ma	Beden, 1985
	West Turkana, Kenya	Upper Lomekwi and Lokalalei Mbs., Nachukui Fm.	2.9–2.3 Ma	Harris et al., 1988a, 1988b

TABLE 15.6 (CONTINUED)

Taxon	Occurrence (Site, Locality)	Stratigraphic Unit	Age	Key References
E. recki shungurensis continued	Omo, Ethiopia (subspecies type)	Mbs. C-lower F, Shungura Fm.	2.9–2.3 Ma	Beden, 1980, 1987a; Feibel et al., 1989; Alemseged, 2003
	Kigagati, Kaiso Village, Uganda	Upper Kaiso Fm., cf. Mbs. C-lower F, Shungura Fm.	Ca. 2.9–2.3 Ma	Cooke and Coryndon, 1970; Sanders, 1990
	Bouri, Ethiopia	Hata Mb., Bouri Fm.	2.5 Ma	Heinzelin et al., 1999
E. recki atavus	Karonga, Uraha, Malawi	Chiwondo Beds	Ca. 2.4–1.8 Ma	Bromage et al., 1995
	Upper Semliki River, Democratic Republic of Congo	Lusso Beds	Ca. 2.36–1.90 Ma (equivalent to Mbs. F, G, Shungura Fm.)	Sanders, 1990
	?West Turkana, Kenya	Kalochoro Mb., Nachukui Fm.	2.35–1.90 Ma	Brugal et al., 2003
	Omo, Ethiopia (subspecies type)	Mbs. upper F–G, Shungura Fm.	2.34–1.90 Ma	Beden, 1980, 1987a; Feibel et al., 1989; Alemseged, 2003
	Fejej FJ-1 Site, Kenya	Unit III	Ca. 2.34–1.90 Ma	Moullé et al., 2001
	Koobi Fora and Ileret, Kenya	Upper Burgi Mb., Koobi Fora Fm. (just under and above the KBS Tuff)	2.0–1.88 Ma	Beden, 1980, 1983; Harris, 1983; Feibel et al., 1989; Brown, 1994
	Konso, Ethiopia	Intervals 1–3, Konso Fm.	Ca. 1.91–1.50 Ma	Suwa et al., 2003
	Olduvai Gorge, Tanzania	Beds I and lower II	~1.87–1.6 Ma (alt. 2.10–1.75 Ma)	Hay, 1976; Beden, 1980, 1985; Brown, 1994; Tamrat et al., 1995
E. recki ileretensis	?West Turkana, Kenya	Kaitio Mb., Nachukui Fm.	1.9–1.6 Ma	Harris et al., 1988a, 1988b
	Omo, Ethiopia	Mbs. ?J, K, and lower L, Shungura Fm.	~?1.74–1.30 Ma	Beden, 1980, 1985, 1987a; Feibel et al., 1989; Alemseged, 2003
	Koobi Fora and Ileret, Kenya (subspecies type)	KBS and Okote Mbs., Koobi Fora Fm. (below and above the Okote Tuff)	~1.7–1.5 Ma	Beden, 1980, 1983, 1985; Harris, 1983; Feibel et al., 1989; Brown, 1994
	Barogali, Republic of Djibouti		Ca. 1.6–1.3 Ma	Berthelet, 2001; Berthelet and Chavaillon, 2001
	Chesowanja, Kenya		>1.42 Ma	Hooker and Miller, 1979; Beden, 1985
	Olduvai Gorge, Tanzania	Upper Bed II	~1..60–1.15 Ma (alt. 1.75 Ma–?)	Hay, 1976; Beden, 1980, 1985; Brown, 1994; Tamrat et al., 1995
E. recki recki	Koobi Fora and Ileret, Kenya	Okote Mb., Koobi Fora Fm. (below the Chari Tuff)	1.64–1.39 Ma	Beden, 1980; 1983; Harris, 1983; Feibel et al., 1989; Brown, 1994
	Konso, Ethiopia	Intervals 4–6, Konso Fm.	Ca. 1.43–1.39 Ma	Suwa et al., 2003
	Omo, Ethiopia	Upper Mb. L, Shungura Fm.	<1.39–1.16 Ma	Beden, 1980, 1987a; Feibel et al., 1989; Alemseged, 2003
	West Turkana, Kenya	Nariokotome Mb., Nachukui Fm.	Ca. 1.3–1.0 Ma	Harris et al., 1988a, 1988b

Taxon	Occurrence (Site, Locality)	Stratigraphic Unit	Age	Key References
E. recki recki continued	Olduvai Gorge, Tanzania (type)	Beds III, IV	1.15–0.60 Ma (alt. ?–1.1 Ma)	Hay, 1976; Beden, 1985; Brown, 1994; Tamrat et al., 1995
	Melka Kunture (Garba IV), Ethiopia	Melka Kunture Fm.	>0.78 Ma, ca. 1.0 Ma	Beden, 1980, 1985; Raynal et al., 2004b
	Olorgesailie, Kenya	Olorgesailie Fm.	Ca. 1.0–0.5 Ma (alt. 0.93–0.70 Ma)	Beden, 1980, 1985; Bye et al., 1987; Deino and Potts, 1990
	Busidima-Telalak region, Afar, Ethiopia	?Wehaietu Fm.	Between 0.8 and 0.2 Ma	Alemseged and Geraads, 2000
	Kathu Pan, South Africa	Acheulean deposit	Middle Pleistocene, sometime in interval of 0.7–0.4 Ma	Klein, 1984, 1988
	Namib IV, Namibia		0.7–0.4 Ma interval (alt. late Lower or early Middle Pleistocene)	Shackley, 1980; Klein, 1984, 1988
	Power's Site, South Africa	lower terrace, Vaal River sequence, Rietputs Fm.	Middle Pleistocene, similar in age to Kathu Pan	Klein, 1988
	Kanjera, Kenya	Apoko Fm.	Ca. 0.5 Ma	Plummer and Potts, 1989; Behrensmeyer et al., 1995
E. recki subsp. indet. (*E. r. shungurensis* or *E. r. brumpti*)	Makapansgat, South Africa	Mb. 4	<3.110 Ma (alt. 2.7–2.5 Ma)	Cooke, 1993; Partridge, 2000; Partridge et al., 2000
	Sterkfontein, South Africa	Mb. 4	2.8–2.4 Ma (alt. 2.5–1.5 Ma)	Cooke, 1993; Berger et al., 2002
(?*E. recki shungurensis*)	Ahl al Oughlam, Morocco		Late Pliocene, ca. 2.5 Ma	Geraads and Metz-Muller, 1999
(?*E. recki shungurensis*)	Salé, Morocco		?Early Pleistocene (alt. late Pliocene)	Arambourg, 1970
	Marsabit Road, Kenya		Early Pleistocene	Gentry and Gentry, 1969
(*Elephas* sp.)	Mansourah (Constantine), Algeria		Early Pleistocene	Chaid-Saoudi et al., 2006
(*Elephas* sp.)	Swartkrans, South Africa	Hanging Remnant Mb. 3	Ca. 1.8–1.5 Ma Ca. 1.5–1.0 Ma	de Ruiter, 2003 Cooke, 1993; Partridge, 2000
	Dandero, northern Danakil Depression, Eritrea	Upper Danakil Fm.	Ca. 1.0 Ma	Shoshani et al., 2001c
(intermediate between *E. r. ileterensis* and *E. r. recki*)	Buia, Eritrea	Upper Danakil Fm.	Ca. 1.0 Ma	Ferretti et al., 2003
Elephas iolensis	Sidi Abderrahmane, Algeria		?Close to middle Pleistocene/late Pleistocene boundary	Arambourg, 1960; Maglio, 1973; Coppens et al., 1978; Raynal et al., 2004
	Omo, Ethiopia	Kibish Fm.	Close to middle Pleistocene/late Pleistocene boundary	Maglio, 1973; Coppens et al., 1978; McDougall et al., 2005
	Vaal River, South Africa	Lower (younger) terraces	Middle–late Pleistocene	Dart, 1927, 1929; Osborn, 1928; Maglio, 1973; Coppens et al., 1978
	Beausejour Farm, Algeria (type)		?Late Pleistocene	Maglio, 1973; Coppens and Gaudant, 1976
	Port de Mastaganem, Algeria		Late Pleistocene	Maglio, 1973; Coppens et al., 1978

TABLE 15.6 (CONTINUED)

Taxon	Occurrence (Site, Locality)	Stratigraphic Unit	Age	Key References
E. iolensis continued	Zouerate, Mauritania		Late Pleistocene	Maglio, 1973; Coppens et al., 1978
	El Douira, Tunisia		90–75,000 y BP	Coppens and Gaudant, 1976
	Kaiso Village and Behanga I, Western Rift, Uganda	?Rwebishengo Beds	Latest Pleistocene	Cooke and Coryndon, 1970; Sanders, 1990; Pickford et al., 1993
	Natodameri, Sudan		35,000 y BP	Maglio, 1973; Coppens et al., 1978

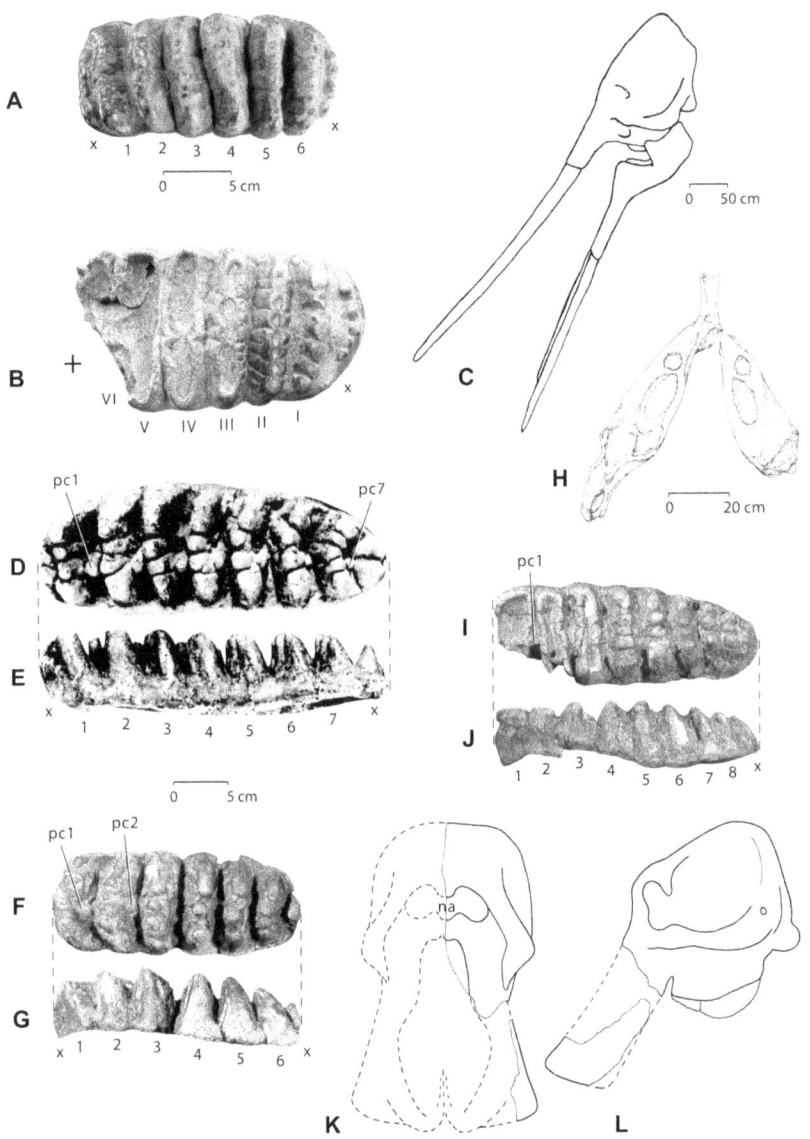

FIGURE 15.17 Aspects of stegodont and elephant cranial, mandibular, and dental morphology. Anterior is to the left except where speci-fied. A) Occlusal view, M2, M15408 (Kazinga Channel, Uganda), *Stegodon kaisensis*. B) Occlusal view, M3, NK92'88 (Nkondo Fm., Uganda), *Stegodon kaisensis*. C) Reconstruction of skull, *Stegotetrabelodon syrticus* (cover illustration, Garyounis Scientific Bulletin, Special Issue No. 4, 1982). D) Occlusal view (reversed), m3, unnumbered (Sahabi, Libya), *Stegotetrabelodon syrticus* (Maglio, 1973: plate I, figure 3). E) Left lateral view (reversed), m3, unnumbered (Sahabi, Libya), *S. syrticus* (Maglio, 1973: plate I, figure 3). F) Occlusal view, M3, KNM-LT 354, part of holotype of *Stegotetrabelodon orbus*. G) Right lateral view, M3, KNM-LT 354, part of holotype of *S. orbus*. H) Dorsal view, mandible, L176-1, *Primelephas korotorensis*, originally attributed to *Stegodibelodon schneideri* (Kalb and Mebrate, 1993: figure 23); anterior is to the top. I) Occlusal view, m3, KNM-LT 351, *P. korotorensis*. J) Left lateral view, m3, KNM-LT 351, *P. korotorensis*. K) Anterior view, reconstruction of cranium of *Loxodonta adaurora*, based on KNM-LT 353 and KNM-KP 385 (Maglio, 1970b: figure 2A). L) Left lateral view, reconstruction of cranium of *L. adaurora*, based on KNM-LT 353 and KNM-KP 385 (Maglio, 1970b: figure 2). C, courtesy of Noel Boaz. D, E, H, permission of the American Philosophical Society. K, L, permission of the Museum of Comparative Zoology, Harvard University.

Africa have been attributed to *S. kaisensis*, previous diagnoses (e.g., Hopwood, 1939; Cooke and Coryndon, 1970) did not distinguish the species from other congeners, or were based on erroneous identification of molars. Tassy's (1995) separation of specimens from Uganda into two stages, without specifying the synapomorphies uniting these forms, further complicated the situation. Thus, the justification for this species has been largely geographic, rather than morphological.

Nonetheless, the species appears to be distinguished by apically anteroposteriorly compressed loph(id)s; deep grooves separating apical digitations (mammillae) that reach the bottom of transverse valleys (figures 15.17A, 15.17B); and strong steplike wear configuration of enamel wear figures combined with weak enamel folding.

Description No skulls of African stegodonts have been conserved (see MacInnes, 1942:85). The species is best documented from the Western Rift, where it has been chronostratigraphically subdivided into two morphs, a primitive "Nkondo stage"

and a more progressive "Warwire stage," based on dental differences (Tassy, 1995). The Nkondo stage includes all of the older collections of Western Rift material attributed to the species (Cooke and Coryndon, 1970; Sanders, 1990), as well as a more recent sample from this region (Tassy, 1995) and specimens of similar age from Chad (table 15.6; Brunet, 2001). Molars of this stage are wide (M3 W = 100–114 mm), with a thin covering of cementum over the plate sides; enamel is thick (ET ≈ 4.6–5.1 mm); lamellar frequency is low (M3 LF = 3.0–4.0); and crown height is very low (HI ≈ 55) (Tassy, 1995). There are no accessory central conules, and apically loph(id)s are anteroposteriorly narrow. Except for thicker enamel and more massive plates, a partial M3 from Kenya (KNM-MP 46; Sanders, 1999) is otherwise morphometrically similar to molars of the Nkondo stage of *S. kaisensis*.

The best preserved specimen of this morph is a complete right M2 from the Kazinga Channel, Uganda (BMNH 15408; Fuchs, 1934), originally the type of *S. fuchsi* (MacInnes, 1942).

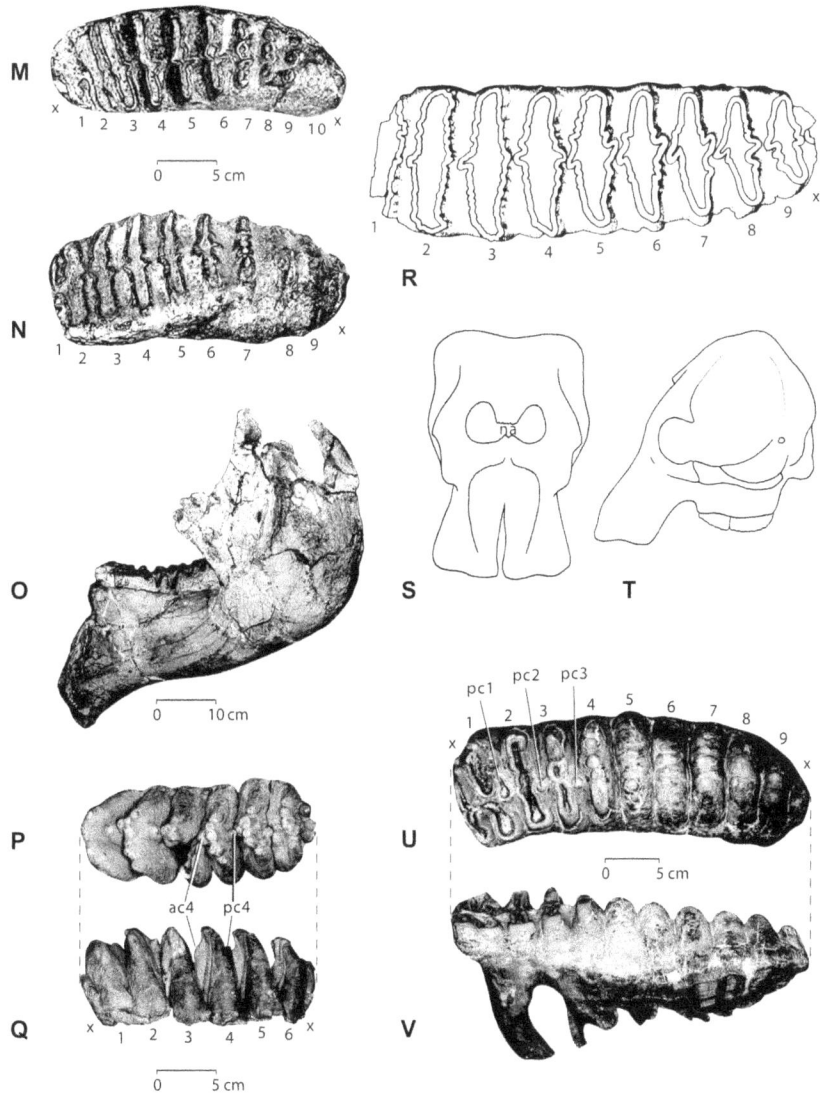

FIGURE 15.17 (CONTINUED)

M) Occlusal view, m3, KNM-KP 385, part of holotype of *L. adaurora*. N) Occlusal view, M3, KNM-KP 385, part of holotype of *L. adaurora*. O) Left lateral view, mandible, KNM-KP 385, part of holotype of *L. adaurora*. P) Occlusal view, M2, SAM-PQ-L45627, holotype of *Loxodonta cookei*. Q) Right lateral view, M2, SAM-PQ-L45627, holotype of *L. cookei*. R) Occlusal view, M3, L 161.19a (Omo Shungura, Ethiopia), *Loxodonta atlantica angammensis* (Beden, 1987a: figure 8). S) Anterior view, cranium, *Loxodonta africana* (Maglio, 1970b: figure 2B). T) Left lateral view, cranium, *L. africana* (Maglio, 1970b: figure 2B). U) Occlusal view, m3, SAM-PQ-L12723, *Mammuthus subplanifrons*. V) Left lateral view, m3, SAM-PQ-L12723, *M. subplanifrons*. R, permission of Editions du Centre National de la Recherche Scientifique. S, T, permission of the Museum of Comparative Zoology, Harvard University.

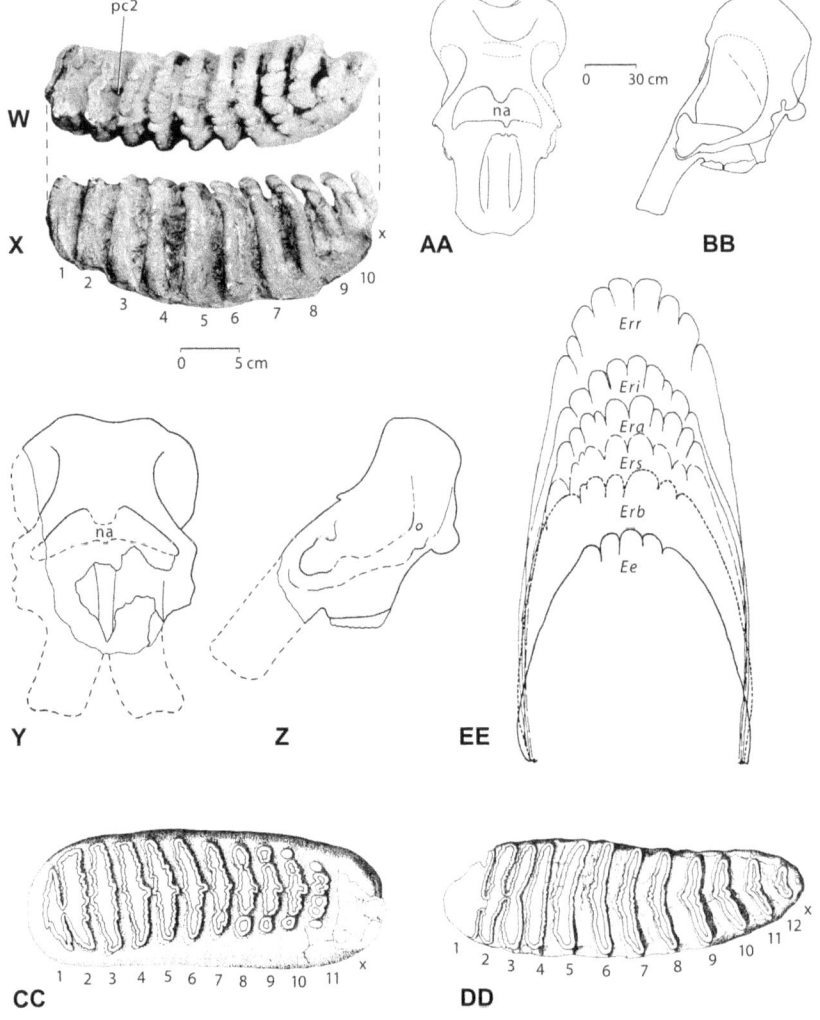

FIGURE 15.17 (CONTINUED)

W) Occlusal view, m3, 1950-1: 12 (Lac Ichkeul, Tunisia), *Mammuthus africanavus*. X) Left lateral view, m3, 1950-1: 12 (Lac Ichkeul, Tunisia), *M. africanavus*. Y) Anterior view, reconstruction of cranium of *Mammuthus meridionalis*, based on Geological Institute of Florence Nos. 1049, 1051, and 1054 (Maglio, 1970b: figure 3A). Z) Left lateral view, reconstruction of cranium of *M. meridionalis*, based on Geological Institute of Florence Nos. 1049, 1051, and 1054 (Maglio, 1970b: figure 3A'). AA) Left lateral view, reconstruction of cranium of *Elephas recki atavus*, KNM-ER 5711 (Beden, 1983: figure 3.17A). BB) Anterior view, reconstruction of cranium of *E. recki atavus*, KNM-ER 5711 (Beden, 1983: figure 3.17B). CC) Occlusal view, m2, lectotype of *E. recki recki* (Beden, 1980: plate IE). DD) Occlusal view, m3, *E. recki atavus* (Beden, 1980: plate IC). EE) Transverse profiles of *E. ekorensis* and *E. recki* molars, showing progessive, time-successive increases in hypsodonty within the lineage (Beden, 1987a: figure 22). Y, Z, permission of the Museum of Comparative Zoology, Harvard University. AA, BB, permission of Oxford University Press, Inc. CC, DD, copyright © 2010 Elsevier, B.V. EE, permission of Editions du Centre National de la Recherche Scientifique.

ABBREIVATIONS: ac, anterior accessory central conule; *Ee, Elephas ekorensis*; *Era, Elephas recki atavus*; *Erb, E. recki brumpti*; *Eri, E. recki ileretensis*; *Err, E. recki recki*; *Ers, E. recki shungurensis*; na, nasal aperture; pc, posterior accessory central conule; x, anterior or posterior cingulum(id); 1, 2, 3, . . ., first second, third, . . . plate.

It has a plate formula of x6x, with 9–11 apical digitations per plate; plates are transversely straight and separated by Y-shaped transverse valleys, wide (W = 92.8) and low (H = 54 mm), and are not closely spaced (LF = 3.75); enamel is thick (5.0 mm); a trace of cementum coats the valleys and is invested around plate apices; and the crown is rectangular in occlusal view (figure 15.17A). Other molars of Nkondo stage stegodonts are morphologically similar, with as many as 13 mammillae per plate (see Tassy, 1995: figure 2), and weakly folded enamel accompanied by pronounced steplike wear configurations of enamel. Plate formulae dp4 = 6; m1 = x6x; M1 = x6x; M2 = x6x; M3 = 7x (Tassy, 1995).

"Warwire stage" molars are also wide (M3/m3 W = 99–105 mm), with numerous bilaterally compressed apical digitations per plate (10–13) (Tassy, 1995). They are primarily distinguished from "Nkondo stage" molars by slighly greater hypsodonty (HI = 63–66) and evidently by more plates, but otherwise are quite similar. Plate formula: m3 = 9 (Tassy, 1995).

Remarks The greater representation of *Stegodon kaisensis* in the Western Rift may be related to humid conditions and proximity of lowland rainforest locally during the late Miocene–Pliocene (Pickford et al., 1993; Boaz, 1994). In overall morphology, *S. kaisensis* Nkondo stage most closely resembles the penecontemporaneous Asian species *S. zdanskyi*, and "Warwire stage" is similar to Javan *S. trigonocephalus* (see Saegusa et al., 2005). *Stegodon kaisensis* is dentally convergent on the primitive elephant *Primelephas korotorensis*, and where one is common, the other is rare or absent (table 15.6).

Further sampling of dental isotopes needs to be undertaken to more widely assess habitat and dietary similarities for these taxa. They both have extremely brachyodont crowns, a low number of plates, few or no accessory central conules in third molars, transversely rectilinear, pyramidal-shaped plates, V-shaped transverse valleys that may be "pinched" at their bases, cementum not completely infilling transverse valleys, thick enamel, and numerous apical digitations in deciduous premolars. This has led to some confusion in identification of specimens. Molars referred to various stegodont species from the Middle Awash, Ethiopia (Kalb and Mebrate, 1993; Tassy, 1995), and Koulà and Kolinga, Chad (Coppens, 1965, 1967) instead belong in *P. korotorensis* (Haile-Selassie, 2001).

Family ELEPHANTIDAE Gray, 1821

Arising in Afro-Arabia during the late Miocene, by the late Pliocene elephants had dispersed throughout the Old World and into North America, and for most of their existence were considerably more taxonomically diverse and widespread than today (Todd and Roth, 1996; Shoshani and Tassy, 1996). The decline of mammutids and gomphotheriids, and increased distribution of C_4 plants (grasses and sedges) (Cerling et al., 1993) created ecological opportunities that were exploited by several successive Mio-Pliocene adaptive radiations of elephants. The first radiation of elephants, in the interval 9.0–7.0 Ma, included *Stegotetrabelodon*, *Primelephas*, and archaic loxodonts, and was followed in the early to mid-Pliocene by a cohort of more progressive species of *Loxodonta* and the first appearance of unequivocal *Mammuthus* and *Elephas* (Maglio, 1973; Sanders, 2004; *contra* Tassy, 2003). These last three genera endured to the Recent as the main, crown lineages of Elephantidae (Maglio, 1973; Todd and Roth, 1996).

At the heart of the rise of elephants was a reconfiguration of craniodental anatomy, correlated with a shift from the grinding-shearing mastication of gomphotheres to a dedicated fore-aft power-shearing translation of the jaws during feeding (Maglio, 1972, 1973) and an emphasis on grazing (Cerling et al., 1999). Anatomical changes included re-organization of molar loph(id)s into transverse lamellae (plates), forward displacement of temporalis muscles, and elevation of parietals and occipitals (Maglio, 1972, 1973; Coppens et al., 1978). The efficiency of this power-shearing adaptation was subsequently improved by loss of lower tusks, anterior shortening of the mandible, increased number of plates and lamellar frequency, enhanced hypsodonty, greater complexity of enamel folding, increased cranial elevation, and delayed serial appearance of molars, all or nearly all of which occurred independently in multiple lineages of elephants (Sikes, 1967; Aguirre, 1969; Maglio, 1973; Froehlich and Kalb, 1995; Todd and Roth, 1996).

Subfamily STEGOTETRABELODONTINAE Aguirre, 1969

Stegotetrabelodonts originated in Afro-Arabia during the late Miocene, and are primarily known from that region. These primitive elephants retained long mandibular symphyses and impressive, projecting lower tusks. In addition, their molars exhibit a host of plesiomorphic features, including strong median longitudinal molar sulci, or clefts; few, low-crowned lamellae, each formed of a small number of conelets; thick, unfolded enamel; and prominent posterior

accessory central conules, often throughout the extent of the crown (Maglio, 1973; Coppens et al., 1978). However, their molars also possess traits common to elephant cheek teeth, such as true plates, loss of trefoil wear patterns, and obliteration of longitudinal sulci by the formation of complete transverse enamel loops in moderate occlusal wear. As well, stegotetrabelodontine crania evidently had a raised, anteroposteriorly compressed skull profile typical of more advanced elephants (Maglio, 1973; see also Tassy, 1999: figure 18.1; *contra* Maglio, 1972).

Despite the low-crowned condition of their molars, stegotetrabelodontines were grazers, a preference shared with other early elephants (Cerling et al., 1999, 2003). This behavior is linked with a late Miocene worldwide pattern of increased seasonality and aridity (Pagani et al., 1999) and widespread expansion of C_4 ecosystems (Cerling et al., 1993). Domination of open grasslands at the expense of heterogenous environments, however, appears to have been more of a Pliocene phenomenon in Africa (Cerling et al., 1993, Kingston et al., 1994), synchronous with further alterations to the elephant masticatory system beyond that of the first radiation of the family.

Often portrayed as broadly ancestral to elephantine elephants as a whole (e.g., Maglio, 1973; Coppens et al., 1978; Beden, 1983), it is alternatively possible that the phylogenetic role of stegotetrabelodontines was limited to being part of an early, side branching of elephants which also included *Primelephas* and *Stegodibelodon* (Tassy and Debruyne, 2001; Sanders, 2004).

Genus *STEGOTETRABELODON* Petrocchi, 1941
STEGOTETRABELODON SYRTICUS Petrocchi, 1941
Figures 15.17C–15.17E

Partial Synonymy Stegotetrabelodon lybicus, Petrocchi, 1943, 1954; *Stegolophodon sahabianus*, Petrocchi, 1943, 1954; *Stegodon syrticus*, Petrocchi, 1954; *Stegotetrabelodon lybicus*, Maglio, 1973; *Stegolophodon sahabianus*, Gaziry, 1982; *Stegotetrabelodon lybicus*, Gaziry, 1982, 1987b; *Stegotetrabelodon syrticus*, Tassy, 1995; *Stegotetrabelodon syrticus*, Sanders, 2008b.

Age and Occurrence Latest Miocene or earliest Pliocene, northern Africa (table 15.6).

Diagnosis Based in part on Maglio (1973). Large elephant with long, straight upper and lower tusks (length reaches over 2,000 mm for both; i2 projecting length >50% of jaw length) that are relatively slender; lower tusks closely appressed to one another; mandibular symphysis elongate and downturned; median longitudinal sulci divide molars; prominent posterior accessory conules may extend nearly throughout the extent of crowns (figures 15.17C, 15.17D).

Description The lone recovered cranium of *S. syrticus* is heavily damaged dorsally, but its preserved morphology suggests an elephantine-like configuration (figure 15.17C; Petrocchi, 1954; Maglio, 1973; Coppens et al., 1978). Upper tusks are rounded in cross section, and lower tusks have ovoid cross sections that are higher than wide. Both lack enamel.

Third and fourth permanent premolars were retained (Gaziry, 1987b), and the dental formula is 1-0-2-3/1-0-2-3 for upper and lower tooth quadrants, respectively, likely preceded by three deciduous premolars in each. Molars are low-crowned, and organized into rows of lamellae, or plates, each composed of a small number of conelets (four–six). Enamel is thick and unfolded, and plates are coated with cementum. Plate formulae dp4 = x4x; m1 = x4x; m3 = x7x; P4 = x2x; M2 = x5–x5x; M3 = x6x.

Lamellar frequency is low (m3 = 2.7; M3 = 3.0). Although unworn molar crowns are longitudinally divided by median clefts and superficially resemble molars of gomphotheriids, with moderate wear plates occlusally form complete transverse enamel loops. In lateral view, plates are pyramidal in shape, and transverse valleys are correspondingly V-shaped (figure 15.17E).

STEGOTETRABELODON ORBUS Maglio, 1970
Figure 15.17F

Partial Synonymy *Primelephas gomphotheroides* (in part), Coppens et al., 1978; cf. *Primelephas gomphotheroides*, Tassy, 1986; Elephantidae gen. et sp. indet. (in part), Tassy, 2003; Elephantidae gen. et sp. incertae sedis A (in part), Tassy, 2003; *Elephas nawataensis* (in part), Tassy, 2003: figure 8.8, plate 4; *Elephas* cf. *E. ekorensis* (in part), Tassy, 2003: figure 8.8, plate 5; Elephantidae gen. and sp. indet., Tassy, 2003: figure 8.5, plates 4–5; Elephantidae gen. et sp. indet., Saegusa and Hlusko, 2007.

Age and Occurrence Late Miocene–early Pliocene, eastern Africa (table 15.6).

Diagnosis Based in part on Maglio (1970b, 1973); Maglio and Ricca (1977); Coppens et al. (1978). Smaller than *S. syrticus*, with less massive mandible; lower tusk length ≤1,000 mm and projecting length less than 50% of overall jaw size; M3 with posterior central accessory conules usually limited to first two plates (figure 15.17F); crown height low but slightly greater than in *S. syrticus* (figure 15.17E, G).

Description The cranium is unknown for this species. Lower tusks relatively slender (KNM-LT 354 L = 1000 mm; H = 70 mm; W = 58 mm), closely parallel to one another, and strongly downturned in their alveoli. Mandibular symphysis extensive, with surprisingly thin alveolar bone for support of such elongate i2s.

Inferred presence of second-fourth deciduous premolars and upper and lower third and fourth premolars; the adult dental formula is 1-0-2-3/1-0-2-3. Plate formulae: dp3 = x3x; p3 = x3x; p4 = x3x; m2 = 5x-x6; m3 = x7x; dP2 = 3x; dP3 = x3x; P4 = x2x; M2 = x5x; M3 = x6x. Molar morphology similar to that in *S. syrticus*, with thick enamel (5.0–8.0 mm in third molars), low hypsodonty indices (66–71) and lamellar frequency (2.75–3.0), and median clefts in unworn crowns.

Remarks Proposed incorporation of *S. orbus* into *S. syrticus* (Gaziry, 1987b) merits consideration, since morphological differences between the species are slight and tusk size distinctions may be ontogenetic. In addition to the type series from Libya, fossils from Abu Dhabi have also been added to *S. syrticus*, including a partial skeleton with a cranium, mandible, and tusks, and a modest sample of molars (Andrews, 1999; Tassy, 1999). The cranium exhibits typical elephant features (see Tassy, 1999: figure 18.1), and the upper and lower tusks are elongate and relatively thin (I2 L = 1,020 mm; i2 projecting length = 530 mm; Tassy, 1999). The m3, however, has eight massive, low plates formed of three to four conelets each, with anterior and posterior pretrite accessory central conules closely appressed to several of them, suggesting that the Abu Dhabi elephant belongs to a different species (Sanders, 2004). An isolated lower molar (?m2) from Jebel Barakah, Abu Dhabi, was allocated by Madden et al. (1982) to *Stegotetrabelodon grandincisivus*, and later was placed by Tassy (1999) in Elephantoidea indet. ?"*Mastodon*" *grandincisivus*. The taxon "*grandincisivus*" appears to be an amebelodont (Tassy, 1985, 1986, 1999); however, the occlusal morphology of the Jebel Barakah specimen exhibits no amebelodont features, and likely does not belong to this taxon. Contrary to Tassy's (1999) assertion, the Jebel Barakah molar does possess features in common with the stegotetrabelodont documented by other specimens at Jebel Barakah and at Shuwaihat: it is very brachyodont, lophids are composed of few conelets (four), a median longitudinal sulcus is persistent with wear, cementum coats the plates and partially fills the transverse valleys, and posterior accessory central conules were apparently present throughout the pretrite side of the crown (see Tassy, 1999: figure 18.15).

A stegotetrabelodont (not *Primelephas*, *contra* Coppens et al., 1978; Beden, 1985; Pickford, 1987a) is present in the heterogeneous elephant assemblage from Kanam, Kenya (MacInnnes, 1942; Maglio, 1973), evidenced by molars (M 15409, 15410, 15411) with low lamellar frequencies, wide plates, small number of robust conelets, presence of accessory conules nearly throughout m3, and persistence of median clefts. M3s in this sample have seven plates. A stegotetrabelodont is also present in the Oluka Fm. of the Western Rift, Uganda (Tassy, 1995), with eight plates in m3. This elephant apparently succeeded locally a more primitive species of *Stegotetrabelodon* from the Kakara Fm. (Tassy, 1995). While an isolated m2 (KNM-NK 42396) from Lemudong'o, Kenya, has been described as belonging to a new, indeterminate genus and species of elephant (Saegusa and Hlusko, 2007), one of us (W.J.S.) does not think the evidence is sufficient to warrant a new taxon and prefers to place the specimen in *Stegotetrabelodon* for now. It has a plate formula of 5x, pyramidal plate shape in lateral view, and posterior accessory conules throughout the crown, all features found in *Stegotetrabelodon*; the value of features such as uniformity of width of the crown and appression of conules on plates for diagnosing a new taxon may be overstated. Similarly, Tassy (2003:346) has assigned a small number of gnathic and dental remains from the Lower and Upper Nawata Fm. at Lothagam, Kenya to "Elephantidae gen. and sp. indet.," but their morphological departure from the type seems acceptable for a normally variable, dimorphic species, and they should be retained in *S. orbus*.

Deriving crown elephant genera (*Loxodonta, Elephas*, and *Mammuthus*) from this primitive subfamily requires a reversal to the dental morphology of gomphotheres, which have free anterior as well as posterior accessory conules associated with their loph(id)s. These are present to some degree in early species of each crown genus, but not in stegotetrabelodonts and other primitive elephant taxa. In nearly all other ways, however, *Stegotetrabelodon* is sufficiently primitive to represent a good model for the ancestral elephant (Coppens et al., 1978).

Subfamily ELEPHANTINAE Gray, 1821

More advanced elephants lacking lower tusks and appreciable median sulci, and with less prominent mandibular symphyses than in stegotetrabelodonts. Minimum number of third molar plates is seven (Kalb and Mebrate, 1993; Kalb et al., 1996a). Tendency for molar crowns to become hypsodont, enamel thinner and folded, plates more closely spaced and numerous, and for accessory conules to become incorporated into plates, except in the earliest forms. Trends in these features are strongly directional over time and occurred independently within the main elephantine lineages (Maglio, 1973).

This subfamily had its origins in Africa in the late Miocene, contemporaneous with stegotetrabelodonts, and includes the last extant representatives of the Proboscidea. As pointed out by Maglio (1973), the rapid evolution and wide geographic distribution of elephantine species make them particularly valuable for biochronological correlation and the study of evolutionary phenomena.

Genus *PRIMELEPHAS* Maglio, 1970
PRIMELEPHAS KOROTORENSIS (Coppens, 1965)
Figures 15.17H–15.17J, 15.18A, and 15.18B

Partial Synonymy Stegodon korotorensis, Coppens, 1965:343; *Stegodon kaisensis* Coppens, 1967:I, figure 2; *Mammuthus (Archidiskodon) subplanifrons* (in part), Cooke and Coryndon, 1970:123; *Primelephas gomphotheroides*, Maglio, 1970b:10; *Primelephas korotorensis*, Maglio, 1970b:12, and Maglio, 1973:22; *Stegotrabelodon orbus*, Kalb and Mebrate, 1993:40–44; cf. *Stegodibelodon schneideri*, Kalb and Mebrate, 1993:44–46; *Mammuthus subplanifrons* (in part), Kalb and Mebrate, 1993:59–63; *Elephas nawataensis* (in part), Tassy, 2003:343–345: figure 8.8, plates 1–3; *Primelephas korotorensis*, Mackaye et al., 2008:227.

Age and Occurrence Late Miocene-early Pliocene, East and Central Africa (table 15.6).

Diagnosis Monospecific genus. Distinguished from other elephants by extreme brachyodonty and from contemporaneous elephant species by narrowness of cheek teeth (East African sample; figures 15.18A, 15.18B) third molar mean W < 100 mm). Differs from *Stegotetrabelodon* by absence of lower tusks and superficial expression or absence of median sulci in molar crowns, less robust molar crown features, and restriction of accessory central conules to the anterior end of the crown in m3. In occlusal view, plates usually more rectilinear than the convex-convex plates of *Stegodibelodon*.

Description Based on Maglio (1970b, 1973); Maglio and Ricca (1977); Coppens et al. (1978); Tassy (1986, 1995, 2003); Sanders (1997, 2004); Haile-Selassie (2001). A partial cranium from the Oluka Fm. of the Western Rift, Uganda, has a convex forehead with small concavities at the lateral corners, weakly divergent, narrow tusk alveoli (W = 80 mm), a large and bulging nasal process of the premaxillary, and massive zygomatic process of the frontal. In addition, the face is anteroposteriorly short, the orbits are situated above the anterior end of the M1s, and the basicranium is semivertical and raised well above the height of the palate (see Tassy, 1995: figure 4, plate III). In lateral profile, the cranium is rounded and the forehead appears convex.

Adult dental formula 1-0-2-3/0-0-2-3; deciduous formula 1-0-3/0-0-3. Mandibular symphysis a distinct, projecting "spout" with no incisor alveoli (figure 15.17H). Molars very low crowned (third molar HI < 70) and composed of few plates. Plate formulae: dp2 = x3x; dp3 = x4x; dp4 = x5 to 6x; p3 = x3; p4 = x3x; m1 = x5x; m2 = x6x; m3 = 7x to 8x; dP2 = x3x; dP3 = x4x; dP4 = x5x; P4 = x3x; M1 = x5x; M2 = x5x; M3 = x7x. Accessory central conules only on the posterior side of plates. Enamel thick, unfolded. Cementum coats plates to apices, but does not completely fill the transverse valleys. In lateral view, plates are pyramidal in shape and widely spaced by V-shaped transverse valleys (figure 15.17J). In occlusal view, plates are usually transversely rectilinear or anteriorly convex (figure 15.17I). Molar plates typically superficially subdivided into four-eight apical digitations. The East African sample of this species has relatively narrow molars; width of third molars is usually < 100 mm (figures 15.18A, 15.18B). Mackaye et al. (2008) state that the new molars from Chad assigned to *Primelephas* generally fit metrically with the East African sample; however, they range to greater width (figures 15.18A, 15.18B).

Remarks Originally erected as a temporal and morphological intermediate between *Stegotetrabelodon* and elephantine

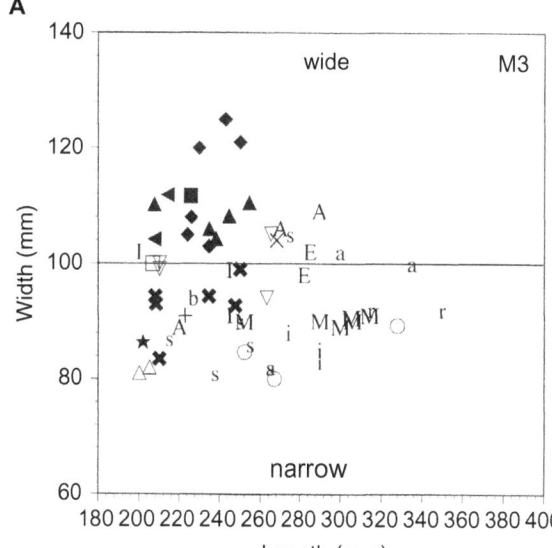

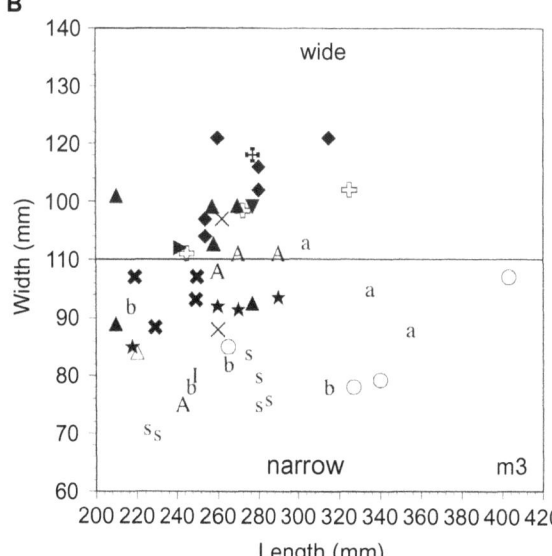

FIGURE 15.18 Bivariate plots of M3 (A) and m3 (B) crown length versus width in African tetralophodonts, stegodonts and elephants. Comparative dimensions supplementing original measurements are from Scott (1907), MacInnes (1942), Petrocchi (1954), Coppens (1965, 1972), Arambourg (1970), Maglio (1973), Beden (1983, 1987a), Harris et al. (1988a, 2003), Kalb and Mebrate (1993), Tassy, (1995, 1999), Sanders (1997, 2007), Haile-Selassie (2001), Tsujikawa (2005b), and Mackaye et al. (2008).

SYMBOLS: closed square, *Tetralophodon* sp. nov.; open square, *Stegodon kaisensis*; closed diamond, *Stegotetrabelodon syrticus*; closed triangle, *S. orbus*; inverted closed triangle, *Stegotetrabelodon* sp., Abu Dhabi; right-facing closed triangle, *Stegotetrabelodon* sp., Kanam; left-facing closed triangle, *Stegotetrabelodon* sp., Nyabusosi, Western Rift; heavy X, *Primelephas korotorensis*; light X, *P. korotorensis*, Toros Menalla, Chad; cross, *Stegodibelodon schneideri*; inverted open triangle, *Loxodonta adaurora adaurora*; open triangle, *L. adaurora kararae*; star, *L. cookei*; +, *L. exoptata*; open circle, *L. atlantica*; open cross, *Mammuthus subplanifrons*; A, *M. africanavus*; M, *M. meridionalis*; E, *Elephas ekorensis*; b, *Elephas recki brumpti*; s, *E. recki shungurensis*; a, *E. recki atavus*; i, *E. recki ileretensis*; r, *E. recki recki*; I, *Elephas iolensis*.

elephants, the taxonomic homogeneity and legitimacy of the taxon have been questioned (Coppens et al., 1978; Beden, 1979, 1983, 1985). Recent addition of considerable new material to the original hypodigm (Mebrate, 1983; Tassy, 1986,

1995, 2003; Kalb and Mebrate, 1993; Sanders, 1996, 1997; Brunet et al., 2000; Haile-Selassie, 2000, 2001; WoldeGabriel et al., 2001; Mackaye et al., 2008) and comprehensive restudy of *Primelephas* (Mundinger and Sanders, 2001; Sanders, 2004), however, have more clearly delineated the taxon and confirmed its validity. Supposed tusk sockets on symphyseal fragment KNM-LT 358 (Maglio, 1970b; Maglio and Ricca, 1977) articulate with KNM-LT 378 to form closed anterior chambers of the mandibular canal (Mundinger and Sanders, 2001; Sanders, 2004). A mandible from the Middle Awash with a projecting symphyseal "spout" and no incisor alveoli (figure 15.17H), a juvenile mandible with a similar symphysis (the type of "*Elephas nawataensis*"; Tassy, 2003), and mandibles from Toros Menalla, Chad (Mackaye et al., 2008) demonstrate the absence of lower tusks in *Primelephas*. The molars of the juvenile mandible differ from other early *Elephas* taxa (e.g., *E. ekorensis, E. recki brumpti*) but closely match molars of *Primelephas* in having posterior but not anterior accessory central conules, and in overall plate construction. In addition, no features were provided to differentiate it from other elephant species in the publication that formally named "*E. nawataensis*" (Tassy, 2003). There is no compelling reason to accept this as a new species. Conversely, an m3 from Lukeino (KNM-LU 7597) with nine plates that was originally placed in *Primelephas* (Tassy, 1986) does not belong in this genus and may be an archaic mammoth (see later discussion).

Maglio (1970b) originally named two species of *Primelephas*, *P. gomphotheroides*, and *P. korotorensis*, the latter based on two distal molar fragments from Koulà and Kolinga, Chad that were originally placed in *Stegodon* by Coppens (1965, 1967). Sanders (1997), Mundinger and Sanders (2001), and Sanders (2004) suggested that *Primelephas korotorensis* should be subsumed into *P. gomphotheroides* as the supposed greater molar hypsodonty of the former is an artefact of measurements taken only on the narrow distal heel of molars—there is no proportional difference between the molars of these species. Mackaye et al. (2008) recently agreed with this assessment and pointed out that in this case, *P. korotorensis* has priority, since Coppens named *(Stegodon) korotorensis* in 1965, well before Maglio's (1970b) erection of *P. gomphotheroides*. Unfortunately, this means that the type reverts to a broken distal molar from Koulà, Chad, rather than the more complete, associated dentition from Lothagam, Kenya, selected as the original type by Maglio (1970b).

Chronostratigraphic refinements at Lothagam, Kenya (Feibel, 2003; McDougall and Feibel, 2003) show that *Primelephas* was contemporaneous with *Stegotetrabelodon* (table 15.6; Tassy, 2003; Sanders, 2004). With new evidence of *Loxodonta* from similar-aged deposits (see later discussion; Tassy, 1995, 2003; Vignaud et al., 2002; Sanders, 2004), it is apparent that *Primelephas* was part of an initial evolutionary radiation of elephants in the late Miocene. Alternately posited as centrally positioned phylogenetically vis-à-vis other elephantines (Todd and Roth, 1996), or as the sister taxon to a more restricted clade of *Elephas* + *Mammuthus* (Beden, 1983; Tassy, 1995), it is more likely that *Primelephas* was part of an early, side branching of elephants separate from the origination of Recent elephant genera (Tassy and Debruyne, 2001; Sanders, 2004).

Genus *STEGODIBELODON* Coppens, 1972
STEGODIBELODON SCHNEIDERI Coppens, 1972

Partial Synonymy Loxodonta schneideri, Beden, 1983, 1985; *Selenetherium kolleensis*, Mackaye et al., 2005.

Age and Occurrence Early Pliocene, Central Africa (table 15.6).

Diagnosis Based in part on Coppens (1972); Sanders (2004); Mackaye et al. (2005). Differs from *Stegotetrabelodon* in absence of lower tusks, shorter symphysis, thinner enamel, higher molar plates, and virtual absence of median sulci. Similar to *Primelephas* in mandibular and dental morphology, but has wider molars and apparently higher-crowned M3s. Distinguished from other elephantine elephants in the length of its mandibular symphysis, low number of plates, enamel thickness (5.0–8.0 mm), and low lamellar frequency (2.1–3.0) in M3/m3.

Description Coppens (1972); Mackaye et al. (2005). Plate formulae: m3 = x7x; M3 = x7x. Poorly known from dental, mandibular, and a small number of postcranial remains, including new specimens from Kollé, Chad. Molars are large, wide, and moderately low crowned (M3 HI = 79). Enamel is thick and unfolded. Plates are well spaced, and pyramidal in lateral view; transverse valleys are correspondingly V shaped and well coated with cementum. In occlusal view, plates of m3 are anteriorly convex, and it is not possible to ascertain if there were accessory central conules present at earlier stages of wear. Upper tusks are large and slightly curved.

Remarks Originally placed in Stegotetrabelodontinae because of the elongation of its symphysis, compared with those of extant elephants (Coppens, 1972). Although exclusion of *Stegodibelodon* from the Elephantinae persists uncritically in the literature (e.g., Kalb et al., 1996b; Shoshani, 1996; Tassy and Shoshani, 1996; Shoshani and Tassy, 2005), there is no reason to continue this arrangement. The lack of lower tusks and reduction of median sulci are synapomorphies of Elephantinae.

In the absence of any other evidence to mark its presence, a mandible lacking teeth from the Middle Awash, Ethiopia, with symphyseal morphology identical to that of the *Stegodibelodon* mandible from Menalla, Chad (Kalb and Mebrate, 1993), is more reasonably attributed to *Primelephas*, which is common there. New specimens from Kollé, Chad, including third molars with *more* than six plates, *contra* Mackaye et al. (2005), are indistinguishable from the modest original *Stegodibelodon* sample, and thus do not warrant erection of a new species. As more becomes known about its overall anatomy, it is possible that *Stegodibelodon* will be recognized as a larger congener of *Primelephas*. Contrary to Beden (1983, 1985), there are no features of the occlusal morphology of *S. schneideri* that would support its placement as a basal member of the *Loxodonta* clade.

Genus *LOXODONTA* Cuvier, 1825
(anonymous emendation, 1827)

The phylogeny of *Loxodonta* includes the living African elephants and extends back to the late Miocene (table 15.6). Isolated molars from Toros Menalla, Chad (Vignaud et al., 2002) and the Lukeino Fm., Tugen Hills, Kenya (including KNM-LU 916) constitute the earliest evidence of the genus: these teeth have as few as eight plates in m3, are low crowned with thick enamel and low lamellar frequencies, and exhibit propeller- or lozenge-shaped enamel wear figures distinctive of the genus.

Compared with *Elephas* and *Mammuthus*, loxodont elephants are morphologically and geographically conservative, with a wholly African evolutionary history (Coppens et al., 1978). For the greater part of the Pleistocene, however, they were supplanted in the eastern part of the continent by

Elephas recki and apparently retreated to northern and southern refugia, and possibly to West–Central African forests. Although Mio-Pliocene species were grazers or mixed feeders with a high proportion of C_4 plants in their diet, in the middle Pleistocene *Loxodonta africana* emerged from exile as a preferential browser (Koch et al., 1995; Cerling et al., 1999, 2003; Zazzo et al., 2000; Harris et al., 2003; Schoeninger et al., 2003; Kingston and Harrison, 2007).

Among the distinctive features shared by most members of the genus, including archaic species, are a globular cranium with a biconvex vault, unexpanded parietals, and distally divergent tusk alveoli; mandibles with long, straight corpora, prominent symphyses, and strongly convex and anteroventrally canted condyles; and "lozenge"-shaped enamel molar occlusal wear figures, incorporating anterior and posterior central accessory conules that are usually present throughout molar crowns (Maglio, 1973; Beden, 1983).

There has been some debate about the relationship of *Loxodonta* to other elephant genera, though morphology based cladistic analyses usually position *Loxodonta* as the sister taxon to an *Elephas-Mammuthus* clade (e.g., Kalb and Mebrate, 1993; Shoshani, 1996; Tassy, 1996c; Shoshani et al., 1998). These results were criticized by Thomas et al. (2000), who felt that features linking *Elephas* and *Mammuthus* are homoplasic. More recent cladistic analyses relying largely on craniodental characters of basal members of elephant genera (Mundinger and Sanders, 2001; Sanders, 2004), however, support the traditional view. Mitochondrial DNA sequence analysis, using the American mastodon as an outgroup, concluded that *Loxodonta* is the sister taxon to *Elephas* + *Mammuthus* (Rohland et al., 2007), a result supported by other analyses (e.g., Yang et al., 1996; Ozawa et al., 1997; Krause et al., 2005; Rogaev et al., 2006), but some genetic studies are contradictory, alternatively finding evidence for a *Loxodonta-Mammuthus* clade (Hagelberg et al., 1994; Noro et al., 1998; Barriel et al., 1999; Thomas et al., 2000; Thomas and Lister, 2001; Debruyne et al., 2003). The phylogenetic connection of *Loxodonta* to archaic elephants remains unresolved: while molars of the earliest loxodonts retain the anterior and posterior accessory conules of the gomphothere "trefoil" throughout their crowns, the molars of *Stegotetrabelodon*, *Primelephas*, and *Stegodibelodon* lack anterior accessory conules, and therefore in this feature seem unsuitable for the ancestry of these crown elephants (Sanders, 2004).

LOXODONTA ADAURORA Maglio, 1970
Figures 15.17K–15.17O, 15.17S, 15.17T, 15.17Z, and 15.17BB

Partial Synonymy *Archidiskodon planifrons nyanzae* (in part), MacInnes, 1942; *Archidiskodon* cf. *meridionalis*, MacInnes, 1942; *Elephas* cf. *planifrons*, Arambourg, 1947; *Loxodonta* cf. *africanava* (in part), Cooke and Coryndon, 1970: plate I, figures D, E; *Loxodonta exoptata*, Kalb and Mebrate, 1993: figure 30.

Age and Occurrence Early–late Pliocene, eastern and southern Africa (table 15.6).

Diagnosis Based in part on Maglio (1970b); Maglio and Ricca (1977); Beden (1983); Harris et al. (2003). Cranium similar to that of *Loxodonta africana* but with a longer frontal, broader forehead, more elongate prenasal region, more flaring tusk alveoli, and prominent frontoparietal ridges lateral to the narial opening (figures 15.17K, 15.17L, 15.17S, 15.17T). As in modern African elephants but differing from *Elephas* spp., the occipital condyles are located below the level of the glenoid fossa. Mandible also similar to the lower jaw of the extant

African elephant, except for a more prominent symphyseal "beak" and broader corpus to accommodate wide molars (figure 15.17O). Molars mesodont to slightly hypsodont, with unfolded or coarsely undulating enamel, and modest number of well spaced plates. Anterior and posterior accessory central conules present throughout molar crowns and incorporated into enamel loops with wear, but do not form prominent median loxodont sinuses ("lozenges"; figures 15.17M, 15.17N); lateral arms of plates may remain anteroposteriorly compressed well into occlusal wear (Tassy, 2003). Posterior accessory conules may be doubled, and crowns may sport marginal accessory conules as well (Kalb and Mebrate, 1993).

Molars more massive than in *Elephas ekorensis* and *L. exoptata*, and with less developed median sinuses than in the latter.

Description *Loxodonta adaurora* resembles *L. africana* in cranial morphology in its convex frontoparietal surface, unexpanded parietals and occipital region, and slight temporal constriction. As in the extant species, the cranium of *L. adaurora* is only slightly compressed anteroposteriorly, and in lateral view has a more rounded profile than in either *Elephas* or *Mammuthus* (figures 15.17L, 15.17Z, 15.17BB; Maglio, 1970b). The nasal aperture is exceptionally large, the occipital condyles are posteriorly prominent, and the incisor alveoli are widely separated and flare distally. The alveoli held massive upper tusks that curved gently upward with no longitudinal torque (Maglio, 1973).

There are no lower tusks or vestigial incisor cavities in the mandible (*contra* Maglio, 1970b, 1973; Beden, 1983). Rather, these cavities form the antechambers of the mandibular canal for nerve plexuses and arteries that supplied the face. The symphyseal gutter is narrow and very deep. The ramus is tall, with a narrow neck, transversely elongated condyle, and coronoid process that leans outward and exhibits a capacious masseteric fossa (Beden, 1983). These features are typical of modern African elephants.

Molars are broadest basally, with crown height slightly less than width (Beden, 1983). In lateral view, plates are subparallel to one another and separated by U-shaped valleys that may be filled with cementum. In the third molars, lamellar frequency ranges from 3.2 to 5.0, and enamel thickness varies from 2.8 to 4.5 mm (Beden, 1983). Plates are comprised of five to six conelets (five to eight in advanced forms), and taper laterally and anteroposteriorly toward the apex. The adult dental formula appears to have been 1-0-0-3/0-0-0-3, with three deciduous precursors in each quadrant of the upper and lower jaws. No permanent premolars are known for *L. adaurora*.

Plate formulae: (*L. adaurora adaurora*) dp2 = 4x; dm3 = 6x; m1 = 6–7; m3 = x10x; dP2 = 3x; dP4 = x8x; M2 = x6x; M3 = x9x–10x; (*L. adaurora kararae*) m3 = 11–12; M2 = x9x; M3 = x10x.

The species is well represented by a nearly complete skeleton from Kanapoi (Maglio, 1973), which anatomically resembles the postcranium of the modern African elephant and is distinct from that of *Elephas* spp. (Maglio and Ricca, 1977). There are at least 18 thoracic, 3 or 4 lumbar, and 4 sacral vertebrae; the long bones exhibit numerous graviportal adaptations and indicate an animal of very large size (femur L = 136.5 cm; humerus L = 116.2 cm); manus and pes elements are especially *Loxodonta africana*–like, except for being larger and more robust (Maglio and Ricca; 1977).

Remarks While skull morphology links *Loxodonta adaurora* with other loxodont elephants (Maglio, 1970b, 1973; Beden, 1983), it is an atypical member of the genus because its enamel wear figures do not develop strong median sinuses (Tassy, 2003). For this reason, it is unlikely that this species was part of the main loxodont lineage leading to *L. africana*.

Beden (1983) divided the species into early to mid-Pliocene *L. adaurora adaurora*, and late Pliocene *L. a. kararae*, based on slight differences in enamel thickness, number of plates in m3, and lamellar frequency. Compared with *L. a. adaurora*, in *L. a. kararae* the cranium also has a shorter premaxillary, orbits that protrude less laterally, and zygomatic arches that are taller and less massive (Beden, 1983). An M2 from the Matabaietu Fm., Middle Awash, Ethiopia attributed to *L. exoptata* (L77-1; Kalb and Mebrate, 1993: figure 30) appears instead to belong to *L. a. kararae*, based on its number of plates (eight), twinned accessory conules, degree of enamel folding, enamel thickness (ET = 2.0–4.0 mm), and moderate plate spacing (LF = 5.0), in addition to its lack of prominent loxodont sinuses.

These elephants appear to have existed in a mosaic of closed woodlands to open savannas (Harris et al., 2003; Schoeninger et al., 2003), with morphological transformation of their craniodental features during the late Pliocene perhaps a response to increase of open habitats and grazing competition due to cooler, drier climates (Behrensmeyer et al., 1997; Alemseged, 2003). No longer included in *L. adaurora* are specimens returned to *Loxodonta exoptata* (Beden, 1987b; Sanders, 2005, in press), particularly the fossil sample from Laetoli, Tanzania. These specimens were once partitioned among *L. adaurora* and *Elephas recki* (Maglio, 1969b, 1973; Coppens et al., 1978), but the molar variation that led to this division is an artifact of differences in the contribution of accessory conules to occlusal plate shapes at successive wear stages (see later discussion).

LOXODONTA COOKEI Sanders, 2007
Figures 15.17P and 15.17Q

Partial Synonymy Mammuthus subplanifrons, Hendey, 1976, 1981; Elephantinae indet., cf. *Loxodonta*, Tassy, 1986; *Loxodonta* sp. "Lukeino stage," Tassy, 1995; *Loxodonta* sp. nov., Sanders, 2006.

Age and Occurrence Late Miocene-early Pliocene, southern and eastern Africa (table 15.6).

Diagnosis Sanders (2007). Primitive loxodont with permanent third and fourth premolars. Distinguished from *L. adaurora*, *L. exoptata*, *L. atlantica*, and *L. africana* by fewer molar plates, especially in posterior teeth, and lower hypsodonty indices, and from nonloxodont elephants by presence of anterior and posterior accessory conules (figures 15.17P, 15.17Q) and tendency for plates to apically form median sinuses with wear, throughout molar crowns. The only crown elephant species known to retain permanent premolars.

Description Tassy (1995); Sanders (2007). Skull unknown. Morphological details of the entire molar sample are similar: plates are formed of five to eight conelets, converge anteroposteriorly and laterally toward the apex, and are not very closely spaced (m3 LF = 3.4–4.2); crown height is usually less than width (m3 HI = 81–102); transverse valleys are U shaped. In more worn specimens, enamel is thick (m3 ET = 4.1–6.2 mm) and unfolded or coarsely folded, and wear figures form median loxodont sinuses that may touch in the midline. Cementum usually nearly fills transverse valleys. Accessory conules are lower than unworn plates and apically free (figure 15.17Q) and only become incorporated into plates with moderate wear. Deciduous premolars are miniature versions of adult molars, except for dp2s, which are tiny, triangular teeth with three closely appressed plates.

Permanent premolars were in occlusion at the same time as adult molars, without a strong wear gradient between them. In occlusal view, P3 is rounded, of nearly equal length and width, while p3 is more ovoid and longer than wide. Fourth permanent premolars are larger, relatively more elongate, and have better expression of accessory central conules. Cementum is well developed in these teeth.

Plate formulae: dp2 = 3x; dp3 = x4x; dp4 = x6x; p3 = x3x–x4x; p4 = x4x; m1 = x5x; m2 = x7x; m3 = x7x–x8x; DP3 = x4; DP4 = x5x or x6; P3 = x3x–x4; P4 = x4; M1 = 5x–x6; M2 = x5x-x6x; M3 = 7–8.

Remarks This species, best represented at Langebaanweg, South Africa, encompasses late Miocene–early Pliocene specimens that exhibit strong development of anteroposterior median enamel expansions ("loxodont sinuses") and are characterized by mesodont molar crowns with a low number of plates (Sanders, 2006). The morphology of these specimens suggests affinity with a subsequent *L. exoptata–L. africana* lineage (see Tassy, 1995). Possibly the oldest representative of *L. cookei* is an m3 reportedly from the Lukeino Fm., Tugen Hills, Kenya (KNM-LU 67; Tassy, 1986). The actual provenance of this specimen is uncertain (M. Leakey, pers. comm.), and its preservation and morphology are a good match for other specimens referable to *L. cookei* from the younger Chemeron Fm.

LOXODONTA EXOPTATA (Dietrich, 1941)

Partial Synonymy Palaeoloxodon antiquus recki, Hopwood, 1936; *Archidiskodon exoptatus*, Dietrich, 1941, 1942; *Elephas antiquus recki*, Hopwood in Kent, 1941; *A. subplanifrons*, Cooke, 1960; *E. recki*, Leakey, 1965; *Loxodonta exoptata*, Coppens, 1965; *E. (Archidiskodon) exoptatus*, Arambourg, 1969; *Loxodonta* sp. (in part), Maglio, 1969b; *E. recki* (in part), Maglio, 1969b; *Loxodonta* sp. "C" (in part), Maglio, 1970a; *E. recki* (in part), Maglio, 1970a; *L. adaurora* (in part), Maglio, 1970b; *Mammuthus (Archidiskodon) recki* (in part), Cooke and Coryndon, 1970; *L. adaurora* (in part), Maglio, 1973; *L. adaurora* (in part), Coppens et al., 1978; *L. exoptata*, Beden, 1987b.

Age and Occurrence Early–late Pliocene, eastern Africa (table 15.6).

Diagnosis Based in part on Beden (1987b). Molars with anterior and posterior accessory conules that fuse with enamel wear figures of plates to form loxodont median sinuses the length of the crown when in wear, and slightly hypsodont, with maximum width a little above the base of the tooth. Lateral borders of molar crowns converge toward the apex, and the median part of each plate is highest. Cementum thick and fills the transverse valleys. Higher crowned with thinner enamel, more plates, and greater lamellar frequencies than molars of *L. cookei* and Miocene *Loxodonta*. Molars with narrower plates, stronger median sinus development, and less robustly constructed than in contemporaneous *L. adaurora*.

Description Cranium unknown. Adult dental formula is 1-0-0-3/0-0-0-3, with three deciduous premolars preceding the eruption of adult molars in each dental quadrant; there are no known permanent premolars. Molars with parallel-sided plates and U-shaped transverse valleys. Enamel thickness moderate (m3 = 2.54.0 mm; M3 = 2.0–4.0 mm); lamellar frequency ranges from 4.1 to 5.5 in third molars; and third molar crown height is usually slightly greater than width (HI > 100). Accessory conules are lower than plate conelets, closely appressed to plates, and have their greatest diameter at about the midheight of the crown. For this reason, they may not be apparent and do not contribute to the formation of median sinuses until the crown is well in wear, and disappear with heavy occlusal wear. In the early stages of occlusal wear, plate apices may form "propeller" shapes, with the center of the enamel figure prominent and rounded, and the lateral "arms" more anteroposteriorly compressed (Kalb and Mebrate, 1993; Sanders, 1997).

Plate formulae: dp2 = x3x–x4x; dp3 = x6x; dp4 = 7; m1 = x7x or x8; m2 = 8–9x; m3 = 11–12; DP2 = x3x–x4; DP3 = x5x–6x; DP4 = x6 or x7x; M2 = 8x–9x; M3 = 11–12 (Beden, 1983, 1987b; Harris et al., 2003).

The postcranial sample for the species from the Upper Unit of the Laetolil Beds and the Upper Ndolanya Beds at Laetoli is primarily composed of isolated bones of the manus and pes. These closely resemble comparable elements in the extant African elephant (Beden, 1987b).

Remarks Although this species, best known from the Laetolil and Upper Ndolanya Beds (table 15.6), has a long and complicated taxonomic history (Beden, 1987b) and has been subdivided and subsumed into *Loxodonta adaurora* and *Elephas recki* (Maglio, 1969b, 1973; Coppens et al., 1978), it is clear that its constituent specimens belong to a distinct species of *Loxodonta* that is on or close to the evolutionary lineage of the extant African elephant. Chronostratigraphic refinements and an expanded fossil record indicate that *L. exoptata* is an early–late Pliocene collateral rather than descendant species of *L. adaurora*, with antecedents more likely in earliest Pliocene *L. cookei*.

From the absence at Laetoli of *Elephas recki*, presumed to be an habitué of dry wooded savannas, Beden (1987b) inferred that *Loxodonta exoptata* lived in humid/wet wooded savannas. While contrasting paleoecological reconstructions of Pliocene Laetoli varying from semiarid bushland (Kovarovic et al., 2002) to habitats "dominated by closed woodlands with a substratum of C_4 grasses or open woodland interspersed with grassy patches" (Kingston and Harrison, 2007:299) have not helped to refine this hypothesis, isotopic studies showing that *all* early elephant species were grazers or mixed feeders with a strong dietary preference for C_4 plants (see earlier discussion) contradict the idea that *L. exoptata* was singularly adapted for feeding in such a specialized ecological niche (Kingston and Harrison, 2007).

LOXODONTA ATLANTICA (Pomel, 1879)
Figure 15.17R

Partial Synonymy Elephas atlanticus, Pomel, 1879; *E. (Loxodon) zulu*, Scott, 1907; *Loxodonta zulu*, Osborn, 1942; *E. pomeli* (in part), Arambourg, 1947, 1952:413, figures 7 and 8, plate 1, figure 4; *Loxodonta (Palaeoloxodon) antiquus recki*, Singer and Crawford, 1958; *Loxodonta atlantica*, Cooke, 1960.

Age and Occurrence ?late Pliocene–early Pleistocene eastern and Central Africa; ?early–middle Pleistocene, northern and southern Africa (table 15.6).

Diagnosis Based on Maglio (1973); Coppens et al. (1978). Largest and most advanced loxodont elephants, subdivided into at least two geographically separate subspecies. Enamel wear figures do not include median sinuses as prominent as those in modern African elephants, and enamel may be finely folded or crimped (Scott, 1907; Osborn, 1942). *Loxodonta atlantica atlantica*: median molar plate loops often bifurcated. Irregular, small enamel loops may occur around worn lozenge-shaped molar plate wear figures. Distinguished from other loxodont elephants by greater number of plates, higher crowns, and plicated enamel. *Loxodonta atlantica zulu*: differentiated from *L. a. atlantica* by thinner enamel.

Description The species is best known from northern Africa. The skull of *L. a. atlantica* is morphologically similar to that of *L. africana*, except for exceptionally large occipital condyles and a narrower premaxillary region (Coppens et al., 1978). Molars of this subspecies are among the most derived of all loxodont elephants, with hypsodonty indices exceeding 200, coarsely folded, moderately thick enamel (2.2–3.6 mm in third molars), and more plates per molar than in other species of *Loxodonta*. Plate formulae: m1 = 7–8; m2 = 11–12; m3 = 10–15; M1 = 8–9; M2 = 9–11; M3 = 14 (Maglio, 1973).

Molars of *L. a. zulu*, from southern Africa, have even thinner enamel (2.0–2.8 in third molars) and are also high crowned (range of HI > 200). Plate spacing is similar to that of the northern subspecies (third molar LF = 3.4–5.4). Enamel is folded, and a right m3 described by Scott (1907) has bifurcated accessory conules that are much more prominent posteriorly, as is the case in molars from Elandsfontein, South Africa (Maglio, 1973). Plates may have trifoliate lateral tips and are anteriorly concave and posteriorly convex (unlike the condition in *L. a. atlantica*, in which m3 plates occlusally exhibit narrow median sinuses and are slightly anteriorly and posteriorly convex). The occlusal morphology of *L. a. zulu* recalls that of *Elephas recki*, particularly in the greater expression of the posterior median plate projections. Plate formulas: m1 = 10; m3 = x11x–13x; M2 = 12; M3 = 12 (Scott, 1907; Osborn, 1942; Maglio, 1973).

Remarks Coppens (1965) named a third subspecies of *L. atlantica*, *L. a. angammensis*, based on craniodental material from the early Pleistocene of Chad, to which Beden (1987a) later added molar specimens from the late Pliocene of the Omo, Ethiopia (see figure 15.17R). Morphometrically, the cranial and dental remains of this subspecies are quite similar to those of *L. africana*, and except for some gross enamel folding, exhibit no special similarity to those of *L. atlantica atlantica* or *L. a. zulu*. Plate formulae are less advanced than those of *L. a. atlantica* and *L. a. zulu*: dp3 = 5x; dp4 = x8; DP3 = 5x; M2 = x8x–9; M3 = x12x. Third molar LF = 3.5–5.0 and HI = 154; enamel thickness ranges from 2.3 to 3.1 mm (Coppens, 1965; Beden, 1987a). It is debatable whether this fossil material belongs in *L. atlantica* (Maglio, 1973; Coppens et al., 1978).

Too derived to be considered a lineal precursor of *Loxodonta africana* (Maglio, 1973), *L. atlantica atlantica/zulu* appear to have been grazing specialists whose demise paralleled that of *Elephas recki* at the end of the middle Pleistocene, attributed at least in part to global environmental change (see Klein, 1988). The relationship between the geographically discontinuous subspecies *L. a. atlantica* and *L. a. zulu* requires further exploration.

LOXODONTA AFRICANA (Blumenbach, 1797)
Figures 15.17S and 15.17T

Partial Synonymy Elephas africana, Blumenbach, 1797; *E. capensis*, Cuvier, 1798; *Loxodonta africana*, Gray, 1843; *E. (Loxodonta) oxyotis*, Matschie, 1900; *E. africanus capensis*, Lydekker, 1907; *E. a. cyclotis*, Lydekker, 1907; *E. a. oxyotis*, Lydekker, 1907; *E. a. knochenhaueri*, Lydekker, 1907; *Loxodonta africana africana*, Heller and Roosevelt, 1914; *L. prima*, Dart, 1929; *L. africana* var. *obliqua*, Dart, 1929; *L. africana pharaohensis*, Deraniyagala, 1948; ?*L. atlantica angammensis*, Coppens, 1965; ?*L. atlantica angammensis*, Beden, 1987a.

Age and Occurrence Early Pleistocene–present, widespread throughout the central, eastern, and southern regions of sub-Saharan Africa (table 15.6).

Diagnosis Based in part on Maglio (1973). Distinguished from *L. adaurora* by cranium with more reduced premaxillaries and tusks, by higher, narrower molar crowns, and by substantial development of anteroposterior median expansions of molar plate wear figures ("loxodont sinuses"). Greater number of molar plates than in *L. adaurora* and *L. cookei*; intermediate molars with more plates than in *L. exoptata*.

Description Based in part on Osborn (1942); Sikes (1971); Maglio (1973). Cranium is rounded in lateral profile and highly pneumatized, with comparatively widely spaced, distally divergent premaxillae, a biconvex frontoparietal surface, no median parietal depression, and with a nearly vertical occipital region (figures 15.17S, 15.17T). Mandible with a long, relatively slender horizontal corpus and forward-leaning ramus. Condyles are convex and usually wider transversely than anteroposteriorly; masseteric fossae are capacious. *Contra* Maglio (1973), the mandibular canal terminates in an antechamber lateral to the symphysis (see Sikes, 1971: figure 43).

The most distinctive feature of *L. africana* is the formation with wear of enamel figures with strong anteroposteriorly projecting median sinuses (anteriorly ∧ shaped and posteriorly ∨ shaped), throughout the occlusal surface of molars. Plates are moderately well spaced (third molar LF = 4.0–5.0). Molar enamel is without plications and has moderately thick borders. HI > 100, but not as extremely hypsodont as in more derived species of *Elephas* and *Mammuthus*, and lower than in *L. atlantica*.

Molars in each quadrant may be numbered M1–M6 or MI-MVI (e.g., Laws, 1966; Sikes, 1967) but are actually DP2–DP4, M1–M3, with no permanent premolars (Roth, 1992); upper tusks have a small deciduous precursor, and curve gently upward and slightly inward, without longitudinal torque.

Plate formulae: dp2 = 3–5; dp3 = 6–8; dp4 = 8–10; m1 = 7–10; m2 = 9–12; m3 = 10–14; DP2 = x3x–5; DP3 = 6–7; DP4 = 8–10; M1 = 8–9; M2 = 11–12; M3 = 11–13. This is very conservative compared with plate formulae of Recent species of *Elephas* and *Mammuthus*.

The postcranium is that of an immense animal that must bear the heft of considerable daily forage (up to 5%–6% of body weight, along with as much as 50 gallons of water; Sikes, 1971). The neck is short; there are 20–21 strong, compact thoracic vertebrae and only 3–4 short lumbars with massive centra, linking the thorax closely to the pelvis; the scapular glenoid and acetabulum of the innominate face downward; the pelvic ilia flare broadly and are united by a sturdy sacrum of four elements; the feet are pseudoplantigrade with thick pads; the legs are thick and pillarlike; and the long anterior thoracic spinous processes and high pelvic girdle anchor a dorsal vertebral suspensory bridge ("swayback") for support of the substantial viscera (Sikes, 1971).

LOXODONTA CYCLOTIS (Matschie, 1900)

Partial Synonymy *E. cyclotis*, Matschie, 1900; *E. africanus pumilio*, Noack, 1906; *E. a. albertensis*, Lydekker, 1907; *L. cyclotis*, Morrison-Scott, 1947; *L. cyclotis*, Grubb et al., 2000; *L. cyclotis*, Roca et al., 2001.

Age and Occurrence Recent, western half of sub-Saharan Africa up to lake region of the Western Rift Valley.

Diagnosis In comparison with the African bush elephant *L. africana*, the African forest elephant *L. cyclotis* varies to appreciably smaller body size and has paler, more finely wrinkled skin, relatively more slender, straighter tusks, smaller, more rounded ears, a greater number of toenails, a more flattened, relatively broader cranium with more widely separated temporal ridges, shorter, less divergent tusk alveoli, and less arched zygomatics, and a mandible with lower rami and a more prominent, narrow spoutlike symphysis (Lydekker, 1907; Allen,

1936; Morrison-Scott, 1947; Grubb et al., 2000; but see Backhaus, 1958).

Description Except for the differences listed in the diagnosis, African forest elephants are similar morphologically to African bush elephants.

Remarks African elephants are the largest living terrestrial animals, exceeding 6,000 kg and 4 m in shoulder height in some individuals (Laursen and Bekoff, 1978). Once inhabitants of all Africa except the most arid areas of the Sahara, competition with humans and poaching for ivory has severely reduced the distribution of these last proboscideans on the continent, and they are now endangered (Kingdon, 1997). The geographic range of *L. cyclotis* extends from western Africa (Sierra Leone) across the Congo forest belt to the lake region of the Western Rift Valley, and *L. africana* is unevenly dispersed throughout sub-Saharan areas of Central, East, and southern Africa (Allen, 1936; Kingdon, 1997).

Although these species have been considered subspecies with hybrid zones, and study of mitochondrial genes suggested gene flow between forest and savanna populations and incomplete speciation (Eggert et al., 2002; Debruyne, 2005), sequence analysis of nuclear genes in African elephants revealed that their mitochondrial and nuclear genomes have very different evolutionary histories and that there is deep, species-level genetic disparity partitioning the forest and bush forms, interpreted as a consequence of strong reproductive isolation between *L. africana* and *L. cyclotis*; this supports their separation at the species level (Roca et al., 2001, 2005; Comstock et al., 2002). Dissociation between mitochondrial and nuclear gene patterns in African elephants is evidence of ancient episodes of hybridization between forest females and savanna males, and a long history of backcrossing of female hybrids to savanna males has obliterated the forest nuclear genome in savanna elephant populations (Roca et al., 2005).

Modern African elephants are mixed feeders whose diet is generally dominated by C$_3$ browse (Cerling et al., 1999); however, this may vary considerably and include heavy reliance on grazing, reflecting local and seasonal availability of foods (Kingdon, 1979; Buss, 1990; Tchamba and Seme, 1993; White et al., 1993; Koch et al., 1995; Cerling et al., 2006; Codron et al., 2006). Forest elephants may consume a greater amount of fruit than their bush cousins, but there is no evidence that they consistently rely on a higher percentage of browse across their range (see Tchamba and Seme, 1993; White et al., 1993). The suggestion that modern elephants may be filling a niche vacated by other browsing ungulates during the Pleistocene (Kingston and Harrison, 2007) is intriguing.

The antiquity of the species is uncertain. Craniodental specimens from the Omo, Ethiopia and Angamma-Yayo, Chad may extend the temporal span of *L. africana* beyond middle Pleistocene East and Central African occurrences (table 15.6) to as early as the late Pliocene–early Pleistocene, or could belong to an ancestral segment of the African elephant lineage, such as *L. cookei* (table 15.6). Originally placed in *L. atlantica angammensis* (Coppens, 1965; Beden, 1987a), these specimens differ from *L. africana* only in subtle features and are not especially like typical middle Pleistocene examples of *L. atlantica* (Maglio, 1973; Coppens et al., 1978). There is no fossil record of forest elephants.

Genus *MAMMUTHUS* Brookes, 1828

Mammuthus encompasses the mammoths, popularly exemplified by the woolly mammoth, *M. primigenius* (Lister and

Bahn, 2007). Following an early to mid-Pliocene African origin, mammoths migrated to Eurasia during the mid-late Pliocene and reached North America by the early Pleistocene (Dudley, 1996; Fisher, 1996; Lister, 1996; Lister and van Essen, 2003), where they thrived until the end of the epoch. In Eurasia and North America, mammoths underwent multiple episodes of speciation and profound evolutionary changes in craniodental anatomy (Lister, 2001; Lister et al., 2005; Lister and Bahn, 2007). Many of these changes paralleled progressive modifications of the skull and molars in *Elephas*, particularly in features correlated with increased hypsodonty and shearing efficiency during mastication, presumably in association with availability of grasslands and adaptation to grazing (Maglio, 1973; Dudley, 1996; Lister and Sher, 2001; but see Koch, 1991; Fisher, 1996). The later species from the Northern Hemisphere were the most derived elephants of all, with more than 25 plates crowded together in third molars whose crown heights exceed twice their width. In addition, these elephants had very raised, extremely foreshortened crania festooned with prodigious, spirally twisted and curved tusks (Maglio, 1973).

Mammoths did not fare as well in Africa, apparently going extinct there before the end of the early Pleistocene (Maglio, 1973). *Mammuthus* skulls are characterized by dorsally expanded parietals and vertically concave, transversely convex frontoparietal surfaces (lacking a midsagittal depression), strong temporal constriction, widely separated orbits, and proximally closely spaced premaxillary sheaths that curve outward distally (Maglio, 1973). Most of these features are also present in crania of early species, which has proven invaluable for positive identification of the genus in Africa.

MAMMUTHUS SUBPLANIFRONS (Osborn, 1928)
Figures 15.17 U and 15.17V

Partial Synonymy Archidiskodon subplanifrons, Osborn, 1928; *A. andrewsi*, Dart, 1929; *A. proplanifrons*, Osborn, 1934; *Mammuthus (Archidiskodon) scotti*, Meiring, 1955; *Stegolophodon* sp., Singer and Hooijer, 1958; *Mammuthus (Archidiskodon) subplanifrons*, Cooke and Coryndon, 1970; *Mammuthus subplanifrons*, Maglio and Hendey, 1970; *M. subplanifrons*, Maglio, 1973; *Primelephas gomphotheroides* (in part), Tassy, 1986.

Age and Occurrence Latest Miocene, early Pliocene, eastern and southern Africa (table 15.6).

Diagnosis Large elephant with broad, brachyodont to mesodont molars, low number of thick plates (leading to low LF = 3.2–3.75), moderately thick (ET = 4.0–5.8 mm), unfolded enamel, and accessory conules limited to the posterior side of plates in the mesial half of the crown. There are no loxodont sinuses formed by the enamel loops with wear.

Description Based in part on Maglio and Hendey (1970); Sanders (2006, 2007). The cranium is unknown for this species. The corpus of a mandible (SAM-PQ-L 12723) from Langebaanweg, South Africa, is relatively long and not heavily constructed. The ramus is high and has a more restricted masseteric fossa than in loxodont elephants. Anterior chambers of the mandibular canal open externally via mandibular foramina; these are lateral to where incisive alveoli would be if lower tusks existed, though there are no signs of these. The anterior chambers never contain incisive tooth buds in proboscideans that do have i2s, so there is no reason to speculate that they once contained vestigial tusk buds in *M. subplanifrons* (*contra* Maglio and Hendey, 1970).

A large tusk from Virginia, South Africa, associated with a molar comparable morphometrically with others assigned to *M. subplanifrons*, displays the spiral twisting typical of the genus (Meiring, 1955; Maglio, 1973). An enigmatic isolated upper tusk from Langebaanweg, South Africa (SAM-PQ-L 40430), with a length of >830 mm and cross sectional height of 103 mm near its midpoint, although flattened throughout, also retains a distinct longitudinal torque and curves upward at its distal tip. Longitudinal torque is typical of mammoths but not of the other proboscidean taxa found at Langebaanweg (*Anancus* and *Loxodonta*); however, the tusk is unusual for an elephant in that it also has lateral sulci that run its length. In the absence of crania, tentative reliance must be placed on these tusks to justify allocating this species to *Mammuthus*.

The m3s of the Langebaanweg mandibular specimen are low crowned (HI = 67–69) with nine robustly built plates composed of three to five conelets (figures 15.17U, 15.17V). The anterior and posterior cingulids are prominent. Posterior accessory central conules are limited to the first three plates, and plates are transversely straight with a dominant central conelet. The plates are less pyramidal in cross section than in *Primelephas,* and the transverse valleys are sub–U shaped. Cementum coats the plates but does not fill the transverse valleys. Several specimens from the Middle Awash, Ethiopia (table 15.6) are morphologically similar, but with one less plate in m3, and greater crown height (m3 HI = 76; M3 HI = 89).

An m3 from the Lukeino Fm., Tugen Hills, Kenya that was originally assigned to *P. gomphotheroides* (KNM-LU 7597A; Tassy, 1986) has nine plates and a distinctive dominant central conelet in each plate, typical of early mammoths. Morphometrically, though well worn, this molar resembles specimens in the Langebaanweg + Middle Awash sample of *M. subplanifrons*. Dated between 6.2 and 5.6 Ma (table 15.6), the Lukeino molar may be the oldest known mammoth fossil.

Plate formulae: m3 = x8x–x9x; M2 = x6x; M3 = 8x–9 (Maglio and Hendey, 1970; Kalb and Mebrate, 1993; Haile-Selassie, 2001).

Remarks As traditionally composed, this species is quite heterogeneous morphologically (Maglio, 1973), and likely is a wastebasket taxon. The holotype, a partial m3 from the Vaal River, South Africa (MMK 3920, "*Archidiskodon subplanifrons*"; Osborn, 1928), and molar specimens of other, synonymized taxa from the Vaal River ("*Archidiskodon proplanifrons*," "*A. andrewsi*"; Dart, 1929; Osborn, 1934; Maglio, 1973) differ in important occlusal details from the Middle Awash + Langebaanweg sample and more closely resemble primitive *Loxodonta* (Sanders, 2006, 2007). At the same time, the Middle Awash + Langebaanweg sample cannot be fit into any other existing proboscidean taxon, and anatomically anticipates at least part of the younger *M. africanavus* hypodigm. For these reasons, the species *M. subplanifrons* is maintained. In the absence of associated crania, however, there is no certainty that this species is a mammoth.

MAMMUTHUS AFRICANAVUS (Arambourg, 1952)
Figures 15.17U–15.17X

Partial Synonymy Elephas meridionalis, Pomel, 1895; *E. planifrons*, Deperet and Mayet, 1923; *E. africanavus*, Arambourg, 1952, 1970; *Loxodonta africana*, Cooke, 1960; *L. africana*, Coppens, 1965; *Mammuthus africanavus*, Maglio, 1973.

Age and Occurrence Mid- to late Pliocene, northern and Central Africa (table 15.6).

Diagnosis Based in part on Arambourg (1970); Maglio (1973). Primitive species of *Mammuthus* with a low number of third molar plates, occasionally undulating but unfolded, moderately thick enamel, modest plate spacing, and retention in anterior half of molar crowns of accessory central conules. Sides of molars taper strongly toward the apex of the crown (figure 15.17W). Cranium and I2s typical for the genus.

Description Based in part on Arambourg (1970); Maglio (1973); Coppens et al. (1978). The upper tusks are massive in cross section (Garet et Tir specimen, W = 136 mm; H = 140 mm), long (L = +2,310 mm), and recurved upward and inward distally (Arambourg, 1970). An associated cranium is reportedly morphologically similar to that of *M. meridionalis* (Maglio, 1973).

Isolated molars are difficult to distinguish from those of archaic *Elephas*. Accessory conules are retained in the anterior portion of the crown and may be particularly prominent posterior to plates. These do not contribute to loxodont sinuses, however. Plates are formed of five to seven conelets, are parallel to one another in lateral view, and are separated by U-shaped transverse valleys that are abundantly filled with cementum (figure 15.17X). In third molars, lamellar frequency varies from 3.0 to 5.2; enamel thickness ranges from 2.6 to 4.3 mm; and crowns are modestly high, reaching hypsodonty indices of 120 (Maglio, 1973). Plate formulae: dp3 = x6x; dp4 = 6–x7x; m1 = 7–x8x; m2 = 8–9; m3 = 10–13; DP2 = x5; DP3 = 5–6; DP4 = 6; M1 = 6–7; M2 = 8–9; M3 = 9.

Remarks The earliest unambiguous evidence of the genus *Mammuthus* in Africa, the age and morphology of this species are close to those for the oldest European mammoths (*M. rumanus*, dated to ca. 3.5–2.5 Ma; Lister and van Essen, 2003; Lister et al., 2005). If there is a connection between this species and its putative precursor *M. subplanifrons*, it might be evidenced by the similarity of specimens such as m3 1950-1:12 from Lac Ichkeul, Tunisia to m3s from early Pliocene Langebaanweg, South Africa, in restriction of accessory conules to the posterior of the first few plates, and relatively rectilinear, simple plates, though with one more plate (10) and higher crowned (figures 15.17U–15.17X). In addition, mammoth dental specimens from Hadar, Ethiopia, of mid-Pliocene age (White et al., 1984) are reportedly morphometrically intermediate between *M. subplanifrons* and *M. africanavus* (Beden, 1985).

MAMMUTHUS MERIDIONALIS (Nesti, 1825)
Figures 15.17Y and 15.17Z

Partial Synonymy Elephas meridionalis, Nesti, 1825; *E. planifrons*, Doumergue, 1928; *Elephas* aff. *meridionalis*, Arambourg, 1952; *E. moghrebiensis*, Arambourg, 1970; *Mammuthus meridionalis*, Maglio, 1973; "*E. moghrebiensis*" = *E. recki ileretensis*, Geraads and Metz-Muller, 1999.

Age and Occurrence Early Pleistocene, northern Africa (table 15.6).

Diagnosis Based on Maglio (1973). Species with characteristic mammoth cranium showing dorsally expanded occipital and parietals, a strongly anteriorly concave frontoparietal surface that is flat to convex transversely, without parietal crests (figures 15.17Y, 15.17Z). Molars moderately derived for the genus, with hypsodont crowns lacking significant development of accessory conules, and slightly more plates than in *M. africanavus*.

Description Molars of this species from Africa are morphologically similar to those from Europe, but with greater hypsodonty (third molar HI = 157–176) and more plates (Arambourg, 1970). In the African specimens, lamellar frequency ranges from 4.0 to 5.0 and enamel thickness from 2.0 to 3.5 mm

(Arambourg, 1970). Molars are long but not particularly wide. Greatest width of the crown is located one-third to halfway above the cervix. Enamel may be slightly folded to undulating, and enamel loops are simple and comprised of five to seven conelets. There are no appreciable accessory central conules. Plate formulae (Africa): ?dp4 = ?12; m3 = 16; DP4 = 9; ?M2 = ?15; M3 = 14–16 (Arambourg, 1970); (Europe): dp2 = 3–4; dp3=5-6; dp4 = 8–9; m1 = 9–10; m2 = 8–10; m3 = 10–14; DP2 = 3–4; DP3 = 5–6; DP4 = 7–8; M1 = 8–10; M2 = 9–11; M3 = 12–14 (Maglio, 1973).

Remarks This species is best known from Europe, and only tentatively documented in North Africa. Geraads and Metz-Muller (1999) place the specimens from Aïn Hanech, Algeria, in *Elephas recki ileretensis*, and it is possible that Arambourg's (1970) "*E. moghrebiensis*" may not be synonymous with *M. meridionalis* (G. Markov, pers. comm.). Because the African specimens are more advanced in crown height and plate number, it is possible that they are derived from the European deme of the species, which in turn almost certainly descended from *M. rumanus* (see Lister and van Essen, 2003; Lister et al., 2005). Nonetheless, the progressive quality of molar morphology and younger geological age of African "*M. meridionalis*" in comparison to *M. africanavus* suggest that these species might be useful for future biochronological sequencing of North African sites, especially if considered along with *Loxodonta atlantica atlantica*, which replaced *Mammuthus* in North Africa during the middle Pleistocene.

Genus ELEPHAS Linnaeus, 1758
Figures 15.17AA–15.17EE

Now endangered, in terms of biogeography, longevity, diversity, and impact on faunas, *Elephas* was the most successful Old World elephant taxon. The genus originated in Africa in the early Pliocene and by the late Pliocene migrated out of the continent into more temperate zones (Maglio, 1973; Todd and Roth, 1996). Once out of Africa, these proboscideans diversified quickly across the Near East, Europe, Asia, and South Asia to become the most speciose of the elephant genera (Coppens et al., 1978). Today, the genus is represented only by the Asian elephant, *E. maximus*, which is widely distributed across Asia and South Asia, though in increasingly fragmented areas and declining numbers (~55,000 individuals; Shoshani and Eisenberg, 1982; Sukumar and Santiapillai, 1996; Fleischer et al., 2001; Blake and Hedges, 2004). The species has been sorted into three subspecies (Shoshani and Eisenberg, 1982; Sukumar and Santiapillai, 1996), with genetic variability evidencing two major clades that appear to have experienced extensive gene flow between populations in the past (Fernando et al., 2000; Fleischer et al., 2001; Vidya et al., 2005).

Even primitive species of *Elephas* are readily recognizable from their cranial morphology, which clearly contrasts with that of *Loxodonta*, and to a lesser degree with that of *Mammuthus*: the skull is high and anteroposteriorly compressed; the frontoparietal surface is flat to concave; there are usually distinct parietooccipital bosses; and the upper edges of the temporal fossae are bordered by sharp, prominent ridges (figures 15.17AA, 15.17BB; Maglio, 1973; Coppens et al., 1978). In derived species, the molars are very high crowned, may have a large number of plates, thin, very plicated enamel, thick cementum, and accessory conules are absent or persist only as larger folds in enamel loops (figures 15.17CC, 15.17DD). These features are convergent on molar structure in advanced forms of *Mammuthus*. In more primitive species, the greater expression of accessory conules is shared by a number of elephant genera, and may cause difficulty for identification of isolated specimens.

There is no consensus on how to more finely partition the African *E. ekorensis–E. recki–E. iolensis* lineage (see Maglio, 1973; Beden, 1980; Todd, 2005). Nevertheless, the geographic and temporal extent of this lineage, particularly *E. recki*, its occurrence in radiometrically well-dated sites, and its progressive morphometric changes over time (figure 15.17EE), make this one of the most useful African mammalian taxa for biochronological correlation. More difficult to understand is the precipitous disappearance of this lineage, after nearly three million years of dominating East African faunas.

ELEPHAS EKORENSIS Maglio, 1970

Partial Synonymy Elephas africanavus, Arambourg et al., 1969; primitive *E. recki*, Howell et al., 1969; *Loxodonta adaurora*, Coppens and Howell, 1974:2275; *E. recki,* Coppens and Howell, 1974:2275.

Age and Occurrence Early to mid–Pliocene, eastern Africa (table 15.6).

Diagnosis Based in part on Maglio (1970b); Beden (1987a). Less pronounced expression of typical *"Elephas"* features in the cranium than other congeners, with only modest bossing of the parietooccipital region and a flatter forehead without much anterior expansion of the parietals. Prominent, widely separated tusk sockets. Large external nasal opening. Differs from *Loxodonta* in having a more anteroposteriorly compressed cranium with a flat rather than rounded forehead, and distally less flaring tusk sockets.

Description This is a primitive member of the genus, as evidenced by the low number of molar plates, moderate hypsodonty (HI ranges from 100 to 175), intermediate plate spacing (third molar LF = 3.8–4.8), and moderately thick, unfolded or coarsely undulating enamel (third molar ET = 3.3–4.0 mm; Maglio, 1973). In addition, molar plates retain anterior and larger posterior accessory conules throughout much of the crown. These are apically free though incorporated as mesiodistal median projections in enamel wear figures, but they do not form strong median sinuses as in *Loxodonta*. Only a small number of conelets (four to six) form each plate, which are broadest near the base of the crown. Third molars are broadest anteriorly, taper drastically posteriorly, and are less massive than in sympatric *L. adaurora*. In lateral view, plates are parallel-sided and separated by cementum-filled, U-shaped transverse valleys.

Plate formulae: m1 = 8; m2 = x9; m3 = 12; DP2 = 3x; DP3 = 6x; DP4 = 8x; M1 = 7–7x; M2 = 9; M3 = 11–11x (Coppens et al., 1978; Beden, 1980, 1987a).

Remarks This is the most ancient unequivocal representative of the genus *Elephas*. Although an older putative congener, *Elephas nawataensis*, was named from the Upper Mb. of the Nawata Fm. and Apak Mb. of the Nachukui Fm. at Lothagam, Kenya (Tassy, 2003; see also Tassy and Debruyne, 2001), its holotype is more sensibly synonymized with *Primelephas korotorensis* and the rest of its type series with *Stegotetrabelodon orbus* (see above; Mundinger and Sanders, 2001; Sanders, 2004).

Morphologically, *E. ekorensis* seems a good ancestral model from which to derive *Elephas recki* and the first Eurasian representative of the genus, *E. planifrons* (Maglio, 1970b; Coppens et al., 1978).

ELEPHAS RECKI Dietrich, 1915
Figures 15.17AA–15.17EE

Partial Synonymy See Beden (1983) for a more complete synonymy. *Elephas antiquus recki,* Dietrich, 1915; *E. zulu,* Hopwood, 1926; *E. recki,* Arambourg, 1942; *Palaeoloxodon antiquus recki,* MacInnes, 1942:42, plate 8, figures 4–5; *Palaeoloxodon recki,* Osborn, 1942; *Omoloxodon,* Deraniyagala, 1955; *Elephas Palaeoloxodon recki,* Beden, 1983.

Age and Occurrence Early Pliocene–middle Pleistocene, primarily eastern Africa (rare occurrences in northern, Central, and southern Africa; table 15.6).

Diagnosis Based in part on Maglio (1973); Coppens et al. (1978); Beden (1980). Medium-sized elephant with hypsodont molars that in later forms have finely folded, thin enamel, and a greater number of closely spaced plates than in *Loxodonta*. Unlike *Mammuthus*, tusks are not spirally twisted, and the forehead is demarcated from the temporal fossae by sharp, acute ridges. Frontoparietal surface more vertical than in *E. ekorensis*.

Description Based in part on Maglio (1973); Coppens et al. (1978); Beden (1980). Cranium raised and anteroposteriorly flat or concave, with a strong frontal crest, large external nasal opening, rectangular prenasal region, deep, wide incisive fossa, nearly parallel, and drawn out zygomatic processes of the frontal; massive incisor alveoli that are closely proximate at their openings; and parietooccipital bosses (which are profound in more advanced subspecies) (figures 15.17AA, 15.17BB). Orbits widely spaced but small. Tusks are gently curved upward in a single plane.

Mandible short, massive, very brevirostrine, with a more rocker-shaped ventral corpus than in *Loxodonta*. The ramus is broad, and the condyles are rounded and set on a short condylar neck. There are no lower tusks.

Differences in molar proportions, plate spacing, enamel thickness, hypsodonty indices, enamel folding, and plate number in different stages or subspecies of *E. recki* are enumerated by Maglio (1973) and Beden (1980, 1983, 1987a). Generally, in earlier forms (e.g., *E. r. brumpti*) enamel is thicker (M3/m3 = 2.8–4.0 mm) and unfolded to coarsely folded, hypsodonty is modest (M3/m3 = 101–116), plates are not particularly closely spaced (M3/m3 LF = 4.0–5.5), and accessory central conules may be retained, particularly in the anterior half to two-thirds of the crown, though median sinuses are absent or only weakly developed (figures 15.17CC, 15.17DD). In more advanced subspecies (e.g., *E. r. recki*), accessory conules are completely absorbed into the plate loops, plate spacing is closer (M3/m3 LF = 4.6–6.0), enamel is thinner (M3/m3 = 1.8–3.0 mm) and well plicated, and molar crowns are relatively higher (M3/m3 HI = 161–200)(Beden, 1980).

Plate formulae: *(E. r. brumpti)* dp2 = x3x–x4; dp3 = 6x–x6x; dp4 = x7x–9; m3 = x11x–14; DP2 = 4x; DP3 = 6x; DP4 = x7x; *(E. r. shungurensis)* dp2 = x3–x4; dp3 = x6–7x; dp4 = 9–10x; m1 = 10x; m2 = 10–x10x; m3 = x12x–15; DP2 = x4–5; DP3 = x5x–x7x; DP4 = x9; M1 = 8x; M2 = 11x; M3 = x12x–15; *(E. r. atavus)* dp2 = 4–4x; dp3 = 6x–x8; dp4 = x8x–10; m1 = 10–11; m2 = 10–12x; m3 = 13x–x17; DP2 = 5–x5; DP3 = 6x–7x; DP4 = 9–9x; M1 = 9–x11x; M2 = 9–11x; M3 = 14x–17; *(E. r. ileretensis)* dp3 = 7x–8x; m1 = x10x; m2 = 11–11x; m3 = x14x; M2 = 12; M3 = x15x–x16; *(E. r. recki)* dp2 = 3–x3x; dp3 = 7x–8; m2 = x12; m3 = 14x–18x; M2 = x10; M3 = x13x–19 (Beden, 1980, 1983, 1987a).

Remarks A highly successful species of great longevity, *Elephas recki* was the dominant elephant in East Africa during the late Pliocene--middle Pleistocene (Beden, 1985) and apparently evolved anagenetically for a period of over three million years. Phyletic transformation of the dentition in this species was directional, and involved increases in molar hypsodonty (figure 15.17EE), number of plates, and enamel folding, accompanied by closer spacing of plates, thinner enamel, and complete incorporation of accessory conules into plates (Maglio, 1973; Coppens et al., 1978; Beden, 1980, 1985). Although the species continued to evolve progressively into the late Pleistocene in the form of *E. iolensis* (see later discussion), by the middle Pleistocene *Loxodonta* had reappeared in East Africa and begun to replace *Elephas* there (Beden, 1985).

The morphological continuum across successive generations of *E. recki* resembles that of a ring species rolled out over time, with the end members as distinct from one another as any two living species, but with morphological changes between intervening generations nearly imperceptible. Although overall differences could warrant partitioning the lineage into a number of species (see Todd, 2005), serial phases have been subdivided into time-successive stages (1–4; Maglio, 1970a, 1973; Coppens et al., 1978) or subspecies (*E. r. brumpti, shungurensis, atavus, ileretensis,* and *recki;* Beden, 1980), whose divisions are largely governed by chronostratigraphic unit boundaries, most notably in the Omo Shungura Formation (table 15.6; Beden, 1980, 1987a). It has been suggested that temporal overlap of these subspecies (see table 15.6) invalidates the hypothesis of anagenetic change (e.g., Todd, 2005). These taxonomic subdivisions are artificial, however, and the arbitrary partitioning of specimens may have typologically overemphasized subspecific or stage demarcations and downplayed variability. When the names or stages are ignored, the emergent pattern is one of a continuously, directionally evolving lineage with robust variation and substantial morphometric overlap between successive generations.

The disappearance of this once widespread, abundant elephant is unlikely to have resulted from direct competition with reemergent loxodonts. There is evidence for the decline of the species toward the end of the Acheulean industrial phase, and absence from subsequent Middle Stone Age faunas, though not necessarily because of overhunting by humans. It is alternatively possible that shifts in temperature and rainfall patterns due to changes in the intensity and periodicity of glacials and interglacials may have upset competitive balances among grazers and given other ungulate taxa an edge over these elephants (Klein, 1988). If the *E. recki* lineage is considered to have terminated in *E. iolensis,* then its extinction prior to the late Pleistocene is illusory.

ELEPHAS IOLENSIS Pomel, 1895

Partial Synonymy Archidiskodon sheppardi, Dart, 1927; *A. transvaalensis,* Dart, 1927; *A. broomi,* Osborn, 1928; *A. hanekomi,* Dart, 1929; *A. yorki,* Dart, 1929; *Pilgrimia yorki,* Dart, 1929; *P. wilmani,* Dart, 1929; *P. kuhni,* Dart, 1929; *P. archidiskodontoides,* Haughton, 1932; *P. subantiqua,* Haughton, 1932; *Elephas pomeli* (in part), Arambourg, 1952; *E. iolensis,* Arambourg, 1960.

Age and Occurrence Late Pleistocene, northern, eastern, and southern Africa (table 15.6).

Diagnosis Based in part on Maglio (1973); Coppens and Gaudant (1976); Coppens et al. (1978). Medium- to large-sized species, with more hypsodont molars than *E. recki,* lacking significant development of median loops or sinuses in molar enamel wear figures.

Description Based in part on Dart (1929); Maglio (1973); Coppens and Gaudant (1976); Coppens et al. (1978). The skull is unknown. The molars are more hypsodont than those of other African elephants (HI ranges to nearly 300). Despite this advanced condition, molars have only a modest number of plates (m1 = 8; m2 = 12; M3 = 13–14). Sectioned molars show that anterior and posterior accessory central conules are completely "captured" by the enamel loops, producing little anteroposterior midline expansion of the wear figures (see Coppens and Gaudant, 1976: plate 3). Enamel is only moderately thin (third molar ET = 2.0–3.5), and plates are thick in lateral view, yet they are crowded together, yielding lamellar frequencies of 5.0–6.3. Enamel is irregularly but strongly folded. Plate breadth is greatest about midheight and may exceed 100 mm in third molars. Plates are parallel sided in lateral view and separated by very narrow, U-shaped transverse valleys that are abundantly filled with cementum.

Remarks Elephas iolensis occurred widely across Africa but is not abundant in the fossil record (Maglio, 1973). Nonetheless, it is temporally well constrained between the close of the middle Pleistocene to nearly the end of the epoch (table 15.6), and it constitutes the closing phase of the *E. ekorensis–E. recki* lineage (Maglio, 1973). With its demise, *Loxodonta africana* and *L. cyclotis* were left as the lone proboscidean inhabitants of Africa. Presumably a grazer, *E. iolensis* might have become extinct for causes ecologically linked with the reasons that the modern African elephant survived as a mixed-feeder/browser.

Summary

EVOLUTIONARY PHASES

Due to their robust fossil record, dynamic course of evolution, and ability to traverse great distances, proboscideans are among the most useful of African mammals for correlative dating of fossil sites and refining the chronology and geographic pattern of regional migratory events. There is now considerably more evidence of their phylogeny, from a greater reach of geological history, than there was at the time of the last major review of African proboscideans (in Maglio and Cooke, 1978). Since then, the proboscidean fossil record has been extended back more than 20 million years, to >55 Ma, and has yielded many new taxa (table 15.1). Temporal range distributions of genera indicate that proboscideans underwent at least eight major phylogenetic diversification events (figure 15.19). These may prove useful for subdividing African mammalian faunas into biochronological stages (see Pickford, 1981), but the relationship between these episodes and biotic, physical, and climatic phenomena is still being investigated.

The earliest documented phase of African proboscidean evolution occurred at the end of the Paleocene and produced the oldest known members of the order, including phosphatheres and daouitheres. This phase coincided with the Paleogene thermal maximum, the warmest period of the Cenozoic (Kennett, 1995; Denton, 1999; Feakins and deMenocal, this volume, chap. 4). At this time, Africa and southern Arabia constituted an island continent separated from Eurasia by the Tethys Sea, and its mammalian fauna was strongly endemic (Cooke, 1968; Coryndon and Savage, 1973; Maglio, 1978; Krause and Maas, 1990; Holroyd and Maas, 1994; Gheerbrant, 1998). Regional differences in climate and ecology were far smaller than they are today (Denton, 1999), and forests were probably widespread across the continent. The connection between this climatic event and proboscidean origins is uncertain, as the diversity of these archaic taxa suggests an even older

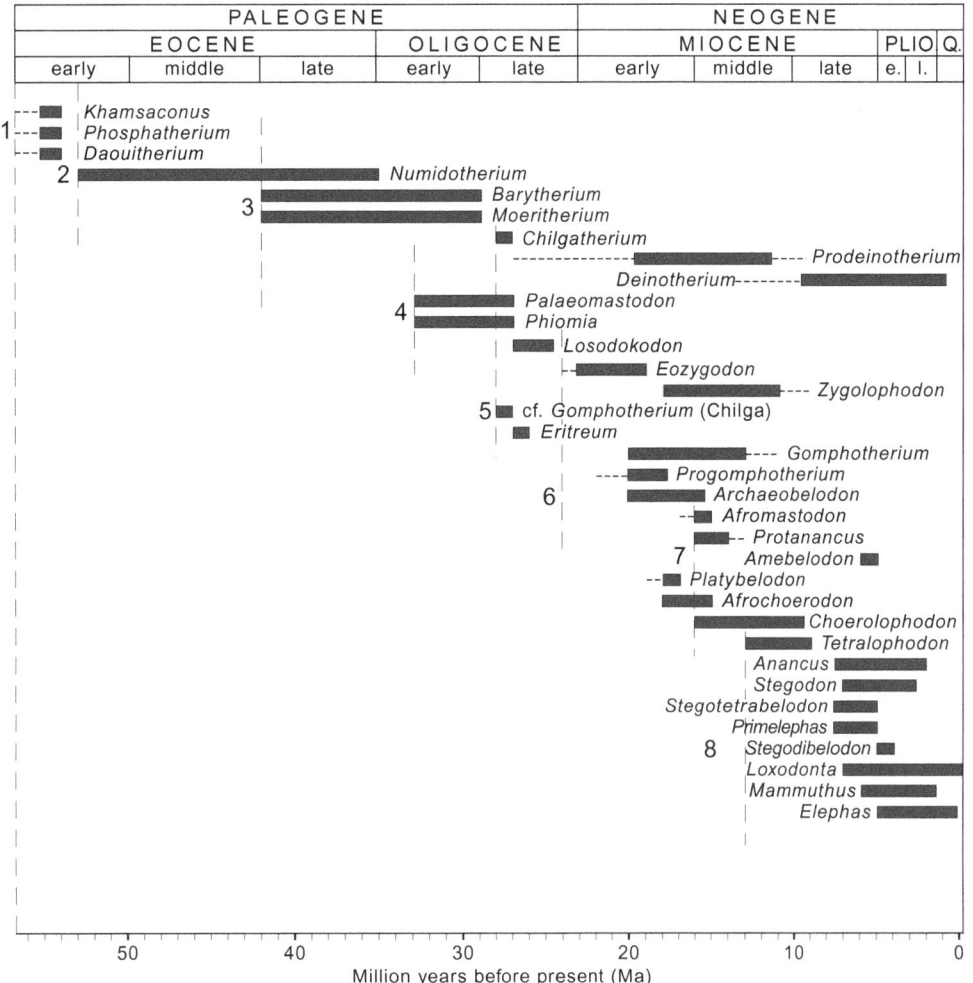

PALEOGENE					NEOGENE				
EOCENE			OLIGOCENE		MIOCENE			PLIO	Q.
early	middle	late	early	late	early	middle	late	e.	l.

FIGURE 15.19   Temporal distribution of African proboscidean genera, based on tables 15.2–15.6. Bars represent known chronological ranges of taxa; horizontal dotted lines represent uncertain dates of occurrence. Numbers 1–8 indicate major proboscidean evolutionary episodes.

divergence from other paenungulates. Subsequent to their first appearance, proboscideans remained indigenous to Afro-Arabia until the end of the Paleogene (Antoine et al., 2003). It is likely that the predominantly North African distribution of proboscideans from nearshore and shallow water environments during the Eocene is an artefact of poor preservation of Paleogene sub-Saharan sites (Sanders et al., 2004).

By the end of the early Eocene (ca. 50 Ma), phosphatheres and daouitheres had been supplanted by numidotheres, of greater body size and more derived skeletal anatomy. The timing of this replacement coincided with the start of a cooling trend, but climate was still quite equable (Denton, 1999; Feakins and deMenocal, this volume, chap. 4), and reasons for the succession are obscure.

In the latter half of the Eocene (ca. 40–37 Ma), a greater diversity of proboscidean taxa came to coexist, including more advanced numidotheres, barytheres, and moeritheres. These taxa were considerably larger in body size, with more specialized, outsized anterior dentitions than their predecessors, and with adaptations for a semiaquatic existence. Habitat specializations likely helped to ensure their survival into the Oligocene.

Barytheres and moeritheres were joined at the start of the Oligocene by palaeomastodonts, the first of the elephant-like proboscideans, possessing trunks, projecting tusks, and terrestrial graviportal postcranial adaptations. This proboscid-

ean assemblage, along with taxa such as creodonts, early anthropoids, saghatheriid hyraxes, anthracotheriid artiodactyls, and arsinoitheres, comprised the typical mammalian "Fayumian" fauna of the African Oligocene (Simons, 1968; Gagnon, 1997). Although in the early Oligocene the collision of the Indian plate with Asia closed off the eastern Tethys Sea, Africa remained separated from Eurasia by the western Tethys and Paratethys Seas (Rögl, 1998), and its fauna continued to be isolated. Global temperatures also declined precipitously (Denton, 1999), but palaeomastodonts were shielded in low latitudes from the extreme effects of global cooling, and they existed in warm, well-watered forested and woodland conditions (Wight, 1980; Bown et al., 1982; Bown and Kraus, 1988; Jacobs et al., 2005).

Distinct global warming trends occurred during the late Oligocene and early Miocene (Miller et al., 1987; Kennett, 1995; Denton, 1999), the former accompanied by the first appearance of elephantoids and chilgatheriine deinotheres, in the Horn of Africa. Palaeomastodonts appear to have been unaffected by these changes, but by the early part of the Miocene and the second warming episode, they and many of the other Fayumian mammals had vanished and the most significant phase of African proboscidean evolution had begun, with the diversification of elephantoids into mammutids, gomphotheriines, amebelodonts, and choerolophodonts, and replacement of chilgatheriines by deinotheriines. Distribution of early Miocene proboscidean sites is more extensive throughout Africa than are Paleogene occurrences (tables 15.1–15.5; Pickford, 2003) and shows that some species had epicontinental ranges.

During this time, Africa and the Arabian plate rotated northward to contact the Anatolian plate, establishing a land bridge between Afro-Arabia and Eurasia (Rögl, 1999). Gomphotheriines, deinotheres, amebelodontines, and mammutids exploited the new intercontinental connection to immigrate to Eurasia in the 20–18 Ma interval (Tassy, 1989), and made their way even earlier to South Asia (Bernor et al., 1987; Antoine et al., 2003). At the same time, a host of Eurasian mammals (rhinos, fissiped carnivores, suids, insectivores, chalicotheres, rodents) invaded Afro-Arabia (Andrews and Van Couvering, 1975; Bernor et al., 1987; Agustí and Antón, 2002; Guerin and Pickford, 2003). Competition with new herbivorous ungulate taxa may have played an important role in the early Miocene morphological specialization and phyletic diversification of elephantoids, and in the demise of the palaeomastodonts (Sanders et al., 2004).

In the early middle Miocene, ca. 16.5–16.0 Ma, at the climax of Neogene warming (Kennett, 1995; Denton, 1999), archaic amebelodonts and choerolophodonts were replaced by more advanced subfamilials, perhaps catalyzed by the immigration into Africa of a second wave of Eurasian mammals (horned bovids, antlered giraffoids, and listriodont suids; Pickford, 1981). There is evidence of continued, progressive evolutionary change, particularly by these taxa, throughout phases of subsequent middle Miocene global cooling in the interval of 15.6–12.5 Ma (Kennett, 1995; Denton, 1999).

The most recent major proboscidean evolutionary episode in Africa occurred in the late Miocene. This involved the local extinction of most gomphotheres and mammutids, immigration into the continent of stegodonts, anancine gomphotheres, and tetralophodonts, perhaps made easier by the beginning of the Messinian Crisis, or closing off of the Mediterranean Sea, which enhanced land connections between Africa and Eurasia via the Gibralter Strait and the Gulf of Aden (Rögl, 1999), and the origin of elephants. Around this time, strong uplift of rift shoulders in eastern Africa began to affect local climate, enhancing seasonal temperature variability, and producing more arid conditions through rainshadow effects (Partridge et al., 1995a). Simultaneously, uplift of the Tibetan Plateau changed wind patterns and also contributed to drier conditions, and global decrease in the worldwide CO_2 content of the atmosphere favored the spread of C_4 plants, including grasses (Cerling et al., 1993; Partridge et al., 1995b). Elephants were among the first African mammals to exploit these new circumstances by evolving craniodental adaptations specialized for grazing. Increased fragmentation and heterogeneity of ecosystems due to climatic deterioration and greater geomorphological relief created conditions favorable to speciation (Partridge et al., 1995a) and may be linked with the initial radiation of archaic elephants ca. 7.0–5.0 Ma. Elephants underwent a series of subsequent diversifications as they continued to refine these adaptations against the pressures of increased competition for C_4 resources (see Cerling et al., 2003). The wide distribution of fossil elephants, and rapid pace of progressive alterations of their craniodental grazing adaptations have proven especially useful for biochronological correlation of sites from the late Miocene to the present, and for the study of evolutionary processes (Cooke and Maglio, 1972; Maglio, 1973).

PHYLOGENY

Most major phylogenetic events in proboscidean evolution occurred in Africa, including the first appearance of the order and the origin of most subsequent major taxa (barytherioids, moeritheres, deinotheres, palaeomastodonts, mammutids, gomphotheres, and elephants). Phylogenetic analyses have now established that moeritheres, barytheres, and deinotheres belong in the Proboscidea (Tassy, 1979b, 1981, 1982, 1985, 1996c; Shoshani et al., 1996; Gheerbrant et al., 2005), linked by a small series of unremarkable traits such as anteroposterior flattening of the femur, loss of the first lower premolar, and hypertrophy of second incisors (Shoshani and Tassy, 1996). Recent discoveries of older, Paleocene and Eocene taxa from North Africa such as *Eritherium, Phosphatherium, Daouitherium,* and *Numidotherium* (Mahboubi et al., 1986; Gheerbrant et al., 1996, 2002, 2005; Court, 1995) have more clearly delineated the primitive condition for the order: relatively small animals lacking graviportal adaptations, with low-slung crania, nearly full dentitions, no trunks or projecting tusks, and bilophodont molars. Addition of these taxa to phylogenetic analyses has reconfigured proboscidean relationships (figure 15.20). As a result, moeritheres, once posited as basal proboscideans (Tassy and Shoshani, 1988; Shoshani et al., 1996; Tassy, 1996c), have been replaced in this position by phosphatheres and barytherioids, and they are now hypothesized to be the sister taxon to Deinotheriidae + Elephantiformes (Gheerbrant et al., 2005). Recovery of more ancient moerithere fossils from Algeria (Delmer et al., 2006) suggests that they are descended from lophodont proboscideans, supporting this hypothesis.

While their relationships are now clearer and better supported (figure 15.20; Shoshani et al., 2001; Shoshani, 1996; Tassy, 1996c; Sanders, 2004; Gheerbrant et al., 2005), problems remain for the interpretation of African proboscidean phylogeny, partly because of strong tendencies for homoplasy within the order, as illustrated by several notable examples. First, the nature of the connection between palaeomastodonts, mammutids, and gomphotheres requires further investigation. Although cladistic analysis has indicated that *Palaeomastodon* is the sister taxon to *Phiomia* + Elephantoidea (Tassy, 1988, 1990; Shoshani, 1996), new fossils of late Oligocene palaeomastodonts and mammutids (Sanders et al., 2004; Gutiérrez and Rasmussen, 2007) suggest instead that *Palaeomastodon* and mammutids have an ancestor-descendant relationship. This would necessitate drastic reclassification of Palaeomastodontidae and Elephantoidea.

A second phylogenetic problem concerns stegodont relationships. *Stegodon* is highly convergent craniodentally on elephants, and has been placed in Elephantidae by some (Arambourg, 1942; Kalb and Mebrate, 1993; Kalb et al., 1996a). Nonetheless, advances in stegodont biogeography and chronostratigraphy show this to be unlikely (Saegusa et al., 2005). Alternatively, stegodonts are often designated as close sister taxa to elephants (e.g., Shoshani, 1996), but new fossil material from Kenya (Tassy, 1995; Tsujikawa, 2005a) indicates a derivation of elephants from *Tetralophodon* and a more immediate relationship of those taxa. Despite their eventual evolution of elephant-like features, it is possible that stegodontids diverged from other elephantoids as long ago as the early Miocene. If so, they would instead be a sister taxon of Gomphotheriidae (figure 15.20).

Third, the relationships of deinotheres remain poorly understood. New fossil finds from Ethiopia (Sanders et al., 2004)

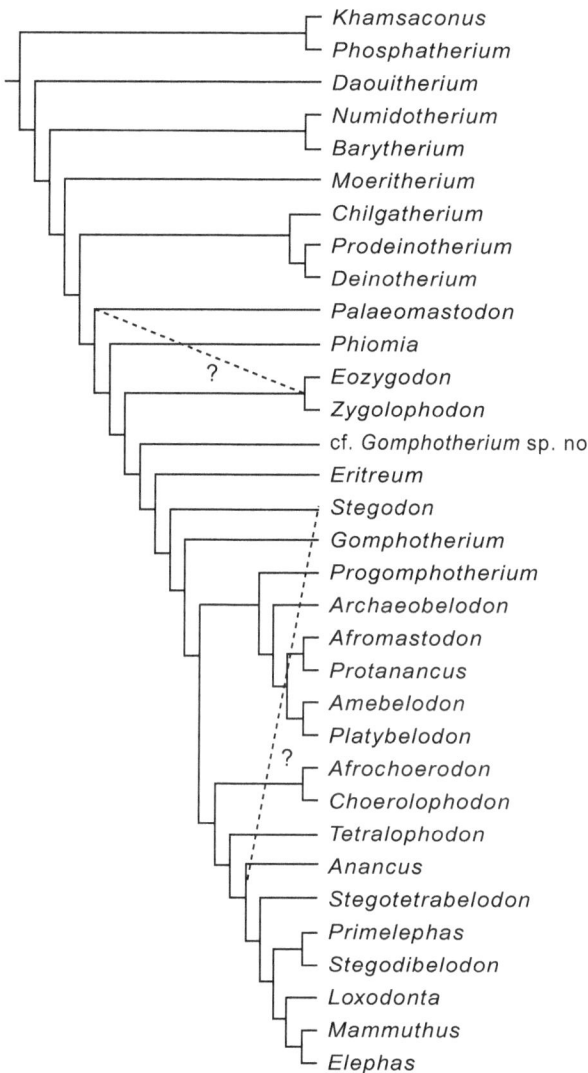

FIGURE 15.20 Cladogram of African proboscidean genera, based in part on Shoshani (1996), Tassy (1996c), Sanders (2004), and Gheerbrant et al. (2005). Dotted lines indicate alternative hypotheses about sister group relationships of taxa.

support an earlier hypothesis of derivation from a moerithere-like ancestor (Harris, 1969, 1975, 1978). Cladistic treatment indicates that while deinotheriines are strongly convergent in cheek tooth morphology with barytheres (Harris, 1978), they are more closely related to Elephantiformes (Gheerbrant et al., 2005). However, recovery of chilgatheriine skulls is critical for more informative testing of these hypotheses.

Finally, despite thorough efforts at description and diagnosis (e.g., Maglio, 1973; Maglio and Ricca, 1977; Kalb and Mebrate, 1993; Tassy, 1995; Sanders, 1997, 2007), elephant relationships remain tangled. The traditional separation of *Loxodonta* from *Elephas* + *Mammuthus* (Coppens et al., 1978; Kalb and Mebrate, 1993; Shoshani, 1996; Tassy, 1996c) has been challenged on morphological and molecular grounds (Hagelberg et al., 1994; Noro et al., 1998; Barriel et al., 1999; Thomas et al., 2000; Thomas and Lister, 2001; Debruyne et al., 2003), and proposed ancestral-descendant relationships between late Miocene–early Pliocene archaic genera and these crown elephantines seems more tenuous now (Sanders, 2004) than they were thought to be 30 years ago (Coppens et al., 1978). Nonetheless, the taxonomic and temporal continuity of crown elephantines within their own lineages is now much better documented, and for *Loxodonta* can be traced back nearly continuously from the present to at least 7 Ma (Sanders, 2007). Genetic studies provide support for the division of modern African elephants into distinct forest and savanna species (Roca et al., 2001, 2005; Comstock et al., 2002).

TRENDS

The pattern of African proboscidean evolution is widely branching, as a result of repeated adaptive radiations, with the greatest variety of taxa in the Miocene (Shoshani and Tassy, 1996). From their inception, proboscideans exhibited surprising diversity, with phosphatheres and more advanced daouitheres found together. In the early and middle Miocene, it was not unusual for multiple species of proboscideans to co-occur at the same localities, as for example at Wadi Moghara, where the fauna includes deinotheres, mammutids, gomphotheriines, amebelodontines, and a species of choerolophodont (Sanders and Miller, 2003). Given the impact of *Loxodonta africana* on modern African ecosystems, it is difficult to imagine the richness of an environment that could support so many mega-herbivores, and yet this seems to have been fairly common in Miocene times. As recently as the late Pliocene, multiple species of elephants, stegodonts, deinotheres, and anancine gomphotheres still shared much of the African landscape, but were finally reduced to a single surviving species, largely because of the ecological ascendance of hominids, and increasing competition with other herbivores in the Pleistocene.

Throughout most of their existence, proboscideans have been the largest or among the largest animals in African terrestrial faunas, with repeated tendencies for gigantism. Although the earliest known members of the order were small, by the latter part of the early Eocene numidotheres had become pig sized, and soon thereafter barytherioids reached elephantine proportions. Most Miocene proboscideans probably weighed several tons and were at least the size of small elephants, with late Miocene deinotheres being the most immense terrestrial mammals to have inhabited the continent (Christiansen, 2004). The long trunk, tusks, and serially replaced dental battery of 2- to 7-ton modern elephants give them the ability to eat a varied and impressive daily amount of forage, thereby maintaining their large mass, and were likely the key adaptations that also maintained the impressive early–middle Miocene radiation of gomphotheres. Common to such huge animals is a suite of gravitportal adaptations such as short, stout feet; pillarlike, elongated long bones (particularly proximal elements) with vertically facing articular surfaces, broad innominates with downward facing acetabulae; and shortened lumbar vertebral regions that bring the thorax in close approximation with the pelvis. These adaptations were present in even the first elephantiforms, by the beginning of the Oligocene.

An important factor in the evolution of horizontal, serial emplacement of cheek teeth was loss of teeth, probably associated with timing of tooth development and their rotation into occlusion, so that most lineages of large-bodied proboscideans exhibited a reduced dental formula in comparison with the first members of the order. The first phases of tooth loss in proboscideans, however, appear to have been linked with rostral elongation and/or specialization of incisors for acquisition of forage, and not with horizontal tooth succession. In these phases, canines and anteriormost premolars were diminished in importance or lost, and some incisors were hypertrophied, while the remaining premolars became more molariform. Starting with a dental formula of I3/2-C1/1-P4/3-M3/3 in *Phosphatherium*, the tooth complement in the sirenian- and hippolike moeritheres and barytheres was modestly reduced to I3/2-C1/0-P3/3-M3/3 and I2/2-C0/0-P3/3-M3/3, respectively.

With the evolution of the first of the elephant-like proboscideans, the palaeomastodonts, the rostrum and symphysis were further elongated, in concert with greater hypertrophy of second incisors, leaving no room or need for additional anterior teeth, and the tooth formula was again reduced, to I1/1-C0/0-P3/2-M3/3. Elephants underwent a secondary reduction of the anterior mandible and lost their lower incisors, as part of a mechanical reorganization of the skull for greater effectiveness of fore-aft mastication, and although more archaic elephant species retained permanent premolars, most crown elephants have a very reduced tooth formula of I1/0-C0/0-P0/0-M3/3, with three deciduous premolars preceding the emergence of molars (Laws, 1966; Sikes, 1967; Roth, 1992). The convergent loss of the lower tusks in choerolophodonts, anancine gomphotheres, and *Stegodon* suggests that they also could have been lost multiple times among elephantines.

The proboscidean tendency for convergent or parallel development of features is best documented in the dentition. The earliest proboscideans, including phosphatheres, daouitheres, and barytherioids, had lophodont cheek teeth with chisellike crests, employed in tapirlike vertical shearing; this masticatory mechanism was evidently separately evolved by deinotheres, and also to some extent by mammutids. Gomphotheres developed a different, rotary grinding and shearing system for chewing that became progressively more effective independently in different lineages through the addition of accessory conules, acquisition of cementum, increase in number of conelets per loph(id), and multiplication of loph(id)s per tooth. Mechanisms to enhance locking precision of molars in occlusion, such as lateral offset of half-loph(id)s, also were developed multiple times by different proboscidean taxa (e.g., choerolophodonts, *Protanancus*, *Anancus*). Crown elephant lineages responded to selective pressure for greater efficiency in grazing by independently evolving molars with more plates, greater lamellar frequency, higher crowns, and thicker cementum, perhaps the most compelling example of parallelism in the development of proboscidean dentitions. The temporally coordinated and progressive, directional pattern of this change across multiple lineages during a time of increasingly widespread open conditions suggests that these elephants evolved via anagenesis.

EPILOGUE OR EPITAPH?

The surviving African elephant are among the most recognizable mammals on the continent, sharing with humans the traits of great intelligence and complex social behavior (Sikes, 1971), and are likely the terminal members of an order that dominated Paleogene and Neogene ecosystems. Maintenance of open woodlands and savannas, and associated herbivore assemblages, is critically dependent on the presence of elephants (Eltringham, 1992), whose absence would likely permanently alter the biotic composition of habitats throughout sub-Saharan Africa. Despite their ecological versatility, however, there is no guarantee that these representatives of one of the most successful orders of African mammals will long survive into the future. Poaching of elephants for ivory has exacted a terrible toll on elephant populations, and the encroachment of human settlements and domestic livestock on their ranges looms as an even greater threat to their survival (Kingdon, 1979; Buss, 1990). Elephants in particular may have helped ensure the initial fortunes of hominids by opening up ecosystems, and it would be tragically ironic if this most African of mammalian orders is brought unnecessarily to extinction, after surviving the rigors of physical, biotic, and climatic upheavals for over 55 million years, by unbounded human fecundity.

ACKNOWLEDGEMENTS

We thank the following individuals and institutions for access to fossil specimens in their care: Jerry Hooker (The Natural History Museum, London), Meave Leakey and Emma Mbua (Kenya National Museums, Nairobi), Philip Gingerich and Gregg Gunnell (University of Michigan Museum of Paleontology, Ann Arbor), Pascal Tassy (Muséum National d'Histoire Naturelle, Paris), Graham and Margaret Avery (Iziko South African Museum, Cape Town), Lyndon Murray and Daniel Brinkman (Peabody Museum, Yale University, New Haven), Mohammed el-Bedawi, Fathi Ibrahim Imbabi, and the late Yusry Attia (Geological Museum, Cairo), Elwyn Simons (Duke University Primate Center), Noel Boaz (International Institute for Human Evolutionary Research, Integrative Centers for Science and Medicine, Ashland, Oregon), Elmar Heizmann (Staatliches Museum für Naturkunde, Stuttgart), Michael Mbago and Amandus Kweka (Tanzanian National Museums, Dar es Salaam), Ezra Musiime (Ugandan Museum, Kampala), Mohammed Arif (Geological Survey of Pakistan, Islamabad), Muluneh Mariam (National Museum of Ethiopia, Addis Ababa), and Bernard Marandat and Jean-Jacques Jaeger (Montpellier II University, Montpellier). We are especially grateful to Bonnie Miljour (University of Michigan Museum of Paleontology) for her expert production of the figures. Illustrations of *Phosphatherium* and *Daouitherium* were aided by the photographic work of D. Serrette and P. Loubry. Their efforts benefited from collaboration with the Cherifian Office of Phosphates (OCP) and Ministry of Energy and Mines of Morocco. We are appreciative of Martin Pickford's contribution of proboscidean specimen images and helpful discussion about them. Financial support for this project was generously provided by several Turner Grants from the Department of Geological Sciences, University of Michigan and through grants to Terry Harrison, John Kappelman, and Laura MacLatchy (to W.J.S.), and by the "Fondation des Treilles," the Sysresource and Synthesys Programs (EU), the Département Histoire de la Terre, USM 203/UMR 5143, Muséum national d'histoire naturelle, Paris, and the "Société des Amis du Muséum" (to C.D.). Finally, we are grateful for many kindnesses and expertise extended by the late Hezy Shoshani, for thorough review of the manuscript by Georgi Markov and Al Roca, and for Lars Werdelin's editorial guidance.

After submission of this chapter, new information on features of the skeleton, dentition, and enamel microstructure of *Numidotherium savagei* was published (Delmer, 2009), detailing distinctions between this species and both *Numidotherium koholense* and *Barytherium* spp. As a result, *N. savagei* has now been placed in a new genus, *Arcanotherium* Delmer, 2009 and assigned a systematic position intermediate between lophodont Eocene proboscideans and bunolophodont moeritheres and elephantiforms.

Literature Cited

Aguirre, E. E. 1969. Evolutionary history of the elephant. *Science* 164:1366–1376.

Aguirre, E. E., and M. T. Alberdi. 1974. *Hipparion* remains from the northern part of the Rift Valley (Kenya). *Proceedings Koninklijke Nederlandse Akademie van Wetenschappen* 77:146–156.

Aguirre, E. E., and P. H. Leakey. 1974. Nakali: Nueva fauna de *Hipparion* del Rift Valley de Kenya. *Estudios Geológicos* 30:219–227.

Agustí, J., and M. Antón. 2002. *Mammoths, Sabertooths, and Hominids: 65 Million Years of Mammalian Evolution in Europe.* Columbia University Press, New York, 313 pp.

Alemseged, Z. 2003. An integrated approach to taphonomy and faunal change in the Shungura Formation (Ethiopia) and its implication for hominid evolution. *Journal of Human Evolution* 44:451–478.

Alemseged, Z., and D. Geraads. 2000. A new middle Pleistocene fauna from the Busidima-Telalak region of the Afar, Ethiopia. *Comptes Rendus de l'Académie des Sciences, Paris: Sciences de la Terre et des Planètes* 331:549–556.

Allard, M. W., R. L. Honeycutt, and M. J. Novacek. 1999. Advances in higher level mammalian relationships. *Cladistics* 15:213–219.

Allen, G. M. 1936. Zoological results of the George Vanderbilt African expedition of 1934: Part II, The forest elephant of Africa. *Proceedings of the Academy of Natural Sciences, Philadelphia* 88:15–44.

Ambrose, S. H., L. J. Hlusko, D. Kyule, A. Deino, and M. Williams. 2003. Lemudong'o: A new 6 Ma paleontological site near Narok, Kenya Rift Valley. *Journal of Human Evolution* 44:737–742.

Andrews, C. W. 1901a. Fossil vertebrates from Egypt. *Zoologist* (4) 5:318–319.

———. 1901b. A new name for an ungulate. *Nature* 64:577.

———. 1901c. Preliminary notes on some recently discovered extinct vertebrates from Egypt. Part I. *Geological Magazine*, Series IV, 8:400–409.

———. 1901d. Über das Vorkommen von Proboscidiern in untertertiären Ablagerungen Aegyptens. *Tagesblatt des V Internationalen Zoologischen-Kongresses*, Berlin 6:4–5.

———. 1902. Über das Vorkommen von Proboscidiern in untertertiären Ablagerungen Aegyptens. In *Verhandlungen des V Internationalen Zoologischen-Kongresses*, Berlin, 528 pp.

———. 1904a. Further notes on the mammals of the Eocene of Egypt. *Geological Magazine*, Series V, 1:109–115.

———. 1904b. Note on the Barypoda, a new order of ungulate mammals. *Geological Magazine*, Decade V, n.s., 1:481–482.

———. 1905. Note on the species of *Palaeomastodon. Geological Magazine*, Series V, 2:512–563.

———. 1906. *A Descriptive Catalogue of the Tertiary Vertebrata of the Fayum, Egypt.* British Museum of Natural History, London, 324 pp.

———. 1908. On the skull, mandible, and milk dentition of *Palaeomastodon*, with some remarks on the tooth change in the Proboscidea in general. *Philosophical Transactions of the Royal Society of London B* 199:393–407.

———. 1911. On a new species of *Dinotherium* from British East Africa. *Proceedings of the Zoological Society of London* 1911:943–945.

Andrews, C. W., and H. J. L. Beadnell. 1902. A preliminary note on some new mammals from the Upper Eocene of Egypt. *Survey Department, Public Works Ministry, Cairo* 1902:1–9.

Andrews, P. 1999. Taphonomy of the Shuwaihat proboscidean, late Miocene, Emirate of Abu Dhabi, United Arab Emirates; pp. 338–353 in P. J. Whybrow and A. Hill (eds.), *Fossil Vertebrates of Arabia.* Yale University Press, New Haven.

Andrews, P., G. E. Meyer, D. R. Pilbeam, J. A. Van Couvering, and J. A. H. Van Couvering. 1981. The Miocene fossil beds of Maboko Island, Kenya: Geology, age, taphonomy, and palaeontology. *Journal of Human Evolution* 10:35–48.

Andrews, P., and J. A. H. Van Couvering. 1975. Palaeoenvironments in the East African Miocene; pp. 62–103 in F. S. Szalay (ed.), *Approaches to Primate Paleobiology: Contributions to Primatology*, vol. 5. Karger, Basel.

Anonymous. 1976. Miocene vertebrates from South West Africa. *South African Journal of Science* 72:355.

Anthony, R., and M. Friant 1940. Remarques sur le *Mastodon borsoni* Hays et les autres Mastodontes Zygolophodontes de l'Europe. *Bulletin du Museum National d'Histoire Naturelle, Paris*, Série II, 12:449–455.

Antoine, P.-O., J.-L. Welcomme, L. Marivaux, I. Baloch, M. Benammi, and P. Tassy. 2003. First record of Paleogene Elephantoidea (Mammalia, Proboscidea) from the Bugti Hills of Pakistan. *Journal of Vertebrate Paleontology* 23:977–980.

Arambourg, C. 1934a. Le *Dinotherium* des Gisements de l'Omo. *Comptes Rendus de la Societé Géologique de France* 1934:86–87.

———. 1934b. Le *Dinotherium* des Gisements de l'Omo (Abyssinie). *Bulletin de la Societé Géologique de France* 4:305–309.

———. 1942. L'*Elephas recki* Dietrich: Sa position systématique et ses affinités. *Bulletin de la Société Géologique de France*, Série 5, 12:73–87.

———. 1945. *Anancus osiris*, un mastodonte nouveau du Pliocène inférieur d'Egypte. *Bulletin de la Société Géologique de France* 15:479–495.

———. 1947. Contribution à l'étude géologique et paléontologique du bassin du lac Rudolf et de la basse vallée de l'Omo. II, Paléontologie. *Editions du Muséum, Paris*, 1947:231–562.

———. 1959. Vertébrés continentaux du Miocène Supérieur de l'Afrique du Nord. *Service de la Carte Géologique de l'Algérie, Paléontologie Mémoire* 4:1–159.

———. 1960. Au sujet de *Elephas iolensis* Pomel. *Bulletin d'Archéologie Marocaine* 3:93–105.

———. 1961. Note préliminaire sur quelques Vertébrés nouveaux du Burdigalien de Libye. *Comptes Rendus Sommaire des Seances de la Société Géologique de France* 1961:107–109.

———. 1970. Les Vertébrés du Pléistocène de l'Afrique du Nord. *Archives du Muséum National d'Histoire Naturelle, Paris*, 7e Série, 10:1–126.

Arambourg, C., and P. F. Burollet. 1962. Restes de Vertébrés oligocènes en Tunisie centrale. *Compte Rendu Sommaire des Séances de la Société Géologique de France* 2:42–43.

Arambourg, C., J. Chavaillon, and Y. Coppens. 1969. Résultats de la nouvelle mission de l'Omo (2e campagne 1968). *Comptes Rendus de l'Académie des Sciences, Paris* 268:759–762.

Arambourg, C., and P. Magnier. 1961. Gisements de Vertébrés dans le bassin tertiaire de Syrte (Libye). *Comptes Rendus Hebdomadaires des Séances de l'Académie des Sciences* 252:1181–1183.

Asher, R. J., and T. Lehmann. 2008. Dental eruption in afrotherian mammals. *BMC Biology* 2008, 6:14; doi:10.1186/1741-7007-6-14.

Asher, R. J., M. J. Novacek, and J. H. Geisler. 2003. Relationships of endemic African mammals and their fossil relatives based on morphological and molecular evidence. *Journal of Mammalian Evolution* 10:131–194.

Assefa, Z., S. Yirga, and K. E. Reed. 2008. The large mammal fauna from the Kibish Formation. *Journal of Human Evolution* 55:501–512.

Bachmann, I. 1875. Beschreibung eines Unterkiefers von *Dinotherium bavaricum* H. v. Meyer. *Abhandlungen der Schweizerischen Paläontologischen Gesellschaft* 2:1–20.

Backhaus, V. D. 1958. Zur Variabilität der äußeren systematischen Merkmale des afrikanischen Elefanten (*Loxodonta* Cuvier, 1825). *Säugetierkundliche Mitteilungen* 6:166–173.

Barbour, E. H. 1930. *Amebelodon sinclairi* sp nov. *Nebraska State Museum Bulletin* 17:155–158.

Barriel, V., E. Thuet, and P. Tassy. 1999. Molecular phylogeny of Elephantidae: Extreme divergence of the extant forest African elephant. *Comptes Rendus de l'Académie des Sciences, Paris* 322:447–454.

Barry, J. C., and L. J. Flynn. 1989. Key biostratigraphic events in the Siwalik sequence; pp. 557–571 in E. H. Lindsay, V. Fahlbusch, and P. Mein (eds.), *European Neogene Mammal Chronology.* Plenum Press, New York.

Beden, M. 1975. A propos des Proboscidiens Plio-Quaternaires des Gisements de l'Omo (Ethiopie); pp. 693–705 in *Colloque International C. N. R. S. no 218* (Paris 4–9 juin 1973): Problèmes Actuels de Paléontologie-Évolution des Vertébrés.

———. 1976. Proboscideans from Omo Group Formations; pp. 193–208 in Y. Coppens, F. C. Howell, G. L. Isaac, and R. F. Leakey (eds.), *Earliest Man and Environments in the Lake Rudolf Basin.* University of Chicago Press, Chicago.

———. 1979. Données récentes sur l'évolution des Proboscidiens pendant le Plio-Pléistocène en Afrique Orientale. *Bulletin de la Société Géologique de France* 21:271–276.

———. 1980. *Elephas recki* Dietrich, 1915 (Proboscidea, Elephantidae): Évolution au cours du Plio-Pléistocène en Afrique orientale. *Géobios* 13:891–901.

———. 1983. Family Elephantidae; pp. 40–129 in J. M. Harris (ed.), *Koobi Fora Research Project*, vol. 2. Clarendon Press, Oxford.

———. 1985. Les Proboscidiens des Grands Gisements à Hominidés Plio-Pléistocènes d'Afrique Orientale; pp. 21–44 in *L'Environnement des Hominidés au Plio-Pléistocène.* Fondation Singer-Polignac, Masson, Paris.

———. 1987a. *Les Faunes Plio-Pléistocènes de la Vallée de l'Omo (Éthiopie). Tome 2. Les Elephantidés (Mammalia, Proboscidea).* Éditions du Centre National de la Recherche Scientifique, Paris, 162 pp.

———. 1987b. Fossil Elephantidae from Laetoli; pp. 259–294 in M. D. Leakey and J. M Harris (eds.), *Laetoli: A Pliocene Site in Northern Tanzania.* Clarendon Press, Oxford.

Behrensmeyer, A. K. 1976. Lothagam Hill, Kanapoi, and Ekora: A general summary of stratigraphy and faunas; pp. 163–170 in Y. Coppens, F. C.

Howell, G. L. Isaac, and R. Leakey (eds.), *Earliest Man and Environment in the Lake Rudolf Basin*. University of Chicago Press, Chicago.

Behrensmeyer, A. K, A. L. Deino, A. Hill, J. D. Kingston, and J. J. Saunders. 2002. Geology and geochronology of the middle Miocene Kipsaramon site complex, Muruyur Beds, Tugen Hills, Kenya. *Journal of Human Evolution* 42:11–38.

Behrensmeyer, A. K, R. Potts, T. Plummer, L. Tauxe, N. Opdyke, and T. Jorstad. 1995. The Pleistocene locality of Kanjera, Western Kenya: Stratigraphy, chronology and paleoenvironments. *Journal of Human Evolution* 29:247–274.

Behrensmeyer, A. K, N. E. Todd, R. Potts, and G. E. McBrinn. 1997. Late Pliocene faunal turnover in the Turkana Basin, Kenya and Ethiopia. *Science* 278:1589–1594.

Berger, L. R., R. Lacruz, and D. J. de Ruiter. 2002. Brief communication: Revised age estimates of *Australopithecus*-bearing deposits at Sterkfontein, South Africa. *American Journal of Physical Anthropology* 119:192–197.

Bergounioux, F. M., and F. Crouzel. 1956. Présence de *Tetralophodon longirostris* dans le Vindobonien inférieur de Tunisie. *Bulletin de la Société Géologique de France* 6:547–557.

———. 1959. Nouvelles observations sur un petit Mastodonte du Cartennien de Kabylie. *Comptes Rendus Sommaire de Séances de la Société Géologique de France* 1959:101–102.

Bernor, R. L., and P. Pavlakis. 1987. Zoogeographic relationships of the Sahabi large mammal fauna (early Pliocene, Libya); pp. 349–383 in N. T. Boaz, A. El-Arnauti, A. W. Gaziry, J. de Heinzelin, and D. D. Boaz (eds.), *Neogene Paleontology and Geology of Sahabi*. Liss, New York.

Bernor, R. L., M. Brunet, L. Ginsburg, P. Mein, M. Pickford, F. Rögl, S. Sen, F. Steininger, and H. Thomas. 1987. A consideration of some major topics concerning Old World Miocene mammalian chronology, migrations and paleogeography. *Geobios* 20:431–439.

Bernor, R. L., and R. S. Scott. 2003. New interpretations of the systematics, biogeography and paleoecology of the Sahabi hipparions (latest Miocene)(Libya). *Geodiversitas* 25:297–319.

Berthelet, A. 2001. L'outillage lithique du site de dépecage à *Elephas recki ileretensis* de Barogali (République de Djibouti). *Comptes Rendus de l'Académie des Sciences, Paris*, Série II, 332:411–416.

Berthelet, A., and J. Chavaillon. 2001. The early Palaeolithic butchery site of Barogali (Republic of Djibouti); pp. 176–179 in G. Cavarretta, P. Gioia, M. Mussi, and M. R. Palombo (eds.), *Proceedings of the First International Congress of La Terra degli Elefanti: The World of Elephants*. Consiglio Nazionale delle Ricerche, Rome.

Bishop, W. W. 1964. More fossil primates and other Miocene mammals from north-east Uganda. *Nature* 203:1327–1331.

———. 1967. The later Tertiary in East Africa: Volcanics, sediments and faunal inventory; pp. 31–56 in W. W. Bishop and J. D. Clark (eds.), *Background to Evolution in Africa*. University of Chicago Press, Chicago.

———. 1972. Stratigraphic succession "versus" calibration in East Africa; pp. 219–246 in W. W. Bishop and J. A. Miller (eds.), *Calibration of Hominoid Evolution*. Scottish Academic Press, Edinburgh.

Bishop, W. W., G. R. Chapman, A. Hill, and J. A. Miller. 1971. Succession of Cainozoic vertebrate assemblages from the northern Kenya Rift Valley. *Nature* 233:389–394.

Bishop, W. W., J. A. Miller, and F. J. Fitch. 1969. New potassium-argon date determinations relevant to the Miocene fossil mammal sequence in East Africa. *American Journal of Science* 267:669–699.

Bishop, W. W., and F. Whyte. 1962. Tertiary mammalian faunas and sediments in Karamoja and Kavirondo, East Africa. *Nature* 196:1283–1287.

Blake, S., and S. Hedges. 2004. Sinking the flagship: The case of forest elephants in Asia and Africa. *Conservation Biology* 18:1191–1202.

Boaz, N. T. 1994. Significance of the Western Rift for hominid evolution; pp. 321–343 in R. S. Corruccini and R. L. Ciochon (eds.), *Integrative Paths to the Past. Paleoanthropological Advances in Honor of F. Clark Howell*. Prentice Hall, Englewood Cliffs, N.J.

Boeuf, O. 1992. *Anancus arverneneis chilhiacensis* nov. subsp. (Proboscidea, Mammalia), un Mastodonte du Plio-Pléistocène de Haute-Loire, France. *Geobios Mémoire Spécial* 14:179–188.

Bonaparte, C. L. 1845. *Catalogo Metodico dei Mammiferi Europei*. Valenciennes, Milan, 36 pp.

Bonnefille, R., R. Potts, F. Chalié, D. Jolly, and O. Peyron. 2004. High resolution vegetation and climate change associated with Pliocene *Australopithecus afarensis*. *Proceedings of the National Academy of Sciences, USA* 101:12125–12129.

Borissiak, A. A. 1929. On a new direction in the adaptive radiation of mastodonts. *Palaeobiologica* 2:19–33.

Boschetto, H. B., F. H. Brown, and I. McDougall. 1992. Stratigraphy of the Lothidok Range, northern Kenya, and K/Ar ages of its Miocene primates. *Journal of Human Evolution* 22:47–71.

Bown, T. M., and M. J. Kraus. 1988. Geology and paleoenvironment of the Oligocene Jebel Qatrani Formation and adjacent rocks, Fayum Depression, Egypt. *U.S. Geological Survey, Professional Paper* 1452:1–60.

Bown, T. M., M. J. Kraus, S. L. Wing, J. G. Fleagle, B. H. Tiffney, E. L. Simons, and C. F. Vondra. 1982. The Fayum primate forest revisited. *Journal of Human Evolution* 11:603–632.

Brives, A. 1919. Sur la découverte d'une dent de *Dinotherium* dans la sablière du Djebel Konif, près Tébessa. *Bulletin de la Société d'Histoire Naturelle d'Afrique du Nord* 10: 90–93.

Bromage, T. G., F. Schrenk, and Y. M. Juwayeyi. 1995. Paleobiogeography of the Malawi Rift: Age and vertebrate paleontology of the Chiwondo Beds, northern Malawi. *Journal of Human Evolution* 28: 37–57.

Brown, F. H. 1994. Development of Pliocene and Pleistocene chronology of the Turkana Basin, East Africa, and its relation to other sites; pp. 285–312 in R. S. Corruccini and R. L. Ciochon (eds.), *Integrative Paths to the Past. Paleoanthropological Advances in Honor of F. Clark Howell*. Prentice Hall, Englewood Cliffs, N.J.

Brown, F. H., I. McDougall, I. Davies, and R. Maier. 1985. An integrated Plio-Pleistocene chronology for the Turkana Basin; pp. 83–90 in E. Delson (ed.), *Ancestors: The Hard Evidence*. Liss, New York.

Brugal, J.-P., H. Roche, and M. Kibunjia. 2003. Faunes et paléoenvironnements des principaux sites archéologiques plio-pléistocènes de la formation de Nachukui (Ouest-Turkana, Kenya). *Comptes Rendus Palevol* 2:675–684.

Brunet, M. 2001. Chadian australopithecines: Biochronology and environmental context; pp. 103–106 in P. V. Tobias, M. A. Raath, J. Moggi-Cecchi, and G. A. Doyle (eds.), *Humanity from African Naissance to Coming Millennia*. Florence University Press, Florence.

Brunet, M., A. Beauvilain, D. Billiou, H. Bocherens, J.-R. Boisserie, L. de Bonis, P. Branger, A. Brunet, Y. Coppens, R. Daams, J. Dejax, C. Denys, P. Duringer, V. Eisenmann, F. Fanoné, P. Fronty, M. Gayet, D. Geraads, F. Guy, M. Kasser, G. Koufos, A. Likius, N. Lopez-Martinez, A. Louchart, L MacLatchy, H. T. Mackaye, B. Marandat, G. Mouchelin, C. Mourer-Chauviré, O. Otero, S. Peigné, P. P. Campomanes, D. Pilbeam, J. C. Rage, D. de Ruitter, M. Schuster, J. Sudre, P. Tassy, P. Vignaud, L Viriot, and A. Zazzo. 2000. Chad: Discovery of a vertebrate fauna close to the Mio-Pliocene boundary. *Journal of Vertebrate Paleontology* 20: 205–209.

Brunet, M., A. Beauvilain, Y. Coppens, E. Heintz, A. H. E. Moutaye, and D. Pilbeam. 1995. The first australopithecine 2,500 kilometres west of the Rift Valley (Chad). *Nature* 378:273–275.

Brunet, M., A. Beauvilain, D. Geraads, F. Guy, M. Kasser, H. T. Mackaye, L. M. MacLatchy, G. Mouchelin, J. Sudre, and P. Vignaud. 1998. Tchad: Découverte d'une faune de mammifères du Pliocène inférieur. *Comptes Rendus de l'Académie des Sciences, Paris, Sciences de la Terre et des Planètes* 326:153–158.

Buss, I. O. 1990. *Elephant Life: Fifteen Years of High Population Density*. Iowa State University Press, Ames, 191 pp.

Butzer, K. W. 1971. The Lower Omo Basin: Geology, fauna and hominids of Plio-Pleistocene formations. *Naturwissenschaften* 58:7–16.

Bye, B. A., F. H. Brown, T. E. Cerling, and I. McDougall. 1987. Increased age estimate for the Lower Palaeolithic hominid site of Olorgesailie, Kenya. *Nature* 329:237–239.

Cerling, T. E., J. M. Harris, S. H. Ambrose, M. G. Leakey, and N. Solounias. 1997. Dietary and environmental reconstruction with stable isotope analyses of herbivore tooth enamel from the Miocene locality of Fort Ternan, Kenya. *Journal of Human Evolution* 33:635–650.

Cerling, T. E., J. M. Harris, and M. G. Leakey. 1999. Browsing and grazing in elephants: The isotope record of modern and fossil proboscideans. *Oecologia* 120:364–374.

———. 2003. Isotope paleoecology of the Nawata and Nachukui Formations at Lothagam, Turkana Basin, Kenya; pp. 605–624 in M. G. Leakey and J. M. Harris (eds.), *Lothagam: The Dawn of Humanity in Eastern Africa*. Columbia University Press, New York.

———. 2005. Environmentally driven dietary adaptations in African mammals; pp. 258–272 in J. R. Ehleringer, M. D. Dearing, and T. E. Cerling (eds.), *History of Atmospheric CO_2 and its Effects on Plants, Animals, and Ecosystems*. Springer, New York.

Cerling, T. E., J. M. Harris, B. J. MacFadden, M. G. Leakey, J. Quade, V. Eisenmann, and J. R. Ehleringer. 1997. Pattern and significance of global ecologic change in the Late Neogene. *Nature* 389:153–158.

Cerling, T. E., Y. Wang, and J. Quade. 1993. Expansion of C_4 ecosystems as an indicator of global ecological change in the late Miocene. *Nature* 361:344–345.

Cerling, T. E., G. Wittemyer, H. B. Rasmussen, F. Vollrath, C. E. Cerling, T. J. Robinson, and I. Douglas-Hamilton. 2006. Stable isotopes in elephant hair document migration patterns and diet changes. *Proceedings of the National Academy of Sciences, USA* 103:371–373.

Chaid-Saoudi, Y., D. Geraads, and J.-P. Raynal. 2006. The fauna and associated artefacts from the Lower Pleistocene site of Mansourah (Constantine, Algeria). *Comptes Rendus Palevol* 5:963–971.

Christiansen, P. 2004. Body size in proboscideans, with notes on elephant metabolism. *Zoological Journal of the Linnean Society* 140:523–549.

Clift, W. 1828. On the fossil remains of two new species of *Mastodon*, and of other vertebrated animals, found on the left bank of the Irawadi. *Transactions of the Geological Society of London* (2) II, Part III:369–375.

Codron, J., J. A. Lee-Thorp, M. Sponheimer, D. Codron, R. C. Grant, and D. J. de Ruiter. 2006. Elephant *(Loxodonta africana)* diets in Kruger National Park, South Africa: Spatial and landscape differences. *Journal of Mammalogy* 87:27–34.

Comstock, K. E., N. Georgiadis, J. Pecon-Slattery, A. L. Roca, E. A. Ostrander, S. J. O'Brien, and S. K. Wasser. 2002. Patterns of molecular genetic variation among African elephant populations. *Molecular Ecology* 11:2489–2498.

Cooke, H. B. S. 1968. The fossil mammal fauna of Africa. *Quarterly Review of Biology* 43:234–264.

Cooke, H. B. S. 1993. Fossil proboscidean remains from Bolt's Farm and other Transvaal cave breccias. *Palaeontologia Africana* 30:25–34.

Cooke, H. B. S., and S. Coryndon. 1970. Fossil mammals from the Kaiso Formation and other related deposits in Uganda; pp. 107–224 in L. S. B. Leakey and R. G. J. Savage (eds.), *Fossil Vertebrates of Africa*, vol. 2. Academic Press, Edinburgh.

Cooke, H. B. S., and V. J. Maglio. 1972. Plio-Pleistocene stratigraphy in East Africa in relation to proboscidean and suid evolution; pp. 303–329 in W. W. Bishop and J. A. Miller (eds.), *Calibration of Hominoid Evolution: Recent Advances in Isotopic and Other Dating Methods Applicable to the Origin of Man*. Wenner-Gren Foundation for Anthropological Research, Scottish Academic Press, New York.

Coppens, Y. 1965. Les Proboscidiens du Tchad. *Actes du Vᵉ Congrés Panafricain de Préhistoire et de l'Étude du Quaternaire* (Santa Cruz de Tenerife), I, 5:331–387.

———. 1967. Les Faunes de Vertebres Quaternaires du Tchad; pp. 89–97 in W. W. Bishop and J. D. Clark (eds.), *Background to Evolution in Africa*. University of Chicago Press, Chicago.

———. 1972. Un nouveau Proboscidien du Pliocène du Tchad, *Stegodibelodon schneideri* nov. gen. nov. sp., et le phylum Stegotetrabelodontinae. *Comptes Rendus de l'Académie des Sciences*, Série D, 274:2962–2965.

Coppens, Y., and M. Beden. 1978. Moeritherioidea; pp. 333–335 in V. J. Maglio and H. B. S. Cooke (eds.), *Evolution of African Mammals*. Harvard University Press, Cambridge.

Coppens, Y., and M. Gaudant. 1976. Découverte d'*Elephas iolensis* Pomel dans le Tyrrhénien de Tunisie. *Bulletin de la Société Géologique de France* 18:171–177.

Coppens, Y., V. J. Maglio, C. T. Madden, and M. Beden. 1978. Proboscidea; pp. 336–367 in V. J. Maglio and H. B. S. Cooke (eds.), *Evolution of African Mammals*. Harvard University Press, Cambridge.

Corvinus, G., and Q. B. Hendey. 1978. A new Miocene vertebrate locality at Arrisdrift in South West Africa (Namibia). *Neues Jahrbuch für Geologie und Paläontologie, Monatshefte* 1978(4):193–205.

Coryndon, S. C., and R. J. G. Savage. 1973. The origin and affinities of African mammal faunas. *Special Papers in Paleontology* (The Palaeontological Association, London) 12:121–135.

Court, N. 1990. The periotic of *Arsinoitherium* and its phylogenetic implications. *Journal of Vertebrate Paleontology* 10:170–182.

———. 1992. Cochlea anatomy of *Numidotherium koholense*: Auditory acuity in the oldest known proboscidean. *Lethaia* 25:211–215.

———. 1993. A dental peculiarity in *Numidotherium koholense*: Evidence of feeding behaviour in a primitive proboscidean. *Zeitschrift für Säugetierkunde* 58:194–196.

———. 1994a. Limb posture and gait in *Numidotherium koholense*, a primitive proboscidean from the Eocene of Algeria. *Zoological Journal of the Linnean Society* 111:297–338.

———. 1994b. The periotic of *Moeritherium* (Mammalia, Proboscidea): Homology or homoplasy in the ear region of Tethytheria McKenna, 1975? *Zoological Journal of the Linnean Society* 112:13–28.

———. 1995. A new species of *Numidotherium* (Mammalia: Proboscidea) from the Eocene of Libya and the early phylogeny of the Proboscidea. *Journal of Vertebrate Paleontology* 15:650–671.

Court, N., and J.-J. Jaeger. 1991. Anatomy of the periotic bone in *Numidotherium koholense*, an example of parallel evolution in the inner ear of tethytheres. *Comptes Rendus de l'Académie des Sciences*, Série II, 312:559–565.

Dart, R. 1927. Mammoths and man in the Transvaal. *Nature*, suppl., no. 3032:1–8.

———. 1929. Mammoths and other fossil elephants of the Vaal and Limpopo watersheds. *South African Journal of Science* 26:698–731.

Debruyne, R. 2005. A case study of apparent conflict between molecular phylogenies: The interrelationships of African elephants. *Cladistics* 21:31–50.

Debruyne, R., V. Barriel, and P. Tassy. 2003. Mitochondrial cytochrome *b* of the Lyakhov mammoth (Proboscidea, Mammalia): New data and phylogenetic analyses of Elephantidae. *Molecular Phylogenetics and Evolution* 26:421–434.

Deino, A., and R. Potts. 1990. Single-crystal ⁴⁰Ar/³⁹Ar dating of the Olorgesailie Formation, southern Kenya Rift. *Journal of Geophysical Research* 95:8453–8470.

Deino, A. L., L. Tauxe, M. Monaghan, and A. Hill. 2002. ⁴⁰Ar/³⁹Ar geochronology and paleomagnetic stratigraphy of the Lukeino and lower Chemeron Formations at Tabarin and Kapcheberek, Tugen Hills, Kenya. *Journal of Human Evolution* 42:117–140.

Delmer, C. 2005. Les premières phases de différenciation des proboscidiens (Tethytheria, Mammalia): Le rôle du *Barytherium grave* de Libye. Unpublished PhD dissertation, Muséum National d'Histoire Naturelle, Paris, 470 pp.

———. 2009. Reassessment of the generic attribution of *Numidotherium savagei* and the homologies of lower incisors in proboscideans. *Acta Palaeontologica Polonica* 54:561–580.

Delmer, C., M. Mahboubi, R. Tabuce, and P. Tassy. 2006. A new species of *Moeritherium* (Proboscidea, Mammalia) from the early late Eocene of Algeria: New perspectives on the ancestral morphotype of the genus. *Palaeontology* 49:421–434.

Denton, G. H. 1999. Cenozoic climate change; pp. 94–114 in T. G. Bromage and F. Schrenk (eds.), *African Biogeography, Climate Change, and Human Evolution*. Oxford University Press, New York.

Depéret, C. 1897. Découverte du *Mastodon angustidens* dans l'étage Cartennien de Kabylie. *Bulletin de la Société Géologique de France* 25:518–521.

Depéret, C., L. Lavauden, and M. Solignag. 1925. Sur la découverte du *Mastodon arvernensis* dans le Pliocène de Ferryville (Tunisie). *Comptes Rendus Sommaire des Séances, Société Géologique de France* 1–2:21–22.

Deraniyagala, P. E. P. 1955. Some extinct elephants, their relatives and the two living species. *Ceylon National Museum Publication, Colombo* 1955:1–161.

De Ruiter, D. J. 2003. Revised faunal lists for Members 1–3 of Swartkrans, South Africa. *Annals of the Transvaal Museum* 40:29–41.

Ditchfield, P., J. Hicks, T. Plummer, L. C. Bishop, and R. Potts. 1999. Current research on the late Pliocene and Pleistocene deposits north of Homa Mountain, southwestern Kenya. *Journal of Human Evolution* 36:123–150.

Doumergue, F. 1928. Découverte de l'*Elephas planifrons* Falconer à Rachgoun (Dépt. d'Oran). *Bulletin du cinquantenaire de la Société de géographie et d'archéologie d'Oran* 1928:114–132.

Drake, R., and G. H. Curtis. 1987. K-Ar geochronology of the Laetoli fossil localities; pp. 48–61 in M. D. Leakey and J. M. Harris (eds.), *Laetoli: A Pliocene Site in Northern Tanzania*. Clarendon Press, Oxford.

Drake, R., J. A. Van Couvering, M. Pickford, G. H. Curtis, and J. A. Harris. 1988. New chronology for the early Miocene mammalian fauna of Kisingiri, western Kenya. *Journal of the Geological Society, London* 145:479–491.

Dubrovo, I. A. 1974. Some new data on mastodonts from Western Mongolia; pp. 64–73 in N. M. Kramarenko (ed.), *Mesozoic and Cenozoic Faunas and Biostratigraphy of Mongolia*. Izdatelstro "Nauka," Moscow.

Dudley, J. P. 1996. Mammoths, gomphotheres, and the Great American Faunal Interchange; pp. 289–295 in J. Shoshani and P. Tassy (eds.), *The Proboscidea: Evolution and Palaeoecology of Elephants and Their Relatives*. Oxford University Press, Oxford.

Eggert, L. S., C. A. Rasner, and D. S. Woodruff. 2002. The evolution and phylogeography of the African elephant inferred from mitochondrial

DNA sequence and nuclear microsatellite markers. *Proceedings of the Royal Society of London, B* 269:1993–2006.

Éhik, J. 1930. *Prodinotherium hungaricum* n.g., n. sp. *Geologica Hungarica series palaeontologica* 6:1–24.

Eizirik, E., W. J. Murphy, and S. J. O'Brien. 2001. Molecular dating and biogeography of the early placental mammal radiation. *Journal of Heredity* 92:212–219.

El-Khashab, B. 1979. A brief account on Egyptian Paleogene Proboscidea. *Annals of the Geological Survey of Egypt* 9:245–260.

Eltringham, S. K. 1992. Ecology and behavior; pp. 124–127 in J. Shoshani (ed.), *Elephants*. Simon and Schuster, London.

Errington de la Croix, J. 1887. Le géologie du Cherichera (Tunisie Centrale). *Comptes Rendus de l'Académie des Sciences, Paris* 105:321–323.

Evans, E. M. N., J. A. H. Van Couvering, and P. Andrews. 1981. Palaeoecology of Miocene Sites in Western Kenya. *Journal of Human Evolution* 10:99–116.

Falconer, H. 1845. Description of some fossil remains of *Dinotherium*, giraffe, and other Mammalia form the Gulf of Cambray, western coast of India. *Quarterly Journal of the Geological Society of London* 1:356–372.

Fara, E., A. Likius, H. T. Mackaye, P. Vignaud, and M. Brunet. 2005. Pliocene large-mammal assemblages from northern Chad: Sampling and ecological structure. *Naturwissenschaften* 92:537–541.

Feibel, C. S. 1999. Tephrostratigraphy and geological context in paleoanthropology. *Evolutionary Anthropology* 8:87–100.

———. 2003. Stratigraphy and depositional setting of the Pliocene Kanapoi Formation, Lower Kerio Valley, Kenya; pp. 9–20 in J. M. Harris and M. G. Leakey (eds.), *Geology and Vertebrate Paleontology of the Early Pliocene Site of Kanapoi, Northern Kenya*. Contributions in Science, Natural History Museum of Los Angeles County, 498.

Feibel, C. S., and F. H. Brown. 1991. Age of the primate-bearing deposits on Maboko Island, Kenya. *Journal of Human Evolution* 21:221–225.

Feibel, C. S., Brown, F. H., and I. McDougall. 1989. Stratigraphic context of fossil hominids from the Omo Group Deposits: northern Turkana Basin, Kenya and Ethiopia. *American Journal of Physical Anthropology* 78:595–622.

Fernando, P., M. E. Pfrender, S. E. Encalada, and R. Lande. 2000. Mitochondrial DNA variation, phylogeography and population structure of the Asian elephant. *Heredity* 84:362–372.

Ferretti, M. P., and R. V. Croitor. 2001. Functional morphology and ecology of Villafranchian proboscideans from Central Italy; pp. 103–108 in G. Cavarretta, P. Gioia, M. Mussi, and M. R. Palombo (eds.), *Proceedings of the First International Congress of La Terra degli Elefanti: The World of Elephants*. Consiglio Nazionale delle Ricerche, Rome.

———. 2005. Functional morphology and ecology of Villafranchian proboscideans from central Italy. *Quaternary International* 126–128:103–108.

Ferretti, M. P., G. Ficcarelli, Y. Libsekal, T. M. Tecle, and L. Rook. 2003. Fossil elephants from Buia (northern Afar Depression, Eritrea) with remarks on the systematics of *Elephas recki* (Proboscidea, Elephantidae). Journal of Vertebrate Paleontology 23:244–257.

Fisher, D. C. 1996. Extinction of proboscideans in North America; pp. 296–315 in J. Shoshani and P. Tassy (eds.), *The Proboscidea: Evolution and Palaeoecology of Elephants and Their Relatives*. Oxford University Press, Oxford.

Fleischer, R. C., E. A. Perry, K. Muralidharan, E. S. Stevens, and C. M. Wemmer. 2001. Phylogeography of the Asian elephant *(Elephas maximus)* based on mitochondrial DNA. *Evolution* 55:1882–1892.

Forster Cooper, C. 1922. Miocene Proboscidia [sic] from Baluchistan. *Proceedings of the Zoological Society of London* 42:606–626.

Fournet, A. 1971. Les Gisements à Faune Villafranchienne de Tunisie. *Notes du Service Géologique Tunisie* 34:53–69

Fourtau, R. 1918. *Contribution a l'Étude Vertébrés Miocènes de l'Égypte.* Survey Department, Ministry of Finance, Egypt, Government Press, Cairo, 121 pp.

Franz-Odendaal, T. A., J. A. Lee-Thorp, and A. Chinsamy. 2002. New evidence for the lack of C_4 grassland expansions during the early Pliocene at Langebaanweg, South Africa. *Paleobiology* 28:378–388.

Frick, C. 1933. New remains of trilophodont-tetrabelodont mastodons. *Bulletin of the American Museum of Natural History* 59:505–652.

Froehlich, D. J., and J. E. Kalb. 1995. Internal reconstruction of elephantid molars: Applications for functional anatomy and systematics. *Paleobiology* 21:379–392.

Fuchs, V. E. 1934. The geological work of the Cambridge expedition to the East African lakes. *Geological Magazine* 71: 97–116, 145–166.

Gagnon, M. 1997. Ecological diversity and community ecology in the Fayum sequence (Egypt). *Journal of Human Evolution* 32:133–160.

Gaudry, A. 1891. Quelques remarques sur les Mastodontes à propos de l'animal du Cherichira. *Mémoires de la Société Géologique de France, Paléontologie* 8:1–6.

Gaziry, A. W. 1976. Jungtertiäre Mastodonten aus Anatolien (Türkei). *Geologisches Jahrbuch* 22:3–143.

———. 1982. Proboscidea from the Sahabi Formation. *Garyounis Scientific Bulletin*, Special Issue No. 4:101–108.

———. 1987a. New mammals from the Jabal Zaltan site, Libya. *Senckenbergiana Lethaea* 68:69–89.

———. 1987b. Remains of Proboscidea from the early Pliocene of Sahabi, Libya; pp. 183–203 in N. T. Boaz, A. El-Arnauti, A. W. Gaziry, J. de Heinzelin, and D. D. Boaz (eds.), *Neogene Paleontology and Geology of Sahabi*. Liss, New York.

Gebo, D., L. L. MacLatchy, R. Kityo, A. Deino, J. Kingston, and D. Pilbeam. 1997. A hominoid genus from the early Miocene of Uganda. *Science* 276:401–404.

Gentry, A. W. 1987. Mastodons from the Miocene of Saudi Arabia. *Bulletin of the British Museum (Natural History)* 41:395–407.

Gentry, A.W., and A. Gentry. 1969. Fossil camels in Kenya and Tanzania. *Nature* 222:898.

Geraads, D. 1982. Paléobiogéographie de l'Afrique du Nord depuis de Miocène terminal, d'après des grand mammifères. *Geobios Mémoire Spécial* 6:473–481.

———. 1989. Vertébrés fossiles du Miocène superieur du Djebel Krechem el Artsouma (Tunisie Centrale): Comparisons biostratigraphiques. *Geobios* 22:777–801.

———. 2002. Plio-Pleistocene mammalian biostratigraphy of Atlantic Morocco. *Quaternaire* 13:43–53.

Geraads, D., Z. Alemseged, and H. Bellon. 2002. The late Miocene mammalian fauna of Chorora, Awash Basin, Ethiopia: Systematics, biochronology and the ^{40}K-^{40}Ar ages of the associated volcanics. *Tertiary Research* 21:113–122.

Geraads, D., and F. Amani. 1998. Bovidae (Mammalia) du Pliocène final d'Ahl al Oughlam, Casablanca, Maroc. *Paläontologisches Zeitschrift* 72:191–205.

Geraads, D., and F. Metz-Muller. 1999. Proboscidea (Mammalia) du Pliocène final d'Ahl al Oughlam (Casablanca, Maroc). *Neues Jahrbuch für Geologie und Paläontologie, Monatshefte* 1999(1):52–64.

Geraads, D., J.-P. Raynal, and V. Eisenmann. 2004. The earliest human occupation of North Africa: a reply to Sahnouni et al. (2002). *Journal of Human Evolution* 46:751–761.

Gheerbrant, E. 1998. The oldest known proboscidean and the role of Africa in the radiation of modern orders of placentals. *Bulletin of the Geological Society of Denmark* 44:181–185.

———. 2009. Paleocene emergence of elephant relatives and the rapid radiation of African ungulates. *Proceedings of the National Academy of Sciences, USA* 106:10717–10721.

Gheerbrant, E., J. Sudre, and H. Cappetta. 1996. A Paleocene proboscidean from Morocco. *Nature* 383:68–70.

Gheerbrant, E., J. Sudre, H. Cappetta, and G. Bignot. 1998. *Phosphatherium escuilliei* du Thanétien du bassin des Ouled Abdoun (Maroc), plus ancien proboscidien (Mammalia) d'Afrique. *Geobios* 30:247–269.

Gheerbrant, E., J. Sudre, H. Cappetta, M. Iarochène, M. Amaghzaz, and B. Bouya. 2002. A new large mammal from the Ypresian of Morocco: Evidence of surprising diversity of early proboscideans. *Acta Palaeontologica Polonica* 47:493–506.

Gheerbrant, E., J. Sudre, H. Cappetta, C. Mourer-Chauvire, E. Bourdon, M. Iarochène, M. Amaghzaz, and B. Bouya. 2003. Les localités à mammifères des carrières de Grand Daoui, Bassin des Ouled Abdoun, Maroc, Yprésien: Premier état des lieux. *Bulletin de la Société Géologique de France* 174:279–293.

Gheerbrant, E., J. Sudre, P. Tassy, M. Amaghzaz, B. Bouya, and M. Iarochène. 2005. Nouvelles données sur *Phosphatherium escuilliei* (Mammalia, Proboscidea) de l'Éocène inférieur du Maroc, apports à la phylogénie des Proboscidea et des ongulés lophodontes. *Geodiversitas* 27:239–333. Figures © Publications Scientifiques du Muséum national d'Histoire naturelle, Paris.

Gingerich, P. D., D. E. Russell, and N. A. Wells. 1990. Astragalus of *Anthracobune* (Mammalia, Proboscidea) from the early–middle Eocene of Kashmir. *Contributions from the Museum of Paleontology* (University of Michigan) 28:71–77.

Glennie, K. W., and B. D. Evamy. 1968. Dikaka: Plants and plant-root structures associated with aeolian sand. *Palaeogeography, Palaeoclimatology, Palaeoecology* 4:77–87.

Göhlich, U. B. 1998. Elephantoidea (Proboscidea, Mammalia) aus dem Mittel- und Obermiozän der Oberen Süßwassermolasse

Süddeutschlands: Odontologie und Osteologie. *Münchner schaftliche Abhandlungen A* 36:1–245.

———. 1999. Order Proboscidea; pp. 157–168 in G. Rössner and K. Heissig (eds.), *The Miocene Land Mammals of Europe*. Pfeil, Munich.

Gräf, I. E. 1957. Die prinzipien der Arbestimmung bei *Deinotherium*. *Palaeontographica Abt. A* 180:131–185.

Grubb, P., C. P. Groves, J. P. Dudley, and J. Shoshani. 2000. Living African elephants belong to two species: *Loxodonta africana* (Blumenbach, 1797) and *Loxodonta cyclotis* (Matschie, 1900). *Elephant* 2:1–4.

Guan, J. 1996. On the shovel-tusked elephantoids from China; pp. 124–135 in J. Shoshani and P. Tassy (eds.), *The Proboscidea: Evolution and Palaeoecology of Elephants and Their Relatives*. Oxford University Press, Oxford.

Guerin, C. and M. Pickford. 2003. *Ougandatherium napakense* nov. gen. nov. sp., le plus ancien Rhinocerotidae Iranotheriinae d'Afrique. *Annales de Paléontologie* 89:1–35.

Gutiérrez, M., and D. Rasmussen. 2007. Late Oligocene mammals from northern Kenya. *Journal of Vertebrate Paleontology* 27, suppl. to no. 3:85A.

Hagelberg, E., M. G. Thomas, C. E. Cook, Jr., A. V. Sher, G. F. Barishnikov, and A. M. Lister. 1994. DNA from ancient mammoth bones. *Nature* 370:333–334.

Haile-Selassie, Y. 2000. A newly discovered early Pliocene hominid-bearing paleontological site in the Mulu Basin, Ethiopia. *American Journal of Physical Anthropology*, suppl. 30:170.

———. 2001. Late Miocene mammalian fauna from the Middle Awash Valley, Ethiopia. Unpublished PhD dissertation, University of California, Berkeley, 425 pp.

Haile-Selassie, Y., G. WoldeGabriel, T. D. White, R. L. Bernor, D. Degusta, P. R. Renne, W. K. Hart, E. Vrba, S. Ambrose, and F. C. Howell. 2004. Mio-Pliocene mammals from the Middle Awash, Ethiopia. *Geobios* 37:536–552.

Hailwood, E. A., and P. J. Whybrow. 1999. Palaeomagnetic correlation and dating of the Baynunah and Shuwaihat Formations, Emirate of Abu Dhabi, United Arab Emirates; pp. 75–87 in P. J. Whybrow and A. Hill (eds.), *Fossil Vertebrates of Arabia*. Yale University Press, New Haven.

Hamilton, W. R. 1973. A Lower Miocene mammalian fauna from Siwa, Egypt. *Palaeontology* 16:275–281.

Hamilton, W. R., P. J. Whybrow, and H. A. McClure. 1978. Fauna of fossil mammals from the Miocene of Saudi Arabia. *Nature* 274:248–249.

Harris, J. M. 1969. *Prodeinotherium* from Gebel Zelten, Libya. Unpublished PhD dissertation, University of Bristol, Bristol, 303 pp.

———. 1973. *Prodeinotherium* from Gebel Zelten, Libya. *Bulletin of the British Museum (Natural History), Geology* 23:285–350.

———. 1975. Evolution of feeding mechanisms in the Family Deinotheriidae. *Zoological Journal of the Linnean Society* 56:332–362.

———. 1976. Cranial and dental remains of *Deinotherium bozasi* (Mammalia, Proboscidea) from East Rudolf, Kenya. *Journal of Zoology* 178:57–75.

———. 1977a. Deinotheres from southern Africa. *South African Journal of Science* 73:282–282.

———. 1977b. Mammalian faunas from East African hominid-bearing localities; pp. 21–48 in T. H. Wilson (ed.), *A Survey of the Prehistory of Eastern Africa*. Pan-African Congress for Prehistory and Quaternary Studies, Nairobi.

———. 1978. Deinotherioidea and Barytherioidea; pp. 315–332 in V. J. Maglio and H. B. S. Cooke (eds.), *Evolution of African Mammals*. Harvard University Press, Cambridge.

———. 1983. Background to the study of the Koobi Fora fossil faunas; pp. 1–21 in J. M. Harris (ed.), *Koobi Fora Research Project: Volume 2. The Fossil Ungulates: Proboscidea, Perissodactyla, and Suidae*. Clarendon Press, Oxford.

———. 1987a. Fossil Deinotheriidae from Laetoli; pp. 294–297 in M. D. Leakey, and J. M. Harris (eds.), *Laetoli: A Pliocene site in northern Tanzania*. Clarendon Press, Oxford.

———. 1987b. Summary; pp. 524–531 in M. D. Leakey and J. M. Harris (eds.), *Laetoli: A Pliocene Site in Northern Tanzania*. Clarendon Press, Oxford.

———. 2003. Deinotheriidae from Lothagam; pp. 359–361 in M. G. Leakey, and J. M. Harris (eds.), *Lothagam: The Dawn of Humanity in Africa*. Columbia University Press, New York.

Harris, J. M., F. H. Brown, and M. G. Leakey. 1988a. Geology and palaeontology of Plio-Pleistocene localities west of Lake Turkana, Kenya. *Contributions in Science, Natural History Museum of Los Angeles County* 399:1–128.

Harris, J. M., F. H. Brown, M. G. Leakey, A. C. Walker, and R. E. Leakey. 1988b. Pliocene and Pleistocene hominid-bearing sites from west of Lake Turkana, Kenya. *Science* 239:27–33.

Harris, J. M., M. G. Leakey, and T. E. Cerling. 2003. Early Pliocene tetrapod remains from Kanapoi, Lake Turkana Basin, Kenya; pp. 39–113 in J. M. Harris and M. G. Leakey (eds.), *Geology and Vertebrate Paleontology of the Early Pliocene Site of Kanapoi, Northern Kenya*. Contributions in Science, Natural History Museum of Los Angeles County, 498.

Harris, J. M., and R. Watkins. 1974. New early Miocene vertebrate locality near Lake Rudolf, Kenya. *Nature* 252:576–577.

Harrison, T. 2002. The first record of fossil hominins from the Ndolanya Beds, Laetoli, Tanzania. *American Journal of Physical Anthropology* 119:83.

Harrison, T., and E. Baker. 1997. Paleontology and biochronology of fossil localities in the Manonga Valley, Tanzania; pp. 361–393 in T. Harrison (ed.), *Neogene Paleontology of the Manonga Valley, Tanzania*. Plenum Press, New York.

Harzhauser, M., W. E. Piller, and F. F. Steininger. 2002. Circum-Mediterranean Oligo-Miocene biogeographic evolution: The gastropods' point of view. *Palaeogeography, Palaeoclimatology, Palaeoecology* 183:103–133.

Haug, É. 1911. II. Les périodes géologiques; pp. 539–2024 in E. Haug, *Traité de Géologie*. Paris.

Hay, R. L. 1976. *Geology of the Olduvai Gorge: A Study of Sedimentation in a Semiarid Basin*. University of California Press, Berkeley, 203 pp.

———. 1987. Geology of the Laetoli area; pp. 23–47 in M. D. Leakey and J. M. Harris (eds.), *Laetoli: A Pliocene Site in Northern Tanzania*. Clarendon Press, Oxford.

Heinzelin, J. de. 1983. *The Omo Group: Volume 1*. Text. Archives of the International Omo Research Expedition. Musée Royal de l'Afrique Centrale, Tervuren, Belgique, Annales Série in-8°, Sciences Géologiques n° 85:1–365.

Heinzelin, J. de, J. D. Clark, T. White, W. Hart, P. Renne, G. WoldeGabriel, Y. Beyene, and E. Vrba. 1999. Environment and behavior of 2.5-million-year-old Bouri hominids. *Science* 284:625–629.

Heinzelin, J. de, and A. El-Arnauti. 1987. The Sahabi Formation and related deposits; pp. 1–22 in N. T. Boaz, A. El-Arnauti, A. W. Gaziry, J. de Heinzelin, and D. D. Boaz (eds.), *Neogene Paleontology and Geology of Sahabi*. Liss, New York.

Hendey, Q. B. 1978a. The age of the fossils from Baard's Quarry, Langebaanweg, South Africa. *Annals of the South African Museum* 75:1–24.

———. 1978b. Preliminary report on the Miocene vertebrates from Arrisdrift, South West Africa. *Annals of the South African Museum* 76:1–41.

———. 1981. Palaeoecology of the late Tertiary fossil occurrences in "E" Quarry, Langebaanweg, South Africa, and a reinterpretation of their geological context. *Annals of the South African Museum* 84:1–104.

Hill, A. 1994. Late Miocene and early Pliocene hominoids from Africa; pp. 123–145 in R. S. Corruccini and R. L. Ciochon (eds.), *Integrative Paths to the Past*. Prentice Hall, Englewood Cliffs, N.J.

———. 1999. The Baringo Basin, Kenya: from Bill Bishop to BPRP; pp. 85–97 in P. Andrews and P. Banham (eds.), *Late Cenozoic Environments and Hominid Evolution: A Tribute to Bill Bishop*. Geological Society, London.

———. 2002. Paleoanthropological research in the Tugen Hills, Kenya. *Journal of Human Evolution* 42:1–10

Hill, A., G. Curtis, and R. Drake. 1986. Sedimentary stratigraphy of the Tugen Hills, Baringo, Kenya; pp. 285–295 in L. E. Frostick, R. W. Renaut, I. Reid, and J.-J. Tiercelin (eds.), *Sedimentation in the African Rifts*. Blackwell and Geological Society of London Special Publication 25, Oxford.

Hill, A., R. Drake, L. Tauxe, M. Monaghan, J. C. Barry, A. K. Behrensmeyer, G. Curtis, B. Fine Jacobs, L. Jacobs, N. Johnson, and D. Pilbeam. 1985. Neogene palaeontology and geochronology of the Baringo Basin, Kenya. *Journal of Human Evolution* 14:759–773.

Holroyd, P. A., and M. C. Maas. 1994. Paleogeography, paleobiogeography, and anthropoid origins; pp. 297–334 in J. G. Fleagle and R. F. Kay (eds.), *Anthropoid Origins*. Plenum Press, New York.

Holroyd, P. A., E. L. Simons, T. M. Bown, P. D. Polly, and M. J. Kraus. 1996. New records of terrestrial mammals from the Upper Eocene Qasr el Sagha Formation, Fayum Depression, Egypt. *Palaeovertebrata* 25:175–192.

Hooijer, D.A. 1963. Miocene Mammalia of Congo. *Koninklijk Museum voor Midden-Afrika, Tervuren, Belgie, Annalen, Series in 8°, Geologische Wetenschappen* 46:1–77.

Hooker, P. J., and J. A. Miller. 1979. K-Ar dating of the Pleistocene fossil hominid site at Chesowanja, North Kenya. *Nature* 282:710–712.

Hopwood, A. T. 1939. The mammalian fossils; pp. 308–316 in T. P. O'Brien, *The Prehistory of Uganda Protectorate*. Cambridge University Press, Cambridge.

Hormann, K. 1963. Note on a mastodontoid from Libya. *Zeitschrift für Säugetierkunde* 28:88–93.

Huttunen, K. J. 2000. Deinotheriidae (Proboscidea, Mammalia) of the Miocene of Lower Austria, Burgenland and Franzensbad, Czech Republic: Systematics, odontology and osteology. Unpublished thesis, University of Vienna, 76 pp.

Ishida, H., and S. Ishida. 1982. *Study of the Tertiary Hominoids and Their Paleoenvironments: Vol. 1. Report of Field Survey in Kirimun, Kenya, 1980*. Osaka University Press, Osaka. 181 pp.

Jacobs, B. F., and C. H. S. Kabuye. 1987. A middle Miocene (12.2 my old) forest in the East African Rift Valley, Kenya. *Journal of Human Evolution* 16:147–155.

Jacobs, B. F., N. Tabor, M. Feseha, A. Pan, J. Kappelman, T. Rasmussen, W. Sanders, M. Wiemann, J. Crabaugh, and J. L. G. Massini. 2005. Oligocene terrestrial strata of northwestern Ethiopia: A preliminary report on paleoenvironments and paleontology. *Palaeontologia Electronica* 8 (1), 25A:19 pp.

Joleaud, L. 1928. Eléphants et dinothériums pliocènes de l'Ethiopie: Contribution à l'etude paléogéographique des proboscidiens africains. *Comptes Rendus du Congress Géologique International*, 14th Session:1001–1007.

Joleaud, L., and J. Lombard. 1933a. Conditions de fossilisation et de gisement des mammifères quaternaires d'Ounianga Kebir. *Bulletin de la Société Géologique de France* 3:239–243.

———. 1933b. Mammifères quaternaires d'Ounianga Kebir (Tibesti sud-oriental). *Comptes Rendus Hebdomadaires des Séances de l'Académie des Sciences* 196:497–499.

Kalandadze, N. N., and S. A. Rautian. 1992. The system of mammals and historical zoogeography [in Russian]; pp. 44–152 in O. L. Rossolimo (ed.), *Filogenetika Mlekopitaûsih: Issledovaniâ po Faune*. Sbornik Trudov Zoologiceskogo Muzeâ Moskovoskogo Gosudarstvennorgo Universiteta, 29.

Kalb, J. E. 1995. Fossil elephantoids, Awash paleolake basins, and the Afar triple junction, Ethiopia. *Palaeogeography, Palaeoclimatology, Palaeoecology* 114:357–368.

Kalb, J. E., D. J. Froehlich, and G. L. Bell. 1996a. Phylogeny of African and Eurasian Elephantoidea of the late Neogene; pp. 101–116 in J. Shoshani and P. Tassy (eds.), *The Proboscidea: Evolution and Palaeoecology of Elephants and Their Relatives*. Oxford University Press, Oxford.

———. 1996b. Palaeobiogeography of late Neogene African and Eurasian Elephantoidea; pp. 117–123 in J. Shoshani and P. Tassy (eds.), *The Proboscidea: Evolution and Palaeoecology of Elephants and Their Relatives*. Oxford University Press, Oxford.

Kalb, J. E., C. J. Jolly, A. Mebrate, S. Tebedge, C. Smart, E. B. Oswald, D. L. Cramer, P. F. Whitehead, C. B. Wood, G. C. Conroy, T. Adefris, L. Sperling, and B. Kana. 1982. Fossil mammals and artefacts from the Middle Awash Valley, Ethiopia. *Nature* 298:25–29.

Kalb, J. E., and A. Mebrate. 1993. Fossil elephantoids from the hominid-bearing Awash Group, Middle Awash Valley, Afar Depression, Ethiopia. *Transactions of the American Philosophical Society* 83:1–114.

Kappelman, J., D. T. Rasmussen, W. J. Sanders, M. Feseha, T. Bown, P. Copeland, J. Crabaugh, J. Fleagle, M. Glantz, A. Gordon, B. Jacobs, M. Maga, K. Muldoon, A. Pan, L. Pyne, B. Richmond, T. Ryan, E. Seiffert, S. Sen, L. Todd, M. C. Wiemann, and A. Winkler. 2003. Oligocene mammals from Ethiopia and faunal exchange between Afro-Arabia and Eurasia. *Nature* 426:549–552.

Kennett, J. P. 1995. A review of polar climatic evolution during the Neogene, based on the marine sediment record; pp. 49–64 in E. S. Vrba, G. H. Denton, T. C. Partridge, and L. H. Burckle (eds.), *Paleoclimate and Evolution, with Emphasis on Human Origins*. Yale University Press, New Haven.

Kent, P. E. 1942. The Pleistocene beds of Kanam and Kanjera, Kavirondo, Kenya. *Geological Magazine* 79:72–77.

Kingdon, J. 1979. *East Africa Mammals: An Atlas of Evolution in Africa: Volume IIIB. Large Mammals*. University of Chicago Press, Chicago, 436 pp.

———. 1997. *The Kingdon Field Guide to African Mammals*. Academic Press, San Diego, 464 pp.

Kingston, J., B. F. Jacobs, A. Hill, and A. Deino. 2002. Stratigraphy, age and environments of the late Miocene Mpesida Beds, Tugen Hills, Kenya. *Journal of Human Evolution* 42:95–116.

Kingston, J., B. D. Marino, and A. Hill. 1994. Isotopic evidence for Neogene hominid paleoenvironments in the Kenya Rift Valley. *Science* 264:955–959.

Kingston, J. D., and T. Harrison. 2007. Isotopic dietary reconstructions of Pliocene herbivores at Laetoli: Implications for early hominin paleoecology. *Palaeogeography, Palaeoclimatology, Palaeoecology* 243:272–306.

Klein, R. G. 1984. The large mammals of southern Africa: Late Pliocene to Recent; pp. 107–146 in R. G. Klein (ed.), *Southern African Prehistory and Paleoenvironments*. Balkema, Rotterdam.

———. 1988. The archaeological significance of animal bones from Acheulean sites in southern Africa. *African Archaeological Review* 6:3–25.

Klein, R. G., G. Avery, K. Cruz-Uribe, and T. E. Steele. 2007. The mammalian fauna associated with an archaic hominin skullcap and later Acheulean artifacts at Elandsfontein, Western Cape Province, South Africa. *Journal of Human Evolution* 52:164–186.

Klein, R. G., and K. Cruz-Uribe. 1991. The bovids from Elandsfontein, South Africa, and their implications for the age, palaeoenvironment, and origins of the site. *African Archaeological Review* 9:21–79.

Koch, P. L. 1991. The isotopic ecology of Pleistocene proboscideans. *Journal of Vertebrate Paleontology* 11 (suppl. to no. 3):40A.

Koch, P. L., J. Heisinger, C. Moss, R. W. Carlson, M. L. Fogel, and A. K. Behrensmeyer. 1995. Isotopic tracking of change in diet and habitat use in African elephants. *Science* 267:1340–1343.

Kovarovic, K., P. Andrews, and L. Aiello. 2002. The palaeoecology of the Upper Ndolanya Beds at Laetoli, Tanzania. *Journal of Human Evolution* 43:395–418.

Krause, J., P. H. Dear, J. L. Pollack, M. Slatkin, H. Spriggs, I. Barnes, A. M. Lister, I. Ebersberger, S. Pääbo, and M. Hofreiter. 2005. Multiplex amplification of the mammoth mitochondrial genome and the evolution of Elephantidae. *Nature* 439:724–727.

Krause, D. W., and M. C. Maas. 1990. The biogeographic origins of late Paleocene-early Eocene mammalian immigrants to the Western Interior of North America; pp. 71–106 in T. M. Bown and K. D. Rose (eds.), *Dawn of the Age of Mammals in the Northern Part of the Rocky Mountains Interior, North America*. Geological Society of America Special Paper 243, Geological Society of American, Boulder, Colo.

Kullmer, O., L. Sandrock, T. B. Viola, W. Hujer, H. Said, and H. Seidler. 2008. Suids, elephantoids, paleochronology, and paleoecology of the Pliocene hominid site Galili, Somali Region, Ethiopia. *Palaios* 23:452–464.

Kumar, K. 1991. *Anthracobune aijiensis* nov. sp. (Mammalia: Proboscidea) from the Subathu Formation. Eocene from NW Himalaya, India. *Geobios* 24:221–239.

Lambert, W. D. 1990. Rediagnosis of the genus *Amebelodon* (Mammalia, Proboscidea, Gomphotheriidae), with a new subgenus and species, *Amebelodon (Konobelodon) britti*. *Journal of Paleontology* 64:1032–1040.

———. 1992. The feeding habits of the shovel-tusked gomphotheres: Evidence from tusk wear patterns. *Paleobiology* 18:132–147.

———. 1996. The biogeography of the gomphotheriid proboscideans of North America; pp. 143–148 in J. Shoshani and P. Tassy (eds.), *The Proboscidea: Evolution and Palaeoecology of Elephants and Their Relatives*. Oxford University Press, Oxford.

Lambert, W. D., and J. Shoshani. 1998. Proboscidea; pp. 606–621 in C. M. Janis, K. M. Scott, and L. L. Jacobs (eds.), *Evolution of Tertiary Mammals of North America: Volume 1. Terrestrial Carnivores, Ungulates, and Ungulatelike Mammals*. Cambridge University Press, Cambridge.

Laursen, L., and M. Bekoff. 1978. *Loxodonta africana*. *Mammalian Species* 92:1–8.

Lavocat, R. 1961. Le gisement de Vertébrés miocènes de Beni Mellal (Maroc): Etude systematique de la faune de Mammifères. *Notes et Mémoires du Service Géologique du Maroc, Rabat* 155:29–94.

Laws, R. M. 1966. Age criteria for the African elephant. *East African Wildlife Journal* 4:1–37.

Leakey, L. S. B. 1965. *Olduvai Gorge 1951–1961: A Preliminary Report on the Geology and Fauna*. Vol. 1. Cambridge University Press, Cambridge, 118 pp.

Leakey, M. G., C. S. Feibel, R. L. Bernor, J. M. Harris, T. E. Cerling, K. M. Stewart, G. W. Storrs, A. Walker, L. Werdelin, and A. J. Winkler. 1996. Lothagam: A record of faunal change in the Late Miocene of East Africa. *Journal of Vertebrate Paleontology* 16:556–570.

Leakey, R. E., and M. G. Leakey. 1986. A new Miocene hominoid from Kenya. *Nature* 324:143–146.

Leakey, M. G., P. S. Ungar, and A. Walker. 1995. A new genus of large primate from the late Oligocene of Lothidok, Turkana District, Kenya. *Journal of Human Evolution* 28:519–531.

LeBlanc, J. 2000. A guide to macrofossil localities of Libya, Africa: With notes on mineral sites. http://sites.google.com/site/leblancjacques/fossilhome.

Lehmann, U. 1950. Über Mastodontenreste in der Bayerischen Staatssammlung in München. *Palaeontographica, Abt. A* 99:121–232.

Lepersonne, J. 1959. Données géologiques et stratigraphiques et comparaison entre le lac Albert et le lac Edouard; in W. Adam, *Mollusques Pléistocènes de la Region du Lac Albert et de la Semliki*, Annales, Musee Royal du Congo Belge, series in 8°, Sciences Géologiques 25:78–119.

Lister, A. 1996. Evolution and taxonomy of Eurasian mammoths; pp. 203–213 in Shoshani and P. Tassy (eds.), *The Proboscidea: Evolution and Palaeoecology of Elephants and Their Relatives*. Oxford University Press, Oxford.

———. 2001. "Gradual" evolution and molar scaling in the evolution of the mammoth; pp. 648–651 in G. Cavarretta, P. Gioia, M. Mussi, and M. R. Palombo (eds.), *Proceedings of the First International Congress of La Terra degli Elefanti: The World of Elephants*. Consiglio Nazionale delle Ricerche, Rome.

Lister, A., and P. Bahn. 2007. *Mammoths: Giants of the Ice Age*. 3rd ed. University of California Press, Berkeley, 168 pp.

Lister, A., and A. V. Sher. 2001. The origin and evolution of the woolly mammoth. *Science* 294:1094–1097.

Lister, A., A. V. Sher, H. van Essen, and G. Wei. 2005. The pattern and process of mammoth evolution in Eurasia. *Quaternary International* 126–128:49–64.

Lister, A., and H. van Essen. 2003. *Mammuthus rumanus* (Stefanescu), the earliest mammoth in Europe; pp. 47–52 in A. Petculescu and E. Stiuca (eds.), *Advances in Vertebrate Paleontology "Hen to Panta."* Romanian Academy "Emile Racovitza" Institute of Speleology, Bucharest.

Liu, A. G. S. C., E. R. Seiffert, and E. L. Simons. 2008. Stable isotope evidence for an amphibious phase in early proboscidean evolution. *Proceedings of the National Academy of Sciences, USA* 105:5786–5791.

Ludwig, K. R., and P. R. Renne. 2000. Geochronology on the paleoanthropological time scale. *Evolutionary Anthropology* 9:101–110.

Lydekker, R. 1907. The ears as a race-character in the African elephant. *Proceedings of the Zoological Society, London* 1907:380–403.

Ma, A., and H. Tang. 1992. On discovery and significance of a Holocene *Ailuropoda-Stegodon* fauna from Jinhua, Zhejiang. *Vertebrata PalAsiatica* 30:295–312.

MacInnes, D. G. 1942. Miocene and post-Miocene Proboscidia [sic] from East Africa. *Transactions of the Zoological Society of London* 25:33–106.

Mackaye, H. T., M. Brunet, and P. Tassy. 2005. *Selenetherium kolleensis* nov. gen. nov. sp.: un nouveau Proboscidea (Mammalia) dans le Pliocène tchadien. *Géobios* 38:765–777.

Mackaye, H. T., Y. Coppens, P. Vignaud, F. Lihoreau, and M. Brunet. 2008. De nouveaux restes de *Primelephas* dans le Mio-Pliocène du Nord du Tchad et revision du genre *Primelephas*. *Comptes Rendus Palevol* 7:227–236.

MacLatchy, L., A. Deino, and J. Kingston. 2006. An updated chronology for the early Miocene of NE Uganda. *Journal of Vertebrate Paleontology* 26 (suppl. to no. 3):93A.

Madden, C. T. 1972. Miocene mammals, stratigraphy, and environment of Muruarot Hill, Kenya. *PaleoBios* 14:1–12.

———. 1977. *Tetralophodon* (Proboscidea, Gomphotheriidae) from Subsaharan Africa. *Revue de Zoologie Africaine* 91:153–160.

———. 1980. *Zygolophodon* from Subsaharan Africa, with observations on the systematics of palaeomastodontid proboscideans. *Journal of Paleontology* 54:57–64.

———. 1982. Primitive *Stegotetrabelodon* from latest Miocene of Subsaharan Africa. *Revue de Zoologie Africaine* 96:782–796.

Madden, C. T., K. W. Glennie, R. Dehm, F. C. Whitmore, R. J. Schmidt, R. J. Ferfoglia, and P. J. Whybrow. 1982. Stegotetrabelodon *(Proboscidea, Gomphotheriidae) from the Miocene of Abu Dhabi*. United States Geological Survey, Jiddah. 22 pp.

Maglio, V. J. 1969a. A shovel-tusked gomphothere from the Miocene of Kenya. *Breviora* 310:1–10.

———. 1969b. The status of the East African elephant *"Archidiskodon exoptatus"* Dietrich 1942. *Breviora* 336:1–24.

———. 1970a. Early Elephantidae of Africa and a tentative correlation of African Plio-Pleistocene deposits. *Nature* 225:328–332.

———. 1970b. Four new species of Elephantidae from the Plio-Pleistocene of northwestern Kenya. *Breviora* 341:1–43.

———. 1972. Evolution of mastication in the Elephantidae. *Evolution* 26:638–658.

———. 1973. Origin and evolution of the Elephantidae. *Transactions of the American Philosophical Society* 63:1–149.

———. 1974. A new proboscidean from the late Miocene of Kenya. *Palaeontology, London* 17:699–705.

———. 1978. Patterns of faunal evolution; pp. 603–619 in V. J. Maglio and H. B. S. Cooke (eds.), *Evolution of African Mammals*. Harvard University Press, Cambridge.

Maglio, V. J., and H. B. S. Cooke. 1978. *Evolution of African Mammals*. Harvard University Press, Cambridge, 641 pp.

Maglio, V. J., and Q. B. Hendey. 1970. New evidence relating to the supposed stegolophodont ancestry of the Elephantidae. *South African Archaeological Bulletin* 25:85–87.

Maglio, V. J., and A. B. Ricca. 1977. Dental and skeletal morphology of the earliest elephants. *Verhandelingen der Koninklijke Nederlandse Akademie van Wetenschappen, Afd. Natuurkunde Eerste Reeks, Deel* 29:1–51.

Mahboubi, M., R. Ameur, J. Y. Crochet, and J.-J. Jaeger. 1984. Earliest known proboscidean from early Eocene of north-west Africa. *Nature* 308:543–544.

———. 1986. El Kohol (Saharan Atlas, Algeria): a new Eocene mammal locality in North-West Africa. *Palaeontographica, Abt. A* 192:15–49.

Makinouchi, T., S. Ishida, Y. Sawada, N. Kuga, N. Kimura, Y. Orihashi, B. Bajope, M. wa Yemba, and H. Ishida. 1992. Geology of the Sinda-Mohari Region, Haut-Zaire Province, eastern Zaire. *African Study Monographs*, suppl. 17:3–18.

Matsumoto, H. 1922. Revision of *Palaeomastodon* and *Moeritherium*. *Palaeomastodon intermedius*, and *Phiomia osborni*, new species. *American Museum Novitates* 51:1–6.

———. 1923. A contribution to the knowledge of *Moeritherium*. *Bulletin of the American Museum of Natural History* 48:97–140.

———. 1924. A revision of *Palaeomastodon* dividing it into two genera, and with descriptions of two new species. *Bulletin of the American Museum of Natural History* 50:1–58.

Mawby, J. E. 1970. Fossil vertebrates from northern Malawi: Preliminary report. *Quaternaria* 13:319–324.

Mazo, A. V. 1996. Gomphotheres and mammutids from the Iberian Peninsula, pp. 136–142 in J. Shoshani and P. Tassy (eds.), *The Proboscidea: Evolution and Palaeoecology of Elephants and Their Relatives*. Oxford University Press, Oxford.

McDougall, I., F. H. Brown, and J. G. Fleagle. 2005. Stratigraphic placement and age of modern humans from Kibish, Ethiopia. *Nature* 433:733–736.

McDougall, I., and C. S. Feibel. 2003. Numerical age control for the Miocene-Pliocene succession at Lothagam, a hominoid-bearing sequence in the northern Kenya Rift; pp. 43–64 in M. G. Leakey and J. M. Harris (eds.), *Lothagam: The Dawn of Humanity in Eastern Africa*. Columbia University Press, New York.

McDougall, I., and R. T. Watkins. 1985. Age of hominoid-bearing sequence at Buluk, northern Kenya. *Nature* 318:175–178.

McKenna, M. C., and S. K. Bell. 1997. *Classification of Mammals above the Species Level*. Columbia University Press, New York, 631 pp.

Mebrate, A. 1983. Late Miocene–middle Pleistocene proboscidean fossil remains from the Middle Awash Valley, Afar Depression, Ethiopia. Unpublished master's thesis, Department of Systematics and Ecology, University of Kansas, 119 pp.

Mebrate, A., and J. E. Kalb. 1985. Anancinae (Proboscidea: Gomphotheriidae) from the Middle Awash Valley, Afar, Ethiopia. *Journal of Vertebrate Paleontology* 5:93–102.

Meiring, A. J. D. 1955. Fossil proboscidean teeth and ulna from Virginia, O.F.S. *Navorsinge van die Nasionale Museum, Bloemfontein* 1(8):187–201.

Metz-Muller, F. 1995. Mise en évidence d'une variation intra-spécifique des caractères dentaires chez *Anancus arvernensis* (Proboscidea, Mammalia) du gisement de Dorkovo (Pliocène ancien de Bulgarie, Biozone MN14). *Geobios* 28:737–743.

Meyer, H. von. 1831. Note on *Dinotherium giganteum* and *D. bavaricum*. *Neues Jahrbuch für Mineralogie* 1831:296–297.

Miller, E. R. 1996. Mammalian paleontology of an Old World monkey locality, Wadi Moghara, early Miocene, Egypt. Unpublished PhD dissertation, Washington University, St. Louis, 372 pp.

———. 1999. Faunal correlation of Wadi Moghara, Egypt: Implications for the age of *Prohylobates tandyi*. *Journal of Human Evolution* 36:519–533.

Miller, K. G., R. G. Fairbanks, and G. S. Mountain. 1987. Tertiary oxygen isotope synthesis, sea level history, and continental margin erosion. *Paleoceanography* 2:1–19.

Moody, R. T. J., and P. J. C. Sutcliffe. 1993. The sedimentology and palaeontology of the Upper Cretaceous-Tertiary deposits of central West Africa. *Modern Geology* 18:539–554.

Morrison-Scott, T. C. S. 1947. A revision of our knowledge of African elephants' teeth, with notes on forest and "pygmy" elephants. *Proceedings of the Zoological Society of London* 117:505–527.

Moullé, P. E., A. Eichassoux, Z. Alemseged, and E. Desclaux. 2001. On the presence of *Elephas recki* at the Oldowan prehistoric site of Fejej FJ-1 (Ethiopia); pp. 122–125 in G. Cavarretta, P. Gioia, M. Mussi, and M. R. Palombo (eds.), *Proceedings of the First International Congress of La Terra degli Elefanti: The World of Elephants*. Consiglio Nazionale delle Ricerche, Rome.

Moustafa, Y. S. 1974a. The Oligocene African Proboscidea: Part I. Introduction. *Annals of the Geological Survey of Egypt* 4:385–415.

———. 1974b. The Oligocene African Proboscidea: Part II. The *Phiomia-Palaeomastodon* question. *Annals of the Geological Survey of Egypt* 4:417–432.

Mundinger, G. S., and W. J. Sanders. 2001. Taxonomic and systematic re-assessment of the Mio-Pliocene elephant *Primelephas gomphotheroides*. In *Scientific Programme and Abstracts*, 8th International Theriological Congress, Sun City, South Africa, Additional Abstracts, 2.

Murphy, W. J., Eizirik, E., W. E. Johnson, Y. P. Zhang, O. A. Ryder, and S. J. O'Brien. 2001. Molecular phylogenetics and the origins of placental mammals. *Nature* 409:614–618.

Nakatsukasa, M., Y. Kunimatsu, H. Nakaya, Y. Sawada, T. Sakai, H. Hyodo, T. Itaya, and E. Mbua. 2007. Late Miocene fossil locality Nakali in Kenya and its paleoenvironment. *American Journal of Physical Anthropology*, suppl. 44:177.

Nakaya, H. 1993. Les Faunes de Mammifères du Miocène supérieur de Samburu Hills, Kenya, Afrique de l'est et l'environnement des Pré-Hominidés. *L'Anthropologie (Paris)* 97:9–16.

Nakaya, H., M. Pickford, Y. Nakano, and H. Ishida. 1984. The late Miocene large mammal fauna from the Namurungule Formation, Samburu Hills, northern Kenya. *African Study Monographs*, suppl. 2:87–131.

Nakaya, H., M. Pickford, K. Yasui, and Y. Nakano. 1987. Additional large mammalian fauna from the Namurungule Formation, Samburu Hills, northern Kenya. *African Study Monographs*, suppl. 5:79–129.

Noro, M., R. Masuda, I. A. Dubrovo, M. C. Yoshida, and M. Kato. 1998. Molecular phylogenetic inference of the woolly mammoth *Mammuthus primigenius*, based on complete sequences of mitochrondrial cytochrome *b* and 12S ribosomal RNA genes. *Journal of Molecular Evolution* 46:314–326.

O'Leary, M. A., E. M. Roberts, M. Bouare, F. Sissoko, and L. Tapanila. 2006. Malian Paenungulata (Mammalia: Placentalia): New African afrotheres from the early Eocene. *Journal of Vertebrate Paleontology* 26:981–988.

Olson, S. L., and D. T. Rasmussen. 1986. Paleoenvironments of the earliest hominoids: New evidence from the Oligocene avifauna of the Fayum Depression, Egypt. *Science* 233:1202–1204.

Osborn, H. F. 1909. The feeding habit of *Moeritherium* and *Palaeomastodon*. *Nature* 81:139–140.

———. 1919. *Palaeomastodon*, the ancestor of the long-jawed mastodons only. *Proceedings of the National Academy of Sciences, USA* 5:265–266.

———. 1921. The evolution, phylogeny and classification of the Proboscidea. *American Museum Novitates* 1:1–15.

———. 1928. Mammoths and man in the Transvaal. *Nature* 121:672–673.

———. 1934. Primitive *Archidiskodon* and *Palaeoloxodon* of South Africa. *American Museum Novitates* 741:1–15.

———. 1936. *Proboscidea: A Monograph of the Discovery, Evolution, Migration and Extinction of the Mastodonts and Elephants of the World: Vol. I. Moeritherioidea, Deinotherioidea, Mastodontoidea*. American Museum Press, New York, 802 pp. Figures courtesy The American Museum of Natural History.

———. 1942. *Proboscidea: A Monograph of the Discovery, Evolution, Migration and Extinction of the Mastodonts and Elephants of the World: Vol. II. Stegodontoidea, Elephantoidea*. American Museum Press, New York, 828 pp.

Osborn, H. F., and W. Granger. 1931. The shovel-tuskers, Amebelodontinae, of Central Asia. *American Museum Novitates* 470:1–12.

———. 1932. *Platybelodon grangeri*, three growth stages, and a new serridentine from Mongolia. *American Museum Novitates* 537:1–13.

Ozawa, T., S. Hayashi, and V. M. Mikhelson. 1997. Phylogenetic position of mammoth and Steller's sea cow within Tethytheria demonstrated by mitochondrial DNA sequences. *Journal of Molecular Evolution* 44:406–413.

Pagani, M., K. H. Freeman, and M. A. Arthur. 1999. Late Miocene atmospheric CO_2 concentrations and the expansion of C_4 grasses. *Science* 285:876–879.

Palmer, R. W. 1924. An incomplete skull of *Dinotherium*, with notes on the Indian forms. *Palaeontologia Indica* 7:1–14.

Partridge, T. C. 2000. Hominid-bearing cave and tufa deposits; pp. 100–125 in T. C. Partridge and R. R. Maud (eds.), *The Cenozoic of Southern Africa*. Oxford University Press, Oxford.

Partridge, T. C., G. C. Bond, C. J. H. Hartnody, P. B. deMenocal, and W. F. Ruddiman. 1995b. Climatic effects of late Neogene tectonism and volcanism; pp. 8–23 in E. S. Vrba, D. H. Denton, T. C. Partridge, and L. H. Burckle (eds.), *Paleoclimate and Evolution, with Emphasis on Human Origins*. Yale University Press, New Haven.

Partridge, T. C., A. G. Latham, and D. Heslop. 2000. Appendix on magnetostratigraphy of Makapansgat, Sterkfontein, Taung and Swartkrans; pp. 126–129 in T. C. Partridge and R. R. Maud (eds.), *The Cenozoic of Southern Africa*. Oxford University Press, Oxford.

Partridge, T. C., B. A. Wood, and P. B. deMenocal. 1995a. The influence of global climatic change and regional uplift on large-mammalian evolution in East and southern Africa; pp. 331–355 in E. S. Vrba, D. H. Denton, T. C. Partridge, and L. H. Burckle (eds.), *Paleoclimate and Evolution, with Emphasis on Human Origins*. Yale University Press, New Haven.

Patterson, B., A. K. Behrensmeyer, and W. D. Sill. 1970. Geology of a new Pliocene locality in northwestern Kenya. *Nature* 256:279–284.

Patterson, C., and A. E. Longbottom. 1989. An Eocene amiid fish from Mali, West Africa. *Copeia* 4:827–836.

Petrocchi, C. 1943. Il giacimento fossilifero di Sahabi. In *Collezione Scientifica e Documentaria a Cura del Ministero dell' Africa Italiana, Verbania*, 12, 169 pp.

———. 1954. Paleontologia di Sahabi: Parte I. Probosidati di Sahabi. *Rendiconti Accademia nazionale dei XL* 4–5:8–74.

Petronievics, B. 1923. Remarks upon the skull of *Moeritherium* and *Palaeomastodon*. *Annals and Magazine of Natural History, London*, series 9, 12:55–61.

Pickford, M. 1981. Preliminary Miocene mammalian biostratigraphy for western Kenya. *Journal of Human Evolution* 10:73–97.

———. 1982. The tectonics, volcanics and sediments of the Nyanza Rift Valley, Kenya. *Zeitschrift für Geomorphologie*, Neue Folge Supplement-Band 42:1–33.

———. 1985. A new look at *Kenyapithecus* based on recent discoveries in Western Kenya. *Journal of Human Evolution* 14:113–144.

———. 1986a. Cainozoic paleontological sites of Western Kenya. *Münchner Geowissenschaftliche Abhandlungen A* 8:1–151.

———. 1986b. Première découverte d'une faune mammalienne terrestre paléogène d'Afrique sub-saharienne. *Comptes Rendus de l'Académie des Sciences, Paris*, Série II, 19:1205–1210.

———. 1987a. The geology and palaeontology of the Kanam Erosion Gullies (Kenya). *Mainzer Geowissenschaftliche Mitteilungen* 16:209–226.

———. 1987b. Recognition of an early Oligocene or late Eocene mammal fauna from Cabinda, Angola. *Musée Royal de l'Afrique Centrale (Belgique), Rapport Annuel du Département de Géologie et de Mineralogie* 1985–1986:89–92.

———. 1991a. Biostratigraphic correlation of the middle Miocene mammal locality of Jabal Zaltan, Libya; pp. 1483–1490 in M. J. Salem, O. S. Hammuda, and B. A. Eliagoubi (eds.), *The Geology of Libya*, vol. 4. Academic Press, New York.

———. 1991b. Revision of Neogene Anthracotheriidae of Africa; pp. 1491–1525 in M. J. Salem and M. T. Busrewil (eds.), *The Geology of Libya*. Academic Press, New York.

———. 2001. *Afrochoerodon* nov. gen. *kisumuensis* (MacInnes)(Proboscidea, Mammalia) from Cheparawa, middle Miocene, Kenya. *Annales de Paléontologie* 87:99–117.

———. 2003. New Proboscidea from the Miocene strata in the lower Orange River Valley, Namibia. *Memoir Geological Survey of Namibia* 19:207–256.

———. 2004. Partial dentition and skeleton of *Choerolophodon pygmaeus* (Deperet) from Ngenyin, 13 Ma, Tugen Hills, Kenya: Resolution of a century old enigma. *Zona Arqueologica: Miscelànea en Homenaje a Emiliano Aguirre, Palaeontologia*, Madrid, Museo Arqueologica Regional 2:429–463.

————. 2005a. *Choerolophodon pygmaeus* (Proboscidea, Mammalia) from the middle Miocene of southern Africa. *South African Journal of Science* 101:175–177.

————. 2005b. Preliminary report on a proboscidean tooth from Daberas (middle Miocene), Sperrgebiet, Namibia. Unpublished report, 2 pp.

————. 2007. New mammutid proboscidean teeth from the Middle Miocene of tropical and southern Africa. *Palaeontologia Africana* 42:29–35.

Pickford, M., and P. Andrews. 1981. The Tinderet Miocene sequence in Kenya. *Journal of Human Evolution* 10:11–33.

Pickford, M., H. Ishida, Y. Nakano, and K. Yasui. 1987. The middle Miocene fauna from the Nachola and Aka Aiteputh Formations, northern Kenya. *African Study Monographs, Supplementary Issue* 5:141–154.

Pickford, M., and P. Mein. 2006. Early middle Miocene mammals from Moroto II, Uganda. *Beiträge zur Paläontologie* 30:361–386.

Pickford, M., and B. Senut. 2000. Geology and palaeobiology of the Namib Desert, Southwestern Africa. *Memoir Geological Survey of Namibia* 18:1–155.

Pickford, M., B. Senut, and D. Hadoto. 1993. *Geology and Palaeobiology of the Albertine Rift Valley, Uganda-Zaire. Volume I: Geology.* Occasional Publication 24, Centre International pour la Formation et les Echanges Géologiques, Orleans (France), pp. 1–190.

Pickford, M., B. Senut, D. Gommery, and E. Musiime. 2003. New catarrhine fossils from Moroto II, early middle Miocene (ca 17.5 Ma) Uganda. *Comptes Rendus Palevol* 2:649–662.

Pickford, M., B. Senut, and D. Hadoto, J. Musisi, and C. Kariira. 1986. Découvertes récentes dans les sites miocènes de Moroto (Ouganda Oriental): aspects biostratigraphiques et paléoécologiques. *Comptes Rendus de l'Académie des Sciences, Paris,* Série II 9:681–686.

Pickford, M., and P. Tassy. 1980. A new species of *Zygolophodon* (Mammalia, Proboscidea) from the Miocene hominoid localities of Meswa Bridge and Moroto (East Africa). *Neues Jahrbuch für Geologie und Paläontologie, Monatshefte* 4:235–251.

Plummer, T. W., and R. Potts. 1989. Excavations and new findings at Kanjera, Kenya. *Journal of Human Evolution* 18:269–276.

Pomel, A. 1879. Ossements d'Éléphants et d'Hippopotames découvertes dans une station préhistorique de la plaine d'Eglis (Province d'Oran). *Bulletin de la Société Géologique de France,* Série 3 7:44–51.

Qiu, Z.-X., B.-Y. Wang, H. Li, T. Deng, and Y. Sun. 2007. First discovery of deinotheres in China. *Vertebrata PalAsiatica* 45:261–277.

Rasmussen, D. T., and M. Gutiérrez. 2009. A mammalian fauna from the late Oligocene of northwestern Kenya. *Palaeontographica, Abt. A* 288:1–52.

Raynal, J.-P., F. Z. S. Alaoui, L. Magoga, A. Mohib, and M. Zouak. 2004a. The Lower Palaeolithic sequence of Atlantic Morocco revisited after recent excavations at Casablanca. *Bulletin d'Archéologie Marocaine* 20:44–76.

Raynal, J.-P., G. Kieffer, and G. Bardin. 2004b. Garba IV and the Melka Kunture Formation: A preliminary lithostratigraphic approach; pp. 137–166 in J. Chavaillon and M. Piperno (eds.), *Studies on the Early Paleolithic Site of Melka Kunture, Ethiopia.* Istituto Italiano di Preistoria e Protostoria, Florence.

Raza, S. M., and G. E. Meyer. 1984. Early Miocene geology and paleontology of the Bugti Hills, Pakistan. *Memoirs of the Geological Survey of Pakistan* 11:43–63.

Remy, J.-A. 1976. Presence de *Deinotherium* sp. Kaup (Proboscidea, Mammalia) dans la faune miocène de Beni Mellal (Maroc). *Géologie Méditerranéenne* 3:109–114.

Renne, P. R., G. WoldeGabriel, W. K. Hart, G. Heiken, and T. D. White. 1999. Chronostratigraphy of the Miocene-Pliocene Sagantole Formation, Middle Awash Valley, Afar Rift, Ethiopia. *Geological Society of America Bulletin* 111:869–885.

Richmond, B. G., J. G. Fleagle, J. Kappelman, and C. C. Swisher III. 1998. First hominoid from the Miocene of Ethiopia and the evolution of the catarrhine elbow. *American Journal of Physical Anthropology* 105:257–277.

Robinson, P. 1974. The Beglia Formation of Tunisia. *Memoires, Bureau de Recherches Géologiques et Minières (France)* 78:235.

Robinson, P., and C. C. Black. 1973. A small Miocene faunule from near Testour, Beja Gouvernorat, Tunisia. *Annales des Mines et de la Géologie* 26:445–449.

————. 1974. Vertebrate faunas from the Neogene of Tunisia. *Annals of the Geological Survey of Egypt* 4:319–332.

Roca, A. L., N. Georgiadis, and S. J. O'Brien. 2005. Cytonuclear genomic dissociation in African elephant species. *Nature Genetics* 37:96–100.

Roca, A. L., N. Georgiadis, J. Pecon-Slattery, and S. J. O'Brien. 2001. Genetic evidence for two species of elephant in Africa. *Science* 293:1473–1477.

Rogaev, E. I., Y. K. Moliaka, B. A. Malyarchuk, F. A. Kondrashov, M. V. Derenko, I. Chumakov, and A. P. Grigorenko. 2006. Complete mitochondrial genome and phylogeny of Pleistocene mammoth *Mammuthus primigenius*. *PLoS Biology* 4(3): e73; doi 10.1371/journal. pbio.0040073, 22 pp.

Roger, O. 1886. Ueber *Dinotherium bavaricum*. *Palaeontographica* 32:215–226.

Roger, J., M. Pickford, H. Thomas, F. de Lapparent de Broin, P. Tassy, W. van Neer, C. Bourdillon-de-Grissac, and S. Al-Busaidi. 1994. Découverte de vertébrés fossiles dans le Miocène de la région du Huqf au Sultanat d'Oman. *Annales de Paléontologie* 80:253–273.

Rögl, F. 1998. Palaeogeographic considerations for Mediterranean and Paratethys seaways (Oligocene to Miocene). *Annalen des Naturhistorischen Museums in Wien* 99A:279–310.

————. 1999. Oligocene and Miocene palaeogeography and stratigraphy of the circum-Mediterranean Region; pp. 485–500 in P. J. Whybrow and A. Hill (eds.), *Fossil Vertebrates of Arabia.* Yale University Press, New Haven.

Rohland, N., A.-S. Malaspinas, J. L. Pollock, M. Slatkin, P. Matheus, and M. Hofreiter. 2007. Proboscidean mitogenomics: chronology and mode of elephant evolution using mastodon as outgroup. *PLoS Biology* 5:1–9.

Roth, V. L. 1992. Quantitative variation in elephant dentitions: implications for the delimitation of fossil species. *Paleobiology* 18:184–202.

Sach, V. V. J., and E. P. J. Heizmann. 2001. Stratigraphie und Säugetierfaunen der Brackwassermolasse in der Umgebung von Ulm (Südwestdeutschland). *Stuttgarter Beiträge zur Naturkunde Serie B (Geologie und Paläontologie)* 310:1–95.

Saegusa, H. 1987. Cranial morphology and phylogeny of the stegodonts. *The Compass* 64:221–243.

————. 1996. Stegodontidae: evolutionary relationships; pp. 178–190 in J. Shoshani and P. Tassy (eds.), *The Proboscidea: Evolution and Palaeoecology of Elephants and Their Relatives.* Oxford University Press, Oxford.

Saegusa, H., and L. J. Hlusko. 2007. New late Miocene elephantoid (Mammalia: Proboscidea) fossils from Lemudong'o, Kenya. *Kirtlandia* 56:140–147.

Saegusa, H., Y. Thasod, and B. Ratanasthien. 2005. Notes on Asian stegodontids. *Quaternary International* 126–128:31–48.

Sahni, M. R., and C. Tripathi. 1957. A new classification of the Indian deinotheres and description of *D. orlovii* sp. nov. *Memoirs of the Geological Society of India, Paleontologia India* 33:1–33.

Sahnouni, M., and J. de Heinzelin. 1998. The site of Aïn Hanech revisited: New investigations at this Lower Pleistocene site in northern Algeria. *Journal of Archaeological Science* 25:1083–1101.

Sahnouni, M., D. Hadjouis, J. van der Made, A. Derradji, A. Canals, M. Medig, H. Belahrech, Z. Harichane, and M. Rabhi. 2002. Further research at the Oldowan site of Aïn Hanech, north-eastern Algeria. *Journal of Human Evolution* 43:925–937.

————. 2004. On the earliest human occupation in North Africa: A response to Geraads et al. *Journal of Human Evolution* 46:763–775.

Said, R. 1962. Über das Miozän in der westlichen wuste Ägyptens. *Geologisches Jahrbuch* 80:349–366.

Sánchez-Villagra, M. R., Y. Narita, and S. Kuratani. 2007. Thoracolumbar vertebral number: the first skeletal synapomorphy for afrotherian mammals. *Systematics and Biodiversity* 5:1–7.

Sanders, W. J. 1990. Fossil Proboscidea from the Pliocene Lusso Beds of the Western Rift, Zaïre; pp. 171–187 in N. T. Boaz (ed.), *Evolution of Environments and Hominidae in the African Western Rift Valley.* Virginia Museum of Natural History Memoir No. 1, Martinsville, Va.

————. 1996. Fossil proboscideans of the Manonga Valley, Tanzania. *Journal of Vertebrate Paleontology* 16:62A–63A.

————. 1997. Fossil Proboscidea from the Wembere-Manonga Formation, Manonga Valley, Tanzania; pp. 265–310 in T. Harrison (ed.), *Neogene Paleontology of the Manonga Valley, Tanzania.* Plenum Press, New York.

————. 1999. Oldest record of *Stegodon* (Mammalia: Proboscidea). *Journal of Vertebrate Paleontology* 19:793–797.

————. 2003. Proboscidea; pp. 202–219 in M. Fortelius, J. Kappelman, S. Sen, and R. L. Bernor (eds.), *Geology and Paleontology of the Miocene Sinap Formation, Turkey.* Columbia University Press, New York.

———. 2004. Taxonomic and systematic review of Elephantidae based on late Miocene–early Pliocene fossil evidence from Afro-Arabia. *Journal of Vertebrate Paleontology* 24 (suppl. to no. 3):109A.

———. 2005. New Pliocene fossil proboscidean specimens from Laetoli, Tanzania. *Journal of Vertebrate Paleontology* 25 (suppl. to no. 3):109A.

———. 2006. Comparative description and taxonomy of proboscidean fossils from Langebaanweg, South Africa. *African Natural History* 2:196–197.

———. 2007. Taxonomic review of fossil Proboscidea (Mammalia) from Langebaanweg, South Africa. *Transactions of the Royal Society of South Africa* 62:1–16.

———. 2008a. Review of fossil Proboscidea from the early–middle Miocene site of Jabal Zaltan, Libya. *Garyounis Scientific Bulletin*, Special Issue No. 5:217–239.

———. 2008b. Review of fossil Proboscidea from the late Miocene–early Pliocene site of As Sahabi, Libya. *Garyounis Scientific Bulletin*, Special Issue No. 5:245–260.

———. In press. Proboscidea; in T. Harrison (ed.), *Laetoli Revisited*, vol. 2. Karger, New York.

Sanders, W. J., and J. Kappelman. 2001. A new late Oligocene proboscidean fauna from Chilga, Ethiopia. P. 120 in *Scientific Programme and Abstracts, 8th International Theriological Congress, Sun City, South Africa*.

Sanders, W. J., J. Kappelman, and D. T. Rasmussen. 2004. New large-bodied mammals from the late Oligocene site of Chilga, Ethiopia. *Acta Palaeontologica Polonica* 49:365–392.

Sanders, W. J., and E. R. Miller. 2002. New proboscideans from the early Miocene of Wadi Moghara, Egypt. *Journal of Vertebrate Paleontology* 22:388–404.

Saunders, J. J. 1996. North American Mammutidae; pp. 271–279 in J. Shoshani and P. Tassy (eds.), *The Proboscidea: Evolution and Palaeoecology of Elephants and Their Relatives*. Oxford University Press, Oxford.

Savage, R. J. G. 1967. Early Miocene mammal faunas of the Tethyan region. *Systematics Association Publication* 7:247–282.

———. 1969. Early Tertiary mammal locality in southern Libya. *Proceedings of the Geological Society of London* 1657:167–171.

———. 1971. Review of the fossil mammals of Libya. *Symposium on the Geology of Libya, Faculty of Science, University of Libya*, pp. 217–225.

———. 1976. Review of early Sirenia. Systematic Zoology 25:344–351.

Savage, R. J. G., and W. R. Hamilton. 1973. Introduction to the Miocene mammal faunas of Gebel Zelten, Libya. *Bulletin of the British Museum (Natural History), Geology* 22:515–527.

Savage, R. J. G., and M. E White. 1965. Two mammal faunas from the early Tertiary of Central Libya. *Proceedings of the Geological Society of London* 1623:89–91.

Savage, R. J. G., and P. G. Williamson. 1978. The early history of the Turkana depression; pp. 375–394 in W. W. Bishop (ed.), *Geological Background to Fossil Man*. Scottish Academic Press, Edinburgh.

Sawada, Y., M. Pickford, T. Itaya, T. Makinouchi, M. Tateishi, K. Kabeto, S. Ishida, and H. Ishida. 1998. K-Ar ages of Miocene Hominoidea (*Kenyapithecus* and *Samburupithecus*) from Samburu Hills, northern Kenya. *Comptes Rendus de l'Académie des Sciences, Paris, Sciences de la Terre et des Planètes* 326:445–451.

Scally, M., O. Madsen, C. J. Douady, W. W. de Jong, M. J. Stanhope, and M. S. Springer. 2001. Molecular evidence for the major clades of placental mammals. *Journal of Mammalian Evolution* 8:239–277.

Schlosser, M. 1905. Review of Andrews' "Notes on an Expedition to the Fayum, Egypt, with descriptions of some new mammals" in *Geological Magazine* (V) 10: 337–343, 1903. *Neues Jahrbuch für Mineralogie, Geologie und Palaeontologie*, I, 1:156–157.

———. 1911. Beiträge zur Kenntnis der Oligozänen Landsäugetiere aus dem Fayum: Ägypten. *Beiträge zur Paläontologie und Geologie Österreich-Ungarns* 24:51–167.

Schoeninger, M. J., H. Reeser, and K. Hallin. 2003. Paleoenvironment of *Australopithecus anamensis* at Allia Bay, East Turkana, Kenya: Evidence from mammalian herbivore enamel stable isotopes. *Journal of Anthropological Archaeology* 22:200–207.

Scott, W. B. 1907. A collection of fossil mammals from the coast of Zululand. *Geological Survey of Natal and Zululand*, Third and Final Report:259–262.

Seiffert, E. R. 2006. Revised age estimates for the later Paleogene mammal faunas of Egypt and Oman. *Proceedings of the National Academy of Sciences, USA* 103:5000–5005.

———. 2007. A new estimate of afrotherian phylogeny based on simultaneous analysis of genomic, morphological, and fossil evidence. *BMC Evolutionary Biology* 7:224.

Semaw, S., S. W. Simpson, J. Quade, P. R. Renne, R. F. Butler, W. C. McIntosh, N. Levin, M. Dominguez-Rodrigo, and M. J. Rogers. 2005. Early Pliocene hominids from Gona, Ethiopia. *Nature* 433:301–304.

Sénégas, F., and D. M. Avery. 1998. New evidence for the murine origins of the Otamyinae and the age of Bolt's Farm. *South African Journal of Science* 94:503–507.

Senut, B., M. Pickford, M. de Wit, J. Ward, R. Spaggiari, and J. Morales. 1996. Biochronology of sediments at Bosluis Pan, Northern Cape Province, South Africa. *South African Journal of Science* 92:249–251.

Shackley, M. 1980. An Acheulean industry with *Elephas recki* fauna from Namib IV South West Africa (Namibia). *Nature* 284:340–341.

Shipman, P. 1986. Paleoecology of Fort Ternan reconsidered. *Journal of Human Evolution* 15:193–204.

Shipman, P., A. Walker, J. A. Van Couvering, P. J. Hooker, and J. A. Miller. 1981. The Fort Ternan hominoid site, Kenya: Geology, age, taphonomy and paleoecology. *Journal of Human Evolution* 10:49–72.

Shoshani, J. 1996. Para- or monophyly of the gomphotheres and their position within Proboscidea; pp. 149–177 in J. Shoshani and P Tassy (eds.), *The Proboscidea.. Evolution and Palaeoecology of Elephants and Their Relatives*. Oxford University Press, Oxford.

Shoshani, J., and J. F. Eisenberg. 1982. *Elephas maximus*. *Mammalian Species* 182:1–8.

Shoshani, J., E. M. Golenberg, and H. Yang. 1998. Elephantidae phylogeny: Morphological versus molecular results. *Acta Theriologica*, Suppl. 5:89–122.

Shoshani, J., W. J. Sanders, and P. Tassy. 2001a. Elephants and other proboscideans: a summary of recent findings and new taxonomic suggestions; pp. 676–679 in G. Cavarretta, P. Gioia, M. Mussi, and M. R. Palombo (eds.), *Proceedings of the First International Congress of la Terra degli Elefanti: The World of Elephants*. Consiglio Nazionale delle Ricerche, Rome.

Shoshani, J., R. J. G. Savage, and R. M. West. 1989. A new species of *Barytherium* (Mammalia, Proboscidea) from Africa, and a discussion of early proboscidean systematics and paleoecology. In *Abstracts of Papers and Posters, Fifth International Theriological Congress, Rome*, p. 160.

Shoshani, J., and P. Tassy. 1992. Classifying elephants; pp. 22–23 in J. Shoshani (ed.), *Elephants*. Weldon Owen, Sydney.

———. 1996. Summary, conclusions, and a glimpse into the future; pp. 335–348 in J. Shoshani and P. Tassy (eds.), *The Proboscidea: Evolution and Palaeoecology of Elephants and Their Relatives*. Oxford University Press, Oxford.

———. 2005. Advances in proboscidean taxonomy and classification, anatomy and physiology, and ecology and behavior. *Quaternary International* 126–128:5–20.

Shoshani, J., R. C. Walter, M. Abraha, S. Berhe, R. T. Buffler, and B. Negassi. 2001c. *Elephas recki* from Dandero, northern Danakil Depression, Eritrea; pp. 143–147 in G. Cavarretta, P. Gioia, M. Mussi, and M. R. Palombo (eds.), *Proceedings of the First International Congress of La Terra degli Elefanti: The World of Elephants*. Consiglio Nazionale delle Ricerche, Rome.

Shoshani, J., R. C. Walter, M. Abraha, S. Berhe, P. Tassy, W. J. Sanders, G. H. Marchant, Y. Libsekal, T. Ghirmai, and D. Zinner. 2006. A proboscidean from the late Oligocene of Eritrea, a "missing link" between early Elephantiformes and Elephantimorpha, and biogeographic implications. *Proceedings of the National Academy of Sciences, USA* 103:17296–17301.

Shoshani, J., R. C. Walter, Y. Libsekal, M. Abraha, and S. Berhe. 2001b. The oldest gomphothere in Africa: A latest Oligocene proboscidean mandible from the Dogali Formation, eastern Eritrea. *Abstracts of Papers and Posters, Eighth International Theriological Congress, Sun City, South Africa, 12–17 August 2001*, Abstract No. 337:128–129.

Shoshani, J., R. M. West, N. Court, R. J. G. Savage, and J. M. Harris. 1996. The earliest proboscideans: general plan, taxonomy, and palaeoecology; pp. 57–75 in J. Shoshani and P. Tassy (eds.), *The Proboscidea: Evolution and Palaeoecology of Elephants and Their Relatives*. Oxford University Press, Oxford.

Sickenberg, O., and M. Schönfeld. 1975. The Chorora Formation: Lower Pliocene limnical sediments in the southern Afar (Ethiopia); pp. 277–284 in A. Pilger and A. Rösler (eds.), *Afar Depression of Ethiopia*, vol. 1. E. Schweizerbart'sche Verlagsbuchhandlung (Nägele und Obermiller), Stuttgart.

Sikes, S. K. 1967. The African elephant, *Loxodonta africana*: A field method for the estimation of age. *Journal of Zoology, London* 154:235–248.

———. 1971. *The Natural History of the African Elephant*. Weidenfeld and Nicolson, London, 397 pp.

Simons, E. L. 1968. Early Cenozoic mammalian faunas. Fayum Province, Egypt: Part I. African Oligocene mammals: Introduction, history of study, and faunal succession. *Peabody Museum of Natural History, Yale University Bulletin* 28:1–21.

Simpson, G. G. 1945. The principles of classification and a classification of mammals. *Bulletin of the American Museum of Natural History* 85:1–350.

Singer, R., and D. A. Hooijer. 1958. A *Stegolophodon* from South Africa. *Nature* 182:101–102.

Springer, M. S., G. C. Cleven, O. Madsen, W. W. de Jong, V. G. Waddell, H. M. Amrine, and M. J. Stanhope. 1997. Endemic African mammals shake the phylogenetic tree. *Nature* 388:61–64.

Springer, M. S., W. J. Murphy, E. Eizirik, and S. J. O'Brien. 2003. Placental mammal diversification and the Cretaceous Tertiary boundary. *Proceedings of the National Academy of Sciences, USA* 100:1056–1061.

Stanhope, M. J., V. G. Waddell, O. Madsen, W. de Jong, S. B. Hedges, G. C. Cleven, D. Kao, and M. S. Springer. 1998. Molecular evidence for multiple origins of Insectivora and for a new order of endemic African insectivore mammals. *Proceedings of the National Academy of Sciences, USA* 95:9967–9972.

Sudre, J., J.-J. Jaeger, B. Sigé, and M. Vianey-Liaud. 1993. Nouvelles données sur les condylarthres du Thanétien et de l'Yprésien du Bassin d'Ouarzazate (Maroc). *Geobios* 26:609–615.

Sukumar, R., and C. Santiapillai. 1996. *Elephas maximus*: Status and distribution; pp. 327–331 in J. Shoshani and P. Tassy (eds.), *The Proboscidea: Evolution and Palaeoecology of Elephants and Their Relatives*. Oxford University Press, Oxford.

Suwa, G., H. Nakaya, B. Asfaw, H. Saegusa, A. Amzaye, R. T. Kono, Y. Beyene, and S. Katoh. 2003. Plio-Pleistocene terrestrial mammal assemblage from Konso, southern Ethiopia. *Journal of Vertebrate Paleontology* 23:901–916.

Suwa, G., T. White, B. Asfaw, G. WoldeGabriel, and T. Yemane. 1991. Miocene faunal remains from the Burji-Soyama area, Amaro Horst, southern sector of the main Ethiopian rift. *Palaeontologia Africana* 28:23–28.

Symeonidis, N. K. 1970. Ein *Dinotherium*-Fund in Zentralmakedonien (Griechenland). *Annales Géologiques des Pays Helléniques* 3:1153–1165.

Tabuce, R., R. J. Asher, and T. Lehmann. 2008. Afrotherian mammals: A review of current data. *Mammalia* 72:2–14.

Tabuce, R., C. Delmer, and E. Gheerbrant. 2007a. Evolution of the tooth enamel microstructure in the earliest proboscideans (Mammalia). *Zoological Journal of the Linnean Society* 149:611–628.

Tabuce, R., L. Marivaux, A. Adaci, M. Bensalah, J.-L. Hartenberger, M. Mahboubi, F. Mebrouk, P. Tafforeau, and J.-J. Jaeger. 2007b. Early Tertiary mammals from North Africa reinforce the molecular Afrotheria clade. *Proceedings of the Royal Society of London, B* 274:1159–1166.

Taieb, M., D. C. Johanson, Y. Coppens, and J. L. Aronson. 1976. Geological and paleontological background of Hadar hominid site, Afar, Ethiopia. *Nature* 260:289–293.

Takai, F., H. Saegusa, Thaung-Htike, and Zin-Maung-Maung-Thein. 2006. Neogene mammalian fauna in Myanmar. *Asian Paleoprimatology* 4:143–172.

Tamrat, E., N. Thouveny, M. Taieb and N. D. Opdyke. 1995. Revised magnetostratigraphy of the Plio-Pleistocene sedimentary sequence of the Olduvai Formation (Tanzania). *Palaeogeography, Palaeoclimatology, Palaeoecology* 114:273–283.

Tassy, P. 1977a. Découverte de *Zygolophodon turicensis* (Schinz) (Proboscidea, Mammalia) au Lieu-Dit Malartic a Simorre, Gers (Vindobonien Moyen): Implications paléoécologiques et biostratigraphiques. *Geobios* 10:655–659.

———. 1977b. Présence du genre *Choerolophodon* Schlesinger (Proboscidea, Mammalia) dans le Miocène est-africain. *Comptes Rendus de l'Académie des Sciences, Paris*, Série D, 284:2487–2490.

———. 1979a. Les Proboscidiens (Mammalia) du Miocène d'Afrique orientale: résultats préliminaires. *Bulletin de la Société Géologique de France* 21:265–269.

———. 1979b. Relations phylogénétiques du genre *Moeritherium* Andrews, 1901 (Mammalia). *Comptes Rendus de l'Académie des Sciences de Paris*, Série D, 189:85–88.

———. 1981. Le crâne de *Moeritherium* (Proboscidea, Mammalia) de l'Eocène de Dor el Talha (Libye) et le problème de la classification phylogénétique du genre dans les Tethytheria McKenna, 1975. *Bulletin du Muséum National d'Histoire Naturelle de Paris*, 4e Série, 3, Section C, 1:87–147.

———. 1982. Les principales dichotomies dans l'histoire des Proboscidea (Mammalia): Une approche phylogénétique. *Geobios Mémoire Spécial* 6:225–245.

———. 1983a. Les Elephantoidea Miocènes du Plateau du Potwar, Groupe de Siwalik, Pakistan: Ie Partie. Introduction: Cadre chronologique et géographique, Mammutidés, Amébélodontidés. *Annales de Paléontologie* 69:96–136.

———. 1983b. Les Elephantoidea Miocènes du Plateau du Potwar, Groupe de Siwalik, Pakistan. IIe Partie: Choerolophodontes et Gomphothères: Les Elephantoidea Miocènes du Plateau du Potwar, Groupe de Siwalik, Pakistan. *Annales de Paléontologie* 69:235–297.

———. 1984. Le mastodonte à dents étroites, le grade trilophodonte et la radiation initiale des Amebelodontidae; pp. 459–473 in E. Buffetaut, J.-M. Mazin, and E. Salmon (eds.), *Actes du Symposium Paléontologique*. Cuvier, Montbeliard.

———. 1985. La place des mastodontes miocènes de l'ancien monde dans la phylogénie des Proboscidea (Mammalia): Hypothèses et conjectures. Unpublished PhD dissertation, Université Pierre et Marie Curie, Paris, vols. 1–3, 861 pp.

———. 1986. *Nouveaux Elephantoidea (Mammalia) dans le Miocène du Kenya*. Cahiers de Paleontologie, Éditions du Centre de la Recherche Scientifique, Paris, 135 pp.

———. 1988. The classification of Proboscidea: How many cladistic classifications? *Cladistics* 4:43–57.

———. 1989. The "Proboscidean Datum Event": How many proboscideans and how many events? pp. 237–252 in E. H. Lindsay, V. Fahlbusch, and P. Mein (eds.), *European Neogene Mammal Chronology*. Plenum Press, New York.

———. 1990. Phylogénie et classification des Proboscidea (Mammalia): Historique et actualité. *Annales de Paléontologie* 76:159–224.

———. 1994a. Gaps, parsimony, and early Miocene elephantoids (Mammalia), with a re-evaluation of *Gomphotherium annectens* (Matsumoto, 1925). *Zoological Journal of the Linnean Society* 112:101–117.

———. 1994b. Origin and differentiation of the Elephantiformes (Mammalia, Proboscidea). *Verhandlungen des Naturwissenschaftlichen Vereins in Hamburg* 34:73–94.

———. 1995. Les Proboscidiens (Mammalia) Fossiles du Rift Occidental, Ouganda; pp. 217–257 in M. Pickford and B. Senut (eds.), *Geology and Palaeobiology of the Albertine Rift Valley, Uganda-Zaire. Vol. II: Palaeobiology*. CIFEG Occasional Publications, 1994/29, Orléans.

———. 1996a. Dental homologies and nomenclature in the Proboscidea; pp. 21–25 in J. Shoshani and P. Tassy (eds.), *The Proboscidea: Evolution and Palaeoecology of Elephants and Their Relatives*. Oxford University Press, Oxford.

———. 1996b. The earliest gomphotheres; pp. 89–91 in J. Shoshani and P. Tassy (eds.), *The Proboscidea: Evolution and Palaeoecology of Elephants and Their Relatives*. Oxford University Press, Oxford.

———. 1996c. Who is who among the Proboscidea? pp. 39–48 in J. Shoshani and P. Tassy (eds.), *The Proboscidea: Evolution and Palaeoecology of Elephants and Their Relatives*. Oxford University Press, Oxford.

———. 1999. Miocene elephantids (Mammalia) from the Emirate of Abu Dhabi, United Arab Emirates: Palaeobiogeographic implications; pp. 209–233 in P. J. Whybrow and A. Hill (eds.), *Fossil Vertebrates of Arabia*. Yale University Press, New Haven.

———. 2003. Elephantoidea from Lothagam; pp. 331–358 in M. G. Leakey and J. M. Harris (eds.), *Lothagam: The Dawn of Humanity in Eastern Africa*. Columbia University Press, New York.

Tassy, P., P. Anupandhanant, L. Ginsburg, P. Mein, B. Ratanasthien, and V. Suteethorn. 1992. A new *Stegolophodon* (Proboscidea, Mammalia) from the early Miocene of northern Thailand. *Geobios* 25:511–523.

Tassy, P., and P. Darlu. 1986. Les Elephantidae: Nouveau regard sur les analyses de parcimonie. *Geobios* 20:487–494.

Tassy, P., and R. Debruyne. 2001. The timing of early Elephantinae differentiation: The palaeontological record, with a short comment on molecular data; pp. 685–687 in G. Cavarretta, P. Gioia, M. Mussi, and M. R. Palombo (eds.), *Proceedings of the First International Congress of La Terra degli Elefanti: The World of Elephants*. Consiglio Nazionale delle Ricerche, Rome.

Tassy, P., and M. Pickford. 1983. Un nouveau Mastodonte zygolophodonte (Proboscidea, Mammalia) dans le Miocène Inférieur d'Afrique Orientale: Systématique et Paléoenvironnement. *Geobios* 16:53–77.

Tassy, P., and J. Shoshani. 1988. The Tethytheria; elephants and their relatives; pp. 283–315 in M. J. Benton (ed.), *The Phylogeny and Classification of the Tetrapods: Vol. 2. Mammals.* The Systematics Association, Special Volume No. 35B, Clarendon, Oxford.

———. 1996. Historical overview of classification and phylogeny of the Proboscidea; pp. 3–8 in J. Shoshani and P. Tassy (eds.), *The Proboscidea. Evolution and Palaeoecology of Elephants and Their Relatives.* Oxford University Press, Oxford.

Tchamba, M. N., and P. M. Seme. 1993. Diet and feeding behaviour of the forest elephant in the Santchou Reserve, Cameroon. *African Journal of Ecology* 31:165–171.

Thomas, M. G., and A. M. Lister. 2001. A statistical appraisal of molecular and morphological evidence for mammoth-elephant relationships; pp. 688–692 in G. Cavarretta, P. Gioia, M. Mussi, and M. R. Palombo (eds.), *Proceedings of the First International Congress of La Terra degli Elefanti: The World of Elephants.* Consiglio Nazionale delle Ricerche, Rome.

Thomas, H., and G. Petter. 1986. Révision de la faune de mammifères du Miocène supérieur de Menacer (ex-Marceau), Algérie: Discussion sur l'âge du gisement. *Geobios* 19:357–373.

Thomas, H., J. Roger, S. Sen, C. Bourdillon-de-Grissac, and Z. Al-Sulaimani. 1989. Découverte de vertébrés fossiles dans l'Oligocène inférieur du Dhofar (Sultanat d'Oman). *Geobios* 22:101–120.

Thomas, H., J. Roger, S. Sen, M. Pickford, E. Gheerbrant, Z. Al-Sulaimani, and S. Al-Busaidi. 1999. Oligocene and Miocene terrestrial vertebrates in the southern Arabian Peninsula (Sultanate of Oman) and their geodynamic and palaeogeographic settings; pp. 430–442 in P. J. Whybrow and A. Hill (eds.), *Fossil Vertebrates of Arabia.* Yale University Press, New Haven.

Thomas, H., S. Sen, M. Khan, B. Battail, and G. Ligabue. 1982. The Lower Miocene fauna of Al-Sarrar (Eastern province, Saudi Arabia). *Atlal* 5:109–136.

Thomas, M. G., E. Hagelberg, H. B. Jones, Z. Yang, and A. M. Lister. 2000. Molecular and morphological evidence on the phylogeny of the Elephantidae. *Proceedings of the Royal Society of London, B* 267:2493–2500.

Tiercelin, J.-J., J. Michaux, and Y. Bandet. 1979. Le Miocène supérieur du Sud de la Dépression de l'Afar, Éthiopie: Sédiments, faunes, âges isotopiques. *Bulletin de la Société Géologique de France* 21:255–258.

Tiffney, B. H., J. G. Fleagle, and T. M. Bown. 1994. Early to middle Miocene angiosperm fruits and seeds from Fejej, Ethiopia. *Tertiary Research* 15:25–42.

Tobien, H. 1962. Über carpus und tarsus von *Deinotherium giganteum* Kaup. *Paläontologische Zeitschrift,* H. Schmidt-Festband 1962:231–238.

———. 1971. *Moeritherium, Palaeomastodon, Phiomia* aus dem Paläogen Nordafrikas und die Abstammung der Mastodonten (Proboscidea, Mammalia). *Mitteilungen aus dem Geologischen Institut der Technischen Universität Hannover* 10:141–163. Figure © 200X The Society of Vertebrate Paleontology.

Tobien, H. 1973a. On the evolution of mastodonts (Proboscidea, Mammalia). Part I: The bunodont trilophodont group. *Notizblatt des Hessischen Landesamtes für Bodenforschung zu Wiesbaden* 101:202–276.

———. 1973b. The structure of the mastodont molar (Proboscidea, Mammalia): Part 1. The bunodont pattern. *Mainzer schaftliche Mitteilungen* 2:115–147.

———. 1975. The structure of the mastodont molar (Proboscidea, Mammalia): Part 2. The zygodont and zygobunodont patterns. *Mainzer Geowissenschaftliche Mitteilungen* 4:195–233.

———. 1978. The structure of the mastodont molar (Proboscidea, Mammalia). Part 3: The Oligocene mastodont genera *Palaeomastodon, Phiomia* and the Eo/Oligocene Paenungulate *Moeritherium. Mainzer Geowissenschaftliche Mitteilungen* 6:177–208.

———. 1980. A note on the skull and mandible of a new choerolophodont mastodont (Proboscidea, Mammalia) from the middle Miocene of Chios (Aegean Sea, Greece), pp. 299–307 in L. L. Jacobs (ed.), *Aspects of Vertebrate History: Essays in Honor of Edwin Harris Colbert.* Museum of Northern Arizona Press, Flagstaff.

———. 1986. Mastodonts (Proboscidea, Mammalia) from the Late Neogene and early Pleistocene of the People's Republic of China. *Mainzer Geowissenschaftliche Mitteilungen* 15:119–181.

———. 1988. Les Proboscidiens Deinotheriidae. *Palaeovertebrata, Mémoire Extraordinaire* 1988:135–175.

Tobien, H. 1996. Evolution of zygodons with emphasis on dentition; pp. 76–85 in J. Shoshani and P. Tassy (eds.), *The Proboscidea: Evolution and Palaeoecology of Elephants and Their Relatives.* Oxford University Press, Oxford.

Tobien, H., G. Chen, and Y. Li. 1986. Mastodonts (Proboscidea, Mammalia) from the late Neogene and early Pleistocene of the People's Republic of China: Part 1. Historical account: The genera *Gomphotherium, Choerolophodon, Synconolophus, Amebelodon, Platybelodon, Sinomastodon, Mainzer Geowissenschaftliche Mitteilungen* 15:119–181.

———. 1988. Mastodonts (Proboscidea, Mammalia) from the late Neogene and early Pleistocene of the People's Republic of China: Part 2. The genera *Tetralophodon, Anancus, Stegotetrabelodon, Zygolophodon, Mammut, Stegolophodon:* Some generalities on the Chinese Mastodonts. *Mainzer Geowissenschaftliche Mitteilungen* 17:95–220.

Todd, N. E. 2005. Reanalysis of African *Elephas recki*: Implications for time, space and taxonomy. *Quaternary International* 126–128:65–72.

———. 2006. Trends in proboscidean diversity in the African Cenozoic. *Journal of Mammalian Evolution* 13:1–10.

Todd, N. E., and V. L. Roth. 1996. Origin and radiation of the Elephantidae; pp. 193–202 in J. Shoshani and P. Tassy (eds.), *The Proboscidea: Evolution and Palaeoecology of Elephants and Their Relatives.* Oxford University Press, Oxford.

Tsoukala, E. S., and J. K. Melentis. 1994. *Deinotherium giganteum* Kaup (Proboscidea) from Kassandra Peninsula (Chalkidiki), Macedonia, Greece. *Geobios* 27:633–640.

Tsujikawa, H. 2005a. The palaeoenvironment of *Samburupithecus kiptalami* based on its associated fauna. *African Study Monographs,* suppl. 32:51–62.

———. 2005b. The updated late Miocene large mammal fauna from Samburu Hills, northern Kenya. *African Study Monographs,* suppl. 32:1–50.

Vacek, M. 1877. Über österreichische Mastodonten und ihre Beziehungen zu den Mastoden-Arten Europas. *Abhandlungen der Kaiserlich-Köninglichen geologischen Reichsanstalt* 7:1–45.

Van Couvering, J. A. H., and J. A. Van Couvering. 1976. Early Miocene mammal fossils from East Africa: aspects of geology, faunistics and palaeoecology; pp. 155–207 in G. L. Isaac and E. R. McCown (eds.), *Human Origins: Louis Leakey and the East African Evidence.* Benjamin, Menlo Park.

Vanoverstraeten, M., J. Van Gysel, P. Tassy, B. Senut, and M. Pickford. 1990. Découverte d'une molaire éléphantine dans le Pliocène de la région d'Ishasha, Parc national des Virunga, sud du lac Édouard, Province du Kivu, Zaïre. *Comptes Rendus de l'Académie des Sciences, Paris,* Série II, 311:887–892.

Vialli, V. 1966. Sul rinvenimento di Dinoterio (*Deinotherium* cf. *hobleyi* Andrews) nelle ligniti di Ath Ugri (Eritrea). *Giornale di Geologia (Bologna)* 33:447–458.

Vidya, T. N. C., P. Fernando, D. J. Melnick, and R. Sukumar. 2005. Population differentiation within and among Asian elephant (*Elephas maximus*) populations in southern India. *Heredity* 94:71–80.

Vignaud, P., P. Duringer, H. T. Mackaye, A. Likius, C. Blondel, J.-R. Boisserie, L. de Bonis, V. Eisenmann, M.-E. Etienne, D. Geraads, F. Guy, T. Lehmann, F. Lihoureau, N. Lopez-Martinez, C. Mourer-Chauviré, O. Otero, J.-C. Rage, M. Schuster, L. Viriot, A. Zazzo, and M. Brunet. 2002. Geology and palaeontology of the Upper Miocene Toros-Menalla hominid locality, Chad. *Nature* 418:152–155.

Walker, A. 1969. Fossil mammal locality on Mount Elgon, eastern Uganda. *Nature* 223:591–593.

Weinsheimer, O. 1883. Über *Dinotherium giganteum* Kaup. *Palaeontologische Abhandlungen* 1:205–282.

Wells, N. A., and P. D. Gingerich. 1983. Review of Eocene Anthracobunidae (Mammalia, Proboscidea) with a new genus and species, *Jozaria palustris,* from the Kuldana Formation of Kohat (Pakistan). *Contributions from the Museum of Paleontology* (University of Michigan) 26:117–139.

West, R. M. 1983. South Asian middle Eocene moeritheres (Mammalia: Tethytheria). *Annals of the Carnegie Museum* 52:359–373.

———. 1984. A review of the South Asian middle Eocene Moeritheriidae (Mammalia: Tethytheria). *Mémoires de la Société Géologique de France* 147:183–190.

White, L. J. T., C. E. G. Tutin, and M. Fernandez. 1993. Group composition and diet of forest elephants, *Loxodonta africana cyclotis* Matschie 1900, in the Lope Reserve, Gabon. *African Journal of Ecology* 31:181–199.

White, T. D., R. V. Moore, and G. Suwa. 1984. Hadar biostratigraphy and hominid evolution. *Journal of Vertebrate Paleontology* 4:575–583.

White, T. D., G. WoldeGabriel, B. Asfaw, S. Ambrose, Y. Beyene, R. L. Bernor, J.-R. Boisserie, B. Currie, H. Gilbert, Y. Haile-Selassie, W. K. Hart, L. J. Hlusko, F. C. Howell, R. T. Kono, T. Lehmann, A. Louchart, C. O. Lovejoy, P. R. Renne, H. Saegusa, E. S. Vrba, H. Wesselman, and G. Suwa. 2006. Asa Issie, Aramis and the origin of *Australopithecus*. *Nature* 440:883–889.

Whybrow, P. J., and D. Clements. 1999. Arabian Tertiary fauna, flora, and localities; pp. 460–473 in P. J. Whybrow and A. Hill (eds.), *Fossil Vertebrates of Arabia*. Yale University Press, New Haven.

Whybrow, P. J., McClure, H. A., and Elliott, G. F. 1987. Miocene stratigraphy, geology and flora (Algae) of eastern Saudi Arabia and the Ad Dabtiyah vertebrate locality. *Bulletin of the British Museum (Natural History)* 41:371–382.

Wight, A. W. R. 1980. Paleogene vertebrate fauna and regressive sediments of Dur at Talhah, southern Sirt Basin, Libya; pp. 309–325 in M. J. Salem and M. T. Busrewil (eds.), *The Geology of Libya*, vol. 1. Academic Press, London.

WoldeGabriel, G., Y. Haile-Selassie, P. R. Renne, W. K. Hart, S. H. Ambrose, B. Asfaw, G. Helken, and T. White. 2001. Geology and palaeontology of the late Miocene Middle Awash valley, Afar rift, Ethiopia. *Nature* 412:175–178.

WoldeGabriel, G., T. D. White, G. Suwa, P. Renne, J. de Heinzelin, W. K. Hart, and G. Helken. 1994. Ecological and temporal placement of early Pliocene hominids at Aramis, Ethiopia. *Nature* 371:330–333.

Wynn, J. G., Z. Alemseged, R. Bobe, D. Geraads, D. Reed, and D. C. Roman. 2006. Geological and palaeontological context of a Pliocene juvenile hominin at Dikika, Ethiopia. *Nature* 443:332–336.

Yang, H., E. M. Golenberg, and J. Shoshani. 1996. Phylogenetic resolution within the Elephantidae using fossil DNA sequence from the American mastodon *(Mammut americanum)* as an outgroup. *Proceedings of the National Academy of Sciences, USA* 93:1190–1194.

Yasui, K., Y. Kunimatsu, N. Kuga, B. Bajope, and H. Ishida. 1992. Fossil mammals from the Neogene strata in the Sinda Basin, Eastern Zaire. *African Study Monographs*, suppl. 17:87–107.

Zazzo, A., H. Bocherens, M. Brunet, A. Beauvilain, D. Billiou, H. T. Mackaye, P. Vignaud, and A. Mariotti. 2000. Herbivore paleodiet and paleoenvironmental changes in Chad during the Pliocene using stable isotope ratios of tooth enamel carbonate. *Paleobiology* 26:294–309.

SIXTEEN

Paleogene "Insectivores"

ERIK R. SEIFFERT

Paleogene Afro-Arabian placentals of "insectivoran" grade—reluctantly referred to here, for sake of brevity and lack of a solid taxonomic framework, simply as Paleogene African "insectivores"—have thus far been placed either in endemic African genera *(Chambilestes, Eochrysochloris, Garatherium,*[1] *Jawharia, Todralestes, Widanelfarasia)* or in genera that have also been documented in the fossil records of Europe *(Aboletylestes, Afrodon)* and North America *(Cimolestes, Palaeoryctes).* Whereas Paleogene insectivores from northern continents are increasingly known from well-preserved cranial remains (e.g., Thewissen and Gingerich, 1989; Asher et al., 2002, 2005; Bloch et al., 2004), the African assemblage is comparatively very limited, being composed of isolated teeth, a few partial maxillae and mandibles, and possibly two distal humeri. These scanty records nevertheless bear directly (though not yet conclusively) on some central outstanding issues in early placental mammalian evolution, such as the time and place of origin of the enigmatic Afrosoricida (the afrotherian clade containing tenrecs [Tenrecoidea] and golden moles [Chrysochloridae]) and the nature of Afro-Arabia's biogeographic isolation through the Late Cretaceous and early Cenozoic. Note that Afrosoricida as used here is equivalent to the Tenrecoidea of Asher (this volume, ch. 9).

Interpretation and classification of the Paleogene African "insectivores" has become extremely problematic since the recognition of the endemic Afro-Arabian clade Afrotheria (Seiffert and Simons, 2000; Gheerbrant and Rage, 2006; Asher and Seiffert, this volume, chap. 46), because molecular data suggest a diphyletic origin of Lipotyphla (e.g., Roca et al., 2004), and hence evolution of detailed dental convergences in the Laurasian (eulipotyphlan) and Afro-Arabian (afrosoricid) radiations. Placement of early fossil "insectivores" into either Eulipotyphla or Afrosoricida purely on the basis of geography would be ill advised, but it is not inconceivable that all of the taxa considered in this chapter might be afrotherians aligned with Afrosoricida. For most taxa the evidence is equivocal, but there are a few nonmolar features (discussed further later) shared by Paleocene *Todralestes,* Eocene *Widanelfarasia,* and Miocene

Protenrec that are interpreted as providing limited support for these taxa being consecutive sister taxa of crown Tenrecoidea, while dental features of early Oligocene *Eochrysochloris* align that genus with golden moles (Seiffert et al., 2007).

Systematic Paleontology

Infraclass PLACENTALIA Owen, 1837
Genus *"ABOLETYLESTES"* Russell, 1964
"?ABOLETYLESTES HYPSELUS" Russell, 1964
Figures 16.1D–16.1F

Age and Occurrence Late Paleocene (late Thanetian), Adrar Mgorn 1 and Ihadjamène, Jbel Guersif Formation, Ouarzazate Basin, Morocco.

Diagnosis Bases of para- and metacone not fused on M1; small conules and weak or absent postparaconule cristae on M1–3; small parastyle on M1; metastylar lobe decreases in size distally.

Description Only upper molars are known (figures 16.1D–16.1F). The parastyle on M1 (IDJ 13) is small but is approximately equal in height to the stylocone on M2–3. The preparacrista is long and meets the stylocone labially. Meta- and paracone are separated at their bases on M1 and become more fused on M2–3. Minute conules are present on M1, lacking on M2–3. Metastylar lobes decrease in size distally. Pre- and postcingula are absent.

Remarks African "?A. hypselus" differs from late Paleocene European *A. hypselus* in having a more narrow M1 with a reduced parastyle and a longer and more buccally oriented preparacrista, and in lacking well-developed conules, internal cristae, cusp "D", and a deep M2 ectoflexus. Phylogenetic analysis with or without a chronobiogeographic character (see the discussion) aligns "?A. hypselus" with other African taxa to the exclusion of European *A. hypselus* (see figure 16.4, later).

"ABOLETYLESTES" ROBUSTUS Gheerbrant, 1992
Figures 16.1A and 16.1B

Age and Occurrence Late Paleocene (late Thanetian), Adrar Mgorn 1, Jbel Guersif Formation, Ouarzazate Basin, Morocco.

---

[1] Gheerbrant and Rage (2006) consider the species identified by Kappelman et al. (1996) as "Herpetotheriinae gen. nov., sp. nov." from the Kartal Formation, Turkey, to be a possible representative of *Garatherium.*

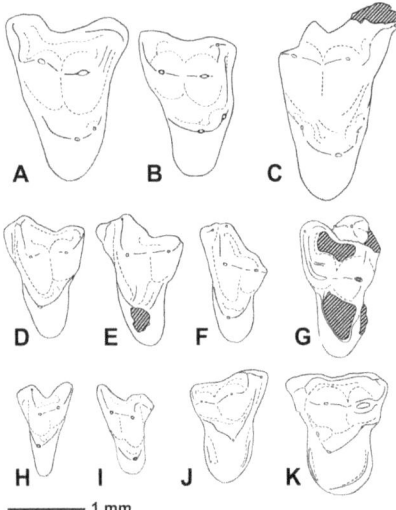

FIGURE 16.1 Upper molars of late Paleocene "insectivores" from the Ouarzazate Basin, Morocco (Adrar Mgorn 1 and Ihadjamène), based on camera lucida drawings of original specimens. A) THR 184, "*Aboletylestes*" *robustus* M2?; B) THR 126, "*Aboletylestes*" *robustus* M1?; C) THR 130, "*Cimolestes*" *cuspulus* M2?; D) IDJ 13, "?*Aboletylestes hypselus*" M1?; E) THR 165, "?*Aboletylestes hypselus*" M2?; F) THR 163, "?*Aboletylestes hypselus*" M3; G) THR 195, possible afrosoricid (chrysochlorid?) M1 or M2; H) THR 159, "*Palaeoryctes*" *minimus* M2?; I) THR 173, "*Palaeoryctes*" *minimus* M1?; J) M2 from THR 134, type of *Todralestes variabilis*; K) M2 from THR 140, also placed in the *T. variabilis* hypodigm.

Diagnosis Bases of robust para- and metacone fused; poorly developed parastylar region; long postmetacristae with carnassial notch present on the M2 postmetacrista.

Description THR 126 is here considered to be an M1, and THR 184 an M2 (figures 16.1A, 16.1B). The primary cusps are robust, and the bases of the para- and metacones are fused. The parastylar region is poorly developed, but the postmetacrista is long and buccally oriented, with a distinct carnassial notch on M2. A small paraconule is present on the relatively mesially oriented preprotocrista; there are no postparaconule cristae and only a faint metaconule on M2. The postprotocristae are more distally oriented than in contemporaneous African species. Pre- and postcingula are absent.

Remarks "*A.*" *robustus* is radically different from *A. hypselus* in the shape and greater development of the stylar region, the relatively long postmetacrista, and the shape of the trigon, and it is certainly not a close relative of the European species. The poorly developed preparacrista and parastylar region, long postmetacrista, and distinct carnassial notch on M2 suggest the possibility of a distant relationship with hyaenodontids. If THR 126 is in fact an M1, then "*A.*" *robustus* would also have had a relatively small M1 when compared with M2, also as in hyaenodontids. "*A.*" *robustus* could lend additional support to the African origin of Hyaenodontidae proposed by Gheerbrant et al. (2006) and Solé et al. (2009).

Genus "*CIMOLESTES*" Marsh, 1889i
"*CIMOLESTES*" *CUSPULUS* Gheerbrant, 1992
Figure 16.1C

Age and Occurrence Late Paleocene (late Thanetian), Adrar Mgorn 1, Jbel Guersif Formation, Ouarzazate Basin, Morocco.

Diagnosis Accessory conules present along molar pre- and postprotocristae (figure 16.1C); metastylar cusp and cusp "D" present; preparacrista courses mesially toward the parastyle; carnassial notch on the M3 preparacrista.

Description Accessory conules along molar pre- and postprotocristae lack internal cristae. Paracone is slightly taller than the metacone and the bases of the cusps are partially fused. A deep ectoflexus, distinct metastylar lobe, metastylar cusp, and cusp "D" are present. The preparacrista is oriented mesially toward the parastyle. Pre- and postcingula are absent. An isolated m3 has robust protoconid and metaconid cusps, the latter of which is placed slightly distal to the former. The cristid obliqua meets the distal face of the trigonid midway between the proto- and metaconid.

Remarks "*C.*" *cuspulus* does not provide sufficient morphological information to convincingly confirm or refute the placement of this species in the otherwise Late Cretaceous and earliest Paleocene North American genus *Cimolestes*, which itself is poorly known, differs little from primitive Late Cretaceous placentals, and is not known to occur on landmasses intermediate between Afro-Arabia and North America.[2] A more appropriate placement for this species is currently Placentalia *incertae sedis*.

Genus "*PALAEORYCTES*" Matthew, 1913
"*PALAEORYCTES*" *MINIMUS* Gheerbrant, 1992
Figures 16.1H–16.1I

Age and Occurrence Late Paleocene (latest Thanetian), Adrar Mgorn 1, Jbel Guersif Formation, Ouarzazate Basin, Morocco.

Diagnosis No P4 metacone; small P4 protocone placed far mesial to the apex of the paracone; relatively short P4–M2 pre- and postprotocristae; M1–2 preprotocristae terminate near the base of the paracone; no conules; buccolingually restricted M1–2 trigons; M1–2 metacones placed buccal to paracones; small M1–2 parastylar lobes; deep M1–2 ectoflexi.

Description M1–2 (figures 16.1H, 16.1I) are very broad, with an acute angle between pre- and postprotocristae. No conules are present. The ectoflexus is deep on ?M2 (THR 159) but relatively small on ?M1 (THR 173), and the stylar area is enclosed by tall prepara- and postmetacrista on ?M2 (the preparacrista is not well developed on ?M1). Parastyles are weak; on ?M2 the parastyle and stylocone are present, but the former is much lower than the latter. Buccal cusps are extensively fused and the metacone is buccally placed, forming a continuous shearing surface. Pre- and postcingula are absent. P4 shows no clear development of a parastyle; the paracone is tall and has a long, buccally curving postparacrista. The P4 protocone is small, situated close to the base of the paracone, and supports small para- and metaconules.

Distinct precingulids are present on lower molars. Molar metaconids are approximately equal in height to the protoconids, and paraconids are relatively small. The cristid obliqua and preentocristids of the talonid basin are tall and oriented at a sharp angle relative to the distal face of the trigonid; the talonid is open lingually and the basin is strongly canted lingually. The cristid obliqua meets the distal face of the metaconid. There is no clear development of an entoconid and the hypoconulid and hypoconid meet to form a tall crest enclosing the buccal aspect of the basin. A larger tooth from Adrar Mgorn 1 (THR 217) was described

[2] Gheerbrant (1992) notes that *Didelphodus* cf. *absarokae* (Godinot, 1981) from Rians, France, might represent *Cimolestes*.

by Gheerbrant (1992) as *"Palaeoryctes* cf. *minimus"*; it has a larger, more mesially oriented paraconid and less specialized talonid basin morphology/orientation than the other *"P." minimus* lower molars.

Remarks "Palaeoryctes" minimus bears only a superficial resemblance to Holarctic palaeoryctids, and Fox (2004) has argued that its placement in Palaeoryctidae "is almost certainly erroneous" (p. 612). *"Palaeoryctes" minimus* is more derived toward zalambdodonty than any other African species known from before the early Miocene, and could be a stem afrosoricid. If zalambdodonty evolved convergently in tenrecs and golden moles (Seiffert et al., 2007), *"P." minimus* could be nested within Afrosoricida, possibly as a stem chrysochlorid (though the limited phylogenetic analysis presented here places the species outside crown Afrosoricida; see figure 16.4, later). The larger unnamed taxon represented by THR 195 (figure 16.1G) might also be a primitive chrysochlorid.

Genus AFRODON Gheerbrant, 1988
AFRODON CHLEUHI Gheerbrant, 1988

Age and Occurrence Late Paleocene (late Thanetian), Adrar Mgorn 1 and Ihadjamène, Jbel Guersif Formation, Ouarzazate Basin, Morocco.

Diagnosis Stylar shelves broad; small ectoflexi; long and buccally oriented preparacristae; no pre- or postcingula; linear centrocristae; para- and metaconules with weak internal cristae. On m1–2, paraconid is relatively small and the metaconid is placed distal to the protoconid.

Description M1 (THR 168) is longer and narrower than M2, and it has a poorly developed parastylar region, much as in "?A. hypselus." M2 is also similar to that of "?A. hypselus" in having a reduced metastylar lobe and a long preparacrista that meets the stylocone. Metaconules are relatively well developed. Pre- and postcingula are absent. The p4 has a small paraconid and a larger metaconid that is about half the height of the protoconid. A crest runs lingually from the p4 hypoconid to connect with the postmetacristid. Lower molar metaconids are well developed and slightly lower than the protoconids; paraconids are relatively small. The cristid obliqua meets the trigonid midway between the proto- and metaconid and delimits well-developed hypoflexids. The hypoconulid varies from being centrally placed to slightly more lingual, and entoconids vary from being cuspidate to cristiform.

Remarks Gheerbrant (1995) suggested that *A. chleuhi* is the sister group of all other known adapisoriculids. Phylogenetic analysis of morphological characters alone supports this hypothesis (figure 16.4A) and implies an African origin for Adapisoriculidae, but analysis following inclusion of a chronobiogeographic character aligns *A. chleuhi* with other African taxa to the exclusion of undoubted European adapisoriculids (figure 16.4B). *Afrodon chleuhi* is here left outside Adapisoriculidae and treated as Placentalia *incertae sedis*. Storch (2008) recently argued that isolated humeri and femora from the late Paleocene of Walbeck (Germany) belong to either *Afrodon germanicus* or *Bustylus* cf. *cernaysi*, two undoubted adapisoriculids that occur at that site. The humero-ulnar articulations of the specimens from Walbeck bear at least a superficial resemblance to those of THR 364 and 365, two distal humeri from Adrar Mgorn 1 that Gheerbrant (1994) preliminarily assigned to *Todralestes variabilis*. If these specimens belong to the slightly larger *A. chleuhi*, they could reasonably be interpreted as providing additional support for

that species' alleged adapisoriculid affinities. Storch (2008) argued that the specimens from Walbeck support plesiadapiform affinities for Adapisoriculidae, but the teeth of these taxa bear no special resemblance to either plesiadapiforms or crown primates.

AFRODON TAGOURTENSIS Gheerbrant, 1993

Age and Occurrence Middle early Eocene (middle Ypresian), N'Tagourt 2, Aït Ouarithane Formation, Ouarzazate Basin, Morocco.

Diagnosis M1 relatively long and narrow, with distinct para- and metaconule; paraconule relatively lingually situated with respect to the metaconule; narrow stylar region without distinct stylar cusps; m1 relatively small in comparison to m2. Distinct p4 entoconid, with no cristid obliqua.

Description Two probable M1s are known (NTG 2-18 and 2-23) and are more similar to those of European adapisoriculids than to that of *A. chleuhi* in having relatively elongate stylar regions. Unlike European species, the parastyle and stylocone are very small or absent, and there is no development of other stylar cusps. *Afrodon tagourtensis* is also more similar to European taxa such as *Adapisoriculus minimus* in having a paraconule that is relatively lingual in position with respect to the metaconule. The preparacrista is short and buccally oriented and meets the stylocone, the ectoflexus is very faint or absent, and there are no pre- or postcingula. As in *A. chleuhi*, the molar metaconids are placed distal to the protoconids and the cristid obliqua meets the trigonid midway between the proto- and metaconid; the angle between the cristid obliqua and hypocristid is acute on m2. The p4 talonid is bicuspid, with a distinct entoconid and hypoconid, but no cristid obliqua is present.

Remarks Phylogenetic analysis of morphological features alone places *A. tagourtensis* as the sister taxon of European adapisoriculids (figure 16.4A, later), while inclusion of a chronobiogeographic character places the species as the sister taxon of *A. chleuhi*, with no special relationship to European taxa (figure 16.4B).

Genus GARATHERIUM CROCHET, 1984
GARATHERIUM MAHBOUBII Crochet, 1984

Age and Occurrence Middle early Eocene (middle Ypresian), El Kohol, Algeria.

Diagnosis Dilambdodont arrangement of buccal crests; stylar cusps well developed, including mesostyle and cusp "D"; preparacrista meets the stylocone; small conules present, with internal cristae.

Description Only a single upper molar of *G. mahboubii* is known, possibly an M1 on the basis of the mesiodistally elongate stylar region. The buccal crests are W shaped, with preparacrista meeting the stylocone and postpara- and premetacristae meeting the mesostyle. Cusp "D" is present just mesial to the buccal terminus of the postmetacrista. Conules are small and have internal cristae, with paraconule situated mesial (rather than lingual) to the metaconule. Pre- and postcingula are absent.

Remarks Garatherium mahboubii was originally described as a peradectine marsupial (Crochet, 1984); adapisoriculid affinities were later suggested by Gheerbrant (Gheerbrant, 1991, 1995). More recently McKenna and Bell (1997) placed *Garatherium* among didelphimorph herpetotheriine (= herpethotheriid) marsupials. *Garatherium* is here considered to be a placental; the genus is very poorly known but interestingly shares a number of M1 features with latest Eocene *Widanelfarasia* (most notably a dilambdodont arrangement of the buccal cusps). If tenrecoid zalambdodonty evolved from dilambdodonty (Seiffert et al., 2007), then *Garatherium* could be a stem tenrecoid.

?*GARATHERIUM TODRAE* Gheerbrant, 1998

Age and Occurrence Late Paleocene (latest Thanetian), Adrar Mgorn 1 and Ihadjamène, Jbel Guersif Formation, Ouarzazate Basin, Morocco.

Diagnosis Slightly larger than *G. mahboubii*.

Description The figured holotype of ?*G. todrae* (Gheerbrant et al., 1998) is likely to be an M2, but it is very similar in morphology to the probable M1 of *G. mahboubii* in having a dilambdodont arrangement of the buccal crests and a distinct stylocone, mesostyle, and cusp D. Gheerbrant et al. (1998) describe an M3 with a straight centrocrista (THR-MFSP 35) as possibly belonging to ?*G. todrae*, but the postparacrista and premetacrista are buccally oriented on another probable M3 that was placed in the ?*G. todrae* hypodigm (THR 273; see Gheerbrant, 1995).

Family CHAMBILESTIDAE Gheerbrant and Hartenberger, 1999
Genus *CHAMBILESTES* Gheerbrant and Hartenberger, 1999
CHAMBILESTES FOUSSANENSIS Gheerbrant and Hartenberger, 1999

Age and Occurrence Early or early middle Eocene (Ypresian or early Lutetian), Chambi, Kasserine Plateau, Tunisia.

Diagnosis Differs from other Paleogene African species in combining the following features: pre- and postcingula present and distinct; linear centrocrista; para- and metacone separated at their bases; conules and internal cristae are present; stylar shelves relatively narrow. P4 with well-developed protocone and paraconule, without postprotocrista.

Description The type and only specimen is a maxilla with P4-M3. The molars are narrow and broad, without well-developed stylar shelves. Para- and metaconules are present and bear internal cristae. Pre- and postcingula are present on M1–2, with small hypocones present on M1–2. The P4 has a well-developed, mesially situated protocone that lacks a postprotocrista but has a small paraconule on the preprotocrista. P4 has a small parastyle and distinct buccal cingulum.

Remarks Gheerbrant and Hartenberger (1999) suggested that *Chambilestes* might have affinities with Laurasian "erinaceomorphs," particularly *Scenopagus*, but their phylogenetic analyses did not support such a relationship. The phylogenetic analyses presented here (figure 16.4) place *Chambilestes* in either a very basal position (when morphological characters alone are considered), or as the sister taxon of *Todralestes* (when a chronobiogeographic character is included).

Cohort ?AFROTHERIA Stanhope et al., 1998
Order ?AFROSORICIDA Stanhope et al., 1998
Family TODRALESTIDAE Gheerbrant, 1991
Genus *TODRALESTES* Gheerbrant, 1991
TODRALESTES VARIABILIS Gheerbrant, 1991
Figures 16.1J, 16.1K, and 16.2

Age and Occurrence Late Paleocene (Thanetian), Adrar Mgorn 1, Ihadjamène, Ilimzi, and Timadriouine, Ouarzazate Basin, Morocco.

Diagnosis Differs from other African species in combing the following features: low and lingually situated P4 protocone; M1–2 with pre- and postcingula, small hypocones, narrow stylar shelves, and small para- and metaconules; p4 paraconid almost as tall as metaconid; molar talonids only slightly narrower than trigonids.

Description There is a considerable amount of variation within the *T. variabilis* hypodigm, and at least one specimen (THR 140; figure 16.1K) is arguably different enough from the holotype (figure 16.1J) to warrant generic distinction. In the type and similar material, P3 is two rooted, and the distal alveolus is placed mesial to the lingual root of P4, as in *Protenrec* and *Widanelfarasia* (Seiffert et al., 2007). The P4 paracone is tall and robust, and the protocone is low, close to the base of the paracone. Upper molar stylar shelves are relatively narrow, paracones are taller than metacones, conules are poorly developed, and weak pre- and postcingula are present. THR 151 and THR 349 demonstrate that *T. variabilis* had a two-rooted p2 but no p1. The p3 is simple, with a dominant, mesially oriented protoconid, while the p4 has a small metaconid and paraconid and a deep hypoflexid defined by a crest that links the hypoconid to the postvallid. Lower molars are similar in having low talonids that are narrower than the tall trigonids, with shallow hypoflexids. Two distal humeri have been referred to *Todralestes*, and both have globular capitula with short tails, entepicondylar foramina, and dorsoepitrochlear fossae; however, as noted earlier, these specimens might belong to the slightly larger species *Afrodon chleuhi*. See Gheerbrant (1994) for a detailed treatment of this material.

Remarks The molar morphology of *Todralestes* is quite primitive (figure 16.2), but the robust P4 paracone, two-rooted and "inset" P3, and presumably stepped transition from P4 to P3 are apomorphic features shared with *Protenrec* and *Widanelfarasia*. *Todralestes* also shares with these and other afrosoricids the apomorphic loss of p1. *Todralestes* could be a stem or crown afrosoricid (Seiffert, 2003; Seiffert et al., 2007).

TODRALESTES BUTLERI Gheerbrant, 1993

Age and Occurrence Middle early Eocene (middle Ypresian), N'Tagourt 2, Aït Ouarithane Formation, Ouarzazate Basin, Morocco.

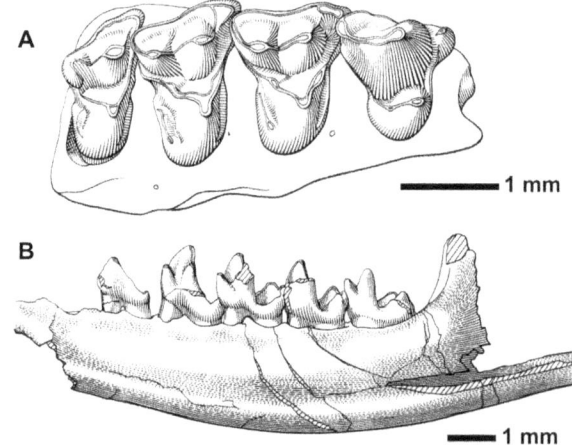

FIGURE 16.2 Upper and lower dentition of late Paleocene *Todralestes variabilis*. A) THR 134, holotype right maxilla with P4-M3; B) THR 90, right mandible with p3-m3. From Gheerbrant, 1991. Permission granted, Copyright 2010 © Elsevier.

Diagnosis Differs from *T. variabilis* in having relatively large conules with postparaconule and premetaconule cristae.

Description On M1, metacone is smaller than the paracone, and the two cusps show some basal fusion. Conules are relatively well developed and have internal cristae. Pre- and postcingula are present and distinct; on one specimen (NTG 2-16), a complete lingual cingulum is present. The stylar shelf is narrow, without a clear ectoflexus. P4 paracone is tall, with a small style on the buccally curving postparacrista; buccal cingulum is incomplete.

Remarks The features that distinguish *T. butleri* from *T. variabilis* are shared with *Chambilestes*, and the former might be a member of Chambilestidae rather than Todralestidae.

Order AFROSORICIDA Stanhope et al., 1998
?Suborder TENRECOMORPHA Butler, 1972
Genus *WIDANELFARASIA* Seiffert and Simons, 2000
WIDANELFARASIA BOWNI Seiffert and Simons, 2000
Figure 16.3

Age and Occurrence Late Eocene (latest Priabonian), Quarry L-41, lower sequence of Jebel Qatrani Formation, northern Egypt.

Diagnosis Differs from other African species in combining the following features: in the upper dentition, P4 has a robust paracone, low protocone, and ectocrista; buccal cusps of M1–2 are situated internally, and preparacristae are long and buccally oriented; buccal crests on M1 arranged in the dilambdodont pattern; pre- and postcingula are absent; conules are minute or absent. In the lower dentition, p1 is absent; p4 talonid is bicuspid, with a crest linking the hypoconid to the lingual part of the postvallid; molar trigonids are tall, talonids are relatively narrow, with deep hypoflexids.

Description In the upper dentition (figure 16.3A), P2–3 are two rooted, and P3 is inset. The P4 has a robust paracone with a buccal ectocrista that delimits an ectofossa, and a low protocone situated close to the base of the paracone. Upper molars have broad stylar shelves and long, buccally oriented preparacristae, and minute conules. M1 is dilambdodont, whereas M2 is quasi-zalambdodont. In the lower dentition (figure 16.3B), i2 is enlarged and bears a distal basal cusp. The canine is large and single-rooted. The p2–3 are double-rooted with mesially inclined protoconids. The p4 has a low paraconid and a larger metaconid; there is a

deep hypoflexid on the talonid basin. The lower molars have tall trigonids and low talonids, and protocristids are transversely oriented; molar hypoflexids are deep, increasing in depth distally.

Remarks Nonmolar features align *Widanelfarasia* with tenrecs via early Miocene *Protenrec* (Seiffert et al., 2007); the latter also shares numerous derived features with the extant Malagasy genus *Geogale*, and more material of *Protenrec* is needed to test its placement as either a stem tenrecoid or a nested member of the Malagasy tenrecid clade. Regardless, *Widanelfarasia*'s well-developed metacones are almost certainly plesiomorphic features that exclude the genus from crown Tenrecoidea. *Widanelfarasia* provides evidence for an intermediate stage between moderate dilambdodonty and zalambdodonty.

WIDANELFARASIA RASMUSSENI Seiffert and Simons, 2000

Age and Occurrence Late Eocene (latest Priabonian), Quarry L-41, lower sequence of Jebel Qatrani Formation, northern Egypt.

Diagnosis Differs from *W. bowni* in its smaller size and in having relatively narrow talonids.

Description Morphology of the lower p4–m3 is very similar to that of *W. bowni*; upper dentition is not known.

Genus *JAWHARIA* Seiffert et al., 2007
JAWHARIA TENRECOIDES Seiffert et al., 2007

Age and Occurrence Early Oligocene (early Rupelian), Quarry E, lower sequence of Jebel Qatrani Formation, northern Egypt.

Diagnosis Differs from *Widanelfarasia* in having a relatively narrow m3 talonid, deeper hypoflexids on m2–3, and a faint ectocristid on m2.

Description Only known from a single jaw with m3 and part of m2. Morphology of m3 is similar to that of *Widanelfarasia*, but the talonid is relatively narrow. The m2 hypoconid is placed buccally but bears a faint ectocristid on its buccal face that defines the distal wall of a well-developed hypoflexid. The m2 cristid obliqua is concave.

Suborder CHRYSOCHLOROIDEA Broom, 1915
Genus *EOCHRYSOCHLORIS* Seiffert et al., 2007
EOCHRYSOCHLORIS TRIBOSPHENUS Seiffert et al., 2007

Age and Occurrence Early Oligocene (early Rupelian), Quarry E, lower sequence of Jebel Qatrani Formation, northern Egypt.

Diagnosis Differs from other Paleogene African taxa in having a one-rooted p3 with a cingular talonid, p4 with tall metaconid and paraconid, a relatively narrow m2 talonid with low talonid cusp relief and a more mesially oriented cristid obliqua, and molar trigonids that are strongly canted lingually with respect to the dorsoventral axis of the mandibular corpus.

Description The p3 is single rooted and has a distal cingulum, as in the Miocene chrysochlorid *Prochrysochloris*. The p4 is molariform, with a large paraconid and metaconid and a well-developed talonid basin. The only known lower molar (m2) has a tall trigonid, narrow trigonid, and, when compared with *Widanelfarasia* or *Protenrec*, a relatively shallow hypoflexid and mesially oriented cristid obliqua. The molar trigonid is canted strongly lingually.

Remarks The occlusal surface of *Eochrysochloris*'s only known molar talonid differs in detail from those of *Widanelfarasia*

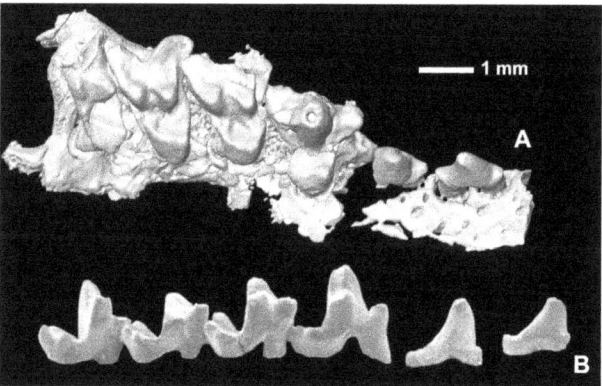

FIGURE 16.3 Upper and lower dentition of late Eocene *Widanelfarasia bowni*. A) DPC 21845, a right maxilla with P2-M3; B) teeth of holotype (CGM 83698), right mandible with p2-m3.

and *Jawharia* in having a relatively poorly developed hypoconid and shallow hypoflexid, suggesting a different arrangement of the occluding upper molar cusps and crests.

Discussion

Paleontologists have often assumed that Africa served as something of a biogeographic cul-de-sac and not a major center of origin for placental clades; as such, new discoveries from Africa have generally been interpreted within the context of the much more complete Laurasian record of placental evolution. Perhaps not surprisingly, most of the late Paleocene and early Eocene African "insectivores" have, at one time or another, been closely aligned with species, genera, or families documented on northern continents, and published taxonomies and phylogenetic interpretations imply at least seven Paleocene or Eocene exchanges of "insectivore" taxa between Afro-Arabia and Eurasia (Gheerbrant and Rage, 2006). There is no evidence for a land bridge connecting Afro-Arabia to Eurasia in the latest Cretaceous or early Paleogene, however, and if the taxa considered here are nonaquatic and nonvolant, such exchanges would have been chance events involving overwater dispersal and/or

island hopping. Morphological evidence for phylogenetic hypotheses that imply such widespread dispersal is, however, very weak—largely being based on a few highly homoplasious or plesiomorphic dental characters.

Figure 16.4 presents the results of phylogenetic analyses of the more complete fossil taxa considered in this chapter (i.e., those that could be scored for at least 50% of the morphological characters sampled), alongside members of the European adapisoriculid radiation that have been identified as close relatives of early African taxa such as *Afrodon* (Gheerbrant, 1995; Gheerbrant and Rage, 2006). The matrix includes 57 morphological characters, primarily from the dentition. A molecular scaffold constraining the monophyly of crown Tenrecoidea and of Malagasy Tenrecidae (cf. Asher and Hofreiter, 2006) was enforced by scoring 10 additional characters as either "0" or "1," first scoring only crown tenrecoids as "1" for characters 59–63, and then scoring only Malagasy tenrecids as "1" for characters 64–67; stem placentals and other extant taxa were scored as "0" for all of these characters.

Analysis of morphological characters alone places African members of the genus *Afrodon* as consecutive sister taxa of European adapisoriculids (figure 16.4A), but most other

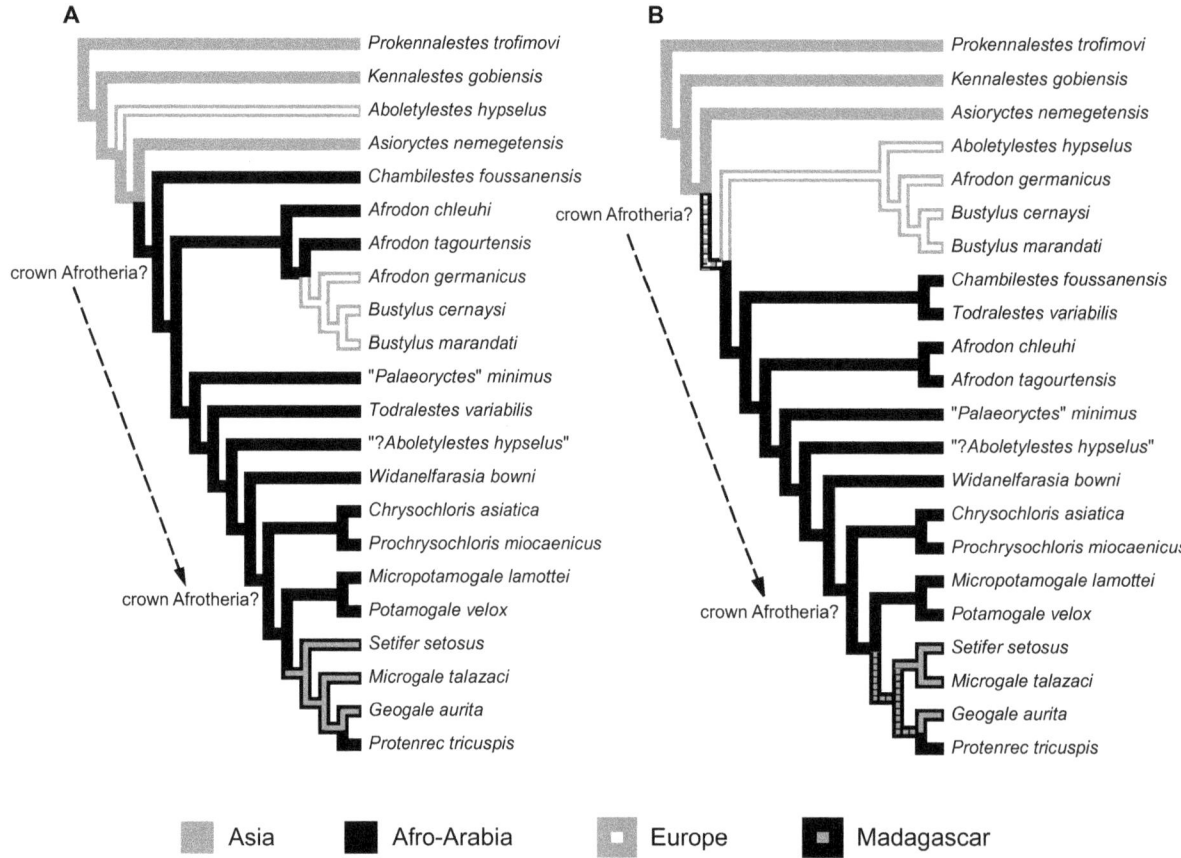

FIGURE 16.4 Strict consensus trees derived from parsimony analysis of 57 morphological characters without (A) and with (B) a chronobiogeographic character. Crown Tenrecoidea and the Malagasy tenrecid clade were constrained to be monophyletic (see text), but fossil taxa were free to fall inside or outside those clades. All multistate characters treated as ordered were scaled, and polymorphisms were scored as an intermediate state. See Rossie and Seiffert (2006) for methodological details of chronobiogeographic analysis. Character matrix is available on request from the author. Tree in (A) has a length (TL) of 162.22 steps, a consistency index (CI) of 0.42, a retention index (RI) of 0.62, and a rescaled consistency index (RCI) of 0.261. The tree in (B) was recovered by first searching (in PAUP 4.0b10) for morphological trees that were equal to or shorter than the shortest tree recovered by heuristic search (in Mesquite v. 2.5) of the matrix that included the chronobiogeographic character (essentially the "debt ceiling" method employed in stratocladistic analysis). The >10,000,000 trees recovered in PAUP that satisfied this criterion were then filtered (using the "Filter trees from other source" command in Mesquite) to determine if any of the trees had a shorter overall length when the chronobiogeographic character was included. Of the >10,000,000 trees, only a single most parsimonious tree was found (that figured in [B]), which was of length 168.33.

African taxa aside from *Chambilestes*, including "?*Aboletylestes hypselus*" are more closely aligned with afrosoricids; early Miocene *Protenrec* is nested within the Malagasy tenrec radiation as the sister taxon of extant *Geogale*, implying a dispersal back across the Mozambique Channel to account for its presence in Africa (cf. Asher and Hofreiter, 2006). A more rigorous test of the dispersals implied by this cladogram is made possible by adding a chronobiogeographic character for which each taxon is assigned a "time/space" state, and which conservatively adds a single step to tree length for each overwater dispersal required by the cladogram (using a step matrix that takes into account reconstructed paleogeography at each "time slice" represented). A complete description of this methodology is beyond the scope of this chapter, but the approach is described in detail in Rossie and Seiffert (2006). The shortest trees recovered by parsimony analysis following inclusion of the chronobiogeographic character reveal that the morphological evidence supporting overwater dispersal is weak, for *Afrodon* is placed as the sister group of all other Afro-Malagasy taxa. *Protenrec* is still placed as the sister taxon of *Geogale*, but a back-migration would not be required by the cladogram in figure 16.4B. *Todralestes* and *Chambilestes* are placed as sister taxa. It is important to note that in both of these trees it is not

known where Afrosoricida would join other afrotherians, and so it is not clear that taxa placed outside crown Afrosoricida on these trees are stem afrosoricids, stem members of Afroinsectivora (Afrosoricida + Macroscelidea), or stem afrotherians.

This analysis incorporates too few characters and taxa to be seen as anything other than a very preliminary assessment of the relationships among Paleogene Afro-Arabian "insectivores." However, based on these results and other direct observations on taxa not included in the phylogenetic analysis, a number of testable hypotheses can be proposed: (1) "*Aboletylestes*" *robustus* is aligned with the endemic Afro-Arabian placental radiation that gave rise to Hyaenodontidae; (2) "?*Aboletylestes hypselus*" (Gheerbrant, 1992) is not a close relative of European *Aboletylestes* and is aligned with other African taxa, possibly as a stem or crown afrosoricid; (3) *Afrodon*, *Garatherium*, "*Palaeoryctes*" *minimus*, *Todralestes*, *Widanelfarasia*, and possibly *Chambilestes* are stem or crown afrosoricids. "*Cimolestes*" *cuspulus* is too poorly known to be considered anything other than Placentalia *incertae sedis*. As with so many other early Paleogene placentals, the superordinal (e.g., afrotherian vs. laurasiatherian) affinities of most early African "insectivores" in fact remain very much open to debate and can only be convincingly tested with much more complete material (see also table 16.1).

TABLE 16.1
Occurrences of Paleogene Afro-Arabian "insectivores"

Taxon	Occurrence (Site, Locality)	Stratigraphic Unit	Age	Key References
"?*Aboletylestes hypselus*"	Adrar Mgorn 1, Ihadjamène (Morocco)	Jbel Guersif Fm.	Late Paleocene	Gheerbrant, 1992
"*Aboletylestes*" *robustus*	Adrar Mgorn 1	Jbel Guersif Fm.	Late Paleocene	Gheerbrant, 1992
"Adapisoriculidae gen. et sp. indet."	N'Tagourt 2 (Morocco)	Aït Ouarithane Formation	Early Eocene	Gheerbrant et al., 1998
Afrodon chleuhi	Adrar Mgorn 1, Ihadjamène	Jbel Guersif Fm.	Late Paleocene	Gheerbrant, 1988, 1995
Afrodon cf. *chleuhi*	Adrar Mgorn 1bis (Morocco)	Jbel Guersif Fm.	Late Paleocene	Gheerbrant, 1995
Afrodon tagourtensis	N'Tagourt 2	Aït Ouarithane Formation	Early Eocene	Gheerbrant, 1993; Gheerbrant et al., 1998
Afrodon sp.	Ihadjamène	Jbel Guersif Fm.	Late Paleocene	Gheerbrant, 1995
Chambilestes foussanensis	Chambi (Tunisia)		Early or middle Eocene	Gheerbrant and Hartenberger, 1999
"*Cimolestes*" *cuspulus*	Adrar Mgorn 1	Jbel Guersif Fm.	Late Paleocene	Gheerbrant, 1992
"*Cimolestes*" cf. *incisus*	Adrar Mgorn 1	Jbel Guersif Fm.	Late Paleocene	Gheerbrant, 1992
"Didelphodontinae, gen. et sp. nov."	N'Tagourt 2	Aït Ouarithane Formation	Early Eocene	Gheerbrant, 1993; Gheerbrant et al., 1998
"Didelphodontinae, gen. et sp. indet. 1"	Adrar Mgorn 1	Jbel Guersif Fm.	Late Paleocene	Gheerbrant, 1992
"Didelphodontinae, gen. et sp. indet. 2"[1]	Adrar Mgorn 1, N'Tagourt 2	Jbel Guersif and Aït Ouarithane Formation Fms.	Late Paleocene and early Eocene	Gheerbrant, 1992; Gheerbrant et al., 1998
Eochrysochloris tribosphenus	Quarry E (Egypt)	Jebel Qatrani Fm.	Early Oligocene	Seiffert et al., 2007
"?*Garatherium* n. sp."	Adrar Mgorn 1	Jbel Guersif Fm.	Late Paleocene	Gheerbrant, 1995
Garatherium mahboubii	El Kohol (Algeria)	El Kohol Fm.	Early Eocene	Crochet, 1984; Mahboubi et al., 1986
?*Garatherium todrae*	Adrar Mgorn 1, Ihadjamène	Jbel Guersif Fm.	Late Paleocene	Gheerbrant et al., 1998
"Insectivora, at least four species"	Taqah (Oman)	Ashawq Fm.	Early Oligocene	Thomas et al., 1999
Jawharia tenrecoides	Quarry E	Jebel Qatrani Fm.	Early Oligocene	Seiffert et al., 2007
"Lipotyphla indet."	El Kohol	El Kohol Fm.	Early Eocene	Mahboubi et al., 1986
"Proteutheria or Lipotyphla indet. 1, 2, 3"	Adrar Mgorn 1	Jbel Guersif Fm.	Late Paleocene	Gheerbrant, 1992

TABLE 16.1 (CONTINUED)

Taxon	Occurrence (Site, Locality)	Stratigraphic Unit	Age	Key References
"Nyctitheriidae gen. et sp. indet."	Aznag (Morocco)	Jbel Tagount Fm.	Middle Eocene	Tabuce et al., 2005
"Soricomorpha gen. et sp. indet."	Aznag	Jbel Tagount Fm.	Middle Eocene	Tabuce et al., 2005
Todralestes variabilis	Adrar Mgorn 1, Ihadjamène	Jbel Guersif Fm.	Late Paleocene	Gheerbrant, 1992
Todralestes butleri	N'Tagourt 2	Aït Ouarithane Formation	Early Eocene	Gheerbrant, 1993
Widanelfarasia bowni	Quarry L-41 (Egypt)	Jebel Qatrani Fm.	Late Eocene	Seiffert and Simons, 2000
Widanelfarasia rasmusseni	Quarry L-41	Jebel Qatrani Fm.	Late Eocene	Seiffert and Simons, 2000

[1] "Didelphodontinae, gen. et sp. indet. 3" in Gheerbrant, 1992.

ACKNOWLEDGMENTS

E. Gheerbrant, B. Marandat, and R. Tabuce provided access to fossils and casts, and L. Gordon and P. Jenkins provided access to osteological material. This research was funded by the U.S. National Science Foundation and the Leakey Foundation.

Literature Cited

Asher, R. J., and M. Hofreiter. 2006. Tenrec phylogeny and the noninvasive extraction of nuclear DNA. *Systematic Biology* 55:181–194.

Asher, R. J., R. J. Emry, and M. C. McKenna. 2005. New material of *Centetodon* (Mammalia, Lipotyphla) and the importance of (missing) DNA sequences in systematic paleontology. *Journal of Vertebrate Paleontology* 25:911–923.

Asher, R. J., M. C. McKenna, R. J. Emry, A. R. Tabrum, and D. G. Kron. 2002. Morphology and relationships of *Apternodus* and other extinct, zalambdodont, placental mammals. *Bulletin of the American Museum of Natural History* 273:1–117.

Bloch, J. I., R. Secord, and P. D. Gingerich. 2004. Systematics and phylogeny of late Paleocene and early Eocene Palaeoryctinae (Mammalia, Insectivora) from the Clarks Fork and Bighorn Basins, Wyoming. *Contributions from the Museum of Paleontology* (University of Michigan) 31:119–154.

Crochet, J.-Y. 1984. *Garatherium mahboubii* nov. gen., nov. sp., marsupial de l'Eocène inférieur d'El Kohol (Sud Oranais, Algérie). *Annales de Paléontologie* 70:275–294.

Fox, R. C. 2004. A new palaeoryctid (Insectivora: Mammalia) from the Late Paleocene of Alberta, Canada. *Journal of Paleontology* 78:612–616.

Gheerbrant, E. 1988. *Afrodon chleuhi* nov. gen., nov. sp., "insectivore" (Mammalia, Eutheria) lipotyphlé (?) du Paléocène marocain: Données préliminaires. *Comptes Rendus de l'Académie des Sciences, Paris*, Série II, 307:1303–1309.

———. 1991. *Bustylus* (Eutheria, Adapisoriculidae) and the absence of ascertained marsupials in the Paleocene of Europe. *Terra Nova* 3:586–592.

———. 1992. Les mammifères paléocenes du Bassin d'Ouarzazate (Maroc): I. Introduction général et palaeoryctidae. *Palaeontographica, Abt. A* 224:67–132.

———. 1993. Premières données sur les mammifères "insectivores" de l'Yprésien du Bassin d'Ouarzazate (Maroc: site de N'Tagourt 2). *Neues Jahrbuch für Geologie und Paläontologie, Abhandlungen* 187:225–242.

———. 1994. Les mammifères paléocènes du Bassin d'Ouarzazate (Maroc): II. Todralestidae (Proteutheria, Eutheria). *Palaeontographica, Abt. A* 231:133–188.

———. 1995. Les mammifères paléocènes du Bassin d'Ouarzazate (Maroc): III. Adapisoriculidae et autres mammifères (Carnivora, ?Creodonta, Condylarthra, ?Ungulata et *incertae sedis*). *Palaeontographica, Abt. A* 237:39–132.

Gheerbrant, E., and J.-L. Hartenberger. 1999. Nouveau mammifère insectivore (?Lipotyphla, ?Erinaceomorpha) de l'Eocène inférieur de Chambi (Tunisie). *Paläontologisches Zeitschrift* 73:143–156.

Gheerbrant, E., M. Iarochène, M. Amaghzaz, and B. Bouya. 2006. Early African hyaenodontid mammals and their bearing on the origin of the Creodonta. *Geological Magazine* 143:475–489.

Gheerbrant, E., and J. C. Rage. 2006. Paleobiogeography of Africa: How distinct from Gondwana and Laurasia? *Palaeogeography, Palaeoclimatology, Palaeoecology* 241:224–246.

Gheerbrant, E., J. Sudre, S. Sen, C. Abrial, B. Marandat, B. Sigé, and M. Vianey-Liaud. 1998. Nouvelles données sur les mammifères du Thanétien et de l'Yprésien du Bassin d'Ouarzazate (Maroc) et leur contexte stratigraphique. *Palaeovertebrata* 27:155–202.

Kappelman, J., M. C. Maas, S. Sen, B. Alpagut, M. Fortelius, and J.-P. Lunkka. 1996. A new early Tertiary mammalian fauna from Turkey and its paleobiogeographic significance. *Journal of Vertebrate Paleontology* 16:592–595.

Mahboubi, M., R. Ameur, J.-Y. Crochet, and J.-J. Jaeger. 1986. El Kohol (Saharan Atlas, Algeria): A new Eocene mammal locality in northwestern Africa. *Palaeontographica, Abt. A* 192:15–49.

McKenna, M. C., and S. K. Bell. 1997. *Classification of Mammals above the Species Level*. Columbia University Press, New York, 631 pp.

Roca, A. L., G. K. Bar-Gal, E. Eizirik, K. M. Helgen, R. Maria, M. S. Springer, S. J. O'Brien, and W. J. Murphy. 2004. Mesozoic origin for West Indian insectivores. *Nature* 429:649–651.

Rossie, J. B., and E. R. Seiffert. 2006. Continental paleobiogeography as phylogenetic evidence; pp. 461–514 in J. G. Fleagle and S. Lehman (eds.), *Primate Biogeography*. Plenum, New York.

Seiffert, E. R. 2003. A phylogenetic analysis of living and extinct Afrotherian placentals. Unpublished PhD dissertation, Duke University, Durham, N.C., 239 pp.

Seiffert, E. R., and E. L. Simons. 2000. *Widanelfarasia*, a diminutive placental from the late Eocene of Egypt. *Proceedings of the National Academy of Sciences, USA* 97:2646–2651.

Seiffert, E. R., E. L. Simons, T. M. Ryan, T. M. Bown, and Y. Attia. 2007. New remains of Eocene and Oligocene Afrosoricida (Afrotheria) from Egypt, with implications for the origin(s) of afrosoricid zalambdodonty. *Journal of Vertebrate Paleontology* 27:963–972.

Solé, F., E. Gheerbrant, M. Amaghzaz, and B. Bouya. 2009. Further evidence of the African antiquity of hyaenodontid ('Creodonta', Mammalia) evolution. *Zoological Journal of the Linnean Society* 156:827–846.

Storch, G. 2008. Skeletal remains of a diminutive primate from the Paleocene of Germany. *Naturwissenschaften* 95:927–930.

Tabuce, R., S. Adnet, H. Cappetta, A. Noubhani, and F. Quillevere. 2005. Aznag (bassin d'Ouarzazate, Maroc), nouvelle localité à sélaciens et mammifères de l'Eocène moyen (Lutétien) d'Afrique. *Bulletin de la Société Géologique de France* 176:381–400.

Thewissen, J. G. M., and P. D. Gingerich. 1989. Skull and endocranial cast of *Eoryctes melanus*, a new palaeoryctid (Mammalia: Insectivora) from the early Eocene of western North America. *Journal of Vertebrate Paleontology* 9:459–470.

Thomas, H., J. Roger, S. Sen, M. Pickford, E. Gheerbrant, Z. Al-Sulaimani, and S. Al-Busaidi. 1999. Oligocene and Miocene terrestrial vertebrates in the southern Arabian peninsula (Sultanate of Oman) and their geodynamic and palaeogeographic settings; pp. 430–442 in P. J. Whybrow and A. Hill (eds.), *Fossil Vertebrates of Arabia*. Yale University Press, New Haven.

PART THREE

EUARCHONTOGLIRES

SEVENTEEN

Rodentia

ALISA J. WINKLER, CHRISTIANE DENYS,
AND D. MARGARET AVERY

Carleton and Musser (2005:745) state, "Rodentia is the largest order of living Mammalia, encompassing 2,277 species . . . or approximately 42% of worldwide mammalian biodiversity." The extant African rodent fauna is tremendously diverse, reflecting the wide variety of habitats present on the continent, from desert to tropical rain forest. It is likely that such an extensive fauna was also present in the past. Certainly our understanding of past diversity has expanded greatly since Lavocat's summary of the African Rodentia in 1973. Lavocat recorded a minimum of 54 genera (excluding extant murines) dating from the Oligocene through the Pleistocene. In this contribution we record about 133 named genera (again excluding extant murines and many genera reported as new, but not yet named); recording individual species would turn this chapter into a book. Unfortunately, the number of fossil rodent specimens far outweighs the number of specialists who study these animals, so past rodent diversity is likely to be underestimated.

This chapter will begin by summarizing the distribution and ecology of the extant families of African rodents. It will then focus on the fossil record, including discussion of systematics, biochronology, and paleobiogeography. Most discussion will of necessity be at the family level; notable exceptions include the subfamilies of the extremely diverse Nesomyidae and Muridae. Summary sections will focus on general geographic regions: northern Africa (countries bordering the Mediterranean Sea and including Chad), eastern Africa (Sudan, Ethiopia, Uganda, Kenya, Tanzania, Democratic Republic of the Congo [= Zaire; although situated in Central Africa], and Malawi), and southern and south-central Africa (figure 17.1; Angola, Zambia, Mozambique, and countries farther south). Concluding statements follow the chronologic history of African rodents, from the earliest records in the early to middle Eocene to the latest Pleistocene.

Methods

The provisional systematic classification followed here (table 17.1) primarily follows specific accounts for extant rodents (with some discussion of fossil forms) in Wilson and Reeder (2005). This is supplemented by a classification of fossil forms by McKenna and Bell (1997).

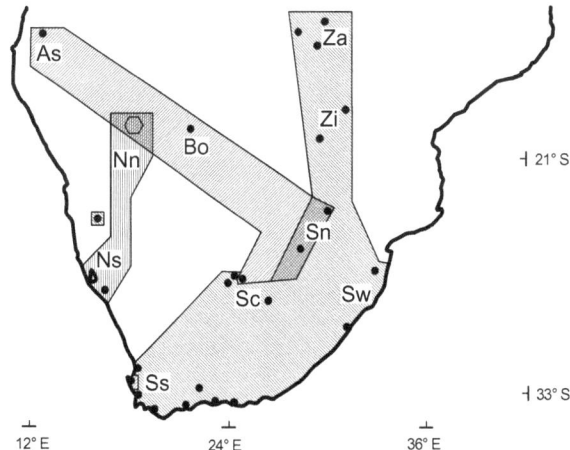

FIGURE 17.1 Southern and south-central African regions used in the text to describe the distribution of fossil rodents.

ABBREVIATIONS FOR REGIONS: As, Angola, southern; Bo, Botswana; Nn, Namibia, northern; Ns, Namibia, southern; Sc, South Africa, central; Sn, South Africa, northern; Sw, South Africa, western; Za, Zambia; Zi, Zimbabwe.

Rodents are often classified into higher-level groups based on skull or jaw structure related to mastication. Sciurognathy and hystricognathy refer to the orientation of the angle (angular process) of the mandible relative to the horizontal process of the mandible (see discussion in Korth, 1994). In the sciurognathous condition (e.g., in the sciurid, *Xerus*), the angle is in the same plane as the horizontal process. In the hystricognathous condition (e.g., in the mole rat, *Bathyergus*), the angle is not in the same plane as the horizontal process. Sciuromorphy, hystricomorphy, and myomorphy refer to the main types of zygomasseteric structure observed in rodents. The following definitions are extremely superficial, but a more thorough discussion is given in Korth (1994). In sciuromorphy (e.g., in the sciurid, *Xerus*), the masseter medialis does not pass through the infraorbital foramen, which is small. In hystricomorphy (e.g., in the cane rat, *Thryonomys*), the masseter medialis is expanded and passes through an enlarged infraorbital foramen. In myomorphy (e.g., in the

Order . . . . Rodentia Bowdich, 1821
  Suborder . . . . Sciuromorpha Brandt, 1855
    Family . . . . Sciuridae Fischer de
              Waldheim, 1817
    Family . . . . Gliridae Muirhead, 1819
  Suborder . . . . Myomorpha Brandt, 1855
    Superfamily . . . . Dipodoidea Fischer de
              Waldheim, 1817
    Family . . . . Dipodidae Fischer de
              Waldheim, 1817
    Superfamily . . . . Muroidea Illiger, 1811
    Family . . . . Spalacidae Gray, 1821
    Family . . . . Nesomyidae Forsyth
              Major, 1897
    Family . . . . Cricetidae Fischer de
              Waldheim, 1817
    Family . . . . Muridae Illiger, 1811
  Suborder . . . . Anomaluromorpha Bugge,
              1974
    Superfamily . . . . Anomaluroidea Gervais,
              1849
    Family . . . . Zegdoumyidae Vianey-Liaud
              et al., 1994
    Family . . . . Anomaluridae Gervais,
              1849
    Superfamily . . . . Pedetoidea Gray, 1825
    Family . . . . Pedetidae Gray, 1825
  Suborder . . . . Hystricomorpha Brandt,
              1855
    Infraorder . . . . Ctenodactylomorpha
              Chaline and Mein, 1979
    Family . . . . Ctenodactylidae Gervais,
              1853
    Infraorder . . . . Hystricognathi Brandt,
              1855
    Family . . . . Bathyergidae Waterhouse,
              1841
    Family . . . . Hystricidae Fischer de
              Waldheim, 1817
    Family . . . . Myophiomyidae Lavocat,
              1973
    Family . . . . Phiomyidae Wood, 1955
    Family . . . . Kenyamyidae Lavocat, 1973
    Family . . . . Petromuridae Wood, 1955
    Family . . . . Thryonomyidae Pocock,
              1922
    Family . . . . *incertae sedis*
      Subfamily . . . . Phiocricetomyinae Lavocat,
              1973
    Family . . . . *incertae sedis*
      Subfamily . . . . nov. of Holroyd, 1994

SOURCE: Adapted from Holroyd (1994), McKenna and Bell (1997), Carleton and Musser (2005), and Woods and Kilpatrick (2005).

root rat, *Tachyoryctes*), the masseter medialis passes through the infraorbital foramen, which is not as large as it is in the hystricomorphous condition, and is usually placed more dorsally: often the infraorbital foramen is shaped like a "keyhole."

As much as possible, the Paleogene time scale follows Luterbacher et al. (2004), the Neogene time scale Lourens et al. (2004), and the Pleistocene time scale Gibbard and Kolfschoten (2004). Note, however, that the "cutoff dates" for these time scales have changed with different authors, and the geologic age of taxa/localities used here may not match those given in the original publications. Complicating the chronology, age assignments for many faunas are based on biostratigraphy (often based on rodents), which may add yet another layer of imprecision when those relative ages are converted to an "absolute" age.

Table 17.1 gives a provisional classification of African fossil and extant Rodentia to the family level. Tables 17.2–17.4 provide a list of the temporal and geographic (mainly by country) occurrence of fossil African rodent genera from their earliest record in the early to middle Eocene through the late Pleistocene. The tables are separated into general geographic areas: table 17.2 for northern Africa (including Chad), table 17.3 for eastern Africa (including the Democratic Repubic of the Congo), and table 17.4 for southern and south-central Africa. Tables 17.5–17.7 provide the details of rodent collections, such as locality information and references, for all sites that are included in this study. As for the tables, the appendices are separated geographically: table 17.5 for northern Africa (including Chad), table 17.6 for eastern Africa (including the Democratic Repubic of the Congo), and table 17.7 for southern and south-central Africa.

ABBREVIATIONS

BUMP, Boston University/Uganda Museum/Makerere University Paleontology Expeditions, Uganda; KNM-TH, National Museums of Kenya, Tugen Hills localities, Kenya; WM, Wembere-Manonga localities, Tanzania.

African Distribution and Ecology of Extant Families

The following is a brief summary of the major areas of distribution and the ecology of the 14 families of rodents currently found in Africa. Most of these families, many of the genera, and some of the species are known also from outside Africa. This discussion will, however, cover only their occurrence in Africa. The geographic distribution of extant rodents provided here is based primarily on political boundaries (i.e., countries), following Wilson and Reeder (2005). For distribution based on African biotic zonation, the reader is referred to Denys (1999).

For more information on extant African rodents, several summary resources are available, for example, Wilson and Reeder (2005; systematics and distribution), Nowak (1999; overviews), Kingdon (1997; "field guide"), and Happold (in press). More regional coverage, specific to rodents, is provided by Rosevear (1969; western Africa), Kingdon (1974; eastern Africa), De Graaff (1981; southern Africa), and Bronner et al. (2003; southern Africa).

Suborder SCIUROMORPHA Brandt, 1855
Family SCIURIDAE Fischer de Waldheim, 1817

All the modern African Sciuridae have been grouped into the Subfamily Xerinae Osborn, 1910, whose monophyly is

TABLE 17.2

Temporal and geographic occurrence of African rodent genera in northern Africa

Abbreviations for countries: Al, Algeria; Ch, Chad; Eg, Egypt; Li, Libya; Mo, Morocco; Tu, Tunisia. Names in parentheses are alternative or previous.

Family	Subfamily	Genus	Eocene Early to middle	Eocene Middle to late	Eocene Late	Early Oligo.	Miocene Early	Miocene Middle	Miocene Late	Pliocene Early	Pliocene Middle	Pliocene Late	Pleistocene Early	Pleistocene Middle	Pleistocene Late
Sciuridae Hemprich, 1820	Xerinae Osborn, 1910	Atlantoxerus Forsyth Major, 1893 (Heteroxerus Stehin & Schaub, 1951; Getuloxerus Lavocat, 1961)	—	—	—	—	—	Mo	Li Al Mo Tu Eg cf.	Mo[1] Al	Al	—	Mo	—	—
		Xerus Hemprich & Ehrenberg, 1833	—	—	—	—	—	—	Ch	Ch	—	—	—	—	—
	?Petauristinae	Genus indet.	—	—	—	—	—	—	—	—	—	—	—	—	—
Gliridae Muirhead, 1819	Leithiinae Lydekker, 1896	Genus indet.	—	—	—	—	—	Mo Al	Al	—	Al	—	—	—	—
		Microdyromys de Bruijn, 1966 (Afrodyromys, Jaeger, 1975)	—	—	—	—	—	Mo Al	Mo Al Tu	—	—	—	—	—	—
		Dryomys Thomas, 1906	—	—	—	—	—	—	Eg	—	—	—	—	—	—
		Eliomys Wagner, 1840	—	—	—	—	—	—	—	Mo[1]	Al	—	—	Mo Tu Al	—
		Genus undet.	—	—	—	—	—	—	—	—	—	—	—	—	—
Dipodidae Fischer de Waldheim, 1817	Allactaginae Vinogradov, 1925	Protalactaga Young, 1927	—	—	—	—	—	Tu Al Mo	Al Tu	—	—	—	—	—	—
	Dipodinae Fisher de Waldheim, 1817	Jaculus Erxleben, 1777	—	—	—	—	—	—	—	—	—	Tu	Mo	—	—
	Sicistinae Allen, 1901	Heterosminthus Schaub, 1930 (Zapodidae)	—	—	—	—	—	Li	—	—	—	—	—	—	—
Dipodidae?		(Genus indet.)	—	—	—	—	Li	—	—	—	—	—	—	—	—
Spalacidae Gray, 1821	Rhizomyinae Winge, 1887	Prokanisamys de Bruijn, Hussain, & Leinders, 1981	—	—	—	—	—	Al	—	—	—	—	—	—	—
		Genus indet.	—	—	—	—	—	—	—	Ch	—	—	—	—	—

TABLE 17.2
(CONTINUED)

Family	Subfamily	Genus	Eocene			Early Oligo.	Miocene			Pliocene			Pleistocene		
			Early to middle	Middle to late	Late		Early	Middle	Late	Early	Middle	Late	Early	Middle	Late
Nesomyidae Forsyth Major, 1897	Dendromurinae G. M. Allen, 1939	Ternania Tong & Jaeger, 1993	—	—	—	—	—	—	Eg	—	—	—	—	—	—
		Senoussimys Jaeger & Ameur-Chabbar, 1978	—	—	—	—	—	—	Al	—	—	—	—	—	—
		Dendromus A. Smith, 1829	—	—	—	—	—	—	Al Eg cf.	—	—	—	—	—	—
		Steatomys Peters, 1846	—	—	—	—	—	—	Eg cf.	—	—	—	—	—	—
Cricetidae Fischer de Waldheim, 1817	Arvicolinae Gray, 1821	Gen. nov. Coiffait-Martin, 1991	—	—	—	—	—	Al	—	—	—	—	—	—	—
		Genus indet.	—	—	—	—	—	—	—	—	—	—	—	—	—
		Ellobius Fischer, 1814	—	—	—	—	—	Li	—	—	—	—	—	Al Mo Tu	Tu
	Cricetinae Fischer de Waldheim, 1817	Apocricetus Freudenthal et al., 1998	—	—	—	—	—	—	Eg	—	—	—	—	—	—
		Cricetus Leske, 1779	—	—	—	—	—	—	Mo Eg	Mo[1]	—	—	—	—	—
	Lophiomyinae Milne-Edwards, 1867	Protolophiomys Aguilar & Thaler, 1987	—	—	—	—	—	—	—	—	Al	—	—	—	—
		Lophiomys Milne-Edwards, 1867	—	—	—	—	—	—	—	Mo[1]	—	—	—	—	—
Muridae Illiger, 1815	Myocricetodontinae Lavocat, 1961	Potwarmus Lindsay, 1988	—	—	—	—	Li cf.	Li	—	—	—	—	—	—	—
		Myocricetodon Lavocat, 1952	—	—	—	—	—	Mo Al	Li Eg Mo Al Tu	Mo[1]	—	—	—	—	—
		Mellalomys Lavocat, 1961	—	—	—	—	Li	Mo Tu Li Al	—	—	—	—	—	—	—
		Cricetodon Lartet, 1857	—	—	—	—	—	—	—	—	—	—	—	—	—
		Dakkamys Jaeger, 1977b	—	—	—	—	—	Mo Al	—	—	—	—	—	—	—
		Aissamys Coiffait-Martin, 1991	—	—	—	—	—	Al	—	—	—	—	—	—	—

Taxon													
Cricetodontinae Schaub, 1925	—	—	—	—	—	—	Mo Al Tu	Al	—	—	—	—	—
Zramys Jaeger & Michaux, 1973	—	—	—	—	—	Mol Tu cf.	Al	—	—	—	—	—	—
Ruscinomys Depéret, 1890	—	Mo	—	—	—	—	—	—	—	—	—	—	—
Gerbillinae Gray, 1825	—	—	—	—	—	—	—	—	—	—	—	—	—
Pseudomeriones Schaub, 1934	—	—	—	—	—	Al	—	—	—	—	—	—	—
Protatera Jaeger, 1977b	—	—	—	—	—	Mol	Al Li	—	—	—	—	—	—
Mascaramys Tong, 1986	—	Al	—	Tu	—	—	—	—	—	—	—	—	—
Gerbillus Demarest, 1804	—	Al Mo Tu	Tu Mo	Mo	—	—	—	—	—	—	—	—	—
Meriones Illiger, 1811	Tu	Al Mo Tu	Mo	—	—	—	—	—	—	—	—	—	—
Genus indet.	—	—	Mo	—	Al Tu	—	—	—	—	—	—	—	—
Progonomys Schaub, 1938	—	—	—	—	—	—	Al / Li Mo / Al Eg	—	—	—	—	—	—
Progonomys and/or *Preacomys*	—	—	—	—	—	—	Eg	—	—	—	—	—	—
Karnimata Jacobs, 1978	—	—	—	—	—	—	Al	—	—	—	—	—	—
Castillomys Michaux, 1969 (*Occitanomys* Michaux, 1969)	—	—	—	—	—	—	Mo Al	—	—	—	—	—	—
Stephanomys Schaub, 1938	—	—	—	—	—	—	Al	—	—	—	—	—	—
Saidomys James & Slaughter, 1974	—	Mo	—	—	—	—	Eg	—	—	—	—	—	—
Paraethomys F. Petter, 1968	Tu	Al Tu / Mo	Tu Mo / Al	Mo Al / Tu	Al Tu	Mol Tu / Al	Mo Al	—	—	—	—	—	—
Apodemus Kaup, 1829	—	Mo	—	Tu	—	Mol	Mo Al	—	—	—	—	—	—
Arvicanthis Lesson, 1842	—	Al	Tu	—	Al	—	—	—	—	—	—	—	—
Pelomys Peters, 1852	—	—	—	—	—	Al	—	—	—	—	—	—	—
Golunda Gray, 1837	—	—	—	—	Al	—	—	—	—	—	—	—	—
Mus Linnaeus, 1758	—	Mo Tu / Al	Tu Mo / Tu Mo	Mo Al Tu / Mo Al Tu	Tu	Mol Tu	—	—	—	—	—	—	—
Praomys Thomas, 1915	—	Al Mo Tu	—	—	—	Mol	—	—	—	—	—	—	—

TABLE 17.2
(CONTINUED)

Family	Subfamily	Genus	Eocene Early to middle	Eocene Middle to late	Eocene Late	Early Oligo.	Miocene Early	Miocene Middle	Miocene Late	Pliocene Early	Pliocene Middle	Pliocene Late	Pleistocene Early	Pleistocene Middle	Pleistocene Late
Zegdoumyidae Vianey-Liaud et al., 1994		Genus undet.	—	—	—	—	—	—	—	—	—	—	—	—	—
		Zegdoumys Vianey-Liaud et al., 1994	Al Tu	—	—	—	—	—	Ch	Ch	Al	—	—	—	—
		Glibia Vianey-Liaud et al., 1994	Al	—	—	—	—	—	—	—	—	—	—	—	—
		Glibemys Vianey-Liaud et al., 1994	Al	—	—	—	—	—	—	—	—	—	—	—	—
Anomaluridae Gervais, 1849		Nementchamys Jaeger et al., 1985	—	Al	—	—	—	—	—	—	—	—	—	—	—
Pedetidae Gray, 1825		Megapedetes MacInnes, 1957	—	—	—	—	—	Mo	—	—	—	—	—	—	—
Ctenodactylidae Gervais, 1853		Metasayimys Lavocat, 1961	—	—	—	—	—	Mo	—	—	—	—	—	—	—
		Sayimys Wood, 1937	—	—	—	—	Li	Li	Li	—	—	—	—	—	—
		(Africanomys)	—	—	—	—	—	—	—	—	—	—	—	—	—
		Africanomys Lavocat, 1961	—	—	—	—	—	Mo Tu Al	Mo Eg	—	—	—	—	—	—
		Irhoudia Jaeger, 1971	—	—	—	—	—	—	Li Mo Al	Mo[1]	—	Mo Tu	Mo	—	—
		Testouromys Robinson & Black, 1973	—	—	—	—	—	Tu	—	—	—	—	—	—	—
		Genus indet.	—	—	—	—	—	Al	Tu	—	Al	Al	—	—	—
Hystricidae Fischer de Waldheim, 1817		Atherurus F. Cuvier, 1817	—	—	—	—	—	—	Eg	—	—	—	—	—	—
		Hystrix Linnaeus, 1758	—	—	—	—	—	—	Ch Al	—	—	Mo	—	Mo	—
Phiomyidae Wood, 1955		Protophiomys Jaeger, Denys, & Coiffait, 1985	—	Al	—	—	—	—	—	—	—	—	—	—	—

Family / Subfamily	Genus											
Diamantomyinae Schaub, 1958	*Phiomys* Osborn, 1908	—	—	—	—	—	—	—	—	—	Li Eg	Eg	—
	Metaphiomys Osborn, 1908	—	—	—	—	—	—	—	—	—	Li Eg	—	—
	n. gen. Holroyd, 1994 (*Phiomys* part)	—	—	—	—	—	—	—	—	—	—	Eg	—
Thryonomyidae Pocock, 1922	*Paraphiomys* Andrews, 1914	—	—	—	—	—	—	—	Mo	—	—	—	—
	n. gen. Holroyd, 1994 (*Paraphiomys* in part)	—	—	—	—	—	—	—	—	—	Eg	—	—
	n. gen. Wessels et al., 2003	—	—	—	—	—	—	—	—	Li	—	—	—
Family *incertae sedis* — Phiocricetomyinae Lavocat, 1973	*Phiocricetomys* Wood, 1968	—	—	—	—	—	—	—	—	—	Eg	—	—
	n. gen. Holroyd, 1994 (*Phiomys* in part)	—	—	—	—	—	—	—	—	—	—	Eg	—
Family *incertae sedis*	*Gaudeamus* Wood, 1968	—	—	—	—	—	—	—	—	—	—	Eg	—

[1]Includes sites reported as dating close to the Miocene-Pliocene boundary.

TABLE I7.3

Temporal and geographic occurrence of African rodent genera in eastern Africa

Abbreviations for countries: Et, Ethiopia; Ke, Kenya; Ug, Uganda; Ta, Tanzania; Zr, Zaire (Democratic Republic of the Congo). Names in parentheses are alternative or previous.

Family	Subfamily	Genus	Late Oligocene	Miocene Early	Miocene Middle	Miocene Late	Pliocene Early	Pliocene Middle	Pliocene Late	Plio-Pleist.	Pleistocene Early	Pleistocene Middle	Pleistocene Late
Sciuridae Hemprich, 1820	Xerinae Osborn, 1910	Vulcaniscurus Lavocat, 1973	—	Ke Ug	Ke	—	—	—	—	—	—	—	—
		Xerus Hemprich and Ehrenberg, 1833 (Paraxerus)	—	—	—	Ke Et	Ke Ta	Et Ta	Et	—	—	—	—
		Kubwaxerus Cifellei et al., 1986	—	—	—	Ke	—	—	—	—	—	—	—
		Heliosciurus Trouessart, 1880	—	—	—	—	Ke	—	—	—	—	—	—
		Paraxerus Forsyth Major, 1893	—	—	—	Ke	Ke Ta	Et	Et	—	—	—	—
		Genus indet.	—	—	—	—	Et	—	—	—	—	—	—
Gliridae Muirhead, 1819	Graphiurinae Winge, 1887	Graphiurus Smuts, 1832	—	—	—	—	—	—	—	—	Ta	Ke	—
Dipodidae Fisher de Waldheim, 1817	Dipodinae Fisher de Waldheim, 1817	Jaculus Erxleben, 1777	—	—	—	—	—	—	Et	Ke	—	Ta	—
Family Spalacidae Gray, 1821	Rhizomyinae Winge, 1887	Pronakalimys Tong & Jaeger, 1993	—	—	Ke	—	—	—	—	—	—	—	—
		Nakalimys Flynn & Sabatier, 1984	—	—	—	Ke Et	—	—	—	—	—	—	—
		Tachyoryctes Rüppell, 1835	—	—	—	Et	Et	Et	Zr	—	—	—	Et
Pseudocricetodontidae Ünay-Bayraktar, 1989[1]		?Pseudocricetodon Thaler, 1969	—	Ug	—	—	—	—	—	—	—	—	—
Nesomyidae Forsyth Major, 1897	Afrocricetodontinae Lavocat, 1973	Afrocricetodon Lavocat, 1973	—	Ke Ug	—	—	—	—	—	—	—	—	—
		Notocricetodon Lavocat, 1973	—	Ke Ug	Ke	—	—	—	—	—	—	—	—
		Protarsomys Lavocat, 1973	—	Ke Ug	—	—	—	—	—	—	—	—	—
	Cricetomyinae Roberts, 1951	Cricetomys Waterhouse, 1840	—	—	—	—	—	—	—	—	—	Ke	—
		Saccostomus Peters, 1846	—	—	—	Ke	Ta[2] Ta	Ta	—	Ke	Ta	Ke	—
		Genus indet.	—	—	Ke	—	—	—	—	—	—	—	—
	Dendromurinae G. M Allen, 1939	Ternania Tong & Jaeger, 1993	—	—	Ke	—	—	—	—	—	—	—	—

Taxon											
Cricetidae Fisher de Waldheim, 1817	—	—	—	—	—	—	—	—	—	—	—
Lophiomyinae Milne-Edwards, 1867	—	—	—	—	—	—	—	—	—	—	—
Mabokomys Winkler, 1998	—	—	—	—	—	—	—	—	Ke	—	—
Dendromus A. Smith, 1829 (Dendromys)	—	Ta	—	—	—	—	Ta	Ke Et aff.	—	—	—
Steatomys Peters, 1846	—	—	—	—	—	—	Ke Ta	Ke Et	Ke	—	—
cf. "Dendromus" gen. nov. (Saccostomus)	—	—	—	—	—	—	—	—	—	—	—
Lophiomys Milne-Edwards, 1867	—	—	—	—	—	—	—	Et	—	—	—
Muridae Illiger, 1815	—	—	—	—	—	—	—	—	—	—	—
Myocricetodontinae Lavocat, 1961	—	—	—	—	—	—	—	—	—	—	—
Leakeymys Lavocat, 1964	—	—	—	—	—	—	—	—	Ke	—	—
Dakkamys Jaeger, 1977b	—	—	—	—	—	—	—	—	Ke	—	—
Megacricetodontinae Fahlbusch, 1964 or Myocricetodontinae Lavocat, 1961	—	—	—	—	—	—	—	—	—	—	—
Genus indet.	—	—	—	—	—	—	—	—	Ke	—	—
Cricetodontinae Schaub, 1925	—	—	—	—	—	—	—	—	—	—	—
Afaromys Geraads, 1998a	—	—	—	—	—	—	—	Et	—	—	—
Democricetodon Fahlbusch, 1964	—	—	—	—	—	—	—	—	Ke	—	—
Genus indet.	—	—	—	—	—	—	—	—	Ke	—	—
Deomyinae Lyddeker, 1889	—	—	—	—	—	—	—	—	—	—	—
Preacomys Geraads, 2001	—	—	—	—	—	—	—	Et Ke	—	—	—
Tectonomys Winkler, 1997	—	—	—	—	—	—	Ke Ta[2]	Et cf.	—	—	—
Gerbillinae Gray, 1825	—	—	—	—	—	—	—	—	—	—	—
Acomys I. Geoffroy, 1838 (Millardia Thomas, 1911)	—	—	—	—	Et	Et	Ke Et	Ke Ug cf.	—	—	—
Abudhabia de Bruijn & Whybrow, 1994	—	—	—	—	—	—	—	Ke	—	—	—
Gerbilliscus Thomas, 1897 (Tatera Lataste, 1897)	—	Ke	Ta	Ke	Et	Et Ta	Ke Ta Et	Ke Et	—	—	—
Gerbillus Demarest, 1804	—	—	Ta	—	Et	—	Ke	—	—	—	—
Murinae Illiger, 1815	—	—	—	—	—	—	—	—	—	—	—
Progonomys Schaub, 1938 (Karnimata Jacobs, 1978)	—	—	—	—	—	—	—	Ke	—	—	—
cf. Parapelomys Jacobs, 1978	—	—	—	—	—	—	—	Et	—	—	—
aff. Stenocephalemys Frick, 1914 (Stenocephalomys)	—	—	—	—	—	—	—	Et	—	—	—
Lukeinomys Mein & Pickford, 2006	—	—	—	—	—	—	—	Ke	—	—	—

271

TABLE 17.3
(CONTINUED)

Family	Subfamily	Genus	Late Oligocene	Miocene			Pliocene			Plio-Pleist.	Pleistocene		
				Early	Middle	Late	Early	Middle	Late		Early	Middle	Late
		Saidomys James & Slaughter, 1974	—	—	—	Ke	Ke Ta[2]	Et	Et	—	—	—	—
		Aethomys Thomas, 1915	—	—	—	Ke	Ke	—	Et	Ke	Ta	Ke	—
		Arvicanthis Lesson, 1842	—	—	—	Ke	Ug	Et Ta	Et	Ke	Ta	Ug Ke	—
		Lemniscomys Trouessart, 1881	—	—	—	Ke Et	Ke Et	Et	Et	—	—	—	—
		Pelomys Peters, 1852 (Golunda[3] Gray, 1837)	—	—	—	—	Et	Et	Et	—	—	—	—
		Zelotomys Osgood, 1910	—	—	—	—	Ke	Et	Et	Ke	Ta	Ke	—
		Mus Linnaeus, 1758	—	—	—	—	—	Et	Et	—	Ta	Ke	—
		Oenomys Thomas, 1904	—	—	—	—	Ke	Et	Et	Ke	Ta	Ke	—
		Thallomys Thomas, 1920	—	—	—	—	Ta	Et Ta	Et	—	Ta	—	—
		Thamnomys Thomas, 1907 (Grammomys)	—	—	—	—	—	—	—	—	Ta	Ke	—
		Thamnomys Thomas, 1907 (Grammomys), or Thallomys	—	—	—	—	Ke	—	—	—	—	—	—
		Mastomys Thomas, 1915	—	—	—	Ke	Ke Ta	—	Ta	—	—	—	—
		Praomys Thomas, 1915 (Mastomys)	—	—	—	—	—	Et	Et	Ke	Ta	Ke	—
		Genus indet.	—	—	—	Ke	—	—	—	—	—	—	—
	Otomyinae Thomas, 1897	Otomys F. Cuvier, 1824	—	—	—	Ke	—	—	Zr	—	—	Ke	—
Anomaluridae Gervais, 1849		Paranomalurus Lavocat, 1973	—	Ke Ug	Ke	—	—	—	—	—	—	—	—
		Anomalurus Waterhouse, 1843	—	—	Ke	—	—	—	—	—	—	—	—
		Zenkerella Matschie, 1898	—	Ke Ug	—	—	—	—	—	—	—	—	—
Pedetidae Gray, 1825		Megapedetes MacInnes, 1957	—	Ke Ug	Ke	—	Ta	—	—	—	—	—	—
		Pedetes Illiger, 1811	—	—	—	Ke	—	Ta	—	—	—	Ke	—
		Genus indet.	—	Ke	Ke	Ke	—	—	—	—	—	—	—

Taxon	1	2	3	4	5	6	7	8	9	10	11
Bathyergidae Waterhouse, 1841											
Bathyergoides Stromer, 1924	—	—	—	—	—	—	—	—	—	Ke Ug	—
Geofossor Mein & Pickford, 2003	—	—	—	—	—	—	—	—	—	Ug	—
Richardus Lavocat, 1988	—	—	—	—	—	—	—	—	Ke	—	—
Bathyerginae Waterhouse, 1841											
Proheliophobius Lavocat, 1973	—	—	—	—	—	—	—	—	—	Ke	—
Heliophobius or *Cryptomys*	—	Ke	—	—	—	—	—	—	—	—	—
Heterocephalinae Landry, 1957											
Heterocephalus Rüppell, 1842	—	—	—	—	Et	Ta	Ta	—	—	—	—
Genus indet.	—	—	—	—	—	—	—	—	—	—	—
Hystricidae Fischer de Waldheim, 1817											
Xenohystrix Greenwood, 1955	—	—	—	Ke	—	Et	Ta Et	Ke Et	Ke	—	—
Hystrix Linnaeus, 1758	—	Ug Ke	—	Et Ke	Et Ke	Et Ta	Ke Ta Et Et	Ke Et	—	—	—
Atherurus F. Cuvier, 1829	—	—	—	—	—	—	Et	Ke Et	—	—	—
Genus indet.	—	—	—	—	—	—	—	—	—	—	—
Myophiomyidae Lavocat, 1973											
Myophiomys Lavocat, 1973	—	—	—	Ke	—	—	—	—	—	Ke Ug	—
Elmerimys Lavocat, 1973	—	—	—	—	—	—	—	—	Ke	Ke	—
Phiomyidae Wood, 1955											
Andrewsimys Lavocat, 1973	—	—	—	—	—	—	—	—	—	Ke Ug	—
Ugandamys Winkler et al., 2005	—	—	—	—	—	—	—	—	—	Ug	—
Diamantomyinae Schaub, 1958											
Metaphiomys Osborn, 1908	—	—	—	—	—	—	—	—	—	—	Ta
Diamantomys Stromer, 1922	—	—	—	—	—	—	—	—	Ke	Ke Ug	Ke
Kenyamyidae Lavocat, 1973											
Kenyamys Lavocat, 1973	—	—	—	—	—	—	—	—	—	Ke	—
Simonimys Lavocat, 1973	—	—	—	—	—	—	—	—	—	Ke Ug	—
Petromuridae Tullberg, 1899											
Petromus A. Smith, 1831	—	—	—	—	—	—	—	Ke	—	—	—
Thryonomyidae Pocock, 1922											
Apodecter Hopwood, 1929	—	—	—	—	—	—	—	Ke	Ke	Ug	—
Paraphiomys Andrews, 1914	—	—	—	—	—	—	—	Ke Et	Ke	Ke Ug	—
Epiphiomys Lavocat, 1973	—	—	—	—	—	—	—	—	—	Ke Ug	—

273

TABLE 17.3
(CONTINUED)

Family	Subfamily	Genus	Late Oligocene	Miocene			Pliocene			Plio-Pleist.	Pleistocene		
				Early	Middle	Late	Early	Middle	Late		Early	Middle	Late
		Paraulacodus Hinton, 1933	—	—	Ke	Ke Et	Ta[2] Ke Et	—	—	—	—	—	—
		Thryonomys Fitzinger, 1867	—	—	—	Ke Et	Ke Et	Et Ta	Zr Et	Ke	—	Ug Ke	—
		Genus indet.	—	—	Ke	—	—	—	—	—	—	—	—
Incertae sedis		*Kahawamys* Stevens et al. In press	Ta	—	—	—	—	—	—	—	—	—	—
		Lavocatomys Holroyd & Stevens, 2009 (*Phiomys*)	—	Ke	—	—	—	—	—	—	—	—	—

[1]This tentative record is not discussed in the main text.

[2]The Manonga Valley sites, Tanzania, are of late Miocene (based on overall fauna) to early Pliocene age (based on rodents).

[3]Musser (1987) suggested assigning *Golunda gurai* from the Hadar Fm (Sabatier, 1982) and Omo Valley, Ethiopia (Wesselman, 1984), to *Pelomys. Golunda* from Aramis (WoldeGabriel et al., 1994) is here referred to *Pelomys.*

TABLE I7.4

Temporal and geographic occurrence of African rodent genera from the Miocene, Pliocene, and Pleistocene in southern and south-central Africa

Abbreviations for regions: As, Angola, southern; Bo, Botswana; Nn, Namibia, northern; Ns, Namibia, southern; Sc, South Africa, central; Sn, South Africa, northern; Sw, South Africa, western; Za, Zambia; Zi, Zimbabwe. Names in parentheses are alternative or previous.

Family	Subfamily	Genus	Miocene			Pliocene		Plio-Pleist.[1]	Pleistocene		
			Early	Middle	Late	Early	Late		Early	Middle	Late
Sciuridae Hemprich, 1820	Subfamily Xerinae Osborn, 1910	Heteroxerus Stehlin & Schaub, 1951 (Vulcanisciurus Lavocat, 1973)	—	—	Nn	—	—	—	—	—	—
		Xerus Hemprich and Ehrenberg, 1833	—	—	—	—	—	—	—	—	Sc
		Paraxerus Forsyth Major, 1893	—	—	Nn	—	—	—	—	—	Sw
		Genus indet.	—	Ns	—	—	—	—	—	—	—
Gliridae Muirhead, 1819	Graphiurinae Winge, 1887	Otaviglis Mein et al., 2000a	—	—	Nn	—	—	—	—	—	—
		Graphiurus Smuts, 1832	—	—	—	—	Sn Nn	—	Nn Sn	Za	Sn Ss Za
Family Spalacidae Gray, 1821	Rhizomyinae Winge, 1887	Nakalimys Flynn & Sabatier, 1984	—	—	Nn	—	—	—	—	—	—
		Harasibomys Mein et al., 2000a (cf. Brachyuromys Major, 1896)	—	—	Nn	—	—	—	—	—	—
Nesomyidae Forsyth Major, 1897	Afrocricetodontinae Lavocat, 1973	Notocricetodon Lavocat, 1973	—	Nn	—	—	—	—	—	—	—
	Cricetomyinae Roberts, 1951	Protarsomys Lavocat, 1973	—	Nn Ns	Nn	—	—	—	—	—	—
		Saccostomus Peters, 1846	—	—	Nn cf.	—	Nn	—	Nn Sn Za	Sc Za	Sn Sw Za
	Otavimyinae Mein et al., 2004	Otavimys Mein et al., 2004	—	—	Nn	—	—	—	—	—	—
		Boltimys Sénégas & Michaux, 2000	—	—	—	Sn	—	—	—	—	—
	Dendromurinae G. M. Allen, 1939	Dendromus A. Smith, 1829 (Dendromys J. B. Fischer, 1830)	—	—	Nn	Ss Sn	Nn Sn	AS Bo	Nn Sc Sn	Sc Za	Sn Ss Sw Za Bo
		Malacothrix Wagner, 1843	—	—	—	Sn	Nn Sn	Bo	Nn Sc Sn	Sc	Sn Sw
		Steatomys Peters, 1846	—	Nn cf.	Nn	Ss Sn	Nn Sn	As	Nn Sn Za	Sc Za	Sn Sw Za Bo
	Mystromyinae Vorontsov, 1966	Proodontomys Pocock, 1987 (Mystromys)	—	—	—	Sn	Sn	—	Sc	—	—

TABLE 17.4
(CONTINUED)

Family	Subfamily	Genus	Miocene			Pliocene		Plio-Pleist.[1]	Pleistocene		
			Early	Middle	Late	Early	Late		Early	Middle	Late
		Mystromys Wagner, 1841	—	—	—	Ss Sn	Nn Sn	—	Nn Sc Sn	Sc Ss	Sn Ss Sw Bo
	Petromyscinae Roberts, 1951	*Stenodontomys* Pocock, 1987 (*Mystromys*)	—	—	Nn	Ss	Nn	—	Nn	—	—
		Harimyscus Mein et al., 2000b (*Petromyscus*)	—	—	Nn	—	—	—	—	—	—
		Petromyscus Thomas, 1926	—	—	Nn	Ss	—	—	Nn	Sc	—
Muridae Illiger, 1815	Myocricetodontinae Lavocat, 1961	*Myocricetodon* Lavocat, 1952	—	Nn	Nn	—	—	—	—	—	—
		Dakkamyoides Lindsay, 1988	—	Nn	Nn	—	—	—	—	—	—
		Mioharimys Mein et al., 2000b (cf. *Mystromys*)	—	—	Nn	—	—	—	—	—	—
		?*Afaromys* Geraads, 1998a	—	—	Nn	—	—	—	—	—	—
	Cricetodontinae Schaub, 1925	*Democricetodon* Fahlbusch, 1964	—	Nn cf.	—	—	—	—	—	—	—
	Namibimyinae Mein et al., 2000b	*Namibinys* Mein et al., 2000b	—	—	Nn	—	—	—	—	—	—
	Deomyinae Lyddeker, 1889	*Preacomys* Geraads, 2001	—	—	Nn	—	—	—	—	—	—
		Acomys I. Geoffroy, 1838 (*Millardia* Thomas, 1911)	—	—	—	Ss Sn	Nn Sn	As Bo	Nn Sc Sn	Sc Za Ss	Ss Za
		Uranomys Dollman, 1909	—	—	—	—	Nn	As	—	Ss	—
	Gerbillinae Gray, 1825	*Desmodillus* Thomas & Schwann, 1904	—	—	—	Ss Sn?	Nn Sn	—	Nn Sc Sn	Sc	—
		Gerbillurus Shortridge, 1942 (*Gerbillus* Damarest, 1804, ?*Taterillus* Thomas, 1910)	—	—	—	Sn	Nn Sn	Bo	Nn	Sc Ss Za	Ns Ss Za Bo
		Desmodillus / Gerbillurus	—	—	—	Ss	Nn Sn	—	—	—	—
		Gerbilliscus Thomas, 1897 (*Tatera* Lataste, 1882)	—	—	—	Ss Sn	Nn Sn	As Bo	Nn Sc Sn Za	Sc Za Ss	Sn Ss Sw Za Bo
	Murinae Illiger, 1815	*Aethomys* Thomas, 1915 (*Aethomys* [*Micaelamys*])	—	—	Nn	Ss Sn	Nn Sn	As	Nn Sn Za	Sc Za Ss	Sn Sw Za
		Micaelamys Ellerman, 1941 (*Aethomys*)	—	—	Nn	Sn	Nn	—	Nn Sn	Sc	Ss Sw
		Arvicanthis Lesson, 1842	—	—	—	—	—	—	Sn Za?	Za	Za
		Lemniscomys Trouessart, 1881	—	—	—	Sn	Sn	—	Nn Sn Za	Sc Za	Sw

		Pelomys Peters, 1852	—	—	—	Sn	—	As Bo	Sn Za	Za	Sw Za
		Rhabdomys Thomas, 1916	—	—	—	Ss Sn	Nn Sn	—	Nn Sc Sn	Sc Ss	Sn Ss Sw Za
		Zelotomys Osgood, 1910	—	—	—	Sn?	Nn Sn	As Bo	Nn Sn	Sc Za Ss	Sn Za?
		Dasymys Peters, 1875	—	—	—	Sn	Nn Sn	As	Sc Sn	Sc Ss Za	Sn Ss Sw Za
		Malacomys Milne-Edwards, 1877	—	—	—	—	—	As	—	—	—
		Mus Linnaeus, 1758 (*Leggada* Gray, 1837)	—	—	—	Ss Sn	Nn Sn	As Bo	Nn Sc Sn Za	Sc Za	Sn Ss Sw Za
		Grammomys Thomas, 1915 (*Thamnomys* Thomas, 1907)	—	—	—	Sn	Sn	As	Sn	—	Ss Sw
		Thallomys Thomas, 1920	—	—	—	Sn cf.	Nn	As	Nn Sc Sn Za	Sc Za	Sn Ss Za
		Mastomys Thomas, 1915 (*Myomys, Praomys, Rattus*)	—	—	—	Sn Ss?	Nn Sn	As	Nn Sc Sn Za	Sc Za	Sn Sw Za
		Myomyscus Shortridge, 1942 (*Myomys, Rattus, Praomys*)	—	—	—	Sn	Sn	—	Sn	Ss	Ss
	Otomyinae Thomas, 1897	*Euryotomys* Pocock, 1976	—	—	—	Ss Sn	—	—	—	—	—
		Palaeotomys Broom, 1937 (*Otomys*)	—	—	—	Sn	Sn	Bo	Sc Sn	Sn	—
		Prootomys Broom, 1948 (*Myotomys*)	—	—	—	Sn	—	Bo	Nn Sc Sn	—	—
		Myotomys Thomas, 1918 (*Otomys*)	—	—	—	Sn	Sn	—	Sn	Sc Ss	Sn Ss
		Otomys F. Cuvier, 1824 (*Myotomys*)	—	—	—	Sn	Nn Sn	As Bo	Nn Sc Sn Za	Sc Ss Za	Sn Ss Sw Za Bo
		Parotomys Thomas, 1918	—	—	—	—	—	—	—	Ss	Ns
Pedetidae Gray, 1825		*Parapedetes* Stromer, 1924	Ns	Nn	—	—	—	—	—	—	—
		Megapedetes MacInnes, 1957	Ns cf.	Ns	—	—	—	—	—	—	—
		Pedetes Illiger, 1811	—	—	—	Sn	Sn	Bo	Sc Sn	Sc	Sc Zi Bo
		Genus indet.	—	—	Nn	—	—	—	—	—	—
Bathyergidae Waterhouse, 1841		*Bathyergoides* Stromer, 1924	Ns	Ns	—	—	—	—	—	—	—
		Gypsorhychus Broom, 1934	—	—	—	Sn	—	—	Sc Sn	—	—
		Geofossor Mein & Pickford, 2003 (*Paracryptomys* Lavocat, 1973)	Ns	Ns	—	—	—	—	—	—	—
	Bathyerginae Waterhouse, 1841	*Proheliophobius* Lavocat, 1973, or *Richardus* Lavocat, 1988	—	—	Nn	—	—	—	—	—	—
		Bathyergus Illiger, 1811	—	—	—	Ss	—	—	—	Ss	Ss
		Georychus Illiger, 1811	—	—	—	—	—	Bo	Sn	Ss	Sn Ss

TABLE 17.4
(CONTINUED)

Family	Subfamily	Genus	Miocene			Pliocene		Plio-Pleist.[1]	Pleistocene		
			Early	Middle	Late	Early	Late		Early	Middle	Late
Hystricidae Fischer de Waldheim, 1817		Cryptomys Gray, 1864	—	—	—	Ss Sn	Nn Sn	As	Nn Sc Sn	Sc Ss Za	Sn Ss Sw Za
		Xenohystrix Greenwood, 1955	—	—	—	Sn	—	—	—	—	—
		Hystrix Linnaeus, 1758	—	—	—	Ss Sn	Sn	As Bo	Sc Sn	Sc Ss	Sn Ss Sw Za Zi
Myophiomyidae Lavocat, 1973	Myophiomyinae Lavocat, 1973	Phiomyoides Stromer, 1926	Ns	—	—	—	—	—	—	—	—
Phiomyidae Wood, 1955		Pomonomys Stromer, 1922	Ns	—	—	—	—	—	—	—	—
	Diamantomyinae Schaub, 1958	Diamantomys Stromer, 1922	Ns	Ns	—	—	—	—	—	—	—
Petromuridae Tullberg, 1899		Petromus A. Smith, 1831 (Petromys Smith, 1834)	Ns	—	—	Sn	—	—	Sc Sn	—	Ns
Thryonomyidae Pocock, 1922		Apodecter Hopwood, 1929a (Paraphiomys)	Ns	—	Nn	—	—	—	—	—	—
		Paraphiomys Andrews, 1914 (Apodecter part)	—	Sn Ns	—	—	—	—	—	—	—
		Neosciuromys Stromer, 1922	Ns	—	—	—	—	—	—	—	—
		Phthinylla Hopwood, 1929a	Ns	—	—	—	—	—	—	—	—
		Paraulacodus Hinton, 1933	—	—	Nn	—	—	—	—	—	—
		Thryonomys Fitzinger, 1867	—	—	—	—	—	—	—	—	Za Zi

[1]The samples from Botswana and Angola have not yet been dated more precisely.

supported by molecular and morphologic studies (Mercer and Roth, 2003; Steppan et al., 2004; Denys et al., 2003). In northern Africa, they are represented by *Atlantoxerus*, while tropical Africa hosts *Xerus, Epixerus, Funisciurus, Heliosciurus, Myosciurus, Paraxerus*, and *Protoxerus*. African squirrels inhabit tropical forest to Sudanian zones along the margins of the Sahara. Only *Atlantoxerus* and *Xerus* (both in the tribe Xerini) are ground-dwelling forms; the other African squirrels (tribe Protoxerini) are arboreal.

Family GLIRIDAE Muirhead, 1819

Representatives of two subfamilies of glirids are found currently in Africa: Graphiurinae (*Graphiurus*) and Leithiinae (*Eliomys*). *Graphiurus*, with 14 species, is restricted to sub-Saharan Africa (Holden, 2005) in forests or in rocky areas in dry tableland, often along waterways (Nowak, 1999). The genus is arboreal, but often found on the ground. Two species of *Eliomys* are known currently from northern Africa (Holden, 2005). In its African and non-African distribution, *Eliomys* may be found in a variety of habitats including extensive forests, swamps, rocky areas, cultivated fields, steppe deserts, and mountains (Nowak, 1999).

Suborder MYOMORPHA Brandt, 1855
Superfamily DIPODOIDEA Fischer de Waldheim, 1817
Family DIPODIDAE Fischer de Waldheim, 1817

Two subfamilies of dipodids are found currently in Africa: Dipodinae (*Jaculus*) and Allactaginae (*Allactaga*). *Jaculus* (two species) and *Allactaga* (one species) are known from desert and semidesert areas of northern Africa (Nowak, 1999). *Jaculus* is reported from Senegal, northeastern Nigeria, Niger, southern Mauritania to Morocco, then east to Somalia (Holden and Musser, 2005). It lives in sandy and saline deserts, rocky valleys, and meadows (Nowak, 1999). *Allactaga* is found on coastal gravel plains in Egypt and eastern Libya (Holden and Musser, 2005).

Superfamily MUROIDEA Illiger, 1811
Family SPALACIDAE Gray, 1821

Only the subfamilies Spalacinae and Tachyoryctinae are found in Africa. Extant spalacids are highly specialized fossorial and subterranean muroids. The spalacines are represented exclusively by *Spalax ehrenbergi*, whose primarily Middle Eastern range also includes the Mediterranean coastal areas of Libya and Egypt (Musser and Carleton, 2005). *Spalax ehrenbergi* is found in deep sandy or loamy soils in a variety of habitats (Nowak, 1999). Musser and Carleton (2005) recognize 13 species of the exclusively African genus *Tachyoryctes*, which is the only extant member of the Tachyoryctinae. *Tachyoryctes* is found in eastern Africa, with most species occurring at high altitude in the East African Rift mountains, generally in areas with >500 mm annual rainfall (Nowak, 1999). It is most common in wet uplands. Preferred habitats include open grassland, thinly treed savanna, moorland, and cultivated areas (Nowak, 1999).

Family NESOMYIDAE Forsyth Major, 1897

Composition of the Nesomyidae as used here follows Musser and Carleton (2005:930), who include the subfamilies Cricetomyinae, Delanymyinae, Dendromurinae, Mystromyinae, Nesomyinae, and Petromyscinae. Representatives of all these subfamilies are currently found only in Africa (Nesomyinae only in Madagascar). The Delanymyinae and Nesomyinae are unknown from the fossil record.

The Cricetomyinae include three extant genera: *Beamys*, *Cricetomys*, and *Saccostomus*. Cricetomyines are found in savanna to forest habitats in sub-Saharan Africa (Nowak, 1999). Extant dendromurines are found also in sub-Saharan Africa (Nowak, 1999). The monophyly of Dendromurinae was established based on molecular evidence; they include the genera *Dendromus, Megadendromus, Malacothrix, Dendroprionomys, Prionomys*, and *Steatomys* (Musser and Carleton, 2005). *Dendromus* and *Steatomys* are widely distributed. *Dendromus* is found in a wide range of habitats from sea level to 4,300 m. *Steatomys* occurs in dry savanna and subtropical-tropical dry lowland grasslands. *Megadendromus* is restricted to the Ethiopian highlands. *Dendroprionomys* and *Prionomys* are endemic to central African forest blocks, while *Malacothrix* is found only in dry areas of southern Africa (ecology of dendromurine genera from, e.g., Nowak, 1999, and Musser and Carleton, 2005). Two genera originally considered to belong to the Dendromurinae, *Deomys* and *Leimacomys*, have been transferred to Muridae (Deomyinae and Leimacomyinae).

Phylogenetic allocation of the monotypic Mystromyinae has been controversial (see discussion in Musser and Carleton, 2005), but recent phylogenetic analysis of two nuclear protein-coding genes allies *Mystromys* with the Nesomyinae, Cricetomyinae, and Dendromurinae (Michaux et al., 2001). *Mystromys* is found currently in South Africa, Lesotho, and south Swaziland (Musser and Carleton, 2005), where it inhabits the Fynbos, Succulent Karoo, Nama Karoo, Grassland, Arid Savanna, and Savanna Woodland Biomes (Mugo et al. in Musser and Carleton, 2005).

The Petromyscinae include only the genus *Petromyscus*, which inhabits Namibia, Angola, and South Africa. *Petromyscus* lives in rocky habitats in dry barren mountains (Nowak, 1999).

Family CRICETIDAE Fischer de Waldheim, 1817

Musser and Carleton (2005:955) briefly summarize the history of the "cricetid-murid question," as to the correct familial assignment for the different subfamilies of muroids that have been variously assigned to the Cricetidae or the Muridae. Musser and Carleton's proposal, which they note is provisional and which we follow here, finds general support for a monophyletic Cricetidae clade based on phylogenetic evaluation of genetic sequence data (Musser and Carleton, 2005:955). Diagnostic morphological characters for the Cricetidae are often ambiguous because of the large size and heterogeneity of the group, and because of parallel evolution of derived characters within the Cricetidae and among the Muroidea. Even so, Musser and Carleton (2005:956) list several characters shared by all or many cricetids (those most relevant to fossils are given in the diagnosis for the family in the Systematic Paleontology section).

Musser and Carleton (2005) recognize six subfamilies of cricetids from around the world, of which only the Lophiomyinae are currently found in Africa. The subfamily Arvicolinae, tribe Ellobiusini, is known from Africa as fossils but is absent from the modern fauna.

The Lophiomyinae includes one genus and species, *Lophiomys imhausi*, the maned rat. *Lophiomys* is known from eastern Sudan to Somalia, northeastern Uganda, Kenya, and western Tanzania (Musser and Carleton, 2005). It is generally believed to inhabit dense montane forests, but appears to be tolerant of a wide range of habitats (Nowak, 1999).

TABLE 17.5
Details of fossil rodent collections from northern Africa
Abbreviations for countries: Al, Algeria; Ch, Chad; Eg, Egypt; Li, Libya; Mo, Morocco; Tu, Tunisia.

Epoch	Region	Locality	Level/Subunit	References
Early to middle Eocene	Al	Gour Lazib (Glib Zegdou)	—	Vianey-Liaud et al., 1994
	Tu	Chambi	—	Vianey-Liaud et al., 1994
Middle to late Eocene	Al	Bir El Ater (= Nementcha)	—	Jaeger et al., 1985
Late Eocene to early Oligocene	Eg	Fayum Depression	Qasr el Sagha and Jebel Qatrani fms	Wood, 1968; Holroyd, 1994
	Li	Dor el Talh (Dur et Tallah)	—	Holroyd, 1994
Early Oligocene	Li	Zallah Oasis (Zella)	—	Fejfar, 1987
Early Miocene	Li	Jebel Zelten	Lowermost fossiliferous units (A, B)	Wessels et al., 2003; Savage, 1990
Middle Miocene	Tu	Testour	—	Robinson & Black, 1973
	Mo	Beni Mellal	—	Lavocat, 1961; Jaeger, 1977a
	Mo	Azdal	—	Benammi et al., 1995
	Li	Jebel Zelten	Measured Section 2; Wadi Atírát	Wessels et al., 2003; Savage, 1990
	Mo	Jebel Rhassoul	—	Benammi, 1997a
	Al	Numerous sites[1]	—	Coiffait-Martin, 1991
	Mo	Pataniak 6	—	Jaeger, 1977b
Late Miocene	Al	Bou Hanifia	Bou Hanifia Fm	Ameur, 1976, 1984
	Mo	Oued Tabia	—	Benammi et al., 1995
	Eg	Sheikh Abdallah (Farafra)	—	Heissig, 1982; Coiffait-Martin, 1991; Pickford et al., 2008
	Eg	Gabal et Muluk, Wadi el Natrun	Stromer's Profile C	James & Slaughter, 1974; Slaughter & James, 1979
	Ch	Tm 266, Toros-Menalla	anthracotherid unit	Vignaud et al., 2002
	Mo	Afoud, Aït Kandoula Bassin	—	Benammi, 1997b, 2001; Benammi et al., 1995; Remy & Benammi, 2006
	Li	Sahabi	Sahabi Fm	Munthe, 1987; Agusti et al, 2004
	Al	Amama 2	—	Jaeger, 1977b
	Mo	Numerous sites[2]	—	Jaeger, 1977b
	Tu	Numerous sites[3]	—	Jaeger, 1977b; Robinson et al., 1982
	Al	Numerous sites[4]	—	Arambourg, 1959; Ameur-Chabbar, 1988; Coiffait-Martin, 1991
Early Pliocene (late Miocene-early Pliocene)[5]	Mo	Aïn Guettara	—	Brandy & Jaeger, 1980
	Mo	Lissasfa	—	Geraads, 1998b
	Ch	Kossom, Bougoudi	Lower & upper green sandstones	Brunet et al., 2000; Denys et al., 2003
Early Pliocene	Mo	Aghouri	—	Benammi et al., 1995
	Mo	Saïz[6]	—	Jaeger, 1975
	Tu	Lac Ichkeul (= Garaet Ichkeul)	—	Jaeger, 1971b
	Al	Amama 3[6]	—	Jaeger, 1975
Middle Pliocene	Al	Oued Athmenia 1	—	Coiffait-Martin, 1991
	Al	Oued Smendou	—	Coiffait-Martin, 1991
	Tu	Djebel Mellah[6]	—	Jaeger, 1975
Late Pliocene	Mo	Ahl al Oughlam	—	Geraads, 1995, 2006
	Al	Argoub Kemellal 2	—	Coiffait-Martin, 1991
	Tu	Aïn Brimba[6] & Bulla Regia I[6]	—	Jaeger, 1975; Mein & Pickford, 1992
Early Pleistocene	Al	Aïn Rouina[6]	—	Jaeger, 1975
	Mo	Sidi Abdallah[6]	—	Jaeger, 1975
	Mo	Thomas Quarry	Level L	Geraads, 2002
Early to late Pleistocene	Tu	Djebel Ressas	—	Mein & Pickford, 1992
Early to middle Pleistocene	Mo	Jebel Irhoud	—	Jaeger, 1970, 1971a, 1971b

Epoch	Region	Locality	Level/Subunit	References
Middle Pleistocene	Al	Tighenif (Ternifine)	—	Jaeger, 1969; Tong, 1986
	Mo	Grotte des Rhinocéros, Oulad Hamida I	—	Geraads, 1994
	Mo	Thomas Quarry	Th1-G, Th1-ABCE	Geraads, 2002
	Al	Aïn Mefta,[6] Tadjera,[6] & Jebel Filfila[6]	—	Jaeger, 1975
	Mo	Sidi Abderrahman[6] & Salé[6]	—	Jaeger, 1975
	Tu	Bulla Regia II[6]	—	Jaeger, 1975

[1]Middle Miocene of Algeria: Chouf Aissa, Fedj el Besbes, Oued Metlili, Polygone 1 & 2, Sidi Messaoud.

[2]Late Miocene of Morocco: Amama 1 & 2, Asif Assermo, Khendek-el-Ouaich, Oued Zra.

[3]Late Miocene of Tunisia: Araguib Kammra C, Jebel Semmene, Sidi Ounis-MDM.

[4]Late Miocene of Algeria: Argoub Kemellal 1, Bab el Ahmar, Beni Mrahim, Bou Adjeb, Dra Temedlet, El Allaiga, El Hiout, Guergour Ferroudi, Maatgua, Mekhancha, Ouled el Arbi, Smendou 6, Zighout Youcef, Fedj el Attauch (= Koudiet et Tine or Oued el Atteuch), Sidi Salem, Menacer (= Marceau).

[5]Fauna from these sites, dated at or near the Miocene-Pliocene boundary, are listed as early Pliocene in table 17.2.

[6]Epoch subdesignation approximate.

TABLE 17.6

Details of fossil rodent collections from eastern Africa

Abbreviations for countries: Et, Ethiopia; Ke, Kenya; Ug, Uganda; Ta, Tanzania; Zr, Zaire (Democratic Republic of the Congo).

Epoch	Country	Locality	Level/Subunit	References
Late Oligocene	Tn	Mbeya Region	TZ-01; Red Sandstone Group	Stevens et al., 2004, 2006; Stevens et al., 2009
	Ke	Losodok	Eragaleit beds	Rasmussen & Gutiérrez, 2009
Early Miocene	Ug	Bukwa	Green clay unit	Winkler et al., 2005
	Ke	Meswa Bridge, Songhor, Koru, Rusinga, Mfwangano, Karunga	Various	Lavocat, 1973; Odhiambo Nengo & Rae, 1992; Holroyd and Stevens, 2009
	Ug	Napak	—	Bishop, 1962; Lavocat, 1973; Pickford et al., 1986; MacLatchy et al., 2007
	Ug	Moroto II	Kagole Beds	Pickford & Mein, 2006
	Ke	Kalakol (Kalodirr)	—	Leakey & Leakey, 1986
Middle Miocene	Ke	Fort Ternan	—	Lavocat, 1964, 1988, 1989 ; Denys & Jaeger, 1992
	Ke	Kipsaramon	Muruyur Beds	Pickford, 1988; Winkler, 1992
	Ke	Maboko	—	Winkler, 1998
	Ke	Kabarsero	Ngorora Fm	Winkler, 2002
Late Miocene	Ke	Nakali		Flynn & Sabatier, 1984
	Ke	Samburu Hills	Namurungule Fm	Kawamura & Nakaya, 1984
	Et	Chorora	Chorora Fm	Jaeger et al., 1980; Geraads, 1998a, 2001
	Ug	NY 45	Kakara Fm	Mein, 1994
	Ke	Lothagam	Lower Mb, Nawata Fm	Cifelli et al., 1986; Winkler, 2003
	Ke	Aragai, Kapcheberek, Kapsomin	Lukeino Fm	Winkler, 2002; Mein & Pickford, 2006
	Ke	Lemudong'o 1, Lemudong'o	Speckled Tuff (primarily)	Ambrose et al., 2003; Hlusko, 2007; Manthi, 2007
	Et	Middle Awash localities	Adu-Asa Fm	Haile-Salassie et al., 2004; Wesselman et al., 2009
	Ke	Lothagam	Upper Mb, Nawata Fm	Winkler, 2003
Late Miocene to early Pliocene[1]	Ta	Inolelo 1, Shoshamagai 2, Manonga Valley	Ibole Mb, Wembere Manonga Fm	Winkler, 1997
Early Pliocene	Et	Middle Awash localities (e.g., Aramis)	Haradaso Mb, Sagantole Fm	WoldeGabriel et al., 1994; Haile-Salassie et al., 2004; Wesselman et al., 2009
	Ke	Kanapoi	Kanapoi Fm	Harris et al., 2003; Manthi, 2006

TABLE 17.6
(CONTINUED)

Epoch	Country	Locality	Level/Subunit	References
	Ke	Tabarin	Chemeron Fm	Winkler, 2002
	Ta	Laetoli	Kakesio Beds	Denys, 1990c
	Ug	Kazinga Channel	Kazinga Beds	Mein, 1994
Early to middle Pliocene	Ta	Lothagam	Kaiyumung Mb, Nachukui Fm	Winkler, 2003
	Ta	Laetoli	Upper Laetolil Beds	Denys, 1985, 1987a; Davies, 1987
Middle Pliocene	Et	Hadar	Sidi Hakoma, Denen-Dora, Kada Hadar mbs, Hadar Fm	Sabatier, 1978, 1979, 1982
Middle to late Pliocene	Et	Omo Valley	Shungura Fm	Wesselman, 1984
Late Pliocene	Zr	Semliki Valley	Lusso Beds	Boaz et al., 1992
Middle to late Pliocene	Ta	Laetoli	Upper Ndolanya Beds	Denys, 1987a; Ndessokia, 1989
Late Pliocene & Plio-Pleistocene	Ke	West Lake Turkana	Nachukui Fm	Harris et al., 1988
Plio-Pleistocene (late Pliocene to early Pleistocene)	Ke	East Lake Turkana	Koobi Fora Fm	Black & Krishtalka, 1986; Denys, 1999
Early Pleistocene	Ta	West Natron (Peninj)	—	Denys, 1987b
	Ta	Olduvai Gorge	Bed I	Jaeger, 1976; Denys, 1989a–d, 1990b, 1992
Middle Pleistocene	Ta	Olduvai Gorge	Masek Beds	Jaeger, 1979
	Ke	Tugen Hills	"Hominid level" Kapthurin Fm	McBrearty, 1999; Denys, 1999
	Et	Asbole, lower Awash Valley	Awash Group	Alemseged & Geraads, 2000; Geraads et al., 2004
	Ke	Isenya	—	Brugal & Denys, 1989
Late Pleistocene	Et	Melka Kunture	—	Sabatier, 1979
	Et	Porc-Epic Cave	—	Assefa, 2006

[1]The Manonga Valley sites are of late Miocene (based on overall fauna) to early Pliocene age (based on rodents); they are recorded as early Pliocene in table 17.3.

Family MURIDAE Illiger, 1811

Composition of the Muridae as used here follows Musser and Carleton (2005:901), who include the subfamilies Leimacomyinae, Deomyinae, Gerbillinae, Murinae, and Otomyinae. Representatives of all these Subfamilies are found currently in Africa. Musser and Carleton (2005) note that phylogenetic analyses of mitochondrial and nuclear genetic sequences support the monophyly of various subsets of these subfamilies.

The enigmatic *Leimacomys*, sole member of the Leimacomyinae and known from only two museum specimens, is unknown from the fossil record. There is a complicated history of phylogenetic assignments/proposed relationships of members of the Deomyinae, summarized by Musser and Carleton (2005). Members of the Deomyinae range in their distribution and ecology from a widespread geographic distribution in mainly arid habitats *(Acomys)*, to savannas *(Uranomys)*, to a more restricted distribution in tropical forests from equatorial Guinea to Uganda *(Deomys and Lophuromys)*(Nowak, 1999; Musser and Carleton, 2005).

The Gerbillinae are defined clearly by both morphological and molecular attributes (Musser and Carleton, 2005). Fourteen genera are currently found in Africa. Based on anatomi-

cal study, Musser and Carleton (2005) split the genus *Tatera* and include in this genus only the Indian *T. indica*. All the African representatives are transferred to *Gerbilliscus*, which becomes a sub-Saharan African endemic genus. Gerbils are found in almost all of Africa, generally in deserts and semideserts (Tong, 1989) and never in tropical forests.

As defined by Carleton and Musser (1984), the Murinae can be characterized by a cluster of external, cranial, postcranial, dental, reproductive, and arterial characteristics, and, in particular, by a derived molar morphology. Extant African murines represent more than 25% of worldwide murine diversity with 32 endemic genera (Musser and Carleton, 2005). They are found throughout the continent in a wide variety of habitats. Recent molecular work confirms some morphological divisions defined previously by Misonne (1969), such as the Arvicanthine, *Rattus*, *Praomys*, and *Mus* divisions. Lecompte et al. (2005) identified three distinct groups of African murines: the Arvicanthini (sensu Ducroz et al., 2001) and a *Praomys* group (*sensu* Lecompte et al., 2002), with *Malacomys* isolated in its own "group." Lecompte et al. (2005) also added the *Mus (Nannomys)* clade. Lecompte et al.'s (2008) analysis of one mitochondrial and two nuclear gene sequences lead them to propose dividing all murines into 10 formal tribes. The endemic African murines would include members

TABLE 17.7
Details of rodent collections from southern and south-central Africa

Abbreviations for regions: As, Angola, southern; Bo, Botswana; Nn, Namibia, northern; Ns, Namibia, southern; Sc, South Africa, central; Sn, South Africa, northern; Sw, South Africa, western; Za, Zambia; Zi, Zimbabwe.

Epoch	Region	Locality	Level/Subunit	References
Early Miocene	Ns	Diamond Fields (various); Auchas	—	Hopwood, 1929b; Lavocat, 1973; Hamilton & Van Couvering, 1977; Stromer, 1924, 1926; Mein & Pickford, 2003; Mein & Senut, 2003
Early to basal middle Miocene[1]	Ns	Arrisdrift	Pit2/AD 8	Hendey, 1978; Mein & Pickford, 2003; Mein & Senut, 2003
Middle Miocene	Nn	Berg Aukas	Blocks BA1, BA47 & BA63	Conroy et al., 1992; Pickford et al., 1994; Senut et al., 1992
Late Miocene	Nn	Berg Aukas	Blocks BA31 & BA90	Conroy et al., 1992; Senut et al.; 1992; Mein et al., 2000a, b, 2004
	Nn	Harasib 3a	Block ARI	Senut et al., 1992; Mein et al. 2000a, b, 2004
Early Pliocene	Sn	Bolt's Farm	Waypoint 160, unspecified	Cooke, 1963; Sénégas & Avery, 1998; Sénégas & Michaux, 2000; Sénégas, 2004
	Sn	Makapansgat	Dumps, EXQRM, MLWD/III, MRCIS, Rodent Cave, unspecified	Cooke, 1963; De Graaff, 1960; Denys, 1999; Lavocat, 1956, 1957, 1978; Pocock, 1987; Turner et al., 1999, Sénégas 2000
	Ss	Langebaanweg	QSM, PPM, unspecified	Denys, 1990a, c, d, 1991, 1994a, b, 1998; Hendey, 1976, 1981, 1984; Pocock, 1976
Late Pliocene	Nn	Jägersquelle	—	Denys, 1999; Senut et al., 1992; Turner et al., 1999
	Nn	Nosib	Blocks NOS1 & NOS2, unspecified	Denys, 1999; Senut et al., 1992; Turner et al., 1999
	Sn	Sterkfontein	Sts/Dumps 1, 2 & 8; Stw/H2; Mb 2 & 4, Type Site, Ext, M4	Avery, 2000a, 2001; Cooke, 1963; Denys, 1999; Pocock, 1987; Turner et al., 1999
	Sn	Drimolen	Cave, lower level Dl, upper level Du	Sénégas et al. 1999
Plio-Pleistocene	As	Humpata	2, unspecified	Denys, 1999; Pickford et al., 1990, 1992, 1994; Turner et al., 1999
	Bo	NW Botswana (Ngamiland: Gcwihaba & Nqumtsa)	—	Denys, 1999; Pickford, 1990; Pickford & Mein, 1988; Pickford et al., 1994; Turner et al., 1999
Early Pleistocene	Nn	Aigamas	Blocks AIG1 & AIG2	Senut et al., 1992
	Nn	Berg Aukas	Blocks BA8 & BA54	Conroy et al., 1992; Senut et al., 1992
	Nn	Friesenberg	—	Sénégas, 1996
	Nn	Uisib	Block UIS	Senut et al., 1992
	Sc	Taung	—	Broom, 1934, 1939, 1948a, b; Broom & Schepers, 1946; Cooke, 1963, 1990; Denys, 1999; Lavocat, 1967, 1978; Turner et al., 1999
	Sn	Gladysvale	—	Avery, 1995a; Cooke, 1963
	Sn	Kromdraai	A, B, unspecified	Cooke, 1963; De Graaff, 1961; Denys, 1999; Pocock, 1985, 1987; Turner et al., 1999
	Sn	Plover's Lake	—	Thackeray and Watson, 1994
	Sn	Schurveberg	—	Broom, 1937a, b
	Sn	Sterkfontein	M5E	Avery, 2000a, 2001

TABLE 17.7
(CONTINUED)

Epoch	Region	Locality	Level/Subunit	References
	Sn	Swartkrans	Mb 1-3	Avery, 1998, 2001; Brain et al., 1988; Denys, 1999; Turner et al., 1999; Watson, 1993
	Za	Kabwe	—	Avery, 2003; Chubb, 1909; Hopwood, 1929a; Mennell & Chubb, 1907
Middle Pleistocene	Sc	Wonderwerk	Acheulean levels	Avery, 1995b, unpubl.; Klein, 1988
	Sn	Cave of Hearths	—	Cooke, 1963; De Graaff, 1960, 1988
	Ss	Duinefontein	DFT2	Klein, 1976b; Klein et al., 1999; Cruz-Uribe et al., 2003
	Ss	Elandsfontein	Main	Klein & Cruz-Uribe, 1991
	Za	Mumbwa	XIV-VIII	Avery, 2000b; Klein & Cruz-Uribe, 2000
	Za	Twin Rivers	—	Avery, 2003
	Ss	Hoedjiespunt 1	ROOF & hominid sands	Matthews et al., 2005, 2006
Late Pleistocene	Ns	Zebrarivier	MSA levels	Avery, 1984
	Sc	*Equus* Cave	Pleistocene levels	Klein et al., 1991
	Sc	Florisbad	Spring, unspecified	Brink, 1987; Scott & Brink, 1992
	Sn	Sterkfontein	M6	Avery, 2000a, 2001
	Ss	Blombos	MSA levels	Henshilwood et al., 2001
	Ss	Boomplaas	Levels BRL-LOH	Avery, 1982b
	Ss	Byneskranskop	Levels 15–19	Avery, 1982b; Klein, 1981; Schweitzer & Wilson, 1982
	Ss	Die Kelders	MSA levels	Klein, 1975; Avery, 1982b; Avery et al., 1997; Grine et al., 1991
	Ss	Elands Bay Cave	Pleistocene levels	Klein & Cruz-Uribe, 1987
	Ss	Klasies River Mouth	1 & 1A Pleistocene levels	Avery, 1987; Klein, 1975, 1976a
	Ss	Nelson Bay Cave	YSL & YGL (LGM levels)	Avery, 1982b; Klein, 1972, 1974
	Sw	Border Cave	Pleistocene levels	Avery, 1982a, 1992; Klein, 1977
	Sw	Umhlatuzana	Layers 5 & 6	Avery, 1991
	Za	Mumbwa	VII & III	Avery, 2000b; Klein & Cruz-Uribe, 2000
	Zi	Pomongwe	ESA & MSA levels	Brain, 1981
	Zi	Redcliff	—	Klein, 1978
	Bo	Drotsky's Cave	—	Robbins et al., 1996

[1]Listed as middle Miocene in table 17.4.

of five tribes: Murini, Praomyini, Malacomyini, Otomyini, and Arvicanthini.

Musser and Carleton (2005) follow a provisional classification for the Otomyinae that includes the extant genera *Myotomys*, *Otomys*, and *Parotomys*. Recent molecular work confirms that these rodents belong in the Murinae (Lecompte et al., 2008; see previous discussion) and are very close to the Arvicanthini, but we follow the traditional classification of considering them a separate subfamily. The otomyines are indigenous to sub-Saharan Africa. They often live in grassy areas in the vicinity of water but may also occur in rocky or otherwise more arid habitats, and some species are endemic to cold high altitudes.

Suborder ANOMALUROMORPHA Bugge, 1974
Superfamily ANOMALUROIDEA Gervais, 1849
Family ANOMALURIDAE Gervais, 1849

This is an exclusively African group. Dieterlen (2005a) divides the anomalurids into two subfamilies, Anomalurinae and Zenkerellinae. The Anomalurinae includes the genus *Anomalurus*, with four species, and the Zenkerellinae includes two genera, *Idiurus* (two species) and *Zenkerella* (one species). Schunke (2005) also recognizes seven species, but considers *"Anomalurus" beecrofti* to be the only representative of the genus *Anomalurops*. Anomalurids are found in rain forest, montane forests, and gallery forests in an essentially west to east band from Senegal to Tanzania (Schunke, 2005). Excepting *Zenkerella*, all anomalurids can perform gliding flight (Schunke, 2005).

Superfamily PEDETOIDEA Gray, 1825
Family PEDETIDAE Gray, 1825

This exclusively African group includes only the genus *Pedetes*, which is fossorial and inhabits open semiarid to arid environments. *Pedetes capensis* occurs in central and southern Africa, while *P. surdaster* is found in Tanzania and Kenya (Dieterlen, 2005c).

Suborder HYSTRICOMORPHA Brandt, 1855
Infraorder CTENODACTYLOMORPHI Chaline and Mein, 1979
Family CTENODACTYLIDAE Gervais, 1853

Although known also from the fossil record of Asia, the ctenodactylids are restricted currently to Africa. There are four genera (*Ctenodactylus*, *Felovia*, *Massoutiera*, and *Pectinator*) and five species, which occur in northern Africa eastward, to Somalia in eastern Africa (Dieterlen, 2005b). Ctenodactylids are found in desert or semidesert habitats (Nowak, 1999).

Infraorder HYSTRICOGNATHI Brandt,1855
Family BATHYERGIDAE Waterhouse, 1841

This group is endemic to Africa south of the Sahara. It includes six genera and 17 species (Woods and Kilpatrick, 2005; Knock et al., 2006). Both morphological and molecular evidence support division into two subfamilies. The subfamily Heterocephalinae is monogeneric, represented by *Heterocephalus* from Kenya, Ethiopia, and Somalia. The subfamily Bathyerginae includes *Bathyergus*, *Georychus*, *Cryptomys*, *Fukomys*, and *Heliophobius*. *Bathyergus* and *Georychus* are restricted mainly to southern South Africa, although *Bathyergus* is known also from southern Namibia and *Georychus* has relict populations in KwaZulu-Natal and Mpumalanga (Nowak, 1999; Friedmann and Daly, 2004). *Cryptomys* occurs from southern Africa to Sudan and west to Ghana (Woods and Kilpatrick, 2005), but this genus is in need of revision. On molecular grounds, Knock et al. (2006) erected a new genus, *Fukomys*, from the Zambezian savannas. The type species is *Cryptomys damarensis*, and the genus includes some *Georychus* and *Cryptomys* species fround north of Limpopo River, and possibly several undescribed cryptic species (Van Daele et al., 2007). *Heliophobius* is known from Zimbabwe, Zambia, and Mozambique to the Democratic Republic of Congo and east to Kenya and Tanzania (Woods and Kilpatrick, 2005). Bathyergids are fossorial and usually live in areas of loose sandy soil; some of them are eusocial to varying degrees (Nowak, 1999).

Family HYSTRICIDAE Fischer de Waldheim, 1817

African hystricids are represented by two genera and three species. *Atherurus africanus* is known from the Guineo-Congolese forests of west-central Africa to Kenya, Uganda, and southern Sudan, and it can reach elevations of 3,000 m (Nowak, 1999; Woods and Kilpatrick, 2005). *Hystrix cristata* ranges from Morocco to Egypt and from Senegal to Ethiopia and northern Tanzania (Woods and Kilpatrick, 2005). *Hystrix africaeaustralis* occurs from the mouth of the Congo River to Rwanda, Uganda, Kenya, and into southern Africa (De Graaff, 1981; Woods and Kipatrick, 2005). The two species of *Hystrix* are sympatric in Central and East Africa (De Graaff, 1981). *Hystrix* is found most commonly in hilly, rocky habitats but is adaptable to almost any habitat except swampy areas, extensive moist forest, and the most barren deserts (Kingdon, 1974; De Graaff, 1981).

Family PETROMURIDAE Wood, 1955

Several authors (although not McKenna and Bell, 1997) group the Petromuridae with the Thryonomyidae into the superfamily Thryonomuroidea (see discussion in Woods and Kilpatrick, 2005). Extant petromurids are known only from southern Africa. This family includes one extant genus and species, *Petromus typicus*. The dassie rat is known from western

South Africa and Namibia to southwestern Angola (Woods and Kilpatrick, 2005). It is found in the South West Arid and marginally in the Southern Savanna biotic zones, and lives in narrow rock crevices and among large boulders in rocky hills and mountainous areas (De Graaff, 1981).

Family THRYONOMYIDAE Pocock, 1922

The thryonomyids include one genus with two species, *Thryonomys gregorianus* (lesser cane rat) and *T. swinderianus* (greater cane rat). *Thryonomys gregorianus* occurs in sub-Saharan Africa approximately in a band from Cameroon in the west to Ethiopia in the east, and south to Mozambique. *Thryonomys swinderianus* is more widely distributed in "Africa, south of the Sahara" (Woods and Kilpatrick, 2005:1545). Although both species are dependent on grass for cover and food, they occupy distinct ecological niches: *T. gregorianus* is more terrestrial and lives in moist savanna, while *T. swinderianus* prefers semiaquatic habitats such as marshes and reed beds (Kingdon, 1974).

Systematic Paleontology

The diagnoses given here include those characters most applicable to fossil remains: soft tissue and/or molecular characters diagnostic of taxa with extant representatives are not included. The "Geologic Age" of taxa refers only to the African record.

Order RODENTIA Bowdich, 1821
Suborder SCIUROMORPHA Brant, 1855
Family SCIURIDAE Fischer de Waldheim, 1817
Figure 17.2

Diagnosis 1/1, 0/0, 1-2/1, 3/3; sciurognathus, sciuromorphous: skull usually with well-developed postorbital processes; presence of DP4/P4 (dp4/p4) and in some cases DP3/P3; molars bunodont, cuspidate with cusps sometimes connected by lophs; lower cheek teeth similar to uppers or basined.

Description Fossil African squirrels are generally known from isolated teeth or incomplete dentitions. The most complete material (i.e., partial skeletons) is from the early Pliocene at Kossom Bougoudi, Chad (Denys et al., 2003), and from the late Miocene at Lothagam, Kenya (Cifelli et al., 1986; Winkler, 2003).

Geologic Age Early Miocene to Recent.

African Distribution Ubiquitous.

Remarks The oldest fossil sciurid from Africa is *Vulcanisciurus* (figure 17.2A), which is reported from the early (ca. 20 Ma; Rusinga, Songhor, and Koru, Kenya, and Napak, Uganda; Lavocat, 1973) middle (14 Ma; Fort Ternan, Kenya; Denys and Jaeger, 1992), and earliest late Miocene (12.5 Ma; Ngorora Formation, Tugen Hills, Kenya; Winkler, 1990, 2002). This is the only described genus from eastern Africa during this time period. From southern Africa, Mein et al. (2000a) describe *Heteroxerus* (previously listed as *Vulcanisciurus*; Senut et al., 1992) from Harasib 3a, Namibia (ca. 10 Ma). Mein et al. (2000a) consider the undetermined sciurid from the middle Miocene Muruyur Beds, Kenya (Winkler, 1992), to pertain also to *Heteroxerus*. Mein et al. (2000a) note that late middle Miocene sciurids from Berg Aukas, Namibia, most closely resemble *Vulcanisciurus*. *Atlantoxerus*, which even now is known only from northern Africa, is first reported from the middle Miocene (ca. 15 Ma) at Beni Mellal, Morocco (Jaeger, 1977a), but is better known from younger deposits (e.g., in the late Miocene from Libya, Algeria, Morocco, Tunisia, and tentatively from Egypt).

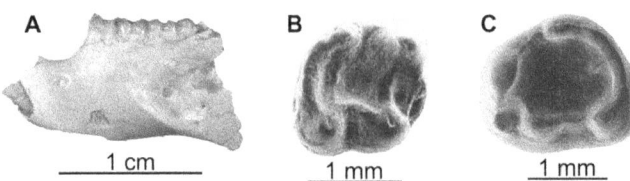

FIGURE 17.2 Examples of African Sciuridae. A) *Vulcanisciurus*, left mandible with p4–m3, uncatalogued specimen, early Miocene, Rusinga, Kenya. B) Subfamily Xerinae, *Paraxerus*, right m1 or m2, KNM-TH 19473, early Pliocene, Tabarin, Chemeron Formation, Kenya. C) Subfamily Xerinae, *Heliosciurus*, KNM-TH 19484, early Pliocene, Tabarin, Chemeron Formation, Kenya.

The first occurrences of *Xerus* are from Chorora, Ethiopia (ca. 11 Ma; Geraads, 2001); Toros-Menalla, Chad (ca. 7–6 Ma; Vignaud et al. 2002); Lemudong'o, Kenya (6 Ma; Manthi, 2007); and the Middle Awash, Ethiopia (5.7 Ma; Wesselman et al., 2009, listed as *Paraxerus* by Haile-Selassie et al., 2004). The extinct exclusively African squirrel, *Kubwaxerus*, is known only from the late Miocene (7.4–6.5 Ma), lower Nawata Formation, Lothagam, Kenya: it shares affinities with modern *Protoxerus* (Cifelli et al. 1986, Winkler, 2003). The earliest occurrences of *Paraxerus* are from Lemudong'o, Kenya (6 Ma; Manthi, 2007); the Tabarin locality, Chemeron Formation, Kenya (figure 17.2B; 4.5–4.4 Ma; Winkler, 1990, 2002); and the Upper Laetolil Beds, Laetoli, Tanzania (ca. 3.8–3.5 Ma; Denys, 1987a). The earliest record of *Heliosciurus* is at the Tabarin locality, Chemeron Formation, Kenya (figure 17.2C; 4.5–4.4 Ma; Winkler, 1990, 2002).

Family GLIRIDAE Muirhead, 1819

Diagnosis 1/1, 0/0, 1/1, 3/3; sciurognathus, myomorphous except *Graphiurus*, which is hystricomorphous; cheek teeth extremely bunodont, shallowly concave, with transverse (often numerous) crests.

Geologic Age Middle Miocene to Recent.

African Distribution Northern (Morocco, Algeria, Tunisia), eastern (Kenya, Tanzania), and southern (South Africa, Namibia, Zambia) Africa.

Remarks The fossil record of glirids in Africa is sparse. The earliest report is of the extinct genus *Microdyromys* from the middle Miocene of Morocco (Jaeger, 1977a, 1977b), and Algeria (Coiffait-Martin, 1991). Senut et al. (1992) and Mein et al. (2000a; extinct genus *Otaviglis*) record late Miocene "graphiurines" from South Africa and Namibia. *Graphiurus* is known from the early Pliocene at Langebaanweg, South Africa (Hendey, 1981; Pocock, 1976); the early Pleistocene at West Natron, Tanzania (Denys, 1987b); and the middle Pleistocene Kapthurin Formation, Kenya (Denys, 1999; McBrearty, 1999). Geraads (1994) reports *Eliomys* from the middle Pleistocene at Oulad Hamida 1, Morocco. Dobson (1998: in Holden, 2005:832) suggests that *Eliomys* dispersed to northern Africa at least twice: during the Messinian from Iberia and during the late Pleistocene from the eastern Mediterranean.

Suborder MYOMORPHA Brandt, 1855
Superfamily DIPODOIDEA Fischer de Waldheim, 1817
Family DIPODIDAE Fischer de Waldheim, 1817

Diagnosis 1/1, 0/0, 0-1/0, 3/3; sciurognathus, myomorphous; much enlarged infraorbital foramen, masseteric plate reduced and strictly ventral, cheek teeth rooted, generally high crowned, and cuspidate; saltatorial postcranial adaptations.

Geologic Age Middle Miocene to Recent.

African Distribution Northern Africa (Morocco, Tunisia, Libya) and eastern Africa (Kenya, Tanzania, Ethiopia).

Remarks The earliest African records of the Dipodidae are from the middle Miocene. *Protalactaga moghrebiensis* is reported by Jaeger (1977b) from the middle Miocene of Morocco. Zazhigin and Lopatin (2000), in their review of the Allactaginae, exclude this species from *Protalactaga* and suggest that it should probably be referred to a new genus. *Heterosminthus* is known from middle Miocene deposits at Jebel Zelten, Libya (Wessels et al., 2003). The earliest report of the extant genus *Jaculus* is from the late Pliocene of Tunisia (Jaeger, 1975). The Dipodidae are first known as fossils in eastern African in the latest Pliocene of the Omo Valley, Ethiopia (Omo Member F, 2.08–1.98 Ma; Wesselman, 1984), and they survive in Kenya and Tanzania, with their last record in Tanzania at 0.6 Ma. The Dipodidae are currently not found in Kenya or Tanzania.

Superfamily MUROIDEA Illiger, 1811
Family SPALACIDAE Gray, 1821

Diagnosis 1/1, 0/0, 0/0, 3/3; sciurognathus, myomorphous; cheek teeth lamelliform, infraorbital foramen high and rather small (from Lavocat, 1978).

Geologic Age Late Miocene to Recent.

African Distribution Northern (Libya), eastern (Kenya, Ethiopia, Zaire), and southern Africa (Namibia).

Remarks We follow Musser and Carleton (2005) in considering the "Rhizomyidae" one of six subfamilies (including Tachyoryctinae) within the family Spalacidae. The earliest reported rhizomyine is *Prokanisamys* from the early Miocene at Jebel Zelten, Libya (Wessels et al., 2003). Tong and Jaeger (1993) describe *Pronakalimys* from the middle Miocene, Fort Ternan, Kenya. Flynn and Sabatier (1984) describe *Nakalimys* from the late Miocene of Kenya and place it in "Rhizomyidae." *Nakalimys* is also reported from Chorora, Ethiopia (ca. 11 Ma; Geraads, 1998a). Mein (2000a:385) allocates *Nakalimys* and *Harasibomys* to the "family Rhizomyidae or Spalacidae" and suggests that the early Miocene spalacine *Debruijnia* from Turkey (ca. 20 Ma) "could represent an ancestral form for the African burrowing rodents."

The oldest record of *Tachyoryctes (T. makooka)* is from the Adu-Asa and lower Sagantole formations, Middle Awash,

Ethiopia (at sites dated at 5.7, 5.6, 5.2, 4.85 Ma; Wesselman et al., 2009; as *Tachyoryctes* sp. in Haile-Selassie et al., 2004). As yet undescribed *Tachyoryctes* are known from Aramis localities in the lower Sagantole formation (Wesselman et al., 2009). Sabatier (1979) describes *T. pliocaenicus* from the Pliocene at Hadar (Hadar Formation), Ethiopia. Both Haile-Selassie et al. (2004) and Sabatier (1979) suggest their material is most similar to the Asian genus *Kanisamys*. *Tachyoryctes* is reported from the late Pliocene Lusso Beds, Zaire (Boaz et al., 1992), and from several Pleistocene localities in eastern Africa, but much of the material is not yet described (Flynn, 1990). Sabatier (1979) studied late Pleistocene *Tachyoryctes* from Melka Kunture, Ethiopia, and suggests that two species are present, including possibly *T. macrocephalus*.

Family NESOMYIDAE Forsyth Major, 1897
Figure 17.3

Diagnosis 1/1, 0/0, 0/0, 3/3; myomorphous, sciurognathus.

Geologic Age Early Miocene to Recent.

African Distribution Afrocricetodontinae: eastern and southern Africa. Cricetomyinae: eastern and southern Africa. Dendromurinae: northern, eastern, and southern Africa. Otavimyinae, Mystromyinae, and Petromyscinae: southern Africa.

Remarks Fossil remains of the Nesomyidae include the subfamilies Afrocricetodontinae (extinct), Cricetomyinae, Otavimyinae (extinct), Dendromurinae, Mystromyinae, and Petromyscinae. Although the Nesomyinae and Delanymyinae are unknown from the fossil record, extinct members of some other subfamilies, such as the Afrocricetodontinae and Petromyscinae *(Stenodontomys)*, may have phylogenetic affinities with the Nesomyinae and Delanymyinae, respectively (e.g., see discussion in Musser and Carleton, 2005).

The Afrocricetodontinae date from the early to middle Miocene of eastern Africa and the middle to late Miocene of Namibia. They include the genera *Afrocricetodon*, *Notocricetodon* (figure 17.3A), and *Protarsomys*.

The earliest record of a cricetomyine is an isolated tooth from the middle Miocene Ngorora Formation of Kenya (genus indet.; Winkler, 1990, 2002). Mein et al. (2004) consider the dendromurine described by Geraads (2001; cf. *"Dendromus"* gen. nov.) from the late Miocene of Ethiopia and by Winkler (1990, 2002) from the middle Miocene of Kenya, to pertain to *Saccostomus*. *Saccostomus* is known securely in eastern Africa from the late Miocene (Kenya; Mein and Pickford, 2006) and the late Miocene/early Pliocene (Tanzania; Winkler, 1997). In southern Africa, *Saccostomus* is identified tentatively from the late Miocene, and confidently from the late Pliocene (both records from northern Namibia; Senut et al., 1992). The only fossil *Cricetomys* is from the middle Pleistocene of Kenya (Kapthurin Formation; McBrearty, 1999; Denys, 1999). Fossils of *Beamys* are unknown.

The Otavimyinae include two genera, *Otavimys* and *Boltimys*. They are known only from Namibia and South Africa, and date from the late Miocene to early Pliocene (Mein et al., 2004).

Fossil dendromurines are relatively common and date back to the middle Miocene of Kenya (*Ternania* from Fort Ternan; Tong and Jaeger, 1993). *Ternania* is also reported from the late Miocene at Sheikh Abdallah, Egypt (Pickford et al., 2008): this is the only record of this genus other than from its type locality. *Senoussimys* is known only from the late Miocene of Algeria (Coiffait-Martin, 1991). Some other genera from the middle and late Miocene, assigned originally to the Dendromurinae (e.g., *Mabokomys*, *Dakkamys*, cf. *"Dendromus"* gen. nov., *Potwarmus*), are considered by other authors to belong in other subfamilies (see discussion in Musser and Carleton, 2005; Mein et al., 2004). The earliest records of the extant genera *Dendromus* and *Steatomys* (figure 17.3C) are from the

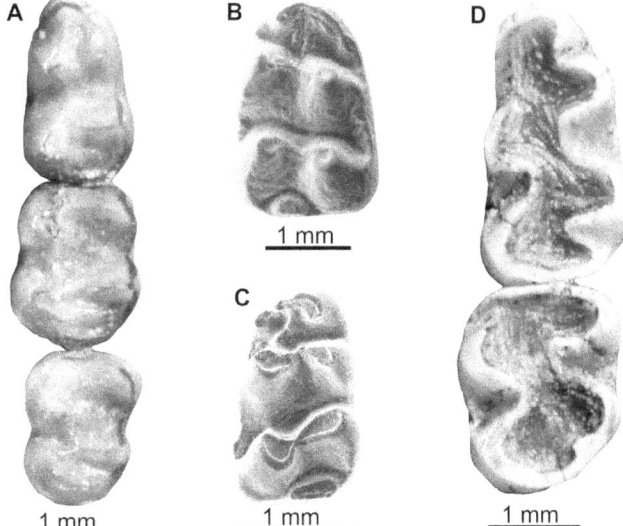

FIGURE 17.3 Examples of African Nesomyidae. A) Subfamily Afrocriceto-dontinae, *Notocricetodon*, left mandible with m1–3, BUMP 1272, early Miocene, Napak, Uganda. B) Subfamily Cricetomyinae, *Saccostomus*, WM 1343/92, right m1, late Miocene–early Pliocene, Manonga Valley, Tanzania. C) Subfamily Dendromurinae, *Steatomys*, left m1, KNM-TH 19467, late Miocene, Kapcheberek, Lukeino Formation, Kenya. D) Subfamily Mystromyinae, *Mystromys*, left mandible with m1–m2, uncatalogued specimen, Drotsky's Cave, Botswana.

late Miocene of northern Africa (e.g., *Dendromus* from Algeria [Coiffait-Martin, 1991], and both genera, tentatively, from Sheikh Abdallah, Egypt [Pickford et al., 2008]) and eastern Africa (Kenya; Mein and Pickford, 2006; *Dendromus* possibly also from Ethiopia; Geraads, 2001). In southern Africa, *Steatomys* is reported tentatively from the middle Miocene of Namibia, and *Dendromus* and *Steatomys* are known from the late Miocene of Namibia (Mein et al., 2004). *Dendromus* is reported from the early Pliocene at Langebaanweg, South Africa (Denys, 1994b), while *Malacothrix* is seen first in the middle Pliocene at Makapansgat, South Africa (Denys, 1999).

Fossil mystromyines, like their extant representatives, are exclusively from southern Africa. The earliest record of the subfamily is from the early Pliocene at Langebaanweg (Pocock, 1987). Fossil Mystromyinae include the extant genus *Mystromys* (figure 17.3D; including three extinct species, *M. hausleitneri* Broom, 1937b, *M. antiquus* Cooke, 1963, and *M. pocockei* Denys, 1991) and the extinct genus *Proodontomys* from the Plio-Pleistocene of South Africa. *Proodontomys* is believed to have become extinct ca. 1–0.7 Ma (Avery, 1998).

The fossil record of the Petromyscinae is also restricted to southern Africa. The earliest forms are from the late Miocene of Namibia, and include the extinct genera *Stenodontomys* (Pocock, 1987) and *Harimyscus* (Mein et al., 2000b). *Stenodontomys* is also reported from South Africa (Cape and Gauteng Provinces) and is last known in the early Pleistocene. *Harimyscus* is known only from the single late Miocene report. The extant genus *Petromyscus* is first known from the early Pliocene of South Africa.

Family CRICETIDAE Fischer de Waldheim, 1817

Diagnosis 1/1, 0/0, 0/0, 3/3; myomorphous, sciurognathus; biserial arrangement of molar cusps with retention of a mure; presence of a discrete anterocone(id) on first molars (from Musser and Carleton, 2005).

Geologic Age Middle Miocene to Recent.
African Distribution Northern Africa (Morocco, Algeria, Tunisia, Egypt) and eastern Africa (Ethiopia).
Remarks Only three of the six extant subfamilies of cricetids are represented as fossils in Africa. Although no longer found in Africa, *Ellobius* (Arvicolinae) is reported from the middle (Morocco, Tunisia, Algeria) and late (Tunisia) Pleistocene (Tong, 1986; Jaeger, 1988). The cricetine, *Apocricetus*, is listed only from Sheikh Abdallah, Egypt (Pickford et al., 2008). The Lophiomyinae are represented by two genera. *Protolophiomys* (extinct) is reported from the middle Pliocene of Algeria (Coiffait-Martin, 1991) and Morocco (Aguilar and Michaux, 1990) and also from the late Miocene of southern Spain (Aguilar and Thaler, 1987). The extant genus *Lophiomys* is first reported from Africa in the Middle Awash, Adu-Asa Formation, Ethiopia, at 5.7 Ma (Wesselman et al., 2009). This is the only eastern African record of this genus. *Lophiomys* is known also from Lissasfa, Morocco (the only northern African record), in the early Pliocene (Miocene-Pliocene boundary; Geraads, 1998b).

Family MURIDAE Illiger, 1811
Figure 17.4

Diagnosis 1/1, 0/0, 0/0, 3/3; myomorphous, sciurognathus; cusps arranged in chevrons in murines and some deomyines.
Geologic Age Early Miocene to Recent.
African Distribution Ubiquitous.
Remarks Three extinct subfamilies of Muridae are known: Myocricetodontinae, Cricetodontinae, and Namibimyinae. Two extinct genera, *Leakeymys* and *Potwarmus*, are of uncertain assignment to subfamily. All extant subfamilies of Muridae are known from the fossil record except for the Leimacomyinae.

The Myocricetodontinae make their first appearance in the late early Miocene of Libya at Jebel Zelten (Wessels et al., 2003).

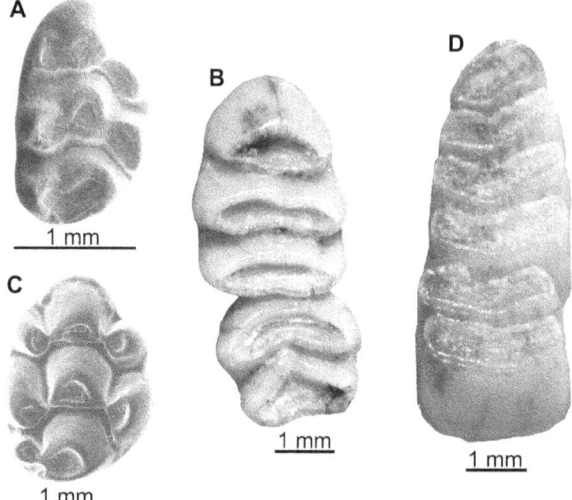

FIGURE 17.4 Examples of African Muridae. A) Subfamily Deomyinae, *Acomys*, right M1, uncatalogued specimen, early Pliocene, Langebaanweg, South Africa. B) Subfamily Gerbillinae, *Gerbilliscus*, right M1–M2, uncatalogued specimen, Drotsky's Cave, Botswana. C) Subfamily Murinae, *Saidomys*, right M1, KNM-TH 18478, early Pliocene, Tabarin, Chemeron Formation, Kenya. D) Subfamily Otomyinae *Otomys*, left m1–m2, uncatalogued specimen, Drotsky's Cave, Botswana. Figure 17.4A, from Xu et al., 1996, © University of Toronto Press Incorporated 1996. Reprinted with permission of the publisher. Figure 17.4C from Winkler, 2002, © Copyright 2002 with permission from Elsevier.

This group becomes relatively common in northern Africa, especially in the middle and late Miocene. Myocricetodontines are more speciose and abundant in northern (four genera) than in eastern (minimum one genus) or southern (three genera) Africa. The earliest reports of the Cricetodontinae are from the middle Miocene. In northern Africa they are from Algeria and Tunisia (Jaeger, 1977b; Robinson et al., 1982), in eastern Africa from Kenya (Winkler, 2002), and in southern Africa from Namibia (cf. *Democricetodon*; Pickford et al., 1994). The Cricetodontinae are best known from the late Miocene of northern Africa. The Namibimyinae are known only from the genus *Namibimys* from the late Miocene of Namibia (Mein et al., 2000b). The enigmatic African endemic genus *Leakeymys* is described from the middle Miocene Fort Ternan locality, Kenya (Lavocat, 1964; Tong and Jaeger, 1993). Coiffait-Martin (1991) lists *Leakeymys* from Farafra (= Sheikh Abdallah), Egypt, but Pickford et al. (2008) do not report this genus. *Potwarmus* is reported only from northern Africa: from the middle Miocene at Jebel Zelten, Libya (Wessels et al., 2003), and, tentatively, from the late Miocene at Sheikh Abdallah, Egypt (Pickford et al., 2008).

Extinct fossil genera of Deomyinae date from the late Miocene to early Pliocene (about 11–4 Ma), and have been described from both eastern and southern Africa. They display size and morphological variability and are both speciose and numerically abundant. *Preacomys* is found at Chorora, Ethiopia (Geraads, 2001), and Harasib, Namibia (Mein et al., 2004) at around 11–9 Ma (possibly Sheikh Abdallah, Egypt [*Progonomys* and/or *Preacomys*], Pickford et al., 2008), and later in the late Miocene in the Lukeino Formation, Kenya (Mein and Pickford, 2006). Although Geraads (2001) has reported cf. *Tectonomys* from Chorora, it is known with certainty only from the early Pliocene sites of Tabarin, Kenya (Winkler, 2002), and the Manonga Valley, Tanzania (Winkler, 1997). Within the Deomyinae, *Acomys* (figure 17.4A), like *Uranomys*, is characterized by a distinctive M3 pattern. *Preacomys* and *Tectonomys* may be synonymous with *Acomys*, based on their M3 pattern. The M3 is currently unknown for *Preacomys* from Chorora and for any sample of *Tectonomys,* but the M3 of the Harasib *Preacomys* does display the *Acomys* pattern.

Acomys is the only extant genus of Deomyinae that is well known as a fossil. It is reported from eastern and southern Africa, and is often numerically abundant. Its oldest occurrences are at Lemudong'o, Kenya (late Miocene; Manthi, 2007), Langebaanweg, South Africa (early Pliocene; Denys, 1990d), and, tentatively, at Kakara, Uganda (late Miocene; Mein, 1994). This genus includes some material referred previously to the murine *Millardia* (Sabatier, 1982; Pickford and Mein, 1988; see discussion in Southern African Rodents, Taxonomic Issues). The extant genus *Uranomys* is listed from the Plio-Pleistocene of southern Angola (Pickford et al., 1992).

The Gerbillinae are generally considered to have evolved from the Myocricetodontinae (Jaeger, 1977b; Tong, 1989; other references in Musser and Carleton, 2005). Extinct genera include *Pseudomeriones*, *Abudhabia*, *Protatera*, and *Mascaramys*. The only record of *Pseudomeriones* is from the early Pliocene of Algeria (Jaeger, 1975). The only African records of *Abudhabia* are from the late Miocene of Kenya (Lothagam: Winkler, 2003; Lukeino Fm.: Mein and Pickford, 2006), although *Protatera yardangi* from Sahabi, Libya (Munthe, 1987), may be referrable to this genus (see discussion in Flynn et al., 2003). *Protatera* is named from Amama 2, late Miocene of Algeria (Jaeger, 1977b), and is also reported from the early Pliocene (Miocene-Pliocene boundary) of Morocco (Coiffait-

Martin, 1991; Geraads, 1998b). *Mascaramys* is from the late Pliocene of Tunisia (Mein and Pickford, 1992) and the middle Pleistocene of Algeria (Tong, 1986). Extant genera are first reported from the late Miocene of Kenya and Ethiopia (*Gerbilliscus* [formerly *Tatera*]; figure 17.4B), early Pliocene of Kenya (*Gerbillus*) and South Africa (*Desmodillus*, *Gerbillurus*), and the early Pleistocene of Morocco (*Meriones*). Tong (1989) provides a thorough study of the origin, evolution, and phylogenetic relationships of northern African gerbils based on examination of Plio-Pleistocene specimens from 15 localities.

The earliest records of murines in Africa are in the late Miocene, ca. 11–10 Ma. In northern Africa, *Progonomys* (extinct) is reported from several sites such as Bou Hanifia, Algeria (Ameur, 1976), Sahabi, Libya (Munthe, 1987), possibly Sheikh Abdallah, Egypt (*Progonomys* and/or *Preacomys*: Pickford et al., 2008), and Oued Tabia (Benammi et al., 1995) and Oued Zra, Morocco (Jaeger, 1977b). Several other extinct late Miocene murines are known from northern Africa including *Paraethomys*, *Stephanomys*, *Karnimata*, *Castillomys*, and *Saidomys*. The earliest report of murines in eastern Africa is from Chorora, Ethiopia (aff. *Stenocephalomys* [sic, *Stenocephalemys*] and cf. *Parapelomys*; Geraads, 2001; members of the tribes Praomyini and Arvicanthini, respectively). *Progonomys*, *Saidomys*, and another extinct murine, *Lukeinomys*, are also known from the late Miocene of eastern Africa. In southern Africa, the earliest murine is from Harasib, Namibia (*Aethomys*; Mein et al., 2004; the only southern African representative of the Arvicanthini). The high diversity of African Murinae is established early, in the late Miocene to early Pliocene. For example, in northern Africa, there are seven first appearances in the late Miocene and three in the early Pliocene. In eastern Africa, there are nine in the late Miocene and three in the early Pliocene. Only 2 genera are reported in the late Miocene of southern Africa and 10 in the early Pliocene, but this is likely related to the paucity of late Miocene southern African faunas. Throughout Africa, there are relatively fewer first appearances after the early Pliocene.

Molecular studies of murines (e.g., Lecompte et al., 2008) suggest that both faunal interchange with Eurasia and in situ diversification played a role in the history of African murines. By and large, conclusions from molecular work agree well with the fossil record. Molecular studies suggest an initial colonization event at around 11 Ma, with subsequent diversification. A second period of diversification (e.g., of modern Arvicanthini + Otomyini and of modern Praomyini) is suggested after about 9–7 Ma. A second period of interchange between Africa and Eurasia is seen about 6.5–5 Ma. Genera involved in this second period of dispersal include, for example, the first occurrences of *Mus* (of southern Asian origin) in the early Pliocene of northern, eastern, and southern Africa. Another genus, the extinct arvicanthine *Saidomys* (figure 17.4C), is first known in northern and eastern Africa in the late Miocene (not known from southern Africa). *Saidomys* is also reported from the early Pliocene of Afghanistan, suggesting dispersal from Africa to southern Asia in this time period (Winkler, 2003).

Based on paleontological, anatomical, molecular, and other data, otomyines are considered to have had their origins in African murines, in particular arvicanthine-like forms (e.g., *Euryotomys;* see discussion in Musser and Carleton, 2005; also Pocock, 1976; Sénégas, 2001). This split was likely to have been around 8.5–7 Ma. Otomyines are absent from northern Africa, present but restricted to the genus *Otomys* in eastern and central Africa, and taxonomically diverse in southern Africa, which is probably their center of origin. Outside of southern Africa, *Otomys* is first known from the late Pliocene of Zaire (Boaz et al., 1992), and it arrives in the Rift Valley only in the late Pleistocene (Denys, 2003). Six genera of otomyines (*Euryotomys, Palaeotomys, Prototomys, Myotomys, Otomys, Parotomys*; figure 17.4D illustrates *Otomys*) are reported from southern Africa, but none from Namibian sites. In southern Africa, otomyines are found exclusively in Cape and Gauteng Province sites, and two of the three modern genera (*Otomys* and *Myotomys*) first occur only around 3 Ma, while *Parotomys* is not recorded until the middle Pleistocene.

Suborder ANOMALUROMORPHA Bugge, 1974
Superfamily ANOMALUROIDEA Gervais, 1849
Family ZEGDOUMYIDAE Vianey-Liaud et al., 1994

Diagnosis Original diagnosis, translated from French. 1/1, 0/0, 1/1, 3/3; [mandibular and cranial structure unknown]; teeth brachydont with tendency toward lophodonty. Upper molars quadrangular, well developed hypocone at same level as protocone; transverse crests present (at least protoloph and metaloph); longitudinal crest generally not clearly defined. On lower molars, no difference in height between trigonid and talonid; summit of external and internal tubercles [cusps] at about same level: only metaconid slightly more elevated than the others; anterior cingulum well developed transversely and always present: its surface at the same height as end of talonid; mesoconid individualized, clearly separated from protoconid and hypoconid, often elongated by a mesolophid; longitudinal crest generally absent; hypoconulid absent. Enamel pauciserial with tendency to uniserial.

Geologic Age Early to middle Eocene.

African Distribution Northern Africa (Algeria, Tunisia).

Remarks The extinct family Zegdoumyidae was erected on the basis of isolated teeth from Algeria (*Zegdoumys, Glibia, Glibemys*) and Tunisia (*Zegdoumys*). Assignment of a few P4/p4 to members of this family suggests there was premolar replacement. Vianey-Liaud et al. (1994) and Vianey-Liaud and Jaeger (1996) consider the Ischyromyidae or Sciuravidae to be the most likely ancestors of the Zegdoumyidae, which in turn are the most likely ancestors of the Anomaluridae and the Graphiurinae. The position of the Zegdoumyidae as ancestral to the Anomaluridae is questioned by Dawson et al. (2003) and Marivaux et al. (2005).

Family ANOMALURIDAE Gervais, 1849
Figure 17.5

Diagnosis Based on Lavocat (1978). 1/1, 0/0, 1/1, 3/3; sciurognathus, myomorphous; cheek teeth bunodont with transverse crests; masseter muscle passes through greatly enlarged infraorbital foramen to insert broadly on muzzle; ascending ramus of orbital arch weak; palatine bone contributing noticeably to orbitotemporal floor; pterygoid fossa not open anteriorly; middle ear with globular promontory.

Geologic Age Middle to late Eocene; early Miocene to Recent.

African Distribution Northern Africa (Algeria), eastern Africa (Kenya, Uganda). Lavocat (1973:196) reports anomalurid remains from the Oligocene of Fayum, Egypt, but this material was never formally described and has not been relocated (P. A. Holroyd, pers. comm.).

Remarks Isolated teeth of the middle to late Eocene genus *Nementchamys* were described by Jaeger et al. (1985) from Bir El Ater (= Nementcha), eastern Algeria. The morphology of these teeth was considered too derived in some respects for this taxon to have been the ancestor of early Miocene and later species. The best known fossil anomalurid is *Paranomalurus* (including *P. bishopi, P. soniae,* and *P. walkeri*; figures 17.5A, 17.5C) from the early Miocene of Kenya and Uganda (Lavocat, 1973). *Paranomalurus* cf. *soniae* is known from the middle Miocene Fort Ternan locality, Kenya (Denys and Jaeger, 1992). A single isolated tooth from the middle Miocene Muruyur Beds, Kipsaramon, Kenya, is assigned to *Anomalurus parvus* (figure 17.5B; Winkler, 1992). *Zenkerella wintoni* is poorly known from the fossil record. The holotype is a single mandible from the early Miocene of Songhor, Kenya (Lavocat, 1973). *Zenkerella* is also reported from Moroto, Uganda (Pickford and Mein, 2006). Fossil anomalurids are unknown from southern Africa and from eastern African sites after the middle Miocene.

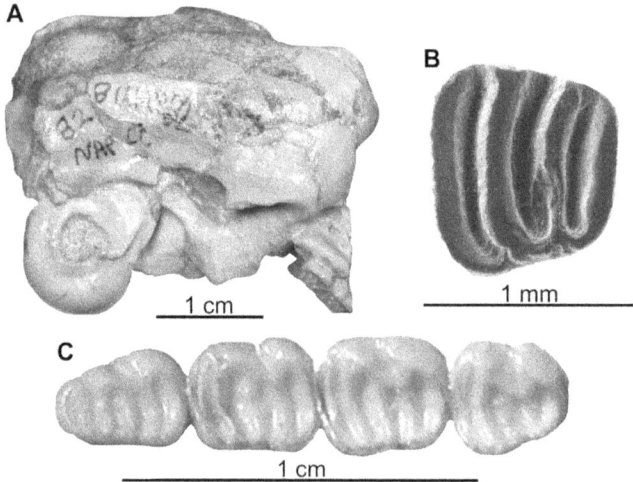

FIGURE 17.5   Examples of Family Anomaluridae. A) *Paranomalurus*, skull, BUMP 82, early Miocene, Napak, Uganda. B) *Anomalurus*, left M3, KNM-TH 19439, middle Miocene, Kipsaraman, Muruyur Beds, Kenya. C) *Paranomalurus*, right p4–m3, BUMP 82, early Miocene, Napak, Uganda. Figure 17.5B from Winkler, 1992, © Copyright 1992 The Society of Vertebrate Paleontology. Reprinted and distributed with the permission of the Society of Vertebrate Paleontology.

Superfamily PEDETOIDEA Gray, 1825
Family PEDETIDAE Gray, 1825

Synonymy *Parapedetidae* Stromer, 1926:128, 143.

Diagnosis 1/1, 0/0, 1/1, 3/3; sciurognathus, sciuromorphous; cheek teeth bunodont (e.g., *Megapedetes*) to highly hypsodont (e.g., *Pedetes; Parapedetes* also hypsodont) with simplified bilobate occlusal pattern.

Description The cheek teeth of the extant genus, *Pedetes*, are ever growing, but those of the extinct taxon *Megapedetes* are rooted as are the deciduous teeth of *Parapedetes*. Many fossil specimens of pedetids are isolated teeth, which may be extremely difficult, if not impossible, to assign to tooth position.

Geologic Age Early Miocene to Recent.

African Distribution Northern Africa (Morocco only), eastern and southern Africa.

Remarks The position of the Pedetidae among rodents has been controversial. For example, Huchon et al. (2000) suggest that the Pedetidae form an independent, early diverging lineage. However, Montgelard et al. (2002) confirm that the Anomaluridae is the sister clade to Pedetidae. McKenna and Bell (1997) created the family Parapedetidae, including the monogeneric *Parapedetes* (early–middle Miocene, Namibia); however, we consider *Parapedetes* to belong to the Pedetidae.

The earliest African records of pedetids are from the early Miocene of Kenya and Uganda (*Megapedetes*; MacInnes, 1957; Lavocat, 1973) and Namibia (the poorly known genus *Parapedetes* Stromer, 1926, and an isolated phalanx of *Megapedetes* from Namibia; Mein and Senut, 2003). *Megapedetes* is also present in the middle Miocene of Kenya at Fort Ternan (Denys and Jaeger, 1992) and Maboko (Winkler, 1998), and in Namibia (Mein and Senut, 2003). Many eastern African specimens of fossil pedetids have not yet been described formally and there are likely to be new taxa, including a smaller form (Lavocat, 1973; Winkler, 2002; Mein and Pickford, 2006).

Megapedetes is found most commonly in eastern Africa. Reports of this genus in Morocco (Beni Mellal, middle Miocene; Lavocat, 1961), Saudi Arabia (early Miocene, date uncertain; Sen in Thomas et al., 1982), Turkey (middle Miocene; Sen, 1977), the Isle of Chios, Greece (middle Miocene; Tobien, 1968), and Israel (late Miocene, but date uncertain; Tchernov et al., 1987; early Miocene; Wood and Goldsmith, 1998), suggest dispersal of *Megapedetes* from eastern to northern Africa and the eastern Mediterranean region in the early Miocene.

The modern genus *Pedetes* appears for the first time in the Laetolil Beds, Tanzania (3.7 Ma: Davies, 1987). In southern Africa, *Pedetes* is first known from the early Pliocene. It is found at a number of sites including Taung (early Pleistocene; Broom, 1934) and Florisbad, South Africa (Dreyer and Lyle, 1931), and Drotsky's Cave, Botswana (late Pleistocene; Robbins et al., 1996).

Suborder HYSTRICOMORPHA Brandt, 1855
Infraorder CTENODACTYLOMORPHI Chaline and Mein, 1979
Family CTENODACTYLIDAE Gervais, 1853

Diagnosis 1/1, 0/0, 1–2/1–2, 3/3; sciurognathus, hystricomorphous; cheek teeth high-crowned, tetralophodont pattern, simplified in extant taxa; P3/p3 may be present. Enlarged infraorbital foramen; masseteric plate reduced and ventral, pterygoid fossa deep but blind; lacrimal large and contacts vertical process of jugal (from Lavocat, 1978).

Geologic Age Early Miocene to Recent.

African Distribution Northern Africa (Libya, Morocco, Tunisia, Algeria, Egypt).

Remarks In Africa, fossil ctenodactylids are known exclusively from the north, where five extinct genera are reported (*Sayimys, Metasayimys, Africanomys, Irhoudia,* and *Testouromys*). Extant genera are not recorded from the fossil record prior to the Holocene. The earliest record of an African ctenodactylid is *Sayimys* from the early Miocene of Libya (Jebel Zelten; Wessels et al., 2003); the genus is last reported in the late Miocene. *Metasayimys* occurs only in the middle Miocene of Morocco (e.g., Beni Mellal; Lavocat, 1961). *Africanomys* is well-known from the middle to late Miocene: it has been recorded from Libya, Morocco, Tunisia, Algeria, and Egypt. *Irhoudia* is first encountered in the late Miocene of Libya, Morocco, and Algeria, and it is last reported from the early Pleistocene of Morocco (Jebel Irhoud; Jaeger, 1971a). *Testouromys* has been found only in the middle Miocene of Tunisia.

African ctenodactylids likely dispersed from Eurasia to northern Africa in the early Miocene. The group is also known from the late Oligocene to late Miocene of Pakistan (Lindsay et al., 2005), the early Miocene of Turkey and the middle Miocene of Israel (see summary in Wessels et al., 2003), and the middle Miocene of Chios Island, Greece (López-Antoñanzas et al., 2004).

Infraorder HYSTRICOGNATHI Brandt, 1855
Family BATHYERGIDAE Waterhouse, 1841

Synonymy Bathyergoididae Lavocat, 1973:109.

Diagnosis 1/1, 0/0, 0/0–2, 3/3; hystricognathus, hystricomorphous; infraorbital foramen primitively large, but may be secondarily reduced (much reduced in extant forms); number of cheek teeth variable in different genera; high-crowned cylindrical molars with simple design (concave dentine base surrounded by wide enamel border) in modern representatives; remains of cusps in early Miocene representatives; most members of family have molars with two sinuses with internal one more distinct, and well-fused roots; incisors with multiserial enamel.

Geologic Age Early Miocene to Recent.

African Distribution Eastern and southern Africa.

Remarks Lavocat (1973) divided the fossil and extant mole rats into two families, the Bathyergoididae and Bathyergidae, within the superfamily Bathyergoidea. His Bathyergoididae, which included only the early Miocene taxon *Bathyergoides neotertiarius*, was diagnosed as "family of Bathyergoidea in which the structure of the jugal teeth is very conservative" (translated from French; Lavocat, 1973:109). Other bathyergoids (except *Paracryptomys* [= *Geofossor* of Mein and Pickford, 2003] from the lower Miocene of Namibia) were assigned to the Bathyergidae, diagnosed as "family of Bathyergoidea in which the structure of jugal teeth is very simplified" (translated from French; Lavocat, 1973:139). *Paracryptomys* was assigned to "family uncertain."

The fossil record shows that this group was more widely distributed and diversified in the past than it is today, although it has never occurred in northern Africa. The oldest representatives of the group (all extinct genera) are in the early Miocene deposits of Namibia (*Bathyergoides* and *Geofossor;* e.g., Stromer, 1924; Mein and Pickford, 2003) and eastern Africa (*Bathyergoides, Geofossor,* and *Proheliophobius;* e.g., Lavocat, 1973; Mein and Pickford, 2003). *Richardus* is present in the middle Miocene of Fort Ternan, Kenya (Lavocat, 1988,

1989; Denys and Jaeger, 1992). As yet undetermined genera are reported from the middle Miocene at Maboko, and from the Muruyur Beds and Ngorora Formation, Kenya (Winkler, 2002). In the late Miocene, *Proheliophobius* or *Richardus* is present at Harasib, Namibia (Mein et al., 2000a). Another extinct genus, *Gypsorhychus*, is first reported from the early Pliocene at Taung, South Africa (Broom, 1948a). The affinities of this genus are not well known due to its highly derived dental pattern.

The first occurrences of modern genera (*Bathyergus* and *Cryptomys*) are in the early Pliocene at Langebaanweg, South Africa (ca. 5 Ma; Denys, 1998). In eastern Africa, *Heterocephalus* makes its first appearance at ca. 4.3 Ma in the Lower Laetolil Beds, Tanzania (Denys, 1987a). The first record of *Georychus* is in the Plio-Pleistocene at Ngamiland, Botswana (Pickford and Mein, 1988). *Heliophobius* is not known definitively from the fossil record, but may be present in the middle Pleistocene at Isenya, Kenya (Brugal and Denys, 1989).

Family HYSTRICIDAE Fischer de Waldheim, 1817
Figure 17.6

Diagnosis Based on Lavocat (1978); Nowak (1999). 1/1, 0/0, 1/1, 3/3: hystricognathus, hystricomorphous; occlusal surface of teeth flat with multiple crests, dP4 replaced by P4, teeth moderately to strongly high crowned; skull strongly domed in *Hystrix* due to inflation of nasal sinuses; anterior palatine foramina very small.

Geologic Age Late Miocene to Recent.

African Distribution Northern, eastern, and southern Africa.

Remarks Fossil hystricids are rare from northern Africa, but better known, especially from the Plio-Pleistocene record, from eastern and southern Africa. There are three genera, *Atherurus*, *Hystrix*, and *Xenohystrix*; only the last named is extinct.

The oldest African record of *Atherurus* is ca. 11–10 Ma from Sheikh Abdallah, Egypt (Pickford et al., 2008). The

1 cm

FIGURE 17.6 Family Hystricidae. *Hystrix*, left mandible with m1–m3, LAET 1368, early to middle Pliocene, Upper Laetolil Beds, Laetoli, Tanzania.

earliest records from eastern Africa are from 6 Ma at Lemudong'o, Kenya (Hlusko, 2007), and from the Adu-Asa Formation, Middle Awash, Ethiopia (5.7 Ma; Wesselman et al., 2009). Wesselman et al. (2009) also mention that *Atherurus* is present (as yet undescribed) from the Pliocene at Aramis, Ethiopia. *Atherurus* is not known from southern Africa.

The earliest occurrences of *Hystrix* are from several late Miocene localities. *Hystrix* is known from the late Miocene at Toros-Menalla, Chad (7–6 Ma; Vignaud et al., 2002), and Menacer (= Marceau), Algeria (Arambourg, 1959), although the geologic age of the Algerian material is uncertain. In eastern Africa, *Hystrix* is recorded from Lothagam (possibly >7.44 Ma; Winkler, 2003) and Lemudong'o (6 Ma; Hlusko, 2007), Kenya, and from the Adu-Asa Formation, Ethiopia (Haile-Selassie et al., 2004; Wesselman et al., 2009). The earliest reports of *Hystrix* from southern Africa are from the early Pliocene at Langebaanweg and Makapansgat, South Africa (De Graaff, 1960; Hendey, 1984). Most fossil *Hystrix* are not referred to species. Specific identification of isolated teeth (the most common fossil) is difficult because the occlusal pattern changes with wear and tooth size varies with crown height. In fact, Van Weers (2005) considered the morphology of the cheek teeth unusable for the distinction of extant subgenera and species of *Hystrix*.

The oldest records of the large, lower crowned *Xenohystrix* are from the late Miocene of Lemudong'o (Hlusko, 2007) and from the Adu-Asa and lower Sagantole formations (Haile-Selassie et al., 2004). The earliest report of *Xenohystrix* from southern Africa is in the early Pliocene at Makapansgat, South Africa (De Graaff, 1960).

The earliest known hystricid, *Atherurus karnuliensis* (allocation by Van Weers, 2005; described originally as *Sivacanthion complicatus* by Colbert, 1933), is a relatively lower-crowned (primitive) taxon from the middle Miocene of northern India. The oldest known remains of *Hystrix* are from the Vallesian (early late Miocene) of Hungary, and *Hystrix* is also known from the late Miocene of China (Van Weers, 2005). Since the earliest African records of hystricids are in the late Miocene, they are likely immigrants, dispersing from Eurasia to Africa in the late Miocene.

Family MYOPHIOMYIDAE Lavocat, 1973

Diagnosis From Lavocat (1973, 1978). 1/1, 0/0, 1–2/1–2, 3/3; hystricognathus, hystricomorphous; small size; cheek teeth with relatively prominent cusps and lower crests; dP3 may be replaced by P3 in *Myophiomys*.

Geologic Age Early to middle Miocene.

African Distribution Eastern Africa (Kenya, Uganda) and southern Africa (Namibia).

Remarks Lavocat (1973) erected the family Myophiomyidae to include the genera *Phiocricetomys*, *Myophiomys*, *Elmerimys*, and *Phiomyoides*. *Phiocricetomys* is reported only from the Fayum (early Oligocene; Wood, 1968). *Myophiomys* (early Miocene) and *Elmerimys* (early to middle Miocene) are known only from eastern Africa, and *Phiomyoides* is recorded only from the early Miocene of Namibia (Stromer, 1926). Holroyd (1994) removed the extremely bunodont *Phiocricetomys* (which also has only three cheek teeth per quadrant) from the Myophiomyidae and considered it family *incertae sedis*. Myophiomyids are relatively uncommon.

Family PHIOMYIDAE Wood, 1955

Figures 17.7A and 17.7B

Diagnosis From revised diagnosis of Holroyd (1994). 1/1, 0/0, 1/1, 3/3; hystricognathus, hystricomorphous. Differs from Thryonomyidae in usually smaller size and dp4 replaced by p4 primitively; DP4 variably replaced by P4; dp4 metaconid less anteriorly placed; relatively stronger hypolophid; and less lingually placed ectolophid.

Geologic Age Middle or late Eocene to early middle Miocene.

African Distribution Northern Africa (Algeria, Libya, Egypt), eastern Africa (Uganda, Kenya, Tanzania), and southern Africa (Namibia).

Remarks Composition of the Phiomyidae follows the proposed revision of the family by Holroyd (1994) and includes the genera *Protophiomys*, *Phiomys*, *Andrewsimys*, *Ugandamys*, and *Pomonomys*, plus the Diamantomyinae (*Metaphiomys*, *Diamantomys*, and a new genus of Holroyd, 1994). The earliest records of the Phiomyidae are isolated teeth of *Protophiomys algeriensis* from the middle or late Eocene of Algeria (Jaeger et al., 1985).

Holroyd (1994) reexamined collections of late Eocene to early Oligocene rodents from the Fayum Province, Egypt, studied previously by Wood (1968) and others. One of the most significant results of her study was the reevaluation of the composition of the genus *Phiomys* and the suggestion that Miocene material from Kenya attributed to *Phiomys* likely pertained to other taxa (Holroyd, 1994). This suggestion is formalized in Holroyd and Stevens (2009) where some of these early Miocene specimens of *"Phiomys andrewsi"* from Kenya are placed in a new genus, *Lavocatomys*, considered family *incertae sedis* pending further work on eastern African rodent interrelationships. Holroyd and Stevens (2009) note that Stromer (1926) reported cf. *P. andrewsi* from the early to middle Miocene of Namibia. They have not examined this material, but suggest it may not be attributable to *Phiomys*.

In northern Africa, *Phiomys* is known from the late Eocene of Egypt and the early Oligocene of Egypt and Libya (Wood, 1968; Holroyd, 1994; Fejfar, 1987; see discussion in Holroyd and Stevens, 2009). Other phiomyids from eastern Africa include *Andrewsimys* (Lavocat, 1973; Pickford and Mein, 2006) and *Ugandamys* (figure 17.7A; Winkler et al., 2005). *Pomonomys* is reported only from the early Miocene of Namibia (Stromer, 1922). *Pomonomys* was originally considered by several authors (e.g., Lavocat, 1973; but not Holroyd, 1994) to be in the Subfamily Diamantomyinae.

The subfamily Diamantomyinae includes the genera *Metaphiomys*, *Diamantomys*, and a new genus of Holroyd (1994; reported as *Acritophiomys* spp. nomina nuda in Lewis and Simons, 2007) from the late Eocene-early Oligocene of the Fayum. *Metaphiomys* has been recovered from the early Oligocene of Egypt and Libya (Wood, 1968; Holroyd, 1994; Fejfar, 1987). It has also been reported from the late Oligocene of Tanzania (Stevens et al., 2006). *Diamantomys* (figure 17.7B) is a large and often numerically abundant rodent first recovered from the late Oligocene at Losodok, Kenya (Rasmussen and Gutiérrez, 2009), and more commonly from the early to middle Miocene. It is now known from several species in eastern (e.g., Lavocat, 1973; Winkler, 1992; Pickford and Mein, 2006) and southern Africa (Mein and Pickford, 2003).

Family KENYAMYIDAE Lavocat, 1973

Diagnosis Emended from Lavocat (1978). 1/1, 0/0, 1/1, 3/3; hystricognathus, hystricomorphous; relatively small size; short masseteric insertion; four brachydont cheek teeth with long narrow crests, individual cusps indistinct. Replacement of DP4/dp4 by P4/p4 uncertain.

Geologic Age Early Miocene.

African Distribution Eastern Africa (Kenya, Uganda).

Remarks This relatively poorly known family, which is not reported from northern or southern Africa, includes only two genera, *Kenyamys* and *Simonimys* (Lavocat, 1973).

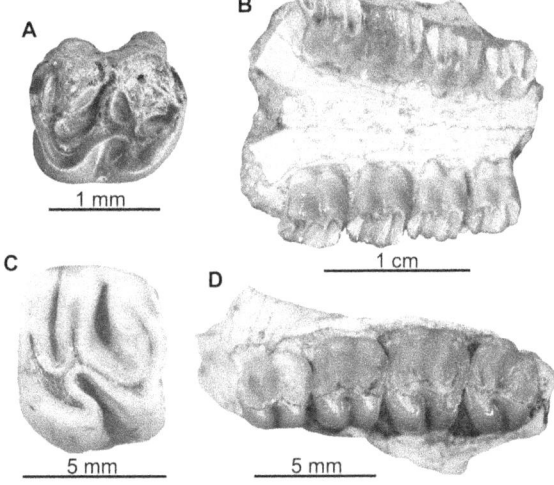

FIGURE 17.7 A) Family Phiomyidae, *Ugandamys*, right M1 or M2, BUMP 1023, early Miocene, Bukwa IIB, Uganda. B) Family Phiomyidae, subfamily Diamantomyinae, *Diamantomys*, palate with DP4–M3, BUMP 13, early Miocene, Napak, Uganda. C) Family Thryonomyidae, *Thryonomys*, left M1 or M2, LU2-24, late Pliocene, Upper Semliki Valley, Zaire. D) Family Thryonomyidae, *Paraphiomys*, left dp4–m3, uncatalogued specimen, early Miocene, Rusinga, Kenya. Figure 17.7A from Winkler et al., 2005, © Copyright 2005 The Society of Vertebrate Paleontology. Reprinted and distributed with the permission of the Society of Vertebrate Paleontology.

Family PETROMURIDAE Wood, 1955

Diagnosis 1/1, 0/0, 1/1, 3/3; hystricognathous, hystrico-morphous; cheek teeth rooted and hypsodont with deep infoldings of the enamel on the lingual side of maxillary cheek teeth and labial side of mandibular cheek teeth.

Geologic Age Late Miocene to Recent.

African Distribution Southern Africa (South Africa, Namibia) and eastern Africa (Kenya).

Remarks The earliest record of the only described genus, *Petromus*, and the only record from outside southern Africa, is from the late Miocene Lukeino Formation, Kenya (*P.* cf. *P. antiquus*; Mein and Pickford, 2006). The oldest species from southern Africa is the extinct *P. antiquus* from the early Pliocene (Waypoint 160; Sénégas, 2004) of northern South Africa. *Petromus* is also reported from the early Pleistocene of northern and central South Africa (*P. minor* from Taung; Broom, 1939), and from the late Pleistocene of southern Namibia (Pickford et al., 1994). Holroyd (1994) considers *Petromus* to belong in the family Thryonomyidae.

Family THRYONOMYIDAE Pocock, 1922
Figures 17.7C and 17.7D

Diagnosis Emended from Lavocat (1978). 1/1, 0/0, 1/1, 3/3; hystricognathus, hystricomorphous; muzzle of normal proportions; where known, masseter muscle insertion extending far in front of infraorbital foramen; anterior palatine foramina well developed; semihypsodont molars with well-developed crests, the number of crests reduced in several forms; DP4 and dp4 not replaced.

Geologic Age Early Miocene to Recent.

African Distribution Northern Africa (Morocco, Egypt, Libya), eastern Africa (Kenya, Uganda, Ethiopia, Tanzania, Zaire), and southern Africa (Namibia, South Africa, Zambia, Zimbabwe).

Remarks Today, the thryonomyids are represented by a single genus, *Thryonomys*, which is found throughout much of Africa and in some areas is relatively abundant. In the past, the thryonomyids were not only numerically abundant but also much more speciose, particularly in eastern and southern Africa. Their record in northern Africa is relatively sparse. Holroyd (1994) reassigned *Paraphiomys simonsi* Wood, 1968, from the early Oligocene of Egypt to a new, as yet unpublished, genus. *Gaudeamus* Wood, 1968, from the late Eocene of Egypt (and Oman) was placed in the family Thryonomyidae by Lavocat, 1973, but placed as family *incertae sedis* by Holroyd (1994). *Paraphiomys occidentalis* is known from the middle Miocene of Morocco at Beni Mellal (Lavocat, 1961) and Azdal (Benammi et al., 1995). A new, unnamed genus has been reported from the early Miocene at Jebel Zelten, Libya (Wessels et al., 2003). The Libyan material is considered by López-Antoñanzas et al. (2004) possibly to pertain to their new species of *Paraphiomys* from Saudi Arabia.

Thryonomyids from eastern and southern Africa are speciose and known from a number of localities. *Apodecter* is small in size and reported from the early (Hopwood, 1929b) and late Miocene of Namibia (Mein et al., 2000a), the early Miocene of Uganda (Pickford and Mein, 2006), and the middle to late Miocene of Kenya (Mein and Pickford, 2006). *Apodecter* is reported from the early Pliocene of South Africa (Sénégas, 2000, in Mein and Pickford, 2006), but this is a late record and has not been verified. *Paraphiomys* (figure 17.7D) is geographically wide-spread, speciose, and numerically abundant. It was first described as including two species from numerous early Miocene sites in Kenya and Uganda (Lavocat, 1973). Since 1973, the number of species in the genus has expanded, however, the

generic assignment of many of these species has been controversial, as has the generic assignment of many other species of Thryonomyidae (see, e.g., Mein and Pickford, 2006; López-Antoñanzas et al., 2004). *Neosciuromys africanus* (reported only from the early Miocene of Namibia [Stromer 1922, 1926]) is similar to *Paraphiomys*, but *Neosciuromys* has higher-crowned cheek teeth and a simpler loph pattern on the lower dentition. *Neosciuromys* was synonymized by Lavocat (1973) with *Paraphiomys pigotti* but is considered a valid genus by other authors (see discussion in López Antoñanzas et al., 2004). *Phthinylla* is a very poorly known genus from the early Miocene of Namibia (Hopwood, 1929b). It has been considered a junior synonym of *Paraphiomys pigotti* and, more likely, of *Neosciuromys* (López Antoñanzas et al., 2004). The genus *Epiphiomys* is yet another poorly known thryonomyid, reported only from the early Miocene of Kenya and Uganda (Lavocat, 1973).

Paraulacodus is derived (has simplified loph morphology) compared to *Paraphiomys*. *Paraulacodus* has been recovered from the middle to late Miocene of eastern Africa (Jaeger et al., 1980; Geraads, 1998a, 2001; Winkler, 2002, 2003), and the late Miocene of southern Africa (Mein et al., 2000a). It is found in Pakistan, where its presence is considered a result of dispersal from eastern Africa about 13 Ma (Flynn and Winkler, 1994).

Thryonomys is first reported in eastern Africa at Lemudong'o, Kenya (6 Ma; Manthi, 2007), and the Middle Awash, Adu-Asa Formation, Ethiopia (5.7–5.6 Ma; Wesselman et al., 2009; Haile-Selassie et al., 2004). Wesselman et al.'s (2009) new species, *T. asakomae*, is significant in having a four-lophed dp4 (an apomorphy of *Thryonomys*; *Paraulacodus* has three lophs), but upper incisors with two grooves (an apomorphy of *Paraulacodus*). One specimen of *T. asakomae* has an incisor with a hint of a third groove (*Thryonomys* has three grooves). After the late Miocene, *Thryonomys* (figure 17.7C) is relatively common. *Thryonomys* has not been reported from northern Africa, and is not known in southern Africa until the late Pleistocene (e.g., Brain, 1981).

Discussion

This overview has necessarily been compiled from the literature, but it has become increasingly clear that many of the collections require reexamination. Accepted taxonomies have changed repeatedly since many of the original descriptions were published, so that it is becoming difficult to update species lists accurately without going back to the original material. Apart from periodic generally accepted changes to synonymies, there are also disagreements between contemporary workers and some controversies between molecular and morphological phylogenies. The situation can be further complicated by a tendency for some names to reappear or to be elevated and demoted cyclically, without reanalysis of the original material. As a result, original lists must be examined in context, and integrated lists require recompilation from original lists in the light of a new taxonomic authority, such as Wilson and Reeder (2005), or the latest results from molecular phylogenetic studies. In the present case, an attempt has been made to reflect differences and changes in generic names (tables 17.2–17.4, names in parentheses), but there are instances where the position is very involved. One such concerns a group of murine rodents, the tribe Praomyini (Lecompte et al., 2008), which includes, in part, *Myomyscus*, *Myomys* (synonymized under *Mastomys*), *Mastomys*, *Praomys*, *Zelotomys*, and *Stenocephalemys*. Attribution of published fossil records to these genera requires attention to be paid to both the history of the naming of each taxon and its distribution before the modern equivalence of original identifications can be assessed. It is, however, very unlikely that one will be able to determine with confidence which genus is represented without reexamining the material. Even then, identification of these genera is very difficult, especially based on dental pattern, because morphological divergence is low, although molecular divergence is high (Lecompte et al., 2005, 2008). The situation is probably less acute at higher taxonomic levels but nevertheless exists and will particularly affect specimens that have not been identified to genus.

NORTHERN AFRICAN RODENTS

Taxonomic Issues

As discussed in more detail for southern African rodents, it would be interesting to reexamine unusual taxonomic records (see table 17.2). For example, the middle Pleistocene occurrence of *Saidomys* in Morocco is very late compared to other records of the genus. The only other record of the genus in northern Africa is in the late Miocene of Egypt, and eastern African records range from the late Miocene to late Pliocene. Outside Africa, *Saidomys* is known from the early Pliocene of Afghanistan (*Saidomys* from Thailand likely belongs to a different genus; Winkler, 2003). Other rare occurrences in northern Africa (e.g., *Paraphiomys*, middle Miocene; and *Megapedetes*, middle Miocene) likely represent range extensions by genera that occur in greater numerical abundance in eastern Africa since the early Miocene.

It may also be of interest to reexamine records of the murine *Golunda* from Africa. At present, *Golunda* is known only as a single species from southern Asia. It is known also in the fossil record of southern Asia since the early Pliocene (Musser and Carleton, 2005). In Africa, *G. jaegeri* is reported from three middle Pliocene localities in Algeria (Oued Athmeneia 1, Oued Smendou, Amama 3; Coiffait and Coiffait, 1981; Coiffait-Martin, 1991) and one in the early Pliocene of Morocco (Azid; Benammi et al., 1995). In Ethiopia, *Golunda gurai* is described from the middle (Hadar; Sabatier, 1979; 1982) and late (Omo; Wesselman, 1984) Pliocene, and is listed from the early Pliocene (Aramis; WoldeGabriel et al., 1994). Musser (1987), using descriptions and illustrations of *Golunda gurai* from Hadar and Omo, suggested this material was not *Golunda* but should probably be referred to *Pelomys*. However, Musser was not able to examine *Golunda* from northern Africa, Aramis, or Omo. It is noteworthy that *G. jaegeri* was originally assigned to *Pelomys europaeus*. The relationship of *Golunda* to other murines has not been resolved, nor has its biogeographic history (Musser and Carleton, 2005), and a better understanding of the African material referred to this genus may help clarify these issues.

Temporal Distribution

Probably the greatest contribution of the northern African faunas to our knowledge of the history of African rodents is that they provide the best sample of Paleogene species: the record from the Fayum, Egypt, is exceptional. Although samples from other localities in northern Africa are still relatively small, there is still more material than from southern Africa, where there are no published records, or from eastern Africa, where only two areas of late Oligocene age (Mbeya Region, Tanzania; and Losodok, northern Kenya) have produced material from the Paleogene. Our understanding of the phylogenetic relationships and early evolutionary history of

several major, almost exclusively African groups (Phiomyidae, Thryonomyidae, Anomaluridae, and Zegdoumyidae) is dependent on continued research in northern Africa.

Because of its geographic location, northern Africa seems also to be the theatre of the first murine radiation, and to have been in contact at several times with Eurasia during the Miocene-Pliocene. During the Neogene, some extant families or subfamilies have their African range restricted to only northern Africa, such as the Leithiinae (Gliridae), Allactaginae (Dipodidae), Sicistinae (Dipodidae), Arvicolinae (Cricetidae), Cricetinae (Cricetidae), and Ctenodactylidae. In total, one finds about 38 extinct genera and only 20 extant genera in the fossil (pre-Holocene) record of this region.

EASTERN AFRICAN RODENTS

Taxonomic Issues

Again, unexpected taxonomic records (table 17.3), such as a possible pseudocricetodontid from the early Miocene of Uganda (Pickford and Mein, 2006), the only record of this group in Africa, should be reexamined. The earliest and only record of the Dassie rat, *Petromus*, from outside southern Africa (Lukeino Formation, late Miocene, Kenya; Mein and Pickford, 2006) is also interesting and warrants further investigation.

Late Miocene–earliest Pliocene records attributed to the cane rat, *Thryonomys*, need to be examined in detail, as some support the hypothesis (see discussion in Winkler, 2003) that *Paraulacodus* is the sister taxon to *Thryonomys*. Wesselman et al.'s (2009) new species, *T. asakomae* (Middle Awash, Ethiopia), provides the best evidence so far for this relationship. *Thryonomys asakomae* has character states apomorphic for both *Thryonomys* (a four-lophed dp4) and *Paraulacodus* (an upper incisor with two grooves). There is also one incisor with a hint of a third groove, as seen in *Thryonomys*. An unnamed small *Thryonomys* from the late Miocene Upper Nawata Formation, Kenya, also has a four-lophed dp4, but the dP4 is similar morphologically to *Paraulacodus*. An upper incisor with two grooves, assigned to *Paraulacodus* (and proportionally smaller than the Upper Nawata specimens), is geologically slightly older (Lower Nawata Formation). Two mandibles of *Thryonomys* from the late Miocene at Lemudong'o, Kenya (Manthi, 2007), have not yet been studied in detail.

An extremely important area for future study is the interrelationships of early eastern African hystricognaths, in particular the phiomyids, with those from the Eocene-Oligocene of northern Africa. This work has begun in conjunction with newly recovered late Oligocene rodents from Tanzania (Holroyd and Stevens, 2009; Stevens et al., 2004, 2006, 2009) and Kenya (Rasmussen and Gutiérrez, 2009). One new genus described recently from the late Oligocene of Tanzania, *Kahawamys*, has a mixture of characters seen in both northern Paleogene and eastern early Miocene hystricognaths (Stevens et al., 2009). Continued study of the Paleogene eastern African material should help to clarify taxonomic relationships among early northern and sub-Saharan African taxa. This will provide clues to the history of small mammal dispersal between these two areas that can be tested by looking at the dispersal history of the large mammals. It is also hoped that these Oligocene eastern African faunas will eventually yield the remains of bathyergids and anomaluroids, groups whose early history is very poorly known and whose taxonomic relationships remain controversial.

Temporal Distribution

Two factors contributing to our understanding of the temporal distribution of many eastern African rodent faunas are of special significance. First, many highly fossiliferous localities are found within the still active East African rift system. As a consequence of a long history of tectonic/volcanic activity, many fossils are not only well preserved but also preserved in radioisotopically datable sediments. This is in contrast to many localities in northern and southern Africa where remains are often found, for example, preserved in infillings in karst deposits. Dating of northern and southern African fossil specimens is thus often dependent on biochronology, though magnetostratigraphic age determination may occasionally be possible.

A second factor (dependent in part on the first) is the presence in eastern Africa of a few highly fossiliferous, relatively continuous, well-calibrated (by absolute dates) stratigraphic sections. A prime example is the Tugen Hills sequence, Baringo Basin, Kenya, which includes numerous faunas dating from the early middle Miocene into the Pleistocene. Within this sequence, one can follow evolutionary trends through time that are relatively free of possible geographic biases (e.g., Winkler, 1990, 2002).

Eastern Africa is at the crossroads between northern and southern Africa, and it has yielded a remarkable Neogene rodent record with about 78 genera and all extant families represented. There is no endemic family in this region. Eastern Africa is characterized by a diversity of rodents slightly higher during the Neogene than it is today: the Neogene record includes about 42 extinct and 36 extant genera. The highest fossil diversity is observed among the Sciuridae, Murinae, Bathyergidae, Phiomyidae, and Thryonomyidae, which may have diversified in response to formation of the East African Rift Valley.

SOUTHERN AFRICAN RODENTS

Taxonomic Issues

As is usual, information that is more obviously out of place or unexpected first focuses attention on potential problems. One such example is *Millardia* (a murine), which has been reported from the Plio-Pleistocene in northwestern Botswana (Pickford and Mein, 1988). This currently Asian genus was first reported from Africa by Sabatier (1982), who named two new species from the Pliocene of Ethiopia. Subsequently, however, Denys (1990d) determined that at least one of these species should be referred to the African genus *Acomys* (a deomyine), a new species of which she described from the early Pliocene at Langebaanweg, southern South Africa (1990a). Under the circumstances, it seems most likely that the specimens from Botswana should also be referred to *Acomys*, and this is reflected in table 17.4.

Two less obviously unexpected genera are the gerbil (tateril) *Taterillus* and the naked mole rat *Heterocephalus*, both of which today are exclusively African genera. At present, they are found only north of the equator. *Taterillus* has been listed as present at Makapansgat in northern South Africa (Pocock, 1987) and in northwestern Botswana (Pickford and Mein, 1988). While it is not impossible that this genus formerly extended much farther south, it seems far more likely that the genus involved is the southern African genus *Gerbillurus* or, possibly, *Desmodillus*. De Graaff (1960, quoting Lavocat, 1957) listed the eastern African bathyergid

Heterocephalus as present at Makapansgat, but Pocock (1987) did not mention this genus. Lavocat (1957) provided no compelling evidence that the specimens were *Heterocephalus*, rather than one of the southern African taxa. The presence of deep indentations on worn teeth of the Makapansgat material suggests assignment to *Georychus*, which has been reported from the Sterkfontein Valley during the early Pleistocene (Avery, 2001). The Sterkfontein record constitutes a northward extension to the current range of *Georychus*, and it would not be surprising for it to have occurred still further north during the Pliocene. An alternative possibility would be *Heliophobius*, whose present distribution lies in Zambia and Mozambique (Kingdon, 1997), in which case its occurrence at Makapansgat would constitute a southward range extension. The material assigned to *Taterillus* and *Heterocephalus* needs reexamination to resolve this issue in the light of current understandings of these genera.

Temporal Distribution

In southern Africa, the distribution of rodent-bearing deposits of different geologic ages is generally mutually exclusive (figure 17.1). Miocene material has been reported only from Namibia: early and middle Miocene samples from southern Namibia, and middle and late Miocene samples from northern Namibia (table 17.4). Pliocene sites occur in a broad band from southern Angola to northern South Africa, with an early Pliocene outlier at Langebaanweg, in southern South Africa. Conversely, and with the exception of Zebrarivier in southern Namibia, Pleistocene sites occur to the south and east of the earlier sites. Thus, apart from the Otavi Mountains in northern Namibia and the Sterkfontein Valley and Makapansgat in northern South Africa (figure 17.1), there is no temporal continuity in any one area. This makes it impossible to follow evolutionary trends that are free of geographic biases in this part of Africa.

None of the early Miocene genera and only one of the 12 named middle Miocene genera is extant, but nearly a third of the 23 late Miocene genera are thought to be extant. Whereas many of the extinct genera are from the Suborder Myomorpha, none of these appeared before the middle Miocene. Apart from one member of the Anomaluromorpha, *Parapedetes*, all the early Miocene rodents are Hystricognathi. From the middle Miocene onward, the Myomorpha assumed numerical dominance, beginning with eight genera in various subfamilies of the Nesomyidae, and continuing with the appearance of many genera in various subfamilies of the Muridae in the early Pliocene. Only two extant genera (the sciurid *Paraxerus* and the nesomyid *Petromyscus*) have so far been identified definitively before the Pliocene, although others, notably several dendromurines, have been identified tentatively (Pickford et al., 1994; Senut et al., 1992; see also table 17.4). In more recent times, only three extinct genera (*Stenodontomys*, *Proodontomys*, and *Gypsorhychus*) continued undoubtedly into the early Pleistocene. Two Otomyinae (*Palaeotomys* and *Prototomys*) have been reported from the middle Pleistocene (De Graaff, 1960) and early Pleistocene, respectively. Whether these genera are separable from the extant *Otomys* and *Myotomys* has not been demonstrated conclusively.

Many of the represented genera, such as the murine *Aethomys* and otomyine *Otomys*, today have wide distributions and, as such, could be found in fossil form both north and south of the equator. In these cases, differences in distribution over time may only be detectable at the species level. In other, presently more restricted genera, range extensions or differences in distribution over time are more likely to be detectable, but problems of identification, such as those mentioned above, need to be solved before these can be determined definitely. Examples of genera that may previously have extended further south than they do today include the tropical murine *Malacomys*, which has been reported from Humpata in southern Angola (Pickford et al., 1992). If correct, this would constitute a considerable southward extension for this genus, which currently reaches no further south than northeastern Angola (Musser and Carleton, 2005). Identification of this unique record of *Malacomys* is in doubt because the tooth morphology of this genus is similar to that of *Praomys*, making generic distinction of fossil remains difficult (see discussion in Wilson and Reeder, 2005). *Uranomys* is a second genus whose reported occurrence in the Humpata deposits (Pickford et al., 1992) may imply a previously wider distribution than that of the present. However, the recorded distribution of this genus is very disjointed, and Kingdon (1997) notes that the genus is poorly known. Less dramatic is the discovery at Border Cave in western South Africa that the murine *Pelomys* occurred several degrees south of its present range during the late Pleistocene (Avery, 1982a). Likewise, the presence of a bathyergid in northern South Africa that is apparently not the widespread *Cryptomys* must constitute a range extension, whichever genus is represented (Friedmann and Daly, 2004).

These examples suggest that examination of the geographic distribution patterns of other genera will almost certainly show change through time. In some instances there has possibly been long-term unidirectional change. One such case is the endemic nesomyid *Mystromys*, which is now listed as endangered in South Africa (Friedmann and Daly, 2004) but was obviously extremely common in the Sterkfontein Valley (northern South Africa) during the late Pliocene and Pleistocene (Avery, 2001). It also occurred in northern Namibia during the late Pliocene and possibly even the late Miocene (Pickford et al., 1992), and in Botswana during the late Pleistocene (Robbins et al., 1996), but is apparently absent from those countries today (Stuart and Stuart, 2001). The only other mystromyine is the extinct *Proodontomys*, so far known only from northern and central (Taung) South Africa during the Pliocene and early Pleistocene.

Southern Africa seems to have acted as the diversification center for the Graphiurinae, Otavimyinae, Mystromyinae, Namibimyinae, Petromyscinae, Otomyinae, and Petromuridae. Nearly all extant African rodent families are represented except the Anomaluridae and Dipodidae. The Rhizomyinae are no longer found in southern Africa. All these pecularities still confirm the relative isolation of southern African rodent faunas during the end of the Miocene.

Summary and Conclusions

The earliest record of African rodents is in the early to middle Eocene in Algeria and Tunisia. Eocene and Oligocene rodents, from Algeria, Tunisia, Egypt, and Libya, include the families Zegdoumyidae, Phiomyidae, Thryonomyidae, and Anomaluridae. The Zegdoumyidae may be derived from the Ischyromyidae, which had a Holarctic distribution beginning in the early Eocene, or from North American (or possibly Asian) Sciuravidae (also dating from the early Eocene). The Phiomyidae and Thryonomyidae have been described as originating within Africa from a single immigration event, probably from

Asia or the Indian Subcontinent (Holroyd, 1994). Mahboubi et al. (1997), however, suggest that Paleogene rodents and other taxa originated from several Paleogene terrestrial interchanges between Africa and the northern Tethyan regions, since, unlike Holroyd (1994), they did not consider the Afro-Arabian continent to have been geographically isolated during the Eocene. The Anomaluridae are considered by some authors to be derived from the Zegdoumyidae. To date, there are only two reported sub-Saharan Paleogene rodent faunas, from the Mbeya Region, Tanzania (Stevens et al., 2004, 2006, 2009), and Losodok, northern Kenya (Rasmussen and Gutiérrez, 2009): both these faunas are still under study.

New rodent families and subfamilies are found in the early Miocene. These include the Sciuridae, Spalacidae, Nesomyidae (Afrocricetodontinae), Muridae (Myocricetodontinae), Pedetidae, Ctenodactylidae, Bathyergidae, Myophiomyidae, and Kenyamyidae (found only in the early Miocene). There is also a single tentative record of the family Pseudocricetodontidae in the early Miocene of Uganda (Pickford and Mein, 2006). The Sciuridae likely immigrated from Europe, the Spalacidae from southern Asia, and the Ctenodactylidae, Myocricetodontinae, and Afrocricetodontinae from Eurasia. The Pedetidae, Bathyergidae, Myophiomyidae, and Kenyamyidae are found exclusively, or almost exclusively, in Africa, and they likely evolved within Africa. The sister group relationships of the Pedetidae and Bathyergidae are controversial. The middle Miocene records the first appearance of the Gliridae and Dipodidae. The earliest glirid, *Microdyromys*, is known from the middle Miocene of northern Africa (Algeria, Morocco). It is a likely immigrant from Europe, where it is first known in the late Oligocene (Jaeger, 1977a). "*Protalactaga*" and *Heterosminthus* are the earliest African dipodids: they are reported from the middle Miocene of Morocco and Libya. African dipodids are likely immigrants from Asia. In the late Miocene, the Cricetidae, Hystricidae, and Petromuridae first occur. The Cricetidae are likely immigrants from Europe (and have a circum-Mediterranean distribution), the Hystricidae from Eurasia, and the Petromuridae probably evolved within Africa.

The late Eocene through early Miocene record is dominated both in number of species and in numerical abundance by the Hystricognathi, particularly the families Phiomyidae and Thryonomyidae. In the early Miocene the Muroidea made their appearance and eventually overshadowed the Hystricognathi. In the early late Miocene, the subfamily Murinae first occurred in Africa. The timing of their appearance is coincident with the dispersal of this group into Europe and northern Asia, after its likely derivation in southern Asia (Jacobs, 1985). In Africa, the murines diversified rapidly beginning in the late Miocene and early Pliocene (likely including additional immigration events), and are currently the most speciose and numerically abundant group of African rodents.

ACKNOWLEDGMENTS

We thank W. J. Sanders and L. Werdelin for the invitation to contribute to this volume. Our chapter benefited from the input and critiques of many individuals, in particular P. Holroyd and an anonymous reviewer. H. Wesselman, T. Rasmussen, M. Pickford, P. Holroyd, and D. C. D. Happold kindly provided information about newly described faunas and/or access to publications still in press. We appreciate the help of D. Winkler for photographing most of the specimens illustrated in this chapter. We thank N. Boaz, T. Harrison, L. MacLatchy, and L. Robbins for permission to include photographs of specimens they collected.

Literature Cited

Aguilar, J. P., and J. Michaux. 1990. Un *Lophiomys* (Cricetidae, Rodentia) nouveau dans le Pliocène du Maroc: Rapport avec les Lophiomyinae fossils et actuels. *Paleontologia i Evolució* 23:205–211.

Aguilar, J. P., and L. Thaler. 1987. *Protolophiomys ibericus* nov. gen. nov. sp. (Mammalia, Rodentia) du Miocène supérieur de Salobrena (Sud de l'Espagne). *Comptes Rendus des Séances de l'Académie des Sciences de Paris*, Série II, 304:859–863.

Agusti, J., A. El-Arnauti, A. Galobart, R. Gaete, and M. Lienas. 2004. Results of a field campaign in the late Miocene of the Sahabi Formation, Libya; p. 6 in *Abstracts of the Sedimentary Basins of Libya, Third Symposium, Geology of East Libya, November 21–23, 2004, Binghazi, Libya*.

Alemseged, Z., and D. Geraads. 2000. A new middle Pleistocene fauna from the Busidima-Telalak region of the Afar, Ethiopia. *Comptes Rendus de l'Académie des Sciences, Paris, Sciences de la Terre et des Planètes* 331:549–556.

Ambrose, S. H., L. J. Hlusko, D. Kyule, A. Deino, and M. Williams. 2003. Lemudong'o: a new 6 Ma paleontological site near Narok, Kenya Rift Valley. *Journal of Human Evolution* 44:737–742.

Ameur, R. C. 1976. Radiometric age of early *Hipparion* fauna in northwest Africa. *Nature* 261:38–39.

———. 1984. Découverte de nouveaux rongeurs dans la formation Miocène de Bou Hanifia (Algérie occidentale). *Geobios* 17:167–175.

Ameur-Chabbar, R. 1988. Biochronologie des formations continentales du Néogène et du Quaternaire de l'Algérie. Contribution des Micromammifères. Unpublished PhD dissertation, Université de Oran, Algeria, 480 pp.

Arambourg, C. 1959. Vertébrés continentaux du Miocène supérieur de l'Afrique du Nord. Publications du Service de la Carte Géologique de l'Algérie (nouvelle série). *Paléontologie, Mémoire* 4:1–161.

Assefa, Z. 2006. Faunal remains from Porc-Epic: Paleoecological and zooarchaeological investigations from a Middle Stone Age site in southeastern Ethiopia. *Journal of Human Evolution* 51:50–75.

Avery, D. M. 1982a. The micromammalian fauna from Border Cave, KwaZulu, South Africa. *Journal of Archaeological Science* 9:187–204.

———. 1982b. Micromammals as palaeoenvironmental indicators and an interpretation of the late Quaternary in the southern Cape Province, South Africa. *Annals of the South African Museum* 85:183–374.

———. 1984. Micromammals and environmental change at Zebrarivier, central Namibia. *Journal of the South West African Scientific Society* 38:79–86.

———. 1987. Late Pleistocene coastal environment of the southern Cape Province of South Africa: Micromammals from Klasies River Mouth. *Journal of Archaeological Science* 14:405–421.

———. 1991. Late Quaternary incidence of some micromammals in Natal. *Durban Museum Novitates* 16:1–11.

———. 1992. The environment of early modern humans at Border Cave, South Africa: micromammalian evidence. *Palaeogeography, Palaeoclimatology, Palaeoecology* 91:71–87.

———. 1995a. The preliminary assessment of the micromammalian remains from Gladysvale Cave, South Africa. *Palaeontologia Africana* 32:1–10.

———. 1995b. Southern savannas and Pleistocene hominid adaptations: the micromammalian perspective; pp. 459–478 in E. S. Vrba, G. H. Denton, T. C. Partridge, and L. H. Burckle (eds.), *Paleoclimate and Evolution with Emphasis on Human Origins*. Yale University Press, New Haven.

———. 1998. An assessment of the lower Pleistocene micromammalian fauna from Swartkrans Members 1–3, Gauteng, South Africa. *Geobios* 31:393–414.

———. 2000a. Notes on the systematics of micromammals from Sterkfontein, Gauteng, South Africa. *Palaeontologia Africana* 36:83–90.

———. 2000b. Past and present ecological and environmental information from micromammals; pp. 63–72 in L. Barham (ed.), *The Middle Stone Age of Zambia, South Central Africa*. Western Academic and Specialist Press, Bristol, England.

———. 2001. The Plio-Pleistocene vegetation and climate of Sterkfontein and Swartkrans, South Africa, based on micromammals. *Journal of Human Evolution* 41:113–132.

———. 2003. Early and middle Pleistocene environments and hominid biogeography: Micromammalian evidence from Kabwe, Twin Rivers, and Mumbwa Caves in central Zambia. *Palaeogeography, Palaeoclimatology, Palaeoecology* 189:55–69.

Avery, G. P., K. Cruz-Uribe, P. Goldberg, F. E. Grine, R. G. Klein, M. J. Lenardi, C. W. Marean, W. J. Rink, H. P. Schwarcz, A. I. Thackeray, and M. L. Wilson. 1997. The 1992–1993 excavations at the Die Kelders Middle and Later Stone Age cave site, South Africa. *Journal of Field Archaeology* 24:263–291.

Benammi, M. 1997a. Magnétostratigraphie du Miocène supérieur continental du Bassin d'Afoud, Bassin d'Aït Kandoula); pp. 285–291 in J. -P. Aguilar, S. Legendre, and J. Michaux (eds.), *Actes du Congrès BiochroM '97*. Mémoires et Travaux de l'Ecole Pratique des Hautes Etudes, l'Institut de Montpellier 21.

———. 1997b. Nouveaux rongeurs du Miocène continental du Jebel Rhassoul (moyenne Moulouya, Maroc). *Geobios* 30:713–721.

———. 2001. Découverte de deux nouvelles espèces du genre *Myocricetodon* dans le Miocène supérieur du basin d'Aït Kandoula (Maroc). *Comptes Rendus de l'Académie des Sciences, Paris, Sciences Terre et des Planètes* 333:187–193.

Benammi, M., B. Orth, M. Vianey-Liaud, Y. Chaimanee, V. Suteethorn, G. Feraud, J. Hernandez, and J.-J. Jaeger. 1995. Micromammifères et biochronologie des formations néogènes du flanc sud du Haut-Atlas Marocain: Implications biogéographiques, stratigraphiques et tectoniques. *Africa Geoscience Review* 2:279–310.

Bishop, W. W. 1962. The mammalian fauna and geomorphological relations of the Napak volcanics, Karamoja. *Records of the Geological Survey, Uganda* 1957–1958:1–18.

Black, C. C., and L. Krishtalka. 1986. Rodents, bats, and insectivores from the Plio-Pleistocene sediments to the east of Lake Turkana, Kenya. *Contributions in Science, Natural History Museum of Los Angeles County* 372:1–15.

Boaz, N. T., R. L. Bernor, A. S. Brooks, H. B. S. Cooke, J. de Heinzelin, R. Dechamps, E. Delson, A. W. Gentry, J. W. K. Harris, P. Meylan, P. P. Pavlakis, W. J. Sanders, K. M. Stewart, J. Verniers, P. G. Williamson, and A. J. Winkler. 1992. A new evaluation of the significance of the Late Neogene Lusso Beds, Upper Semliki Valley, Zaire. *Journal of Human Evolution* 22:505–517.

Brain, C. K. 1981. *The Hunters or the Hunted?* University of Chicago Press, Chicago, 365 pp.

Brain, C. K., C. S. Churcher, J. D. Clark, F. E. Grine, P. Shipman, R. L. Susman, A. Turner, and V. Watson. 1988. New evidence of early hominids, their culture and environment from the Swartkrans cave, South Africa. *South African Journal of Science* 84:828–835.

Brandy, L. D., and J.-J. Jaeger. 1980. Les échanges de faunes terrestres entre l'Europe et l'Afrique nord-occidentale au Messinien. *Comptes Rendus de l'Académie des Sciences, Paris*, Série D, 291:465–468.

Brink, J. 1987. The archaeozoology of Florisbad, Orange Free State. *Memoirs of the National Museum* (Bloemfontein) 24:1–151.

Bronner, G. N., M. Hoffmann, P. J. Taylor, C. T. Chimimba, P. B. Best, C. A. Matthee, and T. J. Robinson. 2003. A revised systematic checklist of the extant mammals of the southern African subregion. *Durban Museum Novitates* 28:56–96.

Broom, R. 1934. On the fossil remains associated with *Australopithecus africanus*. *South African Journal of Science* 31:471–480.

———. 1937a. Notes on a few more new fossil mammals from the caves of the Transvaal. *Annals and Magazine of Natural History, London* 20(10):509–514.

———. 1937b. On some new Pleistocene mammals from limestone caves of the Transvaal. *South African Journal of Science* 33:750–768.

———. 1939. The fossil rodents of the limestone cave at Taungs. *Annals of the Transvaal Museum* 19:315–317.

———. 1948a. The giant rodent mole, *Gypsorhychus*. *Annals of the Transvaal Museum* 21:47–49.

———. 1948b. Some South African Pliocene and Pleistocene mammals. *Annals of the Transvaal Museum* 21:1–38.

Broom, R., and G. W. H. Schepers. 1946. The South African fossil ape-men the Australopithecinae. *Transvaal Museum Memoir* 2:1–272.

Brugal, J. P., and C. Denys. 1989. Vertébrés du site acheuléen d'Isenya (Kenya, district de Kajiado): Implications paléoécologiques et paléobiogéographiques. *Comptes Rendus de l'Académie des Sciences, Paris*, Série II, 308:1503–1508.

Brunet, M., and M. P. E. T. 2000. Chad: Discovery of a vertebrate fauna close to the Mio-Pliocene boundary. *Journal of Vertebrate Paleontology* 20:205–209.

Carleton, M. D., and G. G. Musser. 1984. Chapter 11. Muroid rodents; pp. 289–397 in S. Anderson and J. Knox Jones, Jr. (eds.), *Orders and Families of Recent Mammals of the World*. Wiley, New York.

————. 2005. Order Rodentia; pp. 745–752 in D. E. Wilson and D. M. Reeder (eds.), *Mammal Species of the World*. 3rd ed. Johns Hopkins University Press, Baltimore.

Chubb, E. C. 1909. List of vertebrate remains; pp. 21–25 in F. White, Notes on a cave containing fossilized bones of animals, worked pieces of bone, stone implements and quartzite pebbles, found in a kopje or small hill, composed of zinc and lead ores, at Broken Hill, north-western Rhodesia. *Proceedings of the Rhodesia Scientific Association*, 7.

Cifelli, R. L., A. K. Ibui, L. L. Jacobs, R. W. Thorington, Jr. 1986. A giant tree squirrel from the late Miocene of Kenya. *Journal of Mammalogy* 67:274–283.

Coiffait, B., and P. E. Coiffait. 1981. Découverte d'un gisement de micromammifères d'âge Pliocène dans le basin de Constantine (Algérie). Présence d'un muridé nouveau *Paraethomys athmenia*. Palaeovertebrata 11:1–15.

Coiffait-Martin, B. 1991. Contribution des rongeurs du Néogène d'Algérie à la biochronologie mammalienne d'Afrique nord-occidentale. Unpublished PhD dissertation, Université Nancy I, Nancy, France, 400 pp.

Colbert, E. H. 1933. Two new rodents from the lower Siwalik Beds of India. *American Museum Novitates* 633:1–6.

Conroy, G. C., M. Pickford, B. Senut, J. Van Couvering, and P. Mein. 1992. *Otavipithecus namibensis*, first Miocene hominoid from southern Africa. *Nature* 356:144–148.

Cooke, H. B. S. 1963. Pleistocene mammal faunas of Africa with particular reference to southern Africa; pp. 65–116 in F. C. Howell and F. Bourlière (eds.), *African Ecology and Human Evolution*. Aldine Press, Chicago.

————. 1990. Taung fossils in the University of California collections; pp. 119–134 in G. H. Sperber (ed.), *Apes to Angels: Essays in Anthropology in Honor of Phillip V. Tobias*. Wiley-Liss, New York.

Cruz-Uribe, K., R. G. Klein, G. Avery, M. Avery, D. Halkett, T. Hart, R. G. Milo, C. G. Sampson, and T. P. Volman. 2003. Excavation of buried Late Acheulean (mid-Quaternary) land surfaces at Duinefontein 2, Western Cape Province, South Africa. *Journal of Archaeological Science* 30:559–575.

Davies, C. 1987. Fossil Pedetidae (Rodentia) from Laetoli; pp. 171–190 in M. D. Leakey, and J. M. Harris (eds.), *Laetoli: A Pliocene Site in Tanzania*. Clarendon Press, Oxford.

Dawson, M. R., T. Tsubamoto, M. Takai, N. Egi, S. T. Tun, and C. Sein. 2003. Rodents of the Family Anomaluridae (Mammalia) from southeast Asia (middle Eocene, Pondaung Formation, Myanmar). *Annals of the Carnegie Museum* 72:203–213.

De Graaff, G. 1960. A preliminary investigation of the mammalian microfauna in Pleistocene deposits in the Transvaal System. *Palaeontologia Africana* 7:59–118.

————. 1961. On the fossil mammalian microfauna collected at Kromdraai by Draper in 1895. *South African Journal of Science* 57:259–260.

————. 1981. *The Rodents of Southern Africa*. Butterworths, Durban, South Africa, 267 pp.

————. 1988. The smaller mammals of the Cave of Hearths from basal guano underlying the Acheulean deposits (ca. 200,000 BP); pp. 535–548 in R. J. Mason (ed.), *Cave of Hearths, Makapansgat*. Transvaal Occasional Paper No. 21.

Denys, C. 1985. Paleoenvironmental and paleobiogeographical significance of the fossil rodent assemblages of Laetoli (Pliocene, Tanzania). *Palaeogeography, Palaeoclimatology, Palaeoecology* 52:77–97.

————. 1987a. Fossil rodents (other than Pedetidae) from Laetoli; pp. 118–170 in M. D. Leakey and J. M. Harris (eds.), *Laetoli: A Pliocene Site in Tanzania*. Clarendon Press, Oxford.

————. 1987b. Micromammals from the West Natron Pleistocene deposits (Tanzania): Biostratigraphy and paleoecology. *Bulletin des Sciences Géologiques* 40:185–201.

————. 1989a. Implications paléoécologiques et paléobiogéographiques de la présence d'une gerboise (Rodentia, Mammalia) dans le rift est africain au Pléistocène moyen. *Comptes Rendus de l'Académie des Sciences, Paris*, Série II, 309:1261–1266.

————. 1989b. A new species of Bathyergid rodent from Olduvai Bed I (Tanzania, lower Pleistocene). *Neues Jahrbuch für Geologie und Paläontologie, Monatshefte* 1989:257–264.

————. 1989c. Phylogenetic affinities of the oldest East African *Otomys* (Rodentia, Mammalia) from Olduvai Bed I (Pleistocene, Tanzania). *Neues Jahrbuch für Geologie und Paläontologie, Monatshefte* 1989:705–725.

————. 1989d. Two new Gerbillids (Rodentia, Mammalia) from Olduvai Bed I (Pleistocene, Tanzania). *Neues Jahrbuch für Geologie und Paläontologie, Abhandlungen* 178:243–265.

————. 1990a. Deux nouvelles espèces d'*Aethomys* (Rodentia, Muridae) à Langebaanweg (Pliocène, Afrique du Sud): Implications phylogénétiques. *Annales de Paléontologie* 76:41–69.

————. 1990b. First occurrence of *Xerus* cf. *inauris* (Rodentia, Sciuridae) at Olduvai Bed I (Lower Pleistocene, Tanzania). *Paläontologische Zeitschrift* 64:359–365.

————. 1990c. Implications paléoécologiques et paléobiogéographiques de l'étude de rongeurs plio-pleistocènes d'Afrique orientale et australe. Unpublished PhD dissertation, Université Pierre et Marie Curie (Paris VI), Paris, 406 pp.

————. 1990d. The oldest *Acomys* (Rodentia, Muridae) from the Lower Pliocene of South Africa and the problem of its murid affinities. *Palaeontographica, Abteilung A* 210:79–81.

————. 1991. Un nouveau rongeur *Mystromys pocockei* sp. nov. (Cricetinae) du Pliocène inférieur de Langebaanweg (Region du Cape, Afrique du Sud). *Comptes Rendus de l'Académie des Sciences, Paris*, Série II, 313:1335–1341.

————. 1992. Présence de *Saccostomus* (Rodentia, Mammalia) à Olduvai Bed I (Tanzania, Pléistocène inférieur). Implications phylétiques et paléobiogéographiques. *Geobios* 25:145–154.

————. 1994a. Affinités systématiques de *Stenodontomys* (Mammalia, Rodentia) rongeur Muroidea du Pliocène de Langebaanweg (Afrique du Sud). *Comptes Rendus de l'Académie des Sciences, Paris*, Série II, 318:411–416.

————. 1994b. Nouvelles espèces de *Dendromus* (Rongeurs, Muroidea) à Langebaanweg (Pliocène, Afrique du Sud). Conséquences stratigraphiques et paléoécologiques. *Palaeovertebrata* 23:153–176.

————. 1998. Phylogenetic implications of the existence of two modern genera of Bathyergidae (Mammalia, Rodentia) in the Pliocene site of Langebaanweg (South Africa). *Annals of the South African Museum* 105:265–286.

————. 1999. Of mice and men: Evolution in East and South Africa during Plio-Pleistocene times; pp. 226–252 in T. G. Bromage and F. Schrenk (eds.), *African Biogeography, Climate Change, and Human Evolution*. Oxford University Press, Oxford.

————. 2003. Evolution du genre *Otomys* (Rodentia, Muridae) au Plio-Pléistocène d'Afrique orientale et australe; pp. 75–84 in A. Petrulescu and E. Stiuca (eds.), *Advances in Paleontology, "Hen to Pantha," Volume in Honor of Constantin Radulescu and Petre Mihai Samson*. Romanian Academy, "Emil Racovitză," Institute of Speleology, Bucharest, Romania.

Denys, C., and J.-J. Jaeger. 1992. Rodents of the Miocene site of Fort Ternan (Kenya), First Part: Phiomyids, Bathyergids, Sciurids and Anomalurids. *Neues Jahrbuch für Geologie und Paläontologie, Abhandlungen* 185:63–84.

Denys, C., L. Viriot, R. Daams, P. Pelaez-Campomanes, P. Vignaud, L. Andossa, and M. Brunet. 2003. A new Pliocene Xerine sciurid (Rodentia) from Kossom Bougoudi, Chad. *Journal of Vertebrate Paleontology* 23:676–687.

Dieterlen, F. 2005a. Family Anomaluridae; pp. 1532–1534 in D. E. Wilson and D. M. Reeder (eds.), *Mammal Species of the World*. 3rd ed. Johns Hopkins University Press, Baltimore.

————. 2005b. Family Ctenodactylidae; p. 1536 in D. E. Wilson and D. M. Reeder (eds.), *Mammal Species of the World*. 3rd ed. Johns Hopkins University Press, Baltimore.

————. 2005c. Family Pedetidae; p. 1535 in D. E. Wilson and D. M. Reeder (eds.), *Mammal Species of the World*. 3rd ed. Johns Hopkins University Press, Baltimore.

Dobson, M. 1998. Mammal distributions in the western Mediterranean: The role of human intervention. *Mammal Review* 28:77–88.

Dreyer, T. F., and A. Lyle. 1931. *New Fossil Mammals and Man from South Africa*. Nasionale Pers, Bloemfontein, South Africa, 60 pp.

Ducroz, J. -F., V. Volobouev, and L. Granjon. 2001. An assessment of the systematics of arvicanthine rodents using mitochondrial DNA sequences: Evolutionary and biogeographical implications. *Journal of Mammalian Evolution* 8:173–206.

Fejfar, O. 1987. Oligocene rodents from Zallah Oasis, Libya. *Münchner Geowissenschaftliche Abhandlungen A* 10:265–268.

Flynn, L. J. 1990. The natural history of rhizomyid rodents; pp. 155–183 in E. Nevo and O. A. Reig (eds.), *Evolution of Subterranean Mammals at the Organismal and Molecular Levels*. Liss, New York.

Flynn, L. J., and M. Sabatier. 1984. A muroid rodent of Asian affinity from the Miocene of Kenya. *Journal of Vertebrate Paleontology* 3:160–165.

Flynn, L. J., and A. Winkler. 1994. Dispersalist implications of *Paraulacodus indicus*: A South Asian rodent of African affinity. *Historical Biology* 9:223–235.

Flynn, L. J., A. J. Winkler, L. L. Jacobs, and W. R. Downs III. 2003. Tedford's gerbils from Afghanistan; pp. 603–624, in L. J. Flynn (ed.), *Vertebrate Fossils and Their Context: Contributions in Honor of R. H. Tedford*. American Museum of Natural History Bulletin 279.

Friedmann, Y., and B. Daly (eds.). 2004. *Red Data Book of the Mammals of South Africa: A Conservation Assessment*. CBSG Southern Africa, Conservation and Specialist Breeding Group (SSC/IUCN), Endangered Wildlife Trust, Johannesburg, 722 pp.

Geraads, D. 1994. Rongeurs et Lagomorphes du Pleistocène moyen de la "Grotte des Rhinocéros," Carrière Oulad Hamida 1 à Casablanca, Maroc. *Neues Jahrbuch für Geologie und Paläontologie, Abhandlungen* 191:147–172.

———. 1995. Rongeurs et insectivores (Mammifères) du Pliocène final d'Ahl al Oughlam (Casablanca, Maroc). *Geobios* 28:99–115.

———. 1998a. Rongeurs du Miocène supérieur de Chorora (Ethiopie): Cricetidae, Rhizomyidae, Phiomyidae, Thryonomyidae, Sciuridae. *Paleovertebrata* 27:203–216.

———. 1998b. Rongeurs du Mio-Pliocène de Lissasfa (Casablanca, Maroc). *Geobios* 31:229–245.

———. 2001. Rongeurs du Miocène supérieur de Chorora (Ethiopie): Murinae, Dendromurinae et conclusions. *Paleovertebrata* 30:89–109.

———. 2002. Plio-Pleistocene mammalian biostratigraphy of Atlantic Morocco. *Quaternaire* 13:43–53.

———. 2006. The late Pliocene locality of Ahl al Oughlam, Morocco: Vertebrate fauna and interpretation. *Transactions of the Royal Society of South Africa* 61:97–101.

Geraads, D., Z. Alemseged, D. Reed, J. Wynn, and D. C. Roman. 2004. The Pleistocene fauna (other than Primates) from Asbole, lower Awash Valley, Ethiopia, and its environmental and biochronological implications. *Geobios* 37:697–718.

Gibbard, P., and T. Van Kolfschoten. 2004. The Pleistocene and Holocene Epochs; pp. 441–452 in F. Gradstein, J. Ogg, and A. Smith (eds.), *A Geologic Time Scale 2004*. Cambridge University Press, Cambridge.

Grine, F. E., R. G. Klein, and T. P. Volman. 1991. Dating, archaeology and human fossils from the Middle Stone Age levels of Die Kelders, South Africa. *Journal of Human Evolution* 21:363–395.

Haile-Selassie, Y., G. Woldegabriel, T. D. White, R. L. Bernor, D. Degusta, P. R. Renne, W. K. Hart, E. Vrba, A. Stanley, and F. C. Howell. 2004. Mio-Pliocene mammals from the Middle Awash, Ethiopia. *Geobios* 37:536–552.

Hamilton, W. R., and J. A. Van Couvering. 1977. Lower Miocene mammals from South West Africa. *Bulletin of Namib Desert Ecological Research Unit* 2:9–11.

Happold, D. C. D. (ed.). In press. *The Mammals of Africa: Vol. 3. Rodents and Lagomorphs*. University of California Press, Berkeley.

Harris, J. M., F. H. Brown, and M. G. Leakey. 1988. Stratigraphy and paleontology of Pliocene and Pleistocene localities west of Lake Turkana, Kenya. *Contributions in Science, Natural History Museum of Los Angeles County* 399:1–128.

Harris, J. M., M. G. Leakey, and T. E. Cerling; appendix by A. J. Winkler. 2003. Early Pliocene tetrapod remains from Kanapoi, Lake Turkana Basin, Kenya. *Contributions in Science, Natural History Museum of Los Angeles County* 498:39–113.

Heissig, K. 1982. Kleinsäuger aus einer obermiozänen (Vallesium) Karstfüllung Ägyptens. *Mitteilungen der Bayerischen Staatssammlung für Paläontologie und Historische Geologie* 22:97–101.

Hendey, Q. B. 1976. The Pliocene fossil occurrences in "E" Quarry, Langebaanweg, South Africa. *Annals South African Museum* 69:215–247.

———. 1978. Preliminary report on the Miocene vertebrates from Arrisdrift, South West Africa. *Annals of the South African Museum* 76:1–41.

———. 1981. Palaeoecology of the late Tertiary fossil occurrences in "E" Quarry, Langebaanweg, South Africa, and a reinterpretation of their geological context. *Annals of the South African Museum* 84:1–104.

———. 1984. Southern African late Tertiary vertebrates; pp. 81–106 in R. G. Klein (ed.), *Southern African Prehistory and Paleoenvironments*. Balkema, Rotterdam.

Henshilwood, C. S., J. C. Sealy, R. Yates, K. Cruz-Uribe, P. Goldberg, F. E. Grine, R. G. Klein, C. Poggenpoel, K. Van Niekerk, and I. Watts. 2001. Blombos Cave, southern Cape, South Africa: Preliminary report on the 1992–1999 excavations of the Middle Stone Age levels. *Journal of Archaeological Science* 28:421–448.

Hlusko, L. J. 2007. Earliest evidence for *Atherurus* and *Xenohystrix* (Hystricidae, Rodentia) in Africa, from the late Miocene site of Lemudong'o, Kenya. *Kirtlandia* 56:86–91.

Holden, M. E. 2005. Family Gliridae; pp. 819–841 in D. E. Wilson and D. M. Reeder (eds.), *Mammal Species of the World*. 3rd ed. Johns Hopkins University Press, Baltimore.

Holden, M. E., and G. G. Musser. 2005. Family Dipodidae; pp. 871–893 in D. E. Wilson and D. M. Reeder (eds.), *Mammal Species of the World*. 3rd ed. Johns Hopkins University Press, Baltimore.

Holroyd, P. A. 1994. An examination of dispersal origins for Fayum Mammalia. Unpublished PhD dissertation, Duke University, Durham, N.C., 328 pp.

Holroyd, P. A., and N. J. Stevens. 2009. Differentiation of *Phiomys andrewsi* from *Lavocatomys aequatorialis* (n. gen., n. sp.) (Rodentia: Thryonomyoidea) in the Oligo-Miocene interval on continental Africa. *Journal of Vertebrate Paleontology* 29:1331–1334.

Hopwood, A. T. 1929a. Mammalia; pp. 70–73 in W. P. Pycraft, G. E. Smith, M. Yearsley, J. T. Carter, R. A. Smith, A. T. Hopwood, D. M. A. Bate, and W. E. Swinton (eds.), *Rhodesian Man and Associated Remains*. Trustees of the British Museum (Natural History), London.

———. 1929b. New and little-known mammals from the Miocene of Africa. *American Museum Novitates* 344:1–9.

Huchon, D., F. Catzeflis, and E. J. P. Douzery. 2000. Variance of molecular datings, evolution of rodents and the phylogenetic affinities between Ctenodactylidae and Hystricognathi. *Proceedings of the Royal Society of London B* 267:393–402.

Jacobs, L. L. 1985. The beginning of the age of murids in Africa. *Acta Zoologica Fennica* 170:149–151.

Jaeger, J.-J. 1969. Les rongeurs du Pléistocène moyen de Ternifine (Algérie). *Comptes Rendus de l'Académie des Sciences, Paris*, Série D, 269:1492–1495.

———. 1970. Découverte au Jebel Irhoud des premières faunes de rongeurs du Pléistocène inférieur et moyen du Maroc. *Comptes Rendus de l'Académie des Sciences, Paris*, Série D, 270:920–923.

———. 1971a. Un Cténodactylidé (Mammalia, Rodentia) nouveau, *Irhoudia bohlini* n. g., n. sp. du Pléistocène inférieur du Maroc: Rapports avec les formes actuelles et fossiles. *Notes du Service Géologique du Maroc* 31:113–140.

———. 1971b. Les micromammifères du "Villafranchien" inférieur du lac Ichkeul (Tunisie): Données stratigraphiques et biogéographiques nouvelles. *Comptes Rendus de l'Académie des Sciences, Paris* 273:562–565.

———. 1975. Les Muridae (Mammalia, Rodentia) du Pliocène et du Pléistocène du Maghreb. Origine, évolution, données biogéographiques et paléoclimatiques. Unpublished PhD dissertation, Université de Montpellier, Montpellier, France, 124 pp.

———. 1976. Les Rongeurs (Mammalia, Rodentia) du Pléistocène inférieur d'Olduvai Bed I (Tanzania): Ière partie. Les Muridés; pp. 57–120 in R. J. G. Savage and S. C. Coryndon (eds.), *Fossil Vertebrates of Africa*, vol. 4. Academic Press, New York.

———. 1977a. Rongeurs (Mammalia, Rodentia) du Miocène de Beni Mellal. *Palaeovertebrata* 7:91–125.

———. 1977b. Les rongeurs du Miocène moyen et supérieur du Maghreb. *Palaeovertebrata* 8:1–166.

———. 1979. Les faunes de rongeurs et de lagomorphs du Pliocène et du Pléistocène d'Afrique orientale. *Bulletin de la Société Géologique France* (7) 21:301–308.

———. 1988. Origine et évolution du genre *Ellobius* (Mammalia, Rodentia) en Afrique Nord-Occidentale. *Folia Quaternaria* 57:3–50.

Jaeger, J.-J., C. Denys, and B. Coiffait. 1985. New Phiomorpha and Anomaluridae from the late Eocene of north-west Africa: Phylogenetic implications; pp. 567–588 in W. P. Luckett and J. -L. Hartenberger (eds.), *Evolutionary Relationships among Rodents: A Multidisciplinary Analysis*. NATO ASI Series, Series A: Life Sciences, Vol. 92. Plenum Press, New York.

Jaeger, J.-J., J. Michaux, and M. Sabatier. 1980. Premières données sur les rongeurs de la formation de Ch'orora (Ethiopie) d'âge Miocène supérieur. I. Thryonomyidés. *Palaeovertebrata, Mémoire Jubilée R. Lavocat*: 365–374.

James, G. T., and B. H. Slaughter. 1974. A primitive new middle Pliocene murid from Wadi el Natrun, Egypt. *Annals of the Geological Survey of Egypt, Cairo* 4:333–362.

Kawamura, Y., and H. Nakaya. 1984. Thryonomyid rodent from the Late Miocene Namurungule Formation, Samburu Hills, Northern Kenya. *African Studies Monograph*, suppl. 2:133–139.

Kingdon, J. 1974. *East African Mammals: Vol. IIB. Hares and Rodents.* University of Chicago Press, Chicago, 704 pp.

———. 1997. *The Kingdon Field Guide to African Mammals.* Academic Press, San Diego, 464 pp.

Klein, R. G. 1972. The Late Quaternary mammalian fauna of Nelson Bay Cave (Cape Province, South Africa): Its implications for megafaunal extinctions and for environmental and cultural change. *Quaternary Research* 2:135–142.

———. 1974. Environment and subsistence of prehistoric man in the southern Cape Province. World Archaeology 5:249–284.

———. 1975. Middle Stone-Age man-animal relationships in southern Africa: Evidence from Die Kelders and Klasies River Mouth. *Science* 190:265–267.

———. 1976a. The mammalian fauna of the Klasies River Mouth sites, southern Cape Province, South Africa. *South African Archaeological Bulletin* 31:75–98.

———. 1976b. A preliminary report on the "Middle Stone Age" open-air site of Duinefontein 2 (Melkbosstrand, south-western Cape Province, South Africa). *South African Archaeological Bulletin* 31:12–20.

———. 1977. The mammalian fauna from the Middle and Later Stone Age (Later Pleistocene) levels of Border Cave, Natal Province, South Africa. *South African Archaeological Bulletin* 32:14–27.

———. 1978. Preliminary analysis of the mammalian fauna from the Redcliff Stone Age site, Rhodesia. *Occasional Papers of the National Museums and Monuments of Rhodesia, Series A, Human Sciences* 4:74–80.

———. 1981. Later Stone Age subsistence at Byeneskranskop Cave, South Africa; pp. 166–190 in R. S. O. Harding, and G. Teleki (eds.), *Omnivorous Primates: Gathering and Hunting in Human Evolution.* Columbia University Press, New York.

———. 1988. The archaeological significance of animal bones from Acheulean sites in southern Africa. *African Archaeological Review* 6:3–25.

Klein, R. G., G. Avery, K. Cruz-Uribe, D. Halkett, T. Hart, R. G. Milo, and T. P. Volman. 1999. Duinefontein 2: An Acheulean site in the Western Cape Province of South Africa. *Journal of Human Evolution* 37:153–190.

Klein, R. G., and K. Cruz-Uribe. 1987. Large mammal and tortoise bones from Elands Bay Cave and nearby sites, western Cape Province, South Africa; pp. 132–163 in J. Parkington and M. Hall (eds.), *Papers in the Prehistory of the Western Cape, South Africa.* British Archaeological Reports, Oxford, S332.

———. 1991. The bovids from Elandsfontein, South Africa, and their implications for the age, palaeoenvironment, and origins of the site. *African Archaeological Review* 9:21–79.

———. 2000. Macromammals and reptiles; pp. 51–61 in L. Barham (ed.), *The Middle Stone Age of Zambia, South Central Zambia.* Western Academic and Specialist Press, Bristol, England.

Klein, R. G., K. Cruz-Uribe, and P. B. Beaumont. 1991. Environmental, ecological and paleoanthropological implications of the Late Pleistocene mammalian fauna from *Equus* Cave, northern Cape Province, South Africa. *Quaternary Research* 36:94–119.

Knock, D., C. M. Ingram, L. J. Frabotta, R. L. Honeycutt, and H. Burda. 2006. On the nomenclature of Bathyergidae and *Fukomys* n. gen. (Mammalia: Rodentia). *Zootaxa* 1142:51–55.

Korth, W. W. 1994. *The Tertiary Record of Rodents in North America.* Topics in Geobiology, Vol. 12. Plenum Press, New York, 319 pp.

Lavocat, R. 1956. La faune des rongeurs des grottes à Australopithèques. *Palaeontologia Africana* 4:69–75.

———. 1957. Sur l'âge des faunes de rongeurs des grottes à Australopithèques; pp. 133–134, in J. D. Clark (ed.), *Proceedings of the Third Panafrican Congress of Prehistory, Livingston, 1955.* Chatto and Windus, London.

——— R. 1961. Le gisement de vertébrés fossiles de Beni Mellal (Maroc): Étude systématique de la faune de mammifères et conclusions générales. *Notes et Mémoires du Service Géologique du Maroc* 155:1–144.

———. 1964. Fossil rodents from Fort Ternan, Kenya. *Nature* 202:1131.

———. 1967. Les microfaunes du Quaternaire ancien d'Afrique orientale et australe; pp. 67–72 in W. W. Bishop and J. D. Clark (eds.), *Background to Evolution in Africa.* University of Chicago Press, Chicago.

———. 1973. Les Rongeurs du Miocène d'Afrique Orientale: 1. Miocène inferieur. *Mémoires et Travaux de l'Institut de Montpellier* 1:1–284.

———. 1978. Rodentia and Lagomorpha; pp. 69–89 in V. J. Maglio and H. B. S. Cooke (eds.), *Evolution of African Mammals.* Harvard University Press, Cambridge.

———. 1988. Un rongeur bathyergidé nouveau remarquable du Miocène de Fort Ternan (Kenya). *Comptes Rendus de l'Académie des Sciences, Paris* 306:1301–1304.

———. 1989. Osteologie de la tete de *Richardus excavans* Lavocat, 1988. *Palaeovertebrata* 19:73–80.

Leakey, R. E., and M. G. Leakey. 1986. A new Miocene hominoid from Kenya. *Nature* 324:143–146.

Lecompte, E., K. Aplin, C. Denys, F. Catzeflis, M. Chades, and P. Chevret. 2008. Phylogeny and biogeography of African Murinae based on mitochondrial and nuclear gene sequences, with a new tribal classification of the subfamily. *BMC Evolutionary Biology* 8:199; doi:10.1186/1471-2148-8-199.

Lecompte, E., C. Denys, and E. L. Granjon. 2005. Confrontation of morphological and molecular data: the The *Praomys* group (Rodentia, Murinae) as a case of adaptive convergences and morphological stasis. *Molecular Phylogenetics and Evolution* 37:899–919.

Lecompte, E., L. Granjon, J. Kerbis Peterhans, and C. Denys. 2002. Cytochrome-*b* based phylogeny of the *Praomys* group (Rodentia, Murinae): A new African radiation? *Comptes Rendus Biologies* 325:827–840.

Lewis, P. J., and E. L. Simons. 2007. Morphological trends in the molars of fossil rodents from the Fayum Depression, Egypt. *Palaeontologia Africana* 42:37–42.

Lindsay, E. H., L. J. Flynn, I. U. Cheema, J. C. Barry, K. Downing, A. Rahim Rajpar, and S. Mahmood Raza. 2005. Will Downs and the Zinda Pir Dome. *Palaeontologia Electronica* 8.1.19A:1–18.

López-Antoñanzas, R., S. Sen, and P. Mein. 2004. Systematics and phylogeny of the cane rats (Rodentia: Thryonomyidae). *Zoological Journal of the Linnean Society* 142:423–444.

Lourens, L., F. Hilgen, N. J. Shackleton, J. Laskar, and D. Wilson. 2004. The Neogene Period; pp. 409–440 in F. Gradstein, J. Ogg, and A. Smith (eds.), *A Geologic Time Scale 2004.* Cambridge University Press, Cambridge.

Luterbacher, H. P., J. R. Ali, H. Brinkhuis, F. M. Gradstein, J. J. Hooker, S. Monechi, J. G. Ogg, J. Powell, U. Röhl, A. Sanfilippo, and B. Schmitz. 2004. The Paleogene Period; pp. 384–408 in in F. Gradstein, J. Ogg, and A. Smith (eds.), *A Geologic Time Scale 2004.* Cambridge University Press, Cambridge.

MacInnes, D. G. 1957. A new Miocene rodent from East Africa. *British Museum (Natural History), Fossil Mammals of Africa* 12:1–35.

MacLatchy, L., S. Cote, A. Kingston, A. Winkler, and J. Rossie. 2007. Early Miocene localities at Napak, Uganda. *Journal of Vertebrate Paleontology* 27 (suppl. to no. 3):109A.

Mahboubi, M., F. Mebrouk, and J.-J. Jaeger. 1997. Conséquences paléobiogéographiques tirées à partir de l'étude de quelques gisements paléogènes du Maghreb (Mammifères, Gastéropodes, Charophytes); pp. 275–284 in J.-P. Aguilar, S. Legendre, and J. Michaux (eds.), *Actes du Congrès BiochroM '97.* Mémoires et Travaux de l'Ecole Pratique des Hautes Etudes, l'Institut de Montpellier 21.

Manthi, F. K. 2006. The Pliocene micromammalian fauna from Kanapoi, northwestern Kenya, and its contribution to understanding the environment of *Australopithecus anamensis.* Unpublished PhD dissertation, University of Cape Town, Cape Town, South Africa, 231 pp.

———. 2007. Preliminary review of the rodent fauna from Lemudong'o, southwestern Kenya and its implication to the Late Miocene palaeoenvironments. *Kirtlandia* 56:92-105.

Marivaux, L. S. Ducrocq, J.-J. Jaeger, B. Marandat, J. Sudre, Y. Chaimanee, S. T. Tun, W. Htoon, and A. N. Soe. 2005. New Remains of *Pondaungimys anomaluropsis* (Rodentia, Anomaluroidea) from the latest middle Eocene Pondaung Formation of Central Myanmar. *Journal of Vertebrate Paleontology* 25:214–227.

Matthews, T., C. Denys, and J. E. Parkington. 2005. The palaeoecology of the micromammals from the late middle Pleistocene site of Hoedjiespunt 1 (Cape Province, South Africa). *Journal of Human Evolution* 49:432–451.

Matthews, T., J. E. Parkington, and C. Denys. 2006. The taphonomy of the micromammals from the late middle Pleistocene site of Hoedjiespunt 1 (Cape Province, South Africa). *Journal of Taphonomy* 4:1–16.

McBrearty, S. 1999. The archaeology of the Kapthurin Formation; pp. 143–156 in P. Andrews and P. Banham (eds.), *Late Cenozoic Environments and Hominid Evolution: A Tribute to Bill Bishop.* Geological Society of London, London.

McKenna, M. C., and S. K. Bell. 1997. *Classification of Mammals above the Species Level*. Columbia University Press, New York, 631 pp.

Mein, P. 1994. Micromammifères du Miocène supérieur et du Pliocène du rift occidental, Ouganda; pp. 187–193 in B. Senut, and M. Pickford (eds.), *Geology and Paleobiology of the Albertine Rift Valley, Uganda-Zaïre. Vol. II. Palaeobiology*. CIFEG Occasional Publications, Orléans 29.

Mein, P., and M. Pickford. 1992. Gisements karstiques pléistocènes au Djebel Ressas, Tunisie. *Comptes Rendus de l'Académie des Sciences, Paris*, Série II, 315:247–253.

———. 2003. Rodentia (other than Pedetidae) from the Orange River deposits, Namibia. *Memoirs of the Geological Survey of Namibia* 19:147–160.

———. 2006. Late Miocene micromammals from the Lukeino Formation (6.1 to 5.8 Ma), Kenya. *Bulletin et Mémoires de la Société Linnéen de Lyon* 75:183–223.

Mein, P., and B. Senut. 2003. The Pedetidae from the Miocene site of Arrisdrift (Namibia). *Memoirs of the Geological Survey of Namibia* 19:161–170.

Mein, P., M. Pickford, and B. Senut. 2000a. Late Miocene micromammals from the Harasib karst deposits, Namibia: Part 1. Large muroids and non-muroid rodents. *Communications of the Geological Survey of Namibia* 12:375–390.

———. 2000b. Late Miocene micromammals from the Harasib karst deposits, Namibia. Part 2a–Myocricetodontinae, Petromyscinae and Namibimyinae (Rodentia, Gerbillidae). *Communications of the Geological Survey of Namibia* 12:391–401.

———. 2004. Late Miocene micromammals from the Harasib karst deposits, Namibia: Part 2b. Cricetomyidae, Dendromuridae and Muridae, with an addendum on the Myocricetodontinae. *Communications of the Geological Survey of Namibia* 13:43–61.

Mennell, F. P., and E. C. Chubb. 1907. On an African occurrence of fossil Mammalia associated with stone implements. *Geological Magazine* 5:443–448.

Mercer, J. M., and V. L. Roth. 2003. The effects of Cenozoic global change on squirrel phylogeny. *Science* 299:1568–1572.

Michaux, J., A. Reyes, and F. Catzeflis. 2001. Evolutionary history of the most speciose mammals: molecular phylogeny of muroid rodents. *Molecular Biology and Evolution* 18:2017–2031.

Misonne, X. 1969. African and Indo-Australian Muridae evolutionary trends. *Annales du Musée Royal d'Afrique Centrale, Tervuren, Belgique* 172:1–219.

Montgelard, C., S. Bentz, C. Tirard, O. Verneau, and F. M. Catzeflis. 2002. Molecular systematics of Sciurognathi: the mitochrondrial cytochrome b and 12S rRNA genes support the Anomaluroidea (Pedetidae and Anomaluridae). *Molecular Phylogenetics and Evolution* 22:220–233.

Munthe, J. 1987. Small-mammal fossils from the Pliocene Sahabi Formation of Libya; pp. 135–144 in N. T. Boaz, A. El-Arnauti, A. W. Gaziry, J. de Heinzelin, and D. D. Boaz (eds.), *Neogene Paleontology and Geology of Sahabi*. Liss, New York.

Musser, G. G. 1987. The occurrence of *Hadromys* (Rodentia: Muridae) in early Pleistocene Siwalik strata in northern Pakistan and its bearing on biogeographic affinities between Indian and northeastern African murine faunas. *American Museum Novitates* 2883:1–36.

Musser, G. G., and M. D. Carleton. 2005. Superfamily Muroidea; pp. 894–1531 in D. E. Wilson and D. M. Reeder (eds.), *Mammal Species of the World*. 3rd ed. Johns Hopkins University Press, Baltimore.

Ndessokia, P. N. S. 1989. The mammalian fauna and archaeology of the Ndolanya and Olpiro Beds, Laetoli, Tanzania. Unpublished PhD dissertation, University of California, Berkeley, 203 pp.

Nowak, R. M. 1999. *Walker's Mammals of the World*. 6th ed. Johns Hopkins University Press, Baltimore, 1936 pp.

Odhiambo Nengo, I., and T. C. Rae. 1992. New hominoid fossils from the early Miocene site of Songhor, Kenya. *Journal of Human Evolution* 23:423–429.

Pickford, M. 1988. Geology and fauna of the middle Miocene hominoid site at Muruyur, Baringo District, Kenya. *Human Evolution* 3:381–390.

———. 1990. Some fossiliferous Plio-Pleistocene cave systems of Ngamiland, Botswana. *Botswana Notes and Records* 22:1–15.

Pickford, M., T. Fernandes, and S. Aço. 1990. Nouvelles découvertes de remplissages de fissures à primates dans le Planalto da Humpata, Huila, Sud de l'Angola. *Comptes Rendus de l'Académie des Sciences, Paris* 310:843–848.

Pickford, M., and P. Mein. 1988. The discovery of fossiliferous Plio-Pleistocene cave fillings in Ngamiland, Botwana. *Comptes Rendus de l'Académie des Sciences, Paris* 307:1681–1686.

———. 2006. Early middle Miocene mammals from Moroto II, Uganda. *Beiträge zur Paläontologie* 30:361–386.

Pickford, M., P. Mein, and B. Senut. 1992. Primate bearing Plio-Pleistocene cave deposits of Humpata, Southern Angola. *Human Evolution* 7:17–33.

———. 1994. Fossiliferous Neogene karst fillings in Angola, Botswana, and Namibia. *South African Journal of Science* 90:227–230.

Pickford, M., B. Senut, D. Hadoto, J. Musisi, and C. Kaiira. 1986. Nouvelles découvertes dans le Miocène inférieur de Napak Ouganda Oriental. *Comptes Rendus de l'Académie des Sciences, Paris*, Série II, 302:47–52.

Pickford, M., H. Wanas, P. Mein, and H. Soliman. 2008. Humid conditions in the Western Desert of Egypt during the Vallesian (Late Miocene). *Bulletin, Tethys Geological Society, Cairo* 3:63–79.

Pocock, T. N. 1976. Pliocene mammalian microfauna from Langebaanweg: a new fossil genus linking the Otomyinae with the Murinae. *South African Journal of Science* 72:58–60.

———. 1985. Plio-Pleistocene mammalian microfauna in southern Africa. *Annals of the Geological Survey of South Africa* 19:65–67.

———. 1987. Plio-Pleistocene mammalian microfauna in southern Africa: A preliminary report including description of two new fossil muroid genera (Mammalia: Rodentia). *Palaeontologia Africana* 26:69–91.

Rasmussen, D. T., and M. Gutiérrez. 2009. A mammalian fauna from the Late Oligocene of northwestern Kenya. *Palaeontographica Abt. A* 288:1–52.

Remy, J.-A., and M. Benammi. 2006. Présence d'un Gomphotheriidae indet. (Proboscidea, Mammalia) dans la faune vallésienne d'Afoud AF6 (Bassin d'Aït Kandoula, Maroc), établie d'après la microstructure de l'émail d'un fragment de molaire. *Geobios* 39:555–562.

Robbins, L. H., M. L. Murphy, N. J. Stevens, G. A. Brook, A. H. Ivester, K. A. Haberyan, R. G. Klein, R. Milo, K. M. Stewart, D. G. Matthiesen, and A. J. Winkler. 1996. Paleoenvironment and archaeology of Drotsky's Cave: Western Kalahari Desert, Botswana. *Journal of Archaeological Science* 23:7–22.

Robinson, P., and C. C. Black. 1973. A small Miocene faunule from near Testour, Beji gouvernorat, Tunisia. *Annales des Mines et de la Géologie, Tunis* 26:445–449.

Robinson, P., C. C. Black, L. Krishtalka, and M. R. Dawson. 1982. Fossil small mammals from the Kechabta Formation, northwestern Tunisia. *Annals of Carnegie Museum* 51:231–249.

Rosevear, D. R. 1969. *The Rodents of West Africa*. The Trustees of the British Museum (Natural History), London, 604 pp.

Sabatier, M. 1978. Un nouveau *Tachyoryctes* (Mammalia, Rodentia) du basin Pliocène de Hadar (Éthiopie). *Géobios* 11:95–99.

———. 1979. Les Rongeurs des sites à hominidés de Hadar et Melka-Kunture (Ethiopie). *Thèse 3è cycle, Université des Sciences et techniques du Languedoc, Montpellier II, France*:66–85.

———. 1982. Les rongeurs du site Pliocène à hominidés de Hadar (Ethiopie). *Paleovertebrata* 12:1–56.

Savage, R. J. G. 1990. The African dimension in European early Miocene mammal faunas; pp. 587–600 in E. H. Lindsay, V. Fahlbusch, and P. Mein (eds.), *European Neogene Mammal Chronology*. NATO ASI Series, Series A, Life Sciences. Plenum Press, New York.

Schunke, A. C. 2005. Systematics and biogeography of the African scaly-tailed squirrels (Mammalia: Rodentia: Anomaluridae). Unpublished PhD dissertation, Mathematisch-Naturwissenschaftlichen Fakultat der Rheinischen Friedrich-Wilhelms-Universität Bonn, Germany, 171 pp.

Schweitzer, F. R., and M. L. Wilson. 1982. Byneskranskop 1: A late Quaternary living site in the southern Cape Province, South Africa. *Annals of the South African Museum* 88:1–203.

Scott, L., and J. S. Brink 1992. Quaternary palaeoenvironments of pans in central South Africa: Palynological and palaeontological evidence. *South African Geographer* 19:22–34.

Sen, S. 1977. *Megapedetes aegaeus* nov. sp. (Pedetidae) et à propos d'autres rongeurs africains dans le Miocène d'Anatolie. *Geobios* 10:983–986.

Sénégas, F. 1996. Introduction à l'étude des faunes de rongeurs du Plio-Pleistocène du sud de l'Afrique: Analyse d'un échantillon de brèches fossilifères de gisements du Transvaal (S. Af.) et du site de Friesenberg (Namibie). Diplôme d'Etudes Aprofondies (DEA) Paléontologie, Monpellier II, Montpellier, France.

———. 2000. Les faunes de rongeurs (Mammalia) plio-pléistocène de la province de Gauteng (Afrique du Sud): Mises au point et apports systématiques, biochronologiques et précisions paléoenvironnementales. Unpublished PhD dissertation, Université de Montpellier II, Montpellier, France, 2 vols., 231 pp.

———. 2001. Interpretation of the dental pattern of the South African fossil *Euryotomys* (Rodentia, Murinae, Otomyini) and origin of otomyine dental morphology; pp. 151–160 in C. Denys, A. Poulet, and L. Granjon (eds.), *African Rodents: Proceedings of the 8th ASM Symposium, Collection Colloques et Seminaires*. IRD Editions, Paris.

———. 2004. A new species of *Petromus* (Rodentia, Hystricognatha, Petromuridae) from the early Pliocene of South Africa and its paleoenvironmental implications. *Journal of Vertebrate Paleontology* 24:757–763.

Sénégas, F., and D. M. Avery. 1998. New evidence for the murine origin of the Otomyinae (Mammalia, Rodentia) and the age of Bolt's Farm. *South African Journal of Science* 94:503–507.

Sénégas, F., F. Laudet, and J. Michaux. 1999. Recent and fossil micromammal faunas from the area of Drimolen (Gauteng, South Africa): Paleoenvironmental implications. INQUA 1999, poster session P6, abstract.

Sénégas, F., and J. Michaux. 2000. *Boltimys broomi* gen. nov. sp. nov. (Rodentia, Mammalia), nouveau Muridae d'affinité incertaine du Pliocène inférieur d'Afrique du Sud. *Comptes Rendus de l'Académie des Sciences, Paris, Sciences de la Terre et des Planètes* 330:521–525.

Senut, B., M. Pickford, P. Mein, G. Conroy, and J. Van Couvering. 1992. Discovery of 12 new Late Cainozoic fossiliferous sites in paleokarst of the Otavi Mountains, Namibia. *Comptes Rendus de l'Académie des Sciences, Paris*, Série II, 314:727–733.

Slaughter, B. H., and G. T. James. 1979. *Saidomys natrunensis*, an arvicanthine rodent from the Pliocene of Egypt. *Journal of Mammalogy* 60:421–425.

Steppan, S. J., B. L. Storz, and R. S. Hoffmann. 2004. Nuclear DNA phylogeny of the squirrels (Mammalia: Rodentia) and the evolution of arboreality from c-myc and RAG1. *Molecular Phylogenetics and Evolution* 30:703–719.

Stevens, N. J., M. D. Gottfried, P. M. O'Connor, E. M. Roberts, and S. D. Ngasala. 2004. A new Paleogene fauna from the East African Rift, southwestern Tanzania. *Journal of Vertebrate Paleontology* 24 (suppl. 3):118A.

Stevens, N. J., P. A. Holroyd, E. M. Roberts, P. M. O'Connor, and M. D. Gottfried. 2009. *Kahawamys mbeyaensis* (n. gen., n. sp.) (Rodentia: Thryonomyoidea) from the late Oligocene Rukwa Rift Basin, Tanzania. *Journal of Vertebrate Paleontology* 29:631–634.

Stevens, N. J., P. M. O'Connor, M. D. Gottfried, E. M. Roberts, and S. Ngasala, and M. R. Dawson. 2006. *Metaphiomys* (Rodentia: Phiomyidae) from the Paleogene of southwestern Tanzania. *Journal of Paleontology* 80:407–410.

Stromer, E. 1922. Erste mitteilung über tertiäre Wirbeltier-Reste aus Deutsch-Südwestafrika. *Sitzungsberichte der Bayerischen Akademie der Wissenschaften* 1921:331–340.

———. 1924. Ergebnisse der Bearbeitung mitteltertiärer Wirbeltier-Reste aus Deutsch-Südwestafrika. *Sitzungsberichte der Bayerischen Akademie der Wissenschaften München* 1923:253–270.

———. 1926. Reste land- und süsswasser-bewohnender Wirbeltiere aus den Diamantfeldern Deutsch-Südwestafrikas; pp. 107–153 in E. Kaiser (ed.), *Die Diamantenwuste Südwest-Afrikas*, vol. 2. Dietrich Reimer (Ernst Volsen), Berlin.

Stuart, C., and T. Stuart. 2001. *Field Guide to the Mammals of Southern Africa*. 3rd ed. Struik Publishers, Cape Town, South Africa, 272 pp.

Tchernov, E., L Ginsburg, P. Tassy, and N. F. Goldsmith. 1987. Miocene mammals of the Negev (Israel). *Journal of Vertebrate Paleontology* 7:284–310.

Thackeray, J. F., and V. Watson. 1994. A preliminary account of faunal remains from Plover's Lake. *South African Journal of Science* 90:231–233.

Thomas, H., S. Sen, M. Khan, B. Battail, and G. Ligabue. 1982. Part IIIa. The Lower Miocene fauna of Al-Sarrar (Eastern province, Saudi Arabia). *ATLAL, Journal of Saudi Arabian Archaeology* 5:109–136.

Tobien, H. 1968. Paläontologische Ausgrabungen nach jungtertiären Wirbeltieren auf der Insel Chios (Griechenland) und Maragheh (NW-Iran). *Jahrbuch, Vereinigung Freunde der Universität Mainz* 1968:51–58.

Tong, H. 1986. The Gerbillinae (Rodentia) from Tighenif (Pleistocene of Algeria) and their significance. *Modern Geology* 10:197–214.

———. 1989. Origine et évolution des Gerbillidae (Mammalia, Rodentia) en Afrique du Nord. *Mémoires de la Société Géologique de France*, n.s., 155:1–120.

Tong, H., and J.-J. Jaeger. 1993. Muroid rodents from the middle Miocene Fort Ternan locality (Kenya) and their contribution to the phylogeny of muroids. *Palaeontographica, Abt. A* 229:51–73.

Turner, A., L. C. Bishop, C. Denys, and J. K. McKee. 1999. Appendix: A locality-based listing of African Plio-Pleistocene mammals; pp. 369–399 in T. G. Bromage and F. Schrenck (eds.), *African Biogeography, Climate Change, and Human Evolution*. Oxford University Press, New York.

Van Daele, P. A. A. G., E. Verheyen, M. Brunain, and D. Adriaens. 2007. Cytochrome *b* sequence analysis reveals differential molecular evolution in African mole-rats of the chromosomally hyperdiverse genus *Fukomys* (Bathyergidae, Rodentia) from the Zambezian region. *Molecular Phylogenetics and Evolution* 45:142–157.

Van Weers, D. J. 2005. A taxonomic revision of the Pleistocene *Hystrix* (Hystricidae, Rodentia) from Eurasia with notes on the evolution of the family. *Contributions to Zoology* 74:301–312.

Vianey-Liaud, M., and J.-J. Jaeger. 1996. A new hypothesis for the origin of African Anomaluridae and Graphiuridae (Rodentia). *Palaeovertebrata* 25:349–358.

Vianey-Liaud, M., J.-J. Jaeger, J.-L. Hartenberger, and M. Mahboubi. 1994. Les rongeurs de l'Eocene d'Afrique nord-occidentale-Glib Zegdou (Algérie) et Chambi (Tunesie) et l'origine des Anomaluridae. *Palaeovertebrata* 23:93–118.

Vignaud, P., P. Duringer, H. T. Mackaye, A. Likius, C. Blondel, J. -R. Boisserie, L. de Bonis, V. Eisenmann, M. -E. Etienne, D. Geraads, F. Guy, T. Lehmann, F. Lihoreau, N. Lopez-Martinez, C. Mourer-Chauviré, O. Otero, J. -C. Rage, M. Schuster, L. Viriot, A. Zazzo, and M. Brunet. 2002. Geology and palaeontology of the Upper Miocene Toros-Menalla hominid locality, Chad. *Nature* 418:152–155.

Watson, V. 1993. Composition of the Swartkrans bone accumulations, in terms of skeletal parts and animals represented; pp. 35–73 in C. K. Brain (ed.), *Swartkrans: A Cave's Chronicle of Early Man*. Transvaal Museum Monograph 8.

Wesselman, H. B. 1984. The Omo micromammals: Systematics and paleoecology of early man sites from Ethiopia. *Contributions to Vertebrate Evolution* 7:1–219.

Wesselman, H. B., M. T. Black, and M. Asnake. 2009. Small mammals; pp. 105–133 in Y. Haile-Selassie and G. WoldeGabriel (eds.), *Ardipithecus kadabba: Miocene Evidence from the Middle Awash, Ethiopia*. University of California Press, Berkeley.

Wessels, W., O. Fejfar, P. Peláez-Campomanes, A. van der Meulen, and H. de Bruijn. 2003. Miocene small mammals from Jebel Zelten, Libya. *Coloquios de Paleontología*, suppl. 1:699–715.

Wilson, D. E., and D. M. Reeder (eds.). 2005. *Mammal Species of the World: A Taxonomic and Geographic Reference*. 3rd ed, vol. 1. Johns Hopkins University Press, Baltimore, 743 pp.

Winkler, A. J. 1990. Systematics and biogeography of Neogene rodents from the Baringo District, Kenya. Unpublished PhD dissertation, Southern Methodist University, Dallas, 172 pp.

———. 1992. Systematics and biogeography of middle Miocene rodents from the Muruyur Beds, Baringo District, Kenya. *Journal of Vertebrate Paleontology* 12:236–249.

———. 1997. Systematics, paleobiogeography, and paleoenvironmental significance of rodents from the Ibole Member, Manonga Valley, Tanzania; pp. 311–332 in T. Harrison (ed.), *Neogene Paleontology of the Manonga Valley, Tanzania: A Window into the Evolutionary History of East Africa*. Plenum Press, New York.

———. 1998. A new dendromurine (Rodentia: Muridae) from the middle Miocene of western Kenya; pp. 91–104 in Y. Tomida, L. J. Flynn, and L. L. Jacobs (eds.), *Papers in Vertebrate Paleontology and Chronology in Honor of Everett H. Lindsay*. National Science Museum Monographs, Tokyo.

———. 2002. Neogene paleobiogeography and East African paleoenvironments: Contributions from the Tugen Hills rodents and lagomorphs. *Journal of Human Evolution* 42:237–256.

———. 2003. Rodents and lagomorphs from the Miocene and Pliocene of Lothagam, northern Kenya; pp. 169–198 in M. G. Leakey and J. M. Harris (eds.), *Lothagam, the Dawn of Humanity in Eastern Africa*. Columbia University Press, New York.

Winkler, A. J., L. MacLatchy, and M. Mafabi. 2005. Small rodents and a lagomorph from the early Miocene Bukwa locality, Eastern Uganda. *Palaeontologia Electronica* 8, issue 1, 24A:1–12.

WoldeGabriel, G., T. D. White, G Suwa, P. Renne, J. De Heinzelin, W. K. Hart, and G. Heiken. 1994. Ecological and temporal placement of early Pliocene hominids at Aramis, Ethiopia. *Nature* 371:330–333.

Wood, A. E. 1968. Early Cenozoic mammalian faunas, Fayum Providence, Egypt. Part II: The African Oligocene Rodentia. *Bulletin Yale Peabody Museum Natural History* 28:29–105.

Wood, A. E., and N. F. Goldsmith. 1998. Early Miocene rodents and lagomorphs from Israel. *Journal of Vertebrate Paleontology* 18 (suppl. to no. 3):57A.

Woods, C. A., and C. W. Kilpatrick. 2005. Infraorder Hystricognathi; pp. 1538–1600 in D. E. Wilson and D. M. Reeder (eds.), *Mammal Species of the World*. 3rd ed. Johns Hopkins University Press, Baltimore.

Xu, X., A. J. Winkler, and L. L. Jacobs. 1996. Is the rodent *Acomys* a murine? An evaluation using morphometric techniques; pp. 660–675 in K. M. Stewart and K. L. Seymour (eds.), *Palaeoecology and Palaeoenvironments of Late Cenozoic Mammals*. Tributes to the Career of C. S. (Rufus) Churcher. University of Toronto Press, Toronto.

Zazhigin, V. S., and A. V. Lopatin. 2000. The history of the Dipodoidea (Rodentia, Mammalia) in the Miocene of Asia: 3. Allactaginae. *Paleontological Journal* 34:553–565 (translated from *Paleontologicheskii Zhurnal* 5:82–94).

EIGHTEEN

Lagomorpha

ALISA J. WINKLER AND D. MARGARET AVERY

Our knowledge of fossil African lagomorphs has expanded tremendously since Lavocat's one-paragraph summary of the group in Maglio and Cooke's (1978) *Evolution of African Mammals*. With few exceptions, the first sentence of Lavocat's summary succinctly stated what was known at that time: "There is little to be said about the history of the Lagomorpha" (Lavocat, 1978:84). Although there have been many new discoveries since 1978, the group has still received relatively little attention. In many cases, even with Pleistocene remains (especially see table 18.3, occurrences for southern Africa), specimens are not described and are often identified only to higher taxonomic levels (e.g., Leporidae). The earliest records of leporids, in particular, are sparse and greatly need detailed comparative study.

Methods

Premolar morphology is usually considered the most important criterion for taxonomic assignment of fossil lagomorphs. Thus, dental terminology for p3 and P3 for ochotonids and p3 and P2 for leporids is illustrated in figure 18.1. Tooth terminology is after López-Martínez and Thaler (1975) for ochotonids and White (1991) and López-Martínez et al. (2007; see their summary of the correspondence in tooth terminology among different authors) for leporids.

A phylogenetic analysis of the sister group relationships within the Lagomorpha is outside the scope of this chapter, especially since the relatively poorly studied African fossil lagomorphs can, at this point, add little to prior hypotheses. Systematics of the Lagomorpha is still controversial, and the reader is referred to relationships proposed by other authors, for example, based on morphology (e.g., Hibbard, 1963; Dawson, 1981; Corbet, 1983; Averianov, 1999 [using morphological, one geographic, and one karotypic character]), supertree analysis (Stoner et al., 2003), and molecular supermatrix analysis (Matthee et al., 2004; Robinson and Matthee, 2005). Of special pertinence to African leporids, it is noteworthy that morphological studies support a sister taxon relationship between *Pronolagus* and *Bunolagus* (e.g., Corbet, 1983; Averianov, 1999), but supermatrix analysis (Matthee et al., 2004; Robinson and Matthee, 2005) suggests phylogenetic affinities among *Poelagus* (considered the sister species to *Caprolagus* by Averianov, 1999) and *Pronolagus* (and the Asian *Nesolagus*), with *Bunolagus*, *Oryctolagus*, *Caprolagus*, and *Pentalagus* the

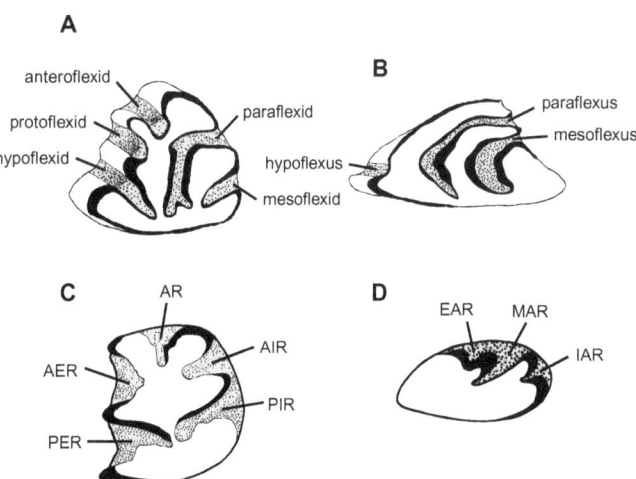

FIGURE 18.1 Tooth terminology for ochotonids and leporids. A and B) ochotonid (after López-Martínez and Thaler, 1975); A, left p3; B, left P3. C and D) leporid (after White, 1991; López-Martinez et al., 2007); C, left p3; D, right P2.

ABBREVIATIONS: for P2, EAR, external anterior reentrant (= mesoflexus); IAR, internal anterior reentrant (= hypoflexus); MAR, main anterior reentrant (= paraflexus). For p3, AER, anteroexternal reentrant (= protoflexid); AIR, anterointernal reentrant (= paraflexid); AR, anterior reentrant (= anteroflexid); PER, posteroexternal reentrant (= hypoflexid); PIR, posterointernal reentrant (= mesoflexid).

derived species within a clade also including *Brachylagus* and *Sylvilagus*. Supermatrix analysis does not provide support for the Palaeolaginae (*sensu* Dice, 1929, or Corbet, 1983).

As much as possible, the Neogene time scale used here follows Lourens et al. (2004), and the Pleistocene time scale follows Gibbard and Kolfschoten (2004). Note, however, that the "cutoff dates" for these time scales have changed with different authors, and the geologic age of taxa/localities used here may not match those given in the original publications.

Tables 18.1–18.3 give the known occurrences of fossil lagomorphs from northern (including Chad), eastern, and southern Africa, respectively.

305

Lagomorpha from northern Africa, Miocene–Pleistocene
Abbreviations: AL, Algeria; MO, Morocco; TU, Tunisia. E, early; M, middle; L, late.

Taxon	Synonymy	Locality	Stratigraphic horizon	Age	References
OCHOTONIDAE					
Kenyalagomys mellalensis	*Austrolagomys mellalensis* Mein and Pickford, 2003	Beni Mellal, MO	—	M. Miocene, ca. 14 Ma	Janvier and de Muizon, 1976
?*Kenyalagomys* sp.	*Austrolagomys* sp. Mein and Pickford, 2003	ATH7A3, Jebel Zelten, Libya	—	M. E. Miocene, 19–18 Ma	Wessels et al., 2003
Alloptox sp.	Ochotonidae indet. Savage, 1990	Measured Section 2, Jebel Zelten, Libya	Lower Marádah Fm	M. Miocene, 15–14 Ma	Wessels et al., 2003
Ochotonidae, gen. and sp. undet.	—	Testour, TU	—	M. Miocene, ca. 14 Ma	Robinson and Black, 1973
PROLAGIDAE					
Prolagus michauxi	—	Argoub Kemellal 1, AL	—	L. Miocene	Coiffait-Martin, 1991
P. michauxi and *P.* cf. *P. michauxi*	—	Afoud, MO	—	L. Miocene	Benammi, 1997; Benammi et al., 1995
P. cf. *P. michauxi*	—	Aghouri, MO	—	E. Pliocene, ca. 3 Ma	Benammi et al., 1995
Prolagus sardus	—	Lac Ichkeul and Bulla Regia I, TU	—	Plio-Pleistocene (Lac Ichkeul E. Pliocene; Benammi et al., 1995)	Mein and Pickford, 1992
Prolagus "sardus"	—	Djebel Ressas NE1, 5, 6, 8, TU	—	E. Pleistocene, ca. 1.6 Ma and < 1.6 Ma	Mein and Pickford, 1992
Prolagus sp.	—	Oued Mellague, TU	Kechabla Fm	L. Miocene	Robinson et al., 1982
		Ahl al Oughlam, MO	—	L. Pliocene, ca. 2.5 Ma	Geraads, 2006
LEPORIDAE					
Serengetilagus tchadensis	—	Toros Menalla, Chad	—	L. Miocene	Lpez-Martínez et al., 2007
Serengetilagus aff. *S. praecapensis*	—	Kossom Bougoudi, Chad	—	E. Pliocene, ca. 5 Ma	Brunet et al. 2000
Trischizolagus raynali	*Serengetilagus raynali* Geraads, 1994	Grotte des Rhinocéros, Oulad Hamida I, MO	—	M. Pleistocene	Geraads, 1994
Serengetilagus or *Trischizolagus* group	—	Ahl al Oughlam, MO	—	L. Pliocene, ca. 2.5 Ma	Geraads, 2006
Lepus cf. *L. capensis*	—	Djebel Ressas, Aïn Bahya, El Mahah, Ternifine, TU	—	Pleistocene, < 1.6 Ma	Mein and Pickford, 1992
	—	Grotte des Rhinocéros, Oulad Hamida I, MO	—	M. Pleistocene	Geraads, 1994
Lepus sp.	—	Ahl al Oughlam, MO	—	L. Pliocene, ca. 2.5 Ma	Geraads, 2006
Lagomorpha indet.	—	Chouf Aïssa, AL	—	M. Miocene	Coiffait-Martin, 1991
		Oued Smendou, AL	—	M. Miocene	
		Polygone 1, AL	—	L. Pliocene	
	—	Gabal el Muluk, Wadi el Natrun, Egypt	Stromer's Profile C	L. Miocene	James and Slaughter, 1974

TABLE 18.2
Lagomorpha from eastern Africa, Miocene–Pleistocene
Abbreviations: ET, Ethiopia; KE, Kenya; UG, Uganda; TA, Tanzania. E, early; M, middle; L, late.

Taxon	Synonymy	Locality	Stratigraphic horizon	Age	References
OCHOTONIDAE					
Kenyalagomys rusingae	*Austrolagomys rusingae* Mein and Pickford, 2003	Rusinga, KE	—	E. Miocene, 17.8 Ma	MacInnes, 1953
K. minor	*Austrolagomys minor* Mein and Pickford, 2003	Rusinga, Mfwangano, Karunga, KE	—	E. Miocene, ca. 17.8 Ma	MacInnes, 1953
	—	Kalodirr (West Kalokol), KE	—	E. Miocene, 18–16 Ma	Leakey and Leakey, 1986
Ochotonidae, gen. and sp. undet.	—	Bukwa, UG	Green clay unit	E. Miocene, 19.5-19.1 Ma	Winkler et al., 2005
LEPORIDAE					
Alilepus sp.	—	Lothagam, KE	Lower member, Nawata Fm	L. Miocene, 6.57–6.54 Ma	Winkler, 2003
	Leporidae, gen. and sp. nov. Winkler, 2002	Tugen Hills, KE	Lukeino Fm	L. Miocene, 6.2–5.6 Ma	Winkler, 2003
	?*Alilepus* sp. Haile-Salassie et al., 2004	Worku Hassan, ET	Haradaso Member, Sagantole Fm	E. Pliocene, 4.85 Ma	Wesselman, et al., 2008
cf. *Alilepus* sp.	—	Lemudong'o, KE	—	Late Miocene, ca. 6.0 Ma	Darwent, 2007
Serengetilagus praecapensis	—	BIK, Saiture Dora, Alayla, ET	Adu-Asa Fm	Late Miocene, 5.8–5.2 Ma	Wesselman, et al., 2008
	—	Lothagam, KE	Apak Member, Nachukui Fm	E. Pliocene, 4.22–4.20 Ma	Winkler, 2003
	—	Laetoli, TA	Upper Ndolanya Beds, Laetolil Beds	M.-L. Pliocene, 3.5–2.4 Ma E.-M. Pliocene, 3.8–3.5 Ma	Dietrich, 1941, 1942; Erbaeva and Angermann, 1983; Davies, 1987
Serengetilagus sp.	—	Aragai, Kapcheberek, Kapsomin, Tugen Hills, KE	Lukeino Fm	L. Miocene, 6.1–5.8 Ma	Mein and Pickford, 2006
cf. *Serengetilagus* sp.	—	Kanam West, KE	—	Pliocene	Flynn and Bernor, 1987
Serengetilagus sp. and/or *Lepus* sp.	—	Olduvai Gorge, TA	Bed I	E. Pleistocene	Leakey, 1965; Leakey, 1971
Lepus capensis	—	Omo Valley, ET	Member E, lower Members F, G; Shungura Fm.	L. Pliocene; ca. 2.08–1.98 Ma	Wesselman, 1984
	—	Porc-Epic Cave, ET	—	L. Pleistocene	Assefa, 2006
Lepus veter	—	Chianda-Uyoma and Kanjera, KE	—	Upper M. Pleistocene	MacInnes, 1953
"*Lepus*" sp.	—	Kanapoi, KE	Kanapoi Fm	E. Pliocene, 4.17–4.07 Ma	Manthi, 2006
	—	Laetoli, TA	Upper Ndolanya Beds	M. Pliocene, 2.6 Ma	Ndessokia, 1990
"?*Lepus*" sp.	—	Middle Awash, ET	Adu-Asa Fm	L. Miocene, 5.8–5.2 Ma	Haile-Salassie et al., 2004
Leporidae, gen. and sp. indet.	—	Tugen Hills, KE	Mpesida Beds	L. Miocene, 7–6.2 Ma	Winkler, 2002
Lagomorpha indet.	—	Koobi Fora, KE	Koobi Fora Fm	L. Pliocene	Harris, 1978
	—	Natoo Member assemblage, KE	Natoo Member, Nachukui Fm	E. Pleistocene, ca. 1.6–1.33 Ma	Harris et al., 1988

TABLE 18.3
Lagomorpha from southern Africa, Miocene–Pleistocene
Abbreviations: AN, Angola; BO, Botswana; NA, Namibia; SA, South Africa; ZA, Zambia; ZI, Zimbabwe. E, early; M, middle; L, late.

Taxon	Synonomy	Locality	Stratigraphic horizon	Age	References
OCHOTONIDAE					
Austrolagomys hendeyi	Kenyalagomys sp. nov. Hendey, 1978	Arrisdrift, NA	Pit 2/ AD8	M. Miocene, 18–14 Ma	Mein and Pickford, 2003
A. inexpectatus	A. simpsoni Hopwood, 1929; K. simpsoni Hendey, 1978	Elisabethfeld, Langental, Grillental, NA	—	E. Miocene	Stromer 1924, 1926; Hopwood, 1929; Mein and Pickford, 2003
LEPORIDAE					
Pronolagus cf. P. randensis	—	Makapansgat, SA Cave of Hearths, SA	—	E. and M. Pleistocene	Cooke, 1963
Pronolagus sp.	—	Langebaanweg, SA	Varswater Fm	E. Pliocene	Hendey, 1981
	—	Makapansgat, SA	EXQRM, MRCIS	Pliocene, ca. 3 Ma	Pocock, 1987
	—	Sterkfontein, SA	STS/Dumps 1, 2 and 8, STW/H2	E. Pleistocene	Pocock, 1987
Serengetilagus sp.	—	Cangalongue 1, AN	—	Plio-Pleistocene; ca. 1.8–1.3 Ma	Pickford et al., 1992
	—	Molo, AN	—	Plio-Pleistocene	Pickford et al., 1992
	—	Tchiua, AN	pink breccia	Plio-Pleistocene	Pickford et al., 1992
Lepus capensis	L. cf. L. capensis Klein, 1976b	Duinefontein 2, SA	Horizon 2	M. Pleistocene	Klein et al., 1999
	—	Blombos Cave, SA	Phases 1–3	L. Pleistocene, <128 ka	Henshilwood et al., 2001
	—	Die Kelders Cave, SA	MSA levels	L. Pleistocene	Grine et al., 1991
	—	Klasies River Mouth 1, SA	Horizon 14	L. Pleistocene	Klein, 1976a
L. cf. L. capensis	—	Bolt's Farm, SA	—	Plio-Pleistocene	Cooke, 1963
	—	Cave of Hearths, SA	—	M. Pleistocene	Cooke, 1963
	Lepus sp. Brink, 1987	Florisbad, SA	Spring, MSA	M. Pleistocene	Scott and Brink, 1992
	—	Border Cave, SA	passim	L. Pleistocene	Klein, 1977
	Lepus sp. Klein, 1972	Nelson Bay Cave, SA	—	L. Pleistocene, 12–14 ka	Klein, 1974
L. saxatilis	—	Blombos Cave, SA	Phases 1–3	L. Pleistocene, <128 ka	Henshilwood et al., 2001
L.? saxatilis	?Lepus crawshayi	Border Cave, SA	passim	L. Pleistocene	Klein, 1977
Lepus sp.	—	Gcwihaba and Nqumtsa, BO	—	Plio-Pleistocene	Pickford, 1990
	—	Rocky II, Kaokoland, NA	—	Plio-Pleistocene	Pickford et al., 1993
	—	Elandsfontein, SA	—	M. Pleistocene	Klein, 1974
cf. Lepus	—	Kromdraai B, SA	Layers 1 and 3	Plio-Pleistocene, 1–2 Ma	Brain, 1981
Gen. et sp. indet.	—	Jägersquelle 1, NA	—	Plio-Pleistocene	Senut et al., 1992
	L. cf. L. capensis Cooke, 1963	Taung, SA	—	Plio-Pleistocene	Cooke, 1990
	—	Kromdraai, SA	—	E. Pleistocene	Cooke, 1990
	L. cf. L. capensis Cooke, 1963	Swartkrans, SA	passim	E. Pleistocene	Watson, 1993
	—	Equus Cave, SA	—	M. Pleistocene	Klein et al., 1991
	—	Wonderwerk Cave, SA	—	M. Pleistocene	Klein, 1988
	—	Mumbwa Caves, ZA	—	M. Pleistocene, <130 ka	Klein and Cruz-Uribe, 2000

Taxon	Synonomy	Locality	Stratigraphic horizon	Age	References
Gen. et sp. indet. continued	—	Pomongwe, ZI	—	M. and L. Pleistocene	Brain, 1981
	Lepus sp. Klein, 1981	Byneskranskop, SA	—	L. Pleistocene	Schweitzer and Wilson, 1982
	—	Redcliff Cave, ZI	—	L. Pleistocene	Klein, 1978
	—	Zebra Cave, NA	MSA levels	L. Pleistocene, >48 ka	Cruz-Uribe and Klein, 1983

ANATOMICAL ABBREVIATIONS

For P2, EAR, external anterior reentrant (= mesoflexus); IAR, internal anterior reentrant (= hypoflexus); MAR, main anterior reentrant (= paraflexus). For p3, AER, anteroexternal reentrant (= protoflexid); AIR, anterointernal reentrant (= paraflexid); AR, anterior reentrant (= anteroflexid); PER, posteroexternal reentrant (= hypoflexid); PIR, posterointernal reentrant (= mesoflexid).

Distribution of Extant Taxa

Ochotonids are not present in the extant African fauna. Worldwide, this family is represented today by only one genus, *Ochotona*, restricted to Eurasia and western North America (Hoffmann and Smith, 2005). Prolagids are also not present in the extant African fauna, nor are they found currently anywhere else in the world. They may have survived *(Prolagus sardus)* until the late 18th century on Mediterranean islands (Dawson, 1969, and see summary in Hoffmann and Smith, 2005). Leporids, however, are known at present from most of Africa (Kingdon, 1974). There are five genera: *Bunolagus* (bushman or riverine rabbit), *Lepus* (hare), *Oryctolagus* (European, Old World, or domestic rabbit), *Poelagus* (bunyoro rabbit, Central African rabbit, or Uganda grass-hare), and *Pronolagus* (red hare or red rock hare).

BUNOLAGUS THOMAS, 1929

This monotypic genus includes only *B. monticularis*, which is known from the central Karoo, almost exclusively in the Northern Cape Province, South Africa (Duthie and Robinson, 1990; Friedmann and Daly, 2004; Hoffmann and Smith, 2005). *Bunolagus* are very rare and have an extremely restricted distribution: they occur in dense bush along seasonal rivers (Duthie and Robinson, 1990; Nowak, 1999). The presence of *Bunolagus* in the fossil record is uncertain.

PRONOLAGUS LYON, 1904

Six species of this genus have been described, but three, *P. crassicaudatus*, *P. randensis*, and *P. rupestris* (Hoffmann and Smith, 2005), or four (the three listed plus *P. saundersiae* for southern Africa; Bronner et al., 2003) are now generally recognized. The genus is reported from southern Africa and the eastern Rift Valley, as far north as Kenya (Kingdon, 1974; Hoffmann and Smith, 2005). *Pronolagus* prefer rocky habitat with mixed dense bush and grass (Kingdon, 1974).

LEPUS LINNAEUS, 1758

The taxonomy of *Lepus* is complicated and remains highly controversial (Hoffmann and Smith, 2005). In Africa, the genus is widespread and is considered by Hoffmann and Smith (2005) to include at least six species. African *Lepus* are found in a variety of generally open habitats. Some species may be found in open, arid habitats, while others inhabit somewhat moister, more wooded savannas (Kingdon, 1974).

ORYCTOLAGUS LILLJEBORG, 1874

This genus is mentioned by Gibb (1990) after Lever (1985) as occurring in north-western Africa. *Oryctolagus* have not been reported formally from the fossil record of Africa.

POELAGUS ST. LEGER, 1932

Poelagus are restricted to parts of eastern and central Africa, with a disjunct population in Angola (Hoffmann and Smith, 2005). Only one species, *P. marjorita*, is recognized. It is found in moist savanna within "grasslands and *Isoberlina* woodlands in association with rocky outcrops and, to a lesser degree, with forest" (Duthie and Robinson, 1990:124). *Poelagus* are currently unknown from the fossil record.

Systematic Paleontology

Eight to ten genera of lagomorphs (the number depends on the status of *Kenyalagomys* and the fossil record of *Bunolagus*) are known from the fossil record of Africa. Of these, six to seven genera are extinct. The following section provides an overview of the systematics, known remains, and distribution in time and space of these fossils. For historical clarity, *Austrolagomys* and *Kenyalagomys* are discussed as distinct genera, although they are synonymized by Mein and Pickford (2003).

Order LAGOMORPHA Brandt, 1855
Family OCHOTONIDAE Thomas, 1897
Subfamily SINOLAGOMYINAE Gureev, 1960
Genus *AUSTROLAGOMYS* Stromer, 1926

Synonymy Austrolagomys Stromer, 1924 (original informal description).
Original diagnosis Translated from Stromer (1926:128). Size a little larger than *Ochotona alpina* (Pallas, 1773); lower incisor a circular equilateral triangle in cross section with enamel also on the sides; premolar and molar formula 3-2/2-3; P2 a simple "pencil" shape; P3 broadly triangular with two weak folds,

one external, one internal; P4–M2 becoming smaller toward M2, of similar structure—broadly square with long narrow middle inside fold; p3 square, broader than long, narrower anteriorly, small fold internally; p4–m2 formed by two transversely oval pillars separated inside and outside by a deep fold; barely broader than long; m3 a narrow transversely oval pillar.

Emended Diagnosis P2 reduced; P3 with single flexus (mesoflexus) and deep enamel fold anteroexternally; P4 molariform; lower incisor root extends backward to terminate as swelling on lingual surface of mandible near root end of m1; m3 unilobed; dental formula 2-0-3-2/1-0-2-3 (possibly a second genus with lower formula 1-0-3-3) (from emended diagnosis of Mein and Pickford, 2003:171, with *Kenyalagomys* synonymized with *Austrolagomys*).

Included Species A. *inexpectatus*, A. *hendeyi*.

African Age and Occurrence Early to middle Miocene, Namibia.

Material *Austrolagomys inexpectatus* and *A. hendeyi* are known from numerous incomplete cranial remains.

Remarks Stromer (1924, 1926) identified the first reported African ochotonid, *A. inexpectatus*, from the early Miocene of Elisabethfeld, Namibia. Hopwood (1929) described a second species, *A. simpsoni*, based on two incomplete mandibles from the same general area. Without stating their reasons, Hamilton and Van Couvering (1977) concluded that the two species of *Austrolagomys* were synonymous, thereby reversing their previous acceptance that there were two species (Van Couvering and Van Couvering, 1976). With access to larger samples, Mein and Pickford (2003) concluded that Hamilton and Van Couvering (1977) were correct that *A. simpsoni* is most likely synonymous with *A. inexpectatus*, but they upheld Hendey's (1978) finding that material from Arrisdrift, Namibia (middle Miocene), should be assigned to a separate species, which they named *A. hendeyi* (Mein and Pickford, 2003).

Martin (2004) studied the evolution of incisor enamel microstructure in lagomorphs. A lower incisor of *A. "simpsoni"* (the paratype; apparently from the same locality as the holotype) was included in his study. The incisor schmelzmuster of this specimen closely matched that of a specimen of *Sinolagomys* sp. indet. (another sinolagomyine ochotonid), except that the Hunter-Schreger bands of *A. "simpsoni"* were more distinct (had a higher angle of decussation; Martin, 2004).

Genus *KENYALAGOMYS* MacInnes, 1953

Diagnosis Dental formula 2, 0, 3, 2/1, 0, 2-3, 3; P3 trapezoidal in section with a deep V-shaped enamel fold on anteroexternal border, occupying most of the crown; p3 with deep enamel fold on anteroexternal border and shallow fold on anterior border; m3 simple.

Included Species K. *rusingae*, K. *minor*, and K. *mellalensis*.

African Age and Occurrence Early–middle Miocene, northern Africa (Morocco, possibly Libya and Tunisia); eastern Africa (Kenya).

Material Numerous, mostly fragmentary, cranial remains of *K. rusingae* and *K. minor* are known. There is a small sample of cranial and postcranial remains of *K. mellalensis*. Three isolated teeth of ?*Kenyalagomys* sp. from Libya and three isolated and broken teeth from Tunisia are reported.

Remarks Based on specimens from Kenya, MacInnes (1953) erected the genus *Kenyalagomys*, which he distinguished from *Austrolagomys* (represented only by *A. inexpectatus*) by five differences in the dentition. However, when examining new

material from Arrisdrift, Hendey (1978) noted that, among the characters used by MacInnes (1953) to distinguish *Kenyalagomys* from *Austrolagomys*, two of the five (a deep external fold on p3 and a marked median angulation of the posterior walls of the anterior lobes of p4–m2) were also used by Hopwood (1929) to distinguish what he considered to be two species of *Austrolagomys* (*A. inexpectatus* and *A. "simpsoni"*). MacInnes (1953) made no mention of the slightly smaller *A. "simpsoni"* Hopwood, 1929, of which he was apparently unaware. Hendey (1978) therefore determined that *A. inexpectatus* should remain the sole species of *Austrolagomys*, and that *A. "simpsoni"* should be transferred to *Kenyalagomys*. The Arrisdrift material, which was considered to be closest to *K. minor* and *K. simpsoni*, but more advanced than either, was identified as *Kenyalagomys* sp. nov. (Hendey, 1978). Whereas Hendey's (1978) acceptance of *Kenyalagomys* led him to assign the Arrisdrift material to this genus, Mein and Pickford (2003) believe that none of the characters used by MacInnes (1953) to separate the two genera are valid. Although they do not completely rule out the possibility that there may be a second genus of sinolagomyine in eastern Africa (because *K. minor* is diagnosed, in part, as sometimes having a rudimentary p2), they propose that *Kenyalagomys* should be considered a synonym of *Austrolagomys*. They provide an emended diagnosis for *Austrolagomys* based on a larger sample, (including new material from the type locality) than that available to Stromer (1926). *Kenyalagomys* sp. nov. Hendey, 1978, becomes synonymous with their new species, *A. hendeyi* Mein and Pickford, 2003.

Remains of sinolagomyines from northern Africa are fragmentary. ?*Kenyalagomys* sp. is from the middle early Miocene Gebel Zelten locality, Libya (Wessels et al., 2003). Mein and Pickford (2003) examined a cast of the P3 from this sample and assigned it to *Austrolagomys*. *Kenyalagomys mellalensis* is from Beni Mellal, Morocco (ca. 14 Ma; Janvier and de Muizon, 1976). Mein and Pickford (2003) suggest that this taxon should be transferred to *Austrolagomys*. Robinson and Black (1973) suggest that the isolated ochotonid teeth from Testour, Tunisia (ca. 14 Ma) may be related to *Kenyalagomys* or *Austrolagomys*.

Subfamily OCHOTONINAE Thomas, 1897
Genus *ALLOPTOX* Dawson, 1961

Diagnosis Dental formula (2), 0, 3, 2/1, 0, 2, 3; cheek teeth hypsodont; P3 crescentic fold connecting to anterobuccal wall; long internal hypostria on P4, M1, M2; p3 anteroexternal fold shallower than in *Ochotona*; long anterointernal fold extending posteroexternally and reaching farther posteriorly than in *Ochotona*; enamel on posterior and buccal walls of anterointernal fold thicker than on anterior wall. Trigonid and talonid of p4, m1, m2 each about equal in width; single column forming m3; two persistent, well-developed anterior folds on P2; posterior process from posteroloph of M2 absent.

Included African Taxa Ochotonidae indet. Savage, 1990 (original identification).

African Age and Occurrence Middle Miocene, Libya.

Material Isolated P3.

Remarks The only African record of *Alloptox* is an isolated P3, identified as *Alloptox* sp. It is considered most similar to *A. anatoliensis* from Turkey (late Miocene) based on presence of a wide paraflexus, well-pronounced metastyle, and a weak hypoflexus (Wessels et al., 2003). *Alloptox* are known primarily from China, Mongolia, and Kazakhstan (Erbajeva, 1994).

Family PROLAGIDAE Gureev, 1960
Genus *PROLAGUS* Pomel, 1853

Diagnosis Emended diagnosis from López-Martínez and Thaler (1975). Teeth with continuous crests lacking valleys; hypoconulid only on m2; p3 with three external grooves, from the addition of a protoconulid, and two internal grooves

Further Informal Attributes Continuously growing dentition; M3/m3 lacking; trilobed m2; p3 with a protoconulid and a rounded, generally isolated anteroconid (López-Martínez, 2001:216).

Included African Species *P. michauxi* and *P. sardus*.

African Age and Occurrence Late Miocene to the early Pleistocene (ca. 1.6 Ma), northern Africa (Morocco, Algeria, Tunisia).

Remarks For consistency, we follow Hoffmann and Smith (2005; who follow Erbajeva, 1988, and Gureev, 1960) in considering the prolagids a distinct family. Other authors place the prolagids within the family Ochotonidae (e.g., López-Martínez, 2001). The African record of the Prolagidae (i.e., *Prolagus*) is restricted to northern Africa: López-Martínez (2001) reports *Prolagus* from 18 (unspecified) localities. There are no formal published studies of African *Prolagus*.

Family LEPORIDAE Fischer de Waldheim, 1817
Subfamily PALAEOLAGINAE Dice, 1917
Genus *ALILEPUS* Dice, 1931
Figures 18.2A and 18.2B

Synonymy *Allolagus* Dice, 1929 (preoccupied).

Diagnosis Revised diagnosis after White (1991). Leporid of medium to large size with fully modernized cranium and dentary; P2 with deeply incised MAR and shallow EAR; p3 with PIR as deep or shallower than PER, PIR often pinched off to form an enamel lake, AIR shallower (usually missing) than PIR, AR absent, TH (thick enamel in the PER) smooth to slightly folded, AER shallow with smooth thin enamel.

African Age and Occurrence Late Miocene (Lothagam, 6.57–6.54 Ma; Tugen Hills, 6.2–5.6 Ma; Lemudong'o, ca. 6.0 Ma; Worku Hassan, 4.85 Ma), Kenya, Ethiopia.

Material Two incomplete skeletons of *Alilepus* sp. are reported from Lothagam, Kenya (figure 18.2B; Winkler, 2003). An isolated p3 of *Alilepus* sp. is from the Tugen Hills, Kenya (figure 18.2A). These remains may all belong to the

same species. The p3 of these specimens differs from those of typical *Alilepus* in having a deeper AER. There is a fragmentary maxilla from Lemudong'o, Kenya, and a specimen with maxillary and mandibular dentition from Worku Hassan, Middle Awash, Ethiopia.

Remarks The Lothagam and Tugen Hills specimens have not been described fully; preliminary discussion of the Lothagam specimens is in Winkler (2003). Haile-Selassie et al. (2004) list (but do not describe or illustrate) ?*Alilepus* in a composite faunal list from the late Miocene Adu-Asa and lower Sagantole formations, Middle Awash, Ethiopia. Upon reexamination of this material, Wesselman et al. (2009) assign a specimen with maxillary and mandibular dentition from the lower Sagantole Formation, Worku Hassan locality, to *Alilepus* sp. Assignment is based on p3 morphology, although the specimen has a small AR. Darwent (2007) attributes a maxillary fragment with P2–P4 from Lemudong'o, Kenya, to cf. *Alilepus*. Possession of a P2 with a deep MAR and shallow EAR suggests this assignment. The Lemudong'o fauna includes an additional 50 dental and (predominantly) postcranial remains of lagomorphs, but these are not specifically identifiable (Darwent, 2007).

Genus *BUNOLAGUS* Thomas, 1929

Diagnosis Thomas (1929). Limbs short, thick, heavily furred; tail bushy, cylindrical, uniformly colored; skull lacks "projecting shoulders" on front corners of zygoma; incisors barely visible from above in front of premaxillae; feet do not expand terminally but are of equal thickness throughout. Duthie and Robinson (1990) also note informally the presence of a dark brown stripe extending from the lower border of the jaw toward the base of the ear.

Included Species (Extant) *B. monticularis*.

African Age and Occurrence Recent; possibly middle Pleistocene (0.4 Ma), South Africa (Klein, 1984).

Remarks R. G. Klein (pers. comm., 2007) now considers that specimens he previously assumed to be *Bunolagus* (Klein, 1984) might be small representatives of *Lepus*, although positive identification has not yet been possible. This would have been the only fossil record of *Bunolagus*. It seems likely that *Bunolagus* has long, if not always, had a very restricted distribution in central South Africa.

Genus *PRONOLAGUS* Lyon, 1904

Diagnosis Lyon (1904) in part. Externally similar to *Lepus*; skull and teeth essentially like those of *Romerolagus* except that auditory bullae are smaller than foramen magnum and anterior face of p3 has two reentrant angles. P3 with all five reentrants present.

Included Species (Fossil Record Only) Tentative referral to *P. randensis*.

African Age and Occurrence Early Pliocene–Recent, South Africa.

Material There are numerous specimens from Langebaanweg, South Africa, representing most, if not all, body parts, which were ascribed to *Pronolagus* by Hendey (1981). For other reports, the material represented is not given.

Remarks Fossil *Pronolagus* is currently known only from southern Africa, where it is the earliest represented genus of leporid (early Pliocene). Hendey (1981) lists it from the early Pliocene at Langebaanweg, but the specimens have not been described. The only mention of a species is a tentative

FIGURE 18.2 Camera lucida drawings of *Alilepus* sp. from the late Miocene of Kenya. Drawings were made from casts of the specimens. A) KNM-TH 967/21219, right p3, from the Lukeino Formation, Tugen Hills, Kenya. B) KNM-LT 22999, right p3–p4, from the Lower Nawata Formation, Lothagam, Kenya.

referral to *P. randensis* from Cave of Hearths, South Africa (Pleistocene, Cooke, 1963). None of the other currently recognized species are recorded from the fossil record, although this may reflect the lack of work on the genus or a paucity of material, rather than a real absence of these species in the past. No extinct species of *Pronolagus* have been recognized.

Subfamily ARCHAEOLAGINAE Dice, 1917
Genus *SERENGETILAGUS* Dietrich, 1942
Figure 18.3

Synonymy Serengetilagus Dietrich, 1941:121 *(nomen nudum)*.

Diagnosis From López-Martínez et al. (2007). Archaeolaginae leporids with short snout, large incisive foramen, narrow choanae and relatively developed tympanic bullae; dentary increasing in height backwards; P2 with deep MAR (paraflexus), weak EAR (mesoflexus) and shallow IAR (hypoflexus) in advanced species; upper molariform teeth with well crenulated hypostria; p3 crescentic in shape with two main, constant external folds (PER [hypoflexid] extending

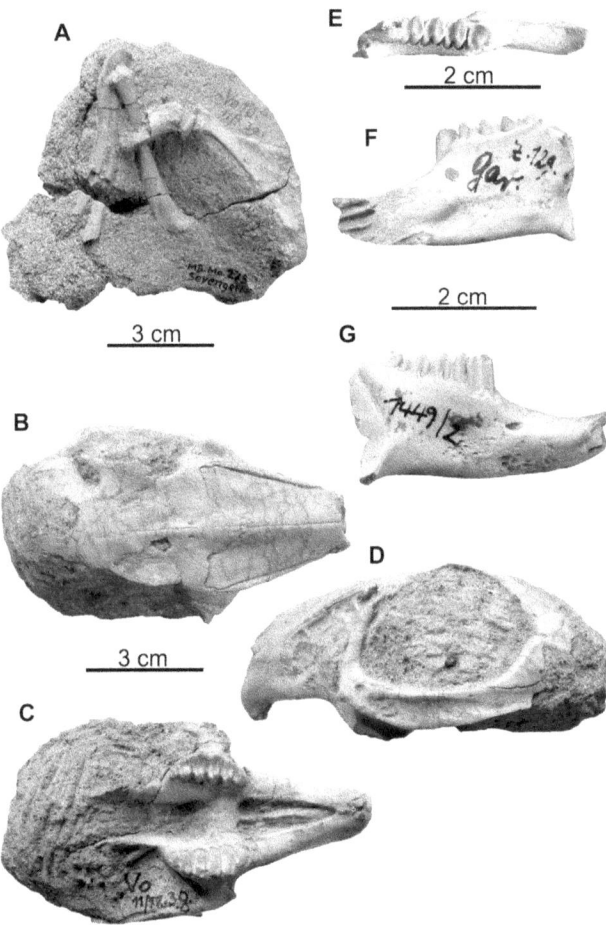

FIGURE 18.3 *Serengetilagus praecapensis* from the early Pliocene of Laetoli, Tanzania. Specimens from the Dietrich collection, Museum für Naturkunde, Berlin, Germany. A) MB.Ma 27543, block with articulated scapula and humerus, and articulated humerus, radius and ulna. B–D) MB.Ma 1447.1, skull, paralectotype of Erbaeva and Angermann (1983), in B, dorsal; C, ventral; and D, lateral views. E–G) MB.Ma 1449/2, right mandible with p3–m3, lectotype of Erbaeva and Angermann (1983), in E, occlusal; F, internal; and G. external views.

about halfway across the crown and shallow AER [protoflexid]) and up to three additional folds variably present (AR [anteroflexid] variably developed, weak AIR [paraflexid] and exceptionally a PIR-enamel lake [mesoflexid-mesofossetid], mainly in young individuals); when an AR (anteroflexid) is present, lingual anteroconid is weaker than the labial one.

Included African Species Serengetilagus praecapensis Dietrich, 1942 (type species); *S. tchadensis* (López-Martínez et al., 2007).

African Age and Occurrence Late Miocene to Pliocene, possibly to early Pleistocene, Chad, Kenya, Tanzania, Ethiopia, Angola.

Material The sample of *S. praecapensis* from Laetoli (type locality) includes thousands of specimens: isolated elements to incomplete skeletons are known (figure 18.3). *Serengetilagus praecapensis* from Lothagam is an incomplete left dentary with i–m3. Three isolated teeth and a mandible of *S. praecapensis* are from the Middle Awash, Ethiopia. A lower jaw of *S. aff. S. praecapensis* is reported from Kossom Bougoudi, Chad. *S. tchadensis* from Toros Menalla, Chad, is known from 18 numbered specimens, some including multiple elements, and some probably referable to the same individuals. Cranial and limited postcranial remains are present. An isolated p3 of cf. *Serengetilagus* sp. is from Kanam West, Kenya (there is also an isolated upper cheek tooth of a leporid from this locality). A minimum of 17 specimens of *Serengetilagus* sp. is reported from the Tugen Hills, Kenya. The referred material of *Serengetilagus* sp. from Olduvai Gorge and Angola is not listed.

Remarks The remains of *Serengetilagus* from Laetoli, Tanzania, are the most numerous and complete of any known fossil lagomorph from Africa. The type species, *S. praecapensis*, was originally described by Dietrich (1941, 1942) based on specimens collected by Kohl-Larsen in 1939 (housed at the Museum für Naturkunde, Humbolt University, Berlin). Unfortunately, Dietrich did not designate a holotype. MacInnes (1953) described material collected by Louis and Mary Leakey in 1935 and housed in the Museum of Natural History, London. Erbaeva and Angermann (1983) provided additional descriptions of Dietrich's cranial specimens and designated a lectotype (figures 18.3E–18.3G). They noted and illustrated that the p3 of this species shows much morphologic variability. Davies (1987) did a preliminary study of *S. praecapensis* collected from Laetoli (Upper Ndolanya and Laetolil Beds) by Mary Leakey and colleagues from 1974 to 1982, now housed at the National Museum, Dar es Salaam, Tanzania. Many additional remains of *Serengetilagus* have been collected from Laetoli by T. Harrison and colleagues and are currently under study.

Four specimens of *S. praecapensis* are illustrated and briefly described from the Adu-Asa Formation (5.8–5.2 Ma), Ethiopia (Wesselman et al., 2009). Winkler (2003) described a mandible of *S. praecapensis* from the early Pliocene (4.22–4.20 Ma) at Lothagam, Kenya. *S. aff. S. praecapensis* is reported from Kossom Bougoudi, northern Chad (ca. 5 Ma; Brunet et al., 2000). This material is described as being primitive with respect to *S. praecapensis* from Laetoli (Brunet et al., 2000).

Serengetilagus tchadensis is considered to have some of the more primitive character states of the genus such as a simpler p3 with only two main external reentrants and upper cheek teeth strongly widened transversely with wear (López-Martínez et al., 2007). These authors assign *Serengetilagus* to the Archaeolaginae and suggest that the lineage may be derived from *Hypolagus* stock (e.g., something like *H. gromovi*). López-Martínez et al.

(2007) suggest a mixed fossorial-cursorial mode of life for both species of the genus based on morphological traits and taphonomic indicators. Davies (1987) had suggested that *S. praecapensis* might have been a burrower.

Serengetilagus sp. has been reported from several other African localities. Flynn and Bernor (1987) suggested that an isolated leporid p3 from the Pliocene Kanam West locality, Kenya, was likely referable to *Serengetilagus*. Unstudied lagomorphs of the "*Serengetilagus-Trischizolagus* group" are reported from Ahl al Oughlam, Morocco (ca. 2.5 Ma; Geraads, 2006). In southern Africa, *Serengetilagus* is listed as occurring in the Plio-Pleistocene of southern Angola by Pickford et al. (1992), although these authors cast some doubt on the identification (Pickford et al., 1992:20).

Serengetilagus sp. was reported from the late Miocene of the Tugen Hills, Kenya, by Mein and Pickford (2006). Assignment of this material to *Serengetilagus* is, however, not definitive based on the descriptions given (there are no illustrations). A p3 from a young individual and a dp3 are described: the authors note that the PER of the young individual crosses the midline of the tooth. Two adult p3s are reported by Mein and Pickford (2006), but neither is described. A single dP2 described by Mein and Pickford (2006: the adult P2 is not described) has a single anterior reentrant. The generic diagnosis provided by López-Martínez et al. (2007; see previous discussion) includes P2 having two to three reentrants (the condition for juveniles is not discussed).

Leakey (1965) lists, and very briefly discusses, lagomorphs from the 1951–1961 excavations at Olduvai Gorge, Tanzania (early Pleistocene). Lagomorphs are reported only from Bed I, and although present in this unit they are described as uncommon. Leakey (1965:19) reports "a very large lagomorph, apparently belonging to the genus *Lepus*, as well as a lagomorph closely resembling *Serengetilagus*." The specimens are not described, and the only illustration is of a distal tibia. Leakey (1971) lists *Serengetilagus* sp. from the 1960–1963 excavations in Bed I; there is no discussion or illustration of the material.

Genus *TRISCHIZOLAGUS* Radulesco and Samson, 1967

Diagnosis Lower p3 divided unequally by two opposed posterior sinuses, the external is longer; isthmus of dentine separating the two rather wide sinuses; trigonid almost two times longer than talonid and showing anterior, anteroexternal, and anterointernal grooves, conferring a characteristic four-lobed form; enamel smooth. Lower incisor with median longitudinal groove on anterior face; posterior groove absent.

Included African Species Trischizolagus raynali.

African Age and Occurrence Middle Pleistocene, Grotte des Rhinocéros, Oulad Hamida I, Morocco.

Material Some P2s, mandible with p3–m3, 4 p3s, about 30 isolated molars, some mandibular fragments.

Remarks Geraads (1994) noted that this material belonged to the *Serengetilagus-Trischizolagus* group: he assigned it to *Serengetilagus* because *Serengetilagus* was known only from Africa, whereas *Trischizolagus* was known only from the late Miocene to late Pliocene of Eurasia. We agree with Sen and Erbajeva (1995), Averianov and Tesakov (1997), and López-Martínez et al. (2007) that the Moroccan specimens pertain to *Trischizolagus*. These authors note that p3 of the Moroccan form has a PIR in the form of an enamel

island and the shape of the p3 (including a large lingual anteroconid) is similar to that of *Trischizolagus* (with an angular rhombic shape) rather than that of *Serengetilagus*, which is more rounded.

Subfamily LEPORINAE Dice, 1917
Genus *LEPUS* Linnaeus, 1758

Diagnosis Lyon (1904). Pelage soft, a patch on throat different in color and texture from surrounding fur; ears as long as or longer than head; tail short but plainly evident. Hind feet long, heavily furred, claws not conspicuous. Sutures of interparietal obliterated in the adult; postorbital processes large and triangular, with distinct anterior and posterior limbs. Palate short, its least length 2.5 or less times the length of M1; choanae wide, about 4 times the length of M1. Teeth "normal." Precocial young (Corbet, 1983). Leporine p3 pattern with PER extending almost completely across the tooth and lacking a PIR.

Included African Species (Fossil Record Only) Lepus veter (extinct), *L. capensis, L. saxatilis.*

African Age and Occurrence Late Pliocene to Recent, northern (Morocco), eastern (Ethiopia, Kenya, Tanzania), and southern Africa (South Africa, Namibia).

Material The fossil record of *L. capensis* includes six isolated teeth from the Omo Valley, Ethiopia. There are illustrations of two p3s, referred to *L. cf. L. capensis*, from Oulad Hamida 1, Morocco. *Lepus veter* is known from an incomplete maxilla, two mandibles, and four postcranial elements. Most reports of fossil *Lepus* do not include what elements are present.

Remarks There are late Miocene–early Pliocene reports of *Lepus* from Africa, but none of this material has been adequately described and these assignments should be considered extremely tentative. These would be very early dates for the presence of *Lepus* in Africa, considering that the earliest records of *Lepus* elsewhere in the world are early (Europe) and late (North America) Pliocene (McKenna and Bell, 1997). Haile-Selassie et al. (2004) tentatively assign isolated leporid teeth from the late Miocene Adu-Asa Formation, Middle Awash, Ethiopia, to *Lepus*. There is no description or illustration of the material. Wesselman et al. (2009) assign Adu-Asa leporids to *Serengetilagus*. *Lepus* has also been reported, but neither described nor adequately figured, from the early Pliocene at Kanapoi, Kenya (4.17–4.07 Ma, Winkler in Harris et al., 2003; Manthi, 2006).

Ndessokia (1990: table 9.1) reports *Lepus* sp. from the late Pliocene (2.6 Ma), Upper Ndolanya Beds, Laetoli, Tanzania. The specimens listed by Ndessokia are not illustrated, but are described as "virtually identical in size and morphology to those of contemporary *Lepus capensis* (Ndessokia, 1990: 90). P2 has two anterior reentrants, but the description of p3 (listed as p2) is not adequate to say conclusively whether it is *Lepus* or *Serengetilagus*.

Lepus sp. is reported (not yet studied) in the late Pliocene from Ahl al Oughlam, Morocco (ca. 2.5 Ma; Geraads, 2006). Wesselman (1984) reports *L. capensis* from Member E, and lower Members F and G, Shungura Formation, Omo Valley, Ethiopia (ca. 2.08–1.98 Ma). An unstudied incomplete skull and skeleton of a 'Lagomorpha' is the only lagomorph reported from the Koobi Fora Formation, Kenya (Harris, 1978).

Although rarely described, *Lepus* is often reported from the Pleistocene. In northern Africa, *Lepus* cf. *L. capensis* is known from the middle Pleistocene Grotte de Rhinocéros, Oulad

Hamida I, Morocco (Geraads, 1994). Geraads (1994) notes that *Lepus* cf. *L. capensis* is less abundant than *"Serengetilagus" raynali* at this locality. In eastern Africa, *Lepus* is reported from Olduvai Bed I, Tanzania (early Pleistocene; Leakey, 1965), but note the discussion on lagomorphs from this locality given above under *Serengetilagus*. *Lepus capensis* is listed from the late Pleistocene of Ethiopia (Assefa, 2006). The only extinct species of African *Lepus*, *L. veter*, is from the middle Pleistocene of Kenya. It is diagnosed as being comparable in size to a large *L. capensis*, but more robust, with teeth having a more complex enamel pattern, and with the anterior flange of the angular process of the mandible more pronounced and displaced laterally (MacInnes, 1953).

In southern Africa, *Lepus* is first known with certainty in the middle Pleistocene (Duinefontein 2 [Klein et al., 1999]; Cave of Hearths [Cooke, 1963]; Elandsfontein [Klein, 1974]; Florisbad [Brink, 1987]). *Lepus* from Rocky II, Kaokoland, Namibia, and Gcwihaba and Nqumtsa, Botswana, may be older, but the sites are not dated securely (Pickford, 1990; Pickford et al., 1993). The *Lepus* from the Plio-Pleistocene (2–1 Ma) at Kromdraai B in South Africa is not confidently identified (Brain, 1981); neither is the *Lepus capensis* from Bolt's Farm (Cooke, 1963), Swartkrans (Watson, 1993), nor Taung (Cooke, 1990). Currently, only two species of *Lepus* (*L. capensis* and *L. saxatilis*) are recognized in southern Africa (Bronner et al., 2003), although each species may include a second cryptic species (Robinson and Matthee, 2005). *L. capensis* is by far the most commonly identified species in middle and late Pleistocene archaeological sites (e.g., Klein 1976a, 1976b, 1984; Klein et al., 1999; Grine et al., 1991; Henshilwood et al., 2001). Klein (1984) suggested that this species might have been present in the middle to late Pliocene, although it is not included in any of the faunal lists for the sites of that age. In the late Pleistocene, *L. saxatilis* has been recorded from Blombos Cave (Henshilwood et al., 2001) and *L.? saxatilis* (as *?L. crawshayi*) from Border Cave (Klein, 1977).

Paleobiogeographic History of African Lagomorphs and the Origins of the Major Genera

The fossil record of African lagomorphs is relatively sparse, but enough is known to make generalizations about the paleobiogeographic history of the group and the origins of the extinct and extant genera. These can be tested with additional collecting. Sinolagomyine ochotonids, the primitive subfamily that includes *Austrolagomys/Kenyalagomys*, are known in Africa in the early and middle Miocene. The origin of the subfamily is probably in the late early Oligocene of Asia (Erbajeva, 1994), and the African record is likely the result of immigration at least by the early Miocene. Pre–early Miocene faunas are rare in Africa, although some, such as those from the Fayum Basin, Egypt, have produced many specimens. So far none of these pre–early Miocene sites have produced any lagomorphs. The apparent extinction of *Austrolagomys/Kenyalagomys* in Africa in the middle Miocene coincides with the worldwide extinction of the Sinolagomyinae by the end of the middle Miocene/beginning of the late Miocene (Erbajeva, 1994). The ochotonine *Alloptox* immigrated from Eurasia, perhaps through Turkey, to northern Africa later in the middle Miocene. There are no African reports of ochotonids after the middle Miocene.

The earliest record of the prolagid *Prolagus* is from the early Miocene of Europe (ca. 20 Ma). About 22 species of the genus have been reported from Europe, Anatolia, and northern

Africa (López-Martínez, 2001). The appearance of *Prolagus* in the late Miocene of northern Africa is likely due to immigration from Europe as part of major faunal interchange among Europe, Asia, and Africa associated with the Messinian salinity crisis (López-Martínez, 2001). *Prolagus* is last reported in Africa in the early Pleistocene.

Based on fossil and molecular data, the Leporidae likely diverged from the Ochotonidae in the late Eocene, or possibly in the early Eocene (Rose et al., 2008). The earliest African leporids are from the late Miocene. In eastern Africa, *Alilepus* is from sites in Kenya dated between 6.6 and 5.6 Ma and from Ethiopia at 4.85 Ma. An isolated leporid cheek tooth, unidentifiable to genus, is from the Mpesida Beds, Kenya, dated at 7–6.2 Ma (Winkler, 2002). The appearance of leporids in the late Miocene of eastern Africa is associated with significant faunal change in the region, affecting both small and large mammals (e.g., Hill, 1995). *Serengetilagus tchadensis*, from Toros Menalla, northern Chad (López-Martínez et al., 2007), is the only leporid from outside eastern Africa described from the late Miocene.

Other than *Alilepus*, *Serengetilagus* and *Trischizolagus* are the only extinct African leporid genera. *Serengetilagus* is known definitively from eastern Africa and from Chad. Outside Africa, *Serengetilagus* sp. may be present in the early Pliocene at Pul-e Charkhi, Afghanistan (Sen and Erbajeva, 1995; considered to be *Trischizolagus* by Averianov and Tesakov, 1997). Topachevsky (1987) described *S. orientieuropaeus* from the lower Pliocene of Ukraine (also considered to be *Trischizolagus* by Averianov and Tesakov, 1997). A *Serengetilagus*-like leporid is listed from the early Miocene of Kazakhstan (Erbajeva and Tyutkova, 1997). A middle Pleistocene sample from Morocco was originally assigned to *Serengetilagus* (Geraads, 1994) but belongs to *Trischizolagus*: this is a geologically young record for either genus. *Trischizolagus* is a genus known well from Eurasia from sites dating from the late Miocene–late Pliocene (Averianov and Tesakov, 1997). The interrelationship between these two genera is unresolved (e.g., see Averianov and Tesakov, 1997; López-Martínez et al., 2007).

Five genera of leporids are currently known from Africa. *Oryctolagus* may have been introduced by man. *Poelagus* is unknown from the fossil record. *Bunolagus* was previously reported as having a single Pleistocene record, but this identification has been questioned recently. The fossil record of *Pronolagus* is confined exclusively to southern Africa, with the oldest report in the early Pliocene (ca. 5 Ma) at Langebaanweg. None of the fossil *Pronolagus* material has been described. Some of the *Alilepus* specimens from Kenya (Winkler, 2003) show similarities in p3 morphology with extant *Pronolagus* and *Bunolagus*, suggesting that these may be sister taxa.

The fifth genus of leporid currently known from Africa is *Lepus*: it is the most speciose, numerically abundant, and geographically widespread of the extant African leporids. Leporids with a p3 pattern similar to that of *Lepus* (leporine pattern with PER extending almost completely across the tooth and lacking a PIR; Dice, 1929) are first reported from Africa (usually as *Lepus* sp.) in the early Pliocene of eastern Africa. None of this material has been described formally and it remains to be determined if this early material actually pertains to *Lepus*. Outside of Africa, the earliest reports of *Lepus* are from the late Pliocene of North America and the early Pliocene of Europe (McKenna and Bell, 1997). Suchentrunk et al. (2006) state that the origin of *Lepus* may be circa 2.5 Ma. Thus, the earliest records of *"Lepus"* from Africa would be among, or would be, the oldest of the genus. *Lepus* is not definitively reported from Africa until the late Pliocene. The early African records of *"Lepus"* may represent another genus, which also has the leporine pattern on p3. For example, among extant African leporids, *Oryctolagus* and *Poelagus* also have this pattern. Certainly, the early African reports of *"Lepus"* are significant, but the specimens need to be studied formally before definitive assignment to *Lepus* should be accepted.

Molecular Evidence for the Origins of Extant African Genera

Matthee et al. (2004) developed a phylogenetic framework for the 11 extant genera of leporids using five nuclear and two mitochondrial gene fragments to construct a molecular supermatrix. Dispersal-vicariance analysis using these data suggested that the current geographic distribution of the group (with regard to Africa) reflects three independent dispersal events from Asia into Africa. Use of a relaxed Bayesian molecular clock provided a chronological framework for these dispersal events. The first event (prior to 11.3 Ma) gave rise to *Pronolagus* and *Poelagus*. A second dispersal event (sometime between 11.3 and 7 Ma) brought *Bunolagus*. *Lepus* arrived most recently (perhaps closer to 5 Ma; Robinson and Matthee, 2005). The arrival of *Lepus* may be correlated with the continued expansion of grasslands in the late Miocene (Robinson and Matthee, 2005). At present, fossil evidence does not support the immigration of leporids into Africa prior to the late Miocene (ca. 7–6 Ma). The earliest African leporids, such as the Kenyan and Ethiopian *Alilepus* (dating from 6.57–4.85 Ma), are members of the Palaeolaginae, as are the extant genera *Pronolagus* and *Bunolagus* (although these extant genera are not sister taxa based on molecular evidence; Matthee et al., 2004). The lack of the AR on the Kenyan *Alilepus* (although present on the Ethiopian specimen; present on extant *Pronolagus* and *Bunolagus*) is the primitive state for leporids. The earliest African fossil record for taxa with a leporine p3 pattern (early Pliocene) roughly correlates with the timing of dispersal of *Lepus* into Africa based on molecular data. The question is, of course, whether these specimens really are attributable to *Lepus*.

Conclusions

Our knowledge of the history of African lagomorphs has increased greatly since Lavocat's summary of the group in *Evolution of African Mammals* (Maglio and Cooke, 1978). Of particular interest are the earliest records of the leporids and how these specimens may elucidate the origins of the extant African genera. What is apparent, however, from the current summary is that in spite of the tremendous increase in the number of specimens recorded, much of that material still remains to be described. Based on the number of specimens, we know that there is a rich history of lagomorphs in Africa. However, the significance of much of that material to our understanding of the evolution, paleoecology, and paleobiogeography of the Lagomorpha remains to be discovered.

ACKNOWLEDGMENTS

We thank W. J. Sanders and L. Werdelin for the invitation to contribute to this volume. Our chapter benefited from the input and critiques of many individuals; we especially thank M. Erbajeva, N. López-Martínez, H. Wesselman, and an anonymous reviewer. Dale Winkler drew figure 18.1. Camera lucida drawings of *Alilepus* sp. from Kenya (figure 18.2) are courtesy of Y. Tomida. We are grateful to T. Harrison for permission to illustrate specimens of *Serengetilagus praecapensis* (figure 18.3; collected by T. Harrison with funding provided to him by the National Geographic Society, the Leakey Foundation, and the National Science Foundation [Grants BCS-9903434 and BCS-0309513]. Figure 18.1C and 18.1D © Copyright 1991 The Society of Vertebrate Paleontology. Reprinted and distributed with permission of the Society of Vertebrate Paleontology.

Literature Cited

Assefa, Z. 2006. Faunal remains from Porc-Epic: Paleoecological and zooarchaeological investigations from a Middle Stone Age site in southeastern Ethiopia. *Journal of Human Evolution* 51:50–75.

Averianov, A. O. 1999. Phylogeny and classification of Leporidae (Mammalia, Lagomorpha). *Vestnik Zoologii* 33:41–48.

Averianov, A. O., and A. S. Tesakov. 1997. Evolutionary trends in Mio-Pliocene Leporinae, based on *Trischizolagus* (Mammalia, Lagomorpha). *Paläontologische Zeitschrift* 71:145–153.

Benammi, M. 1997. Magnétostratigraphie du Miocène supérieur continental du Maroc (Coupe d'Afoud, Bassin d'Aït Kandoula); pp. 285–291 in J. -P. Aguilar, S. Legendre, and J. Michaux (eds.), *Actes du Congrès BiochroM '97*. Mémoires et Travaux de l'Ecole Pratique des Hautes Etudes, l'Institut de Montpellier 21.

Benammi, M., B. Orth, M. Vianey-Liaud, Y. Chaimanee, V. Suteethorn, G. Feraud, J. Hernandez, and J. J. Jaeger. 1995. Micromammifères et biochronologie des formations néogènes du flanc sud du Haut-Atlas Marocain: Implications biogéographiques, stratigraphiques et tectoniques. *Africa Geoscience Review* 2:279–310.

Brain, C. K. 1981. *The Hunters or the Hunted?* University of Chicago Press, Chicago, 365 pp.

Brink, J. 1987. The archaeozoology of Florisbad, Orange Free State. *Memoirs of the National Museum* (Bloemfontein) 24:1–151.

Bronner, G. N., M. Hoffmann, P. J. Taylor, C. T. Chimimba, P. B. Best, C. A. Matthee, and T. J. Robinson. 2003. A revised systematic checklist of the extant mammals of the southern African subregion. *Durban Museum Novitates* 28:56–96.

Brunet, M., and M. P. E. T. 2000. Chad: discovery of a vertebrate fauna close to the Mio-Pliocene boundary. *Journal of Vertebrate Paleontology* 20:205–209.

Coiffait-Martin, B. 1991. Contribution des Rongeurs du Néogène d'Algérie à la biochronologie mammalienne d'Afrique nord-occidentale. PhD dissertation, Université Nancy I, Nancy, France, 400 pp.

Cooke, H. B. S. 1963. Pleistocene mammal faunas of Africa with particular reference to southern Africa; pp. 65–116 in F. C. Howell and F. Bourlière (eds.), *African Ecology and Human Evolution*. Aldine Press, Chicago, Illinois.

———. 1990. Taung fossils in the University of California collections; pp. 119–134 in G. H. Sperber (ed.), *Apes to Angels: Essays in Anthropology in Honor of Phillip V. Tobias*. Wiley-Liss, New York.

Corbet, G. B. 1983. A review of classification in the family Leporidae. *Acta Zoologica Fennica* 174:11–15.

Cruz-Uribe, K., and R. G. Klein. 1983. Faunal remains from some Middle and Later Stone Age sites in South West Africa. *Journal of the South West Africa Scientific Society* 36/37:91–114.

Darwent, C. M. 2007. Lagomorphs (Mammalia) from late Miocene deposits at Lemudong'o, southern Kenya. *Kirtlandia* 56:112–120.

Davies, C. 1987. Note on the fossil Lagomorpha from Laetoli; pp. 190–193 in M. D. Leakey and J. M. Harris (eds.), *Laetoli: A Pliocene Site in Northern Tanzania*. Oxford University Press, New York.

Dawson, M. R. 1969. Osteology of *Prolagus sardus*, a Quaternary ochotonid (Mammalia, Lagomorpha). *Palaeovertebrata* 2:157–190.

———. 1981. Evolution of modern leporids; pp. 1–8 in K. Myers and C. D. MacInnes (eds.), *Proceedings of the World Lagomorph Conference*. University of Guelph Press, Ontario.

Dice, L. R. 1929. The phylogeny of the Leporidae, with description of a new genus. *Journal of Mammalogy* 10:340–344.

Dietrich, W. O. 1941. Die Säugetierpaläontologischen Ergebnisse der Kohl-Larsen'schen Expedition, 1937–1939, in nördlichen Deutsch-Ostafrika. *Neues Jahrbuch für Minerologie, Geologie und Paläontologie* B8:217–223.

Dietrich, W. O. 1942. Ältestquartäre Säugetiere aus der Südlichen Serengeti, Deutsch-Ostafrika. *Palaeontographica, Abt. A* 94:43–133.

Duthie, A. G., and T. J. Robinson. 1990. The African rabbits; pp. 121–127 in J. A. Chapman and J. E. C. Flux (eds.), *Rabbits, Hares and Pikas*. Status Survey and Conservation Action Plan. IUCN/SSC Lagomorph Specialist Group, Gland, Switzerland.

Erbaeva, M. A., and R. Angermann. 1983. Das Originalmaterial von *Serengetilagus praecapensis* Dietrich, 1941—ergänzende Beschreibung und vergleichende Diskussion. *Schriftenreihe für Geologische Wissenschaften, Berlin* 19/20:39–60.

Erbajeva, M. A. 1988. *Cenozoic Pikas (Taxonomy, Systematics, Phylogeny)* [in Russian]. Nauka, Moscow, Russia, 224 pp.

——. 1994. Phylogeny and evolution of Ochotonidae with emphasis on Asian ochotonids; pp. 1–13 in Y. Tomida, C. K. Li, and T. Setoguchi (eds.), *Rodent and Lagomorph Families of Asian Origins and Diversification*. National Science Museum Monographs 8, Tokyo.

Erbajeva, M. A., and L. A. Tyutkova. 1997. Paleogene and Neogene lagomorphs from Kazakhstan; pp. 209–214 in J. -P. Aguilar, S. Legendre, and J. Michaux (eds.), *Actes du Congrès BiochroM '97*. Mémoires et Travaux de l'Ecole Pratique des Hautes Etudes, l'Institut de Montpellier 21.

Flynn, L. J., and R. L. Bernor. 1987. Late Tertiary mammals from the Mongolian People's Republic. *American Museum Novitates* 2872:1–16.

Friedmann, Y., and B. Daly (eds.). 2004. *Red Data Book of the Mammals of South Africa: A Conservation Assessment*. CBSG Southern Africa, Conservation and Specialist Breeding Group (SSC/IUCN), Endangered Wildlife Trust, Johannesburg, South Africa, 722 pp.

Geraads, D. 1994. Rongeurs et Lagomorphes du Pleistocène moyen de la "Grotte des Rhinocéros," Carrière Oulad Hamida 1 à Casablanca, Maroc. *Neues Jahrbuch für Geologie und Paläontologie, Abhandlungen* 191:147–172.

——. 2006. The late Pliocene locality of Ahl al Oughlam, Morocco: Vertebrate fauna and interpretation. *Transactions of the Royal Society of South Africa* 61:97–101.

Gibb, J. A. 1990. The European rabbit *Oryctolagus cuniculus*; pp. 116–120 in J. A. Chapman and J. E. C. Flux (eds.), *Rabbits, Hares and Pikas: Status Survey and Conservation Action Plan*. IUCN/SSC Lagomorph Specialist Group. IUCN, Gland, Switzerland.

Gibbard, P. and T. Van Kolfschoten. 2004. The Pleistocene and Holocene epochs; pp. 441–452 in F. Gradstein, J. Ogg, and A. Smith (eds.), *A Geologic Time Scale 2004*. Cambridge University Press, Cambridge.

Grine, F. E., R. G. Klein, and T. P. Volman. 1991. Dating, archaeology and human fossils from the Middle Stone Age levels of Die Kelders, South Africa. *Journal of Human Evolution* 21:363–395.

Gureev, A. A. 1960. Lagomorphs from the Oligocene of Mongolia and China [in Russian]; pp. 5–34 in K. K. Flerov (ed.), *Neogene Mammals*. Transactions of the Paleontological Institute, AN USSR, Academia Nauk USSR Publishing 77(4).

Haile-Selassie, Y., G. Woldegabriel, T. D. White, R. L. Bernor, D. Degusta, P. R. Renne, W. K. Hart, E. Vrba, A. Stanley, and F. C. Howell. 2004. Mio-Pliocene mammals from the Middle Awash, Ethiopia. *Geobios* 37:536–552.

Hamilton, W. R., and J. A. Van Couvering. 1977. Lower Miocene mammals from South West Africa. *Bulletin of Namib Desert Ecological Research Unit* 2:9–11.

Harris, J. M. 1978. Paleontology; pp. 32–63 in M. G. Leakey and R. E. Leakey (eds.), *Koobi Fora Research Project*, vol. 1. Clarendon Press, Oxford.

Harris, J. M., F. H. Brown, and M. G. Leakey. 1988. Stratigraphy and paleontology of Pliocene and Pleistocene localities west of Lake Turkana, Kenya. *Natural History Museum of Los Angeles County, Contributions in Science* 399:1–128.

Harris, J. M., M. G. Leakey, and T. E. Cerling. Appendix by A. J. Winkler. 2003. Early Pliocene tetrapod remains from Kanapoi, Lake Turkana Basin, Kenya. *Natural History Museum of Los Angeles County, Contributions in Science*, 498:39–113.

Hendey, Q. B. 1978. Preliminary report on the Miocene vertebrates from Arrisdrift, South West Africa. *Annals of the South African Museum* 76:1–41.

——. 1981. Palaeoecology of the late Tertiary fossil occurrences in "E" Quarry, Langebaanweg, South Africa, and a reinterpretation of their geological context. *Annals of the South African Museum* 84:1–104.

Henshilwood, C. S., J. C. Sealy, R. Yates, K. Cruz-Uribe, P. Goldberg, F. E. Grine, R. G. Klein, C. Poggenpoel, K. Van Niekerk, and I. Watts. 2001. Blombos Cave, southern Cape, South Africa: Preliminary report on the 1992–1999 excavations of the Middle Stone Age levels. *Journal of Archaeological Science* 28:421–448.

Hibbard, C. W. 1963. The origin of the P3 pattern of *Sylvilagus, Caprolagus, Oryctolagus* and *Lepus*. *Journal of Mammalogy* 44:1–15.

Hill, A. 1995. Faunal and environmental change in the Neogene of East Africa: evidence from the Tugen Hills sequence, Baringo District, Kenya; pp. 178–193 in E. S. Vrba, G. H. Denton, T. C. Partridge, and L. H. Burckle (eds.), *Paleoclimate and Evolution: With Emphasis on Human Origins*. Yale University Press, New Haven.

Hoffmann, R. S., and A. T. Smith. 2005. Order Lagomorpha; pp. 185–211 in D. E. Wilson and D. M. Reeder (eds.), *Mammal Species of the World*. Johns Hopkins University Press, Baltimore.

Hopwood, A. T. 1929. New and little-known mammals from the Miocene of Africa. *American Museum Novitates* 344:1–9.

James, G. T., and B. H. Slaughter. 1974. A primitive new middle Pliocene murid from Wadi el Natrun, Egypt. *Annals of the Geological Survey of Egypt, Cairo* 4:333–362.

Janvier, P., and C. de Muizon. 1976. Les Lagomorphes du Miocène de Béni Mellal, Maroc. *Géologie Méditerranéenne* 3:87–90.

Kingdon, J. 1974. *East African Mammals: Vol. IIB. Hares and Rodents*. University of Chicago Press, Chicago, 704 pp.

Klein, R. G. 1972. The Late Quaternary mammalian fauna of Nelson Bay Cave (Cape Province, South Africa): Its implications for megafaunal extinctions and for environmental and cultural change. *Quaternary Research* 2:135–142.

——. 1974. Environment and subsistence of prehistoric man in the southern Cape Province. *World Archaeology* 5:249–284.

——. 1976a. The mammalian fauna of the Klasies River Mouth sites, southern Cape Province, South Africa. *South African Archaeological Bulletin* 31:75–98.

——. 1976b. A preliminary report on the "Middle Stone Age" open-air site of Duinefontein 2 (Melkbosstrand, south-western Cape Province, South Africa). *South African Archaeological Bulletin* 31:12–20.

——. 1977. The mammalian fauna from the Middle and Later Stone Age (Later Pleistocene) levels of Border Cave, Natal Province, South Africa. *South African Archaeological Bulletin* 32:14–27.

——. 1978. Preliminary analysis of the mammalian fauna from the Redcliff Stone Age site, Rhodesia. *Occasional Papers of the National Museums and Monuments of Rhodesia, Series A, Human Sciences* 4:74–80.

——. 1981. Later Stone Age subsistence at Byneskranskop Cave, South Africa; pp. 166–190 in R. S. O. Harding, and G. Teleki (eds.), *Omnivorous Primates: Gathering and Hunting in Human Evolution*. Columbia University Press, New York.

——. 1984. The large mammals of southern Africa: Late Pliocene to Recent; pp. 107–146 in R. G. Klein (ed.), *Southern African Prehistory and Paleoenvironments*. Balkema Press, Rotterdam.

——. 1988. The archaeological significance of animal bones from Acheulean sites in southern Africa. *African Archaeological Review* 6:3–25.

Klein, R. G., G. Avery, K. Cruz-Uribe, D. Halkett, T. Hart, R. G. Milo, and T. P. Volman. 1999. Duinefontein 2: An Acheulean site in the Western Cape Province of South Africa. *Journal of Human Evolution* 37:153–190.

Klein, R. G., and K. Cruz-Uribe. 2000. Macromammals and reptiles; pp. 51–61 in L. Barham (ed.), *The Middle Stone Age of Zambia, South Central Zambia*. Western Academic and Specialist Press, Bristol, England.

Klein, R. G., K. Cruz-Uribe, and P. B. Beaumont. 1991. Environmental, ecological and paleoanthropological implications of the Late Pleistocene mammalian fauna from *Equus* Cave, northern Cape Province, South Africa. *Quaternary Research* 36:94–119.

Lavocat, R. 1978. Rodentia and Lagomorpha; pp. 69–89 in V. J. Maglio and H. B. S. Cooke (eds.), *Evolution of African Mammals*. Harvard University Press, Cambridge.

Leakey, L. S. B. 1965. *Olduvai Gorge 1951–1961: Volume 1*. Cambridge University Press, Cambridge, 118 pp.

Leakey, M. D. 1971. *Olduvai Gorge: Vol. 3. Excavations in Beds I and II, 1960–1963*. Cambridge University Press, Cambridge, 306 pp.

Leakey, R. E., and M. G. Leakey. 1986. A new Miocene hominoid from Kenya. *Nature* 324:143–146.

Lever, C. 1985. *Naturalized Mammals of the World*. Longman, London, 487 pp.

————. 2001. Paleobiogeographical history of *Prolagus*, a European ochotonid (Lagomorpha). *Lynx (Praha)*, n.s. 32:215–231.

López-Martínez, N., A. Likius, H. T. Mackaye, P. Vignaud, and M. Brunet. 2007. A new lagomorph from the Late Miocene of Chad (Central Africa). *Revista Española de Paleontología* 22:1–20.

López-Martínez, N., and L. Thaler. 1975. Biogéographie, évolution et compléments à la systématique du groupe d'Ochotonidés *Piezodus-Prolagus* dans le cénozoïque d'Europe Sud-Occidentale. *Bulletin de la Société Géologique de France* 7:850–866.

Lourens, L., F. Hilgen, N. J. Shackleton, J. Laskar, and D. Wilson. 2004. The Neogene period; pp. 409–440 in F. Gradstein, J. Ogg, and A. Smith (eds.), *A Geologic Time Scale 2004*. Cambridge University Press, Cambridge.

Lyon, M. W., Jr. 1904. Classification of hares and their allies. *Smithsonian Miscellanous Collections* 45:321–447.

MacInnes, D. G. 1953. The Miocene and Pleistocene Lagomorpha of East Africa. *Fossil Mammals of Africa* 6:1–30.

Maglio, V. J., and H. B. S. Cooke (eds.). 1978. *Evolution of African Mammals*. Harvard University Press, Cambridge, 641 pp.

Manthi, F. K. 2006. The Pliocene micromammalian fauna from Kanapoi, northwestern Kenya, and its contribution to understanding the environment of *Australopithecus anamensis*. Unpublished PhD dissertation, University of Cape Town, Cape Town, South Africa, 231 pp.

Martin, T. 2004. Evolution of incisor enamel microstructure in Lagomorpha. *Journal of Vertebrate Paleontology* 24:411–426.

Matthee, C. A., B. J. van Vuuren, D. Bell, and T. J. Robinson. 2004. A molecular supermatrix of the rabbits and hares (Leporidae) allows for the identification of five intercontinental exchanges during the Miocene. *Systematic Biology* 53:433–447.

McKenna, M. C., and S. K. Bell. 1997. *Classification of Mammals above the Species Level*. Columbia University Press, New York, 631 pp.

Mein, P., and M. Pickford. 1992. Gisements karstiques pléistocènes au Djebel Ressas, Tunisie. *Comptes Rendus de l'Académie des Sciences, Paris*, Série II, 315:247–253.

————. 2003. Fossil picas (Ochotonidae, Lagomorpha, Mammalia) from the basal Middle Miocene of Arrisdrift, Namibia. *Memoirs of the Geological Survey of Namibia* 19:171–176.

————. 2006. Late Miocene micromammals from the Lukeino Formation (6.1 to 5.8 Ma), Kenya. *Bulletin et Mémoires de la Société Linnéen de Lyon* 75:183–223.

Ndessokia, P. N. S. 1990. The mammalian fauna and archaeology of the Ndolanya and Olpiro Beds, Laetoli, Tanzania. Unpublished PhD dissertation, University of California, Berkeley, 203 pp.

Nowak, R. M. 1999. *Walker's Mammals of the World*. 6th ed. Johns Hopkins University Press, Baltimore, 1936 pp.

Pickford, M. 1990. Some fossiliferous Plio-Pleistocene cave systems of Ngamiland, Botswana. *Botswana Notes and Records* 22:1–15.

Pickford, M., P. Mein, and B. Senut. 1992. Primate bearing Plio-Pleistocene cave deposits of Humpata, Southern Angola. *Human Evolution* 7:17–33.

Pickford, M., B. Senut, P. Mein, and G. C. Conroy. 1993. Premiers gisements fossilifères post-miocènes dans le Kaokoland, nord-ouest de la Namibie. *Comptes Rendus de l'Académie des Sciences, Paris*, Série II, 317:719–720.

Pocock, T. N. 1987. Plio-Pleistocene mammalian microfauna in southern Africa: A preliminary report including description of two new fossil muroid genera (Mammalia: Rodentia). *Palaeontologia Africana* 26:69–91.

Robinson, P., and C. C. Black. 1973. A small Miocene faunule from near Testour, Beji gouvernorat, Tunisia. *Annals des Mines et de la Géologie, Tunis* 26:445–449.

Robinson, P., C. C. Black, L. Krishtalka, and M. R. Dawson. 1982. Fossil small mammals from the Kechabta Formation, northwestern Tunisia. *Annals of the Carnegie Museum* 51:231–249.

Robinson, T. J., and C. A. Matthee. 2005. Phylogeny and evolutionary origins of the Leporidae: A review of cytogenetics, molecular analyses and a supermatrix analysis. *Mammal Review* 35:231–247.

Rose, K. D., V. B. DeLeon, P. Missiaen, R. S. Rana, A. Sahni, L. Singh, and T. Smith. 2008. early Eocene lagomorpha (Mammalia) from Western India and the early diversification of Lagomorpha. *Proceedings of the Royal Society of London, B* 275:1203–1208.

Savage, R. J. G. 1990. The African dimension in European early Miocene mammal faunas; pp. 587–600 in E. H. Lindsay, V. Fahlbusch, and P. Mein (eds.), *European Neogene Mammal Chronology*. NATO ASI Series, Series A, Life Sciences. Plenum Press, New York.

Schweitzer, F. R., and M. L. Wilson. 1982. Byneskranskop 1: A late Quaternary living site in the southern Cape Province, South Africa. *Annals of the South African Museum* 88:1–203.

Scott, L., and J. S. Brink 1992. Quaternary palaeoenvironments of pans in central South Africa: Palynological and palaeontological evidence. *South African Geographer* 19:22–34.

Sen, S. and M. Erbajeva. 1995. Early Pliocene leporids (Mammalia, Lagomorpha) from Afghanistan. *Comptes Rendus de l'Académie des Sciences, Paris*, Série II, 320:1225–1231.

Senut, B., M. Pickford, P. Mein, G. Conroy, and J. Van Couvering. 1992. Discovery of 12 new Late Cainozoic fossiliferous sites in paleokarst of the Otavi Mountains, Namibia. *Comptes Rendus de l'Académie des Sciences, Paris*, Série II, 314:727–733.

Stoner, C. J., O. R. P. Bininda-Emonds, and T. Caro. 2003. The adaptive significance of colouration in lagomorphs. *Biological Journal of the Linnean Society* 79:309–328.

Stromer, E. 1924. Ergebnisse der Bearbeitung mitteltertiärer Wirbeltier-Reste aus Deutsch-Südwestafrika. *Sitzungsberichte der Bayerische Akademie der Wissenschaften, München* 1923:253–270.

Stromer, E. 1926. Reste land- und süsswasser-bewohnender Wirbeltiere aus den Diamantfeldern Deutsch-Südwestafrikas; pp. 107–153 in E. Kaiser (ed.), *Die Diamantenwuste Südwest-Afrikas*, vol. 2. Dietrich Reimer (Ernst Volsen), Berlin.

Suchentrunk, H. B. Slimen, C. Stamatis, H. Sert, M. Scandura, M. Apollonio, and Z. Mamuris. 2006. Molecular approaches revealing prehistoric, historic, or Recent translocations and introductions of hares (genus *Lepus*) by humans. *Human Evolution* 21:151–165.

Thomas, O. 1929. On mammals from the Kaoko-Veld, South-West Africa, obtained during Captain Shortridge's fifth Percy Sladen and Kaffrarian Museum Expedition. *Proceedings of the Zoological Society London* 1929:99–111.

Topachevsky, I. V. 1987. The first find of the genus *Serengetilagus* representative (Lagomorpha, Leporidae) from Pliocene deposits of the Eastern Europe. *Vestnik Zoologii, Kiev* 6:48–51.

Van Couvering, J. A. H., and J. A. Van Couvering. 1976. Early Miocene mammal fossils from East Africa: aspects of geology, faunistics and paleo-ecology; pp. 155–207 in G. L. Isaac and E. R. McCown (eds.), *Human Origins*. W. A. Benjamin Series in Anthropology, Benjamin, Menlo Park, California.

Watson, V. 1993. Composition of the Swartkrans bone accumulations, in terms of skeletal parts and animals represented; pp. 35–73 in C. K. Brain (ed.), *Swartkrans. A Cave's Chronicle of Early Man*. Transvaal Museum Monograph 8.

Wesselman, H. B. 1984. The Omo Micromammals, Systematics and paleoecology of early man sites from Ethiopia. *Contributions to Vertebrate Evolution* 7:1–219.

Wesselman, H. B., M. T. Black, and M. Asnake. 2009. Small mammals; pp. 105–133 in Y. Haile-Selassie and G. WoldeGabriel (eds.), *Ardipithecus kadabba in Africa: Miocene Evidence from the Middle Awash, Ethiopia*. University of California Press, Berkeley.

Wessels, W., O. Fejfar, P. Peláez-Campomanes, A. van der Meulen, and H. de Bruijn. 2003. Miocene small mammals from Jebel Zelten, Libya. *Coloquios de Paleontología*, suppl. 1:699–715.

White, J. A. 1991. North American Leporinae (Mammalia: Lagomorpha) from Late Miocene (Clarendonian) to latest Pliocene (Blancan). *Journal of Vertebrate Paleontology* 11:67–89.

Winkler, A. J. 2002. Neogene paleobiogeography and East African paleoenvironments: Contributions from the Tugen Hills rodents and lagomorphs. *Journal of Human Evolution* 42:237–256.

————. 2003. Rodents and lagomorphs from the Miocene and Pliocene of Lothagam, northern Kenya; pp. 169–198 in M. G. Leakey and J. M. Harris (eds.), *Lothagam: The Dawn of Humanity in Eastern Africa*. Columbia University Press, New York.

Winkler, A. J., L. MacLatchy, and M. Mafabi. 2005. Small rodents and a lagomorph from the early Miocene Bukwa locality, Eastern Uganda. *Palaeontologia Electronica* 8, issue 1, 24A:1–12.

NINETEEN

Paleogene Prosimians

MARC GODINOT

The term *prosimians* is a grouping of all the primates that are outside the anthropoidean, or simian, clade. Because it is gradistic, paraphyletic, it is used here as an informal, but very convenient, group. It can also be used as a formal suborder Prosimii (e.g., Fleagle, 1999). I prefer to use as a formal taxon the suborder Strepsirrhini, which includes two infraorders, Adapiformes and Lemuriformes (as redefined by Hoffstetter, 1977), to which now needs to be added the informal stem lemuriforms, genera that are more closely related to lemuriforms than to adapiforms, but that do not possess the defining character of lemuriforms. All the living African and Asian Lorisoidea and Malagasy Lemuroidea possess a shared derived dental structure, composed of the two lower incisors and the canine, which are procumbent and closely appressed into an anterior tooth comb. This allows the convenient distinction between the lemuriforms (all living Strepsirrhines with a tooth comb) and the extinct adapiforms. Primate genera lying outside Strepsirrhini and Anthropoidea are variously referred to omomyiformes or tarsiiformes, and to the controversial Asiatic families Eosimiidae and Amphipithecidae.

Resolving the affinities of the most primitive African primates has been a long and difficult task. It is far from being finished, as shown later by taxa of uncertain affinities (see also Seiffert et al., this volume, chap. 22, concerning the Paleocene *Altiatlasius*). Primitive and fragmentary fossils are often controversial. In addition, similar molar structures can be reached by convergence between widely divergent groups. For example, the lower molars of *Oligopithecus savagei* resemble those of the adapiform *Hoanghonius*, leading to the idea that they bridged a gap and the suggestion of an adapoid ancestry for anthropoids (Gingerich, 1977), a hypothesis further developed by Rasmussen and Simons (1988) and Rasmussen (1990). Discovery of its skull showed it to be definitely an anthropoid. Conversely, *Plesiopithecus teras* was described as an anthropoid, based on its complete mandible and lower dentition (Simons, 1992; Simons et al., 1994), and discovery of its skull showed it to be a Strepsirrhine (Simons and Rasmussen, 1994). It is more difficult to identify more fragmentary material. The isolated teeth of small Oman strepsirrhines were considered as "adapiforms (?Anchomomyini)," although striking resemblances with lemuriforms were also reported (Gheerbrant et al., 1993). Other small Fayum primates, such as *Wadilemur* and *"Anchomomys" milleri*, were referred to the adapiforms, following the identification of anchomomyins in Oman, and the type-mandible of *Djebelemur martinezi* was also attributed to an adapiform, reinforcing the alleged close phylogenetic relationships between adapiforms and anthropoids (Hartenberger and Marandat, 1992). In this context, the first recovery of teeth documenting a tooth-combed strepsirrhine in the Paleogene beds of the Fayum was a crucial step forward (Seiffert et al., 2003). More material has subsequently allowed the identification of a series of tooth-combed lemuriforms and the realization that two other taxa probably document African stem lemuriforms (Seiffert et al., 2005; Godinot, 2006a). A further advance concerning the Egyptian Fayum sequence is its redating by Seiffert (2006); the new ages are used here.

While these discoveries clarified important points of systematics and phylogeny, others continue to add controversial taxa. This is especially the case with the description of *"Dralestes" hammadaensis* from Algeria and discussion of the affinities of Azibiidae by Tabuce et al. (2004). This has recently been followed by the description of new material, several new synonymies, and a very different view of Azibiidae (discussed later). The case of *Afrotarsius*, known since 1985, is also vexing. There are even a few isolated teeth that will not be formally described and studied here because their affinities are too controversial. An M3 from Fayum Quarry I was described as a lorisid tooth and later suspected to belong to a plesiopithecid (Rasmussen and Nekaris, 1998) or maintained in the lorisids (Simons, 1998). Even after the discovery of a lorisid-like genus in the Fayum, the authors are unable to refute any of the hypotheses offered about this tooth (Seiffert et al., 2003). Similarly, isolated teeth have been taken as indicative of the presence of Omomyidae in Africa. Simons et al. (1986) referred two isolated teeth, a p4 and a P4, from Fayum Quarry E to the Omomyidae. Thomas et al. (1988) referred two other isolated teeth from Thaytiniti (Oman), an m1 and an M3, to "Omomyidae?" These Omani teeth were later attributed to new oligopithecids (Gheerbrant et al., 1995). The isolated teeth from the Fayum do not appear diagnostic enough to

TABLE 19.1
List of the taxa included in this chapter, with their localities and ages
Note that *Afrotarsius* and Azibiidae, having controversial affinities, are not undoubted prosimians (see text).

Family	Genus and Species	Locality	Country	Age
Indet.	*Afrotarsius chatrathi*	Fayum Quarry M	Egypt	Early Oligocene
Azibiidae	*Azibius trerki*	Gour Lazib and Glib Zegdou	Algeria	Late early or early middle Eocene
	Algeripithecus minutus	Glib Zegdou	Algeria	Late early or early middle Eocene
Adapidae	*Aframonius dieides*	Fayum L 41	Egypt	Latest Eocene
	Afradapis longicristatus	Fayum BQ 2	Egypt	Late Eocene (37 Ma)
Djebelemuridae	*Djebelemur martinezi*	Chambi	Tunisia	Late early or early middle Eocene
	Anchomomys milleri	Fayum L 41	Egypt	Latest Eocene
Plesiopithecidae	*Plesiopithecus teras*	Fayum L 41	Egypt	Latest Eocene
Galagidae	*Wadilemur elegans*	Fayum L 41	Egypt	Latest Eocene
	Saharagalago misrensis	Fayum Birket Qarun 2	Egypt	Late Eocene (37 Ma)
Lemuriformes Fam.Indet	*Karanisia clarki*	Fayum Birket Qarun 2	Egypt	Late Eocene (37 Ma)
	Omanodon minor	Taqah	Oman	Early Oligocene
	Shizarodon dhofarensis	Taqah	Oman	Early Oligocene

infer that the family Omomyidae, otherwise known only on the northern continents, was present in the Paleogene of Africa. However, they reveal a diversity of as yet unnamed tiny prosimians or anthropoids. The list of taxa given here (table 19.1) is provisional.

On the whole, these fossils, added to the variety of primates already found in Africa, have sometimes been used as evidence for the origin and diversification of primates on the African continent. However, as will become clear later in this chapter, the African fossil record of prosimians is better interpreted as one group of old African endemics and one or two later immigrant lineages. This view agrees with the probable Asiatic origin of Primates, as indicated by the Asiatic distribution of their euarchontan sister groups.

Most illustrations in this chapter are scanning electron micrographs of casts, indicated as SEM in the figure captions.

Systematic Paleontology

Order PRIMATES Linnaeus, 1758
Infraorder indet.
Family indet.
Genus *AFROTARSIUS* Simons and Bown, 1985
AFROTARSIUS CHATRATHI Simons and Bown, 1985
Figure 19.1

Age and Occurrence Fayum locality Quarry M, Egypt, 249-m level of Jebel Qatrani Formation, early Oligocene.

Diagnosis Small primate; p4–m3 length 8.70 mm. Differs from all other small primates by the following combination of characters: similar-shaped trigonid and paraconid on m1–3; the paraconid is somewhat crestiform, smaller and more labially placed than in Eosimiidae; m1 > m2 > m3, and m3 is lacking a well-formed third lobe.

Description Dental remains of *Afrotarsius* are restricted to the type specimen, a right mandible bearing m1–m3, the posterior extremity of a broken p4, and the root of p3 (Simons and Bown, 1985). None of the molar trigonids is intact, that of m1 is damaged on the protoconid and metaconid, that of m2 is lacking the labial part of the protoconid and the metaconid summit is damaged on m3 (figure 19.1). Only m2 has an intact

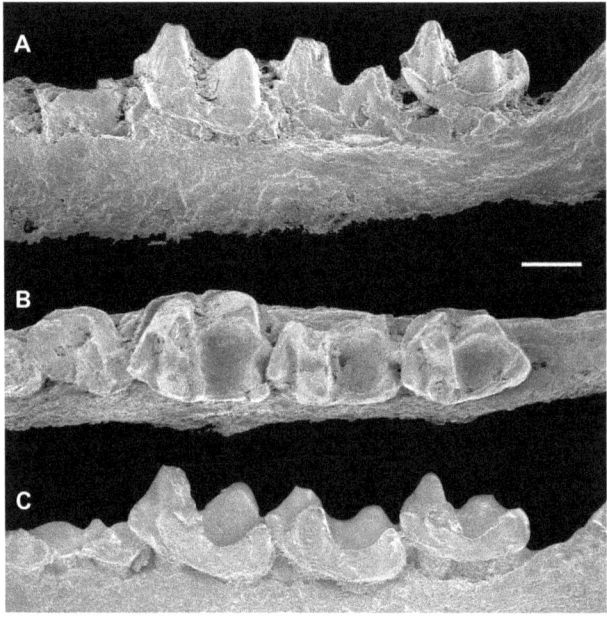

FIGURE 19.1 The type mandible of *Afrotarsius chatrathi* from the Oligocene of the Fayum, Egypt, in labial (A), occlusal (B), and lingual (C) views. SEM of a cast; scale bar is 1 mm.

metaconid, and only m3 has an intact protoconid, making detailed comparisons with other taxa difficult. Molar size decreases from m1 to m3. The trigonids are tall. From what is preserved, it appears that the trigonids of the three molars were of similar morphology and had a similar, almost crestiform paraconid, labially shifted relative to the metaconid, well separated from the latter cusp by a deep groove, but with a ventrolingual base extending in a cingular fashion below the anterior part of the metaconid. The hypoconid is large and tall on m1, decreasing in size from m1 to m3. There is a relatively tall lingual crest joining the metaconid with a poorly defined and posteriorly placed entoconid (undifferentiated on m3). On the back of the metaconids, this crest is more abrupt than in most other primates. The talonid basin is broad, but less so than in *Tarsius*. There is a small median hypoconulid on m1–2.

Remarks *Afrotarsius* must have included a high proportion of insects in its diet. A left proximal tibiofibula found in the same quarry as the mandible has been ascribed to *Afrotarsius* on the basis of overall similarity with *Tarsius* and size fitting with *Afrotarsius* (Rasmussen et al., 1998). However, in the absence of the more diagnostic distal part, this bone could pertain to another small mammal. Further comparisons and discussions in relation to other faunal components (especially rodents) must be made before concluding that *Afrotarsius* shared the leaping specialization of the living tarsier. Concerning the type mandible, an overall similarity to the living *Tarsius* has been emphasized (Simons, 1995, 1998) and is suggested by the name. However, in the initial description the species was referred to "Tarsiiformes, Tarsiidae?" and in fact various affinities have since been suggested for *Afrotarsius*. It is clear that this genus differs from *Tarsius* in having a less bulbous, more crestiform, paraconid, in having m1 > m2 > m3, and in lacking a well-formed third lobe on m3. If there are tarsiids having a well-formed third lobe in the Eocene, as implied by *Tarsius eocaenus* from Shanghuang (Beard et al., 1994), the m3 of *Afrotarsius* implies that this genus is outside of Tarsiidae. *Afrotarsius* differs from Omomyidae in having a similar-shaped trigonid and a similar-sized paraconid on m1–3. The very tall trigonids and shelflike medially placed paraconid, the m3 without a third lobe, and m1 > m2 > m3 are even reminiscent of some insectivores (Godinot, 1994). However, the broad talonid basins and the apparently broad and short p4 more likely indicate a primate. Kay and Williams (1994) suggested that *Afrotarsius* might be the sister group of anthropoids, because it shares with them several derived characters. However, several of the characters, such as the m1 cristid obliqua oriented mesially toward the protoconid, presence of small hypoconulids, and m2 > m3, could well be primitive in Primates. Such a polarity is especially suggested by the morphology of Eosimiidae and outgroup comparison (with, e.g., *Ptilocercus* and not with derived and convergent plesiadapiforms). In any case, dentally primitive insectivorous primates known from fragmentary dental remains are very difficult to assess phylogenetically. A recent phylogenetic study based on 359 morphological characters found it again closely related to *Tarsius* (Seiffert et al., 2005), which is surprising in view of the characters mentioned, which would suggest a position outside presumed tarsiids. *Afrotarsius* shares a number of characters with *Eosimias* and *Xanthorhysis* from the middle Eocene of China, two taxa placed by Beard in two different families (Beard et al., 1994; Beard, 1998) but that should be united in one. *Afrotarsius* appears close to eosimiids in its small m3 with a poorly differentiated third lobe (probably primitive); it differs from, and is probably more derived than, *Eosimias* in the smaller and more labially placed paraconid, a broader talonid basin on m1, and m1 > m2 > m3. In trigonid characters, *Afrotarsius* would appear to be more easily derivable from a genus like *Xanthorhysis*. This would imply an eosimiid-tarsiid dispersal from Asia to Africa during the Paleogene. Alternatively, *Afrotarsius* also shares many similarities with the isolated m1 of Thaytiniti now referred to an oligopithecid (Thomas et al., 1988; Gheerbrant et al., 1995). It could thus turn out to be an especially insectivorous oligopithecid or protoanthropoid. Only complementary material will allow a better understanding of this controversial genus.

Suborder STREPSIRRHINI or ANTHROPOIDEA?
Family AZIBIIDAE Gingerich, 1976
Genus *AZIBIUS* Sudre, 1975

Synonymy Tabelia Godinot and Mahboubi, 1994 (see Tabuce et al., 2009)

AZIBIUS TRERKI Sudre, 1975
Figure 19.2E

Age and Occurrence Locus 1 of the Gour Lazib ("Gouiret el Azib" in Sudre, 1975), and Glib Zegdou, eastern extremity of the Hammada du Drâ, southwest Algeria, late Ypresian or early Lutetian.

Diagnosis Small azibiid primate; p4–m2 length around 7.5 mm. Upper molars bunodont with very large hypocone; low protocone with long lingual slope; no lingual cingulum; paracone and metacone with long labial slopes. Molarized P4 with paracone and metacone forming a high ectoloph; P4 differs from that of *Algeripithecus* by the presence of a preprotocrista, a paraconule, and a parastyle. Lower molars bunodont; crown height rising from m3 to m1; elongated trigonid on m1, with anteriorly directed preprotocristid and posteriorly offset metaconid; talonid basin shallow and lingually opened on m1–2, shorter on m1 than on m2, posteriorly elongated on m3. p3–4 elongated and bladelike, differing from those of *Algeripithecus* by lower metaconids.

Description The ramus of the type mandible is not very tall. A piece of sandstone bearing the original imprint of the specimen shows the beginning of the ascending ramus (Sudre, 1975). The three preserved teeth show apical wear, accentuated on m1, added to erosion and missing enamel in several places (figure 19.2E). This state of preservation is partly responsible for the difficulty in interpreting this fossil. Tabuce et al. (2004) underlined that published drawings of *Azibius* oversimplified the morphology of m1. The p4 is very tall, bladelike, formed by two successive cusps, protoconid and hypoconid, coalescent almost to their tips and recognizable only by extremely shallow depressions. Its crown is posterolingually extended in occlusal view, reflecting a high degree of molarization in the talonid extension, modified by its integration into the anteroposterior blade.

Despite the wear and erosion, one can see that the m1 had an elevated trigonid with an anteriorly directed preprotocristid, and a metaconid that lies posterior and lingual relative to the protoconid. The talonid extends along half of the tooth length. The cristid obliqua runs toward the summit of the metaconid. In lingual view, the highly placed notch between metaconid and entoconid suggests that the talonid basin was relatively shallow.

The m2 has more usual proportions. It is moderately high crowned and quite bunodont. The metaconid is especially massive. A continuous protocristid, not reflected in the drawings of Sudre (1975, 1979), is present. The metaconid seems to have had a distinct, posteriorly extended postmetacristid, apparently continuous with a low preentocristid; in lingual view, both form a rounded notch between metaconid and entoconid, very close to the lowest level of the talonid basin, which appears shallow. The separation between metaconid and entoconid is underlined on the lingual side by a vertical groove lingual and ventral relative to the preceding notch.

Remarks The recent paper by Tabuce et al. (2009) documents new material from Glib Zegdou attributed to *Azibius trerki*,

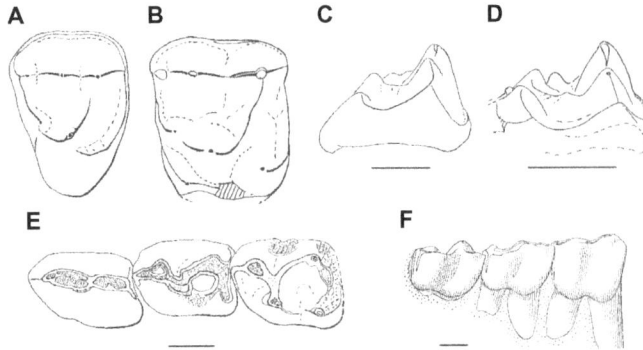

FIGURE 19.2 Drawings of teeth of azibiids from the early–middle Eocene of Algeria. An upper M1 of *Algeripithecus* in occlusal (B) and posterior (D) views at the same scale. *Azibius trerki*, P4, in occlusal (A) and posterior (C) views at the same scale. The p4–m2 on the type mandible of *Azibius trerki*: an interpretation of wear facets in occlusal view (E), and an interpretation of the labial view as if not eroded (F). Note on the posterior views (C) and (D) the similar, very tall and steep paraconc in comparison with the protocone, similar basal exodaenodonty, similar morphology of the crista obliqua. (F) is from Sudre, 1975. Scale bars are 1 mm. 2f, with permission of the Academie des Sciences, Paris.

a species earlier described from Gour Lazib, a 15-km-distant locality (at the same geological level). This material includes upper teeth and a mandible bearing p4–m3. The M2 and m3 are very similar to those described earlier from Glib Zegdou as "*Tabelia hammadae*" (Godinot and Mahboubi, 1994), which appears as a junior synonym of *A. trerki*. Furthermore, its P4 is similar to the tooth described as an M2 and made the type of "*Dralestes hammadaensis*" by Tabuce et al. (2004), making the latter another junior synonym of *A. trerki*. What had been unexpected was the co-occurrence in the same species of the bizarre autapomorphic p3–4 of *Azibius* with molars having an anthropoid appearance. The upper molars of *Azibius* are transversely elongated, bunodont with long lingual and labial slopes. The trigon basin is transversely short; there is a paraconule and no metaconule. The hypocone is large and bears a short prehypocrista. A fragment of maxilla shows a large canine alveolus, a small alveolus for a single-rooted P2, and P3 and P4. P3 has a high paracone and a very small and low protocone. P4 is molarized, with a metacone isolated from the paracone by shallow grooves, a large protocone with preprotocrista interrupted by a paraconule, postprotocrista almost reaching the metacone, and a salient posterior cingulum with an incipient hypocone (figure 19.2A). This maxillary fragment preserves a small surface of the orbital floor, two infraorbital foramina above P2–3, and the narrow ventral part of a groove interpreted by Tabuce et al. (2009) as a part of the lacrimal canal. Azibiids are discussed later.

Suborder STREPSIRRHINI E. Geoffroy Saint-Hilaire, 1812
Infraorder ADAPIFORMES Hoffstetter, 1977
Family ADAPIDAE Trouessart, 1879
Genus *AFRAMONIUS* Simons, Rasmussen and
Gingerich, 1995
AFRAMONIUS DIEIDES Simons, Rasmussen and
Gingerich 1995
Figure 19.3

Age and Occurrence Fayum locality L 41, Fayum Province, Egypt; 45–47 m above the base of the Jebel Qatrani Formation, latest Eocene.

Diagnosis Medium-sized adapiform; length P2–M3 is 19.7 mm. Small incisors, large canine and relatively small premolar series, no P1/p1, small unicuspid P2/p2, moderately sized P3–4 and p3–4. Upper molars with high cusps and crests, labially deflected centrocrista, large hypocone (small on M3) linked to a salient posterior cingulum, broad trigon basin with crenulated enamel, small paraconule; M1–2 with lingually shifted metacone and long metastylar crest. p3 and p4 broad and bearing several crests on the posterior wall of their protoconid. Lower molars with anteroposteriorly short trigonids, no paraconid, long paralophid, long and broad talonids, high hypoconid, cristid obliqua ending below the postvallid on m2–3, postmetacristid lingually salient; posteriorly extended m3 with narrow hypoconulid.

Description Several mandibular and a few maxillary specimens of *A. dieides* have been reported (Simons et al., 1995; Simons and Miller, 1997). The mandibular rami are shallow and show an intraspecific variability concerning symphyseal fusion. The ascending ramus is steep on its anterior side. The articular condyle lies clearly above the tooth row. Anteriorly, the incisor alveoli reflect small, crowded, and vertically implanted incisors. The canines of both sides are relatively close to each other. They appear large and vertically implanted. The strong canine size difference between individuals suggests that *Aframonius* had marked sexual dimorphism (Simons et al., 1995). The small, single rooted p2 is simple and has a lingual cingulid. Its anterior side can be affected by wear from the upper canine. Both p3 and p4 are moderate in size, especially broad in their posterior half in occlusal view, with a rounded posterior outline (figure 19.3). Their protoconid is anteriorly placed. The preprotocristid joins an anterolingual paracingulid, without cusp. The postprotocristid joins a prominent talonid cusp, the hypoconid. A posteroventral crest descends along the lingual side of the protoconid of p3, joining the posterolingual cingulid. This crest is even

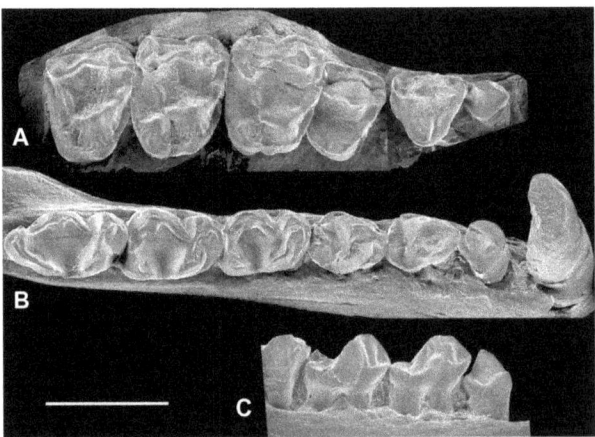

FIGURE 19.3 The dentition of *Aframonius dieides* from the early Oligocene of the Fayum, Egypt. SEM of casts. A) Maxillary teeth M3 to P2 in occlusal view; made from three different views to preserve the occlusal orientation of the teeth. B) Mandibular teeth from m3 to the canine in occlusal view. C) Lower p4 to p2 in lingual view. Scale bar is 5 mm.

stronger, and more anterior, on p4, joining a metaconid (variable, not well defined at its summit) and descending posteriorly as a postmetaconid cristid, which also joins the lingual cingulid, turning around an incipient talonid basin (lingual fovea of Simons et al., 1995). Although this lingual fovea is broad in occlusal view, it is inclined, and there is no real talonid basin. A short labial cingulid runs from the tip of the hypoconid. The p3 and p4 are unusual in showing a marked posterolabial protrusion in occlusal outline, whereas the posterolingual corner of these teeth is often extended in notharctids, including cercamoniines.

The three lower molars have a talonid that is longer and broader than the trigonid (figure 19.3). There is no paraconid. The preprotocristid is steep and short, resulting in a very short trigonid having the protoconid very anteriorly situated. The paralophid is subhorizontal, narrowing lingually, often continued by a thinner and posteroventrally inclined cingulid on the lingual side of these teeth (reminiscent of true adapines). The protoconid and metaconid are relatively close, joined by a short protocristid. A salient postmetacristid is steep, long, and lingually inclined in occlusal view, ending at the deep lingual notch that opens the talonid basin: deep and broad U-shaped opening on m3, slightly less deep V-shaped opening on m1, and intermediate morphology on m2. The hypoconid is almost merged in tall crests: the anterior cristid obliqua, which reaches the trigonid wall below the postvallid on m2–3 (not visible on m1 due to wear), and the long postcristid that joins the entoconid. The entoconid is more lingual than the metaconid and more distal than the hypoconid, leaving space for a very broad talonid basin. The m3 is slightly longer than the m2, and its third lobe is relatively small and narrow, slightly delineated on the labial but not the lingual side. A labial cingulid is present only on the median part of the lower molars. Some enamel wrinkling is present on the lower molars but can be detected only on unworn molars (Simons and Miller, 1997).

A fragment of maxilla shows the partial alveolus for a large upper canine, and such a large canine is also implied by the honing facets of p2 and p3. The P2 is reduced in size, single rooted, bearing strong anterior and posterior crests and a lingual cingulum. The P3 and P4 have a main labial cusp, the paracone, bearing strong anterior and posterior crests; the anterior one joins a cuspidate parastyle arising from a labial and a lingual cingulum; the anterior crest curves labially in a crestiform metastyle joining the cingula. On P4, a low protocone and preprotocrista isolate a trigon "basin," open posteriorly; anterior and posterior cingula are well formed. The P4 might be described as "subrectangular," having a straight labial border, whereas the lingual border is rounded in outline. The P3 has a more reduced lingual part, and an almost triangular outline. The posterolingual side shows a weak concavity in occlusal view, while the posterior cingulum curves anterolingually and ascends toward the base of the paracone, and then descends toward the parastyle, without a protocone cusp differentiated.

The three upper molars are subequal in size (figure 19.3). They are mesiodistally broad. The M1 and M2 are slightly broader posteriorly than anteriorly, due to their lingually salient hypocone, while M3 has a much smaller hypocone and a more triangular, lingually narrower, outline. The paracone and the metacone are tall, crested, and relatively narrow labiolingually; they are linked by a salient centrocrista, labially deflected in occlusal view, showing a longer premetacrista and shorter postparacrista on M2. The metacone appears lingually shifted relative to other cusps; correlatively, the postmetacrista is strongly curved labially on M2 (probably also on M1), forming a high and salient metastylar crest. The preparacrista is straight; a parastyle is barely differentiated on the labial side. The labial cingulum is well formed, and the anterior and posterior cingula are even stronger. The hypocone of M1–3 is linked to the posterior cingulum; the small M3 hypocone is clearly linked to a continuous lingual cingulum. The lingual cingulum is almost complete on M1. It is linked to the base of the (broken) hypocone. The lingual cingulum is more discontinuous and attenuated on M2, which possibly had a slightly larger hypocone than M1. The protocone is voluminous. The steep preprotocrista joins a small, well-formed paraconule (decreasing in size from M1 to M3 at the same time as the steepness of the preprotocrista decreases). The postprotocrista starts with a posterior orientation and turns labially, making an almost complete and low crista obliqua joining the protocone to the metacone. The metaconule is reduced to a slight thickening of the crest on M1–2, without a cuspule (M2 of DPC 15190). Wrinkling of the enamel is present on the broad trigon basin and other parts, and more accentuated on the lingual slopes of the protocone and M1–2 hypocones.

Remarks *Aframonius* has molars emphasizing shearing crests, which suggests a diet including a large proportion of foliage. Its body weight was estimated at 1,600 g by Simons et al. (1995). The marked sexual dimorphism suggests differentiated social structures involving competition between males, and indirectly suggests diurnal habits.

Aframonius is, together with *Afradapis*, the only consensual adapiform of the African fossil record. It was described as a "cercamoniine adapid" by Simons et al. (1995). It is reminiscent of some cercamoniines through its reduction of the anterior premolar series, premolariform p4, and the tall crests and relatively large hypocones of its upper molars. Some *Europolemur* specimens have a similarly labially deflected centrocrista (Godinot, 1988). However, *Aframonius* also has p3 and p4 "unusually complex for a cercamoniine" and p4 "shorter anteroposteriorly and broader labiolingually than many adapids" (Simons et al., 1995), and sexual dimorphism and symphyseal fusion, which have not yet been found in cercamoniines. Similarities with the North American *Mahgarita* have been noted. Among them are several characters found in adapines—for example, broad trigon basin on the upper molars, long lower molars with posteriorly placed entoconid, and broad talonid basin. However, *Mahgarita* differs from *Aframonius* through a series of characters (smaller M3, straight centrocrista, taller and less angled crista obliqua, labiolingually shorter M1, symmetrical protocone of P4, shorter paralophid on m1–2, much simpler and narrower p3–4). The highest overall phenetic similarity is between *Aframonius* and *Caenopithecus*. The larger *C. lemuroides* is generally more massive and more advanced through its well-developed metastylids and mesostyle as well as closer protoconid and metaconid. They otherwise share the adapid characters mentioned, plus large M3 and long, lingually narrowing paralophid, salient postmetacristid with almost incipient metastylid, and talonid basin with deep lingual notch. *Aframonius* has its own autapomorphies in p3–4 morphology and in the lingual shift of the metacone of M1–2 and associated long metastylar crest. *Caenopithecus* is an immigrant in Europe toward the end of the Lutetian, and it is probable that the same adapid stock of Asiatic origin also reached Africa and gave rise to the *Aframonius* lineage. These primitive adapids of Asiatic origin are distinct from cercamoniine notharctids and were referred to as caenopithecines (Godinot, 1998).

Other Asiatic adapiforms have been described since, including early Eocene taxa (Bajpai et al., 2005), so that an extensive Asiatic radiation can be expected.

In sum, it is difficult to exclude the hypothesis that *Aframonius* could be a cercamoniine, having developed convergent adapine-like characters through a more folivorous adaptation. However, it seems more parsimonious to place it within caenopithecines, a group of Asiatic adapids outside the European adapine radiation, which subsequently dispersed to North America, Europe, and Africa. New nonsival-adapid adapiforms recently described from Asia (Marivaux et al., 2001, 2006) lend credit to this scenario.

Genus *AFRADAPIS* Seiffert et al., 2009
AFRADAPIS LONGICRISTATUS Seiffert et al., 2009

Age and Occurrence Birket Qarun locality 2, Fayum Province, Egypt; Birket Qarun Formation, basal Priabonian, around 37 Ma.

Diagnosis Large adapiform close in molar morphology to *Caenopithecus* and *Aframonius*. Differs from other caenopithecines by the following combination of characters: absence of p2; tall P3; tall, very elongated, and trenchant p3; mesostyle on M1–2; and no mesostylid on m1–m3; M3 transversely short with posteriorly elongated talon.

Comments This recently described taxon extends the adapiform diversity in Africa, documents rarely found teeth, and adds one more marked convergence in the anterior dentition between an adapid and some anthropoids (Seiffert et al., 2009). The lower incisors bear lingual keels, and their distal crests were probably aligned into a cropping mechanism. The upper canines have a deep medial groove, and there is no indication of sexual dimorphism in canine size. The lower jaws are fused and have a deep and short corpus. The p2 is lost, and the elongated and tranchent p3 forms a honing mechanism for the upper canine. Such a mechanism is convergent with much younger catarrhines. The high development of crests on the molars of *Afradapis* shows that this species was folivorous. It is by far the largest primate in the late Eocene level of Birket Qarun, occupying a niche that would be reached by anthropoids only several million years later (Seiffert et al., 2009). *Afradapis* is closely related to the younger latest Eocene *Aframonius*, also from the Fayum, Egypt.

Infraorder unnamed, plesion stem Lemuriformes
Family DJEBELEMURIDAE Hartenberger and Marandat, 1992
Genus *ANCHOMOMYS* Stehlin, 1916
ANCHOMOMYS MILLERI Simons, 1997
Figures 19.4B and 19.4D

Age and Occurrence Fayum locality L 41, Fayum Province, Egypt; 45–47 m above the base of the Jebel Qatrani Formation, latest Eocene.

Diagnosis Small primate, p3–m2 length 6.8 mm, known only from its type mandible. Premolariform canine smaller

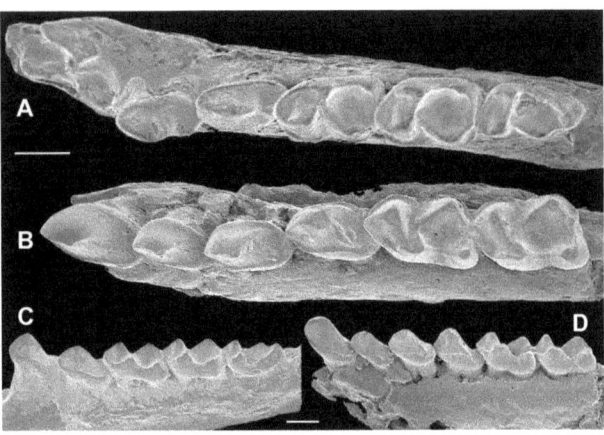

FIGURE 19.4 Mandibles of *Djebelemur martinezi* from the early–middle Eocene of Chambi, Tunisia (A, C) and *Anchomomys milleri* from the early Oligocene of the Fayum, Egypt (B, D). Occlusal (A, B) and lingual (C, D) views. SEM of casts. Scale bars are 1 mm.

than in cercamoniine adapiforms; no p1; p2–p4 relatively low and simple, with continuous lingual cingulum; lower molars crested, without paraconid, with anteriorly directed preprotocristid, narrow paralophid continuous from preprotocristid to base of metaconid, long cristid obliqua. Differs from *Djebelemur* in its larger size, canine with continuous lingual cingulum, p4 shorter and broader, with two broadly distant posterior crests on the posterior wall of its protoconid, somewhat broader lower molars, slightly more mesially extended preprotocristid, and narrower anterior extremity of the paralophid, cristid obliqua more extended on m2, directed toward the labial part of the metaconid (very close to the postvallid).

Description The single rooted p2 is similar in shape to the canine, differing only in its smaller size and more salient postprotocristid joining the cingulum posterolabially (figures 19.4B, 19.4D). The double-rooted p3 is similar but more elongated, with a more accentuated dorsal convexity of its lingual cingulum, and an incipient posterolingual crest on the posterior wall of the protoconid. The p4 is again slightly larger and broader, with a very small cingular talonid, a more salient posterolingual crest of the protoconid, lacking a metaconid. The m1 and m2 are quite similar and show salient crests, including on both sides of the entoconid; the anterior joins the postmetacristid, lingually closing the talonid basin. They have an almost continuous labial cingulum, from its posterior junction with the postcristid (small cingular hypoconulid) to the anterior part of the trigonid. The m1 differs from the m2 in having a slightly more distal position of its metaconid, and correlatively slightly more distal orientation of the paralophid and more lingually shifted direction of the part of the cristid obliqua ascending toward the summit of the metaconid. On m2, the postcristid also shows a marked median ventral concavity.

Genus *DJEBELEMUR* Hartenberger and Marandat, 1992
DJEBELEMUR MARTINEZI Hartenberger and Marandat, 1992
Figures 19.4A and 19.4C

Age and Occurrence Chambi locality in Tunisia, late early or early middle Eocene.

Diagnosis Small primate, p3–m2 length 6.5 mm. The type mandible shows p3 to m3. An associated lower canine is premolariform and shows a short posterior cingulid but no lingual cingulum. The p3 and p4 are remarkably narrow and simple; the posterior side of the protoconid of p4 shows only one median crest, the postprotocristid, which joins the small median cingular talonid cusp (figures 19.4A, 19.4C). In labial view, the molars of *Djebelemur* appear more anteriorly ascending; the anterior cingulum is more abrupt, and the protoconid appears taller relative to the hypoconid than in *A. milleri*. The lower molars of *Djebelemur* differ from those of *A. milleri* in the more abrupt preprotocristid, which is shorter in occlusal view, and an anterior broadening of the paralophid. *Djebelemur* is unusual in the increasingly labial direction of its preprotocristid from m1 to m3. This, added to a narrower paralophid on m3, results in the very narrow transverse trigonid fovea on m3, as opposed to the nearly anteroposterior fovea on m1. On m2 and m3, the cristid obliqua curves toward the summit of the protoconid. In lingual view, the metaconid appears especially low on m1. The crest linking metaconid and entoconid is high in *Djebelemur*, which produces a deep talonid basin and decreases the height of the entoconid in lingal view. In *Djebelemur*, the entoconid is merged into the crest surrounding the talonid basin, which is not the case in *A. milleri*. The m3 is smaller than the m2, being narrower and slightly shorter, despite its third lobe; its entoconid is not differentiated.

Comments on the Djebelemurids The affinities of these two taxa have been difficult to decipher. *Djebelemur* was described as an "Adapidae, ?Cercamoniinae" in line with the views of Gingerich and others favoring an African adapid ancestry for anthropoids (Hartenberger and Marandat, 1992). However, it has been shown that the mode of premolar and molar compaction in *Djebelemur* is markedly different from the morphologies found in European adapiforms (Godinot, 1994). As its name indicates, *A. milleri* was described as a Cercamoniinae, Anchomomyini (Simons, 1997), despite the fact that its mandible shows the low and premolariform canine that reinforces the distinctness of these fossils from the European cercamoniins (Godinot, 2006a). The two species *A. milleri* and *D. martinezi* are very close morphologically, despite their considerable difference in age, and the differences mentioned here that justify two genera. *Djebelemur* is more primitive in its narrow and elongated p4. The remarkably simple and narrow p3 and p4 of these genera seem to typify the group to which they pertain. Similarly narrow premolars are found in the lemuriform *Wadilemur*, and indicate their close relationships, as opposed to adapiforms. The many similarities between djebelemurids and early lemuriforms, such as the anteroposteriorly narrow trigonid with long transverse paralophid joining the metaconid and delimiting a transverse anterior fovea, are responsible for the close phylogenetic relationships found between the two groups in the cladistic analyses of many characters (Seiffert et al., 2003, 2005) as well as in a more qualitative search for shared derived dental characters (Godinot, 2006a). Both types of analyses have led to the proposal that djebelemurids are the primitive sister group of lemuriforms ("crown strepsirrhines" in the terminology of Seiffert et al., 2003, 2005). This is very important because the presence of stem lemuriforms, morphologically preceding the differentiation of the lemuriform tooth comb, in the Eocene of Africa, is the almost definitive proof of the differentiation of lemuriforms on the African landmass.

Two small upper molars from Chambi were described by Hartenberger and Marandat (1992) and attributed to *Djebelemur*. They are extremely bunodont and anthropoid-like (see Seiffert et al., this volume, chap. 22). It seems difficult to associate them with the mandible, especially in view of the crests developed on the m3, which differ markedly from the bunodont aspect of the m3 of *Algeripithecus*. An undescribed and less bunodont isolated M3 from Chambi lends credence to this interpretation, though we should not forget that at present the upper teeth of djebelemurids are unknown, and could provide some surprises.

Infraorder LEMURIFORMES Gregory, 1915 (sensu Hoffstetter, 1977)
Family PLESIOPITHECIDAE Simons and Rasmussen, 1994
Genus *PLESIOPITHECUS* Simons, 1992
PLESIOPITHECUS TERAS Simons, 1992
Figure 19.5

Age and Occurrence Fayum locality L 41, Egypt, base of Jebel Qatrani Formation, latest Eocene.

Diagnosis Medium-sized Strepsirrhine (skull length around 53 mm) having relatively large orbits, a very high muzzle and a marked angle between the palate and the basicranium. Upper canines very large, straight, bilaterally compressed; upper incisors unknown. Lower anterior tooth enlarged, procumbent; three premolars above and below. Upper molars simple, with continuous lingual cingulum and no hypocone. Lower molars relatively broad and having a low relief; mesiodistally compressed trigonid, shelflike paralophid and no paraconid, broad and shallow talonid basin. Molar size decreasing from first to third.

Description A relatively complete and partially crushed cranium and three mandibles of *P. teras* have been reported (Simons, 1992; Simons and Rasmussen, 1994). The cranium shows a very high rostrum, linked to very large and straight upper canines with long roots in the maxilla (figure 19.5).

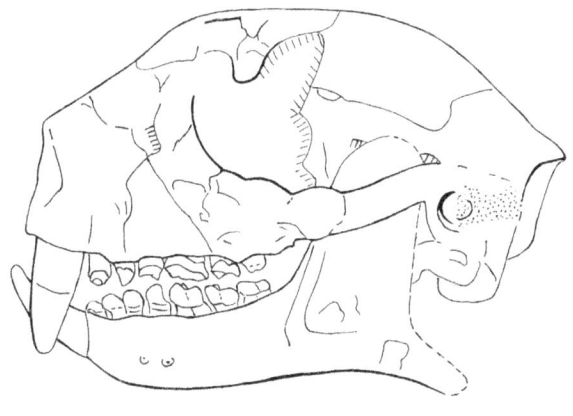

FIGURE 19.5 The skull of *Plesiopithecus teras,* from locality L 41, latest Eocene, Fayum, Egypt. Left side; schematic drawing based on several photographs in Simons and Rasmussen (1994) and Simons (1998). Cranium and mandible were not associated. Scale bar is 1 cm. Courtesy of the National Academy of Science, USA.

The vault is markedly curved, in association with a posteriorly inclined basicranium and a partly ventrally facing nuchal plane. There is a marked angle beween the muzzle and the basicranium (klinorhynchy). The two temporal lines converge and form a sagittal crest at least at the back of the vault. The broken rim of the mastoid on one side suggested to Simons and Rasmussen (1994) that the mastoid probably was large and inflated as in lorisoids. The zygomatic arches appear thin. The orbits are large. Their anterior edge is quite anteriorly placed (above P4 in the text and the figure of the right side of the skull, but not in the reconstruction, of Simons and Rasmussen, 1994). There is a small lacrimal foramen at the margin of the orbit. The infraorbital foramen is ventrally placed above P3.

The large canines are pointed, bilaterally compressed, and bear an anterior and a posterior crest. It is unknown if incisors were present or not, because the premaxillae are broken off. There is no P1, and P2 is small, anteroposteriorly elongated, and bears small anterior and posterior cuspules. The P3 is single cusped, and shows an anterior cingulum and a narrow lingual lobe. The P4 is larger, labiolingually extended, and bears a large protocone and a higher labial cusp. The molars decrease in size from M1 to M3. The three molars are simple, with the three main cusps, a well-formed postprotocrista reaching the metacone on M1–2, and a clear metaconule on M1. M1 and M2 have a posterior and lingual cingulum that is interrupted on the anterior side. The M3 is very small and simple.

The mandibles have an horizontal ramus that increases in height anteriorly (maximum below p3–4), in relation to the very large anterior tooth. The angular process is thin and very salient posteriorly. The ascending ramus is posteriorly inclined, and the broad and flat condyle lies above the dental row. Two mandibles have three premolars between the molars and the large anterior tooth. One mandible has a fourth small anterior tooth, which is either a p1 or a diminutive canine. This small tooth has a narrow, anteriorly inclined crown, overlapping the posterior part of the anterior tooth. The anterior tooth is very large and procumbent. On its posterior side there is a crest, salient and lateral in its distal part, which becomes lower and more median in its proximal part. In medial view, one can see a curved anterior crest, a curved middle crest descending toward the root, and a posterior crest starting proximally as a cingular shelf, ascending and disappearing into the low medial convexity. Such a morphology is reminiscent of that of enlarged anterior incisors in other primates or plesiadapiforms. All three lower premolars are unicuspid; p2 is labiolingually narrow and elongated, whereas p3 and p4 are very round in occlusal outline and bear a well-formed lingual cingulum. The lower molars decrease in size from m1 to m3. They are labiolingually broad, with a relatively lingual protoconid and a correlatively long labial slope (especially on m1, slightly less on m2, and still less on m3). Their trigonids are very short, with a simple paralophid (no paraconid) turning around the base of the metaconid on m1, joining the summit of the metaconid on m3, and intermediate in morphology on m2. All have a broad and shallow talonid basin. The cristid obliqua is anterolabial, being directed toward the summit of the protoconid in occlusal view. The metaconids are low. The postmetaconid cristid is well formed, extended distally, and bears a cuspule ("metastylid" of Simons, 1974) from which a low crest descends into the talonid basin. In lingual view, the cusps appear very low.

Remarks The very unusual and specialized characters of *Plesiopithecus* make this genus an interesting enigma. Its flat and broad lower molars led to its first identification as a bizarre anthropoid (Simons, 1992). Discovery of its skull showed that it had a postorbital bar and was a prosimian, but such an unusual one that it was included in a new superfamily, Plesiopithecoidea (Simons and Rasmussen, 1994). Its large orbits indicate that it was probably a nocturnal species. Its other peculiarities were scrutinized for possible phylogenetic ties but not much in terms of adaptation. The whole cranium and the mandibles show a number of similarities with *Daubentonia*: high muzzle, rounded vault, marked klinorhynchy, orbits anteriorly placed. These characters seem to be correlated with the enlarged upper canine and lower incisor, as in the living aye-aye (Cartmill, 1974). This functional analogy suggests that *Plesiopithecus* probably was engaged in some kind of wood-boring adaptation in search of insect larvae. This would fit with the very low relief of the lower molars. This specialization is not carried as far as in *Daubentonia*, which has evergrowing incisors, a diastema, and further reduced molars, but in many ways *Plesiopithecus* looks as if it were "on the way" toward a daubentoniid-like adaptation. The upper canine is described as vertically implanted in *Plesiopithecus*. However, it looks slightly anteroventrally oriented in the figures and reconstruction of Simons and Rasmussen (1994). This canine is reported to have a small wear facet at its tip, as well as evidence of wear on the distal edge. This should be further studied to evaluate the possibilities of some bark or wood gnawing. Even with these differences in mind, the similarities are enough to suggest some kind of wood processing and soft-bodied insect eating in *Plesiopithecus*.

The phylogenetic relationships of *Plesiopithecus* are unclear. Affinities with living Strepsirrhines were early recognized, leading Simons and Rasmussen (1994) to refer it to "Infraorder cf. Lorisiformes." However, of the seven characters listed by these authors as possibly being shared derived between *Plesiopithecus* and living lorisoids, the similarity in upper molar morphology was hard to understand (lorisoids have a large hypocone lobe and *Plesiopithecus* no hypocone at all), and four others are ambiguous because they can be found in lemuroids as well. Because *Plesiopithecus* also lacks some of the shared dental characters of living lorisoids (molarized P4 and p4, caniniform p2 associated with the tooth comb), Simons and Rasmussen (1994) concluded that *Plesiopithecus* was either the sister taxon of Lorisoidea or the sister taxon of Lemuriformes. Rasmussen and Nekaris (1998) reached similar conclusions, making Plesiopithecidae the sister group of living Lemuriformes. They suggested that the heavily built and low-vaulted cranium and short and deep muzzle could be ancestral lemuriform characters. One might wonder, however, if some of these characters are not linked to the peculiar adaptation of *Plesiopithecus*. Two more recent, comprehensive phylogenetic analyses reached similar conclusions. In the first, molecular characters and one biogeographical character were added to the morphological evidence (Seiffert et al., 2003). *Daubentonia* is grouped with Malagasy lemuroids and *Plesiopithecus* appears as the sister group of all living and fossil lemuriforms. The second analysis is based on 359 morphological characters and recovers the same topology, with *Plesiopithecus* as the sister group to all lemuriforms. However, *Daubentonia* is absent from the sample, and several nodes (Malagasy lemuroids) are constrained by genetic signals (Seiffert et al., 2005). There is a general consensus that *Plesiopithecus* lies close to the base of the lemuriform radiation. However, it could be closer to tooth-combed lemuriforms than to *Djebelemur* and *A. milleri* (Seiffert et al., 2003, 2005), or it might have some significance for daubentoniid origins (Godinot, 2006a).

The identification of the lower anterior tooth is important for an assessment of homology and phylogenetic interpretation. Simons and Rasmussen (1994) and Rasmussen and Nekaris (1998) favored an identification of the large lower anterior tooth of *Plesiopithecus* as a canine, suggesting that it might be derived from a tooth-combed canine. I would rather consider that the small procumbent tooth present on one mandible is a remnant of a disappearing canine, possibly a tooth-combed canine because it has the marked anterior angulation of its crown (hypothesis mentioned by Simons and Rasmussen, 1994), and that the very large anterior tooth, broadly curved as it is, and having an incisiform crown (see prior discussion), is more likely an incisor. In this case, *Plesiopithecus* would be a lemuriform, descended from a tooth-combed ancestor, as often suspected; however, it would also come closer to a possible daubentoniid relative. Further exploration of convergence versus affinity between *Plesiopithecus* and *Daubentonia* is exciting, because a close affinity would have a strong impact on scenarios of Madagascar colonization. Such affinities depend, on one hand, on the homology of the enlarged anterior teeth of *Daubentonia*, which has been contentious, and, on the other hand, on a better understanding of *Plesiopithecus*'s anterior dental morphology and adaptation.

Family GALAGIDAE Mivart, 1864
Genus *WADILEMUR* Simons, 1997
WADILEMUR ELEGANS Simons, 1997
Figures 19.6A and 19.6B

Age and Occurrence Fayum locality L 41, Fayum Province, Egypt; 45–47 m above the base of the Jebel Qatrani Formation, latest Eocene.

Diagnosis Small primate; length of p3–m2 is 8.2 mm. Lower molars with relatively transverse trigonids, continuous paralophid reaching the metaconid and encircling an elongated anterior fovea. The cristid obliqua runs to the middle of the wall between the protoconid and the metaconid on m1–3. The p4 is elongated with a well-formed talonid; p3 elongated and raised anteriorly, also with distinct talonid cusp; elongated single-rooted p2 with cingular talonid and elongated crown raised above the preceding tooth. The upper molars are strongly waisted (emargination of the posterior side), have the labial expansion of the posterior part of the crown linked to the labially deflected postmetacrista, a parastyle, a well-formed and broadly rounded hypocone lobe less salient than in *Saharagalago*, a large cuspidated hypocone linked to two cingular crests, and no prehypocrista. The P4 has a broad protocone lobe and an extreme posterior waisting of its crown. Its low protocone bears a labial preprotocrista and a short posterior postprotocrista nearly linked to a small hypocone, and a posterior cingulum.

Description No tooth-combed incisor or canine has been attributed to *W. elegans;* however, the anteriorly quite complete

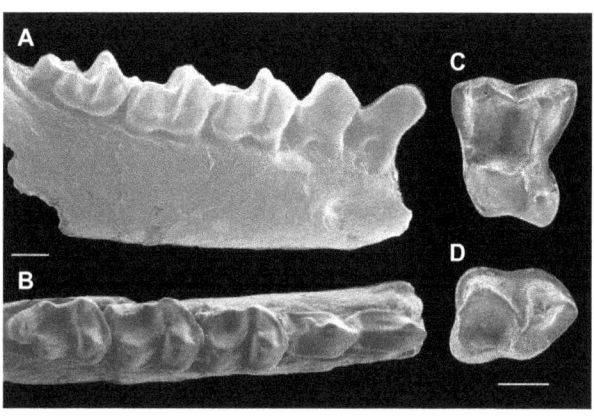

FIGURE 19.6 Dental remains of *Wadilemur elegans* from locality L 41, earliest Oligocene (A, B) and *Saharagalago misrensis* from the late Eocene (C, D), both in the Fayum, Egypt. The mandible of *Wadilemur* is in labial (A) and occlusal (B) views. The M1 (C) and holotype m1 (D) are occlusal views. (A) and (B), courtesy of E. Simons; (C) and (D) are SEM of casts. Scale bars are 1 mm. A, B, with permission of the National Academy of Sciences and courtesy of E. Simons.

mandible DPC 16872 bears a p2 and preserves the alveoli for the canine and the incisors. The p2 and p3 are simple in morphology, having their main cusp anteriorly projecting and their talonid bearing a simple cusp and a posterolingual cingular crest. The p4 shows a transversely developed talonid basin that is lingually inclined, limited by a high hypoconid, and a long cristid obliqua joining the posterior wall of the protoconid. The lower molars show a marked waisting of their labial outline (figures 19.6A, 19.6B). They are of subequal length and narrow posteriorly. The m3 is narrower than the m2 and has a small talonid basin. There is possibly variability in the trigonid, which seems anteroposteriorly narrower on the type specimen (Simons, 1997) than on the specimen illustrated by Seiffert et al. (2005). In figure 2 of Simons (1997), the cristid obliqua of m1 seems to reach the metaconid, contrary to what is written in the text and what can be seen in figure 1E of Seiffert et al. (2005). This appearance is probably the result of wear, and the relatively labial direction of the cristid obliqua in occlusal view seems typical of *Wadilemur*.

On the upper molars, the trigon basin is as broad on M2 as on M1 and is limited by a trenchant postprotocrista without metaconule, and an arcuate preprotocrista bearing a small paraconule. There is almost no labial cingulum on the M1, though a relatively well-differentiated one is present on a referred M2. Both show a labial part of the posterior cingulum.

A partial left femur from L 41, preserving the proximal extremity and a significant portion of the shaft, has been attributed to *W. elegans*. From size alone, it could also pertain to *A. milleri*, but because *Wadilemur* is more common at this quarry and the femur presents several galagid characters, its attribution to *Wadilemur* appears more likely. A series of characters shared with modern galagids include a cylindrical head, a short neck oriented almost perpendicular to the shaft, a large greater trochanter overhanging the anterior aspect of the shaft, and a straight shaft with an elliptical, laterally compressed cross section. However, there are also some differences from living galagids: there is a small but distinct third trochanter, and the angle of the lesser trochanter is greater than in extant galagids and more similar to the angle in *Microcebus murinus*. On the whole, these characters indicate a high percentage of leaping in the locomotor repertoire of *Wadilemur*, but a less specialized locomotion than in typical vertical clingers and leapers.

Remarks Wadilemur was first described as a cercamonine adapid (Simons, 1997); however, the more complete material subsequently found suggests that it is very likely a lemuriform. The p3 and p4 are relatively elongated and labiolingually narrow. Their anterior part and that of m1 overlaps the posterior part of the preceding tooth, revealing a marked crowding of the lower teeth, as opposed to the more horizontal succession found in small adapiforms. The anteriorly inclined and very procumbent p2 indicates a procumbent canine, as in a tooth comb. The morphology of the anterior part of the most complete mandible is reported to be "fully consistent" with the presence of a tooth comb (Seiffert et al., 2005). Moreover, the characters of p4 strongly suggest a galagid line of evolution. The Galagidae is the only living lemuriform family consistently showing a molarized p4 (Maier, 1980). Furthermore, the molarization of the p4 of *Wadilemur* is clearly on the same evolutionary path (Seiffert et al., 2005: figure 1). This, added to other derived galagid characters such as the strong posterior waisting of P4 and M1, clearly ally *Wadilemur* with living galagids. It shows that the family Galagidae was differentiated from other early lemuriforms by the latest Eocene.

Genus *SAHARAGALAGO* Seiffert, Simons and Attia, 2003
SAHARAGALAGO MISRENSIS Seiffert, Simons and
Attia, 2003
Figures 19.6C and 19.6D

Age and Occurrence Birket Qarun locality 2, Fayum Province, Egypt; Birket Qarun Formation, basal Priabonian, around 37 Ma.

Diagnosis Small primate known from two isolated molars. Lower molar close in size to m1–2 of *A. milleri*. "Differs from other living and extinct galagids in having upper molars that are relatively less broad buccolingually, a trenchant prehypocrista coursing from the hypocone to the postprotocrista, and continuous buccal cingulids on lower molars" (Seiffert et al., 2003).

Description and Comments Among the two teeth used to name *Saharagalago*, the upper shows characters strongly evocative of galagids. Seiffert et al. (2003) cite the greater emargination of the distal crown margin, lingual distension of the hypocone lobe with respect to the lingual face of the protocone, absence of lingual and buccal cingula, long and buccally directed postmetacrista, flaring of the molar margin buccal to the metacone, and presence of a prehypocrista (figure 19.6C). However, these characters are not all found in galagids: *Wadilemur* has no prehypocrista, and its M2 has a labial cingulum. The upper molar of *Saharagalago* looks galagid-like, but it would be important to have confirmation of this affinity through other teeth, especially a p4, before accepting the far-reaching conclusion that galagids were differentiated in the basal Priabonian.

Family *Incertae sedis*
Genus *KARANISIA* Seiffert, Simons and Attia, 2003
KARANISIA CLARKI Seiffert, Simons and Attia, 2003
Figure 19.7

Age and Occurrence Birket Qarun locality 2, Fayum Province, Egypt; Birket Qarun Formation, basal Priabonian, around 37 Ma.

Diagnosis Small lemuriform primate. Differs from all lorisoids by the following combination of characters: lower molars with continuous labial cingulid, m3 longer than m2, with a large third lobe. Very short and simple p4 with continous cingulid all around the crown. Upper molars relatively short labiolingually, decreasing in size from M1 to M3, high and continuous crista obliqua reaching the metacone, encircling a mesiodistally broad trigon basin on M1 (with sharp angle of the crista on the posterior wall of the protocone), and narrower trigon basin on M2 and M3. Small crestiform hypocone on a well-formed hypocone lobe, small distal concavity of the crown, continuous lingual cingulum on upper molars and P4. Relatively simple P4 and P3 with one labial cusp and a well-developed protocone.

Remarks The type specimen of *Karanisia clarki* is a mandible bearing m1–m3 (figure 19.7G). The crucial elements that proved the presence of a tooth-combed lemuriform in this early late Eocene locality are another mandibular fragment bearing p4 and showing alveoli for p3, a single rooted p2, a canine and i2, and an isolated canine with an elongated crown very similar to a tooth-combed canine and showing similar traces of wear. Despite the published figure being misleading in suggesting a small canine alveolus (Seiffert et al., 2003: figure 1b), the authors report that the isolated canine would perfectly fit the canine alveolus on the mandible, were they from the same side. The shape of the upper molars of *Karanisia*, especially their broad and rounded hypocone lobe with crestiform hypocone, are tantalizingly similar to some living lorisids. The first phylogenetic analysis of Seiffert et al. (2003) found *Karanisia* nested within living lorisids, as sister group of *Arctocebus*, a living lorisid that possesses a continuous lingual cingulum on its upper molars. This was taken as evidence of an early divergence of lorises and galagos, although the support for the lorisid clade was weak. The subsequent more comprehensive

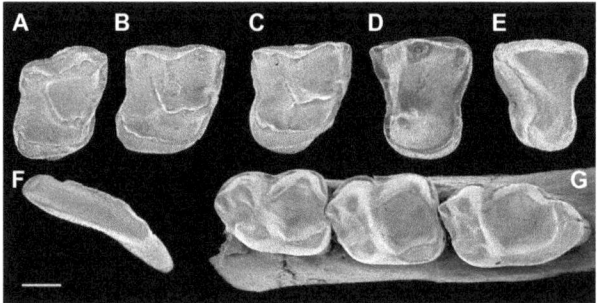

FIGURE 19.7 Teeth and mandible of *Karanisia clarki*, from the late Eocene of the Fayum, Egypt, all in occlusal views. Upper teeth are M3 (A), M2 (B), M1 (C), P4 (D), and P3 (E); lower teeth are a canine (F) and m1–m3 (G). All are SEM of casts; the images in (F) and (G) are inverted left to right to facilitate comparison. Scale bar is 1 mm.

phylogenetic analysis of Seiffert et al. (2005) places *Karanisia* in an unresolved position between lorisoids and lemuroids (i.e., as a basal lemuriform).

Genus *OMANODON* Gheerbrant et al., 1993
OMANODON MINOR Gheerbrant et al., 1993
Figure 19.8

Age and Occurrence Taqah locality, Dhofar Province, Sultanate of Oman, early Oligocene.

Diagnosis Very small primate (m1: 1.75 × 0.95 mm). Upper molars simple, relatively short labiolingually, with continuous crista obliqua and similar broad trigon basin on M1–3, without labial cingulum; small hypocone on M1, very reduced on M2 (very slight swelling on the posterolingual cingulum), lacking on M3. On M1, the hypocone is smaller than in *Wadilemur* and *Saharagalago*, probably close in size to that of *Karanisia* but placed on a shelf that is less salient posteriorly and without a continuous lingual cingulum. In occlusal view, the outline of M1 shows the hypocone lobe salient posterolingually, but the lingual part is narrower than the labial half, and the posterior concavity much less expressed, than in *Wadilemur*, *Karanisia*, and *Saharagalago*. The lower molars are very similar to those of other small African lemuriforms. The paralophid is not continuous until the summit of the metaconid, however, and shows some variability; m1 has a very broad talonid basin and almost an entoconid lobe.

Description The material ascribed to *Omanodon minor* consists of isolated teeth. Variation between the lower molars can be partly due to tooth position (figure 19.8). The type specimen TQ 39 might be an m2. TQ 40 has a cristid obliqua oriented more anteriorly before bifurcating toward the summit of the metaconid (figure 19.8F). Its more discontinuous paralophid, without a well-differentiated premetacristid, indicates a marked variability of the trigonid. A p4 was tentatively attributed to *Omanodon* by Gheerbrant et al. (1993). This attribution is probably correct, and this tooth is very important: it is a very short p4, molarized through the development of a large metaconid, lower than the protoconid, to which it is related by a sharp, only slightly notched protocristid (figure 19.8E). In occlusal view, the protocristid is quite transverse, at an angle of somewhat less than 90° to the preprotocristid.

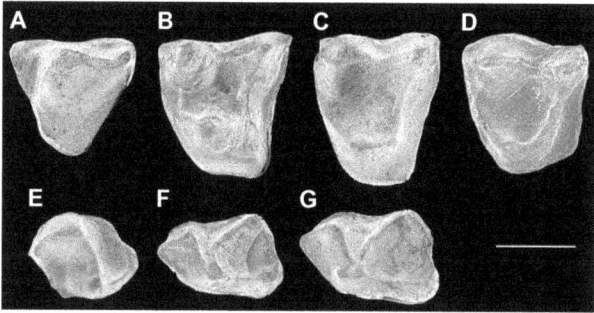

FIGURE 19.8 Teeth of *Omanodon minor*, from the early Oligocene of Oman, all in occlusal view. The upper teeth are interpreted as P3 (A), M1 (B), M2 (C), and M3 (D); the lower teeth are p4 (E), m1 (F), and possibly m2 (G; TQ 39, holotype). All are SEM of casts. Images of (C), (F), and (G) are inverted left to right to facilitate comparison. Scale bar is 1 mm.

Erosion of the anterior border precludes being certain regarding the presence of a paraconid, but the remaining groove suggests that there was probably only a paralophid, almost parallel to the protocristid. A thin, continuous labial cingulid surrounds the protoconid and continues as a small posterior loop around a very small talonid basin, before ascending into a postmetacristid joining the metaconid.

Other isolated teeth referred to *Omanodon* include a probable P3, which has a very large parastyle, a reduced protocone lobe, and a possible upper canine with a very slight lingual cingulum. A tooth ascribed to *Omanodon* as a possible P2, seems not to fit well beside the P3 and will not be commented on further.

Remarks *Omanodon* was described as an anchomomyin cercamoniin. However, Gheerbrant et al. (1993) were aware of the significance of the premolars that they tentatively referred to this genus as possibly indicating an African endemic group. Comparison with subsequently described lemuriforms suggests that it pertains to the same group, but a more precise affinity is not straightforward. The lower molars share with *Saharagalago* the very broad talonid basin, the lingually salient entoconid, and the extent and orientation of the trigonid basin. However, these characters are only slightly less developed in the djebelemurid *A. milleri*. The upper molars have a hypocone that is much smaller than in *Saharagalago* and *Wadilemur*, and only a very slight posterior emargination, so that they appear much less galagid-like. They could be primitive for these characters, and this raises the idea of djebelemurid affinities, though the p4 is, on the contrary, very advanced through its peculiar type of molarization. The p4 was favorably compared with that of *Microcebus* by Gheerbrant et al. (1993), but this tooth in *Microcebus* is much simpler and lacks a metaconid, and a similar short p4 with metaconid can be seen in the living *Loris* (Maier, 1980). Among fossil taxa, the P3 can be compared only with that of *Karanisia*. Its lingual part is clearly more reduced, which could be evocative of cheirogaleids (except *Cheirogaleus*). On the whole, lemuriform affinities may appear more likely, but djebelemurid affinities cannot be excluded without a better knowledge of the upper dentition in this family.

Genus *SHIZARODON* Gheerbrant et al., 1993
SHIZARODON DHOFARENSIS Gheerbrant et al., 1993

Age and Occurrence Taqah locality, Dhofar Province, Sultanate of Oman, early Oligocene.

Diagnosis Very small primate. Lower molars close to those of *Omanodon*, differing from it through a marked labial concavity in the outline, a talonid broader in comparison with the labiolingually narrow trigonid, a larger crestiform paraconid, and a hypoconulid on the postcristid.

Description and Comments Two teeth only, the type specimen and a trigonid, are known of *S. dhofarensis*. The differences from the lower molars of *Omanodon* are not enormous but may suggest a distinct genus.

General Discussion

The fossil record of African Paleogene prosimians is much more interesting now that it allows for a discussion of lemuriform and galagid origins (Seiffert et al., 2003). However, it remains on the whole quite poor (table 19.1). Only one genus, *Plesiopithecus*, is known from cranial remains, and only one postcranial element is attributed to another genus, *Wadilemur*. Four genera are known only from their lower teeth, and four others are known only from isolated teeth. This explains why several genera are still of uncertain phylogenetic affinities.

Despite this poor knowledge, it seems possible to distinguish among these African prosimians two different groups: old African endemics, and a few immigrants having dispersed from Eurasia some time during the Eocene. The latter group includes the ancestor of the only consensual African adapiforms the *Aframonius* and *Afradapis*, and *Afrotarsius* if this taxon is not related to anthropoids. *Afrotarsius* is found in the Oligocene, whereas the other two come from the late Eocene. They could be the result of a dispersal event close or prior to the early late Eocene. Indeed, they are more similar to late middle and early late Eocene Eurasian forms than to late Eocene European or early Oligocene Asian forms. They appear clearly separated from their closest northern relatives by a substantial amount of evolution, which might have taken place either in Africa or in an unknown, probably Asian intermediate area. Despite European cercamoniines having many times been cited in the literature as the probable source of African prosimians, these hypotheses are not confirmed. Dispersal from Asia appears more likely, and this also applies to the source of the other group, the old African endemics: a dispersal of the earliest African strepsirrhines the first stem lemuriforms (possibly including azibiid ancestors), might have occurred at the same time as the dispersal of *Altiatlasius* ancestors during the Paleocene (see Seiffert et al., this volume, chap. 22), or it might have been more recent, in the early Eocene.

The origin of the tooth-combed strepsirrhines the Lemuriformes, has been a matter of debate for a long time, due to the lack of fossils. An Asiatic origin was suggested by Marivaux et al. (2001) in connection with their interpretation of *Bugtilemur*. The discovery of true lemuriforms in the Fayum was a remarkable advance, which led to the subsequent realization that djebelemurids appear as a likely stem group for them (Seiffert et al., 2003, 2005; Godinot, 2006a). The differentiation of tooth-combed lemuriforms in Africa during the Eocene seems now quite probable. This proposition, which agrees with the general view of primate evolution issuing from the fossil record, markedly contrasts with molecular estimates implying that tooth-combed lemuriforms differentiated in the Cretaceous or earliest Paleocene (Martin, 2000; Yoder and Yang, 2004). Despite the incompleteness of the fossil record, these molecular estimates can be refuted (Godinot, 2006a).

However, if the origin of Lemuriformes can now be bracketed in Africa, their phylogeny is far from clear. The only aspect that seems secure is the differentiation of stem galagids in the latest Eocene, as indicated by the affinities of *Wadilemur*. It is tempting to infer that stem galagids were already differentiated at the time of BQ-2, around 37 Ma, because *Saharagalago* consistently clusters with them in the analyses of Seiffert et al. (2003, 2005). However, *Saharagalago* is only known from two isolated teeth. Some characters listed by Seiffert et al. (2003) to support a galagid affinity for *Saharagalago*, like the prehypocrista and the lingual flaring of the hypocone lobe, are also found, less expressed, in lorisids, and the prehypocrista is absent in *Wadilemur*. The place of *Karanisia* moved from stem lorisid (Seiffert et al., 2003) to stem lorisoid (Seiffert et al., 2005), and these analyses have until now

omitted *Omanodon*. Further study of dental characters is needed. *Omanodon* could be linked to stem lemuroids through its lingually reduced P3 and absence of posterior emargination. However, this is very conjectural. The latter character might simply be primitive. Knowledge of the upper teeth of djebelemurids will be critical to assess character polarities within lemuriforms, and probably also to narrow down azibiid affinities. In any case, assessment of the primitive lemuroid dental morphotype is difficult because the Malagasy radiation is broad. A complex, bushy evolution of early lemuriforms can be expected on the African landmass. It will be especially important to identify stem lemuroids—that is, genera closely related to the dispersers into Madagascar. The latter likely belonged to a group of successful African early lemuriforms. Why would that group be unlikely to be found (Seiffert et al., 2005)? In fact, if *Plesiopithecus* really is allied with daubentoniids, some lemuroid history is already documented in Africa (unless the divergence of *Daubentonia* preceded the lemuroid-lorisoid split, as in some molecular studies).

The record of African Paleogene prosimians reveals an interesting ecological aspect in comparison with anthropoids. In the living world, anthropoids are much larger than prosimians. However, the reverse may have been true during a large part of the Paleogene. In one of the richest localities, Fayum L 41, the two largest taxa are prosimians, *Aframonius* and *Plesiopithecus*, weighing approximately 1.6 and 1 kg, respectively (Simons et al., 1995). This is even more conspicuous at Birket Qarun 2, where *Afradapis* is by far the largest species, weighing 2–3 kg (Seiffert et al., 2009). It is only during the Oligocene that anthropoids continued to grow and reached larger sizes, whereas surviving prosimians remained small. Were there larger primate species in the middle and late Eocene, as on the northern continents (5–8 kg), that are not yet sampled? The African fossil record of these periods is still very poor. After their successful colonization of Madagascar, strepsirrhines too, were later able to reach large sizes.

ACKNOWLEDGMENTS

Casts of many of the taxa studied here were provided over the years by E. Simons and more recently by E. Seiffert. E. Gheerbrant commented on a first draft of this chapter, and G. Gunnell provided helpful criticism and suggestions to improve the English of the final draft. E. Seiffert discussed azibiids. C. Chancogne helped the author, and we together took most of the SEM pictures. E. Simons provided those of *Wadilemur*. E. Louis helped in the final preparation of the figures.

Literature Cited

Bajpai, S., V. V. Kapur, J. G. M. Thewissen, D. P. Das, B. N. Tiwari, R. Sharma, and N. Saravanan. 2005. Early Eocene primates from Vastan Lignite Mine, Gujarat, Western India. *Journal of the Palaeontological Society of India* 50(2):12 pp.

Beard, C. K. 1998. A new genus of Tarsiidae (Mammalia: Primates) from the middle Eocene of Shanxi province, China, with notes on the historical biogeography of tarsiers; pp. 260–277 in K. C. Beard and M. Dawson (eds.), *Dawn of the Age of Mammals in Asia*. Bulletin of the Carnegie Museum of Natural History 34.

Beard, C. K., T. Qi, M. R. Dawson, B. Wang, and C. Li. 1994. A diverse new primate fauna from middle Eocene fissure-fillings in southeastern China. *Nature* 368:604–609.

Cartmill, M. 1974. *Daubentonia, Dactylopsila*, woodpeckers and klinorhynchy; pp. 655–670 in R. D. Martin, G. A. Doyle and A. Walker (eds.), *Prosimian Anatomy, Biochemistry and Evolution*. Duckworth, London.

Fleagle, J. G. 1999. *Primate Adaptation and Evolution*. Academic Press, San Diego, 486 pp.

Gheerbrant, E., H. Thomas, J. Roger, S. Sen, and Z. Al-Sulaimani. 1993. Deux nouveaux primates dans l'Oligocène inférieur de Taqah (Sultanat d'Oman): Premiers adapiformes (? Anchomomyini) de la Péninsule Arabique. *Palaeovertebrata* 22:141–196.

Gheerbrant, E., H. Thomas, S. Sen, and Z. Al-Sulaimani. 1995. Nouveau primate Oligopithecinae (Simiiformes) de l'Oligocène inférieur de Taqah, Sultanat d'Oman. *Comptes Rendus de l'Académie des Sciences, Paris*, Série I, 321:425–432.

Gingerich, P. D. 1976. Cranial anatomy and evolution of early Tertiary Plesiadapidae (Mammalia, Primates). *University of Michigan, Papers on Paleontology* 15:1–141.

———. 1977. Radiation of Eocene Adapidae in Europe; in Faunes de Mammifères du Paléogène d'Eurasie. *Geobios, Mémoire Spécial* 1:165–182.

Godinot, M. 1988. Les primates adapidés de Bouxwiller (Eocène Moyen, Alsace) et leur apport à la compréhension de la faune de Messel et à l'évolution des Anchomomyini; pp. 383–407 in J. L. Franzen and W. Michaelis (eds.), *Der eozäne Messelsee—Eocene Lake Messel*. Courier Forschungsinstitut Senckenberg 107.

———. 1994. Early North African primates and their significance for the origin of Simiiformes (= Anthropoidea); pp 235–295 in J. G. Fleagle and R. F. Kay (eds.), *Anthropoid Origins*. Plenum Press, New York.

———. 1998. A summary of adapiform systematics and phylogeny; pp. 218–249 in C. S. Harcourt, R. H. Crompton and A. T. C. Feistner (eds.), *Biology and Conservation of Prosimians. Folia Primatologica* 69, suppl. 1.

———. 2006a. Lemuriform origins as viewed from the fossil record. *Folia Primatologica* 77:446–464.

———. 2006b. Primate origins: A reappraisal of historical data favoring tupaiid affinities; pp. 83–142 in M. Ravosa and M. Dagosto (eds.), *Primate Origins: Adaptations and Evolution*. Springer, New York.

Godinot, M. and M. Mahboubi. 1992. Earliest known simian primate found in Algeria. *Nature* 357:324–326.

———. 1994. Les petits primates simiiformes de Glib Zegdou (Eocène inférieur à moyen d'Algérie). *Comptes Rendus de l'Académie des Sciences, Paris*, Série II, 319:357–364.

Hartenberger, J.-L., and B. Marandat. 1992. A new genus and species of an early Eocene Primate from North Africa. *Human Evolution* 7:9–16.

Hoffstetter, R. 1977. Phylogénie des primates: Confrontation des résultats obtenus par les diverses voies d'approche du problème. *Bulletins et Mémoires de la Société d'Anthropologie de Paris*, Série XIII, 4:327–346.

Kay, R. F., and B. A. Williams. 1994. Dental evidence for anthropoid origins; pp. 361–445 in J. G. Fleagle and R. F. Kay (eds.), *Anthropoid Origins*. Plenum Press, New York.

Maier, W. 1980. Konstruktionsmorphologische Untersuchungen am Gebiss der rezenten Prosimiae (Primates). *Abhandlungen der Senckenbergischen Naturforschenden Gesellschaft* 538:1–158.

Marivaux, L., Y. Chaimanee, P. Tafforeau, and J.-J. Jaeger. 2006. New Strepsirrhine primate from the late Eocene of Peninsular Thailand (Krabi Basin). *American Journal of Physical Anthropology* 130: 425–434.

Marivaux, L., J.-L. Welcomme, P.-O. Antoine, G. Métais, I. M. Baloch, M. Benammi, Y. Chaimanee, S. Ducrocq, and J.-J. Jaeger. 2001. A fossil lemur from the Oligocene of Pakistan. *Science* 294:587–591.

Martin R. D. 2000. Origins, diversity and relationships of lemurs. *International Journal of Primatology* 21:1021–1049.

Rasmussen, D. T. 1990. The phylogenetic position of *Mahgarita stevensi*: protoanthropoid or lemuroid? *International Journal of Primatology* 11:439–468.

Rasmussen, D. T., G. C. Conroy, and E. L. Simons. 1998. Tarsier-like locomotor specializations in the Oligocene primate *Afrotarsius*. *Proceedings of the National Academy of Sciences, USA* 95:14848–14850.

Rasmussen, D. T., and K. A. Nekaris 1998. Evolutionary history of lorisiform primates; pp. 250–285 in C. S. Harcourt, R. H. Crompton and A. T. C. Feistner (eds.), *Biology and Conservation of Prosimians. Folia Primatologica* 69, suppl. 1.

Rasmussen, D. T., and E. L. Simons. 1988. New specimens of *Oligopithecus savagei*, early Oligocene primate from the Fayum, Egypt. *Folia Primatologica* 51:182–208.

Seiffert, E. R. 2006. Revised age estimates for the later Paleogene mammal faunas of Egypt and Oman. *Proceedings of the National Academy of Sciences, USA* 103:5000–5005.

Seiffert, E. R., J. M. G. Perry, E. L. Simons, and D. M. Boyer. 2009. Convergent evolution of anthropoid-like adaptations in Eocene adapiform primates. *Nature* 461:1118–1121.

Seiffert, E. R., E. L. Simons, and Y. Attia. 2003. Fossil evidence for an ancient divergence of lorises and galagos. *Nature* 422:421–424.

Seiffert, E. R., E. L. Simons, T. M. Ryan, and Y. Attia. 2005. Additional remains of *Wadilemur elegans*, a primitive stem galagid from the late Eocene of Egypt. *Proceedings of the National Academy of Sciences, USA* 102:11396–11401.

Simons E. L. 1974. Notes on early Tertiary prosimians; pp. 415–433 in R. D. Martin, G. A. Doyle and A. Walker (eds.), *Prosimian Biology*. Duckworth, London.

———. 1992. Diversity in the early Tertiary anthropoidean radiation in Africa. *Proceedings of the National Academy of Sciences, USA* 89:10743–10747.

———. 1995. Egyptian Oligocene primates: A review. *Yearbook of Physical Anthropology* 38:199–238.

———. 1997. Discovery of the smallest Fayum Egyptian primates (Anchomomyini, Adapidae). *Proceedings of the National Academy of Sciences, USA* 94:180–184.

———. 1998. The prosimian fauna of the Fayum Eocene/Oligocene deposits of Egypt; pp. 286–294 in C. S. Harcourt, R. H. Crompton and A. T. C. Feistner (eds.), *Biology and Conservation of Prosimians. Folia Primatologica* 69, suppl. 1.

Simons, E. L., and T. M. Bown. 1985. *Afrotarsius chatrathi*, first tarsiiform primate (?Tarsiidae) from Africa. *Nature* 313:475–477.

Simons, E. L., T. M. Bown, and D. T. Rasmussen. 1986. Discovery of two additional prosimian primate families (Omomyidae, Lorisidae) in the African Oligocene. *Journal of Human Evolution* 15:431–437.

Simons, E. L., and E. R. Miller. 1997. An upper dentition of *Aframonius dieides* (Primates) from the Fayum, Egyptian Eocene. *Proceedings of the National Academy of Sciences, USA* 94:7993–7996.

Simons, E. L., and D. T. Rasmussen. 1994. A remarkable cranium of *Plesiopithecus teras* (Primates, Prosimii) from the Eocene of Egypt. *Proceedings of the National Academy of Sciences, USA* 91:9946–9950.

Simons, E. L., D. T. Rasmussen, T. M. Bown, and P. S. Chatrath. 1994. The Eocene origin of anthropoid primates: Adaptation, evolution, and diversity; pp. 179–201 in J. G. Fleagle and R. F. Kay (eds.), *Anthropoid Origins*. Plenum Press, New York.

Simons, E. L., D. T. Rasmussen, and P. D. Gingerich. 1995. New cercamoniine adapid from Fayum, Egypt. *Journal of Human Evolution* 29:577–589.

Sudre, J. 1975. Un prosimien du Paléogène ancien du Sahara nord-occidental: *Azibius trerki* n. g. n. sp. *Comptes Rendus de l'Académie des Sciences, Paris*, Série D, 280:1539–1542.

———. 1979. Nouveaux mammifères éocènes du Sahara occidental. *Palaeovertebrata* 9:83–115.

Szalay, F. S., and E. Delson. 1979. *Evolutionary History of the Primates*. Academic Press, New York, 580 pp.

Tabuce, R., M. Mahboubi, P. Tafforeau, and J. Sudre. 2004. Discovery of a highly-specialized plesiadapiform primate in the early–middle Eocene of northwestern Africa. *Journal of Human Evolution* 47: 305–321.

Tabuce, R., L. Marivaux, R. Lebrun, M. Adaci, M. Bensalah, P.-H. Fabre, E. Fara, H. G. Rodrigues, L. Hautier, J.-J. Jaeger, V. Lazzari, F. Mebrouk, S. Peigné, J. Sudre, P. Tafforeau, X. Valentin, M. Mahboubi. 2009. Anthropoid versus strepsirhine status of the African Eocene primates *Algeripithecus* and *Azibius*: craniodental evidence. *Proceedings of the Royal Society B* 276:4087–4094.

Thomas, H., J. Roger, S. Sen, and Z. Al-Sulaimani. 1988. Découverte des plus anciens "Anthropoïdes" du continent arabo-africain et d'un Primate tarsiiforme dans l'Oligocène du Sultanat d'Oman. *Comptes Rendus de l'Académie des Sciences, Paris*, Série II, 306: 823–829.

Yoder A. D., and Z. Yang. 2004. Divergence dates for Malagasy lemurs estimated from multiple gene loci: Geological and evolutionary context. *Molecular Ecology* 13:757–773.

TWENTY

Later Tertiary Lorisiformes

TERRY HARRISON

The lorisiforms are a group of strepsirrhine primates, comprising the extant galagos and lorisids, that are included together in the superfamily Lorisoidea. They share with other crown strepsirrhines the possession of a specialized tooth comb, comprising the lower canines and incisors, reduced upper incisors with a broad central diastema, and a toilet claw on the second pedal digit. Molecular, karyological, and anatomical studies confirm that galagos and lorisids are monophyletic with respect to lemuriforms from Madagascar (Le Gros Clark, 1971; Sarich and Cronin, 1976; Yoder et al., 1996; Roos et al., 2004; Yoder and Yang, 2004; Masters et al., 2007; Horvath et al., 2008). Cranially, they share an ectotympanic ring that is fused to the lateral wall of the auditory bulla, the development of an enlarged ascending pharyngeal artery in the carotid system, and an ethmoid plate (os planum) in the orbital wall (Le Gros Clark, 1971; Groves, 1974; Cartmill, 1975). Lorisoids are relatively small primates, with an average body mass ranging from approximately 60 g to over 1 kg, but most species weigh under 300 g. They are all nocturnal, with a diet consisting of a combination of invertebrates (mainly insects), exudates, and fruits.

Earlier molecular studies produced contradictory results concerning the monophyly of lorisids and galagids respectively (Dene et al., 1976; Sarich and Cronin, 1976; Yoder, 1997; Porter et al., 1997), but support for this inference has comes from a combination of morphological, behavioral, karyological, and recent molecular studies (Dutrillaux, 1988; Zimmermann, 1990; Rasmussen and Nekaris, 1998; Masters and Brothers, 2002; Roos et al., 2004; Masters et al., 2005; Horvath et al., 2008).

The extant lorisids are characterized by a suite of features that include a generalized strepsirrhine dentition with simple nonmolariform p4 and P4; upper molars with a small hypocone, usually lacking a prehypocone crista, without a distinct distolingual lobe or a deep notch in the distal margin of the crown, short and distally directed postmetacrista; lower molars with short subrectangular trigonid; relatively robustly constructed cranium; marked degree of convergence of the orbits; orbits with raised margins; deep mandible and lower face; auditory bulla poorly inflated, with only weak pneumatization of the mastoid region (except for *Arctocebus*); small rounded external ear; first manual digit relatively long, with a wide angle of abduction; vestigial or highly reduced second manual and pedal digits; highly mobile wrist and ankle joints; presence of specialized vascular bundles in the limbs (i.e., retia mirabilia) to facilitate endurance grasping; forelimb less than 15% shorter than the hindlimb, associated with slow cautious quadrupedal climbing; a long torso with an increased number of thoracolumbar vertebrae; transpedicular foramina in the postcervical vertebrae; ribs relatively broad (except *Loris*); relatively short calcaneus and navicular; increased number of sacral vertebrae; and an external tail very short to vestigial (less than 30% of head and body length in potto; less than 6% in other lorisids).

Extant galagids have the following features: a specialized dentition with molariform p4 and P4; upper molars with large hypocone on an expanded distolingual lobe, well-developed prehypocone crista, deeply notched distal margin, long and distobuccally directed postmetacrista; lower molars with elongated, subtriangular trigonid with beaklike mesial margin; relatively lightly constructed cranium; orbits lacking strong frontation and raised margins; shallow mandible and lower face; very inflated auditory bulla with pneumatization extending into the mastoid region; large, mobile external ear; unreduced second manual and pedal digit; more parasagittally constrained wrist and ankle joints; forelimb more than 30% shorter than the hindlimb, associated with specialized leaping behaviors; relatively short thoracolumbar region; ribs relatively narrow; elongated calcaneus and navicular that articulate at an anterior synovial joint; and external tail long and bushy (at least 125% of head and body length)(Hill, 1953; Le Gros Clark, 1956; Napier and Napier, 1967; Cartmill and Milton, 1977; Walker, 1978; Schwartz and Tattersall, 1985; Gebo, 1989; Phillips and Walker, 2002).

The galagos are included together in a single family, the Galagidae, restricted to sub-Saharan Africa. There are at least 24 species currently recognized, belonging to five genera (Kingdon, 1997; Bearder, 1999; Masters and Bragg, 2000; Groves, 2001; Grubb et al., 2003; table 20.1). As for

TABLE 20.1

Classification of the extant and extinct genera of lorisoids

Order	Primates Linnaeus, 1758
Suborder	Strepsirrhini Geoffroy, 1812
Infraorder	Lorisiformes Gregory, 1915
Superfamily	Lorisoidea Gray, 1821
Family	Lorisidae Gray, 1821
Subfamily	Perodicticinae Gray, 1870
Genus	Perodicticus Bennett, 1831
Genus	Arctocebus Gray, 1863
Genus	Pseudopotto Schwartz, 1996
Subfamily	Lorisinae Gray, 1821
Genus	Loris Geoffroy, 1796
Genus	Nycticebus Geoffroy, 1812
Genus	Nycticeboides Jacobs, 1981*
Genus	Microloris Flynn and Morgan, 2005*
Subfamily	Mioeuoticinae (new family-group name)
Genus	Mioeuoticus Leakey, 1962*
Family	Galagidae Gray, 1825
Subfamily	Galaginae Gray, 1825
Genus	Galago Geoffroy, 1796
Genus	Galagoides Smith, 1833
Genus	Otolemur Coquerel, 1859
Genus	Euoticus Gray, 1863
Genus	Sciurocheirus Gray, 1873
Subfamily	Kombinae (new family-group name)
Genus	Komba Simpson, 1967*
Subfamily	incertae sedis
Genus	Progalago MacInnes, 1943*

Asterisk (*) denotes extinct genera.

the less speciose lorisids, nine species are generally recognized, belonging to five genera (Groves, 2001; Brandon-Jones et al., 2004), although the validity of the newly recognized and poorly known *Pseudopotto martini* Schwartz, 1996 has been questioned (Sarmiento, 1998; Grubb et al., 2003; table 20.1). The extant African lorisids, the angwantibos *(Arctocebus)*, potto *(Perodicticus)* and false potto *(Pseudopotto)*, occur in Western and Central Africa, while the Asian lorisids, the slender lorises *(Loris)*, and the slow lorises *(Nycticebus)* are distributed throughout South and Southeast Asia. The nature of the relationships between members of the Lorisidae has proved difficult to resolve (Groves, 1974, 2001; Schwartz and Tattersall, 1985, 1987; Rumpler et al., 1987, 1989; Schwartz, 1992; Yoder, 1997; Masters and Brothers, 2002; Masters et al., 2005), and a number of alternative taxonomic schemes have been proposed. However, the consensus view is that the African and Asian lorisids are separate clades (Yoder, 1997; Rasmussen and Nekaris, 1998; Groves, 2001; Roos et al., 2004; Masters et al., 2005, 2007). The taxonomy of the extant lorisiforms followed here (Grubb et al., 2003; Brandon-Jones et al., 2004) is summarized in table 20.1.

The later Tertiary fossil record of Lorisiformes is quite poor, being restricted to material from the middle to late Miocene of the Siwalik Group of northern Pakistan and to Miocene and Plio-Pleistocene localities in Africa. *Nycticeboides simpsoni* from the late Miocene Dhok Pathan Formation of Pakistan is known from a partial skeleton dated to 8–9 Ma. It is inferred

to be a crown lorisid, probably closely related to *Nycticebus* (Jacobs, 1981; MacPhee and Jacobs, 1986; Flynn and Morgan, 2005). Other isolated teeth from Pakistan, belonging to additional species of lorisids, have been recovered from sediments dating from ~8–16 Ma, including the recently described *Microloris pilbeami* (9–10 Ma; Flynn and Morgan, 2005). A possible crown lorisine, ?*Nycticebus linglom*, is known from the early Miocene of Thailand (17–18 Ma; Mein and Ginsburg, 1997). The fossil lorisiforms from the Miocene of East Africa are currently referred to seven different species belonging to three extinct genera, *Progalago*, *Komba*, and *Mioeuoticus*. An additional species of *Komba* is named here, based on previously described material from the early Miocene of Rusinga and Mfangano Islands in western Kenya. Additional species and genera will eventually need to be recognized in order to accommodate the morphological diversity represented. Isolated teeth and postcranial remains of galagids from Namibia and Egypt are the only specimens known from Miocene localities outside East Africa.

The phylogenetic relationships of the Miocene lorisoids have been the subject of some debate. Walker (1974, 1978) suggested that *Mioeuoticus* is a stem African lorisid or a stem perodicticine, while *Komba* and *Progalago* represent stem galagids. Subsequent authors have generally followed this scheme (Szalay and Delson, 1979; Martin, 1990; Fleagle, 1999; Phillips and Walker, 2000, 2002). However, McCrossin (1992) has argued that *Progalago* is a stem lorisid rather than a galagid. Molecular evidence indicates that galagids and lorisids diverged during the Eocene (Yoder, 1997; Yang and Yoder, 2003; Roos et al., 2004; Horvath, 2008; but see Porter et al., 1997, for a younger divergence date of ~23 Ma), and this is supported by the occurrence of fossil stem galagids (*Saharagalago misrensis* and *Wadilemur elegans*) and a possible lorisid or stem lorisiform *(Karanisia clarki)* from the later Eocene (~35–41 Ma) of Egypt (Seiffert et al., 2003, 2005; Godinot, this volume, chap. 19). One might expect, therefore, to find crown lorisids and galagids among early Miocene faunas, although the possibility of the occurrence of contemporaneous stem group members or basal lorisoids should not be discounted. The taxonomy of the extinct lorisiforms from the later Tertiary shown in table 20.1 is based on their inferred phylogenetic relationships as presented in this chapterr. *Mioeuoticus* and *Komba* are recognized as lorisids and galagids, respectively, but they are sufficiently distinct from extant members of these clades to merit being included in separate subfamilies. The phylogenetic and taxonomic position of *Progalago* is less certain, but the balance of evidence favors it being recognized as a stem galagid.

In addition to Miocene lorisoids, there is a growing number of finds of fossil galagids from the Plio-Pleistocene of Africa. Unfortunately, the fossil record for lorisids from this period is entirely unknown. Much of the Plio-Pleistocene material is rather fragmentary, so it has proved difficult to establish their precise relationships to extant taxa, but they are unequivocally closely related to crown galagids. Only two Plio-Pleistocene species have been named: *Otolemur howelli* from the Omo in Ethiopia and *"Galago" sadimanensis* from Laetoli, Tanzania, and Kapchebrit, Kenya (Wesselman, 1984; Walker, 1987).

History of Research

The first fossil lorisoids in Africa were discovered by A. T. Hopwood at the early Miocene site of Koru in Kenya in 1931, but these specimens were not described for over three decades

(Simpson, 1967). The first description of a Miocene lorisoid was published by MacInnes (1943), based on a single specimen from Songhor in Kenya. This provided the basis for the description of a new genus and species, *Progalago dorae*. The British-Kenya Miocene Expeditions recovered a sizeable collection of fossil lorisoids from the early Miocene sites of Songhor and Rusinga, and the material was described in a monograph published by Le Gros Clark and Thomas (1952). Two new species of *Progalago* were recognized, *P. robustus* and *P. minor*, and these were considered to be primitive galagids. The most complete specimen (KNM-RU 1940) was a partial cranium and associated endocast of an undetermined species of *Progalago*. In 1956, Le Gros Clark described a relatively complete cranium (KNM-RU 2052), referred to *Progalago* sp., that had been found on Rusinga Island by Mary Leakey in 1952 (Le Gros Clark, 1956). An additional partial cranium was later recovered by W. W. Bishop in 1958 at the early Miocene site of Napak in Uganda, and this was designated the holotype of a new genus and species, *Mioeuoticus bishopi* (Leakey, 1962).

Simpson (1965) briefly described the first fossil galagids from the Plio-Pleistocene of East Africa. A collection of jaw fragments, isolated teeth, and postcranial remains, referable to the extant species *Galago senegalensis*, was recovered from FLK I in Bed I, Olduvai Gorge (~1.8 Ma) in northern Tanzania. A taxonomic revision of the Miocene lorisoids from East Africa was undertaken by Simpson (1967). *Progalago dorae* was subdivided into two species, with *Progalago songhorensis* being erected to accommodate a slightly smaller species. *Progalago robustus and P. minor* were transferred to a new genus, *Komba*. Finally, Simpson (1967) described what he erroneously believed to be a new species of lorisid, *Propotto leakeyi*, but the material was later demonstrated to belong to a pteropid bat (Walker, 1969; see Gunnell, this volume, chap. 30).

Walker's (1974, 1978) review of the phylogenetic relationships of the Miocene lorisoids indicated that *Progalago* and *Komba* were galagids, while *Mioeuoticus* was a lorisid. The KNM-RU 2052 cranium was suggested to be a distinct species of *Mioeuoticus* and was subsequently named as the holotype of *M. shipmani* by Phillips and Walker (2000). Following initial studies by Walker (1970, 1974), Gebo (1986, 1989) published a detailed comparative and functional study of the lorisoid postcranial material from the Miocene of East Africa. Additional postcranial remains of fossil lorisoids have been reported from the early Miocene locality of Napak in Uganda (Gebo et al., 1997; MacLatchy and Kityo, 2002).

Further discoveries of Plio-Pleistocene galagids from East African localities were reported in the 1980s. Wesselman (1984) described material from the Shungura Formation (Member B), Omo Valley, Ethiopia, as a new species of galagid, *Galago howelli*. He also identified two isolated teeth as belonging to the extant taxa *Galago demidoff* (now *Galagoides zanzibaricus)* and *Galago senegalensis*. Walker (1987) described a new species of fossil galagid, *Galago sadimanensis*, from the Pliocene localities of Laetoli, Tanzania and Kapchebrit, Kenya. Denys (1987) reported an isolated upper canine of *Galago* sp. indet. from the early Pleistocene Humbu Formation (~1.3–1.7 Ma), Peninj, West Natron, Tanzania.

In 1992, McCrossin described a new species of galagid, *Komba winamensis*, from the middle Miocene locality of Maboko Island in Kenya. As part of his study, McCrossin (1992) reassessed the phylogenetic relationships of the Miocene lorisoids of East Africa and concluded that *Progalago*

was probably a stem lorisid rather than a galagid. However, Rasmussen and Nekaris (1998), who critically reinterpreted the evidence linking *Progalago* with extant lorisids, concluded that these characters were either primitive features of crown strepsirrhines or characters of questionable significance.

In 1994 a mandibular fragment of a galagid was discovered at Kanapoi (~4.1–4.2 Ma) and subsequently referred to cf. *Galago* sp. indet. (Harris et al., 2003). Two upper cheek teeth of a small fossil galagid from the late Miocene (~9–10 Ma) locality of Harasib 3a in Namibia have been figured but not yet described (Conroy et al., 1993, 1996; Rasmussen and Nekaris, 1998). In 1999, McCrossin announced the discovery of a new species of fossil galagid of small size from Maboko Island, based on a mandibular fragment and a calcaneus, but this taxon awaits formal description (McCrossin, 1999). Most recently, Pickford and colleagues have recovered the remains of late Miocene galagids from breccias in the Western Desert of Egypt dated to ~10–11 Ma (Pickford et. al., 2006) and from the Lukeino Formation at Kapsomin in Kenya dated to 5.8–6.1 Ma (Pickford and Senut, 2001; Mein and Pickford, 2006).

In the most recent review of the fossil lorisoids from Africa, Phillips and Walker (2002) reiterate Walker's earlier interpretation (Walker, 1974, 1978) that *Mioeuoticus* is a primitive lorisid, while *Progalago* and *Komba* are galagids. The reader is directed to Rasmussen and Nekaris (1998) and Phillips and Walker (2002) for critical reviews of the history of discoveries and debates regarding the Miocene lorisoids from East Africa.

ABBREVIATIONS

BC, Chemeron Formation; BUMP, Boston Uganda Makerere Paleontology; CA, Chamtwara; FT, Fort Ternan; KNM, National Museum of Kenya; KO, Koru; LAET, Laetoli; MB, Maboko Island; MW, Mfangano Island; MUZM, Makerere University Zoology Museum; RU, Rusinga Island; SO, Songhor. Dental terminology follows that of Szalay and Delson (1979).

Systematic Paleontology

An emended diagnosis of each species of lorisoid from the Miocene and Plio-Pleistocene of Africa is presented below. In addition, brief remarks about the main characteristics, taxonomy, and phylogenetic affinities are given for each genus. The major taxonomic changes adopted here include the recognition of a new species of *Komba* and the identification of *Progalago songhorensis* Simpson, 1967 as a junior synonym of *Mioeuoticus bishopi* Leakey, 1962.

Genus *PROGALAGO* MacInnes, 1943

Distribution Early Miocene (~19 Ma). Songhor and Koru in Kenya, and Napak in Uganda.

Diagnosis Emended from MacInnes (1943); Le Gros Clark and Thomas (1952); Simpson (1967). Differs from *Komba* in the following respects: Mandibular corpus relatively deeper, increasing in depth posteriorly, with a slender flange below and posterior to m3. p4 unicuspid with vestigial metaconid, broad talonid, and small tubercles on distal margin. Lower molars relatively broad, with low and bunodont cusps, trigonid only slightly more elevated than talonid, trigonid mesiodistally short, subquadrate, lacking a prominent mesiobuccal stylid or beaklike projection of mesial margin, and with a buccolingually expanded talonid. M1 and M2 more elongated, with less well-developed hypoparacrista and crista obliqua, no metaconule or paraconule, a less buccodistally directed postmetacrista, a shallower distal marginal notch, relatively smaller hypocone, a less distinct distolingual lobe for the hypocone, and a greater differential in elevation between the buccal and lingual cusps. Differs from *Mioeuoticus* in the following respects: Mandibular corpus with a slender flange below and posterior to m3. Lower molars relatively broad. M1 and M2 with relatively broader crowns, more receding mesiolingual margin, more elevated buccal cusps, large parastyle and metastyle, less well-developed crista obliqua, weakly developed buccal cingulum, distal margin of crown with slight degree of emargination (rather than convex); hypocone positioned more strongly distolingually on a distinct lobe of the crown.

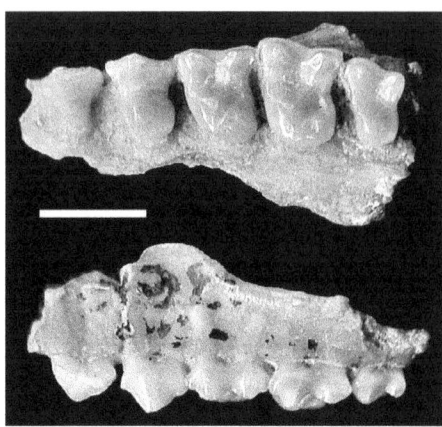

FIGURE 20.2 *Progalago dorae*. KNM-SO 1361, left maxillary fragment with P3–M3. (Top) Occlusal view; (bottom) lateral view. Scale, 5 mm.

PROGALAGO DORAE MacInnes, 1943
Figures 20.1 and 20.2

Holotype KNM-SO 379, left mandibular fragment with p4 and m2 (figure 20.1). Songhor, Kenya.

Distribution Early Miocene (~19 Ma). Songhor and Koru (Legetet Carbonatite Formation, Chamtwara Member of the Kaputtay Nephelinite Agglomerate), Kenya, and Napak IV in Uganda (see table 20.2)

Diagnosis As for genus. See figures 20.1 and 20.2.

Remarks Simpson (1967) recognized two species of *Progalago*, *P. dorae* and *P. songhorensis*. Both species are based on holotypes that are partial mandibles, so details of the lower molars were used as the primary basis for differentiating the taxa. *Progalago songhorensis* was differentiated from the type species by its slightly smaller size, the shape and proportions of the lower molars, and the presence of a buccal cingulum on the trigonid. However, Simpson (1967) was

FIGURE 20.1 *Progalago dorae*. KNM-SO 379, holotype, left mandibular fragment with p4 and m2. (Top) Lateral view; (middle) occlusal view; (bottom) medial view. Scale, 5 mm.

acutely aware of the problem of attributing cranial remains and upper teeth without the advantage of direct association. This was borne out by the fact that Simpson misidentified KNM-RU 2052 as *Progalago dorae* (now recognized as *Mioeuoticus shipmani*). Comparisons of this latter specimen with the upper dentition of *P. songhorensis* provided numerous morphological differences between the two species, which Simpson (1967) suggested were perhaps sufficient to merit generic separation. With the removal of KNM-RU 2052 to *Mioeuoticus* and the subsequent reshuffling of the hypodigms, the upper teeth and maxilla of *Progalago* were sorted into two species based mainly on size (Walker, 1974, 1978). This resulted in *Progalago songhorensis* having upper teeth that were very similar to those of *Progalago dorae*. The only distinction in the upper teeth given by Phillips and Walker (2002) is that the cheek teeth are relatively wider in *P. songhorensis*.

However, my own assessment of the material indicates that all of the upper teeth and maxilla previously attributed to *P. songhorensis* are probably too large to be assigned to this species, and that they should be referred to *P. dorae*. Several observations support this conclusion. First, the range of variation in the upper cheek teeth of the combined samples previously attributed to *P. songhorensis* + *P. dorae* is well within the limits of any single species of extant lorisoid of comparable dental size. Second, the mesiodistal lengths of M2 and m2 are typically subequal in extant strepsirrhines, whereas the upper molars attributed to *P. songhorensis* + *P. dorae* are longer (3.4–4.0 mm) than the lower molars (3.1–3.3 mm) definitively attributed to *P. songhorensis*. Several additional pieces of evidence allow for the correct assignment of *P. songhorensis* specimens. First, an isolated upper molar from Songhor (KNM-SO 1312), of the appropriate size to belong to *P. songhorensis*, is morphologically quite different from that of *P. dorae*. Second, a mandibular fragment of *P. songhorensis* from Rusinga is associated with an M2 that is very similar in size and morphology to KNM-SO 1312. Third, these two upper molars match the M2 in the type specimen of *Mioeuoticus bishopi* from Napak. They share a subrectangular crown that tapers slightly lingually, a relatively narrow crown, a well-developed cingulum along the entire buccal margin, no lingual cingulum, a weakly developed hypoparacrista, a well-developed hypocone connected to the protocone by a short prehypocone crista, the hypocone is positioned distal and slightly lingual to the protocone, and the distal margin is

Species	Localities	Age	Primary References
Mioeuoticus bishopi	Songhor and Rusinga (Hiwegi Fm.), Kenya; Napak I, Uganda	Early Miocene, ~18–19 Ma	Leakey, 1962; Simpson, 1967
M. shipmani	Rusinga (Hiwegi Fm.), Kenya	Early Miocene, ~18 Ma	Le Gros Clark, 1956; Walker, 1974, 1978; Phillips and Walker, 2000
Progalago dorae	Songhor and Koru (Chamtwara Mb. and Legetet Fm.), Kenya; Napak IV, Uganda	Early Miocene, ~19 Ma	MacInnes, 1943; Simpson, 1967; Walker 1969
K. minor	Songhor, Koru (Koru Fm., Legetet Fm., Chamtwara Mb.), and Rusinga (Hiwegi Fm.), Kenya; Napak IV, Uganda	Early Miocene, ~17–20 Ma	Le Gros Clark and Thomas, 1952; Simpson, 1967
K. robustus	Songhor and Koru (Koru Fm., Chamtwara Mb.), Kenya; Napak IV, Uganda	Early Miocene, ~19–20 Ma	Le Gros Clark and Thomas, 1952; Simpson, 1967
K. walkeri sp. nov.	Rusinga (Wayondo, Hiwegi, Kulu Fm.) and Mfangano (Makira Beds), Kenya	Early Miocene, ~16.5–18.0 Ma	Le Gros Clark and Thomas, 1952; Simpson, 1967
Komba sp.	Moroto II, Uganda	Early Miocene, ~17.0–17.5 Ma	Pickford and Mein, 2006
Komba winamensis	Maboko Island, Kenya	Middle Miocene, ~15 Ma	McCrossin, 1992
Galagidae gen. et sp. nov.	Maboko Island, Kenya	Middle Miocene, ~15 Ma	McCrossin, 1999
Lorisidae gen. et sp. nov.	Fort Ternan, Kenya	Middle Miocene, ~14 Ma	Walker, 1978
Galago farafraensis	Sheikh Abdallah, Egypt	Late Miocene, ~10–11 Ma	Pickford et al., 2006
Galagidae indet.	Harasib 3a, Namibia	Late Miocene, ~9–10 Ma	Conroy et al., 1993; Rasmussen and Nekaris, 1998
Galagidae indet.	Kapsomin, Lukeino Fm, Kenya	Late Miocene, ~6 Ma	Pickford and Senut, 2001
Galagidae indet.	Kanapoi, Kenya	Early Pliocene, ~4.1–4.2 Ma	Harris et al., 2003
"*Galago*" *sadimanensis*	Upper Laetolil Beds, Laetoli, Tanzania; Mabaget Fm., Kapchebrit, Kenya	Pliocene, ~3.5–5.0 Ma	Walker, 1987
Otolemur howelli	Mb. B, Shungura Fm., Omo, Ethiopia	Pliocene, ~3.0-3.2 Ma	Wesselman, 1984
Galagoides cf. *zanzibaricus*	Upper Mb. B, Shungura Fm., Omo, Ethiopia	Pliocene, ~3.0 Ma	Wesselman, 1984
Galagidae indet.	Lower Mb. G, Shungura Fm., Omo, Ethiopia	Late Pliocene, ~2.0 Ma.	Wesselman, 1984
Galago senegalensis	Bed I, Olduvai Gorge, Tanzania	Early Pleistocene, ~1.8 Ma	Simpson, 1965
Galagidae indet.	Humbu Fm., Peninj, Tanzania	Early Pleistocene, ~1.3–1.7 Ma	Denys, 1987

convex. Finally, the type specimen of *P. songhorensis* and the type specimen of *M. bishopi* occlude perfectly. The evidence supports the conclusion that the mandibular specimens and isolated lower teeth of *P. songhorensis* from Songhor and Rusinga belong to the same species as the partial cranium that represents the holotype (and previously only recorded specimen) of *Mioeuoticus bishopi* from Napak. Given these findings, *Progalago songhorensis* Simpson, 1967 is here formally considered to be a junior synonym of *Mioeuoticus bishopi* Leakey, 1962.

The phylogenetic relationships of *Progalago* have proved difficult to resolve. MacInnes (1943), Le Gros Clark and Thomas (1952), Walker (1974, 1978), and Phillips and Walker (2002) have argued that it represents a stem galagid, while McCrossin (1992) regards it as a stem lorisid based on the possession of a mandibular corpus that deepens posteriorly, the development of an inferior mandibular flange, and lorisid-like

p4 and upper molars. Simpson (1967) and Rasmussen and Nekaris (1998) contend that there is inadequate basis to link *Progalago* with either the extant lorisids or galagids. As noted by Rasmussen and Nekaris (1998), the features listed by McCrossin (1992) are of questionable significance for assessing the phylogenetic relationships of *Progalago*. A mandibular corpus that deepens posteriorly and the inferior flange are traits that occur variably in extant lorisids. The p4 of *Progalago* does resemble most extant lorisids in being a relatively small ovoid tooth with a well-developed protoconid, and a subsidiary metaconid, but this is the primitive condition for crown strepsirrhines, seen also in cheirogaleids. Similarly, the lorisid-like features of the upper molars of *Progalago* are best interpreted as being closer to the primitive lorisoid condition, rather than synapomorphies shared with extant lorisids. However, *Progalago* does share several features that serve to link it with the galagids: P3 is an elongated triangular tooth,

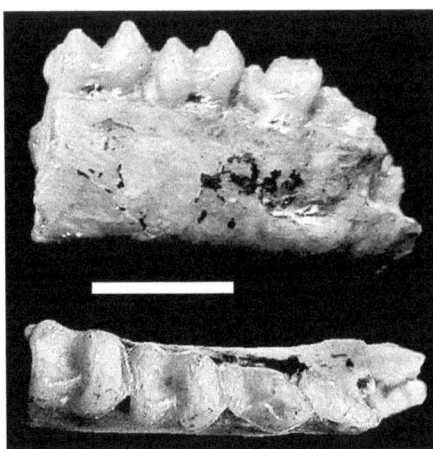

FIGURE 20.3 *Komba robustus.* KNM-SO 501, holotype, right mandibular fragment with p4–m2. (Top) Lateral view; (bottom) occlusal view. Scale, 5 mm.

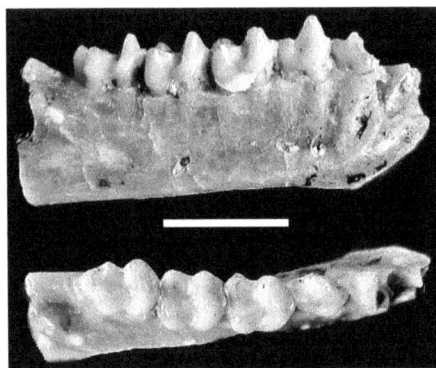

FIGURE 20.4 *Komba robustus.* KNM-SO 1329, right mandibular fragment with p3–m3. (Top) Lateral view; (bottom) occlusal view. Scale, 5 mm.

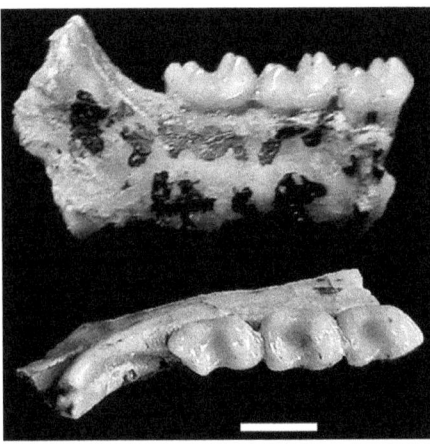

FIGURE 20.5 *Komba minor.* KNM-SO 438, holotype, right mandibular fragment with m1–m3. (Top) Lateral view; (bottom) occlusal view. Scale, 2 mm.

with a well-developed protocone and paracone; P4 shows a tendency toward greater molarization (i.e., incipient development of a hypocone and larger size in relation to M1); M1 and M2 are relatively broad, with a distinct distal notch, a

distolingually offset hypocone, a moderately well-developed hypocone lobe, and a receding mesiolingual margin. Although not as derived as extant galagids or *Komba*, these features are best interpreted as synapomorphies that unite them as a clade (although it should be noted that many of these features have been independently acquired in *Loris tardigradus*). Moreover, further support for this hypothesis might come from the postcranial specimens attributed to *P. dorae*, which are more like those of *Komba* and extant galagids than the lorisid-like postcranials attributed to *Mioeuoticus* (Gebo, 1986, 1989). In conclusion, *Progalago* is provisionally considered a primitive galagid representing the sister taxon of *Komba* + extant galagids.

Genus *KOMBA* Simpson, 1967

Distribution Early to middle Miocene (~15–20 Ma) of Kenya and Uganda (see table 20.2).

Diagnosis Emended from Simpson (1967); McCrossin (1992). Cranium with highly inflated auditory bulla. M1 and M2 with large hypocone, distended distolingual lobe, strong emargination of distal margin between metacone and hypocone, and long distobuccally directed postmetacrista. Mandibular corpus relatively shallow and of constant depth below molar series. Lacks crest or flange on inferior margin of corpus. p4 with distinct metaconid. Lower molars with subtriangular trigonid that has a well-developed mesiobuccal stylid and a protuberant, beaklike mesial margin. Differs from extant galagids in: inflation of auditory bulla not extending into the petromastoid region; P4 bicuspid, lacking metacone and hypocone; p4 with relatively small talonid and lacking entoconid; anterior portion of calcaneus only moderately elongated and lacking synovial joint with navicular.

KOMBA ROBUSTUS (Le Gros Clark and Thomas, 1952)
Figures 20.3 and 20.4

Holotype KNM-SO 501, right mandibular fragment with p4–m2 (figure 20.3). Songhor, Kenya.

Distribution Early Miocene (~19–20 Ma). Songhor and Koru (Koru Formation and Chamtwara Member of the Kapurtay Nephelinite Agglomerate) in Kenya, and Napak IV in Uganda (see table 20.2).

Diagnosis Emended from Le Gros Clark and Thomas (1952); Simpson (1967); McCrossin (1992); Phillips and Walker (2002). A species of *Komba* approximately the size of the extant Allen's galago *(Sciurocheirus alleni)*. Mandibular corpus relatively very shallow. Anterior face of mandibular symphysis with only moderate degree of inclination. Mental foramen single or double, but most frequently double. Molar cusps moderately high and sharp. Buccal cingulum of lower molars highly reduced or absent. m3 is relatively slightly larger than in *K. minor*. (Note: The distinctive features of the calcaneus listed by Phillips and Walker [2002] in their species definition of *K. robustus* do not apply to this species since the specimen [KNM-SO 1364] on which these features were based has been reassigned to *Mioeuoticus bishopi*; see later discussion]. See figures 20.3 and 20.4.

KOMBA MINOR (Le Gros Clark and Thomas, 1952)
Figure 20.5

Holotype KNM-SO 438, right mandibular fragment with m1–m3 (figure 20.5). Songhor, Kenya.

Distribution Early Miocene (~17–20 Ma). Songhor, Koru (Koru Formation, Legetet Carbonatite Formation, and Chamtwara Member of the Kaputray Nephelinite Agglomerate) and Rusinga Island (Hiwegi Formation) in Kenya; Napak IV in Uganda (see table 20.2).

Diagnosis Emended from Le Gros Clark and Thomas (1952); Simpson (1967); McCrossin (1992); Phillips and Walker (2002). A species of *Komba*, smaller than *K. robustus*, approximately the size of the extant Zanzibar galago *(Galagoides zanzibaricus)*. Mandibular corpus relatively very shallow. Anterior face of mandibular symphysis relatively strongly inclined. Mental foramen single, rarely double. Molar cusps very high and sharp. Buccal cingulum of lower molars generally present on trigonid and in median buccal cleft. m3 only slightly greater in area than m2.

KOMBA WINAMENSIS McCrossin, 1992
Figure 20.6

Holotype KNM-MB 20200, right mandibular fragment with p2–m1 (figure 20.6). Maboko Island, Kenya.

Distribution Middle Miocene, ~15 Ma. Maboko Main, Bed 3, Maboko Island, Kenya (see table 20.2).

Diagnosis Emended from McCrossin (1992). A species of *Komba*, larger than *K. robustus* and *K. minor*, approximately the size of the extant greater galago *(Otolemur crassicaudatus)*. Mandibular corpus relatively deep. Multiple mental foramina (3 in the holotype). p2 relatively small in relation to the size of p4. Lower molar cusps moderately low and blunt, with buccal cingulum reduced to inconspicuous vestige on the median buccal cleft, and distobuccal invagination of the postcristid.

KOMBA WALKERI sp. nov.
Figures 20.7–20.9

Etymology Named in honor of Alan Walker for his important contributions to the study of extant and fossil lorisoids.

Holotype KNM-MW 100 (Field Number 44'56). Right mandibular fragment with m1–3 (figure 20.7). Mfangano Island, Kenya.

Referred Specimens KNM-RU 1834, left maxilla with M2–M3; KNM-RU 1930, left maxilla with P4–M2; KNM-RU 1940, partial neurocranium and natural endocast (figure 20.8); KNM-RU 1983, right mandible with m2–m3; KNM-RU 2055, right mandible with m2–m3; KNM-RU 2056, right mandible with m1–m3; KNM-RU 3417, left P3; KNM-RU 3425, left maxilla with M1–M3, right maxilla with M1, left mandible fragment with m1; KNM-RU 3429, right maxilla with M2.

Type Locality Locality A3, Makira Beds, Mfangano Island, Lake Victoria, western Kenya (see Pickford, 1986, for maps of type locality).

Distribution Early Miocene, ~16.5–18 Ma. Rusinga Island (Wayondo, Hiwegi, and Kulu Formations) and Mfangano Island (Makira Beds), Kenya (see table 20.2).

Diagnosis A species of *Komba* similar in overall dental size to *K. robustus* and to the extant *Sciurocheirus alleni*. Mandibular corpus relatively deep. Anterior face of mandibular symphysis only moderately inclined. Molar cusps moderately high and sharp. Buccal cingulum of lower molars highly reduced or absent. m1 and m2 with distobuccal invagination of the postcristid. Differs from *K. robustus* in lower molars slightly larger in overall dental size (average linear dimensions 8.8% greater);

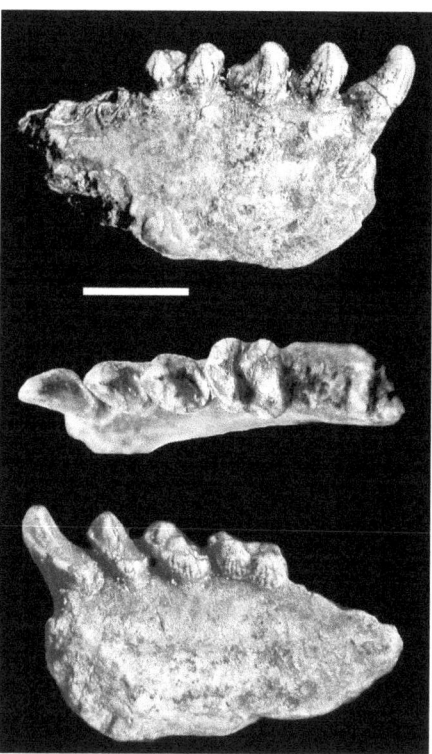

FIGURE 20.6 *Komba winamensis*. KNM-MB 20200, holotype, right mandibular fragment with p2–m1. (Top) Lateral view; (middle) occlusal view; (bottom) medial view. Scale, 5 mm.

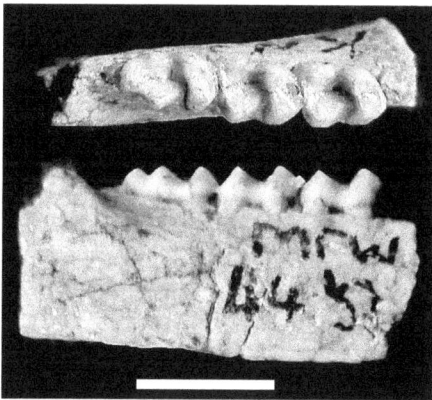

FIGURE 20.7 *Komba walkeri*. KNM-MW 100, holotype, right mandibular fragment with m1–m3. (Top) Occlusal view; (bottom) lateral view. Scale, 5 mm.

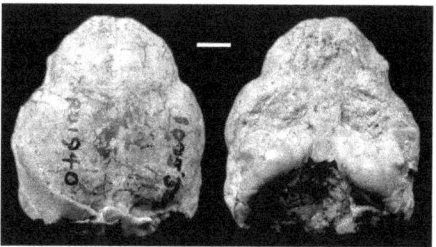

FIGURE 20.8 *Komba walkeri*. KNM-RU 1940. Partial neurocranium and natural endocast. (Top) Dorsal view; (bottom) ventral view. Scale 5 mm.

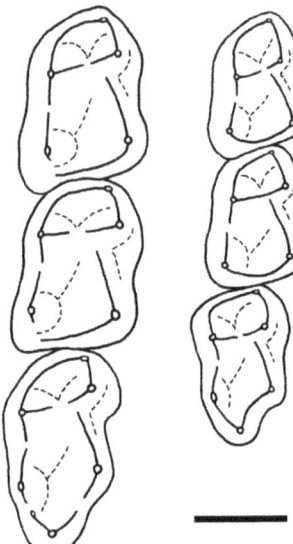

FIGURE 20.9 Comparison of the lower molars of *Komba walkeri* and *Komba robustus*. (Left) *K. walkeri*, KNM-MW 100, right m1–m3. (Right) *K. robustus*, KNM-SO 1329, right m1–m3. Scale, 2 mm. Note the larger size of *K. walkeri*, as well as the entoconid-postcristid invagination, and the narrower m3 with a longer distal heel.

mandibular corpus relatively deeper; m1 and m2 with trigonid not as elevated relative to the talonid (but similar degree of relative cusp height); the postcristid runs from the hypocone to terminate at the distal margin of the entoconid, producing a small but distinct invagination of the postcristid, with a shallow groove separating it from the distobuccal face of the entoconid (as in *K. winamensis*, whereas in *K. robustus* and *K. minor* the crest passes directly to the base of the entoconid); m3 has a relatively narrower crown and a longer distal heel; upper molars are relatively slightly narrower (see figure 20.9). Differs from *K. minor* in being much larger in overall dental size (average linear dimensions of lower molars 37.5% greater); mandibular corpus relatively deeper; mandibular symphysis less inclined; lower molars with trigonid not as elevated relative to the talonid, the cusps not as elevated or sharp, and the buccal cingulum absent to vestigial (more strongly developed in *K. minor*); presence of postcristid invagination (lacking in *K. minor*); m3 relatively larger in relation to the size of m2, with a relatively narrower crown and a longer distal heel. Differs from *K. winamensis* in being much smaller in overall size (average linear dimensions of m1 are 32.4% smaller); lower molars with trigonid that is more elevated relative to the talonid, and cusps higher and sharper.

Remarks In addition to the three previously recognized species of *Komba*, a new species is described here based on material from the early Miocene of Rusinga and Mfangano. Walker (1974, 1978) previously noted that the lower molars of *K. robustus* from Rusinga and Mfangano could be differentiated from those from the type locality of Songhor in being slightly larger, relatively broader, and with less elevated cusps. These and additional features of the dentition and mandible are sufficient to recognize a separate species. Most early Miocene localities have two species of *Komba*, represented by the ubiquitous *K. minor* and either *K. robustus* or *K. walkeri* (see table 20.2). Similarly, *Komba winamensis* from the middle Miocene of Maboko co-occurs with a small species of

galagid (discussed later). Differences in size and dental morphology between coexisting pairs of species imply dietary partitioning, with the smaller species being more specialized for insectivory.

Komba shares a number of derived features with extant galagids, including inflated auditory bulla; P3 elongated and triangular with small paracone; M1 and M2 with distinct distal emargination, receding mesiolingual margin, distolingually displaced hypocone borne on a distinct lobe of the crown, long and distobuccally directed postmetacrista; incipient molarization of p4 with expanded talonid, and small distal tubercles; m1 and m2 with subtriangular trigonid, a distinct mesiobuccal stylid, and a beaklike mesial margin. To these might be added several features of the femur shared exclusively with extant galagids, including a cylindrical head, a straight shaft, and a mediolaterally narrow and elongated distal end (Gebo, 1989). However, extant galagids are more derived than *Komba* in having a more heavily pneumatized mastoid region, more procumbent tooth comb, a p4 that is relatively larger and more molariform bearing prominent metaconid and distal cuspules, P4 molariform with well-developed metacone and hypocone, M3 reduced in size relative to M2, and a greatly elongated calcaneus that articulates with the navicular at an anterior synovial joint. The evidence indicates that *Komba* is the sister taxon of extant galagids. Moreover, since *Komba* is more derived in the degree of molarization of p4 and in the morphology of the upper and lower molars than *Progalago*, it can be deduced that these two Miocene genera represent successive sister taxa to the extant galagids. To reflect these relationships, *Komba* is included here in a separate subfamily, the Kombinae, to distinguish it from the extant galagids, which are included together in the subfamily Galaginae. The type genus of the new family group is designated as *Komba* Simpson, 1967. It is distinguished from the Galaginae in the features listed here.

Genus *MIOEUOTICUS* Leakey, 1962

Distribution Early Miocene (~18–19 Ma). Songhor and Rusinga in Kenya and Napak in Uganda (see table 20.2).

Diagnosis Emended from Leakey (1962); Simpson (1967); Phillips and Walker (2002). Cranium strongly constructed, with distinctly raised temporal ridges. Short and broad snout, with large nasal aperture. Large, upwardly facing orbits with thin orbital floor. Relatively broad interorbital region. Strong maxillary jugum. Shallow lower face, with anterior zygomatic root inferiorly placed. Palate relatively broad, with wide internal nares. The neurocranium is relatively small. Strong nuchal crest that continues laterally to join the suprameatal crest. The bulla is poorly inflated, and pneumatization does not extend into the mastoid region. Mandible relatively deep and increases in depth posteriorly. P2 is unicuspid with two roots and is large in relation to P4. P4 is bicuspid and nonmolariform, with a strong lingual cingulum, at least distolingually. Molars are bunodont with low crowns. M1 and M2 are relatively large, rectangular to nearly square in outline, with convex distal margins. The hypoparacrista is weakly developed. The hypocone is relatively large, and positioned almost directly distal to the protocone. The parastyle is weakly developed and the metastyle is absent. M3 is triangular and lacks a hypocone. p4 unicuspid with vestigial metaconid, broad talonid, and small tubercles on distal margin. Lower molars with trigonid only slightly more elevated than talonid, trigonid mesiodistally short, subquadrate, and lacking a prominent mesiobuccal stylid or beaklike projection of the mesial margin, and with a buccolingually expanded talonid.

MIOEUOTICUS BISHOPI Leakey, 1962
Figures 20.10 and 20.11

Holotype NAP I.3.6/58, anterior portion of the cranium consisting of the face and palate with right P2, P4–M2 and left P4–M3. Napak I, Uganda.

Distribution Early Miocene (~18–19 Ma). Songhor and Rusinga Island (Hiwegi Formation), Kenya; Napak I, Uganda (see table 20.2).

Diagnosis Emended from Leakey (1962); Phillips and Walker (2002). A species of *Mioeuoticus* similar in overall size to the extant potto, *Perodicticus potto*. Single anteriorly facing infraorbital foramen located almost midway between the orbital rim and the alveolar margin of the upper premolars. P3 is three rooted as in *Komba*, *Progalago*, and extant lorisoids. P4 subrectangular and the crown does not taper lingually. It has a cingulum that continues around

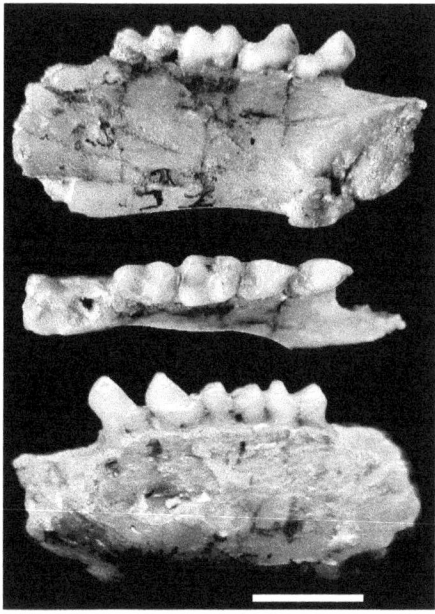

FIGURE 20.11 *Mioeuoticus bishopi.* KNM-SO 1359, left mandibular fragment with p3–m2. (Top) Medial view; (middle) occlusal view; (bottom) lateral view. Scale, 5 mm.

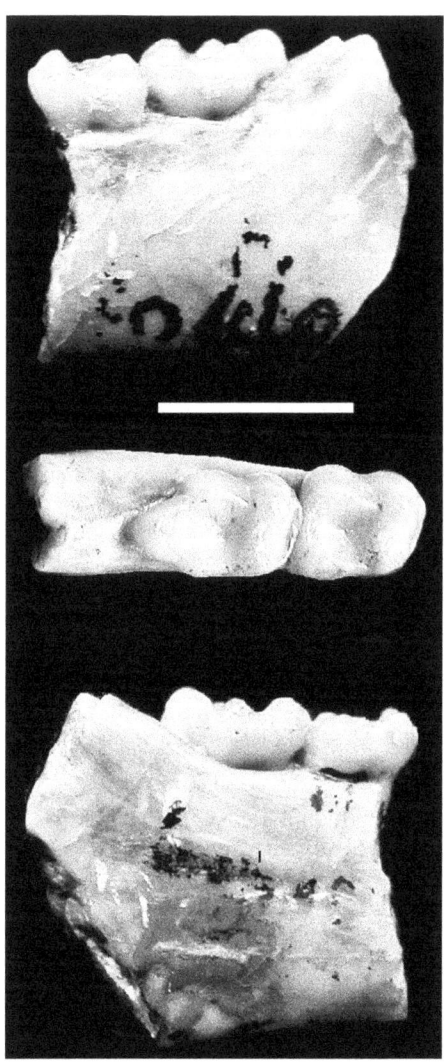

FIGURE 20.10 *Mioeuoticus bishopi.* KNM-SO 469 (holotype of *Progalago songhorensis*), left mandibular fragment with m2–m3. (Top) Lateral view; (middle) occlusal view; (bottom) medial view. Scale, 5 mm.

the distal and lingual margins of the crown. Upper molars smaller and relatively narrower than those of *M. shipmani.* M1 and M2 with hypocone linked to the base of the protocone by a low prehypocone crista. Weakly developed crista obliqua and buccal cingulum. M1 slightly larger in size than M2, and M3 much smaller than either M1 or M2. m2 with protoconid positioned mesial to the metaconid, such that the distal wall of the trigonid is more obliquely aligned, cristid obliqua more obliquely oriented, and moderate development of the buccal cingulum. See figures 20.10 and 20.11.

MIOEUOTICUS SHIPMANI Phillips and Walker, 2000
Figure 20.12

Holotype KNM-RU 2052, partial cranium (figure 20.12). R105b, Rusinga Island, Kenya.

Distribution Early Miocene (~18 Ma). Hiwegi Formation, Rusinga Island, Kenya (see table 20.2).

Diagnosis A species of *Mioeuoticus* similar in overall size to *Mioeuoticus bishopi* and the extant potto, *Perodicticus potto.* Single downward-facing infraorbital foramen nearer the orbital margin than in *M. bishopi.* P3 is two rooted. P4 triangular and the crown tapers lingually. It has a cingulum that is limited to the distal margin and the distolingual corner. Upper molars larger and relatively broader than those of *M. bishopi.* M1 and M2 with hypocone separated from the protocone by a groove. Buccal cingulum and crista obliqua better developed. M1 much smaller than M2, and subequal in size to M3. m2 with protoconid positioned almost transversely opposite the metaconid, the cristid obliqua is more mesially oriented, and no development of a cingulum.

Remarks Although *Mioeuoticus* is less commonly represented at early Miocene sites than *Progalago* and *Komba*, the genus is substantially better known anatomically because of the preservation of a relatively complete cranium of *M. shipmani* (KNM-RU 2052) and a partial cranium of *M. bishopi*

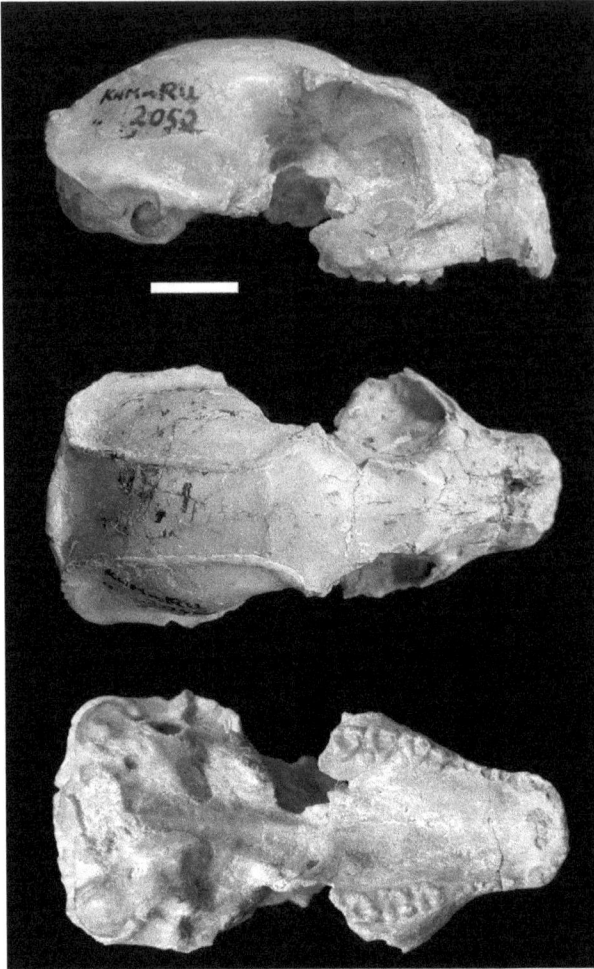

FIGURE 20.12 *Mioeuoticus shipmani.* KNM-RU 2052, holotype, partial cranium. (Top) Lateral view; (middle) dorsal view; (bottom) ventral view. Scale, 5 mm.

(NAP I.3.6/58). The two species of *Mioeuoticus* are similar in overall size and cranial morphology, but are distinguished by features of their upper and lower cheek teeth. The material attributed to *Progalago songhorensis* is here transferred to *Mioeuoticus bishopi*, and the two species are formally synonymized. *Mioeuoticus bishopi* is known primarily from Songhor and Napak, dated at ~19 Ma, but it also occurs in the younger Hiwegi Formation of Rusinga (~17.8 Ma), where it is contemporaneous with *M. shipmani*.

Several features of the cranium confirm the lorisiform affinities of *Mioeuoticus*, including an ethmoid contribution to the medial wall of the orbit and an annular ectotympanic applied to the lateral margin of the bulla. More important, *Mioeuoticus* shares a number of key features linking it with extant lorisids. These include a robustly constructed neurocranium, strongly raised temporal ridges, bulla and mastoid region poorly inflated, posteriorly tilted orbits, wide internal nares, a pronounced suprameatal crest, and relatively narrow upper molars (Walker, 1978; Phillipson and Walker, 2002). Postcranial remains provisionally assigned to *Mioeuoticus* suggest that the specialized slow-climbing adaptations of the extant lorisids may have been incipiently developed in this taxon (Gebo, 1986, 1989; discussed later). This inference is

supported by evidence of relatively small semicircular canals in *M. shipmani* (Walker et al., 2008; discussed later). Nevertheless, *Mioeuoticus* is more primitive than all extant lorisids in having a less inflated bulla and mastoid region, a shallow lower face with a thin orbital floor, a relatively small neurocranium, and a relatively large M3. Based on this evidence, *Mioeuoticus* can be inferred to represent the sister taxon of extant lorisids. As a consequence, it is included here in a separate subfamily, the Mioeuoticinae, to distinguish it from the extant lorisids, which are included in the subfamilies Lorisinae and Perodicticinae. The type genus of the new family-group is designated as *Mioeuoticus* Leakey, 1962. It is distinguished from both Lorisinae and Perodicticinae in the features listed above.

POSTCRANIAL REMAINS FROM THE EARLY MIOCENE OF EAST AFRICA

A number of isolated postcranial elements of lorisoids have been recovered from early Miocene localities in East Africa, but, lacking direct association with craniodental material, their attribution to particular species is uncertain (Walker, 1970, 1974, 1978; Gebo, 1986, 1989; McCrossin, 1992; Gebo et al., 1997; Rasmussen and Nekaris, 1998; MacLatchy and Kityo, 2002). The femora have cylindrical-shaped heads, straight shafts, and anteroposteriorly elongated distal ends as in extant galagids (Walker, 1970; Gebo, 1989). The calcanei have anterior segments that are relatively long compared with those of extant lorisids, but they do not approach the highly derived condition typical of extant galagids, in which the anterior segment is extremely elongated. Broader comparisons show, however, that the fossil calcanei are proportionally most similar to those of cheirogaleids (Szalay, 1976; Gebo 1986), and they probably retain the primitive condition for crown strepsirrhines and lorisiform primates. The fossil tarsals also lack an anterior calcaneonavicular synovial joint, a uniquely derived trait characterizing the extant galagids (Hall-Craggs, 1966; Szalay, 1976; Gebo, 1986). The calcanei are unusual for strepsirrhine primates in having separate anterior calcaneal facets, implying restricted subtalar joint mobility (Gebo, 1986). The navicular facet is associated with a relative deep calcaneal pit, as in extant lorisoids. The tali have moderately long necks and elevated, subparallel trochlear rims. This combination of postcranial features suggest that the early Miocene lorisoids were primarily quadrupedal climbers and leapers, most similar in their locomotor and positional behavior to extant cheirogaleids (Gebo, 1986, 1989). Equally importantly, the postcranial evidence shows that *Komba* and *Progalago* lack important derived features that would link them with either crown lorisids or crown galagids. Possible exceptions are features of the femur (e.g., cylindrical head, straight shaft, and a mediolaterally narrow and elongated distal end), which may represent leaping specializations that would link the early Miocene lorisoids with the extant galagids.

Gebo (1989) attributed a calcaneus from Songhor and a talus from Koru to *Mioeuoticus*, based on their size and the fact that they present more lorisid-like features than the tarsals of *Komba* and *Progalago*. However, until now the genus has not been shown to occur at either of these localities. The calcaneus (KNM-SO 1364) was initially attributed to *K. robustus* (Gebo, 1986) and subsequently reassigned to *Progalago songhorensis* by McCrossin (1992), but should now be assigned to *Mioeuoticus bishopi* (as also indicated by Gebo, 1989). The talus (KNM-KO 159) is possibly attributable to the unnamed lorisid from Koru, which is represented by several isolated teeth of appropriate size (discussed later). A distal humerus from the early Miocene of Napak (MUZM 30), which corresponds in size to either *Progalago dorae* or *Mioeuoticus bishopi*, shares several derived features with extant lorisids, including a cone-shaped trochlear with a downturned medial margin and a deep olecranon fossa (Gebo et al., 1997). Given that it is morphologically distinct from the distal humerus from Songhor attributed to *Progalago* (Gebo, 1989; Gebo et al., 1997), an attribution to *Mioeuoticus* seems reasonable. Similarly, a proximal femur from Napak (BUMP 20) has a large globular head that is elevated above the greater trochanter and suggests a wide range of mobility at the hip as in extant lorisids (MacLatchy and Kityo, 2002). However, the large size of the specimen makes attribution to known species problematic. Nevertheless, these few postcranials confirm that lorisoids with lorisid-like postcranial features, associated with specialized climbing behaviors, had already differentiated by the early Miocene.

Pertinent here is a recent study that has demonstrated a correlation between the radius of curvature of the semicircular canals and the agility of locomotion among extant strepsirrhines (Walker et al., 2008). Compared with lorisids, extant galagids have a relatively much larger canal, which presumably relates to their faster and more agile locomotion. CT scans of the crania of *Komba walkeri* (KNM-RU 1940) and *Mioeuoticus shipmani* (KNM-RU 2052) have shown that the relative size of the semicircular canals are comparable to those of extant galagids and lorisids respectively. This may imply that the acquisition of at least some of the features associated with the specialized modes of locomotion in the modern lorisoid clades was already underway by the early Miocene (Walker et al., 2008).

OTHER MIOCENE LORISOIDS FROM AFRICA

McCrossin (1999) made reference to a mandibular fragment (KNM-MB 18859) of an undescribed galagid from the middle Miocene of Maboko Island. It belongs to a species that is much smaller than *Komba minor*. With an estimated body size of only 30–40 g, about half the size of the smallest extant galagid, *Galagoides demidoff*, it is close in size to the smallest of all extant primates, *Microcebus myoxinus* (Smith and Jungers, 1987; Rowe, 1996). It differs from *Komba* in having higher and more acute cusps, longer occlusal crest, a more elevated trigonid, a deeper and more restricted talonid basin, and no buccal cingulum (McCrossin, 1999). An isolated calcaneus, referred to this species on the basis of size, has the following characteristics: the anterior and medial facets of the calcaneus are well separated, there is no evidence of an anterior calcaneonavicular synovial joint, the anterior calcaneal segment is not as elongated as in extant galagids, but similar to *Komba* and cheirogaleids, and the heel process is straight (McCrossin, 1999). In terms of its positional behavioral, it probably resembled extant cheirogaleids in being a generalized quadrupedal climber, runner, and leaper.

A left maxilla of a lorisoid (KNM-FT 3354) from the middle Miocene (~14 Ma) locality of Fort Ternan in western Kenya belongs to an undescribed species of lorisid about the size of *Perodicticus potto* (figure 20.13). The maxilla below the orbit is relatively deep, and the lateral wall bows out to form a slightly pronounced inferior orbital rim as in extant lorisids. P3 is broken, but it was a triangular tooth with three

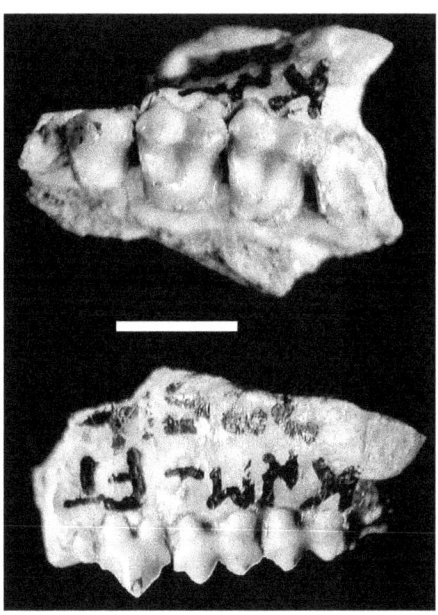

FIGURE 20.13   Lorisidae gen. et sp. nov. KNM-FT 3354, left maxillary fragment with P4–M3. (Top) Occlusal view; (bottom) lateral view. Scale, 5 mm.

roots. P4 is a relatively narrow bicuspid tooth, with a well-developed paracone, much smaller protocone, and prominent mesial and distal styles. M1 and M2 are subrectangular, with lingually tapering crowns, very shallow distal notch, low trigon cusps, well-developed parastyle and metastyle, hypocone lacking (except for a small swelling on the distolingual aspect of lingual cingulum on M2), pronounced shelflike lingual cingulum, well-developed buccal cingulum, especially on M2. M3 is poorly preserved, but it appears to have been a relatively large tooth as in extant lorisids. The Fort Ternan specimen is distinguished dentally from *Mioeuoticus* in the following features: P3 with three roots (as in *M. bishopi*, but only two roots in *M. shipmani*); P4 with taller paracone and less prominent protocone; M1 and M2 with relatively shorter and more lingually tapering crowns, metaconid distinctly smaller than paraconid rather than subequal, mesial and distal margins of crowns slightly concave rather than convex, more strongly developed lingual cingulum, and absence of a hypocone. Clearly the Fort Ternan specimen belongs to a distinct genus and species of lorisid that is more derived than *Mioeuoticus*. Compared with extant and fossil lorisids, it is most similar to *Perodicticus*, and may eventually be referable to the Perodicticinae.

Several upper molars from the early Miocene Chamtwara Member (~19 Ma) of Koru are very similar to those from Fort Ternan, and may belong to the same or a closely related genus of lorisid. They are slightly larger in overall dimensions than the corresponding teeth from Fort Ternan, and are comparable in size to those of *M. shipmani*. The M1 and M2 (KNM-CA 1796 and CA 1797) probably belong to a single subadult individual (figure 20.14). An isolated M3 (KNM-CA 367) is provisionally referred to the same taxon. The M1 and M2 are similar to those from Fort Ternan, and they are distinguished from those of *Mioeuoticus*, in the following features: relatively short and more lingually tapering crowns, metacone much smaller than paracone, mesial and distal margins of crowns slightly concave, strongly developed lingual cingulum, and

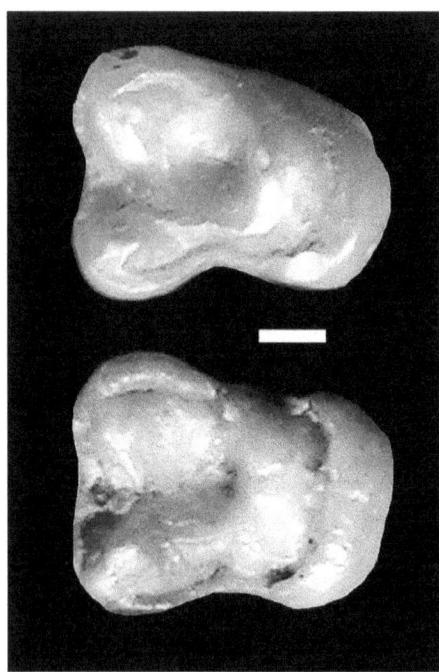

FIGURE 20.14 Lorisidae indet. KNM-CA 1796. (Top) and KNM-CA 1797 (bottom); associated right M1 and M2. Occlusal views. Scale, 1 mm.

Miocene and that suitable ecological conditions prevailed in the western Sahara at this time (Pickford et al., 2006).

Pickford and Senut (2001) and Mein and Pickford (2006) have reported the discovery of a mandible of a galagid, currently undescribed, from the Lukeino Formation (~6 Ma) at the site of Kapsomin in the Tugen Hills in Kenya. Finally, Pickford and Mein (2006) have briefly described an isolated m3 of a galagid from Moroto II, with an estimated age of 17.0–17.5 Ma, which is smaller than *Komba minor*, and may represent a distinct species.

PLIO-PLEISTOCENE LORISOIDS FROM EAST AFRICA

Compared with the Miocene, the Plio-Pleistocene fossil record for lorisoids is relatively much poorer. Only two species of galagids have been formally described, *Otolemur howelli* and *"Galago" sadimanensis*, but a number of fragmentary finds of additional species have been reported from localities in East Africa. There are no recorded fossil lorisids from the Plio-Pleistocene of Africa, and by this time they may already have been restricted to West and Central Africa, where the fossil record is exceedingly poor.

Genus *OTOLEMUR* Coquerel, 1859
OTOLEMUR HOWELLI Wesselman, 1984
Figure 20.15

hypocone absent to vestigial. However, their larger size, broader crowns, and more strongly developed cingulum indicate a species distinction.

An isolated upper molar of a galagid from Harasib 3a in Namibia (Conroy et al, 1993, 1996; Rasmussen and Nekaris, 1998), dated to ~9–10 Ma (Pickford et al., 1994), has the typical characteristics of crown galagids: a distolingually distended hypocone lobe, a distinct prehypocrista, a deeply notched distal margin between the hypocone and metacone, a concave buccal margin with acutely angular mesiobuccal and distobuccal corners, and a long and distobuccally directed postmetacrista. In terms of its overall size and general morphology it is similar to the extant *Galagoides zanzibaricus*, although the crowns are relatively narrower. Today, the only galagid with a range that extends into Namibia is *Galago moholi* (Kingdon, 1997; Groves, 2001), but this species has teeth that are considerably larger than those from Harasib.

Several isolated teeth and postcranials of a small galagid, comparable in size to the extant *Galagoides demidoff*, have recently been recovered from cave breccias at the site of Sheikh Abdallah in the Western Desert of Egypt (Pickford et al., 2006). The associated fauna indicates a late Miocene age (~10–11 Ma). The material formed the basis for the diagnosis of a new species, *Galago farafraensis* Pickford et al., 2006, which is differentiated by its small metaconid on p4 closely associated with the protoconid, reduced metaconules and paraconules on upper molars, and a triangular M3 with a reduced hypocone. Morphologically, it appears to be closest to *G. senegalensis*, although considerably smaller in size, and this has led Pickford et al. (2006) to refer it to the same genus. Today, galagids are restricted to sub-Saharan Africa, with the most northern occurring species, *G. senegalensis*, extending along the Nile Valley as far as central Sudan. The specimens from Egypt demonstrate that galagids occurred more than 1,500 km north of their current distribution during the late

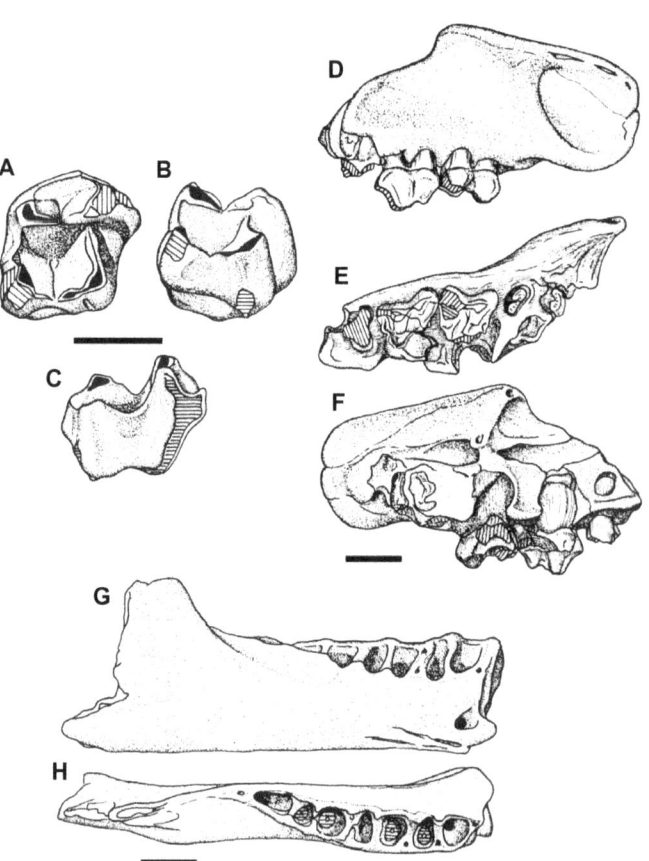

FIGURE 20.15 *Otolemur howelli*. L. 1-378, right m2: A) occlusal view; B) distal view; C) buccal view. L. 1-377, holotype, left maxillary fragment with P4-M1: D) lateral view; E) occlusal view; F) medial view. Omo 229-73-4018, right mandibular fragment, edentulous: G) lateral view; H) dorsal view. Scales, all 2 mm. Reproduced from Wesselman, 1984, with permission from S. Karger AG, Basel.

Holotype L.1–378, left maxillary fragment with P4–M1 (figure 20.15). Member B, Shungura Formation, Omo, Ethiopia.

Distribution Pliocene, ~3.0–3.2 Ma. Member B, Shungura Formation, Omo, Ethiopia (see table 20.2).

Diagnosis Emended from Wesselman (1984). A species of *Otolemur* intermediate in size and morphology between *Otolemur crassicaudatus* and *Sciurocheirus alleni*. It resembles *Otolemur crassicaudatus* in the following features: robust maxilla with a deep, markedly vertical zygoma; square outline of p4 in occlusal view; presence of hypometacrista on m1; mandibular corpus with straight inferior margin and large mental foramen. It resembles *Sciurocheirus alleni* in the following features: mental foramen situated low on the mandibular corpus and located below the posterior root of p4. The taxon is unique in having: mesiodistally short upper cheek teeth that are small in relation to the size of the maxilla; P4 almost square in occlusal outline, with paracone and metacone closely approximated, metacone displaced buccally, hypocone long, and protocone inflated and more mesially situated; M1 mesiodistally shorter than P4, with cusps intermediate in degree of inflation between *S. alleni* and *O. crassicaudatus*; lower molars relatively shorter mesiodistally (as indicated by the alveoli); m2 intermediate in size between *S. alleni* and *O. crassicaudatus*, as well as in the degree of cusp inflation and elevation, depth of basins, and development of occlusal crests.

Remarks This species is known only from a fragmentary maxilla, an isolated m2, and an edentulous mandible from the lower part of the Shungura Formation in the Omo Valley, Ethiopia (Wesselman, 1984). The morphology of the lower face and the molariform P4 are typical of galagids. However, *Otolemur howelli* is unique in a number of features that serve to distinguish it from all extant taxa (see details given earlier). Wesselman (1984) concluded that the fossils are closest in morphology to *Otolemur crassicaudatus* among the extant galagids, and based on these similarities the extinct species is included in the same genus. Currently, the distribution of *Otolemur* extends northward only as far as Uganda and southern Kenya (Kingdon, 1997). The only galagid found in southern Ethiopia today is the widespread Senegal galago, *Galago senegalensis* (Kingdon, 1997).

Genus *GALAGO* Geoffroy, 1796
"GALAGO" SADIMANENSIS Walker, 1987
Figures 20.16–20.17

Holotype LAET 294, right mandibular fragment with p2–m2 (figure 20.16). Laetoli, Tanzania.

Distribution Pliocene, ~3.5–5.0 Ma. Laetoli, Upper Laetolil Beds, Tanzania; Kapchebrit, Mabaget Formation (= Chemeron Formation, Northern Extension), Kenya (table 20.2; figure 20.17).

Diagnosis Emended from Walker (1987). A species of galagid similar in overall dental dimensions to *Galago senegalensis*. It differs from all extant galagids in having the following features: a relatively deeper mandibular corpus with a symphysis that is greater in cross-sectional area, a long axis more vertical, and having a characteristic inverted teardrop (rather than oval) sagittal section; p2 stouter, lower crowned, and more vertically implanted; p4 shorter, more ovoid, and somewhat less molarized, with less well-developed metaconid, shorter and narrower talonid basin, and weakly developed entoconid and hypoconid; lower molars relatively narrower; and m3 relatively small in comparison to m2.

Remarks This species is known from five partial mandibles from the Upper Laetolil Beds at Laetoli (~3.5–3.8 Ma) in

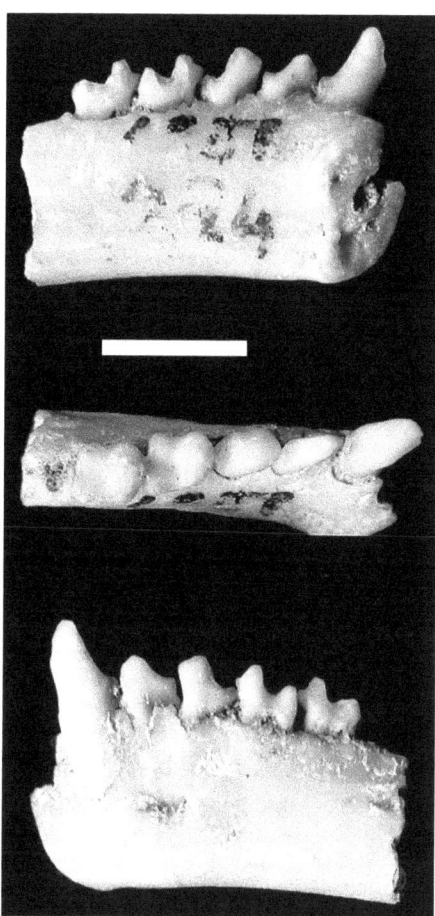

FIGURE 20.16 *"Galago" sadimanensis.* LAET 294, holotype, right mandibular fragment with p2–m2. (Top) Medial view; (middle) occlusal view; (bottom) lateral view. Scale, 5mm.

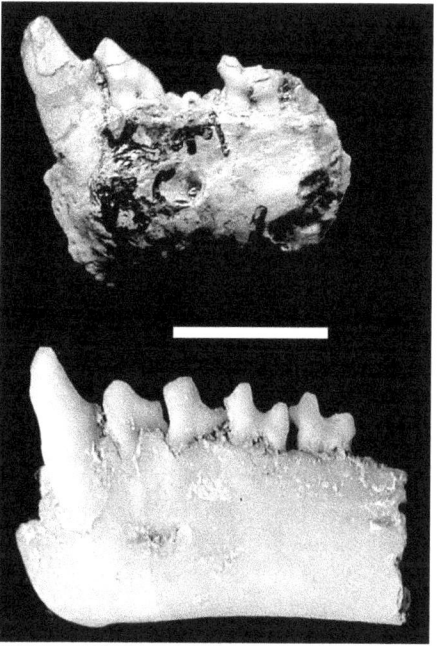

FIGURE 20.17 *"Galago" sadimanensis.* Comparison of KNM-BC 1646 from Kapchebrit. (top) and LAET 294 from Laetoli (bottom); KNM-BC 1646, left mandibular fragment with p2–p3 and m1. LAET 294, left mandibular fragment with p2–m2. Scale, 5 mm.

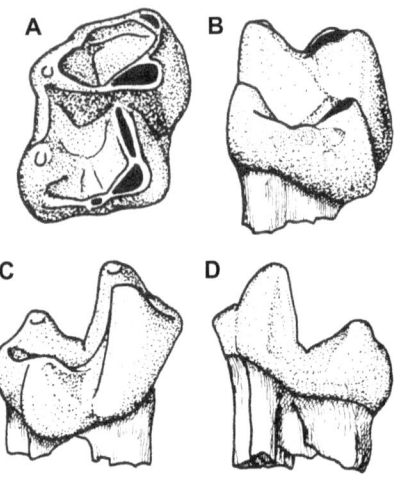

FIGURE 20.18 *Galagoides* cf. *zanzibaricus* from the Omo Valley, Ethiopia. L. 1-521, right m2: A) occlusal view; B) distal view; C) buccal view; D) lingual view. Scale, 1 mm. Reproduced from Wesselman, 1984, with permission from S. Karger AG, Basel.

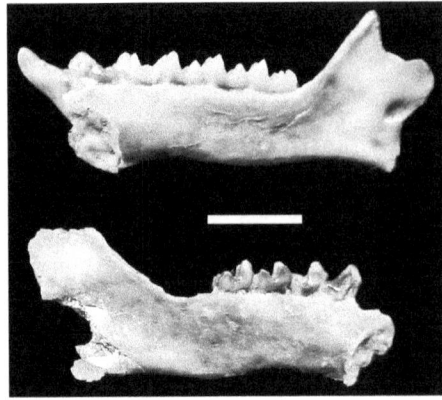

FIGURE 20.19 *Galago senegalensis* from Olduvai Gorge, Tanzania. (Top) Right mandibular fragment with p2–m3, medial view; (bottom) left mandibular fragment with p3–m2, medial view. Scale, 5 mm.

Tanzania (Walker, 1987) and a mandibular fragment from the Mabaget Formation (= Chemeron Formation, Northern Extension; ~5 Ma), in the Tugen Hills of Kenya (Walker, 1987). New material has recently been recovered from Laetoli and awaits description.

The unique features of the mandible and p2, and the reduced size of m3, compared with extant galagids, are best interpreted as autapomorphies (see Walker, 1987). These specialized features are probably functionally linked, and they exhibit some degree of convergence with the derived morphology seen in extant lorisids. However, "*Galago*" *sadimanensis* appears to be more primitive than extant galagids in having a p4 that is less molariform. The p4 is relatively small in relation to m1, with an ovoid crown, small metaconid, and tiny distal cusps. This condition is most closely approximated by *Galago* spp. among extant galagids, although the latter do have more expanded talonid basins. "*Galago*" *sadimanensis* is more derived than *Komba* in the degree of molarization of the p4, but less so than in extant galagids. Thus, it can be inferred to be the sister-taxon of extant galagids (see figure 20.21, later). Its phylogenetic affinities and its unique suite of morphological features negate it being retained in *Galago* sensu lato. A new genus name will be proposed, along with an account of newly recovered material of this species from Laetoli (Harrison, in preparation).

OTHER PLIO-PLEISTOCENE GALAGIDS

A number of isolated teeth from the Plio-Pleistocene of Africa are identifiable as members of Galagidae, but the material is insufficient to assign to a genus or species (see table 20.2). Harris et al. (2003) provided a brief description of a mandibular fragment with m2 of a tiny galagid from Kanapoi in Kenya (dated to ~4.1–4.2 Ma). The size of the molar falls at the lowest end of the size range of *Galagoides zanzibaricus*. However, it differs from the latter in having less elevated molar cusps, a relatively deeper mandibular corpus, and an m3 (judging from the size of its roots) that was longer than m2. This specimen clearly belongs to a previously undescribed species, but formal taxonomic designation will have to await the recovery of better material.

Wesselman (1984) described a fragmentary m2 from lower Member G (~2.0 Ma) of the Shungura Formation, Omo, Ethiopia, which he referred to *Galago senegalensis*. However, the tooth is much smaller than lower molars of *G. senegalensis* and it is even somewhat smaller than those of *Galago moholi*. It is comparable in overall size to *Galagoides zanzibaricus*, although it apparently differs in morphology (Wesselman, 1984). Referral to an extant species does not seem appropriate given these distinctions, and it is best to consider this specimen as Galaginae indet. Wesselman (1984) also described an isolated m2 from upper Member B of the Shungura Formation (~3.0 Ma) that is metrically and morphological very similar to *Galagoides zanzibaricus*, except that the crown is slightly narrower (figure 20.18). It is here tentatively referred to as *Galagoides* cf. *zanzibaricus*. The current geographic distribution of *Galagoides* does not extend as far north as Ethiopia, with *G. zanzibaricus* and *G. thomasi* in Kenya and Uganda being the most northern species.

Simpson (1965) described several mandibular fragments of fossil galagids recovered from excavations at FLK I, Bed I, Olduvai Gorge in northern Tanzania (dated to ~1.8 Ma) (figure 20.19). Additional finds from the same locality include isolated teeth and postcranial remains. These were referred to *Galago senegalensis*. Simpson (1965) compared the specimens to *G. moholi* (which at that time was recognized as a subspecies of *G. senegalensis*), and found that the dimensions of the lower teeth in the Olduvai fossils were 7% larger. My own data confirm that the Olduvai material is identical in size to, or very slightly smaller than, *G. senegalensis braccatus*, the subspecies that inhabits Kenya and northern Tanzania today (Kingdon, 1997; Grubb et al., 2003). Moreover, the morphology of the dentition is identical. The postcranials are also comparable to those of *G. senegalensis* (Szalay and Delson, 1979; Gebo, 1986), and they represent the earliest record of the highly derived hindlimb morphology typical of modern-day galagids.

Finally, Denys (1987) reported an isolated upper canine of a galagid from the Humbu Formation at Peninj, northern Tanzania, dated to ~1.3–1.7 Ma, which is consistent in morphology and only slightly smaller than examples of the corresponding tooth in extant *Galago senegalensis*.

General Discussion

Walker (1974, 1978) and Phillips and Walker (2000, 2002) have argued that *Mioeuoticus* is a stem lorisid or a stem perodicticine, while *Komba* and *Progalago* represent stem galagids. An alternative viewpoint was presented by McCrossin (1992), in which *Mioeuoticus* and *Progalago* were considered to be stem lorisids, while *Komba* represents a stem galagid. Rasmussen and Nekaris (1998) consider the fossil evidence insufficient to resolve the relationships between the Miocene and extant lorisoids. The poor fossil record of lorisoids from the later Tertiary is certainly an impediment to assessing the precise affinities of the currently known extinct taxa, but some tentative conclusions about their phylogenetic relationships can be reached. *Komba* and *Mioeuoticus* from the Miocene of East Africa are best considered as a stem galagid and a stem lorisid, respectively. They are included here in their own subfamilies in order to reflect these relationships at a higher taxonomic level. The phylogenetic relationships of *Progalago* are less evident. As noted by McCrossin (1992), *Progalago* does possess

some specialized features shared with the extant lorisids, and a case could be made to include them as stem members of this clade. However, the balance of evidence favors recognizing them as primitive galagids that are the sister taxon of *Komba* + extant galagids (see figure 20.20).

The mid-Pliocene galagid from the Omo Valley in Ethiopia, identified as *Otolemur howelli*, is provisionally considered to be a crown galagid closely related to the modern greater galagos. However, it is rather poorly known, and better material is needed to confirm its phylogenetic affinities. The fossil galagid from the Pliocene of Laetoli and Kapchebrit, *"Galago" sadimanensis*, is inferred to be the sister taxon of extant galagids and will be formally transferred to a new genus to reflect this relationship (figure 20.21).

ACKNOWLEDGMENTS

I am grateful to Bill Sanders and Lars Werdelin for inviting me to contribute to this volume. Permission to study fossil collections was granted by the governments of Kenya and Tanzania. I thank the curators and staff of the National Museums of Kenya, the National Museum of Tanzania, the Natural History Museum in London, and the American Museum of Natural History for access to collections. The help, advice, and comments of the following individuals are gratefully acknowledged: Peter Andrews, Susanne Cote, Rich Kay, Emma Mbua, Mary Muungu, Martin Pickford, Luca Pozzi, and Alan Walker. I am especially grateful to Martin Pickford for access to unpublished data. Research was supported by grants from the National Geographic Society, the Leakey Foundation, and the National Science Foundation (Grants BCS-9903434 and BCS-0309513).

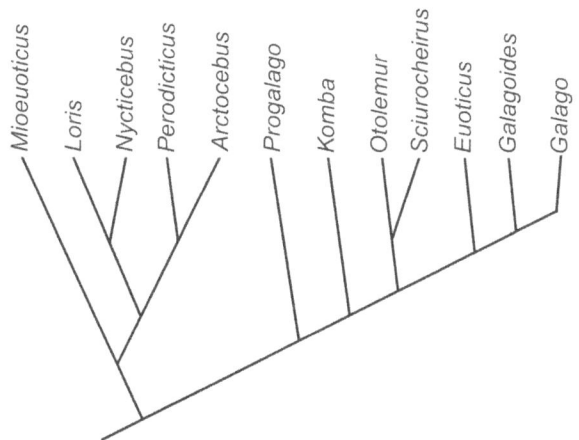

FIGURE 20.20 Cladogram illustrating the inferred phylogenetic relationships between the extant genera of lorisoids (excluding *Pseudopotto*) and the Miocene lorisoids from Africa.

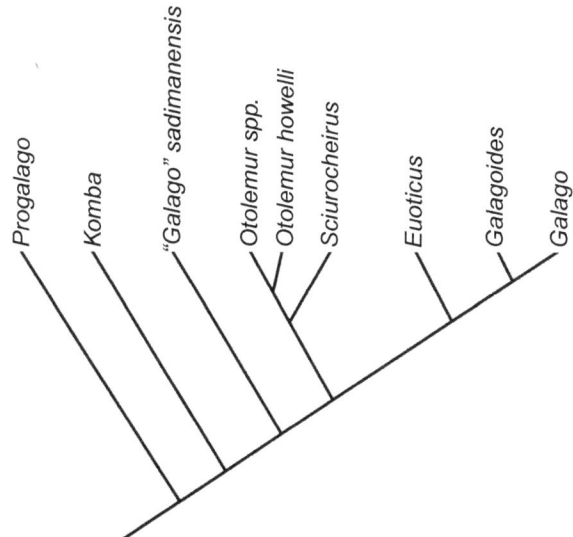

FIGURE 20.21 Cladogram illustrating the inferred phylogenetic relationships between extant and fossil galagids.

Literature Cited

Bearder, S. K. 1999. Physical and social diversity among nocturnal primates: A new view based on long-term research. *Primates* 40:267–282.

Brandon-Jones, D., A. A. Eudey, T. Geissmann, C. P. Groves, D. J. Melnick, J. C. Morales, M. Shekelle, and C.-B. Stewart. 2004. Asian primate classification. *International Journal of Primatology* 25:97–164.

Cartmill, M. 1975. Strepsirhine basicranial structure and the affinities of the Cheirogaleidae; pp. 313–354 in W. P. Luckett and F. S. Szalay (eds.), *Phylogeny of the Primates: A Multidisciplinary Approach*. Plenum Press, New York.

Cartmill, M., and K. Milton. 1997. The lorisiform wrist joint and the evolution of "brachiating" adaptations in the Hominoidea. *American Journal of Physical Anthropology* 47:249–272.

Conroy, G. C., M. Pickford, B. Senut, and P. Mein. 1993. Diamonds in the desert: The discovery of *Otavipithecus namibiensis*. *Evolutionary Anthropology* 2:46–52.

Conroy, G. C., B. Senut, D. Gommery, M. Pickford, and P. Mein. 1996. Brief Communication: New primate remains from the Miocene of Namibia, southern Africa. *American Journal of Physical Anthropology* 99:487–492.

Dene, H., M. Goodman, W. Prychodko, and G. W. Moore. 1976. Immunodiffusion systematics of the primates. III. The Strepsirhini. *Folia Primatologica* 25:35–61.

Denys, C. 1987. Micromammals from the West Natron Pleistocene deposits (Tanzania). Biostratigraphy and paleoecology. *Sciences Géologiques Bulletin* 40:185–201.

Dutrillaux, B. 1988. Chromosome evolution in primates. *Folia Primatologica* 50:134–135.

Fleagle, J. G. 1999. *Primate Adaptation and Evolution*. 2nd ed. Academic Press, San Diego, 596 pp.

Flynn, L. J, and M. E. Morgan, 2005. New lower primates from the Miocene Siwaliks of Pakistan; pp. 81–101 in D. E. Lieberman, R. J. Smith, and J. Kelley (eds.), *Interpreting the Past: Essays on Human, Primate, and Mammal Evolution in Honor of David Pilbeam*. Brill Academic Publishers, Boston.

Gebo, D. L. 1986. Miocene lorisids: The foot evidence. *Folia Primatologica* 47:217–225.

———. 1989. Postcranial adaptation and evolution in Lorisidae. *Primates* 30:347–367.

Gebo, D. L., L. MacLatchy, and R. Kityo. 1997. A new lorisid humerus from the early Miocene of Uganda. *Primates* 38:423–427.

Groves, C. 1974. Taxonomy and phylogeny of prosimians; pp. 449–473 in R. D. Martin, G. A. Doyle, and A. C. Walker (eds.), *Prosimian Biology*. Duckworth, London.

———. 2001. *Primate Taxonomy*. Smithsonian Institution Press, Washington, D.C., 350 pp.

Grubb, P., T. M. Butynski, J. F. Oates, S. K. Bearder, T. R. Disotell, C. P. Groves, and T. T. Struhsaker. 2003. Assessment of the diversity of African primates. *International Journal of Primatology* 24:1301–1357.

Hall-Craggs, E. C. B. 1966. Rotational movements in the foot of *Galago senegalensis*. *Anatomical Record* 154:287–294.

Harris, J. M., M. G. Leakey, and T. E. Cerling. 2003. Early Pliocene tetrapod remains from Kanapoi, Lake Turkana Basin, Kenya. *Contributions in Science* 498:39–113.

Hill, W. C. O. 1953. *Primates: Comparative Anatomy and Taxonomy: Vol. 1, Strepsirhini*. Edinburgh University Press, Edinburgh, 798 pp.

Horvath, J. E., D. W. Weisrock, S. L. Embry, I. Fiorentino, J. P. Balhoff, P. Kappeler. G. A. Wray, H. F. Willard, and A. D. Yoder. 2008. Development and application of a phylogenomic toolkit: Resolving the evolutionary history of Madagascar's lemurs. *Genome Research* 18:489–499

Jacobs, L. L. 1981. Miocene lorisid primates from the Pakistan Siwaliks. *Nature* 289:585–587.

Kingdon, J. 1997. *The Kingdon Field Guide to African Mammals*. Academic Press, San Diego, 464 pp.

Leakey, L. S. B. 1962. Primates; pp. 1–18 in W. W. Bishop (ed.), *The Mammalian Fauna and Geomorphological Relations of the Napak Volcanics, Karamoja*. Records of the Geological Survey of Uganda 1957–58, Kampala, Uganda.

Le Gros Clark, W. E. 1956. *A Miocene Lemuroid Skull from East Africa*. Fossil Mammals of Africa 9. British Museum (Natural History), London, 6 pp.

———. 1971. *The Antecedents of Man*. 3rd ed. Edinburgh University Press, Edinburgh, 394 pp.

Le Gros Clark, W. E., and D. P. Thomas. 1952. *The Miocene Lemuroids of East Africa*. Fossil Mammals of Africa 5. British Museum (Natural History), London, 20 pp.

MacInnes, D. G. 1943. Notes on the East African primates. *Journal of the East Africa and Uganda Natural History Society* 39:521–530.

MacLatchy, L., and R. Kityo. 2002. A Lower Miocene lorid femur from Napak, Uganda. *American Journal of Physical Anthropology* 34 (suppl.):104–105.

MacPhee, R. D. E., and L. L. Jacobs. 1986. *Nycticeboides simpsoni* and the morphology, adaptations, and relationships of Miocene Siwalik Lorisidae. *Contributions to Geology, University of Wyoming, Special Paper* 3:131–161.

Martin, R. D. 1990. *Primate Origins and Evolution: A Phylogenetic Reconstruction*. Princeton University Press, Princeton, 804 pp.

Masters, J. C., N. M. Anthony, M. J. de Wit, and A. Mitchell. 2005. Reconstructing the evolutionary history of the Lorisidae using morphological, molecular and geological data. *American Journal of Physical Anthropology* 127:465–480.

Masters, J. C., M. Boniotto, S. Crovella, C. Roos, L. Pozzi, and M. Delpero. 2007. Phylogenetic relationships among the Lorisoidea as indicated by craniodental morphology and mitochondrial sequence data. *American Journal of Primatology* 69:6–15.

Masters, J. C., and N. P. Bragg. 2000. Morphological correlates of speciation in bush babies. *International Journal of Primatology* 21:793–813.

Masters, J. C., and D. J. Brothers. 2002. Lack of congruence between morphological and molecular data in reconstructing the phylogeny of the Galagonidae. *American Journal of Physical Anthropology* 117:79–93.

McCrossin, M. L. 1992. New species of bushbaby from the middle Miocene of Maboko Island, Kenya. *American Journal of Physical Anthropology* 89:215–233.

———. 1999. Phylogenetic relationships and paleoecological adaptations of a new bushbaby from the middle Miocene of Kenya. *American Journal of Physical Anthropology* 108 (S28):195–196.

Mein, P., and L. Ginsburg. 1997. Les mammifères du gisement miocène inférieur de Li Mai Long, Thaïland: systématique, biostratigraphie et paléoenvironnement. *Geodiversitas* 19:783–844.

Mein, P., and M. Pickford. 2006. Late Miocene micromammals from the Lukeino Formation (6.1 to 5.8 Ma), Kenya. *Bulletin Mensuel de la Société Linnéenne de Lyon* 75:183–223.

Napier, J. R., and P. H. Napier. 1967. *A Handbook of Living Primates: Morphology, Ecology and Behaviour of Nonhuman Primates*. Academic Press, London, 456 pp.

Phillips, E. M., and A. Walker. 2000. A new species of fossil lorisid from the Miocene of East Africa. *Primates* 41:367–372.

———. 2002. Fossil lorisoids; pp. 83–95 in W. C. Hartwig (ed.), *The Primate Fossil Record*. Cambridge University Press, Cambridge.

Pickford, M. 1986. Cainozoic Palaeontological Sites of Western Kenya. *Münchner Geowissenschaftliche Abhandlungen* 8:1–151.

Pickford, M., and P. Mein. 2006. Early middle Miocene mammals from Moroto II, Uganda. *Beiträge zur Paläontologie* 30:361–386.

Pickford, M., P. Mein, and B. Senut. 1994. Fossiliferous Neogene karst fillings in Angola, Botswana and Namibia. *South African Journal of Science* 90:227–230.

Pickford, M., and B. Senut. 2001. The geological and faunal context of Late Miocene hominoid remains from Lukeino, Kenya. *Comptes Rendus de l'Académie des Sciences, Paris, Sciences de la Terre et des Planètes* 332:145–152.

Pickford, M., H. Wanas, and H. Soliman. 2006. Indications for a humid climate in the Western Desert of Egypt 11–10 Myr ago: Evidence from Galagidae (Primates, Mammalia). *Comptes Rendus Palevol* 5:935–943.

Porter, C. A., S. L. Page, J. Czelusniak, H. Schneider, M. C. Schneider, I. Sampaio, and M. Goodman. 1997. Phylogeny and evolution of selected primates as determined by sequences of the ε-globin locus and 5′ flanking regions. *International Journal of Primatology* 18:261–295.

Rasmussen, D. T., and K. A. Nekaris, 1998. Evolutionary history of lorisiform primates. *Folia Primatologica* 69 (suppl. 1):250–285.

Roos, C., J. Schmitz, and H. Zischler. 2004. Primate jumping genes elucidate strepsirrhine phylogeny. *Proceedings of the National Academy of Sciences, USA* 101:10650–10654.

Rowe, N. 1996. *The Pictorial Guide to the Living Primates*. Pogonias Press, East Hampton, N.Y., 263 pp.

Rumpler, Y., S. Warter, B. Ishak, and B. Dutrillaux. 1989. Chromosomal evolution in primates. *Human Evolution* 4:157–170.

Rumpler, Y., S. Warter, B. Meier, H. Preuschoft, and B. Dutrillaux. 1987. Chromosomal phylogeny of three Lorisidae: *Loris tardigradus, Nycticebus coucang*, and *Perodicticus potto*. *Folia Primatologica* 48: 216–220.

Sarich, V. M., and J. E. Cronin. 1976. Molecular systematics of the primates; pp. 141–170 in M. Goodman and R. E. Tashian (eds.), *Molecular Anthropology*. Plenum Press, New York.

Sarmiento, E. E. 1998. The validity of "*Pseudopotto martini*." *African Primates* 3:44–45.

Schwartz, J. H. 1992. Phylogenetic relationships of African and Asian lorisids; pp. 65–81 in S. Matano, R. H. Tuttle, H. Ishida, and M. Goodman (eds.), *Topics in Primatology: Volume 3: Evolutionary Biology, Reproductive Endocrinology, and Virology*. University of Tokyo Press, Tokyo.

Schwartz, J. H., and I. Tattersall. 1985. Evolutionary relationships of living lemurs and lorises (Mammalia, Primates) and their potential affinities with European Eocene Adapidae. *Anthropological Papers of the American Museum of Natural History* 60:1–100.

———. 1987. Tarsiers, adapids and the integrity of the Strepsirhini. *Journal of Human Evolution* 16:23–40.

Seiffert, E. R., E. L. Simons, and Y. Attia. 2003. Fossil evidence for an ancient divergence of lorises and galagos. *Nature* 422: 421–424.

Seiffert, E. R., E. L. Simons, T. M. Ryan, and Y. Attia. 2005. Additional remains of *Wadilemur elegans*, a primitive stem galagid from the late Eocene of Egypt. *Proceedings of the National Academy of Sciences, USA* 102:11396–11401.

Simpson, G. G. 1965. Family: Galagidae; pp. 15–16 in L. S. B. Leakey (ed.), *Olduvai Gorge 1951–61: Volume 1. A Preliminary Report on the Geology and Fauna*. Cambridge University Press, Cambridge.

———. 1967. The Tertiary lorisiform primates of Africa. *Bulletin of the Museum of Comparative Zoology* 136:39–62.

Smith, R. J., and W. L. Jungers. 1997. Body mass in comparative primatology. *Journal of Human Evolution* 32:523–559.

Szalay, F. S. 1976. Systematics of the Omomyidae (Tarsiiformes, Primates): Taxonomy, phylogeny and adaptations. *Bulletin of the American Museum of Natural History* 156:157–450.

Szalay, F. S., and E. Delson. 1979. *Evolutionary History of the Primates*. Academic Press, New York, 580 pp.

Walker, A. C. 1969. True affinities of *Propotto leakeyi* Simpson 1967. *Nature* 223:647–648.

———. 1970. Post-cranial remains of the Miocene Lorisidae of East Africa. *American Journal of Physical Anthropology* 33:249–262.

———. 1974. A review of the Miocene Lorisidae of East Africa; pp. 435–447. in R. D. Martin, G. A. Doyle, and A. C. Walker (eds.), *Prosimian Biology*. Duckworth, London.

———. 1978. Prosimian primates; pp. 90–99 in V. J. Maglio and H. B. S. Cooke (eds.), *Evolution of African Mammals*. Harvard University Press, Cambridge.

———. 1987. Fossil Galaginae from Laetoli; pp. 88–90 in M. D. Leakey and J. M. Harris (eds.) *Laetoli: A Pliocene Site in Northern Tanzania*. Clarendon Press, Oxford.

Walker, A., T. M. Ryan, M. T. Silcox, E. L. Simons, and F. Spoor. 2008. The semicircular canal system and locomotion: The case of extinct lemuroids and lorisoids. *Evolutionary Anthropology* 17:135–145.

Wesselman, H. B. 1984. *The Omo Micromammals: Systematics and Paleoecology of Early Man Sites from Ethiopia*. Contributions to Vertebrate Evolution, Vol. 7. Karger, Basel, Switzerland, 165 pp.

Yang, Z., and A. D. Yoder. 2003. Comparison of likelihood and Bayesian methods for estimating divergence times using multiple gene loci and calibration points, with application to a radiation of cute-looking mouse lemur species. *Systematic Biology* 52:705–716.

Yoder, A. D. 1997. Back to the future: A synthesis of strepsirrhine systematics. *Evolutionary Anthropology* 6:11–22.

Yoder, A. D., M. Cartmill, M. Ruvolo, K. Smith, and R. Vilgalys, 1996. Ancient single origin for Malagasy primates. *Proceedings of the National Academy of Sciences, USA* 93:5122–5126.

Yoder, A. D., and Z. Yang. 2004. Divergence dates for Malagasy lemurs estimated from multiple gene loci: geological and evolutionary context. *Molecular Ecology* 13:757–773.

Zimmermann, E. 1990. Differentiation of vocalizations in bushbabies (Galaginae, Prosimiae, Primates) and the significance for assessing phylogenetic relationships. *Zeitschrift für Zoologische Systematik und Evolutionsforschung* 28:217–239.

TWENTY-ONE

Subfossil Lemurs of Madagascar

LAURIE R. GODFREY, WILLIAM L. JUNGERS,
AND DAVID A. BURNEY

Madagascar's living lemurs (order Primates) belong to a radiation recently ravaged by extirpation and extinction. There are three extinct and five extant families (two with extinct members) of lemurs on an island of less than 600,000 km^2. This level of familial diversity characterizes no other primate radiation. The remains of up to 17 species of recently extinct (or subfossil lemurs) have been found alongside those of still extant lemurs at numerous Holocene and late Pleistocene sites in Madagascar (figure 21.1, table 21.1). The closest relatives of the lemurs are the lorisiform primates of continental Africa and Asia; together with the lemurs, these comprise the suborder Strepsirrhini.

Most researchers have defended an ancient Gondwanan (African or Indo-Madagascan) origin for lemurs. On the basis of molecular data, some posit an origin of primates between 85 and 90 Ma (e.g., Martin, 2000; Eizirik et al., 2001; Springer et al., 2003; Yoder and Yang, 2004; Miller et al., 2005), and of lemurs on Madagascar by ~80 Ma (e.g., Yoder and Yang, 2004). Using nuclear genes only, Poux et al. (2005) place the origin of primates at ~80 Ma and the colonization of Madagascar at between 60 and 50 Ma, with a 95% credibility interval of 70–41 Ma. There is general agreement that lemurs must have been established on Madagascar by the middle Eocene (Roos et al., 2004). Actual putative fossil primates (e.g., *Altiatlasius*, plesiadapiforms) first appear ~60 Ma in Algeria (Hooker et al., 1999; Tabuce et al., 2004). Seiffert et al. (2003, 2005) identify the earliest definitive strepsirrhines (*Karanisia, Saharagalago*, and *Wadilemur*) as primitive lorisoids; they do not appear in the fossil record until after ~40 Ma (in northern Egypt, alongside definitive anthropoids). On the basis of limited dental evidence, Marivaux et al. (2001) have described a possible fossil lemur *(Bugtilemur)*—the only one found outside Madagascar—in 30-million-year-old deposits in Pakistan (see Seiffert et al., 2003, and Godinot, 2006, for critiques).

If primates originated 80–90 million years ago, then the first quarter or third of the primate evolutionary record is missing entirely. An early primate origin can be defended on the basis of the fossil record if existing fossils are used to model the tempo of primate diversification (Tavaré et al., 2004), and 2.5 million years is used as an average species lifespan. An alternative explanation is that the early evolution of the Primates was more explosive than a model based on steady, gradual diversification would suggest. We believe that the latter scenario is more consistent with the fossil record.

The question of how lemurs got to Madagascar is still far from resolved (Godinot, 2006; Masters et al., 2006; Stevens and Heesy, 2006; Tattersall, 2006a, 2006b). It is clear that Madagascar (with the Indian plate) separated from Africa long before primates evolved and that it arrived at its present position relative to Africa by 120–130 Ma (Krause et al., 1997; Roos et al., 2004; Masters et al., 2006; Rabinowitz and Woods, 2006). Most scholars favor chance rafting of an ancestral lemur from continental Africa to Madagascar (Krause et al., 1997; Kappeler, 2000; Roos et al., 2004; Rabinowitz and Woods, 2006), with prior divergence of lemurs and lorises. Reports of floating islands at sea (such as one observed in 1902 some 30 miles off the coast of San Salvador supporting a troop of monkeys and plentiful vegetation including coconut trees) are intriguing in this regard (Van Duzer, 2004). Others are skeptical of long-distance water rafting for terrestrial mammals (Lawlor, 1986; Hedges et al., 1996; de Wit and Masters, 2004; Stankiewicz et al., 2006; Masters et al., 2006) and urge consideration of other models. For example, an early Indo-Madagascan origin for primates might account for the presence of lemurs on Madagascar and a possible cheirogaleid, *Bugtilemur*, in Pakistan 30 million years ago.

But if primates originated on the Indo-Madagascan plate, or if they colonized Madagascar during the Cretaceous, then primates of some sort might be expected to occur in the Cretaceous deposits of Madagascar. The rich Gondwanan fossil record of Madagascar provides no such corroboration (Krause et al., 1997); there were apparently no proto-lemurs on Madagascar during the Cretaceous. Furthermore, if primates originated on Indo-Madagascar instead of Asia or Africa, some sweepstakes mechanism (long-distance water rafting or dispersal over a land bridge or stepping-stones) is still needed to explain their presence in northern Africa 60 million years ago. The breakup of India and Madagascar was underway around 90 million years ago (Storey, 1995), but India did not collide with continental Asia until much later—glancingly at 57 million years ago and fully at 35 million years ago (Ali and Aitchison, 2008). During the critical time period (90–60 million years ago), Madagascar, the Indian plate, and the

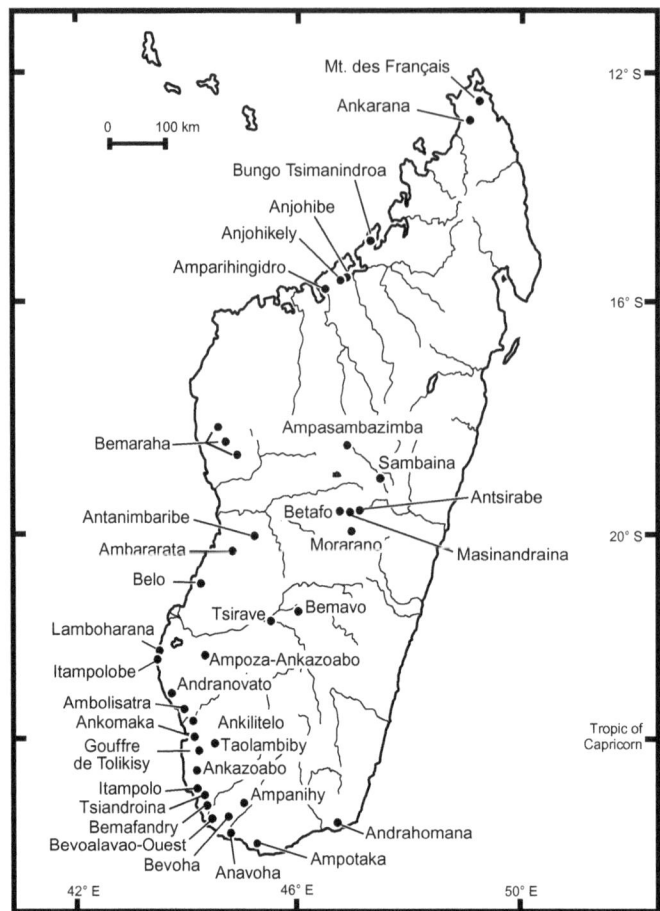

FIGURE 21.1   Map showing subfossil sites.

continental African and Asian plates functioned as isolated landmasses, although passage between southern India and Madagascar may have been facilitated by the Seychelles-Mascarene Plateau and nearby areas of elevated seafloor prior to the K/T catastrophic extinction event (65 million years ago; Ali and Aitchison, 2008). The Kerguelen Plateau, which formed ~118 million years ago and connected the Indo-Malagasy plate to what is now Australia and Antarctica during the mid-Cretaceous, had drowned by ~90 million years ago (Ali and Aitchison, 2008). During the early Cenozoic, there may have been intermittent stepping-stone islands south-southwest of India (along the Deccan-Réunion hot-spot ridge), but by this time, huge stretches of ocean separated India and Madagascar (Ali and Aitchison, 2008). A putative land "bridge" connecting Africa to Madagascar (45–26 Ma; see McCall, 1997) cannot explain the introduction of primates from Madagascar to Africa or the colonization of Madagascar by African primates because it is too recent (Poux et al., 2005); the Mozambique Channel seems to have been sufficiently narrow to allow sporadic independent crossings from the early Cenozoic onward (see also Ali and Huber, 2010).

Relationships among families of lemurs are also problematic. There is strong evidence that the Daubentoniidae were the first to diverge (~60 Ma, perhaps earlier; see Yoder et al., 1996; Yoder, 1997; DelPero et al., 2001, 2006; Pastorini et al., 2001; Poux and Douzery, 2004; Yoder and Yang, 2004; Roos et al., 2004, Horvath et al., 2008), but the ages of divergence and relationships among the remaining four extant families (Cheirogaleidae, Lepilemuridae, Indriidae, and Lemuridae)

are poorly understood; alternative molecular data sets favor different topologies. There is some molecular support for a sister taxon relationship between Indriidae and Lemuridae (DelPero et al., 2001; Roos et al., 2004), with an age of divergence between 32 and 52 Ma (Roos et al., 2004). Poux et al. (2005) support a sister taxon relationship for the Cheirogaleidae and Lepilemuridae, with the Indriidae as the sister to that group, and the Lemuridae diverging from a cheirogaleid-lepilemurid-indriid clade just over 30 Ma. The latter topology was also supported by Horvath et al. (2008) using 11 novel markers from nine chromosomes for 18 extant lemur species, but Orlando et al. (2008), using a taxonomically broader data set (35 lemur species, including six extinct ones) but relying on a smaller set of genes (12S and Cytb), found support for the former (i.e., a sister taxon relationship between the Indriidae plus their extinct relatives and the Lemuridae plus their extinct relatives), with the Lepilemuridae as the first family to diverge after the Daubentoniidae, followed by the Cheirogaleidae, and then the Indriidae and Lemuridae. Using a large, composite data set and a variety of analytical methods, DelPero et al. (2006) found equally strong support for two topologies—one identical to that found by Orlando et al. (2008), and the other with the Indriidae first to diverge after the Daubentoniidae, followed by the Lemuridae, and finally, the Lepilemuridae and Cheirogaleidae. In each of these topologies, the Cheirogaleidae are nested well within the lemur clade. The formerly favored notion (grounded in some remarkable morphological and developmental similarities; Szalay and Katz, 1973; Cartmill, 1975; Schwartz and Tattersall,

TABLE 21.1
Major occurrences and ages of extinct lemurs of Madagascar

Taxon	Occurrence (Site, Region)	Calibrated Range of Dated Specimens at 2σ	Key References
	PALAEOPROPITHECIDAE		
Palaeopropithecus maximus	Ampasambazimba, Itasy, possibly Ankarana	BP 2352–2157	Standing, 1903, 1905; Grandidier, 1899, 1901
P. ingens	Ambolisatra, Ampoza, Anavoha, Andranovato, Ankazoabo-Grotte, Ankilitelo, Ankomaka, Beavoha, Belo-sur-mer, Betioky-Toliara, Itampolobe, Lower Menarandra, Manombo-Toliara, Taolambiby, Tsiandroina, Tsivonohy	BP 2366–2315, AD 640–946, AD 1300–1620	Filhol, 1895; Godfrey and Jungers, 2003
Palaeopropithecus kelyus	Amparihingidro, Anjohibe, Belobaka, Ambongonambakoa, perhaps Ampoza		Gommery et al., 2009 MacPhee et al., 1984
Archaeoindris fontoynontii	Ampasambazimba	BP 2362–2149 BP 2711–2338	Standing, 1909, 1910; Lamberton, 1934a; Vuillaume-Randriamanantena, 1988
Babakotia radofilai	Ankarana, Anjohibe	BP 5290–4840	Godfrey et al., 1990; Jungers et al., 1991; Simons et al., 1992
Mesopropithecus globiceps	Anavoha, Ankazoabo-Grotte, Belo-sur-mer, Manombo-Toliara, Taolambiby, Tsiandroina, Tsirave	BC 354–60, AD 58–247, AD 245–429	Lamberton, 1936; Tattersall, 1971
M. pithecoides	Ampasambazimba	AD 570–679	Standing, 1905; Tattersall, 1971
M. dolichobrachion	Ankarana		Simons et al., 1995
	ARCHAEOLEMURIDAE		
Archaeolemur majori	Ambararata-Mahabo, Ambolisatra, Anavoha, Andrahomana, Ankazoabo-Grotte, Ankilitelo, Beavoha, Belo-sur-mer, Bemafandry, Betioky-Toliara, Itampolobe, Lamboharana, Manombo-Toliara, Nosy-Ve, Taolambiby, Tsiandroina, Tsirave. Possibly Ampasambazimba, Ampoza-Ankazoabo, Ankarana, Bungo-Tsimanindroa.	AD 260–530, AD 410–620, AD 620–700	Filhol, 1895; Forsyth-Major, 1896; Tattersall, 1973; Hamrick et al., 2000; Godfrey et al., 2005
A. edwardsi	Ampasambazimba, Ampoza-Ankazoabo, Ankarana, Belo-sur-mer, Masinandraina, Morarano-Betafo, Sambaina, Vakinanakaratra. Possibly Ambolisatra, Amparihingidro, Anjohibe, Anjohikely	BP 8740–8410, BC 350–AD 80, AD 910–1150	Filhol, 1895; Standing, 1905; Tattersall, 1973; Hamrick et al., 2000; Godfrey et al., 2005
Hadropithecus stenognathus	Southern, southwestern, and central Madagascar: Ambovombe, Ampasambazimba, Anavoha, Andrahomana, Belo-sur-Mer, Tsirave	BP 7660–7490, BP 2344–1998, AD 444–772	Lorenz von Liburnau, 1902; Lamberton, 1938; Godfrey et al., 1997b; Godfrey et al., 2006a
	MEGALADAPIDAE		
Megaladapis edwardsi	Ambolisatra, Ampanihy, Ampoza-Ankazoabo, Anavoha, Andrahomana, Andranovato, Ankomaka, Beavoha, Betioky-Toliara, Itampolobe, Lamboharana, Taolambiby, Tsiandroina	BP 5436–5059, AD 27–412, AD 666–816,	Grandidier, 1899; Lorenz von Liburnau, 1905; Jungers, 1977, 1978; Lamberton, 1934c; Vuillaume-Randriamanantena, et al., 1992

TABLE 21.I
(CONTINUED)

Taxon	Occurrence (Site, Region)	Calibrated Range of Dated Specimens at 2σ	Key References
M. madagascariensis	Ambararata-Mahabo, Ambolisatra, Amparihingidro, Ampoza-Ankazoabo, Anavoha, Andrahomana, Anjohibe, Ankarana, Ankilitelo, Beavoha, Belo-sur-mer, Bemafandry, Itampolobe, Mt. des Français, Taolambiby, Tsiandroina, Tsirave, Tsivonohy.	BP 15670–14380, BP 2870–2760 AD 1280–1420	Forsyth-Major 1893, 1894; Jungers 1977, 1978; Vuillaume-Randriamanantena, et al., 1992
M. grandidieri	Ampasambazimba, Antsirabe, Itasy, Morarano-Betafo	AD 900–1040	Standing, 1903; Lamberton, 1934c; Jungers, 1977, 1978; Vuillaume-Randriamanantena, et al. 1992

LEMURIDAE			
Pachylemur insignis	Ambararata-Mahabo, Ambolisatra, Anavoha, Andrahomana, Ampoza-Ankazoabo, Belo-sur-mer, Bemafandry, Lamboharana, Manombo-Toliara, Taolambiby, Tsiandroina, Tsirave, perhaps Amparihingidro	AD 680–960, BC 110–AD 100	Filhol, 1895; Lamberton, 1948
P. jullyi	Ampasambazimba, Antsirabe, Morarano-Betafo, possibly Ankarana	—	Grandidier 1899 Lamberton 1948

DAUBENTONIIDAE			
Daubentonia robusta	Anavoha, Lamboharana, Tsirave	AD 891–1027	Grandidier 1929 Lamberton 1934b MacPhee and Raholimavo 1988 Simons 1994

NOTE: Source for dates Burney et al., 2004.

1985; Yoder, 1994), that the cheirogaleids are actually primitive lorisiforms invading Madagascar independently of lemurs, is countered by consistent and mounting molecular evidence to the contrary (Yoder, 1994, 1997; Yoder et al. 1996; Yoder and Yang, 2004; Roos et al., 2004; Poux et al., 2005; DelPero et al., 2006).

With regard to extinct lemurs, morphological, developmental and molecular data support a sister taxon relationship for the Palaeopropithecidae (four genera) and the Indriidae (Tattersall and Schwartz, 1974; Godfrey, 1988; Godfrey et al., 2002; Schwartz et al., 2002; Karanth et al., 2005). Morphological data (postcranial characters in particular) suggest that, within the Palaeopropithecidae, *Mesopropithecus* diverged first, then *Babakotia*; *Palaeopropithecus* and *Archaeoindris* share the most recent ancestor (Godfrey, 1988; Godfrey et al., 1990; Jungers et al., 1991; Simons et al., 1992, 1995; Godfrey and Jungers, 2002). Morphological and molecular evidence also favors a sister taxon relationship for the extinct *Pachylemur* and still extant *Varecia*, and their status as the sister to a *Eulemur-Lemur-Hapalemur* clade (Seligsohn and Szalay 1974; Crovella et al. 1994; Wyner et al., 2000; Pastorini et al., 2002). Relationships of the Archaeolemuridae and the Megaladapidae to extant lemurs have been more

controversial. Morphological data support affinity of *Megaladapis* and *Lepilemur* (Tattersall and Schwartz, 1974; Wall, 1997); most existing molecular data fail to support this connection (Yoder et al., 1999; Yoder, 2001; Karanth et al., 2005; Orlando et al., 2008). The latter instead affirm a close relationship of the Megaladapidae to the Lemuridae and suggest that Montagnon et al.'s (2001a, 2001b) molecular support for a link between the megaladapids and *Lepilemur* results from polymerase chain reaction (PCR) contamination. Within the genus *Megaladapis*, two species (*M. madagascariensis* and *M. grandidieri*) are clear sister taxa (Vuillaume-Randriamanantena et al., 1992). The Archaeolemuridae have long been considered the sister to the palaeopropithecid-indriid clade, largely on the basis of molar morphology and the number of teeth in the (modified) tooth comb (Tattersall and Schwartz, 1974; Godfrey, 1988; Godfrey and Jungers, 2002). Other morphological as well as developmental characters suggest closer affinity to the Lemuridae (King et al., 2001; Godfrey et al., 2006a; Lemelin et al., 2008), but recent molecular analysis has accorded support for the former scenario. Orlando et al. (2008) found strong molecular support for a close relationship among the Archaeolemuridae, Palaeopropithecidae, and Indriidae, but their data could not resolve phylogenetic relationships within this group. A sister taxon relationship of *Archaeolemur* and *Hadropithecus*, long supported by morphological evidence (e.g., Tattersall, 1973; Godfrey, 1988; Godfrey and Jungers, 2002), has now received strong molecular support, however (Orlando et al., 2008).

Systematic Paleontology

Family PALAEOPROPITHECIDAE Tattersall, 1973

This is the most speciose of extinct lemur families, with four genera *(Palaeopropithecus, Archaeoindris, Babakotia, Mesopropithecus)* and eight recognized species future revisions might collapse some of these species. All have the same adult dental formula (2.1.2.3/2.0.2.3) as in extant indriids, with only two pairs of premolars and four teeth in the tooth comb. *Mesopropithecus*, like *Babakotia* (and unlike *Palaeopropithecus* and *Archaeoindris*), retains a number of primitive craniodental features, including an inflated auditory bulla with intrabullar ectotympanic ring, and a conventional tooth comb of indriid type, with four teeth. Palaeopropithecids share with indriids accelerated dental crown formation (Schwartz et al., 2002; Godfrey et al., 2006c), but the largest taxa *(Palaeopropithecus* and *Archaeoindris*) differ from indriids in details of the nasal aperture (e.g., paired protuberances) and the petrosal bone (e.g., deflated bulla). The namesake of the family, *Palaeopropithecus*, exhibits the most derived postcranial specializations for hind- and forelimb suspension—inferred behaviors that are correlated with greatly curved proximal phalanges (Jungers et al., 1997) and with high intermembral indices (Jungers, 1980; Jungers et al., 2002). The palaeopropithecids have been dubbed the "sloth lemurs" due to their remarkable postcranial convergences to sloths (Godfrey and Jungers, 2003), and recent research on the semicircular canals of giant lemurs lends support to this argument (Walker et al., 2008).

Genus *PALAEOPROPITHECUS* G. Grandidier, 1899
PALAEOPROPITHECUS MAXIMUS Standing, 1903

Partial Synonymy Palaeopropithecus raybaudii, Standing, 1903.
Age and Occurrence Late Quaternary, central, possibly northern Madagascar.
Diagnosis Largest species of genus; skull length averages 191 mm; orbits small; orbital margin raised to form a bony rim; petrosal elongated to form a tube; bulla not inflated; neurocranium small, frontal region of skull depressed; postorbital constriction strong; large frontal sinuses; facial retroflexion strong; sagittal crest often present; mandible deep (particularly in gonial region) but mandibular corpus thin; dental rows nearly parallel; paroccipital processes large; dorsal portions of the premaxillae (as well as, to a far lesser extent, the lateral termini of the nasals) inflated and bulbous; "tooth comb" with four short, blunt, and slightly separated incisors; molars with low cusps; the hypocone on M1 and M2 is extremely reduced in height, almost shelflike; diastema present between anterior and posterior mandibular premolars; anterior mandibular dentition has been modified from a true tooth comb into four short and stubby procumbent teeth.
Description Body mass estimated at ~46 kg (Jungers et al., 2008); intermembral index 144–145 (Godfrey and Jungers, 2002). This species is about twice the size of the new species

from the northwest, but only minimally larger than *P. ingens*. The cheek teeth resemble those of the indriids (especially *Propithecus*) in cusp configuration and stylar development, but the first and second molars of both upper and lower jaws are more buccolingually compressed and mesiodistally elongated, and the third molars are smaller in relative size. The occlusal enamel tends to be crenulated. The lingual borders of the anterior maxillary molars are elongated so that they roughly equal the lengths of the buccal borders. As in indriids, the lower molars have accentuated trigonid and talonid basins, a strong protoconid, and a low hypoconid. Crests connect the protoconid and metaconid, as well as the hypoconid and entoconid. A paraconid is present on m1 and m2, separated from the metaconid by a moderately deep groove.

The hands and feet bear long, strongly curved metapodials and phalanges with deep flexor grooves; the metacarpo- and metatarsophalangeal joints are "notched" in a tongue-and-groove manner. The vertebral spinous processes are short and blunt throughout the entire thoracosacral vertebral column, and the transverse processes of the thoracic and lumbar vertebrae arise from the vertebral arches (Shapiro et al., 2005). The os coxae have a prominent ischial spine, but only a rudimentary anterior inferior iliac spine. The iliac blades flare laterally, and the pubis is long and flattened superoinferiorly. The small sacral hiatus suggests a reduced, if not vestigial, tail. By comparison, the pectoral girdle is poorly known. The humerus is long and robust and carries an entepicondylar foramen; the olecranon process of the ulna is reduced and its styloid process projects distally well beyond the head. The femur is short and anteroposteriorly flattened, with a shallow patellar groove and a reduced greater trochanter; the collodiaphyseal angle approaches 180 degrees, and the large, ball-like femoral head lacks a fovea capitis. The tibia and fibulae have very reduced (essentially absent) medial and lateral malleoli.

PALAEOPROPITHECUS INGENS G. Grandidier, 1899
Figure 21.2A

Partial Synonymy Thaumastolemur grandidieri Filhol, 1895; *Bradytherium madagascariense*, G. Grandidier, 1901.
Age and Occurrence Late Quaternary, southern and western Madagascar.
Diagnosis Skull (at ~184 mm in length) and teeth similar to *P. maximus* but slightly smaller; mandibular symphysis shorter.
Description Body mass estimated at approximately 42 kg (Jungers et al., 2008); intermembral index 135–138 (Godfrey and Jungers, 2002). Because the jaws of *P. ingens* are smaller than *P. maximus*, full adults from the south may lack diastemata separating the mandibular premolars. The length and development of the diastema between the anterior and posterior lower premolars in *Palaeopropithecus* depends on biological age and adult body size; diastemata are absent in all individuals when the premolars first erupt, but they may form and lengthen as the jaw grows. The dental microstructure of an individual belonging to this species was used to derive dental developmental data for *Palaeopropithecus* (Schwartz et al., 2002; Godfrey et al., 2006c); the permanent teeth show extremely accelerated crown formation, and they appear to have erupted when the jaws were still small, very like the condition in extant indriids. Postcrania largely similar to *P. maximus*, but carpal and tarsal bones are also known

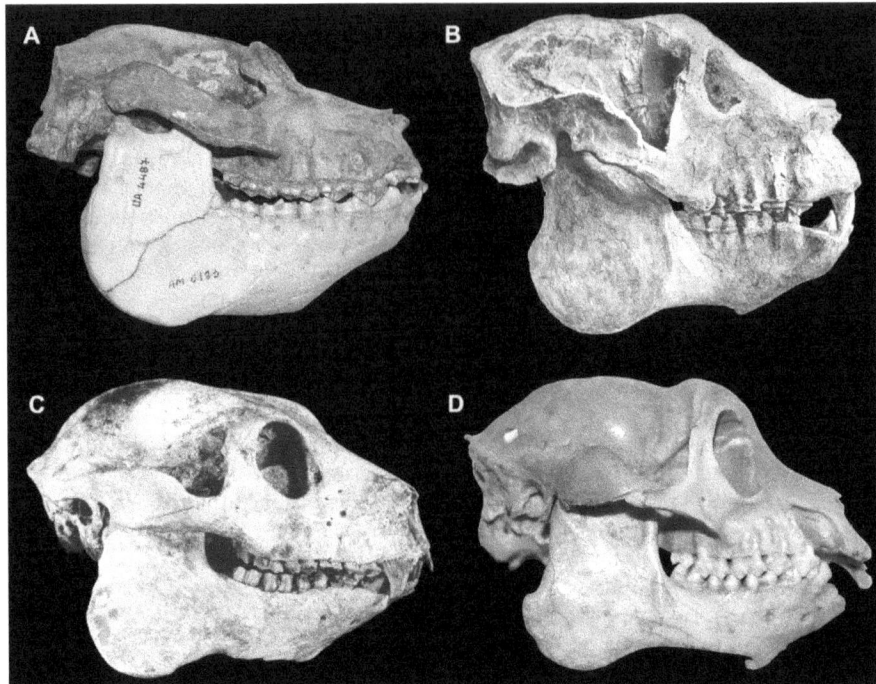

FIGURE 21.2 Lateral views of skulls of palaeopropithecids: A) *Palaeopropithecus ingens* (composite skull from southern Madagascar, collection of the Université d'Antananarivo); B) *Archaeoindris fontoynontii* (from Ampasambazimba, collection of the Académie Malgache); C) *Babakotia radofilai* (from the Ankarana Massif, collection of the Duke Primate Center); D) *Mesopropithecus globiceps* (from southern Madagascar, collection of the Université d'Antananarivo). Not to scale. The photograph of *Archaeoindris* was reproduced from Lamberton, 1934a: Plate I, and mirror-imaged.

for *P. ingens* (e.g., Hamrick et al., 2000; Jungers and Godfrey, 2003). The long ulnar styloid is excluded from the pisiform and articulates like a "mortar in pestle" with the triquetrum; the overall carpus has a flexed set. The hindfoot is reduced, especially the calcaneus, and the talar trochlea is globular; the plantar-flexed talar head articulates uniquely with both the navicular and cuboid.

Remarks on the Genus Palaeopropithecus Of all palaeopropithecids, *Palaeopropithecus* is most specialized for suspension, bearing long, curved phalanges, a very reduced hindfoot, the highest intermembral index (exceeding all living primates except orangutans), and extremely reduced spinous processes on thoracolumbar and sacral vertebrae. The sloth lemurs exhibit suspensory adaptations that imply a life almost entirely in the trees, even in areas that pollen evidence shows were not primarily dense forest, but rather a mosaic of woodland and grassland environments (e.g., Burney, 1987a, 1987b; Matsumoto and Burney, 1994). It is likely that *Palaeopropithecus* is the animal described by Etienne de Flacourt (1658) as the "tretretretre" and represented in Malagasy folklore as an ogre incapable of moving on smooth rocky surfaces (Godfrey and Jungers, 2003). The smallest and most gracile of the *Palaeopropithecus* species, known from two fossil localities in the northwest (Amparihingidro and Anjohibe/Anjohikely), has yet to be formally described but is outside the observed ranges of the other two species in most respects (see MacPhee et al., 1984, for a description of the discovery of a skeleton belonging to this variant). A small but distinct hypocone is manifested on the first and second molars of the new species from the northwest, and its proximal phalanges are extremely curved. Further analysis may make it difficult to maintain the specific distinction between *P. maximus* and *P. ingens*.

Genus *ARCHAEOINDRIS* Standing, 1909
ARCHAEOINDRIS FONTOYNONTII Standing, 1909
Figure 21.2B

Partial Synonymy Lemuridotherium Standing, 1910
Age and Occurrence Late Quaternary, central Madagascar.
Diagnosis Largest of extinct lemurs; length of single known skull 269 mm—shorter (but wider) than that of *Megaladapis*; neurocranium small; sagittal and nuchal crests strong; as in *Palaeopropithecus*: anterior molars buccolingually compressed; upper third molars reduced; postorbital constriction marked, external auditory meatus tubular (probably petrosal in origin); auditory bulla deflated; lower incisors stubby and blunt; diastema separates p2 and p4; palate rectangular; cheek tooth enamel crenulated; paired protuberances over the nasal aperture. The limited postcrania recall those of *Palaeopropithecus* but are much larger; large femoral head lacks fovea capitis; collodiaphyseal angle high; greater trochanter reduced; differs from *Palaeopropithecus* in having: much more massive and extremely robust postcranial bones; relatively deeper skull; orbits less dorsally oriented, and lacking the distinctly thickened rimming that characterizes those of *Palaeopropithecus*; cheek teeth are less wrinkled and slightly higher-crowned.

Description Body mass estimated at ~160 kg (Jungers et al., 2008). Knowledge of this species is based on one complete skull, additional fragmentary jaws, a fragmentary humerus and femur of an adult, and four long bone diaphyses of an immature individual. These bones are sufficient to demonstrate that the intermembral index well exceeded 100 but was probably lower than that of *Palaeopropithecus*.

Remarks on Genus Archaeoindris Adaptations for scansoriality are interesting in light of the massive size of *Archaeoindris*, which has been interpreted as convergent on ground sloths (Lamberton, 1934a; Jungers, 1980). This genus is only known from one site, Ampasambazimba, in the western highlands. Additional details on the postcranial anatomy of *Archaeoindris* are given by Vuillaume-Randriamanantena (1988).

Genus *BABAKOTIA* Godfrey et al., 1990
BABAKOTIA RADOFILAI Godfrey et al., 1990
Figure 21.2C

Age and Occurrence Late Quaternary, northern and northwestern Madagascar.

Diagnosis Skull length averages 114 mm; dentition similar to that of *Propithecus* but with greater mesiodistal elongation of premolars; cheek tooth enamel heavily crenulated; shearing crests well developed; face long as in *Indri*; differs from latter in greater postorbital constriction and more robust mandible. As in extant indriids, tooth comb is of the conventional indriid type, with four elongated teeth (Jungers et al., 2002).

Description Body mass estimated at ~21 kg (Jungers et al., 2008). The auditory bulla is inflated and possesses an intrabullar, ringlike ectotympanic. The postorbital bar is robust. There are no orbital tori or circumorbital protuberances.

Postcranially, *Babakotia* is more specialized for suspension than *Mesopropithecus*, but less so than *Palaeopropithecus*. The intermembral index is 118. There is a moderate degree of spinous process reduction in the thoracolumbar region. The innominate sports an incipient ischial spine, reduced rectus femoris process, and long pubis with some degree of superoinferior flattening. The femoral head is globular and somewhat cranially directed (but not to the extent seen in *Palaeopropithecus* and *Archaeoindris*); the collodiaphyseal angle is high; the femoral shaft is anteroposteriorly compressed; the patellar groove is shallow; the tibial malleolus is reduced; the calcaneus is quite reduced. There is also some reduction in relative lengths of the pollex and hallux; the proximal phalanges are long and curved with marked flexor ridges.

Remarks on the Genus Babakotia Geographically restricted to the north and northwest. First specimens discovered in the late 1980s and described in 1990. *Babakotia* was morphologically intermediate in the morphocline between *Mesopropithecus* and *Palaeopropithecus*. Many features of its axial and appendicular skeleton ally it functionally and phylogenetically with *Palaeopropithecus* (Jungers et al. 1991; Simons et al. 1992).

Genus *MESOPROPITHECUS* Standing, 1905
MESOPROPITHECUS GLOBICEPS Lamberton, 1936
Figure 21.2D

Partial Synonymy Neopropithecus globiceps Lamberton, 1936; *Neopropithecus platyfrons* Lamberton, 1936.

Age and Occurrence Late Quaternary, southern, southwestern, and southeastern Madagascar.

Diagnosis Skull length averages 94 mm; very similar to but slightly smaller than *M. pithecoides*; differs from latter in having a more gracile skull; snout narrows anteriorly to a greater degree; teeth very like (though slightly larger than) those of *Propithecus*, except: upper and lower premolars relatively shorter; M3 moderately buccolingually constricted. Tooth

comb form typical of living indriids. Auditory bulla remains inflated. Both *M. globiceps* and *M. pithecoides* differ from *M. dolichobrachion* in limb proportions: relatively shorter forelimb.

Description Body mass estimated at ~11 kg (Jungers et al., 2008); intermembral index 97. Forelimb relatively conservative (indriid-like); hindlimb and axial skeleton more specialized for suspension (more like *Palaeopropithecus* and *Babakotia*). Forelimbs and hindlimbs approximately equal in length.

MESOPROPITHECUS PITHECOIDES Standing, 1905

Age and Occurrence Late Quaternary, central Madagascar.

Diagnosis Skull length averages 98 mm; very like *M. globiceps*; skull with well-developed sagittal and nuchal cresting; massive zygomatic arches; muzzle broader anteriorly than in *M. globiceps*.

Description Intermembral index 99. Limb proportions are virtually identical to those of *M. globiceps*. Marked craniodental similarities to *M. globiceps*.

MESOPROPITHECUS DOLICHOBRACHION Simons et al., 1995

Age and Occurrence Late Quaternary, northern Madagascar.

Diagnosis Skull length averages 102 mm; *M. dolichobrachion* differs little from congeners craniodentally, except in having a third upper molar with relatively wider trigon and smaller talon. Chief distinctions postcranial: humerofemoral (~104) and intermembral (~113) indices relatively high (hence its specific nomen); humerus substantially longer and more robust than that of either congener; humerus unique among congeners in exceeding length of femur.

Description Largest of the *Mesopropithecus* species at ~14 kg (Jungers et al., 2008). As in other *Mesopropithecus*, the central upper incisor is larger than the lateral, and there is a small gap separating them at prosthion; the upper premolars are short mesiodistally. Tooth comb is present. Sagittal and nuchal crests are evident, orbits are small, postorbital constriction is marked, and the muzzle is wide and squared anteriorly.

M. dolichobrachion has an indriid-like carpus but strongly curved proximal phalanges. Moderately reduced neural spines of lumbar vertebrae, and reduced rectus femoris process. The fovea capitis is reduced, the femoral condyles anteroposteriorly compressed. The value for the brachial index of *M. dolichobrachion* (also ~104) also deviates from those of its congeners (~101). Of all its congeners, *M. dolichobrachion* is most similar to *Babakotia*, *Archaeoindris*, and *Palaeopropithecus*, suggesting greater specializations for suspension.

Remarks on Genus Mesopropithecus Tattersall (1971) considered *Mesopropithecus* the sister taxon to *Propithecus*, but Godfrey (1988) defended a closer relationship to *Palaeopropithecus* and *Archaeoindris*. New discoveries have added evidence in favor of the latter. Two of the three species, *M. globiceps* and *M. pithecoides*, are very alike and allopatric. The latter may be a slightly larger-bodied, geographic variant of the former that should not be accorded separate species status. In comparison to *Propithecus*, *Mesopropithecus* has relatively smaller and more convergent orbits, a steeper facial angle, greater postorbital constriction, a more robust postorbital bar, a relatively wider and anteriorly squared muzzle, and zygoma that are more robust and cranially convex in outline. The temporal lines are anteriorly confluent and may form a sagittal crest; the nuchal ridge is confluent with poste-rior root of the zygoma. *Mesopropithecus dolichbrachion* is the most distinct, and is geographically restricted to the extreme north.

Family ARCHAEOLEMURIDAE G. Grandidier, 1905

This family, dubbed the "monkey lemurs," includes three recognized species in two genera: *Archaeolemur majori* (south-ern and western Madagascar), *A. edwardsi* (central Madagascar), and *Hadropithecus stenognathus* (largely southern and western Madagascar). Variants of *A. majori* and *A. edwardsi* exist in other parts of Madagascar, and a full review of this variation is war-ranted. The archaeolemurids have a dental formula of 2.1.3.3/2.0.3.3; they possess a highly modified tooth comb with four instead of six teeth, likely missing the lower canine. The lower incisors are procumbent but occlude directly with the uppers. The central upper incisors are considerably larger than the lateral, and there is substantial contact of the mesial edges of the two central incisors; they thus lack the typical strepsirrhine interincisal gap. The premolar series is modified into a continuous shearing blade in all species, and P4 is molar-iform (greatly buccolingually expanded with a distally emplaced and distinct protocone). As in most extant lemurs (except cheirogaleids), there is an inflated petrosal bulla with a free intrabullar tympanic ring, and the carotid foramen is located on the posterior wall of the bulla. The neurocranium of archaeolemurids is relatively large (at least by strepsirrhine standards) (see Tattersall, 1973, for detailed descriptions of the craniodental morphology of the archaeolemurids). A number of postcranial features suggest that the archaeolemurids spent considerable time on the ground (Walker, 1974; Godfrey, 1988). These include a posteriorly directed humeral head with greater tubercle projecting above it, a relatively deep olecranon fossa, a reduced and dorsomedially reflected medial epicondyle, and a greater trochanter projecting above the femoral head. In comparison to like-sized cercopithecids, the archaeolemurids have limb bones that are relatively short and robust, and very short metapodials. Hamrick et al. (2000) and Jungers et al. (2005) document newly discovered cheirideal elements of *Archaeolemur*, and the first known cheirideal elements of *Hadropithecus* have recently also been described (Wunderlich et al., 1996; Godfrey et al., 1997b, 2006a; Lemelin et al., 2008).

Genus *ARCHAEOLEMUR* Filhol, 1895
ARCHAEOLEMUR MAJORI Filhol, 1895
Figure 21.3

Partial Synonymy Nesopropithecus australis Forsyth-Major, 1900; *Protoindris globiceps* Lorenz von Liburnau, 1900; *Globil-emur flacourti* Forsyth-Major, 1897; *Bradylemur bastardi*, G. Grandidier, 1900.

Age and Occurrence Late Quaternary, southern and west-ern Madagascar; possibly central and northern Madagascar.

Diagnosis Skull length averages 128 mm; similar in mor-phology to but smaller in size than *A. edwardsi*; differs from latter in having less development of sagittal and nuchal crests; shallower (less steep) facial profile; as in *A. edwardsi*, lower incisors long, slender, and obliquely implanted; their tips wear flat; central upper incisors enormous and spatulate; upper canine is very broad and low crowned; p2 caniniform and robust; molars buccolingually expanded (broader than they are long), the first two with classic bilophodonty; third molars reduced but may exhibit incipient bilophodonty.

Description Body mass estimated at ~18 kg; intermembral index 92. Like its congener, *Archaeolemur majori* has short metapodials and phalanges, and relatively straight proximal phalanges. The pelvic girdle is broad, the scapula is relatively short (along the spine) but broad, with a particularly well-developed infraspinous fossa, and both fore- and hindlimbs are relatively short.

ARCHAEOLEMUR EDWARDSI (Filhol, 1895)

Partial Synonymy Nesopropithecus roberti Forsyth-Major, 1896; *Bradylemur robustus* G. Grandidier, 1899; *Archaeolemur platyrrhinus*, Standing, 1908.

Age and Occurrence Late Quaternary, central Madagascar; possibly western, northern, and southeastern Madagascar.

Diagnosis Skull length averages 147 mm; differs from its congener in having relatively greater postorbital constriction;

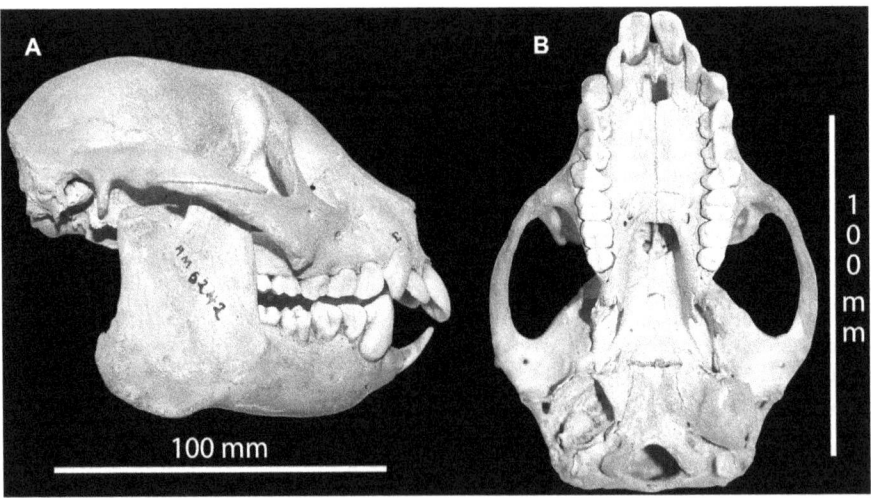

FIGURE 21.3 Archaeolemuridae: A) Lateral view of skull of *Archaeolemur majori* (from Tsirave, collec-tion of the Université d'Antananarivo); B) Ventral view of same skull of *A. majori*.

larger teeth (particularly the molars); relatively broader upper third molars; steeper facial profile; and greater development of sagittal and nuchal crests.

Description This is the larger of the two species of this genus, at ~26.5 kg (Jungers et al., 2008). There are few differences between *A. edwardsi* and *A. majori* other than body size. They are similar in morphology and in proportions; thus, for example, the intermembral index of *A. edwardsi* is 92, just as in *A. majori*. There are minor differences in robusticity, with the larger species tending also to be more robust. The hand of *Archaeolemur* sports a free os centrale, large pisiform, reduced pollex, and hamate with reduced hamulus. On the foot, the calcaneus, cuboid and fifth metatarsal have large tuberosities, and the hallux is reduced. There are enormous apical tufts on the distal phalanges of all digits. Unlike other extinct and extant lemurs, there is no evidence of a grooming claw.

Remarks on Genus Archaeolemur Recent research on dental microstructure has demonstrated that crown formation time was more prolonged in *Archaeolemur* than in *Megaladapis* or *Palaeopropithecus*, but not as prolonged as in *Hadropithecus* (Godfrey et al., 2005). Burney et al. (1997) and Vasey and Burney (unpubl.) found evidence for mollusk and small vertebrate consumption, in addition to herbivory, in fecal pellets that apparently belonged to *Archaeolemur*. The molars of *Archaeolemur* show high prism decussation and relatively thick enamel (Godfrey et al., 2005); this is normally indicative of hard-object feeding.

Genus *HADROPITHECUS* Lorenz von Liburnau, 1899
HADROPITHECUS STENOGNATHUS
Lorenz von Liburnau, 1899
Figure 21.4A

Partial Synonymy Pithecodon sikorae Lorenz von Liburnau, 1899.

Age and Occurrence Late Quaternary, southern, western, and central Madagascar.

Diagnosis See Lorenz von Liburnau (1902); Tattersall (1973). Skull length ~141 mm; face short, facial profile steep, mandible deep and very robust; zygomatic arch and postorbital bar well developed and robust; neurocranium relatively broad; as in *Archaeolemur*, premolar series modified into a continuous shearing blade; unlike *Archaeolemur*, anterior premolars, upper canine, and all incisors diminutive; all upper premolars have protocone developed to some extent; P4 is broader than M1 and completely molariform; lower incisors orthally implanted.

Description Body mass estimated at approximately 35 kg (Jungers et al., 2008). The limb bones of *Hadropithecus* recall those of *Archaeolemur* in many respects but differ in proportions; for example, the humerofemoral index (~103) is considerably higher and the brachial index (ca. 84) considerably lower (Godfrey et al., 2006a). The femur is considerably more robust and its shaft more anteroposteriorly compressed. *Hadropithecus* exhibits a number of traits (especially of the cheiridea) that may reflect greater terrestriality (e.g., virtually no hamulus or hook on the hamate, a more mediolateral orientation of the articular facet of the hamate for the triquetrum; Godfrey et al., 2006a; Lemelin et al., 2008), but the limb bone anatomy (including the greater anteroposterior compression of the femoral shaft, and greater asymmetry of the femoral condyles) suggests that this species was not cursorial. This inference has now gained support from study of the semicircular canals (Walker et al., 2008).

Remarks on genus Hadropithecus Recent discoveries have confirmed that Lamberton's (1938) hindlimb attributions for *Hadropithecus* are incorrect; the actual hindlimb bones of *Hadropithecus* are described by Godfrey et al. (1997b, 2006a). Stable carbon isotope values indicate a diet unlike that of any other lemur, high in C$_4$ and/or CAM plant products (Burney et al., 2004). Recent research on dental microstructure has demonstrated a unique dental developmental pattern, wherein crown formation was prolonged (approaching the developmental timing of chimpanzees), suggesting late molar eruption and prolonged infancy (Godfrey et al., 2005, 2006a, 2006b, 2006c).

Family MEGALADAPIDAE Forsyth-Major, 1894

The family Megaladapidae, or "koala lemurs," includes only one genus, *Megaladapis* with two subgenera, *Megaladapis* and

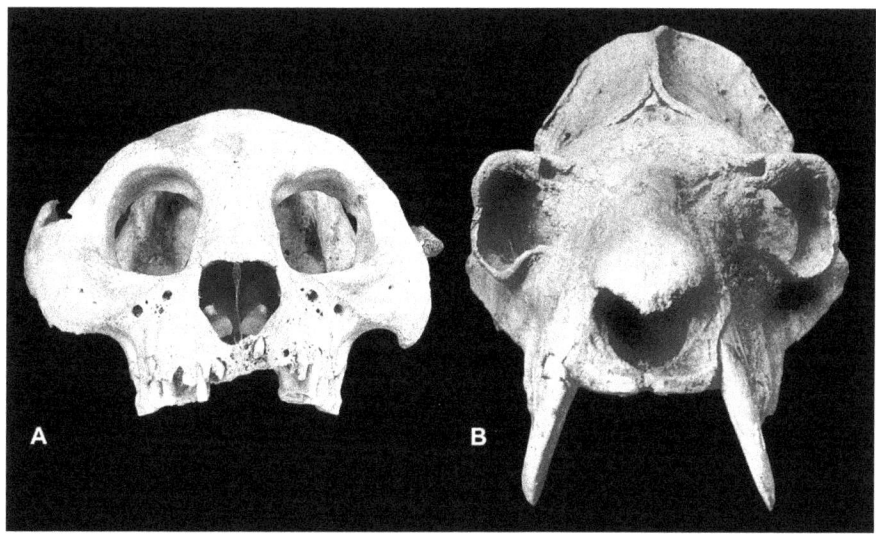

FIGURE 21.4 Frontal views of skulls of *Hadropithecus stenognathus* (A; from Tsirave, collection of the Académie Malgache) and *Megaladapis edwardsi* (B; southern Madagascar, collection of the Université d'Antananarivo). Not to scale.

Peloriadapis (Vuillaume-Randriamanantena *et al.*, 1992). The adult dental formula (0.1.3.3/2.1.3.3) is identical to that of *Lepilemur*. There is a typical strepsirrhine tooth comb comprising six teeth, and no permanent upper incisors. The angled mandibular symphysis fuses completely in adults. A diastema of variable length is present between the upper canine and the first premolar, and between the caniniform lower premolar and p3. Molar size increases from M1 to M3, and molars exhibit mesiodistally long shearing crests or "ectolophs." There is a posterior extension of the mandibular condyle's articular surface (and reciprocal expansion of the postglenoid process), another apparent homoplasy with *Lepilemur*. The skull is narrow, elongate, and bears both sagittal and nuchal crests. A small neurocranium is hafted onto the long and massive facial skeleton via a very large frontal sinus. There is strong postorbital constriction, a large temporal fossa, robust zygomatic arches, and a broad interorbital region. The nuchal plane is vertical and the occipital condyles face posteriorly; the paroccipital processes are long. The orbits are relatively small, laterally divergent and encircled by bony tori. The facial axis is retroflexed (i.e., marked airorhynchy). The autapomorphic nasals are long, projecting beyond prosthion, and flexed downward above the nasal aperture. The olfactory tracts are very long, and the optic foramina are relatively small. The auditory bulla is not inflated, the tympanic ring is fused laterally, and the tubular external auditory meatus is petrosal in origin (MacPhee, 1987). The mandible sports an expanded gonial region and a robust corpus.

The limbs are relatively short and very robust, and the upper limb is longer than the lower one (Jungers et al., 2002). Humerofemoral and intermembral indices are greater than 100. Slow, deliberate locomotion, inferred from postcranial morphology and proportions, has now been confirmed in a study of the semicircular canal system (Walker et al., 2008). Both hands and feet are relatively enormous, with divergent and robust pollex and hallux. Moderately curved proximal phalanges (Jungers et al., 1997). Spinous processes of thoracolumbar vertebrae are blunt and very reduced (but not to the extent seen in *Palaeopropithecus*). Transverse processes arise from the vertebral arch in the thoracolumbar region. The ilium is long, with the gluteal surface facing posteriorly; it broadens cranially and terminates with hooklike anterior superior spines. The sacrum is long and rectangular, and the sacral hiatus is narrow (the tail was no doubt quite short). The olecranon fossa of the humerus is shallow, the olecranon process of the ulna is prominent and retroflexed, and the ulnar styloid process is large and projecting. The femoral head is large and globular, and the knee exhibits an unusual "bowlegged" angle between femur and tibia. The femur is flattened in the anteroposterior plane. The fibula is robust and curved. The pisiform is dorsopalmarly expanded; the scaphoid tubercle and the hamate hamulus are similar in length to modern pronograde lemurs (Hamrick et al. 2000).

Genus *MEGALADAPIS* Forsyth-Major, 1894
MEGALADAPIS (PELORIADAPIS) EDWARDSI
(G. Grandidier, 1899)
Figure 21.4B

Partial Synonymy Peloriadapis edwardsi G. Grandidier, 1899; *Megaladapis insignis* Forsyth-Major 1900; *Megaladapis brachycephalus* Lorenz von Liburnau 1900; *Megaladapis dubius* Lorenz von Liburnau 1900; *Palaeolemur destructus* Lorenz von Liburnau 1900; *Megaladapis destructus* Lorenz von Liburnau 1901.

Age and Occurrence Late Quaternary, southern and southwestern Madagascar.

Diagnosis Largest of the koala lemurs. Lacks upper incisors, variable diastemata, M3 largest cheek tooth, airorhynch facial skeleton. Long projecting upper canines, prominent caniniform lower P3. Functional tooth comb present; mandibular symphysis fused. Bulla is flat and external auditory tube is "tubular." Higher intermembral and humerofemoral indices in comparison to congeners.

Description Very large body size (approximately 85 kg; Jungers et al., 2008). Cranial length averages 296 mm. Absolutely short diastemata. Extremely large molars (e.g., mesiodistal length of M^1 is 18.8 mm on average). Intermembral index ca. 120. Very robust long bones. Relatively straight humeral and radial diaphysis. Extremely varus knee joint. Dominance of medial condyle of proximal tibia and very lateral projection of tibial tuberosity. Relatively small tubercle on fifth metatarsal. Flattened surface of talar trochlea and malleolar facets. Low crural index. Reduced spinous processes, small sacral hiatus. Iliac blades long, broadening superiorly with hooklike anterior superior spines; rugose iliac crest for origin of abdominal musculature.

MEGALADAPIS (MEGALADAPIS) MADAGASCARIENSIS
Forsyth-Major, 1894

Partial Synonymy Megaladapis filholi G. Grandidier, 1899.
Age and Occurrence Late Quaternary, southern and southwestern Madagascar.
Diagnosis Smallest of the koala lemurs, especially in the postcranium. Close phenetic affinities with *M. grandidieri* (Vuillaume-Randriamanantena et al., 1992). Limb bones very robust; humerus broadens distally with large brachioradialis flange. Prominent tuberosity on fifth metatarsal. Large calcaneus with medially projecting tuberosity. Long, robust, and divergent hallux. Talar trochlea less flattened, more grooved.

Description Smallest of the three species of *Megaladapis*, at approximately 46.5 kg (Jungers et al., 2008). Skull length averages 245 mm. Mean length of M^1 is 14.0 mm. Longer diastemata. Intermembral index ca. 114. Humeral head exhibits greater longitudinal curvature. Olecranon fossa is deeper. Broad distal humerus with projecting medial epicondyle and broad brachialis flange. Radial diaphysis quite curved. Relatively large lesser trochanter. Prominent lateral tubercle of fifth metatarsal. Axial skeleton and bony girdles are still poorly known.

MEGALADAPIS (MEGALADAPIS) GRANDIDIERI
Standing, 1903
Figure 21.5

Partial Synonymy Megalindris gallienii Standing, 1908.
Age and Occurrence Late Quaternary, central Madagascar.
Diagnosis "Medium-sized" koala lemur. Teeth are small relative to the postcranium. Strong phenetic affinities with *M. madagascariensis*. Relatively long diastemata. Some limb bones (e.g., tibia and fibula) overlap in length and robusticity with *M. edwardsi*. Shortest mass-adjusted femora and least curved proximal phalanges of the koala lemurs.

Description Body mass estimated at ~74 kg (Jungers et al., 2008). Skull length estimated at 289 mm; M1 length is 15.4 mm. Absolutely and relatively large diastemata. Shearing crests more prominent on molars (Jungers et al., 2002). Larger body size but morphologically similar to *M. madagascariensis*. Intermembral index ca. 115. Postcrania recall those

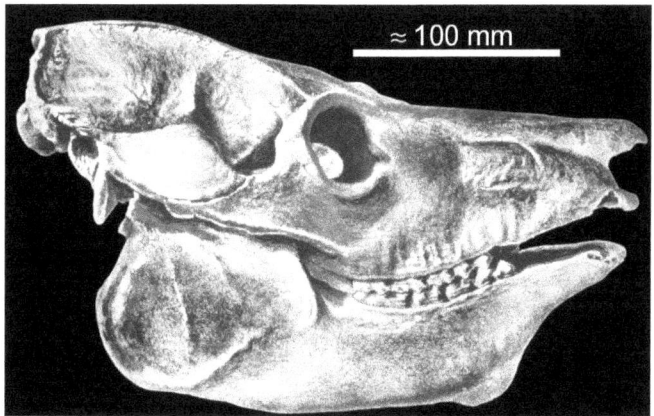

FIGURE 21.5   Lateral view of skull of *Megaladapis grandidieri* from Ampasambazimba, reproduced from Lamberton, 1934c: Plate II.

of *M. madagascariensis*, but are larger overall. One large tibia and apparently associated fibula formerly misidentified as belonging to *Archaeoindris* (Vuillaume-Randriamanantena et al., 1992).

Remarks on the Genus Megaladapis *Megaladapis* sp. cf. *M. grandidieri/madagascariensis* (provisional) from the extreme north and northwest of Madagascar is intermediate in size between *M. madagascariensis* and *M. grandidieri*, and very similar anatomically to both. The variant of *Megaladapis* from Anjohibe has particularly small teeth. Schwartz et al. (2005) demonstrate rapid dental development in *Megaladapis* despite its enormous size, although not as rapid as in *Palaeopropithecus*.

Family LEMURIDAE Gray, 1821

This family is comprised mainly of extant forms as well as the extinct genus *Pachylemur*, which resembles *Varecia* in numerous characteristics but is much larger in body size (Walker, 1974; Seligsohn and Szalay, 1974). The adult dental formula is as in other lemurids (2.1.3.3/2.1.3.3); there is a typical lemurid tooth comb, and the mandibular symphysis remains unfused throughout life. The orbits are relatively small. Separate genus status for *Pachylemur* is supported by differences in the postcranial skeletons and inferred positional behavior of *Pachylemur* vs. other lemurids. Whereas the appendicular skeleton closely resembles that of other lemurids in some morphological details, the limbs are shorter and more robust relative to the vertebral column, and the proportions are different. The two species in the genus *Pachylemur* (*Pachylemur insignis* and *P. jullyi*) are sometimes considered regional variants of the same species.

Genus *PACHYLEMUR* Lamberton, 1948
PACHYLEMUR INSIGNIS (Filhol, 1895)
Figures 21.6A and 21.6B

Partial Synonymy Lemur intermedius Filhol, 1895; *Varecia insignis* Walker, 1974.

Age and Occurrence Late Quaternary, southern and southwestern Madagascar, perhaps northwest.

Diagnosis Skull length averages 117 mm; as in *Varecia*, distinguished from *Lemur* and *Eulemur* by suite of dental traits (elongate talonid basins, protocone fold on the first upper molar, anterior expansion of the lingual cingulum of first and second upper molars); differs from *Varecia* in having

orbits more frontally oriented; broader skull; more massive jaws; larger teeth; differs from *P. jullyi* in having: smaller palate, smaller teeth; mandibular cheek teeth are more buccolingually compressed; superior temporal lines generally do not meet at midline; talonid basins of the lower molars skewed into a rhombus that opens distolingually; buccal cusps positioned mesial to adjacent lingual cusps (Vasey et al., 2005).

Description Adult dental formula (2.1.3.3/2.1.3.3), dental morphology similar in most respects to *Varecia*, but about three to four times larger in body size, and far more robust. This is the smaller of the two species of *Pachylemur* at ~11.5 kg (Jungers et al., 2008). Intermembral index ca. 97. Intermembral index higher than in *Varecia* (the fore- and hindlimbs more equal in length), but mass-adjusted limb lengths are shorter. Greater tubercle and greater trochanter project just proximal to humeral and femoral heads, respectively. Short lumbar vertebral bodies; lumbar spinous processes are somewhat reduced and exhibit less anticliny.

PACHYLEMUR JULLYI (G. Grandidier, 1899)

Partial Synonymy Paleochirogalus jullyi Grandidier, 1899; *Lemur jullyi* Standing, 1904; *L. maxiensis* Standing, 1904; *L. majori* Standing 1908.

Age and Occurrence Late Quaternary, central Madagascar, possibly north.

Diagnosis Similar to congener but larger in skull and tooth size; average skull length ~125 mm; mandibular molars wider, talonid basins squarer with adjacent cusps transversely aligned (Vasey et al., 2005).

Description Larger of the two species of *Pachylemur* at ~13 kg (Jungers et al., 2008); intermembral index ~94. Sagittal and nuchal crests generally occur. Humerofemoral, intermembral, brachial and crural indices are all slightly higher than in *P. insignis*.

Remarks on the Genus Pachylemur DNA confirms a close relationship of *Pachylemur* to *Varecia* (Crovella et al., 1994); dental anatomical data also support such a relationship (Seligsohn and Szalay 1974; Tattersall, 1982). The dominant element in the diet of *Pachylemur*, as in *Varecia*, was probably fruit. It was likely the most important large-seed disperser of the extinct lemurs (Godfrey et al., 2008). Its *Varecia*-like teeth show a rather high incidence of caries and uneven dental wear (Vasey et al., 2005). Seligsohn and Szalay (1974) argue, on basis of molar morphology, that in comparison to *Eulemur*,

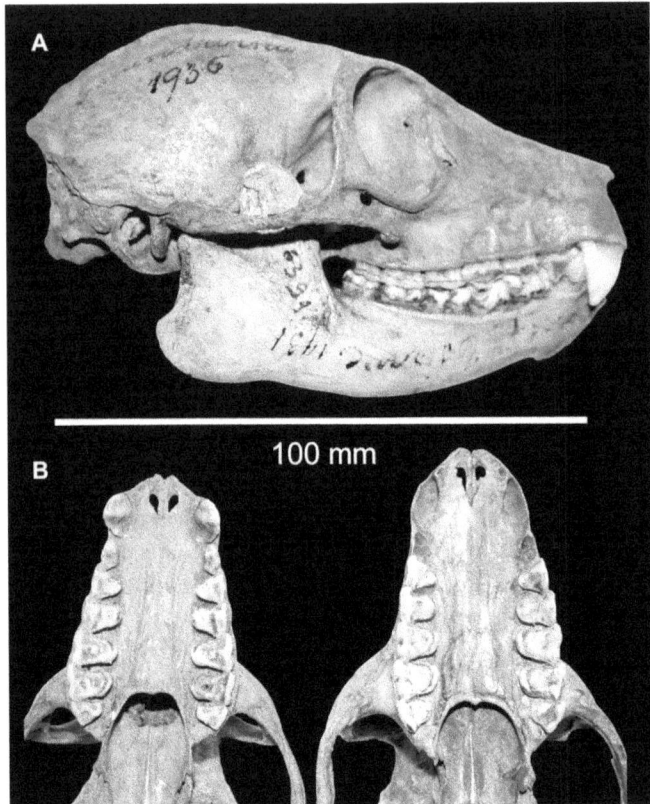

FIGURE 21.6. A) lateral view of a composite skull of *Pachylemur insignis* (cranium from Tsiandroina, mandible from Tsirave, both in southern Madagascar, collection of the Université d'Antananarivo); B) occlusal view of maxillary dentition of *P. insignis* (left, southern Madagascar) and *P. jullyi* (right, central Madagascar), both in the collections of the Université d'Antananarivo. A and B are not to scale; scale bar applies only to A.

Pachylemur would have consumed fewer leaves and more stems and hard fruits. They cite evidence for hard fruit consumption in *Varecia*. *Pachylemur* from the north is not known from a whole skull; materials are insufficient to designate species affinities with any confidence.

Family DAUBENTONIIDAE Gray, 1863

This family comprises a single genus (*Daubentonia*) with two species, the still-extant *D. madagascariensis* and the giant extinct aye-aye, *D. robusta*. Incisors are hypertrophied and curved, chisel-like, with enamel on the anterior surface only. Both upper and lower incisors are laterally compressed and open rooted; the mesial enamel and distal dentine create a sharp cutting edge through differential wear. Whereas the incisors are known for both extinct and extant species, no skull belonging to the extinct form has been found. Existing skeletal remains demonstrate broad similarities. In the extant form (and likely the extinct), a long diastema separates the anterior teeth from the reduced cheek teeth. The cheek teeth are flattened and exhibit indistinct, rounded cusps; there is a single peglike upper premolar. Adult dental formula is 1.0.1.3/1.0.0.3.

Genus *DAUBENTONIA* E. Geoffroy Saint-Hilaire, 1795
DAUBENTONIA ROBUSTA (Lamberton, 1934b)

Partial Synonymy Chiromys Illiger, 1811, *Cheiromys* G. Cuvier 1817, *Chiromys robustus* Lamberton, 1934b

Age and Occurrence Late Quaternary, southwestern Madagascar.

Diagnosis Postcranial skeleton very similar in morphology to that of extant congener but much more robust (Lamberton, 1934b; Simons, 1994); differing in limb proportions (e.g., humerofemoral index higher); limbs short in comparison to body mass; intermembral index ~85. As in congener, femoral head relatively small, ilia narrow and rodlike; incisors hypertrophied, and manual digit III with thin, filiform phalanges and elongated metacarpal; forelimb short and robust in comparison to the hindlimb

Description Body mass estimated at ~14 kg (Jungers et al., 2008), roughly 5 times that of living congener. Of the skull, only the incisors are known (Grandidier, 1929; MacPhee and Raholimavo, 1988). The postcrania exhibit a number of distinctive features (Simons, 1994); for example, the brachialis flange is enormous and winglike, accommodating a massive brachioradialis (Soligo, 2005).

Remarks on the Genus Daubentonia *Daubentonia* possesses a suite of appendicular and especially manual adaptations that facilitate the manual extraction (through bored holes) of nuts, insects, insect larvae, and other foodstuffs. The metacarpophalangeal joint of the third digit of extant aye-ayes allows an extraordinary range of movement; the aye-aye can insert this digit at odd angles into the longitudinal channels created by wood-boring insects (Erickson, 1994, 1995). Aye-ayes exhibit postcranial as well as craniodental convergences to *Dactylopsila* (the striped possum), which forages in a similar manner (Cartmill, 1974; Godfrey et al., 1995). Invertebrates most likely complemented a primary diet of nuts and other plant products. The giant extinct aye-aye and still extant aye-aye appear to have been allopatric, with the former restricted to the drier habitats of the southwest.

General Discussion: The Extinction of the Subfossil Lemurs

The extinct lemurs of Madagascar formed a key portion of a megafauna that was unique in many ways. In the first place, this is the only primate-dominated assemblage among the world's extinct late Quaternary megafaunas. Additionally, the extinction losses here were more severe than on any of the continents and most other large islands. Madagascar lost all of its endemic animals above 10 kg, including not merely the big strepsirrhine primates but also birds, reptiles, and the other large mammals. It is also one of the most recent of the prehistoric megafaunal crashes, so the evidence is relatively fresh.

As we have documented here and elsewhere (Godfrey et al., 1997a, 2006b; Godfrey and Jungers, 2002; Jungers et al., 2002; see also Tattersall, 1982), the adaptive diversity represented by the combination of extant and extinct lemurs is extraordinary. Primate body masses on Madagascar once ranged from roughly 30 g to over 150 kg, and not long ago there were "monkey lemurs," "sloth lemurs," "koala lemurs," and giant aye-ayes to round out the amazing roster of Malagasy primates. Along with elephant birds, giant tortoises and hippos, all lemurs ~10 kg and larger are missing now from the still impressive array of endemic vertebrates. What and/or who killed the giant lemurs and other megafauna of Madagascar?

Various theories have been offered to account for this last of the great megafaunal extinctions (reviewed in Burney et al., 2004; Burney, 2005). Several hypotheses are dramatically unicausal and imply rapid extirpation of the subfossil lemurs and other large-bodied terrestrial vertebrates at the hands of colonizing Indonesians: "great fires" (e.g., Humbert, 1927), "blitzkrieg hunting" (e.g., Martin, 1984), and "hypervirulent diseases" (MacPhee and Marx, 1997). Although the fingerprints of humans are surely present at this Holocene crime scene, a recently compiled ^{14}C chronology for late prehistoric Madagascar is incompatible with the extreme versions of these extinction scenarios (Burney et al., 2004). The anthropogenic "smoking gun" smoldered for a very long time, much too long, in fact, to validate the predictions of any model of overnight eradication of the subfossil lemurs and other megafauna.

The accumulated evidence, backed by 278 age determinations (primarily ^{14}C dating) documents late Pleistocene climatic events as well as the apparently human-caused transformation of the environment in the late Holocene (reviewed in Burney et al., 2004). Multiple lines of evidence (including modified bones of extinct species and the appearance in sediment cores of exotic pollen of introduced *Cannabis*) point to the earliest human presence at ca. 2300 ^{14}C yr BP (350 cal yr BC). A decline in megafauna, inferred from a drastic decrease in spores of the coprophilous fungus *Sporormiella* spp. (a proxy for megafaunal biomass) in sediments at 1720 ± 40 ^{14}C yr BP (230–410 cal yr AD), is followed by large increases in charcoal particles in sediment cores (Burney et al., 2003). This pattern begins in the southwest part of the island and spreads to other coasts and the interior over the next millennium. The record of human occupation is initially sparse but shows large human populations throughout the island by the beginning of the second millennium AD.

Dating of the extinct large lemurs, as well as pygmy hippos, elephant birds, and giant tortoises, demonstrates that most if not all the extinct taxa were still present on the island when humans arrived. Many overlapped chronologically with humans for a millennium or more. Among the extinct lemurs, *Hadropithecus stenognathus, Pachylemur insignis, Mesopropithecus pithecoides,* and *Daubentonia robusta* were still present near the end of the First Millennium AD. *Palaeopropithecus ingens, Megaladapis edwardsi,* and *Archaeolemur* sp. cf. *A. edwardsi* may have survived until the middle of the second millennium AD. The accumulated evidence suggests that humans may have collapsed these ecosystems through a combination of impacts, including overhunting (e.g. MacPhee and Burney, 1991; Perez et al. 2005); landscape modification (e.g., Burney, 1993; Burney et al., 2003) and perhaps other interacting factors, such as invasive species and climatic desiccation (Dewar, 1984; Burney, 1999).

The extinction explanation we favor in lieu of single-cause, very rapid scenarios is thus the "synergy" hypothesis (Burney, 1999). Extinctions are still regarded primarily as the handiwork of humans, but hunting, burning, and habitat transformation and degradation interact in a very slow and mosaic fashion, and the various human impacts may well have differed in significance from region to region across the island. Background climatic change (e.g., dessication in the southwest) and the introduction of domesticated species (e.g., livestock proliferation in the northwest) are regarded as probable contributing factors in the extinction process but cannot serve as stand-alone explanations. For example, in the southwest it seems likely that the open-country, nonprimate grazers and browsers (e.g., tortoises, elephant bird, and hippos) were reduced drastically in density by intense human predation within a few centuries of colonization. As plant biomass increased as a consequence, fires of human origin increased in frequency and ferocity, and this promoted major ecological restructuring, including the loss of wooded savannas and the preferred (arboreal) habitats of subfossil lemurs. Slowly reproducing, large-bodied lemurs, probably already at low population densities, were unable to "bounce back," and extinction proceeded slowly but inexorably. Few places, if any, in Madagascar were untouched by humans as they expanded into other areas at different times, but the result, the "deadly syncopation" (MacPhee and Marx, 1997), was invariably the same. Regrettably, the lethal synergies we have proposed are still in place in Madagascar, and the extinction window remains all too open.

ACKNOWLEDGMENTS

We are deeply indebted to our many colleagues and friends in Madagascar for their assistance and continuing support of our research efforts in their country. Without their kind and generous cooperation, our work there would have been impossible. We also wish to express our admiration for and appreciation of Elwyn Simons, who was instrumental in bringing L.R.G. and W.L.J. back to Madagascar more than two decades ago. We also offer our sincere thanks to the many museum curators who provided access to their skeletal collections of living and extinct mammals. We thank Luci Betti-Nash for her help in preparing the figures. We gratefully acknowledge our numerous collaborators in this work, especially our Malagasy colleagues who have facilitated our field research on subfossil lemurs (Berthe Rakotosamimanana and Gisèle Randria, both recently deceased, and Armand Rasoamiaramanana). Our collaborators include Lida Pigott Burney, Kierstin Catlett, Prithijit Chatrath, Alan Cooper, Brooke Crowley, Frank Cuozzo, Brigitte Demes, Mary Egan, Steve Goodman, Mark Hamrick, Mitchell Irwin, Helen James, A. J. Timothy Jull, Stephen King, Pierre Lemelin, Patrick Mahoney, Malgosia Nowak-Kemp, Robert Paine, Ventura Perez, Andrew Petto, Lydia Raharivony, Berthe Rakotosamimanana, Mirya Ramarolahy, Ramilisonina, Gisèle Randria, Jonah Ratsimbazafy, Jeannette Ravaoarisoa, Brian Richmond, Timothy Ryan, Karen Samonds, Gary Schwartz, Jessica Scott, Rob Scott, Gina Semprebon, Liza Shapiro, Cornelia Simons, Elwyn Simons, Michael Sutherland, Mark Teaford, Peter Ungar, Natalie Vasey, Martine Vuillaume-Randriamanantena, Alan Walker, Christine Wall, William Wheeler, Trevor Worthy, Henry Wright, and Roshna Wunderlich. Support for our laboratory and field research in Madagascar has been provided by the National Science Foundation, the Wenner-Gren Foundation for Anthropological Research, and the National Geographic Society (L.R.G.); the National Science Foundation, the Margot Marsh Biodiversity Fund, and the Stony Brook University Medical Center (W.L.J.); and the National Science Foundation, the National Geographic Society, the Smithsonian Institution, and the NOAA Human Dimensions of Global Change program (D.A.B.).

Literature Cited

Ali, J. R., and J. C. Aitchison. 2008. Gondwana to Asia: Plate tectonics, paleogeography and the biological connectivity of the Indian subcontinent from the Middle Jurassic through latest Eocene (166–35 Ma). *Earth Science Reviews* 88:145–166.

Ali, J. R., and M. Huber. 2010. Mammalian biodiversity on Madagascar controlled by ocean currents. *Nature* 463:653–656.

Burney, D. A. 1987a. Late Holocene vegetational change in central Madagascar. *Quaternary Research* 28:130–143.

———. 1987b. Pre-settlement vegetation changes at Lake Tritrivakely, Madagascar. *Palaeoecology of Africa* 18:357–381.

———. 1993 Late Holocene environmental changes in arid Southwestern Madagascar. *Quaternary Research* 40:98–106

———. 1999. Rates, patterns, and processes of landscape transformation and extinction in Madagascar; pp 145–164 in R. D. E. MacPhee (ed.), *Extinction in Near Time*. Kluwer Academic/Plenum Publishers, New York.

———. 2005. Finding the connections between paleoecology, ethnobotany, and conservation in Madagascar. *Ethnobotany Research & Applications* 3:385–389.

Burney, D. A., L. P. Burney, L. R. Godfrey, W. L. Jungers, S. M. Goodman, H. T. Wright, and A. J. Timothy Jull. 2004. A chronology for late prehistoric Madagascar. *Journal of Human Evolution* 47:25–63.

Burney, D. A., H. F. James, F. V. Grady, J.-G. Rafamantanantsoa, Ramilisonina, H. T. Wright and J. B. Cowart. 1997. Environmental change, extinction, and human activity: Evidence from caves in NW Madagascar. *Journal of Biogeography* 24:755–767.

Burney, D. A., G. S. Robinson, and L. P. Burney. 2003. *Sporormiella* and the late Holocene extinctions in Madagascar. *Proceedings of the National Academy of Sciences, USA* 100:10800–10805.

Cartmill, M. 1974. *Daubentonia, Dactylopsila*, woodpeckers, and klinorhynchy; pp. 655–670 in R. D. Martin, G. A. Doyle, and A. C. Walker (eds.), *Prosimian Biology*. Duckworth, London.

Cartmill, M. 1975. Strepsirhine basicranial structures and the affinities of the Cheirogaleidae; pp. 313–354 in W. P. Luckett and F. S. Szalay (eds.), *Phylogeny of the Primates: A Multidisciplinary Approach*. Plenum Press, New York.

Crovella, S., D. Montagnon, B. Rakotosamimanana, and Y. Rumpler. 1994. Molecular biology and systematics of an extinct lemur: *Pachylemur insignis*. *Primates* 35:519–522.

DelPero, M., J. C. Masters, P. Cervella, S. Crovella, G. Ardito, and Y. Rumpler. 2001. Phylogenetic relationships among the Malagasy lemuriforms (Primates: Strepsirrhini) as indicated by mitochondrial sequence data from the 12S rRNA gene. *Zoological Journal of the Linnean Society* 133:83–103.

DelPero, M., L. Pozzi, and J. C. Masters. 2006. A composite molecular phylogeny of living lemuroid primates. *Folia Primatologica* 77:434–445.

Dewar, R. E. 1984. Extinctions in Madagascar: The loss of the subfossil fauna; pp. 574–593 in P. S. Martin, and R. G. Klein (eds.), *Quaternary Extinctions: A Prehistoric Revolution*. University of Arizona Press, Tucson.

deWit, M., and J. C. Masters. 2004. The geological history of Africa, India and Madagascar, dispersal scenarios for vertebrates. *Folia Primatologica* 75:117.

Eizirik, E., W. J. Murphy, and S. J. O'Brien. 2001. Molecular dating and biogeography of the early placental mammal radiation. *Journal of Heredity* 92:212–219.

Erickson, C. J. 1994. Tap-scanning and extractive foraging in aye-ayes, *Daubentonia madagascariensis*. *Folia Primatologica* 62:125–135.

———. 1995. Feeding sites for extractive foraging by the aye-aye, *Daubentonia madagascariensis*. *American Journal of Primatology* 35.235–240.

Filhol, H. 1895. Observations concernant les mammifères contemporains des *Aepyornis* à Madagascar. *Bulletin du Muséum d'Histoire Naturelle Paris* 1:12–14.

Flacourt, E. de. 1658. *Histoire de la Grande Isle Madagascar composée par le Sieur de Flacourt*. Chez G. de Lynes, Paris.

Forsyth-Major, C. I. 1893. Verbal report on an exhibition of a specimen of a subfossil lemuroid skull from Madagascar. *Proceedings of the Zoological Society of London* 36:532–535.

———. 1894. On *Megaladapis madagascariensis*, an extinct gigantic lemuroid from Madagascar, with remarks on the associated fauna, and on its geologic age. *Philosophical Transactions of the Royal Society of London, B* 185:15–38.

———. 1896. Preliminary notice on fossil monkeys from Madagascar. *Geological Magazine*, n.s., Decade 4, 3:433–436.

Godfrey, L. R. 1988. Adaptive diversification of Malagasy strepsirhines. *Journal of Human Evolution* 17:93–134.

Godfrey, L. R., and W. L. Jungers 2002. Quaternary fossil lemurs; pp. 97–121 in W. C. Hartwig (ed.), *The Primate Fossil Record*. Cambridge University Press, Cambridge.

———. 2003. The extinct sloth lemurs of Madagascar. *Evolutionary Anthropology* 12:252–263.

Godfrey, L. R., W. L. Jungers, D. A. Burney, N. Vasey, Ramilisonina, W. Wheeler, P. Lemelin, L. J. Shapiro, G. T. Schwartz, S. J. King, M. F. Ramarolahy, L. L. Raharivony, and G. F. N. Randria. 2006a. New discoveries of skeletal elements of *Hadropithecus stenognathus* from Andrahomana Cave, southeastern Madagascar. *Journal of Human Evolution* 51:395–410.

Godfrey, L. R., W. L. Jungers, K. E. Reed, E. L. Simons, and P. S. Chatrath. 1997a. Subfossil lemurs: Inferences about past and present primate communities; pp. 218–256 in S. M. Goodman, and B. D. Patterson (eds.), *Natural Change and Human Impact in Madagascar*. Smithsonian Institution Press, Washington, D.C.

Godfrey, L. R., W. L. Jungers, and G. T. Schwartz. 2006b. Ecology and extinction of Madagascar's subfossil lemurs; pp. 41–64 in L. Gould, and M. L. Sauther (eds.), *Lemurs: Ecology and Adaptation*. Springer, New York, New York.

Godfrey, L. R., W. L. Jungers, G. T. Schwartz, and M. T. Irwin. 2008. Ghosts and orphans: Madagascar's vanishing ecosystems; pp. 361–395 in J. G. Fleagle and C. C. Gilbert (eds.), *Elwyn Simons. A Search for Origins*. Springer, New York.

Godfrey, L. R., W. L. Jungers, R. E. Wunderlich, and B. G. Richmond. 1997b. Reappraisal of the postcranium of *Hadropithecus* (Primates, Indroidea). *American Journal of Physical Anthropology* 103:529–556.

Godfrey, L. R., A. J. Petto, and M. R. Sutherland. 2002. Dental ontogeny and life-history strategies: The case of the giant extinct indroids of Madagascar; pp. 113–157 in J. M. Plavcan, R. F. Kay, W. L. Jungers, and C. P. van Schaik (eds.), *Reconstructing Behavior in the Primate Fossil Record*. Kluwer Academic/Plenum Publishers, New York.

Godfrey, L. R., G. T. Schwartz, K. E. Samonds, W. L. Jungers, and K. K. Catlett. 2006c. The secrets of lemur teeth. *Evolutionary Anthropology* 15:142–154.

Godfrey, L.R., G. M. Semprebon, G. T. Schwartz, D. A. Burney, W. L. Jungers, E. K. Flanagan, F. P. Cuozzo, and S. J. King. 2005. New insights into old lemurs: The trophic adaptations of the Archaeolemuridae. *International Journal of Primatology* 26:825–854.

Godfrey, L. R., E. L. Simons, P. S. Chatrath, and B. Rakotosamimanana. 1990. A new fossil lemur (*Babakotia*, Primates) from northern Madagascar. *Comptes Rendus de l'Académie des Sciences, Paris*, Série II, 310:81–87.

Godfrey, L.R., M. R. Sutherland, R. R. Paine, F. L. Williams, D. S. Boy, and M. Vuillaume-Randriamanantena. 1995. Limb joint surface areas and their ratios in Malagasy lemurs and other mammals. *American Journal of Physical Anthropology* 97:11–36.

Godinot, M. 2006. Lemuriform origins as viewed from the fossil record. *Folia Primatologica* 77:446–464.

Gommery, D., N. Ramanivosoa, S. Tombomiadana-Raveloson, H. Randrianantenaina, and P. Kerloc'h. 2009. Une nouvelle espèce de lémurien géant subfossile nu Nord-Ouest de Madagascar (*Palaeopropithecus kelyus,* Primates). *Comptes Rendus Palevol* 8:471–480.

Grandidier, G. 1899. Description d'ossements de lémuriens disparus. *Bulletin du Muséum d'Histoire Naturelle Paris* 5:272–276, 344–348.

———. 1901. Un nouvel édenté subfossile de Madagascar. *Bulletin du Muséum d'Histoire Naturelle Paris* 7:54–56.

———. 1929 (for the year 1928). Une variété du *Cheiromys madagascariensis* actuel et un nouveau *Cheiromys* subfossile. *Bulletin de l'Académie Malgache* (n.s.) 11:101–107.

Hamrick, M. W., E. L. Simons, and W. L. Jungers. 2000. New wrist bones of the Malagasy giant subfossil lemurs. *Journal of Human Evolution* 38:635–650.

Hedges, S. B., P. H. Parker, C. G. Sibley, and S. Kumar. 1996. Continental breakup and the ordinal diversification of birds and mammals. *Nature* 381:226–229.

Hooker, J. J., D. E. Russell, and A. Phelizon. 1999. A new family of Plesiadapiformes (Mammalia) from the Old World lower Paleogene. *Palaeontology* 42:377–407.

Horvath, J. E., D. W. Weisrock, S. L. Embry, I. Fiorentino, J. P. Balhoff, P. Kappeler, G. A. Wray, H. F. Willard, and A. D. Yoder. 2008. Development and application of a phylogenomic toolkit: Resolving the evolutionary history of Madagascar's lemurs. *Genome Research* 18:489–499.

Humbert, H. 1927. Destruction d'une flore insulaire par le feu. *Mémoires de l'Académie Malgache* 5:1–80.

Jungers, W. L. 1977. Hindlimb and pelvic adaptations to vertical climbing and clinging in *Megaladapis*, a giant subfossil prosimian from Madagascar. *Yearbook of Physical Anthropology* 20:508–524.

———. 1978. The functional significance of skeletal allometry in *Megaladapis* in comparison to living prosimians. *American Journal of Physical Anthropology* 19:303–314.

———. 1980. Adaptive diversity in subfossil Malagasy prosimians. *Zeitschrift für Morphologie und Anthropologie* 71:177–186.

Jungers, W. L., B. Demes, and L. R. Godfrey. 2008. How big were the "giant" extinct lemurs of Madagascar? pp. 343–360 in J. G. Fleagle and C. C. Gilbert (eds.), *Elwyn Simons: A Search for Origins.* Springer, New York.

Jungers, W. L., and L. R. Godfrey. 2003. Box 2: Extreme sport. *Evolutionary Anthropology* 12:258.

Jungers, W. L., L. R. Godfrey, E. L. Simons, and P. S. Chatrath. 1997. Phalangeal curvature and positional behavior in extinct sloth lemurs (Primates, Palaeopropithecidae). *Proceedings of the National Academy of Sciences, USA* 94:11998–12001.

Jungers, W. L., L. R. Godfrey, E. L. Simons, P. S. Chatrath, and B. Rakotosamimanana. 1991. Phylogenetic and functional affinities of *Babakotia radofilai*, a new fossil lemur from Madagascar. *Proceedings of the National Academy of Sciences, USA* 88:9082–9086.

Jungers, W. L., L. R. Godfrey, E. L. Simons, R. E. Wunderlich, B. G. Richmond, and P. S. Chatrath. 2002. Ecomorphology and behavior of giant extinct lemurs from Madagascar; pp. 371–411 in J. M. Plavcan, R. F. Kay, W. L. Jungers, and C. P. van Schaik (eds.), *Reconstructing Behavior in the Primate Fossil Record.* Kluwer Academic/Plenum Publishers, New York.

Jungers, W. L., P. Lemelin, L. R. Godfrey, R. E. Wunderlich, D. A. Burney, E. L. Simons, P. S. Chatrath, H. F. James, and G. F. N. Randria. 2005. The hands and feet of *Archaeolemur*: Metrical affinities and their functional significance. *Journal of Human Evolution* 49:36–55.

Kappeler, P. M. 2000. Lemur origins: Rafting by groups of hibernators? *Folia Primatologica* 71:422–425.

Karanth, K. P., T. Delefosse, B. Rakotosamimanana, T. J. Parsons, and A. D. Yoder. 2005. Ancient DNA from giant extinct lemurs confirms single origin of Malagasy primates. *Proceedings of the National Academy of Sciences, USA* 102:5090–5095.

King, S. J., L. R. Godfrey, and E. L. Simons. 2001. Adaptive and phylogenetic significance of ontogenetic sequences in *Archaeolemur*, subfossil lemur from Madagascar. *Journal of Human Evolution* 41:545–576.

Krause, D. W., J. H. Hartman, and N. A. Wells. 1997. Late Cretaceous vertebrates from Madagascar: Implications for biotic change in deep time; pp. 3–43 in S. M. Goodman, and B. D. Patterson (eds.), *Natural Change and Human Impact in Madagascar.* Smithsonian Institution Press, Washington, D.C.

Lamberton, C. 1934a. Contribution à la connaissance de la faune subfossile de Madagascar: Lémuriens et Ratites. *L'Archaeoindris fontoynonti* Stand. *Mémoires de l'Académie Malgache* 17:9–39.

———. 1934b. Contribution à la connaissance de la faune subfossile de Madagascar: Lémuriens et Ratites: *Chiromys robustus* sp. nov. Lamb. *Mémoires de l'Académie Malgache* 17:40–46.

———. 1934c. Contribution à la connaissance de la faune subfossile de Madagascar: Lémuriens et Ratites: Les *Megaladapis*. *Mémoires de l'Académie Malgache* 17:47–105.

———. 1936. Nouveaux lémuriens fossiles du groupe des Propithèques et l'intérêt de leur découverte. *Bulletin du Muséum National d'Histoire Naturelle, Paris*, Série II, 8:370–373.

———. 1938 (for the year 1937). Contribution à la connaissance de la faune subfossile de Madagascar: Note III. Les Hadropithèques. *Bulletin de l'Académie Malgache* (n.s.) 20:127–170.

———. 1948 (for the year 1946). Contribution à la connaissance de la faune subfossile de Madagascar: Note XVII. Les Pachylemurs. *Bulletin de l'Académie Malgache* (n.s.) 27:7–22.

Lawlor, T. E. 1986. Comparative biogeography of mammals on islands. *Biological Journal of the Linnean Society* 28:99–125.

Lemelin, P., M. W. Hamrick, B. G. Richmond, L. R. Godfrey, W. L. Jungers, and D. A. Burney. 2008. New hand bones of *Hadropithecus stenognathus*: Implications for the paleobiology of the Archaeolemuridae. *Journal of Human Evolution* 54:404–413.

Lorenz von Liburnau, L. 1902. Über *Hadropithecus stenognathus* Lz. nebst bemerkungen zu einigen anderen ausgestorbenden Primaten von Madagaskar. *Denkschriften der Mathematisch-Naturwissenschaftliche Klasse der Kaiserlichen Akademie der Wissenschaften zu Wien* 72:243–254.

———. 1905. *Megaladapis edwardsi* G. Grandidier. *Denkschriften der Mathematisch-Naturwissenschaftlichen Klasse der Kaiserlichen Akademie der Wissenschaften zu Wien* 77: 451–490.

MacPhee, R. D. E. 1987. Basicranial morphology and ontogeny of the extinct giant lemur *Megaladapis*. *American Journal of Physical Anthropology* 74:333–355.

MacPhee, R. D. E., and D. A. Burney. 1991. Dating of modified femora of extinct dwarf hippopotamus from southern Madagascar: Implications for constraining human colonization and vertebrate extinction events. *Journal of Archaeological Science* 18:695–706.

MacPhee, R. D. E., and P. A. Marx. 1997. The 40,000-year plague: Humans, hypervirulent diseases, and first-contact extinctions; pp. 169–217 in S. M. Goodman, and B. D. Patterson (eds.), *Natural Change and Human Impact in Madagascar.* Smithsonian Press, Washington, D.C.

MacPhee, R. D. E., and E. M. Raholimavo. 1988. Modified subfossil aye-aye incisors from southwestern Madagascar: Species allocation and paleoecological significance. *Folia Primatologica* 51:126–142.

MacPhee, R. D. E, E. L. Simons, N. A. Wells, and M. Vuillaume-Randriamanantena. 1984. Team finds giant lemur skeleton. *Geotimes* 29:10–11.

Marivaux, L., J.-L. Welcomme, P.-O. Antoine, G. Métais, I. M. Baloch, M. Benammi, Y. Cahimanee, S. Ducrocq, and J.-J. Jaeger. 2001. A fossil lemur from the Oligocene of Pakistan. *Science* 294:587–591.

Martin, P. S. 1984. Prehistoric overkill: The global model; pp. 354–403 in P. S. Martin, P.S. and R. G. Klein, (eds.), *Quaternary Extinctions: A Prehistoric Revolution.* University of Arizona Press, Tucson.

Martin, R. D. 2000. Origins, diversity and relationships of lemurs. *International Journal of Primatology* 21:1021–1049.

Masters, J. C., M. J. de Wit, and R. J. Asher. 2006. Reconciling the origins of Africa, India and Madagascar with vertebrate dispersal scenarios. *Folia Primatologica* 77:399–418.

Matsumoto, K., and D. A. Burney. 1994. Late Holocene environmental changes at Lake Mitsinjo, northwestern Madagascar. *The Holocene* 4:16–24.

McCall, R. A. 1997. Implications of recent geological investigations of the Mozambique Channel for the mammalian colonization of Madagascar. *Proceedings of the Royal Society of London B* 264:663–665.

Miller, E. R., G. F. Gunnell, and R. D. Martin. 2005. Deep time and the search for anthropoid origins. *Yearbook of Physical Anthropology* 48:60–95.

Montagnon, D., B. Ravaoarimanana, B. Rakotosamimanana, and Y. Rumpler. 2001a. Ancient DNA from *Megaladapis edwardsi* (Malagasy subfossil): Preliminary results using partial cytochrome b sequence. *Folia Primatologica* 72:30–32.

Montagnon, D., B. Ravaoarimanana, and Y. Rumpler. 2001b. Ancient DNA from *Megaladapis edwardsi*: Reply. *Folia Primatologica* 72:343–344.

Orlando, L., S. Calvignac, C. Schnebelen, C. J. Douady, L. R. Godfrey, and C. Hänni. 2008. DNA from extinct giant lemurs links archaeolemurids to extant indriids. *BMC Evolutionary Biology* 8:121; http://www.biomedcentral.com/1471-2148/8/121.

Pastorini, J, M. R. J. Forstner, and R. D. Martin. 2002. Phylogenetic relationships among Lemuridae (Primates): Evidence from mtDNA. *Journal of Human Evolution* 43:463–478.

Pastorini, J., U. Thalmann, and R. D. Martin. 2001. Molecular phylogeny of the lemur family Cheirogaleidae (Primates) based on mitochondrial DNA sequences. *Molecular Phylogenetics and Evolution* 19:45–56.

Perez, V. R., L. R. Godfrey, M. Nowak-Kemp, D. A. Burney, J. Ratsimbazafy, and N. Vasey. 2005. Evidence of early butchery of giant lemurs in Madagascar. *Journal of Human Evolution* 49:722–742.

Poux, C., and J. P. Douzery. 2004. Primate phylogeny, evolutionary rate variations, and divergence times: A contribution from the nuclear gene IRBP. *American Journal of Physical Anthropology* 124:1–16.

Poux, C., O. Madsen, E. Marquard, D. R. Vieites, W. W. de Jong, and M. Vences. 2005. Asynchronous colonization of Madagascar by the four endemic clades of primates, tenrecs, carnivores, and rodents as inferred from nuclear genes. *Systematic Biology* 54:719–730.

Rabinowitz, P. D. and S. Woods. 2006. The Africa-Madagascar connection and mammalian migrations. *Journal of African Earth Sciences* 44:270–276.

Roos, C., J. Schmitz, and H. Zischler. 2004. Primate jumping genes elucidate strepsirrhine phylogeny. *Proceedings of the National Academy of Sciences, USA* 101:10650–10654.

Schwartz, G. T., P. Mahoney, L. R. Godfrey, F. P. Cuozzo, W. L. Jungers, and G. F. N. Randria. 2005. Dental development in *Megaladapis edwardsi* (Primates, Lemuriformes): Implications for understanding life history variation in subfossil lemurs. *Journal of Human Evolution* 49:701–721.

Schwartz, G. T., K. E. Samonds, L. R. Godfrey, W. L. Jungers, and E. L. Simons. 2002. Dental microstructure and life history in subfossil Malagasy lemurs. *Proceedings of the National Academy of Sciences, USA* 99:6124–6129.

Schwartz, J. H., and I. Tattersall. 1985. Evolutionary relationships of living lemurs and lorises (Mammalia, Primates) and their potential affinity with European Eocene Adapidae). *Anthropological Papers of the American Museum of Natural History* 60:1–100.

Seiffert, E. R., E. L. Simons, and Y. Attia. 2003. Fossil evidence for an ancient divergence of lorises and galagos. *Nature* 422:388–389.

Seiffert, E. R., E. L. Simons, T. M. Ryan, and Y. Attia. 2005. Additional remains of *Wadilemur elegans*, a primitive stem galagid from the late Eocene of Egypt. *Proceedings of the National Academy of Sciences, USA* 102:11396–11401.

Seligsohn, D., and F. S. Szalay. 1974. Dental occlusion and the masticatory apparatus in *Lemur* and *Varecia*: Their bearing on the systematics of living and fossil primates; pp. 543–561 in R. D. Martin, G. A. Doyle, and A. C. Walker (eds.), *Prosimian Biology*. Duckworth, London.

Shapiro, L. J., C. V. M. Seiffert, L. R. Godfrey, E. L. Simons, and G. F. N. Randria. 2005. Morphometric analysis of lumbar vertebrae in extinct Malagasy strepsirrhines. *American Journal of Physical Anthropology* 128:823–839.

Simons, E. L. 1994. The giant aye-aye *Daubentonia robusta*. *Folia Primatologica* 62:14–21.

Simons, E. L., L. R. Godfrey, W. L. Jungers, P. S. Chatrath, and B. Rakotosamimanana. 1992. A new giant subfossil lemur, *Babakotia*, and the evolution of the sloth lemurs. *Folia Primatologica* 58:197–203.

Simons, E. L., L. R. Godfrey, W. L. Jungers, P. S. Chatrath, and J. Ravaoarisoa, J. 1995. A new species of *Mesopropithecus* (Primates, Palaeopropithecidae) from northern Madagascar. *International Journal of Primatology* 16:653–682.

Soligo, C. 2005. Anatomy of the hand and arm in *Daubentonia madagascariensis*: A functional and phylogenetic outlook. *Folia Primatologica* 76:262–300.

Springer, M. S., W. J. Murphy, E. Eizirik, and S. J. O'Brien. 2003. Placental mammal diversification and the Cretaceous-Tertiary boundary. *Proceedings of the National Academy of Sciences, USA* 100:1056–1061.

Standing, H.-F. 1903. Rapport sur des ossements sub-fossiles provenant d'Ampasambazimba. *Bulletin de l'Académie Malgache* 2:227–235.

———. 1905. Rapport sur des ossements sub-fossiles provenant d'Ampasambazimba. *Bulletin de l'Académie Malgache* 4:95–100.

———. 1909 (for the year 1908). Subfossiles provenant des fouilles d'Ampasambazimba. *Bulletin de l'Académie Malgache* 6:9–11.

———. 1910 (for the year 1909). Note sur les ossements subfossiles provenant des fouilles d'Ampasambazimba. *Bulletin de l'Académie Malgache* 7:61–64.

Stankiewicz, J., C. Thiart, J. C. Masters, and M. J. de Wit. 2006. Did lemurs have sweepstake tickets? An exploration of Simpson's model for the colonization of Madagascar by mammals. *Journal of Biogeography* 33:221–235.

Stevens, N. J., and C. P. Heesy. 2006. Malagasy primate origins: Phylogenies, fossils, and biogeographic reconstructions. *Folia Primatologica* 77:419–433.

Storey, B. C. 1995. The role of mantle plumes in continental breakup: Case histories from Gondwananland. *Nature* 377:301–308.

Szalay, F. S., and C. C. Katz. 1973. Phylogeny of lemurs, galagos and lorises. *Folia Primatologica* 19:88–103.

Tabuce, R., M. Mahboubi, P. Tafforeau, and J. Sudre. 2004. Discovery of a highly specialized plesiadapiform primate in the early middle Eocene of northwestern Africa. *Journal of Human Evolution* 47:305–321.

Tattersall, I. 1971. Revision of the subfossil Indriinae. *Folia Primatologica* 15:257–269.

———. 1973. Cranial anatomy of the Archaeolemurinae (Lemuroidea, Primates). *Anthropological Papers of the American Museum of Natural History* 52:1–110.

———. 1982. *The Primates of Madagascar*. Columbia University Press, New York, 382 pp.

———. 2006a. Historical biogeography of the strepsirhine primates of Madagascar. *Folia Primatologica* 77:477–487.

———. 2006b. Origin of the Malagasy strepsirhine primates; pp. 3–17 in L. Gould, and M. L. Sauther (eds.), *Lemurs: Ecology and Adaptation*. Springer, New York.

Tattersall, I., and J. H. Schwartz. 1974. Craniodental morphology and the systematics of the Malagasy lemurs (Primates, Prosimii). *Anthropological Papers of the American Museum of Natural History* 52:139–192.

Tavaré, S., C. R. Marshall, O. Will, C. Soligo, and R. D. Martin. 2002. Using the fossil record to estimate the age of the last common ancestor of extant primates. *Nature* 416:726–729.

Van Duzer, C. 2004. *Floating Islands: A Global Bibliography with an Edition and Translation of G. C. Munz's* Exercitatio Academica de Insulis Natantibus *(1711)*. Cantor Press, Los Altos Hills, Calif., 404 pp.

Vasey, N., L. R. Godfrey, and V. R. Perez, V.R. 2005. The paleobiology of *Pachylemur*, 2005. *American Journal of Physical Anthropology* (suppl.) 40:212.

Vuillaume-Randriamanantena, M. 1988. The taxonomic attributions of giant subfossil lemur bones from Ampasambazimba: *Archaeoindris* and *Lemuridotherium*. *Journal of Human Evolution* 17:379–391.

Vuillaume-Randriamanantena, M., L. R. Godfrey, W. L. Jungers, and E. L. Simons. 1992. Morphology, taxonomy and distribution of *Megaladapis*: Giant subfossil lemur from Madagascar. *Comptes Rendus de l'Académie des Sciences, Paris*, Série II, 315:1835–1842.

Walker, A. C. 1974. Locomotor adaptations in past and present prosimian primates; pp. 349–381 in F.A. Jenkins, Jr., (ed.), *Primate Locomotion*. Academic Press, New York.

Walker, A. C., Ryan, T. M., Silcox, M. T., Simons, E. L., and Spoor, F. 2008. The semicircular canal system and locomotion: the case of extinct lemuroids and lorisoids. *Evolutionary Anthropology* 17:135–145.

Wall, C. E. 1997. The expanded mandibular condyle of the Megaladapidae. *American Journal of Physical Anthropology* 103:263–276.

Wunderlich, R. E., E. L. Simons, and W. L. Jungers. 1996. New pedal remains of *Megaladapis* and their functional significance. *American Journal of Physical Anthropology* 100:115–138.

Wyner, Y., R. DeSalle, and R. Asher. 2000. Phylogeny and character behavior in the family Lemuridae. *Molecular Phylogenetics and Evolution* 15:124–134.

Yoder, A. D. 1994. Relative position of the Cheirogaleidae in strepsirhine phylogeny: A comparison of morphological and molecular methods and results. *American Journal of Physical Anthropology* 94:25–46.

———. 1997. Back to the future: A synthesis of strepsirrhine systematics. *Evolutionary Anthropology* 6:11–22.

———. 2001. Ancient DNA from *Megaladapis edwardsi*. *Folia Primatologica* 72:342–343.

Yoder, A. D., M. Cartmill, M. Ruvolo, K. Smith, and R. Vilgalys. 1996. Ancient single origin of Malagasy primates. *Proceedings of the National Academy of Sciences, USA* 93:5122–5126.

Yoder, A. D., B. Rakotosamimanana, and T. Parsons. 1999. Ancient DNA in subfossil lemurs: Methodological challenges and their solutions; pp. 1–17 in B. Rakotosamimanana, H. Rasamimanana, J. U. Ganzhorn, and S. M. Goodman (eds.), *New Directions in Lemur Studies*. Plenum Press, New York.

Yoder, A. D., and Z. H. Yang. 2004. Divergence dates for Malagasy lemurs estimated from multiple gene loci: Geological and evolutionary context. *Molecular Ecology* 13:757–773.

TWENTY-TWO

Paleogene Anthropoids

ERIK R. SEIFFERT, ELWYN L. SIMONS,
JOHN G. FLEAGLE, AND MARC GODINOT

Anthropoid primates were among the most common members of Afro-Arabian mammal faunas during the late Paleogene, and they may have been present on that landmass as early as the late Paleocene (Godinot, 1994; Beard and Wang, 2004; Seiffert et al., 2005a; Beard, 2006; Rossie and Seiffert, 2006). Specialists continue to debate the role of Asia in early anthropoid diversification, and whether stem anthropoids originated in Asia or Afro-Arabia (Beard, 2004; Marivaux et al., 2005; Miller et al., 2005; Beard, 2006; Heesy et al., 2006; Tabuce et al., 2009), but the African record provides the oldest purported anthropoid, the most continuous record of early anthropoids, and the greatest diversity of anthropoid taxa (e.g., Beard, 2002; Rasmussen, 2002). Significantly, some early anthropoids from Africa are known from relatively complete cranial, and in some cases postcranial, remains, making identification of their anthropoid status more secure.

Most of this chapter deals with late Eocene and early Oligocene anthropoids, among which three divergent clades can be identified: Parapithecoidea, Proteopithecidae, and Catarrhini. It is generally believed that the stem lineage of Platyrrhini also originated in Afro-Arabia (e.g., Hoffstetter, 1982), but the evidence for this hypothesis is strictly phylogenetic; as yet the remains of stem platyrrhines either have not been recovered in Afro-Arabia or have not yet been correctly identified as such (but see Miller and Simons, 1997; Takai et al., 2000). Ghost lineages such as these should not be unexpected, though, for Paleogene anthropoid fossils have only been recovered from ten areas in Afro-Arabia, five of which are in North Africa.

Aside from the advanced latest Oligocene catarrhine *Kamoyapithecus* from Kenya (Leakey et al., 1995), the only fossil anthropoid species that have been named from the Paleogene of Afro-Arabia are from northern Africa and Oman; otherwise, an isolated canine is known from the Oligocene of Angola (Pickford, 1986), and a distal humerus and dental remains have recently been discovered in the Oligocene of Tanzania (Stevens et al., 2005, 2009). Of the 23 species discussed here, 18 are from the Fayum area in northern Egypt.

In this contribution, we first provide a review of all early anthropoid taxa from the Paleogene of Africa, with a summary of the taxonomic history, morphology, and reconstructed adaptations of each taxon. In the discussion we summarize patterns of phylogenetic and adaptive changes through time in the African record of early anthropoid evolution. The taxonomy proposed here is based on the results of the most comprehensive phylogenetic analyses of Paleogene anthropoids currently available (Seiffert et al., 2005a, 2009). Ages for major primate-bearing quarries in the Jebel Qatrani Formation are from Seiffert (2006; this volume, chap. 2). Taxa referred to as "stem anthropoids" (e.g., *Altiatlasius*, parapithecoids) are those extinct taxa that are more closely related to extant Anthropoidea than to Tarsiidae or Strepsirrhini. "Crown Anthropoidea" refers to the clade containing Platyrrhini, Catarrhini, and all extinct descendants from their last common ancestor; within that clade, "stem catarrhines" (herein oligopithecids and propliopithecids) are those extinct taxa that are more closely related to "crown catarrhines" (the clade containing cercopithecoids, hominoids, and all extinct descendants from their last common ancestor) than to Platyrrhini, but which are not nested within the former clade.

Systematic Paleontology

Order PRIMATES Linnaeus, 1758
Suborder ?ANTHROPOIDEA Mivart, 1864
Family *incertae sedis*
Genus *ALTIATLASIUS* Sigé et al., 1990
ALTIATLASIUS KOULCHII Sigé et al., 1990
Figure 22.1

Age and Occurrence Latest Paleocene (latest Thanetian) part of geomagnetic polarity Chron 24r, <56.67 Ma (Ogg and Smith, 2004), Adrar Mgorn 1, Jbel Guersif Formation, Ouarzazate Basin, Morocco.

Diagnosis Differs from later Afro-Arabian anthropoids in combining the following features: lower molars with well-developed paraconids, small hypoconulids, and relatively short talonids; M2? with very small hypocone and pericone, well-developed metaconule, elongate postmetacrista, and broad stylar shelf and trigon.

Description The *A. koulchii* hypodigm includes isolated upper and lower molars, and one mandibular fragment (THR 136) with the m2 germ and alveoli for m1 and the distal root of p4. There is a considerable amount of morphological

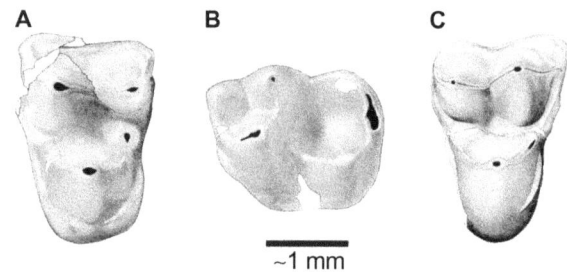

A   B   C

~1 mm

FIGURE 22.1 Isolated teeth of *Altiatlasius koulchii*, late Paleocene of Morocco. A) THR 141, holotype M1 or M2; B) THR 135, m1 or m2; C) THR 144, M1 or M2.

variation in the hypodigm and it is not yet clear that the ?M1 (THR 144; figure 22.1C), M3 (THR 145), and p3 or p4 (THR 128) described by Sigé et al. (1990) belong to *A. koulchii*, the holotype of which (THR 141; figure 22.1A) is a possible M2. The lower molars (m1 and m2) have well-developed paraconids situated slightly labial to the metaconids (figure 22.1B). Talonid basins are shorter than those of later Afro-Arabian anthropoids, particularly on m1, and trigonids are relatively tall. Molar cusps show moderate basal inflation, particularly those of the trigonid. The oblique cristids are oriented toward the protoconid on m1–2. Premetacristids are not present on the lower molars. Molar hypoconulids are small, crestiform, and centrally placed on the distal border of the talonid.

Morphological variation among the upper molars attributed to *A. koulchii* is problematic. The holotype ?M2 (THR 141) differs from the alleged M1 (THR 144) and M3 (THR 145) in the hypodigm in having a complete lingual cingulum, a pericone, a well-developed metaconule without pre- or postmetaconule cristae, and a relatively broad trigon. The morphology of the postprotocrista on THR 144 is very different from that of THR 141 in that the crest does not terminate at an enlarged metaconule but rather curves around the distal aspect of the crown to meet the distal aspect of the metacone. THR 141 is also considerably larger than the other upper molars. Furthermore, the relative heights of the buccal cusps on these upper molars are arguably inconsistent with Sigé et al.'s (1990) attribution to locus: in buccal view the paracone and metacone of primate M1s are generally subequal in size, and the relative size of the metacone decreases distally, with that cusp being smallest on M3. However on the alleged M1 (THR 144) of *A. koulchii* the metacone is smaller than the paracone, whereas on the ?M2 (THR 141) the cusps are about equal in size. Sigé et al.'s (1990) proposed arrangement of the upper molars in the *A. koulchii* hypodigm is also peculiar given their reconstruction of lower molar size as decreasing distally from m1 to m2. This could indicate that THR 141 is, in fact, an M1 while THR 144 is an M2 of *A. koulchii*, but the teeth would still differ markedly in the aforementioned characters. Alternatively, the more waisted distal margin of THR 144 provides some evidence for it being an M1, because the M1s of some primitive primates such as *Teilhardina* and *Eosimias* have a very concave distal margin. On the basis of size, it does appear that the lower molars in the hypodigm likely belong to *A. koulchii* (the holotype upper molar THR 141 is 1.75 mm long; the lower molars range in length from 1.75–2.04 mm; but THR 144 is only 1.55 mm long).

Remarks *Altiatlasius* was originally described as an omomyid with possible relevance for anthropoid origins (Sigé et al., 1990), and a number of authorities have since suggested that *Altiatlasius* might be both the oldest stem anthropoid (or

"proto-simiiform") primate and the oldest known crown primate (or euprimate)(Gingerich, 1990; Godinot, 1994; Beard, 1998; Seiffert et al., 2005a). The Adrar Mgorn 1 site that produced the remains of *A. koulchii* has been tied to Chron 24r, which spans the Paleocene-Eocene boundary, but invertebrates and selachians from the Jbel Guersif Formation make a late Paleocene age more likely (Gheerbrant et al., 1998). *Altiatlasius* combines anthropoid-like features of the lower molars (e.g., moderate bunodonty, laterally placed oblique cristids, paraconids placed labial to the metaconids, large protoconids, and small hypoconulids) and, on the holotype upper molar, a complete lingual cingulum, tiny hypocone and pericone, and a well-developed postprotocrista that meets a metaconule that lacks pre- or postmetaconule cristae.

Altiatlasius has also been interpreted as a toliapinid plesiadapiform by Hooker et al. (1999), but we consider the similarities shared with toliapinids to be less compelling than those shared with anthropoids. Another interesting lower molar (THRbis ?) from the Ouarzazate Basin locality Adrar Mgorn 1bis (also likely to be late Paleocene in age) was described by Gheerbrant et al. (1998) as being either a plesiadapiform or primate, and as being similar to *A. koulchii*. The specimen is larger than *A. koulchii* and has a much broader talonid and more lingually oriented cristid obliqua, but interestingly also a hypoconulid that is situated close to the entoconid, as in many later Afro-Arabian anthropoids. More

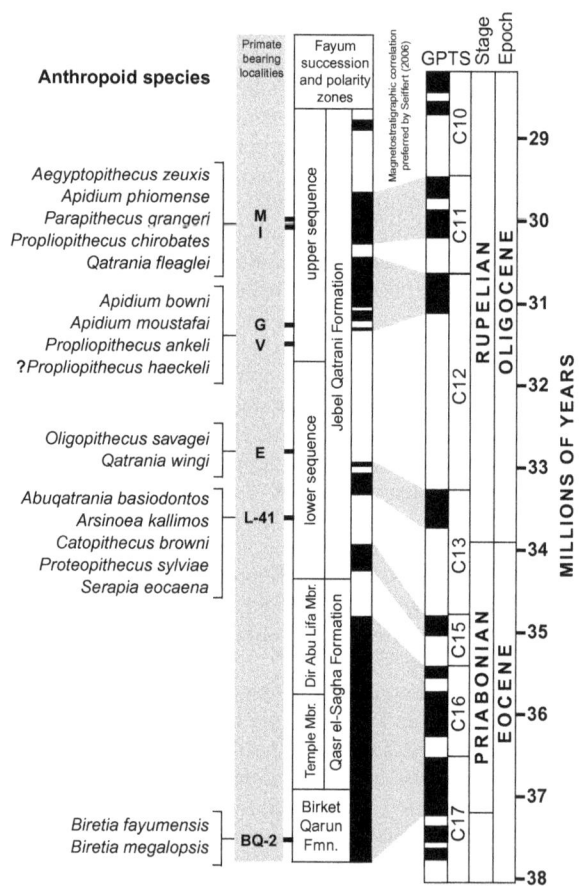

FIGURE 22.2 Ages of major anthropoid-bearing localities in the Fayum Depression. On right, correlation of the local Fayum magnetostratigraphy to the Geomagnetic Polarity Timescale [(GPTS, based on Gradstein et al. (2004)] proposed by Seiffert (2006). On left, anthropoid species represented at each of the major anthropoid-bearing localities.

material will be needed to determine the higher-level affinities of this species, but interestingly this tooth adds to the evidence for primate *(sensu lato)* diversity in the Paleocene of northern Africa.

Suborder ANTHROPOIDEA Mivart, 1864
Family PROTEOPITHECIDAE Simons, 1997
Genus *PROTEOPITHECUS* Simons, 1989
PROTEOPITHECUS SYLVIAE Simons, 1989
Figures 22.3, 22.4, 22.5A, 22.6C, and 22,.6D

Age and Occurrence Latest Eocene (latest Priabonian), Quarry L-41, lower sequence of Jebel Qatrani Formation, northern Egypt (figure 22.2).

Diagnosis Dental formula: 2.1.3.3/2.1.3.3. Differs from other undoubted Afro-Arabian anthropoids in the following combination of dental features: in its lower dentition, *P. sylviae* has a mesiodistally compressed lower canine; a p2 that is as large or larger than p3; an enlarged p4 metaconid situated transverse, or only slightly distal, to the protoconid; moderately bunodont molars with basally inflated cusps and relatively tall trigonids; large paraconid on m1; cristid obliqua on m1 oriented toward protoconid; small paraconids variably present on m2–3; and twinned entoconid and hypoconulid on m1–2 (Miller and Simons, 1997). In the upper dentition (figure 22.4), P2 is single rooted and variably bears a very small protocone; P3–4 have small hypocones and no lingual cingula; M1–2 are transversely broad and have well-developed hypocones, variably incomplete lingual cingula, and well-developed buccal cingula; small para- and metaconules are variably present. Upper and lower third molars are very small in proportion to m2/M2. Postcranial differentiae include a relatively short humeral trochlea and a medial strut of the entepicondylar foramen that is confluent with the medial aspect of the trochlea (Seiffert et al., 2000); gluteal tuberosity on the femur positioned at the same proximodistal level as the lesser trochanter; femoral trochanteric fossa closed distally by an intertrochanteric crest (Gebo et al., 1994; Seiffert et al., 2004); distal femoral articulation that is more shallow, anteroposteriorly, than that of *Apidium* (Seiffert et al., 2004); astragalar fibular facet steep sided; medial and lateral crests of astragalar trochlea roughly equal in height; cotylar fossa poorly developed; astragalar head not mediolaterally expanded as in *Catopithecus* and *Aegyptopithecus*.

Description The lower incisors are very small in proportion to the postcanine teeth; i2 is larger than i1, and the mesial aspect of the i2 crown slightly overlaps the distal aspect of the i1 crown. Both incisors bear weak lingual cingulids. The lower canine is large, sexually dimorphic (Simons et al., 1999), and mesiodistally compressed. The p2 is as large as or larger than p3 (figure 22.3), and both p2 and p3 have well-developed lingual cingulids, no metaconids and no clear development of hypoconids. The p3 protocristid is directed distolingually. The p4 metaconid is large and only slightly smaller than the protoconid; the former cusp is situated lingual (as opposed to distal) to the latter and both cusps are linked by a transverse protocristid. The p4 talonid is low and short, with only a very small hypoconid and a short cristid obliqua that meets the distal trigonid wall between the proto- and metaconid. A large paraconid is present on m1, but there is no premetacristid enclosing the lingual aspect of the trigonid basin (figure 22.3). Small paraconids are variably present

on m2–3. The entoconid and hypoconulid cusps are twinned on m1–m2.

The P2 is much smaller than P3 or P4, is single rooted, and variably bears a tiny protocone (figure 22.4). None of the upper premolars have lingual cingula, but P3–4 have hypocones and well-developed protocones. The M1–2 have large hypocones and lingual cingula that are interrupted lingually by variably developed pericone cusps. Conules vary in development, and the postprotocrista does not reach far up onto the base of the metacone as in some other Paleogene anthropoids. The buccal cingula are generally very distinct. M3 has only a very small metacone and no hypocone.

Craniofacial morphology of *P. sylviae* is known from two dorsoventrally crushed crania (Simons, 1997; figure 22.5A). The orbits are relatively small, suggesting a diurnal activity pattern (Heesy and Ross, 2001), and are enclosed by well-developed postorbital septa. The orbital apertures are more convergent than those of parapithecids such as *Apidium* and *Parapithecus* (Simons, 2008). The lacrimal bone and foramen are positioned within the orbit, as in all other known anthropoids, and there is no zygomatic-lacrimal contact. The rostrum is more pronounced than those of small platyrrhines, and the nasals appear to be relatively long. When compared with *Catopithecus* and *Aegyptopithecus*, the ascending wing of the premaxilla is relatively small. There is a small zygomaticofacial foramen and a single infraorbital foramen positioned over P2 or P3. The metopic suture is completely fused, and the temporal lines are separate through most of their length; posteriorly there is a short sagittal crest on DPC 14095. The basicranium is badly distorted, but it is clear that the carotid foramen is rostrally positioned, the ectotympanic is annular, and the auditory bulla is relatively uninflated, as in platyrrhines and other Fayum anthropoids for which the region is known. The temporomandibular joint is relatively flat, not guttered as in *Tarsius* and omomyids, and there is a small postglenoid foramen just medial and caudal to the postglenoid process, in a position very similar to that of many small platyrrhines. The caudal border of the palate is positioned rostral to M3, there is no postpalatine torus, and a distinct maxillary notch is present lateral to the pterygoids as in *Parapithecus*. There is no contact between the lateral pterygoid plate and the auditory bulla.

The humerus of *P. sylviae* (figures 22.6C, 22.6D) is similar in morphology to those of parapithecids (Seiffert et al., 2000). The trochlea is relatively short and is confluent with the capitulum. The medial epicondyle is oriented slightly posteriorly, with a distinct dorsoepitrochlear fossa on its dorsal aspect. The deltopectoral crest is well developed, and the bicipital groove is narrow and deep. The entepicondylar foramen is more laterally placed than in oligopithecids and propliopithecids, and its medial strut is confluent with the medial aspect of the trochlea. The trochanteric fossa of the femur is "walled off" by an intertrochanteric crest, as in the parapithecids *Apidium* and *Parapithecus*, and there is a small gluteal tuberosity that is placed at about the level of the lesser trochanter (Gebo et al., 1994; Simons and Seiffert, 1999). The femoral head is round and bears a distinct fovea capitis. A crista paratrochanterica is present on the dorsal surface of the femoral neck, as in some platyrrhines. The distal articulation is slightly deeper than it is wide, but is not as deep as that of *Apidium* (Fleagle and Simons, 1995; Seiffert et al., 2004). The tibia is relatively long, and a crural index of 106 can be calculated from an associated tibia and femur (Simons and

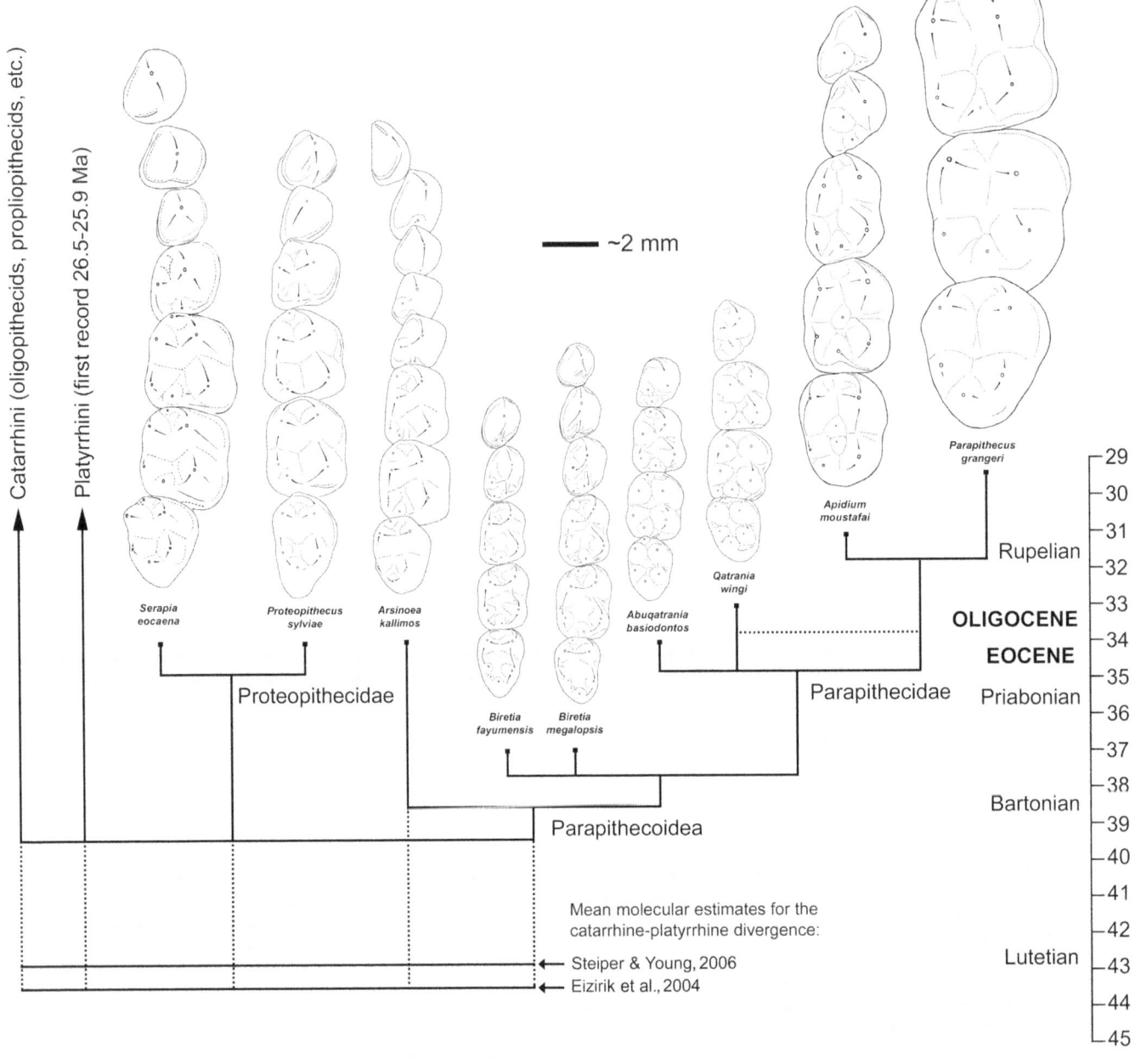

FIGURE 22.3 Phylogenetic relationships among catarrhines, platyrrhines, proteopithecids, and parapithecoids based on the phylogenetic analysis of Seiffert et al. (2005a), with line drawings of the lower dentitions of the latter two clades. Divergences among species are arbitrarily resolved at 1 Ma.

Seiffert, 1999). The proximal articular surface is moderately retroflexed, and the cnemial crest is relatively proximally placed. The distal tibiofibular articulation takes the form of a short syndesmosis, as in many small platyrrhines. The astragalus is similar in shape to those of small platyrrhines in having a tall, relatively straight-sided body with a short medial malleolar articulation and a centrally placed groove for flexor fibularis (Seiffert and Simons, 2001). The astragalar head is narrow when compared with those of *Catopithecus* and *Aegyptopithecus*.

Remarks The genus *Proteopithecus* was originally placed in the subfamily Oligopithecinae, alongside *Catopithecus* and

Oligopithecus, by Simons (1989) on the basis of a maxilla with M2–3, partial M1, and associated P2 (at that time mistakenly identified as P3). Material collected and described since that time has demonstrated that *Proteopithecus* is radically different from *Catopithecus* in its craniodental (Miller and Simons, 1997; Simons, 1997) and postcranial morphology (Seiffert et al., 2004), and it is not closely related to that genus. On the basis of craniodental morphology, Simons (1997) erected the family Proteopithecidae to reflect *Proteopithecus*'s phylogenetic distance from oligopithecids and parapithecids, and this distinction has been upheld and further supported by analyses of postcranial morphology. Phylogenetic analyses

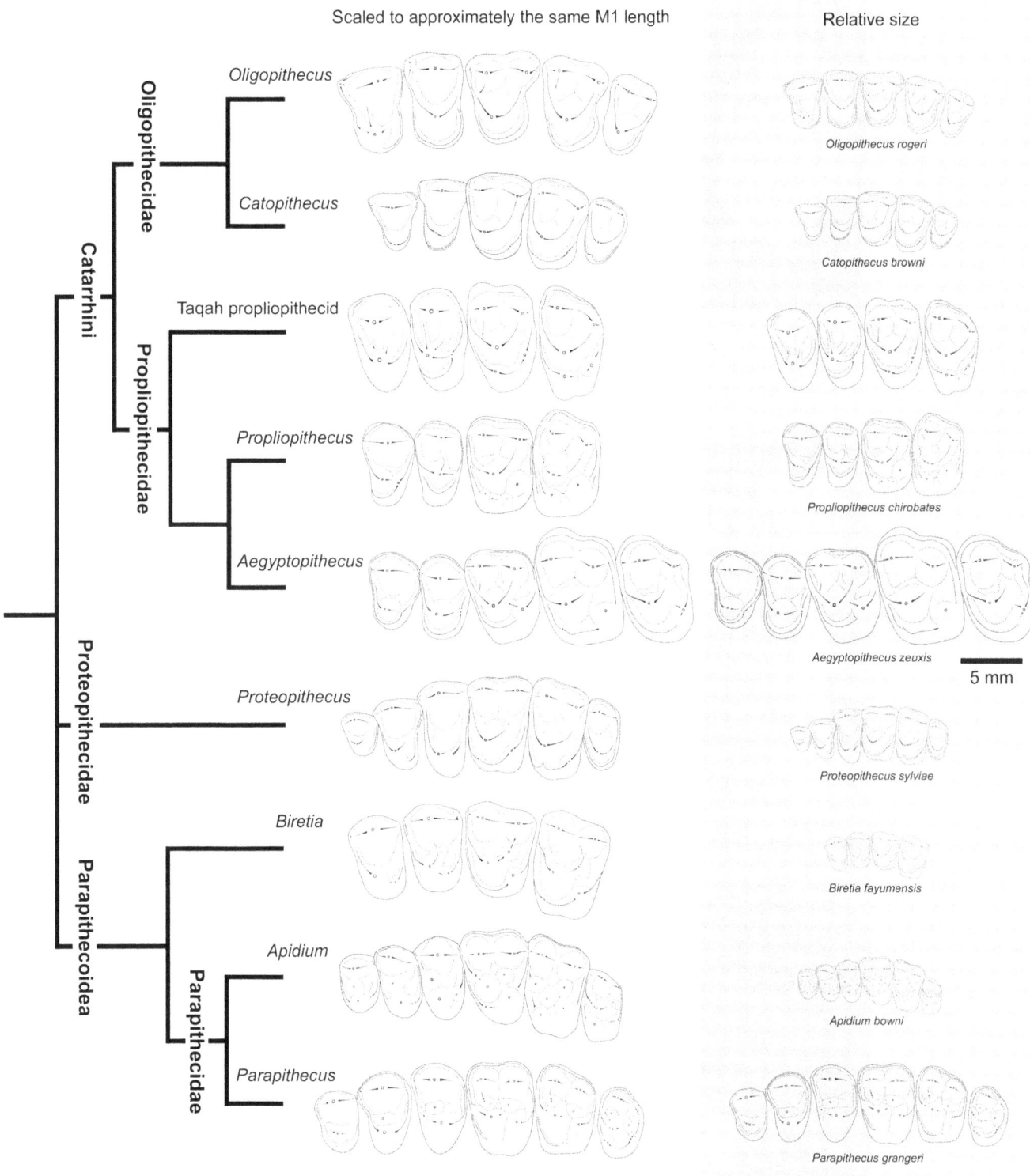

Scaled to approximately the same M1 length

Relative size

Oligopithecus

Catopithecus

Taqah propliopithecid

Propliopithecus

Aegyptopithecus

Proteopithecus

Biretia

Apidium

Parapithecus

Oligopithecidae

Propliopithecidae

Proteopithecidae

Parapithecidae

Catarrhini

Proteopithecidea

Parapithecoidea

Oligopithecus rogeri

Catopithecus browni

Propliopithecus chirobates

Aegyptopithecus zeuxis

5 mm

Proteopithecus sylviae

Biretia fayumensis

Apidium bowni

Parapithecus grangeri

FIGURE 22.4 Line drawings of maxillary postcanine dentitions from representatives of each major anthropoid clade known from the Afro-Arabian Paleogene. Some of these reconstructions are composites from isolated teeth or multiple individuals, and some specimens are reversed for ease of comparison. From top to bottom, *Oligopithecus* is represented by *Oligopithecus rogeri* from the early Oligocene of Oman (Gheerbrant et al., 1995); *Catopithecus* by *Catopithecus browni* from the latest Eocene of Egypt (e.g., Simons, 1995b; Simons and Rasmussen, 1996); *Propliopithecus* by *Propliopithecus chirobates* from the early Oligocene of Egypt (e.g., Kay et al., 1981); *Aegyptopithecus* by *Aegyptopithecus zeuxis* from the early Oligocene of Egypt (e.g., Simons, 1965; Kay et al., 1981); *Proteopithecus* by *Proteopithecus sylviae* from the latest Eocene of Egypt (e.g., Simons, 1989; Simons, 1997); *Biretia* by *Biretia fayumensis* from the earliest late Eocene of Egypt (Seiffert et al., 2005a); *Apidium* by *Apidium bowni* from the early Oligocene of Egypt (Simons, 1995a); *Parapithecus* by *Parapithecus grangeri* (Simons, 1974, 1986).

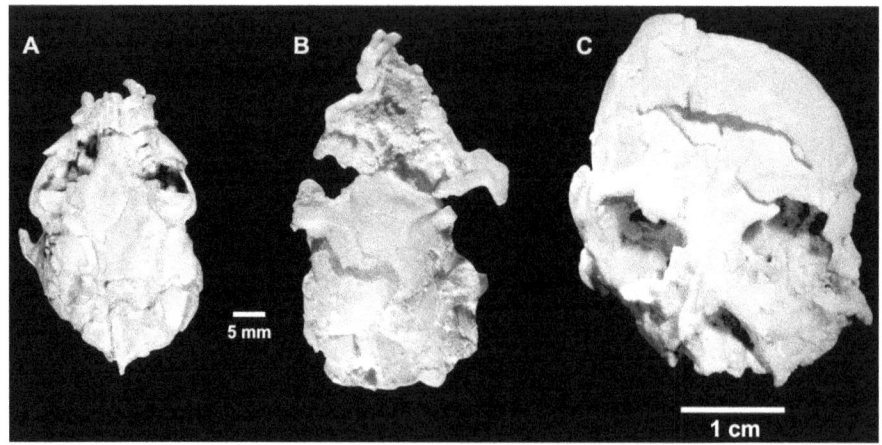

FIGURE 22.5   Cranial remains of *Catopithecus* and *Proteopithecus* from the terminal Eocene Quarry L-41. A) Dorsal view of DPC 14095, crushed cranium of *Proteopithecus sylviae*; B) dorsal view of DPC 11594, crushed cranium of *Catopithecus browni*; C) rostral view of DPC 12367. 5 mm scale bar for A and B; 1 cm scale bar for C.

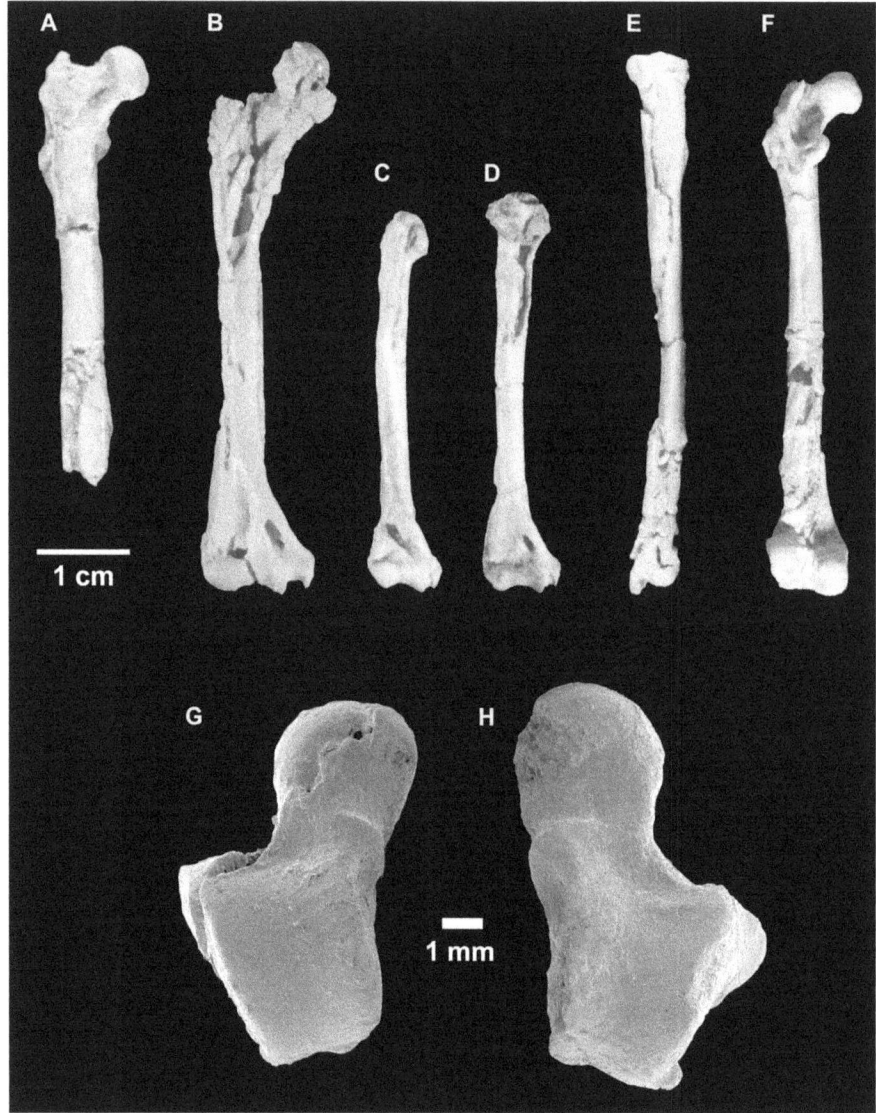

FIGURE 22.6   Postcranial remains of *Catopithecus* and *Proteopithecus* from Quarry L-41. Upper row (A-F) are limb bones: A) DPC 8256, right femur of *Catopithecus browni* (missing distal end) in anterior view; B) DPC 12274, right humerus of *C. browni* in anterior view; C) DPC 18256, right humerus of *Proteopithecus sylviae* in anterior view; D) DPC 20191, right humerus of *P. sylviae* in anterior view; E) DPC 16873, left tibia of *P. sylviae*; F) DPC 17031, left femur of *P. sylviae*. Bottom row are astragali of G) *P. sylviae* (left side, DPC 15417) and H) *C. browni* (right side, DPC 10037). 1 cm scale bar applies to A-F; 1 mm scale bar applies to G and H.

that have taken into account craniodental and postcranial characters have universally rejected a close relationship of *Catopithecus* and *Proteopithecus* (Kay et al., 2004; Seiffert et al., 2004, 2005a; Marivaux et al., 2005). *Proteopithecus* was probably a diurnal and arboreal anthropoid with a mixed insectivorous/frugivorous diet that lived in large social groups and depended largely on rapid quadrupedal locomotion with some pronograde leaping.

Some recently published phylogenetic analyses (Kay et al., 2004; Marivaux et al., 2005; Seiffert et al., 2005b) align proteopithecids with parapithecoids. *Proteopithecus* is similar to *Apidium* and *Parapithecus* in some details of distal humeral, pelvic, and proximal femoral morphology, but the peculiar hindlimb features shared by *Proteopithecus* and parapithecids may be plesiomorphic because a similar morphology is recorded in remains of possible stem anthropoids from the middle Eocene of China (Gebo et al., 2008). Furthermore, *Proteopithecus* has a more crown anthropoid-like p4, with a large, transversely placed metaconid, and has more convergent orbits than those of parapithecids. For these reasons, we suspect that proteopithecids are probably more closely related to crown anthropoids than to parapithecoids, and that the postcranial features shared with the latter are either plesiomorphic or convergent; more complete fossils from horizons older than L-41 will be required to determine whether *Proteopithecus* is a stem anthropoid, a stem platyrrhine, or a basal parapithecoid.

Genus *SERAPIA* Simons, 1992
SERAPIA EOCAENA Simons, 1992

Age and Occurrence Latest Eocene (latest Priabonian), Quarry L-41, lower sequence of Jebel Qatrani Formation, northern Egypt.

Diagnosis Dental formula: /?.1.3.3, upper dentition not known. *Serapia* is larger and more bunodont than *Proteopithecus*; when compared with that genus the p4 metaconid is relatively low and placed slightly more distal.

Description The lower canine is mesiodistally compressed and bears a discontinuous lingual cingulid. The p2 is larger than p3, at least in the type specimen, and both teeth have lingual cingulids and faint distobuccal cingulids; there is a small hypoconid on p3. The p4 is larger than the other premolars and has a small paraconid. A small centrally placed hypoconid is present on the talonid, from which a cristid obliqua courses mesially to terminate near the lingual side of the protoconid. The m1 is only slightly larger than the m2 and differs from that tooth in having a small paraconid. On both m1–2, the cristid obliqua terminates behind the protoconid and the entoconid and hypoconulid cusps are twinned; the hypoconulid is more centrally placed on the relatively short m3. All lower molars lack premetacristids and have faint buccal cingulids.

Remarks *Serapia* was originally placed in the family Parapithecidae (Simons, 1992) but *S. eocaena* is essentially a larger and more bunodont version of *Proteopithecus*; indeed, the similarities between the two taxa are so striking that they are arguably congeneric. Simons et al. (2001) formally transferred *Serapia* from the family Parapithecidae to the family Proteopithecidae; this arrangement was independently suggested by Beard (2002). Proteopithecid monophyly was supported in the phylogenetic analyses of Simons et al. (2001), Seiffert et al. (2004, 2005b), and Gunnell and Miller (2001).

Infraorder PARAPITHECOIDEA Schlosser, 1911
Genus *ARSINOEA* Simons, 1992
ARSINOEA KALLIMOS Simons, 1992
Figure 22.3

Age and Occurrence Latest Eocene (latest Priabonian), Quarry L-41, lower sequence of Jebel Qatrani Formation, northern Egypt.

Diagnosis Dental formula: /2.1.3.3. *Arsinoea* differs from other early anthropoids in combining the following features: large and spatulate i2 that is much longer than i1; small, somewhat premolariform lower canine; large p2; small distally placed metaconids, tall hypoconids, and lingual cingulids on p3–4; m1–2 with lingually positioned paraconids, twinned hypoconulids and entoconids, and relatively low trigonids; metaconid on m1 is distally placed, and the cristid obliqua terminates distolingual to the protoconid; m3 relatively long.

Description The i2 is high crowned and spatulate, and has an occlusal surface that sits only just below the relatively short canine (figure 22.3). The canine is somewhat premolariform in being relatively low crowned and mesially inclined (the only known specimen is, however, likely to be a female); a complete lingual cingulid and a well-developed hypoconid cusp are present. The p2 is subequal in size to the p3 and has a complete lingual cingulid and a small hypoconid cusp placed high on the distal face of the crown. The p3–4 are mesiodistally short and buccolingually broad, and somewhat mesially oriented (p4 overlaps the distal face of p3). Both teeth have short talonids with small hypoconids, small metaconids, and complete lingual cingulids. The hypoconulids are twinned with the entoconids on m1–2, but more centrally placed on m3. On m1 the cristid obliqua terminates lingual to the protoconid, and a large paraconid is present, which is widely spaced from a distally placed metaconid; on m2 the two cusps are closely twinned. Premetacristids are absent on both teeth, but postmetacristids and preentocristids meet to enclose the lingual aspect of the m1–2 talonids. The molar trigonids are relatively low when compared with those of proteopithecids, and the crowns are basally inflated. The m3 lacks a paraconid, and although the hypoconulid does not protrude distally, overall the tooth is not reduced relative to m2 as in proteopithecids and oligopithecids. Despite having a small canine, the mandibular corpus of the holotype is very deep.

Remarks *Arsinoea* preserves a confusing mix of primitive and derived dental characters that leave it as perhaps the most problematic of all the undoubted Jebel Qatrani anthropoids. The premolar dentition is more similar to that of derived parapithecids than that of the older parapithecoid *Biretia* in having a small p3 metaconid, mesially oriented protoconids, and distinct hypoconids on p2–4. However *Arsinoea* also has lingually situated paraconids and twinned entoconid and hypoconulid on m1–3, which are arguably primitive features that are not seen in *Biretia*. This pattern seems to indicate that either the central placement of the hypoconulids in *Biretia* is convergent with the pattern seen in later parapithecoids, or the premolar morphology of *Arsinoea* is convergent on that of parapithecids. If the latter scenario proves to be correct, there would be little reason to place *Arsinoea* within Parapithecoidea. Simons et al. (2001) proposed the monogeneric family Arsinoeidae for the genus, but we here consider *Arsinoea* to be best considered as *incertae sedis* within Parapithecoidea. Ross (2000) placed *Arsinoea* in the otherwise Asian and middle and late Eocene family Pondaungidae (= Amphipithecidae), but subsequent phylogenetic analyses that have sampled *Arsinoea* and Amphipithecidae (Kay et al., 2004; Seiffert et al., 2005a) have not provided support for such a placement.

Genus *BIRETIA* Bonis et al., 1988
BIRETIA FAYUMENSIS Seiffert, Simons, Clyde, et al., 2005
Figures 22.3 and 22.4

Age and Occurrence Earliest late Eocene (early Priabonian), Locality BQ-2, Umm Rigl Member of Birket Qarun Formation, northern Egypt.

Diagnosis Differs from other undoubted Afro-Arabian anthropoids aside from *Qatrania wingi* in its small size. In its lower dentition (figure 22.3), *B. fayumensis* differs from other species in combining the following features: p3 with complete lingual cingulid and no metaconid; a small, distally placed p4 metaconid; p4 talonid with a large hypoconid and small entoconid; relatively low molar trigonids with centrally placed hypoconulids; paraconid and, variably, a premetacristid present on m1. In the upper dentition (figure 22.4), *B. fayumensis* differs from other species in combining: protocones on P2–4 and hypocones on P3–4; no buccal or lingual cingula on the upper premolars; M1–2 with well-developed hypocones, small conules, no hypoparacristae, and no buccal cingula.

Description An isolated p2 (DPC 21757E) attributed to *Biretia megalopsis* could conceivably belong to *B. fayumensis*, however, the root of that tooth is of the correct size and shape to fit into the p2 alveolus of the *B. megalopsis* specimen DPC 21539B. Therefore, more material will be needed to determine whether *Biretia* had a relatively small p2 as in amphipithecids and eosimiids (as would appear to be the case if DPC 21757E belonged to *B. megalopsis*). The premolars of *B. fayumensis* are very similar to those of *B. megalopsis*, but the p3 and p4 have more distally oriented lateral protocristids, and the p4 talonid is slightly shorter. The p4 has a prominent hypoconid and a smaller entoconid, and a trenchant cristid obliqua that terminates behind the protoconid. The first lower molar has a well-developed paraconid and a distally placed metaconid; on all lower molars the hypoconulids are centrally placed, and small distolingual foveae are present. Small paraconids are variably present on m2. Protocristids are present on all lower molars, and postmetacristids meet preentocristids to enclose the lingual aspect of the talonids. The molar trigonids are relatively low when compared with those of proteopithecids.

The upper premolars of *B. fayumensis* are known only from isolated specimens. Seiffert et al. (2005a) suggested that a bicuspid and three-rooted upper premolar (DPC 21759C) was likely a P2 of the species, because the tooth lacks features (such as small hypocones and relatively large protocones) observable in other small anthropoid premolars that likely represent the P3–4 of *B. fayumensis*. More complete specimens will be needed to convincingly demonstrate this, but, if correct, then *Biretia* shares with the later parapithecid *Apidium* an apomorphic complex of having a three-rooted and bicuspid P2. The P3 and P4 both have well-developed protocones, small hypocones, and no lingual cingula. The P4 figured by Seiffert et al. (2005a) has a crest coursing buccally from the protocone apex that terminates in a small cuspule that may be homologous with the parapithecid paraconule. The M1–2 have well-developed lingual cingula, large hypocones, and distinct metaconules; a pericone is present on M2, and a small paraconule is present on M1.

Remarks *B. fayumensis* is similar to *Biretia piveteaui* from Bir el Ater (Nementcha) but is slightly smaller and has a relatively narrow m1 trigonid. The possibility exists that the two are actually conspecific, but the available material of *B. piveteaui* is so limited (one possible m1 and probably an M2—discussed later) that this cannot currently be demonstrated. Nevertheless, the presence of such similar species of *Biretia* at BQ-2 and Bir el Ater lends some support to the hypothesis that the latter is either late middle Eocene or early late Eocene in age (Tabuce et al., 2000) and not of early Oligocene age (Rasmussen et al., 1992). *Biretia fayumensis* is also very similar to *B. megalopsis* (discussed later) but differs in a few details and is smaller and less bunodont. More material may reveal that *Biretia* is actually a paraphyletic genus near the base of radiation of late Eocene and early Oligocene parapithecoids.

BIRETIA MEGALOPSIS Seiffert, Simons, Clyde, et al., 2005
Figure 22.3

Age and Occurrence Early late Eocene (early Priabonian), Locality BQ-2, Umm Rigl Member of Birket Qarun Formation, northern Egypt.

Diagnosis Dental formula: ?.?.?.3/?.1.3.3. Differs from *B. fayumensis* and *B. piveteaui* in being larger and more bunodont. Differs from *B. fayumensis* in having a p3 with a more lingually oriented protocristid and a p4 protocristid that descends from the apex of the protoconid to meet the metaconid (figure 22.3); an m1 cristid obliqua that is more lingually oriented, meeting the trigonid lingual to the protoconid; larger conules on M1–2; and the variable presence of postprotocristae on M1–2.

Description *Biretia megalopsis* is very similar in dental morphology to *B. fayumensis*, aside from the differentiae noted in the diagnosis. A small part of the mandibular symphyseal region is preserved on DPC 21539B and suggests that the symphysis was unfused. The mandibular corpus is quite deep, with the depth under m2 being well over twice the length of that tooth. Otherwise, the species is unique among living and extinct anthropoids in having orbital and palatal laminae of the maxilla fused, leaving a very shallow suborbital region and the lingual roots of the M1–2 exposed within the orbital floor (Seiffert et al., 2005a). This configuration has never been observed in a Paleogene anthropoid and is otherwise only known in nocturnal *Tarsius* among definitive crown haplorhines (the nocturnal anthropoid *Aotus* has lingual molar tooth roots exposed within the orbital floor but no fusion of the orbital and palatal lamina of the maxilla). The morphology of *B. megalopsis* is most parsimoniously explained as being due to orbital hypertrophy on a level similar to that of *Aotus* or even *Tarsius*. If, alternatively, *B. megalopsis* was diurnal and had small orbits, then early parapithecoids had a very different orbitofacial structure than that observable in crown anthropoids, and later parapithecoids such as *Apidium* and *Parapithecus* convergently evolved crown anthropoid-like suborbital morphology.

Remarks The dental morphology of *B. megalopsis* is essentially what would be expected in a primitive parapithecoid of its age, but its suborbital morphology is wholly unexpected. Jaeger and Marivaux (2005) suggested that all of the north African anthropoids of this age might have had enlarged orbits, but an undescribed anthropoid genus from BQ-2 clearly lacks this apparently derived arrangement of the suborbital region. More complete facial material of *B. megalopsis* will be required to definitively establish whether its strange orbital morphology is in fact the direct result of orbital hypertrophy; however, at present this seems to be the best working hypothesis.

BIRETIA PIVETEAUI Bonis et al., 1988

Age and Occurrence Latest middle Eocene (Bartonian) or late Eocene (Priabonian), Bir el Ater (= Nementcha), northern Algeria.

Diagnosis Larger than *B. fayumensis*, smaller than *B. megalopsis*; m1 paraconid apparently somewhat reduced relative to those of other *Biretia* species; cristid obliqua more buccally placed than in *B. megalopsis*. Differs from Fayum species of *Biretia* in having an m1 trigonid that is approximately equal in width to the talonid.

Description The holotype of *B. piveteaui* is an isolated m1 (BRT 84–17). It is similar to *Biretia* species from the Fayum in its small size, centrally placed hypoconulid, postmetacristid, and in being bunodont but not having a crown that is as basally inflated as younger parapithecids. An M2 from Bir el Ater was described by Tabuce et al. (2001) and referred to "cf. *Algeripithecus*." This M2 has many detailed similarities to those of the aforementioned *Biretia* species (Seiffert et al., 2005a), and it very likely belongs to *B. piveteaui*.

Family PARAPITHECIDAE Schlosser, 1911
Genus *ABUQATRANIA* Simons et al., 2001
ABUQATRANIA BASIODONTOS Simons et al., 2001
Figure 22.3

Synonymy "*Qatrania* sp." Kirk and Simons, 2001:204; "*Qatrania* sp." Gunnell and Miller, 2001:178; "*Qatrania*" Rasmussen, 2007:899, 909.

Age and Occurrence Latest Eocene (latest Priabonian), Quarry L-41, lower sequence of Jebel Qatrani Formation, northern Egypt.

Diagnosis Dental formula: /?.1.3.3. Differs from younger species of *Qatrania* in having: the distal face of the m2 entoconid confluent with the distal aspect of the tooth (in *Qatrania* the cusp is mesiodistally short and a very well-defined distolingual fovea is present) (figure 22.3); a p4 with a small paraconid and relatively small hypoconid; relatively tall and narrow molar trigonids with larger paraconids; a more buccally placed m1 cristid obliqua; and relatively well-developed hypocristids. Further differs from *Q. fleaglei* in having a continuous p4 cristid obliqua that meets the metaconid.

Description Only the morphology of p4–m3 are known, but it is clear from alveoli that the p2 was single rooted, and the double-rooted p3 was obliquely implanted in the mandible. The p4 has a small paraconid and a small metaconid placed distal to the protoconid; the cristid obliqua courses mesiolingually from the hypoconid to meet the metaconid. On the lower molars, the hypoconulids are centrally placed, the paraconids are relatively well developed, protocristids and premetacristids are absent, and the trigonids are elevated relative to the talonid. The lower molar cusps are bulbous, becoming more so distally along the toothrow, and most molar cristids are only weakly developed; the cristid obliqua is reduced to a tiny spur on m3. On m1 the metaconid is distally placed and the cristid obliqua meets the trigonid slightly lingual to the protoconid. The mandibular corpus is very shallow, and the almost vertically oriented ascending ramus bears a tall coronoid process that extends high above the level of the mandibular condyle.

Remarks As noted by Simons et al. (2001), *Abuqatrania* does not exhibit any clear autapomorphies, and could conceivably represent an ancestor for all later parapithecids. Such a hypothesis accords well with *Abuqatrania*'s stratigraphic position, as it is probably at least one million years older than the next-oldest parapithecid, *Qatrania wingi*.

Genus *QATRANIA* Simons and Kay, 1983
QATRANIA FLEAGLEI Simons and Kay, 1988

Age and Occurrence Later early Oligocene (late Rupelian), Quarry M, upper sequence of Jebel Qatrani Formation, northern Egypt.

Diagnosis Larger than *Q. wingi*, from which it also differs in having a more prominent m1 paraconid and a reduced p4 cristid obliqua.

Description Dental formula: /?.1.3.3. At least one incisor was present. The p4 has a small metaconid placed distal to the protoconid, and only very slight development of the cristid obliqua; there is no lingual cingulid on that tooth as in *Arsinoea*. The m1 has a large paraconid and a capacious mesial fovea that is bordered by a large and distally placed metaconid. There are no premetacristids or protocristids on m1–2, and both teeth have voluminous, basally inflated cusps, centrally placed hypoconulids, and little or no development of cristids aside from the m1 cristid obliqua, which meets the trigonid about mid-way between the metaconid and protoconid. The m2 entoconid is displaced mesially, and there is a discrete, well-defined distolingual fovea on that tooth. Well-developed precingulids are present on m1–2.

Remarks *Q. fleaglei* documents the only known extension of a lower sequence primate genus into the Jebel Qatrani Formation's upper sequence. *Qatrania* species evidently persisted as rare elements of the Fayum primate fauna for about three to four million years, over the course of which they are only documented by a total of six specimens.

QATRANIA WINGI Simons and Kay, 1983
Figure 22.3

Age and Occurrence Early Oligocene (early Rupelian), Quarry E, lower sequence of Jebel Qatrani Formation, northern Egypt.

Diagnosis Dental formula: /?.1.3.3. differs from *A. basiodontos* and *Q. wingi* in the features already noted in those species' diagnoses.

Description Similar to *Q. fleaglei* but differs in being smaller and having a distinct p4 cristid obliqua (figure 22.3). Cristid obliqua orientation evidently varies within the species from being quite lingually oriented (as in CGM 40240) to being more buccally oriented (DPC 6125). The m3 morphology is similar to that of *Abuqatrania* in that the molar cusps are more bulbous and mammiform than those of m1–2; cristids are poorly developed or absent altogether. A single broken upper molar (DPC 2804) is known; the reconstruction of the tooth provided by Simons and Kay (1988) suggests that the specimen would have been very similar in shape to the M1s of *B. fayumensis*, with which it shares a large hypocone, distinct para- and metaconule, complete lingual cingulum, and no buccal cingulum.

Subfamily PARAPITHECINAE Schlosser, 1911
Genus *APIDIUM* Osborn, 1908
APIDIUM BOWNI Simons, 1995
Figure 22.4

Age and Occurrence Middle early Oligocene (middle Rupelian), Quarry V, upper sequence of Jebel Qatrani Formation, northern Egypt.

Diagnosis Dental formula: ?.1.3.3/2.1.3.3. Smaller than younger *Apidium moustafai* and *Apidium phiomense*, with relatively small and poorly developed lower molar centroconids.

Description *A. bowni* is known from a number of mandibles and a distorted face (Simons, 1995a). The p3 is equal in height to p4 and has a small metaconid, a large hypoconid, and, variably, a weak lingual cingulid. The metaconid on p4 is small and distally positioned as in other parapithecids. The p4 cristid obliqua is variably expressed and courses mesially from a centrally placed hypoconid. A small paraconid is present on m1 and is situated mesiolingually; the cusp is variably present on m2. The m1 metaconid is placed distal to the protoconid and the cristid obliqua meets the trigonid lingual to the protoconid on m1–2. Molar hypoconulids are centrally placed on the talonid, and centroconids are present but often only weakly developed. The m3 lacks the paraconid cusp and is either equal in length to or shorter than m2. The upper canine is oval in cross section and has no lingual cingulum. The three upper premolars are three rooted and bicuspid and lack lingual cingula (figure 22.4); paraconules and hypocones are well expressed on P3–4. M1–2 have distinct lingual cingula, pericones, hypocones, and conules; postprotocristae and buccal cingula are lacking. The M3 is much reduced relative to M2; a hypocone is present on the former, but the metacone is poorly developed. The mandibular corpus is shallow.

The partial cranium of *A. bowni* (DPC 5264) preserves a frontal that does not appear to be distorted; its angle of orbital convergence, using the method proposed by Simons and Rasmussen (1996), is only 90°, which is considerably lower than that reported for *A. phiomense* (115°) and *P. grangeri* (102°). The metopic suture is completely fused. There is a small zygomaticofacial foramen and a single infraorbital foramen that is positioned between P2–3.

Remarks The primitive species *A. bowni* provides some evidence for convergent evolution between *Parapithecus* and later species of *Apidium*. For instance, despite apomorphically sharing centroconids and various other derived dental features with other *Apidium* species, some *A. bowni* individuals primitively retain a distinct paraconid on m1 and a smaller paraconid on m2. The m2 paraconid is absent in *Parapithecus*, *A. moustafai* and *A. phiomense*, and the m1 paraconid, when present in *A. moustafai*, is comparatively small. The combination of plesiomorphic molar paraconids and apomorphic centroconids in *A. bowni* suggests that molar paraconid loss is a possible synapomorphy of a *Parapithecus* clade containing *Parapithecus fraasi* and *Parapithecus grangeri*. Thomas et al. (1999) have reported "cf. *Apidium*" and "Parapithecidae indet." from the Taqah locality in Oman, which is probably of roughly the same age as Quarry V (Seiffert 2006).

APIDIUM MOUSTAFAI Simons, 1962
Figure 22.3

Age and Occurrence Middle early Oligocene (middle Rupelian), Quarries G and L-77, upper sequence of Jebel Qatrani Formation, northern Egypt.

Diagnosis Dental formula: ?.1.3.3/2.1.3.3. differs from *A. bowni* in its larger size and from *A. phiomense* in being smaller, having a more mesiodistally compressed lower canine, no

development of a p2 buccal cingulid, a shorter p4 talonid, relatively trenchant molar cristids, and variable presence of the P4 hypocone.

Description In known specimens the lower canine is mesiodistally compressed. The p2 is one-rooted and taller than p3–4. The p3 is taller than p4, and both teeth are basally inflated and have small, distally placed metaconids and well-developed hypoconids (figure 22.3); paraconids, protocristids, premetacristids, and oblique cristids are lacking. Paraconids and premetacristids are variably present on m1, and the paraconid is invariably absent on m2–3. The m1 metaconid is placed distal to the protoconid and the cristid obliqua meets the trigonid lingual to the protoconid. Centroconids and large, centrally placed hypoconulids are present on all molars, and accessory cusps are variably present in the distolingual foveae. On m2, the trigonid is approximately equal in width to the talonid and the cristid obliqua varies in orientation from meeting the protoconid to terminating lingual to that cusp. Molar entoconids vary in position from being mesial to the hypoconid to being transversely placed relative to that cusp. The m3 is either equal in size to, or longer than the m2. The upper dentition is very similar to that of *A. bowni* aside from having a faint buccal cingulum on the upper molars, and a comparatively small hypocone on M3.

APIDIUM PHIOMENSE Osborn, 1908
Figure 22.7

Age and Occurrence Later early Oligocene (late Rupelian), Quarries I, M, and X, upper sequence of Jebel Qatrani Formation, northern Egypt.

Diagnosis Dental formula: 2.1.3.3/2.1.3.3. Largest species of *Apidium*; further differs from other species of the genus in having a lower canine that is more oval in cross section, no p3 lingual cingulid, a relatively large p4 talonid, a discontinuous buccal cingulid on the lower molars, greater basal inflation of molar cusps, an m2 trigonid that is variably much wider than the talonid, and variable presence of a lingual cingulum on the upper canine. The mandibular corpus tends to be deeper than those of younger *Apidium* species.

Description Aside from the features already discussed in the diagnosis, *A. phiomense*'s dentition is very similar to that of *A. moustafai*. The more complete material of *A. phiomense* reveals some additional information about the lower incisors, which are very small and spatulate; i1 is smaller than i2 and both bear lingual cingulids. Dental eruption sequence is dp3, dp4, m1, m2, p2, p4, (p3, m3), canine (Kay and Simons, 1983). The mandibular symphysis is fused and the masseteric fossa is usually only weakly developed. The coronoid process extends far above the relatively low mandibular condyle.

The cranial morphology of *A. phiomense* is known from one partial cranium and probably also an isolated frontal bone, without locality information, that was collected early in the 20th century (Simons, 1995a). The former specimen confirms that *A. phiomense* had complete postorbital closure. The temporal lines converge near the frontal-parietal suture, and there is only slight development of a sagittal crest near the back of the cranial vault. The shape of the neurocranium suggests that *A. phiomense* had a relatively small brain, perhaps similar in size to those of other Fayum anthropoids such as *Parapithecus* and *Aegyptopithecus* (Simons et al., 2007).

A. phiomense is represented by numerous isolated postcranial bones (Fleagle and Simons, 1983, 1995; Fleagle and Kay, 1987;

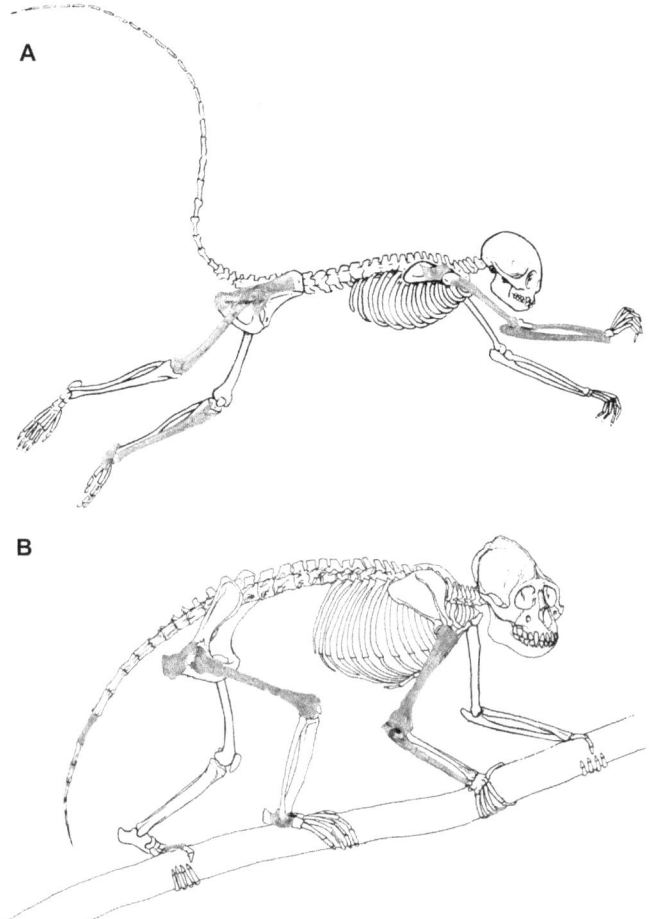

FIGURE 22.7 Postcranial elements and skeletal reconstructions of the parapithecid *Apidium* (A) and the propliopithecid *Aegyptopithecus* (B). Known elements are represented by photographs. Note that these species are represented by cranial material as well (particularly *Aegyptopithecus*; see figure 22.9). Original artwork by Luci Betti-Nash, Stony Brook University.

figure 22.7), and this material has been described in great detail by Fleagle and Simons (1995); their work forms the basis for most of the following. The humerus has a prominent deltopectoral crest and a very simple distal articulation that has little or no distinction between the trochlea and the ovoid capitulum; the trochlea is conical and relatively narrow mediolaterally, when compared with those of oligopithecids and propliopithecids (Seiffert et al., 2000). An entepicondylar foramen is present, and the medial strut is usually confluent with the medial aspect of the trochlea. The radial fossa is invariably deep, and some specimens have a supratrochlear foramen. The medial epicondyle is dorsally oriented at an angle of about 20°, and a dorsoepitrochlear fossa is present on its dorsal surface. The ulnar olecranon process is relatively long and extends proximally along the same axis as the shaft. The lateral anconeal process is well developed, and the floor of the sigmoid cavity is broad and saddle shaped. The articular facet of the projecting ulnar styloid process faces distally. The radial shaft is broad and flat and the distal articulation is shallow, extending onto a sharp styloid process laterally; a distinct tubercle is present just proximal to the styloid. There is no evidence for a synovial joint with the distal ulna. The radial head is oval, dorsolaterally oriented, and has a shallow depression for articulation with the capitulum.

Ossa coxae have a broad gluteal plane and an expanded iliac plane adjacent to the acetabulum that becomes narrower proximally. The lower ilium is short (about 60%–70% of the ischium), and the acetabulum is positioned closer to the auricular surface than in most anthropoids. The ischial tuberosities are expanded dorsally and ventrally but not medially or laterally. The proximal articular surface of the femur extends posteriorly and medially onto the femoral neck, and the femoral head varies from being somewhat spherical to being slightly more oval in dorsal view. A distinct fovea capitis is present. A small crista paratrochanterica is present in some specimens. The proximal surface of the greater trochanter is low, varying in position from being about level with the most proximal aspect of the femoral head to sitting below it. As in *Proteopithecus*, the trochanteric fossa is walled off distally; but unlike *Proteopithecus* there is no development of a third trochanter, and the lesser trochanter is large and platelike. The distal articulation is deeper than it is wide, with femoral condyles that are subequal in size and separated by a correspondingly deep and symmetrical intercondylar notch. The proximal articulation of the tibia has a convex lateral condyle whose posterior surface faces dorsally; the medial condyle is smaller, oval and concave. The proximal part of the tibial shaft is triangular in cross section and mediolaterally compressed, and it has a sharp cnemial crest protruding from the proximal fifth of the shaft. The distal end is more rounded in cross section and has an extensive scar for the tibiofibular syndesmosis that extends across 40% of the shaft. The distal articulation is quite broad mediolaterally and anteroposteriorly, and it has an anterior "beak" and a prominent medial malleolus with a laterally facing articular surface.

The astragalar body is tall and steep sided and has a deeply incised trochlear surface. The cotylar fossa is generally only weakly developed when compared with those of oligopithecids, propliopithecids, and later catarrhines. On average, the astragalar neck diverges from the long axis of the trochlea at an angle of about 20°. The astragalar head varies from being relatively wide to being about as wide as tall, and in some specimens the navicular facet extends onto the dorsolateral surface of the head. The long axis of the astragalar head is obliquely oriented in distal view, with the lateral surface more dorsal to the medial surface. The ectal facet is relatively narrow distally and tends to broaden proximally, which is the reverse of the pattern observable in Fayum catarrhines. The calcaneal sustentacular facet is semicircular and generally restricted, although it occasionally meets the anterior calcaneal facet. The cuboid facet is lunate or fan shaped, and it partially surrounds a deep pit for the calcaneocuboid ligament. There is a distinct peroneal tubercle at about the level of the sustentaculum tali. There is no heel process. The lateral aspect of the cuboid apparently has a facet for metatarsal V, which is a very peculiar feature otherwise seen only in *Tarsius* among extant primates. The navicular has facets for the cuneiform bones that are arranged in an L, as in other living and extinct haplorhines; the bone is not elongated as in some prosimians.

Remarks *Apidium phiomense* is the most common primate in the upper sequence of the Jebel Qatrani Formation, and its postcranial anatomy is the best known of any Paleogene anthropoid. *A. phiomense* was likely a diurnal and largely frugivorous arboreal quadruped that lived in large social groups. The species exhibits some postcranial features seen in extant pronograde leapers, and it was likely a very active species with a mode of locomotion similar to that of *Saimiri*.

Genus *PARAPITHECUS* Schlosser, 1910
PARAPITHECUS FRAASI Schlosser, 1910

Age and Occurrence Holotype mandible is from the Jebel Qatrani Formation, but its stratigraphic position within that formation is unknown. Simons (2001) suggested that two specimens, DPC 2374 (from Quarry M) and DPC 3135 (from Quarry I) might be attributable to *P. fraasi* rather than *P. grangeri*.

Diagnosis Differs from *Apidium* in possibly retaining only one set of deciduous or permanent incisors into adulthood; in having a relatively small p3 (compared to p4), a narrow p4 that is relatively large when compared with m1, less crowding of the lower premolars, no m1–2 paraconids or centroconids, relatively broad m2–3 trigonids, an m2 entoconid placed transverse to (rather than mesial to) the hypoconid, slightly lower molar trigonids with weak postmetacristids, no accessory cusps in the m2 distolingual fovea, and a relatively small m3. Differs from *P. grangeri* in possibly retaining only one set of deciduous or permanent incisors into adulthood, having a larger p3 hypoconid, a narrower p4 with a lower talonid basin, a narrower m1, and larger molar hypoconulids.

Description The morphology of *P. fraasi* is definitively known from only a single specimen collected by Richard Markgraf in the early part of the 20th century. In this type specimen there is only one pair of small incisors that are lighter in color

than the other, clearly permanent, teeth implanted in the jaw. These incisors have been interpreted by Kay and Simons (1983) and Simons (2001) as deciduous incisors; Simons (2001) identified another specimen from Quarry M (DPC 3135) as a possible *P. fraasi*, but this specimen has canine alveoli closely appressed, with no incisor alveoli. Simons (2001) has argued that there is no evidence for additional incisor alveoli in the type of *P. fraasi* and that there would not have been enough space available for more than one pair of deciduous or permanent incisors. Kirk and Kay (2004) have argued that the number of incisor alveoli in the type cannot be determined due to breakage of the mandibular symphysis, and that retention of one pair of incisors into adulthood in one species (*P. fraasi*) and loss of both pairs in another (*P. grangeri*) is sufficient evidence for placing *P. grangeri* in a different genus, *Simonsius*.

The lower canine in the type of *P. fraasi* is tall and has a lingual cingulid. The p2 is large, also has a lingual cingulid, and a small, buccally placed hypoconid cusp. Both p3–4 have bulbous metaconids that are placed distal to protoconids, and large hypoconids that are not linked to the trigonids by oblique cristids. There are no paraconids on m1–3, but faint premetacristids and postmetacristids are present on those teeth. The cristid obliqua meets the trigonid lingual to the protoconid on m1 and more buccally on m2. Most molar cusps are somewhat bulbous and basally inflated; there are large, centrally placed hypoconulids on all lower molars, but no centroconids are present.

Remarks The taxonomy of *Parapithecus* has been quite controversial since Gingerich's (1978) reassessment of the Stuttgart parapithecids, in which he argued that *P. fraasi* was actually *A. phiomense*, and erected the genus *Simonsius* for *Parapithecus grangeri*. *Parapithecus fraasi* is nevertheless still universally recognized as a species very distinct from those of *Apidium*, but *Simonsius* now has priority as a generic replacement name for the species *grangeri* if it can be shown to be sufficiently distinct from *P. fraasi*. Ultimately, more complete material of undoubted *P. fraasi* specimens will be required to convincingly resolve the ongoing debate about its dental formula. The monophyly of a *Parapithecus* clade containing the species *P. fraasi* and *P. grangeri* was recovered in Seiffert et al.'s (2004) phylogenetic analysis, but this clade was not consistently recovered with the larger taxon and character sets compiled by Seiffert et al. (2005a, 2005b); Kay et al. (2004) found *P. fraasi* to be more closely related to *Apidium*, but notably their analysis included only the most specialized species of *Apidium*, *A. phiomense*. As noted earlier, some specimens of *A. bowni* primitively preserve large paraconids on m1, which are absent in *A. phiomense*, *P. fraasi*, and *P. grangeri*; it is therefore possible that m1 paraconid loss is in fact a synapomorphy of *Parapithecus* and acquired independently within Parapithecinae. If *P. fraasi* is more closely related to *Apidium* as suggested by Kay et al. (2004), then m1 paraconids would have been independently lost in three parapithecine lineages (unless the cusp was reacquired in *A. bowni* and, variably, in *A. moustafai*).

PARAPITHECUS GRANGERI Simons, 1974
Figures 22.4 and 22.8

Age and Occurrence Later early Oligocene (late Rupelian), Quarries I and M, upper sequence of Jebel Qatrani Formation, northern Egypt.

Diagnosis Dental formula: 2.1.3.3/0.1.3.3. Differs from *P. fraasi* in the features noted in that species' diagnosis.

Description The morphology of *P. grangeri* is fairly well-known, as the species is represented by a complete cranium

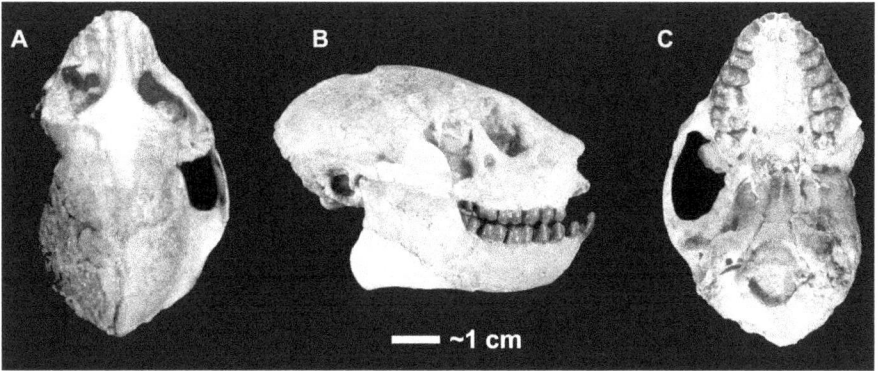

FIGURE 22.8 Cranium of the early Oligocene parapithecid *Parapithecus grangeri* (DPC 18651, from Fayum Quarry I) in dorsal (A), lateral (B, with unassociated mandible), and ventral (C) views.

(Simons, 2001; figure 22.8) as well as numerous mandibles, partial maxillae, and postcranial elements. In its postcranial morphology, there are evidently no notable differences from *A. phiomense* aside from being larger and somewhat more robust. The strangest feature of *P. grangeri* is the complete loss of lower incisors (Simons, 1986) and development of interproximal contact between the lower canines. It is now known that there was no correlated loss of the upper incisors (Simons, 2001). The morphology of the lower teeth is similar to that of *P. fraasi*, but the premolar hypoconids and molar hypoconulids are not as well developed, and the cusps show less basal inflation. Some specimens show little or no development of the p3 metaconid, and there is occasionally a small paraconid on p4. Molar trigonids invariably have very restricted mesial foveae and lack paraconids, and premetacristids are variably present on m1. The m1 is relatively short and broad when compared with other parapithecids (figure 22.3), and cristid obliqua orientation on the tooth varies from terminating lingual to the protoconid to being more buccally placed. Postmetacristids are variably present on the lower molars. The trigonid is roughly equal in height to the talonid. Weak buccal cingulids are variably present on the lower molars. The mandibular symphysis is fused. The p2 and m1–2 erupt before the p3–4, and the lower canine and m3 erupt last (Kay and Simons, 1983).

A well-preserved cranium of *P. grangeri* has four small alveoli for upper incisors; thus, the upper incisors must have occluded directly with the lower canines. The upper canine alveoli are comparatively quite large, and unlike other Fayum anthropoids appear to have been somewhat internally rotated. P2–4 have distinct protocones, lack lingual cingula and hypocones, and have convex (as opposed to waisted) distal margins (figure 22.4); P3–4 have paraconules as in *Apidium*. The upper molars have large hypocones, distinct para- and metaconules, and buccal cingula; postprotocristae are absent. A pericone is variably present on M2 and the lingual cingulum is sometimes interrupted on the lingual aspect of the protocone. The M3 hypocone is absent or very small.

When compared with contemporaneous *Aegyptopithecus*, *P. grangeri* has a less prominent rostrum, a smaller premaxilla, and a more bell-shaped palate, all of which are presumably correlated in part with the reduction of the upper incisors and internal rotation of the upper canine. The facial profile is quite linear, with only a slight angle between the nasals and frontal in lateral view. The nasals broaden rostrally and the nasal aperture, though distorted, was apparently quite broad and not teardrop shaped as in *Aegyptopithecus*. The infraorbital foramen is placed at about the level of the contact between P3 and P4, and there is a large zygomaticofacial foramen. The palate terminates medial to M3 and is incised by a deep notch lateral to the pterygoids. The palatine foramina are large and distinct. The orbits are not as frontated or as convergent as those of crown anthropoids, but postorbital closure is very well developed, with only a small inferior orbital fissure. The temporal lines converge near the back of the frontal and there is a weak sagittal crest. The glenoid fossa is relative flat and broad, with no development of a large entoglenoid process as in *Aegyptopithecus*. The postglenoid foramen is placed posterior to the postglenoid process, which itself is partially fused with the auditory bulla. The lateral pterygoid plate also contacts the lateral aspect of the bulla, and the foramen ovale is placed medial rather than lateral to that plate. The basioccipital broadens caudally, and slightly overlaps the petrosal (though not to the extent seen in *Tarsius* and some omomyids). The carotid foramen is placed medially and rostral to the fenestra cochleae. An anterior accessory cavity is present but does not appear to be extensively trabeculated as in crown anthropoids. The ectotympanic is annular and rims the external auditory meatus.

The brain of *P. grangeri* is small relative to cranium and body size, and has frontal lobes that are relatively unexpanded when compared with those of extant anthropoids. Relative brain size was probably at about the level of extant strepsirrhines (Bush et al., 2004a, 2004b; Simons et al., 2007). Relative orbit size is consistent with a diurnal activity pattern, and the relative size of the optic foramen suggests that *Parapithecus* had high acuity vision like other known anthropoids (Bush et al., 2004b; Kirk and Kay, 2004; Kirk, 2006). The nuchal region is quite prominent, but not as prominent as in known male *Aegyptopithecus*.

Infraorder CATARRHINI Geoffroy, 1812
Family OLIGOPITHECIDAE Kay and Williams, 1994
Genus *CATOPITHECUS* Simons, 1989
CATOPITHECUS BROWNI Simons, 1989
Figures 22.4, 22.5B, 22.5C, 22.6A, 22.6B, 22.6H, and 22.9

Age and Occurrence Latest Eocene (latest Priabonian), Quarry L-41, lower sequence of Jebel Qatrani Formation, northern Egypt.

Diagnosis Dental formula: 2.1.2.3/2.1.2.3. Differs from species of *Oligopithecus* in having a more labially positioned p4 paraconid; a relatively short p4 hypocristid; relatively small P3 protocones; more prominent M1-2 hypocones, and more inflated M1-2 protocones. Differs from *Oligopithecus savagei* in having a relatively shallow mandibular corpus; a more oval (as opposed to mesiodistally compressed) lower canine; relatively well-developed p4 postproto- and postmetacristids; a relatively large m2 hypoconulid; a waisted P3 and no hypocones on P3-4; buccal cingula at least weakly expressed on upper molars, occasionally bearing a small mesostyle nodule; and a short mesial cingulum on M1-2. Differs from *Oligopithecus rogeri* in having a smaller p4 entoconid, relatively short p4-m3, no paraconid on m2, m3 hypoconulids that are twinned with entoconids (rather than being centrally placed), and upper molars that are less waisted.

Description The i2 is larger than i1 and both incisors are relatively high crowned, though not to the degree observable in propliopithecids. Lingual cingulids on the lower incisors are poorly developed or absent. The lower canines are large and sexually dimorphic (Simons et al., 1999) but not mesiodistally compressed as in *Proteopithecus* and *Oligopithecus*. The p2 is absent, and p3 varies in height from being as tall as, or taller than, p4. The p3 has a lingual cingulum, a small paraconid, and a trenchant postprotocristid (figures 22.9A, 22.9B). The p4 has a small paraconid and a large metaconid that is placed transverse to the protoconid and that is

connected to that cusp by a well-developed protocristid. The p4 talonid is tall and broad and extends up to the level of the mesial fovea on m1; its cristid obliqua terminates behind the protoconid. A small paraconid is present on the m1 (but not on m2-3) and is positioned labial to the metaconid. Premetacristids are occasionally present on m1 and are usually present on m2-3. The m1 metaconid is placed approximately transverse to the protoconid and is connected to that cusp by a well-developed protocristid. Postmetacristids are present on all molars; in some specimens these crests take the form of small metastylids. The trigonid cusps are only slightly taller than those of the trigonid. Entoconid and hypoconulid cusps are twinned on all lower molars.

The upper incisors are spatulate, the I1 is larger than I2, and both teeth have weak lingual cingula. The upper canine is oval in cross section, has a deep mesial groove, and lacks a lingual cingulum. The P3 is waisted and bicuspid and has a small protocone (figure 22.4); the protocone on P4 is larger. Both upper premolars have lingual cingula. The upper molars are simple, with small hypocones, no conules, trenchant pre- and postprotocristae, complete lingual cingula, and occasionally parahypocristae. Pericones are variably present on M2. Buccal cingula on the upper molars vary from being weakly developed to bearing small mesostyle nodules. The upper and lower third molars are smaller than the upper and lower second molars, respectively.

As with *Proteopithecus*, the craniofacial morphology of *C. browni* is known only from crushed crania (Simons, 1995b; Simons and Rasmussen, 1996; figure 22.5B, 22.5C). The orbits are relatively small, suggesting a diurnal activity pattern (Heesy and Ross, 2001), and are enclosed by well-developed

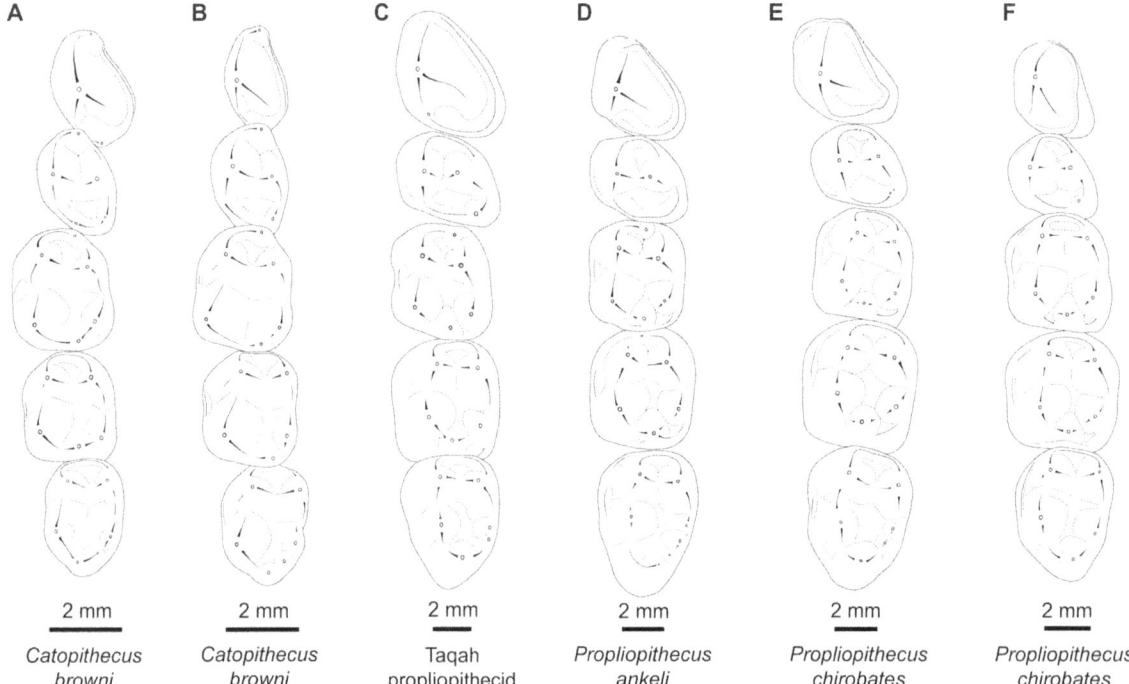

A	B	C	D	E	F
2 mm	2 mm	2 mm	2 mm	2 mm	2 mm
Catopithecus browni	*Catopithecus browni*	Taqah propliopithecid	*Propliopithecus ankeli*	*Propliopithecus chirobates*	*Propliopithecus chirobates*

FIGURE 22.9 Postcanine dentition of stem catarrhines from the latest Eocene and early Oligocene of Egypt and Oman. A) DPC 18271, male *Catopithecus browni* from the latest Eocene Fayum Quarry L-41; B) DPC 12708, female *C. browni* from Quarry L-41; C) TQ4, propliopithecid from Taqah, Oman (probable male); D) DPC 8706, female *Propliopithecus ankeli* from Fayum Quarry V; E) DPC 3836, male *Propliopithecus chirobates* from Fayum Quarry I; F) DPC 9865, female *P. chirobates* from Quarry I. Note sexual dimorphism in the size and shape of the lower third premolar.

postorbital septa. The orbital apertures are more convergent than those of parapithecids and *Proteopithecus*. The lacrimal bone and foramen are positioned within the orbit, and there is no zygomatic-lacrimal contact. When compared with many small platyrrhines, the rostrum is well developed, and the nasals are quite long. The ascending wing of the premaxilla is relatively broad, as in *Aegyptopithecus* and some Miocene catarrhines. There is a small zygomaticofacial foramen and a single infraorbital foramen positioned over P3 or P4. The metopic suture is completely fused, and the temporal lines converge on the parietals to form a short sagittal crest. Basicrania of *Catopithecus* are badly distorted, but the preserved morphology indicates that the position of the carotid foramina is similar to that of *Aegyptopithecus*, the ectotympanic was annular as in other Fayum anthropoids, and there was a large bony canal for the promontorial artery. The mastoid region appears to have been pneumatized. As in *Aegyptopithecus*, there is a well-developed entoglenoid process and a small postglenoid foramen that is positioned posterior, rather than medial, to the postglenoid process. There is no postpalatine torus, and a distinct maxillary notch is present lateral to the pterygoids, as in *Parapithecus* and *Proteopithecus*. There is no contact between the lateral pterygoid plate and the auditory bulla as in *Parapithecus*.

The postcranial morphology of *Catopithecus* is known from a number of associated and isolated elements (figures 22.6A, 22.6B, and 22.6H). The humeral morphology of *Catopithecus* resembles that of the propliopithecids *Aegyptopithecus* and *Propliopithecus* in having long trochleae relative to capitular width, and medially placed entepicondylar foramina (Seiffert et al., 2000). *Catopithecus*'s humerus was evidently long relative to body size, as in later living and extinct catarrhines. The trochlea is confluent with the capitulum, the medial epicondyle is oriented posteriorly, and a dorsoepitrochlear fossa is present. As in *Aegyptopithecus*, the femoral trochanteric fossa is distally open, rather than being "walled off" by an intertrochanteric crest as in *Proteopithecus* and parapithecids, and there is a small gluteal tuberosity that is placed distal to the lesser trochanter (Gebo et al., 1994; Ankel-Simons et al., 1998). The femoral head is round and has a distinct fovea capitis. The astragalus is similar to *Aegyptopithecus* in having a relatively broad head, a deep cotylar fossa, an elevated lateral trochlear keel, a laterally projecting fibular facet, and a well-developed lateral tubercle buttressing the groove for the tendon of flexor fibularis (Seiffert and Simons, 2001). This pattern differs considerably from that observable in omomyids, parapithecids, *Proteopithecus*, and nonateline platyrrhines.

Remarks *Catopithecus* is the most common primate at the Fayum Quarry L-41 and is now represented by numerous mandibles, maxillae, crania, and postcranial remains. The genus has also been reported, but not described, from the early Oligocene Taqah locality in Oman (Thomas et al., 1999). *Catopithecus* was originally described as a propliopithecid catarrhine by Simons (1989), but its retention of various plesiomorphies not observable in crown anthropoids, such as an unfused mandibular symphysis and an m1 paraconid, led Kay et al. (1997) and Ross et al. (1998) to suggest that *Catopithecus* was a stem anthropoid and a close relative of *Proteopithecus*. The recovery of additional postcranial material of *Aegyptopithecus*, *Catopithecus*, and *Proteopithecus* has convincingly demonstrated that the latter two taxa are not closely related and that *Catopithecus* shares with *Aegyptopithecus* a number of apomorphic postcranial features that support their placement as stem catarrhines. The most comprehensive parsimony

analyses of morphological data have since placed *Catopithecus* as a stem catarrhine (Marivaux et al., 2005; Seiffert et al., 2005a), and most authorities now recognize the genus as such (Gunnell and Miller, 2001; Rasmussen, 2002; Beard, 2004; Marivaux et al., 2005; Seiffert et al., 2005a). This placement requires that mandibular symphyseal fusion and m1 paraconid loss evolved independently within crown Anthropoidea.

Catopithecus has well-developed shearing crests on its upper and lower molars and has been interpreted as a mixed frugivore-insectivore that might have also consumed leaves (Kirk and Simons, 2001). Postcranial remains suggest that *Catopithecus* was more likely to have been a relatively slow climber than an active arboreal quadruped. The lower canines are sexually dimorphic, suggesting that the species lived in large social groups with intense male-male competition for access to females.

Genus *OLIGOPITHECUS* Simons, 1962
OLIGOPITHECUS ROGERI Gheerbrant et al., 1995
Figure 22.4

Age and Occurrence Middle early Oligocene (middle Rupelian), Taqah, Shizar Member of the Ashawq Formation, Oman.

Diagnosis Differs from *Catopithecus* in the ways mentioned in the diagnosis of that genus. Differs from *O. savagei* in having a relatively elongate p4 with a well-developed entoconid; relatively long upper and lower molars; more lingually placed molar paraconids, variably present on m2; waisted upper premolars without hypocones; and a more concave distal outline of the upper molars.

Description The anterior dentition is not known. The lower p3 is enlarged relative to p4, has a complete lingual cingulum and trenchant protocristid, and is evidently sexually dimorphic. The p4 is relatively long when compared with those of other oligopithecids, and slightly more molarized, with a distinct entoconid on the talonid. The molars are also relatively long, and the m1 has a large, lingually placed paraconid; this cusp is variably present on m2. The hypoconulid is twinned with the entoconid on m1–2 but is relatively centrally placed on m3. There are no buccal cingulids on the lower teeth. The upper premolars are bicuspid and bear lingual cingula, and the P3 protocone is relatively large when compared with that of *Catopithecus* (figure 22.4). The upper molars are simple, with no conules, a complete lingual cingulum, and a small hypocone, and are relatively long and narrow when compared with those of other oligopithecids.

Remarks *Oligopithecus rogeri* is the youngest and most specialized oligopithecid. Thomas et al. (1999) report three additional "oligopithecine" species from the Taqah locality, including a new species of *Catopithecus*, but none of these have been described.

OLIGOPITHECUS SAVAGEI Simons, 1962

Age and Occurrence Early Oligocene (early Rupelian), Quarry E, lower sequence of Jebel Qatrani Formation, northern Egypt.

Diagnosis Differs from *O. rogeri* in having a relatively short p4 with a less complex talonid, relatively short lower molars, a more labially placed m1 paraconid, and hypocones on P3–4. Differs from *C. browni* in having a relatively deep mandibular corpus, a mesiodistally compressed lower canine, no postmetacristid on p4, a relatively small m2 hypoconulid, hypocones on P3–4, and no buccal cingula or mesial cingula on upper molars.

Description Oligopithecus savagei is known from a single mandible with c-m2 and a number of isolated teeth, all from the Fayum Quarry E (Simons, 1962; Rasmussen and Simons, 1988). The lower canine in the type specimen is tall and mesiodistally compressed. The p3 is larger than p4 and bears a well-defined wear surface for the distal face of the upper canine, a complete lingual cingulum, and a distinct protocristid. As in other oligopithecids, the p4 has a small paraconid and a large metaconid placed approximately transverse to the protoconid. The p4 talonid is broad and tall, but lacks the very distinct entoconid cusp present in *O. rogeri*. The m1 has a small paraconid that is positioned just labial to the metaconid; no paraconid is present on m2, and the mesial fovea on that tooth is mesiodistally restricted. The talonid basins are broad, and hypoconulids are twinned with the entoconid on m1–2. Buccal cingulids are present at the base of the m1–2 hypoflexids. The upper premolars have well-developed protocones, lingual cingula, and small hypocones. The only known upper molar (M1?) has a complete lingual cingulum but no mesial cingulum, buccal cingulum, or conules. The pre- and postprotocristae are well developed, and the hypocone is very small. There is no pericone.

Remarks Oligopithecus savagei has a lower shearing quotient than older *Catopithecus* and is thought to have been frugivorous (Kirk and Simons, 2001). Nothing is known about its postcranial or cranial morphology.

Family PROPLIOPITHECIDAE Straus, 1961
Genus *AEGYPTOPITHECUS* Simons, 1965
AEGYPTOPITHECUS ZEUXIS Simons, 1965
Figures 22.4, 22.7, and 22.10

Synonymy Propliopithecus zeuxis Szalay and Delson 1979:438.

Age and Occurrence Later early Oligocene (late Rupelian), Quarries I and M, upper sequence of Jebel Qatrani Formation, northern Egypt.

Diagnosis Dental formula: 2.1.2.3/2.1.2.3. Differs from *Propliopithecus chirobates* in being larger and in having less distinct lingual cingulids on the lower incisors; paraconid never present on p4 or m1; relatively weak p4 postprotocristids and relatively well-developed postmetacristids; lower molars that are more basally inflated, increase in size posteriorly, and have relatively well-developed buccal cingulids; a weaker lingual cingulum on the upper canine; hypocones and buccal cingula variably present on P3–4; considerable buccal expansion of upper molar paracones. The mandibular corpus is relatively deep, particularly in males. Differs from *Propliopithecus haeckeli* in being larger; having a more lingually positioned hypoconid on p4; more basal inflation of the lower premolars and molars; and relatively large and more centrally placed molar hypoconulids. Differs from *Propliopithecus ankeli* in having more basally inflated lower molars and larger distolingual foveae.

Description Lower incisors bear weak lingual cingulids and are relatively high crowned, narrow, and labiolingually broad when compared with those of *P. chirobates*. The lower canines and p3s are strongly sexually dimorphic (Fleagle et al., 1980), and the p3 is taller than p4 (particularly in males). The p3 bears a complete lingual cingulid and a trenchant protocristid. Paraconids never occur on any lower teeth. The p4 has a large metaconid that is placed transverse to the protoconid, and a well-developed talonid basin that is

often offset lingually. The lower molars are basally inflated and have buccal cingulids, broad talonid basins, low trigonids with mesiodistally short mesial foveae, and centrally placed hypoconulids and large distolingual fovea. Wear facet "X" is present. Molar size generally increases posteriorly; when compared with *P. chirobates* there is minimal overlap in m2/m3 length ratios. The mandibular symphysis is fused, superior and inferior transverse tori are present, and mandibular corpora are consistently very deep when compared with molar length or breadth. Lingual cingula are consistently present, and small hypocones are usually present, on P3–4 (figure 22.4). M1–2 have complete lingual cingula, short mesial cingula, large hypocones, trenchant pre- and postprotocristae, usually have some development of the buccal cingulum, and lack conules. A pericone is occasionally present on M2, and a hypocone is variably present on M3.

The craniofacial morphology of *A. zeuxis* is known from two complete crania (CGM 40237 and 85785) and a number of faces (Simons, 1987; Simons et al., 2007; figure 22.10). There is considerable craniofacial dimorphism within the species—males have well-developed frontal trigons, sagittal crests, nuchal crests, pronounced rostra, and tall maxillae, but these features are absent in the small female cranium CGM 85785. The following features can be observed in both sexes: the premaxilla is quite broad as in early Miocene *Afropithecus* and proconsulids, and it differs from the reduced condition seen in parapithecids, pliopithecids, and most platyrrhines; nasals are relatively long; facial profile is concave in lateral view; an elongate postpalatine spine is present; there is no deep maxillary notch; the origin of the lateral pterygoid plate is positioned close to the lingual margin of M3. The canal for the internal carotid artery enters the petrosal medially and somewhat rostrally and does not shield the fenestra cochleae. The internal carotid takes a perbullar pathway, and the anterior accessory cavity is trabeculated (e.g., Ross, 1994). The ectotympanic is annular, but has a ragged lateral margin ventrally, and extends out into a short process dorsally, suggesting incipient development of the tubular form seen in crown catarrhines. *Aegyptopithecus*'s ratio of endocranial volume to skull length is more strepsirrhine-like than anthropoid-like, and the frontal lobes are not as expanded as in most extant anthropoids. The central sulcus is not clearly present in the small juvenile female (CGM 85785). There is a distinct lunate sulcus delimiting the primary visual cortex.

The postcranial morphology of *Aegyptopithecus* is known from a number of isolated elements that can be assigned to the genus based on size (figure 22.7). The humerus (Fleagle and Simons, 1978, 1982a) is robust and has a well-defined deltopectoral crest and brachialis flange and prominent greater and lesser tuberosities that do not extend proximal to the somewhat elliptical and posteriorly facing humeral head. The bicipital groove is broad and shallow. On the distal articular surface, the trochlea is conical and generally confluent with the ovate capitulum. A dorsoepitrochlear fossa is present, and the medial epicondyle is oriented posteriorly at about 20°. The ulna (Fleagle et al., 1975) has an elongate olecranon process and an anteroposteriorly deep shaft. The femur (Ankel-Simons et al., 1998) has a small and distally positioned gluteal tuberosity, an "open" trochanter fossa, and relatively deep condyles (knee index of 0.83). The astragalus has a very deep cotylar fossa as in later catarrhines, a broad astragalar head, a tall lateral keel of the trochlea, a laterally projecting fibular facet, and a distinct tubercle buttressing the groove for the flexor fibularis—all as in *Catopithecus* and as in later catarrhines (particularly early

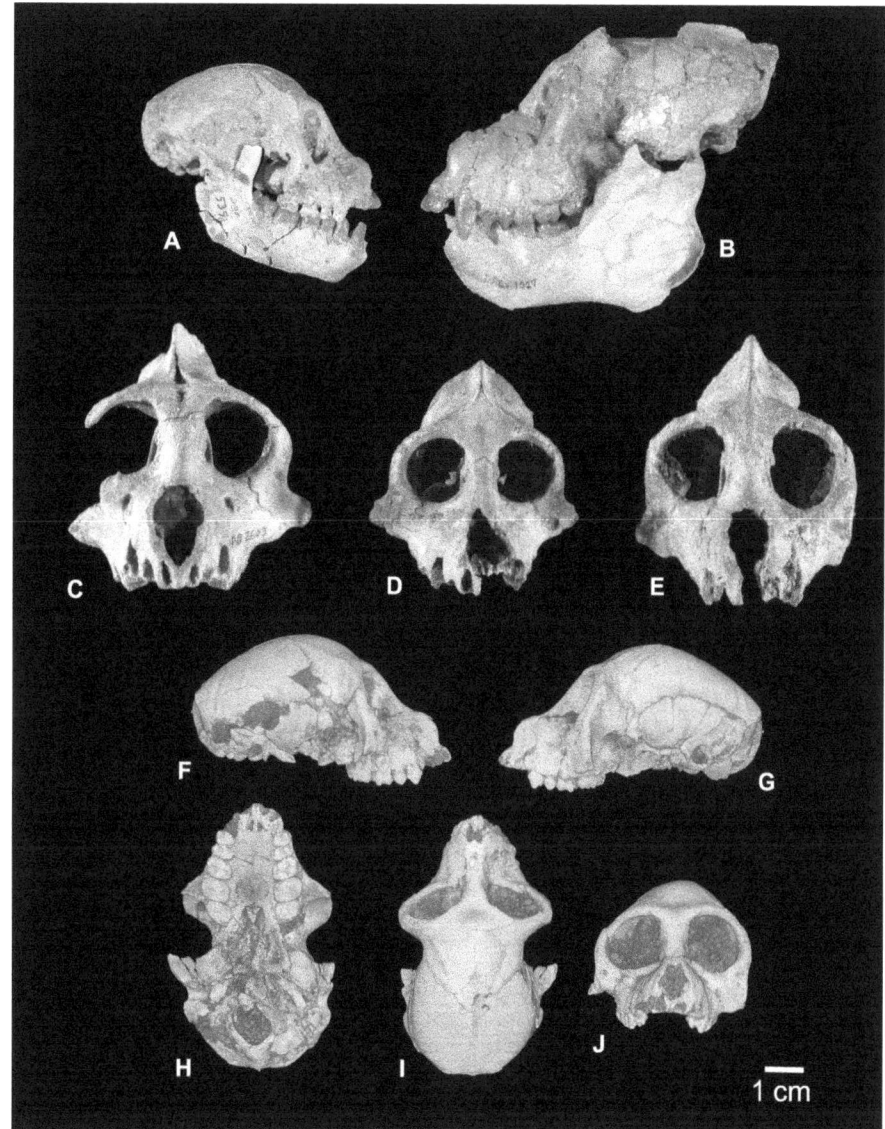

FIGURE 22.10 Cranial remains of *Aegyptopithecus zeuxis* from the early Oligocene of Egypt. A) female cranium (CGM 85785) and B) male cranium (CGM 40237) with unassociated mandibles; rostral views of C) DPC 2803, D) DPC 3161, and E) CGM 42842. F-J, 3D reconstructions of CGM 85785 from high-resolution micro-CT scans, in right lateral (F), left lateral (G), ventral (H), dorsal (I), and rostral (J) views.

Miocene catarrhines and living and extinct hominoids) (Seiffert and Simons, 2001). The calcaneus has a broad sustentaculum, a long posterior calcaneal facet, a relatively elongate distal segment, and a fan-shaped distal facet for the cuboid (Gebo and Simons, 1987). The first metatarsal has a small peroneal process as in other anthropoids (Conroy, 1976a, 1976b). Manual and pedal phalanges are more curved than those of *Apidium*, have well-developed flexor sheath ridges, and well-developed plantar tubercles (Hamrick et al., 1995).

Remarks Aegyptopithecus is the largest of all the Fayum anthropoids and is known from abundant craniodental and postcranial remains. The genus shows a number of similarities to early Miocene catarrhines in its postcranial and cranial anatomy and in a few details is intermediate between more generalized propliopithecids and later crown catarrhines. One of these features is extreme sexual dimorphism in body mass, with size differences between males and females likely being even greater than those which are observable among many dimorphic early Miocene species. Despite exhibiting a level of dimorphism that is otherwise only seen in terrestrial or semi-terrestrial cercopithecoids, in many ways *Aegyptopithecus*'s postcranium is most similar to that of the arboreal platyrrhine *Alouatta* (howler monkeys) among extant anthropoids (Fleagle, 1983), and there is no clear evidence for terrestrial behavior in the genus. *Aegyptopithecus* is also important for exhibiting a brain/body size ratio that is lower than that of extant catarrhines and platyrrhines, demonstrating that anthropoid-like encephalization occurred independently in Africa and in the Neotropics (Simons et al., 2007).

Genus *MOERIPITHECUS* Schlosser, 1910
MOERIPITHECUS MARKGRAFI Schlosser, 1910

Synonymy Propliopithecus markgrafi Simons, 1965; *Propliopithecus haeckeli (partim)* Gingerich, 1978; *Propliopithecus markgrafi* Kay et al., 1981.

Age and Occurrence Provenance unknown.

Diagnosis Differs from other propliopithecids in combining the following features of m1–2: close approximation of the hypoconulid and entoconid and little or no development of distolingual foveae; less basal inflation of the cusps, and more trenchant crests; no paraconid on m1.

Description The species is only known from a single specimen preserving m1–2. Both teeth lack paraconids and have restricted mesial foveae. The hypoconulids are twinned with the entoconids and all of the cusps are inset relative to the crown base, leaving very constricted talonid and trigonid basins. The m2 is broader than long.

Remarks M. markgrafi was recovered from the Jebel Qatrani Formation by the professional collector Richard Markgraf in the early part of the 20th century, and unfortunately its stratigraphic position in that succession is unknown. Thomas et al. (1991) assigned propliopithecid specimens from Taqah, Oman, as well as lower dentitions of *P. ankeli*, to *M. markgrafi*, but the latter appears to be distinct from the former two. *M. markgrafi* is an important transitional taxon in catarrhine phylogeny because the species preserves propliopithecid (or advanced catarrhine) features such as restricted mesial foveae and basal inflation of the molar crowns, while also having closely twinned hypoconulid and entoconid and little or no distolingual foveae (which provide occlusal area for the hypocone) as in more primitive oligopithecid catarrhines. This occlusal pattern suggests that the species likely did not have enlarged hypocones as in other propliopithecids.

Genus *PROPLIOPITHECUS* Schlosser, 1910
PROPLIOPITHECUS ANKELI Simons et al., 1987
Figures 22.5 and 22.9D

Age and Occurrence Middle early Oligocene (middle Rupelian), Quarry V, upper sequence of Jebel Qatrani Formation, northern Egypt; probably also the Shizar Member of the Ashawq Formation, Oman.

Diagnosis Dental formula: ?.1.2.3/2.1.2.3. Two mandibular specimens are known from the Fayum Quarry V (one large, DPC 5392, and one small, DPC 8706; figure 22.9D); the smaller specimen is similar in size to *P. chirobates*, but the larger male specimen is larger than male *P. chirobates*. The larger specimen DPC 5392 (see figures 2B and 3B in Simons et al., 1987) differs from DPC 8706 in having molars that are very short relative to their widths, primarily due to extreme basal inflation of the buccal margins of the teeth. The lower premolars of both specimens are larger and more robust than those of *P. haeckeli*, and are more similar to those of *A. zeuxis*. A worn paraconid is present on m1 of DPC 8706, but not on DPC 5392. Differs from *A. zeuxis* in having more peripheral upper premolar and molar cusps and more restricted distolingual foveae.

Description The lower incisors are not known. The lower canines are very large in the only known (male) specimen preserving these teeth. The lower premolars are very similar to those of *Aegyptopithecus*, and p3s are of different sizes in the large (DPC 5392) and small (DPC 8706) specimens, but the latter has a ratio of p3/m2 area that is more similar to that of male *A. zeuxis* and *P. chirobates*, suggesting that either (1) there is considerable size variation among *P. ankeli* males, or (2) *P. ankeli* females have relatively large p3s when compared with other propliopithecids (figure 22.5). Another possibility is that two propliopithecid species are present at

Quarry V and that DPC 8706 represents a male of an unnamed species other than *P. ankeli*. In addition to differences in basal inflation of the lower molars, as noted in the diagnosis, weak support for the latter hypothesis comes from the fact that DPC 8706 also clearly has a small worn paraconid on m1 that is not present on DPC 5392. With only two propliopithecid mandibular specimens known from Quarry V, we here take the more conservative position that both belong to *P. ankeli*. Cusp placement on the lower molars is otherwise similar to those of other propliopithecids, but the hypoconulids are more lingually positioned than in *A. zeuxis*. The m3s are longer than m2s, and *P. ankeli* falls within the range of variation observable for *A. zeuxis* m2/m3 ratios, but falls outside the range of variation observable for *P. chirobates*. The mandibular corpus is very robust, and the symphysis is fully fused. The upper premolars are more ovate than in other propliopithecid species. The upper molars are not known from unworn specimens but evidently would have had well-developed hypocones and buccal cingula. The only known maxilla of *P. ankeli* was very tall in lateral view as in *Aegyptopithecus*, and would have had a very large canine.

Remarks Propliopithecus ankeli is probably the oldest known propliopithecid, but is already quite specialized. As noted above, lower dentitions of *P. ankeli* were placed in *M. markgrafi* by Thomas et al. (1991), but the type of the latter species appears to be distinct from the Taqah and Quarry V propliopithecids. However, there is very little difference between the propliopithecid mandible from Taqah and those of *P. ankeli*—particularly the small *P. ankeli* specimen DPC 8706, which, like the Taqah specimen, preserves a small worn paraconid on m1 (figures 22.5C, 22.5D). Unfortunately, there are no unworn upper molars of *P. ankeli*, so it is not possible to make meaningful comparisons with the propliopithecid upper teeth from Taqah; but based on lower dental morphology, the Omani specimens are here considered to fit better as members of the species *P. ankeli* than as members of *M. markgrafi*. Thomas et al. (1999) reported an additional, undescribed propliopithecid species from Taqah, Oman.

PROPLIOPITHECUS CHIROBATES Simons, 1965
Figures 22.4, 22.5E, and 22.5F

Synonymy Aeolopithecus chirobates Simons, 1965
Age and Occurrence Later early Oligocene (late Rupelian) Quarry I, upper sequence of Jebel Qatrani Formation, northern Egypt.

Diagnosis Dental formula: ?.1.2.3/2.1.2.3. Similar in size and morphology to *P. haeckeli* but differs from that species in having greater development of the p4 postprotocristid; premolars that are somewhat more oblique in root implantation, crown shape, and cusp placement; molar hypoconulids placed farther from entoconids; relatively long m3; mandibular corpus more shallow (relative to m2 length).

Description Propliopithecus chirobates is known from a few mandibles (figures 22.5E, 22.5F), partial maxillae (figure 22.4), a calcaneus, a humerus, and a tibia. The mandibular symphysis is fused. The lower incisors are somewhat high crowned and relatively narrow, but not as narrow (relative to postcanine teeth) as those of *A. zeuxis*. The i2 is larger than i1 and both have distinct but incomplete lingual cingulids. The lower canines and p3s are large and sexually dimorphic (Fleagle et al., 1980). The latter has a complete lingual cingulum and a well-developed protocristid. As in other propliopithecids, the p4 has a large metaconid placed transverse to the protoconid, and a well-developed and tall talonid basin. Small paraconids and premetacristids are occasionally present on p4. The m1 rarely has a small paraconid, and the mesial foveae are mesiodistally restricted on all molars. The molar talonids have centrally placed hypoconulids and well-developed distal foveae. Lower molar buccal cingulids are generally weak or absent. The m3 is usually slightly longer than m2, but not as long as in *A. zeuxis*.

The upper canine is oval in cross section and has a distinct lingual cingulum. The upper premolars have large protocones and lingual cingula, and lack hypocones and buccal cingula; small parastyles are variably present. The buccal roots of P3 appear to be fused on known specimens. The upper molars are relatively broad when compared with those of *A. zeuxis*, and have well-developed lingual cingula, short mesial cingula, hypocones, and pre- and postprotocristae. Buccal cingula are present but often weakly developed. Conules are absent, and a distinct pericone is consistently present on M2 but not on M1.

The humeral and calcaneal morphology of *P. chirobates* is very similar to that of *A. zeuxis* (Fleagle and Simons, 1982b; Gebo and Simons, 1987), suggesting a similar mode of locomotion. The calcaneus has a heel process, which is found in taxa that often use hindlimb suspensory postures. Propliopithecid tibial morphology is known from only a single specimen that has been attributed to *P. chirobates* (Fleagle and Simons, 1982b). The specimen has a moderately retroflexed proximal articular surface, a very short articulation with the fibula, no compression of the distal part of the shaft, a posteriorly positioned groove for tibialis posterior, and a large medial malleolus with extension of the articular surface well onto the anterior surface of that process (suggesting that, like *A. zeuxis*, *P. chirobates* would have had a deep astragalar cotylar fossa with which the medial malleolus articulated).

Remarks *Propliopithecus chirobates* is somewhat more generalized than *A. zeuxis*, but overall the two species are quite similar in postcanine and postcranial morphology. Both appear to have been sexually dimorphic, but the canine and postcanine dimorphism of *P. chirobates* is much less pronounced than that of *A. zeuxis*. Both were presumably relatively slow moving arboreal quadrupeds, probably with some capacity for hindlimb suspension. It has only recently become clear that *P. chirobates* is restricted to the Fayum Quarry I, whereas *A. zeuxis* is regularly found at Quarries I and M (Simons et al., 2007); this geographic distribution might indicate that *P. chirobates* was more restricted than *A. zeuxis* in its environmental preferences.

PROPLIOPITHECUS HAECKELI Schlosser, 1910

Age and Occurrence Provenance unknown.

Diagnosis Most similar in size and morphology to *P. chirobates*, differing in the features noted in that species' diagnosis (discussed earlier).

Description *Propliopithecus haeckeli* is known from only a single specimen, a mandible preserving c-m3. The mandibular corpus is deeper than that of *P. chirobates* specimens, and the symphysis is fully fused. Premolar and molar morphology is very similar to that of *P. chirobates*.

Discussion

ADAPTIVE CHANGES IN EARLY ANTHROPOID EVOLUTION

There is no doubt that our current understanding of anthropoid evolution in Africa is based on a very limited sampling of the actual diversity that was present on that continent. Ongoing paleontological work in new places and previously unsampled time periods, as well as in areas that have been studied for many decades, continues to produce new and surprising discoveries. Despite the limited geographic and temporal sampling, it seems worthwhile noting broad patterns in the fossil record of the Paleogene anthropoids of Africa. This is especially appropriate and reasonable because in one area, the Fayum region of Egypt, there is a sequence of fossiliferous horizons ranging in age from the earliest late Eocene through latest early Oligocene that indicates considerable change through time in the adaptive and phylogenetic composition of the early anthropoid assemblages. In addition, the more limited record from other sites generally accords with the trends seen in the Fayum anthropoids, suggesting that the patterns observable in the Egyptian sites are not taphonomic artifacts.

One of the most striking features of the record of early anthropoids from the later Eocene and early Oligocene of North Africa is the dramatic and continuous increase in body size through time (figure 22.11). *Biretia*, from the early late Eocene of Egypt and Algeria, is in the size range of the smallest living anthropoids, with estimated body masses less that 200 g. The diverse early anthropoids from the younger late Eocene deposits of the Fayum (Quarry L-41) are (with

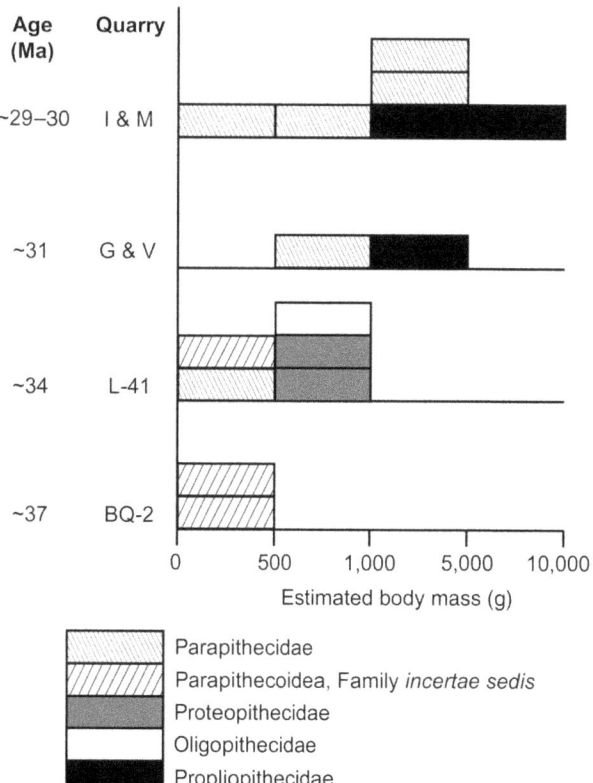

FIGURE 22.11 Changes in anthropoid body mass through the ~8 million years of evolution documented in the Birket Qarun and Jebel Qatrani formations.

the exception of *Abuqatrania*) larger but still relatively small compared with living anthropoids, with all estimated body masses less that 1,000 g. Larger anthropoids first appear in deposits clearly dating to the early Oligocene (Quarry V in the Fayum and the Ashawq Formation in Oman), and become even more common and diverse in the higher parts of the Jebel Qatrani Formation (Quarries I and M). Very large anthropoids comparable in size to many living Old World anthropoids only appear in the latest Oligocene and early Miocene of East Africa (Leakey et al., 1995; Fleagle, 1999; Harrison, 2002).

The size changes recorded in the Fayum sequence are only in part associated with changes in the taxonomic composition of the assemblages through time. The later, more derived propliopithecids are all larger than the earlier and in many ways more primitive oligopithecids. However, the parapithecids span the entire size range, and in the youngest deposits include both very small (*Q. fleaglei*) and large (*P. grangeri*) taxa.

Similarly, the changes in the size and taxonomic composition of the Paleogene anthropoid assemblages through time seem only loosely associated with other adaptive features of these taxa. In contrast with living anthropoids, in which canine and body mass sexual dimorphism seem to be correlated with body mass, there is evidence for canine dimorphism in Paleogene anthropoids of all sizes (Fleagle et al., 1980; Simons et al., 1999, 2007). Sexual dimorphism in canines and body size is common among extant anthropoids, but is generally absent or minimally expressed in extant strepsirrhines and tarsiers, and also very rare among Paleogene primates. Evidence of canine dimorphism has been documented among notharctine adapoids of North America (Krishtalka et al., 1991; Gingerich, 1995) and suggested for the European adapine *Adapis* (Gingerich, 1981). There is no evidence of canine dimorphism among other adapoids, any omomyiforms or among either eosimiids or amphipithecids of Asia (but see Jaeger et al., 2004). In contrast, canine dimorphism (along with likely dimorphism in body size) is common among the early anthropoids from the late Eocene and early Oligocene of Africa. Canine dimorphism has been documented in Proteopithecidae *(Proteopithecus)*, Oligopithecidae *(Catopithecus)*, Parapithecidae *(Apidium)*, and propliopithecids *(Propliopithecus, Aegyptopithecus)*(Fleagle et al., 1980; Simons et al., 1999). In *Aegyptopithecus*, there is also evidence of considerable size dimorphism in the cranium (Simons, 1987; Simons et al., 2007). All of these dimorphic early anthropoids were probably diurnal based on the relative size of their orbits. However, the late Eocene anthropoids *Proteopithecus* and *Catopithecus* are smaller than any extant primate characterized by canine dimorphism. Among living anthropoids, canine sexual dimorphism is associated with polygynous social groups and the evidence of canine dimorphism in Eocene anthropoids from Africa has played a prominent role in reconstructing scenarios of anthropoid origins (Ross, 2000; Plavcan, 2004). The widespread dimorphism in canines and likely body size among Oligocene propliopithecids foreshadows the condition found among later anthropoids of Africa and Eurasia and the attendant problems with the identification of taxonomic diversity within fossil assemblages (Plavcan and Cope, 2002).

The Eocene-Oligocene boundary was a time of global climate change and associated faunal turnover in many parts of the world (Meng and McKenna, 1998; Hooker et al., 2004; Coxall et al., 2005; Zanazzi et al., 2007). However, at present there is no compelling evidence for either taxonomic or adaptive changes among the African early anthropoids associated with this global event (Seiffert, 2006). There is only evidence for broad, size-related temporal changes in diet through time among the Paleogene anthropoids. The smaller, earlier taxa seem to include both frugivorous and insectivorous species, while the larger, later taxa are largely frugivorous (Kay and Simons, 1980; Kirk and Simons, 2001). By the time of deposition of the upper sequence of the Jebel Qatrani Formation (about 31 Ma), parapithecids and propliopithecids had convergently evolved dental and gnathic similarities such as paraconid loss and mandibular symphyseal fusion, but these records occur at least 2.5 million years after the major cooling event. If these features are related to earliest Oligocene climate change in some way, then they are now best interpreted as delayed reactions. Interestingly, the Eocene anthropoids from the ~37 Ma Locality BQ-2 and the ~34 Ma Quarry L-41 appear to have lived alongside much larger, and highly folivorous, adapiforms that disappeared from northern Africa at the Eocene-Oligocene boundary (Seiffert, 2007; Seiffert et al., 2009). Anthropoid encephalization also has no clear link to Oligocene climate change, for even at ~30 Ma the relative brain sizes of parapithecids and propliopithecids were surprisingly small (Simons et al., 2007), being strepsirrhine-like or smaller.

PROSPECTUS FOR FUTURE RESEARCH

Very little is known about Afro-Arabian anthropoid evolution prior to about 37 million years ago (near the middle–late Eocene boundary), when Locality BQ-2 in the Fayum Depression was deposited. Although *Altiatlasius* exhibits a few dental features that (along with biogeography) appear to support affinities with Fayum anthropoids, the material is too fragmentary, and the features in question are too prone to homoplasy, to stand as conclusive evidence for a very early stem anthropoid presence in Africa. Much more complete fossil evidence from the Eocene of Africa and Asia will be needed to provide a convincing test of outstanding hypotheses surrounding early anthropoid biogeography. Key in this regard will be renewed work in the early and middle Eocene of Africa (which, fortunately, is already underway; Adaci et al., 2007; Tabuce et al., 2009) and the recovery of cranial and postcranial material of enigmatic Asian clades such as Eosimiidae and Amphipithecidae. Circumstantial evidence for an early anthropoid presence in Africa is provided by molecular dating, which places the catarrhine-platyrrhine divergence around 43–44 Ma (Eizirik et al., 2004; Steiper and Young, 2006), but the studies that provided these estimates also imply very ancient dates for the strepsirrhine-haplorhine divergence that are not consistent with the available fossil evidence from northern continents. Two other intriguing, but again only fragmentary, pieces of evidence are an isolated talonid and trigonid from the ~45.8–43.6 Ma (middle Eocene) Aznag locality in Morocco (Tabuce et al., 2005), the former of which appears to have crestiform but closely approximated entoconid and hypoconulid, as in some Fayum anthropoids. The presence of *Biretia* at BQ-2 confirms that parapithecoids had diverged from crown anthropoids by the middle–late Eocene boundary, and this evidence leaves it likely that parapithecoid origins extend well back into the middle Eocene.

ACKNOWLEDGEMENTS

P. Chatrath managed the fieldwork in Egypt that produced most of the fossils discussed here, and the late Y. Attia long acted as the primary Egyptian collaborator in this research. B. Marandat and R. Tabuce provided access to fossils from Algeria and Morocco. E.R.S. and E.L.S.'s research for this chapter has been funded by the U.S. National Science Foundation, the Leakey Foundation, Gordon Getty, and Verna Simons.

Literature Cited

Adaci, M., R. Tabuce, F. Mebrouk, M. Bensalah, P.-H. Fabre, L. Hautier, J.-J. Jaeger, V. Lazzari, M. Mahboubi, L. Marivaux, O. Otero, S. Peigné, and H. Tong. 2007. Nouveaux sites à vertébrés paléogènes dans la région des Gour Lazib (Sahara nord-occidental, Algérie). *Comptes Rendus Palevol* 6:535–544.

Ankel-Simons, F., J. G. Fleagle, and P. S. Chatrath. 1998. Femoral anatomy of *Aegyptopithecus zeuxis*, an early Oligocene anthropoid. *American Journal of Physical Anthropology* 106:413–424.

Beard, K. C. 1998. A new genus of Tarsiidae (Mammalia: Primates) from the middle Eocene of Shanxi Province, China, with notes on the historical biogeography of tarsiers. *Bulletin of the Carnegie Museum of Natural History* 34:260–277.

———. 2002. Basal anthropoids; pp. 133–149 in W. C. Hartwig (ed.), *The Primate Fossil Record.* Cambridge University Press, Cambridge.

———. 2004. *The Hunt for the Dawn Monkey: Unearthing the Origins of Monkeys, Apes, and Humans.* University of California Press, Berkeley, 348 pp.

———. 2006. Mammalian biogeography and anthropoid origins; pp. 439–467 in S. M. Lehman and J. G. Fleagle (eds.), *Primate Biogeography: Progress and Prospects.* Springer, New York.

Beard, K. C., and J. Wang. 2004. The eosimiid primates (Anthropoidea) of the Heti Formation, Yuanqu Basin, Shanxi and Henan Provinces, People's Republic of China. *Journal of Human Evolution* 46:401–432.

Bush, E. C., E. L. Simons, and J. M. Allman. 2004a. High-resolution computed tomography study of the cranium of a fossil anthropoid primate, *Parapithecus grangeri*: New insights into the evolutionary history of primate sensory systems. *The Anatomical Record, Part A* 281:1083–1087.

Bush, E. C., E. L. Simons, D. J. Dubowitz, and J. M. Allman. 2004b. Endocranial volume and optic foramen size in *Parapithecus grangeri*; pp. 603–614 in C. F. Ross and R. F. Kay (eds.), *Anthropoid Origins: New Visions.* Kluwer Academic/Plenum Press, New York.

Conroy, G. C. 1976a. Hallucial tarsometatarsal joint in an Oligocene anthropoid, *Aegyptopithecus zeuxis*. *Nature* 262:684–686.

———. 1976b. Primate postcranial remains from the Oligocene of Egypt. *Contributions to Primatology* 8:1–134.

Coxall, H. K., P. A. Wilson, H. Pälike, C. H. Lear, and J. Backman. 2005. Rapid stepwise onset of Antarctic glaciation and deeper calcite compensation in the Pacific Ocean. *Nature* 433:53–57.

Eizirik, E., W. J. Murphy, M. S. Springer, and S. J. O'Brien. 2004. Molecular phylogeny and dating of early primate divergences; pp. 45–64 in C. F. Ross and R. F. Kay (eds.), *Anthropoid Origins: New Visions.* Kluwer Academic Press, New York.

Fleagle, J. G. 1983. Locomotor adaptations of Oligocene and Miocene hominoids and their phyletic implications; pp. 301–324 in R. L. Ciochon and R. Corruccini (eds.), *New Interpretations of Ape and Human Ancestry.* Plenum Press, New York.

———. 1999. *Primate Adaptation and Evolution.* 2nd ed. Academic Press, San Diego, 596 pp.

Fleagle, J. G., and R. F. Kay. 1987. The phyletic position of the Parapithecidae. *Journal of Human Evolution* 16:483–532.

Fleagle, J. G., R. F. Kay, and E. L. Simons. 1980. Sexual dimorphism in early anthropoids. *Nature* 287:328–330.

Fleagle, J. G., and E. L. Simons. 1978. Humeral morphology of the earliest apes. *Nature* 276:705–707.

———. 1982a. The humerus of *Aegyptopithecus zeuxis*: A primitive anthropoid. *American Journal of Physical Anthropology* 59:175–193.

———. 1982b. Skeletal remains of *Propliopithecus chirobates* from the Egyptian Oligocene. *Folia Primatologica* 39:161–177.

———. 1983. The tibio-fibular articulation in *Apidium phiomense*, an Oligocene anthropoid. *Nature* 301:238–239.

———. 1995. Limb skeleton and locomotor adaptations of *Apidium phiomense*, an Oligocene anthropoid from Egypt. *American Journal of Physical Anthropology* 97:235–289.

Fleagle, J. G., E. L. Simons, and G. C. Conroy. 1975. Ape limb bone from Oligocene of Egypt. *Science* 189:135–137.

Gebo, D. L., M. Dagosto, K. C. Beard, and X. Ni. 2008. New primate hind limb elements from the middle Eocene of China. *Journal of Human Evolution* 55:999–1014.

Gebo, D. L., and E. L. Simons. 1987. Morphology and locomotor adaptations of the foot in early Oligocene anthropoids. *American Journal of Physical Anthropology* 74:83–101.

Gebo, D. L., E. L. Simons, D. T. Rasmussen, and M. Dagosto. 1994. Eocene anthropoid postcrania from the Fayum, Egypt; pp. 203–233 in J. G. Fleagle and R. F. Kay (eds.), *Anthropoid Origins.* Plenum Press, New York.

Gheerbrant, E., J. Sudre, S. Sen, C. Abrial, B. Marandat, B. Sigé, and M. Vianey-Liaud. 1998. Nouvelles données sur les mammifères du Thanétien et de l'Yprésien du Bassin d'Ouarzazate (Maroc) et leur contexte stratigraphique. *Palaeovertebrata* 27:155–202.

Gheerbrant, E., H. Thomas, S. Sen, and Z. Al-Sulaimani. 1995. Nouveau primate Oligopithecinae (Simiiformes) de l'Oligocène inférieur de Taqah, Sultanat d'Oman. *Comptes Rendus de l'Académie des Sciences, Paris,* Série IIA, 321:425–432.

Gingerich, P. D. 1978. The Stuttgart collection of Oligocene primates from the Fayum Province, Egypt. *Paläontologisches Zeitschrift* 52:82–92.

———. 1981. Cranial morphology and adaptations in Eocene Adapidae: I. Sexual dimorphism in *Adapis magnus* and *Adapis parisiensis*. *American Journal of Physical Anthropology* 56:217–234.

———. 1990. Primate evolution: African dawn for primates. *Nature* 346:411.

———. 1995. Sexual dimorphism in earliest Eocene *Cantius torresi* (Mammalia, Primates, Adapoidea). *Contributions from the Museum of Paleontology* (University of Michigan) 29:185–199.

Godinot, M. 1994. Early North African primates and their significance for the origin of Simiiformes (= Anthropoidea); pp. 235–296 in J. G. Fleagle and R. F. Kay (eds.), *Anthropoid Origins.* Plenum Press, New York.

Gradstein, F., J. Ogg, and A. Smith. 2004. *A Geological Time Scale 2004.* Cambridge University Press, Cambridge; 589 pp.

Gunnell, G. F., and E. R. Miller. 2001. Origin of Anthropoidea: Dental evidence and recognition of early anthropoids in the fossil record, with comments on the Asian anthropoid radiation. *American Journal of Physical Anthropology* 114:177–191.

Hamrick, M. W., D. J. Meldrum, and E. L. Simons. 1995. Anthropoid phalanges from the Oligocene of Egypt. *Journal of Human Evolution* 28:121–145.

Harrison, T. 2002. Late Oligocene to middle Miocene catarrhines from Afro-Arabia; pp. 311–338 in W. C. Hartwig (ed.), *The Primate Fossil Record.* Cambridge University Press, Cambridge.

Heesy, C. P., and C. F. Ross. 2001. Evolution of activity patterns and chromatic vision in primates: morphometrics, genetics, and cladistics. *Journal of Human Evolution* 40:111–149.

Heesy, C. P., N. J. Stevens, and K. E. Samonds. 2006. Biogeographic origins of primate higher taxa; pp. 419–437 in J. G. Fleagle and S. Lehman (eds.), *Primate Biogeography: Progress and Prospects.* Springer, New York.

Hoffstetter, R. 1982. Les Primates Simiiformes (= Anthropoidea) (Compréhension, phylogénie, histoire biogéographique). *Annales de Paléontologie* 68:241–280.

Hooker, J. J., M. E. Collinson, and N. P. Sille. 2004. Eocene-Oligocene mammalian faunal turnover in the Hampshire Basin, UK: Calibration to the global time scale and the major cooling event. *Journal of the Geological Society, London* 161:161–172.

Hooker, J. J., D. E. Russell, and A. Phélizon. 1999. A new family of Plesiadapiformes (Mammalia) from the Old World lower Paleogene. *Palaeontology* 42:377–407.

Jaeger, J.-J., Y. Chaimanee, P. Tafforeau, S. Ducrocq, A. N. Soe, L. Marivaux, J. Sudre, S. T. Tun, W. Htoon, and B. Marandat. 2004. Systematics and paleobiology of the anthropoid primate *Pondaungia* from the late Middle Eocene of Myanmar. *Comptes Rendus Palevol* 3:243–255.

Jaeger, J.-J., and L. Marivaux. 2005. Shaking the earliest branches of anthropoid primate evolution. *Science* 310:244–245.

Kay, R. F., J. G. Fleagle, and E. L. Simons. 1981. A revision of the Oligocene apes of the Fayum Province, Egypt. *American Journal of Physical Anthropology* 55:293–322.

Kay, R. F., C. Ross, and B. A. Williams. 1997. Anthropoid origins. *Science* 275:797–804.

Kay, R. F., and E. L. Simons. 1980. The ecology of Oligocene African Anthropoidea. *International Journal of Primatology* 1:21–37.

———. 1983. Dental formulae and dental eruption patterns in Parapithecidae (Primates, Anthropoidea). *American Journal of Physical Anthropology* 62:363–375.

Kay, R. F., B. A. Williams, C. F. Ross, M. Takai, and N. Shigehara. 2004. Anthropoid origins: A phylogenetic analysis; pp. 91–135 in C. F. Ross and R. F. Kay (eds.), *Anthropoid Origins: New Visions*. Kluwer Academic/Plenum Press, New York.

Kirk, E. C. 2006. Visual influences on primate encephalization. *Journal of Human Evolution* 51:76–90.

Kirk, E. C., and R. F. Kay. 2004. The evolution of high visual acuity in the Anthropoidea; pp. 539–602 in C. F. Ross and R. F. Kay (eds.), *Anthropoid Origins: New Visions*. Kluwer Academic/Plenum Press, New York.

Kirk, E. C., and E. L. Simons. 2001. Diets of fossil primates from the Fayum Depression of Egypt: A quantitative analysis of molar shearing. *Journal of Human Evolution* 40:203–229.

Krishtalka, L., R. K. Stucky, and K. C. Beard. 1991. The earliest evidence for sexual dimorphism in primates. *Proceedings of the National Academy of Sciences, USA* 87:5223–5226.

Leakey, M. G., P. S. Ungar, and A. Walker. 1995. A new genus of large primate from the Late Oligocene of Lothidok, Turkana District, Kenya. *Journal of Human Evolution* 28:519–531.

Marivaux, L., P.-O. Antoine, S. R. H. Baqri, M. Benammi, Y. Chaimanee, J.-Y. Crochet, D. de Franceschi, N. I.-J. Jaeger, G. Métais, G. Roohi, and J.-L. Welcomme. 2005. Anthropoid primates from the Oligocene of Pakistan (Bugti Hills): Data on early anthropoid evolution and biogeography. *Proceedings of the National Academy of Sciences, USA* 102:8436–8441.

Meng, J., and M. C. McKenna. 1998. Faunal turnovers of Paleogene mammals from the Mongolian Plateau. *Nature* 394:364–367.

Miller, E. R., G. F. Gunnell, and R. D. Martin. 2005. Deep time and the search for anthropoid origins. *Yearbook of Physical Anthropology* 48:60–95.

Miller, E. R., and E. L. Simons. 1997. Dentition of *Proteopithecus sylviae*, an archaic anthropoid from the Fayum, Egypt. *Proceedings of the National Academy of Sciences, USA* 94:13760–13764.

Ogg, J. G., and A. G. Smith. 2004. The geomagnetic polarity time scale; pp. 63–86 in F. M. Gradstein, J. G. Ogg, and A. G. Smith (eds.), *A Geological Time Scale 2004*. Cambridge University Press, Cambridge.

Pickford, M. 1986. Première découverte d'une faune mammalienne terrestre paléogène d'Afrique sub-saharienne. *Comptes Rendus de l'Académie des Sciences, Paris*, Série II, 302:1205–1210.

Plavcan, J. M. 2004. Evidence for early anthropoid social behavior; pp. 383–412 in C. F. Ross and R. F. Kay (eds.), *Anthropoid Origins: New Visions*. Kluwer Academic/Plenum Press, New York.

Plavcan, J. M., and D. A. Cope. 2002. Metric variation and species recognition in the fossil record. *Evolutionary Anthropology* 10:204–222.

Rasmussen, D. T. 2002. Early catarrhines of the African Eocene and Oligocene; pp. 203–220 in W. C. Hartwig (ed.), *The Primate Fossil Record*. Cambridge University Press, Cambridge.

Rasmussen, D. T., T. M. Bown, and E. L. Simons. 1992. The Eocene-Oligocene transition in continental Africa; pp. 548–566 in D. R. Prothero and W. A. Berggren (eds.), *Eocene-Oligocene Climatic and Biotic Evolution*. Princeton University Press, Princeton.

Rasmussen, D. T., and E. L. Simons. 1988. New specimens of *Oligopithecus savagei*, early Oligocene primate from the Fayum, Egypt. *Folia Primatologica* 51:182–208.

Ross, C., B. Williams, and R. F. Kay. 1998. Phylogenetic analysis of anthropoid relationships. *Journal of Human Evolution* 35:221–306.

Ross, C. F. 1994. The craniofacial evidence for anthropoid and tarsier relationships; pp. 469–548 in J. G. Fleagle and R. F. Kay (eds.), *Anthropoid Origins*. Plenum Press, New York.

———. 2000. Into the light: The origin of Anthropoidea. *Annual Review of Anthropology* 29:147–194.

Rossie, J. B., and E. R. Seiffert. 2006. Continental paleobiogeography as phylogenetic evidence; pp. 461–514 in J. G. Fleagle and S. Lehman (eds.), *Primate Biogeography*. Plenum, New York.

Seiffert, E. R. 2006. Revised age estimates for the later Paleogene mammal faunas of Egypt and Oman. *Proceedings of the National Academy of Sciences, USA* 103:5000–5005.

———. 2007. Evolution and extinction of Afro-Arabian primates near the Eocene-Oligocene boundary. *Folia Primatologica* 78:314–327.

Seiffert, E. R., and E. L. Simons. 2001. Astragalar morphology of late Eocene anthropoids from the Fayum Depression (Egypt) and the origin of catarrhine primates. *Journal of Human Evolution* 41:577–606.

Seiffert, E. R., E. L. Simons, W. C. Clyde, J. B. Rossie, Y. Attia, T. M. Bown, P. Chatrath, and M. Mathison. 2005a. Basal anthropoids from Egypt and the antiquity of Africa's higher primate radiation. *Science* 310:300–304.

Seiffert, E. R., E. L. Simons, and J. G. Fleagle. 2000. Anthropoid humeri from the late Eocene of Egypt. *Proceedings of the National Academy of Sciences, USA* 97:10062–10067.

Seiffert, E. R., E. L. Simons, J. M. G. Perry, and D. M. Boyer. 2009. Convergent evolution of anthropoid-like adaptations in Eocene adapiform primates. *Nature* 461:1118–1121.

Seiffert, E. R., E. L. Simons, T. M. Ryan, and Y. Attia. 2005b. Additional remains of *Wadilemur elegans*, a primitive stem galagid from the late Eocene of Egypt. *Proceedings of the National Academy of Sciences, USA* 102:11396–11401.

Seiffert, E. R., E. L. Simons, and C. V. M. Simons. 2004. Phylogenetic, biogeographic, and adaptive implications of new fossil evidence bearing on crown anthropoid origins and early stem catarrhine evolution; pp. 157–181 in C. F. Ross and R. F. Kay (eds.), *Anthropoid Origins: New Visions*. Kluwer Academic/Plenum Press, New York.

Sigé, B., J.-J. Jaeger, J. Sudre, and M. Vianey-Liaud. 1990. *Altiatlasius koulchii* n. gen. n. sp., primate omomyidé du Paléocène du Maroc, et les origines des euprimates. *Palaeontographica, Abt. A* 214:31–56.

Simons, E. L. 1962. Two new primate species from the African Oligocene. *Postilla* 64:1–12.

———. 1965. New fossil apes from Egypt and the initial differentiation of Hominoidea. *Nature* 205:135–139.

———. 1974. *Parapithecus grangeri* (Parapithecidae, Old World Higher Primates): New species from the African Oligocene of Egypt and the initial differentiation of Cercopithecoidea. *Postilla* 64:1–12.

———. 1986. *Parapithecus grangeri* of the African Oligocene: An archaic catarrhine without lower incisors. *Journal of Human Evolution* 15:205–213.

———. 1987. New faces of *Aegyptopithecus* from the Oligocene of Egypt. *Journal of Human Evolution* 16:273–289.

———. 1989. Description of two genera and species of late Eocene Anthropoidea from Egypt. *Proceedings of the National Academy of Sciences, USA* 86:9956–9960.

———. 1992. Diversity in the early Tertiary anthropoidean radiation in Africa. *Proceedings of the National Academy of Sciences, USA* 89:10743–10747.

———. 1995a. Crania of *Apidium*: Primitive anthropoidean (Primates, Parapithecidae) from the Egyptian Oligocene. *American Museum Novitates* 3124:1–10.

———. 1995b. Skulls and anterior teeth of *Catopithecus* (Primates: Anthropoidea) from the Eocene and anthropoid origins. *Science* 268:1885–1888.

———. 1997. Preliminary description of the cranium of *Proteopithecus sylviae*, an Egyptian late Eocene anthropoidean primate. *Proceedings of the National Academy of Sciences, USA* 94:14970–14975.

———. 2001. The cranium of *Parapithecus grangeri*, an Egyptian Oligocene anthropoidean primate. *Proceedings of the National Academy of Sciences, USA* 98:7892–7897.

———. 2008. Convergence and frontation in Fayum anthropoid orbits; pp. 407–429 in C. J. Vinyard, M. J. Ravosa, and C. E. Wall (eds.), *Primate Craniofacial Function and Biology*. Springer, New York.

Simons, E. L., and R. F. Kay. 1988. New material of *Qatrania* from Egypt with comments on the phylogenetic position of the Parapithecidae (Primates, Anthropoidea). *American Journal of Primatology* 15:337–347.

Simons, E. L., J. M. Plavcan, and J. G. Fleagle. 1999. Canine sexual dimorphism in Egyptian Eocene anthropoid primates: *Catopithecus* and *Proteopithecus*. *Proceedings of the National Academy of Sciences, USA* 96:2559–2562.

Simons, E. L., and D. T. Rasmussen. 1996. Skull of *Catopithecus browni*, an early Tertiary catarrhine. *American Journal of Physical Anthropology* 100:261–292.

Simons, E. L., and E. R. Seiffert. 1999. A partial skeleton of *Proteopithecus sylviae* (Primates, Anthropoidea): First associated dental and postcranial remains of an Eocene anthropoidean. *Comptes Rendus de l'Académie des Sciences, Paris*, Série IIA, 329:921–927.

Simons, E. L., D. T. Rasmussen, and D. L. Gebo. 1987. A new species of *Propliopithecus* from the Fayum, Egypt. *American Journal of Physical Anthropology* 73:139–147.

Simons, E. L., E. R. Seiffert, P. S. Chatrath, and Y. Attia. 2001. Earliest record of a parapithecid anthropoid from the Jebel Qatrani Formation, northern Egypt. *Folia Primatologica* 72:316–331.

Simons, E. L., E. R. Seiffert, T. M. Ryan, and Y. Attia. 2007. A remarkable female cranium of the Oligocene anthropoid *Aegyptopithecus zeuxis* (Catarrhini, Propliopithecidae). *Proceedings of the National Academy of Sciences, USA* 104:8731–8736.

Steiper, M. E., and N. M. Young. 2006. Primate molecular divergence dates. *Molecular Phylogenetics and Evolution* 41:384–394.

Stevens, N. J., P. M. O'Connor, M. D. Gottfried, E. M. Roberts, and S. Ngasala. 2005. An anthropoid primate humerus from the Rukwa Rift Basin, Paleogene of southwestern Tanzania. *Journal of Vertebrate Paleontology* 25:986–989.

Stevens, N. J., P. M. O'Connor, E. M. Roberts, M. D. Gottfried, and J. Temba. 2009. New primate fossils from the late Oligocene Nsungwe Formation, Rukwa Rift Basin, Tanzania. *American Journal of Physical Anthropology* (suppl.) 47:384.

Tabuce, R., B. Coiffait, P.-E. Coiffait, M. Mahboubi, and J.-J. Jaeger. 2000. A new species of *Bunohyrax* (Hyracoidea, Mammalia) from the Eocene of Bir el Ater (Algeria). *Comptes Rendus de l'Académie des Sciences, Paris*, Série IIA, 331:61–66.

———. 2001. Knowledge of the evolution of African Paleogene mammals: Contribution of the Bir El Ater locality (Eocene, Algeria); pp. 215–229 in C. Denys, L. Granjon, and A. Poulet (eds.), *African Small Mammals*. IRD Press, Paris.

Tabuce, R., L. Marivaux, R. Lebrun, M. Adaci, M. Bensalah, P.-H. Fabre, E. Fara, H. G. Rodrigues, L. Hautier, J.-J. Jaeger, V. Lazzari, F. Mebrouk, S. Peigné, J. Sudre, P. Tafforeau, X. Valentin, M. Mahboubi. 2009. Anthropoid versus strepsirhine status of the African Eocene primates *Algeripithecus* and *Azibius*: craniodental evidence. *Proceedings of the Royal Society B* 276:4087–4094.

Takai, M., F. Anaya, N. Shigehara, and T. Setoguchi. 2000. New fossil materials of the earliest New World monkey, *Branisella boliviana*, and the problem of platyrrhine origins. *American Journal of Physical Anthropology* 111:263–281.

Thomas, H., J. Roger, S. Sen, M. Pickford, E. Gheerbrant, Z. Al-Sulaimani, and S. Al-Busaidi. 1999. Oligocene and Miocene terrestrial vertebrates in the southern Arabian peninsula (Sultanate of Oman) and their geodynamic and palaeogeographic settings; pp. 430–442 in P. J. Whybrow and A. Hill (eds.), *Fossil Vertebrates of Arabia*. Yale University Press, New Haven.

Thomas, H., S. Sen, J. Roger, and Z. Al-Sulaimani. 1991. The discovery of *Moeripithecus markgrafi* Schlosser (Propliopithecidae, Anthropoidea, Primates), in the Ashawq Formation (early Oligocene of Dhofar Province, Sultanate of Oman). *Journal of Human Evolution* 20:33–49.

Zanazzi, A., M. J. Kohn, B. J. MacFadden, and D. O. Terry. 2007. Large temperature drop across the Eocene-Oligocene transition in central North America. *Nature* 445:639–642.

TWENTY-THREE

Cercopithecoidea

NINA G. JABLONSKI AND STEPHEN FROST

Old World monkeys are some of the most common and visible components of the modern mammalian fauna of Africa, and are the dominant nonhuman primates in Africa today with respect to the overall numbers of species present (from 39 to 44, depending on species definitions) and the number of ecological zones inhabited (from primary evergreen rainforests and swamp forests, to woodlands, savannas, grassy plateaus, arid subdeserts, and steppes). What is rarely appreciated is that Old World monkeys have risen to a position of ecological dominance among primates only recently in geological time. During the early and middle Miocene, the Cercopithecoidea were well established in Africa, but not taxonomically diverse. The absence or near absence of monkey fossils from prolific early Miocene sites like Rusinga Island suggests that the animals were genuinely rare elements of the mammalian fauna at the time.

The earliest African cercopithecoids belong to the Victoriapithecidae, an extinct family from the early to middle Miocene of eastern Africa that exhibit a mosaic of basal catarrhine and modern Old World monkeylike morphological features. The ambiguity of the morphology of the victoriapithecids has persuaded most authorities that the group represents an early radiation, distinct from rather than ancestral to, other Old World monkeys; however, the possibility that they are a series of only loosely related stem cercopithecoids cannot be ruled out. They are often classified in a family equivalent to the Cercopithecidae. The latter is here divided into two subfamilies, the Cercopithecinae (comprising the modern vervets, guenons, mangabeys, mandrills, geladas, common baboons and their fossil relatives) and Colobinae (comprising the green, red, and black-and-white colobus monkeys and their fossil relatives), following Groves (Groves, 2001) and the generally accepted convention for the discipline, but in contrast to the scheme followed in a previous review (Jablonski, 2002).

In this chapter, we refer to fossil-bearing sites within various regions of Africa, such as northern Africa. For our purposes, northern Africa comprises Morocco, Algeria, Libya, Egypt, and the Sudan. Northeastern Africa includes Ethiopia, Somalia, and Eritrea. Eastern includes Kenya, Uganda, and Tanzania. Southern Africa comprises South Africa, while southwestern Africa refers to Angola.

The events that produced the major lineages recognized at the family and subfamily levels within the Cercopithecoidea occurred in the early Miocene of Africa, based on first appearances of Victoriapithecidae in the fossil record of northern and eastern Africa beginning approximately 20 mya (Benefit and McCrossin, 2002) and the absence of fossil cercopithecoids outside of Africa until the very late Miocene (Barry, 1987; Jablonski, 2002). Molecular methods have been used increasingly in the last two decades to aid in the reckoning of cercopithecoid phylogeny in the face of widespread morphological homoplasy and to refine the timing of lineage splitting events in the absence of informative fossils. The split between the Colobinae and Cercopithecinae has been estimated using mtDNA at about 16.2 Ma (with an approximate 95% confidence interval of 14.4–17.9 Ma; Raaum et al., 2005), but detailed knowledge of this milestone and of the early history of the modern subfamilies is still unclear due to a paucity of middle Miocene fossils.

By the late Miocene, clearly differentiated members of the Cercopithecinae and Colobinae are present in the African fossil record, and from this time onward, Old World monkey lineages underwent increasing cladogenesis and dispersal. Dispersal of at least one major lineage each of colobine and cercopithecine monkeys into Eurasia in the late Miocene resulted in the seeding of that continent with the ancestors of *Mesopithecus* (and descendant colobines) and the earliest forms of macaques, respectively. Increased fragmentation of forests and greater habitat heterogeneity resulting from major mountain-building events and their climatic sequelae in Asia, Europe, and Africa created new ecological opportunities for the Cercopithecoidea throughout the Old World from the late Miocene onward. Phyletic and ecological diversification of the group increased through the Plio-Pleistocene of Africa, mirroring an increase in environmental seasonality and habitat heterogeneity on the continent. By the Plio-Pleistocene, Cercopithecoidea were among the most diverse, widespread, and prolific of mammals in Africa, even without taking into account the many taphonomic factors, which biased the fossil record against preservation of small-bodied and forest-dwelling species.

The success of many cercopithecoid species in Africa today is due to their adaptability and behavioral flexibility, which render them less sensitive to increases in climatic variability, environmental seasonality, and other ecological disturbances than strepsirrhines and apes. Cercopithecoids can eat a wider variety of foods and can engage more readily in food switching than other primates (excepting humans). They also exhibit relatively fast life histories typified by early age at first birth, short interbirth intervals, and relatively short weaning periods. These attributes permit successful reproduction under highly seasonal conditions or during times of environmental uncertainty (Jablonski et al., 2000). The parallel evolution of these qualities in the Old World monkey and human lineages has been one of the main reasons sustaining consistently high levels of scholarly interest in the fossil record of African Cercopithecoidea over the last half century. African fossil Cercopithecoidea have been offered as models of hominin differentiation, evolution, and behavior in numerous studies, the most enduring and worthy of these being Clifford Jolly's "seed-eater hypothesis" in which Plio-Pleistocene *Theropithecus* was invoked as a model organism in the study of the adaptive radiation of hominids (Jolly, 1970).

The roots of the adaptability and evolutionary success of the Old World monkeys can be traced not only to their dietary flexibility and life histories, but also to their generalized anatomies. Cercopithecoidea are distinguished from Hominoidea in their dentition and skeleton by bilophodont molars, a prominent developmental sulcus on the buccal surface of the male upper canine, the absence or considerable reduction of the paranasal sinuses, and a prominent medial trochlear keel on the distal humerus (Szalay and Delson, 1979; Strasser and Delson, 1987). Cercopithecoids are agile quadrupeds with grasping cheiridia. While a lack of hooves precludes seasonal migrations and reduces potential day range size, dexterous and sensitive fingers permit extraction and harvesting of a wide variety of foods on the ground and in trees. Foods picked with the fingers are further processed in the mouth by spatulate incisors and bilophondont molars. The dental formula for the group is 2.1.2.3 in both jaws. Bilophodont molars, one of the hallmarks of the Cercopithecoidea, are capable of reducing a variety of foods by cutting and crushing, depending on the thickness of the enamel, height of the cusps, steepness of the shearing surfaces, and total occlusal surface area. Cercopithecines, with a few notable exceptions, are eclectic feeders. They manually harvest and eat fruits, seeds, flowers, rhizomes, insects, and even small vertebrates, which they triturate with thick-enameled molars of generally low relief. The living species have ischial callosities and all have cheek pouches, which can provide temporary storage space for high-quality food items. This capability can be particularly important in light of frequently high levels of within-group scramble competition for food. Colobines concentrate on eating plant foods with generally higher fiber and allelochemical content, including young leaves, seeds, and unripe fruits that are harvested by hand and slowly digested with the assistance of bacterial symbionts in a ruminant stomach. Most colobine species exhibit greatly reduced or absent external thumbs. Although they are often referred to as the leaf-eating monkeys, seeds are important components of the diet of most extant species (Lucas and Teaford, 1994), and seed eating may have been one of the selective forces involved in the original differentiation of the group. Molars with thin enamel, high cusp relief, and tall shearing crests permit colobines to finely shred and pulverize ingested vegetation, thus increasing the surface area of material exposed to salivary enzymes (Stewart et al., 1987; Zhang et al., 2002). Foregut fermentation, including recruitment of lysozyme as a bacteriolytic enzyme in the stomach, evolved independently in colobines and ruminant artiodactyls, and it was a major element of the evolutionary success of both groups in the changing environments of the late Miocene and Pliocene. Colobines do not chew the cud, but their ruminant digestive apparatus permits them to chemically transform the complex carbohydrates in vegetation into fatty acids as a source of energy. The prolonged and chemically complex digestive processes in the colobine gut also break down toxic secondary compounds found in seeds and leaves, a process that renders many "nonprimate foods" suitable fare. Ruminant digestion also reduces animals' dependence on drinking water, thus increasing potential niche breadth (Van Soest, 1982).

Fossils of Old World monkeys in Africa have been recovered in significant numbers from a wide variety of sites in northern and sub-Saharan Africa. Readers are referred to recently published reviews for discussions of the history of discovery of fossil Cercopithecoidea in these regions (Benefit and McCrossin, 2002; Jablonski, 2002). The distribution of fossil sites in space and time in Africa is uneven, however, and our knowledge of the past occurrences, evolutionary histories, and adaptations of African Cercopithecoidea is biased and incomplete. The largest concentrations of monkey-bearing sites occur along the Great Rift Valley of eastern and northeastern Africa, and in the breccia-filled caves of the Transvaal region of South Africa and Angola. Most of these sites are of Pliocene and Pleistocene age, but recent finds of late Miocene monkeys from Ethiopia, Kenya, and Chad (Brunet et al., 2002; Hlusko, 2006, 2007; Frost et al., 2009) have shed light on an earlier and little-known phase of the group's evolution. A scattering of Plio-Pleistocene sites yielding monkey fossils also occurs in Algeria, Morocco, and Egypt. The evolutionary proximity and co-occurrence of hominins and cercopithecoids has meant that, at most sites, monkey fossils have been collected completely and prepared promptly, leaving paleontologists with much to study.

The propinquity of the Old World monkeys to humans has led to intensive study of the phylogeny and systematics of the Cercopithecoidea, and numerous modifications to the classification of the group in recent decades. Controversies have arisen in large part because the recency of the group's radiation has contributed to the existence of considerable homoplasy in features of the dentition, skull, and postcranial skeleton of phyletically distant taxa. The increased use of soft tissue and molecular characters, as mentioned above, has overcome some of these problems, and has yielded some considerable surprises for classification of the group in general. Among these are the recognition that mangabeys are diphyletic (Barnicot and Hewtt-Emmett, 1972; Gilbert, 2007b), that the so-called odd-nosed monkeys of Asia comprise a distinct clade (Sterner et al., 2006), recognition of a single "terrestrial" clade among the guenons (Disotell and Raaum, 2002; Tosi et al., 2002a, 2004; Xing et al., 2005, 2007), and that pronounced elongation of the muzzle evolved independently at least four times within the Papionini (in *Mandrillus*, *Papio*, the *Theropithecus brumpti* lineage; Eck and Jablonski, 1987), and the Plio-Pleistocene baboon-like macacines *Paradolichopithecus* and *Procynocephalus* of Eurasia (Szalay and Delson, 1979). With the introduction of better and more stable molecular phylogenies, classifications of the Cercopithecoidea are now being duly revised (Benefit and McCrossin, 2002; Jablonski, 2002), but efforts to fit fossil species into these schemes are still in their infancy. This is because many African monkey fossil species are not clearly related to living species except by suites of shared ancestral features, and because some higher taxa contain only extinct species (e.g., all of the Victoriapithecidae and six of the eight recognized genera of African Colobinae are known only in the fossil record). For these reasons, no attempt has been made here to present a phylogenetic hypothesis or evolutionary tree summarizing the positions of the fossil monkeys of Africa.

INSTITUTIONAL ABBREVIATIONS

AMNH = American Museum of Natural History, New York; ARA-VP = Aramis, Middle Awash, National Museum of Ethiopia; CGM = Cairo Geological Museum, Cairo, Egypt; KA = Kromdraai A, Department of Palaeontology, Transvaal Museum, Pretoria, South Africa; KNM = National Museums of Kenya, Nairobi, Kenya (suffixes – BC = Baringo Chemeron; –BN = Baringo Ngeringerowa; -ER = East Turkana; -KP = Kanapoi; -LT = Lothagam; -MB = Maboko; -NK = Narok District, including Lemudong'o; WT = West Turkana); M = British Museum (Natural History), London; M (and MP) = Makapansgat, Department of Anatomy, University of the Witwatersrand and Bernard Price Institute, Johannesburg, South Africa; NME = National Museum of Ethiopia, Addis Ababa, Ethiopia; SAM = Iziko South African Museum, Capetown, South Africa; SB = Schurweberg (Skurweberg, Skurveberg), Department of Palaeontology, Transvaal Museum, Pretoria, South Africa; SK = Swartkrans, Department of Palaeontology, Transvaal Museum, Pretoria, South Africa; STS = Sterkfontein (pre-1966), Department of Palaeontology, Transvaal Museum, Pretoria, South Africa; SWP = Sterkfontein (post-1966) = Department of Anatomy, University of the Witwatersrand, Johannesburg, South Africa; and YPM = Yale Peabody Museum.

Systematic Paleontology

The classification of the Cercopithecoidea followed here (table 23.1) is based on that used in previous reviews (Benefit and McCrossin, 2002; Jablonski, 2002), duly updated to reflect recent fossil discoveries and phylogenetic interpretations. Interpretations of the geological contexts and ages of many African fossil Cercopithecoidea have been controversial (table 23.2); a useful review of this information for East African fossils has been published elsewhere and should be consulted (Gundling and Hill, 2000).

No tribes are recognized within the Colobinae because of ongoing study of the phyletic relationships between genera within the subfamily. Old World monkeys and their fossils have been studied by many people over more than a century, and, as a consequence, the synonymies for many species are lengthy and complex. Considerations of length preclude their reproduction in full here, so the reader is commended to other sources for this information (Freedman, 1957; Simons and Delson, 1978; Szalay and Delson, 1979; Groves, 2001). Photographs of specimens are by the authors unless otherwise stated.

Order PRIMATES Linnaeus, 1758
Family VICTORIAPITHECIDAE von Koenigswald, 1969

Victoriapithecidae is an extinct family of basal cercopithecoids. Victoriapithecids differ from other Cercopithecoidea in the possession of a lower and narrow neurocranium, supraorbital costae, a frontal trigon formed in part by the anterior convergence of the temporal lines, orbits that are taller than wide, a deep malar region, variable occurrence of the crista obliqua on the deciduous P4–M3, absence of transverse distal lophs on the upper molars and deciduous upper premolars, and other features (Benefit, 1993, 1999; Benefit and McCrossin, 2002). Although the nominate genus is represented by abundant well-preserved fossils from Maboko Island (Kenya), *Prohylobates* is not, and considerable uncertainty remains over the morphology and taxonomic status of some of the earliest Old World monkeys. Renewed study of victoriapithecids from Egypt and northern Kenya (including fossils originally classified as *Prohylobates* sp. nov. from Buluk, Kenya; Leakey, 1985; Miller et al., 2009) has led to a forthcoming revision of the family that promises to shed light on these problems.

Genus *PROHYLOBATES* Fourtau, 1918

Diagnosis Delson (1979); Miller et al. (2009). Distinguished from *Victoriapithecus* by the possession of lower permanent molars exhibiting incomplete bilophodonty, and a reduced M3 hypoconulid.

Description *Prohylobates* is known from relatively few fossil jaws and teeth from the type locality of Wadi Moghara in Egypt and Jabal Zaltan, Libya.

Age Early–middle Miocene.

African Occurrence Northern Africa.

Remarks Poor preservation of most specimens has hindered detailed comparison with *Victoriapithecus* and assessment of the position of the genus within its family. Despite these difficulties, the molar morphology of available specimens indicates incomplete development of bilophodonty, and the conclusion that the genus is more primitive than *Victoriapithecus* (Benefit, 2009).

Northern Africa was a center of catarrhine diversification from the Oligocene through the middle Miocene. The region's importance in cercopithecoid evolution was great and is increasingly recognized as such as more fossils of middle Miocene age are recovered from sites in Egypt and Libya. Logistical and physical difficulties continue to hamper this effort throughout the region, however.

PROHYLOBATES SIMONSI Delson, 1979

Diagnosis Miller et al. (2009). Differs from other species of *Prohylobates* in its much larger size and possession of M2 and M3 of equal size. Its distinctive morphology may warrant placement in a new genus.

Description The species is known from the holotype, AMNH 17768, a partial mandible.

Age Middle Miocene.

African Occurrence Jabal Zaltan (= Gebel Zelten)(Libya).

Remarks This species shares with other Victoriapithecidae a quadrate arrangement of cusps on a waisted M2, pronounced molar flare, and molar wear that begins as circular depressions on cusp tips (Benefit and McCrossin, 2002).

PROHYLOBATES TANDYI Fourtau, 1918

Diagnosis *Prohylobates tandyi* exhibits smaller molar teeth than most members of its family and differs from other species of its genus in its possession of a P4 which is large relative to the lower molars, and a very small M3 relative to

Order . . . . Primates Linnaeus, 1758

Infraorder . . . . Catarrhini É. Geoffroy
Saint-Hilaire, 1812

Superfamily. . . . . Cercopithecoidea Gray, 1821

Family . . . . Victoriapithecidae von
Koenigwald, 1969

Genus . . . . *Prohylobates* Fourtau, 1918
Prohylobates simonsi Delson,
1979
Prohylobates tandyi Fourtau,
1918

Genus . . . . *Victoriapithecus* von
Koenigswald, 1969
Victoriapithecus macinnesi
von Koenigswald, 1969

Family . . . . Cercopithecidae Gray, 1821

Subfamily . . . . Colobinae Jerdon, 1867

Genus . . . . *Cercopithecoides* Mollett, 1947
Cercopithecoides kerioensis
M. G. Leakey, Teaford, and
Ward, 2003
Cercopithecoides kimeui M. G.
Leakey, 1982
Cercopithecoides meaveae Frost
and Delson, 2002
Cercopithecoides williamsi
Mollett, 1947
Cercopithecoides alemayehui
Gilbert and Frost, 2008
cf. *Cercopithecoides*

Genus . . . . *Colobus* Illiger, 1811
Colobus freedmani Jablonski
and M.G. Leakey, 2008
Colobus guereza Rüppell,
1835 *Colobus* sp. indet.

Genus . . . . *Kuseracolobus* Frost 2001
Kuseracolobus aramisi Frost,
2001
Kuseracolobus hafu Hlusko,
2006

Genus . . . . *Libypithecus* Stromer, 1913
Libypithecus markgrafi
Stromer, 1913

Genus . . . . *Microcolobus* Benefit and
Pickford, 1986
Microcolobus tugenensis Benefit
and Pickford, 1986

Genus . . . . *Paracolobus* R. E. F. Leakey,
1969
Paracolobus chemeroni R. E. F.
Leakey, 1969
Paracolobus enkorikae Hlusko,
2007
Paracolobus mutiwa M. G.
Leakey, 1982
Paracolobus sp. indet.

Genus . . . . *Rhinocolobus* M. G. Leakey,
1982
Rhinocolobus turkanensis
M. G. Leakey, 1982

Subfamily . . . . Cercopithecinae Gray, 1821

Tribe . . . . Cercopithecini Gray, 1821

Genus . . . . *Cercopithecus* Brunnich, 1772
Cercopithecus sp. indet.

Genus . . . . . . *Chlorocebus* Gray, 1879
Chlorocebus cf. *patas*
cf. *Chlorocebus* aff. *aethiops*

Tribe . . . . Papionini Burnett, 1828

Subtribe. . . . . . . Macacina Owen, 1843

Genus . . . . . . *Macaca* Lacépède, 1799
Macaca libyca Stromer,
1920
Macaca sylvanus Linnaeus,
1758

Subtribe. . . . . . . Papionina Burnett, 1828

Genus . . . . . . *Cercocebus* É. Geoffroy
Saint-Hilaire, 1812
Cercocebus sp. indet.

Genus . . . . . . *Dinopithecus* Broom, 1937
Dinopithecus ingens Broom,
1937

Genus . . . . . . *Gorgopithecus* Broom and
Robinson, 1949
Gorgopithecus major Broom,
1940

Genus . . . . . . *Lophocebus* Palmer, 1903
Lophocebus cf. *albigena* Gray,
1850

Genus . . . . . . *Papio* Müller, 1773
Papio hamadryas Linneaus,
1758
Papio izodi Gear, 1926

Genus . . . . . . *Parapapio* Jones, 1937
Parapapio ado Hopwood, 1936
Parapapio broomi Jones, 1937
Parapapio jonesi Broom, 1940
Parapapio lothagamensis M. G.
Leakey, Teaford, and Ward,
2003
Parapapio sp. indet.

Genus . . . . . . *Pliopapio* Frost 2001
Pliopapio alemui Frost,
2001

Genus . . . . . . *Procercocebus* Gilbert, 2007
Procercocebus antiquus
(Haughton, 1925)

Genus . . . . . . *Theropithecus* I. Geoffroy
Saint-Hilaire, 1843

Subgenus . . *Theropithecus* Geoffroy
Saint-Hilaire, 1843
*Theropithecus (Theropithecus)
darti* (Broom and Jensen,
1946)
*Theropithecus (Theropithecus)
oswaldi* Andrews, 1916

Subgenus . . *Omopithecus* Delson, 1993
*Theropithecus (Omopithecus)
baringensis* (R. Leakey, 1969)
*Theropithecus (Omopithecus)
brumpti* (Arambourg, 1947)
*Theropithecus (Omopithecus)
quadratirostris* (Iwamoto,
1982)
Theropithecus (Omopithecus)
sp. indet.

Subgenus . . *Theropithecus* sp. indet.

TABLE 23.2
Summary of major site occurrences and ages for African fossil Cercopithecoidea
Age ranges provided are based on the most recent chronometric estimates available; otherwise age has been estimated only to epoch or subepoch.
See text for details and complete list of site occurrences.

Taxon	Major Site Occurrences	Age	Key References
	VICTORIAPITHECIDAE		
Prohylobates simonsi	Jabal Zaltan [= Gebel Zelten] (Libya)	Middle Miocene	Delson, 1979
Prohylobates tandyi	Wadi Moghra (Egypt)	Early Miocene	
Victoriapithecus macinnesi	Maboko Island (Kenya), Napak (Uganda)	19.5 Ma (Napak); ~15–12.1 Ma (Maboko)	Benefit, 1993, 1999; Benefit and McCrossin, 2002
	CERCOPITHECIDAE: COLOBINAE		
Cercopithecoides kerioensis	Lothagam (Kenya)	5–4.2 Ma	Leakey et al., 2003
Cercopithecoides kimeui	East Turkana and Rawi Gulley (Kenya); Hadar (Ethiopia)	3–1.5 Ma	Frost et al., 2003b; Jablonski et al., 2008a
Cercopithecoides meaveae	Hadar (Ethiopia)	3.4–3.28 Ma	Frost and Delson, 2002
Cercopithecoides williamsi	East Turkana (Kenya); Sterkfontein, Makapansgat, Swartkrans, and Kromdraai (South Africa)	3.3–1.5 Ma	Freedman, 1957; Heaton, 2006; Jablonski et al., 2008a
Colobus freedmani	East Turkana (Kenya)	1.527–1.485 Ma	Jablonski and Leakey, 2008b
Colobus guereza	Wad (Wadi) Medani (Sudan)	Pleistocene	Simons, 1967; Frost and Alemseged, 2007
Colobus sp. indet.	Kanam East (Kenya); Omo Group, Afar, and Middle Awash (Ethiopia); Kazinga (Uganda)	Plio-Pleistocene; late Pleistocene	Frost and Alemseged, 2007
Kuseracolobus aramisi	Middle Awash and Gona (Ethiopia)	5.2–4.4 Ma	Frost, 2001b
Kuseracolobus hafu	Asa Issie (Ethiopia)	4.4–3.75 Ma	Hlusko, 2006
Libypithecus markgrafi	Wadi Natrun (Egypt)	Latest Miocene	Stromer, 1913; Delson, 1975
Microcolobus tugenens	Ngeringerowa (Kenya)	9.5–9.0 Ma	Benefit and Pickford, 1986
Paracolobus chemeroni	Chemeron (Kenya); Middle Awash (Ethiopia)	3–2 Ma	Leakey, 1969; Birchette, 1982
Paracolobus enkorikae	Lemudong'o (Kenya)	6 Ma	Hlusko, 2007
Paracolobus mutiwa	East and West Turkana (Kenya)	3.36–1.88 Ma	Leakey, 1982
Paracolobus sp. indet.	Laetoli (Tanzania); Western Rift (Uganda)	3.8–?2.5 Ma	Leakey and Delson, 1987; Senut, 1994
Rhinocolobus turkanensis	Omo Group (Ethiopia); East Turkana (Kenya)	3.4–1.5 Ma	Leakey, 1982; Jablonski et al., 2008a
	CERCOPITHECIDAE: CERCOPITHECINAE: CERCOPITHECINI		
Cercopithecus sp. indet.	Omo Group (Ethiopia); East Turkana and Kanam East (Kenya)	Pliocene and Pleistocene	Eck and Howell, 1972; Harrison and Harris, 1996; Jablonski et al., 2008b
Chlorocebus cf. *patas* cf. *Chlorocebus* aff. *aethiops*	Asbole and Middle Awash (Ethiopia)	Pleistocene	Frost, 2001a; Frost and Alemseged, 2007
	CERCOPITHECIDAE: CERCOPITHECINAE: PAPIONINI		
Macaca sylvanus	North Africa	Plio-Pleistocene	
Macaca libyca	Wadi Natrun (Egypt)	Latest Miocene	Stromer, 1913; Delson, 1980
Cercocebus sp. indet.	Makapansgat, Kromdraai, and ?Swartkrans (South Africa)	Late Pliocene	Gilbert, 2007b
Dinopithecus ingens	Schurweberg and Swartkrans (South Africa)	Late Pliocene	Broom, 1937; Freedman, 1957; Szalay and Delson, 1979
Gorgopithecus major	Kromdraai (South Africa)	Plio-Pleistocene	Broom and Robinson, 1949; Freedman, 1957; Heaton, 2006
Lophocebus cf. *albigena*	East Turkana (Kenya)	?2.0–1.38 Ma	Jablonski et al., 2008b
Lophocebus sp. indet.	Kanam East (Kenya)	Plio-Pleistocene	Harrison and Harris, 1996
Papio hamadryas	Sterkfontein, Kromdraai, Drimolen, Swartkrans, and Bolt's Farm (South Africa); Olduvai Gorge (Tanzania); Asbole, (Ethiopia)	Pleistocene	Freedman, 1976; Frost and Alemseged, 2007

TABLE 23.2

(CONTINUED)

Taxon	Major Site Occurrences	Age	Key References
Papio izodi	Taung, Sterkfontein Members 2 and 4, Kromdraai, and Coopers, (South Africa)	Plio-Pleistocene	Freedman, 1957; Szalay and Delson, 1979; Heaton, 2006
Parapapio ado	Laetoli (Tanzania); East Turkana (Kenya)	Pliocene	Leakey and Delson, 1987; Jablonski et al., 2008b
Parapapio broomi	Sterkfontein and Bolt's Farm (South Africa)	Pliocene	Jones, 1937; Broom, 1940; Freedman, 1957; Heaton, 2006
Parapapio jonesi	Makapansgat, Sterkfontein (South Africa); Afar and Middle Awash (Ethiopia)	Pliocene	Freedman, 1957; Frost, 2001b; Frost and Delson, 2002
Parapapio lothagamensis	Lothagam (Kenya)	7.4–ca. 5 Ma	Leakey et al., 2003; Jablonski et al., 2008b
Parapapio sp. indet.	East and West Turkana and Lothagam (Kenya)	Plio-Pleistocene	Jablonski et al., 2008b
Pliopapio alemui	Middle Awash (Ethiopia)	?5.7–4.2 Ma	Frost, 2001b; Frost et al., 2009
Procercocebus antiquus	Taung (South Africa)	Plio-Pleistocene	Freedman, 1957; Gilbert, 2007b
Theropithecus (Theropithecus) darti	Omo Group, Middle Awash, and Hadar (Ethiopia); Makapansgat (South Africa)	3.5–2.4 Ma	Freedman, 1957, 1976; Eck, 1993; Frost, 2001b
Theropithecus (Theropithecus) oswaldi	Ain Jourdel and Ternifine (Algeria); Thomas Quarries (Morocco); Omo Group, Middle Awash, Afar Region, and Konso (Ethiopia); East and West Turkana, Olorgesailie, and Kapthurin (Kenya); Olduvai Gorge and Peninj (Tanzania); Kaiso (Uganda); and Swartkrans, Sterkfontein, Hopefield, and Bolt's Farm (South Africa)	2.5–0.25 Ma	Jolly, 1972; Dechow and Singer, 1984; Eck, 1987; Delson, 1993; Delson et al., 1993; Delson and Hoffstetter, 1993; Leakey, 1993; Jablonski et al., 2008b
Theropithecus (Omopithecus) baringensis	Chemeron (Kenya); Leba (Angola)	3–2 Ma	Leakey, 1969; Eck and Jablonski, 1984; Delson and Dean, 1993
Theropithecus (Omopithecus) brumpti	East and West Turkana (Kenya); Omo Group (Ethiopia)	3.4–2.68 Ma	Eck and Jablonski, 1987; Leakey, 1993; Jablonski et al., 2002
Theropithecus (Omopithecus) quadratirostris	Omo Group (Ethiopia)	~3 Ma	Iwamoto, 1982; Eck and Jablonski, 1984; Delson and Dean, 1993
Theropithecus (Omopithecus) sp. indet.	East Turkana (Kenya)	3.94 Ma	Jablonski et al., 2008b
Theropithecus subgen. et sp. indet.	Middle Awash (Ethiopia)	3.9 Ma	Frost, 2001a

M2 and equal to M1 (Benefit and McCrossin, 2002; Miller et al., 2009).

Description This species is known from few specimens, including the holotype, CGM 30936, and a newly referred specimen, DPC 6235, both mandible fragments (Benefit and McCrossin, 2002).

Age Early Miocene.

African Occurrence Wadi Moghra (Egypt).

Remarks Understanding of this important species has been hampered by a small sample of specimens, all of which have abraded teeth.

Genus *VICTORIAPITHECUS* von Koenigswald, 1969
VICTORIAPITHECUS MACINNESI von Koenigswald, 1969
Figure 23.1

Diagnosis Benefit and McCrossin (2002). Distinguished from *Prohylobates* by a cercopithecine-like inferior transverse torus on the mandible, P4 small compared to molars,

presence of metaconid on p3, molar size gradient M1 < M2 > M3 and m1 < m2 < m3.

Description The skull of *Victoriapithecus* is known from many, mostly fragmentary craniodental remains, the best of which are the male cranium, KNM-MB 29100 (figure 23.1), maxillae KNM-MB 18995 and KNM-MB 18996, the mandible KNM-MB 18993.

Age Early–middle Miocene, 19.5 Ma (Napak); and ~15–12.1 Ma (Maboko Formation)(Gundling and Hill, 2000; Benefit and McCrossin, 2002).

African Occurrence Numerous sites in Kenya, especially Maboko Island (Kenya), Napak (Uganda), and possibly Ongoliba (Democratic Republic of the Congo), as enumerated elsewhere (Benefit and McCrossin, 2002) but not universally accepted (Delson, 1979).

Remarks At present, the genus contains the single species, *V. macinnesi*, which is the best known of the Victoriapithecidae. It exhibits a distinctive mosaic of cercopithecine-like and colobine-like traits that are consistent with its age and its status

FIGURE 23.1 *Victoriapithecus macinnesi*. Lateral view of male cranium KNM-MB 29100 with mandible KNM-MB 18993. Photograph courtesy of Brenda Benefit; © National Museums of Kenya.

as a generalized basal monkey. The cranium is narrower and lower relative to length than those of other cercopithecoids and exhibits slight ventral deflection of the cranial vault relative to the basicranium (klinorhynchy)(Benefit and McCrossin, 1997). It resembles those of cercopithecines in its narrow interorbital region and narrow nasal bones, low and narrow nasal aperture, moderately long and anteriorly tapering snout, and moderately long premaxilla (Benefit and McCrossin, 1997, 2002). The molar teeth *Victoriapithecus* are bilophodont but lack most of the derived specializations seen in the molars of Cercopithecinae and Colobinae, as thoroughly described by Benefit (Benefit, 1993; Benefit and McCrossin, 2002). The postcranial elements known for the species suggest that it was a mostly terrestrial quadruped, as reviewed elsewhere (Harrison, 1989).

Family CERCOPITHECIDAE Gray, 1821
Subfamily COLOBINAE Jerdon, 1867

Medium- to very large-sized monkeys with reduced or absent thumbs in most species, long tails, bilophodont molars with long shearing crests and steep sides in most species, and a sacculated stomach accommodating bacterial symbionts for foregut fermentation. The distal humerus, which is often preserved as a fossil, is distinguished in colobines by medial and lateral pillars flanking the olecranon fossa that are approximately equal in width and are relatively flat, giving the dorsal surface of the distal humerus a flattened look.

Genus CERCOPITHECOIDES Mollett, 1947

Diagnosis Distinguished from other colobines by its shallow mandibular corpus with a slightly convex inferior border in most specimens, a shallow and relatively thin mandibular symphysis, a nonexpanded gonion, and a low ramus oriented obliquely relative to the occlusal plane. Differs from other colobines in its squared muzzle and strong terrestrial adaptations in the postcranium, especially the forelimb. Differs from *Rhinocolobus*, *Libypithecus*, and *Nasalis* in its short muzzle and short, rounded calvaria. It differs from Asian colobines, *Paracolobus*, and *Rhinocolobus* in the absence of a protocone on the P3.

Description The earliest occurrences of the genus are fragmentary gnathic and postcranial remains in early Pliocene exposures in the East African Rift Valley recognized as cf. *Cercopithecoides* and *C. kerioensis*. The genus is best known from plentiful remains of *C. williamsi* retrieved from the Plio-Pleistocene limestone breccias of most of the South African cave sites, where it is one of the few fossil colobines known until the late Pleistocene (see *Colobus*, later). In East Africa, *C. williamsi* appears to have been rare but is represented by one partial skeleton in good condition from Koobi Fora, KNM-ER 4420. More common and widespread in the East African Plio-Pleistocene is *C. kimeui*, which is significantly larger. It is represented by several crania and mandibles, most of which bear heavily worn teeth (Jablonski et al., 2008a). *Cercopithecoides meaveae* is smaller than *C. williamsi* and known only from the middle Pliocene of the Afar region. *Cercopithecoides kerioensis* from the early Pliocene of Lothagam is the oldest known member of the genus.

Age Pliocene and early Pleistocene.

African Occurrence Eastern and southern Africa.

Remarks In most species, the face of *Cercopithecoides* is wide, and the large, widely spaced orbits are rectangular. The nasal aperture is small and the nasal bones are moderately long. The supraorbital torus is thick; the postorbital constriction is not marked. Sexual dimorphism appears low to moderate. The molars bear thin enamel and exhibit relatively high, columnar cusps with capacious basins and large foveae. The canines and P3 are sexually dimorphic. The postcranium is fairly well-known and shows many features characteristic of terrestrial cercopithecoids, especially in the forelimb.

Cercopithecoides was an agile and committed terrestrialist, and it shows many of the specializations to mostly terrestrial locomotion seen today in forms of *Papio* baboons in Africa and in temple langurs *(Semnopithecus entellus)* in Asia. The unique shallow but thick mandibular morphology in later species of the genus suggests that these forms of *Cercopithecoides* ate foods of a different consistency than those known for other colobines. In craniodental morphology, early species of *Cercopithecoides* most closely resemble *Colobus*. The similarities are strongest between *Colobus* and the earliest, most primitive forms of *Cercopithecoides*,

C. kerioensis from Lothagam and the specimens of the genus from early Pliocene horizons of the Koobi Fora Formation that share deep and relatively thin mandibles. Although *Colobus* and *Cercopithecoides* differ in limb proportions, they share many similarities in the morphology of the proximal and distal humerus and femur. These morphological resemblances suggest that modern *Colobus* may have descended from an early, primitive species of *Cercopithecoides*. This hypothesis warrants testing.

Cercopithecoides was the dominant colobine of the African Pliocene, but it became less common by the later Pliocene and earlier Pleistocene, and extinct by the middle Pleistocene. From the *Cercopithecoides* lineage arose severally regionally restricted species of different body sizes and locomotor propensities.

CERCOPITHECOIDES KERIOENSIS Leakey et al., 2003

Diagnosis Differs from *C. kimeui* and *C. williamsi* males in its small size, thin supraorbital tori, narrow interorbital width, strong nuchal crests, a sagittal crest close to inion, and a relatively short and deep mandibular corpus (Leakey et al., 2003). Differs from *C. meaveae* in its deeper and more sloping mandibular symphysis (Frost and Delson, 2002).

Description The species is represented by the holotype, KNM-LT 9277, a male partial skull.

Age Early Pliocene, ca. 5–4.2 Ma.

African Occurrence Lothagam (Kenya), probably the Apak Member of the Nachukui Formation.

Remarks Cercopithecoides kerioensis is the earliest known species of its genus. The species shares a deep and relatively thin-bodied mandible with other representatives of the genus from the Lonyumun and Lokochot members of the Koobi Fora Formation, and it is likely that this morphology is primitive for the genus (Leakey, 1982).

CERCOPITHECOIDES KIMEUI Leakey, 1982
Figure 23.2

Diagnosis Larger than *C. williamsi* with a globular calvaria that is flatter than that of *C. williamsi* (Leakey, 1982). The mandible is more robust than that of *C. williamsi*, especially in the thickness of the mandibular corpus and symphysis. Upper molars lower crowned than those of other colobines.

Description This species is represented by the holotype male calvaria and maxillae from middle Bed II Olduvai Gorge, Tanzania, two partial female crania and a series of other cranial and postcranial specimens from the Upper Burgi and KBS members of the Koobi Fora Formation (Jablonski et al., 2008a), a female cranium and mandible from a latest Pliocene to earliest Pleistocene horizon at Hadar, Ethiopia (Frost and Delson, 2002), and a male face and mandible from Rawi Gulley, Kenya (Frost et al., 2003b; figure 23.2). An isolated lower molar from Bed III at Olduvai Gorge, if it represents *C. kimeui*, would extend the known range up to as young as 1.2 Ma.

Age Plio-Pleistocene, 2.6–1.5 Ma.

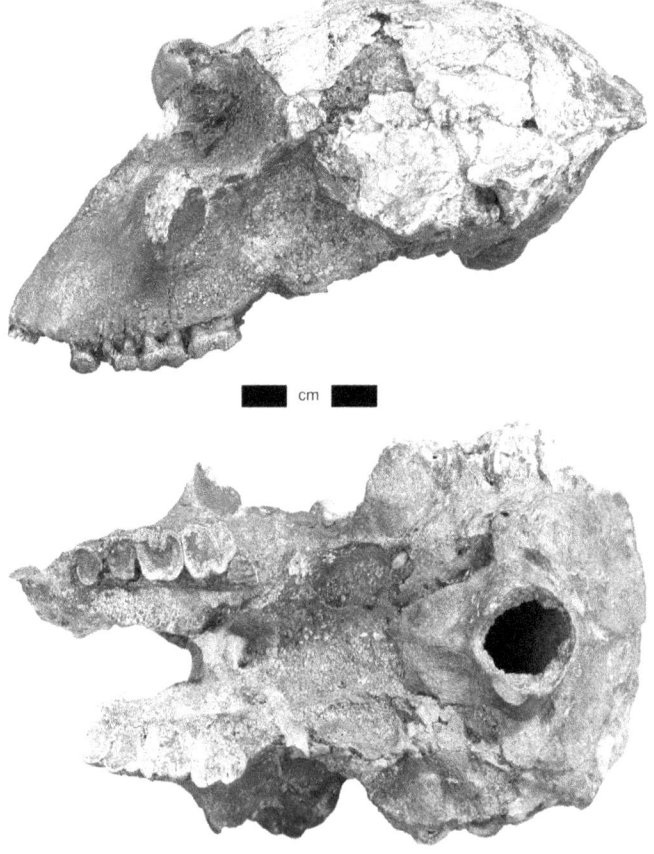

FIGURE 23.2 *Cercopithecoides kimeui*. Lateral and basal views of female cranium KNM-ER 398. The heavy wear on the cheek teeth is characteristic of adults of this species. © National Museums of Kenya.

African Occurrence Olduvai Gorge (Tanzania), East Turkana and Rawi Gulley (Kenya), Hadar (Ethiopia).

Remarks A very large colobine monkey with females and males estimated at approximately 25 and 50 kg, respectively (Delson et al., 2000; Frost and Delson, 2002). The muzzle is relatively narrow and inflated just below the infraorbital margin, as in *Lophocebus*. The mandible is shallow and thick, and supported broad and puffy molar crowns of low relief. The weakly columnar cusps of the molars were covered in thin enamel and wore down quickly to functional obsolescence (Jablonski et al., 2008a). Judging by the nature and degree of tooth wear in most specimens, *C. kimeui* ate a highly abrasive diet. Although its mandible is thick, it lacks the deep and strongly reinforced mandibular corpus and symphysis and large muscles of mastication found in monkeys that chew tough and highly fibrous vegetation or a lot of vegetation. It is possible that *C. kimeui* had a relatively soft but highly abrasive diet that has no analogue among living monkeys but that included high percentages of fruits and leaves covered with grit or containing high concentrations of phytoliths (Benefit, 2000). The elbow joint of *C. kimeui* suggests that the species was highly terrestrial but exhibited considerable forearm flexibility consistent with an adaptation in the species to manipulation of food objects with the hand. This intriguing species is described and discussed at greater length elsewhere (Jablonski et al., 2008a).

CERCOPITHECOIDES MEAVEAE Frost and Delson, 2002

Diagnosis Emended after Frost and Delson (2002). Smaller than either *C. williamsi* or *C. kimeui*, and comparable to the extant species *Nasalis larvatus* in body size. Similar to *C. kerioensis*, but different from *C. kimeui* and *C. williamsi* in the absence of a median mental foramen. The mandibular symphysis is more vertically oriented than that of *C. kerioensis*.

Description This species is known from the holotype partial skeleton from Leadu as well as a maxilla and male mandible from the Sidi Hakoma Member of the Hadar Formation (Frost and Delson, 2002). It exhibits the relatively shallow and robust mandible typical of the genus. The known postcrania (including a tentatively assigned distal humeral fragment from Hadar) exhibit features of the shoulder, elbow, and hip joints associated with terrestrial locomotion (Frost and Delson, 2002).

Age Middle Pliocene, 3.4–3.28 Ma.

African Occurrence Leadu and Hadar (Ethiopia).

Remarks This species is younger than *C. kerioensis* and generally older than most of the material allocated to *C. williamsi* and *C. kimeui*. It is generally similar to *C. williamsi* in many aspects of its cranial, dental, and postcranial morphology, although it is not as extreme in its adaptations to a terrestrial lifestyle as is *C. williamsi* from Koobi Fora (Frost and Delson, 2002). Several isolated teeth of similar size to *C. meaveae* are known from the Omo (Shungura Mbs. B–G), Ethiopia, as well as from the upper Laetolil Beds, Tanzania. These specimens might represent *C. meaveae*, but at this point are best considered of indeterminate affinity.

CERCOPITHECOIDES WILLIAMSI Mollett, 1947
Figure 23.3

Diagnosis Larger in body size than all members of the genus except for *C. kimeui*, being similar in cranial and dental size to *Rhinocolobus*. Molar teeth possess higher crowns than those of *C. kimeui*. Differentiated from *C. meaveae* and *C. kerioensis* by the presence of a medial mental foramen. Mandibular corpus is shallower and thicker than that of *C. kerioensis*.

Description *Cercopithecoides williamsi* is the only fossil colobine known from South Africa and occurs at many Plio-Pleistocene cave sites in the country. Among the most complete cranial specimens are an associated female cranium

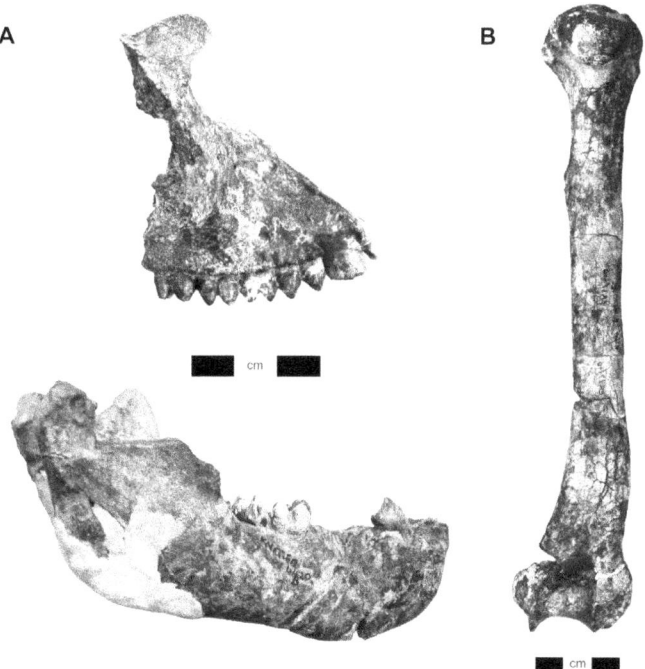

FIGURE 23.3 *Cercopithecoides williamsi*. Selected elements of partial male skeleton KNM-ER 4420. A) Lateral view of right maxilla and mandible; B) dorsal view of humerus. © National Museums of Kenya.

and mandible from Bolt's Farm (BF 56784) a female cranium from Sterkfontein (STS 394A), and two male crania from Makapan (MP 113 [= M2999], and BPI M3055)(Freedman, 1957; Maier, 1971; Jablonski, 2002). In East Africa, the species is recognized only from East Turkana from whence the most complete specimen of the species, the partial skeleton KNM-ER 4420 (figure 23.3), has been recovered (Jablonski et al., 2008a).

Age Plio-Pleistocene, 3.5–1.5 Ma.

African Occurrence Lokochot, upper Burgi, and KBS members of the Koobi Fora Formation, East Turkana (Kenya); Leba (Angola); and Taung, Sterkfontein, Makapansgat, Swartkrans, Kromdraai faunal site, Coopers, Swartkrans II, Graveyard, Bolt's Farm, and Haasgat (South Africa). In South Africa, *C. williamsi* co-occurs with all fossil papionin species (Heaton, 2006; Jablonski et al., 2008a).

Remarks *Cercopithecoides williamsi* is a colobine with a large and rounded calvaria, a short, relatively narrow and rounded muzzle, a wide face, large widely spaced rectangular orbits, and a thick supraorbital torus. Sexual dimorphism in cranial shape is apparent in the relatively long and narrow crania of the males that contrast to the females with shorter crania and more rounded calvariae. The shoulder and elbow joint show features (including a massive, superiorly projecting greater tuberosity of humerus, a broad, flange-bearing humeral trochlea and retroflexed olecranon) associated with stability that are characteristic of terrestrial cercopithecoids (Jablonski et al., 2008a).

The skeleton of *C. williamsi* presents an interesting mixture of features, some of which are similar to living species and others that are not. The cranium and dentition of the species are quite unlike those of other colobines, particularly in the construction of the masticatory apparatus. The gracile construction of the jaws suggests that the species may have subsisted on foods that did not require strong occlusal forces or highly repetitive chewing to be processed, such as unripe fruits and young leaves. The postcranium of *C. williamsi* exhibits similarities to modern large-bodied colobines and cercopithecines that spend most of their time on the ground, but forage and sleep in trees.

A comparative study of the South and East African morphs of the species is long overdue, in order to determine whether they represent distinct species or members of one widespread and geographically variable species such as the modern temple langur, *Semnopithecus entellus*, from southern Asia.

CERCOPITHECOIDES ALEMAYEHUI Gilbert and Frost, 2008

Diagnosis Gilbert and Frost (2008). Smaller in size than *C. williamsi* and *C. kimeui*. Supraorbital torus more projecting than than of *C. kerioensis* or *C. meaveae*. Nasal bones longer than those of all other species of *Cercopithecoides*, extending inferior to the orbital rim.

Description This species is known from a single adult male calvaria and maxilla. In cranial size it is comparable to *C. kerioensis* and *C. meaveae*, but it has relatively larger and squarer upper molars. The holotype shows some evidence of healed cranial trauma.

Age Pleistocene, 1.0 Ma.

African Occurrence Daka Member of the Bouri Formation, Middle Awash, Ethiopia.

Remarks This taxon represents the youngest member of this long-lived genus and possibly the smallest. It may represent a cranially autapomorphic relict.

cf. CERCOPITHECOIDES

Diagnosis Jablonski et al. (2008a). Smaller in overall size and lacking the distinctly thick and shallow mandibular corpora of other species of *Cercopithecoides*, but referred to the genus because of the strong similarities in molar shape and cusp configuration. Postcrania share details of humeral and femoral morphology with *C. williamsi*.

Description A large number of isolated teeth, fragmentary mandibles, and isolated partial long bones, mostly deriving from exposures of the Lonyumun Member at Area 261–A in Allia Bay, east of Lake Turkana are referred to cf. *Cercopithecoides*.

Age Pliocene, 3.95–3.45 Ma.

African Occurrence Upper Lonyumun and lower Lokochot members of the Koobi Fora Formation (Kenya).

Remarks The specimens assigned to cf. *Cercopithecoides* were derived from animals of moderate size, probably closely comparable to modern black-and-white colobus monkeys based on mandibular and dental dimensions. The assemblage may include two species distinguished by male canine morphology, one of which bears a close resemblance to *Cercopithecoides kerioensis* from Lothagam.

Genus COLOBUS Illiger, 1811

Diagnosis Colobine monkeys with small heads and teeth typical of the subfamily. Unlike *Procolobus*, the cranial vault lacks an anterior sagittal crest, the supraorbital ridge is thin with small supraorbital foramina or notches, and the pterygoid fossae are not perforate. The distal lophid of the lower third molar is broader than the mesial. Limbs are relatively long, with greatly reduced or absent thumbs, and shortened tarsus. Body size is larger than *Procolobus (Procolobus)* and *Presbytis*, but smaller than *Nasalis* and larger subspecies of *Semnopithecus*.

Description Remains of *Colobus* similar or identical to living black-and-white colobus monkeys are abundant at several sites in the Afar region of Ethiopia, particularly in the middle Pleistocene (Kalb et al., 1982a; Frost, 2001a, 2001b; Frost and Alemseged, 2007).

Age Pliocene–present.

African Occurrence Northern and eastern Africa.

Remarks Most fossils of *Colobus* are found in Pleistocene strata of Africa where they sometimes co-occur with the mangabey, *Lophocebus* cf. *albigena*, the baboons *Theropithecus oswaldi* and *Papio hamadryas*, and the vervet *Chlorocebus aethiops*. The notable exception to this is the fragment of a right mandibular fragment with teeth from Kanam East, Kenya; the deposits may be of early Pliocene age (Harrison and Harris, 1996). Species of *Colobus* are common in forests in much of sub-Saharan Africa today, including riparian habitats within arid regions. *Colobus guereza* lives in the Afar region today and occurred throughout much of Ethiopia in modern times (Napier, 1985). As discussed, it is possible that *Colobus* descended from an early form of *Cercopithecoides*, on the basis of morphological similarities between the two in the mandible, and proximal and distal humerus and femur.

COLOBUS FREEDMANI Jablonski and Leakey, 2008
Figure 23.4

Diagnosis Swindler and Orlosky (1974); Jablonski and Leakey (2008b). Smaller than modern *Colobus guereza* and

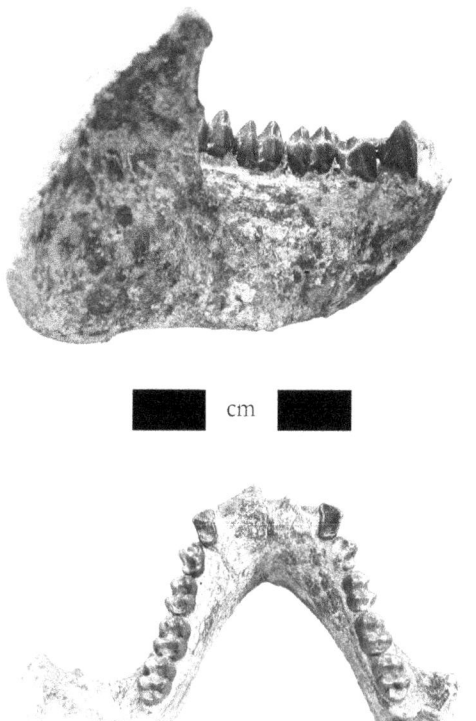

FIGURE 23.4 *Colobus freedmani*. Lateral and occlusal views of probable female mandible KNM-ER 44224 (holotype). © National Museums of Kenya.

distinguished from it by a narrower and slightly more sectorial lower fourth premolar with a small, distally offset lingual cusp. The size and position of the lingual cusp (metaconid) of the P4 distinguishes colobines from cercopithecines, and further distinguishes African from Asian colobines; in *C. freedmani*, the metaconid is small and distally offset, a condition which is uncommon in modern *C. guereza*.

Description The species is best represented by the holotype mandible, KNM-ER 44224 (figure 23.4), and by the partial skeleton, KNM-ER 5896, from Koobi Fora.

Age Early Pleistocene, 1.527–1.485 Ma.

African Occurrence Lower part of the Okote Member of the Koobi Fora Formation, East Turkana (Kenya).

Remarks The mandibular corpus and symphysis of *C. freedmani* are sexually dimorphic, to a slightly greater extent than in modern *Colobus*. In the postcranium, the clavicle is flatter and slightly less superiorly curved at its distal extremity, and the greater tuberosity of the humerus projects slightly above the level of the humeral head, indicating enhanced muscular stabilization of the shoulder than seen in the modern species (Jablonski and Leakey, 2008b). As in modern colobus, the thumb appears to have been markedly reduced. A complete description of this species is available elsewhere (Jablonski and Leakey, 2008b). *Colobus freedmani* was very similar in overall appearance to modern *C. guereza* and was probably an equally lithe and competent arborealist. Its upper limb anatomy suggests that it may have relied slightly less on overhead suspensory locomotion than its modern congener.

COLOBUS GUEREZA Rüppell, 1835

Diagnosis The modern black-and-white colobus monkey.

Description The species is best represented by a nearly complete fossil cranium of a young female individual, YPM 19063.

Age Presumed Pleistocene.

African Occurrence Near Wad (Wadi) Medani, central Sudan.

Remarks The cranial morphology of fossil *C. guereza* was similar to that of modern *C. polykomos*, with a short face and short nasal bones, well-marked supraorbital ridges, and pronounced postorbital constriction. Similar to *C. polykomos* in most features except for the buccally bowed cheek tooth rows and relatively small teeth, which more closely resemble *C. polykomos polykomos*.

The occurrence of modern black-and-white colobus monkeys indicates that during the Pleistocene the Blue Nile supported a lush riparian corridor capable of supporting arboreal monkey populations.

COLOBUS sp. indet.

Diagnosis Similar in size and morphology to modern *Colobus*, but not identifiable at the species level.

Description A heterogeneous group from of mostly Pleistocene age, *Colobus* sp. indet. is represented mostly by isolated mandibles. The sample from Asbole in the Afar Region of Ethiopia is by far the largest, consisting of many partial crania, mandibles, and fragmentary postcrania (Jablonski, 2002; Frost and Alemseged, 2007).

Age Plio-Pleistocene; late Pleistocene.

African Occurrence Plio-Pleistocene occurrences are from East Africa, including Kanam East (Kenya), the Omo Group, Asbole, Afar Region, and Middle Awash (Ethiopia); the late Pleistocene specimen is from Kazinga (Uganda)(Harrison and Harris, 1996; Jablonski, 2002).

Remarks Most specimens closely resemble modern black-and-white colobus monkeys, with some being slightly smaller and some somewhat larger. This group also probably includes forms more closely related to the modern red colobus, *Piliocolobus badius* (Frost, 2001b; Frost and Alemseged, 2007). It is interesting that the middle Pleistocene morph assigned to *Colobus* sp. indet. from the Asbole in the Afar region of Ethiopia is not *C. guereza*. The relationship of this morph to living species of *Colobus* and to other fossils of the genus, such as the roughly contemporaneous cranium from Wad Medani, Sudan, is currently unknown (Frost, 2001b). The fossil Afar species is within the size range of most extant *Colobus*, lacks the large female canines of extant *C. guereza*, and has relatively large central incisors.

Genus KUSERACOLOBUS Frost, 2001

Diagnosis Frost (2001a). The genus is characterized by a short face with anteriorly positioned zygomata and a mandibular corpus that is deep and robust, and deepens posteriorly. The interorbital region is broad, unlike *Nasalis*, *Rhinocolobus*, and *Libypithecus*. Unlike *Cercopithecoides* and distinguished from extant African colobines by well-developed protocones on the p3 and metaconids of the P4. Distal lophids of the M3 are approximately equal in breadth with the mesial lophids; this condition is similar to *Procolobus* but unlike *Colobus*, where the distal lophid is usually broader.

Description The type species, *K. aramisi*, is known from deposits in the Middle Awash ranging in age from 5.2 to 4.4 Ma, with tentative identifications perhaps extending the range back to 5.7 Ma (Frost, 2001b). It is also known from the western margin at Gona dated to 4.5 Ma (Semaw et al., 2005). A second, larger species, *K. hafu*, has recently been described from Asa Issie and is approximately 4.1 Ma (Hlusko, 2006).

Age Latest Miocene and early Pliocene, 5.2–4.1 Ma.

African Occurrence Afar Region (Ethiopia).

Remarks The geologically younger *K. hafu* is considerably larger (Hlusko, 2006; Frost et al., 2009). The presence of *Kuseracolobus* species in the Mio-Pliocene fossil record of northeastern Africa indicates that the colobine radiation was more diverse at an earlier period in Africa than previously recognized. The probable arboreal proclivities of *Kuseracolobus* species (Hlusko, 2006) also indicate that previous generalizations about the mostly terrestrial locomotor preferences of early colobines need to be revised.

KUSERACOLOBUS ARAMISI Frost, 2001
Figure 23.5

Diagnosis Same as for genus and considerably smaller than *K. hafu* (figure 23.5).

Description Described on the basis of a series of mandibles, maxillae, and dentition from the lower Aramis Member of the Sagantole Formation in the Middle Awash dated to 4.4 Ma (Frost, 2001b). A single mandible—originally thought to perhaps be *Libypithecus* (Kalb et al., 1982b)—and several isolated teeth from the 5.2 Ma old Kuseralee Member of the Sagantole Formation have been allocated to this species (Frost et al., 2009). A fragmentary series of fossils from the Adu Asa Formation in the Middle Awash may also represent this species and would extend the chronological range to 5.7 Ma (Frost, 2001b). Finally, a series of mandibles, maxillae, and dentition from Gona are also from this species (Semaw et al., 2005).

Age 5.2–4.4 Ma.

African Occurrence Middle Awash and Gona (Ethiopia).

Remarks *Kuseracolobus aramisi* is similar in size to *Cercopithecoides meaveae* and the extant proboscis monkey of Asia, *Nasalis larvatus* (Hlusko, 2006; Frost et al., 2009). Postcrania, especially of the forelimb, are similar in some aspects of their morphology to those of *Paracolobus*, *Rhinocolobus*, and extant arboreal colobines (Hlusko, 2006).

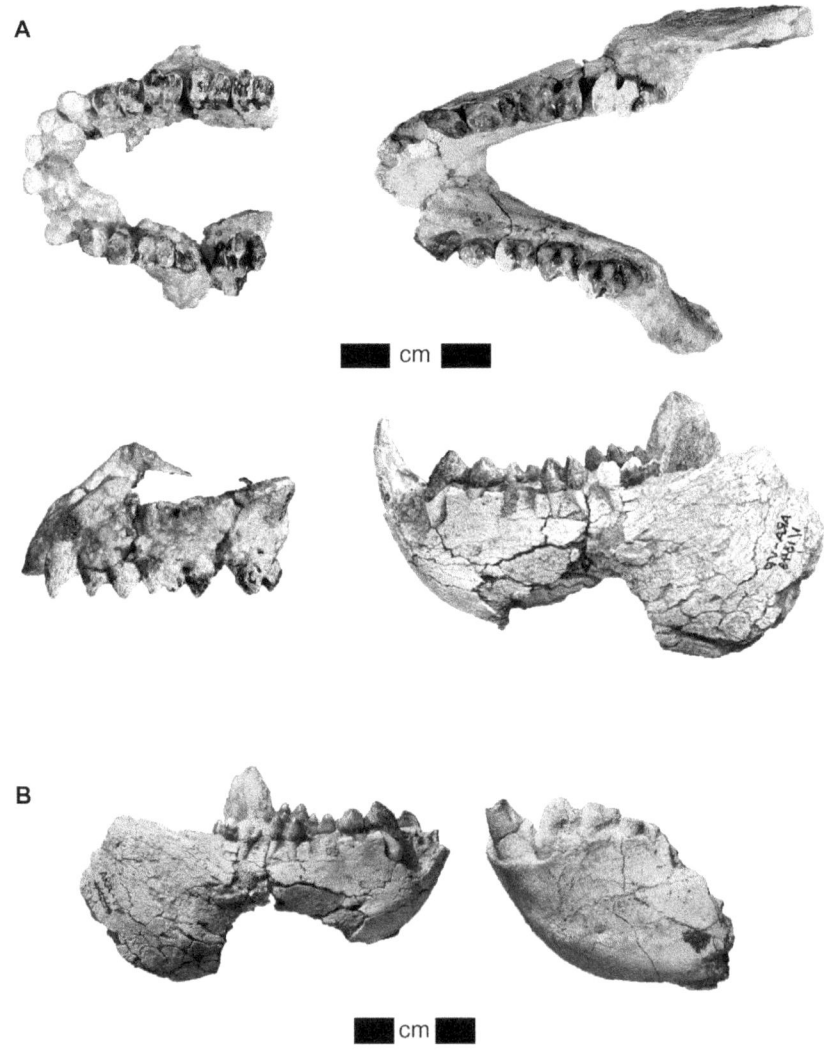

FIGURE 23.5 A) *Kuseracolobus aramisi*. Left: Occlusal and lateral views of maxilla ARA-VP-6/1686; right: Occlusal and lateral views of male mandible ARA-VP-1/87 (holotype), photographically reversed. B) Comparison of lateral views of partial mandibles of *K. aramisi* ARA-VP-1/87 (holotype) with *K. hafu* ASI BVP 2/100. Photograph courtesy of Leslea Hlusko.

KUSERACOLOBUS HAFU Hlusko, 2006
Figure 23.5

Diagnosis Distinguished from *K. aramisi* by its larger size (figure 23.5).

Description The species is known from fragmentary jaws, isolated teeth, and mostly forelimb postcrania as described at length elsewhere (Hlusko, 2006).

Age 4.1 Ma.

African Occurrence Asa Issie area of the Middle Awash Region (Ethiopia).

Remarks Kuseracolobus hafu was larger than any living colobine and larger than most Plio-Pleistocene colobines, with an estimated male body mass between 30 and 40 kg (Hlusko, 2006). The forelimb morphology of the species indicates that it spent a large proportion of its time in the trees, despite its large size.

Genus *LIBYPITHECUS* Stromer, 1913
LIBYPITHECUS MARKGRAFI Stromer, 1913

Diagnosis A primitive colobine, lacking clear apomorphies with most other African and Asian colobines (Strasser and Delson, 1987). *Libypithecus* is also characterized by a maxillary sinus, a feature unknown in extant colobines (Rae et al., 2007). Its long and low cranium, projecting rostrum, and narrow interorbital distance may be symplesiomorphous for cercopithecoids in general, being shared with *Victoriapithecus, Nasalis,* and *Rhinocolobus.*

Description Libypithecus is primarily known by the holotype cranium of *L. markgrafi* and an isolated M1 from Wadi Natrun, Egypt (Stromer, 1914). In addition, a few isolated colobine molars from the late Miocene site of Sahabi, Libya (Meikle, 1987), have been tentatively allocated to *L. markgrafi,* largely on the basis of size. At both Wadi Natrun and Sahabi *L. markgrafi* occurs with the early papionin *Macaca libyca* (Delson, 1973, 1975).

Age Latest Miocene.

African Occurrence Gar Maluk, Wadi Natrun (Egypt).

Remarks The holotype male skull is nearly complete, except that damage renders the join between the middle and lower face and the upper face and vault tenuous, so that their relationship must be estimated (Szalay and Delson, 1979). The premaxilla of the male holotype cranium shows clear evidence of healed trauma. Related to its relatively long rostrum, the calvaria exhibits a marked sagittal crest which increases in height posteriorly. The dentition is typical of colobines with small incisors, the upper lateral incisor being caniniform, and the lowers show the presence of lingual enamel. The protocone on the p3 is reduced as in extant African colobines (Benefit and Pickford, 1986).

Genus *MICROCOLOBUS* Benefit and Pickford, 1986
MICROCOLUBUS TUGENENSIS Benefit and Pickford, 1986

Diagnosis A small colobine with a mandibular symphysis that lacks an inferior transverse torus below the genioglossal fossa and a median symphyseal foramen. The P3 exhibits a well-developed heel but lacks a mesiolingual sulcus; the P4 metaconid and protoconid are of almost equal height. The molar cusps are tall and anteriorly directed, and the lingual notches are deep, but the projection of the M1 and M2 cusps above the lingual notch is low compared to other colobines (Benefit and Pickford, 1986). Mandible slightly deeper below

M3 than M1 as in *Colobus, Rhinopithecus, Nasalis,* and *Mesopithecus,* and absent median symphyseal foramen as in most other colobines. In its dentition, it is most similar to small examples of *Mesopithecus pentelici,* with moderately high molar cusps angled slightly forward, and well-developed mesial and distal shelves.

Description A rare colobine best known from the holotype, KNM-BN 1740, a nearly complete mandible from Ngeringerowa in the Baringo Basin of Kenya.

Age 9.5–9.0 Ma (Gundling and Hill, 2000).

African Occurrence Ngeringerowa (Kenya).

Remarks The morphological similarities between *Microcolobus* and the widespread Eurasian Mio-Pliocene colobine *Mesopithecus* suggest a close phyletic relationship between the two.

Genus *PARACOLOBUS* Leakey, 1969

Diagnosis Based on Hlusko (2007). *Paracolobus* is a medium to large monkey distinguished from *Rhinocolobus* by its moderately long cranium, broad muzzle, and wide face. The nasal bones are short but longer than those of *Rhinocolobus,* and the nasal aperture is long with relatively thick lateral margins. Differs notably from *Cercopithecoides* in its deep and slender mandibular corpus and the absence of a median mental foramen at the symphysis. The premolars are relatively large, and the p3 has a small protocone, lacking in *Cercopithecoides.* The molars are tall with angular cusps separated by deep valleys; the maxillary molars are wide relative to length and flare at the cervix. Most features of the postcranium are typical of arboreal colobines, but some are intermediate between those and the terrestrial cercopithecine condition.

Description Paracolobus enkorikae from the terminal Miocene site of Lemudong'o, Kenya, is represented by many gnathic remains and some forelimb postcrania. *Paracolobus chemeroni* and *P. mutiwa* are of late Pliocene age, the former being recognized from a partial skeleton from Chemeron (Kenya) and possibly a mandible and humerus from the Middle Awash (Ethiopia). *Paracolobus mutiwa* is known from sites within the Omo–Lake Turkana Basin of Kenya and Ethiopia (Leakey, 1982). Examples of the genus not assignable to species are also known from Laetoli, Tanzania (Leakey and Delson, 1987).

Age Terminal Miocene and late Pliocene.

African Occurrence East and northeast Africa.

Remarks Species of *Paracolobus* are some of the most intriguing fossil monkeys to ever have been recovered because they exhibit suites of postcranial characteristics that have no close modern equivalents. They exhibit a range of body sizes from that of modern black-and-white colobus monkeys *(P. enkorikae)* to much larger than any living colobine *(P. chemeroni). Paracolobus enkorikae* and *P. chemeroni* were probably mostly arboreal, but *P. mutiwa* may have been far more terrestrial. The recognition of the terminal Miocene species *P. enkorikae* greatly expands the temporal span of the genus (Hlusko, 2007).

PARACOLOBUS CHEMERONI Leakey, 1969
Figure 23.6

Diagnosis Larger than *P. enkorikae* and lacking the squared rostrum, maxillary ridges, and expanded mandibular gonion of *P. mutiwa.*

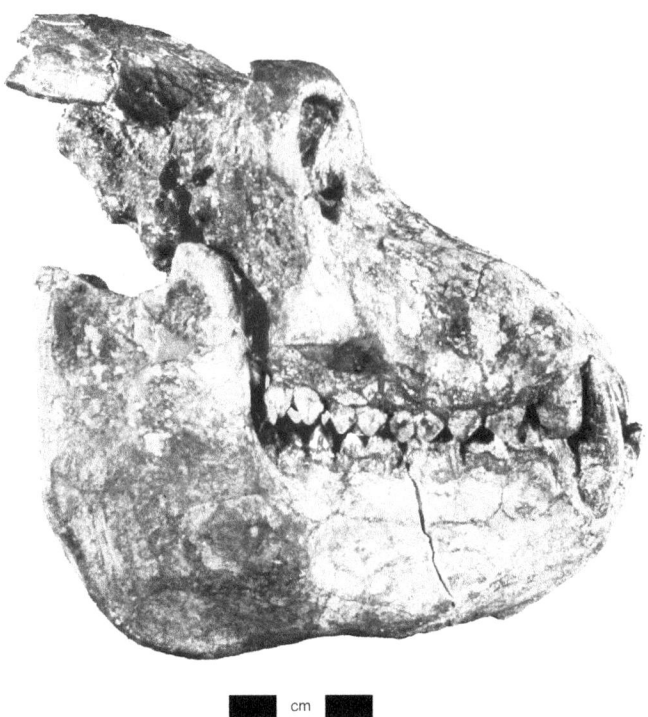

FIGURE 23.6 *Paracolobus chemeroni*. Lateral view of male skull KNM-BC 3. Photograph courtesy of Gerald Eck. © National Museums of Kenya.

Description The species is best known from the holotype, a remarkable and nearly complete male skeleton, KNM-BC 3 (figure 23.6), from Chemeron, Kenya (Leakey, 1969). A single left male mandibular corpus fragment from Matabaietu in the Middle Awash, dated to 2.5 Ma, has been referred to cf. *P. chemeroni* (Frost, 2001a, 2001b).

Age Pliocene, 3–2 Ma.

African Occurrence Chemeron (Kenya) and Middle Awash (Ethiopia).

Remarks The postcranial anatomy of the species was studied in detail by Birchette whose unpublished dissertation is a classic of descriptive paleontology (Birchette, 1982). Its unique features, including a poorly developed deltoid tuberosity of the humerus, shallow humeral trochlea, and anteriorly inclined olecranon process of the ulna, were interpreted as being consistent with a predominantly arboreal lifestyle. The body mass of the species was very great for a monkey, with an estimated range for males of 39–46 kg (Delson et al., 2000; Hlusko, 2007).

PARACOLOBUS ENKORIKAE Hlusko, 2007
Figure 23.7

Diagnosis Hlusko (2007). Distinguished from *P. mutiwa* and from *Kuseracolobus, Microcolobus,* and *Rhinocolobus* by the absence of significant mandibular gonial expansion, and from *Cercopithecoides* by a mandibular cross section that is less robust and rounded; differs from *P. chemeroni* and *P. mutiwa* mainly in its much smaller size.

Description Known from the holotype nearly complete mandible, KNM-NK 44770 (figure 23.7), numerous partial jaws, and isolated teeth (Hlusko, 2007).

Age Terminal Miocene, 6 Ma.

African Occurrence Lemudong'o, Narok District (Kenya).

Remarks *Paracolobus enkorikae* exhibits close affinities with *P. chemeroni* (Leakey, 1982) and may be its ancestor. It

shares overall dental proportions with *Victoriapithecus macinnesi*. On the basis of study of the mandibular morphology of the species, Hlusko has suggested that *Paracolobus* may have closer evolutionary affinities to modern *Colobus* than do the Plio-Pleistocene species of *Rhinocolobus, Cercopithecoides,* and *Kuseracolobus.*

PARACOLOBUS MUTIWA Leakey, 1982

Diagnosis Distinguished from other species of *Paracolobus* by a high muzzle, maxillary fossae and ridges, wide frontal process of the zygoma, narrower interorbital region, and less sharply converging temporal lines. The gonion is greatly expanded.

Description The species is known from the holotype female fragmentary cranium from East Turkana, KNM-ER 3843, and numerous gnathic fragments and isolated teeth (Leakey, 1982). A partial skeleton from West Turkana has not been fully described (Harris et al., 1988).

Age Pliocene, 3.36–1.88 Ma.

African Occurrence East and West Turkana (Kenya), and the Omo Valley (Ethiopia).

Remarks Preliminary study of the partial skeleton from West Turkana indicates that the species shows strong positive allometry of the skull and teeth relative to the postcranium and that it may have been strongly terrestrial (Jablonski, unpublished observations).

PARACOLOBUS sp. indet.

Diagnosis Distinguished from *P. chemeroni* and *P. mutiwa* by the smaller size of the teeth, but remains are insufficient to allow assignment to species.

Description A large number of fragmentary mandibular and maxillary specimens from Laetoli, Tanzania, the best being LAET 247, 259, and 4596, and two femoral fragments,

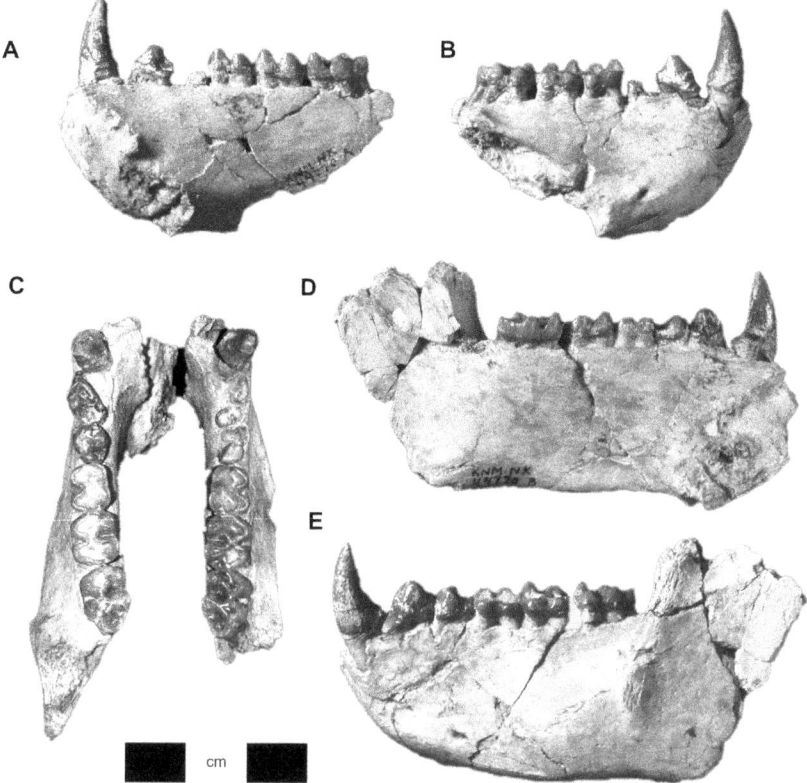

FIGURE 23.7 *Paracolobus enkorikae.* Views of male mandible KNM-NK 44770 (holotype). A) Lingual view of right mandible; B) buccal view of right mandible; C) occlusal view of right and left mandibles; D) lingual view of left mandible; E) buccal view of left mandible. Photograph courtesy of Leslea Hlusko. © National Museums of Kenya.

LAET 247 and LAET 327, and several isolated teeth (Leakey and Delson, 1987); isolated molars and a distal humerus from the Western Rift (Senut, 1994); and, possibly, three fragments, including an isolated M3 from Makapansgat (Eisenhart, 1974).

Age Pliocene, 3.8–?2.5 Ma.

African Occurrence Laetoli (Tanzania), and possibly Western Rift (Uganda), and Makapansgat (South Africa).

Remarks The morphology of the teeth and femur indicate affinities with *Paracolobus*, but it is possible that at least the Laetoli assemblage represents a different taxon altogether (Leakey, 1982; Leakey and Delson, 1987; Jablonski et al., 2008a). The occurrences of the genus in Uganda and South Africa must be viewed with doubt pending further discoveries and study.

Genus *RHINOCOLOBUS* Leakey, 1982

RHINOCOLOBUS TURKANAENSIS Leakey, 1982

Figure 23.8

Diagnosis After Jablonski et al. (2008b). A large colobine, which shares with *Libypithecus, Dolichopithecus,* and modern *Nasalis* but not other colobines a long braincase and muzzle, narrow interorbital distance, marked postorbital constriction, and relatively small orbits. The mandibular corpus is thin and deep and the gonion expanded; a median symphyseal foramen is present. Distinguished from *Libypithecus* by its larger overall size, larger and more rounded calvaria, longer muzzle, and lack of a large sagittal crest, and from *Nasalis* by short nasal bones, wide malar region, postglabellar sul-

cus, and nuchal crests. The near absence of nasal bones in *Rhinocolobus* is remarkable and distinguishes it from all other colobines save the Asian odd-nosed monkey genus, *Rhinopithecus.* In the postcranium, *Rhinocolobus* is distinguished from other large colobines by its large and medially protuberant medial humeral epicondyle and medially inwardly curved distal humeral shaft.

Description The species is represented by abundant remains, including a well-preserved female skull (KNM-ER 1485; figure 23.8), and a partial male skeleton (KNM-ER 1542), which are fully described and reconstructed elsewhere (Jablonski et al., 2008a). Several fragmentary gnathic remains and an isolated humerus from the Sidi Hakoma and Denen Dora members of the Hadar Formation (3.4–3.2 Ma) have been tentatively allocated to *R. turkanaensis* (Frost and Delson, 2002).

Age Plio-Pleistocene, 3.4–1.5 Ma.

African Occurrence Shungura and Hadar Formations (Ethiopia) and the Koobi Fora Formation, East Turkana (Kenya).

Remarks Rhinocolobus was one of the largest known colobines, exceeding extant African colobines and the widespread fossil species, *Cercopithecoides williamsi,* in body size, but smaller in tooth and body size than *C. kimeui, Paracolobus chemeroni,* and *P. mutiwa.* It was highly sexually dimorphic in body mass, with males being an estimated 31 kg and females 17 kg (Delson et al., 2000); its canines and P3 were also highly sexually dimorphic. *Rhinocolobus* was predominantly arboreal, as judged by the similarities of its postcranium to that of large extant species such as the equally sexually dimorphic odd-nosed Asian colobines *Nasalis larvatus* and *Rhinopithecus brelichi.*

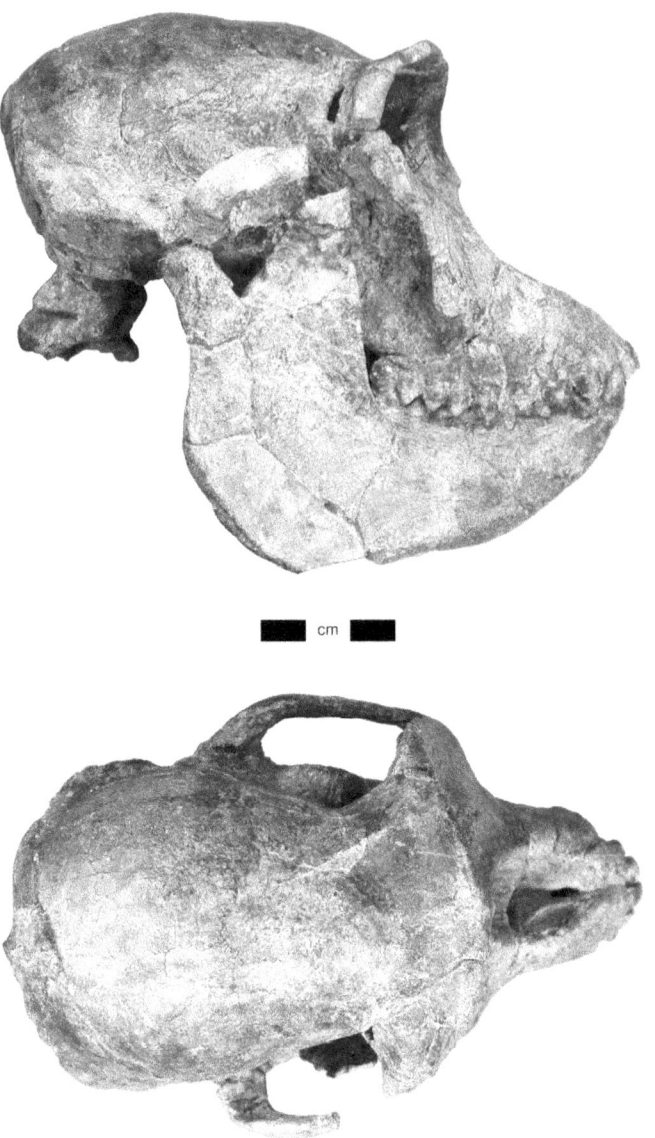

FIGURE 23.8 *Rhinocolobus turkanaensis*. Lateral and vertical views of female skull KNM-ER 1485. © National Museums of Kenya.

Subfamily CERCOPITHECINAE Gray, 1821
Tribe CERCOPITHECINI Gray, 1821
Genus *CERCOPITHECUS* Brunnich, 1772
CERCOPITHECUS sp. indet.

Diagnosis Small cercopithecines comparable in size and morphology to living species of *Cercopithecus*, and characterized by relatively arboreal locomotor behavior compared to *Chlorocebus*. Like all members of the tribe, the M3 lacks a hypoconulid.

Description Cercopithecus has been recognized from gnathic fragments and isolated teeth from the Omo (Eck and Howell, 1972; Eck and Jablonski, 1987), from a left fragmentary mandible from Kanam East (M 15923)(Harrison and Harris, 1996), from fragmentary jaws, isolated teeth, an ulna and a femur as two morphs (*Cercopithecus* sp. indet. A and B) from Koobi Fora (Jablonski et al., 2008b), a jaw fragment from Kuguta (Napier, 1985), isolated teeth from late Pleistocene levels at Olduvai Gorge (Tanzania) and Loboi (Kenya)(Leakey, 1988), and a series of isolated teeth from a later Pliestocene deposit near Taung (South Africa)(Delson, 1984).

Age Pliocene and Pleistocene.

African Occurrence The Omo–Lake Turkana Basin (Ethiopia and Kenya); Kuguta and Kanam East (Kenya); later Pleistocene deposits near Taung (South Africa).

Remarks The scarcity of *Cercopithecus* fossils may be due to their genuine rarity in the past but is equally likely to be due to their small size, which renders them susceptible to destruction by carnivores or scavengers or to rapid decomposition by the elements. For whatever reason, the fossil record of *Cercopithecus* is sparse (Jablonski et al., 2008b). The smaller morph (*Cercopithecus* sp. indet. B) recognized at Koobi Fora may represent the Pliocene ancestor of the modern talapoin, the smallest extant guenon (Jablonski et al., 2008b). Several of the fragmentary specimens allocated to this genus may in fact represent *Chlorocebus* as they are craniodentally similar, and only recently have species (*C. aethiops*, and the *C. lhoesti* species group) previously included in *Cercopithecus* been referred to *Chlorocebus* (Tosi et al., 2002a, 2003b, 2004, 2005; Xing et al., 2007).

Genus CHLOROCEBUS Gray, 1870

Diagnosis Cercopithecin monkeys of average to large size for the tribe. Characterized by adaptations for terrestrial locomotion. Sexual swellings are absent.

Description Best known from a series of fossils from middle Pleistocene deposits from Asbole and the Middle Awash (Ethiopia).

Age Middle Pleistocene, 0.6 Ma and younger.

African Occurrence Eastern Africa.

Remarks Recent molecular analyses have consistently grouped the patas monkeys (formerly *Erythrocebus patas*) with the more terrestrially adapted species traditionally included in *Cercopithecus*: *C. aethiops*, *C. lhoesti*, *C. solatus*, and *C. preussi* (Tosi et al., 2002a, 2002b, 2003b, 2004, 2005). Following the recommendations of Tosi and colleagues (Tosi et al., 2004), these taxa are included in the genus *Chlorocebus*, which has priority over *Erythrocebus* and *Allochrocebus*. Thus, all of the guenon species that are more terrestrially adapted (Gebo and Sargis, 1994) and that occur in more open habitat types, such as *Ch. aethiops* and *Ch. patas* (Napier, 1981), are grouped in this genus. To varying degrees, this arrangement has also received some support from karyotypic analyses and vocalizations (Dutrillaux et al., 1988; Gautier, 1988). Molecular estimates indicate that *Chlorocebus* diverged from the *Miopithecus* + *Cercopithecus* group approximately 8 Ma (Tosi et al., 2005).

Thus far, the only fossil material allocated to this genus derives from the middle Pleistocene of the Afar (Ethiopia) (Kalb et al., 1982b; Frost, 2001a; Frost and Alemseged, 2007), but fragmentary specimens from eastern and southern Africa that have been allocated to *Cercopithecus* in the past are likely to represent this genus as well.

Cf. *CHLOROCEBUS* aff. *AETHIOPS*

Diagnosis A guenon similar in size to the extant vervet monkey, larger than *Miopithecus* and smaller than *Ch. patas*. Rostrum shorter and less squared than that of *Ch. patas*. The i2 is larger relative to the i1 than is the case in *Miopithecus*, *Cercopithecus* or *Ch. lhoesti*. The molar cusps are also not as tall and sharp as are those of *Ch. lhoesti*.

Description This species is best represented by a series of gnathic remains from Asbole, dated to approximately 600 Ka, as well as the younger site of Andalee in the Middle Awash (Ethiopia)(Frost, 2007b; Frost and Alemseged, 2007).

Age Middle Pleistocene, 0.6 Ma and younger.

African Occurrence Asbole and Middle Awash (Ethiopia).

Remarks The material assigned to this taxon shows affinities with the extant vervet monkey, present throughout diverse African habitats. The fossil assemblages derive from middle Pleistocene sites in the Afar region that were possibly closer to a major stream, such as the paleo-Awash, than other contemporary sites, such as Bodo and the other Dawaitoli Formation sites (Kalb et al., 1982b; Frost and Alemseged, 2007).

Cf. *CHLOROCEBUS* cf. *PATAS*

Diagnosis Comparable in size to extant *Ch. patas*, but larger than other known guenons. Rostrum is relatively long and narrow, and the palate is relatively deep.

Description This species is known from a few cranial fragments and an os coxae and femur from Asbole (Ethiopia).

A few specimens from Andalee (Ethiopia) may also represent this taxon (Frost, 2007b).

Age Middle Pleistocene, 0.6 Ma and possibly younger.

African Occurrence Asbole and Middle Awash (Ethiopia).

Remarks If this material belongs to the patas monkey, then it would represent an eastward extension of the species' known range. The femur, tentatively allocated to this taxon, has a relatively low neck-shaft angle and tall greater trochanter, both indicating a relatively terrestrial locomotor mode.

Tribe PAPIONINI Burnett, 1828
Subtribe MACACINA Owen, 1843
Genus *MACACA* Lacépède, 1799

Diagnosis Small to large size monkeys with narrow interorbital regions, rounded muzzles of moderate length, molars with flared sides and relatively short crowns, and moderate to great sexual dimorphism in the canine-premolar complex, and postcranium. Distinguished from other papionins by the general absence of suborbital fossae and maxillary ridges, presence of a maxillary sinus, and a lack of extreme lateral flare of the molars.

Description The representation of this genus in Africa is small compared to that in Asia, where it is the dominant papionin of the past and present. The African fossil species are similar in morphology to the modern Barbary macaque, *Macaca sylvanus*.

Age Late Miocene—Recent.

African Occurrence Northern Africa.

Remarks Today, *Macaca* is the most widely distributed nonhuman primate genus, with *M. sylvanus* occurring in North Africa and between 14 and 19 species distributed across southern and eastern Asia. It also spans the widest range of ecosystems including temperate forests in Japan, high altitude areas of the Himalayan foothills, as well as tropical forests and woodland (Fooden, 1980; Szalay and Delson, 1979; Jablonski, 2002). The genus has an extensive fossil record throughout the Pliocene and Pleistocene of Europe and Asia (Delson, 1980; Jablonski, 2002). Delson proposed that the North African and Eurasian *Macaca* became isolated from the sub-Saharan papionins in the late Miocene, perhaps due to the Sahara becoming more of a biogeographic barrier (Delson, 1975, 1980). The timing and biogeography of this scenario has essentially been corroborated by molecular analyses (Morales and Melnick, 1998; Tosi et al., 2000, 2002b, 2003b; Jablonski, 2002; Xing et al., 2005). Most morphological and molecular analyses agree on the position of this genus as the sister to all other extant papionins, and it is often placed in its own subtribe Macacina, with the remaining genera in the subtribe Papionina (Delson, 1973; Strasser and Delson, 1979, 1987; Jablonski, 2002).

The oldest material that may represent this genus is a series of isolated teeth from Menacer (formerly Marceau, Algeria), where it is found in association with a relatively large species of colobine. This population was initially named *Macaca flandrini*, but the holotype was in fact a specimen representing the co-occurring colobine. Therefore, the papionin material has remained unnamed pending discovery of more diagnostic morphology. Menacer has been faunally dated to the late Miocene (Geraads, 1987). Delson tentatively allocated this material to *Macaca* largely on the basis of geography but noted that it lacks any features distinguishing it from a stem papionin (Delson, 1975, 1980; Szalay and Delson, 1979).

Three isolated teeth from the middle Pliocene site of Ahl al Oughlam (Morocco) have been tentatively assigned to *Macaca*, again based largely on biogeography (Alemseged and Geraads, 1998), as has an isolated distal humeral fragment from Garaet Ichkeul (Tunisia)(Delson, 1975, 1980), but the latter interpretation is not universally accepted (Alemseged and Geraads, 1998).

MACACA LIBYCA (Stromer, 1920)

Diagnosis A medium-sized papionin similar to extant *M. sylvanus* in both size and the relatively few preserved aspects of its morphology.

Description The sample from Wadi Natrun consists of a few mandibular fragments and isolated teeth (Stromer, 1920; Delson, 1975, 1980; Szalay and Delson, 1979). Additional mandibular fragments from Sahabi may also represent this species (Meikle, 1987).

Age Late Miocene—earliest Pliocene.

African Occurrence North Africa, Wadi Natrun (Egypt), and Sahabi (Libya).

Remarks *Macaca libyca* co-occurs with the colobine *Libypithecus markgrafi* (discussed earlier) at Wadi Natrun and possibly Sahabi. Due to the ambiguous nature of the morphology preserved by the relatively fragmentary specimens, it has generally been included in *Macaca* for biogeographic reasons but considered distinct from *M. sylvanus* due to its age, which approximates molecular estimates for the last common ancestor of the genus (Delson, 1975, 1980; Szalay and Delson, 1979; Tosi et al., 2003a).

MACACA SYLVANUS (Linnaeus, 1758)

Diagnosis Medium-sized papionin, with robust supraorbital torus, lacking maxillary or mandibular corpus fossae. Dentition is typical of the tribe. Limbs are relatively robust, and the external tail is reduced to a stub. Lacks the derived reproductive anatomy of many Asian species.

Description Fragmentary jaws and teeth from several sites in North Africa have been assigned to this species.

Age Middle and late Pleistocene, and possibly Pliocene.

African Occurrence *Macaca sylvanus* is known from Ain Mefta and Tamar Hat (Algeria), as well as from two isolated teeth tentatively included from the Pliocene site from Ain Brimba (Tunisia), as *Macaca aff. sylvanus* (Delson, 1980).

Remarks In Europe, this species has a much more extensive chronological range than in Africa, extending back to the Ruscinian, and is recognized in three chronological subspecies: *M. s. prisca*, *M. s. florentina*, and *M. s. pliocena* (Delson, 1980), but the relationship of the mostly younger African material to these European forms is unclear.

Subtribe PAPIONINA Burnett, 1828
Genus CERCOCEBUS Geoffroy Saint-Hilaire, 1812
CERCOCEBUS sp. indet.

Diagnosis Medium-sized monkeys distinguished from *Cercopithecus* and *Macaca* by the presence of suborbital fossae and large incisors relative to molar size; distinguished from *Lophocebus* by relatively shallow maxillary fossae and mandibular molars strongly laterally flared at the cervix.

Description The fossil evidence for the modern genus of *Cercocebus* mangabeys is fragmentary and comprises only a partial cranium from Makapansgat (M3057/8/9, M218), fragmentary jaws from Kromdraai (Eisenhart, 1974), and possibly some other gnathic remains previously assigned to female *Parapapio jonesi* from Swartkrans, such as SK 414, SK 543, and SK573a and b (Gilbert, 2007b). East African specimens previously referred to *Cercocebus* (Jablonski, 2002) are now considered *Lophocebus* cf. *albigena* (Jablonski et al., 2008b) and are described later in this chapter.

Age Late Pliocene.

African Occurrence Makapansgat, Kromdraai, and ?Swartkrans (South Africa).

Remarks The *Cercocebus* lineage has recently been recognized in South Africa, from material originally assigned to *Parapapio antiquus* (Gilbert, 2007b). (See later discussion of *Procercocebus antiquus*.) Molecular and morphological evidence has conclusively demonstrated that the mangabeys are diphyletic and that *Cercocebus* belongs to the *Cercocebus/Mandrillus* clade, while *Lophocebus* belongs to the *Lophocebus/Papio/Theropithecus* clade (Cronin and Sarich, 1976; Page and Goodman, 2001; Gilbert, 2007b). The existence of a fossil representative of the *Cercocebus* lineage in southern Africa suggests that modern *Cercocebus* mangabeys achieved their current equatorial distribution following a Pleistocene dispersal from southern Africa (Gilbert, 2007b), but other biogeographic interpretations are possible, as discussed later.

Genus DINOPITHECUS Broom, 1937
DINOPITHECUS INGENS Broom, 1937
Figure 23.9

Diagnosis Distinguished from other papionins by its very large size, broad interorbital region, the robustness and rugosity of cranial surfaces associated with the masticatory and nuchal musculature, the strength of the temporal lines in both sexes (and sagittal crest in males), strong nuchal crests, and large, broad molars that often show accessory cuspules and short P3s with large anterior foveae in males (Freedman, 1957). Distinguished from *Papio* by its general lack of maxillary and mandibular corpus fossae. *Dinopithecus ingens* was sexually dimorphic.

Description Remains of the species are known from two sites only, the type locality of Schurweberg (Skurweberg; 1 specimen, SB 7) and Swartkrans (over 30 specimens). Among the collection from Swartkrans are three nearly complete crania of females (SK 553, SK 600, and SK 603; figure 23.9) and a fragmentary cranium of a male (SK 599) lacking the muzzle (Simons and Delson, 1978; Szalay and Delson, 1979; Delson and Dean, 1993; Heaton, 2006).

Age Late Pliocene.

African Occurrence Schurweberg (Skurweberg) and Swartkrans (South Africa).

Remarks The phyletic affinities, distribution, and classification of *Dinopithecus* have been a subject of some controversy for many years, partly because the most complete specimens referred to the taxon are female, and the majority of diagnostic papionin apomorphies are expressed only in males. Delson proposed that *Dinopithecus* was best considered an extinct subgenus of *Papio* and that several large fossil papionins from eastern and southwestern Africa could be accommodated within it; this view has been adopted by Frost and Heaton but is not followed here as these similarities may be primitive retentions. Several papionin fossils from Leba, Angola, considered possibly referable to *Dinopithecus*, have been argued by one of us (N.G.J.) to belong to an early, large representative of *Theropithecus* (Jablonski, 1994), as discussed later. The other author (S.F.) does not concur with this allocation. Similarities between the molar teeth of *D. ingens* and *Gorgopithecus major* noted in an early study (Freedman, 1957)

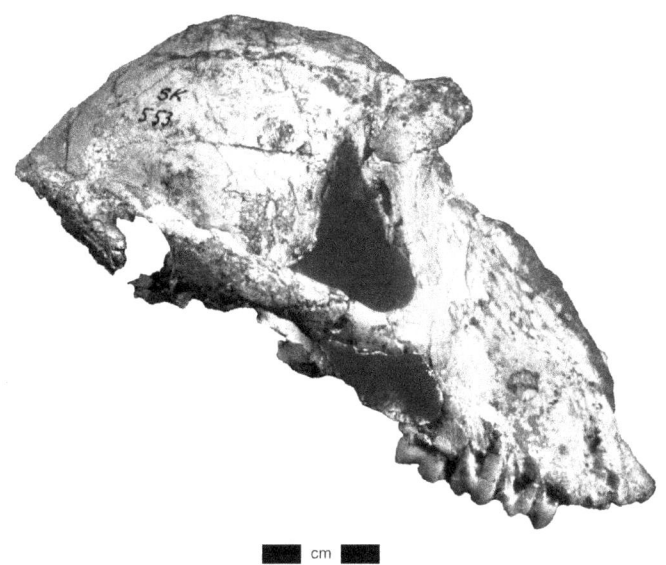

FIGURE 23.9 *Dinopithecus ingens.* Lateral view of female cranium SK 553. Photograph courtesy of Gerald Eck.

may indicate phyletic propinquity between these two large and unusual papionins, but the lack of comparable facial remains precludes necessary study.

Genus *GORGOPITHECUS* Broom and Robinson, 1949
GORGOPITHECUS MAJOR (Broom, 1940)

Diagnosis Distinguished from other papionins by its relatively short, high, and narrow muzzle, deep maxillary and mandibular fossae, great interorbital breadth, and long calvaria.

Description This species is recognized with confidence only from Kromdraai and is best known from a single specimen of crushed male cranium, KA 192, a partial female cranium, KA 153, and two badly damaged mandibles (KA 150 and KA 152). A single palate from Swartkrans has also been tentatively allocated to this species (Delson, 1984).

Age Plio-Pleistocene.

African Occurrence Kromdraai (South Africa).

Remarks Gorgopithecus major was one of two Plio-Pleistocene papionins (along with *Papio angusticeps*) recognized only at Kromdraai; it is probably also present at Coopers A and possibly Coopers B (Delson, 1984). The distinctive facial skeleton of this species features maxillae dropping steeply inferiorly from the sides of the nasal bones and nasal aperture to the alveolar margin. The short nasals are oriented almost horizontally to the posterior edge of the nasal aperture; the muzzle drops steeply from there to the anterior margin of the aperture on the premaxilla (Freedman, 1957). The zygomatic arch in both sexes is heavily built and the bizygomatic breadth is great. As mentioned earlier, the molars of *G. major* exhibit similarities to *Dinopithecus ingens* and may denote a close phyletic relationship between the two.

Genus *LOPHOCEBUS* Palmer, 1903

Description Lophocebus is recognized from a large collection of jaws and some postcrania from the Koobi Fora Formation, from a small but diagnostic collection of teeth and a temporal bone fragment from Kanam East, and tentatively from Olduvai and Omo.

Age Plio-Pleistocene.

African Occurrence Kanam East and East Turkana (Kenya), and possibly Olduvai (Tanzania) and Omo (Ethiopia).

Remarks Recognition of the diphyly of the mangabeys, discussed in connection with *Cercocebus* earlier, has led to renewed interest in the fossil history of both mangabey lineages. *Lophocebus*, as attested by the assemblages at East Turkana and Kanam East, diverged from the common ancestor of *Lophocebus/Papio/Theropithecus* in the late Miocene or early Pliocene in eastern Africa, probably from a species of *Parapapio. Lophocebus* mangabeys occupied the eastern Rift Valley through the middle Pleistocene, apparently favoring the riparian and forest habitats there. When those habitats deteriorated in the face of increasing Pleistocene seasonality, the mangabeys shifted their distribution westward. Today, *Lophocebus* is found only in central and western equatorial Africa, from the western Rift westward. It is significant that *Lophocebus* and *Pan* share this pattern of biogeographic history (McBrearty and Jablonski, 2005).

LOPHOCEBUS cf. *ALBIGENA* (Gray, 1850)
Figure 23.10

Diagnosis Larger in size than the modern gray-cheeked mangabey, *Lophocebus albigena*, and morphologically distinguished from it only in its consistently large and broad M3 hypoconulid.

Description The most diagnostic examples of this morph are from the Koobi Fora Formation, where they comprise mostly gnathic remains (figure 23.10), the best being female left partial maxillae KNM-ER 595 and 44260, male maxillae KNM-ER 827 and 3090, and many mandibles or parts thereof, including female specimens KNM-ER 898 and 6014 and partial male specimens KNM-ER 594 and 6063. A humerus and proximal femur are also included in the sample. A mandible from Upper Bed II at Olduvai (Tanzania) and another mandible, originally identified as possibly *Parapapio* (Leakey and Leakey, 1976) from Shungura Member K (Ethiopia), Omo K6 '70 C146, may also represent this species (Eck, 1976).

Age 2.0–1.38 Ma.

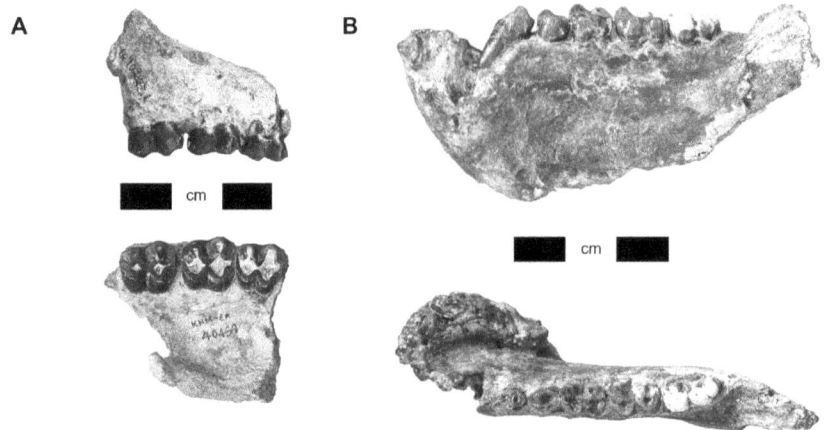

FIGURE 23.10 *Lophocebus* cf. *albigena*. A) Lateral and occlusal views of partial left maxilla KNM-ER 40459; B) lateral and occlusal views of male partial right mandible KNM-ER 594. © National Museums of Kenya.

African Occurrence KBS and Okote members, Koobi Fora Formation, East Turkana (Kenya); Upper Bed II, Olduvai (Tanzania); and, possibly, Shungura Formation, Omo Group (Ethiopia).

Remarks This species was originally identified as *Cercocebus* sp., but at that time, *Lophocebus* was a junior synonym of *Cercocebus*, and the Koobi Fora material was suggested to be most similar to the "*albigena*-group" (Leakey and Leakey, 1976). Like modern *L. albigena*, the fossil form exhibits moderate sexual dimorphism in canine and cranial dimensions. The molars of the species denote a frugivorous diet similar to that of the modern gray-cheeked mangabey, which uses its large incisors for preparing thick-husked fruits and seeds (Hylander, 1975). The postcranial remains of *L.* cf. *albigena* indicate that the species was an arboreal climber, with a hindlimb adapted for stability at the hip joint (Jablonski et al., 2008b). The co-occurrence of the gray-cheeked mangabey and an early form of black-and-white colobus monkey (*Colobus freedmani*) in the Lake Turkana Basin in the early Pleistocene was the first step in the emergence of two of the modern cercopithecoid fauna in East Africa.

LOPHOCEBUS sp. indet.

Diagnosis Same as for genus.

Description A small assemblage from Kanam East, consisting of associated and heavily worn upper central incisors and M2 (M15922) and a fragment of right temporal bone (M18800; Harrison and Harris, 1996), has been referred to *Lophocebus*.

Age Plio-Pleistocene.

African Occurrence Kanam East, Kenya.

Remarks The distinctively robust and convexly curved incisors and the pronounced buccal flare of the isolated molar indicate a probable attribution to *Lophocebus* (Harrison and Harris, 1996).

Genus *PAPIO* Müller, 1773

Diagnosis A papionin of moderate to large body size with elongated muzzles in males; a strong supraorbital torus with a prominent glabella; extreme levels of canine-premolar and somatic sexual dimorphism; and shoulder, elbow, and hip joints strongly adapted for terrestriality. Following Heaton, the genus is distinguished from *Parapapio* by a sharp anteorbital drop near the glabella, leading to a muzzle that slopes gradually toward alveolare, by maxillary fossae in both sexes (which also distinguishes it from *Dinopithecus*), and by a flattened muzzle dorsum and sharp maxillary ridges in males (Heaton, 2006).

Description Taung and Sterkfontein have yielded examples of the small and short-muzzled *P. izodi*, including the type specimen TP 7. Heaton concluded that most of the specimens of *Papio* from late Pliocene sediments at Sterkfontein previously allocated to *P. hamadryas robinsoni*, including female SWP 2946, and male SWP Un2, were in fact *P. izodi* (Heaton, 2006). The fossil morph of the modern baboon, *P. hamadryas robinsoni*, appears to be a valid and widespread form present at most South African Plio-Pleistocene sites between 2.0 and 1.0 Ma except Schurweberg (Heaton, 2006). The modern savanna baboon, *Papio hamadryas*, is represented at the 600 Ka site of Asbole, Ethiopia (Alemseged and Geraads, 2000) and is probably also present at Bodo, in the Middle Awash as attested by two isolated molars (Frost et al., 2009). It is also known by a juvenile cranium from probable middle Pleistocene levels at Olduvai (Remane, 1925).

Age Plio-Pleistocene–Recent.

African Occurrence Northeast and South Africa.

Remarks Classification of extinct and extant species within *Papio* has been unstable for decades due to problems of species recognition. The assignment of all living forms of the genus into several subspecies within the single species *P. hamadryas*, following Groves (Groves, 1989, 2001; Frost et al., 2003a), is advocated by most scholars and is followed here.

The widespread presence of fossil members of the genus in Africa over space and time was assumed by many early paleontologists in light of the ubiquity of modern *Papio* baboons. Recent molecular and paleontological studies have cast doubt on this interpretation, however. *Papio* appears to have arisen in the terminal Pliocene, possibly first in South Africa (Newman et al., 2004; Wildman et al., 2004). Fossils of *Papio hamadryas* do not appear in eastern and northeastern Africa until the middle Pleistocene (Jablonski and Leakey, 2008a; Jablonski et al., 2008b).

Most of the fossil evidence for true *Papio* derives from South Africa, where the history of discovery and naming of fossil representatives of the genus has been long and tortured, partly because of the absence of morphological and metrical distinctions between the molars of *Papio* and *Parapapio* (Freedman, 1957; McKee, 1993; Heaton, 2006). Heaton's recent study has resulted in significant clarification of the morphological distinctions between the two genera, and has shed light on their probable chronologies in South Africa (Heaton, 2006).

Controversy surrounds the presence of *Papio* species outside of South Africa prior to the Pleistocene, partly because most examples identified as such have been fragmentary or immature. A relatively small species of this genus was identified in Ethiopia from the 2.5 Ma Hatayae Member of the Bouri Formation based on a partial female cranium. A relatively large morph from Members E–G of the Shungura Formation is known from two female partial crania and a male rostrum (Eck, 1976; Delson and Dean, 1993; de Heinzelin et al., 1999; Frost, 2001b). This series has been attributed by Delson and Dean to *Dinopithecus* and associated with the significantly older holotype of *Theropithecus quadratirostris* from the Usno Formation (Delson and Dean, 1993). Eck and Jablonski considered the Usno holotype to be an unquestioned *Theropithecus* and did not discuss the status of the other large papionin from the Shungura Formation (Eck and Jablonski, 1984). In this chapter, the Usno holotype is discussed under *Theropithecus*, but the large Shungura papionin is considered to be a large species of *Papio* or, possibly, *Dinopithecus*. A distal humerus and isolated deciduous premolar from Laetoli, Tanzania were referred tentatively to *Papio* (Leakey and Delson, 1987). In the absence of clear *Papio* synapomorphies, all of these cases are more parsimoniously considered as *Parapapio*. Specimens from high in the Koobi Fora succession reported as close in size to modern *Papio* (Harris et al., 1988) have been referred to *Parapapio*, as discussed later in this chapter and elsewhere (Jablonski et al., 2008b).

PAPIO HAMADRYAS (Linnaeus, 1758)

Diagnosis The modern savanna baboon, distinguished from other species of the genus by sharp maxillary ridges in males.

Description This heterogeneous group of modern *Papio* baboons comprises two morphs from South Africa: *P. hamadryas robinsoni* Freedman, 1957 and *P. hamadryas ursinus* Kerr, 1782, and at least two from eastern Africa: *P. hamadryas* subsp. indet., and *P. h. hamadryas* or *P. h. anubis*. None save *P. h. robinsoni* are known from significant samples.

Age Latest Pliocene and Pleistocene.

African Occurrence *Papio hamadryas robinsoni*: Sterkfontein Member 5, Kromdraai A and B, Drimolen, Swartkrans, and Bolt's Farm; *Papio hamadryas ursinus*: Pretoria, Transvaal; *Papio hamadryas* ssp. indet.: Olduvai Gorge (Tanzania); Asbole, Afar Region, (Ethiopia).

Remarks The South African origin of *Papio hamadryas* in the late Pliocene is attested to by molecular evidence (Newman et al., 2004; Wildman et al., 2004), and by the absence of any unambiguous fossils assignable to the taxon before that time. The middle Pleistocene fossils of the species from Asbole is probably either *P. h. hamadryas* or *P. h. anubis* (Frost and Alemseged, 2007). The specimen of *Papio* from an unknown level at Olduvai (Remane, 1925) appears to be a modern *Papio* because, despite its juvenile status, the concave profile of the orbital region and sloping muzzle clearly identify the specimen as *Papio* as opposed to *Parapapio* and its bulbous and low-crowned molars distinguish it from *Theropithecus*.

PAPIO IZODI Gear, 1926

Diagnosis Following Heaton, the species exhibits well-developed supraorbital tori in both sexes and a sharp anteorbital drop from glabella, and is distinguished from other *Papio* species and *Parapapio* by its lightly developed maxillary ridges, and well-defined maxillary fossae sited over the infraorbital foramina (Heaton, 2006). Delson showed that it had larger orbits and teeth than either modern *P. h. kindae* or fossil *P. [h.] angusticeps* of similar size (Delson, 1988).

Description Represented by the lectotype TP 7 (formerly AD 992) and numerous other good specimens from Taung and from many mostly fragmentary cranial remains from Sterkfontein, including the partial male skull SWP 29a and b, and the female SWP 2946 (Freedman, 1965; McKee, 1993; Heaton, 2006).

Age Plio-Pleistocene.

African Occurrence Taung, Sterkfontein Members 2 and 4, Kromdraai, and Coopers, (South Africa).

Remarks *Papio izodi* appears to be the most ancient species of true *Papio*, and may be ancestral or at least phyletically close to *P. hamadryas*. *Papio izodi* is conceived here as including most specimens originally assigned to *P. angusticeps* (Szalay and Delson, 1979), but Heaton (2006) recognizes the latter species as distinct and endemic to Kromdraai, along with *Gorgopithecus major*.

Genus PARAPAPIO Jones, 1937

Diagnosis Freedman (1957); Eisenhart (1974); Heaton (2006); Gilbert (2007b). A papionin of medium to large size distinguished by a straight or only slightly concave facial profile from nasion to rhinion, a lightly built and nonprojecting supraorbital torus, weak or absent maxillary ridges, poorly excavated or absent maxillary and mandibular fossae, and moderate sexual dimorphism in the cranium and canine-premolar complex.

Description Many morphs of this genus are known, with the earliest being those in eastern Africa. The greatest diversification of *Parapapio* took place in southern Africa, after a presumed southward dispersal event. Species of the genus were probably ancestral to all of the papionins that emerged in the Pliocene and Pleistocene—that is, to the earliest forms of *Theropithecus*, *Lophocebus*, *Papio*, *Mandrillus*, and *Cercocebus*. The precise nature and timing of this diversification have not been illumined by the fossil record, but molecular evidence for the timing of the major cladogenetic events within the Papionina leaves little doubt that the African radiation of the subtribe derived from *Parapapio* ancestry.

Age Late Miocene–early Pleistocene.

African Occurrence Eastern and southern Africa.

Remarks The lack of unambiguous and well-defined apomorphies has resulted in the repeated reassessment and reallocation of species between and among *Parapapio* morphs, and between *Parapapio* and *Papio*. This has resulted in a confusing literature and multiple conflicting synonymies (Leakey and Delson, 1987; Jablonski et al., 2008b). Recent studies have shed light on the diversity of *Parapapio* in southern and eastern Africa, respectively, but Pan-African comparisons are urgently needed in order to determine which species are shared between regions and the timing and nature of dispersal or vicariant events.

The earliest and most primitive members of the genus are known from East Africa, in the form of the late Miocene species, *P. lothagamensis*, and the early Pliocene species *P. ado* from Laetoli and *P.* cf. *ado* from Allia Bay at Koobi Fora and Kanapoi. *Parapapio* appears to have undergone its first diversification event in the Mio-Pliocene of eastern Africa. One possible scenario is as follows: In the early to middle Pliocene, an East African *Parapapio* species, possibly *P. jonesi*, dispersed to southern Africa, where it or a descendant form underwent regional diversification to give rise to the many species of *Parapapio* known from the Plio-Pleistocene cave sites of South Africa. The ancestor of modern *Papio* almost certainly arose from *Parapapio*, but the fossil record is not informative as to whether that evolutionary step occurred in northeastern or southern Africa. The molecular evidence suggests the latter (Newman et al., 2004).

PARAPAPIO ADO (Hopwood, 1936)

Diagnosis Leakey and Delson (1987). A small species of *Parapapio* distinguished from *Papio* and larger species of *Parapapio* by its more sloping mandibular symphysis, with a projecting alveolar margin, straight-sided molars, an absence of cuspules in the median lingual clefts of the upper molars, and slight buccal flexures between the M3 lophids that slightly kink the tooth.

Description Most of the specimens referred to this species are fragmentary jaws, the most complete being the mandible of a young adult female, LAET 1209 (Leakey and Delson, 1987). A mandible from Kanapoi, originally identified as possibly *P. jonesi* (Patterson, 1968), was tentatively reassigned to this species (Leakey and Delson, 1987; Frost, 2001b), and collections of recent mandibles and dentition have confirmed this (Harris et al., 2003). A collection of isolated molars and fragmentary jaws including the male partial maxilla KNM-ER 37118 and partial mandible KNM-ER 37060 from East Turkana has been identified tentatively as *Parapapio* cf. *ado* (Jablonski et al., 2008b).

Age Early Pliocene, 4.1–3.49 Ma.

African Occurrence Laetoli (Tanzania); Kanapoi, and Lokochot and Lonyumun members of the Koobi Fora Formation, East Turkana (Kenya).

Remarks The poor preservation of many of the remains referred to this species, and—especially—the absence of diagnostic facial material, precludes the clear morphological definition of the taxon. As discussed in connection with *Cercopithecus*, the lack of relevant fossils is probably due to taphonomic bias adversely affecting the preservation of small-bodied monkeys.

PARAPAPIO BROOMI Jones, 1937
Figure 23.11

Diagnosis Heaton (2006). A species of *Parapapio* of moderate size distinguished from *Parapapio jonesi* by more strongly developed maxillary ridges and maxillary fossae, from *Parapapio whitei* by smaller size, and shorter rostrum, and from *Papio izodi* by less marked development of the same features; shares with *Parapapio jonesi* but not with *P. izodi* anterolaterally "fleeting" zygomatic arches.

Description The species is known from a large number of partial crania (figure 23.11) from Sterkfontein including the female STS 255 and the STS 253, as well as a slightly larger morph from Bolt's Farm, BF 43.

Age Plio-Pleistocene.

African Occurrence Members 2 and 4, Sterkfontein; Members 2–4 at Makapansgat; and Bolt's Farm (South Africa).

Remarks Heaton's research has suggested that *P. whitei* (discussed later) is a junior synonym of *P. broomi*, with the expanded concept of the species being more sexually dimorphic. If his analysis is correct, this would result in significant expansion of the sample size the specimens assigned to *P. broomi* (Heaton, 2006).

PARAPAPIO JONESI Broom, 1940
Figure 23.12

Diagnosis Following Heaton (2006), a small *Parapapio* morph distinguished from *Parapapio broomi* and *Papio izodi* by a straight nasal profile and the least marked development of maxillary ridges and fossae (Frost, 2001b; Frost and Delson, 2002).

Description Best known from South Africa, where it is represented by cranial and abundant gnathic remains including female cranium STS 565 and male cranium SWP 2947, and craniodental remains from Hadar and the Middle Awash (Frost, 2001b; Frost and Delson, 2002).

Age Middle Pliocene.

African Occurrence Makapansgat and Members 2 and 4, Sterkfontein (South Africa); tentatively from Kada Hadar Member, Hadar Formation (Ethiopia), and below the Sidi Hakoma Tuff in the Middle Awash (Ethiopia).

Remarks The identification of *P. jonesi* from the Hadar Formation (ca. 3.0 Ma) and from Middle Awash sediments (3.5 Ma) is based on a nearly complete male skull, female face, and several mandibular fragments (figure 23.12; Frost and Delson, 2002). Morphologically, this material is similar to that from

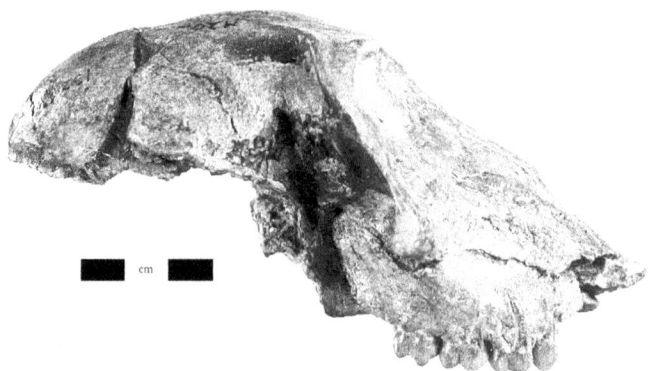

FIGURE 23.11 *Parapapio broomi*. Lateral view of male cranium MP 2 (M 202). Photograph courtesy of Leonard Freedman.

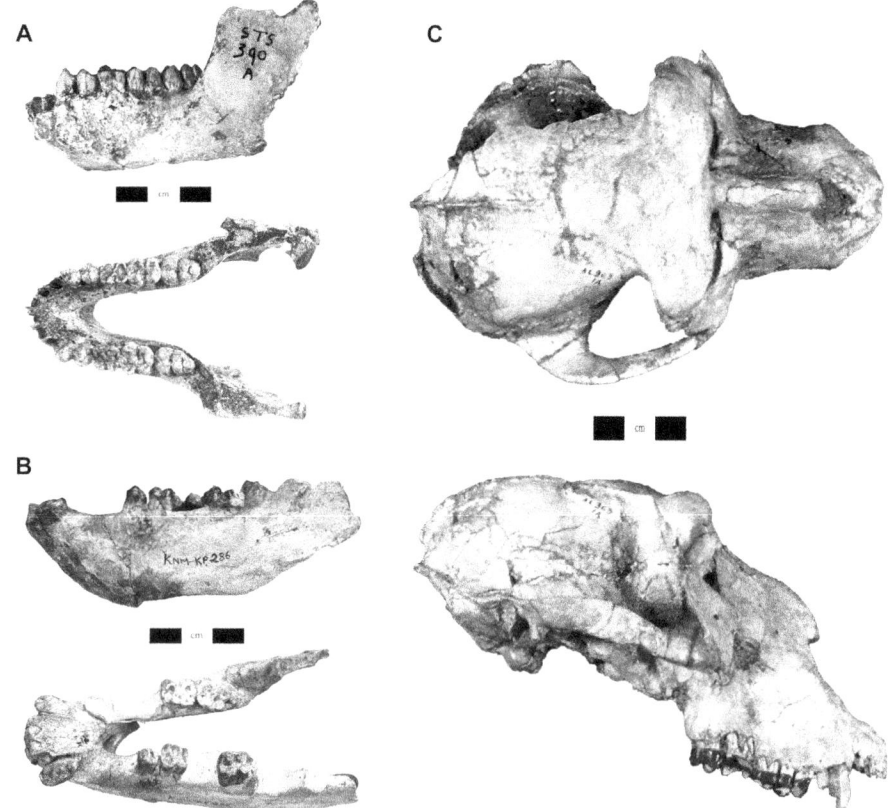

FIGURE 23.12 *Parapapio jonesi.* Comparison of specimens allocated or tentatively allocated to *P. jonesi* from different parts of Africa. A) Lateral and occlusal views of female mandible STS 390A from Sterkfontein; photograph courtesy of Leonard Freedman. B) Lateral and occlusal views of male mandible KNM-KP 286 from Kanapoi. © National Museums of Kenya. C) Vertical and lateral views of male cranium A.L.363-1a from Hadar.

Makapansgat and Sterkfontein Member 4, but it differs in the prominent and rounded nasal bones and robust and thickened brow ridge of the male cranium. If the remains assigned to *P. jonesi* from northeastern and southern Africa are conspecific or closely related, it would establish the species as the earliest papionin with a nearly Pan-African distribution. If, as is argued later, *Parapapio* is originally a northeastern or eastern African genus, then *P. jonesi* or a closely related morph may represent the base of the genus' radiation in South Africa (Heaton, 2006).

PARAPAPIO LOTHAGAMENSIS Leakey et al., 2003

Diagnosis Leakey et al. (2003). Distinguished from all other *Parapapio* species by its small size; long, narrow, and proclined mandibular symphysis; broad P3; and deciduous P4 lacking a transverse crest. Shares with *Victoriapithecus* but not most other cercopithecoids considerable molar flare, distally constricted m3, and other characteristics.

Description Known from mostly fragmentary gnathic and postcranial remains and isolated teeth, the most complete specimen of the species is the holotype mandible, KNM-LT 23091.

Age Late Miocene, 7.4–ca. 5 Ma.

African Occurrence Lower and Upper members, Nawata Formation, Lothagam (Kenya).

Remarks This species is the oldest, smallest, and most primitive of the recognized papionins. Its presence at Lothagam, and the existence of other Mio-Pliocene morphologically similar forms of *Parapapio* at nearby sites in the Omo–Lake Turkana Basin, suggests that the region was a center of diversification for *Parapapio.*

PARAPAPIO WHITEI Broom, 1940

Diagnosis Larger than other species of *Parapapio.* Rostrum is relatively longer than that of *P. broomi.*

Description The species is known from a series of crania from Makapansgat and more fragmentary remains from Sterkfontein.

Age Plio-Pleistocene.

African Occurrence Members 2 and 4, Sterkfontein, and Members 2–4 Makapansgat (South Africa).

Remarks Freedman (1957) recognized three species of *Parapapio* at Sterkfontein and Makapansgat, largely on the basis of dental size. In response to the poorly defined boundaries among these species, Heaton undertook an important reevaluation of *Parapapio* (Heaton, 2006). He synonymized *P. broomi* and *P. whitei*, and if this is correct, it would result in the definition of a more sexually dimorphic *P. broomi.*

PARAPAPIO sp. indet.

Diagnosis Includes small-, medium-, and large-sized examples of the genus that are too fragmentary or that are represented by too few specimens to diagnose to species.

Description This sample includes partial and fragmentary jaws, and a large number of isolated papionin postcrania that lack the apomorphies of *Papio* or *Theropithecus.*

Age Plio-Pleistocene.

African Occurrence Apak and Kaiyumung members, Nachukui Formation, Lothagam (Kenya); Lonyumun, Lokochot, Upper Burgi, and KBS members, Koobi Fora Formation, East Turkana (Kenya)(Leakey et al., 2003; Jablonski et al., 2008b).

Remarks This collection almost certainly includes one or more new species, but the parlous nature of the material has precluded the erection of formal taxa to date. Prior to 3 Ma, *Parapapio* was the dominant papionin in the Turkana Basin. After that time, the genus slips into the background of the Turkana Basin monkey fauna, seemingly having been eclipsed by species of *Theropithecus*, while continuing to thrive in southern Africa. *Parapapio* does not disappear altogether from East Africa, however, because it continues to be represented by a series of morphologically intriguing, but poorly known morphs that are distinct from their *Theropithecus* contemporaries or from modern *Papio* baboons (Frost, 2001b).

Genus *PLIOPAPIO* Frost 2001
PLIOPAPIO ALEMUI Frost, 2001

Diagnosis Frost (2001a). A medium-sized papionin similar in cranial and dental size to larger species of *Macaca* but smaller than *Mandrillus*, most subspecies of modern *Papio*, *Gorgopithecus*, *Paradolichopithecus*, *Dinopithecus*, or extinct species of *Theropithecus*. Distinguished from *Papio* by a rounded muzzle dorsum and weak or absent maxillary ridges and facial fossae, and from *Parapapio* by a clear anteorbital drop and ophryonic groove.

Description *Pliopapio alemui* is best known from the male holotype cranium, ARA-VP-6/933, and a large sample of mandibles, maxillae, and dentition from Aramis and contemporary sites (Frost, 2001b); also present at Gona (Semaw et al., 2005), and known from a female maxilla and mandible, along with more fragmentary remains from the 5.2-Ma Kuseralee Member of the Sagantole Formation, and several isolated teeth from the Adu Asa Formation may extend its range back to 5.7 Ma (Frost et al., 2009).

Age Latest Miocene–early Pliocene, 5.2–4.2 Ma.

African Occurrence Aramis and Kuseralee members of the Sagantole Formation, Middle Awash; and Gona (Ethiopia).

Remarks *Pliopapio* is a medium-sized papionin from the Afar region of Ethiopia. Its phylogenetic position is not known, but Frost (2001b) hypothesized that it may represent a stem member of the *Papio*, *Lophocebus*, *Theropithecus* clade, although a position closer to *Parapapio* or a stem African papionin cannot be ruled out. The obliquely oriented (proclined) morphology of the mandibular symphysis that *Pliopapio* shares with early species of *Parapapio* such as *P. ado* and *P. lothagamensis* suggests either a close phyletic relationship or similar patterns of ingestion and mastication, or both.

Genus *PROCERCOCEBUS* Gilbert, 2007
PROCERCOCEBUS ANTIQUUS (Haughton, 1925)

Diagnosis Gilbert (2007b). A small- to medium-sized papionin distinguished from other papionins by an extremely straight nasal profile, widely divergent anterior temporal lines, and upturned nuchal crest, and other features; and from *Cercocebus* primarily in its more widely divergent temporal lines and smaller premolars.

Description Recognized only from Taung, the best specimens of the species are the lectotype, SAM 5384, a partial

male cranium, a nearly complete adult male cranium, UW-AS TP9, and a partial adult female cranium, UW-AS TP8 (Gilbert, 2007b).

Age Plio-Pleistocene.

African Occurrence Taung (South Africa).

Remarks As mentioned in connection with *Lophocebus* earlier, the recognition of *Procercocebus* in South Africa lends weight to the hypothesis that the two lineages of extant mangabeys evolved from geographically distant forebears in eastern and southern Africa.

Genus *THEROPITHECUS* I. Geoffroy Saint-Hilaire, 1843

Diagnosis Delson (1993); Jablonski (2002); Jablonski et al. (2008b). Large, heavily built, and highly sexually dimorphic papionins distinguished from other papionins by high-crowned and columnar-cusped cheek teeth with deep foveae and pronounced infoldings of thick enamel, mandibular teeth arranged in an anteroposteriorly convex curve (Curve of Spee) in most species, anterior union of temporal lines and long sagittal crest, a high opposability index produced by its elongated pollex and abbreviated index finger, and many other features.

Description Following Delson, *Theropithecus* is recognized here as consisting of two subgenera, *T. (Omopithecus)* and *T. (Theropithecus)*, which represent the two major lineages within the genus as they are now understood (Delson, 1993). An early, if not the earliest *Theropithecus*, is recognized from early Pliocene sequences of the Koobi Fora Formation in the Omo–Lake Turkana Basin (described later under *Theropithecus* sp. indet.) and appears to represent an early member of the *T. (Omopithecus)* lineage. Both subgenera may appear at similar times in the fossil record, however, with *T. (Omopithecus)* in early Pliocene sequences of the Koobi Fora Formation, and *T. (Theropithecus)* potentially from Wee-ee in the Middle Awash. The *T. (Omopithecus)* lineage was associated with well-watered environments, including riparian forests, in the Omo–Lake Turkana and Baringo basins during the Pliocene. The *T. (Theropithecus)* lineage, by contrast, dominated more open, grassland-dominated environments throughout the Pleistocene, and it is known from localities in northern, northeastern, eastern, and southern Africa. The sole living representative of the genus, *T. (T.) gelada*, is found only in the grass-covered highlands of central Ethiopia.

Age Pliocene–Recent.

African Occurrence Pan-African, except for equatorial rain forests.

Remarks Fossils of *Theropithecus* occur more widely and in larger numbers than those of any other primate genus in Africa, and the genus appears to have been one of the most successful primates of all time. This was due to a remarkable combination of anatomical features of the hand, dentition, and masticatory apparatus that permitted theropiths to harvest, ingest, and chew large quantities of vegetation, especially grass. These specializations allowed open-country-dwelling forms of the genus, especially *T. (T.) oswaldi*, to successfully compete in savanna and savanna-woodland environments against hooved mammals until the middle Pleistocene, when a combination of factors, notably increasing environmental seasonality and an inability to engage in seasonal migrations, led to the extirpation of the genus from all but the most remote and competition-free grassland environments. The genus is represented today by only one relict species in the Ethiopian highlands, *T. gelada*. The subgeneric definitions advanced by Delson for division of extinct *Theropithecus* species (Delson, 1993) are followed here, as in a previous review (Jablonski, 2002). The first comprehensive description of fossil *Theropithecus* by Jolly (1972) is a classic work in vertebrate paleontology, to which modern studies of fossil primate functional anatomy and ecomorphology owe their inspiration.

Subgenus *THEROPITHECUS* I. Geoffroy Saint-Hilaire, 1843

Diagnosis Distinguished from *Theropithecus (Omopithecus)* by tendencies toward reduction of muzzle length, weak development of maxillary ridges and corresponding rounding of the muzzle cross section, reduction of incisors, and molarization of premolars.

Description After the tentatively identified material from Wee-ee, Middle Awash, already mentioned, the earliest members of the subgenus, represented by *T. (T.) darti*, are known from abundant remains at Hadar and the Middle Awash (including Maka and contemporary sites) in Ethiopia, and Makapansgat in South Africa. *Theropithecus (T.) oswaldi* is recognized from almost all Plio-Pleistocene fossil sites in Africa, with the largest samples deriving from sites in the Omo–Lake Turkana Basin and Kanjera and Olorgesailie in Kenya and Swartkrans in South Africa.

Age Plio-Pleistocene.

African Occurrence North, northeastern, eastern, and southern Africa (including Algeria, Ethiopia, Kenya, Tanzania, and South Africa).

Remarks The fossils of *T. (Theropithecus)* form a geographically and temporally extensive array of medium- to large-bodied papionins. Some students have lumped constituent taxa into a single species, *T. oswaldi*, with three slightly morphologically differentiated chrono-subspecies: *T. oswaldi darti*, *T. oswaldi oswaldi*, and *T. oswaldi leakeyi* (Leakey, 1993). *Theropithecus (Theropithecus) darti* is here retained as a separate species, following Eck and Jablonski (1987), but questions remain over the validity of this species and its certain identification from widely geographic distant sites in northeastern, eastern, and southern Africa. One of us (S.F.) considers the *darti* morph to be the earliest subspecies of *T. (T). oswaldi*.

THEROPITHECUS (THEROPITHECUS) DARTI
Broom and Jensen, 1946
Figure 23.13

Diagnosis Distinguished from *T. (T.) oswaldi*, by its smaller overall size, narrow and ovoid piriform aperture, a large infratemporal fossa contained by a broadly bowed and smoothly curved zygomatic arch, and a deeply excavated triangular depression on the mandibular ramus.

Description The largest sample of this species is from Hadar, which has produced a partial skull of a male (AL205-1a–1c; figure 23.13) as well as many partial crania of juveniles and subadults (Eck, 1993). The species was first recognized from Makapansgat in South Africa, where a male mandible with broken cheek teeth, MP 1 (M201), represents the holotype (Freedman, 1957; Maier, 1972). Makapansgat has also produced a complete female juvenile skull with occluded cranium and mandible, M3073, and the adult female cranium MP 222. Fragmentary specimens of *T. (T.). darti* are also known from lower Omo Group and Middle Awash deposits (Eck, 1987, 1993; Frost, 2001b), and from East Turkana, where the species

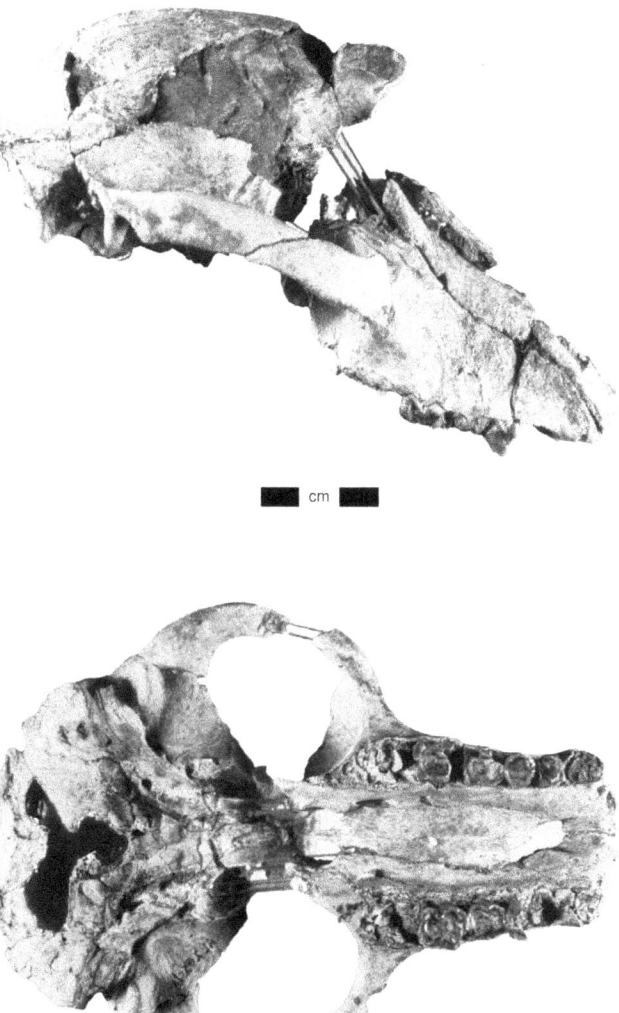

FIGURE 23.13 *Theropithecus (Theropithecus) darti.* Lateral and basal views of male cranium AL 205-1. Photographs courtesy of Gerald Eck.

is represented by a fragmented male cranium, KNM-ER 3025 (Leakey, 1993).

Age Pliocene, 3.5–2.4 Ma.

African Occurrence Upper Lokochot Member, Koobi Fora Formation, East Turkana (Kenya); lower members of the Shungura Formation, Omo Group (Ethiopia), Hadar, Formation "W" below the Sidi Hakoma Tuff, Middle Awash (Ethiopia); and Makapansgat (South Africa).

Remarks *Theropithecus (T.) darti* was geographically widespread, but was only reasonably common in northeastern Africa. Although the taxonomic equivalence of the different geographic representatives of the species is still debated, the occurrence of *T. (T.) darti* in South Africa appears to be the result of a dispersal event near the end of the species' history. Like *T. (T.) oswaldi*, *T. (T.) darti* appeared to enjoy limited success in southern Africa probably because of competition from species and descendants of *Parapapio*. As this species grades into *T. (T.) oswaldi*, it is often regarded as an early chrono-subspecies *T. (T.) oswaldi darti* (Leakey, 1993; Frost, 2001a; Frost and Delson, 2002; Frost and Alemseged, 2007).

THEROPITHECUS (THEROPITHECUS) OSWALDI
Andrews, 1916
Figure 23.14

Diagnosis Same as for genus and subgenus. Distinguished from other papionins, especially *T. darti*, by its larger size, robust postglenoid processes and greater relative reduction of incisors and canines. This long-lived species exhibited considerable variation through time.

Description Most students recognize at least two subspecies, an earlier and smaller *T. (T.) oswaldi oswaldi* and a later and larger *T. (T.) oswaldi leakeyi*. Both morphs are represented by abundant remains, including partial skeletons, which are enumerated elsewhere (Eck, 1987; Delson et al., 1993; Frost, 2001b; Frost and Delson, 2002; Frost and Alemseged, 2007). Occurrences of the former subspecies include partial crania from the Middle Awash, material from the upper part of the Hadar Formation, Konso, and large numbers of specimens from upper members of the Shungura Formation from Ethiopia (Jablonski et al., 2008b); from plentiful cranial and postcranial remains from the upper Burgi and KBS members of the Koobi Fora Formation (Delson et al., 1993), West Turkana and Kanjera in Kenya; and from Olduvai Gorge in Tanzania (Freedman, 1970; Delson et al., 1993; Heaton, 2006; Frost and Alemseged, 2007; Gilbert, 2007a; Frost et al., 2009; figure 23.14). *Theropithecus (T.) oswaldi leakeyi* is known from the Daka Member of the Bouri Formation, the Dawaitoli Formation, Andalee in the Middle Awash, Asbole in the Afar, the southern Ethiopian site of Konso, and Member L of the Shungura Formation. It is also recognized from the Okote Member of the Koobi Fora Formation, West Turkana, and Olorgesailie in Kenya, from Olduvai Gorge and Peninj in Tanzania, and from several South African cave sites (Delson and Hoffstetter, 1993). It is also known from the northern African sites of Ain Jourdel and Ternifine, Algeria, and Thomas Quarries, Morocco (Delson et al., 2000). The largest and probably last representative of the species appears to have occurred at Kapthurin in the Baringo Basin.

Age Plio-Pleistocene, 2.5–0.25 Ma.

African Occurrence Ain Jourdel and Ternifine (Algeria); Thomas Quarries (Morocco); Upper Kada Hadar Member, Hadar Formation (Ethiopia); Matabaietu Formation, and Hata and Daka members, Bouri Formation, Dawaitoli Formation, Middle Awash (Ethiopia); Asbole, Afar Region (Ethiopia); Melka Kontoure (Ethiopia); Konso (Ethiopia); Members D–L, Shungura Formation (Ethiopia); upper Burgi, KBS, and Okote members, Koobi Fora Formation, East Turkana (Kenya); Nachukui Formation, West Turkana (Kenya); Olorgesailie (Kenya); Kapthurin (Kenya); Beds I–IV and Masek Beds, Olduvai Gorge and Peninj (Tanzania); Kaiso (Uganda); and Swartkrans, Sterkfontein (Member 5), Gladysvale, Hopefield, Coopers, and Bolt's Farm (South Africa).

Remarks *Theropithecus oswaldi* was a chronospecies, and the subspecies recognized by some workers represent specimen allocations based on size, enamel folding complexity, anterior dental reduction, or occurring on either side of depositional

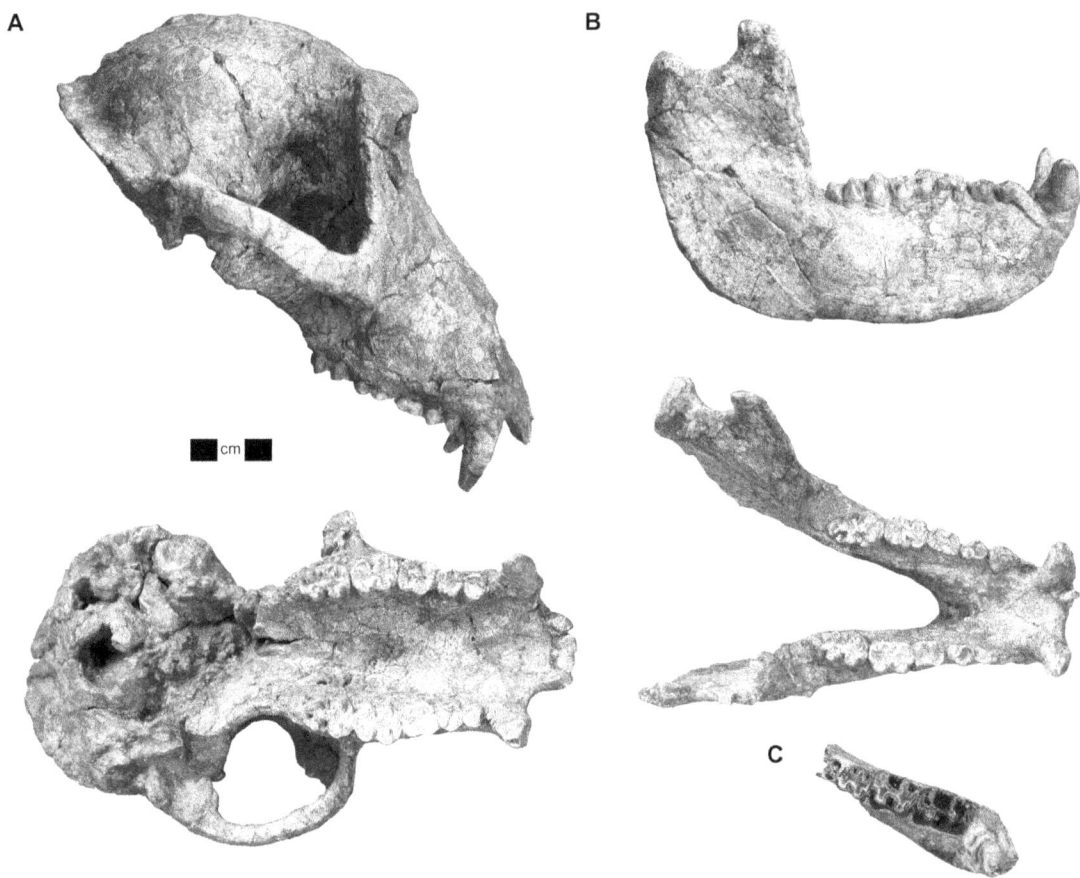

FIGURE 23.14 *Theropithecus (Theropithecus) oswaldi*. A) Lateral and basal views of male cranium KNM-ER 18925; B) lateral and occlusal views of associated mandible KNM-ER 18925; C) occlusal view of partial left mandible KNM-ER 880. Note the crown height and complex infoldings of enamel that characterize the late examples of this species.

hiatuses. Through the long tenure of the species, the animals became considerably larger, with estimates for male body mass ranging from 42 kg in early representatives of the species to >85 kg by the time the species went extinct (Delson et al., 2000). *Pari passu*, the postcanine dentition underwent increases in occlusal surface area and in the length and complexity of molar enamel ridges, while the anterior dentition underwent significant reduction (Jolly, 1972; Jablonski et al., 2008b). *Theropithecus oswaldi* was one of the most highly specialized primates that ever lived. The species was large bodied, highly sexually dimorphic, mostly—if not exclusively—terrestrial, and capable of chewing prodigious amounts of manually harvested vegetation (mostly grass) with its powerful jaws and ungulate-like teeth (Benefit and McCrossin, 1990; Jablonski, 1993). *Theropithecus oswaldi* was the only primate other than *Homo* known to have dispersed out of Africa in the Plio-Pleistocene; fragmentary but convincing fossil remains are known from Spain and India (Delson et al., 1993).

Subgenus *OMOPITHECUS* Delson, 1993

Diagnosis Jablonski (1994). Distinguished from *Theropithecus (Theropithecus)* by an elongated muzzle with a flat dorsum and well-developed maxillary ridges, a large and anteriorly expanded zygoma, a robust zygomatic arch that is triangular in cross section, a robust mandibular symphysis bearing sinusoidal mental ridges, and features of the postcranial skeleton that reflect adaptations for elbow stability and shoulder flexibility.

Description This subgenus contains three recognized species, *T. (O.) baringensis*, *T. (O.) brumpti*, and *T. (O.) quadratirostris*, from the Omo–Lake Turkana Basin of Kenya. Remains referred to *T. (O.) baringensis* have also been described from Angola (Jablonski, 2002). All species are represented by complete crania (Jablonski et al., 2002), and *T. (O.) brumpti* from extensive fossils including a partial skeleton (Jablonski et al., 2008b)

Age Pliocene, 3.94–2.0 Ma.

African Occurrence Chemeron (Kenya); Tugen Hills (Kenya); Lonyumun, Lokochot and Tulu Bor members, Koobi Fora Formation, East Turkana (Kenya); Nachukui Formation, West Turkana (Kenya), Members B–G, Shungura Formation, Omo Group (Ethiopia); Usno Formation, Omo Group (Ethiopia).

Remarks *Theropithecus (Omopithecus)* contains the earliest recognized fossils of *Theropithecus*, as represented by isolated molar teeth (Benefit and McCrossin, 1990). *Theropithecus (O.) quadratirostris* and *T. (O.) baringensis* exhibit broad, boxy muzzles that lack well-defined maxillary ridges and zygomatic arches that show anterior thickening, not flare. Their molar teeth are low crowned and lack the pinched, columnar cusps typical of most *T. (O.) brumpti*. The dietary preferences of the species in the *T. (Omopithecus)* lineage are not well-known but do not appear to be closely comparable to those of living primates (Jablonski, 1986; Jablonski et al., 2002).

Postcranial fossils are only known from *T. (O.) brumpti*, and these indicate that the species was mainly terrestrial but that it had a high opposability index and was capable of undertaking the same kind of dexterous manipulation of food and other items that the gelada does today (Jablonski et al., 2002). *Theropithecus (Omopithecus) brumpti* may have been quite similar to the living mandrill in appearance and behavior (Jablonski, 1994).

THEROPITHECUS (OMOPITHECUS) BARINGENSIS
(Leakey, 1969)
Figure 23.15

Diagnosis Same as for the subgenus. *Theropithecus (Omopithecus) baringensis* shows the most conservative facial morphology of the subgenus, with a broad muzzle lacking well-defined maxillary ridges and zygomatic arches showing anterior thickening but no anterior flare. The molars are somewhat lower crowned and exhibit less pinching of the cusps than in *T. (O.) brumpti*.

Description The species is best known from the holotype, a complete skull, KNM-BC 2, from Chemeron (figure 23.15). It is also known from a mandibular specimen, KNM-BC 1647, from the same site, and a series of cranial remains from the Angolan site of Leba including a complete juvenile cranium, TCH 38 '90, and a partial cranium of a male, TCH 25 '90 (Delson and Dean, 1993). Other remains referable to this species are discussed and illustrated as *Papio quadratirostris* by Delson and Dean (1993).

Age Pliocene, 3–2 Ma (for Chemeron)(Gundling and Hill, 2000).

African Occurrence Chemeron (Kenya); Leba (Angola).

Remarks Originally classified as *Papio baringensis* (Leakey, 1969), the classification and phyletic position of this species continues to be debated in the literature (Eck and Jablonski, 1984, 1987; Delson and Dean, 1993; Leakey, 1993; Jablonski et al., 2002). The holotype and paratype are best accommodated within *Theropithecus (Omopithecus)* because of the relatively small size and weakly developed columnar cusp morphology of the molars that presage the conditions seen in *T. (O.) brumpti*.

THEROPITHECUS (OMOPITHECUS) BRUMPTI
(Arambourg, 1947)
Figure 23.16

Diagnosis Same as for the subgenus, with accentuation of facial features, specifically those of the maxillae and zygomatic arch. *Theropithecus (O.) brumpti* is distinguished from other species of the subgenus by very high-crowned, straight-sided, and columnar-cusped molars; more deeply excavated fossae of the mandibular corpus; more pronounced maxillary ridges; and anterolaterally flaring zygomatic arches present in both sexes.

Description Good examples of this species (figure 23.16) have been retrieved from deposits of the Omo and Turkana basins and include a partial skeleton, KNM-WT 39368, two nearly complete crania of adult males (L345-287 and KNM-WT 16828), partial crania of females (L32-155 and L122-34), and nearly complete mandibles of males (L576-8, KNM-ER 2015) (Eck and Jablonski, 1987).

Age Pliocene, 3.4–2.0 Ma.

African Occurrence Lokochot and Tulu Bor members, Koobi Fora Formation, East Turkana (Kenya); Nachukui Formation, West Turkana (Kenya); Members B–G, Shungura Formation, Omo Group (Ethiopia).

Remarks This species would have been one of the most extraordinary-looking monkeys that ever lived. Its large and anterolaterally flaring zygomatic arches are unique among mammals and supported the attachment of a greatly enlarged superficial masseter muscle, which permitted a wide gape for canine displays (Jablonski, 1986; Jablonski et al., 2002). The shape of the maxillary ridges and the shape and orientation

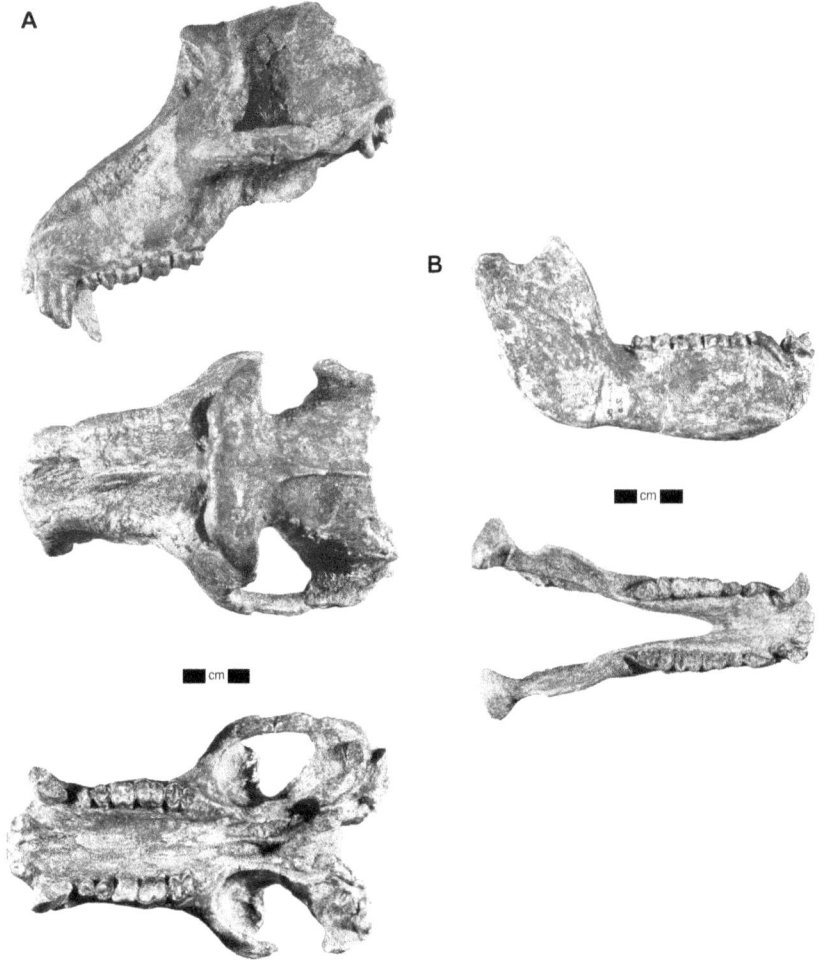

FIGURE 23.15 *Theropithecus (Omopithecus) baringensis*. Views of male skull, KNM-BC 2 (holotype). A) Lateral, vertical, and basal views of cranium; B) lateral and occlusal views of mandible (lateral is a mirror image). © National Museums of Kenya.

of the zygomatic flare vary considerably between individuals of the same sex. Remains of the hand skeletons of two individuals exhibit the elongated pollex and abbreviated index digit that contributed to excellent opposability, as seen in extant *Theropithecus gelada* (Iwamoto, 1982). The discovery of this manual architecture in *T. (O.) brumpti* indicates that the configuration probably was primitive for the genus and that it may have played an important role facilitating food harvesting and manipulation.

THEROPITHECUS (OMOPITHECUS) QUADRATIROSTRIS (Iwamoto, 1982)

Diagnosis Same as for the subgenus, except that diagnostic features of the cranium appear intermediate in state between those of *T. (O.) baringensis* and *T. (O.) brumpti*. The broad, long muzzle bears rounded maxillary ridges, the robust zygomatic arches are triangular in cross section at the zygomaticomaxillary suture and thickened anteriorly, and the broad frontal process of the zygoma lends depth to the malar region, as in other *Theropithecus*. The species is distinguished from *T. (O.) baringensis* by its elongated neurocranium, more posteriorly oriented temporalis musculature, and posterior union of the temporal lines and formation of sagittal crest posterior to the bregma in males.

Description The species is known only from the holotype, a nearly complete cranium of a male (unnumbered; Delson and Dean, 1993)

Age Pliocene, 3.3 Ma.

African Occurrence Usno Formation, Omo Group (Ethiopia).

Remarks This species is only recognized with certainty from the type specimen; specimens allocated by Delson and Dean (Iwamoto, 1982) to *Papio quadratirostris* from Angola are assigned to *T. (O.) baringensis* (Jablonski, 1994). The material from the Shungura Formation is discussed under *Papio*. As in the case of *T. (O.) baringensis*, this species was originally assigned to *Papio* (Eck and Jablonski, 1984) and was subsequently referred to *Theropithecus* (Delson and Dean, 1993). This referral has been questioned (including by S.F.) (Eck and Jablonski, 1984) but is retained here because of the synapomorphies that unite it with both *T. (O.) baringensis* and *T. (O.) brumpti*, as discussed elsewhere (Jablonski et al., 2008b).

THEROPITHECUS (OMOPITHECUS) sp. indet.

Diagnosis Molars with moderately pinched, columnar cusps covered with thick enamel; distinguished from *Theropithecus (Theropithecus)* by molars with lower crowns, lower cusp relief, and less pronounced enamel infoldings.

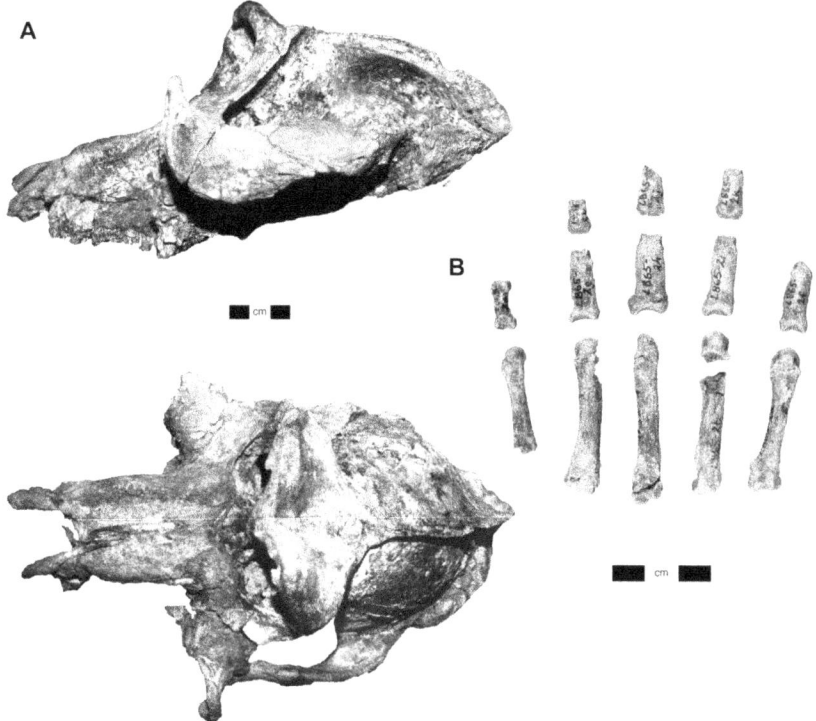

FIGURE 23.16 *Theropithecus (Omopithecus) brumpti.* A) Lateral and vertical views of male cranium KNM-WT 16828. Note strongly anteriorly expanded and flaring zygomatic arches. © National Museums of Kenya. B) Dorsal view of the hand skeleton of *T. (O.) brumpti* (L865-2) showing the derived condition of elongated metacarpals and abbreviated proximal phalanges that contribute to a high opposability index.

Description Specimens assigned to this morph include mostly isolated teeth and a few highly weathered jaw and postcranial fragments that have been described from Allia Bay at Koobi Fora and that were recently assigned to *Theropithecus* sp. indet. or cf. *Theropithecus* (Kalb et al., 1982b).

Age Pliocene, 3.94 Ma.

African Occurrence Lokochot and Lonyumun members, Koobi Fora Formation, East Turkana (Kenya).

Remarks These remains constitute one of the oldest, if not the oldest, occurrences of the genus. At Allia Bay, *Theropithecus* appears to have been the rarest of the monkeys, with its remains greatly outnumbered by those of *Parapapio* and *Cercopithecoides*.

THEROPITHECUS subgen. et sp. indet.

Diagnosis This small group comprises fragmentary specimens probably belonging to *Theropithecus*, but lacking sufficient apomorphies to permit more specific diagnosis.

Description A small female mandible from the Afar has been tentatively assigned to *Theropithecus* (Frost, 2001a). A male mandible from the Middle Awash may also belong to *Theropithecus* but is is more equivocal (Frost, 2001a; Benefit and McCrossin, 2002; Benefit, 2009).

Age Pliocene, 3.9 Ma.

African Occurrence Wee-ee in the Middle Awash, Afar Region (Ethiopia).

Remarks These occurrences of *Theropithecus*, if confirmed, would be the earliest known and would point to an origin of the genus in northeastern Africa. An isolated molar from

Kanapoi dated to 4.1 has also been tentatively included in *Theropithecus* (Frost, 2001a; Harris et al., 2003).

General Discussion

The African fossil record of the Cercopithecoidea defines the history of the superfamily from its origins in the middle Miocene through the latest Pleistocene. From the late Miocene onward, monkeys were a major element of the African mammalian fauna, but the composition of the monkey fauna was not stable through time. Changes in the taxonomic and morphological composition of the monkey fauna occurred as a consequence of climatic and environmental change, and because of competition with other mammals. This discussion will touch on only a small number of the most important topics brought to light by the fossil record of the African Cercopithecoidea. These are, first, the nature and meaning of the fossil record of the Victoriapithecidae and of the earliest representatives of the modern subfamilies Colobinae and Cercopithecinae. The second topic to be explored is that of the Pliocene radiations of *Cercopithecoides* and *Parapapio*. This will pave the way for discussions of the evolution of the large, terrestrial colobines and *Theropithecus*, and for examinations of the origin of the modern cercopithecoid fauna of Africa that highlight our state of knowledge concerning, especially, the evolution of the mangabeys and *Papio* baboons.

The earliest clearly defined Old World monkeys belong to the family Victoriapithecidae, with the northeastern African species *Prohylobates tandyi* generally considered the most primitive known cercopithecoid species. The circum-Mediterranean region has long been considered a geographic focus for early cercopithecoid evolution (Delson, 1973, 1974, 1975). The presence of the middle Miocene *P. tandyi* in Egypt and of the earliest representatives of the Colobinae and Cercopithecinae in late Miocene and earliest Pliocene deposits of Greece, Algeria, Libya, and Kenya considerably strengthens this surmise. Primate paleontologists have generally discounted the possibility that the ancestors of any African late Miocene or Plio-Pleistocene cercopithecids dispersed from Eurasia, but this possibility cannot be excluded at least for colobines because of the strong craniodental morphological similarities between Eurasian *Mesopithecus* and the African Mio-Pliocene genera *Libypithecus*, *Paracolobus*, and *Rhinocolobus* (Jablonski, 1998; Stewart and Disotell, 1998).

The nature of the phyletic relationships of victoriapithecids to later cercopithecids is still a topic of much debate (Benefit and McCrossin, 2002), because the constituent species do not display patterns of synapomorphies that permit their convincing assignment to one or another of the modern cercopithecid subfamilies. The conditions leading to the emergence of the Colobinae and Cercopithecinae thus remain unclear but were no doubt related to the evolution of divergent dietary preferences. The earliest colobines presumably exhibited specializations for gut fermentation and allelochemical detoxification consistent with a diet consisting mostly of seeds and leaves (Lucas and Teaford, 1994). Many such foods, especially leaves, are not patchily distributed and are not usually contested. On the other hand, the earliest cercopithecines would have possessed cheek pouches, which afforded the animals the ability to temporarily store high-quality and highly contestable food items that might require incisal or manual preparation before ingestion (Lambert, 2005; Smith et al., 2008).

The Cercopithecoides of the latest Miocene and earliest Pliocene in Africa included two large and idiosyncratic genera endemic to Ethiopia, *Kuseracolobus* and *Pliopapio*, as well as early representatives of the colobine genera *Paracolobus* and *Cercopithecoides* and the papionin genus *Parapapio*. These fossils have been recorded mostly from closed environments, and the animals have been interpreted as being exclusively arboreal or at least not exclusively terrestrial in their modes of locomotion (Hlusko, 2006, 2007). These findings denote that arboreality was the rule, not the exception, for late Miocene and early Pliocene monkeys, and that the evolution of predominantly terrestrial forms occurred only later in the history of both cercopithecid subfamilies in Africa.

Major changes in African environments and mammalian faunas occurred by the middle Pliocene. Marine and terrestrial paleoclimatic records indicate that a shift from warmer and wetter conditions to more seasonally contrasted, cooler, drier, and more variable conditions occurred after about 3 Ma (deMenocal, 2004; Feakins and deMenocal, this volume, chap. 4). From this time onward, the predominant trend was toward increased seasonality of rainfall, intensification of seasonal pulses of ecosystem productivity, and increasing climatic variability, with steplike increases in savanna vegetation apparent at 1.8 Ma, 1.2 Ma, and 0.6 Ma (deMenocal, 2004). These changes impacted the hydrological regimes of most of the continent and the vegetation on which mammalian herbivores, including the monkeys, depended (Bobe et al., 2002; Bobe and Behrensmeyer, 2004; deMenocal, 2004). From the middle Pliocene onward, the African cercopithecoid fauna underwent marked changes in numbers and kinds of species, distributions of body sizes, and dietary

and locomotor specializations (Elton, 2007). Monkeys are particularly sensitive indicators of ecosystem productivity because they require foods of relatively high nutritional yield with respect to amino acid, fatty acid, and sugar content. Monkeys are not as narrowly restricted as apes in their dietary preferences (Jablonski et al., 2000), but they require higher-quality foods than other mammals of similar body size because of their relatively large and metabolically costly brains. They are also limited in their physical ability to range widely for preferred foods because their prehensile hands and feet are no match for wear-resistant hooves when it comes to covering long distances. Thus, monkeys—like other primates, but unlike many ungulates—cannot and do not undertake long-distance migrations in order to track seasonally available food and water resources. For these and other reasons, monkeys were vanguard indicators of environmental change in the African Plio-Pleistocene. They were sensitive to environmental perturbations and were among the first to suffer during periods of climatic deterioration. There is no evidence, however, that they underwent significant pulses of speciation and extinction in coordination with specific climatic swings (Frost, 2007a).

By the middle Pliocene the geographic center for cercopithecid lineage diversification had shifted from northern and northeastern Africa to eastern Africa. Against the backdrop of prevailing environmental change, species of the colobine *Cercopithecoides* and the papionin *Parapapio* appear to have been the most successful, at least in those habitats preserved in the fossil record.

The fossil record of the Omo–Lake Turkana Basin is particularly instructive in shedding light on cercopithecid evolution from the middle Pliocene through middle Pleistocene (Jablonski and Leakey, 2008a). With the onset of increasingly seasonal rainfall regimes in the middle Pliocene, the fauna of this region became dominated by one terrestrial cercopithecine apparently restricted to riparian forest floors (*Theropithecus [O.] brumpti*) and to large-bodied colobines capable of exploiting ecotonal woodland savanna habitats. The latter included one mostly arboreal species, *Rhinocolobus turkanaensis*, and two highly terrestrially adapted forms, *Cercopithecoides williamsi* and *C. kimeui* (Jablonski et al., 2008a). The success that large-bodied colobines enjoyed from the middle Pliocene through early Pleistocene relative to cercopithecines probably had much to do with their capacity for fermentative, ruminant digestion, which would have permitted them to subsist on vegetation of lower quality and with lower requirements for fresh drinking water (Van Soest, 1982). Their command of these niches was short-lived, however, as competition from more highly mobile and better dentally equipped ruminant artiodactyls intensified. By the Plio-Pleistocene boundary, environments were considerably more heterogeneous than in the Pliocene, and pronounced differences in habitat quality and ecosystem productivity existed between areas adjoining permanent sources of water and those not. Under these conditions, monkeys probably suffered from increasing competition with ruminant artiodactyls and suids, which were becoming ever-more taxonomically and ecologically diverse (Gagnon and Chew, 2000; Cerling et al., 2003).

The earliest cercopithecoids in South Africa are from the middle Pliocene. At that time, the monkey fauna included *Cercopithecoides williamsi*, several species of *Parapapio* and *Theropithecus (T.) darti*. By the later Pliocene, *Papio izodi* was present in southern Africa. Although the details of latest

Tertiary and early Quaternary climatic and environmental change are less well-known for southern than for eastern Africa, available records indicate progressive cooling since 3.2 Ma and increasing dominance of arid-adapted vegetation (deMenocal, 2004). For reasons that are still not clear, *Parapapio* and its putative regional descendants were more diverse in southern Africa than were theropiths and colobines. *Parapapio* underwent considerable regional diversification and gave rise to the first species of *Papio* and *Procercocebus*, the probable ancestor to the *Cercocebus* lineage of mangabeys. At the same time, colobines and theropiths in South Africa continued to be represented by one ecological generalist species each: *Cercopithecoides williamsi* and *Theropithecus (T.) darti* (and later *T. [T.] oswaldi*), respectively. During the late Pliocene and early Pleistocene, *Papio* and the *Cercocebus* mangabeys appear to have remained confined to southern Africa. Judging from its Pleistocene distribution in southern Africa, *Papio* evolved in markedly seasonal environments where it developed the ecological opportunism that was eventually to propel it to greater success and continental hegemony by the late middle and late Pleistocene.

In the late Pliocene environments of eastern Africa, *Theropithecus (O.) brumpti* disappeared from its riparian forest-floor niche, probably supplanted by suines (such as *Kolpochoerus* spp.) and dry-adapted tragelaphine bovids, which colonized their habitats and more successfully exploited subcanopy food resources (Flagstad et al., 2001; Cerling et al., 2003). *Theropithecus (T.) oswaldi* became the dominant cercopithecine by the Plio-Pleistocene boundary, not by ecologically displacing *T. (O.) brumpti* but by establishing a successful niche as a terrestrial, primarily graminivorous habitué of ecotonal and open habitats (Bobe et al., 2002). In this environment, *T. (T.) oswaldi* joined the large colobine, *Cercopithecoides kimeui*, while *Rhinocolobus turkanaensis* maintained a more arboreal niche. By the early middle Pleistocene, woodland environments became rarer, the trend toward increasing seasonality of precipitation had intensified, habitats had become highly heterogeneous, and monkeys had become less common and less taxonomically diverse. *Theropithecus (T.) oswaldi* appears to have been the most successful of the monkeys in competing with ungulates and suids for grazing niches in mostly open habitats, largely because of its large, high-crowned, and occlusally complex molars. Most of the other monkey species recognized at that time were species closely comparable to modern forms and include close relatives of the gray-cheeked mangabey, *Lophocebus albigena*, colobus monkeys (*Colobus* spp.) and various guenons (*Cercopithecus* and *Chlorocebus* spp.). These species occupied uniquely primate niches, which were unassailable by hoofed competitors (Jablonski and Leakey, 2008a).

In eastern Africa, the middle and late Pleistocene were characterized by inexorably increasing seasonality of rainfall and exaggerated habitat heterogeneity. This period saw the extinction of the last of the large Plio-Pleistocene monkeys, epitomized by the demise of the colobine *Cercopithecoides kimeui* and the giant papionin *Theropithecus (T.) oswaldi*. Representatives of *Papio* in eastern Africa begin to be detected by the later middle Pleistocene. There are few sites in which *Theropithecus (T.) oswaldi* and *Papio* appear to have coexisted, at least until the very end of the middle Pleistocene, and it is unlikely that it was ecological competition with *Papio* that drove *T (T.) oswaldi* to extinction. The largely graminivorous theropith became extinct because it could not survive as a highly encephalized and relatively immobile ecological specialist. *Papio* did not take over this niche but succeeded in a completely different one, that of the intensive, highly opportunistic, and omnivorous forager.

Through time, the centers of diversification of the Cercopithecoidea in Africa shifted from the north and northeast in the middle and late Miocene to east Africa in the middle Pliocene to southern Africa in the terminal Pliocene and Pleistocene. Mostly endemic monkey faunas evolved in the these regions, but generalist forms did succeed in dispersing, first from northeast to eastern Africa in the latest Miocene, then from eastern to southern African in the later Pliocene. The enormous latitudinal expanse of Africa meant that the monkeys evolving in different regions faced markedly different climatic and environmental challenges. It is telling that, by the end of the Pleistocene, the dominant cercopithecids of open African environments—baboons of the genus *Papio*— underwent their early evolution under the highly seasonal conditions at the southernmost extremity of the continent. With few exceptions, the remaining elements of the modern African monkey fauna are committed arborealists who established and maintained highly folivorous or frugivorous niches in woodlands and forests, far from the hoofbeats and claws of the savanna.

Summary

All evidence at hand indicates that Old World monkeys originated in Africa in the early Miocene. The earliest cercopithecoids were recovered from sites in Egypt, Libya, and Kenya and have been assigned to species of Victoriapithecidae, a family lacking the anatomical specializations of later Cercopithecidae. Although one site, Maboko Island, preserves large numbers of *Victoriapithecus* fossils, victoriapithecids appear to have been genuinely rare elements of the African mammalian fauna. Until around 10 mya, hominoids were the dominant catarrhines throughout the Old World. The lack of diversity of early cercopithecoids can readily be appreciated in figure 23.17, which depicts the temporal ranges of the genera of Africa fossil Cercopithecoidea.

In the interpretation of the history of Old World monkeys in Africa, it is difficult to differentiate taphonomic effects from biogeographic trends. The fossil record indicates that the prevailing direction of dispersal of cercopithecoids in Africa was from north (or northeast) to south and began in the middle Miocene. By the late Miocene, species representing the two modern cercopithecid subfamilies, Colobinae and Cercopithecinae, were present in the fossil records of Ethiopia and Kenya; by the early Pliocene, colobine and cercopithecine fossils were plentiful in the Omo–Lake Turkana Basin. The plentiful and well-preserved monkey fossils from the basin have been tacitly considered to represent the archetypal fossil monkeys of Africa and putative ancestors for later monkey lineages. Evidence provided by detailed studies of the northeastern and southern African monkey faunas suggests, however, that the Omo–Lake Turkana Basin was geographically isolated during intervals of the Pliocene and Pleistocene and that some of its most famous fossil monkeys (e.g., *Rhinocolobus turkanaensis* and *Theropithecus brumpti*) were strictly endemic. With respect to overall biogeographic trends, the most important south to north dispersal of monkeys was that of *Papio* in the middle Pleistocene.

Morphological and molecular evidence indicates that the earliest true colobines possessed foregut fermentation (Stewart et al., 1987; Stewart, 1999), a key innovation that allowed members of the subfamily to evolve a variety of

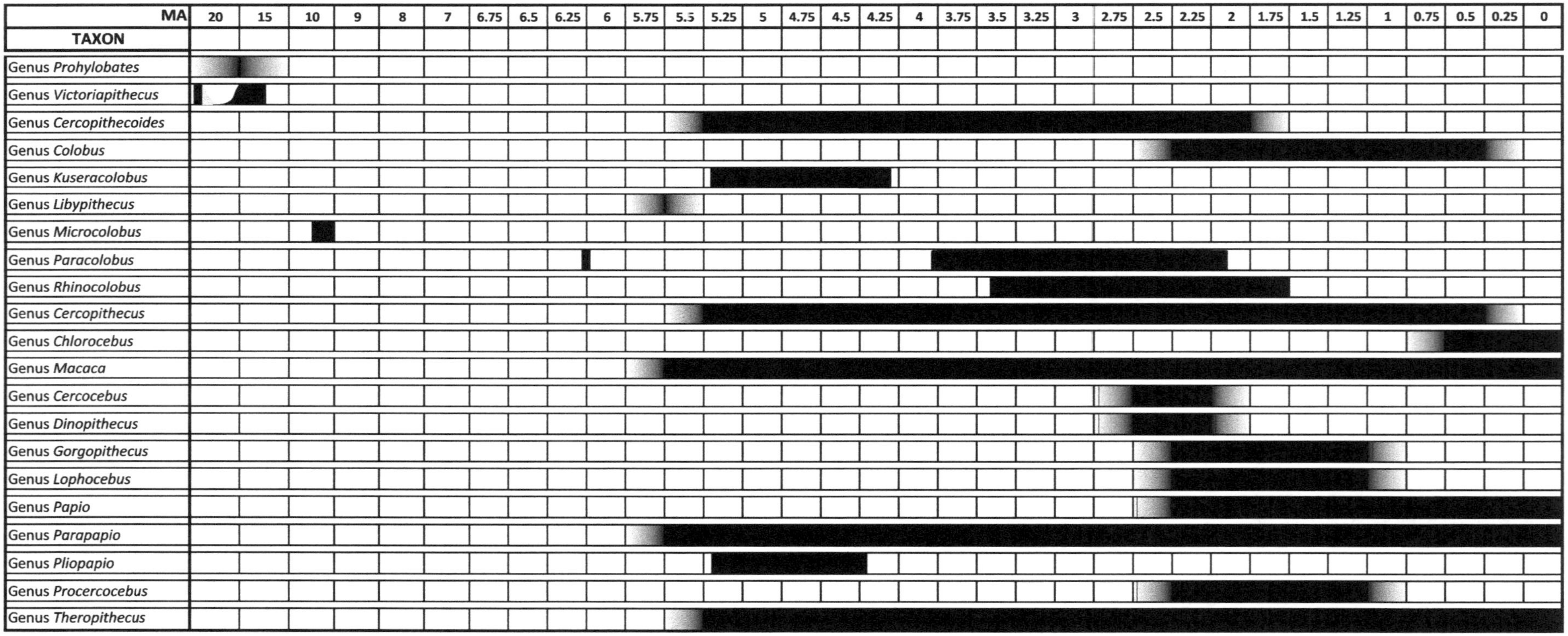

FIGURE 23.17   The temporal ranges of African fossil monkey genera. Shading at ends of range bars indicates uncertainty about dates of first and last appearances.

herbivore niches in forests, ecotonal woodland-grasslands, and open savannas through the Pliocene and Pleistocene. The Pliocene colobines were large bodied (>10 kg), and many were terrestrial or semiterrestrial. Species of the genus *Cercopithecoides* were particularly numerous and widespread. They competed favorably with ungulates in woodland-grassland and ecotonal woodland ecosystems through the Plio-Pleistocene, and they only became extinct in these habitats when increased environmental seasonality throughout Africa led to deterioration of food quality and critical shortfalls of food at important milestones in animal life histories. The modern colobine fauna of Africa consists of exclusively arboreal species, and probably includes descendants of some semiterrestrial or terrestrial Plio-Pleistocene forms.

Three genera, *Parapapio, Theropithecus*, and *Cercopithecus (sensu lato)* dominated cercopithecine evolution in Africa through the Pliocene and Pleistcene. The radiation of *Parapapio* species is still not well understood because few complete fossils are known, but most species appear to have been macaque-like in size, morphology, and habitus. Regional species or populations of *Parapapio* were probably the ancestors of *Theropithecus* and *Lophocebus* in eastern Africa, and of *Papio* and *Cercocebus* in southern Africa; in due course, the taxonomy of the genus may have to be revised to accommodate these cladogenetic events. The evolution of *Theropithecus* is well documented, with the species *T. oswaldi* being its most important representative. This widespread and highly adaptable grass eater was the only primate other than early *Homo* that dispersed out of Africa in the Plio-Pleistocene, and continues to be a well-deserved subject of study. Its extinction occurred later than that of the large-bodied colobines but was probably precipitated by a similar course of events. The evolution of the guenons (*Cercopithecus* and close relatives) is the least well documented in the African fossil record probably because of taphonomic factors that mostly prevented the preservation of forest-dwellers of smaller body size (5–10 kg). Nonetheless, the group's radiation rivaled that of papionins with regard to the total number of species produced and the total area of the continent they occupied.

The most visible and widespread of the nonhuman primates of Africa today, the baboons of the genus *Papio*, are relative newcomers whose unquestioned fossil record began only in the middle Pleistocene. These baboons have achieved success because their eclectic, opportunistic foraging strategies made it possible for them to create intensivist niches that were different from those of most of the open-country-dwelling primates who preceded them. In this way, *Papio* baboons were similar to another remarkable primate lineage, that of *Homo sapiens*.

ACKNOWLEDGMENTS

We thank Bill Sanders and Lars Werdelin for inviting us to contribute to this volume and for being extremely patient while we prepared this chapter. Discussions with Brenda Benefit, Eric Delson, Todd Disotell, Leslea Hlusko, Cliff Jolly, Meave Leakey, Ellen Miller, and Tony Tosi, especially in the context of the Cercopithecoid Analytical Working Group of the Revealing Hominid Origins Initiative, were useful in helping us organize our thoughts and keeping us current on new fossil discoveries. Eric Delson's comments on the original version of this chapter greatly improved the text. We are extremely grateful to Brenda Benefit, Gerald Eck, Leonard Freedman, and Leslea Hlusko for providing us with photos of various fossil monkey species.

Bonnie Warren of the Department of Anthropology of the California Academy of Sciences and Tess Wilson of the Department of Anthropology of The Pennsylvania State University prepared the figures for publication. Tess Wilson is thanked for maintaining the bibliographic database and for preparing the final draft for submission.

Literature Cited

Alemseged, Z., and D. Geraads. 1998. *Theropithecus atlanticus* (Thomas, 1884)(Primates: Cercopithecidae) from the late Pliocene of Ahl al Oughlam, Casablanca, Morocco. *Journal of Human Evolution* 34:609–621.

———. 2000. A new Middle Pleistocene fauna from the Busidima-Telalak region of the Afar, Ethiopia. *Comptes Rendus de l'Académie des Sciences*, Série IIA, 331:549–556.

Barnicot, N. A., and D. Hewtt-Emmett. 1972. Red cell and serum proteins of *Cercocebus, Presbytis, Colobus*, and certain other species. *Folia Primatologica* 17:442–457.

Barry, J. C. 1987. The history and chronology of Siwalik cercopithecids. *Human Evolution* 2:47–58.

Benefit, B. R. 1993. The permanent dentition and phylogenetic position of *Victoriapithecus* from Maboko Island, Kenya. *Journal of Human Evolution* 25:83–172.

———. 1999. *Victoriapithecus*: The key to Old World monkey and catarrhine origins. *Evolutionary Anthropology* 7:155–174.

———. 2000. Old World monkey origins and diversification: An evolutionary study of diet and dentition; pp. 133–179 in P. F. Whitehead and C. J. Jolly (eds.), *Old World Monkeys*. Cambridge University Press, Cambridge.

———. 2009. The biostratigraphy and paleontology of fossil cercopithecoids from eastern Libya; pp. 247–266 in M. J. Salem et al. (eds.) Geology of East Libya, Vol. 3. Tripoli, Libya. *Geology of East Libya*.

Benefit, B. R., and M. L. McCrossin. 1990. Diet, species diversity and distribution of African fossil baboons. *Kroeber Anthropological Society Papers* 71–72:77–93.

———. 1997. Earliest known Old World monkey skull. *Nature* 388:368–371.

———. 2002. The Victoriapithecidae, Cercopithecoidea; pp. 241–253 in W. C. Hartwig (ed.), *The Primate Fossil Record*. Cambridge University Press, Cambridge.

Benefit, B. R., and M. Pickford. 1986. Miocene fossil cercopithecoids from Kenya. *American Journal of Physical Anthropology* 69:441–464.

Birchette, M. G. 1982. The postcranial skeleton of *Paracolobus chemeroni*. Unpublished PhD dissertation, Harvard University, Cambridge, 494 pp.

Bobe, R., and A. K. Behrensmeyer. 2004. The expansion of grassland ecosystems in Africa in relation to mammalian evolution and the origin of the genus *Homo*. *Palaeogeography, Palaeoclimatology, Palaeoecology* 207:399–420.

Bobe, R., A. K. Behrensmeyer, and R. E. Chapman. 2002. Faunal change, environmental variability and late Pliocene hominin evolution. *Journal of Human Evolution* 42:475–497.

Broom, R. 1937. On some new Pleistocene mammals from limestone caves of the Transvaal. *South African Journal of Science* 33:750–768.

———. 1940. The South African Pleistocene cercopithecid apes. *Annals of the Transvaal Museum* 20:89–100.

Broom, R., and J. T. Robinson. 1949. A new type of fossil baboon, *Gorgopithecus major*. *Proceedings of the Zoological Society of London* 119:379–387.

Brunet, M., F. Guy, D. Pilbeam, H. T. Mackaye, A. Likius, D. Ahounta, A. Beauvilain, C. Blondel, H. Bocherens, J.-R. Boisserie, L. De Bonis, Y. Coppens, J. Dejax, C. Denys, P. Duringer, V. Eisenmann, G. Fanone, P. Fronty, D. Geraads, T. Lehmann, F. Lihoreau, A. Louchart, A. Mahamat, G. Merceron, G. Mouchelin, O. Potero, P. P. Campomanes, M. Ponce de Leon, J.-C. Rage, M. Sapanet, M. Schuster, J. Sudre, P. Tassy, X. Valentin, P. Vignaud, L. Viriot, A. Zazzo, and C. Zollikofer. 2002. A new hominid from Upper Miocene of Chad, Central Africa. *Nature* 418:145–151.

Cerling, T. E., J. M. Harris, and B. H. Passey. 2003. Dietary preferences of East African Bovidae based on stable isotope analysis. *Journal of Mammalogy* 84:456–471.

Cronin, J. E., and V. M. Sarich. 1976. Molecular evidence for dual origin of mangabeys among Old World monkeys. *Nature* 260:700–702.

Dechow, P. P. C., and R. R. Singer. 1984. Additional fossil *Theropithecus* from Hopefield, South Africa: A comparison with other African sites and a reevaluation of its taxonomic status. *American Journal of Physical Anthropology* 63:405–35.

de Heinzelin, J., J. D. Clark, T. White, W. Hart, P. Renne, G. Wolde-Gabriel, Y. Beyene, and E. Vrba. 1999. Environment and behavior of 2.5-million-year-old Bouri hominids. *Science* 284:625–635.

Delson, E. 1973. Fossil colobine monkeys of the Circum-Mediterranean region and the evolutionary history of the Cercopithecidae (Primates, Mammalia). Unpublished PhD dissertation, Columbia University, 856 pp.

———. 1974. Preliminary review of cercopithecid distribution in the circum-Mediterranean region. *Mémoires du Bureau des Recherches Géologiques et Minières (France)* 78:131–135.

———. 1975. Evolutionary history of the Cercopithecidae. *Contributions to Primatology.* 5:167–217

Delson, E. 1979. *Prohylobates* (Primates) from the Early Miocene of Libya: A new species and its implications for cercopithecid origins. *Geobios* 12:725–733.

Delson, E. 1980. Fossil macaques, phyletic relationships and a scenario of deployment; pp. 10–30 in D. G. Lindburg (ed.), *The Macaques: Studies in Ecology, Behavior and Evolution.* Van Nostrand Reinhold, New York.

———. 1984. Cercopithecid biochronology of the African Plio-Pleistocene: Correlation among eastern and southern hominid-bearing localities. *Courier Forschungs-Institut Senckenberg* 69:199–218.

———. 1988. Chronology of South African australopith site units; pp. 317–324 in F. E. Grine (ed.), *Evolutionary History of the "Robust" Australopithecines.* Aldine de Gruyter, New York.

———. 1993. *Theropithecus* fossils from Africa and India and the taxonomy of the genus; pp. 157–189 in N. G. Jablonski (ed.), Theropithecus: *The Rise and Fall of a Primate Genus.* Cambridge University Press, Cambridge.

Delson, E., and D. Dean. 1993. Are *Papio baringensis* R. Leakey, 1969, and *P. quadratirostris* Iwamoto, 1982, species of *Papio* or *Theropithecus*? pp. 125–156 in N. G. Jablonski (ed.), Theropithecus: *The Rise and Fall of a Primate Genus.* Cambridge University Press, Cambridge.

Delson, E., G. G. Eck, M. G. Leakey, and N. G. Jablonski. 1993. A partial catalogue of fossil remains of *Theropithecus*; pp. 499–525 in N. G. Jablonski (ed.), Theropithecus: *The Rise and Fall of a Primate Genus.* Cambridge University Press, Cambridge.

Delson, E., and R. Hoffstetter. 1993. *Theropithecus* from Ternifine, Algeria; pp. 191–208 in N. G. Jablonski (ed.), Theropithecus: *The Rise and Fall of a Primate Genus.* Cambridge University Press, Cambridge.

Delson, E., C. J. Terranova, W. L. Jungers, E. J. Sargis, N. G. Jablonski, and P. C. Dechow. 2000. Body mass in Cercopithecidae (Primates, Mammalia): Estimation and scaling in extinct and extant taxa. *Anthropological Papers of the American Museum of Natural History* 83:1–159.

deMenocal, P. B. 2004. African climate change and faunal evolution during the Pliocene-Pleistocene. *Earth and Planetary Science Letters* 220:3–24.

Disotell, T. R., and R. L. Raaum. 2002. Molecular timescale and gene tree incongruence in the Guenons; pp. 27–36 in M. E. Glenn and M. Cords (eds.), *The Guenons: Diversity and Adaptation in African Monkeys.* Kluwer Academic Publishers, New York.

Dutrillaux, B., M. Muleris, and J. Couturier. 1988. Chromosomal evolution of Cercopithecinae; pp. 150–159 in A. Gautier-Hion, F. Bourlière, J. P. Gautier, and J. Kingdon (eds.), *A Primate Radiation: Evolutionary Biology of the African Guenons.* Cambridge University Press, New York.

Eck, G. G. 1976. Cercopithecoidea from Omo Group deposits; pp. 332–344 in Y. Coppens, F. C. Howell, G. L. Isaac, and R. E. F. Leakey (eds.), *Earliest Man and Environments in the Lake Rudolf Basin.* University of Chicago Press, Chicago.

———. 1987. *Theropithecus oswaldi* from the Shungura Formation, Lower Omo Basin, southwestern Ethiopia; pp. 123–140 in *Les Faunes Plio-Pleistocenes de la Vallée de l'Omo (Ethiopie): Cercopithecidae de la Formation de Shungura.* Centre National de la Recherche Scientifique, Paris.

———. 1993. *Theropithecus darti* from the Hadar Formation, Ethiopia; pp. 15–83 in N. G. Jablonski (ed.), Theropithecus: *The Rise and Fall of a Primate Genus.* Cambridge University Press, Cambridge.

Eck, G., and F. C. Howell. 1972. New fossil *Cercopithecus* material from the Lower Omo Basin, Ethiopia. *Folia Primatologica* 18:325–355.

Eck, G. G., and N. G. Jablonski. 1984. A reassessment of the taxonomic status and phyletic relationships of *Papio baringensis* and *Papio quadratirostris* (Primates: Cercopithecidae). *American Journal of Physical Anthropology* 65:109–134.

———. 1987. The skull of *Theropithecus brumpti* as compared with those of other species of the genus *Theropithecus*; pp. 11–122 in *Les Faunes Plio-Pleistocenes de la Vallée de l'Omo (Ethiopie): Cercopithecidae de la Formation de Shungura.* Centre National de la Recherche Scientifique, Paris.

Eisenhart, B. 1974. The Fossil Cercopithecoids of Makapansgat and Sterkfontein. Unpublished senior honors thesis, Harvard University, Cambridge.

Elton, S. 2007. Environmental correlates of the cercopithecoid radiations. *Folia Primatologica* 78:344–364.

Flagstad, Ø., P. O. Syvertsen, N. C. Stenseth, and K. S. Jakobsen. 2000. Environmental change and rates of evolution: The phylogeographic pattern within the hartebeest complex as related to climatic variation. *Proceedings of the Royal Society of London, B* 268:667–677.

Fooden, J. 1980. Classification and distribution of living macaques (*Macaca* Lacepede, 1799); pp. 1–9 in D. Lindburg, G (ed.), *The Macaques: Studies in Ecology, Behavior and Evolution.* Van Nostrand Reinhold, Los Angeles,.

Freedman, L. 1957. The fossil Cercopithecoidea of South Africa. *Annals of the Transvaal Museum* 23:8–262.

———. 1965. Fossil and subfossil primates from the limestone deposits at Taung, Bolt's Farm and Witkrans, South Africa. *Palaeontologia Africana* 9:19–48.

———. 1970. A new check-list of fossil Cercopithecoidea of South Africa. *Palaeontologia Africana* 13:109–110.

———. 1976. South African fossil Cercopithecoidea: A re-assessment including a description of new material from Makapansgat, Sterkfontein and Taung. *Journal of Human Evolution* 5:297–315.

Frost, S. R. 2001a. Fossil Cercopithecidae of the Afar Depression, Ethiopia: Species systematics and comparison to the Turkana Basin. Unpublished PhD dissertation, City University of New York, New York, 463 pp.

———. 2001b. New Early Pliocene Cercopithecidae (Mammalia: Primates) from Aramis, Middle Awash Valley, Ethiopia. *American Museum Novitates* 3350:1–36.

———. 2007a. Fossil Cercopithecidae from the Middle Pleistocene Dawaitoli Formation, Middle Awash Valley, Afar Region, Ethiopia. *American Journal of Physical Anthropology* 134:460–467.

———. 2007b. African Pliocene and Pleistocene cercopithecid evolution and global climatic change; pp. 51–76 in R. Bobe, Z. Alemseged, and A. K. Behrensmeyer (eds.), *Hominid Environments in the East African Pliocene: An Assessment of the Faunal Evidence.* Springer, New York.

Frost, S. R., and Z. Alemseged. 2007. Middle Pleistocene fossil Cercopithecidae from Asbole, Afar Region, Ethiopia. *Journal of Human Evolution* 53:227–259.

Frost, S. R., and E. Delson. 2002. Fossil Cercopithecidae from the Hadar Formation and surrounding areas of the Afar Depression, Ethiopia. *Journal of Human Evolution* 43:687–748.

Frost, S. R., Y. Haile-Selassie, and L. G. Hlusko. 2009. Cercopithecidae; pp. 135–158 in Y. Haile-Selassie (ed.), Ardipithecus kadabba. *Late Miocene evidence from the Middle Awash, Ethiopia.* University of California Press, Berkeley.

Frost, S. R., L. F. Marcus, F. L. Bookstein, D. P. Reddy, and E. Delson. 2003a. Cranial allometry, phylogeography, and systematics of large-bodied papionins (Primates: Cercopithecinae) inferred from geometric morphometric analysis of landmark data. *The Anatomical Record Part A* 276A:1048–1072.

Frost, S. R., T. Plummer, L. C. Bishop, P. Ditchfield, J. Ferraro, and J. Hicks. 2003b. Partial cranium of *Cercopithecoides kimeui* Leakey, 1982 from Rawi Gully, southwestern Kenya. *American Journal of Physical Anthropology* 122:191–199.

Gagnon, M., and A. E. Chew. 2000. Dietary preferences in extant African Bovidae. *Journal of Mammalogy* 81:490–511.

Gautier, J. P. 1988. Interspecific affinities among guenons as deduced from vocalizations; pp. 194–226 in A. Gautier-Hion, F. Bourlière, J. P. Gautier, and J. Kingdon (eds.), *A Primate Radiation: Evolutionary Biology of the African Guenons.* Cambridge University Press, New York.

Gebo, D. L., and E. J. Sargis. 1994. Terrestrial adaptations in the postcranial skeletons of guenons. *American Journal of Physical Anthropology* 93:341–371.

Geraads, D. 1987. Dating the Northern African cercopithecid fossil record. *Human Evolution* 2:19–27.

Gilbert, C. C. 2007a. Craniomandibular morphology supporting the diphyletic origin of mangabeys and a new genus of the *Cercocebus/Mandrillus* clade, *Procercocebus. Journal of Human Evolution* 53:69–102.

———. 2007b. Identification and description of the first *Theropithecus* (Primates: Cercopithecoidae) material from Bolt's Farm, South Africa. *Annals of the Transvaal Museum* 44:1–10.

Gilbert, W. H., and S. R. Frost. 2008. Cercopithecidae; pp. 115–132 in W. H. Gilbert and B. Asfaw (eds.), Homo erectus: *Pleistocene Evidence from the Middle Awash, Ethiopia.* University of California Press, Berkeley.

Groves, C. P. 1989. *A Theory of Human and Primate Evolution.* Clarendon Press, Oxford, 384 pp.

———. 2001. *Primate Taxonomy.* Smithsonian Institution Press, Washington, D.C., 350 pp.

Gundling, T., and A. Hill. 2000. Geological context of fossil Cercopithecoidea from eastern Africa; pp. 180–213 in P. F. Whitehead and C. J. Jolly (eds.), *Old World Monkeys.* Cambridge University Press, Cambridge.

Harris, J. M., F. H. Brown, and M. G. Leakey. 1988. Stratigraphy and paleontology of Pliocene and Pleistocene localities west of Lake Turkana, Kenya. *Contributions in Science* 399:1–128.

Harris, J. M., M. G. Leakey, and T. E. Cerling. 2003. Early Pliocene tetrapod remains from Kanapoi, Lake Turkana Basin, Kenya. *Contributions in Science, Natural History Museum of Los Angeles County, Los Angeles* 498:39–114.

Harrison, T. 1989. New postcranial remains of *Victoriapithecus* from the middle Miocene of Kenya. *Journal of Human Evolution* 18:3–54.

Harrison, T., and E. E. Harris. 1996. Plio-Pleistocene cercopithecids from Kanam, western Kenya. *Journal of Human Evolution* 30:539–561.

Heaton, J. L. 2006. Taxonomy of the Sterkfontein fossil Cercopithecinae: The Papionini of members 2 and 4 (Gauteng, South Africa). Unpublished PhD dissertation, Indiana University, 503 pp.

Hlusko, L. J. 2006. A new large Pliocene colobine species (Mammalia: Primates) from Asa Issie, Ethiopia. *Geobios* 39:57–69.

———. 2007. A new late Miocene species of *Paracolobus* and other Cercopithecoidea (Mammalia: Primates) fossils from Lemudong'o, Kenya. *Kirtlandia* 56:72–85.

Hylander, W. L. 1975. Incisor size and diet in anthropoids with special reference to Cercopithecidae. *Science* 189:1095–1098.

Iwamoto, M. 1982. A fossil baboon skull from the lower Omo basin, southwest Ethiopia. *Primates* 23:533–541.

Jablonski, N. G. 1986. The hand of *Theropithecus brumpti*; pp. 173–182 in *Selected Proceedings of the Tenth Congress of the International Primatological Society: Volume 1. Primate Evolution.* Cambridge, Cambridge University Press.

———. 1993. The evolution of the masticatory apparatus in *Theropithecus*; pp. 299–329 in N. G. Jablonski (ed.), Theropithecus: *The Rise and Fall of a Primate Genus.* Cambridge University Press, Cambridge.

———. 1994. New fossil cercopithecid remains from the Humpata Plateau, southern Angola. *American Journal of Physical Anthropology* 94:435–64.

———. 1998. Primate evolution—in and out of Africa: Comments from Nina G. Jablonski. *Current Biology* 9:119–122.

———. 2002. Fossil Old World monkeys: The late Neogene radiation; pp. 255–299 in W. C. Hartwig (ed.), *The Primate Fossil Record.* Cambridge University Press, Cambridge.

Jablonski, N. G., and M. G. Leakey. 2008a. The importance of the Cercopithecoidea from the Koobi Fora Formation in the context of primate and mammalian evolution; pp. 397–416 in N. G. Jablonski and M. G. Leakey (eds.), *Koobi Fora Research Project: Volume 6. The Fossil Monkeys.* California Academy of Sciences, San Francisco.

———. 2008b. Systematic paleontology of the small colobines; pp. 12–30 in N. G. Jablonski and M. G. Leakey (eds.), *Koobi Fora Research Project: Volume 6. The Fossil Monkeys.* California Academy of Sciences, San Francisco.

Jablonski, N. G., M. G. Leakey, and M. Antón. 2008a. Systematic paleontology of the cercopithecines; pp. 103–300 in N. G. Jablonski and M. G. Leakey (eds.), *Koobi Fora Research Project: Volume 6. The Fossil Monkeys.* California Academy of Sciences, San Francisco.

Jablonski, N. G., M. G. Leakey, C. Kiarie, and M. Antón. 2002. A new skeleton of *Theropithecus brumpti* (Primates: Cercopithecidae) from Lomekwi, West Turkana, Kenya. *Journal of Human Evolution* 43:887–923.

Jablonski, N. G., M. G. Leakey, C. V. Ward, and M. Antón. 2008b. Systematic paleontology of the large colobines; pp. 31–102 in N. G. Jablonski and M. G. Leakey (eds.), *Koobi Fora Research Project: Volume 6. The Fossil Monkeys.* California Academy of Sciences, San Francisco.

Jablonski, N. G., M. J. Whitfort, N. Roberts-Smith, and Q. Xu. 2000. The influence of life history and diet on the distribution of catarrhine primates during the Pleistocene in eastern Asia. *Journal of Human Evolution* 39:131–157.

Jolly, C. J. 1970. The seed-eaters: A new model of hominid differentiation based on a baboon analogy. *Man* 5:5–26.

———. 1972. The classification and natural history of *Theropithecus (Simopithecus)*(Andrews, 1916), baboons of the African Plio-Pleistocene. *Bulletin of the British Museum (Natural History), Geology* 22:1–123.

Jones, T. R. 1937. A new fossil primate from Sterkfontein, Krugersdorp, Transvaal. *South African Journal of Science* 33:709–728.

Kalb, J. E., M. Jaegar, C. J. Jolly, and B. Kana. 1982a. Preliminary geology, paleontology and paleoecology of a Sangoan site at Andalee, Middle Awash Valley, Ethiopia. *Journal of Archaeological Science* 9:349–363.

Kalb, J. E., C. J. Jolly, S. Tebedge, A. Mebrate, C. Smart, E. B. Oswald, P. F. Whitehead, C. B. Wood, T. Adefris, and V. Rawn-Schatzinger. 1982b. Vertebrate faunas from the Awash Group, Middle Awash Valley, Afar, Ethiopia. *Journal of Vertebrate Paleontology* 2:237–258.

Lambert, J. E. 2005. Competition, predation, and the evolutionary significance of the cercopithecine cheek pouch: The case of *Cercopithecus* and *Lophocebus. American Journal of Physical Anthropology* 126:183–192.

Leakey, M. G. 1982. Extinct large colobines from the Plio-Pleistocene of Africa. *American Journal of Physical Anthropology* 58:153–172.

———. 1985. Early Miocene cercopithecids from Buluk, northern Kenya. *Folia Primatologica* 44:1–14.

———. 1988. Fossil evidence for the evolution of the guenons; pp. 7–12 in A. Gautier-Hion, F. Bourlière, J.-P. Gautier, and J. Kingdon (eds.), *A Primate Radiation: Evolutionary Biology of the African Guenons.* Cambridge University Press, Cambridge.

———. 1993. Evolution of *Theropithecus* in the Turkana Basin; pp. 85–123 in N. G. Jablonski (ed.), Theropithecus: *The Rise and Fall of a Primate Genus.* Cambridge University Press, Cambridge.

Leakey, M. G., and E. Delson. 1987. Fossil Cercopithecidae from the Laetolil Beds; pp. 91–107 in M. D. Leakey and J. M. Harris (eds.), *Laetoli: A Pliocene Site in Northern Tanzania.* Clarendon Press, Oxford.

Leakey, M. G., and R. E. F. Leakey. 1976. Further Cercopithecinae (Mammalia, Primates) from the Plio/Pleistocene of East Africa. *Fossil Vertebrates of Africa* 4:121–146.

Leakey, M. G., M. F. Teaford, and C. V. Ward. 2003. Cercopithecidae from Lothagam; pp. 201 - 248 in M. G. Leakey and J. M. Harris (eds.), *Lothagam: The Dawn of Humanity in Eastern Africa.* Columbia University Press, New York.

Leakey, R. E. F. 1969. New Cercopithecidae from the Chemeron Beds of Lake Baringo, Kenya. *Fossil Vertebrates of Africa* 1:53–69.

Lucas, P. W., and M. F. Teaford. 1994. Functional morphology of colobine teeth; pp. 173–203 in A. G. Davies and J. F. Oates (eds.), *Colobine Monkeys: Their Ecology, Behaviour and Evolution.* Cambridge University Press, Cambridge.

Maier, W. 1971. Two new skulls of *Parapapio antiquus* from Taung and a suggested phylogenetic arrangement of the genus *Parapapio. Annals of the South African Museum* 59:1–16.

———. 1972. The first complete skull of *Simopithecus darti* from Makapansgat, South Africa, and its systematic position. *Journal of Human Evolution* 1:395–405.

McBrearty, S., and N. G. Jablonski. 2005. First fossil chimpanzee. *Nature* 437:105–108.

McKee, J. K. 1993. Taxonomic and evolutionary affinities of *Papio izodi* fossils from Taung and Sterkfontein. *Palaeontologia Africana* 30:43–49.

Meikle, W. E. 1987. Fossil Cercopithecidae from the Sahabi Formation; pp. 119–127 in N. T. Boaz, A. El-Arnauti, A. W. Gaziry, J. de Heinzelin, and D. D. Boaz (eds.), *Neogene Paleontology and Geology of Sahabi.* Liss, New York.

Miller, E. R., B. R. Benefit, M. L. McCrossin, J. M. Plavcan, M. G. Leakey, A. N. El-Barkooky, M. A. Hamdan, M. K. Abdel Gawad, S. M. Hassan, and E. L. Simons. 2009. Systematics of early and middle Miocene Old World monkeys. *Journal of Human Evolution* 57:195–211.

Morales, J. C., and D. J. Melnick. 1998. Phylogenetic relationships of the macaques (Cercopithecidae: *Macaca*), as revealed by high resolution restriction site mapping of mitochondrial ribosomal genes. *Journal of Human Evolution* 34:1–23.

Napier, P. H. 1981. *Catalogue of Primates in the British Museum (Natural History) and Elsewhere in the British Isles: Part II. Family Cercopithecidae, Subfamily Cercopithecinae.* British Museum (Natural History), London, 203 pp.

———. 1985. *Catalogue of Primates in the British Museum (Natural History) and Elsewhere in the British Isles: Part III. Family Cercopithecidae, Subfamily Colobinae.* British Museum (Natural History), London, 111 pp.

Newman, T. K., C. J. Jolly, and J. Rogers. 2004. Mitochondrial phylogeny and systematics of baboons *(Papio)*. *American Journal of Physical Anthropology* 124:17–27.

Page, S. L., and M. Goodman. 2001. Catarrhine phylogeny: Noncoding DNA evidence for a diphyletic origin of the mangabeys and for a human-chimpanzee clade. *Molecular Phylogenetics and Evolution* 18:14–25.

Patterson, B. 1968. The extinct baboon, *Parapapio jonesi*, in the early Pleistocene of northwestern Kenya. *Breviora* 282:1–4.

Raaum, R. L., K. N. Sterner, C. M. Noviello, C.-B. Stewart, and T. R. Disotell. 2005. Catarrhine primate divergence dates estimated from complete mitochondrial genomes: Concordance with fossil and nuclear DNA evidence. *Journal of Human Evolution* 48:237–257.

Rae, T. C., O. Rohrer-Ertl, C.-P. Wallner, and T. Koppe. 2007. Paranasal pneumatization of two late Miocene colobines: *Mesopithecus* and *Libypithecus* (Cercopithecidae: Primates). *Journal of Vertebrate Paleontology* 27:768–771.

Remane, A. 1925. Der Fossile Pavian (*Papio* Sp.) von Oldoway nebst Bemerkungen über die Gattung *Simopithecus* C. W. Andrews; pp. 83–90 in H. Reck (ed.), *Wissenschaftliche Ergebnisse der Oldoway-Expedition 1913*, vol. 2.

Semaw, S., S. W. Simpson, J. Quade, P. R. Renne, R. F. Butler, W. C. McIntosh, N. Levin, M. Dominguez-Rodrigo, and M. J. Rogers. 2005. Early Pliocene hominids from Gona, Ethiopia. *Nature* 433:301–305.

Senut, B. 1994. Cercopithecoidea Neogénes et Quaternaires du Rift Occidental (Ouganda); pp. 195–205 in M. Pickford and B. Senut (eds.), *Geology and Palaeobiology of the Albertine Rift Valley, Uganda-Zaire: Vol. II. Palaeobiology.* CIFEG, Orléans.

Simons, E. L. 1967. A Fossil *Colobus* Skull from the Sudan (Primates, Cercopithecidae). *Postilla* 111:1–12.

Simons, E. L., and E. Delson. 1978. Cercopithecidae and Parapithecidae; pp. 100–119 in V. J. Maglio and H. B. S. Cooke (eds.), *Evolution of African Mammals.* Harvard University Press, Cambridge.

Smith, L. W., A. Link, and M. Cords. 2008. Cheek pouch use, predation risk, and feeding competition in blue monkeys (*Cercopithecus mitis stuhlmanni*). *American Journal of Physical Anthropology* 137:334–341.

Sterner, K. N., R. L. Raaum, Y.-P. Zhang, C.-B. Stewart, and T. R. Disotell. 2006. Mitochondrial data support an odd-nosed colobine clade. *Molecular Phylogenetics and Evolution* 40:1–7.

Stewart, C.-B. 1999. The colobine Old World monkeys as a model system for the study of adaptive evolution at the molecular level; pp. 29–38 in P. Dolhinow and A. Fuentes (eds.), *The Nonhuman Primates.* Mayfield, London.

Stewart, C.-B., and T. R. Disotell. 1998. Primate evolution—in and out of Africa. *Current Biology* 8:R583–R588.

Stewart, C.-B., J. W. Schilling, and A. C. Wilson. 1987. Adaptive evolution in the stomach lysozymes of foregut fermenters. *Nature* 330:401–404.

Strasser, E., and E. Delson. 1987. Cladistic analysis of cercopithecid relationships. *Journal of Human Evolution* 16:81–99.

Stromer, E. 1914. Mitteilungen über Wirbeltierreste aus dem Mittelpliocän des Natrontales (Ägypten). *Zeitschrift der Deutschen Geologischen Gesellschaft* 65:350–372.

———. 1920. Mitteilungen über Wirbeltierreste aus dem Mittelpliocän des Natrontales (Ägypten) 5. Nachtrag zur Affen. *Sitzungsberichte der Bayerischen Akademie der Wissenschaften* 1920:345–370.

Swindler, D. R., and F. J. Orlosky. 1974. Metric and morphological variability in the dentition of colobine monkeys. *Journal of Human Evolution* 3:135–160.

Szalay, F. S., and E. Delson. 1979. *Evolutionary History of the Primates.* Academic Press, New York, 580 pp.

Tosi, A. J., P. J. Buzzard, J. C. Morales, and D. J. Melnick. 2002a. Y-chromosomal window onto the history of terrestrial adaptation in the Cercopithecini; pp. 13–24 in M. E. Glenn and M. Cords (eds.), *The Guenons: Diversity and Adaptation in African Monkeys.* Kluwer Academic Publishers, New York.

Tosi, A. J., K. M. Detwiler, and T. R. Disotell. 2005. X-chromosomal window into the evolutionary history of the guenons (Primates: Cercopithecini). *Molecular Phylogenetics and Evolution* 36.58–66.

Tosi, A. J., T. R. Disotell, J. C. Morales, and D. J. Melnick. 2003a. Y-chromosome data provide a test of competing morphological evolutionary hypotheses. *Molecular Phylogenetics and Evolution* 27:510–521.

Tosi, A. J., D. J. Melnick, and T. R. Disotell. 2004. Sex chromosome phylogenetics indicate a single transition to terrestriality in the guenons (tribe Cercopithecini). *Journal of Human Evolution* 46:223–237.

Tosi, A. J., J. C. Morales, and D. J. Melnick. 2000. Comparison of Y chromosome and mtDNA phylogenies leads to unique inferences of macaque evolutionary history. *Molecular Phylogenetics and Evolution* 17:133–144.

———. 2002b. Y-chromosome and mitochondrial markers in *Macaca fascicularis* indicate introgression with Indochinese *M. mulatta* and a biogeographic barrier in the Isthmus of Kra. *International Journal of Primatology* 23:161–178.

———. 2003b. Paternal, maternal, and biparental molecular markers provide unique windows onto the evolutionary history of macaque monkeys. *Evolution* 57:1419–1435.

Van Soest, P. J. 1982. *Nutritional Ecology of the Ruminant.* Your Town Press, Salem, Mass., 476 pp.

Wildman, D. E., T. J. Bergman, A. al-Aghbari, K. N. Sterner, T. K. Newman, J. E. Phillips-Conroy, C. J. Jolly, and T. R. Disotell. 2004. Mitochondrial evidence for the origin of hamadryas baboons. *Molecular Phylogenetics and Evolution* 32:287–296.

Xing, J., H. Wang, K. Han, D. A. Ray, C. H. Huang, L. G. Chemnick, C. B. Stewart, T. R. Disotell, O. A. Ryder, and M. A. Batzer. 2005. A mobile element based phylogeny of Old World monkeys. *Molecular Phylogenetics and Evolution* 37:872–80.

Xing, J., H. Wang, Y. Zhang, D. A. Ray, A. J. Tosi, T. R. Disotell, and M. A. Batzer. 2007. A mobile element-based evolutionary history of guenons (Tribe Cercopithecini). *BMC Biology* 5:5.

Zhang, J., Y.-p. Zhang, and H. F. Rosenberg. 2002. Adaptive evolution of a duplicated pancreatic ribonuclease gene in a leaf-eating monkey. *Nature Genetics* 30:411–415.

TWENTY-FOUR

Dendropithecoidea, Proconsuloidea, and Hominoidea

TERRY HARRISON

Catarrhine primates of modern aspect closely related to extant hominoids and cercopithecoids originated in Afro-Arabia during the late Oligocene (Harrison, 1982, 1987, 2002, 2005; Andrews, 1985, 1992b; Fleagle, 1986, 1999; Rasmussen, 2002). These taxa share key derived features with extant catarrhines, such as a tubular ectotympanic and loss of the entepicondylar foramen of the distal humerus (Harrison, 1987). Such features are not found in primitive catarrhines, such as propliopithecids from the early Oligocene of Egypt and Oman (see Seiffert et al. this volume, chap. 22) or plio-pithecids from the Miocene of Eurasia, which primitively retain an annular ectotympanic and/or an entepicondylar foramen (Harrison, 1987, 2002, 2005; Andrews et al., 1996). Three superfamilies of noncercopithecoid catarrhines are recognized in Africa from the late Oligocene onward: Den-dropithecoidea, Proconsuloidea, and Hominoidea (Harrison, 2002; Ward and Duren, 2002; see table 24.1). The Dendro-pithecoidea contains a single family of three closely related genera, *Micropithecus*, *Dendropithecus*, and *Simiolus* (table 24.1). They are all relatively small catarrhines, with primitive dental and postcranial features that indicate that they are the sister taxon to Proconsuloidea + (Hominoidea + Cercopithecoidea). The Proconsuloidea contains a single family, the Proconsuli-dae, divided into three subfamilies: Proconsulinae, Afropith-ecinae, and Nyanzapithecinae (table 24.1). Given the level of taxonomic and adaptive diversity in the Proconsulidae, it may prove desirable at some later date to elevate these sub-families to family rank. The Proconsulidae are medium- to large-sized catarrhines that are more derived postcranially than dendropithecoids, implying a closer relationship to crown catarrhines. The proconsulids are recognized here as the sister taxon to cercopithecoids + hominoids, based on the retention of a number of primitive cranial and postcranial features that are more derived in extant catarrhines (Harrison, 1987, 1988, 1993, 2002, 2005; Harrison and Gu, 1999; Rossie et al., 2002). However, it should be noted that most scholars prefer to recognize the proconsuloids as stem hominoids (the sister group of Hylobatidae + Hominidae; see, e.g., Rose, 1983, 1992, 1997; Andrews, 1985, 1992b; Andrews and Martin, 1987a; Begun et al., 1997; Kelley, 1997; Rae, 1997, 1999; Ward, 1997; Ward et al., 1997; Fleagle, 1999; Singleton, 2000; Pickford and Kunimatsu, 2005) or even stem hominids

(the sister to great apes + humans; see Walker and Teaford, 1989; Walker, 1997). Regardless of their precise phylogenetic affinities, it is evident from the general similarity of their craniodental and postcranial anatomy that the proconsuloids occupy an evolutionary grade that is close to the initial radia-tion of all recent catarrhines (see Harrison, 1987, 1988, 1993, 2002, 2005).

In addition to the species attributed with some confidence to either the Proconsuloidea or Dendropithecoidea, there are several problematic taxa that are difficult to classify, primarily because they are poorly known. *Otavipithecus* is most likely a proconsulid, possibly with affinities to the afropithecines (Andrews, 1992a, 1992b; Singleton, 2000), but its precise rela-tionships cannot be determined at this time. *Kalepithecus*, *Limnopithecus*, *Kogolepithecus*, *Lomorupithecus*, and *Kamoyapith-ecus* are not well enough known to classify them with any confidence. It is probable, however, that most of these taxa are members of the Proconsuloidea or the Dendropithecoidea. *Lomorupithecus* has been suggested to be a member of the Plio-pithecidae (Rossie and McLatchy, 2006), but the inferred syn-apomorphies are contradicted by features used to link the Eur-asian members of this clade (Andrews et al., 1996; Harrison and Gu, 1999), and it is more likely that *Lomorupithecus* repre-sents a dendropithecoid. Based on the primitive morphology of the upper molars of *Kamoyapithecus*, this taxon may be the sister taxon of all other catarrhines from the Miocene and later, including the Proconsuloidea and Dendropithecoidea (Harrison, 2002). Currently, 29 species of stem catarrhines are known from late Oligocene and Miocene localities in East Africa, and they clearly represented a taxonomically and adaptively diverse radiation (tables 24.1 and 24.2).

During the early Miocene, the crown catarrhines—homi-noids and cercopithecoids—diverged, although the fossil record documenting the earliest representatives of these two groups is rather poorly known. Cercopithecoids are first rep-resented in the fossil record by an isolated tooth from Napak in Uganda dated to ~19 Ma. However, Old World monkeys do not become common until after ~17 Ma, and even then their taxonomic diversity remains low until the late Miocene when they begin a major radiation that continues into the Plio-Pleistocene (Benefit and McCrossin, 2002; Jablonski, 2002; see Jablonski and Frost, this volume, chap. 23). The hominoid

Infraorder . . . . Catarrhini E. Geoffroy, 1812

Superfamily . . . . Dendropithecoidea,
  Harrison, 2002

Family . . . . Dendropithecidae, Harrison,
  2002

Genus . . . . *Dendropithecus* Andrews and
  Simons, 1977

Dendropithecus macinnesi
  (Le Gros Clark and Leakey,
  1950)

Genus . . . . *Micropithecus* Fleagle and
  Simons, 1978

Micropithecus clarki Fleagle
  and Simons, 1978

Micropithecus leakeyorum
  Harrison, 1989

Genus . . . . *Simiolus* Leakey and Leakey,
  1987

Simiolus enjiessi Leakey and
  Leakey, 1987

Simiolus cheptumoae Pickford
  and Kunimatsu, 2005

Simiolus andrewsi sp. nov.

Superfamily . . . . Proconsuloidea Leakey, 1963

Family . . . . Proconsulidae Leakey, 1963

Subfamily . . . . Proconsulinae Leakey, 1963

Genus . . . . *Proconsul* Hopwood, 1933a

Proconsul africanus Hopwood,
  1933a

Proconsul nyanzae Le Gros
  Clark and Leakey, 1950

Proconsul major Le Gros Clark
  and Leakey, 1950

Proconsul heseloni Walker,
  Teaford, Martin and
  Andrews, 1993

Proconsul gitongai (Pickford
  and Kunimatsu, 2005)

Subfamily . . . . Afropithecinae Andrews,
  1992a

Genus . . . . *Afropithecus* Leakey and
  Leakey, 1986a

Afropithecus turkanensis
  Leakey and Leakey,
  1986a

Genus . . . . *Heliopithecus* Andrews and
  Martin, 1987b

Heliopithecus leakeyi Andrews
  and Martin, 1987b

Genus . . . . *Nacholapithecus* Ishida et al.,
  2004

Nacholapithecus kerioi Ishida
  et al., 2004

Genus . . . . *Equatorius* S. Ward et al.,
  1999

Equatorius africanus (Le Gros
  Clark and Leakey, 1950)

Subfamily . . . . Nyanzapithecinae, Harrison,
  2002

Genus . . . . *Nyanzapithecus* Harrison,
  1986

*Nyanzapithecus
  vancouveringorum*
  (Andrews, 1974)

Nyanzapithecus pickfordi
  Harrison, 1986

Nyanzapithecus harrisoni
  Kunimatsu, 1997

Genus . . . . *Mabokopithecus* Von
  Koenigswald, 1969

Mabokopithecus clarki Von
  Koenigswald, 1969

Genus . . . . *Rangwapithecus* Andrews,
  1974

Rangwapithecus gordoni
  Andrews, 1974

Genus . . . . *Turkanapithecus* Leakey and
  Leakey, 1986b

Turkanapithecus kalakolensis
  Leakey and Leakey,
  1986b

Genus . . . . *Xenopithecus* Hopwood,
  1933a

Xenopithecus koruensis
  Hopwood, 1933a

Family . . . . *incertae sedis*

Genus . . . . *Otavipithecus* Conroy, Pickford,
  Senut and Mein, 1992

Otavipithecus namibiensis
  Conroy, Pickford, Senut
  and Mein, 1992

Superfamily . . . . *incertae sedis*

Family . . . . *incertae sedis*

Genus . . . . *Limnopithecus* Hopwood,
  1933a

Limnopithecus legetet
  Hopwood, 1933a

Limnopithecus evansi
  MacInnes, 1943

Genus . . . . *Lomorupithecus* Rossie and
  MacLatchy, 2006

Lomorupithecus harrisoni
  Rossie and MacLatchy,
  2006

Genus . . . . *Kalepithecus* Harrison, 1988

Kalepithecus songhorensis
  (Andrews, 1978)

Genus . . . . *Kamoyapithecus* Leakey,
  Ungar and Walker, 1995

Kamoyapithecus hamiltoni
  (Madden, 1980a)

Genus . . . . *Kogolepithecus* Pickford,
  Senut, Gommery and
  Musiime, 2003

Kogolepithecus morotoensis
  Pickford et al., 2003

Superfamily . . . . Hominoidea

Family . . . . Hylobatidae Gray, 1870

Family . . . . Hominidae Gray, 1825

Subfamily . . . . Kenyapithecinae Andrews,
  1992a

Genus . . . . *Kenyapithecus* Leakey, 1962
  (Leakey, 1961 reference)

Kenyapithecus wickeri Leakey,
  1962 (Leakey, 1961
  reference)

Subfamily . . . . Homininae Gray, 1825

Tribe . . . . Gorillini Frechkop, 1943

Genus . . . . *Gorilla* Geoffroy, 1852
    Gorilla gorilla (Savage and
       Wyman, 1847)
Tribe . . . . Hominini Gray, 1825
  Subtribe . . . . Panina Delson, 1977
    Genus . . . . *Pan* Oken, 1816
      Pan troglodytes Gmelin, 1788
      Pan paniscus Schwarz, 1929
  Subtribe . . . . Hominina Gray, 1825
    Genus . . . . *Australopithecus* Dart, 1925
    Genus . . . . *Paranthropus* Broom, 1938
    Genus . . . . *Homo* Linnaeus, 1758
  Subtribe . . . . Hominina?
    Genus . . . . *Ardipithecus* White et al.,
      1995
    Genus . . . . *Orrorin* Senut et al., 2001

Genus . . . . *Sahelanthropus* Brunet et al.,
  2002
Subfamily . . . . *incertae sedis*
  Genus . . . . *Samburupithecus* Ishida and
    Pickford, 1997
    Samburupithecus kiptalami
    Ishida and Pickford,
    1997
  Genus . . . . *Chororapithecus* Suwa et al.,
    2007
    Chororapithecus abyssinicus
    Suwa et al., 2007
  Genus . . . . *Nakalipithecus* Kunimatsu
    et al., 2007
    Nakalipithecus nakayamai
    Kunimatsu et al., 2007

SOURCE: Harrison (2002); Andrews and Harrison (2005).

TABLE 24.2

Geographic and temporal distribution of fossil dendropithecoids,
proconsuloids, and hominoids from the late Oligocene and Miocene of Afro-Arabia

Age	Kenya-Ethiopia	Uganda	Southern Africa	Other Regions
Late Miocene (10–5 Ma)	*Orrorin tugenensis* *Ardipithecus kadabba* *Samburupithecus kiptalami* *Chororapithecus abyssinicus* *Nakalipithecus nakayamai*			*Sahelanthropus tchadensis* [Chad]
Middle Miocene (16–10 Ma)	*Micropithecus leakeyorum* *Simiolus cheptumoae* *Simiolus andrewsi* *Proconsul* sp. (Fort Ternan) *Proconsul gitongai* *Nacholapithecus kerioi* *Equatorius africanus* *Nyanzapithecus pickfordi* *Nyanzapithecus harrisoni* *Mabokopithecus clarki* Nyanzapithecinae indet. (Fort Ternan, Kapsibor) *Kenyapithecus wickeri*		*Otavipithecus namibiensis* [Namibia]	*Heliopithecus leakeyi* [Saudi Arabia]
Early Miocene, late (18–16 Ma)	*Dendropithecus macinnesi* *Simiolus enjiessi* *Proconsul heseloni* *Proconsul nyanzae* *Afropithecus turkanensis* *Nyanzapithecus vancouveringorum* *Turkanapithecus kalakolensis*	*Afropithecus turkanensis* *Kogolepithecus morotoensis*	Nyanzapithecinae indet. (Ryskop) [South Africa]	
Early Miocene, early (23–18 Ma)	*Dendropithecus macinnesi* *Micropithecus clarki* *Proconsul africanus* *Proconsul major* *Proconsul* sp. (Meswa Bridge) *Rangwapithecus gordoni* *Xenopithecus koruensis* *Limnopithecus legetet* *Limnopithecus evansi* *Kalepithecus songhorensis*	*Micropithecus clarki* *Proconsul major* *Lomorupithecus harrisoni* *Limnopithecus legetet* *Limnopithecus evansi*		
Late Oligocene (27–23 Ma)	*Kamoyapithecus hamiltoni*			

fossil record is equally sparse during the early part of the Miocene (if one excludes all of the proconsuloids; Harrison, 2002). Until recently, the best contender for an early hominoid was *Morotopithecus*, dating to ~21 Ma (Gebo et al., 1997). However, the associated fauna indicates a much younger age, and recent comparisons of *Morotopithecus* support the contention that it is a junior synonym of *Afropithecus* (Pickford, 2002; see also Andrews and Martin, 1987b). If *Morotopithecus* is excluded from the Hominoidea and placed in the Proconsuloidea, then the next oldest contenders for hominoid status are *Nacholapithecus*, *Equatorius*, and *Otavipithecus* from the middle Miocene. However, these taxa have no definitive synapomorphies linking them with extant hominoids, and they seem to have their closest affinities with afropithecine proconsulids (Andrews, 1992b; Singleton, 2000; Kelley et al., 2002; Ward and Duren, 2002). The current evidence indicates that *Kenyapithecus wickeri* from the middle Miocene (~14 Ma) of Fort Ternan is the earliest African hominoid. This species is known only from a handful of fragmentary fossils from a single locality in western Kenya, but it does appear to be more derived than both *Equatorius* and *Nacholapithecus* in aspects of its dentition and facial anatomy (Pickford, 1985, 1986c; Harrison, 1992; S. Ward et al., 1999; Kelley et al., 2002).

Although the later Miocene record is quite sparse, the available material demonstrates that Africa supported a relatively high diversity of crown hominoids during this period. Teeth of indeterminate large catarrhines have been reported from the Ngorora Formation in Kenya (~12.0–12.5 Ma; Bishop and Chapman, 1970; Hill and Ward, 1988; Hill, 1999; Hill et al., 2002), and Pickford and Senut (2005a, 2005b) have recently identified isolated teeth of hominoids from the Ngorora Formation (~12.5 Ma) and the Lukeino Formation, Kenya (Kapsomin and Cheboit, ~5.9 Ma) that they claim are related to gorillas and chimpanzees respectively. Unfortunately, the material is not adequate to confirm their relationships, but it does seems likely that they represent several species of hominids, and some may even prove to be stem hominines. *Samburupithecus* from the late Miocene (~9.5 Ma) of Kenya is probably an early hominine (Ishida and Pickford, 1997; Pickford and Ishida, 1998), but clear-cut evidence linking it to the extant African apes or humans is meager at best. Recently recovered material from Nakali in central Kenya (~9.8–9.9Ma) and from Beticha in the Chorora Formation of Ethiopia (~10.0–10.5 Ma) belong to two new species of hominids that are inferred to be closely related to crown hominines (Nakatsukasa et al., 2006; Kunimatsu et al., 2007; Suwa et al., 2007). *Nakalipithecus* is most similar to, and probably closely related to, *Ouranopithecus* from Greece, but it is slightly older and somewhat more primitive dentally (Kunimatsu et al., 2007). *Chororapithecus* has been suggested to be the sister taxon to *Gorilla*, based on details of its molar morphology (Suwa et al., 2007), but the paucity of the material and the high probability of functional convergence prevent a definitive assessment of its phylogenetic relationships.

Middle to late Pleistocene remains of chimpanzees have been reported from several sites in East Africa, but none is definitively identifiable as belonging to *Pan*. The proximal femur from Kikorongo Crater in southwestern Uganda (De Silva et al., 2006) and the isolated teeth from the Kapthurin Formation of Kenya (McBrearty and Jablonski, 2005) are possibly attributable to *Homo*. The specimens previously described as *Pan* sp. from Mumba Höhle in northern Tanzania (Lehmann, 1957) have been identified as *H. sapiens* (T. Harrison, unpublished data). While the fossil evidence documenting the

evolutionary history of the African apes remains scanty, the homininan record is becoming increasingly well documented with major discoveries from the late Miocene (~7–6 Ma) onward (see MacLatchy et al. this volume, chap. 25).

Systematic Paleontology

Superfamily DENDROPITHECOIDEA Harrison, 2002
Family DENDROPITHECIDAE Harrison, 2002
Genus *DENDROPITHECUS* Andrews and Simons, 1977

Included Species D. macinnesi (Le Gros Clark and Leakey, 1950)(type species).

DENDROPITHECUS MACINNESI
(Le Gros Clark and Leakey, 1950)
Figure 24.1

Distribution Early Miocene (~17–20 Ma). Rusinga Island (Wayando, Hiwegi, and Kulu Formations Mfwangano Island, Angulo (Rangoye Beds), Karungu, Songhor, and Koru (Chamtwara Member) in Kenya (Pickford and Andrews, 1981; Harrison, 1981, 1982, 1988, 2002; Pickford, 1981, 1983, 1986a, 1986b; Pickford et al., 1986b; Drake et al., 1988).

Description A small- to medium-sized catarrhine with estimated body weights of ~9 kg and ~5–6 kg in males and females respectively. The main craniodental characteristics are as follows: incisors high crowned and narrow, and small in relation to the size of the molars; i2 asymmetrical in shape, with a convex distal margin; canines strongly sexually dimorphic in size and morphology; canines high crowned and bilaterally compressed in males, lower crowned and less compressed in females; upper canine in males with double mesial groove; upper premolars broad, with paraconid much more elevated than protoconid; p3 sectorial, with a high and bilaterally compressed crown, and a long mesiobuccal honing face; upper molars broad and rectangular, with high and voluminous cusps, well-developed crests, well-defined mesial and distal foveae and trigon basin, and a broad lingual cingulum; M1 < M3 < M2; lower molars long and quite broad, with high conical cusps, sharp occlusal crests, broad and transverse mesial fovea, well-defined and slightly obliquely oriented distal fovea, broad and deep talonid basin, and moderately well-developed buccal cingulum; marked increase in size from m1 to m3; palate long and narrow; large paired incisive foramina; nasal aperture narrow and tapers inferiorly between the roots of the upper central incisors; short subnasal clivus; maxillary sinus extensive; mandibular corpus low and robust; symphysis buttressed by moderately well-developed superior and inferior transverse tori (Le Gros Clark and Thomas, 1951; Le Gros Clark and Leakey, 1951; Andrews and Simons, 1977; Andrews, 1978; Harrison, 1981, 1982, 1988, 2002).

Dendropithecus macinnesi is known from several partial skeletons from Rusinga Island (Le Gros Clark and Thomas, 1951; Ferembach, 1958; Harrison, 1982; Fleagle, 1983; Rose, 1993). The main features are as follows: long and slender limb bones; proximal humerus lacks torsion; humeral shaft slightly retroflexed; distal humerus with a dorsal epitrochlear fossa, but lacking an entepicondylar foramen; distal articulation of humerus with globular capitulum, spool-shaped trochlea, and low lateral trochlear keel; proximal ulna with well-developed olecranon process; radius with oval head and relatively long neck; tarsals, metapodials, and phalanges generally resemble those of *Proconsul*. *Dendropithecus* was an active, arboreal

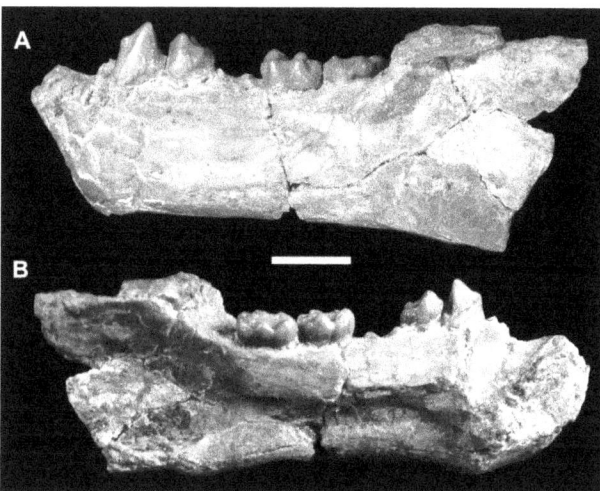

FIGURE 24.1 *Dendropithecus macinnesi*. BM(NH) M 16650 (holotype), left mandibular fragment with p3–p4 and m2–m3: A) lateral view; B) medial view. Scale: 1 cm. Courtesy of P. Andrews.

quadrupedal primate, capable of powerful climbing, and at least some degree of forelimb suspension, most similar in its locomotor capabilities to the larger extant platyrrhines (Le Gros Clark and Thomas, 1951; Harrison, 1982, 2002; Fleagle, 1983; Rose, 1983, 1993; figure 24.1).

Genus *MICROPITHECUS* Fleagle and Simons, 1978

Included Species Mi. clarki Fleagle and Simons, 1978 (type species), *Mi. leakeyorum* Harrison, 1989.

Description Small catarrhines with an estimated body weight of ~4.5 kg and ~3 kg in males and females respectively. Key features are I1 broad and relatively high crowned; I2 almost bilaterally symmetrical; incisors large relative to the size of the cheek teeth; canines high crowned, bilaterally compressed, and markedly sexually dimorphic; upper premolars narrow with well-developed transverse crests; p3 sectorial, with long and narrow crown; p4 ovoid to circular, generally longer than broad; upper molars relatively narrow, with hypocone more lingually placed than protocone, trigon slightly broader than long, large distal fovea, and weak to moderately well-developed lingual cingulum; lower molars ovoid, with low rounded crests, and slightly oblique mesial fovea; lower face very short and broad; premaxilla probably did not make contact with the nasals; nasoalveolar clivus short; nasal aperture broad, and narrows inferiorly between the roots of the central incisors; orbits relatively large, and subcircular in outline; inferior orbital margin overlaps with the nasal aperture; broad interorbital region; inferior orbital fissure extensive; no supraorbital torus or glabellar eminence; weakly developed and widely spaced temporal lines; anterior root of the zygomatic arch originates above M2, close to the alveolar margin, and posteriorly placed in relation to the inferior orbital margin; maxillary sinus extensive; palate broad and shallow; large paired incisive foramina; sulcal pattern on endocranial surface of frontal similar to that of *Proconsul* and extant platyrrhines; mandible high and gracile, with low superior and inferior transverse tori (Pilbeam and Walker, 1968; Fleagle, 1975; Radinsky, 1975; Fleagle and Simons, 1978; Harrison, 1981, 1982, 1988, 1989, 2002).

MICROPITHECUS CLARKI Fleagle and Simons, 1978
Figure 24.2

Distribution Early Miocene (~19–20 Ma). Napak in Uganda and Koru (Koru Formation, Legetet Formation, and Kapurtay Agglomerates, Chamtwara Member) in Kenya (Bishop et al., 1969; Pickford and Andrews, 1981; Pickford, 1983, 1986a, 1986b; Pickford et al., 1986b; Harrison, 1988, 2002).

Description A species distinguished from *Mi. leakeyorum* by the following features: p3 strongly sectorial, with moderately narrow crown; p4 and lower molars relatively broader, with weaker buccal cingulum and more poorly defined mesial and distal foveae; m3 much smaller than m2, with a marked reduction of the cusps and crests distally; upper molars slightly narrower, with narrower trigon and smaller hypocone; M3 relatively smaller with less well-developed cusps distally; M3 ≤ M1 < M2 (Harrison, 1989; figure 24.2). Postcranials from Koru and Napak provisionally referred to this species are smaller, but morphologically similar to those of *D. macinnesi* (T. Harrison, 1982, unpublished data). The frontal bone from Napak (UMP 68–25) was originally attributed to a cercopithecid, but subsequent workers have preferred to ascribe the specimen to *Micropithecus clarki* (Fleagle and Simons, 1978; Harrison, 1982, 1988). However, Rossie and MacLatchy (2006) have recently suggested that the specimen could belong to a cercopithecid after all. Further detailed comparisons are needed to fully resolve this issue, but attribution to *Micropithecus clarki* still seems the most likely taxonomic assignment for the Napak frontal.

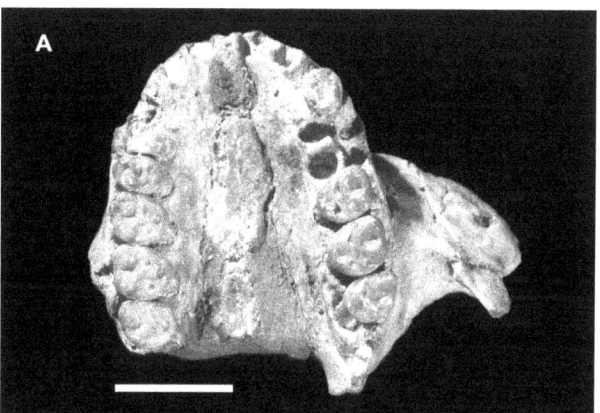

FIGURE 24.2 *Micropithecus clarki*. A) UMP 64-02 (holotype), palate and lower face, occlusal view. Courtesy of J. G. Fleagle. B) KNM-CA 380, mandible, occlusal view. Scale: 1 cm. Courtesy of National Museums of Kenya.

MICROPITHECUS LEAKEYORUM Harrison, 1989
Figure 24.3

Distribution Middle Miocene (~15–16 Ma). Maboko Island and Majiwa, Kenya (Andrews et al., 1981; Pickford, 1981, 1983, 1986a, 1986b; Feibel and Brown, 1991).

Description A species distinguished from *Mi. clarki* by the following features: p3 more bilaterally compressed, with only moderate development of a honing face mesially; p4 relatively longer and narrower; lower molars relatively narrower, with a more pronounced buccal cingulum and better defined mesial and distal fovea; m3 subequal to or slightly larger in occlusal area than m2, and no indication on m3 of marked reduction of the cusps and occlusal crests distally; upper molars slightly broader, with a shorter and more restricted trigon and a larger hypocone; M3 relatively larger with better-developed cusps distally; M1 < M3 < M2 (Harrison, 1989; figure 24.3).

Remarks Based on new finds of *Mi. leakeyorum* from Maboko, Benefit (1991) and Gitau and Benefit (1995) argue that the taxon should be transferred to the genus *Simiolus*. Unfortunately, no detailed descriptions of the material from Maboko have yet been published to substantiate this proposal. An undescribed facial fragment of a male individual apparently differs from *Micropithecus clarki* in having a deeper lower face and a different orientation of the anterior root of the zygomatic arch (Gitau and Benefit, 1995), but these differences could be due to sexual dimorphism. The lower cheek teeth of *Mi. leakeyorum* and *Sim. enjiessi* are generally similar, but the proportions and morphology of the upper cheek teeth are strikingly different. The following features distinguish *Mi. leakeyorum* from *Sim. enjiessi*: cheek teeth smaller in size; p3 with shorter mesiobuccal face and less obliquely aligned crown; p4 lower crowned; lower molars relatively narrower, with smaller mesial and distal foveae, relatively more elongated

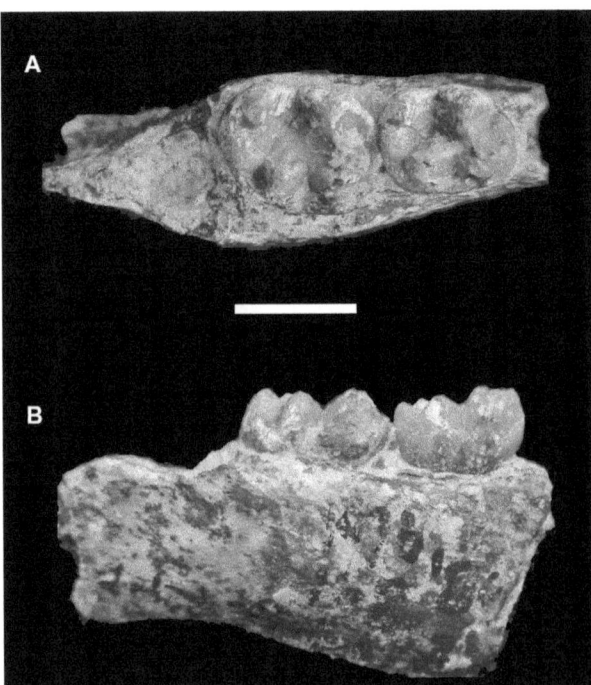

FIGURE 24.3 *Micropithecus leakeyorum.* KNM-MB 14250, right mandibular fragment with m1–m2 (immature). A) Occlusal view; B) lateral view. Scale: 5 mm. Courtesy of National Museums of Kenya.

mesial fovea, a more transversely aligned distal fovea, a hypoconid and hypoconulid that are more closely twinned, and a more strongly developed buccal cingulum; m3 subequal in size to m2 or only slightly larger; buccal cusps on m3 linearly arranged, with the hypoconulid placed more buccally; M2 and M3 relatively much broader; size differential between M1 and M2 much less marked, and lacks the peculiar shape difference between M1 and M2 typical of *Sim. enjiessi*; lingual margin of upper molars more convex, giving the cingulum a C-shaped rather than L-shaped configuration, mesial fovea relatively narrower, crest linking the hypocone and metacone less well developed, lingual cingulum does not continue around the hypocone; superior transverse torus of the mandible may have been more strongly developed than the inferior transverse torus (Harrison, 2002).

These differences between *Mi. leakeyorum* and *Sim. enjiessi* provide adequate justification to include the two species in separate genera. It is interesting to note, however, that the two newly recognized species of *Simiolus* (discussed later), with their more elongated m1–m2 and relatively reduced m3, do narrow the morphological gap between *Micropithecus* and *Simiolus*, at least in their lower dentitions. Nevertheless, based on the strong morphological similarities between *Mi. leakeyorum* and *Mi. clarki*, they are retained here in a single genus. Once the undescribed material from Maboko is fully analyzed, and the relationships between the taxa included in *Micropithecus* and *Simiolus* have been carefully and critically reassessed, it may prove necessary to designate a distinct genus for the species from Maboko.

Genus *SIMIOLUS* Leakey and Leakey, 1987

Included Species *Sim. enjiessi* Leakey and Leakey, 1987 (type species), *Sim. cheptumoae* Pickford and Kunimatsu, 2005, *Sim. andrewsi* sp. nov.

Description Small catarrhine primates with the following combination of features: lower incisors narrow, and small in relation to the size of the molars; upper canine (in females) moderately high crowned and buccolingually compressed; P3 triangular in occlusal outline; upper molars relatively long mesiodistally, with elevated cusps and crests, and a strong transverse crest linking the metacone and hypocone; p3 moderately bilaterally compressed with a long and steep honing face; lower molars relatively long and narrow, ovoid in occlusal outline, with moderately high and sharp cusps and crests, and well-defined basins; the mandible has a high and slender corpus, and well-developed superior and inferior transverse tori. The postcranial remains are comparable in morphology to *Dendropithecus*, and a similar positional behavior can be inferred (Leakey and Leakey, 1987; Rose et al., 1992; Rose, 1993).

SIMIOLUS ENJIESSI Leakey and Leakey, 1987
Figure 24.4

Distribution Early Miocene (~16.8–17.5 Ma). Kalodirr and Locherangan, northern Kenya (Leakey and Leakey, 1987; Anyonge, 1991; Boschetto et al., 1992).

Description A catarrhine primate with an estimated body weight of ~6 kg and ~4 kg in males and females, respectively (Rose et al., 1992; Harrison, 2002). Characteristic features include face relatively short with orbits positioned far anteriorly; incisive foramen large; mandibular symphysis with superior transverse torus subequal to or larger than the inferior transverse torus; incisors relatively narrow and high crowned;

canines buccolingually compressed; p3 high crowned and strongly sectorial, with relatively long mesiobuccal honing face; p4 long and narrow; P3 almost triangular in occlusal outline, with a pronounced degree of extension of enamel onto the buccal root; P4 with limited flare of the cusps, and well-developed lingual cingulum; molars with elevated cusps and sharp occlusal crests; lower molars relatively long and narrow, with poorly developed buccal cingulum, and well-defined, slightly oblique distal fovea; upper molars with large talon basin and well-developed lingual cingulum that continues around the hypocone; M1 much smaller than M2, and differs in shape, being much shorter and more rectangular; M2 and M3 relatively elongated mesiodistally and subequal in size; M3 relatively large without marked reduction of the distal cusps; upper molars with well-developed crest linking the hypocone and metacone (Leakey and Leakey, 1987; Harrison, 2002; figure 24.4).

A number of postcranial specimens of *Simiolus* are known from Kalodirr (Leakey and Leakey, 1987; Rose et al., 1992; Rose, 1993). The most important features are as follows: humerus with slender and slightly retroflexed shaft, distinct dorsal epitrochlear fossa, no entepicondylar foramen, and distal articulation with modest lateral trochlear keel; femur with relatively small head, high neck angle, distinct tubercle on the neck; talus similar to that of other dendropithecids and to proconsulids; metacarpals and phalanges indicate a narrow hand with good flexion-grasping capabilities (Harrison, 1982;

Rose et al., 1992; Rose, 1993). *Simiolus*, like *Dendropithecus*, is inferred to have been an active and agile arboreal quadruped most similar in its positional behavior to the larger extant platyrrhines (Rose et al., 1992; Rose, 1993, 1997).

SIMIOLUS CHEPTUMOAE Pickford and Kunimatsu, 2005

Distribution Middle Miocene (~14.5 Ma). Kipsaraman Main, Muruyur Formation, Tugen Hills, Kenya (Pickford and Kunimatsu, 2005).

Description A species that is slightly smaller than the type species, *Sim. enjiessi*. p4 crown longer than broad, with weak buccal cingulum, and triangular mesial fovea. p4 differs from that of *Sim. enjiessi* in being relatively broader, the long axis of the crown is more obliquely oriented relative to the mesiodistal axis of the crown, and the mesial fovea is more triangular. P4 relatively broad, with a single transverse crest connecting the protocone and paracone. m1 with marked buccal flare, protoconid more mesially positioned than the metaconid, protocristid obliquely oriented, hypoconulid situated just to the buccal side of the midline of the tooth, postmetacristid bears a well-developed mesostylid (bifid metaconid apex of Pickford and Kunimatsu, 2005), buccal cingulum weakly developed and distal fovea small. m3 relatively small (subequal in size to m1), whereas in *Sim. enjiessi* it is much larger than m1 and m2. Lower molars more elongated than those of *Sim. enjiessi* (Leakey and Leakey, 1987; Pickford and Kunimatsu, 2005).

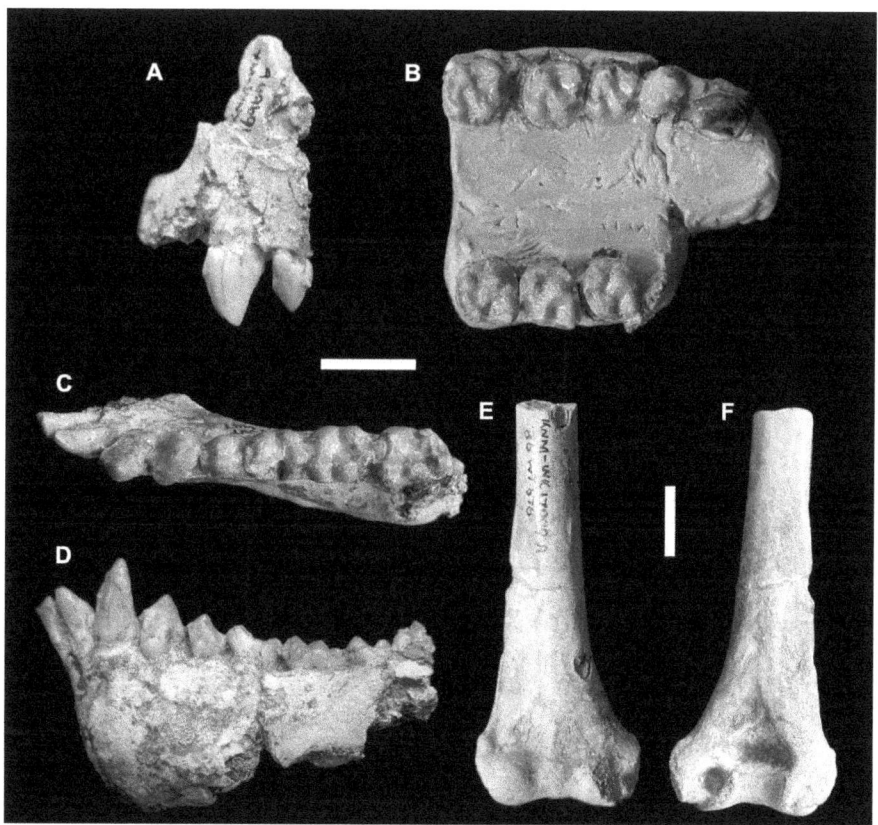

FIGURE 24.4 *Simiolus enjiessi*. A) KNM-WK 16960 (holotype), left premaxilla/maxilla with C–P3, lateral view. B) KNM-WK 16960, right C, P4 and M1–M3, and right M1–M3, occlusal view. C) KNM-WK 16960, left mandible with i1–m3, occlusal view. D) KNM-WK 16960, left mandible with i1–m3, lateral view. E) KNM-WK 17009, right distal humerus, anterior view. F) KNM-WK 17009, right distal humerus, posterior view. Scale: 1 cm. Courtesy of National Museums of Kenya.

SIMIOLUS ANDREWSI sp. nov.
Figure 24.5

Distribution Middle Miocene (~13.7 Ma)(Pickford et al., 2006). Fort Ternan, Kenya.

Holotype Left mandibular corpus with c–m3 (KNM-FT 20), and associated i2 (KNM-FT 25), p4 (KNM-FT 24), m2 (KNM-FT 21), and m3 (KNM-FT 23) from the right side. Two specimens, incorrectly accessioned as having been recovered from Maboko Island (a left i2, KNM-MB 124) and Songhor (a right lower canine, KNM-SO 1102), respectively, are identical in morphology and preservation to KNM-FT 20 and KNM-FT 25, and clearly represent antimeres of the same individual (Andrews and Walker, 1976; Harrison, 1992, 2002; figure 24.5).

Referred Specimens KNM-FT 13, edentulous mandibular symphysis of an immature individual; KNM-FT 14, left mandibular corpus of an immature individual with m1 exposed in its crypt; KNM-FT 19, left M3 (Harrison, 1992).

Etymology Named after Peter Andrews in recognition of his important contributions to the study of Miocene catarrhines.

Diagnosis A species of *Simiolus* similar in overall dental size to *Sim. enjiessi* and slightly larger than *Sim. cheptumoae*. Differs from *Sim. enjiessi* in the following features: i2 relatively higher crowned and slightly broader, with a more distinctly angular distal margin, and a better-developed lingual pillar; lower canine (comparing those of presumed females) is slightly taller and more slender; p3 not as elongated or as bilaterally compressed, with a shorter honing face; p4 slightly broader, with more widely spaced cusps, more oblique transverse crest linking the main cusps, and less well-developed buccal cingulum; m2 relatively narrower (average breadth-length index is 79.4 in *Sim. enjiessi* and 75.7 in *Sim. andrewsi*) with a greater size differential between m1 and m2; m2 with slightly longer mesial fovea, more transversely aligned protocristid, broader distal fovea, somewhat better developed buccal cingulum, and hypoconulid more buccally displaced; m3 smaller than m2,

with a more transversely oriented protocristid, a better-developed buccal cingulum, a relatively larger entoconid, and a smaller distal fovea; M3 mesiodistally shorter, relatively smaller in size, with more markedly reduced distal cusps (Harrison, 1992, 2002). Differs from *Sim. cheptumoae* in the following features: dentition slightly larger in size (lower cheek tooth areas average 17.3% larger); p4 relatively broader, with shorter mesial fovea; m1 with less pronounced buccal flare, more strongly developed buccal cingulum, and lacking a mesostylid on the postmetacristid; lower molars not as elongated (average breadth-length indices for m1 and m3 are 73.1 and 74.6 in *Sim. cheptumoae* and 81.4 and 77.8 in *Sim. andrewsi*); m3 larger than m1.

Description i2 narrow and moderately high crowned, with angular distal margin, rounded lingual cingulum and distinct lingual pillar; lower canine of female individual moderately high crowned and slender, with a short distal heel; p3 is low crowned, mesiodistally elongated and bilaterally compressed, with a relatively short and steeply inclined honing face; p4 is long and narrow, and ovoid in occlusal outline, with a slight trace of a buccal cingulum; lower molars long and narrow, with shallow talonid basin, hypoconulid situated slightly toward the buccal side of the midline of the crown, large and well-defined distal fovea, and well-developed buccal cingulum. m3 slightly smaller than m2, but larger than m1. M3 relatively short and broad, with reduced distal cusps, and a moderately well-developed lingual cingulum (see Andrews and Walker, 1976; Harrison 1982, 1992 for additional illustrations, descriptions, and measurements).

Remarks The small catarrhine primates from the middle Miocene of Fort Ternan were first described by Leakey (1968), who provisionally referred them to *Limnopithecus* sp. Andrews and Walker (1976) presented a more detailed description of the material, and tentatively assigned the specimens to *Limnopithecus legetet*. Later, Andrews (1980) suggested that the material had its closest affinities with *Dendropithecus*. Following the description of *Simiolus enjiessi*, I referred the Fort Ternan material to the latter genus but did not assign it to a species (Harrison, 1992, 2002). Given that that the Fort Ternan material is sufficiently distinct from all other previously described species and that the material is adequate to diagnose a separate taxon, the recognition of a new species appears fully justified. Nevertheless, the question remains whether this species should be attributed to *Simiolus*. Pickford and Kunimatsu (2005) have recently argued that the Fort Ternan material should not be included in *Simiolus*, based mainly on the morphological differences in the lower dentition that distinguish it from *Simiolus cheptumoae* from Kipsaraman. However, one could use the same argument to negate the generic attribution of the material from Kipsaraman. The Fort Ternan species and *Simiolus enjiessi* share high-crowned and slender lower canines, bilaterally compressed p3 with relatively long honing faces, elongated lower molars with high conical cusps arranged peripherally and connected by well-developed crests; and a high and slender mandibular corpus. Among all of the East African Miocene catarrhine taxa, the species from Fort Ternan is morphologically closest to *Simiolus enjiessi*, being remarkably similar in many respects, and it is appropriate to include them together in the same genus. Unfortunately, the upper dentition, which is so distinctive in the type species, is largely unknown from Fort Ternan (and Kipsaraman). Additional finds might eventually necessitate the recognition of more than one genus for these three species, but pending such discoveries, it seems appropriate to retain them in the genus *Simiolus*.

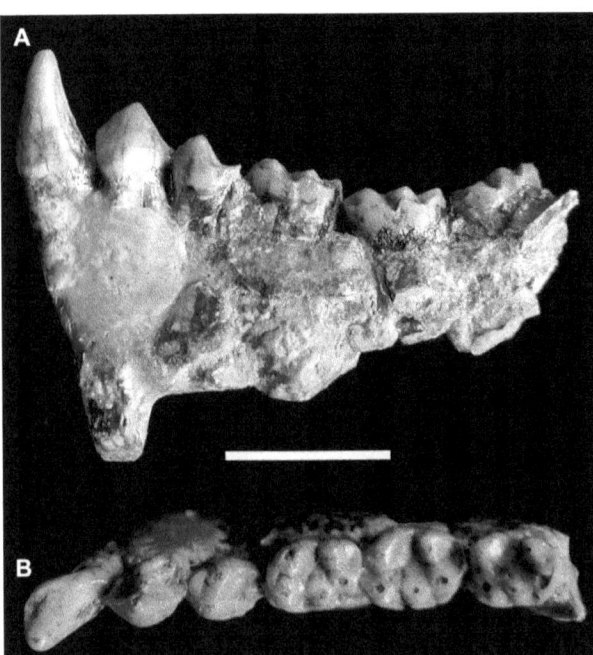

FIGURE 24.5 *Simiolus andrewsi*. KNM-FT 20 (holotype), left mandibular fragment with c–m3, A) lateral view; B) occlusal view. Scale: 1 cm. Courtesy of National Museums of Kenya.

Superfamily PROCONSULOIDEA Leakey, 1963
Family PROCONSULIDAE Leakey, 1963
Subfamily PROCONSULINAE Leakey, 1963
Genus *PROCONSUL* Hopwood, 1933

Included Species *P. africanus* Hopwood, 1933 (type species), *P. heseloni* Walker, Teaford, Martin and Andrews, 1993, *P. nyanzae* Le Gros Clark and Leakey, 1950, *P. major* Le Gros Clark and Leakey, 1950, *P. gitongai* (Pickford and Kunimatsu, 2005).

Description Medium- to large-sized catarrhines. Lower face moderately short and broad. Incisive fossa with large fenestra. C/I2 diastema relatively large in males, but small in females. Nasal aperture relatively broad, rhomboidal in shape, tapering inferiorly between the central incisor roots, and widest just below mid-height. Subnasal clivus short. Nasal bones long and narrow, and supported laterally by premaxillary alae. Premaxilla makes contact with the nasal bones, thereby excluding the maxilla from the margin of pyriform aperture (*contra* Andrews, 1978; Rae, 1999). Prominent canine jugum and shallow canine fossa. Single large infraorbital foramen. Palate long, rectangular and shallow. Maxillary sinus extensive. Zygomatic arch originates relatively low on the face. Articular fossa gutterlike, with well-developed eminence and postglenoid process. Short, broad tubelike external auditory meatus. Nuchal plane short and steeply angled, with a strongly developed external occipital protuberance located high on the neurocranium. Large subarcuate fossa. Frontal process of the zygomatic perforated by multiple zygomaticofacial foramina situated slightly above the inferior margin of the orbits. Low indistinct supraorbital costae and a slightly swollen glabellar region with an extensive frontal sinus. Interorbital region relatively broad. Orbits subrectangular, slightly broader than high, with a distinct angulation of the superolateral margin. Lacrimal duct located within the orbit. Neurocranium relatively large, with slight postorbital constriction. Superior temporal lines strongly marked and converge posteriorly, but do not meet to form a sagittal crest, at least in females. Cortical sulcal pattern generally similar to that of extant platyrrhines. Mandibular symphysis with superior transverse torus moderate to well developed, and inferior transverse torus generally weaker or entirely absent. Single mental foramen located below the premolars (Le Gros Clark and Leakey, 1951; Napier and Davis, 1959; Davis and Napier, 1963; Corruccini and Henderson, 1978; Whybrow and Andrews, 1978; McHenry et al., 1980; Walker and Pickford, 1983; Falk, 1983; Walker et al., 1983; Walker, 1997; Harrison, 2002; Rossie et al., 2002).

Upper incisors slightly procumbent. I1 narrow and high crowned, and much larger than I2. Upper canines relatively stout, moderately bilaterally compressed, with a single mesial groove. Canines strongly sexually dimorphic. Upper premolars mesiodistally short and broad, with a marked height differential between the paracone and protocone, especially on P3, and a weak lingual cingulum variably present on P4. Enamel on molars ranges from thin to thick (Beynon et al., 1998; Smith et al., 2003). Upper molars rectangular to rhomboidal in shape, and buccolingually broader than long. Cusps and crest moderately elevated, and occlusal basins generally well defined. Protoconule usually conspicuous. Lingual cingulum broad, and commonly beaded. Buccal cingulum variably developed. Hypocone large, with poorly developed crests linking it to the protocone or crista obliqua. M3 subequal in size to M2 or slightly larger, with variable regression of the metacone and hypocone. Lower incisors narrow and high crowned. p3 moderately sectorial, with relatively long mesiobuccal face. p4 usually broader than long, with a weak buccal cingulum. Lower molars relatively long and narrow, with simple "crystalline" cusps, and

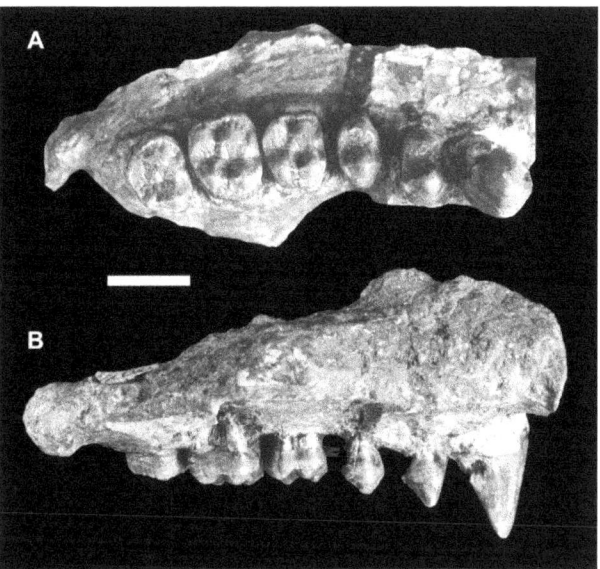

FIGURE 24.6 *Proconsul africanus*. BM(NH) M 14084 (holotype), left maxilla with C–M3, A) occlusal view; B) medial view. Scale: 1 cm. Courtesy of P. Andrews.

few wrinkles. Buccal cingulum variably developed. Mesial and distal foveae generally well defined. m1 < m2 < m3 (Andrews, 1978; Walker et al., 1983, 1993; Andrews and Martin, 1991; Walker, 1997; Beynon et al., 1998; Harrison, 2002).

PROCONSUL AFRICANUS Hopwood, 1933
Figure 24.6

Distribution Early Miocene (~19–20 Ma). Koru (Koru Formation, Legetet Formation, and Chamtwara Member of the Kapurtay Nephelinite Agglomerates), Songhor (Kapurtay Agglomerates), and Mteitei Valley, Kenya (Bishop et al., 1969; Pickford and Andrews, 1981; Pickford, 1981, 1983, 1986a, 1986b; Harrison, 1988, 2002).

Description A medium-sized catarrhine, comparable in overall size to *P. heseloni* (body weight estimates given later), although the teeth tend to be slightly smaller. It differs from *P. heseloni* in the following respects: lower canines more bilaterally compressed; upper premolars narrower, with greater height differential between the paracone and protocone; cingula and occlusal crests better developed on upper molars; hypocone and protocone subequal in size; M1 relatively narrower; M1< M3 < M2; lower molars with better-developed buccal cingula; m3 hypoconulid and hypoconid subequal in size, entoconid relatively larger and more distally placed relative to the hypoconid, distal fovea broader with better-developed crest linking the entoconid and hypoconulid, broader less triangular talonid with more reduced distal cusps; molar enamel thinner; mandibular symphysis with massive superior transverse torus only; mandibular corpus tends to be deeper, and shallows posteriorly more strongly (Andrews, 1978; Walker et al., 1993; Harrison, 2002; Smith et al., 2003; figure 24.6).

PROCONSUL HESELONI Walker et al.,1993
Figure 24.7

Distribution Early Miocene (~17.0–18.5 Ma). Rusinga Island (Wayando, Kiahera, Hiwegi, and Kulu Formations) and Mfwangano Island (Kiahera, Rusinga Agglomerate, and Hiwegi formations)(Drake et al., 1988).

Description A medium-sized catarrhine, similar in dental size to *P. africanus*, with an estimated body weight of ~10 kg and ~20 kg for females and males respectively (Ruff et al., 1989; Rafferty et al., 1995). *Proconsul heseloni* differs from *P. africanus* in the following characteristics: lower canines less bilaterally compressed; upper premolars slightly broader, with reduced height differential between the paracone and protocone; cingula and occlusal crests less well developed on upper molars; hypocone smaller than the protocone; M1 relatively broader; M1 < M2 ≤ M3; lower molars with less well-developed buccal cingulum; m3 hypoconulid larger than hypoconid, entoconid relatively smaller and positioned more directly transversely opposite the hypoconid, distal fovea less transversely aligned with weaker crest linking the entoconid and hypoconulid, and narrower and more triangular talonid with well-developed distal cusps; molar enamel intermediate thick; mandibular symphysis with inferior transverse torus subequal to or less pronounced than the supe-rior transverse torus; mandibular corpus not as deep, and shallows less strongly posteriorly (Andrews, 1978; Walker et al., 1993; Harrison, 2002; Smith et al., 2003).

The skull of *P. heseloni* has been described and analyzed in some detail previously (Le Gros Clark, 1950; Le Gros Clark and Leakey, 1951; Napier and Davis, 1959; Davis and Napier, 1963; Corruccini and Henderson, 1978; McHenry et al., 1980; Walker and Pickford, 1983; Walker et al., 1983; Teaford et al., 1988), and it provides the basis for the description presented for the genus above. Walker et al. (1983) have estimated the cranial capacity of the type specimen at 167.3 cm^3, suggesting that *Proconsul heseloni* was more encephalized than extant cercopithecids of comparable body size. However, Manser and Harrison (1999) have predicted a brain size of only 130.3 cm^3 based on a regression of foramen magnum area, which suggests that the degree of encephalization in *Proconsul* was close to the mean for anthropoids (figure 24.7).

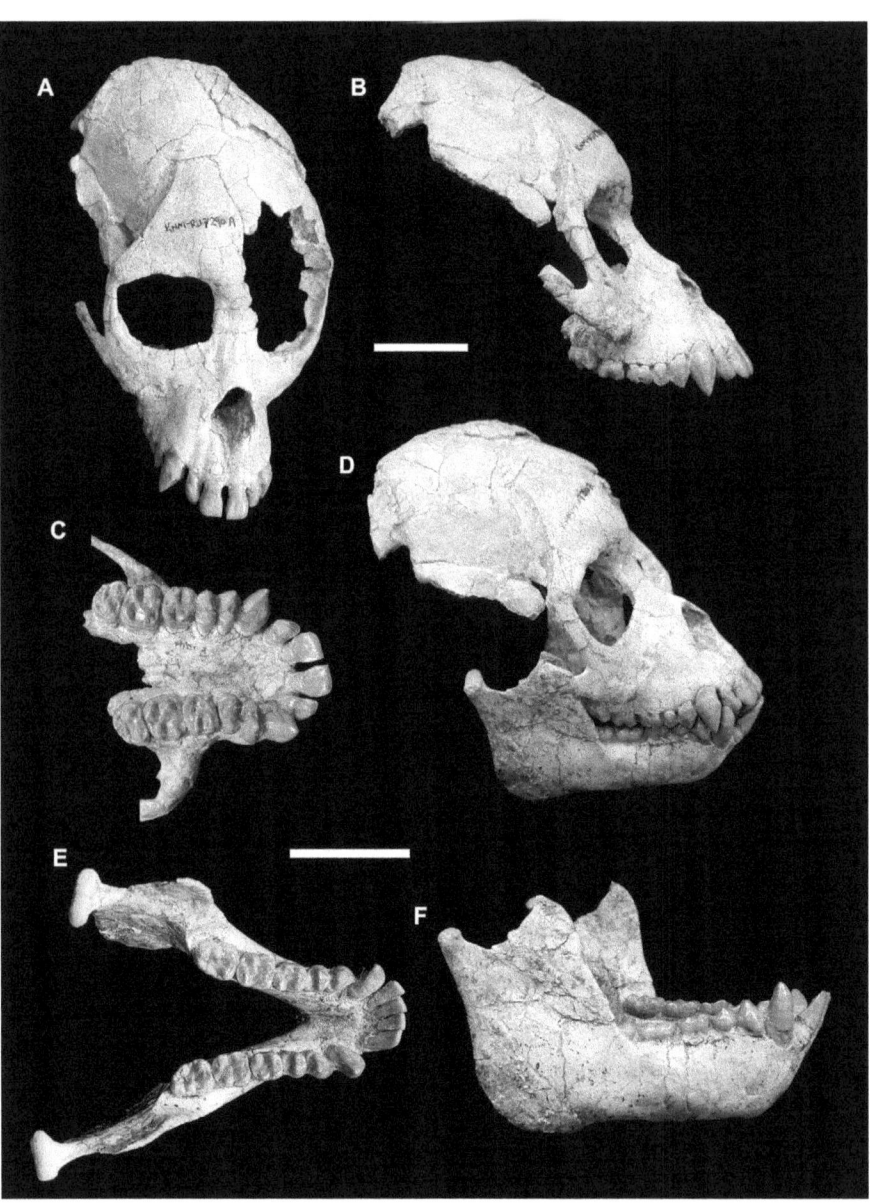

FIGURE 24.7   *Proconsul heseloni.* KNM-RU 7290 (holotype), partial skull: A) cranium, facial view; B) cranium, lateral view; C) cranium, palatal view; D) cranium, oblique superolateral view; E) mandible occlusal view; F) mandible lateral view. Scale: 3 cm—upper bar for A, B, and D; lower bar for C, E, and F. Courtesy of National Museums of Kenya.

The postcranial skeleton of *Proconsul heseloni* is well-known, being represented by at least nine partial skeletons from Rusinga Island (Le Gros Clark and Leakey, 1951; Napier and Davis, 1959; Walker and Pickford, 1983; Walker et al., 1985; Walker and Teaford, 1989). Although the postcranium corresponds closely in almost every respect to the primitive catarrhine morphotype, a few traits have been inferred to be synapomorphies linking *Proconsul* with extant hominoids (Harrison, 1982, 1987, 1993, 2002; Rose, 1988, 1992, 1993, 1997; Walker and Pickford, 1983; Walker et al., 1993; Walker, 1997; C. Ward et al., 1991, 1993, 1995, 1997; Ward, 1993, 1997, 1998; Kelley, 1997). Thorax relatively long and narrow (Ward, 1993, 1997; Ward et al., 1993). Lumbar vertebrae with centra that are long, with relatively small cranial and caudal surface areas (Sanders and Bodenbender, 1994; Harrison and Sanders, 1999) and moderately well-developed ventral keels. Sacrum relatively narrow, with small sacroiliac joint (Rose, 1993). Based on a purported partial sacrum it has been argued that *P. heseloni* did not have a tail, but this inference has been contested (Ward et al., 1991; Ward, 1997; Harrison, 1998; C. Ward et al., 1999). Ilium narrow. Ischial tuberosities lacking as in platyrrhines (Rose, 1993; Harrison and Sanders, 1999). Estimated intermembral, brachial and crural indices of 88, 96, and 92, respectively (Walker and Pickford, 1983; Harrison, 2002). Limb bones relatively robust (Ruff et al., 1989). Scapula most similar to those of colobines and platyrrhines (Rose, 1993, 1997). Humerus with posteriorly directed head, retroflexed shaft, no entepicondylar foramen or dorsiepitrochlear fossa, and distal articulation with distinct lateral keel, globular capitulum, and narrow zona conoidea (Harrison, 1982, 1987; Rose, 1993, 1997). Radial head ovoid, with beveled margin. Ulna with well-developed olecranon process and a styloid process that articulates with the carpus (Napier and Davis, 1959; Harrison, 1982, 1987; Beard et al., 1986). Os centrale unfused. Hand relatively long (Walker et al., 1993), with a well-developed thumb and a mobile trapezium–first metacarpal joint (Rafferty, 1990; Rose, 1992, 1993). Femur with high neck angle (Walker, 1997). Distal end of femur relatively broad, with medial condyle slightly larger than lateral condyle. Fibula stout. Foot similar to those of arboreal quadrupedal primates (Harrison, 1982; Langdon, 1986; Rose, 1993; Strasser, 1993). Phalanges quite stout and slightly curved (Begun et al., 1993). Hallux well developed with a powerful grasping capability. The postcranium indicates that *P. heseloni* was an arboreal, quadrupedal catarrhine, most similar in its locomotor repertoire to extant colobines and larger platyrrhines (Harrison, 1982; Rose, 1983, 1993, 1994; Walker and Pickford, 1983; Walker, 1997; Li et al., 2002).

PROCONSUL NYANZAE Le Gros Clark and Leakey, 1950
Figure 24.8

Distribution Early Miocene (~17.0–18.5 Ma). Rusinga Island (Wayando, Kiahera, Hiwegi and Kulu Formations), Mfangano Island (Kiahera, Rusinga Agglomerate, and Hiwegi Formations), and Karungu, western Kenya (Drake et al., 1988).

Description A large catarrhine, intermediate in dental size between *P. heseloni* and *P. major*. Estimated body weight range of 20–50 kg (Ruff et al., 1989; Rafferty et al., 1995), with males and females probably averaging ~28 kg and ~40 kg, respectively. *Proconsul nyanzae* is morphologically very similar to *P. heseloni*. It differs primarily in its larger size, and the following morphological features: canines less bilaterally compressed; P3 with greater height differential between paracone and protocone;

lower molars relatively broader; greater size differential between m1 and m2; m3 has a larger entoconid connected to the hypoconulid by a well-developed crest; greater size differential between M1 and M2; M1 and M2 relatively narrower, with hypocone subequal in size to metacone, lingual cingulum better developed, and distal transverse crest more pronounced; M3 relatively larger; greater degree of secondary wrinkling on upper and lower molars; slightly thicker molar enamel; no inferior transverse torus on the mandible (Harrison, 2002; Smith et al., 2003). The postcranium of *Proconsul nyanzae* is represented by a partial skeleton from Mfangano Island (Ward et al., 1993), two partial skeletons from the Kaswanga Primate Site, and a number of associated and isolated postcranial elements from Rusinga Island (Le Gros Clark and Leakey, 1951; Le Gros Clark, 1952; Preuschoft, 1973; Harrison, 1982, 2002). Despite the size difference, *Proconsul nyanzae* is remarkably similar in its postcranial morphology to *P. heseloni* (figure 24.8).

PROCONSUL MAJOR Le Gros Clark and Leakey, 1950
Figure 24.9

Distribution Early Miocene (~19–20 Ma). Songhor, Mteitei Valley, and Koru (Koru Formation, Legetet Formation, and Kaputray Agglomerate, Chamtwara Member), Kenya and Napak (I, IV, V, IX, and CC), Uganda (Bishop et al., 1969; Pickford and Andrews, 1981; Pickford, 1981, 1983, 1986a, 1986b; Pickford et al., 1986b; Harrison, 1988, 2002; Senut et al., 2000; MacLatchy and Rossie, 2005). Senut et al. (2000) and Pickford et al. (2003) have referred several isolated teeth and a fragmentary femur from Moroto II (17.0–17.5 Ma) to

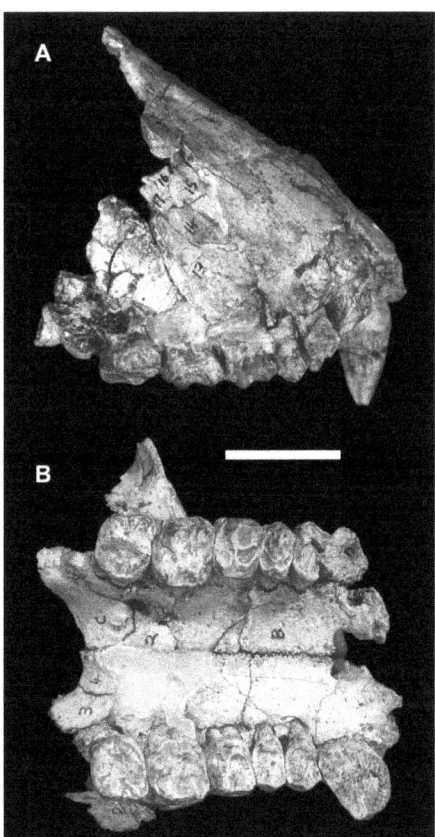

FIGURE 24.8 *Proconsul nyanzae.* BM(NH) M 16647 (holotype), lower face and palate: A) lateral view, B) occlusal view. Scale: 2 cm. Courtesy of P. Andrews.

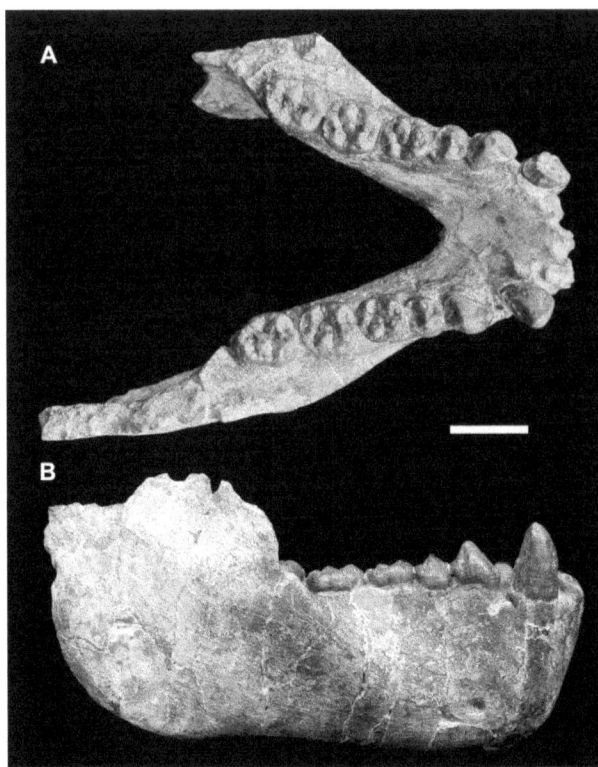

FIGURE 24.9 *Proconsul major*, KNM-LG 452, partial mandible: A) occlusal view, B) lateral view. Scale: 2 cm. Courtesy of National Museums of Kenya.

this taxon, but this attribution is questionable (see also MacLatchy and Rossie, 2005), and the teeth, at least, are consistent in morphology with those of *Afropithecus turkanensis* (T. Harrison, unpublished data).

Description A large catarrhine, similar to or slightly larger in dental size than *Pongo pygmaeus*, with an estimated body weight of 60–90 kg (Harrison, 1982; Rafferty et al., 1995; Gommery et al., 1998). *Proconsul major* (along with *P. gitongai*) is the largest species of *Proconsul*, with average dental dimensions almost 20% larger than those of *P. nyanzae*. It is not as well-known as *P. heseloni* or *P. nyanzae*, being represented by a small number of jaw fragments and isolated teeth (Pilbeam, 1969; Andrews, 1978; Martin, 1981). *Proconsul major* is characterized by the following features: massive superior transverse torus with no inferior torus (as in *P. africanus*); upper and lower incisors relatively broader; upper canines less bilaterally compressed and more tusklike; canines with distinctive sinusoidal curvature of distal crest and a bladelike tip; upper premolars narrower, with cusps more similar in height; lower molar proportions differ from *P. nyanzae* in having a less marked size differential between m1 and m2; m3 tends to be larger relative to m2; lower molars have a stronger buccal cingulum than in *P. nyanzae* and *P. heseloni*; M3 relatively large as in *P. nyanzae* (Andrews, 1978; Martin, 1981; Harrison, 2002). Postcranial remains from Koru, Songhor, and Napak include a scapular fragment, humeral shaft fragments, metapodials, phalanges, a navicular, calcanei, tali, several femoral fragments, and a distal tibia (MacInnes, 1943; Le Gros Clark and Leakey, 1951; Preuschoft, 1973; Harrison, 1982; Conroy and Rose, 1983; Langdon 1986; Rafferty et al., 1995; Gommery et al., 1998, 2002; Senut et al., 2000). Despite their larger size, they are generally similar to the corresponding elements of *P. heseloni* and

P. nyanzae, but distinctions do indicate differences in locomotor and positional behavior (Harrison, 1982; Senut et al., 2000; Gommery et al., 2002). Nengo and Rae (1992) have described a fragmentary distal ulna of *P. major* as being much more hominoid-like than that of *P. heseloni*, but the attribution of this specimen is questionable (Rose, 1997; Walker, 1997; figure 24.9).

Remarks Senut et al. (2000) have argued that *Proconsul major* is distinct enough from other species of *Proconsul* to merit being placed in its own genus, *Ugandapithecus*. Justification for this position is provided by its unique combination of morphological features, especially the specializations of the canines and premolars. However, one could argue with equal justification that these are differences that one would typically expect to distinguish species within a genus. In their diagnosis of *Ugandapithecus*, Senut et al. (2000) include several features of the proximal femur that differentiate it from *P. heseloni* and *P. nyanzae*, but unfortunately comparative material of *P. africanus* is not available.

There is no doubt that *P. major* is closest morphologically to the other species of *Proconsul* when compared to other Miocene taxa, and together they form a tight-knit clade. However, despite the close similarity of these species, one could argue for a separate genus on phylogenetic grounds if it could be demonstrated that *P. major* was the sister group to a clade comprising the other species of *Proconsul*, or that it shared derived features with another taxon. Unfortunately, it has not been possible to establish the relationships among *Proconsul* species to help resolve this matter. *Proconsul heseloni* and *P. nyanzae* are morphologically very similar to each other, and it does seem reasonable to conclude that they are one another's sister taxa. *Proconsul major* shares a well-developed superior transverse torus of the mandible with *P. africanus*, and this specialization may serve to unite these two species as a clade. If this is the case, pairs of sister species of *Proconsul* would co-occur at Kenyan localities in chronological succession, with *P. major* + *P. africanus* at 19–20 Ma, followed by *P. heseloni* + *P. nyanzae* at 17–18 Ma (Harrison, 2002; MacLatchy and Rossie, 2005; Harrison and Andrews, 2009). If confirmed, it would argue against recognizing *P. major* as a separate genus. In fact, a much stronger case might be made for separating the two species pairs as different genera, in which case *africanus* + *major* would be retained in *Proconsul* (the type species being *P. africanus*), and a new genus would have to be recognized for *heseloni* + *nyanzae*. As an alternative interpretation, *P. major* does exhibit a few features that presage the derived morphology seen in Afropithecinae, such as more tusklike canines and narrower upper premolars with cusps more similar in height. This could imply a closer relationship with this latter clade, thereby supporting a generic distinction. However, given the close morphological similarity between species of *Proconsul*, and our current lack of understanding of the precise relationships among them, it is best to group them together in a single genus, *Proconsul*. As a result, *Ugandapithecus* is recognized here as a junior synonym of *Proconsul* (see also MacLatchy and Rossie, 2005).

If these species are included together in *Proconsul*, then an interesting consequence is that the genus contains taxa with at least a fivefold difference in estimated average body mass. Very few modern genera of mammals have species that encompass such a range of body mass, presumably because the ecological and physiological correlates associated with such differences in body size profoundly influence behavior and morphology, which in turn necessitates recognizing different genera. However, exceptions do exist among African large mammals (e.g., *Theropithecus* and *Tragelaphus*). Males of *Theropithecus* spp. (extant and extinct), for example, probably ranged in body mass from about 20 kg to almost 100 kg (Delson et al., 2000). This observation may have implications for interpreting the ecological and behavioral plasticity of *Proconsul*.

PROCONSUL GITONGAI (Pickford and Kunimatsu, 2005)
Figure 24.10

Distribution Early middle Miocene (~14.5 Ma). Kipsaraman, Muruyur Formation, Tugen Hills, Kenya (Hill et al., 1991; Pickford, 1998; Behrensmeyer et al., 2002; Pickford and Kunimatsu, 2005).

Description A species of *Proconsul* similar in size or slightly larger than *P. major*. It is poorly known, being represented only by two associated upper molars (holotype) and seven isolated teeth (Hill et al., 1991; Pickford and Kunimatsu, 2005). The upper molars differs from those of *P. major* in the following respects: cusps with higher relief and more blocky appearance; trigon basin and distal fovea somewhat deeper; broader lingual cingulum, extending distally onto the mesiolingual aspect of the hypocone, with a tendency to develop an accessory cuspule; and enamel more coarsely wrinkled. An upper canine (female) and a germ of a lower canine (male) are known, and these are comparable in all respects to those of *P. major*, including the distinctive bladelike tip in the lower canine. m3 has a narrow mesial fovea, postmetacristid bearing a distinct mesostylid, a tendency to develop accessory cuspules between the main cusps, and a narrow and discontinuous buccal cingulum (Pickford and Kunimatsu, 2005; figure 24.10).

Remarks Pickford and Kunimatsu (2005) included this species in *Ugandapithecus*, but as discussed earlier, this genus is recognized here as a junior synonym of *Proconsul*. Moreover, given the paucity of material, the relatively minor distinctions in the upper and lower molars, and the almost complete overlap in size, I am not convinced that *P. gitongae* is sufficiently different from *P. major* to merit attribution to a separate species. It is telling that Pickford and Kunimatsu (2005) consider an upper molar from Moroto II, attributed to *P. major* (Pickford et al., 1999, 2003), as possibly belonging to *P. gitongai*, based on its very large size (although the specimen is considered here to belong to *Afropithecus turkanensis*). This implies that size is a key factor in discriminating these species, yet the sample sizes of *P. major* are not adequate to determine the range of intraspecific variation in this respect. Certainly, similar morphological and metrical distinctions could be used to separate early Miocene samples of *P. major* (Martin, 1981), and such differences appear to be typical of intraspecific variation between populations of extant species (e.g., Pilbrow, 2006). The overall similarity in the molars and the distinctive morphology of the lower canines does suggest that the *P. major* and *P. gitongai* samples might be conspecific or at least closely related members of a single evolving lineage. The difference in age between the Kipsaraman sample and the youngest specimens of *P. major* (more than 4 myrs—less if specimens from Moroto are included in *P. major*) may have been an important contributing factor in the decision to recognize a new species. A time range of over 5 myrs for a single species of fossil catarrhine would represent a remarkable temporal span, and no other species from the early Miocene of East Africa can be definitively demonstrated to have survived into the middle Miocene. Until larger samples are available to establish its morphological distinctiveness, or

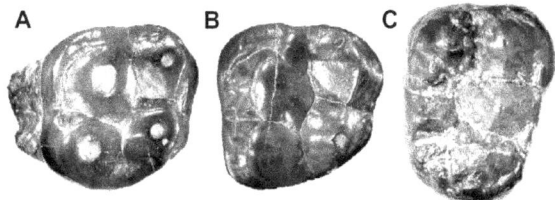

FIGURE 24.10 *Proconsul gitongai*. A) Bar 737'02, left M1 (holotype). B) Bar 210'02, left M2. C) Bar 213'02, right m3. Scale: 1 cm. Courtesy of M. Pickford.

until a detailed taxonomic revision of the larger proconsulids is undertaken, *Proconsul gitongai* is provisionally retained as a separate species.

PROCONSUL MESWAE Harrison and Andrews, 2009

Distribution Early Miocene (~22.5 Ma). Meswa Bridge (Locality 36), Muhoroni Agglomerate, Kenya (Bishop et al., 1969; Pickford, 1981, 1986a; Pickford and Andrews, 1981; Harrison and Andrews, 2009).

Description A species of *Proconsul* that is intermediate in dental size between *P. nyanzae* and *P. major*. It differs from all known species in the following features: incisors relatively low crowned; deciduous canines relatively larger, more robust, and high crowned; molars and deciduous premolars relatively broader and higher crowned, with a more pronounced degree of buccolingual flare, cusps less voluminous and situated farther from the crown margin, and better developed cingula; p4 broader, with better-developed buccal cingulum and smaller distal basin; lower molars less rectangular in occlusal outline, with a longer and narrower mesial fovea, smaller distal fovea, more restricted talonid basin, hypoconid larger than the protoconid, and a tendency for a smaller hypoconulid; size differential among dp4, m1, and m2 not as great; mandibular corpus in infants of comparable dental age relatively more slender with a less prominent development of the superior transverse torus (at least compared with *P. major*); maxilla robust with larger diastema and better developed canine jugum and canine fossa (Andrews et al., 1981; Harrison and Andrews, 2009).

Remarks All of the material comes from a single excavation site at Meswa Bridge, and it comprises at least four individuals ranging in age from infant to late juvenile. While acknowledging the distinctiveness of the material, Andrews et al. (1981) deferred naming a new species because the hypodigm consists entirely of immature individuals. However, sufficient examples of the permanent dentition are available to be able to make comparisons with other known species of *Proconsul* and to distinguish the Meswa Bridge sample as a separate species (Harrison and Andrews, 2009). This material represents the oldest known species of *Proconsul*, and only *Kamoyapithecus* among fossil catarrhines has greater antiquity in the East African fossil record. Like *Kamoyapithecus*, *Proconsul* from Meswa Bridge primitively retains broad molars with marked buccolingual flare and prominent cingula that are reduced in all later proconsulids. It presumably represents the sister taxon of the other five species of *Proconsul* but is close enough morphologically to be included in the same genus.

Subfamily AFROPITHECINAE Andrews, 1992
Genus *AFROPITHECUS* Leakey and Leakey, 1986

Included Species A. *turkanensis* Leakey and Leakey, 1986 (type species).

AFROPITHECUS TURKANENSIS Leakey and Leakey, 1985
Figures 24.11–24.14

Distribution Early Miocene (~17–18 Ma). Kalodirr and Moruorot (Lothidok Formation, Kalodirr Member), Buluk (Bakate Formation, Buluk Member), and Locherangan, northern Kenya, and Moroto I and II, eastern Uganda (Leakey and Walker, 1985, 1997; McDougall and Watkins, 1985; Leakey and Leakey, 1986a; Watkins, 1989; Anyonge, 1991; Boschetto et al., 1992). Radiometric dates indicate an age for Moroto of older than 20.6 Ma (Gebo et al., 1997). However, the fauna correlates best with those from late early Miocene or early middle Miocene localities (Pickford et al., 1986a, 1999, 2003), and it seems likely that Moroto is broadly contemporaneous with the *Afropithecus* localities in northern Kenya.

Description A large catarrhine, comparable in dental size to *Proconsul major*, but possibly smaller in body size, with an estimated body weight of 30–55 kg (Sanders and Bodenbender, 1994; Leakey and Walker, 1997; Gebo et al., 1997; MacLatchy and Pilbeam, 1999). Skull with the following characteristics: long, broad and domed muzzle; palate shallow, long and narrow, with toothrows parallel sided or converging slightly posteriorly; incisive foramen comprising large paired openings; large diastema between C and I2; premaxilla narrow but anteriorly protruding, with contact superiorly with the nasals; steeply inclined frontal; strong postorbital constriction; temporal lines strongly marked and converge in the midline far anteriorly to form a frontal trigon; frontal sinus present in the glabellar region; supraorbital costae slender; supraorbital notch at the medial angle of the orbital margin; broad interorbital region; nasals long and narrow, with midline keeling and concave contour in lateral view; pyriform aperture only slightly higher than broad, and oval in shape; subnasal clivus relatively short; canine jugum prominent, with shallow canine fossa; distinct maxillary fossa just below and anterior to the orbit; double infraorbital foramina; anterior root of the zygomatic arch deep, superiorly sloping, and attaches relatively low on the face; maxillary sinus extensive; orbit broader than high, and asymmetrical in shape; orbital process of frontal narrow; lacrimal fossa extends onto the face just anterior to the margin of the orbit; mandible with very deep corpus, distinct mandibular fossa, single mental foramen, ramus set at an oblique angle to the corpus, symphysis with strong inferior transverse torus and lacking superior transverse torus, and steeply sloping subincisive planum (Allbrook and Bishop, 1963; Pilbeam, 1969; Andrews, 1978; Leakey and Leakey, 1986a; Leakey et al., 1988a; Leakey and Walker, 1997; Pickford, 2002; figures 24.11 and 24.12).

Upper incisors strongly procumbent, and angled obliquely toward the midline; I1 relatively broad, and much larger than I2; lower incisors broad, especially i2; upper canine in males broad and tusklike, with an almost circular basal cross section, a deep mesial groove and a bladelike tip as in *P. major*;

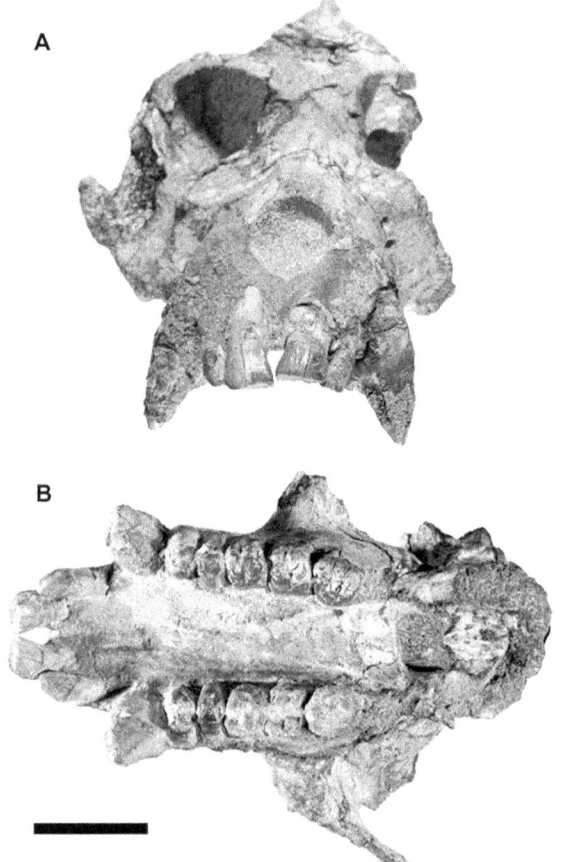

FIGURE 24.11 *Afropithecus turkanensis*. KNM-WK 16999 (holotype), partial cranium: A) frontal view; B) occlusal view. Scale: 3 cm. Courtesy of the National Museums of Kenya.

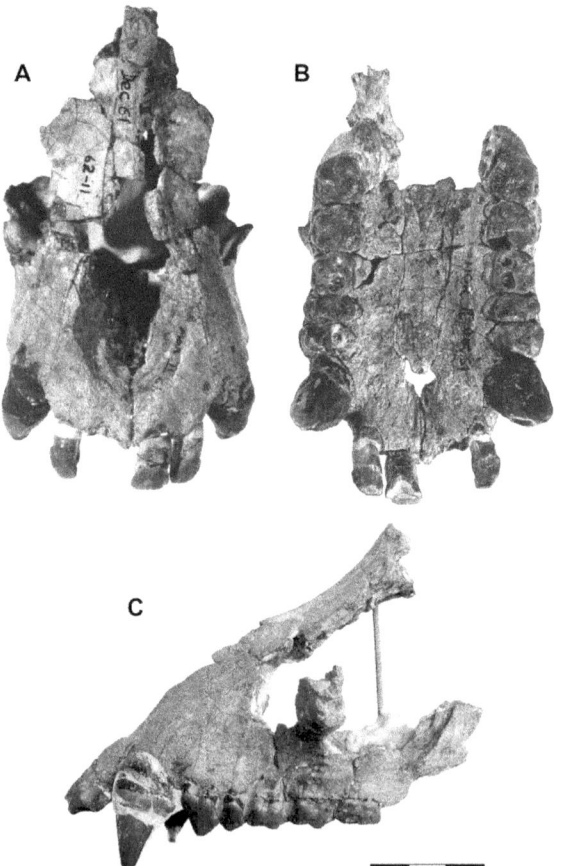

FIGURE 24.12 *Afropithecus turkanensis*, UMP 62-11, palate and lower face: A) facial view; B) palatal view; C) lateral view. Scale: 3 cm. Courtesy of D. Pilbeam and Martin Pickford.

lower canine stout, bilaterally compressed and relatively low crowned; strong sexual dimorphism in canine size; P3 larger than P4; upper premolars broad, with only moderate difference in height between paraconid and protoconid, and lacking a lingual cingulum; upper premolars relatively large in relation to M1; p3 relatively large, narrow and sectorial; p4 generally broader than long; upper premolars and molars have marked buccolingual flare; upper molars relatively narrow, with bunodont cusps, wrinkled enamel, small mesial fovea, moderate to weak development of lingual cingulum, and large hypocone (subequal in size to protocone); M1 < M2 ≤ M3; lower molars relatively broad; m1 < m2 < m3; enamel of cheek teeth thick with heavy wrinkling (Leakey and Leakey, 1986a; Leakey et al., 1988a; Leakey and Walker, 1997; Smith et al., 2003).

Isolated postcranials of *Afropithecus* from Kalodirr and Buluk (Leakey and Leakey, 1986a; Leakey et al., 1988a; Leakey and Walker, 1997) are similar in size and morphology to those of *P. nyanzae* (Leakey et al., 1988a; Rose, 1997; Ward, 1997, 1998). In addition, a small sample of postcranial specimens is known from Moroto. The lumbar vertebrae share some morphological specialization with hominoids, including robust pedicles, lack of anapophyses, reduced ventral keeling, a caudally inclined spinous process, and dorsally oriented transverse process arising from the pedicle (Walker and Rose, 1968; Ward, 1993; Sanders and Bodenbender, 1994; MacLatchy et al., 2000; Nakatsukasa, 2008). The glenoid articular surface of the scapula is rounded and expanded superiorly as in hominoids (MacLatchy and Pilbeam, 1999; MacLatchy et al., 2000), although several authors have argued that this specimen may not belong to a primate (Pickford et al., 1999; Senut et al., 2000; Johnson et al., 2000). The femoral and phalangeal fragments are similar in morphology to those of *Proconsul* (MacLatchy and Bossert, 1996; Gebo et al.,

1997; MacLatchy and Pilbeam, 1999; Pickford et al., 1999; MacLatchy et al., 2000). Taken together the postcranials indicate an arboreal quadrupedal locomotor pattern similar to *Proconsul*, but with a greater emphasis on orthograde climbing and clambering (Leakey and Walker, 1997; Ward, 1998; MacLatchy et al., 2000; Nakatsukasa, 2008; figure 24.13).

Remarks Leakey and Walker (1985) described a small collection of fossil catarrhines from the locality of Buluk in northern Kenya, and assigned part of the material to *Sivapithecus* (an attribution questioned by Delson [1985] and Pickford [1986c]). These specimens were later assigned to *Afropithecus turkanensis*, along with material from Kalodirr, Moruorot, and Locherangan (Leakey and Walker, 1997). Recently, Pickford (in Pickford and Kunimatsu, 2005) suggested that the Buluk material might have closer affinities with *P. gitongai*. In addition to the material from northern Kenya, specimens from Moroto in eastern Uganda, previously assigned to the species *Morotopithecus bishopi*, are included here in *A. turkanensis*.

A close taxonomic and phylogenetic association between *Afropithecus* and *Morotopithecus* has been suggested previously (Andrews and Martin, 1987b; Leakey et al., 1988a; Andrews, 1992b; Leakey and Walker, 1997). However, Pickford's (2002) revised reconstruction of the Moroto lower face (UMP 62-11) and his critical reassessment of the morphological similarities between the Moroto specimen and the partial cranium of *A. turkanensis* (KNM-WK 16999) provide convincing evidence that the two should be included together in a single species (figure 24.14). This interpretation is further supported by the recent findings of Patel and Grossman (2006), who, seemingly unaware of Pickford's (2002) paper, concluded from their comparison of dental metrics that the holotypes of *Morotopithecus* and *Afropithecus* are not sufficiently different to justify a taxonomic distinction. Linking the catarrhine faunas from Buluk, Kalodirr, Locherangan and Moroto

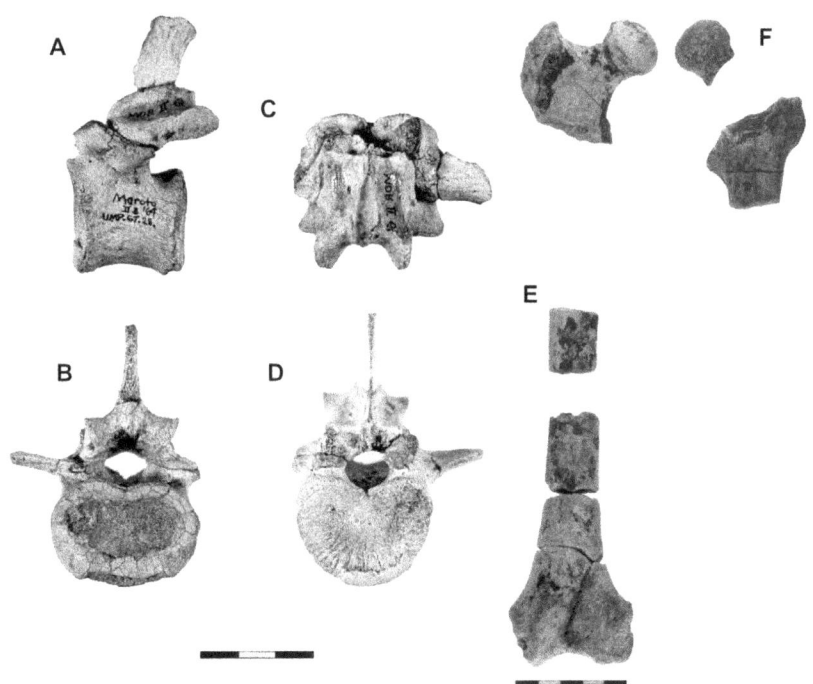

Figure 24.13 *Afropithecus turkanensis*. UMP 67-28, lumbar vertebra A) lateral view; B) caudal view; C) dorsal view; D) cranial view. Scale: 3 cm. MUZM 80, E) right femur; F) left proximal femur. Scale: 5 cm. Courtesy of L. MacLatchy.

further emphasizes the strong provinciality that distinguishes the late early Miocene faunas from northern East Africa and those from western Kenya (Hill et al., 1991; Harrison, 2005).

Previously, *Morotopithecus* was considered to be an early Miocene hominoid based on the presence of shared derived characteristics of the postcranium linking it with extant apes (Ward, 1993; Sanders and Bodenbender, 1994; Gebo et al., 1997; MacLatchy et al., 2000; Young and MacLatchy, 2004). In fact, the most detailed phylogenetic analysis published to date indicates that *Morotopithecus* is a stem hominid (Young and MacLatchy, 2004). The best evidence for this comes from lumbar vertebrae from Moroto II, which share derived characteristics with extant hominoids (Walker and Rose, 1968; Ward, 1993; Sanders and Bodenbender, 1994; MacLatchy et al., 2000; Nakatsukasa, 2008). These specializations are functionally and behaviorally associated with a dorsostable lower back and more orthograde postures, similar to the derived positional pattern seen in modern hominoids (Ward, 1993; Sanders and Bodenbender, 1994; MacLatchy et al., 2000). The lumbar morphology of *Morotopithecus* contrasts with the condition seen in primitive catarrhines, such as *Proconsul*, which have long and flexible lower backs (Ward, 1993; Sanders and Bodenbender, 1994; Nakatsukasa, 2008). It should be noted, however, that Pickford (2002) prefers to assign these vertebrae to *Ugandapithecus*. Other postcranials from Moroto indicate a general similarity to *Proconsul* and to *Afropithecus* from Kalodirr and imply that it may have retained the primitive catarrhine morphology, at least in its appendicular skeleton (Pickford et al., 1999; MacLatchy et al., 2000). Although

I ultimately preferred to recognize *Morotopithecus* as a stem hominoid (Harrison, 2002, 2005), I kept open the possibility that the dental similarities with afropithecines were valid synapomorphies, and that *Morotopithecus* was merely a large orthograde proconsulid that developed its own unique adaptations in the vertebral column in parallel with those of extant hominoids (see also Nakatsukasa, 2008). This latter interpretation now seems to be the most likely alternative. As a result, *Morotopithecus* is considered here to be a junior subjective synonym of *Afropithecus* (following Pickford, 2002) and is included in the Proconsuloidea rather than the Hominoidea.

Genus *HELIOPITHECUS* Andrews and Martin, 1987

Included Species *H. leakeyi* Andrews and Martin, 1987 (type species).

HELIOPITHECUS LEAKEYI Andrews and Martin, 1987
Figure 24.15

Distribution Early middle Miocene. Dam Formation, Ad Dabtiyah, Saudi Arabia (Andrews et al., 1978; Andrews and Martin, 1987b).

Description A large catarrhine intermediate in size between *Proconsul heseloni* and *P. nyanzae*, and somewhat smaller than *Afropithecus turkanensis* (Andrews and Martin, 1987b; Andrews, 1992b). The genus and species is based on a maxillary fragment (the holotype) and four isolated teeth, so knowledge of its anatomy is rather limited (Andrews et al., 1978; Andrews and Martin, 1987b). The main features are as follows: palate relatively shallow and narrow, with parallel tooth rows; large diastema between C and I2 (at least in males); upper premolars large in relation to molars; P3 larger than P4; P3 with marked difference in height between paracone and protocone; P4 with lingual cingulum; upper cheek teeth relatively low crowned with voluminous cusps and relatively thick enamel; upper molars slightly broader than long, with moderate development of the lingual cingulum, and a small buccal cingulum (Andrews et al., 1978; Andrews and Martin, 1987b; figure 24.15).

Remarks *Heliopithecus* differs from *Proconsul* and resembles *Afropithecus* (as well as *Equatorius*) in the following derived characters: upper premolars relatively large, and upper molars narrower with reduced development of the lingual cingulum, more bunodont cusps, and thicker enamel. Several researchers have suggested that *Heliopithecus* may be congeneric with *Afropithecus* (which has priority; see Andrews et al., 1987; Andrews and Martin, 1987b; Leakey et al., 1988a). However, *Heliopithecus* is distinguished from *Afropithecus* in having relatively broader cheek teeth, greater differential between the heights of

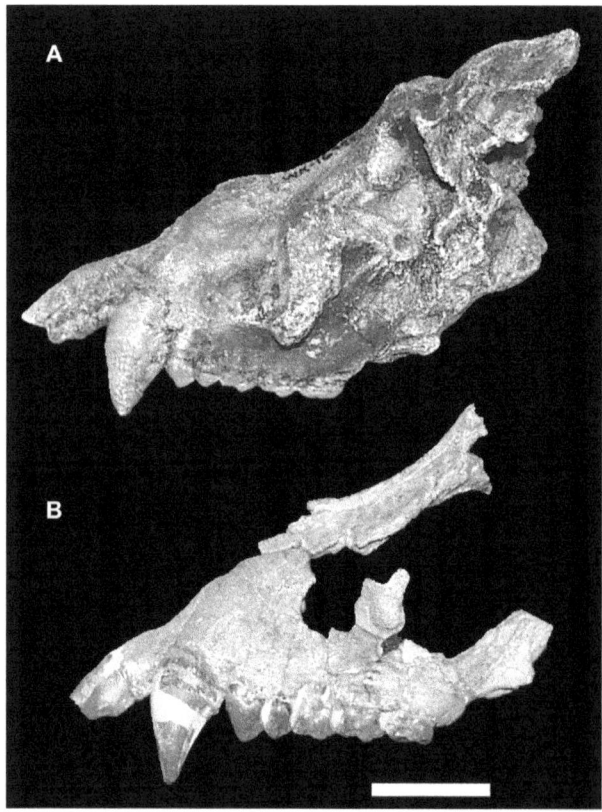

FIGURE 24.14 Comparison of *Afropithecus* and *Morotopithecus*. A) *Afropithecus turkanensis* (holotype) KNM-WK 16999, face in lateral view. B) *Morotopithecus bishopi* (holotype), UMP 62-11, lower face in lateral view. Scale: 3 cm.

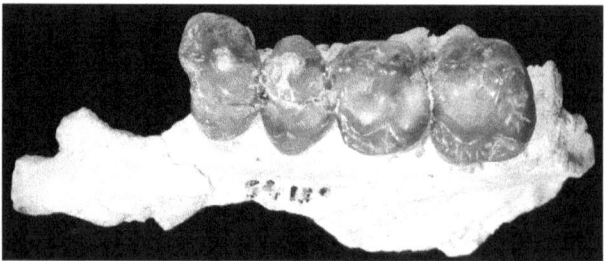

FIGURE 24.15 *Heliopithecus leakeyi*, BM(NH) M 35145 (holotype), left maxilla with P3–M2, occlusal view. Courtesy of P. Andrews.

the paracone and protocone on P3, presence of a lingual cingulum on P4, upper molars with relatively smaller hypocone, and better-developed lingual cingulum. These differences indicate that *Heliopithecus* is more primitive than *Afropithecus* and should be recognized as a distinct genus.

Genus *NACHOLAPITHECUS* Ishida et al., 2004

Included Species Na. kerioi Ishida et al., 2004 (type species).

NACHOLAPITHECUS KERIOI Ishida et al., 2004
Figures 24.16–24.18

Distribution Early middle Miocene (~15–16 Ma). Aka Aiteputh Formation, Nachola, Samburu District, Kenya (Pickford et al., 1984a, 1984b; Makinouchi et al., 1984; Itaya and Sawada, 1987; Sawada et al., 1987, 1998).

Description Nacholapithecus kerioi is represented by a partial skeleton (KNM-BG 35250), a number of isolated postcranial specimens, and a sizable collection of jaw fragments and teeth (Ishida et al., 1984, 2004; Rose et al., 1996; Nakatsukasa et al., 1998, 2002, 2003a; Kunimatsu et al., 2004). It is a medium-sized catarrhine, with an estimated body mass of 20–22 kg and 10 kg in males and females, respectively (Rose et al., 1996; Nakatsukasa et al., 2003a; Ishida et al., 2004). Key features of the skull are as follows: Face relatively short. Nasal aperture tall and narrow, widest above midheight, and tapering inferiorly. Subnasal clivus moderately low. Premaxilla overlaps slightly with the palatine process of maxilla to produce a "stepped" nasal floor and restricted incisive fossa (Ishida et al., 2004; Kunimatsu et al., 2004). Premaxilla slightly protruding, with procumbent upper incisors. Prominent canine jugum bordered posteriorly by a deep canine fossa in males; less well developed in females. Relatively large diastema between I2 and C in male individuals; small in females. Anterior root of zygomatic arch situated low on the face above M1/M2 and laterally projecting. Maxillary sinus not as extensive as in *Proconsul,* terminating anteriorly at M1, and its floor is level with or slightly lower than the apices of the molar roots. Palate relatively shallow. Mandibular corpus moderately deep, with shallow postcanine fossa on the lateral side. Symphysis steeply inclined, with moderately well-developed inferior transverse torus (Ishida et al., 2004; Kunimatsu et al., 2004; figures 24.16 and 24.17).

I1 is narrow, buccolingually stout, with a broad lingual pillar. I2 narrower, with mesiodistal diameter about 75% that of I1. Upper canines in males robust but relatively low crowned. Upper premolars moderately large, and quite broad. P3 with paracone much more elevated than protocone and connected by a pair of transverse crests. P4 ovoid, with paracone and protocone subequal in height. Upper molars rectangular, broader than long, with slightly longer lingual moiety than buccal moiety. Cusps low and voluminous. Large hypocone. Lingual cingulum weakly developed or absent. Upper molars increase in size from M1 to M3. M3 tapers distally, with reduced distal cusps. Lower incisors tall and mesiodistally narrow. Lower canines in males robust, relatively low crowned, with strong bilateral compression. Lower molars rectangular, with moderately low and rounded cusps. Entoconid relatively small. Well-developed transverse crests demarcate the mesial and distal foveae. Buccal cingulum poorly developed. m3 triangular in outline, with reduced entoconid, and large hypoconulid aligned with protoconid and hypoconid. m3 is much larger than m2 (Kunimatsu et al., 2004).

The cervical vertebrae are relatively large. It is not possible to determine the number of thoracic vertebrae, but there are six or seven lumbar vertebrae as in *Proconsul* and in most extant nonhominoid anthropoids (Ward et al., 1993; Nakatsukasa et al., 2007). The lumbar vertebrae have relatively small and elongated centra, transverse processes that originate from the centrum cranially and from the base of the pedicle caudally,

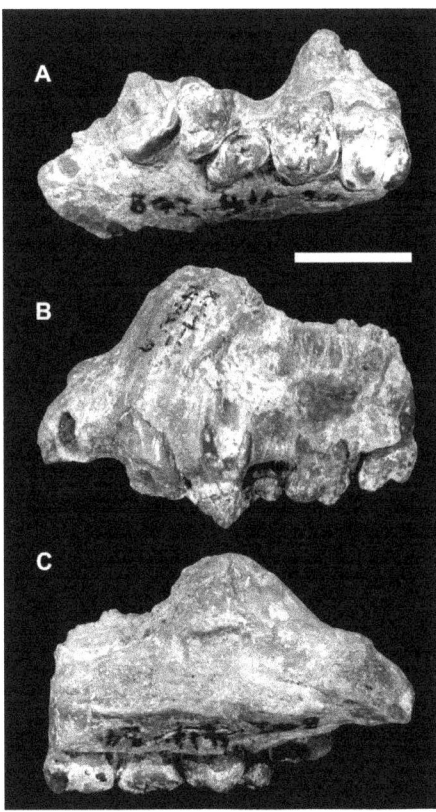

FIGURE 24.16 *Nacholapithecus kerioi*. KNM-BG 14700A, left maxilla with P3-M2. A) occlusal view; B) lateral view; C) medial view. Scale: 2 cm. Courtesy of Y. Kunimatsu.

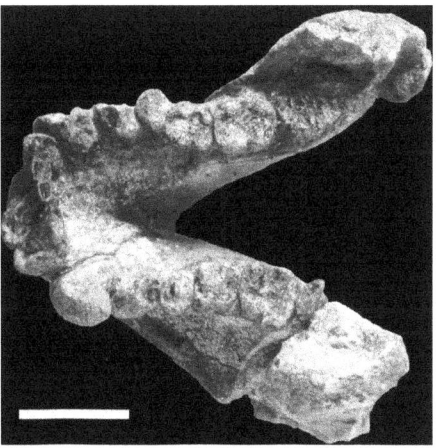

FIGURE 24.17 *Nacholapithecus kerioi*. KNM-BG 35250 (holotype), mandible with right i1–m3 and left c–m2. Scale: 2 cm. Courtesy of Y. Kunimatsu.

retention of a small anapophysis on at least one vertebra, and a strong median ventral keel. They differ from *Proconsul* in having a more caudally positioned lumbar spinous process, suggesting greater stability. The first sacral vertebra indicates that the iliac blades were oriented more parasagittally than in extant hominoids, while the cranial elevation of the centrum relative to the alae and the small size of the lumbosacral joint are more hominoid-like. First coccygeal vertebra indicates loss of external tail (Nakatsukasa et al., 2003b). The vertebral column indicates a long flexible trunk, typical of arboreal palmigrade quadrupeds. Clavicle long and slender, with moderate degree of curvature, possibly implying a dorsally positioned scapula (Senut et al., 2004). Scapula with broad glenoid fossa, and relatively elongated acromion process (Ishida et al., 2004; Senut et al., 2004). Forelimb bones are relatively large compared with those of the hindlimb. Humeral shaft with flat deltoid plane. Distal humerus with well-developed supracondylar crest, massive lateral epicondyle, shallow radial fossa, moderately large and deep coronoid fossa, well-developed medial epicondyle directed posteromedially, deep olecranon fossa, globular capitulum, zona conoidea forming a distinct gutter, and a weakly developed lateral trochlear keel. Proximal ulna with moderately long and nonretroflexed olecranon process, wide and mediolaterally weakly convex trochlear notch, protuberant coronoid process, strong proximolateral extension of the articular surface on the lateral side of the olecranon, small and laterally facing radial notch. Distal radius robust, with a large styloid process, and a straight shaft. Scaphoid with unfused os centrale. Ischium with well-developed ischial spine for attachment of the gemelli muscles, as in *Proconsul* spp. Proximal femur with large globular head, short neck, high neck-shaft angle, and low greater trochanter. Distal femur with square patella groove, and slight asymmetry of the condyles. Patella broad, anteroposteriorly shallow, and almost circular in outline. Tibia with slender shaft, deep fibular notch and prominent medial malleolus. Fibula robust, with large malleolus. Talus with slightly wedged and deeply grooved trochlear surface, lateral trochlear rim more projecting than medial rim, well-developed concavity on the distal margin of the trochlea to receive the anterior margin of the distal tibia, deep malleolar cup, and strong groove for *M. flexor hallucis longus*. Calcaneus with deep pit for the interosseus talocalcaneal ligament, a moderately broad sustentaculum, contiguous anterior and middle

talar articular facets, and deep groove for the tendon of *M. flexor hallucis longus*. Expanded medial process of the heel tuberosity of calcaneus, characteristic of arboreal primates that are adept at pedal grasping (Rose et al., 1996). Medial cuneiform with mediolaterally convex articular facet for the first metatarsal, as in arboreal primates with opposable halluces. Metatarsal I lacking facet for prehallux. Long and well-developed hallux and pollex. Lateral metatarsals and proximal manual and pedal phalanges relatively long and slender, with slight to moderate curvature. Middle phalanges with straight shafts. Terminal phalanges with mediolaterally compressed ungual tufts. Postcranial morphology is generally similar to that of *Proconsul*, and indicates an arboreal quadruped. However, *Nacholapithecus* is more derived than *Proconsul* in having relatively larger forelimb bones, longer clavicle and scapular spine, and longer pedal rays, indicating a greater propensity for vertical climbing, hoisting, quadrumanous clambering and suspension, and bridging behaviors (Rose et al., 1996; Nakatsukasa et al., 1996, 2002, 2003a, 2003b, 2007; Ishida et al., 2004). Similar forelimb dominated behaviors have been inferred for *Equatorius* (McCrossin et al., 1998; figure 24.18).

Remarks The material assigned to this species was originally assigned to *Kenyapithecus* sp. or *Kenyapithecus* cf. *africanus* (Ishida et al., 1984; Rose et al., 1996). Ishida et al. (1999) proposed a new genus and species name, *Nacholapithecus kerioi*, but the purely descriptive diagnosis does not serve to differentiate the species from other taxa as required by ICZN Article 13.1.1 (i.e., to be available, a new name published after 1930 must be accompanied by a description or definition that states in words characters that are purported to differentiate the taxon; International Commission on Zoological Nomenclature, 1999). Technically, the name constituted a *nomen nudum* until it became available when Ishida et al. (2004) provided craniodental and postcranial features that served to differentiate it from other East African Miocene catarrhines. This means that if *Nacholapithecus* and *Equatorius* enter into synonymy, the latter name takes priority.

Craniodentally and postcranially, *Nacholapithecus* is quite similar to *Proconsul*, but it does appear to be more derived in having reduced molar cingula, low-crowned and robust canines in males, a restricted maxillary sinus, possibly a somewhat restricted incisive fossa, a relatively deeper subnasal clivus, a mandibular symphysis dominated by an inferior

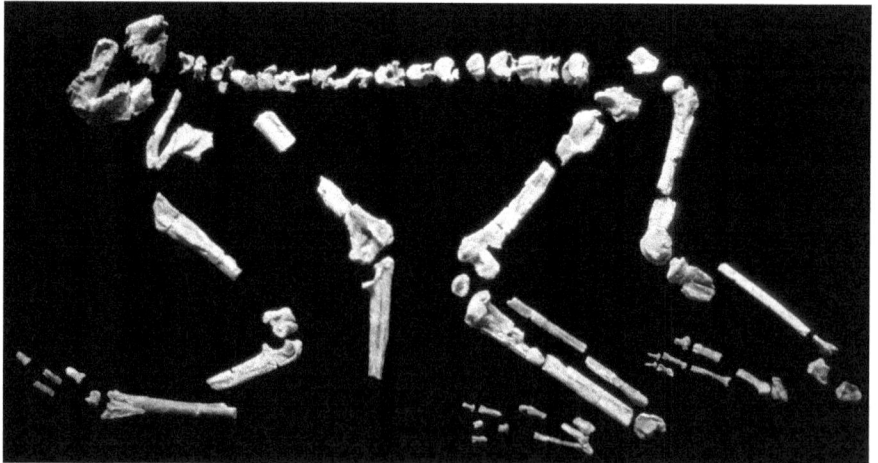

FIGURE 24.18 *Nacholapithecus kerioi*. KNM-BG 35250 (holotype), partial skeleton. Courtesy of M. Nakatsukasa.

transverse torus, a forelimb-dominated skeleton, a short femoral neck with a high angle, and absence of a prehallux facet on the first metatarsal (Kunimatsu et al., 2004; Ishida et al., 2004). In these respects *Nacholapithecus* more closely resembles *Equatorius*. Kunimatsu et al. (2004) and Ishida et al. (2004) have suggested that the subnasal morphology could represent an incipient development of the more derived morphology seen in extant great apes, in which the subnasal clivus overlaps with the palatal process. However, this anatomical region in *Nacholapithecus* is somewhat deformed, and the original configuration of the incisive fossa is difficult to interpret. By contrast, the lack of postcranial specializations shared with crown hominoids would suggest that *Nacholapithecus* is a stem catarrhine or stem hominoid, rather than a hominid. *Nacholapithecus* has a suite of dental, cranial and postcranial specializations not seen in *Proconsul*, but none of these represent clear-cut synapomorphies that would definitively link it with crown hominoids, with the possible exception of the purported loss of the external tail (Nakatsukasa et al., 2003b). However, many of the specializations listed here point to closer affinities with *Afropithecus* and especially *Equatorius*, and as a consequence *Nacholapithecus* is considered here to be a specialized member of the Proconsuloidea, and provisionally included in the Afropithecinae.

Genus *EQUATORIUS* S. Ward et al., 1999

Included Species E. africanus S. Ward et al., 1999 (type species)

EQUATORIUS AFRICANUS (Le Gros Clark and Leakey, 1950)
Figure 24.19

Distribution Middle Miocene (~14.5–16.0 Ma). Maboko Island, Majiwa, Kaloma, and Nyakach (Kaimogool North and Chepetet West) in western Kenya (Andrews and Molleson, 1979; Pickford, 1982, 1986a; Feibel and Brown, 1991; Wynn and Retallack, 2001); Muruyur Beds at Kipsaraman and Cheparawa, Tugen Hills, central Kenya (Behrensmeyer et al., 2002; Pickford and Kunimatsu, 2005). The holotype (BMNH M16649) was originally reported as coming from Rusinga Island (Le Gros Clark, 1950, 1952), but the preservation of the specimen and the adhering matrix are inconsistent with such a provenance (Andrews and Molleson, 1979). It is almost certain that the type specimen derives from Maboko Island (figure 24.19).

Description Medium-sized catarrhine, with an estimated body mass of approximately 20–40 kg, and demonstrating marked sexual dimorphism (McCrossin et al., 1998). Craniodental and postcranial material is well represented by specimens from Maboko Island and Kipsaraman. Maxilla with anterior root of zygomatic arch situated close to the alveolar margin, and maxillary sinus extending anteriorly into the alveolar region of the upper premolars. Mandible with long and strongly proclined symphysis, shelflike inferior transverse torus, weak superior transverse torus, and posteriorly directed genioglossal fossa. The corpus is low and robust. Lower incisors are tall, narrow, buccolingually thick, and strongly procumbent. I1 mesiodistally broad relative to height, with narrow lingual cingulum and small basal lingual tubercle. I1 is much broader than I2. I2 is conical and asymmetrical, with an oblique lingual cingulum. Upper and lower canines of males relatively low crowned and robust. Upper canines with deep mesial groove and weak lingual cingulum. Lower canines with prominent distal heel and moderately developed lingual

cingulum. p3 has a tall protoconid, moderately developed mesiobuccal honing face, a mesiolingual beak, a small metaconid, and a vestigial lingual cingulum. p4 broader than long with two main cusps subequal in size, a pair of distal tubercles (variably developed), a large talonid basin, and buccal cingulum vestigial to absent. Upper premolars relatively large compared with size of molars. P3 with protocone moderately lower than the paracone, cusps separated by mesiodistally oriented fissure, no formation of central fovea, and lingual cingula highly reduced to absent. Molars have thick enamel, low occlusal relief, restricted basins, moderate buccolingual flare, cingula absent to reduced, and crenulated enamel. Lower molars rectangular, relatively broad, and they increase in size posteriorly. The hypoconulid is positioned on the buccal side of the midline of the crown. m3 triangular, with crown tapering distally. Upper molars relatively narrow. M1 ≤ M3 < M2. M3 with reduced hypocone and talon basin (McCrossin and Benefit, 1993, 1997; Benefit and McCrossin, 1995; S. Ward et al., 1999; Kelley et al., 2002).

A number of postcranial remains are known from Maboko Island (Le Gros Clark and Leakey, 1951; Benefit and McCrossin, 1995; McCrossin and Benefit, 1994, 1997; Rose, 1997; McCrossin et al., 1998). The proximal humerus has a posteriorly directed head, large greater tubercle that projects proximally beyond the level of the humeral head, and a shallow bicipital groove. The shaft is markedly anteriorly flexed, with a strong deltopectoral crest and weakly developed supinator crest. The proximal ulna has a well-developed posteriorly reflected olecranon process. The pisiform retains a distinct articular facet for the styloid process of the ulna, as in cercopithecoids. Metacarpal III exhibits a strong transverse dorsal ridge bordering the distal articulation (Benefit and McCrossin, 1995; McCrossin and Benefit, 1997; Allen and McCrossin, 2007). The phalanges are short, relatively stout, and only slightly curved. The femur is slender and slightly longer than the humerus, with an estimated humerofemoral index of 95 (McCrossin and Benefit, 1997). The proximal femur has a small head, a long neck with a high neck angle, and a distinct posterior tubercle. The distal femur is moderately broad and anteroposteriorly shallow, with a broad and low patellar groove. The patella is relatively broad (McCrossin and Allen, 2007). The distal tibia is anteroposteriorly thick, with a well-developed medial keel for articulation with the trochlea of the talus. The talus is comparable to that of *Proconsul*, with a wedged trochlea and a relatively elevated lateral trochlear keel. The medial cuneiform has a relatively flat distal articular surface for the first metatarsal, and a well-developed peroneal tubercle indicating that the hallux was habitually

FIGURE 24.19 *Equatorius africanus*. BM(NH) M 16649 (holotype), left maxilla with P3–M1. Scale: 1 cm. Courtesy of P. Andrews.

adducted as in semiterrestrial cercopithecids. A prehallux facet is lacking. The first metatarsal is robust (McCrossin and Benefit, 1994, 1997; Benefit and McCrossin, 1995; McCrossin et al., 1998).

In addition to the material from Maboko Island, a partial skeleton of *Equatorius* (KNM-TH 28860) is known from Kipsaraman (S. Ward et al., 1999; Sherwood et al., 2002). The main anatomical features are as follows: scapula with robust acromion process that extends well beyond the margin of the glenoid, and axillary border longer than the vertebral border; clavicle robust and relatively straight; humerus with posteriorly directed head, shaft retroflexed with well-developed deltopectoral crest, sharp lateral supracondylar ridge, small posteromedially directed medial epicondyle, and deep olecranon fossa; ulna with long olecranon process, laterally facing radial notch, and long styloid process that articulates with the carpus; radius straight and slender, with circular head, and a bicipital tuberosity located close to the proximal end; scaphoid robust with free os centrale; hamate with deep pits for hamatotriquetral and hamatocapitate ligaments and a deep and spiraled triquetral groove; metacarpals slender with only slight curvature, broad heads ventrally that narrow dorsally; phalanges only slightly curved; thoracic vertebrae with small heart-shaped centra and strong ventral keel. The postcranial morphology indicates that *Equatorius* was an agile, semiterrestrial quadruped (McCrossin and Benefit, 1997; Sherwood et al., 2002).

Remarks The convoluted taxonomic and nomenclatural history of *Equatorius africanus* is reviewed by Andrews and Molleson (1979), Madden (1980a), Pickford (1985), and McCrossin and Benefit (1994). Following S. Ward et al. (1999), this species is here considered generically distinct from *Kenyapithecus wickeri*, although a number of current workers prefer to recognize these taxa as congeneric (e.g., McCrossin and Benefit, 1994; Benefit and McCrossin, 2000; Kunimatsu et al., 2004). Begun (2000, 2002) recognizes *Equatorius* as a junior synonym of *Griphopithecus*, a thick-enameled hominoid from broadly contemporary localities in central Europe and Turkey. However, S. Ward et al. (1999) and Kelley et al. (2002), following earlier studies (i.e., Pickford, 1985, 1986c; Harrison, 1992), provide adequate justification to distinguish *Equatorius* from both *Kenyapithecus* and *Griphopithecus*.

Equatorius and *Nacholapithecus* share a suite of specialized features relative to *Proconsul* that suggest that they are closely related. These include a somewhat more restricted maxillary sinus with an elevated floor in relation to molar root apices; moderately deep subnasal clivus; very well-developed inferior transverse torus of the mandible; mandibular corpus relatively robust; tall and somewhat procumbent incisors; low-crowned and robust canines; relatively narrow premolars that are large in relation to the molars; cheek teeth with thick enamel, low relief of dentine-enamel junction; low and rounded cusps and crests, and strongly reduced cingula; upper molars relatively narrower. Many of these specializations are also characteristic of *Afropithecus* and *Heliopithecus*, and these similarities provide the basis for the inclusion of all four genera in the Afropithecinae. However, *Nacholapithecus* and *Equatorius* are more derived than *Afropithecus* in having a more restricted maxillary sinus, a much more pronounced inferior transverse torus, a more robust mandibular corpus, narrower premolars, and more reduced molar cingula. Although no detailed comparisons of the postcranials of *Nacholapithecus* and *Equatorius* have yet been published, the preliminary accounts suggest that they are remarkably similar in many respects, further emphasizing their close relationship. Their primitive postcranial morphology also provides further support for their inclusion in the Proconsuloidea rather than Hominoidea.

If, as implied in the Introduction, the subfamilies included within Proconsulidae are eventually elevated to family status it might prove worthwhile to include *Equatorius* and *Nacholapithecus* in a separate subfamily within the Afropithecidae. Equatorinae, a taxonomic concept proposed by Cameron (2004), would be available. Although Cameron (2004) used an incorrect stem in the formation of his family group name (Equator- rather than Equatori-), Article 29.4 of the International Code of Zoological Nomenclature (International Commission for Zoological Nomenclature, 1999) allows maintenance of the original spelling as correct for family group names proposed after 1999.

Subfamily NYANZAPITHECINAE Harrison, 2002
Genus *NYANZAPITHECUS* Harrison, 1986

Included Species Ny. vancouveringorum (Andrews, 1974) (type species), *Ny. pickfordi* Harrison, 1986, *Ny. harrisoni* Kunimatsu, 1997.

Description Nyanzapithecus is a small- to medium-sized catarrhine intermediate in size between *P. heseloni* and *D. macinnesi*. Judging from the dentition, *Ny. vancouveringorum* and *Ny. pickfordi* have estimated body weights of ~11 kg and ~8 kg for males and females, respectively, while *Ny. harrisoni* was probably slightly smaller. *Nyanzapithecus* is distinguished from other proconsulids by the following dental features: I1 broad and spatulate, relatively low crowned, and stoutly constructed; I2 broad, moderately low crowned and robust, and approaching I1 in size; lower incisors broad and moderately high crowned; P3 structurally similar to P4; upper premolars ovoid in occlusal outline and relatively long and narrow, with elevated and inflated cusps of similar height, poorly developed occlusal crests, and an inflated lingual cingulum, at least on P4; p3 long and narrow, with only slight extension of enamel onto the buccal aspect of the mesial root; p4 long and narrow with high cusps, and mesial fovea much more elevated than the distal basin; upper molars long and narrow, with low, rounded, and voluminous cusps, buccally displaced protocone, restricted trigon basin and foveae, well-developed lingual cingulum, low and rounded occlusal crests; M1 < M2 ≤ M3; lower molars very long and narrow, with low, rounded, and inflated cusps, short and rounded crests, long and narrow talonid basin, restricted mesial and distal foveae, poorly developed buccal cingulum, and deep lingual notch; m1 < m2 < m3; dP4 longer than broad, with voluminous cusps and relatively restricted occlusal foveae (Harrison, 1986, 2002; Kunimatsu, 1992a, 1992b, 1997).

The fragmentary cranial remains of *Ny. vancouveringorum* and *Ny. pickfordi* indicate that the genus has a relatively short face, low and broad nasal aperture, and robust premaxilla (Harrison, 1986). A proximal humerus from Maboko has been provisionally attributed to *Ny. pickfordi* (McCrossin, 1992), and a proximal humerus from Rusinga, previously attributed by Gebo et al. (1988) to *Dendropithecus macinnesi* or *Proconsul heseloni*, is best assigned to *Ny. vancouveringorum* on the basis of size (Harrison, 2002). These two specimens are morphologically similar to of other proconsulids (Gebo et al., 1988; McCrossin, 1992).

NYANZAPITHECUS VANCOUVERINGORUM (Andrews, 1974)
Figure 24.20

Distribution Early Miocene (~17.0–18.5 Ma). Rusinga Island and Mfangano Island, Kenya (Drake et al., 1988).

Description This species is poorly known, being best represented by a maxilla (holotype) and a mandible from Rusinga. It differs from other species of *Nyanzapithecus* in the following features: both upper premolars have well-developed lingual cingulum; upper molars and dP4 only slightly longer than broad, and generally rectangular to square in occlusal outline; upper molar with moderately inflated cusps that encroach only partially into the occlusal basin, trigon basin and mesial and distal foveae restricted but well defined, hypocone connected to the protocone by short crest, lingual cingulum well developed, both lingually and mesially; lower molars moderately long and narrow, with reduced mesial and distal foveae, relatively expansive talonid basin, and moderately inflated cusps (Harrison, 1986, 2002; figure 24.20).

NYANZAPITHECUS PICKFORDI Harrison, 1986

Distribution Middle Miocene (~15–16 Ma). Maboko Island and Kipsaraman, Kenya (Pickford, 1981, 1983, 1986a, 1986b; Feibel and Brown, 1991; Kelley et al., 2002; Pickford and Kunimatsu, 2005).

Description A species of *Nyanzapithecus* distinguished from *Ny. vancouveringorum* by the following characteristics: P3 lacking a lingual cingulum; upper molars higher crowned, much longer than broad, tending to taper distally and become waisted midway along their length, with inflated cusps that crowd the occlusal basins and restrict the mesial and distal foveae; hypocone connected by a crest to the crista obliqua, but with no direct connection to the protocone; lingual cingulum particularly well developed mesially, but reduced lingually; lower molars longer and narrower with very inflated cusps and extremely restricted occlusal basins (Harrison, 1986).

Remarks Renewed excavations at Maboko Island have led to the recovery of a large sample of additional specimens assigned to this species, including a nearly complete mandible of a subadult female individual (McCrossin, 1992; Benefit and McCrossin, 1997; Gitau et al., 1998), but the specimens have not yet been described. Newly discovered specimens from Kipsaraman, attributed to *N.* cf. *pickfordi* by Pickford and Kunimatsu (2005), are morphologically similar to the material from Maboko Island, but the teeth may be slightly larger.

NYANZAPITHECUS HARRISONI Kunimatsu, 1997
Figure 24.21

Distribution Middle Miocene (~13–15 Ma). Aka Aiteputh Formation, Nachola, Kenya (Kunimatsu, 1992a, 1992b, 1997).

Description A species of *Nyanzapithecus* somewhat smaller than *Ny. vancouveringorum* and *Ny. pickfordi*. It is distinguished from *Ny. vancouveringorum* in having upper molars higher crowned; molar cusps higher and more inflated; occlusal basins and foveae more restricted; upper molars that tend to taper distally; more distinct lingual cingulum; M3 crown shorter; p4 more elongated; lower molars relatively short (especially m3); mandible more slender (Kunimatsu, 1997). It differs from *Ny. pickfordi* in having upper and lower molars less elongated; lingual cingulum on upper molars less well developed mesially and better developed lingually; hypocone linked to the protocone directly, rather than to the crista obliqua; less waisted

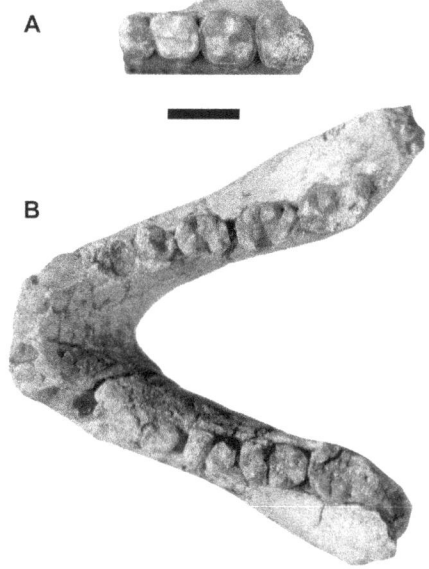

FIGURE 24.20 *Nyanzapithecus vancouveringorum.* A) KNM-RU 2058 (holotype), left maxilla with P4–M3, occlusal view. B) KNM-RU 1855, mandible with right p4–m3 and left m1–m3. Scale: 1 cm. Courtesy of National Museums of Kenya.

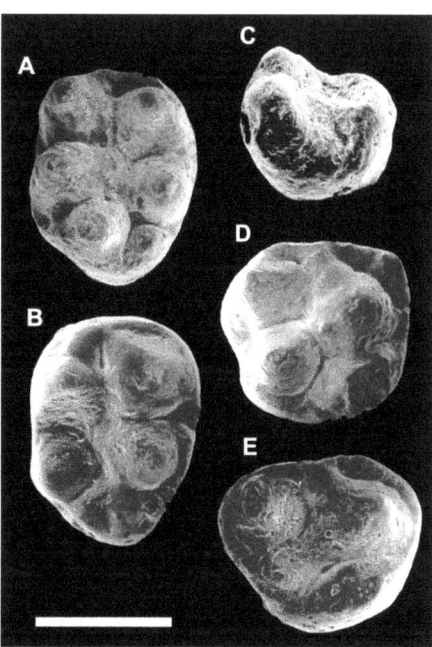

FIGURE 24.21 *Nyanzapithecus harrisoni.* A) KNM-BG 15235, left m2, occlusal view; B) KNM-BG 15227, right m3, occlusal view; C) KNM-BG 15318, left p4, occlusobuccal view; D) KNM-BG 15237 (holotype), right M2, occlusal view; E) KNM-BG 15344, right M3, occlusal view. Scale: 5 mm. Courtesy of Y. Kunimatsu.

upper molars; p3 with weak but continuous buccal cingulum; hypoconulid on m3 tends to be located more medially (Kunimatsu, 1997; figure 24.21).

Remarks *Nyanzapithecus* clearly forms a well-defined and specialized clade of proconsulids. The three species can be arranged in a phyletic series of increasing specialization from the early Miocene *Ny. vancouveringorum* through *Ny. harrisoni* to *Ny. pickfordi* in the middle Miocene (Harrison, 1986; Kunimatsu, 1997).

Genus *MABOKOPITHECUS* Von Koenigswald, 1969

Included Species Ma. clarki Von Koenigswald, 1969 (type species)

MABOKOPITHECUS CLARKI Von Koenigswald, 1969

Distribution Middle Miocene (~15–16 Ma). Maboko Island, Kenya (Pickford, 1981, 1986a, 1986b; Feibel and Brown, 1991).

Description A small- to medium-sized catarrhine comparable in dental size to *Nyanzapithecus pickfordi*. Until recently, this species was known only from two isolated m3s (Von Koenigswald, 1969; Harrison, 1986), so comparisons were very limited. The m3 of *Mabokopithecus clarki* is characterized by the following features: crown long and narrow, with a distinctive curvature; cusps high, conical, and voluminous; protoconid and metaconid large, well-developed and transversely aligned, and separated by a deep median groove that communicates with an elongated mesial fovea; hypoconid lingually displaced, so that the buccal cusps are not aligned, leading to the distinctive concavity along the buccal margin of the crown; low rounded crests descend from the metaconid and protoconid into the talonid basin and converge with a similar crest from the hypoconid at a distinct mesoconid (present in the holotype only); hypoconulid very large; entoconid and hypoconulid linked by a prominent crest, which defines a pitlike distal fovea; subsidiary tubercle located between the metaconid and entoconid; and buccal cingulum narrow and irregular (Harrison, 1986, 2002).

Remarks Renewed excavations at Maboko Island have yielded a nearly complete mandible of a female individual containing an m3 that provides a close match with the holotype of *Mabokopithecus clarki* (Benefit et al., 1998). The rest of the dentition, however, is very similar to *Ny. pickfordi*, and there can be little doubt that they should be included together in the same genus. Since *Mabokopithecus* Von Koenigswald, 1969 has priority over *Nyanzapithecus* Harrison, 1986, all three species of *Nyanzapithecus* should eventually be transferred to the former genus, as I have suggested (Harrison, 2002). Such a taxonomic move awaits the formal description of the new material from Maboko. However, from my own brief comparisons, I favor including all of the nyanzapithecine specimens from Maboko in a single species, *Mabokopithecus clarki*. Benefit et al. (1998), on the other hand, consider that *Mabokopithecus clarki* should be retained as specifically distinct from "*Ny.*" *pickfordi*.

Genus *RANGWAPITHECUS* Andrews, 1974

Included Species R. gordoni Andrews, 1974 (type species).

RANGWAPITHECUS GORDONI Andrews, 1974
Figure 24.22

Distribution Early Miocene (~19–20 Ma). Songhor and Lower Kaputay, Kenya (Bishop et al., 1969; Pickford and Andrews, 1981; Pickford, 1983, 1986a, 1986b; Harrison, 1988; Cote and Nengo, 2007).

Description A medium-sized catarrhine similar in dental size to *P. africanus* and *P. heseloni*. It differs from *Proconsul* in the following respects: upper and lower incisors high crowned and relatively narrow; upper incisors moderately procumbent; canines markedly sexually dimorphic; upper canine strongly bilaterally compressed, with a bladelike distal crest; upper premolars and molars narrow and relatively elongated; upper premolars with more ovoid occlusal outline, paracone and protocone more similar in height, and inflated lingual cingulum on both P3 and P4; P3 < P4; molars with low cusps, well-developed crests, wrinkled occlusal surface, and marked wear differential; upper molars with strong lingual cingulum, enlarged hypocone, and rhomboidal arrangement of cusps and occlusal outline; M1 < M2 < M3; lower canine high crowned and bilaterally compressed in males; p3 elongated and bilaterally compressed; p4 and lower molars long and narrow; lower molars with buccal cingulum represented by deep foveae between the buccal cusps; m1 < m2 < m3, with marked size increase along the series (Andrews, 1978; Nengo and Rae, 1992). Several mandibular specimens have recently been recovered from Songhor and Lower Kaputay (Hill and Odhiambo, 1987; Nengo and Rae, 1992; Cote and Nengo, 2007) that document more fully the morphology of the lower jaw and dentition, but these still await detailed description. Cranial and mandibular specimens with the following features: premaxilla relatively short; diastema small; palate long and narrow, and broadens posteriorly; maxillary sinus deeply excavated between the roots of the upper molars; nasal aperture probably relatively broad; anterior root of the zygomatic arch positioned low on the face above M1–M2; mandibular corpus deep with a strongly developed superior transverse torus (Andrews, 1978). Postcranials are generally similar in morphology to those of *P. heseloni*, and indicate a pronograde arboreal primate (Preuschoft, 1973; Harrison, 1982; Langdon, 1986; Nengo and Rae, 1992). A proximal femur from Songhor has a relatively high femur neck angle compared with *Proconsul*, being most similar to *Pongo* and *Ateles* among living anthropoids, and probably implies a greater degree of specialization for climbing and hindlimb suspension (Harrison, 1982; figure 24.22).

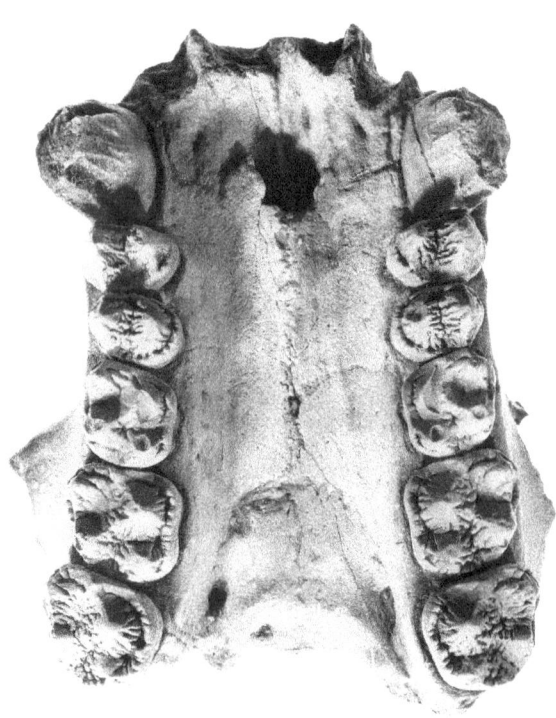

FIGURE 24.22 *Rangwapithecus gordoni*. KNM-RU 700 (holotype), lower face and palate, occlusal view. Courtesy of P. Andrews.

Genus *TURKANAPITHECUS* Leakey and Leakey, 1986

Included Species T. kalakolensis Leakey and Leakey, 1986 (type species)

TURKANAPITHECUS KALAKOLENSIS Leakey and Leakey, 1986
Figure 24.23

Distribution Late early Miocene (~17.0–17.5 Ma). Kalodirr (Lothidok Formation, Kalodirr Member), northern Kenya (Boschetto et al., 1992). Possibly represented at Fejej in southern Ethiopia, dated to older than 16.2 Ma (Richmond et al., 1998).

Description A medium-sized catarrhine in which male specimens are comparable in cranial and postcranial size to female specimens of *P. heseloni*, with an average body weight of ~10 kg. The main characteristics of the skull are as follows: face relatively short, with broad and domed snout; large incisive fossa; narrow palate, with tooth rows that converge posteriorly; premaxillary suture makes contact with the nasals; nasals are relatively broad and expand inferiorly and superiorly; nasal aperture broad, ovoid in shape, and narrows between the central incisor roots; very broad interorbital region; orbits subcircular in outline, with shallow supraorbital notch; slightly thickened supraorbital tori, with depressed glabella region; possibly a large frontal sinus; single large infraorbital foramen, located just below the orbital margin; lacrimal fossa located just anterior to the orbital margin; well-developed canine jugum; canine fossa indistinct; extensive maxillary sinus; anterior root of zygomatic arch situated low on the face; zygomatic arches relatively deep and widely flaring, with a slight upward sweep; postorbital constriction marked, with large temporal fossae; multiple zygomatico-facial foramina located above the inferior margin of the orbit; frontal process of the malar mediolaterally narrow, with a rugose anterior face; temporal lines strongly marked and converge posteriorly, possibly resulting in a sagittal crest posteriorly, at least in males; cranial capacity estimated (from the area of the foramen magnum) to be only 84.3 cm³, and much less encephalized than *P. heseloni* (Manser and Harrison, 1999); inferior orbital fissure large; nuchal plane relatively short, with a heavy nuchal crest; mandibular symphysis with weakly developed superior transverse torus and indistinct inferior transverse torus; corpus shallow and relatively slender, with a constant depth below the molars; ramus anteroposteriorly long, and superoinferiorly low, with its anterior margin sloping posteriorly at an angle of ~120° to the alveolar plane; pronounced inferior expansion of the angular region of the ramus; articular condyle knoblike; multiple mental foramina located below the premolars (Leakey and Leakey, 1986b; Leakey et al., 1988b; Harrison, 2002).

Dental characteristics include upper canine large and strongly bilaterally compressed in males, with a deep mesial groove and a flangelike distal margin; upper premolars relatively narrow, with paracone not much more elevated than protocone; lingual cingulum present on both P3 and P4; P4 < P3; upper molars with elongated rhomboidal-shaped crowns that narrow distally; trigon narrow and dominated by a voluminous protocone; hypocone closely appressed to protocone and linked to it (or the crista obliqua) by a distinct crest; buccal cingulum moderately well developed, with variable development of accessory cuspule; M2 and M3 with distinct paraconule at the termination of the preparacrista; lingual cingulum broad, and continues mesially around the protocone to form a distinct mesial ledge; prominent secondary conule on the mesiolingual margin of the cingulum; M1 <

M3 < M2; lower molars long and narrow, with only slight development of a buccal cingulum; m1 < m2 < m3; m3 only slightly larger than m2 (Leakey and Leakey, 1986b; Leakey et al., 1988b; Harrison, 2002; figure 24.23).

Postcranial specimens from Kalodirr are generally similar to those of *Proconsul* but do differ in a number of respects that indicate locomotor differences (Leakey and Leakey, 1986b; Leakey et al., 1988b; Rose, 1993, 1994; Ward, 1997). The principal features are as follows: proximal ulna with moderately long and proximally directed olecranon process; distal ulna with long styloid process that articulates with carpus; radius with oval head and relatively long neck; metacarpals and phalanges comparable to those of *Proconsul*; trapezium-first metacarpal joint mobile; femur more slender than in *Proconsul*, with high bicondylar angle; distal articular surfaces of femur less expanded mediolaterally than in *Proconsul* (Leakey et al., 1988b; Rose, 1993, 1994; Ward, 1997). *Turkanapithecus* can be inferred to have been an arboreal quadruped similar in its general locomotor repertoire to *Proconsul*, but possibly with enhanced climbing abilities (Rose, 1993). A partial ulna

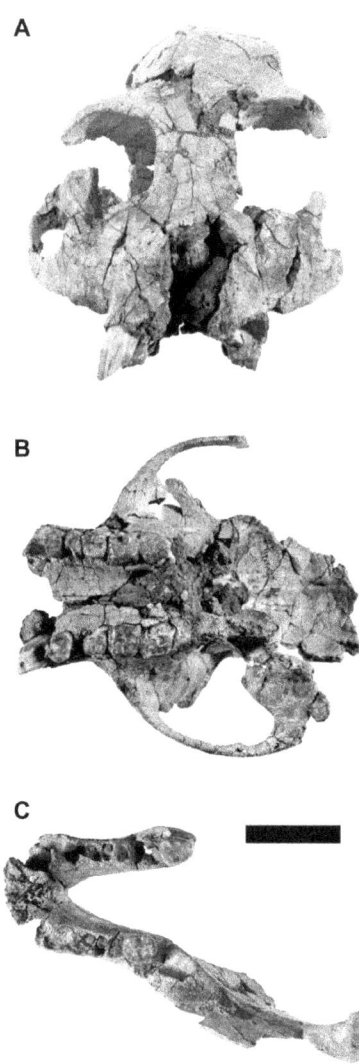

FIGURE 24.23 *Turkanapithecus kalakolensis.* KNM-WK 16950 (holotype), partial skull. A) Cranium, facial view; B) palatal view; C) mandible, occlusal view. Scale: 2 cm. Courtesy of the National Museums of Kenya.

from Fejej in southern Ethiopia, dated to older than 16.2 Ma, is morphologically similar to that of *Turkanapithecus* from Kalodirr (Richmond et al., 1998). Unfortunately, there are no craniodental finds associated with the specimen to definitively identify the taxon to which is belongs, but given the similarity in size and morphology, the correspondence in age, and the close geographic proximity to Kalodirr, a provisional attribution to *Turkanapithecus* seems justified.

Genus *XENOPITHECUS* Hopwood, 1933

Included Species *X. koruensis* Hopwood, 1933 (type species)

XENOPITHECUS KORUENSIS Hopwood, 1933

Distribution Early Miocene (~19–20 Ma). Locality 14 (Maize Crib), Legetet Carbonatite Formation, Koru, Kenya (Pickford, 1986a).

Description Known only from the type specimen, a left maxilla with M1–M2. Upper molars rectangular with convex mesial and distal margins; low, rounded, and voluminous cusps that crowd the trigon basin; crests low and rounded, and relatively short; trigon cusps subequal; hypocone relatively small; mesial fovea narrow; broad lingual cingulum, with a tendency to develop subsidiary tubercles; narrow trigon in relation to the total breadth of the crown; and strong buccolingual flare.

Xenopithecus koruensis can be distinguished from *Proconsul* in the following combination of features: the upper molars are smaller; the size differential between M1 and M2 is not as marked; the molar crowns are more bilaterally symmetrical with the greatest length in the midline (rather than lingually); the cusps are more voluminous and conical (rather than pyramidal), and they crowd the occlusal basins; the cusps are set closer together with a greater degree of buccolingual flare—this results in a narrower mesial fovea and trigon; lingual cingulum more strongly developed, with subsidiary tubercles; well-developed crest linking hypocone to protocone; less pronounced secondary wrinkling of enamel; flatter wear on molars; and less extensive maxillary sinus.

Remarks *Xenopithecus koruensis* was first described by Hopwood (1933a, 1933b), who distinguished it from *Proconsul africanus* by its crowded trigon cusps, small hypocone and well-developed cingula. Le Gros Clark and Leakey (1951), as first revisers, regarded *X. koruensis* as a junior synonym of *P. africanus*, and this has generally been followed by most subsequent workers (Simons and Pilbeam, 1965; Andrews, 1978; Szalay and Delson, 1979). *Xenopithecus* was briefly resurrected as a subgenus by Madden (1980b) to accommodate a new species from Lothidok, *Proconsul (Xenopithecus) hamiltoni*, but this latter taxon was later transferred to a new genus, *Kamoyapithecus* (Leakey et al., 1995). However, Pickford (1986a) and Pickford and Kunimatsu (2005) have recognized the distinctiveness of *X. koruensis*, without formally resurrecting the taxon. The fact that the holotype is still the only known specimen of *X. koruensis*, more than 70 years after its initial description, has provided the greatest impediment to recognizing it as a separate species. Most scholars have preferred to consider it as an aberrant variant of *Proconsul africanus*. However, the morphological and metrical differences between the type specimen of *Xenopithecus koruensis* and the contemporary *Proconsul africanus* provide adequate grounds to recognize the former as a separate species and genus. Presumably, it was an exceedingly rare species in the Koru primate paleocommunity.

Xenopithecus shares several distinctive traits of the upper molars that link it with *Turkanapithecus* and indicate that it might represent a stem nyanzapithecine. These include conical and voluminous cusps, a narrow trigon, and the development of subsidiary conules on the expanded lingual cingulum. Based on this evidence, *Xenopithecus* is considered here to be the primitive sister taxon of the other nyanzapithecines, thereby placing the origin of the clade prior to 20 Ma.

Nyanzapithecine from Ryskop

Distribution Early Miocene (~17.5–18.0). Avontuur Mine, Ryskop near Hondeklip Bay, Namaqualand, South Africa (Senut et al., 1997; Pickford and Senut, 1997).

Description The specimen consists of an incomplete left M1 or M2 (SAM PQ RK 1402), lacking most of the buccal moiety of the crown. It is slightly smaller than upper molars of extant *Gorilla gorilla* and *Samburupithecus kiptalami*. Although incomplete buccally, it is evident from the preserved portion of the metacone that the crown would have been longer than broad (with an estimated length-breadth index of ~107). Cusps low and rounded, with poorly developed occlusal crests. Protocone voluminous, and displaced buccally away from the lingual margin of the crown. Lingual cingulum forms a broad ledge around the mesial and lingual aspects of the protocone, and extends mesially almost to the midline of the crown. Well-developed paraconule on the mesial marginal ridge. Crista obliqua short. Hypocone smaller than protocone, but still relatively large. Crest joins the hypocone to the crista obliqua. Trigon basin narrow. Talon relatively large (Senut et al, 1997; Pickford and Senut, 1997).

Remarks Without additional material not much can be deduced about the relationships of the Ryskop specimen. The proportions of the crown, and the distinctive occlusal morphology certainly indicate that it is a member of the Nyanzapithecinae. However, its size and morphology preclude it from being referred to any of the named taxa from East Africa, and it very likely represents a novel species. Its closest affinities appear to be with the nyanzapithecines from the middle Miocene of Fort Ternan (discussed later). Its most important contribution, perhaps, relates to documenting the biogeographic distribution of catarrhines during the early Miocene. This is the southernmost occurrence of a Miocene catarrhine in Africa, occurring at latitude 30°S, 1,200 km south of Berg Aukas, the only other Miocene locality that has produced a fossil catarrhine from southern Africa. Today, African apes extend southward only to 8°S, although cercopithecids occur at the southern tip of Africa (Kingdon, 1997).

Nyanzapithecine from Fort Ternan and Kapsibor

Distribution Middle Miocene (~13.7 Ma). Fort Ternan and Kapsibor, Kenya (Pickford et al., 2006).

Description Known only from three isolated teeth from Fort Ternan (P4, M1, and m3) and an isolated upper molar from the neighboring locality of Kapsibor (Leakey, 1968; Harrison, 1986, 1992; Pickford, 1986a). P4 ovoid in shape, with a relatively narrow crown. Protocone and paracone elevated, voluminous, closely associated, and subequal in height. Crests poorly developed. Mesial and distal basins shelflike, and continuous with lingual cingulum. Upper molars much longer than broad, and taper strongly distally. M1 and M2 with estimated length-breadth index of 110.8 and 111.3 respectively. Cusps rounded and voluminous, and their bases fill much of the occlusal basins. Protocone displaced from the mesial and lingual margins of the crown. Trigon basin restricted to a narrow C-shaped groove around the mesiolingual aspect of paracone. Lingual cingulum well developed, especially mesially. Preprotocrista long and mesially directed, and terminates at the mesial marginal ridge in a small tubercle. m3 relatively long, with voluminous cusps, low rounded crests, finely wrinkled enamel, large mesial fovea, entoconid and hypoconid transversely aligned, and small tubercle (mesoconid) in the talonid basin at the junction of postprotocristid and prehypocristid. An isolated upper canine of a male individual may also belong to this species. The crown, although missing its apex, is tall, bilaterally strongly compressed, with double mesial groove. It resembles *Dendropithecus* and *Rangwapithecus* in the degree of bilateral compression of the crown, but only *Dendropithecus* typically has a double mesial groove (Harrison, 1986, 1992).

Remarks Leakey (1967, 1968) first noted the presence of a large species of oreopithecid at Fort Ternan. He suggested that the specimens closely resembled *Oreopithecus bambolii* from the late Miocene of Italy, and tentatively referred the material to the same genus. Later, Simons (1972) concluded that the material was conspecific with the European taxon. However, the occurrence of oreopithecids at Fort Ternan was contested by Andrews and Walker (1976), who concluded that the specimens were not primates at all but suids. A reexamination of the material led me to conclude (Harrison, 1986, 1992) that the specimens did indeed belong to an oreopithecid, probably one closely related to, but more derived than, *Nyanzapithecus*. However, subsequent phylogenetic and taxonomic studies have removed all of the East African Miocene taxa from the Oreopithecidae, and included them in the Proconsulidae (Harrison and Rook, 1997; Harrison, 2002). The concordance in size and morphology of the material from Fort Ternan and Kapsibor indicates that they can be referred to a single species. Compared with *Nyanzapithecus*, the specimens are larger and more derived, and they clearly represent a different genus and species, but the available material is not adequate to diagnose a new taxon. The upper molars are generally similar to the specimen from Ryskop (discussed earlier), but they are somewhat smaller in size.

Subfamily *incertae sedis*
Genus *OTAVIPITHECUS* Conroy et al., 1992

Included Species O. namibiensis Conroy et al., 1992 (type species)

OTAVIPITHECUS NAMIBIENSIS Conroy et al., 1992
Figures 24.24 and 24.25

Distribution Middle Miocene (~12–13 Ma). Berg Aukas, Otavi Mountains, Namibia (Conroy et al., 1992; Senut et al., 1992; Conroy, 1996). A proximal ulna is associated with a fauna that indicates a somewhat younger age (late Miocene) than the other specimens attributed to *Otavipithecus* (Senut and Gommery, 1997).

Description The species is represented by a fragmentary mandible preserving the symphyseal region and right corpus with p4–m3 (holotype), a partial frontal, an atlas, a middle manual phalanx, and a proximal ulna (Conroy et al., 1992, 1993a, 1993b, 1996; Conroy, 1996; Pickford et al., 1997; Senut and Gommery, 1997). It is a medium-sized catarrhine. Based on regressions of molar size, Conroy et al. (1992) have estimated the body mass at 14–20 kg, and this is confirmed by the size of the postcranials. Mandibular corpus relatively low, and

shallows only slightly posteriorly. Inferior transverse torus extends somewhat more posteriorly than the superior torus. Single mental foramen below p3. Well-developed postcanine fossa laterally. The anterior root of the ramus does not obscure m3 in lateral view, leaving a small retromolar space. The incisal region is relatively narrow. Canine root relatively large, suggestive of a male individual for the holotype. No diastema between c and p3. p4 ovoid, with oblique long-axis, low, rounded protoconid and metaconid of equal height, and a small, but distinct entoconid. Lower molars short, broad and relatively square in outline, with pronounced basal flare. They have low, rounded, "puffy" cusps, separated by deep grooves. Occlusal crests poorly developed, and no secondary wrinkling of the enamel surface. Mesial and distal foveae relatively small. On m2 and m3 the preprotocristid terminates marginally at a well-developed tubercle. Lingual face of hypoconid expanded into talonid basin. Hypoconulid positioned just buccal to the midline of crown. Remnants of cingulum occur between the buccal cusps. m3 relatively small, intermediate in size between m1 and m2. It does not taper distally, and there is only a short hypoconulid lobe. Molar enamel thin, as in extant African apes (Conroy et al., 1992, 1995; Singleton, 2000). Frontal fragment similar to *Proconsul*, with slender superciliary ridge bordering the superior margin of orbit, an extensive frontal sinus, a broad interorbital region, and widely spaced temporal lines (Pickford et al., 1997; figures 24.24 and 24.25).

The middle manual phalanx is relatively long and slender, with a moderate degree of proximodistal curvature, as in arboreal palmigrade quadrupeds, including *Proconsul* (Conroy et al., 1993a, 1993b; Senut and Gommery, 1997). The atlas displays a mosaic of features intermediate between cercopithecoids and hominoids (Conroy et al., 1996; Senut and Gommery, 1997; Kunimatsu et al., 2004), such as more horizontally oriented superior and inferior facets, relatively small transverse processes, dorsally expanded vertebral canal, and loss of the lateral atlas bridge. The proximal ulna is weathered and poorly preserved. According to Senut and Gommery (1997), the proximal ulna resembles that of *Proconsul* and

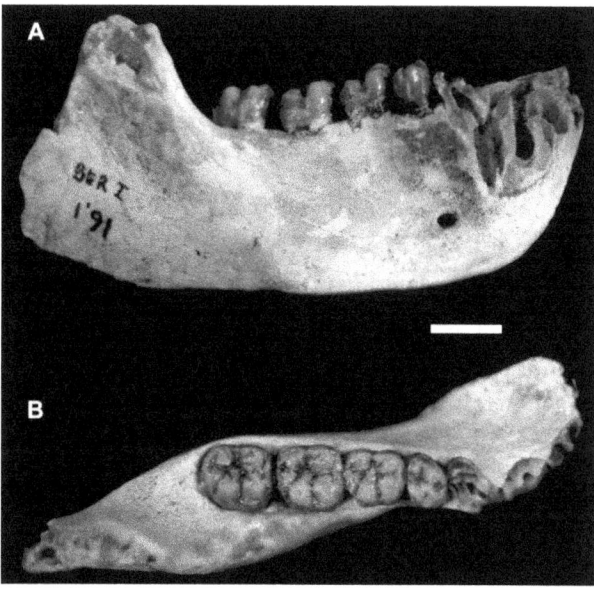

FIGURE 24.24 *Otavipithecus namibiensis*. BER I (holotype), right mandibular fragment with p3–m3. A) Lateral view; B) occlusal view. Scale: 1 cm. Courtesy of G. Conroy.

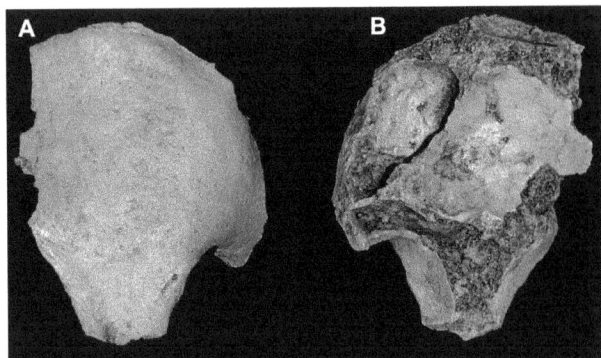

FIGURE 24.25 *Otavipithecus namibiensis*. BA 52'94, frontal. A) Frontal view; B) endocranial view. Courtesy of M. Pickford

extant arboreal quadrupeds. However, the stout shaft and shallow olecranon notch are unusual features for a primate ulna, and they cast doubt on whether this specimen should really be attributed to *Otavipithecus*.

Remarks Fossils from Berg Aukas are derived from breccias accumulated in karstic fissure fillings. Individual blocks of breccia were obtained from past mining operations, and the contained fossils indicate that these blocks vary in age from Miocene to Plio-Pleistocene. Based on correlations of the associated rodents with North African faunas, the age of the block containing the type mandible of *Otavipithecus* is estimated to have been middle Miocene (~12–13 Ma; Conroy et al., 1992; Conroy, 1996). However, comparisons with rodent faunas from East Africa indicate that a somewhat earlier date is possible.

Given the paucity of material, it has proved difficult to establish the phylogenetic relationships of *Otavipithecus*. Alternative analyses have led to the conclusion that it might be closely related to *Afropithecus* or *Heliopithecus* (Andrews, 1992a, 1992b; Singleton, 2000), the sister taxon to extant African apes + humans (Conroy, 1994), or a stem hominid (Begun, 1994; Pickford et al., 1994). Singleton (2000), in the most comprehensive study of the mandibular specimen to date, favored a close relationship with *Afropithecus* but cautiously noted a lack of statistical support for this association and the weakness of the defining synapomorphies. Given that comparisons are limited to the morphology of the mandible and lower cheek teeth, any similarities with *Afropithecus* might be the result of functional convergence. On the basis of the evidence available, including that from the frontal and postcranials, it seems likely that *Otavipithecus* does represent a proconsulid, possibly an afropithecine proconsulid, although its precise affinities cannot be determined with the material available.

Regardless of its phylogenetic status, *Otavipithecus* is important from a biogeographical perspective. It is one of only two samples of Miocene noncercopithecoid catarrhines known from well south of the equator (apart from an isolated tooth of an indeterminate species of nyanzapithecine from South Africa—see earlier discussion). At a latitude of 19.5°S, *Otavipithecus* occurs ~2,700 km southwest of Miocene proconsuloid and hominoid localities in western Kenya, of which Karungu at 0.9°S is the southernmost occurrence (Pickford, 1986a). This implies that proconsuloids had a geographic distribution that may have extended across much of sub-Saharan Africa during the early and middle Miocene.

Superfamily *incertae sedis*
Genus *LIMNOPITHECUS* Hopwood, 1933

Included Species *Lim. legetet* Hopwood, 1933 (type species), *Lim. evansi* MacInnes, 1943.

Description Small catarrhine primates intermediate in dental size between *Simiolus* and *Micropithecus* (probably averaging ~5 kg). *Limnopithecus* is represented primarily by isolated teeth and jaw fragments, and this hampers attempts to adequately determine its phylogenetic relationships. The main characteristics are as follows: lower face short; shallow subnasal clivus; nasal aperture narrow and elliptical; orbits situated low on the face, and positioned relatively far forward, with the anterior margin situated just above the canine root; relatively inflated maxillary sinus that extends posteriorly beyond M3 and laterally into the anterior root of the zygomatic arch; mandible shallow and gracile; prominent superior transverse torus and poorly developed inferior transverse torus *(Lim. evansi)* or possibly absent *(Lim. legetet)*; incisors small in relation to the size of the molars; i2 bilaterally asymmetrical in shape; upper premolars relatively broad with paracone and protocone of approximately equal height; upper molars with high conical cusps, protocone relatively small, trigon only slightly broader than long, and well-developed lingual cingulum; M1 < M3 ≤ M2; M3 only slightly reduced in size compared with M2; lower molars ovoid to rectangular in occlusal outline, with broad talonid basin, and well-developed buccal cingulum; m1 < m2 < m3 (Harrison, 1981, 1982, 1988, 2002). Postcranial remains referred to *Limnopithecus* on the basis of size are morphologically similar to the corresponding elements in *Dendropithecus macinnesi* (Harrison, 1982).

LIMNOPITHECUS LEGETET Hopwood, 1933
Figure 24.26

Distribution Early Miocene (~17–20 Ma). Koru (Koru Formation, Legetet Formation, and Kaputray Agglomerates, Chamtwara Member), Rusinga Island in Kenya, and Napak (IV and V) and Bukwa II in eastern Uganda (Bishop et al., 1969; Andrews and Pickford, 1981; Harrison, 1981, 1982, 1988, 2002; Pickford, 1981, 1983, 1986a, 1986b). The specimen from Bukwa II may extend the temporal range for this species if the published radiometric dates (~22 Ma) are confirmed (Walker, 1968, 1969; Brock and MacDonald, 1969; Bishop et al., 1969), but Pickford (1981, 1986a) has suggested that the fauna is closer in age to the Hiwegi Formation, Rusinga Island (~17.8 Ma). An isolated lower molar referred to *Lim. legetet* from Williams Flat (Loperot) in northern Kenya is of uncertain age (Harrison, 1982).

Description Incisors broad and low crowned; canines relatively small; p3 ovoid to almost circular, with a short mesiobuccal face (*Lim. legetet* may be unique among early Miocene East African catarrhines in having a reduced development of the C/p3 honing complex); p4 broad and ovoid to circular; upper premolars and molars moderately long and broad, with well-defined occlusal crests; lower molars broad and rectangular, with high and sharp cusps and crests, large and well-defined talonid basin and mesial and distal foveae; distal fovea broad and slightly obliquely oriented in m1 and m2, and very obliquely oriented in m3; m3 relatively large, with the entoconid situated transversely opposite the hypoconid (Harrison, 1981, 1982, 1988, 2002; figure 24.26).

LIMNOPITHECUS EVANSI MacInnes, 1943
Figure 24.26

Distribution Early Miocene (~19–20 Ma). Songhor and Mteitei Valley in western Kenya, and Napak in Uganda (Bishop et al.,

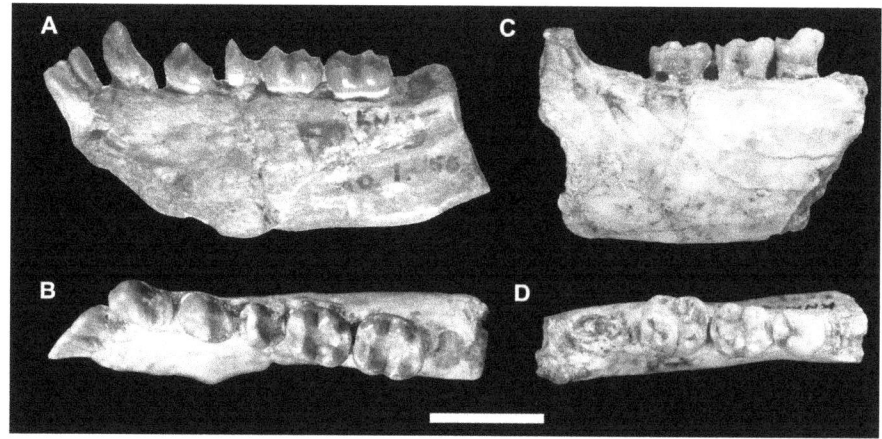

FIGURE 24.26   *Limnopithecus legetet*, KNM-KO 8, right mandibular fragment with i1–m2. A) Medial view;
B) occlusal view. *Limnopithecus evansi*, KNM-SO 385 (holotype), right mandibular fragment with p4–m2.
C) Lateral view; D) occlusal view. Scale: 1 cm. Courtesy of National Museums of Kenya,

1969; Andrews and Pickford, 1981; Harrison, 1981, 1982, 1988, 2002, unpublished data; Pickford, 1981, 1983, 1986a, 1986b).

Description Limnopithecus evansi is distinguished from the type species by the following features: incisors narrower and relatively higher crowned; canines somewhat larger; p3 is narrower with a moderately developed sectorial face on the mesiobuccal aspect of the crown (more typical of other early Miocene catarrhines); p4 long and narrow, with a large mesial fovea; upper premolars and molars relatively broader, with less well-defined occlusal crests; distal cusps on M3 smaller; lower molars have low, rounded cusps and occlusal crests; crest connecting the entoconid and hypoconulid poorly developed or entirely lacking, and as a consequence the distal fovea is ill defined and communicates directly with the talonid basin (a feature unique to *Lim. evansi*); m3 smaller, and entoconid more distally positioned in relation to the hypoconid; mandibular corpus below the cheek teeth slightly higher (Harrison, 1981, 1982, 1988, 2002).

Remarks MacInnes (1943) initially recognized *Limnopithecus evansi* based on new material from Songhor. Later, Le Gros Clark and Leakey (1951) considered that the differences between *Lim. legetet* and *Lim. evansi* were insufficient to recognize separate species, and they formally synonymized the two taxa (although this move was later questioned by Le Gros Clark [1952]). Andrews (1978), following Le Gros Clark and Leakey (1951), attributed additional specimens from Songhor to *Lim. legetet*. Harrison (1982, 1988) resurrected *Lim. evansi*, and included all of the *Limnopithecus* material from Songhor and Mteitei Valley to this species. In addition, Harrison (1982, 1988) also noted that a number of specimens from Napak were morphologically and metrically consistent with *Lim. evansi*, but that a conclusive identification was not possible based on the available material. However, a recent reassessment of the taxonomy of the Ugandan material has now provided confirmation that *Lim. evansi* does definitively occur in the collections from Napak (T. Harrison, unpublished data), and that the type specimen of *Lomorupithecus harrisoni* (discussed later) also belongs to this taxon.

Genus *LOMORUPITHECUS* Rossie and MacLatchy, 2006

Included Species Lom. harrisoni Rossie and MacLatchy, 2006 (type species)

LOMORUPITHECUS HARRISONI Rossie and MacLatchy, 2006
Figures 24.27–24.29

Distribution Early Miocene (~19–20 Ma). Napak IX, Uganda (Rossie and MacLatchy, 2006).

Description A small catarrhine primate with an estimated body weight of 4.3 kg. It has the following morphological characteristics: lower face very short; orbit situated low on the face above P3/P4, and overlaps with the nasal aperture inferiorly; nasal bones short, rhomboidal in shape, and mediolaterally domed; nasal aperture narrow, with a sharply defined margin; inferior margin of nasal aperture V shaped and extends between the roots of the central incisors; premaxilla makes contact with the nasals, and excludes the maxilla from the margin of the nasal aperture; canine jugum prominent, with a deep postcanine fossa; malar surface of zygomatic process slopes postero-inferiorly away from the inferior margin of the orbit; anterior root of the zygomatic arch situated very low on the face, just above M1; frontal sinus present; maxillary sinus extensive and invades the maxillary process between the roots of the cheek teeth; incisive fenestra apparently relatively large; palate shallow and quite narrow, with tooth rows that diverge slightly posteriorly; small I2/C diastema; mandibular corpus slender and shallows posteriorly, with a well-developed superior transverse torus only (figures 24.27 and 24.28).

Upper canine buccolingually compressed; P3 short, broad and triangular in outline, with paracone moderately more elevated than the protocone; P4 ovoid in outline, with a rounded lingual cingulum; m1 relatively broad with a rounded distolingual margin, distally positioned hypocone, and well-developed lingual cingulum; m1 ovoid in outline, tapering mesially, with low rounded cusps and crest, slightly oblique cristid obliqua, large mesial and distal foveae, narrow buccal cingulum, relatively small hypoconulid placed buccal to the midline of the crown; m2 with peripherally placed cusps, long mesial fovea, broad and deep talonid basin and distal fovea, hypoconulid and hypoconid closely appressed, and hypoconid with a weak crest descending into the talonid basin (forming the distal arm of a pliopithecine triangle; figure 24.29).

Remarks Rossie and MacLatchy (2006) presented a phylogenetic analysis from which they deduced that *Lomorupithecus* represents a member of the Pliopithecoidea. If substantiated, this inference would have important biogeographic and

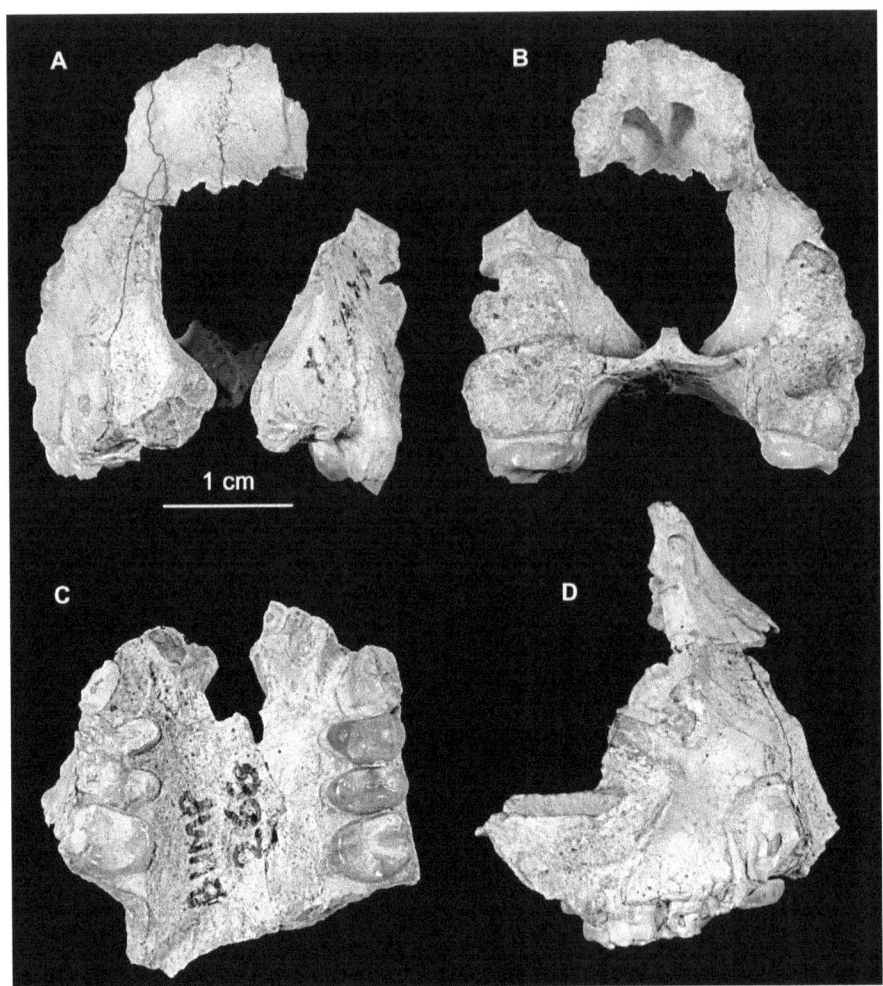

FIGURE 24.27 *Lomorupithecus harrisoni*. BUMP 266 (holotype), lower face with left C–M1 and right P3–M1.
A) frontal view; B) posterior view; C) occlusal view; D) lateral view. Scale: 1 cm. Courtesy of J. Rossie.

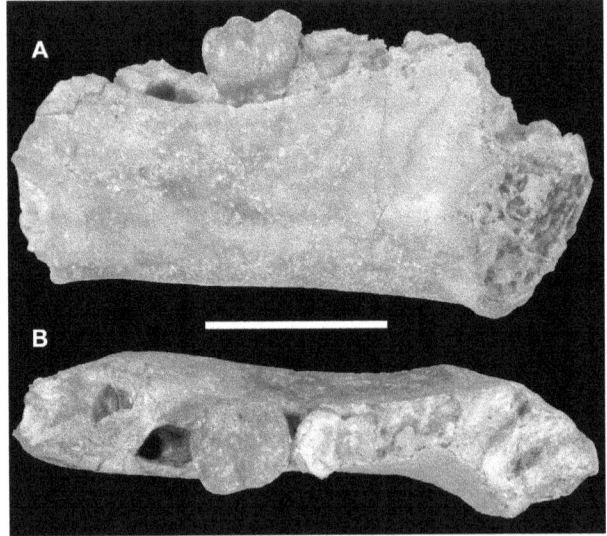

FIGURE 24.28 *Lomorupithecus harrisoni*. BUMP 268, left mandibular fragment with m1 (and m2 in the crypt). A) Medial view; B) occlusal view.
Scale: 1 cm. Courtesy of J. Rossie.

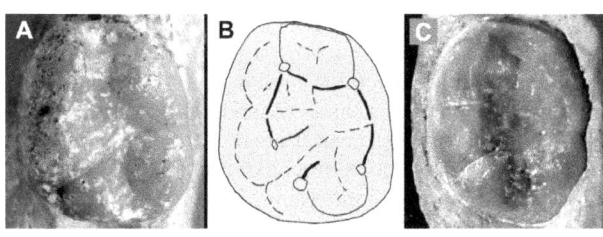

FIGURE 24.29 *Lomorupithecus harrisoni.* BUMP 268, dentition from left mandibular fragment. A) left m1, occlusal view; B) left m1 illustration, occlusal view; C) left m2 (exposed in crypt), occlusal view. Courtesy of J. Rossie.

phylogenetic implications, since pliopithecoids are otherwise entirely unknown from the Miocene of Africa, being strictly a Eurasian clade (Andrews et al., 1996; Harrison and Gu, 1999). The most important features identified by Rossie and MacLatchy (2006) as purported synapomorphies of pliopithecoids are as follows: narrow M1–2 mesial fovea with a prowlike paracristid, narrow and distally projecting distal fovea, and twinned hypoconulid. However, none of these features has previously been identified as diagnostic features of pliopithecoids, and they appear to be of limited utility in determining phylogenetic relationship among Miocene catarrhines. The absence of definitive pliopithecoid autapomorphies in the lower molars of *Lomorupithecus*, as well the lack of evidence for the distinctive incisors and p3 morphology that characterize Eurasian pliopithecids (Andrews et al., 1996; Harrison and Gu, 1999), calls into question to the attribution of this taxon to Pliopithecoidea. The morphology of the face and the cheek teeth are much more similar to those of other stem catarrhines from East Africa than they are to Eurasian pliopithecoids, and, in my view, *Lomorupithecus* is most likely a dendropithecoid or proconsuloid.

Moreover, the upper dentition and lower face are metrically and morphologically indistinguishable from those of *Limnopithecus evansi* from contemporary sites in western Kenya (Harrison, 1982, 1988), a taxon that has previously been identified as possibly occurring at Napak (Harrison, 1982, 1988). Recent comparisons of the Ugandan and Kenyan material by the author have confirmed that *Limnopithecus evansi* does occur at Napak, and that *Lomorupithecus harrisoni* likely represents a junior synonym of *Lim. evansi*. The holotype of *Lom. harrisoni* (BUMP 266) is very similar to material from Songhor attributed to *Lim. evansi*, while the mandibular fragment (BUMP 268) included in the hypodigm of *Lom. harrisoni* by Rossie and MacLatchy (2006) can be identified as belonging to *Micropithecus clarki*. *Lomorupithecus harrisoni* is provisionally retained here as a separate species until a more detailed comparative study can be presented (T. Harrison, unpublished ms.), but ultimately the combined hypodigms will be assigned to a single species, *Lomorupithecus evansi*.

Genus *KALEPITHECUS* Harrison, 1988

Included Species Kal. songhorensis Harrison, 1988 (type species)

KALEPITHECUS SONGHORENSIS (Andrews, 1978)
Figure 24.30

Distribution Early Miocene (~19–20 Ma). Songhor, Mteitei Valley, and Koru (Legetet Formation and Kapurtay Agglomerates, Chamtwara Member) in western Kenya (Harrison, 1988).

Description A small catarrhine primate similar in dental size to *Lim. legetet*, with an estimated body weight of ~5 kg. I1 relatively broader and more spatulate compared with those in *Limnopithecus* or *Dendropithecus*; I2 markedly bilaterally asymmetrical in shape, and relatively much smaller than I1; lower incisors high crowned, slender and relatively bilaterally symmetrical; canines moderately high crowned, with only slight buccolingual compression; upper premolars relatively narrow, with a well-developed transverse crest linking the main cusps; p3 exhibits a moderate degree of sectoriality; p4 relatively large and ovoid, frequently being broader than long; upper molars relatively broad due to the strong development of a lingual cingulum; protocone voluminous and markedly buccally displaced away from the margin of the crown; breadth of the trigon only slightly greater than its length; lower molars short and broad, and rectangular to ovoid in shape; mesial fovea slightly oblique; buccal cingulum broad, but rounded and poorly defined; m1 < m2 ≤ m3; molars have low rounded and voluminous cusps that restrict the extent of the foveae and occlusal basins; occlusal crests low, rounded, and poorly developed; anterior teeth large in relation to the size of the cheek teeth; unlike other early Miocene catarrhines, nasal aperture broad, particularly inferiorly, and nasoalveolar clivus relatively deep (Harrison, 1981, 1982, 1988, 2002; figure 24.30).

Remarks An isolated M1 from the middle Miocene (~14.5 Ma) locality of Kipsaramon in Kenya was provisionally identified as *Kalepithecus* cf. *songhorensis* (Hill et al., 1991). However, Pickford and Kunimatsu (2005) have reassigned this specimen to *Limnopithecus* sp.

Genus *KAMOYAPITHECUS* Leakey et al., 1995

Included Species Kam. hamiltoni Leakey et al., 1995 (type species)

KAMOYAPITHECUS HAMILTONI (Madden, 1980)
Figure 24.31

Distribution Late Oligocene (~24–27 Ma). Erageleit Beds of the Kalakol Basalts, Lothidok, Kenya (Boschetto et al., 1992; Leakey et al., 1995).

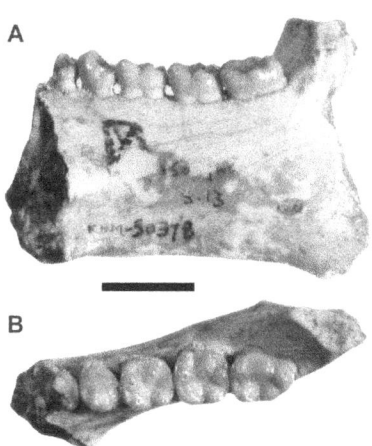

FIGURE 24.30 *Kalepithecus songhorensis.* KNM-SO 378 (holotype), right mandibular fragment with p3–m3. A) Medial view; B) occlusal view. Scale: 2 cm. Courtesy of National Museums of Kenya.

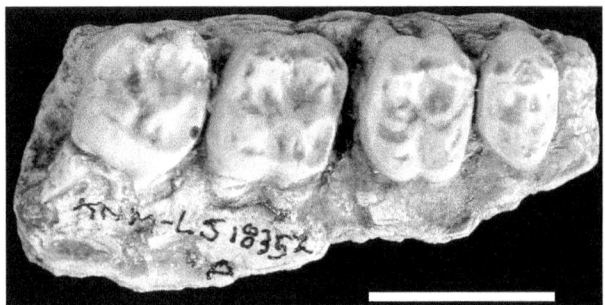

FIGURE 24.31 *Kamoyapithecus hamiltoni.* KNM-LS 18352 (holotype), right maxilla with P4–M3 (cast), occlusal view. Scale: 3 cm. Courtesy of L. Sarlo.

Description A large catarrhine primate approximating *Proconsul nyanzae* in dental size. It differs from other East African catarrhines in the following combination of features: mandibular symphysis with large superior transverse torus and much smaller inferior transverse torus; mental foramen anteriorly facing and positioned just anterior to the mesial root of p3; i2 mesiodistally compressed without tapering at the cervix and with a robust root; canines with stout roots; upper canine with relatively short and bilaterally compressed crown, deep mesial groove, and sharp distal crest; p3 mesiodistally elongate, presumably with a long mesiobuccal honing face; P4 ovoid with moderate degree of lingual flare and slight buccal flare; upper molars very broad and low crowned, with distinct buccolingual flare, conical and bunodont cusps, a relatively narrow trigon, a broad lingual cingulum without crenulation (although subsidiary conules are situated on the lingual cingulum), and a narrow buccal cingulum; M2 with large hypocone set close to the trigon, and crown distinctly shorter in buccal moiety than lingual moiety; M3 with diminutive distal cusps; M1 < M3 ≤ M2 (Madden, 1980b; Leakey et al., 1995; figure 24.31).

Remarks Comparisons with propliopithecids and pliopithecids (i.e., primitive stem catarrhines) suggest that *Kamoyapithecus* may retain features of its dentition that are more primitive than those of all other East African fossil catarrhines, including the dendropithecoids. These include short stout canines with robust roots; upper molars very broad (average length-breadth index of M2 is 77.1, compared with 80.6–115.3 for proconsulids, 82.1–97.1 for dendropithecids, and 66.6–82.8 for propliopithecids + pliopithecids), with strong flare, low rounded cusps, a relatively narrow trigon, a broad lingual cingulum; M2 with large hypocone set close to the trigon, and crown distinctly shorter in its buccal moiety than its lingual moiety (Leakey et al., 1995; Harrison and Gu, 1999). The primitiveness of *Kamoyapithecus* is consistent with its late Oligocene age, and it probably represents the sister taxon of dendropithecoids + (proconsuloids + crown catarrhines), but additional material is needed to settle the question of its phylogenetic and taxonomic status.

Genus *KOGOLEPITHECUS* Pickford et al., 2003

Included Species Kog. morotoensis Pickford et al., 2003 (type species)

KOGOLEPITHECUS MOROTOENSIS Pickford et al., 2003
Figure 24.32

Distribution Late early Miocene (~17.0–17.5 Ma) from Moroto II, eastern Uganda (Pickford et al., 2003).

Description Kogolepithecus is a poorly known taxon, represented by four lower teeth (Pickford et al., 2003). Additional specimens from Moroto II have recently been recovered and await description. It is a small- to medium-sized catarrhine, with teeth slightly larger than those of *Dendropithecus macinnesi*. The genus differs from other East African Miocene catarrhines in the follow combination of features: p4 relatively broad, with a short mesial fovea, widely spaced cusps and a well-developed buccal cingulum; the lower molars are relatively broad, with relatively high and conical cusps, a well-developed buccal cingulum with a small tubercle adjacent to the mesiobuccal margin of the hypoconid, a tendency to develop bifid metaconids (with a prominent mesostylid on the postmetacristid separated from the main cusp by a distinct groove) and entoconids, a very short mesial fovea, hypometacristid weakly developed or lacking with confluence of the mesial fovea and talonid basin, long and lingually inflected crests linking the protoconid and hypoconid, transverse alignment of the entoconid and hypoconid, a subsidiary crest often descends from the hypoconid into the talonid basin, a deep lingual notch, a deep and expansive talonid basin, a very small hypoconulid positioned relatively close to the hypoconid, and m3 smaller than m2 or subequal in size (figure 24.32).

Remarks The morphological and metrical distinctiveness of the lower molars of *Kogolepithecus* provide good evidence to support the recognition of a separate genus and species. However, one has to be cautious about interpreting the significance of some of these features if the lower molars all belong to a single individual, as indicated by Pickford et al. (2003), because the unique traits shared by them could represent individual variation rather than distinctive features of all members of the species. Nevertheless, the combination of size, breadth-length proportions, and occlusal morphology of the lower molars clearly distinguishes this material from all other Miocene catarrhines. With such limited material available for comparison it is impossible to determine the phylogenetic affinities of *Kogolepithecus*, so, following Pickford et al. (2003), the genus is left unassigned at the superfamily level.

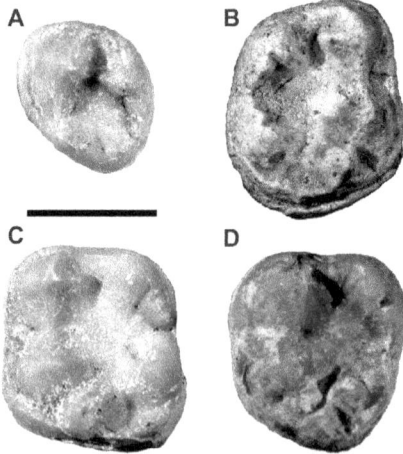

FIGURE 24.32 *Kogolepithecus morotoensis.* A) Mor II 27'03, left p4, occlusal view; B) Mor II 29'03, right m2, occlusal view; C) Mor II 28'03 (holotype), left m2, occlusal view; D) Mor II 10'03, left m3, occlusal view. Scale: 5 mm. Courtesy of M. Pickford.

Superfamily HOMINOIDEA Gray, 1825
Family HOMINIDAE Gray, 1825
Subfamily KENYAPITHECINAE Andrews, 1992
Genus *KENYAPITHECUS* Leakey, 1962

Included Species Ken. wickeri Leakey, 1962 (type species)

KENYAPITHECUS WICKERI Leakey, 1962
Figure 24.33

Distribution Middle Miocene (~13.7 Ma). Fort Ternan, Kenya (Andrews and Walker, 1976; Pickford, 1985; Harrison, 1992; Pickford et al., 2006).

Description Kenyapithecus wickeri is rather poorly known, being represented only by two maxillary fragments, three mandibular fragments, and 10 isolated teeth (Andrews and Walker, 1976; Harrison, 1992). It is a medium-sized catarrhine, similar in dental size to *Proconsul nyanzae* and *Equatorius africanus*. Lower face relatively short and deep, with pronounced canine fossa. Anterior root of zygomatic arch positioned relatively high on the face above M1, and widely flaring and superiorly sweeping. Maxillary sinus extensive, penetrating anteriorly as far as P4 and extending laterally into the anterior root of the zygomatic arch, but it does not invade the alveolar region. Palate arched and moderately deep. Mandible with low and robust corpus, well-developed inferior transverse torus, long subincisive planum, and a single mental foramen situated low on the corpus. I1 low crowned, broad and spatulate, with inflated mesial and distal marginal ridges on the lingual face, a narrow lingual cingulum, but no development of a lingual pillar. I2 relatively bilaterally symmetrical, low crowned, and broad, with a narrow lingual cingulum and low rounded lingual pillar. I2 much smaller than I1. Canines demonstrate marked sexual dimorphism. Upper canines of males relatively high crowned and robust, with a deep mesial groove, and it was probably medially inclined and externally rotated. Upper canines of females low crowned, relatively stout and conical, with a broad and deep mesial groove, a distinct rounded lingual cingulum, and finely wrinkled lingual face. P4 relatively broad and ovoid in shape, paracone only slightly more elevated that the protocone, cusps connected by a pair of transverse crests that delimit a central fovea, and no expression of a lingual cingulum. Molars with thick enamel. Upper molars trapezoidal in shape, slightly broader than long and tapering distally, with low rounded cusps and crests, hypocone well separated from trigon, no lingual cingulum, and minimal secondary wrinkling of the enamel surface. M2 much larger than M1. Lower canine of males high crowned and moderately bilaterally compressed, with narrow lingual cingulum; those of females low crowned and bilaterally compressed, with a weak lingual cingulum. p3 sexually dimorphic, being high crowned in males with long mesiobuccal honing face, and lower crowned in females with short honing face. p4 short and broad, with two main cusps of similar size and elevation, a well-developed transverse crest, a large talonid basin with distinct buccal and lingual tubercles, and vestigial buccal cingulum. Lower molars rectangular and relatively narrow, with broad mesial fovea, low rounded and voluminous cusps and crests, buccal cingulum vestigial, and no secondary wrinkling of the enamel. M3 slightly larger than M1 (Andrews and Walker, 1976; Pickford, 1985, 1986c; Harrison, 1992; McCrossin and Benefit, 1997; Ward and Duren, 2002; figure 24.33).

A distal humerus from Fort Ternan most likely belongs to *Kenyapithecus*, but it cannot be entirely ruled out that it

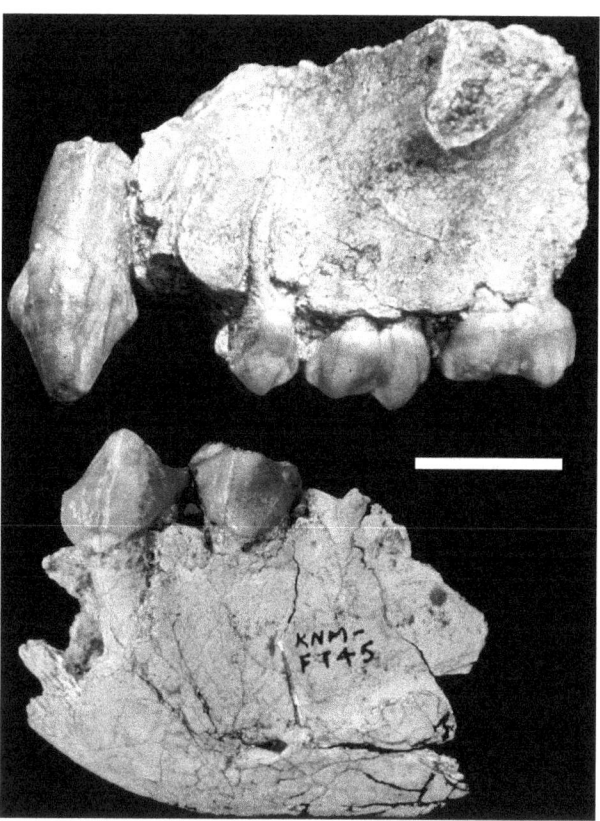

FIGURE 24.33 *Kenyapithecus wickeri.* KNM-FT 45, left mandible with p3–p4; KNM-FT 46a–b (holotype), left maxilla with c, p3–m2, lateral view. Scale: 1 cm. Courtesy of P. Andrews.

belongs to the large nyanzapithecine from this site (Harrison, 1992). It is characterized by a wide spool-shaped trochlea, a well-developed lateral trochlear keel, a distinct zona conoidea, a globular capitulum, a short and strongly posteromedially directed medial epicondyle, a coronoid fossa that is larger than the radial fossa, a deep, well-defined olecranon fossa bordered laterally by a sharp flange, and a strong supinator crest (McCrossin and Benefit, 1994; Rose, 1997). The distal humerus indicates that *Kenyapithecus* was a semiterrestrial quadruped, with ranges of motion at the elbow that match those of extant hominoids.

Remarks A number of researchers consider *Kenyapithecus wickeri* and *Equatorius africanus* to be congeneric and include them together in *Kenyapithecus* (e.g., McCrossin and Benefit, 1994; Benefit and McCrossin, 2000; Kunimatsu et al., 2004). However, a good case has been made to recognize a generic distinction between these two species (Pickford 1985, 1986c; Harrison, 1992; S. Ward et al., 1999; Kelley et al., 2002). *Kenyapithecus* can be distinguished from *Equatorius* in the following respects: anterior root of the zygomatic arch originates higher on the face, and is more markedly laterally flaring and upwardly sweeping; canine fossa more distinct; maxilla much less pneumatized in the region above the premolars; maxillary sinus more elevated, with floor situated well above the apices of the molar roots; palate relatively deeper; upper incisors relatively broader and more robust; I1 with more strongly developed lingual cingulum, well-developed mesial and distal marginal swellings, and lack of a lingual pillar; I2 broader, with a more symmetrical crown; upper canines of females larger and less bilaterally compressed; P4 slightly broader, with stronger transverse crests, no lingual cingulum, and large relative to the size of M1; upper molars taper slightly distally, giving them a more trapezoidal shape, crowns relatively narrower, and greater reduction of the lingual cingulum; lower canines in females slightly larger in size; p3 with stronger lingual cingulum; lower molars relatively narrower; m3 hypoconulid arranged more in line with other buccal cusps (Harrison, 1992; S. Ward et al., 1999; Ward and Duren, 2002; Kelley et al., 2002). As noted by Harrison (1992), many of the derived features listed above are features that *Kenyapithecus* shares uniquely with the Ponginae, but they are likely to represent primitive retentions from the ancestral hominoid or hominid morphotypes. Although this inference needs to be tested with more complete material and with more rigorous phylogenetic analyses, the derived morphology of the lower face and dentition does suggest that *Kenyapithecus* is a stem hominoid or a stem hominid. Recently, a Eurasian species of *Kenyapithecus* has been recognized at the early middle Miocene (~16.0–16.5 Ma) locality of Paşalar in Turkey (Andrews and Kelley, 2007).

Finally, there has been much confusion over the correct nomenclature and authorship of the family group name based on the type genus *Kenyapithecus*. As noted by Pickford and Kunimatsu (2005), the earliest reference to such a family group name is by Andrews (1992b), who included the genus in a new tribe, Kenyapithecini. Begun (2002) and Ward and Duren (2002) incorrectly attributed the family group name to Leakey (1962). In addition, Begun (2002) created a new family, Griphopithecidae, to include the subfamilies Griphopithecinae and Kenyapithecinae, even though the latter has priority as a family group name.

Subfamily *incertae sedis*
Genus *SAMBURUPITHECUS* Ishida and Pickford, 1997

Included Species Sam. kiptalami Ishida and Pickford, 1997 (type species)

SAMBURUPITHECUS KIPTALAMI Ishida and Pickford, 1997
Figure 24.34

Distribution Early late Miocene (~9.5 Ma). Locality SH 22, Lower Member, Namurungule Formation, Samburu Hills, Kenya (Pickford et al., 1984a, 1984b; Makinouchi et al., 1984; Nakaya et al., 1984; Itaya and Sawada, 1987; Sawada et al., 1987, 1998; Tsujikawa, 2005).

Description Samburupithecus kiptalami is known only from the type specimen (KNM-SH 8531), a left premaxilla-maxilla with P3–M3 (Ishida and Pickford, 1997; Pickford and Ishida, 1998). The dentition is similar in size to that of female gorillas (~70–100 kg; Smith and Jungers, 1997). Lower face moderately prognathic. Lateral margin of the nasal aperture sharp edged. Anterior root of zygomatic process pneumatized and situated relatively low on the face. Large, shallow postcanine fossa anterior to the zygomatic process. Extensive maxillary sinus. Palate moderately deep. Upper canine alveolus relatively small, suggesting that this individual was probably female. Upper premolars mesiodistally expanded, and relatively narrow. P3 triangular in outline, while P4 is oval. Molars bunodont, with low rounded and voluminous cusps and poorly developed crests. Molar enamel relatively thick, with a high-relief enamel-dentine junction and no secondary wrinkling. Trigon basin restricted in size. Protocone largest of main cusps; paracone, metacone and hypocone subequal. Well-developed lingual cingulum around the mesiolingual face of protocone. Upper molars increase in size from M1 to M3, with a marked size

differential between successive molars. The very large size of M3 is rare among hominoids (but is typical for *Nyanzapithecus* and *Rangwapithecus*, and does occur occasionally in extant and fossil hominines). In occlusal view the alveolar process is almost straight from C–M3, with the M3 slightly offset medially (Ishida and Pickford, 1997; figure 24.34).

Remarks Samburupithecus has been linked to the African ape + human clade (i.e., Homininae; see Ishida et al., 1984; Andrews, 1992b). Ward and Duren (2002:395) offer a more general interpretation, suggesting that *Samburupithecus* represents a "definite hominoid of modern aspect." However, there are no synapomorphies linking *Samburupithecus* with extant hominoids, and its status as a member of the African hominine clade (i.e., Hominini) is based primarily on its large size and its late Miocene age. *Samburupithecus* is more primitive than all extant hominids in having a relatively inferiorly placed zygomatic process, a nasal aperture that extended inferiorly close to the roots of the incisors resulting in what must have been a short subnasal clivus (based on the remnant of the lateral margin of the nasal aperture), and the occurrence of a well-developed lingual cingulum on the upper molars. These are all primitive features that *Samburupithecus* shares with proconsulids. Moreover, *Samburupithecus* shares a number of peculiar specializations with *Rangwapithecus*, and to a lesser extent with other nyanzapithecines, that might imply a close relationship with this clade. These features include a marked size increase of the upper molars posteriorly, with M3 being much larger than M2, and elongated upper premolars. Given the absence of synapomorphies linking *Samburupithecus* with crown hominoids, the possibility that it represents a late surviving proconsulid cannot be discounted (see Begun, 2001). Unfortunately, there is insufficient material at present to resolve its phylogenetic relationships, but it is provisionally retained here as a hominine.

Genus *CHORORAPITHECUS* Suwa et al., 2007

Included Species C. abyssinicus Suwa et al., 2007 (type species)

CHORORAPITHECUS ABYSSINICUS Suwa et al., 2007
Figure 24.35

Distribution Late Miocene (~10.0–10.5 Ma). Beticha, upper part of Chorora Formation, Ethiopia (Geraads et al., 2002; Suwa et al., 2007).

Description This species is known only from a small collection of isolated teeth, comprising a lower canine and eight

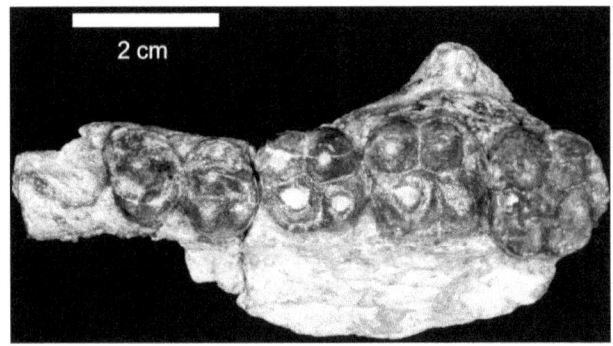

FIGURE 24.34 *Samburupithecus kiptalami*. KNM-SH 8531 (holotype), left maxilla with P3–M3. Scale: 2 cm. Courtesy of Y. Kunimatsu.

upper and lower molars. The dentition is comparable in size to that of *Gorilla gorilla*. The lower canine is low crowned and relatively robust, with a high mesial shoulder and a short distal heel. The morphology and size of the tooth suggests that it belonged to a female individual. The upper and lower molars have low and voluminous cusps, cusp apices arranged peripherally, low and rounded, but distinct occlusal crests, relatively thick enamel (i.e., intermediate thickness), especially on the lingual side in uppers and the buccal side in lowers, and a relatively low topography of the cuspal dentine-enamel junction. CHO-BT 4 (the holotype), an isolated M2, is the best preserved of the upper molars. It has the following characteristics: crown longer than broad (with a length-breadth ratio of 106.1), falling at the upper end of the range for extant *Gorilla gorilla* and *Hylobates syndactylus*, and similar to *Nyanzapithecus* spp., *Samburupithecus kiptalami*, and *Oreopithecus bambolii* (Harrison, 1982, 1986, 2002); crown narrows slightly distally, and exhibits a moderate degree of buccolingual waisting midway along its length; mesial fovea narrow and very short; preprotocrista long and well developed; metacone much smaller than protocone; metacone and protocone linked by a robust crista obliqua; shallow trigon basin; hypocone small, and separated from the protocone and the metacone by a deep L-shaped fissure; and weak lingual cingulum on the mesiolingual margin of the protocone. CHO-BT 5 is described as an M3, but based on the preserved occlusal morphology (the margins of the crown are eroded), it could be a left m2 instead. Suwa et al. (2007) suggest that CHO-BT 4 and CHO-BT 5 are possibly associated. In this case, if it were an M3, it would be much smaller in occlusal area than M2 (only 86% of the area of M2). Other isolated teeth confirm the relatively small size of M3. The m1 is ovoid in shape, with low rounded cusps, a simple Y-shaped groove system, and a small distal fovea. The m3 is a narrow, triangular tooth that tapers distally. It has a narrow mesial fovea, a strong Y-shaped groove system with slight secondary wrinkling, moderately tall metaconid with a long and steep postmetacristid separated from the small entoconid by a deep lingual notch, a relatively large hypoconulid placed in the midline of the crown, and a slight trace of the buccal cingulum between the protoconid and hypoconid (Suwa et al., 2007; figure 24.35).

Remarks With so little anatomical evidence available, the phylogenetic and taxonomic affinities of *Chororapithecus* are difficult to determine. Suwa et al. (2007) have argued that *Chororapithecus* may be the sister taxon to *Gorilla*. Shared specializations include relatively distinct shearing crests on the upper and lower molars (especially the preprotocrista, postmetacristid and distal trigonid crest), as well as aspects of the enamel-dentine junction. As noted by Suwa et al. (2007), however, these can be interpreted as incipient specializations for a fibrous diet that could have been independently acquired in *Chororapithecus* and *Gorilla* as a consequence of functional convergence. Added weight is given to this scenario by the fact that dental adaptations to exploit more fibrous diets are common among later Miocene catarrhines from eastern Africa, such as *Nyanzapithecus*, *Equatorius*, *Kenyapithecus*, and *Samburupithecus*, as a response to increasingly seasonal environments associated with a shift to cooler and drier climatic conditions at this time (Harrison 1989, 1992). However, the upper and lower molars of *Chororapithecus* are readily distinguished from those of *Gorilla* in being higher crowned with thicker enamel, less elevated cusps, lower and shorter occlusal crests, and less lingually inflected postprotocristid and prehypocristid. However, these differences in *Chororapithecus* are plausibly interpreted as primitive hominid features that do not preclude it from being closely related to *Gorilla* (Suwa et al., 2007). In terms of its phylogenetic relationships a close relationship between *Chororapithecus* and *Gorilla* remains a possibility, as supported by the evidence presented by Suwa et al. (2007). However, given the paucity of the material and the high probability of functional convergence, it would seem that such a relationship cannot be convincingly demonstrated. Instead, *Chororapithecus* is best regarded as a stem hominid, or possibly a stem hominine, of uncertain phylogenetic affinities.

Among Miocene hominoids, *Chororapithecus* is most similar to *Samburupithecus* (Suwa et al., 2007). With comparisons limited to the upper molars, few distinguishing features can be identified. According to Suwa et al. (2007), *Chororapithecus* differs from *Samburupithecus* in the following respects: more open occlusal basins, less robust crests, possession of a continuous crista obliqua on M2 (in *Samburupithecus* it is disrupted by a

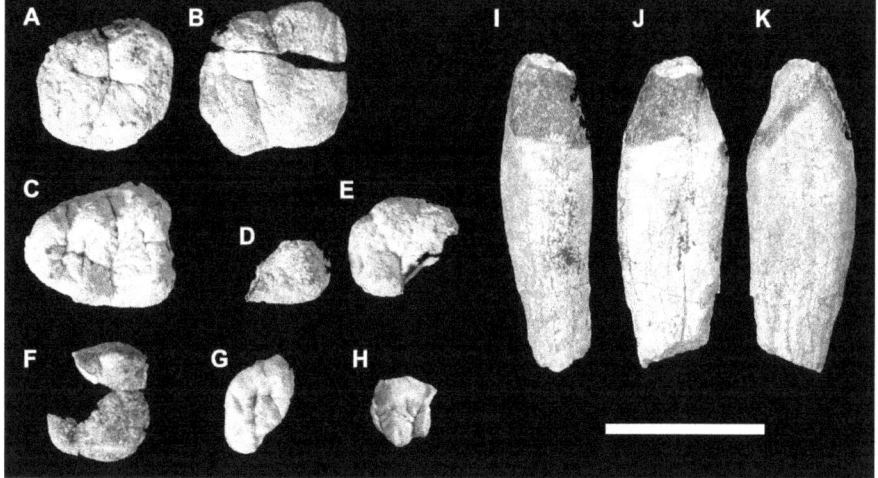

FIGURE 24.35 *Chororapithecus abyssinicus.* A) CHO-BT 5, right M3?, occlusal view; B) CHO-BT 4 (holotype), right M2, occlusal view; C) CHO-BT 7, left m3; D) CHO-BT 9, left lower molar fragment, occlusal view; E) CHO-BT 8, left m1, occlusal view; F) CHO-BT 6, right M3, occlusal view; G) CHO-BT 11, incomplete right M3, occlusal view; H) CHO BT 10, right lower molar fragment, occlusal view; I–K) CHO-BT 3, left lower canine, distal, buccal and lingual views respectively. Scale: 2 cm. Courtesy of G. Suwa.

longitudinal fissure), and weaker development of the lingual cingulum on the mesiobuccal margin of the protocone. Another possible difference is the relative size of M3, which is smaller than M2 in *Chororapithecus* and larger than M2 in *Samburupithecus*. However, these are relatively minor differences given the overall similarity between the two taxa in molar size, proportions, and occlusal morphology. Based on the variation observed in extant hominoids, one could argue that the Chorora and Namurungule samples should be referred to separate species within the same genus or even combined into a single species. The range of metrical variation in the dental remains from both sites combined does not exceed that of a single species of extant great ape. Although synonymy seems likely, until more detailed comparisons are available or until additional specimens are recovered, *Chororapithecus abyssinicus* is provisionally retained here as a separate genus and species.

Genus *NAKALIPITHECUS* Kunimatsu et al., 2007

Included Species Nakalipithecus nakayamai Kunimatsu et al., 2007 (type species)

NAKALIPITHECUS NAKAYAMAI Kunimatsu et al., 2007
Figure 24.36

Distribution Late Miocene (~9.8–9.9 Ma); Nakali, Upper Member of Nakali Formation, Kenya (Kunimatsu et al., 2007).

Description Nakalipithecus is a large hominoid, comparable in dental size to female gorillas. The species is known only from a right mandibular fragment with heavily worn m1–m3 (the holotype) and 11 isolated teeth. The mandibular corpus is relatively high and slender, and shallows slightly posteriorly. It has a moderately well-developed superior transverse torus and a more strongly developed inferior transverse torus that extends posteriorly as far as m1. i2 relatively tall and narrow, with a convex distal margin and lacking a distinct lingual cingulum or pillar. p3 subtriangular in occlusal outline with a broad distal fovea and a prominent lingual cingulum. p4 broader than long, with two peripherally placed cusps and traces of a buccal cingulum. The lower molars exhibit the following features: relatively thick enamel; dentine-enamel junction with low topography; cusps relatively low, voluminous, and peripherally arranged; distinct traces of a buccal cingulum; m1 smaller than m2, but size differential not marked; m3 much larger than m2, at least in the holotype, but isolated teeth indicate a good deal of size variation in this tooth. I1 low crowned and broad, with strongly convex distal margin, elevated lingual cingulum, and strong lingual pillar. The upper canine (of a presumed female individual) is low crowned and robust, with a distinct cuspule perched on the well-developed lingual cingulum, and prominent buccal cingula. Upper premolars relatively narrow, with voluminous cusps, and weak heteromorphy. M1 slightly broader than long, with a distinct remnant of the lingual cingulum on the mesiolingual margin of the protocone (figure 24.36).

Remarks Kunimatsu et al. (2007) present a convincing case that *Nakalipithecus nakayamai* is most similar to, and probably most closely related to, *Ouranopithecus macedoniensis* from the late Miocene (8.7–9.6 Ma) of Greece. In addition to overall size, they share the following suite of distinctive features: mandibular corpus relatively deep and slender; p3 subtriangular, with low protoconid, and broad distal fovea; m1:m2 size differential relatively small; molars with thick enamel and cingula reduced to remnants; and upper canine in presumed females relatively low crowned and robust. However, *Nakalipithecus* differs from

FIGURE 24.36 *Nakalipithecus nakayamai*. KNM-NA 46400 (holotype), right mandible with m1–m3. A) Lateral view; B) occlusal view; C) medial view. Scale: 2 cm. Courtesy of Y. Kunimatsu.

Ouranopithecus in the following respects: upper canine in presumed females higher crowned, with a more elevated mesial shoulder; P3 with more marked buccal cingula and styles, and a better-developed mesial transverse crest linking the two cusps; P4 with less inflated cusps; upper premolars more elongated; molars with thinner enamel, less inflated cusps, larger occlusal basins, and better developed buccal cingula; p3 with more marked buccal and lingual cingula. The less inflated cusps and the better-developed cingula on the cheek teeth indicate that *Nakalipithecus* is more primitive than *Ouranopithecus*. Based on the available evidence, it is reasonable to infer that the lineage leading to *Ouranopithecus* originated in Africa from a species similar in morphology to *Nakalipithecus*. Along with the growing diversity of other possible hominines from the later Miocene of Africa, this provides support for the contention that the hominine clade originated in Africa, and that one or more members subsequently spread to Europe (Cote, 2004; Kunimatsu et al., 2007; Suwa et al., 2007; *contra* Stewart and Disotell, 1998; Begun, 2001, 2005).

Evolutionary Relationships

Fossil catarrhines from the late Oligocene onwards can be grouped into four superfamilies: Dendropithecoidea, Proconsuloidea, Hominoidea, and Cercopithecoidea. The first two clades are stem catarrhines of modern aspect restricted to the Oligo-Miocene of Afro-Arabia. They are derived relative to primitive stem catarrhines (i.e., propliopithecids and pliopithecids) in the development of a fully formed tubular ectotympanic (at least in proconsuloids) and the loss of the entepicondylar foramen in the distal humerus (Harrison, 1987, 2002). Cercopithecoids and hominoids are crown catarrhines that share a suite of cranial and postcranial synapomorphies that are absent in dendropithecoids and proconsuloids (Harrison and Sanders, 1999; Harrison, 1993, 2002; figure 24.37).

The Dendropithecoidea share the following features: upper and lower canines strongly bilaterally compressed; p3 exhibits moderate to strong sectoriality; the limb bones are slender; the humerus has a relatively straight shaft; the distal humerus has a large medially directed medial epicondyle, a well-developed dorsal epitrochlear fossa, a broad and shallow zona conoidea, a weak lateral trochlear keel, a trochlea with minimal spooling, and a shallow olecranon fossa (Harrison, 1987, 1988, 1993; Rose et al., 1992; Rose 1997). Most of these characters can be interpreted as the primitive condition for catarrhines, while the distinctive C/p3 honing complex probably corresponds closely to the primitive condition for catarrhines of modern aspect (Harrison and Gu, 1999). It is possible, therefore, that the dendropithecoids represent a paraphyletic group, but their close morphological similarity, especially in their postcranial morphology, makes it more likely that they represent phyletically closely related taxa (figure 24.37).

Proconsuloidea is more derived than Dendropithecoidea in many aspects of its postcranium, especially the forelimb (see Harrison, 1987; Rose et al., 1992; Rose, 1993, 1997). *Proconsul*, as the best-known early Miocene catarrhine, is of key impor-

tance for interpreting the phylogenetic affinities of the Proconsuloidea. However, several different hypotheses have been proposed concerning the relationships of *Proconsul*. Most researchers contend that *Proconsul* shares key synapomorphies with extant hominoids and consider it a stem hominoid (e.g., Rose, 1983, 1992, 1997; Andrews, 1985, 1992b; Fleagle, 1986, 1999; Senut, 1989; Begun et al., 1997; Kelley, 1997; Ward, 1997; Rae, 1999) or a stem hominid (Rae, 1997; Walker, 1997; Walker and Teaford, 1989). However, I prefer to recognize *Proconsul* as a stem catarrhine, representing the sister-group to cercopithecoids + hominoids (Harrison 1987, 1988, 1993, 2002, 2005; Harrison and Rook, 1997; Harrison and Gu, 1999). This latter conclusion is based on derived features of the face, ear region, and postcranium shared by extant catarrhines, but primitively lacking in *Proconsul* (figure 24.37).

The Proconsulidae contains three subfamilies. The Proconsulinae, comprising at least five species of *Proconsul*, includes the most generalized proconsuloids. *Proconsul heseloni* and *P. nyanzae* are very similar, and they are almost certainly sister taxa. *Proconsul major* appears to be morphologically the most distinctive, but a few derived features may link it with *P. africanus*. Members of the Afropithecinae share a suite of derived features, including tall and procumbent incisors, tusklike upper canines, enlarged upper premolars molars with low cusps and crests, thick enamel, and reduced cingula, narrow upper molars, and moderate to well-developed inferior transverse torus in the mandible (Andrews, 1992b; Leakey and Walker, 1997). The subfamily may also include *Otavipithecus* (Andrews, 1992b; Singleton, 2000), but the phylogenetic relationships of this taxon are difficult to establish. The morphology of the face and postcranium clearly associate *Afropithecus* (and *Otavipithecus*) more closely to *Proconsul* than to extant hominoids (Rose, 1993, 1997; Leakey and Walker, 1997; Ward, 1997, 1998; Harrison, 2002).

The Nyanzapithecinae is distinguished from other proconsuloids by its specialized dentition. *Nyanzapithecus*, the most specialized member of the clade, is characterized by the following suite of derived features: relatively robust premaxilla; stoutly constructed upper incisors; P3 ovoid, with cusps similar in height; P4 with well-developed lingual cingulum; upper premolars narrow; p4 long and narrow with high cusps and elevated mesial fovea; upper molars with elevated and inflated cusps, poorly defined occlusal crests, restricted basins, well-developed lingual cingulum, especially mesially, and crown longer than broad; M3 equal in size to or larger than M2; and lower molars with elevated and inflated cusps, poorly defined occlusal crests, poorly developed buccal cingulum, restricted mesial and distal foveae, and crown very long and narrow. A reassessment of the relationship between *Nyanzapithecus* and *Mabokopithecus* will have to await detailed comparisons of new finds from Maboko Island, but it is evident that the two taxa are closely related and may even be congeneric. *Rangwapithecus* is more primitive than *Nyanzapithecus* in a number of important respects, but they do share the following derived characters: upper canines strongly bilaterally compressed; upper premolars with relatively long and narrow crowns, more ovoid occlusal outline, buccal and lingual cusps more similar in height, inflated lingual cingulum on both P3 and P4 (except *Ny. pickfordi*); P3 structurally similar to P4; upper molars long and relatively narrow, and taper distally; M3 is the largest tooth in the molar series; lower p4 and molars long and narrow. *Turkanapithecus* shares a subset of these synapomorphies (i.e., upper canine strongly bilaterally compressed; upper premolars with relatively long and narrow crowns; buccal and lingual cusps more similar in height; inflated lingual cingulum on both P3 and P4; upper molars long and relatively narrow,

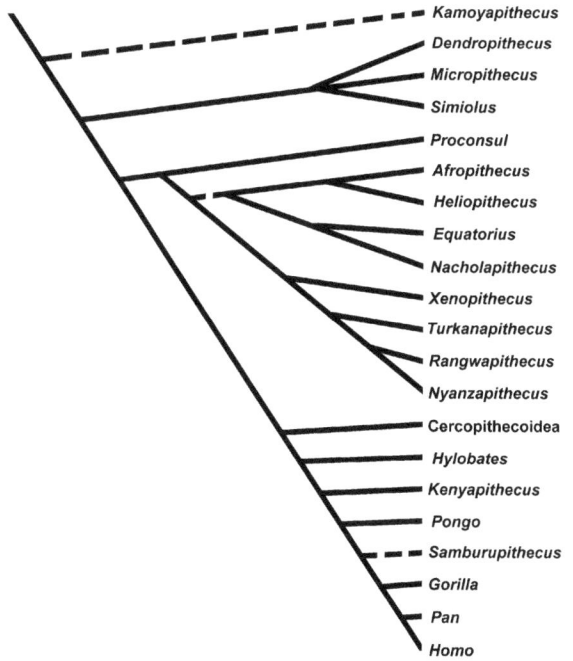

FIGURE 24.37 Cladogram illustrating the inferred phylogenetic relationships between the Afro-Arabian dendropithecoids, proconsuloids and hominoids as presented in this chapter. Broken lines represent uncertain relationships.

and taper distally; upper molars with well-developed mesial ledge and large protoconule; lower P4 and molars long and narrow). It can be inferred to represent the sister taxon of *Rangwapithecus + Nyanzapithecus* (Harrison, 2002). Even though *Proconsul* and *Turkanapithecus* have strikingly different facial characteristics that distinguish them, both taxa share a similar general facial plan that is quite different from that of extant catarrhines. *Turkanapithecus* is the only nyanzapithecine with associated cranial and postcranial material, and the latter is morphologically very similar to that of *Proconsul* and more derived than that of *Dendropithecus*, *Micropithecus*, and *Simiolus* (Rose, 1997; figure 24.37).

Kalepithecus, *Limnopithecus*, *Kamoyapithecus*, *Kogolepithecus*, and *Lomorupithecus* are not well enough known to establish their phylogenetic affinities with any degree of confidence. *Kamoyapithecus* retains several features of its dentition that are more primitive than those of all other East African fossil catarrhines, including the dendropithecoids, and it may represent the sister taxon of dendropithecoids + (proconsuloids + crown catarrhines). Although poorly known, *Kenyapithecus wickeri* from the middle Miocene of Fort Ternan probably represents the earliest known hominoid in Africa. It appears to be more derived than afropithecines, including *Nacholapithecus* and *Equatorius*, in the morphology of its lower face and dentition, and it likely represents a stem hominoid or a stem hominid.

Tantalizing evidence from the later Miocene of East Africa demonstrates that hominoids were quite diverse during this period, although few species have been formally named because of the paucity of the material. *Samburupithecus* from the late Miocene of Kenya is known from a single maxillary specimen, making it difficult to ascertain its phylogenetic relationships, but it probably represents a primitive hominine (figure 24.37). The recently described *Chororapithecus* from the late Miocene of Ethiopia has been interpreted as closely related to *Gorilla* (Suwa et al., 2007), but it more likely represents a stem hominine closely related to *Samburupithecus*. *Nakalipithecus* from the late Miocene of Kenya is a stem hominine that represents the primitive sister taxon of *Ouranopithecus* from Europe (Kunimatsu et al., 2007). Other specimens from later Miocene localities in Kenya (i.e., from Lukeino, Nakali and Ngorora) apparently represent at least three additional unnamed species of hominids, some of which may prove to be stem hominines once better material is recovered (Hill and Ward, 1988; Hill et al., 2002; Pickford and Senut, 2005a, 2005b; Kunimatsu et al., 2007). Given the diversity and apparent relationships of these later Miocene hominids from East Africa, it is likely that hominines originated in Africa, rather than in Eurasia. Pickford et al. (2008, 2009) have recently reported the discovery of a purported fossil hominoid from the late Miocene of Niger, but the very fragmentary nature of the specimen and the uncertain provenance and relative paucity of the associated fauna dictate that any conclusions about its taxonomy or inferred age should be considered tentative at best. Middle to late Pleistocene remains of chimpanzees have been recorded from sites in East Africa (Lehmann, 1957; McBrearty and Jablonski, 2005; De Silva et al., 2006), but none can be definitively attributed to *Pan*. The homininan fossil record is well documented from the late Miocene onward.

ACKNOWLEDGMENTS

I would like to thank Bill Sanders and Lars Werdelin for inviting me to prepare this contribution for this volume. I thank the following colleagues and institutions for allowing me access to the fossil material and casts in their care: E. Mbua, M. G. Leakey, and R. E. Leakey, National Museums of Kenya, Nairobi; P. Andrews and J. Hooker, The Natural History Museum, London; E. Delson, American Museum of Natural History, New York; and J. Fleagle, Stony Brook, New York. The following individuals kindly provided me with photographs of specimens: P. Andrews, G. Conroy, J. Fleagle, Y. Kunimatsu, M. G. Leakey, L. MacLatchy, M. Nakatsukasa, M. Pickford, D. Pilbeam, J. Rossie, L. Sarlo and G. Suwa. Numerous colleagues have contributed to the research and ideas presented here, but the following deserve special mention: P. Andrews, D. Begun, B. Benefit, E. Delson, J. Fleagle, A. Hill, C. Jolly, J. Kelley, Y. Kunimatsu, M. G. Leakey, L. MacLatchy, L. Martin, M. McCrossin, M. Nakatsukasa, M. Pickford, D. Pilbeam, M. Rose, W. Sanders, B. Senut, A. Walker, C. Ward, and S. Ward.

Literature Cited

Allbrook, D., and W. W. Bishop. 1963. New fossil hominoid material from Uganda. *Nature* 197:1187–1190.

Allen, K. L., and M. L. McCrossin. 2007. Functional morphology of the *Kenyapithecus* hand from Maboko Island (Kenya). *American Journal of Physical Anthropology* 132, S44:62

Andrews, P. 1978. A revision of the Miocene Hominoidea of East Africa. *Bulletin of the British Museum (Natural History), Geology* 30:85–225.

———. 1980. Ecological adaptations of the smaller fossil apes. *Zeitschrift für Morphologie und Anthropologie* 71:164–173.

———. 1985. Family group systematics and evolution among catarrhine primates; pp. 14–22 in E. Delson (ed.), *Ancestors: The Hard Evidence*. Liss, New York.

———. 1992a. An ape from the south. *Nature* 356:106.

———. 1992b. Evolution and environment in the Hominoidea. *Nature* 360:641–647.

Andrews, P., W. R. Hamilton, and P. J. Whybrow. 1978. Dryopithecines from the Miocene of Saudi Arabia. *Nature* 274:249–250.

Andrews, P., and T. Harrison. 2005. The last common ancestor of apes and humans; pp. 103–121 in D. E. Lieberman, R. J. Smith, and J. Kelley (eds.), *Interpreting the Past: Essays on Human, Primate, and Mammal Evolution in Honor of David Pilbeam*. Brill Academic Publishers, Boston.

Andrews, P., T. Harrison, E. Delson, R. L. Bernor, and L. Martin. 1996. Distribution and biochronology of European and Southwest Asian Miocene catarrhines; pp. 168–207 in R. L. Bernor, V. Fahlbusch, and H.-W. Mittmann (eds.), *The Evolution of Western Eurasian Neogene Mammal Faunas*. Columbia University Press, New York.

Andrews, P., T. Harrison, L. Martin, and M. Pickford. 1981. Hominoid primates from a new Miocene locality named Meswa Bridge in Kenya. *Journal of Human Evolution* 10:123–128.

Andrews, P., and J. Kelley. 2007. Middle Miocene dispersals of apes. *Folia Primatologica* 78:328–343.

Andrews, P., and L. B. Martin. 1987a. Cladistic relationships of extant and fossil hominoids. *Journal of Human Evolution* 16:101–118.

———. 1987b. The phyletic position of the Ad Dabtiyah hominoid. *Bulletin of the British Museum (Natural History), Geology* 41:383–393.

———. 1991. Hominoid dietary evolution. *Philosophical Transactions of the Royal Society of London, B* 334:199–209.

Andrews, P., and T. I. Molleson. 1979. The provenance of *Sivapithecus africanus*. *Bulletin of the British Museum (Natural History), Geology* 32:19–23.

Andrews, P., and E. Simons. 1977. A new Miocene gibbon-like genus, *Dendropithecus* (Hominoidea, Primates) with distinctive postcranial adaptations: Its significance to origin of Hylobatidae. *Folia Primatologica* 28:161–169.

Andrews, P., and A. Walker. 1976. The primate and other fauna from Fort Ternan, Kenya; pp. 279–304 in G. L. Isaac and E. R. McCown (eds.), *Human Origins*. Benjamin, Menlo Park, Calif.

Anyonge, W. 1991. Fauna from a new lower Miocene locality west of Lake Turkana, Kenya. *Journal of Vertebrate Paleontology* 11:378–390.

Beard, K. C., M. F. Teaford, and A. Walker. 1986. New wrist bones of *Proconsul africanus* and *nyanzae* from Rusinga Island, Kenya. *Folia Primatologica* 47:97–118.

Begun, D. R. 1994. The significance of *Otavipithecus namibiensis* to interpretations of hominoid evolution. *Journal of Human Evolution* 27:385–394.

———. 2000. Middle Miocene hominoid origins. *Science* 287:2375a.

———. 2001. African and Eurasian Miocene hominoids and the origins of the Hominidae; pp. 231–253 in L. de Bonis, G. Koufos and P. Andrews (eds.), *Hominoid Evolution and Climate Change in Europe: Volume 2. Phylogeny of the Neogene Hominoid primates of Eurasia.* Cambridge University Press, Cambridge.

———. 2002. European hominoids; pp. 339–368 in W. C. Hartwig (ed.), *The Primate Fossil Record.* Cambridge University Press, Cambridge.

———. 2005. *Sivapithecus* is east and *Dryopithecus* is west, and never the twain shall meet. *Anthropological Science* 113:53–64.

Begun, D. R., M. F. Teaford, and A. Walker. 1993. Comparative and functional anatomy of *Proconsul* phalanges from the Kaswanga Primate Site, Rusinga Island, Kenya. *Journal of Human Evolution* 26:89–165.

Begun, D. R., C. V. Ward, and M. D. Rose. 1997. Events in hominoid evolution; pp. 389–415 in D. R. Begun, C. V. Ward, and M. D. Rose (eds.), *Function, Phylogeny, and Fossils: Miocene Hominoid Evolution and Adaptation.* Plenum Press, New York.

Behrensmeyer, A. K., A. L. Deino, A. Hill, J. D. Kingston, and J. J. Saunders. 2002. Geology and geochronology of the middle Miocene Kipsaramon site complex, Muruyur Beds, Tugen Hills, Kenya. *Journal of Human Evolution* 42:11–38.

Benefit, B. R. 1991. The taxonomic status of Maboko small apes. *American Journal of Physical Anthropology*, suppl. 12:50–51.

Benefit, B. R., and M. L. McCrossin. 1995. Miocene hominoids and hominid origins. *Annual Review of Anthropology* 24:237–256.

———. 1997. New fossil evidence bearing on the relationship of *Nyanzapithecus* and *Oreopithecus*. *American Journal of Physical Anthropology*, suppl. 24:74.

———. 2000. Middle Miocene hominoid origins. *Science* 287:2375a.

———. 2002. The Victoriapithecidae, Cercopithecoidea; pp. 241–253 in W. C. Hartwig (ed.), *The Primate Fossil Record.* Cambridge University Press, Cambridge.

Benefit, B. R., S. N. Gitau, M. L. McCrossin, and A. K. Palmer. 1998. A mandible of *Mabokopithecus clarki* sheds new light on oreopithecid evolution. *American Journal of Physical Anthropology*, suppl. 26:109.

Beynon, A. D., M. C. Dean, M. G. Leakey, D. J. Reid, and A. Walker. 1998. Comparative dental development and microstructure of *Proconsul* teeth from Rusinga Island, Kenya. *Journal of Human Evolution* 35:163–209.

Bishop, W. W., and G. R. Chapman. 1970. Early Pliocene sediments and fossils from the northern Kenya Rift Valley. *Nature* 226: 914–918.

Bishop, W. W., J. A. Miller, and F. J. Fitch. 1969. New potassium-argon age determinations relevant to the Miocene fossil mammal sequence in East Africa. *American Journal of Science* 267:669–699.

Boschetto, H. B., F. H. Brown, and I. McDougall. 1992. Stratigraphy of the Lothidok Range, northern Kenya, and K-Ar ages of its Miocene primates. *Journal of Human Evolution* 22:47–71.

Brock, P. W. G., and R. MacDonald. 1969. Geological environment of the Bukwa mammalian fossil locality, eastern Uganda. *Nature* 223:593–596.

Cameron, D. W. 2004. *Hominid Adaptations and Extinctions.* University of New South Wales Press, Sydney, 288 pp.

Conroy, G. C. 1994. *Otavipithecus*: Or how to build a better hominid—not. *Journal of Human Evolution* 27:373–383.

———. 1996. The cave breccias of Berg Aukas, Namibia: A clustering approach to mine dump paleontology. *Journal of Human Evolution* 30:349–355.

Conroy, G. C., J. W. Lichtman, and L. B. Martin. 1995. Brief communication: Some observations on enamel thickness and enamel prism packing in the Miocene hominoid *Otavipithecus namibiensis*. *American Journal of Physical Anthropology* 98:595–600.

Conroy, G. C., M. Pickford, B. Senut, and P. Mein. 1993a. Additional Miocene primates from the Otavi Mountains, Namibia. *Comptes Rendus de l'Académie des Sciences, Paris*, Série II, 317:987–990.

———. 1993b. Diamonds in the desert. The discovery of *Otavipithecus namibiensis*. *Evolutionary Anthropology* 2:46–52.

Conroy, G. C., M. Pickford, B. Senut, J. Van Couvering, and P. Mein. 1992. *Otavipithecus namibiensis*, first Miocene hominoid from southern Africa. *Nature* 356:144–148.

Conroy, G. C., and M. D. Rose. 1983. The evolution of the primate foot from the earliest primates to the Miocene hominoids. *Foot and Ankle* 3:342–364.

Conroy, G. C., B. Senut, D. Gommery, M. Pickford, and P. Mein. 1996. Brief communication: New primate remains from the Miocene of Namibia, Southern Africa. *American Journal of Physical Anthropology* 99:487–492.

Corruccini, R. S., and A. M. Henderson. 1978. Palatofacial comparison of *Dryopithecus (Proconsul)* with extant catarrhines. *Primates* 19:35–44.

Cote, S. M. 2004. Origins of the African hominoids: An assessment of the palaeobiogeographic evidence. *Comptes Rendus Palevol* 3:323–340.

Cote, S., and I. Nengo. 2007. Inter- and intra-sexual dimorphism in an early Miocene catarrhine—*Rangwapithecus gordoni*—as demonstrated by new material from western Kenya. *American Journal of Physical Anthropology* 132:S44:91.

Davis, P. R., and J. Napier. 1963. A reconstruction of the skull of *Proconsul africanus* (R.S. 51). *Folia Primatologica* 1:20–28.

Delson, E. 1985. The earliest *Sivapithecus*? *Nature* 318:107–108.

Delson, E., C. J. Terranova, W. L. Jungers, E. J. Sargis, N. G. Jablonski, and P. C. Dechow. 2000. Body mass in Cercopithecidae (Primates, Mammalia): Estimation and scaling in extinct and extant taxa. *Anthropological Papers of the American Museum of Natural History* 83:1–159.

De Silva, J., E. Shoreman, and L. MacLatchy. 2006. A fossil hominoid proximal femur from Kikorongo Crater, southwestern Uganda. *Journal of Human Evolution* 50:687–695.

Drake, R., J. A. Van Couvering, M. Pickford, G. Curtis, and J. A. Harris. 1988. New chronology for the early Miocene mammalian faunas of Kisingiri, western Kenya. *Journal of the Geological Society of London* 145:479–491.

Falk, D. 1983. A reconsideration of the endocast of *Proconsul africanus*: Implications for primate brain evolution; pp. 239–248 in R. L. Ciochon and R. S. Corruccini (eds.), *New Interpretations of Ape and Human Ancestry.* Plenum Press, New York.

Feibel, C. S., and F. H. Brown. 1991. Age of the primate-bearing deposits on Maboko Island, Kenya. *Journal of Human Evolution* 21:221–225.

Ferembach, D. 1958. Les Limnopithèques du Kenya. *Annales de Paléontologie* 44:149–249.

Fleagle, J. G. 1975. A small gibbon-like hominoid from the Miocene of Uganda. *Folia Primatologica* 24:1–15.

———. 1983. Locomotor adaptations of Oligocene and Miocene hominoids and their phyletic implications; pp. 301–324 in R. L. Ciochon and R. S. Corruccini (eds.), *New Interpretations of Ape and Human Ancestry.* Plenum Press, New York.

———. 1986. The fossil record of early catarrhine evolution; pp. 130–149 in B. Wood, L. Martin, and P. Andrews (eds.), *Major Topics in Primate and Human Evolution.* Cambridge University Press, Cambridge.

———. 1999. *Primate Adaptation and Evolution.* 2nd ed. Academic Press, New York, 596 pp.

Fleagle, J. G., and E. L. Simons. 1978. *Micropithecus clarki*, a small ape from the Miocene of Uganda. *American Journal of Physical Anthropology* 49:427–440

Gebo, D. L., K. C. Beard, M. F. Teaford, A. Walker, S. G. Larson, W. L. Jungers, and J. G. Fleagle. 1988. A hominoid proximal humerus from the early Miocene of Rusinga Island, Kenya. *Journal of Human Evolution* 17:393–401.

Gebo, D. L., L. MacLatchy, R. Kityo, A. Deino, J. Kingston, and D. Pilbeam. 1997. A hominoid genus from the early Miocene of Uganda. *Science* 276:401–404.

Geraads, D., Z. Alemseged, and H. Bellon. 2002. The late Miocene mammalian fauna of Chorora, Awash basin, Ethiopia: systematics, biochronology and the ^{40}K-^{40}Ar ages of the associated volcanics. *Tertiary Research* 21:113–122.

Gitau, S. N., and B. R. Benefit. 1995. New evidence concerning the facial morphology of *Simiolus leakeyorum* from Maboko Island. *American Journal of Physical Anthropology*, suppl. 20:99.

Gitau, S. N., B. R. Benefit, M. L. McCrossin, and T. Roedl. 1998. Fossil primates and associated fauna from 1997 excavations at the middle Miocene site of Maboko Island, Kenya. *American Journal of Physical Anthropology*, suppl. 26:87.

Gommery, D., B. Senut, and M. Pickford. 1998. Nouveaux restes postcrâniens d'Hominoidea du Miocène inférieur de Napak, Ouganda. *Annales de Paléontologie* 84:287–306.

Gommery, D., B. Senut, M. Pickford, and E. Musiime. 2002. Les nouveaux restes de squelette d'*Ugandapithecus major* (Miocène inférieur de Napak, Ouganda). *Annales de Paléontologie* 88:167–186.

Harrison, T. 1981. New finds of small fossil apes from the Miocene locality of Koru in Kenya. *Journal of Human Evolution* 10:129–137.

———. 1982. Small-bodied apes from the Miocene of East Africa. Unpublished PhD Dissertation, University of London, 647 pp.

———. 1986. New fossil anthropoids from the middle Miocene of East Africa and their bearing on the origin of the Oreopithecidae. *American Journal of Physical Anthropology* 71:265–284.

———. 1987. The phylogenetic relationships of the early catarrhine primates: A review of the current evidence. *Journal of Human Evolution* 16:41–80.

———. 1988. A taxonomic revision of the small catarrhine primates from the early Miocene of East Africa. *Folia Primatologica* 50:59–108.

———. 1989. A new species of *Micropithecus* from the middle Miocene of Kenya. *Journal of Human Evolution* 18:537–557.

———. 1992. A reassessment of the taxonomic and phylogenetic affinities of the fossil catarrhines from Fort Ternan, Kenya. *Primates* 33:501–522.

———. 1993. Cladistic concepts and the species problem in hominoid evolution; pp. 345–371 in W. H. Kimbel and L. B. Martin (eds.), *Species, Species Concepts, and Primate Evolution*. Plenum Press, New York.

———. 1998. Evidence for a tail in *Proconsul heseloni*. *American Journal of Physical Anthropology*, suppl. 26:93–94.

———. 2002. Late Oligocene to middle Miocene catarrhines from Afro-Arabia; pp. 311–338 in W. C. Hartwig (ed.), *The Primate Fossil Record*. Cambridge University Press, Cambridge.

———. 2005. The zoogeographic and phylogenetic relationships of early catarrhine primates in Asia. *Anthropological Science* 113:43–51.

Harrison, T., and P. Andrews. 2009. The anatomy and systematic position of a new species of early Miocene proconsulid from Meswa Bridge, Kenya. *Journal of Human Evolution* 56:479–496.

Harrison, T., and Y. Gu. 1999. Taxonomy and phylogenetic relationships of early Miocene catarrhines from Sihong, China. *Journal of Human Evolution* 37:225–277.

Harrison, T., and L. Rook. 1997. Enigmatic anthropoid or misunderstood ape? The phylogenetic status of *Oreopithecus bambolii* reconsidered; pp. 327–362 in D. R. Begun, C. V. Ward, and M. D. Rose (eds.), *Function, Phylogeny, and Fossils: Miocene Hominoid Evolution and Adaptation*. Plenum Press, New York.

Harrison, T., and W. J. Sanders. 1999. Scaling of lumbar vertebrae in anthropoid primates: Its implications for the positional behavior and phylogenetic affinities of *Proconsul*. *American Journal of Physical Anthropology*, suppl. 28:146.

Hill, A. 1999. The Baringo Basin, Kenya: from Bill Bishop to BPRP; pp. 85–97 in P. Andrews and P. Banham (eds.), *Late Cenozoic Environments and Hominid Evolution: A Tribute to Bill Bishop*. Geological Society, London.

Hill, A., K. Behrensmeyer, B. Brown, A. Deino, M. Rose, J. Saunders, S. Ward, and A. Winkler. 1991. Kipsaramon: A lower Miocene hominoid site in the Tugen Hills, Baringo District, Kenya. *Journal of Human Evolution* 20:67–75.

Hill, A., M. Leakey, J. D. Kingston, and S. Ward. 2002. New cercopithecoids and a hominoid from 12.5 Ma in the Tugen Hills succession, Kenya. *Journal of Human Evolution* 42:75–93.

Hill, A., and I. Odhiambo. 1987. New mandible of *Rangwapithecus* from Songhor, Kenya. *American Journal of Physical Anthropology* 72:210.

Hill, A., and S. Ward. 1988. Origin of the Hominidae: The record of African large hominoid evolution between 14 My and 4 My. *Yearbook of Physical Anthropology* 31:49–83.

Hopwood, A. T. 1933a. Miocene primates from British East Africa. *Annals and Magazine of Natural History*, series 10, 11:96–98.

———. 1933b. Miocene primates from Kenya. *Linnean Society's Journal, Zoology* 38:437–464.

International Commission on Zoological Nomenclature. 1999. *International Code of Zoological Nomenclature*. 4th ed. International Trust for Zoological Nomenclature, London.

Ishida, H., Y. Kunimatsu, M. Nakatsukasa, and Y. Nakano. 1999. New hominoid genus from the middle Miocene of Nachola, Kenya. *Anthropological Science* 107:189–191.

Ishida, H., Y. Kunimatsu, T. Takano, Y. Nakano, and M. Nakatsukasa. 2004. *Nacholapithecus* skeleton from the middle Miocene of Kenya. *Journal of Human Evolution* 46:69–103.

Ishida, H., and M. Pickford. 1997. A new late Miocene hominoid from Kenya: *Samburupithecus kiptalami* gen. et sp. nov. *Comptes Rendus de l'Académie des Sciences Paris, Sciences de la Terre et des Planètes* 325:823–829.

Ishida, H., M. Pickford, H. Nakaya, and Y. Nakano. 1984. Fossil anthropoids from Nachola and Samburu Hills, Samburu District, Kenya. *African Study Monographs*, suppl. 2:73–85.

Itaya, T., and Y. Sawada. 1987. K-Ar ages of volcanic rocks in the Samburu Hills area, northern Kenya. *African Study Monographs*, suppl. 5:27–45.

Jablonski, N. G. 2002. Fossil Old World monkeys: The late Neogene radiation; pp. 255–299 in W. C. Hartwig (ed.), *The Primate Fossil Record*. Cambridge University Press, Cambridge.

Johnson, K. B., M. L. McCrossin, and B. R. Benefit. 2000. Circular shapes do not an ape make: Comments on interpretation of the inferred *Morotopithecus* scapula. *American Journal of Physical Anthropology*, suppl. 30:189–190.

Kelley, J. 1997. Paleobiological and phylogenetic significance of life history in Miocene hominoids; pp. 173–208 in D. R. Begun, C. V. Ward, and M. D. Rose (eds.), *Function, Phylogeny, and Fossils: Miocene Hominoid Evolution and Adaptation*. Plenum Press, New York.

Kelley, J., S. Ward, B. Brown, A. Hill, and D. L. Duren. 2002. Dental remains of *Equatorius africanus* from Kipsaramon, Tugen Hills, Baringo District, Kenya. *Journal of Human Evolution* 42:39–62.

Kingdon, J. 1997. *The Kingdon Field Guide to African Mammals*. Academic Press, San Diego, California, 464 pp.

Kunimatsu, Y. 1992a. New finds of small anthropoid primate from Nachola, northern Kenya. *African Study Monographs* 14:237–249.

———. 1992b. A revision of the hypodigm of *Nyanzapithecus vancouveringi*. *African Study Monographs* 14:231–235.

———. 1997. New species of *Nyanzapithecus* from Nachola, northern Kenya. *Anthropological Science* 105:117–141.

Kunimatsu, Y., H. Ishida, M. Nakatsukasa, Y. Nakano, Y. Sawada, and K. Nakayama. 2004. Maxillae and associated gnathodental specimens of *Nacholapithecus kerioi*, a large-bodied hominoid from Nachola, northern Kenya. *Journal of Human Evolution* 46:365–400.

Kunimatsu, Y., M. Nakatsukasa, Y. Sawada, T. Sakai, M. Hyodo, H. Hyodo, T. Itaya, H. Nakaya, H. Saegusa, A. Mazurier, M. Saneyoshi, H. Tsujikawa, A. Yamamoto, and E. Mbua. 2007. A new Late Miocene great ape from Kenya and its implications for the origins of African great apes and humans. *Proceedings of the National Academy of Sciences, USA* 104:19220–19225.

Langdon, J. 1986. Functional morphology of the Miocene hominoid foot. *Contributions to Primatology* 22:1–225.

Le Gros Clark, W. E. 1950. New palaeontological evidence bearing on the evolution of the Hominoidea. *Quarterly Journal of the Geological Society of London* 105:225–264.

———. 1952. Report on fossil hominoid material collected by the British-Kenya Miocene Expedition, 1949–1951. *Proceedings of the Zoological Society of London* 122:273–286.

Le Gros Clark, W. E., and L. S. B. Leakey. 1951. *The Miocene Hominoidea of East Africa*. Fossil Mammals of Africa, No. 1. British Museum (Natural History), London, 117 pp.

Le Gros Clark, W. E., and D. P. Thomas 1951. *Associated Jaws and Limb Bones* of Limnopithecus macinnesi. Fossil Mammals of Africa No. 3. British Museum (Natural History), London, 27 pp.

Leakey, L. S. B. 1961 (1962). A new lower Pliocene fossil primate from Kenya. *Annals and Magazine of Natural History*, series 13, 4:689–696.

———. 1967. Notes on the mammalian faunas from the Miocene and Pleistocene of East Africa; pp. 7–29 in W. W. Bishop and J. D. Clark (eds.), *Background to Evolution in Africa*. University of Chicago Press, Chicago.

———. 1968. Upper Miocene primates from Kenya. *Nature* 218:527–528.

Leakey, M. G., P. S. Ungar, and A. Walker. 1995. A new genus of large primate from the late Oligocene of Lothidok, Turkana District, Kenya. *Journal of Human Evolution* 28:519–531.

Leakey, M. G., and A. Walker. 1997. *Afropithecus*: Function and phylogeny; pp. 225–239 in D. R. Begun, C. V. Ward, and M. D. Rose (eds.), *Function, Phylogeny, and Fossils: Miocene Hominoid Evolution and Adaptations*. Plenum Press, New York.

Leakey, R. E., and M. G. Leakey. 1986a. A new Miocene hominoid from Kenya. *Nature* 324:143–146.

———. 1986b. A second new Miocene hominoid from Kenya. *Nature* 324:146–148.

———. 1987. A new Miocene small-bodied ape from Kenya. *Journal of Human Evolution* 16:369–387.

Leakey, R. E., M. G. Leakey, and A. C. Walker. 1988a. Morphology of *Afropithecus turkanensis* from Kenya. *American Journal of Physical Anthropology* 76:289–307.

———. 1988b. Morphology of *Turkanapithecus kalakolensis* from Kenya. *American Journal of Physical Anthropology* 76:277–288.

Leakey, R. E., and A. C. Walker. 1985. New higher primates from the early Miocene of Buluk, Kenya. *Nature* 318:173–175.

Lehmann, U. 1957. Eine jungpleistozäne Wirbeltierfauna aus Ostafrika (Material der Expedition Kohl-Larsen 1938 von der Mumba-Höhle). *Mitteilungen aus dem Geologischen Staatsinstitut in Hamburg* 26:100–140.

Li, Y., R. H. Crompton, M. Günter, W. Wang, and R. Savage. 2002. Reconstructing the mechanics of quadrupedalism in an extinct hominoid. *Zeitschrift für Morphologie und Anthropologie* 83:265–274.

MacInnes, D. G. 1943. Notes on the East African Miocene primates. *Journal of the East Africa and Uganda Natural History Society* 17:141–181.

MacLatchy, L., and W. H. Bossert. 1996. An analysis of the articular surface distribution of the femoral head and acetabulum in anthropoids, with implications for hip function in Miocene hominoids. *Journal of Human Evolution* 31:425–453.

MacLatchy, L., D. Gebo, R. Kityo, and D. Pilbeam. 2000. Postcranial functional morphology of *Morotopithecus bishopi*, with implications for the evolution of modern ape locomotion. *Journal of Human Evolution* 39:59–183.

MacLatchy, L., and D. Pilbeam. 1999. Renewed research in the Ugandan early Miocene; pp. 15–25 in P. Andrews and P. Banham (eds.), *Late Cenozoic Environments and Hominid Evolution: A Tribute to Bill Bishop.* Geological Society, London.

MacLatchy, L., and J. B. Rossie. 2005. The Napak hominoids: Still *Proconsul major*; pp. 15–28 in D. E. Lieberman, R. J. Smith, and J. Kelley (eds.), *Interpreting the Past: Essays on Human, Primate, and Mammal Evolution in Honor of David Pilbeam.* Brill Academic Publishers, Boston.

Madden, C. T. 1980a. East African *Sivapithecus* should not be identified as *Proconsul nyanzae*. *Primates* 21:133–135.

———. 1980b. New *Proconsul (Xenopithecus)* from the Miocene of Kenya. *Primates* 21:241–252.

Makinouchi, T., T. Koyaguchi, T. Matsuda, H. Mitsushio, and S. Ishida. 1984. Geology of the Nachola area and the Samburu Hills, west of Baragoi, northern Kenya. *African Study Monographs*, suppl. 2:15–44.

Manser, J., and T. Harrison. 1999. Estimates of cranial capacity and encephalization in *Proconsul* and *Turkanapithecus*. *American Journal of Physical Anthropology* 28:189.

Martin, L. 1981. New Specimens of *Proconsul* from Koru, Kenya. *Journal of Human Evolution* 10:139–150.

McBrearty, S., and N. Jablonski. 2006. First fossil chimpanzee. *Nature* 437:105–108.

McCrossin, M. L. 1992. An oreopithecid proximal humerus from the middle Miocene of Maboko Island, Kenya. *International Journal of Primatology* 13:659–677.

McCrossin, M. L., and K. L. Allen. 2007. Articular kinematics of the knee of *Kenyapithecus*. *American Journal of Physical Anthropology* 132, S44:167–168.

McCrossin, M. L., and B. R. Benefit. 1993. Recently recovered *Kenyapithecus* mandible and its implications for great ape and human origins. *Proceedings of the National Academy of Sciences, USA* 90:1962–1966.

———. 1994. Maboko Island and the evolutionary history of Old World monkeys and apes; pp. 95–121 in R. S. Corruccini and R. L. Ciochon (eds.), *Integrative Paths to the Past: Paleoanthropological Advances in Honor of F. Clark Howell.* Prentice Hall, Englewood Cliffs, N.J.

———. 1997. On the relationships and adaptations of *Kenyapithecus*, a large-bodied hominoid from the middle Miocene of eastern Africa; pp. 241–267 in D. R. Begun, C. V. Ward, and M. D. Rose (eds.), *Function, Phylogeny, and Fossils: Miocene Hominoid Evolution and Adaptations.* Plenum Press, New York.

McCrossin, M. L., B. R. Benefit, S. N. Gitau, A. K. Palmer, and K. T. Blue. 1998. Fossil evidence for the origins of terrestriality among Old World higher primates; pp. 353–396 in E. Strasser, J. Fleagle, A. Rosenberger, and M. McHenry (eds.), *Primate Locomotion.* Plenum Press, New York.

McDougall, I., and R. Watkins. 1985. Age of hominoid-bearing sequence at Buluk, northern Kenya. *Nature* 318:175–178.

McHenry, H. M., P. Andrews, and R. S. Corruccini. 1980. Miocene hominoid palatofacial morphology. *Folia Primatologica* 33:241–252.

Nakatsukasa, M. 2008. Comparative study of Moroto vertebral specimens. *Journal of Human Evolution* 55:581–588.

Nakatsukasa, M., Y. Kunimatsu, Y. Nakano, and H. Ishida. 2002. Morphology of the hallucial phalanges in extant anthropoids and fossil hominoids. *Zeitschrift für Morphologie und Anthropologie* 83:361–372.

Nakatsukasa, M., Y. Kunimatsu, Y. Nakano, T. Takano, and H. Ishida. 2003a. Comparative and functional anatomy of phalanges in *Nacholapithecus kerioi*, a middle Miocene hominoid from northern Kenya. *Primates* 44:371–412.

———. 2007. Vertebral morphology of *Nacholapithecus kerioi* based on KNM-BG 35250. *Journal of Human Evolution* 52:347–369.

Nakatsukasa, M., Y. Kunimatsu, Y. Sawada, T. Sakai, H. Hyodo, T. Itaya, M. Saneyoshi, H. Tsujikawa, and E. Mbua. 2006. Late Miocene primate fauna in Nakali, central Kenya. *American Journal of Physical Anthropology* 132, 42:136.

Nakatsukasa, M., D. Shimizu, Y. Nakano, and H. Ishida. 1996. Three-dimensional morphology of the sigmoid notch of the ulna in *Kenyapithecus* and *Proconsul*. *African Study Monographs*, suppl. 24:57–71.

Nakatsukasa, M., H. Tsujikawa, D. Shimizu, T. Takano, Y. Kunimatsu, Y. Nakano, and H. Ishida. 2003b. Definitive evidence of tail loss in *Nacholapithecus*, an East African Miocene hominoid. *Journal of Human Evolution* 45:179–186.

Nakatsukasa, M., A. Yamanaka, Y. Kunimatsu, D. Shimizu, and H. Ishida, H. 1998. A newly discovered *Kenyapithecus* skeleton and its implications for the evolution of positional behavior in Miocene East African hominoids. *Journal of Human Evolution* 34:657–664.

Nakaya, H., M. Pickford, Y. Nakano, and H. Ishida. 1984. The late Miocene large mammal fauna from the Namurungule Formation, Samburu Hills, northern Kenya. *African Study Monographs*, suppl. 2:15–44.

Napier, J. R., and P. R. Davis. 1959. *The Fore-limb Skeleton and Associated Remains of* Proconsul africanus. Fossil Mammals of Africa, No. 16. British Museum (Natural History), London, 69 pp.

Nengo, I. O., and T. C. Rae. 1992. New hominoid fossils from the early Miocene site of Songhor, Kenya. *Journal of Human Evolution* 23:423–429.

Patel, B. A., and A. Grossman. 2006. Dental metric comparisons of *Morotopithecus* and *Afropithecus*: Implications for the validity of the genus *Morotopithecus*. *Journal of Human Evolution* 51:506–512.

Pickford, M. 1981. Preliminary Miocene mammalian biostratigraphy for western Kenya. *Journal of Human Evolution* 10:73–97.

———. 1982. New higher primate fossils from the Middle Miocene deposits at Majiwa and Kaloma, western Kenya. *American Journal of Physical Anthropology* 58:1–19.

———. 1983. Sequence and environment of the lower and middle Miocene hominoids of western Kenya; pp. 421–440 in R. L. Ciochon and R. S. Corruccini (eds.). *New Interpretations of Ape and Human Ancestry.* Plenum Press, New York.

———. 1985. A new look at *Kenyapithecus* based on recent discoveries in western Kenya. *Journal of Human Evolution* 14:113–143.

———. 1986a. Cainozoic palaeontological sites in western Kenya. *Münchner Geowissenschaftliche Abhandlungen, Reihe A, Geologie und Paläontologie* 8:1–151.

———. 1986b. The geochronology of Miocene higher primate faunas of East Africa; pp. 19–33 in J. G. Else and P. C. Lee (eds.), *Primate Evolution.* Cambridge University Press, Cambridge.

———. 1986c. Hominoids from the middle Miocene of East Africa and the phyletic position of *Kenyapithecus*. *Zeitschrift für Morphologie und Anthropologie* 76:117–130.

———. 1998. Geology and fauna of the middle Miocene hominoid site at Muruyur, Baringo District, Kenya. *Human Evolution* 3:381–390.

———. 2002. New reconstruction of the Moroto hominoid snout and a reassessment of its affinities to *Afropithecus turkanensis*. *Human Evolution* 17:1–19.

Pickford, M., and P. J. Andrews. 1981. The Tinderet Miocene sequence in Kenya. *Journal of Human Evolution* 10:11–33.

Pickford, M., Y. Coppens, B. Senut, J. Morales, and J. Braga. 2009. Late Miocene hominoid from Niger. *Comptes Rendus Palevol* 8:413–425.

Pickford, M., and H. Ishida. 1998. Interpretation of *Samburupithecus*, an Upper Miocene hominoid from Kenya. *Comptes Rendus de l'Académie des Sciences Paris, Sciences de la Terre et des Planètes* 326:299–306.

Pickford, M., H. Ishida, Y. Nakano, and H. Nakaya. 1984a. Fossiliferous localities of the Nachola-Samburu Hills area, northern Kenya. *African Study Monographs*, suppl. 2:45–56.

Pickford, M., and Y. Kunimatsu. 2005. Catarrhines from the Middle Miocene (ca. 14.5 Ma) of Kipsaraman, Tugen Hills, Kenya. *Anthropological Science* 113:189–224.

Pickford, M., S. Moya Sola, and M. Köhler. 1997. Phylogenetic implications of the first African middle Miocene hominoid frontal bone from Otavi, Namibia. *Comptes Rendus de l'Académie des Sciences Paris, Sciences de la Terre et des Planètes* 325:459–466.

Pickford, M., H. Nakaya, H. Ishida, and Y. Nakano. 1984b. The biostratigraphic analyses of the faunas of the Nachola area and Samburu Hills, northern Kenya. *African Study Monographs*, suppl. 2:67–72.

Pickford, M., Y. Sawada, R. Tayama, Y. Matsuda, T. Itaya, H. Hyodo, and B. Senut. 2006. Refinement of the age of the Middle Miocene Fort Ternan Beds, Western Kenya, and its implications for Old World biochronology. *Comptes Rendus Geoscience* 338:545–555.

Pickford, M., and B. Senut. 1997. Cainozoic mammals from coastal Namaqualand, South Africa. *Palaeontologia Africana* 34:199–217.

——. 2005a. Hominoid teeth with chimpanzee- and gorilla-like features from the Miocene of Kenya: Implications for the chronology of ape-human divergence and biogeography of Miocene hominoids. *Anthropological Science* 113:95–102.

——. 2005b. Implications of the presence of African ape-like teeth in the Miocene of Kenya; pp. 121–133 in F. d'Errico and L. Backwell (eds.), *From Tools to Symbols: From Early Hominids to Modern Humans.* Witwatersrand University Press, Johannesburg.

Pickford, M., B. Senut, G. C. Conroy, and P. Mein. 1994. Phylogenetic position of *Otavipithecus*: questions of methodology and approach; pp. 265–272 in B. Thierry, J. R. Anderson, J. J. Roeder, and N. Herrenschmidt (eds.), *Current Primatology: Volume 1. Ecology and Evolution.* Université Louis Pasteur, Strasbourg.

Pickford, M., B. Senut, and D. Gommery. 1999. Sexual dimorphism in *Morotopithecus bishopi*, an early Middle Miocene hominoid from Uganda, and a reassessment of its geological and biological contexts; pp. 27–38 in P. Andrews and P. Banham (eds.), *Late Cenozoic Environments and Hominid Evolution: A Tribute to Bill Bishop.* Geological Society, London.

Pickford, M., B. Senut, D. Gommery, and E. Musiime. 2003. New catarrhine fossils from Moroto II, early middle Miocene (ca 17.5 Ma) Uganda. *Comptes Rendus Palevol* 2:649–662.

Pickford, M., B. Senut, D. Hadoto, J. Musisi, and C. Kariira. 1986a. Découvertes récentes dans les sites Miocènes de Moroto (Ouganda oriental): Aspects biostratigraphiques et paléoécologiques. *Comptes Rendus de l'Academie des Sciences, Paris* 302:681–686.

——. 1986b. Nouvelles découvertes dans le Miocène inférieur de Napak, Ouganda Oriental. *Comptes Rendus de l'Academie des Sciences, Paris* 302:47–52.

Pickford, M., B. Senut, J. Morales, and J. Braga. 2008. First hominoid from the Late Miocene of Niger. *South African Journal of Science* 104:337–339.

Pilbeam, D. R. 1969. Tertiary Pongidae of East Africa: Evolutionary relationships and taxonomy. *Bulletin of the Peabody Museum of Natural History* 31:1–185.

Pilbeam, D. R., and A. Walker. 1968. Fossil monkeys from the Miocene of Napak, north-east Uganda. *Nature* 220:657–660.

Pilbrow, V. 2006. Population systematics of chimpanzees using molar morphometrics. *Journal of Human Evolution* 51:646–662.

Preuschoft, H. 1973. Body posture and locomotion in some East African Miocene Dryopithecinae; pp. 13–46 in M. Day (ed.), *Human Evolution, Symposium of the Society for the Study of Human Biology*, vol. 11. Taylor and Francis, London.

Radinsky, L. B. 1975. Primate brain evolution. American Scientist 63:656–663.

Rae, T. C. 1997. The early evolution of the hominoid face; pp. 59–77 in D. R. Begun, C. V. Ward, and M. D. Rose (eds.), *Function, Phylogeny, and Fossils: Miocene Hominoid Evolution and Adaptation.* Plenum Press, New York.

——. 1999. Mosaic evolution in the origin of the Hominoidea. *Folia Primatologica* 70:125–135.

Rafferty, K. L. 1990. The functional and phylogenetic significance of the carpometacarpal joint of the thumb in anthropoid primates. Unpublished master's thesis, New York University, New York.

Rafferty, K. L., A. Walker, C. Ruff, M. D. Rose, and P. J. Andrews. 1995. Postcranial estimates of body weights in *Proconsul*, with a note on a distal tibia of *P. major* from Napak, Uganda. *American Journal of Physical Anthropology* 97:391–402.

Rasmussen, D. T. 2002. Early catarrhines of the African Eocene and Oligocene; pp. 311–338 in W. C. Hartwig (ed.), *The Primate Fossil Record.* Cambridge University Press, Cambridge.

Richmond, B. G., J. G. Fleagle. J. Kappelman, and C. C. Swisher. 1998. First hominoid from the Miocene of Ethiopia and the evolution of the catarrhine elbow. *American Journal of Physical Anthropology* 105:257–277.

Rose, M. D. 1983. Miocene hominoid postcranial morphology: Monkey-like, ape-like, neither, or both? pp. 405–420 in R. L. Ciochon and R. S. Corruccini (eds.), *New Interpretations of Ape and Human Ancestry.* Plenum Press, New York.

——. 1988. Another look at the anthropoid elbow. *Journal of Human Evolution* 17:193–224.

——. 1992. Kinematics of the trapezium-1st metacarpal joint in extant anthropoids and Miocene hominoids. *Journal of Human Evolution* 22:255–266.

——. 1993. Locomotor anatomy of Miocene hominoids; pp. 252–272 in D. L. Gebo (ed.), *Postcranial Adaptation in Nonhuman Primates.* Northern Illinois University Press, DeKalb.

——. 1994. Quadrupedalism in some Miocene catarrhines. *Journal of Human Evolution* 26:387–411.

——. 1997. Functional and phylogenetic features of the forelimb in Miocene hominoids; pp. 79–100 in D. R. Begun, C. V. Ward, and M. D. Rose (eds.), *Function, Phylogeny, and Fossils: Miocene Hominoid Evolution and Adaptation.* Plenum Press, New York.

Rose, M. D., M. G. Leakey, R. E. F. Leakey, and A. C. Walker. 1992. Postcranial specimens of *Simiolus enjiessi* and other primitive catarrhines from the early Miocene of Lake Turkana, Kenya. *Journal of Human Evolution* 22:171–237.

Rose, M. D., Y. Nakano, and H. Ishida. 1996. *Kenyapithecus* postcranial specimens from Nachola, Kenya. *African Study Monographs*, suppl. 24:3–56.

Rossie, J. B. and L. MacLatchy. 2006. A new pliopithecoid genus from the early Miocene of Uganda. *Journal of Human Evolution* 50:568–586.

Rossie, J. B., E. L. Simons, S. C. Gauld, and D. T. Rasmussen. 2002. Paranasal sinus anatomy of *Aegyptopithecus*: Implications for hominoid origins. *Proceedings of the National Academy of Sciences, USA* 99:8454–8456.

Ruff, C. B., A. Walker, and M. F. Teaford. 1989. Body mass, sexual dimorphism and femoral proportions of *Proconsul* from Rusinga and Mfangano Islands, Kenya. *Journal of Human Evolution* 18:515–536.

Sanders, W. J., and Bodenbender, B. E. 1994. Morphometric analysis of lumbar vertebra UMP 67–28, Implications for spinal function and phylogeny of the Miocene Moroto hominoid. *Journal of Human Evolution* 26:203–237.

Sawada, Y., M. Pickford, T. Itaya, T. Makinouchi, M. Tateishi, K. Kabeto, S. Ishida, and H. Ishida. 1998. *Comptes Rendus de l'Académie des Sciences, Paris, Sciences de la Terre et des Planètes* 326:445–451.

Sawada, Y., M. Tateishi, and S. Ishida. 1987. Geology of the Neogene system in and around the Samburu Hills, northern Kenya. *African Study Monographs*, suppl. 5:7–26.

Senut, B. 1989. *Le Coude Chez les Primates Hominoïdes: Anatomie, Fonction, Taxonomie et Évolution.* Cahiers de Paléoanthropologie. CNRS, Paris, France, 231 pp.

Senut, B., and D. Gommery. 1997. Squelette postcrânien d'*Otavipithecus*, Hominoidea du Miocène moyen de Namibie. *Annales de Paléontologie* 83:267–284.

Senut, B., M. Nakatsukasa, Y. Kunimatsu, Y. Nakano, T. Takano, H. Tsujikawa, D. Shimizu, M. Kagaya, and H. Ishida. 2004. Preliminary analysis of *Nacholapithecus* scapula and clavicle from Nachola, Kenya. *Primates* 45:97–104.

Senut, B., M. Pickford, D. Gommery, and Y. Kunimatsu. 2000. Un nouveau genre d'hominoïde du Miocène inférieur d'Afrique orientale: *Ugandapithecus major* (Le Gros Clark & Leakey, 1950). *Comptes Rendus de l'Académie des Sciences, Paris, Sciences de la Terre et des Planètes* 331:227–233.

Senut, B., M. Pickford, P. Mein, G. Conroy, and J. Van Couvering. 1992. Discovery of 12 new Late Cainozoic fossiliferous sites in palaeokarst of the Otavi Mountains, Namibia. *Comptes Rendus de l'Academie des Sciences de Paris*, Série II, 314:727–733.

Senut, B., M. Pickford, and D. Wessels. 1997. Panafrican distribution of Lower Miocene Hominoidea. *Comptes Rendus de l'Académie des Sciences, Paris, Sciences de la Terre et des Planètes* 325:741–746.

Sherwood, R. J., S. Ward, A. Hill, D. L. Duren, B. Brown, and W. Downs. 2002. Preliminary description of the *Equatorius africanus* partial skeleton (KNM-TH 28860) from Kipsaramon, Tugen Hills, Baringo District, Kenya. *Journal of Human Evolution* 42:63–73.

Simons, E. L. 1972. *Primate Evolution: An Introduction to Man's Place in Nature.* MacMillan. New York, 322 pp.

Simons, E. L., and D. R. Pilbeam. 1965. Preliminary revision of the Dryopithecinae (Pongidae, Anthropoidea). *Folia Primatologica* 3:81–152.

Singleton, M. 2000. The phylogenetic affinities of *Otavipithecus namibiensis. Journal of Human Evolution* 38:537–573.

Smith, R. J., and W. L. Jungers. 1997. Body mass in comparative primatology. *Journal of Human Evolution* 32:523–559.

Smith, T. M., L. B. Martin, and M. G. Leakey. 2003. Enamel thickness, microstructure and development in *Afropithecus turkanensis. Journal of Human Evolution* 44:283–306.

Stewart, C.-B., and T. R. Disotell. 1998. Primate evolution—in and out of Africa. *Current Biology* 8:R582–R587.

Strasser, E., 1993. Kaswanga *Proconsul* foot proportions. *American Journal of Physical Anthropology,* suppl. 16:191

Suwa, G., R. T. Kono, S. Katoh, B. Asfaw, and Y. Beyene. 2007. A new species of great ape from the late Miocene epoch in Ethiopia. *Nature* 448:921–924.

Szalay, F. S., and E. Delson 1979. *Evolutionary History of the Primates.* Academic Press, New York, 580 pp.

Teaford, M. F., K. C. Beard, R. E. Leakey, and A. Walker. 1988. New hominoid facial skeleton from the early Miocene of Rusinga Island, Kenya, and its bearing on the relationship between *Proconsul nyanzae* and *Proconsul africanus. Journal of Human Evolution* 17:461–477.

Tsujikawa, H. 2005. The updated late Miocene large mammal fauna from Samburu Hills, northern Kenya. *African Study Monographs,* suppl. 32:1–50.

Von Koenigswald, G. H. R. 1969. Miocene Cercopithecoidea and Oreopithecoidea from the Miocene of East Africa; pp. 39–52 in L. S. B. Leakey (ed.), *Fossil Vertebrates of Africa,* vol. 1. Academic Press, London.

Walker, A. 1968. The Lower Miocene fossil site of Bukwa, Sebei. *Uganda Journal* 32:149–156.

———. 1969. Fossil mammal locality on Mount Elgon, Eastern Uganda. *Nature* 223:591–593.

———. 1997. *Proconsul* function and phylogeny; pp. 209–224 in D. R. Begun, C. V. Ward, and M. D. Rose (eds.), *Function, Phylogeny, and Fossils: Miocene Hominoid Evolution and Adaptation.* Plenum Press, New York.

Walker, A., D. Falk, R. Smith, and M. Pickford. 1983. The skull of *Proconsul africanus:* reconstruction and cranial capacity. *Nature* 305:525–527.

Walker, A., and M. Pickford. 1983. New postcranial fossils of *Proconsul africanus* and *Proconsul nyanzae;* pp. 325–351 in R. L. Ciochon and R. S. Corruccini (eds.), *New Interpretations of Ape and Human Ancestry.* Plenum Press, New York.

Walker, A., and M. Rose. 1968. Fossil hominoid vertebra from the Miocene of Uganda. *Nature* 217:980–981.

Walker, A., and M. Teaford. 1989. The hunt for *Proconsul. Scientific American* 260:76–82.

Walker, A., M. F. Teaford, and R. E. Leakey. 1985. New information regarding the R114 *Proconsul* site, Rusinga Island, Kenya; pp. 143–149 in J. Else and P. Lee (eds.) *Primate Evolution.* Cambridge University Press, Cambridge.

Walker, A., M. F. Teaford, L. Martin, and P. Andrews. 1993. A new species of *Proconsul* from the early Miocene of Rusinga/Mfangano Islands, Kenya. *Journal of Human Evolution* 25:43–56.

Ward, C. V. 1993. Torso morphology and locomotion in *Proconsul nyanzae. American Journal of Physical Anthropology* 92:291–328.

———. 1997. Functional anatomy and phyletic implications of the hominoid trunk and hindlimb; pp. 101–130 in D. R. Begun, C. V. Ward, and M. D. Rose (eds.), *Function, Phylogeny, and Fossils: Miocene Hominoid Evolution and Adaptation.* Plenum Press, New York.

———. 1998. *Afropithecus, Proconsul,* and the primitive hominoid skeleton; pp. 337–352 in E. Strasser, J. Fleagle, A. Rosenberger, and H. McHenry (eds.), *Primate Locomotion: Recent Advances.* Plenum Press, New York.

Ward, C. V., D. R. Begun, and M. D. Rose. 1997. Function and phylogeny in Miocene hominoids; pp. 1–12 in D. R. Begun, C. V. Ward, and M. D. Rose (eds.), *Function, Phylogeny, and Fossils: Miocene Hominoid Evolution and Adaptation.* Plenum Press, New York.

Ward, C. V., C. B. Ruff, A. Walker, M. D. Rose, M. F. Teaford, and I. O. Nengo. 1995. Functional morphology of *Proconsul* patellas from Rusinga Island, Kenya, with implications for other Miocene-Pliocene catarrhines. *Journal of Human Evolution* 29:1–19.

Ward, C. V., A. Walker, and M. F. Teaford. 1991. *Proconsul* did not have a tail. *Journal of Human Evolution* 21:215–220.

———. 1999. Still no evidence for a tail in *Proconsul heseloni. American Journal of Physical Anthropology,* suppl. 28:273.

Ward, C. V., A. Walker, M. F. Teaford, and I. Odhiambo. 1993. Partial skeleton of *Proconsul nyanzae* from Mfangano Island, Kenya. *American Journal of Physical Anthropology* 90:77–111.

Ward, S., and D. L. Duren. 2002. Middle and late Miocene African hominoids; pp. 385–397 in W. C. Hartwig (ed.), *The Primate Fossil Record.* Cambridge University Press, Cambridge.

Ward, S., B. Brown, A. Hill, J. Kelley, and W. Downs. 1999. *Equatorius:* A new hominoid genus from the middle Miocene of Kenya. *Science* 285:1382–1386.

Watkins, R. T. 1989. The Buluk Member, a fossil hominoid-bearing sedimentary sequence of Miocene age from Northern Kenya. *Journal of African Earth Sciences* 8:107–112.

Whybrow, P. J., and P. J. Andrews. 1978. Restoration of the holotype of *Proconsul nyanzae. Folia Primatologica* 30:115–125.

Wynn, J. G., and G. J. Retallack. 2001. Paleoenvironmental reconstruction of middle Miocene paleosols bearing *Kenyapithecus* and *Victoriapithecus,* Nyakach Formation, southwestern Kenya. *Journal of Human Evolution* 40:263–288.

Young, N. M., and MacLatchy, L. 2004. The phylogenetic position of *Morotopithecus. Journal of Human Evolution* 46:163–184.

TWENTY-FIVE

Hominini

LAURA M. MACLATCHY, JEREMY DESILVA, WILLIAM J.
SANDERS, AND BERNARD WOOD

The late Miocene–Pliocene emergence of the Hominini coincides with considerable mammalian faunal turnover in Africa, including the endemic radiation of groups such as ruminants, suids, and cercopithecids; the first appearance through immigration of taxa such as *Equus* and *Giraffa*; the demise of previously successful groups such as anthracotheres; and, overall, an increased representation of taxa adapted to more open savanna-mosaic environments (e.g., Behrensmeyer, et al. 1992).

Most, if not all, of the major taxonomic events for the tribe occur in Africa, including the emergence of our own species. Only two hominin species (from one eurytopic genus) are found entirely outside Africa (*Homo neanderthalensis* and *Homo floresiensis*), and all hominins are restricted to the continent until just under 2 million years ago.

The principal adaptive changes of the tribe also occur in Africa. Early hominins are distinguished by a suite of postcranial features showing reliance on terrestrial bipedalism. By the late Pliocene (and possibly sooner), these adaptations gave hominins one of the most efficient forms of locomotion of all mammals. The selective advantages that led to the evolution of terrestrial bipedalism (carrying? tool use? energetics?) are intensively discussed within paleoanthropology. Late Miocene–early Pliocene hominins are further characterized by an increase in the size of the postcanine dentition, a reduction in size of the anterior dentition, and a loss of the upper canine/lower premolar honing complex. Late Pliocene hominins then specialize in one of two ways: the australopiths (discussed later) become megadont, and *Homo* reduces postcanine tooth size. Like striding bipedalism, the selective factors responsible for these dental and masticatory changes are debated, as are the dietary implications.

It is only during the last one-third of the existing record of hominin evolution that brain size expands significantly and stone tools appear. As current evidence stands, the latter occurs prior to the former. These two innovations, likely reflecting complex biocultural feedback involving language, also coincide with a decrease in hominin diversity from the late Pliocene, when as many as five African species occur, to the mid-Pleistocene, after which only one evolving lineage dominates the continent.

Taxonomy in this chapter errs on the side of cutting up samples finely, as lumping may obscure potentially significant differentiation (Groves, 1989). We believe it is also generally easier to amalgamate taxa rather than subdivide, should variation later be shown to be less than, or exceed, that expected within a single species. However, there is also strong opinion against "diversity systematics" within paleoanthropology (e.g., Wolpoff, 1999), and an antidote against this approach has recently been offered by White (2009).

A taxon recognized by many in the paleoanthropological community as a valid species, *Homo ergaster*, is not considered separately here, because we are convinced by Spoor et al.'s (2007) argument that many of the cranial differences between it and *Homo erectus* are size related. Overall, we present a taxonomic scheme that we feel most accurately reflects current knowledge about hominin systematics and is widely used by paleoanthropologists. We are not wholly in agreement with this taxonomy and anticipate that taxonomic changes are imminent. For example, widespread recognition that *Australopithecus* is paraphyletic (e.g., Strait and Grine, 2004; Kimbel, 2007; Collard and Wood, 2007) will likely result in more genera for the australopith grade. Other taxa, currently conceived as potential chronospecies (*Australopithecus anamensis* and *Au. afarensis*; *Paranthropus aethiopicus*, and *P. boisei*) may be lumped; and the possibility exists that certain poorly known taxa may be subsumed within taxa with larger hypodigms, once evidence for more overlapping anatomical regions has been recovered and analyzed (e.g., *Orrorin* may be subsumed within *Ardipithecus*, *Au. bahrelghazali* within *Au. afarensis*). We use the informal term *australopith* instead of the more traditional *australopithecine*, because the latter should only be used for taxa within *Australopithecus* and *Paranthropus* that are thought to belong in their own subfamily, the Australopithecinae. Likewise, we use the informal term *hominin* rather than *hominid* (the family-level colloquialism), since in cladistic taxonomies, which we support, Hominidae includes extant great apes as well as humans. Site abbreviations used throughout the text are defined in table 25.1.

Of previously published taxonomies, we follow Wood and Richmond (2000):

Superfamily Hominoidea
    Family Hominidae
        Subfamily Homininae
            Tribe Hominini ("hominins")
                Subtribe incertae sedis
                    Sahelanthropus
                    Orrorin
                    Ardipithecus
                Subtribe Australopithecina
                    Australopithecus
                    Paranthropus
                Subtribe *incertae sedis*
                    Kenyanthropus
                Subtribe Hominina ("hominans")
                    Homo

Systematic Paleontology

Family HOMINIDAE Gray, 1825
Subfamily HOMININAE Gray, 1825
Tribe HOMININI Gray, 1825
Subtribe INCERTAE SEDIS
Genus *SAHELANTHROPUS* Brunet et al., 2002
SAHELANTHROPUS TCHADENSIS Brunet et al., 2002
Figure 25.1 and Table 25.2

Holotype The type specimen is TM 266-01-060-1 ("Toumaï"), a nearly complete but distorted cranium (flattened dorsoventrally and depressed on the right side). A list of other specimens referred to *S. tchadensis*, including three mandibles from three localities, is provided by Brunet et al. (2002, 2005).

Age and Occurrence Late Miocene, Central Africa (table 25.2).

Diagnosis Based on Brunet et al. (2002). *Sahelanthropus tchadensis* differs from *Pan* and *Gorilla* in its short, more orthognathic subnasal region; smaller canines; short basioccipital and more anterior foramen magnum; flat, long, more horizontally oriented nuchal plane; and downward lipping of the nuchal crest. It further differs from *Gorilla* (which it has been claimed to resemble; Wolpoff et al., 2002) in its smaller-sized cranium, lack of supratoral sulcus, and lower crowned cheek teeth. It differs from the penecontemporaneous *Ardipithecus* in the well-developed crests and cingulum on the lingual aspect of I1, less incisiform upper canines with a low distal shoulder, buccolingually narrower lower canines with stronger distal tubercle and two-rooted p4. Lingual I1 topography also distinguishes *Sahelanthropus* from *Orrorin*, as do the short, apically worn upper canines.

Description A cranium with great ape-sized brain case, U-shaped dental arcade, an orthognathic, anteroposteriorly short subnasal region and relatively vertical upper face, a wide interorbital pillar and large, continuous supraorbital torus, marked postorbital constriction and posteriorly located sagittal crest, a flat, long, approximately horizontally oriented nuchal plane with downward lipping of nuchal crest, anteriorly positioned foramen magnum, and short basioccipital. Dental remains show small, apically worn canines, no lower c/p3 diastema or upper I2/C diastema and the apparent absence of a functional honing C/p3 complex. The lower canine has a large distal tubercle and low shoulders. Mandibular premolars have two roots. Postcanine teeth are bunodont and slightly crenulated and enamel thickness (1.2–1.9 mm) is intermediate between *Pan* and *Australopithecus* (see Brunet et al., 2002, 2005; Zollikofer et al., 2005; Guy et al., 2005).

TABLE 25.1
Site abbreviations found throughout the text
Adapted from Wood and Richmond, 2000.

Abbreviation	Site
ALA-VP	Alaya—Vertebrate Paleontology
AL or A.L.	Afar Locality
AME-VP	Amba East—Vertebrate Paleontology
ARA-VP	Aramis—Vertebrate Paleontology
ASK-VP	Asa Koma—Vertebrate Paleontology
BAR	Baringo
BC	Border Cave
BOU-VP	Bouri—Vertebrate Paleontology
BSN	Busidima Formation
DNH	Drimolen
EP	Eyasi Plateau, Tanzania
ER	East Rudolf (Koobi Fora or East Turkana)
GAM-VP	Gamedah—Vertebrate Paleontology
GWM	Gona Western Margin
Is	Ishango
KGA	Konso Gardula
KNM	Kenya National Museums
KP	Kanapoi
KRM	Klasies River Mouth
KT	Koro Toro
LH or L.H.	Laetoli Hominin
LT	Lothagam
LU	Lukeino Formation
Ma	Mega annum
MLD	Makapansgat Limeworks Dump
OH or O.H.	Olduvai Hominin
OL	Olorgesailie
Omo (L)	Shungura Formation, Ethiopia
SAM-AP	South African Museum, Klasies River Cave
SE	Sterkfontein Extension site
SK, SKW	Swartkrans, or Swartkrans Wits hominin
SKX	Swartkrans Excavation
STD-VP	Saitune Dora—Vertebrate Paleontology
Sts	Sterkfontein type site
Stw, StW, Stw/H, or StW/H	Sterkfontein Wits hominin
TH	Tabarin hominin
TM (in context of South African hominins)	Transvaal Museum
TM (in context of *Sahelanthropus*)	Toros-Menalla
UA	Uadi Aalad
UR	Uraha
WT	West Turkana (including Nariokotome)

Remarks The virtual reconstruction of the holotype cranium TM 266-01-60-1 (Zollikofer et al., 2005) is a crucial although potentially controversial anatomical reference, as the holotype is distorted by displacement of fragments around multiple fractures and by plastic deformation, most notably in the maxilla. Fortunately, there is good anatomical continuity along the sagittal and parasagittal planes of the skull, and no major regions are missing. The 3-D reconstruction of the cranium was

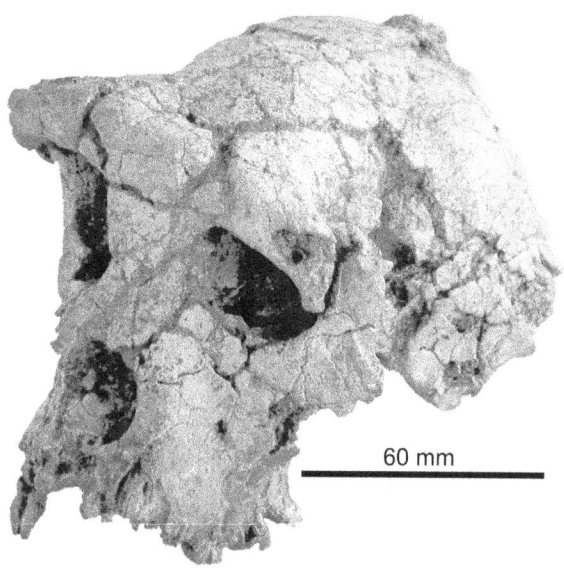

FIGURE 25.1   Holotype of *Sahelanthropus tchadensis*, cranium TM
266-01-60-1. Courtesy of Michel Brunet. © MPFT.

produced by disassembling a CT-generated digital representation along major cracks, removing matrix, and then reassembling the segments while ensuring that the face/neurocranium/basicranium fit together at multiple points (Zollikofer et al., 2005). Compared to the original, the reconstruction is wider and has a sagitally rounder occipital contour, a more horizontally oriented nuchal plane, a taller face and larger, more rounded orbits (Zollikofer et al., 2005).

The cranium presents a mosaic of features that fall into three categories: primitive for African apes, derived relative to *Pan* and *Gorilla* (and some also derived relative to australopiths such as *Au. afarensis*), and autapomorphic features. Primitive features that clearly resemble the condition found in *Pan* and *Gorilla* are found primarily in the neurocranium: the superior contour is long and low, the brain size is small (360–370 cc), and there is pronounced postorbital constriction (Guy et al., 2005). The shape and shallowness of the palate also resemble the conditions found in *Pan* and *Gorilla* (Guy et al., 2005), and the mandibular premolars have two roots and three separate pulp canals (Brunet et al., 2005).

Derived features relative to *Pan* and *Gorilla* support the taxon's claim for hominin status and rest on two complexes of features. The first set of features is basicranial and may reflect habitual orthogrady and therefore bipedalism. The basioccipital is short and the foramen magnum is more anteriorly positioned than is typical of *Pan* and *Gorilla*. In addition, the angle formed between a line delimiting the rim of the foramen magnum in sagittal view and the orbital plane is about 90°, as in extant *Homo*. This angle is acute in *Pan* because although the orbital plane will always tend toward the vertical in lateral view, due to the superiorly sloping posterior aspect of the nuchal plane on which the foramen is situated, the plane of the foramen is oblique in *Pan*, not horizontal as in *Homo*. The condition in *Pan* is thought to reflect the latter's more frequent use of pronograde postures (Zollikofer et al., 2005). However, measurement of this angle is difficult unless specimens are complete, and Wolpoff et al. (2006) have claimed that there is overlap in the angle among *Pan* and early australopiths, poten-

tially diminishing the utility of the character in diagnosing bipeds. TM 266-01-60-1 also has a flat nuchal plane and the nuchal crest exhibits downward lipping; both are similar to the condition in *Australopithecus* and *Homo* and unlike that seen in *Pan* (Zollikofer et al., 2005).

There is also a cluster of features related to the small canines. The subnasal region is short and flat. Both upper and lower canines show apical wear, there are no c/p3 or I2/C diastemae and a honing C/p3 complex is absent.

Autapomorphic features include the remarkably thick supraorbital region, outside the range of male *Gorilla* (Brunet et al., 2002).

If *Sahelanthropus* is not a hominin, it is closely related to them. While Wolpoff et al.'s (2006) argument that convergence may explain the reduced canines, *Ouranopithecus* being a salient example, basicranial features, which are plausibly linked to habitual orthogrady (or more recently to neural reorganization; see Suwa et al., 2009b) would seem less likely to be the result of convergence.

Three localities in the Toros-Menalla (TM) region of Chad (TM 266, 247, and 292) have yielded fossils of *Sahelanthropus*; their biochronological age, based on comparisons with the Lukeino fauna and fauna from the Nawata Fm., Lothagam, is 7.0–6.0 Ma (Vignaud et al., 2002; Brunet et al., 2005). The fauna is described as being more similar to Lothagam (7.4–5.2 Ma) than to Lukeino (ca. 6.0 Ma); for example, *Loxodonta* from Toros-Menalla is more primitive than the Lukeino species. This assessment is supported by recent cosmogenic nuclide dating that constrains the fossils between 7.2 and 6.8 Ma (Lebatard et al., 2008). Existing genetic evidence suggests a *Pan/Homo* split ca. 4–8 Ma (e.g., Steiper et al., 2004; Bradley, 2008) even allowing for subsequent gene flow (Patterson et al., 2006). Given the lack of a good fossil calibration, it is premature to claim *Sahelanthropus* is too old to be a hominin (*contra* Wolpoff et al., 2006); and if the Miocene hominoid *Chororapithecus*, dated at 10.5–10.0 Ma, is a member of the *Gorilla* clade as claimed (Suwa et al., 2007), this would also support a hominin divergence of at least 7.0 Ma.

TABLE 25.2
Major occurrences and ages of African and selected Levantine hominins
? = attribution or occurrence uncertain; abbreviation: alt., alternatively.

Taxon	Location	Unit, Formation, Member or Bed	Age	References
TRIBE HOMININI, LATE MIOCENE–PRESENT				
SUBTRIBE *INCERTAE SEDIS*, LATE MIOCENE–EARLY PLIOCENE				
Sahelanthropus tchadensis	Toros Menalla, Chad (type)	Anthracotheriid Unit	Late Miocene, ca. 7.0–6.0 Ma	Brunet et al., 2002, 2004, 2005; Vignaud et al., 2002; Zollikofer et al., 2005
			6.8–7.2	Lebatard et al., 2008
Orrorin tugenensis	Tugen Hills, Kenya (type)	Lukeino Fm.	6.2–5.6 Ma	Hill et al., 1985, 1986; Senut et al., 2001; Hill, 2002; Sawada et al., 2002
Ardipithecus kadabba	Saitune Dora, Alayla (type), Asa Koma, and Digiba Dora, Middle Awash; Ethiopia	Asa Kona Mb., Adu Asa Fm.	5.8–5.5 Ma	Haile-Selassie, 2001; Haile-Selassie et al., 2004b
	Amba East, Central Awash Complex, Ethiopia	Kuseralee Mb., Sagantole Fm.	5.6–5.2 Ma	Haile-Selassie, 2001
	Gona, Ethiopia	Adu Asa Fm.	>5.4 Ma	Simpson et al., 2007; Levin et al., 2008
Ardipithecus ramidus	As Duma, Gona, Ethiopia	GWM-3 and 5 deposits	4.51–4.32 Ma	Semaw et al., 2005
	Aramis, Middle Awash, Ethiopia (type)	Above the Gàala Vitric Tuff Complex	ca. 4.4 Ma (4.48–4.29 Ma) (alt. 4.39–3.89 Ma)	White et al., 1994, 1995, 2006; WoldeGabriel et al., 1994, 1995, 2009; Kappelman and Fleagle, 1995
	?Lothagam, Kenya		5.0–4.2 Ma	McDougall and Feibel, 1999
	?Tabarin, Baringo Basin, Kenya	Chemeron Fm.	4.48–4.41 Ma	Deino et al., 2002
SUBTRIBE AUSTRALOPITHECINA, EARLY PLIOCENE–EARLY PLEISTOCENE				
Australopithecus anamensis	Kanapoi, Turkana Basin, Kenya		4.17 ± 0.03–4.07 ± 0.02 Ma and < 4.07	Leakey et al., 1998
	Allia Bay, Turkana Basin		3.95 Ma	Leakey et al., 1995
	Aramis, Middle Awash, Ethiopia	Adgantole Mb., Sagantole Fm.	4.2–4.1 Ma	White et al., 2006
	Asa Issie, Middle Awash, Ethiopia		4.2–4.1 Ma	White et al., 2006
	?Fejej, Ethiopia,	Harr Fm.	4.18–>4.0 Ma	Fleagle et al., 1991; Kappelman et al., 1996
	?Belohdelie, Middle Awash	Belohdelie Mb., Sagantole Fm.	3.89–3.86 Ma	Asfaw, 1987; White et al., 1993
Australopithecus afarensis	Laetoli, Tanzania	Upper Laetolil Beds	3.76–3.46 Ma	Drake and Curtis, 1987; Leakey, 1987
	Hadar, Ethiopia	Sidi Hakoma, Denan Dora and Kada Hadar Mbs., Hadar Fm.	3.4–2.95 Ma	Kimbel et al., 1994
	Dikika, Ethiopia	Basal Mb., Hadar Fm.	>3.4 Ma	Walter and Aronson, 1993; Alemseged et al., 2005
	Maka, Middle Awash, Ethiopia	Matabaietu Fm.	3.4 Ma	White et al., 1993, 2000
	East Turkana, Kenya	Tulu Bor Mb., Koobi Fora Fm.	3.4 Ma	Kimbel and White, 1988; Brown, 1994
	West Turkana, Kenya	Lomekwi Mb., Nachukui Fm.	3.35–3.26 Ma	Feibel et al., 1989; Brown et al., 2001

Taxon	Location	Unit, Formation, Member or Bed	Age	References
Australopithecus bahrelghazali	Koro Toro, Tchad		3.5–3.0 Ma 3.6 Ma	Brunet et al., 1995, 1997 Lebatard et al., 2008
Australopithecus africanus	Sterkfontein, South Africa	?Mb. 2	?3.5–3.0 Ma (alt. 4.17 Ma; alt. 2.2 Ma)	Broom, 1936; Schwartz et al., 1994; Clarke and Tobias, 1995; Partridge et al., 1999, 2003; Kuman and Clarke, 2000; Partridge, 2000; Berger et al., 2002; Pickering et al., 2004a; Walker et al., 2006
		Mb. 4	2.8–2.6 Ma (alt. 2.5–1.5 Ma)	
	Makapansgat, South Africa	Mbs. 3, 4	Ca. 3.2–2.9 Ma	Dart, 1948; McKee et al., 1995; Partridge, 2000; Herries, 2003; Latham et al., 2007
	Taung, South African (type)	Dart Deposits	Ca. 2.8–2.6 Ma	Dart, 1925; McKee, 1993
	Gladysvale, South Africa	Eccles Fm.	Ca. 2.5–1.7 Ma	Berger et al., 1993; Schmid, 2002; Pickering et al., 2007
Australopithecus garhi	Middle Awash, Ethiopia (type)	Hatayae Mb., Bouri Fm.	2.5 Ma	Asfaw et al, 1999; de Heinzelin et al., 1999
Paranthropus aethiopicus	Laetoli, Tanzania	Upper Ndolanya Beds	Ca. 2.7–2.5 Ma	Drake and Curtis, 1987; Harrison, 2002
	Omo, Ethiopia (type)	Mbs. C–F, Shungura Fm.	2.6–2.33 Ma	Arambourg and Coppens, 1968; Suwa, 1988; Feibel et al., 1989; Rak and Kimbel, 1991; Wood and Constantino, 2007
	West Turkana, Kenya	Lokalalei Mb., Nachukui Fm.	Ca. 2.5–2.4 Ma	Walker et al., 1986; Harris et al., 1988; Feibel et al., 1989
P. boisei	?Malema, Malawi	Unit 3A, Chiwondo Beds	Ca. 2.5–2.3 Ma	Kullmer et al., 1999
	Omo, Ethiopia	Mbs. G, K, Shungura Fm.	Ca. 2.33–1.39 Ma	Brown and Feibel, 1988; Suwa, 1988; Feibel et al., 1989; Wood et al., 1994
	Koobi Fora, Kenya	Upper Burgi Mb., KBS Mb., Okote Mb., Koobi Fora Fm.	2.0–1.39 Ma	Brown and Feibel, 1988; Wood et al., 1994
	West Turkana, Kenya	Kaitio Mb., Nachukui Fm.	1.87–1.65 Ma	Brown and Feibel, 1988; Harris et al., 1988; Wood et al., 1994
	Oludvai Gorge, Tanzania (type)	Beds I and II	Ca. 1.85–?1.2 Ma (alt. 1.79–1.45 Ma)	Leakey, 1959; Hay, 1976; Wood et al., 1994; Tamrat et al., 1995; Wood and Constantino, 2007
	Peninj, Tanzania	Humbu Fm.	1.70–1.56 Ma (alt. 1.4 Ma)	Leakey and Leakey, 1964; Isaac, 1967; Wood et al., 1994
	Chesowanja, Kenya		>1.42 Ma	Carney et al., 1971; Bishop et al., 1978; Hooker and Miller, 1979; Gowlett et al., 1981
	Konso, Ethiopia	Between Karat Tuff and Trail Bottom Tuff	1.43–1.41 Ma	Suwa et al., 1997; Silverman et al., 2001
P. robustus	Kromdraai B East, South Africa (type)	Mb. 3	Ca. 2.0–1.7 Ma (alt. ca. 1.8–1.7 Ma)	Broom, 1938; McKee et al., 1995; Aiello and Andrews, 2000; Wood and Strait, 2004

TABLE 25.2 (CONTINUED)

Taxon	Location	Unit, Formation, Member or Bed	Age	References
P. robustus (continued)	?Sterkfontein South Africa	Mb. 5	Ca. 2.0–1.7 Ma	Clarke, 1994b; Kuman, 1994; Kuman and Clarke, 2000
	Drimolen, South Africa	Drimolen Main Quarry	Ca. 2.0–1.5 Ma	Keyser, 2000; Keyser et al., 2000
	Cooper's D, South Africa		Ca. 1.5–1.4 Ma	Ruiter et al., 2009
	Gondolin, South Africa		slightly > 1.78 Ma	Kuykendall and Conroy, 1999; Menter et al., 1999;
			(alt. ca. 1.9–1.5 Ma)	Herries et al., 2006
	Swartkrans, South Africa (type, *P. "crassidens"*)	Mbs. 1-3; ?Mb. 5	Ca. 1.8–1.5 Ma	Broom, 1949; Vrba, 1985, 1995; Brain and Watson,
			(alt. ca. 1.8–0.7 Ma)	1992; McKee et al., 1995; Avery, 1998; Curnoe et al.,
			(alt. ca. 1.7–1.0 Ma)	2001; Ruiter, 2003

SUBTRIBE *INCERTAE SEDIS*

Kenyanthropus platyops	Lomekwi, West Turkana, Kenya	Kataboi and Lomekwi Mbs., Nachukui Fm.	3.57–3.3 Ma	Leakey et al., 2001

SUBTRIBE HOMININA, LATE PLIOCENE–PRESENT

Taxon	Location	Unit, Formation, Member or Bed	Age	References
Homo habilis	Olduvai, Tanzania	Bed I	1.87–1.75 Ma	Hay, 1976; Egeland et al., 2007;
		Lower Bed II	<1.75 Ma	Blumenschine et al., 2003; Walter et al., 1991
	East Turkana and Ileret, Kenya	Upper Burgi, KBS and Okote Mbs., Koobi Fora Fm.	1.9–1.44 Ma	Feibel et al., 1989; Gathogo and Brown, 2006; Spoor et al., 2007
	Omo, Ethiopia	Shungura Fm.	2.4–2 Ma	Feibel et al., 1989; Suwa et al., 1996
	Hadar, Ethiopia	Kada Hadar Member, Hadar Fm.	2.33 Ma	Kimbel et al., 1996
	?West Turkana, Kenya	Kalochoro Member, Nachukui Fm.	>1.88	Harris et al., 1988; Feibel et al., 1989
			< 2.34 Ma	Prat et al., 2005
	?Sterkfontein, South Africa	"StW infill" Mb.5 Mb, 4	<2.6–2 Ma ca 2–1.7 Ma 2.8–2.6 Ma (alt. 2.5–1.5 Ma)	Kuman and Clarke, 2000; Berger et al., 2002
	?Swartkrans, South Africa	Mbs. 1 and 2	ca 1.8–1.5	Wood, 1992; Curnoe et al., 2001; de Ruiter, 2003
Homo rudolfensis	?Baringo Basin, Kenya	Chemeron Fm.	2.4 Ma	Hill et al., 1985
	East Turkana, Kenya	Upper Burgi Mb., Koobi Fora Fm.	1.88–1.9 Ma	Feibel et al., 1989; Gathogo and Brown, 2006
	?Omo, Ethiopia	Shungura Fm.	2.4–2 Ma	Feibel et al., 1989; Suwa et al., 1996
	?Uraha, Malawi	?Chiwondo Beds	2.5–2.3 Ma	Bromage et al., 1995a
Homo erectus	East Turkana and Ileret, Kenya	Upper Burgi, KBS and Okote Mbs., Koobi Fora Fm.	1.9–1.5 Ma	Leakey and Walker, 1976; Feibel et al., 1989; Antón, 2003; Gathogo and Brown, 2006; Spoor et al., 2007; Suwa et al., 2007
	Swartkrans, South Africa	Hanging Remnant Mb. 1 Lower Bank, Mb. 1, Mb. 2, ?Mb. 3	Ca. 1.8–1.0 Ma	Broom and Robinson, 1949; Clarke et al., 1970; Vrba, 1985; Delson, 1988; Curnoe et al., 2001
	Olduvai Gorge, Tanzania	Upper Bed II Bed III Bed IV, Olduvai Fm.	1.78–1.47 Ma Ca. 1.47–1.2 Ma 1.2–0.78 Ma	Leakey, 1961; Hay, 1976; Walter et al., 1991; Manega, 1993; Tamrat et al., 1995

Taxon	Location	Unit, Formation, Member or Bed	Age	References
H. erectus (continued)	Gona, Ethiopia	Busidima Fm.	Ca. 1.4–0.9 Ma	Simpson et al., 2008
	Lincoln Cave, South Africa	Lincoln Cave South Deposits	Ca. 1.7–1.4 Ma	Kuman and Clarke, 2000; Reynolds et al., 2007
	Sterkfontein, South Africa	Mb. 5 West	Ca. 1.7–1.4 Ma	Robinson, 1962; Kuman and Clarke, 2000; Curnoe and Tobias, 2006
	Melka Kunturé, Ethiopia	Gomboré IB	Ca. 1.7–1.6 Ma	Chavaillon et al., 1974, 1977; Westphal et al., 1979;
		Garba IV, Level E	Ca. 1.5 Ma	Chavaillon and Coppens, 1986
		Gomboré II, Melka Kunturé Fm.	Ca. 800–700,000 y	Condemi, 2004
	Nariokotome, West Turkana, Kenya	Natoo Mb., Nachukui Fm.	1.6–1.0 Ma	Brown et al., 1985; Feibel et al., 1989; Walker and Leakey, 1993; Brown et al., 2001
	Omo, Ethiopia	Mb. K, Shungura Fm.	1.5–1.4 Ma	Howell, 1976, 1978; Feibel et al., 1989
	Konso-Gardula, Ethiopia	Kayle Mb.	1.45–1.41 Ma	Asfaw et al., 1992; Nagaoka et al., 2005;
		Karat Mb., Konso Fm.	1.41–1.25 Ma	Suwa et al., 2007
	Laetoli, Tanzania	Lower Ngaloba Beds	Ca. 1.2–0.12 Ma	Leakey and Harris, 1987
	Middle Awash, Ethiopia	Dakanihylo Mb., Bouri Fm.	1.0 Ma	Heinzelin et al., 1999; Asfaw et al., 2002
	Uadi Aalad (Buia), Eritrea	Danakil Fm.	1.0 Ma	Abbate et al., 1998
	Olorgesailie, Kenya	Mb. 5 and 6/7 boundary, Olorgesailie Fm.	970–900,000 y	Potts et al., 2004
	Angamma-Yayo, Chad		?Early–?middle Pleistocene	Coppens, 1966; 1967
	Tighenif (Ternifine or Palikao), Algeria		Ca. 700,000 y	Arambourg, 1954, 1955; Howell, 1960; Hublin, 1985; Geraads et al., 1986
Homo heidelbergensis	Olduvai Gorge, Tanzania	Masek Beds	Ca. 780–490,000 y	Leakey, 1969; Hay, 1976; Walter et al., 1991;
		Ndutu Beds, Olduvai Fm.	Ca. 490–62,000 y	Manega, 1993; Tamrat et al., 1995
	Kabwe, Broken Hill Cave, Zambia	No. 1 Kopje outcrop	Ca. 700–400,000 y (alt. ca. 300–125,000 y) (alt. ca. 1.3–0.78 Ma)	Woodward, 1921; Vrba, 1982; Klein, 1994; Rightmire, 1998; McBrearty and Brooks, 2000; Barham et al., 2002
	Bodo, Ethiopia	Bodo Mb., Wehaietu Fm.	640–550,000 y	Conroy et al., 1978; Kalb et al., 1982; Clark et al., 1994
	Saldanha (Hopefield), South Africa	Elandsfontein Main, Elandsfontein Farm	Ca. 600,000 y	Drennen, 1955; Singer, 1954; Klein et al., 2007
	Baringo Basin, Kenya	Middle Silts and Gravels Mb., Kapthurin Fm.	512–510,000 y	Leakey et al., 1969; Wood and Van Noten, 1986; Deino and McBrearty, 2002
	Cave of Hearths, Makapansgat, South Africa	Bed 3	Ca. 500–200,000 y (alt. ca. 700,000 y)	Dart, 1948; Tobias, 1971; Mason, 1988; Latham and Herries, 2004; Beaumont and Vogel, 2006
	Lake Ndutu, Tanzania	?Masek Beds	Ca. 500–300,000 y	Mturi, 1976; Clarke, 1976, 1990
	Littorina Cave, Sidi Abderrahman, Morocco		492–376,000 y	Arambourg and Biberson, 1956; Howell, 1960; Hublin, 1985; Rhodes et al., 2006
	Thomas Quarry, Morocco	Mb. 1, Oulad Hamida Fm.	470–360,000 y	Ennouchi, 1969; Hublin, 1985; Raynal et al., 2001; Rhodes et al., 2006

TABLE 25.2 (CONTINUED)

Taxon	Location	Unit, Formation, Member or Bed	Age	References
H. heidelbergensis (continued)	Salé, Morocco		455–389,000 y	Hublin, 1985, 1991
	Lainyamok, Kenya	Skeleton Hill Patch	393–323,000 y	Shipman et al., 1983; Potts et al., 1988; Potts and Deino, 1995
	?Guomde, Kenya	Chari Mb., Koobi Fora Fm.	Ca. 300–270,000 y (alt. minimum age 160,000 y)	Bräuer et al., 1992b, 1997
	Lake Eyasi, Tanzania	Eyasi Beds	Ca. 300–200,000 y	Kohl-Larsen and Reck, 1936; Leakey, 1936; Mehlman, 1987; Bräuer and Mabulla, 1996; Trinkaus, 2004; Domínguez-Rodrigo et al., 2008
	Melka Kunturé, Ethiopia	Layer B, Garba III	Ca. 300–200,000 y	Chavaillon et al., 1987
	Hoedjiespunt, South Africa	HOMS (Hominid Sands)	Ca. 300–100,000 y	Churchill et al., 2000; Stynder et al., 2001
	Florisbad, South Africa	Peat I layer	Ca. 260,000 y	Dreyer, 1935; Grün et al., 1996
	Wadi Dagadlé, Djibouti	>250,000 y		de Bonis et al., 1988
	Aïn Maarouf (El Hajeb), Morocco		?Early middle Pleistocene	Geraads et al., 1992; Hublin, 1992
	Rabat, Morocco	Khebibat (Mifsud-Giudice) Quarry	?Middle Pleistocene	Marçais, 1934; Howell, 1960; Saban, 1977
	Berg Aukas Mine, Namibia	Level 5, Berg Aukas Fm.	?Middle Pleistocene	Grine et al., 1995
	Loyangalani, Kenya		?Middle–early late Pleistocene	Twiesselmann, 1991
Homo sapiens	Omo, Ethiopia	KHS and PHS, Upper Mb. 1, Kibish Fm.	196,000 ± 2000 y (Middle Stone Age)	Day, 1969; Leakey, 1969; Rightmire, 1976; Stringer, 1978; Day and Stringer, 1982; McDougall et al., 2005; Brown and Fuller, 2008; Feibel, 2008; Fleagle et al., 2008
	Jebel Irhoud, Morocco		190–105,000 y (alt. 125–90,000 y) (Mousterian) (alt. 160–130,000 y)	Ennouchi, 1962. 1963, 1969; Biberson, 1964; Hublin and Tillier, 1981; Hublin et al., 1987; Grün and Stringer, 1991; Hublin, 1991
	Herto, Middle Awash, Ethiopia	Upper Herto Mb., Bouri Fm.	160–154,000 y (early Middle Stone Age or transition from Acheulean to MSA)	Clark et al., 2003; White et al., 2003
	Singa, Sudan	Limestone calcrete, exposed bed of Nile	>133,000 y (alt. 160-140,000 y, Oxygen isotope stage 6) (alt. 160,000 ± 27,000 y; 97,000 ± 15,000 y)	Woodward, 1938; Stringer, 1979; Stringer et al., 1985; Spoor et al., 1998; Grün and Stringer, 1991; McDermott et al., 1996
	Skhul, Israel	Layers B	135–130,000 y (alt. 119,000 ± 18,000 y) (Mousterian)	McCown, 1937; McCown and Keith, 1939; Grün and Stringer, 1991; Mercier et al., 1993; Grün et al., 2005
	Qafzeh, Israel		130–90,000 y (alt. 92,000 ± 5000 y) (Mousterian)	Schwarcz et al., 1988; Valladas et al., 1988; Grün and Stringer, 1991; Yokoyama et al., 1997
	Mumba Rock Shelter, Lake Eyasi, Tanzania	Bed VI	Ca. 130,000 y (Middle Stone Age)	Mehlman, 1979, 1987, 1991; Bräuer and Mehlman, 1988

Taxon	Location	Unit, Formation, Member or Bed	Age	References
H. sapiens (continued)	Laetoli, Tanzania	Ngaloba Beds	120,000 ± 30,000 y (Middle Stone Age) (alt. 130,000 y)	Day et al., 1980; Hay, 1987; Stringer, 1988; Cohen, 1996
	Aduma, Middle Awash, Ethiopia	Ardu Beds	Late Pleistocene; ?105–79,000 y (Middle Stone Age)	Haile-Selassie et al., 2004a
	Bouri, Middle Awash, Ethiopia	Beach deposit, cf. Ardu (B) Beds	Late Pleistocene, cf. Aduma (Middle Stone Age)	Haile-Selassie et al., 2004a
	Diré-Dawa (Porcupine Cave), Ethiopia		?Early late Pleistocene (Middle Stone Age)	Vallois, 1951; Briggs, 1968; Tobias, 1968
	Klasies River Mouth, South Africa		100–80,0000 y (Middle Stone Age I, II) (alt. 110,000 y) 60,000 y (MSA III)	Singer and Wymer, 1982; Deacon and Geleijnse, 1988; Grün et al., 1990; Rightmire and Deacon, 1991; Deacon, 1993; Grine et al, 1998; Grine, 2000; Vogel, 2001; Feathers, 2002
	Oranjemund, Namibia		Ca. 100–50,000 y	Senut et al., 2000
	Die Kelders Cave, South Africa		80–60,000 y (Middle Stone Age)	Tankard and Schweitzer, 1974, 1976; Schweitzer, 1979; Grine et al., 1991; Avery et al., 1997; Grine, 2000
	Haua Fteah, Libya	Interface between layers XXXII and XXXIII	Ca. 75,000 y (Mousterian) (alt. ca. 47,000 y)	McBurney et al., 1953a, 1953b; Tobias, 1967b; Trevor and Wells, 1967 (in McBurney et al., 1953b); Klein and Scott, 1986; Stringer and Brooks, 2000
	Mugharet el Aliya; Témara (Smugglers Cave); Dar-es-Soltane II cave; Zouhara Cave (El Harhoura), Morocco		60–35,000 y (Aterian Industry) (alt. Oxygen isotope stage 5 or 6, late middle Pleistocene or early late Pleistocene) (alt. >70,000 y)	Coon, 1939; Vallois and Roche, 1958; Ferembach 1976a, b; Debénath, 1975, 1980, 1991; Roche and Texier, 1976; Debénath et al., 1982, 1986; Hublin, 1992; Wrinn and Rink, 2003
	Taramsa Hill, Egypt		55,500 ± 3,700 y (late Middle Paleolithic)	Vermeersch et al., 1998
	Border Cave, South Africa		>48,700 y (final Middle Stone Age; Howieson's Poort, BC4) (alt. ca. 80–70,000 y; ca. 65–55,000 y, BC3 and BC5; 60-40,000 y, BC4)	Cooke et al., 1945; Wells, 1950; Villiers, 1973, 1976; Beaumont, 1980; Grün and Stringer, 1991; Grün and Beaumont, 2001; Grün et al., 2003
	Nazlet Khater, Egypt		Ca. 37,570 ± 350 y (alt. 33,000 y)	Thoma, 1984; Pinhasi and Semal, 2000; Crevecoeur and Trinkaus, 2004

TABLE 25.2 (CONTINUED)

Taxon	Location	Unit, Formation, Member or Bed	Age	References
H. sapiens (continued)	Hofmeyr, Eastern Cape Province, South Africa	Dry channel bed, Vlekpoort River	36,200 ± 3,300 y	Grine et al., 2007
	Boskop, South Africa		Late Pleistocene (Middle Stone Age)	Haughton, 1917; Broom, 1918; Galloway, 1937a; Singer, 1961
	Eliye Springs, Kenya		?Late Pleistocene	Bräuer and Leakey, 1986a, b; Bräuer et al., 2004
	Mumbwa, Zambia		19,780 ± 130 y (Late Stone Age)	Jones, 1940; Gabel, 1963; Protsch, 1975a
	Wadi Kubbaniya, Egypt		19–17,000 y (Upper Paleolithic)	Stewart et al., 1986; Wendorf and Schild, 1986; Thorpe, 2003; Shackelford, 2007
	Esna, Egypt	Dune sand unit, Ballana Fm.	18,020 ± 330 y (Upper Paleolithic)	Wendorf et al., 1970; Butler in Lubell, 1974
	Olduvai Gorge, Tanzania	Naisiusiu Beds	16,920 ± 920y	Leakey et al., 1933; Protsch, 1974
	Taforalt, Morocco		16,750 y (alt. 21,000 y)	Ferembach, 1962, 1965
	Kalemba Rockshelter, Zambia		1: ca. 15,000 y; 2–4: 8–7,000 y; 5: 5–4,500 y (Late Stone Age)	Phillipson, 1976
	Wadi Halfa, Sudan		13,740 ± 600 y (Epipaleolithic)	Greene and Armelagos, 1972
	Kom Ombo, Egypt		Ca. 13,000 BP	Reed, 1965; Smith, 1967, 1976; Churcher, 1972
	Afalou, Algeria		12,500–10,500 y (alt. 13,120 ± 370–11,450 ± 230 y)	Chamla, 1978
	Bushman Rockshelter, South Africa		12,500–10,000 y (Late Stone Age)	Vogel, 1969; Protsch and Villiers, 1974
	Jebel Sahaba, Sudan		Ca. 12,000 y (Epipaleolithic)	Anderson, 1968; Wendorf, 1968; Greene and Armelagos, 1972; Thorpe, 2003
	Fish Hoek (Skildergat), South Africa		?Ca. 12,000 y (?Late Stone Age)	Keith, 1931; Protsch, 1975b; Deacon and Wilson, 1992
	Tushka, Egypt	Ballana Fm.	Ca. 12–10,000 y	Wendorf et al., 1970
	Iwo Eleru, Nigeria		11,200 ± 200y (Late Stone Age)	Brothwell and Shaw, 1971
	Tuinplaas (Springbok Flats), South Africa		End of late Pleistocene (?Middle Stone Age) (alt. 5,570 y)	Broom, 1929; Galloway, 1937b; Schepers, 1941; Oakley et al., 1977; Hughes, 1990
	Afalou-Bou-Rhummel, Algeria		Late Pleistocene/ Holocene (Late Stone Age) Post-Pleistocene	Arambourg, 1929; Vallois, 1952; Oakley et al., 1977
	Gamble's Cave II, Kenya		10–8,000 y (Kenya Capsian)	Leakey, 1931, 1935; Rightmire, 1975
	Naivasha Railway Site, Kenya		10,850 300 y (Kenya Capsian)	Leakey, 1942; Rightmire, 1975; Protsch, 1976
	Mechta-el-Arbi, Algeria		Ca. 8,500 y (Epipaleolithic)	Arambourg et al., 1934; Briggs, 1950
	Bromhead's Site (Elmenteita), Kenya		7,410 ± 160 y (Elementeitan)	Leakey, 1927, 1935; Rightmire, 1975

Taxon	Location	Unit, Formation, Member or Bed	Age	References
H. sapiens (continued)	Ishango, Democratic Republic of Congo	Zone Post-Emersion/ Katwe Ash Niveau Fossilifere Principal	6,890 y (Bone harpoons) 30–20,000 y (Bone harpoons)	Brooks and Smith, 1987; Boaz et al., 1990
	Fayum, Egypt	Fayum Lake Beds	Ca. 6,000 y (Neolithic)	Oakley et al., 1977
	Gwisho Hotsprings, Zambia		5–3000 y (Late Stone Age)	Gabel, 1962; Fagan and Van Noten, 1971
	Otijiseva, South West Africa		4,440 ± 70y	Sydow, 1969; Villiers, 1972
	Njoro River Cave, Kenya		3,000 y	Leakey and Leakey, 1950; Rightmire, 1975; Merrick and Monaghan, 1984
	Hyrax Hill, Willey's Kopje, Makalia Burial Site, Nakuru Burial Site, Kenya		1,000 B.C. ("Neolithic" [Gumban A, B] into 2nd millenium AD)	Leakey, 1931; Rightmire, 1975

At Toros-Menalla, the hominin-bearing part of the section for all three hominin localities consists of ~2 m thick sandstones, the Anthracotheriid Unit (AU), which represents a perilacustrine period of deposition. This middle unit is underlain by aeolian sandstones and capped by an upper unit sampling a true lacustrine environment. Within the AU at TM 266, bovids comprise 55% of all mammal remains and include high-crowned grazers, suggesting the presence of open grasslands (Vignaud et al., 2002). Giraffids point to wooded savanna, colobines to the probable presence of gallery forest, and numerous amphibious mammals (hippos, otters, anthracotheres) and shallow and deep-water fish indicate the presence of permanent water bodies. An isotopic study on the TM hippopotamids shows that they were mixed C_3 and C_4 plant eaters, similar to modern *Hippopotamus amphibius*, but with a greater proportion of C_3 plants in their diet (Boisserie et al., 2005). Whether this reflects a habitat difference, less hypsodonty, or a behavioral filter is unknown. Viewed as a whole, the fauna suggests the classic hominin "mosaic environment" reconstruction, in which both closed and open environments are in the vicinity; in this case, proximate habitats include a desert and a lake margin (Vignaud et al., 2002).

Sahelanthropus, like *Au. bahrelghazali* is of high biogeographic importance (discussed later) and points to the need for further work in locations outside the East African rift and the southern African cave sites.

Genus *ORRORIN* Senut et al., 2001
ORRORIN TUGENENSIS Senut et al., 2001
Figure 25.2 and Table 25.2

Holotype The type specimen is BAR 1000'00, a fragmentary mandible. A list of other specimens is provided in Senut et al. (2001).

Age and Occurrence Late Miocene, eastern Africa (table 25.2).

Diagnosis Based on Senut et al. (2001). The postcanine teeth are relatively small, with rectangular m2 and m3, triangular M3, and "thick" enamel. I1 large, C "short" with a shallow and narrow vertical mesial groove and no elevation of crown shoulders. The proximal femur has a large, anteriorly rotated, spherical head and a long neck. *Orrorin tugenensis* differs from *Ardipithecus*, *Pan*, and *Gorilla* in its reportedly thicker enamel (3.1 mm at apex of paraconid of M3). It differs from *Australopithecus* in its smaller, mesiodistally shorter cheek teeth. It differs from *Australopithecus* and *Ardipithecus*, but resembles *Pan* in the presence of a mesial groove on the upper canine.

Description The hypodigm includes dental and postcranial specimens from four localities. Unfortunately, none of the dental specimens are very complete and so comparisons are difficult. The "Lukeino molar" (KNM LU-335), a lower molar germ discovered in 1973 (Pickford, 1975), was reported to resemble chimpanzees in terms of cusp proportions (McHenry and Corruccini, 1980) and australopiths in terms of its buccal flare (Ungar et al., 1994), but interpretations were hampered by lack of knowledge about character polarity and lack of fossils of similar antiquity. The assignment of the specimen to *Orrorin* is reasonable given its location, but similarities between the lower molars in the holotype and the Lukeino molar were not explicitly made in the description of *Orrorin* by Senut et al. (2001). Singleton (2003) has posited that pronounced lower molar flare, a characteristic of KNM-LU 335, is a basal hominin synapomorphy, which would support its hominin status. However, she also notes that *Sahelanthropus* and *Ardipithecus* seem to lack any significant lower molar flare.

A broken m2 and an m3 are partially preserved in the holotype BAR 1000'00. Cusps are low, and enamel is described as "thick"; however, Haile-Selassie (2001) has criticized reports on enamel thickness based on unspecified natural breaks given the highly variable nature of this trait within and between teeth, a view with strong empirical support (Kono, 2004).

An upper central incisor originally allocated to *Orrorin* (BAR 1001'00) has since been reallocated to a nonhominin hominoid, as have two molars (from *"Orrorin"* sites Kapsomin and Cheboit) with purported similarities to *Gorilla* (Senut, 2007). The upper

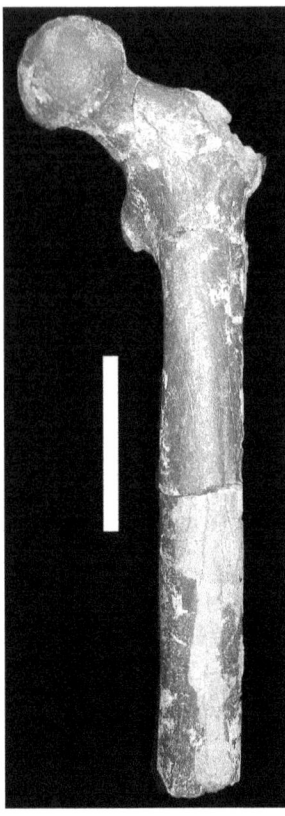

FIGURE 25.2 Anterior view of proximal femur of *Orrorin tugenensis*, BAR 1002'00. Courtesy of Martin Pickford. Scale bar is 4 cm.

canine (BAR 1425'00) resembles that of female *Pan* (i.e., is short with a pointed apex and a shallow mesial groove), and we are unaware of a published rationale for why this specimen is hominin (discussed later).

The most significant postcranial fossil is a relatively complete left proximal femur (BAR 1002'00) preserving the head, neck, lesser trochanter and about two-thirds of the shaft, but missing the greater trochanter and the distal end (Senut et al., 2001; figure 25.2). The specimen is reported to have a long, anteroposteriorly flattened femoral neck (Pickford et al., 2002), and an obturator externus groove (OEG; Pickford et al., 2002). An OEG occurs on most but not all modern human femora, is rare or absent in other primates (Lovejoy et al., 2002; DeSilva et al., 2006), and its presence on this specimen is suggestive of the kind of full hip extension associated with bipedalism. The cortical bone of the femoral neck is reported to be thinner superiorly than inferiorly (Pickford et al., 2002; Galik et al., 2004); however, published computerized tomographic scans have had poor image quality so that the character state of this trait remains ambiguous (Ohman et al., 2005; Richmond and Jungers, 2008). The proximal shaft of the femur has been noted to be relatively wide mediolaterally, and Richmond and Jungers (2008) suggest that this reflects a response to high gluteal muscle bending moments engendered by the longer femoral neck and more widely flaring ilia in early hominin bipeds. In addition, published photographs of the *Orrorin* femur indicate that the articular surface distribution on the femoral head is more extensive anteriorly than posteriorly, whereas in *Pan*, the opposite is true. In modern humans, the more extensive articular surface on the anterior aspect of the head reflects a more ventrally and inferiorly oriented acetabulum, facilitates more medially oriented femoral positions and helps to maintain femoral-acetabular contact during adduction, while in *Pan*, the articular surface distribution reflects a more strictly laterally facing acetabulum and facilitates abduction (MacLatchy, 1996). Femoral head articular surface distribution thus tentatively suggests the same kind of femoral-acetabular relations as those found in hominin taxa that are almost certainly habitually bipedal.

Other postcranial specimens include a humeral shaft fragment BAR (1004'00), with a well-developed brachioradialis crest, similar to the condition found in *Pan* (Senut et al., 2001), and a proximal manual phalanx (BAR 349'00) with a degree of curvature similar to that of *Pan* phalanges (Richmond and Jungers, 2008).

Remarks The Lukeino Formation, Tugen Hills, Baringo Basin, Kenya, has been independently dated by two teams. The dates concur, but those of Deino and colleagues using ^{40}Ar/^{39}Ar dating support a slightly older age (spanning 6.37 ± 0.05 to 5.73 ± 0.05 Ma; Deino et al., 2002; Kingston et al., 2002) than do those of Sawada et al. (2002) using K-Ar dating (6.17 ± 0.15 to 5.66 ± 0.14). Both teams agree that some relevant fossil sites are below a tuff dated at ~5.7 Ma and above a paleomagnetic reversal interval dated at 5.88 Ma (Deino et al., 2002; Sawada et al., 2002), bracketing most referred *Orrorin* fossils between 5.9 and 5.7 Ma, although doubts have been raised as to how well fossiliferous sediments can be traced laterally to dated beds within the Lukeino Formation (Kingston et al., 2002). In addition, two specimens, the Lukeino molar (KNM-LU 335) and a bone fragment attributed to a proximal femur (Senut et al., 2001) from lower in the section are constrained by the underlying trachyte lava that is older than 6.0 Ma.

Specimens are from four localities; Cheboit, Kapsomin, and Kapcheberek are up to 3 kilometers from one another, and Aragai is ~20 km to the south (based on coordinates in Pickford and Senut, 2001). Until more specific chronostratigraphic control is demonstrated for each of the localities or they are correlated laterally, the temporal range of uncertainty for the sites, as described, is on the order of several 100 ky. Thus, it is possible that the fragmentary *Orrorin* hypodigm represents more than one taxon, including nonhominin hominoids, as Senut (2007) has acknowledged. The preponderance of evidence supports bipedality as a significant component of the positional repertoire of the taxon represented by BAR 1002'00, however.

Comparatively little has been written about the teeth assigned to *Orrorin*. Molars are described as *Pan*-sized with thick enamel, although the "thick" versus "thin" designation has been criticized as overly simplistic (Haile-Selassie, 2001; Haile-Selassie et al., 2004b). Senut et al. (2001) have suggested that *Orrorin* gave rise to *Homo* and that there was no megadont phase in this lineage prior to the emergence of *Homo*; that is, all australopiths are a side branch in human evolution, a view with little support (Haile-Selassie, 2001; Richmond and Jungers, 2008). The large canine with mesial groove resembles those of female *Pan* and lacks the elevated crown shoulders found in *Sahelanthropus*, *Ardipithecus*, and hominins from younger strata.

Most of the Lukeino fossils (including *Orrorin*) are reported to be from shallow lake and floodplain deposits. Ruminants, especially impala, are common, suggesting open woodland, as do proboscideans, with five different species reported. The presence of *Colobus* suggests denser trees lining a lake margin (Pickford and Senut, 2001).

Genus *ARDIPITHECUS* White et al., 1995
Table 25.2

Synonymy Australopithecus (White et al., 1994).

Age and Occurrence Late Miocene–early Pliocene, eastern Africa (table 25.2).

Diagnosis Based on White et al., 1994, 1995. A taxon with less postcanine megadontia, relatively larger upper and lower canines, small, narrow dm1 with minimal cuspule development, possibly thinner canine and molar enamel, and p3 and P3 more asymmetrical and with taller buccal cusps than *Australopithecus*. Temporomandibular joint lacks a definable articular eminence. Compared to *Pan*, the taxon has a smaller, more incisiform upper canine (especially *Ar. ramidus*) and smaller lower lateral incisors.

Referred Species Ardipithecus kadabba and *Ardipithecus ramidus*.

Remarks See species' sections.

ARDIPITHECUS KADABBA Haile-Selassie et al., 2004
Figure 25.3 and Table 25.2

Synonymy Ardipithecus ramidus kadabba, Haile-Selassie, 2001.

Holotype The type specimen is ALA-VP-2/10, a right mandibular corpus with right m3, left i2, c, p4, m2, and an m3 root fragment. A list of other referred specimens is provided in Haile-Selassie (2001) and Haile-Selassie et al. (2004b, 2009).

Age and Occurrence Late Miocene, eastern Africa (table 25.2).

Diagnosis Based on Haile-Selassie, 2001; Haile-Selassie et al., 2004b. Sharp m3 lingual cusps; squared distal outline to M3 with four distinct cusps; shallow mesial fovea on P3; tendency for less relief on mesiolingual crown face of the lower canine; mesiolingually to distobuccally compressed lower canines; and clearly defined anterior fovea on p3. It differs from *Pan*, *Gorilla*, and other extant and Miocene hominoids, including *Orrorin*, in its tendency toward more incisiform lower canines with a developed distal tubercle and in its variable expression of mesial crown shoulder height and mesial marginal ridge development. It differs from *Ardipithecus ramidus* in its more basal termination of the mesial and distal apical crests on the upper canine, and by a more asymmetrical crown outline and relatively smaller anterior fovea on p3.

Description Eleven specimens collected between 1997 and 2001 from five middle Awash localities dated to between 5.8 and 5.2 Ma were initially allocated to a subspecies of *Ardipithecus ramidus*, *Ar. r. kadabba* (Haile-Selassie, 2001). In 2004, an enlarged hypodigm adding six more teeth from the Asa Koma locality (5.8–5.6 Ma) resulted in the elevation of all 17 specimens to a new species, *Ardipithecus kadabba* (Haile-Selassie et al., 2004b).

An upper canine and p3 recovered in 2002 were critical to the designation of the new species. Haile-Selassie et al. (2004b) have indicated that the upper canine has a more basal termination of the mesial apical crest and is higher crowned, and that the p3 crown outline of *Ar. kadabba* is more asymmetrical and retains a buccal wear facet, in contrast to *Ar. ramidus*. The right upper canine (ASK-VP-3/400) has little apical wear, and in this respect is more like *Pan* and unlike *Au. afarensis* and *Sahelanthropus*. The specimen also has low mesial shoulders. Upper and lower canines are projecting and interlocking. However, although not consistently expressed, lower canines are more incisiform, with

better-developed distal tubercles and mesial marginal ridges, and elevated mesial crown shoulders compared to those of *Pan*. One lower canine (STD-VP-2/61) is taller with a narrow apex and has a lower mesial crown shoulder and less prominent distal tubercle than the lower canine in the holotype, which is shorter. The holotype canine has a distal tubercle, which has a posteriorly oriented wear facet, as in apes with a C/p3 honing complex; however, the facet is worn horizontally, not diagonally as in apes, suggesting that a fully functioning C/p3 honing complex was not operating (see Haile-Selassie et al., 2004b).

A left i2 is relatively small and has simple lingual morphology (Haile-Selassie et al., 2009). Lower third molar cusps are less rounded than in *Australopithecus*, and the m3 is smaller than those of *Au. anamensis* but larger than in *Pan* (Haile-Selassie, 2001). Maximum radial thicknesses of lateral enamel have recently been published for a number of molars and range between 1.0 and 2.0 mm (Haile-Selassie et al., 2009). Molars have more buccal than lingual wear, with scooped dentin exposure, unlike later hominins, and Haile-Selassie and colleagues (2009) have interpreted this to indicate "erosive" rather than "abrasive" wear. An M1 lacks the occlusal surface crenulation found in *Pan* molars (Haile-Selassie, 2001).

The mandible has been described as resembling those of *Sahelanthropus* and *Ar. ramidus* (including the Tabarin and Lothagam specimens, discussed later) in being smaller and thinner than is typical for early *Australopithecus* (Haile-Selassie et al., 2009).

Forelimb bones include the distal two-thirds of an intermediate hand phalanx (ALA-VP-2/11) that is larger than those of *Au. afarensis,* although it is similar to the latter taxon in terms of the deep fossae for the flexor digitorum superficialis muscle (Haile-Selassie, 2001). A distal humerus fragment (ASK-VP-3/78) has a deep olecranon fossa with steep walls (Haile-Selassie et al., 2009), and a clavicle (STD-VP-2/893) is robust with a strongly marked deltoid insertion (Haile-Selassie, 2001).

A left fourth proximal foot phalanx (AME-VP-1/71, from the chronologically youngest sediments) has a dorsally canted proximal articular surface (Haile-Selassie, 2001), suggesting force was transmitted during dorsiflexion rather than plantarflexion (like *Au. afarensis* and unlike *Pan*). This is provisional evidence for bipedality (Latimer and Lovejoy, 1990a). Phalanx size and curvature correspond to known *Au. afarensis* specimens (Haile-Selassie, 2001) and are shorter and less curved than in *Pan* (Haile-Selassie et al., 2009).

Remarks As noted by Haile-Selassie et al. (2004b:1505): "No known *Ardipithecus ramidus*, *Australopithecus afarensis* or *Australopithecus anamensis* lower p3 exhibits any sign of a mesiobuccally oriented [honing] facet on its buccal crown face." The p3 of *Ar. kadabba* has a buccal wear facet; although this is distinguishable from that of extant apes, it suggests an intermediate stage of C/p3 anatomy not far removed from the presumed honing complex of the last common ancestor of *Pan* and *Homo*. The demonstration of this intermediate stage of C/p3 anatomy places *Ar. kadabba* in a possible basal hominin role.

Haile-Selassie et al. (2004b) have suggested that *Sahelanthropus*, *Orrorin*, and *Ardipithecus* may be congeneric (in which case the genus name would be *Ardipithecus*) and perhaps conspecific. Relatively little evidence contraindicates this for *Orrorin*; distinguishing dental features (e.g., molar flare and enamel thickness) are subtle and rest on small sample sizes. The canine has been distinguished from those of other

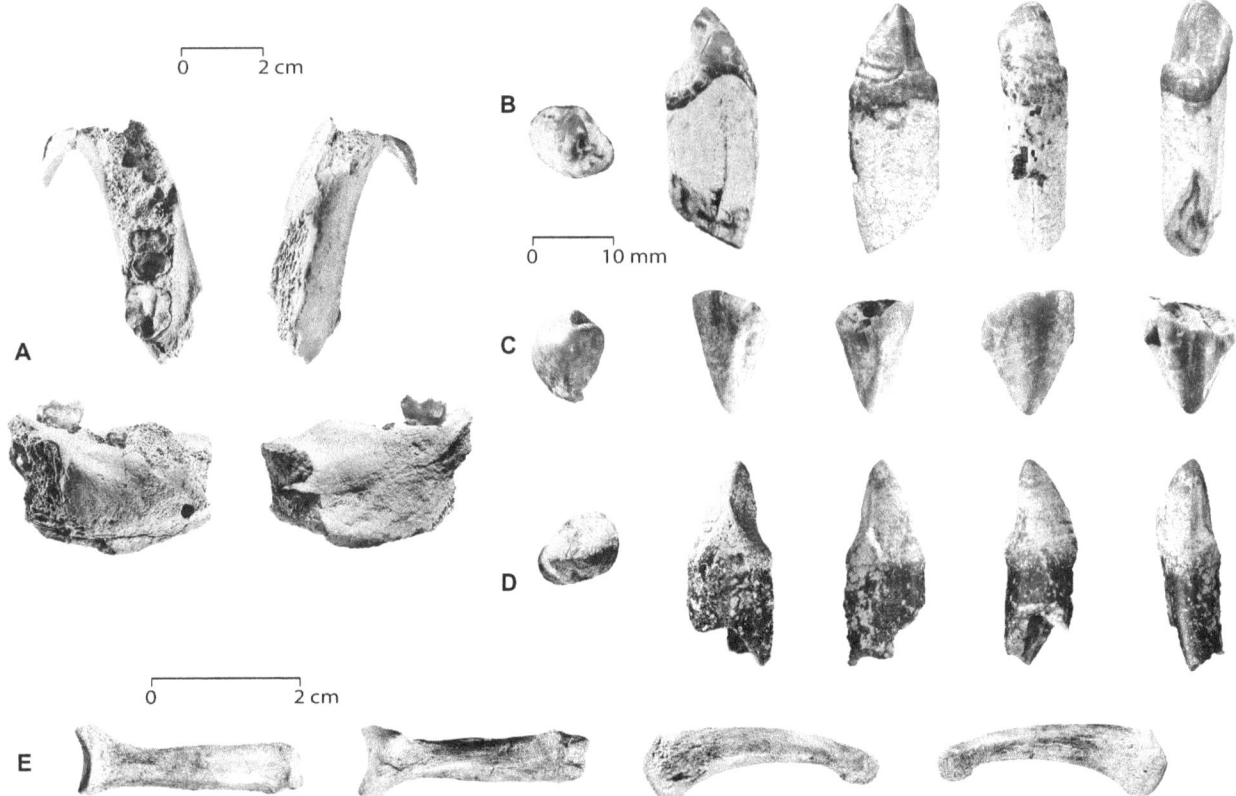

FIGURE 25.3 A) Holotype of *Ardipithecus kadabba*, mandible ALA-VP-2/10; B) holotype of *Ardipithecus kadabba*, canine ALA-VP-2/10; C) isolated right upper canine of *Ardipithecus kadabba* ASK-VP-3/400; D) isolated right lower canine of *Ardipithecus kadabba* STD-VP-2/61; E) proximal foot phalanx of *Ardipithecus kadabba* AME-VP-1/71. Courtesy of Tim White.

hominins but not *Pan*, and as discussed, the *Orrorin* teeth may not represent the same taxon as the postcrania. A greater number of features distinguish *Sahelanthropus* and *Ardipithecus* (e.g., in addition to features in the initial diagnosis, *Sahelanthropus* lacks a diastema between I2/C and has a much thicker browridge, even when sexual dimorphism is taken into account), but the two share derived aspects of basicranial morphology (Suwa et al., 2009b).

In a preliminary report on the environment, WoldeGabriel et al. (2001) note that the Adu-Asa Formation hominin sites (four of five hominin sites yielding 10 of 11 hominin fossils) are part of a vertebrate assemblage that includes reduncine bovids (suggesting open woodland or wooded grassland), *Tachyoryctes* (root rats, found today in upland grasslands), and *Thryonomys* (cane rats, found today at lake and river margins), but few hares (suggesting that open grassland is not well sampled). They contrast this with the slightly younger (5.2 Ma) Kuseralee Member of the Lower Sagantole Formation deposits that have yielded only one hominin fossil as part of an assemblage including unspecified bovids, carnivores, and cercopithecid monkeys. Pedogenic oxygen isotope values have been interpreted as evidence that the Adu-Asa sites are sampling higher-altitude and possibly wetter habitats than the Lower Sagantole sites, but sample size is small. Stable carbon isotopic $\delta^{13}C$ values from the paleosol carbonates do not distinguish between the two members (nine samples from the Adu-Asa Formation range between −7.5‰ and −4.1‰, with a mean of −6.4‰, while two samples for the Lower Sagantole are −5.2‰ and −6.3‰) and are compatible with woodland

to grassy woodland habitats. These authors emphasize that the hominins are most common at the localities sampling higher elevation and possibly more closed habitats. However, the presence of pedogenic carbonates is itself a semi-arid to arid indicator, and the carbon values are relatively enriched. This, along with limited published faunal data, suggests that additional evidence is needed to support this team's contention that hominins older than 4.4 Ma "may have been confined to woodland and forest habitats" (WoldeGabriel et al., 2001:177). An association with closed environments for *Ar. kadabba* has also been questioned by Levin et al. (2008), who find stable carbon isotopic data from fossil herbivore enamel from *Ar. kadabba*– (and *Ar. ramidus*–) bearing deposits at Gona (~ 90 km north of Middle Awash sites) to resemble the signatures found in extant herbivores living in bushland and eating both C_4 and C_3 plants.

ARDIPITHECUS RAMIDUS White et al., 1995
Figure 25.4 and Table 25.2

Synonymy Australopithecus ramidus, White et al., 1994; *Homo antiquus praegens,* Ferguson, 1989, in reference to the Tabarin mandible (see below).

Holotype The type specimen is ARA-VP-6/1, an associated set of teeth including left I1, C, P3, P4, and right I1, C, P4, and M2, and lower right p3–4 (White et al., 1994). A list of other specimens is provided by White et al., 1994, 1995, 2009b, and Semaw et al., 2005.

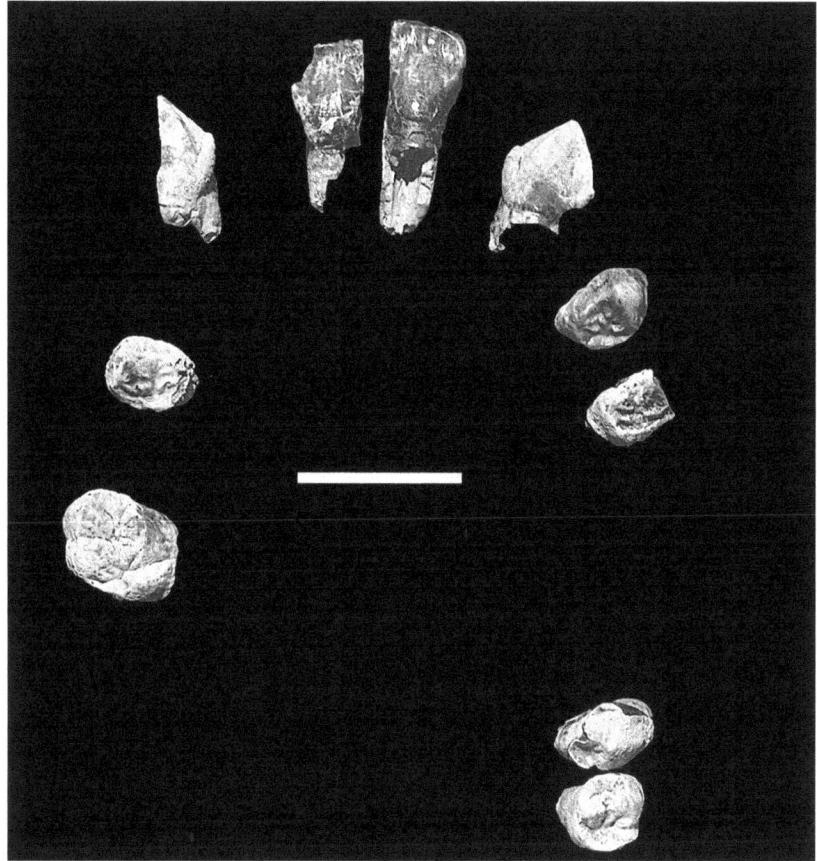

FIGURE 25.4 Holotype of *Ardipithecus ramidus*, associated teeth ARA-VP-6/1. Courtesy of Tim White. Scale = 25 mm.

Age and Occurrence Early Pliocene, eastern Africa (table 25.2).

Diagnosis Based on White et al. (1994). The features distinguishing this taxon from australopiths are listed in the generic diagnosis given earlier. *Ardipithecus ramidus* differs from extant apes in its more incisiform canine morphology and relatively higher canine crown shoulders; relatively small p3 without functional honing facet; relatively broader lower molars; and more anteriorly positioned foramen magnum. It is further distinguishable from *Pan* in its smaller I1, elongate and relatively larger m3, and less crenulated molars.

Description Key to assignment of *Ardipithecus* to Hominini is the upper canine and anterior lower premolar anatomy. The upper canine is less projecting and has more extensive mesial and distal apical crests that that of *Ar. kadabba*, while the p3 is reduced in size and lacks a honing facet. Several authors have noted that *Ar. kadabba–Ar. ramidus–Au. anamensis–Au. afarensis* form an increasingly derived continuum of canine/premolar anatomy (Leakey et al., 1995, 1998; Kimbel et al., 2006; White et al., 2006). Mean canine size in the Aramis sample is comparable to that of female *Pan*, and Suwa and colleagues (2009a) propose that the upper canine is relatively more reduced than the lower canine.

Canine dimorphism is low, with purported males having mean crown diameters only 10–15 % larger than assigned females (Suwa et al., 2009a). Incisors are smaller than those of *Pan* and *Pongo*, while postcanine teeth have been found to be less megadont than in *Australopithecus* or *Pongo*, but larger than in *Pan* (Suwa et al., 2009a). Molars from Aramis are not as broad buccolingually as are those attributed to *Au. afarensis* and they

possess thinner enamel, though not as thin as in *Pan* (Suwa et al., 2009a) *contra* early assessments (White et al., 1994).

The morphology of the dm1 is also considered significant: it is narrow, lacks an anterior fovea, has a large protoconid, a small, distally placed metaconid, and a small, poorly differentiated talonid (White et al., 1994). In these features, it retains the presumed primitive morphology, resembling extant *Pan* and Miocene taxa such as *Dryopithecus*, and differs from the condition seen in australopiths (with the exception of *Au. anamensis*, which is intermediate in morphology between *Ar. ramidus* and *Au. afarensis* [Leakey et al., 1998]), which have buccolingually expanded crowns, larger metaconids, and larger talonids with distinct cusp relief (White et al., 1994).

A partial adult basicranium (ARA-VP-1/500) may also support the hominin status of *Ardipithecus*. The anterior border of the foramen magnum is almost at the level of the carotid foramina, and the skull is described as having a shorter basioccipital region of the cranial base than in extant apes (White et al., 1994; Suwa et al., 2009b), possibly reflecting more habitually erect postures (i.e., during bipedalism) than in these taxa or, alternatively, neural reorganization (Suwa et al., 2009b). This specimen and one other (ARA-VP-1/125) also preserve temporal bone anatomy: there is marked pneumatization of the temporal squama, the temporomandibular joint is flat and lacks an articular eminence, and the tympanic is tubular.

Discovery of a partial skeleton was reported by White et al. in 1995, but until 2009 the only published elements referred to *Ar. ramidus* remained restricted to some incomplete jaws, teeth, cranial fragments, and a few postcranial fossils, mostly of the forelimb. The first additions to the *Ar. ramidus* hypodigm came from Gona (Semaw et al., 2005), but subsequently more fossils, including the ARA-VP-6/500 associated skeleton, have been reported from the Aramis locality (White et al., 2009b), as well as from two other localities, Kuseralee Dora (KUS) and Sagantole (SAG), in the Central Awash Complex (White et al., 2009b). The fragmented skull of ARA-VP-6/500 includes portions of the vault, base, right face, and much of the left side of the mandibular corpus. The maxilla displays a diastema between I2 and C and has weak subnasal prognathism, which is reflected in a zygomatic root positioned above M1, more posterior than in *Australopithecus* but more anterior than in *Pan* (Suwa et al., 2009b). The frontal torus is at the low end of the range of vertical thickness found in *Pan troglodytes* (Suwa et al., 2009b) and is unlike the torus in *Sahelanthropus*, which exceeds that of *Gorilla* in size (Brunet et al., 2002). This character, along with the diastema, distinguish *Ardipithecus* from *Sahelanthropus*; otherwise the two genera share a similar cranial base morphology, a projecting midface, and a low endocranial capacity (Suwa and colleagues [2009b] estimate the endocranial volume of ARA-VP-6/500 to be in the range of 280–350 cc.)

Mandibular specimen ARA-VP-1/401 has a receding symphysis similar to that of *Au. anamensis* but has a less inflated corpus (Suwa et al., 2009b). It also appears the canine was not incorporated into the incisor tooth row, unlike in *Australopithecus* (Suwa et al., 2009b).

Although an associated left humerus, ulna, and radius (ARA-VP-7/2) were part of the original paratype of this taxon (White et al., 1994), ARA-VP-6/500, and other additions to the hypodigm reported in White et al. (2009a) greatly expand our knowledge of the postcranial skeleton of *Ar. ramidus*. The ARA-VP-6/500 individual, like the rest of the hypodigm from sites in the Central Awash Complex, is radiometrically dated at 4.4 Ma (WoldeGabriel et al., 2009). ARA-VP-6/500 has among the smallest canines for the entire Aramis sample, and probabilistic assessments using bootstrapping suggest it is unlikely that the canines could be attributable to a male, even assuming low–moderate dimorphism (Suwa et al., 2009a). The thin supraorbital torus has also been cited as evidence that the skeleton represents a female (Suwa et al., 2009b). However, body size estimates derived using geometric means of measures of the talus and capitate are in the range of 50 kg (Lovejoy et al., 2009b), larger than any female body weight estimates for sufficiently well-sampled australopiths, but similar to estimates of body weight derived from the limited *Au. anamensis* remains. The latter lacks evidence for postcranial dimorphism (i.e., it has been assumed males have been sampled) but does evince canine dimorphism (Ward et al., 2001). Since the Aramis individual is at the high end of the postcranial size distribution, it could be suggestive of low–moderate body size sexual dimorphism, more comparable to that of genus *Pan* than *Gorilla* (Suwa et al., 2009a). Body size estimates from other remains, especially long bones, will be useful in evaluating this claim.

Hand bones of the individual are unusually complete and well preserved and have been interpreted by Lovejoy and colleagues (2009b) as lacking knuckle-walking features. For example, the dorsal surface of the proximal metacarpals do not possess ridges, and the heads are not expanded. Moreover, the hand is also interpreted as not having adaptations for suspensory behavior. The 5th metacarpal-hamate articulation is described as cylindrical/condyloid and therefore mobile, rather than planar and therefore rigid as in large-bodied suspensors

(Lovejoy et al., 2009b). Likewise, the articular relations of metacarpals 2 and 3, and the capitate and trapezoid, are described as lacking the complexity that leads to rigidity in this region in great apes and that may be functionally related to stiffening the wrist for stable manual suspension. *Ardipithecus* is also reported to lack the elongation of the metacarpals that characterizes extant hominoids, though it does have moderately elongated phalanges. Proximal ulnar and distal humeral morphology are suggestive of full elbow extension with considerable loading in this position. For example, the olecranon process is short, and the zona conoidea of the humerus is deep, with a posteriorly extended lateral wall. However, it is proposed that a fully extended elbow joint was used during manipulative foraging rather than during suspension (Lovejoy et al., 2009b).

The pelvis of ARA-VP-6/500 is represented by an almost complete but crushed and distorted left innominate and a portion of the right ilium. If the published reconstruction accurately portrays the superior and lateral extent of the ilium, then the ilium was superoinferiorly shortened relative to all extant nonhuman hominoids, suggesting orthograde postural control in the sagittal plane, and lateral flare, possibly resulting in enhanced abductor function during one-legged stance, an important adaptation for bipedality (Lovejoy et al., 2009d). The ischium, however, is African ape–like with a presumably large ischial tuberosity anchoring the important climbing muscles of the hamstrings (Lovejoy et al., 2009d). The pelvis also preserves an anterior inferior iliac spine. Lovejoy and colleagues' (2009d) reconstruction of the *Ardipithecus* pelvis also suggests that the lower lumbar vertebrae were not stabilized to the ilia with ligaments but instead were "free" and thus able to situationally produce lumbar lordosis during bipedal locomotion. However, this interpretation is likely erroneous, since the lower lumbar vertebrae of modern humans are both lordotic *and* powerfully fixed to the ilium by ligaments, and there is evidence that this was the case for australopiths, as well; furthermore, the lower lumbar vertebrae of all terrestrial mammals have some degree of ligamentous connection to the pelvis (Sanders, 1998). The lower limb of *Ardipithecus* is represented by a fragmentary femoral shaft (ARA-VP-1/701), and parts of the femur, tibia, and fibula of ARA-VP-6/500. Both the fibula and the apparently nearly complete tibia remain essentially undescribed, though the tibia is reconstructed as roughly 262 mm in length. Both the femur and tibia of ARA-VP-6/500 are noted to have been quite damaged. The femur of *Ardipithecus* lacks a lateral spiral pilaster, which delineates the attachment for the vastus lateralis and gluteus maximus muscles in African apes. Instead, this femur possesses a third trochanter and a rugosity that may be homologous to the hypotrochanteric fossa found in australopith femora. These morphologies are interpreted as primitive, with the lateral spiral pilaster being derived in African apes (Lovejoy et al., 2009d).

As was the case with the hands, the foot bones of ARA-VP-6/500 are remarkably complete and well preserved, with 33 individual fossils and even two sesamoids represented (White et al., 2009b). Fifteen other foot fossils attributed to *Ardipithecus* have been found from sites within the Middle Awash study area (navicular, seven metatarsals, and seven phalanges; White et al., 2009b). The foot of *Ardipithecus* displays a unique combination of anatomies not seen in any extant or extinct hominoid. The following description of the foot is from Lovejoy et al. (2009a). The interpretation of this novel foot anatomy is that *Ardipithecus* was capable of both arboreal grasping and terrestrial propulsion. Most notably, the foot of *Ardipithecus* possesses a strongly abducted hallux, indicative of careful arboreal climbing in this taxon. The abduction angle is 68°, similar to the mean in the African apes.

The tarsal region is represented by a talus, cuboid, and cuneiforms. The well-preserved talus (ARA-VP-6/500-023) is African ape–like, possessing a mediolaterally wide distal aspect of the talar trochlea, and a relatively high talar axis angle. This latter angle suggests that *Ardipithecus* lacked the strong bicondylar angle found in later *Australopithecus*. The talus also has a trapezoidal angle forming the groove for the flexor hallucis longus tendon, a feature found more often in African apes than in modern humans and australopiths. However, the talus does have a palpable tubercle for the anterior talofibular ligament, an important ligament for ankle stability found in modern humans but only rarely in the great apes.

The cuboid is a rarely preserved element in the hominin fossil record, yet both the right and a portion of the left are preserved in ARA-VP-6/500 (White et al., 2009b). Unlike modern humans and the OH 8 hominin, the *Ardipithecus* cuboid does not have an eccentrically positioned calcaneal process and therefore may have lacked the derived calcaneocuboid locking mechanism found in obligate bipeds. However, like modern humans, the cuboid is proximodistally elongated, which would increase the lever arm for the plantarflexors during toe-off. A proximodistally enlarged navicular (ARA-VP-6/503) also contributes to the elongated midtarsal region. The cuboid-metatarsal joint is flat, like that found in modern humans, and unlike the joint surface of nonhuman primates. This morphology suggests that *Ardipithecus* may have possessed a rigid midfoot, perhaps unable to produce midfoot flexion. This assertion is supported by the presence of a facet for the os peroneum on the lateral aspect of the cuboid. This sesamoid (perhaps represented by ARA-VP-6/500-093) repositions the peroneus longus tendon out of the cuboid groove to a position more obliquely oriented across the plantar aspect of the foot, helping to stiffen the midfoot. The os peroneum is normally not present in African apes which have more midfoot mobility and are able to conform to arboreal substrates.

Further evidence for plantar rigidity can also be found in the relatively expanded bases of the lateral metatarsals, which are dorsoplantarly tall relative to the length of the metatarsals. The preserved lateral metatarsal head (ARA-VP-6/505-MT3) is domed and possesses a sulcus between the head and the shaft of the bone, consistent with the type of strong phalangeal dorsiflexion that occurs during the toe-off phase of bipedal locomotion. The second metatarsal (ARA-VP-6/1000) displays strong shaft torsion like that found in African apes, consistent with the presence of a grasping first ray. However, the third metatarsal does not possess African ape–like metatarsal torsion, and instead is more modern human–like, suggesting that any toe-off occurred along the oblique axis of the foot. The first metatarsal may have served a balance role during terrestrial travel, rather than a propulsive one, thus the shift to the transverse axis during the bipedal locomotion practiced by later hominins is a more recent adaptation. The pedal phalanges are curved, and are similar in relative length to that in the *Gorilla*, but unlike African ape phalanges, they have a dorsiflexion cant to them, a morphology found in modern human phalanges.

Ardipithecus ramidus specimens from Gona, Ethiopia, described by Semaw and colleagues (2005), include seven partial jaw and dental specimens, three manual phalanges, and a pedal phalanx. Both upper (GWM9n/P51) and lower (GWM9nP50) canines are represented. The upper canine is large and diamond shaped and, as in the Aramis sample, has distinct mesial and distal shoulders. The lower canine has an

elevated mesial shoulder and a distinct distal marginal tubercle. An m1 is small, bunodont, and has enamel thickness (~1 mm) comparable to the Aramis specimens. Overall, wear rates seem low, suggesting a nonabrasive diet. Manual phalanges are reported to resemble those of *Au. afarensis,* except that the base of the proximal articular surface is transversely broad and long. A proximal pedal phalanx preserves the proximal articular surface, and has a dorsally oriented, transversely broad proximal facet, as reported for *Ar. kadabba* (Haile-Selassie et al., 2004b) and *Au. afarensis* (Latimer and Lovejoy, 1990b).

Remarks Two fragmentary hominoid specimens from Kenya are of the appropriate age and may belong to the *Ar. ramidus* hypodigm. White and colleagues were hesitant to refer them to *Ar. ramidus* when the latter taxon was erected because at the time (in 1994) they lacked diagnostic features.

- The Lothagam mandible (KNM-LT 329) is a fragment of the right side of the mandibular corpus preserving the m1 crown and roots of m2–3 (White, 1986a; Hill and Ward, 1988). It is from the Lothagam sequence in northern Kenya and is dated between 5.0–4.2 Ma (McDougall and Feibel, 1999).

- The Tabarin mandible (KNM-TH 13150), is a fragment of the right side of the mandibular corpus with worn m1–2 and portions of the alveoli of p4 and m3 (Ward and Hill, 1987). Although tentatively assigned to *Au. afarensis* (e.g., Ward and Hill, 1987), the publication of *Ar. ramidus* made apparent certain primitive features, including the narrower molars and thinner enamel, shared by the Tabarin specimen and *Ar. ramidus,* relative to *Au. afarensis* (Hill, 1999; Deino et al., 2002). The Tabarin mandible was recovered from the Chemeron Formation, Baringo Basin, Kenya, and is dated at 4.48–4.41 Ma (Deino et al., 2002).

Reported lack of suspensory adaptations in the hand of *Ardipithecus* has been used by Lovejoy and colleagues (2009a, 2009b, 2009c, 2009d; White et al., 2009b) to argue that these features, and therefore forelimb suspensory behavior in general, evolved independently in hylobatids and each great ape genus, rather than being lost in *Ardipithecus*. This perspective is echoed in this team's analysis of the entire postcranium, and shared features related not just to suspension, but vertical climbing and orthogrady, are considered hominoid homoplasies. Their assumption is that *Ardipithecus ramidus* preserves in many respects the attributes of the Last Common Ancestor (LCA) of *Pan* and *Homo*. Unfortunately, there is still considerable uncertainty as to the timing of the *Pan-Homo* split, with estimate ranging from 4–8 Ma (Bradley et al., 2008). Pinpointing this estimate is of importance, for if the younger estimates are true, then if *Ardipithecus* is a hominin, it may indeed preserve many features of the LCA. On the other hand, if the older estimates are true, then *Ardipithecus* may have experienced upward of 3 Ma of independent evolution since the split.

An alternative idea, that *Ardipithecus* is derived from a suspensory, orthograde ancestor, remains to be rigorously tested with more detailed comparative analyses and a more explicit consideration of the polarity of character states that also includes additional information about Miocene hominoids. It is also possible that *Ardipithecus* is not a hominin, in which case, features such as the foramen magnum, reduced canines, and bipedal features in the foot and pelvis would be homoplasies, or that *Ardipithecus* is a hominin but not ancestral to australopiths. For now, phylogenetic continuity among the

Ar. kadabba–Ar. ramidus–Au. anamensis–Au. afarensis series remains well supported biogeographically and by dental evidence, but the postcranial data, in light of the new partial skeleton, are ambiguous and controversial. If *Ar. ramidus* is a hominin, then the scale of homoplasy implied in closely related clades makes the hypothesis that it is a basal hominin a far from the most parsimonious interpretation of this important new evidence for African higher primate evolution.

Despite the detailed functional reconstructions of the Aramis partial skeleton, the overall locomotor profile of ARA-VP-6/500 is enigmatic. It is suggested that this individual neither vertically climbed nor used forelimb suspension to any significant degree, was pronograde and palmigrade in the trees and orthograde and bipedal on the ground (Lovejoy et al., 2009a, 2009b, 2009c, 2009d). No modern or previously described fossil primate occupies such a niche. This functional interpretation raises the intriguing question of how a 50+ kg animal can be a successful arborealist, absent vertical climbing and suspension. Large-bodied extant pronograde monkeys are awkward in the trees (with the possible exception of *Nasalis*), and great apes rely critically on suspension to move arboreally. However, *Ardipithecus* is not reconstructed as a ripe fruit eater like *Pan* or *Pongo* (e.g., as evidenced by narrow incisors), but rather as a generalized frugivore/omnivore (Suwa et al., 2009a). Thus, if the locomotor limitations reconstructed are correct, then while foraging arboreally, it may have moved ponderously on large supports in the lower canopy rather than the upper canopy (Lovejoy et al., 2009b). The arboreal food resources available at this level, and which would have to have been incentive to maintain a grasping hallux, are not detailed, however.

If body size and canine dimorphism are low as suggested, then this may imply weak male-male competition, more comparable to *Pan paniscus* than *Pan troglodytes* (Suwa et al., 2009a). In addition, if *Ardipithecus* is the sister taxon to *Australopithecus*, and if canine reduction was initially a consequence of social change, then postcanine megadonty and thick enamel may have been enabled by a hypothetical decrease in selection to maintain long, interlocking canines. However, since *Ardipithecus* molars are relatively broad, and the enamel thicker than that of *Pan*, it is hard to resolve whether or not dietary selection was already at work. This is in part due the lack of relevant fossils of the African ape and modern human LCAs. Many middle and late Miocene hominoids have thick enamel, so it is also possible that *Ardipithecus* had reduced enamel thickness relative to the LCA.

Seven localities at Gona have yielded fossils attributed to *Ar. ramidus*. The Gona Western Margin sequence is composed of small-scale fluvial, lacustrine, and volcaniclastic elements, and the environment of deposition is reconstructed as lakes, swamps, springs, and streams amid local volcanic centers (Semaw et al., 2005). Faunal remains support the presence of some open habitats; for example, papionines are more common than colobines and there are numerous grazing bovids. Stable carbon isotopic analyses of tooth enamel reveal a high proportion of grazing herbivores (Levin et al., 2008), while paleosol samples yielded mean $\delta^{13}C$ values of -7.5 ‰, indicating mixed habitat, with both C_3 and C_4 plants (Semaw et al., 2005). Semaw and colleagues (2005; Levin et al., 2008) note that although woodland, grassy woodland and C_4 grasslands were presumably present, habitat preferences within this mosaic are as yet unknown.

The Gona paleoenvironmental interpretation is somewhat at odds with the reading of the environment at Aramis, where hominins are localized as occupying "woodland with patches of forest" (White et al., 2009a:92). However, 28 paleosol carbonate values for the Lower Aramis Member (excluding those from the nonhominin-bearing sites to the east interpreted to be more open) have a mean of -4.1‰ (Woldegabriel et al., 2009), indicating a greater C_4 component and potentially more open habitats relative to Gona. Furthermore, $\delta^{13}C$ values from mammal tooth enamel at Aramis show evidence of herbivores with predominantly C_3, predominantly C_4 as well as mixed diets (White et al., 2009a), and include large-bodied grazers such as *Anancus*, which would require significant amounts of grass. Five *Ardipithecus* samples show little variation, with a mean $\delta^{13}C$ of -10.25‰, similar to values obtained for C_3 browsers (White et al., 2009a). Comparisons with published values for other hominoids suggest a diet enriched in ^{13}C plants relative to chimpanzees but less enriched than australopiths (Merwe et al., 2008; Smith et al., 2010; White et al., 2009a).

The Aramis fossils accumulated on a flat plain, and though tooth marks are common on the bone, carnivore activity is not thought to have resulted in bone concentration, and there is no evidence of fluviatile transport (WoldeGabriel et al., 1994; Louchart et al., 2009). Thus, taphonomic bias in terms of the ecomorphological implications of the faunal composition may be minimal, and the finding that woodland-adapted bovids (e.g. *Tragelaphus*) and arboreal cercopithecoids represent over 50 % of all macrovertebrate remains may be significant (White et al., 2009a). Nonetheless, overall, the local vegetative ecosystem appears to have been heterogeneous, like that at Gona, and where the specific ecological niche of *Ardipithecus* fits within this mosaic is not obvious. It could be argued that *Ardipithecus*' possession of a grasping hallux is one of the best indicators that the taxon may have preferentially used the woodland/forest over the grassy woodland savanna habitats.

Subtribe AUSTRALOPITHECINA Gregory and Hellman, 1939
Genus *AUSTRALOPITHECUS* Dart, 1925
Table 25.2

Partial Synonymy Plesianthropus, Broom, 1936; *Paranthropus,* Broom, 1938; *Homo,* Mayr, 1950; *Praeanthropus,* Şenyürek, 1955

Age and Occurrence Early Pliocene–early Pleistocene, eastern, southern, and Central Africa (table 25.2).

Diagnosis Dart's (1925) initial description of the taxon was based on a juvenile cranium, the Taung child, and it referred to several features, such as an enlarged brain size relative to apes, and relatively slight facial prognathism, that would not be given emphasis now. However, the anteriorly positioned foramen magnum and robust mandible with small canines and no diastema remain key diagnostic characters.

Currently, the genus *Australopithecus* is recognized in the following gradistic terms, by a suite of both primitive and derived features, and is considered paraphyletic (see also Kimbel, 2007): extant ape-sized brain; small incisors and canines relative to body weight; lower anterior premolar does not hone the upper canine; postcanine teeth relatively large with thick enamel and bulbous cusps; premolars with more complex occlusal anatomy; robust maxilla, zygomatic bone and mandible; short, vertical midface; subnasal prognathism; anteriorly placed foramen magnum; and postcranial adaptations for bipedality.

Referred Species Australopithecus africanus, Dart, 1925; *Au. afarensis,* Johanson et al., 1978; *Au. anamensis,* Leakey et al., 1995; *Au. bahrelghazali,* Brunet et al., 1995; *Au. garhi,* Asfaw et al., 1999.

Remarks See species' sections.

AUSTRALOPITHECUS ANAMENSIS Leakey et al., 1995
Figure 25.5 and Table 25.2

Holotype The type specimen is KNM-KP 29281, a mandible preserving all teeth but without rami. A partial left temporal bone is likely to be from the same individual and has the same accession number. More specimens are listed in Leakey et al., 1995, 1998 and White et al., 2006.

Age and Occurrence Early Pliocene, eastern Africa (table 25.2).

Diagnosis Based on Leakey et al., 1995, 1998. It differs from all other species of *Australopithecus* in its small, elliptically shaped external auditory meatus; long mandibular bodies; closely spaced, parallel tooth rows in both mandible and maxilla; mental region of mandible not strongly convex; symphysis steeply inclined posteriorly; canines with long, robust roots; trigons of upper molars wider than talons; smoothly continuous lateral nasal aperture; and small medullary cavity of humerus. It can be distinguished from *Ar. ramidus* by its thicker enamel; more buccolingually expanded molars; subequally sized m1 and m2; a tympanic tube that extends only to the medial edge of the glenoid process; somewhat larger dm1; and humerus with weakly developed lateral trochlear ridge.

Description Relatively complete mandibular and maxillary remains referred to both male and female specimens have revealed a consistent suite of dental features that allow this taxon to be distinguished from the geologically more recent *Au. afarensis* (see Leakey et al., 1995, 1998). The canine in the holotype is smaller than two isolated canines and the canine socket in a large mandible (KNM-KP 29287); therefore, the type is presumed to be from a female and canine dimorphism is thought to be substantial; all canines have long, robust roots, and the upper canine has two basal tubercles and a large root. The long axis of the mandibular symphysis is posteriorly inclined, and the three mandibles have strongly receding but smoothly convex symphyseal contours, unlike many *Au. afarensis* specimens (especially those from Hadar). The tooth rows are closely spaced, and the cross-sectional profile of the corpora is unlike that of African apes, whose contours become flatter in section basally.

The i2 is larger, the p3 is more unicuspid and asymmetrical, and the C/p3 complex is more ape-like than in *Au. afarensis*. Molar enamel thickness is similar to that of *Au. afarensis* (1–2 mm), but thicker than in *Ardipithecus ramidus* and thinner than in *Paranthropus* taxa (Ward et al., 2001). Molars are more buccolingually expanded, and canine enamel thicker apically, than in *Ar. ramidus*. The lateral nasal aperture is smoothly continuous with the maxillary bone, as in apes, and unlike the condition in *Au. afarensis*, which has distinct lateral nasal crests.

Part of a left temporal bone is probably associated with the type mandible. The external auditory meatus is small and elliptical in outline, unlike the meatus of *Au. afarensis* but like those in *Pan* and *Ardipithecus*. The tympanic tube is shorter than in *Ardipithecus,* and the articular eminence is less well developed than in other australopiths and *Homo.*

Dental specimens from Asa Issie, Ethiopia (see White et al., 2006), all possess enamel thickness, molar size and canine shape that are similar to known *Au. anamensis* specimens from Kenya. The canines of a maxilla are as large or larger than those of *Au. afarensis* and *Au. anamensis,* and they are mesiodistally long as in *Au. anamensis* from Kenya.

Postcranial specimens attributed to *Au anamensis* have received considerable attention because of their large size and because the hindlimb is derived toward bipedality, while the forelimb retains numerous primitive features. A right tibia

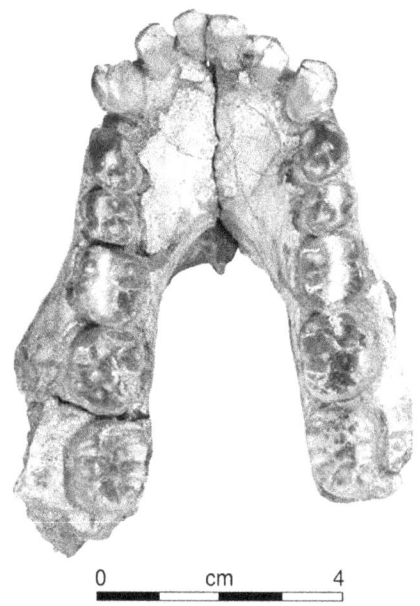

FIGURE 25.5 Holotype of *Australopithecus anamensis*, mandible KNM-KP 29281. Courtesy of National Museums of Kenya.

(KNM-KP 29285) preserving both epiphyses, but missing a connecting portion of the shaft, is larger than the largest Hadar tibia attributed to *Au. afarensis*. Using regression equations based on modern human data, the upper epiphyseal surface area yields a body mass estimate of 55 kg, while the lower epiphyseal surface area yields a body mass estimate of 47 kg (Leakey et al., 1995). Bipedal features of the tibia include rectangular proximal surface with anteroposterior lengthening of articular surfaces, concave condyles of equal area, vertically oriented, straight shaft, and an inferiorly facing, square-shaped distal articular surface (Leakey et al., 1995). Primitive features include a proximal metaphysis that is not expanded, and a strong insertion for the gracilis muscle next to the anterior border of the shaft.

Body mass regressions (using modern human data) for the Kanapoi left distal humerus (KNM-KP 271) first described in 1967 (Patterson and Howells, 1967) suggest a body mass of 58 kg (McHenry, 1992a). In anatomy it closely resembles specimens attributed to *Au. afarensis*. It lacks the lateral extension of the trochlear joint surface found in apes (which is thought to resist loads during hyperextension of the elbow during knuckle walking) and lacks the prominent lateral epicondyle found in *Ar. ramidus* (Ward et al., 2001). It possesses very thick cortical bone, near the maximum of the observed range in African apes and modern humans and more like that found in *Pongo*.

The Kanapoi capitate (KNM-KP 31724) is more primitive than known specimens of *Au. afarensis* in that the articular facet for metacarpal II faces strictly laterally as in apes, suggesting little rotational capacity at the carpometacarpal II joint (Leakey et al., 1998). It has a large, globular head and is larger than two *Au. afarensis* capitates from Hadar. The facet for the lunate is greater than that for the scaphoid, as in other *Australopithecus* specimens.

The Allia Bay radius (KNM-ER 20419) is also large; if proportioned like *Homo*, then stature calculated for this individual ranges between 176 and 183 cm (Ward et al., 2001). Ward et al. (2001) consider it more likely that *Au. anamensis* had relatively long arms. Richmond and Strait (2000) claim the radius retains evidence of knuckle walking, but Ward et al. (2001) dispute this because the former authors did not correct for the missing styloid process in a cast. The contact facet for the lunate is larger than the scaphoid, as in other *Australopithecus* radii.

There is also an *Au. afarensis*–like proximal manual phalanx from Kanapoi (KNM-KP 30503) attributed to *Au. anamensis* (see Ward et al., 2001), as well as a right femoral shaft fragment from Asa Issie (White et al., 2006). The latter does not retain either articular end but has thick cortical bone, a rugose attachment for gluteus maximus, and no linea aspera.

Remarks All but one of the specimens from Kanapoi, Turkana Basin, Kenya, are constrained between 4.17 ± 0.03 and 4.07 ± 0.02 Ma; the exception is a mandible that is slightly younger than 4.07 Ma (Leakey et al., 1998). The Allia Bay, Turkana Basin, Kenya sample is ~3.95 Ma in age (Leakey et al., 1995). The Asa Issie, Aramis, Middle Awash locality in Ethiopia is 4.2–4.1 Ma in age (White et al., 2006).

The history of finds for this taxon is a long one. Bryan Patterson found the distal humerus during an expedition to Kanapoi in 1965. Many of the Allia Bay isolated teeth were found in the 1980s, including one hemimaxilla. The type specimen and tibia were found in 1994/95, and additional fossils (dm1, capitate) were found between 1995 and 1997. Finds from Aramis, Ethiopia were published in 2006. Temporally, however, all the specimens are from a tightly constrained window between 4.2 and 3.9 Ma in the Eastern Rift, and all evince an anatomy widely deemed intermediate between *Ar. ramidus* and *Au. afarensis* (Kimbel et al., 2006; White et al., 2006; Kimbel, 2007).

There are also six worn mandibular teeth and one unworn p4 from Fejej, Ethiopia (4.18–4.0 Ma [Kappelman et al., 1996]) that are contemporaneous with *Au. anamensis,* but the severity of the wear and fragmentary nature of the specimens make it difficult to assign them to *Au. anamensis* versus *Au. afarensis* (Ward et al., 2001). Ward et al. (2001) have also noted that the Belohdelie frontal (3.89–3.86 Ma [White et al., 1993]) may represent *Au. anamensis,* but appropriate comparisons cannot be made because a frontal bone is not yet represented in the *Au. anamensis* sample.

The dietary and adaptive implications of the dental and masticatory features are significant; the increase in masticatory robusticity over *Ar. ramidus* is dramatic and may represent a punctuated event in anatomical adaptation (White et al., 2006). The molar expansion, thicker enamel, reduced anterior tooth wear, somewhat reduced canines, and robust mandible may all reflect the beginning stage of an emphasis on processing harder and/or more abrasive food items than were processed by the earlier purported hominins (Ward et al., 2001). Microwear comparisons among *Ar. ramidus*, *Au. anamensis*, and *Au. afarensis* will be useful in evaluating this hypothesis.

Postcranial implications are of considerable importance, as the tibia is, *Orrorin* apart, presently the oldest uncontested evidence of well-adapted bipedality (Leakey et al., 1995, 1998; Ward et al., 2001). The ankle and knee joints were clearly reorganized relative to those of all extant apes, to facilitate stable movement of flexion and extension, but constrain dorsiflexion and inversion at the ankle, and axial rotation at knee. Primitive features of the upper limb may reflect tree-climbing abilities and perhaps compensated for a hindlimb with restricted joint mobility; alternatively, features such as long arms may have been selectively neutral. It has been argued that the adaptive significance of primitive retentions is not easily testable unless the features have a strong epigenetic component (see Ward et al., 2001; Ward, 2002).

Thus far, recovered postcrania are large, and canine dental dimorphism is high, suggesting attribution of existing postcrania to males. Ward et al. (2001) have posited that large male body size may have conferred reproductive advantages resulting from male-male combat.

Available published evidence indicates that *Ar. ramidus* has no derived features that would preclude it from being ancestral to *Au. anamensis*; likewise, the latter cannot be excluded from ancestry of *Au. afarensis*, and there is evidence (discussed later) that they may be time-successive species. *Australopithecus anamensis* is more similar to (older) Laetoli *Au. afarensis* specimens than to Hadar specimens, and the Allia Bay sample, which is younger than the Kanapoi sample, is most similar to Laetoli. For example, the first lower deciduous molar is larger than the very small, narrow dm1 of *Ar. ramidus,* but smaller than that of *Au. afarensis*. The dm1 also lacks buccal and lingual grooves and basin differentiation found in *Au. afarensis*. Thus, *Au. anamensis* and *Au. afarensis* may have been an evolving chronospecies (Kimbel et al., 2006; White et al., 2006; Kimbel, 2007). In the case of the *anamensis-afarensis* transition, Kimbel and colleagues currently favor retaining the taxonomic status quo because it "helps localize and communicate about the clustering of morphology in time and space." (2007:145) In the case of the *ramidus-anamensis* transition, White et al. (2006) claim that the record is still too sparse to determine whether branching may have occurred between 4.4 and 4.2 Ma.

Faunal reconstructions for Kanapoi are based on over 30 mammalian taxa (Leakey et al., 1995). Cercopithecids outnumber colobines and several bovid species are represented, including *Tragelaphus*, impala and kudu. The most common carnivore is *Parahyaena* (Leakey et al., 1995), and tooth marks on bones are common (Ward et al., 2001). Stable carbon isotopic analysis of paleosols has shown soils that are associated with semiarid vegetational mosaics and a mixed ecosystem including edaphic grasslands, bush/woodland and gallery woodland (Wynn, 2000). Ward et al. (2001) have cautioned that transport of fossils by carnivores is a possibility, and so it is difficult to know if *Au. anamensis* was actually living in the dry environments sampled isotopically. Faunal and isotopic analyses for Allia Bay indicate a mosaic of environments ranging from woodland with extensive canopy to open grasslands (Coffing et al., 1994; Schoeninger et al., 2003). White et al. (2006) have undertaken preliminary analyses of the two Asa Issie localities. (Note, at the Aramis locality, the single *Au. anamensis* specimen [maxilla VP-14] was unaccompanied by significant associated faunal remains and paleoecological reconstructions have not been attempted.) At ASI-VP-2 and ASI-VP-5, hominins are associated with more than 500 other vertebrate fossils, with primates being most common, followed by bovids. Colobines outnumber cercopithecines 57:9 and among bovids, *Tragelaphus* is the most abundant. Stable carbon isotopic analysis of paleosols from these sites have been interpreted as indicative of humid woodland/savannah environments, with ca. 25%–35% C_4 grass. Overall, paleoecological reconstructions for Kanapoi, Allia Bay, and Asa Issie indicate that *Au. anamensis*, like *Au. afarensis*, was associated with habitat heterogeneity. Specific environmental preferences within this variability are unknown and taphonomic factors have yet to be fully accounted for. Currently, we are limited to inferring that early members of the *Au. anamensis*–*Au. afarensis* populations utilized and/or tolerated a wide range of habitats.

AUSTRALOPITHECUS AFARENSIS Johanson et al., 1978
Figure 25.6 and Table 25.2

Partial Synonymy Meganthropus africanus, Weinert, 1950; *Praeanthropus africanus,* Şenyürek, 1955; *Praeanthropus afarensis,* Strait and Grine, 2004.

The Garusi I maxilla from Laetoli was named *Praeanthropus* by Hennig (1948), but no type was designated. Later, Weinert (1950) proposed the specimen should be placed in *Meganthropus africanus* because of similarities to an Indonesian taxon (now *H. erectus*). Johanson et al. (1978) did not recognize either of these taxa and instead placed relevant Laetoli and Hadar material in *Australopithecus afarensis,* but Strait et al. (1997) suggested that *Praeanthropus africanus* be resurrected. In 1999, the International Commission for Zoological Nomenclature ruled *africanus* be suppressed but left *Praeanthropus afarensis* available. Currently, some authors (e.g., Strait and Grine, 2004; Grine et al., 2006) use *Praeanthropus afarensis,* but this usage has not been widely accepted.

Holotype The type specimen is L.H. 4, a mandibular corpus with right broken c, p3–4, m1–3, and left p4, m1–2. For additional specimens, see Johanson et al., 1978 and *American Journal of Physical Anthropology* 57(4), 1982 (Hadar); Leakey and Harris, 1987 (Laetoli); White et al., 2000 (Maka); and Alemseged et al., 2005 (Dikika). An overview of the hypodigm is provided in Kimbel (2007) and Kimbel and Delezène (2009).

Age and Occurrence Early to mid-Pliocene, East Africa (table 25.2).

Diagnosis Based on Johanson et al., 1978; Kimbel, 2007. Compared to *Au. anamensis, Au. afarensis* has a larger external auditory meatus; a less inclined anterior corpus profile of mandible; an asymmetric upper canine with more apically placed mesial crown shoulder; more symmetric, molarized p3 crown with frequent development of the metaconid (second cusp); more molarized dm1; and sharper lateral margins and a more distinct inferior margin to the nasal aperture. Compared to later australopiths, it has relatively large upper central incisors; absolutely smaller postcanine dentition; a strongly protruding, convex subnasal plane that projects beyond the bicanine line; a less vertical anterior corpus profile of mandible; shallower mandibular fossa and less well developed articular eminence; and a horizontally inclined, tubular tympanic. The canines are asymmetrical with considerable size variation and mandibular tooth rows vary from subrectangular to U shaped. Diastemata often occur between I2/C and c/p3, and although there are sometimes vertical wear striae on the buccal face of p3, the C/p3 complex is considered functionally nonhoning. The mandibular corpus is relatively deep anteriorly in large specimens; the ramus of the mandible is broad and relatively low; and there is strong alveolar prognathism. Compared to *Kenyanthropus,* it has larger upper molars (Leakey et al., 2001). See Johanson et al. (1978) for initial diagnosis and Kimbel (2007) and Kimbel and Delezène (2009) for an updated summary. Diagnostic postcranial features are numerous, compatible with bipedality in the hindlimb (see Ward, 2002) and described later.

Description Australopithecus afarensis is perhaps the most thoroughly known australopith. Some 400 specimens are attributed to the taxon, about 90% of which are from Hadar in the Afar depression of Ethiopia (Kimbel et al., 2004). The Hadar remains include the famous "Lucy" skeleton (A.L. 288-1 [3.18 Ma]), a collection of fossils from A.L. 333 (3.2 Ma) known as the "First Family" that likely represents members of a single population, as well as the most complete cranium and mandible of a single adult *Au. afarensis* individual (A.L. 444-2 [3.0 Ma]).

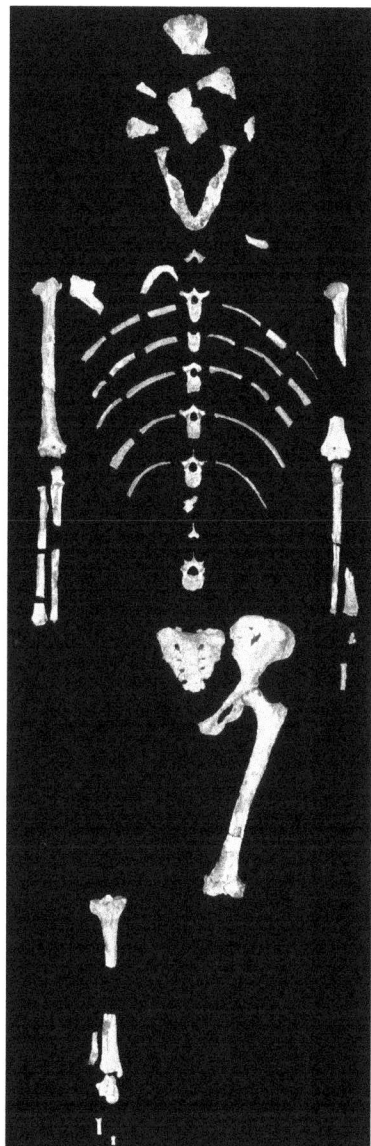

FIGURE 25.6   Partial skeleton of *Australopithecus afarensis*, AL 288-1 "Lucy." Courtesy of the Cleveland Museum of Natural History.

As one of the most intact adult hominin skeletons known, A.L. 288-1 is of special significance (Johanson et al., 1982). It is small (~27 kg; McHenry, 1991b) and for various reasons (overall size, morphology) is presumed to be female (Johanson and White, 1979; Tague and Lovejoy, 1986). Additional postcranial fossils from Hadar have augmented the information obtained from the Lucy skeleton, and they suggest a taxon with an amalgam of primitive and derived features, along with distinctive features that are challenging to interpret functionally as they have no analog among extant hominoids. Primitive features include manual and pedal phalanges that are more curved than those of *Homo* (Stern and Susman, 1983); a cranially oriented glenoid fossa (Stern and Susman, 1983); a funnel-shaped thorax (McHenry, 1991a); a relatively long foot (Jungers and Stern, 1983); lower limbs that are short relative to upper limbs (Jungers, 1982); small vertebral centra including small sacral body (Sanders, 1990); and lesser cranial expansion of the acetabular articular surface (MacLatchy, 1996). Derived features that resemble the condition found in *Homo* include lumbar lordosis and sacral retroflexion (Lovejoy, 2004); expansion of the retroauricular region of the ilium, superoinferiorly short, mediolaterally expanded iliac blades and short ischium (McHenry, 1991a); flattened inferior contour of the lateral femoral condyle, deep patellar groove with high lateral lip and high femoral bicondylar angle (Johanson and Coppens, 1976; Tardieu, 1981); perpendicular orientation of the distal tibial articular surface (Latimer et al., 1987); anteriorly unexpanded distal tibial articular surface (DeSilva, 2008); large calcaneum with well-developed structures to dissipate stress at heel strike (Latimer and Lovejoy 1989); convergent hallux (Latimer and Lovejoy, 1990b); proximal pedal phalanges with dorsally oriented proximal articular surfaces, suggesting dorsi- rather than plantar flexion (Latimer and Lovejoy, 1990a); and relatively short pedal phalanges (White, 1994).

Novel features are particularly evident in the pelvis of A.L. 288-1. The iliac blades are laterally flared but lack the sagittal alignment found in *Homo* pelves. The flaring iliac blades, which are found in other australopiths, have been interpreted to provide sufficient spatial displacement (in conjunction with a long femoral neck) to increase the lever arm length for the lesser gluteals to effect abduction and control body torque during single leg stance on an extended lower limb (e.g., Lovejoy, 2004). However, Stern and colleagues propose that without iliac blades that are oriented in the sagittal plane, the lesser gluteals would not act as abductors but rather as medial rotators, stabilizing the trunk on a flexed thigh (e.g., Stern and Susman, 1981, 1983, 1991). Other disagreements about the best way to reconstruct *Au. afarensis*'s positional behavior (i.e., especially in relation to other novel features, such as limb proportions) are considered later.

The craniodental anatomy is well represented, and the hypodigm includes over 60 mandibles and mandible fragments (Kimbel et al., 2004). The mandible is characterized by a deep mandibular corpus that is rounded and bulbous anteriorly, and hollowed laterally; a low, rounded inferior transverse torus; a weak to moderate superior transverse torus; integration of the canine crown into the pre- rather than postcanine dental arch (Kimbel et al., 2004)(though this is less true of L.H. 4; Kimbel et al., 2006); and a vertical, anteriorly positioned ramus. Like other australopiths, the mandibular corpus is transversely thick, even when considered relative to molar size (Teaford and Ungar, 2000).

The large sample size reveals variation in some characters. Mandibular corpus size increases over time (Lockwood et al., 2000), the slope of the symphyseal axis is highly variable (Kimbel et al., 2004), and the symphyses of some Laetoli mandibles have a convex external surface and recede inferiorly, similar to the condition in *Au. anamensis*, and unlike the straight external contour found in Hadar mandibles (Kimbel et al., 2006). Dental features show stasis overall, but lower canine dimensions vary with regard to degree of mesiodistal compression; and p3 mesiodistal length decreases, and M3 dimensions increase, from the Laetoli to the Hadar sample (Lockwood et al., 2000). Laetoli upper canines also appear to resemble those of *Au. anamensis* and are mesiodistally longer than those from Hadar (Kimbel et al., 1996). As with other australopiths, premolars and molars are large compared to incisors and canines; molars have low, bunodont cusps and thick enamel; and the postcanine tooth area is large (Teaford and Ungar, 2000).

Cranially, *Au. afarensis* shares a number of derived features with later hominins, including reduced upper facial prognathism; anteriorly positioned foramen magnum; short anterior cranial base; distinct lateral margins in the nasal aperture; and a larger auditory meatus (Kimbel et al., 2004; Kimbel, 2007). Plesiomorphic features include compound temporonuchal crests, shallow mandibular fossae with weakly developed articular eminences, somewhat flat, low frontal squama, no frontal trigon, vertical midface and convex subnasal plane, distinct subnasal and intranasal parts of clivus, narrow interorbital and nasal aperture breadths, flat, wide, robust zygomatic region, and weakly flexed cranial base and tubular tympanic (Walker et al., 1986; Kimbel, 2007). Endocranial capacities are known for three adult and one subadult specimen, and fall in the range of extant great apes: A.L. 444-2, ca. 550 cm^3; A.L. 333-45, ca. 500 cm^3; A.L. 162-28, ca. 400 cm^3; A.L. 333-105 (juvenile), ca. 320 cm^3 (Falk, 1985; Kimbel et al., 2004).

A second important partial *Au. afarensis* skeleton from Ethiopia is the Dikika juvenile dated to 3.35–3.31 Ma (Wynn et al., 2006) and described by Alemseged et al. (2006). The M1 crown is fully formed but unerupted; using ape models of development, the individual may have been about 3 years old at time of death. Several features are worthy of note. The hyoid bone was African ape–like; since this bone is highly modified by the time of Neanderthals (Arensburg et al., 1989), its primitive form in *Au. afarensis* suggests that selection for vocal anatomy reorganization had not begun. The manual phalanges of the Dikika child are already curved; if this is an epigenetic trait, then it suggests some climbing was being undertaken. The scapula is also ape-like, with a cranially tilted glenoid fossa, but overall the anatomy is reported to be more like that of *Gorilla* than *Pan*, although with a reduced supraspinous fossa. Derived hindlimb features include a femoral bicondylar angle, robust calcaneum, and a transversely expanded proximal tibia.

Remarks High levels of variability in the *Au. afarensis* sample have been the subject of considerable study, with an emerging, complex picture of geographical separation, phyletic change over time (especially between the Laetoli and Hadar samples) and sexual dimorphism as contributing factors (Lockwood et al., 2000; Kimbel et al., 2004, 2006; Kimbel, 2007). Nonetheless, there is a "paleoanthropological consensus . . . that *Au. afarensis* is, indeed, both biologically and statistically speaking, a 'good' species" (Kimbel et al., 2004:4, 8) and that variation can be attributed to intraspecific anagenesis (Grine et al., 2006).

Although it is now well accepted that Hadar is sampling a single species, the level of dimorphism within this comparatively well-sampled taxon is still debated. The conventional view that *Au. afarensis* had high body size dimorphism, with males (45 kg) estimated to be 50% larger than females (29 kg) (McHenry, 1992a), or even double the mass of females (e.g., Richmond and Jungers, 1995), has been challenged by Reno et al. (2003, 2005) using the A.L. 333 sample of individuals and A.L. 288-1. Using the proportional relationships among skeletal elements within Lucy, they estimated the femoral head size that would correspond to each postcranial element from the A.L. 333 sample. They concluded that the pattern of femoral head size variation is similar to that of modern *Homo*, and thus that intraspecific variation (including sexual dimorphism) is lower than previously thought. The implications of this pattern of sexual dimorphism are potentially far-reaching because certain behavioral characteristics are correlated with higher degrees of body size dimorphism, including polygynous social systems, male-male aggression, and male-driven predator aggression and territoriality (McHenry, 1994a; Ward, 2002). Closer similarity in body sizes is associated with lower reproductive variance in males, more similar operational sex ratios and are compatible with earlier theories linking bipedality and pair bonding (Lovejoy, 1981). Although the view that *Au. afarensis* had low–moderate levels of sexual dimorphism has been vigorously challenged (Plavcan et al., 2005), an emphasis on skeletal over body size dimorphism may prove a useful paleontological approach overall.

The pointed disagreement over the functional interpretation of individual postcranial features, as discussed, has led to difference in opinion in how to synthesize the wealth of information that comes from the postcranial record, with some postulating that *Au. afarensis* was a committed and efficient terrestrial biped (e.g., Latimer et al., 1987; Latimer and Lovejoy, 1989, 1990a, 1990b; Lovejoy, 2004) and others that *Au. afarensis* was an inefficient biped still utilizing arboreal supports (e.g., Stern and Susman, 1983, 1991; Susman et al., 1984). Furthermore, both efficiency and performance style of *Au. afarensis*'s bipedality remain contested. For example, some researchers support a bent knee, bent hip (BKBH) gait on the grounds that it would have minimized oscillations in center of mass, lessened peak vertical reaction forces and increased stride length (Schmitt, 2003) while others suggest that BKBH gaits would be too energetically expensive and would have raised core body temperature to such a degree as to be unsustainable (Crompton et al., 1998; Carey and Crompton, 2005). Furthermore, the Lucy pelvis and other sufficiently well-preserved australopith pelves all show evidence of a ventral pelvic tilt, which places hip extensor muscles in a more favorable position so that they can retract the lower limb when legs are straight. Chimpanzees lack this tilt and must walk with a BKBH gait when bipedal in order to place the extensor insertions anterior to the pelvic origins so that they can retract the leg.

However, as Ward (2002) has summarized, researchers have been approaching the reconstruction of posture and locomotion from different philosophical vantages and with different goals in mind. Stern and Susman advocate the examination of all anatomical features and contend that the current, inferred utility of features is of paramount importance in reconstructing what australopiths were *capable* of doing (e.g., Stern and Susman, 1991). For example, since long fingers are good for grasping arboreal supports and a cranially oriented glenoid fossa would facilitate elevating the forelimb above the head, *Au. afarensis* was likely climbing trees. Latimer and Lovejoy (1989, 1990a, b) weight characters more explicitly and give primitive characters less consideration if there is some evidence of directional selection. They reason that if fingers are less curved than the condition inferred for the LCA of *Pan* and *Homo* (itself difficult to determine), then climbing ability is being compromised and either selected against or having a neutral effect on fitness.

One of the most evocative lines of evidence of the bipedalism of *Au. afarensis* comes from the trail of hominin footprints in Tuff 7 of the Laetolil beds, discovered by Andrew Hill in 1976 (Leakey and Harris, 1987) and dated to 3.66 Ma (Deino, in press). White and Suwa (1987) have made a strong case that these prints were made by *Au. afarensis* on the basis of foot anatomy, rather than on the basis of age and location alone. The tuff preserves three sets of footprints, with one individual stepping "pace for pace" in the footprints of an individual moving ahead, and a third individual walking alongside creating a parallel trackway (Leakey, 1987; Robbins, 1987). The footprints preserve several modern human–like attributes, including heel strike, hallucal toeing off, a lateral to medial shift in weight bearing on the sole of the foot, an adducted hallux, and a longitudinal arch (Tuttle, 1987), but Bennett et al. (2009) take the view that there are significant differences between the Laetoli prints and those of habitually unshod modern humans.

The diet of *Au. afarensis* has also been the subject of much research, but a lack of isotopic sampling leaves a large lacuna in our knowledge. The postcanine tooth crowns of *Au. afarensis* are larger than those of *Au. anamensis*, and the enamel is thicker. Flat teeth with planar surfaces would presumably be inefficient at processing pliant foods such as meat, leaves or tough fruit, but efficient at crushing hard, brittle foods, as well as weak foods (Teaford and Ungar, 2000). Thickly enameled teeth would also resist abrasion and, depending on microstructure, be less likely to fracture under stress when hard objects are being consumed (Lucas et al., 2008).

A relatively small anterior dentition implies that *Au. afarensis* probably did not regularly eat fruit with thick husks, or fruits with flesh adhering to seeds (Teaford and Ungar, 2000). Microwear on anterior teeth include scratches and pits, suggesting incisors may have been used to strip gritty plant parts (Ryan and Johanson, 1989). Recent analysis of molar microwear (Grine et al., 2006) from Hadar and Laetoli shows little variation over time. Teeth possess scratches reminiscent of mountain gorilla teeth and compatible with an abrasive and possibly tough diet, but none of the pitting associated with hard object feeding in extant primates (Grine et al., 2006). This characterization is more similar to microwear results for *Australopithecus africanus* than for *Paranthropus robustus* (Grine and Kay, 1988; Scott et al, 2005). Similarity in microwear between *Gorilla* and *Au. afarensis* contrasts with observed differences in molar topology, the shearing crests of the former contrasting with the relatively flat teeth of the latter. However, Grine et al. (2006) explore the idea that the robust masticatory system of *Au. afarensis* may have evolved to process seasonal hard foods (i.e., the hard foods were critical, fall-back foods, rather than staple foods). Overall, the gross dental anatomy, if not the microwear, suggests that *Au. afarensis* was puncturing, grinding and chewing abrasive, hard, while the robusticity of the mandible would have provided a high resistance to mechanical failure (Teaford and Ungar, 2000).

There has been considerable interest in reconstructing the paleoecology of *Au. afarensis* sites, in large part as they may bear on the selective forces at work in the maintenance of the signature adaptations of the genus: bipedality and megadonty. A consensus is growing that *Au. afarensis* tolerated a wide range of environmental conditions and was broadly distributed across a heterogeneous landscape (e.g., White et al., 1993; Bonnefille et al., 2004; Reed, 2008).

The faunal composition of the Laetolil beds, Laetoli (including invertebrates, chelonians, galagids, cercopithecids, rodents, carnivores, perissodactyls, suids, giraffids, and bovids) was initially reconstructed as indicating dry wooded or bush savanna ecosystems with well-defined wet and dry seasons (see references in Leakey and Harris, 1987, especially Harris, 1987). Its proximity to an active volcano would have produced a soil/vegetation gradient with grassland grading into woodland at increasing distance from the volcano (Andrews, 1989). Harris (1987) noted that because the depositional environment was volcanic, and not fluviatile or lacustrine like most Ethiopian and Kenyan sites, it actually may be more representative of much of East Africa, and the Serengeti has been invoked as a likely modern analog (e.g., Andrews, 1989). Recently, additional research has complicated this picture. Some faunal analyses indicate there was a high proportion of arboreal and frugivorous mammals (Walker, 1987;

Andrews, 1989; Reed, 1997) and isotopic work on Laetoli herbivores indicates dietary guilds dominated by mixed
browsing/grazing or browsing foraging strategies, suggesting
woodland as a more important component of the environment than previously recognized (Kingston and Harrison,
2007). Andrews and Bamford's (2008) topographic reconstructions, and the inferred soils, drainage, and vegetation
that would accompany their topography, also support a significant woodland component. Su and Harrison (2008:678)
support a "predominantly open woodland" environment at
Laetoli, and reason that the low density of hominin remains
relative to Hadar implies less optimal habitats, which they
attribute to the less densely wooded, drier mosaicism of
Laetoli compared to Hadar. Recent syntheses of Hadar paleoecology (Bonnefille et al., 2004; Reed, 2008) do not support a
significant difference between these two localities, however,
and the resolution needed to address ecological and taphonomic sources of differences in species' fossil densities is not
readily available (Cote, 2008). Although there appears to be a
trend toward increasingly open environments in the Hadar
sequence associated with *Au. afarensis* (Reed, 1997, 2008),
detailed pollen analyses suggest a variety of habitats were
present over its stratigraphic range (Bonnefille et al., 2004),
and faunal analyses of over 4,000 mammalian specimens
also support heterogenity, with bushland, open woodland,
shrubland, and edaphic grassland as habitat components
(Reed, 2008). Other *Au. afarensis* sites have yet to be subjected
to such detailed paleoenvironmental analyses. White et al.
(1993) have described the Maka fauna as broadly comparable
to that from the Denen Dora Member, which Reed (2008) has
reconstructed as bushland or woodland/floodplain grassland.
The Tulu Bor Member of Koobi Fora has a depositional environment and faunal list compatible with floodplains (Feibel
et al., 1991; Reed, 1997), and shrubland/wetland/grasslands/
woodlands have also been invoked (Harris, 1991; Reed, 2008).
At Dikika, *Au. afarensis* occurs in the oldest or Basal Member
of the Hadar Formation, and the younger Sidi Hakoma
Member (Alemseged et al., 2005, 2006). Reed (2008) has
looked at faunal composition of these members in deposits
from the other side of the Awash River at the Hadar site and
reconstructs the Basal Member as a woodland/shrubland
mosaic, and Sidi Hakoma as similar, but trending toward
drier and more open conditions. However, Wynn et al. (2006)
have noted that the Sidi Hakoma Member at Dikika preserves
a relatively high proportion of grazing bovids, and so was
perhaps a more open habitat than the more wooded regions
represented at Hadar

Australopithecus afarensis is no longer the most primitive
member of *Australopithecus*, and as additional *Ar. kadabba*,
Ar. ramidus and *Au. anamensis* fossils have been recovered,
the evidence for ancestor-descendant relationships among
the four taxa has strengthened (White et al., 2006; Kimbel
et al., 2006). Bipedalism and moderate megadontia remain
hallmarks of *Au. afarensis*, and it continues to serve as a
key basis of comparison for all other late Miocene through
Pliocene taxa. This is because of the relative richness of its
fossil record as well as a general lack of autapomorphies
that would preclude it from an ancestral relationship with
later taxa (Kimbel et al., 2004; but see Rak et al., 2007,
regarding the gorilla-like ramal anatomy of Hadar specimen A.L. 822-1).

AUSTRALOPITHECUS BAHRELGHAZALI Brunet et al., 1996
Figure 25.7 and Table 25.2

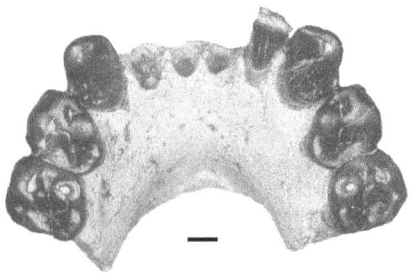

Synonymy Australopithecus afarensis, Brunet et al., 1995.

Holotype The type specimen KT12/H1 is an anterior mandibular corpus with right i1 alveolus and i2–p4 and left i1
alveolus, i2 root and c-p4 (Brunet et al., 1995). An isolated P3
is the only other published specimen (Brunet et al., 1996)
although the presence of *Australopithecus* sp. was reported from
nearby locality KT13 (Brunet et al., 1997).

Age and Occurrence Mid-Pliocene, Central Africa (table 25.2).

Diagnosis Based on Brunet et al., 1995, 1996. Long axis of
mandibular symphysis oriented subvertically; lower canines
and incisors large and canines with high crowns and cingula;
lower premolars have three roots, and are buccolingually broad
with buccal cingula; p3 is bicuspid with a strong metaconid; p4
is molarized with a small talonid; P3 has three roots and an
asymmetrical crown. The taxon differs from *Ar. ramidus* in its
thicker enamel, three-rooted premolars and less asymmetrical;
p3; from *Au. anamensis* in its more vertical symphyseal region,
short planum alveolare, reduced inferior transverse torus and
bicuspid p3 with strong metaconid; from *Au. afarensis* by its
subvertical, relatively flat symphyseal region and three-rooted
lower premolars; from *Au. africanus* by its less vertical posterior
symphysis, less robust corpus, larger anterior dentition and
three-rooted lower premolars (see Brunet et al., 1995, 1996).

Description Brunet and colleagues (1995, 1996) cite the flat,
more vertical orientation of the symphyseal region as distinguishing it from *Au. afarensis*, but metrics supporting this view have yet
to be published. If additional finds support the characterization
of a less prognathic taxon, it would be interesting in light of the
fact the anterior teeth are *Au. afarensis*-sized. The canine is asymmetrical with a long distal cuspule and strong lingual crest. The
upper third premolar has three roots, like most robust australopiths but unlike most *Au. afarensis* and *Au. africanus*, which have
two roots; the lower premolars have three distinct roots.

Remarks The referral of the two published fossils to a new
species has been disputed on the grounds that there is insufficient material (White, 2002), and that the diagnostic features
are represented in the Laetoli, Hadar and Maka *Au. afarensis*
collections (Kimbel, 2007). For example, L.H. 24 has a three-
rooted premolar (White et al., 2000) and A.L. 444-2 has a vertical symphyseal cross section (Kimbel et al., 2004).

The biogeographic importance of the Chadian finds is
nonetheless undiminished, regardless of whether additional
fossils eventually bolster the case for a distinct species, or
confirm the occurrence as a variant of East African *Au. afarensis*. The former sets up a scenario of increasing cladogenesis
in the mid-Pliocene; if *Kenyanthropus platyops* is ultimately
shown to be another, separate taxon from this interval (discussed later; Leakey et al., 2001) then hominins may have up
to three lineages between 3.5 and 3 Ma.

The alternate, single-taxon scenario extends the range of *Au. afarensis* so that it is no longer confined to a 1,500-km swath along a north-south gradient of the East African rift and places it 2,500 km to the northwest. Although *Au. afarensis* might not have been present over this broad a geographic area at any point in time, it would have been sufficiently mobile to thrive at least for intervals across a heterogeneous landscape. This does not necessarily imply a taxon of more generalized niche, however. Many of the KT taxa, both woodland (e.g., *Kolpochoerus afarensis*) and grassland (e.g., *Hipparion* sp. aff. *afarense/hasumense*) "specialists" are the same species as those found at Hadar and Laetoli, indicating that these taxa were also are widely distributed latitudinally in sub-Saharan Africa. At the least, this suggests that dispersal of taxa with either woodland or grassland affinities was not limited by profound habitat homogeneity; rather, habitats were potentially varied enough over a small scale, but over a wide enough geographic area and time depth to permit habitat specialists to attain a wide geographic distribution. Brunet et al. (1995) have characterized sub-Saharan Africa, from the Atlantic to the Indian Ocean, southward to Cape of Good Hope, as a woodland savannah belt. The heterogeneous nature of such a belt, and associated variation in factors such as seasonality, rainfall, and altitude has played a major role in theories of hominin diversification (e.g., Potts, 1998; Kingston, 2007).

The nonhominin fauna at KT 12 include silurid fish, suggesting a lakeside environment, as well as taxa indicating the presence of forest or woodland (e.g., reduncine bovids, *Kolpochoerus afarensis*) and more open habitats (e.g., *Ceratotherium* and *Hipparion*)(Brunet et al., 1995). Locality KT 13 has a similar biochronological age and environmental reconstruction (Brunet et al., 1997).

AUSTRALOPITHECUS AFRICANUS Dart, 1925
Figure 25.8 and Table 25.2

Partial Synonymy Australopithecus transvaalensis, Broom, 1936; *Plesianthropus transvaalensis,* Broom, 1937; *Australopithecus prometheus*, Dart, 1948; *Homo transvaalensis*, Mayr, 1950; *Australopithecus africanus africanus*, Robinson, 1954; *Australopithecus africanus transvaalensis,* Robinson, 1954; *Homo africanus*, Robinson, 1972; Olson, 1978.

Holotype The type specimen is Taung 1, a juvenile skull with a natural endocast.

Age and Occurrence Mid- to late Pliocene, southern Africa (table 25.2).

Diagnosis Dart's (1925) original description of *Australopithecus africanus* differentiated this taxon from modern apes by a slightly enlarged brain, with a posteriorly positioned lunate sulcus caused by enlarged parietal lobes, and an anteriorly positioned foramen magnum suggestive of upright walking. In addition, the Taung mandible is robust, though equipped with small canines, and no diastema. Further diagnosis has been aided by additional discoveries of *Au. africanus* (Lockwood and Tobias, 1999, 2002; Moggi-Cecchi et al., 2006) and by the analyses of White et al. (1981). Relative to *Au. afarensis*, *Au. africanus* has a slightly less prognatic face with a flat nasoalveolar clivus, a deeper palate, a more robust mandibular corpus and increased buttressing of the anterior corpus, larger postcanine dentition, and a deciduous lower molar crown with a twinned medial basin. Pneumatization is restricted to the mastoid region, unlike in *Au. afarensis*, where it extends to the temporal squama. *Australopithecus africanus* crania do not possess the compound temporal nuchal crest present in *Au. afarensis* fossils and have a maxillary furrow lateral to the nasal opening rather than the

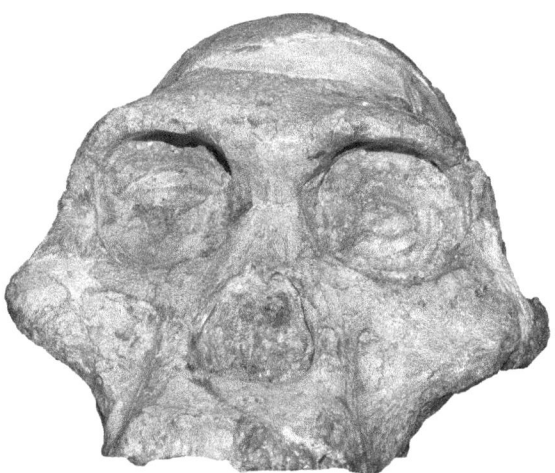

FIGURE 25.8   *Australopithecus africanus* cranium STS 5 "Mrs. Ples." Courtesy of the Transvaal Museum, (Northern Flagship Institution).

canine fossa found in *Au. afarensis*. Though variation exists for all of these features in *Au. afarensis*, *Au. africanus* specimens uniformly have a bicuspid third premolar, canines with apical wear pattern, and no diastema between the maxillary canine and lateral incisors. In *Au. afarensis* and *Paranthropus robustus*, the third molar is typically the largest tooth, whereas in *Au. africanus* the second molar tends to be the largest. Unlike *Paranthropus*, *Au. africanus* fossils have an enlarged anterior dentition, distinct supraorbital morphology often consisting of a supraciliary eminence and strongly pronounced glabellar region, the absence of temporal lines merging with the supraorbital torus, moderate postorbital constriction, only weakly developed sagittal cresting on male crania, an expanded cranial base, and lower fourth premolars with three rather than two cusps. Relative to specimens assigned to *Homo*, *Au. africanus* possesses a shallower temporomandibular fossa and a small cranial capacity.

Description Besides the juvenile Taung skull, *Au. africanus* is craniodentally represented by Sts 5 ("Mrs. Ples"), an almost complete female cranium lacking the maxillary dentition, StW 505, a presumed male, which preserves an almost complete right side of the cranium and parts of the left frontal and maxilla (Lockwood and Tobias, 1999), and Sts 71, which preserves most of the left part of a skull and the right maxilla (note, Sts 71 may be associated with the mandible Sts 36; Wallace, 1972). MLD 37/38 preserves most of the calvaria, though the face has been sheared off. Other relatively complete specimens include the partial cranium StW 252, a distorted partial cranium StW 13, a basicranium Sts 19, and the associated maxilla StW 52a and mandible StW 52b, and perhaps StW 53 (Kuman and Clarke, 2000; though see Curnoe and Tobias, 2006, who consider StW 53 to be early *Homo*).

Lockwood and Tobias (1999) have commented that *Au. africanus* has very few autapomorphies and instead possesses an amalgam of plesiomorphic features found in *Au. afarensis* and derived features found in later *Homo* and *Paranthropus* specimens. Present on most *Au. africanus* craniofacial fossils are prominent columns of bone along the nasal-maxillary junctions, termed the anterior pillars (Rak, 1985). This morphology has been suggested to be an adaptation that resisted bending forces in the facial skeletal skeleton in *Au. africanus* associated with dietary changes and molarization of the premolars (Rak, 1985). The presence of anterior pillars in *Au. africanus* and *P. robustus* may be a shared-derived feature suggestive of an ancestor-descendant relationship (Rak, 1985).

This feature is particularly well developed in specimens such as StW 13, though only weakly developed on others like TM 1512, and perhaps absent altogether in StW 391 (Lockwood and Tobias, 2002). *Australopithecus africanus* crania also have flaring zygomatics with strong zygomatic prominences. Though there is usually not a sagittal crest present (but see Sts 17), the temporal lines are positioned high on the cranium. Furthermore, *Au. africanus* dentitions combine the large anterior teeth found in earlier *Au. afarensis* remains with the large postcanine dentition found in later robust australopiths and in some early *Homo*. In this respect, some *Au. africanus* fossils, including StW 252, are similar to the type specimen of *Au. garhi*.

Average cranial capacity in *Au. africanus* is 463.9 cm^3 ± 51.9 cm^3 (range 400 cm^3–560 cm^3) based on data from eight crania and endocasts (Conroy et al., 1990, 1998, 2000a; Holloway et al., 2004). This is slightly greater than the mean cranial capacity (383.4 cm^3) of the similar-sized chimpanzee (Tobias, 1971). In addition to brain size, brain organization has been studied in detail for *Au. africanus*. Based on CT scans of MLD 37/38, it has been suggested that *Au. africanus* shares with *Homo* expanded anastomotic channels efficient for cooling cranial blood (Falk and Conroy, 1983; Conroy et al., 1990). This is in contrast to *Au. afarensis* and the paranthropines, which typically have an enlarged occipital-marginal sinus (Conroy et al., 1990). However, there is variation in this feature, as the type cranium from Taung has an enlarged occipital-marginal sinus, despite having other endocranial features clearly linking this specimen to *Au. africanus* and not to *Paranthropus*. These include squared-off frontal lobes (Falk and Clarke, 2007), features found in the endocasts of *Au. africanus* and early *Homo*, but not in *Paranthropus* taxa (Falk et al., 2000).

Examples of almost all of the skeletal elements (minus a few tarsal, carpal, and phalangeal elements) of *Au. africanus* have been recovered from the Sterkfontein cave (for an inventory of the 1936–1999 discoveries, see the appendix in Pickering et al., 2004b). Even a fossilized stapes is known for *Au. africanus*, and suggests that this hominin could hear higher frequencies than modern humans (Moggi-Cecchi and Collard, 2002). More complete postcranial remains include the partial skeletons Sts 14 and StW 431, and potentially the StW 573 "Little Foot" remains, though Clarke (2008) suggests that StW 573 may belong to a different *Australopithecus* species. The morphology of the vertebral column, pelvis, and lower limb has clearly demonstrated that *Au. africanus* was an habitual biped (Robinson, 1972; Lovejoy, 1974). In fact, the southern African *Au. africanus* remains led Washburn and Patterson (1951) to propose that instead of encephalization, it was adaptations for upright walking that differentiated the earliest hominins from the apes. Based on postcranial remains, *Au. africanus* males were approximately 1.38 m tall and 41 kg, whereas the females were roughly 1.15 m tall and 30 kg (McHenry, 1992a; McHenry and Coffing, 2000). The level of sexual dimorphism is presumed to have been like that of the common chimpanzee (Lockwood, 1999).

The StW 431 pelvis has modern human–like attachments for the gluteals and latissimus dorsi (Häusler, 2002) and to judge from the rugosity of their attachments the sacrotuberous, dorsal iliac and interosseous ligaments were well developed, powerful, and would have helped maintain the tilt of the sacrum in upright posture (Sanders, 1998). Pelvic remains from StW 441/465 are also reconstructed as well adapted for bipedality (Häusler and Berger, 2001). As in *Au. afarensis*, the ilia flare laterally in Sts 14 and StW 431 (Kibii and Clarke, 2003). Macchiarelli et al. (1999) found that the trabecular

patterns in the ilia Sts 14, Sts 65, StW 431, MLD 7, and MLD 25 differed slightly from the modern human condition and may reflect differences in the magnitude and direction of stress incurred on the ilium during locomotion.

The morphology of the lumbar region of the vertebral columns of StW 14, StW 431, and StW H8/H41 is generally consistent with adaptations for bipedality, though *Au. africanus* had very small centra (Shapiro, 1993; Sanders, 1998; Toussaint et al., 2003). Based on the morphology of the vertebrae, in conjunction with other postcranial features, it has been suggested that *Au. africanus* may have been more versatile in its locomotor capacities and perhaps engaged in both bipedalism and climbing activities (Shapiro, 1993; Sanders, 1998). However, because a short lumbar region (usually three vertebrae) is thought to be an adaptation for orthograde climbing in the hominoids, climbing would have been kinematically different from that practiced by modern apes, as *Au. africanus* had either five or six lumbar vertebrae (Sanders, 1998; Touissant et al., 2003). Whitcome et al. (2007) recently found that the lordosis angle in Sts 14 and StW 431 fits the pattern distinguishing modern male and female lumbar vertebrae, suggesting full bipedality and concomitant adaptations for pregnancy in the presumed female Sts 14.

The femora StW 99, StW 598, and MLD 46 possess a long femoral neck, which would help to increase the mechanical advantage of the lesser gluteals during the single-legged, stance phase of the walking cycle (Reed et al., 1993; Partridge et al., 2003). StW 99 also has a mediolaterally expanded subtrochanteric region, which may have helped to resist bending loads during bipedalism (Richmond and Jungers, 2008). The fragmentary proximal femur StW 522 has a strikingly deep obturator externus groove, which suggests hyperextension at the hip, though this specimen also has a short femoral neck and an ape-like margin around the rim of the femoral head. Distal femora TM 1513 and Sts 34 both possess a strong bicondylar angle, suggesting that the knee of *Au. africanus* was positioned directly under the center of mass. The proximal tibia StW 514 has a curved lateral condyle and a single attachment for the lateral meniscus, leading Berger and Tobias (1996) to suggest that *Au. africanus* may have had chimpanzee-like locomotor capacities. However, Organ and Ward (2006) found that the convexity of the lateral condyle does not discriminate between modern humans and extant African apes. A single point of attachment for the lateral meniscus of StW 514 is similar to what is found in *Au. afarensis*. To judge by the horizontally oriented distal tibial articular surface relative to the long axis of the shaft in StW 358, StW 389, and StW 514b, in these specimens the ankle was also aligned under the knee and thus under the center of mass (DeSilva, 2008). These tibiae and the tali StW 88, StW 102, StW 347, StW 363, and StW 486 also lack adaptations that would allow *Au. africanus* to put its foot in positions of dorsiflexion and inversion, which are important during vertical climbing (DeSilva, 2008). However, the tali of *Au. africanus* are also ape-like in possessing a deep trochlear groove (Harcourt-Smith, 2002; Deloison, 2003). The calcaneum of *Au. africanus* StW 352 is similar to *Au. afarensis* calcanei in having a cross-sectional area in the range of modern humans and larger than that found in African apes (Latimer and Lovejoy, 1989). The large peroneal tubercle of the StW 352 calcaneus resembles the condition seen in *Au. afarensis*.

The morphology of the StW 573 foot has been interpreted as being consistent with a grasping hallux (Clarke and Tobias, 1995), but this hypothesis has been refuted by more detailed studies, including a morphometric analysis of the medial cuneiform and first metatarsal (Harcourt-Smith, 2002; Harcourt-Smith and Aiello, 2004; Kidd and Oxnard, 2005; McHenry and Jones, 2006).

Many *Au. africanus* metatarsals and phalanges await description and functional analysis. An analysis of the fourth metatarsal StW 485 and complete fifth metatarsal StW 114/115 suggests a stable lateral column of the foot in *Au. africanus* (DeSilva and MacLatchy, 2008; DeSilva, 2009; Zipfel et al., 2009).

Despite the many adaptations of the lower limb for habitual bipedality, the morphology of the *Au. africanus* upper limb is suggestive of some degree of arboreality. The associated humerus, radius, and ulna of StW 431 are robust (Toussaint et al., 2003) and exhibit relatively larger upper than lower limb joint surfaces (McHenry and Berger, 1998). A study of upper to lower limb size in multiple *Au. africanus* fossil specimens using a resampling approach found that *Au. africanus* had more ape-like proportions than did *Au. afarensis* (Green et al., 2007). This result suggests that *Au. africanus* may have engaged in more activities that loaded the upper limb, such as arboreality, than *Au. afarensis* (Green et al., 2007). Interestingly, though, preliminary examination of StW 573 suggests that the arms of this *Au. africanus* individual are not long relative to its associated legs (Clarke, 2002).

The metacarpals of *Au. africanus* are modern human–like in their length, which would allow modern human–like manual dexterity, though they lack the robusticity useful for tool-making grips (Green and Gordon, 2008). The distal thumb phalanx (StW 294) of *Au. africanus* is more robust than that found in apes and may be evidence for power gripping in *Au. africanus*. This is consistent with the robust thumb of StW 573 currently being excavated. These morphologies may be evidence for climbing or tool making in *Au. africanus* (Ricklan, 1987; Clarke, 1999, 2002). The hand of StW 573 appears to have modern human–like proportions with short fingers and a long thumb (Clarke, 2002). If *Au. africanus* was still engaged in arboreal locomotion, reduced finger length would indicate that it climbed in a manner different from and perhaps less efficient than modern apes (Ricklan, 1990).

Remarks Pollen data collected from *Au. africanus*–bearing deposits at Makapansgat suggest that this hominin lived along a forest margin. In a synthesis of data from the faunal fossil record, pollen, and geomorphology of the Makapansgat Valley, Rayner et al. (1993) reconstruct the *Au. africanus* habitat as consisting of patches of subtropical forest. These data are consistent with analyses of fossilized wood from Sterkfontein Member 4 identified as *Dichapetalum* cf. *mombuttense*, which grows as a liana in closed forests (Bamford, 1999). Reduced pitting on the molars (Grine, 1986) and microwear patterns on the incisors (Ungar and Grine, 1991) are consistent with a diet rich in soft fruit and leaves. But the diet of *Au. africanus* may have included resources from a drier open woodland or grassland environment. Faunal analysis of Sterkfontein Member 4 reconstructs the paleoenvironment as an open woodland (Reed, 1997). Isotope analysis found that four teeth of *Au. africanus* recovered from Makapansgat have relatively high $\delta^{13}C$ values of −5.6 to −11.3 ‰, suggesting that this species exploited not only leaves and fruits, but also C_4 plant resources from an open woodland or grassland environment (Sponheimer and Lee-Thorp, 1999). Additional isotopic work on 10 teeth from Sterkfontein Member 4 deposits yielded similar results ($\delta^{13}C$ range of −4.4 to −8.8 ‰), suggesting that *Au. africanus* ate a varied diet perhaps consisting of grasses, seeds, underground storage organs, invertebrates, and grazing mammals in addition to the occasional leafy vegetable and soft fruit (van der Merwe et al.,

2003). Elevated Sr/Ca ratios in *Au. africanus* are also consistent with a diet consisting of insects and underground storage organs (Sponheimer et al., 2005). These data suggest that *Au. africanus* was capable of exploiting a range of environments.

Close to 600 individual fossils, including 495 teeth, have been recovered from Member 4 deposits at Sterkfontein. There is enough morphological diversity in the Sterkfontein Member 4 hominin assemblage, however, for some to suggest that two hominin species are represented. This hypothesis has been promoted primarily by Clarke (1988, 1994) who argues that a "pre-*Paranthropus*" hominin, possibly represented by the StW 252 cranium and the StW 573 skeleton, is a distinct species from *Au. africanus* as represented by the Taung skull and crania such as Sts 5 and Sts 17. Clarke (1988) notes in particular the similarity in postcanine tooth size between StW 252 and the *P. robustus* maxilla SK 13/14 (though StW 252 has much larger anterior teeth than any *Paranthropus*). In a thorough description of the craniodental remains from Member 4, Lockwood and Tobias (2002) argued that specimens StW 183, StW 255, and the partial cranium StW 252 are morphologically similar to one another, but distinct from the *Au. africanus* hypodigm, and may be members of a different species, and the similarity of StW 255 to *P. aethiopicus* KNM-WT 17000 has also been noted (Spoor, 1993). Other specimens from Member 4 that have received attention include Sts 19, which was thought not to be *Au. africanus* by some (Kimbel and Rak, 1993) but believed to be part of the normal variation within a species in another study (Ahern, 1998). The juvenile specimen StW 151 has been regarded as *Homo*-like and potentially distinct from *Au. africanus* (Moggi-Cecchi et al., 1998). Partridge et al. (2003) have argued that the variation found in the cranial remains and femora of Member 4 hominins and fossils recovered from the older Jacovec Cave deposits cannot be accommodated within a single species, and Schwartz and Tattersall (2005) recognize two distinct morphs in the Sterkfontein Member 4 assemblage.

Despite these suggestions for a second species besides *Au. africanus* in Member 4, none has yet been named or described. Furthermore, in a recent study of the Sterkfontein Member 4 dental remains, the morphological variation in the sample was less than that known in *P. boisei* or *H. habilis sensu stricto* (Moggi-Cecchi et al., 2006). The troublesome fossil StW 252 has enlarged postcanine dentition like *Paranthropus*, but like other members of *Au. africanus* it retains the enlarged anterior dentition as well. It is possible that the variation seen in *Au. africanus* can be accounted for by variation within a single species, perhaps being sampled across different time periods.

Phylogenetically, *Au. africanus* has been proposed as a sister taxon to *Homo* (e.g., Strait et al., 1997) or a sister taxon to *Paranthropus* (e.g., Johanson and White, 1979; Rak, 1983). Either hypothesis has been regarded as possible in recent work (Asfaw et al., 1999; Kimbel et al., 2004).

AUSTRALOPITHECUS GARHI Asfaw et al., 1999
Figure 25.9 and Table 25.2

Holotype The type specimen is BOU-VP-12/130, an adult male cranium consisting of frontal, parietals, lower face, palate, and upper dentition.

Age and Occurrence Late Pliocene, eastern Africa (table 25.2).

Diagnosis Based on Asfaw et al. (1999). *Australopithecus garhi* differs from *Au. afarensis* in having absolutely larger anterior and postcanine dentition, and in having a more derived upper third premolar morphology consisting of a more oval, symmetrical

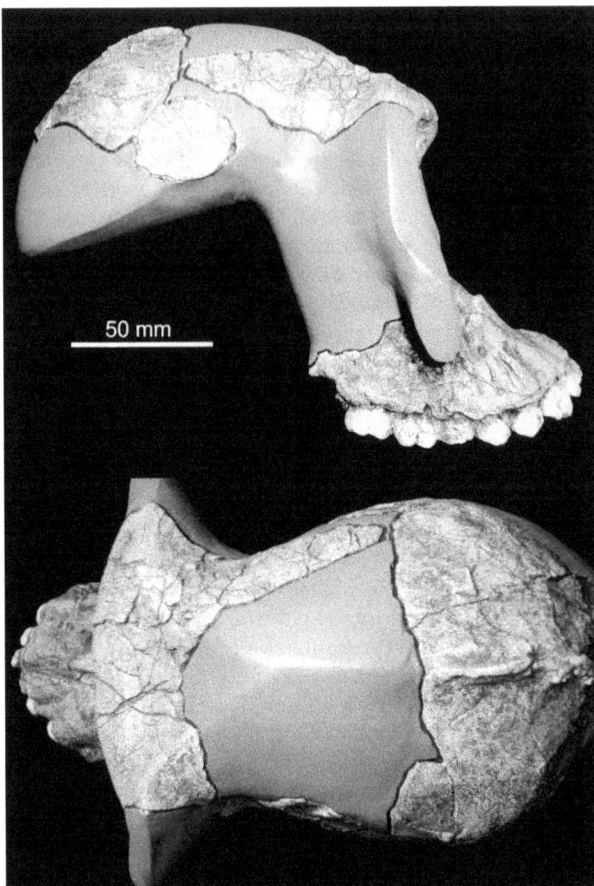

FIGURE 25.9 Holotype of *Australopithecus garhi*, cranium BOU-VP-12/130. Courtesy of Tim White.

occlusal outline and a weaker projection of the mesiobuccal enamel line. *Australopithecus garhi* differs from *Au. africanus* in having a more primitive subnasal region with a convex clivus contour and lacking anterior pillars; and in aspects of its frontal anatomy, such as the presence of a frontal trigon, frontal convergence of the temporal lines, and a strong sagittal crest. *Australopithecus garhi* lacks the derived facial anatomy of *Paranthropus* and also differs from *Paranthropus* in having a relatively larger anterior dentition and thinner tooth enamel. The canine to molar ratio is *Homo*-like, though the prognathic lower face is reminiscent of more primitive hominins like *Au. afarensis*.

Description Based on Asfaw et al. (1999). The type cranium shows a combination of primitive and derived anatomies. BOU-VP-12/130 has a small brain of approximately 450 cm³. The frontal bone shows evidence of a frontal trigon and has strong temporal lines and a marked postorbital constriction. The parietals are complete enough to demonstrate the presence of a strong sagittal crest. The facial anatomy of BOU-VP-12/130 is *Au. afarensis*–like, with a prognathic subnasal region that possesses a convex clivus contour, and canine and lateral incisor roots that are in line with or lateral to the nasal aperture. Unlike *Paranthropus robustus* and most *Au. africanus* specimens, the maxilla is not reinforced with anterior pillars. The dental arch is U-shaped, and a small diastema is present between the upper canine and lateral incisor. Perhaps the most striking feature of the BOU-VP-12/130 cranium is the absolute size of the teeth. The postcanine dentition, and in particular the premolars, are as large and in some cases larger than *Paranthropus* teeth. However, unlike *Paranthropus*, BOU-VP-12/130 has large canines and

incisors as well. Therefore, the relative proportions of the teeth are *Au. africanus* and *Homo*-like, though the absolute sizes of the teeth are some of the largest yet discovered in the hominin fossil record. The large dentition and the presence of strong ectocranial markings, such as a sagittal crest, are evidence for this cranium belonging to a male individual.

For now, only the holotype cranium BOU-VP-12/130 is assigned to *Australopithecus garhi*. However, other specimens found nearby in stratigraphic horizons of similar age to that yielding the type (Asfaw et al., 1999; White et al., 2005), and megadont specimens from Omo and sites in Kenya may be included in the hypodigm in the future. Between 2.7 and 2.3 million years ago, the fossil remains currently known from East Africa can be broadly grouped into those that were moving toward the *Paranthropus* condition, and nonrobust specimens that share morphology found in early *Homo* (Suwa et al., 1996). Craniodental remains primarily from the Omo do not support the hypothesis of multiple nonrobust species between 2.7–2.3 Ma (Suwa et al., 1996). For this interval, the only two named East African hominins are *P. aethiopicus* and *Au. garhi*. Fossils from 2.7–2.3 million years ago not assignable to *Paranthropus* may ultimately be united under the hypodigm of *Au. garhi*. These include BOU-VP-17/1, a 2.5 Ma mandible with dentition, GAM-VP-1/1, an edentulous left mandibular corpus, and GAM-VP-1/2, a parietal fragment, from 3.0–2.0 Ma deposits at the site of Gamedah. White et al. (2005) concluded that nothing about the morphology of the Gamedah fossils would preclude them from being assigned to *Au. garhi* but stopped short of doing so. The two mandibles, based on similar morphologies to the BOU-VP-12/130 type, may represent *Au. garhi* females (Asfaw et al., 1999). The BOU-VP-12/130 *Au. garhi* cranium has postcanine dental arcade length and proportions that are quite similar to the 2.7 Ma associated teeth from Turkana, KNM-ER 5431 (White et al., 2005). The BOU-VP-17/1 and GAM-VP-1/2 mandibles share derived premolar and molar morphology with 2.7–2.5 Ma nonrobust Omo specimens L824-5, L362-14, and L45-2 (Asfaw et al., 1999; White et al., 2005).

There are currently no postcrania assigned to *Au. garhi*. However, in the description of *Au. garhi*, Asfaw et al. (1999) reported the discovery of a femur and associated humerus, ulna, and radius, partial fibula, and foot phalanx (BOU-VP-12/1A-G) in the Bouri Hata sediments. These were described by DeGusta (2004). This partial skeleton was found in the same 2.5 Ma horizon, 278 m away from the BOU-VP-12/130 cranium. The femur, humerus, and radius are complete enough to estimate limb proportions in BOU-VP-12/1, and Asfaw et al. (1999) suggest that this skeleton represents the earliest evidence for modern human–like limb proportions, with a relatively elongated femur. The forearm is still quite long in BOU-VP-12/1, similar to the condition found in the A.L. 288-1 *Au. afarensis* skeleton. These data suggest that femur elongation preceded forearm shortening in the taxon represented by this individual.

One of the few comparable postcranial elements from this time period in East Africa is the KNM-WT 16002 femur from 2.7-million-year-old deposits in the Lomekwi Member in West Turkana, Kenya. The morphology of this femur, however, is reportedly distinct from the femur from the BOU-VP-12/1 skeleton (Lovejoy, pers. comm. in Brown, et al., 2001). Brown et al. (2001) tentatively suggest that the KNM-WT 16002 femur may belong to *P. aethiopicus*.

Remarks There are few fossil hominin remains from eastern Africa between 3.0 and 2.0 Ma. Although this time period is well represented in the southern African fossil record, the discovery of *Au. garhi* provides important insights into the evolutionary trajectory of hominins in East Africa during this part of the late Pliocene. Soon after 2.5 million years ago, the earliest members of the genus *Homo* appeared. Until the description of *Au. garhi* in 1999, *P. aethiopicus* was the only named East African hominin between 2.7 and 2.3 million years ago, and it was evidently more closely related to *P. boisei* than to early members of the genus *Homo*. Although Asfaw et al. (1999) are cautious in assigning phylogenetic significance to BOU-VP-12/130, they regard *Au. garhi* as a "candidate ancestor" for the genus *Homo*. The authors note, for instance, that the length and proportions of the dental arcade in the type cranium are "equivalent" to the 2.15 Ma early *Homo* mandible from the Shunguru Formation Omo 75–14 (Asfaw et al., 1999).

Important in this discussion of craniodental anatomy is a comparison of *Au. garhi* with the temporally contemporaneous southern African hominin *Au. africanus*. Asfaw et al. (1999) list nine features found primarily in the maxillary and frontal regions that differentiate *Au. garhi* and *Au. africanus*. However, many of the distinguishing characters are variably present in the collection of *Au. africanus* specimens from the Member 4 deposits in Sterkfontein Cave. For example, an I2/C diastema is present in the partial cranium StW 252, which also has tooth proportions and overall tooth dimensions that are quite similar to BOU-VP-12/130. A frontal trigon and frontal convergence of the temporal line is present in Sts 17, and anterior pillars are absent in TM 1512 and StW 498, though StW 498 is from an immature individual (Lockwood and Tobias, 1999). StW 391 also lacks strong anterior pillars and has a convex clivus contour, though this specimen may also be from an adolescent (Lockwood and Tobias, 1999, 2002). The apparent close similarities between BOU-VP-12/130 and certain specimens currently assigned to *Au. africanus*, such as StW 252, are worthy of further investigation.

Faunal analysis suggests that the Hata hominins were living along a lake margin rich with grazing bovids, and zooarchaeological remains found in the same horizon as the BOU-VP-12/130 type cranium and the BOU-VP-12/1 skeleton suggest that hominins were utilizing these bovids as a food supply (Heinzelin et al., 1999). The Hata material, in the form of percussion and cut marks on bovid mandible and tibia, and an equid femur provides the earliest direct evidence for meat and marrow acquisition in the hominin fossil record (Heinzelin et al., 1999), and the earliest stone tools have been recovered from 2.6-million-year-old sediments at Gona, roughly 100 km north of the Bouri formation (Semaw, 2000).

With this circumstantial evidence that at least one Bouri hominin species was beginning to modify stone to acquire meat and marrow, and evidence that a species of hominin had evolved elongated lower limbs, the fossils of the Hata deposits preserve evidence of two of the major evolutionary transitions in the hominin lineage (the incorporation of significant amounts of meat and marrow into the diet, and a shift to a more efficient form of bipedal locomotion). Whether these transitions occurred in the species *Au. garhi*, or in another taxon, remains to be established.

Genus *KENYANTHROPUS* Leakey et al., 2001
KENYANTHROPUS PLATYOPS Leakey et al., 2001
Figure 25.10 and Table 25.2

Synonymy *Australopithecus afarensis*, White, 2003.
Holotype The type is KNM-WT 40000, a nearly complete but distorted cranium, and the paratype is KNM-WT 38350, a left partial maxilla (Leakey et al., 2001).

Age and Occurrence Mid-Pliocene, eastern Africa (table 25.2).

Diagnosis Based on Leakey et al. (2001). Cranium with ape-size brain. Facial contour is flat in the transverse plane at a level just below the nasal bones; zygomaticoalveolar crest low and curved; tall malar region; vertically oriented maxillary zygomatic process positioned above P3–P4; nasoalveolar clivus long and transversely and sagittally flat; moderate subnasal prognathism; incisors in line with canine; thin palate; M1 and M2 small with thick enamel; upper incisor roots of similar size, small external auditory meatus (as in *Au. anamensis* and *Ar. ramidus*, and unlike more derived australopiths); and mediolaterally long tympanic element lacking a petrous crest.

Kenyanthropus platyops differs from *Ar. ramidus* in its buccolingually narrow M2, thicker molar enamel, a more cylindrical articular eminence and deeper mandibular fossa; from *Australopithecus* in its reduced subnasal prognathism, more anteriorly positioned maxillary zygomatic process; transversely and sagittally flat nasoalveolar clivus, low and curved zygomaticoalveolar crest, similarly sized upper incisors, and small M1–2 crowns. It differs from *Paranthropus* in its tall malar region, flat midface, thinner palate, stepped entrance to the nasal cavity, small M1–2 crowns, and thinner enamel.

Description Based on Leakey et al. (2001). The holotype cranium KNM-WT 40000 is relatively complete but it is considerably distorted and lacks most of the basicranium and the anterior and premolar tooth crowns.

One of the most striking features of the holotype is the flat transverse facial contour below the nasal bones. The incisor alveoli are situated almost on the bicanine line, contributing to an orthognathic subnasal region. However, unlike *Paranthropus*, the midface is flat, not dished.

The M2 in KNM-WT 40000 is the only tooth whose width and length can be measured, and it falls below the range of known early hominins. The M1 of KNM-WT 38350 is also small, but comparable to the smallest known specimens of *Au. anamensis*, *Au. afarensis* and *H. habilis*. Molar enamel is thick, comparable to that found in *Australopithecus*, but not as thick as in *Paranthropus*.

Remarks There are several interesting evolutionary implications associated with this genus, should its taxonomic validity be substantiated. White (2003, 2009) has disputed its validity because he contends the type has been so altered

FIGURE 25.10 Holotype of *Kenyanthropus platyops*, cranium KNM-WT 40000. Courtesy of National Museums of Kenya.

by expanding matrix distortion (EMD) that its true anatomy cannot be accurately gauged. EMD, a type of postmortem deformation, results when matrix becomes interspersed between adjacent fragments of fossil bone, displacing the fragments relative to one another and altering anatomy in an unpredictable way. White's caution may be warranted in terms of whether the facial part of the diagnosis can differentiate *Kenyanthropus* from *Au. afarensis*. However, the relatively diminutive molar size is unaffected by distortion and remains evidence for consideration of a new taxon given that postcanine megadontia is a defining characteristic of *Australopithecus*.

The implications of the taxon, as currently described, are as follows. First, it suggests modest hominin diversification between 4.0 and 3.0 million years ago, since, as currently interpreted, this interval is occupied by the *Au. anamensis-Au. afarensis* lineage. However, as White (2003) argues, two to three (if *Au. bahrelghazali* is included) taxa between 4.0 and 3.0 Ma may not qualify as the components of an adaptive radiation per se.

A second implication is that it provides a plausible backstory to the enigmatic cranium KNM-ER 1470 attributed to *Homo rudolfensis*, and dated at 1.9 Ma. This taxon has long been troubling to anthropologists because of its large, flat face and big brain, contemporaneous with smaller-brained specimens with relatively delicate faces that are difficult to reconcile as belonging to a single taxon. If *Kenyanthropus* gave rise to *Homo rudolfensis*, but not *Homo habilis*, and later *Homo erectus*, then a possible corollary is that big brains may have evolved more than once.

A third implication concerns the unique combination of a *Paranthropus*-like facial morphology combined with small molars. The zygomatic arch is anteriorly positioned, a configuration typically associated with a more anterior line of action for the masseter and hence more chewing power. An anterior zygomatic and large molars have been thought to be functionally and developmentally linked, but it appears that they can be independent (i.e., an anterior zygomatic is not just the result of large masticatory musculature, driven by large tooth/jaw size; Leakey et al., 2001). Leakey et al. (2001:439) suggest that *Kenyanthropus* occupied a "distinct dietary adaptive zone," but the nature of its dietary niche was not elaborated. One possibility is that the combination of strong masseters and thick enamel with unexpanded molars would enable the jaws to exert higher bite forces per unit area, as would be useful in hard object feeding. It is also tempting to look to the paleoenvironment for clues. Although Leakey et al. (2001) conclude that the Lomekwi paleoenvironment may have been more vegetated and wetter than at Hadar, paleoecological resolution remains too coarse to identify specific ecological factors that might be associated with a novel dietary niche.

Genus *PARANTHROPUS* Broom, 1938
Table 25.2

Partial Synonymy Zinjanthropus, Leakey, 1959; *Australopithecus*, Tobias 1967.

Age and Occurrence Late Pliocene to early Pleistocene, eastern and southern Africa (table 25.2).

Diagnosis Same as for *P. robustus*.

Referred Species Paranthropus aethiopicus Arambourg and Coppens, 1968; *P. boisei* Leakey, 1959, *P. robustus* Broom, 1938.

Remarks The hominins in this genus are often called "robust," but this is a misleading term that should only refer to their heavy jaws, megadont to hypermegadont cheek teeth, and inferred massive masticatory musculature (Grine, 1988; McHenry, 1991b; McCollum, 1999), as body size estimates based on regression analysis of postcranial dimensions to body mass and stature show that they were not heavier or taller than so-called gracile hominin taxa in the genus *Australopithecus* (Jungers, 1988; McHenry, 1988, 1991a, 1991b, 1992a, 1992b). Grine (1988) has recommended that these misleading terms be abandoned. There is some disagreement about the proper genus allocation of *P. aethiopicus*, *P. boisei*, and *P. robustus*, with many authorities preferring to place them in *Australopithecus*. Nonetheless, their derived craniofacial architecture, associated novel alignment of powerful chewing muscles, the disproportion of their small anterior teeth to their large (immense in the case of *P. boisei*) cheek teeth, and analyses of tooth wear and dental isotopic composition suggest that they had substantial trophic differences from *Au. afarensis* and *Au. africanus*, and thus separation at the genus level appears justified (Clarke, 1996). Whether the genus is truly monophyletic remains more uncertain, and depends on the phylogenetic position of *P. robustus* vis-à-vis the East African *Paranthropus* taxa and *Au. africanus* (Aiello and Andrews, 2000; Wood and Constantino, 2007). The reason for their disappearance is also not well understood and may be due less to competition with sympatric early species of *Homo* than to other factors, such as the effect of turnover of carnivore guilds during the early Pleistocene (Walker, 1984; Klein, 1988). Considerable work remains to be done on the dietary and postural and locomotor adaptations of this most unusual group of hominins.

PARANTHROPUS AETHIOPICUS (Arambourg and Coppens, 1968)
Figure 25.11 and Table 25.2

Partial Synonymy Paraustralopithecus aethiopicus, Arambourg and Coppens, 1967, 1968; *Australopithecus africanus* (in part), Howell, 1978; *Australopithecus boisei*, Walker et al., 1986; *Paranthropus aethiopicus*, Chamberlain and Wood, 1987; *Au. aethiopicus*, Kimbel and White, 1988; *Au. aethiopicus*, Kimbel et al., 1988; *Au. walkeri*, Ferguson, 1989; *Paranthropus aethiopicus*, Clarke, 1996; *Au. boisei*, Curnoe, 2001; *Paranthropus aethiopicus*, Wood and Constantino, 2007.

Holotype Omo 18-1967-18, mandible lacking rami and tooth crowns except for a partial left canine (figure 25.11), Mb. C, Shungura Fm., Omo, Ethiopia (Arambourg and Coppens, 1967), dated to 2.6 Ma (Feibel et al., 1989). A list of other specimens referred to *P. aethiopicus* is provided by Wood and Constantino (2007).

Age and Occurrence Late Pliocene, East Africa (table 25.2).

Diagnosis The species was provisionally recognized by Arambourg and Coppens in 1967, and formally named by these authors in 1968, based on a nearly toothless mandible from Mb. C, Unit C8 of the Shungura Fm., Omo, Ethiopia. The original diagnosis purported to differentiate the specimen from other early hominin jaws by its overall morphology, including the general massivity of the specimen, very thick corpus, deep genioglossal fossa, inferred macrodonty of cheek teeth, short, parabolic aspect of the alveolar rows, very reduced size of the canine and incisor region, and a deep, receding symphysis (Arambourg and Coppens, 1968). The referral of cranium KNM-WT 17000 to the species permits the diagnosis to be more meaningfully emended to include, in comparison with other species of *Paranthropus* (and *P. boisei* in particular), more prognathic face; palate less retracted; weaker flexion of the cranial base; longer distance between M1 and the temporomandibular joint; higher inclination of the nuchal plane; shorter postcanine tooth row; shallower mandibular fossa; lower articular eminence; flatter, shallower palate; smaller cranial capacity; parietals low and sloping; maxillary process directed backward; nasomaxillary basin more pronounced; possibly larger incisors; and inferred greater I1–C length relative to buccolingual width of P4 (Walker et al., 1986; Ferguson, 1989; Suwa, 1989; Wood and Richmond, 2000; Wood and Constantino, 2007).

Description Despite initial claims for uniqueness (Arambourg and Coppens, 1967, 1968), the type specimen closely resembles mandibles of *P. boisei* in the robustness of its corpus inferred massiveness of its postcanine teeth, and small size and transversely straight alignment of its canine-incisor row. It also exhibits wide extramolar sulci, and prominent superior and inferior transverse tori (Howell and Coppens, 1976). The V-shaped configuration of its alveolar profile and relatively modest height (33.0 mm) and width (26.0 mm) of the corpora at m2 contrast with the condition seen in many *P. boisei* mandibles (Wood et al., 1994); however, the robusticity index of corpus dimensions (W × 100/H at m2 = 79) fall within the range of indices for *P. boisei* (65–87), and the shape of the jaw and absolute dimensions are close to those of presumed female mandibles of that species (Sanders, 1987; Leakey and Walker, 1988; Walker and Leakey, 1988).

A slightly geologically younger (ca. 2.45 Ma) partial mandible from West Turkana, Kenya, KNM-WT 16005, is larger than Omo 18-1967-18 and presumably from a male individual (Walker et al., 1986; Leakey and Walker, 1988; Walker and Leakey, 1988). It also had relatively small incisors and canines, immense cheek teeth, and massive corpus. The size of the incisors and canines is inferred from their mesiodistally compressed roots (Leakey and Walker, 1988). Only the crowns or partial crowns of left p3–m2 and right p3–m1 are preserved. The mesiodistal and buccolingual dimensions of the left tooth crowns are p3 = 10.7 × 13.8 mm, p4 = est. 12.0 × est. 15.0 mm, m1 = 15.7 × 14.3 mm, and m2 = est. 17.0 × 16.7 mm, respectively (Walker et al., 1986; Leakey and Walker, 1988). The p3s are asymmetrical in occlusal outline, while the p4s are more molarized (Leakey and Walker, 1988). Anteriorly, the symphysis has a mild central keel, with slight concavities between the keel and the canine juga (Leakey and Walker, 1988). The closest similarity in dimensions of this specimen is with the *P. boisei* mandible from Peninj, Tanzania (Walker et al., 1986; Walker and Leakey, 1988).

Undoubtedly the most distinctive specimen documenting *P. aethiopicus* as a valid species (Kimbel et al., 1988; but see Curnoe, 2001) is KNM-WT 17000, the so-called Black Skull (e.g., Bower, 1987; Wilford, 1987), an adult cranium lacking the portion of the frontal posterior to the frontal trigon, the midsections of the zygomatic arches, large sections of the parietals, particularly anteriorly, and fragments from the occipital, pterygoid processes, and maxillae, as well as all of the teeth except for a premolar and half a molar (figure 25.11; Walker et al., 1986; Leakey and Walker, 1988). This specimen is close in geological age (ca. 2.5 Ma) to Omo 18-1967-18, and is remarkable for having the largest sagittal crest and one of the smallest fossil hominin cranial capacities (and most diminutive brain size among "robust" australopiths, 410 cm³; Walker et al., 1986; Falk, 1987; Falk et al., 2000). It also would have accommodated an immense mandible, similar in size to the largest known for *P. boisei* (Walker et al., 1986; Leakey and Walker, 1988). Morphology of the specimen was described in detail by Walker

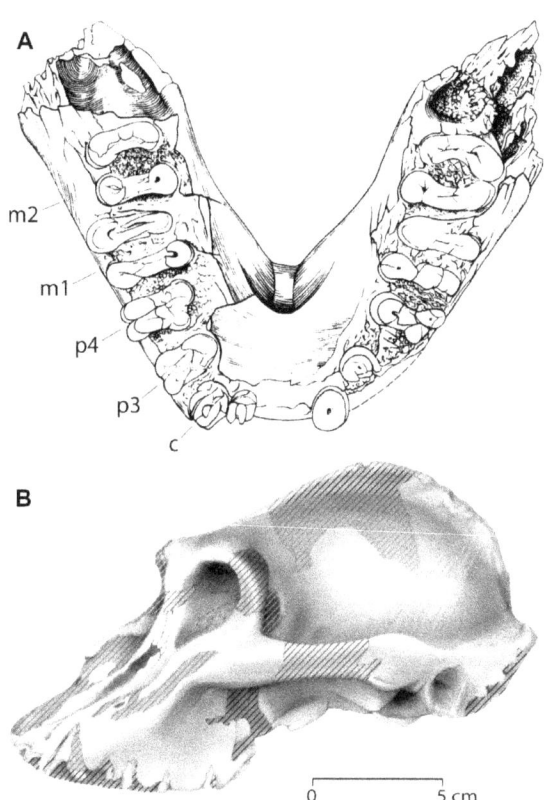

FIGURE 25.11 A) Holotype of *Paranthropus aethiopicus*, mandible Omo 18-1967-18 (after Arambourg and Coppens, 1968: fig. 1); B) *Paranthropus aethiopicus* cranium KNM WT 17000.

et al. (1986) and Leakey and Walker (1988). Only its most salient features are listed here: cranium massively built, with a large, prognathic face; neurocranium relatively very small; incisor and canine roots disproportionately small relative to size of cheek tooth roots; P4 = 16.2 mm long × 11.5 mm wide; palate very large, broad (45.0 mm wide between P3 and 4), and flat; nasal opening pear shaped with slightly everted superolateral margins; low infraorbital foramina; visorlike flare of the zygomatics; triangular nasomaxillary basins bordered laterally by more anteriorly set zygomatic "visors"; root of zygomatic above P3; temporal foramen very large; bar-like supraorbital tori joined by a modestly inflated glabella; strong frontal trigon; sagittal crest most pronounced posteriorly, implying hypertrophy of the posterior fibers of the temporalis mm.; foramen magnum heart shaped; and cerebellar lobes of the endocast not tucked under occipital poles of the cerebrum (Walker et al., 1986; Leakey and Walker, 1988).

Another cranial specimen assigned by some (e.g., Wood and Constantino, 2007) to *P. aethiopicus* is the posterior portion of a juvenile calotte from Mb. E, Shungura Fm., Omo, Ethiopia, L338y-6 (Howell, 1976). This specimen was described in detail by Rak and Howell (1978), who placed it in *Australopithecus boisei*. It is close in age to KNM-WT 16005. The calotte is comprised of the occipital squama, most of the parietals, and a small portion of the frontal and is associated with a basioccipital fragment. A hole in the anterior part of the right parietal bone may be a tooth puncture mark from a large carnivore (Rak and Howell, 1978); a carnivore tooth mark is also present in KNM-WT 17000, just posterior to the right temporal line at the point of the middle of the orbit (Leakey and Walker, 1988). Cranial capacity is low, estimated at 427 cm^3 (Holloway, 1983). Age is estimated by the openness of its sutures, and the prominence of its muscle markings suggests that it is from a male individual (Rak and Howell, 1978). Due to the young age of the individual, postorbital constriction is inferred to still have been moderate at death, the sagittal contour of the calotte is rounded, and the temporal lines, though prominent, had not yet approximated a sagittal crest. The nuchal planum is roughened, and there is a pronounced external occipital protuberance. Paired salient depressions about the midline of the basilar part of the occipital are inferred to have been insertions sites for longus capitis muscles, a condition found in apes and other australopiths, but not humans (Rak and Howell, 1978). The foramen magnum was apparently heart shaped. Ridges on the temporal margins of the parietals indicate substantial overlap between the temporals and parietals, typical of *Paranthropus* (Rak and Kimbel, 1991). The superomedial position of these striae especially resemble their distribution in *P. boisei* crania, where they are located evenly along the arc of the parietotemporal suture, and contrast with their inferred arrangement in KNM-WT 17000, where the area of most substantial striae and overlap between the temporal and parietal is narrowly oriented posteromedially, in alignment with the most rugose segment of the sagittal crest (Rak and Kimbel, 1991; 1993; but see Walker et al., 1993). Variation in this feature between KNM-WT 17000 and L338y-6 could be ontogenetic. Based on cranial capacity, pattern of meningeal branching, cerebellar morphology, and perceived absence of an enlarged occipital-marginal venous sinus, Holloway (1981) concluded that the closest similarity of the endocast of L338y-6 is with either *Au. afarensis* or *Au. africanus*. These features were subsequently shown to be incorrectly interpreted or taxonomically uninformative, and other aspects of the endocast of L338y-6 link it with *Paranthropus* (White and Falk, 1999).

Possibly the oldest specimens of *P. aethiopicus* derive from the Upper Ndolanya Beds at Laetoli, Tanzania (table 25.2; Harrison, 2002). They include a portion of the lower face and palate (EP 1500/01), and left proximal tibia (EP 1000/98) that await comprehensive description. Morphometrically, the tibia resembles that of *Au. afarensis* individual A.L. 288-1, and the cranial fragment is similar to KNM-WT 17000 in the shallowness of its palate and position of the infraorbital foramen.

Remarks Although *Australopithecus* and *Paranthropus* have been used interchangeably in the literature to refer to *P. aethiopicus*, Groves (1999) has pointed out that the first usage of *Au. aethiopicus* was in reference to *Au. afarensis* from Hadar (Tobias, 1980), and that application of this nomen to a species typified by Omo 18–1967–18 would constitute a taxonomic homonym and be invalid. If *Paranthropus* is not deemed monophyletic, the available name for the species would be *Australopithecus walkeri* (Ferguson, 1989).

Specimens of *P. aethiopicus* represent the oldest occurrences of the genus, succeeded closely in time in East Africa by *P. boisei* (table 25.2). Little attention was paid initially to the type mandible from the Omo, but interest in the specimen was rekindled two decades later by the finding of the similar-aged cranium KNM-WT 17000. Although WT 17000 was characterized as an early member of the *"Australopithecus boisei"* lineage by its describers, they cautiously noted that its combination of primitive similarities to *Au. afarensis* and *Paranthropus* "robust" australopith features might warrant assignment to a different species, which because of temporal and geographical connections to the Omo mandible they suggested should be named *"Australopithecus aethiopicus"*

(Walker et al., 1986). In addition, Suwa (1988) showed that *P. aethiopicus* lacked the extreme expansion of p4 talonids, diagnostic of *P. boisei*. This combination of features, dominated as it is by traits held in common with *Au. afarensis*, contrasts with the small number of synapomorphies shared with *P. boisei*, and is evidence supporting the validity of *"Au. aethiopicus"* (Kimbel et al., 1988; Ward, 1991).

More importantly, recognition of *P. aethiopicus* as an early stage of *Paranthropus* served to dramatically reconfigure early hominin systematics. While Rak (1983) had envisioned an evolutionary progression of *Au. africanus* > *P. robustus* > *P. boisei*, temporal and morphological considerations suggest that *P. aethiopicus* was intermediate between *Au. afarensis* and *P. boisei/P. robustus*, with *Au. africanus* essentially left as a side branch (Kimbel et al., 1988) or perhaps antecedent only to *P. robustus* (Walker and Leakey, 1988). Wood and Constantino (2007) felt that *P. aethiopicus* and *P. boisei* could be viewed as chronospecies within an evolving lineage; indeed, the oldest known cranium attributed to *P. boisei*, Omo-323, from the 2.1 Ma–aged Mb. unit G8 of the Shungura Fm., retains traits of *P. aethiopicus* in the morphology of its glabellar region, supraorbital tori, and articular eminence, but otherwise has *P. boisei*-type features (Alemseged et al., 2002). This is just the sort of mosaic change one would expect in a lineage evolving anagenetically (Alemseged et al., 2002). Acceptance of a close phylogenetic relationship between these two species has not been universal, however, and others have posited *P. aethiopicus* as part of a polyphyletic group, separate from a *P. boisei+P. robustus* clade (Skelton and McHenry, 1992).

Almost nothing is known of the postcranial skeleton of this species (Wood and Richmond, 2000), making it difficult to reconstruct its paleobiology. However, it has been suggested that *P. aethiopicus* inhabited more closed environments than *P. boisei* (Reed, 1997).

PARANTHROPUS BOISEI (Leakey, 1959)
Figure 25.12 and Table 25.2

Partial Synonymy Zinjanthropus boisei, Leakey, 1959; *Paranthropus boisei*, Robinson, 1960; *Australopithecus (Zinjanthropus) boisei*, Tobias, 1967.

Holotype OH 5, subadult male cranium, Olduvai Gorge, Tanzania (Leakey, 1959; figure 25.12).

Age and Occurrence Late Pliocene–early Pleistocene, East Africa (table 25.2).

Diagnosis The type cranium was discovered by Mary Leakey in July, 1959 at locality FLK 1 in Olduvai Gorge, near the bottom of Bed I. The original diagnosis of the type cranium placed it in a new genus, *"Zinjanthropus,"* which was distinguished from other australopiths (including southern African *Paranthropus*) by greater reduction of the canines; extent of muscular attachment area on the malars; a deeper palate; coincidence of nasion with the most anterior aspect of the glabellar region; thinness of the parietals; development of the nuchal crest as a continuous ridge across the occipital in males; high vaulted posterior region of the cranium; less elongate foramen magnum; occurrence of a massive horizontal torus above the mastoids; extensive pneumatization of the mastoid region of the temporals; development of keeled anterior margins of the sagittal crest for attachment of anterior temporalis muscle fibers; great interorbital width; shape and position of the external orbital angle elements of the frontal bone; m2 > m3; and greater overall massiveness of the cranium, especially the face (Leakey, 1959).

Later detailed description of *P. boisei* crania shows that they can be further discriminated from those of *P. robustus* by presumed merging (and consequent loss) of anterior pillars into the maxillary infraorbital surface; absence of maxillary fossula; expansion of the lateral infraorbital region into broad "visors" (in some cases); development of a nasomaxillary basin; blunt lateral margins of the pyriform aperture; extent of palatal retraction and extreme forward extension of the masseter muscle attachments; very wide flare of the zygomatic arches; and extreme postorbital constriction and very capacious extent of the temporal foramen (Rak, 1983). Also distinctive of *Paranthropus boisei* is the robustness of its mandibles and its hypermegadont posterior dentition, with premolars more "molarized" and cheek tooth crown areas larger than those of other hominins (Kimbel and White, 1988; Suwa, 1988).

Description Based in part on Leakey (1959); Tobias (1967a); Howell (1978); Rak (1983); Leakey and Walker (1988); Walker and Leakey (1988); Brown et al. (1993); Suwa et al. (1997); McCollum (1999); Alemseged et al. (2002). The history of recovery of fossils attributed to *Paranthropus boisei* is summarized in Wood and Constantino (2007). Crania, mandibles, and teeth, particularly of male individuals, are relatively abundant in the fossil sample of this species, due in part to their heavy construction. The crania recovered of *P. boisei* include the type OH 5, from Olduvai, Omo 323-1976-896, the oldest of the species, KNM-ER 23000, KNM-ER 406, KNM-ER 13750, and the presumed female crania KNM-ER 407 and KNM-ER 732 from East Turkana, KNM-WT 17400, from West Turkana, and the geologically relatively young KGA10-525 specimen from

Konso, Ethiopia (table 25.2). Although Rak (1983) depicted an idealized cranial morphology for *P. boisei* that was starkly contrasted with the cranial anatomy of *P. robustus*, more recent work, especially by Brown et al. (1993), Suwa et al. (1997), and Alemseged et al. (2002), reveals a great degree of polymorphism beyond that expressed via strong sexual dimorphism in the *P. boisei* sample, and shows that morphological distinctions between the two species are not always clear-cut.

Nonetheless, in general, the facial mask of *P. boisei* is less complicated than that of *P. robustus*, with no anterior pillars or maxillary fossulae. The face of *P. boisei* is broader and longer than in *Australopithecus* (Bilsborough and Wood, 1988), and the lateral infraorbital region sweeps out into the shape of a "visor," but this is less pronounced in some specimens (Brown et al., 1993). Other features exhibiting variability in presumed male crania are the position of the greatest projection of the sagittal crest, the shape of the supraorbital tori, the degree of massiveness of the glabella, the size and projection of the lower face, and the size of the temporal foramen. The midface comprises a sunken, nasomaxillary basin, or "dished" face. The nasal aperture has blunt lateral margins, and the nasoalveolar "gutter" and clivus resemble those of *P. robustus* (discussed later). Nasion is coincident with the glabella. Nasal bones are tucked well under the glabella and are narrow inferiorly. Supraorbital tori vary widely in thickness. Postorbital constriction is strong, and the temporal lines bound a small, concave frontal trigon. The forehead is flattened. Zygomatic arches are high, strongly constructed (Corruccini and Gill, 1993), and flare widely. The mastoids are large in males, with a strong crest above, and the cranium

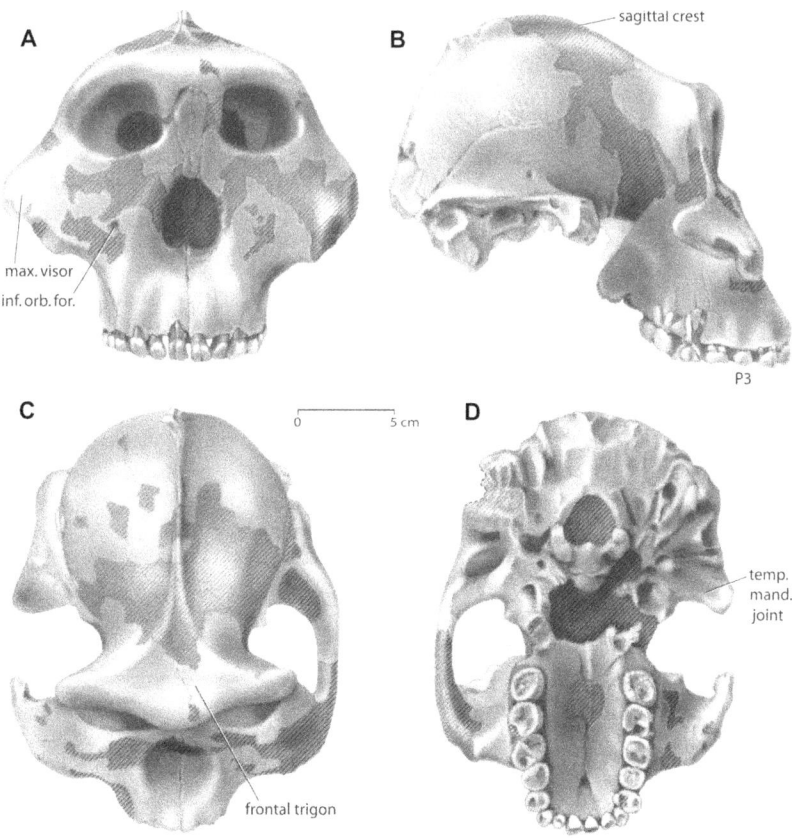

FIGURE 25.12   Holotype of *Paranthropus boisei*, cranium OH 5.

is pneumatized in this region. The palate is very retracted and the massester muscle attachment sites are extended forward to a greater degree than in any other australopith. The palate may be very deep and is bounded by a dentition in a parabolic arcade that has disproportionately large premolars and molars, compared with the very reduced canines and incisors (which are crowded in a transverse line at the anterior end of the tooth row). Temporomandibular joints tend to be large. There may be a compound temporal-nuchal crest. The foramen magnum is heart shaped and situated well forward of the bitympanic line (Dean and Wood, 1982). Occipital condyles are diminutive. Venous drainage of the cranium in *P. boisei* occurred primarily through an enlarged occipital-marginal sinus system (Falk, 1986, 1988); however, in KNM-ER 23000, drainage on the left side of the cranium occurred via a transverse-sigmoid sinus system (Brown et al., 1993). Cranial capacity ranges slightly higher than in other australopiths; Falk et al. (2000) estimated 500 cm^3 for OH 5, and 438 cm^3 and 466 cm^3 for the presumed female specimens KNM-ER 407 and KNM-ER 732, respectively. In addition, Brown et al. (1993) estimated endocranial volumes of 490 cm^3 for Omo 323-1976-896, 491 cm^3 for KNM-ER 23000, and 500 cm^3 for KNM-WT 17400. The greatest cranial capacity estimated for a specimen of *P. boisei* is that of KGA10-525 from Konso, 545 cm^3 (Suwa et al., 1997). These estimates are substantially lower than those for sympatric early *Homo*.

Examples of polymorphism in male crania of *P. boisei* include greater prognathism in KNM-ER 406 than in OH 5 (Rak, 1983), which is not an expression of regional intraspecific differences, since KNM-WT 17400 is more like OH 5 in this regard (Leakey and Walker, 1988). The zygomatic processes of OH 5 and KNM-ER 23000 are less visorlike than those of KNM-ER 406 and KNM-ER 13750, and their sides parallel the cranium reminiscent of the manner in *P. robustus* (Brown et al., 1993). KGA10–525 departs the most from the "ideal" cranial morphology envisioned for *P. boisei* by Rak (1983). Its zygomatic processes are configured similarly to those of *P. robustus*; it has a short lower face like that of KNM-ER 406, but more orthognathic than that of OH 5; its sagittal crest is more posteriorly developed, as in the *P. aethiopicus* cranium KNM-WT 17000; it has a high placement of the infraorbital foramen; and its palate is autapomorphically broad, shallow, and anteroposteriorly short, with a *Homo*-like shape (Suwa et al., 1997). Suwa et al. (1997) felt that the shape of the zygomatic processes and several other features in this individual strengthened the case for monophyly of *Paranthropus*. The morphological uniqueness of KGA 10–525, however, occurs primarily in features that have low heritability and are strongly liable to mechanical strains from mastication (Wood and Lieberman, 2001), which suggests that cranial polymorphism in *P. boisei* may be associated with a wide range of dietary habits.

Gorilla-like levels of sexual dimorphism in cranial morphology are also observed in *P. boisei*: female crania (e.g., KNM-ER 732, KNM-ER 407) are much smaller than those of males and lack ectocranial cresting, though they do exhibit features consistent with their placement in the species, such as depressed nasal bones, advanced placement of the malar region, and coronally oriented petrous temporal bones (Dean and Wood, 1982; Bilsborough and Wood, 1988). Despite its geographic range and temporal extent, in nearly all aspects of its skull anatomy, *P. boisei* does not exceed the degree of variation observed in extant hominoids (Silverman et al., 2001; Wood and Lieberman, 2001), although the very derived mor-

phology of *P. boisei* means that detailed comparisons cannot be carried out.

There is a relative abundance of mandibles in the species sample, particularly from East Turkana, where some 20 specimens representing both sexes have been recovered. In contrast to their crania, other than the less robust corpora and smaller postcanine dentition in females (Walker and Leakey, 1988), mandibles of *P. boisei* exhibit only a small degree of variability (Wood and Lieberman, 2001). In addition, there appears to have been little mandibular or dental morphological change over geological time in *P. boisei* (Wood et al., 1994). Mandibles of this species have corpora that are absolutely and relatively very broad and high (almost rounded in cross section in some individuals), with tall, thick symphyses, wide extramolar sulci, and tall rami that are rooted as anterior as m2 and that have extensive muscle attachment areas, particularly for mm. masseter and the medial pterygoid muscles. The symphysis is posteriorly inclined, and has prominent superior and inferior transverse tori; the superior torus may continue as far posterior as the premolars, and bounds a deep median fossa. As with the upper dentition, the lower dentition exhibits a gross imbalance between diminutive anterior teeth and massive cheek teeth, including strongly molarized premolars. Deciduous premolars are very large and molarized. The p4 of *P. boisei* exceeds that of all other hominins in expansion of the talonid (Suwa, 1988), and in size and occlusal area of its cheek teeth the species is unmatched (Kimbel and White, 1988; Suwa et al., 1994). While M3 may be smaller or equal in size to M2, m3 is usually larger than m2.

Studies of tooth development and emergence indicate that in *Paranthropus*, permanent incisors and first molars formed their crowns at about the same time (similar to modern humans), but that they came into occlusion earlier in the life of an individual than in modern humans (Bromage and Dean, 1985; Smith, 1986; Dean, 1987a, 1988). However, there is some disagreement about the timing and speed of crown formation in *P. boisei*, with competing claims of rapid development (Beynon and Wood, 1987; Dean, 1987b, based on samples from East Turkana and Olduvai) versus rapid enamel differentiation and secretion, but longer overall crown formation time than in modern humans, associated with hyperthick enamel (Ramirez-Rozzi, 1993, based on samples from Omo).

For the most part, postcranial remains are only tentatively attributed to *P. boisei* (Wood and Richmond, 2000). Most of these are limb bones (Howell, 1978; McHenry, 1994b; McHenry et al., 2007), which yield mean body mass and height estimates of 49 kg and 137 cm for males and 34 kg and 124 cm for females (McHenry, 1991a, 1991b, 1992a, 1992b), with body mass ranging from 33.0 kg to 88.6 kg, depending on the regression used (Jungers, 1988). Feldesman and Lundy's (1988) estimates of stature for *P. boisei* are slightly greater than McHenry's (1991a) calculations; nonetheless, it is clear that this species was very similar in stature and body mass to other australopith species, and that sexual dimorphism in body size was likely considerable. A partial skeleton associated with a mandibular fragment, KNM-ER 1500, has been assigned to *P. boisei* (Grausz et al., 1988; but see Wood, 2005). This specimen, presumably of a female, has relatively large forelimbs and small hindlimbs, in comparison with modern humans, similar to the condition in *Au. afarensis* (McHenry, 1994b). Many of the forelimb bones attributed to *P. boisei* males are relatively even larger, which could be due to sexual selection (McHenry, 1994b) or retained adaptation to climbing. Distal tibial morphology is consistent with bipedality (McHenry, 1994b; DeSilva, 2009), but astragalar morphology indicates that the distal tibiofibular articulation with the ankle differed from the arrangement in modern humans (Grausz et al., 1988).

Remarks Given the extraordinary morphology of its masticatory system and its broad sympatry with early *Homo*, the most important questions about *P. boisei* involve diet and parameters of its ecological niche. The relative and absolute massiveness of the mandibular corpora and molar hypermegadonty of *P. boisei* are unmatched among primates, and have been referred to as "super-simian" by Wood and Aiello (1998). Biomechanical analysis of cross-sectional dimensions of the corpus in *P. boisei* indicates adaptation to resist transverse bending ("wishboning") and torsion during the power stroke of mastication (Daegling, 1989), possibly caused in part by the lateral position of the masseter muscles, relative to the occlusal plane, and large horizontal movements of the cheek teeth during chewing (Hylander, 1988). This is consistent with a need to powerfully crush and grind food, probably plant materials that required prolonged and extensive chewing (Hylander, 1988). Although it has been estimated that *P. boisei* could have generated much greater bite force across its cheek teeth than apes and modern humans, this force scales to the size of the occlusal platform in a manner similar to that in other hominids (Demes and Creel, 1988). Along with light construction of its facial skeleton (Ward, 1991), this evidence indicates its masticatory adaptations may not have been particularly suitable to break down hard, gritty food items, but could have processed tough or fibrous vegetation.

It has been argued, from an analysis of many factors of its lifeways and morphology, that *P. boisei* was a dietary eurytope, capable of eating a wide range of foods while maintaining the ability to access tough or hard specialized, seasonal fallback plant parts (Wood and Strait, 2004). It is also possible that *P. boisei* was capable of exploiting a wide range of habitats. The study of Shipman and Harris (1988) on faunal association of *P. boisei* concluded that it occupied primarily closed, wet habitats, while other studies (e.g., Schrenk et al., 1995; Reed, 1997; Suwa et al., 1997; Wood and Strait, 2004) interpret *P. boisei* sites to have included grassland or open woodland near dependable water sources. Dental microwear and isotopic analyses have not yet resolved these questions. Carbon isotope analysis shows that *P. boisei* had a diet rich in C_4 food items (van der Merwe et al., 2008)—grasses, sedges (including tough papyrus), or animals that eat these plants—but, in contrast to findings made on *P. robustus* (discussed later), microscopic examination of *P. boisei* teeth have not yielded results consistent with a constant diet of hard or tough food items (Ungar et al., 2008). That meat might have been incorporated in the diet of *P. boisei* was suggested by the results of dental strontium-calcium analysis (Boaz and Hampel, 1978). Between these findings and interpretations of its morphology and habitat preferences, it seems reasonable to at least tentatively conclude that *P. boisei* was the "higher primate equivalent of a bushpig" (Wood and Richmond, 2000:38).

PARANTHROPUS ROBUSTUS Broom, 1938
Figure 25.13 and Table 25.2

Partial Synonymy Paranthropus crassidens Broom, 1949; *Australopithecus crassidens*, Howell, 1978; *Australopithecus robustus*, Howell, 1978.

Holotype TM 1517, young adult male cranium, Site B, Kromdraai, South Africa (Broom, 1938).

Age and Occurrence Early Pleistocene, southern Africa (table 25.2).

Diagnosis The type specimen was discovered at Kromdraai, South Africa by a schoolboy and subsequently named by Broom in 1938. As the East African species of *Paranthropus* were not yet known, Broom (1938) focused on differentiating *P. robustus* from "*Plesianthropus transvaalensis*" (*Australopithecus africanus*), on the criteria of more diminutive anterior teeth, and the large size and morphology of the cheek teeth. Later, Broom (1949) attributed several upper anterior teeth and a mandible with cheek teeth from Swartkrans, South Africa to a new species of *Paranthropus*, "*P. crassidens*," based on the larger size of its premolars and molars than in the type specimen of *P. robustus*. Comparison of *Paranthropus* crania shows that *P. robustus* can be readily distinguished from *P. aethiopicus* and *P. boisei* by the morphology of its midfacial region and usually by configuration of its zygomatic prominence (see earlier discussion). Compared with *Au. africanus* and *Au. afarensis*, the splanchnocranium of *P. robustus* is deeper vertically (Bilsborough and Wood, 1988), cheek teeth are larger and more disproportionate relative to the size of the anterior dentition, and the midface is more sunken.

Description Based in part on Broom (1939); Broom and Robinson (1952); Howell (1978); Rak (1983); McKee (1989); Susman (1989); Grine and Daegling (1993); Grine and Strait (1994); Keyser (2000); Susman et al. (2001); de Ruiter et al. (2006). Craniodental and mandibular specimens of this species are well represented in the combined sample from Swartkrans, Kromdraai, and Drimolen, South Africa (Grine, 1989; Lockwood et al., 2007). Males appear to comprise a disproportionate percentage of the sample. Similar to *P. boisei*, cranial morphology of *P. robustus* is highly derived and distinctive. The high degree of morphometric variation in the sample has indicated to some the presence of two *Paranthropus* species, *P. crassidens* from Swartkrans and *P. robustus* from Kromdraai (e.g., Broom, 1949; Broom and Robinson, 1952; Howell, 1978; Grine, 1982). Recent discoveries at Drimolen (Keyser, 2000) and interpretive work by Lockwood et al. (2007), however, show that this variation can be accommodated within a single species (see also Kimbel and White, 1988) and accounted for in part by strong sexual dimorphism and bimaturism, in which males continue skeletal growth long after eruption of M3.

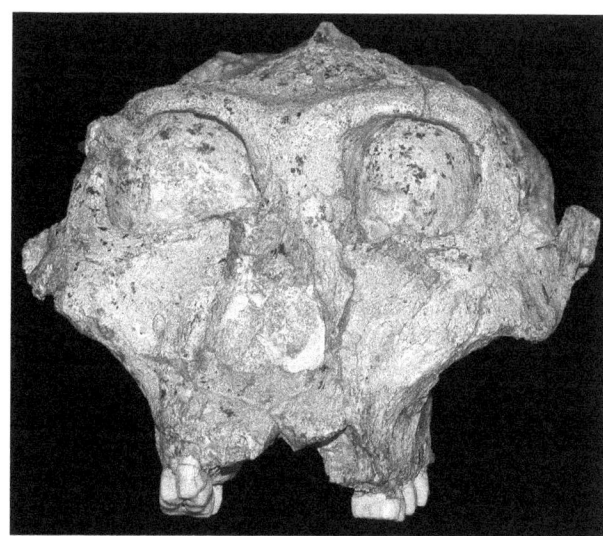

FIGURE 25.13 *Paranthropus robustus* cranium SK 48. Courtesy of the Transvaal Museum (Northern Flagship Institution).

The most complete male cranium of *P. robustus* is SK 48; other notable cranial specimens include the type TM 1517, SK 83, SK 52, SK 46, SK 79, and SKW 18 (de Ruiter et al., 2006; Lockwood et al., 2007). Considered together, these are characterized by a face with a unique maxillary trigon on each side, bordered medially by an anterior pillar, laterally by a zygomaticomaxillary "step," and inferiorly by an obliquely angled zygomaticoalveolar crest. The anterior pillars frame the pyriform aperture and nasoalveolar clivus, which curves smoothly into the nasoalveolar "gutter" at the base of the aperture. The infraorbital foramina open moderately low on the face, and are separated on each side by a subforamen divide from maxillary fossulae.

As noted by Broom (1938, 1939), the *P. robustus* midface is "dished," with its central area depressed relative to the anterior projection of the laterally situated zygomatic prominences. These structures do not flare into visors as in *P. boisei*, but angle sharply posteriorly into the zygomatic arches. The nasals are nearly flat and in the same plane as the maxillae. Glabella is massive, rectangular shaped, and anteriorly prominent, widely separating the orbits. The supraorbital margin on each side is arched into a modest, riblike torus, or costa supraorbitalis (Clarke, 1977); temporal lines running from the midline to these tori enclose a concave frontal trigon that is bounded anteriorly by the glabella. In superior view, postorbital constriction is severe; in lateral view, the cranium does not rise much above the height of the supraorbital tori, and there is no perceptible forehead. In addition, the root of the zygomatic arch and attachment for the masseter muscle are very high and anteriorly placed, relative to the tooth row.

Bizygomatic width in *P. robustus* is great, compared with biorbital breadth, resulting in a large temporal foramen on either side of the cranium. The temporal lines unite posterior to the frontal trigon and are raised in male crania to form a sagittal crest, necessitated by attachment of presumably thick temporalis muscles on a relatively small (and thin-vaulted) braincase. Cranial capacity for SK 1585 is estimated at only 476 cm^3, and features of the endocast such as shape of the frontal and temporal lobes resemble those of apes and other robust australopiths, rather than those of modern humans, early *Homo*, and *Australopithecus* (Falk et al., 2000).

Flexion of the cranial base in *P. robustus* parallels that of modern humans and is greater than in *Australopithecus*, and the foramen magnum is situated anterior to the bitympanic line (Dean and Wood, 1981, 1982; Wood and Richmond, 2000). The occipital condyles are relatively small, but the mastoid processes are large, heavily pneumatized, and are inflated laterally such that they constitute the widest points of the cranial base. The nuchal planum is weakly inclined and bordered by a well-developed external occipital crest; however, there is no compound temporal-nuchal crest. The mandibular fossa is deep and backed by a small postglenoid process that is closely adjacent to a cone-shaped tympanic. Intracranially, bony markings suggest that venous blood drained via a supplementary occipitomarginal pathway (Wood and Richmond, 2000).

The palate is shallow anteriorly and deepens posteriorly, is constructed of bone that is thick in cross section (McCollum, 1997; Strait et al., 2007), posteriorly retracted, and is bordered by teeth that form an elongate, U-shaped arcade, with incisors and canines nearly in a straight line transversely. McCollum (1999) linked development of the thickened palate in *Paranthropus* with growth of a tall mandibular ramus, which she felt was functionally integrated with the expanded occlusal area

of the cheek teeth. Compared with the anterior teeth, the premolars and molars are very large—the premolars are somewhat "molarized"—and have bulbous cusps and thick enamel (Grine and Martin, 1988; Zilberman et al., 1990; Conroy, 1991); comprehensive descriptions and information about *P. robustus* tooth morphology are provided by Broom and Robinson (1952) and Grine (1989). In size, the cheek teeth are modally intermediate between those of *Au. africanus* and *P. boisei*, but overlap the range of each (Kimbel and White, 1988). In fact, an m2 from Gondolin, South Africa, probably an outlier of *P. robustus*, is at the upper end of the size range for *P. boisei* (Menter et al., 1999). The permanent tooth formation sequence was apparently I1, M1, I2, C, P3, M2, P4, and M3, in contrast to the modern human pattern of I1, M1, I2, P3, C, P4, M2, and M3 (Broom and Robinson, 1952; Dean, 1985; but see Grine, 1987), and study of enamel apposition in *P. robustus* indicates crown formation timing similar to that in apes, rather than to the developmental rate of modern humans (Dean et al., 1993).

In contrast, female crania, exemplified by the extraordinarily complete DNH 7 specimen from Drimolen, are considerably smaller and lack the cresting seen in adult male crania. Anterior pillars are weakly developed, there are no incisal eminences, and the midface is not as deeply "dished" as in male crania. Although glabella is prominent and borders a depressed frontal trigon, and the postorbital constriction is strong, there are no supraorbital tori. In addition, the mastoid process is small. The anterior teeth (including blunt-tipped canines) are very reduced relative to the size of the cheek teeth, which include molarized, three-rooted premolars. Molar size progression follows the usual pattern seen in *Paranthropus*, with M3 > M2 > M1, and molars are wider than long. These teeth have thick enamel, expanded distal cusps, and are worn flat. When articulated with its mandible, the skull of DNH 7 exhibits a pronounced underbite. Other, less complete female *P. robustus* cranial specimens include SK 21, SK 821, and SKW 8 (Lockwood et al., 2007).

Mandibles of *P. robustus* are characterized by high, anteroposteriorly extensive rami with prominent coronoid processes, a robustly constructed symphysis with heavy inferior and superior transverse tori (but no simian shelf), and a deep genioglossal pit. The anterior margin of the mandible slopes gently backward inferiorly, and is higher than the posterior region of the corpus on each side. The corpora are well buttressed internally and externally, and they have huge extramolar sulci. As in the upper jaw, the lower cheek teeth are disproportionately larger than the anterior dentition, and molar size progression is m3 > m2 > m1. Although *Au. africanus* m2 and m3 approach those of *P. robustus* in size, m1 is generally much larger and molar talonids are relatively expanded in the latter (Suwa et al., 1994). Markings on the mandible for m. masseter, m. temporalis, and the pterygoid muscles are very pronounced. Evidence for substantial sexual dimorphism can be seen in comparison of the mandible from the female skull DNH 7 with male mandibles DNH 8 and SK 12, which have considerably larger dimensions, more massive teeth, and heavier buttressing of tori (Keyser, 2000).

Hominin postcranial elements are well represented in the *P. robustus* sites of Kromdraai and particularly Swartkrans, but specific attribution at the latter site is problematic because of the co-occurrence there of early *Homo* (Brain, 1976, 1988; Wood and Richmond, 2000). Because of the great numerical disparity between the craniodental remains of hominin taxa at Swartkrans, it has been argued that the overwhelming probability is that most postcrania from the site belong to *P. robustus* (Susman, 1988). Hominin postcranial remains from Swartkrans have been enumerated by Broom and Robinson (1952), Robinson (1970, 1972), Howell (1978), Grine and Susman (1991), McHenry (1994b), Susman et al. (2001), and Susman (1989), and include elements from all regions of the skeleton.

The vertebrae are much smaller than those of modern humans and are similar dimensionally to those of *Au. africanus*. A last lumbar vertebra, SK 3981b, is dorsally wedged, indicative of lumbar lordosis, its pedicles are as robust as those in modern humans, and it has a massive accessory tuberosity on the transverse process for attachment of powerful iliolumbar ligaments, all adaptations to frequent bipedal posture and locomotion (Sanders, 1998). Similarly, configuration of the innominate of *P. robustus* (e.g., SK 3155b; but see Brain et al., 1974) suggests effectiveness in extending the leg and maintaining balance in upright posture. The iliac blade is broad and low, and reflected posteriorly, providing expanded surface area for gluteal muscles and positioning them for better extensor muscle action, as well as lowering the center of gravity and moving it closer in line with the vertebral column and legs (Robinson, 1972). Differences exist between the innominates of this hominin and modern humans. Some of these differences are likely to be primitive retentions in *P. robustus*: the acetabulum is deep but relatively small (e.g., SK 50, SK 3155b), as is the auricular surface for the sacrum; the iliac blades flare more laterally; the well-developed anterior superior iliac spine projects more laterally; and the ischial tuberosity projects farther from the acetabular rim (McHenry, 1975, 1994b). Robinson (1972) felt that this last feature correlated with an ape-like, power-oriented propulsive mechanism and incomplete adaptation to striding bipedalism. The hip joint of *P. robustus* is small, which is an australopith trait, as is the elongation and anteroposterior flattening of the femoral neck (e.g., SK 3121, SKW 19; SK 82, SK 97)(Robinson, 1972; McHenry, 1994b; Susman et al., 2001). In addition, cross-sectional buttressing of *P. robustus* (and *P. boisei*) femoral diaphyses is significantly greater mediolaterally relative to the condition in *Homo* (Ruff et al., 1999). These traits suggest that *P. robustus* may have differed kinematically or mechanically from modern humans in its bipedality. Support for this notion is found in the morphology of the first metatarsal (e.g., SKX 5017, SK 45690, SK 1813), which is not configured for human-like toe-off (Susman and Brain, 1988; Susman, 1989; Susman and de Ruiter, 2004).

Although Susman (1989) generally allocated postcranial fossils from Swartkrans with close similarity to modern humans to *Homo*, manual fossils with a number of derived, modern human-like features from Member 1 were attributed to *P. robustus* (Susman, 1988). These include a pollical distal phalanx (SKX 5016) with a broad apical tuft and muscle marking for a large m. flexor pollicis longus, a first metacarpal (SKX 5020) with a modern human–like lateral marginal crest for a strong opponens pollicis muscle, and a manual proximal phalanx (SKX 5018) with modern human–like shaft curvature. Combined with the proportions of digits II–V, these features indicate modern human–like capabilities for precision grip; based on this interpretation, Susman (1988, 1991a) suggested that the Oldowan stone tools and bone and horn implements found in Member 1 (where *Homo* is very poorly represented) and elsewhere could have been manufactured by *Paranthropus*. This interpretation has not been embraced without some reservations (e.g., Hamrick and Inouye, 1995; Ohman et al., 1995), and has been rejected by others as overly reliant on taphonomic, as opposed to morphological, criteria (e.g., Trinkaus and Long, 1990; but see Susman, 1991b, 1995), and remains controversial. It seems illogical to assume that *Paranthropus*, with its emphasis on craniofacial adaptations to heavy mastication, was the primary maker of stone tools, while *Homo* exhibited a progressive and probably functionally related association of tooth size diminution and improved stone tool manufacture that continued after the demise of its evolutionary cousins.

Remarks The degree of craniodental size differences between females and males suggests gorilla-like levels of sexual dimorphism (Lockwood et al., 2007). Along with the suggestion that older males with higher social rank may have had more exaggerated development of diagnostic features such as anterior pillars, this indicates a social system in which "male reproductive success is concentrated in a period of dominance resulting from intense male-male competition" (Lockwood et al., 2007:1444). Body size in *P. robustus* has been estimated to have ranged from 37.1 to 57.5 kg, or 42.2 to 88.6 kg, depending on the regression employed (Jungers, 1988), and averages between 40.2–49.8 kg for males and 31.9–40.3 kg for females, again depending on the regression used (McHenry, 1992a, 1992b, 1994a). These contrasts are somewhat less than the dimorphism observed in gorillas, so the issue of body size and social structure in *P. robustus* requires further study.

New studies indicate that the derived masticatory apparatus of *P. robustus* may not have been correlated with a narrow dietary specialization. Though most paleoenvironments associated with *P. robustus* are open grasslands, correspondence analysis of faunal assemblages that include this species indicates that it had a woodland habitat preference (de Ruiter et al., 2008). Moreover, carbon isotope analysis of enamel shows that *P. robustus* had a mixed diet primarily of C_3 foods, supplemented by a significant amount of C_4 sources, either grasses, sedges, or animals that consume these plants, and that this diet varied interannually and seasonally (Lee-Thorp et al., 1994, 2000; Sponheimer et al., 2005, 2006; van der Merwe et al., 2008). Thus, it appears that *P. robustus* was a dietary and habitat generalist, perhaps periodically venturing from woodland settings to acquire fallback foods in more open settings, and relying on its powerful occlusal platform to process tough, critical resources.

The dietary adaptations and functional anatomy of the southern African *Paranthropus*, however, are not yet clearly understood. Although bone and horn fragments from Swartkrans have been interpreted as tools for digging up tubers (Brain et al., 1988), subsequent microwear analysis of these tools suggests that they were used to forage for termites (Backwell and d'Errico, 2000). Tooth wear studies do not show microwear features on *P. robustus* teeth consistent with grazing (Grine, 1981; Grine and Kay, 1988), so it is possible that the source of C_4 in these hominins was termites and small vertebrates (Lee-Thorp et al., 2000). The greater incidence of pits, broader wear features, and heterogeneity of scratches observed in dental microwear indicates that *P. robustus* had a different diet than *Au. africanus*, primarily of hard food items (Grine, 1986; Kay and Grine, 1988; Scott et al., 2005). Strontium-calcium ratios in *P. robustus* samples are quite low, however, which is inconsistent with a diet specialized in seeds, roots, and rhizomes. They do fit either with a preference for leaves and shoots of forbs and woody plants, or with omnivory, with substantial intake of animals that graze (Sillen, 1992), including termites, though Sponheimer et al. (2005) have argued that C_4 food other than sedges and termites might have been important in the diet of *P. robustus*. Nonetheless, it is difficult to imagine hard food items that *P. robustus* might have consumed if not seeds and nuts, and processing such food items is consistent with their heavy jaws, postcanine megadonty, and thick enamel (Lee-Thorp et al., 2000; Lucas et al., 2008).

The relationship of *P. robustus* to East African *Paranthropus* and other australopiths requires further investigation, and it remains possible that the unique extracranial cresting, dished midfacial configuration, heavy jaws and immense cheek teeth are convergent adaptive responses to similar environmental changes, rather than a shared derived complex (Wood and Constantino, 2007, and references therein); for example, the facial pillars of *P. robustus* have more in common with *Au. africanus* cranial morphology than with the East African *Paranthropus* lineage (if they were not merged into the facial architecture in this group) and could reflect an independent, endemic southern African derivation. Additionally, it has been noted that molar talonid expansion in *P. robustus* occurred via enlargement of the entoconid, whereas it occurred by enlargement of the hypoconid in *P. boisei* (Suwa et al., 1994), suggesting convergence rather than synapomorphy for molar size increase.

Subtribe HOMININA Gray, 1825
Genus *HOMO* Linnaeus, 1758
Table 25.2

Partial Synonymy Anthropopithecus, Dubois, 1893; *Pithecanthropus*, Dubois, 1894; *Sinanthropus*, Black, 1927; *Meganthropus*, Weidenreich, 1944; *Atlanthropus*, Arambourg, 1954; *Telanthropus*, Broom and Robinson, 1949.

Age and Occurrence Late Pliocene to Recent (first appearance, eastern and southern Africa; increasingly cosmopolitan following the late Pliocene; table 25.2).

Diagnosis When Leakey et al. (1964) erected the species *Homo habilis*, they presented a diagnosis of *Homo* as follows: postcranium adapted to erect posture and bipedal gait; low intermembral index; fully opposable pollex with well-developed precision and power grips; cranial capacity variable but larger on average than those of australopiths and ranging between ~600 and 1,600 cc; temporal lines do not reach to midline; less postorbital constriction than in australopiths; no concavity in facial profile although degree of orthognathism varies; variation in supraorbital torus development and symphyseal contour; dental arcade parabolic and usually lacking diastema; bicuspid p3; smaller and buccolingually narrower molars than in australopiths; small canines relative to most other hominoids. Wood (1992) published an explicitly cladistic list of eight *Homo* synapomorphies: increased cranial vault thickness; reduced postorbital constriction; increased contribution of the occipital bone to cranial sagittal arc length; increased cranial vault height; more anteriorly positioned foramen magnum; reduced lower face prognathism; buccolingually narrow tooth crowns, especially lower premolars; and shorter molar tooth row. More recently, Wood and Collard (1999; Collard and Wood, 2007) have distilled the criteria for allocation to *Homo* to distinctive features that are adaptively relevant and reliably inferable from the paleontological record: trend toward absolutely larger body size; relatively longer lower limbs; larger brain size relative to body size; prolonged ontogeny; fully committed terrestrial bipedalism; and more gracile masticatory apparatus relative to body size. In addition, they suggest that to be referred to *Homo*, species must be shown to be more closely related to the type species, *Homo sapiens,* than to the type species of any other hominin genus.

Referred Species (partial list) Homo habilis, Homo rudolfensis, Homo erectus, Homo ergaster, Homo antecessor, Homo heidelbergensis, Homo neanderthalensis, Homo floresiensis, Homo sapiens.

Remarks See species' sections.

HOMO HABILIS Leakey et al., 1964
Figure 25.14 and Table 25.2

Partial Synonymy Homo ergaster, Groves and Mazak 1975; *Homo rudolfensis*, Alexeev 1986; *Homo microcranous*, Ferguson 1995; *Australopithecus habilis*, Wood and Collard, 1999.

Holotype The type specimen OH 7 includes both parietals, partial mandible and hand bones of a juvenile (but see below as to whether the cranial and postcranial remains can be reliably associated). Paratypes referred to *H. habilis* by Leakey et al. (1964) include OH 4, 6, 8, 13; OH 14 and OH 16 were also referred to the species. For additional Olduvai and Koobi Fora specimens, see Wood (1992), Groves (1989), and Schrenk et al. (2007).

Age and Occurrence Late Pliocene to early Pleistocene, East and southern Africa (table 25.2).

Diagnosis Of the characters listed by Leakey et al. (1964) in the original diagnosis, the following remain widely supported: mean cranial capacity greater than that of *Australopithecus* but smaller than *H. rudolfensis* or *H. erectus*; smaller maxillae and mandibles than those of *Australopithecus*, and within the range of *Homo erectus*; premolars that are buccolingually narrower than those of *Australopithecus*, and tendency toward buccolingual narrowness and mesiodistal elongation of all teeth, especially lower premolars and molars; reduced subnasal prognathism compared to *Australopithecus*; and relatively thin molar enamel (see Dunsworth and Walker, 2002; Kimbel et al., 1997). *Homo habilis* also lacks derived features found in *H. erectus* including: frontal and sagittal keeling; mediolaterally narrow temporomandibular joint; angled tympanic-petrous; less postorbital constriction; thick cranial vault; and opisthocranion positioned high on the occipital profile (Spoor et al., 2007).

Description The time period between about 2.4 and 1.8 Ma shows the earliest evidence of major trends in the *Homo*

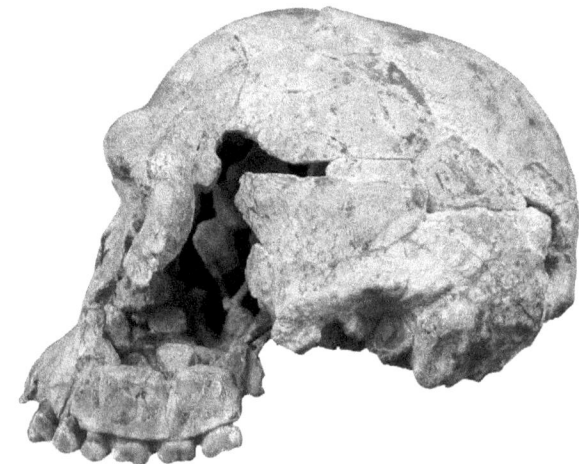

FIGURE 25.14 *Homo habilis* cranium KNM ER 1813. Courtesy of National Museums of Kenya.

lineages: increase in brain size and decrease in tooth size. However, the non-*Paranthropus* hominins during this interval have high morphological variability in absolute and relative brain size and postcanine occlusal area, and in cranial and facial architecture. Postcranial morphology is also highly variable and further confounded by lack of associations with craniodental material. Consequently, this period is best viewed as a transitional time with a poor fossil record. Nonetheless, current evidence best supports the presence of more than one species of non-*Paranthropus* hominin at Koobi Fora, and possibly elsewhere in Africa.

The following descriptions apply to those Olduvai and Koobi Fora specimens allocated by Wood (1992) to *Homo habilis*. Groves' (1989) allocations coincide with those of Wood, with the exception that the Koobi Fora small forms (KNM-ER 1813, 1805) are considered different from both *H. habilis sensu stricto* and *H. rudolfensis*. Some prefer to refer all or most of these specimens to "early *Homo*" (e.g., Suwa et al., 1996; Asfaw et al., 1999), "habilines" (e.g., Wolpoff, 1999), or species of *Australopithecus* (e.g., Wood and Collard, 1999; see later discussion).

Although 600 cc was the endocranial threshold cited for admittance into *Homo* by Leakey and colleagues (1964), this was lowered considerably from earlier such "cerebral Rubicons." It is now apparent that an absolute endocranial threshold is unworkable and biologically irrelevant in the absence of reliable estimates of body size (Wood and Collard, 1999). Furthermore, small Koobi Fora crania have endocranial volumes below 600 cm^3 [KNM-ER 1813 = 510 cm^3; KNM-ER 1805 = 582 cm^3 (Falk, 1987)], although some Olduvai specimens are larger (OH 7 = 674 cm^3; OH 13 = 673 cm^3; OH 16 = 638 cm^3; Tobias, 1971). Attempts to assess relative brain size using postcranial referents (e.g., McHenry, 1994a; but see below regarding postcranial attributions) and cranial proxies such as orbital area (Wood and Collard, 1999) find that *H. habilis* is only modestly encephalized relative to australopiths. However, although a large brain is correlated with slower maturation, the life history pattern of early *Homo* may have been like that seen in the australopiths. Dean et al. (2001) have shown that the timing of tooth development events resembles those of modern and fossil African apes.

Overall, there is reduction in tooth row length, jaw size, and absolute size of the postcanine dentition, but molars fall within the lower range of *Australopithecus* and the upper range of *Homo erectus* (Dunsworth and Walker, 2002). A more rectangular tooth shape (i.e., buccolingually narrow and mesiodistally elongated) is a consistent feature of the taxon, as is thinner (compared with *Australopithecus* and *Paranthropus*) molar enamel.

Supraorbital torus development is variable, but may be described as "incipient" in most specimens. The coronal chord is greater than the sagittal chord in the parietals, upper facial breadth exceeds midface breadth, and the nasal margins are sharp, with an everted nasal sill (Dunsworth and Walker, 2002).

The OH 65 specimen is a maxilla thought by Blumenschine et al. (2003) to have affinities with the KNM-ER 1470 *H. rudolfensis* lectotype, in particular in terms of its broad, flattened naso-alveolar clivus. This is disputed by Spoor et al. (2007), who note similarities between OH 65 and KNM-ER 42703, the youngest known specimen assigned to *H. habilis* (1.44 Ma). For example, both are of similar size and lack the anteriorly placed and forward-sloping zygomatic process found in KNM-ER 1470.

Southern African specimens that may represent *Homo habilis* come from Sterkfontein and Swartkrans, the most complete of which are crania Stw 53 and SK 847 (Grine et al., 1993, 1996; Curnoe and Tobias, 2006), although the former has also been attributed to *Australopithecus africanus* (Kuman and Clarke, 2000) and the latter to *H. erectus* (Kimbel et al., 1997). These two crania have been found to resemble one another more than East African *Homo* specimens KNM-ER 1813, 1470, 3733, OH 24, and KNM-WT 15000, raising the possibility that they represent a geographic variant of *H. habilis* or even a separate species of *Homo* not sampled in East Africa (Grine et al., 1993, 1996). However, the bony labyrinths of the two crania differ. While the semicircular canals of SK 847 resemble those of *Homo sapiens* and *Homo erectus*, suggesting similarity in movement perception to well adapted bipeds, the semicircular canals of Stw 53 were found to have a much less derived configuration (Spoor et al., 1994).

Three *Homo* specimens predating *Homo habilis* from Olduvai and Koobi Fora are securely dated radiometrically. The oldest is a temporal bone from Lake Baringo, Kenya, dated at 2.4 Ma (Hill et al., 1992). It is attributed to *Homo* on the basis of plausible synapomorphies absent in *Australopithecus* and *Paranthropus* temporal bones (although as noted by Asfaw et al. [1999] and Sherwood et al. [2002], comparisons with penecontemporaneous *Au. garhi* are not yet possible), including a medially positioned mandibular fossa, a mandibular fossa containing an anteromedial recess and a flange of tegmen tympani, a reduced temporomandibular tubercle and a sharp petrous crest (Hill et al., 1992; Sherwood et al., 2002). However, given the fragmentary nature of the specimen, it was not assigned to a particular species of *Homo*. A second *Homo* specimen is AL 666-1, a well-preserved maxilla with right P3–M1 crowns and left I2–M2 dated at 2.33 Ma (Kimbel et al., 1996). It has been referred to *Homo* on the basis of 10 characters including relatively broad palate, reduced subnasal prognathism, flat nasoalveolar clivus sharply angled to the floor of nasal cavity, and to a male of *H. habilis* on the basis of dental size and morphology, overall phenetic similarity, large size and lack of derived features found in *H. erectus* (e.g., inclined nasoalveolar clivus) or *H. rudolfensis* (e.g., remodeled subnasal region)(Kimbel et al., 1997). The third specimen is an isolated lower molar of a juvenile from the Nachukui Formation, West Turkana, which was found just above a tuff dated at 2.34 Ma (Prat et al., 2005).

The only postcranial specimens securely associated with early *Homo* craniodental material are OH 62 and KNM-ER 3735, dated to ~1.8 Ma and 1.9, respectively (Häusler and McHenry, 2004). OH 62 has been assigned to *H. habilis* by most workers, but KNM-ER 3735 is referred to *Homo* sp. Remains for OH 62 include maxillary, mandibular, radial, ulnar, humeral, tibial, and femoral fragments, while KNM-ER 3735 is represented by temporal and zygomatic, distal humerus, proximal radius, femoral, tibial and sacral fragments. Although initially thought to possess more primitive limb proportions than A.L. 288-1 (Johanson et al., 1982), it has since been emphasized that femur length cannot be reliably estimated for OH 62 (Asfaw et al., 1999; Dunsworth and Walker, 2002; Reno et al., 2005). Moreover, Häusler and McHenry (2004) have claimed that the OH 62 femur is overall more similar in proportion to an Olduvai specimen (OH 34)(of uncertain age [either Bed II or Bed III] and taxonomic assignment [*Homo* sp.]) than to australopith femora, although the OH 62 femur is small, with a reconstructed body mass of 33 kg (McHenry, 1992a). This finding would be significant as it would undermine reconstructions of *Homo habilis* as being more primitive in limb proportions, and smaller and more dimorphic than later *Homo erectus* samples, but Ruff (2008) has since provided further evidence that the morphology of OH 62 is not consistent with it being an obligate biped. At present, there is a substantial range of morphology represented in hominin postcranial remains recovered from 1.9 to 1.5 Ma. In the Lake Turkana Basin, *H. erectus* (1.9–1.5 Ma), *H. habilis* (1.9–1.44), *H. rudolfensis* (1.9 Ma), and *P. boisei* (2–1.39 Ma) co-occur. Large, derived postcrania such as the innominate KNM-ER 3228, femora KNM-ER 1481A and 1472 and talus KNM-ER 813 could plausibly belong to any of these taxa. Thus the nature of the transition from an *Australopithecus*-like postcranial grade to the tall, long-legged, and large hindlimb jointed morph (exemplified by the aforementioned postcrania and by *H. erectus* partial skeletons KNM-ER 1808 and KNM-WT 15000) cannot be reliably reconstructed.

The OH 7 hand, OH 35 tibia, and fibula and OH 8 foot, all from Bed I, may represent *Homo habilis* or *P. boisei* (e.g., Gebo and Schwartz, 2006; Susman, 2008; Moyà-Solà et al., 2008). The hand bones are of a juvenile and previous researchers have noted the following: robust, but otherwise modern humanlike distal phalanges, robust, slightly curved middle and proximal phalanges and a broad, flattened carpometacarpal (CM) joint on the trapezium (Susman and Stern, 1982). These features are compatible with strong grasping, a powerful, mobile thumb, and powerful fingertips. However, Robinson (1972), Dunsworth and Walker (2002) and Moyà-Solà et al. (2008) all doubt whether the hand elements can be reliably associated with the cranial remains, the main grouping argument being the juvenile status of all specimens. The OH 35 tibia has an articular surface that faces inferiorly and limited ability for either dorsiflexion or plantarflexion (Susman and Stern, 1982; DeSilva, 2009). The OH 8 foot has several derived attributes, including a human-like pattern of metatarsal robusticity (i.e., a robust fifth metatarsal, indicating the lateral to medial weight transfer that occurs in modern humans), a lack of abductory capabilities in the hallux and a stiff lateral column, suggestive of a longitudinal arch (Susman and Stern, 1982; DeSilva, 2009). Crocodile and leopard bite marks are present on both the OH 8 talus, and the OH 35 tibia (Njau and Blumenschine, 2007). Citing a "perfect" fit between the OH 8 talus and the OH 35 tibia, Stern and Susman (1982) argued that these bones are not only both from *H. habilis* but possibly from the same individual. Their recovery in different geological horizons makes this hypothesis unlikely (Hay, 1976). Nevertheless, the association of OH 8 and OH 35 was tested by examining the congruence of the talar and tibial articular surfaces of associated human and ape skeletons using a 3-D laser scanner (Aiello et al., 1998; Wood et al., 1998). The results suggested that the articular surfaces of OH 8 and OH 35 were incongruent, and perhaps not only from different individuals but from different species as well (Aiello et al., 1998; Wood et al., 1998). Susman (2008) has recently reiterated the claim (Susman and Stern, 1982) that the OH 8 foot and the OH 7 type mandible belonged to the same, adolescent individual. However, the arthritic lateral metatarsals, and the obliterated epiphyseal line on the base of the first metatarsal indicate an older age for the OH 8 individual than the OH 7 *H. habilis* individual (DeSilva, 2008). Body weight estimates are 32 kg for the tibia and ~31 kg for the talus (McHenry, 1992a). Body weight estimates from orbital area of *Homo habilis* are comparable, ca. 30–35 kg (Aiello and Wood, 1994; Kappelman, 1996).

Remarks *Homo habilis* has been a controversial taxon since its inception—first, either because inclusion of relatively small-brained specimens into the genus (e.g., OH 7) was thought to be unjustifiable (e.g., Holloway, 1965) or because specimens were thought to be subsumable within *Homo erectus* (e.g., Brace et al., 1973). Later, controversy centered on whether *H. habilis* represented either one, highly variable, or two species (see Wood, 1992), with many researchers finding the degree and pattern of variation in *H. habilis sensu lato* to be unlike intraspecific variation found in extant *Homo, Pan* or *Gorilla* (Wood, 1991). Troubling to many was the co-occurrence of KNM-ER 1813 and KNM-ER 1470 at 1.9 Ma at Koobi Fora (e.g., Wood, 1985; Lieberman et al., 1988): the former has a small endocranial volume, a small face and teeth, and incipient browridges, while the latter has a larger endocranial volume but a larger facial skeletal, and presumably, dentition, and a transversely flat facile profile. However, an endocranial volume range of 510–750 cc does not exceed the level of variation found in dimorphic extant primates (Miller, 1991), and early *Homo* crania from Dmanisi, Republic of Georgia (~1.77 Ma) have endocranial capacities with almost as wide a range (from 600 to 775 cc)(Gabunia et al., 2000, 2002; Vekua et al., 2002).

Gathogo and Brown (2006) have recently suggested a new age for KNM-ER 1813 of 1.65 Ma, and proposed that this may remove some objections about whether the Koobi Fora sample can be accommodated within one pre-*erectus Homo* taxon (i.e., *Homo habilis sensu lato*). However, the stratigraphic revision on which the new age is based is disputed by Feibel et al. (2007) and is not widely accepted.

While it is very possible that more than one early *Homo* taxon is represented in the Turkana Basin, the same cannot be said for Olduvai, where there is general agreement that only *Homo habilis* has been sampled.

The most significant development in the interpretation of early *Homo* in the last decade is the proposal by Wood and Collard (1999) that inclusion of *Homo habilis* and *H. rudolfensis* in *Homo* produces such poor adaptive coherence that they should be removed, and transferred to genus *Australopithecus*. These authors made the case that the hypodigms for these taxa correspond to an ecological niche or adaptive grade that, overall, more closely resembles those of taxa belonging to *Australopithecus* than to *Homo* (see also Collard and Wood, 2007), but demonstration that these taxa are more closely related to *Australopithecus* than to *H. sapiens* is more equivocal (e.g., Strait et al., 1997; Strait and Grine, 2004). An alternative to placing *H. habilis* and *H. rudolfensis* in *Homo* (which would expand the definition of the genus beyond acceptable limits for many) or *Australopithecus* (already paraphyletic) would be to transfer these taxa to a different genus, or genera (Collard and Wood, 2007). Strait and Grine (2004) advocate leaving these taxa in *Homo*, as they believe the genus retains monophyly with their inclusion, and cladistic arguments should take precedence over gradistic arguments. However, gradistic arguments to remove *H. habilis* and *H. rudolfensis* from *Homo* can still be made, provided the evidence for niche separation is compelling and provided the remaining members retained in *Homo* constitute a monophyletic and holophyletic group. If *H. habilis* and *H. rudolfensis* are shown to be sister taxa (e.g., as in Wood, 1992, Box 4) they could be placed in the same genus. If they are not sister taxa but represent two divergence events that predate the divergence of *Homo erectus* (e.g., as in Strait and Grine, 2004: figures 4 and 5), then they could be removed from *Homo*, but two new genus names would have to be implemented. It has been suggested that facial similarities between *Kenyanthropus* and KNM-ER 1470 could reflect a close phylogenetic relationship (Leakey at el., 2001). If future finds bear this out, *Homo rudolfensis* may be transferred to *Kenyanthropus rudolfensis*. Given the lack of data to resolve phylogenetic issues, and the likelihood that the degree of niche separation to be inferred from even expanded hypodigms may remain low, many authors prefer to retain *H. habilis* and *H. rudolfensis* within *Homo*, at least for the time being.

Egeland et al. (2007) provide a recent overview of Olduvai Basin Bed I paleoecology. The landscape was dominated by a saline, alkaline paleolake with fluctuating levels; streams drained from volcanoes to the south and east, and the east contained an alluvial fan and plain (Hay, 1976). Paleosol carbonates from trenches in Upper Bed I, between ~1.845 and 1.785 Ma indicate that C_4 plants were a major component of the vegetation, perhaps 40–60% (Sikes and Ashley, 2007). Wooded grasslands/grassy woodlands dominated the edges of the paleolake, and prior to 1.76 Ma, the Olduvai Basin is reconstructed as supporting mixed habitats. Significant aridification takes place between 1.76 and 1.75 ma (Sikes and Ashley, 2007; Egeland et al., 2007).

The Omo Group Plio-Pleistocene deposits of the Turkana Basin are exposed in East and West Lake Turkana, Kenya and Omo Valley, Ethiopia (Feibel et al., 1989; Bobe and Behrensmeyer, 2004). Sediments were deposited by fluvial, lacustrine, and deltaic activity, and the landscape was variously dominated by a large paleolake (between 4 and 2 Ma) or the Omo River. Faunal records of the Turkana Basin indicate that species with adaptations for a continuum of habitats from closed to open persisted between 4.0 and 1.0 Ma (Behrensmeyer et al., 1997) but periods of high faunal turnover occur in intervals from 3.4–3.2, 2.8–2.6, 2.4–2.2, and 2.0–1.8 Ma (Bobe and Behrensmeyer, 2004).

Bobe and Behrensmeyer (2004) demonstrate that large cyclical shifts in the fauna begin at 2.5 Ma in the Turkana Basin, at about the presumed time of the origin of *Homo*, and attempt to link it with environmental change. They note: "The fundamental importance of grasslands [for hominin evolution] may lie in the complexity and heterogeneity they added to the range of habitats available to the early species of the genus *Homo*" (399). Indeed, one of the most seductive environmental scenarios in paleoanthropology has been the idea that increasing seasonality and aridification associated with the late Pliocene was a potent selective force in hominin evolution, linked not just to the origins of genus *Homo* but to encephalization, stone tool manufacture, and a concomitant increase in manual dexterity, and greater commitment to terrestrial bipedality. For example, Vrba's (1985) documentation of a turnover in bovid taxa between 2.7 and 2.5 Ma was thought to occur in synchrony with environmentally driven extinction (*Au. africanus*) and speciation (*P. robustus* and *H. habilis*) among hominins in southern Africa. Support for a pan-African biotic turnover event has not materialized, however (Behrensmeyer et al., 1997), although Reed (1997) has shown that in East Africa, as in the south, *Homo* co-occurs with taxa adapted to more open, arid environments than do australopiths. Against this backdrop of evidence for increasingly frequent associations between hominins and more open environments over time, stands the recognition that persistent heterogeneity and, particularly, instability (albeit cyclical) in habitat due to factors such as short-term orbitally forced wet/dry oscillations may be a more dominant selective force in hominin evolution (Potts, 1998; Kingston and Harrison, 2007; Kingston, 2007).

The technical skills associated with tool making were long thought to be linked with brain expansion; however, this picture has been complicated by provisional evidence for tool use and manufacture among small-brained australopiths such as *Au. garhi* and *P. robustus*. The oldest Oldowan stone tools from Gona, Ethiopia (2.6 Ma), are not associated with hominin remains and slightly predate the earliest record of *Homo* (table 25.2). Nonetheless, early *Homo* and all subsequent members of the genus are consistently associated with stone tools. Oldowan stone tool manufacture and animal butchery reflect a significant shift in hominin foraging patterns (see Plummer [2004] for a review of the Oldowan sites) and signal the dawn of an ever-increasing dependence on culture as an adaptive strategy. Although an increased dependence on processing animal and possibly plant tissue with stone tools plays a role in the transition from the *Australopithecus* to the *Homo* grade, the nature of this dietary change as inferred from the anatomical, biomechanical, microwear, and isotopic evidence remains ambiguous (Ungar et al., 2006).

HOMO RUDOLFENSIS Alexeev, 1986
Figure 25.15 and Table 25.2

Lectotype This taxon was not formally diagnosed by Alexeev (1986) or later by Groves (1989) and no holotype was assigned, but KNM-ER 1470, an edentulous adult cranium is the lectotype (Wood 1992). Crania KNM-ER 1590 and 3732 and mandibles 1802 and UR 501 are included by some researchers, as are 1.9 Ma large-sized, derived postcrania from Koobi Fora such as KNM-ER 3228 and KNM-ER 1481 (Groves, 1989; Wood, 1992; Dunsworth and Walker, 2002; Schrenk et al., 2007, but see Wood and Richmond, 2000:41).

Partial Synonymy Pithecanthropus rudolfensis, Alexeev 1986; *Homo habilis*, Leakey et al., 1964; *Homo ergaster*, Groves and Mazak 1975; *Australopithecus rudolfensis*, Wood and Collard, 1999.

Age and Occurrence Late Pliocene to early Pleistocene, eastern and southern Africa (table 25.2).

Diagnosis Larger endocranial volume (752 cc for KNM-ER 1470 [Wood, 1991], but see the suggestion in Bromage et al. 2008 that the endocranial volume may be smaller than the estimate given here) than *Australopithecus*, *Paranthropus*, and the mean for *H. habilis*; very weak supraorbital tori; moderate

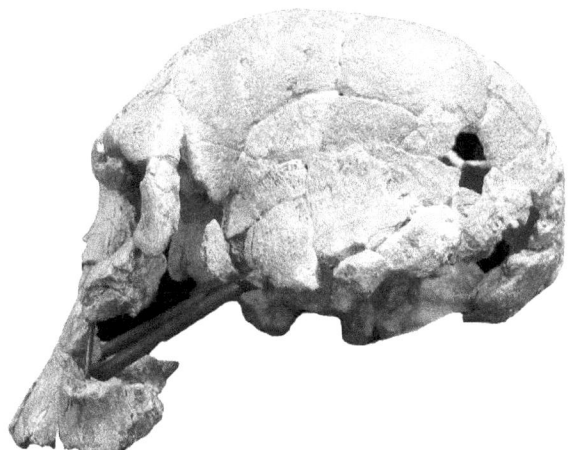

FIGURE 25.15 *Homo rudolfensis* cranium KNM-ER 1470. Courtesy of National Museums of Kenya.

postorbital constriction; midface broad relative to upper face; prognathic overall but with an orthognathic lower face; relatively broad, short palate compared to *H. habilis*; anteriorly placed and forward sloping zygomatic process; superior surface of posterior zygoma flat; less everted nasal margins than *H. habilis*; no nasal sill; rounded mandibular symphysis with no internal buttressing; anterior and posterior dentition inferred to be large; more complex premolar root system. Can be distinguished from *H. erectus* by a larger face and dentition and lack of well-developed supraorbital tori (Wood, 1992; Groves, 1989; Dunsworth and Walker, 2002).

Description KNM-ER 1470 was initially referred to *Homo* sp. indet. by Leakey (1973). Although the endocranial capacity clearly aligned it with *Homo*, the face was noted to have similarities with *Australopithecus* and even *Paranthropus* (e.g., Leakey, 1973; Walker, 1976; Wood, 1991); moreover, the orientation of the face was recognized to be uncertain because of expanding matrix distortion of the frontal base (Leakey, 1973; Bromage, 2008). In addition to the features detailed in the diagnosis here, KNM-ER 1470 exhibits anteriorly positioned glenoid fossae and external auditory meati, and weakly developed muscle markings on the occipital and temporal bones (Leakey, 1973). Orbital area and orbital height have been used to reconstruct a body mass of ca. 49 kg (Kappelman, 1996) and 53 kg (Aiello and Wood, 1994) respectively.

Wood (1991) recognized that early *Homo* mandibles from Koobi Fora sort into two types. Those attributed to *H. rudolfensis* (KNM-ER 1482, 1483, 1801, 1802) are noted for their robust corpi, large postcanine crown areas, broad postcanine teeth, p3 molarization including developed talonids, and roots of p3 and p4 that are plate-like (Bromage et al., 1995b).

Wood (1992) tentatively allocated large-sized (and derived) postcrania from Koobi Fora to *H. rudolfensis* but later noted that no postcranial fossils can be reliably linked to *H. rudolfensis* (Wood and Collard, 1999; Wood and Richmond, 2000). KNM-ER 1470 is not directly associated with any postcrania; however, higher in the stratigraphic section in "area 131" where KNM-ER 1470 was found, three separate femora were recovered (KNM-ER 1472, 1475, 1481), one of which (1481) is associated with a tibia and fibula. All four specimens come from below the KBS tuff and are considered to be 1.89 ± 0.05 Ma (Feibel et al., 1989). KNM-ER 1475 is quite fragmentary, but 1472 and 1481 are well preserved. KNM-ER 1481 has some features that resemble australopith femora; for example, its neck is relatively long, and its shaft is anteroposteriorly flattened. However, it is much longer and has an absolutely large femoral head diameter (body weight based on head size is estimated to be 57 kg; McHenry, 1992b), and a similar distribution of femoral subchondral bone as modern humans (MacLatchy, 1996). KNM-ER 1481 may also be from *H. erectus* (Kennedy, 1983), though others find it more likely that this femur is from early *Homo* (i.e., *H. habilis sensu lato*)(Trinkaus, 1984a).

KNM-ER 3228 is 1.95 ± 0.05 Ma (Feibel et al., 1989) and as such is the oldest well-dated postcranial fragment whose size (i.e., body mass based on acetabulum size is estimated to be 62 kg; McHenry, 1992b), and morphology resemble those of *Homo sapiens*. Based on its size, it has been argued that KNM-ER-3228 may represent early *H. erectus* (Antón, 2003), but the only purported *H. erectus* specimen of this antiquity is the occipital fragment KNM-ER 2598 (discussed later) that is contemporaneous with KNM-ER 1470. The enlarged acetabulum on KNM-ER 3228 suggests high joint reaction forces at the hip perhaps as a result of large body mass. There is also a prominent iliac pillar reinforcing the bone and resisting the bending forces that would be imposed on the laterally flaring ilium during bipedal locomotion, and in this way it is similar to the *H. erectus* OH 28 pelvis (Rose, 1984; Aiello and Dean, 1990).

Remarks Other than the Koobi Fora material, the only relatively complete specimens that have been referred to *Homo rudolfensis* by at least some authors are two mandibles: Omo 75-14 from Ethiopia and UR 501 from Malawi. In addition to the mandible, the Omo collection includes upward of 20 isolated teeth of *Homo* affinity (Suwa et al., 1996). Suwa and colleagues have noted that the sizes of the teeth tend to fall above the mean for *H. habilis sensu stricto* and correspond in some respects (e.g., p3 molarization) to the *H. rudolfensis* morphological pattern laid out by Wood (1991, 1992). However, these authors posit that the more robust dentition of *H. rudolfensis* may represent the primitive condition for *Homo*, with rapid gracilization occurring within this lineage (during Upper Burgi Member time), to yield *H. habilis sensu stricto*. Under this model of anagenetic change, the hypodigms of the two early *Homo* taxa would be subsumed into *H. habilis*. However, it has also been suggested that the *H. habilis* morphotype may represent the primitive condition for *Homo* (Kimbel et al., 1997).

The Malawi mandible is of biogeographic significance in that it is associated with a mostly East African, rather than South African, endemic fauna (Bromage et al., 1995b). Schrenk et al. (2002) report a faunal age of 2.5–2.3 Ma for UR 501, largely on the basis that the form of the suid *Notochoerus scotti* from Uraha is reportedly more advanced than those from Member C of the Shungura Formation, which is dated at ca. 2.8 Ma (Feibel et al., 1989) but less advanced than those from Member G, below the KBS tuff at ca. 2.0 Ma (Feibel et al., 1989). Given the ~2,500 km separating Omo and Uraha, a less precise faunal age for the mandible is probably warranted, in the range of 2.7–2.0 Ma. Hill (1995) has also suggested caution in attaching too narrow a faunally-based date to this specimen.

HOMO ERECTUS (Dubois, 1893), Weidenreich, 1940
Figure 25.16 and Table 25.2

Partial Synonymy Anthropopithecus erectus, Dubois, 1892; *Pithecanthropus erectus*, Dubois, 1894; *Sinanthropus pekinensis*, Black, 1927; *Homo (Javanthropus) soloensis*, Oppenoorth, 1932; *Homo primigenius asiaticus*, Weidenreich, 1933; *Homo neanderthalensis soloensis*, von Koenigswald, 1934; *Homo soloensis*, Dubois, 1936; *Homo erectus javensis*, Weidenreich, 1940; *Homo erectus pekinensis*, Weidenreich, 1940; *Pithecanthropus robustus*, Weidenreich, 1945; *Meganthropus palaeojavanicus*, Weidenreich, 1945; *Pithecanthropus pekinensis*, Boule and Vallois, 1946; *Telanthropus capensis*, Broom and Robinson, 1949; *Pithecanthropus modjokertensis*, von Koenigswald, 1950; *Paranthropus palaeojavanicus*, Robinson, 1954; *Atlanthropus mauritanicus*, Arambourg, 1954; *Australopithecus capensis*, Oakley, 1954; *Pithecanthropus capensis*, Simonetta, 1957; *Pithecanthropus palaeojavanicus*, Piveteau, 1957; *Pithecanthropus sinensis*, Piveteau, 1957; *Homo leakeyi*, Heberer, 1963; *Homo sapiens soloensis*, Campbell, 1964; *Sinanthropus lantianensis*, Woo, 1964; *Tchadanthropus uxoris*, Coppens, 1966; *Homo ergaster*, Groves and Mazák, 1975; *Homo modjokertensis*, von Koenigswald, 1975; *Pithecanthropus soloensis*, Jacob, 1978; *Homo erectus trinilensis*, Sartono, 1982; *Homo palaeojavanicus sangiranensis*, Sartono, 1982; *Homo palaeojavanicus mojokertensis*, Sartono, 1982; *Homo palaeojavanicus robustus*, Sartono, 1982; *Homo erectus ngandongensis*, Sartono, 1982; *Homo georgicus*, Gabounia et al., 2002.

Holotype Trinil 2. Calotte discovered along the Solo River, Java in 1891.

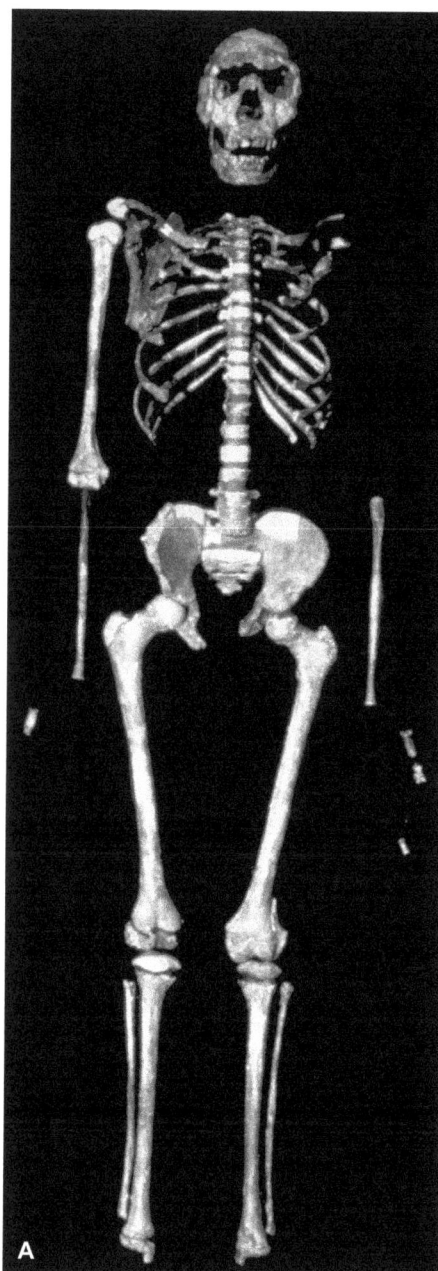

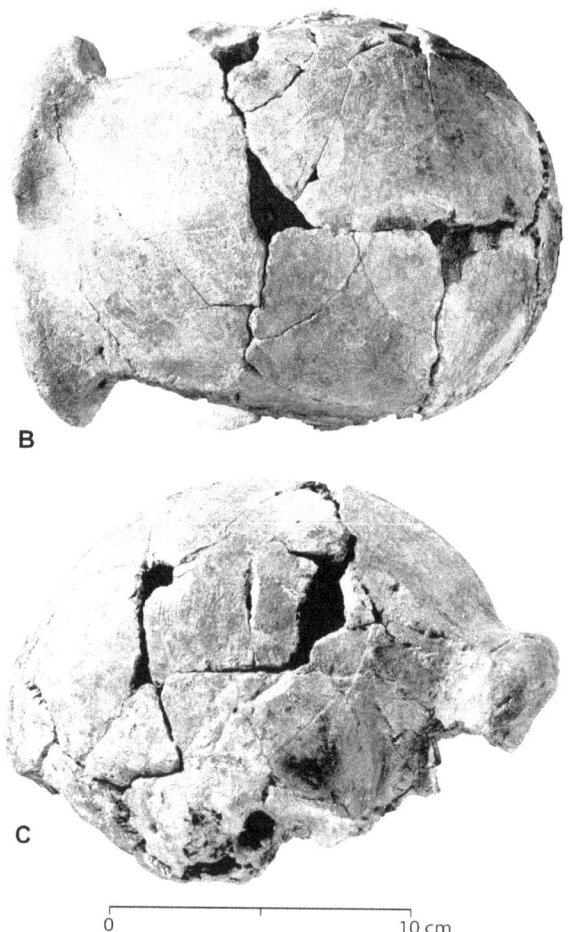

FIGURE 25.16   A) *Homo erectus* partial skeleton KNM-WT
15000. Courtesy of National Museums of Kenya.
B, C) *Homo erectus* calvaria BOU VP-2/66 in B) dorsal,
and C) right lateral, view. Courtesy of Tim White.

Age and Occurrence Early to middle Pleistocene Africa,
Asia, Europe. Perhaps into later Pleistocene in sites in China
and Indonesia (table 25.2).

Diagnosis The Trinil calotte was the first fossil to demonstrate
the existence of a small-brained hominin in the human fossil
record. Compared to *H. sapiens*, the type specimen has a smaller
cranial capacity (~850 cc); a low, sloping frontal bone with a
thick, continuous supraorbital torus; moderate postorbital constriction; a midline keel; a strongly angled occipital with a thick
transverse occipital torus. Fossils from Swartkrans, South Africa
(Broom and Robinson, 1949), Olduvai Gorge, Tanzania (Heberer,
1963), Lake Turkana, Kenya (Leakey and Walker, 1976; Walker
and Leakey, 1993), and Middle Awash, Ethiopia (Asfaw et al.,
2002; Gilbert and Asfaw, 2008), have greatly expanded our

knowledge of *H. erectus*. Compared to *H. sapiens*, *H. erectus* has a
wider face; moderate subnasal prognathism; does not possess a
mental eminence though the mandibular corpus is more robust,
the ramus is mediolaterally wide, and the bicondylar breadth
large; and a relatively larger third molar. Postcranially, *H. erectus*
had six lumbar vertebrae; a longer femoral neck with associated
broader pelvis; and thicker cortical bone in long bone midshafts.
Compared to *Australopithecus* and earlier *Homo*, *H. erectus* has a
larger average cranial capacity; vertically oriented parietals;
thicker supraorbitals; overall thicker cranial bones especially in
inner and outer tables; a strongly angled occipital region;
reduced temporal fossa; a narrow but deep temporomandibular
fossa; smaller postcanine teeth (especially M3) relative to body
size; mesiodistally reduced upper M3; more platymeric femora;
thicker cortical bone; more modern human–like intermembral
index; and an overall larger body height and weight.

Description *Homo erectus* crania tend to be quite broad relative to
their height, with parallel-sided parietals when viewed posteriorly, a
robustly built occipital region often with an occipital torus, and a
supraorbital torus that varies in projection and thickness, perhaps
as a function of sexual dimorphism. Above the supraorbital torus is
often a shelflike supratoral sulcus. The crania are typically thick and
possess keeling along the midline and often a postbregmatic eminence. There is usually a degree of subnasal prognathism. The
robusticity of the occipital and supraorbital region of the cranium
may scale allometrically (Spoor et al., 2007). The thickened cranial
vaults, expanded nuchal plane, and prominent supraorbitals may
be a suite of characters functionally related to the increased anterior
loading of the skull during mastication (Wolpoff, 1999). *Homo*

erectus is more modern human–like in craniodental morphology and postcranial anatomy than earlier *Homo* or australopiths.

The average cranial capacity in eight African fossils assigned to *H. erectus* is 870 cm^3 ± 129 cm^3 (range 691 cm^3–1067 cm^3; Holloway et al., 2004). These fossils include the relatively complete crania KNM-ER 3733, KNM-ER 3883, KNM-WT 15000, KNM-ER 42700, OH 9, UA 31 (Buia), BOU-VP-2/66 (Daka), and OH 12. Perhaps the earliest evidence for cranial expansion is the 1.9 Ma KNM-ER 2598 occipital fragment, which has a wide posterior cranial fossa, an angled occipital with a transverse occipital torus, but some claim this fossil may not be 1.9 Ma, and instead may have weathered from more recent deposits (White, 1995). *Homo erectus* has midfacial anatomy different from earlier hominins, including the presence of larger orbits and larger nasal regions. The large nasal regions may have been selected to increase the volume of air, and to retain water during expiration (Franciscus and Trinkaus, 1988). Other important craniodental fossils of *H. erectus* include KNM-ER 730, an associated mandible, frontal, and occipital of an older adult female. Consistent with other presumed female *H. erectus* fossils, the supraorbital is not markedly tall, and there is a weak nuchal torus. Craniodental remains of *H. erectus* may also be present in the later Sterkfontein cave deposits (SE 1508 and SE 1937) and at Swartkrans cave (SK 15 mandible and SK 847 partial cranium preserving part of the face)(Clarke, 1994a; Curnoe and Tobias, 2006).

The craniodental remains of *H. erectus* show a substantial range of variation, perhaps related to a persistence of *Australopithecus*-like levels of sexual dimorphism. A roughly 950,000-year-old frontal and temporal fossil (KNM-OL 45500) from the archaeologically rich site of Olorgesailie is quite gracile (Potts et al., 2004). Though the glabella region is prominent, the frontal breadth is reduced and the supraorbital thinner than all known *H. erectus* specimens except perhaps OH 12 (Potts et al., 2004). Recently described fossils from Ileret, Kenya, are consistent with the hypothesis that *H. erectus* displayed marked sexual dimorphism (Spoor et al., 2007). The 1.55 Ma calvaria KNM-ER 42700 has the smallest cranial capacity (691 cm^3) of any definitive *H. erectus* from Africa, and like in KNM-OL 45500, the supraorbitals are thin. The cranial vault is thinner than most *H. erectus* fossils, and the occipital not as strongly angled and lacks a strong occipital torus. A recent morphometric study of KNM-ER 42700 found it to be quite distinct from known *H. erectus* crania and perhaps not attributable to that species (Baab, 2008a; but see reply in Spoor et al., 2008). Despite this substantial range of variation in cranial morphology, Suwa et al. (2007) have found morphological continuity in the dental remains of *H. erectus* from 1.65 to 1.0 Ma, with a slight tendency toward dental gracility after 1.4 Ma. This may be correlated with the appearance of Acheulean tool technology, first preserved in 1.4 Ma deposits at the *H. erectus* site of Konso-Gardula (Asfaw et al., 1992).

The postcranial anatomy of *H. erectus* is known primarily from the remarkably complete skeleton of the young male from Nariokotome, KNM-WT 15000 (Brown et al., 1985; Walker and Leakey, 1993). Most of the skeleton is preserved; the specimen lacks some of the cervical and thoracic vertebrae, the left humerus, both radii, and hand and foot bones. Readers are referred to the Nariokotome volume (Walker and Leakey, 1993) for a detailed treatment of this specimen. The morphology of the ribs suggests that *H. erectus* was as barrel chested as modern humans (Jellema et al., 1993). The femora and tibiae are elongated, the proximal femur has a long femoral neck and the ilia flare laterally, though the ilia are poorly preserved in this specimen. Latimer and Ward (1993) believe that the vertebrae and sacrum have relatively small, australopith-like centra (but see Sanders, 1998), a modern human–like lumbar lordosis, and by inference strong erector spinae. The shoulder of the Nariokotome Boy possesses a combination of derived morphologies, including a modern human–like scapula, with a less cranially oriented glenoid than those found in apes and australopiths; primitive morphologies including a short clavicle and humerus with reduced torsion (Larson et al., 2007).

Brown et al. (1985) estimate that KNM-WT 15000 was roughly 12 years old at death, but others use perikymata to suggest that he was only 8 years old (Dean et al., 2001). Nevertheless, at the age of 8–12, he already had a thicker supraorbital torus and more robust facial morphology than the adult KNM-ER 3733. He was also already 1.66 m and roughly 48 kg (Ruff and Walker, 1993). Dean et al. (2001) have suggested that if this young *H. erectus* male had already attained this size within only 8 years, then *H. erectus* may have had an accelerated life history relative to modern humans, including an earlier weaning age, a rapid period of growth, and an earlier age of first reproduction.

Other possible postcranial remains from *H. erectus* have been described from Koobi Fora, Olduvai Gorge, and from sites in southern Africa. These include KNM-ER 1808, a pathological skeleton of a tall *H. erectus*. Walker et al. (1982) found that the pattern of bone formation on the KNM-ER 1808 skeleton was similar to skeletal material from individuals who had consumed large quantities of raw liver, and in consequence had suffered from an overdose of vitamin A. The authors concluded that the morphology of the KNM-ER 1808 skeleton was evidence not only for the consumption of meat in *H. erectus*, but for conspecific care, as well (Walker and Shipman, 1996). Skinner (1991) argued that hypervitaminosis A could also result from eating too much honey, and Rothschild et al. (1995) most recently argued that the 1808 skeleton is more consistent with this individual suffering from yaws, not from hypervitaminosis A. Nevertheless, both KNM-ER 1808 and KNM-WT 15000 display modern human body proportions. This differs from earlier australopiths, which tend to have a relatively higher intermembral index (Aiello and Dean, 1990). Another specimen, KNM-ER 803 preserves parts of both upper and lower limb morphology, though this skeleton is quite fragmentary.

Other postcranial remains suggested to be from *H. erectus* include femora KNM-ER 736, KNM-ER 737, BOU-VP-1/15, BOU-VP-2/15, BOU-VP-19/63; tibiae KNM-ER 741, KNM-ER 19700, BOU-VP-1/109, StW 567; tali KNM-ER 5428 and BOU-VP-2/95, and the OH 28 femur and pelvis (Walker, 1994; Antón, 2003; Gilbert and Asfaw, 2008). These fossils collectively suggest that *H. erectus* was a large, muscular hominin. Body size estimates from the tibia and femora range from 45–68 kg (Antón, 2003). Using modern human-based regression equations, the large talus KNM-ER 5428 would be from an 86.7-kg individual (McHenry, 1992b). *Homo erectus* femora are characterized by thick midshaft cortical bone, subtrochanteric platymerism, and a distal position of the minimum shaft breadth (Kennedy, 1983; Gilbert and Asfaw, 2008). The lower limb postcranial anatomy of *H. erectus* is consistent with a bipedal locomotor gait similar, if not indistinguishable, from that of modern humans. This assertion has recently been supported by an analysis of 1.52-Ma footprints presumably left by *H. erectus* at Ileret, Kenya (Bennett et al., 2009).

An *H. erectus* female pelvis and lumbar vertebra (BSN49/P27a-d) have recently been described from 0.9- to 1.4-Ma deposits in Gona, Ethiopia (Simpson et al., 2008). The pelvis is from a small, presumably female, individual (1.2–1.46 m in height) and retains the laterally flaring ilia characteristic of the australopith pelvis. However, the dimensions of the birth canal suggest that *H. erectus* females were capable of delivering infants with large (300- to 315-cc) brains, suggesting that *H. erectus* had evolved a modern human–like prenatal brain growth pattern.

Remarks Readers are advised to consult Antón (2003) for a more detailed treatment of the biology and evolution of *Homo erectus*. As already discussed, the allocation of unassociated postcrania is problematic; however the partial skeletons KNM-ER 15000 and 1808 are evidence that *H. erectus* is a consistently larger hominin than the australopiths or perhaps early *Homo*.

Bramble and Lieberman (2004) have suggested that evolution of the body proportions and anatomies first seen at 1.9 Ma and present in KNM-WT 15000 and -ER 1808 are adaptations for long-distance running and may have been selected for to increase hunting or scavenging success. Relatively long legs provide an elongated stride and energy conserving tendons increase the efficiency of long-distance travel. Modern humans have important physiological differences when compared to chimpanzees that are related to heat dispersal, such as sweat glands and reduced body hair. Adoption of diurnal hunting, scavenging, and long-distance travel would impose such a selection pressure against body hair. However, with the removal of body hair, selection would act fiercely on modern human skin color. Jablonski and Chaplin (2000) have elegantly shown that under an equatorial African sun, light skin color would result in folic acid destruction, whereas dark skin pigmentation would protect folic acid, while still allowing penetration of enough ultraviolet radiation to maintain sufficient vitamin D production. Rogers et al. (2004) sequenced the MC1R gene in modern humans and other primates and found that this gene, which helps regulate pigmentation, coalesced in hominins at roughly 1.5 Ma. These data suggest a selective sweep in a gene partially responsible for skin color variation in hominins near the time that *H. erectus* appeared. *Homo erectus* may therefore have been the first hominin with reduced body hair, perhaps as a result of long-distance diurnal travel related to hunting and scavenging.

Homo erectus specimens also indicate a shift in dietary strategies compared to earlier hominins. Australopith postcanine teeth are both relatively and absolutely larger than either early *Homo* or early *H. erectus* teeth. However, there is an increase in incisor size in *H. erectus*, suggesting a greater emphasis on anterior tooth loading. The evidence for an increase in meat consumption around 2 million years ago is supported by genetic studies on the tapeworm, which presumably evolved a relationship with hominins after being consumed as part of an animal carcass (Hoberg et al., 2001). Finally, a species cannot become reliant on meat if that food source is not present in the environment. The evolution of *H. erectus* and evidence that this species began to consume more meat tissue than its predecessors is supported by paleoecological evidence for faunal evolution in Africa between 2.5 and 1.8 million years ago during a time of variable climates with the trend toward drier and a greater variety of habitats (Behrensmeyer et al., 1997). These conditions would support the evolution of many of the prey animals found in *H. erectus* assemblages.

The pattern of stone tool sites on the African landscape changes during the early evolution of *H. erectus* (Cachel and Harris, 1998). These authors have noted that at that time archaeological sites begin to increase in volume, and the distances that hominins traveled to obtain the raw material for their stone tools increased. The patterns of stone tools thus indicate an increase in the home range occupied by *H. erectus*. This has also been reported in the later *H. erectus* locality of Olorgesailie, which reflects a shift toward greater use of the landscape and a more deliberate selection of stone tool raw materials (Potts, et al.

1999). These archaeological data are consistent with work by Antón et al. (2002) who have shown that an increase in body size and a change in diet correlate with an increase in home ranges across primates. Critically, the increase in home range is not just within Africa, but *H. erectus* is presumably the first hominin species to migrate out of Africa. This occurred shortly after the first fossil evidence for *H. erectus* (~1.95 Ma), as fossils likely assignable to this species have been found in 1.77-Ma sites in Dmanisi, Georgia (Gabunia and Vekua, 1995; Gabunia et al., 2000; Vekua et al., 2002; Lordkipanidze et al., 2006, 2007), and in 1.8-Ma sites in Indonesia (Swisher et al., 1994).

Related to an increase in big-game hunting, long-distance travel, and stone-tool sophistication is the tantalizing, but difficult-to-test, question of whether *H. erectus* possessed language. Based on the reduced size of the thoracic vertebral canals in the Nariokotome skeleton, MacLarnon (1993) suggested that *H. erectus* might have lacked the precise motor control of the intercostal and abdominal muscles necessary for modern human–like speech. However, the vertebrae of the Nariokotome skeleton may be pathological (Latimer and Ohman, 2001), and vertebrae from other *H. erectus* skeletons suggest that the size of the vertebral canals are within the modern human range and do not preclude *H. erectus* from possessing language (Meyer, 2006).

Based on morphological differences between the African and Asian *Homo* fossils from the early Pleistocene, some have suggested that *H. erectus* be reserved for fossils from Asia, and the majority of African fossils from this time period be allocated to *H. ergaster* (Groves and Mazák, 1975). This hypothesis of taxonomic diversity in the *H. erectus* sample has other supporters (e.g., Wood and Richmond, 2000; Schwartz and Tattersall, 2003). However, fossils of *H. erectus* crania from Eritrea dated to 1.0 Ma (Abbate et al., 1998) and remains from 1.4- to 1.0-Ma sites in Ethiopia (Asfaw et al., 2002; Suwa et al., 2007) overlap in variation with Asian and African *H. erectus* specimens, suggesting that *H. erectus* is a single, morphologically diverse taxon, as suggested earlier (Rightmire, 1993). Results of a 3-D geometric morphometric study of variation in *H. erectus* found that a single-species hypothesis best fit the data (Baab, 2008b).

Recently, a 1.44 Ma maxilla KNM-ER 42703 was assigned to *H. habilis* (Spoor et al., 2007). If the taxonomic assignment is correct, then *H. habilis* and *H. erectus* were contemporaries, challenging the view that *H. habilis* evolved into *H. erectus* via anagenesis.

HOMO HEIDELBERGENSIS Schoetensack, 1908
Figure 25.17 and Table 25.2

Partial Synonymy Palaeanthropus heidelbergensis, Bonarelli, 1909; *Homo rhodesiensis*, Woodward, 1921; *Cyphanthropus rhodesiensis*, Pycraft, 1928; *Homo (Africanthropus) helmei*, Dreyer, 1935; *Homo florisbadensis (helmei)*, Drennan, 1935; *Paleoanthropus njarensis*, Kohl-Larsen and Reck, 1936; *Homo steinheimensis*, Berckhemer, 1936; *Africanthropus njarasensis*, Weinert, 1939; *Homo marstoni*, Paterson, 1940; *Homo swanscombensis*, Kennard, 1942; *Homo saldanensis*, Drennan, 1935; *Homo sapiens rhodesiensis*, Campbell, 1964; *Homo sapiens steinheimensis*, Campbell, 1964; *Homo sapiens steinheimensis*, Campbell, 1964; *Homo erectus petraloniensis*, Murrill, 1983; *Homo antecessor*, Bermúdez de Castro et al., 1997; *Homo cepranensis*, Mallegni et al., 2003

Holotype Mauer Mandible, complete adult mandible from Rösch sandpit in the village of Mauer, near Heidelberg, Germany (Schoetensack, 1908).

Age and Occurrence Middle Pleistocene, Europe, Asia, Africa (table 25.2).

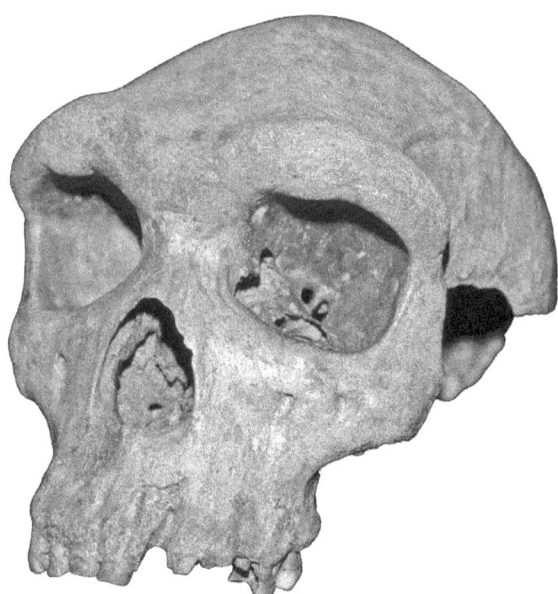

FIGURE 25.17 *Homo heidelbergensis* cranium from Kabwe, or "Broken Hill." Courtesy of Philip Rightmire.

Diagnosis A nearly complete mandible from Germany described by Schoetensack (1908) and reevaluated by Howell (1960) and more recently by Mounier et al. (2009) differentiated *H. heidelbergensis* from *H. erectus* and from *H. sapiens*. Compared to *H. erectus* mandibles, the Mauer mandible has a broader ramus; a taller anterior corpus; a posteriorly positioned mental foramen; a truncated gonial angle; an enlarged buccal cusp on the third premolar; and taurodontism of the molar pulp cavities. Unlike *H. sapiens*, the Mauer mandible has a thick symphysis with no projecting mental eminence; an extended planum alveolare; and the second molar is larger than the first. Similarities between the Mauer type mandible and mandibles and a cranium from the Arago site in France has led Rightmire to amend the diagnosis to include features shared in common by the Arago specimen, and fossils from Africa including Kabwe, Bodo, and Ndutu (Rightmire, 1998, 2008). In comparison with *H. erectus*, *H. heidelbergensis* possesses a larger cranial capacity achieved through an expanded parietal region and reduced postorbital constriction; a longer, more vertical occiput and shorter, more horizontally oriental nuchal plane; increased flexion of the anterior cranial base; larger frontal sinuses; a thinner tympanic plate; discontinuous supraorbital tori with a shallower supratoral sulcus; a shallower mandibular fossa; an anteriorly positioned incisive canal; and a more vertically oriented nasal margin. In comparison with *H. sapiens*, *H. heidelbergensis* possesses less parietal expansion; superior-inferiorly thicker and more projecting supraorbital tori; thicker cranial bones; midline keeling of a less vertically oriented frontal bone; an angular torus on the parietals; and a large, broad face.

Description Based on Rightmire (2008). *Homo heidelbergensis* possesses an interesting mixture of primitive features found in *H. erectus* and more derived features found in later *H. sapiens* specimens. These include a large, broad face, with a brain size that is within the range of modern humans. The brain is encased in a differently shaped and more robust cranium. The frontal bone is low, and there is a distinctive sagittal keel. The cranial vault is thick, particularly in the occipital region. There is also some subnasal prognathism. The supraorbital tori are projecting, superoinferiorly tall and discontinuous. The tori achieve their maximum thickness in the mid-orbit region, and appear everted or twisted laterally (Wolpoff, 1999). However, relative to earlier *H. erectus* crania, *H. heidelbergensis* has expanded parietals, a broader frontal, and a more rounded occiput, all features consistent with a larger brain volume. Postcranially, *H. heidelbergensis* shares with *H. sapiens* the same limb proportions; however, the long bones are more robustly built. The three most complete crania from Bodo, Kabwe, and Ndutu will be discussed here.

The earliest African specimen that may belong to *H. heidelbergensis* is the 600 ka (Clarke et al., 1994) Bodo cranium recovered in the Middle Awash, Ethiopia in 1976 (Conroy et al., 1978). The Bodo cranium consists of most of the face, and 41 cranial fragments pieced together to form most of the frontal bone, parietals, some of the anterosuperior aspect of the temporals, and some of the right aspect of the superior occipital. There is also a missing portion of the left maxilla and zygomatic. None of the teeth are preserved well enough to discern any occlusal detail. Bodo possesses a very large face, with massive zygomatics, a broad nasal opening, and a robustly built arched supraorbital torus. The supraorbital height is approximately 17.5 mm, slightly less than the 21 mm thick supraorbitals on the Kabwe skull. The breadth of the face (15.8 cm) is matched only by the large Indonesian *H. erectus* skull Sangiran 17. Bodo possesses a gently sloping frontal bone, with limited postorbital constriction. There is also a keel running along the sagittal aspect of the cranium. The cranial bones of the Bodo specimen are extremely thick, approaching 13 mm at bregma—greater than in any known *H. erectus* specimen (Conroy et al., 1978). Using a CT reconstruction of the skull, Conroy et al. (2000b) estimated a cranial capacity of 1,250 cc. Distinct cutmarks on the frontal and maxilla may be evidence of the deliberate defleshing and potential cannibalism of Bodo (White, 1986b).

In many ways, the Bodo cranium is similar to another large Middle Pleistocene cranium from Kabwe (or "Broken Hill"). Both skulls are considered to be from males (Rightmire, 1998; Wolpoff, 1999). The Kabwe cranium was the first major discovery of a fossil human on the continent of Africa and thus holds important historical significance. It was discovered in 1921 in the Broken Hill Mine in Zambia. Kabwe is nearly complete, missing only a region consisting of the right temporal and the right side of the basicranium (Woodward, 1921). Like Bodo, the Kabwe cranium is robustly built, with a broad face, and a thick, arched supraorbital torus. Also like Bodo, Kabwe has a gently receding forehead, and a midline keel. However, Kabwe possesses more gracile zygomatics than those found on the Bodo cranium. All of the maxillary teeth are preserved and they show considerable wear, with many of the teeth littered with cavities. The cranial capacity of Kabwe is estimated to be ca. 1,300 cm^3 (Holloway et al., 2004). Postcranial remains recovered from the Broken Hill Cave include several femora, a complete tibia (E 691), and an innominate (E 719) that all show modern human proportions, though are more robustly built than modern human lower limbs (Pearson, 2000). This postcranial robusticity is evident as well in a large Middle Pleistocene femur from Berg Aukas, Namibia (Grine et al., 1995), and KNM-ER 999, a large femur from Koobi Fora, Kenya (Day and Leakey, 1974; Trinkaus, 1993b).

Similar in morphology to Bodo and Kabwe, though slightly more gracile, the fragmentary Ndutu cranium is probably from a female (Rightmire, 1998; Wolpoff, 1999). Ndutu has a projecting supraorbital torus, though it is thinner than that found on either Bodo or Kabwe. The Ndutu cranium also has expanded parietals and a long vertical occiput (Clarke, 1976, 1990). The cranial capacity is estimated by Holloway et al. (2004) to be 1,100 cm^3.

Temporally younger specimens, such as those from Guomde, Florisbad, and Lake Eyasi, are more similar morphologically to modern *Homo sapiens* than early members of *H. heidelbergensis,* and thus it is difficult to confidently assign these fossils to a taxon, and the absence of accurate and precise information about the age of these fossils (Millard, 2008) currently limits our ability to accurately assess the tempo of evolutionary change from a *H. heidelbergensis*–like ancestor to the earliest definitive *H. sapiens.*

Remarks The legitimacy of *Homo heidelbergensis* as a taxonomic unit is controversial. Some regard *Homo heidelbergensis* as a late version of the evolving lineage *Homo erectus* that ultimately gave rise to our own species *Homo sapiens* in Africa (White et al., 2003). This view necessarily evokes transitional fossils with intermediate morphologies. Formerly, these fossils were regarded as "archaic" *Homo sapiens,* but most paleontologists now refer to them as *H. heidelbergensis.* Two studies have recently tested the distinctiveness of *H. heidelbergensis* and both have supported its taxonomic validity (Rightmire, 2008; Mounier et al., 2009).

Since *H. erectus, H. heidelbergensis,* and *H. sapiens* may represent a single evolving lineage, some have argued that the former two taxonomic distinctions should be sunk into *H. sapiens sensu lato* (Tobias, 1995; Wolpoff, 1999). However, this approach is untenable given overwhelming morphological and genetic (e.g., Stringer, 2002; White et al., 2003; Green et al., 2006; Wall and Kim, 2007) evidence that Neanderthals are a distinct lineage of extinct hominins. The most recent common ancestor of *H. sapiens* and *H. neanderthalensis* requires a name that is neither *H. sapiens* nor *H. neanderthalensis.* Some have given that distinction to *H. erectus* (e.g., White et al., 2003), while others suggest that middle Pleistocene hominins are sufficiently and consistently different from earlier *H. erectus* to warrant a separate species, that would be *H. heidelbergensis* (e.g., Tattersall, 1986; Rightmire, 2008). The taxonomic murkiness in the middle Pleistocene is further complicated by suggestions that European *H. heidelbergensis* fossils (represented perhaps by the crania from Petralona, Arago, and Atapuerca) are morphologically distinct from African fossils from Bodo and Kabwe, and that the European specimens of *H. heidelbergensis* form a chronospecies with *H. neanderthalensis* (Stringer, 1996). Nonetheless, the coefficient of variation for over 20 features of all crania assigned to *H. heidelbergensis* is within the expected range for a single species (Rightmire, 2008). If future studies find the European and African *H. heidelbergensis* fossils different enough to warrant species distinction, the African fossils will be regarded as *Homo rhodesiensis,* with the Kabwe cranium as the type specimen (Rightmire, 2008).

HOMO SAPIENS Linnaeus, 1758
Figures 25.18 and 25.19; Table 25.2

Partial Synonymy Homo capensis, Broom, 1918; *Palaeoanthropus palestinensis,* McCown and Keith, 1939; *Homo sapiens neanderthalensis,* Howell, 1978; *Homo sapiens afer,* Howell, 1978; *Homo sapiens capensis,* Galloway, 1937a, 1937b; *Homo helmei,* McBrearty and Brooks, 2000; *Homo sapiens idaltu,* White et al., 2003; *Homo idaltu,* Basell, 2008.

Diagnosis Because of the gradual accumulation of features that today characterize our own species, *Homo sapiens* (considered here to be synonymous with "humans"), it has proven difficult to construct a diagnosis applicable to all members of the lineage, or to identify the point at which the species began (Howell, 1978). If a criterion for inclusion in *H. sapiens* was the possession of features described in some literature as "anatomically modern," this would exclude some extant humans (Lieberman

et al., 2002) and much of the late middle through late Pleistocene African hominin sample many would attribute to *H. sapiens.* Although Howell (1978) provided a thorough summary of modern human skeletal anatomy, including many features uniquely found in *Homo sapiens,* but not in *H. neanderthalensis,* or in other species of *Homo,* he emphasized that he was not offering a diagnosis of *H. sapiens.* Subsequent studies have attempted to identify and quantify autapomorphies of *Homo sapiens* (e.g., Day and Stringer, 1982; Stringer et al., 1984; Lieberman, 1995; Lieberman et al., 2002); however, because these features arose sequentially, it is preferable to adopt a lineage-based definition of the species, in which descendants of *H. heidelbergensis* subsequent to the separation from the *H. neanderthalensis* lineage are considered modern humans, and to view the accumulation of autapomorphies in this context (Lieberman et al., 2002).

Among the autapomorphic traits identified as characteristic of, or unique to, the *Homo sapiens* lineage are limb bones with thin cortical bone and small articular surfaces, presence of a canine fossa, large endocranial capacity, cerebral asymmetry, elevated cranial vault with a high, vertical forehead and greatest width biparietally, and inferred associated expansion of the prefrontal cortex and parietal lobes, with parietal bossing and loss of sagittal keeling and parasagittal flattening, high frontal angle, narrow, high, rounded occipital planum of the occipital bone, strong basicranial flexion with the foramen magnum tucked well under the braincase, expanded middle cranial fossa, associated with inferred expansion of lateral and inferior areas of the temporal lobes that relate to language, reduced, orthognathic face, separation of the supraorbital region from glabella and subdivision of the superior orbital margin into supraorbital and supraciliary portions, inferior orbital plane tilted down and back from the inferior orbital margin, extreme lateral placement of the styloid processes, reduced dental crown size and concomitant reduction in size of alveolar processes of the upper and lower jaws, reduction of cranial robusticity, including thinner cranial bones, and expansion of the mental trigon and mental fossae of the mandible to form a bony chin (Howell, 1978; Day and Stringer, 1982; Arsuaga et al., 1999; Lieberman, 1998; Spoor et al., 1999; Lieberman et al., 2002; Schwartz and Tattersall, 2003; Bastir et al., 2008; Pearson, 2008).

Description Survey of the fossil record shows that the features characteristic of anatomically modern humans accumulated progressively in a mosaic fashion, beginning in the late middle Pleistocene and reaching full expression only by the end of the late Pleistocene (Howell, 1978; Habgood, 1989; Stringer, 2002; Trinkaus, 2005; Bräuer, 2008; Pearson, 2008). These features appeared first in Africa and the geographically closely linked Levant at the same time that the distinctive Neanderthal morphological pattern was developing in Europe and western Asia (Stringer, 2002) and when more archaic hominins (i.e., *H. erectus*) still inhabited eastern Asia (Klein, 1995). The fossil record of *Homo sapiens* in Africa is copious, particularly from localities dated to the end of the late Pleistocene and Holocene, making it impossible to comprehensively list and describe all the relevant specimens within the scope of this overview. Table 25.2 provides instead a representative sampling of fossil *Homo sapiens* occurrences on the continent and several from the Levant. Specimens are described as exemplars of archaic, near-modern, and modern human categories, with greatest emphasis on the earlier phases of the lineage, though it should be noted that these phases grade into one another without clear demarcations and that there was considerable morphological heterogeneity at any particular time (Foley and Lahr, 1992).

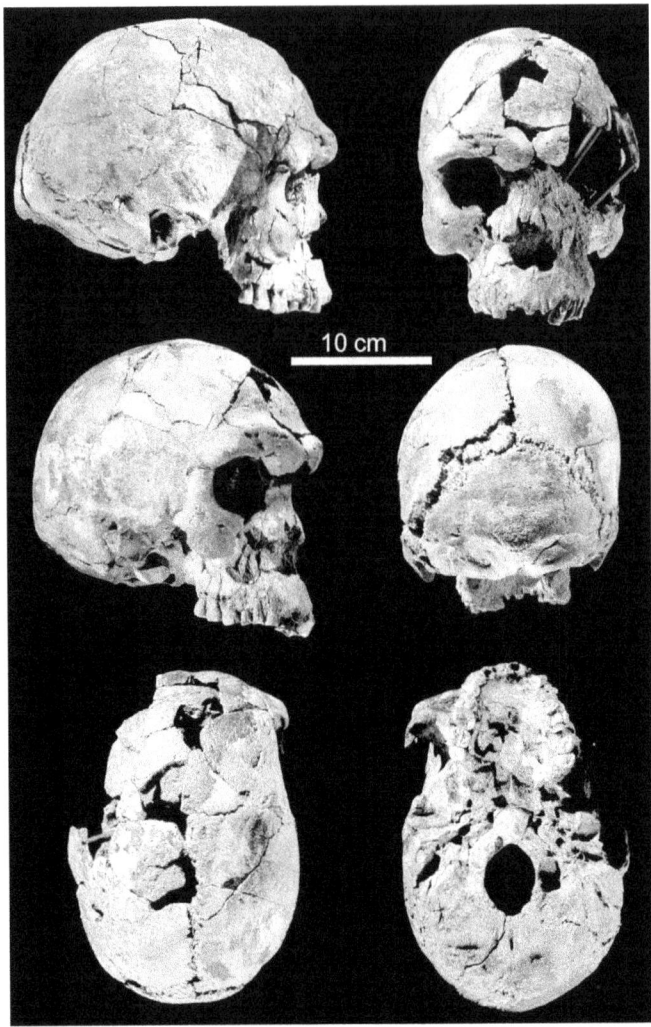

FIGURE 25.18 Holotype of *Homo sapiens idaltu*, cranium BOU-VP-16/1. Courtesy of Tim White.

Archaic humans (early fossil *H. sapiens*) typically exhibit some of the morphological configuration of extant humans, while retaining varying degrees of structural primitiveness. They date from the late middle to early late Pleistocene, and their cultural context is usually Mousterian or Middle Stone Age (McBrearty and Brooks, 2000; Basell, 2008). The oldest of these hominins may be from the Omo Kibish deposits (table 25.2). Omo I, comprised of parts of the skull, dentition, and postcranial skeleton, has a cranium that is robust in comparison with modern human crania, with a prominent glabella, slightly receding forehead, prominent supraorbital torus, and large teeth, accompanied by more derived features such as a rounded occipital profile, contracted nuchal planum with modest muscle markings, a relatively high vault, expanded parietal region and widest point high on the vault, and absence of a sagittal keel and parasagittal flattening (Day, 1969). Its skeleton is morphometrically within the modern human range (but see Pearson, 2000), though it is robustly built with strong muscle markings (Day, 1969; Rightmire, 1976; Stringer, 1978; Day and Stringer, 1982). In contrast to the condition of the Omo I cranium, the Omo II calvaria has greater resemblance to specimens of *Homo heidelbergensis* and is more heavily constructed, with strong muscle markings; it has a receding forehead, large, flat nuchal plane, greater occipital angulation, modest sagittal keel, and shallow parasagittal

depressions, and a massive occipital torus accompanied by a transverse supratoral sulcus; nonetheless, its cranial capacity is estimated to be 1,435 cm³ (Day, 1969; Day and Stringer, 1982). The Omo I postcranials and other remains from Omo Kibish indicate that individuals from the site were of medium to tall stature (ca. 162–182 cm; Pearson, 2000; Pearson et al., 2008a, 2008b). Though Omo I clearly belongs in *Homo sapiens* and neither specimen has anatomical affinities with Neanderthals (Day and Stringer, 1982; *contra* Brose and Wolpoff, 1971), its relationship to Omo II and the phylogenetic position of the latter remain unclear (Fleagle et al., 2008).

An extraordinary set of archaic *H. sapiens* crania, penecontemporaneous with the Omo fossils (table 25.2), was recovered from Herto, Ethiopia. The most complete of these is BOU-VP-16/1, which has a long, high vault. Its more archaic morphology includes a modestly receding forehead, strongly flexed occipital with a prominent external occipital protuberance, large teeth, large, flared pterygoid plates, broad, deep glenoid fossa, very well-developed temporal lines, robust supraorbital region, and great distance between the articular eminence and occlusal plane; however, it also exhibits the advanced features of a divided supraorbital torus, greatest breadth high on the vault, relatively little prognathism, and modest-sized orbits and malars (White et al., 2003). Typical of *H. sapiens* from this time period, BOU-VP-16/1 and the other adult (BOU-VP-16/2; BOU-VP-16/43) and immature (BOU-VP-16/5) cranial remains from Herto have no special morphometric affinity with any regional modern *H. sapiens* population, but they demonstrate that modern human morphology was developing in Africa prior to the disappearance of Neanderthals from Europe and western Asia (White et al., 2003). Cut marks on these crania reveal the earliest evidence for nonutilitarian defleshing and mortuary practices (Clark et al., 2003).

Also from this time period are hominin remains from Jebel Irhoud, Morocco (table 25.2), including two partial crania (Irhoud 1 and 2), and a juvenile mandible and humerus (Irhoud 3 and 4)(Hublin and Tillier, 1981; Hublin et al., 1987). Although the Irhoud specimens have been considered by some to have Neanderthal affinities (e.g., Ennouchi, 1962, 1963, 1969; Mann and Trinkaus, 1973), this has been largely discounted as only a superficial resemblance (e.g., Briggs, 1968; Hublin and Tillier, 1981; Hublin et al., 1987; Hublin, 1992, 2001). Instead, the Irhoud specimens are typical of other late middle Pleistocene *H. sapiens* from Africa, in exhibiting a mix of archaic and advanced features that anticipates the morphology of modern humans. Irhoud 1 has a long, low cranial vault and large upper face. It also possesses a weak occipital torus and moderately elongate nuchal planum. The interorbital distance is broad. Cranial capacity was recalculated at a modest 1,305 cm^3 (Holloway, 2000), after an initial estimate of 1,480 cm^3 by Anthony (1966). However, the forehead is only slightly receding, the frontal attains a great vertical dimension at bregma, and the lower face is gracile. In addition, although the supraorbital tori are arched, robust, and continuous across glabella, they thin out laterally. The parietals rise vertically and are expanded superiorly (Hublin, 1992), so the greatest width is high on the cranium, and in posterior view the cranial vault has a pentagonal profile (Hublin, 2001). Alveolar prognathism is pronounced but not outside the modern human range, and there is no midfacial prognathism (Hublin, 1992). Irhoud 2 has an even more modern-looking frontal profile, and its supraorbital tori are more separated by glabella than in Irhoud 1. Conversely, Irhoud 2 appears more primitive in the posterior outline of the cranial vault and extent of the nuchal planum (Hublin, 1992). X-ray synchrotron microtomography of the teeth in Irhoud 3 shows that dental development and tooth eruption were like that of modern humans, the oldest evidence of modern life history parameters such as prolonged growth and a correlated increased juvenile learning period (Smith et al., 2007). This mandible has a true chin, small condyle, and the height of the corpus decreases posteriorly, but primitively it has large teeth, a genioglossal fossa, and a planum alveolare (Hublin, 2001).

Hominin remains from the Levantine sites of Jebel Qafzeh and Skhul (Israel) provide further evidence of the development of *H. sapiens* features in the late middle Pleistocene–earliest late Pleistocene (table 25.2), and represent the first known migration of *Homo sapiens* out of Africa proper. At Skhul, at least 10 individuals were recovered (Schwartz and Tattersall, 2003), most from intentional burials (Garrod and Bate, 1937). The best-preserved adult skull is Skhul V, which exhibits a high, rounded vault, vertical forehead, diminished nuchal planum (compared with *Homo heidelbergensis* crania), expanded parietal eminences, a posteriorly placed lateral origin of the petro-tympanic crest, and large mastoid processes, in combination with more archaic features such as barlike supraorbital tori that continue across an anteriorly prominent glabella, large teeth, and a very broad interorbital area (McCown and Keith, 1939; Howells, 1970; Harvati, 2003; Schwartz and Tattersall,

2003). The mandible of this individual has no incisive alveolar planum, the corpus decreases in height posteriorly, and although it has a projecting "chin," the jaw lacks a proper mental trigon or mental tubercles (Schwartz and Tattersall, 2003), its ramus is quite high and vertical, and the coronoid process is very high. Other crania (e.g., Skhul II, IV) also show archaic features, with greater development of the supraorbital region, thicker bones, and lower crania with longer nuchal planes and more receding foreheads. The juvenile cranium Skhul I has a comparatively more modern appearance: it has a vertical forehead, raised cranial vault, and parietal expansion producing a pentagonal outline in posterior view. Estimated brain sizes of the more complete adult crania are impressive, ranging from 1,520 to 1,590 cm^3 (Holloway, 2000).

Fourteen hominin individuals have been recovered from Jebel Qafzeh, most from intentional graves in Mousterian contexts (Vandermeersch, 1981). There is some variation in this sample in the degree of development of supraorbitals and mental trigons, but the overall expression of anatomically modern *H. sapiens* features is unmistakable (Vandermeersch, 1981; Stringer, 1974; Trinkaus, 1984b; *contra* Brose and Wolpoff, 1971). The crania are generally long, high vaulted, with vertical to near-vertical frontals and rounded occipital profiles. Parietal expansion and bossing is obvious, as is reduction of the lower face. In some individuals (Qafzeh 9, 11), the supraorbital region is bipartite, the mastoid process is large and juxtamastoid eminence small, and the mandible exhibits a true mental trigon or chin (Harvati, 2003; Schwartz and Tattersall, 2003). Endocranial volume is capacious, calculated as 1,568 cm^3 for Qafzeh 6 and 1,508 cm^3 for Qafzeh 9 (Vandermeersch, 1981; Holloway, 2000). In contrast to penecontemporaneous Neanderthals from the region, postcranial features of the Skhul and Qafzeh hominins are far more like those of modern humans (e.g., higher neck-shaft angles of the femur, short, stout superior pubic rami, position of the external obturator groove, lower limb cross-sectional anatomy; Ben-Itzhak et al., 1988; Smith et al., 1983, 1984; Rak, 1990; Trinkaus, 1992, 1993a). The differences in femoral angles have a high correlation with varying activity levels during development, and from this it can be implied that adults endured less femoral strain and juvenile individuals from the Qafzeh-Skhul population(s) underwent lower levels of locomotor activity and greater age-grade division of activities than Neanderthal juveniles from the region (Ruff and Hayes, 1983; Trinkaus, 1993a). In addition, principal components analysis of crania demonstrate that Qafzeh 6 and Skhul 5 fall within a grouping of *Homo sapiens* crania from northern Africa dated between 35,000 and 5,000 y, and not with the Neanderthal sample (Bräuer and Rimbach, 1990). Dental analysis of prey species shows that these Levantine *H. sapiens* may have had a more efficient strategy of resource exploitation than Neanderthals from the same region and hunted more seasonally, demonstrating that similarities in stone tool cultures did not necessarily correlate with identical behaviors (Lieberman and Shea, 1994). Measurements estimated from the ratio of femoral length to stature indicate that the Qafzeh-Skhul hominins were reasonably tall, with adult heights of between 164 and 193 cm (Feldesman et al., 1990).

The persistence of archaic features and considerable morphological heterogeneity in the African late Pleistocene (Stringer, 2002) is evidenced in specimens such as those from Klasies River Mouth in South Africa, and L.H. 18 from the Ngaloba Beds at Laetoli, Tanzania. Though not as old geologically as the Herto hominins (table 25.2), L.H. 18 has a more primitive appearance, with a relatively low vault, small mastoid process, marked recession of the forehead, slight keeling of the frontal, inferred facial prognathism, occipitomastoid crest, thick cranial bones, central occipital torus, and low cranial capacity (1,200 cm³)(Day et al., 1980; Rightmire, 1984). More advanced traits in the specimen include a rounded occipital profile, low position of inion, parietal bossing, presence of a nasal spine, canine fossa, absence of parasagittal flattening, and a divided supraorbital torus (Day et al., 1980; Rightmire, 1984).

Fossils from Klasies River Mouth, South Africa, are close in age to the Ngaloba hominin (table 25.2), and derive from Middle Stone Age contexts (Singer and Wymer, 1982). These have been prominent in the debate about the antiquity of the emergence of modern humans in Africa, with some workers stressing their modern human features (e.g., Singer and Wymer, 1982; Rightmire and Deacon, 1991; Bräuer et al., 1992) and others denying their modernity (e.g., Wolpoff et al., 1994). The debate is fueled in part by the degree of variation in the sample, particularly in mandibular morphology (Grine et al., 1998). Mandibular specimen KRM 41815 (SAM-AP 6222) is small but robust, with remnants of evidently modest-sized teeth. The ramus has an anteriorly projecting expansion of the coronoid process, and a broad, shallow sigmoid notch. The corpus decreases in height posteriorly. Although the chin is not anteriorly prominent, nonetheless it is well demarcated, with a clear mental trigon flanked by shallow depressions. In contrast, the anterior profile of the symphyseal region of mandible SAM-AP 6223 is nearly vertical, and its mental trigon is more weakly demarcated (Schwartz and Tattersall, 2003). Detailed comparative morphological examinations of the malar (KRM 16651 = SAM-AP 6098) and temporal (SAM-AP 6269) specimens from the site show that they are within the range of variation observed in modern humans (Bräuer and Singer, 1996; Grine et al., 1998). The supraorbital region of the frontal fragment from Klasies River Mouth (KRM 16425) is divided into supraorbital and superciliary portions and is essentially modern (Grine et al., 1998). Variability in the hominin sample from this site extends to the postcranium, which is described as exhibiting a mix of modern and archaic features (Churchill et al., 1996; Pearson and Grine, 1997; Pearson, 2000).

The partial cranium (M.A.R. 89.4.1.3, or Dar es Soltane 5) from Dar es Soltane II, Morocco is reminiscent of the Ngaloba specimen, though it may be much younger geologically (table 25.2). It clearly is not anatomically modern in all respects, attesting to the persistence of archaic morphology in *H. sapiens* well into the late Pleistocene. Its frontal is slightly receding and bounded by thick, arched supraorbital tori, which project more anteriorly than glabella, it has a broad interorbital area, the nasals are deeply set under glabella, the articular fossa is deep, and cranial bones are moderately thick, imparting a primitive aspect to the cranium (Ferembach, 1976a; Bräuer, 1984). Nonetheless, the supraorbitals are each faintly subdivided into medial and lateral segments, it has canine fossae, and the frontal vault rises to impressive height near bregma (Schwartz and Tattersall, 2003; Trinkaus, 2005).

By the latter half of the late Pleistocene in Africa, hominins were near-modern human anatomically, but usually still relatively robust in build and dimensions and generally not morphometrically affiliated with a particular modern human population. The skeleton from Nazlet Khater, Egypt (table 25.2), is typical of this group of hominins. The postcranial anatomy of this specimen is indistinguishable from that of modern humans, but the skull exhibits strong alveolar prognathism, a robust mandibular corpus, very great breadth of the mandibular ramus, and "does not display clear affinities with modern Negroid

populations" (Thoma, 1984; Pinhasi and Semal, 2000, p. 282; Trinkaus, 2005). Nonetheless, in principal components analysis of cranial variables, Nazlet Khater and Dar es Soltane 5 fall within the range or closer to anatomically modern humans than to Neanderthals, or closer than Neanderthals are to modern humans, including Upper Paleolithic Europeans (Bräuer and Rimbach, 1990; Crevecoeur, 2008). Some of the features found in the Nazlet Khater skull (anteriorly positioned zygomatic; exceptionally wide mandibular ramus) are shared with the oldest known early modern human in Europe, Pestera cu Oase 2, from Romania, dated to ca. 40,000 y (Rougier et al., 2007).

A cranium from Hofmeyr, South Africa, is of similar antiquity to Nazlet Khater (table 25.2). This specimen is large, robust, and retains primitive features such as a broad nasal opening, glabellar prominence, a continuous, moderately well-developed supraorbital torus, large molars, and a broad frontal process of the maxilla (Grine et al., 2007). Though it also exhibits many modern human features, such as a steeply vertical frontal, high, rounded braincase with parietal expansion and greatest width high on the parietals, and has an associated mandibular fragment lacking a retromolar gap, its overall construction does not match that of crania from extant African populations; however, 3-D geometric and linear morphometric analyses show a close affinity between the Hofmeyr specimen and Upper Paleolithic European crania (Grine et al., 2007). This supports the idea that the ancestry of Upper Paleolithic Eurasians was rooted in Africa, as previously indicated by the work of Bräuer and Rimbach (1990). While Wolpoff (1989) has argued to the contrary that early humans from Africa exhibit features linking them closely with contemporary African populations, the evidence for this is unconvincing (Habgood, 1989).

Hominin fossils from Border Cave, South Africa also belong in this group of near moderns. A very fragmentary cranium, BC 1, has thick, arched supraorbital tori that are not subdivided and that project anterior to glabella, resembling the Dar es Soltane II cranium in this regard. It has a wide interorbital region, and prominent mastoid and supramastoid crests. However, the frontal rises steeply vertically and is "bulging," glabella is little developed, and the vault is large (cranial capacity ca. 1,510 cm³; Holloway, 2000) and high (Cooke et al., 1945, de Villiers, 1973; Rightmire, 1979; Habgood, 1989). The mandibles have moderately developed chins, corpora that recede in height posteriorly, and lack retromolar gaps, but they have anteroposteriorly expanded rami with high, shallow notches between the coronoid processes and condyles. Statistical analyses showing close affinity between BC 1 and modern African populations such as southern African Negro and Khoisan (e.g., de Villiers 1976; Rightmire, 1979; de Villiers and Fatti, 1982) are statistically suspect and flawed by comparatively including only modern African populations that it was assumed a priori had a phylogenetic relationship with the Border Cave hominin (Campbell, 1980; Habgood, 1989).

A slightly older *H. sapiens* specimen is the juvenile skeletal burial from Taramsa Hill, Egypt (table 25.2). The long bones are slender, and the cranium exhibits a number of modern features such as a high, vertical forehead, rounded occipital, divided supraorbital, and expanded parietals, but it retains a relatively large, prognathic face and large teeth (Vermeersch et al., 1998). This appears to be the oldest known intentional burial north of the equator in Africa (Vermeersch et al., 1998).

By the end of the late Pleistocene–early Holocene, most hominins (the exceptions are *H. neanderthalensis* in Europe, and late-surviving *H. erectus* and *H. floresiensis*, both in Indonesia) were essentially anatomically modern, and their accompanying archeological record of Late Stone Age or Upper Paleolithic and Epipaleolithic cultures exhibits the signs of modern human–like cognitive skills and behaviors (including, at Wadi Kubbaniya, Egypt, and Jebel Sahaba, Sudan, evidence of murder or warfare; Wendorf, 1968; Wendorf and Schild, 1986; Thorpe, 2003). Hominins from this time period include a number of specimens from the Upper Semliki Valley, Democratic Republic of Congo, dated to the late Pleistocene (e.g., Is 11 fossils) and Holocene (e.g., Is 1-1 and 1-2; Ky 2), respectively (table 25.2; Boaz et al., 1990). The most complete of these, Is 1-1, includes a cranium (figure 25.19), mandible, and partial skeleton from an adult male. In all respects, the morphology of this individual matches that of anatomically modern humans, and the results of multivariate discriminant analysis show that it closely resembles modern Bantu or Central African Negroid populations in its cranial anatomy (Boaz et al., 1990). This fits a pattern in which other African fossil *Homo sapiens* from this interval (table 25.2) routinely have strong morphometric affinities with extant African populations (Rightmire, 1975).

Remarks A number of models have been advanced to explain the origin and phylogeny of *Homo sapiens*, including the African Replacement Model, which states that *H. sapiens* first arose in Africa, migrated to other regions of the Old World, and replaced archaic, indigenous populations in these regions with little or no interbreeding; the African Hybridization and Replacement Model, which allows for some degree of genetic exchange between African emigrants and populations being replaced; the Assimilation Model, which posits an African origin for humans and subsequent significant gene flow between regions, but denies an important role for migratory replacement; and the Multiregional Evolution Model, which denies a recent African origin for modern humans, emphasizing instead regional genetic and morphological continuity over time and gene flow between regions, with humans emerging contemporaneously in different regions (Aiello, 1993; Stringer, 2002).

As shown, the African (and Levantine) fossil record supports an African origin for *Homo sapiens*. Skeletal traits associated with modern humans first appeared in Africa during the late middle Pleistocene–early late Pleistocene, long before evidence for this morphology in other parts of the Old World (Foley and Lahr, 1992), and well before the disappearance of

Neanderthals in Europe (Aiello, 1993). In addition, the earliest *H. sapiens* populations outside Africa resemble contemporaneous African *H. sapiens* cranially and postcranially (Bräuer and Rimbach, 1990; Holliday and Trinkaus, 1991; Ruff, 1994; Holliday, 1997, 1998, 2000; Pearson, 2000; Grine et al., 2007), suggesting that migration "out of Africa" played an important role in regional populational transformations from archaic hominins to *Homo sapiens* (McBrearty and Brooks, 2000).

The central importance for Africa in the establishment of *H. sapiens* throughout the Old World in the Pleistocene is further supported by genetic studies, which indicate that all modern humans share a late Pleistocene African ancestor (e.g., Wainscoat et al., 1986; Cann et al., 1987, 1994; Mountain et al., 1993; Stoneking, 1993; Stoneking et al., 1993; Bowcock et al., 1994; Cavalli-Sforza et al., 1994; Nei, 1995; Tischkoff et al., 1996; Ingman et al., 2000; Pearson, 2004; but see Templeton, 1993; Relethford, 1995). Study of the mitochondrial (mt) haplogroup M, originally thought to be an ancient marker of East Asian origin, demonstrated that this haplogroup is rooted in eastern Africa; its distribution and variation indicate migration of *H. sapiens* from Africa to Asia via western India around 50,000 years ago (Quintana-Murci et al., 1999). In contrast, mtDNA studies of Neanderthals reveal an ancient separation time of their lineage from the one leading to modern humans, within the interval 741,000–317,000 years ago (Krings et al., 2000; Ovchinnikov et al., 2000). Furthermore, the amount of gene flow needed to spread modern human morphology among small peripheral populations is incompatible with the maintenance over time of regional features in those populations (Stringer, 2002). Thus, the balance of evidence does not support the Assimilation and Multiregional Evolution Models. Genetic evidence also does not generate much support for the Hybridization and Replacement Model (Pearson, 2004).

The *Homo sapiens* lineage in Africa forms a good paleospecies: it is morphologically more advanced than *Homo heidelbergensis* and *Homo erectus*, differs anatomically substantially from the penecontemporaneous Neanderthal lineage, and exhibits progressive accumulation of features that characterize modern humans. As with many other basal segments of mammalian clades, however, recognition of the earliest members of the lineage is difficult because of the retention of a great number of plesiomorphies. Moreover, taxonomic "splitters" may prefer to emphasize the primitive features in these early modern humans by subdividing the lineage formally (cf. McBrearty and Brooks's [2000] use of "*Homo helmei*" and "*Homo sapiens*"). Nonetheless, it appears that the *Homo sapiens* lineage emerged in the latter part of the middle Pleistocene, close in time to the beginnings of Middle Stone Age (MSA) culture, which was distinguished by prepared core technology, and flake and blade tools including unifacial or bifacial projectile points (McBrearty and Brooks, 2000). Although the first appearance of the MSA is dated by ^{40}Ar/^{39}Ar in the Ethiopian Rift to >276,000 years (Morgan and Renne, 2008), older than the corpus of fossil evidence for archaic *Homo sapiens* (table 25.2), specimens such as Florisbad (Dreyer, 1935; Grün et al., 1996), and possibly the Guomde hominin, KNM-ER 3884 (Bräuer et al., 1992b)(closer in age to the beginning of the MSA; table 25.2), seem more advanced than other specimens assigned to *Homo heidelbergensis* and could represent the beginnings of the *Homo sapiens* lineage.

While the origin of *Homo sapiens* may have been coincident with the beginning of MSA culture, the connection between these cultural and anatomical changes is obscure, since Mousterian and MSA stone tool kits found with early *H. sapiens* are technologically identical to stone tool industries of Neanderthals (Klein, 1995). There is ongoing debate about the

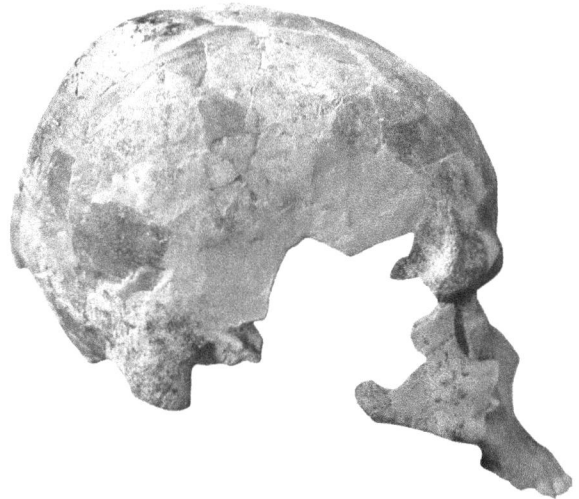

FIGURE 25.19 Lateral view of anatomically modern *Homo sapiens* cranium Is 1-1 from Ishango, Democratic Republic of Congo. Courtesy of Noel Boaz.

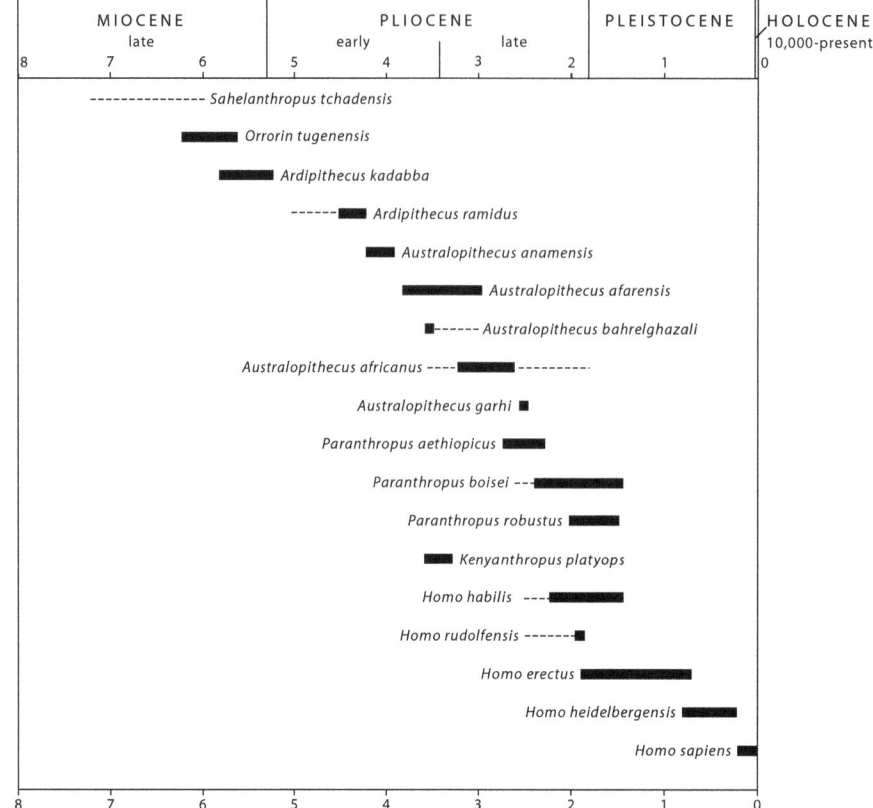

MIOCENE
late

PLIOCENE
early   late

PLEISTOCENE

HOLOCENE
10,000-present

8   7   6   5   4   3   2   1   0

-------------- *Sahelanthropus tchadensis*

Orrorin tugenensis

Ardipithecus kadabba

------ *Ardipithecus ramidus*

Australopithecus anamensis

Australopithecus afarensis

■------ *Australopithecus bahrelghazali*

Australopithecus africanus ---- ---------·

Australopithecus garhi ■

Paranthropus aethiopicus

Paranthropus boisei --

Paranthropus robustus

Kenyanthropus platyops

Homo habilis ---

Homo rudolfensis ------

Homo erectus

Homo heidelbergensis

Homo sapiens

8   7   6   5   4   3   2   1   0

FIGURE 25.20   Time line of hominin species' ranges.

relationship between the emergence of modern human features and cultural change. Changes in African *H. sapiens* cranial anatomy seem to have started with increased brain size and expansion of the frontal, parietal, and perhaps temporal regions, implying a significant reorganization of the brain and cognitive abilities, as well as with diminution of the lower face. This was followed by "modernization" of the occipital profile, lessening of midfacial robusticity, and thinning and bipartite division of supraorbital tori. Tooth size reduction and lessening of mandibular robustness occurred more recently, toward the end of the late Pleistocene. Some of the most recent transformations of human morphology were probably associated with new methods of processing food. If the acquisition of modern human morphology was not causally related to equally unique and modern behaviors, what was the reason for anatomical transformation into *H. sapiens*, and when did modern human behavior begin?

In the current debate, the Human Revolution Model posits that the first unequivocal signs of fully modern cognitive and communicative abilities occurred in the African archeological record relatively late, around 50–40,000 years ago, driven by largely undetectable reorganization of human neurological networks (e.g., Klein, 1992, 1995, 2000). The evidence for this relatively recent cognitive and technological shift is found in cultural factors associated with the Late Stone Age (LSA) and Upper Paleolithic, including customary shaping of bone, antler, shell, and ivory into formal artifact types; expression of ritual in art and elaborate graves; spatial organization of camp floors; greater diversity and standardization of artifact types; capability of blade and microlithic production; increased geographic range and widespread trade networks; personal ornamentation; and capability of fishing (Klein, 1992, 1995, 2000). In contrast, others (e.g., Lahr and Foley, 1998; McBrearty and Brooks, 2000) have argued

that the African Middle Stone Age was not just a regional variant of Mousterian culture, and that modern cultural features had been gradually accumulating throughout the duration of the MSA, in phase with the mosaic development of modern human morphology. If this view is correct, *H. sapiens* cognitive changes, increased utilization of coastal resources (e.g., Walter et al., 2000), and migratory patterns may have been driven by cycles of cooling and aridity, correlated with Northern Hemisphere glacial cycles (Carto et al., 2009). These climatic pulses are connected to episodic emigrations of humans from Africa throughout the late Pleistocene, leading to establishment of *Homo sapiens* as a global species by the end of the epoch (Carto et al., 2009).

Summary

The three earliest purported hominin species (*Sahelanthropus tchadensis, Orrorin tugenensis*, and *Ardipithecus kadabba*) are characterized by subtly modified canines relative to fossil and extant hominoids, molars as large or slightly larger than those of *Pan*, slightly thicker enamel than is found in extant African apes, and provisional evidence for bipedality. Paleoenvironmental contexts suggest at least some heterogeneity with grassland, woodland and forest represented at hominin-bearing sites. The specific ecological niches of hominins within these mosaic environments, however, remain unknown.

Haile-Selassie et al. (2004a, 2004b) have suggested that only one genus may be sampled thus far in the late Miocene, and it is not unexpected that the systematics of the late Miocene taxa are debated, given such sparse material. The record in the early Pliocene is less ambiguous taxonomically and craniodental evidence supports three apparently time successive species with ancestor-descendant relationships possible: *Ardipithecus ramidus* (itself a

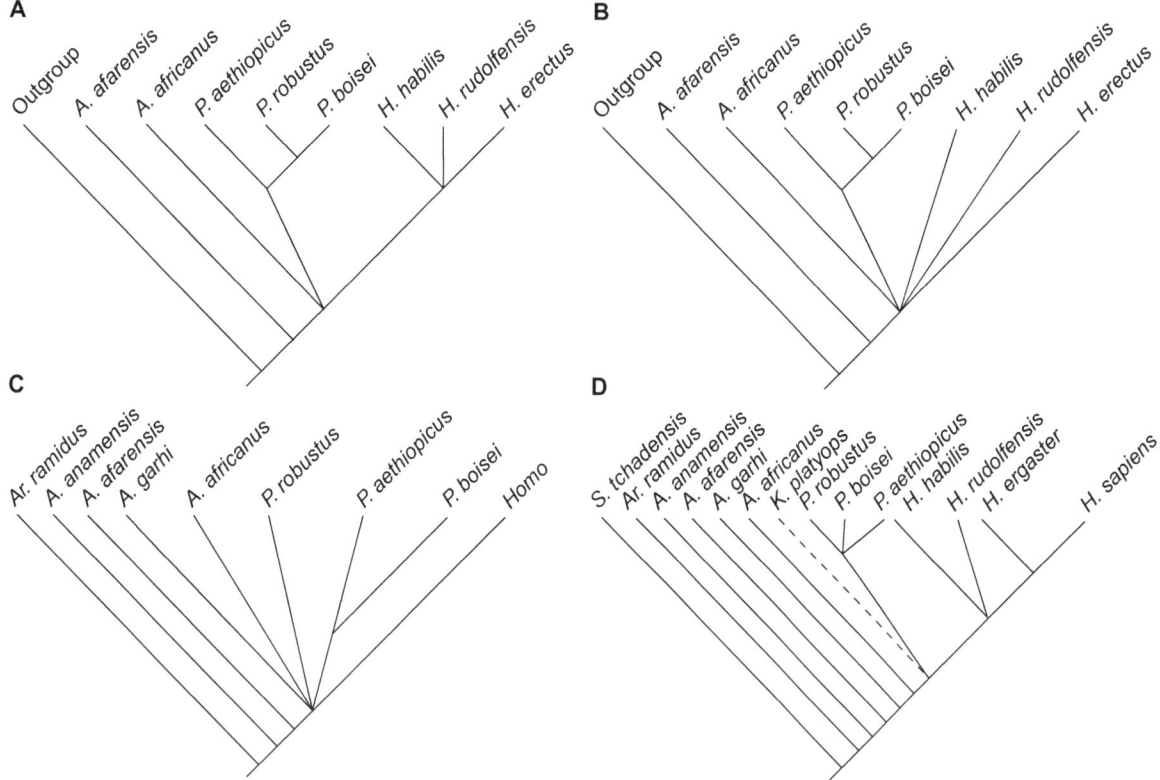

FIGURE 25.21   A) Strict consensus of the three shortest trees found by PAUP's branch and bound algorithm in phylogenetic analysis published by Kimbel et al., 2004. B) Strict consensus of the five shortest trees found by PAUP's branch and bound algorithm in phylogenetic analysis published by Kimbel et al., 2004. C) Cladogram published by Asfaw et al., 1999, showing an unresolved polychotomy as a major feature. D) Strict consensus of the most parsimonious cladograms published by Strait and Grine, 2004. The dashed line reflects these authors' uncertainty as to whether *Kenyanthropus* is a distinct species.

likely descendant of *Ardipithecus kadabba*), *Australopithecus anamensis*, and *Australopithecus afarensis*. If these taxa are related, it suggests that over time there was directional selection for postcanine megadonty, and a concomitant change in canine function and morphology (i.e., canines become more incisiform) as the premolars became molarized. Postcranially, *Ardipithecus* can now be distinguished in numerous respects from *Australopithecus*, raising the possibility of rapid aquisition of bipedal features. Alternatively, *Ardipithecus* may not be the sister taxon to *Australopithecus*.

The mid-Pliocene finds the first record of hominins in southern Africa. *Australopithecus africanus* is craniodentally derived relative to *Au. afarensis* in several respects, and shares features of the jaw, face, and basicranium with later hominins. Both *Au. afarensis* and *Au. africanus* are well represented postcranially and are incontrovertibly terrestrial bipeds, although the details of their gait and degree of commitment to bipedalism are subjects of ongoing discussion. The mid-Pliocene also has two controversial taxa, *Kenyanthropus platyops* and *Au. bahrelghazali*, considered by some to be conspecific with *Au. afarensis*.

As figure 25.20 reveals, the late Pliocene/early Pleistocene is the only part of the African hominin record where multiple lineages clearly co-occur in time and possibly space, in southern and East Africa. This runs counter to prevailing notions of the hominin diversification being rather bushy (see also White, 2003, 2009) although future finds may increase evidence for cladogenesis. Overlap of multiple taxa makes attribution of fossil postcrania and stone tools with dental remains problematic, however, it is apparent that hominins of this time were experimenting with new and different adaptive pathways.

Both *Au. garhi* and *Paranthropus* evolve even larger masticatory apparatuses than found in earlier australopiths, allowing these taxa to puncture, crush and grind food with abrasive and/or hard mechanical properties. Encephalization and reduction in tooth size also evolve during this period, and fossils exhibiting these traits are almost invariably associated with stone tool manufacture and placed in one of three recognized species of early *Homo* that, somewhat ambiguously, co-occur as early as 1.9 Ma. More long-legged and bipedally efficient body plans also appear in the late Pliocene, but while *Homo erectus* is known to possess derived postcrania due to associated skeletal material, *Homo habilis*, *Homo rudolfensis*, and *Paranthropus* have few postcrania definitively assigned to them.

Environmental mosaicism is associated with all hominins during this period, although there is a general trend of increasingly frequent associations between hominins and more open environments over time. The instability and variability of such habitats is itself a likely selective force in hominin evolution contributing to behavioral plasticity and a generalist strategy associated with the tribe's success in Africa and worldwide.

Phylogenetic hypotheses abound for this period (figure 25.21). Monophyly of South and East African *Paranthropus* species is supported by most parsimony-based analyses but it remains plausible that the forms evolved independently. There is no clear candidate among known australopith-grade hominins for an ancestral relationship to genus *Homo*, although *Au. africanus* and *Au. garhi* seem more closely related than *Au. afarensis*. *Paranthropus* is deemed too derived to have played an ancestral role, but some analyses (figure 25.21) place *Paranthropus* as the sister clade to *Homo*. Finally, if *Homo habilis* and

Homo rudolfensis are two different species, then it is presently unclear which is more closely related to *Homo erectus*.

There is general agreement that *Homo erectus*, *Homo heidelbergensis*, and *Homo sapiens* represent an evolving lineage within Africa, although the species boundaries on either side of *H. heidelbergensis* are obscure. Behaviorally, this lineage became ever more complex, as brain size increased dramatically and tools became more sophisticated. Many have argued that behavioral evolution took a punctuated leap sometime after the origin of *H. sapiens* ca. 200,000 years ago, so that humans were acquiring anatomically modern features before exhibiting fully modern behavior. Archeological evidence in Africa suggests that modern behavior may have accreted slowly initially, then exploded after a critical threshold was reached. Africa was thus the crucible that selected for virtually all of the adaptations that allowed one hominin taxon to spread into every biome on Earth.

ACKNOWLEDGMENTS

The authors are grateful to the following people and institutions for providing photographs: Michel Brunet, Yohannes Haile-Selassie, and the Cleveland Museum of Natural History; Emma Mbua and the National Museums of Kenya, Martin Pickford, Noel Boaz, Philip Rightmire, Carol Ward, Francis Thackeray, and the Transvaal Museum (Northern Flagship Institution); Tim White; and the University of California Press. We thank Meave Leakey and Emma Mbua (National Museums of Kenya), Phillip Tobias and Ron Clarke (University of the Witwatersrand), Francis Thackeray (Transvaal Museum), Michael Mbago and Amandus Kweka (National Museums of Tanzania), and Mamitu Yilma (Ethiopian National Museum) for access to specimens in their care. John Kingston and Tim White provided helpful comments and insights on portions of the manuscript. Bonnie Miljour expertly produced figures 25.11, 25.12, 25.19, 25.20, and 25.21, and Grace Holliday helped with manuscript revisions. Funding for W.J.S.'s research was generously provided by the Wenner-Gren Foundation, L. S. B. Leakey Foundation, and several Turner Grants from the Department of Geological Sciences, University of Michigan. LM was supported by the Department of Anthropology, University of Michigan. BW acknowledges support from the GW Vice-President for Academic Affairs and from GWs Selective Excellence Program. We are also grateful to Lars Werdelin for his editorial guidance and Francisco Reinking and Chuck Crumly of the University of California Press for their patience and assistance.

Literature Cited

Abbate, E., A. Albianelli, A. Azzaroli, M. Benvenuti, B. Tesfamariam, P. Bruni, N. Cipriani, R. J. Clarke, G. Ficcarelli, R. Macchiarelli, G. Napoleone, M. Papini, L. Rook, M. Sagri, T. Madhin Tecle, D. Torre, and I. Villa. 1998. A one-million-year-old *Homo* cranium from the Danakil (Afar) Depression of Eritrea. *Nature* 393:458–460.

Ahern, J. C. M. 1998. Underestimating intraspecific variation: The problem with excluding Sts 19 from *Australopithecus africanus*. *American Journal of Physical Anthropology* 105:461–480.

Aiello, L. C. 1993. The fossil evidence for modern human origins in Africa: A revised view. *American Anthropologist* 95:73–96.

Aiello, L. C., and P. Andrews. 2000. The australopithecines in review. *Human Evolution* 15:17–38.

Aiello, L. C., and C. Dean. 1990. *An Introduction to Human Evolutionary Anatomy*. Academic Press, Harcourt Brace, London, 596 pp.

Aiello, L. C. and B. A. Wood. 1994. Cranial variables as predictors of hominine body mass. *American Journal of Physical Anthropology* 95:409–426.

Aiello, L. C., B. A. Wood, C. Key, and C. Wood. 1998. Laser scanning and palaeoanthropology: An example from Olduvai Gorge, Tanzania. pp. 223–226 in E. Strasser, J. Fleagle, A. Rosenberger, and H. McHenry (eds.), *Primate Locomotion: Recent Advances*. Plenum Press: New York.

Alekseev, V. P. 1986. *The Origin of the Human Race*. Progress Publishers, Moscow, 336 pp.

Alemseged, Z., Y. Coppens, and D. Geraads. 2002. Hominid cranium from Omo: Description and taxonomy of Omo-323–1976–896. *American Journal of Physical Anthropology* 117:103–112.

Alemseged, Z., F. Spoor, W. H. Kimbel, R. Bobe, D. Geraads, D. Reed, and J. G. Wynn. 2006. A juvenile early hominin skeleton from Dikika, Ethiopia. *Nature* 443:296–301.

Alemseged, Z., J. G. Wynn, W. H. Kimbel, D. Reed, D. Geraads, and R. Bobe. 2005. A new hominin from the Basal Member of the Hadar Formation, Dikika, Ethiopia, and its geological context. *Journal of Human Evolution* 49:499–514.

Anderson, J. E. 1968. Late Palaeolithic skeletal remains from Nubia; pp. 996–1040 in F. Wendorf (ed.), *The Prehistory of Nubia*, vol. 2. Southern Methodist University, Dallas.

Andrews, P., and M. Bamford. 2008. Past and present vegetation ecology of Laetoli, Tanzania. *Journal of Human Evolution* 54:78–98.

Andrews, P. J. 1989. Lead review: Palaeoecology of Laetoli. *Journal of Human Evolution* 18:173–181.

Anthony, J. 1966. Premières observations sur le moulage endocrânien des hommes fossiles du Jebel Irhoud (Maroc). *Comptes Rendus Hebdomadaires des Séances de l'Académie des Sciences, Paris, Série D* 262:556–558.

Antón, S. C. 2003. Natural history of *Homo erectus*. *Yearbook of Physical Anthropology* 46:126–170.

Antón, S. C., W. R. Leonard, and M. L. Robertson. 2002. An ecomorphological model of the initial hominid dispersal from Africa. *Journal of Human Evolution* 43:773–785.

Arambourg, C. 1929. Découverte d'un ossuaire humain du Paléolithique supérieur en Afrique du Nord. *Anthropologie* 39:219–221.

———. 1954. L'hominien fossile de Ternifine (Algérie). *Comptes Rendus des Séances de l'Academie des Sciences, Paris* 239:893–895.

———. 1955. Une nouvelle mandibule *d'Atlanthropus* du gisement de Ternifine. *Comptes Rendus Hebdomadaires des Séances de l'Académie des Sciences, Paris* 241:895–897.

Arambourg, C., and P. Biberson. 1956. The fossil human remains from the Paleolithic site of Sidi Abderrahman (Morocco). *American Journal of Physical Anthropology* 14:467–490.

Arambourg, C., M. Boule, H. V. Vallois, and R. Verneau. 1934. Les grottes paléolithiques des Beni-Segoual (Algérie). *Archives de l'Institut de Paléontologie Humaine* 13:1–36.

Arambourg, C., and Y. Coppens. 1967. Sur la découverte dans le Pléistocène inférieur de la Vallée de l'Omo (Éthiopie) d'une mandibule d'Australopithécien. *Comptes Rendus des Séances de l'Académie des Sciences* 265:589–590.

———. 1968. Découverte d'un Australopithécien nouveau dans les gisements de l'Omo (Éthiopie). *South African Journal of Science* 64:58–59.

Arensburg, B., A. M. Tiller, B. Vandermeersch, H. Duday, L.A. Schepartz, and Y. Rak. 1989. A Middle Paleolithic human hyoid bone. *Nature* 338:758–760.

Arsuaga, J. L., I. Martínez, C. Lorenzo, A. Gracia, A Muñoz, C. Alonso, and J. Gelego. 1999. The human cranial remains from Gran Dolina Lower Pleistocene site (Sierra de Atapuerca, Spain). *Journal of Human Evolution* 37:431–458.

Asfaw, B. 1987. The Behlodelie frontal: New evidence of early hominid cranial morphology from the Afar of Ethiopia. *Journal of Human Evolution* 16:611–624.

Asfaw, B., Y. Beyene, G. Suwa, R. C. Walter, T. D. White, G. WoldeGabriel, G., and T. Yemane. 1992. The earliest Acheulian from Konso-Gardula. *Nature* 360:732–735.

Asfaw, B., W. H. Gilbert, Y. Beyene, W. K. Hart, P. R. Renne, G. WoldeGabriel, E. Vrba, and T. D. White. 2002. Remains of *Homo erectus* from Bouri, Middle Awash, Ethiopia. *Nature* 416:317–320.

Asfaw, B., T. White, O. Lovejoy, B. Latimer, S. Simpson, and G. Suwa. 1999. *Australopithecus garhi*: A new species of early hominid from Ethiopia. *Science* 284:629–635.

Avery, D. M. 1998. An assessment of the Lower Pleistocene micromammalian fauna from Swartkrans Members 1–3, Gauteng, South Africa. *Geobios* 31:393–414.

Avery, G., K. Cruz-Uribe, P. Goldberg, F. E. Grine, R. G. Klein, M. J. Lenardi, C. W. Marean, W. J. Rink, H. P. Schwarcz, A. I. Thackeray, and M. L. Wilson. 1997. The 1992–1993 excavations at the Die Kelders Middle and Later Stone Age cave site, South Africa. *Journal of Field Archaeology* 24:263–291.

Baab, K. L. 2008a. A re-evaluation of the taxonomic affinities of the early *Homo* cranium KNM-ER 42700. *Journal of Human Evolution* 55:741–746.

————. 2008b. The taxonomic implications of cranial shape variation in *Homo erectus*. *Journal of Human Evolution* 54:827–847.

Backwell, L. R., and F. d'Errico. 2001. Evidence of termite foraging by Swartkrans early hominids. *Proceedings of the National Academy of Sciences, USA* 98:1358–1363.

Bamford, M. 1999. Pliocene fossil woods from an early hominid cave deposit, Sterkfontein, South Africa. *South African Journal of Science* 95:231–237.

Barham, L. S., A. Pinto, and C. Stringer. 2002. Bone tools from Broken Hill (Kabwe) cave, Zambia, and their evolutionary significance. *Before Farming* 2:1–16.

Basell, L. S. 2008. Middle Stone Age (MSA) site distributions in eastern Africa and their relationship to Quaternary environmental change, refugia and the evolution of *Homo sapiens*. *Quaternary Science Reviews* 27:2484–2498.

Bastir, M., A. Rosas, D. E. Lieberman, and P. O'Higgins. 2008. Middle cranial fossa anatomy and the origin of modern humans. *The Anatomical Record* 291:130–140.

Beaumont, P. 1980. On the age of Border Cave hominids 1–5. *Palaeontologia Africana* 23:21–33.

Beaumont, P. B. and J. C. Vogel. 2006. On a timescale for the past million years of human history in central South Africa. *South African Journal of Science* 102:217–228.

Behrensmeyer, A. K., J. D. Damuth, W. A. DiMichele, R. Potts, H.-D. Sues, and S. L. Wing eds. 1992. *Terrestrial Ecosystems through Time*. University of Chicago Press, Chicago, 568 pp.

Behrensmeyer, A. K., N. E. Todd, R. Potts, and G. E. McBrinn. 1997. Late Pliocene faunal turnover in the Turkana Basin, Kenya and Ethiopia. *Science* 278:1589–1594.

Ben-Itzhak, S., P. Smith, and R. A. Bloom. 1988. Radiographic study of the humerus in Neanderthals and *Homo sapiens sapiens*. *American Journal of Physical Anthropology* 77:231–242.

Bennet M. R., J. W. K. Harris, B. G. Richmond, D. R. Braun, E. Mbua, P. Kiura, D. Olago, M. Kibunjia, C. Omuombo, A. K. Behrensmeyer, D. Huddart, and S. Gonzales. 2009. Early hominin foot morphology based on 1.5 million-year-old footprints from Ileret, Kenya. *Science* 323:1197–1201.

Beynon, A. D., and B. A. Wood. 1987. Patterns and rates of enamel growth in the molar teeth of early hominids. *Nature* 326:493–496.

Berger, L. R., D. J. de Ruiter, C. M. Steininger, and J. Hancox. 2003. Preliminary results of excavations at the newly investigated Coopers D deposit, Gauteng, South Africa. *South African Journal of Science* 99:276–278.

Berger, L. R., A. W. Keyser, and P. V. Tobias. 1993. Gladysvale: first early hominid site discovered in South Africa since 1948. *American Journal of Physical Anthropology* 92:107–111.

Berger, L. R., R. Lacruz, and D. J. de Ruiter. 2002. Brief communication: Revised age estimates of *Australopithecus*-bearing deposits at Sterkfontein, South Africa. *American Journal of Physical Anthropology* 119:192–197.

Berger, L. R., and P. V. Tobias. 1996. A chimpanzee-like tibia from Sterkfontein, South Africa and its implications for the interpretation of bipedalism in *Australopithecus africanus*. *Journal of Human Evolution* 30:343–348.

Biberson, P. 1964. La place des hommes du Paléolithique marocain dans la chronologie du Pléistocène atlantique. *Anthropologie* 68:475–526.

Bilsborough, A., and B. A. Wood. 1988. Cranial morphometry of early hominids: facial region. *American Journal of Physical Anthropology* 76:61–86.

Bishop, W. W., A. Hill, and M. Pickford. 1978. Chesowanja: A revised geological interpretation; pp. 309–327 in W. W. Bishop (ed.), *Geological Background to Fossil Man*. Geological Society of London, Scottish Academic Press, Edinburgh.

Blumenschine, R. J., C. R. Peters, F. T. Masao, R. J. Clarke, A. L. Deino, R. L. Hay, C. C. Swisher, I. G. Stanistreet, G. M. Ashley, L. J. McHenry, N. E. Sikes, N. J. van der Merwe, J. C. Tactikos, A. E. Cushing, D. M. Deocampo, J. K. Njau, and J. I. Ebert. 2003. Late Pliocene *Homo* and hominid land use from western Olduvai Gorge, Tanzania. *Science* 299:1217–1221.

Boaz, N. T., and J. Hampel. 1978. Strontium content of fossil tooth enamel and diet of early hominids. *Journal of Paleontology* 52:928–933.

Boaz, N. T., P. P. Pavlakis, and A. S. Brooks. 1990. Late Pleistocene–Holocene human remains from the Upper Semliki, Zaire. *Virginia Museum of Natural History Memoir* 1:273–299.

Bobe, R., and A. K. Behrensmeyer. 2004. The expansion of grassland ecosystems in Africa in relation to mammalian evolution and the origin of the genus *Homo*. *Paleogeography, Palaeoclimatology, Palaeoecology* 207:399–420.

Boisserie, J.-R., A. Zazzo, G. Merceron, C. Blondel, P. Vignaud, A. Likius, H. T. Mackaye, and M. Brunet. 2005. Diets of modern and late Miocene hippopotamids: Evidence from carbon isotope composition and micro-wear of tooth enamel. *Palaeogeography, Palaeoclimatology, Palaeoecology* 221:153–174.

Bonarelli, G. 1909. *Palaeoanthropus* (n.g.) *heidelbergensis* (Schoet.). *Rivista Italiana di Paleontologia* 15:26–31.

Bonnefille, R., R. Potts, F. Chalié, D. Jolly, and O. Peyron. 2004. High resolution vegetation and climate change associated with Pliocene *Australopithecus afarensis*. *Proceedings of the National Academy of Sciences, USA* 101:12125–12129.

Boule, M., and H. V. Vallois. 1946. *Les Hommes Fossiles*. Masson et Cie, Paris, 583 pp.

Bowcock, A. M., A. Ruiz-Linares, J. Tomfohrde, E. Minch, K. R. Kidd, and L. L. Cavalli-Sforza. 1994. High resolution of human evolutionary trees with polymorphic microsatellites. *Nature* 368:455–457.

Bower, B. 1987. Family feud: Enter the "Black Skull." *Science News* 131:58–59.

Brace, C. L., P. E. Mahler and R. E. Rosen. 1973. Tooth measurement and the rejection of the taxon *"Homo habilis."* *Yearbook of Physical Anthropology* 16:50–68.

Bradley, B. 2008. Reconstructing phylogenies and phenotypes: A molecular view of human evolution. *Journal of Anatomy* 212:337–353.

Brain, C. K. 1976. A re-interpretation of the Swartkrans site and its remains. *South African Journal of Science* 72:141–146.

————. 1988. New information from the Swartkrans Cave of relevance to "robust" australopithecines; pp. 311–316 in F. E. Grine (ed.), *Evolutionary History of the "Robust" Australopithecines*. Aldine de Gruyter, New York.

Brain, C. K., C. S. Churcher, J. D. Clark, F. E. Grine, P. Shipman, R. L. Susman, A. Turner, and V. Watson. 1988. New evidence of early hominids, their culture and environment from the Swartkrans cave, South Africa. *South African Journal of Science* 84:828–835.

Brain, C. K., E. S. Vrba, and J. T. Robinson. 1974. A new hominid innominate bone from Swartkrans. *Annals of the Transvaal Museum* 29:55–63.

Brain, C. K., and V. Watson. 1992. A guide to the Swartkrans early hominid cave site. *Annals of the Transvaal Museum* 35:343–365.

Bramble, D. M, and D. E. Lieberman. 2004. Endurance running and the evolution of *Homo*. *Nature* 432:345–352.

Bräuer, G. 1984. A craniological approach to the origin of anatomically modern *Homo sapiens* in Africa and implications for the appearance of modern Europeans; pp. 327–410 in F. Smith and F. Spencer (eds.), *The Origin of Modern Humans*. Liss, New York.

————. 2008. The origin of modern anatomy: by speciation or intraspecific evolution? *Evolutionary Anthropology* 17:22–37.

Bräuer, G., H. J. Deacon, and F. Zipfel. 1992a. Comment on the new maxillary finds from Klasies River, South Africa. *Journal of Human Evolution* 23:419–422.

Bräuer, G., C. Groden, F. Groning, A. Kroll, K. Kupczik, E. Mbua, A. Pommert, and T. Schiemann. 2004. Virtual study of the endocranial morphology of the matrix-filled cranium from Eliye Springs, Kenya. *The Anatomical Record, Part A* 276A:113–133.

Bräuer, G., and R. E. Leakey. 1986a. The ES-11693 cranium from Eliye Springs, West Turkana, Kenya. *Journal of Human Evolution* 15:289–312.

————. 1986b. A new archaic *Homo sapiens* cranium from Eliye Springs, West Turkana, Kenya. *Zeitschrift für Morphologie und Anthropologie* 76:245–252.

Bräuer, G., R. E. Leakey, and E. Mbua. 1992b. A First Report on the ER-3884 Cranial Remains from Ileret/East Turkana, Kenya; pp. 111–119 in G. Bräuer and F. H. Smith (eds.), *Continuity or Replacement: Controversies in Homo sapiens Evolution*. Balkema, Rotterdam.

Bräuer, G., and A. Z. P. Mabulla. 1996. New hominid fossil from Lake Eyasi, Tanzania. *Anthropos (Brno)* 34:47–53.

Bräuer, G., and M. J. Mehlman. 1988. Hominid molars from a Middle Stone Age level at the Mumba Rock Shelter, Tanzania. *American Journal of Physical Anthropology* 75:69–76.

Bräuer, G., and K. W. Rimbach. 1990. Late archaic and modern *Homo sapiens* from Europe, Africa, and southwest Asia: craniometric comparisons and phylogenetic implications. *Journal of Human Evolution* 19:789–807.

Bräuer, G., and R. Singer. 1996. The Klasies zygomatic bone: archaic or modern? *Journal of Human Evolution* 30:161–165.

Bräuer G., Y. Yokoyama, C. Falguères, and E. Mbua. 1997. Modern human origins backdated. *Nature* 386:337–338.

Briggs, L. C. 1950. On three skulls from Mechta-el-Arbi, Algeria. *American Journal of Physical Anthropology* 8:305–313.

———. 1968. Hominid evolution in northwest Africa and the question of the North African "Neanderthaloids." *American Journal of Physical Anthropology* 29:377–386.

Bromage, T. G., and M. C. Dean. 1985. Re-evaluation of the age at death of Plio-Pleistocene fossil hominids. *Nature* 317:525–528.

Bromage, T. G., J. M. McMahon, J. F. Thackeray, O. Kullmer, R. Hogg, A. L. Rosenberger, F. Schrenk, and D.H. Enlow. 2008. Craniofacial architectural constraints and their importance for reconstructing the early *Homo* skull KNM-ER 1470. *Journal of Clinical Pediatric Dentistry* 33:43–54.

Bromage, T. G., F. Schrenk, and Y. M. Juwayeyi. 1995a. Paleobiogeography of the Malawi Rift: Age and vertebrate paleontology of the Chiwondo Beds, northern Malawi. *Journal of Human Evolution* 28:37–57.

Bromage, T. G., F. Schrenk, and F. W. Zonneveld. 1995b. Paleoanthropology of the Malawi Rift: An early hominid mandible from the Chiwondo Beds, northern Malawi. *Journal of Human Evolution* 28:71–108.

Brooks, A. S., and C. C. Smith. 1987. Ishango revisited: new age determinations and cultural interpretations. *African Archaeological Review* 5:65–78.

Broom, R. 1918. The evidence afforded by the Boskop skull of a new species of primitive man (*Homo capensis*). *Anthropological Papers of the American Museum of Natural History* 23:63–79.

———. 1929. The Transvaal fossil human skeleton. *Nature* 123:415–416.

———. 1936. A new fossil anthropoid skull from South Africa. *Nature* 138:486–488.

———. 1938. The Pleistocene anthropoid apes of South Africa. *Nature* 142:377–379.

———. 1939. A restoration of the Kromdraai skull. *Annals of the Transvaal Museum* 19:327–329.

———. 1949. Another new type of fossil ape-man (*Paranthropus crassidens*). *Nature* 163:57.

Broom, R., and J. T. Robinson. 1949. A new type of fossil man. *Nature* 164:322–323.

———. 1952. Swartkrans ape-man. *Paranthropus crassidens*. *Transvaal Museum Memoir* 6:1–123.

Brose, D. S., and M. H. Wolpoff. 1971. Early Upper Paleolithic man and late Middle Paleolithic tools. *American Anthropologist* 73:1156–1194.

Brothwell, D. R., and T. Shaw. 1971. A late Upper Pleistocene proto-West African Negro from Nigeria. *Man* 6:221–227.

Brown, B., F. H. Brown, and A. Walker. 2001. New hominids from the Lake Turkana Basin, Kenya. *Journal of Human Evolution* 41:29–44.

Brown, B., A. Walker, C. V. Ward, and R. E. F. Leakey. 1993. New *Australopithecus boisei* calvaria from East Lake Turkana, Kenya. *American Journal of Physical Anthropology* 91:137–159.

Brown, F. H. 1994. Development of Pliocene and Pleistocene chronology of the Turkana Basin, East Africa, and its relation to other sites; pp. 285–312 in R. S. Corruccini and R. L. Ciochon (eds.), *Integrative Paths to the Past*. Prentice Hall, Upper Saddle River, N.J.

Brown, F. H. and C. S. Feibel. 1988. "Robust" hominids and Plio-Pleistocene paleogeography of the Turkana Basin, Kenya and Ethiopia; pp. 325–341 in F. E. Grine (ed.), *Evolutionary History the "Robust" Australopithecines*. Aldine de Gruyter, New York.

Brown, F. H., and C. R. Fuller. 2008. Stratigraphy and tephra of the Kibish Formation, southwestern Ethiopia. *Journal of Human Evolution* 55:366–403.

Brown, F. H., J. Harris, R. Leakey, and A. Walker. 1985. Early *Homo erectus* skeleton from west Lake Turkana, Kenya. *Nature* 316:788–792.

Brunet, M., A. Beauvilain, Y. Coppens, E. Heintz, A. H. E. Moutaye, and D. Pilbeam. 1995. The first australopothecine 2,500 kilometers west of the Rift Valley (Chad). *Nature* 378:273–275.

———. 1996. *Australopithecus bahrelghazali*, une nouvelle espèce d'Hominidé ancien de la région de Koro Toro (Tchad). *Comptes Rendus de l'Académie des Aciences, Série II, Sciences de la Terre et des Planètes* 322:907–913.

Brunet, M., A. Beauvilain, D. Geraads, F. Guy, M. Kasser, H. T. Mackaye, L. M. MacLatchy, G. Mouchelin, J. Sudre, and P. Vignaud. 1997. Tchad: Un nouveau site à Hominidés Pliocènes. *Comptes Rendus de l'Académie des Aciences, Série II, Sciences de la Terre et des Planètes* 324:341–345.

Brunet, M., F. Guy, J.-R. Boisserie, A. Djimdoumalbaye, T. Lehmann, F. Lihoreau, A. Louchart, M. Schuster, P. Tafforeau, A. Likius, H. T. Mackaye, C. Blondel, H. Bocherens, L. De Bonis, Y. Coppens, C. Denis, P. Duringer, V. Eisenmann, A. Flisch, D. Geraads, N. Lopez-Martinez, O. Otero, P. P. Campomanes, D. Pilbeam, M. Ponce de León, P. Vignaud, L. Viriot, and C. Zollikofer. 2004. "Toumaï," Miocène supérieur du Tchad, le nouveau doyen du rameau humain. *Comptes Rendus Palevol* 3:277–285.

Brunet, M., F. Guy, D. Pilbeam, D. E. Lieberman, A. Likius, H. T. Mackaye, M. S. Ponce de León, C. P. E. Zollikofer, and P. Vignaud. 2005. New material of the earliest hominid from the Upper Miocene of Chad. *Nature* 434:752–755.

Brunet, M., F. Guy, D. Pilbeam, H. T. Mackaye, A. Likius, D. Ahounta, A. Beauvilain, C. Blondel, H. Bocherens, J.-R. Boisserie, L. De Bonis, Y. Coppens, J. Dejax, C. Denys, P. Duringer, V. Eisenmann, G. Fanone, P. Fronty, D. Geraads, T. Lehmann, F. Lihoreau, A. Louchart, A. Mahamat, G. Merceron, G. Mouchelin, O. Otero, P. P. Campomanes, M. Ponce de Léon, J.-C. Rage, M. Sapanet, M. Schuster, J. Sudre, P. Tassy, X. Valentin, P. Vignaud, L. Viriot, A. Zazzo, and C. Zollikofer. 2002. A new hominid from the Upper Miocene of Chad, Central Africa. *Nature* 418:145–151.

Cachel, S., and J. W. K. Harris. 1998. The lifeways of *Homo erectus* inferred from archaeological and evolutionary ecology: a perspective from East Africa; pp. 108–132 in M. D. Petraglia and R. Korisettar (eds.), *Early Human Behavior in a Global Context: The Rise and Diversity of the Lower Paleolithic Record*. Routledge, New York.

Campbell, N. A. 1980. On the study of the Border Cave remains: Statistical comments. *Current Anthropology* 21:532–535.

Cann, R. L., O. Richards, and J. K. Lum. 1994. Mitochondrial DNA and human evolution: Our one Lucky Mother; pp. 135–148 in M. H. Nitecki and V. Nitecki (eds.), *Origins of Anatomically Modern Humans*. Plenum Press, New York.

Cann, R. L., M. Stoneking, and A. C. Wilson. 1987. Mitochondrial DNA and human evolution. *Nature* 325:31–36.

Carey, T. S., and R.W. Crompton. 2005. The metabolic costs of "bent-hip, bent-knee" walking in humans. *Journal of Human Evolution* 48:25–44.

Carney, J., A. Hill, J. A. Miller, and A. Walker. 1971. Late australopithecine from Baringo District, Kenya. *Nature* 230:509–514.

Carto, S. L., A. J. Weaver, R. Hetherington, Y. Lam, and E. C. Wiebe. 2009. Out of Africa and into an ice age: On the role of global climate change in the late Pleistocene migration of early modern humans out of Africa. *Journal of Human Evolution* 56:139–151.

Cavalli-Sforza, L. L., P. Menozzi, and A. Piazza. 1994. *The History and Geography of Human Genes*. Princeton University Press, Princeton, N.J., 1088 pp.

Chamla, M. C. 1978. Le peuplement de l'Afrique du Nord de épipaléolithique à l'époque actuelle. *L'Anthropologie* 82:385–430.

Chavaillon, J., C. Brahimi, and Y. Coppens. 1974. First discovery of hominid in one of Acheulian sites of Melka-Kunturé (Ethiopia). *Comptes Rendus Hebdomadaires des Séances de l'Académie des Sciences, Paris, Série D* 278:3299–3302.

Chavaillon, J., N. Chavaillon, Y. Coppens, and B. Senut. 1977. Hominid in Oldowan site of Gombore-1, Melka-Kunturé, Ethiopia. *Comptes Rendus Hebdomadaires des Séances de l'Académie des Sciences, Paris, Série D* 285:961–963.

Chavaillon, J., and Y. Coppens. 1986. New discovery of *Homo erectus* in Melka Kunturé (Ethiopia). *Comptes Rendus de l'Académie des Sciences, Paris, Série II*, 303:99–104.

Chavaillon, J., F. Hours, and Y. Coppens. 1987. Discovery of hominid fossil remains in association with a late Acheulian assemblage in Melka Kunturé (Ethiopia). *Comptes Rendus de l'Académie des Sciences, Paris, Série II*, 304:539–542.

Churcher, C. S. 1972. Late Pleistocene vertebrates from archaeological sites in the plain of Kom Ombo, Upper Egypt. *Life Sciences Contributions of the Royal Ontario Museum* 82:1–172.

Churchill, S. E., L. R. Berger, and J. E. Parkington. 2000. A middle Pleistocene human tibia from Hoedjiespunt, Western Cape, South Africa. *South African Journal of Science* 96:367–368.

Churchill, S. E., O. M. Pearson, F. E. Grine, E. Trinkaus, and T. W. Holliday. 1996. Morphological affinities of the proximal ulna from Klasies River Main Site: Archaic or modern? *Journal of Human Evolution* 31:213–237.

Clark, J. D., Y. Beyene, G. WoldeGabriel, W. K. Hart, P. R. Renne, H. Gilbert, A. Defleur, G. Suwa, S. Katoh, K. R. Ludwig, J.-R. Boisserie, B. Asfaw, and T. D. White. 2003. Stratigraphic, chronological and behavioural contexts of Pleistocene *Homo sapiens* from Middle Awash, Ethiopia. *Nature* 423:747–752.

Clark, J. D., J. de Heinzelin, D. K. Schick, W. K. Hart, T. D. White, G. WoldeGabriel, R. C. Walter, G. Suwa, B. Asfaw, E. Vrba, and Y. Haile-Selassie. 1994. African *Homo erectus*: old radiometric ages and young oldowan assemblages in the Middle Awash Valley, Ethiopia. *Nature* 264:1907–1910.

Clarke, R. J. 1976. New cranium of *Homo erectus* from Lake Ndutu, Tanzania. *Nature* 262:485–487.

———. 1977. The cranium of the Swartkrans hominid, SK 847, and its relevance to human origins. Unpublished PhD dissertation, Department of Anatomy, University of the Witwatersrand, Johannesburg.

———. 1988. A new *Australopithecus* cranium from Sterkfontein and its bearing on the ancestry of *Paranthropus*, pp. 285–292 in F. E. Grine (ed.), *Evolutionary History of the "Robust" Australopithecines*. Aldine de Gruyter, New York.

———. 1990. The Ndutu cranium and the origin of *Homo sapiens*. *Journal of Human Evolution* 19:699–736.

———. 1994a. Advances in understanding the craniofacial anatomy of South African early hominids, pp. 205–222 in R. S. Corruccini and R. L. Ciochon (eds.), *Integrative Paths to the Past: Paleoanthropological Advances in Honor of F. Clark Howell*. Prentice Hall, Upper Saddle River, N.J.

———. 1994b. On some new interpretations of Sterkfontein stratigraphy. *South African Journal of Science* 90:211–214.

———. 1996. The genus *Paranthropus*: What's in a name?; pp. 93–104 in W. E. Meikle, F. C. Howell, and N. G. Jablonski (eds.), *Contemporary Issues in Human Evolution*. Memoir 21, California Academy of Sciences, San Francisco.

———. 1999. Discovery of complete arm and hand of the 3.3 million-year-old *Australopithecus* skeleton from Sterkfontein. *South African Journal of Science* 95:477–480.

———. 2002. Newly revealed information on the Sterkfontein Member 2 *Australopithecus* skeleton. *South African Journal of Science* 98:523–526.

———. 2008. Latest information on Sterkfontein's *Australopithecus* skeleton and a new look at *Australopithecus*. *South African Journal of Science* 104:443–449.

Clarke, R. J., F. C. Howell and C. K. Brain. 1970. More evidence of an advanced hominid at Swartkrans. *Nature* 225:1219–1222.

Clarke, R. J., and P. V. Tobias. 1995. Sterkfontein Member 2 foot bones of the oldest South African hominid. *Science* 269:521–524.

Coffing, K., C. Feibel, M. Leakey, and A. Walker. 1994. Four-million-year-old hominids from east Lake Turkana, Kenya. *American Journal of Physical Anthropology* 93:55–65.

Cohen, P. 1996. Fitting a face to Ngaloba. *Journal of Human Evolution* 30:373–379.

Collard, M., and B. Wood. 2007. Defining the genus *Homo*; pp. 1575–1610 in W. Henke and I. Tattersall (eds.), *Handbook of Paleoanthropology*, vol. 3. Springer, New York.

Condemi, S. 2004. Studies on Melka Kunturé. The Oldowan and the Developed Oldowan. pp. 60–78 in J. Chavaillon and M. Piperno (eds.), *Studies on the early Paleolithic Site of Melka Kunturé, Ethiopia*. Istituto Italiano di Preistoria e Protostoria, Florence.

Conroy, G. C. 1991. Enamel thickness in South African australopithecines: noninvasive evaluation by computed tomography. *Palaeontologia Africana* 28:53–59.

Conroy, G. C., D. Falk, J. Guyer, G. W. Weber, H. Seidler, W. Recheis. 2000a. Endocranial capacity in Sts 71 (*Australopithecus africanus*) by three-dimensional computed tomography. *The Anatomical Record* 258:391–396.

Conroy, G. C., C. J. Jolly, D. Cramer, and J. E. Kalb. 1978. Newly discovered fossil hominid skull from the Afar depression, Ethiopia. *Nature* 276:67–70.

Conroy, G. C., M. W. Vannier, and P. V. Tobias. 1990. Endocranial features of *Australopithecus afarensis* revealed by 2– and 3–D computed tomography. *Science* 247:838–841.

Conroy, G. C., G. W. Weber, H. Seidler, W. Recheis, D. Z. Nedden, and J. H. Mariam. 2000b. Endocranial capacity of the Bodo cranium determined from three-dimensional computed tomography. *American Journal of Physical Anthropology* 113:111–118.

Conroy, G. C., G. W. Weber, H. Seidler, P. V. Tobias, A. Kane, and B. Brunsden. 1998. Endocranial capacity in an early hominid cranium from Sterkfontein, South Africa. *Science* 280:1730–1731.

Cooke, H. B. S, B. D. Malan, and L. H. Wells. 1945. Fossil man in the Lebombo mountains, South Africa: The "Border Cave," Ingwavuma District, Zululand. *Man* 45:6–13.

Coon, C. 1939. *The Races of Europe*. Macmillan, New York; 739 pp.

Coppens, Y. 1966. An early hominid from Chad. *Current Anthropology* 7:584–585.

———. 1967. Les Faunes de Vertébrés Quaternaires du Tchad; pp. 89–97 in W. W. Bishop and J. D. Clark (eds.), *Background to Evolution in Africa*. University of Chicago Press, Chicago.

Corruccini, R. S., and P. S. Gill. 1993. Multivariate allometry of the robust australopithecine zygomatic foramen: bootstrap approach to confidence limits. *Human Evolution* 8:11–15.

Cote, Susanne. 2008. Sampling and ecology in three early Miocene catarrhine assemblages from East Africa. Unpublished PhD dissertation, Harvard University.

Crevecoeur, I. 2008. *Étude Anthropologique du Squelette du Paléolithique Supérieur de Nazlet Khater 2 (Égypte)*. Leuven University Press; 318 pp.

Crevecoeur, I., and E. Trinkaus. 2004. From the Nile to the Danube: A comparison of the Nazlet Khater 2 and Oase 1 early modern human mandibles. *Anthropologie* 42:203–213.

Crompton, R. H., Y. Li, W. Wang, M. Gunther, and R. Savage. 1998. The mechanical effectiveness of erect and "bent-knee, bent-hip" bipedal walking in *Australopithecus afarensis*. *Journal of Human Evolution* 35:55–74.

Curnoe, D. 2001. Cranial variability in East African "robust" hominins. *Human Evolution* 16:168–198.

Curnoe, D., R. Grün, L. Taylor, and F. Thackeray. 2001. Direct ESR dating of a Pliocene hominin from Swartkrans. *Journal of Human Evolution* 40:379–391.

Curnoe, D., and P. V. Tobias. 2006. Description, new reconstruction, comparative anatomy, and classification of the Sterkfontein Stw 53 cranium, with discussions about the taxonomy of other southern African early *Homo* remains. *Journal of Human Evolution* 50:36–77.

Daegling, D. J. 1989. Biomechanics of cross-sectional size and shape in the hominoid mandibular corpus. *American Journal of Physical Anthropology* 80:91–106.

Dart, R. A. 1925. *Australopithecus africanus*: The man-ape of South Africa. *Nature* 115:195–199.

———. 1948. The Makapansgat proto-human *Australopithecus prometheus*. *American Journal of Physical Anthropology* 6:259–284.

Day, M. H. 1969. Omo human skeletal remains. *Nature* 222: 1135–1138.

Day, M. H., and R. E. F. Leakey. 1974. New evidence of the genus *Homo* from East Rudolf, Kenya (III). *American Journal of Physical Anthropology* 41:367–380.

Day, M. H., M. D. Leakey, and C. Magori. 1980. A new hominid fossil skull (L.H. 18) from the Ngaloba Beds, Laetoli, northern Tanzania. *Nature* 284:55–56.

Day, M. H., and C. B. Stringer. 1982. A reconsideration of the Omo Kibish remains and the *erectus-sapiens* transition; pp. 814–846 in *1er Congrès Internationale de Paléontologie Humaine*. Prétirage, Nice.

Deacon, H. J. 1993. Southern Africa and modern human origins; pp. 104–117 in M. J. Aitken, C. B. Stringer, and P. A. Mellars (eds.), *The Origin of Modern Humans and the Impact of Chronometric Dating*. Princeton University Press, Princeton, N.J.

Deacon, H. J., and V. B. Geleijnse. 1988. The stratigraphy and sedimentology of the main site sequence, Klasies River, South Africa. *South African Archaeological Bulletin* 43:5–14.

Deacon, J., and M. Wilson. 1992. Peers Cave: The "cave the world forgot." *Digging Stick* 9:2–5.

Dean, M. C. 1985. The eruption pattern of the permanent incisors and first permanent molars in *Australopithecus (Paranthropus) robustus*. *American Journal of Physical Anthropology* 67:251–257.

———. 1987a. The dental development status of six East African juvenile fossil hominids. *Journal of Human Evolution* 16:197–213.

———. 1987b. Growth layers and incremental markings in hard tissues: A review of the literature and some preliminary observations about enamel structure in *Paranthropus boisei*. *Journal of Human Evolution* 16:157–172.

———. 1988. Growth of teeth and development of the dentition in *Paranthropus*; pp. 43–53 in F. E. Grine (ed.), *Evolutionary History of the "Robust" Australopithecines*. Aldine de Gruyter, New York.

Dean, M. C., A. D. Beyon, J. F. Thackeray, and G. A. Macho. 1993. Histological reconstruction of dental development and age at death of a juvenile *Paranthropus robustus* specimen, SK 63, from Swartkrans, South Africa. *American Journal of Physical Anthropology* 91:401–419.

Dean, C., M. G. Leakey, D. Reid, F. Schrenk, G. T. Schwartz, C. Stringer, and A. Walker. 2001. Growth processes in teeth distinguish modern humans from *Homo erectus* and earlier hominins. *Nature* 414:628–630.

Dean, M. C., and B. A. Wood. 1981. Metrical analysis of the basicranium of extant hominoids and *Australopithecus*. *American Journal of Physical Anthropology* 54:63–71.

———. 1982. Basicranial anatomy of Plio-Pleistocene hominids from East and South Africa. *American Journal of Physical Anthropology* 59:157–174.

Debénath, A. 1975. Découverte de restes humains probablement atériens à Dar es Soltane (Maroc). *Comptes Rendus de l'Académie des Sciences, Paris, Série D* 281:875–876.

———. 1980. New human Aterian remains from Morocco. *Comptes Rendus Hebdomadaires des Séances de l'Académie des Sciences, Paris, Série D* 290:851–852.

———. 1991. Les atériens du Maghreb. *Les Dossiers d'Archéologie* 161:52–57.

Debénath, A., J. P. Raynal, J. Roche, and J. P. Texier. 1986. Position, habitat, typologie et devenir de l'atérien marocain: données récentes. *Anthropologie* 90:233–246.

Debénath, A., J. P. Raynal, and J. P. Texier. 1982. Position stratigraphique des restes humains paléolithiques marocains sur la base des travaux recents. *Comptes Rendus de l'Académie des Sciences, Paris* 294:1247–1250.

de Bonis, L., D. Geraads, J.-J. Jaeger, and S. Sen. 1988. Vertébrés du Pléistocène de Djibouti. *Bulletin de la Societe Géologique de France* 4:323–334.

DeGusta, D. 2004. Pliocene hominid postcranial fossils from the Middle Awash, Ethiopia. Unpublished PhD dissertation, University of California, Berkeley, 593 pp.

Deino, A. L. In press. ^{40}Ar/^{39}Ar dating of Laetoli, Tanzania; in T. Harrison (ed.), *Paleontology and Geology of Laetoli: Human Evolution in Context: Volume 1. Geology, Geochronology, Paleoecology and Paleoenvironment.* Springer, Dordrecht.

Deino, A. L., and S. McBrearty. 2002. ^{40}Ar/^{39}Ar dating of the Kapthurin Formation, Baringo, Kenya. *Journal of Human Evolution* 42:185–210.

Deino, A. L., L. Tauxe, M. Monaghan, and A. Hill. 2002. ^{40}Ar/^{39}Ar geochronology and paleomagnetic stratigraphy of the Lukeino and lower Chemeron Formations at Tabarin and Kapcheberek, Tugen Hills, Kenya. *Journal of Human Evolution* 42:117–140.

Deloison, Y. 2003. Anatomie des os fossiles des pieds des hominides d'Afrique du sud dátes entre 2,4 et 3,5 millions d'annés. Interprétation quant à leur mode de locomotion. *Biométrie Humaine et Anthropologie* 21:189–230.

Delson, E. 1988. Chronology of South African australopith site units; pp. 317–325 in F. E. Grine (ed.), *Evolutionary History of the "Robust" Australopithecines.* Aldine de Gruyter, New York.

Demes, B., and N. Creel. 1988. Bite force, diet, and cranial morphology of fossil hominids. *Journal of Human Evolution* 17:657–670.

de Ruiter, D. J., R. Pickering, C. M. Steininger, J. D. Kramers, P. J. Hancox, S. E. Churchill, L. R. Berger, and L. Backwell. 2009. New *Australopithecus robustus* fossils and associated U-Pb dates from Cooper's Cave (Gauteng, South Africa). *Journal of Human Evolution* 56:497–513.

DeSilva, J. M. 2008. Vertical climbing adaptations in the anthropoid ankle and midfoot: implications for locomotion in Miocene catarrhines and Plio-Pleistocene hominins. Unpublished PhD dissertation, University of Michigan, Ann Arbor, 363 pp.

———. 2009. Functional morphology of the ankle and the likelihood of climbing in early hominins. *Proceedings of the National Academy of Sciences, USA* 106:6567–6572.

DeSilva, J. M., and L. M. MacLatchy. 2008. Revisiting the midtarsal break. *American Journal of Physical Anthropology* (suppl.) 135(S46):89.

DeSilva, J., E. Shoreman and L. MacLatchy. 2006. A fossil hominoid proximal femur from Kikorongo Crater, Southwestern Uganda. *Journal of Human Evolution* 50:687–695.

Domínguez-Rodrigo, M., A. Mabulla, L. Luque, J. W. Thompson, J. Rink, P. Bushozi, R. Díez-Martin, and L. Alcala. 2008. A new archaic *Homo sapiens* fossil from Lake Eyasi, Tanzania. *Journal of Human Evolution* 54:899–903.

Drake, R., and G. H. Curtis. 1987. K-Ar geochronology of the Laetoli fossil localities; pp. 48–52 in M. D. Leakey and J. M. Harris (eds.), *Laetoli: A Pliocene Site in Northern Tanzania.* Clarendon Press, Oxford.

Drennan, M. R. 1935. The Florisbad skull. *South African Journal of Science* 32:601–602.

———. 1955. The special features and status of the Saldanha skull. *American Journal of Physical Anthropology* 13:625–634.

Dreyer, T. 1935. A human skull from Florisbad. *Proceedings of the Academy of Sciences, Amsterdam* 38:119–128.

Dubois, E. 1893. Palaeontologische onderzoekingen op Java. *Verslag van het Mijnwesen* 3:10–14.

———. 1894. Pithecanthropus erectus, *eine Menschenaehnliche Übergangsform aus Java.* Landesdruckerei, Batavia.

———. 1936. Racial identity of *Homo soloensis* Oppenoorth (including *Homo modjokertensis*, von Koenigswald) and *Sinanthropus pekinensis*, Davidson Black. *Proceedings of the Academy of Sciences, Amsterdam* 39:1180–1185.

Dunsworth, H., and A. Walker. 2002. Early genus *Homo*; pp. 419–435 in W. C. Hartwig (ed.), *The Primate Fossil Record.* Cambridge University Press, Cambridge.

Egeland, C. P., M. Domínguez-Rodrigo, and R. Barba. 2007. Geological and paleoecological overview of Olduvai Gorge; pp. 33–38 in M. Domínguez-Rodrigo, R. Barba, and C. P. Egeland (eds.), *Deconstructing Olduvai: A Taphonomic Study of the Bed I Sites.* Springer, New York.

Ennouchi, E. 1962. Un crâne d'homme ancien au Jebel Irhoud (Maroc). *Comptes Rendus de l'Académie des Sciences, Paris* 254:4330–4332.

———. 1963. Les Néanderthaliens du Jebel Irhoud (Maroc). *Comptes Rendus de l'Académie des Sciences, Paris* 256:2459–2460.

———. 1969. Présence d'un enfant néanderthalien au Jebel Irhoud (Maroc). *Annales de Paléontologie (Vert.-Invert.)* 55:251–255.

Fagan, B. M., and F. L. Van Noten. 1971. *The Hunter-Gatherers of Gwisho.* Musée Royal de l'Afrique Centrale, Tervuren, 228 pp.

Falk, D. 1985. Hadar AL 162–28 endocast as evidence that brain enlargement preceded cortical reorganization in hominid evolution. *Nature* 313:45–47.

———. 1986. Evolution of cranial blood drainage in hominids: Enlarged occipital/marginal sinuses and emissary foramina. *American Journal of Physical Anthropology* 70:311–324.

———. 1987. Hominid paleoneurology. *Annual Review of Anthropology* 16:13–30.

———. 1988. Enlarged occipital/marginal sinuses and emissary foramina: Their significance in hominid evolution; pp. 85–96 in F. E. Grine (ed.), *Evolutionary History of the "Robust" Australopithecines.* Aldine de Gruyter, Hawthorne, New York.

Falk, D., and R. J. Clarke. 2007. Brief communication: new reconstruction of the Taung endocast. *American Journal of Physical Anthropology* 134:529–534.

Falk, D. and G. C. Conroy. 1983. The cranial venous sinus system in *Australopithecus afarensis. Nature* 306:779–781.

Falk, D., J. C. Redmond Jr., J. Guyer, G. C. Conroy, W. Recheis, G. W. Weber, and H. Seidler. 2000. Early hominid brain evolution: A new look at old endocasts. *Journal of Human Evolution* 38:695–717.

Feathers, J. K. 2002. Luminescence dating in less than ideal conditions: case studies from Klasies River Mouth and Duinefontein, South Africa. *Journal of Archaeological Science* 29:177–194.

Feibel, C. S. 2008. Microstratigraphy of the Kibish hominin sites KHS and PHS, Lower Omo Valley, Ethiopia. *Journal of Human Evolution* 55:404–408.

Feibel, C. S., F. H. Brown, and I. McDougall. 1989. Stratigraphic context of fossil hominids from the Omo Group deposits: northern Turkana Basin, Kenya and Ethiopia. *American Journal of Physical Anthropology* 78:595–622.

Feibel, C. S., J. M. Harris, and F. H. Brown. 1991. Paleoenvironmental context for the Late Neogene of the Turkana Basin; pp. 321–370 in J. M. Harris (ed.), *Koobi Fora Research Project*, vol. 3. Clarendon Press, Oxford.

Feibel, C., C. Lepre, and R. Quinn. 2007. Integrated stratigraphic approaches: Evolving perspectives on time, facies, and paleoenvironmental systems in the Plio-Pleistocene of the Turkana Basin. *In East Africa Paleoanthropology Society Annual Meeting, Abstracts.* http://www.paleoanthro.org/meeting.htm.

Feldesman, M. R., J. G. Kleckner, and J. K. Lundy. 1990. Femur/stature ratio and estimates of stature in mid- and late-Pleistocene fossil hominids. *American Journal of Physical Anthropology* 83:359–372.

Feldesman, M. R., and J. K. Lundy. 1988. Stature estimates for some African Plio-Pleistocene fossil hominids. *Journal of Human Evolution* 17:583–596.

Ferembach, D. 1962. La nécropole épipaléolithique de Taforalt: Étude des squelettes humains. Centre National pour la Recherche Scientifique, Rabat, Appendix:123.

———. 1965. Diagrammes crâniens sagittaux, et mensurations individuelles des squelletes Iberomaurusiens de Taforalt (Maroc Oriental). *Travaux du Centre de Recherches Anthropologiques, Préhistoriques, et Ethnographiques, Alger.* Arts et Métiers Graphiques, Paris, 124 pp.

———. 1976a. Les restes humains de la grotte de Dar es Soltane 2 (Maroc): Campagne 1975. *Bulletins et Mémoires de la Société d'Anthropologie de Paris* 3:183–193.

———. 1976b. Les restes humains de Temara (Campagne 1975). *Bulletins et Mémoires de la Société d'Anthropologie de Paris* 3:175–180.

Ferguson, W. W. 1989. A new species of the genus *Australopithecus* (Primates: Hominidae) from Plio/Pleistocene deposits west of Lake Turkana in Kenya. *Primates* 30:223–232.

Fleagle, J. G., Z. Assefa, F. H. Brown, and J. J. Shea. 2008. Paleoanthropology of the Kibish Formation, southern Ethiopia: Introduction. *Journal of Human Evolution* 55:360–365.

Fleagle, J. G., D. T. Rasmussen, S. Yirga, T. M. Bown, and F. E. Grine. 1991. Current events: New hominid fossils from Fejej, Southern Ethiopia. *Journal of Human Evolution* 21:145–152.

Foley, R. A., and M. M. Lahr. 1992. Beyond "out of Africa": reassessing the origins of *Homo sapiens*. *Journal of Human Evolution* 22:523–529.

Franciscus, R. G., and E. Trinkaus. 1988. Nasal morphology and the emergence of *Homo erectus*. *American Journal of Physical Anthropology* 75:517–527.

Gabel, C. 1962. Human crania from the Late Stone Age of the Kafue Basin, northern Rhodesia. *South African Journal of Science* 58:307–314.

———. 1963. Further human remains from the Central African Later Stone Age. *South African Archaeological Bulletin* 18:40–48.

Gabunia, L., M.-A. de Lumley, A. Vekua, D. Lordkipanidze, and H. de Lumley. 2002. Découvert d'un nouvel hominidé à Dmanissi (Transcaucasie, Georgie). *Comptes Rendus Palevol* 1:243–253.

Gabunia, L., and A. Vekua. 1995. A Plio-Pleistocene hominid from Dmanisi, East Georgia, Caucasus. *Nature* 373:509–512.

Gabunia, L., A. Vekua, D. Lordkipanidze, C. C. Swisher III, R. Ferring, A. Justus, M. Nioradze, M. Tvalchrelidze, S. C. Antón, G. Bosinski, O. Jöris, M-A. de Lumley, G. Majsuradze, and A. Mouskhelishvili. 2000. Earliest Pleistocene hominid cranial remains from Dmanisi, Republic of Georgia: taxonomy, geological setting, and age. *Science* 288:1019–1025.

Galik, K., B. Senut, M. Pickford, D. Gommery, J. Treil, A. J. Kuperavage, and R. B. Eckhardt. 2004. External and internal morphology of the BAR 1002'00 *Orrorin tugenensis* femur. *Science* 305:1450–1453.

Galloway, A. 1937a. The characteristics of the skull of the Boskop physical type. *American Journal of Physical Anthropology* 23:31–47.

———. 1937b. Man in Africa in the light of recent discoveries. *South African Journal of Science* 34:89–120.

Garrod, D., and D. Bate. 1937. *The Stone Age of Mount Carmel, Volume I: Excavations at the Wadi el-Mughara*. Clarendon Press, Oxford, 240 pp.

Gathogo, P. N., and F. H. Brown. 2006. Revised stratigraphy of Area 123, Koobi Fora, Kenya, and new age estimates of its fossil mammals, including hominins. *Journal of Human Evolution* 51:471–479.

Gebo, D. and G. T. Schwartz. 2006. Foot Bones from Omo: implications for hominid evolution. *American Journal of Physical Anthropology* 129:499–511.

Geraads, D., F. Amani, and J.-J. Hublin. 1992. Le gisement pléistocène moyen de l'Aïn Maarouf près de El Hajeb, Maroc: présence d'un hominidé. *Comptes Rendus de l'Académie des Sciences, Paris, Série II*, 314:319–323.

Geraads, D., J.-J. Hublin, J.-J. Jaeger, S. Sen, H. Tong, and P. Toubeau. 1986. The Pleistocene hominid site of Ternifine, Algeria: New results on the environment, age and human industries. *Quaternary Research* 25:380–386.

Gilbert, W. H., and B. Asfaw. 2008. *Homo erectus: Pleistocene Evidence from the Middle Awash, Ethiopia*. University of California Press, Berkeley, 480 pp.

Gowlett, J. A. J., J. W. K. Harris, D. Walton, and B. Wood. 1981. Early archaeological sites, hominid remains and traces of fire from Chesowanja, Kenya. *Nature* 294:125–129.

Grausz, H. M., R. E. Leakey, A. C. Walker, and C. V. Ward. 1988. Associated cranial and postcranial bones of *Australopithecus boisei*; pp. 127–132 in F. E. Grine (ed.), *Evolutionary History of the "Robust" Australopithecines*. Aldine de Gruyter, New York.

Green, D. J., and A. D. Gordon. 2008. Metacarpal proportions in *Australopithecus africanus*. *Journal of Human Evolution* 54:705–719.

Green, D. J., A. D. Gordon, and B. G. Richmond. 2007. Limb-size proportions in *Australopithecus afarensis* and *Australopithecus africanus*. *Journal of Human Evolution* 52:187–200.

Green, R. E., J. Krause, S. E. Ptak, A. W. Briggs, M. T. Ronan, J. F. Simons, L. Du, M. Egholm, J. M. Rothberg, M. Paunovic, and S. Pääbo, S. 2006. Analysis of one million base pairs of Neanderthal DNA. *Nature* 444:330–336.

Greene, D. L., and G. J. Armelagos. 1972. *Mesolithic Populations from Wadi Halfa*. Department of Anthropology Research Reports No. 1, University of Massachusetts, Amherst, 136 pp.

Grine, F. E. 1981. Trophic differences between "gracile" and "robust" australopithecines; a scanning electron microscope analysis of occlusal events. *South African Journal of Science* 77:828–835.

———. 1982. A new juvenile hominid (Mammalia: Primates) from Member 3, Kromdraai Formation, Transvaal, South Africa. *Annals of the Transvaal Museum* 33:165–239.

———. 1986. Dental evidence for dietary differences in *Australopithecus* and *Paranthropus*: A quantitative analysis of permanent molar microwear. *Journal of Human Evolution* 15:783–822.

———. 1987. On the eruption pattern of the permanent incisors and first permanent molars in *Paranthropus*. *American Journal of Physical Anthropology* 72:353–359.

———. 1988. Evolutionary history of the "robust" australopithecines: A summary and historical perspective; pp. 509–520 in F. E. Grine (ed.), *Evolutionary History of the "Robust" Australopithecines*. Aldine de Gruyter, Hawthorne, New York.

———. 1989. New hominid fossils from the Swartkrans Formation (1979–1986 excavations): craniodental specimens. *American Journal of Physical Anthropology* 79:409–450.

———. 2000. Middle Stone Age human fossils from Die Kelders Cave 1, Western Cape Province, South Africa. *Journal of Human Evolution* 38:129–145.

Grine, F. E., R. M. Bailey, K. Harvati, R. P. Nathan, A. G. Morris, G. M. Henderson, I. Ribot, and A. W. G. Pike. 2007. Late Pleistocene human skull from Hofmeyr, South Africa, and modern human origins. *Science* 315:226–229.

Grine, F. E., and D. J. Daegling. 1993. New mandible of *Paranthropus robustus* from Member 1, Swartkrans Formation, South Africa. *Journal of Human Evolution* 24:319–333.

Grine, F. E., B. Demes, W. L. Jungers, and T. M. Cole III. 1993. Taxonomic affinity of the early Homo cranium from Swartkrans, South Africa. *American Journal of Physical Anthropology* 92:411–426.

Grine, F. E., W. L. Jungers, and J. Schultz. 1996. Phenetic affinities among early *Homo* crania from East and South Africa. *Journal of Human Evolution* 30:189–225.

Grine, F. E., W. L. Jungers, P. V. Tobias, and O. M. Pearson. 1995. Fossil *Homo* femur from Berg Aukas, Northern Namibia. *American Journal of Physical Anthropology* 97:151–185.

Grine, F. E., and R. F. Kay. 1988. Early hominid diets from quantitative image analysis of dental microwear. *Nature* 333:768–770.

Grine, F. E., R. G. Klein, and T. P. Volman. 1991. Dating, archaeology and human fossils from the Middle Stone Age levels of Die Kelders, South Africa. *Journal of Human Evolution* 21:363–395.

Grine, F. E., and L. B. Martin. 1988. Enamel thickness and development in *Australopithecus* and *Paranthropus*; pp. 3–42 in F. E. Grine (ed.), *Evolutionary History of the "Robust" Australopithecines*. Aldine de Gruyter, New York.

Grine, F. E., and D. S. Strait. 1994. New hominid fossils from Member 1 "Hanging Remnant," Swartkrans Formation, South Africa. *Journal of Human Evolution* 26:57–75.

Grine, F. E., and R. L. Susman. 1991. Radius of *Paranthropus robustus* from Member 1, Swartkrans Formation, South Africa. *American Journal of Physical Anthropology* 84:229–248.

Grine, F. E., O. M. Pearson, R. G. Klein, and G. P. Rightmire. 1998. Additional human fossils from Klasies River Mouth, South Africa. *Journal of Human Evolution* 35:95–107.

Grine, F. E., P. S. Ungar, M. F. Teaford, and S. El-Zaatari. 2006. Molar microwear in *Praeanthropus afarensis*: Evidence for dietary stasis through time and under diverse paleoecological conditions. *Journal of Human Evolution* 51:297–319.

Groves, C. P. 1989. *A Theory of Human and Primate Evolution*. Clarendon Press, Oxford, 375 pp.

———. 1999. Nomenclature of African Plio-Pleistocene hominins. *Journal of Human Evolution* 37:869–872.

Groves, C. P., and V. Mazák. 1975. An approach to the taxonomy of the Hominidae: gracile Villafranchian hominids of Africa. *Casopis pro Mineralogii Geologii* 20:225–247.

Grün, R., and P. Beaumont. 2001. Border Cave revisited: A revised ESR chronology. *Journal of Human Evolution* 40:467–482.

Grün, R., P. Beaumont, P. V. Tobias, and S. Eggins. 2003. On the age of Border Cave 5 human mandible. *Journal of Human Evolution* 45:155–167.

Grün R., J. S. Brink, N. A. Spooner, L. Taylor, C. B. Stringer, R. G. Franciscus, A. S. Murray. 1996. Direct dating of Florisbad hominid. *Nature* 382:500–501.

Grün, R., N. J. Shackleton, and H. Deacon. 1990. Electron-spin-resonance dating of tooth enamel from Klasies River Mouth Cave. *Current Anthropology* 31:427–432.

Grün, R., and C. Stringer. 1991. Electron spin resonance dating and the evolution of modern humans. *Archaeometry* 33:153–199.

Grün, R., C. Stringer, F. McDermott, R. Nathan, N. Porat, S. Robertson, L. Taylor, G. Mortimer, S. Eggins, and M. McCulloch. 2005. U-series and ESR analyses of bones and teeth relating to the human burials from Skhul. *Journal of Human Evolution* 49:316–334.

Guy, F., D. E. Lieberman, D. Pilbeam, M. Ponce de León, A. Likius, H. T. Mackaye, P. Vignaud, C. Zollikofer, and M. Brunet. 2005. Morphological affinities of the *Sahelanthropus tchadensis* (Late Miocene hominid from Chad) cranium. *Proceedings of the National Academy of Sciences, USA* 102:18836–18841.

Habgood, P. J. 1989. An examination of regional features on middle and early late Pleistocene sub-Saharan African hominids. *South African Archaeological Bulletin* 44:17–22.

Häusler, M. 2002. New insights into the locomotion of *Australopithecus africanus* based on the pelvis. *Evolutionary Anthropology* S1:53–57.

Häusler, M., and L. R. Berger. 2001. StW 441/465: a new fragmentary ilium of a small-bodied *Australopithecus africanus* from Sterkfontein, South Africa. *Journal of Human Evolution* 40:411–417.

Haeusler, M., and H. M. McHenry. 2004. Body proportions of *Homo habilis* reviewed. *Journal of Human Evolution* 46:433–465.

Haile-Selassie, Y. 2001. Late Miocene hominids from the Middle Awash, Ethiopia. *Nature* 412:178–181.

Haile-Selassie, Y., B. Asfaw, and T. D. White. 2004a. Hominid cranial remains from Upper Pleistocene deposits at Aduma, Middle Awash, Ethiopia. *American Journal of Physical Anthropology* 123:1–10.

Haile-Selassie, Y., G. Suwa, and T. D. White. 2004b. Late Miocene teeth from Middle Awash, Ethiopia, and early hominid dental evolution. *Science* 303:1503–1505.

———. 2009. Hominidae; pp. 159–236 in Y. Haile-Selassie and G. WoldeGabriel (eds.), *Ardipithecus kadabba: Late Miocene Evidence from the Middle Awash, Ethiopia.* University of California Press, Berkeley.

Hamrick, M. W., and Inouye, S. E. 1995. Thumbs, tools, and early humans. *Science* 268:586–587.

Harcourt-Smith, W. E. H. 2002. Form and function in the hominoid tarsal skeleton. Unpublished PhD dissertation, University College London.

Harcourt-Smith, W. E. H., and L. Aiello. 2004. Fossils, feet and the evolution of human bipedal locomotion. *Journal of Anatomy* 204:403–416.

Harris, J. M. 1987. Summary; pp. 524–532 in M. D. Leakey and J. M. Harris (eds.), *Laetoli: A Pliocene Site in Northern Tanzania.* Clarendon Press, Oxford.

———. 1991. *Koobi Fora Research Project: Volume 3. The Fossil Ungulates: Geology, Fossil Artiodactyls, and Paleoenvironments.* Clarendon Press, Oxford.

Harris, J. M., F. H. Brown, and M. G. Leakey. 1988. Stratigraphy and paleontology of Pliocene and Pleistocene localities west of Lake Turkana, Kenya. *Contributions in Science, Natural History Museum of Los Angeles County* 399:1–128.

Harrison, T. 2002. The first record of fossil hominins from the Ndolanya Beds, Laetoli, Tanzania. *American Journal of Physical Anthropology* 32(suppl.):83.

Harvati, K. 2003. Quantitative analysis of Neanderthal temporal bone morphology using three-dimensional geometric morphometrics. *American Journal of Physical Anthropology* 120:323–338.

Haughton, S. H. 1917. Preliminary note on the ancient human skull remains from the Transvaal. *Transactions of the Royal Society of South Africa* 6:1–14.

Hay, R. L. 1976. *Geology of the Olduvai Gorge: A Study of Sedimentation in a Semiarid Basin.* University of California Press, Berkeley, 203 pp.

———. 1987. Geology of the Laetoli area; pp. 23–47 in M. Leakey and J. Harris (eds.), Results of the Laetoli Expeditions 1975–1981. Oxford University Press, Oxford.

Heberer, G. 1963. Über einen neuen neuen archanthropinin Typus aus der Oldoway Schlucht. *Zeitschrift für Morphologie und Anthropologie* 53:171–177.

Heinzelin, J. de, J. D. Clark, T. White, W. Hart, P. Renne, G. WoldeGabriel, Y. Beyene, and E. Vrba. 1999. Environment and behavior of 2.5–million-year-old Bouri hominids. *Science* 284: 625–635.

Hennig, E. 1948. Quartärfaunen und Urgeschichte Ostafrikas, *Naturwissenschaftliche Rundschau.* 1(5):212–217.

Herries, A. I. R. 2003. Magnetostratigraphy of the South African hominid palaeocaves. *American Journal of Physical Anthropology* 36(suppl.):113.

Herries, A. I. R., J. W. Adams, K. L. Kuykendall, and J. Shaw. 2006. Speleology and magnetobiostratigraphic chronology of the GD 2 locality of the Gondolin hominin-bearing paleocave deposits, North West Province, South Africa. *Journal of Human Evolution* 51:617–631.

Hill, A. 1995. Faunal and environmental change in the Neogene of East Africa: evidence from the Tugen Hills Sequence, Baringo District, Kenya; pp. 178–193 in E. S. Vrba, G. H. Denton, T. C. Partridge, and L. H. Burkle (eds.), *Paleoclimate and Evolution, with Emphasis on Human Origins.* Yale University Press, New Haven.

———. 1999. The Baringo Basin, Kenya: from Bill Bishop to BPRP; pp. 85–97 in P. Andrews and P. Banham (eds.), *Late Cenozoic Environments and Hominid Evolution: A tribute to Bill Bishop.* Geological Society, London.

———. 2002. Paleoanthropological research in the Tugen Hills, Kenya: Introduction. *Journal of Human Evolution* 42:1–10.

Hill, A., G. Curtis, and R. Drake. 1986. Sedimentary stratigraphy of the Tugen Hills, Baringo District, Kenya. pp. 285–295 in L. E. Frostick, R. W. Renaut, I. Reid, and J.-J. Tiercelin (eds.), *Sedimentation in the African Rifts.* Geological Society of London Special Publication 25. Blackwell, Oxford.

Hill, A., R. Drake, L. Tauxe, M. Monaghan, J. C. Barry, A. K. Behrensmeyer, G. Curtis, B. Fine Jacobs, L. Jacobs, N. Johnson, and D. Pilbeam. 1985. Neogene palaeontology and geochronology of the Baringo Basin, Kenya. *Journal of Human Evolution* 14:759–773.

Hill, A., and S. Ward. 1988. Origin of the Hominidae: The record of African large hominoid evolution between 14 my and 4 my. *Yearbook of Physical Anthropology* 31:49–83.

Hill, A., S. Ward, A. Deino, G. Curtis, and R. Drake. 1992. Earliest *Homo*. *Nature* 355:719–722.

Hoberg, E. P., N. L. Alkire, A. De Queiroz, and A. Jones. 2001. Out of Africa: origins of the *Taenia* tapeworms in humans. *Proceedings of the Royal Society of London, B* 268:781–787.

Holliday, T. W. 1997. Body proportions in late Pleistocene Europe and modern human origins. *Journal of Human Evolution* 32:423–447.

———. 1998. Brachial and crural indices of European late Upper Paleolithic and Mesolithic humans. *Journal of Human Evolution* 36:549–566

———. 2000. Evolution at the crossroads: Modern human emergence in western Asia. *American Anthropologist* 102:54–68.

Holliday, T. W., and E. Trinkaus. 1991. Limb-trunk proportions in Neandertals and early anatomically modern humans. *American Journal of Physical Anthropology* (suppl.) 12:93–94.

Holloway, R. L. 1965. Cranial capacity of the hominine from Olduvai Bed I. *Nature* 208:205–206.

———. 1981. The endocast of the Omo L338y-6 juvenile hominid: Gracile or robust *Australopithecus*? *American Journal of Physical Anthropology* 54:109–118.

———. 1983. Human paleontological evidence relevant to language behavior. *Human Neurobiology* 2:105–114.

———. 2000. Brain; pp. 141–149 in E. Delson, I. Tattersall, and J. Van Couvering (eds.), *Encyclopedia of Human Evolution and Prehistory.* Garland Publishing, New York.

Holloway, R. L., D. C. Broadfield, M. S. Yuan, J. H. Schwartz, and I. Tattersall. 2004. *The Human Fossil Record: Volume 3. Brain Endocasts: The Paleoneurological Evidence.* Wiley-Liss, New York. 315 pp.

Hooker, P. J., and J. A. Miller. 1979. K-Ar dating of the Pleistocene hominid site at Chesowanja, North Kenya. *Nature* 282:710–712.

Howell, F. C. 1960. European and Northwest African Middle Pleistocene hominids. *Current Anthropology* 1:195–232.

———. 1976. Overview of the Pliocene and earlier Pleistocene of the Lower Omo Basin, southern Ethiopia; pp. 227–268 in G. Ll. Isaac and E. R. McCown (eds.), *Human Origins: Louis Leakey and the East African Evidence.* Benjamin, Menlo Park, Calif.

———. 1978. Hominidae; pp. 154–248 in V. J. Maglio and H. B. S. Cooke (eds.), *Evolution of African Mammals.* Harvard University Press, Cambridge.

Howell, F. C., and Y. Coppens. 1976. An overview of Hominidae from the Omo Succession, Ethiopia; pp. 522–532 in Y. Coppens, F. C. Howell, G. L. Isaac, and R. E. F. Leakey (eds.), *Earliest Man and Environments in the Lake Rudolf Basin.* University of Chicago Press, Chicago.

Howells, W. 1970. Mount Carmel Man: Morphological relationships. *Proceedings of the VIIIth Congress of Anthropological and Ethnological Sciences, Tokyo and Kyoto 1968* 1:269–272. Science Council of Japan, Tokyo.

Hublin, J.-J. 1985. Human Fossils of the North African Middle Pleistocene and the origin of *Homo sapiens*; pp. 283–288 in E. Delson (ed.), *Ancestors: The Hard Evidence*. Liss, New York.

———. 1991. L'émergence des *Homo sapiens* archaiques: Afrique du Nord-Ouest et Europe occidentale. Unpublished thesis, Université de Bordeaux.

———. 1992. Recent human evolution in northwestern Africa. *Philosophical Transactions of the Royal Society London, B* 227:185–191.

———. 2001. Northwestern African middle Pleistocene hominids and their bearing on the emergence of *Homo sapiens*; pp. 99–121 in L. Barham and K. Robson-Brown (eds.), *Human Roots. Africa and Asia in the Middle Pleistocene*. Western Academic & Specialist Press Limited, Bristol.

Hublin, J.-J., and A. M. Tillier. 1981. The Mousterian juvenile mandible from Irhoud (Morocco): A phylogenetic reinterpretation; pp. 167–185 in C. B. Stringer (ed.), *Aspects of Human Evolution*. Taylor and Francis, London.

Hublin, J.-J., A. M. Tillier, and J. Tixier. 1987. L'humérus d'enfant moustérien (Homo 4) de Jebel Irhoud (Maroc) dans son contexte archéologique. *Bulletins et Mémoires de la Société d'Anthropologie de Paris* 4:115–142.

Hughes, A. 1990. The Tuinplaas human skeleton from the Springbok Flats, Transvaal; pp. 197–214 in G. H. Sperber (ed.), *From Apes to Angels: Essays in Honour of Philip V. Tobias*. Wiley-Liss, New York.

Hylander, W. L. 1988. Implications of *in vivo* experiments for interpreting the functional significance of "robust" australopithecine jaws; pp. 55–83 in F. E. Grine (ed.), *Evolutionary History of the "Robust" Australopithecines*. Aldine de Gruyter, New York.

Ingman, M., H. Kaessmann, S. Pääbo, and U. Gyllensten. 2000. Mitochrondrial genome variation and the origin of modern humans. *Nature* 408:708–713.

Isaac, G. L. 1967. The stratigraphy of the Peninj Group: Early middle Pleistocene formations west of Lake Natron, Tanzania; pp. 229–258 in W. W. Bishop and J. D. Clark (ed.), *Background to Evolution in Africa*. University of Chicago Press, Chicago.

Jablonski, N. G., and G. Chaplin. 2000. The evolution of human skin coloration. *Journal of Human Evolution* 39:57–106.

Jellema, L. M., B. Latimer, and A. Walker. 1993. The rib cage, pp. 294–325 in A. Walker and R. Leakey (eds.), *The Nariokotome* Homo erectus *Skeleton*. Harvard University Press, Cambridge.

Johanson, D. C., and Y. Coppens. 1976. A preliminary anatomical diagnosis of the first Plio/Pleistocene hominid discoveries in the Central Afar, Ethiopia. *American Journal of Physical Anthropology* 45:217–234.

Johanson, D. C., C. O. Lovejoy, W. H. Kimbel, T. D. White, S. C. Ward, M. E. Bush, B. M. Latimer, and Y. Coppens. 1982. Morphology of the Pliocene partial hominid skeleton (A.L. 288-1) from the Hadar Formation, Ethiopia. *American Journal of Physical Anthropology* 57:403–451.

Johanson, D. C., and T. D. White. 1979. A systematic assessment of early African hominids. *Science* 203:321–330.

Johanson, D. C., T. D. White, and Y. Coppens. 1978. A new species of the genus *Australopithecus* (Primates: Hominidae) from the Pliocene of eastern Africa. *Kirtlandia* 28:2–14.

Jones, T. R. 1940. Human skeletal remains from the Mumbwa Cave, northern Rhodesia. *South African Journal of Science* 37:313–319.

Jungers, W. L. 1982. Lucy's limbs: skeletal allometry and locomotion in *Australopithecus afarensis*. *Nature* 297:676–678.

———. 1988. New estimates of body size in australopithecines; pp. 115–125 in F. E. Grine (ed.), *Evolutionary History of the "Robust" Australopithecines*. Aldine de Gruyter, New York.

Jungers, W. L., and J. T. Stern Jr. 1983. Body proportions, skeletal allometry and locomotion in the Hadar hominids: a reply to Wolpoff. *Journal of Human Evolution* 12:673–684.

Kalb, J. E., C. J. Jolly, S. Tebedge, A. Mebrate, C. Smart, E. B. Oswald, P. F. Whitehead, C. B. Wood, T. Adefris, and V. Rawn-Schatzinger. 1982. Vertebrate faunas from the Awash Group, Middle Awash Valley, Afar, Ethiopia. *Journal of Vertebrate Paleontology* 2:237–258.

Kappelman, J. 1996. The evolution of body mass and relative brain size in fossil hominids. *Journal of Human Evolution* 30:243–276.

Kappelman, J., and J. G. Fleagle. 1995. Age of early hominids. *Nature* 376:558–559.

Kappelman, J., C. C. Swisher III, J. G. Fleagle, S. Yirga, T. M. Bown, and M. Feseha. 1996. Age of *Australopithecus afarensis* from Fejej, Ethiopia. *Journal of Human Evolution* 30:139–146.

Kay, R. F., and F. E. Grine. 1988. Tooth morphology, wear and diet in *Australopithecus* and *Paranthropus* from southern Africa; pp. 427–447 in F. E. Grine (ed.), *Evolutionary History of the "Robust" Australopithecines*. Aldine de Gruyter, New York.

Keith, A. 1931. *New Discoveries Relating to the Antiquity of Man*. Norton, New York., 512 pp.

Kennedy, G. E. 1983. A morphometric and taxonomic assessment of a hominine femur from the lower member, Koobi Fora, Lake Turkana. *American Journal of Physical Anthropology* 61:429–436.

Keyser, A. W. 2000. The Drimolen skull: the most complete australopithecine cranium and mandible to date. *South African Journal of Science* 96:189–193.

Keyser, A. W., C. G. Menter, J. Moggi-Cecchi, T. R. Pickering, and L. R. Berger. 2000. Drimolen: a new hominid-bearing site in Gauteng, South Africa. *South African Journal of Science* 96:193–197.

Kibii, J. M., and R. J. Clarke. 2003. A reconstruction of the StW 431 *Australopithecus* pelvis based on newly discovered fragments. *South African Journal of Science* 99:225–226.

Kidd, R., and C. Oxnard. 2005. Little foot and big thoughts: A re-evaluation of the Stw573 foot from Sterkfontein, South Africa. *Homo* 55:189–212.

Kimbel, W. H. 2007. The species and diversity of australopiths; pp. 1539–1573 in W. Henke and I. Tattersall (eds.), *Handbook of Paleoanthropology*, vol. 3. Springer, New York.

Kimbel, W. H., D. C. Johanson, and Y. Rak. 1994. The first skull and other new discoveries of *Australopithecus afarensis* at Hadar, Ethiopia. *Nature* 368:449–451.

———. 1997. Systematic assessment of a maxilla of *Homo* from Hadar, Ethiopia. *American Journal of Physical Anthropology* 103:235–262.

Kimbel, W. H., C. A. Lockwood, C. V. Ward, M. G. Leakey, Y. Rak, and D. C. Johanson. 2006. Was *Australopithecus anamensis* ancestral to *A. afarensis*? A case of anagenesis in the hominin fossil record. *Journal of Human Evolution* 51:134–152.

Kimbel, W. H., and L. K. Delezene. 2009. "Lucy" redux: A review of research on *Australopithecus afarensis*. *Yearbook of Physical Anthropology* 52:2–48.

Kimbel, W. H., and Y. Rak. 1993. The importance of species taxa in paleoanthropology and an argument for the phylogenetic concept of the species category; pp. 461–484 in W. H. Kimbel, and L. B. Martin (eds.), *Species, Species Concepts and Primate Evolution*. Plenum, New York.

Kimbel, W. H., Y. Rak, and D. C. Johanson. 2004. *The Skull of* Australopithecus afarensis. Oxford University Press, New York, 272 pp.

Kimbel, W. H., R. C. Walter, D. C. Johanson, K. E. Reed, J. L. Aronson, Z. Assefa, C. W. Marean, G. G. Eck, R. Bobe, E. Hovers, Y. Rak, C. Vondra, T. Yemane, D. York, Y. Chen, N. M. Evensen, and P. E. Smith. 1996. Late Pliocene *Homo* and Oldowan tools from the Hadar Formation (Kada Hadar Member), Ethiopia. *Journal of Human Evolution* 31:549–561.

Kimbel, W. J., and T. D. White. 1988. Variation, sexual dimorphism and the taxonomy of *Australopithecus*; pp. 175–192 in F. E. Grine (ed.), *Evolutionary History of the "Robust" Australopithecines*. Aldine de Gruyter, New York.

Kimbel, W. J., T. D. White, and D. C. Johanson. 1988. Implications of KNM-WT 17000 for the evolution of "robust" *Australopithecus*; pp. 259–268 in F. E. Grine (ed.), *Evolutionary History of the "Robust" Australopithecines*. Aldine de Gruyter, New York.

Kingston, J. D. 2007. Shifting adaptive landscapes: Progress and challenges in reconstructing early hominid environments. *Yearbook of Physical Anthropology* 50:20–58.

Kingston, J. D., and T. Harrison. 2007. Isotopic dietary reconstructions of Pliocene herbivores at Laetoli: Implications for early hominin paleoecology. *Palaeogeography, Palaeoclimatology, Palaeoecology* 243:272–306.

Kingston, J. D., B. F. Jacobs, A. Hill, and A. Deino. 2002. Stratigraphy, age and environments of the late Miocene Mpesida Beds, Tugen Hills, Kenya. *Journal of Human Evolution* 42:95–116.

Klein, R. G. 1988. The causes of "robust" australopithecine extinction; pp. 499–505 in F. E. Grine (ed.), *Evolutionary History of the "Robust" Australopithecines*. Aldine de Gruyter, New York.

———. 1992. The archeology of modern human origins. *Evolutionary Anthropology* 1:5–14.

———. 1994. Southern Africa before the Iron Age. pp. 471–519 in R. S. Corruccini and R. Ciochon (eds.), *Integrative Paths to the Past: Paleoanthropological Advances in Honor of F. Clark Howell*. Prentice Hall, Englewood Cliffs, N.J.

———. 1995. Anatomy, behavior, and modern human origins. *Journal of World Prehistory* 9:167–198.

————. 2000. Archeology and the evolution of human behavior. *Evolutionary Anthropology* 9:17–36.

Klein, R. G., G. Avery, K. Cruz-Uribe, and T. E. Steele. 2007. The mammalian fauna associated with an archaic hominin skullcap and later Acheulean artifacts at Elandsfontein, Western Cape Province, South Africa. *Journal of Human Evolution* 52:164–186.

Klein, R. G., and K. Scott. 1986. Re-analysis of faunal assemblages from the Haua Fteah and other late Quaternary archaeological sites in Cyrenaican Libya. *Journal of Archaeological Science* 13:515–542.

Kohl-Larsen, K., and H. Reck. 1936. Ersten Ueberblick über die Jungdiluvialen Tier und Menschenfunde Dr Kohl-Larsen's im Nordöstlichen Teil des Njarasa-Grabens (Ostafrika). *International Journal of Earth Sciences* 27:401–441.

Kono, R. T. 2004. Molar enamel thickness and distribution patterns in extant great apes and humans: new insights based on a 3-dimensional whole crown perspective. *Anthropological Science* 112:121–146.

Krings, M., C. Capelli, F. Tschentscher, H. Geisert, S. Meyer, A. von Haesler, K. Grossenschmidt, G. Possnert, M. Paunovic, and S. Pääbo. 2000. A view of Neanderthal genetic diversity. *Nature Genetics* 26:144–146.

Kullmer, O., O. Sandrock, R. Abel, F. Schrenk, T. G. Bromage, and Y. M. Juwayeyi. 1999. The first *Paranthropus* from the Malawi Rift. *Journal of Human Evolution* 37:121–127.

Kuman, K. 1994. The archaeology of Sterkfontein—past and present. *Journal of Human Evolution* 27:471–495.

Kuman, K., and R. J. Clarke. 2000. Stratigraphy, artefact industries and hominid associations for Sterkfontein, Member 5. *Journal of Human Evolution* 38:827–847.

Kuykendall, K. L. and G. C. Conroy. 1999. Description of the Gondolin teeth: Hyper-robust hominids in South Africa? *American Journal of Physical Anthropology* 28(suppl.):176–177.

Lahr, M. M., and R. A. Foley. 1998. Towards a theory of modern human origins: Geography, demography, and diversity on recent human evolution. *Annual Review of Anthropology* 27:137–176.

Larson, S. G., W. L. Jungers, M. J. Morwood, T. Sutkna, Jatmiko, E. W. Saptomo, R. Awe Due, and T. Djubiantono. 2007. *Homo floresiensis* and the evolution of the hominin shoulder. *Journal of Human Evolution* 53:718–731.

Latham, A. G., J. K. McKee, and P. V. Tobias. 2007. Bone breccias, bone dumps, and sedimentary sequences of the western Limeworks, Makapansgat, South Africa. *Journal of Human Evolution* 52:388–400.

Latham, A.G. and A. I. R. Herries. 2004. On the formation and sedimentary infilling of the Cave of Hearths and Historic Cave Complex, Makapansgat, South Africa. *Geoarchaeology* 19:232–342.

Latimer, B., and C. O. Lovejoy. 1989. The calcaneus of *Australopithecus afarensis* and its implications for the evolution of bipedality. *American Journal of Physical Anthropology* 78:369–386.

————. 1990a. Hallucal tarsometatarsal joint in *Australopithecus afarensis*. *American Journal of Physical Anthropology* 82:125–133.

————. 1990b. Metatarsophalangeal joints of *Australopithecus afarensis*. *American Journal of Physical Anthropology* 83:13–23.

Latimer B., and J. C. Ohman. 2001. Axial dysplasia in *Homo erectus*. *Journal of Human Evolution* 40:12.

Latimer, B., J. C. Ohman, and C. O. Lovejoy. 1987. Talocrural joint in African Hominoids: Implications for *Australopithecus afarensis*. *American Journal of Physical Anthropology* 74:155–175.

Latimer B., and C. V. Ward. 1993. The thoracic and lumbar vertebrae; pp. 266–293 in A. Walker and R. Leakey (eds.), *The Nariokotome* Homo erectus *Skeleton*. Harvard University Press, Cambridge.

Leakey, L. S. B. 1927. Stone age man in Kenya Colony. *Nature* 120:85–86.

————. 1931. *Stone Age Cultures of Kenya Colony*. Cambridge University Press, Cambridge, 287 pp.

————. 1935. *Stone Age Races in Kenya*. Oxford University Press, London, 150 pp.

————. 1936. A new fossil skull from Eyassi, East Africa: Discovery by a German expedition. *Nature* 138:1082–1084.

————. 1942. The Naivasha fossil skull and skeleton. *Journal of the East Africa Natural History Society* 16:169–177.

————. 1959. A new fossil skull from Olduvai. *Nature* 184:491–493.

————. 1961. New finds at Olduvai Gorge. *Nature* 189:649–650.

Leakey, L. S. B. and M. D. Leakey. 1964. Recent discoveries of fossil hominids in Tanganyika—at Olduvai + near Lake Natron. *Nature* 202:5–6.

Leakey, L. S. B., H. Reck, P. G. H. Boswell, A. T. Hopwood, and J. D. Solomon. 1933. The Oldoway human skeleton. *Nature* 131:397–398.

Leakey, L. S. B., P. V. Tobias, and J. R. Napier. 1964. A new species of the genus *Homo* from Olduvai Gorge. *Nature* 202:7–9.

Leakey, M. D. 1987. Introduction; pp. 490–523 in M. D. Leakey and J. M. Harris (eds.), *Laetoli: A Pliocene Site in Northern Tanzania*. Clarendon Press, Oxford.

Leakey, M. D., and J. M. Harris. 1987. *Laetoli: A Pliocene Site in Northern Tanzania*. Clarendon Press, Oxford, 561 pp.

Leakey, M. D., and L. S. B. Leakey. 1950. *Excavations at Njoro River Cave*. Oxford University Press, Oxford, 78 pp.

Leakey, M. D., P. V. Tobias, J. E. Martyn, and R. E. F. Leakey. 1969. An Acheulean industry with prepared core technique and the discovery of a contemporary hominid mandible at Lake Baringo, Kenya. *Proceedings of the Prehistoric Society* 3:48–76.

Leakey, M. G., C. S. Feibel, I. McDougall, and A. Walker. 1995. New four-million-year-old hominid species from Kanapoi and Allia Bay, Kenya. *Nature* 376:565–571.

Leakey, M. G., C. S. Feibel, I. McDougall, C. Ward, and A. Walker. 1998. New specimens and confirmation of an early age for *Australopithecus anamensis*. *Nature* 393:62–66.

Leakey, M. G., F. Spoor, F. H. Brown, P. N. Gathogo, C. Kiarie, L. N. Leakey, and I. McDougall. 2001. New hominin genus from eastern Africa shows diverse middle Pliocene lineages. *Nature* 410:433–440.

Leakey, R. E. F. 1969. Early *Homo sapiens* remains from the Omo River region of south-west Ethiopia. *Nature* 222:1132–1133.

————. 1973. Evidence for an advanced Plio-Pleistocene hominid from East Rudolf, Kenya. *Nature* 242:447–450.

Leakey, R. E. F., and A. C. Walker. 1976. *Australopithecus, Homo erectus* and the single species hypothesis. *Nature* 261:572–574.

————. 1988. New *Australopithecus boisei* specimens from East and West Lake Turkana, Kenya. *American Journal of Physical Anthropology* 76:1–24.

Lebatard, A-E., D. L. Bourlès, P. Duringer, M. Jolivet, R. Braucher, J. Carcaillet, M. Schuster, N. Arnaud, P. Monié, F. Lihoreau, A. Likius, H. T. Mackaye, P. Vignaud, and M. Brunet. 2008. Cosmogenic nuclide dating of *Sahelanthropus tchadensis* and *Australopithecus bahrelghazali*: Mio-Pliocene hominids from Chad. *Proceedings of the National Academy of Sciences, USA* 105:3226–3231.

Lee-Thorp, J. A., J. F. Thackeray, and N. J. van der Merwe. 2000. The hunters and the hunted revisited. *Journal of Human Evolution* 39:565–576.

Lee-Thorp, J. A., N. J. van der Merwe, and C. K. Brain. 1994. Diet of *Australopithecus robustus* at Swartkrans from stable carbon isotopic analysis. *Journal of Human Evolution* 27:361–372.

Levin, N. E., S. W. Simpson, J. Quade, T. Cerling, S. R. Frost. 2008. Herbivore enamel carbon isotopic composition and the environmental context of *Ardipithecus* at Gona, Ethiopia; pp. 215–234 in J. Quade and J. G. Wynn (eds.), *The Geology of Early Humans in the Horn of Africa*. Geological Society of America Special Paper 446, Boulder, Colorado.

Lieberman, D. E. 1995. Testing hypotheses about recent human evolution from skulls: Integrating morphology, function, development, and phylogeny. *Current Anthropology* 36:159–197.

————. 1998. Sphenoid shortening and the evolution of modern human cranial shape. *Nature* 393:158–162.

Lieberman, D. E., B. M. McBratney, and G. Krovitz. 2002. The evolution and development of cranial form in *Homo sapiens*. *Proceedings of the National Academy of Sciences, USA* 99:1134–1139.

Lieberman, D. E., D. R. Pilbeam and B. A. Wood. 1988. A probabilistic approach to the problem of sexual dimorphism in *Homo habilis*: A comparison of KNM-ER 1470 and KNM-ER 1813. *Journal of Human Evolution* 17:503–511.

Lieberman, D. E., and J. J. Shea. 1994. Behavioral differences between archaic and modern humans in the Levantine Mousterian. *American Anthropologist* 96:300–332.

Lockwood, C. A. 1999. Sexual dimorphism in the face of *Australopithecus africanus*. *American Journal of Physical Anthropology* 108:97–127.

Lockwood, C. A., W. H. Kimbel, and D. C. Johanson. 2000. Temporal trends and metric variation in the mandibles and dentition of *Australopithecus afarensis*. *Journal of Human Evolution* 39:23–55.

Lockwood, C. A., C. G. Menter, J. Moggi-Cecchi, and A. W. Keyser. 2007. Extended male growth in a fossil hominin species. *Science* 318:1443–1446.

Lockwood, C. A., and P. V. Tobias. 1999. A large male hominin cranium from Sterkfontein, South Africa, and the status of *Australopithecus africanus*. *Journal of Human Evolution* 36:637–685.

—————. 2002. Morphology and affinities of new hominin cranial remains from Member 4 of the Sterkfontein Formation, Gauteng Province, South Africa. *Journal of Human Evolution* 42:389–450.

Lordkipanidze, D., T. Jashashvili, A. Vekua, M. S. Ponce de León, C. P. E. Zollikofer, G. P. Rightmire, H. Pontzer, R. Ferring, O. Oms, M. Tappen, M. Bukhsianidze, J. Agusti, R. Kahlke, G. Kiladze, B. Martinez-Navarro, A. Mouskhelishvili, M. Nioradze, and L. Rook. 2007. Postcranial evidence from early *Homo* from Dmanisi, Georgia. *Nature* 449:305–310.

Lordkipanidze, D., A. Vekua, R. Ferring, G. P. Rightmire, C. P. E. Zollikofer, M. S. Ponce de Léon, J. Agustí, G. Kiladze, A. Mouskhelishvili, M. Nioradze, and M. Tappen. 2006. A fourth hominin skull from Dmanisi, Georgia. *Anatomical Record, A* 288:1146–1157.

Louchart, A., H. Wesselman, R. J. Blumenschine, L. J. Hlusko, J. K. Njau, M. T. Black, M. Asnake, and T. D. White. 2009. Taphonomic, avian, and small-vertebrate indicators of *Ardipithecus ramidus* habitat. *Science* 326:66e1–66e4.

Lovejoy, C. O. 1974. The gait of australopithecines. *Yearbook of Physical Anthropology* 17:147–161.

—————. 1975. Biomechanical perspectives on the lower limb of early hominids; pp. 291–326 in R. H. Tuttle (ed.), *Primate Functional Morphology.* Aldine, Chicago.

—————. 1981. The origin of man. *Science* 211:341–350.

—————. 2004. Review: The natural history of human gait and posture: Part 1. Spine and pelvis. *Gait and Posture:* 1–17.

Lovejoy, C. O., B. Latimer, G. Suwa, B. Asfaw, and T. D. White. 2009a. Combining prehension and propulsion: The foot of *Ardipithecus ramidus. Science* 326:72e1–72e8.

Lovejoy, C. O., R. S. Meindl, J. C. Ohman, K. G. Heiple, and T. D. White. 2002. The Maka femur and its bearing on the antiquity of human walking: Applying contemporary concepts of morphogenesis to the human fossil record. *American Journal of Physical Anthropology* 119:97–133.

Lovejoy, C. O., S. W. Simpson, T. D. White, B. Asfaw, and G. Suwa. 2009b. Careful climbing in the Miocene: The forelimbs of *Ardipithecus ramidus* and humans are primitive. *Science* 326: 70e1–70e8.

Lovejoy, C. O., G. Suwa, S. W. Simpson, J. H. Matternes, and T. D. White. 2009c. The Great Divides: *Ardipithecus ramidus* reveals the postcrania of our last common ancestors with African apes. *Science* 326:100–106.

Lovejoy, C. O., G. Suwa, L. Spurlock, B. Asfaw, and T. D. White. 2009d. The pelvis and femur of *Ardipithecus ramidus:* The emergence of upright walking. *Science* 326:71e1–71e6.

Lubell, D. 1974. The Fakhurian, a late Paleolithic industry from Upper Egypt. *Geological Survey of Egypt* 58:176–183.

Lucas, P., P. Constantino, B. Wood, and B. Lawn. 2008. Dental enamel as a dietary indicator in mammals. *BioEssays* 30:374–385.

Macchiarelli, R., L. Bondioli, V. Galichon, and P. V. Tobias. 1999. Hip bone trabecular architecture shows uniquely distinctive locomotor behavior in South African australopithecines. *Journal of Human Evolution* 36:211–232.

MacLarnon, A. 1993. The vertebral canal, pp. 359–390 in A. Walker and R. Leakey (eds.), *The Nariokotome* Homo erectus *Skeleton.* Harvard University Press, Cambridge.

MacLatchy, L. M. 1996. Another look at the australopithecine hip. *Journal of Human Evolution* 31:455–476.

Manega, P.C., 1993. Geochronology, geochemistry, and isotopic study of the Plio-Pleistocene hominid sites and the Ngorongoro Volcanic Highland in northern Tanzania. Unpublished PhD dissertation, University of Colorado, Boulder.

Mann, A., and E. Trinkaus. 1973. Neandertals and Neandertal-like fossils from the Upper Pleistocene. *Yearbook of Physical Anthropology* 17:169–193.

Marcais, J. 1934. Découverte de restes humains fossiles dans les Ores quaternaires de Rabat (Maroc). *L'Anthropologie* 44:579–583.

Mason, R. J. 1988. *Cave of Hearths, Makapansgat, Transvaal.* Archaeology Research Unit Occasional Paper No 21. University of the Witwatersrand Press, Johannesburg.

McBrearty, S., and A. S. Brooks. 2000. The revolution that wasn't: a new interpretation of the origin of modern human behavior. *Journal of Human Evolution* 39:453–563.

McBurney, C. B. M., J. C. Trevor, and L. H. Wells. 1953a. A fossil human mandible from a Levalloiso-Mousterian horizon in Cyrenaica. *Nature* 172:889–891.

—————. 1953b. The Haua Fteah fossil jaw. *Journal of the Royal Anthropological Institute* 83:71–85.

McCollum, M. A. 1997. Palatal thickening and facial form in *Paranthropus:* examination of alternative developmental models. *American Journal of Physical Anthropology* 103:375–392.

—————. 1999. The robust australopithecine face: a morphogenetic perspective. *Science* 284:301–305.

McCown, T. D. 1937. Mugharet Es-Skhul. Description and excavations; pp. 91–107 in D. A. Garrod and D. M. A. Bate (eds.), *The Stone Age of Mount Carmel: Volume 1. Excavations at the Wady El-Mughara.* Clarendon Press, Oxford.

McCown, T. D., and A. Keith. 1939. *The Stone Age of Mount Carmel: Volume 2. The Fossil Human Remains from the Levalloiso-Mousterian.* Clarendon Press, Oxford, 390 pp.

McDermott, F., C. Stringer, R. Grün, G. T. Williams, V. K. Din, and C. J. Hawkesworth. 1996. New late Pleistocene uranium-thorium and ESR dates for the Singa hominid (Sudan). *Journal of Human Evolution* 31:507–516.

McDougall, I., F. H. Brown, and J. G. Fleagle. 2005. Stratigraphic placement and age of modern humans from Kibish, Ethiopia. *Nature* 433:733–736.

McDougall, I., and C. S. Feibel. 1999. Numerical age control for the Miocene-Pliocene succession at Lothagam, a hominoid-bearing sequence in the northern Kenya Rift. *Journal of the Geological Society, London* 156:731–745.

McHenry, H. M. 1975. A new pelvic fragment from Swartkrans and the relationship between the robust and gracile australopithecines. *American Journal of Physical Anthropology* 43:245–261.

—————. 1988. New estimates of body weight in early hominids and their significance to encephalization and megadontia in "robust" australopithecines; pp. 133–148 in F. E. Grine (ed.), *Evolutionary History of the "Robust" Australopithecines.* Aldine de Gruyter, New York.

—————. 1991a. Femoral lengths and stature in Plio-Pleistocene hominids. *American Journal of Physical Anthropology* 85:149–158.

—————. 1991b. Petite bodies of the "robust" australopithecines. *American Journal of Physical Anthropology* 86:445–454.

—————. 1992a. Body size and proportions in early hominids. *American Journal of Physical Anthropology* 87:407–431.

—————. 1992b. How big were early hominids? *Evolutionary Anthropology* 1:15–20.

—————. 1994a. Behavioral ecological implications of early hominid body size. *Journal of Human Evolution* 27:77–87.

—————. 1994b. Early hominid postcrania: Phylogeny and function; pp. 251–268 in R. S. Corruccini and R. L. Ciochon (eds.), *Integrative Paths to the Past. Paleoanthropological Advances in Honor of F. Clark Howell.* Prentice Hall, Englewood Cliffs, N.J.

McHenry, H. M., and L. R. Berger. 1998. Body proportions in *Australopithecus afarensis* and *A. africanus* and the origin of the genus *Homo. Journal of Human Evolution* 35:1–22.

McHenry, H. M., C. C. Brown, and L. J. McHenry. 2007. Fossil hominin ulnae and the forelimb of *Paranthropus. American Journal of Physical Anthropology* 134:209–218.

McHenry, H. M., and K. Coffing. 2000. *Australopithecus* to *Homo:* Transformation in body and mind. *Annual Review of Anthropology* 29:125–146.

McHenry, H. M., and R. S. Corruccini. 1980. Late Tertiary hominoids and human origins. *Nature* 285:397–398.

McHenry, H. M., and A. L. Jones. 2006. Hallucial convergence in early hominins. *Journal of Human Evolution* 50:534–539.

McKee, J. K. 1989. Australopithecine anterior pillars: Reassessment of the functional morphology and phylogenetic relevance. *American Journal of Physical Anthropology* 80:1–9.

—————. 1993. Faunal dating of the Taung hominid fossil deposit. *Journal of Human Evolution* 25:363–378.

McKee, J. K., J. F. Thackeray, and L. R. Berger. 1995. Faunal assemblage seriation of southern African Pliocene and Pleistocene fossil deposits. *American Journal of Physical Anthropology* 96:235–250.

Mehlman, M. J. 1979. Mumba-Höhle revisited: the relevance of a forgotten excavation to some current issues in East African prehistory. *World Archaeology* 11:80–94.

—————. 1987. Provenience, age and associations of archaic *Homo sapiens* crania from Lake Eyasi, Tanzania. *Journal of Archaeological Science* 14:133–162.

—————. 1991. Context for the emergence of modern man in Eastern Africa: some new Tanzanian evidence; pp. 177–196 in J. D. Clark (ed.), *Cultural Beginnings: Approaches to Understanding Early Hominid Lifeways in the African Savanna.* Monograph 19. Forschungsinstitut für Vor- und Frühgeschichte, Römisch-Germanisches Zentralmuseum, Bonn.

Menter, C. G., K. L. Kuykendall, A. W. Keyser, and G. C. Conroy. 1999. First record of hominid teeth from the Plio-Pleistocene site of Gondolin, South Africa. *Journal of Human Evolution* 37:299–307.

Mercier, N., H. Valladas, O. Bar-Yosef, B. Vandermeersch, C. B. Stringer, and J.-L. Joron. 1993. Thermoluminescence date for the Mousterian burial site of Es-Skhul, Mt. Carmel. *Journal of Archaeological Science* 20:169–174.

Merrick, H. V., and M. C. Monaghan. 1984. The date of the cremated burials in Njoro River Cave. *Azania* 19:7–11.

Merwe, N. J. van der, F. T. Masao, and M. K. Bamford. 2008. Isotopic evidence for contrasting diets of early hominins *Homo habilis* and *Australopithecus boisei* of Tanzania. *South African Journal of Science* 104:153–155.

Merwe, N. J. van der, J. F. Thackeray, J. A. Lee-Thorpe, and J. Luyt. 2003. The carbon isotope ecology and diet of *Australopithecus africanus* at Sterkfontein, South Africa. *Journal of Human Evolution* 44:581–597.

Meyer, M. 2006. Evidence for the anatomical capacity for spoken language in *Homo erectus*. *American Journal of Physical Anthropology* 41(suppl.):130.

Millard, A. R. 2008. A critique of the chronometric evidence for hominid fossils: I. Africa and the Near East 500–50 ka. *Journal of Human Evolution* 54:848–874.

Miller, J. A. 1991. Does brain size variability provide evidence of multiple species in *Homo habilis*? *American Journal of Physical Anthropology* 84:385–398.

Moggi-Cecchi, J., and M. Collard. 2002. A fossil stapes from Sterkfontein, South Africa, and hearing capabilities of early hominids. *Journal of Human Evolution* 42:259–265.

Moggi-Cecchi, J., F. E. Grine, and P. V. Tobias. 2006. Early hominid dental remains from Members 4 and 5 of the Sterkfontein Formation (1966–1996 excavations): Catalogue, individual associations, morphological descriptions and initial metrical analysis. *Journal of Human Evolution* 50:239–328.

Moggi-Cecchi, J., P. V. Tobias, and A. D. Beynon. 1998. The mixed dentition and associated skull fragments of a juvenile fossil hominid from Sterkfontein, South Africa. *American Journal of Physical Anthropology* 106:425–466.

Morgan, L. E., and P. R. Renne. 2008. Diachronous dawn of Africa's Middle Stone Age: New ^{40}Ar/^{39}Ar ages from the Ethiopian Rift. *Geology* 36:967–970.

Mounier A., F. Marchal, and S. Condemi. 2009. Is *Homo heidelbergensis* a distinct species? New insight on the Mauer mandible. *Journal of Human Evolution* 56:219–246.

Mountain, J. L., A. A. Lin, A. M. Bowcock, and L. L. Cavalli-Sforza. 1993. Evolution of modern humans: evidence from nuclear DNA polymorphisms; pp. 69–83 in M. J. Aitken, C. B. Stringer, and P. A. Mellars (eds.), *The Origin of Modern Humans and the Impact of Chronometric Dating.* Princeton University Press, Princeton, N.J.

Moyà-Solà, S., M. Köhler, D. M. Alba, and S. Almécija. 2008. Taxonomic attribution of the Olduvai Hominid 7 manual remains and the functional interpretation of hand morphology in robust australopithecines. *Folia Primatologia* 79:215–250.

Mturi, A. A. 1976. New hominid from Lake Ndutu, Tanzania. *Nature* 262:484–485.

Nagaoka S., S. Katoh, G. WoldeGabriel, H. Sato, H. Nakaya, Y. Beyene, and G. Suwa. 2005. Lithostratigraphic and sedimentary environments of the hominid-bearing Plio–Pleistocene Konso Formation in the southern Main Ethiopian Rift, Ethiopia. *Paleogeography, Paleoclimatology, Paleoecology* 216:333–357.

Nei, M. 1995. Genetic support of the out-of-Africa theory of human evolution. *Proceedings of the National Academy of Sciences, USA* 92:6720–6722.

Njau, J. K. and Blumenschine, R. J. 2007. A diagnosis of crocodile feeding traces on larger mammal bone, with fossil examples from the Plio-Pleistocene Olduvai Basin, Tanzania. *Journal of Human Evolution* 50:142–162.

Oakley, K. P., B. G. Campbell, and T. I. Molleson. 1977. *Catalogue of Fossil Hominids. Part I. Africa.* 2nd ed. British Museum (Natural History), London, 223 pp.

Ohman, J. C., C. O. Lovejoy and T. White. 2005. Questions about *Orrorin* femur. *Science* 307:845.

Ohman, J. C., M. Slanina, G. Baker, and R. P. Mensforth. 1995. Thumbs, tools, and early humans. *Science* 268:587–588.

Organ, J. M., and C. V. Ward. 2006. Contours of the hominoid lateral tibial condyle with implications for *Australopithecus. Journal of Human Evolution* 51:113–127.

Ovchinnikov, I., G. Anders, A. Götherström, G. Romanova, V. Kharitonov, K. Lidén, and W. Goodwin. 2000. Molecular analysis of Neanderthal DNA from the northern Caucasus. *Nature* 404:490–493.

Partridge, T. C. 2000. Hominid-bearing cave and tufa deposits; pp. 100–130 in T. C. Partridge and R. R. Maud (eds.), *The Cenozoic of Southern Africa.* Oxford University Press, Oxford.

Partridge, T. C., D. E. Granger, M. W. Caffee, and R. J. Clarke. 2003. Lower Pliocene hominid remains from Sterkfontein. *Science* 300:607–612.

Partridge, T. C., J. Shaw, D. Heslop, and R. J. Clarke. 1999. The new hominid skeleton from Sterkfontein, South Africa: Age and preliminary assessment. *Journal of Quaternary Science* 14:293–298.

Patterson, B., and W. W. Howells. 1967. Hominid humeral fragment from early Pleistocene of North-western Kenya. *Science* 156:64–66.

Patterson, N., D. J. Richter, S. Gnerre, E. S. Lander, and D. Reich. 2006. Genetic evidence for complex speciation of humans and chimpanzees. *Nature* 441:1103–1108.

———. 2000. Postcranial remains and the origin of modern humans. *Evolutionary Anthropology* 9:229–247.

———. 2004. Has the combination of genetic and fossil evidence solved the riddle of modern human origins? *Evolutionary Anthropology* 13:145–159.

———. 2008. Statistical and biological definitions of "anatomically modern" humans: Suggestions for a unified approach to modern morphology. *Evolutionary Anthropology* 17:38–48.

Pearson, O. M., J. G. Fleagle, F. E. Grine, and D. F. Royer. 2008a. Further new hominin fossils from the Kibish Formation, southwestern Ethiopia. *Journal of Human Evolution* 55:444–447.

Pearson, O. M., and F. E. Grine. 1997. Re-analysis of the hominid radii from Cave of Hearths and Klasies River Mouth, South Africa. *Journal of Human Evolution* 32:577–592.

Pearson, O. M., D. F. Royer, F. E. Grine, and J. G. Fleagle. 2008b. A description of the Omo I postcranial skeleton, including newly discovered fossils. *Journal of Human Evolution* 55:421–437.

Phillipson, D. W. 1976. *The Prehistory of Eastern Zambia.* British Institute in Eastern Africa, Memoir No. 6, Thames & Hudson, Nairobi, 229 pp.

Pickering, R., R. J. Clarke, and J. L. Heaton. 2004a. The context of Stw 573, an early hominid skull and skeleton from Sterkfontein Member 2: Taphonomy and paleoenvironment. *Journal of Human Evolution* 46:277–295.

Pickering, R., P. J. Hancox, J. A. Lee-Thorp, R. Grün, G. E. Mortimer, M. McCulloch, and L. R. Berger. 2007. Stratigraphy, U-Th chronology, and paleoenvironments at Gladysvale Cave: insights into the climate control of South African hominin-bearing cave deposits. *Journal of Human Evolution* 53:602–619.

Pickering, T. R., R. J. Clarke, and J. Moggi-Cecchi. 2004b. Role of carnivores in the accumulation of the Sterkfontein Member 4 hominid assemblage: A taphonomic reassessment of the complete hominid fossil sample (1936–1999). *American Journal of Physical Anthropology* 125:1–15.

Pickford, M. 1975. Late Miocene sediments and fossils from the Northern Kenya Rift Valley. *Nature* 256:279–284.

Pickford, M., and B. Senut. 2001. The geological and faunal context of Late Miocene hominid remains from Lukeino, Kenya. *Comptes Rendus de l'Académie des Sciences, Série IIA, Earth and Planetary Sciences* 332:145–152.

Pickford, M., B. Senut, D. Gommery, and J. Treil. 2002. Bipedalism in *Orrorin tugenensis* revealed by its femora. *Comptes Rendus Palevol* 1:1–13.

Pinhasi, R., and P. Semal. 2000. The position of the Nazlet Khater specimen among prehistoric and modern African and Levantine populations. *Journal of Human Evolution* 39:269–288.

Plavcan, J. M., C. A. Lockwood, W. H. Kimbel, M. R. Lague and E. H. Harmon. 2005. Sexual dimorphism in *Australopithecus afarensis* revisited: How strong is the case for a human-like pattern of dimorphism? *Journal of Human Evolution* 48:313–320.

Plummer, T. 2004. Flaked stones and old bones: Biological and cultural evolution at the dawn of technology. *Yearbook of Physical Anthropology* 47:118–164.

Potts, R. 1998. Environmental hypotheses of hominin evolution. *Yearbook of Physical Anthropology* 41:93–136.

Potts, R., A. K. Behrensmeyer, A. Deino, P. Ditchfield, and J. Clark. 2004. Small mid-Pleistocene hominin associated with East African acheulean technology. *Science* 305:75–78.

Potts, R., A. K. Behrensmeyer, and P. Ditchfield. 1999. Paleolandscape variation and early Pleistocene hominid activities: Members 1 and 7, Olorgesailie Formation, Kenya. *Journal of Human Evolution* 37:747–788.

Potts, R., and A. Deino. 1995. Mid-Pleistocene change in large mammal faunas of eastern Africa. *Quaternary Research* 43:106–113.

Potts, R., P. Shipman, and E. Ingall. 1988. Taphonomy, paleoecology and hominids of Lainyamok, Kenya. *Journal of Human Evolution* 17:597–614.

Prat, S., J-P. Brugal, J-J. Tiercelin, J-A. Barrat, M. Bohn, A. Delagnes, S. Harmand, K. Kimeu, M. Kibunjia, P-J. Texier, and H. Roche. 2005. First occurrence of early *Homo* in the Nachukui Formation (West Turkana, Kenya) at 2.3–2.4 Myr. *Journal of Human Evolution* 49:230–240.

Protsch, R. 1974. The age and stratigraphic position of Olduvai hominid I. *Journal of Human Evolution* 3:379–385.

———. 1975a. The absolute dating of Upper Pleistocene sub-Saharan fossil hominids and their place in human evolution. *Journal of Human Evolution* 4:297–322.

———. 1975b. The Kohl-Larsen Eyasi and Garusi hominid finds in Tanzania and their relation to *Homo erectus*; pp. 217–226 in B. Sigmon and J. Cybulski (eds.), Homo erectus*: Papers in Honor of Davidson Black*. University of Toronto Press, Toronto.

———. 1976. The Naivasha hominid and its confirmed late Upper Pleistocene age. *Anthropologischer Anzei*ger 35:97–102.

Protsch, R., and H. de Villiers. 1974. Bushman rock shelter, Origstad, eastern Transvaal, South Africa. *Journal of Human Evolution* 3:387–396.

Pycraft, W. P., G. E. Smith, M. Yearsley, J. T. Carter, R. A. Smith, A. T. Hopwood, D. M. A. Bate, and W. E. Swinton. 1928. *Rhodesia Man and Associated Remains*. British Museum of Natural History, London.

Quintana-Murci, L., O. Semino, H.-J. Bandelt, G. Passarino, K. McElreavey, and A. S. Santachiara-Benerecetti. 1999. Genetic evidence of an early exit of *Homo sapiens sapiens* from Africa through eastern Africa. *Nature Genetics* 23:437–441.

Rak, Y. 1983. *The Australopithecine Face*. Academic Press, New York. 169 pp.

———. 1985. Australopithecine taxonomy and phylogeny in light of facial morphology. *American Journal of Physical Anthropology* 66:281–287.

———. 1990. On the differences between two pelvises of Mousterian context from the Qafzeh and Kebara Caves, Israel. *American Journal of Physical Anthropology* 81:323–332.

Rak, Y., and F. C. Howell. 1978. Cranium of a juvenile *Australopithecus boisei* from the Lower Omo Basin, Ethiopia. American Journal of Physical Anthropology 48:345–366.

Rak, Y., and W. H. Kimbel. 1991. On the squamosal suture of KNM-WT 17000. *American Journal of Physical Anthropology* 85:1–6.

Rak, Y., and W. H. Kimbel. 1993. Reply to Drs. Walker, Brown, and Ward. *American Journal of Physical Anthropology* 90:506–507.

Rak, Y., A. Ginzburg, and E. Geffen. 2007. Gorilla-like anatomy on *Australopithecus afarensis* mandibles suggests *Au. afarensis* link to robust australopiths. *Proceedings of the National Academy of Sciences, USA* 104:6568–6572.

Ramirez-Rozzi, F. V. 1993. Tooth development in East African *Paranthropus*. *Journal of Human Evolution* 24:429–454.

Raynal, J.-P., F.-Z. Sbihi-Alaoui, D. Geraads, L. Magoga, and A. Mohib. 2001. The earliest occupation of North-Africa: the Moroccan perspective. *Quaternary International* 75:65–75.

Rayner, R. J., B. P. Moon, and J. C. Masters. 1993. The Makapansgat australopithecine environment. *Journal of Human Evolution* 24:219–231.

Reed, C. A. 1965. A human frontal bone from the late Pleistocene of the Kom Ombo Plain, Upper Egypt. *Man* 65:101–104.

Reed, K. E., J. M. Kitching, F. E. Grine, W. L. Jungers, and L. Sokoloff. 1993. Proximal femur of *Australopithecus africanus* from Member 4, Makapansgat, South Africa. *American Journal of Physical Anthropology* 92:1–15.

Reed, K. E. 1997. Early hominid evolution and ecological change through the African Plio-Pleistocene. *Journal of Human Evolution* 32:289–322.

———. 2008. Paleoecological patterns at the Hadar hominin site, Afar Regional State, Ethiopia. *Journal of Human Evolution* 54:743–768.

Relethford, J. H. 1995. Genetics and modern human origins. *Evolutionary Anthropology* 4:53–63.

Reno, P. L., R. S. Meindl, M. A. McCollum, and C. O. Lovejoy. 2003. Sexual dimorphism in *Australopithecus afarensis* was similar to that of modern humans. *Proceedings of the National Academy of Sciences. USA* 100:9404–9409.

———. 2005. The case is unchanged and remains robust: *Australopithecus afarensis* exhibits only moderate skeletal dimorphism. *Journal of Human Evolution* 49:279–288.

Reynolds, S. C., R. J. Clarke, and K. A. Kuman. 2007. The view from the Lincoln Cave: mid- to late Pleistocene fossil deposits from Sterkfontein hominid site, South Africa. *Journal of Human Evolution* 53:260–271.

Rhodes, E. J., J. S. Singarayer, J.-P. Raynal, K. E. Westeway, and F. Z. Sbihi-Alaoui. 2006. New age estimates for the Palaeolithic assemblages and Pleistocene succession of Casablanca, Morocco. *Quaternary Science Reviews* 25:2569–2585.

Richmond, B. G., and W. L. Jungers. 1995. Size variation and sexual dimorphism in *Australopithecus afarensis* and living hominoids. *Journal of Human Evolution* 29:229–245.

———. 2008. *Orrorin tugenensis* femoral morphology and the evolution of hominin bipedalism. *Science* 319:1662–1665.

Richmond, B. G. and Strait, D. S. 2000. Evidence that humans evolved from a knuckle-walking ancestor. *Nature* 404:382–385.

Ricklan, D. E. 1987. Functional anatomy of the hand of *Australopithecus africanus*. *Journal of Human Evolution* 16:643–664.

———. 1990. The precision grip in *Australopithecus africanus*: anatomical and behavioral correlates; pp. 171–183 in G. H. Sperber (ed.), *From Apes to Angels: Essays in Anthropology in Honor of Phillip V. Tobias*. Wiley-Liss, New York.

Rightmire, G. P. 1975. New studies of post-Pleistocene human skeletal remains from the Rift Valley, Kenya. *American Journal of Physical Anthropology* 42:351–370.

———. 1976. Relationships of Middle and Upper Pleistocene hominids from sub-Saharan Africa. *Nature* 260:238–240.

———. 1979. Implications of Border Cave skeleton remains for later Pleistocene human evolution. *Current Anthropology* 20:23–35.

———. 1984. *Homo sapiens* in sub-Saharan Africa; pp. 85–115 in F. H. Smith, and F. Spencer (eds.), *The Origins of Modern Humans: A World Survey of the Fossil Evidence*. Liss, New York.

———. 1993. *The Evolution of* Homo erectus. Cambridge University Press, New York, 276 pp.

———. 1998. Human evolution in the Middle Pleistocene: The role of *Homo heidelbergensis*. *Evolutionary Anthropology* 6:218–227.

———. 2008. *Homo* in the Middle Pleistocene: hypodigms, variation, and species recognition. *Evolutionary Anthropology* 17:8–21.

Rightmire, G. P., and H. J. Deacon. 1991. Comparative studies of late Pleistocene human remains from Klasies River Mouth, South Africa. *Journal of Human Evolution* 20:131–156.

Robbins, L. M. 1987. Hominid footprints from Site G; pp. 496–502 in M. D. Leakey and J. M. Harris (eds.), *Laetoli: A Pliocene Site in Northern Tanzania*. Clarendon Press, Oxford.

Robinson, J. T. 1962. Sterkfontein stratigraphy and the significance of the Extension Site. *South African Archaeological Bulletin* 17:87–107.

Robinson, J. T. 1970. Two new early hominid vertebrae from Swartkrans. *Nature* 225:1217–1225.

———. 1972. *Early Hominid Posture and Locomotion*. University of Chicago Press, Chicago, 361 pp.

Roche, J., and J.-P. Texier. 1976. Découverte de restes humains dans un niveau atérien supérieur de la grotte des Contrebandiers, à Temara (Maroc). *Comptes Rendus de l'Académie des Sciences, Paris* 282:45–47.

Rogers, M. D. Iltis, and S. Wooding. 2004. Genetic variation at the MC1R locus and the time since loss of human body hair. *Current Anthropology* 45:105–108.

Rose, M., D. 1984. A hominine hip bone, KNM-ER 3228, from east Lake Turkana, Kenya. American *Journal of Physical Anthropology* 63:371–378.

Rothschild, B. M., I. Herskovitz, and C. Rothschild. 1995. Origin of yaws in the Pleistocene. *Nature* 378:343–344.

Rougier, H., S. Milota, R. Rodrigo, M. Gherase, L. Sarcina, O. Moldovan, J. Zilhão, S. Constantin, R. G. Franciscus, C. P. E. Zollikofer, M. Ponce de Léon, and E. Trinkaus. 2007. Pestera cu Oase 2 and the cranial morphology of early modern Europeans. *Proceedings of the National Academy of Sciences* 104:1165–1170.

Ruff, C. 1994. Morphological adaptation to climate in modern and fossil hominids. *Yearbook of Physical Anthropology* 37:65–107.

———. 2008. Relative limb strength and locomotion in *Homo habilis*. *American Journal of Physical Anthropology* 138:90–100.

Ruff, C., and W. C. Hayes. 1983. Cross-sectional geometry of Pecos-Pueblo femora and tibiae: A biomechanical investigation: 1. *American Journal of Physical Anthropology* 60:359–381.

Ruff, C. B., H. M. McHenry, and J. F. Thackeray. 1999. Cross-sectional morphology of the SK 82 and 97 proximal femora. *American Journal of Physical Anthropology* 109:509–521.

Ruff, C. B., and A. Walker. 1993. Body size and body shape; pp. 234–265 in A. Walker and R. Leakey (eds.), *The Nariokotome* Homo erectus *Skeleton*. Harvard University Press, Cambridge.

Ruiter, D. J. de. 2003. Revised faunal lists for Members 1–3 of Swartkrans, South Africa. *Annals of the Transvaal Museum* 40:29–41.

Ruiter, D. J. de, M. Sponheimer, and J. A. Lee-Thorp. 2008. Indications of habitat associations of *Australopithecus robustus* in the Bloubank Valley, South Africa. *Journal of Human Evolution* 55:1015–1030.

Ruiter, D. J. de, C. M. Steininger, and L. R. Berger. 2006. A cranial base of *Australopithecus robustus* from the hanging remnant of Swartkrans, South Africa. *American Journal of Physical Anthropology* 130:435–444.

Ryan, A. S., and Johanson, D. C. 1989. Anterior dental microwear in *Australopithecus afarensis*: comparisons with human and nonhuman primates. *Journal of Human Evolution* 18:235–268.

Saban, R. 1977. Place of Rabat man (Kebibat, Morocco) in human evolution. *Current Anthropology* 18:518–524.

Sanders, W. J. 1987. A review of the initial interpretations of KNM-WT 17000. New York University *Journal of Anthropology* 2:24–33.

———. 1990. Weight transmission through the lumbar vertebrae and sacrum in Australopithecines. *American Journal of Physical Anthropology* 81:289.

———. 1998. Comparative morphometric study of the australopithecine vertebral series Stw-H8/H41. *Journal of Human Evolution* 34:249–302.

Sawada, Y., M. Pickford, B. Senut, T. Itaya, M. Hyodo, T. Miura, C. Kashine, T. Chujo, and H. Fujii. 2002. The age of *Orrorin tugenensis*, an early hominid from the Tugen Hills, Kenya. *Comptes Rendus Palevol* 1: 293–303.

Schepers, G. W. H. 1941. The mandible of the Transvaal fossil human skeleton from Springbok Flats. *Annals of the Transvaal Museum* 20:253–271.

Schmitt, D. 2003. Insights into the evolution of human bipedalism from experimental studies of humans and other primates. *Journal of Experimental Biology* 206:1437–1448.

Schmid, P. 2002. The Gladysvale project. *Evolutionary Anthropology* 11:45–48.

Schoeninger, M. J., H. Reeser, and C. Hallin. 2003. Paleoenvironment of *Australopithecus anamensis* at Allia Bay, East Turkana, Kenya: Evidence from mammalian herbivore enamel stable isotopes. *Journal of Anthropological Archeology* 22:200–207.

Schoetensack, O. 1908. *Der Unterkiefer des* Homo heidelbergensis *aus den Sanden von Mauer bei Heidelberg*. Englemann, Leipzig. 67 pp.

Schrenk, F., T. G. Bromage, A. Gorthner, and O. Sandrock. 1995. Paleoecology of the Malawi Rift: vertebrate and invertebrate faunal contexts of the Chiwondo Beds, northern Malawi. *Journal of Human Evolution* 28:59–70.

Schrenk, F., O. Kullmer, and T. Bromage. 2007. The earliest putative *Homo* fossils; pp. 1611–1631 in W. Henke and I. Tattersall (eds.), *Handbook of Paleoanthropology*, vol. 3. Springer, New York.

Schrenk, F., O. Kullmer, O. Sandrock, and T. G. Bromage. 2002. Early hominid diversity, age and biogeography of the Malawi-Rift. *Human Evolution* 17:113–122.

Schwartz, H. P., R. Grün, and P. V. Tobias. 1994. ESR dating of the australopithecine site of Sterkfontein, South Africa. *Journal of Human Evolution* 26:175–181.

Schwartz, H. P., R. Grün, B. Vandermeersch, O. Bar-Yosef, H. Valladas, and E. Tchernov. 1988. ESR dates for the hominid burial site of Qafzeh in Israel. *Journal of Human Evolution* 17:733–737.

Schwartz, J. H., and I. Tattersall. 2003. *The Human Fossil Record. Volume Two. Craniodental Morphology of Genus* Homo *(Africa and Asia)*. Wiley-Liss, New York, 603 pp.

———. 2005. *The Human Fossil Record: Volume 4, Craniodental Morphology of Early Hominids (Genera* Australopithecus, Paranthropus, Orrorin*), and Overview*. Wiley-Liss, New York, 616 pp.

Schweitzer, F. 1979. Excavations at Die Kelders, Cape Province, South Africa: the Holocene deposits. *Annals of the South African Museum* 78:101–233.

Scott, R. S., P. S. Ungar, T. S. Bergstrom, C. A. Brown, F. E. Grine, M. F. Teaford, and A. Walker. 2005. Dental microwear texture analysis shows within-species diet variability in fossil hominins. *Nature* 436:693–695.

Semaw, S. 2000. The world's oldest stone artifacts from Gona, Ethiopia: Their implications for understanding stone technology and patterns of human evolution between 2.6–1.5 million years ago. *Journal of Archaeological Science* 27:1197–1214.

Semaw, S., S. W. Simpson, J. Quade, P. R. Renne, R. F. Butler, W. C. McIntosh, N. Levin, M. Dominguez-Rodrigo, and M. J. Rogers. 2005. Early Pliocene hominids from Gona, Ethiopia. *Nature* 433:301–305.

Senut, B. 2007. The Earliest putative hominids; pp. 1519–1538 in W. Henke and I. Tattersall (eds.), *Handbook of Paleoanthropology*, vol. 3. Springer, New York.

Senut, B., M. Pickford, J. Braga, D. Marais, and Y. Coppens. 2000. Découverte d'un *Homo sapiens* archaïque à Oranjemund, Namibie. *Comptes Rendus de l'Academie des Sciences, Paris, Sciences de la Terre et des Planètes* 330:813–819.

Senut, B., M. Pickford, D. Gommery, P. Mein, K. Cheboi, and Y. Coppens. 2001. First hominid from the Miocene (Lukeino formation, Kenya). *Comptes Rendus de l'Académie des Sciences, Series IIA, Earth and Planetary Science* 332:137–144.

Şenyürek, M. 1955. A note on the teeth of *Meganthropus africanus* Weinert from Tanganyika Territory. *Belleten (Ankara)* 19:1–54.

Shackelford, L. L. 2007. Regional variation in the postcranial robusticity of late Upper Paleolithic humans. *American Journal of Physical Anthropology* 133:655–668.

Shapiro, L. 1993. Evaluation of "unique" aspects of human vertebral bodies and pedicles with a consideration of *Australopithecus africanus*. *Journal of Human Evolution* 25:433–470.

Sherwood, R. J., S. C. Ward, and A. Hill. 2002. The taxonomic status of the Chemeron temporal (KNM-BC 1). *Journal of Human Evolution* 42:153–184.

Shipman, P., and J. M. Harris. 1988. Habitat preference and paleoecology of *Australopithecus boisei* in eastern Africa; pp. 343–381 in F. E. Grine (ed.), *Evolutionary History of the "Robust" Australopithecines*. Aldine de Gruyter, New York.

Shipman, P., R. Potts, and M. Pickford. 1983. Lainyamok, a new middle Pleistocene hominid site. *Nature* 306:365–368.

Sikes, N. E., and G. M. Ashley. 2007. Stable isotopes of pedogenic carbonates as indicators of paleoecology in the Plio-Pleistocene (upper Bed I), western margin of the Olduvai Basin, Tanzania. *Journal of Human Evolution* 53:574–594.

Sillen, A. 1992. Strontium-calcium rations (Sr/Ca) of *Australopithecus robustus* and associated fauna from Swartkrans. *Journal of Human Evolution* 23:495–516.

Silverman, N., B. Richmond, and B. Wood. 2001. Testing the taxonomic integrity of *Paranthropus boisei sensu stricto*. *American Journal of Physical Anthropology* 115:167–178.

Simpson, S. W., J. Quade, L. Kleinsasser, N. Levin, W. MacIntosh, N. Dunbar, and S. Semaw. 2007. Late Miocene hominid teeth from Gona Project Area, Ethiopia. *American Journal of Physical Anthropology* 44 (suppl.):219.

Simpson, S. W., J. Quade, N. E. Levin, R. Butler, G. Dupont-Nivet, M. Everett, and S. Semaw. 2008. A female *Homo erectus* pelvis from Gona, Ethiopia. *Science* 322:1089–1092.

Singer, R. 1954. The Saldanha skull from Hopefield, South Africa. *American Journal of Physical Anthropology* 12:345–362.

———. 1961. Pathology in the temporal bone of the Boskop skull. *South African Archaeological Bulletin* 16:103–104.

Singer, R., and J. Wymer. 1982. *The Middle Stone Age at Klasies River Mouth in South Africa*. University of Chicago Press, Chicago, 234 pp.

Singleton, M. 2003. Functional and phylogenetic implications of molar flare variation in Miocene hominoids. *Journal of Human Evolution* 45:57–79.

Skelton, R. R., and H. M. McHenry. 1992. Evolutionary relationships among early hominids. *Journal of Human Evolution* 23:309–349.

Skinner, M. 1991. Bee brood consumption: an alternative explanation for hypervitaminosis A in KNM-ER 1808 (*Homo erectus*) from Koobi Fora, Kenya. *Journal of Human Evolution* 20:493–503.

Smith, B. H. 1986. Dental development in *Australopithecus* and early *Homo*. *Nature* 323:327–330.

Smith, C. C., M. E. Morgan, and D. Pilbeam. 2010. Isotopic ecology and dietary profiles of Liberian chimpanzees. *Journal of Human Evolution* 58:43–55.

Smith, P., R. A. Bloom, and J. Berkowitz. 1983. Bone morphology and biomechanical efficiency in fossil hominids. *Current Anthropology* 24:662–663.

———. 1984. Diachronic trends in humeral cortical thickness of Near Eastern populations. *Journal of Human Evolution* 13:603–611.

Smith, P. E. L. 1967. New investigations in the late Pleistocene archaeology of the Kom Ombo Plain (Upper Egypt). *Quaternaria* 9:141–152.

———. 1976. Stone-age man on the Nile. *Scientific American* 235:30–38.

Smith, T. M., P. Tafforeau, D. J. Reid, R. Grün, S. Eggins, M. Boutakiout, and J.-J. Hublin. 2007. Earliest evidence of modern human life history in North African early *Homo sapiens*. *Proceedings of the National Academy of Sciences, USA* 104:6128–6133.

Sponheimer, M., D. de Ruiter, J. A. Lee-Thorp, and A. Späth. 2005a. Sr/Ca and early hominin diets revisited: new data from modern and fossil tooth enamel. *Journal of Human Evolution* 48:147–156.

Sponheimer, M., and J. A. Lee-Thorp. 1999. Isotopic evidence for the diet of an early hominid, *Australopithecus africanus*. *Science* 283:368–370.

Sponheimer, M., J. Lee-Thorp, D. de Ruiter, D. Codron, J. Codron, A. T. Baugh, and F. Thackeray. 2005b. Hominins, sedges, and termites: New carbon isotope data from the Sterkfontein valley and Kruger National Park. *Journal of Human Evolution* 48:301–312.

Sponheimer, M., B. H. Passey, D. J. de Ruiter, D. Guatelli-Steinberg, T. E. Cerling, and J. A. Lee-Thorp. 2006. Isotopic evidence for dietary variability in the early hominin *Paranthropus robustus*. *Science* 314:980–982.

Spoor, C. F. 1993. The comparative morphology and phylogeny of the human bony labyrinth. Unpublished PhD dissertation, University of Utrecht.

Spoor, F., C. Stringer, and F. Zonneveld. 1998. Rare temporal bone pathology of the Singa Calvaria from Sudan. *American Journal of Physical Anthropology* 107:41–50.

Spoor, F., B. Wood, and F. Zonneveld. 1994. Implications of early hominid labyrinthine morphology for evolution of human bipedal locomotion. *Nature* 369:645–648.

Spoor, F., M. G. Leakey, S. C. Antón, and L. N. Leakey. 2008. The taxonomic status of KNM-ER 42700: A reply to Baab (2008a). *Journal of Human Evolution* 55:747–750.

Spoor, F., P. O'Higgins, C. Dean, and D. E. Lieberman. 1999. Anterior sphenoid in modern humans. *Nature* 397:572.

Spoor, F., M. G. Leakey, P. N. Gathogo, F. H. Brown, S. C. Antón, I. McDougall, C. Kiarie, F. K. Manthi, and L. N. Leakey. 2007. Implications of new early *Homo* fossils from Ileret, east of Lake Turkana, Kenya. *Nature* 448:688–691.

Steiper, M.E., N. M. Young, and T. Y. Sukarna. 2004. Genomic data support the hominoid slowdown and an early Oligocene estimate for the hominoid-cercopithecoid divergence. *Proceedings of the National Academy of Sciences, USA* 101:17021–17026.

Stern, J. T. Jr., and R. L. Susman. 1981. Electromyography of the gluteal muscles in *Hylobates, Pongo,* and *Pan*: Implications for the evolution of hominid bipedality. *American Journal of Physical Anthropology* 55:153–166.

———. 1983. The locomotor anatomy of *Australopithecus afarensis*. *American Journal of Physical Anthropology* 60:279–317.

———. 1991. "Total morphological pattern" versus the "magic trait": Conflicting approaches to the study of early hominid bipedalism; pp. 99–111 in Y. Coppens and B. Senut (eds.), *Origine(s) de la Bipédie chez les Hominidés*. Centre National de la Recherche Scientifique, Paris.

Stewart T. D., M. Tiffany, J. L. Angel, and J. O. Kelley. 1986. Description of the human skeleton. pp. 49–70 in F. Wendorf, R. Schild, and A. E. Close (eds.), *The Wadi Kubbaniya Skeleton: A Late Paleolithic Burial from Southern Egypt*. Southern Methodist University, Dallas.

Stoneking, M. 1993. DNA and recent human evolution. *Evolutionary Anthropology* 2:60–73.

Stoneking, M., S. T. Sherry, A. J. Redd, and L. Vigilant. 1993. New approach to dating suggests a recent age for the human mtDNA ancestor; pp. 84–103 in M. J. Aitken, C. B. Stringer, and P. A. Mellars (eds.), *The Origin of Modern Humans and the Impact of Chronometric Dating*. Princeton University Press, Princeton, N.J.

Strait, D. S., and F. E. Grine. 2004. Inferring hominoid and early hominid phylogeny using craniodental characters: the role of fossil taxa. *Journal of Human Evolution* 47:399–452.

Strait, D. S., F. E. Grine, and M.A. Moniz. 1997. A reappraisal of early hominid phylogeny. *Journal of Human Evolution* 32:17–82.

Strait, D. S., B. G. Richmond, M. A. Spencer, C. F. Ross, P. C. Dechow, and B. A. Wood. 2007. Masticatory biomechanics and its relevance to early hominid phylogeny: An examination of palatal thickness using finite-element analysis. *Journal of Human Evolution* 52:585–599.

Stringer, C. B. 1974. Population relationships of later Pleistocene hominids: Multivariate study of available crania. *Journal of Archaeological Science* 1:317–342.

———. 1978. Some problems in Middle and Upper Pleistocene hominid relationships; pp. 395–418 in D. J. Chivers and K. Joysey (eds.), *Recent Advances in Primatology*. Academic Press, London.

———. 1979. A re-evaluation of the fossil human calvaria from Singa, Sudan. *Bulletin of the British Museum (Natural History), Geology* 32:77–83.

———. 1988. Archaic *Homo sapiens*; pp. 49–54 in I. Tattersall, E. Delson, and J. van Couvering (eds.), *Encyclopedia of Human Evolution and Prehistory*. Garland Press, New York.

———. 1996. Current issues in modern human origins; pp. 115–134 in E. Meikle, F. C. Howell, and N. Jablonski (eds.), *Contemporary Issues in Human Evolution*. California Academy of Sciences, San Francisco, Memoir 21.

———. 2002. Modern human origins: progress and prospects. *Philosophical Transactions of the Royal Society London, B* 357:563–579.

Stringer, C. B., and A. S. Brooks. 2000. Haua Fteah; p. 305 in E. Delson, I. Tattersall, J. A. Van Couvering and A. S. Brooks (eds.), *Encyclopedia of Human Evolution and Prehistory*. Garland Press, New York.

Stringer, C. B., L. Cornish, and P. Stuart-Macadam. 1985. Preparation and further study of the Singa skull from Sudan. *Bulletin of the British Museum (Natural History), Geology* 38:347–358.

Stringer, C., J.-J. Hublin, and B. Vandermeersch. 1984. The origin of anatomically modern humans in Western Europe; pp. 51–135 in F. H. Smith, and F. Spencer (eds.), *The Origins of Modern Humans: A World Survey of the Fossil Evidence*. Liss, New York.

Stynder, D., J. Moggi-Cecchi, L. R. Berger, and J. E. Parkington. 2001. Human mandibular incisors from the late Middle Pleistocene locality of Hoedjiespunt 1, South Africa. *Journal of Human Evolution* 41:369–383.

Su, D. F., and T. Harrison. 2008. Ecological implications of the relative rarity of fossil hominins at Laetoli. *Journal of Human Evolution* 55:672–681.

Susman, R. L. 1988. Hand of *Paranthropus robustus* from Member 1, Swartkrans: fossil evidence for tool behavior. *Science* 240:781–784.

———. 1989. New hominid fossils from the Swartkrans Formation (1979–1986): postcranial specimens. *American Journal of Physical Anthropology* 79:451–474.

———. 1991a. Species attribution of the Swartkrans thumb metacarpals: reply to Drs. Trinkaus and Long. *American Journal of Physical Anthropology* 86:549–552.

———. 1991b. Who made the Oldowan tools? Fossil evidence for tool behavior in Plio-Pleistocene hominids. *Journal of Anthropological Research* 47:129–151.

———. 1995. Thumbs, tools, and early humans: Response. *Science* 268:589.

———. 2008. Evidence Bearing on the Status of *Homo habilis* at Olduvai Gorge. *American Journal of Physical Anthropology* 137:356–361.

Susman, R. L., and T. M. Brain. 1988. New first metatarsal (SKX 5017) from Swartkrans and the gait of *Paranthropus robustus*. *American Journal of Physical Anthropology* 77:7–15.

Susman, R. L., and D. J. de Ruiter. 2004. New hominin first metatarsal (SK 1813) from Swartkrans. *Journal of Human Evolution* 47:171–181.

Susman, R. L., D. de Ruiter, and C. K. Brain. 2001. Recently identified postcranial remains of *Paranthropus* and early *Homo* from Swartkrans Cave, South Africa. *Journal of Human Evolution* 41:607–629.

Susman, R. L., and J. T. Stern. 1982. Functional morphology of *Homo habilis*. *Science* 217:931–934.

Susman, R. L., J. T. Stern Jr., and W. L. Jungers. 1984. Arboreality and bipedality in the Hadar hominids. *Folia Primatologica* 43:113–156.

Suwa, G. 1988. Evolution of the "robust" australopithecines in the Omo succession: Evidence from mandibular premolar morphology; pp. 199–222 in F. E. Grine (ed.), *Evolutionary History of the "Robust" Australopithecines*. Aldine de Gruyter, New York.

———. 1989. The premolar of KNM-WT 17000 and relative anterior to posterior dental size. *Journal of Human Evolution* 18:795–799.

Suwa, G., B. Asfaw, Y. Beyene, T. D. White, S. Katoh, S. Nagaoka, H. Nakaya, K. Uzawa, P. Renne, and G. WoldeGabriel. 1997. The first skull of *Australopithecus boisei*. *Nature* 389:489–492.

Suwa, G., B. Asfaw, Y. Haile-Selassie, T. D. White, S. Katoh, G. WoldeGabriel, W. K. Hart, H. Nakaya, and Y. Beyene. 2007. Early Pleistocene *Homo erectus* fossils from Konso, southern Ethiopia. *Anthropological Science* 115:133–151.

Suwa, G., B. Asfaw, R. T. Kono, D. Kubo, C. O. Lovejoy, and T. D. White. 2009b. The *Ardipithecus ramidus* skull and its implications for hominid origins. *Science* 369:68e1–68e7.

Suwa, G., R. T. Kono, S. W. Simpson, B. Asfaw, C. O. Lovejoy, and T. D. White. 2009a. Paleobiological implications of the *Ardipithecus ramidus* dentition. *Science* 369:94–99.

Suwa, G., T. D. White, and F. C. Howell. 1996. Mandibular postcanine dentition from the Shungura Formation, Ethiopia: Crown morphology, taxonomic allocations, and Plio-Pleistocene hominid evolution. *American Journal of Physical Anthropology* 101:247–282.

Suwa, G., B. A. Wood, and T. D. White. 1994. Further analysis of mandibular molar crown and cusp areas in Pliocene and early Pleistocene hominids. *American Journal of Physical Anthropology* 93:407–426.

Swisher, C. C., G. H. Curtis, T. Jacob, A. G. Getty, A. Suprijo, and Widiasmoro. 1994. Age of the earliest known hominids in Java, Indonesia. *Science* 263:1118–1121.

Sydow, W. 1969. Discovery of a Boskop skull at Otjiseva, near Windhoek, South West Africa. *South African Journal of Science* 65:77–82.

Tague, R. G., and C. O. Lovejoy. 1986. The obstetric pelvis of A. L. 288-1 (Lucy). *Journal of Human Evolution* 15:237–255.

Tamrat, E., N. Thouveny, M. Taïeb, and N. D. Opdyke. 1995. Revised magnetostratigraphy of the Plio-Pleistocene sedimentary sequence of the Olduvai Formation (Tanzania). *Palaeogeography, Palaeoclimatology, Palaeoecology* 114:273–283.

Tankard, A. J., and F. R. Schweitzer. 1974. The geology of Die Kelders Cave and environs: a palaeoenvironmental study. *South African Journal of Science* 70:365–369.

———. 1976. Textural analysis of cave sediments: Die Kelders, Cape Province, South Africa; pp. 289–316 in D. A. Davidson and M. L. Shackley (eds.), *Geoarchaeology*. Duckworth, London.

Tardieu, C. 1981. Morpho-functional analysis of the articular surfaces of the knee-joint in primates; pp. 68–80 in A. B. Chiarelli and R. S. Corruccini (eds.), *Primate Evolutionary Biology*. Springer, Berlin.

Tattersall, I. 1986. Species recognition in human paleontology. *Journal of Human Evolution* 15:165–175.

Teaford, M. F., and P. S. Ungar. 2000. Diet and the evolution of the earliest human ancestors. *Proceedings of the National Academy of Sciences, USA* 97:13506–13511.

Templeton, A. R. 1993. The "Eve" hypothesis: A genetic critique and reanalysis. *American Anthropologist* 95:51–72.

Thoma, A. 1984. Morphology and affinities of the Nazlet Khater Man. *Journal of Human Evolution* 13:287–296.

Thorpe, I. J. N. 2003. Anthropology, archaeology, and the origin of warfare. *World Archaeology* 35:145–165.

Tishkoff, S. A., E. Dietzsch, W. Speed, A. J. Pakstis, J. R. Kidd, K. Cheung, B. Bonné-Tamir, A. S. Santachiara-Benerecetti, P. Moral, M. Krings, S. Pääbo, E. Watson, N. Risch, T. Jenkins, and K. K. Kidd. 1996. Global patterns of linkage disequilibrium at the CD4 locus and modern human origins. *Science* 271:1380–1387.

Tobias, P. V. 1967a. *The Cranium and Maxillary Dentition of* Australopithecus (Zinjanthropus) boisei. *Olduvai Gorge*, vol. 2. Cambridge University Press, Cambridge. 264 pp.

———. 1967b. The hominid skeletal remains of Haua Fteah; pp. 337–352 in C. B. M. McBurney (ed.), *Haua Fteah and the Stone Age of the South East Mediterranean*. Cambridge University Press, Cambridge.

———. 1968. Middle and early Upper Pleistocene members of the genus *Homo* in Africa; 176–194 in G. Kurth (ed.), *Evolution and Hominisation*. 2nd ed. Fischer, Stuttgart.

———. 1971. *The Brain in Hominid Evolution*. Columbia University Press, New York. 170 pp.

———. 1980. *"Australopithecus afarensis"* and *A. africanus*: Critique and an alternative hypothesis. *Palaeontologia Africana* 23:1–17.

———. 1995. The place of *Homo erectus* in nature with a critique of the cladistic approach, pp. 31–46 in J. R. F. Bower, and S. Sartono (eds.), *Paleoanthropology. Human Evolution in its Ecological Context*, vol. 1. Pithecanthropus Centennial Foundation, Leiden.

Toussaint, M., G. A. Macho, P. V. Tobias, T. C. Partridge, and A. R. Hughes. 2003. The third partial skeleton of a late Pliocene hominin (StW 431) from Sterkfontein, South Africa. *South African Journal of Science* 99:215–223.

Trinkaus, E. 1984a. Does KNM-ER 1481A establish *Homo erectus* at 2.0 Myr BP? *American Journal of Physical Anthropology* 64:137–139.

———. 1984b. On affinities of the Forbes Quarry (Gibraltar-1) cranium. *Current Anthropology* 25:687–688.

———. 1992. Morphological contrasts between the Near Eastern Qafzeh-Skhul and late archaic human samples: Grounds for a behavioral difference; pp. 277–294 in T. Akazawa, K. Aoki, and T. Kimura (eds.), *The Evolution and Dispersal of Modern Humans in Asia*. Hokusen-Sha, Tokyo.

———. 1993a. Femoral neck-shaft angles of the Qafzeh-Skhul early modern humans, and activity levels among immature Near Eastern Middle Paleolithic hominids. *Journal of Human Evolution* 25:393–416.

———. 1993b. A note on the KNM-ER 999 hominid femur. *Journal of Human Evolution* 24:493–504.

———. 2004. Eyasi 1 and the suprainiac fossa. *American Journal of Physical Anthropology* 124:28–32.

———. 2005. Early modern humans. *Annual Review of Anthropology* 34:207–230.

Trinkaus, E., and J. C. Long. 1990. Species attribution of the Swartkrans Member 1 first metacarpals: SK 84 and SKX 5020. *American Journal of Physical Anthropology* 83:419–424.

Tuttle, R. H. 1987. Kinesiological inferences and evolutionary implications from Laetoli bipedal trails G-1, G-2,3, and A; pp. 503–523 in M. D. Leakey and J. M. Harris (eds.), *Laetoli: A Pliocene Site in Northern Tanzania*. Clarendon Press, Oxford.

Twiesselmann, F. 1991. La mandibule et le fragment de maxillaire supérieur de Loyangalani (rive est du lac Turkana, Kenya). *Anthropologie et Préhistoire* 102:77–95.

Ungar, P. S., and F. E. Grine. 1991. Incisor size and wear in *Australopithecus africanus* and *Paranthropus robustus*. *Journal of Human Evolution* 20:313–340.

Ungar, P. S., F. E. Grine, and M. F. Teaford. 2008. Dental microwear and diet of the Plio-Pleistocene hominin *Paranthropus boisei*. *PLoS One* 3:e2044. doi 10.1371/journal.pone.0002044.

———. 2006. Diet in early *Homo*: A review of the evidence and a new model of adaptive versatility. *Annual Review of Anthropology* 35:209–228.

Ungar, P. S., A. Walker, and K. Coffing. 1994. Reanalysis of the Lukeino Molar (KNM-LU 335). *American Journal of Physical Anthropology* 94:165–173.

Valladas, H., J. L. Reys, J. L. Joron, G. Valladas, O. Bar-Yosef, and B. Vandermeersch. 1988. Thermoluminescence dating of Mousterian "proto-Cro-Magnon" remains from Israel and the origin of modern man. *Nature* 331:614–616.

Vallois, H. V. 1951. La mandibule humaine fossile de la grotte du Porc-Épic, près Diré-Daoua (Abyssinie). *Anthropologie* 55:231–238.

———. 1952. Diagrammes sagittaux et mensurations individuelles des hommes fossiles d'Afalou-bou-Rhummel. *Travaux du Laboratoire d'Anthropologie et Archéologie Préhistorique du Musée du Bardo* 5:1–134.

Vallois, H. and J. Roche. 1958. The Acheulian mandible from Temara, Morocco. *Comptes Rendus Hebdomadaires des Séances de l'Académie des Sciences, Paris* 246:3113–3116.

Vandermeersch, B. 1981. *Les Hommes Fossiles de Qafzeh (Israel)*. Cahiers de Paléontologie. CNRS, Paris, 319 pp.

Vekua, A., D. Lordkipandize, G. P. Rightmire, J. Agusti, R. Ferring, G. Maisuradze, A. Mouskhelishvili, M. Nioradze, M. Ponce de Léon, M. Tappen, M. Tvalchrelidze, and C. Zollikofer. 2002. A new skull of early *Homo* from Dmanisi, Georgia. *Science* 297:85–89.

Vermeersch, P. M., E. Paulissen, S. Stokes, C. Charlier, P. Van Peer, C. Stringer, and W. Lindsay. 1998. A Middle Palaeolithic burial of a modern human at Taramsa Hill, Egypt. *Antiquity* 72:475–484.

Vignaud, P., P. Duringer, H. T. Mackaye, A. Likius, C. Blondel, J-R. Boisserie, L. de Bonis, V. Eisenmann, M-E. Etienne, D. Geraads, F. Guy, T. Lehmann, F. Lihoreau, N. Lopez-Martinez, C. Mourer-Chauviré, O. Otero, J-C. Rage, M. Schuster, L. Viriot, A. Zazzo, and M. Brunet. 2002. Geology and palaeontology of the Upper Miocene Toros-Menalla hominid locality, Chad. *Nature* 418:152–155.

Villiers, H. de. 1972. The first fossil human skeleton from South West Africa. *Transactions of the Royal Society of South Africa* 40:187–196.

———. 1973. Human skeletal remains from Border Cave, Ingwavuma District, KwaZulu, South Africa. *Annals of the Transvaal Museum* 28:229–256.

———. 1976. A second adult human mandible from Border Cave, Ingwavuma District, KwaZulu, South Africa. *South African Journal of Science* 72:212–215.

Villiers, H. de, and L. P. Fatti. 1982. The antiquity of the Negro. *South African Journal of Science* 72:212–215.

Vogel, J. C. 1969. Radiocarbon dating of Bushman rockshelter, Ohrigstad District. *South African Archaeological Bulletin* 24:56.

———. 2001. Radiometric dates for the Middle Stone Age in South Africa; pp. 261–268 in P. V. Tobias, M. A. Raath, J. Moggi-Cecchi, and G. A. Doyle (eds.), *Humanity from African Naissance to Coming Millenia: Colloquia in Human Biology and Palaeoanthropology*. Florence University Press, Florence.

Vrba, E. S. 1982. Biostratigraphy and chronology, based particularly on Bovidae, of southern hominid-associated assemblages: Makapansgat, Sterkfontein, Taung, Kromdraai, Swartkrans; also Elandsfontein (Saldanha), Broken Hill (now Kabwe) and Cave of Hearths; pp. 707–752

in H. de Lumley and M-A. de Lumley (eds.), Prétirage, 1er Congrès
International de la Palèontologie Humaine. Centre National pour la
Recherche Scientifique, Nice.
———. 1985. Ecological and adaptive changes associated with early
hominid evolution; pp. 63–71 in E. Delson (ed.), *Ancestors: The Hard
Evidence*. Alan R. Liss, New York.
———. 1995. The fossil record of African antelopes (Mammalia, Bovidae) in relation to human evolution and paleoclimate; pp. 385–424
in E. S. Vrba, G. H. Denton, T. C. Partridge, and L. H. Burckle (eds.),
Paleoclimate and Evolution, with Emphasis on Human Origins. Yale Uni-
versity Press, New Haven.
Wainscoat, J. S., A. V. S. Hill, A. L. Boyce, J. Flint, M. Hernandez, S. L.
Thein, J. M. Old, J. R. Lynch, A. G. Falusi, D. J. Weatherall, and J. B.
Clegg. 1986. Evolutionary relationships of human populations from
an analysis of nuclear DNA polymorphisms. *Nature* 319:491–493.
Walker, A. 1976. Remains attributable to *Australopithecus* in the East
Rudolf succession; pp.484–489 in Y. Coppens, F. C. Howell, G. Ll.
Isaac, and R. L. Leakey (eds.), *Earliest Man and Environments in the Lake
Rudolf Basin*. University of Chicago Press, Chicago.
———. 1984. Extinction in hominid evolution; pp. 119–152 in M. H.
Nitecki (ed.), *Extinctions*. University of Chicago Press, Chicago.
———. 1987. Fossil Galaginae from Laetoli; pp. 88–90 in M. D. Leakey
and J. M. Harris (eds.), *Laetoli: A Pliocene Site in Northern Tanzania*.
Clarendon Press, Oxford.
———. 1994. Early *Homo* from 1.8–1.5 million year deposits at Lake
Turkana, Kenya, pp. 167–173 in J. F. Franzen (ed.), *100 Years of Pith-
ecanthropus: The* Homo erectus *Problem*. Courier Forschunginstitut
Senckenberg, Frankfurt.
Walker, A., and R. Leakey. 1993. *The Nariokotome* Homo erectus *Skeleton*.
Harvard University Press, Cambridge, Mass. 468 pp.
Walker, A., R. E. Leakey, J. M. Harris, and F. H. Brown. 1986. 2.5-Myr
Australopithecus boisei from west of Lake Turkana, Kenya. *Nature*
322:517–522.
Walker, A., M. R. Zimmerman, and R. E. F. Leakey. 1982. A possible case
of hypervitaminosis A in *Homo erectus*. *Nature* 296:248–250.
Walker, A., and P. Shipman. 1996. *The Wisdom of the Bones*. Knopf, New
York. 368 pp.
Walker, A. C., and C. B. Ruff. 1993. The reconstruction of the pelvis; pp.
221–233 in A. Walker and R. Leakey (eds.), *The Nariokotome* Homo
erectus *Skeleton*. Harvard University Press, Cambridge, Mass. 468 pp.
Walker, A. C., B. Brown, and S. C. Ward. 1993. Squamosal suture of
cranium KNM-WT 17000. *American Journal of Physical Anthropology*
90:501–505.
Walker, A. C., and R. E. Leakey. 1988. The evolution of *Australopithecus
boisei*; pp. 247–258 in F. E. Grine (ed.), *Evolutionary History of the
"Robust" Australopithecines*. Aldine de Gruyter, New York.
Walker, J., R. A. Cliff, and A. G. Latham. 2006. U-Pb isotopic age of
the Stw 573 hominid from Sterkfontein, South Africa. *Science*
314:1592–1594.
Wall, J. D. and S. K. Kim. 2007. Inconsistencies in Neanderthal genomic
DNA sequences. *PLOS Genetics* 3:1862–1866.
Wallace, J. 1972. The dentition of South African early hominids: A study
of form and function. Unpublished PhD dissertation, University of
Witwatersrand, Johannesburg.
Walter, R. C., and J. L. Aronson. 1993. Age and source of the Sidi
Hakoma Tuff, Hadar Formation, Ethiopia. *Journal of Human Evolution*
25:229–240.
Walter, R. C., R. T. Buffler, J. H. Bruggemann, M. M. M. Guillaume, S. M.
Berhe, B. Negassi, Y. Libeskal, H. Cheng, R. L. Edwards, R. von Cosel,
D. Néraudeau, and M. Gagnon. 2000. Early human occupation of the
Red Sea coast of Eritrea during the last interglacial. *Nature*
405:65–69.
Walter, R. C., P. C. Manega, R. L. Hay, R. E. Drake, and G. H. Curtis.
1991. Laser-fusion ^{40}Ar/^{39}Ar dating of Bed I, Olduvai Gorge, Tanzania.
Nature 385:145–149.
Ward, C. V. 2002. Interpreting the posture and locomotion of *Australo-
pithecus afarensis*: Where do we stand? *Yearbook of Physical Anthropol-
ogy* 45:185–215.
Ward, C. V., M. G. Leakey, and A. Walker. 2001. Morphology of *Aus-
tralopithecus anamensis* from Kanapoi and Allia Bay, Kenya. *Journal of
Human Evolution* 41:255–368.
Ward, S., and A. Hill. 1987. Pliocene hominid partial mandible from
Tabarin, Baringo, Kenya. *American Journal of Physical Anthropology*
72:21–37.
Ward, S. C. 1991. Taxonomy, paleobiology, and adaptations of the
"robust" australopithecines. *Journal of Human Evolution* 21:469–483.

Washburn, S. L., and B. Patterson. 1951. Evolutionary importance of the
South African "man-apes." *Nature* 167:650–651.
Weinert, H. 1950. Über die neuen Vor-und Frühmenschenfunde aus
Afrika, Java, China und Frankreich. *Zeitschrift für Morphologie und
Anthropologie* 42:113–148.
Wells, L. 1950. The Border cave skull, Inguwavuma District, Zululand.
American Journal of Physical Anthropology 8:241–243.
Wendorf, F. 1968. Site 117: a Nubian final Palaeolithic graveyard near
Jebel Sahaba, Sudan; pp. 954–1040, in F. Wendorf (ed.), *The Prehistory
of Nubia*. Southern Methodist University Press, Dallas.
Wendorf, F., R. Said, and R. Schild. 1970. Egyptian prehistory: some new
concepts. *Science* 169:1161–1171.
Wendorf, F., and R. Schild. 1986. *The Wadi Kubbaniya Skeleton: A Late
Paleolithic Burial from Southern Egypt*. Southern Methodist University
Press, Dallas. 85 pp.
Westphal, M., J. Chavaillon, and J.-J. Jaeger 1979. Magnétostratigraphie
des dépôts pléistocènes de Melka-Kunturé (Ethiopie): Premières données. *Bulletin de la Société Géologique de France* 21:237–241.
Whitcome, K. K., L. J. Shapiro, and D. E. Lieberman. 2007. Fetal load
and the evolution of lumbar lordosis in bipedal hominins. *Nature*
450:1075–1080.
White, D. D., and D. Falk. 1999. A quantitative and qualitative reanalysis of the endocast from the juvenile *Paranthropus* specimen L338y-6
from Omo, Ethiopia. *American Journal of Physical Anthropology*
110:399–406.
White, T. D. 1986a. *Australopithecus afarensis* and the Lothagam Mandible: Fossil man—new facts, new ideas. *Anthropos (Brno)* 23: 79–90.
———. 1986b. Cut marks on the Bodo cranium: A case of prehistoric
defleshing. *American Journal of Physical Anthropology* 69:503–509.
———. 1994. Ape and hominid limb length. *Nature* 369:194.
———. 1995. African omnivores: global climatic change and Plio-
Pleistocene hominids and suids; pp. 369–384 in E. S. Vrba, G. H.
Denton, T. C. Partridge, and L. H. Burckle (eds.), *Paleoclimate and
Evolution, with Emphasis on Human Origins*. Yale University Press,
New Haven.
———. 2002. Earliest hominids; pp. 407–417 in W. C. Hartwig (ed.), *The
Primate Fossil Record*. Cambridge University Press, Cambridge.
———. 2003. Early hominids: Diversity or distortion? *Science* 299:1994–
1997.
———. 2009. Ladders, bushes, punctuations, and clades: Hominid paleobiology in the late twentieth century; pp. 122–148 in D. Sepkoski
and M. Ruse (eds.), *The Paleobiological Revolution*. University of Chicago Press, Chicago.
White, T. D., S. H. Ambrose, G. Suwa, D. F. Su, D. DeGusta, R. L. Bernor,
J.-R. Boisserie, M. Brunet, E. Delson, S. Frost, N. Garcia, I. X. Giaourtsakis, Y. Haile-Selassie, M. Teaford, and E. Vrba. 2009a. Macrovertebrate paleontology and the Pliocene habitat of *Ardipithecus ramidus*.
Science 369:87–93.
White, T. D., B. Asfaw, Y. Beyene, Y. Haile-Selassie, C. O. Lovejoy, G.
Suwa, and G. WoldeGabriel. 2009b. *Ardipithecus ramidus* and the
paleobiology of early hominids. *Science* 369:75–86.
White, T. D., B. Asfaw, D. DeGusta, H. Gilbert, G. D. Richards, G. Suwa,
and F. C. Howell. 2003. Pleistocene *Homo sapiens* from Middle Awash,
Ethiopia. *Nature* 423:742–747.
White, T. D., B. Asfaw, and G. Suwa. 2005. Pliocene hominid fossils from
Gamedah, Middle Awash, Ethiopia. *Transactions of the Royal Society of
South Africa* 60:79–83.
White, T. D., D. C. Johanson, and W. H. Kimbel. 1981. *Australopithecus
africanus*: its phyletic position reconsidered. *South African Journal of
Science* 77:445–470.
White, T. D., and G. Suwa. 1987. Hominid footprints at Laetoli: Facts and
interpretations. *American Journal of Physical Anthropology* 72:485–514.
White, T. D., G. Suwa, and B. Asfaw. 1994. *Australopithecus ramidus*, a
new species of early hominid from Aramis, Ethiopia. *Nature*
371:306–312.
———. 1995. Corrigendum. *Australopithecus ramidus*, a new species of
early hominid from Aramis, Ethiopia. *Nature* 375:88.
White, T. D., G. Suwa, W. K. Hart, R. C. Walter, G. WoldeGabriel, J. de
Heinzelin, J. D. Clark, B. Asfaw, and E. Vrba. 1993. New discoveries of
Australopithecus at Maka in Ethiopia. *Nature* 366: 261–265.
White, T. D., G. Suwa, S. Simpson, and B. Asfaw. 2000. Jaws and teeth of
Australopithecus afarensis from Maka, Middle Awash, Ethiopia. *Ameri-
can Journal of Physical Anthropology* 111:45–68.
White, T. D., G. WoldeGabriel, B. Asfaw, S. Ambrose, Y. Beyene, R. L.
Bernor, J-R. Boisserie, B. Currie, H. Gilbert, Y. Haile-Selassie, W. K.
Hart, L. J. Hlusko, F. C. Howell, R. T. Kono, T. Lehmann, A. Louchart,

C. O. Lovejoy, P. R. Renne, H. Saegusa, E. S. Vrba, H. Wesselman, and G. Suwa. 2006. Asa Issie, Aramis and the origin of *Australopithecus. Nature* 440:883–889.

Wilford, J. N. 1987. New fossil is forcing family tree revisions. *New York Times*, April 14:C1–C2.

WoldeGabriel, G., S. Ambrose, D. Barboni, R. Bonnefille, L. Bremond, B. Currie, D. DeGusta, W. H. Hart, A. M. Murray, P. R. Renne, M. C. Jolly-Saad, K. M. Stewart, and T. D. White. 2009. The geological, isotopic, botanical, invertebrate, and lower vertebrate surroundings of *Ardipithecus ramidus. Science* 326:65e1–65e5.

WoldeGabriel, G., Y. Haile-Selassie, P. R. Renne, W. K. Hart, S. H. Ambrose, B. Asfaw, G. Heiken, and T. White. 2001. Geology and palaeontology of the Late Miocene Middle Awash valley, Afar Rift, Ethiopia. *Nature* 412:175–178.

WoldeGabriel, G., P. Renne, T. D. White, G. Suwa, J. de Heinzelin, W. K. Hart, and G. Heiken. 1995. Reply. Age of early hominids. *Nature* 376:559.

WoldeGabriel, G., T. D. White, G. Suwa, P. Renne, J. de Heinzelin, W. K. Hart, and G. Heiken. 1994. Ecological and temporal placement of early Pliocene hominids at Aramis, Ethiopia. *Nature* 371:330–333.

Wolpoff, M. H. 1989. Multiregional evolution: the fossil alternative to Eden; pp. 62–108 in P. Mellars and C. B. Stringer (eds.), *The Origin and Dispersal of Modern Humans: Behavioural and Biological Perspectives.* Edinburgh University Press, Edinburgh.

———. 1999. *Paleoanthropology.* McGraw-Hill, Boston, 878 pp.

Wolpoff, M. H., J. Hawks, B. Senut, M. Pickford, and J. Ahern. 2006. An ape or *the* ape: Is the Toumaï Cranium TM 266 a hominid? *PaleoAnthropology* 2006:36–50.

Wolpoff, M. H., B. Senut, M. Pickford, and J. Hawks. 2002. *Sahelanthropus* or "*Sahelpithecus*"? *Nature* 419:581–582.

Wolpoff, M. H., A. G. Thorne, F. H. Smith, D. W. Frayer, and G. G. Pope. 1994. Multiregional evolution: A world-wide source for modern human populations; pp. 175–199 in M. Nitecki and D. Nitecki (eds.), *Origins of Anatomically Modern Humans.* Plenum Press, New York.

Woo, J. K. 1964. A newly discovered mandible of the *Sinanthropus* type: *Sinanthropus lantianensis. Scientia Sinica* 13:891–911.

Wood, B. 1985. Early *Homo* in Kenya, and its systematic relationships; pp. 206–214 in E. Delson (ed.), *Ancestors: The Hard Evidence.* Liss, New York.

———. 1991. *Koobi Fora Research Project: Volume 4. Hominid Cranial Remains.* Clarendon Press, Oxford.

———. 1992. Origin and evolution of the genus *Homo. Nature* 355:783–790.

———. 2002. Hominid revelations from Chad. *Nature* 418:133–135.

———. 2005. A tale of two taxa. *Transactions of the Royal Society of South Africa* 60:91–94.

Wood, B., and L. C. Aiello. 1998. Taxonomic and functional implications of mandibular scaling in early hominins. *American Journal of Physical Anthropology* 105:523–538.

Wood, B., and M. Collard. 1999. The human genus. *Science* 284:65–71.

Wood, B., and P. Constantino. 2007. *Paranthropus boisei*: Fifty years of evidence and analysis. *Yearbook of Physical Anthropology* 50:106–132.

Wood, B., and D. E. Lieberman. 2001. Craniodental variation in *Paranthropus boisei*: A developmental and functional perspective. *American Journal of Physical Anthropology* 116:113–25.

Wood, B., and B. G. Richmond. 2000. Human evolution: taxonomy and paleobiology. *Journal of Anatomy* 196:19–60.

Wood, B., and D. Strait. 2004. Patterns of resource use in early *Homo* and *Paranthropus. Journal of Human Evolution* 46:119–162.

Wood, B. A., and F. Van Noten. 1986. Preliminary observations on the BK 8518 mandible from Baringo, Kenya. *American Journal of Physical Anthropology* 69:117–127.

Wood, B., C. Wood, and L. Koningsberg. 1994. *Paranthropus boisei*: An example of evolutionary stasis? *American Journal of Physical Anthropology* 95:117–136.

Wood, B. A., L. C. Aiello, C. Wood, and C. Key (1998) A technique for establishing the identity of 'isolated' fossil hominin limb bones. *Journal of Anatomy* 193:61–72.

Woodward, A. 1921. A new cave man from Rhodesia, South Africa. *Nature* 108:371–372.

———. 1938. A fossil skull of an ancestral Bushman from the Anglo-Egyptian Sudan. *Antiquity* 12:193–195.

Wrinn, P. J., and W. J. Rink. 2003. ESR dating of tooth enamel from Aterian levels at Mugharet el 'Aliya (Tangier, Morocco). *Journal of Archaeological Science* 30:123–133.

Wynn, J. G. 2000. Paleosols, stable carbon isotopes, and paleoenvironmental interpretation of Kanapoi, Northern Kenya. *Journal of Human Evolution* 39:411–432.

Wynn, J. G., Z. Alemseged, R. Bobe, D. Geraads, D. Reed, and D. C. Roman. 2006. Geological and palaeontological context of a Pliocene juvenile hominin at Dikika, Ethiopia. *Nature* 443:332–336.

Yokoyama, Y., C. Falguères, and M.-A. de Lumley. 1997. Datation directe d'un crâne Proto-Cro-Magnon de Qafzeh par la spéctrometrie gamma non descructive. *Comptes Rendus de l'Académie des Sciences, Paris, Série IIA*, 324:773–779.

Zilberman, U., P. Smith, and G. H. Sperber. 1990. Components of australopithecine teeth: A radiographic study. *Human Evolution* 5:515–529.

Zipfel, B., J. M. DeSilva, R. S. Kidd. 2009. Earliest complete hominin fifth metatarsal- implications for the evolution of the lateral column of the foot. *American Journal of Physical Anthropology* 140:532–545.

Zollikofer, C. P. E., M. S. Ponce de León, D. E. Lieberman, F. Guy, D. Pilbeam, A. Likius, H. T. Mackaye, P. Vignaud, and M. Brunet. 2005. Virtual cranial reconstruction of *Sahelanthropus tchadensis. Nature* 434:755–759.

PART FOUR

LAURASIATHERIA

TWENTY-SIX

Creodonta

MARGARET E. LEWIS AND MICHAEL MORLO

The order Creodonta was first named by Cope (1875) and removed from its original placement in the order Carnivora (Cuvier, 1822). Some researchers still maintain that creodonts and carnivorans (or carnivoramorphans) are sister taxa (e.g., Tedford, 1976; Savage, 1977; Novacek and Wyss, 1986; Wozencraft, 1989; Flynn and Wesley-Hunt, 2005), though not all agree (Fox and Youzwyshyn, 1994; Polly, 1996). While both creodonts and carnivorans possess carnassials, creodont carnassials are either P4/M1 and m1/m2 (Oxyaenidae, limnocyonine hyaenodontids) or P4/M1/M2 and m1/m2/m3 (other Hyaenodontidae). Members of the order Carnivora all have carnassials at P4 and m1. While creodonts and carnivorans share an ossified tentorium and a few basicranial and tarsal features, their possible synapomorphies are few (Flynn et al., 1988; Wyss and Flynn, 1993; Rose, 2006).

Like carnivorans, creodonts vary greatly in their postcranial adaptations. Unlike carnivorans, most creodonts have fissured terminal phalanges, with the exception of the European Proviverrinae, and an unfused scaphoid and lunate. In general, hyaenodontid limbs are relatively short with respect to body size in comparison to carnivorans (Mellett, 1977), but more elongate and gracile than in the oxyaenids. A more general description of creodont craniodental and postcranial features can be found elsewhere (e.g., Jenkins and Camazine, 1977; Gingerich and Deutsch, 1989; Rose, 1990; Gebo and Rose, 1993; Polly, 1996; Gunnell, 1998; Morlo and Habersetzer, 1999; Morlo and Gunnell, 2003).

All known Afro-Arabian creodonts belong to the family Hyaenodontidae and range in age from the Eocene (and possibly Paleocene) to the middle Miocene. While the earliest African members of this family are found in the north, this family eventually spread southward into eastern and southern Africa. In Africa, as elsewhere, this family is diverse both in body size and morphology, with body size ranging from some of the smallest of forms to the largest hyaenodontid known *(Megistotherium)*. A brief list of the Afro-Arabian hyaenodontid subfamilies and genera can be found in table 26.1, while the distribution of species can be found in tables 26.2 and 26.3 and, later, in figure 26.8.

ABBREVIATIONS

AMNH, American Museum of Natural History, New York; BMNH, The Natural History Museum, London; CGM, Cairo Geological Museum, Egypt; KNM, Kenya National Museums, Nairobi, Kenya; UM, Uganda Museum, Kampala, Uganda; YPM, Peabody Museum of Natural History, Yale University, New Haven.

Systematic Paleontology

Order CREODONTA Cope, 1875
Family HYAENODONTIDAE Leidy, 1869

Diagnosis Tritubercular to sectorial molars with carnassial blades in P4, M1, M2, and m1, m2, m3 (except in Limnocyoninae). M3 transverse or absent.

TABLE 26.1
Afro-Arabian genera and subfamilies

Subfamily	Genus
Apterodontinae	*Apterodon*
	Quasiapterodon
Hyainailourinae	*Akhnatenavus*
	Anasinopa
	Buhakia
	Dissopsalis
	Hyainailouros
	Isohyaenodon
	Leakitherium
	Masrasector
	Megistotherium
	Metapterodon
	Metasinopa
	Pterodon
Koholiinae	*Koholia*
Proviverrinae	*Boualitomus*
	Tinerhodon
Teratodontinae	*Teratodon*

TABLE 26.2
Distribution of Afro-Arabian hyaenodontid creodonts from the Paleocene through the Oligocene

Abbreviations: LO = late Oligocene, EO = early Oligocene, E/O = Eocene/Oligocene, LE = late Eocene, EE = early Eocene, LP = late Paleocene, U = Upper, L = Lower

Time	Site	Akhnatenavus leptognathus	Apterodon altidens	Apterodon macrognathus	Apterodon saghensis	Apterodon sp.	Boualitomus marocanensis	Hyainailouros sp.	Koholia atlasense	Masrasector aegypticum	Masrasector ligabuei	Masrasector nov. sp.	Metapterodon brachycephalus	Metapterodon markgrafi	Metapterodon schlosseri	Metapterodon sp.	Metasinopa ethiopica	Metasinopa fraasi	"Pterodon" africanus	Pterodon phiomensis	Pterodon syrtos	Quasiapterodon minutus	Tinerhodon disputatum	Hyaenodontidae indet.
LO	Chilga, Eth.							x																
EO	Jebel Qatrani (U), Eg.		x							x	x	x	?	x				x			x	x		
E/O	Dur el Talhah, Lib.																							x
LE	Jebel Qatrani (L), Eg.	x	x	x						?	x		?	x				x		x	x			
	Taqah, Oman										x													
	Qasr el-Sagha, Eg.				x	x										x								
EE	El Kohol, Alg.								x															
	Gour Lazib, Alg.																							x
	Grand Daoui, Mor.						x																	
LP	Adrar Mgorn, Mor.																						x	

Remarks For many years, the earliest African representatives of this family dated to the late Eocene (Savage, 1978). Recent finds push the earliest occurrence back to at least the earliest Eocene and possibly the late Paleocene (e.g., Gheerbrant, 1995; Gheerbrant et al., 2006).

Subfamily APTERODONTINAE Szalay, 1967

Diagnosis Lower molars with paraconid, protoconid, and more or less complex talonid. Upper molars triangular, as in Tritemnodon and Sinopa, with protocone prominent, subequal paracone, metacone, and styles. Teeth tubercular rather than sectorial (Osborn, 1909:417).

Age and Occurrence Late Eocene to Oligocene; Europe and Egypt.

Remarks The genus Apterodon was originally placed within the subfamily Hyaenodontinae. Szalay (1967) suggested that this genus should be placed in a separate tribe from other hyaenodontines. Van Valen (1967) agreed but wanted to abandon the use of formal tribes until the subfamily was better known. Holroyd (1994) later elevated the tribe to subfamily status.

Genus APTERODON Fischer, 1880

Synonymy ?Apterodon sp. nov. Simons, 1868; Dasyurodon Andreae, 1887; Pterodon Andrews, 1904 (P. macrognathus only).

Diagnosis As for subfamily.

Type Species Apterodon gaudryi Fischer, 1880.

Age and Occurrence Late Eocene to Oligocene; Europe and Egypt.

Afro-Arabian Species Apterodon altidens Schlosser, 1910; A. macrognathus Andrews, 1904; A. saghensis Simons and Gingerich, 1976.

Remarks Van Valen (1966) removed Apterodon from the Creodonta and placed it in the Mesonychidae. Szalay (1967) later reaffirmed the creodont affinities of Apterodon and was supported by Van Valen (1967).

The oldest member of this genus is found in northern Africa in the late Eocene (Holroyd et al., 1996). Simons and Gingerich (1976) suggest that Apterodon probably evolved in isolation in Africa and dispersed around the Tethys Sea to Europe in the early Oligocene. They based this on the relative frequency of Apterodon in the Fayum deposits in contrast to the relatively low number of species and specimens (one per species) from Europe. A. macrognathus may have been ancestral to later species in the Fayum, as well as to European taxa (Tilden et al., 1990; Holroyd, 1994). Others have suggested that Apterodon originated in Asia and crossed into northern Africa before continuing to Europe (Lange-Badré and Böhme, 2005).

Genus QUASIAPTERODON Lavrov, 1999

Synonymy Apterodon Fischer, 1880 (partim).

Diagnosis After Lavrov (1999). Small, with skull length about 100 mm. Preorbital opening ahead of P4 posterior edge. Protocone developed on P4 and P3. Paraconids of m1 and m2 strongly displaced medially from axis of dental row; their talonids retain rudimentary interior basin.

Type Species Apterodon minutus Schlosser, 1910.

Age and Occurrence Late Eocene to Oligocene; Egypt.

Afro-Arabian Species Quasiapterodon minutus (Schlosser, 1910).

Remarks Quasiapterodon minutus originally was placed in Apterodon Schlosser, 1910. After several authors had stated that it probably did not belong in this genus (Tilden et al., 1990; Holroyd, 1994), Lavrov (1999) formally created the genus Quasiapterodon. There is some doubt, however, as to whether

TABLE 26.3
Distribution of Afro-Arabian hyaenodontid creodonts during the Miocene

Site	Anasinopa haasi	Anasinopa leakeyi	Buhakia moghraensis	Dissopsalis pyroclasticus	Hyainailouros fourtaui	Hyainailouros napakensis	Hyainailouros nyanzae	Hyainailouros sulzeri	Hyainailouros sp.	Isohyaenodon andrewsi	Isohyaenodon matthewi	Isohyaenodon pilgrimi	Leakitherium hiwegi	Leakitherium sp.	Megistotherium osteothlastes	Metapterodon kaiseri	Metapterodon stromeri	Metapterodon zadoki	Metasinopa napaki	Teratodon enigmae	Teratodon spekei	cf. Teratodon	Hyaenodontidae indet.
MIDDLE MIOCENE																							
Gafsa (L), Tun.																							X
Baringo-L2/56, Ken.					X																		
Grillental, Namib.								X															
Arrisdrift, Namib.								X															
Tugen Hills-Kabarsero, Ken.															X								
Tugen Hills-Bartule, Ken.															X								
Kaboor, Ken.					X																		
Fort Ternan, Ken.		?							X														
Baringo-Chaparawa, Ken.															X								
EARLY MIOCENE																							
Wadi Moghra, Eg.			X		X										X							X	
Negev, Israel	X																						
Gebel Zelten, Lib.	?														X								
Maboko, Ken.		X																					
Muruarot 2, Ken.		X								X													
Ombo, Ken.							X			X													
Mfwanganu, Ken.		X																					
Rusinga 2, Ken.								X															
R. 106,31, Ken.		X																					
Rusinga, Ken.							X			X	X	X				X		X					
R.1,1a,3,12,18 Ken.							X			X			X										
Napak, Ug.						X	X					X		X				X	X		X		
Moroto I, Ug.															X								
Koru, Ken.																					X		
Songhor, Ken.										X	X	X						X		X	X		
Karungu, Ken.		X														X							
Langental, Namib.																	X						
Elisabethfeld, Namib.																X							

it belongs in Apterodontinae or should instead be placed close to *"Metasinopa" ethiopica* (B. Lange-Badré, pers. comm.; and M.M., pers. obs.).

Subfamily HYAINAILOURINAE Pilgrim, 1932 (emend.)

Synonymy Pterodontidae Polly, 1996; Hyaenaelurinae Egi et al., 2007.

Type Genus Hyainailouros Biedermann, 1863.

Diagnosis Emended after Holroyd (1999). Hyaenodontids with metaconid small to absent, with distinct but unbasined talonid on lower molars, connate metacone and paracone present on M1–M2, a weak to absent P3 lingual cingulum, P4 lacking continuous lingual cingulum, relatively large anterior keels on lower molars, m3 talonid reduced relative to that of m1–m2, lower molar protoconids and paraconids subequal in length, circular subarcuate fossa present on petrosal, and nuchal crest not extending laterally to mastoid processes. Emended to allow for the presence of a small metaconid on m3 as partly present in *Dissopsalis* and related taxa. *Orienspterodon* is excluded from this subfamily due to the fact that the talonid of m3 is less reduced than the talonid of m2 (discussed later).

Age and Occurrence Early Eocene to late middle Miocene; northern and eastern Africa, Europe, questionable from Asia and North America.

Remarks Pilgrim named this subfamily using the Latinized form "Hyaenaelurinae," even though he correctly cited the type genus as *Hyainailouros* (the Greek version), as created by Biedermann (1863). This has led to much nomenclatural confusion. The ICZN 35.4.1 (1999) states that a familial name has to be corrected if based on an incorrect emendation of the type genus name. As *Hyainailouros* is correct, the subfamily has to be based on this name and not on its (incorrectly) Latinized version *Hyaenaelurus*. We thus emend "Hyaenaelurinae" Pilgrim, 1932 to the correct form Hyainailourinae Pilgrim, 1932.

The genera included in this subfamily have long been considered to be members of the Hyaenodontinae despite the fact that Pilgrim (1932) placed them in their own subfamily. Pilgrim's taxonomy was not followed by subsequent authors until Polly (1996) provided a phylogenetic analysis supporting Pilgrim. Holroyd (1999) revised Polly's original diagnosis of this taxon, which he had named Pterodontinae, as she noted that his diagnosis is based almost solely upon upper dentitions and crania, while the sole feature of the lower dentition, the loss of the metaconid, is known to have evolved independently in other hyaenodontid lineages.

Hyainailourine origins may lie with the early Eocene "proviverrine" genus *Arfia* (Van Valen, 1967; Polly, 1996; Peigné et al., 2007), which is now known to have occurred in all northern continents (Gingerich, 1989; Smith and Smith, 2001; Lavrov and Lopatin, 2004; Smith et al., 2006). The supposed earliest true hyainailourine is *Francotherium* from the late early Eocene (MP 10) of France (Polly, 1996). Holroyd (1999) also includes the middle Eocene Asian taxon *"Pterodon" dakhoensis* in the subfamily. It is clear, however, that the generic assignment of this species does not lie within *Pterodon* (Lavrov, 1996), leading Egi et al. (2007) to rename this species *Orienspterodon dakhoensis*. More importantly, assignment of *Orienspterodon* to Hyainailourinae is unlikely due to the fact that the talonid is less reduced in m3 than in m2. Instead, the genus may be close to the South Asian "proviverrines" *Paratritemnodon*, *Kyawdawia*, and *Yarshea* that have been interpreted as of separate descent from *Arfia* and thus as a sister group of Hyainailourinae (Peigné et al., 2007).

Details of the tooth eruption sequence that distinguish hyainailourines from proviverrines as well as from hyaenodontines place *Dissopsalis* and *Buhakia* into this subfamily (Morlo et al., 2007). The consequence of this move is that the undisputed close relatives of these genera, *Metasinopa* and *Anasinopa*, should be interpreted as hyainailourines as well. Questionable is *Masrasector*, which has been interpreted either as an ancestor of *Anasinopa* (Tchernov et al., 1987) or as a possible descendant of an early Eocene prototomine-like proviverrine (Peigné et al., 2007). Only in the first case would *Masrasector* belong to the Hyainailourinae (discussed later).

Genus *AKHNATENAVUS* Holroyd, 1999

Synonymy Pterodon Osborn, 1909:419 *(partim)*; *Pterodon* Simons, 1968:18.

Diagnosis Differs from *Pterodon* and *Metapterodon* in its smaller size and in having more gracile and relatively smaller premolars; presence of diastemata between p1 and p2, p2 and p3; smaller and shorter talonids on m1 and m2; lingual bulge on p3; slender canines; paraconid smaller relative to the protoconid. Differs from *Metapterodon* in retaining a m3 talonid.

Type Species Pterodon leptognathus Osborn, 1909.

Age and Occurrence Late Eocene/early Oligocene; northern Africa.

Afro-Arabian Species Akhnatenavus leptognathus (Osborn, 1909).

Genus *ANASINOPA* Savage, 1965
Figure 26.1

Diagnosis Hyainailourine with dental formula 3.1.?4.3/ 3.1.4.3; skull elongate and jaws slender; two-rooted p1; lower premolars compressed, crowded posteriorly, length slightly greater than height; p4 with distinct talonid; P4 tubercular, parastyle smaller than metacone; M1 + 2 tritubercular, triangular; metacone and paracone close together but not connate; metastyle shearing; metaconule and paraconule present; protocone V shaped; M3 present; m1 smallest and m3 largest; protoconid and paraconid subequal with height approximately equal to trigonid length; metaconid much smaller; talonid basined; m1 and m2 talonid length slightly less than trigonid; m3 talonid much reduced.

Type Species Anasinopa leakeyi Savage, 1965.

Age and Occurrence Early to middle? Miocene; eastern and northern Africa, Israel.

Afro-Arabian Species Anasinopa haasi Tchernov et al., 1987; *A. leakeyi* Savage, 1965.

Remarks This genus was first named from numerous fragmentary dental and mandibular specimens from Kenya (figure 26.1). Members of this genus are relatively small for hyaenodontids, with *A. leakeyi* being about the size of the European wolf *(Canis lupus)*(Savage, 1965) and *A. haasi* being much smaller. Savage (1965) notes that the presence of metaconids on the lower molars places this genus and species firmly in the proviverrines (as understood at that time).

Savage (1965) suggests that this genus represents an intermediate stage between African "*Sinopa*" (now *Metasinopa ethiopica*) and *Tritemnodon* from the middle Eocene of North

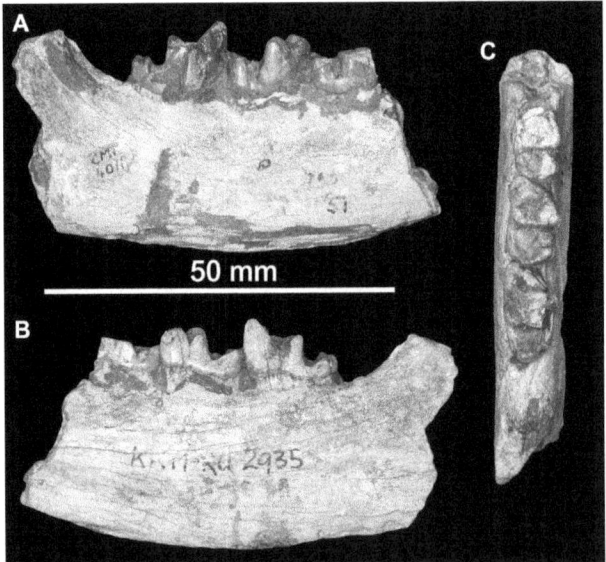

FIGURE 26.1 KNM RU 2935 (= CMF.4018), *Anasinopa leakeyi*, right mandible fragment with m1–m3 from Rusinga Island, Kenya (Savage, 1965:260): A) buccal, B) lingual, and C) occlusal views.

America. However, he also notes that it is less derived than specimens of *Metasinopa fraasi*. Tchernov et al. (1987) suggest that *Anasinopa* is a direct descendant of the Oligocene genus *Masrasector* from the Fayum. This genus was not known to Savage in 1965.

Van Valen (1967), however, transferred *Anasinopa* to *Paracynohyaenodon* due to what he believed was Savage's reliance on incorrect drawings in the original description of *Paracynohyaenodon*. This view has not been supported in the literature and *Anasinopa* has remained as a distinctly Afro-Arabian genus.

Genus *BUHAKIA* Morlo et al., 2007

Diagnosis From Morlo et al. (2007). Differs from *Pterodon* and *Metapterodon* in having larger talonids on m1 and m2 that point posteriorly instead of posterolingually. Differs from all other African hyainailourines in lacking a metaconid on m2, but with relatively large talonids.

Type Species Buhakia moghraensis Morlo et al., 2007.

Age and Occurrence Early Miocene; northern Africa.

Afro-Arabian Species Buhakia moghraensis Morlo et al., 2007.

Remarks The holotype of the sole species of the genus is a juvenile mandibular fragment (DPC 8994) from Wadi Moghra (= Moghara), Egypt. This species represents a hypercarnivorous member of a pre-Miocene diversification of middle- to large-sized hyaenodontids (Morlo et al., 2007). The interpretation of hypercarnivory is based on the reduction of the metaconids and trenchant character of the cristid obliquum on the lower molars.

Genus *DISSOPSALIS* Pilgrim, 1910

Diagnosis After Colbert (1933); Savage (1965); Barry (1988). Dental formula ?.1.4.3./?.1.4.3. Mandible slightly bowed and with distinct swelling under p4. Protocone prominent, anterior and remote from paracone; paracone much lower and smaller than metacone; metaconule and paraconule present but reduced; parastyle small or absent; very long, trenchant, posteriorly directed metastyle. Premolars robust with well-developed cingulum; P3 with slight posterolingual expansion; P4 with large lobate and bulbous lingual cusp (protocone), slightly inflated paracone, and parastyle small or absent; P4 almost as large as M1; M1 and M2 with long and very narrow, anterolingually directed protocone: paracone and metacone closely appressed, but not connate (*contra* Savage, 1965); M3 small, narrow, and transverse; p3 with small anterior accessory cusp, weak talonid, and salient entocristid; p4 with tall and robust protoconid and distinct talonid with a basin formed by an entocristid; p4 larger than m1; molars trenchant; m1 with three cusped trigonid and basined talonid; metaconid small; m2 with trenchant trigonid; greatly reduced metaconid; short, basined talonid; m3 with trenchant trigonid; metaconid vestigial or absent; talonid only a small trenchant tubercle.

Type Species Dissopsalis carnifex Pilgrim, 1910.

Age and Occurrence Middle Miocene; Asia and eastern Africa.

Afro-Arabian Species Dissopsalis pyroclasticus Savage, 1965.

Remarks This genus was first described by Pilgrim (1910) from fragmentary material from the Chinji Formation in the Siwalik sequence of northern Pakistan. Savage (1965) later named an additional species, *D. pyroclasticus*, from Kenya.

The relationship of *Dissopsalis* to other taxa has been highly disputed, as the degree of carnassial specialization in this genus approaches that of hyaenodontines. While this genus traditionally has been placed in the proviverrines due to the possession of a separate paracone and metacone on the upper molars and retention of reduced metaconids on the lowers (Pilgrim, 1914; Colbert, 1933, 1935), these features are primitive (Barry, 1988). Barry (1988) concludes that *Dissopsalis* and *Anasinopa* may be sister taxa, with *Anasinopa leakeyi* being a plausible ancestor for *Dissopsalis*.

Despite the derived nature of *Dissopsalis*, Barry (1988) rejected the placement of this genus within the Hyaenodontinae. He notes that both taxa share the following synapomorphies: development of an oblique shearing paracristid, suppression of the metaconids and talonids, and near fusion of the paracones and metacones. Barry argued that *Dissopsalis* retained many primitive features and thus did not have all of the hyaenodontine synapomorphies.

In contrast, Van Valen (1965) suggested a special relationship to *Arfia*, a relationship supported by Polly (1996). Polly's cladistic analysis of cranial and postcranial characters suggested the following relationship: (*Arfia* (*Dissopsalis* (*Pterodon, Hyainailouros*))). Thus, *Dissopsalis* can be interpreted as a hyainailourine. As Polly found hyaenodontines (*sensu latu*) to be diphyletic (Hyaenodontinae and Hyainailourinae), the importance of the supposed hyaenodontine (*sensu latu*) synapomorphies listed by Barry (1988) can be questioned. Moreover, tooth eruption patterns in *Dissopsalis* are unlike those of hyaenodontines but fit with hyainailourines (M.M., pers. obs.). For these reasons, we follow Morlo et al. (2007) and remove *Dissopsalis* from the proviverrines and place it in the Hyainailourinae.

The African species, *D. pyroclasticus*, is apparently easily confused with other hyaenodontids and with carnivorans and suids (Barry, 1988). Although it is recorded from Fort Ternan, Maboko, Moroto, and Napak (Bishop, 1967; Andrews et al., 1981; Pickford, 1981; Shipman et al., 1981; Pickford et al., 1986), Barry considers only the type specimen (BMNH M.19082) from Kaboor, Kenya and a single upper molar from the Ngorora Formation of the Baringo Basin (Locality 2/56; KNM-BN 1191) to belong to this species with any assurance. Barry refers the upper molar from Baringo to this species due to the small paracone, narrow protocone, and trenchant metastyle, which are a suite of characters found together only in *Dissopsalis*. Morlo et al. (2007) implied a generic separation of *D. pyroclasticus* from *Dissopsalis* by stating that *Buhakia* was closer to *D. carnifex* than to *D. pyroclasticus*. This separation, however, was not formally published; therefore, we follow the traditional assignment of this taxon.

Genus *HYAINAILOUROS* Biedermann, 1863

Synonymy Hyaenailurus Rütimeyer, 1867; *Hyaenaelurus* Stehlin, 1907; *Hyaenaelurus* Helbing, 1925; *Hyaenaelurus* Koenigswald, 1947; *Hyainolouros* Holroyd, 1994; *Hyainolouros* Holroyd, 1999

Diagnosis Very large to giant hyainailourines with P3–4 three rooted, diastema between single-rooted p1 and p3, and reduced talonids in m1–3.

Type Species Hyainailouros sulzeri Biedermann, 1863.

Age and Occurrence Early to middle Miocene; northern and eastern Africa, western Europe, and Pakistan.

Afro-Arabian Species Hyainailouros fourtaui Koenigswald, 1947; *H. napakensis* Ginsburg, 1980; *H. nyanzae* (Savage, 1965); *H. sulzeri* Biedermann, 1863

Remarks Fourtau (1920) identified a P3 from Moghra, Egypt as *Hyaena* sp. indet. The species *Hyainailouros fourtaui* was named for this specimen by Koenigswald (1947), and it is now considered to be a P4 (Ginsburg, 1980; Morlo et al., 2007).

Despite the fact that this genus has been known longer than any other hyainailourine except *Pterodon*, its diagnosis is more than unclear. The relationship of *Hyainailouros* to Miocene African "*Pterodon*" ("*P*." *nyanzae*, *P. africanus*), as well as to *Megistotherium osteothlastes* (which may be conspecific with *H. bugtiensis*), are unresolved and need further study. Here we follow the discussion of the genus provided by Morlo et al. (2007).

Until recently, *H. sulzeri* was only known from western Europe (Switzerland, France, Germany, and Spain)(Ginsburg, 1999). This taxon has been identified at Arrisdrift, Namibia, from a maxillary fragment and a juvenile canine and mandible (Morales et al., 2003) and has also been reported from Grillental, Namibia (Morales and Pickford, 2005), although the Grillental material was later referred to *Megistotherium* (Morales and Pickford, 2008). Morlo et al. (2007) suggest that all described African species of *Hyainailouros* may ultimately be recognized as conspecifics of the type species, given how poorly known this genus is within Africa. They suggest that all known taxa may be reducible to three: *H. bugtiensis* (= *Megistotherium osteothlastes*), *H. sulzeri* (all African and European specimens), and *Sivapterodon lahirri* (India). *H. sulzeri* in this scheme would be highly sexual dimorphic, as Miocene "*Pterodon*" (*P. nyanzae* including Miocene *P. africanus*, see Holroyd 1994), *H. fourtaui*, and *H. napakensis* are clearly smaller than *H. sulzeri* (sensu stricto). Hence, Turner and Antón (2004: figure 5.2) accept the presence of two African species of *Hyainailouros* that are separated by size. Undescribed material from the lower Miocene of Meswa Bridge, Kenya, including a mandibular ramus, various upper and lower teeth, humerus, radius, ulna, and tibia, may help to resolve some of these issues.

Genus *ISOHYAENODON* Savage, 1965
Figures 26.2 and 26.3

Diagnosis After Morales et al. (1998a). Differs from *Hyaenodon* in upper and lower molars more robust, lower molars with subequal paraconid and protoconid; m2 less reduced with respect to m3, and upper molars with better developed protocone.

Type Species Isohyaenodon andrewsi Savage, 1965.

Age and Occurrence Oligocene through Lower Miocene; northern and eastern Africa.

Afro-Arabian Species Isohyaenodon andrewsi Savage, 1965 (status questioned); *I. matthewi* Savage, 1965 (status questioned); *I. pilgrimi* Savage, 1965.

Remarks This genus was originally proposed as a subgenus of *Hyaenodon* by Savage (1965), with the suggestion that it might merit generic distinction. Morales et al. (1998a) elevated *Isohyaenodon* to generic status. They recognized two species, *I. andrewsi* and *I. pilgrimi*, and referred *Metapterodon zadoki* to this genus.

The status of the species *I. andrewsi* (figure 26.2) is in question. Van Valen (1967) suggests that the type specimen may, in fact, be a mandible of *Leakitherium hiwegi* despite being slightly smaller than the holotype of *L. hiwegi*. He also suggests that other species of *Isohyaenodon* may belong in *Leakitherium*. Morales et al. (2007) suggest that *L. hiwegi* and *I. andrewsi* may be upper and lower dentitions, respectively, of the same species.

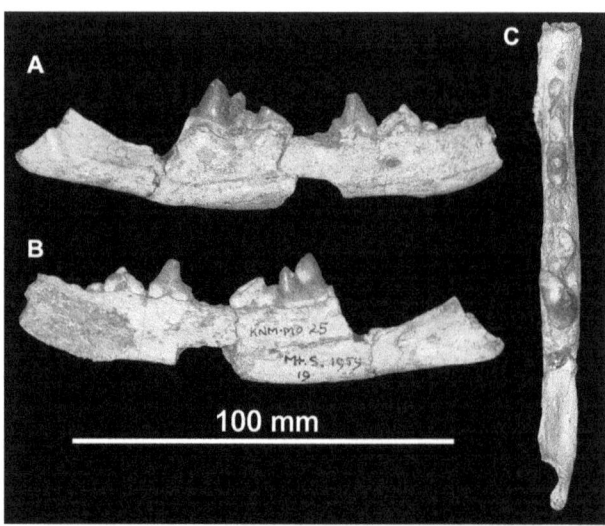

FIGURE 26.2 KNM MO 25, *Isohyaenodon andrewsi*, right mandible fragment with p1–p3, m2, alveolus for p4, broken m1 from Maboko, Kenya: A) buccal B) lingual, and C) occlusal views.

They note that if this true, then *Leakitherium* would have priority. In contrast, Holroyd (1999) considers *I. andrewsi* to be the previously unknown lower dentition of *Metapterodon kaiseri*, currently known only from the upper dentition, or a closely related species to *M. kaiseri*.

The status of the second species, *I. matthewi*, is also in question. Morales et al. (1998a) noted that *Metapterodon zadoki*, which they referred to *Isohyaenodon*, had priority in Savage's publication. They then synonymized *I. matthewi* with *I. zadoki* as they believed that "*H*." *matthewi* and *M. zadoki* are lower and upper dentitions, respectively, of the same species. They also refer one specimen of *I. matthewi* (KNM-CMF 4060) to *I. andrewsi* as this specimen conjoins with an *I. andrewsi* mandible (KNM-CMF 4023). Further material from Napak has been described as *Isohyaenodon zadoki* (Morales et al., 2007).

The third species, *I. pilgrimi* (figure 26.3), is the only one of Savage's (1965) three eastern African (former) *Hyaenodon* species whose status has not been questioned. Perhaps this is due to the fact that it is the only species of *Isohyaenodon* to include material of both upper and lower dentitions. New material has recently been described from Napak, Uganda (Morales et al., 2007).

As *I. andrewsi* is the type species of the genus, synonymizing this species with *Leakitherium hiwegi* or *Metapterodon kaiseri* will require a taxonomic revision of all members of the genus. Morales et al. (2007:74) has noted that the m2 of *I. pilgrimi*, *M. zadoki*, and a possible specimen of *Leakitherium* from Napak (UM NAP I 44'99) all share a "mesial valley on the anterior margin." The enamel surface of the m2 in *M. zadoki*, *L. hiwegi*, and UM NAP I 44'99 has "vertical, rounded parallel ridges or swellings" in contrast to the usual wrinkled creodont enamel (2007:74). Thus, even if *Isohyaenodon* is synonymized with *Leakitherium* and all current taxa are placed within that genus, the relationship of *Leakitherium* and *Metapterodon* must be clarified and the placement of *M. zadoki* within a genus addressed. Synonymy of *Isohyaenodon* with *Metapterodon* requires evaluating the relationships among current *Isohyaenodon* species and *L. hiwegi* to determine if they all belong to this genus, as well. Clearly,

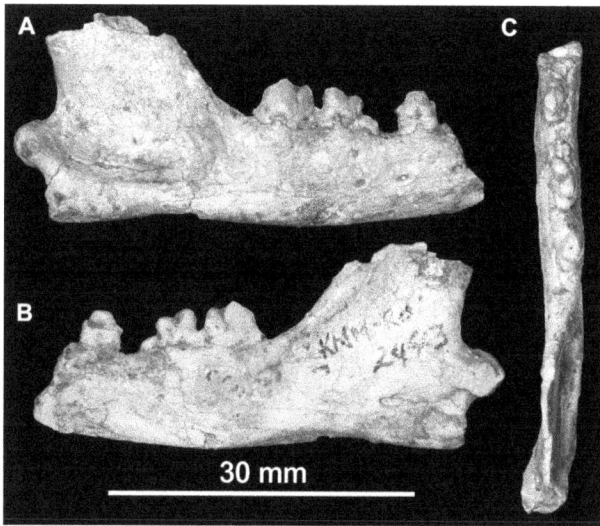

FIGURE 26.3   KNM RU 2943 (= CMF.4062), *Isohyaenodon pilgrimi*, right mandible fragment with p4, m2–m3 from Rusinga Island, Kenya (Savage, 1965:284): A) buccal, B) lingual and C) occlusal views.

more associated upper and lower dentitions are necessary to resolve these issues.

Note that there are currently no African members of the genus *Hyaenodon*. Other African taxa originally referred to *Hyaenodon* include *H. brachycephalus*. Holroyd (1999) removed *H. brachycephalus* from *Hyaenodon* and placed it in *Metapterodon*. Clearly, the confusion over what taxa should be incorporated within *Isohyaenodon* or *Metapterodon* suggests that both taxa are in need of further revision.

Genus *LEAKITHERIUM* Savage, 1965

Synonymy Leakeytherium Morales et al., 2007.

Diagnosis Hyainailourine without M3; M1+2 highly sectorial, protocone greatly reduced on M2; molars with connate paracone and metacone and shearing metastyle; P4 with protocone and prominent parastyle, central paracone, metacone and trenchant metastyle.

Type Species Leakitherium hiwegi Savage, 1965.

Age and Occurrence Miocene; eastern Africa.

Afro-Arabian Species Leakitherium hiwegi Savage, 1965.

Remarks The species *L. hiwegi* has strongly sectorial molars like *Pterodon*. This species is roughly similar in size to *Isohyaenodon andrewsi* and is about the size of a leopard (Savage, 1965; Van Valen, 1967; Morales et al., 2007).

The type species is only known from upper dentitions. Van Valen (1967) and Morales et al. (2007) have suggested that this material may belong to the same species as *Isohyaenodon andrewsi*, which is only known from lower dentitions. Morales et al. (2007) refer a lower molar fragment from Napak to this genus noting that the morphology is also similar to other species of *Isohyaenodon*, but roughly 50% larger (see remarks in *Isohyaenodon* section).

Genus *MASRASECTOR* Simons and Gingerich, 1974
Figure 26.4

Diagnosis Emended from Holroyd (1994); Peigné et al. (2007). Differs from all other hyaenodontids, including all Fayum taxa, Oligocene European *Quercitherium*, and Miocene

African *Teratodon*, in its small length/breadth index of m1, combined with the low, short, blunt p3–4 (Peigné et al., 2007). *Masrasector* further differs strongly from *Metasinopa* and *Anasinopa* (where known) in its generally smaller size and in having lower crowned teeth; less pronounced molar size increase from front to back; less reduced metaconid; slight buccal cingulum on lower molars and well-developed buccal cingulum on upper molars; more poorly developed paracristid shearing; entoconid and hypoconid distinct and separate on molars; p3 with posterior accessory cusp; p4 with distinct entoconid, entocristid and hypoconulid and a talonid basin; small mesostyle module on upper molars; lower molars with hypoconid more labially placed and distinct from the hypoconulid and entoconid; molar hypoconulid more posteriorly projecting; relatively larger molar talonids that are subequal in length to trigonid; large, bulbous P4 protocone; shorter, less obliquely oriented metastyle; a smaller paraconule bearing a stronger postparaconule crista; relatively shorter and wider talon basin on M1–2; paracone and metacone more connate, less appressed to one another; shorter M3 parastyle, narrower M3 talon basin; less anteriorly positioned protocone relative to the buccal edge of the tooth; P3 metastyle lacking; weaker P3 buccal and lingual cingula; shorter and wider P3; less angled anterior face of P3.

Type Species Masrasector aegypticum Simons and Gingerich, 1974.

Age and Occurrence Late Eocene to early Oligocene; northern Africa and Arabian Peninsula.

Afro-Arabian Species Masrasector aegypticum Simons and Gingerich, 1974; *M. ligabuei* Crochet et al., 1990; *Masrasector* n.sp. Holroyd, 1994.

Remarks Simons and Gingerich (1974) believed *Masrasector* to be a stage between the Eocene *Sinopa* or *Proviverra* and the Oligocene and Miocene *Metasinopa*. In this scenario, the paraconids, metaconids, and protoconids become progressively larger, resulting in a change in shear, where *Masrasector* is intermediate between the early group *(Sinopa and Proviverra)* and *Metasinopa*. Simons and Gingerich note that the crowns of the upper and lower premolars are typically worn flat at an early age and that talonid basin wear indicates a propalinal component at the end of the masticatory stroke.

Simons and Gingerich (1974) note that a specimen (YPM 20944) that is about half the size of *M. aegypticum* is present at Quarry G (Jebel Qatrani Formation, Upper Sequence) as well, suggesting a second species present at the same level. Holroyd (1994) demonstrated that this specimen has deciduous teeth and referred it to *M. aegypticum*. However, a short review of the specimen reveals a clear difference from *M. aegypticum* in the relative proportions of m1, which is relatively less broad in the juvenile (see figure 26.4). In our view, YPM 20944 does not belong to *Masrasector aegypticum* and may not be a *Masrasector* at all.

The relationships of *Masrasector* are completely unresolved as it differs strongly from other hyaenodontids. Its supposed ancestorship to *Anasinopa* (Simons and Gingerich, 1974)—and thus its relationship to Hyainailourinae—is only weakly justified, as is the alternative interpretation of *Masrasector* being a descendant of a prototomine-like proviverrines from the early Eocene of Africa (Peigné et al., 2007). The hypothesized relationship to South Asian "proviverrines" (Egi et al., 2005), however, has been convincingly falsified (Peigné et al., 2007). The blunt premolars of *Masrasector* suggest a lifestyle similar to that of the hypothesized molluscivores *Quercitherium* and *Teratodon*, even if molar morphology separates all

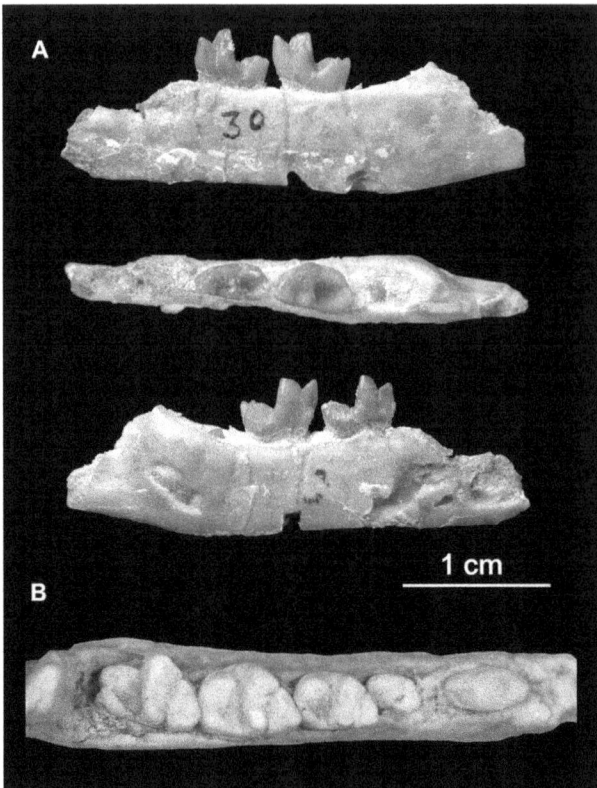

FIGURE 26.4 A) Juvenile mandible (YPM 20944) from Quarry G, Jebel Qatrani Formation Upper Sequence, Fayum Province, Egypt, in, from top to bottom, buccal, occlusal, and lingual views. B) Cast of type specimen of *Masrasector aegypticum* (CGM 30978) in occlusal view also from Quarry G. Note the differences in the relative proportions of m1 between the two specimens that lead us to conclude that YPM 20944 belongs to a different species, and possibly different genus, than CGM 30978.

three genera phylogenetically from each other. This lifestyle has therefore evolved independently.

Genus *MEGISTOTHERIUM* Savage, 1973

Diagnosis Rasmussen et al. (1989). Gigantic hyainailourine, single large upper incisor; large upper canine, laterally placed with respect to other teeth; palate constricted near P1–2; P3–4 three rooted, P4 width greater than length (unlike *Hyainailouros*); M1–2 trenchant; M3 small, transverse (Savage, 1973). Mandible deep and bowed, forming thick horizontal torus along inferior edge; p4 differs from that of other hyaenodontines in being obliquely oriented with respect to the axis of the mandible; p4 further differs from that of *Hyainailouros* in bearing a small, trenchant hypoconid; m1 differs from that of *Pterodon* in its extreme size reduction, and the lack of differentiated cusps and crests; m2–3 resemble those of most hyaenodontines but differ from *Apterodon* in having long, bladelike paraconids and (at least on m2) small hypoconids.

Type Species Megistotherium osteothlastes Savage, 1973.

Age and Occurrence Early Miocene; northern and eastern Africa, possibly Pakistan.

Afro-Arabian Species Megistotherium osteothlastes Savage, 1973.

Remarks This is the largest known hyaenodontid. Savage (1973) named the only species based on a skull from Libya

(BMNH M26173) and referred a mandible from the Bugti Hills, Pakistan (BMNH M 12049), to this genus. The referred material includes a partial mandible (DPC 14557) from Egypt and an upper canine, damaged m2 and partial left humerus from Kenya (Rasmussen et al., 1989; Morales and Pickford, 2005, 2008). Morales and Pickford (2008) refer the Grillental *Hyainailouros* to *Megistotherium* and mention unpublished material of this genus from Moroto I, Uganda.

Although originally considered a hyaenodontine, *Megistotherium* was later placed in Pterodontinae by Holroyd (1999). Morales and Pickford (2005) have noted that this genus has never been clearly demonstrated to differ from *Hyainailouros*, while Morlo et al. (2007) have suggested that *M. osteothlastes* may be referable specifically to *H. bugtiensis*.

Savage (1973) notes the remarkably large size of the holotype of *M. osteothlastes*, including an enormous sagittal crest, paranasal sinuses, and canines. The premolars are described as "heavy" and "blunt," as in *Crocuta*. Both by its name ("greatest beast—bone crusher") and through comparison to *Crocuta*, Savage indicates that he views this species as equivalent, at the least, to extant spotted hyenas in bone-cracking capabilities.

Unlike many African hyaenodontids, this species has attributed postcranial material. Savage (1973) states that these specimens are distinguished by their large size from all other species in the same beds at Gebel Zelten, Libya. The humerus from the Ngorora Formation, Kenya, is similar in morphology to the Gebel Zelten humerus but is smaller in size (Morales and Pickford, 2005). The Kenyan humerus, along with the damaged m2, represents the latest possible occurrence of this genus (13–12 Ma; Morales and Pickford, 2005, 2008).

Genus *METAPTERODON* Stromer, 1926
Figure 26.5

Synonymy Hyaenodon Andrews, 1906:218; *Hyaenodon* Osborn, 1909:423; *Pterodon* Osborn, 1909:423 *(partim)*; ?*Metasinopa* Osborn, 1909:423; *Metapterodon* Stromer, 1926:110; *Hyaenodon* Savage, 1965:268 *(partim)*; *Metapterodon* Savage, 1965:268 *(partim)*; *Hyaenodon (Isohyaenodon)* Savage, 1965:280 *(partim)*; *Metasinopa*(?) sp. Savage, 1965:264.

Diagnosis Revised after Holroyd (1999). Differs from *Pterodon* and *Akhnatenavus* in having P4 postparacrista developed into a shearing crest; narrower lower molars; narrower upper molar metastyles; undulant labial face on upper molars; p2 with anterior accessory cuspid; bladelike P4 metastyle; double-rooted P3. Differs from *Pterodon* in smaller size and in having longer P4 metastyles; smaller protocones; narrower upper molars; more labially placed and more salient anterior accessory cuspids on p2–p3; p1 lacking; more sharply angled posterior trigonid wall; much shorter m1–m2 talonids; m3 talonid lacking. Differs from *Hyaenodon* in having upper molar protocones primitively present; lower molar posterior paracristid not markedly longer than anterior paracristid.

Type Species Metapterodon kaiseri Stromer, 1926.

Age and Occurrence Late Eocene to early Miocene; northern, eastern, and southern Africa.

Afro-Arabian Species Metapterodon brachycephalus (Osborn, 1909); *M. kaiseri* Stromer, 1926; *M. markgrafi* Holroyd, 1999; *M. schlosseri* Holroyd, 1999; *M. stromeri* Morales et al., 1998a; *M. zadoki* Savage, 1965 (*M. zadoki* may belong in *Isohyaenodon*).

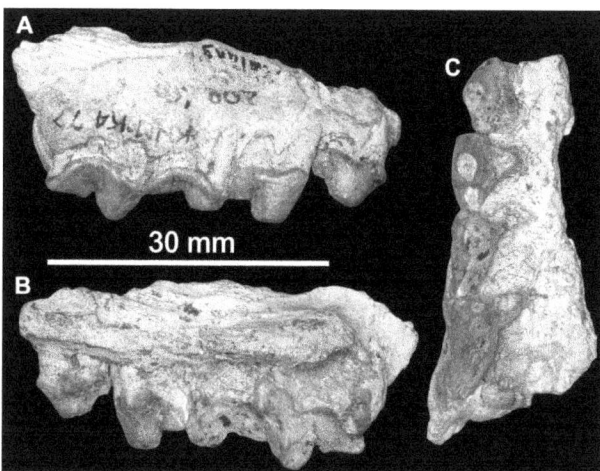

FIGURE 26.5 KNM KA 77 (= CMF.4038), *Metapterodon kaiseri*, right maxilla fragment with P3–M3 from Karungu, Kenya (Savage, 1965:268, figure 28, plate 4, figure 2) in A, buccal; B, lingual; C, occlusal views.

Remarks The genus *Metapterodon* has proven to be quite problematic. Although described originally from Namibia, additional fragmentary Kenyan material was added to the Namibian species *M. kaiseri* by Savage (1965). Savage provided the first revision of the genus based on his disagreement with the meaning of characters cited by Stromer (1926). Van Valen (1967) synonymized *Metapterodon* with *Pterodon* but noted that the latter genus might be polyphyletic. Savage (1978) appears to have agreed with Van Valen, as *M. kaiseri* and *M. zadoki* were placed in *Pterodon* in this later work. Dashzeveg (1985) synonymized *M. zadoki* with *M. kaiseri*, a view that has not been supported by later researchers (e.g., Morales et al., 1998a; Holroyd, 1999; Morales et al., 2007). Morales et al. (1998a) resurrected *Metapterodon* for the Namibian material and placed *M. zadoki* and the eastern Africa specimens of *M. kaiseri* in the genus *Isohyaenodon* (see remarks in the *Isohyaenodon* section.) This change supports Van Valen's (1967) view that the Kenyan material assigned to *M. kaiseri* differs from the Namibian holotype and should be placed in another taxon. Morales et al. (1998a) did not address the Fayum material.

Holroyd (1999), apparently unaware of the work of Morales et al. (1998a), independently revived *Metapterodon*. She recognized *M. kaiseri* and *M. zadoki* as valid species, referred *Hyaenodon brachycephalus* to *Metapterodon*, and named two new species. Like Van Valen (1967) and Morales et al. (1998a), Holroyd (1999) recognized that *M. kaiseri* might be different from later taxa assigned to this genus. She noted that her two new species may ultimately prove to belong to a separate genus from *M. kaiseri*, but she still placed them in this genus to demonstrate their uniqueness from both *Isohyaenodon* and *Pterodon*. Note that Holroyd considered "*H. (Isohyaenodon) andrewsi*" to be the previously unknown lower dentitions of *M. kaiseri* or a closely related species. She also tentatively placed two specimens previously referred to *Hyaenodon* (Andrews, 1906; Osborn, 1909; Savage, 1965) in her new species *M. schlosseri*: AMNH 13262 (formerly ?*Hyaenodon* from Quarry B) and BMNH C8812-13 (formerly *Hyaenodon (Isohyaenodon) andrewsi*, possibly from the lower sequence of the Jebel Qatrani Formation).

The relationships between African forms of *Isohyaenodon*, *Pterodon*, and *Metapterodon* are in need of further clarification. While *M. zadoki* may well belong in *Isohyaenodon*, we have chosen the conservative approach and left it in *Metapterodon*

for the moment until the relationships among the aforementioned genera are clarified.

Genus *METASINOPA* Osborn, 1909

Synonymy Sinopa Andrews, 1906 (*S. ethiopica* only); *Sinopa* Schlosser, 1911.

Diagnosis Revised after Holroyd (1994). p1 absent; p3 and m3 present; small lower premolars; basal talonid present, persistent metaconid on m2 and m3; small trenchant heels on lower molars (Osborn, 1909). Differs from Miocene *Anasinopa* in having more poorly developed metaconids; strong preparacrista; smaller p4 talonid; narrower mandibular cheek teeth; narrower protoconid and talonid; smaller m2 relative to m3; more mediolaterally oriented paracristid; better separated paracone and metacone.

Type Species Metasinopa fraasi Osborn, 1909.

Age and Occurrence Late Eocene to ?Miocene; northern and eastern Africa.

Afro-Arabian Species ?*Metasinopa ethiopica* (Andrews, 1906); *M. fraasi* Osborn, 1909; *M. napaki* Savage, 1965.

Remarks This genus was originally named from material in the Fayum. *Metasinopa* is distinguished from *Pterodon* and *Apterodon* by a persistent metaconid on m2 and m3 that Osborn believed related it to *Sinopa* and *Tritemnodon*. The genus is distinguished from *Hyaenodon* by having a basal talonid, as in *Pterodon* and *Apterodon* (Osborn, 1909).

Metasinopa fraasi and *M. ethiopica* both retain a metaconid on the lower molars but differ in the greater breadth of the talonid and overall greater size of *M. fraasi* (Osborn, 1909). Analyses by Barry (1988) suggest that at least *M. fraasi* is related to *Anasinopa* and *Dissopsalis*. Barry has even suggested that *M. fraasi*, and possibly all of *Metasinopa*, may belong within the Miocene genera *Anasinopa* or *Dissopsalis*, or even the older European Eocene genus *Prodissopsalis*. Holroyd (1994) kept this species in the genus *Metasinopa*. The type specimen is still the only known specimen of *M. fraasi*.

"*M.*" *ethiopica* was originally described as "*Sinopa*" *ethiopica* and later transferred to *Metasinopa* by Savage (1965). In her unpublished dissertation, Holroyd (1994), however, argues for a generic separation of this taxon from *Metasinopa*. It should be noted, that *Quasiapterodon* is very close to this species as well.

M. napaki was described from a fragmentary mandible (BMNH M19097) found in northeast Uganda. Savage placed this new species provisionally in *Metasinopa* "largely for convenience" (1965:264). This species is smaller overall than *M. fraasi* and larger than *M. ethiopica*. The m3 talonid of *M. napaki* is relatively longer than in the other members of this genus. Van Valen (1967) questioned the referral of *M. napaki* to *Metasinopa* due to the moderate size of the metaconid on m3 (i.e., larger than that of *M. fraasi*). Instead, Van Valen proposed moving this species to *Paracynohyaenodon*; a move that has not been accepted in the literature.

The only known upper dentition of *Metasinopa* is a fragmentary maxilla (BMNH M19096) from Napak that Savage (1965) provisionally assigned to *M. napaki*. Savage believed that the absence of a parastyle on P4 prevented the inclusion of this specimen in *Sinopa*, *Anasinopa*, *Dissopsalis*, or *Prodissopsalis*. Barry (1988), however, noted that the absence of a parastyle has also been used as a diagnostic character for *Dissopsalis*. Barry, therefore, suggested that this specimen probably belongs to *Anasinopa leakeyi* or a small, second African species of *Dissopsalis* due to the possession of a bulbous and anteriorly directed protocone, inflated paracone, and absence of the parastyle on

P4. While the type of *A. leakeyi* has a small parastyle, Barry believed that the strength of the parastyle, or parastylar cingulum, varies in *A. leakeyi* much as it does in *D. carnifex*.

Holroyd (1994) has suggested that *M. napaki* as a whole is actually a member of *Anasinopa* or an unnamed, closely related genus. If either Barry (1988) or Holroyd (1994) is correct, then the known range of *Metasinopa* would be restricted to northern Africa during the late Eocene/early Oligocene.

Genus *PTERODON* (Blainville, 1839)

Diagnosis Emended after Holroyd (1999). Differs from *Metapterodon* in its larger size and in having m3 talonid present; relatively larger m1–m2 talonids; more lingually placed and less salient anterior accessory cuspulids on p2–p3; narrower upper molars; relatively shorter upper molar metastyles; incomplete fusion of paracone and metacone; a single, pronounced upper molar ectoflexus. Differs from *Akhnatenavus* in having molar talonids as wide as trigonids; molar protoconid and paraconid subequal in size; anterior "bulge" on molars lacking; unreduced premolars; premolar diastemata lacking. Differs from *Megistotherium* and *Hyainailouros* in its smaller size and in having relatively larger talonids and larger anterior keels on lower molars.

Type Species Pterodon dasyuroides Blainville, 1839

Age and Occurrence Late Eocene through early Miocene; northern and eastern Africa, Europe.

Afro-Arabian Species Pterodon africanus Andrews, 1906; *P. phiomensis* Osborn, 1909; *P. syrtos* Holroyd, 1999.

Remarks Holroyd (1994, 1999) notes that *Pterodon* has been used as a "wastebasket taxon" for middle to late Paleogene and Miocene hyaenodontines lacking the derived features of *Hyaenodon*. As such, she restricted *Pterodon* via the above diagnosis to species that share derived characters with the type species, *P. dasyuroides*. Egi et al. (2007) removed the last remaining Asian species of *Pterodon*, *P. dakhoensis*, from this genus and placed it in the newly erected genus *Orienspterodon*.

Savage (1965) suggested that all species vary mainly in size, while others disagree (e.g., Holroyd, 1994, 1999). Nonetheless, the members of this genus do vary in size, with *P. grandis* being "two-thirds as large again" as *P. africanus* (Savage, 1965:272) and *P. phiomensis* being two-thirds the size of *P. africanus*. *P. syrtos* is the smallest member of this genus.

Like most African creodonts, this genus is not without controversy. Van Valen (1967) believed that Napak material included in *P. africanus* by Savage (1965) should be placed in *Hyainailouros*. Holroyd (1999) suggested that both of Savage's *P. africanus* specimens (BMNH M19090 and KNM-CMF 4024) should be referred to *P. nyanzae* (= *Hyainailouros nyanzae*) as the type of *P. nyanzae* is not significantly larger than his *P. africanus* specimens (a feature used by Savage to distinguish the two). She also discounts his perception of a greater anterior keel on the P4 and M1 of *P. nyanzae* relative to *P. africanus*. Interestingly, although Morales et al. (2007) list both both taxa as having been described from Napak, they list only *P. africanus* with no explanation in their final creodont taxon list. As has been discussed before, however, it is unclear how small *Hyainailouros* can be separated from large Miocene "*Pterodon*." In our view, Miocene "*Pterodon*" more probably belongs to *Hyainailouros*.

While the late Eocene European form, *P. dasyuroides*, has been suggested to be ancestral to African *Pterodon* (Savage, 1978), the redating of the earliest portion of the Fayum to the late Eocene makes this much less likely. Lange (1967) believed *P. africanus* to be the most primitive *Pterodon*. A brief discussion of the history of *Pterodon* can be found in Holroyd (1999).

Subfamily KOHOLIINAE Crochet, 1988
Genus *KOHOLIA* Crochet, 1988

Diagnosis Hyaenodontids possessing a P4 with a particularly slender principal cusp, a relatively long and trenchant posterior crest and a very small protocone. M1 with a tall parastyle, a large stylar shelf and the paracone and metacone equal in distance from the labial edge of the crown, and a paracone that is more developed and taller than the metacone.

Age and Occurrence Lower Eocene; northern Africa.

Afro-Arabian Species Koholia atlasense Crochet, 1988.

Remarks The morphology of P4, the fact that M2 is larger than M1, the size of the protocone, and the fact that the paracone and metacone are nearly connate place this taxon in the Hyaenodontidae (Crochet, 1988). Crochet placed this genus in its own subfamily, the Koholiinae, based on three features that differ from hyaenodontines: "1. P4 possesses a very slender principal cusp, a very elongated and trenchant metastylar crest, and a very developed lingual cingulum (but much smaller than in proviverrines); 2. The paracone of M1 is very developed and taller than the metacone. Such a character is seen in certain proteutherian families (namely Palaeoryctidae), condylarths (mainly Mesonychia), and fissipeds (canids, for example). Among the hyaenodontine creodonts, *Pterodon dasyuroides* (Ludian, MP 18 to 20, western Europe) and *Pterodon africanus* (Savage, 1965) possess M1 and M2 paracones slightly larger than the metacones and well joined to the latter. For the latter species, Savage (1965) noted that the M2 possesses a tall paracone. Of the other hyaenodontids, only the proviverrine *Tritemnodon* (North American Eocene) has a paracone that is significantly more developed than the metacone (Matthew, 1909). 3. The stylar shelf of the M1 is particularly developed (Fig. 1 and 2). This character distinguishes the Kohol form from every described hyaenodontid." (See Crochet 1988:1796; translated from the French). Crochet notes elsewhere in his description that the stylar shelf is particularly wide.

Subfamily PROVIVERRINAE Schlosser, 1886

Content Includes North American and Asian Limnocyoninae, Morlo and Gunnell, 2003, 2005; Morlo et al., 2007.

Diagnosis After Matthew (1909); Gunnell (1998); Morlo and Habersetzer (1999). Narrow skull with long face; M1–3 tritubercular; m1–3 tuberculosectorial; metaconids present on lower molars; less derived carnassial specializations; limbs mostly unspecialized, but scansorial and cursorial specialists also occur.

Age and Occurrence Eocene, but possibly also late Paleocene. Oligocene and Miocene occurrences disputed. North America and northern Africa (Eocene), Egypt (Oligocene), and Indo-Pakistani region and eastern Africa (Miocene).

Remarks Barry (1988), Polly (1996), Morlo and Habersetzer (1999), and subsequent authors have suggested that this subfamily is paraphyletic. Some researchers (Barry, 1988; Egi et al., 2005) have tried to support a relationship between the African genera *Dissopsalis*, *Anasinopa*, and *Metasinopa*, possibly *Masrasector*, and even *Teratodon* and European and/or Asian proviverrines based on postulated derived features. However, *Dissopsalis* (and, consequently, its relatives *Anasinopa*, *Metasinopa*, and *Buhakia*) is now assigned to Hyainailourinae (and thus distinct from the subfamily Proviverrinae) based on tooth morphology (Lange-Badré, 1979; Peigné et al., 2007), basicranial morphology (Polly, 1996), and dental ontogeny (Morlo et al., 2007). *Teratodon* (Savage, 1965) has also been placed in its own subfamily

(Van Valen, 1967; Morlo and Habersetzer, 1999). This leaves the Moroccan genera *Tinerhodon* (Paleocene) and *Boualitomus* (early Eocene) as the only undisputed proviverrines of Africa, while the assignment of *Masrasector* awaits clarification.

Genus *BOUALITOMUS* Gheerbrant in Gheerbrant et al., 2006

Diagnosis From Gheerbrant et al. (2006). Dental morphology close to that of proviverrine hyaenodontids (m1 smaller than m2–3, m2 and m3 similar in size, large paraconids and sharp paracristid, protoconids high and pointed, metaconid not reduced, trigonid moderately compressed, talonid narrow with weak cusps, entoconid distal, premolars sharp and elongated, p2–3 asymmetric, diastemata between anterior premolars). Dental morpholog=y closest to *Prototomus* among proviverrines but differs in the absence of p1. Small size, close to that of *P. minimus*. Differs from *P. minimus* in having m3 unreduced with respect to m2, the cusps less differentiated on the talonid of the molars, the p4 and molars slightly narrower, and the talonid narrower in m1. The talonid of p4 bears at least two accessory cusps.

Type Species Boualitomus marocanensis Gheerbrant in Gheerbrant et al., 2006.

Age and Occurrence Earliest Eocene; northern Africa.

Afro-Arabian Species Boualitomus marocanensis Gheerbrant in Gheerbrant et al., 2006.

Remarks Gheerbrant et al. (2006) link this genus from the Ouled Abdoun Basin in Morocco with *Prototomus*, a primitive "proviverrine," and related forms within the subfamily (see Peigné et al., 2007, and references cited therein for a discussion suggesting that prototomine-like "proviverrines" are not related to *Proviverra*-like proviverrines). In fact, these authors note that the only autapomorphic feature is the loss of p1, which may be related to the shortened anterior portion of the dentary. Prototomine-like "proviverrines" have a single-rooted p1, while Proviverrinae *sensu stricto* have a double-rooted p1 (Morlo and Habersetzer, 1999). Estimated body mass ranges from 300 to 570 g, indicating that this is a very small form.

Genus *TINERHODON* (Gheerbrant, 1995)

Diagnosis Emended after Gheerbrant (2006). Dentition showing affinities with the proviverrine hyaenodontids, and especially with *Boualitomus*: m3 not reduced; paraconids lingual and enlarged, only slightly smaller than metaconid; paracristid and protocristid sharp; carnassial notches on paracristid, protocristid and cristid obliqua; trigonid moderately compressed mesiodistally; talonid narrower than trigonid and bearing cusps of similar height; talonid elongated and oblique with respect to the longitudinal axis; entoconid distal and close to the hypoconulid; premolars simple and sharp; p2–3 laterally compressed, elongated and with asymmetric lateral profile; diastemata between anterior premolars; large mental foramina below p4 and p2. *Tinerhodon* especially resembles *Boualitomus* in the morphology of the talonid of p4 that bears several accessory susps, including a bulbous protostylid. *Tinerhodon* differs from *Boualitomus* and other proviverrines in some unusual primitive features: (1) smaller size (half the size of *Boualitomus*); (2) molars with wider talonid, with more cuspidate talonid cusps (hypoconulid especially larger), and with variable accessory cusps; (3) p4 with occlusal outline more inflated transversely and with talonid more molarized (lingual accessory cusps more developed, protostylid more inflated, postfossid distinct, and hypoflexid more developed). It also

differs from *Boualitomus* in having the protostylid closer to the protoconids and the presence of a metaconid ridge on p4, and the distally more recurved protoconids on the molars.

Type Species Tinerhodon disputatum Gheerbrant, 1995.

Age and Occurrence Latest Paleocene; northern Africa.

Afro-Arabian Species Tinerhodon disputatum Gheerbrant, 1995.

Remarks This genus and species is known from Adrar Mgorn 1 and Ihadjamene in the Ouarzazate Basin of Morocco (Gheerbrant, 1995). It was described by Gheerbrant (1995) as being most similar to creodonts, but also similar to basal Carnivora and to pantolestids.

Later work (Gheerbrant et al., 2006) indicates that this genus is related to the primitive Eocene genus *Boualitomus*, also from Morocco. The material from the Grand Daoui Quarries (lowermost Eocene) in the Ouled Abdoun Basin of Morocco has been suggested to be part of a lineage leading from the Adrar Mgorn 1 material (Gheerbrant et al., 2003, 2006). The Ouled Abdoun material is larger and more derived than the Adrar Mgorn material. The two genera share the primitive talonid morphology of p4, with *Tinerhodon* showing particularly cimolestid-like features including "small size, p4 less simplified and more inflated transversely, talonid of molars wider and bearing more developed cusps" (Gheerbrant et al., 2006:486). It should be noted, however, that *Tinerhodon*—if it is a hyaenodontid—is closer to Proviverrinae *sensu stricto* than to *Prototomus*-like "proviverrines" (to which *Boualitomus* belongs). This is evident in the structure of its talonids, where the hypoconids, hypoconulids, and entoconids are strong and similarly separated from each other while in *Boualitomus* (and other *Prototomus*-like "proviverrines") the hypoconulid is placed much closer to the entoconid and all talonid cusps are much weaker.

Subfamily TERATODONTINAE Savage, 1965

Diagnosis M3/m3 present, M3 transverse, M2/m3 main carnassial pair, M1/m2 less functional as carnassials. Premolars large, bunodont, tubercular with thick enamel: P4 larger than M1. Lower molars with small talonid and metaconid present; m2 larger than m1. Jaw relatively short. M1 and M2 metacone slightly larger than and connate with paracone; elongate metastyle; M2 slightly larger than M1; protocone almost as large as paracone. Lower molars with well-developed metaconid, trigonid cusps high, talonid small, paraconid-protoconid shear very oblique. p3 large with low single cusp.

Age and Occurrence Miocene; eastern Africa.

Remarks Savage placed this taxon at the family level as a member of the Oxyaenoidea. Teratodontidae was later reduced to a subfamily by Van Valen (1967), a stance that has been viewed as a possibility by other workers (e.g., Polly, 1996), fully supported (Morlo and Habersetzer, 1999), or disregarded (Morales et al., 2007).

Genus *TERATODON* Savage, 1965
Figures 26.6 and 26.7

Diagnosis As for subfamily.

Type Species Teratodon spekei Savage, 1965.

Age and Occurrence Miocene; eastern Africa and possibly Egypt.

Afro-Arabian Species Teratodon spekei Savage, 1965; *T. enigmae* Savage, 1965.

Remarks The species *T. spekei* is about the size of *Vulpes vulpes*, while *T. enigmae* has more robust jaws and dentition (figures 26.6 and 26.7). A poorly preserved mandibular fragment from Wadi Moghara, Egypt, has been referred to cf. *Teratodon* due to similarities in size and the presence of a metaconid and short talonid on m1 (Morlo et al., 2007). *Teratodon* shares its blunt, durophagous premolars with *Masrasector* and the European *Quercitherium*. However, molar morphology separates all three from each other, implying that durophagy evolved independently in all three hyaenodontids.

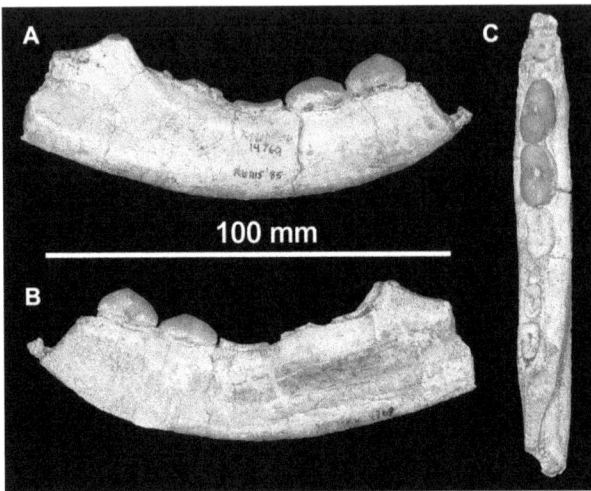

FIGURE 26.6   KNM RU 14769B, *Teratodon* sp., right mandible fragment with p2–p3, broken p4–m3 in A) buccal, B) lingual, and C) occlusal views.

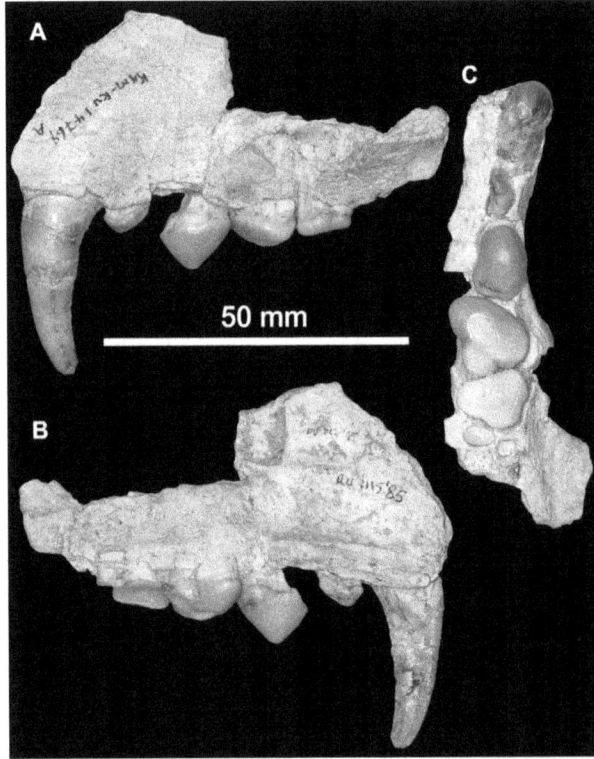

FIGURE 26.7   KNM RU 14769A, *Teratodon* sp., left maxilla fragment with C, P1–P4, roots of M1 in A) buccal, B) lingual, and C) occlusal views.

General Discussion

SYSTEMATICS AND TAXONOMY

The fossil record of creodonts extends from the Paleocene to the Miocene. As noted by Polly (1996), this large expanse of geological time has contributed to the paucity of systematic revisions of the order and its families since Matthew's (e.g., 1901, 1909) original taxonomic work.

While the number of creodont families recognized has differed from author to author, there are at least two that are generally recognized: the Oxyaenidae and the Hyaenodontidae. The earliest creodonts are oxyaenids from the late Paleocene (Tiffanian LMA) of North America, suggesting a North American origin for this family (Gunnell, 1998; Eberle and McKenna, 2002). The primary radiation of oxyaenids occurs in North America (Gunnell, 1998), although they do appear in the Eocene in Eurasia. No oxyaenids have been found in Africa.

Hyaenodontids appear first in Asia and Africa and eventually are found in Europe and North America. The relationship between the hyaenodontids and oxyaenids and the question of whether they should be elevated to unrelated orders has long been debated (Van Valen, 1967; Gingerich, 1980; Polly, 1996; Gunnell, 1998; Flynn and Wesley-Hunt, 2005; Gheerbrant et al., 2006). A brief taxonomic history of the hyaenodontids has been presented by Polly (1996).

A third family, the Limnocyonidae, has been recognized in North America (e.g., Gazin, 1946; Gunnell, 1998). This group has also been placed as a subfamily within the Oxyaenidae (e.g., Wortman, 1902; Matthew, 1909), but should be considered a subfamily within the Hyaenodontidae (e.g., Denison, 1938; Simpson, 1945; Van Valen, 1966; Morlo and Gunnell, 2003; Peigné et al., 2007).

Savage (1965) proposed an additional family, the Teratodontidae (including *Teratodon* and *Quercitherium*), as a derived descendant of the proviverrine hyaenodontids present in Africa and Europe. Van Valen (1967) thought that this was a viable taxon, although he believed it should be at a lower taxonomic level. In this chapter, we have placed *Teratodon* within its own subfamily as the dentally derived condition of this genus cannot be ignored.

As with creodont families, various formulations of subfamilies have also appeared. Hyaenodontids have generally been subdivided into the Proviverrinae and Hyaenodontinae following Matthew's (1909) suggestion that the first was ancestral to the latter. Two additional subfamilies, Limnocyoninae and Machaeroidinae, have been placed variably within this family or the Oxyaenidae, as noted earlier. More recently, Crochet (1988) named a new subfamily, Koholiinae, from Algeria.

The two traditional subfamilies, Proviverrinae and Hyaenodontinae, are probably paraphyletic. Barry (1998) undertook a phylogenetic analysis to explore the relationship of *Dissopsalis* to other taxa. As a result, he placed some proviverrines in a monophyletic group termed "advanced proviverrines" composed of the "Neogene proviverrines" (*Dissopsalis*, *Anasinopa*, and *Metasinopa*), as well as *Prodissopsalis*, *Allopterodon*, *Cynohyaenodon*, and probably *Paratritemnodon* and *Paracynohyaenodon*. *Prodissopsalis* and *Allopterodon* are the sister group of the "Neogene proviverrines." The sister group of the "advanced proviverrines" includes the European species of *Proviverra* and possibly *Masrasector aegypticum*. Together, all of these taxa form an "Old World proviverrine" assemblage that Barry states can be distinguished from hyaenodontines, limnocyoninines, and the "undifferentiated residue of proviverrines" (1998:43), although the interrelationships of these groups is unclear. Barry concludes that *Dissopsalis* either was African in origin or came from a lineage common to both Africa and southern Asia.

Holroyd (1994) took a different tack and proposed that some taxa found in both Asia and Africa (*Masrasector, Dissopsalis,* and *Anasinopa*) along with *Paratritemnodon* and *Metasinopa* form a monophyletic group with respect to other hyaenodontids. In fact, she suggests that the Indo-Pakistani genus *Paratritemnodon* had an African origin. Egi et al. (2004) added the new genus *Yarshea* from Myanmar to this clade, extending its geographic range farther east.

While Barry assumed that hyaenodontines were monophyletic, Polly (1996) suggested that the subfamily Hyaenodontinae is diphyletic and thus split this taxon into the Pterodontinae (*Pterodon* and possibly *Hyainailouros*) and the Hyaenodontinae (*Hyaenodon,* "*Pterodon*" *hyaenoides,* and possibly *Oxyaenoides*) with several genera unassigned to subfamily. In this scenario, the metaconid is lost at least twice. Holroyd (1999) later revised the diagnosis for Pterodontinae, as Polly's diagnosis was based almost exclusively on upper dentition and cranial features. Holroyd notes that the metaconid was lost independently in several hyaenodontid lineages but retains Polly's general definition of pterodontines as hyaenodontids lacking a metaconid that are more closely related to *Pterodon* than *Hyaenodon*. We must note, however, that Pilgrim's Hyainailourinae has priority over Polly's Pterodontinae.

Holroyd (1994, 1999) noted that Paleogene hyainailourines (= her pterodontines) appear to be monophyletic, at least when compared to selected European and Asian proviverrines and hyaenodontines. She suggests that African hyainailourines form a clade with European *Pterodon dasyuroides* possibly united by a "weak to absent P3 lingual cingulum, relatively larger anterior keels on the lower molars, m3 talonid reduction, and lower molar protoconids and paraconids subequal in length" (1999:17). Among the Neogene hyainailourines, Holroyd (1999) suggests that the larger taxa ("*Megistotherium, Hyainailouros,* and *P. nyanzae*") are most closely related to the Fayum species of *Pterodon*. Smaller Neogene hyainailourines may be more closely related to *Metapterodon* or a new small species from Locality 41. However, *Metapterodon* is itself a problematic taxon (see Morales et al., 1998a; Holroyd, 1999). Holroyd resisted formally recognizing these relationships within hyainailourines as she notes that morphological similarities could be due to convergence.

Based on new finds in the Pondaung Formation in Myanmar, Peigné et al. (2007) proposed a new set of relationships among proviverrines, hyainailourines, and limnocyonines, thereby rendering any previously hypothesized relationships between Asian and African taxa invalid. Proviverrines are broken into (1) *Proviverra*-like Proviverrinae, (2) *Arfia*-like South Asian Proviverrinae, and (3) *Prototomus*-like Proviverrinae. Briefly, *Prototomus*-like proviverrines (including *Sinopa*) form a clade with *Masrasector* and the limnocyonines (including *Prolimnocyon*). This clade is the sister group to a clade formed by hyainailourines and *Arfia*-like South Asian proviverrines (*Paratritemnodon* and *Kyawdawia*). All of these taxa then form a sister group to *Proviverra*-like proviverrines. *Arfia* is considered the ancestor of the hyainailourines and the *Arfia*-like South Asian proviverrines. This has implications for the biogeography of these taxa, as hyainailourines were not present in Asia before the early Miocene. The South Asian taxa now can be interpreted as an endemic clade that evolved independently in the Eocene/Oligocene hyainailourine radiation of Africa and Europe, including some taxa with morphologies similar to hyainailourines (*Orienspterodon*).

Attempts have been made to place hyaenodontines into tribes. Szalay (1967) created the tribe Apterodontini consisting solely of the genus *Apterodon*. However, Holroyd (1994) elevates this tribe to the subfamily level with Apterodontinae and includes not only *Apterodon* but also what is now known as *Quasiapterodon* (based on "*Apterodon*" *minutus*). This genus is derived from material assigned to *Apterodon* aff. *A. macrognathus* (Holroyd, 1994: figure 5.1). The European species *A. gaudryi* is descended from the *A. macrognathus* lineage (see also Lange-Badré and Böhme, 2005, for a slightly different approach).

It is clear that a large-scale study of the African Creodonta needs to be undertaken given the key nature of this group for understanding biogeography and dispersal issues worldwide.

BIOGEOGRAPHY OF THE HYAENODONTIDAE

Throughout most of hyaenodontid evolution, the joint land mass of Africa and the Arabian Peninsula (i.e., the Afro-Arabian continent) was separated from Eurasia. Short-lived dispersal routes permitting Simpsonian sweepstakes dispersal or island hopping may have occurred throughout this time (see Holroyd, 1994, for a discussion of some of these routes), but definitive connections between Eurasia and Africa were submerged in the Tethys at the end of the Cretaceous and did not occur again until the Miocene or latest Oligocene (see Morlo et al., 2007, and references therein).

Despite this relative isolation, early workers posited that the first African creodonts were not endemic, but were closely related to Eurasian forms that had migrated into Africa at some point during the late Paleocene to late Eocene (e.g., Andrews, 1906; Schlosser, 1911; Cooke, 1968; Coryndon and Savage, 1973; Maglio, 1978). Gingerich (1980, 1989; Gingerich and Deutsch, 1989), on the other hand, suggested that hyaenodontids, and creodonts overall, originated in Africa based on the diversity of the Fayum taxa. Redating of the Fayum along with recent African discoveries have indicated that the time of dispersal, regardless of direction, must have been before the late Eocene.

More recently, a specifically Asian origin for all hyaenodontids has been suggested. The oldest member of this taxon is often considered to be *Prolimnocyon chowi* from the Bayan Ulan fauna of Inner Mongolia (Meng et al., 1998; Beard and Dawson, 1999). In this model, hyaenodontids and other mammalian taxa (e.g., artiodactyls, perissodactyls, and primates) disperse from Asia near the Paleocene/Eocene boundary (Beard, 1998; Beard and Dawson, 1999; Eberle and McKenna, 2002). Note that this model subsumes the family Limnocyonidae, of which *Prolimnocyon* is a member, into the Hyaenodontidae such that *Prolimnocyon* is presumably the ancestor of Limnocyonidae and a sister-group of prototomine-like "proviverrines." Due to its already small, reduced m3, Bumbanian (lower Eocene of Asia), *Prolimnocyon* cannot be the ancestor of all Hyaenodontidae. In any case, prototomine-like proviverrines and *Prolimnocyon* appear in North America by the earliest Eocene (Wasatchian NALMA) and are already widely dispersed (Gunnell, 1998). In Europe, hyaenodontids, including the presumed earliest hyainailourine *Francotherium*, appear in the early Eocene (Dormaal, Belgium, and Le Quesnoy, France; Smith and Smith, 2001), although they appear to be limited to western Europe.

An alternate model has been proposed by Gheerbrant (1990). Gheerbrant hypothesized that hyaenodontids, along with omomyid primates, dispersed northward from Africa near the Paleocene/Eocene boundary. The earliest supposed creodont in Africa is *Tinerhodon disputatum* from the latest Paleocene site of Adrar Mgorn 1 in Morocco (Cappetta et al., 1978, 1987; Gheerbrant, 1995; Gheerbrant et al., 2006). Gheerbrant et al. (2006) proposed that *Tinerhodon* was a primitive proviverrine with some cimolestid-like features, suggesting to them that even earlier African proviverrines are the ancestors of all other creodonts. They dismiss the Mongolian *P. chowi* as being too derived to be ancestral to all other creodonts. Their identification of *Cimolestes* (a taxon that has been variously regarded as the ancestor of carnivorans, creodonts, or both) in the late Paleocene of Africa suggests to them that the earliest creodonts evolved in northern Africa from African cimolestids that then underwent a trans-Tethyan dispersal into Europe and eventually into North America. Gheerbrant and Rage (2006) envisioned five to seven faunal dispersals between Africa and Laurasia from the Late Cretaceous to the Eocene/Oligocene boundary. Several of these dispersals involve hyaenodontids moving northward from Africa.

Holroyd (1994) challenged Gheerbrant's original model, as a simple northward dispersal does not explain the close relationships between late Eocene and early Oligocene mammals of the Fayum and other areas at the genus and family level. Her research indicated that creodonts engaged in several intercontinental dispersals, with proviverrine dispersal events occurring in the early Eocene and hyaenodontine dispersal events occurring from the early through middle Eocene. Egi et al. (2004) supported the idea of an Afro-Asian proviverrine group that was widely distributed around the Tethys Sea and roamed between northern Africa and southeast Asia during the Eocene. Holroyd (1994), however, postulated a later dispersal of Fayum taxa (e.g., *Apterodon*) from northern Africa into Europe, a hypothesis that was supported by Morlo et al. (2007) for *Hyainailorous*.

At present, it is not possible to rule out either Asia or Africa as the origin of hyaenodontids. Creodont phylogeny is poorly understood, and the late Paleocene/early Eocene record of hyaenodontids is relatively poor. Phylogenetic relationships between African, Asian, and European taxa are disputed (e.g., Egi et al., 2005; Peigné et al., 2007). The presence of taxa such as *Tinerhodon* and *Boualitomus* from the Paleocene and Eocene of Morocco, respectively, and *Koholia* from the Algerian Eocene provide intriguing glimpses into the early Tertiary carnivore guilds of this region.

However, the model of Peigné et al. (2007) may simplify these problems by interpreting "proviverrine" relationships in a different way. If prototomine-like "Proviverrinae" (including limnocyonines and *Arfia*) are not or only roughly related to Proviverrinae s. str., then the origin of both groups may have lain in different continents: Asia for prototomine-like taxa (with *Prolimnocyon chowi* as the oldest known representative) and Africa for Proviverrinae s. str. (with *Tinerhodon* as the oldest known representative). The late early Eocene African *Boualitomus* could then easily be interpreted as an African prototomine-like "proviverrine" that migrated to Africa from Europe in one of the several early Eocene faunal exchanges between Africa and Europe (Gheerbrant and Rage, 2006). In Europe, as in all Northern continents, the group was well established in the early Eocene (e.g., Smith and Smith, 2001). On the other hand, Proviverrinae s. str. may have migrated from Africa to Europe later than the earliest

Eocene as their first European representatives are known not before MP 8 (*"Proviverra" eisenmanni* and *Parvagula*; see Morlo and Habersetzer, 1999).

Within Africa, hyaenodontids began to diversify and disperse during the Eocene. In the early Eocene, hyaenodontids are known from the Ouled Abdoun Basin in Morocco (Gheerbrant et al., 2006) and two sites in Algeria: El Kohol and Gour Lazib (Crochet, 1988; Adaci et al., 2007). By the late Eocene and early Oligocene, hyaenodontids had dispersed across northern Africa and the Arabian Peninsula (e.g., Osborn, 1909; Schlosser, 1910, 1911; Simons, 1968; Simons and Gingerich, 1974; Crochet et al., 1990; Holroyd, 1994, 1999), which was part of the African continent at the time (Smith et al., 1994).

By the late Oligocene, hyaenodontids had reached Ethiopia (Sanders et al., 2004), yet they are poorly represented elsewhere in the African fossil record. This is in contrast to the situation in Eurasia and North America where only one genus, *Hyaenodon*, occurred. In Europe and North America, *Hyaenodon* survives until shortly before the end of this epoch (Mellett, 1977; Lange-Badré, 1979; Gunnell, 1998; Tedford et al., 2004) and lasts until the earliest Miocene in Asia (Wang et al., 2005). Hyaenodontids disappear from Europe during the late Oligocene (Ginsburg, 1999) but reimmigrated in the latest early Miocene.

In the early Miocene, hyaenodontids are well-known components of the African fossil record. They ranged from the Arabian Peninsula and northern Africa to eastern and even southern Africa (Stromer, 1926; Koenigswald, 1947; Savage, 1965; Bishop, 1967; Savage, 1973; Andrews et al., 1981; Pickford et al., 1986; Tchernov et al., 1987; Rasmussen et al., 1989; Morales et al., 1998a; Morlo et al., 2007).

At the beginning of the Miocene, the Arabian plate collided with Asia. It is not surprising, therefore, that hyaenodontids reappear in Europe during the Miocene. Made (1999) defines the "Creodont Event" (17.5? Ma) as the point where the hyaenodontids *Hyainailouros* disperses from Africa to Eurasia, possibly in the company of other African taxa during a relatively warm interval in Europe. Morlo et al. (2007), however, suggest that *Hyainailouros* may have migrated twice out of Africa: once to Asia (19.6 Ma, MN 3, Bugti fauna) and once to Europe (MN 4). This hypothesis is based on the greater similarity between African *Hyainailouros* and European *H. sulzeri* than between *H. sulzeri* and *H. bugtiensis*. *Hyainailouros* then persists through the early Miocene of the Indo-Pakistani region and into the earliest middle Miocene (MN 5) of Europe. Citations of *Hyainailouros* being present in MN 7/8 (Ginsburg, 1980, 1999; Welcomme et al., 1997) are based on its occurrence at La Grive, which incorporates not only a mammalian fauna from MN 7/8 but also faunal elements of MN 5. Besides *Hyainailouros*, the carnivorans *Paralutra jaegeri* (Heizmann and Morlo, 1998) and *Martes munki*, both typical faunal elements for MN 5 (Nagel et al., 2009), are also found in La Grive.

During the middle Miocene, hyaenodontids are found across northern, eastern, and southern Africa (e.g., Savage, 1978; Shipman et al., 1981; Barry, 1988; Morales et al., 1998b, 2003). By the end of the Miocene, however, hyaenodontids disappear from both Asia and Africa.

The appearance of modern forms of Carnivora has been suggested to be the cause of the extinction of creodonts (e.g., Gunnell, 1998; Ginsburg, 1999). Ginsburg (1999) proposed that once migrations between Eurasia and Africa were established in the early Miocene (MN 3), hyaenodontids declined rapidly as members of the order Carnivora dispersed through Africa. However, Morlo et al. (2007) provide evidence for the immigration of carnivorans into Africa as early as the late Oligocene, which implies a long time of co-occurrence. In fact, these early carnivoran immigrants include only small stenoplesictids that did not compete with hyaenodontids due to their smaller size and different diet. In the early Miocene, amphicyonids and the sabre-toothed barbourofelids are also present in African faunas. As with stenoplesictids, both amphicyonids and barbourofelids differ clearly in their respective ecomorphologies from the co-occurring medium-sized and giant hyainailourines. Moreover, studies of changes in carnivore guilds after migration events have not necessarily shown competition to be an important factor. Instead, the new arriving taxa were incorporated into the fauna (Van Valkenburgh, 1999; Morlo et al., in press). This is also evident from studies on the arrival of single carnivorans on islands (Phillips et al., 2007).

While it is therefore tempting to blame the disappearance of the creodonts solely on carnivorans, one must remember that dramatic reversals in diversity (and presumably abundance) occur in other groups (e.g., hominoids) during the Miocene of Africa. Presumably, climatic changes and changes in the diversity and abundance of preferred prey species, in concert with the dispersal and diversification of carnivorans within Africa, led to the demise of the original carnivores of Africa.

Conclusions

Hyaenodontids are found throughout the Cenozoic of this continent (figure 26.8) and are undeniably successful until the end of the Miocene. Although no formal studies have been carried out on the structure of specific African hyaenodontid guilds for a specific place and time (and, indeed, there is often not enough material of individual species for such a study), it is clear that members of this group ranged from small, insectivorous forms to the largest carnivores in Africa.

The taxonomy of hyaenodontid creodonts is fairly complex. We support the presence of five subfamilies within the Afro-Arabian region: Apterodontinae, Hyainailourinae, Koholiinae, prototomine-like Proviverrinae, and Teratodontinae. Unfortunately, analysis of the relationships between these subfamilies is beyond the scope of this chapter. Material in northern Africa and Asia demonstrate that revision of the taxonomy of African hyaenodontids cannot be carried out in isolation.

ACKNOWLEDGMENTS

We would like to thank Bill Sanders and Lars Werdelin for the invitation to participate in this volume and the anonymous reviewers and editors for providing helpful comments and suggestions.

NOTE ADDED IN PROOF

After submission of this manuscript, significant new discoveries of African Paleogene hyaenodontids have been reported (not listed in Table 26.2). Pickford et al. (2008) describe *Pterodon* sp. and an indet. genus and species of proviverrine from the middle-late Eocene of Namibia. Rasmussen and Gutiérrez (2009) describe material from the late Oligocene of northern Kenya including the new genus and species *Mlanyama sugu* (referred to Proviverrinae), and *Hyainailouros* sp. from Nakwai and Losodok, respectively. Finally, Solé et al. (2009) describe the new koholiine *Lahimia selloumi* from the Late Paleocene of Sidi Chennane in the Ouled Abdoun Basin, Morocco. This find is particularly significant as it indicates that *Boualitomus* also belongs in the Koholiinae, rather than the Proviverrinae, as previously thought, and provides strong evidence for the African origin of hyaenodontids. This discovery will precipitate a fundamental shift in our understanding of hyaenodontids, a topic that will be pursued further elsewhere.

Literature Cited

Adaci, M., R. Tabuce, F. Mebrouk, M. Bensalah, P. H. Fabre, L. Hautier, J.-J. Jaeger, V. Lazzari, M. H. Mahboubi, L. Marivaux, O. Otero, S. Peigné, and H. Tong. 2007. Nouveaux sites à vertébrés paléogènes dans la région des Gour Lazib (Sahara nord-occidental, Algérie). *Comptes Rendus Palevol* 6:535–544.

Andrews, C. W. 1906. *A Descriptive Catalogue of the Tertiary Vertebrata of the Fayum, Egypt.* British Museum (Natural History), London, 324 pp.

Andrews, P. J., G. E. Meyer, D. R. Pilbeam, J. A. Van Couvering, and J. A. H. Van Couvering. 1981. The Miocene fossil beds of Maboko Island, Kenya: Geology, age, taphonomy, and paleontology. *Journal of Human Evolution* 10:35–48.

Barry, J. C. 1988. *Dissopsalis*, a middle and late Miocene proviverrine creodont (Mammalia) from Pakistan and Kenya. *Journal of Vertebrate Paleontology* 8:25–45.

Beard, K. C. 1998. East of Eden: Asia as an important center of taxonomic origination in mammalian evolution; pp. 5–39 in K. C. Beard and M. R. Dawson (eds.), *Dawn of the Age of Mammals in Asia.* Bulletin of the Carnegie Museum of Natural History, 34.

Beard, K. C., and M. R. Dawson. 1999. Intercontinental dispersal of Holarctic land mammals near the Paleocene/Eocene boundary: Paleogeographic, paleoclimatic, and biostratigraphic implications. *Bulletin de la Société Géologique de France* 170:697–706.

Biedermann, W. G. A. 1863. *Petrefacten aus der Umgegend von Winterthur: II. Die Braunkohlen von Elgg. Anhang:* Hyainailouros sulzeri. Bleuler-Hausheer, Wintherthur, 23 pp.

Bishop, W. W. 1967. The later Tertiary in East Africa: Volcanics, sediments, and faunal inventory; pp. 31–56 in W. W. Bishop and J. D. Clark (eds.), *Background to Evolution in Africa.* University of Chicago Press, Chicago.

Cappetta, H., J.-J. Jaeger, M. Sabatier, B. Sigé, J. Sudre, and M. Vianey-Liaud. 1978. Découverte dans le Paléocène du Maroc des plus anciens Mammifères euthériens d'Afrique. *Geobios* 11:257–263.

FIGURE 26.8 Map of sites with creodonts in Africa and the Arabian Peninsula. Cross hatching indicates a region with a large number of early Miocene sites.

● Paleocene
★ Eocene
✪ Eocene/Oligocene
   boundary
● Oligocene
× Early Miocene
▲ Middle Miocene

Cappetta, H., J.-J. Jaeger, B. Sigé, J. Sudre, and M. Vianey-Liaud. 1987. Compléments et précisions biostratigraphiques sur la faune paléocène à Mammifères et Sélaciens du bassin d'Ouarzazate (Maroc). *Tertiary Research* 8:147–157.

Colbert, E. H. 1933. The skull of *Dissopsalis carnifex* Pilgrim, a Miocene creodont from India. *American Museum Novitates* 603:1–8.

———. 1935. Siwalik mammals in the American Museum of Natural History. *Transactions of the American Philosophical Society* 26:1–401.

Cooke, H. B. S. 1968. Evolution of mammals on southern continents. II. The fossil mammal fauna of Africa. *Quarterly Review of Biology* 43:234–264.

Cope, E. D. 1875. On the supposed Carnivora of the Eocene of the Rocky Mountains. *Proceedings of the Academy of Natural Sciences, Philadelphia* 27:444–448.

Coryndon, S. C. and R. J. G. Savage. 1973. Origins and affinities of African mammal faunas. *Special Papers in Palaeontology* (Palaeontological Association, London) 12:121–135.

Crochet, J.-Y. 1988. Le plus ancien Créodonte africain: *Koholia atlasense* nov. gen., nov. sp. (Eocène inférieur d'El Kohol, Atlas saharien, Algérie). *Comptes Rendus de l'Académie des Sciences, Paris*, Série II, 307:1795–1798.

Crochet, J.-Y., H. Thomas, J. Roger, S. Sen, and Z. Al-Sulaimani. 1990. Première découverte d'un créodonte dans la péninsule Arabique: *Masrasector ligabuei* nov. sp. (Oligocène inférieur de Taqah, Formation D'Ashawq, Sultanat d'Oman. *Comptes Rendus de l'Académie des Sciences, Paris*, Série II, 311:1455–1460.

Cuvier, G. 1822. Portions de tête et de machoire d'une grande espece appartenant à un genre de la famille des Coatis, des Ratons, etc.; pp. 269–272 in G. Cuvier (ed.), *Recherches sur les Ossemens Fossiles, ou l'on Retablit les Charactères de Plusieurs Animaux, Dont les Révolutions du Globe ont Detruit les Espèces*. New ed. Dufour and d'Ocagne, Paris.

Dashzeveg, D. 1985. Nouveaux Hyaenodontinae (Creodonta, Mammalia) du Paléogène de Mongolie. *Annales de Paléontologie* 71:233–256.

Denison, R. H. 1938. The broad-skulled Pseudocreodi. *Annals of the New York Academy of Sciences* 37:163–257.

Eberle, J. J., and M. C. McKenna. 2002. Early Eocene Leptictida, Pantolesta, Creodonta, Carnivora, and Mesonychidae (Mammalia) from the Eureka Sound Group, Ellesmere Island, Nunavut. *Canadian Journal of Earth Sciences* 39:899–910.

Egi, N., P. A. Holroyd, T. Tsubamoto, N. Shigehara, M. Takai, Soe Thura Tun, Aye Ko Aung, and Aung Naing Soe. 2004. A new genus and species of hyaenodontid creodont from the Pondaung Formation (Eocene, Myanmar). *Journal of Vertebrate Paleontology* 24:502–506.

Egi, N., P. A. Holroyd, T. Tsubamoto, A. N. Soe, M. Takai, and R. L. Ciochon. 2005. Proviverrine hyaenodontids (Creodonta: Mammalia) from the Eocene of Myanmar and a phylogenetic analysis of the proviverrines from the Para-Tethys Area. *Journal of Systematic Paleontology* 3:337–358.

Egi, N., T. Tsubamoto, and M. Takai. 2007. Systematic status of Asian "*Pterodon*" and early evolution of hyaenaelurine hyaenodontid creodonts. *Journal of Paleontology* 81:770–778.

Flynn, J. J., N. A. Neff, and R. H. Tedford. 1988. Phylogeny of the Carnivora; pp. 73–116 in M. J. Benton (ed.), *The Phylogeny and Classification of the Tetrapods: Volume 2. Mammals*. The Systematics Association Special Volume Number 35B. Clarendon Press, Oxford.

Flynn, J. J., and G. D. Wesley-Hunt. 2005. Carnivora; pp. 175–198 in K. D. Rose and J. D. Archibald (eds.), *The Rise of Placental Mammals: Origins and Relationships of the Major Extant Clades*. Johns Hopkins University, Baltimore.

Fourtau, R. 1920. *Contribution a l'Étude des Vertébrés Miocènes de l'Égypte*. Egypt Survey Department, Cairo, 121 pp.

Fox, R. C., and G. P. Youzwyshyn. 1994. New primitive carnivorans (Mammalia) from the Paleocene of western Canada, and their bearing on relationships of the order. *Journal of Vertebrate Paleontology* 14:382–404.

Gazin, C. L. 1946. *Machaeroides eothen* Matthew, the saber-toothed creodont of the Bridger Eocene. *Proceedings of the United States National Museum* 96:335–347.

Gebo, D. L. and K. D. Rose. 1993. Skeletal morphology and locomotor adaptation in *Prolimnocyon atavus*, an early Eocene hyaenodontid creodont. *Journal of Vertebrate Paleontology* 13:125–144.

Gheerbrant, E. 1990. On the early biogeographical history of African placentals. *Historical Biology* 4:107–116.

———. 1995. Les mammifères paléocènes du bassin d'Ouarzazate (Maroc). III. Adapisoriculidae et autres mammifères (Carnivora, ?Creodonta, Condylarthra, ?Ungulata et *incertae sedis*). *Palaeontographica, Abt. A* 237:39–132.

Gheerbrant, E., M. Iarochène, M. Amaghzaz, and B. Bouya. 2006. Early African hyaenodontid mammals and their bearing on the origin of the Creodonta. *Geological Magazine* 143:475–489.

Gheerbrant, E., and J.-C. Rage. 2006. Paleobiogeography of Africa: How distinct from Gondwana and Laurasia? *Palaeogeography, Palaeoclimatology, Palaeoecology* 241:224–246.

Gheerbrant, E., J. Sudre, H. Cappetta, C. Mourer-Chauviré, E. Bourdon, M. Iarochène, M. Amaghzaz, and B. Bouya. 2003. Les localités à mammifères des carrières de Grand Daoui, bassin des Ouled Abdoun, Maroc, Yprésien: Premier état des lieux. *Bulletin de la Société Géologique de France* 174:279–293.

Gingerich, P. D. 1980. *Tytthaena parrisi*, oldest known oxyaenid (Mammalia, Creodonta) from the late Paleocene of western North America. *Journal of Paleontology* 54:570–576.

———. 1989. New Earliest Wasatchian mammalian fauna from the Eocene of northwestern Wyoming: Composition and diversity in a rarely sampled high-floodplain assemblage. *University of Michigan, Papers on Paleontology* 28:1–97.

Gingerich, P. D., and H. A. Deutsch. 1989. Systematics and evolution of early Eocene Hyaenodontidae (Mammalia, Creodonta) in the Clarks Fork Basin, Wyoming. *Contributions of the Museum of Paleontology* (University of Michigan) 27:327–391.

Ginsburg, L. 1980. *Hyainailouros sulzeri*, mammifère Créodonte du Miocène d'Europe. *Annales de Paléontologie (Vertébrés)* 66:19–73.

———. 1999. Order Creodonta; pp. 105–108 in G. E. Rössner and K. Heissig (eds.), *The Miocene Land Mammals of Europe*. Pfeil, Munich.

Gunnell, G. F. 1998. Creodonta; pp. 91–109 in C. M. Janis, K. M. Scott, and L. L. Jacobs (eds.), *Evolution of Tertiary Mammals of North America: Volume 1. Terrestrial Carnivores, Ungulates, and Ungulatelike Mammals*. Cambridge University Press, Cambridge.

Heizmann, E. P. J. and M. Morlo. 1998. Die semiaquatische *Lartetictis dubia* (Mustelina, Carnivora, Mammalia) von Goldberg/Ries (Baden-Württemberg). *Mainzer naturwissenschaftliches Archiv, Beiheft* 21:141–153.

Holroyd, P. A. 1994. An examination of dispersal origins for Fayum Mammalia. Unpublished PhD dissertation, Duke University, Durham, N.C.

———. 1999. New Pterodontinae (Creodonta: Hyaenodontidae) from the late Eocene–early Oligocene Jebel Qatrani Formation, Fayum province, Egypt. *Paleobios* 19:1–18.

Holroyd, P. A., E. L. Simons, T. M. Bown, P. D. Polly, and M. J. Kraus. 1996. New records of terrestrial mammals from the upper Eocene Qasr el Sagha Formation, Fayum depression, Egypt. *Paleovertebrata* 25:175–192.

ICZN. 1999. *International Code of Zoological Nomenclature*. 4th ed. International Trust for Zoological Nomenclature, London, 306 pp.

Jenkins, F. A., and S. M. Camazine. 1977. Hip structure and locomotion in ambulatory and cursorial carnivores. *Journal of Zoology, London* 181:351–370.

Koenigswald, G. H. R. v. 1947. Ein *Hyaenaelurus* aus dem Miocaen Nordafrikas. *Société Paléontologique Suisse, Contribution à l'Étude des Vertébrés miocènes de l'Egypte* 292–294.

Lange, B. 1967. Créodontes des phosphorites du Quercy, *Apterodon gaudryi*. *Annales de Paléontologie* 53:79–90.

Lange-Badré, B. 1979. Les créodontes (Mammalia) d'Europe occidentale de l'Éocène supérieur à l'Oligocène supérieur. *Mémoires du Muséum National d'Histoire Naturelle, Série C, Sciences de la Terre* 42:1–249.

Lange-Badré, B., and M. Böhme. 2005. *Apterodon intermedius*, sp. nov., a new European creodont mammal from MP22 of Espenhain (Germany). *Annales de Paléontologie* 91:311–328.

Lavrov, A. V. 1996. A new species of *Neoparapterodon* (Creodonta, Hyaenodontidae) from the Khaichin-Ula 2 Locality (the Khaichin Formation, Middle-Upper Eocene, Mongolia) and the phylogeny of Asiatic representatives of *Pterodon*. *Paläontologisches Zeitschrift* 4:95–107.

———. 1999. New material on the Hyaenodontidae (Mammalia, Creodonta) from the Ergiliyn Dzo Formation (late Eocene of Mongolia) and some notes on the system of the Hyaenodontidae. *Paleontological Journal* 33:321–329.

Lavrov, A. V. and A. V. Lopatin. 2004. A new species of *Arfia* (Hyaenodontidae, Creodonta) from the Basal Eocene of Mongolia. *Paleontological Journal* 38:448–457.

Made, J. van der 1999. Intercontinental relationship Europe-Africa and the Indian Subcontinent; pp. 457–472 in Rössner, G. E. and K. Heissig (eds.), *The Miocene Land Mammals of Europe*. Pfeil, Munich.

Maglio, V. J. 1978. Patterns of faunal evolution; pp. 603–620 in V. J. Maglio and H. B. S. Cooke (eds.), *Evolution of African Mammals*. Harvard University Press, Cambridge.

Matthew, W. D. 1901. Additional observations on the Creodonta. *Bulletin of the American Museum of Natural History* 14:1–38.

———. 1909. The Carnivora and Insectivora of the Bridger Basin, Middle Eocene. *Memoirs of the American Museum of Natural History* 9:289–567.

Mellett, J. S. 1977. Paleobiology of North American *Hyaenodon* (Mammalia, Creodonta). *Contributions to Vertebrate Evolution* 1:1–134.

Meng, J., R. Zhai, and A. R. Wyss. 1998. The Late Paleocene Bayan Ulan Fauna of Inner Mongolia, China; pp. 148–185 in K. C. Beard and M. R. Dawson (eds.), *Dawn of the Age of Mammals in Asia*. Bulletin of the Carnegie Museum of Natural History, 34.

Morales, J., and M. Pickford. 2005. Carnivores from the middle Miocene Ngorora Formation (13–12 Ma), Kenya. *Estudios Geológicos* 61:271–284.

———. 2008. Creodonts and carnivores from the Middle Miocene Muruyur Formation at Kipsaraman and Cheparawa, Baringo District, Kenya. *Comptes Rendus Palevol* 7:487–497.

Morales, J., M. Pickford, S. Fraile, M. J. Salesa, and D. Soria. 2003. Creodonta and Carnivora from Arrisdrift, early Middle Miocene of Southern Namibia. *Memoir of the Geological Survey of Namibia* 19:177–194.

———. 2007. New carnivoran material (Creodonta, Carnivora and Incertae sedis) from the early Miocene of Napak, Uganda. *Paleontological Research* 11:71–84.

Morales, J., M. Pickford, and D. Soria. 1998a. A new creodont *Metapterodon stromeri* nov. sp. (Hyaenodontidae, Mammalia) from the early Miocene of Langental (Sperrgebiet, Namibia). *Comptes Rendus de l'Académie des Sciences, Paris, Série Sciences de la Terre et des Planètes* 327:633–638.

Morales, J., M. Pickford, D. Soria, and S. Fraile. 1998b. New carnivores from the basal Middle Miocene of Arrisdrift, Namibia. *Eclogae Geologicae Helvetiae* 91:27–40.

Morlo M. and G. F. Gunnell 2003. Small Limnocyoninae (Hyaenodontidae, Mammalia) from the Bridgerian, middle Eocene of Wyoming: Thinocyon, *Iridodon* n. gen., and *Prolimnocyon*. Contributions from the Museum of Paleontology, The University of Michigan 31:43–78.

Morlo, M., and G. F. Gunnell. 2005. New species of *Limnocyon* (Mammalia, Creodonta) from the Bridgerian (Middle Eocene). *Journal of Vertebrate Paleontology* 25:251–255.

Morlo, M., G. F. Gunnell, and D. Nagel. In press. Ecomorphological analysis of carnivore guilds in the Eocene through Miocene of Laurasia; in A. Goswami and A. Friscia (eds.), *Carnivoran Evolution: New Views on Phylogeny, Form and Function*. Cambridge University Press, Cambridge.

Morlo, M., and J. Habersetzer. 1999. The Hyaenodontidae (Creodonta, Mammalia) from the lower Middle Eocene (MP 11) of Messel (Germany) with special remarks to new x-ray methods. *Courier Forschungs-Institut Senckenberg* 216:31–73.

Morlo, M., E. R. Miller, and A. N. El-Barkooky. 2007. Creodonta and Carnivora from Wadi Moghra, Egypt. *Journal of Vertebrate Paleontology* 27:145–159.

Nagel, D., C. Stefen, and M. Morlo. 2009. The carnivoran community from the Miocene of Sandelzhausen, Germany. *Paläontologische Zeitschrift* 83:151–174.

Novacek, M. J., and A. Wyss. 1986. Higher-level relationships of the Recent eutherian orders: Morphological evidence. *Cladistics* 2:257–287.

Osborn, H. F. 1909. New carnivorous mammals from the Fayûm Oligocene, Egypt. *Bulletin of the American Museum of Natural History* 26:415–424.

Peigné, S., M. Morlo, Y. Chaimanee, S. Ducrocq, S. T. Tun, and J.-J. Jaeger. 2007. New discoveries of hyaenodontids (Creodonta, Mammalia) from the Pondaung Formation, middle Eocene, Myanmar: Paleobiogeographic implications. *Geodiversitas* 29:441–458.

Phillips, R. B., C. S. Winchell, and R. H. Schmidt. 2007. Dietary overlap of an alien and native carnivore on San Clemente Island, California. *Journal of Mammalogy* 88:173–180.

Pickford, M. 1981. Preliminary Miocene mammalian biostratigraphy for western Kenya. *Journal of Human Evolution* 10:73–97.

Pickford, M., B. Senut, D. Hadoto, J. Musisi, and C. Kariira. 1986. Nouvelles découvertes dans le Miocène inférieur de Napak, Ouganda Oriental. *Comptes Rendus de l'Académie des Sciences, Paris*, Série II 302:47–52.

Pickford, M., B. Senut, J. Morales, P. Mein, and I. M. Sanchez. 2008. Mammalia from the Lutetian of Namibia. *Memoir of the Geological Survey of Namibia* 20:465–514.

Pilgrim, G. E. 1910. Notices of new mammalian genera and species from the Tertiaries of India. *Records of the Geological Survey of India* 40:63–71.

———. 1914. Description of teeth referable to the lower Siwalik creodont genus *Dissopsalis*, Pilgrim. *Records of the Geological Survey of India* 44:265–279.

———. 1932. The fossil Carnivora of India. *Memoir of the Geological Survey of India, Paleontologica Indica*, n.s. 18:1–232.

Polly, P. D. 1996. The skeleton of *Gazinocyon vulpeculus* gen. et comb. nov. and the cladistic relationships of Hyaenodontidae (Eutheria, Mammalia). *Journal of Vertebrate Paleontology* 16:303–319.

Rasmussen, D. T., C. D. Tilden, and E. L. Simons. 1989. New specimens of the gigantic creodont, *Megistotherium*, from Moghara, Egypt. Journal of Mammalogy 70:442–447.

Rasmussen, D. T. and M. Gutiérrez, 2009. A mammalian fauna from the late Oligocene of northwestern Kenya. *Palaeontographica Abt. A* 288:1–52.

Rose, K. D. 1990. Postcranial skeletal remains and adaptations in early Eocene mammals from the Willwood Formation, Bighorn Basin, Wyoming; pp. 107–133 in T. M. Bown and K. D. Rose (eds.), *Dawn of the Age of Mammals in the Northern Part of the Rocky Mountain Interior, North America*. Geological Society of America, Special Paper 243, Boulder, Colo.

———. 2006. *The Beginning of the Age of Mammals*. Johns Hopkins University Press, Baltimore, 428 pp.

Sanders, W. J., J. Kappelman, and D. T. Rasmussen. 2004. New large-bodied mammals from the late Oligocene site of Chilga, Ethiopia. *Acta Palaeontologica Polonica* 49:365–392.

Savage, R. J. G. 1965. Fossil mammals of Africa: 19. The Miocene Carnivora of East Africa. *Bulletin of the British Museum (Natural History), Geology* 10:239–316.

———. 1973. *Megistotherium*, gigantic hyaenodont from Miocene of Gebel Zelten, Libya. *Bulletin of the British Museum (Natural History), Geology* 22:485–511.

———. 1977. Evolution in carnivorous mammals. *Palaeontology* 20:237–271.

———. 1978. Carnivora; pp. 249–267 in V. J. Maglio and H. B. S. Cooke (eds.), *Evolution of African Mammals*. Harvard University Press, Cambridge.

Schlosser, M. 1910. Über einige fossile Säugetiere aus dem Oligocän von Ägypten. *Zoologischer Anzeiger* 35:500–508.

———. 1911. Beiträge zur Kenntnis der oligozänen Landsäugetiere aus dem Fayûm, Ägypten. *Beiträge zur Paläontologie und Geologie Österreich-Ungarns, Wien* 14:51–167.

Shipman, P., A. Walker, J. A. Van Couvering, P. J. Hooker, and J. A. Miller. 1981. The Fort Ternan hominoid site, Kenya: Geology, age taphonomy and paleoecology. *Journal of Human Evolution* 10:49–72.

Simons, E. L. 1968. Early Cenozoic mammalian faunas, Fayum Province, Egypt. Part I. African Oligocene mammals: Introduction, history of study, and faunal succession. *Bulletin of the Peabody Museum of Natural History* (Yale University) 28:1–21.

Simons, E. L., and P. D. Gingerich. 1974. New carnivorous mammals from the Oligocene of Egypt. *Annals of the Geological Survey of Egypt* 4:157–166.

———. 1976. A new species of *Apterodon* (Mammalia, Creodonta) from the Upper Eocene Qasr el Sagha Formation of Egypt. *Postilla* 168:1–9.

Simpson, G. 1945. The principles of classification and a classification of mammals. *Bulletin of the American Museum of Natural History* 85:1–350.

Smith, T., K. D. Rose, and P. D. Gingerich. 2006. Rapid Asia-Europe-North America geographic dispersal of earliest Eocene primate *Teilhardina* during the Paleocene-Eocene Thermal Maximum. *Proceedings of the National Academy of Science, USA* 103:11223–11227.

Smith, T., and R. Smith. 2001. The creodonts (Mammalia, Ferae) from the Paleocene-Eocene transition in Belgium (Tienen Formation, MP7). *Belgian Journal of Zoology* 131:117–135.

Smith, A. G., D. G. Smith, and B. M. Funnell. 1994. *Atlas of Mesozoic and Cenozoic Coastlines*. Cambridge University Press, Cambridge, 99 pp.

Solé, F., E. Gheerbrant, M. Amghzaz, and B. Bouya. 2009. Further evidence of the African antiquity of hyaenodontid ('Creodonta', Mammalia) evolution. *Zoological Journal of the Linnean Society* 156:827–846.

Stromer, E. 1926. Reste land-und süsswasser-bewohnender Wirbeltiere aus dem Diamentenfelden Deutsch-Südwestafrikas; pp. 107–153 in E. Kaiser (ed.), *Die Diamantenwüste Südwestafrikas*, vol. 2. Ditrich Reimer, Berlin.

Szalay, F. S. 1967. The affinities of *Apterodon* (Mammalia, Deltatheridia, Hyaenodontidae). *American Museum Novitates* 2293:2–17.

Tchernov, E., L. Ginsburg, P. Tassy, and N. F. Goldsmith. 1987. Miocene mammals of the Negev (Israel). *Journal of Vertebrate Paleontology* 7:284–310.

Tedford, R. H. 1976. Relationship of pinnipeds to other carnivores (Mammalia). *Systematic Zoology* 25:363–374.

Tedford, R. H., L. B. Albright, III, A. D. Barnosky, I. Ferrusquía-Villafranca, R. M. Hunt, Jr., J. E. Storer, C. C. Swisher, III, M. R. Voorhies, S. D. Webb, and D. P. Whistler. 2004. Mammalian biochronology of the Arikareean through Hemphillian interval (Late Oligocene through early Pliocene Epochs); pp. 169–231 in M. O. Woodburne (ed.), *Late Cretaceous and Cenozoic Mammals of North America*. Columbia University Press, New York.

Tilden, C. D., P. A. Holroyd, and E. L. Simons. 1990. Phyletic affinities of *Apterodon* (Hyaenodontidae, Creodonta). *Journal of Vertebrate Paleontology* 10 (suppl. to no. 3):46A.

Turner, A., and M. Antón. 2004. *Evolving Eden: An Illustrated Guide to the Evolution of the African Large-Mammal Fauna*. Columbia University Press, New York, 269 pp.

Van Valen, L. 1965. Some European Proviverrini (Mammalia, Deltatheridia). *Paleontology* 8:638–665.

———. 1966. Deltatheridia, a new order of mammals. *Bulletin of the American Museum of Natural History* 132:1–176.

———. 1967. New Paleocene insectivores and insectivore classification. *Bulletin of the American Museum of Natural History* 135:217–284.

Van Valkenburgh, B. 1999. Major patterns in the history of carnivorous mammals. *Annual Review of Earth and Planetary Sciences* 27:463–93.

Wang, X., Z. Qiu, and B. Wang. 2005. Hyaenodonts and carnivorans from the early Oligocene to early Miocene of Xianshuihe Formation, Lanzhou Basin, Gansu Province, China. *Palaeontologia Electronica* 8:1–14.

Welcomme, J.-L., P.-O. Antoine, F. Duranthon, P. Mein, and L. Ginsburg. 1997. Nouvelles découvertes de vertébrés miocènes dans le synclinal de Dera Bugti (Balouchistan, Pakistan). *Comptes Rendus de l'Académie des Sciences* 325:532–536.

Wortman, J. L. 1902. Studies of Eocene Mammalia in the Marsh Collection, Peabody Museum, Part I: Carnivora. *American Journal of Science*. 11:333–348, 437–450; 12:143–154, 193–206; 281–296, 377–382, 421–432; 13:39–46, 115–128, 197–206, 433–448; 14:17–23.

Wozencraft, W. 1989. Appendix: classification of the recent Carnivora; pp. 569–594 in J. Gittleman (ed.), *Carnivore Behavior, Ecology, and Evolution*. Cornell University Press, New York.

Wyss, A. R., and J. J. Flynn. 1993. A phylogenetic analysis and definition of the Carnivora; pp. 32–52 in F. S. Szalay, M. J. Novacek, and M. C. McKenna (eds.), *Mammal Phylogeny: Placentals*. Springer, New York.

TWENTY-SEVEN

Prionogalidae (Mammalia *Incertae Sedis*)

LARS WERDELIN AND SUSANNE M. COTE

Priongale breviceps is a very small mammal with carnivorous adaptations described by Schmidt-Kittler and Heizmann (1991) on the basis of fragmentary craniodental material from a number of early Miocene localities in Kenya and Uganda. The dental homologies as reconstructed are unique among mammals, and therefore the taxon is placed in Mammalia *incerta sedis*. It has recently been accompanied in the family Prionogalidae by *Namasector soriae*, from Namibia (Morales et al., 2008) and the family was suggested by those authors to belong in the Creodonta.

Systematic Paleontology

Family PRIONOGALIDAE Morales et al., 2008
Genus *PRIONOGALE* Schmidt-Kittler and Heizmann, 1991
PRIONOGALE BREVICEPS Schmidt-Kittler and Heizmann, 1991

Emended Diagnosis Hypercarnivorous mammal as small as or smaller than any extant carnivoran. Tooth row greatly reduced and snout foreshortened. Dental formula I?/2, C?/1, P3/2, M1/2. Two pairs of carnassial teeth present: P4/m1 and M1/m2. P4 molarized as in extant carnivorans with enlarged, anterolingually situated protocone, buccally placed paracone and prominent metacone. P4 with mesiodistally long metastyle. M1 with broad stylar shelf and reduced paracone situated far from buccal margin of tooth, metacone large with metastyle extending at >30° to the direction of the pre- and postmetacristae. p4 and m1 very similar in morphology, with three cusps corresponding to paraconid, protoconid, and hypoconid. The paraconid/protoconid complex is trenchant. The postprotocristid leads directly to the prehypocristid and a trenchant hypoconid. There is no trace of a metaconid. The lingual margin of the talonid has a shallow but wide basin. m2 has only two cusps, paraconid and protoconid, which form a single cutting blade as in extant hypercarnivores. The talonid is absent or reduced to a minute distal cusplet.

Holotype KNM-SO 1431, right maxilla fragment with P3–M1 (Schmidt-Kittler and Heizmann, 1991: figures 1a–1b; erroneously listed as left maxilla fragment KNM-SO 1413 by Schmidt-Kittler and Heizmann, 1991:6).

Type Locality Songhor, Kenya (early Miocene, ca. 19.5 Ma).

Age and Occurrence Songhor, Legetet, Chamtwara, and Rusinga, Kenya; Napak IV, Uganda (all early Miocene, ca. 20–17 Ma).

Additional Material Songhor: KNM-SO 1380 left mandible fragment with p4–m1 (Schmidt-Kittler and Heizmann, 1991: figures 3a–3c); KNM-SO 1698, left mandible fragment with broken m2; SO 5056, left mandible fragment with p4–m1 (Schmidt-Kittler and Heizmann, 1991: figure 6); KNM-SO 8334, right mandible fragment with p4–m1 (Schmidt-Kittler and Heizmann, 1991: figures 5a–5b); KNM-SO 15976, right mandible fragment with broken m1, m2 fragment; KNM-SO 15977, right mandible fragment with m2 fragment; KNM-SO 15979, right p4 (listed as SO 3160 in Schmidt-Kittler and Heizmann, 1991: table 1); KNM-SO 15980, left P4 (listed as SO 3120 in Schmidt-Kittler and Heizmann, 1991: table 1, figures 8a–8c, figure 15); KNM-SO 22192, left mandible fragment with m2; KNM-SO 22371, right mandible fragment with broken m1, m2; KNM-SO 22856, left mandible fragment with p4, m1 fragment (Schmidt-Kittler and Heizmann, 1991: figures 4a–4c), specimen currently missing; KNM-SO 22858, left mandible fragment with m2 fragment (Schmidt-Kittler and Heizmann, 1991: figures 7a–7c); KNM-SO 22857, left mandible fragment with p4, fragment of c (listed as KNM SO 22957 by Schmidt-Kittler and Heizmann, 1991: table 1, figures 2a–2c); KNM-SO 22859, right mandible fragment with m2 ; KNM-SO 22860, left mandible fragment with p4–m1 in Schmidt-Kittler and Heizmann (1991:7, figures 14a–14d), specimen currently missing; KNM-SO 22946, right mandible fragment with m2, broken p4–m1; Sgr 4168.66, isolated right m1. Legetet: KNM-LG 1554, right mandible fragment with m1–m2; KNM-LG 2380, right mandible fragment with m2 (Schmidt-Kittler and Heizmann, 1991: figures 12a–12c); KNM-LG 2389, right mandible fragment with broken m1; KNM-LG 2395, right mandible fragment with broken m2; KNM-LG 2398, right P4 (Schmidt-Kittler and Heizmann, 1991: figure 11); KNM-LG 2402, ?P4 fragment. Chamtwara: KNM-CA 301, right mandible fragment with broken m1 and m2; KNM-CA 302, right mandible fragment with m2 fragment; KNM-CA 2083, left edentulous mandible fragment; KNM-CA 2731, left mandible fragment with m2 (Schmidt-Kittler and Heizmann, 1991: figures 10a–10c); KNM-CA 2800, left mandible fragment with broken m2; KNM-CA

3164, right mandible fragment with p4 (Schmidt-Kittler and Heizmann, 1991: figures 9a–9b). Rusinga: KNM-RU 15928, right mandible fragment with broken m1 and m2; KNM-RU 19690, left mandible fragment with broken p4, m1, roots of m2; ?KNM-RU 16759, broken ?m2. Napak IV: right mandible fragment with m1–m2 (Schmidt-Kittler and Heizmann, 1991: figures 13a–13c); left mandible fragment with m1–m2; left P4 (all Napak specimens unnumbered and presently missing).

Remarks As noted, there is some confusion regarding the identity of at least two of these specimens. On the basis of present knowledge, we cannot explain the discrepancies. The specimens currently listed as KNM-SO 22856 and SO 22860 do not match any of the other specimens listed by Schmidt-Kittler and Heizmann (1991), nor do either of the specimens illustrated under these numbers by those authors match any other specimens in the specimen list. In addition, several of the specimens originally described by Schmidt-Kittler and Heizmann (1991) cannot at present be located in the National Museums of Kenya, Nairobi, or the Uganda Museum, Kampala.

The dental homologies given here are dependent on acceptance of the arguments presented by Schmidt-Kittler and Heizmann (1991). In their view, the first upper molariform tooth must be P4 because the difference between this tooth and the subsequent one is too great for them to be M1 and M2. Among the differences demonstrated by these authors, the position of the paracone stands out. In the putative P4, this cusp is situated close to the buccal edge of the tooth, while in the putative M1 the paracone is located close to the lingual margin and is flanked on the buccal side by a wide stylar shelf. The authors also call attention to similarities between the ?P4 of *Prionogale* and the P4 of some primitive Paleocene "insectivores" such as *Acmeodon*. Given this upper cheek tooth homology, the posterior-most lower cheek tooth must be m2.

Against this interpretation stands the very similar p4 and m1. Like the appearance of very dissimilar M1 and M2, very similar p4 and m1 is rare or nonexistent among carnivorous mammals. If these teeth are instead m1 and m2, the posterior-most molar must be m3. In such a scenario, the upper sectorial cheek tooth pair would be M1 and M2, regardless of morphology. In this case, *Prionogale* could represent a highly derived hyaenodontid creodont, as suggested by Morales et al. (2008).

Both of these scenarios seem possible, although the weight of evidence tends to favor the original interpretation of Schmidt-Kittler and Heizmann (1991). However, accepting this interpretation leaves unresolved the question of the relationships of *Prionogale*. Schmidt-Kittler and Heizmann (1991: figure 17) indicate a closer relationship of *Prionogale* with Leptictoidea than with Carnivora or Creodonta, while in their abstract (but not in their text) they suggest that *Prionogale* is a relic of the endemic African paleofauna (i.e., an afrotherian). No characters can at present be used to support such an interpretation, but it must surely be considered on biogeographic grounds and because *Prionogale* is morphologically very distant from contemporaneous carnivorous animals, suggesting a long separate evolution for the genus.

Genus *NAMASECTOR* Morales et al., 2008
NAMASECTOR SORIAE Morales et al., 2008

Diagnosis Morales et al. (2008). Very small hypercarnivorous creodont, comparable in size to *Thereutherium* and *Prionogale*, M1–m2 and P4–m1 functioning as highly specializsed carnassials. P3 elongated with a strong linguobasal cuspid, P4 and p4 elongated with morphology similar to M1 and m1, respectively.

Holotype EF 118′04, right maxilla with P3–M1 (apparently erroneously given as EF 118′01 in the text but correctly in the figure captions; Morales et al., 2008: plate 1:2, figure 2:1).

Type Locality Elisabethfeld (Tortoise site), Namibia (early Miocene, ca. 20–19 Ma).

Age and Occurrence Type locality only.

Additional Material EF 50′01, left mandible, EF 60′01, right mandible, EF 118′01, right P4, EF 118′01 right maxilla fragment.

Remarks All this material is suggested by Morales et al. (2008) to belong to a single individual. As in the case of *Prionogale*, the dental homologies given here are entirely based on arguments presented by the authors of the taxon. In this case the homologies are much better established as the teeth are more "orthodox" in morphology than those of *Prionogale*. However, there is a slim possibility that the anteriormost of the preserved (P3 and p4 above) teeth might be deciduous. However, this should not affect the suggested homologies of the remaining teeth and thus at present has no bearing on the relationships of *Namasector*.

Discussion

Morales et al. (2008) find a number of similarities between *Namasector* and *Prionogale*, such as the diminutive size, the development of two carnassial pairs, the small m1, and the reduced premolar row. They further consider *Namasector* (and by extension *Prionogale*) to belong to the Creodonta due to the similarities in the carnassials with that group of carnivores. Because of differences from all known creodonts, however, they erect the new family Prionogalidae, encompassing the nominal genus *Prionogale* and the referred genus *Namasector*, and place it as *incertae sedis* within the order Creodonta.

We agree that *Namasector* does, indeed, show probable affinities with the Creodonta, and specifically with the smaller Hyaenodontidae such as *Isohyaenodon* (see Lewis and Morlo, this volume, chap. 26). We also agree that there are features that are shared between *Namasector* and *Prionogale* that may be indicative of relationship, such as the two carnassial pairs. However, the major difference between *Prionogale* and *Namasector* (and hyaenodontid creodonts) lies in the morphology of M1. In *Prionogale*, as stated earlier, the paracone is lingually placed in M1 and buccally placed in P4. *Namasector* shows a pattern more typical of creodonts, with the paracone in both teeth relatively buccally placed, though it is slightly more lingual in P4, which is the opposite of the condition in *Prionogale*. This issue has yet to be addressed in detail.

For this reason, we have here preferred to retain the Prionogalidae as Mammalia *incertae sedis* rather than place them with the Creodonta as suggested by Morales et al. (2008). We have placed *Namasector* here as well, rather than with the Creodonta, to highlight the ongoing discussion regarding the affinities of this genus to *Prionogale*.

Literature Cited

Morales, J., M. Pickford, and M. Salesa. 2008. Creodonta and Carnivora from the early Miocene of the northern Sperrgebiet, Namibia. *Memoir of the Geological Survey of Namibia* 20:291–310.

Schmidt-Kittler, N., and P. J. Heizmann. 1991. *Prionogale breviceps* n.gen. n.sp.: Evidence of an unknown major clade of eutherians in the Lower Miocene of East Africa. *Münchner Geowissenschaftliche Abhandlungen* 19:5–16.

TWENTY-EIGHT

Primitive Ungulates ("Condylarthra" and Stem Paenungulata)

EMMANUEL GHEERBRANT

Africa is probably the most poorly known paleobiogeographic province in the evolutionary history of the primitive ungulates collectively called "Condylarthra." This is related to the more general problem of the very poor fossil record of mammals in the early Paleogene of Africa. Consequently, knowledge of the origin and early evolution of modern African ungulates relies mostly on molecular phylogenetic analyses (e.g., Madsen et al., 2001; Murphy et al. 2001).

The few described African condylarths have been found in the Paleocene and early Eocene of Morocco, and in the middle Eocene of Senegal (table 28.3). These sites have yielded limited mammalian material represented mostly by isolated teeth. The best-preserved condylarth material comes from the Ouled Abdoun Basin, Morocco (Gheerbrant et al., 2001). Consequently, the systematic and phylogenetic position of African condylarth taxa remains unresolved. Besides the few described taxa, several condylarth occurrences have been inferred in Africa from phylogenetic hypotheses of the origin of African ungulates. "Hyopsodontids" and phenacodonts have been especially advanced as potential stem groups of the endemic modern ungulates of Africa. The most robust current hypotheses involve a probable Laurasian "hyopsodontid" origin for the macroscelideans (Hartenberger, 1986; Tabuce et al., 2001, 2007; Zack et al., 2005a, 2005b). The ungulate affinity of macroscelideans is supported by molecular studies that include them in the superclade Afrotheria (e.g., Madsen et al., 2001; Murphy et al., 2001). The study by Tabuce et al. (2001) of the macroscelidid *Nementchatherium* from the Eocene of Algeria supports close relationships between macroscelideans and African ungulates such as hyracoids and proboscideans. The recent analysis of Tabuce et al. (2007) also supports a relationship between paenungulates and macroscelideans and European louisinines. "Hyopsodontidae" are, however, still nearly unknown in Africa, perhaps with the exception of two damaged teeth reported by Tabuce et al. (2005) from the Lutetian Aznag locality (Ouarzazate Basin, Morocco).

We here follow Archibald's (1998) systematics of condylarths, with the Zack et al. (2005b) emendations for the "Hyopsodontidae," "Mioclaenidae," and Apheliscidae.

"Mioclaenidae" is considered a junior synonym of Hyopsodontidae, and includes Mioclaeninae, Kollpaniinae, Pleuraspidotheriinae, and Hyopsodontinae, which is restricted to *Hyopsodus*. Classical "Hyopsodontidae" are referred to Apheliscidae, which includes Apheliscinae, Louisininae and several genera *incertae sedis* (see Zack et al., 2005b: table 4).

ABBREVIATIONS

CPSGM, OCP DEK/GE: Collections of the Office Chérifien des Phosphates, Khouribga, Morocco; TZT, THR, and NTG2: material respectively from Talazit, Adrar Mgorn 1 and N'Tagourt 2, Ouarzazate Basin, Morocco, collection of the University Montpellier II; BD, M'Bodione Dadere, Senegal, collection of the University Montpellier II; NHN, collection of the Muséum National d'Histoire Naturelle, Paris. In measurements: L = length, W = width, H = height.

Systematic Paleontology

Superorder PAENUNGULATA Simpson, 1945
Family *Incertae sedis* (nov. 1)
Genus *ABDOUNODUS* Gheerbrant and Sudre, 2001 (in Gheerbrant et al., 2001)
ABDOUNODUS HAMDII Gheerbrant and Sudre, 2001 (in Gheerbrant et al., 2001)
Figures 28.1 and 28.2

Age and Occurrence Probably Thanetian of the Ouled Abdoun Basin, Morocco (see table 28.3, later). This species was described from the earliest Ypresian, as for *Phosphatherium escuilliei* (Gheerbrant et al., 2001), but the exact level and locality of the holotype, previously the only known specimen, remain unknown. We here report on new specimens from more southern localities than those of *Phosphatherium* (quarries of Meraa El Arech and Sidi Chennane) and from older levels called "phosphates bed II," of Thanetian age (Gheerbrant et al., 2003). Further details on the age of the mammal levels from Sidi Chennane and Meraa El Arech will be published separately.

Material Holotype: MNHN PM21, a left dentary with p3–4, m1–2, diastema and alveoli for p?, c or i?.

Referred new material:

- OCP DEK/GE 308, fragment of left dentary with m3 (figures 28.1D–28.1F), from Sidi Chennane, section B7, loc. 32°39′68 N, 6°44′02 W, local phosphates bed II, Thanetian;

- MNHN PM35, fragment of left dentary with m2? (figures 28.1A–28.1C), loc. Sidi Chennane?

- OCP DEK/GE 310, fragment of right dentary with m1–3 (figures 28.2A–28.2C), from Meraa El Arech, section S4, loc. N 32° 44, 044′ W 06° 47,603′, local phosphates bed II, Thanetian.

- Two other specimens with m2–3 from private collections (PM67, and PM68) are also referred here to *A. hamdii.*

Diagnosis Modified from Gheerbrant et al. (2001). Dental morphology close to "mioclaenids" in the strong bunodonty and in the simplified and inflated premolars. Differs from "mioclaenids," and especially from kollpaniines, by the following combination of derived features: (1) dental row: presence of a diastema in front of p3; (2) molars: talonid short and narrow, narrower than the trigonid; postfossid partially filled by the convex and confluent internal flanks of the hypoconid and entoconid, which bear a small but distinct hypolophid; entoconid voluminous; cingula reduced; weak cristid obliqua joining the trigonid in its labial midwidth on m1–2; hypoconulid small and labial. Kollpaniines differ, moreover, in the more distal metaconid, which invades the postfossid, in the less rectilinear protocristid, and in having the hypoconulid close to the entoconid. *Abdounodus hamdii* differs from all known condylarths, including "mioclaenids" and apheliscids (= "Hyopsodontidae"), in the incipient hypolophid, the typically low postcristid (cingulum-like), especially on m1–2, and bearing a small labial hypoconulid and an incipient postentoconulid.

Description Abdounodus hamdii, described by Gheerbrant et al. (2001) based on a single partial lower jaw retaining damaged p3–4, m1–2 (holotype), is one of the most poorly known mammals from the Ouled Abdoun basin. We refer to this species recently discovered material from Ouled Abdoun that is nearly identical to the holotype. This new material significantly enhances our knowledge of the lower molar morphology of *Abdounodus* and indicates that *A. hamdii* is from older beds than *Phosphatherium* (i.e. from phosphates bed II of Thanetian age; see Gheerbrant et al., 2003).

Abdounodus hamdii is a small species, the size of a large species of *Hyopsodus.* The teeth are very bunodont, although the crests are still developed, especially on the molar protoconid. There is a distinct diastema in front of p3 on the single known specimen (holotype) preserving this area.

The p3–4 are not enlarged. They are extended mesiodistally but also inflated labiolingually. The morphology is typically simplified: paraconid and metaconid reduced or absent, talonid reduced, postfossid absent. Cingula are absent.

The molars are very bunodont: the cusps are bulbous and low; the crests are weak; the basins are shallow; the crown is low with a talonid that is slightly narrower than the trigonid; the crown is inflated laterally, including on the lingual side. The m2 is slightly wider than m1 and m3. The mesiodistal compression of the trigonid increases posteriorly. The cingula are vestigial. The paraconid is bulbous. The paracristid is typically mesially extended and joins the paraconid on its mesiolabial flank (OCP DEK/GE 310, PM67). The paraconid is more labial than the metaconid, especially on anterior molars. On m2 and m3, there is a thin premetacristid, but no postmetacristid. The protoconid is selenodont-like: the paracristid and protocristid are widely concave and crescentic, the paracristid is mesially convex, and the protocristid is long, sharp and underlined by a semilunar wear facet, indicating an incipient functional protolophid. The metaconid is slightly distal to the protoconid. The talonid of m1–2 is short. It is as wide or wider than the trigonid on m2, and narrower than the trigonid on m1 and m3. The cristid obliqua ends labially against the trigonid, at the transverse level of the protoconid

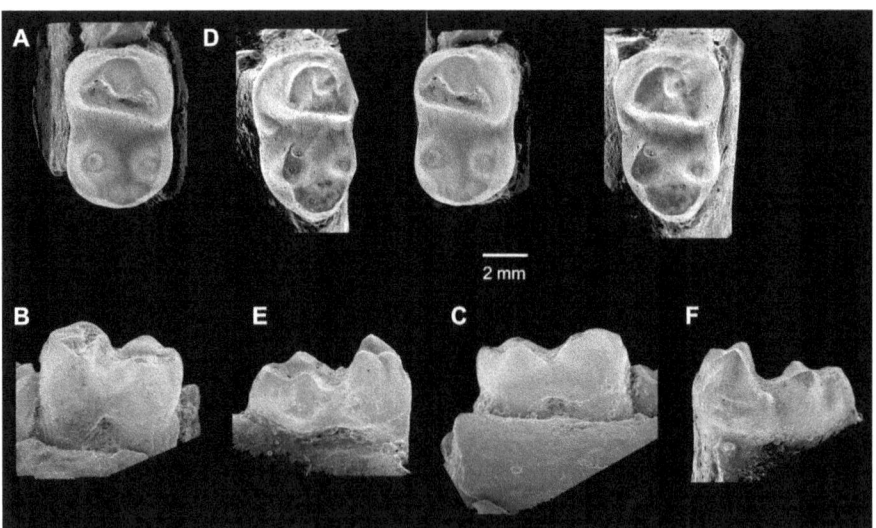

FIGURE 28.1 *Abdounodus hamdii* from the Thanetian of the Ouled Abdoun Basin, Morocco. A–C) MNHN PM35, fragment of left dentary with m2? (loc. Sidi Chennane?) in occlusal (stereopair), labial, and lingual views. D–F) OCP DEK/GE 308, fragment of left dentary with m3 (loc. Sidi Chennane), in occlusal (stereopair), lingual, and labial views.

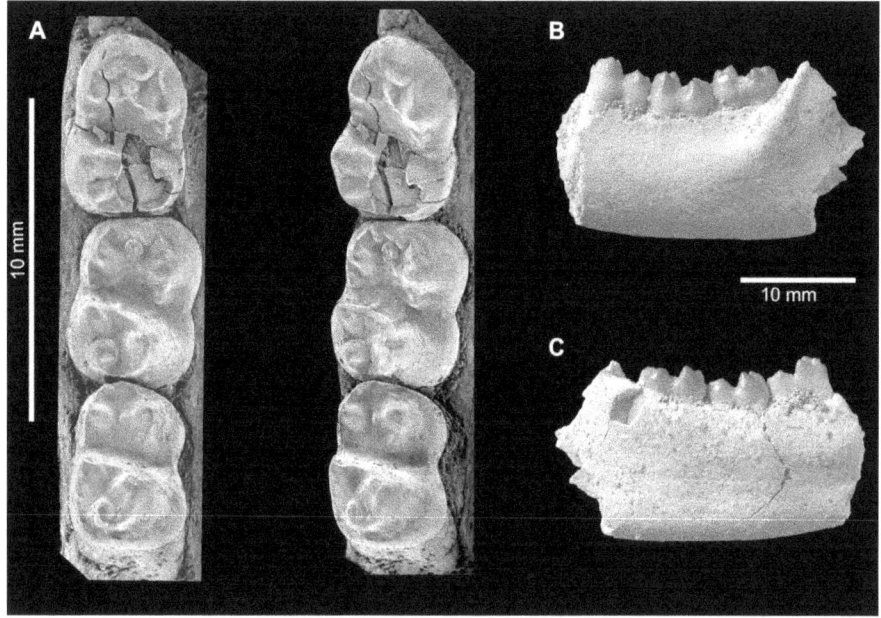

FIGURE 28.2 *Abdounodus hamdii* from the Thanetian of the Ouled Abdoun Basin, Morocco. A–C) OCP DEK/GE 310, fragment of left lower jaw with m1–3 (loc. Meraa El Arech) in occlusal (s.e.m. stereopair), labial, and lingual views.

apex (m2–3) or more labially (m1). The cristid obliqua and the entocristid are weak anteriorly, so that the postfossid is open laterally near the trigonid. The hypoconid and entoconid are well developed and their internal flanks are inflated and confluent, showing a slight but distinct hypolophid crest on the unworn m2 and m3 (figure 28.2). In the postfossid, there is an occasional small accessory cusp that is close to the hypoconulid and linked to the hypolophid (OCP DEK/GE 310). The postcristid is cingulum-like—that is, located low and distal and weakly linked to the entoconid and hypoconid, on worn specimens such as OCP DEK/GE 308. It bears a hypoconulid and a postentoconulid. The hypoconulid is labial, close to the hypoconid, but separated from it by a

notch. It is large on m3, and small on m1–2. The mesoconid and an entoconulid are more or less distinct. m3 is two rooted and not enlarged, and has a weak posterior lobe bearing a well-developed hypoconulid and a small postentoconulid (OCP DEK/GE 310, OCP DEK/GE 308). However, the distal root of m3 is oblique and shows a developed distal lobe that is compressed laterally and salient distally (PM67).

The corpus of the dentary is moderately high. The ramus rises labially to m3 (talonid) on OCP DEK/GE 308 and 310. There is a very small coronoid foramen located behind m3, on the concave anterior side of the ramus. Dimensions of the described material of *Abdounodus hamdii* are given in table 28.1.

TABLE 28.1
Measurement data for *Abdounodus hamdii* in mm, * = estimate.

DENTITION

Specimen	Lp3	Lp4	Lm1	Wm1	Lm2	Wm2	Lm3	Wm3	Lm1–2	Lm1–3
MNHN PM21	4.4*	4.9*	4.9*	3.85*	5.2*	—	—	—	10*	—
OCP DEK/GE 308	—	—	—	—	—	—	6.3	4.2	—	—
MNHN PM35	—	—	—	—	5.7	4.7	—	—	—	—
OCP DEK/GE 310	—	—	5.5	4	5.7	4.5	6.1	4.3	11	17
PM67	—	—	—	—	5.7	4.4	5.5	3.8	11.1	—
PM68	—	—	—	—	5.4	3.7	5.3	3.3	10.8	—

DENTARY (CORPUS)

	MNHN PM21	MNHN PM33	OCP DEK/GE 310	OCP DEK/GE 308
Height below m2	—	—	11	—
Transverse width below m1	6.05*	—	6.5	—

Remarks In *Abdounodus hamdii*, the strong bunodonty, simplified, inflated, and small p3–4, bulbous paraconid, short and narrow talonid, large talonid notch and mesially convex paracristid are all features of "Mioclaenidae," a family of small condylarths that has been renamed Hyopsodontidae by Zack et al. (2005b) to include Mioclaeninae, Kollpaniinae, and *Hyopsodus*. *Abdounodus* was indeed tentatively referred to "Mioclaenidae" (Gheerbrant et al., 2001). However, none of these "mioclaenid" taxa show clear affinity with *Abdounodus*.

Abdounodus is, moreover, distinguished by several significant characters, some of which are elucidated by the new material reported here, such as OCP DEK/GE 310:

1. the inflated and confluent internal flank of the hypoconid and entoconid, which bears an incipient hypolophid; in "mioclaenids," the entoconid is laterally compressed;

2. the typically low, more or less cingulum-like postcristid, especially on m1–2;

3. the small, labially positioned hypoconulid and the presence of an incipient postentoconulid;

4. the occurrence of a diastema in front of p3.

Feature 4, if representative of the species, is autapomorphic, at least at the generic level. The presence of an incipient hypolophid (feature 1) is reminiscent of *Ocepeia* (discussed later), although the construction differs in detail (in *Ocepeia* the internal crest is absent on the hypoconid of m1–2). Several differences (see *Ocepeia*) indicate that *Abdounodus* and *Ocepeia* belong to distinct genera and families. However, feature (1) and the general resemblance of *Abdounodus* and *Ocepeia* to each other—for instance, in the strong bunodonty, simplified premolars, bulbous paraconid, crescentic, and concave paracristid and protocristid, hypolophid, and reduced cingula—might suggest a relationship between the two genera at a high systematic level relative to other early ungulates; that is, they belong to the same major African ungulate group. Features 1, 2, and 3 all are reminiscent of an archaic lophodont ungulate. This is in accordance with the development of the protocristid with its concave shape and crescentic wear pattern, suggesting a functional protolophid. The labial hypoconulid is known in proboscideans and the presence of a postentoconulid is known in tethytheres (Gheerbrant et al., 2005). The incipient hypolophid (feature 1), low postcristid (feature 2), and bunodonty are especially reminiscent of primitive hyracoids such as *Seggeurius*. As a whole, these features are more likely indicative of a relationship between *Abdounodus* and paenungulates, rather than with north-Tethyan condylarths such as "mioclaenids" (Gheerbrant et al., 2001). However, the structure is much more primitive than in any known paenungulate (e.g., strong bunodonty vs. very incipient lophodonty, paraconid bulbous, trigonid not compressed in m1–2, m3 only slightly longer than m2, and with a short hypoconulid lobe). *Abdounodus* might be representative of a primitive lineage of paenungulates, although it seems autapomorphic in features such as the occurrence of diastema in front of p3 (holotype). Determination of its exact relationships with and within paenungulates awaits description of more complete material. In any case, *Abdounodus* probably belongs to a new and primitive family of African ungulates.

Superorder ?PAENUNGULATA Simpson, 1945
Family *Incertae sedis* (nov. 2)
Genus *OCEPEIA* Gheerbrant and Sudre, 2001 (in Gheerbrant et al., 2001)
OCEPEIA DAOUIENSIS Gheerbrant and Sudre, 2001 (in Gheerbrant et al., 2001)
Figure 28.3

Age and Occurrence Thanetian? of the Ouled Abdoun Basin, Morocco (table 28.3). This species was described from the earliest Ypresian as for *P. escuilliei* (Gheerbrant et al., 2001), but the exact level and locality of the described material are unknown (two specimens recovered from a commercial source). Some new specimens referred to the species *O. daouiensis* are probably from the Thanetian of Sidi Chennane (Phosphates bed II).

Material Holotype: CPSGM-MA1, fragment of right dentary with p4, m1; MNHN PM20, fragment of left dentary with m2–3. New referred material: MNHN PM41, left dentary with p3–4, m1–3 (figures 28.3A–28.3C); MNHN PM49, fragment of right dentary bearing p3–4, m1, and alveoli for the canines and two or three incisors (figure 28.3D); OCP DEK/GE 309, fragment of left dentary with c1, p3–4, m1–2 (figure 28.3E).

Diagnosis Modified from Gheerbrant et al. (2001). Morphology close to that of loxolophine arctocyonids, especially in the bulbous and median paraconid on m3, but more derived in several features, some of which are known in phenacodontids such as *Ectocion* (a), while others are probably autapomorphic (b):

(a) selenodont trend with well developed labial crests (especially the postcristid); postmetacristid and metastylid developed; mesoconid large; entoconulid present; entoconid strong; hypoconulid located very lingually.

(b) anterior dentition remarkably shortened, with p1–2 lost and diastema absent or reduced; p3–4 morphology simplified and trenchant, with development of a long mesiodistal crest; p4 crown inflated labially; cingula absent or only vestigial; labial flank of molars inflated; hypoconulid reduced in m1–2; horizontal ramus of dentary transversely inflated.

Ocepeia is more primitive than phenacodontids in several respects, including the bulbous paraconid of the molars and the simple p4. The occurrence of a lower molar entolophid is also distinctive with respect to arctocyonids and phenacodontids; it recalls lophodont ungulates such as paenungulates.

Description The new material establishes the lower dental formula of *O. daouiensis* as i1–2–3?, c1, p2–3, m1–2–3.

The anterior dentition is very short, with loss of p1 and p2, diastema absent or very short, small and compressed lower incisors (length of i1–3 less than the length of c1), and short symphysis. The root of i1–2 is mesiodistally compressed and labiolingually wide; i2 was larger than i1. A small, circular alveolus suggests the probable presence of i3; it was considerably smaller than i1–2 and possibly peglike. The canine is a large and stout tooth. It is somewhat primate-like, with a lingual cingulum and an asymmetrical labiolingual profile (flat lingual flank). It is elliptic in cross section, with an oblique (mesiolabial to distolingual) long axis. The lower canine has mesiodistal crests that are linked to the lingual cingulum.

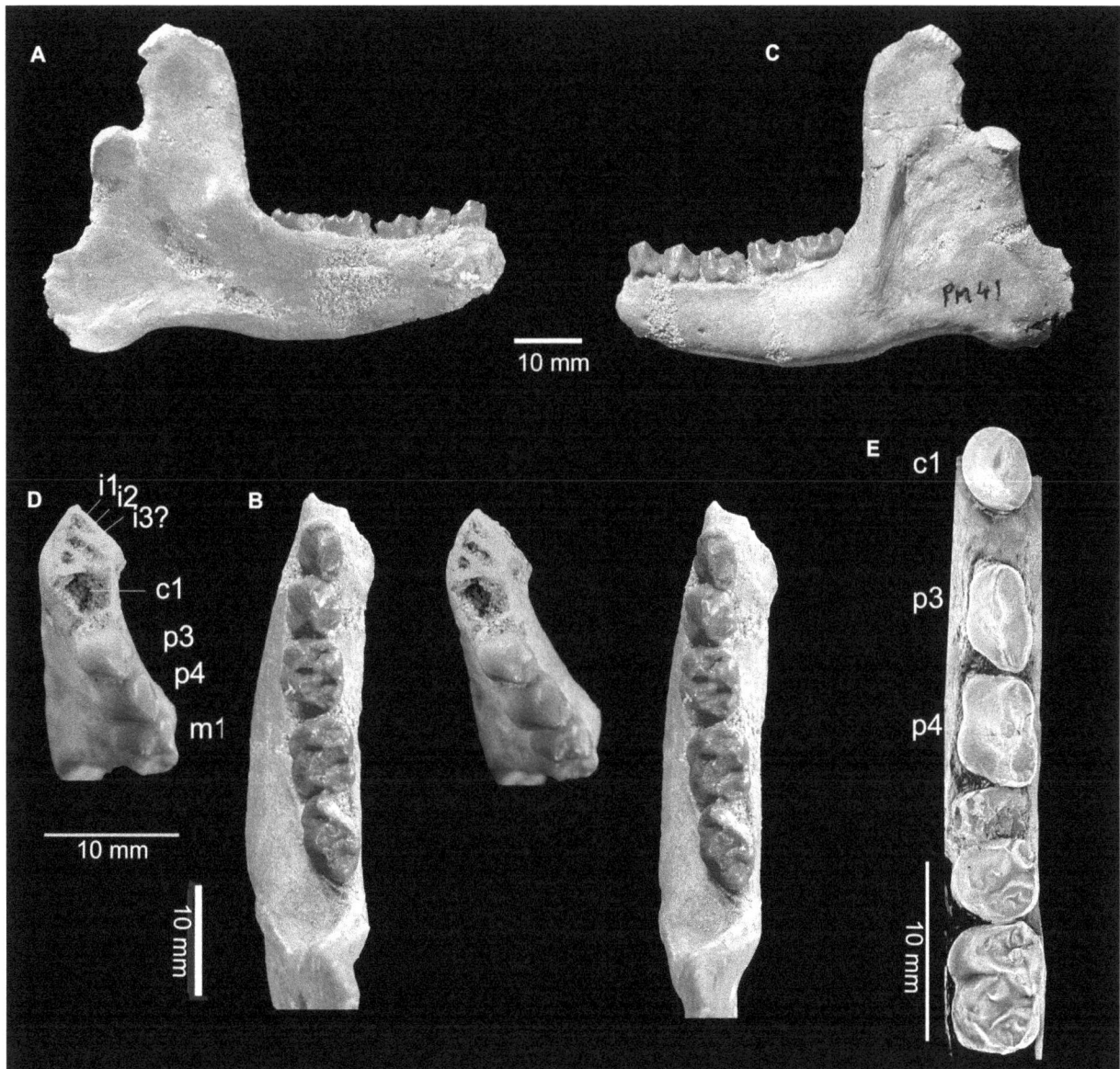

FIGURE 28.3 *Ocepeia daouiensis* from the Thanetian? of the Ouled Abdoun Basin, Morocco. A–C) MNHN PM 41, lower jaw preserving p3–4, m1–3, in lingual, occlusal (stereopair), and labial views. D) MNHN PM49, fragment of right dentary bearing p3–4, m1, and alveoli for the canine and two or three incisors, in occlusal stereophotographic view; E) OCP DEK/GE 309, fragment of left dentary with c1, p3–4, m1–2, in occlusal s.e.m. view.

The p3 and p4 are similar, with a typically simplified morphology, longer than wide, and trenchant. Although elongated, the crown remains broad and bunodont. It is exodaenodont below the talonid and p4 is inflated mesiolabially. The paraconid is very reduced. The talonid is reduced. There is a variable and more or less continuous lingual cingulum.

The molars are bunodont and low, but also selenodont (long and sharp labial crests). Their length increases from m1 to m3. The labial flank is noticeably inflated, and it lacks a cingulum. The talonid is large. It is much longer, as wide or wider, and lower than the trigonid. The trigonid is moderately compressed. The paracristid and protocristid are broadly concave and bear a semilunar wear facet, suggesting a selenodont trend. The paraconid is well developed and bulbous on all molars. It is only slightly lingual to the metaconid in m1 and m2, and much more lingual in m3. The metaconid is

slightly more distal than the protoconid. A premetacristid and postmetacristid are present; the postmetacristid base is inflated as a more or less distinct metastylid. The postfossid is relatively large and complicated. The mesoconid is large and inflated, and the entoconulid is small. The hypoconid is slightly more mesial than the entoconid. It bears long and concave crests with well-developed wear facets 3 and 4. The cristid obliqua contacts the trigonid at about its midwidth. The postcristid is crenulated and bears several small accessory cusps, at least one of which is labial and close to the hypoconid, and one lingual and close to the entoconid. The hypoconulid is weak on m1–2 (smaller than the mesoconid), but it forms a large and distally salient lobe on m3. Comparison and homology with the m3 hypoconulid lobe suggests, that on m1–2, the hypoconulid corresponds to the lingual-most accessory cusp of the postcristid (close to the entoconid). The well-developed entoconid shows a typical lingual crest called an entolophid by Gheerbrant et al. (2001); on m1–2 it is linked to the labial-most accessory cusp on the postcristid, and on m3 (MNHN PM20, MNHN PM41) to the hypoconid. The entocristid is not reduced. Abrasive wear is weak, whereas the shearing attrition wear facets are well developed.

The dentary is robust, with a labially remarkably inflated corpus. The corpus is high nearly the entire length of the tooth row, even anteriorly at the symphysis, providing a robust construction to the whole anterior lower dentition. The anterolingual bony crest of the ramus is less concave posteriorly than in *Phosphatherium*, where the ramus is more inclined anteriorly. The coronoid process is vertical, narrow and very high (higher than in *Phosphatherium*). The articular condyle is moderately high above the tooth row. It is expanded as a transverse cylinder, but with distinct lateral and medial articular surfaces. The articular surface is extended ventrally on the medial side. The mandibular angular process is well developed and strongly protruding ventrodistally. It is narrower than in *Phosphatherium*. There are at least two large mental foramina, below p3 and the posterior part of p4. The short symphysis ends below p3 and is unfused. Dimensions

of the described material of *Ocepeia daouiensis* are given in table 28.2.

Remarks The most striking feature of *Ocepeia daouiensis* is its very short (e.g., loss of p1–2 and reduced diastema) and robust (e.g., deep and inflated dentary, stout c1) anterior dentition. The corresponding short-snouted morphology is a remarkable and unusual ungulate convergence with primates, especially with anthropoids. It suggests a peculiar oral ingestion pattern in *Ocepeia*, probably characterized by a greater strength of the anterior dentition for the gripping and/or cutting of food (e.g., Butler, 1983). We might speculate that this is linked to peculiar ecological adaptations, e.g., with a possible primate-like arboreal life. The robust and short-snouted morphology, the small incisors, the selenodont (incipiently lophodont) molar pattern, and the wear pattern, illustrating a predominant shearing function (phase I of mastication), suggest a peculiarly specialized folivorous ungulate. *Ocepeia* has a distinct adaptive dental function and diet compared to the noticeable crushing trend (premastication phase) of *Abdounodus*, which is more bunodont and shows significant abrasive tooth wear; *Abdounodus* was possibly more frugivorous.

Ocepeia shows several other familial differences from *Abdounodus*. It differs especially in its larger size, less bunodont

TABLE 28.2
Measurement data for *Ocepeia daouiensis*

DENTITION

Specimen	c1 L	c1 W	c1 H	p3 L	p3 W	p4 L	p4 W	m1 L	m1 W	m2 L	m2 W	m3 L	m3 W
CPSGM-MA1	—	—	—	—	—	6.4	5.6	7.3	5.4	—	—	—	—
MNHN PM20	—	—	—	—	—	—	—	—	—	7	5.3	8.3	5
MNHN PM41	—	—	—	5.5	4	6.1	4.5	7.1	4.8	—	—	—	—
MNHN PM49	—	—	—	5.6	4.3	5.6	5.2	6.8	6	7.2	6.2	8.3	—
OCP DEK/GE 309	4.3	4.4	7.7	5.9	4.5	6.1	5.5	—	—	7.4	6.4	—	—

TOOTH ROW

Specimen	Lm1–3	Lm1–2	Lp2–m3	Lp3–4
MNHN PM41	22	14.3	32.2	10.8
MNHN PM49	—	—	—	11.4
OCP DEK/GE 309	—	14.4	—	12.2

DENTARY (HORIZONTAL RAMUS)

	CPSGM-MA1	MNHN PM20	MNHN PM41	MNHN PM49	OCP DEK/GE 309
Height below m2	—	17	16.4	—	17.6
Transverse width of the below m1	—	9.6	9.1	—	9.1

- Height of the coronoid-angular apophyses: 48.2 mm
- Height of the articular condyle–alveolar border: 11 mm.
- Height of the coronoid apophysis–articular condyle: 15.4 mm
- Transverse width of the articular condyle: 13 mm
- Width of the angular apophysis: 13 mm
- Maximal length of the dentary: 63.4 mm; estimated total length of the dentary: 72–73 mm

TABLE 28.3
Major occurrences and ages of African condylarths
? = attribution uncertain.

Taxon	Occurrence (Site, Locality)	Stratigraphic Unit	Age
"CONDYLARTHRA" / STEM PAENUNGULATA, PALEOCENE–EOCENE			
Paenungulata *Abdounodus hamdii* (5)	Sidi Chennane, Meraa El Arech, Ouled Abdoun Basin, Morocco	Phosphates bed II	Thanetian, about 60 Ma
?Paenungulata *Ocepeia daouiensis* (5)	Sidi Chennane and Grand Daoui?, Ouled Abdoun Basin, Morocco	Phosphates bed II?	Thanetian?, 60? Ma
?Condylarthra"indet. 1 (1)	Talazit, Ouarzazate Basin, Morocco	Fm Jebel Guersif	Late Thanetian, 55–58 Ma
"Condylarthra" indet. 2 (THR 100) (2)	Adrar Mgorn 1, Ouarzazate Basin, Morocco	Fm Jebel Guersif	Late Thanetian, 55–58 Ma
"Condylarthra" indet. 3 (THR 303) (2)	Adrar Mgorn 1, Ouarzazate Basin, Morocco	Fm Jebel Guersif	Late Thanetian, 55–58 Ma
"Condylarthra" indet. 4 (NTG-52) (3)	N'Tagourt 2, Ouarzazate Basin, Morocco	Fm Aït Ouarithane or Jbel Ta'louit	Ypresian, 55 Ma
Condylarthra indet. 5 (4)	M'Bodione Dadere, Senegal	?	Lutetian, middle Eocene

NOTES: (1) Sudre et al. 1993; (2) Gheerbrant 1995; (3) Gheerbrant et al. (1998); (4) Sudre (1979); (5) Gheerbrant et al. (2001) and this chapter.

molars (especially less inflated lingually) that are more shearing than crushing, larger m3 with a developed hypoconulid lobe, larger talonid, entolophid not extended on the hypoconid of m1–2, hypoconulid lingual, paraconid more lingual on m1 and more labial (median) on m3, and strong postmetacristid. However, as discussed for *Abdounodus*, there is a general resemblance, including the selenodont, incipiently lophodont molar pattern that suggests their probable affinity at a high systematic level (i.e., within Paenungulata).

Among known "condylarths," the basic molar and premolar pattern of *Ocepeia* is especially reminiscent of the loxolophine arctocyonids, known from the Paleocene of North America. The resemblance includes the paraconid, which is bulbous, and located at the trigonid mid-width in m3 (in contrast to "Mioclaenidae"), the lingual hypoconulid, the large talonid, and the inflated premetacristid. My colleagues and I (Gheerbrant et al., 2001) mentioned *Lambertocyon* (late Paleocene) as the closest loxolophine. Loxolophines differ in the smaller m3 (but not so in *Lambertocyon*), the reduced entocristid, the slightly shorter and more piercing premolars, and several primitive features (Gheerbrant et al., 2001: table 4), including in the anterior dentition. *Ocepeia* was also favorably compared to phenacodontids based on their very similar molar patterns, which is also supported by the large and stout lower canine. However, new data on the anterior dentition serve to distinguish *Ocepeia* from any known phenacodontids. The lower canine differs in its primate-like morphology. The simplified premolars of *Ocepeia*, previously considered to be a secondary feature with respect to phenacodontids (Gheerbrant et al., 2001), are now viewed as more probably primitive, in accordance with several other plesiomorphic features of the Moroccan genus (unreduced paraconid, hypoconulid lobe of m3 developed, cristid obliqua more lingual on the trigonid, no protostylid, trigonid of m1 less compressed). The molarized p4 of phenacodontids is probably a significant specialized difference from *Ocepeia*. In this regard, *Ocepeia* remains phenetically closer to arctocyonid loxolo-

phines such as *Lambertocyon* than to phenacodontids. As stated previously, the arctocyonid loxolophines seem to represent the best known structural ancestral morphotype for *Ocepeia*. The new morphological data on *Ocepeia* would suggest that the median paraconid in m3 is a significant shared derived feature with Loxolophinae.

The shortened and robust anterior dentition is a remarkable specialized morphology of *Ocepeia*, indicating an African ungulate lineage that diverged early from the ancestral generalized condylarth pattern retained in known arctocyonids. The occurrence of an entolophid suggests, as for *Abdounodus*, a possible paenungulate affinity. The primitive molar pattern associated with the autapomorphic anterior dentition indicates a basal and divergent paenungulate lineage. A recent cladistic analysis indicated a primitive basal position for *Ocepeia* (Gheerbrant et al., 2005) with respect to "Taxeopoda" and Paenungulata. However, *Ocepeia* was very poorly known at that time, and cladistic analysis incorporating the new material will certainly help to further resolve its position. More strikingly than for *Abdounodus*, it appears that *Ocepeia* cannot be referred to any known condylarth family.

Undetermined African Condylarthra

OUARZAZATE BASIN, MOROCCO

TZT 1, a right isolated p3 or p4 (figures 28.4E and 28.4F) from the Thanetian of Talazit, was described by Sudre et al. (1993) as an undetermined arctocyonid. This is one of the very few ungulate teeth and one of the largest mammal teeth (L = 5.0 mm; W = 3.2 mm) discovered in the Ouarzazate basin, which has yielded mostly small insectivorous or carnivorous species. The premolariform crown is dominated by a large inflated protoconid, which is flanked distally by a small posterior bulbous cusp. The protoconid apex is extensively worn by abrasion through most of its height. The paraconid was absent or very small. Sudre et al. (1993) stressed resemblances with

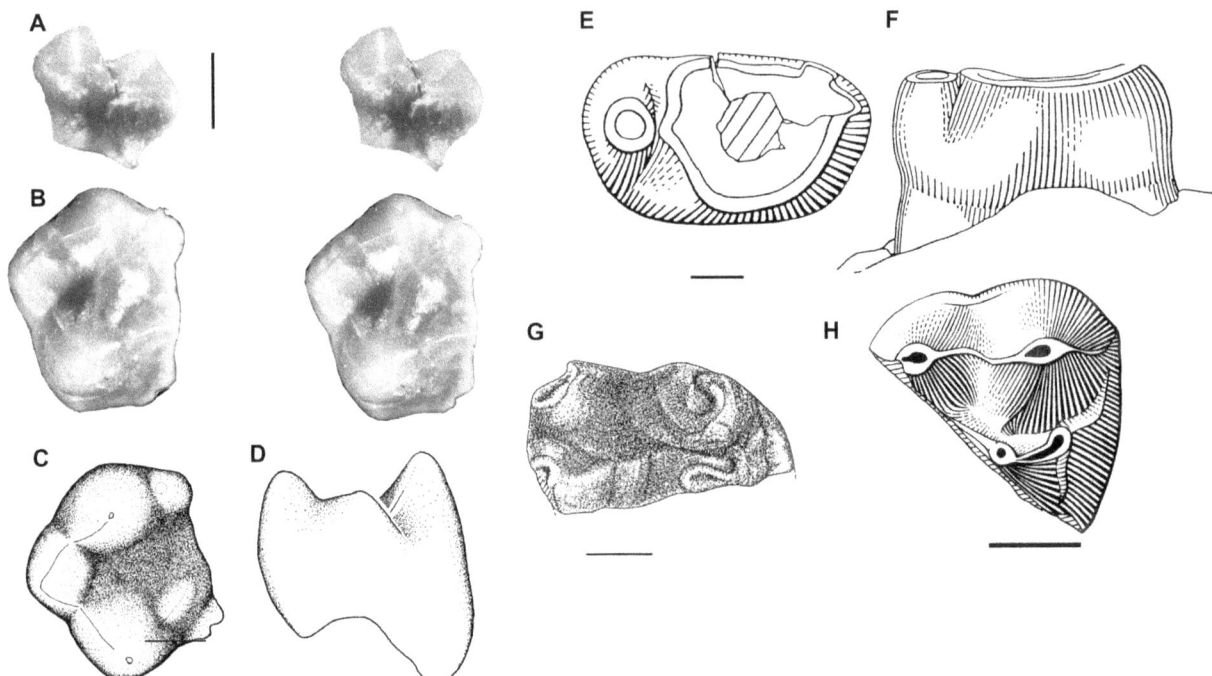

FIGURE 28.4   Undetermined African Condylarthra. A) THR 100, a fragment of a left lower molar from the Thanetian of the Ouarzazate Basin (Morocco), in labial stereo view (Gheerbrant 1995: plate 5, figure 8); B–D) THR 303, an isolated talonid of a right m1 or m2 from the Thanetian of the Ouarzazate Basin (Morocco) in occlusal and distal views (Gheerbrant 1995: figures 36 and plate 6, figure 3); E–F) TZT 1, a right isolated p3 or p4 from the Thanetian of the Ouarzazate Basin (Morocco), in occlusal and labial views (Sudre et al. 1993: figure 1); G) NTG2–52, a damaged left lower molar from the Ypresian of the Ouarzazate Basin (Morocco) in occlusal view (Gheerbrant et al. (1998: figure 7); H) BD323, right dP4/? from the middle Eocene of M'Bodione Dadere (Senegal) in occlusal view (Sudre 1979: figure A). Scale bars = 1 mm.

North American triisodontine arctocyonids such as *Goniacodon*, especially in the morphology of the distal cusp. A resemblance is also noted with the p3 of Hyopsodontidae *sensu* Zack et al. (2005b)(= "Mioclaenidae"), which additionally share the strong apical abrasion of the protoconid. This wear pattern and the poorly developed crests indicate a predominant vertical orthal chewing movement typical of primitive early Paleogene ungulates. The Apheliscidae also shares the predominant development of the protoconid and the reduced talonid. The systematic position of this form is here considered undetermined.

THR 100, a fragment of a lower molar (figure 28.4A) from the Thanetian of the Adrar Mgorn 1, was described by Cappetta et al. (1987: figure 3) as an undetermined condylarth. This is a small (L max = 2.2 mm) and bunodont, molariform, two-rooted tooth. The trigonid is slightly inclined and weakly compressed mesiodistally. The protoconid is robust and low. The talonid is wider than the trigonid. The hypoconid is large and robust. The cristid obliqua joins the trigonid in its labial part and does not rise on it. The enamel is thick. The affinity of this tooth remains enigmatic. More complete material would justify a comparison with small European louisinine apheliscids.

THR 303, an isolated talonid (W = 3.4 mm; L = 3.4 mm) of a right m1 or m2 (figures 28.4B–28.4D) from the Thanetian of the Adrar Mgorn 1, was described by Gheerbrant (1995) as an indeterminate placental of possible condylarth affinity. The crown is high (H = 3.9 mm) labially where the enamel extends far below the hypoconid, as an original structure. The cusps are robust and bulbous, but also slightly crested. The postfossid is relatively large. The entoconid and hypoconid are large and subequal, and their internal flanks are inflated and confluent. The hypoconulid is well developed and salient distally. The entoconid and hypoconulid are approximated with respect to the hypoconid, from which they are separated by a wide notch extending distolabially as a vestigial postcingulid. The mesoconid and entoconulid are distinct. The general construction, including the accessory cusps, is similar to phenacodontids, but THR 303 is less bunodont and has a larger hypoconulid. Gheerbrant (1995) also made favorable comparisons with ptolemaiids. However, THR 303 is original in the strong exodaenodonty of the tooth below the (tall) hypoconid, and the confluent entoconid and hypoconid. Although the condylarth affinity of this species is likely, it exact systematic position remains unknown.

NTG2-52 (figure 28.4G), a damaged lower molar from the Ypresian of N'Tagourt 2, was described earlier (Gheerbrant et al., 1998) as an indeterminate condylarth. This is among the largest teeth discovered at this locality (L = 4.1 mm), although it belongs to a smaller species than *Khamsaconus bulbosus* (Sudre et al. 1993). It is bunodont. The precingulid is inflated and robust. The protocristid is strongly reduced and a deep notch separates the paraconid and metaconid. The entoconid and hypoconid are large and bulbous, and they are also deeply separated, the postcristid being strongly reduced. The cristid obliqua joins the trigonid very labially. The precise systematic affinity of this form is unknown.

BD323 (figure 28.4H), an upper molariform tooth from the middle Eocene locality of M'Bodione Dadere was interpreted by Sudre (1979) as a dP4 of a condylarth. The tooth is small (L = 2.9 mm; W = 2.8 mm) and square in occlusal outline. The crown is low, with weak crests. The paracone and metacone are large, low, and blunt, and their labial flank is expanded. There is no mesostyle. A very small mesial paracrista is separated from the paracone by a small notch. The centrocrista is rectilinear, in line with the preparacrista. The protocone is slightly distal to the paracone. There is a trace of a crest at the lingual base of the paracone, but no clear evidence of a true protoloph. The precingulum is linked to the mesiolabial end of the preprotocrista, which is inflated and strongly abraded at this level (paraconule?). From this site, Sudre (1979) also described a distal part of an upper molariform tooth, BD1, similar in size to BD323, but of uncertain taxonomic position. This very bunodont tooth bears a broad postcingulum delineating a posterior fovea, and large hypocone. The condylarth affinity of BD323 is supported by the overall bunodont (not lophodont) morphology. The morphology of this tooth might be suggestive of relationships with louisinine apheliscids and primitive macroscelidids according to current study by R. Tabuce (pers. comm.). Its exact systematic position remains to be determined.

Conclusions

The best-known African condylarths are *Abdounodus* and *Ocepeia* from the late Paleocene (and early Eocene?) of the Ouled Abdoun Basin, Morocco. They show clearly derived features with respect to the generalized primitive ungulates called "condylarths." Some are striking autapomorphies indicating an old, at least Paleocene, African history; others, such as the incipient lophoselenodonty, suggest possible relationships to the Paenungulata.

Early Paleogene African localities (including M'Bodione Dadere) may also provide key data related to the recent hypothesis of a "condylarth" (i.e., "hyopsodontid") origin of the macroscelideans (Hartenberger, 1986; Tabuce et al., 2001; Zack et al., 2005a). The origin and initial radiation of the African endemic ungulates and, in fact, of the whole African placental fauna are among the major challenges in contemporary mammalian paleontology.

ACKNOWLEDGEMENTS

I thank the editors, L. Werdelin and W. J. Sanders, for their kind invitation to contribute to this book, and for the editorial corrections that improved the chapter. The study of the mammal material from the Ouled Abdoun basin benefited from the collaboration of the paleontological Convention with the Ministère de l'Energie et des Mines (Direction de la Géologie) and the Office Chérifien des Phosphates (OCP) of Morocco. Casts of the material from the Ouarzazate Basin and from M'Bodione Dadere were kindly provided by R. Tabuce (University Montpellier II). The photographs were taken by P. Loubry (UMR 5143) and C. Chancogne (UMR 5143).

Literature Cited

Archibald, J. D. 1998. Archaic ungulates ("Condylarthra"); pp. 292–331 in C. M. Janis, K. M. Scott, and L. L. Jacobs (eds.), *Evolution of Tertiary Mammals of North America.* Cambridge University Press, Cambridge.

Butler, P. M. 1983. Evolution and mammalian dental morphology. *Journale de Biologie Buccale* 11:285–302.

Cappetta, H., J.-J. Jaeger, B. Sigé, J. Sudre, and M. Vianey-Liaud. 1987. Compléments et précisions biostratigraphiques sur la faune paléocène à Mammifères et Sélaciens du bassin d'Ouarzazate (Maroc). *Tertiary Research* 8:147–157.

Gheerbrant, E. 1995. Les mammifères Paléocènes du Bassin d'Ouarzazate (Maroc) : III. Adapisoriculidae et autres mammifères (Carnivora, ?Creodonta, Condylarthra, ?Ungulata et Incertae Sedis). *Palaeontographica, Abt. A* 237:39–132.

Gheerbrant, E., J. Sudre, H. Cappetta, C. Mourer-Chauviré, E. Bourdon, M. Iarochène, M. Amaghzaz, and B. Bouya. 2003. Les localités à mammifères des carrières de Grand Daoui, Bassin des Ouled Abdoun, Maroc, Yprésien: Premier état des lieux. *Bulletin de la Société Géologique de France* 174:279–293.

Gheerbrant, E., J. Sudre, M. Iarochène, and A. Moumni. 2001. First ascertained African "Condylarth" mammals (primitive ungulates: cf. Bulbulodentata and cf. Phenacodonta) from the earliest Ypresian of the Ouled Abdoun Basin, Morocco. *Journal of Vertebrate Paleontology* 21:107–118.

Gheerbrant, E., J. Sudre, S. Sen, C. Abrial, B. Marandat, B. Sigé, and M. Vianey-Liaud 1998. Nouvelles données sur les mammifères du Thanétien et de l'Yprésien du bassin d'Ouarzazate (Maroc) et leur contexte stratigraphique. *Palaeovertebrata* 27:155–202.

Gheerbrant, E., J. Sudre, P. Tassy, M. Amaghzaz, B. Bouya, and M. Iarochène. 2005. Nouvelles données sur *Phosphatherium escuilliei* (Mammalia, Proboscidea) de l'Eocène inférieur du Maroc, apports à la phylogénie des Proboscidea et des ongulés lophodontes. *Geodiversitas* 27:239–333.

Hartenberger, J.-L. 1986. Hypothèse paléontologique sur l'origine des Macroscelidea (Mammalia). *Comptes Rendus de l'Académie des Sciences, Paris,* Série II 302:247–249.

Madsen, O., M. Scally, C. J. Douady, D. J. Kao, R. W. DeBry, R. Adkins, H. M. Amrine, M. J. Stanhope, W. W. De Jong, and M. S. Springer. 2001. Parallel adaptive radiations in two major clades of placental mammals. *Nature* 409:610–614.

Murphy, W. J., E. Eizirik, W. E. Johnson, Y.-P. Zhang, O. R. Ryder, and S. J. O'Brien. 2001. Molecular phylogenetics and the origins of placental mammals. *Nature* 409:614–618.

Sudre, J. 1979. Nouveaux Mammifères éocènes du Sahara occidental. *Palaeovertebrata* 9:83–115.

Sudre, J., J.-J. Jaeger, B. Sigé, and M. Vianey-Liaud. 1993. Nouvelles données sur les condylarthres du Thanétien et de l'Yprésien du Bassin d'Ouarzazate (Maroc). *Geobios* 26:609–615.

Tabuce, R., S. Adnet, H. Cappetta, A. Noubhani, and F. Quillevere. 2005. Aznag (bassin d'Ouarzazate, Maroc), nouvelle localité à sélaciens et mammifères de l'Eocène moyen (Lutétien) d'Afrique. *Bulletin de la Société de Géologie de France* 176:381–400.

Tabuce, R., B. Coiffait, P. E. Coiffait, M. Mahboubi, and J.-J. Jaeger. 2001. A new genus of Macroscelidea (Mammalia) from the Eocene of Algeria: A possible origin for elephant shrews. *Journal of Vertebrate Paleontology* 21:535–546.

Tabuce, R., L. Marivaux, M. Adaci, M. Bensalah, J.-L. Hartenberger, M. Mahboubi, F. Mebrouk, P. Tafforeau, and J.-J. Jaeger. 2007. Early Tertiary mammals from North Africa reinforce the molecular Afrotheria clade. *Proceedings of the Royal Society of London B,* 274:1159–1166.

Zack, S., T. A. Penkrot, J. Bloch, and K. D. Rose. 2005a. Affinities of the "hyopsodontids" to elephant shrews and a holarctic origin of Afrotheria. *Nature* 434:497–501.

Zack, S., T. A. Penkrot, D. W. Krause, and M. C. Maas. 2005b. A new apheliscine "condylarth" mammal from the late Paleocene of Montana and Alberta and the phylogeny of the "hyopsodontids." *Acta Palaeontologica Polonica* 50:809–830.

TWENTY-NINE

Neogene Insectivora

PERCY M. BUTLER

The order Insectivora was formerly used (e.g., Simpson, 1945) to comprise a miscellany of eutherians with primitive characters, thought to be relatively little modified descendants from the ancestral eutherian stock. Subsequent investigations of eutherian phylogeny have resulted in the removal of a number of insectivoran families to separate orders. Of the African families, the Macroscelididae (order Macroscelidea) and the Tenrecidae and Chrysochloridae (order Tenrecoidea) are treated in separate chapters in this book, leaving only the Erinaceidae and Soricidae to be the subject of the present chapter. These two families have a wide distribution outside Africa, with a fossil record going back to the Eocene, but they appear in Africa only in the Miocene, as immigrants (see tables 29.1 and 29.2).

Systematic Paleontology

Order ERINACEOMORPHA Gregory, 1910
Family ERINACEIDAE Fischer, 1814

Recent erinaceids are divided into two subfamilies: the spiny hedgehogs (Erinaceinae), widely distributed in Eurasia and Africa, and the moonrats (Echinosoricinae = Hylomyinae of Frost et al., 1991), of Southeast Asia. Echinosoricines similar to the modern forms have been found in the early Miocene of Thailand (Mein and Ginsburg, 1997). The extinct genus *Galerix* and related genera from the Oligocene and Miocene of Europe are generally regarded as members of the same subfamily, called Galericinae, and are divided into the tribes Galericini and Echinosoricini (Butler, 1948). The earliest member of the Galericinae is *Eogalericius*, from the middle Eocene of Mongolia (Lopatin, 2004). Doubt has been cast on the special relationship between the Galericini and Echinosoricini by Gould (1995), who carried out a cladistic analysis of fossil and living Erinaceidae. Gould failed to find any derived characters shared by *Galerix* and the Echinosoricini; the differences between them are due to *Galerix* having more primitive characters. The database of the analysis, however, is very incomplete: of 94 skeletal characters used, only 29 are available for *Galerix*, and these are mostly dental. Until the relationship is clarified, Galericinae should be regarded as a paraphyletic taxon. Erinaceinae and Galericinae appeared in Europe in the early Oligocene, and both entered Africa in the early Miocene.

African Fossil Record Two genera of Galericinae occurred in Africa, both of them represented in Europe and Asia. *Galerix africanus* is known by about 30 specimens from the early Miocene of Kenya, including mandibles, maxillae, and isolated teeth (Butler, 1956, 1969, 1984). The best specimen, from Rusinga, is a maxilla with associated mandibles. It resembles the European *Galerix exilis* but is larger, with a proportionately shorter premolar series. On the upper molars, the lingual root is divided in some specimens, and the metaconule has no posterior crest; in these characters *G. africanus* approaches the Echinosoricini (Butler, 1984). The report of *Galerix* from the middle Miocene of the Otavi Mountains, Namibia (Conroy et al., 1992; Senut et al., 1992), shows that it spread widely over Africa. An isolated incisor from the middle Miocene of Beni Mellal, Morocco (Lavocat, 1961), is of uncertain identification.

A related genus, *Schizogalerix*, is represented by some isolated molars from four middle to late Miocene localities in Algeria and Morocco (Engesser, 1980). *Schizogalerix* is distinguished from *Galerix* by the enlargement and division of the mesostyle of upper molars. It has been found in Turkey, Greece, and Austria, but not in France or Spain, and it must have invaded Africa from the east, rather than by the Iberian route (Engesser, 1980).

The Erinaceinae are distinguished from the Galericinae by their characteristic dentition. The anterior incisors I1 and i2 are enlarged to act as forceps. The i1 is absent, and I3 is larger than I2 and usually two rooted. Of the premolars, P1, p1, and p3 are missing, P3 is small, and p4 has a tall, upright paraconid. The last molars are simplified: M3 has no metacone and m3 has no talonid. The genus *Amphechinus* (= *Palaeoerinaceus*), of which there are several species in the Oligocene and Miocene of Europe and Asia, reached Africa by the early Miocene. It is specialized in the greater enlargement of the anterior incisors: the root of i2 reaches back below p4, and the premaxilla is lengthened to accommodate the root of I1. Also, the second and third molars are reduced in size in comparison with the first. These specializations are absent in modern erinaceines. At the same time, *Amphechinus* has retained primitive characters lost in modern forms; for example, the lachrymal foramen opens in the orbit instead of on

TABLE 29.1
African fossil Erinaceidae

Taxon	Distribution	Reference
	GALERICINAE	
Galerix africanus Butler, 1956	E. Miocene, East Africa (Koru, Legetet, Songhor, Rusinga, etc.)	Butler, 1956, 1969 (as *Lanthanotherium*), 1984
Galerix sp.	M. Miocene, SW Africa (Otavi)	Conroy et al., 1992; Senut et al., 1992
?Galerix sp.	M. Miocene, NW Africa (Beni Mellal)	Lavocat, 1961
Schizogalerix	M.–L. Miocene, NW Africa	Engesser, 1980
	ERINACEINAE	
Amphechinus rusingensis Butler, 1956	E.–M. Miocene, E. Africa (Legetet, Chamtwara, Songhor, Maboko, etc.), SW Africa (Arrisdrift)	Butler, 1956, 1984; Mein and Pickford, 2003
Amphechinus sp.	M. Miocene, E. Africa (Fort Ternan)	Butler, 1984
Gymnurechinus leakeyi Butler, 1956	E. Miocene, E. Africa (mainly Hiwegi Fm., Rusinga)	Butler, 1956, 1984
Gymnurechinus camptolophus Butler, 1956	E. Miocene, E. Africa (Songhor, Rusinga)	Butler, 1956, 1969, 1984 (includes *G. songhorensis*)
Protechinus salis Lavocat, 1961	M. Miocene, NW Africa (Beni Mellal)	Lavocat, 1961
Erinaceus (Atelerix) broomi Butler and Greenwood, 1973	E. Pleistocene, S. Africa (Bolt's Farm), E. Africa (Olduvai)	Broom, 1937, 1948 (as *Atelerix major*); Butler and Greenwood, 1973
Erinaceus (Atelerix) sp.	L. Pliocene, NW Africa (Ahl al Oughlam)	Geraads, 1995

TABLE 29.2
African fossil Soricdae

Taxon	Distribution	Reference
	CROCIDOSORICINAE	
Lartetium dehmi africanum (Lavocat, 1961)	M. Miocene, NW Africa (Beni Mellal)	Lavocat, 1961
	CROCIDURINAE	
Myosorex robinsoni Meester, 1955	L. Pliocene, E. Africa (Omo); E. Pleistocene, E. Africa (Olduvai); L. Pliocene–E. Pleistocene, S. Africa (Makapansgat, Bolt's Farm, Sterkfontein, etc.)	Butler and Greenwood, 1979; Meester, 1955; Meester and Meyer, 1972
Myosorex sp.	M. Miocene, NW Africa (Tunisia); E. Pliocene, S. Africa (Langebaanweg)	Robinson and Black, 1974; Pocock, 1976; Hendey, 1981
Sylvisorex granti (Thomas, 1907)	E. Pleistocene, E. Africa (Olduvai)	Butler and Greenwood, 1979
Sylvisorex olduvaiensis Butler and Greenwood, 1979	E. Pleistocene, E. Africa (Olduvai)	Butler and Greenwood, 1979
Suncus varilla (Thomas, 1895)	L. Pliocene–E. Pleistocene, S. Africa (Makapansgat, Bolt's Farm, Sterkfontein, etc.); M. Pleistocene, E. Africa (Isenya)	Meester and Meyer, 1972; Brugal and Denys, 1989
Suncus varilla meesteri Butler and Greenwood, 1979	E. Pleistocene, E. Africa (Olduvai)	Butler and Greenwood, 1979
Suncus infinitesimus Heller, 1912	L. Pliocene—E. Pleistocene, S. Africa (Sterkfontein, Sterkfontein extension)	Meester and Meyer, 1972
Suncus leakeyi Butler and Greenwood, 1979	E. Pleistocene, E. Africa (Olduvai)	Butler and Greenwood, 1979
Suncus shungurensis Wesselman, 1984	L. Pliocene, E. Africa (Omo)	Wesselman, 1984

TABLE 29.2 (CONTINUED)

Taxon	Distribution	Reference
Suncus barbarus Geraads, 1993	L. Pliocene, NW Africa (Ahl al Oughlam)	Geraads, 1995
Suncus lixus (Thomas, 1908)	L. Pliocene, E. Africa (Omo); M. Pleistocene, E. Africa (Isenya—cf.)	Wesselman, 1984; Brugal and Denys, 1989
Suncus hesaertsi Wesselman, 1984	L. Pliocene, E. Africa (Omo)	Wesselman, 1984
Suncus sp.	L. Miocene, E. Africa (Lukeino). Pliocene, S. Africa (Langebaanweg)	Mein and Pickford, 2006. Pocock, 1976; Hendey, 1981
Crocidura kapsominensis Mein and Pickford, 2006	L. Miocene, E.Africa (Lukeino)	Mein and Pickford, 2006
Crocidura aithiops Wesselman, 1984	L. Pliocene, E Africa (Omo)	Wesselman, 1984
Crocidura dolichura Peters, 1876	L. Pliocene, E. Africa (Omo—cf.); E. Pleistocene, E. Africa (Koobi Fora—cf.)	Wesselman, 1984; Black and Krishtalka, 1986
Crocidura cyanea (Duvernoy, 1838)	L. Pliocene, E. Africa (Laetoli—cf.)	Butler, 1987
Crocidura balsaci Butler and Greenwood, 1979	E. Pleistocene, E. Africa (Olduvai)	Butler and Greenwood, 1979
Crocidura nana Dobson, 1890	E. Pleistocene, E. Africa (Koobi Fora—cf.)	Black and Krishtalka, 1986
Crocidura hildegardeae Thomas, 1904	M. Pleistocene, E. Africa (Isenya)	Brugal and Denys, 1989
Crocidura yankariensis Hutterer and Jenkins, 1980	M. Pleistocene, E. Africa (Isenya)	Brugal and Denys, 1989
Crocidura taungsensis Broom, 1948	L. Pliocene, S. Africa (Taung); L. Pleistocene, S. Africa (Cave of Hearths)	Broom , 1948; Meester, 1955; De Graaff, 1960
Crocidura fuscomurina (Heuglin, 1865)	E. Pleistocene, S. Africa (Bolt's Farm—cf.)	Davis and Meester, unpublished; De Graaff, 1960
Crocidura hirta Peters, 1852	E. Pleistocene, S. Africa (Bolt's Farm—cf.); L. Pleistocene, S. Africa (Witkrans Cave, Cave of Hearths—cf.)	Butler, unpublished; Davis and Meester, unpublished; De Graaff, 1960
Crocidura jaegeri Rzebik-Kowalska, 1988	Plio-Pleistocene, NW Africa (Irhoud Ochre)	Rzebik-Kowalska, 1988
Crocidura marocana Rzebik-Kowalska, 1988	M. Pleistocene, NW Africa (Irhoud Derbala Virage)	Rzebik-Kowalska, 1988
Crocidura maghrebiana Hutterer, 1991	M. Pleistocene, NW Africa (Irhoud Derbala Virage; Oulad Hamida)	Rzebik-Kowalska, 1988 (as *C.* cf. *viaria*); Hutterer, 1991; Geraads, 1993 (as *C. darelbeidae*)
Crocidura tarfayensis Vesmanis and Vesmanis, 1980	M. Pleistocene, NW Africa (Oulad Hamida—cf.)	Geraads, 1993
Crocidura whitakeri de Winton, 1898	M. Pleistocene, NW Africa (Ain Mefta—cf.)	Rzebik-Kowalska, 1988
Crocidura russula (Hermann, 1780)	M. Pleistocene, NW Africa (Ain Mefta)	Rzebik-Kowalska, 1988
Crocidura canariensis Hutterer, Lopez-Jurado and Vogel, 1987	L. Pleistocene, Canary Islands	Michaux et al., 1991
Crocidura spp.	L. Pleistocene, S. Africa (Kabwe); Plio-Pleistocene, SW Africa (Otavi Mountains)	Hopwood, 1928; Senut et al., 1992
Diplomesodon fossorius Repenning, 1965	L. Pliocene, S. Africa (Makapansgat)	Repenning, 1965, 1967

SORICIDAE

Taxon	Distribution	Reference
Asioriculus maghrebiensis (Rzebik-Kowalska, 1988)	L. Pliocene–E. Pleistocene, NW Africa (Irhoud Ochre, Ahl al Oughlam)	Rzebik-Kowalska, 1988; Geraads, 1995 (as *Episoriculus*)

the face, and the condyle and posterior part of the jaw are less elevated, so that the coronoid process is inclined rather than vertical. *Amphechinus rusingensis* (figure 29.1B) is represented by several partial skulls and jaw fragments from Kenya, ranging in age from 20 Ma (Legetet) to 15 Ma (Maboko)(Butler, 1956, 1969, 1984), and it has also been found at Arrisdrift, Namibia (Mein and Pickford, 2003). It is less advanced than the European species in that the roots of p2 are not united and the metaconule of upper molars is less reduced, implying an Asiatic rather than a European ancestry.

Gymnurechinus, the second early Miocene erinaceine genus, is unknown outside Africa. In the dentition, it lacks the specializations of *Amphechinus* and resembles modern Erinaceinae; thus, the anterior incisors are less enlarged and the second and third molars less reduced. At the same time, it shares with *Amphechinus* primitive characters such as the orbital lachrymal foramen and the low elevation of the condyle. Two species are recognized, *G. leakeyi* (figure 29.1A) and *G. camptolophus* (figure 29.1C), distinguished by the pattern of rugosity on the skull roof, by the shape of the molar teeth,

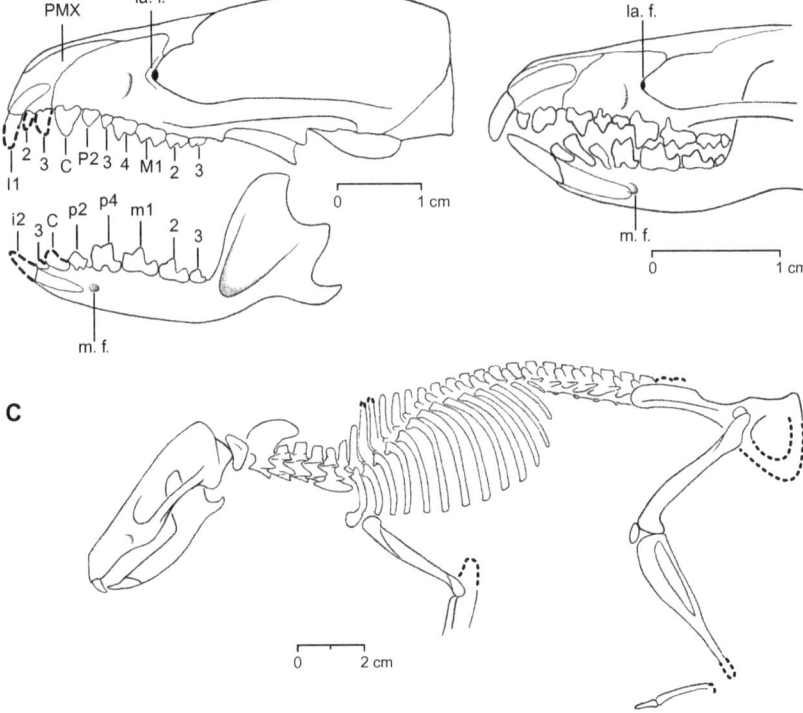

FIGURE 29.1 A) *Gymnurechinus leakeyi.* B) *Amphechinus rusingensis.* C) *Gymnurechinus camptolophus.* After Butler, 1956.

ABBREVIATIONS: la.f., lachrymal foramen; m.f., mental foramen; pmx, premaxilla.

and by size. Both species are represented by well-preserved skulls, permitting a detailed description of the structure, including the endocast (Butler, 1956). Compared with modern hedgehogs, the skull is less flattened, with a higher occiput and a longer snout. An incomplete skeleton of *G. camptolophus* from Rusinga Island indicates an agile animal with a strong neck, high thoracic neural spines, and flexible lumbar region. It has greater resemblance to *Echinosorex* or *Hylomys* than to *Erinaceus*.

Modern erinaceines are specialized in developing spines on their backs, accompanied by complex cutaneous musculature. This forms a sphincter at the margin of the spine-covered area, which by contracting flexes the back and pulls the skin over the head and hindquarters. This mechanism of rolling up explains several features of the skeleton, including the short snout, flattened skull, short neck, low neural spines, and other muscle processes, that distinguish *Erinaceus* from *Gymnurechinus*. Thus, *Gymnurechinus* seems to be a plesiomorphic erinaceine that had not developed the spiny pelage.

Protechinus salis, based on fragmentary material from the middle Miocene of Morocco (Lavocat, 1961), might be derived from *Gymnurechinus*. It is advanced in the facial position of the lachrymal foramen and the union of the roots of p2, but it remains primitive in the low elevation of the condyle and coronoid process. Erinaceinae of modern type were present at the same time in Europe and Turkey (Engesser, 1980).

All the Miocene genera became extinct, and sub-Saharan Africa is now inhabited by species of *Erinaceus*, of which they form the subgenus *Atelerix*. Two molars from the late Pliocene (2.5 Ma) of Morocco (Geraads, 1995) are very similar to *E. (A.) algirus*, which now lives in the Mediterranean zone. The facial part of a skull from Bolt's Farm, South Africa (early Pleisto-

cene), was described by Broom (1937) as *Atelerix major* (now changed on priority grounds to *Erinaceus (Atelerix) broomi*). Fragmentary material from Olduvai, including limb bones, was studied by Butler and Greenwood (1973), who concluded that *E. (A.) broomi* was related to the extant *E. (A.) albiventris* but had primitive characters shared with *E. europaeus*. The existing species of *E. (Atelerix)* inhabit grassland and savanna; *E. (A.) albiventris* has lost the hallux, a cursorial adaptation.

Probably the most recent erinaceid invaders, not found as fossils, are *Hemiechinus auritus*, which ranges from Egypt to the European steppes, and *Paraechinus aethiopicus* in the Sahara and Arabia, the most desert-adapted hedgehog, with close relatives in India. The successive erinaceid invasions reflect increasing aridity in the regions of entry, from forest in the Miocene, to savanna in the Pliocene, to desert today.

Order SORICOMORPHA Gregory, 1910
Family SORICIDAE Fischer, 1814

Shrews form an important part of the small-mammal fauna of Africa at the present time. With about 150 species, they are second only to the rodents in diversity; however, their paleontological record is comparatively poor. Except for two fossil species, all African shrews belong to the subfamily Crocidurinae ("white-toothed shrews"), and they show little morphological diversity. In most cases, the species are distinguished by combinations of small differences, so that fairly complete jaws or maxillae are required for identification. Nearly all the material is fragmentary and identification to species, and sometimes even to genus, may be uncertain. Differences between primitive and derived characters in the living species of Crocidurinae were investigated by Heim de Balsac and

Lamotte (1956, 1957), Butler and Greenwood (1979), Butler et al. (1989), McLellan (1994), and Jenkins et al. (1998).

African Fossil Record Shrews are characterized by specialized anterior incisors (I1, i1) and double articulation of the mandibular condyle. The two extant subfamilies, Soricinae and Crocidurinae, differ in the form of p4 and the mandibular condyle (Repenning, 1967; Reumer, 1994). Both subfamilies are believed to be derived from the more primitive Crocidosoricinae, which are known from the Oligocene of Mongolia and Europe, and reached their maximum diversity in Europe in the early Miocene (Reumer, 1994). A crocidosoricine, *Lartetium dehmi africanum*, was present in Morocco in the middle Miocene (Lavocat, 1961); it belongs to a genus found in the early and middle Miocene of Europe and Turkey. *Lartetium* has also been recorded from the late Miocene of Egypt (Pickford et al., 2008). The diversity of Crocidurinae in Africa today strongly indicates that the subfamily originated there, but evidence from before the Pliocene is very sparse. A jaw of *Crocidura* from Rusinga (Butler and Hopwood, 1957) is probably not early Miocene but modern; it resembles large Recent shrews of the *C. olivieri* group. No other shrews are known from the early Miocene deposits of East Africa, though many specimens of Tenrecidae of similar size to shrews have been found. If the Rusinga specimen is rejected, the earliest record of Crocidurinae is undescribed material of *Myosorex* from the middle Miocene (ca. 12 Ma) of Tunisia (Robinson and Black, 1974; Robinson et al., 1982). From the late Miocene Lukeino Formation of Kenya (ca. 6 Ma), Mein and Pickford (2006) described *Crocidura kapsominensis* and *Suncus* sp., based on fragmentary material. Two more species of *Crocidura*, represented by single specimens, occur in the Middle Awash (5.7 Ma), Ethiopia (H. B. Wesselman, pers. comm.). By the early Pliocene, crocidurines had reached South Africa, and appeared in Europe (the late Miocene crocidurine teeth from Turkey [Engesser, 1980] have been reidentified by Reumer [1994] as crocidosoricine). The basic radiation of the Crocidurinae thus probably took place in Africa in the middle to late Miocene. In the absence of fossil evidence, the pattern of branching must be inferred from the comparison of living species.

Myosorex, together with the extant *Congosorex* and *Surdisorex*, differs widely from other crocidurines and shares some characters with Soricinae (Heim de Balsac, 1967). It has an additional minute tooth in the lower jaw between the unicuspid i2 and p4. The comparatively small anterior incisors and condyle resemble the condition in Crocidosoricinae. Isozyme analysis (Maddalena and Bronner, 1982) and mitochondrial rRNA (Quérouil et al., 2001) place *Myosorex* in an intermediate position between Crocidurinae and Soricinae. It may have entered Africa independently. *Myosorex robinsoni* is common in the South African Pleistocene breccias and at Olduvai. It is related to the extant *M. cafer* and *M. varius* (Meester, 1955; Avery, 2000). These species inhabit moist habitats (forest and river valleys), and other species are in tropical mountain forest. They have fossorial adaptations.

The affinity between the closely related genera *Sylvisorex*, *Suncus*, and *Crocidura* has been much discussed. The first two genera differ from *Crocidura* in the presence of a small fourth upper unicuspid tooth anterior to P4 (probably P3). This is present in *Myosorex* and is presumably a primitive trait. *Sylvisorex* differs from *Suncus* in having a higher proportion of primitive characters—for example, smaller i1, mental foramen under p4 rather than m1, additional cusps on p4, basined talonid on m3, narrower ventral condyle. These are, however, not present in all the species. Heim de Balsac and Lamotte

(1957) regarded *Sylvisorex* as broadly ancestral to the other genera. Mitochondrial rRNA (Quérouil et al., 2001) shows that *Sylvisorex* and *Suncus* are paraphyletic, on multiple branches from the stem of the *Crocidura* crown group.

When the upper dentition is not available, generic identification must depend on resemblance to living species. Two species of *Sylvisorex* were identified by Butler and Greenwood (1979) from the early Pleistocene of Olduvai. One is very similar to *S. granti*, though it differs in some details, and has some resemblance to *S. megalura*. The other, *S. olduvaiensis*, is distinct from the living species, though it shares some derived characters with *S. johnsoni*, which is much smaller, and it also has some resemblance to *Suncus lixus*. It is the commonest shrew in the lower part of Bed I but is rare in the upper part, perhaps due to greater aridity at that time. Existing species of *Sylvisorex* are inhabitants of tropical mountain forest, except *S. megalura*, which extends to savanna.

As 11 of 16 living species of *Suncus* occur in southern Asia, it has been suggested that the genus originated there (Butler, 1978), from a *Sylvisorex*-like ancestor and that the African species represent an invasion from Asia. The most primitive species, *Suncus dayi* from India, is very similar to *Sylvisorex* (Jenkins et al., 1998). However, the mitochondrial rRNA analysis of Quérouil et al. (2001) makes *S. dayi* the sister species of *Sylvisorex megalura*, within a clade that also includes *Suncus etruscus* and the African *S. infinitesimus* and *S. remyi*. Therefore, it seems probable that *Suncus* originated in Africa, and more than one species invaded Asia. As no fossil *Suncus* has been found outside Africa, the date of the invasion(s) remains speculative.

The oldest evidence of *Suncus* is a mandible fragment and a molar from the late Miocene (ca. 6 Ma) of Kenya. It is the size of the extant *S. varilla*. In South Africa, *Suncus* is recorded, but not described, from the early Pliocene of Langebaanweg, South Africa (Hendey, 1981). *Suncus varilla*, similar to the living species, is present in the Plio-Pleistocene breccias of South Africa (Meester and Meyer, 1972). A primitive form, *S. varilla meesteri* (figure 29.2), occurs at Olduvai and probably Makapansgat (Butler and Greenwood, 1979). *Suncus lixus*, which is larger than *S. varilla*, has not been recorded from the South African deposits, though some material from Bolt's Farm might belong to it. Wesselman (1984) identified as *Suncus* aff. *lixus* a fragment of mandible from the late Pliocene (3.0 Ma) of Omo, but the ramus is shallower than in *S. lixus*. *Suncus infinitesimus*, distinguished from *S. varilla* by smaller size, occurs at Sterkfontein, South Africa (Meester and Meyer, 1972). It is represented at Olduvai by a more primitive species, *S. leakeyi*. Probably related is *S. shungurensis*, from Omo (3.0 Ma), known by three mandibular fragments with molars (Wesselman, 1984). The generic identity of *S. haessertsi*, from the same place, is uncertain; it is known only by a fragment with m3 of a large shrew that might belong to *Crocidura*. *Suncus barbarus*, from the late Pliocene (2.5 Ma) of Morocco (Geraads, 1995) is a distinctive species, known by upper and lower dentitions. *Crocidura jaegeri*, from the late Pliocene of Algeria (Rzebik-Kowalska, 1988) is a small shrew that is much like *Suncus leakeyi* from Olduvai, and it might be a *Suncus*. A maxillary fragment shows a small tooth anterior to P4 that might be the fourth unicuspid.

Heim de Balsac and Lamotte (1957) regarded *Crocidura* as diphyletic, partly derived from *Sylvisorex* and partly from *Suncus*, with independent loss of P3. Butler et al. (1989) also postulated a multiple origin of *Crocidura*, with different groups of species derived from *Sylvisorex*-like ancestors. However, the

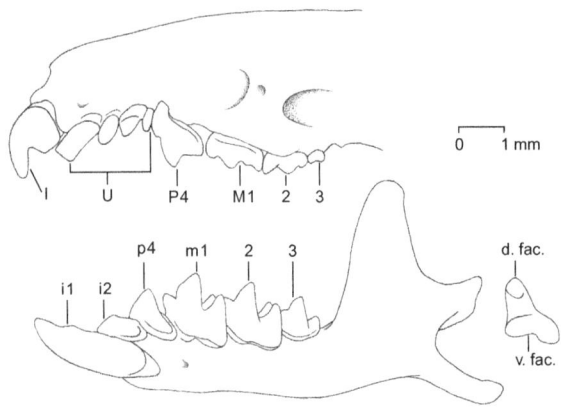

FIGURE 29.2 *Suncus varilla meesteri.*

ABBREVIATIONS: d. fac. and v. fac., dorsal and ventral articular facets of condyle; U, unicuspids. After Butler and Greenwood, 1979.

allozyme analysis of Maddalena (1990) and the mitochondrial rRNA analysis by Quérouil et al. (2001) indicate a single origin, with *Sylvisorex* and *Suncus* on earlier branches from the *Crocidura* stem. Of 19 species and subspecies of *Crocidura* included in the cladistic analysis of McLellan (1994), based on morphology, 17 are in a single clade, but *C. dolichura* and *C. nilotica* are in more basal positions.

The oldest record of *Crocidura*, if the Rusinga jaw is rejected, is of *C. kapsominensis,* from the late Miocene (ca. 6.0 Ma) of the Lukeino Formation, Kenya (Mein and Pickford, 2006). The first upper unicuspid is exceptionally large. From the late Pliocene (3.0 Ma) of Omo, Ethiopia, Wesselman (1984) described *Crocidura aithiops*, a large, advanced form in the *C. olivieri* group, and *C.* aff. *dolichura*, which is small and primitive. Forms similar to these occur in the late Miocene (5.7 Ma) and early Pliocene (4.4 Ma) of the Middle Awash, Ethiopia (H. B. Wesselman, pers. comm.). Also from the late Pliocene is a jaw fragment from Laetoli (ca. 2.7 Ma; Upper Ndolanya Beds), unnamed but of the size of *C. cyanea* (Butler, 1987). Though based on very limited material, these show that diversification of *Crocidura* was already advanced in the Pliocene. In South Africa, the oldest species is *C. taungsensis*, from the late Pliocene of Taung (Broom, 1948; Meester, 1955), based on the anterior part of a skull. It is a shrew that resembles *C. fuscomurina* in size, but it differs in the small parastyle of P4 and the more posterior position of the infraorbital foramen. It survived into the late Pleistocene (De Graaff, 1960).

The largest collection of shrews from the early Pleistocene is from Olduvai Bed I, although *Crocidura* is poorly represented (Butler and Greenwood, 1979). The only species described is *C. debalsaci*, based on 19 mandibular specimens out of a total of 848. Multivariate analysis indicates that it is related to *C. voi* (Butler et al., 1989), a savanna species included in the subgenus *Afrosorex* by Hutterer (1986). In South Africa, *Crocidura* has not been recorded from Makapansgat, Sterkfontein (prior to 0.1 Ma), Swartkrans, and other localities that have produced many specimens of *Suncus*. *Crocidura fuscomurina* (distinct from *C. taungsensis*) and *C. hirta* appear at Bolt's Farm; *C. silacea* occurs in the most recent deposits at Sterkfontein (Avery, 2000). Perhaps the comparative scarcity of *Crocidura* reflects environmental conditions; the African species of *Suncus* are inhabitants of savanna, while the more primitive species of *Crocidura* are mainly forest animals, and

the genus spread to savanna rather late (Hutterer and Happold, 1983; Hutterer, 1986; McLellan, 1994).

In northwest Africa, after *C. jaegeri* (1.7 Ma), mentioned as a possible *Suncus*, the record of *Crocidura* goes back only to the middle Pleistocene (ca. 0.5 Ma). Most of the fossils are close to species still living in the area (see Rzebik-Kowalska, 1988; Hutterer, 1991; Geraads, 1993), but names have been given to *C. marocana* and *C. maghrebiana* (synonym *C. darelbeidae*), thought to be related to *C. whitakeri* and *C. viaria*, respectively.

Maddalena (1990) investigated the relationships between species of *Crocidura* by electrophoretic isozyme analysis. After the separation of the primitive *C. luna* and *C. bottegi* on basal branches, the species divided into two clades, containing the Palaearctic and Afrotropical species, respectively. The two groups also differ karyotypically; the primitive diploid chromosome number of 36–40 is retained or reduced in the Palaearctic and Oriental species, and increased in the Afrotropical species (Maddalena and Ruedi, 1994). Thus, the non-African species appear to have a unique common ancestor that migrated from Africa probably near the Miocene-Pliocene boundary.

A reverse migration also occurred. The only African soricine in Morocco in the late Pliocene, *Asioriculus maghrebiensis*, belongs to a genus that ranged from the late Miocene to Pleistocene in Europe and Asia Minor. It probably entered Africa from the east. *Crocidura russula*, now living in northwest Africa as well as western Europe (Catzeflis et al., 1985), may have crossed the Strait of Gibraltar from Europe to Africa at a time of low sea level during the Pleistocene. This would also account for the close relationship between *Crocidura sicula*, in Sicily, and *C. canariensis* on the Canary Islands (Sará, 1995).

The most problematic fossil African shrew is *Diplomesodon fossorius*, known only from Makapansgat (Repenning, 1965), where it is represented by several mandibles and partial skulls. It differs from *Crocidura* in having only two upper unicuspids between I1 and P4. In this it resembles *D. pulchellum*, the only living species of the genus, from Central Asia east of the Aral Sea. *Diplomesodon fossorius* is larger and more specialized, with proportionately larger anterior incisors, transversely wider molars, more reduced last molars, and a transversely extended ventral condyle. Several of its derived characters are approached by or shared with *Crocidura macarthuri*, living in East Africa. Butler (1978) suggested that *D. fossorius* and *D. pulchellum* had been independently derived from *Crocidura*, in Africa and Asia, respectively, and that their resemblance is due to parallel evolution. However, protein electrophoresis shows that *D. pulchellum* is not a member of the clade to which the Palaearctic species of *Crocidura* belong but rather is an early offshoot of the Afrotropical clade (Ruedi, 1998). This implies that *Diplomesodon* emigrated from Africa independently from *Crocidura*, and supports the generic assignment of *D. fossorius*.

The primary habitat of the Crocidurinae was probably tropical rain forest, where at the present time the species of *Sylvisorex* and most of the more primitive species of *Crocidura* occur. The African species of *Suncus* and many species of *Crocidura* have adapted to drier conditions and spread to savanna (Hutterer and Happold, 1983; Hutterer, 1986; McLellan, 1994). This process may have occurred earlier in *Suncus* than in *Crocidura*, to judge by the relative scarcity of *Crocidura* in Olduvai Bed I and the older deposits of South Africa. *Diplomesodon* may be another early example. Changes of climate have undoubtedly had a major influence on the distribution of species, such as the spread of *Myosorex* into southern Africa (Meester, 1968). With retreat of the forest belt due to greater aridity, populations were isolated in forest islands and on mountains, where they developed into separate species (Heim de Balsac, 1967; McLellan, 1994).

ACKNOWLEDGMENTS

I thank Alisa Winkler for help with the literature search and for editorial improvements of the text, Bonnie Miljour for setting up the figures, and Hank Wesselman for permission to cite unpublished research on fossil shrews.

Literature Cited

Avery, D. M. 2000. Notes on the systematics of micromammals from Sterkfontein, Gauteng, South Africa. *Palaeontologia Africana* 38:83–90.

Black, C. C., and L. Krishtalka. 1986. Rodents, bats and insectivores from the Plio-Pleistocene to the east of Lake Turkana. *Contributions in Science, Natural History Museum of Los Angeles County* 372:1–15.

Broom, R. 1937. Notices of a few more fossil mammals from the caves of the Transvaal. *Annals and Magazine of Natural History* (10) 28:509–514.

———. 1948. Some South African Pliocene and Pleistocene mammals. *Annals of the Transvaal Museum* 21:1–38.

Brugal, J. P., and C. Denys. 1989. Vertébrés du site acheuléen d'Isenya (Kenya, District de Kajiado). Implications paléoécologiques et paléobiogeographiques. *Comptes Rendus Hebdomadaires des Séances de l'Académie des Sciences, Paris*, Série II, 308:1503–1508.

Butler, P. M. 1948. On the evolution of the skull and dentition in the Erinaceidae, with special reference to fossil material in the British Museum. *Proceedings of the Zoological Society of London* 118:446–500.

———. 1956. Erinaceidae from the Miocene of East Africa. *Fossil Mammals of Africa* 11:1–75.

———. 1969. Insectivores and bats from the Miocene of East Africa; pp. 1–37 in L. S. B. Leakey (ed.), *Fossil Vertebrates of Africa*, vol. 1. Academic Press, London.

———. 1978. Insectivora and Chiroptera; pp. 56–68 in V. J. Maglio and H. B. S. Cooke (eds.), *Evolution of African Mammals*. Harvard University Press, Cambridge.

———. 1984. Macroscelidea, Insectivora and Chiroptera from the Miocene of East Africa. *Palaeovertebrata* 14:117–200.

———. 1987. Fossil insectivores from Laetoli; pp. 85–87 in M. D. Leakey and J. M. Harris (eds.), *Laetoli: A Pliocene Site in Northern Tanzania*. Clarendon Press, Oxford.

Butler, P. M., and M. Greenwood. 1973. The early Pleistocene hedgehog from Olduvai, Tanzania; pp. 7–42 in L. S. B. Leakey, R. J. G. Savage, and S. C. Coryndon (eds.), *Fossil Vertebrates of Africa*, vol. 3. Academic Press, London.

———. 1979. Soricidae (Mammalia) from the early Pleistocene of Olduvai Gorge, Tanzania. *Zoological Journal of the Linnean Society* 67:329–379.

Butler, P. M., and A. T. Hopwood. 1957. Insectivora and Chiroptera from the Miocene Rocks of Kenya Colony. *Fossil Mammals of Africa* 13:1–35.

Butler, P. M., R. S. Thorpe, and M. Greenwood. 1989. Interspecific relations of African crocidurine shrews (Mammalia, Soricidae) based on multivariate analysis of mandibular data. *Zoological Journal of the Linnean Society* 96:373–412.

Catzeflis, F., T. Maddalena, S. Hellwing, and P. Vogel. 1985. Unexpected findings on the taxonomic status of East Mediterranean *Crocidura russula* auct. (Mammalia, Insectivora). *Zeitschrift für Säugetierkunde* 30:185–201.

Conroy, G. C., M. Pickford, B. Senut, J. Van Couvering, and P. Mein. 1992. *Otavipithecus namibiensis*, first Miocene hominoid from southern Africa. *Nature* 356:144–148.

Davis, H. D. S., and J. Meester. Report on the microfauna in the University of California collections from the South African cave breccias. Unpublished report.

De Graaff, G. 1960. A preliminary investigation of the mammalian microfauna in Pleistocene deposits of caves in the Transvaal system. *Palaeontologia Africana* 7:59–116.

Engesser, B. 1980. Insectivora and Chiroptera (Mammalia) aus dem Neogen der Türkei. *Schweizerischen Paläontologischen Abhandlungen* 102:45–149.

Frost, D. R., W. C. Wozencraft, and R. S. Hoffmann. 1991. Phylogenetic relationships of hedgehogs and gymnures (Mammalia, Insectivora, Erinaceidae). *Smithsonian Contributions to Zoology* 518:1–69.

Geraads, D. 1993. Middle Pleistocene *Crocidura* (Mammalia, Insectivora) from Oulad Hamida I, Morocco, and their phylogenetic relationships. *Proceedings Koninklijke Nederlandse Akademie van Wetenschappen* 96:281–294.

———. 1995. Rongeurs et insectivores (Mammalia) du Pliocene final de Ahl al Oughlam (Casablanca, Maroc). *Geobios* 28:99–115.

Gould, G. C. 1995. Hedgehog phylogeny (Mammalia, Erinaceidae)—the reciprocal illumination of the quick and the dead. *American Museum Novitates* 3131:1–45.

Heim de Balsac, H. 1967. Faits nouveaux concernant les *Myosorex* (Soricidae) de l'Afrique orientale. *Mammalia* 31:610–628.

Heim de Balsac, H., and M. Lamotte. 1956. Évolution et phylogénie des Soricidés africains: I. La lignée *Myosorex-Surdisorex*. *Mammalia* 20:140–167.

———. 1957. Évolution et phylogénie des Soricidés africains: II. La lignée *Sylvisorex-Suncus-Crocidura*. *Mammalia* 21:15–49.

Hendey, Q. B. 1981. Palaeoecology of the late Tertiary fossil occurrences in "E" Quarry, Langebaanweg, South Africa, and reinterpretation of their geological context. *Annals of the South African Museum* 84:1–104.

Hopwood, A. T. 1928. Mammalia; pp. 70–73 in W. P. Pycraft, G. E. Smith, M. Yearsley, J. T. Carter, R. A. Smith, A. T. Hopwood, D. M. A. Bate, W. E. Swinton, and F. A. Bather: *Rhodesian Man and Other Associated Remains*. British Museum, London.

Hutterer, R. 1986. African shrews allied to *Crocidura fischeri*: Taxonomy, distribution and relationship. *Cimbebasia* (A)4:23–35.

———. 1991. Variation and evolution of the Sicilian shrew: Taxonomic conclusions and description of a possibly related species from the Pleistocene of Morocco (Mammalia: Soricidae). *Bonner Zoologische Beiträge* 42:241–251.

Hutterer, R., and D. C. D. Happold. 1983. The shrews of Nigeria. Bonner Zoologische Monographien 18:1–79.

Jenkins, P., M. Ruedi, and F. M. Catzeflis. 1998. A biochemical and morphological investigation of *Suncus dayi* (Dobson, 1888) and a discussion of relationships in *Suncus* Hemprich & Ehrenberg, 1833, *Crocidura* Wagler, 1832, and *Sylvisorex* Thomas, 1904 (Insectivora: Soricidae). *Bonner Zoologische Monographien* 47:257–276.

Lavocat, R. 1961. Le gisement de Vertébrés miocènes de Beni Mellal (Maroc): Étude systématique de la faune de Mammifères et conclusions génerales. *Notes et Mémoires du Service Géologique de Maroc* 155:1–145.

Lopatin, A. V. 2004. A new genus of the Galericinae (Erinaceidae, Insectivora) from the middle Eocene of Mongolia. *Palaeontological Journal* 38:319–326.

Maddalena, T. 1990. Systematics and biogeography of Afrotropical and Palaearctic shrews of the genus *Crocidura* (Insectivora, Soricidae): An electrophoretic approach; pp. 297–308 in G. Peters and R. Hutterer (eds.), *Vertebrates in the Tropics*. Museum Alexander Koenig, Bonn.

Maddalena, T., and G. Bronner. 1992. Biochemical systematics of the endemic African genus *Myosorex* Gray, 1838. *Israel Journal of Zoology* 38:245–252.

Maddalena, T., and M. Ruedi. 1994. Chromosomal evolution in the genus *Crocidura* (Insectivora: Soricidae). *Special Publication of the Carnegie Museum of Natural History* 18:335–344.

McLellan, L. J. 1994. Evolution and phylogenetic affinities of the African species of *Crocidura*, *Suncus*, and *Sylvisorex* (Insectivora: Soricidae). *Special Publication of the Carnegie Museum of Natural History* 18:379–391.

Meester, J. 1955. Fossil shrews of South Africa. *Annals of the Transvaal Museum* 22:271–278.

———. 1968. The origins of the southern African mammal fauna. *Zoologia Africana* 1:87–95.

Meester, J., and I. J. Meyer. 1972. Fossil *Suncus* (Mammalia, Soricidae) from southern Africa. *Annals of the Transvaal Museum* 27:269–277.

Mein, P., and L. Ginsburg. 1997. Les mammifères du gisement miocène inférieur de Li Mae Long, Thailand: Systematique, biostratigraphie et paléoenvironnement. *Geodiversitas* 19:783–844.

Mein, P., and M. Pickford. 2003. Insectivora from Arrisdrift, a basal middle Miocene locality in southern Namibia. *Memoir of the Geological Survey of Namibia* 19:141–146.

Mein, P., and M. Pickford. 2006. Late Miocene micromammals from the Lukeino Formation (6.1 to 5.8 Ma), Kenya. *Bulletin et Mémoires de la Société Linnéen de Lyon* 75:183–223.

Michaux, J., R. Hutterer, and N. Lopez Martinez. 1991. New fossil fauna from Fuerteventura, Canary Islands: evidence for a Pleistocene age of endemic rodents and shrews. *Comptes Rendus Hebdomadaires des Séances de l'Académie des Sciences, Paris*, Série III 312:801–806.

Pickford, M., H. Wanas, P. Mein, and H. Soliman. 2008. Humid conditions in the Western Desert of Egypt during the Vallesian (Late Miocene). *Bulletin of the Tethys Geological Society, Cairo* 3:63–79.

Pocock, T. N. 1976. Pliocene mammalian microfauna from Langebaanweg: A new fossil genus linking the Otomyinae with the Murinae. *South African Journal of Science* 72:58–60.

Quérouil, S., R. Hutterer, P. Barrière, M. Colyn, J. C. K. Peterhaas, and E. Verheyen. 2001. Phylogeny and evolution of African shrews (Mammalia: Soricidae) inferred from 16S rRNA sequences. *Molecular Phylogenetics and Evolution* 20:185–195.

Repenning, C. A. 1965. An extinct shrew from the early Pleistocene of South Africa. *Journal of Mammalogy* 46:189–196.

———. 1967. Subfamilies and genera of the Soricidae. *United States Geological Survey Professional Paper* 565:1–74.

Reumer, J. W. F. 1994. Phylogeny and distribution of the Crocidosoricinae (Mammalia: Soricidae). *Special Publication Carnegie Museum of Natural History* 18:345–355.

Robinson, P., and C. C. Black. 1974. Vertebrate faunas from the Neogene of Tunisia. *Annals of the Geological Survey of Egypt* 4:319–332.

Robinson, P., C. C. Black, L. Krishtalka, and M. R. Dawson. 1982. Fossil small mammals from the Kechabta Formation, northeastern Tunisia. *Annals of the Carnegie Museum* 51:231–249.

Ruedi, M. 1998. Protein evolution in shrews; pp. 269–294 in M. Wolsan and J. M. Wojcik (eds.), *Evolution of Shrews*. Mammal Research Institute, Polish Academy of Sciences, Bialowieza.

Rzebik-Kowalska, B. 1988. Soricidae (Mammalia, Insectivora) from the Plio-Pleistocene and middle Quaternary of Morocco and Algeria. *Folia Quaternaria* 57:51–90.

Sará, M. 1995. The Sicilian *(Crocidura sicula)* and the Canary *(C. canariensis)* shrew (Mammalia, Soricidae): Peripheral isolate formation and geographic variation. *Bollettino di Zoologia* 62:173–182.

Senut, B., M. Pickford, P. Mein, G. Conroy, and J. Van Couvering. 1992. Discovery of 12 new Late Cainozoic fossiliferous sites in palaeokarsts of the Otavi Mountains, Namibia. *Comptes Rendus Hebdomadaires des Séances de l'Académie des Science, Paris*, Série II 314:727–733.

Simpson, G. G. 1945. The principles of classification and a classification of mammals. *Bulletin of the American Museum of Natural History* 85:1–350.

Wesselman, H. B. 1984. *The Omo Micromammals: Systematics and Paleoecology of Early Man Sites from Ethiopia*. Contributions to Vertebrate Evolution, vol. 7. Karger, Basel, 219 pp.

THIRTY

Chiroptera

GREGG F. GUNNELL

The fossil record of bats is relatively poor (Gunnell and Simmons, 2005), although there are places (e.g., the Quercy karst deposits in France) where bat fossils can be quite common. Except for some exceptional preservation in lagerstät-ten such as Messel in Germany (Habersetzer and Storch, 1987; Simmons and Geisler, 1998; Storch, 2001) and the Green River Formation (Jepsen, 1966; Simmons et al., 2008) in Wyoming (USA), most bat fossils consist of fragmentary skulls and dentitions. The African record of bats is no different.

Three separate areas preserve fossil bats on the mainland African continent—North, East, and South Africa (figure 30.1). The oldest records are from North (early Eocene) and East (middle Eocene) Africa, while East and South Africa have the best records of Plio-Pleistocene bats. Additionally, there is a restricted sample of late Oligocene bats from Taqah, Oman (Sigé et al., 1994), on the nearby Arabian Peninsula and good samples of subfossil bats from Madagascar (Samonds, 2006, 2007).

The African fossil bat record includes scant records of pteropodids (Old World fruit bats) from the Miocene and Pliocene and from Pleistocene and subfossil samples from Kenya and Madagascar. With the possible exception of *Tanzanycteris* (Gunnell et al., 2003), which may be most closely related to *Hassianycteris* (an archaic bat from Messel; Smith and Storch, 1981), all known African bats represent modern superfamilies if not modern bat families.

Systematic Paleontology

Order CHIROPTERA Blumenbach, 1779
Family PTEROPODIDAE Gray, 1821
Genus *PROPOTTO* Simpson, 1967
PROPOTTO LEAKEYI Simpson, 1967
Figures 30.2A and 30.2B

Diagnosis Lower dental formula (2?).1.3.3, lower canine large, p2 small and peglike, lower molars rounded, low crowned with central basin surrounded by low cusps and lacking central anteroposterior groove typical of extant fruit bat molars, m3 small and circular in outline, dentary deepens anteriorly and possesses a small inferior, symphyseal torus (see Walker, 1969).

Age and Occurrence Early Miocene, Burdigalian (18–20 Ma); Koru, Songhor, Chamtwara, and Rusinga Island, Kenya.

Remarks There are six described specimens of *Propotto leakeyi*, all represented by lower jaws and teeth. The type (KNM-SO 508; this was Simpson's [1967] specimen "R" and was also numbered CMM 421A) and an additional specimen (KNM-RU 2084; this was Simpson's [1967] specimen "T" and was also numbered CMM 745), originally described as possibly coming from Rusinga Island (Simpson, 1967:46), are from Songhor. KNM-KO 101 is from Koru Locality 25, KNM-CA 1999 is from Chamtwara, KNM-RU 1879 (this was Simpson's [1967] specimen "S") is questionably from Rusinga locality R1, and KNM-RU 3690 is from Rusinga locality R3A (Butler, 1984). When originally described by Simpson (1967), *Propotto* was thought to be a lorisid primate, but Walker (1969) noted the lack of a tooth comb and the anteriorly deep dentary and argued that the true affinities of these specimens were with fruit bats (see Sigé and Aguilar, 1987, for a contrary opinion).

Except for a single, enigmatic tooth from the late Eocene of Thailand (Ducrocq et al., 1993) and pteropodid records from the late Oligocene and early Miocene of southern France (Aguilar et al., 1986; Sigé and Aguilar, 1987; Sigé et al., 1997), *Propotto* remains the earliest and best record yet known for fossil fruit bats. A previously described partial skeleton of a possible Oligocene pteropodid from Italy (Meschinelli, 1903; Dal Piaz, 1937) has been suggested to be an echolocating bat instead (Russell and Sigé, 1970; Schutt and Simmons, 1998).

Genus *EIDOLON* Rafinesque, 1815
EIDOLON aff. *E. HELVUM* (Kerr, 1792)
Figure 30.2C

Age and Occurrence Late Pliocene (2.95 Ma); Omo Locality 1, Member B, Shungura Formation, Ethiopia.

Remarks A single upper second molar of *Eidolon* was described by Wesselman (1984) from Omo Locality 1. Other than being somewhat larger and lacking a bulge along its lingual ridge, this tooth is nearly identical to M2 in the extant straw-colored fruit bat, *Eidolon helvum*. *Eidolon* is distributed across Arabia, sub-Saharan Africa and Madagascar today and is among the most common of African fruit bats (Nowak, 1994).

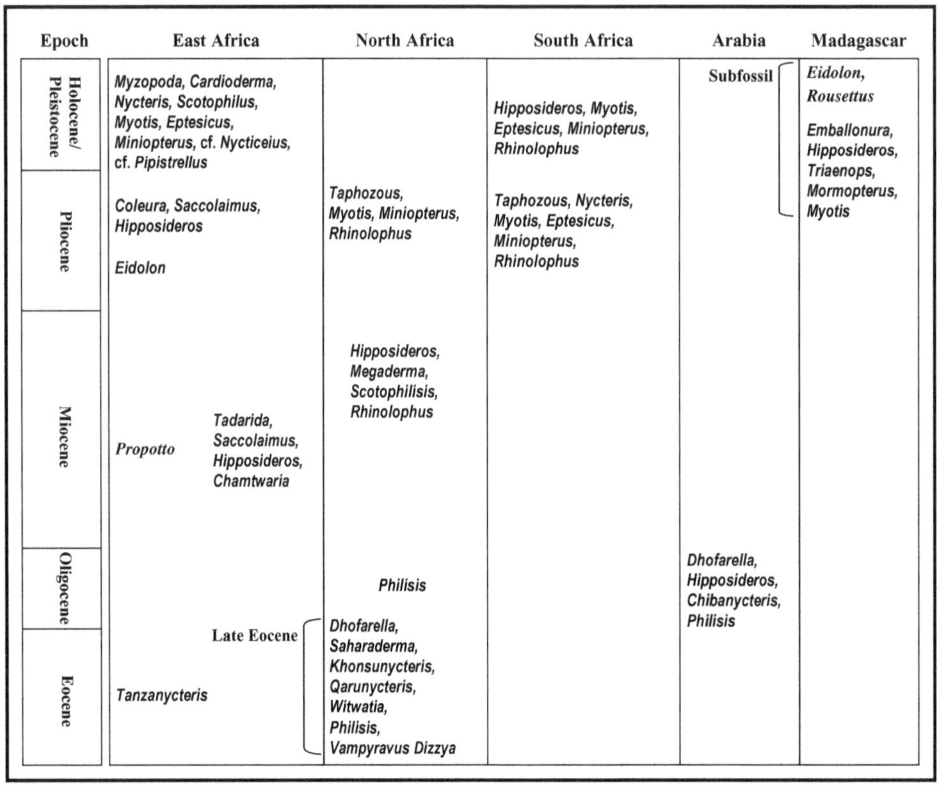

Epoch	East Africa	North Africa	South Africa	Arabia	Madagascar
Holocene/ Pleistocene	*Myzopoda, Cardioderma, Nycteris, Scotophilus, Myotis, Eptesicus, Miniopterus,* cf. *Nycticeius,* cf. *Pipistrellus*	*Taphozous, Myotis, Miniopterus, Rhinolophus*	*Hipposideros, Myotis, Eptesicus, Miniopterus, Rhinolophus*	Subfossil	*Eidolon, Rousettus* — *Emballonura, Hipposideros, Triaenops, Mormopterus, Myotis*
Pliocene	*Coleura, Saccolaimus, Hipposideros* — *Eidolon*		*Taphozous, Nycteris, Myotis, Eptesicus, Miniopterus, Rhinolophus*		
Miocene	*Propotto* — *Tadarida, Saccolaimus, Hipposideros, Chamtwaria*	*Hipposideros, Megaderma, Scotophilisis, Rhinolophus*			
Oligocene		*Philisis*		*Dhofarella, Hipposideros, Chibanycteris, Philisis*	
Eocene	Late Eocene — *Tanzanycteris*	*Dhofarella, Saharaderma, Khonsunycteris, Qarunycteris, Witwatia, Philisis, Vampyravus Dizzya*			

FIGURE 30.1 Stratigraphic distribution of Cenozoic chiropteran genera in East, North, and South Africa, the Arabian Peninsula, and Madagascar. Relative position of taxon name does not necessarily indicate precise placement within any epoch.

EIDOLON DUPREANUM (Pollen, 1866, in Schlegel and Pollen, 1866)

Age and Occurrence Subfossil, late Pleistocene-Holocene, ≤ 10,000 BP; Old SE Locality, Anjohibe Cave, Madagascar.

Remarks The presence of Malagasy subfossil *Eidolon dupreanum* is documented by several isolated dental and postcranial remains described by Samonds (2007).

Genus *ROUSETTUS* Gray, 1821
ROUSETTUS MADAGASCARIENSIS G. Grandidier, 1928

Age and Occurrence Subfossil, late Pleistocene, exact age uncertain; SS2 Locality, Anjohibe Cave, Madagascar.

Remarks *Rousettus madagascariensis* is represented by a single distal right humerus that was described by Samonds (2007).

ROUSETTUS cf. *R. MADAGASCARIENSIS*

Age and Occurrence Subfossil, late Pleistocene, 69,600–86,800 BP; NCC-1 Locality, Anjohibe Cave, Madagascar.

Remarks *Rousettus* cf. *R. madagascariensis* is represented by a single fragmentary right m3 that was described by Samonds (2007).

PTEROPODID sp.

Age and Occurrence Early Miocene, Burdigalian (19 Ma) and early Pleistocene (1.6 Ma); Songhor, Chamtwara, and Area 130A, Okote Member, Koobi Fora Formation, Kenya.

Remarks Six partial humeri from Chamtwara and two from Songhor represent pteropodids but cannot be assigned to any particular genus (Butler, 1984). A single lower first molar of an indeterminate pteropodid from Koobi Fora Area 130A was described by Black and Krishtalka (1986). They noted similarities to extant *Myonycteris* (little collared fruit bat), *Epomops* (epauleted bat), and *Epomophorus* (epauleted fruit bat) but were not able to determine which, if any, of these genera the tooth represents.

Family EMBALLONURIDAE Gervais, 1855
Genus *DHOFARELLA* Sigé et al., 1994
DHOFARELLA THALERI Sigé et al., 1994

Diagnosis Sigé et al. (1994). Possesses upper molars with moderately notched labial border, parastyle projecting mesially, large and distally projecting hypocone shelf (talon), protocone lacking a postprotocrista, lower molars with relatively straight cristid obliqua, strong hypoconid and entoconid, and high and sharply defined entocristid.

Description The sample of *Dhofarella thaleri* from Taqah is small, being composed only of five specimens, two partial lower jaw fragments, and three isolated teeth (Sigé et al., 1994).

Age and Occurrence Early Oligocene, Rupelian (31.5 Ma); Taqah, Oman.

Remarks Among living African emballonurids, *Dhofarella* is most similar to *Coleura afra* (African sheath-tailed bat) but differs in having even shorter and broader lower molars and straight cristid obliquae that produce shallow hypoflexids.

DHOFARELLA SIGEI Gunnell, Simons and Seiffert, 2008

Diagnosis Differs from *Dhofarella thaleri* in being 30% smaller in comparable tooth dimensions, and in having a relatively shorter m1 trigonid and talonid.

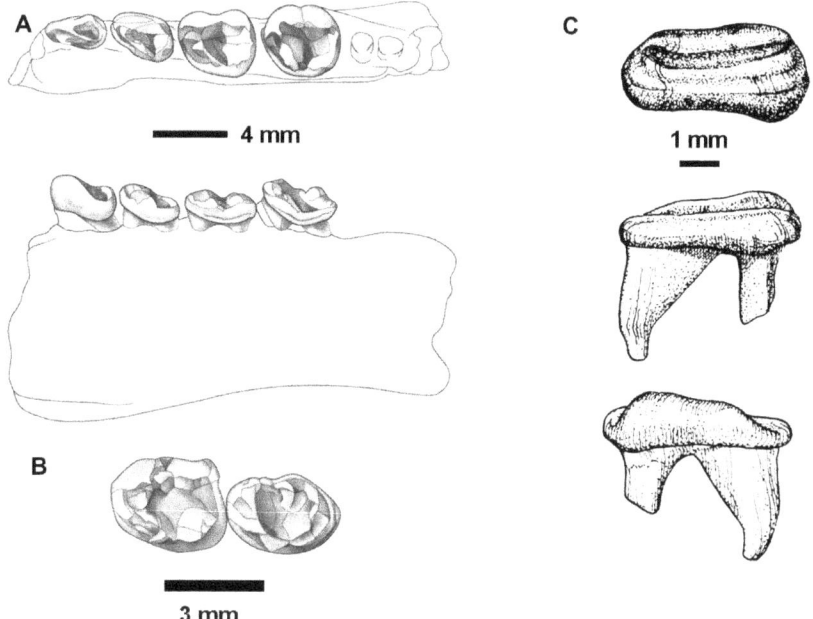

FIGURE 30.2   African Pteropodidae. A) Holotype (KNM-SO 508) of *Propotto leakeyi*, right dentary with p3–m2 in occlusal (top) and medial (bottom) views. B) Referred specimen (KNM-RU 2084) of *Propotto leakeyi*, left dentary with m2–3. C) Right M2 of *Eidolon* cf. *E. helvum* from Omo Locality 1 in occlusal (top), lingual (middle), and labial (bottom) views. A and B adapted from plate 1, figures 6 and 8 in Simpson (1967); C adapted from figure 14 in Wesselman (1984). A, B reproduced by permission of the Museum of Comparative Zoology, Harvard University. C reproduced by permission of S. Karger AG, Basel.

Description Like the sample from Taqah, *Dhofarella* is poorly known from the Fayum as well being represented only by the holotype specimen (CGM 83670), a left dentary m1–3.

Age and Occurrence Latest Eocene, Priabonian (34 Ma); Fayum Quarry L-41, Egypt

Remarks *Dhofarella sigei* establishes the presence of emballonurids in Africa by the late Eocene, indicating that the family has had a relatively long presence on the continent.

Genus *SACCOLAIMUS* Temminck, 1838
SACCOLAIMUS INCOGNITA Butler and Hopwood, 1957

Diagnosis Butler and Hopwood (1957). Similar in size to *Saccolaimus nudiventris* (= *Taphozous nudiventris*) but differs in having a relatively larger P2, less divergent frontal crests, and a zygomatic root originating opposite M2.

Description The holotype (BMNH M14222) of *Saccolaimus incognita* is a left maxillary fragment preserving portions of P4 and M2 and alveoli for C1–P2, M1, and the anterior roots of M3. Judging from the alveolus, P2 was less reduced than in extant species of *Saccolaimus* and M3 was relatively reduced unlike in most other African emballonurid genera. The hard palate terminates at M2 unlike in *Emballonura* and *Coleura* where it extends to the posterior margin of M3. There is a well-developed postorbital process (broken) present, a characteristic typical of emballonurids.

Age and Occurrence Early Miocene, Legetet Formation (20 Ma and 17.5 Ma), Burdigalian; "Maize Crib," Koru, Kenya, and Moroto II, Uganda.

Remarks Extant *Saccolaimus* has often been considered a subgenus of *Taphozous* and the species *nudiventris* is now placed in *Taphozous*. However, the characters of *S. incognita* described by Butler and Hopwood (1957) support placing this species in *Saccolaimus* as they originally suggested.

An additional specimen of *S. incognita* (KNM-LG 1514), consisting of a palate with partial left and right dentitions preserved, was described by Butler (1984) as *Taphozous incognita*. Pickford and Mein (2006) describe a proximal humerus from Moroto II that they assign to *Taphozous incognita*.

SACCOLAIMUS ABITUS (Wesselman, 1984)
Figure 30.3A

Diagnosis Relatively large species, intermediate in size between *Saccolaimus peli* and *Taphozous nudiventris*, but differing from these extant species in having a more mesiodistally compressed tooth row, an anteriorly deep dentary, and relatively shorter and broader p2–m1; characteristics shared with *S. peli* include lower molars with heavy labial cingulum, cristid obliqua joining postvallid labially, and lacking a high entocristid.

Description The holotype (L1–375) is a right dentary with p2–m1 from Omo Locality 1. There is an additional isolated left m2 from Omo Locality 28.

Age and Occurrence Late Pliocene (2.1–2.95 Ma); Omo Locality 1, Member B, and Omo Locality 28, member F, Shungura Formation, Ethiopia.

Remarks Wesselman (1984) included *Saccolaimus* as a subgenus of *Taphozous* and placed *T. abitus* in that subgenus. *Saccolaimus* is here recognized as a valid genus distinct from *Taphozous* (following Simmons, 2005) so the species *abitus* is recognized as *Saccolaimus abitus*.

Genus *TAPHOZOUS* E. Geoffroy, 1818
TAPHOZOUS sp.

Age and Occurrence Plio-Pleistocene (1.8–2.3 Ma); Sterkfontein, Cave Breccia, South Africa.

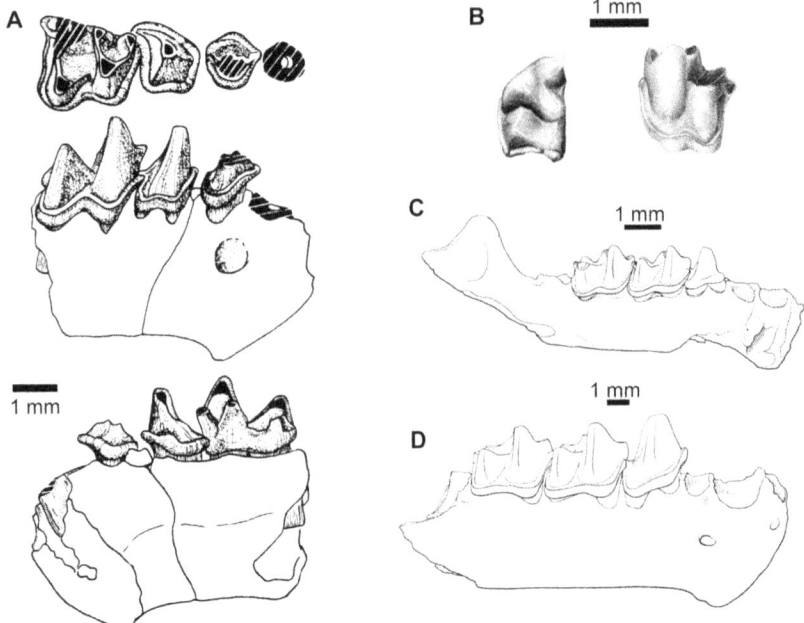

FIGURE 30.3 African Plio-Pleistocene echolocating bats. A) Holotype (L1-375) of *Saccolaimus abitus*, right dentary with p2–m1 in occlusal (top), lateral (middle), and medial (bottom) views. B) Left m1 (KNM-ER 5955) of *Scotophilus* sp. in occlusal (left) and labial (right) views. C) Right dentary (TSM 93) with p4–m2 of *Triaenops furculus* in lateral view. D) Right dentary (TSM 29) with p4–m2 of *Hipposideros commersoni* in lateral view. A adapted from figures 15A–15C in Wesselman (1984); B adapted from figures 4A–4B in Black and Krishtalka (1986); C and D adapted from figures 2 and 3 in Sabatier and Legendre (1985). A reproduced by permission of S. Karger AG, Basel. B reproduced by permission of The Natural History Museum of Los Angeles County.

Remarks *Taphozous* sp. was included in a faunal list from Sterkfontein (Pocock, 1987) but was not mentioned in the text of that paper. *Taphozous* sp. is also known from the late Pliocene (2.5 ma) at Ahl al Oughlam in Morocco (Eiting et al., 2006).

Genus *COLEURA* PETERS, 1867
COLEURA MUTHOKAI Wesselman, 1984

Diagnosis Similar to extant *Coleura afra* (African sheath-tailed bat) but differs in being somewhat smaller, in having labiolingually narrower molars, and in having a somewhat lower entocristid.

Description The holotype (L28–211) and only specimen is a left m1. This tooth has a strong labial cingulid, is nyctalodont, and has a relatively basined talonid that is somewhat broader than the trigonid.

Age and Occurrence Late Pliocene (2.1 Ma); Omo Locality 28, Member F, Shungura Formation, Ethiopia.

Genus *EMBALLONURA* Temminck, 1838
EMBALLONURA ATRATA Peters, 1874

Age and Occurrence Pleistocene? Lac Tsimanampetsotsa, Karst Breccia, Madagascar.

Remarks The presence of extant *Emballonura atrata* (Old World sheath-tailed bat) in the Lac Tsimanampetsotsa faunal sample was noted by Sabatier and Legendre (1985). These authors suggested that the Lac Tsimanampetsotsa deposits were at least Pleistocene in age if not older. Recently (Samonds, 2006; pers. comm.) has questioned this age assessment and has suggested that this represents a subfossil Holocene assemblage instead. *Emballonura atrata* is only found on Madagascar today.

EMBALLONURID gen. et sp. indet.

Age and Occurrence Early Miocene, Burdigalian (18–20 Ma); Rusinga Locality R3, Rusinga Island, Songhor, and Locality 10, Legetet Formation, Kenya.

Remarks Several isolated humerus fragments have been assigned to indeterminate emballonurids by Butler (1969, 1984). These include KNM-SO 5523 and BMNH M 34152, proximal and distal humeri, respectively, from Songhor, and two distal humeri, KNM-LG 1507 from Legetet Locality 10 and an unnumbered specimen in the Kenya National Museum collection from Rusinga locality R3.

Family HIPPOSIDERIDAE Lydekker, 1891
Genus *HIPPOSIDEROS* Gray, 1831
HIPPOSIDEROS (BRACHIPPOSIDEROS) OMANI
Sigé et al., 1994

Diagnosis Teeth smaller than in any other known species of *Hipposideros*.

Description H. (B.) omani is represented by a few isolated teeth from Taqah. The holotype (TQ 72, left M1) measures only 0.9 × 1.15 mm (Sigé et al., 1994).

Age and Occurrence Early Oligocene, Rupelian (31.5 Ma); Taqah, Oman.

HIPPOSIDEROS KAUMBULUI Wesselman, 1984

Diagnosis Very similar to extant *Hipposideros commersoni* (Old World leaf-nosed bats) but differs in having slightly smaller tooth dimensions, an m1 with a broader talonid, a much more massive hypoconid, a shallower and more lingually closed trigonid fovea, an m3 with a lower and less distinct entocristid, and an M3 with a strong distal cingulum.

Description The holotype (L28–167) is an m1. There are two other teeth, an M3 and a fragmentary m3, that also represent this species from the type locality.

Age and Occurrence Late Pliocene (2.1 Ma); Omo Locality 28, Member F, Shungura Formation, Ethiopia.

HIPPOSIDEROS CYCLOPS (Temminck, 1853)

Age and Occurrence Late Pliocene (2.95 Ma); Omo Locality 1, Member B, Shungura Formation, Ethiopia.

Remarks This extant species is represented at Omo Locality 1 by two isolated teeth (Wesselman, 1984).

HIPPOSIDEROS COMMERSONI E. Geoffroy, 1813
Figure 30.3D

Age and Occurrence Middle to late Pleistocene (266,000–69,600 BP) and Pleistocene? Twin Rivers Cave, Zambia; NCC-1 Locality, Anjohibe Cave, and Lac Tsimanampetsotsa, Karst Breccia, Madagascar.

Remarks The presence of extant *Hipposideros commersoni* is documented from the middle Pleistocene Twin Rivers Cave in Zambia (Avery, 2003), the late Pleistocene NCC-1 Locality in Anjohibe Cave (Samonds, 2007), and in the Lac Tsimanampetsotsa faunal sample (Sabatier and Legendre, 1985). *Hipposideros commersoni* is known from much of sub-Saharan African and Madagascar today (Nowak, 1994).

HIPPOSIDEROS cf. H. COMMERSONI

Age and Occurrence Late Pleistocene to Holocene, exact age uncertain; Old SE Locality (≤10,000 BP) and SS2 Locality (age uncertain), Anjohibe Cave, Madagascar.

Remarks The presence of *Hipposideros* cf. *H. commersoni* is documented by isolated teeth from these two Anjohibe Cave localities (Samonds, 2007).

HIPPOSIDEROS BESAOKA Samonds, 2007

Diagnosis Samonds (2007). Differs from *H. commersoni* in having larger and more robust teeth and in having upper molars that are broader relative to their length.

Description A feature shared between *H. besaoka* and extant *H. commersoni* is a great deal of variability in the depth and thickness of the mandibular corpus. *Hipposideros besaoka* has the typical *Hipposideros* dental formula of 1/2, 1/1, 2/2, 3/3, and it also shares a laterally shifted and reduced P2 and a relatively large p2 with other species of *Hipposideros*.

Age and Occurrence Late Pleistocene to Holocene (≤10,000 BP); TW-10 Locality, Anjohibe Cave, Madagascar.

Remarks The hypodigm of *H. besaoka* is easily the largest of any fossil hipposiderine yet described from Africa. In addition to the holotype maxilla (UA 9478), there are over 600 referred specimens ranging from relatively complete dentaries to isolated teeth (Samonds, 2007).

HIPPOSIDEROS (SYNDESMOTIS) VETUS (Lavocat, 1961)

Diagnosis Lavocat (1961); Legendre (1982). Larger than extant *H. (S.) megalotis*; angular process long and laterally deflected, coronoid process low and relatively short; C1 with well-developed secondary cusp; P2 absent; M3 reduced with elongate mesostyle but lacking metacone and metastyle.

Description Three specimens were assigned to this species by Lavocat (1961). The holotype (BML 1124) is a maxilla with P4–M3, while the other two specimens are an upper canine (BML 1197) and a lower jaw with m1–3 (BML 779). Legendre (1982) lists three other referred specimens including two lower jaws (BNM 1 and 2) and an upper canine (BNM 3).

Age and Occurrence Early late Miocene, Tortonian; Beni Mellal, Morocco.

Remarks Lavocat (1961) noted some resemblances of his new species to *Hipposideros caffer* but chose to place the new species questionably within *Asellia* pending discovery of more complete specimens. Sigé (1976) retained these specimens in *Asellia* but Legendre (1982) moved them to *Hipposideros* in the subgenus *Syndesmotis*.

There is also a late Miocene (10–11 Ma) record of *Hipposideros (Syndesmotis)* sp. from Sheikh Abdallah, in the Western Desert of Egypt (Wanas et al., 2009).

HIPPOSIDEROS sp.

Remarks *Hipposideros* sp. has been reported from Kanapoi (late Pliocene) in Kenya (Winkler, 1998, 2003)(two species), Lac Tsimanampetsotsa (?Pleistocene) in Madagascar and questionably from Songhor (early Miocene) in Kenya (Butler, 1969, 1984; Sabatier and Legendre, 1985).

Genus TRIAENOPS Dobson, 1871
TRIAENOPS GOODMANI Samonds, 2007

Diagnosis Samonds (2007). Larger than any other known species of *Triaenops*; molar crowns narrow with labially rounded protoconid and hypoconid, preentocristid weak to absent, talonid wider than trigonid, and m2 with small hypoflexid shelf.

Description *Triaenops goodmani* is only represented by three partial lower dentitions, two of which contain worn m2–3 only. The holotype (UA 9010) preserves relatively unworn p4–m2 and provides most of the morphological information pertaining to this species. The p4 is tall and simple; the molars have labial cusps taller than lingual cusps, talonids slightly lower than trigonids, and are nyctalodont.

Age and Occurrence Late Pleistocene to Holocene (≤ 10,000 BP); Old SE Locality, Anjohibe Cave, Madagascar.

Remarks Samonds (2007) described *T. goodmani* as being the largest known species of *Triaenops*, but her measurements of m2 seem to contradict this statement. *T. goodmani* is reported to have an average m2 length and width of 1.56 × 1.00 (see Samonds, 2007: table 3). Comparable measurement means of *T. auritus* and *T. furculus* for m2 length and width are 2.14 × 1.99 and 2.52 × 2.13 mm, respectively (note that these measurements apparently were reversed in Samonds, 2007: table 3), while the same measurements for *T. rufus* are 1.45 × 0.95. This suggests that *T. goodmani* at least in m2 size is one of the smaller species of *Triaenops*.

TRIAENOPS FURCULUS Trouessart, 1906
Figure 30.3C

Age and Occurrence Pleistocene? Lac Tsimanampetsotsa, Karst Breccia, Madagascar.

Remarks The presence of extant *Triaenops furculus* (triple nose-leaf bats) in the Lac Tsimanampetsotsa faunal sample was recorded by Sabatier and Legendre (1985). Today *Triaenops furculus* is restricted to Madagascar and certain islands in the Seychelles.

TRIAENOPS sp.

Age and Occurrence Late Pleistocene (69,600–86,800 BP) and late Pleistocene-Holocene (≤10,000 BP); NCC-1 Locality and Old SE Locality, Anjohibe Cave, Madagascar.

Remarks The presence of Malagasy subfossil *Triaenops* sp. at Anjohibe was noted by Samonds (2006, 2007).

HIPPOSIDERINE Gen. et sp. indet.

Remarks Ten teeth of an indeterminate hipposiderine were described from the late early Oligocene locality of Taqah (31.5 Ma) in Oman by Sigé et al. (1994). Three specimens of an indeterminate hipposiderine are also present in the late Pliocene (2.5 Ma) Ahl al Oughlam sample from Morocco (Eiting et al., 2006). Butler (1984) noted the presence of humeral fragments of Miocene hipposiderines, a proximal humerus from Chamtwara and three distal humeri from Songhor.

Family MYZOPODIDAE Thomas, 1904
Genus *MYZOPODA* Milne-Edwards and Grandidier, 1878
MYZOPODA sp.

Age and Occurrence Early Pleistocene (1.8–1.9 Ma); Olduvai Bed I, Olduvai Gorge, Tanzania.

Remarks Butler (1978) mentions the presence of a humerus from Olduvai that closely resembles that of extant *Myzopoda aurita* (Old World sucker-footed bat) except for its larger size. *Myzopoda aurita*, the sole extant member of the family, is today only known from Madagascar.

Family MEGADERMATIDAE H. Allen, 1864
Genus *SAHARADERMA* Gunnell et al., 2008
SAHARADERMA PSEUDOVAMPYRUS Gunnell et al., 2008

Diagnosis Possesses m1 longer than m2, distinct paraconids, metaconids, and hypoconulids on lower molars (especially m1), less bilaterally compressed molar talonids (especially m3), relatively long and narrow p4, and a relatively shallow mandibular horizontal ramus.

Description The holotype (CGM 83672) and only known specimen is represented by a right dentary with p4–m3.

Age and Occurrence Latest Eocene, Priabonian (34 Ma); Fayum Quarry L-41, Egypt.

Remarks The recognition of *Saharaderma pseudovampyrus* as a megadermatid in the late Eocene of North Africa is an important record of this Old World family because it extends the temporal range of megadermatids by at least 16 million years—the next oldest sample is an enigmatic record from Rusinga Island (Butler and Hopwood, 1957).

Genus *CARDIODERMA* Peters, 1873
CARDIODERMA SP.

Age and Occurrence Early Pleistocene (1.8–1.9 Ma); Olduvai Bed I, Olduvai Gorge, Tanzania.

Remarks Butler and Greenwood (1965) mentioned the presence of a mandible and maxilla of an extinct megadermatid from Olduvai Bed I. Butler (1978) later suggested that these specimens represented a species similar to the extant *Cardioderma cor* (African false vampire bat).

Genus *MEGADERMA* E. Geoffroy, 1810
MEGADERMA GAILLARDI (Trouessart, 1898)

Description Seven teeth from Beni Mellal were assigned to this taxon by Sigé (1976). These include three teeth—BM 1200, a relatively large m3 (3.5 × 3.0 mm), BM 1201, an M3 (2.0 × 4.2 mm), and BM 1202, a p2—that were originally placed in a new genus and species, *Afropterus gigas*, by Lavocat (1961).

Age and Occurrence Early late Miocene, Tortonian; Beni Mellal, Morocco.

Remarks When Lavocat (1961) described *Afropterus gigas*, he compared it with the New World phyllostomid *Vampyrum* (*Vampiris* in Lavocat, 1961), although he declined to place it in a specific family, indicating that he didn't believe it was a phyllostomid. Russell and Sigé (1970) suggested that *A. gigas* was a megadermatid, a position supported by Butler (1978). Sigé (1976) felt that *Afropterus gigas* was not distinct enough from *Megaderma gaillardi* to warrant specific separation and synonymized the African genus and species, a view that is accepted here.

MEGADERMA JAEGERI Sigé, 1976

Diagnosis Relatively small size; upper canine with well differentiated anterior cuspule; m1 with open trigonid that is basally more narrow than the talonid.

Description The hypodigm consists of the holotype (BML Br 522, a left m1) and four other isolated teeth (an upper canine, another m1, and two M1's).

Age and Occurrence Early late Miocene, Tortonian; Beni Mellal, Morocco.

Remarks One of the specimens referred to this species by Sigé (1976), a left M1 (BML La 1178), was figured by Lavocat (1961) and referred to family indeterminate. One other referred specimen was noted in the Sigé (1976) hypodigm as BML (La) 3000 (right M1) but in his figure 8 (p. 78) was labeled as BML (La) 300 (also a right M1) instead.

MEGADERMA sp.

Age and Occurrence Late Miocene (10–11 Ma); Sheikh Abdallah, Western Desert, Egypt.

Remarks *Megaderma* sp. occurs in western Egypt in the late Miocene (Wanas et al., 2009).

"MEGADERMIDAE" Indet.

Age and Occurrence Early Miocene, Burdigalian (18 Ma); Kasawanga, Rusinga Island, Kenya.

Remarks A single specimen (BMNH M 34141, represented by a left dentary with m1–3, was assigned to "Megadermidae" genus and species indeterminate by Butler and Hopwood (1957). Butler (1984) noted that this specimen was from Kasawanga, not Rusinga Locality R106 as originally stated.

MEGADERMATIDAE gen. et sp. nov.

Age and Occurrence Late Pliocene (2.7–3.1 Ma); Makapansgat, Cave Breccia, South Africa.

Remarks Megadermatidae, genus nov. was included in a faunal list from Makapansgat (Pocock, 1987) but was not mentioned in the text of that paper.

Family MOLOSSIDAE Gervais, in de Castelnau, 1855
Genus *TADARIDA* Rafinesque, 1814
TADARIDA RUSINGAE Arroyo-Cabrales et al., 2002

Diagnosis Largest species of genus, sagittal crest present near lambdoid crest, palate deeply domed, M2 postprotocrista strong, M1–2 with hypocone developed and connected to postprotocrista by a small crest.

Description The holotype (KNM-RU 14357) and only specimen of *T. rusingae* is a nearly complete skull (length = 30.49 mm).

Age and Occurrence Early Miocene, Burdigalian (17.5–18 Ma); Rusinga Locality R106, Hiwegi Formation, Rusinga Island, Kenya.

Remarks The skull of *T. rusingae* represents one of only two African fossil bats known by relatively complete skulls (the other being *Tanzanycteris*).

Genus *MORMOPTERUS* Peters, 1865
MORMOPTERUS JUGULARIS Peters, 1865

Age and Occurrence Pleistocene? Lac Tsimanampetsotsa, Karst Breccia, Madagascar.

Remarks The presence of extant *Mormopterus jugularis* (Little Goblin bat) in the Lac Tsimanampetsotsa faunal sample was noted by Sabatier and Legendre (1985). *Mormopterus jugularis* is today restricted to Madagascar.

MOLOSSIDAE indet.

Remarks Indeterminate molossids have been recorded from Beni Mellal in Morocco (Lavocat, 1961), from Olduvai Bed I in Tanzania (Butler, 1978), and from Napak, Uganda (Butler, 1984).

Family NYCTERIDAE Van der Hoeven, 1855
Genus *CHIBANYCTERIS* Sigé et al., 1994
CHIBANYCTERIS HERBERTI Sigé et al., 1994

Diagnosis Sigé et al. (1994). Upper molar protocone not clearly separated from bases of labial cusps, M2 with protofossa closed off by protocristae, M2 talon relatively small, p2 simple with little or no differentiation of the trigonid.

Description The holotype (TQ 77, right M2) of *Chibanycteris herberti* is of relatively small size (2.4 × 2.7 mm). Five other teeth (four p2's and one m3) were also assigned to this species by Sigé et al. (1994).

Age and Occurrence Early Oligocene, Rupelian (31.5 Ma); Taqah, Oman.

Remarks This is the earliest record of a nycterid known from Africa. The family Nycteridae (slit-faced or hollow-faced bats) is represented by a single genus (*Nycteris*) today, and of the 13 living species, 10 are restricted to Africa (Nowak, 1994). Simmons (2005:391) discusses the nomenclature of the family name, which is often (and perhaps more technically correctly) spelled Nycterididae.

NYCTERIS sp.

Age and Occurrence Late Pliocene (2.7–3.1 Ma); Makapansgat, Cave Breccia, South Africa.

Remarks *Nycteris* also has been tentatively identified from Area 130A, early Pleistocene (1.6 Ma), Okote Member, Koobi Fora Formation, Kenya (Black and Krishtalka, 1986). Butler

(1978), citing a personal communication from Bryan Patterson, stated that there are some undescribed specimens of a nycterid from Kanapoi (late Pliocene, 2.5 Ma, Kenya). It is not clear if these specimens are among those later referred by Winkler (1998, 2003) to *Hipposideros* sp.

NYCTERIDAE indet.

Age and Occurrence Early Miocene, Burdigalian (19 Ma); Chamtwara, Kenya.

Remarks A single distal humerus from Chamtwara (KNM. CA 2198) was described by Butler (1984). He noted that it was very similar to extant *Nycteris thebaica*.

Family RHINOPOMATIDAE Bonaparte, 1838
Genus *QARUNYCTERIS* Gunnell et al., 2008
QARUNYCTERIS MOERISAE Gunnell et al., 2008

Diagnosis Differs from *Rhinopoma* in being larger, in having paracone and metacone of nearly equal height, a less well-developed parastyle, and a relatively smaller and less labially extended mesostyle.

Description The holotype (CGM 83671) and only known specimen is a right M2 from Quarry BQ-2.

Age and Occurrence Latest Eocene, Priabonian (37 Ma); Fayum Quarry BQ-2, Egypt.

Remarks Despite being only a single tooth, *Qarunycteris* is an extremely important record because it is the only African fossil rhinopomatid known.

Family VESPERTILIONIDAE Gray, 1821
Genus *KHONSUNYCTERIS* Gunnell,
Simons and Seiffert, 2008
KHONSUNYCTERIS AEGYPTICUS Gunnell,
Simons and Seiffert, 2008

Diagnosis Possesses relatively large and robust c1 with relatively short posterior shelf and lacking anterior cusplet, p2 larger than p3 which is much smaller than p4, p3 double rooted, p4 relatively elongate with an anterior cusplet, molars relatively short and broad, robust, tall and straight ascending ramus (not angled anteriorly).

Description The holotype (CGM 83673) and only known specimen is a left dentary c1–m2.

Age and Occurrence Latest Eocene, Priabonian (34 Ma); Fayum Quarry L-41, Egypt.

Remarks Khonsunycteris is by far the oldest vespertilionid known from Africa, predating the next oldest species by 15 million years.

Genus *CHAMTWARIA* Butler, 1984
CHAMTWARIA PICKFORDI Butler, 1984

Diagnosis Possesses three upper premolars with P3 smaller than P2, but P3 not as reduced as in *Myotis*; differs from *Kerivoula* in having a longer, narrower face, nasals extending above canines, infraorbital foramen close to tooth row, a sharp infraorbital crest, and a lower P4 paracone.

Description The holotype (KNM-CA 2237) is a rostrum with left C1–M1.

Age and Occurrence Early Miocene (19.5 Ma), Burdigalian; Chamtwara, Kenya.

Remarks Among fossil vespertilionids, *Chamtwaria* most closely resembles *Stehlinia* from Europe. It is intermediate in premolar characters between *Stehlinia* and extant *Kerivoula* and is grouped with the latter genus as the only two members of the subfamily Kerivoulinae by McKenna and Bell (1997).

CHAMTWARIA? sp.

Age and Occurrence Early Miocene (17.5 Ma), Burdigalian; Moroto II, Uganda.

Remarks Pickford and Mein (2006) describe a left upper canine from Moroto II that they questionably assign to *Chamtwaria*.

Genus *SCOTOPHILUS* Leach, 1821
SCOTOPHILUS SP.
Figure 30.3B

Age and Occurrence Early Pleistocene (1.6 Ma); Area 130A, Okote Member, Koobi I Formation, Kenya.

Remarks Black and Krishtalka (1986) describe an upper and a lower molar from Koobi Fora as *Scotophilus* sp.

Genus *MYOTIS* Kaup, 1829
MYOTIS TRICOLOR (Temminck, 1832)

Age and Occurrence Pleistocene (1.3–1.9 Ma); Swartkrans, Cave Breccia, South Africa.

Remarks *Myotis* cf. *M. tricolor* is also known from the Plio-Pleistocene (1.7–2.0 Ma), Kromdraai B Cave Breccia in South Africa (Pocock, 1987).

MYOTIS GOUDOTI (A. Smith, 1834)

Age and Occurrence Late Pleistocene, exact age uncertain; SS2 Locality, Anjohibe Cave, Madagascar.

Remarks *Myotis goudoti* from Anjohibe cave is represented by a single lower canine that can be identified by its small size, single, sharp cusp, and uniquely spade-shaped lingual surface (Samonds, 2007).

MYOTIS cf. *M. WELWITSCHII* (Gray, 1866)

Age and Occurrence Plio-Pleistocene (1.7–2.0 Ma); Kromdraai B Cave Breccia in South Africa (Pocock, 1987).

Remarks *Myotis* cf. *M. welwitschii* was included in a faunal list from Kromdraai B by Pocock (1987). No mention was made of this taxon in the text of that paper.

MYOTIS sp.

Occurrence *Myotis* species indet. is known from Ahl al Oughlam, Morocco (Eiting et al., 2006); Bolt's Farm, South Africa (Broom, 1948); and Olduvai Bed I, Tanzania (Butler, 1978).

Genus *EPTESICUS* Refinesque, 1820
EPTESICUS cf. *E. BOTTAE* (Peters, 1896)

Age and Occurrence Late Pliocene (2.7–3.1 Ma); Makapansgat Cave Breccia in South Africa (Pocock, 1987).

EPTESICUS cf. *E. HOTTENTOTUS* (A. Smith, 1833)

Age and Occurrence Plio-Pleistocene (1.7–3.1 Ma); Makapansgat and Kromdraai B Cave Breccias in South Africa (Pocock, 1987) and Olduvai Bed I, Tanzania (Butler, 1978).

EPTESICUS sp.

Age and Occurrence Early Pliocene (5.0 Ma) and Pleistocene (1.3–1.9 Ma); Langebaanweg, Varswater Formation, South Africa (Hendey, 1981); and Swartkrans Cave Breccia, South Africa (Avery, 1998).

Genus *MINIOPTERUS* Bonaparte, 1837
MINIOPTERUS SCHREIBERSI (Kuhl, 1817)

Age and Occurrence Middle Pleistocene (170–266 Ka); Twin Rivers Cave, Zambia (Avery, 2003).

Remarks *Miniopterus schreibersi* (Schreiber's long-fingered bat) is widespread across much of the Old World (Simmons, 2005).

MINIOPTERUS cf. *M. SCHREIBERSI* (Kuhl, 1817)

Age and Occurrence Plio-Pleistocene (1.8–2.3 Ma); Sterkfontein Cave Breccia in South Africa (Pocock, 1987).

Remarks *Miniopterus* cf. *M. schreibersi* has also been recorded from Olduvai Bed I, Tanzania (Butler, 1978).

MINIOPTERUS sp.

Age and Occurrence Late Pliocene (2.5–3.1 Ma); Makapansgat Cave Breccia in South Africa (Pocock, 1987).

Remarks A new species of *Miniopterus* is present in the late Pliocene (2.5 Ma) of Ahl al Oughlam, Morocco (Eiting et al., 2006).

Genus *NYCTICEIUS* Rafinesque, 1819
Cf. *NYCTICEIUS (SCOTEINUS) SCHLIEFFENI* (Peters, 1859)

Age and Occurrence Early Pleistocene (1.8–1.9 Ma); Olduvai Bed I, Tanzania (Butler and Greenwood, 1965).

Genus *PIPISTRELLUS* Kaup, 1829
Cf. *PIPISTRELLUS (SCOTOZOUS) RUEPPELLI* (Fischer, 1859)

Age and Occurrence Early Pleistocene (1.8–1.9 Ma); Olduvai Bed I, Tanzania (Butler and Greenwood, 1965).

VESPERTILIONIDAE indet.

Remarks An indeterminate vespertilionid has been noted from the Mio-Pliocene, Beni Mellal, Morocco (Lavocat, 1961). Indeterminate partial humeri of vespertilionids have been described from Chamtwara and Songhor by Butler (1984).

Family PHILISIDAE Sigé, 1985
Genus *WITWATIA* Gunnell et al., 2008
WITWATIA SCHLOSSERI Gunnell et al., 2008

Diagnosis Larger than other philisids, upper molars with robust and distinct parastyles, relatively tall protocones higher and more steeply descending postprotocristae, p2 and p4 robust, p4 with distinct cuspule developed on posterolingual margin, lower molars with robust anterior cingulids, c1 lacks complete cingulid and lacks descending groove along posterior flank.

Description The holotype of *W. schlosseri* (CGM 83668) is a left dentary with c1–m3. The hypodigm consists of 13 other specimens, all from Quarry BQ-2.

Age and Occurrence Latest Eocene, Priabonian (37 Ma); Fayum Quarry BQ-2, Egypt.

Remarks Witwatia schlosseri is among the largest fossil bats known, similar in size to *Afropterus gigas* (Lavocat, 1961). Both of these fossil taxa rival extant *Macroderma gigas* (Australian ghost bat), one of the largest known microbats.

WITWATIA EREMICUS Gunnell et al., 2008

Diagnosis Differs from *W. schlosseri* in being smaller; similar in size to *Philisis sphingis* but differs in having upper molars with more parallel postpara- and premetacristae resulting in notched mesostyle, ectoloph narrow and extending to labial margin, preprotocrista terminates at base of paracone and does not extend anteriorly to the parastyle, trigon basin narrower and more steeply sloping, talon more distinctly defined by postprotocingulum that is distinct and separated from the postprotocrista; lower molars have more lingually angled cristid obliqua and deeper ectoflexid as in *W. schlosseri*, paraconid lower, less distinct and more appressed to base of protoconid, trigonid generally more anteroposteriorly compressed, anterior cingulid more robust, and hypoconulid relatively smaller.

Description Witwatia eremicus is represented by upper and lower molars and some canines. Like all philisids, *W. eremicus* has myotodont molars.

Age and Occurrence Latest Eocene, Priabonian (37 Ma); Fayum Quarry BQ-2, Egypt.

Genus *PHILISIS* Sigé, 1985
PHILISIS SPHINGIS Sigé, 1985
Figures 30.4C and 30.4D

Diagnosis Size intermediate, infraorbital canal short, P4 with relatively large talon and small, low protocone, upper molars without hypocone or talon development, ectoloph long and narrow extending to mesostyle that may be partially open (doubled), preprotocrista extends to parastyle, postprotocrista terminates before reaching metacone, p4 relatively simple, no para- or metaconid developed, lower molars myotodont, cristid obliquae only weakly angled lingually and flexed, hypoflexid relatively shallow, hypoconulid large.

Description The holotype (YPM 34488) of *P. sphingis* is a right maxilla P4–M3. There is also a referred lower jaw p4–m2 from the type locality. Gunnell et al. (2008) note the presence of two other specimens from Quarry I as well.

Age and Occurrence Early Oligocene, Rupellian (30 Ma); Fayum Quarry I, Egypt.

PHILISIS SEVKETI Sigé et al., 1994

Diagnosis Differs from other species of *Philisis* in being smaller.

Description The holotype is a left m1 (TQ 51). In addition there are seven other fragmentary specimens from Taqah that constitute the hypodigm of *P. sevketi*.

Age and Occurrence Early Oligocene, Rupelian (31.5 Ma); Taqah, Oman.

CF. *PHILISIS* sp.

Description Two broken upper molars were assigned to cf. *Philisis* sp. by Sigé et al. (1994).

Age and Occurrence Early Oligocene, Rupelian (31.5 Ma); Taqah, Oman.

Remarks An additional specimen from an unknown level in the Jebel Qatrani Formation was referred to *Philisis* sp. by Sigé, 1985.

Genus *DIZZYA* Sigé, 1991
DIZZYA EXSULTANS Sigé, 1991
Figures 30.4A and 30.4B

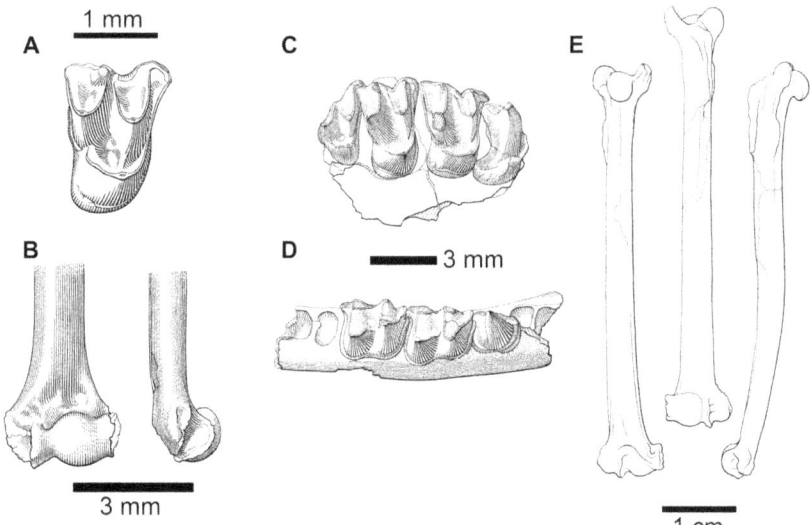

FIGURE 30.4 African Eocene-Oligocene echolocating bats. A) Holotype (CB 1–17) of *Dizzya exsultans*, right M1 or M2 in occlusal view. B) Referred specimen (CB 1–18) of *Dizzya exsultans*, left distal humerus in lateral (left) and anterior (right) views. C) Holotype (YPM 34488) of *Philisis sphingis*, right maxilla with P4–M3 in occlusal view. D) Referred specimen (YPM 34489) of *Philisis sphingis*, right dentary with p4–m2 in occlusal view. E) Holotype (Unnumbered) of *Vampyravus orientalis*, right humerus in medial (left), lateral (middle), and anterior (right) views. A and B adapted from figures 2 and 5 in Sigé (1991), http://www.schweizerbart.de; C–E adapted from figures 3B, 4B, and 5 in Sigé (1985). A, B reproduced by permission of Schweizerbart Publishers.

Diagnosis Sigé (1991). Small size, upper molars with metacone slightly stronger than paracone, mesostyle weakly open, weak para- and metalophs developed; postprotocrista and metacingulum continuous; lower molars nyctalodont with moderately high hypoconids.

Description The holotype of *Dizzya exsultans* is an upper M1 or M2 (CB 1–17). Sigé (1991) also attributed a lower dentary fragment with broken m1–2 (CB 1–15) and a distal humerus fragment (CB 1–18) to *D. exsultans*.

Age and Occurrence Early Eocene, Ypresian (50 Ma); Chambi, Tunisia.

Remarks The presence of both rhinolophoids and vespertilionoids (in the form of the philisid *Dizzya*) at Chambi in the early Eocene led Sigé (1991) to postulate a southern origin for modern bat groups noting that only archaic bats (Eochiroptera) were known at this time in the northern continents.

Genus *SCOTOPHILISIS* Horáček et al., 2006
SCOTOPHILISIS LIBYCUS Horáček et al., 2006

Diagnosis Molars with heavily built protoconid, high lingual crown wall, and shallow protofossid; p3 absent; symphysis robust and oval and not extending beyond p2, mandibular angle indistinct.

Description The holotype (NMPC/OF.JZel./Chi 1) is a right dentary with m1–2.

Age and Occurrence Early middle Miocene, Langhian (16.0 to 14.0 Ma); Jebel Zelten, Libya.

Remarks *Scotophilisis* has been proposed as an intermediate taxon between Oligocene *Philisis* and extant *Scotophilus* (Horáček et al., 2006; see also Wessels et al., 2003).

Family RHINOLOPHIDAE Gray, 1825
Genus *RHINOLOPHUS* Lacépède, 1799
RHINOLOPHUS HILDEBRANDTII Peters, 1878

Age and Occurrence Middle Pleistocene (170,000–266,000 BP); Twin Rivers Cave, Zambia (Avery, 2003).

Remarks *Rhinolophus hildebrandtii* (Hildebrandt's horseshoe bat) is relatively common in southern and central Africa today (Simmons, 2005).

RHINOLOPHUS CF. *R. CAPENSIS* Lichtenstein, 1823

Age and Occurrence Plio-Pleistocene (1.7–3.1 Ma); Kromdraai B Cave Breccia in South Africa (Pocock, 1987).

Remarks *Rhinolophus* cf. *R. capensis* is also known from Makapansgat Cave Breccia, South Africa (De Graaff, 1960).

RHINOLOPHUS CLIVOSUS Cretzschmar, 1828

Age and Occurrence Late Pleistocene (300,000 to 15,540 BP); Hoedjiespunt 1 and Saldanha Bay Yacht Club localities, west coast of South Africa (Matthews et al., 2007).

RHINOLOPHUS CF. *R. CLIVOSUS* Cretzschmar, 1828

Age and Occurrence Plio-Pleistocene (1.8–3.1 Ma); Makapansgat and Sterkfontein Cave Breccias in South Africa (Pocock, 1987).

RHINOLOPHUS CF. *R. DARLINGI* K. Anderson, 1905

Age and Occurrence Plio-Pleistocene (1.8–3.1 Ma); Makapansgat and Sterkfontein Cave Breccias in South Africa (Pocock, 1987).

RHINOLOPHUS FERRUMEQUINUM MELLALI Lavocat, 1961

Diagnosis Lavocat (1961). Possesses c1–m1 dental series longer than *Rhinolophus ferrumequinum*, teeth narrower and less robust, premolars relatively longer, molars relatively shorter, upper molars subrectangular with para-, meta-, and mesostyles aligned instead of being concave at the mesostyle.

Description *Rhinolophus ferrumequinum mellali* is based on two specimens, a holotype dentary (BM 707) with c1–m1, and a referred maxilla (BM 1216) with P4–M2.

Age and Occurrence Early late Miocene, Tortonian; Beni Mellal, Morocco (Lavocat, 1961).

RHINOLOPHUS sp.

Age and Occurrence Late Miocene (10–11 Ma) and Pleistocene (1.3–1.9 Ma); Sheikh Abdallah, Western Desert, Egypt (M. Pickford, pers. comm. Wanas et al., 2009), and Swartkrans Cave Breccia in South Africa (Avery, 1998).

Remarks *Rhinolophus* sp. is present in the Egyptian Western Desert (Wanas et al., 2009), while a new species of *Rhinolophus* is present in the late Pliocene (2.5 ma) of Ahl al Oughlam, Morocco (Eiting et al., 2006).

RHINOLOPHOID sp.

Age and Occurrence Early Eocene (50 Ma); Chambi, Tunisia (Sigé, 1991).

Remarks A broken first or second upper molar was assigned to an indeterminate rhinolophoid by Sigé (1991). Nothing else can be added to Sigé's original discussion of this tooth.

Family TANZANYCTERIDAE Gunnell et al., 2003
Genus *TANZANYCTERIS* Gunnell et al., 2003
TANZANYCTERIS MANNARDI Gunnell et al., 2003

Diagnosis Possesses very large cochlear diameter relative to basicranial width, first rib relatively broader than other ribs, clavicle articulating with coracoid, trochiter (= greater tuberosity) of humerus extending proximally beyond humeral head, anterior laminae on ribs present, manubrium with bilaterally compressed ventral keel, and a narrow, dual-faceted scapular infraspinous fossa.

Description *Tanzanycteris mannardi* is known from a single, partial skeleton. It is the oldest placental mammal known from sub-Saharan Africa (Gunnell et al., 2003).

Age and Occurrence Middle Eocene, Lutetian (46 Ma); Mahenge, Tanzania.

Remarks *Tanzanycteris* is unique among Eocene bats in possessing very enlarged cochleae. This suggests that sophisticated forms of echolocation were already being employed by bats as early as the middle Eocene (Gunnell et al., 2003).

FAMILY indet.
Genus *VAMPYRAVUS* Schlosser, 1910
VAMPYRAVUS (= PROVAMPYRUS) ORIENTALIS
Schlosser, 1910
Figure 30.4E

Diagnosis The humerus is robust, distally curved, trochiter and lesser tuberosity extend beyond humeral head, head relatively large, round, and robust, distal articular facets offset laterally, capitulum with distinct lateral tail, entepicondyle relatively robust and extended.

Description The holotype (unnumbered, in the Staatliches Museum für Naturkunde, Stuttgart) is a relatively large right humerus (49 mm in length) similar in length to living *Rousettus aegyptiacus* (Egyptian rousette fruit bat, body weight approximately 140 g). The bone is very robust and is fully described in Sigé (1985).

Age and Occurrence Early Oligocene, Rupelian; Jebel Qatrani Formation, Egypt (precise locality unknown).

Remarks Schlosser originally mentioned the type humerus in 1910 (p. 507), noting that it was similar to *Vampyrus* (= *Vampyrum*) and *Stenoderma*, two New World phyllostomid bats, but larger than both, being almost twice as large as the humerus of *Stenoderma*. He stated that this specimen was the basis for *Vampyravus orientalis* n. g. n. sp. ("Ich basiere hierauf *Vampyravus orientalis* n. g. n. sp."). However, in the following year (Schlosser, 1911), he published a formal and much more extensive description, with figures, of the same humerus, this time proposing the name *Provampyrus orientalis* for the specimen. Sigé (1985) argued that *Vampyravus* should be considered as a *nomen oblitum* ("forgotten name") because it had not been used in publication between 1910 and 1968, therefore rendering *Vampyravus* still valid and available. McKenna and Bell (1997) considered *Vampyravus* to be a likely *nomen nudum* and therefore not available as a valid name.

According to the International Code of Zoological Nomenclature (ICZN), Article 12 states that in order "to be available, every new name published before 1931 must satisfy the provisions of Article 11 [which *Vampyravus orientalis* does] and must be accompanied by a description or a definition of the taxon that it denotes, or by an indication." The issue at hand seems to be whether Schlosser's (1910) cryptic description of *Vampyravus* constitutes enough of a taxon definition to stand as valid. It would not constitute a valid diagnosis and description today, but since it was published in 1910, it seems as though it does. At least, there is no apparent reason to invali-

date the senior synonym without petitioning the ICZN. Therefore, I follow Sigé (1985) and continue to recognize *Vampyravus* as valid. A further complicating factor may be that the humerus of *Vampyravus orientalis* could well represent the same taxon as *Philisis sphingis*, a presumably similar sized bat from the Jebel Qatrani Formation (Sigé, 1985) known only from teeth. This determination awaits more complete material from the Fayum.

MICROCHIROPTERAN indet.

Remarks Black and Krishtalka (1986) describe a worn lower molar from Area 131A, Okote Member, Koobi Fora Formation as an indeterminate chiropteran. K. Muldoon (pers. comm.) reports the present of three bat genera (*Miniopterus*, *Otomops*, and *Mormopterus*) from the late Holocene (500 BP) locality of Ankilitelo Cave in southwestern Madagascar.

General Discussion and Summary

The fossil record of African bats begins in the early Eocene in Tunisia (table 30.1). With the possible exception of *Tanzanycteris* (Gunnell et al., 2003), all African Eocene bats are members of extant superfamilies, if not families. Sigé (1991) suggested that perhaps modern bats had a southern origin because of the lack of modern bat families on northern continents in the Eocene. However, since 1991, emballonurids have been described from the early (Hooker, 1996; although see Storch et al., 2002) and middle (Storch et al., 2002) Eocene of Europe, and vespertilionoids have been described from the early Eocene of North America (Beard et al., 1992). There does, however, remain a curious lack of archaic bats from the African Eocene suggesting that, if not a center of origin, Africa (and Gondwana) may have played a significant role as a center of diversification for many modern groups of bats.

TABLE 30.1
Occurrences and ages of African Chiroptera
Bold = first described in primary reference. South African Cave Site dates interpreted based on Vrba (1982), Delson (1984), and Partridge (2000).

Taxon	Occurrence (Site, Location)	Stratigraphic Unit	Age	Primary Reference
		PTEROPODIDAE		
Propotto leakeyi	Koru, Songhor, Chamtwara, Rusinga Island, Kenya		Early Miocene (18.0–20.0 Ma)	Simpson, 1967
Eidolon aff. *E. helvum*	Omo Loc. 1, Ethiopia	Shungura Fm., Member B	Late Pliocene (2.95 Ma)	Wesselman, 1984
Eidolon dupreanum	Anjohibe Cave, Madagascar	Old SE Locality	Subfossil (≤10,000 BP)	Samonds, 2007
Rousettus madagascariensis	Anjohibe Cave, Madagascar	SS2 Locality	Subfossil	Samonds, 2007
Rousettus cf. *R. madagascariensis*	Anjohibe Cave, Madagascar	NCC-1 Locality	Late Pleistocene (69,600–86,800 BP)	Samonds, 2007
Pteropodid sp.	Songhor and Chamtwara, Kenya		Early Miocene (19 Ma)	Butler, 1984
Pteropodid sp.	Loc. 130-A, Kenya	Koobi Fora Fm., Okote Member	Early Pleistocene (1.6 Ma)	Black & Krishtalka, 1986

TABLE 30.1 (CONTINUED)

Taxon	Occurrence (Site, Location)	Stratigraphic Unit	Age	Primary Reference
		EMBALLONURIDAE		
Dhofarella thaleri	Taqah, Oman		Early Oligocene (31.5 Ma)	Sigé et al., 1994
Dhofarella sigei	Fayum Quarry L-41, Egypt	Jebel Qatrani Fm.	Late Eocene (34.0 Ma)	Gunnell et al., 2008
Saccolaimus incognita	Koru, Kenya	Legetet Fm., Kenya	Early Miocene (20.0 Ma)	Butler & Hopwood, 1957
Taphozous (= Saccolaimus) incognita	Moroto II, Uganda		Early Miocene (17.5 Ma)	Pickford & Mein, 2006
Saccolaimus abitus	Omo Locs. 1 & 28, Ethiopia	Shungura Fm., Members B and F	Late Pliocene (2.1–2.95 Ma)	Wesselman, 1984
Taphozous sp.	Sterkfontein, South Africa	Cave Breccia	Plio-Pleistocene (1.8–2.3 Ma)	Pocock, 1987
Taphozous sp.	Ahl al Oughlam, Morocco	Cave Breccia	Late Pliocene (2.5 Ma)	Eiting et al., 2006
Coleura muthokai	Omo Loc. 28, Ethiopia	Shungura Fm., Member F	Late Pliocene (2.1 Ma)	Wesselman, 1984
Emballonura atrata	Tsimanampetsotsa, Madagascar	Karst Breccia	Pleistocene or Subfossil	Sabatier & Legendre, 1985
Emballonurid sp.	Rusinga Loc. R3, Kenya		Early Miocene (18.0 Ma)	Butler, 1969
Emballonurid sp.	Songhor, Legetet Locality 10	Legetet Fm.	Early Miocene (20.0 Ma)	Butler, 1984
		HIPPOSIDERIDAE		
Hipposideros (Brachipposideros) omani	Taqah, Oman		Early Oligocene (31.5 Ma)	Sigé et al., 1994
Hipposideros kaumbului	Omo Loc. 28, Ethiopia	Shungura Fm., Member F	Late Pliocene (2.1 Ma)	Wesselman, 1984
Hipposideros cyclops	Omo Loc. 1, Ethiopia	Shungura Fm., Member B	Late Pliocene (2.95 Ma)	Wesselman, 1984
Hipposideros commersoni	Twin Rivers Cave, Zambia		Middle Pleistocene (170 – 266 Ka)	Avery, 2003
Hipposideros commersoni	Anjohibe Cave, Madagascar	NCC-1 Locality	Late Pleistocene (69,600 – 86,800 BP)	Samonds, 2007
Hipposideros commersoni	Tsimanampetsotsa, Madagascar	Karst Breccia	Pleistocene or Subfossil	Sabatier & Legendre, 1985
Hipposideros cf. H. commersoni	Anjohibe Cave, Madagascar	Old SE Locality	Subfossil (≤10,000 BP)	Samonds, 2007
Hipposideros cf. H. commersoni	Anjohibe Cave, Madagascar	SS2 Locality	Subfossil	Samonds, 2007
Hipposideros besaoka	Anjohibe Cave, Madagascar	TW-10 Locality	Subfossil (≤ 10,000 BP)	Samonds, 2007
Hipposideros (Syndesmotis) vetus	Beni Mellal, Morocco		Early late Miocene	Lavocat, 1961
Hipposideros (Syndesmotis) sp.	Sheikh Abdallah, Egypt	Cave Breccia	Late Miocene (10-11 Ma)	Wanas et al., 2009
Hipposideros sp.	Songhor, Kenya		Early Miocene (20.0 Ma)	Butler, 1969
Hipposideros spp.	Kanapoi, Kenya		Early Pliocene (4 Ma)	Winkler, 2003
Hipposideros sp.	Tsimanampetsotsa, Madagascar	Karst Breccia	Pleistocene or Subfossil	Sabatier & Legendre, 1985
Triaenops goodmani	Anjohibe Cave, Madagascar	Old SE Locality	Subfossil (≤ 10,000 BP)	Samonds, 2007
Triaenops furculus	Tsimanampetsotsa, Madagascar	Karst Breccia	Pleistocene or Subfossil	Sabatier & Legendre, 1985

Taxon	Occurrence (Site, Location)	Stratigraphic Unit	Age	Primary Reference
Triaenops sp.	Anjohibe Cave, Madagascar	NCC-1 Locality	Late Pleistocene (69,600 – 86,800 BP)	Samonds, 2007
Triaenops sp.	Anjohibe Cave, Madagascar	Old SE Locality	Subfossil (≤ 10,000 BP)	Samonds, 2007
Hipposiderine sp.	Taqah, Oman		Early Oligocene (31.5 Ma)	Sigé et al., 1994
Hipposiderine sp.	Ahl al Oughlam, Morocco	Cave Breccia	Late Pliocene (2.5 Ma)	Eiting et al., unpub. ms.
Hipposiderine sp.	Chamtwara, Kenya		Early Miocene (19.0 Ma)	Butler, 1984
Hipposiderine sp.	Songhor, Kenya		Early Miocene (20.0 Ma)	Butler, 1984

MYZOPODIDAE

Taxon	Occurrence (Site, Location)	Stratigraphic Unit	Age	Primary Reference
Myzopoda sp.	Olduvai Bed I, Tanzania		Early Pleistocene (1.8-1.9 Ma)	Butler, 1978

MEGADERMATIDAE

Taxon	Occurrence (Site, Location)	Stratigraphic Unit	Age	Primary Reference
Saharaderma pseudovampyrus	Fayum Quarry L-41, Egypt	Jebel Qatrani Fm.	Late Eocene (34.0 Ma)	Gunnell et al., 2008
Cardioderma sp.	Olduvai Bed I, Tanzania		Early Pleistocene (1.8-1.9 Ma)	Butler & Greenwood, 1965
Megaderma gaillardi	Beni Mellal, Morocco		Early late Miocene	Sigé, 1976
Megaderma jaegeri	Beni Mellal, Morocco		Early late Miocene	Sigé, 1976
Megaderma sp.	Sheikh Abdallah, Egypt	Cave Breccia	Late Miocene (10-11 Ma)	Wanas et al., 2009
Megadermatid sp.	Rusinga Loc. R106, Kenya	Hiwegi Fm.	Early Miocene (18.0 Ma)	Butler & Hopwood, 1957
Megadermatid sp.	Makapansgat, South Africa	Cave Breccia	Late Pliocene (2.7-3.1 Ma)	Pocock, 1987

MOLOSSIDAE

Taxon	Occurrence (Site, Location)	Stratigraphic Unit	Age	Primary Reference
Tadarida rusingae	Rusinga Loc. R106, Kenya	Hiwegi Fm.	Early Miocene (18.0 Ma)	Arroyo-Cabrales et al., 2002
Mormopterus jugularis	Tsimanampetsotsa, Madagascar	Karst Breccia	Pleistocene or Subfossil	Sabatier & Legendre, 1985
Molossid Indet.	Beni Mellal, Morocco		Early late Miocene	Lavocat, 1961
Molossid Indet. (4 species)	Olduvai Bed I, Tanzania		Early Pleistocene (1.8-1.9 Ma)	Butler, 1978
Molossid Indet.	Napak, Uganda		Early Miocene (19 Ma)	Butler, 1984

NYCTERIDAE

Taxon	Occurrence (Site, Location)	Stratigraphic Unit	Age	Primary Reference
Chibanycteris herberti	Taqah, Oman		Early Oligocene (31.5 Ma)	Sigé et al., 1994
Nycteris sp.	Makapansgat, South Africa	Cave Breccia	Late Pliocene (2.7–3.1 Ma)	Pocock, 1987
Nycteris sp.	Loc. 130-A, Kenya	Koobi Fora Fm., Okote Member	Early Pleistocene (1.6 Ma)	Black & Krishtalka, 1986
?Nycterid Indet.	Kanapoi, Kenya		Late Pliocene (4 Ma)	Butler, 1978
Nycterid Indet.	Chamtwara, Kenya		Early Miocene (19 Ma)	Butler, 1984

TABLE 30.1 (CONTINUED)

Taxon	Occurrence (Site, Location)	Stratigraphic Unit	Age	Primary Reference
		RHINOPOMATIDAE		
Qarunycteris moerisae	Fayum Quarry BQ-2, Egypt	Birket Qarun Fm.	Late Eocene (37.0 Ma)	Gunnell et al., 2008
		VESPERTILIONIDAE		
Khonsunycteris aegypticus	Fayum Quarry L-41, Egypt	Jebel Qatrani Fm.	Late Eocene (34.0 Ma)	Gunnell et al., 2008
Chamtwaria pickfordi	Chamtwara, Kenya		Early Miocene (19.5 Ma)	Butler, 1984
Chamtwaria?	Moroto II, Uganda		Early Miocene (17.5 Ma)	Pickford & Mein, 2006
Scotophilus sp.	Loc. 130-A, Kenya	Koobi Fora Fm., Okote Member	Early Pleistocene (1.6 Ma)	Black & Krishtalka, 1986
Myotis tricolor	Swartkrans, South Africa	Cave Breccia	Pleistocene (1.3–1.9 Ma)	Avery, 1998
Myotis cf. M. tricolor	Kromdraai B, South Africa	Cave Breccia	Plio-Pleistocene (1.7–2.0 Ma)	Pocock, 1987
Myotis goudoti	Anjohibe Cave, Madagascar	SS2 Locality	Subfossil	Samonds, 2007
Myotis cf. M. welwitschii	Kromdraai B, South Africa	Cave Breccia	Plio-Pleistocene (1.7–2.0 Ma)	Pocock, 1987
Myotis sp.	Ahl al Oughlam, Morocco	Cave Breccia	Late Pliocene (2.5 Ma)	Eiting et al., 2006
Myotis sp.	Bolt's Farm, South Africa	Cave Breccia	Early Pliocene (4.5 Ma)	Broom, 1948
Myotis sp.	Olduvai Bed I, Tanzania		Early Pleistocene (1.8–1.9 Ma)	Butler, 1978
Eptesicus cf. E. bottae	Makapansgat, South Africa	Cave Breccia	Late Pliocene (2.7–3.1 Ma)	Pocock, 1987
Eptesicus cf. E. hottentotus	Makapansgat, South Africa	Cave Breccia	Late Pliocene (2.7–3.1 Ma)	Pocock, 1987
Eptesicus cf. E. hottentotus	Kromdraai B, South Africa	Cave Breccia	Plio-Pleistocene (1.7–2.0 Ma)	Pocock, 1987
Eptesicus cf. E. hottentotus	Olduvai Bed I, Tanzania		Early Pleistocene (1.8–1.9 Ma)	Butler, 1978
Eptesicus sp.	Langebaanweg, South Africa	Varswater Fm., QSM Member	Early Pliocene (5.0 Ma)	Hendey, 1981
Eptesicus sp.	Swartkrans, South Africa	Cave Breccia	Pleistocene (1.3–1.9 Ma)	Avery, 1998
Miniopterus schreibersii	Twin Rivers Cave, Zambia		Middle Pleistocene (170–266 Ka)	Avery, 2003
Miniopterus cf. M. schreibersii	Sterkfontein, South Africa	Cave Breccia	Plio-Pleistocene (1.8–2.3 Ma)	Pocock, 1987
Miniopterus cf. M. schreibersii	Olduvai Bed I, Tanzania		Early Pleistocene (1.8–1.9 Ma)	Butler, 1978
Miniopterus sp.	Makapansgat, South Africa	Cave Breccia	Late Pliocene (2.7–3.1 Ma)	Pocock, 1987
Miniopterus n. sp.	Ahl al Oughlam, Morocco	Cave Breccia	Late Pliocene (2.5 Ma)	Eiting et al., 2006
Cf. Nycticeius (Scoteinus) schlieffeni	Olduvai Bed I, Tanzania		Early Pleistocene (1.8–1.9 Ma)	Butler & Greenwood, 1965
Cf. Pipistrellus (Scotozous) rueppelli	Olduvai Bed I, Tanzania		Early Pleistocene (1.8–1.9 Ma)	Butler & Greenwood, 1965

Taxon	Occurrence (Site, Location)	Stratigraphic Unit	Age	Primary Reference
Vespertilionid Indet.	Beni Mellal, Morocco		Early late Miocene	Lavocat, 1961
Vespertilionid Indet.	Songhor & Chamtwara, Kenya		Early Miocene (19 Ma)	Butler, 1984

PHILISIDAE

Taxon	Occurrence (Site, Location)	Stratigraphic Unit	Age	Primary Reference
Witwatia schlosseri	Fayum Quarry BQ-2, Egypt	Birket Qarun Fm.	Late Eocene (37.0 Ma)	Gunnell et al., 2008
Witwatia eremicus	Fayum Quarry BQ-2, Egypt	Birket Qarun Fm.	Late Eocene (37.0 Ma)	Gunnell et al., 2008
Philisis sphingis	Fayum Quarry I, Egypt	Jebel Qatrani Fm.	Early Oligocene (30.0 Ma)	Sigé, 1985
Philisis sevketi	Taqah, Oman		Early Oligocene (31.5 Ma)	Sigé et al., 1994
Cf. *Philisis* sp.	Taqah, Oman		Early Oligocene (31.5 Ma)	Sigé et al., 1994
Philisis sp.	Fayum, Egypt	Jebel Qatrani Fm.	Early Oligocene (30.0 Ma)	Sigé, 1985
Dizzya exsultans	Chambi, Tunisia		Early Eocene (50.0 Ma)	Sigé, 1991
Scotophilisis libycus	Jebel Zelten, Libya		Early Middle Miocene (14.0–16.0 Ma)	Horáček et al., 2006

RHINOLOPHIDAE

Taxon	Occurrence (Site, Location)	Stratigraphic Unit	Age	Primary Reference
Rhinolophus hildebrandtii	Twin Rivers Cave, Zambia		Middle Pleistocene (170–266 Ka)	Avery, 2003
Rhinolophus cf. *R. capensis*	Kromdraai B, South Africa	Cave Breccia	Plio-Pleistocene (1.7–2.0 Ma)	Pocock, 1987
Rhinolophus cf. *R. capensis*	Makapansgat, South Africa	Cave Breccia	Late Pliocene (2.7–3.1 Ma)	De Graaf, 1960
Rhinolophus clivosus	Hoedjies-punt 1	Fossil hyena lair	Late Pleistocene (300,000 BP)	Matthews et al., 2007
Rhinolophus clivosus	Saldanha Bay Yacht Club	Fossil owl accumulation	Late Pleistocene (15,540 BP)	Matthews et al., 2007
Rhinolophus cf. *R. clivosus*	Makapansgat, South Africa	Cave Breccia	Late Pliocene (2.7–3.1 Ma)	Pocock, 1987
Rhinolophus cf. *R. clivosus*	Sterkfontein, South Africa	Cave Breccia	Plio-Pleistocene (1.8–2.3 Ma)	Pocock, 1987
Rhinolophus cf. *R. darlingi*	Makapansgat, South Africa	Cave Breccia	Late Pliocene (2.7–3.1 Ma)	Pocock, 1987
Rhinolophus cf. *R. darlingi*	Sterkfontein, South Africa	Cave Breccia	Plio-Pleistocene (1.8–2.3 Ma)	Pocock, 1987
Rhinolophus ferrumequinum mellali	Beni Mellal, Morocco		Early late Miocene	Lavocat, 1961
Rhinolophus sp.	Sheikh Abdallah, Egypt	Cave Breccia	Late Miocene (10–11 Ma)	Wanas et al., 2009
Rhinolophus sp.	Swartkrans, South Africa	Cave Breccia	Pleistocene (1.3–1.9 Ma)	Avery, 1998
Rhinolophus n. sp.	Ahl al Oughlam, Morocco	Cave Breccia	Late Pliocene (2.5 Ma)	Eiting et al., 2006
Rhinolophoid sp.	Chambi, Tunisia		Early Eocene (50.0 Ma)	Sigé, 1991

TANZANYCTERIDAE

Taxon	Occurrence (Site, Location)	Stratigraphic Unit	Age	Primary Reference
Tanzanycteris mannardi	Mahenge, Tanzania		Middle Eocene (46.0 Ma)	Gunnell et al., 2003

TABLE 30.I (CONTINUED)

Taxon	Occurrence (Site, Location)	Stratigraphic Unit	Age	Primary Reference
		FAMILY ?		
Vampyravus (= *Provampyrus*) *orientalis*	Fayum, Egypt	Jebel Qatrani Fm.	Early Oligocene	Schlosser, 1910
Microchiropteran Indet.	Loc. 131-A, Kenya	Koobi Fora Fm., Okote Member	Early Pleistocene (1.6 Ma)	Black & Krishtalka, 1986

ACKNOWLEDGMENTS

I thank Lars Werdelin and Bill Sanders for the invitation to participate in this volume. I thank Bernard Sigé for his many helpful insights and for his thorough review of the manuscript. I have benefited greatly from discussions with Thierry Smith, Nancy Simmons, Erik Seiffert, Thomas Eiting, Kaye Reed, Martin Pickford, Laura MacLatchy, Karen Samonds, Kathleen Muldoon, Bill Sanders, and Philip Gingerich. Alan Walker kindly provided a cast of the holotype of *Tadarida rusingae*, and Marilyn Fox and Mary Ann Turner produced and provided casts of the type and referred specimens of *Philisis sphingis*.

Literature Cited

Aguilar, J.-P., M. Calvet, J.-Y.Crochet, S. Legendre, J. Michaux, and B. Sigé. 1986. Première occurrence d'un Mégachiroptère Ptéropodidé dans le Miocène moyen d'Europe (Gisement de Lo Fournas-II, Pyrénées-Orientales, France). *Palaeovertebrata* 16:173–184.

Arroyo-Cabrales, J., R. Gregorin, D. A. Schlitte, and A. Walker. 2002. The oldest African molossid bat cranium (Chiroptera: Molossidae). *Journal of Vertebrate Paleontology* 22:380–387.

Avery, D. M. 1998. An assessment of the lower Pleistocene micromammalian fauna from Swartkrans Members 1–3, Gauteng, South Africa. *Geobios* 31:393–414.

———. 2003. Early and Middle Pleistocene environments and hominid biogeography; micromammalian evidence from Kabwe, Twin Rivers, and Mumbwa Caves in central Zambia. *Palaeogeography, Palaeoclimatology, Palaeoecology* 189:55–69.

Beard, K. C., B. Sigé, and L. Krishtalka. 1992. A primitive vespertilionoid bat from the early Eocene of central Wyoming. *Comptes Rendus des Séances de l'Académie des Sciences, Paris* 314:735–741.

Black, C. C., and L. Krishtalka. 1986. Rodents, bats, and insectivores from the Plio-Pleistocene sediments to the east of Lake Turkana, Kenya. *Contributions in Science, Natural History Museum of Los Angeles County* 372:1–15.

Broom, R. 1948. Some South African Pliocene and Pleistocene mammals. *Annals of the Transvaal Museum* 21:1–38.

Butler, P. M. 1969. Insectivores and bats from the Miocene of East Africa: New material; pp. 1–38 in L. S. B. Leakey (ed.), *Fossil Vertebrates of Africa*, vol. 1. Academic Press, London.

———. 1978. Insectivora and Chiroptera; pp. 56–68 in V. J. Maglio and H. B. S. Cooke (eds.), *Evolution of African Mammals*. Harvard University Press, Cambridge.

———. 1984. Macroscelidea, Insectivora and Chiroptera from the Miocene of East Africa. *Palaeovertebrata* 14:117–200.

Butler, P. M., and M. Greenwood. 1965. Order Insectivora; pp. 13–14 in L. S. B. Leakey (ed.), *Olduvai Gorge 1951–1961: Volume 1. Fauna and Background*. Cambridge University Press, Cambridge.

Butler, P. M., and A. T. Hopwood. 1957. Insectivora and Chiroptera from the Miocene rocks of Kenya Colony. *British Museum (Natural History) Fossil Mammals of Africa* 13:1–35.

Dal Piaz, G. 1937. I mammiferi dell'Oligocene Veneto *Archaeopteropus transiens*. *Memorie degli Istituti di Geologia e Mineralogia dell'Università di Padova* 11:1–8.

De Graaff, G. 1960. A preliminary investigation into the mammalian microfauna in Pleistocene deposits of caves in the Transvaal System. *Palaeontologia Africana* 7:79–118.

Delson, E. 1984. Cercopithecoid biochronology of the African Plio-Pleistocene: Correlation among eastern and southern hominid-bearing localities. *Courier Forschungsinstitut Senckenberg* 69:199–218.

Ducrocq, S., J.-J. Jaeger, and B. Sigé. 1993. Un mégachiroptère dans l'Eocène supérieur de Thaïlande: Incidence dans la discussion phylogénique du groupe. *Neues Jahrbuch für Mineralogie, Geologie, und Paläontologie, Monatshefte* 9:561–575.

Eiting, T. P., D. Geraads, and G. F. Gunnell. 2006. New late Pliocene bats (Chiroptera) from Ahl al Oughlam, Casablanca, Morocco. *Journal of Vertebrate Paleontology* 26(suppl. to no. 3):58A.

Gunnell, G. F., B. F. Jacobs, P. S. Herendeen, J. J. Head, E. Kowalski, C. P. Msuya, F. A. Mizambwa, T. Harrison, J. Habersetzer, and G. Storch. 2003. Oldest placental mammal from sub-Saharan Africa: Eocene microbat from Tanzania—evidence for early evolution of sophisticated echolocation. *Palaeontologia Electronica* 5:1–10.

Gunnell, G. F., and N. B. Simmons. 2005. Fossil evidence and the origin of bats. *Journal of Mammalian Evolution* 12:209–246.

Gunnell, G. F., E. L. Simons, and E. R. Seiffert. 2008. New bats (Mammalia: Chiroptera) from the late Eocene and early Oligocene, Fayum Depression, Egypt. *Journal of Vertebrate Paleontology* 28:1–11.

Habersetzer, J., and G. Storch. 1987. Klassifikation und funktionelle Flügelmorphologie paläogener Fledermäuse (Mammalia, Chiroptera). *Courier Forschungsinstitut Senckenberg* 91:11–150.

Hendey, Q. B. 1981. Palaeoecology of the Late Tertiary fossil occurrences in "E" Quarry, Langebaanweg, South Africa, and a reinterpretation of their geological context. *Annals of the South African Museum* 84:1–104.

Hooker, J. J. 1996. A primitive emballonurid bat (Chiroptera, Mammalia) from the earliest Eocene of England. *Palaeovertebrata* 25:287–300.

Horáček, I., O. Fejfar, and P. Hulva. 2006. A new genus of vespertilionid bat from early Miocene of Jebel Zelten, Libya, with comments on *Scotophilus* and early history of vespertilionid bats (Chiroptera). *Lynx (Praha)* 37:131–150.

Jepsen, G. L. 1966. Early Eocene bat from Wyoming. *Science* 154:1333–1339.

Lavocat, R. 1961. Le gisement des vertébrés miocènes de Beni Mellal (Maroc): Etude systématique de la faune de mammifères et conclusions générales. *Notes et Mémoires du Service Géologique du Maroc* 155:29–94.

Legendre, S. 1982. Hipposideridae (Mammalia: Chiroptera) from the Mediterranean middle and late Neogene, and evolution of the genera *Hipposideros* and *Asellia*. *Journal of Vertebrate Paleontology* 2:372–385.

Matthews, T., C. Denys, and J. E. Parkington. 2007. Community evolution of Neogene micromammals from Langebaanweg 'E' Quarry and other west coast fossil sites, south-western Cape, South Africa. *Palaeogeography, Palaeoclimatology, Palaeoecology* 245:332–352.

McKenna, M. C., and S. K. Bell. 1997. *Classification of Mammals above the Species Level*. Columbia University Press, New York, 631 pp.

Meschinelli, L. 1903. Un nuovo Chirottero fossile (*Archaeopteropus transiens* Mesch.) delle lignite di Monteviale. *Atti Reale Istituto Veneto di Scienze, Lettere ed Arti* 62:1329–1344.

Nowak, R. M. 1994. *Walker's Bats of the World.* Johns Hopkins University Press, Baltimore, 287 pp.

Partridge, T. C. 2000. Hominid-bearing cave and tufa deposits; pp. 100–125 in T. C. Partridge and R. R. Maud (eds.), *The Cenozoic of Southern Africa.* Oxford University Press, Oxford.

Pickford, M., and P. Mein. 2006. Early Middle Miocene mammals from Moroto II, Uganda. *Beiträge zur Paläontologie* 30:361–386.

Pocock, T. N. 1987. Plio-Pleistocene fossil mammalian microfauna of southern Africa: A preliminary report including description of two new fossil muroid genera (Mammalia: Rodentia). *Palaeontologia Africana* 26:69–91.

Russell, D. E., and B. Sigé. 1970. Révision des chiroptères lutétiens de Messel (Hesse, Allemagne). *Palaeovertebrata* 3:83–182.

Sabatier, M., and S. Legendre. 1985. Une faune à rongeurs et chiroptères Plio-Pleistocènes de Madagascar. *Actes du 100 Congrés National des Sociétés Savants, Montpellier, Sciences* 4:21–28.

Samonds, K. E. 2006. The origin and evolution of Malagasy bats: Implications of new Late Pleistocene fossils and cladistic analyses for reconstructing biogeographic history. Unpublished PhD dissertation, Stony Brook University, Stony Brook, N.Y., 403 pp.

———. 2007. Late Pleistocene bat fossils from Anjohibe Cave, northwestern Madagascar. *Acta Chiropterologica* 9:39–65.

Schlosser, M. 1910. Über einige fossile Säugetiere aus dem Oligocän von Ägypten. *Zoologischen Anzeiger* 35:500–508.

———. 1911. Beiträge zur Kenntnis der oligozänen Landsäugetiere aus dem Fayum, Ägypten. *Beiträge zur Paläontologie und Geologie Österreich-Ungarns und des Orients* 24:51–167.

Schutt, W. A., Jr., and N. B. Simmons. 1998. Morphology and homology of the chiropteran calcar, with comments on the phylogenetic relationships of *Archaeopteropus. Journal of Mammalian Evolution* 5:1–32.

Sigé, B. 1976. Les Megadermatidae (Chiroptera, Mammalia) miocènes du Beni Mellal, Maroc. *Géologie Méditerranéenne* 3:71–86.

———. 1985. Les chiroptères oligocènes du Fayum, Egypte. *Geologica et Palaeontologica* 19:161–189.

———. 1991. Rhinolophoidea et Vespertilionoidea (Chiroptera) du Chambi (Eocène inférieur de Tunisie). Aspects biostratigraphiques, biogéographiques et paléoécologiques de l'origine des chiroptères modernes. *Neues Jahrbuch für Geologie und Paläontologie, Abhandlungen* 182:355–376.

Sigé, B., and J.-P. Aguilar. 1987. L'extension stratigraphique des mégachiroptères dans le Miocène d'Europe méridionale. *Comptes Rendus des Séances de l'Académie des Sciences, Paris* 304:469–474.

Sigé, B., J.-Y. Crochet, J. Sudre, J.-P. Aguilar, and G. Escarguel. 1997. Nouveaux sites d'âges varies dans les Remplissages Karstiques du Miocène inférieur de Bouzigues (Hérault, Sud de la France): Partie I. Sites et faunes 1 (Insectivores, Chiroptères, Artiodactyles). *Geobios* 20:477–483.

Sigé, B., H. Thomas, S. Sen, E. Gheerbrant, J. Roger, and Z. Al-Sulaimani. 1994. Les chiroptères de Taqah (Oligocène inférieur, Sultanat d'Oman): Premier inventaire systématique. *Münchner Geowissenschaftliche Abhandlungen* 26:35–48.

Simmons, N. B. 2005. Order Chiroptera; pp. 312–529 in D. E. Wilson and D. M. Reeder (eds.), *Mammal Species of the World: A Taxonomic and Geographic Reference.* 3rd ed. Johns Hopkins University Press, Baltimore.

Simmons, N. B., and J. H. Geisler. 1998. Phylogenetic relationships of *Icaronycteris, Archaeonycteris, Hassianycteris,* and *Palaeochiropteryx* to extant bat lineages, with comments on the evolution of echolocation and foraging strategies in Microchiroptera. *Bulletin of the American Museum of Natural History* 235:1–182.

Simmons, N. B., K. L. Seymour, J. Habersetzer, and G. F. Gunnell. 2008. Primitive early Eocene bat from Wyoming and the evolution of flight and echolocation. *Nature* 451:818–821.

Simpson, G. G. 1967. The Tertiary lorisiform primates of Africa. *Bulletin of the Museum of Comparative Zoology* 136:39–62.

Smith, J. D., and G. Storch. 1981. New Middle Eocene bats from "Grube Messel" near Darmstadt, W-Germany (Mammalia: Chiroptera). *Senckenbergiana Biologica* 61:153–167.

Storch, G. 2001. Paleobiological implications of the Messel mammalian assemblage; pp. 215–235 in G. F. Gunnell (ed.), *Eocene Biodiversity: Unusual Occurrences and Rarely Sampled Habitats.* Kluwer Academic/Plenum Publishers, New York.

Storch, G., B. Sigé, and J. Habersetzer. 2002. *Tachypteron franzeni* n. gen., n. sp., earliest emballonurid bat from the middle Eocene of Messel (Mammalia, Chiroptera). *Paläontologische Zeitschrift* 76:189–199.

Vrba, E. S. 1982. Biostratigraphy and chronology, based particularly on Bovidae, of southern hominid-associated assemblages: Makapansgat, Sterkfontein, Taung, Kromdraai, Swartkrans; also Elandsfontein (Saldanha), Broken Hill (now Kabwe) and Cave of Hearths; pp. 707–752 in H. de Lumley and M. A. de Lumley (eds.), *Prétirage 1er Congrés International de Paléontologie Humaine.* Union Internationale des Sciences Prehistoriques et Protohistoriques 2, Nice, France.

Walker, A. C. 1969. True affinities of *Propotto leakeyi* Simpson, 1967. *Nature* 223:647–648.

Wanas, H. A., M. Pickford, P. Mein, H. Soliman, and L. Segalen. 2009. Late Miocene karst system at Sheikh Abdallah, between Bahariya and Farafra, Western Desert, Egypt: Implications for palaeoclimate and geomorphology. *Geologica Acta* 7:475–487.

Wesselman, H. B. 1984. The Omo micromammals—Systematics and paleoecology of early man sites from Ethiopia; pp. 1–219 in M. K. Hecht and F. S. Szalay (eds.), *Contributions to Vertebrate Evolution,* vol. 7. Karger, Basel, Switzerland.

Wessels, W., O. Fejfar, P. Peláez-Campomanes, A. van der Meulen, and H. de Bruijn. 2003. Micromamíferos miocenos de Jebel Zelten, Libia; pp. 699–715 in N. López-Martínez, P. Peláez-Campomanes, and M. Hernández Fernández (eds.), *Coloquios de Paleontología,* volume 1 in honor of Dr. Remmert Daams.

Winkler, A. J. 1998. New small mammal discoveries from the early Pliocene at Kanapoi, West Turkana, Kenya. *Journal of Vertebrate Paleontology* 18 (suppl. to no. 3):87A.

———. 2003. Rodents and lagomorphs from the Miocene and Pliocene of Lothagam, northern Kenya; pp. 169–198 in M. G. Leakey and J. M. Harris (eds.), *Lothagam: The Dawn of Humanity in Eastern Africa.* Columbia University Press, New York.

THIRTY-ONE

Pholidota

TIMOTHY J. GAUDIN

The family Manidae includes all the living pangolins, or scaly anteaters. Pangolins are a small group, encompassing eight extant species that are distributed across sub-Saharan Africa, the Indian subcontinent, southern China, Southeast Asia, and the East Indies eastward to the Phillipines (Barlow, 1984; Feiler, 1998; Nowak, 1999; Gaubert and Antunes, 2005; Schlitter, 2005). Four of these species presently occur in Africa: *Smutsia gigantea*, the giant pangolin; *S. temmincki*, the ground pangolin; *Phataginus tetradactyla*, the long-tailed pangolin; and *P. tricuspis*, the tree pangolin. All are found in the forests of western and central Africa except *S. temmincki*, which occurs in drier and more open habitats in eastern and southern Africa (Kingdon, 1997). Pangolins are generally rare (although some species may be locally common; Kingdon, 1997) and very unusual mammals recognized most readily by their extraordinary external carapace of large, overlapping, keratinous epidermal scales that form a protective cover for the back, tail, head, and limbs. Pangolins are toothless, with smooth conical skulls and greatly reduced mandibles (Grassé, 1955; Kingdon, 1974, Heath, 1992a, 1992b, 1995). They feed almost exclusively on ants and termites, capturing these insects with a tremendously elongated tongue covered with copious amounts of sticky saliva and masticating them using a keratinized stomach lining (Grassé, 1955; Kingdon, 1974, 1997; Heath, 1992a, 1992b, 1995; Nowak, 1999; Swart et al., 1999). Most pangolins are characterized by powerful forelimbs and strong claws, used to break open ant and termite nests (Grassé, 1955; Kingdon, 1974, 1997; Heath, 1992a, 1992b, 1995; Nowak, 1999). They range in habitus from terrestrial to semiarboreal to more fully arboreal forms. The two arboreal species from Africa have elongated, prehensile tails (Grassé, 1955; Kingdon, 1974, 1997; Nowak, 1999).

The fossil record of pangolins is sparse. This can be attributed in part to their ecology—their preference for forested habitats, their solitary lifestyles and concomitantly low population densities. It can also be attributed to their morphology, and in particular their lack of teeth, the most commonly preserved and readily identifiable portion of the skeleton in most mammals. Indeed, because of the absence of teeth, their fossils may often be overlooked by mammalian paleontologists even when present (Pickford and Senut, 1991; Gaudin et al., 2006).

The family Manidae is placed in the order Pholidota. Pholidota also includes at least two extinct families of pango-

lins, the Patriomanidae and Eomanidae (Szalay and Shrenk, 1998; Storch, 2003; Gaudin, 2004; Gaudin et al., 2006, 2009). Additionally, several authors (Emry, 1970; McKenna and Bell, 1997; Rose et al., 2005) have advocated close ties between pangolins and an extinct group of mostly North American fossorial mammals with reduced dentitions, the palaeanodonts. Emry (1970) and McKenna and Bell (1997) actually include palaeanodonts formally within the order Pholidota. I will follow Gaudin et al. (2009) and refer to the order Pholidota as including living and fossil pangolins but not palaeanodonts, and employ Pholidotamorpha to refer to the group including both pangolins and palaeanodonts. The Paleogene record of the Pholidota *sensu stricto* is largely confined to Laurasian continents. The oldest pangolins are *Eomanis waldi* and *Euromanis krebsi* from the middle Eocene (roughly 45 Ma) Messel deposits of Germany (Storch, 1978; Storch and Martin, 1994; Gaudin et al., 2009). Somewhat younger taxa are known from the late Eocene of Central Asia (*Cryptomanis gobiensis*; Gaudin et al., 2006) and western North America (*Patriomanis americana*; Emry, 1970, 2004), and even younger pangolins derive from the Oligocene-Miocene of France and Germany (various species of *Necromanis*; Koenigswald, 1969, 1999; Koenigswald and Martin, 1990). By contrast, the record from Africa and from those parts of Asia where pangolins occur today is confined to the Pliocene and Pleistocene with one exception (discussed later) and is represented by less complete material (Guth, 1958; Emry, 1970; Botha and Gaudin, 2007).

The first published cladistic study of pangolin phylogenetic relationships was conducted by Gaudin and Wible (1999). We based our study on the cranial anatomy of the seven extant pangolins plus the one fossil form represented by well-preserved cranial material, *Patriomanis*. This study confirmed the monophyly of the extant forms relative to *Patriomanis*. Szalay and Shrenk (1998) suggested that all well-known Laurasian fossil pangolins could be placed in the family Patriomanidae. However, Gaudin et al. (2006) noted that such a family was likely paraphyletic, and Storch (2003) suggested that *Eomanis* should be assigned to its own family, Eomanidae. A comprehensive phylogenetic study of pangolin interrelationships has been conducted (Gaudin, 2004; Gaudin et al., 2009). The phylogenetic results of this study are illustrated in figure 31.1 and are consistent with the allocation of extant pangolins (perhaps plus closely allied Plio-Pleistocene forms) to the family Manidae, the allocation of *Patriomanis* and *Cryptomanis* to the Patriomanidae, and *Eomanis* to the Eomanidae.

As discussed elsewhere (Gaudin and Wible, 1999; Gaudin et al., 2009), uncertainty has existed historically not only concerning the phylogenetic relationships among living and fossil pangolins, but also concerning the appropriate taxonomic arrangement of the living taxa, in particular the number of genera to which they should be assigned. The eight extant species have been placed in as many as six different genera (Pocock, 1924), although most recent taxonomies have placed them in a single genus, *Manis* (Barlow, 1984; Nowak, 1999; Schlitter, 2005). However, Patterson (1978) recognized two separate genera, and McKenna and Bell (1997) recognized four. Studies by Emry (1970) and Gaudin and Wible (1999) agreed that extant pangolins ought to be placed in multiple genera, but they did not make any formal taxonomic pronouncements pending more thorough systematic study. The phylogenetic study reported elsewhere (i.e., Gaudin, 2004; Gaudin et al., 2009; see also figure 31.1) supports the allocation of the African ground pangolins to the genus

Smutsia, the African arboreal pangolins to the genus *Phataginus*, and the Asian pangolins to the genus *Manis*. These generic assignments are followed in the present study.

Systematic Paleontology

The fossil record of pangolins in Africa is scant, although there is evidence that pangolins have occupied the continent since at least Oligocene times. Indeed, as noted by Pickford and Senut (1991), their record might be more substantive if paleontologists were better able to recognize the postcranial elements of these animals in the field and in existing collections. The published record is summarized here.

Order PHOLIDOTA Weber, 1904
UNNAMED TAXON Gebo and Rasmussen, 1985
Figure 31.2

Age and Occurrence Fayum province, Egypt; Jebel Qatrani Formation (Gebo and Rasmussen, 1985). The Jebel Qatrani Formation is assigned an early Oligocene age by Gingerich (1992).

Discussion Remains include two manual ungual phalanges with deeply divided, bifid tips like those characteristic of modern and most fossil pangolins (Gebo and Rasmussen, 1985). Although Pickford and Senut (1991) question this allocation, the overall morphology of these unguals resembles that of pangolins much more closely than other digging mammals with bifid unguals, and the identification thus seems reasonable based on the limited evidence available.

Family MANIDAE Gray, 1821
Genus *SMUTSIA* Gray, 1865
SMUTSIA GIGANTEA (Illinger, 1815)
Figure 31.3

Age and Occurrence Langebaanweg, South Africa; Varswater Formation, early Pliocene (ca. 5 Ma; Hendey, 1973, 1976; Botha and Gaudin, 2007). Also Lake Albert Basin, Uganda; Warwire formation, late Pliocene (3.6–3.45 Ma; Pickford and Senut, 1991, 1994).

Discussion The older remains from Langebaanweg are the more complete, consisting of a partial skeleton that includes portions of the skull, vertebral column, and limbs (Botha and Gaudin, 2007). Indeed, apart from the isolated unguals just discussed, this is the oldest pangolin skeletal material from the African fossil record and by far the most complete. The specimens exhibit a number of pathologies but otherwise closely resemble the extant pangolin species *S. gigantea*. The younger remains from Uganda consist of a single left radius, also assigned to *S. gigantea* (Pickford and Senut, 1991). Both the South African and Ugandan specimens are slightly smaller than extant *S. gigantea*.

SMUTSIA cf. *S. TEMMINCKII* (Smuts, 1832)

Age and Occurrence Late Pleistocene (12–18 ka; Klein, 1972); Nelson Bay Cave, South Africa.

Discussion Klein (1972) records the presence of a single individual that he ascribes to *S.* cf. *temminckii*. However, as noted by Botha and Gaudin (2007), this assignment has subsequently been questioned.

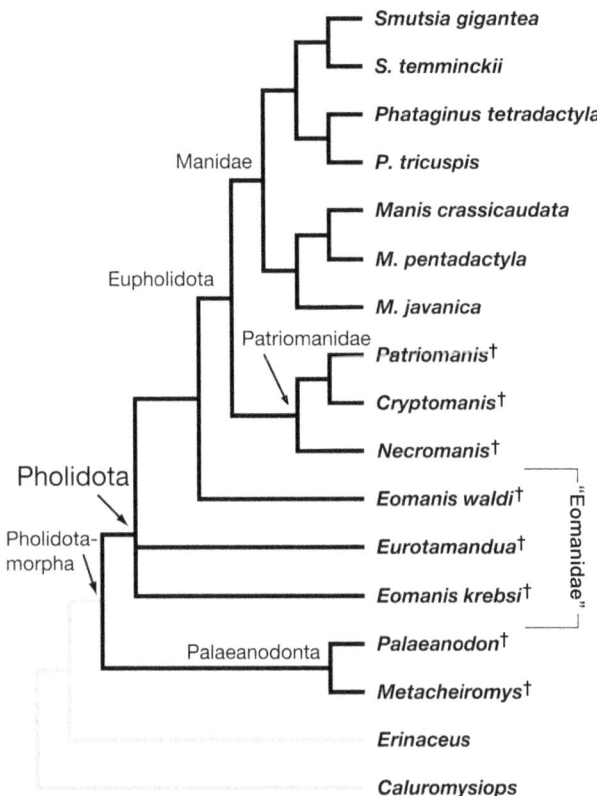

FIGURE 31.1 Phylogeny of the Pholidota from Gaudin (2004) based on analysis of 395 morphological characters drawn from the cranial and postcranial skeleton, scored for 15 in-group taxa. The results shown are a consensus of two most parsimonious trees derived from different character weighting schemes. Basal nodes for the families Manidae and Patriomanidae are labeled. Eomanidae is paraphyletic in this analysis and includes the putative anteater *Eurotamandua*. The order Pholidota includes all fossil pangolins, including eomanids and *Eurotamandua*. The Pholidotamorpha is a clade including both pholidotans and palaeanodonts, and it is equivalent to the "Order Pholidota" as defined by Emry (1970) and McKenna and Bell (1997). This phylogeny differs in several details from that of Gaudin et al. (2009).

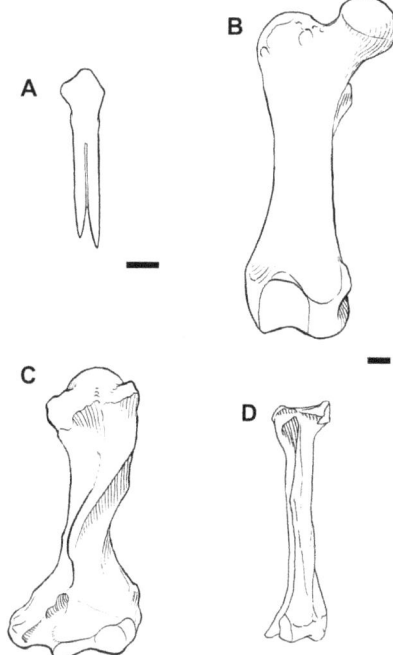

FIGURE 31.2 A) Isolated ungual phalanx in dorsal view, unnamed taxon from the Oligocene of Egypt (modified from Gebo and Rasmussen, 1985). B) Extant specimen of *Smutsia gigantea*, AMNH 53858, right femur, anterior view. C) *S. gigantea*, AMNH 53858, left humerus, anterior view. D) *S. gigantea*, AMNH 53858, left radius, anterior view. Scale bars equal 5 mm in A, 1 cm in B-D.

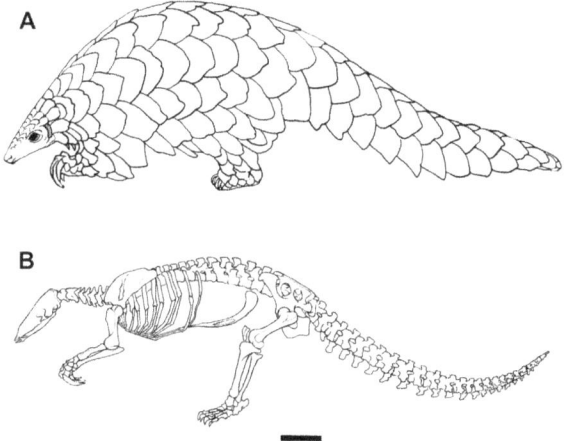

FIGURE 31.3 A) *Smutsia temmincki*. B) Skeleton of *S. gigantea* in left lateral view. Scale bars =1 10 cm. B reprinted with permission from Rose et al., 2005.

Discussion

With only four published records, the African fossil record for pangolins is extremely poor. Yet the Oligocene Fayum record would appear to indicate that pangolins have been present in Africa for some 30 million years or more. Based on this record, Gebo and Rasmussen (1985) suggest that pangolins form part of the characteristic "old African" assemblage of mammals

that includes Proboscidea, Hyracoidea, and anthropoid primates. As noted by Gaudin et al. (2006), the presence of ancient pangolins in the early Cenozoic of Africa would accord with morphology-based supraordinal phylogenies of placental mammals (e.g., Novacek and Wyss 1986), which support a sister-group relationship between Pholidota and the Gondwanan (South American) group Xenarthra. However, the fossil pangolin record as it is currently known better matches molecular-based phylogenies that link Pholidota to Laurasian groups, especially the Carnivora, within the clade Laurasiatheria (e.g., Delsuc et al., 2002; Springer et al., 2004). The oldest fossil pangolins, the greatest number of fossil pangolin taxa, and the best-preserved fossil pangolins are all to be found on the northern continents. Moreover, the putative sister group to the Pholidota, the Palaeanodonta, is also a Laurasian clade (Rose et al., 2005; Gaudin et al., 2009). Thus the bulk of the evidence points to a Laurasian origin for pangolins, and implies that the absence of pangolins from the early Cenozoic of Africa is not simply an artifact of poor preservation or collection biases.

Nevertheless, improvement of the Oligocene and Neogene record of pangolins should contribute significantly to the resolution of outstanding questions concerning the phylogenetic diversification and biogeographic origin of the Manidae itself. The biogeographic range of extant manids lies entirely outside the range of the Paleogene fossil taxa. The fossil record of undisputed manids in sub-Saharan Africa, where pangolins occur today, extends back only to the early Pliocene, as noted earlier. In Southeast Asia, the record does not extend back further than the Pleistocene (Emry, 1970). All but one of these Plio-Pleistocene fossils, the giant Asian pangolin *Manis palaeojavanica*, has been assigned to modern species. It remains unclear whether modern pangolins originated in Africa and subsequently migrated to Southeast Asia or the reverse. Pangolins persist in Europe until the Pliocene (Kormos, 1934), raising the possibility that both the extant African and Asian clades trace their origin back to Europe.

Finally, substantial morphological differences exist between extant manids and fossil pangolins such as *Cryptomanis* (Gaudin et al., 2006) and *Patriomanis* (Emry, 1970, 2004; Rose and Emry, 1993; Gaudin and Wible, 1999). For example, Gaudin et al. (2006) point out that the proximal limb elements are more robust but the distal limb elements less robust in *Cryptomanis* when compared to modern forms. Furthermore, there is substantial diversity in the locomotor habits among modern pangolins. For example, members of the genus *Phataginus* are small arboreal taxa with elongate, prehensile tails, *Smutsia gigantea* is a terrestrial quadruped and a strong digger, whereas *S. temminckii* often walks bipedally and does very little digging (Kingdon, 1974, 1997). Therefore, improvements in the African fossil record for pangolins should provide insight into the evolution of morphological diversity within Manidae, and potentially the origin of derived morphologies that characterize the family as a whole.

ACKNOWLEDGMENTS

I thank the editors for inviting me to contribute to this volume. I thank my artist, Julia Morgan Scott, for her always excellent work in preparing figures 31.2 and 31.3. Work on this manuscript was supported by National Science Foundation RUI Grant DEB 0107922.

Literature Cited

Barlow, J. C. 1984. Xenarthrans and pholidotes; pp. 219–239 in S. Anderson and J. Knox Jones, Jr. (eds.), *Orders and Families of Recent Mammals of the World.* Wiley, New York.

Botha, J., and T. J. Gaudin. 2007. A new pangolin (Mammalia: Pholidota) from the Pliocene of Langebaanweg, South Africa. *Journal of Vertebrate Paleontology* 27:484–491.

Delsuc, F., M. Scally, O. Madsen, M. J. Stanhope, W. W. de Jong, F. M. Catzefelis, M. S. Springer, and E. J. P. Douzery. 2002. Molecular phylogeny of living xenarthrans and the impact of character and taxon sampling on the placental tree rooting. *Molecular Biology and Evolution* 19:1656–1671.

Emry, R. J. 1970. A North American Oligocene pangolin and other additions to the Pholidota. *Bulletin of the American Museum of Natural History* 142:459–510.

———. 2004. The edentulous skull of the North American pangolin, *Patriomanis americanus. Bulletin of the American Museum of Natural History* 285:130–138.

Feiler, A. 1998. Das Philippinen-Schuppentier, *Manis culionensis* Elera, 1915, eine fast vergessene Art (Mammalia: Pholidota: Manidae). *Zoologische Abhandlungen, Staatliche Naturhistorische Sammlungen, Dresden Museum für Tierkunde* 50:161–164.

Gaubert, P., and A. Antunes. 2005. Assessing the taxonomic status of the Palawan pangolin *Manis culionensis* (Pholidota) using discrete morphological characters. *Journal of Mammalogy* 86:1068–1074.

Gaudin, T. J. 2004. Phylogenetic analysis of extinct and extant pangolins (Mammalia, Pholidota) and related taxa using postcranial data. *Journal of Vertebrate Paleontology* 24(suppl. to no. 3):63A.

Gaudin, T. J., R. J. Emry, and B. Pogue. 2006. A new genus and species of pangolin (Mammalia, Pholidota) from the late Eocene of Inner Mongolia, China. *Journal of Vertebrate Paleontology* 26:146–159.

Gaudin, T. J., R. J. Emry, and J. R. Wible. 2009. The phylogeny of living and extinct pangolins (Mammalia, Pholidota) and associated taxa: A morphology-based analysis. *Journal of Mammalian Evolution* 16:235–305.

Gaudin, T. J., and J. R. Wible. 1999. The entotympanic of pangolins and the phylogeny of the Pholidota (Mammalia). *Journal of Mammalian Evolution* 6:39–65.

Gebo, D. L., and D. T. Rasmussen. 1985. The earliest fossil pangolin (Pholidota: Manidae) from Africa. *Journal of Mammalogy* 66:538–540.

Gingerich, P. D. 1992. Marine mammals (Cetacea and Sirenia) from the Eocene of Gebel Mokattam and Fayum, Egypt: Stratigraphy, age and paleoenvironments. *University of Michigan Papers on Paleontology* 30:1–84.

Grassé, P. P. 1955. Ordre de Pholidotes; pp. 1267–1282 in P. P. Grassé (ed.), *Traité de Zoologie,* vol. 17, Mammifères. Masson et Cie, Paris.

Guth, C. 1958. Pholidota; pp. 641–647 in J. Piveteau (ed.), *Traité de Paléontologie,* vol. 2, no. 6, Mammifères Évolution. Masson et Cie, Paris.

Heath, M. E. 1992a. *Manis pentadactyla. Mammalian Species* 414:1–6.

———. 1992b. *Manis temminckii. Mammalian Species* 415:1–5.

———. 1995. *Manis crassicaudata. Mammalian Species* 513:1–4.

Hendey, Q. B. 1973. Fossil occurrences at Langebaanweg, Cape Province. *Nature* 244:13–14.

———. 1976. The Pliocene fossil occurrences in "E" quarry, Langebaanweg, South Africa. *Annals of the South African Museum* 69:215–247.

Kingdon, J. 1974. *East African Mammals,* vol. 1. University of Chicago Press, Chicago, 446 pp.

———. 1997. *The Kingdon Field Guide to African Mammals.* Princeton University Press, Princeton, 476 pp.

Klein, R. G. 1972. The Late Quaternary mammalian fauna of Nelson Bay Cave (Cape Province, South Africa): Its implications for megafaunal extinctions and environmental and cultural change. *Quaternary Research* 2:135–142.

Koenigswald, W. V. 1969. Die Maniden (Pholidota, Mamm.) des europäischen Tertiärs. *Mitteilungen der Bayerischen Staatssammlung für Paläontologie und Historische Geologie* 9:61–71.

———. 1999. Order Pholidota; pp. 75–80 in G. E. Rössner and K. Heissig (eds.), *The Miocene Land Mammals of Europe.* Pfeil, Munich.

Koenigswald, W. von, and T. Martin. 1990. Ein Skelett von *Necromanis franconica,* einem Schuppentier (Pholidota, Mammalia) aus dem Aquitan von Saulcet im Allier-Becken (Frankreich). *Eclogae Geologicae Helvetiae* 83:845–864.

Kormos, T. 1934. *Manis hungarica,* n. sp., das erste Schuppentier aus dem europäischen Oberpliozän. *Folia Zoologia et Hydrobiologia* 6:87–94.

McKenna, M. C., and S. K. Bell. 1997. *Classification of Mammals above the Species Level.* Columbia University Press, New York, 631 pp.

Novacek, M. J., and A. R. Wyss. 1986. Higher-level relationships of the recent eutherian orders: Morphological evidence. *Cladistics* 2:257–287.

Nowak, R. M. 1999. *Walker's Mammals of the World.* 6th ed. Johns Hopkins University Press, Baltimore, 1936 pp.

Patterson, B. 1978. Pholidota and Tubulidentata; pp. 268–278 in V. J. Maglio and H. B. S. Cooke (eds.), *Evolution of African Mammals.* Harvard University Press, Cambridge.

Pickford, M., and B. Senut. 1991. The discovery of a giant pangolin in the Pliocene of Uganda. *Comptes Rendus de l'Académie des Sciences, Paris,* Serie II, 313:827–830.

———. 1994. Fossil Pholidota of the Albertine Rift Valley, Uganda; pp. 259–260 in M. Pickford and B. Senut (eds.), *Geology and Palaeobiology of the Albertine Rift Valley, Uganda-Zaire, Palaeobiology II.* CIFEG Occasional Papers, Orleans.

Pocock, R. I. 1924. The external characters of the pangolins (Manidae). *Proceedings of the Zoological Society of London* 1924:707–723.

Rose, K. D., and R. J. Emry. 1993. Relationships of Xenarthra, Pholidota, and fossil "edentates": The morphological evidence; pp. 81–102 in F. S. Szalay, M. J. Novacek, and M. C. McKenna (eds.), *Mammal Phylogeny: Placentals.* Springer, New York.

Rose, K. D., R. J. Emry, T. J. Gaudin, and G. Storch. 2005. Xenarthra and Pholidota; pp. 106–126 in K. D. Rose and J. D. Archibald (eds.), *The Rise of Placental Mammals: Origins and Relationships of the Major Extant Clades.* Johns Hopkins University Press, Baltimore.

Schlitter, D. A. 2005. Order Pholidota; pp. 530–531 in D. E. Wilson and D. M. Reeder (eds.), *Mammal Species of the World.* 3rd ed. Johns Hopkins University Press, Baltimore.

Springer, M. S., M. J. Stanhope, O. Madsen, and W. W. de Jong. 2004. Molecules consolidate the placental mammal tree. *Trends in Ecology and Evolution* 19:430–438.

Storch, G. 1978. *Eomanis waldi,* ein Schuppentier aus dem Mittel-Eozän der "Grube Messel" bei Darmstadt (Mammalia: Pholidota). *Senckenbergiana Lethaea* 59:503–529.

———. 2003. Fossil Old World "edentates"; pp. 51–60 in R.A. Fariña, S. F. Vizcaíno, and G. Storch (eds.), *Morphological Studies in Fossil and Extant Xenarthra (Mammalia).* Senckenbergiana Biologica 83.

Storch, G., and T. Martin. 1994. *Eomanis krebsi,* ein neues Schuppentier aus dem Mittel-Eozän der Grube Messel bei Darmstadt (Mammalia: Pholidota). *Berliner Geowissenschaftliche Abhandlungen* E 13:83–97.

Swart, J. M., P. R. K. Richardson, and J. W. H. Ferguson. 1999. Ecological factors affecting the feeding behavior of pangolins *(Manis temminckii). Journal of Zoology* 247:281–292.

Szalay, F. S., and F. Schrenk. 1998. The middle Eocene *Eurotamandua* and a Darwinian phylogenetic analysis of "edentates." *Kaupia: Darmstädter Beiträge zur Naturgeschichte* 7:97–186.

THIRTY-TWO

Carnivora

LARS WERDELIN AND STÉPHANE PEIGNÉ

The order Carnivora has a shorter history in Africa than on any other continent except Australasia and South America. The definite record of the order on the African continent extends back to the Lower Miocene, though some earlier records may exist (discussed later). During this time, the order has diversified enormously, first as a result of migrations from Eurasia and later as a result of in situ speciation. Despite this, our knowledge of the history of African Carnivora still is poorer than for most continents, mainly due to the geographically biased fossil record on the African continent. For the Plio-Pleistocene, only parts of northern, eastern, and southern Africa have an adequate Carnivoran fossil record, and for the Miocene the situation is much worse, as only some time slices of this epoch have an adequate record in some parts of eastern Africa, with most of the rest of the continent simply a white spot on the map.

Nevertheless, this review encompasses more than 100 genera and about twice that many species. The organization is by family (in standard order) and genus (in alphabetical order), with a series of subheadings providing the bulk of the information. These are as follows:

Diagnosis We have tried to provide reasonable diagnoses of all genera. In most cases these have been taken from the original publications or from subsequent revisions. Many extant genera are diagnosed on the basis of soft-tissue characters, and for these we have tried to present provisional diagnoses based on craniodental information. These diagnoses should be treated as general indications only.

African Species A list of the species of each genus that are known from the African fossil record. When "*Genus* sp." is listed, this means that an unnamed species is known to differ significantly from all named species, or at least cannot comfortably be included in the named species.

Age The approximate first and last appearance datums for each African genus (not including extra-African occurrences).

Geographic Occurrence An alphabetical list of the African countries in which each genus has been found (again, extra-African occurrences are not included). Locality data are provided in the tabular material.

Remarks Any comments of a mainly taxonomic nature that we have found to be relevant in our study of the various taxa.

We conclude with short sections on biogeography and migration patterns, based on the data we have collected in creating the review.

Systematic Paleontology

Family AMPHICYONIDAE Haeckel, 1866
Table 32.1
Genus *AFROCYON* Arambourg, 1961
Figure 32.1

Diagnosis Revised from Arambourg (1961). Amphicyonid of large size, comparable to *Amphicyon giganteus* or *A. shahbazi*; p4 simple with slightly enlarged talonid and distal accessory cuspid; m1 voluminous with posteriorly located metaconid, relatively short talonid, large hypoconid in buccal position; m2 with well-developed protoconid and hypoconid, entoconid reduced; m3 longer than wide, bilobed and two-rooted; mandibular corpus very tall and narrow.

African Species *A. burolleti* Arambourg, 1961.

Age Ca. 19–15 Ma.

Geographic Occurrence Libya.

Remarks The single species of the genus is known from a fragmentary and poorly preserved left hemimandible with

FIGURE 32.1 *Afrocyon burolleti*, type specimen (MNHN, no number) in buccal and occlusal views.

TABLE 32.1
Occurrences of Amphicyonidae species

Sites	Afrocyon burolleti	Agnotherium antiquum	Agnotherium kiptalami	Agnotherium sp.	Amphicyon giganteus	Amphicyonidae indet.	Amphicyonidae sp. A	Bonisicyon illacabo	Cynelos euryodon	Cynelos macrodon	Cynelos minor	Cynelos sp.	Myacyon dojambir	Ysengrinia ginsburgi	Ysengrinia sp.
Arrisdrift					X									X	
Beni Mellal		cf.													
Bled Douarah Beglia		cf.													
Buluk						X									
Chamtwara									X						
Elisabethfeld															X
Escarpment (Gona)								X							
Fejej						X									
Fiskus															X
Fort Ternan				X											
Grillental															X
Hamadi Das (Gona)								X							
Hiwegi R 1									X						
Hiwegi R 3									X						
Hiwegi R 5									X	X					
Hondeklip Bay				X											
Jebel Zelten	X				cf.										
Kalodirr									X						
Kiahera Hill									X						
Kipsaraman			cf.						X	X					
Koru									X						
Kulu									X						
Langental															X
Legetet												X			
Lemudong'o								X							
Lothagam Lower Nawata							X	X							
Lothagam Upper Nawata								X							
Maboko						X									
Malembe						X									
Mfwangano									X						
Moroto II												?			
Napak									X						
Ngorora Member D			X												
Nyamsingula									X						
Oued Mya 1													X		
Rusinga									X	X					
Samburu Hills Namurungule						X									
Songhor									X						
Toros Menalla						X									
Wadi Moghra												X			

p4–m3. Its main distinctive feature is the presence of a double-rooted m3, which is unique among Amphicyoninae. Due the fragmentary nature of the holotype, detailed comparison with other taxa must await future discoveries.

Genus *AGNOTHERIUM* Kaup, 1833

Diagnosis Revised from Kurtén (1976). Amphicyonid of medium to large size, with felinoid characters: short snout, elongated upper canines, reduced anterior premolars, small, double-rooted P3, large P4 with small parastyle and reduced protocone; large M1–2; high-crowned, large p4, large m1 with trenchant trigonid, no metaconid, and reduced talonid with only a trenchant hypoconid crest; m2 reduced relative to m1 and lacking the paraconid; jaw deep.

African Species A. cf. *antiquum* Kaup, 1833, *A. kiptalami* Morales and Pickford, 2005, *A.* sp.

Age Ca. 14.8–9.5 Ma.

Geographic Occurrence Kenya, Morocco, South Africa, Tunisia.

Remarks This genus shows a derived condition toward hypercarnivory, including reduced premolars and m2–3, m1 lacking metaconid and with a talonid formed solely by the hypoconid, and extreme reduction of the P4 protocone. The diagnosis proposed by Kurtén (1976) includes the absence of m3. However, material from Steinheim (Helbing, 1929: figure 1) and Frohnstetten (Kuss, 1962) clearly shows the alveolus for m3. In addition, the isolated m2 from Beni Mellal (Morocco) assigned to a form close to *A. antiquum* by Ginsburg (1977) has a small facet on its distal face that indicates the presence of an m3. There is considerable size variation in African *Agnotherium*, from the early and small species from Fort Ternan to later, larger forms such as *Agnotherium kiptalami* from Ngorora D or *Agnotherium* sp. cf. *A. antiquum* from Bled Douarah. The taxonomic status of the species from Fort Ternan is not yet clear. It is smaller than other *Agnotherium* and the m1 has a small metaconid (Morales and Pickford, 2005a; Werdelin and Simpson, 2009, fig. 2), whereas the metaconid is completely absent in all other known specimens of *Agnotherium*.

Genus *AMPHICYON* Lartet, 1836

Diagnosis Translated and revised from Kuss (1965). Medium- to large-sized amphicyonids with dental formula I 3/3, C 1/1, P4/4, M 3/3; P1–3/p1–3 rounded, short, lacking posterior accessory cusps and with strong basal cingulum; p4 with posterior accessory cusps and sometimes a weak anterior bump; P4 somewhat shorter than m1, generally with a parastyle and a retracted and reduced protocone; m1 relatively low-crowned, paraconid somewhat truncated and with an anterior crest, metaconid reduced but always present, talonid wide with a tall hypoconid; M1 triangular or distally somewhat concave, lingual cingulum mostly present only posterior to the protocone; M2 and M3 relatively large; M2 enlarged and as wide or slightly wider than M1; m2 more or less rectangular, longer than wide, with vestigial paraconid and talonid shorter and narrower than trigonid.

African Species A. giganteus (Schinz, 1825).

Age Ca. 19–15 Ma.

Geographic Occurrence Libya, Namibia.

Remarks The genus is known from complete skeletons and many dental remains from North America and Europe. It is also known from many poorly characterized Asian species, the generic assignment of which requires confirmation (see review in Peigné et al., 2006). Although rare in Africa, the genus potentially has a Pan-African distribution, since it is present in northern (Jebel Zelten; Ginsburg and Welcomme, 2002) and southern (Arrisdrift; Morales et al., 2003) Africa. In both cases, remains have been assigned to *Amphicyon giganteus* or a closely related species. *Amphicyon giganteus* is a large-sized generalized species based on dental remains from the middle Miocene of France. Specimens from Libya (distal humerus and astragalus) are assigned to *Amphicyon* sp. cf. *A. giganteus* on the basis of their overall size only, as these remains are not diagnostic at the species level in *Amphicyon*. The presence of *A. giganteus* at Arrisdrift is well supported, with a subcomplete mandible with p4–m2 and some metapodials (Morales et al., 2003).

Genus *BONISICYON* Werdelin and Simpson, 2009

Diagnosis From Werdelin and Simpson (2009). Amphicyonidae of small size; carnassial shear on m1 entirely mesiodistal; m1 hypoconid an elongated crest, separated from trigonid by a narrow postvallid notch and effectively a part of the carnassial shear; m1 metaconid in evidence only as a bulge on the lingual side of the protocond; m1 relatively wide and bulbous at the base of the crown; m2 broad and short.

African Species B. illacabo Werdelin and Simpson, 2009.

Age Ca. 7.5–5.5 Ma.

Geographic Occurrence Ethiopia, Kenya.

Remarks This genus and species brings together material of a small amphicyonid from a number of sites. It includes the material from the Upper Nawata Fm., Lothagam described by Werdelin (2003) as Amphicyonidae sp. B as well as the tooth described as *Simocyon* sp. from Lemudong'o by Howell and García (2007). Isolated teeth from Gona, Ethiopia, bear witness to the uniqueness of this taxon.

Genus *CYNELOS* Jourdan, 1862

Diagnosis Translated and revised from Kuss (1965) and Hunt (1998a). Small- to large-sized amphicyonids with dental formula I 3/3, C 1/1, P 4/4, M 3/3; incisors tend to have accessory cusps; strong canines; long and slender premolars; p4 with posterior accessory cuspid only; M2–3 and m2–3 enlarged, with M2 only slightly more reduced than M1, and with the paracone as large or only slightly larger than the metacone; lower jaw slender.

African Species C. euryodon (Savage, 1965), *C. macrodon* (Savage, 1965), *C. minor* (Morales and Pickford, 2008), *C.* sp.

Age Ca. 20.5–16 Ma.

Geographic Occurrence Egypt, Kenya, Uganda.

Remarks Cynelos is the most diverse amphicyonine genus with at least six species in North America and up to nine in Europe during the late Oligocene and Miocene, although there is no consensus about the generic assignment of some European species (Hunt, 1998a; Peigné and Heizmann, 2003). In Africa, *Cynelos* is by far the most common Amphicyonidae with two species. *C. macrodon* is known only from isolated teeth but *C. euryodon*, the smallest African species, is well-known from several early Miocene localities in Uganda and Kenya. The arrival of *Cynelos* coincides with the first wave of migrations of Carnivora to Africa. Recently, Morales et al. (2007) proposed resurrecting *Hecubides* for the African species. In our opinion, these authors demonstrate the distinction between *Cynelos euryodon* and *C. lemanensis* only. Given the fragmentary nature of the known material, we see no strong support for a generic distinction of African *Cynelos*.

Genus *MYACYON* Sudre and Hartenberger, 1992

Diagnosis Translated and modified from Sudre and Hartenberger (1992). Amphicyonid of large size characterized by its sectorial molars; m1 large with an elongated talonid, strong protoconid with tall, strong trenchant anterior crest, paraconid indistinct and separated from protoconid by a very weak notch (visible on the buccal margin of the anterior crest), metaconid reduced and situated slightly posteriorly, talonid short, with a strong, crested hypoconid and a smaller, poorly developed entoconid situated far distally and lacking crest; m2 short and oblong, with protoconid trenchant, no paraconid, poorly developed metaconid situated at the level of the protoconid, and talonid short and narrower than the trigonid, with a strong hypoconid but no entoconid; trigonid of m1 and m2 with a strong buccal cingulum.

African Species M. *dojambir* Sudre and Hartenberger, 1992.

Age Ca. 11.2–9 Ma.

Geographic Occurrence Algeria.

Remarks This species is represented by a fragmentary right hemimandible with m1–m2 (m3 not yet erupted). This is a very large species that reached the size of the largest species of *Amphicyon*, A. *ingens* from North America (Hunt, 2003). As previously pointed out (Sudre and Hartenberger, 1992), *Myacyon* has nothing to do with any of the known Amphicyoninae. It remains a geographically and morphologically isolated species in northern Africa.

Genus *YSENGRINIA* Ginsburg, 1965

Diagnosis Modified from Hunt (1998a). Medium- to large-sized amphicyonid with dental formula I 3/3, C 1/1, P 4/4, M 3/3. P1–3/p1–3 low, reduced and lacking accessory cuspids; p4 tall with well-developed posterior accessory cuspid; robust, massive m1 trigonid with strongly reduced metaconid, talonid dominated by a centrally to buccally placed, prominent hypoconid crest and a reduced entoconid; M2–3/m2–3 not enlarged relative to M1/m1 as in, for example, *Amphicyon*; m2 elliptical in occlusal view, short, with large trigonid comprising a vestigial paraconid, strong protoconid, reduced metaconid, and low, short, posteriorly tapering talonid with a prominent hypoconid crest but no entoconid; mandibular corpus robust and tall, especially anteriorly.

African Species Y. *ginsburgi* Morales et al., 1998; *Ysengrinia* sp.

Age Ca. 20–17 Ma.

Geographic Occurrence Namibia, South Africa.

Remarks The only African species of *Ysengrinia* is known through dental and postcranial remains from Arrisdrift (Morales et al., 1998, 2003). There are many morphological differences between this species and the type species Y. *gerandiana* (Heizmann and Kordikova, 2000; Peigné and Heizmann, 2003), notably the more reduced size of p4 relative to m1 in the African species. The species assigned to *Ysengrinia* do not really form a homogeneous group (especially with the inclusion of poorly known species such as Y. *depereti* and Y. *valentiana*), and a detailed analysis shows many differences between them (Peigné and Heizmann, 2003).

AMPHICYONIDAE indet.

Age Ca. 20.5–5.5 Ma.

Geographic Occurrence Angola, Chad, Ethiopia, Kenya, Uganda.

Remarks The earliest and the latest occurences of the Amphicyonidae in Africa are documented by indeterminate remains.

An isolated incisor, particularly difficult to assign precisely, is known from Malembe (Hooijer, 1963). Though assigned to cf. *Amphicyon*, this tooth could equally belong to *Cynelos*. Aside from *Bonisicyon illacabo*, Lothagam includes a large amphicyonid (size of A. *giganteus*) from the Lower Nawata Fm., known from an upper molar and fragmentary postcranial elements.

Other records of Amphicyonidae are mentioned in faunal lists that require confirmation. Thus, ?*Cynelos* sp. may be present in the early middle Miocene site of Moroto II, Uganda (Pickford et al., 2003), but here we consider it an undetermined amphicyonid. One exception is the recent description of new material from the Namurungule Fm. (Samburu Hills) that includes lower teeth of a large Carnivore assigned to Amphicyonidae or Ursidae (Tsujikawa, 2005). Illustrations of the specimens show that they belong to an amphicyonid. The author suggests close relationships to *Agnotherium*, but in our opinion, the m2 is much more similar to that of *Ysengrinia* spp. (especially Y. *gerandiana* and Y. *americana*) in having an elliptical outline, a posteriorly tapering talonid with a prominent, laterally placed hypoconid, and a distinct buccal cingulum. The m2 from Samburu Hills is, however, larger in size and more elongated than in *Ysengrinia*, and, above all, it comes from geologically much younger strata. In addition, the p4 of the same individual differs from that of *Ysengrinia* in lacking accessory cuspids. The absence of the posterior accessory cuspid on p4 is a derived feature of *Pseudarctos bavaricus* and *Ictiocyon socialis*, two much smaller species with an m2 that is morphologically distinct from the Samburu Hills amphicyonid. The species from Namurungule may represent a new species, but additional remains are necessary to confirm this hypothesis.

Family URSIDAE Fischer, 1817
Table 32.2
Genus *AGRIOTHERIUM* Wagner, 1837

Diagnosis Modified from Hunt (1998c). Large-sized ursine with dental formula I 3/3, C 1/1, P 3–4/2–4, M 2/3; sexually dimorphic; short-snouted robust skull, somewhat brachycephalic; palate wide; premolar toothrow much shortened with anterior premolars reduced in size, single rooted, and low crowned; P4 robust, with strong parastylar cusp and protocone shelf; M2 with only rudimentary talon; m1–2 cusp pattern variable; premasseteric fossa present; symphyseal region of lower jaw ventrally produced as a "chin"; long-footed, plantigrade limbs.

African Species A. *africanum* Hendey, 1980, A. *aecuatorialis* Morales et al., 2005, A. sp.

Age Ca. 6.3–4 Ma.

Geographic Occurrence ?D. R. Congo, Ethiopia, Kenya, Libya, South Africa, Uganda.

Remarks This genus has a stratigraphic range in Africa from the latest Miocene to the early Pliocene. Though generally rare, it was successful and spread through the entire continent. The Muishondfontein Pelletal Phosphorite Mb. of the Varswater Fm. of Langebaanweg has yielded a large number of cranial, dental and postcranial specimens of *Agriotherium africanum*, representing a minimum of 14 individuals. The ursid from Sahabi, previously identified as *Indarctos* sp. (Howell, 1987), is probably a close relative of this species (Morales et al., 2005). The possible record of the genus from the upper member of the Sinda Beds of the Democratic Republic of Congo (late Miocene to Pliocene) is speculative given the available material (Yasui et al., 1992). Additional material is also known from early Pliocene sites in Ethiopia (L.W., pers. obs.).

TABLE 32.2
Occurrences of Ursidae species

Sites	*Agriotherium aecuatorialis*	*Agriotherium africanum*	*Agriotherium* sp.	*Hemicyon* sp.	*Indarctos arctoides*	Ursidae indet.	*Ursus arctos*	*Ursus etruscus*	*Ursus* sp.
Afalou bou Rhummel							X		
Ahl al Oughlam								cf.	
Ain Bahya III									X
Ain Rouina								cf.	
Ali Bacha							X		
Aramis		cf.							
Babors							X		
Bouknadel							X		
Boulhaut							X		
Brèche entre Oran et Mers-el-Kébir							X		
Djebel Thaya							X		
Doukkala I							X		
Douar Debagh							X		
El Khenzira							X		
El Ksiba							X		
Fort Bourdonneau							X		
Grotte de l'Akouker							X		
Grotte d'Os (Djurdjura)							X		
Grotte de l'Ours (Djurdjura)							X		
Grotte des Ours (Constantine)							X		
Grotte des Ours G0 (Casablanca)							X		
Grotte du Mouflon (Constantine)							X		
Hadar Denen Dora						X			
Hadar Kada Hadar						X			
Khifan bel Ghomari							X		X
Koobi Fora Lonyumun						X			
Koobi Fora Okote						?			
Koobi Fora Tulu Bor						X			
L'Anou Tenechiji							X		
L'Ifri en Terga Roumi							X		
La Madeleine							X		
La Pointe Pescade							X		
Langebaanweg PPM		X							
Les Falaises							X		
Menacer					aff.				
Nachukui Lower Lomekwi						X			
Nkondo			X						
Oulad Hamida 1—rhino cave							X		
Oulad Hamida Th III—*H. erectus*							X		
Rusinga				X					
Sahabi		cf.							
Sagantole Fm.			X						
Sidi Abderrahmane							X		
Sinda			X						
Takouatz Guerrissène									X

TABLE 32.2

(CONTINUED)

Sites	*Agriotherium aecuatorialis*	*Agriotherium africanum*	*Agriotherium* sp.	*Hemicyon* sp.	*Indarctos arctoides*	Ursidae indet.	*Ursus arctos*	*Ursus etruscus*	*Ursus* sp.
Tamar Hat							X		X
Ternifine							X		
Thomas I—*H. erectus* cave							X		
Thomas II							X		
Thomas III							X		
Tighenif						X			
Toulkine							X		
Tugen Mabaget	X								

Genus *HEMICYON* Lartet, 1851

Diagnosis Translated and modified from Ginsburg and Morales (1998). Mid- to large-sized species of Hemicyoninae with dental formula I 3/3, C 1/1, P 4/4, M 2/3; m1 talonid simple and nearly symmetrical, talonid groove shallow and roughly axial; P4 with paracone anterobuccally inflated and without anterobuccal cingulum; M1 subrectangular; M2 tends to be oval in shape; P4 protocone elongated; lower premolars simple; m1 with a strong metaconid well separated from the protoconid; talonid shallow, with low hypoconid and lingual tubercles of the talonid poorly or not developed; m2 generally broader than m1, with protoconid and metaconid very prominent; m3 small, low, with only protoconid and metaconid developed.

African Species H. sp.

Age Ca. 18–17 Ma.

Geographic Occurrence Kenya.

Remarks Hemicyon belongs to the Hemicyoninae, traditionally considered a subfamily of Ursidae. It is known in Africa from a single tooth, a P4 from the early Miocene of Rusinga (Schmidt-Kittler, 1987), which is only slightly younger than the earliest, much smaller, Eurasian species of the genus, *H. gargan* (Ginsburg and Morales, 1998). Schmidt-Kittler (1987) rightly casts doubts on the generic assignment of the Rusinga tooth, arguing that his specimen compares well with an isolated P4 from Wintershof-West (early Miocene, Germany) that is now assigned to another hemicyonine, *Phoberocyon dehmi* (Ginsburg and Morales, 1998).

Genus *INDARCTOS* Pilgrim, 1913

Diagnosis Modified from Hunt (1998c). Mid- to large-sized ursid, sexually dimorphic with dental formula I 3/3, C 1/1, P 4/4, M 2/3; skull dolichocephalic and snout short; P1/ p1–P3/p3 well developed in earlier species and reduced in size but nearly always present and single rooted in advanced species; P4 robust with parastylar cusp present but usually not as developed as in *Agriotherium*; molars low crowned; M2 with elongate talon; protoconid-metaconid-entoconid-entoconulid of m1 aligned in smooth descending curve; premasseteric fossa absent; anterior part of lower jaw tapers forward, not as squared off and blunt as in *Agriotherium*; "chin" present only in old individuals; plantigrade.

African Species I. aff. *arctoides* (Depéret, 1895).

Age Ca. 7.1–5.3 Ma.

Geographic Occurrence Algeria.

Remarks Apart from the *Hemicyon* sp. from Rusinga, *Indarctos* is the earliest ursid in Africa. It is known through a single record from Menacer, late Miocene of Algeria (Petter and Thomas, 1986). The genus has also been identified from Sahabi (Howell, 1987) but assignment of this material is debatable, and this record is here assigned to *Agriotherium*. According to Howell (1987), two genera, *Agriotherium* and *Indarctos*, are present at Sahabi, mainly based on the size variability of the sample. However, size alone cannot be used to distinguish the strongly sexually dimorphic species of the two genera. It is therefore most probable that only one genus (*Agriotherium*) is present at the Libyan locality.

Genus *URSUS* Linnaeus, 1758

Diagnosis Modified from Hunt (1998c). Mid- to large-sized ursid; sexually dimorphic, with dental formula I 3/3, C 1/1, P 2–4/1–4, M 2/3; skull dolichocephalic; no premasseteric fossa; m2 shorter than m1 in earlier forms but m2 later increasing to exceed m1 length; great elongation of m2-3/ M1-2; Pleistocene forms exhibit twinning of metaconid of m1–m2 and considerable widening of m2-3 relative to m1.

African Species U. cf. *etruscus* Cuvier, 1823, *U. arctos* Linnaeus, 1758, *U.* sp.

Age Ca. 2.5 Ma–Holocene.

Geographic Occurrence Algeria, Morocco, Tunisia.

Remarks Ursus originated in Eurasia and reached Africa in the mid-Pliocene. It is only known from the Maghreb, where it became extinct around the mid-19th century (Servheen et al., 1998; Nowak, 2005, p. 124). The earliest record of the genus, *Ursus* sp. cf. *U. etruscus*, is from the mid-Pliocene site of Ahl al Oughlam. The presence of a form close to this typical Villafranchian Eurasian bear in northwestern Africa supports a migration event from Europe, possibly through Spain and the Gibraltar Strait, where the species is known from as early as the early Pliocene of Layna (Fraile et al., 1997). A large number of Pleistocene to Holocene records of *Ursus* spp. are known from northern Africa, especially from Morocco and Algeria (Hamdine et al., 1998).

URSIDAE indet.

Age Ca. 4.4–0.7 Ma.
Geographic Occurrence Algeria, Ethiopia, Kenya.
Remarks At least one species of undescribed ursine is represented by postcranial material from several members at Hadar, Koobi Fora and West Turkana (Werdelin and Lewis, 2005).

Family CANIDAE Fischer, 1817
Table 32.3
Genus *CANIS* Linnaeus, 1758

Diagnosis Modified after Munthe (1998). Canids of large size; shorter, broader face than *Vulpes*; wide zygomatic arches; width of skull across widest part of arches equal to at least half the length of the skull; frontal sinuses enlarged and invading the postorbital processes; incisors with accessory cusps and I3 enlarged; P4 about the same length as M1, no parastyle; m1 talonid with crest uniting hypoconid and entoconid.
African Species C. adustus Sundevall, 1847, *C. atrox* Broom, 1948, *C. aureus* Linnaeus, 1758, *C. brevirostris* Ewer, 1956, *C. falconeri* Forsyth Major, 1877, *C. mesomelas* Schreber, 1776, *C. pictus* (Temminck, 1820), *C.* sp.
Age Ca. 3.5 Ma–Recent.
Geographic Occurrence Algeria, Ethiopia, Kenya, Morocco, South Africa, Tanzania, Zambia.
Remarks The genus *Canis* (here including *Lycaon*) is widely distributed in Africa today. Modern *Canis* are cursorial, open-habitat adapted taxa. This is likely to have been true in the past as well and may explain their relative scarcity as fossils, since such habitats seem to be underrepresented in the carnivoran fossil record of Africa. The record of *Canis* sp. from South Turkwel is, at ca. 3.5 Ma, one of the oldest for the genus outside North America (Werdelin and Lewis, 2000).

Genus *EUCYON* Tedford and Qiu, 1996
Figure 32.2A

Diagnosis Modified after Tedford and Qiu (1996). Skull with frontal sinus invading the base of the postorbital process, usually removing the "vulpine-crease" on the dorsal surface of the process; paroccipital process expanded posteriorly, usually with a salient tip; mastoid process enlarged into a knob or ridgelike prominence; lacking foxlike lateral flare and eversion of the dorsal border of the orbital part of the zygoma; lacking a transverse cristid connecting the hypoconid and entoconid of the m1 talonid; second posterior cusplet present on p4.
African Species E. intrepidus Morales et al., 2005, *E. minimus* Haile-Selassie et al., 2009, *E. wokari* García, 2008, *E.* sp. (figure 32.2A).

Age Ca. 6.1–2.6 Ma.
Geographic Occurrence Ethiopia, Morocco, Kenya, South Africa.
Remarks The genus *Eucyon* was originally erected for some Eurasian and North American species transitional between the primitive *Leptocyon* spp. and derived *Canis* (Tedford and Qiu, 1996). *Eucyon intrepidus* from the Lukeino Fm., Kenya, recently described by Morales et al. (2005), is the oldest known representative of the genus in Africa. It is based on isolated teeth, but these are diagnostic for the genus. A second possible record of this species from Lemudong'o in Kenya has recently been described by Howell and García (2007). Recently, García (2008) has described the new species *E. wokari* from the early Pliocene of Ethiopia (Aramis and Kuseralee). More complete material of *Eucyon* sp. is known from Langebaanweg (figure 32.2A), in the form of a skull, mandibles, and postcranial elements of more than one individual (Spassov and Rook, 2006). *Eucyon* is also known from the much younger site of Ahl al Oughlam, where material described by Geraads (1997) as *Canis* aff. *aureus* at least in part is attributable to *Eucyon* sp. (but see Geraads, 2008).

Genus *NYCTEREUTES* Temminck, 1838
Figures 32.2B and 32.2C

Diagnosis Modified after Ward and Wurster-Hill (1990). Skull small, greatest length <130 mm with short, narrow muzzle and low forehead; parietals with rugose surface and slight interparietal crest; mandible with a distinct rounded subangular lobe on the posterior margin; dental formula I 3/3, C 1/1, P 4/4, M 2/3, total 42, but an extra upper molar is common. Carnassial blades reduced and molars relatively large.
African Species N. abdesalami Geraads, 1997, *N. terblanchei* (Broom in Broom and Schepers, 1946; figure 32.2C), *N.?* sp. nov. Werdelin and Dehghani, in press (figure 32.2B), *N.* sp.
Age Ca. 3.1–1.3 Ma.
Geographic Occurrence Morocco, South Africa, Tanzania.
Remarks Nyctereutes is mainly a Eurasian taxon and its lineage was one of the first canids to evolve after the family dispersed into Eurasia near the end of the Miocene (Tedford and Qiu, 1991). Barry (1987) described a medium-sized canid from the Laetolil Beds, Upper Unit, as aff. *Canis brevirostris*. Tedford (pers. comm., 1999) has suggested that this taxon should be placed in *Nyctereutes*, and on the basis of dental characteristics we concur (Werdelin and Dehghani, in press). It should be noted, however, that the Laetoli form lacks the subangular lobe of the mandible that is characteristic of *Nyctereutes*. There is also a possible specimen of *Nyctereutes* from the latest Miocene of Lissasfa in Morocco (Geraads, 2008). Specimens definitely belonging to *Nyctereutes* have been recorded from Ahl al Oughlam in Morocco (Geraads, 1997) and Kromdraai and Elandsfontein in South Africa (Hendey, 1974; Ficcarelli et al., 1984).

Genus *OTOCYON* Müller, 1836

Diagnosis Small-sized Canidae; dentition with supernumerary molars, dental formula I 3/3, C 1/1, P 4/4, M 3–4/4–5; premolars small, with mesial and distal diastemata; upper molars similar in size, multicusped; P4 and m1 hypocarnivorous; m1 with low trigonid, long talonid; P4 triangular, with very short metastyle.
African Species O. megalotis (Desmarest, 1822).

TABLE 32.3
Occurrences of Canidae species

Sites	Canidae indet.	Canis adustus	Canis atrox	Canis aureus	Canis brevirostris	Canis falconeri	Canis mesomelas	Canis pictus	Canis sp.	Eucyon intrepidus	Eucyon minimus	Eucyon sp.	Nyctereutes abdesalami	Nyctereutes terblanchei	Nyctereutes sp. nov.	Nyctereutes sp.	Otocyon megalotis	Prototocyon recki	Prototocyon sp.	Vulpes chama	Vulpes pattisoni	Vulpes pulcher	Vulpes riffautae	Vulpes rueppellii	Vulpes vulpes	Vulpes sp.
Ahl al Oughlam												X	X											aff.		
Ain Bahya				X																					X	
Ain Boucherit				X																					X	
Ain Hanech						cf.																			X	
Asbole				cf.																						
Awash 7 (Kalb)							cf.																			
Bouknadel				X																					X	
Busidima-Telalak									X																	
Cap Achakar									X																	
Cooper's									X																	
Dar es Soltane									X																	
Diepkloof Rock Shelter							X																			
Djebel Irhoud																									X	
Doukkala I									X																	
Doukkala II				X																					X	
Drimolen																				X						
Duinefontein 2							X																			
Equus Cave								X																		
Elands Bay Cave							X															X				
Elandsfontein Main							X	X														X				
El Harhoura 1				X																						
Florisbad								X																		
Gladysvale									X																	
Grotte des félins				X																						X
Hadar Denen Dora									X																	
Hadar Kada Hadar									X																	
Hadar Sidi Hakoma	X																									
Hopefield							X	X																		
Isenya							aff.																			
Kabwe		X																								
Kifan Bel Gohmari				X																						
Konso-Gardula 2									X																	
Koobi Fora KBS							cf.																			
Koobi Fora Upper Burgi									X																	
Kossom Bougoudi	X																									
Kromdraai Member A			X				cf.		X					X								X				
Laetolil Beds Upper Unit									cf.						X				X							
Lainyamok		X					X		X								X									
Langebaanweg Baard's Quarry									X																	
Langebaanweg PPM												X														
Lemudong'o									aff.																	
Lissasfa																?										

Sites	Canidae indet.	Canis adustus	Canis atrox	Canis aureus	Canis brevirostris	Canis falconeri	Canis mesomelas	Canis pictus	Canis sp.	Eucyon intrepidus	Eucyon minimus	Eucyon sp.	Nyctereutes abdesalami	Nyctereutes terblanchei	Nyctereutes sp. nov.	Nyctereutes sp.	Otocyon megalotis	Prototocyon recki	Prototocyon sp.	Vulpes chama	Vulpes pattisoni	Vulpes pulcher	Vulpes riffautae	Vulpes rueppellii	Vulpes vulpes	Vulpes sp.
Lothagam Kaiyumung									X																	
Makapansgat Member 3		cf.																		X						
Makapansgat Member 4																				X						
Melkbos							X																			
Melkbosstrand							X																			
Mursi Fm.	X																									
Nachukui Kalochoro									X																	
Nachukui Natoo									X																	
Olduvai Bed I							cf.											X								
Olduvai Bed II						X																				
Olduvai Bed IV							?																			
Oulad Hamida 1—rhino cave		aff.						aff.																		
Oulad Hamida Th III—*H. erectus*		aff.																								
Plovers Lake	X						X	X									X			X						
Sagantole Fm.											X															
Saldanha Lime Quarry							X																			
Saldanha Sea Harvest							X													X						
Sibudu HP	X																									
Slangkop							X																			
South Turkwel									X																	
Sterkfontein Member 4						X	cf.																			
Sterkfontein Member 5							cf.		X																	
Sterkfontein Jacovec Cave									X											X						
Swartklip 1							X	X												X						
Swartkrans Member 2							cf.		X																	
Swartkrans Member 3							cf.		X											X						
Taung Dart deposits																					X					
Thomas 1—*H. erectus* cave		aff.							cf.																	
Tighenif		cf.																								
Toros Menalla																							X			
Tugen Lukeino										X																
Ysterfontein							X																			

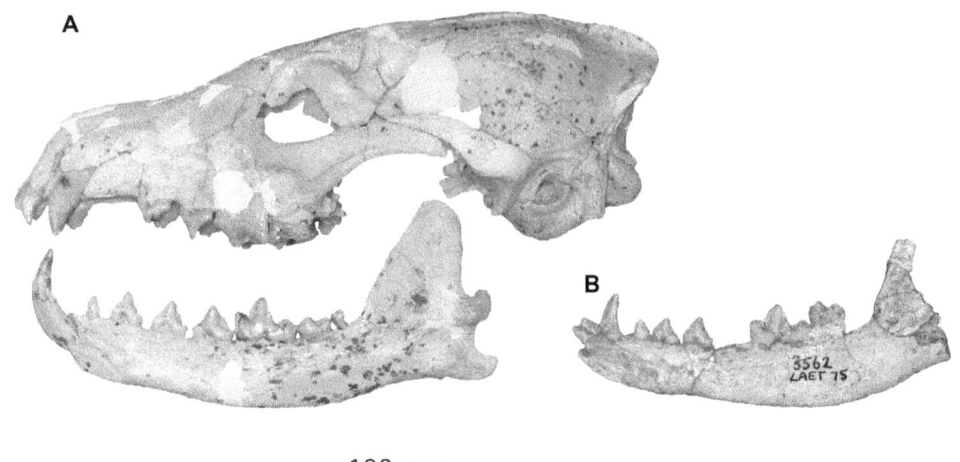

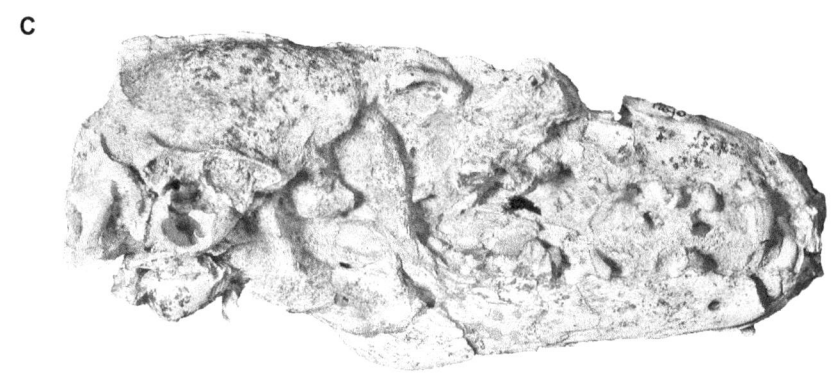

FIGURE 32.2   African Canidae. A) *Eucyon* sp. PQ-L 31272 from the Muishondfontein Pelletal Phosphorite Mb., Varswater Fm., Langebaanweg, cranium in left lateral view and right hemimandible in lingual view. B) *?Nyctereutes* sp. nov. (part) LAET 75-3522 (specimen incorrectly labeled) from the Upper Laetolil Beds, Laetoli, right hemimandible in lingual view. C) *Nyctereutes terblanchei* type specimen KA 1290 from Kromdraai Mb. A skull in right lateral view.

Age Ca. 0.3 Ma–Recent.
Geographic Occurrence Kenya.
Remarks The extant species *O. megalotis* is in the fossil state only known from Lainyamok in Kenya at ca 300 ka (Potts and Deino, 1995). Earlier records of bat-eared foxes are referred to *Prototocyon* (see next section).

Genus *PROTOTOCYON* Pohle, 1928

Diagnosis Translated and modified from Pohle (1928). Small, foxlike genus of dogs, close to *Otocyon*; frontal sinuses and sagittal crest absent; the weak lambdoid crests form a central invagination; dental formula I 3/3, C 1/1, P 4/4, M 2/3; premolars small; carnassial with a weak parastyle, width more than 60% of length; molars relatively large and very similar to each other, resembling those of *Otocyon*; lower jaw with a processus subangularis.
African Species *P. recki* Pohle, 1928, *P.* sp.
Age Ca. 2.0–1.7 Ma (possibly as early as 3.7 Ma).
Geographic Occurrence Tanzania.

Remarks This genus, known through material of several individuals from Olduvai, Bed I (Petter, 1973), is only doubtfully distinct from the modern genus, *Otocyon*. The main differences lie in the somewhat more primitive dentition of *P. recki*. It is a typical bat-eared fox and probably had the same ecological habits as its modern counterpart. Remains of a canid in the *Otocyon* lineage have been recorded from Laetoli (Werdelin and Dehghani, in press), but whether they can truly be referred to the genus *Prototocyon* remains to be seen.

Genus *VULPES* Frisch, 1775

Diagnosis Modified after Munthe (1998). Skull with long, sharp muzzle; postorbital process thin and concave dorsally; frontal bones less convex than in *Canis*; simple incisors; canine teeth slender, relatively longer than in *Canis*; cheek teeth slender with sharp, well-defined cusps and crests.
African Species *V. chama* (Smith, 1833), *V. pattisoni* Broom, 1948, *V. pulcher* Broom, 1939, *V. riffautae* Bonis et al., 2007, *V. rueppellii* (Schinz, 1825), *V. vulpes* (Linnaeus, 1758).

Age Ca. 7 Ma–Recent.

Geographic Occurrence Algeria, Chad, Kenya, Morocco, South Africa, Tanzania.

Remarks Most finds of foxes in Africa are from the southern part of the continent. However, some of the oldest finds thus far (e.g., an unidentified ?Vulpini from the Mursi Fm., Ethiopia) have come from eastern Africa, suggesting that much of the group's history on the continent remains to be discovered. Recent discoveries have demonstrated that a small-sized fox is present in Toros Menalla, late Miocene of Chad, which implies a much earlier occurrence for the genus in Africa, and for the whole tribe Vulpini in the Old World (Bonis et al., 2007b).

CANIDAE indet.

Age Ca. 5.3–4 Ma.

Geographic Occurrence Chad, Ethiopia, South Africa.

Remarks As noted earlier, a probable vulpine has been recovered from the Mursi Fm. Since it may not belong to any of the listed genera, we prefer to list it here. Indeterminate remains of canids are also known from Kossom Bougoudi, Chad (Bonis et al., 2008).

Family MUSTELIDAE Fischer, 1817
Table 32.4
Genus *AONYX* Lesson, 1827

Diagnosis After Willemsen (1992). Dentition more robust than in *Lutra*; teeth broad; P4 with large talon, protocone elongated, extending distally nearly to the metacone, no hypocone; m1 with talonid broader than trigonid, hypoconid present, entoconid absent or smaller than hypoconid, buccal cingulum of talonid strong.

African Species *A. capensis* (Schinz, 1821).

Age Ca. 4.3 Ma–Recent.

Geographic Occurrence Ethiopia, Kenya, South Africa, Tanzania.

Remarks Compared to other Lutrinae, members of the Aonyxini are rare in the African fossil record. The oldest record is from the Laetolil Beds, Lower Unit.

Genus *DJOURABUS* Peigné et al., 2008

Diagnosis After Peigné et al. (2008a). Large-sized species; tall and thick mandibular corpus; m1 bunodont, short, and very broad with a trigonid broader than the talonid, transversely oriented paraconid shelf, bases of the metaconid and paraconid connate, short; m1 talonid molarised and short, with squared-off distal margin and individualized and low entoconid and hypoconid, and cingulid very reduced and restricted to the mesial margin of the paraconid.

African Species *D. dabba* Peigné et al., 2008.

Age Ca. 7 Ma.

Geographic Occurrence Chad.

Remarks This species is based on a fragmentary mandible on which the canine and the m1 are the only preserved teeth. Despite the fragmentary nature of the material, the morphology of the mandible and carnassial of this bunodont lutrine is easily distinguished from that of the other bunodont genera (*Enhydriodon, Enhydritherium, Paludolutra,* and *Sivaonyx*). As it is one of the earliest-known bunodont lutrines, the origin of *Djourabus* remains unclear.

Genus *EKORUS* Werdelin, 2003

Diagnosis After Werdelin (2003). Mustelidae of gigantic size; dentition highly modified for slicing, with narrow canines and slender premolars; m1 lacking metaconid, talonid reduced to a single, tall cusp placed directly posterior to the trigonid blade; m2 small and peg shaped; M1 very reduced and much broader than long; appendicular skeleton modified, with relatively long limbs but short, broad, semiplantigrade feet; humerus lacking entepicondylar foramen; vertebral column slender; tail long.

African Species *E. ekakeran* Werdelin, 2003.

Age Ca. 7.4–6.5 Ma.

Geographic Occurrence Kenya.

Remarks This form was described on the basis of a nearly complete skeleton from the Lower Nawata Mb. at Lothagam (Werdelin, 2003). It is by far the most hypercarnivorous and cursorial mustelid known. Its origin is unknown but may lie with the Eurasian Miocene *Ischyrictis* group, which also includes large, hypercarnivorous forms. If so, the *Ekorus* lineage has a long, independent history in Africa.

Genus *EOMELLIVORA* Zdansky, 1924

Diagnosis Modified after Wolsan and Semenov (1996). Musteline mustelids of very large size; P3 with one or two posterior accessory cusps; P4 with a subconical protocone, and with paracone-protocone and paracone-parastyle crests; M1 with a vestigial metacone, an arched, ridge-shaped protocone continuing into the anterior protocone crest, and a talon about equally expanded anteriorly and posteriorly; p4 with a posterior accessory cusp; m1 metaconid absent and replaced by a distinct crest, talonid with lingual talonid crest enclosing a vestigial basin, with single but strong, nearly centrally positioned hypoconid; m2 elongated anteroposteriorly, with a low crown surrounded by a cingulum and a very small, almost centrally placed protoconid, linked with the cingulum by crests anteriorly, posteriorly, and lingually.

African Species *E. tugenensis* Morales and Pickford, 2005.

Age Ca. 12–11.5 Ma.

Geographic Occurrence Kenya.

Remarks Morales and Pickford (2005a) report material from the Ngorora Fm. that they refer to a new species of *Eomellivora*. This material is morphologically similar to the Eurasian *E. wimani* but much smaller, being approximately the size of the extant *Mellivora capensis*.

Genus *EROKOMELLIVORA* Werdelin, 2003

Diagnosis After Werdelin (2003). Small-sized Mellivorinae with slender mandibular horizontal ramus; premolars long and slender; m1 low and long with small metaconid; m2 present and single rooted.

African Species *E. lothagamensis* Werdelin, 2003.

Age Ca. 6.5–5.5 Ma.

Geographic Occurrence Kenya.

Remarks This taxon resembles *Mellivora* dentally but lacks several of the derived features of the extant genus, such as the loss of m2. It is possible that *Mellivora* is a direct descendant of *Erokomellivora*.

Genus *HOWELLICTIS* Bonis et al., 2009

Diagnosis After Bonis et al. (2009). Small-sized Mellivorinae without P1/p1 differing from *Mellivora* in its smaller size, less imbricated and narrower premolars, lack of a sagittal crest, and presence of a reduced m2.

African Species *H. valentini* Bonis et al., 2009.

Age Ca. 7 Ma.

Geographic Occurrence Chad.

Remarks The species is represented by a fairly good sample from Toros-Menalla, Chad. It is morphologically different from *Mellivora* and could belong the ancestral stock of this genus, along with *Erokomellivora*.

Genus *ICTONYX* Kaup, 1835

Diagnosis Small-sized Mustelidae; dental formula I 3/3, C 1/1, P 3/3, M 1/2; premolars small and slender; m1 with a low trigonid and broad and flat talonid about as long as the trigonid; P4 slender; M1 short with reduced metastyle wing.

African Species *I. libyca* (Hemprich and Ehrenberg, 1833), *I. striatus* (Perry, 1810).

Age Ca. 2.5 Ma–Recent.

Geographic Occurrence Angola?, Morocco, South Africa.

Remarks *Ictonyx* is only known as a fossil from two records of extant species, since *I. bolti* has been transferred to the genus *Prepoecilogale* (Petter and Howell, 1985).

Genus *KENYALUTRA* Schmidt-Kittler, 1987

Diagnosis After Schmidt-Kittler (1987). Lutrine of the size of, or somewhat smaller than, the Recent clawless otter; trigonid of the lower carnassial as low as the talonid, metaconid and protoconid equal in size, talonid markedly broader that the trigonid, and hypoconid situated far backward.

African Species *K. songhorensis* Schmidt-Kittler, 1987.

Age Ca. 20–19 Ma.

Geographic Occurrence Kenya.

Remarks *Kenyalutra* was erected by Schmidt-Kittler (1987) on the basis of two lower carnassials, one of them fragmentary, from Songhor, Kenya. He assigned them to the Lutrinae mainly on the basis of two features: trigonid and talonid of nearly equal height and talonid very broad, considerably broader than the trigonid. However, these are characters of questionable significance given the age difference between *Kenyalutra* and other African Lutrinae, the very small size of *Kenyalutra* compared to comparable Lutrinae, and the highly derived nature of some features, such as the breadth of the talonid and the posteriorly placed hypoconid. Morales et al. (2000) suggested that the *Kenyalutra* specimens might be the unknown lower dentition of *Kelba quadeemae*, a mammal of uncertain affinities first described by Savage (1965) on the basis of isolated upper molars. They further suggested that this combination might be referred to the hemigaline viverrids. However, recently described material clearly shows that *Kelba* is not a carnivoran (Cote et al., 2007), whereas the *Kenyalutra* specimens almost certainly are. *Kenyalutra* might still be a viverrid, but more material is needed to make any sort of assessment of its phylogenetic relationships. Thus, for the time being we retain it in the Mustelidae.

Genus *LUOGALE* Schmidt-Kittler, 1987

Diagnosis After Schmidt-Kittler (1987). Mandibular ramus very strong: m1 shorter and taller than in the *Laphictis-Ischyrictis* group, with small but clearly developed metaconid and short talonid with a single blunt cusp; P4 with strong carnassial blade and well-separated rounded protocone, small parastyle present.

African Species *L. rusingensis* Schmidt-Kittler, 1987.

Age Ca. 18–17 Ma.

Geographic Occurrence Kenya.

Remarks Schmidt-Kittler (1987) described this new genus and species on the basis of some tooth and mandible fragments from Rusinga, Kenya. It shows similarities with less derived forms of the *Ischyrictis* group in the structure of m1. Schmidt-Kittler (1987) suggests that it may have originated from the European *Paragale-Plesiogale* group.

Genus *LUTRA* Brisson, 1762

Diagnosis After Willemsen (1992). Lutrine with wide, flat head; fingers and toes extensively webbed and with well-developed claws; tail tapering evenly, somewhat flattened. Skull long and depressed, facial part rather long, intertemporal region not swollen; dental formula I 3/3, C 1/1, P 4/3, M 1/2; teeth with sharp cutting edges, never blunt and not very robust; P4 with a sharp cutting blade on the buccal side, formed by paracone and metacone, talon not covering entire lingual side of trigon; m1 with sharp cuspids, hypoconid forming a sharp edge, external cingulum of talonid not strongly developed.

African Species *L. fatimazohrae* Geraads, 1997, *L. libyca* Stromer, 1914.

Age Ca. 7.1 Ma–Recent.

Geographic Occurrence Egypt, Morocco.

Remarks The species listed here are all quite similar to each other and to *Torolutra* (discussed later). The number of valid species and their relationship to each other and to other species of *Lutra* (e.g., *L. simplicidens* Thenius, 1948) remains obscure at present.

Genus *MARTES* Pinel, 1792

Diagnosis After Anderson (1970). Skull ranging in basilar length from 60 to 115 mm; facial angle slight; auditory bullae elongated, thin walled, moderately inflated, not in close contact with paroccipital processes; auditory meatus short but distinct; palate extends beyond last upper molar; dental formula I 3/3, C 1/1, P 4/4, M 1/2; P4 trenchant with small, well-developed protocone; m1 with small metaconid, trigonid longer than talonid, talonid semibasined.

African Species *M. khelifensis* Ginsburg, 1977.

Age Ca. 12.5–11.2 Ma.

Geographic Occurrence Morocco.

Remarks Pre-Pleistocene *Martes* is in need of revision. This genus, like *Mustela*, has become a wastebasket nomen for various small mustelids of uncertain relationship to each other. *Martes khelifensis* is one of these. It is known from a single M1 germ that Ginsburg (1977) compares with the homologous element of the Asian *M. anderssoni*, another small *Martes*-like mustelid of uncertain affinity. Nevertheless, *M. khelifensis* is the only such taxon in the African fossil record.

Genus *MELLALICTIS* Ginsburg, 1977

Diagnosis Translated from Ginsburg (1977). Mellivorine with elongated face; p4 tall, lacking anterior accessory cusp; m1 short with well-developed metaconid and short talonid where the hypoconid is the largest cusp; M1 with strongly developed parastyle, metacone reduced to a narrow crest posterior to the paracone, paraconule isolated, protocone elongated surrounded by an extensive cingulum.

African Species M. mellalensis Ginsburg, 1977.
Age Ca. 12.5–11.2 Ma.
Geographic Occurrence Morocco.
Remarks Mellalictis is known from a number of isolated teeth and postcranial fragments from Beni Mellal, Morocco. It has similarities with both ischyrictines and mellivorines, but Ginsburg (1977) suggests that its relationships probably lie with the former, particularly with the genera *Ischyrictis* and *Hadrictis*. If this is correct, it may also be related to the much later and larger *Ekorus* from Kenya (see earlier discussion).

Genus *MELLIVORA* Storr, 1780

Diagnosis Large-sized Mustelidae; molar dentition reduced, dental formula I 3/3, C 1/1, P 3/3, M 1/1; premolars slender, m1 with short, trenchant talonid, metaconid low; M1 antero-posteriorly shortened, paracone and metacone small.
African Species M. benfieldi Hendey, 1978, *M. capensis* (Schreber, 1776), *M. carolae* Michel, 1988.
Age Ca. 5.8 Ma–Recent.
Geographic Occurrence Ethiopia, Kenya, Morocco, South Africa, Tanzania.
Remarks The oldest published record of *Mellivora* is from Langebaanweg, based on material referred to *M. benfieldi* (Hendey, 1974). However, Petter (1987) suggests that this species may simply be the ancestral starting point of the *M. capensis* lineage, in which case the two would be conspecific. Only better material can elucidate the intraspecific variation of these species. *M. carolae* from the Pleistocene of Morocco (Michel, 1988) does not appear to differ from the living species, though this is difficult to evaluate due to the absence of illustrations of the relevant specimens.

Genus *MUSTELA* Linnaeus, 1758

Diagnosis Modified from Bryant et al. (1993). Small-sized Mustelidae; skull with anterior opening of palatine canal in maxilla, inflated bulla, and no paroccipital process; dental formula I 3/3, C 1/1, P 3/3, M 2/2; no accessory cuspid on p4; M1 transversely elongated and with buccal cusps reduced; no metaconid on m1, m1 talonid not basined, with a strong central hypoconid, an entoconid, and a buccal cingulum.
African Species M. putorius Linnaeus, 1758.
Age Ca. 0.04 Ma.
Geographic Occurrence Morocco.
Remarks The only fossil record of the genus in Africa is *Mustela putorius* from the Upper Pleistocene of El Harhoura 1 cave, Morocco (Aouraghe, 2000). The species is currently absent from the continent.

Genus *NAMIBICTIS* Morales et al., 1998

Diagnosis After Morales et al. (1998). Mustelinae with hypercarnivorous dentition; lower canine with very high crown; lower premolars and m1 mediolaterally compressed; p4 high crowned; m1 with a vertical paraconid, a residual metaconid, a small talonid comprised of a bevelled hypoconid.
African Species N. senuti Morales et al., 1998.
Age Ca. 17.5–17 Ma.
Geographic Occurrence Namibia.
Remarks Namibictis is a medium-sized mustelid with hypercarnivorous dentition known from Arrisdrift, Namibia. It is similar in carnassial morphology to ischyrictines, especially

Mellalictis, and is probably related to the latter taxon, as suggested by Morales et al. (1998).

Genus *PLESIOGULO* Zdansky, 1924

Diagnosis After Kurtén (1970). Wolverine with relatively large anterior premolars, carnassials relatively shorter than in *Gulo*; upper molar larger and more expanded anteroposteriorly than in *Gulo*, with inner lobe produced posterad; lower carnassial with a long, weakly basined talonid, and with or without a metaconid; m2 less reduced than in *Gulo*.
African Species P. botori Haile-Selassie et al., 2004, *P. monspessulanus* Viret, 1939, *P. praecocidens* Kurtén, 1970.
Age Ca. 6.1–5 Ma.
Geographic Occurrence Ethiopia, Kenya, South Africa.
Remarks P. monspessulanus, from South Africa (Hendey, 1978) and *P. botori* from Kenya and Ethiopia (Haile-Selassie et al., 2004a) are both large species of the genus, while *P. praecocidens* from Kenya is a smaller species. Whether *P. monspessulanus* and *P. praecocidens* are correctly assigned to species is at present a moot point, but it is clear that *Plesiogulo* had a short-lived presence in Africa at the end of the Miocene–beginning of the Pliocene.

Genus *PREPOECILOGALE* Petter and Howell, 1985

Diagnosis After Cooke (1985). Musteline slightly smaller than *Ictonyx striatus*, resembling *Poecilogale albinucha* in size, skull proportions, and in the extent of the lambdoid crest; P2 present as in *Ictonyx*; P4 intermediate between *Ictonyx* and *Poecilogale* in protocone development; tympanic bullae less inflated than in *Ictonyx striatus*.
African Species P. bolti (Cooke, 1985).
Age Ca. 3.7–2.6 Ma.
Geographic Occurrence Morocco, South Africa, Tanzania.
Remarks P. bolti is known from Bolt's Farm in South Africa, where it was originally described as *Ictonyx bolti* (Cooke, 1985), from Ahl al Oughlam, and from Laetoli, indicating a wide distribution over a relatively short time span.

Genus *SIVAONYX* Pilgrim, 1931
Figure 32.3

Diagnosis After Morales et al. (2005). Lutrinae with P4 with continuous cutting blade, with no incision separating the paracone and metastyle, protocone moderate, hypocone moderate, well separated from the protocone and the paracone-metastyle; M1 somewhat wider than long, with peripheral cusplets and a wide, flat central valley; m1 with relatively long trigonid, and unmolarised talonid with a wide basin.
African Species S. africanus (Stromer, 1931), *S. beyi* Peigné et al., 2008, *S. hendeyi* Morales et al., 2005, *S. ekecaman* (Werdelin, 2003)(figure 32.3D), *S. kamuhangirei* Morales and Pickford, 2005, *S. soriae* Morales and Pickford, 2005, *S.* sp.
Age Ca. 7.1–1.6 Ma.
Geographic Occurrence Chad, Egypt, Ethiopia, Kenya, South Africa, Uganda.
Remarks The majority of the taxa listed above have in the past been referred to *Enhydriodon*, but Morales and Pickford (2005b) and Morales et al. (2005) have transferred them to *Sivaonyx*. Neither taxon is particularly well characterized. The lineage includes an impressive array of African species, the geologically younger of which were among the most massive carnivorans of all time (figures 32.3A–C, E).

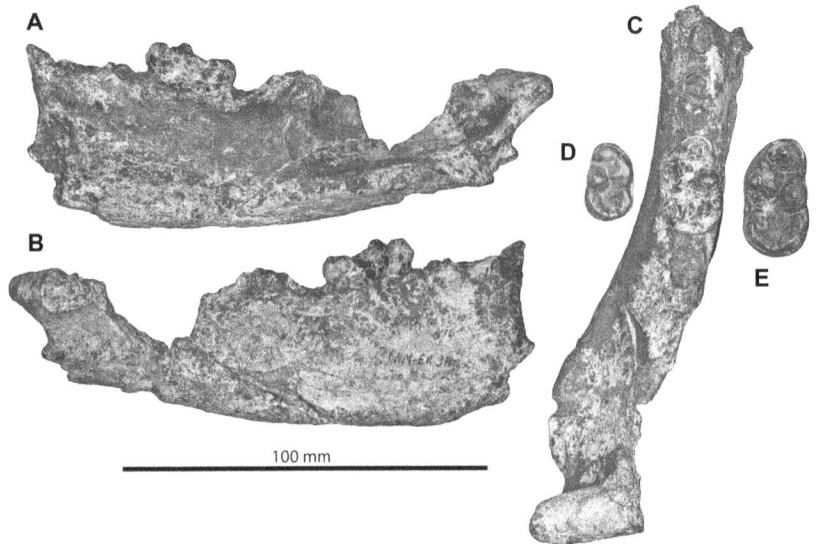

FIGURE 32.3 *Sivaonyx* spp. A–C) *Sivaonyx* n. sp. A, KNM-ER 3110 left ramus from the Tulu Bor Mb., Koobi Fora Fm. in buccal (A), lingual (B), and occlusal (C) views. D) *S. ekecaman* type specimen KNM-KP 10034 (part), right m1 from Kanapoi in occlusal view. E) *Sivaonyx* n. sp. B, L56-1 left from Mb. C, Shungura Fm in occlusal view.

Genus *TOROLUTRA* Petter et al., 1991

Diagnosis Translated from Petter et al. (1991). Mandible with relatively deep masseteric fossa, separated from the ventral margin of the ramus and with anterior border behind alveolus for m2; p4 robust, oval in occlusal view, surrounded by a thick, continuous cingulum; m1 with tall and sharp trigonid cusps, the protoconid being the tallest trigonid cusp, talonid narrow, with flat and obliquely oriented lingual face of the hypoconid.

African Species T. ougandensis Petter et al., 1991, *T.* sp.

Age Ca. 7.1–1.6 Ma.

Geographic Occurrence Egypt, Ethiopia, Kenya, Uganda.

Remarks The taxonomy of small and medium-sized lutrines is in a state of confusion at present (see *Lutra*), and where *Torolutra* might fit into this picture is currently not determinable.

Genus *VISHNUONYX* Pilgrim, 1932

Diagnosis Modified after Pilgrim (1932). Lutrinae of rather small size; P4 triangular, buccal anteroposterior diameter greater than lingual, and also much exceeding transverse diameter, with high pointed paracone, metacone lower but elongated, parastyle weak, protocone and hypocone both much lower than paracone, protocone situated rather far forward; internal cingulum slight; M1 rather small; mandible with deep ramus; p4 elongate, only slightly broader posteriorly, with strong posterior accessory cusp, not situated so much to the outside of the main cusp as in *Sivaonyx*, without anterior accessory cusp, with broad cingulum; m1 with talonid broader and shorter than trigonid, surrounded by a crenulated rim, entoconid as strong as hypoconid; m2 elongate oval, rather longer than in *Sivaonyx*.

African Species V. chinjiensis Pilgrim, 1932, *V. angololensis* Werdelin, 2003.

Age Ca. 12–6.5 Ma.

Geographic Occurrence Kenya.

Remarks The *V. chinjiensis* specimen from Ngorora matches that described from India by Pilgrim (1932) in size and morphology, while *V. angololensis* is rather larger. Morales and Pickford (2005a) suggest that *V. angololensis* might belong to *Torolutra* and this merits further consideration, as none of these taxa is particularly well characterized.

LUTRINAE indet.

Age Ca. 7.4–2.3 Ma.

Geographic Occurrence Chad, Ethiopia, Kenya, Tanzania.

Remarks Lutrinae that do not fit into the aforementioned genera or that are indeterminate are found at a number of localities, but none of these are outside the stratigraphic or geographic range of the more complete material. In addition to *Djourabus dabba* and *Sivaonyx beyi*, the late Miocene locality of Toros-Menalla, Chad, has also yielded at least two additional species known from fragmentary material. One species displays piscivorous dental adaptations and is related to *Torolutra ougandensis* and the other species is represented by a complete femur very similar to that of *Aonyx capensis* (Peigné et al., 2008a). A partial juvenile skeleton with maxilla fragment from the lower unit of the Laetolil beds is allied to *Aonyx*.

MUSTELIDAE indet.

Age Ca. 11–1 Ma.

Geographic Occurrence Chad, Ethiopia, Kenya, Tanzania, South Africa, Uganda.

Remarks Mustelidae of uncertain affinity are known from a number of localities (table 32.4). Most of these are known from fragmentary material, and several represent small species that may belong to known genera but have not been sufficiently studied. In a faunal list, Pickford et al. (1990) record ? *Poecilictis* from Cangalungue Cave in Angola. The generic attribution of this material needs verification, though clearly a small mustelid is indicated.

Family NANDINIIDAE Pocock, 1929
Table 32.5
Genus *NANDINIA* Gray, 1843

TABLE 32.4
Occurrences of Mustelidae species

Sites	Vishnuonyx chinjiensis	Vishnuonyx angololensis	Torolutra sp.	Torolutra ougandensis	Sivaonyx sp.	Sivaonyx soriae	Sivaonyx kamuhangirei	Sivaonyx hendeyi	Sivaonyx ekecaman	Sivaonyx beyi	Sivaonyx africana	Prepoecilogale bolti	Plesiogulo praecocidens	Plesiogulo monspessulanus	Plesiogulo botori	Namibictis senuti	Mustelidae indet.	Mustela putorius	Mellivora carolae	Mellivora capensis	Mellivora benfieldi	Mellalictis mellalensis	Martes khelifensis	Lutrinae indet.	Lutra libyca	Lutra fatimazohrae	Luogale rusingensis	Kenyalutra songhorensis	Ictonyx sp.	Ictonyx striatus	Ictonyx libyca	Howellictis valentini	Erokomellivora lothagamensis	Eomellivora tugenensis	Ekorus ekakeran	Djourabus dabba	Aonyxini indet.	Aonyx capensis
Adu-Asa Fm.											cf.				X						aff.			X		X					X							
Ahl al Oughlam				cf.	X							cf.								cf.																		
Allia Bay																	X																					
Aramis																																						
Arrisdrift																X				X																		
Asbole																																						
Awash 10 (Kalb)																																						
Awash 7 (Kalb)					X							X																										cf.
Beni Mellal																						X	X															
Bolt's Farm																																						
Bouknadel																																						
Cangalongue																													?									
Duinefontein 2																			X											X								
Elands Bay Cave																		X		X										X								
Elandsfontein Main																				X										X								X
El Harhoura																																						
Equus Cave																				X																		X
Eshoa Kakurongori					X												X																					
Florisbad																																						X
Hadar Denen Dora					X																X																	
Hadar Pinnacle					X																																	

TABLE 32.4
(CONTINUED)

Sites	Aonyx capensis	Aonyxini indet.	Djourabus dabba	Ekorus ekakeran	Eomellivora tugenensis	Erokomellivora lothagamensis	Howellictis valentini	Ictonyx libyca	Ictonyx striatus	Ictonyx sp.	Kenyalutra songhorensis	Luogale rusingensis	Lutra fatimazohrae	Lutra libyca	Lutrinae indet.	Martes khelifensis	Mellalictis mellalensis	Mellivora benfieldi	Mellivora capensis	Mellivora caroli	Mustela putorius	Mustelidae indet.	Namibictis senuti	Plesiogulo botori	Plesiogulo monspessulanus	Plesiogulo praecocidens	Prepoecilogale bolti	Sivaonyx africana	Sivaonyx beyi	Sivaonyx ekecaman	Sivaonyx hendeyi	Sivaonyx kamuhangirei	Sivaonyx soriae	Sivaonyx sp.	Torolutra ougandensis	Torolutra sp.	Vishnuonyx angololensis	Vishnuonyx chinjiensis
Hadar Sidi Hakoma																																		X				
Hopefield																			X											X								
Kanapoi	X																															X						
Kazinga																			X																			
Klasies River Mouth																						X																
Konso-Gardula 2		cf.																																	cf.			
Koobi Fora KBS																																		X				
Koobi Fora Lokochot																																		X				
Koobi Fora Tulu Bor																																						
Koobi Fora Upper Burgi																		cf.				X													cf.			
Koro Toro															X																			X				
Laetolil Beds Lower Unit																			cf.								X											
Laetolil Beds Upper Unit																																						
Langebaanweg Baard's Quarry																		X	cf.																			
Langebaanweg PPM																															X							
Langebaanweg QSM																									aff.													

Lemudong'o																							X														
Lothagam Lower Nawata				X																	X																X	
Lothagam Upper Nawata						X															X																	
Manonga 1																					X																	
Nachukui Lower Lomekwi																																	X	cf.				
Nachukui Upper Lomekwi															X																							
Nakali																					X																	
Nakoret																																	X					
Napak																					X																	
Ngorora Member D					X																																	X
Nkondo																															X			X				
Omo Shungura B															X																		X					
Omo Shungura C																																	X					
Omo Shungura E + F															X																		X					
Omo Shungura H																																	X		X			
Omo Usno																																	X	cf.				
Plovers Lake	X								X										X			X																
Rusinga												X																										
Sagantole Fm.															X			aff.																				
Saldanha Sea Harvest																			X																			
Sibudu HP									cf.													X																
Songhor											X																											
South Turkwel															X																							
Swartklip 1	X								X										X																			
Toros Menalla		aff.	X				X								X														X									
Tugen Lukeino																									X							X		X	aff.			
Tugen Mabaget																																						
Tygerfontein																			X																			
Wadi Natrun													X														X								cf.			
Warwire																																		X				

TABLE 32.5
Occurrences of Nandiniidae and Viverridae species

Sites	NANDINIIDAE	Nandinia sp.	VIVERRIDAE	Africanictis hyaenoides	Africanictis meini	Africanictis schmidtkittleri	Civettictis civetta	Civettictis howelli	Civettictis sp.	Genetta genetta	Genetta tigrina	Genetta sp.	Herpestides aegypticus	Herpestides aequatorialis	?Herpestides afarensis	Kanuites lewisae	Ketketictis solida	Mioprionodon pickfordi	Moghradictis nedjema	Orangictis gariepensis	Pseudocivetta ingens	Sahelictis korei	Stenoplesictis muhoronii	Tugenictis ngororaensis	Vishnuictis africana	Viverra howelli	Viverra leakeyi	Viverridae indet.
Adu-Asa Fm.												X																
Ahl al Oughlam												X															X	
Allia Bay																												X
Aramis																												X
Arrisdrift					X	X														X								
As Duma																												X
Awash 10 (Kalb)												X																
Awash 7 (Kalb)									X																			
Beni Mellal												X																X
Chamtwara				X																								
Chorora															X													
Duinefontein 2												X																
Elands Bay Cave												X																
Elandsfontein Main							X																					
Elisabethfeld																												X
Fort Ternan																X												X
Hadar Denen Dora																												X
Hadar Pinnacle																												X
Hadar Sidi Hakoma																												X
Kanapoi												X																
Kipsaraman																									X			
Klein Zee												X																
Konso-Gardula 1												X																X
Konso-Gardula 2												X																X
Koobi Fora Chari																												X
Koobi Fora KBS																					X							X
Koobi Fora Lokochot																												X
Koobi Fora Okote									X																			
Koobi Fora Tulu Bor																												X
Koobi Fora Upper Burgi									X												X							X
Koru																												
Laetolil Beds Upper Unit												X															X	
Langebaanweg PPM																											X	X
Langebaanweg QSM												X															X	X
Legetet																												
Lemudong'o												X																
Lothagam Lower Nawata												X														X	cf.	X
Lothagam Upper Nawata												X																
Manonga 2																												X
Melka Kunture Garba IV																					X							

Sites	NANDINIIDAE	*Nandinia* sp.	VIVERRIDAE	*Africanictis hyaenoides*	*Africanictis meini*	*Africanictis schmidtkittleri*	*Civettictis civetta*	*Civettictis howelli*	*Civettictis* sp.	*Genetta genetta*	*Genetta tigrina*	*Genetta* sp.	*Herpestides aegypticus*	*Herpestides aequatorialis*	*?Herpestides afarensis*	*Kanuites lewisae*	*Ketketictis solida*	*Mioprionodon pickfordi*	*Moghradictis nedjema*	*Orangictis gariepensis*	*Pseudocivetta ingens*	*Sahelictis korei*	*Stenoplesictis muhoronii*	*Tugenictis ngororaensis*	*Vishnuictis africana*	*Viverra howelli*	*Viverra leakeyi*	Viverridae indet.
Nachukui Kalochoro																												X
Nachukui Lower Lomekwi																												X
Nachukui Upper Lomekwi																												X
Napak																												
Ngorora Member A																								X				
Ngorora Member B																												X
Olduvai Bed I																					X							
Olduvai Bed II																					X							
Omo Shungura B											X																	
Omo Shungura C											cf.																X	X
Omo Shungura D									cf.																			X
Omo Shungura E+F																					X							X
Omo Shungura G																					X							X
Omo Shungura L																												X
Plovers Lake										X																		
Rusinga														X									X					
Sagantole Fm.											X																cf.	
Sahabi																									X		cf.	
Saldanha Sea Harvest												cf.																
Sibudu HP												cf.																X
Songhor														X				X					X					
South Turkwel									cf.																			
Sterkfontein Jacovec Cave										X																		
Swartkrans Member 3										X																		
Toros Menalla																						X						X
Tugen Lukeino		X									X																aff.	
Tugen Mabaget								X																				
Wadi Moghra													X				X		X									

Geographic Occurrence Kenya.

Remarks Nandinia is an arboreal forest taxon and as such less likely to be represented in the fossil record than terrestrial taxa. However, it is possible that postcranial remains, especially femora, of *Nandinia* are mistaken for monkey postcrania (B. Senut, pers. comm. to L.W., 2006) and that the family has a more extensive record than a single tooth from the Lukeino Fm. suggests.

Family VIVERRIDAE Gray, 1821
Table 32.5
Genus *AFRICANICTIS* Morales et al.,1998

Diagnosis After Morales et al. (1998). Feloid carnivoran of the size of the European Miocene hyaenid *Protictitherium crassum*; p1 reduced; premolars tall and narrow; m1 with tall trigonid and short talonid with a small hypoconid; P4 narrow and elongate, with strong parastyle and conical protocone; M2 small; M1 with well-developed parastylar area, shortened in the type species.

African Species A. hyaenoides Morales et al., 2003, *A. meini* Morales et al., 1998, *A. schmidtkittleri* Morales et al., 1998.

Age Ca. 20–17 Ma.

Geographic Occurrence Kenya, Namibia.

Remarks This genus of small carnivorans, at present known only from Arrisdrift, Namibia, and Chamtwara, Kenya (Schmidt-Kittler, 1987; Morales et al., 1998, 2003), is a morphological link between *Stenoplesictis* and the percrocutids. Its dental morphology is similar to primitive hyaenids such as *Protictitherium* (Morales et al., 1998), and *Tungurictis* (Colbert, 1939; Werdelin and Solounias, 1991). Of the three species, the older *A. schmidtkittleri* is the smallest and most primitive, while *A. meini* and *A. hyaenoides* differ in the structure of the m1. The latter is similar in many respects to primitive *Percrocuta* spp.

Genus *CIVETTICTIS* Pocock, 1915

Diagnosis Large-sized Viverridae; dental formula I 3/3, C 1/1, P 4/4, M 2/2; P4 protocone enlarged, metastyle short; m1 with short trigonid and broad, three-cusped talonid; premolars robust but slender; m1 and M1–2 broad and with crushing adaptations.

African Species C. civetta (Schreber, 1776), *C. howelli* Morales, Pickford and Soria, 2005, *C.* sp.

Age Ca. 5.1 Ma–Recent.

Geographic Occurrence Ethiopia, Kenya, South Africa.

Remarks Civettictis spp. are rare members of eastern African Pliocene faunas. Most fossil finds are fragmentary.

Genus *GENETTA* Cuvier, 1817

Diagnosis From Gaubert et al. (2004). Small-sized viverrid with dental formula I 3/3, C 1/1, P 4/4, M 2/2; plesiomorphic dental anatomy; M2 present; m1 talonid developed; m2 well developed.

African Species G. genetta (Linnaeus, 1758), *G. tigrina* (Schreber, 1776), *G.* sp.

Age Ca. 14 Ma–Recent.

Geographic Occurrence Ethiopia, Kenya, Morocco, South Africa, Tanzania.

Remarks Fossils of *Genetta* spp. are not uncommon in the African Upper Miocene-Pleistocene, but referral to specific species is hampered by the fragmentary nature of most remains, especially the earliest records (Beni Mellal, Lothagam, Lemudong'o). The first definite record of the genus is from Kanapoi. The oldest material that is indistinguishable from modern *G. genetta* on diagnostic features is from the Upper Burgi Mb. at Koobi Fora.

Genus *HERPESTIDES* Beaumont, 1967

Diagnosis Small-sized viverrid with dental formula I 3/3, C 1/1, P 4/4, M 2/2; skull with strongly developed sagittal crest; ossified auditory bulla with inflated ectotympanic, bilaminar septum, and so forth (see Hunt, 1991); P3 with third root more or less well individualized and supporting a small cuspid or a prominent lingual cingulum; P4 with protocone enlarged and parastyle developed and trenchant; M1 with crescent-shaped protocone much larger than buccal cusps; hypoconid prominent on m1; single-rooted m2 lacking the paraconid.

African Species H. aegypticus Morlo et al., 2007, *H. aequatorialis* Schmidt-Kittler, 1987, ?*H. afarensis* Geraads et al., 2002.

Age Ca. 18–?10 Ma.

Geographic Occurrence Egypt, ?Ethiopia, Kenya.

Remarks The European *H. antiquus* is known through abundant material from the Lower Miocene of St-Gérand-le-Puy, France (Beaumont, 1967). Hunt (1991) has shown this taxon to be a true viverrid. In Africa, the genus is known only from dental remains, of which those of *H. afarensis* from Chorora (Geraads et al., 2002) cannot be considered diagnostic at the generic level. This species may belong to *Kanuites* (see next section).

Genus *KANUITES* Dehghani and Werdelin, 2008

Diagnosis From Dehghani and Werdelin, 2008. Viverridae with long and narrow skull and moderately hypercarnivorous dentition; dental formula (upper dentition only): I ?3, C, P 4, M 2; P1 simple, peglike; P2–3 tall, slender; P4 robust, protocone slightly mesial to paracone, metastyle robust; M1 mesiodistally reduced, triangular, buccal cingulum strongly developed; M2 small, triangular.

African Species K. lewisae Dehghani and Werdelin, 2008.

Age Ca. 14–13.7 Ma.

Geographic Occurrence Kenya.

Remarks Known from the middle Miocene of Fort Ternan, Kenya, *Kanuites* is the oldest known viverrid from Africa with intact basicranium. Its dentition and skull morphology are quite generalized, but dentally it shows resemblances to *Herpestides* and may be close to that taxon.

Genus *KETKETICTIS* Morlo et al., 2007

Diagnosis Modified after Morlo et al. (2007). Differs from other African Miocene Viverridae s. l. except *Herpestides* and *Kichechia* in having a very broad and plesiomorphic m1; differs from *Herpestides* and *Kichechia* in having the trigonid cingulid surrounding the anterior tip of the tooth, the metaconid placed more posteriorly, and the low talonid equipped with extremely low cuspids and low crista obliqua; differs from *Herpestides* in being broader and having a less trenchant trigonid; differs from *Kichechia* in being much larger, having a smaller entoconid, and the cristid obliqua oriented more buccally.

African Species K. solida Morlo et al., 2007.

Age Ca. 18–17 Ma.

Geographic Occurrence Egypt.

Remarks Ketketictis is known from a single, isolated m1 from Wadi Moghara initially identified as *Herpestides* sp. indet. A (Miller, 1999). Due to the fragmentary nature of this species, and its resemblance to *Kichechia*, an assignment to Herpestidae cannot be ruled out. It should be noted that the familial allocation of poorly known early Miocene African taxa to either Viverridae s. l. or Herpestidae is strongly subjective and generally not based on apomorphies.

Genus *MIOPRIONODON* Schmidt-Kittler, 1987

Diagnosis Revised after Schmidt-Kittler (1987). Comparable in general features of the dentition to *Stenoplesictis* but smaller and differing from that genus in: m1 trigonid lower and metaconid less reduced; p1 two rooted; P4 less elongate; M1 relatively larger.

African *Species* M. *pickfordi* Schmidt-Kittler, 1987.
Age Ca. 20–19 Ma.
Geographic Occurrence Kenya.
Remarks *Mioprionodon* is a genus of small, relatively hypercarnivorous viverrids known from a number of mandible fragments with teeth from Songhor, Kenya. It resembles *Stenoplesictis*, *Semigenetta*, and *Palaeoprionodon* in the general morphology of its dentition, but it is considerably less derived in its hypercarnivorous features. *Mioprionodon* is distinguished from the aforementioned taxa on the basis of plesiomorphic characters and probably is a basal member of that clade.

Genus *MOGHRADICTIS* Morlo et al., 2007

Diagnosis After Morlo et al. (2007). Large, hypercarnivorous "stenoplesictid," differing from all others in having an m1 with small metaconid combined with long talonid, which is unbasined due to the presence of a tall hypoconid and no entoconid; differs from European "stenoplesictids" in having a P4 paracone that protrudes backwards; differs from "*Stenoplesictis*" *muhoronii*, in being much larger and having the m1 postparacristid pointing more anteriorly.
African Species M. *nedjema* Morlo et al., 2007.
Age Ca. 18–17 Ma.
Geographic Occurrence Egypt.
Remarks This recently described carnivoran from Wadi Moghara is represented only by a few dental remains. The species has been previously listed as *Herpestides* sp. indet. B (Miller, 1999). Morlo et al. (2007) consider assignment of "*Stenoplesictis*" *muhoronii* from Kenya to this genus possible.

Genus *ORANGICTIS* Morales et al., 2001

Diagnosis After Morales et al. (2001a). Primitive viverrine intermediate in size between *Viverricula indica* and *Viverra zibetha*; dentition robust; p4 with greatly reduced anterior cusplet; m1 short with high and closed trigonid, in which the metaconid is important and the paraconid is in a very lingual position, talonid small with very well-developed entoconid, attaining the height of the hypoconid; m2 relatively large, open trigonid with small paraconid and metaconid slightly taller than protoconid, talonid deeply excavated like that of m1, but with hypoconulid higher than entoconid and separated from it and the hypoconid.
African Species O. *gariepensis* Morales et al., 2001.
Age Ca. 17.5–17 Ma.
Geographic Occurrence Namibia.
Remarks This genus is based on two mandibular specimens from Arrisdrift, Namibia (Morales et al., 2003). It shows some dental affinities with *Herpestides* and, like it, may be close to the basal stock of the Viverridae and/or Viverrinae. It also bears resemblances to *Tugenictis* (discussed later), though it is much smaller than that taxon. Morales and Pickford (2005a) suggest that they may belong to the same lineage.

Genus *PSEUDOCIVETTA* Petter, 1967

Diagnosis Modified and translated from Petter (1973). Civet of large size with bunodont dentition; M1 and M2 rectangular, with low, rounded cusps; p4 with large, rounded main cusp followed by a talonid with bunodont cuspids.
African Species P. *ingens* Petter, 1967.
Age Ca. 2.4–0.8 Ma.

Geographic Occurrence Ethiopia, Kenya, Tanzania.
Remarks *Pseudocivetta* is by far the most aberrant viverrid, though it bears some resemblances to *Tugenictis* and *Orangictis* (see these genera). Its cheek dentition is very bunodont, with extremely low relief and with a p4 that has a low, rounded main cusp and a posterior area that supports a number of low, bunodont cusps. *Pseudocivetta* was a short-lived but successful taxon that was present over a relatively large geographic area for about 1 Ma near the Plio-Pleistocene boundary.

Genus *SAHELICTIS* Peigné et al., 2008

Diagnosis Modified from Peigné et al. (2008b). Viverrid of large size. Compared to *Viverra leakeyi*, V. *howelli*, V. *pepratxi*, V. *peii*, m1 with reduced metaconid; compared to these species and to *Megaviverra carpathorum*, m1 talonid narrower, with a reduced hypoconid, no entoconid, no or greatly reduced hypoconulid, and a very low mesiolingual rim; compared to *Viverra leakeyi*, V. *howelli*, V. *pepratxi*, V. *peii*, V. (?) *chinjiensis*, and *Vishnuictis salmontanus*, m1 talonid shorter; compared to V. *leakeyi* and V. *bakerii*, M1 more reduced relative to P4 (or less shortened P4 relative to M1); compared to V. *leakeyi*, M1 transversely less elongated; compared to *Vishnuictis durandi*, long rostrum, skull lacking expanded frontal regions and marked postorbital constriction; compared to *Viverra howelli*, V. *pepratxi*, V. (?) *chinjiensis*, and *Vishnuictis salmontanus*, much larger size.
African Species S. *korei* Peigné et al., 2008.
Age Ca. 7 Ma.
Geographic Occurrence Chad.
Remarks *Sahelictis korei* is thus far known only from Toros-Menalla, Chad. However, the large viverrid described from Langebaanweg and previously assigned to V. *leakeyi*, could be closely related to the Chadian species and belong to the same genus. With V. *leakeyi* and P. *ingens*, S. *korei* is one of the rare fossil viverrids from Africa to reach a size larger than the extant C. *civetta*.

Genus *STENOPLESICTIS* Filhol, 1880

Diagnosis Modified from Peigné and de Bonis (1999) and Peigné (2000). Small feloid with dental formula I 3/3, C 1/1, P 3–4/3–4, M 2/2; snout narrow; auditory region with inflated ectotympanic that widely contacts the basicranium; caudal entotympanic not ossified; septum bullae bilaminar for half of its height; lateral rims of basicranium ventrally depressed; P3 with small but distinct posterior accessory cusp; M2 reduced but double rooted in earlier species; p3–4 with distinct posterior accessory cuspid; posterior cingulum of lower premolars not trenchant; m1 metaconid reduced but relatively larger and less retracted than in *Palaeoprionodon* and *Haplogale*; m1 talonid basined, with a tall buccal crest on which the hypoconid may be prominent; m2 reduced and single rooted, with trigonid including a dominant protoconid.
African Species "S." *muhoronii* Schmidt-Kittler, 1987.
Age Ca. 20–17 Ma.
Geographic Occurrence Kenya.
Remarks Like *Africanictis* and *Moghradictis* (discussed earlier), *Stenoplesictis* is included in the family Stenoplesictidae in recent contributions (e.g., Morales et al., 1998; Morlo et al., 2007). However, this name used as a family-group taxon is paraphyletic, as pointed out as early as 1931 by Pilgrim, and is supported only by dental convergences. Therefore, it should not be used as valid (see also Hunt, 1989, 1998b). It

should be noted, however, that integration of these three genera within the Viverridae s. l., which is also paraphyletic, is not satisfying, either. A second alternative would be to include these genera in the Percrocutidae. The sole African species of *Stenoplesictis* is known from a maxillary fragment from Songhor and a mandibular fragment from Rusinga (Schmidt-Kittler, 1987). As previously pointed out (Peigné and de Bonis, 1999), the generic assignment of the species is debatable. It may be closer to *Semigenetta* (Hunt, 1996).

Genus *TUGENICTIS* Morales and Pickford, 2005

Diagnosis After Morales and Pickford (2005a). Viverridae the size of *Civettictis civetta*; m1 robust, with high, bunodont cusps, paraconid located anteriorly, metaconid smaller that other trigonid cusps, talonid relatively short, with hypoconid and entoconid as tall as trigonid, hypoconid pyramidal, with an anterolingual crest running obliquely to the axis of the tooth as far as the contact between protoconid and metaconid, entoconid high and crescentiform; m2 with reduced trigonid, formed of a voluminous protoconid, small residual paraconid, and metaconid somewhat narrower than protoconid, and talonid dominated by a strong, rounded hypoconid well separated from the protoconid and a hypoconid joined to the entoconid, which is relatively small and well separated from the metaconid.

African Species T. ngororaensis Morales and Pickford, 2005.

Age Ca. 13–12.5 Ma.

Geographic Occurrence Kenya.

Remarks T. ngororaensis is a large, bunodont viverrid known from a single lower carnassial from Member A of the Ngorora Formation (Morales and Pickford, 2005a). This tooth is reminiscent of the lower carnassial of *Orangictis* (Morales et al., 2001a) but is much larger. It also bears similarities to the lower carnassial of *Pseudocivetta* (Petter, 1967) but is overall far less derived. Nevertheless, Morales and Pickford (2005a) suggest that these three genera might be related as part of a lineage of bunodont viverrids culminating with the highly derived *Pseudocivetta*.

Genus *VISHNUICTIS* Pilgrim, 1932

Diagnosis Modified after Pilgrim (1932). Medium- to large-sized viverrids with elongate, high, and narrow skull; contracting gradually to a considerable distance behind the postorbital processes, also contracting in front of the postorbital processes; slender muzzle; sides of face steeply descending from a nasal maxillary ridge; teeth not compressed laterally and with rather blunt cusps; braincase exceptionally narrow; upper molars rather large, relatively broad lingually; P3 without a lingual cusp; premolar series rather spaced, premolars simple; mandible rather stout but shallow; m1 with relatively long trigonid, relatively short talonid; m2 rather large, oblong.

African Species V. africana Morales and Pickford, 2008.

Age Ca. 15.8–15.4 Ma.

Geographic Occurrence Kenya.

Remarks The only record of this Asian genus in Africa is a fragmentary mandible with p4–m1 from the middle Miocene of Kipsaraman, Tugen Hills, Kenya (Morales and Pickford, 2008). According to these authors, this species is very similar to *Viverra* (?) *chinjiensis*, compared to which it is slightly smaller and differs in the more gracile dentition and a larger distal accessory cuspid in p4. They provisionally assign *V.* (?) *chinjiensis* and its African relative to the genus *Vishnuictis*.

Genus *VIVERRA* Linnaeus, 1758
Figures 32.4A and 32.4B

Diagnosis Modified after Veron (1995). Medium-sized viverrids with dental formula I 3/3, C 1/1, P 4/4, M 2/2; skull with posterior rim of nasals across anterior rim of orbits, posterior opening of the carotid canal located in the middle of the medial rim of the bulla; anteromedial ventral process of the promontorium of the petrosal flattened and located level with the basisphenoid-basioccipital suture; paroccipital process very thin in the lower part, thicker in the upper part.

African Species V. howelli Rook and Martínez-Navarro, 2004, *V. leakeyi* Petter, 1963 (figures 32.4A and 32.4B).

Age Ca. 7.4–1.88 Ma.

Geographic Occurrence Ethiopia, Kenya, Libya, Morocco, South Africa, Tanzania.

Remarks Small *Viverra* species are difficult to positively identify. Such records are hidden among the Viverridae indet. For reasons stated in Werdelin (2003), we are inclined to place *V. leakeyi* in the extant genus rather than in *Megaviverra* as in Morales et al. (2005)(but see also *Sahelictis*, earlier).

VIVERRIDAE indet.
Figure 32.4C

Age Ca. 12.5–0.7 Ma.

Geographic Occurrence Chad, Ethiopia, Kenya, Morocco, Namibia, South Africa, Tanzania.

Remarks The numerous mentions of Viverridae indet. are, except for those from the Ngorora Fm., Fort Ternan, and Lothagam, and Toros-Menalla, Pliocene to Pleistocene in age. Most can be assigned to extant genera, though others, including one from the late Pliocene of Koobi Fora (figure 32.4C), represent extinct taxa, in some cases much larger than any living viverrid. The specimen from Member B of the Ngorora Formation (Morales and Pickford, 2005a) represents a small, *Genetta*-like species. A tooth from Beni Mellal, identified by Ginsburg (1977) as *Felis* sp. is more likely to belong to a viverrid.

Family HERPESTIDAE Bonaparte, 1845
Table 32.6
Genus *ATILAX* F. Cuvier in Geoffroy and Cuvier, 1826
Figures 32.5G–32.5I

Diagnosis Skull broad, supraoccipital crest large, sagittal crest well developed; mandible strong, ventral surface curved; coronoid and angular processes large; dental formula I 3/3, C 1/1, P 3–4/3–4, M 2/2; upper canine straight, lower curved; premolars robust; m1 with robust, long trigonid, short talonid; m2 with trigonid and talonid of similar length; P4 robust with large protocone; M1 with large metastyle wing; M1 and M2 bunodont.

African Species A. mesotes (Ewer, 1956) (see figures 32.5G–32.5I), *A. paludinosus* (Cuvier, 1829), *A. sp.*

Age Ca. 1.7 Ma–Recent.

Geographic Occurrence South Africa, Tanzania.

Remarks A. mesotes, known only from Kromdraai Mb. A, was described as a species of *Herpestes*, but following suggestions by Ewer (1956) that this species is on the lineage to the

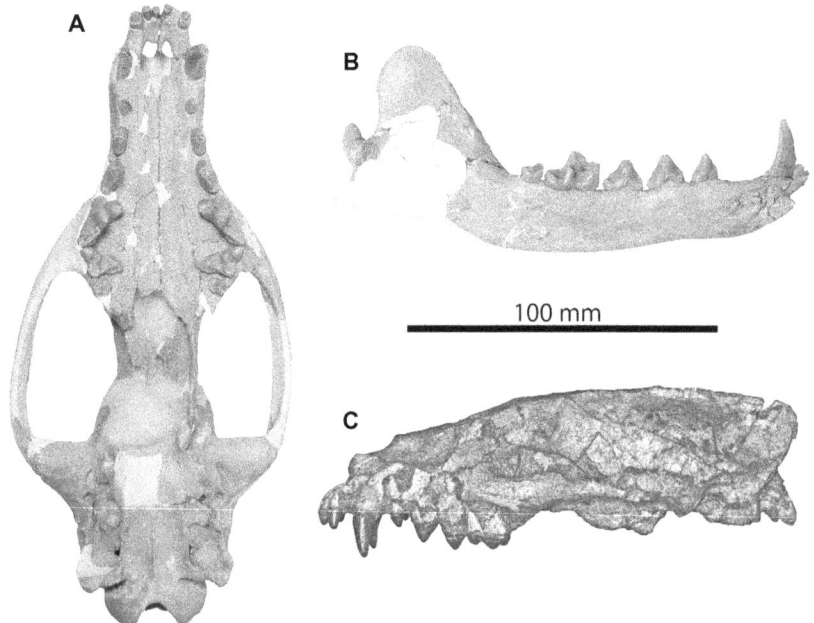

FIGURE 32.4   African Viverridae. A–B) *Viverra leakeyi*, PQ-L 51590, unknown member, Varswater Fm., Langebaanweg, skull in ventral view and left hemimandible in lingual view; C) Viverridae n. gen. and sp., KNM-ER 5339, KBS Mb., Koobi Fora Fm., in left lateral view.

extant marsh mongoose, we here transfer it to *Atilax*. In addition, the genus is known from Olduvai, Bed II.

Genus *CROSSARCHUS* F. Cuvier in Geoffroy and Cuvier, 1825

Diagnosis Modified after Goldman (1987). Small-sized herpestids; dental formula I 3/3, C 1/1, P 3/3, M 2/2; skull with elongated snout, triangular palate, ectotympanic developed and greatly inflated, but much less than caudal entotympanic; P1/p1 absent; P4 wide, with short metastylar shearing surface, well-developed parastyle, and prominent, slender protocone; M1 wide, paracone and metacone subequal, buccally projecting parastyle and metastyle; M2 similar to M1; p3 with posterior accessory cusp projecting buccally; m1 with sharp trigonid, paraconid and metaconid well defined, talonid widened; m2 well developed, trigonid low, paraconid not visible, talonid with basin and hypoconulid rim.

African Species *C. transvaalensis* Broom, 1937.

Age Ca. 1.6 Ma–Recent.

Geographic Occurrence South Africa.

Remarks Fossil *Crossarchus* is only known from Bolt's Farm and possibly from Kromdraai Mb. A. The relationships of *C. transvaalensis* to extant cusimanses, none of which are known from southern Africa, are obscure.

Genus *CYNICTIS* Ogilby, 1833

Diagnosis Modified after Veron (1995) and Taylor and Meester (1993). Small-sized herpestids with dental formula I 3/3, C 1/1, P 4/4, M 2/2; skull relatively tall; orbit completely surrounded by bony ring; anterior part of bulla as large as posterior; anterior chamber of bulla more inflated than posterior; upper incisors disposed on a transverse line; parastyle of P4 enlarged, nearly the size of the protocone; carnassial notch of m1 deep.

African Species *C. penicillata* Cuvier, 1829.

Age Ca. 3.3 Ma–Recent.

Geographic Occurrence South Africa.

Remarks *Cynictis* was tentatively reported from Laetoli by Petter (1987), but this record cannot be considered valid (Werdelin and Dehghani, in press). The only fossil records of the genus are from South Africa.

Genus *GALERELLA* Gray, 1865

Diagnosis Small-sized Herpestidae; skull smaller than *Herpestes*; dental formula I 3/3, C 1/1, P 4/3, M 2/2; p2 relatively smaller than in *Herpestes*.

African Species *G. debilis* Petter, 1973, *G. palaeoserengetensis* (Dietrich, 1942), *G. pulverulenta* (Wagner, 1839), *G. sanguinea* (Rüppell, 1835), *G.* sp.

Age Ca. 7.4 Ma–Recent.

Geographic Occurrence Chad, South Africa, Tanzania.

Remarks For convenience, we retain *Galerella* as a separate genus despite evidence that it does not constitute a monophyletic group (Veron et al., 2004). One of us (LW) is inclined to doubt the specific identification of *G. sanguinea* from the Miocene of Chad, though this material clearly does not belong to either of the previously named fossil species of *Galerella*.

Genus *HELOGALE* Gray, 1861
Figures 32.5A–32.5C

Diagnosis Small-sized herpestid with dental formula I 3/3, C 1/1, P 3/3, M 2/2; skull with no postorbital bar; premolars tall and sharp; P4 protocone very large and tall, much larger than the parastyle; no accessory cusps on P2–3/p2–3, a small distal one on p4; carnassial notches shallow; m1 trigonid tall with paraconid and protoconid of roughly equal height, and metaconid tall but more reduced than the paraconid.

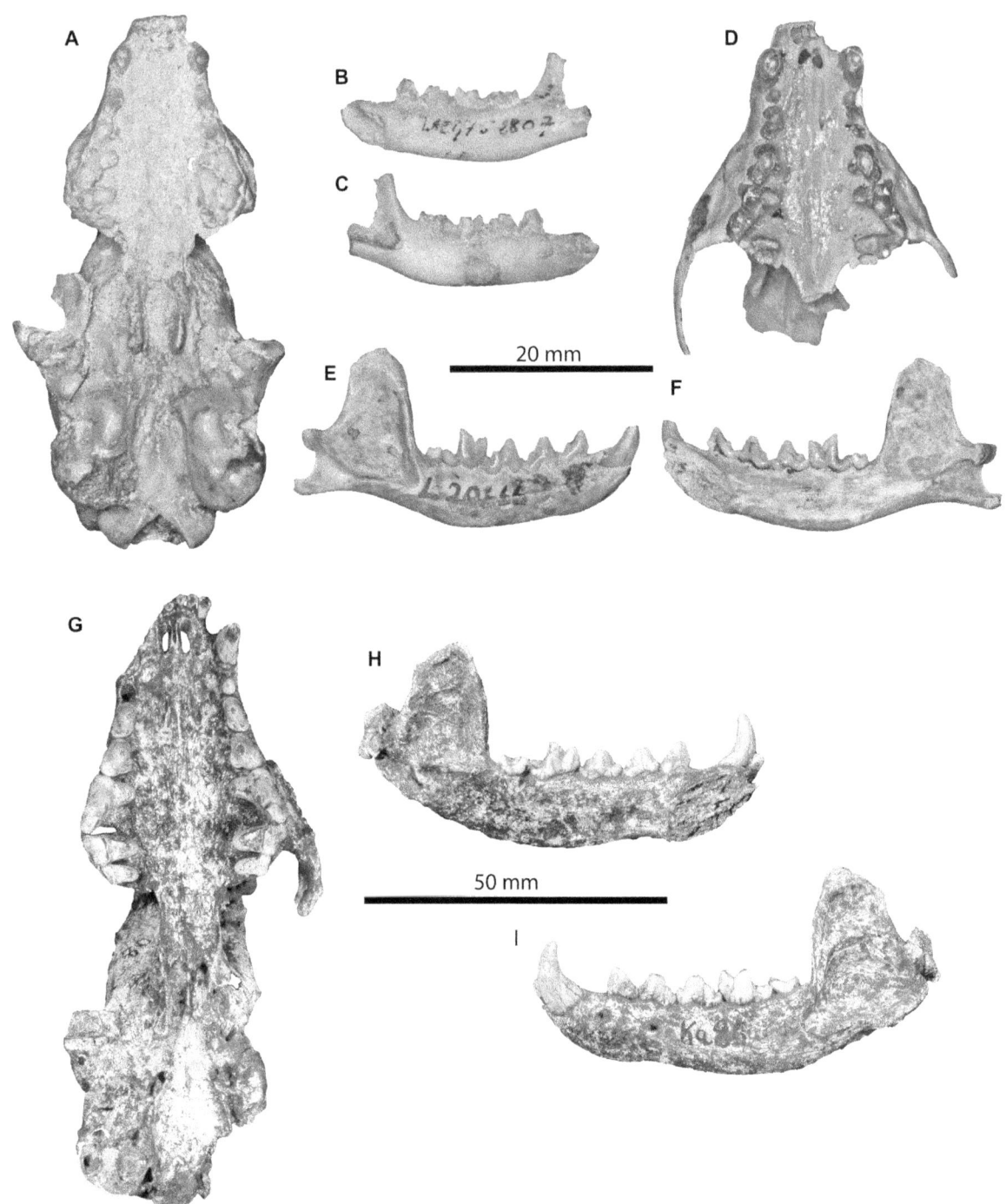

FIGURE 32.5   African Herpestidae. A–C) *Helogale palaeogracilis,* LAET 75-2807 from the Upper Laetolil Beds, Laetoli: A) cranium in ventral view, B–C) right ramus in lingual and buccal view; D) *Herpestes* sp., PQ-L 20735 from the Langeberg Quartzose Sand Mb., Varswater Fm., Langebaanweg, snout in ventral view; E–F) *Herpestes* sp., PQ-L 20666 from the Langeberg Quartzose Sand Mb., Varswater Fm., Langebaanweg, right mandible in buccal and lingual view; G–I) *Atilax mesotes* type specimen KA 86 from Kromdraai Mb. A, skull in ventral view and left mandible in lingual and buccal view.

African Species *H. hirtula* Thomas, 1904, *H. kitafe* Wesselman, 1984, *H. palaeogracilis* (Dietrich, 1942) (figures 32.5A–32.5C), *Helogale* sp.

Age Ca. 6 Ma–Recent.

Geographic Occurrence Ethiopia, Kenya, Tanzania.

Remarks *Helogale* spp. are present at a number of Pliocene localities in eastern Africa, with the youngest record being from the Shungura Fm., Member G. No *Helogale* has been reported from the Pleistocene of Africa. Whether this represents a bias

due to collecting methods or a bias in the sampled environments is not clear.

Genus *HERPESTES* Illiger, 1811
Figures 32.5D–32.5F

Diagnosis Modified after Veron (1995). Small- to medium-sized herpestids with dental formula I 3/3, C 1/1, P 4/4, M 2/2; skull elongated with palate triangular, lacrymal and jugal not

TABLE 32.6

Occurrences of Herpestidae species

Species \ Sites	Adu-Asa Fm., Asa Koma Mb.	Ahl al Oughlam	Aramis	Asbole	Awash 7 (Kalb)	Bolt's Farm	Busidima-Telalak	Cave of Hearths	Chamtwara	Diepkloof Rock Shelter	Drimolen	Duinefontein 2	Elands Bay Cave	Elandsfontein Main	El Harhoura	Elisabethfeld	Equus Cave	Florisbad	Grillental	Hadar Denen Dora	Hadar Sidi Hakoma	Hiwegi R 1	Hiwegi R 106	Hiwegi R 3
Ugandictis napakensis																								
Suricata sp.											X													
Suricata suricatta														?			X							
Suricata major														?										
Mungos sp.																								
Mungos mungo																								
Mungos minutus																								
Mungos dietrichi																								
Leptoplesictis senutae																		X						
Leptoplesictis rangwai																								
Leptoplesictis namibiensis																								
Leptoplesictis mbitensis																								
Legetetia nandii									X															
Kichechia sp.																								
Kichechia zamanae																						X	X	X
Ichneumia sp.								cf.																
Ichneumia nims		X																						
Ichneumia albicauda			aff.											X										
Herpestidae indet.															X									
Herpestes sp.					X										X									
Herpestes ichneumon			cf.						X	X	X	X	X					X						
Herpestes alaylaii	X																							
Herpestes abdelalii		X																						
Helogale sp.	X																							
Helogale palaeogracilis																								
Helogale kitafe		cf.																						
Helogale hirtula																								
Galerella sp.																								
Galerella sanguinea																								
Galerella pulverulenta										X			X											
Galerella palaeoserengetensis																								
Galerella debilis																								
Cynictis penicillata											X													
Crossarchus transvaalensis						X																		
Atilax sp.																								
Atilax paludinosus												X					X	X						
Atilax mesotes																								

TABLE 32.6
(CONTINUED)

Sites	Atilax mesotes	Atilax paludinosus	Atilax sp.	Crossarchus transvaalensis	Cynictis penicillata	Galerella debilis	Galerella palaeoserengetensis	Galerella pulverulenta	Galerella sanguinea	Galerella sp.	Helogale hirtula	Helogale kitafe	Helogale palaeogracilis	Helogale sp.	Herpestes abdelalii	Herpestes alaylaii	Herpestes ichneumon	Herpestes sp.	Herpestidae indet.	Ichneumia albicauda	Ichneumia nims	Ichneumia sp.	Kichechia zamanae	Kichechia sp.	Leggetia nandii	Leptoplesictis mbitiensis	Leptoplesictis namibiensis	Leptoplesictis rangwai	Leptoplesictis senutae	Mungos dietrichi	Mungos minutus	Mungos mungo	Mungos sp.	Suricata major	Suricata suricatta	Suricata sp.	Ugandictis napakensis
Hiwegi R 5																	X						X														
Kabwe																																					
Kalodirr																								X													
Karungu (Nira & Kachuka)																								X													
Kiahera R 114																							X														
Kipsaraman		X															X	X										X									
Klasies River Mouth																			X																		
Klein Zee																			X																		
Koobi Fora KBS																									X												
Koru																																					
Kromdraai Member A	X			?																																	
Laetolil Beds Lower Unit							X						cf.				X	X												X							
Laetolil Beds Upper Unit													X					X																			
Langebaanweg PPM																		X																			
Langebaanweg QSM																																					
Langental														X													X										
Legetet																									X			X									
Lemudong'o												aff.						X	X	aff.																	
Mfwangano																								X													
Middle Awash Adu Asa																																					

	Moruorot Hill	Nachukui Middle Lomekwi	Nachukui Upper Lomekwi	Napak	Ndolanya Beds Upper Unit	Ngorora Member A	Olduvai Bed I	Olduvai Bed II	Omo Shungura B	Omo Shungura C	Omo Shungura E+F	Omo Shungura G	Plovers Lake	Rusinga	Saldanha Sea Harvest	Sibudu HP	Songhor	Sterkfontein Member 2	Sterkfontein Member 4	Sterkfontein Member 5	Sterkfontein Jacovec Cave	Swartklip 1	Thomas I	Toros-Menalla	Tugen Lukeino	Tugen Mabaget
				X													X									
																				X						
													X					X	X							
																				X						
													cf.													
							X																			
	X		X		X		X																			
														X												
														X												
																	X									
	X			?									X				X									
																										cf.
						aff.																				
						X																			X	
																		X	X					X	
						aff.											X							X	X	
				X																						
								X		X																
											X	X														
																X										
												cf.	X			X		X						X		

in contact; anterior rim of orbits formed by the frontal, maxillary, and jugal; auditory meatus somewhat triangular in shape; anterior chamber of the bulla less inflated than the posterior one; dentition trenchant; upper incisors disposed on a straight line; three-rooted P3; carnassial notch on m1 deep.

African Species *H. abdelalii* Geraads, 1997, *H. alaylaii* Haile-Selassie and Howell, 2009, *H. ichneumon* (Linnaeus, 1758), *H.* sp. (figures 32.5D–32.5F)

Age Ca. 15.8 Ma–Recent.

Geographic Occurrence Chad, Ethiopia, Kenya, Morocco, South Africa, Tanzania, Zambia.

Remarks *Herpestes* sp. indet. was recently described from Lemudong'o in Kenya by Howell and García (2007). *H. abdelalii* may be close to *Galerella pulverulenta* (and thus not a species of *Herpestes*), but this requires further analysis. Toros-Menalla, Chad, has yielded the earliest definite African record of the genus, though older material has been attributed to it (Morales and Pickford, 2008).

Genus *ICHNEUMIA* I. Geoffroy St.-Hilaire, 1837

Diagnosis Modified after Taylor (1972). Frontal region of skull expanded and more elevated than parietal region; anterior orbital margin above P4; postorbital processes well developed, forming complete postorbital bridge in fully adult skulls; sagittal and lambdoidal crests well developed; posterior chamber of bulla strongly inflated; dentition heavy, especially P4, with inner lobe of tooth occupying considerably more than half the anteroposterior diameter of the crown and metastyle reduced; M1 fairly symmetrical, parastyle slightly projecting; anterior root of zygomatic arch far behind P4; m1 with connate paraconid and metaconid; m2 elongate with six cusps, lower canines recurved, upper canines only slightly recurved; dental formula I 3/3, C 1/1, P 4/4, M 2/2.

African Species *I. albicauda* (Cuvier, 1829), *I. nims* Geraads, 1997, *I.* sp.

Age Ca. 2.5 Ma–Recent.

Geographic Occurrence Ethiopia, Kenya, Morocco, Tanzania.

Remarks White-tailed mongooses are known only from a pair of records that probably pertain to the extant species (Petter, 1973; Geraads et al., 2004) and the single tooth of *I. nims* from Ahl al Oughlam (Geraads, 1997). Howell and García (2007) describe as *Ichneumia* aff. *albicauda* a right mandibular corpus with p4 from Lemudong'o in Kenya. In view of the much greater age of this specimen than other known *Ichneumia*, this record requires additional taxonomic scrutiny.

Genus *KICHECHIA* Savage, 1965

Diagnosis Modified after Savage (1965). Herpestid with upper dental formula I 3/?3, C 1/1, P 4/4, M 2/2; teeth not compressed; canine long and slender; parastyle present only on P4; upper molars without conules and without hypocone, protocone crescentic and without anterior and posterior wings.

African Species *K. zamanae* Savage, 1965, *K.* sp.

Age Ca. 20.5–16.4 Ma.

Geographic Occurrence Kenya, Uganda.

Remarks *K. zamanae* is the most common small carnivoran of the eastern African Lower Miocene and is known from a considerable number of localities.

Genus *LEGETETIA* Schmidt-Kittler, 1987

Diagnosis After Schmidt-Kittler (1987). Equal in size to *Kichechia* or somewhat smaller. Differs from *Kichechia* in the following characters: all cones of the dentition more acute; m2 two-rooted and very elongate bearing a complete trigonid and strong hypoconulid; p4 with strongly developed anterior accessory cusp; p1 two rooted; M1 and M2 with broader shelf buccally to the external cusps; M2 with well-developed lingual cingulum; P4 more elongate.

African Species *L. nandii* Schmidt-Kittler, 1987.

Age Ca. 20–19 Ma.

Geographic Occurrence Kenya.

Remarks *Legetetia* is known from a large number of isolated teeth and mandible fragments from several west Kenyan Lower Miocene localities. It is notable for its hypocarnivorous dentition and particularly the elongated, two-rooted m2. Schmidt-Kittler (1987) notes similarities with the European *Sivanasua-Euboictis* group and suggests that they and *Legetetia* may be derived from a common stock.

Genus *LEPTOPLESICTIS* Forsyth Major, 1903

Diagnosis After Roth (1988). Small-sized carnivoran; dental formula (lower dentition only) I 3, C 1, P 4, M 2; premolars with tall cusps; p4 posterior accessory cusp very large; m1 postvallid notch less deep than in *Herpestes*; m2 trigonid and talonid distinct.

African Species *L. mbitensis* Schmidt-Kittler, 1987, *L. namibiensis* Morales et al., 2008, *L. rangwai* Schmidt-Kittler, 1987, *L. senutae* Morales et al., 2008.

Age Ca. 20.5–16.4 Ma.

Geographic Occurrence Kenya, Namibia.

Remarks *Leptoplesictis* is poorly defined, as the characters cited by Roth (1988) to distinguish it from *Herpestes* are likely to be plesiomorphic. It is generally considered to be a herpestid because of its similarity to *Herpestes*, though definitive proof from basicrania is still lacking. It is an occasional member of the European carnivoran guild in the early and middle Miocene (Roth, 1988). In Africa the genus is known from the early Miocene of Kenya and Namibia.

Genus *MUNGOS* E. Geoffroy St.-Hilaire and Cuvier, 1795

Diagnosis Modified after Petter (1973) and Veron (1995). Small-sized herpestids with dental formula I 3/3, C 1/1, P 3/3, M 2/2; skull with orbits closed or nearly closed by a postorbital bar; posterior rim of nasals across the anterior rim of orbits; anterior rim of orbits formed by the frontal, maxillary, jugal, and lacrymal; auditory meatus wider than tall; P1/p1 absent (p1 present in *M. dietrichi*); P3 with three roots; P4 molariform, broader than long; talonid of m1 present but simple and with no cuspids.

African Species *M. dietrichi* Petter, 1963, *M. minutus* Petter, 1973, *Mungos* sp.

Age Ca. 3.7 Ma–Recent.

Geographic Occurrence Kenya, South Africa, Tanzania.

Remarks All fossil records of *Mungos* are of extinct species. *Mungos dietrichi* accounts for most of these records with five occurrences. The record of this species from the Laetolil Beds, Upper Unit, is the earliest appearance of *Mungos* in Africa. In Olduvai, Bed I it is sympatric with another *Mungos*, *M. minutus*, which differs from it in the loss of p1 (as in the extant species) and in being markedly smaller, although Petter later (1987) synonymized the two species.

Genus *SURICATA* Desmarest, 1804

Diagnosis Partly modified after van Staaden (1994). Small-sized Herpestidae; dental formula I 3/3, C 1/1, P 3/3, M 2/2; skull high and rounded, with large closed orbits and interorbital space greater than or equal to postorbital constriction; incisor rows slightly curved; all cheek teeth have tall cusps adapted for puncture-crushing of small prey; P1/p1 absent; protocones of P4–M2 large, metastyles very short; m1 trigonid short, talonid long.

African Species *S. suricatta* (Schreber, 1776), *S. major* Hendey, 1974, *S.* sp.

Age Ca. 1.8 Ma–Recent.

Geographic Occurrence South Africa.

Remarks All fossil meerkat records are from southern Africa and all except one are referable to the extant species *S. suricatta*. *S. major* was described by Hendey (1974) on the basis of material from Elandsfontein. It differs from *S. suricatta* mainly in size.

Genus *UGANDICTIS* Morales et al., 2007

Diagnosis Adapted from Morales et al. (2007). Small size; m2 medium sized; m1 with v-shaped trigonid, metaconid lower than paraconid, which is located anterolingually, talonid smaller than trigonid with strong hypoconid, subdivided hypoconulid and entoconid; p4 narrow with anterior, central and posterior cuspids and a basal platform in a posterolingual position; p3 and p2 relatively short with posterior cuspids; P4 with short metastyle, paracone similar in size to the metastyle, conical in form, prominent parastyle, protocone strong, located anterior to parastyle, and posterior to the basal cingulum of the metastyle; P3 and P4 elongated, with weak basal cingula; upper molars with three roots.

African Species *U. napakensis* Morales et al., 2007.

Age 20–18.5 Ma.

Geographic Occurrence Kenya, Uganda.

Remarks Some of the material recently assigned to this new taxon by Morales et al. (2007) was previously referred to *Leptoplesictis rangwai* and *Kichechia zamanae*.

HERPESTIDAE indet.

Age Ca. 13–0.07 Ma.

Geographic Occurrence Ethiopia, Kenya, Namibia, South Africa.

Remarks Indeterminate Herpestidae are known from a variety of sites, the oldest of which is Ngorora Mb. A (Morales and Pickford, 2005a).

Family PERCROCUTIDAE Werdelin and Solounias, 1991
Table 32.7
Genus *DINOCROCUTA* Schmidt-Kittler, 1976

Diagnosis Modified from Howell and Petter (1985). Percrocutids of large to very large size; P3 without internal root; p4 long relative to p3; p2 hypertrophied with a high robusticity index.

African Species *D. algeriensis* (Arambourg, 1959), *D. senyureki* (Ozansoy, 1957).

Age Ca. 11.2–4.5 Ma.

Geographic Occurrence Algeria, Libya.

Remarks *Dinocrocuta* is mainly distributed in Eurasia, where it was prominent until some time before the Miocene-Pliocene boundary (Werdelin, 1996). In Africa, the genus

TABLE 32.7
Occurrences of Percrocutidae species

Sites	*Dinocrocuta algeriensis*	*Dinocrocuta senyureki*	*Percrocuta* sp.	*Percrocuta leakeyi*	*Percrocuta tobieni*
Bled Douarah Beglia			X		
Bou Hanifia	X				
Fort Ternan			X		
Menacer	X				
Nakali				X	X
Ngorora Member B					X
Ngorora Member D					X
Ngorora Member E					X
Sahabi		X			
Samburu Hills Namurungule			X		

is only known from North Africa (Bou Hanifia, Menacer, Sahabi). It may be found to be present elsewhere, but the absence of *Dinocrocuta* from Ngorora suggests that they cannot have been common members of the late middle–early late Miocene fauna of sub-Saharan Africa.

Genus *PERCROCUTA* Kretzoi, 1938

Diagnosis Modified from Howell and Petter (1985). Percrocutids of relatively small size. Last molars (M2/m2) lost; m1 metaconid absent or vestigial; tendency toward shortening of the talonid and elongation of the trigonid, accentuated in more evolved representatives; P3 with or without an internal root; P4 with a reduced protocone, situated more or less posterior to the anterior margin of the parastyle; anterior premolars hypertrophied; p2 and p3 short (relative to p4 and m1) and broad with high robusticity indices.

African Species *P. tobieni* Crusafont Pairó and Aguirre, 1971, *P. leakeyi* (Howell and Petter, 1985), *P.* sp.

Age Ca. 14.8–9.7 Ma.

Geographic Occurrence Kenya, Tunisia.

Remarks *Percrocuta tobieni* is a fairly generalized, medium-sized *Percrocuta*. It is only known with certainty from eastern Africa, but the specimen described as *Capsatherium luciae* from Bled Douarah, Tunisia (Kurtén, 1976), likely also belongs to this species. More than one species may be involved, as material from the Namurungule Formation, Kenya (Nakaya et al., 1984), differs from the type specimen of *P. tobieni*, but the necessary comparisons have yet to be made. Aguirre and Leakey (1974) mention the presence of "*Hiperhyaena leakeyi*" from Nakali, but as pointed out by Morales and Pickford (2006), this is a *nomen nudum* and the name dates from Howell and Petter (1985). Morales and Pickford (2006) transfer the species from *Dinocrocuta*, in which it was placed by Howell and Petter (1985), to *Percrocuta*, and this is probably correct.

Family HYAENIDAE Gray, 1821
Table 32.8
Genus *ADCROCUTA* (Kretzoi, 1938)

Diagnosis Emended after Kretzoi (1938). Hyaenidae of large size; teeth massive; premolars powerful with weak accessory cusps; m1 talonid composed of entoconid and hypoconid; metaconid poorly developed or absent; P4 protocone very reduced.

African Species A. *eximia* (Roth and Wagner, 1854).

Age Ca. 6.1–5.2 Ma.

Geographic Occurrence Libya.

Remarks Adcrocuta is nearly ubiquitous in Eurasian faunas from MN 10–13, but in Africa is known only from Sahabi. No derived, bone-cracking hyaenid (type 6 of Werdelin and Solounias, 1996) is known from sub-Saharan Africa.

Genus *CHASMAPORTHETES* Hay, 1921
Figure 32.6D

Diagnosis Emended after Kurtén and Werdelin (1988). Medium- to large-sized hyaenids with long, slender limbs; rostrum broad; frontal profile stepped; cheek teeth slender, trenchant; dental formula I 3/3, C 1/1, P 3–4/3–4, M 1/1; P1 large, with a tendency to be shed early in life; anterior premolars with strong posterior accessory cusps, anterior cusps variable but usually strong; P4 with large protocone, short paracone, and elongated parastyle; p4 elongate, with large accessory cusps; m1 without metaconid and with single, laterally compressed, blade-like talonid cusp.

African Species C. *australis* (Hendey, 1974), C. *darelbeidae* Geraads, 1997, C. *nitidula* (Ewer, 1954) (figure 32.6D), C. sp.

Age Ca. 7.4–1 Ma.

Geographic Occurrence Chad, Ethiopia, Kenya, Libya, Morocco, South Africa, Tanzania.

Remarks Chasmaporthetes is a common member of the carnivoran guild in Pliocene Old World faunas and is the only hyaenid to migrate to North America (Kurtén and

Werdelin, 1988). Its earliest African appearance is at Toros-Menalla (Bonis et al., 2007a), but it is best known from South Africa, where the species C. *nitidula* is a characteristic member of the Transvaal cave assemblages (Ewer, 1954b; Werdelin and Turner, 1996). We here consider C. *darelbeidae* from Ahl al Oughlam (Geraads, 1997) to be a valid species.

Genus *CROCUTA* Kaup, 1828
Figures 32.6A–32.6C

Diagnosis Emended after Kretzoi (1938). Hyaenidae of large size; premolars massive, P4 paracone weak, metastyle blade greatly elongated; m1 trigonid elongated, metaconid vestigial or lost, talonid greatly reduced.

African Species C. *crocuta* (Erxleben, 1777), C. *dietrichi* Petter and Howell, 1989 (figures 32.6A–32.6C), C. *eturono* Werdelin and Lewis, 2008, C. *ultra* Ewer, 1954, C. sp.

Age Ca. 3.8 Ma–Recent.

Geographic Occurrence Ethiopia, Kenya, Morocco, South Africa, Tanzania.

Remarks We consider C. *dbaa* (Geraads, 1997) to be a synonym of C. *dietrichi*. The history of *Crocuta* in Africa is complex and has only been presented in outline (Lewis and Werdelin, 1997, 2000). Werdelin and Lewis (2008) have recently described the new and aberrant C. *eturono* from the Pliocene of the Turkana Basin. Still more recently, Gilbert et al. (2008) have described a new subspecies, C. *crocuta yangula* from the Daka Mb. of the Bouri Fm., Middle Awash, Ethiopia.

Genus *HYAENA* Brisson, 1762
Figure 32.6E

Diagnosis Large-sized Hyaenidae with moderate bone-cracking adaptations; limbs slender with elongated metapodials; skull robust: dental formula I 3/3, C 1/1, P 4/3, M 1/1; premolars with some bone-cracking adaptations; P4 short, with metastyle shorter than paracone; m1 with short but well-formed talonid, metaconid usually present.

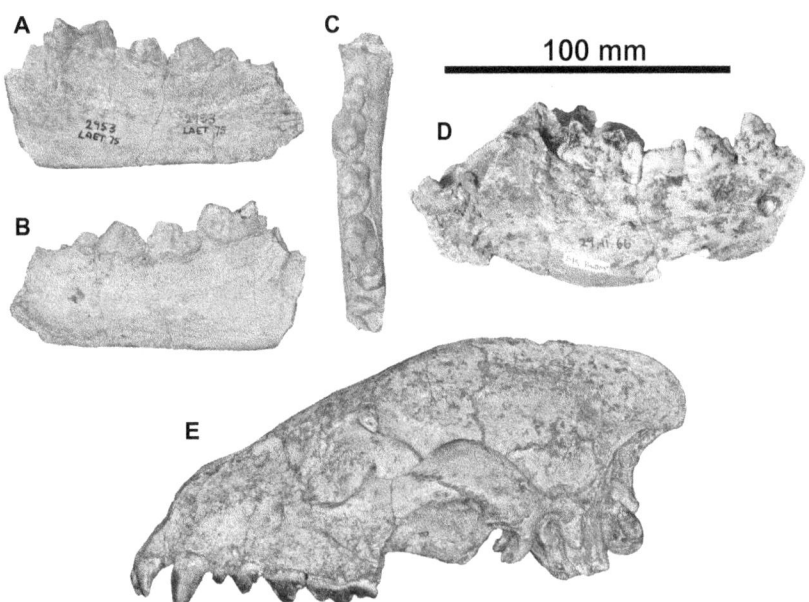

FIGURE 32.6 African Hyaenidae. A–C) *Crocuta dietrichi* type specimen LAET 75-2953 from the Upper Laetolil Beds, Laetoli, left ramus in lingual, buccal, and occlusal view; D) *Chasmaporthetes nitidula* SK 14.005 from Member 1, Swartkrans, right and left rami in right lateral view; E) *Hyaena* sp. KNM-ER 1548, Upper Burgi Mb., Koobi Fora Fm., cranium in left lateral view.

TABLE 32.8
Occurrences of Hyaenidae species

Sites	*Adcrocuta eximia*	*Chasmaporthetes australis*	*Chasmaporthetes nitidula*	*Chasmaporthetes daartbeidae*	*Chasmaporthetes* sp.	*Crocuta crocuta*	*Crocuta dietrichi*	*Crocuta eturono*	*Crocuta ultra*	*Crocuta* sp.	*Hyaena hyaena*	*Hyaena makapani*	*Hyaena* sp.	*Hyaenictis graeca*	*Hyaenictis hendeyi*	*Hyaenictis wehaietu*	*Hyaenictis* sp.	*Hyaenictitherium minimum*	*Hyaenictitherium namaquensis*	*Hyaenictitherium parvum*	*Hyaenictitherium* sp.	Hyaenidae indet.	*Ictitherium ebu*	*Ikelohyaena abronia*	*Lycyaena crusafonti*	*Lycyaenops silberbergi*	*Pachycrocuta brevirostris*	*Pachycrocuta* sp.	*Parahyaena brunnea*	*Parahyaena howelli*	*Pliocrocuta perrieri*	*Proteles amplidentus*	*Proteles* sp.	*Protictitherium crassum*	*Protictitherium punicum*	*Protictitherium* sp.
Adu-Asa Fm.				X			X									X					X			cf.							X					
Ahl al Oughlam											X																				cf.					
Ain Boucherit						X																														
Ain Brimba					X	X																								X						
Ain Hanech																						X														
Ain Tit Mellil																						X														
Allia Bay			X		X					X			X	X								X									cf.			cf.		
Aramis						X																			X										X	
As Duma						cf.																														
Awash 4 (Kalb)						X																							X							
Awash 7 (Kalb)											cf.																									
Beni Mellal						X																							X							
Bled Douarah Beglia						X																							X							
Bloembos						X																							X							
Busidima-Telalak										X	X																									
Doukkala II																																				
Drimolen																																				
Duinefontein 2																																				
Elandsfontein Main																																				
Elandsfontein Wes																																				
El Harhoura																																				
Equus Cave																																				
Grès de Rabat																																				
Grotte des félins																																				
Hadar Denen Dora					cf.		X																								cf.					
Hadar Sidi Hakoma					X		X																	X							cf.					

TABLE 32.8
(CONTINUED)

Sites	Adcrocuta eximia	Chasmaporthetes australis	Chasmaporthetes nitidula	Chasmaporthetes dartbeidae	Chasmaporthetes sp.	Crocuta crocuta	Crocuta dietrichi	Crocuta eturono	Crocuta ultra	Crocuta sp.	Hyaena hyaena	Hyaena makapani	Hyaena sp.	Hyaenictis graeca	Hyaenictis hendeyi	Hyaenictis wehaietu	Hyaenictis sp.	Hyaenictitherium minimum	Hyaenictitherium namaquensis	Hyaenictitherium parvum	Hyaenictitherium sp.	Hyaenidae indet.	Ictitherium ebu	Ikelohyaena abronia	Lycyaena crusafonti	Lycyaenops silberbergi	Pachycrocuta brevirostris	Pachycrocuta sp.	Parahyaena brunnea	Parahyaena howelli	Pliocrocuta perrieri	Proteles amplidentus	Proteles sp.	Prototitherium crassum	Prototitherium punicum	Prototitherium sp.
Hopefield						X																							X							
Kabwe						X																								X						
Kanapoi																																				
Kifan Bel Ghomari																													X							
Klasies River Mouth																																				
Klein Zee																			X																	
Koobi Fora KBS							X		X		X																									
Koobi Fora Lokochot							X	X			X		X																							
Koobi Fora Okote									X																											
Koobi Fora Tulu Bor							X		X			X																								
Koobi Fora Upper Burgi																																				
Koro Toro													cf.									X														
Kosia																						X		X			X		X	X						
Kossom Bougoudi																																				
Kromdraai Member A							X		X																					X						
Laetolil Beds Lower Unit																								cf.			cf.	cf.								
Laetolil Beds Upper Unit																																				
Langebaanweg Baard's Quarry																																				
Langebaanweg PPM		X													X				X					X												
Langebaanweg QSM		X													aff.									X												
Lemudong'o																																				
Lothagam Apak																	cf.						X	cf.												

Lothagam Lower Nawata																cf.											
Lothagam Upper Nawata									cf.	cf.		cf.	cf.														
Makapansgat Member 3				X													X											
Melkbos		X																										
Middle Awash Adu Asa																							X					
Nachukui Kataboi				X			X										X											
Nachukui Lower Lomekwi			X	X									X	X														
Nachukui Middle Lomekwi							X					X	X	X														
Nakoret					X			X				X	X															
Nkondo					X			X				X	X															
Olduvai Bed I								X																				
Olduvai Bed II					X			X																				
Olduvai Bed IV						X																						
Omo Shungura K								X	X			X	X															
Plovers Lake							X																					
Sagantole Fm.								X																				
Sahabi	X									X																		
Saldanha Sea Harvest																												
Samburu Hills Namurungule			cf.																					cf.				
South Turkwel								X																				
Sterkfontein Member 2	X				X				X									X	X			X		X				
Sterkfontein Member 4	X				X				X									X	X			X	X	X				
Sterkfontein Member 5	X								X										X	X				X				
Sterkfontein Jacovec Cave	X				X															X				X				
Swartklip 1																												
Swartkrans Member 2	X				X				X									X	X			X		X				
Swartkrans Member 3	X				X				X									X	X			X	X	X				
Toros-Menalla										X	X		X	X		X							X					
Tugen Lukeino																												
Tugen Mabaget									X		X		X															

African Species *H. hyaena* (Linnaeus, 1758), *H. makapani* Toerien, 1952, *H.* sp. (figure 32.6E).

Age Ca. 3.6 Ma–Recent.

Geographic Occurrence Chad, Ethiopia, Kenya, South Africa, Tanzania, Tunisia.

Remarks The evolutionary history of *Hyaena* in Africa after the possible divergence of the genus from *Ikelohyaena* or some closely related form is fairly straightforward. *H. makapani* is present until ca. 2 Ma, while the extant species *H. hyaena* appears somewhat earlier than this, suggesting a moderate temporal overlap. The record of *H. hyaena precursor* from Aïn Brimba, Tunisia requires further analysis.

Genus *HYAENICTIS* Gaudry, 1861

Diagnosis Emended after Werdelin et al. (1994). Medium- to large-sized Hyaenidae; mandible long and slender; dental formula I 3/3, C 1/1, P 4/3–4, M 1/1–2; p2–3 slender, with weakly developed anterior accessory cusps; P4 with metastyle slightly longer than paracone; m1 with short, two- or three-cusped talonid; m2 generally present.

African Species H. graeca Gaudry, 1861, *H. hendeyi* Werdelin et al., 1994, *H. wehaietu* Haile-Selassie and Howell, 2009, *H.* sp.

Age Ca. 14–4.5 Ma.

Geographic Occurrence Ethiopia, Kenya, Morocco, South Africa.

Remarks *Hyaenictis* is present at a number of African localities, of which Langebaanweg has by far the best sample. Ginsburg (1977) referred a *Hyaenictis* from Beni Mellal to the Eurasian *H. graeca*, but the specific referral must be considered doubtful.

Genus *HYAENICTITHERIUM* Zdansky, 1924

Diagnosis Modified after Kretzoi (1938). Medium- to large-sized hyaenids; dentition more crushing than in *Ictitherium*, with shorter muzzle, stronger mandibular ramus, shorter and more massive canines; m1 talonid with two cusps; m2 and M2 present but reduced.

African Species H. minimum Bonis, Peigné, Likius, Mackaye, Vignaud and Brunet, 2005, *H. namaquensis* (Stromer, 1931), *H.* cf. *parvum* (Khomenko, 1914).

Age Ca. 7.4–4.5 Ma.

Geographic Occurrence Chad, Ethiopia, Kenya, Libya, South Africa.

Remarks Most records of *Hyaenictitherium* in Africa are referable to *H. namaquensis*, but some are smaller. It may turn out that the material referred to *H.* cf. *parvum* by Werdelin (2003) should be placed in the approximately coeval species from Chad, *H. minimum*.

Genus *ICTITHERIUM* Roth and Wagner, 1854

Diagnosis Modified after Kretzoi (1938). Medium-sized forms with generalized doglike hyaena morphology; m1 talonid large, low; P4 with metastyle shorter than paracone and protocone reduced; m2 and M2 present and unreduced; premolars with incipient crushing morphology.

African Species I. ebu Werdelin, 2003.

Age Ca. 7.4–4.7 Ma.

Geographic Occurrence Kenya.

Remarks *Ictitherium* is a common member of the later Miocene carnivoran guild of Eurasia but is rare in Africa, with the only record being the aberrant, long-limbed *I. ebu* from

Lothagam (Werdelin, 2003), though Semenov (2008) has questioned the generic attribution of this species.

Genus *IKELOHYAENA* Werdelin and Solounias, 1991

Diagnosis Modified from Werdelin and Solounias (1991) Slightly smaller than extant *H. hyaena* in size; maxillary contribution to zygomatic arch large; premolars, especially P3 and p3, enlarged, but not strongly conical, and anterior edge only slightly convex; anterior accessory cusps not appressed to main cusp of premolars; infraorbital foramen positioned above midline of P3; P4 metastyle short; M2 and m2 generally present.

African Species I. abronia (Hendey, 1974).

Age Ca. 6.5–2.5 Ma.

Geographic Occurrence Ethiopia, Kenya, Morocco, South Africa, Tanzania.

Remarks *Ikelohyaena abronia*, known from both southern and eastern Africa, is very similar to the extant striped hyaena except for the retention of the second molars. It is not clear whether *I. abronia* should be placed before or after the split between *Hyaena* and *Parahyaena*, and we therefore retain the genus *Ikelohyaena* here.

Genus *LYCYAENA* Hensel, 1863

Diagnosis Modified after Kretzoi (1938). Large-sized forms with slender premolars with moderately strong accessory cusps; m1 talonid short; m2 very reduced; M2 lost.

African Species L. crusafonti Kurtén, 1976.

Age Ca. 12.5–10.5 Ma.

Geographic Occurrence Tunisia.

Remarks The Eurasian hyaenid genus *Lycyaena*, a primitive member of the hunting hyaena lineage (Werdelin and Solounias, 1991, 1996), is only known from a single African record; a lower jaw fragment from Bled Douarah, Tunisia, and apparently never reached sub-Saharan Africa.

Genus *LYCYAENOPS* Kretzoi, 1938

Diagnosis Emended after Kretzoi (1938). Closely related to *Lycyaena* but with lower and longer m1; significantly more massive premolars with well-developed accessory cusps; p2–p3 posteriorly expanded, with posterior end squared off.

African Species L. silberbergi (Broom in Broom and Schepers, 1946).

Age Ca. 4–1.8 Ma.

Geographic Occurrence South Africa.

Remarks This species is commonly placed in *Chasmaporthetes* (see Werdelin and Turner, 1996), but Werdelin (1999b) noted similarities between it and the European *L. rhomboideae* and suggested that they might be congeneric.

Genus *PACHYCROCUTA* Kretzoi, 1938

Diagnosis Modified after Kretzoi (1938). Very large, robust Hyaenidae; premolars very massive; p2–3 lacking anterior accessory cusps; P4 with massive paracone and protocone; m1 short and broad lacking metaconid; talonid consisting only of the hypoconid.

African Species P. brevirostris (Aymard, 1846), *P.* sp.

Age Ca. 4.1–1.3 Ma.

Geographic Occurrence Kenya, South Africa, Tanzania.

Remarks This highly distinctive species has long been known from South Africa as *P. bellax* (Ewer, 1954a). More recently, it has been identified from eastern Africa as well (Werdelin, 1999a).

Genus *PARAHYAENA* Hendey, 1974

Diagnosis Large-sized Hyaenidae with moderate to strong bone-cracking adaptations; limbs slender with elongated metapodials; premolars robust; P3 and p3 with strong bone-cracking ability; dental formula I 3/3, C 1/1, P 4/3, M 1/1; P4 with large protocone, metastyle longer than paracone; m1 short with short but well-formed talonid, metaconid commonly lost.

African Species *P. brunnea* (Thunberg, 1820), *P. howelli* Werdelin, 2003.

Age Ca. 4.3 Ma–Recent.

Geographic Occurrence Kenya, South Africa, Tanzania, Zambia.

Remarks Living brown hyenas are restricted to southern Africa, but *P. brunnea* has been recorded as a fossil in eastern Africa (Werdelin and Barthelme, 1997). The presence of *P. howelli* at several eastern African localities suggests that the origin of the species may lie there rather than in southern Africa.

Genus *PLIOCROCUTA* Kretzoi, 1938

Diagnosis Modified after Kretzoi (1938). Hyaenidae of large size; with unreduced P4 protocone; short, massive m1 lacking metaconid but with a wide talonid bearing ento- and hypoconid.

African Species *P. perrieri* (Croizet and Jobert, 1828).

Age Ca. 3.4–2 Ma.

Geographic Occurrence Ethiopia, Morocco.

Remarks The only certain African record of this species (*P. perrieri latidens* Geraads, 1997) is from Ahl al Oughlam. The taxon has been reported from other North African sites, such as Aïn Boucherit, but the only eastern African record that may be valid is from the Sidi Hakoma and Denen Dora Mbs at Hadar, though this material also requires renewed evaluation.

Genus *PROTELES* I. Geoffroy St.-Hilaire, 1824

Diagnosis Hyaenidae with cheek teeth reduced to pegs; dental formula I 3/3, C 1/1, P+M 3–4/2–4, sometimes less; canines not reduced; skull slender; palate long and rectangular.

African Species *P. amplidentus* Werdelin and Solounias, 1991, *P. cristatus* (Sparrman, 1783), *P. sp.*

Age Ca. 2.5 Ma–Recent (possibly as old as 4 Ma).

Geographic Occurrence South Africa.

Remarks Although *Proteles* is easy to recognize from craniodental remains, it is only known from scarce records in South Africa. *P. amplidentus* differs from the extant species in its slightly less reduced cheek teeth. Postcranial remains of a very small hyaenid from the Laetolil Beds, Lower Unit may possibly belong to *Proteles* or a related form (Werdelin and Dehghani, in press).

Genus *PROTICTITHERIUM* Kretzoi, 1938

Diagnosis Modified after Kretzoi (1938). Small, slender Hyaenidae; well-developed P4 protocone; m1 with very tall talonid cusps, tall, well-developed metaconid, talonid relatively short; upper molars relatively reduced.

African Species *P. crassum* (Depéret, 1892), *P. punicum* (Kurtén, 1976), *P. sp.*

Age Ca. 14–7.5 Ma.

Geographic Occurrence Kenya, Morocco, Tunisia.

Remarks The African representation of this primitive hyaenid genus is limited. The only certain records are from Bled Douarah (see Werdelin and Solounias [1991] for a discussion of this material) and Beni Mellal, described as *Ictitherium* cf. *arambourgi*, a synonym of *P. crassum*, cf. Schmidt-Kittler (1976). A single sub-Saharan record also exists, described as *Ictitherium* sp. by Nakaya et al. (1984).

HYAENIDAE indet.

Age Ca. 7–1.4 Ma.

Geographic Occurrence Chad, Ethiopia, Kenya, Libya, South Africa, Uganda.

Remarks Unidentified and indeterminate Hyaenidae are known from a number of localities of late Miocene to early Pleistocene age. With the exception of "hyaenid sp. E" from Langebaanweg (Hendey, 1974) and a specimen from Chad showing similarities to it, none appear to represent taxa other than those already listed.

Family BARBOUROFELIDAE Schultz et al., 1970
Table 32.9
Genus *AFROSMILUS* Kretzoi, 1929
Figures 32.7C–32.7E

Diagnosis Modified after Morlo et al. (2004). Medium-sized barbourofelids with upper canines very compressed; P2 absent; P3 with anterior accessory cusp present; P4 protocone about the size of the parastyle; M1 reduced with P4 length/M1 width ratio 1.6–1.8; M1 with protocone very reduced but still distinct and metacone completely reduced and not projected posteriorly; m1 with talonid vestigial or absent; mandibular flange poorly developed.

African Species *A. africanus* (Andrews, 1914), *A. turkanae* Schmidt-Kittler, 1987 (figures 32.7C–32.7E), *A. sp.*

Age Ca. 20–15.4 Ma Ma.

Geographic Occurrence Kenya, Namibia.

Remarks The genus is the best known of the family in Africa, although our knowledge remains rather limited. This is apparently the only Lower Miocene African genus that migrates out of the continent, with *A. hispanicus* known from the Lower Miocene of Spain (Morales et al., 2001b); this constitutes evidence of a first migration of the family to Europe.

Genus *GINSBURGSMILUS* Morales et al., 2001
Figures 32.7F–32.7K

Diagnosis Barbourofelids with upper canine transversely compressed, with crenulated anterior and posterior crests, and vertical grooves on both faces; P2 present but very reduced; p3 with anterior accessory cuspid present.

African Species *G. napakensis* Morales et al., 2001 (figures 32.7F–32.7K).

Age Ca. 20–19 Ma.

Geographic Occurrence Kenya, Uganda.

Remarks The single species assigned to the genus is the oldest and most primitive of the family, the origin of which remains unknown. The available material of *Ginsburgsmilus* does not display any derived character of the upper dentition; consequently, those mentioned in the revised diagnosis

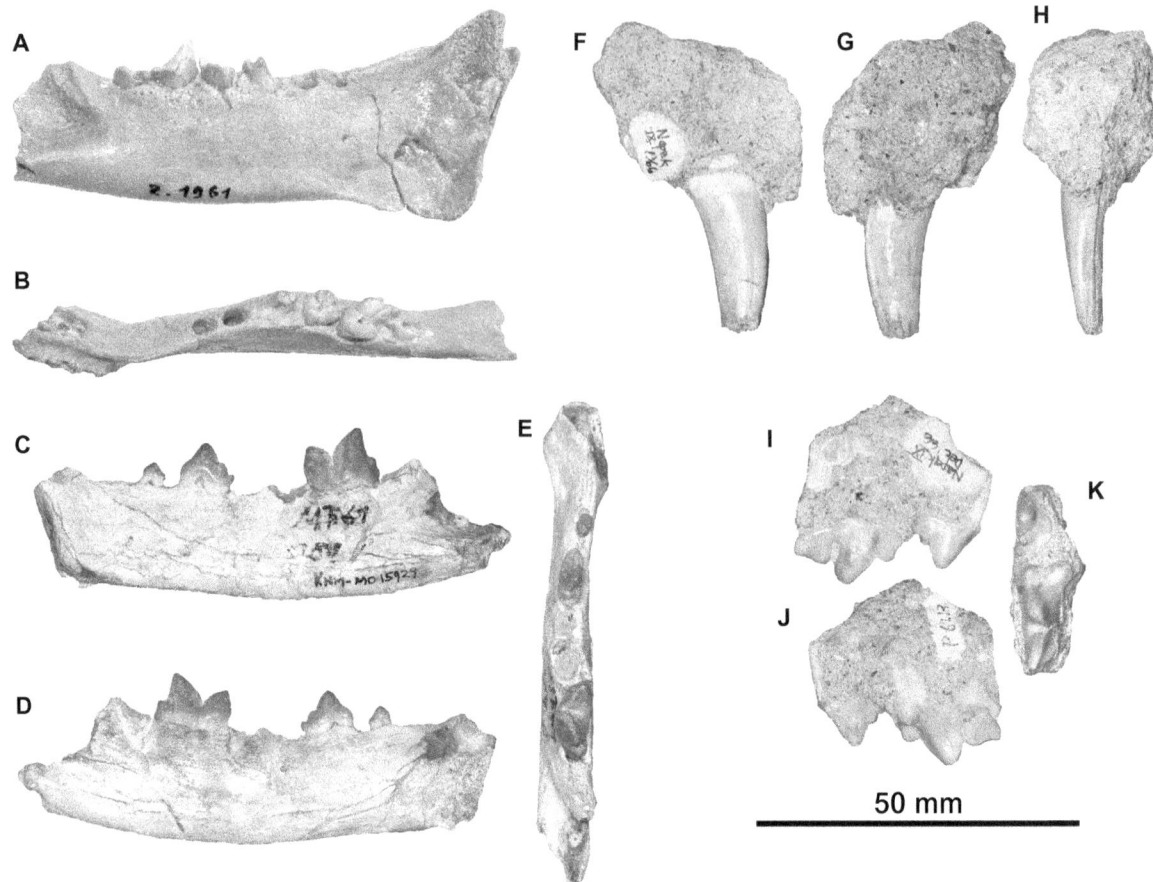

FIGURE 32.7 African Barbourofelidae. A–B) *Syrtosmilus syrtensis* type specimen MNHN-2-1961 right ramus from Jebel Zelten in buccal and occlusal view; C–E) *Afrosmilus turkanae* type specimen KNM-MO 15292 left ramus from Moruorot in buccal, lingual, and occlusal views; F–K) *Ginsburgsmilus napakensis* type specimen UM-P67-13 right C and P3-P4, from Napak IX: F–H) right upper canine in lateral, medial, and anterior views, I–K) right P3–P4 in buccal, lingual, and occlusal views.

proposed here could apply to any basal barbourofelid (see Morlo et al., 2004). Morales et al. (2001b) also mentioned diagnostic characters of the lower dentition, based on a specimen initially described by Schmidt-Kittler (1987, p. 118) as *Afrosmilus turkanae*. The only significant character may be the presence of a small anterior accessory cuspid on p3 (not "p2" as specified in the original diagnosis [Morales et al., 2001b: 98]). If correctly assigned, the calcaneum of this species recently described from Napak I (Morales et al., 2007) indicates plantigrade hindlimb posture.

Genus *SYRTOSMILUS* Ginsburg, 1978
Figures 32.7A and 32.7B

Diagnosis Medium-sized barbourofelid, smaller than *Sansanosmilus palmidens*; p1, p2, and m2 absent; mandibular flange little developed and less pronounced than in *Sansanosmilus*; mandibular corpus strongly curved; diastema between c and p3 longer; p3 less reduced than in *Sansanosmilus*.

African Species S. syrtensis Ginsburg, 1978 (figures 32.7A and 32.7B).

Age Ca. 19–15 Ma.

Geographic Occurrence Libya.

Remarks The single and holotype specimen from Jebel Zelten is a fragmentary right hemimandible with p4 and m1 crown base. The absence of p2, the length of the c–p3 diastema and the strong curvature (in dorsal view) of the dentary indicate that this is an advanced species of the family, more derived than *Ginsburgsmilus* and *Afrosmilus*.

Genus *VAMPYRICTIS* Kurtén, 1976

Diagnosis Derived from Morlo et al. (2004). Large-sized barbourofelid with crenulated deciduous upper canine; lower carnassial elongated, extremely compressed, with strongly curved blade, no talonid and vestigial metaconid.

African Species V. vipera Kurtén, 1976.

Age Ca. 12.5–10.5 Ma.

Geographic Occurrence Tunisia.

Remarks This species was the youngest member of the family in Africa until the recent discovery of cranial material from Kenya (discussed later). Although poorly known (fragment of upper canine and m1), the familial assignment is well supported. The m1 of *Vampyrictis* is much more derived than the contemporary machairodontine felid *Machairodus aphanistus* and has a barbourofelid dental character: the reduction of the talonid (absent in *Vampyrictis*) before that of the metaconid (vestigial). The species has reached the same evolutionary grade as the latest European *Sansanosmilus* in terms of sabertooth adaptation. From what we know of the

TABLE 32.9
Occurrences of Barbourofelidae species

Sites	Afrosmilus africanus	Afrosmilus turkanae	Afrosmilus sp.	Barbourofelidae indet.	Ginsburgsmilus napakensis	Syrtosmilus syrtensis	Vampyrictis vipera
Bled Douarah Beglia							X
Fiskus	X						
Grillental	X						
Jebel Zelten						X	
Karungu (Nira & Kachuka)	X						
Langental	X						
Locherangan	X						
Moruorot Hill		X					
Napak					X		
Rusinga	X	X					
Samburu Hills Namurungule				X			
Songhor			X		X		

dental eruption sequence in barbourofelids and given the young age of the holotype individual, the fragment of upper canine is probably deciduous (Bryant, 1988; pers. obs.).

BARBOUROFELIDAE indet.

Age Ca. 10–7.5 Ma.

Geographic Occurrence Kenya.

Remarks Tsujikawa (2005) recently described a nearly complete skull with damaged/worn dentition from the Namurungule Formation and assigned it to the felid subfamily Machairodontinae. However, this specimen undoubtedly belongs to a derived barbourofelid species. From the available illustrations (Tsujikawa, 2005: figure 2), the skull is ca. 20 cm long from the posterior margin of the occipital condyle to the anteriormost border of the premaxilla. In addition, the general morphology of the skull (short, wide snout, strong maxillary constriction just in front of P3, short palate, strongly lowered postglenoid process, reduced interbullar space, dorsally located occipital condyles, etc.), and the upper dentition (anteroposteriorly short but strongly compressed upper canine, absence of P2, short postcanine diastema, reduced P3 and much more elongated P4) distinguishes this specimen from contemporary machairodontines and indicates that it probably belongs to *Sansanosmilus* (size of *S. vallesiensis*) or to a new, closely related genus. The published information is too limited to allow for a precise taxonomic allocation of the Namurungule specimen.

Family FELIDAE Fischer, 1817
Table 32.10
Genus *ACINONYX* Brookes, 1828

Diagnosis Large-sized Felidae; dental formula I 3/3, C 1/1, P 3/2, M 1/1; skull rounded with high vault, splanchnocranium nearly vertical; limbs slender, claws semiretractile, claw sheaths reduced; cheek dentition sectorial; premolars with tall accessory cusps.

African Species A. aicha Geraads, 1997, *A. jubatus* (Schreber, 1776), *A.* sp.

Age Ca. 4 Ma–Recent.

Geographic Occurrence Ethiopia, Kenya, Morocco, South Africa, Tanzania.

Remarks The fossil history of the cheetah in Africa is poorly known. The earliest *Acinonyx* are recorded from Sterkfontein Mb. 2 and the Laetolil Beds, Upper Unit, but these and most later records cannot be definitely assigned to the living species. The modern species is, however, present at Swartkrans, Mb. 3, in the latest Pliocene or early Pleistocene.

Genus *AMPHIMACHAIRODUS* Kretzoi, 1929

Diagnosis Modified from Sardella (1994). Felids of large size; skull massive and elongated; symphyseal region of mandible robust; ascending ramus of mandible low; teeth crenulated; incisors robust; lower incisors elevated relative to cheek teeth; upper canine compressed and elongated; lower canine reduced in height; P2 present but very reduced; P3, p3, and p4 reduced relative to P4 and m1 but with well-developed cusps; upper carnassial with small preparastyle and moderately reduced protocone; m1 with very small talonid.

African Species A. kabir Peigné et al., 2005, *A.* sp.

Age Ca. 7–4.5 Ma.

Geographic Occurrence Chad, Ethiopia, Libya, South Africa, Tanzania.

Remarks *Amphimachairodus* spp. are not uncommon elements in late Upper Miocene faunas in Africa, but the remains are nearly always very limited and the only named species is *A. kabir* from Chad (Peigné et al., 2005). The African record of *Amphimachairodus* was recently reviewed by Sardella and Werdelin (2007).

TABLE 32.10
Occurrences of Felidae species

Sites	Acinonyx jubatus	Acinonyx aicha	Acinonyx sp.	Amphimachairodus kabir	Amphimachairodus sp.	Caracal caracal	Caracal sp.	Diamantofelis ferox	Dinofelis aronoki	Dinofelis barlowi	Dinofelis darti	Dinofelis diastemata	Dinofelis petteri	Dinofelis piveteaui	Dinofelis sp.	Felis margarita	Felis silvestris	Felis sp.	Felidae indet.	Homotherium hadarensis	Homotherium problematicum	Homotherium africanum	Homotherium sp.	Leptailurus serval	Leptailurus sp.	Lokotunjailurus emagaritus	Lynx sp.	Lynx thomasi	Machairodus aphanistus	Machairodus robinsoni	Machairodus sp.	Megantereon ekidoit	Megantereon ekidoit/whitei	Megantereon whitei	Metailurus obscurus	Metailurus sp.	Namafelis minor	Panthera leo	Panthera pardus	Panthera sp.
Adu-Asa Fm.		X													X		X						X				X				X								X	
Ahl al Oughlam																		X	X			X																		
Ain Bahya																																								
Ain Brimba																																								
Allia Bay													X										X										X							
Aramis																																					X			
Arrisdrift								X																																
As Duma																			X																					
Asbole																	X																					cf.		
Awash 10 (Kalb)																									X															
Awash 5 (Kalb)																																								
Awash 7 (Kalb)			X												X				X				X										X							X
Awash 8 (Kalb)															X																		X							
Awash Bouri Daka																		cf.			cf.																	X	cf.	
Bled Douarah Beglia																			X											X										
Bolt's Farm										X																								X						
Bouknadel																																						X		
Busidima-Telalak																	cf.																					cf.		
Chorora																													cf.											
Cooper's							X								X										X															
Dahkleh Oasis																																								
Diepkloof Rock Shelter																	X		X																					
Doukkala I	X																X																					X	X	
Doukkala II	X																X																							
Drimolen																	X		X																				X	
Duinefontein 2																	X		X																			X		
Elands Bay Cave																	X		X																				X	X
Elandsfontein Main																	X		X														X					X	?	

El Harhoura						X										X	X																					X		
Equus Cave																																							X	
Eshoa Kakurongori																							X																	
Fish Hoek																																							X	
Gladysvale																																						X		
Grotte des félins																	X																					X	X	
Hadar Denen Dora													X					X		X																				
Hadar Kada Hadar									X										X	X																				
Hadar Lip																			X																					
Hadar Pinnacle																			X	X																				
Hadar Sidi Hakoma													X						X																					
Hopefield																																						X		
Inolelo 1																			X																					
Isenya	cf.																																							
Jebel Zelten																			X																					
Kabwe																								X														X	X	
Kanam East														X																										
Kanapoi													X					X					X																	
Kifan Bel Ghomari																	X																					X		
Kipsaraman																			X																					
Klasies River Mouth																																							X	
Konso-Gardula 1														X																										X
Konso-Gardula 2														X				X																						X
Koobi Fora KBS			X												X								X																X	X
Koobi Fora Lonyumun																			X				X																	
Koobi Fora Okote			X											X									X											X				X		X
Koobi Fora Tulu Bor													X						X				X																	
Koobi Fora Upper Burgi			X						X														X																cf.	X
Koro Toro																		X																						
Kosia																			X																					
Kossom Bougoudi															X																									
Kromdraai Member A							X							X									X											X				X	X	
Laetolil Beds Lower Unit																																								
Laetolil Beds Upper Unit			?				X						X					X					X		X														cf.	X
Langebaanweg PPM					X																														X					
Langebaanweg QSM					X										X				X																					
Leba																			X																					

TABLE 32.10
(CONTINUED)

Sites	Acinonyx jubatus	Acinonyx aicha	Acinonyx sp.	Amphimachairodus kabir	Amphimachairodus sp.	Caracal caracal	Caracal sp.	Diamantofelis ferox	Dinofelis aronoki	Dinofelis barlowi	Dinofelis darti	Dinofelis diastemata	Dinofelis petteri	Dinofelis piveteaui	Dinofelis sp.	Felis margarita	Felis silvestris	Felis sp.	Felidae indet.	Homotherium hadarensis	Homotherium problematicum	Homotherium africanum	Homotherium sp.	Leptailurus serval	Leptailurus sp.	Lokotunjailurus emageritus	Lynx sp.	Lynx thomasi	Machairodus aphanistus	Machairodus robinsoni	Machairodus sp.	Megantereon ekidoit	Megantereon ekidoit/whitei	Megantereon whitei	Metailurus obscurus	Metailurus sp.	Namafelis minor	Panthera leo	Panthera pardus	Panthera sp.
Lemudong'o																									cf.	X										X				
Lothagam Apak									X						X				X																					
Lothagam Kaiyumung															X											X														
Lothagam Lower Nawata															X											X														
Lothagam Upper Nawata																																								
Maboko																																								
Makapansgat Member 3			X				X				X								X		X				X															
Melka Kunture Gombore IB																																								
Melkbos					X																												X					X		
Middle Awash Adu Asa															X								X																	
Nachukui Kaitio																							X																	
Nachukui Kataboi																							X																	
Nachukui Lower Lomekwi													X																											
Nachukui Middle Lomekwi													X																											
Nachukui Upper Lomekwi																							X																	
Nakoret									X										X				X																	
Ndolanya Beds Upper Unit																							X																cf.	X
Nkondo					X																															cf.				
Olduvai Bed I															X																							X	X	
Olduvai Bed II															X																							X	X	
Olduvai Bed IV																	X																					X	X	

Site																																
Omo Shungura A	X	cf.									cf.										X				X							cf.
Omo Shungura B	X	cf.			X							X				cf.			X		X	X							X			
Omo Shungura C	cf.	cf.			X				cf.			X										X									cf.	
Omo Shungura D	X	cf.			X							X										X										
Omo Shungura E+F	cf.	cf.			X														X													
Omo Shungura G	cf.																															
Omo Usno	cf.	X	X	cf.	X	cf.		cf.	X			X				cf.	cf.		X		X		X								cf.	
Oulad Hamida 1—rhino cave	X						X									X																
Oulad Hamida III—H. erectus	X																															
Plovers Lake																																
Sagantole Fm.																																
Sahabi	aff.						X																									
Saldanha Sea Harvest	X																															
Samburu Hills Namurungule	X	X				X																										
Shoshamagai			X						X																							
Sibudu HP																																
Sidi Abderrahman																																
Songhor								X		X		X																				
South Turkwel	X	X	X																													
Sterkfontein Member 2	X	X			X	X					X	X																				
Sterkfontein Member 4	X	X		X	X						X	X																				
Sterkfontein Member 5	X	X		X	X						X	X																				
Sterkfontein Jacovec Cave	X									X																						
Swartklip 1	X	X	X	X								X																				
Swartkrans Member 2	X											X																				
Swartkrans Member 3	X	X										X																				
Taung Hrdlicka deposits	X											X																				
Thomas 1—H. erectus cave	X		X																													
Tighenif	X		X																													
Toros-Menalla			X						X																							
Tugen Lukeino		?																														

Genus *CARACAL* Gray, 1843

Diagnosis Derived from Salles (1992). Medium-sized felids; skull with deep pterygoid fossa with marked lateral pterygoid process; subarcuate fossa of the petrosal residual; internal auditory meatus with distinctive, salient border; marked angle formed by the nasal and frontal parts of the profile (45° or higher); rostral constriction limited to the posterior frontonasal region; anterolateral projection of the jugal over the infraorbital foramen present.

African Species *C. caracal* (Schreber, 1776), *C.* sp.

Age Ca. 4–Recent.

Geographic Occurrence Ethiopia, Morocco, South Africa, Tanzania.

Remarks Throughout the Plio-Pleistocene, there are a number of finds of medium-sized felids. These come in two size groups. One of these matches *C. caracal* in size and the other matches *L. serval*. On the assumption that these size groups at least represent the modern genera, if not species, the oldest *Caracal* are from Sterkfontein, Mb. 2 and from the Usno Fm, Ethiopia.

Genus *DIAMANTOFELIS* Morales et al., 1998

Diagnosis Emended after Morales et al. (1998). Medium-sized Felidae with rounded mandibular symphysis; p2 absent; short postcanine diastema; premolars narrow and high; m1 with talonid reduced and lacking a metaconid.

African Species *D. ferox* Morales et al., 1998.

Age 17.5–17 Ma.

Geographic Occurrence Namibia.

Remarks This early felid, known from a partial mandibular corpus and some tentatively referred postcrania from Arrisdrift, bears similarities in the cheek teeth to *Afrosmilus* (Morales et al., 1998) but is clearly not of the sabertoothed type. Its affinities are obscure, except for an apparent close relationship to *Namafelis minor* (discussed later).

Genus *DINOFELIS* Zdansky, 1924
Figure 32.8B

Diagnosis After Werdelin and Lewis (2001). Machairodontinae of moderate to large size; cranium rounded, with large to very large sagittal crest; lower jaw stout; coronoid process large compared to other Machairodontinae; no anterior flange present but anterior margin of ramus strong and flat; upper canine short and only very slightly mediolaterally compressed; no longitudinal grooves; posterior crest well developed, but no serrations present; P2 lost except in *D. cristata*; P3 not or slightly reduced and slender, but less so than in other Machairodontinae; P4 with somewhat reduced protocone; lower canine reduced but not incisiform; p2–p4 slender; m1 elongated, but less so than in other Machairodontinae.

African Species *D. aronoki* Werdelin and Lewis, 2001, *D. barlowi* (Broom, 1937), *D. darti* (Toerien, 1955), *D. diastemata* (Astre, 1929), *D. petteri* Werdelin and Lewis, 2001, *D. piveteaui* (Ewer, 1955) (see also figure 32.8B), *D.* sp.

Age Ca. 7.4–1 Ma.

Geographic Occurrence Chad, Ethiopia, Kenya, Morocco, South Africa, Tanzania.

Remarks Dinofelis is perhaps the most common fossil felid taxon in the African Neogene. There was an extensive endemic radiation of the genus in Africa that is still only

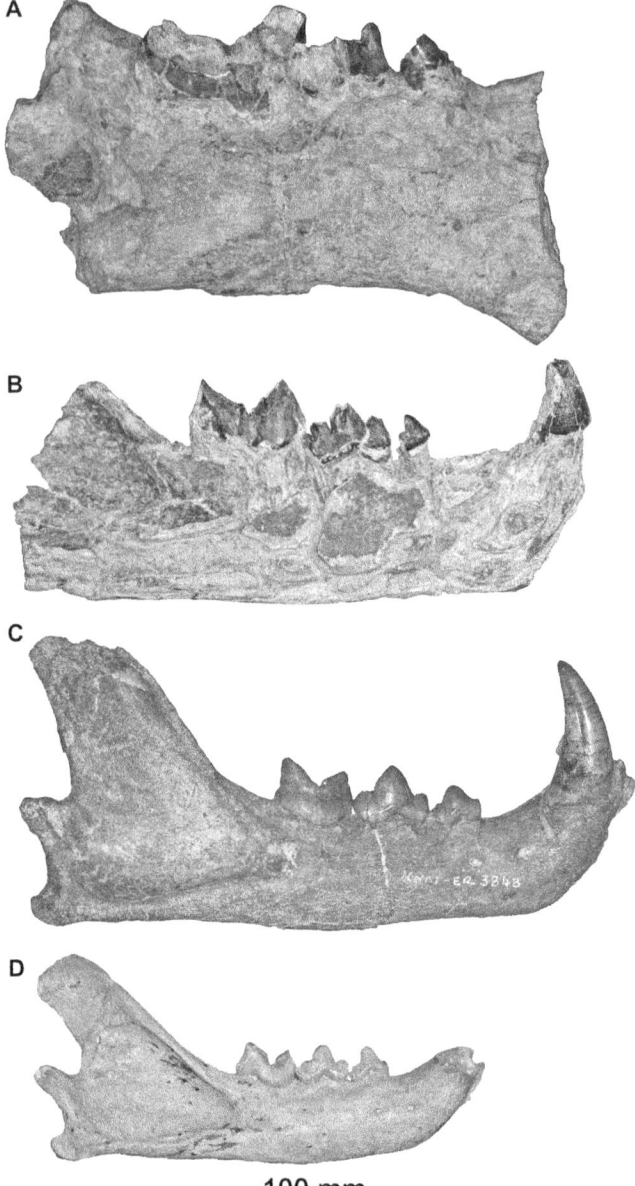

FIGURE 32.8 African Felidae. A) *Homotherium* sp. KNM-ER 931 right ramus from the KBS Mb., Koobi Fora Fm., in buccal view; B) *Dinofelis piveteaui* KNM-ER 40482 left ramus (reversed) from the Okote Mb., Koobi Fora Fm., in buccal view; C) *Panthera pardus* KNM-ER 3848 right hemimandible from the Upper Burgi Mb., Koobi Fora Fm., in buccal view; D) Felidae indet. PQ-L 40321 left hemimandible (reversed), unknown member, Varswater Fm., Langebaanweg in buccal view.

partly explored (Werdelin and Lewis, 2001). The referral of material from Lukeino to the European species *D. diastemata* (Morales et al., 2005) is in our opinion based on material that is not diagnostic at the species level, and in any case the referral of material from Langebaanweg to that species cannot be maintained (Hendey, 1974; Werdelin and Lewis, 2001).

Genus *FELIS* Linnaeus, 1758

Diagnosis Modified after Salles (1992). Small-sized Felidae; dental formula I 3/3, C 1/1, P 3/2, M 1/1; premolar cusps tall, trenchant; m1 metaconid and talonid absent or vestigial;

except *F. chaus*, skull with deep pterygoid fossa with marked lateral pterygoid process, marked angle formed by the nasal and frontal parts of the profile (45° or higher), and modified flat, enlarged, and medially inflected lower rim surface of the orbit; great lateral expansion of the frontal bone, frontal breadth measuring more than half the condylobasal length; anterolateral projection of the jugal over the infraorbital foramen present.

African Species F. margarita (Loche, 1858), *F. silvestris* Schreber, 1775, *F.* sp.

Age Ca. 7 Ma–Recent.

Geographic Occurrence Chad, Egypt, Ethiopia, Kenya, Morocco, South Africa, Tanzania.

Remarks Felid remains of the size of *Felis* spp. are known from many localities in Africa. The oldest of these is Toros-Menalla (Peigné et al., 2008b). The fragmentary nature of most of the material makes referral to a species difficult. The oldest material referred to a specific species of *Felis* is *F. silvestris lybica* from Ahl al Oughlam. The tooth from Beni Mellal identified by Ginsburg (1977) as *Felis* sp. is more likely to belong to a viverrid.

Genus *HOMOTHERIUM* Fabrini, 1890
Figure 32.8A

Diagnosis Emended after Sardella (1994). Machairodont of large size; dentition serrated and crenulated; incisors robust, I3 caniniform; incisors placed in an arch; C flattened and elongated, crenulations large; c small, subequal to i3; p3 reduced, sometimes lost in derived forms; P4 with completely reduced protocone, preparastyle present; p4 reduced, main cusp large; m1 elongated, metaconid and talonid absent; skull elongated, especially the splanchnocranium; sagittal and lambdoid crests large and robust.

African Species H. africanum Petter and Howell, 1987, *H. hadarensis* Petter and Howell, 1988, *H. problematicum* (Collings, 1972), *H.* sp. (figure 32.8A)

Age Ca. 4.2–0.7 Ma.

Geographic Occurrence Algeria, Ethiopia, Kenya, Morocco, South Africa, Tanzania.

Remarks Homotherium is one of the most common carnivorans in African Plio-Pleistocene sites, though it is rarely present in large numbers. Because of the fragmentary nature of much of the material, the taxonomy and evolution of *Homotherium* in Africa is poorly understood.

Genus *LEPTAILURUS* Severtzow, 1858

Diagnosis Derived from Salles (1992). Medium-sized felid; skull with marked lateral expansion of the frontal bone; marked angle formed by the nasal and frontal parts of the profile (45° or higher); rostral constriction limited to the posterior frontonasal region; anterolateral projection of the jugal over the infraorbital foramen present; protocone of P4 markedly reduced.

African Species L. serval (Schreber, 1776), *L.* sp.

Age Ca. 4 Ma–Recent (possibly as old as 6 Ma).

Geographic Occurrence Ethiopia, Kenya, South Africa, Tanzania, Zambia.

Remarks If the comments provided under *Caracal* apply, the earliest *Leptailurus* is from Makapansgat Mb. 3 and from the Laetolil Beds, Upper Unit, unless the record from Lemudong'o (Howell and García, 2007) can be verified as coming from this genus.

Genus *LOKOTUNJAILURUS* Werdelin, 2003

Diagnosis After Werdelin (2003). Felidae of large size; mandibular horizontal ramus slender, ascending ramus relatively tall; upper canine strongly laterally compressed, with serrations present on both anterior and posterior edges; P2 present but small and peglike; upper carnassial long and slender with completely reduced protocone; p3 small and single rooted, though not peglike; lower carnassial long, slender, and low; metaconid-talonid complex absent; appendicular skeleton relatively slender, lacking extreme machairodont features; first digit of manus very robust, with extremely large claw, more than twice the size of the claws on the other digits.

African Species L. emageritus Werdelin, 2003.

Age Ca. 7.4–5.5 Ma.

Geographic Occurrence Kenya.

Remarks This form, known from a partial skeleton and other remains from Lothagam, shows a mixture of derived and primitive features (Werdelin, 2003). Its relationship to other African machairodont cats is unknown at present. It is most similar to a medium-sized machairodont from Langebaanweg, but that taxon is considerably less dentally derived (Hendey, 1974; Sardella and Werdelin, 2007). A second possible record of the species was recently reported from Lemudong'o by Howell and García (2007).

Genus *LYNX* Kerr, 1792

Diagnosis Small- to medium-sized felids with short face; P2 absent; three grooves on upper canine; limbs long, especially posterior limb; tail short.

African Species L. thomasi Geraads, 1980, *L.* sp.

Age Ca. 2.5–0.6 Ma.

Geographic Occurrence Morocco.

Remarks Fossils of *Lynx* spp. are rare in Africa and have thus far only been identified in North Africa. *Lynx thomasi* from Oulad Hamida is a clearly distinct species, but its relationships to other members of the genus remain obscure.

Genus *MACHAIRODUS* Kaup, 1833

Diagnosis After Antón et al. (2004). Lion-sized felid; skull narrow across the zygoma, with moderately convex dorsal profile, well-developed sagittal crest, zygomatic arch low and gently curved in side view, temporal fossa elongated, paroccipital process well developed and projecting inferiorly beyond the relatively small mastoid process, nasofrontal suture intermediate between pantherine (pointed) and evolved machairodontine (straight) condition, postorbital processes large but low; small lower incisors arranged in a straight row; large lower canines with flattened roots and an oval cross section to the crown; small diastema between c and p3; large lower premolars with a complete set of accessory cusps and p3 large relative to p4; well-developed metaconid-talonid complex on m1; mandibular horizontal ramus thick and high, with an undeveloped mandibular flange, coronoid process high and posteriorly inclined; relatively large upper incisors set in a shallow arc; high-crowned and very flattened upper canines; P2 variably present; P3 with posterior expansion; upper carnassial with a distinct protocone and preparastyle.

African Species ?M. aphanistus Kaup, 1833, *M. robinsoni* Kurtén, 1976.

Age Ca. 12.5–10 Ma.

Geographic Occurrence ?Ethiopia, Tunisia.

Remarks Antón et al. (2004) restrict *Machairodus* to *M. aphanistus* only, but we feel that *M. robinsoni*, though poorly known, is so similar to *M. aphanistus* in its known features that it is difficult to distinguish them at the specific, let alone generic level. The record of *M. aphanistus* from Chorora (Geraads et al., 2002) must at present be considered non-diagnostic and pertains to a machairodont of unknown affinities.

Genus *MEGANTEREON* Croizet and Jobert, 1828

Diagnosis Medium-sized sabertooth felids; powerful forequarters, relatively short limbs; dental formula I 3/3, C 1/1, P 2/2, M 1/1; C very long, lacking serrations; m1 not elongated, metaconid and talonid lost.

African Species M. ekidoit Werdelin and Lewis, 2000, *M. whitei* (Broom, 1937).

Age Ca. 4.4–1.3 Ma.

Geographic Occurrence Ethiopia, Kenya, South Africa.

Remarks The specific identity of African *Megantereon* has been intensively debated over the past 20 years (Turner, 1987; Martínez Navarro and Palmqvist, 1995, 1996; Palmqvist, 2002; Werdelin and Lewis, 2002) but now appears to be resolved, at least so far as *M. whitei* is concerned (Palmqvist et al., 2007).

Genus *METAILURUS* Zdansky, 1924

Diagnosis Medium-sized machairodont felids with rounded skull; dental formula I 3/3, C 1/1, P 2/2, M 1/1; C straight, uncompressed, lacking longitudinal grooves, lacking serrations; cheek teeth sectorial; P4 protocone reduced but present; M1 very small; m1 with vestigial metaconid.

African Species M. major Zdansky, 1924, *M. obscurus* (Hendey, 1974), *M.* sp.

Age Ca. 6.1–1.2 Ma.

Geographic Occurrence Kenya, South Africa, Tanzania.

Remarks Hendey (1974) suggested relationships of his *F. obscura* to *Sivapanthera* Kretzoi, 1929, while Turner et al. (1999) refer Langebaanweg material (presumably *F. obscura*) to the North American genus *Adelphailurus*. Recently, Morales et al. (2005) placed the species in *Megantereon*, but this seems in part to be based on a mistaken belief in the presence of *Megantereon* in the Upper Miocene of Baode, China (Zdansky, 1924). In fact, the Chinese specimen can be referred to *Paramachaerodus*. Instead, the Lukeino material shows no appreciable differences from *Metailurus* spp. Petter (1973) tentatively attributes a specimen from Olduvai, Bed II to *Metailurus*, but, though compelling, this requires further material for confirmation. Most recently, Howell and García (2007) report *M. major* from Lemudong'o, Kenya on the basis of a mandible fragment with p3 and p4. Though referral to *Metailurus* seems indicated, we doubt that this material is diagnostic at the species level.

Genus *NAMAFELIS* Morales et al., 2003

Diagnosis After Morales et al. (2003). Felidae the size of *Caracal caracal*; lower dental series p2–m1; symphysis rounded; m1 trigonid lacking a metaconid, talonid formed of a well-defined cusp surrounded posteriorly by a low cingulum.

African Species N. minor Morales, Pickford, Fraile, Salesa and Soria, 2003.

Age Ca. 17.5–17 Ma.

Geographic Occurrence Namibia.

Remarks Like *Diamantofelis* (see above), this is a felid of pseudaelurine grade with no clear affinities to any particular Eurasian *Pseudaelurus*.

Genus *PANTHERA* Oken, 1816
Figure 32.8C

Diagnosis Large-sized Felidae; dental formula I 3/3, C 1/1, P 3/2, M 1/1; hyoid ossified; premolar accessory cusps, especially the posterior, generally large; small talonid present on m1; canines short and robust.

African Species P. leo (Linnaeus, 1758), *P. pardus* (Linnaeus, 1758) (see also figure 32.8C), *P.* sp.

Age Ca. 4 Ma–Recent.

Geographic Occurrence Algeria, Ethiopia, Kenya, Morocco, South Africa, Tanzania, Zambia.

Remarks The early history of *Panthera* in Africa is very fragmentary and although some commentators (e.g., Turner, 1990) have suggested that the extant species *P. leo* and *P. pardus* are present in the Laetolil Beds, Upper Unit (ca. 3.7 Ma), no *Panthera* material older than ca. 2 Ma is diagnostic at the species level. From about this time onwards the extant species are the only species of *Panthera* in Africa.

FELIDAE indet.
Figure 32.8D

Age Ca. 18 Ma–Recent

Geographic Occurrence Angola, Chad, Ethiopia, Kenya, Libya, Morocco, South Africa, Tunisia, Uganda.

Remarks Records of undetermined felids in Africa are numerous (figure 32.8D). Most of these records are based on fragmentary material that does not allow for a better assignment. Other records, from faunal lists, need to be confirmed. This is the case for early and middle Miocene records such as Maboko (Pickford, 1986) and Jebel Zelten (Savage and Hamilton, 1973; Ginsburg, 1979). The earliest undoubted record of Felidae in Africa is an undescribed small species from Songhor (Kenya), represented by a partial hemimandible. Some records are based on recent finds, like the rich samples from Toros-Menalla (Chad), Koobi Fora (Kenya), and the Middle Awash (Ethiopia). A first glance at the Chadian material indicates that the Felidae is the most diverse family there, with a minimum of seven species, from the size of a domestic cat (*Felis* sp., discussed earlier) to a size larger than a tiger (*Amphimachairodus kabir*). The diversity and richness of fossil Felidae in Africa could therefore substantially increase in the next few years.

Family OTARIIDAE Gray, 1825
Table 32.11
Genus *ARCTOCEPHALUS* Geoffroy and Cuvier, 1826

Diagnosis Modified from Berta and Wyss (1994); Nowak (2003). Otariid characterized by facial angle of the skull always greater than 125°, P3 and M1 single rooted, and calcaneal secondary shelf present. Differs from *Callorhinus* in having tibia and fibula fused proximally; differs from Otariinae (i.e., all other extant genera except *Callorhinus*) in retaining an I3 incisiform and oval in cross-section.

African Species A. pusillus (Schreber, 1775).

Age 120 000 yrs BP–Recent;

Geographic Occurrence South Africa.

Remarks The Cape fur seal is known from large samples from several coastal archeological sites in South Africa.

TABLE 32.11
Occurrences of pinniped species

Sites	ODOBENIDAE	Ontocetus emmonsi	OTARIIDAE	Arctocephalus pusillus	PHOCIDAE	Homiphoca capensis	Lobodon carcinophaga	Messiphoca mauretanica	Mirounga leonina	Mirounga sp.	Monachus monachus	Pliophoca etrusca	Monachinae indet.
Ahl al Oughlam		X											
Die Kelders 1				X					X				
Duinefontein 2				X									
Klasies River Mouth Main Site				X					X				
Langebaanweg Baard's Quarry						X							
Langepaanweg PPM						X							
Langebaanweg indet. Member						X							
Raz-el-Aïn								X					X
Ryskop										?			
Sahabi													X
Sea Harvest Site				X			X						
Thomas Quarry 1 level G											aff.		
Wadi Natrun												X	

Family ODOBENIDAE Allen, 1880
Table 32.11
Genus ONTOCETUS Leidy, 1859

Diagnosis Modified from Kohno and Ray (2008). Odobenine odobenid distinguished from other genera by having tusklike canine with thin cementum and orthodentine layers and trefoiled incisive foramina. Differs from *Aivukus* and *Protodobenus* by having tusklike upper canine with well-defined core of globular osteodentine. Differs from *Valenictus* and *Odobenus* by retention of the tusk-like upper canine with strong curvature, taper, lateral compression, and longitudinal surface fluting; two upper incisors not in line with the cheek teeth; and two well-developed lower incisors. Functional dental formula I 2–3/2, C 1/1, P 4/4, M 1–0/1–0.

African Species *O. emmonsi* Leidy, 1859.

Age Ca. 2.5 Ma.

Geographic Occurrence Morocco.

Remarks The African material of this species was described by Geraads (1997) as *Alachtherium africanum*. It was recently synonymized with the Plio-Pleistocene *O. emmonsi* by Kohno and Ray (2008). The Ahl al Oughlam material represents the southernmost record of a walrus in the Old World.

Family PHOCIDAE Gray, 1821
Table 32.11
Genus HOMIPHOCA Muizon and Hendey, 1980

Diagnosis Modified after Muizon and Hendey (1980). A monachine phocid. Dental formula: I 2/2, C 1/1, P 4/4, M 1/1. Differs from *Monachus* in having a relatively large rostrum, wide posteriorly and narrow anteriorly. The premaxillae terminate against the nasals and have prominent anterior tuberosities. The coronoid process is relatively high as in Lobodontini and not strongly medially recurved as in *Monachus*. Premolars similar to *Monachus* but differ in being lower crowned, relatively narrower, and having a pronounced posterolingual expansion of the cingulum. Small but distinct premolar accessory cusps. M1 distinct in having strongly recurved, sharp principal cusp. M1 largest of cheek teeth with posteriorly slanted principal cusp. Interorbital region broad, tapering posteriorly as in *Lobodon* but unlike all other monachines. Tympanic bulla covers petrosal; mastoid forms a lip overlapping the posterior border of the bulla. Humerus lacking entepicondylar foramen; tibia and fibula anteriorly fused.

African Species *H. capensis* (Hendey and Repenning, 1972).

Age Ca. 5.2–5 Ma.

Geographic Occurrence South Africa.

Remarks Muizon and Hendey (1980) consider this species to be related to *Lobodon*, taking a morphologically intermediate position between that genus and *Monachus*. The Langebaanweg material, which is extensive, is the only sub-Saharan record of a pre–middle Pleistocene pinniped.

Genus LOBODON Gray, 1844

Diagnosis Modified after Bininda-Emonds and Russell (1996). Characterized by the following features: anterior nasal bones trident shaped, with lateral prongs shorter than the median prong; size of preorbital process of maxilla small; medium or great degree of invagination of frontal (posterodorsal) edge of widened maxillofrontal suture; tendency to reduce the orbitosphenoid; least interorbital width located distinctly posterior to middle of interorbital region;

greatest zygomatic width at level with glenoid fossa (i.e., at squamosal); weak degree of interlock between jugal and dorsal process of squamosal process of zygomatic arch; outline of palatine bones "butterfly-shaped" in ventral view; shape of posterior edge of palatine roughly triangular; medial portion of caudal entotympanic inflated; posterior opening of carotid canal directed roughly 45° medially; mastoid lip in region of external cochlear foramen present; dorsal region of petrosal expanded; intermediate size and shape of petrosal apex; incisors caniniform; outermost lower incisor about equal in size to remaining incisors; postcanine teeth multicuspate; size of accessory cusps in postcanines large, distinct from major cusp; first postcanines smaller than rest, which are subequal; slanted (relative to vertical) implantation of postcanines; lingual face of mandible concave at middle postcanines; distinct medially directed flange present along ventral edge of jaw located posterior to mandibular symphysis and ventral to posterior postcanines; gluteal fossa on ilium shallow; no curvature of pelvis around long axis (i.e., pelvis straight); ischiatic spine located in posterior postacetabular region; medium depth of trochanteric fossa on femur.

African Species L. carcinophaga (Hombron and Jacquinot, 1842).

Age <50 000 yrs BP–Recent (occasional).

Geographic Occurrence South Africa.

Remarks A single specimen of crabeater seal has been reported from the late Pleistocene Sea Harvest Site, South Africa.

Genus *MESSIPHOCA* Muizon, 1981

Diagnosis Translated and modified from Muizon (1981). Monachinae with greatly developed ulna olecranon and radial facet anteriorly oriented. Differs from *Monachus* and *Pliophoca* in the curvature of the ulna diaphysis and in the shape of the articular facet with the pyramidal. The radius and ulna are greatly flattened. The humerus lacks an entepicondylar foramen.

African Species M. mauretanica Muizon, 1981.

Age Ca. 7–6 Ma.

Geographic Occurrence Algeria.

Remarks This poorly known taxon from Raz-el-Aïn, Algeria, is of particular interest for the light it may shed on the state of the Mediterranean during the Messinian salinity crisis (Muizon, 1981).

Genus *MIROUNGA* Gray, 1827

Diagnosis Modified after Bininda-Emonds and Russell (1996). Basal monachine phocid characterized by: nasal processes of maxilla extending along maxilla only part way of nasals; nasal bones not trident shaped; presence of a deep fossa on the ventrolateral side of premaxilla; preorbital process of maxilla large; incisive foramina absent; pterygoid hamuli directed medially; basioccipital-basisphenoid region concave; hypomastoid fossa absent; median lacerate foramen in the auditory bulla absent; stylomastoid and auricular foramina confluent; parietal contributing to the bony falx; one lower incisor per hemimandible; upper incisors rounded in cross section; postcanine teeth peglike (unicuspate); first and fifth lower postcanine teeth noticeably smaller and first and fifth upper postcanine teeth noticeably larger than the other postcanines, which are subequal; tendency to single-rooting of lower postcanines; reversely arched upper postcanine tooth row; lingual

face of mandible at middle postcanines concave; scapular spine of medium size; gluteal fossa on ilium shallow; no distinct trochanteric fossa on femur; claw semicircular in cross section.

African Species M. leonina (Linnaeus, 1758), M. sp.

Age Ca. 0.1 Ma-Recent (possibly present as early as 18 Ma).

Geographic Occurrence South Africa.

Remarks Pickford and Senut (2000) refer remains from Ryskop in Namibia to ?*Mirounga* sp. These remains are by far the oldest pinniped remains in Africa. Remains of the extant southern elephant seal have been found at the middle Pleistocene archeological sites of Die Kelders and Klasies River Mouth in South Africa.

Genus *MONACHUS* Fleming, 1822

Diagnosis Modified after Bininda-Emonds and Russell (1996). Derived monachine phocid characterized by: Broad contact between nasal processes of premaxilla and nasals; fossa on ventrolateral side of premaxilla shallow; interorbital septum anterior to optic foramina present; incisive foramina of medium size; foramen ovale entirely in the squamosal; medial portion of caudal entotympanic not inflated; petrosal not covered by the auditory bulla; hypomastoid fossa deep; external cochlear foramen open; petrosal apex unexpanded and pointed; incisors not procumbent; no tendency to form or lose additional cusps in triconodont postcanine teeth; accessory cusps in postcanine teeth small; postcanine teeth touching or overlapping; anterior/posterior end of postcanine teeth directed laterally; upper postcanine tooth row kinked between postcanine 1 and 2, otherwise straight; gluteal fossa on the wing of ilium absent; no ridges in anterior portion of obturator foramen.

African Species M. monachus (Hermann, 1779).

Age Ca. 200 000 yrs BP–Recent?

Geographic Occurrence Morocco.

Remarks There is a single record of Mediterranean monk seal from the Thomas Quarry Level G, Morocco.

Genus *PLIOPHOCA* Tavani, 1941

Diagnosis Modified from Koretski and Ray (2008). Medium-sized phocid with relatively broad and robust teeth; P1 and p1 single rooted, the other postcanine teeth double rooted; in the maxilla, anterior alveoli smaller than posterior alveoli, and slightly oblique to the labial side, while the posterior alveoli are slightly oblique to the lingual side; the opposite is true in the dentary; retromolar space elongated, mandibular body not high, ramus very strong and high; p2–m1 five cusped, with basal cingula very well developed; alveoli of m1, M1 shorter than those of p4, P4; diastemata present in maxilla, absent in the mandible; proximal part of deltoid crest of humerus located higher than the head, but lower than the minor tubercle; deltoid crest strongly developed, terminates slightly lower than the middle of diaphysis; greater breadth of deltoid crest located on its middle part; greater trochanter of femur rectangular, placed at the same level as head; trochanteric fossa shallow and caudally elongated; intertrochanteric crest not developed; least width of femur diaphysis is at the middle of the bone; femoral condyles placed wide apart and flattened, with greatest distance between then being 0.15 of the bone length; distal end of femur wider than proximal end by 0.81.

African Species P. etrusca Tavani, 1941.

Age Ca. 6–4 Ma.

Geographic Occurrence Egypt.

Remarks The single African record of this species is based on a fragmentary mandible with one postcanine tooth and an indeterminate deciduous tooth from Wadi Natrun, Egypt (Stromer, 1907, 1913). The species was first assigned to *Pristiphoca* sp. aff. *P. occitana* by Stromer (1913). According Koretski and Ray (2008), however, this species is based on a canine tooth of an indeterminate carnivore, which is presumably lost. Pending rediscovery of the holotype and availability of associated material, these authors regard the species as a nomen nudum. They also consider that the fragmentary right hemimandible with two postcanines referred to as the same species by Gervais (1853) cannot be confidently assigned to the same species and assign it to *Pliophoca etrusca*. Following Howell and Garcia (2007), we then assign the species from Wadi Natrun to *Pliophoca etrusca*.

MONACHINAE indet.

Age Ca. 7–6 Ma.

Geographic Occurrence Algeria, Libya.

Remarks Muizon (1981) and Howell (1987) record additional monachines from Raz-el-Aïn and Sahabi, respectively, although the material is too fragmentary to allow for a more precise diagnosis.

Biogeography

Two uncertain occurrences may attest to a pre-Miocene presence of the order Carnivora in Africa. A Carnivora indet. is represented by several isolated premolars from the Ouarzazate Basin (Paleocene of Morocco; Gheerbrant, 1995), while the second Paleogene record, *Glibzegdouia tabelbalaensis*, is based on an isolated m1 from Glib Zegdou (Eocene of Algeria; Crochet et al., 2001). In both cases, the ordinal assignment is uncertain. Lower molars of recently described proviverrine creodonts from the early late Paleocene and early Eocene of Morocco (Gheerbrant et al., 2006) show some resemblance to those of *G. tabelbalaensis*. More precise comparisons of these teeth may therefore indicate that these putative carnivorans belong to the Creodonta.

Like many nonafrotherian mammalian orders, the presence of carnivorans in Africa is correlated with the collision between the Afroarabian and Eurasian plates near the Oligocene-Miocene boundary. Unambigous remains of Carnivora in Africa are from early Miocene strata in Kenya, Uganda, Congo, and Namibia (but see Note added in proof). Soon after their arrival, the Carnivora successfully and rapidly expanded across the entire continent.

The quality of the fossil record is very heterogeneous in Africa, which renders the distribution of Neogene carnivorans difficult to interpret. The Neogene fossil record of carnivorans is at best extremely poor in western Africa (all countries south of Mauritania, Mali, Niger), central and west-central Africa (Central African Republic, Sudan, Cameroon, Gabon, Congo-Brazzaville, Democratic Republic of Congo), and some southern countries (Angola, Botswana, Mozambique, Zambia). Some regions in Africa have, in contrast, yielded extremely rich faunas for certain times and/or have a good and continuous fossil record for a long time. Eastern Africa is the best "hot spot" for Neogene paleomammalogy in Africa, and in particular for the Plio-Pleistocene (Werdelin and Lewis, 2005); southern Africa also has rich and diversified Plio-Pleistocene faunas but not as many Miocene mammal sites as eastern Africa. This does not imply that carnivorans were absent from southernmost Africa, since they are diverse and numerous, from, for example, the early–middle Miocene of Namibia. With the increasing amount and intensity of field research in Africa, some regions or countries are showing their scientific potential, demonstrating that eastern African dominance may not only be due to its unique geographic and tectonic position, which created exceptional conditions of fossilization, but also to the near absence of a fossil record in most other African regions. One of these emerging regions is northern Chad, the fossiliferous sites of which to date have yielded thousands of Mio-Pliocene fossils.

Over 30 years ago, Savage (1978) concluded his contribution to Maglio and Cooke's *Evolution of African Mammals* about creodonts and carnivorans thus: "This account of origins and migrations is unhappily more conjecture than fact. Only many more facts will substantially improve it. Pliny wrote nearly 2,000 years ago, 'Ex Africa semper aliquid novi': it is still true today." Although Pliny's statement will be true in paleontology for many years to come, 30 years after this confession, the late R. J. G. Savage would certainly acknowledge that the paleontology of carnivores in Africa has been much improved, especially in the last decade, thanks to a much better, and still improving, fossil record. A comparison between the state of the art of Savage (1978) and the present account shows this (see table 32.12). In particular, the total number of extinct genera and occurrences is much greater. All the families are involved, but the Ursidae (Savage did not really provide a synthesis of the numerous Quaternary records), Hyaenidae, and Mustelidae (in particular the Lutrinae) display a tremendous improvement in their fossil record (number of species with fossil record and occurrences).

Hunt (1996) provided a detailed analysis of the biogeography of each family incorporating the fossil record, while Werdelin and Lewis (2005) discussed the fossil record of Plio-Pleistocene carnivorans in Africa. Therefore, we here summarize the main biogeographic trends of each family, especially for the Miocene. Many recent papers have described new or unpublished material from this epoch that has revealed a much greater richness and diversity than previously known. Significant examples include, for example, Arrisdrift (eight species; Morales et al., 1998, 2003), Ngorora Fm. (seven species; Morales and Pickford, 2005a), Lukeino and Mabaget Fms (19 species; Morales et al., 2005), Nawata Fm. (18 species; Werdelin, 2003), Adu-Asa and lower Sagantole Fms (at least 14 species; Haile-Selassie and Howell, 2009), and Lemudong'o (12 species; Howell and García, 2007).

In Africa, there are no fossils assigned to the Miacidae, Viverravidae, Nimravidae, Procyonidae, or Ailuridae (with the *Simocyon* sp. of Howell and García [2007] reassigned to Amphicyonidae). Despite the lack of a fossil record in many African countries, these families of carnivores, the three former being typical of the Paleogene, apparently never reached Africa. More surprising is the apparent absence of a fossil record for the Eupleridae (including all the Malagasy carnivorans), although they had an African ancestry with a divergence time extending back to the early Miocene (24–18 Ma; Yoder et al., 2003). The extant carnivore fauna of Africa is composed of five feliform (Felidae, Hyaenidae, Viverridae, Nandiniidae, Herpestidae) and two caniform (Canidae, Mustelidae) families, including about 80 species (Sunquist and Sunquist, 2002; Gaubert et al., 2004; Sillero-Zubiri et al., 2004; Nowak, 2005). In addition, Amphicyonidae,

TABLE 32.12

Extant and fossil records as reported by Savage and this study

The table only reports the extant and fossil records for Africa. Savage listed the known total number of extinct and living genera for all Carnivora, which is not provided here.

Family	Extinct Genera		Genera with Fossil Record		Living Genera[a]		Total of Genera		Species with Fossil Record		Occurrences	
	Savage	This Study	Savage	This Study	Savage	This Study	Savage	This Study	Savage	This Study	Savage	This Study
Amphicyonidae		7		7		0		7		11		46
Canidae	2	3	7	6	5	3	7	6	23	22	59	110
Ursidae	1	4	1	4	1	0	2	4	2	7	2	55
Mustelidae	1	15	6	20	8	7	9	22	16	33	33	114
Nandiniidae		0		1		1		1		1		1
Viverridae	2	13	13	15	16	3	19	16	30	24	64	96
Herpestidae		4		13		14		18		27		115
Percrocutidae		2		2		0		2		4		11
Hyaenidae	5	11	7	15	3	4	8	15	17	29	56	151
Barbourofelidae		4		4		0		4		5		14
Felidae	5	10	8	15	3	6	8	16	28	29	105	279
Otariidae	0	0	0	1	1	1	1	1	0	1	0	4
Odobenidae	0	1	0	1	0	0	0	1	0	1	0	1
Phocidae	2	5	2	6	2	1	4	6	2	7	2	12
Total	18	79	44	110	39	40	58	119	118	201	321	1,009

[a]Living genera include the genera currently present in Africa, with or without a fossil record. Differences observed between our study and Savage's in the living genera are due to the classification used. See appendix for further explication.

Barbourofelidae, and Ursidae, the former two wholly extinct, have fossil records in Africa. Except for Nandiniidae, the fossil record in Africa for each extant family is continuous or nearly continuous since its first occurrence.

The Barbourofelidae and Amphicyonidae are common but not abundant members of predator communities of the early and middle Miocene of Africa. Amphicyonids in particular are diverse in eastern Africa, where the fossil record of the family is more or less continuous from the early to the late Miocene. They are also found in Mediterranean countries (Morocco, Tunisia, Algeria, Libya, Egypt) and southern Africa (Namibia and South Africa). The extinction of the Amphicyonidae (ca. 5.5 Ma) is slightly later than the arrival of large ursines (*Indarctos* and *Agriotherium*) and canids. No known locality documents the sympatry of these three families, however. Some sites have either Ursidae and Canidae (e.g., Lukeino) or Amphicyonidae and Canidae (e.g., Toros-Menalla, Lemudong'o), but there is no known late Miocene locality documenting a potential sympatry of Amphicyonidae and Ursidae in Africa. Evidence supporting competitive exclusion is lacking, but competition for large prey was possible between large amphicyonids (*Agnotherium*) and large ursids (*Agriotherium*). Early African canids (late Miocene-early Pliocene) were much smaller species and not able to compete with any of the contemporary amphicyonids or ursids (though competition with the small amphicyonid *Bonisicyon* may have occurred). The last Barbourofelidae are probably *Vampyrictis vipera* from Bled Douarah and the Barbourofelidae indet. from Namurungule. Both have reached the evolutionary grade of *Sansanosmilus*. Felids, and in particular machairodontines, were their main competitors. The extinction of this already rare family in the late Miocene of Africa may be a result of the arrival of Machairodontinae on the continent.

Remains of Ursidae are rare but they had a pan-African distribution (from Libya to South Africa). A single tooth from Rusinga (early Miocene, Kenya) indicates an early expansion of this originally Holarctic family southwards. However, this record is separated from the next oldest record of the family (*Indarctos* from Menacer, Algeria) by a 10-million-year hiatus. *Ursus*, the only extant genus with a fossil record in Africa was restricted to northwestern Africa but is now extinct there (Servheen et al., 1998).

Mustelids are among the first carnivoran immigrants to Africa, with otterlike forms from the early Miocene of East Africa. The radiation of otters during the late middle and late Miocene is one of the most significant traits of the family in Africa. Lutrines are by far the most common mustelids, with about two-thirds of the total occurrences of the family. Also worth noting is the great size range of the species, spanning from the size of the living river otter, *Lutra lutra*, to a much greater size than any living otter (>>100 kg). These giant otters were distributed from Egypt to South Africa, including a great diversity in eastern Africa.

The Canidae are exclusive to North America until the late Miocene, after which *Eucyon*- and *Vulpes*-like taxa appear in western Europe, Asia, and Africa. Our knowledge of the Canidae in Africa is still limited though rapidly improving. Recently, Morales et al. (2005) described from Lukeino (ca. 6 Ma) what was then the oldest canid of Africa, *Eucyon intrepidus*. Even more recently, however, a foxlike canid has been discovered from older sediments in Chad (Toros-Menalla, ca. 7 Ma; Bonis et al., 2007b). The family reaches southern Africa in the early

Pliocene and northwestern Africa in the mid-Pliocene. The final radiation of the Canidae (Caninae) in the Pleistocene is seen mainly in eastern Africa, but also in South Africa and Algeria.

The Viverridae and the Herpestidae are among the first immigrants to Africa during the early Miocene and are first known from the same eastern African localities (e.g., Koru, Napak, Rusinga, Songhor). Viverrids are found in northern (Egypt) and southern (Namibia) parts of the continent soon after, but remain rare or absent during the middle Miocene, with a few records in Kenya (Fort Ternan, Ngorora), Morocco (Beni Mellal), and Namibia (Arrisdrift), and are only slightly more abundant during the late Miocene in Ethiopia (Chorora, Adu-Asa), Kenya (Lothagam, Lukeino), and Chad (Toros-Menalla). After the early Miocene records, herpestids are nearly absent in Africa until the end of the Miocene. Solitary mongooses (*Galerella* and *Ichneumia*) are first found in the late Miocene of Chad (only record for central Africa) and Kenya. Social mongooses (*Helogale*) appear in Africa in the latest Miocene of Ethiopia (Adu Asa Fm.). Plio-Pleistocene taxa show greater diversity than today and are known mainly from eastern (Ethiopia, Kenya, Tanzania) and South Africa. In addition to these areas, modern mongooses are known only from a single northern African site (mid-Pliocene, Morocco).

The Hyaenidae has a particularly rich and diverse fossil record in Africa, which strongly contrasts with the present low diversity (four extant species). In contrast to many families, the oldest hyaenids in Africa are from northern Africa, with *Hyaenictis ?graeca*, *Protictitherium punicum*, *P. crassum*, and *Lycyaena crusafonti*, all from roughly contemporaneous, late middle–early late Miocene localities of Morocco (Beni Mellal) and Tunisia (Bled Douarah). *Protictitherium* expands its range southwards to Kenya (Namurungule). A hyaenid fauna composed of *Chasmaporthetes*, *Ikelohyaena*, *Hyaenictis*, *Hyaenictitherium*, *Ictitherium*, and *Adcrocuta*, of which the latter never reaches sub-Saharan Africa, characterizes the late Miocene–earliest Pliocene. Except for *Chasmaporthetes* and *Ikelohyaena*, all these genera become extinct ca. 5 mya ago, while bone-cracking species of *Crocuta*, *Pliocrocuta*, *Pachycrocuta*, *Parahyaena*, and *Hyaena* appear soon after this date.

The Percrocutidae is represented in Africa by two genera, *Dinocrocuta* and *Percrocuta*. Many records are late middle Miocene in age, with the oldest being from Kenya (*Percrocuta* sp. from Fort Ternan and *P. tobieni* from Ngorora). Slightly younger records of *Percrocuta* are from Tunisia (Bled Douarah) and Kenya (Nakali, Namurungule). *Dinocrocuta* is slightly younger and is found at early late Miocene localities of Algeria and Kenya. The last records of the family are from the latest Miocene.

The Felidae was a very diverse family, with at least 15 genera and 35 taxa. Including undetermined material would certainly raise the total number of felid species to over 40. The Felidae first appears in Africa during the early Miocene. The family therefore has a long history on this continent though the distribution is, as for other families, very heterogeneous. About 90% of the occurences are in eastern or southern Africa. The subfamily Machairodontinae, which includes the sabertooth taxa, first appears in the late middle Miocene and becomes extinct during the middle Pleistocene. For more than 10 million years, sabertooth cats were diverse (they account for about half of the total number of African species of Felidae) and significant members of the large predator guild. They had their greatest diversity during the late Miocene–early Pliocene with about 10 species, most of which were large-sized taxa. Extant genera and species have a fossil record restricted to the Pliocene (after ca. 4.3 Ma).

Migration Patterns

As noted, the Carnivora arrives in Africa in the Lower Miocene or slightly earlier, which is quite late in the history of the order. Clearly, their arrival in Africa was due to a migration event, from either Europe or Asia, or both. Throughout the history of Carnivora in Africa, there has been a series of such events, both into and out of Africa. Although the fossil record biases the data, a number of these events are distinct and broadly associated with tectonic and climatic trends of the Neogene

The first such event occurred some 20 Ma ago. It involved members of the families Amphicyonidae, Ursidae, Felidae, Mustelidae, Herpestidae, Viverridae, and Barbourofelidae. The latter two families may have migrated from Asia via Africa to Europe, while the other five most likely are immigrants from Europe. There is no evidence for size sorting during this event, as it involves both very small (e.g., Herpestidae) and very large (e.g., Amphicyonidae) taxa. This event is broadly correlative with the first land passage between Africa and Eurasia, the so-called *Gomphotherium* land bridge (Rögl, 1998), a result of the isolation of the Indo-Pacific from the Paratethys during the later part of sea level cycle TB2.1 (Haq et al., 1987, 1988).

The next event occurs prior to 14 Ma and involves immigration of the family Percrocutidae, the subfamily Mustelinae (earlier mustelids are of uncertain subfamily, but possibly Lutrinae), the amphicyonid genus *Agnotherium*, and the lutrine *Vishnuonyx*, the latter of which almost certainly arrived from the Indian subcontinent. The paleogeography of this time is very similar to the time of the *Gomphotherium* land bridge, with the Indo-Pacific and Paratethys once again isolated from one another during sea level cycle TB2.4, which is dated ca. 15–13.8 Ma (Berggren et al., 1995). Again, no clear size sorting is discernible during this event.

The third event occurred prior to 10 Ma and involved the family Hyaenidae and the genus *Machairodus*, with the latter appearing nearly simultaneously in Eurasia and Africa. This event seems to have involved the immigration to Africa of mainly medium- to large-sized species and is associated with sea-level cycles TB2.6 (ca. 12.5–11.3 Ma) and TB3.1 (ca. 11.3–8.9 Ma). It should be noted that as presently reconstructed, there is no direct land connection between Africa and Europe at this time, and exchange between these continents may have occurred via an easterly route around the Paratethys.

The next event occurred prior to ca 7.5 Ma and involves a series of genus-level taxa: the ursids *Indarctos* and *Agriotherium*, the mustelids *Plesiogulo* and *Sivaonyx*, the hyaenids *Adcrocuta*, *Chasmaporthetes*, and *Hyaenictis*, and the felid *Metailurus*. All these are relatively large taxa, indicating significant size sorting. *Agriotherium* and *Sivaonyx* appear more or less simultaneously in Africa and Eurasia. This event may also, if recent data are accurate, involve the family Canidae as the last carnivoran family to migrate into Africa. It is broadly correlative with sea-level cycle TB3.3, ca. 7.2–5.3 Ma.

The Pliocene also saw a series of carnivoran migration events, but this time mainly out of Africa. The first such event occurred ca. 4.5–4 Ma and involved migration into Africa of the canid genus *Nyctereutes* and migration out of Africa of the felid genera *Megantereon* and *Homotherium*. All these are medium- to large-sized taxa. The event may be correlated with sea level cycle TB3.4, ca. 4.1–3.6 Ma, but it is more likely that at this point migrations were climatically rather than tectonically mediated, as the paleogeography changes very little after the middle Upper Miocene.

An additional set of genera of medium- to large sized carnivorans migrates out of Africa ca. 3.5 Ma: the canid *Canis*, the hyaenid *Pachycrocuta*, and the felid *Acinonyx*. A final event occurs before ca. 2 Ma and involves migration out of Africa of the hyaenid genera *Hyaena* and *Crocuta* and the felid genus *Panthera*. At this time the mustelid genus *Aonyx* may have migrated to Africa, but this is very uncertain. Except *Aonyx* these are all large-sized taxa.

In summary, carnivoran migrations in the Miocene appear to be mainly into Africa and to be tectonically controlled. The earliest migrations show no size sorting, but gradually migrations appear to involve mostly medium- to large-sized taxa. In the Pliocene, migrations are mainly out of Africa and are climatically controlled. They involve mostly medium- to large-sized taxa.

Conclusions

Despite their late appearance, the Carnivora have diversified greatly on the African continent, as shown in the present compilation. At the same time, the geographic and temporal biases in the material show just how much remains to be discovered. The Plio-Pleistocene, up to about one million years ago, is reasonably well-known for some African regions, but the origin of the extant fauna is obscure (Werdelin and Lewis, 2005). The latest Miocene has seen a tremendous upswing in interest and discoveries in the last few years (Werdelin, 2003; Haile-Selassie et al., 2004b; Morales et al., 2005; Howell and García, 2007), and we hope that in the next decade a similar upswing will occur for the middle Miocene, which is very poorly known.

One aspect of carnivoran evolution in Africa that is becoming evident as new discoveries accumulate is the importance of Africa to the evolution of carnivorans elsewhere. Though we would perhaps not go quite as far as Pickford (2004), it is becoming clear that important Plio-Pleistocene carnivoran taxa such as *Canis*, *Homotherium*, *Acinonyx*, *Panthera*, and others may have had an African origin, or at least that Africa was important to their gobal diversification. This makes the factors allowing intermittent migration of mammals out from Africa of critical importance for understanding the origin and composition of modern carnivoran guilds in the Old World and in some cases North America. Carnivorans are also important in the evolutionary history of our own genus (Turner, 1984), and hence understanding carnivoran evolution will help in understanding human evolution as well.

ACKNOWLEDGMENTS

We would like to thank all those many people who collected, curated, and initially described the specimens on which the taxa incorporated in this chapter are based. We would also like to thank many colleagues for fruitful discussions over the years about carnivores and African paleontology. L.W. would like to thank Susanne Cote, Meave Leakey, Margaret Lewis, Jorge Morales, and Martin Pickford for help that significantly improved the chapter, and the Swedish Research Council for a series of grants that made the work possible. S.P. would like to thank the Swedish Museum of Natural History and financial support through the Synthesys project (SE-TAF 1380), which was made available by the European Community's Research Infrastructure Action under the FP6 "Structuring the European Research Area" Program.

APPENDIX

This appendix contains details of the analysis presented in table 32.12.

For Canidae, we include *Fennecus* in *Vulpes* and *Lycaon* in *Canis* (see Werdelin and Lewis, 2005); for Ursidae, we consider *Ursus* as extinct in Africa; for Mustelidae, we consider, following current use, *Paraonyx* as used by Savage a junior synonym of *Aonyx* and *Poecilictis* to be a junior synonym of *Ictonyx*; Viverridae in Savage (1978) are now separated into four families (Viverridae, Herpestidae, Nandiniidae, and Eupleridae); Stenoplesictidae is not regarded as valid here. The differences from Savage concern the Herpestidae, in which we recognize *Galerella*, *Paracynictis*, *Liberiictis*, and *Dologale* as valid genera and *Xenogale* as a junior synonym of *Herpestes*, and the African Viverridae, in which we follow current usage by considering *Osbornictis* a junior synonym of *Genetta*. Differences among the Hyaenidae are due to the recognition of *Parahyaena* as distinct from *Hyaena* and the recognition of the Percrocutidae as a distinct family. For Felidae, we recognize Barbourofelidae as a distinct family. For extant genera, we follow recent systematists and recognize *Caracal*, *Leptailurus*, and *Profelis* as valid genera distinct from *Felis*; species number may be difficult to establish because of the many indeterminate records. Like Savage (1978), we attempt here to include some of them. Species numbers are therefore a lowest estimate that will increase when better material is found to document indeterminate or poorly known taxa. We report evidence that supports the recognition of new taxa in addition to the named species. For Amphicyonidae, we have 10 species and consider that Amphicyonidae indet. (including sp. A) includes at least one additional species. For Ursidae, in addition to the five named species, there are at least two additional species: (1) a new ursine to be described from East Africa; (2) the hemicyonine species from Rusinga. For Canidae, in addition to the 20 named species, we have added two species, considering (1) that *Canis* sp. includes at least one new species (from Turkwel and possibly Laetoli); other, more recent mentions may be referred to either of the previously described species; (2) that Canidae indet. from the Mursi Formation probably represents a species new for the continent. For Mustelidae, in addition to the 32 named species, we consider that Mustelidae indet. includes at least one additional species. For Viverridae, some undetermined mentions certainly represent new taxa (especially from Koobi Fora). The exact number is not possible to establish, but we consider that Viverridae indet. comprises at least two additional species. For Herpestidae, we have included only the named species. For Percrocutidae, we did not consider *Percrocuta* from Namurungule Fm. as a distinct species. For Hyaenidae, in addition to the 27 named species, we have considered as distinct species (1) in Hyaenidae indet., Hyaenidae indet. E from Langebaanweg; and (2) that at least one record of *Proteles* sp. may correspond to the extant species *Proteles cristatus*. For Barbourofelidae, we did not recognize Barbourofelidae indet. from Namurungule as a distinct species because of a possible identity with *Vampyrictis* of the same age. For Felidae, in addition to the 27 named species (*Megantereon ekidoit/whitei* does not correspond to a species but represents uncertainty of assignment), two species have been added, considering that (1) *Felis* sp. certainly includes at least one species distinct from *Felis silvestris lybica* and (2) that Felidae indet., especially the record from Songhor, certainly includes another one. For pinnipeds, "living genera" include those that are frequently observed along the African coasts that is, *Monachus monachus*, *Arctocephalus pusillus*, and *A. tropicalis* (Nowak, 2003). In addition to the six named species, we have considered that ?*Mirounga* sp. from Ryskop certainly represents a species distinct from *M. leonina*.

NOTE ADDED IN PROOF

Rasmussen and Gutiérrez (2009) recently described the carnivoran *Mioprionodon hodopeus* from the late Oligocene of Nakwai, northwestern Kenya. This represents the first record of the Carnivora in Africa and suggests dispersal opportunities from Eurasia earlier than previously thought (see also Morlo et al. (2007)).

Literature Cited

Adam, P. J. 2005. *Lobodon carcinophaga. Mammalian Species* 772:1–14.

Aguirre, E., and P. Leakey. 1974. Nakali: Nueva fauna del *Hipparion* de Rift valley de Kenya. *Estudios Geológicos* 30:219–227.

Anderson, E. 1970. Quaternary evolution of the genus *Martes* (Carnivora, Mustelidae). *Acta Zoologica Fennica* 130:1–132.

Antón, M., M. J. Salesa, J. Morales, and A. Turner. 2004. First known complete skulls of the scimitar-toothed cat *Machairodus aphanistus* (Felidae, Carnivora) from the Spanish Late Miocene site of Batallones: 1. *Journal of Vertebrate Paleontology* 24:957–969.

Aouraghe, H. 2000. Les carnivores fossiles d'El Harhoura 1, Temara, Maroc. *L'Anthropologie* 104:147–171.

Arambourg, C. 1961. Note préliminaire sur quelques Vertébrés nouveaux du Burdigalien de Libye. *Compte Rendu sommaire des séances de la Société Géologique de France* 1961(4):107–108.

Barry, J. C. 1987. Large carnivores (Canidae, Hyaenidae, Felidae) from Laetoli; pp. 235–258 in M. D. Leakey and J. M. Harris (eds.), *Laetoli: A Pliocene Site in Northern Tanzania.* Clarendon Press, Oxford.

Beaumont, G. de. 1967. Observations sur les Herpestinae (Viverridae, Carnivora) de l'Oligocène supérieur avec quelques remarques sur des Hyaenidae du Néogène. *Archives des Sciences, Genève* 20:79–108.

Berggren, W. A., D. V. Kent, C. C. Swisher, III, and M.-P. Aubry. 1995. A revised Cenozoic geochronology and chronostratigraphy; pp. 129–212 in W. A. Berggren, D. V. Kent, M.-P. Aubrey, and J. Hardenbol (eds.), *Geochronology, Time Scales and Global Stratigraphic Correlation.* SEPM Special Publication 54.

Berta, A., and A. R. Wyss. 1994. Pinniped phylogeny. *Proceedings of the San Diego Society of Natural History* 29:33–56.

Bininda-Emonds, O. R. P., and A. P. Russell. 1996. A morphological perspective on the phylogenetic relationships of the extant phocid seals (Mammalia: Carnivora: Phocidae). *Bonner Zoologische Monographien* 41:1–256.

Bonis, L. de, S. Peigné, A. Likius, H. T. Mackaye, M. Brunet, and P. Vignaud. 2007a. First occurence of the "hunting hyaena" *Chasmaporthetes* in the Late Miocene fossil bearing localities of Toros-Menalla, Chad (Africa). *Bulletin de la Société Géologique de France* 178:317–326.

Bonis, L. de, S. Peigné, F. Guy, A. Likius, H. T. Mackaye, P. Vignaud, and M. Brunet. 2009. A new mellivorine (Carnivora, Mustelidae) from the late Miocene of Toros Menalla, Chad. *Neues Jahrbuch für Geologie und Paläontologie* 252:33–54.

Bonis L. de, S. Peigné, A. Likius, H. T. Mackaye, P. Vignaud, and M. Brunet. 2007b. The oldest African fox (*Vulpes riffautae* n. sp., Canidae, Carnivora) recovered in late Miocene deposits of the Djurab desert, Chad. *Naturwissenschaften* 94:575–580.

Bonis, L. de, S. Peigné, H. T. Mackaye, A. Likius, P. Vignaud, and M. Brunet. 2008. The fossil vertebrate locality Kossom Bougoudi, Djurab desert Chad: A window in the distribution of the carnivoran faunas at the Mio-Pliocene boundary in Africa. *Comptes Rendus Palevol* 7:571–581.

Bryant, H. N. 1988. Delayed eruption of the deciduous upper canine in the sabertoothed carnivore *Barbourofelis lovei* (Carnivora, Nimravidae). *Journal of Vertebrate Paleontology* 8:295–306.

Bryant, H. N., A. P. Russell, and W. D. Fitch. 1993. Phylogenetic relationships within the extant Mustelidae (Carnivora): Appraisal of the cladistic status of the Simpsonian subfamilies. *Zoological Journal of the Linnean Society* 108:301–334.

Colbert, E. H. 1939. Carnivora of the Tung Gur Formation of Mongolia. *Bulletin of the American Museum of Natural History* 76:47–81.

Cooke, H. B. S. 1985. *Ictonyx bolti*, a new mustelid from cave breccias at Bolt's Farm, Sterkfontein area, South Africa. *South African Journal of Science* 81:618–619.

Cote, S., L. Werdelin, E. R. Seiffert, and J. C. Barry. 2007. Additional material of the enigmatic early Miocene mammal *Kelba* and its relationship to the order Ptolemaiida. *Proceedings of the National Academy of Sciences, USA* 104:5510–5515.

Crochet, J.-Y., S. Peigné, and M. Mahboubi. 2001. Ancienneté des Carnivora (Mammalia) en Afrique; pp. 91–100 in C. Denys, L. Grangeon, and A. Poulet (eds.), *Proceedings of the 8th International Symposium on African Small Mammals, Paris, 4–9 July 1999*. IRD Editions, Collection Colloques et Séminaires, Orléans.

Dehghani, R., and L. Werdelin. 2008. A new small carnivoran from the Middle Miocene of Fort Ternan, Kenya. *Neues Jahrbuch für Geologie und Paläontologie* 248:233–244.

Ewer, R. F. 1954a. The fossil carnivores of the Transvaal caves: The Hyaenidae of Kromdraai. *Proceedings of the Zoological Society of London* 124:565–585.

———. 1954b. The fossil carnivores of the Transvaal caves: The lycyaenas of Sterkfontein and Swartkrans, together with some general considerations of the Transvaal fossil hyaenids. *Proceedings of the Zoological Society of London* 124:839–857.

———. 1956. The fossil carnivores of the Transvaal caves: Two new viverrids, together with some general considerations. *Proceedings of the Zoological Society of London* 126:259–274.

Ficcarelli, G., D. Torre, and A. Turner. 1984. First evidence for a species of racoon dog, *Nyctereutes* Temminck, 1838, in South African Plio-Pleistocene deposits. *Bollettino della Società Paleontologica Italiana* 23:125–130.

Fraile, S., B. Pérez, I. de Miguel, and J. Morales. 1997. Revisión de los carnívoros presentes en los yacimientos del Neógeno español; pp. 77–80 in J. P. Calvo and J. Morales (eds.), *Avances en el Conocimiento del Terciario Ibérico*. III Congreso del Grupo Español del Terciario, Cuenca.

García, N. 2008. New *Eucyon* remains from the Pliocene Aramis Member (Sagantole Formation), Middle Awash Valley (Ethiopia). *Comptes Rendus Palevol* 7:583–590.

Gaubert, P., C. A. Fernandes, M. W. Bruford, and G. Veron. 2004. Genets (Carnivora, Viverridae) in Africa: an evolutionary synthesis based on cytochrome b sequences and morphological characters. *Biological Journal of the Linnean Society* 81:589–610.

Geraads, D. 1997. Carnivores du Pliocène terminal de Ahl al Oughlam (Casablanca, Maroc). *Geobios* 30:127–164.

———. 2008. Plio-Pleistocene Carnivora of northwestern Africa: a short review. *Comptes Rendus Palevol* 7:591–599.

Geraads, D., Z. Alemseged, and H. Bellon. 2002. The Late Miocene mammalian fauna of Chorora, Awash Basin, Ethiopia: Systematics, biochronology and the ^{40}K-^{40}Ar ages of the associated volcanics. *Tertiary Research* 21:113–122.

Geraads, D., Z. Alemseged, D. H. Reed, and J. G. Wynn. 2004. The Pleistocene fauna (other than Primates) from Asbole, lower Awash Valley, Ethiopia, and its environmental and biochronological implications. *Geobios* 37:697–718.

Gervais, P. 1853. Description de quelques ossements de phoques et de cétacés. *Mémoires de la section des Sciences, Académie des sciences et lettres de Montpellier* 2:307–314.

Gheerbrant, E. 1995. Les mammifères paléocènes du Bassin d'Ouarzazate (Maroc): III. Adapisoriculidae et autres mammifères (Carnivora, ?Creodonta, Condylarthra, ?Ungulata et incertae sedis). *Palaeontographica, Abt. A* 237:39–132.

Gheerbrant, E., M. Iarochène, M. Amaghzaz, and B. Bouya. 2006. Early African hyaenodontid mammals and their bearing on the origin of the Creodonta. *Geological Magazine* 143:475–489.

Gilbert, W. H., N. García, and F. C. Howell. 2008. Carnivora; pp. 95–113 in W. H. Gilbert and B. Asfaw (eds.), *Homo erectus: Pleistocene Evidence from the Middle Awash, Ethiopia*. University of California Press, Berkeley.

Ginsburg, L. 1977. Les carnivores du Miocène de Beni Mellal (Maroc). *Géologie Méditerranéenne* 4:225–240.

———. 1979. Les migrations de mammifères carnassiers (Créodontes + Carnivores) et le problème des relations intercontinentales entre l'Europe et l'Afrique au Miocène inférieur. *Annales Géologiques des Pays Helléniques H.S.* 1979:461–466.

Ginsburg, L., and J. Morales. 1998. Les Hemicyoninae (Ursidae, Carnivora, Mammalia) et les formes apparentées du Miocène inférieur et moyen d'Europe occidentale. *Annales de Paléontologie* 84:71–123.

Ginsburg, L., and J.-L. Welcomme. 2002. Nouveaux restes de Créodontes et de Carnivores des Bugti (Pakistan). *Symbioses* (n.s.) 7:65–68.

Goldman, C. A. 1987. *Crossarchus obscurus*. *Mammalian Species* 290:1–5.

Haile-Selassie, Y., and F. C. Howell. 2009. Carnivora; pp. 237–275 in Haile-Selassie, Y. and WoldeGabriel, G. (eds.), *Ardipithecus kadabba: Late Miocene Evidence from the Middle Awash, Ethiopia*. University of California Press, Berkeley.

Haile-Selassie, Y., L. J. Hlusko, and F. C. Howell. 2004a. A new species of *Plesiogulo* (Mustelidae: Carnivora) from the Late Miocene of Africa. *Palaeontologia Africana* 40:85–88.

Haile-Selassie, Y., G. WoldeGabriel, T. D. White, R. L. Bernor, D. Degusta, P. R. Renne, W. K. Hart, E. Vrba, S. Ambrose, and F. C. Howell. 2004b. Mio-Pliocene mammals from the Middle Awash, Ethiopia. *Geobios* 37:536–552.

Hamdine, W., M. Thévenot, and J. Michaux. 1998. Histoire récente de l'ours brun au Maghreb. *Comptes Rendus de l'Académie des Sciences, Paris, Sciences de la Vie* 321:565–570.

Haq, B. U., J. Hardenbol, and P. R. Vail. 1987. Chronology of fluctuating sea levels since the Triassic (250 million years ago to Present). *Science* 235:1156–1167.

———. 1988. Mesozoic and Cenozoic chronostratigraphy and cycles of sea level change; pp. 71–108 in C. K. Wilgus, B. J. Hastings, H. Posamentier, J. C. van Wagoner, C. A. Ross, and C. G. S. C. Kendall (eds.), *Sea-Level Change: An Integrated Approach*. SEPM Special Publication 42.

Heizmann, E. P. J., and E. G. Kordikova. 2000. Zur systematischen Stellung von *"Amphicyon" intermedius* H. v. Meyer, 1849 (Carnivora, Amphicyonidae). *Carolinea* 58:69–82.

Helbing, H. 1929. *Pseudocyon sansaniensis* Lartet von Steinheim am Albuch. *Eclogae Geologicae Helvetiae* 22:180–184.

Hendey, Q. B. 1974. The late Cenozoic Carnivora of the south-western Cape Province. *Annals of the South African Museum* 63:1–369.

———. 1978. Late Tertiary Mustelidae (Mammalia, Carnivora) from Langebaanweg, South Africa. *Annals of the South African Museum* 76:329–357.

Hooijer, D.A. 1963. Miocene Mammalia of Congo. *Musée Royal de l'Afrique Centrale, Tervuren, Belgique, Annales, Série in 8°, Sciences Géologiques* 46:1–77.

Howell, F. C. 1987. Preliminary observations on Carnivora from the Sahabi Formation (Libya); pp. 153–181 in N. T. Boaz, A. El-Arnauti, A. W. Gaziry, J. de Heinzelin, and D. D. Boaz (eds.), *Neogene Paleontology and Geology of Sahabi*. Liss, New York.

Howell, F. C., and N. García. 2007. Carnivora (Mammalia) from Lemudong'o (Late Miocene, Narok District, Kenya). *Kirtlandia* 56:121–139.

Howell, F. C., and G. Petter. 1985. Comparative observations on some Middle and Upper Miocene hyaenids. Genera: *Percrocuta* Kretzoi, *Allohyaena* Kretzoi, *Adcrocuta* Kretzoi (Mammalia, Carnivora, Hyaenidae). *Geobios* 18:419–476.

Hunt, R. M. Jr. 1989. Evolution of the aeluroid Carnivora: significance of the ventral promontorial process of the petrosal, and the origin of basicranial patterns in the living families. *American Museum Novitates* 2930:1–32.

———. 1991. Evolution of the aeluroid Carnivora: Viverrid affinities of the Miocene carnivoran *Herpestides*. *American Museum Novitates* 3023:1–34.

———. 1996. Biogeography of the order Carnivora; pp. 485–541 in J. L. Gittleman (ed.), *Carnivore Behavior, Ecology, and Evolution*, vol. 2. Cornell University Press, Ithaca, N.Y.

———. 1998a. Amphicyonidae; pp. 196–227 in C. M. Janis, K. M. Scott, and L. L. Jacobs (eds.), *Evolution of Tertiary Mammals of North America: Volume 1. Terrestrial Carnivores, Ungulates, and Ungulatelike Mammals*. Cambridge University Press, Cambridge.

———. 1998b. Evolution of the aeluroid Carnivora: diversity of the earliest aeluroids from Eurasia (Quercy, Hsanda-Gol) and the origin of felids. *American Museum Novitates* 3252:1–65.

————. 1998c. Ursidae; pp. 174–195 in C. M. Janis, K. M. Scott, and L. L. Jacobs (eds.), *Evolution of Tertiary Mammals of North America: Volume 1. Terrestrial Carnivores, Ungulates, and Ungulatelike Mammals.* Cambridge University Press, Cambridge.

————. 2003. Intercontinental migration of large mammalian carnivores: Earliest occurrence of the Old World beardog *Amphicyon* (Carnivora, Amphicyonidae) in North America. *Bulletin of the American Museum of Natural History* 279:77–115.

Kohno, N. and C. E. Ray. 2008. Pliocene walruses from the Yorktown Formation of Virginia and North Carolina, and a systematic revision of the North Atlantic Pliocene walruses. *Virginia Museum of Natural History Special Publication* 14:39–80.

Koretski, I. A., and C. E. Ray. 2008. Phocidae of the Pliocene of Eastern USA. *Virginia Museum of Natural History Special Publication* 14:81–140.

Kretzoi, M. 1938. Die Raubtiere von Gombaszög nebst einer Übersicht der Gesamtfauna. *Annales Historico-Naturales Musei Nationalis Hungarici* 31:89–157.

Kurtén, B. 1970. The Neogene wolverine *Plesiogulo* and the origin of *Gulo* (Carnivora, Mammalia). *Acta Zoologica Fennica* 131:1–22.

————. 1976. Fossil Carnivora from the Late Tertiary of Bled Douarah and Cherichira, Tunisia. *Notes du Service Géologique de Tunisie* 42:177–214.

Kurtén, B., and L. Werdelin. 1988. A review of the genus *Chasmaporthetes* Hay, 1921 (Carnivora, Hyaenidae). *Journal of Vertebrate Paleontology* 8:46–66.

Kuss, S. E. 1962. Problematische Caniden des europäischen Tertiärs. *Berichte der Naturforschenden Gesellschaft zu Freiburg im Brisgau* 52:123–172.

————. 1965. Revision der europäischen Amphicyoninae (Canidae, Carnivora, Mamm.) ausschliesslich der voroberstampischen Formen. *Sitzungsberichte der Heidelberger Akademie der Wissenschaften, Mathematisch-Naturwissenschaftlige Klasse* 1965:5–168.

Lewis, M. E., and L. Werdelin. 1997. Trends in the evolution and ecology of the genus *Crocuta*. *Journal of Vertebrate Paleontology* 17 (suppl. to number 3):60A.

————. 2000. The evolution of spotted hyaenas *(Crocuta)*. *Hyaena Specialist Group Newsletter* 7:34–36.

Martínez Navarro, B., and P. Palmqvist. 1995. Presence of the African machairodont *Megantereon whitei* (Broom, 1937)(Felidae, Carnivora, Mammalia) in the lower Pleistocene site of Venta Micena (Orce, Granada, Spain), with some considerations on the origin, evolution and dispersal of the genus. *Journal of Archaeological Science* 22:569–582.

————. 1996. Presence of the African saber-toothed felid *Megantereon whitei* (Broom, 1937)(Mammalia, Carnivora, Machairodontinae) in Apollonia-1 (Mygdonia Basin, Macedonia, Greece). *Journal of Archaeological Science* 23:869–872.

Michel, P. 1988. Un nouveau Mellivorinae (Carnivora, Mustelidae) du Pléistocène de Buknadel (région de Rabat-Maroc) = *Mellivora carolae* n. sp. *Comptes Rendus de l'Académie des Sciences, Paris*, Série II, 306:935–938.

Miller, E. R. 1999. Faunal correlation of Wadi Moghara, Egypt: Implications for the age of *Prohylobates tandyi*. *Journal of Human Evolution* 36:519–533.

Morales, J., and M. Pickford. 2005a. Carnivores from the Middle Miocene Ngorora Formation (13–12 Ma), Kenya. *Estudios Geológicos* 61:271–284.

————. 2005b. Giant bunodont Lutrinae from the Mio-Pliocene of Kenya and Uganda. *Estudios Geológicos* 61:233–246.

————. 2006. A large percrocutid carnivore from the Late Miocene (ca. 10–9 Ma) of Nakali, Kenya. *Annales de Paléontologie* 92:359–366.

————. 2008. Creodonts and carnivores from the Middle Miocene Muruyur Formation at Kipsaraman and Cheparawa, Baringo District, Kenya. *Comptes Rendus Palevol* 7:487–497.

Morales, J., M. Pickford, S. Fraile, M. Salesa, and D. Soria. 2003. Creodonta and Carnivora from Arrisdrift, early Middle Miocene of southern Namibia. *Memoirs of the Geological Survey of Namibia* 19:177–194.

Morales, J., Pickford, M., and Salesa, M. J. 2008. Creodonta and Carnivora from the early Miocene of the northern Sperrgebiet, Namibia. *Memoirs of the Geological Survey of Namibia* 20:291–310.

Morales, J., M. Pickford, M. Salesa, and D. Soria. 2000. The systematic status of *Kelba*, Savage, 1965, *Kenyalutra*, Schmidt-Kittler, 1987 and *Ndamathaia* Jacobs et al, 1987 (Viverridae, Mammalia) and a review of early Miocene mongoose-like carnivores of Africa. *Annales de Paléontologie* 86:243–251.

Morales, J., M. Pickford, and D. Soria. 2005. Carnivores from the Late Miocene and basal Pliocene of the Tugen Hills, Kenya. *Revista de la Sociedad Geológica de España* 18:39–61.

————. 2007. New carnivoran material (Creodonta, Carnivora and Incertae sedis) from the early Miocene of Napak, Uganda. *Paleontological Research* 11:71–84.

Morales, J., M. Pickford, D. Soria, and S. Fraile. 1998. New carnivores from the basal Middle Miocene of Arrisdrift, Namibia. *Eclogae Geologicae Helvetiae* 91:27–40.

————. 2001a. New Viverrinae (Carnivora: Mammalia) from the basal Middle Miocene of Arrisdrift, Namibia. *Palaeontologia Africana* 37:99–102.

Morales, J., M. J. Salesa, M. Pickford, and D. Soria. 2001b. A new tribe, new genus and two new species of Barbourofelinae (Felidae, Carnivora, Mammalia) from the early Miocene of East Africa and Spain. *Transactions of the Royal Society of Edinburgh, Earth Sciences* 92:97–102.

Morlo, M., E. R. Miller, and A. N. El-Barkooky. 2007. Creodonta and Carnivora from Wadi Moghra, Egypt. *Journal of Vertebrate Paleontology* 27:145–159.

Morlo, M., S. Peigné, and D. Nagel. 2004. A new species of *Prosansanosmilus*: Implications for the systematic relationships of the family Barbourofelidae new rank (Carnivora, Mammalia). *Zoological Journal of the Linnean Society* 140:43–61.

Muizon, C. de. 1981. Premier signalement de Monachinae (Phocidae, Mammalia) dans le Sahélien (Miocène supérieur) d'Oran (Algérie). *Palaeovertebrata* 11:181–194.

Muizon, C. de, and Q. B. Hendey. 1980. Late Tertiary seals of the South Atlantic Ocean. *Annals of the South African Museum* 82:91–128.

Munthe, K. 1998. Canidae; pp. 124–143 in C. M. Janis, K. M. Scott, and L. L. Jacobs (eds.), *Evolution of Tertiary Mammals of North America: Volume 1. Terrestrial Carnivores, Ungulates, and Ungulatelike Mammals.* Cambridge University Press, Cambridge.

Nakaya, H., M. Pickford, Y. Nakano, and H. Ishida. 1984. The Late Miocene large mammal fauna from the Namurungule Formation, Samburu Hills, Northern Kenya. *African Study Monographs*, Supplementary Issue 2:87–131.

Nowak, R. M. 2003. *Walker's Marine Mammals of the World.* Johns Hopkins University Press, Baltimore, 264 pp.

————. 2005. *Walker's Carnivores of the World.* Johns Hopkins University Press, Baltimore, 328 pp.

Palmqvist, P. 2002. On the presence of *Megantereon whitei* at the South Turkwel hominid site, northern Kenya. *Journal of Paleontology* 76:928–930.

Palmqvist, P., V. Torregrosa, J. A. Pérez-Claros, B. Martínez-Navarro, and A. Turner. 2007. A re-evaluation of the diversity of *Megantereon* (Mammalia, Carnivora, Machairodontinae) and the problem of species identification in extinct carnivores. *Journal of Vertebrate Paleontology* 27:160–175.

Peigné, S. 2000. Systématique et évolution des Feliformia (Mammalia, Carnivora) du Paléogène d'Eurasie. Unpublished PhD dissertation, Université de Poitiers, Poitiers, France, 396 pp.

Peigné, S., Y. Chaimanee, C. Yamee, P. Tian, and J.-J. Jaeger. 2006. A new amphicyonid (Mammalia, Carnivora, Amphicyonidae) from the late Middle Miocene of northern Thailand and a review of the amphicyonid record in Asia. *Journal of Asian Earth Sciences* 26:519–532.

Peigné, S., and L. de Bonis. 1999. The genus *Stenoplesictis* Filhol (Mammalia, Carnivora) from the Oligocene deposits of the Phosphorites of Quercy, France. *Journal of Vertebrate Paleontology* 19:566–575.

Peigné, S., L. de Bonis, A. Likius, H. T. Mackaye, P. Vignaud, and M. Brunet. 2005. A new machairodontine (Carnivora, Felidae) from the Late Miocene hominid locality of TM266, Toros-Menalla, Chad. *Comptes Rendus Palevol* 4:243–253.

————. 2008a. Late Miocene Carnivora from Chad: Lutrinae (Mustelidae). *Zoological Journal of the Linnean Society* 152:793–846.

Peigné, S., L. de Bonis, H. T. Mackaye, A. Likius, P. Vignaud and M. Brunet. 2008b. Late Miocene Carnivora from Chad: Herpestidae, Viverridae, and small-sized Felidae. *Comptes Rendus Palevol* 7:499–527.

Peigné, S., and E. P. J. Heizmann. 2003. The Amphicyonidae (Mammalia: Carnivora) from Ulm-Westtangente (MN 2, early Miocene), Baden-Württemberg, Germany: Systematics and ecomorphology. *Stuttgarter Beiträge zur Naturkunde, Serie B (Geologie und Paläontologie)* 343:1–133.

Petter, G. 1967. Petits carnivores villafranchiens du Bed I d'Oldoway (Tanzanie); pp. 529–538 in *Problèmes actuels de Paléontologie (Évolution des Vertébrés)*. Colloques Internationaux du CNRS, Paris.

———. 1973. Carnivores pléistocènes du ravin d'Olduvai (Tanzanie); pp. 44–100 in L. S. B. Leakey, R. J. G. Savage, and S. C. Coryndon (eds.), *Fossil Vertebrates of Africa*, vol. 3. Academic Press, London.

———. 1987. Small carnivores (Viverridae, Mustelidae, Canidae) from Laetoli; pp. 194–234 in M. D. Leakey and J. M. Harris (eds.), *Laetoli: A Pliocene Site in Northern Tanzania*. Clarendon Press, Oxford.

Petter, G. and F. C. Howell. 1985. Diversité des Carnivores (Mammalia, Carnivora) dans les faunes du Pliocène moyen et supérieur d'Afrique orientale. Indications paléoécologiques; pp. 133–149 in Fondation Singer-Polignac (ed.), *L'environnement des Hominidés au Plio-Pléistocène*. Masson, Paris.

Petter, G., M. Pickford, and F. C. Howell. 1991. La loutre piscivore du Pliocène de Nyaburogo et de Nkondo (Ouganda, Afrique occidentale): *Torolutra ougandensis* n. g., n. sp. (Mammalia, Carnivora). *Comptes Rendus de l'Académie des Sciences, Paris* 312:949–955.

Petter, G., and H. Thomas. 1986. Les Agriotheriinae (Mammalia, Carnivora) néogènes de l'ancien monde. Présence du genre *Indarctos* dans la faune de Menacer (ex-Marceau), Algérie. *Geobios* 19:573–586.

Pickford, M. 1986. Cainozoic Palaeontological Sites of Western Kenya. *Münchner Geowissenschaftliche Abhandlungen A* 8:1–151.

———. 2004. Southern Africa: A cradle of evolution. *South African Journal of Science* 100:205–214.

Pickford, M., T. Fernandes, and S. Aço. 1990. Nouvelles découvertes de remplissages de fissures à primates dans le "Planalto da Humpata," Huíla, Sud de Angola. *Comptes Rendus de l'Académie des Sciences* 310:843–848.

Pickford, M., and B. Senut. 2000. Geology and paleobiology of the Namib Desert, southwestern Africa. *Memoirs of the Geological Survey of Namibia* 18:1–155.

Pickford, M., B. Senut, D. Gommery, and E. Musiime. 2003. New Catarrhine fossils from Moroto II, early Middle Miocene (ca. 17.5 Ma) Uganda. *Comptes Rendus Palevol* 2:649–662.

Pilgrim, G. E. 1931. *Catalogue of the Pontian Carnivora of Europe in the Department of Geology*. British Museum (Natural History), London, 174 pp.

———. 1932. The fossil Carnivora of India. *Palaeontologia Indica* 18:1–232.

Pohle, H. 1928. Die Raubtiere von Oldoway. *Wissenschaftliche Ergebnisse der Oldoway-Expedition 1913* (N.F.) 3:45–54.

Potts, R., and A. Deino. 1995. Mid-Pleistocene change in large mammal faunas of eastern Africa. *Quaternary Research* 43:106–113.

Rasmussen, D. T. and M. Gutiérrez. 2009. A mammalian fauna from the Late Oligocene of Northwestern Kenya. *Palaeontographica Abt. A* 288:1–52.

Rögl, F. 1998. Palaeogeographic considerations for Mediterranean and Paratethys seaways (Oligocene to Miocene). *Annalen des Naturhistorisches Museums in Wien* 99:279–310.

Roth, C. 1988. *Leptoplesictis* Major 1903 (Mammalia, Carnivora, Viverridae) aus dem Orleanium und Astaracium/Miozän von Frankreich und Deutschland. *Paläontologische Zeitschrift* 62:333–343.

Salles, L. O. 1992. Felid phylogenetics: Extant taxa and skull morphology (Felidae, Aeluroidea). *American Museum Novitates* 3047:1–67.

Sardella, R. 1994. Sistematica e distribuzione stratigrafica dei Macairodontini dal Miocene Superiore al Pleistocene. Unpublished PhD dissertation, Department of Geology, University of Florens, Florens, Italy, 137 pp.

Sardella, R., and L. Werdelin. 2007. *Amphimachairodus* (Felidae, Mammalia) from Sahabi (latest Miocene-earliest Pliocene, Libya), with a review of African Miocene Machairodontinae. *Rivista Italiana di Paleontologia e Stratigrafia* 113:67–77.

Savage, R. J. G. 1965. Fossil mammals of Africa: 19. The Miocene Carnivora of East Africa. *Bulletin of the British Museum (Natural History): Geology* 10:241–316.

———. 1978. Carnivora; pp. 249–267 in V. J. Maglio and H. B. S. Cooke (eds.), *Evolution of African Mammals*. Harvard University Press, Cambridge.

Savage, R. J. G., and W. R. Hamilton. 1973. Introduction to the Miocene mammal faunas of Gebel Zelten, Libya. *Bulletin of the British Museum (Natural History), Geology* 22:513–527.

Schmidt-Kittler, N. 1976. Raubtiere aus dem Jungtertiär Kleinasiens. *Paleontographica, Abt. A* 155:1–131.

———. 1987. The Carnivora (Fissipedia) from the Lower Miocene of East Africa. *Palaeontographica, Abt. A* 197:85–126.

Servheen, C., S. Herrero, and B. Peyton eds. 1998. *Bears*. Status Survey and Conservation Action Plan. IUCN/SSC Bear and Polar Bear Specialist Groups. IUCN, Gland, Switzerland.

Sillero-Zubiri, C., M. Hoffmann, and D. W. Macdonald. 2004. *Canids: Foxes, Wolves, Jackals and Dogs*. Status Survey and Conservation Action Plan. IUCN, Gland, Switzerland.

Spassov, N., and L. Rook. 2006. *Eucyon marinae* sp. nov. (Mammalia, Carnivora), a new canid species from the Pliocene of Mongolia, with a review of forms referable to the genus. *Rivista Italiana di Paleontologia e Stratigrafia* 112:123–133.

Staaden, M. J. van 1994. *Suricata suricatta*. *Mammalian Species* 483:1–8.

Stromer, E. 1907. Fossile Wirbeltier-Reste aus dem Uadi Fâregh und Uadi Natrûn in Ägypten. *Abhandlungen Herausgegeben von der Senckenbergischen Naturforschenden Gesellschaft* 29:99–132.

———. 1913. Mitteilungen über Wirbeltierreste aus dem Mittelpliocän des Natrontales (Ägypten). *Zeitschrift der Deutschen Geologischen Gesellschaft, Abhandlungen A* 65:350–372.

Sudre, J., and J.-L. Hartenberger. 1992. Oued Mya 1, nouveau gisement de mammifères du Miocène supérieur dans le sud Algérien. *Geobios* 25:553–565.

Sunquist, M., and F. Sunquist. 2002. *Wild Cats of the World*. University of Chicago Press, Chicago, 452 pp.

Taylor, M. E. 1972. *Ichneumia albicauda*. *Mammalian Species* 12:1–4.

Taylor, P. J., and J. A. J. Meester. 1993. *Cynictis penicillata*. *Mammalian Species* 432:1–7.

Tedford, R. H., and Z. Qiu. 1991. Pliocene *Nyctereutes* (Carnivora: Canidae) from Yushe, Shanxi, with comments on Chinese fossil raccoon-dogs. *Vertebrata PalAsiatica* 29:179–189.

———. 1996. A new canid genus from the Pliocene of Yushe, Shanxi Province. *Vertebrata PalAsiatica* 34:27–40.

Tedford, R. H., X. Wang, and B. E. Taylor. 2009. Phylogenetic systematics of the North American fossil Caninae (Carnivora: Canidae). *Bulletin of the American Museum of Natural History* 325:1–218.

Tsujikawa, H. 2005. The updated Late Miocene large mammal fauna from Samburu Hills, northern Kenya. *African Study Monographs* 32:1–50.

Turner, A. 1984. Hominids and fellow travellers: Human migration into high latitudes as part of a large mammal community; pp. 193–217 in R. Foley (ed.), *Hominid Evolution and Community Ecology*. Academic Press, London.

———. 1987. *Megantereon cultridens* (Cuvier)(Mammalia, Felidae, Machairodontinae) from Plio-Pleistocene deposits in Africa and Eurasia, with comments on dispersal and the possibility of a New World origin. *Journal of Paleontology* 61:1256–1268.

———. 1990. The evolution of the guild of larger terrestrial carnivores during the Plio-Pleistocene in Africa. *Geobios* 23:349–368.

Turner, A., L. Bishop, C. Denys, and J. K. McKee. 1999. Appendix: A locality-based listing of African Plio-Pleistocene mammals; pp. 369–399 in T. G. Bromage and F. Schrenk (eds.), *African Biogeography, Climate Change, and Human Evolution*. Oxford University Press, Oxford.

Veron, G. 1995. La position systématique de *Cryptoprocta ferox* (Carnivora) : Analyse cladistique des caractères morphologiques de carnivores Aeluroidea actuels et fossiles. *Mammalia* 59:551–582.

Veron, G., M. Colyn, A. E. Dunham, P. Taylor, and P. Gaubert. 2004. Molecular systematics and origin of sociality in mongooses (Herpestidae, Carnivora). *Molecular Phylogenetics and Evolution* 30:582–598.

Ward, O. G., and D. H. Wurster-Hill. 1990. *Nyctereutes procyonoides*. *Mammalian Species* 358:1–5.

Werdelin, L. 1996. Carnivores, exclusive of Hyaenidae, from the later Miocene of Europe and Western Asia; pp. 271–289 in R. L. Bernor, V. Fahlbusch, and H.-W. Mittmann (eds.), *The Evolution of Western Eurasian Neogene Mammal Faunas*. Columbia University Press, New York.

Werdelin, L. 1999a. *Pachycrocuta* (hyaenids) from the Pliocene of east Africa. *Paläontologische Zeitschrift* 73:157–165.

———. 1999b. Studies of fossil hyaenas: Affinities of *Lycyaenops rhomboideae* Kretzoi from Pestlörinc, Hungary. *Zoological Journal of the Linnean Society* 126:307–317.

———. 2003. Mio-Pliocene Carnivora from Lothagam, Kenya; pp. 261–328 in M. G. Leakey and J. D. Harris (eds.), *Lothagam: Dawn of Humanity in Eastern Africa*. Columbia University Press, New York.

Werdelin, L., and J. Barthelme. 1997. Brown hyena (*Parahyaena brunnea*) from the Pleistocene of Kenya. *Journal of Vertebrate Paleontology* 17:758–761.

Werdelin, L., and R. Dehghani. In press. Carnivora. In T. Harrison (ed.), *Paleontology and Geology of Laetoli, Tanzania: Human Evolution in Context.* Vertebrate Paleobiology and Paleoanthropology Series. Springer, New York.

Werdelin, L., and M. E. Lewis. 2000. Carnivora from the South Turkwel hominid site, northern Kenya. *Journal of Paleontology* 74:1173–1180.

———. 2001. A revision of the genus *Dinofelis* (Mammalia, Felidae). *Zoological Journal of the Linnean Society* 132:147–258.

———. 2002. Species identification in *Megantereon*: A reply to Palmqvist. *Journal of Paleontology* 76:931–933.

———. 2005. Plio-Pleistocene Carnivora of eastern Africa: Species richness and turnover patterns. *Zoological Journal of the Linnean Society* 144:121–144.

———. 2008. New species of *Crocuta* from the early Pliocene of Kenya, with an overview of early Pliocene hyenas of eastern Africa. Journal *of Vertebrate Paleontology* 28:1162–1170.

Werdelin, L., and N. Solounias. 1991. The Hyaenidae: Taxonomy, systematics and evolution. *Fossils and Strata* 30:1–104.

———. 1996. The evolutionary history of hyaenas in Europe and western Asia during the Miocene; pp. 290–306 in R. L. Bernor, V. Fahlbusch, and H.-W. Mittmann (eds.), *The Evolution of Western Eurasian Neogene Mammal Faunas.* Columbia University Press, New York.

Werdelin, L., and A. Turner. 1996. The fossil and living Hyaenidae of Africa: Present status; pp. 637–659 in K. Stewart and K. Seymour (eds.), *Palaeoecology and Palaeoenvironments of Late Cenozoic Mammals: Tributes to the Career of C. S. (Rufus) Churcher.* University of Toronto Press, Toronto.

Werdelin, L., A. Turner, and N. Solounias. 1994. Studies of fossil hyaenids: The genera *Hyaenictis* Gaudry, and *Chasmaporthetes* Hay, with a reconsideration of the Hyaenidae of Langebaanweg, *South Africa. Zoological Journal of the Linnean Society* 111:197–217.

Willemsen, G. F. 1992. A revision of the Pliocene and Quaternary Lutrinae from Europe. *Scripta Geologica* 101:1–115.

Wolsan, M., and Y. A. Semenov. 1996. A revision of the Late Miocene mustelid carnivoran *Eomellivora. Acta Zoologica Cracoviensia* 39:593–604.

Yasui, K., Y. Kunimatsu, N. Kuga, B. Bajope, and H. Ishida. 1992. Fossil mammals from the Neogene strata in the Sinda Basin, eastern Zaire. *African Study Monographs* 17:87–107.

Yoder, A. D., M. M. Burns, S. Zehr, T. Delefosse, G. Veron, S. M. Goodman, and J. J. Flynn. 2003. Single origin of Malagasy Carnivora from an African ancestor. *Nature* 421:734–737.

Zdansky, O. 1924. Jungtertiäre Carnivoren Chinas. *Palaeontologia Sinica,* serie C, 2:1–149.

THIRTY-THREE

Chalicotheriidae

MARGERY C. COOMBS AND SUSANNE M. COTE

Chalicotheres are an unusual group of extinct fossil perissodactyls that, despite a dentition suited for a herbivorous diet, had claws on their digits instead of hooves. The group first appeared in the Eocene, reached its highest diversity in the Miocene, and went extinct in the Pleistocene. Chalicotheres seem to have undergone much of their diversification in Asia but are also found in Europe, Africa, and North America.

Oligocene and later chalicotheres belong to the family Chalicotheriidae (*sensu* Coombs, 1989), which includes two subfamilies, the Chalicotheriinae and Schizotheriinae. The Chalicotheriinae have relatively short, low-crowned cheek teeth and strangely proportioned, gorilla-like bodies in which the forelimbs are much longer than the hindlimbs (Zapfe, 1979); they appear to have lived primarily in moist forested environments and never reached North America. Schizotheriinae have longer, higher-crowned cheek teeth and more typical ungulate (somewhat okapi-like) body proportions; they are found in a variety of depositional environments from moist forests to drier, more open, treed areas. They reached North America via Bering connections in the Miocene.

There has been much debate about the use of the claws in chalicotheres. It is likely that chalicotheres browsed in an upright bipedal position. They may have used their clawed digits to hook down branches and bring them within reach of the mouth (Zapfe, 1979; Coombs, 1983), tear off branches, cut up fruit, or debark trees (Koenigswald, 1932; Geraads et al., 2006), or uproot trees (Schaub, 1943). Feeding appears to have involved browsing (in the broad sense) on some combination of leaves, fruit, bark, and twigs, depending on the taxon and prevailing environmental conditions. The presence of chalicotheres in a fauna is usually thought to be an indicator of some trees and shrubs in the vicinity. Dental mesowear studies of several European Miocene chalicotheres by Schulz et al. (2007) suggested an abrasive dietary component. These authors interpreted the abrasive part of the diet in terms of bark and twigs. Coombs and Semprebon (2005) suggested on the basis of low-magnification stereoscopic dental microwear that both bark/twig feeding and fruit consumption were significant abrasive factors in various chalicothere diets; a more complete microwear study is pending.

African Chalicotheriidae, represented sequentially by both the Chalicotheriinae and Schizotheriinae, are found from the early Miocene into the Pleistocene (figure 33.1). Chalicothere fossils are rare and in Africa are known mostly from fragmentary remains. Fortunately, many of their skeletal elements are distinctive enough to allow them to be identified from isolated bones, particularly those of the manus and pes.

Virtually all known African members of the Chalicotheriinae are restricted to the early Miocene and belong to a single, relatively basal species, *Butleria rusingensis*. Later African chalicotheres belong to the Schizotheriinae and have been referred to three species: *Ancylotherium hennigi, Ancylotherium cheboitense*, and *"Chemositia" tugenensis*. *Ancylotherium hennigi* was the last surviving member of the Schizotheriinae worldwide.

ABBREVIATIONS

BMNH, British Museum of Natural History (now The Natural History Museum), London; BPI, Bernard Price Institute for Paleontological Research, Johannesburg; BSPG, Bayerischen Staatssammlung für Paläontologie und historische Geologie, Munich; CMK, Community Museums of Kenya, Nairobi; IPMH, Institut für Paläontologie und Museum der Humboldt-Universität, Berlin; KNM, Kenya National Museums, Nairobi; NMT, National Museum of Tanzania, Dar es Salaam; MNHN, Muséum National d'Histoire Naturelle, Paris; NMAA, National Museum of Ethiopia, Addis Ababa; UMP, Department of Paleontology, Uganda Museum, Kampala; Mc, metacarpal; Mt, metatarsal.

Systematic Paleontology

Order PERISSODACTYLA Owen, 1848
Superfamily CHALICOTHERIOIDEA Gill, 1872
Family CHALICOTHERIIDAE Gill, 1872
Table 33.1
Subfamily CHALICOTHERIINAE Gill, 1872
Genus *BUTLERIA* Bonis et al., 1995
BUTLERIA RUSINGENSIS (Butler, 1965)
Figure 33.2

Synonymy Chalicotherium rusingense Butler, 1965; *Butleria rusingensis* (Butler) Bonis et al., 1995; *Butleria rusingensis* (Butler) Anquetin et al., 2007.

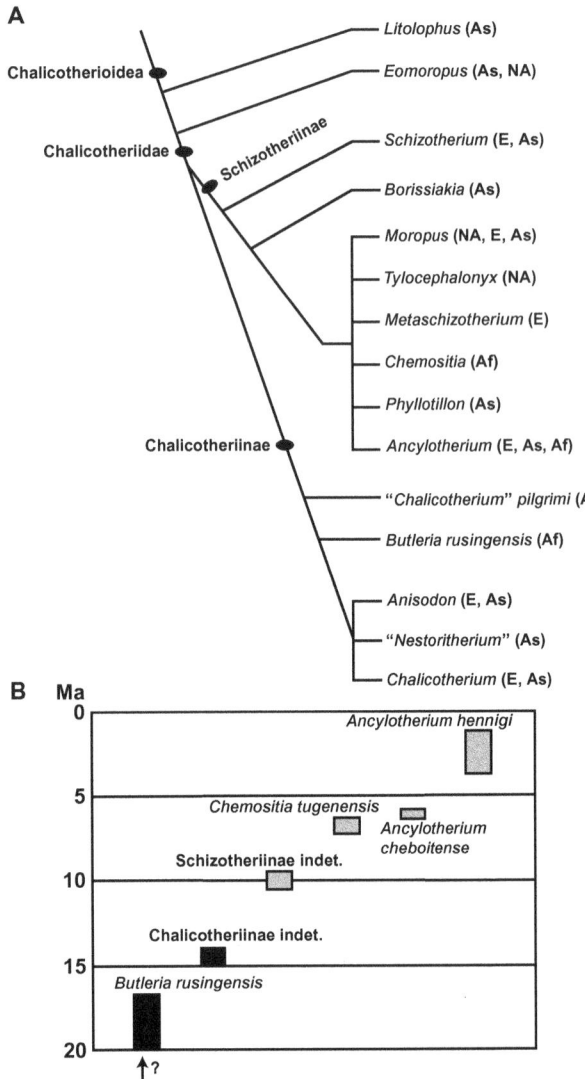

A

Litolophus (As)

Eomoropus (As, NA)

Chalicotherioidea

Schizotheriinae

Schizotherium (E, As)

Chalicotheriidae

Borissiakia (As)

Moropus (NA, E, As)

Tylocephalonyx (NA)

Metaschizotherium (E)

Chemositia (Af)

Phyllotillon (As)

Chalicotheriinae

Ancylotherium (E, As, Af)

"Chalicotherium" pilgrimi (As)

Butleria rusingensis (Af)

Anisodon (E, As)

"Nestoritherium" (As)

Chalicotherium (E, As)

B Ma

0

Ancylotherium hennigi

5

Chemositia tugenensis

Ancylotherium cheboitense

Schizotheriinae indet.

10

Chalicotheriinae indet.

15

Butleria rusingensis

20

↑?

FIGURE 33.1 Relationships and geologic time distributions of African chalicotheres. A) Cladogram of potential relationships of selected chalicotheres, including those discussed in the text (based on Coombs, 1989, and Bonis et al., 1995); B) time distributions of African chalicotheres, based on table 33.1.

ABBREVIATIONS: Af, Africa; E, Europe; As, Asia; NA, North America. Chalicotheriinae: black boxes; Schizotheriinae: gray boxes.

Holotype BMNH M25270, a left maxilla with P2–M3, from site R107, Rusinga Island, Kenya.

Diagnosis Small, basal chalicotheriine with relatively long face and no perinasal depression (maxilla elongated, upper border of nasal opening above P2–P3); mandibular symphysis short (just to the level of p2); i1–i3 present; diastema relatively long; paracone and metacone of upper molars displaced somewhat lingually on M1–M3; "metastylid" distinct on lower molars; astragalus depressed but less so than in *Anisodon grande* (previously placed in *Macrotherium* or *Chalicotherium*; Butler, 1965; Bonis et al., 1995; Anquetin et al., 2007).

Description Like other members of the Chalicotheriinae, *Butleria* has well-developed upper and lower canines and has apparently lost the upper incisors. Three small lower incisors are present, as in *Anisodon grande*, in contrast to their

reduction and loss in more derived members of the genus *Anisodon* (Xue and Coombs, 1985; Anquetin et al., 2007). A crista and crochet are variably present on upper molars. *B. rusingensis* is like other Chalicotheriinae in having distinctive postcranials in which the hindfoot is shorter than the forefoot, as reflected in proportions of the wide, low astragalus and short metatarsals. It is not clear whether the astragalus had an articulation with the cuboid, but if a facet was present, it must have been smaller than that in *A. grande* and *Chalicotherium goldfussi* (Butler, 1965). Zapfe (1979) provided measurement tables showing that at least Mt II and Mt III (Mt IV has not been described) are somewhat longer compared to width than their counterparts in *A. grande*; proportions for Mt II of *B. rusingensis* are closer to those of *"Chalicotherium" pilgrimi* (BMNH M12168). Fusion of proximal and middle phalanges does not occur in *B. rusingensis* or in any other chalicotheriine.

Age Early Miocene (definitively known from deposits from ~19.7 to ~16.8 Ma; see table 33.1).

Occurrence East Africa: Kenya and Uganda.

Remarks Butler (1965) considered *Ch. rusingense* to be a relatively primitive species within *Chalicotherium* and included only two genera, *Chalicotherium* and *Nestoritherium*, within the Chalicotheriinae. Later, Bonis et al. (1995) began an evaluation of species previously referred to *Chalicotherium* and split off *Butleria* as a new basal genus of the Chalicotheriinae. Anquetin et al. (2007), in a more recent revision of chalicotheriine phylogeny, retained the name *Butleria rusingensis* and upheld its position as a basal member of the Chalicotheriinae. At the same time, they encouraged a more complete and detailed evaluation of *B. rusingensis* and comparison with early Miocene Chalicotheriinae from Asia. Currently, *Butleria* includes only African *B. rusingensis*. Its relatively basal morphology does, however, resemble in many ways that of *"Ch." pilgrimi* from the Bugti Beds of Pakistan; Bonis et al. (1995) and Anquetin et al. (2007) did not evaluate *"Ch." pilgrimi*, so its generic assignment is still unsettled and beyond the scope of this chapter.

Butleria rusingensis is the best-known African chalicothere. Significant additional material has been added to collections of the Kenya National Museum since Butler's description of the original specimens in 1965. The most abundant material comes from deposits at Rusinga, but there are also associated foot elements from Mfwangano (figure 33.2). New finds from Legetet and Chamtwara in the Tinderet region of Western Kenya are roughly the same age as the P3 designated by Butler (1965) from Koru. Butler (1965) observed significant size variation in the Rusinga sample, recognizing common "small" and much rarer "large" forms, as is also seen in more recently collected material. These two forms likely represent two sexes, as sexual dimorphism is common in chalicotheres (Coombs, 1975).

All the occurrences listed here are based on clearly identifiable elements, except for Meswa Bridge, which is represented by an isolated canine tooth (KNM-ME 10508). This canine is small for *B. rusingensis* but compares reasonably well morphologically with chalicotheriine lower canines. If confirmed, this occurrence would be the earliest known appearance of Chalicotheriinae in Africa.

Hooijer (1963) identified an isolated incisor and lower molar crown fragment as *Macrotherium* (?) spec. from Miocene deposits of Malembe, near the Atlantic Coast in western Congo (now Cabinda, Angola). Butler (1965, 1978) argued that both specimens were too large to belong to *B. rusingensis*,

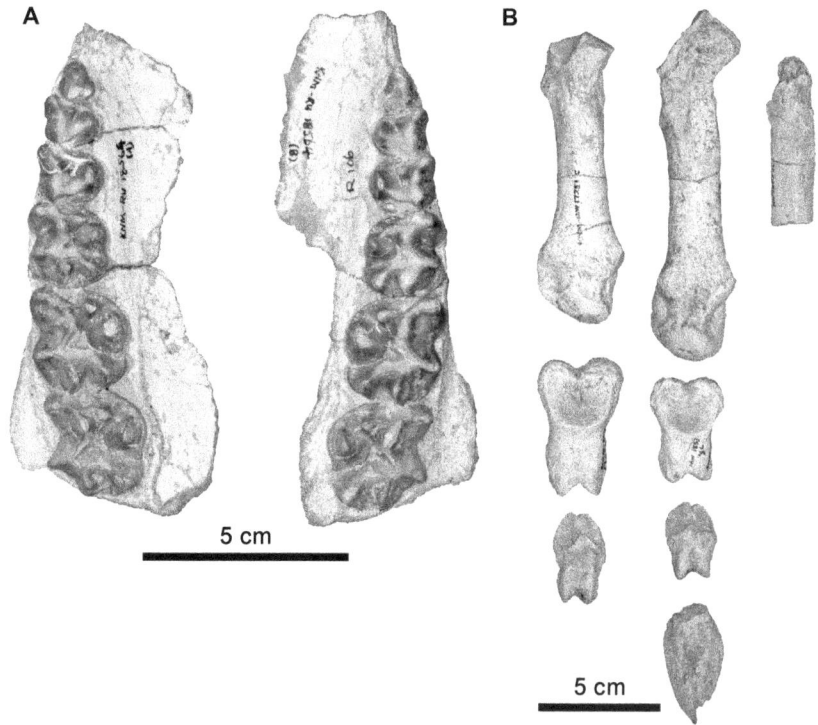

FIGURE 33.2 *Butleria rusingensis:* A) KNM-RU 18564, palate with P2–M3, both sides, from R106, Rusinga; B) part of KNM-MW 17251, forefoot with Mc II, Mc III, Mc IV, and phalanges, from Mfwangano.

and questioned whether they belonged to a chalicothere. Pickford has attributed these fragments to *Arsinoitherium* (Guérin and Pickford, 2005). In any case, they should not be included in any consideration of chalicothere temporal and geographic ranges.

Subfamily SCHIZOTHERIINAE Holland and Peterson, 1914
Genus *ANCYLOTHERIUM* Gaudry, 1862

Diagnosis Slightly modified from Geraads et al. (2007). Large schizotheriine chalicothere. Metaloph of upper molars short, ectoloph relatively flat between very prominent styles, crochet often present. Second lobe of m3 short. Manus as a whole, and each individual metacarpal concave dorsally; trapezium fused to Mc II or (more likely) lost; proximal carpal row shifted in the volar direction in respect to the distal row; scaphoid contacting Mc II in extreme flexion; lunate and magnum with volar processes much reduced or absent. Mc V lost. Mt III the longest metatarsal in a clearly mesaxonic hindfoot.

ANCYLOTHERIUM HENNIGI (Dietrich, 1942)
Figure 33.3

Synonymy Chalicothere, Andrews, 1923; Chalicotheriidae, Hopwood, 1926; *Metaschizotherium hennigi* Dietrich, 1942; *Metaschizotherium hennigi* Hopwood, 1951; *Metaschizotherium (?) transvaalensis* George, 1950; *Metaschizotherium (?) transvaalensis* Webb, 1965; *Ancylotherium hennigi* (Dietrich) Thenius, 1953; *Ancylotherium hennigi* (Dietrich) Butler, 1965.

Type Material Dietrich (1942) did not designate a single type for this species but described and figured the following specimens: Vo 330 18/9–Okt. 1938, right M2; Deturi 2/39, proximal phalanx of the manus; Vo 11/12.1.39, proximal part

of left Mc III; and Gar. Kor 1/39, middle phalanx of digit II of the pes. All are in the collections of the Institut für Paläontologie und Museum der Humboldt-Universität (Berlin) and were collected from the Laetolil beds, Vogel River, Garussi area, northwest of Lake Eyasi, Tanzania. Cooke and Coryndon (1970) listed all four specimens as syntypes. Among the syntypes, the best candidates to be lectotype are the M2 and the partial Mc III. Dietrich's description was unfortunately quite brief, so most of our knowledge of this species relies on subsequent work, especially that of Butler (1965), on additional material. Because the manus is one of the most diagnostic parts of *A. hennigi*, it seems most logical to choose Dietrich's Mc III fragment as the species lectotype.

Diagnosis Metacarpals less flattened and hollowed out dorsally and scaphoid proportionally taller (proximal to distal) than in *Ancylotherium pentelicum* (Butler, 1965); astragalus broad and low compared with that of *A. pentelicum* and *Ancylotherium cheboitense* (Guérin, 1987; Guérin and Pickford, 2005); body size smaller than *A. pentelicum*.

Description A. hennigi has the dental and postcranial specializations of *Ancylotherium*, but is less robust and shows a more moderate degree of postcranial modification than both *A. pentelicum* and *A. cheboitense*, despite its later age. The proportionally low astragalus is, however, a derived character.

Age Plio-Pleistocene (definitively known from deposits from 3.7 to 1.33 Ma).

Occurrence East Africa: Ethiopia, Kenya, Uganda, Tanzania. South Africa: Republic of South Africa (table 33.1).

Remarks The species of *Ancylotherium* with the most complete available material is *A. pentelicum*, a late Miocene species best known from Pikermi (Greece) but distributed through southeast Europe and into Asia. It reached the largest size of any chalicothere and had the highest-crowned molars. Skulls

TABLE 33.1
Summary of African chalicothere occurrences, ages, and distributions
Starred references cite occurrences; unstarred are used for age determinations only.
Abbreviations: K, Kenya; E, Ethiopia; T, Tanzania; U, Uganda; SA, South Africa.
Specimens in boldfaced museums were examined by the authors.

Taxon	Locality	Formation	Age	Material	Museums	References
CHALICOTHERIINAE						
Butleria rusingensis	?Meswa Bridge (K)	Muhoroni agglomerates	~20–23.5 Ma	Possible canine	**KNM**	*This chapter; Bishop et al., 1969; Pickford and Andrews, 1981
	Koru, Tinderet (K)	Koru or Legetet Fm.	~19.5–19.6 Ma	P3, M3	**BMNH**	*Butler, 1965; Bishop et al., 1969; Pickford and Andrews, 1981
	Legetet (Tinderet locality 10) (K)	Legetet Fm.	~19.5–19.6 Ma	dP3, P3	**KNM**	*This chapter; Bishop et al., 1969; Pickford and Andrews, 1981
	Songhor (K)	Kapurtay Agglomerates	~19.6–19.7 Ma	Mt II, phalanges, various isolated teeth	**KNM**	*Butler, 1965; *Bishop, 1967; Bishop et al. 1969; Pickford and Andrews, 1981
	Chamtwara (Tinderet locality 34) (K)	Chamtwara Member (lateral equivalent of Kapurtay Agglomerates)	~19.6–19.7 Ma	Canine, phalanges	**KNM**	*This chapter; Bishop et al., 1969; Pickford and Andrews, 1981
	Napak I, IV, V, IX (U)	"Napak Volcanics"	~19.5 Ma	Proximal phalanx, distal Mc, Mt, various teeth and jaw fragments, other phalanges	**UMP, BMNH**	*Bishop, 1967; Bishop et al., 1969; *Pickford, 1979
	Rusinga (K): Kiahera, Hiwegi, Kiakanga, Kaswanga, Gumba	Wayando Fm, Rusinga Agglomerate, Hiwegi Fm., Kulu Fm., Gumba redbeds	17–18 Ma	Holotype, plus numerous additional teeth and bones of the manus and pes	**BMNH, KNM**	*Butler, 1965; *Bishop, 1967; *Drake et al., 1988
	Mfwangano (K)	Lateral equivalent of Kiahera Fm. and Rusinga Agglomerate	17.9 Ma	Mt II, astragalus, phalanges, associated manus elements	**BMNH, KNM**	*Butler, 1965; *Bishop, 1967; Drake et al., 1988
	Moruorot (K)	Kalodirr Member, Lothidok Fm.	16.8–17.8 Ma	Ungual phalanx	**KNM**	*Pickford, 1979, Drake et al., 1988; Boschetto et al., 1992
Undetermined	Nachola, near Baragoi (K)	Aka Aiteputh Fm.	14–15 Ma	Metatarsal	**KNM**	*Tsujikawa and Nakaya, 2005; *Tsujikawa, pers. comm.; Sawada et al.; 1998
SCHIZOTHERIINAE						
Undetermined	Awash Basin (E)	Chorora Fm.	~10.5 Ma	Proximal phalanx	NMAA	Geraads et al., 2002, as *Ancylotherium* sp. cf. *A. tugenense*
	Samburu Hills, Locality 14 (K)	Upper Member, Namurungule Fm.	9.5 Ma	Proximal phalanx	**KNM**	*Nakaya et al., 1984; *Nakaya, 1994; Sawada et al., 1998; *Tsujikawa, 2005;
	Nkondo (U)	Nkondo Fm.	~5–6 Ma	Scaphoid, broken calcaneum	UMP?	Guérin, 1994, as Chalicotherioidea indet.

Taxon	Locality	Formation	Age	Material	Museums	References
"Chemositia" tugenensis	Koitugum, Tugen Hills (K)	Mpesida Fm.	6.37–7.2 Ma	Holotype broken femur and pes elements (broken tooth KNM MP214 in holotype is a suid)	**KNM**	*Pickford, 1979; Kingston et al., 2002
Ancylotherium cheboitense	Cheboit, Tugen Hills (K)	Lower Lukeino Fm.	5.9–6.1 Ma	Holotype pes elements, right lower molar	CMK	*Guérin and Pickford, 2005
?Ancylotherium cheboitense	Kapcheberek, Tugen Hills (K)	Lukeino Fm.	5.7–6.5 Ma	Proximal phalanx, ungual phalanx, lower molar	**KNM**	*Pickford, 1975; *Pickford, 1979 (as A. hennigi); Deino et al., 2002
?Ancylotherium cheboitense	Sagatia, Tugen Hills (K)	Mabaget Fm.	4.4–5 Ma	Damaged metapodial	CMK	*Guérin and Pickford, 2005
Ancylotherium hennigi	Laetoli (T)	Laetolil Beds	3.6–3.7 Ma	Type material (M2, proximal phalanx of manus, partial Mc III, middle phalanx), astragalus, calcaneum, two duplexes, distal metapodial, two track imprints	IPMH, NMT	*Dietrich, 1942; *Guérin, 1987; *Guérin in Leakey, 1987; Drake and Curtis, 1987
	Tugen Hills/ Baringo JM 511/K001/ BPRP#1, K015/ BPRP#15 (K)	Chemeron Fm.	3.2–3.6 Ma	P4, proximal phalanx, upper molar, talonid	**KNM**	*Hooijer, 1972; *Hooijer, 1973; A. Hill, pers. comm.
	Makapansgat, Transvaal (SA)	Makapan Limeworks Quarry	~3 Ma	Numerous (86+), but typically broken specimens, including many adult and juvenile teeth	BPI	*George, 1950; *Webb, 1965; Guérin and Pickford, 2005
	Kaiso (U)	Probably "Kaiso Village"	2.3 Ma	Proximal phalanx	**BMNH**	*Andrews, 1923; *Hopwood, 1926; *Butler, 1965; *Cooke and Coryndon, 1970; Guérin, 1994
	Lower Omo Basin (E)	Members D, G, ?C, Shungura Fm., Omo Group	1.9–?2.85 Ma	Left P4, right m1 or m2, calcaneum, left Mt IV, ?skull piece	MNHN, NMAA	*Hooijer, 1975; *Guérin, 1976; *Guérin, 1985; Feibel et al., 1989; Guérin and Pickford, 2005
	Bed I, Olduvai (T)	Olduvai Beds	1.71–2.1 Ma	Scaphoid, lunate, cuneiform, Mc II, Mc III, proximal phalanx, 2 middle phalanges	**BMNH**	*Hopwood, 1951; Leakey, 1951; *Butler, 1965; Guérin and Pickford, 2005
	Konso (E)	Intervals 1, 4, and 5, Konso Formation	~1.33–1.91 Ma	16 dental and manus/ pes specimens, most from youngest Interval 5	NMAA	*Suwa et al., 2003; Suwa, pers. comm.

CHALICOTHERIIDAE

Taxon	Locality	Formation	Age	Material	Museums	References
Undetermined	Maboko (K)	Maboko Fm.	>14.7 Ma (~15 Ma)	Abraded middle phalanx	**KNM**	*This chapter; Feibel and Brown, 1991

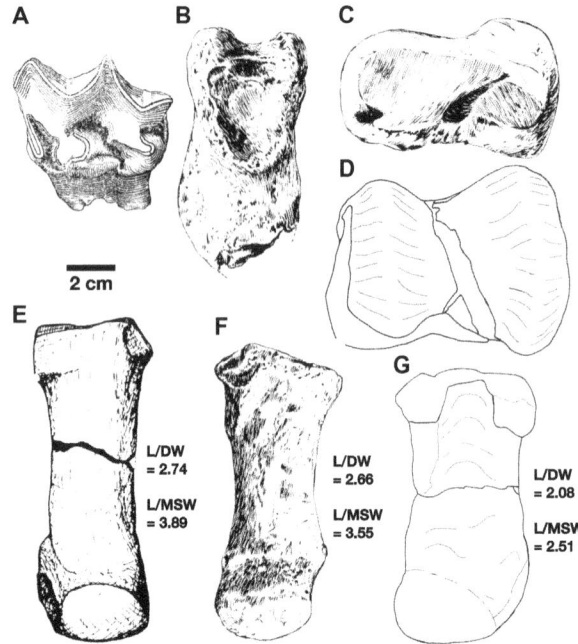

FIGURE 33.3 Morphology of African Schizotheriinae: A) left M2 of *Ancylotherium hennigi* from Makapansgat; B) dorsal view of broken duplex of *A. hennigi* from Laetoli ; C) volar (= palmar) view of astragalus of *A. hennigi* from Laetoli; D) dorsal view of astragalus of holotype of *A. cheboitense*; E) right Mt IV of holotype of *"Chemositia" tugenensis*; F) left Mt IV of *A. hennigi* from Shungura (Omo); and G) left Mt IV of holotype of *A. cheboitense*. A from George (1950), courtesy of South African Journal of Science; B, C from Guérin (1987), courtesy of Oxford University Press; D, G from Guérin and Pickford (2005); E from Pickford (1979), courtesy of Cainozoic Research; F from Guérin (1985).

ABBREVIATIONS FOR THE PROPORTIONS OF MT IV: L = length; DW = distal width; MSW = minimum shaft width.

show inflation of the frontal bone forming a modest-sized skull dome (Geraads et al., 2007). As Schaub (1943) described, *A. pentelicum* had a highly modified forelimb, in which strong flexion of the carpus and hyperextension of the clawed phalanges was possible. The hindlimbs were less strongly modified but still distinctive (see Roussiakis and Theodorou, 2001). Geraads et al. (2006) discussed variability of certain characters within *A. pentelicum*; this information is useful in comparisons with the African species. *A. hennigi* is clearly recognizable as *Ancylotherium*, but most known parts of the skeleton are more conservative in morphology.

Like other derived members of the Schizotheriinae, *Ancylotherium* fused the proximal and middle phalanges of digit II of the manus to form a bone called a duplex. Specimens of *A. hennigi* from Laetoli show this character well (Guérin, 1987; see figure 33.3B). However, the supposed fused phalanges figured by Dietrich (1942, Deturi 2/39) as one of the syntypes is not a duplex; Schaub (1943) identified it as a proximal phalanx belonging to digit IV of the manus. The first discovered specimen of *A. hennigi* was a proximal phalanx from Kaiso, Uganda (Andrews, 1923; Hopwood, 1926; Butler, 1965), still the only described chalicothere specimen from Kaiso. Butler (1965, 1978) thought that this phalanx (BMNH M12693) was the broken proximal part of a duplex and referred it to digit II of the manus; personal observation of this specimen (MCC) suggests that it is not a duplex, because the broken distal end

shows a remnant of the articular facet. A variety of other phalanges of *A. hennigi* have also been identified.

A. hennigi had a widespread distribution in eastern and southern Africa but is a very rare species in several otherwise abundant fossil faunas. It is still understood primarily from single isolated and often broken elements, and much of its anatomy remains unknown. The most complete forefoot material of *A. hennigi* is that from Olduvai Bed I described by Butler (1965). George (1950) and Webb (1965) reported numerous (86), but typically broken, specimens from Makapansgat, South Africa. Most of these are craniodental remains, which provide good evidence for a crochet on upper molars (figure 33.3A), a character common in *Ancylotherium*. Webb (1965) also illustrated a well-developed, spatulate lower incisor, which is better developed than any of the reduced lower incisors in BSPG AS II 147, the Pikermi specimen of *A. pentelicum* in which this part of the jaw is best preserved. George (1950) originally made an M2 (figure 33.3A) from Makapansgat the type of a new species, *Metaschizotherium (?) transvaalensis*, but Butler saw no reason to separate it from *A. hennigi*, whose syntypes include a very similar upper molar (Dietrich, 1942). Known hindfoot elements include an astragalus (Laetoli; figure 33.3C) and two calcanea (Laetoli, Shungura). The relatively low astragalus nonetheless shows the typical (and probably plesiomorphic) schizotheriine character of articulating only with the navicular (Guérin, 1987), in contrast to astragali of derived Chalicotheriinae, which have both navicular and cuboid articulations. The Mt IV from Shungura (Guérin, 1985; see figure 33.3F) is extremely useful for comparisons with the same element in *A. cheboitense* and *"Chemositia" tugenensis*.

Guérin (in Leakey, 1987), named a new ichnotaxon, *Ancylotheriopus tanzaniae*, for two unusual, large footprints from Site C of the Laetoli footprint tuff and attributed them to *A. hennigi*. The clear, deep prints are tridactyl and clawed, with the middle digit longest, all characters that might be expected in footprints of *Ancylotherium*, and they seem to corroborate the idea that the phalanges were held hyperextended, having little contact with the ground. Guérin interpreted one print as a left hindfoot and the other a right forefoot, but with only two separated prints it is hard to be certain that this is correct. In particular, Guérin's figure 12.14B, which he interpreted as a metacarpograde forefoot print, looks more like a heeled plantigrade print. In sticky mud a hindfoot might more likely imprint in a plantigrade posture than a forefoot.

ANCYLOTHERIUM CHEBOITENSE
Guérin and Pickford, 2005
Figure 33.3C and 33.D

Holotype Kipsaraman Museum (CMK) BAR 323"01, a left articulated pes, including the astragalus, a calcaneum fragment, cuboid, coossified navicular and ectocuneiform, mesocuneiform, and Mt II–IV, from Cheboit in the Lukeino Formation.

Diagnosis Robust, proportionally short, metatarsals; all metatarsals with a strongly depressed anterior surface of the shaft.

Description Guérin and Pickford (2005) reported that the ectocuneiform and navicular are fused in the holotype. Whether this condition was widespread throughout the species is unknown, but it has not been reported in any other chalicothere. The species diagnosis of *A. cheboitense* by Guérin and Pickford (2005) noted a large tall astragalus. The astragalus (figure 33.3D) is indeed large (near the high end but within the known size range of *A. pentelicum*), but its height/width proportions are not remarkable, falling near the middle of the range for *A. pentelicum* (Geraads et al., 2006; Coombs, 2009). In contrast, the astragalus of *A. hennigi* (figure 33.3C) is proportionally shorter than in both *A. pentelicum* and *A. cheboitense*.

Age Late Miocene (holotype age 5.9–6.1 Ma).

Occurrence East Africa: Kenya.

Remarks Guérin and Pickford (2005) also referred a lower molar to this species, noting that it was high crowned, as in other species of *Ancylotherium*. They also considered a very damaged metapodial from the Mabaget Formation (Sagatia) as possibly referable to this species.

Interestingly, Guérin and Pickford (2005) did not consider the possibility that KNM-LU 845, a chalicothere proximal phalanx, and KNM-LU 929, a broken ungual phalanx, both referred to *A. hennigi* by Pickford (1979), might instead be referable to *A. cheboitense*. At present it is not possible to assign the two Lukeino phalanges unequivocally, because there are no phalanges associated with the type of *A. cheboitense*. However, in our opinion it is more parsimonious to refer them tentatively to *A. cheboitense* on the basis of geographic and age proximity. Because the Lukeino phalanges are the only previous basis for extending the range of *A. hennigi* down into the late Miocene, reassigning them as possible *A. cheboitense* narrows the confirmed range of *A. hennigi* to the Plio-Pleistocene. A broken lower molar, KNM-LU 844, has also been found at the same locality (2/225) in the Lukeino Formation.

The relatively sparse material of *A. cheboitense* provides a fascinating perspective on hindfoot modifications in African schizotheriine chalicotheres. It is curious, for example, that *A. cheboitense* has shortened metatarsals without a similarly modified astragalus, while *A. hennigi* shortened the astragalus but not the metatarsals. While assignment of *A. cheboitense* to *Ancylotherium* is justified on the basis of current evidence, we can only hope for additional remains to help elucidate the affinities and morphology of this species.

Genus *CHEMOSITIA* Pickford, 1979
CHEMOSITIA TUGENENSIS Pickford, 1979
Figures 33.3E–33.3G

Holotype KNM-MP 229, parts of a right hindlimb, including a femur fragment, partial astragalus and calcaneum, complete Mt IV (figure 33.3E), proximal and middle phalanx, from the Mpesida Formation of the Tugen Hills. KNM-MP 214, including a molar fragment that was originally included in the holotype, belongs to a suid (A. Hill, pers. comm.).

Diagnosis A medium-large schizotheriine chalicothere in which Mt IV is longer compared to width than in both *A. hennigi* and *A. cheboitense*; Mt IV apparently lacking any articulation with the ectocuneiform.

Age Late Miocene (holotype age 6.37–7.2 Ma).

Occurrence East Africa; Kenya.

Remarks No material other than the holotype is unequivocally attributed to this species. Coombs (1989) briefly reviewed "*Chemositia*" *tugenensis*, queried the "volar facets" of the proximal phalanx that Pickford (1979) had considered diagnostic, and expressed doubt about the validity of the genus *Chemositia*. At the same time, the Mt IV of "*C.*" *tugenensis* (figure 33.3E) does differ from that of both the other known African schizotheriines, *Ancylotherium hennigi* (figure 33.3F) and

A. cheboitense (figure 33.3G); its morphology and proportions are closer to those of Mt IV of *A. hennigi* but are unlikely to be confused with the latter. Thus the species may well be valid, and the generic name provides a convenient placeholder until the affinities of "*C.*" *tugenensis* become clearer. The most likely generic assignment would be either *Ancylotherium* or *Metaschizotherium*.

Thenius (1953) considered the European genus *Metaschizotherium* to be synonymous with *Ancylotherium*. Coombs (1974, 1989) continued this synonymy and treated *Metaschizotherium* as a basal subgenus within *Ancylotherium*. More recently, however, a detailed study of *Metaschizotherium bavaricum* (Coombs, 2009) has shown that synonymy of *Metaschizotherium* with *Ancylotherium* is incorrect. Although *Metaschizotherium* and *Ancylotherium* are clearly separate genera, Mt IV of "*C.*" *tugenensis* has proportions and morphology resembling those of *Metaschizotherium* from the middle Miocene of Europe. At the same time it could be a basal species within *Ancylotherium*, whose occurrence in the late Miocene of Eurasia is well documented. For now, the known remains of "*C.*" *tugenensis* are too fragmentary for assignment, so decisions on the synonymy of *Chemositia* must await additional, more diagnostic material.

OTHER AFRICAN CHALICOTHERIIDAE

The best-known African chalicotheres, *B. rusingensis* (Chalicotheriinae) and *A. hennigi* (Schizotheriinae), are well separated temporally. The schizotheriines "*C.*" *tugenensis* and *A. cheboitense* are useful additions to our knowledge of chalicotheres in the intervening time gap. Several additional finds provide tantalizing evidence of African chalicotheres but cannot be clearly assigned to a given species. Tsujikawa and Nakaya (2005) referred one such specimen to *B. rusingensis*, a metatarsal (H. Tsujikawa, pers. comm.) from the 14–15 Ma (Sawada et al., 1998) Aka Aiteputh Formation of Nachola, Samburu Hills (Kenya). A highly abraded, probable middle phalanx (KNM-MB 28161), which represents the only plausible chalicothere element from Maboko (Kenya), also extends the range of chalicotheres into the middle Miocene, but its affinity is difficult to determine.

Nakaya et al. (1984) referred a proximal phalanx (KNM-SH 12138) from the Namurungule Formation of the Samburu Hills, Kenya (9.5–9.6 Ma; Sawada et al., 1998), to *Ancylotherium* sp., while Geraads et al. (2002) referred a proximal phalanx (NMAA CHO1–10), from the Chorora Formation, Ethiopia (10.5 ± 0.5 Ma), to *Ancylotherium* cf. *tugenense*. Both phalanges are referable to the Schizotheriinae and expand the early temporal range of Schizotheriinae in Africa.

Pickford (1979) suggested that the Schizotheriinae appeared even earlier in Africa than is usually indicated and identified three chalicothere specimens from the early Miocene of Napak, Uganda, as possible schizotheriines. One of us (SMC) has reviewed this material and concluded that the proximal phalanx and distal metacarpal (UMP Nap V'61 and Nap V'61B) are referable to the chalicotheriine *B. rusingensis*, already identified from Napak. The proximal metapodial (UM Nap 1 58) does not seem to be chalicothere at all but corresponds to the Mt II of a rhinocerotid, possibly *Ougandatherium napakense* Guérin and Pickford, 2003. It articulates well with a rhinocerotid entocuneiform with the same collection number. The "abnormal premolar" (BMNH M21832) that was figured by Butler (1965; figures 33.3G and 33.3H) and cited by Pickford (1979) as a possible schizotheriine is not convincingly a member of this subfamily either. Therefore we conclude that the hypothesis of early Miocene Schizotheriinae in Africa is not supported.

Discussion and Conclusions

Butleria rusingensis, a member of the Chalicotheriinae, first appears in the African fossil record in the early Miocene, about 20 Ma (figure 33.1). It seems to have had an Asian origin, inasmuch as undoubted Chalicotheriinae are not known in Europe until about 15–16 Ma (late MN5; Heissig, 1999), while the basal chalicotheriine *"Ch." pilgrimi* occurs in the Oligo-Miocene Bugti Beds of Pakistan. It is unclear exactly how closely *B. rusingensis* and *"Ch." pilgrimi* are related, since *"Ch." pilgrimi* is known from few remains, and the characters shared by these two species are plesiomorphic for the subfamily. The deposits in which *B. rusingensis* is found range from subaerial to fluviatile to lacustrine and lake margin (Pickford, 1981), and paleoenvironments are generally interpreted as warm, humid forests in a volcanic terrain. There is no clear evidence that *B. rusingensis* survived past the early Miocene. Middle Miocene chalicothere fossils from Maboko and Nachola are either too poorly preserved or do not match perfectly with earlier material of *B. rusingensis*.

It is not altogether clear when the Schizotheriinae arrived in Africa, but the presence of a phalanx in Ethiopia at 10.5 Ma (Geraads et al., 2002) suggests that at least one representative was present by that time. Geographic origin is hard to pinpoint because related contemporaneous Miocene schizotheriines occur in both Europe and Asia. The well-known Eurasian schizotheriine species *Ancylotherium pentelicum*, typically found in MN 12–13 faunas (such as Pikermi and Samos) at ~8–5.5 Ma, has not been found in Africa, and the split (or splits) leading to the African taxa most likely preceded the appearance of that species.

Very little is known of the early history of the Schizotheriinae in Africa. The earliest named species *"C." tugenensis* (~ 7 Ma) and *A. cheboitense* (~ 6 Ma) seem to have lived in relatively moist environments with available forest. In particular, the Mpesida Beds include abundant silicified remnants of substantial tree trunks (Kingston et al., 2002), and the Cheboit locality in the Lukeino Formation has lateritic paleosols and preserves numerous colobine monkeys and forest-adapted bovids (Guérin and Pickford, 2005). In contrast, *A. hennigi* seems to have lived in a drier environment. Guérin (1985, 1987, 1994) has suggested a savanna environment with bushes and shrubs or gallery forest.

The rarity of chalicothere fossils makes it difficult to use their absence as an environmental indicator. No chalicothere fossils have been reported so far in some very abundant fossil faunas, such as East and West Turkana (Koobi Fora and Nachukui Formations, Kenya), Langebaanweg (South Africa), and the late Miocene and Pliocene deposits of Chad, but the reason for this is not clear.

By the end of the Miocene, chalicotheres were extinct everywhere except for two specialized lineages. Derived species of *Anisodon* (previously known as *Nestoritherium*; see Anquetin et al., 2007), a member of the Chalicotheriinae, survived into the early Pleistocene in forested environments of the Siwaliks and eastern Asia. In Africa, *Ancylotherium hennigi*, a member of the Schizotheriinae, also persisted into the Pleistocene, living in more open environments. Sometime after 1.33 Ma, *A. hennigi* also disappeared.

ACKNOWLEDGMENTS

We thank W. Sanders and L. Werdelin for their invitation to submit this chapter. E. Mbua and M. Mungu at the KNM and E. Kamuhangire and N. Abiti at the Uganda Museum kindly allowed access to collections under their care. We also thank A. Hill, T. Kunimatsu, D. Geraads, C. Guérin, T. White, G. Suwa, J.-R. Boisserie, L. MacLatchy, L. Hlusko, F. Guy, M. Pickford, and H. Tsujikawa for photographs or information about faunal ages and unpublished fossil material. D. Geraads reviewed the manuscript, providing some useful insights and suggestions. P. Getty and W. Coombs contributed helpful insights concerning trackways. S.M.C.'s research in East Africa was supported by the National Science Foundation (BCS-0524944), the Leakey Foundation, the Quaternary Association's Bill Bishop Award, and the Department of Anthropology at Harvard University.

Literature Cited

Andrews, C. W. 1923. An African chalicothere. *Nature* 112:696.

Anquetin, J., P.-O. Antoine, and P. Tassy. 2007. Middle Miocene Chalicotheriinae (Mammalia, Perissodactyla) from France, with a discussion of chalicotheriine phylogeny. *Zoological Journal of the Linnean Society* 151:577–608.

Bishop, W. W. 1967. The later Tertiary in east Africa: Volcanics, sediments and faunal inventory; pp. 31–56 in W. W. Bishop and J. D. Clark (eds.), *Background to Evolution in Africa*. University of Chicago Press, Chicago.

Bishop, W. W., J. A. Miller, and F. J. Fitch. 1969. New potassium-argon determinations relevant to the Miocene fossil mammal sequence in east Africa. *American Journal of Science* 267:669–699.

Bonis, L. de, G. Bouvrain, G. Koufos, and P. Tassy. 1995. Un crâne de chalicothère (Mammalia, Perissodactyla) du Miocène supérieur de Macédoine (Grèce): Remarques sur la phylogénie des Chalicotheriinae. *Palaeovertebrata* 24:135–176.

Boschetto, H. B., F. H. Brown, and I. McDougall. 1992. Stratigraphy of the Lothidok Range, northern Kenya and K/Ar ages of its Miocene primates. *Journal of Human Evolution* 22:47–71.

Butler, P. M. 1965. Fossil mammals of Africa No. 18: East African Miocene and Pleistocene chalicotheres. *Bulletin of the British Museum (Natural History), Geology* 10:165–237.

———. 1978. Chalicotheriidae; pp. 368–370 in V. J. Maglio and H. B. S. Cooke (eds.), *Evolution of African Mammals*. Harvard University Press, Cambridge.

Cooke, H. B. S., and S. C. Coryndon. 1970. Pleistocene mammals from the Kaiso Formation and other related deposits in Uganda; pp. 107–224 in L. S. B. Leakey and R. J. G. Savage (eds.), *Fossil Vertebrates of Africa*, vol. 2. Academic Press, London.

Coombs, M. C. 1974. Ein Vertreter von *Moropus* aus dem europäischen Aquitanien und eine Zusammenfassung der europäischen postoligozänen Schizotheriinae (Mammalia, Perissodactyla, Chalicotheriidae). *Sitzungsberichte der österreichische Akademie von Wissenschaften, Mathematik-naturwissenschaftlich Klasse, Abteilung I* 182:273–288.

———. 1975. Sexual dimorphism in chalicotheres (Mammalia, Perissodactyla). *Systematic Zoology* 24:55–62.

———. 1983. Large mammalian clawed herbivores: A comparative study. *Transactions of the American Philosophical Society* 73(7):1–96.

———. 1989. Interrelationships and diversity in the Chalicotheriidae; pp. 438–457 in D. R. Prothero and R. M. Schoch (eds.), *The Evolution of Perissodactyls*. Clarendon Press, New York.

———. 2009. The chalicothere *Metaschizotherium bavaricum* (Perissodactyla, Chalicotheriidae, Schizotheriinae) from the Miocene (MN 5) Lagerstätte of Sandelzhausen (Germany): description, comparison, and paleoecological significance. *Paläontologische Zeitschrift* 83:85–129.

Coombs, M. C., and G. M. Semprebon. 2005. The diet of chalicotheres (Mammalia, Perissodactyla) as indicated by low magnification stereoscopic microwear analysis. *Journal of Vertebrate Paleontology* 25(suppl. to no. 3):47A.

Deino, A. L., L. Tauxe, M. Monaghan, and A. Hill. 2002. ^{40}Ar/^{39}Ar geochronology and paleomagnetic stratigraphy of the Lukeino and lower Chemeron Formations at Tabarin and Kapcheberek, Tugen Hills, Kenya. *Journal of Human Evolution* 42:117–140.

Dietrich, W. O. 1942. Ältestquartäre säugetiere aus der südlichen Serengeti, Deutsch-Ostafrika. *Palaeontographica, Abt. A*, 94A:43–130.

Drake, R. E., and G. H. Curtis. 1987. K-Ar geochronology of the Laetoli fossil localities; pp. 48–52 in M. D. Leakey and J. M. Harris (eds.), *Laetoli: A Pliocene Site in Northern Tanzania*. Clarendon Press, Oxford.

Drake, R. E., J. A. Van Couvering, M. H. Pickford, G. H. Curtis, and J. A. Harris. 1988. New chronology for the early Miocene faunas of Kisingiri, western Kenya. *Journal of the Geological Society, London* 145:479–491.

Feibel, C. S., and F. H. Brown. 1991. Age of the primate-bearing deposits on Maboko Island, Kenya. *Journal of Human Evolution* 21:221–225.

Feibel, C. S., F. H. Brown, and I. McDougall. 1989. Stratigraphic context of fossil hominids from the Omo Group deposits: Northern Turkana Basin, Kenya and Ethiopia. *American Journal of Physical Anthropology* 78:595–622.

George, M. 1950. A chalicothere from the Limeworks Quarry of the Makapan Valley, Potgietersrust District. *South African Journal of Science* 46:241–242.

Geraads, D., Z. Alemseged, and H. Bellon. 2002. The late Miocene mammalian fauna of Chorora, Awash basin, Ethiopia: Systematics, biochronology and the ^{40}K-^{40}Ar ages of the associated volcanics. *Tertiary Research* 21:113–122.

Geraads, D., N. Spassov, and D. Kovachev. 2006. The Bulgarian Chalicotheriidae (Mammalia): an update. *Revue de Paléobiologie, Genève* 25:429–437.

Geraads, D., E. Tsoukala, and N. Spassov. 2007. A skull of *Ancylotherium* (Chalicotheriidae, Mammalia) from the late Miocene of Thermopigi (Serres, N. Greece) and the relationships of the genus. *Journal of Vertebrate Paleontology* 27:461–466.

Guérin, C. 1976. Rhinocerotidae and Chalicotheriidae (Mammalia, Perissodactyla) from the Shungura Formation, Lower Omo Basin; pp. 214–221 in Y. Coppens, F. C. Howell, G. L. I. Isaac, and R. E. F. Leakey (eds.), *Earliest Man and Environments in the Lake Rudolf Basin: Stratigraphy, Paleoecology, and Evolution*. University of Chicago Press, Chicago.

———. 1985. Les rhinoceros et les chalicothères (Mammalia, Perissodactyla) des gisements de la Vallée de l'Omo en Éthiopie; pp. 67–95 in Y. Coppens and F. C. Howell (eds.), *Les Faunes Plio-Pléistocènes de la Basse Vallée de l'Omo (Éthiopie) : Tome 1. Les Périssodactyles, Les Artiodactyles (les Bovidae), Expédition internationale 1967–1976*. Cahiers de Paléontologie, Travaux de Paléontologie Est-africaine, Éditions du Centre National de la Recherche Scientifique, Paris.

———. 1987. Chalicotheriidae (Mammalia, Perissodactyla) remains from Laetoli; pp. 315–320 in M. D. Leakey and J. M. Harris (eds.), *Laetoli: A Pliocene Site in Northern Tanzania*. Clarendon Press, Oxford University Press, Oxford.

———. 1994. Les Chalicotheriidae (Mammalia, Perissodactyla) du Néogène de l'Ouganda. pp. 281–287 in B. Senut and M. Pickford (eds.), *Geology and Palaeobiology of the Albertine Rift Valley, Uganda-Zaire: Vol. II. Palaeobiology*. CIFEG Occasional publications, Orléans.

Guérin, C., and M. Pickford. 2003. *Ougandatherium napakense* nov. gen. nov. sp., le plus ancien Rhinocerotidae Iranotheriinae d'Afrique. *Annales de Paléontologie* 89:1–35

———. 2005. *Ancylotherium cheboitense* nov. sp., nouveau Chalicotheriidae (Mammalia, Perissodactyla) du Miocène supérieur des Tugen Hills (Kénya). *Comptes Rendus Palevol* 4:225–234.

Heissig, K. 1999. Family Chalicotheriidae; pp. 189–192 in G. E. Rössner and K. Heissig (eds.), *The Miocene Land Mammals of Europe*. Pfeil, Munich.

Hooijer, D. A. 1963. Miocene Mammalia of Congo. *Annales Musée Royal de l'Afrique Centrale, Tervuren, Belgium, Series In 8°, Sciences Géologiques* 46:1–77.

———. 1972. A late Pliocene rhinoceros from Langebaanweg, Cape Province. *Annals of the South African Museum* 59:151–191.

———. 1973. Additional Miocene to Pliocene rhinoceroses of Africa. *Zoologische Mededelingen, Leiden* 46:149–178.

———. 1975. Note on some newly found perissodactyl teeth from the Omo Group deposits, Ethiopia. *Proceedings Koninklijke Nederlandse Akademie van Wetenschappen, series B, Physical Sciences* 78:186–190.

Hopwood, A. T. 1926. Fossil Mammalia; pp. 13–26 in E. J. Wayland (ed.), *The Geology and Paleontology of the Kaiso Bone-beds*. Occasional Papers of the Geological Survey of Uganda, vol. 2, Entebbe.

———. 1951. The Olduvai fauna; pp. 20–24 in L. S. B. Leakey (ed.), *Olduvai Gorge: A Report on the Evolution of the Hand-axe Culture in Beds I–IV*. Cambridge University Press, Cambridge.

Kingston, J. D., B. F. Jacobs, A. Hill, and A Deino. 2002. Stratigraphy, age and environments of the late Miocene Mpesida Beds, Tugen Hills, Kenya. *Journal of Human Evolution* 42:95–116.

Koenigswald, G. H. R. von. 1932. *Metaschizotherium fraasi* n. g. n. sp., ein neuer Chalicotheriide aus dem Obermiocän von Steinheim a. Albuch. Bemerkungen zur Systematik der Chalicotheriiden. *Palaeontographica* (suppl.) 8:1–24.

Leakey, L. S. B. 1951. The significance of the geological and palaeontological evidence; pp. 25–33 in L. S. B. Leakey (ed.), *Olduvai Gorge: A Report on the Evolution of the Hand-axe Culture in Beds I–IV*. Cambridge University Press, Cambridge.

Leakey, M. D. 1987. Animal prints and trails; pp. 451–489 in M. D. Leakey and J. M. Harris (eds.), Laetoli: A Pliocene Site in Northern Tanzania. Clarendon Press, Oxford.

Nakaya, H. 1994. Faunal change of late Miocene Africa and Eurasia: Mammalian fauna from the Namurungule Formation, Samburu Hills, Northern Kenya. *African Study Monographs* (suppl.) 29:1–112.

Nakaya, H., M. Pickford, Y. Nakano, and H. Ishida. 1984. The late Miocene large mammal fauna from the Namurungule Formation, Samburu Hills, northern Kenya. *African Study Monographs* (suppl.) 2:87–131.

Pickford, M. 1975. Another African chalicothere. *Nature* 253:85.

———. 1979. New evidence pertaining to the Miocene Chalicotheriidae (Mammalia, Perissodactyla) of Kenya. *Tertiary Research* 2:83–91.

———. 1981. Preliminary Miocene mammalian biostratigraphy for Western Kenya. *Journal of Human Evolution* 10:73–97.

Pickford, M., and P. Andrews. 1981. The Tinderet Miocene sequence in Kenya. *Journal of Human Evolution* 10:11–33.

Roussiakis, S. J., and G. E. Theodorou. 2001. *Ancylotherium pentelicum* (Gaudry & Lartet, 1856)(Perissodactyla, Mammalia) from the classic locality of Pikermi (Attica, Greece), stored in the Palaeontological and Geological Museum of Athens. *Geobios* 34:563–584.

Sawada, Y., M. Pickford, T. Itaya, T. Makinouchi, M. Tateishi, K. Kabeto, S. Ishida, and I. Ishida. 1998. K-Ar ages of Miocene Hominoidea (*Kenyapithecus* and *Samburupithecus*) from Samburu Hills, northern Kenya. *Comptes Rendus de l'Académie des Sciences Paris, Sciences de la Terre et des Planètes* 326:445–451.

Schaub, S. 1943. Die Vorderextremität von *Ancylotherium pentelicum* Gaudry und Lartet. *Schweizerischen Palaeontologischen Abhandlungen* 64:1–36.

Schulz, E., J. M. Fahlke, G. Merceron, and T. Kaiser. 2007. Feeding ecology of the Chalicotheriidae (Mammalia, Perissodactyla, Ancylopoda). Results from dental micro- and mesowear analyses. *Verhandlungen des naturwissenschaftlichen Vereins Hamburg* (N.F.) 43:5–31.

Suwa, G., H. Nakaya, B. Asfaw, H. Saegusa, A. Amzaye, R. Kono, Y. Beyene, and S. Katoh. 2003. Plio-Pleistocene terrestrial mammalian assemblage from Konso, southern Ethiopia. *Journal of Vertebrate Paleontology* 23:901–916.

Thenius, E. 1953. Studien über fossile Vertebraten Griechenlands. III. Das Maxillargebiss von *Ancylotherium pentelicum* Gaudry und Lartet. *Annales Géologiques des Pays Hélléniques* 5:97–106.

Tsujikawa, H. 2005. The updated late Miocene large mammal fauna from Samburu Hills, northern Kenya. *African Study Monographs* (suppl.) 32:1–50.

Tsujikawa, H., and H. Nakaya. 2005. [Geologic age and palaeoenvironments of mammalian faunas from Samburu Hills, northern Kenya.] *Earth Monthly* 27:603–611. [In Japanese]

Webb, G. L. 1965. Notes on some chalicothere remains from Makapansgat. *Palaeontologia Africana* 9:49–73.

Xue, X.-X., and M. C. Coombs. 1985. A new species of *Chalicotherium* from the upper Miocene of Gansu Province, China. *Journal of Vertebrate Paleontology* 5:336–344.

Zapfe, H. 1979. *Chalicotherium grande* (Blainv.) aus der miozänen Spaltenfüllung von Neudorf an der March (Devinská Nová Ves), Tschechoslowakei. *Neue-Denkschriften des Naturhistorischen Museums in Wien* 2:1–282.

THIRTY-FOUR

Rhinocerotidae

DENIS GERAADS

Among the Perissodactyla, Rhinocerotidae have traditionally been allied with tapirs because they lack a mesostyle, even though other primitive perissodactyls may also lack it (Hooker and Dashzeveg, 2004). The upper cheek teeth are π-shaped (figure 34.1) except M3, which is triangular. The incisors are separated from the cheek teeth by a diastema, as there is no canine; they consist of a chisel-shaped I1, borne by a slender premaxilla, a tusk-shaped i2, plus much smaller I2 and i1. However, I1, or both I1 and i2, become reduced or disappear in several lineages. Nasal and sometimes frontal horns, consisting of agglomerated hair (thus rarely fossilized), grow on more or less recognizable skull bosses in many genera; they are usually inserted behind one another but may rarely sit side by side. Although extensively pneumatized, the skull is robust, with thick bone and sutures fused in adulthood, and this certainly accounts for the good fossil record of the family. The temporal fossa is long, but the cranial base is shortened. The mandible has a transversely elongated condyle, plus an extra articular facet for the postglenoid process. The latter may be united with the posttympanic process beneath the auditory foramen. Horned forms (roughly the Rhinocerotini of Prothero et al. [1986], Rhinocerotinae of Cerdeño [1995], or Rhinocerotina of Antoine [2002]) lack a mastoid exposure, but it may have been present (as in the tapirs), in some hornless forms. They have three digits in the posterior limb, and three or four (the fifth digit being reduced but functional) in the anterior one. Dental terminology is shown in figure 34.1. The various stages of premolar molarisation are shown in figure 34.1C.

During the past two decades, various attempts have been made to resolve the phyletic relationships within the family. Almost every author agrees that this is a difficult task, mainly owing to the dearth of clearly identifiable synapomorphies, and the broad divergences in the published cladograms confirm this. The most parsimonious recent cladistic analyses, using no less than 282 characters (Antoine, 2002; Antoine et al., 2003), unite under the Rhinocerotini (which includes the bulk of the Rhinocerotinae) as an unresolved trichotomy, the Teleoceratina (Old and New World brachypotheres), the Aceratheriina (Old World aceratheres and related forms), and the Rhinocerotina (nonelasmothere Old World horned rhinos); the Elasmotheriini are the sister

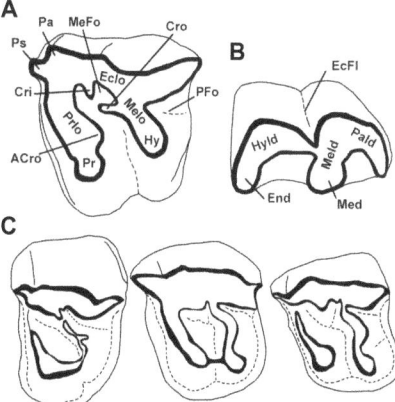

FIGURE 34.1 A) Terminology of upper tooth elements: ACro: antecrochet; Cri: crista; Cro: crochet; Eclo: ectoloph; Hy: hypocone; MeFo: medifossette; Melo: metaloph; Pa: paracone fold; PFo: post-fossette; Pr: protocone; Prlo: protoloph; Ps: parastyle. B) Terminology of lower tooth elements: EcFl: ectoflexid; End: entoconid; Hyld: hypolophid; Med: metaconid; Meld: metalophid; Pald: paralophid. C) Morphology of upper premolars; from left to right: submolariform, semimolariform, molariform (from Heissig, 1969).

group of the Rhinocerotinae. Even though many criticisms can be made of parsimony analysis (choice of characters, of coding, of number of states, of equal weighing, subjectivity of state control, etc.), which leads to significantly different results even when performed by renowned specialists, the phylogenies proposed by Antoine et al. (2002, 2003) can be used as working hypotheses.

There are five living species, all of them seriously threatened or even close to extinction. The small two-horned *Dicerorhinus sumatrensis*, found in Sumatra and the Malaysian peninsula, numbers at most a few hundred surviving individuals. Of the two single-horned species of *Rhinoceros*, *R. sondaicus* and *R. unicornis*, also from southeastern Asia, the former is the most seriously threatened, with perhaps 60 animals remaining in the wild. The African forms, *Ceratotherium* simum and *Diceros bicornis*, are closely related (infra-tribe Dicerotini). Some morphological cladistic analysis (Groves, 1983; Prothero et al., 1986; Cerdeño, 1995; but not Geraads, 1988 and Antoine et al., 2003) and mitochondrial gene sequencing (Tougard et al., 2001) suggest that, among living forms, African rhinos are the sister group of *Dicerorhinus + Rhinoceros*, but more molecular analyses would be welcome.

During the Miocene, African rhinos underwent a diversification comparable to those of the northern continents, but they have received much less attention than the Eurasian forms, especially from systematic and phylogenetic aspects. A number of specific studies, especially by Guérin and Hooijer, have appeared in the last decades, but the last broad review is 30 years old (Hooijer, 1978). As a result, the commonly used taxonomy in Africa is one which was in use a long time ago in Europe, where the meanings of the generic names *Brachypotherium*, *Aceratherium* and *Dicerorhinus* are now much more restricted than they used to be (Heissig, 1999). As in Eurasia, where many species have been wandering through several genera, the phylogeny and systematics of African rhinos are still confused. Much new material, a large part of it still unpublished, has come to light in recent decades, and there is little doubt that serious revisions of the African rhinos are needed. The present account takes a rather conservative view; I have tried to update the systematics, and raise a few phyletic issues, but this account should not be considered as more than preliminary.

ABBREVIATIONS

BMNH, Natural History Museum, London; FSL, Faculté des Sciences, Lyon; KNM, Kenya National Museums, Nairobi; MNHN, Muséum National d'Histoire Naturelle, Paris; NME, National Museum of Ethiopia, Addis Ababa.

Systematic Paleontology

Family RHINOCEROTIDAE Gray, 1821
Subfamily RHINOCEROTINAE Gray, 1821
Tribe RHINOCEROTINI Gray, 1821
Subtribe TELEOCERATINA Hay, 1902
Genus *BRACHYPOTHERIUM* Roger, 1904

Type Species Brachypotherium goldfussi (Kaup, 1834), from the early late Miocene (Vallesian) of Eppelsheim, Germany.

Diagnosis Large rhinos with broad and low skull, short hornless nasals, orbit far forward, powerful anterior dentition and especially large I1s with a short root, brachyodont cheek teeth and short but broad premolars. Upper and lower molars tend to have flattened labial walls and the latter have shallow ectoflexids. Short massive terminal limb segments, with a characteristically low talus.

BRACHYPOTHERIUM nov. sp.?
Figure 34.2

Some fossils from Buluk (= West Stephanie) in northern Kenya, collected and kindly made available to me by E. Miller, apparently belong to a new species. The best specimen is a relatively complete skull, KNM-WS-46072 (figure 34.2), which is low and broad, especially in the occipital area, with an almost flat cranial profile, a deep zygoma, and short hornless nasals. The short and broad premolars match those of the brachypotheres, and there are several typical brachypothere

FIGURE 34.2 *Brachypotherium* nov.sp. ?, skull KNM-WS 46072 from the lower Miocene of Buluk, Kenya. © Publications Scientifiques dyu Muséum national d'Histoire naturelle, Paris (Cerdeño, 1993).

upper incisors and tali in the Buluk collection. The size is that of the small European brachypothere *Prosantorhinus*, but this genus has a saddle-shaped skull, with a sagittal crest and a small nasal horn (Heissig, 1972). The Buluk skull is more like *Brachypotherium brachypus* from Europe (e.g., Cerdeño, 1993: plate 4, figure 11), and, pending detailed study, I tentatively include it in this genus, although the Kenyan material is certainly younger.

BRACHYPOTHERIUM SNOWI Fourtau, 1918
Figure 34.3

Synonymy Aceratherium campbelli Hamilton, 1973.
Type Maxilla figured by Fourtau (1920, fig. 26); housed in the Cairo Geological Museum.
Type Locality Wadi Moghara, Egypt, ca. 17–18 Ma. (Miller, 1999).
Diagnosis A *Brachypotherium* of large size (length of cheek tooth row about 270 mm); skull low and wide, nasals rather long, probably carrying a small (pair of) horn(s), very broad zygomatic arches, temporal lines almost fused into a sagittal crest, dorsal profile strongly concave, occipital rounded, nasal notch above front of P3, anterior border of orbit above M2.
Remarks Brachypotherium snowi was established by Fourtau on the basis of a maxilla with worn teeth and the socket of the upper incisor, plus a fragment of mandible and some teeth. He pointed out the large size of the animal, the shortness and great width of the upper premolars, their lack of a labial cingulum and the reduction of the lingual one (a difference from European brachypotheres), and the moderate development of the antecrochet on all teeth. On the lower teeth he noticed the lack of cingula and of labial flattening, the large size of i2, and the presence of i1. A referred third metatarsal is stout, but not extremely so.

Several specimens from Jebel Zelten, Libya, a set of localities probably mostly dating to about 16 Ma., were referred to this species by Hamilton (1973). The i2s are large and separated by minute i1s; the cingulum is reduced on the upper teeth; P2 is much narrower than P3, which is broad. A third metacarpal is smaller than the Mt III from Moghara. Most of the specimens described by Hamilton as *Aceratherium campbelli* also belong here, as first recognized by Gentry (1987: 430). The holotype skull of the latter species, as well as another, uncollected skull (Hamilton, 1973: plate 3) are clearly from brachypotheres, as shown by their large size, skull regularly broadening from front to rear, with very robust zygomatic arches and posteriorly very broad, low rounded occipital surface, short, broad upper teeth with flat labial walls, and large upper incisor. Several of the diagnostic features of the species are based upon these specimens. The nasals, if correctly identified by Hamilton, decrease in width toward the anterior end, but are rather thick and broad. Paired dorsal swellings suggest that some kind of horn may have been present.

In East and South Africa, this species has been called *B. heinzelini*, but this name should be restricted to the type specimen (discussed later). It is represented by sparse remains from Rusinga (Hooijer, 1966), and has been reported from a few other sites. The most complete specimen is an unpublished mandible from Mwiti (Kajong), Kenya, dated to ca. 16–17 Ma (figure 34.3). It has a straight ventral edge and a widely expanded angular area, as in the European *B. brachypus* (Cerdeño, 1993: plate 5, figure 12); the large i2s are followed by a long diastema; the cheek teeth are brachyodont and have a shallow ectoflexid; the premolars are short, and the missing p2 was certainly small.

As in other brachypotheres, the talus may be very characteristic in its broad and low proportions at Jebel Zelten (Hamilton, 1973: plate 6, figure 7), but, as in Eurasia, the distinction from other rhinos may not always be so clear-cut. The talus from Gumba (Hooijer, 1966: plate 14, figure 3) is high and might not belong to this genus.

Brachypotherium snowi shows some resemblances to the contemporaneous European *Prosantorhinus* (Heissig, 1972; Cerdeño, 1996), but the latter has well-marked terminal horn bosses on the nasals, probably a lower and broader skull, and metapodials that are still shorter.

BRACHYPOTHERIUM LEWISI Hooijer and Patterson, 1972

Synonymy ?*Brachypotherium heinzelini* Hooijer, 1963
Type Skull KNM-LT 88.
Type Locality Lower member of the Nawata Formation at Lothagam, Kenya (Hooijer and Patterson, 1972).
Diagnosis Mostly from Hooijer and Patterson (1972). Size very large: condylobasal length of type skull over 70 cm, anterotransverse diameters of M1–2 some 90 mm as opposed to 70 mm in *B. snowi*. Nasals hornless, slender, not very long, deepest point of nasomaxillary notch above P4, anterior border of orbit above anterior end of M2, frontals flat and hornless, inferior squamosal processes united below sub-

aural channel. Upper incisors very large, upper cheek teeth brachyodont, ectoloph flattened behind paracone style, antecrochet moderate, protocone constriction slight, external cingula often present. Lower i2s of small to moderate size, brachyodont cheek teeth, external cingula often developed. Trochanter tertius of femur strongly developed.

Differs from *B. snowi* in its larger size, straight dorsal cranial profile, dorsal orbital border at least as high as the skull roof, V-shaped choanae, nasal notch deeper, shorter diastemas, lack of i1, smaller i2s.

Remarks The material from Lothagam (Hooijer and Patterson, 1972) includes a rather complete but crushed skull, and a second, less deformed skull lacking most of the teeth; a few more specimens were added more recently (Harris and Leakey, 2003). The material is basically similar to that of *B. snowi* but differs in the characters mentioned in the diagnosis.

Metacarpals from Lothagam are larger than those of *B. snowi* from Jebel Zelten or Rusinga, but not significantly different in their proportions; a molar from Sahabi, Libya (d'Erasmo, 1954), probably of similar age, is truly gigantic. The talus (Hooijer 1963: plate 5, figure 10) is larger and more trapezoidal than that of *B. snowi*.

Brachypotherium heinzelini was established on a P4 from Sinda-Ongoliba (Zaire), as well as on some tooth fragments and a talus, all supposed by Hooijer (1963) to be of early Miocene age. The P4 was mainly characterized by the presence of a labial cingulum, flattened ectoloph, and weak antecrochet, the first of these features being the main distinction from *B. snowi*. It has been shown since (Pickford et al., 1993) that Sinda is probably of latest Miocene age; thus, *B. heinzelini* should rather be compared with *B. lewisi*, and Pickford et al. (1993: 109) suggested that these names may be synonymous. However, the labial cingulum is "virtually absent" on the type of *B. lewisi* (Hooijer and Patterson, 1972: 5), while another difference between them is size, *B. lewisi* being larger, though if the type of *B. heinzelini* is a P3, not a P4, this difference would vanish. If the two names are synonymous, *B. heinzelini* has priority, and some confusion would arise, as this name as hitherto been widely given to early and middle Miocene forms. To avoid confusion, this name should be restricted to the type specimen, while other specimens hitherto called *B. heinzelini* can be referred to *B. snowi*.

Brachypotherium lewisi is best known from the late Miocene, and the transition from *B. snowi* is poorly documented (table 34.1). The latest definite record of the genus is from the Upper Member of the Nawata Formation of Lothagam, dated to ca. 6 Ma, but a possible later record is from the Apak Member, dated to ca. 4.2 Ma (Harris and Leakey, 2003); the extinction of *Brachypotherium* therefore took place about 2 to 4 Ma later than in Europe.

Subtribe ACERATHERIINA Dollo, 1885
Genus *PLESIACERATHERIUM* Young, 1937

Type Species *Plesiaceratherium gracile* Young, 1937.
Diagnosis Modified from Yan and Heissig (1986). Medium-sized to large Aceratheriini with primitive type of skull and dentition. Upper incisors reduced but still shearing against the lower ones in some species. Lower i2 flattened, horizontal and weakly curved. Skull hornless, with deep nasal notch and narrow braincase. Upper cheek teeth with

20 cm

Figure 34.3 *Brachypotherium* cf. *snowi*, mandible KNM-MI 3 from the lower Miocene of Mwiti, Kenya.

TABLE 34.1
List of the main African fossil localities with Rhinocerotidae

Many ages are estimates, not necessarily supported by absolute dating.

Site	Country	Age (Ma)	Key References	Published Identifications	Present Identifications
Haua Fteah	Libya	0.1	Klein and Scott, 1986	*C. simum*; *D. mercki*	*C. simum*; *S. mercki*
Bouknadel	Morocco	0.1	Michel, 1992	*C. simum*; *D. hemitoechus*	*Ceratotherium* sp.; *D. hemitoechus*
Aïn Bahya, Doukkala	Morocco	0.2	Michel, 1992	*C. simum*; *D. hemitoechus*	*C. mauritanicum?*
Isenya	Kenya	0.5	Brugal and Denys, 1989	Rhinocerotidae	Rhinocerotidae indet.
Grotte des Rhinocéros	Morocco	0.5	Raynal et al., 1993	*C. mauritanicum*	*C. mauritanicum*
Duinefontein	South Africa	0.5	Klein et al., 1999	*D. bicornis*; *C. simum*	*D. bicornis*; *C. simum*
Asbole	Ethiopia	0.6	Geraads et al, 2004	*Diceros* sp.	*Diceros* sp.; *C. simum*
Elandsfontein (Hopefield)	South Africa	0.6	Hooijer and Singer, 1960	*D. bicornis*; *C. simum*	*D. bicornis*; *C. simum*
M. Awash-Bodo	Ethiopia	0.7	Kalb et al., 1980	Rhinocerotidae	Rhinocerotidae
Tighenif	Algeria	0.7	Geraads et al., 1986	*C. simum*	*C. mauritanicum*
Olorgesailie	Kenya	0.9	Hooijer, 1969	*C. simum*	*C. simum*
Buia	Eritrea	1	Martinez-Navarro et al., 2004	*C. simum*	*C. simum*
Bouri Daka	Ethiopia	1	Asfaw et al., 2002	*Ceratotherium* sp.	*Ceratotherium* sp.
Kanjera Fm. (N)	Kenya	1	Pickford, 1986; Ditchfield et al., 1999	*D. bicornis*; *C. simum*	*D. bicornis*; *C. simum*
Olduvai upper Bed II, III, IV	Tanzania	1	Hooijer, 1969	*D. bicornis*; *C. simum*	*D. bicornis*; *C. simum*
Aïn Hanech	Algeria	1.4	Arambourg, 1970	*C. simum germanoafricanum*	*C. mauritanicum*
Chemoigut	Kenya	1.5	Bishop et al., 1975	*Ceratotherium* sp.	*Ceratotherium* sp.
Anabo Koma	Djibouti	1.6	Bonis et al., 1988	*Ceratotherium* sp.	*C. mauritanicum*
Peninj	Tanzania	1.7	Geraads, 1987	*C. simum*	*C. simum*
Konso Fm.	Ethiopia	1.8	Suwa et al., 2003	*D. bicornis*; *C. simum*	*D. bicornis*; *C. simum*
Olduvai Bed I, lower Bed II	Tanzania	1.8	Hooijer, 1969	*C. simum*	*C. simum*
Nyabusosi	Uganda	1.8	Guérin, 1994b	*D. bicornis*	Dicerotini
Aïn Boucherit	Algeria	2.0	Arambourg, 1970	*C. simum mauritanicum*	*C. mauritanicum*
Baard's Quarry lower levels	South Africa	2.0	Hendey, 1978	*D. bicornis*; *Ceratotherium* sp.	*D. bicornis*; *Ceratotherium* sp.
Semliki—Lusso	Congo	2.1	Boaz et al., 1992	cf. *Ceratotherium* sp.	*B. lewisi?*; Rhinocerotidae indet.
Koobi Fora	Kenya	2.5	Harris, 1983	*D.bicornis*; *C.praecox*; *C. simum*	*D. praecox*; *D. bicornis*; *C. mauritanicum*
Ahl al Oughlam	Morocco	2.5	Geraads, 2006	*C. mauritanicum*	*C. mauritanicum*
Laetoli—Upper Ndolanya	Tanzania	2.6	Kovarovic et al., 2002	*C. simum*	*C. mauritanicum*
Hohwa	Uganda	2.6	Guérin, 1994b	*C. praecox*	*C. mauritanicum?*
Rawi Fm.	Kenya	2.8	Ditchfield et al. 1999	*C. simum*	*C. mauritanicum?*
Omo	Ethiopia	3.0	Hooijer, 1973; Guérin,1985; Hooijer and Churcher, 1985	*D. bicornis*; *C. simum*	*Diceros* sp.; *C. mauritanicum*
Hadar—Kada Hadar	Ethiopia	3.0	Geraads, 2005	*D. praecox*; *C. mauritanicum*	*D. praecox*; *C. mauritanicum*
West Turkana	Kenya	3.0	Harris et al, 1988	*D. bicornis*; *Ceratotherium* sp.; *C. simum*	*Diceros* sp.; *C. mauritanicum*
Lothagam-Kaiyumung	Kenya	3.0	Harris and Leakey, 2003	*C. praecox*	*D. praecox*
Makapansgat	South Africa	3.0	Hooijer, 1958	*D. bicornis*; *C. simum*	*Diceros* sp.; *Ceratotherium* sp.
Aïn Brimba	Tunisia	3.0	Arambourg, 1970	*C. simum germanoafricanum*	*C. mauritanicum*
Koro Toro 13	Chad	3.2	Likius, 2002	*D.* cf. *bicornis*; *C. praecox*; *Stephanorhinus* sp.	*C. mauritanicum*; *Stephanorhinus* sp.
Hadar—Denen Dora	Ethiopia	3.2	Geraads, 2005	*D. praecox*; *C. mauritanicum*	*D. praecox*; *C. mauritanicum*
Turkwell South	Kenya	3.2	Ward et al., 1999	Rhinocerotidae	Rhinocerotidae indet.
Hadar—Sidi Hakoma	Ethiopia	3.3	Geraads, 2005	*D. praecox*; *C. mauritanicum*	*D. praecox*; *C. mauritanicum*

Site	Country	Age (Ma)	Key References	Published Identifications	Present Identifications
Ekora	Kenya	3.5	Hooijer and Patterson, 1972	*C. praecox*	*D. praecox*
Laetoli	Tanzania	3.6	Guérin, 1987	*D. bicornis*; *C. praecox*	*D.* cf. *praecox*; *C. mauritanicum*
Kanapoi	Kenya	4.0	Hooijer and Patterson, 1972	*C. praecox*	*D. praecox*
Ichkeul	Tunisia	4.0	Arambourg, 1970	*C. simum*; *D. africanus*	*C. mauritanicum*; *Stephanorhinus* sp.
Kanam East and West	Kenya	4.3	Hooijer, 1969; Pickford, 1987	*D. bicornis*; *C. simum germanoafricanum*	*Diceros* sp.?; *C. mauritanicum?*
Lothagam-Apak	Kenya	4.3	Harris and Leakey, 2003	*D. bicornis*; *C. praecox*; *B. lewisi*	*D. praecox*; *Ceratotherium* sp.; *B. lewisi*
Manonga-Kiloleli	Tanzania	4.3	Harrison and Baker, 1997	*C. praecox*	*Ceratotherium* sp.
M. Awash-Aramis	Ethiopia	4.4	WoldeGabriel et al., 1994	*C.* cf. *praecox*	Rhinocerotidae indet.
Chemeron	Kenya	4.5	Hooijer, 1973; Guérin, 2000	*C. simum, C.praecox*; *B. heinzelini*; *D. leakeyi*; *A. acutirostratum*	*C. mauritanicum*
Kollé	Chad	4.5	Likius, 2002	*D.* cf. *bicornis*; *C. praecox*	*C. mauritanicum*; Rhinocerotidae indet.
Warwire	Uganda	4.5	Guérin, 1994b	*D. bicornis*; *C. praecox*	Dicerotini
Kossom Bougoudi	Chad	5.0	Likius, 2002	*D.* cf. *bicornis*	Rhinocerotidae indet.
Hamada Damous	Tunisia	5.0	Coppens, 1971	*C. simum*	Rhinocerotidae indet.
Nkondo	Uganda	5.0	Guérin, 1994b	*D. bicornis*; *C. praecox*	Dicerotini
Langebaanweg PPM	South Africa	5.1	Hooijer, 1972; Hendey, 1981	*C. praecox*	*Ceratotherium* sp.
Langebaanweg QSM	South Africa	5.2	Hooijer, 1972; Hendey, 1981	*C. praecox*	*Ceratotherium* sp.
M. Awash-late Miocene	Ethiopia	5.5	Giaourtsakis et al., 2009	*Diceros* sp.; *D. douariensis*	*Diceros?* sp.
Lukeino A-B	Kenya	6.0	Pickford and Senut, 2001	*Diceros?*; *C. praecox*	*Brachypotherium* sp.?
Lissasfa	Morocco	6.0	Raynal et al., 1999	Rhinocerotidae	*Ceratotherium* sp.
Hondeklip	Namibia	6.0	Pickford and Senut, 1997	*C. praecox*	*Ceratotherium* sp.?
Lothagam-upper Nawata	Kenya	6.5	Harris and Leakey, 2003	*D. bicornis*; *C. praecox*; *B. lewisi*	*Ceratotherium* sp.; *B. lewisi*
Mpesida	Kenya	6.5	Hooijer, 1973 ; Kingston et al., 2002	*C. praecox*; *B. lewisi*	*Ceratotherium* sp.; *B. lewisi*
Menacer (Marceau)	Algeria	7.0	Thomas and Petter, 1986	Rhinocerotidae	Rhinocerotidae indet.
Sinda	Congo	7.0	Hooijer, 1966; Guérin, 2000	*B. heinzelini*; *A. acutirostratum*	Rhinocerotidae indet.
Lothagam-lower Nawata	Kenya	7.0	Hooijer and Patterson, 1972; Harris and Leakey, 2003	*C. praecox*; *B. lewisi*	*Ceratotherium* sp.; *B. lewisi*
Sahabi	Libya	7.0	d'Erasmo, 1954; Bernor et al., 1987	*D. neumayri*; *Brachypotherium* sp.	*Brachypotherium* sp.; *C. douariense?*
Douaria	Tunisia	7.0	Guérin, 1966	*D. douariensis*	*C. douariense*; Rhinocerotidae indet.
Karugamania	Congo	8.0	Guérin, 2000	*B. heinzelini*; *A. acutirostratum*	*B. snowi?*; Rhinocerotidae indet.
Oued-Mya-1	Algeria	9.0	Sudre and Hartenberger, 1992	*Aceratherium* sp.	Dicerotini?
Ngeringerowa	Kenya	9.0	Pickford, 1983	Rhinocerotidae	Rhinocerotidae indet.
Namurungule	Kenya	9.5	Nakaya et al., 1987; Nakaya, 1993	*Paradiceros* sp.; *Kenyatherium bishopi*; *Chilotheridium* sp.	*Ceratotherium* sp.?; *Kenyatherium bishopi?*
Nakali	Kenya	9.5	Aguirre and Guérin, 1974; Antoine, 2002	*Kenyatherium bishopi*	*Kenyatherium bishopi*
Bou Hanifia	Algeria	10.0	Arambourg, 1959	*D. primaevus*	*C.* cf. *primaevum*
Ngorora E	Kenya	10.0	Hooijer, 1971; Guérin, 2000	*B. lewisi*; *Aceratherium* or *Dicerorhinus*; *C. pattersoni*	*Brachypotherium* sp.; Elasmotheriinae?
Bled Douarah (upper Beglia Fm.)	Tunisia	10.0	Robinson and Black, 1974	Rhinocerotidae	Rhinocerotidae indet.
Djebel Krechem	Tunisia	10.0	Geraads, 1989	*D.* cf. *douariensis*; *B.* cf. *lewisi*	*C. douariense?*; Rhinocerotidae indet.

TABLE 34.I
(CONTINUED)

Site	Country	Age (Ma)	Key References	Published Identifications	Present Identifications
Chorora	Ethiopia	10.5	Geraads et al., 2002	Dicerotini	*Ceratotherium* sp.*?*
Ngorora A-D	Kenya	12.0	Nakaya, 1993; Guérin, 2000	*B. lewisi* ; *C. pattersoni*	Rhinocerotidae indet.
Kabasero	Kenya	12.5	Hill et al., 2002	Rhinocerotidae	Rhinocerotidae indet.
Beni Mellal	Morocco	12.5	Guérin, 1976	cf. *Paradiceros mukirii*	cf. *P. mukirii*
Alengerr	Kenya	13.0	Hooijer, 1973; Guérin, 2000	*D. leakeyi*; *A. acutirostratum*	Rhinocerotidae indet.
Fort Ternan	Kenya	13.0	Hooijer, 1968	*Paradiceros mukirii*	*P. mukirii*
Kisegi	Uganda	13.5	Guérin, 1994b; Guérin, 2000	*Paradiceros mukirii*	Rhinocerotidae indet.
Muruyur-Kipsaramon	Kenya	13.7	Pickford, 1988	*A. acutirostratum*	Rhinocerotidae indet.
Kirimun	Kenya	15.0	Hooijer, 1971; Guérin, 2000	*Dicerorhinus* or *Aceratherium*; *Chilotheridium pattersoni*	*B. snowi?*; Rhinocerotidae indet.
Nyakach	Kenya	15.0	Pickford, 1986	*Brachypotherium* sp.	*Brachypotherium* sp.; *Plesiaceratherium* sp.*?*
Maboko-Ombo	Kenya	15.5	Hooijer, 1973; Pickford, 1986	*B. heinzelini*; *D. leakeyi*; *A. acutirostratum*; *C. pattersoni*	Elasmotheriinae?; *Chilotheridium* sp.*?*
Nachola	Kenya	15.5	Pickford et al., 1987	Rhinocerotidae	Elasmotheriinae?
Moroto I and II	Uganda	16.0	Pickford et al., 1986	Rhinocerotidae	Rhinocerotidae indet.
Mwiti (Kajong)	Kenya	16.5	Savage and Williamson, 1978	Rhinocerotidae	*Brachypotherium* cf. *snowi*
Buluk (W. Stephanie)	Kenya	16.5	Leakey and Walker, 1985	*D. leakeyi*; *A. acutirostratum*; *Chilotheridium pattersoni*	*Brachypotherium* nov. sp.?; Rhinocerotidae indet.
Jebel Zelten	Libya	16.5	Hamilton, 1973	*B. snowi*; *A. campbelli*	*B. snowi*
Loperot	Kenya	17.0	Hooijer, 1971; Guérin, 2000	*Chilotheridium pattersoni*	*Chilotheridium pattersoni*
Langental	Namibia	17.0	Heissig, 1971; Hooijer, 1973; Guérin, 2000	*B. heinzelini*	*Brachypotherium* sp.
Ryskop	Namibia	17.0	Pickford and Senut, 1997	Rhinocerotidae	Rhinocerotidae indet.
Moruorot	Kenya	17.2	Deraniyagala, 1951; Hooijer, 1968	*A. acutirostratum*	*T. acutirostratum*; Rhinocerotidae indet.
Moghara	Egypt	17.5	Fourtau 1920, Miller, 1999	*B. snowi*; *Aceratherium* sp.	*B. snowi*; Rhinocerotidae indet.
Karungu	Kenya	17.5	Hooijer, 1966; Pickford, 1986	*B. heinzelini*; *D. leakeyi*; *A. acutirostratum*;	*Brachypotherium* sp.; *R. leakeyi?*
Bukwa	Uganda	17.5	Walker, 1968; Hooijer, 1971	*B. heinzelini*; *C. pattersoni*	Elasmotheriinae?;
Kulu Fm. (Rusinga)	Kenya	17.7	Hooijer, 1966; Pickford, 1986	*D. leakeyi*; *A. acutirostratum*; *Chilotheridium pattersoni*	*R. leakeyi*; *T. acutirostratum*
Uyoma peninsula	Kenya	17.7	Pickford, 1986; Guérin, 2000	*B. heinzelini*; *Dicerorhinus* or *Aceratherium*	*B. snowi?*
Hiwegi Fm. (Rusinga)	Kenya	17.8	Hooijer, 1966, 1968	*B. heinzelini*; *D. leakeyi*; *A. acutirostratum*	*B. snowi*; *R. leakeyi*
Mfwangano	Kenya	17.8	Pickford, 1986	*B. heinzelini*; *D. leakeyi*	*B. snowi*; *R. leakeyi*
Wayando Fm. (Rusinga)	Kenya	18.0	Pickford, 1986; Hooijer, 1966	*B. heinzelini*; *D. leakeyi*; *A. acutirostratum*; *C. pattersoni?*	*B. snowi*; *R. leakeyi*
Arrisdrift	Namibia	18.0	Guérin, 2000; Guérin, 2003	*D. australis*; cf. *C. pattersoni*	*"D." australis*; *Chilotheridium* sp.*?*
Auchas	Namibia	19.5	Pickford and Senut 2003; Guérin, 2000	Rhinocerotidae	Rhinocerotidae indet.
Napak-Napak	Uganda	19.5	Pickford et al., 1986; Guérin and Pickford, 2003	*D. leakeyi*; *A. acutirostratum*; *Ougandatherium napakense*	*Ougandatherium napakense*; Rhinocerotidae indet.
Koru-Songhor	Kenya	20.0	Pickford, 1986; Hooijer, 1966	*D. leakeyi*	Rhinocerotidae indet.
Napak-Iriri	Uganda	20.0	Pickford et al., 1986; Hooijer, 1966, 1973	*B. heinzelini*; *D. leakeyi*	*Brachypotherium* sp.; Rhinocerotidae indet.

faint constriction of the inner cusps. Premolars with high lingual cingulum. Lower premolars long and narrow, with shallow labial groove and protoconid flattened labially. Vertical rugosities on the labial wall are common. Limbs long and slender, mainly their distal segments. Manus tetradactyl.

PLESIACERATHERIUM sp.

I provisionally refer to this genus two incomplete skulls from Nyakach, Kenya, found by M. Pickford, numbered KNM-NC 10486 and KNM-NC 10510, and dated to about 15 Ma. The nasals are remarkably long, straight, and hornless. The nasal notch is deep, and its bottom is U shaped, almost rectangular. The dorsal skull profile is concave, and the orbit is elevated, with an inflated and rounded supraorbital tuberosity. The premaxillae are slender, but probably carried incisors (an isolated upper incisor of medium size could be of the same species). The cheek teeth have a simple morphology (quite distinct from that of the next genus), with slightly pinched protocones on the molars, weak or absent crochet, but the premolars are reduced in size and their lingual cingulum is weak.

These skulls resemble those of early–middle Miocene Eurasian forms included in *Plesiaceratherium*, but these have larger premolars with a strong, continuous lingual cingulum (Heissig, 1972; Antunes and Ginsburg, 1983). The related genus *"Hoploaceratherium,"* best known from *"H." tetradactylum* from Sansan, France, has teeth more like the Nyakach ones, but there are small terminal horns (a minor difference), and it is said to have lost its upper incisors (Heissig, 1999).

The Nyakach rhino is probably a member of this *Plesiaceratherium-Hoploaceratherium* group, but more detailed evidence, especially relating to its upper incisors and postcranials, are still needed to ascertain its phyletic position. It may well be that some specimens from various sites previously referred to *Aceratherium* or *Dicerorhinus* belong here.

Genus TURKANATHERIUM Deraniyagala, 1951

Type Species Turkanatherium acutirostratum Deraniyagala, 1951.

Diagnosis Skull dolichocephalic, occiput vertical, frontoparietal profile concave, temporal lines meet to form a sagittal crest, nasals elongate, nasal notch U shaped and shallow (bottom above front of P3), anterior orbital margin above front of M2. Premolars with long transverse lophs, vestigial bridge between protocone and hypocone, molars without crista, antecrochet strong, at least on M1.

TURKANATHERIUM ACUTIROSTRATUM Deraniyagala, 1951
Figure 34.4

Type Skull (housed in the Colombo Museum, Sri Lanka).
Type Locality Moruorot, Kenya, about 17 to 17.5 Ma.
Diagnosis As for genus.
Remarks The type skull is preserved in the Sri Lanka National Museum but seemingly has never been examined by western researchers, who have had to rely mostly on the descriptions and figures of Deraniyagala (1951). The skull (figure 34.4) is high and narrow, the dorsal profile is concave, the condyles much higher than the tooth row, the temporal lines meet to form a long sagittal crest, the nasal notch has the shape of a wide U, and its bottom is above the middle of P3, and thus rather far from the anterior orbital border, which is above the front of M2. The nasals carry no horn,

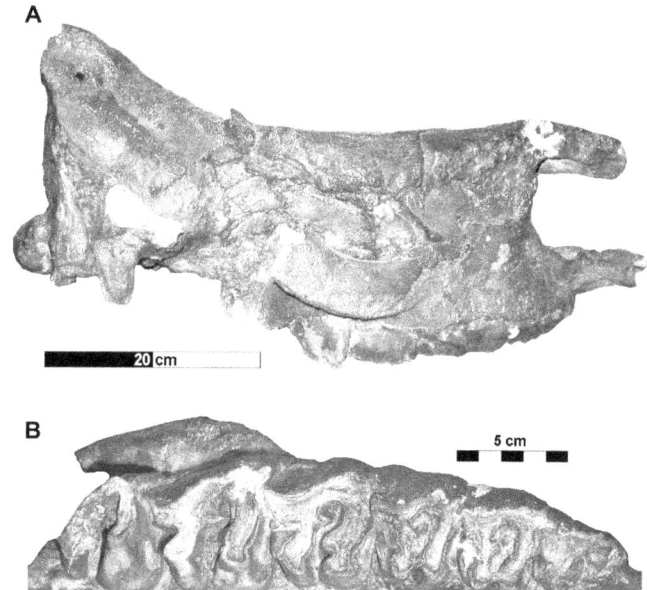

FIGURE 34.4 *Turkanatherium acutirostratum*, holotype skull from Moruorot; lateral view (A) and occlusal view (B) of teeth. Courtesy of Ceylon Journal of Science.

but they are long and slender, extending forward well beyond the level of P2. The long premaxillae were said by Deraniyagala (1951) to be edentulous; this is very unlikely, as noted by Hooijer (1966), but the size of I1 is unknown. The protoloph is constricted on the molars, especially M1, which has a strong antecrochet, but the crochet is weak. The premolars are small, broad, with lophs converging lingually, a lingual connection between them, an incomplete internal cingulum and the postfossette is transversely elongated.

An incomplete mandible from the same locality (MT-66 in Hooijer 1968, now KNM-MO 43) shows, from the shape of their alveoli, that the lateral incisors had long roots and some outward curvature. Arambourg (1933) described from Losodok in the same area two slender metatarsals, but they might not belong to the same taxon.

Arambourg (1959) and Hooijer (1963, 1966) referred *T. acutirostratum* to *Aceratherium*, without much discussion, and this generic attribution has been accepted ever since, but this was done at a time when the latter genus had a very broad meaning, including most middle and many of the late Miocene nonbrachypothere hornless rhinos from Eurasia. The type species of *Aceratherium* is *A. incisivum* from the early late Miocene of Germany, and recent revisions (Heissig, 1999) favor restriction of the generic name to this species. Even though this can be disputed, *T. acutirostratum* clearly differs from *A. incisivum*, which has a shorter skull, an almost flat cranial profile, a very robust zygomatic arch, a deeper nasal notch extending closer to the orbit, and larger and more molariform premolars with a lingual connection occurring only in late wear. *Turkanatherium* can thus be retained as a valid generic name, because its type species differs considerably from that of *Aceratherium*.

The cheek teeth of a lower jaw KNM-RU 3012 (850-47 in Hooijer, 1966) do not differ from those of the sympatric *"Dicerorhinus"*; the associated nasals were relatively long and broad, but not bowed anteroposteriorly, and certainly carried no large horn, but identification of all these remains is uncertain.

Hooijer (1966) found it difficult, if not impossible, to tell apart the limb bones of *"Dicerorhinus"* and *"Aceratherium,"* in spite of the occurrence at Rusinga of two skeletons that he referred to each of these genera; both had primitive limb proportions. *Turkanatherium* (and *Plesiaceratherium*) might be expected to differ in the retention of the fifth digit of the manus, but it has not yet been found.

"Aceratherium" has been reported from a number of African sites, often on the basis of fragmentary remains, but identification of isolated or even incomplete teeth of rhinos is seldom reliable. For instance, identification of teeth from Karugamania, Democratic Republic of Congo as *A. acutirostratum* by Hooijer (1963) was made on the assumption of an early/middle Miocene age of the deposits. Their reassignment to the late Miocene (Pickford et al., 1993) rules out their belonging to this species (especially as the P4 lacks the remarkable transverse broadening of the type specimen). It is likely that *T. acutirostratum* was a common species in the early and middle Miocene of Africa but, besides the type, it is hard to refer any specimen to this species with certainty.

Genus *CHILOTHERIDIUM* Hooijer, 1971

Type Species Chilotheridium pattersoni Hooijer, 1971.

Diagnosis Slightly modified from Hooijer (1971). Small single nasal horn in both sexes; frontals and parietals pneumatized; orbit not placed as near upper contour of skull as in *Chilotherium;* cranium and occiput rather narrow; parietal crests not widely separated; inferior squamosal processes not united below; symphysial portion of mandible narrow, slightly expanding anteriorly. Cheek teeth fully hypsodont as in *Chilotherium* and with the same pattern: uppers with paracone style fading away basally and posterior portion of ectoloph flattened; protocone well set off by folds and flattened internally; anterior fold in metaloph, marking off hypocone; antecrochet prominent basally, curving inward to medisinus entrance; crochet usually well developed, and crista weak or absent; metacone bulge at base in M3; anterior cingulum strong, internal cingulum weak and usually forming cusp at medisinus entrance. i2 subtriangular in cross section, depressed dorsoventrally, internal edge sharpened by wear, outer lower edge rounded, and outer upper edge ridged. Scapula low and wide; limb and foot bones not much shortened; radius and ulna, and tibia and fibula not ankylosed; radius with pyramidal facet; metacarpal V present, three-fifths the length of metacarpal IV; lateral metapodials somewhat divergent posteriorly; femur with small third trochanter; calcaneum without tibia facet; talus with trochlea markedly shifted laterally; navicular nearly rectangular; cuboid wider than high; metatarsal III with small cuboid facet.

CHILOTHERIDIUM PATTERSONI Hooijer, 1971

Type Skull figured by Hooijer (1971: plate 1); numbered 70-64K, B12 in KNM.
Type Locality Loperot, Kenya, ca. 17 Ma.
Diagnosis As for genus.
Remarks The species was erected on a large collection of fossils, but they are much fragmented and distorted. The skulls are made up of a mosaic of fragments that make their actual shape hard to figure out, although tooth features and limb bone proportions are certainly correct.

Upper incisors were said to be lacking, but the premaxillaries are broken off on both skulls from Loperot, and the absence of isolated upper incisors in the Loperot collection is not a strong argument against their actual presence (there are only two isolated i2s).

The main feature of the postcranial skeleton is the retention of a functional fifth metacarpal, but since even a vestigial Mc V articulates with a similar facet on the Mc IV, occurrence of this fifth digit is hard to demonstrate at other localities. Although not noted by Hooijer, the tali are characteristic, with a low medial lip of the trochlea, and a very salient distomedial tuberosity with a slanting proximal border. This morphology is absent from sites other than Loperot, showing that *Chilotheridium* is certainly a rare form.

However, Hooijer (1971) identified *Chilotheridium* from a few other sites, mostly on the basis of isolated teeth. An i2 of large size from Kirimun would perhaps better match *Brachypotherium*. A few upper cheek teeth from Bukwa (Walker, 1968) also referred to *Chilotheridium* by Hooijer, are much worn but are remarkable in the depth of the grooves that tend to isolate pillars: the antecrochet is strong, the protocone is double, and the hypocone is sharply set off from the metaloph. All these features perhaps better fit an elasmothere. Two tooth fragments from Rusinga were assigned to *Chilotheridium* mostly because of their hypsodonty. An upper tooth series from Ngorora (Hooijer, 1971: plate 11, figure 1; now KNM-BN 133) is too worn to be reliably identified. Some isolated teeth from the Samburu Hills (Nakaya et al., 1987) are also hardly identifiable.

Chilotheridium was assumed by Hooijer to be close to the mainly Asiatic late Miocene genus *Chilotherium*, but resemblances concern mostly cheek tooth morphology, estimated depth of the nasal notch, some shortening of the metapodials, and the presence of an articulation between radius and pyramidal. The latter feature is primitive, and the others are prone to parallelism. On the other hand, *Chilotherium* differs considerably from the Kenyan genus in its broad skull, flat frontals, high orbits, short hornless nasals, broadened mandibular symphysis with large i2s, and much shortened metapodials, and the two genera are probably not closely related.

Subtribe RHINOCEROTINA Gray, 1821
Genus *RUSINGACEROS* nov. gen.

Type Species Dicerorhinus leakeyi Hooijer, 1966.
Diagnosis Simplified from Hooijer (1966) for *Dicerorhinus leakeyi*. A rhino of medium size, with a long and low skull. Frontal and nasal horns present; nasal notch very shallow; long, slanting premaxilla bearing moderate-sized incisors; small i1s present, i2s parallel, medium sized; occiput as highly elevated as in *Lartetotherium*. Upper premolars with protoloph and metaloph united lingually up to at least 15 mm from crown base, cingulum weak. Upper molars with internal cingulum very weak or absent, protocone not or hardly constricted off, antecrochet absent, ectoloph depressed between the roots, crochet and crista weak or absent, M3 bulging out at junction of ectoloph and metaloph.

RUSINGACEROS LEAKEYI (Hooijer, 1966)

Type Skull and associated mandible, KNM-RU 2821 (Hooijer, 1966: plates 1 and 2, figures 1 and 2).
Type Locality Rusinga, precise locality unknown.
Diagnosis As for genus.
Remarks This species was described by Hooijer (1966) on the basis of the type, plus another associated maxilla and mandible (now KNM-RU 2822). It was originally referred to the genus *Dicerorhinus* Gloger, 1841, of which the modern

D. sumatrensis is the type species. There are some similarities in the cranial profile, shape and orientation of the nasals and premaxillae, size of the main incisors, but the Rusinga type skull is longer and lower with a longer facial portion, the orbit is more posterior, the zygomatic arch is extremely robust, and the cheek teeth are much more primitive, with submolariform premolars, weaker cristae on the molars, and no metacone fold (the very strong metacone fold of the premolars of the Sumatran rhino is certainly a derived feature), and the posterior limb is relatively longer. There is no evidence of a close relationship between the Rusinga material and the modern species, and including the Rusinga form in Dicerorhinus would expand the content of this genus to virtually every two-horned rhino with front teeth.

Several authors (Groves, 1983; Geraads, 1988; Cerdeño, 1995) have included the Rusinga species in Lartetotherium, a genus based on L. sansaniense from the late middle Miocene site of Sansan in France, thus much later in age than Rusinga. The resemblances include a high occiput, the size of the front teeth, and probably (the Rusinga teeth are highly worn) the molarisation stage of the upper premolars, but the skull of L. sansaniense is much higher and shorter, the zygomatic arch is weaker, the antorbital part shorter (the anterior orbital margin is above the anterior end of M2), the nasal notch is deeper (bottom above P2–P3), the symphysis is broader and shorter. An earlier form of L. sansaniense, from Sandelzhausen in Germany, has a nasal notch situated farther rostrally (Heissig, 1972), and the skull looks relatively longer than that from Sansan (but both are crushed). It partly bridges the chronological and morphological gaps between the Rusinga and Sansan rhinos, but the lack of a frontal horn is a difference from both.

Hooijer (1966) reported R. leakeyi from various sites at Rusinga, and from Songhor and Napak, and some other occurrences were added more recently (see table 34.1) but, besides Rusinga, most of the identifications are based on isolated teeth. Regardless, R. leakeyi is the earliest rhino of modern type, i.e., with a strong nasal and smaller frontal horn. In Eurasia, the earliest "Dicerorhinus" is documented by a few isolated teeth (of doubtful generic attribution) from Baigneaux in France (Ginsburg and Bulot, 1984), a locality dated to late MN4, i.e., somewhat later than Rusinga.

Genus STEPHANORHINUS Kretzoi, 1942

Type Species Stephanorhinus etruscus (Falconer in Murchison, 1868) from the Plio-Pleistocene of Italy.

This is a mostly European genus, the limits of which are controversial. It includes several Pliocene and Pleistocene species previously referred to Dicerorhinus and perhaps dates back to the late Miocene; the whole genus is in need of revision.

STEPHANORHINUS ? AFRICANUS (Arambourg, 1970)

Type M3, MNHN-1948-2-21 (Arambourg, 1970: plate 15, figure 1).
Type Locality Lake Ichkeul, Tunisia, early to middle Pliocene.
Diagnosis Translated from Arambourg (1970). Intermediate in size between S. etruscus and D. sumatrensis, with molars more brachyodont but morphologically similar to those of the living species.
Remarks The type locality yielded only the type, a mandible fragment, and an atlas. Given its age, it is unlikely to be of African origin, since only Dicerotini and brachypotheres survive in the rich East African sites after the middle Miocene.

The large mammals from Ichkeul are mostly of African affinity, but some genera are known on both sides of the Mediterranean at that time, and S. ? africanus is probably of northern origin. This is confirmed by the occurrence of the genus at Koro Toro 13 in Chad, a locality dated at about 3–3.5 Ma. (Likius, 2002), but not in East Africa.

STEPHANORHINUS HEMITOECHUS
(Falconer in Murchison, 1868)

Synonymy Rhinoceros subinermis Pomel, 1895: 21.
Type Skull figured by Falconer in Murchison (1868: plate 15); BMNH M27836.
Type Locality Clacton, Essex, Great Britain; middle Pleistocene.
Remarks The species has been revised by Guérin (1980) and Fortelius et al. (1993). No rhino related to European forms is known in the late Pliocene or early and middle Pleistocene of North Africa, and the Stephanorhinus found in the late Pleistocene of Morocco and Algeria must be an immigrant from the North, together with Sus and cervids. Long referred to S. mercki, it is now believed to belong to S. hemitoechus, the last species of the genus, with a large nasal horn supported by wide nasals buttressed by a robust nasal septum.

Genus PARADICEROS Hooijer, 1968

Type Species Paradiceros mukirii Hooijer, 1968.
Diagnosis Mostly from Hooijer (1968). Two horns, placed on nasals and frontals, respectively. Inferior squamosal processes separate. Lower orbital border rounded. Bottom of nasal notch above front of P3. Mandibular symphysis abbreviated but not widened; edentulous in the adult. Cheek teeth brachyodont, protocone constricted, antecrochet prominent in milk and first molars rather than in last and premolars. Last upper molar subtriangular. Upper molars with wide and low medisinus entrance, upper premolars with high internal pass. Limbs and some of the foot bones more shortened than in Aceratherium or Dicerorhinus though not to the extent seen in Brachypotherium or Chilotherium.

PARADICEROS MUKIRII Hooijer, 1968
Figure 34.5

Type Juvenile skull, figured by Hooijer (1968, pl. 1); KNM-FT 2866.
Type Locality Fort Ternan, ca. 13–14 Ma.
Diagnosis As for genus.
Remarks The species has also been reported from Kisegi in Uganda (Guérin, 1994b) and Beni Mellal in Morocco (Guérin,

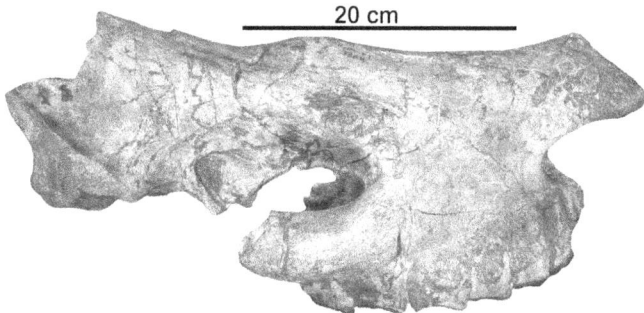

FIGURE 34.5 Paradiceros mukirii, skull KNM-FT 3328 from the middle Miocene of Fort Ternan, Kenya.

1976) on the basis of very poor remains; these identifications are likely but may have been influenced by the age of these sites.

An unpublished, almost complete skull, KNM-FT 3328 (figure 34.5), shows further features. The occiput is inclined backward. There is no true postorbital process. The frontal horn forms a conspicuous median boss, which is much more posterior than that of the African living forms, of *Rusingaceros*, and even of *Dicerorhinus sumatrensis*, casting some doubt on their homology. Behind the orbits, the temporal lines remain far apart before turning medially, suggesting that this frontal horn had a very broad base. The few metapodials, which exhibit marked size variation, are similar in robustness to those of the Dicerotini.

Many features of *Paradiceros* would fit a primitive Dicerotini, but none definitively supports a close relationship, and the peculiar features of the posterior horn seem to speak against it.

Genus *CERATOTHERIUM* Gray, 1868

Synonymy Serengeticeros Dietrich, 1942.
Type Species Ceratotherium simum (Burchell, 1817), living African "white" or square-lipped rhino.
Diagnosis Nasal and frontal horns; nasal bones rounded and short, not contacting lacrimal; lower border of orbit sloping downward; weak postorbital process; broad nuchal crest; premaxilla much reduced; upper and lower incisors vestigial or absent; paracone fold weak; antecrochet absent (Geraads, 2005).
Remarks Ceratotherium obviously shares a common ancestry with *Diceros*, but there is some disagreement about what should be included in either genus. It had long been assumed that modern *Diceros* is closer to the ancestral morphology, but I have argued (Geraads, 2005) that its cranial morphology is in fact derived and that Miocene forms should rather be placed in *Ceratotherium*. I follow this classification here, although the affinities of the incompletely known African Miocene forms are certainly debatable.

CERATOTHERIUM ? PRIMAEVUM (Arambourg, 1959)

Type Incomplete juvenile skull, MNHN 1951–9–222 (Arambourg, 1959: plate 6).
Type Locality Oued el Hammam (= Bou Hanifia), Algeria, early late Miocene.
Diagnosis A two-horned rhino, the anterior horn on a strongly convex nasal boss, no postorbital process on the frontal, lower orbital floor inclined. Incisors reduced or lost, strong parastyle but no cristae on the molars, protocone slightly pinched. Metapodials rather slender.
Remarks This species has been described only from the type locality. The sample is large, but cranial and dental specimens are mainly from juvenile individuals. Arambourg (1959) viewed it as a relative of the Sumatran rhino, but I showed (Geraads, 1986) that its skull displays some apomorphic features of the Dicerotini (see diagnosis). Many important elements of the adult skull, front dentition, and premolars, are absent from the type locality, so that the precise phyletic position of this species is unclear, but it is certainly valid, and I provisionally include it in the paraphyletic genus *Ceratotherium*.

A P2 from the late Miocene of Chorora, Ethiopia (Geraads et al., 2002) has its lingual lobes almost free from the ectoloph, as in the earlier *Paradiceros mukirii*, but it probably also belongs to an early *Ceratotherium*. A maxilla from the late Miocene Namurungule Fm. (Nakaya et al., 1987: plate 6, figure 1), assigned to *Paradiceros mukirii*, is more likely to belong here.

CERATOTHERIUM DOUARIENSE (Guérin, 1966)

Type Associated partial skull and mandible, FSL-16750 and 16751, figured by Guérin, 1966: figures 1, 3, 4 (top), 5, 7–10.
Type Locality Douaria, Tunisia. Age not known with precision, but almost certainly late Miocene.
Diagnosis Translated and simplified from Guérin (1966). Large two-horned skull; nasal notch at the level of P2–P3, anterior orbital border above M1–M2; strong lacrimal processes directed posteroventrally; strong postglenoid process; posterior border of symphysis at the level of p3. Upper premolars with strong lingual cingulum, strong crochet, weak antecrochet. Upper molars with strong crochet.
Remarks The occipital region is unknown, but the rest of the skull does not display the derived features of *Diceros*, and it seems better to leave this species in the paraphyletic genus *Ceratotherium*, pending discovery of a more complete specimen. It is doubtfully distinct from *C. neumayri* from the late Miocene of the Balkano-Iranian province, but more North African specimens would be welcome. It has also been reported from Jebel Krechem, mostly on the basis on geographic proximity (Geraads, 1989), but not outside Tunisia. A tooth from the latest Miocene of Sahabi, Libya, identified as *Diceros neumayri* by Bernor et al. (1987), can be included here, too.

CERATOTHERIUM sp.

The rhino from the early Pliocene of Langebaanweg, South Africa, was described by Hooijer (1972) as *Ceratotherium praecox*. This species should be included in *Diceros* (discussed later), but the Langebaanweg rhino displays derived features of the Pliocene *Ceratotherium* clade, such as a flattened ectoloph, more plagiolophodont teeth, and tendency to close the medi- and postfossettes (Geraads, 2005).

Some other specimens are difficult to fit into the evolution of *Ceratotherium*. A maxilla from the Mio-Pliocene of Lissasfa, Morocco, is unusual in its high premolars but lingually fused protocone and metacone, reminiscent of the primitive Vallesian *Ceratotherium* from Pentalophos, Greece (Geraads and Koufos, 1990), but also of modern *C. simum*.

CERATOTHERIUM MAURITANICUM (Pomel, 1888)
Figure 34.6

Synonymy Serengeticeros efficax Dietrich, 1942.
Type M2, MNHN no. TER-2261, figured by Pomel (1895: plate 1, figures 1 and 2).
Type Locality Tighenif (= Ternifine, = Palikao), Algeria, lower/middle Pleistocene.
Diagnosis Size larger than in *C. neumayri*; nuchal crest stretched more caudally; nasal notch shallower; premolar row shortened; transverse lophs of upper teeth long and narrow; metaloph extending distolingually into distal cingulum, closing postfossette (Geraads, 2005).
Remarks North African *Ceratotherium mauritanicum* (figure 34.6) is clearly distinct from *C. simum*, and Guérin (1994a) and Guérin and Faure (2007) recognized its specific distinctness; in this region it survives until the late middle Pleistocene. I observed (Geraads, 2005) that its main features can also be found in East Africa in most of the specimens usually referred to *C. simum germanoafricanum* (discussed later), and many of those called *Ceratotherium praecox* (Geraads, 2005: table 4), and accordingly referred these East African

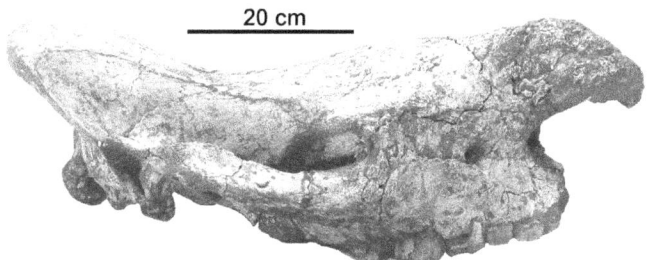

FIGURE 34.6 *Ceratotherium mauritanicum,* skull from the middle Pleistocene of "Grotte des Rhinocéros," Casablanca, Morocco.

specimens to *C. mauritanicum.* Again, precise delimitation of the species may be difficult, as there is little doubt that, in East Africa, it is directly ancestral to the living *C. simum.*

CERATOTHERIUM SIMUM (Burchell, 1817)

Diagnosis Geraads (2005). Strong postorbital constriction; nuchal crest narrow; postglenoid process weak; craniomandibular articulation horizontal; upper cheek teeth hypsodont, with very long narrow lophs and enlarged fossettes; labial walls sinuous in the upper part of crown; occlusal surface flat; premolar row shortened, DP1 shed before adulthood, P2 small; on molars, curved protoloph, oblique narrow metaloph, closed medifossette, post and prefossettes closed in advanced wear; lower cheek teeth rectangular, with closed fossettids on worn teeth; metapodials short and stout.

Remarks In historical times, the "white" rhino had a more restricted range than its "black" cousin, but in the Pleistocene it extended as far as the Mediterranean. Of the two living subspecies, *C. simum cottoni,* which was the more common 100 years ago, is now restricted to a few individuals in northeastern Congo. The southern *C. simum simum,* whose numbers had plummeted to 20 in 1895, now includes about 12,000 individuals, almost all of them in South Africa (International Rhino Foundation data). Late Holocene records of white rhinos in Kenya suggest that their distributions in the recent past were less widely separated.

The fossil form *Rhinoceros simus germanoafricanus* Hilzheimer, 1925, whose subspecific name is often used as a species name for the Pleistocene form, is based on a lost skull from Olduvai, probably from its upper levels, where definite *C. simum* are known. Since it also shows an oblique metaloph, like the modern form, there is no reason to separate them. In any case, using a species name based on a type that survives only through a sketch of two teeth could easily lead to confusion.

Ceratotherium simum is clearly descended from *C. mauritanicum.* The transition, which took place near the Plio-Pleistocene boundary in East Africa, is mainly marked by features associated with an increasingly grazing diet, with some convergences with other grazing species, such as *Coelodonta* and *Elasmotherium,* in tooth morphology. Except for a somewhat larger size, early Pleistocene forms are already identical with the modern one.

Genus *DICEROS* Gray, 1821

Type Species Diceros bicornis Linnaeus, 1758, living African "black" rhino.
Diagnosis Geraads (2005). Premaxilla absent or vestigial. Cranium short and relatively broad. Neurocranium tilted anterodorsally relative to the splanchnocranium, resulting in more vertically oriented occipital plane or even one inclined anterodorsally, nuchal crest less expanded posteriorly, more deeply concave cranial profile, basioccipital angled relative to basisphenoid, shortened face with orbits more anteriorly positioned and closer to nasal notch, and often nasolacrimal contact.

"*DICEROS*" *AUSTRALIS* Guérin, 2000

Type Left third metacarpal, AD 52′97, figured by Guérin (2000: figures 5.3–5.4), housed in the Geological Survey of Namibia, Windhoek.
Type Locality Arrisdrift, Namibia, ca. 17 Ma.
Diagnosis Guérin (2000). A very large cursorial rhinoceros of the Dicerotine type. Upper cheek teeth brachyodont, with a more or less continuous crenulated inner cingulum, and a crochet as the only or main internal fold. Ectoloph of upper premolars with a strong parastyle, paracone fold thick but not very prominent, and no mesostyle or metacone fold. Upper molars have a large paracone fold on their ectoloph, with a weak vertical bulge in the middle of it, and a protocone weakly constricted on its anterior face. Tall and slim but sturdy limb bones. Lateral and medial metapodials very long with respect to the central one.
Remarks This species is known only from isolated teeth and limb bones. The upper incisors are unknown; the lower ones are smaller than in *R. leakeyi,* although not vestigial. The P4 has no constricted protocone, a rather flat ectoloph behind the paracone fold, and no lingual connection between the lophs. These features, plus the large size for this age, led Guérin to include the Arrisdrift species in the Dicerotini, hence in *Diceros,* as he considers this genus as the earliest and most primitive member of this tribe. Indeed, Tougard et al. (2001) rooted the tribe into the late Oligocene, but the characters of *D. australis* are not exclusive of it. This is certainly a valid species, but only cranial material would shed more light on its affinities.

DICEROS PRAECOX (Hooijer and Patterson, 1972)

Type Poorly preserved incomplete skull, KNM-KP 36 (Hooijer and Patterson, 1972: figure 9A).
Type Locality Kanapoi, Kenya, about 4 Ma.
Diagnosis Geraads (2005). The original diagnosis consists entirely of plesiomorphic characters. This species has only a few apomorphic features with respect to its likely ancestor *C. neumayri*: orbit more anterior with respect to tooth row; skull profile more concave; occipital plane more vertical; nuchal crest less extended posteriorly.
Remarks This species had long been included in *Ceratotherium,* but I showed (Geraads, 2005) that the type and a referred skull from Ekora, which formed the basis of the original description, are both closer to *Diceros* in their tooth morphology, concave cranial profile, and occiput more vertical than in *Ceratotherium.* The distinction from *D. bicornis* may not be easy. I referred (Geraads, 2005) a fragmentary skull from the base of the Sidi Hakoma member of the Hadar Fm to *D. praecox,* and an as yet uncollected skull from higher up in the sequence at Dikika appears transitional but is less derived than *D. bicornis* in its larger size, less shortened skull, less anteriorly shifted orbit, and wide nuchal crest. I also referred (Geraads, 2005) to this species several specimens previously called either *C. praecox* or *D. bicornis,* but most of the material previously reported under the name "*Ceratotherium praecox*" belongs to what is called here *C. mauritanicum.*

DICEROS BICORNIS (Linnaeus, 1758)

Diagnosis Geraads (2005). Size smaller than *D. praecox*; face more angled on neurocranium; nuchal crest not expanded; cheek teeth narrower, lophs more transverse; premolar row shortened.

Remarks The living "black" rhino was once widespread outside dense forest in sub-Saharan Africa, but it has never been reported north of the present-day Sahara. Numbers sharply declined with the introduction of firearms, and several subspecies have recently become extinct. A minimum was reached in 1995, with only 2400 remaining wild individuals, but a slight rise since then has brought the number to about 3700 (IRF data), scattered from Kenya to South Africa and Namibia, with three remaining subspecies.

This species has been recorded from as early as the late Miocene (e.g., Lothagam), but I preferred (Geraads, 2005) to regard these pre-Pleistocene forms as *D. praecox*. The best early representative of the living species is skull KNM-ER 636, from the KBS member of Koobi Fora (Harris, 1983).

Subfamily ELASMOTHERIINAE Bonaparte, 1845
Genus *KENYATHERIUM* Aguirre and Guérin, 1974

Type Species Kenyatherium bishopi Aguirre and Guérin, 1974.

Diagnosis Translated from Aguirre and Guérin (1974). A medium-sized Elasmotheriinae; upper premolar hypsodont with regularly convex ectoloph and a very weak paracone fold. Opening of the median valley fully blocked by a wall uniting protocone to hypocone. Medifossette lacking true folds but with localized microfolds. Protocone constricted by a groove on the mesial side of the protoloph.

KENYATHERIUM BISHOPI Aguirre and Guérin, 1974

Type Upper premolar, probably P4, KNM-NA 198.
Type Locality Nakali, Kenya, early late Miocene.
Diagnosis As for genus.
Remarks The holotype and an incomplete molar are indeed remarkable in the features mentioned in the diagnosis, plus the presence of cement, large postfossette, distally closed by a high cingulum, and small protocone and hypocone well set off from the lophs. The authors viewed *Kenyatherium* as close to the Eurasiatic Miocene genera *Iranotherium*, *Hispanotherium*, and *Caementodon*. In the cladistic analysis of Antoine (2002), it occupies a basal position among the elasmotheres because of its transverse metaloph on P4, long metaloph on M1 or M2, lack of cristae, presence of lingual cingulum on upper teeth, and hypocone fully merged into the metaloph on the molar, though the latter two characters are disputable. In any case, the material from Nakali is too poor to precisely determine its systematic position.

The species has also been reported from the roughly contemporaneous lower member of the Namurungule Formation (Nakaya et al., 1987); this is the more likely identification, as the antecrochet is stronger than in *Chilotheridium*, and the grooves isolating the protocone and hypocone on the molars are deeper.

Genus *OUGANDATHERIUM* Guérin and Pickford, 2003

Type Species Ougandatherium napakense Guérin and Pickford, 2003.

Diagnosis Translated from Guérin and Pickford (2003). Small Elasmotheriinae with short, hornless nasals. Hypsodont upper cheek teeth, mesial width greater than distal width; medifossette filled with cement, and with crochet as single fold. Upper premolars small, with a lingual wall connecting protocone to hypocone; ectoloph with well-marked folds; slanting lingual cingulum; constricted protocone. Upper molars with much folded labial wall; protocone strongly constricted; no lingual or labial cingulum, but a strong anterior cingulum encloses a prefossette in the protocone groove. Cursorial limbs, with lengthened second and third segments. Long slender metapodials, the central ones with broadened distal diaphysis, the lateral ones relatively long.

OUGANDATHERIUM NAPAKENSE Guérin and Pickford, 2003

Type Guérin and Pickford (2003) listed as holotype two third metacarpals and two third metatarsals, but the whole material from Napak belongs to only two individuals (Guérin and Pickford, 2003: 8), and these authors did not state why they assumed that these four bones, and only they, belong to one of these individuals. Stored in the Uganda Museum, Kampala.
Type Locality Napak I, Uganda, ca. 19 Ma.
Diagnosis As for genus.
Remarks This species is known only from the lower level of Napak from remains of two incomplete skeletons and skull fragments. The nasals are not fused, triangular, and quite short; the premaxillae look long, but whether they carried an incisor is not known. The upper premolars resemble the P4 of *Kenyatherium*, with a transversely elongated postfossette, but the hypocone is more reduced, especially on P3, and more closely apressed to the protocone, so that the teeth are more premolariform (i.e., more primitive).

Evolution

Like that of several other mammalian groups, the Miocene record of African Rhinocerotidae is relatively good between 18 and 15 Ma, and after 7 Ma, but more patchy between these periods, and before 18 Ma. The late early to early middle Miocene is the period of greatest diversity, with at least four contemporaneous genera in Kenya. Of these, only the brachypotheres are clearly linked to later forms, although it is likely that *Rusingaceros* is related to later *Paradiceros* and Dicerotini, despite the significant time gap.

Chilotheridium remains a mysterious genus, partly owing to the poor preservation of the Loperot material. Unfortunately, the holotype of *Turkanatherium acutirostratum* is not available, but it may play a central role in the evolution of African rhinos, as one may suspect that it is in fact an elasmothere. The systematic status of this group is not fully settled; a recent revision (Cerdeño, 1995) considered it diphyletic, but the latest ones (Antoine, 2002; Antoine et al., 2003) viewed it as a valid clade. The main features suggesting that *Turkanatherium* belongs here are the transversely elongated postfossette on P3–4, and the lingual connection, through a high narrow bridge, of the lophs on these teeth. This is the "semimolariform" morphology of Heissig (1969). *Turkanatherium* would then document part of the ghost lineage of African elasmotheres leading to *Kenyatherium*, as postulated by Antoine (2002).

In fact, at the time of its description, *Kenyatherium* was clearly separated from the other known African middle

Miocene forms, which had mainly been described and illustrated by Hooijer. Now that more of them have been described or discovered, the distinction dwindles, and several African rhinos have premolars reminiscent of the elasmotheres. A complete skull and parts of a skeleton found at Nachola (Baragoi), a site dated to 15–16 Ma., by a Japanese team led by H. Ishida, are now under study by H. Tsujikawa, who kindly allowed me to mention them. The premolars are similar to those from Moruorot, while the molars with strongly pinched hypocone and protocone, and the high zygomatic arches and very long nasals (but without lateral flange) recall those of the elasmothere *Procoelodonta*. Well-preserved skulls from Maboko, kindly made available by B. Benefit, also have strongly pinched protocones and hypocones on the (much worn) molars, antecrochet almost connecting the hypocone and transversely elongated postfossette on a premolar, plus a protruding orbital border and a strong nasal horn.

The question is whether features of the cheek teeth suffice to identify a rhino as an elasmothere. One of the strongest synapomorphies of this group according to Antoine (2002; also Antoine et al., 2003) is the purportedly "submolariform" morphology of P3–P4 but, while this is probably true of *Ougandatherium*, the type of *Kenyatherium* fully matches the description of the "semimolariform" morphology, distinct from the former one by "protocone and hypocone wider apart, bridge between them longer and shifted labially, so that the protocone becomes lingually separated from it by the protocone grooves" (Heissig, 1969: 16, my translation, and figure 4c therein). If both *Ougandatherium* and *Kenyatherium* are elasmotheres, molarisation does increase in this group, and it is hard to keep premolar morphology as a major distinctive feature of it, especially as its members have diverse cranial morphologies. Furthermore, non-molariform premolars are common in Oligocene and early Miocene rhinos, and it may be difficult to distinguish the elasmothere morphology from the primitive condition. Full study of the recently collected material may clarify these issues.

Biogeography

Rhinos are absent from the Oligocene sites, and the African fauna of that time is so clearly endemic that it is unlikely that an Oligocene African rhino will ever be discovered. By contrast, their absence from the earliest Miocene of Meswa Bridge might be due to incomplete sampling, as the fauna is poor but contains a Eurasian immigrant *(Dorcatherium)*; the earliest African rhinos, obviously of northern origin, may thus prove to be older than those presently recorded at Songhor and Napak, at about 20 Ma.

For the rest of the Miocene, uncertainties about real affinities hinder the reconstruction of past ranges and migration routes. *Rusingaceros* predates all Eurasian two-horned rhinos, which may have immigrated from Africa together with the Proboscideans; if the Nyakach rhino really belongs to *Plesiaceratherium*, this genus must also be part of the pre-Langhian exchange, together with the brachypotheres. The period between 15 and 10 Ma. is very poorly sampled; if *Paradiceros*, unknown in Eurasia, is not ancestral to the Dicerotini, the next exchange concerns *Ceratotherium* or its immediate ancestors, at the beginning of the late Miocene. Later immigrations from the North are those of the two *Stephanorhinus* species.

Conclusions

The diversity of African Miocene Rhinocerotidae is clearly greater than was assumed by Hooijer (1978). An undesirable consequence is that it becomes impossible to identify them by their teeth only, as different genera may share similar dental morphology (e.g., the cheek teeth of *Paradiceros mukirii* are almost identical to those of the ? *Plesiaceratherium* from Nyakach, although the skulls are quite distinct). It follows that many previous identifications, based upon fragmentary remains, must be treated with the utmost caution. Table 34.1 lists the main rhino-bearing Cenozoic African localities, with both published and revised identifications. The latter are usually more conservative, and often tentative, because the material is incomplete, or not described and not seen by me. The great number of "Rhinocerotidae indet." gives an idea of what remains to be done.

ACKNOWLEDGMENTS

I am eager to thank L. Werdelin and W. Sanders for having invited me to contribute to this volume. I am especially grateful to E. Mbua, who granted me access to the invaluable collections of the National Museums of Kenya, through the kind help of M. Muungu. A. Currant, A. Prieur, C. Sagne, and M. Yilma also gave me access to collections in their care in the BMNH, FSL, MNHN, and NME, respectively. B. Benefit, Y. Kunimatsu, E. Mbua, E. Miller, M. Pickford, and H. Tsujikawa kindly allowed me to mention a lot of unpublished material. M. Goonatilake kindly sent me photos of the *Turkanatherium* skull in the Sri Lanka National Museum. Thanks also to I. Giaourtsakis for his help and fruitful discussions, and to C. Guérin for reviewing this chapter.

Literature Cited

Aguirre, E., and C. Guérin. 1974. Première découverte d'un Iranotheriinae (Mammalia, Perissodactyla, Rhinocerotidae) en Afrique: *Kenyatherium bishopi* nov. gen. sp. de la formation vallésienne (Miocène supérieur) de Nakali (Kenya). *Estudios Geológicos* 30:229–233.

Antoine, P.-O. 2002. Phylogénie et évolution des Elasmotheriina (Mammalia, Rhinocerotidae). *Mémoires du Muséum National d'Histoire Naturelle* 188:1–359.

Antoine, P.-O., F. Duranthon, and J.-L. Welcomme. 2003. *Alicornops* (Mammalia, Rhinocerotidae) dans le Miocène supérieur des collines Bugti (Balouchistan, Pakistan): Implications phylogénétiques. *Geodiversitas* 25:575–603.

Antunes, M. T., and L. Ginsburg. 1983. Les rhinocérotidés du Miocène de Lisbonne: Systématique, écologie, paléobiogéographie, valeur stratigraphique. *Ciências da Terra (UNL)* 7:17–98.

Arambourg, C. 1933. Mammifères miocènes du Turkana. *Annales de Paléontologie* 22:123–146.

———. 1959. Vertébrés continentaux du Miocène supérieur de l'Afrique du Nord. *Service de la Carte géologique de l'Algérie, Mémoire* 4:5–159.

———. 1970. Les vertébrés du Pléistocène de l'Afrique du Nord. *Archives du Muséum National d'Histoire Naturelle, Série 7*, 10:1–127.

Asfaw, B., W. H. Gilbert, Y. Beyene, W. K. Hart, P. R. Renne, G. Wolde-Gabriel, E. S. Vrba, and T. D. White. 2002. Remains of *Homo erectus* from Bouri, Middle Awash, Ethiopia. *Nature* 416:317–320.

Bernor, R. L., K. Heissig, and H. Tobien. 1987. Early Pliocene Perissodactyla from Sahabi, Libya; pp. 233–254 in N. T. Boaz, A. El-Arnauti, A. W. Gaziry, J. de Heinzelin, and D. D. Boaz (eds.), *Neogene Paleontology and Geology of Sahabi*. Liss, New York.

Bishop, W. W., M. Pickford, and A. Hill. 1975. New evidence regarding the Quaternary, geology, archaeology and hominids of Chesowanja, Kenya. *Nature* 258:204–208.

Boaz, N. T., R. L. Bernor, A. S. Brooks, H. Cooke, J. de Heinzelin, R. Dechamps, E. Delson, A. W. Gentry, J. W. K. Harris, P. A. Meylan, P. P. Pavlakis, W. J. Sanders, K. M. Stewart, J. Verniers, P. G. Williamson, and A. J. Winkler. 1992. A new evaluation of the significance of the Late Neogene Lusso Beds, Upper Semliki Valley, Zaire. *Journal of Human Evolution* 22:505–517.

Bonis, L. de, D. Geraads, J.-J. Jaeger, and S. Sen. 1988. Vertébrés du Pléistocène de Djibouti. *Bulletin de la Société Géologique de France* 4:323–334.

Brugal, J.-P., and C. Denys. 1989. Vertébrés du site acheuléen d'Isenya (Kenya, District de Kajiado). Implications paléoécologiques et paléobiogéographiques. *Comptes Rendus de l'Académie des Sciences, Paris* 308:1503–1508.

Cerdeño, E. 1993. Etude sur *Diaceratherium aurelianense* et *Brachypotherium brachypus* (Rhinocerotidae, Mammalia) du Miocène moyen de France. *Bulletin du Muséum National d'Histoire Naturelle C* 15:25–77.

———. 1995. Cladistic analysis of the family Rhinocerotidae (Perissodactyla). *American Museum Novitates* 3143:1–25.

———. 1996. *Prosantorhinus*, the small teleoceratine rhinocerotid from the Miocene of Western Europe. *Geobios* 29:111–124.

Coppens, Y. 1971. Les Vertébrés villafranchiens de Tunisie: Gisements nouveaux, signification. *Comptes Rendus de l'Académie des Sciences, Paris* 273:51–54.

Deraniyagala, P. E. P. 1951. A hornless rhinoceros from the Mio-Pliocene deposits of East Africa. *Spolia Zeylanica* 26:133–135.

d'Erasmo, G. 1954. Sopra un molare di *Teleoceras* del giacimento fossilifero di Sahabi in Cirenaica. *Rendiconti della Accademia nazionale dei Quaranta* 4:89–102.

Ditchfield, P., J. Hicks, T. W. Plummer, L. Bishop, and R. Potts. 1999. Current research on the late Pliocene and Pleistocene deposits north of Homa Moutain, southwestern Kenya. *Journal of Human Evolution* 36:123–150.

Fortelius, M., P. Mazza, and B. Sala. 1993. *Stephanorhinus* (Mammalia, Rhinocerotidae) of the Western European Pleistocene, with a revision of *S. etruscus* (Falconer, 1868). *Palaeontographia Italica* 80:63–155.

Fourtau, R. 1920. *Contribution à l'Étude des Vertébrés Miocènes de l'Egypte.* Government Press, Cairo, 121 pp.

Gentry, A. W. 1987. Rhinoceroses from the Miocene of Saudi Arabia. *Bulletin of the British Museum (Natural History), Geology* 41:409–432.

Geraads, D. 1986. Sur les relations phylétiques de *Dicerorhinus primaevus* Arambourg, 1959, Rhinocéros du Vallésien d'Algérie. *Comptes Rendus de l'Académie des Sciences, Paris*, Série II 302:835–837.

———. 1987. La faune des dépôts pléistocènes de l'Ouest du lac Natron (Tanzanie); interprétation biostratigraphique. *Sciences Géologiques, Bulletin* 40:167–184.

———. 1988. Révision des Rhinocerotidae (Mammalia) du Turolien de Pikermi: Comparaison avec les formes voisines. *Annales de Paléontologie* 74:13–41.

———. 1989. Vertébrés du Miocène supérieur du Djebel Krechem el Artsouma (Tunisie centrale). Comparaisons biostratigraphiques. *Geobios* 22:777–801.

———. 2005. Pliocene Rhinocerotidae (Mammalia) from Hadar and Dikika (Lower Awash, Ethiopia), and a revision of the origin of modern African rhinos. *Journal of Vertebrate Paleontology* 25:451–461.

———. 2006. The late Pliocene locality of Ahl al Oughlam, Morocco: Vertebrate fauna and interpretation. *Transactions of the Royal Society of South Africa* 61:97–101.

Geraads D., Z. Alemseged, and H. Bellon. 2002. The late Miocene mammalian fauna of Chorora, Awash basin, Ethiopia: Systematics, biochronology and ⁴⁰K-⁴⁰Ar age of the associated volcanics. *Tertiary Research* 21:113–122.

Geraads D., Z. Alemseged, D. Reed, J. Wynn, and D. C. Roman. 2004. The Pleistocene fauna (other than Primates) from Asbole, lower Awash Valley, Ethiopia, and its environmental and biochronological implications. *Geobios* 37:697–718.

Geraads, D., J.-J. Hublin, J.-J. Jaeger, H. Tong, S. Sen, and P. Toubeau. 1986. The Pleistocene Hominid site of Ternifine, Algeria: New results on the environment, age and human industries. *Quaternary Research* 25:380–386.

Geraads, D., and G. D. Koufos. 1990. Upper Miocene Rhinocerotidae (Mammalia) from Pentalophos-1, Macedonia, Greece. *Palaeontographica, Abt. A* 210:151–168.

Giaourtsakis, I. X., C. Pehlevan, and Y. Haile-Selassie. 2009. Rhinocerotidae; pp. 429–468 in Y. Haile-Selassie and G. WoldeGabriel (eds.), *Ardipithecus kadabba. Late Miocene evidence from the Middle Awash, Ethiopia.* University of California Press, Berkeley.

Ginsburg, L., and C. Bulot. 1984. Les Rhinocérotidés (Perissodactyla, Mammalia) du Miocène de Bézian à La Romieu (Gers). *Bulletin du Muséum National d'Histoire Naturelle C*, 4ème série, 6:353–377.

Groves, C. P. 1983. Phylogeny of the living species of Rhinoceros. *Zeitschrift für zoologische Systematik und Evolutionsforschung* 21:293–313.

Guérin, C. 1966. *Diceros douariensis* nov. sp., un Rhinocéros du Mio-Pliocène de Tunisie du Nord. *Documents des Laboratoires de Géologie de la Faculté des Sciences de Lyon* 16:1–50.

———. 1976. Les restes de Rhinocéros du gisement miocène de Beni Mellal, Maroc. *Géologie Méditerranéenne* 3:105–108.

———. 1980. Les Rhinocéros (Mammalia, Perissodactyla) du Miocène terminal au Pleistocène supérieur en Europe occidentale: Comparaison avec les espèces actuelles. *Documents des Laboratoires de Géologie de la Faculté des Sciences de Lyon* 79:1–1185.

———. 1985. Les rhinocéros et les chalicothères (Mammalia, Perissodactyla) des gisements de la vallée de l'Omo en Ethiopie; pp. 67–89 in Y. Coppens and F. C. Howell (eds.), *Les Faunes Plio-Pléistocènes de la Basse Vallée de l'Omo (Ethiopie).* Cahiers de Paléontologie-Travaux de paléontologie est-africaine. CNRS, Paris.

———. 1987. Fossil Rhinocerotidae (Mammalia, Perissodactyla) from Laetoli; pp. 320–348 in M. D. Leakey and J. M. Harris (eds.), *Laetoli: A Pliocene site in Northern Tanzania.* Clarendon Press, Oxford.

———. 1994a. Le genre *Ceratotherium* (Mammalia, Rhinocerotidae) dans le Plio-Pléistocène d'Ethiopie et son évolution en Afrique. *Études Éthiopiennes: Vol. 1. Actes Xème Conférence Internationale des Études Éthiopiennes* (Paris, August 1988):13–29.

———. 1994b. Les Rhinocéros (Mammalia, Perissodactyla) du Néogène de l'Ouganda; pp. 263–280 in B. Senut and M. Pickford (eds), *Geology and Palaeobiology of the Albertine Rift Valley, Uganda-Zaire.* Centre International pour la Formation et les Echanges Géologiques. Publication Occasionnelle 29, Orléans.

———. 2000. The Neogene rhinoceroses of Namibia. *Palaeontologia Africana* 36:119–138.

———. 2003. Miocene Rhinocerotidae of the Orange River Valley, Namibia. *Memoir, Geological Survey of Namibia* 19:257–281.

Guérin, C., and M. Faure. 2007. Etude paléontologique des mammifères du Pléistocène supérieur de l'oued El Akarit; pp. 365–390 in J. P. Roset and M. Harbi-Riahi (eds.), *El Akarit: Un Site Archéologique du Paléolithique Moyen dans le Sud de la Tunisie.* Editions Recherches sur les Civilisations, Paris.

Guérin, C., and M. Pickford. 2003. *Ougandatherium napakense* nov. gen. nov. sp., le plus ancien Rhinocerotidae Iranotheriinae d'Afrique. *Annales de Paléontologie* 89:1–35.

Hamilton, W. R. 1973. North African lower Miocene rhinoceroses. *Bulletin of the British Museum (Natural History), Geology* 24:349–395.

Harris, J. M. 1983. Family Rhinocerotidae; pp. 130–156 in J. M. Harris (ed.), *Koobi Fora Research Project: Volume 2. The Fossil Ungulates: Proboscidea, Perissodactyla and Suidae.* Clarendon Press, Oxford.

Harris, J. M., F. H. Brown, and M. G. Leakey. 1988. Stratigraphy and paleontology of Pliocene and Pleistocene localities west of Lake Turkana, Kenya. *Contributions in Science, Natural History Museum of Los Angeles County* 399:1–128.

Harris, J. M., and M. G. Leakey. 2003. Lothagam Rhinocerotidae; pp. 371–385 in M. G. Leakey and J. M. Harris (eds.), *Lothagam: The Dawn of Humanity in Eastern Africa.* Columbia University Press, New York.

Harrison, T., and E. Baker. 1997. Paleontology and Biochronology of fossil localities in the Manonga Valley, Tanzania; pp. 361–393 in T. Harrison (ed.), *Neogene Paleontology of the Manonga Valley, Tanzania.* Topics in Geobiology 14. Plenum Press, New York.

Heissig, K. 1969. Die Rhinocerotidae (Mammalia) aus der oberoligozänen Spaltenfüllung von Gaimersheim bei Ingolstadt in Bayern und ihre phylogenetische Stellung. *Bayerische Akademie der Wissenschaften, Mathematisch-Naturwissenschaftliche Klasse, Abhandlungen* (N.F.) 138:1–133.

———. 1971. *Brachypotherium* aus dem Miozän von Südwestafrika. *Mitteilungen der Bayerischen Staatsammlung für Paläontologie und Historische Geologie* 11:125–128.

———. 1972. Die obermiozäne Fossil-Lagerstätte Sandelzhausen. 5. Rhinocerotidae (Mammalia), Systematik und Ökologie. *Mitteilungen der Bayerischen Staatssammlung für Paläontologie und Historische Geologie* 12:57–81.

———. 1999. Family Rhinocerotidae; pp. 175–188 in G. Rössner and K. Heissig (eds.), *The Miocene Land Mammals of Europe.* Pfeil, Munich.

Hendey, Q. B. 1978. The age of the fossils from Baard's Quarry, Langebaanweg, South Africa. *Annals of the South African Museum* 75:1–24.

———. 1981. Palaeoecology of the Late Tertiary fossil occurrences in "E" Quarry, Langebaanweg, South Africa, and a reinterpretation of their geological context. *Annals of the South African Museum* 84:1–104.

Hill, A., M. G. Leakey, J. D. Kingston, and S. Ward. 2002. New cercopithecoids and a hominoid from 12.5 Ma in the Tugen Hills succession, Kenya. *Journal of Human Evolution* 42:75–93.

Hooijer, D. A. 1958. Fossil rhinoceroses from the Limeworks Cave, Makapansgat. *Palaeontologia Africana* 6:1–13.

———. 1963. Miocene Mammalia of Congo. *Musée Royal de l'Afrique Centrale: Tervuren, Belgique, Annales*, Series in 8° Sciences Géologiques 46:1–77.

———. 1966. Miocene rhinoceroses of East Africa. *Bulletin of the British Museum (Natural History), Geology* 13:119–190.

———. 1968. A Rhinoceros from the late Miocene of Fort Ternan, Kenya. *Zoologische Mededelingen* 43:77–92.

———. 1969. Pleistocene East African rhinoceroses; pp. 71–98 in L. S. B. Leakey (ed.), *Fossil Vertebrates of Africa*, vol. 1. Academic Press, London.

———. 1971. A new rhinoceros from the late Miocene of Loperot, Turkana district, Kenya. *Bulletin of the Museum of Comparative Zoology* 142:339–392.

———. 1972. A late pliocene rhinoceros from Langebaanweg. *Annals of the South African Museum* 59:151–191.

———. 1973. Additional Miocene to Pleistocene rhinoceroses of Africa. *Zoologische Mededelingen* 46:149–178.

———. 1978. Rhinocerotidae; pp. 371–378 in V. J. Maglio and H. B. S. Cooke (eds.), *Evolution of African Mammals*. Harvard University Press, Cambridge.

Hooijer, D. A., and C. S. Churcher. 1985. Perissodactyla of the Omo Group Deposits, American Collections; pp. 99–117 in *Les Faunes Plio-Pléistocènes de la Basse Vallée de l'Omo (Ethiopie)*. Cahiers de Paléontologie-Travaux de Paléontologie Est-Africaine. CNRS, Paris.

Hooijer, D. A., and B. Patterson. 1972. Rhinoceroses from the Pliocene of Northwestern Kenya. *Bulletin of the Museum of Comparative Zoology* 144:1–26.

Hooijer, D. A., and R. Singer. 1960. Fossil Rhinoceroses from Hopefield, South Africa. *Zoologische Mededelingen* 37:113–128.

Hooker, J. J., and D. Dashzeveg. 2004. The origin of Chalicotheres (Perissodactyla, Mammalia). *Palaeontology* 47:1363–1386.

Kalb, J. E., C. B. Wood, C. Smart, E. B. Oswald, A. A. Mabrete, S. Tebedge, and P. Whitehead. 1980. Preliminary geology and palaeontology of the Bodo D'Ar Hominid Site, Afar, Ethiopia. *Palaeogeography, Palaeoclimatology, Palaeoecology* 30:197–120.

Kingston, J. D., B. F. Jacobs, A. Hill, and A. L. Deino. 2002. Stratigraphy, age and environments of the late Miocene Mpesida Beds, Tugen Hills, Kenya. *Journal of Human Evolution* 42:95–116.

Klein, R. G., G. Avery, K. Cruz-Uribe, D. Halkett, T. Hart, R. G. Milo, and T. P. Volman. 1999. Duinefontein 2: An Acheulean site in the Western Cape Province of South Africa. *Journal of Human Evolution* 37:153–190.

Klein, R. G., and K. Scott. 1986. Re-analysis of faunal assemblages from the Haua Fteah and other Late Quaternary Archaeological Sites in Cyrenaican Libya. *Journal of Archaeological Science* 13:515–542.

Kovarovic, K., P. Andrews, and L. C. Aiello. 2002. The palaeoecology of the Upper Ndolanya Beds at Laetoli, Tanzania. *Journal of Human Evolution* 43:395–418.

Leakey, R. E. F., and A. Walker. 1985. New higher primates from the early Miocene of Buluk, Kenya. *Nature* 318:173–108.

Likius, A. 2002. Les grands ongulés du Mio-Pliocène du Tchad (Rhinocerotidae, Giraffidae, Camelidae): systématique, implications paléobiogéographiques et paléoenvironnementales. Unpublished PhD dissertation, Université de Poitiers, France, 193 pp.

Martínez-Navarro, B., L. Rook, A. Segid, D. Yosief, M. P. Ferretti, J. Shoshani, T. M. Tecle, and Y. Libsekal. 2004. The large fossil mammals from Buia (Eritrea): Systematics, biochronology and paleoenvironments. *Rivista Italiana di Paleontologia e Stratigrafia* (suppl.) 110:61–88.

Michel, P. 1992. Contribution à l'étude paléontologique de Vertébrés fossiles du Quaternaire marocain à partir de sites du Maroc atlantique, central et oriental. Unpublished PhD dissertation, Muséum National d'Histoire Naturelle, Paris, 1,152 pp.

Miller, E. 1999. Faunal correlation of Wadi Moghara, Egypt: Implications for the age of *Prohylobates tandyi*. *Journal of Human Evolution* 36:519–533.

Nakaya, H. 1993. Les faunes de mammifères du Miocène supérieur de Samburu Hills, Kenya, Afrique de l'Est, et l'environnement des pré-Hominidés. *Anthropologie* 97:9–16.

Nakaya, H., M. Pickford, K. Yasui, and Y. Nakano. 1987. Additional large mammalian fauna from the Namurungule formation, Samburu Hills, Northern Kenya. *African Study Monographs* (suppl.) 5:79–129.

Pickford, M. 1983. Sequence and environments of the lower and middle Miocene hominoids of Western Kenya; pp. 421–439 in R. Ciochon, and R. Corruccini (eds), *New Interpretations of Ape and Human Ancestry*. Plenum, New York.

———. 1986. Cainozoic paleontological sites of Western Kenya. *Münchner Geowissenschaftliche Abhandlungen* A 8:1–151.

———. 1987. The geology and palaeontology of the Kanam erosion gullies (Kenya). *Mainzer Geowissenschaftliche Mitteilungen* 16:209–226.

———. 1988. Geology and Fauna of the middle Miocene hominoid site at Muruyur, Baringo District, Kenya. *Human Evolution* 3:381–390.

Pickford, M., H. Ishida, Y. Nakano, and K. Yasui. 1987. The middle Miocene fauna from the Nachola and Aka Aiteputh Formations, Northern Kenya. *African Study Monographs* (suppl.) 5:141–154.

Pickford, M., and B. Senut. 1997. Cainozoic mammals from coastal Namaqualand, South Africa. *Palaeontologia Africana* 34:199–217.

———. 2001. The geological and faunal context of Late Miocene hominid remains from Lukeino, Kenya. *Comptes Rendus de l'Académie des Sciences, Paris, Sciences de la Terre et des Planètes* 332:145–152.

———. 2003. Miocene Palaeobiology of the Orange River Valley, Namibia. *Memoirs of the Geological Survey of Namibia* 19:1–22.

Pickford, M., B. Senut, and D. Hadoto. 1993. Geology and Paleobiology of the Albertine rift valley, Uganda-Zaire: Volume I. Geology. *Centre International pour la Formation et les Études Géologiques, Publication Occasionnelle* 24:1–190.

Pickford, M., B. Senut, D. Hadoto, J. Musisi, and C. Kariira. 1986. Découvertes récentes dans les sites miocènes de Moroto (Ouganda oriental): Aspects biostratigraphiques et paléoécologiques. *Comptes Rendus de l'Académie des Sciences, Paris* 302:681–686.

Pomel, A. 1895. *Les Rhinocéros Quaternaires*. Carte Géologique de l'Algérie, Paléontologie, Monographies, 49 pp.

Prothero, D. R., E. Manning, and C. B. Hanson. 1986. The phylogeny of the Rhinocerotoidea (Mammalia, Perissodactyla). *Zoological Journal of the Linnean Society* 87:341–366.

Raynal, J.-P., D. Geraads, L. Magoga, A. El Hajraoui, J.-P. Texier, D. Lefevre and P.-Z. Sbihi-Alaoui. 1993. La grotte des Rhinocéros (carrière Oulad Hamida 1, anciennement Thomas III, Casablanca), nouveau site acheuléen du Maroc atlantique. *Comptes Rendus de l'Académie des Sciences, Paris*, Série II, 316:1477–1483.

Raynal J.-P., D. Lefevre, D. Geraads, and M. El Graoui. 1999. Contribution du site paléontologique de Lissasfa (Casablanca, Maroc) à une nouvelle interprétation du Mio-Pliocène de la Meseta. *Comptes Rendus de l'Académie des Sciences, Paris, Sciences de la Terre et des Planètes* 329:617–622.

Robinson, P., and C. C. Black. 1974. Vertebrate faunas from the Neogene of Tunisia. *Annals of the Geological Survey of Egypt* 4:319–332.

Savage, R., and P. G. Williamson. 1978. The early history of the Turkana depression; pp. 375–394 in W. W. Bishop (ed.), *Geological Background to Fossil Man*. Scottish Academic Press, London.

Sudre, J., and J.-L. Hartenberger. 1992. Oued Mya 1, nouveau gisement de mammifères du Miocène supérieur dans le sud Algérien. *Geobios* 25:553–565.

Suwa, G., H. Nakaya, B. Asfaw, H. Saegusa, A. Amzaye, R. T. Kono, Y. Beyene, and S. Katoh. 2003. Plio-Pleistocene terrestrial mammal assemblage from Konso, southern Ethiopia. *Journal of Vertebrate Paleontology* 23:901–916.

Thomas, H., and G. Petter. 1986. Révision de la faune de mammifères du Miocène supérieur de Menacer (ex-Marceau), Algérie: discussion sur l'âge du gisement. *Geobios* 19:357–373.

Tougard, C., T. Delefosse, C. Hänni, and C. Montgelard. 2001. Phylogenetic relationships of the five rhinoceros species (Rhinocerotidae, Perissodactyla) based on mitochondrial cytochrome *b* and 12S rRNA genes. *Molecular Phylogenetics and Evolution* 19:34–44.

Walker, A. 1968. The lower Miocene site of Bukwa, Sebei. *Uganda Journal* 32:149–156.

Ward, C. V., M. G. Leakey, B. Brown, F. Brown, J. Harris, and A. Walker. 1999. South Turkwell: a new Pliocene hominid site in Kenya. *Journal of Human Evolution* 36:69–95.

Wolde Gabriel, G., T. D. White, G. Suwa, P. Renne, J. de Heinzelin, W. K. Hart, G. Heiken. 1994. Ecological and temporal placement of early hominids at Aramis, Ethiopia. *Nature* 371:330–334.

Yan, D., and K. Heissig. 1986. Revision and autopodial morphology of the Chinese-European Rhinocerotid genus *Plesiaceratherium* Young, 1937. *Zittelliana* 14:81–109.

THIRTY-FIVE

Equidae

RAYMOND L. BERNOR, MIRANDA J. ARMOUR-CHELU,
HENRY GILBERT, THOMAS M. KAISER, AND ELLEN SCHULZ

Representatives of the Equidae in Africa are known from localities of late Miocene to Recent age, approximately 10.5 Ma to the present. Three-toed equids of the tribe Hipparionini first occur in the early late Miocene, and persist to about 0.5 Ma. The first appearance of the genus *Equus* in eastern Africa is in the Omo Shungura sequence (lower Member G), ca. 2.33 Ma. This is somewhat late compared to Eurasia, where *Equus* first occurred at 2.6 Ma (Lindsay et al., 1980).

We recognize a diverse assemblage of African hipparionine horses, including at the supraspecific rank: *"Cormohipparion," ?"Sivalhippus," Eurygnathohippus, Cremohipparion,* and possibly *Hipparion s.s.* (the last two in northern Africa only). *"Cormohipparion"* is a very early form that is an evolutionary derivative of the North American genus of the same name that first occurred in the Old World 11.2 Ma. Later Vallesian age (ca. 9 Ma) taxa believed to be related to Siwalik hipparionines of the same and younger age are referred to as *?"Sivalhippus"*. The Eurasian genus *Cremohipparion* apparently made a successful range extension into northern Africa during the later portion of the late Miocene (Bernor and Scott, 2003), while *Hipparion s.s.* may also occur at Sahabi (Bernor et al., 2008). *Eurygnathohippus* is a genus of African Hipparionini that first appear in the late Miocene Nawata Formation, Kenya, and successfully spread throughout nontropical forest Africa in the Pliocene and early Pleistocene. It is plausible that *Eurygnathohippus* last occurred in the middle Pleistocene of Africa, but, if present, it was very rare.

We follow Churcher and Richardson (1978) closely in taxonomic allocation of African species of *Equus*. We adopt Groves (2002) in general for the systematics of extant African *Equus* species and develop arguments based on the recent molecular literature where inconsistencies have arisen with that taxonomy. Finally, we provide an update on the dietary behavior of extant zebras and emergent work on African hipparion paleodiet.

Systematic Paleontology

Family EQUIDAE Gray, 1821

Diagnosis Modified from Churcher and Richardson (1978). Unguligrade perissodactyls exhibiting progressive skeletal and dental adaptations for cursorial locomotion and grazing diet respectively. Dentition 3:1–0:4–3:3 (three incisors, one or no canine, four or three premolars, and three molars) in both the maxillary and mandibular dentitions. Incisors in later forms often possessing infundibula (cusps or marks). Canines usually small, sometimes absent or unerupted in females. Premolars 2–4 molarized; dP1's/dp1's primitively being large and sustained into adult life, not replaced by permanent P1's/p1's, and absent in more advanced forms. Upper molars, except in earliest forms, with three lophs: a mesiodistal ectoloph comprised of two arcuate portions; and two obliquely transverse buccolingual lophs (protoloph and metaloph) becoming J-shaped and elaborated with small folds or plis in later forms. Lower molars with progressively enlarged metaconid, entoconid, and metastylid at the junction of both lophids. Body size ranges from small to large. Extremities range from short and robust to elongate and slender. Hipparionine horses persistently tridactyl, *Equus* species always monodactyl.

Age Early late Miocene to Recent in Africa.

Subfamily EQUINAE Steinmann and Doederlein, 1890

Diagnosis Modified from Churcher and Richardson (1978). Skull with postorbital bar. Upper incisors with more or less developed infundibula; in later forms on some or all lower incisors also. Cheek teeth hypsodont with species maximum crown heights ranging from 34 to 90+ mm, being equal to or much greater than mesiodistal length; cementum in fossettes and in later forms over all of crown; small enamel folds or plis present on the pre- and postfossettes and frequently with multiple pli caballins. Premolars 2–4 fully molarized; dP1's/dp1's as for the family Equidae diagnosis. Maxillary molar protolophs and metalophs join ectolophs even in early stages of wear, but in premolars the transverse lophs may remain free until a late wear stage; fossettes usually close with wear. In middle stage of adult wear, protocone is isolated in Hipparionini, while being connected to the protoloph in *Equus*. In mandibular cheek teeth metaconids and metastylids vary from being rounded, to elongate, to squared, with a pointed aspect lingually. Ulna greatly reduced in shaft length sometimes incomplete, fused to radius in nearly all forms. Fibula narrow, reduced, sometimes with only proximal and distal ends remaining; fused distally with tibia in nearly all forms. Metapodial III's vary from short and robust to elongate and slender. First phalanges III vary from being relatively short and massive to elongate and slender. In hipparionines the lateral hoofs are laterally compressed and do not extend to the plantar surface of the central digit (III). In *Equus*, lateral phalanges are lost and lateral metapodials reduced to splint bones fused to the shaft of metapodial III.

Genus *HIPPARION* Christol, 1832

Diagnosis After Bernor et al. (1990, 1996). Medium sized, preorbital bar (POB) length reduced compared to *Hippotherium*, but with the anterior edge of the lacrimal placed still more than half the distance from the anterior orbital rim to the posterior rim of the preorbital fossa (POF). Nasal notch is consistently at or near the anterior border of P2. POF progressively reduced in dorsoventral and anteroposterior dimensions, posterior pocketing, medial depth, and peripheral rim expression. Infraorbital foramen usually encroaches on the anteroventral border of the POF. DP1s/dp1s absent in adults. Maxillary cheek teeth with a maximum crown height of 50–60 mm, pre- and postfossettes moderately complex, often having many plis, but of markedly shortened amplitude, posterior wall of postfossette distinct, pli caballins variably double or single, hypoglyphs deeply to moderately deeply incised, protocones oval shaped to round, isolated from protoloph until very late wear stage, protocone spurs very rare to absent. Mandibular dentition with elongate P2/p2 anterostyle/paraconid, metaconids and metastylids mostly rounded, protoconids reduced to absent and most often covered by cementum, ectostylids absent in adult cheek teeth, premolar and molar linguaflexids variably V- to shallow U-shaped. Metapodials and first phalanges III elongate and slender.

Remarks We adopt the concept for *Hipparion s.s.* previously defined by Bernor et al. (1990, 1996). *Hipparion s.s.* is based on the species *H. prostylum* Christol, 1832 from the middle Turolian locality of Mt. Luberon (= Cucuron = Mt. Leberon, province of Vaucluse, France; MacFadden, 1980; Woodburne and Bernor, 1980; Bernor et al., 1989, 1996). Christol noted that *H. prostylum* was characterized by having three toes on each foot and an isolated protocone. In 1849, Gervais reiterated these two characteristics and further mentioned the occurrence of additional stylids on lower deciduous premolars. Gervais (1849) designated three Mt. Luberon species: *H. prostylum*, *H. diplostylum*, and *H. mesostylum*. Later, Gervais (1859) united all of these species into *H. prostylum*, recognizing the ontogenetic variability of the characters. Gaudry (1873), Osborn (1918), and in later years Sondaar (1974), Skinner and MacFadden (1977), Woodburne and Bernor (1980), MacFadden (1980, 1984), and Bernor (1985; Bernor et al., 1990, 1996) have all agreed in referring at least the majority of the Mt. Luberon material to *H. prostylum*. Sondaar (1974) stated that the specimen figured by Gervais (1859: plate 19, figure 2; maxilla fragment with P3–M2) is the holotype of *H. prostylum*. Bernor (1985) noted that a holotype was never designated and rectified the problem by assigning the most complete specimen, BMNH M33603 to *H. prostylum* as a suitable lectotype. This proposal also maintained the original concept of *H. prostylum* as much as possible.

Sondaar (1974) stated that some postcranial elements in the Mt. Luberon hipparion sample suggest the presence of a larger rare species. Following an unpublished manuscript by Woodburne, Bernor (1985) noted the existence of two cranial morphologies, one a rare form with a more developed POF (BMNH M26617) that was referred to *H.* aff. *prostylum*, and the other, more common form that he referred to *H. prostylum*. Bernor et al. (1990) concluded that the best explanation is that the majority of the Mt. Luberon sample is of *H. prostylum*, while the rarer larger horse is probably of a different clade (Group 1 of Woodburne and Bernor, 1980), simply referred to "*H.*" aff. *prostylum*. Bernor et al. (1996) recognize a number of Eurasian taxa as belonging to *Hipparion s.s.*, including *H. melendezi* (MN10, Spain), *H. gettyi* (MN10/11, Iran), *H. prostylum* (MN12 of France and possibly MN11/12 of Iran), *H. dietrichi* (MN12 of Greece), *H. campbelli* (MN12 of Iran), and *H. concudense* (MN12 of Spain). Bernor et al. (1990) also recognized *H. hippidiodus* (Turolian of China) as a likely member of the *Hipparion s.s.* lineage. Bernor et al. (1989, 1996) referred the Siwalik late Miocene species "*H.*" *antelopinum* to this clade, but later, Bernor and Scott (2003) stated their preference for referring this species to *Cremohipparion antelopinum*.

There is no certain record of *Hipparion s.s.* as defined herein in Africa. Bernor et al. (2008) recently assessed a growing assemblage of hipparionine postcrania from Sahabi, Libya and suggested the possibility that some of this material could relate closely to *H. s.s.* Later in this chapter, we refer all early late Miocene African hipparionines to either "*Cormohipparion*" or ?"*Sivalhippus*" largely due to differences in the skull or postcrania. The previously persistent use of the genus *Hipparion* for advanced African Hipparionini has been rejected by Bernor and Harris (2003), Bernor et al. (2005), and Bernor and Kaiser (2006), except for Sahabi specimens. Eisenmann's and Geraads's (2007) recent referral of a middle Pliocene hipparion sample from Ahl al Oughlam, Morocco to *H. pomeli* is contradicted by the occurrence of well-developed ectostylids on the adult cheek teeth and metapodial III proportions that compare closely with eastern African members of the *Eurygnathohippus* clade (discussed later). It is certainly possible that members of *Hipparion s.s.* could be identified in northern African late Miocene localities such as Sahabi, but this would require virtually complete skull and postcranial material.

Genus *CORMOHIPPARION* Skinner and MacFadden, 1977

Diagnosis After Woodburne (1996, 2007). Species of *Cormohipparion* have mean cheek tooth occlusal lengths from 116 to 140 mm. The unworn or little-worn M1 mesostyle height is reported to range from about 34 to 66 mm (Woodburne, 2007: table 3B). The dorsal POF is prominent with a relatively well-developed and usually continuous anterior rim, with the IOF (infraorbital foramen) alternatively located above the P2/P3 boundary, above the P3, or above the P3/P4 boundary, and consistently very close to the anteroventral limit of the POF. Posteriorly, this fossa has a well-developed rim and deeply recessed pocket. In general, the fossa is oval or teardrop shaped in outline and situated far anterior to the orbit, resulting in a wide POB. The anterior tip of the lacrimal bone enters into the rear of the dorsal POF in primitive species and becomes placed posterior to the POF in advanced species. The POF is lost or severely reduced in advanced species of *Cormohipparion* (*Notiocradohipparion*), such as *C.* (*N.*) *emsliei* (Hulbert, 1987). The dP1 is relatively large and persistent into adulthood, except in more derived species, where it tends to be smaller and not to persist into adult wear, or it tends to be limited to females. The protocone is isolated (except in P2) until very late wear, with a spur in plesiomorphic taxa. The protocone becomes ovate to

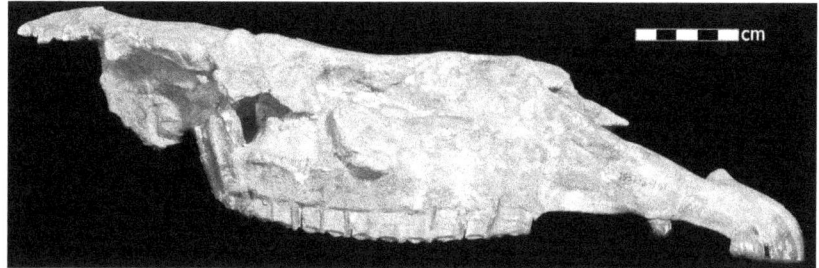

FIGURE 35.1 *"Cormohipparion" africanum* Type MNHN 1951-9-141 cranium, right lateral view.

elongate-oval in shape in advanced species. The P2 antero-style is well developed; P2 is slightly (plesiomorphically) longer to much longer than other cheek teeth. The pli caballin is prominent; usually has multiple plis in premolars and a single one in molars. Fossette borders are moderately to very complex (especially the opposing borders of pre- and postfossettes). The anterior border of the prefossette is increasingly complex in derived taxa. The lower cheek teeth generally possess protostylids (except in *C. goorisi*). Woodburne (2007) modified his diagnosis of *Cormohipparion* following Skinner and MacFadden (1977), MacFadden (1984), and Hulbert (1987). *Cormohipparion* is a genus with several species variously having mesodont to hypsodont cheek teeth. Woodburne (2007) believes that there are no recognized species of *C. s.s.* outside North America.

Remarks Woodburne (2007) recognizes the following 16–10 Ma series of North American *Cormohipparion* species, in order of both their geologic age and phyletic branching pattern, which are congruent: *C. goorisi* MacFadden and Skinner, 1981 (early Barstovian, Texas); *C. quinni* Woodburne, 1996 (late Barstovian, Nebraska and Colorado); *C. merriami* Woodburne, 2007 (early Clarendonian, Nebraska); *C. johnsoni* Woodburne, 2007 (early Clarendonian, Nebraska); *C. fricki* Woodburne, 2007 (early middle Clarendonian, Texas and Nebraska); *C. skinneri* Woodburne, 2007 (late middle Clarendonian, Texas); *C. matthewi* Woodburne, 2007 (late middle Clarendonian, Nebraska), and *C. occidentale* Woodburne, 2007 (late middle Clarendonian, Nebraska and South Dakota). Materials allocated to *C.* sp. occur in El Paso Basin faunas of middle Clarendonian age in California (Woodburne, 2005: figure 2) and in the Devil's Punchbowl, Valyermo, California (Woodburne, 2005: figure 1), also likely of middle Clarendonian age (Woodburne, 2005).

Bernor et al. (2003) recognized a new species of Old World *"Cormohipparion,"* *C. sinapensis*, from the early Vallesian (MN9) of Turkey based on low cheek tooth crown height and a variety of skull and postcranial dimensions and characters. This species allocation was supported by both discrete and continuous variables, and the latter were subjected to univariate, bivariate, multivariate, and log₁₀ ratio analyses. Bernor et al. (2003) likewise assigned postcranial material from another MN9 Turkish locality, Esme Akçaköy, to *C. sinapensis*. Bernor et al. (2003) found consistent morphologic and morphometric differences of the dentition and postcrania between these Turkish *Cormohipparion* and central European *Hippotherium primigenium*. Turkish *C. sinapensis* also lacked essential derived characters typical of more advanced Eurasian clades such as *Cremohipparion* and *Hipparion s.s.* that supported the assignment of the Turkish material to *Cormohipparion*. Woodburne (2007) favored a referral of *C. sinapensis* to the genus *Hippotherium*. This

remains an open issue, but clearly the low crown heights and postcranial morphometrics of *C. sinapensis* are more similar to North American *Cormohipparion* than central European *H. primigenium*.

Bernor et al. (2004) referred a small sample of cheek teeth from the early late Miocene (10.7–10.0 Ma, likely 10.5 Ma) locality of Chorora, Ethiopia to *"Cormohipparion"* sp. This hipparion has a primitive occlusal pattern and apparently increased crown height relative to Sinap *"Cormohipparion"* and Bou Hanifia *"C." africanum*, discussed later. There is no postcranial material, but characters of the cheek teeth generally support this referral. Finally, Bernor et al. (2004) determined that *"C."* sp. from Chorora had the dietary spectrum of an intermediate feeder, very similar to central European Vallesian and Turolian members of the genus *Hippotherium* (Kaiser, 2003; Kaiser et al., 2003), but differed in that it ate C₄ grass.

"CORMOHIPPARION" AFRICANUM (Arambourg, 1959)
Figures 35.1 and 35.2

Diagnosis Modified from Woodburne and Bernor (1980). A medium-sized hipparionine horse with a very long POF (85.6 mm in MNHNM 1951-9-141, type specimen) whose anterior limit is placed above P2. POB is wide, known to range between 42.0 mm (MNHNM 1951-9-141) and 47.7 mm (MNHNM 1951-9-116). The POF is posteriorly pocketed, medially deep, and apparently has an anterior rim (diagnostic for *Cormohipparion* [*sensu* Skinner and MacFadden, 1977]). Maxillary cheek tooth row maximum length is 154 mm. Maximum measured crown height is, on a slightly worn M3, 54.6 mm (MNHN-1951-9-97), suggesting that absolute maximum crown height on an unworn P4 or M1 was about 60 mm. Cheek teeth are complexly plicated, but perhaps not to the degree of *H. primigenium* (Kaiser and Fortelius, 2003). Metapodials have modest length, are slender and well developed craniocaudally at the midshaft, are very similar in size and proportions to *C. sinapensis*, and are primitive relative to *H. primigenium s.s.*

Remarks Arambourg (1959) described a new species of hipparion from the Algerian late Miocene (early Vallesian, MN9, 10.5 Ma; Sen, pers. comm., 1990) locality of Bou Hanifia (figure 35.1). Woodburne and Bernor (1980) and Bernor et al. (1980) studied this material in the context of initial proposals on the supraspecific identity of Old World late Miocene Hipparionini and related *"Hipparion" africanum* to their primitive (Group 1) hipparionines. Bernor et al. (1996) referred the Bou Hanifia hipparion to *Hippotherium africanum* largely based on its plesiomorphic characters.

Study of the early Pliocene hipparion from Langebaanweg E Quarry, *Eurygnathohippus hooijeri* (Langebaanweg; Bernor and Kaiser, 2006), found that *"Hipparion" africanum* (Bou

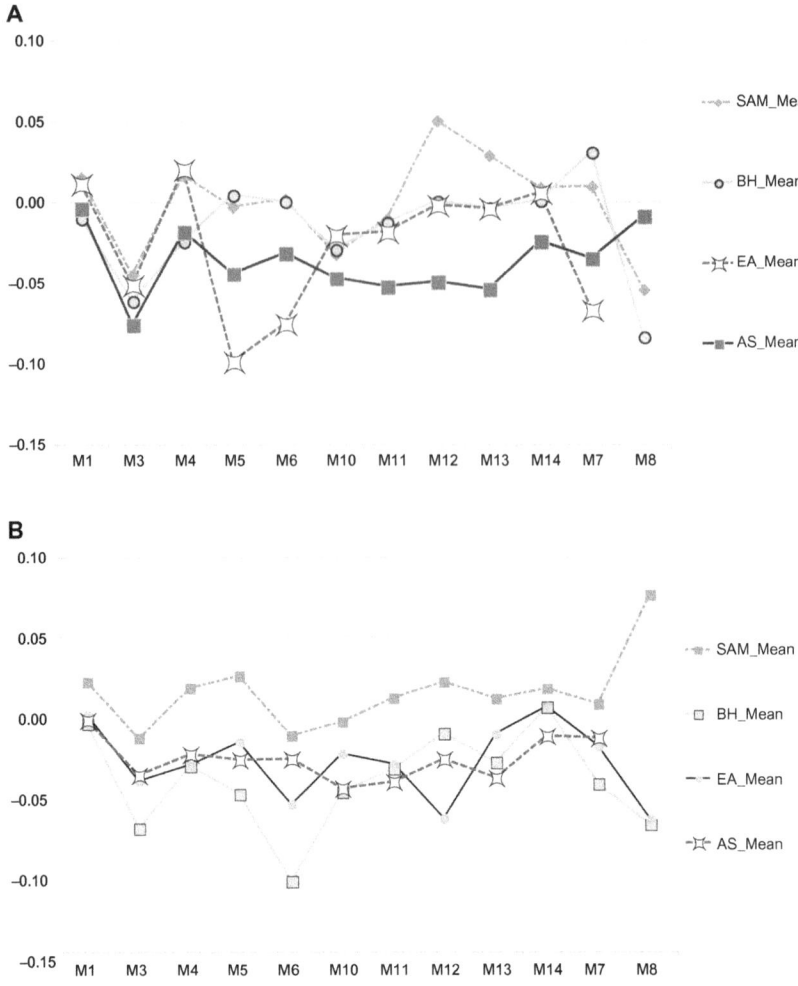

FIGURE 35.2 Log$_{10}$ ratio diagram comparing Bou Hanifia, Sinap, Esme Akçaköy, and Langebaanweg metapodials to the Höwenegg assemblage mean. A) MC III; B) MT III.

Hanifia) had remarkably similar metapodial proportions to *Cormohipparion sinapensis* (Esme Akçaköy and Sinap). Figures 35.2A and 35.2B plot log$_{10}$ ratios of Turkish *C. sinapensis*, Bou Hanifia (Algeria, MN9, ca. 10 Ma) "*C.*" *africanum*, and Langebaanweg (ca. 5.2 Ma) *Eu. hooijeri* metacarpal III's and metatarsal III's (after Bernor and Kaiser, 2006) against the Höwenegg standard. A notable difference was found between the Bou Hanifia metapodials and Turkish *C. sinapensis* metapodials. This difference leads us to provisionally assign the Bou Hanifia hipparion to "*C.*" *africanum*; this sample clearly does not relate to central European *Hippotherium* postcranially and those facial and dental characters they share may be plesiomorphic for Old World hipparionines. "*Cormohipparion*" *africanum* appears to be derived in its elongate fossa (although the type specimen does not exhibit great length on both sides of the skull due to dorsoventral crushing); however, the preorbital fossa's great dorsoventral extent provides the functional constraint on maximum crown height, meaning that it would be primitive for this critical character (Woodburne and Bernor, 1980; Bernor et al., 1980, 1996). The maximum crown height for this species would appear to have been about 60 mm. We cite the further use of the provisional nomen "*Cormohipparion*" as it applies to Siwalik taxa "*C.*" *theobaldi* and "*C.*" cf. *nagriensis* discussed later.

Genus *CREMOHIPPARION* Qiu et al., 1988

Diagnosis Modified after Bernor and Tobien (1989). Hipparionine horses ranging from small to large size with length of tooth row 105–170 mm. POB short, with lacrimal usually touching or invading the posterior limit of the POF. When present, POF is subtriangular in shape and mostly anteroventrally oriented; posterior pocketing slight to absent; medial depth great to slight, medial internal pits occur only in the most derived species, *C. licenti*; peripheral outline strong to weakly defined, anterior rim distinct to absent. Infraorbital foramen placed inferior to and encroaching on the anteroventral border of the POF. Buccinator fossa distinct and unpocketed except in *C. licenti*. Canine fossa present in some but not all species. Malar fossa lacking except in *C. licenti*. Nasal notch tends to become retracted in this lineage and may or may not curve inward in species included within the group. No persistent and functional dP1. In adult, middle stage-of-wear individuals' maxillary cheek teeth with maximum crown height of 40–50 mm, fossette ornamentation complex to simple, posterior wall of postfossette always distinct, pli caballins double or single, hypoglyphs deeply to shallowly incised. Also in adult, middle stage-of-wear individuals' protocones possibly exhibiting some lingual flattening, but tending to become rounded and clearly always isolated from protoloph, usually with no noticeable protoconal spur, and lingually placed relative to the hypocone. P2 anterostyle/paraconid usually elongate, but becomes shortened in some species. Mandibular teeth insufficiently known across all species, but in no case is there reported evidence of ectostylids in the adult dentition. Metapodials, when known, are elongate, to very elongate and slender.

Remarks Qiu et al. (1988) nominated the subgenus *Hipparion (Cremohipparion)* for two Chinese species *H. (Cr.) forstenae* and *H. (Cr.) licenti* from the late Miocene and early Pliocene of China distinguished by their retracted and recurved nasals. Bernor and Tobien (1989) and Bernor et al. (1989, 1996) raised the rank of this taxon to the genus *Cremohipparion* and applied it to the concept of Woodburne and Bernor's "Group 2" hipparionines. *Cremohipparion* is a group of hipparionines known to range from the circum Mediterranean region, through southwestern Asia, southern and eastern Asia. This genus is species rich, and composed of two morphotypes. The first to occur are members of the medium to large morphotype: *Cr. moldavicum* (MN10), and later derived forms *Cr. mediterraneum* (MN11) and *Cr. proboscideum* (MN11 and 12), and Chinese Turolian and Ruscinian age correlative forms *Cr. forstenae* and *Cr. licenti*. These species all have very large and medially deep POFs (although they are secondarily reduced in *Cr. forstenae*) with short POBs; *Cr. proboscideum, Cr. forstenae* and *Cr. licenti* have sharply retracted nasals (Bernor et al., 1996). Having dorsoventrally extensive POFs means that they would have had relatively low cheek tooth crowns. These taxa are most abundantly represented in the "Subparatethyan Province" of Bernor (1983, 1984): eastern Mediterranean to southwestern Asia and the Ukraine. The earliest appearing smaller *Cremohipparion* is *Cr. macedonicum,* first appearing in MN10 of Greece (Koufos, 1984; Bernor et al., 1996). Smaller members of this clade occur in MN12–13 horizons of the Subparatethyan Province (specifically, Iran, Turkey, Greece, Italy, and Spain) and have been referred to the small form *Cr. matthewi* and the very tiny forms *Cr. nikosi* and *Cr. periafricanum. Cr. matthewi* and *Cr. nikosi* are represented by skulls, dentitions and postcrania, and other than their very small size they are remarkable for their extremely slender and elongate metapodials. *Cr. nikosi* also has sharply retracted nasals, which would appear to be convergent on *Cr. proboscideum, Cr. forstenae,* and *Cr. licenti* (from China).

Bernor and Scott (2003) recognized the Siwalik small hipparion with slender elongate metapodials and first phalanges III as being best referred to *Cremohipparion antelopinum*, and metapodial material from Sahabi as being referable to *"Cr."* aff. *matthewi*. The Sahabi postcrania correspond most closely to those from the Samos Main Bone Beds, and as such they are supportive of a late MN12 to early MN13 age for Sahabi. *Cr.* aff. *matthewi* is not reported from any other northern African locality but could relate to the type material of *"Hipparion" sitifense* Pomel, 1897, from St. Arnaud Cemetery, Algeria. The nomen *"H." sitifense* has been applied to a number of African small mammal samples, but as pointed out by Bernor and Harris (2003) and Bernor and Scott (2003), this is inappropriate since there was no type material ever nominated for this nomen and, according to Eisenmann (pers. comm.) the type assemblage cannot be located. What Pomel (1897) figured and reported was insufficient for species recognition and genus assignment. Bernor and Scott (2003) proposed that *"H." sitifense* be considered a *nomen dubium*. That being said, *Cremohipparion* would

FIGURE 35.3 *"Cormohipparion" theobaldi* AMNH 98728 cranial fragment, lateral view.

appear to only be represented in northern Africa, with no current evidence of this genus occurring in sub-Saharan Africa.

Genus ? *"SIVALHIPPUS"* Lydekker, 1877
Figures 35.3–35.5

Diagnosis Large-size hipparionine horses with POF restricted in its dimensions and placed dorsoventally high on the face. POB long, to very long in more advanced forms such as *?"Sivalhippus" perimense*. Cheek teeth high crowned, being 65–75+ mm in maximum crown height. In middle adult wear maxillary cheek teeth complexly ornamented usually with bifid pli caballins, protocones ovate and often flattened lingually, hypoglyphs frequently deeply incised and sometimes encircling the hypocone. Ectostylids absent on adult mandibular cheek teeth. Metapodial III's slender in primitive forms, becoming robust to massive in more advanced forms. First phalanges III robust to massive.

Remarks Bernor and Hussain (1985) recognized three taxa of *Sivalhippus* from the Indian subcontinent, *"Cormohipparion" (Sivalhippus) theobaldi, "C." (S.)* sp. and *"C." (S.) perimense.* MacFadden and Woodburne (1982) earlier referred *"C." (S.) theobaldi* simply to *Cormohipparion theobaldi* and some of the *"C." (S.) perimense* of Bernor and Hussain (1985) to *"Hipparion" feddeni* (skull fragment GSI C349 from Perim Island).

"C." (S.) theobaldi is rare. The type specimen, GSI C153 is a left juvenile maxilla fragment with dP2–4 in early wear. In this specimen, the POF extends very low down on the face. A very large adult skull fragment, AMNH 98728, has a POF that is immense in size, being dorsoventrally extensive, anteroposteriorly elongate, anteroventrally oriented, very deep medially and deeply pocketed posteriorly (figure 35.3).

MacFadden and Woodburne (1982) referred a left skull fragment (YGSP 12507; locality YGSP 330, 9.623 Ma), with P4–M3 and a dorsoventrally extensive POF to *Cormohipparion* cf. *nagriensis*. Specimen YGSP 12507 is not as large as AMNH 98728, and it has a rather short POB length estimate (32 mm est.; MacFadden and Woodburne, 1982: table 1). The cheek teeth have complex plications of the fossettes and lingually flattened protocones on P4–M2, with that of M3 being ovate (MacFadden and Woodburne 1982: figure 13, with associated caption given as figure 16). *"Cormohipparion"* cf. *nagriensis* would appear to be a distinct and more primitive species, and may be derived from North American

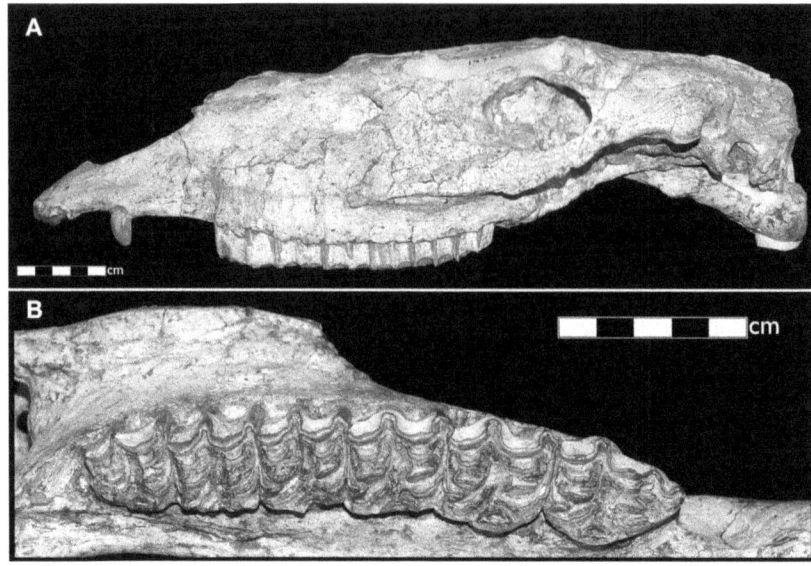

FIGURE 35.4 *"Sivalhippus" perimense* AMNH 19761 cranium. A) Lateral view; B) occlusal view.

Cormohipparion. However, following Woodburne's (2007) revision of the *"C." occidentale* complex and its phylogenetic relationships, neither *S. theobaldi* nor *"C." nagriensis* strictly conform to the concept of *Cormohipparion s.s.* However, AMNH 98728 does have a large POF with a strongly continuous rim, and relatively low crown height that supports its attribution to *Cormohipparion* (*sensu* Skinner and MacFadden, 1977 and MacFadden and Woodburne, 1982). Likewise, *"C."* cf. *nagriensis* has a dorsoventrally extensive POF, with resulting lack of derived high crowns that supports a referral to *Cormohipparion.* While it is possible that the juvenile type specimen of *S. theobaldi* may define a taxon that includes the adult skull fragments that we refer to, this is a somewhat tenuous taxonomic referral. If we chose to restrict the nomen *S. theobaldi* to the immature type, we need a generic name to apply to the adult skull, AMNH 98728. We therefore provisionally apply the nomen *"C." theobaldi* to AMNH 98728 and *"C."* cf. *nagriensis* to YGSP 12507, in broad agreement with MacFadden and Woodburne (1982) and Bernor and Hussain (1985). We underline the importance of further systematic study on Siwalik Hipparionini to better resolve this nomenclature.

Eisenmann (1994) assigned a right P2–P4 and M3, NY 256'90 from the Kakara Formation of the Kisegi-Nyabusosi region, Toro, Uganda (Western Rift) to a new species *Hipparion macrodon.* The assemblage is believed to be early late Miocene. Eisenmann (1994: figures 2 and 3) demonstrated that *"H." macrodon* was much larger than central European *Hippotherium primigenium* (*sensu* Bernor et al., 1990, 1996, 1997) as well as *"C." africanum* from Bou Hanifia. This would mean that *"H." macrodon* was much larger, than any species of *Hipparion s.s.* and thus likely would have had robust postcrania. Eisenmann's figure (1994: plate 1, figure 1) of the occlusal surface of type specimen P2–P4 reveals a species with complex plications of the pre- and postfossette, ovate protocones, and variable pli caballins. In size, *"H." macrodon* most closely compares to *"C." theobaldi* as represented by AMNH 98728. The material is too limited to be referred to another species-level taxon. It is unlikely, however, that *"H." macrodon* is referable to *Cormohipparion, Hipparion s.s., Hippotherium,* or *Cremohipparion.*

The type specimen of *"S." perimense* is GSI C349, a skull fragment from uncertain time horizons at Perim Island, off the coast of the Indian subcontinent. The POB of this specimen is long (61.6 mm) with a moderate POF length (40.2 mm) and dorsoventral height (33.1 mm). Bernor and Hussain (1985) referred an extensive series of GSI skulls (GSI K13/123, GSI C277 [K13/121], GSI C275, and GSI C151), a Munich specimen (GSM H690), and a beautifully preserved skull, AMNH 19761, to this species. Their assessment was based on similarities in skull and dental morphology: a very long POB, a POF reduced in length and dorsoventral height placed far anteriorly on the face—often medially deep with sharply reduced posterior pocketing. Lacrimal's anterior limit positioned far posterior to the POF's distal rim, cheek teeth similar in their complex enamel plications—often with double pli caballins and a protocone that is frequently flattened lingually. AMNH 19761 (figure 35.4) is the most advanced specimen referred to *"S." perimense,* with a long POB (55.8 mm; but not so long as the type specimen), a short POF (43.1 mm), and a very restricted dorsoventral height (26.8 mm). The species hypodigm of *"S." perimense* (*sensu* Bernor and Hussain, 1985) is likely somewhat extended, but still coherent as a clade. If the type specimen of *Sivalhippus theobaldi* (the juvenile maxilla fragment GSI C153) is truly relevant to *"Cormohipparion" theobaldi* as seems possible (dorsoventally extensive POF, low crown heights), then the nomen *Sivalhippus* should be restricted to that taxon. We cannot currently be assured of this assignment. This means that *Sivalhippus* may prove to be inapplicable to the species *"S." perimense.* For the purpose of this paper, and until this problem can be resolved with a thorough phylogenetic analysis, we will apply the nomen *"Sivalhippus"* to *"S." perimense* and apparently related assemblages of taxa in Africa.

Nakaya and Watabe (1990) reported on a modest-sized but important assemblage of early late Miocene hipparionines from the Namurungule Formation, Samburu Hills, Kenya, believed to be 9.0 Ma (Watabe, pers. comm.). A cranium from the site, KNM-SH 15683, was referred to *Hipparion* aff. *africanum* (figure 35.5). As a basis of reference to the type assemblage of *"Hipparion"* (= *"Cormohipparion"* herein) aff. *africanum* we follow Bernor et al.'s (1990, 1996) report on the type assemblage from Bou Hanifia, Algeria, herein. Essential benchmark statistics are: POB length = 44–49 mm; POF length (not secure due to skulls' dorsoventral distortion) = 71–79 mm; dorsoventral height extensive (not measurable due to crushing of skull specimens).

The Samburu Hills skull, KNM-SH 15683, has a well-preserved, ovate, dorsoventrally extensive POF that recalls the type specimen (in particular) of "S." perimense in its overall morphology, rather more than "C." africanum. Remarkable is the POB, which is reported to be 48.9 mm long on the left side and 53.6 mm long on the right side (Nakaya and Watabe, 1990). This compares closely with POBs in the Hoewenegg sample of H. primigenium, whose mean is 48.96 mm (n = 5; Bernor et al., 1997), and "C." africanum cited earlier. It is not as long as the POB seen in the type specimen of "S." perimense (= 61.6 mm) but approaches AMNH 19761 (length = 55.8 m). The POF is not dorsoventrally extensive (42.3 mm) and is moderately elongate (60.3 mm; Nakaya and Watabe, 1990). Overall, the facial morphology of KNM-SH 15683, as exhibited on the left side (Nakaya and Watabe, 1990: plate 1, figure 1A), recalls Siwalik members of "S." perimense and betrays precociously elevated crown heights. In fact, the authors describe a left P4, KNM-SH 12271, with a measured crown height of 66.7 mm and a mandibular p3, KNM-SH 12250, with a measured crown height of 61.0 mm. It is reasonable to estimate that the Samburu Hills hipparion had a maximum crown height of 70+ mm. This crown height exceeds by a margin of >25% any member of the H. primigenium clade, and is similar to early members of the "Sivalhippus" Complex in South Asia and Africa (Bernor and Harris, 2003; Kaiser et al. 2003; Bernor and Kaiser, 2006; Bernor and Haile-Selassie, 2009). The maxillary cheek teeth are also similar to Siwalik hipparionines in their plication complexity and protocone morphology. A single metacarpal III, KNM-SH 12288, has been figured by Nakaya and Watabe (1990) and their measurements suggest affinities with "C." sinapensis and "C." africanum. The same can be said for the first phalanges III (Nakaya and Watabe, 1990: figure 18), which visually appear primitive and less massive in their dimensions.

In summary, the Samburu Hills hipparion may be an early derivative of the "Sivalhippus" Complex that is related to both the Siwalik "S." perimense and African Eurygnathohippus clades. It cannot be referred to Eurygnathohippus due to the lack of any evidence of ectostylids on the adult mandibular cheek teeth (Nakaya and Watabe, 1990; Bernor and Harris, 2003; Bernor and Kaiser, 2006). We follow Bernor and White (in press) in referring this taxon to "C." aff. africanum, being mindful of the advances seen in this taxon's crown height and POF morphology. In fact, the Samburu Hills hipparion could prove to be the sister taxon of the "Sivalhippus" and Eurygnathohippus clades.

Bernor and Harris (2003) have discussed the occurrence of Hipparionini with large and heavily built first phalanges III. In eastern Africa, these are represented best by the Lothagam Nawata Formation form Eurygnathohippus turkanense, and also in proportions, a species of Hipparionini from the Ngorora Formation, Kenya, represented by KNM-BN 1202 and KNM-BN 1598, ca. 9 Ma and similar in proportion to Lothagam Eu. turkanense (Bernor and Harris, 2003) and Sahabi "Sivalhippus" sp. as represented by 2P111A (Bernor et al., 1987, 2005; Bernor and Haile-Selassie, 2009). The Sahabi equid fauna has thus far yielded no adult mandibular cheek teeth with ectostylids, hence our referral of this phalanx to "Sivalhippus" sp. here.

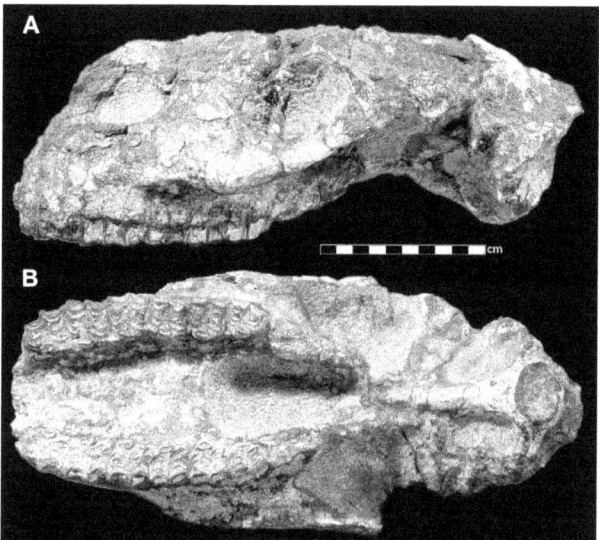

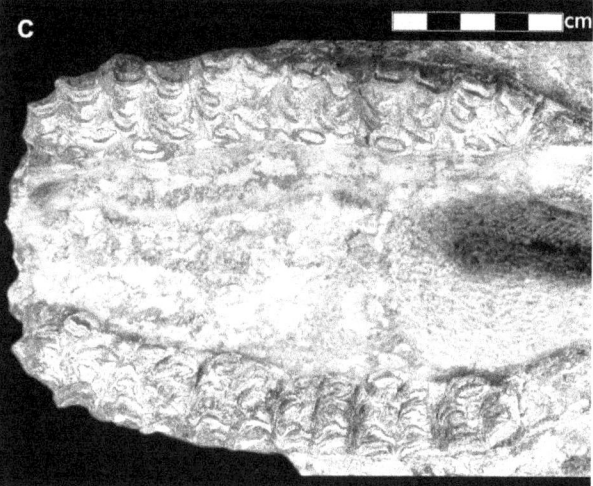

FIGURE 35.5 "Hipparion" (= "Cormohipparion" in this chapter) aff. africanum KNM-SH 15683 cranium. A) Lateral view; B) ventral view; C) occlusal view.

However, the paleobiogeographic relationships between Lothagam and Sahabi are substantial (Bernor et al., 2008), and this phalanx could equally relate to Nawata Formation Eu. turkanense.

Our discussion of Eurasian and African "Cormohipparion," Hippotherium, and "Sivalhippus" should reflect considerable uncertainty as to the generic status of Eurasian and African early late Miocene (ca. 11.2–9 Ma) hipparionines. What is clear is that Hipparionini began to diversify across its Eurasian and African geographic range, in most major clades, within this interval of time. "Sivalhippus" and Eurygnathohippus species are likely sister taxa, united by their sharply increased crown height (the apparent exception being Eurygnathohippus feibeli from Kenya and Ethiopia), with resulting restriction of the POF occurring high and far anterior on the face. Postcrania of these "Sivalhippus" Complex clades were diverse, with some evolving more elongate-slender morphologies of the metapodials and first phalanges III (apparently "C." aff. africanum from the Samburu Hills, Eurygnathohippus feibeli and Eurygnathohippus hooijeri from South Africa), and others exhibiting more robust, and even massive proportions ("S." perimense from the Siwaliks and Eu. turkanense from Kenya and Ethiopia).

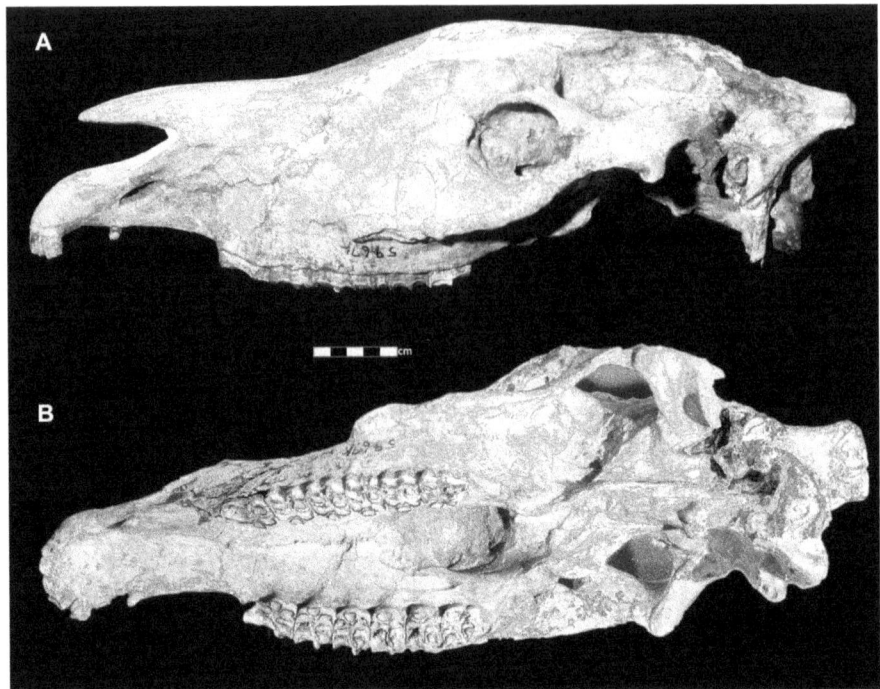

FIGURE 35.6   *Eurygnathohippus turkanense* type cranium, KNM-LT 136. A) Lateral view; B) ventral view.

Genus *EURYGNATHOHIPPUS* Van Hoepen, 1930

Synonymy Stylohipparion Van Hoepen, 1932

Diagnosis Revised from Bernor and Harris (2003). All African hipparionines of the genus *Eurygnathohippus* are united by the synapomorphy of ectostylids occurring on the permanent mandibular cheek teeth. Adult ectostylids are initially small and do not rise high on the labial side of the crown. However, in Africa they progressively evolve in length, width, and height over their chronological range, which is believed to be between 6.5 and less than 1.0 Ma. There is a report of an isolated occurrence of a southern Asian specimen with an ectostylid on permanent cheek teeth (Forsten, 1997) that requires further investigation. Primitive central European hipparionines have small, variably expressed ectostylids. However, to date, no species of Eurasian *Hippotherium* (other than rare occurrences in some MN9 populations), *Hipparion s.s.*, *Cremohipparion*, *Plesiohipparion*, or *Proboscidhipparion* are reported to have ectostylids on the permanent cheek teeth. No North American hipparion species has been reported to have this character either (Bernor and Harris, 2003; Bernor and Kaiser, 2006; Woodburne, 2007).

EURYGNATHOHIPPUS TURKANENSE
(Hooijer and Maglio, 1973)
Figures 35.6–35.8

Diagnosis After Bernor and Harris (2003); modified from and expanded beyond Hooijer and Maglio (1973). A large species of hipparionine equid with a short, moderately broad snout posterior to a line intersecting the two I3s. POF vestigial with no clearly distinguished outline, lacking a posterior rim, very shallow in its medial depth, lacking internal pits, and lacking a peripheral outline and anterior rim. POB nearly indistinguishable due to strong reduction of the preorbital fossa. Orbital surface of lacrimal bone with a distinct and large foramen. Infraorbital foramen inferior to the remnant depression of the POF. Buccinator fossa distinct from the canine fossa and not pocketed. Nasal notch incised just anterior to P2. Maxillary cheek teeth with dP1 strongly reduced. Maxillary cheek teeth believed to be moderately curved, maximum crown height 65+ mm, fossette ornamentation moderately complex, posterior wall of postfossette mostly distinct, pli caballin usually double, hypoglyph moderately deeply to more deeply incised, protocone subtriangular shaped and lingually flattened, protoconal spur generally absent, protocone more lingually placed than hypocone in both premolars and molars. There is no mandible associated with the type cranium, but the larger mandibular teeth from the Upper and Lower Members of the Nawata Formation at Lothagam that have been referred to this species have the following characteristics: p2 paraconid elongate; metaconid and metastylid generally rounded to elongate; metastylid spurs usually lacking on premolars and molars; ectoflexids usually separate, metaconid and metastylid on premolars but not molars; pli caballinid usually absent, rarely single or complex; protostylid often not expressed on occlusal surface, but when it occurs it has a looplike shape projected posterolabially; on permanent cheek teeth ectostylids are variable and usually being small and short in height; linguaflexids are generally V shaped on the premolars, and a deeper and broader U shape on the molars; preflexids and postflexids have enamel margins that vary in their complexity; protoconid enamel band is usually rounded. Postcrania are relatively massive for a hipparionine, metapodials are short and robust with broad proximal and distal articular surfaces; anterior first phalanges III are also stoutly built with broad anterior and posterior articular surfaces.

Remarks The holotype of *Eurygnathohippus turkanense*, KNM-LT 136, is a skull of an old adult female with the premaxillary and maxillary dentition (Hooijer and Maglio, 1973; figure 35.6). The type specimen originates from the Upper Member of the Nawata Formation, Kenya. Hooijer (1975) attributed a limited dental sample from the Mpesida beds, the Lukeino Formation of Kenya (circa 7–6 Ma), and the early Pliocene Yellow Sands of the Mursi Formation, Ethiopia, to *"Hipparion" turkanense*. Hooijer (1975) believed that *"H." turkanense* was similar to *Eurygnathohippus hooijeri* from Langebaanweg, South Africa in lacking ectostylids. Hooijer and Maglio (1973) and later Hooijer (1975) believed that *"H." turkanense* was closely related to *Hipparion hippidiodum* Sefve, 1927 (*sensu* Bernor et al., 1990) and cited its stratigraphic range as being 7–4 Ma. Churcher and Richardson (1978) believed that *"H." turkanense* had affinities with European *"Hipparion"* (= *Hippotherium*) *primigenium* based on incisor morphology and nasal notch incision, while being distinct from *"H." africanum* from Bou Hanifia.

Bernor and Harris (2003) reported a moderate but skeletally diverse dental and postcranial sample from both the Lower and Upper Members of the Nawata Formation, Lothagam that was greatly augmented by collections undertaken by Meave Leakey and colleagues (Leakey et al., 1996; Leakey and Harris, 2003). While the chronologic range of this assemblage is on the order of 7.4–5.5 Ma, most of the sample is derived from the 6.5–6.0 Ma range (C. Feibel, pers. comm.). Bernor and Harris (2003) found that ectostylids variably occur and, when present, are small, rising vertically only a short distance on the labial margin of adult cheek teeth. This morphology is very similar to the Vienna Basin early Vallesian sample from Gaiselberg, Austria (ca. 11.2 Ma; figure 35.7), as well as new, hitherto undescribed late Miocene material from the Middle Awash, Ethiopia. As such, this structure is believed to be rare in the first Old World Hipparionini and becomes a prominent feature only in the *Eurygnathohippus* clade, which thus far is recognized as occurring only in Africa (although Forsten, 1997, has reported a single specimen from South Asia). Bernor and Haile-Selassie (2009) have found it virtually impossible to distinguish *Eu. turkanense* on teeth alone.

Remarkable, and unique for African Hipparionini, is the very robust postcranial skeleton referred to *Eu. turkanense* (Bernor and Harris, 2003). Bernor et al. (2005) reported a strong similarity in metapodial and first phalanx III size and proportions between *Eu. turkanense* and Siwalik late Miocene (circa 8 Ma, J. Barry, pers. comm.) *"Sivalhippus" perimense*. Bernor et al. (2005) further noted that massive first phalanges III are also known from the Baringo Basin (circa 9 Ma, A. Hill, pers. comm.), Kenya, and Sahabi (circa 7 Ma), Libya. Figures 35.8A and 35.8B are log$_{10}$ ratio diagrams of Siwalik *"S." perimense* and *Eu. turkanense* (*sensu lato*) exhibiting the characteristic short and robust morphology that is common between the South Asian and African sister taxa. An interesting collateral issue is whether we can identify ectostylids in the Siwalik members of this group to further strengthen our hypothesis of phylogenetic relationships between *"Sivalhippus"* and *Eurygnathohippus*. We support Hooijer's (1975) observation of a 7–4 Ma chronologic range for *Eu. turkanense*, and note that it occurs only in Kenya and Ethiopia.

EURYGNATHOHIPPUS FEIBELI Bernor and Harris, 2003
Figures 35.9 and 35.10

Diagnosis After Bernor and Harris (2003). A small hipparionine equid with gracile limbs. Metacarpal III elongate and slender with midshaft depth substantially greater than width. Anterior first phalanx III long with very narrow midshaft width. Maxillary cheek teeth with thin parastyle and mesostyle; labiolingually moderately curved to straight, maximum crown height believed to be between 50 and 60 mm; mostly moderate to

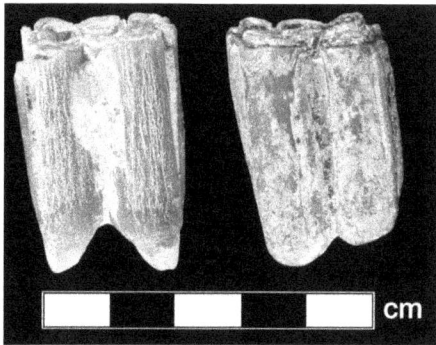

FIGURE 35.7 Gaiselberg *"Hippotherium" primigenium* lower cheek teeth. A) NHMW3, right p4, labial view; B) NHMW3540-206, left p4, labial view.

simple complexity of the pre- and postfossettes; posterior wall of postfossette mostly separated from posterior wall of the tooth; pli caballin mostly single or poorly defined double; hypoglyph variable with wear; protocone tending to be elongate and compressed; protoconal spur usually absent but may appear as a small, vestigial structure; premolar and molar protocone placed lingually to hypocone. Mandibular cheek teeth having premolar metaconid/metastylid mostly rounded, molar metaconid/metastylid mostly rounded to elongate; metastylid spur absent; ectoflexid not separating metaconid/metastylid in premolars, variably separating metaconid/metastylid in molars; pli caballinid mostly absent; when expressed, protostylid is most often presented as a posteriorly directed, open loop; ectostylids are variably expressed and when present are diminutive structures that do not rise high on the labial side of the tooth; premolar and molar linguaflexid shallow V shape; preflexid and postflexid enamel margins generally with simple complexity; protoconid enamel band rounded.

Remarks Bernor and Harris (2003) assigned KNM-LT 139, a partial right forelimb including fragmentary radius, metacarpal III, anterior first phalanx III, anterior second phalanx III, partial metacarpal II, first, second, and third phalanx II and partial metacarpal IV to *Eu. feibeli* (figure 35.9). This specimen has been figured by Hooijer and Maglio (1974: plate 5, figure 7), who assigned it to *"Hipparion"* cf. *sitifense*, considered by Bernor and Scott (2003) to be a *nomen dubium*. The type specimen originates from the Upper Member of the Nawata Formation at Lothagam. In addition, Bernor and Harris (2003) referred a modest sample of Lower and Upper Nawata cheek teeth and postcranial remains to this species.

Eurygnathohippus feibeli has also been identified based on a first phalanx III (JAB-VP-1/1) from the Middle Awash locality of Jara-Borkana, Ethiopia (6.0 Ma; Bernor et al., 2005; Bernor and Haile-Selassie, 2009). Beyond the Lothagam and Jara-Borkana records, the stratigraphic range of *Eu. feibeli* is not certain. There is a record of a small hipparion in the Middle Awash sequence that is secure at 5.7 Ma from the locality of Bilta that is likely *Eu. feibeli*. The best Bilta specimen is an adult mandible, with a short symphysis that preserves adult cheek teeth with small but distinct ectostylids (Bernor and Haile-Selassie, 2009). There are other smaller Hipparionini from the Middle Awash as young as 5.4 Ma that may also be referable to *Eu. feibeli*. There is good material of a larger hipparion with somewhat more robust proportions from Amba West and Amba East that Bernor and Haile-Selassie (2009) referred to *Eu.* aff. *feibeli*, which may yet prove to be a distinct species.

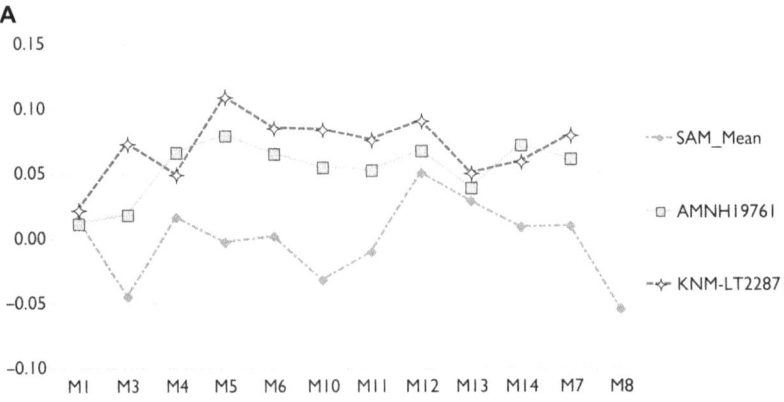

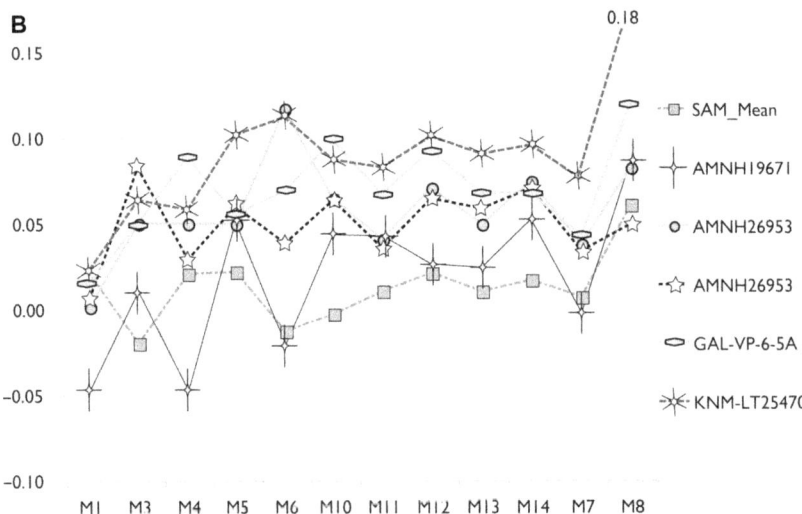

FIGURE 35.8 Log$_{10}$ ratio diagram of robust limbed African and Eurasian *"Sivalhippus"* complex first Phalanges III.

Another important specimen, possibly referable to *Eu. feibeli*, is an early Pliocene (ca. 4.0 Ma) skull from Ekora, Kenya (KNM-EK 4; Hooijer and Maglio, 1974:13–15, plate IV, figures 1–3, plate 5, figure 1; figure 35.10). This specimen is comprised of a partial cranium with deciduous dentition (dP2–4) and M1 clearly exposed in its crypt (Bernor and Harris, 2003). Hooijer and Maglio (1974) referred this specimen to *"Hipparion" primigenium* largely because of the well-developed POF. While the preorbital fossa is large for a sub-Saharan African Pliocene hipparion, it is not as well developed as *H. primigenium s.s.* (Bernor et al., 1988, 1989; Bernor et al., 1997). Moreover, the overall cranial size is less than that of *H. primigenium s.s.*, while the measured crown height of the unerupted M1 (= 55.0 mm) is similar to *H. primigenium s.s.* (Bernor and Franzen 1997; Bernor et al., 1997; Bernor and Harris, 2003; Kaiser et al. 2003; Woodburne, 2007).

As reported by Bernor and Harris (2003), KNM-EK 4, contrasts sharply with the two skulls of *Eu. turkanense*. The Ekora skull is smaller, with a shorter POB and lacrimal situated closer to the POF. The POF is large, subtriangular in shape, anteroposteriorly oriented with moderate posterior pocketing, significant medial depth, and has a peripheral outline that is moderately well delineated. The nasal notch is not preserved, but would not have been highly retracted. Specimen EK 4 is clearly a different species from *Eu. turkanense* but its phylogenetic position is uncertain. The cranial size and cheek tooth morphology of KNM-EK 4 suggest an affinity with Lower and Upper Members, Nawata Formation *Eu. feibeli*, and it was provisionally referred to *Eu.* aff. *feibeli* by Bernor and Harris (2003). In their morphometric comparisons of the larger Langebaanweg hipparion, *Eu. hooijeri*, Bernor and Kaiser (2006) found that while larger, its metapodial proportions exhibited remarkable, and distinctive similarities to the type material of *Eu. feibeli*. It is plausible that *Eu. feibeli* and *Eu. hooijeri* are sister taxa.

EURYGNATHOHIPPUS HOOIJERI Bernor and Kaiser, 2006
Figure 35.11

Diagnosis After Bernor and Kaiser (2006). A large species of hipparionine equid with a moderately elongate snout and arcuate incisor arcade. POB moderately long (37.1 mm) with lacrimal extending near to, but distinctly posterior to, the posterior rim of the POF. Preorbital fossa distinct, but reduced, being unpocketed posteriorly, moderately deep medially, and having a strong dorsoventral orientation. Nasal notch unretracted, extending to near the mesial limit of P2. Infraorbital foramen is situated high on the face indicating great crown height.

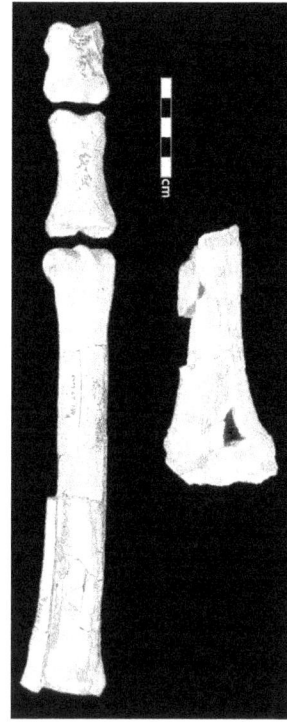

FIGURE 35.9 *Eurygnathohippus fei-beli* type KNM-LT 139 anterior foot, dorsal view.

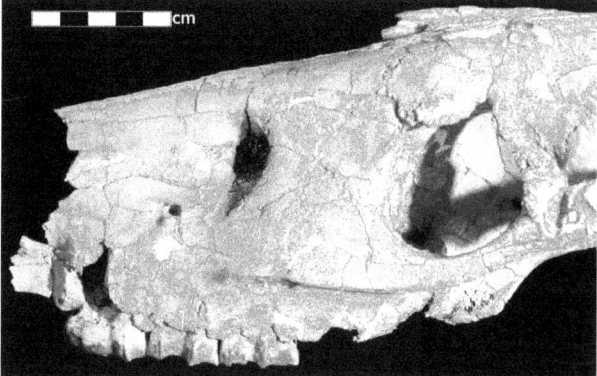

FIGURE 35.10 *Eurygnathohippus* aff. *feibeli* KNM-EK 4 cranium, lateral view.

Maximum crown height of the cheek teeth is 77.5+ (likely 80) mm. Maxillary I3 exhibits slight mesial lengthening with a distinct taper. Maxillary cheek teeth have elongate to oval-shaped protocones, fossette plications are only moderately complex, and pli caballins may vary from single to double. Mandibular cheek teeth have rounded to elongate metaconids, metastylids that may be elongate or pointed in earlier wear and more squared in later wear. Ectostylids are rare but observed on the labial wall of some teeth. Postcrania are mostly at the slender end of the size range of the central European species *Hippotherium primigenium*, exhibiting advances in metapodial III morphology, including lengthening, midshaft that is slender and deep, wider proximal and distal articular surfaces, and distal sagittal keel with increased diameter.

Remarks The holotype of *Eu. hooijeri* is an old adult female skull, SAM PQ-L22187 (figure 35.11) from the Langebaanweg E Quarry (early Pliocene, ca. 5 Ma) originally described and figured by Hooijer (1976: figures 1.1–1.3) under the nomen *Eu.* cf. *baardi*. The Langebaanweg E Quarry hipparion assemblage is extensive and includes the type skull, other partial skull material, mandible, cheek tooth and postcranial material. Some of this postcranial material is associated and has proven to be very valuable for comparisons with other Old World and North American hipparion species.

Bernor and Kaiser (2006) found that the Langebaanweg hipparion exhibits a mosaic of primitive and advanced attributes. The skull has characters that are relatively primitive, such as the shallow incision of the nasal notch, and the moderately long POB with lacrimal closely approaching, but not invading, the POF. Advanced skull characters include: loss of preorbital fossa posterior pocketing, reduced medial depth and infraorbital foramen placed high on the face. The height of the skull, particularly in the region from the anterior cheek tooth row to the posterior nasals, is very great.

The maxillary and mandibular cheek tooth dentition is most remarkable for its great crown height, 77.5 mm being recorded by Bernor for a slightly worn m2. Bernor and Kaiser agree with Hooijer's (1976) initial estimation of a maximum crown height of 80 mm in *Eu. hooijeri*. This crown height is more than 25% greater than in any of the late Miocene Eurasian Hipparionini of the *Hippotherium*, *Cremohipparion*, or *Hipparion s.s.* clades. Such high crowns may be found among advanced members of the *Sivalhippus* Complex in southern and eastern Asia (Bernor and Wolf, in progress). Bernor has not observed cheek tooth crown heights of this magnitude in Ethiopian or Kenyan samples of *Eu. turkanense* and *Eu. feibeli*. In fact, an unworn and unerupted M1 of *Eu.* aff. *feibeli* from Ekora (KNM-EK 4) has a crown height of 55.0 mm. (Bernor and Harris, 2003), and there is no evidence in the Ethiopian or Kenyan late Miocene–early Pliocene record of this clade that contradicts this observation. This evidence supports differences in postcranial size and proportions in asserting that *Eu. feibeli* and *Eu. hooijeri* should be recognized as distinctly separate species.

Bernor and Kaiser (2006) also reported that ectostylids only rarely occur in the Langebaanweg assemblage. They suggest that this rarity is both because ectostylids are not well developed at this early time and are always weakly expressed when present, and because many of the lower cheek teeth have the cementum eroded away, along with any small ectostylid that may have existed. The fact that specimens with an ectostylid exist, and that they are also present in South African *Eurygnathohippus* "namaquense" (after *Notohipparion namaquense* Haughton, 1932 and *Hipparion namaquense* of Churcher and Richardson, 1978) are indicators that they do occur in early Pliocene South African hipparionines.

The E Quarry *Eu. hooijeri* postcrania are generally at the slender end of the size range for Höwenegg *Hippotherium primigenium*. They are, at the same time, strikingly similar to *Cormohipparion sinapensis* from Turkey, and even more similar in their fundamental proportions to *"Hipparion" africanum* from Bou Hanifia, Algeria. The metacarpal III's and metatarsal III's are remarkable for their lengthening, their sharp contrast between midshaft width (M3) versus midshaft depth (M4) and their elevated dimensions for proximal (M5) and distal articular (M11) width and distal sagittal keel depth (M12)(see figure 35.2). These features of the metapodials provide empirical support for the hypothesis that *Eu. hooijeri* undertook functional shifts in its postcranial skeleton that facilitated more effective cursorial locomotion.

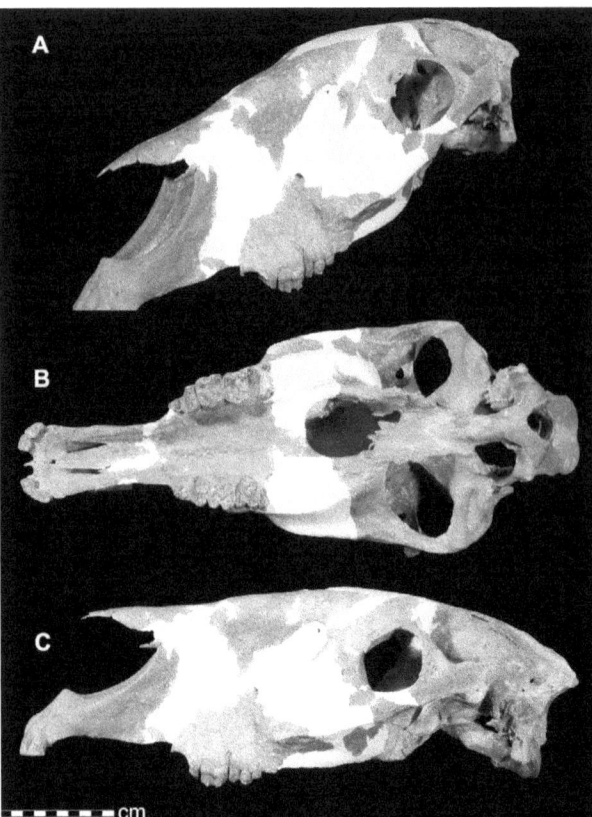

FIGURE 35.11 *Eurygnathohippus hooijeri* type cranium, SAM PQ-L22187. A) Oblique view; B) ventral view; C) lateral view.

Bernor and Kaiser (2006) abandoned the use of Hooijer's (1976) nomen *Hipparion* cf. *baardi*, and alternatively "*Eurygnathohippus*" cf. *baardi* (Franz-Odendaal et al. 2003; Bernor 2006), because the type specimen of *Hipparion (Hipparion) albertense baardi* Boné and Singer, 1965 is insufficient to relate to the Langebaanweg E Quarry hipparion. Bernor and Kaiser (2006) have contended that the type specimen of this species could be referred to any of a number of Old World hipparionines: it is undiagnostic at the species and even the genus level. Finally, Bernor and Kaiser (2006) followed Franz-Odendaal et al. (2003) in determining that *Eu. hooijeri*'s diet is at the grazing end of the dietary spectrum, similar to living zebras.

Eisenmann and Geraads (2007) provided a brief comment on the Langebaanweg hipparion and proposed a new species name, *Hipparion hendeyi*, with the same type specimen selected by Bernor and Kaiser (2006; SAM PQ-L22187; figure 35.11 herein). Eisenmann and Geraads (2007) nomination of "*H.*" *hendeyi* is unsupported by any analysis, morphologic or metric, and contains no figures, no developed diagnosis, and no meaningful systematic comparisons. Moreover, it is superseded by Bernor and Kaiser's (2006) nomination of *Eu. hooijeri*; Eisenmann and Geraads (2007) nomen "*H.*" *hendeyi* is thus a junior objective synonym of *Eu. hooijeri*. We retain the generic designation *Eurygnathohippus*, given the morphological support noted above and in Bernor and Kaiser (2006), and we reemphasize that species of *Eurygnathohippus* are neither closely related to, nor can be objectively included in, the genus *Hipparion* s.s. (Woodburne and Bernor, 1980; Bernor et al., 1996; Bernor and Harris, 2003, Zouhri and Bensalmia, 2005; *contra* Eisenmann and Geraads, 2007).

EURYGNATHOHIPPUS AFARENSE (Eisenmann, 1976)
Figure 35.12

Diagnosis Modified from Eisenmann (1976). A large hipparionine with length basion-P2 being 377 mm. Occipital fossa large. Glenoid processes (= postglenoid processes) large and flat. Vomerine notch acute (V shaped), not arcuate as in *Equus* (and most hipparionines), vomerine index high (14). POF strongly reduced with slight depression and slight posterior rim. POB long (53.2 mm). Maxillary cheek teeth with complex plications of the fossettes, ovate protocones, single to bifid pli caballins. I3 not atrophied (as in *Eu. cornelianus*). Mandibular cheek teeth with ectostylids (albeit not large), preflexid and postflexid enamel borders not complexly plicated, wide linguaflexids. Incisors large, with lingual grooving. Postcrania not certainly known.

Remarks The type specimen of *Eurygnathohippus afarense* is a partial skull, AL363-18, from the Kada Hadar Member of Hadar, Ethiopia (KH3, ca. 3.0 Ma; Eisenmann, 1976). Eisenmann (1976) also referred a mandible, AL177-21, to *Eu. afarense* (figure 35.12). The large incisors of the mandible were an important aspect of Eisenmann's (1976) diagnosis. However, Bernor and Armour-Chelu (1999) noted that the large size compared to *Eu. hasumense* could possibly be due to AL177-21's young age versus *Eu. hasumense* AL340-8's greater age. Eisenmann and Geraads (2007) have subsequently agreed with Bernor and Armour-Chelu (1997) and transferred AL177-21 to the hypodigm of Hadar *Eu. hasumense*. There is no assemblage other than Hadar that has *Eu. afarense* identified. With only a partial skull and dentition and no referable postcrania, the efficacy of *Eu. afarense* is in question. Yet Eisenmann (1976) stated the view that "*Hipparion*" *afarense* represented a "pre-ethiopicum" stage of evolution. She believed that *Eu. afarense* represents the most likely immediate ancestor of African Pleistocene Hipparionini, differing from them in the "nonreduction of the third incisors, the slighter development of ectostylids, and the inferior height of the mandibular ramus, which denotes a lesser degree of hypsodonty" (translated by Churcher and Richardson, 1978).

Churcher and Richardson (1978) considered the cheek teeth of *Eu. afarense* as being close to those of *Hippotherium primigenium*, on the one hand, and more advanced Pleistocene Hipparionini, on the other. In fact, the cheek teeth of *Eu. afarense* are distinctly African Plio-Pleistocene in character, being virtually 50% higher crowned than in *H. primigenium* and having ectostylids on the lower adult cheek teeth along with more ovate protocones and less complexity of the pre- and postfossettes. The origin of *Eu. afarense* may prove to be autochthonous for the northern rift, with its sister taxon plausibly being *Eu. hasumense*. The enlarged incisor teeth of *Eu. afarense*'s were likely an adaptation to increased dependence on grazing, although neither isotopic nor paleodietary analyses have yet been conducted on this sample. This species has not been identified outside Hadar, Ethiopia.

EURYGNATHOHIPPUS HASUMENSE (Eisenmann, 1983)
Figures 35.13 and 35.14

Diagnosis A large hipparion. Skull with little to no preorbital fossa, long POB, elongate narrow snout, with nasal notch incised to mesial border of P2, maxillary cheek teeth with moderate plication complexity and elongate oval protocones, maxillary incisors known to be rather small. Mandible with long and narrow symphysis and rounded incisor arcade. Metapodial III's are elongate and robustly built. First phalanges III robustly built.

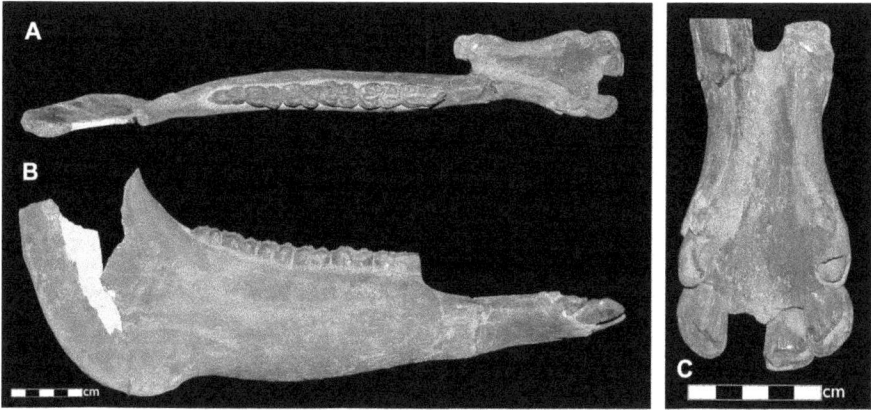

FIGURE 35.12   AL 177–21. *Eurygnathohippus afarense* (*sensu* Eisenmann 1976) AL177–21, mandible.
A) Occlusal view; B) lateral view.

Remarks Eisenmann (1983) recognized *"Hipparion"* (= *Eurygnathohippus*) *hasumense* based on a right p4–m2 cheek tooth row (holotype, KNM-ER 2776) from zones B and C of the Kubi Algi Formation (underneath the Hasuma Tuff). She has included in the hypodigm of *Eu. hasumense* cheek teeth of common morphology from the Chemeron Formation (Kenya) and the Denen Dora Member of the Hadar Formation. Included in this hypodigm was the partial skeleton, including cheek teeth, AL155-6 from DD2 (ca. 3.2 Ma). The AL155 postcrania were analyzed, along with other Hadar and Ethiopian metapodial and first phalangeal material (6–2.9 Ma) by Bernor et al. (2005). Denen Dora 2 has also produced a beautifully preserved skull with an associated mandible of *Eu. hasumense* (*sensu* Bernor et al., 2005), AL340-8 (figure 35.13).

The Hadar sample of *Eu. hasumense* can be characterized from a number of well-preserved specimens. The cranium and mandible (AL340-8) are large and of an old, probably female, adult. The premaxilla is elongate and narrow with an arcuate dental arcade. The nasal notch extends to the mesial boundary of P2 with the anterior nasal opening being long and V shaped anteriorly. The mandible lacks the anterior incisor dentition and symphysis, and only the left p2–p4 are preserved. The preserved teeth exhibiting modestly sized ectostylids on p2–p4; cheek teeth have rounded to elongate metaconids and metastylids that are rounded to pointed linguodistally; pre- and postflexids are labiolingually compressed and lack complex plications. Another mandible, AL425-1 appears to have weathered out from the beneath the uppermost Denen Dora sandstone which caps the member. This mandible is remarkable for its very long symphysis and very narrow, rounded incisor arcade (figure 35.14). It is, in fact, longer in this dimension than the AL177-21 mandible of *Eu. afarense* (now transferred to *Eu. hasumense* by Eisenmann and Geraads, 2007) and does not have the large incisors found in that specimen (although they are quite worn due to advanced age). The AL425-1 mandible closely corresponds to the AL340–8 skull, particularly in its elongate snout portion and narrow, rounded dental arcade.

Bernor et al. (2005) analyzed the large sample of metapodials III and first phalanges III from Hadar and found that it was very advanced among African Pliocene hipparionines. The postcranial skeleton is very large for a hipparionine, with the metapodials III and first phalanges III being robustly built and elongate. The authors concluded that most of the Hadar postcranial material was homogeneous in this regard (except one metapodial III from KH3, which was smaller) and that the pre-

dominant horse at Hadar was *Eu. hasumense*. Bernor et al. (2005) found that species diversity in Ethiopian horizons between 6 and 2.9 Ma was low, with the vast majority of specimens being allocated to the *Eu. feibeli–Eu. hasumense* lineage.

Bernor and Armour-Chelu (1997) reported a partial skull from Manonga Valley Beredi 3, WM1958/92, as being morphologically very similar to the AL340-8 skull from Hadar. The Beredi postcrania are currently under study by Bernor and Armour-Chelu and should prove useful for further comparisons with the Ethiopian *Eu. feibeli–Eu. hasumense* lineage.

Eu. hasumense would appear to have a chronologic range of at least 3.5–2.9 Ma, and is known to range geographically from Ethiopia to Tanzania. As cited, it is a member of the *Eu. feibeli–Eu. hasumense* lineage, which is sister to *Eu. afarense*.

EURYGNATHOHIPPUS CORNELIANUS van Hoepen, 1930

Diagnosis A medium-sized hipparion with cranium lacking POF and having an elongate premaxilla with hypertrophied first and second incisors and atrophied third incisors. All maxillary incisors are strongly grooved on the lingual surface. Mandibular symphysis very broad with hypertrophied, procumbent i1's and i2's with heavily developed ribs on the lingual surface. Mandibular i3's are atrophied and placed adjacent and immediately behind the i2's. Maxillary cheek teeth have moderately complex plications of the pre- and postfossettes and have a maximum crown height as high as 90 mm in their later stratigraphic occurrence. Postcrania referred to this taxon are similar in size, but are mostly somewhat more elongate than *Hippotherium primigenium*.

Remarks Van Hoepen (1930:23, plates 20–22) described an anterior mandibular dentition with very large, hypertrophied i1's and i2's and atrophied i3's placed immediately posterior to the i2's. As pointed out by Cooke (1950), Eisenmann (1983) and Bernor and Armour-Chelu (1997), Van Hoepen (1930) mistook the atrophied i3's for canines. Dietrich (1942:97) is credited as being the first author to suggest a synonymy between *Eurygnathohippus* and *Stylohipparion* (Eisenmann, 1983). Leakey (1965: plate 20, 4 figures) reported the occurrence of *Stylohipparion* (= *Eurygnathohippus*) *albertense* from Bed II of Olduvai Gorge, figuring three mandibular symphyses and one premaxilla complete (lower left corner of figure) with their incisor dentitions. The mandibular dentitions were identical to the type specimen of Van Hoepen from the locality of Cornelia, (Orange) Free State.

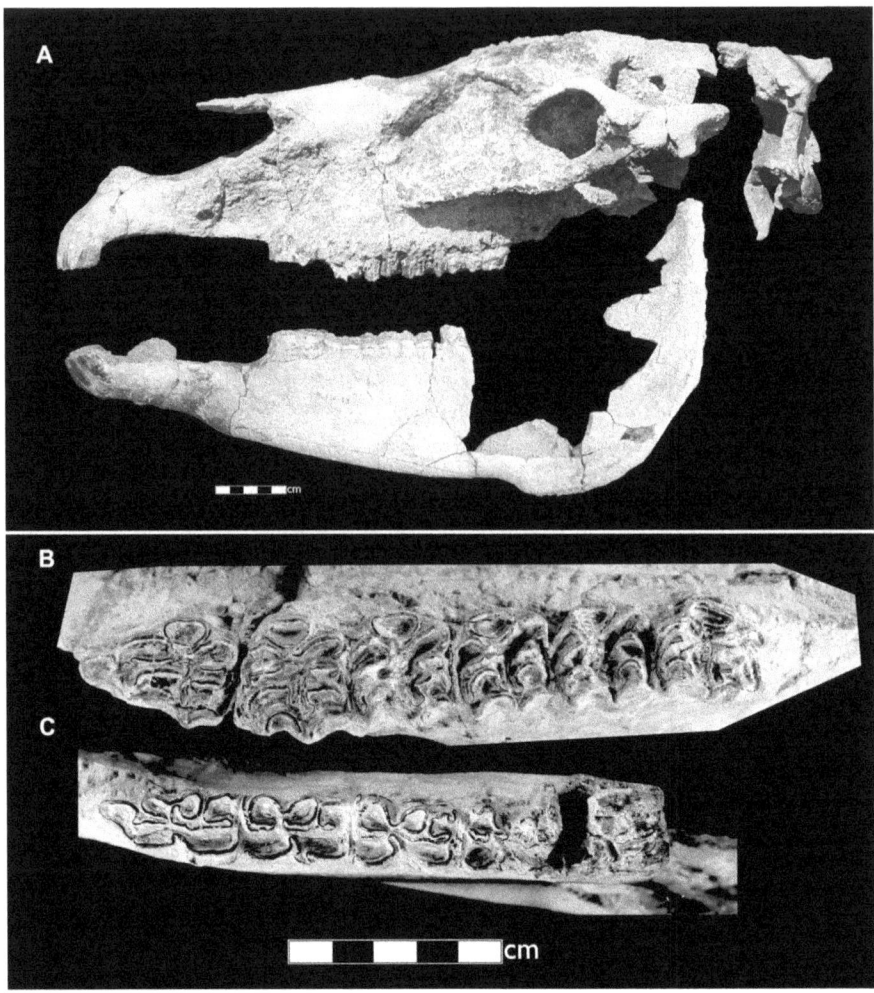

FIGURE 35.13 *Eurygnathohippus hasumense* AL340-8. A) Cranium and mandible, lateral view; B) right maxillary cheek teeth occlusal view; C) left mandibular cheek teeth, occlusal view. Courtesy of Vera Eisenmann.

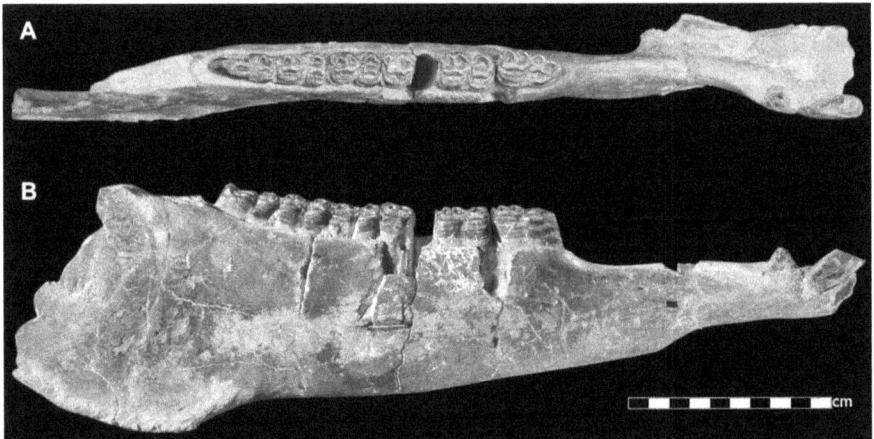

FIGURE 35.14 *Eurygnathohippus hasumense* A.L. 425-1 mandible. A) Occlusal view; B) buccal view; C) anterior dentition.

Hooijer (1975: plates 7–8) reported an adult skull from Olduvai BK II (plate 6, nos. 2845–2846), which he referred to *Hipparion* cf. *ethiopicum*. The skull is nearly complete, albeit distorted due to crushing. This skull is of an older individual with M3 appearing well worn. Hooijer (1975) reported that there is no POF. The snout is relatively long, but there is apparently some portion of the maxilla missing anterior to P3 that disallows an appreciation of its true length. The nasal notch is preserved on the left side and is not retracted, apparently incised a short distance anterior to P2. The incisors are worn. Hooijer (1975:30) reported, "The incisors of the Olduvai skull are of the type of *Eurygnathohippus cornelianus* Van Hoepen (1930): the first and second large, anteriorly flattened, with thick enamel in front and thin enamel lingually, and large cusps completely filled with cement. There is a shallow longitudinal groove along the center of the labial surface of the first incisor, and the second has two such anterior grooves. The lingual surface (of the incisors) is grooved also. The third incisor is much reduced." We add that the right I3 would appear to conform to other Olduvai BK II specimens cited earlier as referable to *Eu. cornelianus.* Hooijer (1975) also cites the lack of a canine in this Olduvai specimen.

Hooijer (1975) reported on another cranium, OLD/63 BK II no. 283 (from the Channel Sand) that includes P2–M2 (P4 erupting and M3 still in the crypt). The specimen lacks the snout and cranium. This specimen provides a suitable contrast to the older adult cranium (nos. 2845–2846) for viewing cheek tooth morphology. These cheek teeth exhibit moderate plications of the fossettes, protocones that are elongate earlier in wear, shorter in later wear and nearly uniformly flattened lingually and rounded labially. Hooijer (1975: 33) reported that the barely worn M2 has a crown height of 75 mm, and the unworn M3 has a chord (not arc) measure of 75 mm. These data imply a maximum crown height of over 75 mm.

Eisenmann (1983: plate 5-3 A–C) did not recognize the existence of *Eu. cornelianus* at Olduvai following Hooijer's (1975) description. She did, however, refer an immature cranium, KNM-ER 3539 with dP2–4, M1 to *Hipparion* (= *Eurygnathohippus*) *cornelianum* (Eisenmann, 1983; however, note that the same specimen is described by Eisenmann [1976: plate 3] as *H.* cf. *ethiopicum* following Hooijer, 1975). This is a well-preserved cranium lacking a POF. Eisenmann (1983) claims that the incisors are large and grooved, but none are figured (neither are they figured by Eisenmann, 1976): the premaxilla in its figured dorsal view is devoid of teeth. This juvenile skull has a moderately long and rather narrow snout. Eisenmann (1983) reports similarities in vomer and incisor morphology between the Koobi Fora skull and Hadar material of *Eu. afarense.* Eisenmann (1983) has further reported *Eu. cornelianus*–type incisors from Omo Shungura D, E, F, G, and L. We have not seen these.

Eisenmann (1983) recognized a second taxon from Koobi Fora and the Omo Shungura Formation, distinguished by its smaller incisor teeth, and referred it to *Hipparion ethiopicum.* She reports a mandible from the same stratigraphic level at Koobi Fora as the *Eu. cornelianus* skull (collecting area 105, *Notochoerus scotti* zone) that has a symphysis notably smaller than the typical *H. cornelianum.* She characterized this taxon as having "large molars relative to small premolars, large anteroposterior development of the ectostylid (high ectostylid index), a great vestibulo-lingual development of the ectostylid (great occlusal width). In addition, the ectostylids are complicated, with accessory pillars and some molars have shallow vestibular grooves."

Bernor and Armour-Chelu (1999) reviewed the problematic classification of *"Hipparion" ethiopicum* and *"H." cornelianum* and determined that it was difficult to recognize a single taxon given the diverse size and morphologies of the sample. Eisenmann's (1983) referral of cranial specimen KNM-ER 3539 to *"H." cornelianum* was believed to be questionable because of its juvenile stage of development, absent

incisors, assertion of atrophied third incisors (third incisors are atrophied in the mandible, not maxilla; see Leakey: 1965, plate 20, lower left maxilla vs. other three mandibular incisors). Moreover, the KNM-ER 3539 skull is placed at 1.89 Ma, while the Olduvai BK II assemblage is >1.2 Ma, (possibly near 1.3 Ma) meaning that considerable evolution could have occurred within this interval of time. Eisenmann and Geraads (2007) refigured KNM-ER 3539, but with a previously unpublished view of the dorsal aspect of the premaxillary incisal region. In this photograph, one can see a very large right and left I1 just beginning to emerge from the crypt. However, this specimen is so immature that one should not be caught in the trap of recognizing a taxon based upon enlarged incisors at this stage of development as was done by Eisenmann (1976) with the A.L. 177-21 specimen of *Eurygnathohippus "afarense."* Our sense is that the Koobi Fora cranium does not have an extraordinarily broad incisor region and cannot readily be separated from Eisenmann's sense of *H. ethiopicum.* More fossils are necessary to address this issue. Particularly convincing would be an adult mandibular symphysis region with hypertrophied i1–i2 and atrophied i3 as found in the holotype *Eu. cornelianus* and Olduvai Bed II specimens figured by Leakey (1965).

Armour-Chelu et al. (2006) have reviewed Hooijer's concept of *H. ethiopicum* and, by extension, *Eu. cornelianus.* They argue that the first evidence of this clade may be from the Upper Ndolanya Beds, Tanzania, circa 2.6 Ma (Ndessokia, 1990). It is also likely represented from Omo Shungura F, dated to 2.36 Ma. Armour-Chelu et al. (2006) have further pointed out that Churcher and Richardson (1978) referred collections with the fundamental cheek tooth morphology (high crowned, large ectostylids) to three regional subspecies: *Hipparion libycum libycum, Hipparion libycum ethiopicum,* and *Hipparion libycum steytleri* for assemblages from northern Africa, eastern Africa, and southern Africa, respectively. Armour-Chelu et al. (2006) have pointed out three problems with this taxonomic solution: first, a detailed morphological comparison of all these populations has not been made; second, crucial statistical analysis of postcranial elements has not been made; third, the diagnostic anterior premaxillary and mandibular symphysial dentitions are lacking in most African Plio-Pleistocene assemblages (outside of Olduvai Bed II and the type material of *Eu. cornelianus* from Cornelia, South Africa). We also add here that stratigraphically younger Olduvai material (possibly from Bed IV) has measured maximum crown heights of 90 mm (Bernor, pers. obs.), which is substantially higher than the 75-mm crown heights reported by Hooijer (1975) for BK II.

Our current understanding of *Eu. cornelianus* is that this species is a member of an evolving lineage that occurred between ca. 2.4 and < 1 Ma. As a result of their statistical analysis on Olduvai Bed II metapodials and astragali, Armour-Chelu et al. (2006) concluded that Olduvai Beds I and II had a hipparion referable to *Eu. cornelianus s.s.,* which is related, at least in part, to Hooijer's hypodigm of *H.* cf. *ethiopicum.* Specimen KNM-ER 3539 may be a member of this lineage, however, its referral to *Eu. cornelianus s.s.* cannot be determined because of its juvenile status. Also, their analysis revealed the likelihood that there is a second, smaller species of *Eurygnathohippus* known from Olduvai Bed II, as well as Omo Shungura F, H, and K that is not referable to *Eu. cornelianus.* Armour-Chelu et al. (2006)

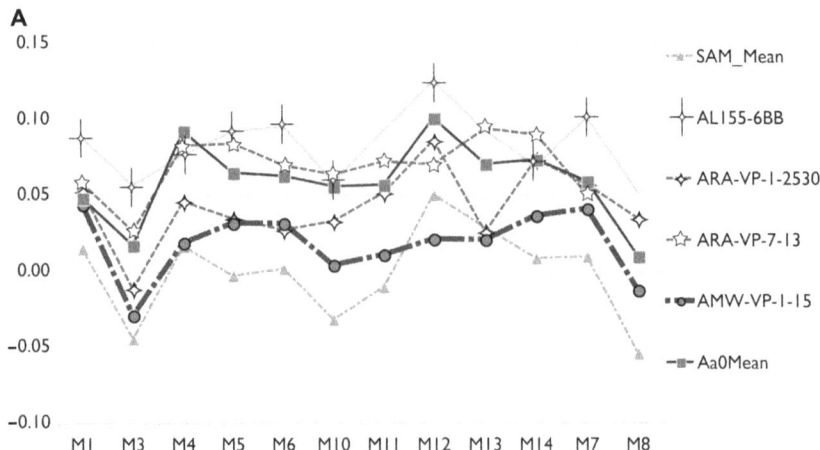

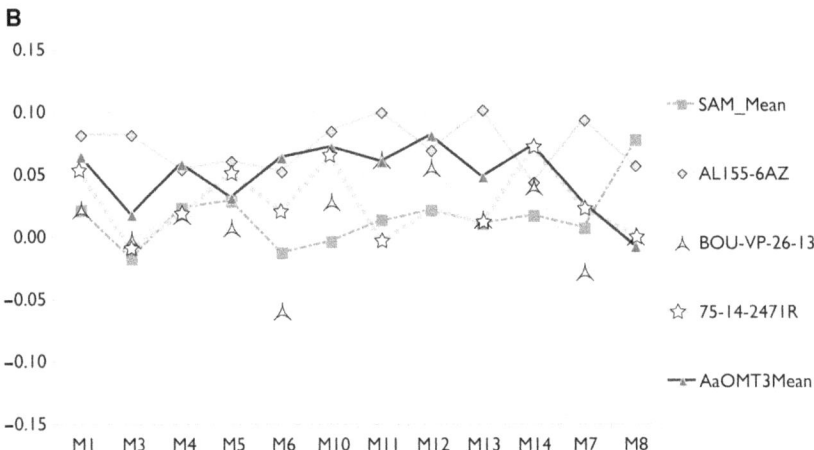

FIGURE 35.15 Log₁₀ ratio diagram comparing Langebaanweg, Hadar, Aramis, Amba West, Daka, Gona, and Ahl al Oughlam MT III's and MC III's. A) MC III; B) MT III.

followed Bernor and Armour-Chelu (1999) in provisionally recognizing the nomen *Eu.* "*ethiopicus*" for some Omo Shungura F, G, and H hipparionines. Clearly, demonstrating the highly derived, hypertrophied incisor structure and accompanying broad mandibular symphysis of *Eu. cornelianus* is important for referral of an assemblage to that species.

Gilbert and Bernor (2008) identified cheek teeth, metapodials and astragali from the 1 Ma Daka fauna of Ethiopia that related well to the Olduvai BK II *Eu. cornelianus* assemblage. As a result, they referred the Daka hipparion assemblage to *Eu.* cf. *cornelianus*.

EURYGNATHOHIPPUS POMELI
(Eisenmann and Geraads, 2007)
Figures 35.15 and 35.16

Diagnosis Modified from Eisenmann and Geraads (2007). A species of *Eurygnathohippus* with moderately elongate and moderately wide muzzle; preorbital fossa reduced, with long POB (50 mm); incisor arcade relatively narrow and tightly arcuate; cheek teeth large and hypsodont, with protocones rather elongate and ovate; lower adult cheek teeth with well-developed and persistent ectostylids rising high on the crown; metapodials elongate and similar in proportions to

early and middle Pliocene *Eurygnathohippus* sp. from Ethiopia and Tanzania, not being as large and derived as the Hadar hipparion.

Remarks Eisenmann and Geraads (2007) reported a well-preserved assemblage of hipparionini Ahl al Oughlam near Casablanca. This assemblage has yielded a very rich micro- and macromammal fauna together with fishes, reptiles, and birds. Eisenmann and Geraads (2007) argue that the sample is homogeneous and biochronologically correlative with eastern African faunas that are ca. 2.5 Ma. in age, roughly contemporaneous with Omo Shungura D.

Eisenmann and Geraads's (2007) analysis of the skull has been largely focused on the shape of the vomer. They claim that *"Hipparion" pomeli* is a member of the "*H.*" *hasumense* group because the basion to vomer distance is short and the cheek teeth are relatively large. They distinguish this "group" from the "*H.*" *afarense* group which, based on a single specimen, has a V-shaped vomer. They relate "*H.*" *afarense* to the juvenile Koobi Fora skull, KNM-ER 3539 (discussed earlier), which they refer to "*H.*" *cornelianum*, along with a series of Olduvai skull fragments (Olduvai BKII-264, BKII-283, BKII-067/5465). They also relate a number of other Hadar, Omo, East Turkana, and Cornelia specimens to this "group" of hipparionines. In fact, Hadar-aged and younger African hipparions are consistent in the one character that best

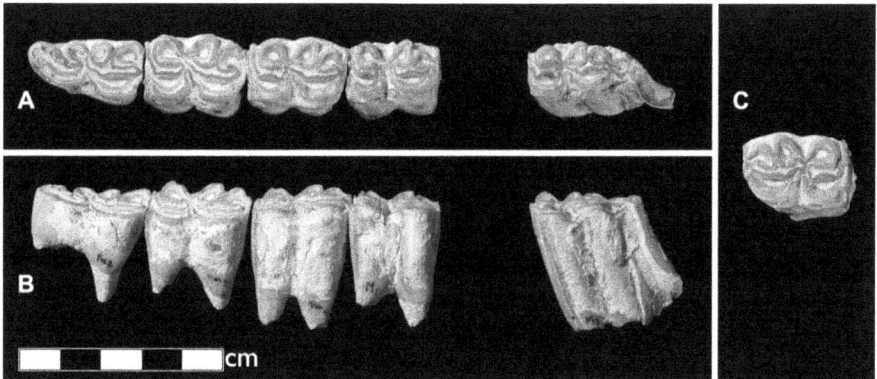

FIGURE 35.16 *Notohipparion namaquense* SAM PQ-9982 type mandibular dentition discovered between Langebaanweg C and E quarries.

unites the *Eurygnathohippus* clade: the presence of ectostylids on the permanent cheek teeth. The ectostylid may not be present on the occlusal surface of a fresh crown, but it is always present on the side of the crown and becomes apparent with wear. Later Pliocene-Pleistocene hipparions, including *Eu. pomeli*, have well-developed ectostylids on the permanent cheek teeth.

Eisenmann and Geraads (2007: figure 8) provide a view of Ahl al Oughlam mandibular cheek teeth that clearly have well-developed, enlarged ectostylids on the permanent cheek teeth. The evolutionary stage of these cheek teeth is consistent with a middle Pliocene correlation and confirms referral to the genus *Eurygnathohippus*. There is likewise an excellent sample of metacarpals III and metatarsals III. Figures 35.15A and 35.15B are $\log_{10}$ ratio diagrams comparing the mean figures given by Eisenmann and Geraad's (2007) for metacarpal III and metatarsal III to a number of more slender-limbed African *Eurygnathohippus* species. Figure 35.15A compares the Ahl al Oughlam metacarpal III mean measurements (AaO) with *Eu. hooijeri* from Langebaanweg (SAM, ca. 5.2 Ma), *Eu. aff. feibeli* from Amba West (Middle Awash, 5.2 Ma; remarkably similar to the Langebaanweg hipparion), and two specimens from Aramis (Middle Awash, 4.2–4.0 Ma); and AL155–6BB of *Eu. hasumense* from Hadar (3.2 Ma). This comparison reveals that the metacarpal III's are most similar to the Aramis specimens, and most definitely not as elongate, not as wide at the midshaft, and without such pronounced keel development as Hadar *Eu. hasumense*. Figure 35.15B compares the Ahl al Oughlam (AaO) metatarsal III mean with Hadar *Eu. hasumense*, Daka (Middle Awash, 1.0 Ma) *Eu. cf. cornelianus*, and Laetoli *Eurygnathohippus* sp. (75–14–2471B, Upper Ndolanya Beds, 2.6 Ma). The Ahl al Oughlam metatarsal III is most like the Laetoli specimen, particularly in length and midshaft width, and is unlike Hadar *Eu. hasumense*. A similarity with Laetoli is consistent with Eisenmann's and Geraads' ca. 2.5 Ma correlation.

There is no basis to deny inclusion of *"Hipparion" pomeli* in the genus *Eurygnathohippus*. The large, well-developed ectostylids on the lower cheek teeth secure this referral. The highly reduced POF, elongate oval protocones and metacarpal III and metatarsal III proportions offer congruent morphological support for this referral. *"H. Hipparion" pomeli* is thus not referable to *Hipparion s.s.* Also, *Eu. pomeli* is not closely related to *Eu. hasumense* but rather has limb proportions more similar to other species of African *Eurygnathohippus*

ranging from 5–1 Ma, suggesting that Hadar *Eu. hasumense* may have been an evolutionary side branch of *Eurygnathohippus* evolution.

REMARKS ON THE TAXONOMY OF HIPPARIONINI

African hipparionine horses have a taxonomic history that is best described as mind numbing. The group's taxonomy requires a major review of specimens, associated geologic context, and taxonomic history that is beyond the scope of our current report here. Churcher and Richardson (1978) made a valiant effort to unravel the alpha taxonomy of the African Hipparionini; however, the time and place in Neogene Old World equid systematics predicated referring all hipparionines to the genus *"Hipparion."* In the last 30 years, Old World *Hipparion* has been demonstrated to be a highly paraphyletic group better segregated into genus-level lineages (cf. Bernor et al., 1996; Bernor and Armour-Chelu, 1999).

Contrary to Eisenmann and Geraads (2007), the vast majority of African hipparionines, and in particular the Plio-Pleistocene forms, are not referable to the genus *Hipparion s.s.* The continued practice of referral of African hipparionines to the Eurasian genus *Hipparion* (with the exception of some Sahabi specimens) is scientifically regressive. Bernor and Armour-Chelu (1999) gave an overview of Africa's diverse group of hipparionines and followed Woodburne and Bernor (1980) and Bernor et al. (1990, 1996) in offering a revisionary sketch. Since then, Bernor and Harris (2003), Bernor and Scott (2003), Bernor et al. (2004, 2005), Franz-Odendaal et al. (2003), Bernor and Kaiser (2006), Armour-Chelu et al. (2006), Bernor and Kaiser (2006), Gilbert and Bernor (2008), and Bernor and Haile-Selassie (2009) have undertaken specimen-based studies of eastern and southern African hipparionines. We update Bernor and Armour-Chelu's (1999: table 14-2) list of African *Hipparion* taxa in table 35.1, continuing the progressive review of this complex problem.

In this contribution we no longer support Churcher and Richardson's (1978) recognition of the genus *Hipparion s.s.* throughout Africa, although the genus plausibly occurs at Sahabi (Bernor et al., 2008). Churcher and Richardson's (1978) application of the nomen *"Hipparion"* (= *Hippotherium*) *primigenium* is likewise unsubstantiated at the species level for localities in Algeria, Ethiopia, Uganda, Kenya, and South Africa. We have cited herein our preference for the use

TABLE 35.1
Summary of the biogeographic distribution of African Neogene

Taxon	Chronology	Geographic Range
HIPPARIONINI		
Hipparion s.s.	9.7–6.5 Ma	Eurasia and northern Africa
Cremohipparion	9.7–4 Ma	Eurasia and northern Africa
Cremohipparion aff. *matthewi*	6.5	Sahabi
Cormohipparion	16–8 Ma	Predominantly North America, in Eurasia and possibly northern Africa 11.2–9.7 Ma
"Cormohipparion" theobaldi	8–7 Ma	South Asia and related form "?*C.*" *megadon* in East Africa
"Cormohipparion" africanum	10.5 Ma	Northern Africa
"Sivalhippus" spp.	9–6.5 Ma	South Asia (*"S." perimense*), East Africa (*"S."* sp. from Samburu Hills) and Sahabi
Eurygnathohippus	6.5–0.5	Africa
Eurygnathohippus turkanense	6.5–4.0 Ma	Ethiopia and Kenya
Eurygnathohippus feibeli	6.0–4.0 Ma	Ethiopia and Kenya
Eurygnathohippus hooijeri	5.2 Ma	South Africa
Eurygnathohippus afarense	3.0 Ma	Ethiopia
Eurygnathohippus hasumense	3.5–2.9 Ma	Ethiopia, Kenya, and Tanzania
Eurygnathohippus cornelianus lineage; includes *Eu.* "*ethiopicus*"	2.5–0.5 Ma	Ethiopia, Kenya, Tanzania, and South Africa
Eurygnathohippus pomeli	2.5 Ma (or older?)	Northern Africa
EQUINI		
Equus	2.33–present	Africa
Equus koobiforensis	2–1.0 Ma	Kenya
Equus oldowayensis	2.3–1.0 Ma	Tanzania
Equus capensis	2–0.001 Ma	South African
Equus numidicus	1.9–1.2 Ma	Northern Africa
Equus tabeti	1.9–1.2 Ma	Northern Africa
Equus melkiensis	Late Pleistocene	Northern Africa
Equus algericus	Late Pleistocene	Northern Africa
Equus grevyi	Late Pleistocene–recent	Central Asia–East Africa
Equus quagga	?1Ma–recent	East and South Africa
Equus zebra	?Middle Pleistocene–recent	South Africa, Namibia, Angola
Equus africanus	1.2 Ma–recent	East Africa

of *"Cormohipparion" africanum* for the Bou Hanifia material (10.5 Ma), ably described by Arambourg (1959). The remainder of Churcher and Richardson's (1978) hypodigm is difficult to assign, but the Ethiopian, Kenyan, and South African material is likely best referable to *Eurygnathohippus* sp.

Hipparion albertense Hopwood, 1926 was correctly restricted to the type specimen by Churcher and Richardson (1978) because it was insufficient for species recognition: the buccal two-thirds of an upper second molar from the Plio-Pleistocene of Uganda. Hooijer (1975) asserted that the nomen *H. albertense* should be considered a *nomen vanum*. He further elaborated on the taxonomic muddle invoked by the application of this "species name" for African assemblages, while Churcher and Richardson (1978) avoided substantive comment. Churcher and Richardson (1978) did address Plio-Pleistocene material from Kaiso, which Cooke and Coryndon (1970) referred to *H. (Hipparion) albertense*.

Hipparion baardi of Boné and Singer, 1965 was described from a modest assemblage of isolated teeth from Langebaanweg Baard's Quarry. These authors remarked on the rather hypsodont teeth of this assemblage (ca. 70 mm), and assigned it to *H. (H.) albertense* as a new subspecies *baardi*. Hooijer

(1975) evaluated the status of *H. (H.) albertense baardi* and concluded that this sample should have its own name, *H. baardi*. Churcher and Richardson (1978) followed this recommendation. Hooijer (1976) acknowledged likely stratigraphic differences between Langebaanweg Baard's Quarry and E Quarry, referring the much better material from Langebaanweg E Quarry to *H.* cf. *baardi*. Prior to the actual study of the Langebaanweg E Quarry hipparion material described by Hooijer (1976), Bernor and Armour-Chelu (1999), and later Franz-Odendaal et al. (2003) provisionally referred this sample to *"Eurygnathohippus"* cf. *baardi*. Most recently, Bernor and Kaiser (2006) undertook a detailed study of the Langebaanweg E Quarry sample and referred that material to *Eurygnathohippus hooijeri*, citing its extraordinary representation of the skeleton (see description given earlier).

The nomen *Notohipparion namaquense* Haughton, 1932 is based on a well-worn mandibular dentition discovered between Langebaanweg C and E Quarries, Langebaanweg, South Africa (Churcher and Richardson, 1978; figure 35.16). Churcher and Richardson (1978) noted that Haughton (1932) neither designated a holotype nor provided a formal diagnosis. Churcher and Richardson (1978) further accurately described the specimen as exhibiting a variable expression of ectostylids in the worn (less than 35 mm in height) crown. Besides expressing ectostylids, the pre- and postflexids were found to have a wavy complexity. Hooijer (1975) recognized a resemblance between *"N." namaquense* and a partial mandible recovered from Shungura Member B11 of the Omo group. He assigned the Omo specimen to *"Hipparion"* sp. B, citing that it was clearly less advanced than *"H." libycum* and indistinguishable from *"Hipparion" primigenium*. In fact, the mandible of *"N." namaquense* conforms closely to known members of the *Eurygnathohippus* clade in its stage of ectostylid evolution: ectostylids are variably expressed and do not normally rise high on the labial side of the teeth. This fundamental morphology is found in both the Lothagam (Bernor and Harris, 2003) and Middle Awash (Bernor and Haile-Selassie, 2009) latest Miocene/earliest Pliocene Hipparionini. Its provenience between the C and E Quarries of Langebaanweg, commensurate with what Bernor and Kaiser have observed for the E Quarry *Eu. hooijeri* assemblage, suggests a referral of *"N." namaquense* to *Eu.* cf. *hooijeri*.

Churcher and Richardson (1978) recognized *Hipparion sitifense* as a valid taxon occurring in northern, central, and eastern Africa. They further synonymized, in part, *H. (H.) albertense* (after Cooke and Coryndon, 1970) with their concept of *Hipparion sitifense*. Churcher and Richardson (1978) cited small size as the most remarkable feature of *H. sitifense* but noted the lack of a type specimen and suitable diagnosis of the material. Bernor and Harris (2003) and Bernor and Scott (2003) reported that the original material described by Pomel (1897) could not be located. Furthermore, Bernor and Harris (2003) and Bernor and Scott (2003) noted that the originally described assemblage lacked sufficient preserved fossil material to accurately assign it to any genus or species based on our current understanding of phylogenetically meaningful features. Currently, late Miocene African hipparionines are assigned to four different genera: *"Cormohipparion"* (*"C." africanum*, Bou Hanifia; *"C."* aff. *africanum*, Samburu Hills; *"Cormohipparion"* sp., Chorora), *Hipparion s.s.* (Sahabi, Libya), *Cremohipparion* aff. *matthewi* (Sahabi, Libya), and *Eurygnathohippus* (*Eu. feibeli* and *Eu. turkanense* (Lothagam, Kenya, and Middle Awash, Ethiopia). As demonstrated by Armour-Chelu et al. (2006), there are likely multiple small species of *Eurygnathohippus* recorded in the Plio-Pleistocene of eastern Africa. Unfortunately, much of the limited material of these hipparionines is easily confused with small, undescribed species of *Equus*.

Hooijer and Maglio (1973) originally described the type material of *Hipparion turkanense* from the late Miocene of Lothagam based on a beautifully preserved skull. Churcher and Richardson (1978) supported Hooijer's view that *H. turkanense* co-occurred with *Hipparion primigenium* at Lothagam. *"H." primigenium* was also cited as occurring at Bou Hanifia (= *"C". africanum* here) and Ekora (EK4, *Eu.* aff. *feibeli* here). Churcher and Richardson (1978) did recognize however that *"H." turkanense* had a dorsoventrally higher maxilla with likely higher crowned teeth. In fact, Bou Hanifia *"Cormohipparion" africanum* has a dorsoventrally deep POF, which reflects substantially lower maximum crown heights than *Eu. turkanense*, *Eu. hooijeri*, and *Eu. hasumense* (figures 35.6, 35.11, and 35.13) and metapodial morphology is dramatically different between these taxa (Bernor et al., 2005, Bernor and Kaiser, 2006). Since Hooijer and Maglio's (1973) original description, and Churcher and Richardson's (1978) minor revision, very important metapodial and first phalangeal

material of *Eu. turkanense* has been collected from Lothagam by Meave Leakey and her team. Bernor and Harris (2003) reported the material as being very robust and relatively massive for a hipparionine. Bernor et al. (2005) found striking similarities between the postcrania of *Eu. turkanense* and Siwalik *"S." perimense* material, and also found rare occurrences of this postcranial morphology from the early Pliocene of Ethiopia.

Churcher and Richardson (1978) chose to recognize *Hipparion libycum* as a single species of latest Pliocene to late Pleistocene hipparionine that ranged across the African continent. Their resulting synonymy (Churcher and Richardson, 1978, p. 399) was extensive. They defined these hipparionines as being extremely hypsodont (over 70 mm unworn height in adult cheek teeth), with strong development of ectostylids, and with marked reduction and medial migration of the lateral incisors in both jaws. Unfortunately, the crucial maxillary and mandibular incisor morphology is only recorded in the Olduvai Bed II, BK II locality and the type specimen of *Eu. cornelianus* is from Cornelia, South Africa (Bernor and Armour-Chelu, 1997; Armour-Chelu et al., 2006). While Churcher and Richardson (1978) are correct that the essential morphology of late Plio-Pleistocene hipparionines is similar, crown height evolves, potentially convergently, through the African hipparion sequence (Bernor and Armour-Chelu, 1997). Striking species-level comparisons between late Pliocene and Pleistocene pan-African hipparionine assemblages will have to address anterior dentitions and postcranial remains to secure species-level identities.

Genus *EQUUS* Linnaeus, 1758

Diagnosis After Churcher and Richardson (1978). Facial region usually shallower than in hipparionine horses, with lacrimal fossa absent or rudimentary. Sagittal crest absent. Fossae on occiput for nuchal attachment small, and those directly above the occipital condyles poorly developed. Bony auditory meatus variable in length and orientation. Basicranial region with or without low longitudinal crest. Coronoid process of the mandible lower than hipparionine horses, and ascending ramus with obliquely posterior orientation. Grooves on mandibular incisors variably developed. Canines usually absent in females. Cheek teeth very hypsodont. Upper cheek teeth with protocone connected to protolophs by a narrow isthmus and few, to no enamel plis on the pre- and postfossette mesial and distal borders. Lower cheek teeth with well-developed metaconid and metastylid, always lacking parastylids. Limbs straighter than in Hipparionini, ulna often with a discontinuous shaft; extremities always monodactyl, lateral digits absent, and lateral metapodials reduced to shortened splints that exhibit contact for no more than two-thirds of the length of metapodial III. Lacking a rudiment of MCV. Hoofed phalanges lack developed median slits on the anterior margin and are more arcuate shaped in their outline than in Hipparionini.

Age Late Pliocene (2.6 Ma)–Recent in the Old World.

Remarks Churcher and Richardson (1978) provided an extensive review of African *Equus* taxonomy. They utilized all information available at that time, which was essentially restricted to living species size, coat color and pattern, and osteological features of the skeleton. Generally speaking, they distinguished species by proportions of the long bones: onagers and half-asses being slender, donkeys small, zebras heavily built or stocky. They noted the considerable variability in cheek tooth size and occlusal morphology. However, Churcher and

Richardson (1978) noted that zebras and asses usually lacked a pli caballin in the upper cheek teeth (contrary to caballine *Equus*) and the usual occurrence of a V-shaped linguaflexid.

Churcher and Richardson (1978) reported remains of *Equus* spp. from the late Pliocene of northern and eastern Africa and the late Pleistocene of eastern and southern Africa. They claimed that these lineages became extinct at the latest by the end of the early Pleistocene in northern Africa, and late Pleistocene or early Holocene in sub-Saharan Africa. Introduction of true horses, *Equus caballus,* is reported to have taken place in northern Africa following the invasion of Egypt by the Hyksos kings around 1580 B.C. (Zeuner, 1963; Churcher and Richardson, 1978). Subsequently, *E. caballus* dispersed along the southern Mediterranean coastline and across the Sahara by human introduction. A later introduction of *E. caballus* took place in South Africa along the Cape Province by Dutch settlers in the late 17th century (Churcher and Richardson, 1978).

Churcher and Richardson (1978) recognized that many Pliocene *Equus* fossils were assigned to *Equus numidicus* and to two Pleistocene species, *Equus oldowayensis* in eastern Africa and *Equus capensis* in southern Africa. They believed that these taxa were likely antecedents of the extant Grevy's zebra, *Equus (Dolichohippus) grevyi,* which is now restricted to the arid and semiarid regions of the Horn of Africa. They (1978:403) asserted that *E. (D.) grevyi* occurred in the middle to late Pliocene of Africa.

Churcher and Richardson (1978) recognized two subgenera of zebra, *Equus (Dolichohippus)* and *Equus (Hippotigris),* but not *Equus (Quagga)*: plains and mountain zebras were considered to be of the subgenus *(Hippotigris).* Churcher and Richardson (1978) reported that there are fossil records of *E. zebra* and *E. quagga* from southern Africa. They speculated that the *E. zebra* and *E. quagga* clades evolved as a result of the early speciation of southern African populations of *E. burchellii*: *E. zebra* being adapted to broken montane countryside, and *E. quagga* presumably evolving in response to life in semiarid regions of southern Africa. We find this to be an entirely credible hypothesis.

EQUUS (DOLICHOHIPPUS) KOOBIFORENSIS
(Eisenmann, 1983)

Diagnosis After Eisenmann (1983)—A large species of *Equus* approaching the size of *Equus sanmeniensis* of China. Palate relatively long with respect to the muzzle. Upper cheek teeth with deep postprotoconal valleys and relatively small protocones. Maxillary P2 with somewhat shallower postprotoconal valleys. Mandibular p2 at least occasionally bearing a protostylid; stenonine double knot (metaconid-metastylid) on the lower cheek teeth; vestibular grooves (ectoflexids) at least occasionally shallow in m3.

Remarks The type skull, KNM-ER 1484, was recovered from the *Notochoerus scotti* zone, Area 130, below the KBS tuff of Koobi Fora. *Equus koobiforensis* is an exceptionally large horse (Eisenmann, 1983). Beyond Eisenmann's diagnosis, other characters include the following: a protostylid is present on dp2, cheek teeth show deep linguaflexids and shallow ectoflexids, and muzzle width is similar to that found in *E. oldowayensis.* Eisenmann (1983) reported a number of close dental similarities shared by *E. koobiforensis* and European *E. stenonis,* including deep postprotoconal valleys, short protocones, and similar proportions of the lengths and protoconal indices of the upper cheek tooth series. She further noted that the P2 and M3 are relatively elongate and the protoconal index increases from P2 to M3.

Eisenmann (1983) was unsettled about the hypodigm of *E. koobiforensis.* She reported a great range of variability in the size of the Koobi Fora *Equus* assemblage, believing that the sample must include two species. The type skull, KNM-ER 1484, is that of a large, young adult mare. Eisenmann (1983: figure 5-4) argued that, compared to several other species of *Equus, E. koobiforensis* is very large and most similar to the Chinese Pleistocene form, *E. sanmeniensis.* Eisenmann (1983) further indicated that there is a mismatch between the very large *E. sanmeniensis* skull, and proportionally smaller metapodials; only the lower cheek tooth series KNM-ER 4051 matches that of the holotype cranium.

Eisenmann (1983) reported that at least the following two characters of the holotype cranium indicate that *Equus koobiforensis* is distinct from, and probably more primitive than, any modern species of *Equus*: (1) the relatively long palate associated with a short snout, also seen in *Dinohippus interpolatus, D. leidyanus, E. simplicidens, E. stenonis vireti,* and *E. stenonis senezensis*; (2) the low protoconal indices, quite similar to *E. stenonis.* With regard to other African large *Equus* species, Eisenmann (1983) reported difficulty making comparisons to both *E. capensis* and *E. oldowayensis* due to the lack of precise data for the skull, cheek teeth, and limb bones, but admitted that *E. koobiforensis* could become the junior synonym of *E. oldowayensis* following further study.

EQUUS (DOLICHOHIPPUS) OLDOWAYENSIS Hopwood, 1937
Figure 35.17

Diagnosis After Churcher and Hooijer (1980). A large horse overlapping in size with *E. grevyi.* The dentition of the skull has a large and broad incisor arcade with incisor teeth having infundibula on I1 and I2 and which may be absent on lower third incisor; males possess a large canine, while the canine is vestigial in females; in middle wear, premolar protocones are generally shorter and rounder on P2 and P3 than on P4, while in the molars, the protocones are persistently more elongate; the mesial portion of the protocone has a strong connection to the protoloph; pli caballins are vestigial or absent; fossette plications are simple. Mandibular incisors are as in the premaxillary incisors, having distinct infundibula; lower cheek teeth have a diminutive metaconid on p2 and rounded to slightly elongate metaconid on p3–m3; metastylid is generally rounded to square-shaped distally on p3–m3 giving a distinctly straight linguomesial border; linguaflexids are very shallow in p2 and V shaped on p3–m3. Metapodials referred to this taxon vary in length and slenderness with those from the Omo being rather gracile, while material from Bed I Olduvai is more robust. The diversity of postcranial measurements alone suggests that there may be more than a single species included within this diagnosis.

Remarks Churcher and Hooijer (1980) reviewed the taxonomy of *E. oldowayensis,* which we follow closely herein. Hopwood (1937: figures 1 and 2) designated a lower jaw from an animal about 2 years old (Catalogue Number VIII, 353, in the Bayerische Paläontologische Staatssammlung, Munich) as the holotype of *E. oldowayensis.* Hopwood (1937) also designated a lower incisive region with the left incisors and right first incisor (BMNH M14199) as the paratype. The original Olduvai collection deposited in Munich, which included the type of *E. oldowayensis,* was destroyed, together with its catalogue, during WW II (K. Heissig, pers. comm. to Churcher

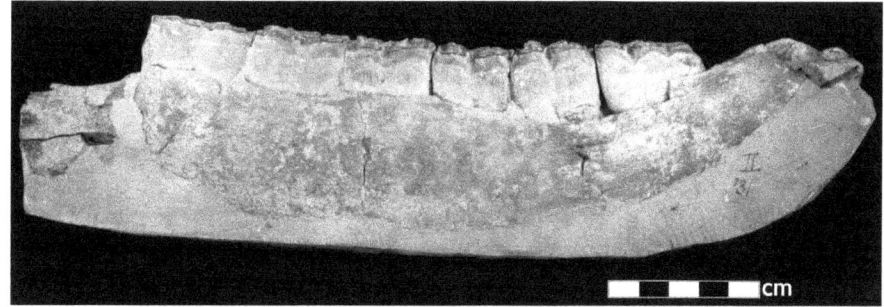

FIGURE 35.17 *Equus oldowayensis* BMNH M14134 left mandible, buccal view.

and Hooijer as cited in 1980:266). Cooke (1963: table 4) assigned a large equid from the Omo deposits to *E. oldowayensis* Hopwood, 1937, or possibly *E.* aff. *grevyi*, but without discussion. Hooijer (1976) reported a large *Equus* from the Omo that he considered to be indeterminate to species in the absence of skulls, and accepted a great deal of variation in cheek tooth characters, including variably long or short protocones and V- or U- shaped linguaflexids. Coppens (1971) noted that "true *Equus* appears in Member G together with *Hipparion ethiopicum*." Churcher and Richardson (1978: 381, table 20.1) recognized *E. (Dolichohippus) oldowayensis* from the Omo Shungura Formation Members F through J, explicitly recognizing it as being a member of the Grevy's zebra lineage. In order to stabilize the taxon *E. oldowayensis* Churcher and Hooijer (1980) selected the mandible, BMNH M14184, as the neotype of *E. oldowayensis* (see figure 35.17 for representative specimen). Churcher and Hooijer (1980) reported on the extensive Olduvai *Equus* material currently housed (on loan) in the Rijksmuseum van Natuurlijke Historie, Leiden, and currently under study by Armour-Chelu.

Equus oldowayensis was equipped with a broad muzzle and horizontal incisive row. This morphology suggests an adaptation to short grass feeding. *Equus oldowayensis* is the most widespread and abundant horse in Plio-Pleistocene deposits of Ethiopia, Kenya, and Tanzania (Churcher 1981; Hooijer and Churcher, 1985). The earliest reported stratigraphic occurrence is from the Shungura Member G (2.33 Ma), and its latest reported occurrence may be Olorgesailie (1.0 Ma) and perhaps the Daka 1.0 Ma horizons. Schulz and Kaiser have studied a small sample of Olduvai *E. oldowayensis* and found that they are at the most abrasion-dominated end of the grazing spectrum.

EQUUS CAPENSIS Broom, 1909

Diagnosis A large bodied horse estimated to be 150 cm at the withers and with a body mass of approximately 400 kg (Eisenmann, 2000) being similar in size to *E. oldowayensis*. Teeth are large and muzzle broader than found in *E. oldowayensis*. Cheek teeth generally with simple plication frequencies. The protocone is moderately long with a lingual depression.

Remarks Churcher (1970) provided a useful review of the nomen *E. capensis* that we follow in part here. *E. capensis* was founded by Broom (1909: plates 18, 19, 22) based on a right mandible, embedded within a block of calcrete, containing p2–m3 (figure 35.18). It was recovered from Table Bay, Ysterfontein, Maitland, South Africa (Broom, 1909; Churcher, 1970). Broom's (1909) original description did not include a photograph, but Churcher (1970) was able to locate the type specimen in the SAM collections. According to

Churcher (1970), Broom's (1928) figuring of the teeth was inaccurate. Cooke (1950) illustrated the original. Wells (1959) discarded the nomen *E. capensis* as being indeterminate and a *nomen nudum*. Churcher (1970) believed that *E. helmei, E. cawoodi, E. kubmi, E. zietsmani*, and specimens of *E. harrisi* and *E. plicatus* were all referable to *Equus capensis*. Churcher and Richardson's (1978:495–496) synonymy was even more extensive.

Widely distributed in South African sites dating from the Plio-Pleistocene, *E. capensis* became extinct during the terminal Pleistocene, along with other members of the megafauna (Klein, 1974). The complex synonymy of this taxon was reviewed in Churcher and Richardson (1978). They concluded that *E. oldowayensis* and *E. capensis* are similar to one another, implying that they may be conspecific. This would give *E. capensis* priority over *E. oldowayensis*. While *E. capensis* and *E. oldowayensis* are very similar in several respects, indicating that they are closely related, we have chosen to retain them as separate taxa for the present. Based on mesowear analysis of the middle Pleistocene Elandsfontein (South Africa) sample, Kaiser and Franz-Odendaal (2004) interpreted *E. capensis*'s paleodiet to indicate a mixed feeding niche. This departure from penecontemporaneous *E. oldowayensis* paleodiet is attributed to the unique fynbos vegetation of the Cape Province, South Africa.

EQUUS (DOLICHOHIPPUS) NUMIDICUS Pomel, 1897

Diagnosis After Churcher and Richardson (1978). A medium-sized horse about the size of a large zebra. The maxillary cheek teeth have a mesiodistally short protocone, especially distal to its junction with the protoloph; the plis

FIGURE 35.18 *Equus capensis* right mandible embedded in calcrete.

are small and simple and the pli caballin is absent; the hypoglyph is shallow, simple, and V shaped, and the styles are rounded, slightly set off at angles from the ectolophs, which are shallowly concave buccally. The cementum is not thick.

Remarks Equus numidicus was erected by Pomel (1897) based on cheek teeth from Ain Boucherit. Eisenmann (1980) attributed three associated molars and 16 isolated teeth from Ain Boucherit and 10 upper cheek teeth from Ain Hanech to *E. numidicus*. There are no reported skulls or postcrania of this species, making its species-level status difficult to defend.

Arambourg (1970) correlated Ain Boucherit with the Lower Villafranchian. Coppens (1971) correlated it with Members A–D of the Omo Shungura Formation. Eisenmann (1980), however, pointed out that *Equus* does not occur in the Omo until Member G. Sahnouni et al. (2002) estimate Ain Boucherit's age as being younger than the Ahl al Oughlam paleontological site (ca. 2.5 Ma) and slightly older than Ain Hanech. Ain Boucherit can be correlated to between 2.4 and 2.0 Ma based on the presence of *Equus*.

EQUUS TABETI Arambourg, 1970

Diagnosis After Eisenmann (1983). A moderate sized species of *Equus* with asinine upper cheek teeth, stenonine lower cheek teeth, and slender MCIII and first phalanges III.

Remarks Equus tabeti is based on material from Ain Hanech, Algeria (Arambourg, 1970). Churcher and Richardson (1978:408) attributed to these deposits a Pleistocene age and synonymised *E. tabeti* with *E. (Hippotigris) burchelli*. Geraads et al. (2004) estimated the age of Ain Hanech to be ca. 1.2 Ma. *Equus tabeti* is largely known from northern Africa, although Eisenmann (1983) tentatively recorded its presence, as *E.* cf. *tabeti*, at Koobi Fora, Kenya. The Koobi Fora assemblage includes a modest sample of cranial, dental and postcranial material. The most remarkable specimens are a fragmentary skull of a young adult with the M3 (KNM-ER 1211), a fragmentary metacarpal III of a medium-sized form (*E.* cf. *tabeti*, KNM-ER 2069), a very small and exceedingly slender metacarpal III (KNM-ER 2067), which Eisenmann (1983) reports to be unlike any *Equus* that she has ever seen and likely pathological.

Upper cheek teeth are reported to be ass-like in their protoconal index, while lower cheek teeth are typically stenonine. Furthermore, the size of cheek teeth are quite similar to northern African *E. tabeti*: in the type material, the P2–P4 ranges from 82 to 95 mm in length, while it is 91 mm in KNM-ER 1211; the molar length of *E. tabeti* ranges from 70 to 81 mm in length, while being 78 mm in KNM-ER 1211 and 73 mm in KNM-ER 325. Also, a single known m3 has a deep ectoflexid, like *E. tabeti,* and unlike modern asses (Eisenmann, 1983). The proportions of the partial metacarpal III (KNM-ER 2069) as well as the first phalanges III, KNM-ER 2069, are similar to *Equus tabeti*.

Eisenmann (1983) believes that the Koobi Fora *E.* cf. *tabeti* is a primitive ass and may be derived from *E. numidicus,* but it has more gracile metapodials and phalanges. Ecomorphological analyses of postcranial elements have concluded that *E. tabeti* most likely lived in arid environments like modern hemiones (Eisenmann and Karchoud, 1982).

EQUUS MELKIENSIS Bagtache, Hadjouis and Eisenmann, 1984

Diagnosis After Bagtache et al. (1984). An *Equus* of probably *Asinus* affinities measuring 1.35–1.40 m. at the withers.

FIGURE 35.19 *Equus algericus* type IPH 61-803 right lower m2.

Metapodials are of moderate robusticity with elevated proximal craniocaudal dimensions.

Remarks The species *E. melkiensis* was nominated by Bagtache et al. (1984) based on a metacarpal III recovered from Allobroges, Algeria, dating to the Aterian (sub-latest Pleistocene). It was reported to co-occur with the caballine horse, *E. algericus* (discussed later). The type specimen is a metacarpal III, I.P.H. Allo. 61-1314, measuring 211 mm in length. Bagtache et al. (1984) nominated two paratypes: a left m2, I.P.H. Allo. 61-1969 and a left metatarsal III, I.P.H. Allo. 61-1834.

The left m2 is cited by Bagtache et al. (1984) as being typical "stenonine" *Equus* with the following characters (reconstructed from their figure 2): a somewhat elongate-rounded metaconid and rounded-square metastylid, V-shaped linguaflexid, short preflexid with a strong mesiolingual pli; postflexid small and labiolingually constricted, shallow ectoflexid not invading between pre- and postflexid. The reconstructed height and metapodial proportions suggest asinine affinities to Bagtache et al. (1984). *E. melkiensis* is also reported from Morocco (Eisenmann, 1995).

EQUUS ALGERICUS Bagtache, Hadjouis and Eisenmann, 1984
Figure 35.19

Diagnosis After Bagtache et al. (1984). A caballine species of *Equus* with a height at the whithers of about 1.44 m. Metapodials are stockily built (trapus).

Remarks The type specimen of *E. algericus* is reported to be a lower second molar, (IPH 61-803), from Allobroges, Algeria (Bagtache et al. 1984: figure 1). As figured, the type specimen would appear to be of a left, not a right cheek tooth. Not enough information is given to verify that it is in fact an m2 rather than a p3, p4, or m1, although its size and occlusal morphology could be that of a premolar rather than a molar. The type specimen has a rounded metaconid connected by a long isthmus to the squared metastylid, giving the linguaflexid a shallow and wide character (figure 35.19). The preflexid is very long, with a distinct, mesiolabial pli directed labially. The postflexid is shorter than the preflexid. There is a small pli caballinid.

There are a dozen metapodials, complete and fragmentary, from the type assemblage. The complete metacarpals III have a length of about 225 mm, while the metatarsals III are about 271 mm in length, and are more lightly built than zebras and *Equus mauritanicus*. The metacarpals III are reported to resemble those of *E. caballus* cf. *gallicus* from Jaurens and Solutré, France (late Pleistocene). *E. algericus* is also reported from Morocco (Aouraghe and Debenath, 1999). *E. algericus* has several caballine features including an asymmetrical double knot (= metaconid-metastylid), but not in the type specimen.

Metapodials are stockily built (robust), and Bagtache et al. (1984) reconstructed this species height at the withers as being 144 cm.

We question the utility of using a single lower cheek tooth as a type specimen because of the degree to which they can vary ontogenetically, both in size and morphology.

EXTANT SPECIES OF *EQUUS*

EQUUS (DOLICHOHIPPUS) GREVYI (Oustalet, 1882)

Diagnosis Modified after Williams (in press). The largest living nondomestic equid. Its withers height is 140–160 cm. Weight in males 353–431 kg (mean 386 kg). It is large headed and long legged, with extremely large ears with rounded tips. The skull is very elongate, exceeding the cervical spine in length (Groves, 2002). Other cranial characters are a raised regio-occipitalis, a deep postorbital constriction, long vomer and muzzle, and a rounded nasal end of the premaxilla, which is wedged between the nasals (Williams, in press). Protostylids are often well developed on p2 (Eisenmann, 1976). The narrow muzzle (Eisenmann, 1980) is grey to tan with a white margin before fine black and white stripes. Sexual dimorphism is slight, with males being 10% heavier than females (King, 1965). Males and females are easily distinguished by the female's black labia. Males have large upper and lower canines, which are absent in females. The mane stands tall and erect on the nape of the neck and is striped black and white continuously with the body. The neck is broad with thickest stripes on the body. The narrow, closely set stripes cover most of the head and body. Flank stripes are fine and vertical, tapering out at a level above the elbow, leaving the underside unstriped. The white belly coloration extends partway up the sides. A wide black hairline passes down the spine and continues onto the tail. The tail ends in a tuft of hair. There is a narrow white zone on either side of this broad black dorsal stripe. Stripes on the hindquarters curl down from the dorsal stripe and taper out leaving the buttocks white. Fine horizontal striping extends down the legs to the hooves and merges at the fetlock. Those stripes on the hindquarters remain vertical until above the hind legs (rather than being primarily horizontal as in other zebra species). In foals, stripes are brown and later turn black, starting with neck and ears, followed by head and limbs, and finally ending with flanks. The dorsal stripe remains brown-black and fluffy for up to 2 years.

Remarks Forsten (1992) believed that the early stenonine *Equus,* which first occurred in the Old World 2.6 Ma (Lindsay et al., 1981), may have been ancestral to *Equus grevyi.* Churcher (1981) referred *Equus* material of Member L of the Omo Shungura Formation to *E. oldowayensis,* which Eisenmann (1985) later described as the earliest *E. grevyi.*

E. grevyi is currently distributed in the arid regions of Ethiopia, northern Kenya, and has recently vanished from Somalia, Djibouti, and Eritrea. Churcher and Richardson (1978) referred *E. numidicus, E. oldowayensis* and *E. capensis* to the subgenus *Dolichohippus.* They further recognized *E. (D.) grevyi* as occurring in the early Pleistocene to Recent of Ethiopia and Kenya (1.8 Ma–Recent). Like Churcher and Richardson (1978), Eisenmann (1983) recognized *E. numidicus* and *E. oldowayensis.* Furthermore, Eisenmann (1983) recognized *E.* cf. *grevyi* from the *Metridiochoerus compactus* zone, the Guomde Formation and the Galana Boi beds of Kenya, based both on cheek tooth dentitions and postcranial remains.

Marean and Gifford-Gonzalez (1991) argue that *E. grevyi* formerly ranged as far south as Tanzania (Lake Manyara) during late Pleistocene times and also occurs in Neolithic levels at Dakhleh Oasis, Egypt (Churcher 1986). Kingdon (1997) has reported ancestral or proto-Grevy's Zebra forms, extended from central Asia to southern Africa. Historically, *E. grevyi* ranged west of the Rift Valley in Kenya (Stigand, 1913; Stewart and Stewart, 1963) to western Somalia, and from the Danakil desert in Eritrea, through the Awash Valley and Ethiopian Rift Valley, the Ogaden, and northeast of Lake Turkana in Ethiopia to north of Mt. Kenya (Williams, in press).

EQUUS (QUAGGA) QUAGGA (Boddaert, 1785)

Diagnosis Modified after Klingel (in press). The shoulder height of an adult male plains zebra (*E. quagga*) is 123–133 cm (mean 128 cm) and 115–126 in females (mean 123 cm). Weight 220–284 kg in the male (mean 248 kg) and 175–241 kg in the female (mean 219 kg). Forehead distinctly convex, zygomatic arches relatively robust. Ear bullae are comparatively small in size; paroccipital processes are well developed. Postorbital constriction is relatively pronounced. The mandible is massive, and the premaxillae curve downward below the level of the alveolar line of the cheek teeth (Grubb, 1981; Groves, 2002). After some initial wear, upper incisors have infundibula that change their shape with wear. In advanced wear, infundibula disappear. In the lower incisors, infundibula are absent or reduced to a simple circular shape in most northern populations, but more usually present in the south. Molars are hypsodont. The tooth eruption sequence is similar to the horse. The first molar is the first permanent cheek tooth to erupt, while the third incisor is the last. Forelock and mane are thin to absent in the northern-most populations, and well developed in the southern populations. When present, the mane may vary widely from thick and long to short and thin. The stripes of plains zebra are rather broad, especially towards its rump. However, plains zebras vary in color and pattering across their range. Stripe patterns are distinct in the northern subspecies, but less prominent in the south. Shadowing on the flank and rump between the dark and white stripes is thin or even absent in the north and wider in the southern populations. The ground color varies from white (north) to buff (south). The basic pattern characteristics, stripe width, complexity, branching, stripe length, is individually unique and inherited. Northern populations have legs striped to the hooves, southern populations usually do not. Rump stripes do not meet the dorsal midline, while in the northern populations, and often also in the southern populations, the stripes extend to the ventral midline. Body stripes are broader than in Grevy's zebra and the mountain zebra, although less so in Cape mountain zebra *E. zebra zebra* than in Hartmann's mountain zebra *E. zebra hartmannae.* The stripe patterns vary with subspecies and geographic location. The neck is relatively short and legs are robust. Ears are pointed, the muzzle is relatively blunt and cropped. The plains zebra has one pair of mammae. There is little sexual dimorphism, although males are usually about 10% larger than females, but there is no colour variation between the sexes. Males also have large canines, whereas in mares the canines are lacking or vestigial. Patterns from various subspecies and populations are published in Cabrera (1936), Antonius (1951), Rau (1974, 1978), and Kingdon (1979, 1997).

Remarks E. quagga is one of the most widely distributed African ungulates, ranging from southern Sudan and southern

Ethiopia to northern Namibia and northern South Africa. Geographic variation is pronounced and includes several morphotypes, often recognized as subspecies: *E. q. crawshaii, E. q. borensis, E. q. boehmi, E. q. chapmanni, E. q. burchellii,* and *E. q. quagga.* From the latest Pliocene and early Pleistocene to the Holocene the species extended over almost the whole African continent. Fossil remains are reported from northern Africa (Morocco, Algeria, Tunisia, Mali, Libya) to the Cape of South Africa (Churcher, 1970; Churcher and Richardson, 1978). Opinions as to whether the quagga represents part of a cline with plains zebra remain divided. Groves and Ryder (2000), Groves and Bell (2004), and Rau (1978) place them in a single taxon that gives the nomen *E. quagga* priority. Leonard et al. (2005) suggest that these taxa differentiated between 120,000 and 290,000 years ago. Fossil *E.* cf. *burchellii* was described by Mendrez (1966) from the Sterkfontein Extension site. Fossil remains of quagga have been reported by Cooke (1941) from the Wonderwerk Cave. The eruption sequence and dental wear patterns have been used for aging (Erz, 1964; Klingel and Klingel, 1966; Smuts 1972, 1974).

Wild extant *Equus* is highly adaptable to locally available plant foods (Berger, 1986). Smuts (1972, 1975) mentions over 65 species of plants ingested by the plains, among which are 50 species of grass as well as 9 taxa of trees and bushes. The choice of grass species broadly reflects what is available, but zebras do show some selectivity. Of seven major grass species, one, *Panicum maximum,* contributed 40% of the intake, and the same preferences were shown all through the year (Ben-Shahar, 1991). Other favored grass species in studies include *Themeda triandra, Cynodon dactylon,* and *Eragrostis superba* (Grubb, 1981). In extremely dry periods they also take browse and even rhizomes (Pienaar, 1963; Gwynne and Bell, 1968). The variety of food items ingested by *Equus quagga* under given habitat conditions indicates that these horses are not to be regarded as specialized grazers, but grass may be the only source of food during some seasons. Based on tooth wear equilibria, Kaiser and Schulz (2006) have demonstrated that the diets of plains zebras vary in abrasiveness and reflect the water availability in a given habitat. In dry habitats the zebra's diet has been shown to be more abrasive than in moister habitats. They further suggest that *E. quagga* may be a suitable indicator for subtle differences in habitat and climate.

Equus mauritanicus occurs in the middle and later Pleistocene of northern Africa (Maghreb region). Churcher and Richardson (1978) referred the material to *E. burchellii mauritanicus* of the subgenus *Hippotigris,* thus recognizing it as a member of the mountain zebra, quagga, and plains zebra clade. Eisenmann (1980, 1983) recognized *E. mauritanicus* as a distinct species, claiming similarities to the more primitive *E. stenonis* in the dentition, and at the same time a close resemblance to *E. burchellii.* Armour-Chelu (pers. obs.) has studied the *E. mauritanicus* material in Paris and concurs with Eisenmann that this material is best recognized as a species other than *E. burchellii.* We recognize *E. mauritanicus* as a species needing further study, analysis, comparison, and formal diagnosis

Eisenmann (1983) also introduced the notion of cross breeding between *E. mauritanicus* and quaggas. Churcher and Richardson (1978) reported an extensive series of late Pliocene to Recent occurrences of *E.* ("*H.*") *burchellii* from Morocco, Algeria, Kenya, Tanzania, Zambia, Zimbabwe and the Republic of South Africa. We follow Eisemann's stance that *E. mauritanicus* is a potentially valid species based on observations by Armour-Chelu, but a great deal of descriptive work is still necessary.

Eisenmann (1983) recognized *Equus* cf. *burchellii* from a sample of upper and lower cheek teeth, anterior and posterior first phalanges from the *Metridiochoerus compactus* zone of Koobi Fora, Kenya.

EQUUS (HIPPOTIGRIS) ZEBRA Linnaeus, 1758

Diagnosis Modified after Penzhorn (in press). Medium-sized, long-legged zebra. The smallest living zebra. The shoulder height of an adult male Cape Mountain Zebra is 127 cm and is 116–129 cm females (mean 124 cm). Weight ca 250–260 kg in males and 204–257 kg in females (mean 234 kg). Joubert (1971) noted that Hartmann's mountain Zebra males (mean=298 kg) are heavier than females (mean = 276 kg). The mountain Zebra is a stocky equid with a relatively short head, which is also striped. Skull dimensions of *E. zebra hartmannae* are greater than those found in *E. zebra zebra* (Lundholm, 1952). According to Penzhorn (in press), the mountain zebra, as Grevy's zebra, has the regio-occipitalis placed high and the postorbital constriction is pronounced. The orbit is placed slightly behind the posterior border of the third molar, and the dorsoventral diameter of the orbit is greater than in the plains zebra (*E. quagga*). The maxillary tuberosity does not extend as far back as in plains zebra, so that the pterygo-palatine fossa is visible from below; the external auditory meatus is large (3% of the basal length of the skull) and directed horizontally (instead of upward or backward); the nasofrontal suture is almost straight, and the temporal lines diverge more rapidly rostrally than in plains zebra, and at a wider angle (Smuts and Penzhorn, 1988; Groves 2002). Canine teeth are prominent in adult males, while those of the females are rudimentary and normally do not cut through the gums (Joubert, 1971).

Remarks Two subspecies, *E. z. hartmannae* (Hartmann's mountain zebra) and *E. z. zebra* (Cape mountain zebra) are recognized, although Groves and Ryder (2000) consider them to be distinct at the species rank: *E. hartmannae* and *E. zebra.* Recent studies on 15 microsatellite loci and 445 base pairs of the mitochondrial control sequences of *E. zebra* (Moodley and Harley 2005; Moodley et al. 2006) support the classification in two subspecies.

Lundholm (1952) described a fossil subspecies, *E. z. greatheadi,* from Vanwyksfontein, close to the Orange River. Today, natural populations of the Cape mountain zebra occur on the Bankberg, Gamka Mountain Reserve, and the Kamanassie Mountains. Historically, mountain zebras occurred from the southern parts of South Africa through Namibia and into extreme southwestern Angola. In the Eastern and Western Cape Provinces of South Africa, they were widely distributed along mountain ranges forming the southern and western edge of the central plateau. Hartmann's mountain zebras occur in the mountainous zone between the Namib Desert and the central plateau of Namibia. The subspecies is structured into four distinct populations (Penzhorn, in press).

In studies of cranial morphology, sexual dimorphism was evident in *E. z. zebra,* where males are smaller than females, while sexes are the same size in *E. z. hartmannae* (Groves and Bell 2004); however, no sexual dimorphism was found in *E. z. zebra* skulls by Smuts and Penzhorn (1988). Age determination, primarily based on incisor wear, has been discussed by Joubert (1972), Penzhorn (1982, 1987), and Penzhorn and Grimbeek (1987).

The muzzle is tan to dark gray between the nostrils and black at the tip. The mane is stiff, upright on the nape and is striped continuously with the neck. They also have a tufted tail with a flowing tassel of white and black hair. The rear spine and upper tail are marked with a gridiron pattern. The long rounded ears are white tipped with black patches. The most diagnostic feature of the mountain zebras is the dewlap, a square flap of skin on this zebra's throat, most developed in males. Stripes are black to deep chocolate brown on a white to buff background. The stripes on the head are narrowest, followed by those on the neck and body. Stripes on lower part of the face are suffused chestnut to orange colored.

The mountain zebra is distinguishable from other zebra species by the thin and relatively closely spaced vertical black stripes on its neck and rump, which form a gridiron pattern and are narrower and more numerous than those of the plains zebra. The horizontal to oblique stripes on the hindquarters are clearly much broader. Also distinctive are the wide, horizontal bands on its haunches, which are broader than both those of Grevy's zebra and the plains zebra. Unlike the plains zebra, the mountain zebra lacks shadow stripes, and the stripes do not meet under the creamy to white belly, which sometimes is interrupted by the first few stripes behind the forelegs that may extend the length of the belly. The black stripes of Hartmann's mountain zebra are thin with much wider white interspaces, while this is the opposite in Cape mountain zebra (Penzhorn, in press). The distinct striping on the legs may encircle the entire limb to the hooves.

Churcher and Richardson (1978) report a relatively small sample of ?middle Pleistocene to Recent fossil remains of *E. zebra* from South Africa.

EQUUS (ASINUS) AFRICANUS (Heuglin and Fitzinger, 1866)

Diagnosis After Churcher and Richardson (1978). Equines of small size and stocky build. Maxillary molars square. Ectoloph of relatively "low relief," with styles lower and valleys shallower than in the subgenera *Equus* and *Hippotigris*. Ectoloph valleys flat or slightly convex laterally; parastyle and mesostyle squared and tilted mesially. Pli caballins typically lacking or reduced. Cheek tooth pre- and postfossettes small with simple plis. Protocone with mesial arm comprising approximately 40% of the length. Lower molars with a pli caballinid. Ectoflexid shallow except in earliest wear; ptychostylid small. Metaflexid with sharp buccal angle. Metaconid elongate, lingually flattened, directed mesiolingually; metastylid rounded, directed distolingually; entoconid rounded and constricted at base; hypoconulid and hypostylid prominent. Linguaflexid open and V shaped.

Remarks Churcher and Richardson (1978) synonymized *E. africanus* and *Equus hydruntinus* within the nomen *Equus (Asinus) asinus*. Churcher (1982) reported the earliest occurrence of this taxon from the middle of Bed II, Olduvai Gorge (>1.2 Ma). This identification was based on a single metatarsal III, which was short (231 mm) and slender, although not to the extent of *E. tabeti*.

Following the classification of Groves (1986), Denzau and Denzau (1999), and Groves and Smeenk (2007), two subspecies are accepted: *E. (A.) africanus africanus* von Heuglin and Fitzinger, 1866 (Nubian wild ass) and *E. (A.) africanus somaliensis* Noak, 1884 (Somali wild ass). In historical times the Nubian wild ass was distributed in eastern Sudan and northern Eritrea, while the known range of the Somali wild ass was southern Eritrea, northeastern Ethiopia, Djibouti and northern Somalia (Yalden et al., 1986; Denzau and Denzau, 1999; Moehlman, 2002; Groves and Smeenk, 2007). According to

Groves (1974), *E. africanus* has the highest and narrowest hooves compared with all extant species of *Equus* indicating an evolutionary adaptation to rocky habitats (Klingel, 1970, 1971). The subspecies *E. a. somalicus* has a whithers height of about 1.25 m while *E. a. africanus* is smaller, having a whithers height of about 1.15 m (Clark, 1985). *E. a. somalicus* is characterized by the following: a rosy gray coat; white belly, legs, and muzzle; well-defined black leg stripes; and a faint, or completely absent shoulder stripe. The second subspecies *E. a. africanus* is rose-gray, too, but a bit lighter colored. Belly and muzzle are also white, but no stripes are observed on the legs, while the shoulder stripe is well defined. The African wild ass *E. africanus* is widely accepted as the ancestor of the domestic donkey *E. asinus* Linnaeus, 1758. The colors and markings of the wild form can be found in domestic donkeys too (Clark, 1985; Denzau and Denzau, 1999). In general, *E. africanus* is much larger than the domestic conspecific *E. asinus*. A huge variability in the domestic breeds is also observed (Poitou asses and Spanish giants often measure 1.5 m at the withers)(Clark, 1985).

REMARKS ON THE TAXONOMY OF *EQUUS*

The taxonomy and systematics of the extant Equidae is currently being intensively debated. Key equid systematics publications include: Groves and Mazak (1967), Groves and Willoughby (1981), Kingdon (1979), Eisenmann (1997), Groves (2002), and Groves and Bell (2004). There is considerable debate about the number of genera, subgenera, and species of zebra that should be recognized. Markers and methodology used range from morphological and ethological approaches (Groves, 1986 et seq.) to molecular genetics. Molecular studies have increasingly flourished since the 1980s (re: Higuchi et al., 1984). Table 35.1 follows Groves (2002), which we accept as a basic template, on which we develop further discussion, with some differences of opinion given later here (cf. tables 35.2 and 35.3).

Classification Based on Morphology

The first inventories of extant *Equus,* and consideration of the taxonomic position of fossil *Equus* occurred around the turn of the 19th to 20th centuries (Gray, 1825; Gidley, 1901; Allen, 1909; Lydekker, 1916). Our predecessors, Churcher and Richardson (1978), provided a monumental documentation of African Neogene equid systematics. Their insights on the taxonomy of African *Equus* are still extensively cited. Churcher and Richardson (1978) recognize three subgenera of *Equus*: *E. (Dolichohippus)*, *E. (Asinus)*, and *E. (Hippotigris)*. They did not consider Eurasian or American *Equus* in their classification. Within the subgenus *Hippotigris*, Churcher and Richardson (1978) recognized three species of zebra: *E. (H.) burchellii*, *E. (H.) quagga*, and *E. (H.) zebra*. *Equus grevyi* is placed in the subgenus *E. (Dolichohippus)*.

Bennett (1980) undertook a cladistic analysis on 96 skulls and 57 postcranial skeletons of members of Pliocene to Recent *Equus*. She recognized only two subgenera of *Equus*: *E. (Equus)* and *E. (Asinus)*. With reference to Willoughby (1974), Azzaroli (1979), and Groves and Willoughby (1981), Eisenmann (1986) recognized the three subgenera of *Equus*: *E. (Equus)*, *E. (Hemionus)*, and *E. (Asinus)*. Eisenmann (1986) agreed that the formerly proposed zebra subgenera *E. (Hippotigris)*, *E. (Quagga)*, and *E. (Dolichohippus)* deserved further study. Craniological studies by Eisenmann and Turlot

TABLE 35.2

Extant African *Equus* species recognized in key literature

Acronyms refer to subgenera as discussed in key literature: A, *Asinus*; D, *Dolichohippus*; E, *Equus*; Hi, *Hippotigris*; Q, *Quagga*.

Reference	*E. africanus* Heuglin and Fitzinger, 1866	*E. africanus africanus* Heuglin and Fitzinger, 1866	*E. africanus somaliensis* Noack, 1884	*E. africanus f. asinus* Linnaeus, 1758	*E. grevyi* Oustalet, 1882	*E. quagga* Boddaert, 1785	*E. quagga quagga* Boddaert, 1785	*E. quagga burchellii* Gray, 1824	*E. zebra* Linnaeus, 1758	*E. zebra zebra* Linnaeus, 1758	*E. zebra hartmannae* Matschie, 1898
Groves and Mazak, 1967	X (A)				X (Hi)	X (Hi)		X (Hi)	X (Hi)		
Rau, 1978, 1986						X (Q)					
Churcher and Richardson, 1978				X		X (Hi)					
Bennett, 1980				X	X (E)	X (Q)		X (E)	X (E)		
Groves and Willoughby, 1981									X (Hi)		
Higuchi et al. 1984						X (Q)			X		
Loewenstein and Ryder, 1985			X	X			X	X			
George and Ryder, 1986			X		X	X		X			X
Eisenmann, 1986											
Yalden et al., 1986					X (D)						
Churcher, 1993					X (D)						
Ishida et al., 1995				X				X	X		
Oakenfull and Clegg, 1998				X	X			X	X		
Oakenfull and Ryder, 1998											
Eisenmann and Mashkour, 1999											
Klein and Cruz-Uribe, 1999						X (Q)		X	X	X	
Groves and Ryder, 2000											
Oakenfull et al., 2000								X			
Groves, 2002	X	X	X			X (Q)	X (Q)	X (Q)	X (Hi)	X	X
Hack et al., 2002							X (Hi)	X (Hi)			
Burke et al., 2003											
Groves and Bell, 2004					X (D)	X (Q)		X (Q)		X	X
Leonard et al., 2005						X (Q)	X (Q)	X (Q)			
Orlando et al., 2006								X			

SOURCES: Original authors as follows: *E. africanus* (Heuglin and Fitzinger, 1866), *E. africanus africanus* (Heuglin and Fitzinger, 1866), *E. africanus somaliensis* (Noack, 1884), *E. africanus f. asinus* (Linnaeus, 1758), *E. grevyi* (Oustalet, 1882), *E. quagga* (Boddaert, 1785), *E. quagga quagga* (Boddaert, 1785), *E. quagga burchellii* (Gray, 1824), *E. zebra* (Linnaeus, 1758), *E. zebra zebra* (Linnaeus, 1758), and *E. zebra hartmannae* (Matschie, 1898).

(1978), Eisenmann (1979), and Skinner (1996), as well as work based on skins only (Rau, 1974) support the hypothesis that the quagga (*E. quagga quagga*) and the plains zebra (*E. quagga burchellii*) are closely related, but clearly distinct from the mountain zebra (*E. zebra zebra*). Based on multivariate analysis of 340 *Equus* skulls, Klein and Cruz-Uribe (1999) challenged these studies and underscored the possibility that the quagga and the plains zebra have to be distinguished at the species level. But Eisenmann and Brink (2000) argue that Klein and Cruz-Uribe (1999) used only a subset of variables instead of the entire set of cranial variables given in the measurement convention by Eisenmann (1980, 1986). Therefore, Eisenmann and Brink (2000) emphasized that the choice of variables was critical to the distinction of equid skulls and agreed with Klein and Cruz-Uribe (1999) that genetic studies were needed to clarify this issue. Groves and Bell (2004) revised the taxonomy of the zebras and considered *E. q. quagga* to grade as a morphocline into *E. q. burchellii*.

Groves (2002) *Equus* taxonomy (table 35.1) is based on extended morphological studies, ethology, and reproductive biology that attempt to bridge to recent molecular results. There are broad areas of agreement between Eisenmann (1986) and Groves (2002). In all of his studies, Groves and his coauthors have embraced the "phylogenetic species concept" of Cracraft (1983), which basically asserts that if a population can be shown to have distinct morphological characters and geographic separation, it should be recognized as a distinct operational taxonomic unit (OTU). If we accept that African zebras are a monophyletic group, then following Groves (2002), there would have to be a taxonomic rank between the genus *Equus* and the recognized subgenera *E. (Equus)*, *E. (Hippotigris)*, and *E. (Quagga)* to taxonomically unite zebras. Groves (2002) differs from the International Union for Conservation of Nature and Natural Resources (IUCN, 2006) somewhat and reflects both previous and subsequent data and conclusions given by Groves and Ryder (2000), Groves (2002), and Groves and Bell (2004).

TABLE 35.3

Extant Eurasian *Equus* species recognized in key literature

Acronyms refer to subgenera as discussed in key literature: A, *Asinus*; He, *Hemionus*.

Reference	*E. ferus* Boddaert, 1785	*E. ferus f. caballus* Linnaeus, 1758	*E. ferus przewalskii* Poliakov, 1881	*E. hemionus* Pallas, 1775	*E. hemionus kulan* Groves and Mazak, 1967	*E. hemionus onager* Boddaert, 1785	*E. kiang* Lydekker, 1916	*E. khur* Lesson, 1827
Groves and Mazak, 1967		X	X	X (A)	X (A)		X (A)	
Rau, 1978, 1986								
Churcher and Richardson, 1978								
Bennett, 1980		X		X (A)		X (A)	X (A)	
Groves and Willoughby, 1981								
Higuchi et al., 1984								
Loewenstein and Ryder, 1985		X	X		X	X		
George and Ryder, 1986		X	X		X	X		
Eisenmann, 1986		X	X	X (He)	X	X	X (He)	
Yalden et al., 1986								
Churcher, 1993								
Ishida et al., 1995		X	X					
Oakenfull and Clegg, 1998		X	X	X				
Oakenfull and Ryder, 1998			X					
Eisenmann and Mashkour, 1999				X				
Klein and Cruz-Uribe, 1999		X						
Groves and Ryder, 2000								
Oakenfull et al., 2000	X		X	X	X	X	X	
Groves, 2002				X (He)		X (He)	X (He)	X (He)
Hack et al., 2002								
Burke et al., 2003		X						
Groves and Bell, 2004								
Leonard et al., 2005								
Orlando et al., 2006		X		X				

SOURCES: Original authors as follows: *E. ferus* (Boddaert, 1785), *E. ferus f. caballus* (Linnaeus, 1758), *E. ferus przewalskii* (Poliakov, 1881), *E. hemionus* (Pallas, 1775), *E. hemionus kulan* (Groves and Mazak, 1967), *E. hemionus onager* (Boddaert, 1785), *E. kiang* (Lydekker, 1916), and *E. khur* (Lesson, 1827).

Classification Using Genetic and Immunological Markers

George and Ryder (1986) analyzed mitochondrial DNA restriction-endonuclease maps of *E. africanus, E. burchellii, E. grevyi, E. hemionus,* and *E. zebra* suggesting that there are at least three major clades in modern *Equus*: the zebras, the wild asses, and the true horses. Ishida et al. (1995) investigated the mitochondrial DNA sequence of the control region and support the interpretation by George and Ryder (1986), though they consider *E. zebra* as the outgroup to all other zebras. Oakenfull and Clegg (1998), the first investigators to study the DNA sequences of a globular protein gene (α and θ globin), also found a marked divergence between the zebras, wild asses, and true horses. Furthermore, these data suggest that *E. ferus przewalskii* and *E. ferus caballus* diverged from all other species of *Equus* around 1.3 Ma. The relationship between zebras and asses, however, cannot be resolved based on the molecular data presently available.

At the species level, the first genetic studies aimed at posing hypotheses of phylogenetic relationships of extant *Equus* were conducted by Higuchi et al. (1984). They analyzed 229 base pairs of mitochondrial DNA (mtDNA) to resolve the question of the origin of the quagga. According to their results, the cladogenetic event that separated the maternal lineages of the quagga and mountain zebra should have occurred between 3 and 4 Ma, based on the calibration of a molecular clock using the split of the primate and ungulate clades dated to 80 Ma. Higuchi et al.'s (1984) analysis therefore suffers from three limitations: from the calibration of the molecular clock, from the use of limited data that included the skin of only a single quagga (Museum Mainz, Germany, without number), and from the comparatively small number of 229 base pairs analyzed. Lowenstein and Ryder (1985) extended the sample size to three quaggas and employed radioimmunoassay techniques to study proteins from the skin. These were compared to serum proteins of three extant zebras (*E. burchellii, E. zebra, E. grevyi*), two Asian wild asses (*E. hemionus onager* and *E. hemionus kulan*) and two horses (*E. ferus przewalskii* and *E. caballus*). These immunological data have suggested that the quagga is best considered to be a subspecific variant of the plains zebra, *E. burchellii*. It is further problematic that neither Higuchi et al. (1984) nor Lowenstein and Ryder (1985) have provided precise geographic provenance and museum specimen ID's of the quagga material they investigated, and that the remainder of their samples were zoo specimens. Recent, more extensive studies of quagga mtDNA (Leonard et al., 2005) confirm earlier morphology-based data (Rau, 1978; Groves, 2002; Groves and Bell, 2004), suggesting that the quagga and the plains zebra should be synonymized.

Oakenfull and Ryder (1998), who conducted the first study of mtDNA (control region and 12S rRNA) markers in true horses (*E. f. przewalskii*), found no significant variation in the mtDNA lineages of Przewalski's horse and domestic horse. However, they did not address the question of the ancestry of the two taxa.

Forstén (1992) attempted to edify the origin of the wild asses by relating mtDNA data (George and Ryder, 1986) with paleontological data from various sources. She concluded that wild asses first occurred in the late Pleistocene of India, the Levant, and northern Africa. Aranguren-Mendez et al. (2001) used 15 microsatellite loci to resolve the ancestry of Spanish domesticated donkey breeds, which previously were considered to derive from the African wild ass, *Equus africanus*. Subsequently, Aranguren-Mendez et al. (2004) analyzed 313 base pairs of the cytochrome *b* gene of 79 individuals and 383 base pairs of the control region of 91 individuals. Aranguren-Mendez et al. (2004) confirmed the existence of two divergent maternal lineages of African origin (*E. asinus africanus* and *E. a. somaliensis*). In conclusion, the ancestry of the African ass lineage still remains unresolved.

The majority of the aforementioned studies were not intended as syntheses of molecular data with morphological, ethological, and biogeographic evidence. The first synthetic approach combining molecular (mtDNA) with morphological data, as well as biogeographic aspects, is by Oakenfull et al. (2000), who analyzed a large sample of several species (including the Asian kiang *E. kiang*). Until now, molecular-based research has neither included all recognized extant *Equus* species nor a sufficient number of wild-shot individuals and thus has contributed minimal insights into the phylogeny of *Equus*. Since most studies (Oakenfull and Ryder, 1998; Oakenfull et al., 2000; Orlando et al., 2006) investigated only one (mtDNA) genetic marker, the bias caused by introgression cannot be known. Further studies should acknowledge that Sakagami et al. (1999) regard mtDNA as being a less suitable single marker system for perissodactyl systematics because of an increased nucleotide substitution rate compared to other groups (Brown et al. 1979). Great potential can be seen in combined approaches using different molecular markers (Hillis et al., 2002).

Evolutionary Biogeography of African Equidae

Figure 35.20 portrays the evolution of African Equidae graphically. The first occurring hipparionine in Africa was previously claimed to be of the genus *Hippotherium*, best and most certainly represented in the central European Vallesian and Turolian ages (Bernor et al., 1996, 1997). Woodburne (2007) undertook a major revision of North American *Cormohipparion* and concluded that there are no bona fide *Cormohipparion* in Eurasia. However, we follow Bernor and White (in press) in recognizing *C. sinapensis* at the early late Miocene locality of Sinap, Turkey, as well as a close relative from Bou Hanifia, northern Africa, *"C." africanum*. The skull morphologies of the Sinap and Bou Hanifia horses are not greatly divergent

from North American *Cormohipparion*, while the postcrania are primitive as in *Cormohipparion*, being distinctly different from the larger and more robustly built European lineage, *Hippotherium*. Rather than erect a new genus for this African hipparion, we prefer to refer it to *"C." africanum* to distinguish it from European *Hippotherium*. The occurrence of this clade also in East Africa at Chorora (Ethiopia) and the Samburu Hills (Kenya) Ethiopia (Chorora) during the earliest part of the African equid record, 10.5 Ma, suggests that they were derived from the founding group of African Hipparionini. The retention of similar, albeit measurably derived, metapodial proportions in the Langebaanweg hipparionines suggests that this founding population may have been at the base of the African hipparionine radiation.

We recognize chronologically early attributes of the *Sivalhippus* Complex (*sensu* Bernor and Hussain, 1985) in late Vallesian correlative (ca. 9 Ma) Samburu Hills hipparionines. Nakaya and Watabe (1990) have reported similarities of the Samburu hipparion to Bou Hanifia *"Hipparion" africanum*, and there are in fact distinct similarities in skull and postcranial morphology. However, there are important morphological features in the facial morphology, and a precocious increase in crown height (to about 70 mm) that suggest the Samburu Hills hipparion shares an affinity with the Siwalik clade referred to *"S." perimense*. Bernor and Harris (2003) have previously argued that the latest Miocene Lothagam form *Eurygnathohippus turkanense* is likewise derived from *"S." perimense*. The Samburu Hills hipparion however is relatively primitive in its postcranial features, having more elongate and slender metapodials, and it is therefore more comparable to *"C." africanum*, than either *"S." perimense* or *Eu. turkanense*. The Samburu Hills hipparion could prove to be a somewhat derived step in the ancestry of both *"Sivalhippus"* and *Eurygnathohippus*, having neither derived massive metapodials like *"S." perimense*, *S. theobaldi*, and *Eu. turkanense*, nor apparent ectostylids like species of *Eurygnathohippus*. Its slender metapodials could prove to have proportions very similar to *"C." africanum* and ancestral to *Eu. hooijeri* (Bernor and Kaiser, 2006) and *Eu. feibeli* (Bernor and Harris, 2003). As such, it could share phylogenetic relationships between African and Asian Hipparionini.

The Baringo Basin has yielded a ca. 9 Ma hipparionine with massive first phalanges III. Bernor et al. (2005), and we here show, that these are very similar in proportions to Siwalik and eastern African massive-limbed forms. Altogether, there are multiple lines of evidence that support a hypothesis of extended, perhaps ephemeral, biogeographic relationships between southern Asia and eastern Africa through much of the late Miocene among species belonging to the *"Cormohipparion,"* *"Sivalhippus,"* and *Eurygnathohippus* clades. Study of the phylogenetic relationships and biogeographic history of these hipparionines is currently being undertaken by Bernor and Wolf.

The *Hipparion s.s.* and *Cremohipparion* lineages are evident from the latest Miocene locality of Sahabi and are not known to occur in sub-Saharan Africa. Both of these lineages are identified on the basis of postcrania only, and are compared with material from Turolian localities in Greece, Turkey, and Iran. Their limited chronologic and geographic relationships suggests a middle-late Turolian biogeographic extension from the eastern Mediterranean into northern Africa.

The distinct divergence of African Hipparionini from Eurasian Hipparionini by the earliest Pliocene suggests a strong vicariant biogeographic pattern. Compared to the Eurasian

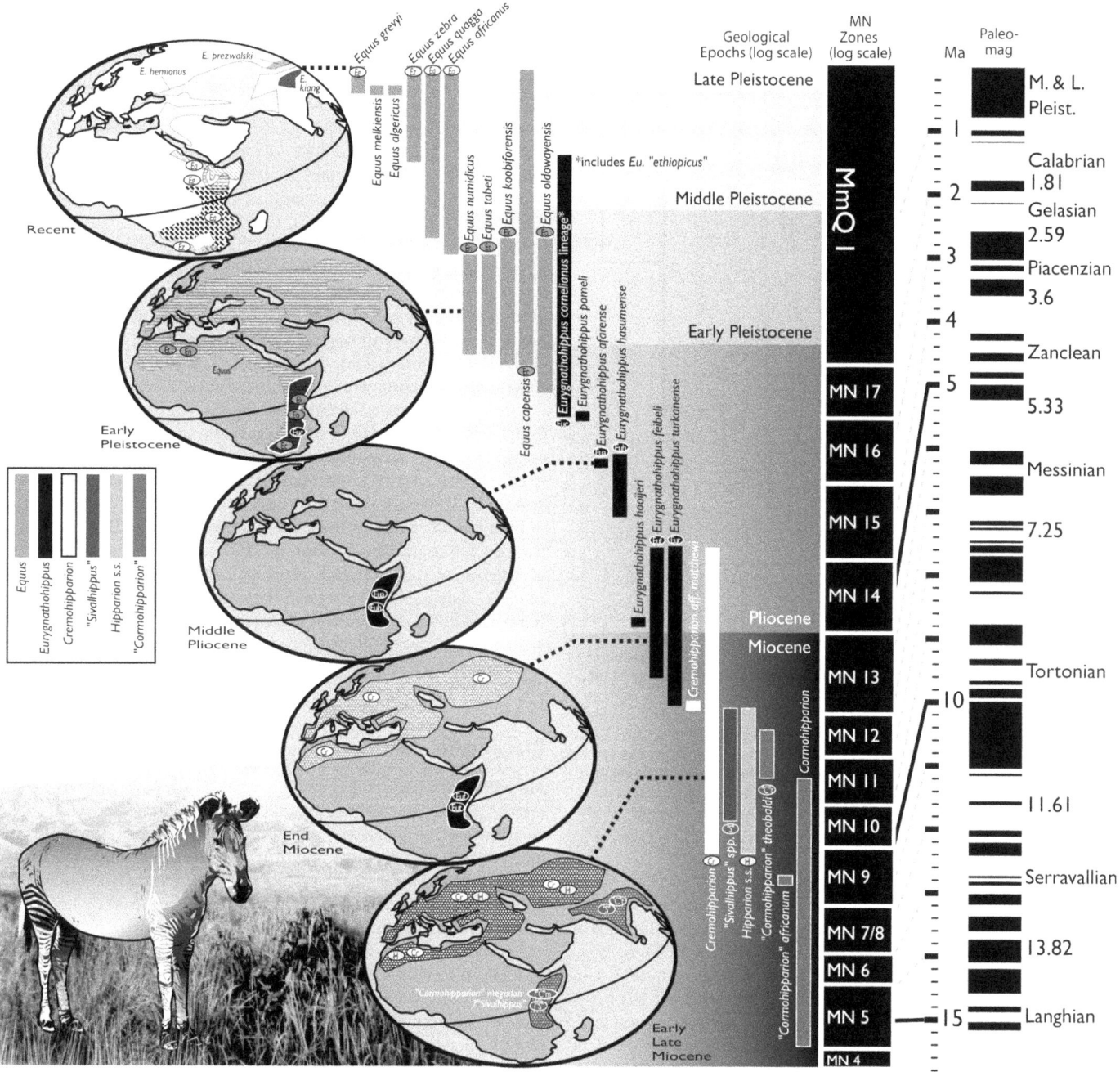

FIGURE 35.20 Chronological and geographic distribution of Neogene and Quaternary African Equidae. The oldest African hipparions thus far known *("Cormohipparion")* are from the early late Miocene. The almost contemporaneous central European *Hippotherium primigenium* behaved as a browser at Höwenegg (Germany); data after Kaiser (2003). In the Turolian of Dorn-Dürkheim (Germany), *H.* aff. *primigenium* and *H. kammerschmittea* are mixed feeders and browsers, respectively, demonstrating the ecological variability of late Miocene European hipparions (data from Kaiser et al., 2003). *Eurygnathohippus hooijeri* from Langebaanweg, South Africa (ca. 5.2 Ma), is a grazer classifying at the more attrition dominated end of the mesowear continuum (data from Franz-Odendaal et al. 2003). The early Pleistocene *E. oldowayensis* (Olduvai Gorge, Tanzania) also classifies at the attrition-dominated end of the grazer spectrum, while *E. capensis* from the middle Pleistocene of Elandsfontein (South Africa) was foraging a mixed diet (Kaiser and Franz-Odendaal, 2004).

late Miocene and Pliocene (Bernor et al., 1996), the diversity of African Hipparionini is quite low. The *Eu. feibeli/Eu. hasumense* clade would appear to dominate in eastern Africa between 6 and 2.9 Ma (Bernor et al., 2005). This clade shares a number of characters, in particular postcranial proportions, with the early Pliocene Langebaanweg form, *Eu. hooijeri*. Middle Pliocene *Eu. pomeli* (Eisenmann and Geraads, 2007) shows affinities with the Kenyan and Ethiopian earlier Pliocene Hipparionini. At 2.5 Ma there appears to be a replacement of earlier forms with a new group of Hipparionini, perhaps derived from southern African populations, that evolve very great crown heights, and at least in one lineage, *E. cornelianus*, a very broad gape with hypertrophied incisors. These Hipparionini no doubt became dedicated short-grass feeders from the later Pliocene until their extinction in the middle Pleistocene (Bernor and Armour-Chelu, 1999).

The evolutionary biogeography of African *Equus* is engulfed in controversy. This problem is exacerbated by a lack of

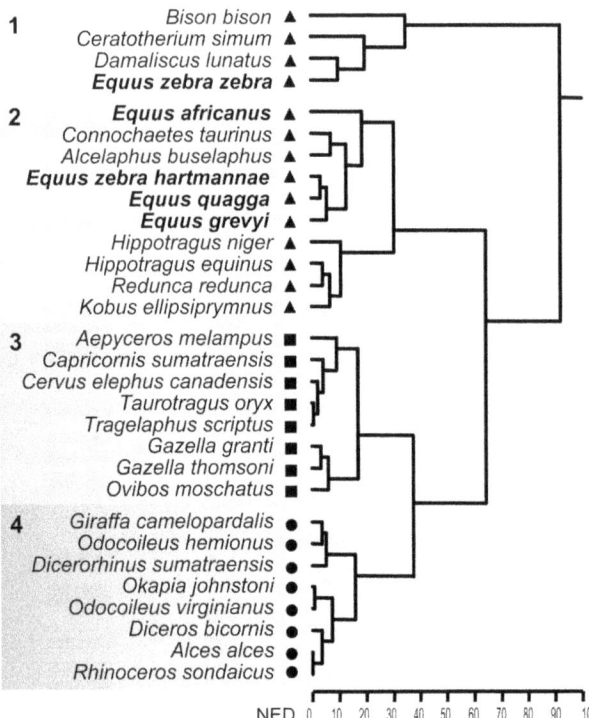

1 Bison bison ▲
Ceratotherium simum ▲
Damaliscus lunatus ▲
Equus zebra zebra ▲

2 **Equus africanus** ▲
Connochaetes taurinus ▲
Alcelaphus buselaphus ▲
Equus zebra hartmannae ▲
Equus quagga ▲
Equus grevyi ▲
Hippotragus niger ▲
Hippotragus equinus ▲
Redunca redunca ▲
Kobus ellipsiprymnus ▲

3 Aepyceros melampus ■
Capricornis sumatraensis ■
Cervus elephus canadensis ■
Taurotragus oryx ■
Tragelaphus scriptus ■
Gazella granti ■
Gazella thomsoni ■
Ovibos moschatus ■

4 Giraffa camelopardalis ●
Odocoileus hemionus ●
Dicerorhinus sumatraensis ●
Okapia johnstoni ●
Odocoileus virginianus ●
Diceros bicornis ●
Alces alces ●
Rhinoceros sondaicus ●

NED 0 10 20 30 40 50 60 70 80 90 100

FIGURE 35.21 Hierarchical dendrogram based on a set of 30 comparative species taken from Fortelius and Solounias (2000). The closer the species are in their mesowear signature, the smaller is the normalized Euclidean distance (root mean-squared difference; NED) at the branching point. Cluster 1 contains those grazers (triangles) with the most abrasion-dominated diet, cluster 2 contains the grazers with less abrasive diet, cluster 3 four intermediate feeders squares, and cluster 4 the browsers (circles). Feeding traits of extant taxa are according to the "conservative" (CONS) classification by Fortelius and Solounias (2000). Note that there is debate on the classification of Tragelaphus scriptus, which is classified a browser elsewhere (see Kaiser and Rössner, 2007).

congruence between osteological data, soft anatomical data, and molecular data. While there are substantial collections of fossil *Equus* from across the African continent, there is an absolute dearth of detailed morphological description, consistent analysis, and rigorous comparison. We have left unresolved whether many of the African fossil species such as *E. mauritanicus* and *E. lylei* are valid species or referable to extant zebra species. We have erred on the side of conservatism in these cases and followed Churcher and Richardson (1978) because there is simply too little information to support additional nomina.

Equus is first securely recorded in Africa in Omo Shungura Formation Member G, 2.33 Ma (Geraads et al., 2004). Late Pliocene to early Pleistocene *Equus*, such as *E. koobiforensis*, shows affinities with European *E. stenonis* and undoubtedly reflects a proximate phylogenetic relationship. Likewise, *E. koobiforensis* shows morphological and size similarities to Chinese Nihowan age *E. sanmeniensis* and *E. oldowayensis* (Eisenmann, 1983). However, the late Pliocene/early Pleistocene record of *Equus* would appear to have a recorded species diversity, which, however, has yet to be documented sufficiently to recognize more than one species at any particular locality.

Churcher and Richardson (1978) explicitly supported the notion that the Olduvai *Equus* was a dolichohippine and directly related to the *E. grevyi* lineage. In turn, Churcher (1981) and Hooijer and Churcher (1985) asserted that

E. oldowayensis was the most widespread and abundant horse in Plio-Pleistocene deposits of Ethiopia, Kenya, and Tanzania and ranging in age from the Omo Shungura G member to Olorgesaile (2.33–1.0 Ma). Churcher and Richardson (1978) further believed that *E. capensis* was a broadly distributed southern African *Equus* that was closely related to *E. oldowayensis*. Also, Churcher and Richardson (1978) identified the large northern African *Equus*, *E. numidicus*, as a member of the dolichohippines. Its correlation is best considered to be with Bed I Olduvai (Eisenmann, 1980).

The earliest occurrences of asses may be documented by *E. tabeti*. While Churcher and Richardson (1978) synonymised *E. tabeti* with *E. burchellii*, Eisenmann (1983) made cogent arguments for its recognition as a primitive ass. She further reports *E. tabeti* in both northern and eastern Africa. Bagtache et al. (1984) reported a late Pleistocene ass, *E. melkiensis*, from Algeria, co-occurring with a caballine species, *E. algericus*.

The fossil record would appear to support the occurrence of dolichohippine *Equus* prior to quaggas, mountain zebras, and asses. The molecular evidence is far from clear and needs more work. The ancestry of *E. africanus* is still unknown (Oakenfull and Ryder, 1998; Aranguren-Mendez et al., 2001). The study of microsatellites and mtDNA of *Equus zebra* by Moodley and Harley (2005) and Moodley et al. (2006) indicates that the divergence of the mountain zebra (*E. zebra*) and the plains zebra (*E. quagga*) is still unresolved because of an underestimation caused by constraints on allele size and mutation rate. *E. przewalskii/E. caballus* would appear to have diverged from all other species of *Equus* by 1.3 Ma (Ishida et al., 1995; Oakenfull and Clegg, 1998; Lowenstein and Ryder, 1985).

Diet and Rainfall

All extant equids are widely recognized as being typical grazers (Nowak, 1991), but habitat conditions such as differences in mean annual precipitation, temperature, and evapotranspiration may affect food availability and thus shift the dietary signal as inferred from dental wear proxies (Kaiser and Schulz, 2006). Tooth wear in equids has been shown to be correlated with subtle habitat differences (Schulz et al., 2007).

The mesowear method was applied to evaluate tooth wear in relation to climatic data. Four upper tooth positions (premolars 4, and molars 1, 2, and 3) were scored applying the mesowear method developed by Fortelius and Solounias (2000), first tested by Kaiser et al. (2000), and modified by extension to several upper and lower tooth positions by Kaiser and Solounias (2003) and Kaiser and Fortelius (2003). For each locality where extant *Equus* data were sampled, the mean annual precipitation was calculated using ArcView 3.1 software based on climate data from the IIASA database (International Institute for Applied System Analyses), a reliable source of precipitation data compiled by the United Nations (Leemans and Kramer, 1981). Mean weighted values for climate variables were calculated for each extant *Equus* species.

Museum specimens comprising 501 individuals of extant African *Equus* that died between 1886 and 2000 were used in this investigation. These specimens are housed at the following institutions: American Museum of Natural History, New York; National History Museum, London; Etosha Ecological Institute, Okaukuejo; Muséum National d'Histoire Naturelle, Paris; Museum für Naturkunde, Berlin; National Museum of Namibia, Windhoek; Naturmuseum Senckenberg, Frankfurt am Main; Naturhistorisches Museum, Bern; Nico van Rooyen Taxidermy, Pretoria; Russian Academy of Science, St. Petersburg;

TABLE 35.4
Frequencies of mesowear variables of P4 and M1–3

N, number of dental tooth specimens scored (number of individuals in brackets); h%, percent high relief parameters; s%, percent sharp cusps; b%, percent blunt cusps; MAP, mean annual precipitation in millimeters, weighted average for the distribution range of each species.

Species	N	h%	s%	b%	MAP
Equus africanus	24(10)	66.7	29.2	12.5	224.5
Equus grevyi	205(54)	52.7	6.6	8.6	618.5
Equus quagga	944(289)	49.2	7.2	13.3	718.8
Equus zebra hartmanae	387(117)	48.8	2.6	16.5	378
Equus zebra zebra	86(31)	31.4	12.9	30.6	462.6

h, s, b / *E. hemionus, E. quagga* and *E. grevyi,* chi-square 41.030, *p* < 0.001
h, s, b / *E. africanus* and *E. zebra zebra,* chi-square 7.179, *p* < 0.05
h, s, b / *E. zebra* and *E. hemionus,* chi-square 26.417, *p* < 0.001

Smithsonian National Museum of Natural History, Washington, D.C.; Transvaal Museum, Pretoria; Zoologisches Museum, Hamburg; Zoologische Staatssammlung, Munich.

The mesowear variable frequencies were calculated for each species, and cluster statistics were performed in order to test these equids versus a set of 25 extant ungulate species used as dietary references by Fortelius and Solounias (2000; see figure 35.21). The resulting cluster diagram ranks taxa relative to one another so that the closer the data sets plot, the greater their similarity as measured by the normalized Euclidean distance (NED) at the branching point. Mesowear variable frequencies of species and mean annual precipitation in their range of distribution are listed per species in table 35.4. In three species, low occlusal reliefs prevail over high reliefs. Cusp shape scoring ranges from 3% sharp to 29% sharp, while blunt cusps never exceed 31%.

Cluster analysis classified the 30 species including 5 equid taxa into four major clusters (figure 35.21): one containing those grazers with the most abrasion dominated diet (cluster 1), one containing the grazers with a less abrasive diet (cluster 2), one containing all intermediate feeders (cluster 3), and one containing all the browsers (cluster 4). The equids are classified in clusters 1 and 2. *E. zebra zebra* is the only species in cluster 1, where it is closely linked to *Damaliscus lunatus,* a grazing African antelope. Cluster 2 contains three zebras, *E. zebra hartmannae, E. quagga* and *E. grevyi,* which classify close to the most abrasion dominated end of Cluster 2. All equid species are significantly different in their mesowear signatures (chi-square probabilities *p* < .05; table 35.4). With the exception of *E. zebra zebra,* an abrasion dominated signature is related to aridity (small MAP). The natural range of *E. zebra zebra* is not the driest environment in this comparison, however, the species has the most abrasion dominated feeding trait (Schulz, 2008).

For *E. quagga,* a positive correlation is found between the humidity (mean annual precipitation = MAP) and dental mesowear variables, indicating a less abrasion-dominated feeding niche, as has been suggested by Kaiser and Schulz (2006). This relationship also seems to apply to *E. africanus, E. zebra hartmannae,* and *E. grevyi,* and it supports the observation that equid feeding behaviour reflects environmental parameters. *E. africanus,* which is associated with the driest natural distribution range in Africa (224.5 mm rain fall per annum), shows the second most abrasion dominated feeding regime. The extreme abrasion dominated trait of the Cape mountain zebra *(E. zebra zebra)* is thus interpreted as a result of

overhunting, habitat fragmentation and early conservation management, which forced this species into a restricted population at the end of the 19th century (Novellie et al., 2002). The Cape mountain zebra was subsequently reintroduced into environments with poorer-quality food resources that were more abrasive. In conclusion, feeding traits of extant equids indicate climate related environmental parameters such as humidity but also reflect habitat fragmentation and other effects of human management and competition for resources.

Conclusions

The African later Neogene record includes two tribes of Equidae, Hipparionini, and Equini. Hipparionini, a tribe of tridactyl horses with an isolated protocone, are first recorded in northern and eastern Africa circa 10.5 Ma. The richest sample of these early hipparionines is from Bou Hanifia, Algeria and includes a taxon that we refer to *"Cormohipparion" africanum.* *"C." africanum* is similar in cranial morphology to Central European *Hippotherium primigenium* and, for that matter, North American *C. occidentale* (Bernor et al., 2003), but is more primitive in its slender, elongate postcranial morphology. At the same time it is advanced in its maximum crown height. We believe that it was related to *"C." sinapensis,* which in turn is a plausible descendant of North American *Cormohipparion.* The penecontemporaneous equid sample from Chorora Ethiopia, *"Cormohipparion"* sp. is meager in comparison, but similar in its dental stage of evolution. The slightly younger, ca. 9.0 Ma Samburu Hills hipparion *"Sivalhippus"* sp. could conceivably be derived from *"C." africanum,* but it is derived in its facial morphology and maximum crown height, while retaining primitive proportions of the metapodials.

Samburu Hills *"Sivalhippus"* sp., younger Sahabi *"Sivalhippus"* sp., and Lothagam *Eurygnathohippus turkanense* exhibit plausible evolutionary relationships to Siwalik (India) hipparionines. We review here and update the literature on the taxonomic tangle that best describes the literature enveloping Siwalik Hipparionini. We recognize *"Cormohipparion" nagriense* as a valid species related to the immigration of *Cormohipparion* into southern central Eurasia. *"Cormohipparion" theobaldi* is a very large hipparion with a dorsoventrally enormous, medially and posteriorly deep POF and relatively low cheek tooth crown height. Its uniquely large size and low crown height make it very similar to *"H." megadon,* and we refer that taxon to *"C." megadon.* We believe that *"C." theobaldi* may be legitimately referred to the type *S. theobaldi* deciduous

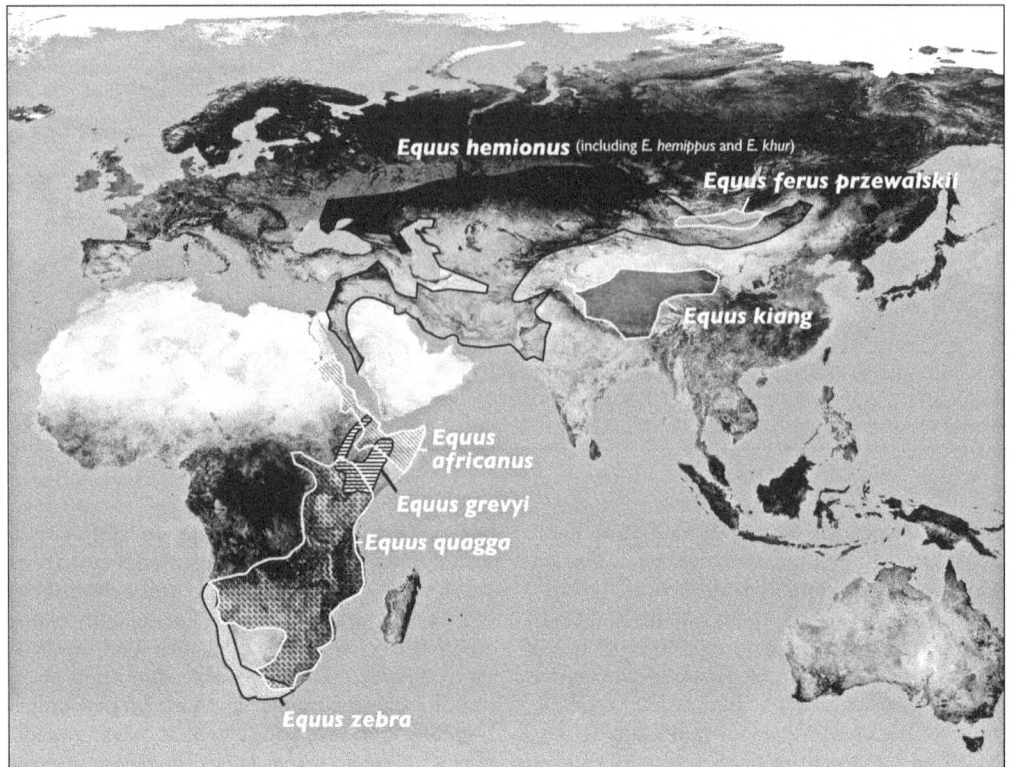

FIGURE 35.22 Reference map of historic ranges of extant *Equus* species (after Moehlman, 2002; Heptner et al., 1966; Denzau and Denzau, 1999). See tables 35.2 and 35.3 for taxonomic citations. Background image credit: NASA Visible Earth; http://veimages.gsfc.nasa.gov/2430/land_ocean_ice.8192.tif

dentition; and if future study verifies this observation, we suggest the formal recognition of this genus, with the species *S. theobaldi* and perhaps *S. megadon*.

The most prevalent of Siwalik hipparionines is "*S.*" *perimense*. This is a large hipparion with its POF reduced and placed dorsally high and far anteriorly on the face, an increased crown height, and massive metapodials. It would appear to be related to *Eu. turkanense* from the Lower Nawata, Lothagam Hill, differing from *Eu. turkanense* in its lack of ectostylids. However, *Eu. turkanense* shares remarkably similar metapodial and first phalangeal proportions with "*S.*" *perimense* and "*S.*" *sp.* from Ngorora and Sahabi. If *S. theobaldi* proves to be a valid genus-rank taxon, then "*S.*" *perimense* and "*Sivalhippus*" sp. need a new genus-level nomen.

There is a modest evolutionary radiation of *Eurygnathohippus* species in Africa. *Eurygnathohippus hooijeri,* a large, precociously hypsodont species is reported only from the early Pliocene of South Africa, ca. 5.2 ma. It would appear to be related to latest Miocene *Eu. feibeli,* a smaller form reported from Kenya and Ethiopia. In the northern portion of the East African Rift, there are two closely related species, *Eu. hasumense* (Ethiopia, Kenya, and Tanzania) and *Eu. africanum* (Ethiopia only). Beginning at 2.5 Ma, the *Eu. cornelianus* lineage appears in Tanzania and Ethiopia, and presumably Kenya, and is characterized by moderate size, elongate, slender metapodials, hypertrophied I1s and I2s, atrophied I3s, and progressively higher-crowned cheek teeth with a climax of 90 mm crown height. It may prove to be appropriate to recognize *Eu. "ethiopicus"* in earlier, pre-Olduvai Bed II eastern African horizons. This depends largely on identifying mandibular symphyses with incisor and canine dentitions, and better records of maximum crown heights in

2.5–0.5 Ma eastern and southern African hipparionines. There may well have been more than one species of hipparionine living in this chronologic interval.

The Eurasian genera *Hipparion s.s.* and *Cremohipparion* are only thus far reported from Sahabi Libya and date to about 7–6 Ma. These genera are prominent in eastern Mediterranean and Southwest Asian faunas and no doubt made their geographic extension in the later portion of the late Miocene.

Equus makes its first appearance in eastern Africa in Omo Shungura Formation Member G, 2.33 Ma. This age of occurrence is delayed relative to Eurasia, where it is 2.6 Ma (Lindsay et al., 1980). First occurring African *Equus* is apparently related to European *E. stenonis* and Chinese *E. sanmeniensis*. The eastern African species *E. koobiforensis* and *E. oldowayensis* are evidently closely related to Eurasian *E. stenonis* (= "stenonine" horses) and in fact may be conspecific. *E. capensis* is a related South African species. A detailed morphologic comparison of *E. stenonis, E. sanmeniensis, E. koobiforensis,* and *E. oldowayensis* has not been undertaken, but it would appear to be necessary to resolve their relationships and perhaps eventually their taxonomic status.

There are four northern African species of *Equus* which we retain herein: *E. numidicus, E. tabeti, E. melkiensis,* and *E. algericus*. *E. numidicus* is large-sized (about the size of a large zebra) and is considered to be a close relative of *E. grevyi*. *E. tabeti* is a smaller (moderate-sized) *Equus* believed to be related to *Equus (Asinus)* and occurs in the Pleistocene of northern and eastern Africa. Eisenmann (1983) reported that it was adapted to warm, dry environments. *E. melkiensis* is a smaller late Pleistocene *Equus* with asinine characters that co-occurs with the caballine horse, *E. algericus*. Figure 35.22 plots the historic distribution of nondomesticated Eurasian *Equus* species.

The first occurrence of zebra species *E. grevyi*, *E. quagga*, and *E. zebra* is widely debated. Churcher and Richardson prefer to recognize *E. grevyi* and *E. quagga* within the Pleistocene, whereas Eisenmann prefers to recognize the distinction of these and several other species that Churcher and Richardson synonymize into extant African zebra species. The molecular evidence would not seem to offer conclusive evidence to support one argument or the other. Sufficient fossil material exists to test these hypotheses with morphologic data and will provide insight for molecular based hypotheses.

All African equids were adapted to eating grass. African Hipparionini would appear to have become progressively more dedicated to eating grass with the recorded increase of maximum crown height from 60 to 90 mm across the 10.5–1.0 Ma timeframe. Zebras also eat grass and are adapted to severe drought conditions in many circumstances. Recent work by Kaiser and Schulz here demonstrates that mesowear is an excellent predictor of environment and in particular rainfall.

ACKNOWLEDGMENTS

The authors thank Bill Sanders and Lars Werdelin for inviting us to participate in this volume and their extraordinary patience in our need for more time than anticipated to complete this chapter. We thank Vera Eisenmann for many of the equid figures included herein. We thank Denis Geraads for the figure of *Eurygnathohippus pomeli* and Hideo Nakaya and Mahito Watabe for the figure of *"Cormohipparion"* aff. *africanum*. We also thank the National Science Foundation (grant number EAR-125009) for supporting the hipparion research (awards to R.L.B.) and the Revealing Hominid Origins Initiative (NSF grant BCS-0321893) to F. C. Howell and T. D. White for supporting R. L. Bernor, M. J. Armour-Chelu, W. H. Gilbert, and T. M. Kaiser, who are members of the Perissodactyl Research Group (R. L. Bernor, working group leader). T.M.K. and E.S. thank the American Museum of Natural History, New York; the British National History Museum, London; the Etosha Ecological Institute, Okaukuejo; the Muséum National d'Histoire Naturelle, Paris; the Museum für Naturkunde, Berlin; the National Museum of Namibia, Windhoek; the Naturmuseum Senckenberg, Frankfurt am Main; the Naturhistorisches Museum, Bern; the Nico van Rooyen Taxidermy, Pretoria; the Russian Academy of Science, St. Petersburg; the Smithsonian National Museum of Natural History, Washington, D.C.; the Transvaal Museum, Pretoria; the Zoologisches Museum der Universität, Hamburg; and the Zoologische Sammlung, Munich, for access to the specimens investigated in this study. The work of T.M.K. and E.S. was partly supported by the Deutsche Forschungsgemeinschaft (AZ: KA 1525-4-1/4-2, KA 1525 6-1). T.M.K. further thanks Tamara Franz-Odendaal (currently Department of Biology, Dalhousie University, Halifax, Nova Scotia, Canada) for providing access to specimens of *Equus capensis* and *Eurygnathohippus hooijeri*. We also wish to thank the Deutsche Akademische Austauschdienst (DAAD) for a scholarship awarded to T. Franz-Odendaal.

Literature Cited

Allen, J. A. 1909. Mammals from British East Africa, collected by the Tjader Expedition of 1906. *Bulletin of the American Museum of Natural History* 26:147–175.

Antonius, O. 1951. *Die Tigerpferde die Zebras*. Monographien der Wildsäugetiere 11, Schöps, Frankfurt am Main, 148 pp.

Arambourg, C. 1959. Vertébrés continentaux du Miocène supérieur de l'Afrique du Nord. *Publications du Service de la Carte Géologique de l'Algérie, n.s. Paléontologie, Mémoire* 4:1–161.

———. 1970. Les Vertébrés du Pléistocène de l'Afrique du Nord. *Archives du Muséum national d'Histoire Naturelle* 10:1–127.

———. 1979. *Vertébrés Villafranchiens d'Afrique du Nord (Artiodactyles, Carnivores, Primates, Reptiles, Oiseaux)*. Fondation Singer-Polignac, Paris, 141 pp.

Aranguren-Mendez, J., A. Beja-Pereira, R. Avellanet, K. Dzama, and J. Jordana. 2004. Mitochondrial DNA variation and genetic relationships in Spanish donkey breeds *(Equus asinus)*. *Journal of Animal Breeding and Genetics* 121:319–330.

Aranguren-Mendez, J., J. Jordana, and M. Gomez. 2001. Genetic diversity in Spanish donkey breeds using microsatellite DNA markers. *Genetics, Selection, Evolution* 33:433–442.

Armour-Chelu, M., R. L. Bernor, and H.-W. Mittmann. 2006. Hooijer's hypodigm for *"Hipparion"* cf. *ethiopicus* (Equidae, Hipparionini) from the East African Plio-Pleistocene; pp. 15–24 in L. W. van den Hoek Ostende, D. Nagel, and M. Harzhauser (eds.), *Beiträge zur Paläontologie 30*. Festschrift für Gudrun Daxner-Höck (Verein zur Förderung der Paläontologie), Vienna.

Aouraghe, H., and A. Debenath. 1999. Les équidés du Pléistocène supérieur de la Grotte Zouhrah à El Harhoura, Maroc. *Quaternaire* 10:283–292.

Azzaroli, A. 1979. On a late Pleistocene ass from Tuscany, with notes on the history of asses. *Palaeontographica Italica* 71:27–47.

Bagtache, B., D. Hadjouis, and V. Eisenmann. 1984. Presence d'un *Equus* caballine (*E. algericus* n. sp.) et d'une autre espece nouvelle d'*Equus* (*E. melkiensis* n. sp.) dans l'Atérien des Allbroges, Algerie. *Comptes Rendus des Séances de l'Académie des Sciences, Paris*, Série II 298:609–612.

Bennett, D. K. 1980. Stripes do not a zebra make: Part I. A cladistic analysis of *Equus*. *Systematic Zoology* 29:272–287.

Ben-Shahar, R. 1991. Abundance of trees and grasses in a woodland savanna in relation to environmental factors. *Journal of Vegetation Science* 2:345–350.

Berger, J. (ed.). 1986. *Wild Horses of the Great Basin: Social Competition and Population Size*. University of Chicago Press, Chicago, 326 pp.

Bernor, R. L. 1983. Geochronology and zoogeographic relationships of Miocene Hominoidea; pp. 21–64 in R. L. Ciochon and R. Corruccini (eds.), *New Interpretations of Ape and Human Ancestry*. Plenum Press, New York.

———. 1984. A zoogeographic theater and biochronologic play: The time/biofacies phenomena of Eurasian and African Miocene mammal provinces. *Paléobiologie Continentale* 14:121–142.

———. 1985. Systematics and evolutionary relationships of the hipparionine horses from Maragheh, Iran. *Paleovertebrata* 15:173–269.

Bernor, R. L., and M. J. Armour-Chelu. 1997. Later Neogene Hipparions from the Manonga Valley, Tanzania; pp. 219–264 in T. Harrison (ed.), *Neogene Paleontology of the Manonga Valley, Tanzania*. Topics in Geobiology Series. Plenum Press, New York.

———. 1999. The family Equidae; pp. 193–202 in G. E. Rössner and K. Heissig (eds.), *The Miocene Land Mammals of Europe*. Pfeil, Munich.

Bernor, R. L., and J. Franzen. 1997. The equids (Mammalia, Perissodactyla) from the Late Miocene (early Turolian) of Dorn-Dürkheim 1 (Germany, Rheinhessen). *Courier Forschungsinstitut Senckenberg* 30: 117–185.

Bernor, R. L., and Y. Haile-Selassie. 2009. Equidae; pp. 397–428 in Y. Haile-Selassie and G. WoldeGabriel (eds.), *Ardipithecus kadabba*. Late Miocene evidence from the Middle Awash, Ethiopia. University of California Press, Berkeley.

Bernor, R. L., and J. Harris. 2003. Systematics and evolutionary biology of the Late Miocene and early Pliocene Hipparionine horses from Lothagam, Kenya; pp. 387–438 in J. M. Harris and M. Leakey (eds.), *Lothagam: The Dawn of Humanity in Eastern Africa*. Columbia University Press, New York.

Bernor, R. L., K. Heissig, and H. Tobien. 1987. Early Pliocene perissodactyla from Sahabi, Libya; pp. 233–254 in N. T. Boaz, J. de Heinzelin, W. Gaziry, and A. El-Arnauti (eds.), *Neogene Geology and Paleontology of Sahabi*. Liss, New York.

Bernor, R. L., and S. T. Hussain. 1985. An assessment of the systematic, phylogenetic and biogeographic relationships of Siwalik hipparionine horses. *Journal of Vertebrate Paleontology* 5:32–87.

Bernor, R. L., and T. M. Kaiser. 2006. Systematics and paleoecology of the earliest Pliocene equid, *Eurygnathohippus hooijeri* n. sp. from Langebaanweg, South Africa. *Mitteilungen aus dem Hamburgischen Zoologischen Museum und Institut* 103:149–186.

Bernor, R. L., T. M. Kaiser, and S. V. Nelson. 2004. The Oldest Ethiopian *Hipparion* (Equinae, Perissodactyla) from Chorora: Systematics, Paleodiet and Paleoclimate. *Courier Forschungsinstitut Senckenberg* 246:213–226.

Bernor, R. L., T. M. Kaiser, and D. Wolf. 2008. Revisiting Sahabi equid species diversity, biogeographic patterns and diet preferences. *Gharyounis Bulletin Special Issue* 5:159–167.

Bernor, R. L., G. D. Koufos, M. O. Woodburne, and M. Fortelius. 1996. The evolutionary history and biochronology of European and southwest Asian late Miocene and Pliocene hipparionine horses; pp. 307–338 in R. L. Bernor, V. Fahlbusch, and H.-W. Mittmann (eds.), *The Evolution of Western Eurasian Neogene Mammal Faunas*. Columbia University Press, New York.

Bernor, R. L., J. Kovar, D. Lipscomb, F. Rögl, and H. Tobien. 1988. Systematic, stratigraphic and paleoenvironmental contexts of first appearing *Hipparion* in the Vienna Basin, Austria. *Journal of Vertebrate Paleontology* 8:427–452.

Bernor, R. L., Z. Qiu, and L.-A. Hayek. 1990. Systematic revision of Chinese *Hipparion* species described by Sefve, 1927. *American Museum Novitates* 2984:1–60.

Bernor, R. L., and R. S. Scott. 2003. New interpretations of the systematics, biogeography and paleoecology of the Sahabi Hipparions (latest Miocene), Libya. *Geodiversitas* 25:297–319.

Bernor, R. L., R. S. Scott, M. Fortelius, J. Kappelman, and S. Sen. 2003. Systematics and evolution of the Late Miocene hipparions from Sinap, Turkey; pp. 220–281 in M. Fortelius, J. Kappelman, S. Sen, and R. L. Bernor (eds.), *The Geology and Paleontology of the Miocene Sinap Formation, Turkey*. Columbia University Press, New York.

Bernor, R. L., R. S. Scott, and Y. Haile-Selassie. 2005. A contribution to the evolutionary history of Ethiopian hipparionine horses: Morphometric evidence from the postcranial skeleton. *Geodiversitas* 27:133–158.

Bernor, R. L., and H. Tobien. 1989. Two small species of *Cremohipparion* (Equidae, Mamm.) from Samos, Greece. *Mitteilungen der Bayerischen Staatssammlung für Paläontologie und Historische Geologie* 29:207–226.

Bernor, R. L., H. Tobien, L.-A. Hayek, and H.-W. Mittmann. 1997. The Höwenegg hipparionine horses: Systematics, stratigraphy, taphonomy and paleoenvironmental context. *Andrias* 10:1–230.

Bernor, R. L., H. Tobien, and M. O. Woodburne. 1989. Patterns of Old World hipparionine evolutionary diversification and biogeographic extension; pp. 263–319 in E. H. Lindsay, V. Fahlbusch, and P. Mein (eds.), *European Neogene Mammal Chronology*. Plenum Press, New York.

Bernor, R. L., and T. White. In press. Systematics and biogeography of "*Cormohipparion*" *africanum*, early Vallesian (MN9, ca. 10.5 Ma) of Bou Hanifia, Algeria. *Transactions of the American Philosophical Society*.

Bernor, R. L., M. O. Woodburne, and J. A. Van Couvering. 1980. A contribution to the chronology of some Old World Miocene faunas based on hipparionine horses. *Geobios* 13:705–739.

Broom, R. 1909. On evidence of a large horse recently extinct in South Africa. *Annals of the South African Museum* 7:281–282.

———. 1928. On some new mammals from the diamond gravels of the Kimberley district. *Annals of the South African Museum* 22:439–444.

Brown, W. M., M. George, and A. C. Wilson. 1979. Rapid evolution of animal mitochondrial DNA. *Proceedings of the National Academy of Sciences, USA* 76:1967–1971.

Cabrera, A. 1936. Subspecific and individual variation in the Burchell zebras. *Journal of Mammalogy* 17:89–112.

Churcher, C. S. 1970. The fossil equidae from the Krugersdorp Caves. *Annals of the Transvaal Museum* 26:145–168.

———. 1981. Zebras (genus *Equus*) from nine Quaternary sites in Kenya, East Africa. *Canadian Journal of Earth Sciences* 18:330–341.

———. 1982. Oldest ass recovered from Olduvai Gorge, Tanzania, and the origin of asses. *Journal of Paleontology* 56:1124–1132.

———. 1986. Equid remains from Neolithic horizons at Dakheh Oasis, Western Desert of Egypt; pp. 413–421 in R. H. Meadow and H. P. Uerpmann (eds.), *Equids in the Ancient World*. Reichert, Wiesbaden, Germany.

Churcher, C. S., and D. A. Hooijer. 1980. The Olduvai zebra *(Equus oldowayensis)* from the later Omo beds, Ethiopia. *Zoologische Mededelingen* 55:265–280.

Churcher, C. S., and M. L. Richardson. 1978. Equidae; pp. 379–422 in V. J. Maglio and H. S. Cooke (eds.), *Evolution of African Mammals*. Harvard University Press, Cambridge.

Clark, B. 1985. *Equus africanus*. Convention on International Trade in Endangered Species of Wild Fauna and Flora. *Appendix* 1:1–29.

Cooke, H. B. S. 1941. A preliminary account of the Wonderwerk Cave, Kuruman District: Section II. The fossil remains. *South African Journal of Science* 37:300–312.

———. 1950. Quaternary fossils from northern Rhodesia; pp. 137–152 in J. D. Clark, *The Stone Age Cultures of Northern Rhodesia*. South African Archaeological Society, Cape Town.

———. 1963. Pleistocene mammal faunas of Africa, with particular reference to southern Africa; pp. 65–116 in F. C. Howell and F. Bourlière (eds.), *African Ecology and Human Evolution*. Viking Fund Publications in Anthropology, New York.

Cooke, H. B. S., and S. Coryndon. 1970. Pleistocene mammals from the Kaiso Formation and other related deposits in Uganda; pp. 107–224 in L. S. B. Leakey and R. J. G. Savage (eds.), *Fossil Vertebrates of Africa*, vol. 2. Academic Press, London.

Coppens, Y. 1971. Les vértèbres villafanchiens de Tunisie: Gisements nouveaux, signification. *Comptes Rendus des Séances de l'Académie des Sciences, Paris, Série D, Sciences Naturelles* 273:51–54.

Cracraft, J. 1983. Species concept and speciation analysis; pp. 159–187 in R. F. Johnston (ed.) *Current Ornithology*. Plenum Press, New York.

Denzau, G., and H. Denzau (eds.). 1999. *Wildesel*. Thorbecke, Stuttgart, 221 pp.

Dietrich, W. O. 1942. Ältestquartäre Säugetiere aus der südlichen Serengeti, Deutsch-Ostafrika. *Palaeontographica, Abt. A* 94:43–133.

Eisenmann, V. 1976. Le protostylide: Valeur systematique et signification phylétique chez les espèces actuelles et fossiles du genre *Equus* (Perissodactyla, Mammalia). *Zeitschrift für Säugetierkunde* 41:349–365.

———. 1979. Le genre *Hipparion* (Mammalia, Perissodactyla) et son intérêt biostratigraphique en Afrique. *Bulletin de la Société gèologique de France* (7) 21:277–281.

———. 1980. Les chevaux *(Equus sensu lato)* fossiles et actuels: Crânes et dents jugales supérieures. *Cahiers de Paleontologie*. CNRS, Paris, 186 pp.

———. 1983. Family Equidae; pp. 156–214 in J. M. Harris (ed.), *Koobi Fora Research Project: Volume 2: The Fossil Ungulates: Proboscidea, Perissodactyla, and Suidae*. Clarendon Press, Oxford.

———. 1985. Les équidés des gisements de la Vallée de l'Omo en Éthiopie; pp. 13–55 in *Les Faunes Plio-Pléistocènes de la Basse Vallée de l'Omo Éthiopie*. Cahiers de Paleontologie. CNRS, Paris.

———. 1986. Comparative osteology of modern and fossil horses, halfasses and asses; pp. 67–116 in R. H. Meadow and H. R. Uerpmann (eds.), *Equids in the Ancient World: Beihefte zum Tübinger Atlas des Vorderen Orients*. Reihe A, Reichert, Wiesbaden, Germany.

———. 1994. Equidae of the Albertine Rift Valley, Uganda; pp. 289–307 in B. Senut and M. Pickford (eds.), *Geology and Palaeobiology of the Albertine Rift Valley Uganda—Zaire: Volume II. Palaeobiology/Paléobiologie*. CIFEG Occasional Publication 1994/29, Orléans, France.

———. 1995. What metapodial morphometry has to say about some Miocene hipparions; pp. 148–163 in E. S. Vrba, G. H. Denton, T. C. Partridge, and L. H. Burckle (eds.), *Paleoclimate and Evolution: With Emphasis on Human Origins*. Yale University Press, New Haven.

———. 1997. Does the taxonomy of the quagga really need to be reconsidered? *South African Journal of Science* 93:66–68.

———. 2000. *Equus capensis* (Mammalia, Perissodactyla) from Elandsfontein. *Palaeontologia Africana* 36:91–96.

Eisenmann, V., and J. S. Brink. 2000. Koffiefontein quaggas and true Cape quaggas: The importance of basic skull morphology. *South African Journal of Science* 96:529–533.

Eisenmann, V., and D. Geraads. 2007. *Hipparion pomeli* sp. nov from the late Pliocene of Ahl al Oughlam, Morocco, and a revision of the relationships of Pliocene and Pleistocene African hipparions. *Palaeontologia Africana* 42:51–98.

Eisenmann, V., and A. Karchoud. 1982. Analyses multidimensionnelles de métapodes d'*Equus* (Mammalia, Perissodactyla). *Bulletin du Muséum Nationale d'Histoire Naturelle*, 4e Série, Sect. C 4:75–103.

Eisenmann, V., and J. C. Turlot. 1978. Sur la taxonomie du genre *Equus* (Equides). *Les Cahiers de l'Analyse des Donnees* 3:179–201.

Erz, W. 1964. Tooth eruption and replacement in Burchell's Zebra, *Equus burchellii* Gray 1825. *Arnoldia* 22:1–8.

Forsten, A. 1992. Mitochondrial-DNA time-table and the evolution of *Equus*: Comparison of molecular and paleontological evidence. *Annales Zoologici Fennici* 28:301–309.

———. 1997. Caballoid hipparions (Perissodactyla, Equidae) in the Old World. *Acta Zoologica Fennica* 205:27–51.

Fortelius, M., and N. Solounias. 2000. Functional characterization of ungulate molars using the abrasion-attrition wear gradient: A new method for reconstructing paleodiets. *American Museum Novitates* 3301:1–36.

Franz-Odendaal, T. A., T. M. Kaiser, and R. L. Bernor. 2003. Systematics and dietary evaluation of a fossil equid from South Africa. *South African Journal of Science* 99:453–459.

Gaudry, A. 1873. *Animaux Fossiles du Mont Léberon (Vaucluse): Étude sur les Vertébrés*. Savy, Paris, 180 pp.

George, M., and O. Ryder. 1986. Mitochondrial DNA evolution in the genus *Equus*. *Molecular Biology and Evolution* 3:535–546.

Geraads, D., J.-P. Raynal, and V. Eisenmann. 2004 The earliest human occupation of North Africa: A reply to Sahnouni et al. (2002). *Journal of Human Evolution* 46: 751–761.

Gervais, F. 1849. Note sur la multiplicité des espèces d'hipparions (genre de chevaux à trois doigts) qui sont enfouis a Cucuron (Vaucluse). *Comptes Rendus de l'Académie des Sciences, Paris* 29:284–286.

———. 1859. *Zoologie et Paléontologie Françaises (Animaux Vertébrés)*. 2nd ed. Paris, 544 pp.

Gidley, J. W. 1901. Tooth characters and revision of the North American species of the genus *Equus*. *Bulletin of the American Museum of Natural History* 14:91–142.

Gilbert, W. H., and R. L. Bernor. 2009. Equidae; pp. 133–166 in W. H. Gilbert and B. Asfaw, (eds.), Homo erectus: *Pleistocene Evidence from the Middle Awash, Ethiopia*. University of California Press, Berkeley.

Gray, J. E. 1825. A revision of the family Equidae. *Zoological Journal* 1:241–248.

Groves, C. P. (ed.). 1974. *Horses, Asses, and Zebras in the Wild*. David and Charles, Newton Abbot, London, 192 pp.

———. 1986. The taxonomy, distribution, and adaptations of recent equids; pp. 11–65 in R. H. Meadow and H.-P. Uerpmann (eds.), *Equids in the Ancient World*. Reichert, Wiesbaden, Germany.

———. 2002. Taxonomy of living Equidae. *Report* 8:94–107. Gland, Switzerland, and Cambridge, IUCN/SCC Equid Specialist Group, IUCN.

Groves, C. P., and C. H. Bell. 2004. New investigations on the taxonomy of the zebra genus *Equus*, subgenus *Hippotigris*. *Mammalian Biology* 69:182–196.

Groves, C. P., and V. Mazak. 1967. On some taxonomic problems of Asiatic wild asses: With the description of a new subspecies (Perissodactyla; Equidae). *Zeitschrift für Säugetierkunde* 32:321–335.

Groves, C. P., and O. A. Ryder. 2000. Systematics and phylogeny of the horse; pp. 1–24 in A. T. Bowling and A. Ruvinsky (eds.), *The Genetics of the Horse*. CABI Publishing, Oxford.

Groves, C. P., and C. Smeenk. 2007. The nomenclature of the African wild ass. *Zoologische Mededelingen* 81:121–135.

Groves, C. P., and D. R. Willoughby. 1981. Studies on the taxonomy and phylogeny of the Genus *Equus*: 1. Subgeneric classification of the recent species. *Mammalia* 45:321–354.

Grubb, P. 1981. *Equus burchellii*. *Mammalian Species* 157:1–9.

Gwynne, M. D., and R. H. V. Bell. 1968. Selection of vegetation components by grazing ungulates in the Serengeti National Park. *Nature* 220:390–393.

Haughton, S. H. 1932. The fossil Equidae of South Africa. *Annals of the South African Museum* 28:407–427.

Heptner, V. G., A. A. Nasimovic, and A. G. Bannikov. 1966. *Die Säugetiere der Sowjetunion, Band I: Paarhufer und Unpaarhufer*. VEB Gustav Fischer, Jena, Germany, 758 pp.

Higuchi, R., B. Bowman, M. Freiberger, O. A. Ryder, and A. C. Wilson. 1984. DNA sequences from the quagga, an extinct member of the horse family. *Nature* 312:282–284.

Hillis, D. M., B. K. Mable, and C. Moritz. 1996. Applications of molecular systematics: the state of the field and a look to the future; pp. 515–543 in D. M. Hillis, C. Moritz, and B. K. Mable (eds.), *Molecular Systematics*. 2nd ed. Sinauer Associates, Sunderland, Mass.

Hoepen, E. C. N. van. 1930. Fossiele Perde van Cornelia, O. V. S. *Paleontologiese Navorsing van die Nasionale Museum, Bloemfontein* 2:13–34.

Hooijer, D. A. 1975. Miocene to Pleistocene Hipparions of Kenya, Tanzania and Ethiopia. *Zoologische Verhandelingen* 142:1–80.

———. 1976. The late Pliocene Equidae of Langebaanweg, Cape Province, South Africa. *Zoologische Verhandelingen* 148:3–39.

Hooijer, D. A., and C. S. Churcher. 1985. Perissodactyla of the Omo Group deposits, American collections; pp. 97–117 in Y. Coppens and F. C. Howell (eds.), *Les Faunes Plio-Pléistocènes de la Basse Vallée de l'Omo (Éthiopie): Tome 1: Périssodactyles—Artiodactyles (Bovidae)*. CNRS, Paris.

Hooijer, D. A., and V. J. Maglio. 1973. The earliest *Hipparion* south of the Sahara, in the late Miocene of Kenya. *Proceedings Koninklijke Nederlandse Akademie van Wetenschappen B* 76:311–315.

———. 1974. Hipparions from the Late Miocene and Pliocene of northwestern Kenya. *Zoologische Verhandelingen* 134:1–34.

Hopwood, A. T. 1937. Die fossilen Pferde von Oldoway. *Wissenschaftliche Ergebnisse der Oldoway-Expedition* 1913 (N.F.) 4:112–136.

Hulbert, R. C. 1987. A new *Cormohipparion* (Mammalia, Equidae) from the Pliocene (latest Hemphillian and Blancan) of Florida. *Journal of Vertebrate Paleontology* 7:451–468.

Ishida, N., T. Oyunsuren, S. Mashima, H. Mukoyama, and N. Saitou. 1995. Mitochondrial DNA sequences of various species of the genus *Equus* with special references to the phylogenetic relationship between Przewalskii's wild horse and domestic horse. *Journal of Molecular Evolution* 41:180–188.

IUCN. 2006. *Red List of Threatened Species*. Downloaded January 10, 2007, from www.iucn.org.

Joubert, E. 1971. Ecology, behaviour and population dynamics of the Hartmann zebra, *Equus zebra hartmannae* Matschie, 1898 in South West Africa. Unpublished PhD dissertation, University of Pretoria, South Africa.

———. 1972. Tooth development and age determination in the Hartmann zebra *Equus zebra hartmannae*. *Madoqua* 6:5–16.

Kaiser, T. M. 2003. The dietary regimes of two contemporaneous populations of *Hippotherium primigenium* (Perissodactyla, Equidae) from the Vallesian (upper Miocene) of Southern Germany. *Palaeogeography, Palaeoclimatology, Palaeoecology* 198:381–402.

Kaiser, T. M., R. L. Bernor, J. Franzen, R. S. Scott, and N. Solounias. 2003. New Interpretations of the Systematics and Palaeoecology of the Dorn-Dürkheim 1 Hipparions (Late Miocene, Turolian Age [MN11]), Rheinhessen, Germany. *Senckenbergiana Lethaea* 83:103–133.

Kaiser, T. M., and M. Fortelius. 2003. Differential mesowear in occluding upper and lower molars: Opening mesowear analysis for lower molars and premolars in hypsodont equids. *Journal of Morphology* 258:67–83.

Kaiser, T. M., and T. A. Franz-Odendaal. 2004. A mixed feeding *Equus* species from the Middle Pleistocene of South Africa. *Quaternary Research* 62:316–323.

Kaiser, T. M., and E. Schulz. 2006. Tooth wear gradients in zebras as an environmental proxy: A pilot study. *Mitteilungen aus dem Hamburgischen Zoologischen Museum und Institut* 103:187–210.

Kaiser, T. M., and N. Solounias. 2003. Extending the tooth mesowear method to extinct and extant equids. *Geodiversitas* 25:321–345.

Kaiser, T. M., N. Solounias, M. Fortelius, R. L. Bernor, and F. Schrenk. 2000. Tooth mesowear analysis on *Hippotherium primigenium* from the Vallesian Dinotheriensande (Germany): A blind test study. *Carolinea* 58:103–114.

King, J. M. 1965. A field guide to the reproduction of the Grant's zebra and the Grevy's zebra. *East African Wildlife Journal* 3:99–117.

Kingdon, J. 1979. *East African Mammals: An Atlas of Evolution in Africa: Volume 3B. Large Mammals*. Academic Press, London, 434 pp.

———. 1997. *The Kingdon Field Guide to African Mammals*. Academic Press, San Diego, 476 pp.

Klein, R. 1974. A provisional statement on terminal Pleistocene mammalian extinctions in the Cape biotic zone (Southern Cape Province, South Africa). *South African Archaeological Society* 2:39–45.

Klein, R. G., and K. Cruz-Uribe. 1999. Craniometry of the genus *Equus* and the taxonomic affinities of the extinct South African quagga. *South African Journal of Science* 95:81–86.

Klingel, H. 1972. *Somali Wild Ass: Status Survey in the Danakil Region*. WWF Field Survey. WWF, Braunschweig, 9 pp.

———. In press. *Equus quagga*; in J. S. Kingdon and M. Hoffmann (eds.), *The Mammals of Africa: Volume 5. Carnivores, Pangolins, Rhinos and Equids*. Academic Press, Amsterdam.

Klingel, H., and U. Klingel. 1966. Tooth development and age determination in the plains zebra (*Equus quagga boehmi* Matschie). *Der Zoologische Garten* 33:34–54.

Koufos, G. D. 1984. A new hipparion (Mammalia, Perissodactyla) from the Vallesian (Late Miocene) of Greece. *Paläontologische Zeitschrift* 58:307–317.

Leakey, L. S. B. (ed.). 1965. *Olduvai Gorge 1951–1961: Volume 1. A Preliminary Report on the Geology and Fauna*. Cambridge University Press, Cambridge, 118 pp.

Leakey, M. G., C. S. Feibel, R. L. Bernor, T. E. Cerling, J. M. Harris, I. McDougall, K. M. Stewart, A. Walker, L. Werdelin, and A. J. Winkler.

1996. Lothagam: A record of faunal change in the late Miocene of East Africa. *Journal of Vertebrate Paleontology* 16:556–570.

Leakey, M. G., and J. Harris (eds.). 2003. *Lothagam: The Dawn of Humanity in Eastern Africa*. Columbia University Press, New York, 678 pp.

Leemans, R., and W. Cramer. 1991. *The IIASA Climate Database for Mean Monthly Values of Temperature, Precipitation and Cloudiness on a Terrestrial Grid*. Research Report RR-91-18. International Institute for Applied Systems Analysis, Laxenburg, Austria, 62 pp.

Leonard, J. A., N. Rohland, S. Glaberman, R. C. Fleischer, A. Caccone, and M. Hofreiter. 2005. A rapid loss of stripes: The evolutionary history of the extinct quagga. *Biology Letters* 1:291–295.

Lindsay, E. H., N. D. Opdyke, and N. M. Johnson. 1980. Pliocene dispersal of the horse *Equus* and late Cenozoic mammalian dispersal events. *Nature* 287:135–138.

Lowenstein, J. M., and O. A. Ryder. 1985. Immunological systematics of the extinct quagga (Equidae). *Experientia* 41:1192–1193.

Lundholm, B. 1952. *Equus zebra greatheadi* n. subsp., a new South African fossil zebra. *Annals of the Transvaal Museum* 22:25–27.

Lydekker, R. (ed.). 1913–1916. *Catalogue of the Ungulate Mammals in the British Museum (Natural History): Volume 5. Perissodactyla (Horses, Tapirs, Rhinoceroses), Hyracoidea (Hyraxes), Proboscidea (Elephants)*. Trustees of the British Museum (Natural History), London, 207 pp.

MacFadden, B. J. 1980. The Miocene horse *Hipparion* from North America and from the type locality in southern France. *Palaeontology* 23:617–635.

———. 1984. Systematics and phylogeny of *Hipparion*, *Neohipparion*, *Nannippus*, and *Cormohipparion* (Mammalia, Equidae) from the Miocene and Pleistocene of the New World. *Bulletin of the American Museum of Natural History* 179:1–195.

MacFadden, B. J., and M. O. Woodburne. 1982. Systematics of the Neogene Siwalik hipparions (Mammalia, Equidae) based on cranial and dental morphology. *Journal of Vertebrate Paleontology* 2:185–218.

Marean, C. W., and D. Gifford-Gonzalez. 1991. Late Quaternary extinct ungulates of East Africa and palaeoenvironmental implications. *Nature* 350:418–420.

Mendrez, C. 1966. On *Equus (Hippotigris)* cf. *burchellii* (Gray) from "Sterkfontein Extension," Transvaal, South Africa. *Annals of the Transvaal Museum* 25:91–97.

Moehlman, P. D. (ed.). 2002. *Equids: Zebras, Asses and Horses*. Status Survey and Conservation Action Plan. IUCN/SSC Equid Specialist Group. IUCN, Gland, Switzerland, and Cambridge, 184 pp.

Moodley, Y., and E. H. Harley. 2005. Population structuring in mountain zebras *(Equus zebra)*: The molecular consequences of divergent demographic histories. *Conservation Genetics* 6:953–968.

Moodley, Y., I. Baumgarten, and E. H. Harley. 2006. Horse microsatellites and their amenability to comparative equid genetics. *Animal Genetics* 37:258–261.

Nakaya, H., and M. Watabe. 1990. *Hipparion* from the Upper Miocene Namurungule Formation, Samburu Hills, Kenya: Phylogenetic significance of the newly discovered skull. *Geobios* 23:195–219.

Ndessokia, P. N. S. 1990. Mammalian Fauna and Archaeology of the Ndolanya and Olpiro Beds, Laetoli, Tanzania. Unpublished PhD dissertation, University of California, Berkeley.

Novellie, P., M. Lindeque, P. Lindeque, P. H. Lloyd, and J. Koen. 2002. Chapter 3; pp. 28–42 in *Status and Action Plan for the Mountain Zebra (Equus zebra)*. IUCN/SCC Equid Specialist Group. IUCN, Gland, Switzerland.

Nowak, R. M. (ed.). 1991. *Walker's Mammals of the World*. 5th ed., vol. 2. Johns Hopkins University Press, Baltimore, 1,629 pp.

Oakenfull, E. A., and J. B. Clegg. 1998. Phylogenetic relationships within the genus *Equus* and the evolution of alpha and theta globin genes. *Journal of Molecular Evolution* 47:772–783.

Oakenfull, E. A., H. N. Lim, and O. A. Ryder. 2000. A survey of equid mitochondrial DNA: Implications for the evolution, genetic diversity and conservation of *Equus*. *Conservation Genetics* 1:341–355.

Oakenfull, E. A., and O. A. Ryder. 1998. Mitochondrial control region and 12S rRNA variation in Przewalski's horse *(Equus przewalskii)*. *Animal Genetics* 29:456–459.

Orlando, L., M. Mashkour, A. Burke, C. J. Douday, V. Eisenmann, and C. Hänni. 2006. Geographic distribution of an extinct equid *(Equus hydruntinus*: Mammalia, Equidae) revealed by morphological and genetical analyses of fossils. *Molecular Ecology* 15:2083–2093.

Osborn, H. F. 1918. Equidae of the Oligocene, Miocene and Pliocene of North America: Iconographic type revision. *Memoirs of the American Museum of Natural History* (n.s.) 2:1–217.

Penzhorn, B. L. 1982. Age determination in Cape mountain zebras *Equus zebra zebra* in the Mountain Zebra National Park. *Koedoe* 25:89–102.

———. 1987. Descriptions of incisors of known-age Cape mountain zebras, *Equus zebra zebra*, from the Mountain Zebra National Park. *Onderstepoort Journal of Veterinary Research* 54:135–141.

———. In press. *Equus zebra*; pp. in J. S. Kingdon and M. Hoffmann (eds.), *The Mammals of Africa: Volume 5. Carnivores, Pangolins, Rhinos and Equids*. Academic Press, Amsterdam.

Penzhorn, B. L., and R. J. Grimbeek. 1987. Incisor wear in free-ranging Cape mountain zebras. *South African Journal of Wildlife Research* 17:99–102.

Pienaar, U. de V. 1963. The large mammals of the Kruger National Park: Their distribution and present-day status. *Koedoe* 6:1–37.

Pomel, A. 1897. Les equidés: Carte géologie de l'Algérie. *Monographies de Paléontologie* 12:1–44.

Qiu, Z., W. Huang, and Z. Guo. 1988. Chinese hipparionines from the Yushe Basin. *Palaeontologica Sinica*, Series C 175:1–250.

Rau, R. 1974. Revised list of the preserved material of the extinct Cape Colony quagga, *Equus quagga quagga* (Gmelin). *Annals of the South African Museum* 65:41–87.

———. 1978. Additions to the revised list of preserved material of the extinct Cape colony quagga and notes on the relationship and distribution of southern plains zebras. *Annals of the South Africa Museum* 77:27–45.

Sahnouni, M., D. Hadjouis, J. van der Made, A. Derradji, A. Canals, M. Medig, and H. Belahrech. 2002. Further research at the Oldowan site of Ain Hanech, north-eastern Algeria. *Journal of Human Evolution* 43:925–937.

Sakagami, M., K. Hiromura, L. G. Chemnick, and O. A. Ryder. 1999. Distribution of the ERE-1 family in Perissodactyla. *Mammalian Genome* 10:930–933.

Schulz, E. 2008. Habitatabhängige Usurgleichgewichte in der Bezahnung verschiedener Arten der Gattung *Equus* LINNAEUS, 1758, als Instrument zur Rekonstruktion der Lebensraumparameter pleistozäner Equidae (Perissodactyla) Mitteleuropas. Unpublished PhD dissertation, University of Hamburg, Biocenter Grindel and Zoological Museum, Hamburg.

Schulz, E., T. M. Kaiser, A. Stubbe, M. Stubbe, R. Samjaa, N. Batsajchan, and J. Wussow. 2007. Comparative demography and dietary resource partitioning of two wild ranging Asiatic equid populations. *Exploration into the Biological Resources of Mongolia* 10:77–90.

Skinner, J. 1996. Further light on the speciation in the quagga. *South African Journal of Science* 92:301–302.

Skinner, M. F., and B. J. MacFadden. 1977. *Cormohipparion* n. gen. (Mammalia, Equidae) from the North American Miocene (Barstovian-Clarendonian). *Journal of Paleontology* 51:912–91.

Smuts, G. L. 1972. Seasonal movements, migration and age determination of Burchell's zebra *(Equus burchellii antiquorum*, H. Smith, 1841) in the Kruger National Park. Unpublished master's thesis, University of Pretoria, Pretoria.

———. 1974. Age determination in Burchell's zebra *(Equus burchellii antiquorum)* from the Kruger National Park. *Equus* 41:103–115.

———. 1975. Home range sizes for Burchell's zebra, *Equus burchellii antiquorum* from the Kruger National Park. *Koedoe* 18:139–146.

Smuts, M. M. S., and B. L. Penzhorn. 1988. Description of anatomical differences between skulls and mandibles of *Equus zebra* and *E. burchellii* from Southern Africa. *South African Journal of Zoology* 23:328–336.

Sondaar, P. Y. 1974. The *Hipparion* of the Rhone Valley. *Geobios* 7:289–306.

Stewart, D. R. M., and J. Stewart. 1963. The distribution of some large mammals in Kenya. *Journal of the East African Natural History Society and Coryndon Museum* 24:1–52.

Stigand, C. H. 1913. *The Game of British East Africa*. 2nd ed. Cox, London, 310 pp.

Wells, L. H. 1959. The nomenclature of South African fossil equids. *South African Journal of Science* 55:64–66.

Williams, S. D. In press. *Equus grevyi*; pp. in J. S. Kingdon and M. Hoffmann (eds.), *The Mammals of Africa: Volume 5. Carnivores, Pangolins, Rhinos and Equids*. Academic Press, Amsterdam.

Willoughby, D. P. (ed.). 1974. The Empire of *Equus*. Barnes, Cranbury, N.J., 475 pp.

Woodburne, M. O. 1996. Reappraisal of the *Cormohipparion* from the Valentine Formation, Nebraska. *American Museum Novitates* 3163:1–56.

———. 2005. A new occurrence of *Cormohipparion*. *Journal of Vertebrate Paleontology* 25:252–253.

———. 2007. Phyletic diversification of the *Cormohipparion occidentale* Complex (Mammalia; Perissodactyla, Equidae), late Miocene, North America, and the origin of the Old World *Hippotherium* Datum. *Bulletin of the American Museum of Natural History* 306:1–138.

Woodburne, M. O., and R. L. Bernor. 1980. On superspecific groups of some Old World hipparionine horses. *Journal of Paleontology* 54:1319–1348.

Yalden, D. W., M. J. Largen and D. Kock. 1986. Catalogue of the mammals of Ethiopia 6. Perissodactyla, Proboscidea, Hyracoidea, Lagomorpha, Tubulidentata, Sirenia and Cetacea. *Monitore Zoologico Italiano*, suppl. 21:31–103.

Zeuner, F. E. 1963. *A History of Domesticated Animals*. Hutchinson, London, 560 pp.

Zouhri, S., and A. Bensalmia. 2005. Révision systématique des *Hipparion sensu lato* (Perissodactyla, Equidae) de l'Ancien Monde. *Estudios Geológicos* 61:61–99.

THIRTY-SIX

Tragulidae

DENIS GERAADS

Present-day tragulids are small ruminants restricted to some humid environments of the Old World intertropical zone. They had a wider distribution in the past, with a good record in the European, East African, and southern Asian Miocene. There are five living species, only one of them, *Hyemoschus aquaticus*, being African. *Moschiola*, with at least three species, lives in Sri Lanka and India, and the two species of *Tragulus*, *T. javanicus*, and *T. napu*, in Southeast Asia (see Meijaard and Groves, 2004, for a revision of *Tragulus*, and Groves and Meijaard, 2005, for a revision of *Moschiola*). The question of their relationships has never been addressed in detail, but *Moschiola* is usually associated with *Tragulus*, *M. meminna* being often considered a species of the latter, mainly on the basis of metacarpal fusion. However, *Moschiola* shares with *H. aquaticus* a number of mostly primitive features, absent in *Tragulus*, which show that *Moschiola* is an early offshoot of the Asiatic branch.

Besides the living genera, tragulids include *Dorcabune* and *Siamotragulus* from the Miocene of southern Asia, and the more widespread *Dorcatherium*, the only genus to have been reported from Africa. *Archaeotragulus* from the late Eocene of Thailand, although described as a Tragulidae, is here not included in this family.

Systematic Paleontology

Order ARTIODACTYLA Owen, 1848
Suborder RUMINANTIA Scopoli, 1777
Family TRAGULIDAE Milne Edwards, 1864

The Tragulidae share with other Ruminantia a complex stomach and the ability to ruminate, as well as a few skeletal characters that help in defining this group, such as selenodonty, the lack of upper incisors and P1, a postcanine diastema, an incisiform lower canine, the fusion of capitate with trapezoid, cuboid with navicular, and ecto- with mesocuneiform, and the reduction of the fibula to a malleolar bone. Most of the present-day Ruminantia belong to the Pecora, which have a four-chambered stomach; usually some kind of cranial appendages; a crescentic articulation between atlas and axis; completely fused third and fourth metapodials; and extremely reduced lateral digits. The Tragulidae are more primitive than the Pecora in the lack of skull appendages, functionally replaced by long, strongly curved upper canines in males, cheek teeth that are not fully selenodont, no postglenoid process in the squamosal, very narrow to absent mastoid exposure, noncrescentic odontoid apophysis of axis vertebra, unfused radius and ulna and incompletely fused central metapodials, distal keel absent on the dorsal part of the metapodials, proximal and distal trochleae of the talus usually not parallel, and persistence of complete lateral metapodials and digits. The main primitive characters of the soft anatomy are the absence of the omasum in the stomach, and a diffuse placenta, as opposed to the cotylodenous one in Pecora.

Derived characters are the enlarged orbits and merging of the optic foramina, cancellous bulla, spaced central incisors, and several characters of the cheek teeth. The lower molars have a peculiar structure: the distal half of the first lobe consists of four cristids forming a Σ (or M). They are, from lingual to labial, the *Dorcatherium* fold, the distal half of the metaconid, the distal half of the protoconid, and the *Tragulus* fold. The latter differs from the *Palaeomeryx* fold of some Pecora in its more lingual position, as its tip connects with the mesial tip of the hypoconid, instead of its flank. The premolar row is long; in p4, and sometimes also in p3, two parallel cristids usually descend distally from the main cuspid. This structure is also found is several Oligocene ruminants such as *Bachitherium* and *Iberomeryx*. On upper molars, the distal lobe is smaller than the mesial one, especially in M3, and strongly shifted labially.

As a whole, the postcranial skeleton loses flexibility (Morales et al., 2003, have shown that flexibility decreases from Miocene to Recent). The cubonavicular is frequently fused with the ectomesocuneiform, and the tibia with the malleolar bone; grooves for tendons are deeply incised on the radius and tibia, the articulation between fibula and calcaneus is relatively flat (obviously a derived state, as camels, suids, and even early cetaceans have an articulation more like that of the Pecora). A more or less ossified dorsal shield is present in the modern forms; it is likely that it was also present in some fossils, although Thomas et al. (1990) did not mention it in their description of the associated skeletal elements of *Siamotragulus*.

Virtually all African fossil tragulids come from the Miocene of western Kenya; none is known within the range of the living form, and the African fossil record is therefore very patchy. Arambourg (1933) was the first to mention a fossil tragulid in Africa; Whitworth (1958) added several new species. More recent additions to African tragulid systematics and evolution were provided by Pickford (2001, 2002). However, no comprehensive revision of the family has been undertaken. All Miocene forms were included in *Dorcatherium*, a genus well-known in many European and (mostly southern) Asian sites, but it is by no means certain that all African forms belong to the same genus. Most fossil tragulids are known from teeth and postcrania only, and in Africa, reasonably complete tooth rows are far fewer than in Europe. This makes specific distinctions based on morphological characters still harder, and they have consequently been based mostly on size, which is clearly not satisfactory. As this review does not intend to be a revision, I shall follow the currently accepted taxonomy; larger and better-preserved samples are needed before it can be reliably improved.

Genus *DORCATHERIUM* Kaup, 1833
Table 36.1

Diagnosis Cingulum present on upper molars, but variably developed; metacarpals not fused. The controlled diagnosis of *Dorcatherium* is very short, but it is likely that many of the characters listed later for *D. pigotti* are indeed valid for the whole genus (or at least its African representatives). The fibula is free in all known African specimens but may be fused with the tibia in European forms.

DORCATHERIUM MORUOROTENSIS Pickford, 2001

Type Locality and Age Moruorot. 17.2 to 17.5 Ma, according to Pickford.
Diagnosis Modified from Pickford (2001). Tiny *Dorcatherium*, lower molar row ca 16–16.5 mm long. Differs from *D. minimum* from Pakistan in its slightly smaller size and in the presence of a cingulum on the anterior and lingual aspects of the upper molar protocones, and lesser buccal flare of the paracone and metacone.
Range See table 36.1.
Description This species is based on two upper molars and a lower molar row (that I have not seen). For the lower teeth, Pickford (2001) mentioned, besides the lack of an ectostylid and the presence of a mesial cingulum, the absence of a *Dorcatherium* fold, which would make it unlike other tragulids. The mesial and distal cristids of the lower molars look tall (Pickford, 2001: figure 1), and *D. moruorotensis* is certainly quite distinct from other African *Dorcatherium*.

DORCATHERIUM PARVUM Whitworth, 1958
Figure 36.1

Type Locality and Age Rusinga R39; formation unknown (Pickford, 1986).
Diagnosis Small *Dorcatherium*, lower molar row ca. 20 mm long. Some individuals lacked p1 (Whitworth, 1958).
Range This species is best known from Rusinga (figure 36.1), but Pickford (2002) described it also from Napak, Uganda. In the KNM, mandible SO 1345 from Songhor has the right size (L m1–m3 = 19.5; L m3 = 8.2) for this species, confirming Whitworth's (1958:42) identification based on limb bones. This species is also present at Maboko, and was even listed at Ngorora, a much younger site, by Pickford (2001: table 7); if these identifications are correct, the species range might cover most of the early and middle Miocene (table 36.1).

Description According to Whitworth (1958), this species has less bunodont teeth than other Miocene ones, no ectostylid, a p2 shorter than p3, and the lingual and labial crests of the Σ are weak. It is indeed true that in most cases the *Tragulus* fold is weak and too short to reach the tip of the hypoconid; this is reminiscent of the morphology of the smallest modern tragulids, and could be a consequence of smaller size. The morphology of p4 is variable: there may be two parallel cristids descending distally from the main cuspid or a morphology closer to that of the Pecora, with short transverse cristids arising from a main longitudinal cristid.

DORCATHERIUM SONGHORENSIS Whitworth, 1958

Type Locality and Age Songhor, about 20 Ma.
Diagnosis Medium-sized *Dorcatherium*, lower molar row ca. 24 mm long.
Range This is the earliest tragulid in Africa, as it occurs at Meswa Bridge (M. Pickford, pers. comm.), but is best known from the early Miocene of Songhor; some specimens from Rusinga are of the right size for this species. It was also reported from Langental in Namibia (Pickford, 2001).
Description Whitworth (1958) noticed that this species is less bunodont than *D. pigotti*, and the specimens from Songhor are very selenodont, with high tubercles. The ectostylid is not pillar shaped, but the mesial and labial cingula are strong. The p4 is relatively longer than in *D. pigotti*. A definitely tragulid cubonavicular from Songhor is not fused with the ecto-meso-cuneiform. The metatarsal is longer than in *Hyemoschus*, but much more slender.

DORCATHERIUM PIGOTTI Whitworth, 1958

Synonymy *D. libiensis* Hamilton, 1973?
Type Locality and Age Rusinga, perhaps R106 (Whitworth, 1958); R106 is in the Hiwegi Fm (Pickford, 1986).
Diagnosis In part after Whitworth (1958). Medium-sized *Dorcatherium*, lower molar row ca. 30 mm long. Ethmoidal vacuity much reduced; large orbit; mandible with a gently convex lower border; occipital narrow, auditory bulla spherical, mastoid not exposed, teeth less bunodont than in *Hyemoschus*, p3 longer than p2 and p4.
Range See table 36.1. *Dorcatherium libiensis* Hamilton, 1973, known from a few remains from Jebel Zelten, is included in the synonymy, following Pickford (2001).
Description *D. pigotti* is by far the best known African *Dorcatherium* (hereunder abbreviated as *D.*), thanks to an associated skull and mandible (KNM-RU 46441) found at Rusinga by A. Walker, who kindly made it available to me. It is the best-preserved known skull of *Dorcatherium*. Although similar in general shape, it differs from *Hyemoschus* (abbreviated as *H.*) in a number of points and in many of them is more similar to *Moschiola* (*M.*):

- Some bone is missing in the lacrymal area, but it is fairly certain that the ethmoidal vacuity is much smaller, as in *M.*, and perhaps even absent, as in the Eppelsheim *D.*; all bones in this area probably had an X-contact, rather than the broad fronto-maxillary suture of *Tragulus*.

- It is impossible to tell whether the premaxilla was separated from the nasal, as in *H.*, or if there was a contact, as in all other Tragulidae, including the Eppelsheim *D.*

TABLE 36.1

Distribution of species of *Dorcatherium,* and of *Hyemoschus aquaticus,* in African Miocene localities, listed by increasing age (many of these ages are biochronological estimates)

Location	Locality	Ma	Main reference	*Dorcatherium moruorotensis*	*D. parvum*	*D. songhorensis*	*D. pigotti*	*D. chappuisi*	*D. iririensis*	*Hyemoschus aquaticus*
1	Mabaget, Kenya*	5.0	Pickford et al., 2004							+
2	Ngeringerowa, Kenya*	9.0	Pickford, 1991				+			
	MIDDLE MIOCENE									
3	Ngorora B, Kenya	12.0	Hill et al., 2002		?		+	+		
4	Fort Ternan, Kenya*	13.0	Pickford, 2001					+		
5	Serek, Kenya*	14.0	Pickford, 2002				+			
6	Kirimon-Mbagathi, Kenya	15.0	Pickford, 2001		+			+		
7	Nyakach, Kenya*	15.0	Pickford, 2001				+	+		
8	Maboko-Ombo, Kenya*	15.5	Pickford, 2001		+		+	+		
9	Nachola, Kenya*	15.5	Pickford et al., 1987				+	+		
10	Kipsaraman, Kenya*	15.7	Behrensmeyer et al., 2002				+	+		
11	Moghara, Egypt*	16.0	Pickford, 2001 (but see other contributions in this volume)				+			
12	Jebel Zelten, Libya*	16.5	Hamilton, 1973				+			
13	Kajong (Mwiti), Kenya	17.0	Pickford, 2001				?			
	EARLY MIOCENE									
14	Kalodirr, Kenya	17.0	Leakey and Leakey, 1986		+		+	+		
15	Langental, Namibia*	17.0	Pickford, 2001			+				
16	Loperot, Kenya	17.0	Pickford, 2001		+					
17	Buluk, Kenya*	17.2	Pickford, 2001				+	+		
18	Moruorot, Kenya*	17.2	Pickford, 2001	+				+		
19	Arrisdrift, Namibia*	17.5	Morales et al., 2003				+			
20	Bukwa, Uganda	17.5	Pickford, 2001		+		+			
21	Karungu, Kenya*	17.5	Pickford, 2001	+	+		+	+		
22	Locherangan, Kenya*	17.5	Anyonge, 1991		+		+			
23	Uyoma peninsula, Kenya*	17.7	Pickford, 2001	?	?	?	?			
24	Rusinga (Kulu Fm), Kenya*	17.7	Pickford, 2001	+	+		+	+		
25	Rusinga (Hiwegi Fm)*	17.8	Pickford, 2001	+	+	+	+	+		
26	Rusinga (Kiahera Fm)*	17.9	Pickford, 2001		+	+	+	+		
27	Rusinga (Wayando Fm)*	18.0	Pickford, 2001		+	+	+	+		
28	Mfwangano, Kenya*	18.0	Pickford, 1986	+			+	+		
29	Napak, Napak Mb, Uganda*	19.5	Pickford, 2002		+		+		+	
30	Napak, Iriri Mb, Uganda*	20.0	Pickford, 2002				+			
31	Koru-Legetet, Kenya*	20.0	Pickford, 2001				+			
32	Songhor-Mteitei, Kenya*	20.0	Pickford, 2001		+		+			
33	Meswa Bridge, Kenya	22.0	Pickford, 2001				+			

SOURCE: Data from Anyonge (1991), Behrensmeyer et al. (2002), Leakey and Leakey (1986), Pickford (1991, 2001, 2002), Pickford et al. (1987, 2004), Retallack et al. (2002).

NOTE: Some of these records are mere mentions in faunal lists that could not be checked; more detailed descriptions and/or checked occurrences of *Dorcatherium* are indicated by an asterisk in the first column.

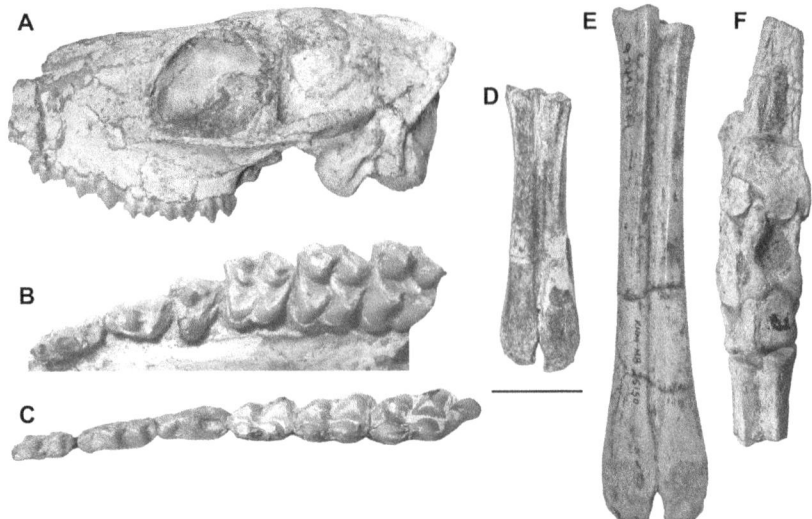

FIGURE 36.1 *Dorcatherium pigotti*, skull KNM-RU 46441 from Rusinga in NMK. A) lateral view; B) upper teeth; C) lower teeth. D) *D*. cf. *pigotti*, left metatarsal from Maboko, KNM-MB 22063. E) *D. chappuisi*, left metatarsal from Maboko, KNM-MB-25150. F) *D*. cf. *parvum*, right ankle from Rusinga, KNM-RU 15136; note the free fibula, but cubonavicular fused with ectocuneiform. Scale bar equals 2 cm for figures 36.1A, 36.1D, 36.1E; 1 cm for figures 36.1B, 36.1C, 36.1F.

- There is no preorbital fossa, as in *H.*

- In *H.*, the auditory region is often extended dorsoventrally, and the auditory foramen opens very high, so that the zygomatic arch is directed dorsocaudally; the Rusinga skull has a normal horizontal arch, as in the other living tragulids.

- Correlatively, the lower border of the mandible has a strong convexity below m3 in *H.*, whereas it is weaker and less localized in *D.*

- The occipital of *H.* is much broader than that of the other tragulids, including the Rusinga skull; in the latter it is quite narrow, with dorsoventrally elongated condyles, more like those of *M.*

- The auditory bulla is large and rounded in *H.* and *M.*, more flattened in *Tragulus*. In the Rusinga skull it is almost spherical, but not very large.

- Living tragulids have a small mastoid exposure, and an open slit between the squamosal and occipital; the mastoid is not visible on the Rusinga skull, and there is no gap between squamosal and occipital.

- The teeth, especially the lower premolars, are less bunodont; in *H.*, after moderate wear, the cristids soon become quite short, whereas they remain well distinct in *D.*

Discussion The Rusinga skull is important to the generic distinction between *Hyemoschus* and *Dorcatherium*, which has been discussed more than once (e.g., Gentry, 1978). Both names are usually retained, mainly because of the geographic and chronological gap between them, but these are of course not valid arguments. Unfortunately, the only known skull of *D. naui*, type species of the genus *Dorcatherium*, from the early late Miocene of Eppelsheim in Germany, is badly crushed and too imperfect for details to be compared. Given the possible amount of morphological differences between two species of Tragulidae, generic distinction between the Rusinga skull and *Hyemoschus* is probably warranted on a phenetic basis, but there is no strong evidence that all Miocene African forms (let alone the European ones) belong to one and the same genus. This is why I prefer not to include the features of the Rusinga skull in the diagnosis of *Dorcatherium*.

DORCATHERIUM IRIRIENSIS Pickford, 2002
Table 36.1

Type Locality and Age Napak V, Napak Member, Uganda.
Diagnosis A species of *Dorcatherium* intermediate in size between *D. pigotti* and *D. chappuisi* (Pickford, 2002).
Range Table 36.1. The species is only known from Napak.
Description The species is defined on the basis of size; if Napak is earlier than all Rusinga sites, it could well be ancestral to *D. chappuisi*, which is only slightly larger.

DORCATHERIUM CHAPPUISI Arambourg, 1933
Table 36.1

Type Locality and Age Moruorot, Kenya.
Range See table 36.1.
Diagnosis Large *Dorcatherium*, lower molar row ca. 17–20 mm long; p1 retained. Shallow mandibular ramus. Whitworth (1958) thought that the cuneiform remained separate from the cubonavicular, on the basis of purported tragulid bones from Rusinga.
Description The species was erected for a mandible with complete cheek teeth from Moruorot. The teeth are strongly weathered, which may increase their apparent bunodonty, but the posterior arm of the hypoconid is very short lingually and connects only to the labial arm of the hypoconulid on m3. There is no ectostylid (Gentry, 1978). The posterolabial cristid of p4 also fails to curve lingually at its distal end. Specimens of the right size to belong to *D. chappuisi* are known

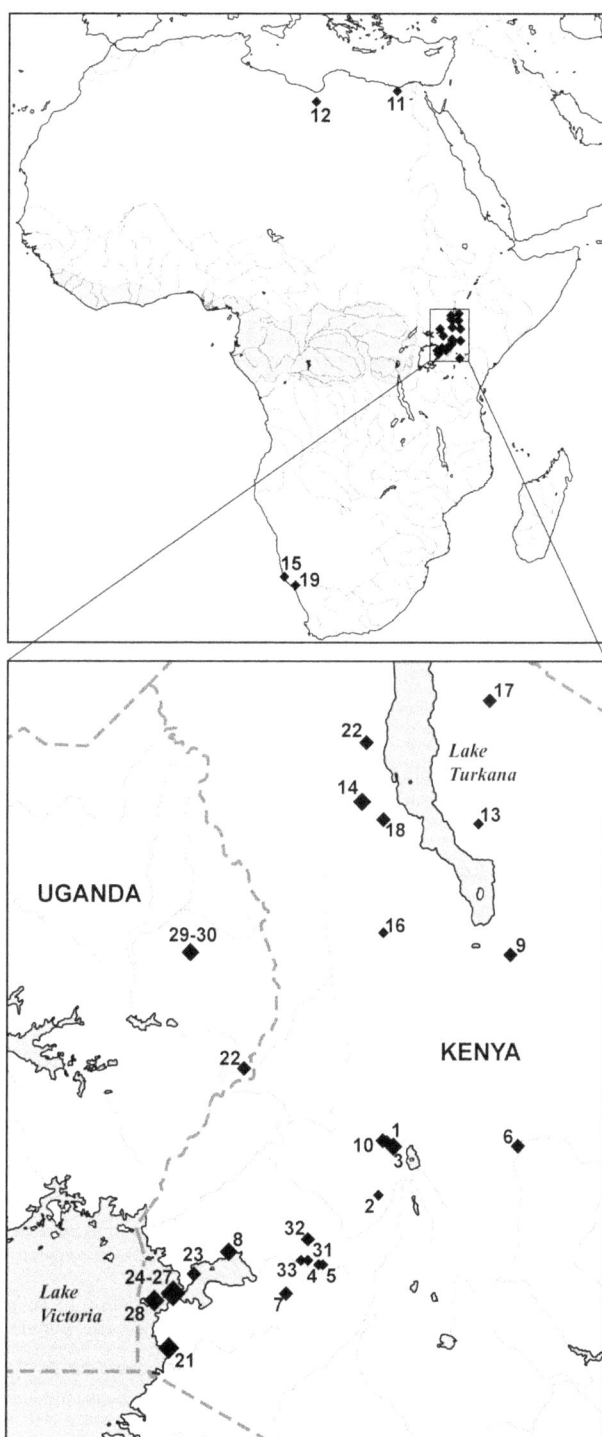

FIGURE 36.2 A) African localities with *Dorcatherium*, and range of living *Hyemoschus aquaticus*. B) Detail of western Kenya; diamond size represents the number of *Dorcatherium* species in each locality. Numbers refer to table 36.1.

that more than one species may have been collected under this name.

A complete metatarsal from Maboko (figure 36.1E) is more slender than those of other species, and much more than that of the living *Hyemoschus*.

Genus *HYEMOSCHUS* Gray, 1845
HYEMOSCHUS AQUATICUS (Ogilby, 1841)
Figure 36.2

Diagnosis Differs from the Miocene *Dorcatherium* by cingulum usually weak or absent on upper molars; cristids of lower cheek teeth short and quickly obscured by wear, plus some details of the postcranial skeleton (Morales et al., 2003) to which should be added the shortness of the metapodials and reduction of MT V. *Hyemoschus* differs from *Tragulus* + *Moschiola* in its larger size, broad occipital, thick premolars, unfused metacarpals, fibula not fused with tibia, and further differs from *Tragulus* in its ethmoidal fissure and coat pattern. Thus, the branching pattern of the modern forms is certainly *Hyemoschus (Moschiola (Tragulus))*, but *Hyemoschus* shares no synapomorphy with other living tragulids to the exclusion of *Dorcatherium*.

Age Recent, but perhaps present in the early Pliocene.

Range The living water chevrotain is usually found under heavy cover, close to permanent water, in most of the rain forests of tropical Africa, but has a disjunct distribution in two areas, one extending from Sierra Leone to Ghana, another from Nigeria and Congo to Western Uganda (figure 36.2). It does not seem to be seriously threatened.

Description Pickford et al. (2004) recorded the species from the early Pliocene of the Mabaget Formation, Kenya, on the basis of an incomplete upper molar and an incisor. Both specimens are indeed similar to those of the living water chevrotain, but the selenodont upper molar has a stronger cingulum than most modern *Hyemoschus*; and, with its labially curved postprotocrista, it is also similar to those of the late Miocene *Dorcatherium* (which are mostly European, such as *D. jourdani* or *D. puyhauberti*); unfortunately, upper teeth are less distinctive than lower ones. The Mabaget specimens provide good evidence for the derivation of *Hyemoschus* from *Dorcatherium*, but the poor African late Miocene record precludes any hypothesis about its geographic origin.

Phylogeny

The interrelationships of the selenodont Artiodactyls have been much debated in recent years (Barry et al., 2005, and references therein), but no consensus has been reached; and when fossil forms (such as the Lophiomerycidae, *Iberomeryx*, Leptomerycidae, and *Bachitherium*) are taken into account, the position of the Tragulidae varies strongly. Many of these primitive hornless Ruminants have often been clustered with tragulids to make up the "Tragulina," but derived characters to support this grouping are hard to find. As already stated (Geraads et al., 1987) the complete closure of the first lobe of the lower molars makes the Tragulidae definitely more derived than the Leptomerycidae, *Cryptomeryx*, and *Iberomeryx*.

The earliest record of the family is at Meswa Bridge (the material has not been described, but M. Pickford, pers. comm., confirms its occurrence there). It is only about 2 Ma later that they appear in Europe and southern Asia, while becoming more diverse in Africa (Pickford, 2001). No possible

from the early and middle Miocene, and as late as Ngorora, but they are more selenodont than the type, with longer cristids, and there may be an ectostylid; even at Rusinga the *Tragulus* fold is longer, showing that selenodonty does not merely increase with age. Size of m3 also seems to vary erratically, but a single p4 from Ngorora is relatively shorter and more complex than earlier ones. All this variation suggests ancestor is known in Africa, as there is no Ruminant known at Chilga or the Fayum (but the Oligocene African record is very poor) and close relatives must be sought elsewhere, but purported early Tragulidae, such as *Archaeotragulus* Métais et al. (2001) despite the reported occurrence of a *Tragulus*-fold, look to be closer to *Iberomeryx*. The biogeographic history of tragulids in the Miocene is wholly unknown, because the lack of cranial appendages, extreme rarity of cranial remains, and uniformity of most characters have prevented the deciphering of their phylogenetic relationships. In Africa, most of the current taxonomy is based on size, which is far from satisfactory. Pickford (2001) recognized five species on the basis of talus and m3 size. His figure 2 suggests that size within each species remained rather stable for almost 5 Ma, but my own measurements of a sample about as large as that of Pickford in the NMK suggest somewhat different conclusions. Most m3s from Maboko (Pickford's faunal set III) are intermediate between *D. songhorensis* and *D. pigotti*; if the former species went extinct earlier than Maboko, they must belong to a *D. pigotti* smaller than at Rusinga. The same is true of *D. chappuisi*, which is smaller at Maboko than in earlier (Rusinga) or later (Fort Ternan) sites. Identifications based on size are probably valid at Rusinga, but using the same size groups in the late middle Miocene may be misleading. Morphological differences do exist, but they are hard to relate to species. For instance, most p4s from early sites have the typical tragulid morphology, with two parallel cristids descending distally from the main cuspid, but several specimens from Maboko, the size of small *D. pigotti*, have only short transverse cristids, as in *Hyemoschus*. A p4 from Ngorora, instead, has a complex p4, with both transverse and longitudinal cristids. Lower molars from Maboko are remarkable in the very labial position of the preentocristid (mesial arm of the entoconid), which almost reaches the *Tragulus*-fold. In Europe, late Miocene *Dorcatherium* are more selenodont than most middle Miocene forms (Gentry, 1990), although precociously selenodont teeth, known as *D. guntianum*, are already present in the early middle Miocene. A parallel can perhaps be found in Africa, with the early *D. songhorensis* being more selenodont than many later forms.

Ecology

Data on the ecology of living water chevrotain in Gabon were provided by Dubost (1978). It lives in the rain forest, where it prefers rather dense shelter providing shade and safety from predators. It feeds mostly on fruit, losing weight during dry seasons. Despite its name, it normally lives on dry ground, entering water only for refuge, being a very good swimmer, and this seems to be the main reason why it is never found far from water. The male is a solitary animal, usually avoiding encounters with its congeners, but it is not territorial. Population densities are relatively low (about 10 individuals per square kilometer). By contrast, the high number of fossils found in some sites (in Europe, Asia, and Africa) suggests that they aggregated at least in family groups. However, similarities in anatomy between *Hyemoschus* and *Dorcatherium* make it unlikely that the latter had a very different diet or lived outside dense cover, although differences in metapodial proportions between species (figures 36.1D and 36.1E) probably indicate some habitat partitioning. In the Miocene of Europe, the distribution in time and space of *Dorcatherium* follows those of tapirs, *Deinotherium* and dryopithecines. It is nearly absent from late Miocene open environments.

At Maboko, tragulids are more than twice as common in the riparian woodland ("dhero") as in grassland or bushland (Retallack et al., 2002). Their almost complete disappearance with the late Miocene in Africa is certainly linked with the expansion of grasslands, even if they survived in some places (Pickford et al., 2004).

ACKNOWLEDGMENTS

I thank L. Werdelin and W. Sanders for having invited me to contribute to this volume. I am especially grateful to B. Benefit and A. Walker, who kindly allowed me to mention here some of the unpublished material that they found at Maboko and Rusinga, respectively; to A. W. Gentry for his helpful comments and information on *Dorcatherium*; and to an anonymous reviewer whose detailed comments and original information on many Kenyan Tragulidae greatly improved the manuscript. Best thanks also, for granting access to collections in their care, to E. Mbua and M. Muungu (Kenya National Museum, Nairobi), J. Cuisin, C. Lefèvre, M. Pickford and C. Sagne (Muséum National d'Histoire Naturelle, Paris), G. Gruber (Hessisches Landesmuseum, Darmstadt), and W. Wendelen (Musée Royal de l'Afrique Centrale, Tervuren).

Literature Cited

Anyonge, W. 1991. Fauna from a new Lower Miocene locality west of Lake Turkana, Kenya. *Journal of Vertebrate Paleontology* 11:378–390.

Arambourg, C. 1933. Mammifères miocènes du Turkana. *Annales de Paléontologie* 22:123–146.

Barry, J. C., S. Cote, L. MacLatchy, E. H. Lindsay, R. Kityo, and A. R. Rajpar. 2005. Oligocene and early Miocene ruminants (Mammalia, Artiodactyla) from Pakistan and Uganda. *Palaeontologia Electronica* 8(1):1–29.

Behrensmeyer, A. K., A. L. Deino, A. Hill, J. D. Kingston, and J. J. Saunders. 2002. Geology and geochronology of the middle Miocene Kipsaramon site complex, Muruyur beds, Tugen Hills, Kenya. *Journal of Human Evolution* 42:11–38.

Dubost, G. 1978. Un aperçu sur l'écologie du chevrotain africain *Hyemoschus aquaticus* Ogilby, Artiodactyle Tragulidé. *Mammalia* 42:1–62.

Gentry, A. W. 1978. Tragulidae and Camelidae; pp. 536–539 in V. J. Maglio and H. B. S. Cooke (eds.), *Evolution of African Mammals*. Harvard University Press, Cambridge.

———. 1990. Ruminant artiodactyls of Paşalar, Turkey. *Journal of Human Evolution* 19:529–550.

Geraads, D., G. Bouvrain, and J. Sudre. 1987. Relations phylétiques de *Bachitherium* Filhol, Ruminant de l'Oligocène d'Europe occidentale. *Palaeovertebrata* 17:43–73.

Groves, C. P., and E. Meijaard. 2005. Interspecific variation in *Moschiola*, the Indian chevrotain. *Raffles Bulletin of Zoology* (suppl.) 12:413–420.

Hamilton, W. R. 1973. The Lower Miocene ruminants of Gebel Zelten, Libya. *Bulletin of the British Museum (Natural History), Geology* 21:75–150.

Hill, A., M. G. Leakey, J. D. Kingston, and S. Ward. 2002. New cercopithecoids and a hominoid from 12.5 Ma in the Tugen Hills succession, Kenya. *Journal of Human Evolution* 42:75–93.

Leakey, R. E. F., and M. G. Leakey. 1986. A new Miocene hominoid from Kenya. *Nature* 124:143–146.

Meijaard, E., and C. P. Groves. 2004. A taxonomic revision of the *Tragulus* mouse-deer (Artiodactyla). *Zoological Journal of the Linnean Society* 140:63–102.

Métais, G., Y. Chaimanee, J.-J. Jaeger, and S. Ducrocq. 2001. New remains of primitive ruminants from Thailand: Evidence of the early evolution of the Ruminantia in Asia. *Zoologica Scripta* 30:231–248.

Morales, J., D. Soria, I. M. Sanchez, V. Quiralte, and M. Pickford. 2003. Tragulidae from Arrisdrift, basal Middle Miocene, southern Namibia. *Memoirs of the Geological Survey of Namibia* 19:359–369.

Pickford, M. 1986. Cainozoic palaeontological sites of Western Kenya. *Münchner Geowissenschaftliche Abhandlungen A* 8:1–151.

———. 1991. Biostratigraphic correlation of the Middle Miocene mammal locality of Jabal Zaltan, Libya. pp. 1483–1490 in: M. J. Salem, O. S. Hammuda, and B. A. Eliagoubi, (eds.), *Third Symposium on the Geology of Libya (Tripoli, 1987)*. Elsevier, Amsterdam.

Pickford, M. 2001. Africa's smallest ruminant: A new tragulid from the Miocene of Kenya and the biostratigraphy of East African Tragulidae. *Geobios* 34:437–447.

———. 2002. Ruminants from the early Miocene of Napak, Uganda. *Annales de Paléontologie* 88:85–113.

Pickford, M., H. Ishida, Y. Nakano, and K. Yasui. 1987. The middle Miocene fauna from the Nachola and Aka Aiteputh formations, northern Kenya. *African Study Monographs* (suppl.) 5:141–154.

Pickford, M., B. Senut, and C. Mourer-Chauviré. 2004. Early Pliocene Tragulidae and peafowls in the Rift Valley, Kenya: Evidence for rainforest in East Africa. *Comptes Rendus Palevol* 3:179–189.

Retallack, G. J., J. G. Wynn, B. R. Benefit, and M. L. McCrossin. 2002. Paleosols and paleoenvironments of the Middle Miocene, Maboko Formation, Kenya. *Journal of Human Evolution* 42:659–703.

Thomas, H., L. Ginsburg, C. Hintong, and V. Suteethorn. 1990. A new tragulid *Siamotragulus sanayathanai* n.g.n.sp. (Artiodactyla, Mammalia) from the Miocene of Thailand (Amphoe Pong, Phayao Province). *Comptes Rendus de l'Académie des Sciences, Paris* 310:989–995.

Whitworth, T. 1958. Miocene ruminants of East Africa. *Fossil Mammals of Africa* 15:1–50.

THIRTY-SEVEN

Pecora *Incertae Sedis*

SUSANNE M. COTE

The Ruminantia is commonly divided into two infraorders: the Tragulina and the Pecora. Pecoran monophyly is well accepted with five modern families: the Giraffidae, Bovidae, Moschidae (musk deer in Asia), Antilocapridae (pronghorns of North America), and Cervidae (deer). These are commonly placed in three superfamilies: Bovoidea, Cervoidea, and Giraffoidea (e.g., Flower, 1883; Gentry, 1994; Hernández Fernández and Vrba, 2005). Of modern pecorans, only the Bovidae and Giraffidae are widespread in Africa (cervids dispersed into North Africa in the early late Pleistocene; see Gentry, this volume, chapter 40). In the African Miocene, there are several named taxa of pecoran ruminants that are not easily assigned to the modern families or superfamilies. Some of these taxa have been assigned to either extant or extinct families, but their status is disputed. These taxa are the subject of this chapter. Taxa that are clearly attributable to the bovids, giraffids, and cervids are fully discussed in separate chapters.

The interfamilial relationships of the Pecora have been the subject of continual debate, and almost all possible phylogenetic arrangements have been suggested in the literature. Morphological analyses have often suggested a close relationship between the Bovidae and Giraffidae (Morales et al., 1986; Gentry and Hooker, 1988; Gentry, 1994, 2000), and the position of the Moschidae has been particularly controversial (see Hassanin and Douzery, 2003: figure 1 for a review). There is a growing consensus from molecular data and new "supertree" analyses that Bovidae and Cervidae together are the sister group of the Giraffidae (Hassanin and Douzery, 2003 and references therein; Hernández Fernández and Vrba, 2005; Price et al., 2005). Moschidae are part of the clade including Bovidae and Cervidae, normally considered more closely related to cervids, although one study has linked them more closely with bovids (Hassanin and Douzery, 2003). It is also suggested that Giraffidae and Antilocapridae may be sister taxa (Hassanin and Douzery, 2003; Hernández Fernández and Vrba, 2005). Although most nodes in these new phylogenetic trees are resolved, consistency indices are sometimes low. Hernández Fernández and Vrba (2005) point out that new phylogenies are best viewed as "working hypotheses" and also draw attention to the potential problem of long-branch attraction—particularly for the Giraffidae and Antilocapridae. The basal position of Giraffidae, far removed from the Bovidae, along with a potential sister taxon relationship between Giraffidae and Antilocapridae, significantly complicates scenarios of biogeography.

Divergence of the Pecora and Tragulina is estimated at approximately 50 Ma (Hernández Fernández and Vrba, 2005; 54–37 Ma using the Bayesian relaxed molecular clock analysis of Hassanin and Douzery, 2003), and diversification of pecoran families appears to have occurred around 30 Ma (36–26 Ma in Hassanin and Douzery, 2003; 32–28 Ma in Hernández Fernández and Vrba, 2005). Importantly, these molecular dating studies give rather early divergence dates (early Miocene) for the tribes within Bovidae and Cervidae, as well as for the giraffid lineage (late Oligocene), suggesting the presence of long "ghost lineages" in these groups. The divergence of pecoran families occurs just after extensive global cooling at the Eocene/Oligocene boundary. It has been suggested that the simultaneous radiation of pecoran families, adapting to the same type of climate change, may have led to parallel evolution of several traits, and contributes greatly to the difficulties of resolving pecoran phylogeny (Janis and Scott, 1988; Hernández Fernández and Vrba, 2005).

In addition to the living families of ruminants, there are numerous names given to extinct alleged families of both the infraorders Tragulina and Pecora, largely from Eurasia. The status of many of these (including to which infraorder they should be assigned) is controversial. Only three extinct families are relevant here: (1) Gelocidae Schlosser, 1886. These are small early pecorans of the Eocene–Oligocene of Europe and perhaps Asia, variably considered basal pecorans (Webb and Taylor, 1980; Blondel, 1997), tragulines (Vislobokova, 2001), or a paraphyletic assemblage (Janis and Scott, 1987). (2) Palaeomerycidae Lydekker, 1883. Miocene pecorans of Eurasia and (as Dromomerycinae) North America. They are generally larger than other contemporaneous pecorans, and most bear cranial appendages and are usually regarded as cervoids. (3) Climacoceratidae Hamilton, 1978. African Miocene ruminants well accepted as a family within the Giraffoidea, bearing branched cranial appendages.

ABBREVIATIONS

AMNH—American Museum of Natural History, New York; BSP—Bayerische Staatssammlung für Paläontologie und historische Geologie, Munich; BU—Department of Geology, Bristol University, Bristol; CGM—Cairo Geological Museum, Cairo; DLC—Duke Lemur Center, North Carolina; FT—Fort Ternan; GSN—Geological Survey of Namibia, Windhoek; KNM—National Museums of Kenya, Department of Paleontology, Nairobi; NHM—The Natural History Museum, Department of Palaeontology, London; SAM—South African Museum, Cape Town; UMP—Uganda Museum, Paleontology, Kampala.

Terminology for dental morphology, including the critical *Palaeomeryx*-fold, follows that of Janis and Scott (1987: figure 5). Dental measurements presented in this chapter were taken by the author as maximum lengths and breadths, and they cannot be compared with those of Hamilton (1973, 1978a, 1978b), which appear to be occlusal surface measurements.

Systematic Paleontology

Order ARTIODACTYLA Owen, 1848
Suborder RUMINANTIA Scopoli, 1777
Infraorder PECORA Flower, 1883

Diagnosis Adapted from Webb and Taylor (1980); Janis and Scott (1987); Gentry and Hooker (1988). Ruminants with a compact, parallel-sided astragalus and an axis vertebra with a spout-shaped odontoid process. A four-chambered stomach with a well-developed omasum is present. Metapodial characters include: complete metapodial keels; fusion of metacarpals III and IV; and metatarsals II and V greatly reduced (or absent). Cranial characters include loss of the stapedial artery; an enlarged fossa for the stapedial muscle, shallow subarcuate fossa, and loss of the promontorium on the petrosal; and a broadened basiooccipital with strong flexion stops on condyles. Metastylids present on lower molars.

Discussion The Gelocidae is sometimes considered the most basal family of pecorans (e.g., Webb and Taylor, 1980; Gentry, 1994). Conversely, Janis and Scott (1987, 1988) exclude the Gelocidae (which they consider paraphyletic) from the Pecora on the basis that metastylids are absent in *Gelocus*. However, small metastylids are clearly present on the lower molars of *Gelocus communis* from Ronzon in the NHM (A. Gentry, pers. comm.; pers. obs.), although metacarpals III and IV are unfused in this taxon (Webb and Taylor, 1980). Lower molar metastylids are also absent in Antilocapridae (Hassanin and Douzery, 2003: table 2), though this may be a secondary loss.

In addition to the synapomorphies in the diagnosis, pecorans are characterized by selenodont cheek teeth with reduced (or absent) lingual cingulum in the upper molars. Upper canines are generally absent but are present in moschids and some cervids. Numerous features of the pecoran limb (fore- and hindlimbs of roughly equal length; elongated limbs, complete distal metapodial keels, fusion of metapodials, and loss of side toes) are adaptations to life in more open habitats, and it is possible that these features evolved in parallel (Janis and Scott, 1988).

Pecorans have also traditionally been characterized by the possession of cranial appendages. Cranial appendages are absent in the Moschidae and in the modern cervid *Hydropotes* (a secondary loss). Cranial appendages are also absent in several fossil taxa of clear pecoran status. Abundant evidence now indicates that cranial appendages evolved in parallel in the pecoran families and cannot be used as a character to unite the Pecora (or Eupecora, *sensu* Webb and Taylor, 1980). Further inquiry is necessary, but the current understanding of the ontogeny of cranial appendages shows that they have different developmental bases, ontogenetic trajectories, and microstructure (e.g., Janis and Scott, 1987; Azanza et al., 2003). The parallel evolution of cranial appendages may be related to the loss of upper canines, a common feature of pecorans that may also be nonhomologous.

Genus *WALANGANIA* Whitworth, 1958

Type Species Walangania africanus Whitworth, 1958.

This is here regarded as the only species in the genus. Hamilton (1973) synonymized *Palaeomeryx africanus* Whitworth, 1958 with *Walangania gracilis* Whitworth, 1958 as *W. africanus*, as the name *Palaeomeryx* (a European genus) does not apply to this material.

WALANGANIA AFRICANUS Whitworth, 1958
Figure 37.1 and Table 37.1

Synonymy Walangania gracilis Whitworth, 1958; *Palaeomeryx africanus*, Whitworth, 1958; *Kenyameryx africanus*, Ginsburg and Heintz, 1966.

Holotype NHM M 21358 right mandible with p3–m3 from Songhor, Kenya.

Occurrence Early Miocene, East Africa (table 37.1).

Diagnosis Emended from Whitworth (1958); Barry et al. (2005). Medium-sized pecoran (m1–m3 length ~40 mm). Cranial appendages unknown and likely absent; p1 at least variably present; p4 variably bifurcated anteriorly, with transverse entostylid and well-developed metaconid that sometimes has a posterior flange. Lower molars with metaconid and entoconid slightly oblique and compressed; small metastylid situated lingual to the posterior end of postmetacristid. *Palaeomeryx*-fold variably present. Upper molars with large

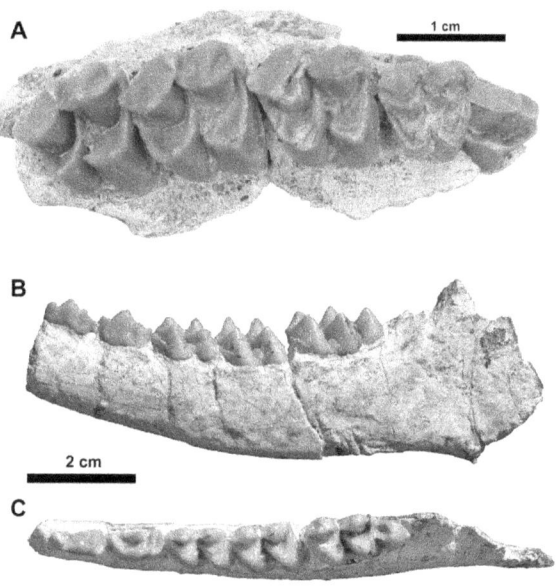

FIGURE 37.1 A) UMP BUMP 274. *Walangania africanus* right maxilla with erupting P3, dP4–M3, occlusal view. B–C) KNM-SO 1627; *W. africanus* left mandible with p3–m3, occlusal (B) and buccal (C) views.

TABLE 37.1
Distribution of Pecoran ruminants

Abbreviation: T, type locality.

Specimens in boldfaced museums were examined by the author; museum abbreviations that appear in italics are casts only.

Starred references cite occurrences, while nonstarred references are used for age determinations only.

Taxon	Occurrence	Age	Museum	References
Walangania africanus	Moroto, Uganda	>20.6 Ma	**UMP**	*Barry et al., 2005; *Cote, 2004; *Pickford and Mein, 2006; Gebo et al., 1997
	Songhor, Kenya (T)	19.5 Ma	**KNM, NHM**	*Whitworth, 1958; Bishop et al., 1969; Pickford and Andrews, 1981
	Koru, Kenya	19–20 Ma	**KNM, NHM**	*Whitworth, 1958; Bishop et al., 1969; Pickford and Andrews, 1981
	Chamtwara	19–20 Ma	**KNM**	*This chapter; Bishop et al., 1969; Pickford and Andrews, 1981
	Legetet	19–20 Ma	**KNM**	*This chapter; Bishop et al., 1969; Pickford and Andrews, 1981
	Napak, Uganda	19–20 Ma	**UMP, NHM**	*Barry et al., 2005, *Cote, 2004, *Pickford, 2002, MacLatchy et al., 2006
	Rusinga	17–18.3 Ma	**KNM, NHM**	*Whitworth, 1958; Drake et al., 1988
	Mfwanganu	17.9 Ma	**KNM, NHM**	*Whitworth, 1958; Drake et al., 1988
	Morourot	16.8–17.9 Ma	**KNM, NHM**	*Whitworth, 1958; Hamilton, 1973; Boschetto et al., 1992
	Kalodirr	16.8–17.5 Ma	**KNM**	*This chapter; Boschetto et al., 1992
	Nachola	14–15 Ma	**KNM**	*Nakaya, 1994; Sawada et al., 1998
	Fort Ternan	14 Ma	**KNM**	*Gentry, 1970
"Gelocus" whitworthi	Songhor (T)	19.5 Ma	**NHM**	*Hamilton, 1973; Bishop et al., 1969; Pickford and Andrews, 1981
Propalaeoryx austroafricanus	Elisabethfeld (T)	18–20 Ma	GSN, BSP, *NHM*	*Stromer, 1926; *Hamilton and Van Couvering, 1977; *Morales et al., 1999; Pickford and Senut, 2003
	Langental	18–20 Ma	AMNH, GSN	*Hamilton and Van Couvering, 1977; Morales et al., 1999
Propalaeoryx nyanzae	Rusinga (T)	17–18.3 Ma	**NHM**	*Whitworth, 1958; Drake et al., 1988
	Mfwanganu	17.9 Ma	**NHM**	*Whitworth, 1958; Drake et al., 1988
Prolibytherium magnieri	Gebel Zelten (T)	~18–16 Ma	**NHM**, MNHN, BU	*Arambourg, 1961, *Hamilton, 1973; *Pickford et al., 2001; Miller and Simons, 1996
	Wadi Moghara	~18–16 Ma	DLC, CGM	*Miller and Simons, 1996, *Pickford et al., 2001
Sperrgebietomeryx wardi	Elisabethfeld (T)	~18–20 Ma	GSN, SAM, BSP, *NHM*	*Morales et al., 1999; Pickford and Senut, 2003
	Langental	~18–20 Ma		*Stromer, 1926; *Hamilton and Van Couvering, 1977; *Morales et al., 1999; Pickford and Senut, 2003
Orangemeryx hendyi	Arrisdrift (T)	~17.5 Ma	GSN	*Morales et al., 1999; *Morales et al., 2003; Pickford and Senut, 2003
Namibiomeryx senuti	Elisabethfeld (T)	18–20 Ma	GSN, AMNH, *NHM*	*Hopwood, 1929; *Morales et al., 1995; Pickford and Senut, 2003

metaconules on M1 and M2; paracone with strong labial rib, metacone rib weak or absent; strong parastyle, weaker metastyle; and subsidiary crests present in the anterior fossette and separate from posteriorly directed postprotocrista. Limbs of advanced pecoran type, with closed metatarsal gullies.

Description Whitworth named two similar pecoran taxa in his 1958 review of East African ruminants: *Walangania gracilis* was named for a juvenile maxilla and mandible with associated partial skeleton from Mfwanganu, and *Palaeomeryx africanus* from numerous specimens from several East African localities. Features that differentiated *W. gracilis* and *P. africanus* included the presence of a p1 and a *Palaeomeryx*-fold on the lower molars of the later. Hamilton (1973) combined *Palaeomeryx africanus* and *Walangania gracilis* into a single taxon, *Walangania africanus*, stating that the *Palaeomeryx*-fold

is present in the *W. gracilis* type and that the p1 is absent in *P. africanus*. Subsequent researchers have tended to follow this synonymy (but see Janis and Scott, 1987). In fact, a p1 is present in some specimens attributed to *Walangania africanus*, but the condition in the old *W. gracilis* type from Mfwanganu cannot be observed.

Walangania africanus is a widespread and common taxon in East African early Miocene localities, and there is a great deal of morphological and size variation encompassed within it, as indicated in the emended diagnosis presented here. Important morphological characters that vary include bifurcation of the p4 paraconid; the size and configuration of the p4 metaconid; and the presence or absence of *Palaeomeryx*-folds. *Palaeomeryx*-folds exhibit inter- and intraindividual variation, with some specimens showing a *Palaeomeryx*-fold on m3, but not on m1 or m2, and also differ greatly in how strong the fold is (which may be a factor of tooth wear). This is significant, as the *Palaeomeryx*-fold is often considered a critical feature for phylogenetic relationships. There is also a considerable amount of size variation present in the collections. In particular, there are small postcrania and a few dental specimens that seem too small to belong within *W. africanus*, and it seems likely that a second, unnamed taxon is present in the sample currently attributed to *W. africanus*.

Discussion Ginsburg and Heintz (1966) transferred *Palaeomeryx africanus* out of the Palaeomerycidae, citing the presence of p1 and the primitive nature of the premolars as nonpalaeomerycid characters, but did not assign it to any other family. Hamilton (1973, 1978a) united the species with *Walangania gracilis* under the name *W. africanus* and believed it to be a likely bovid. In contrast, Janis and Scott (1987) placed *W. africanus* and *W. gracilis* as separate taxa within the Cervoidea, a suggestion adopted by Barry et al. (2005). Gentry (1978) originally suggested that *Walangania* might be a bovid, but subsequently indicated that it is more cervid-like (Gentry, 1994, 2000). Gentry has also pointed out similarities to *Dremotherium* (1994; pers. comm. cited in Hendey, 1978), a primitive ruminant from the late Oligocene of Europe normally placed in the Cervoidea (Janis and Scott, 1987) and sometimes within Moschidae (Webb and Taylor, 1980).

Cervoid-like characters of *Walangania* include the presence of a *Palaeomeryx*-fold, closed metatarsal gullies, and an enlarged metaconid on p4 (Janis and Scott, 1987; Gentry, 1994). The metatarsal gullies of specimens assigned to *Walangania* are closed in all specimens where the character can be observed (isolated specimens not associated with dental material). The bridge tends to be short and does not extend as far up the metatarsus as in deer, and it is possible that this character evolved independently (Janis and Scott, 1987). Currently, the bulk of evidence would suggest that *Walangania* has cervoid affinities, but the character support for this is not strong and rests largely on the variably present *Palaeomeryx*-fold and the closed metatarsal gully. Similarities to *Dremotherium* are certainly apparent and deserving of further investigation.

Younger Occurrences of Walangania Hamilton (1973) transferred a maxilla fragment from Moruorot to *Propalaeoryx nyanzae*, which Whitworth (1958) had previously included in *Palaeomeryx (Walangania) africanus*. Additional material from Moruorot and the nearby locality of Kalodirr is very similar to Whitworth's material and may belong in *Walangania*. The specimens are larger and possess stronger metastyles and weaker paracone ribs than most *Walangania* upper molars, but they are within the range of size variation observed in the Songhor collection.

Walangania africanus has also been reported from the Aka Aiteputh formation, dated to 14–15 Ma (Pickford et al., 1987; Nakaya, 1994; Sawada et al., 1998). Three left lower molars in the KNM do indeed resemble *Walangania*. The size and morphology of the complete lower molar are a good match for *W. africanus* m1s from Songhor and Napak, but there is no trace of a *Palaeomeryx*-fold.

Gentry (1970) mentioned several teeth from Fort Ternan that might represent *Walangania*. Two upper molars (KNM-FT 927 and KNM-FT 3143 [Gentry's Ft 61.702]) are very similar to *Walangania*, except for being buccolingually narrower and having a more triangular outline. Gentry interpreted KNM-FT 3143 as an M3, which would make it very small for *W. africanus*. If these teeth were assigned to *Walangania*, then they would be its youngest record. It is likely that the long temporal range of *Walangania* is an artifact of the limited material available from younger localities, and that additional material would show that many or all of them are not *Walangania*.

Genus indet.
"GELOCUS" WHITWORTHI Hamilton, 1973
Table 37.1

Synonymy Palaeomeryx africanus Whitworth, 1958 *(partim)*.
Holotype NHM M 26692, left mandible fragment with m2–m3 from Songhor, Kenya (listed as Sgr 365.1949 in Hamilton, 1973).
Occurrence Early Miocene, East Africa (table 37.1).
Diagnosis Emended from Hamilton (1973). Medium-sized pecoran with a rounded metaconid on the lower molars; median valley of lower molars very open lingually; length of the lower molar row ~33mm. The posterior end of the entoconid (postentocristid) is forked, and the m3 hypoconulid loop has an entostylid.

Description Hamilton (1973) named *Gelocus whitworthi* based only on the holotype and four isolated lower molars, one of which had been previously described as *Walangania africanus* by Whitworth (1958). Although the *"G." whitworthi* molars are only slightly smaller than *Walangania* (and overlap with the range of variation seen in *Walangania*), they show several distinctive features, most notably the forked posterior end of the entoconid, the presence of an anterolingual cingulum high up on the tooth crown that meets with the mesostylid (Hamilton's "anterior crest curving anterolingually"), and the entostylid on m3, that make it clear that this is a separate taxon. *Palaeomeryx*-folds are present. No further material of *"G." whitworthi* has been recognized, despite extensive collecting. It is possible that some of the variation in premolar morphology and postcranial size in material currently attributed to *Walangania* may in fact represent *"G." whitworthi*, but no additional examples of the distinctive lower molars have ever been recovered. Janis and Scott (1987) report that *"G." whiworthi* teeth are known from Maboko, but there are no published records of this occurrence. Hamilton (1978b) includes one m3 from Rusinga in the hypodigm of *"G." whitworthi*, but this specimen is unconvincing (there is no entostylid; pers. obs.), although it also seems unlikely to belong to *Walangania africanus*.

Discussion While the dental features of *G. whitworthi* are distinctive enough to warrant separation from *Walangania africanus*, it is less clear that this taxon should be included in *Gelocus*, leading Janis and Scott to refer to it as *"Gelocus" whitworthi*, a convention that has been followed by other authors (Gentry, 1994; Barry et al., 2005; this chapter), although no new generic name has been provided. *Gelocus* is either a basal pecoran (Gentry, 1994; Webb and Taylor, 1980) or a nonpecoran ruminant (Janis and Scott, 1987; Vislobokova, 2001). The former seems more likely, as Janis and Scott (1987) report that it is the lack of metastylids that would bar gelocids from inclusion in the Pecora, whereas metastylids are clearly visible on material of *Gelocus communis* from Ronzon in the NHM (pers. obs.).

Hamilton (1973) thought that *"G." whitworthi* showed similarities to *Gelocus communis*, including the forked entoconid and the configuration of the metastylid. However, the entoconid is not forked in *Gelocus communis* specimens in the NHM, and forking similar to that described by Hamilton

(1973) occurs in *Palaeomeryx* from Sansan (pers. obs.). The *Palaeomeryx*-fold also suggests that this species does not belong in *Gelocus*, and led Janis and Scott (1987) to believe that it might have cervoid affinities.

Genus *PROPALAEORYX* Stromer, 1926

Type Species Propalaeoryx austroafricanus Stromer, 1926.
Diagnosis Emended from Stromer (1926). Pecora of medium size, with shallow mandible and rather brachyodont, selenodont lower cheek teeth, closed from p2 to m3. p1 present. Lower molars with strong metastylid and entostylid; pronounced median rib on lingual surface of metaconid, similar rib on entoconid; accessory stylid in median external valley. *Palaeomeryx*-fold absent. No horns are known, although there are also no known cranial remains.

PROPALAEORYX AUSTROAFRICANUS Stromer, 1926
Table 37.1

Holotype BSP 1926–507 right mandible with alveoli p1, p2–m3 from Elisabethfeld.
Occurrence Early Miocene, southern Africa (table 37.1).
Diagnosis As for genus. Larger than *Propalaeoryx nyanzae*.
Description The alveoli for p1 demonstrate that it is a two-rooted tooth. Stromer's (1926) diagnosis (reproduced in Whitworth, 1958) states that no *Palaeomeryx*-fold is present. Morales et al (1999:237–238) state that "a moderate *Palaeomeryx*-fold, most marked in the m1" is present "as in the holotype"; however, a cast of the holotype in London shows no evidence of a *Palaeomeryx*-fold. *Propalaeoryx austroafricanus* is differentiated from its contemporary *Sperrgebietomeryx* only by its larger size, presence of p1 and more complex lower premolar morphology. The small collection of casts of this taxon present in the NHM suggests that there is a great deal of size variability present in the material currently assigned to *P. austroafricanus*.
Discussion Stromer (1926) originally suggested that *Propalaeoryx* was a bovid, whereas Arambourg (1933), followed by Whitworth (1958), thought that the dentition was cervoid-like. Janis and Scott (1987) state that there are no definitive cervoid characters present in *Propalaeoryx*, and they instead suggest that the bifurcated posterior crest of the metaconule indicates that it may be related to giraffoids. Gentry (1994) also tentatively links *Propalaeoryx* with giraffoids. Morales et al. (1999) place *Propalaeoryx* in their new subfamily Sperrgebietomerycinae (discussed later) within Climacoceratidae. The general consensus that *Propalaeoryx* belongs within the Giraffoidea is likely correct, although the state of the canine is unknown. Stromer (1926) tentatively assigned two anterior teeth to *P. austroafricanus*, neither of which was described as being bifurcated, but these teeth are now missing (Hamilton, 1978a). The relationships of *Propalaeoryx* (and other possible early Miocene African giraffoids) to the potential giraffoids *Teruelia* and *Lorancameryx* from the early Miocene of Spain (Moyà-Solà, 1987; Morales et al., 1993) have not been considered.

PROPALAEORYX NYANZAE Whitworth, 1958
Table 37.1

Holotype NHM M 21368 (Ru 324.47) left mandible fragment with m1–m2 from the Lower Hiwegi Beds, Rusinga Island.
Occurrence Early Miocene, East Africa (table 37.1).

Diagnosis From Whitworth (1958). m1–m3 series measuring ~45 mm. Lower molars with prominent accessory tubercle in median external valley. Teeth smaller and lower crowned than in *P. austroafricanus*. *Palaeomeryx*-fold absent.
Description Propalaeoryx nyanzae was named by Whitworth for a small number of lower molars from Rusinga and Mfwanganu. It is smaller than the type species, and not much larger than *W. africanus* from Songhor and Napak. *Propalaeoryx nyanzae* has weaker internal stylids (metastylid, ectostylid) and a stronger median pillar than the type species, but the internal stylids are stronger than is common in *Walangania*. Whitworth (1958) also assigned a distal metatarsal with an open gully for the extensor tendon to *Propalaeoryx nyanzae*.

Hamilton (1973) identified several teeth as the upper molars of *Propalaeoryx nyanzae* (including a maxilla from Moruorot that Whitworth [1958] assigned to *W. africanus*). These teeth are similar to those of *Walangania*, differing only in that they are slightly larger; have a stronger anterior cingulum and weaker posterior cingulum; and have a stronger mesostyle, parastyle, metastyle, and paracone rib. Some of these features are variable in *Walangania*, and it is unclear to what degree some of these characters may be size related and therefore of little phylogenetic significance.

Discussion While agreeing with Arambourg (1933) that the dentition of *Propalaeoryx* indicated cervid affinities, Whitworth (1958) assigned a metatarsal from Rusinga (Ru 1635'50) with an open gully for the extensor tendon to this taxon. He then suggested that early bovids may have "differed little in their dental characteristics from contemporaneous cervids" and left *Propaleoryx* of uncertain position within Pecora.

Propalaeoryx nyanzae is so poorly understood that it is difficult to make any definitive statements about its morphology or evolutionary relationships. One of the more interesting questions about *P. nyanzae* is how it is related to *P. austroafricanus*. Whitworth (1958) stated with confidence that this species belongs in *Propalaeoryx*, even suggesting that further material might eliminate the species-level difference. New material from Namibia (Morales et al., 1999) indicates that there probably is a specific difference and in some ways makes *P. nyanzae* appear more similar to *Walangania*. The relationship between *P. nyanzae* and *W. africanus* has not been considered and surely must be investigated when further material of *P. nyanzae* becomes available.

Genus *PROLIBYTHERIUM* Arambourg, 1961

Type Species Prolibytherium magnieri Arambourg, 1961. Only species of the genus.

PROLIBYTHERIUM MAGNIERI Arambourg, 1961
Table 37.1

Holotype Set of ossicones numbered 1961–5–1 (Arambourg, 1961).
Occurrence Early-middle Miocene, North Africa (table 37.1).
Diagnosis After Arambourg (1961); Hamilton (1973). The skull bears large, flat, wing-shaped ossicones that are fused to the frontal and parietal bones with no visible sutures. These appendages extend anteriorly and posteriorly over the entire skull roof and show signs of vascularization. Occipital condyles large with thickened bone.
Description Arambourg (1961) named *Prolibytherium* on the basis of a set of ossicones and a separate skull fragment. Hamilton (1973) provides a description of the dentition,

additional cranial material, and some postcranial elements from Gebel Zelten. Most of the known dentitions are very worn, and consequently the dental anatomy of this taxon is not well-known. The teeth are relatively hypsodont, and the lower molars have flat lingual walls with no *Palaeomeryx*-fold. The cervical vertebrae are short and robust. Hamilton (1973) suggests that a thick, short neck and short, stocky forelimbs were adaptations to support the weight of the frontal appendages, possibly used in fighting. Pickford et al. (2001) described additional material from Wadi Moghara that is likely conspecific with the material from Gebel Zelten, although there has been no detailed analysis.

Discussion Arambourg (1961) and Hamilton (1973) both viewed *Prolibytherium* as an early sivatheriine giraffid. However, Hamilton later (1978a, 1978b) stated that there is no evidence to identify *Prolibytherium* as a member of the Sivatheriinae or even the Giraffoidea, as the lower canine is unknown, and lists it as Pecora *incertae sedis*. Conversely, in the same volume, Churcher (1978) continued to include *Prolibytherium* in the Sivatheriinae.

More recently, several authors have suggested that *Prolibytherium* may belong within the Climacoceratidae. Gentry (1994) cites the flat lingual walls of the lower molars, m3 hypoconulid lobe without central fossette, and crown height as characters shared with *Climacoceras*, and places *Prolibytherium* in the Climacoceratidae in his classification (1994:147). Others follow suit (Pickford et al., 2001; Morales et al., 2003), although they do not provide any direct character support for this relationship, but cite a general similarity of the dentitions of *Prolibytherium* and *Climacoceras* as well as similarities between the skull of *Prolibytherium* and those of *Sperrgebietomeryx* and *Orangemeryx* (taxa they consider to belong to the Climacoceratidae). If *Prolibytherium* is a climacoceratid, then specialized elongation of the neck vertebrae would not be a feature uniting Climacoceratidae (Pickford et al., 2001; Morales et al., 2003; but *contra* Morales et al., 1999).

Attribution of *Prolibytherium* to the Climacoceratidae is hampered because the canine is unknown (following the view that all giraffoids must exhibit a bilobed canine; Hamilton 1978; Harris et al, this volume, chapter 39). In addition, the odd cranial appendages of *Prolibytherium* are in no way analogous to the tined appendages of *Climacoceras* (Azanza et al., 2003). Barry et al. (2005) point out that *Prolibytherium* has a reduced intercondyloid notch (the occipital condyles are continuous ventrally), similar to material that they assign to *Progiraffa exigua* from the Vihowa Formation of Pakistan. The phylogenetic significance of this feature is unclear.

At present, the balance of evidence would suggest that *Prolibytherium* is a member of the Giraffoidea, although strict assignment to this group, or to the Climacoceratidae in particular, must await discovery of additional material.

Genus *SPERRGEBIETOMERYX* Morales et al., 1999

Type Species Sperrgebietomeryx wardi Morales et al., 1999. Only species in genus.

SPERRGEBIETOMERYX WARDI Morales et al., 1999
Table 37.1

Synonymy cf. *Strogulognathus sansaniensis* Filhol (Stromer, 1926).
Holotype EF 37'93 skull, mandible, and associated vertebrae and hindlimbs from Elisabethfeld, Namibia.

Occurrence Early Miocene, southern Africa (table 37.1).
Diagnosis From Morales et al. (1999). Medium-sized ruminant with long and gracile premolar series. Lower p4 with simple metaconid, directed posteriorly, anterior wing without bifurcation. P2 and p2 nearly the same size as P3 and p3.
Description The holotype is a partial skeleton including a skull and mandible. The skull has strong sagittal and nuchal crests, but no cranial appendages. The holotype dentition is rather worn, but the upper dentition is supposed to be characterized by strong styles and late fusion of the internal lobes. There is a *Palaeomeryx*-fold on the lower molars, and the p3 and p4 are simple with unbifurcated anterior ends and simple metaconids directed posterolingually. The lower canine of *Sperrgebietomeryx* is unknown. Several other isolated postcranial elements from Elisabethfeld, identified largely on the basis of size, are also included in this taxon.
Discussion Stromer's cf. *Strogulognathus sansaniensis* material was from Langental. Hamilton and Van Couvering (1977; Hamilton, 1978a) recovered additional ruminants from Langental and Elisabethfeld and synonymized cf. *S. sansaniensis* with *Propalaeoryx austroafricanus*. Morales et al. (1999) erected *Sperrgebietomeryx* based on the new specimen from Elisabethfeld and included Stromer's original material, but they did not comment on which specimens from the Hamilton and Van Couvering collections are included in this taxon, nor did they comment on its similarities to or differences from *Strogulognathus*. *Sperrgebietomeryx* is differentiated from *P. austroafricanus* by its slightly smaller size, more primitive premolar morphology, the absence of p1 (Morales et al., 1999), and presumably the *Palaeomeryx*-fold.

Morales et al. (1999) include *Sperrgebietomeryx* as a member of the family Climacoceratidae. Hamilton (1978b:168) originally diagnosed the Climacoceratidae (= Climacoceridae in Hamilton 1978b; see Gentry 1994 for ICZN correction to family name) as "Giraffoids having large ossicones carrying many tines." In order to accommodate *Sperrgebietomeryx* within the Climacoceratidae, Morales et al. (1999:232) give a greatly altered diagnosis for the family and also name Sperrgebietomerycinae as a subfamily of climacoceratids characterized by the absence of cranial protuberances. *Climacoceras* does not have a *Palaeomeryx*-fold, and the presence of this feature in a purported climacoceratid (or any member of the Giraffoidea) is problematic.

Genus *ORANGEMERYX* Morales et al., 1999

Type Species Orangemeryx hendeyi Morales et al., 1999.

ORANGEMERYX HENDEYI Morales et al., 1999
Table 37.1

Synonymy Climacoceras sp. Hendey, 1978.
Holotype AD 595'94 left frontal fragment with apophysis.
Occurrence Early middle Miocene or latest early Miocene, southern Africa (table 37.1).
Diagnosis From Morales et al. (1999). Having supraorbital apophyses that are conical in outline, elongated, slightly compressed, ornamented at the base with rounded tubercles, and possessing bifurcated or trifurcated upper terminations.
Description All material assigned to *Orangemeryx* is isolated; no associated cranial apophyses and dentitions are known, nor is any of the postcranial material assigned to *Orangemeryx* associated. Morales et al. (2003) note a great deal of morphological and size variation within the assemblage, but they

conclude that it can presently be considered a single, sexually dimorphic taxon. Conversely, Hendey (1978:29) thought that a second pecoran taxon may have been present at Arrisdrift, based on variation in the size of postcrania and also variation in size and morphology of dental remains.

The cranial appendages of *Orangemeryx* are thought to be apophyseal in origin, like those of *Climacoceras*, but are said to differ from this genus in their short and conical, rather than long and cylindrical outline. The dentition is also similar to *Climacoceras* and is characterized by upper molars with strong styles, a relatively high degree of hypsodonty, and lower molars without *Palaeomeryx*-folds. The lower premolars are more complex than those of *Sperrgebietomeryx*, with a bifurcated anterior end and more lingually oriented transverse crests. The limbs and cervical vertebrae are elongated. Morales et al. (2003) describe a tooth that they consider to be the upper canine of *Orangemeryx*. Numerous isolated incisiform lower teeth representing either the canine or incisors have been assigned to *Orangemeryx*. None of these teeth is bilobed, but neither is it certain that the canine is represented in this assemblage.

Discussion Orangemeryx shares numerous features, most notably its branched apophyseal cranial appendages, with *Climacoceras*, and the attribution of this genus to the Climacoceratidae is more secure than is the case for *Sperrgebietomeryx*, *Propalaeoryx*, or *Prolibytherium*. However, the fact that the canine is not definitively known will lead some authors (e.g., Harris et al, this volume, chapter 39) to leave *Orangemeryx* out of the Giraffoidea at present. The discovery of an associated canine could confirm the status of *Orangemeryx* as a climacoceratid, or perhaps more interestingly, shed doubt on the status of the bilobed canine as a synapomorphy of the Giraffoidea.

Genus *NAMIBIOMERYX* Morales et al., 1995

Type Species Namibiomeryx senuti Morales et al., 1995. Only species in genus.

NAMIBIOMERYX SENUTI Morales et al., 1995
Table 37.1

Synonymy ? "small tragulid" Hopwood, 1929.
Holotype Left mandible with m1–m3; no specimen number is provided.
Occurrence Early Miocene, southern Africa (table 37.1).
Diagnosis From Morales et al. (1995). Dentition hypsodont; lower molars without *Palaeomeryx*-fold, moderate stylids, and hypoconid separated from entoconid; upper molars with strong para- and metastyles and internal lobes high crowned with early fusion; premolar series elongated; p4 with simple metaconid directed posteriorly and anterior end not bifurcated.
Description Namibiomeryx is a small ruminant, about the size of a dik-dik. Morales et al. (1995) describe it as a hornless ruminant, though only dental material has been published. *Namibiomeryx* is placed in the Bovoidea on the basis of its lack of *Palaeomeryx*-fold and lower molars with moderate stylids, as well as early fusion of the upper molar lingual cusps in wear. They further place *Namibiomeryx* in the Bovidae, solely on the basis of its moderately hypsodont dentition.

Several specimens described by Hopwood (1929) as a "smaller tragulid" are in fact not tragulid and may represent *Namibiomeryx*. In size and morphology they are a good match

to published drawings and measurements of *Namibiomeryx*, though the hypoconid is not separated anteriorly to the same degree. Hopwood's specimens are in the AMNH (Nos. 22525 and 22526) with casts preserved in the NHM. If this material can be considered *Namibiomeryx*, it indicates that a p1 was not present in this taxon.

Discussion Morales et al. (1995) point out that the earliest true bovids are likely to be hornless and that *Namibiomeryx* cannot be excluded from the family on this basis. Unfortunately, it is difficult to find dental characters that definitively unite the Bovidae, hypsodonty being an unsatisfactory character due to the likelihood that it has developed independently in several lineages. Further recovery and description of cranial and postcranial material may shed further light on the affinities of *Namibiomeryx*.

ADDITIONAL MIOCENE PECORA OF UNCERTAIN STATUS

Rusinga

In addition to "*Palaeomeryx*" *africanus* (now considered *Walangania africanus* and not a representative of the Palaeomerycidae), Whitworth (1958:24) listed five specimens from Rusinga as "*Palaeomeryx* sp." Hamilton (1978a) transferred all but one of these teeth to *Canthumeryx*, a synonymy that is generally recognized (e.g., Harris et al, this volume, chapter 41). The specimen that Hamilton (1978a) did not transfer is NHM M 35250 (Ru 442'51), an m3 that Hamilton thought was too brachydont to be attributed to *Canthumeryx*, but also could not belong in *Palaeomeryx*. The advanced wear stage of this tooth makes identification difficult, but the tooth is of an appropriate size to belong with the rest of the Rusinga *Canthumeryx* material.

Gebel Zelten

Hamilton (1973) assigned two fragmentary third lower molars to "*Palaeomeryx* sp." (NHM M 26691 and BU 20112). He later (1978a) said that these are not attributable to *Palaeomeryx*, while maintaining that they cannot belong to either *Canthumeryx* or *Prolibytherium*. The specimen in London (BU specimen not seen) is very worn, but similar in size to *Prolibytherium*, although Hamilton thought this tooth was too brachydont for this taxon.

NHM M 26690 is a set of cranial appendages that Hamilton (1973: plate 1, figure 6) considered "Palaeomerycidae indet." They are long, straight, and unbranching, with their bases set close together. Hamilton (1978a) later suggested that this specimen might represent the female or juvenile condition of *Prolibytherium magneri*.

Fort Ternan

Gentry (1970) listed a lower molar (now numbered KNM-FT 3136) in his review of Fort Ternan ruminants, noting that it was similar in morphology to *Propalaeoryx austroafricanus*, but was larger. This tooth is similar in size to the m1 of *Climacoceras* but cannot be assigned to this taxon. Compared with *Climacoceras*, this tooth is lower crowned and more bunodont, with a less flattened lingual surface; has a central valley that is more open lingually with a more strongly developed metastylid and weaker parastylid; and possesses stronger anterior and posterior cingula. Its size and the presence of a faint but clearly visible *Palaeomeryx*-fold preclude attribution to any of the named African early Miocene pecoran taxa, although it is similar in size to published measurements of m1 and m2 of *Orangemeryx*.

These few published specimens, along with several other fragmentary unpublished remains from East Africa provide tantalizing evidence of further pecoran diversity in the African early and middle Miocene.

Discussion

The taxa discussed in this chapter represent the oldest known pecoran ruminants in Africa. The origins of the Pecora are unclear, though the group certainly originated in Eurasia, only dispersing into Africa in the late Oligocene or earliest Miocene. Although the Tragulidae are the closest living sister group of Pecora, the true sister taxa of pecorans are likely found among the extinct families of the Tragulina (Webb and Taylor, 1980; Gentry and Hooker, 1988; Gentry, 1994, 2000). Morales et al. (1995) state that as the earliest bovid, *Namibiomeryx* demonstrates that the bovoid-bovid transition occurred in Africa and not in Eurasia. However, it is not certain that *Namibiomeryx* is a bovid (see also Gentry, this volume, chapter 38) and definitive bovids of equal antiquity occur in western Asia (e.g., Barry et al., 2005). It is likely that neither bovids nor giraffids originated in Africa but migrated into Africa from separate ancestral stocks in Eurasia.

Several of the taxa discussed here have been included in the Giraffoidea *(Prolibytherium, Propalaeoryx, Sperrgebietomeryx, Orangemeryx).* If the presence of a bilobed lower canine can be considered a synapomorphy of the Giraffoidea (Hamilton, 1978b; Harris et al., this volume, chapter 39), these taxa must be left as Pecora *incertae sedis*, as the state of their lower canines is unknown. The bilobed canine is given great importance in giraffoid systematics, and though it is never ideal to use a single synapomorphy to define a group, it is difficult to find other characters that unite the superfamily Giraffoidea. The reported presence of *Palaeomeryx*-folds in *Sperrgebietomeryx*, and possibly some *Propalaeoryx austroafricanus* (Morales et al., 1999:237), would also be unusual for a giraffoid.

Morales and colleagues (1999, 2003; Pickford et al., 2001) have included *Propalaeoryx*, *Sperrgebietomeryx*, and *Orangemeryx* in the Climacoceratidae, largely based on a very different concept of what climacoceratids are. A more traditional classification would exclude *Sperrgebietomeryx* and *Propalaeoryx* from the Climacoceratidae due to their apparent lack of tined cranial appendages (Hamilton, 1978b). *Orangemeryx* and *Prolibytherium* might be better candidates for inclusion in Climacoceratidae as they do bear cranial appendages, but these may not all be homologous. *Prolibytherium* does not possess an elongated neck, part of the diagnosis of Climacoceratidae presented by Morales et al. (1999:232). Other characters of this diagnosis, including open metatarsal gullies and hypsodonty, are likely evolved in parallel in several ruminant lineages. Morales et al. (2003) also question whether a bilobed canine is truly present in *Climacoceras*, following Churcher (1990:193), who had suggested that the bilobed canine present in the type specimen of *Climacoceras gentryi* does not belong to this individual. Reexamination of the *C. gentryi* type shows that while the mandible was broken anteriorly, there is an excellent fit between the dentary and the small anterior fragment containing the incisor and canine and that the bilobed nature of this tooth is represented very faithfully by Hamilton (1978b: figure 3). There can be little doubt that at least *Climacoceras gentryi* did possess a bilobed canine,

although this does not necessarily resolve the significance of this feature for giraffoid systematics.

The presence of a *Palaeomeryx*-fold in taxa that some believe to belong within Giraffoidea is problematic, as this feature is often viewed as a character of Cervoidea (Janis and Scott, 1987, 1988; Gentry, 1994). Either the *Palaeomeryx*-fold must be viewed as a primitive feature common to several lineages of ruminants (e.g., Hamilton, 1973:148) or less likely, a feature that evolved in parallel, or these purported giraffoids must be excluded from this subfamily. The variability in the presence, strength, and morphology of the *Palaeomeryx*-fold in *Walangania* is disconcerting. Variation along the molar row in a single individual, as well as variation in the same tooth between different individuals, indicates that the *Palaeomeryx*-fold may not be as phylogenetically meaningful as has often been assumed.

Conclusions

New material of late Oligocene and early Miocene pecorans may help to clarify the origins of the modern pecoran families. However, as numerous authors have pointed out, the rapid divergence of pecoran groups in the Oligocene (based on molecular dating) may obscure pecoran familial relationships. Clear attribution of all pecoran taxa to a modern family is not possible until the late Miocene. Another problem is that attribution of early taxa into one of the modern groups rests largely on single synapomorphies, such as the bilobed canine for giraffoids, or *Palaeomeryx*-fold for cervoids, and further inquiry into the validity of these synapomorphies, and identification of new characters, is important.

Similarities between several of these African early Miocene pecorans and Eurasian taxa such as *Dremotherium* suggest that the African material cannot be studied in isolation and that intercontinental connections are vital to furthering our understanding of the phylogenetic relationships of these taxa. As African pecorans are immigrant taxa, a greater understanding of the interrelationships and origins of the Pecora may likely come from outside the African continent.

ACKNOWLEDGMENTS

E. Mbua and M. Muungu at the KNM; E. Kamuhangire, E. Musimae, and N. Abiti at the Uganda Museum; and J. Hooker and A. Currant at the NHM kindly allowed access to collections under their care. I would also like to thank E. Miller, J. Morales, M. Pickford, and H. Tsujikawa for helpful information. This research was supported by the National Science Foundation (BCS-0524944), the Leakey Foundation, the Quaternary Research Association, and the Department of Anthropology at Harvard University.

Personal thanks to Lars Werdelin and Bill Sanders for including these "forgotten" taxa in this volume, Laura MacLatchy for permission to study material she collected in Uganda, and John Barry for first introducing me to ruminants. I extend my most particular regards to Alan Gentry not only for suggesting to the editors that I write this chapter, but for believing that I could.

Literature Cited

Arambourg, C. 1933. Mammifères miocènes du Turkana (Afrique orientale). *Annales de Paléontologie* 22:121–148.
———. 1961. *Prolibytherium magnieri*, un Velléricorne nouveau du Burdigalien de Libye (Note préliminaire). *Comptes Rendus des Séances de la Société Géologique de France, Paris* 3:61–63.

Azanza, B., J. Morales, and M. Pickford. 2003. On the nature of the multibranched cranial appendages of the climacoceratid *Orangemeryx hendeyi*. *Memoirs of the Geological Survey of Namibia* 19:345–357.

Barry, J. C., S. Cote, L. MacLatchy, E. Lindsay, R. Kityo, and A. R. Rajpar. 2005. Oligo-Miocene Ruminants from Pakistan and Uganda. *Paleontologia Electronica* 8(1):22A

Bishop, W. W., J. A. Miller, and F. J. Fitch. 1969. New potassium-argon determinations relevant to the Miocene fossil mammal sequence in east Africa. *American Journal of Science* 267:669–699.

Blondel, C. 1997. Les ruminants de Pech Desse et de Pech du Fraysse (Quercy; MP 28): Évolution des ruminants de l'Oligocène d'Europe. *Geobios* 30:573–591.

Boschetto, H. B., F. H. Brown, and I. McDougall. 1992. Stratigraphy of the Lothidok Range, northern Kenya and K/Ar ages of its Miocene primates. *Journal of Human Evolution* 22:47–71.

Churcher, C. S. 1978. Giraffidae; pp. 509–535 in V. J. Maglio and H. B. S. Cooke (eds.), *Evolution of African Mammals*. Harvard University Press, Cambridge.

———. 1990. Cranial appendages of Giraffoidea; pp. 180–194 in G. Bubenik and A. Bubenik (eds.), *Horns, Pronghorns and Antlers*. Springer, New York.

Cote, S. 2004. New ruminant specimens from the early Miocene of Karamoja District, Uganda. *Journal of Vertebrate Paleontology* 24 (suppl. to no. 3):48A

Drake, R. E., J. A. Van Couvering, M. H. Pickford, G. H. Curtis, and J. A. Harris. 1988. New chronology for the early Miocene faunas of Kisingiri, western Kenya. *Journal of the Geological Society, London* 145:479–491.

Flower, W. H. 1883. On the arrangement of the orders and families of existing Mammalia. *Proceedings of the Zoological Society of London* 1883:1–10.

Gebo, D., L. MacLatchy, R. Kityo, A. Deino, J. Kingston, and D. Pilbeam. 1997. A hominoid genus from the early Miocene of Uganda. *Science* 276:401–404.

Gentry, A. W. 1970. The Bovidae (Mammalia) of the Fort Ternan Fossil Fauna; pp. 243–323 in L. S. B. Leakey and R. J. G. Savage (eds.), *Fossil Vertebrates of Africa: Volume 2*. Academic Press, London.

———. 1978. Bovidae; pp. 540–572 in V. J. Maglio and H. B. S. Cooke (eds.), *Evolution of African Mammals*. Harvard University Press, Cambridge.

———. 1994. The Miocene differentiation of Old World Pecora (Mammalia). *Historical Biology* 7:115–158.

———. 2000. The ruminant radiation; pp. 11–25 in E. S. Vrba and G. B. Schaller (eds.), *Antelopes, Deer and Relatives: Fossil Record, Behavioral Ecology, Systematics and Conservation*. Yale University Press, New Haven.

Gentry, A. W., and J. J. Hooker. 1988. The phylogeny of the Artiodactyla; pp. 235–272 in M. J. Benton (ed.), *The Phylogeny and Classification of the Tetrapods: Volume 2. Mammals*. Clarendon Press, Oxford.

Ginsburg, L., and E. Heintz. 1966. Sur les affinités du genre *Palaeomeryx* (Ruminant du Miocène européen). *Comptes Rendus Hebdomadaires de l'Académie des Sciences, Paris* 262:979–982.

Hamilton, W. R. 1973. The lower Miocene ruminants of Gebel Zelten, Libya. *Bulletin of the British Museum (Natural History), Geology* 21:75–150.

———. 1978a. Cervidae and Palaeomerycidae; pp. 496–508 in V. J. Maglio and H. B. S. Cooke (eds.), *Evolution of African Mammals*. Harvard University Press, Cambridge.

———. 1978b. Fossil giraffes from the Miocene of Africa and a revision of the phylogeny of the Giraffidae. *Philosophical Transactions of the Royal Society of London B* 283:165–229.

Hamilton, W. R., and J. A. Van Couvering. 1977. Lower Miocene mammals of South West Africa. *Namibia Bulletin* (suppl. 2, *Transvaal Museum Bulletin*):9–11.

Hassanin, A., and E. Douzery. 2003. Molecular and morphological phylogenies of Ruminantia and the alternative position of the Moschidae. *Systematic Biology* 52:206–228.

Hendey, Q. B. 1978. Preliminary report on the Miocene vertebrates from Arrisdrift, South West Africa. *Annals of the South African Museum* 76:1–41.

Hernández Fernández, M., and E. S. Vrba. 2005. A complete estimate of the phylogenetic relationships in Ruminantia: A dated species-level supertree of the extant ruminants. *Biological Reviews* 80:269–302.

Hopwood, A. T. 1929. New and little-known mammals from the Miocene of Africa. *American Museum Novitates* 344:1–9.

Janis, C., and K. Scott. 1987. The interrelationships of higher ruminant families with special emphasis on the members of the Cervoidea. *American Museum Novitates* 2893:1–85.

———. 1988. The phylogeny of the Ruminantia (Artiodactyla, Mammalia); pp. 273–282 in M. J. Benton (ed.), *The Phylogeny and Classification of the Tetrapods: Volume 2. Mammals*. Clarendon Press, Oxford.

MacLatchy, L., A. Deino, and J. Kingston. 2006. An updated chronology for the early Miocene of Uganda. *Journal of Vertebrate Paleontology* 26 (suppl. to no. 3):93A.

Miller, E., and E. Simons. 1996. Relationships between the mammalian fauna from Wadi Moghara, Qattara Depression, Egypt, and other early Miocene faunas. *Proceedings of the Geological Survey of Egypt, Centennial Conference*:547–580.

Morales, J. L. Ginsburg, and D. Soria. 1986. Los Bovoidea (Artiodactyla, Mammalia) del Mioceno inferior de España: Filogenia y biogeografía. *Paleontología y Evolució* 20:259–265.

Morales, J., M. Pickford, and D. Soria. 1993. Pachyostosis in a lower Miocene giraffoid from Spain, *Lorancameryx pachyostoticus* nov. gen. nov. sp. and its bearing on the evolution of bony appendages in artiodactyls. *Geobios* 26:207–230.

Morales, J., D. Soria, M. Nieto, P. Pelaez-Campomanes, and M. Pickford. 2003. New data regarding *Orangemeryx hendeyi* Morales et al., 2000, from the type locality, Arrisdrift, Namibia. *Memoirs of the Geological Survey of Namibia* 19:305–344.

Morales, J., D. Soria, and M. Pickford. 1995. Sur les origines de la famille des Bovidae (Artiodactyla, Mammalia). *Comptes Rendus de l'Académie des Sciences, Paris* 321:1211–1217.

———. 1999. New stem giraffoid ruminants from the early and middle Miocene of Namibia. *Geodiversitas* 21:229–253.

Moyà-Solà, S. 1987. Los ruminantes (Cervoidea y Bovoidea, Artiodactyla, Mammalia) del Ageniense (Mioceno inferior) de Navarrete del Río (Teruel, Espana). *Paleontología y Evolució* 21:247–269.

Nakaya, H. 1994. Faunal change of late Miocene Africa and Eurasia: Mammalian fauna from the Namurungule Formation, Samburu Hills, Northern Kenya. *African Study Monographs* (suppl.) 29:1–112.

Pickford, M. 2002. Ruminants from the early Miocene of Napak, Uganda. *Annales de Paléontologie* 88:85–113.

Pickford, M., and P. Andrews. 1981. The Tinderet Miocene sequence in Kenya. *Journal of Human Evolution* 10:11–33.

Pickford, M., Y. Attia, and M. Abd El Ghany. 2001. Discovery of *Prolibytherium magnieri* Arambourg, 1961 (Artiodactyla, Climacoceratidae) in Egypt. *Geodiversitas* 23:647–652.

Pickford, M., H. Ishida, Y. Nakano, and K. Yasui. 1987. The Middle Miocene fauna from the Nachola and Aka Aiteputh Formations, Northern Kenya. *African Study Monographs* (suppl.) 5:141–154.

Pickford, M., and P. Mein. 2007. Early Middle Miocene Mammals from Moroto II, Uganda. *Beiträge zur Paläontologie* 30:361–386.

Pickford, M., and B. Senut. 2003. Miocene Paleobiology of the Orange River Valley, Namibia. *Memoir of the Geological Survey of Namibia* 19:1–22.

Price, S., O. Bininda-Emonds, and J. Gittleman. 2005. A complete phylogeny of the whales, dolphins and even-toed hoofed mammals (Cetartiodactyla). *Biological Reviews* 80:445–473.

Sawada, Y., M. Pickford, T. Itaya, T. Makinouchi, M. Tateishi, K. Kabeto, S. Ishida, and I. Ishida. 1998. K-Ar ages of Miocene Hominoidea (*Kenyapithecus* and *Samburupithecus*) from Samburu Hills, northern Kenya. *Comptes Rendus de l'Académie des Sciences, Paris, Sciences de la Terre et des Planètes* 326:445–451.

Stromer, E. 1926. Reste Land- und Süsswasser-Bewohnender Wirbeltiere aus den Diamantfeldern Deutsch-Südwestafrikas; pp. 107–153 in E. Kaiser (ed.), *Die Diamantenwüste Südwest-africas*, vol. 2. Reimer, Berlin.

Vislobokova, I. 2001. Evolution and classification of Tragulina (Ruminantia, Artiodactyla). *Paleontological Journal* 35(S2):S69–S145.

Webb, S. D., and B. E. Taylor. 1980. The phylogeny of higher ruminants and a description of the cranium of *Archaeomeryx*. *Bulletin of the American Museum of Natural History* 167:117–158.

Whitworth, T. 1958. Miocene ruminants of East Africa. *Fossil Mammals of Africa* 15:1–50.

THIRTY-EIGHT

Bovidae

ALAN W. GENTRY

Bovidae contain the cattle, sheep, goats, and antelopes. The word *antelope* is used for bovids outside Europe, mostly in Africa, or not domesticated before Linnaeus' lifetime. It does not correspond with a formal taxonomic category. Most phylogenies postulate bovids being closer to cervids than to giraffids (Marcot, 2007). Unlike the cervoid *Moschus* in relation to Cervidae, there is no living hornless pecoran thought to be a bovoid (member of a superfamily Bovoidea including Bovidae and any related families, the latter as yet unknown). Table 38.1 shows an overall classification of Bovidae, and figure 38.1 shows their evolutionary relationships.

Bovid Attributes and Their History

Bovidae are defined by their hollow horns (hence the old name Cavicornia) in which a keratinized epidermal sheath fits over a bony core. The bovid horn core consists of spongy bone except in some advanced bovids with internal sinuses. Neither sheath nor core is branched or seasonally shed. Some species have horns in both sexes. Upper incisors are lacking and only a few early species retain minute upper canines. Enlarged upper canines have never been found in early bovids, but they have been in the possibly bovoid *Hispano-meryx* (Moyá Solá, 1986). First premolars have disappeared. The cheek teeth are selenodont and the crescentic cusps join to one another earlier in wear than in cervids or giraffids. The cheek teeth are nearly always more hypsodont and with smoother enamel than in cervids or giraffids. Styles, stylids and ribs are not very bulky and cingula are weak, all unlike most deer. Metapodials lateral and medial to the main cannon bone are absent or splintlike and more reduced than in cervids. The metatarsal has an open groove distally on its anterior surface. Compared with cervids or giraffids, many bovids show strong cursorial characters in their limb bones. The majority of them show territorial behaviour.

Horn cores are an extra help in identification but can be more variable within species than is useful. The first, early Miocene stage in their evolution (figure 38.2A) looks as if it were short, stumpy spikes, widely separated and above the back of the orbits (Morales et al., 2003: plate 1, figures 2–3; Moyá Solá, 1983: plate 1, figures 1–2). In the middle Miocene horn cores began to lengthen, their insertions often became closer and more posterior, and keels appeared in Boselaphini. Lengthened, anteroposteriorly enlarged and backwardly curved horn cores could have been successive, overlapping or simultaneous advances during the middle Miocene. Backward curvature appeared in *Tethytragus, Gazella,* early *Hippotragus* and others; in longer horn cores this prevents tips becoming too anteriorly pointed. Figure 38.2 shows that lengthening alone may have taken place in the ancestors of smaller antelopes such as Neotragini, while *Pelea* is a further example of a larger antelope with small-diameter horn cores. From the late Miocene onward, many changes, parallels and reversals became possible in inclination, divergence, curvature, and torsion. The result is the well-known diversity of bovid horn core shapes.

Horn cores of some bovids show clockwise or anticlockwise torsion. The direction can be decided by viewing a right horn core from either end and assuming the torsion to be away from and not toward the observer. In two other conventions clockwise torsion on the right is called *homonymous* or *inverse* (antonyms: *heteronymous* or *normal*). *Homonymous* and *heteronymous* are preferable to *clockwise* and *anticlockwise* in that the same word applies to the horn cores of both sides in an individual animal. Whenever used in English texts, the words need careful initial definition since they are inadequately defined or absent (in this meaning) in dictionaries. Torsion can be manifested as twisting of the axis and need not disturb the overall straightness of the horn core, as in an eland, *Taurotragus oryx.* In other species it can combine with changing divergence along the course of the horn core to produce lyration as in impala, *Aepyceros melampus,* or ultimately the open spiraling of the greater kudu, *Tragelaphus strepsiceros.* Kingdon (1982) gives an ontogenetic interpretation of tragelaphine horn cores that involves different adjectives for the lyrated and spiraled states as used in this chapter.

Primitive states of skull characters are not known for sure, and all the following are tentative to varying degrees. Sinuses within the frontals and horn pedicels were of small extent, and the frontals were not elevated between the horn core bases in front view. Supraorbital foramina were small and without surrounding pits; the preorbital fossa large; the back of the tooth row lay below the orbit; the infraorbital foramen was placed low and anteriorly; the median indent at the back

TABLE 38.1
A possible classification of Bovidae into subfamilies and tribes

Horizontal lines separate possible larger groupings: A, the early and perhaps
diphyletic Hypsodontinae; B, Boselaphini and allied tribes; C, a cluster centred around Antilopini;
D, the caprine-alcelaphine group.

Taxon	*Characteristics*
GROUP A	
Subfamily †Hypsodontinae	
†Hypsodontini	Middle Miocene, perhaps diphyletic to other bovids
GROUP B	
Subfamily Bovinae	
Boselaphini	Nilgai and four-horned antelope
Tragelaphini	Kudu, bushbuck group
Bovini	Cattle, buffalo
GROUP C	
Subfamily Antilopinae	
Cephalophini	Duikers
Neotragini	Dik dik, steenbok and other small antelopes
Antilopini	Impala, blackbuck, saiga antelope, gazelles
†*Criotherium*	Late Miocene, plus †*Palaeoreas*, coming from Antilopini
Peleini	Vaal rhebok, *Pelea capreolus*.
Subfamily Reduncinae	
Reduncini	Waterbuck and reedbuck group, perhaps originating from near *Pelea*
Subfamily †Oiocerinae	
†Oiocerini	Late Miocene, includes †*Urmiatherium*
GROUP D	
Subfamily Hippotraginae	
Hippotragini	Roan, sable antelope, oryx, addax
Alcelaphini	Hartebeest and wildebeest group. This and the preceding tribe arose near the complicated base of the Caprinae.
?Subfamily of Its Own	
†*Tethytragus*	Middle Miocene. Relationship with Oiocerini, Hippotragini, *Pantholops*, or Caprinae still to be decided.
?Subfamily of Its Own	
Pantholops	Chiru, one genus near the origin of Caprinae
Subfamily Caprinae	
Rupicaprini or Naemorhedini	Chamois, serow, goral
Budorcas	Takin, not in the Ovibovini
Ovibovini	Muskox
Caprini	Goats, but tribe for sheep still to be decided

SOURCE: Based on Gentry (1992), Gatesy et al. (1997), Hassanin and Douzery (1999), Vrba and
Schaller (2000), Matthee and Davis (2001), Hernández Fernández and Vrba (2005), and Marcot
(2007). Note that if subfamily Antilopinae were regarded as the cladistic sister group to subfamily
Bovinae, all ranks in C and D would need downgrading. *Criotherium*, Oiocerini, *Tethytragus*,
Pantholops, and Rupicaprini are not represented in Africa.

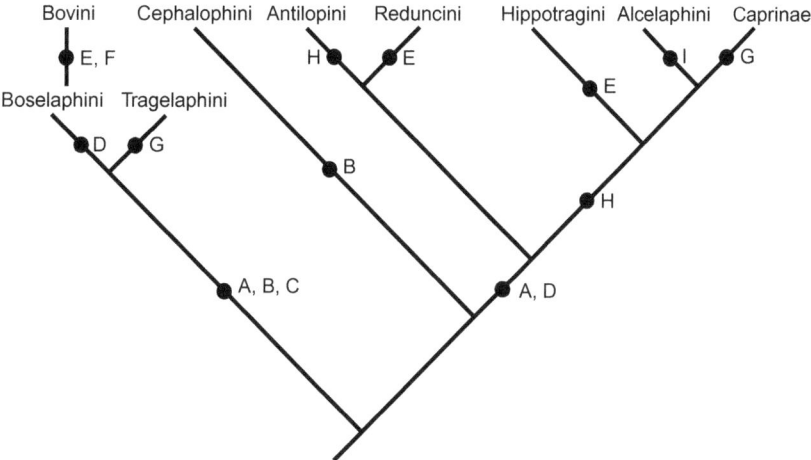

FIGURE 38.1 A simplified scheme of possible subfamily and tribal-level evolutionary relationships among Bovidae. The ancestor would have had short, nearly upright horn core spikes and brachyodont or mesodont teeth with occlusal surfaces as in figure 38.4A. A) Horn cores lengthen, B) horn cores become inclined in side view, C) horn cores acquire keels, D) teeth become higher crowned, E) teeth eventually become boödont, F) much enlarged body size, G) teeth become aegodont, H) horn cores become backwardly curved, and I) long faces, teeth with rounded lobes, no basal pillars.

of the palate was at a more posterior level than the lateral ones. The braincase roof was slightly inclined, but later it could shorten and become more inclined. Temporal lines approached each other closely toward the back of the cranial roof. Each side of the occipital surface faced partly laterally as well as backward. The basioccipital was small and narrow anteriorly, later becoming quadrangular.

Probably the original state of bovid teeth was like the middle Miocene *Eotragus* and *Tethytragus* with slightly rugose enamel and long premolar rows. Since then, various tribes have evolved in their own ways either toward occlusal complexity or simplicity (figures 38.3, 38.4). Many opportunities for parallelisms have been taken up, but no tribe has ended with a suite of characters coinciding with any other tribe. In the past, a rather unclear dichotomy was often seen between the Boodontia and Aegodontia of Schlosser (1911) or "millers" and "cutters" of Köhler (1993). It is paradoxical that although first noticed by palaeontologists, boödonty and aegodonty only become well manifest in middle Pleistocene and later bovids and are not very helpful for most of bovid history. Boödont attributes are large occlusal area, enlarged basal pillars and a complicated outline of the central fos-

settes, and strong vertical ribs on the labial walls of upper molars and lingual walls of lower molars. Aegodont attributes are loss of basal pillars, simple and less curved central fossettes, and flat or slightly concave walls in place of the ribs of boödonts.

Conventions

Authors and dates of subfamily and tribe names are given in Grubb (2001) and Wilson and Reeder (2005). The treatment of species does not include complete locality or synonym citations because (1) many species are widespread; (2) many identifications are dubious, debated, and often based on fragmentary material; (3) authors are not agreed on the morphological or chronostratigraphical levels of transitions to living species; (4) late Pleistocene and later sites are very numerous and lead into an enormous archaeological literature (see Plug and Badenhorst, 2001, for South Africa alone). A good deal of information about earlier finds of fossil bovids of interest is given in Gentry and Gentry (1978), which is indexed. Early, middle, and late temporal subdivisions are used according to normal practices for the Miocene and Pleistocene epochs. The Pliocene is divided into three equal parts, which means that the middle subdivision runs from approximately 4.0 to 3.0 Ma. Localities mentioned in the text are listed in table 38.2. The often-mentioned Siwaliks are a range of sediments associated with the Himalayan orogeny in India and Pakistan and containing a rich sequence of Miocene, Pliocene, and early Pleistocene vertebrate faunas (Barry, 1995).

ABBREVIATIONS

AMNH = American Museum of Natural History, New York; BMNH = Natural History Museum, London; ICZN = International Commission on Zoological Nomenclature, * = type species or type locality, Fm = formation, Mb = member. DAP × DT = horn core index = anteroposterior and transverse diameters at the base of a horn core. Skull orientations are described with the tooth row visualized as horizontal.

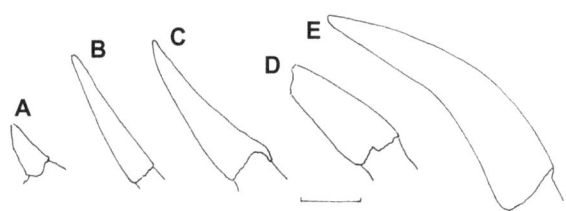

FIGURE 38.2 Side views of bovid horn cores showing early stages in their evolution, anterior to the right. A) Short, stumpy spikes, based on *Namacerus* and early *Eotragus*. B) Increase in length, based on modern *Raphicerus campestris*. C) Increased anteroposterior girth as in *Eotragus clavatus* or the *Pseudoeotragus seegrabensis* of Made (1989). D) Beginning of backward curvature, based on supposed *Homoiodorcas* sp. (Thomas, 1983). E) Longer horn cores with more definite backward curvature, based on late Miocene *Gazella*. Scale = 25 mm.

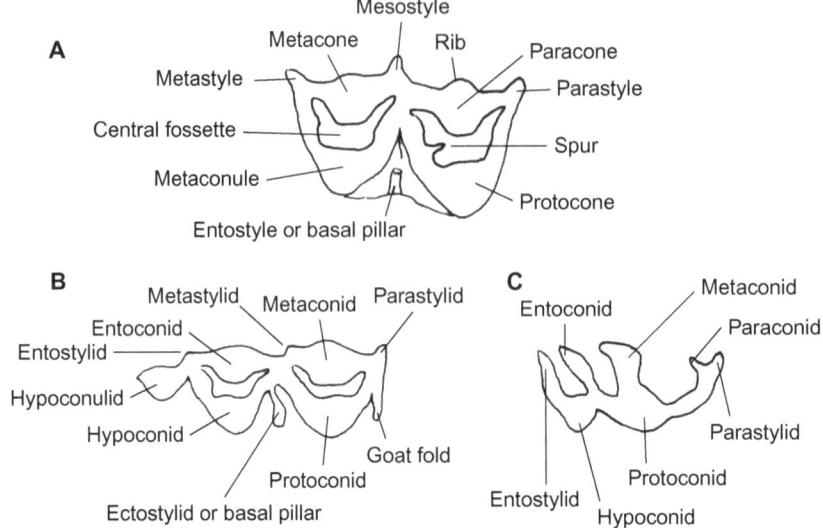

FIGURE 38.3 Nomenclature of bovid right cheek teeth: A) upper molar; B) m3; C) p4. Anterior to the right.

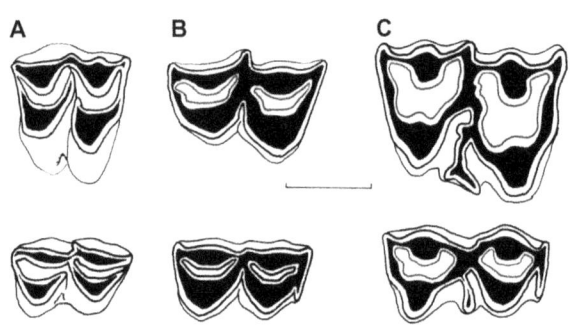

FIGURE 38.4 Aegodont and boödont teeth of the right side in Bovidae. Top row, upper molars; lower row, lower molars. A) Early bovid based on *Eotragus*. B) Aegodont bovid based on *Capra*. C) Boödont bovid based on modern *Hippotragus*. Anterior to the right. Boödont features are large occlusal area, strong ribs, large basal pillars, and a complicated outline of the central fossettes. Aegodont features are flatter or straighter labial walls of upper molars and lingual walls of lower molars, absence of basal pillars, and straighter or more simple central fossettes. Goat folds can occur in either type. Scale = 25 mm.

Fossil Bovidae in Africa

In Eurasia tiny bovid-like dental remains are known well back to the early Oligocene of Mongolia (Dmitrieva, 2002), but nothing is known of pre-Miocene ruminants in Africa. Pecorans such as *Walangania*, *Propalaeoryx*, and *Namibiomeryx* do appear in the early Miocene (Cote, this volume, chapter 37), and the last has been claimed to be a bovid by Morales et al. (1995). From the middle Miocene onward, faunas with two or more bovid species can be found. Four such localities are Gebel Zelten and Maboko, both close to the start of the middle Miocene ~16.0 Ma; Fort Ternan ~14.0 Ma; and the Ngorora Formation lasting from about 12.0–8.0 Ma. At the first three of these localities, one or more giraffoids accompany the bovids. The bovids of these and other faunas are recognisably in different subfamilies and tribes (table 38.3). Much information about subsequent bovid evolution in Africa comes from the Awash area and the Shungura Formation in Ethiopia, the Koobi Fora and Nachukui Formations in Kenya, and Olduvai Gorge in Tanzania, all of which have analyzed temporal spans. The Sterkfontein Valley and Makapansgat Limeworks cave sites (formerly the "Transvaal cave sites"), Chad and several north African localities are also informative.

Family BOVIDAE Gray, 1821
Subfamily and Tribe INDET.
Genus *NAMACERUS* Morales, Soria, Pickford and Nieto, 2003

Type Species Namacerus gariepensis Morales et al., 2003.

NAMACERUS GARIEPENSIS Morales et al., 2003

Synonymy Namacerus gariepensis Morales et al., 2003:372, plates 1–4. *Arrisdrift. Middle Miocene, ca. 17.0 Ma.

Remarks This species is notably informative as an African bovid on the threshold of the middle Miocene. It is about the size of a steenbok, *Raphicerus campestris*, and does have horn cores, very short and stumpy like those of early *Eotragus* at Buñol, Spain (Moyá Solá, 1983), but only about three-quarters of their size. The teeth are said to differ from the early Miocene (ca. 18.0 Ma) *E. artenensis* in Europe by being slightly more hypsodont, and the lower molars by lacking metastylids and having continuous lingual walls. The length of the lower premolar rows relative to the molar rows is about 70%, which is like *Eotragus* and probably primitive for bovids. The metapodials and tibia were not very long, and the limb proportions were thereby like those of a *Cephalophus* or a muntjac deer.

Subfamily HYPSODONTINAE Köhler,1987

Type Genus Hypsodontus Sokolov, 1949.

The now extinct Hypsodontinae reached their zenith and their end in the middle Miocene. By this time, they had acquired horns and become widespread. In the early Miocene they had lived in China (Chen, 1988) and possibly Arabia (Whybrow et al., 1982; Thomas et al., 1982; Gentry, 1987b) and, most probably, had not possessed horns. Some hypsodontines had hypsodont teeth, and bovid-like hypsodont and nonhypsodont teeth go back into the Paleogene of eastern Asia (Wang, 1992; Dmitrieva, 2002).

TABLE 38.2

Adu-Asa Formation, Middle Awash, Ethiopia, late Miocene
(Haile-Selassie et al., 2004); 5.8–5.7 Ma

Ahl al Oughlam, Morocco, late Pliocene
(Geraads and Amani, 1998); ~ 2.5 Ma

Aïn Boucherit, Algeria, late Pliocene (Arambourg, 1979); ~2.0 Ma

Aïn Brimba, Algeria, late Pliocene; ~3–2.5 Ma

Aïn Hanech, Algeria, early Pleistocene; ~1.4–1.2 Ma

Aïn Jourdel, Algeria, late Pliocene; ~2.2 Ma

Aïn Maarouf, Morocco, middle Pleistocene (Geraads and
Amani, 1997); ~0.8 Ma, slightly younger than Tighenif

Alayla, Middle Awash, Ethiopia, latest Miocene
(Haile-Selassie et al., 2004); 5.8–5.5 Ma

Algiers, gisement des Phacochères, Algeria, late Pleistocene
(Hadjouis, 2002)

Anabo Koma, Djibouti, early Pleistocene
(Bonis et al., 1988) ; ~1.6 Ma

Aramis, Middle Awash, Sagantole Fm, Ethiopia, early Pliocene
(Vrba, 1997); ~4.4 Ma

Arrisdrift, Namibia, middle Miocene
(Morales et al., 2003) ; ~17.0 Ma

Asbole, Ethiopia, middle Pleistocene
(Geraads et al., 2004) ; 0.8–0.6 Ma

Beglia Formation, Tunisia, (?middle-) late Miocene
(Robinson, 1972); 11.0–10.0 Ma

Beni Mellal, Morocco, middle Miocene (Choubert and
Faure-Muret, 1961)

Bizerte, Tunisia, late Pleistocene

Bodo 1, Awash, Ethiopia, middle Pleistocene (Vrba, 1997); 0.64 Ma

Bolt's Farm, South Africa, late Pliocene or middle Pleistocene
(Cooke, 1991; Vrba, 1978)

Bou Hanifia (formerly Oued el Hammam), Algeria, early late
Miocene (Arambourg, 1959); 10.0–9.0 Ma

Bouri 1–2, Ethiopia, early Pleistocene (Vrba, 1997); ~1.0 Ma

Buffalo Cave, Limpopo Province, South Africa. Plio-Pleistocene
(Kuykendall et al., 1995)

Buia, Eritrea, early Pleistocene
(Martínez-Navarro et al., 2004); 1.0 Ma

Casablanca , Morocco, late Pleistocene (Arambourg, 1939)

Chelmer, Zimbabwe, late Pleistocene (Cooke and Wells, 1951)

Chiwondo Beds, Malawi, middle Pliocene–early Pleistocene
(Sandrock et al., 2007); three levels: 4.0 Ma or older, 3.76–2.00
Ma or younger, and 1.6–1.5 Ma

Chorora Formation, Ethiopia, late Miocene (Geraads et al., 2002)

Cornelia, South Africa, middle Pleistocene
(Cooke, 1974); 0.8–0.6 Ma

Djebel Krechem, Tunisia, late Miocene (Geraads, 1989); ~9.0 Ma

Djebel-Thaya, Algeria, late Pleistocene (Bourguignat, 1870)

Elandsfontein, South Africa, middle Pleistocene (Klein et al.,
2007); 0.7(- 0.4?) Ma

El Hamma du Djérid, Tunisia, late Miocene (Thomas, 1979);

Florisbad, Free State Province, South Africa, late Pleistocene
(Brink, 1987);

Fort Ternan, middle Miocene, Kenya (Gentry, 1970); 14.0 Ma

Gamedah, Ethiopia, late Pliocene (Vrba, 1997); 2.6 Ma

Garba IV, Ethiopia, early Pleistocene
(Geraads et al., 2004b); ~1.3 Ma

Gebel Zelten, Libya, near the early-middle Miocene transition
(Hamilton, 1973); ~16.0 Ma

Gladysvale, South Africa, middle Pleistocene
(Lacruz et al., 2002); ~0.7 Ma

Hadar Formation, Ethiopia, middle Pliocene
(Alemseged et al., 2005); 3.4–2.9 Ma

Isimila, Tanzania, middle Pleistocene (Howell et al., 1972)

Kabwe, Zambia, middle Pleistocene; ~0.2 Ma

Kaiso Formation deposits of unknown age in the Kazinga
Channel, Uganda (Cooke and Coryndon, 1970; Geraads and
Thomas, 1994)

Kaiso Village Beds, Uganda, late Pliocene
(Geraads and Thomas, 1994); ?(2.6–2.3) Ma

Kanam, Kenya, late Pliocene – early Pleistocene
(Ditchfield et al., 1999)

Kanapoi, Kenya, early Pliocene (Harris et al., 2003); 4.17–4.07 Ma

Kanjera, Kenya, late Pliocene – middle Pleistocene
(Ditchfield et al., 1999)

Karmosit beds, Kenya, Pliocene (Bishop et.al., 1971); ?3.5 Ma

Katwe Ashes, Semliki, DR Congo, late Pleistocene (Boaz, 1990)

Klasies River Mouth, South Africa, late Pleistocene (Klein, 1994)

Kom Ombo, Egypt, late Pleistocene (Churcher, 1972)

Koobi Fora Formation, Kenya, late early Pliocene–middle
Pleistocene (Harris, 1991)

Koro Toro, Chad, middle Pliocene
(Geraads et al., 2001); 3.5–3.0 Ma

Kromdraai A (= Kromdraai Faunal Site), South Africa, early
(-middle?) Pleistocene (Vrba, 1978, 1995; Herries et al., 2009);
1.9–1.5 Ma (–0.7 Ma?)

Laetoli, Laetolil Beds, Tanzania, middle Pliocene (Su and
Harrison, 2007; Harrison, in press); 3.7–3.5 Ma

Laetoli, Upper Ndolanya Beds, Tanzania, late Pliocene
(Kovarovic et al., 2002); 2.6–2.5 Ma

Lainyamok, Kenya, middle Pleistocene
(Potts and Deino, 1995); ~0.36 Ma

Langebaanweg, Varswater Formation, South Africa, early
Pliocene (Gentry, 1980); ~5.0 Ma

Lothagam, Kenya, Nawata Formation to Kaiyumung Member of
the Nachukui Formation, late Miocene–middle Pliocene
(Leakey and Harris, 2003); 7.5–3.6 Ma

Lower Terrace Complex, Semliki, DR Congo, middle Pleistocene
(Boaz, 1990)

Lukeino, Kenya, late Miocene (Thomas, 1980; Deino et al.,
2002); 6.0–5.7 Ma

Lukenya Hill, Kenya, late Pleistocene (Marean, 1992)

Maboko, Kenya, middle Miocene (Feibel and Brown, 1991);
16.0–15.0 Ma

Maka, Middle Awash, Ethiopia, middle Pliocene
(Vrba, 1997); 3.2 Ma

Makapansgat Limeworks, Limpopo Province, South Africa,
late Pliocene–early Pleistocene (Reed, 1998; Vrba, 1995);
Mb 3 = 2.8–2.6 Ma, Mb 4 = 2.7–2.5 Ma, and Mb 5 = 1.8–1.6 Ma

Manonga (= Wembere), Tanzania, late Miocene–early Pliocene
(Harrison and Baker, 1997)

Mansourah, Algeria, early Pleistocene
(Chaîd-Saoudi et al., 2006); ~1.7 Ma

Marceau, Algeria. See Menacer.

Matabaietu, Ethiopia, late Pliocene (Vrba, 1997); ~2.5 Ma

Melka Kunturé, Ethiopia, early–middle Pleistocene
(Geraads et al., 2004b); includes Garba

Melkbos, South Africa, late Pleistocene (Hendey, 1968)

TABLE 38.2

(CONTINUED)

Menacer (formerly Marceau), Algeria, late Miocene
 (Thomas and Petter, 1986)
Middle Awash, Ethiopia, late Miocene–early Pliocene
 (Haile-Selassie et al., 2004)
Modder River, South Africa. See Vaal River Gravels.
Mpesida, Kenya, late Miocene (Thomas, 1980;
 Kingston et al., 2002); 7.0–6.0 Ma
Mugharet el Aliya, Morocco, late Pleistocene (Arambourg, 1957)
Mumba Cave, Eyasi, Tanzania, late Pleistocene (Lehmann, 1957)
Mursi Formation, Omo, Ethiopia, middle Pliocene
 (Brown, 1994); ~4.0 Ma

Nachukui Formation, Kenya. See Lothagam.
Namurungule Formation, Kenya, late Miocene (Nakaya, 1994);
 10.0–8.5 Ma
Ngorora Formation, Kenya, middle–late Miocene
 (Thomas, 1981); ~13.0–10.0 Ma
Nkondo Formation, Uganda, early Pliocene
 (Geraads and Thomas, 1994); 5.0–4.0 Ma
Nyakach, Kenya, middle Miocene

Olduvai Gorge, Tanzania, late Pliocene–Pleistocene (Gentry and
 Gentry, 1978); ~2.0–0.5 Ma. Top Bed I = 1.8 Ma, II = 1.15 Ma,
 III = 0.8 Ma, and IV = 0.6 Ma
Oued el Hammam. See Bou Hanifia.

Peninj, Tanzania, early Pleistocene (Gentry and Gentry, 1978)

Quarry 8, Morocco. See Rabat.

Rabat, Morocco, middle Pleistocene (Ennouchi, 1953)
Rusinga Island, Kenya, late Pleistocene or Holocene deposits
 (Pickford and Thomas, 1984)

Sagantole Formation, Middle Awash, Ethiopia,
 latest Miocene Pliocene
Sahabi, Libya, late Miocene (Bernor and Rook, 2008; Boaz, 2008)

Saint Arnaud (now El Eulma), Algeria, late Pliocene. Close to
 Aïn Boucherit and stratigraphically equivalent.
Setif, Algeria, ?late Pliocene
Shungura Formation, Omo, Ethiopia, middle Pliocene–early
 Pleistocene (Gentry, 1985)
Simbiro, Ethiopia, early Pleistocene
 (Geraads et al., 2004b); ~1.3 Ma
Singa, Sudan, late Pleistocene (Bate, 1951)
Sterkfontein, Gauteng Province, South Africa, middle–late
 Pliocene (Clarke, 2006; Partridge et al. 2000, 2003);
 Mb 2 = 4.2–3.3 Ma, Mb 4 (type site) = 2.6–2.2 Ma, and
 Mb 5E = ?1.0 Ma
Swartkrans, Gauteng Province, South Africa, early Pleistocene
 (De Ruiter, 2003); Sk1 = ~1.6 Ma, Sk2–3 = ?1.4 Ma ("broadly
 contemporaneous with Sk1")

Taung, North West Province, South Africa, late Pliocene
 (Day, 1986); 2.2 Ma. Also late Pleistocene (Broom, 1934)
Tighenif (formerly Palikao, then Ternifine), Algeria, early
 Pleistocene (Geraads, 1981); ~0.8 Ma
Tihodaïne, Algeria, middle Pleistocene (Thomas, 1977)
Toros-Menalla, Chad, late Miocene
 (Vignaud et al., 2002); ~7.0 Ma

Usno Formation (= White Sands and Brown Sands), Omo,
 Ethiopia, middle Pliocene (Gentry, 1985)

Vaal River Gravels, Free State Province, South Africa, middle-late
 Pleistocene (Cooke, 1949)

Wadi Derna, Libya, late Pleistocene (Bate, 1955)
Wadi Natrun, Egypt, early Pliocene (Stromer, 1907)
Warwire Formation, Uganda, middle Pliocene
 (Geraads and Thomas, 1994); 4.0–3.0 Ma
Wee-ee, Middle Awash, Ethiopia, middle Pliocene
 (Vrba, 1997); ~3.7–3.6 Ma

Hypsodontine horn cores have rather upright insertions and weak homonymous torsion. Species differ in horn core length, but the degree of curvature is likely to be both individually variable and to be strong in shorter horn cores. The torsion of the horn cores led to much discussion of possible relationship to late Miocene *Oioceros* and to Caprinae, starting with Pilgrim (1934) and Gentry (1970), but all this may now be ignored. The premolar rows may be short and the molars hypsodont especially in larger species. Lower molars have flat lingual walls. Many teeth are found in later wear and with misshapen occlusal surfaces, suggesting a difficult feeding regime. Morales et al. (2003) discussed the implications of the prominent, but presumably primitive, sagittal crest in Hypsodontinae. Several generic names (*Kubanotragus*, *Turcocerus*, and others) are in use, but only *Hypsodontus* is used in this chapter.

Genus *HYPSODONTUS* Sokolov, 1949

Type Species Hypsodontus miocenicus Sokolov, 1949:1103, text figure 3. *Belometschetskaya, Georgia. Middle Miocene (see appendix note 1, at the end of this chapter);

HYPSODONTUS spp.

Localities and Age Gebel Zelten and Beglia Fm. Middle (?and earliest late) Miocene.

Remarks The earliest hypsodontine in Africa may be a Gebel Zelten horn core and unassociated mandible BMNH M26688 and M26685 cited as *Eotragus* sp. and *Gazella* sp. by Hamilton (1973, plate 13, figure1 [right], figures 2–3), but recognized as hypsodontine by Morales et al. (2003). It is a bigger antelope than *Namacerus gariepensis*. The association of *Gazella negevensis* Tchernov et al., 1987, in Israel with this hypsodontine is unlikely. High-crowned teeth in the Beglia Formation (Robinson, 1986) may be of *Hypsodontus* (Solounias, 1990). If they had survived into the levels with hipparionine horses, they would be the last of the Hypsodontinae.

HYPSODONTUS PICKFORDI (Thomas, 1984)

Synonymy Nyanzameryx pickfordi Thomas, 1984, *Palaeontographica A*, 183:73, plate 2, figures 1–2, text figure 5. *Maboko, Nyakach. Middle Miocene.

Remarks An early middle Miocene cranium BMNH M15543 was described as the type species of a new giraffoid climacoceratid *Nyanzameryx* but reidentified as Hypsodontinae by Morales et al. (2003). It has upright horn cores seemingly like the Chios (Greece) *Hypsodontus* cf. *gaopense* of Bonis et al. (1998: figure 1) and unlike later *Hypsodontus tanyceras*.

TABLE 38.3
Bovid species at Gebel Zelten, Maboko, Fort Ternan and Ngorora: the beginning of bovid diversification in Africa

Taxon	Gebel Zelten	Maboko	Fort Ternan	Ngorora Fm.
SUBFAMILY HYPSODONTINAE				
Hypsodontus sp.	X			
Hypsodontus pickfordi (Thomas, 1984)		X		
Hypsodontus tanyceras (Gentry, 1970)			X	
SUBFAMILY BOVINAE, TRIBE BOSELAPHINI				
?Eotragus sp.			X	
Kipsigicerus labidotus (Gentry, 1970)			X	?
Boselaphini, larger sp.				X
?Sivoreas sp.				X
SUBFAMILY ANTILOPINAE				
Homoiodorcas sp.	X	X		
Homoiodorcas tugenium Thomas, 1981				X
SUBFAMILY ANTILOPINAE, TRIBE ANTILOPINI				
Gazella sp. (1)			X	
SUBFAMILY ANTILOPINAE, TRIBE ?REDUNCINI				
*?Reduncini, sp. or spp				X
SUBFAMILY INDET.				
Gentrytragus thomasi Azanza and Morales, 1994			X	
Gentrytragus gentryi (Thomas, 1981)				X

HYPSODONTUS TANYCERAS (Gentry, 1970)

Synonymy Oioceros tanyceras Gentry, 1970:262, plates 5–7, plate 8, figures 3–4, etc. *Fort Ternan. Middle Miocene.

Remarks A male and a female skull and abundant other remains were recognized from the outset as being related to (?)*Oioceros grangeri* and *noverca* (Pilgrim, 1934) of Tungur, Mongolia, but Gentry failed to sever Hypsodontinae from Caprinae or *Oioceros* until after Köhler (1987) had done so. Male *Hypsodontus tanyceras* have uniquely divergent and curved long horn cores but the female is hornless. Premolar rows are unusually short in relation to molar rows, even for a hypsodontine, and p2 is lost in some individuals. The limb bones are cursorially adapted. The species diverges somewhat from Eurasian hypsodontines but does not need a new generic name. Thomas (1984a) discusses sex dimorphism in hypsodontines.

ASSESSMENT OF HYPSODONTINAE

The Paleogene bovid-like hypsodont and nonhypsodont teeth of eastern Asia have been accepted by Dmitrieva (2002), Wang (1992), and others as Bovidae and are often mentioned in papers dealing with Hypsodontinae. The authors have been aware of the difficulties of fitting them in with any likely story of bovid evolution derived from fossils elsewhere in the Old World, but they have remained circumspect on the subject. Janis and Manning (1998) noted a possible connection from the Oligocene teeth of Mongolia to the early Miocene Antilocapridae (Merycodontinae) of North America. Inevitably the idea arises that Hypsodontinae could be an extinct Old World family, Hypsodontidae, allied with the Antilocapridae rather than with the Bovidae. However Métais et al. (2003) were still able to doubt that the main Oligocene hypsodontine, *Palaeohypsodontus*, was pecoran at all.

Subfamily BOVINAE Gray, 1821
Tribe BOSELAPHINI Knottnerus-Meyer, 1907

Type Genus Boselaphus Blainville, 1816.
Boselaphini excluding *Eotragus* appeared around the middle of the middle Miocene in the Siwaliks, East Africa, and Georgia, but only near its end in Europe and in the late Miocene in China. The two extant species are the nilgai, *Boselaphus tragocamelus*, and the small four-horned antelope, *Tetracerus quadricornis*, both in the Indian subcontinent. Bovini or bovine-like taxa probably evolved on several occasions from boselaphines. Boselaphines are known in Africa until the start of the Pliocene.

The horn cores have keels, often show heteronymous torsion, and often a sharp diminution in anteroposterior diameter (a demarcation) along their course. They usually show signs of anterior growth or modification at their horn core bases and pedicels (Solounias, 1990, figures 12–15). In *Tetracerus* this extends to having an extra anterior pair of horns. Postcornual fossae are of good size but shallow depth. Strong temporal ridges on the cranial roof brace the back of the horn cores, and a rugose surface often develops between the temporal ridges. Many boselaphine characters remain primitive: the braincase roof as a whole does not become inclined; horn core insertions often remain wide apart; supraorbital pits are small; the preorbital fossa large; the infraorbital foramen positioned low and anteriorly; the median indent at the back of the palate lies behind the lateral ones; tooth enamel is rugose for bovids; the premolar rows are long.

Genus *EOTRAGUS* Pilgrim, 1939:137

Type Species Eotragus clavatus (Gervais, 1850) in Gervais, 1848–52. *Sansan, France. Middle Miocene.

Remarks Eotragus, the most familiar and longest-known early bovid, is not accepted by all as a boselaphine. It is well-known in Europe, and there are some likely records in Asia but not Africa. Its horn cores are upright spikes, widely separated above the back of the orbits. The closest plausible record to Africa comes from Negev, Israel (Tchernov et al., 1987).

?EOTRAGUS sp.

Localities and Age Fort Ternan. Middle Miocene.
A single horn core (Gentry, 1970: plate 15, figure 6) is not a satisfactory record.

Genus *KIPSIGICERUS* Thomas, 1984

Type Species Kipsigicerus labidotus (Gentry, 1970).

**KIPSIGICERUS LABIDOTUS* (Gentry, 1970)

Synonymy Protragocerus labidotus Gentry, 1970:247, plates 1, 3–4, etc. *Fort Ternan. Middle Miocene.
? (*Protragocerus labidotus* Thomas, 1981:343, text figures 4–5, plates 1, 2, figures 2–5. Ngorora Fm, Mbs A–D. Middle and early late Miocene).

Remarks This is the only species of *Kipsigicerus*. It is known from a male and a female skull and abundant other remains. The male, but less clearly the female, has a low and wide cranium. The horn cores are strongly compressed mediolaterally. The steady diminution in their degree of divergence from base to tip resembles the little-known European *Protragocerus chantrei*. Pronounced forward growth of the lower parts of the horn cores and pedicels is linked with the prominent demarcation shortly above the base of the horn core. These are distinctive features for a species well back in the middle Miocene. The length of the premolar row relative to the molar row (59%, lowers) is less than in *Eotragus* and similar to the nonboselaphine *Tethytragus* in Europe, so it could be significant that no *Tethytragus*-like antelopes are known from Fort Ternan. *Kipsigicerus* has minute upper canines, nowadays a rare occurrence in individual bovids (Dekeyser and Derivot, 1956). It is well distinct ecologically from *Hypsodontus tanyceras* at Fort Ternan. Thomas (1983: plate 2,

figures 1–5) recorded probable boselaphine teeth from Al Jadidah, Saudi Arabia, close in age to Fort Ternan. The supposed *K. labidotus* in the Ngorora Formation (Gentry, 1978; Thomas, 1981) shows constant divergence and less or no anterior extension of the horn core base, and it is not unlike *Sivoreas* or some *Helicoportax* in the Siwaliks.

Genus *SIVOREAS* Pilgrim, 1939

Type Species Sivoreas eremita Pilgrim, 1939:131, plate 4, figures 1, 1a. *Near Chinji, Siwaliks. Middle Miocene.
The horn cores have front and back keels and tight torsion of their nonspiraled axes. Pilgrim placed the type species in the Tragelaphini, but it looks more like a boselaphine.

?SIVOREAS sp.

Synonymy Sivoreas eremita Thomas, 1981:359, text figures 6–8, plate 3 figures 6–7. Ngorora Fm, Mbs A–D.

Remarks A few horn core pieces are less narrow and with a tighter torsion than the questionable Ngorora *Kipsigicerus labidotus*. It is still unknown whether they belong to a larger and mostly later (Thomas, 1981: table 25) boselaphine represented by a female hornless cranium BN 1078 and other specimens. (The mandible BN 1235 [Thomas, 1981, text figure 21] is of a second large Ngorora species later in the sequence, but probably not a boselaphine.) Material in the late Miocene Namurungule Formation formerly attributed to the southeast European *Palaeoreas* or *Ouzocerus* (Nakaya et al., 1984; Nakaya 1994) may be the species later thought by Tsujikawa (2005) to be more similar to *Sivoreas*. Thus, the case for *Sivoreas* in Africa is reasonable but as yet undecided. If present, its dates would come near or after the end of its likely Siwaliks span of 14.0–11.9 Ma.

Genus *MIOTRAGOCERUS* Stromer, 1928

Type Species Miotragocerus monacensis Stromer 1928:3b, figures 1a, 1b. *Oberföhring, Munich, Germany. Late Miocene (Vallesian).

This widespread Eurasian late Miocene genus entered Africa at least once during that epoch. Herein *Miotragocerus* is used to include most species of *Tragoportax* Pilgrim, 1937. Both names have been widely used in recent decades in place of the well-understood *Tragocerus* Gaudry, 1861, discovered to be a junior homonym of a beetle (Kretzoi, 1968). Other generic names have also been used, and species often transferred from one to another. The nomenclature is a complicated and much discussed subject (Kostopoulos, 2005, and earlier references). Horn cores have an anterior keel, are mediolaterally compressed, and often have demarcations or steps on the anterior edge. Horn insertions become close, frontals are raised between the horn bases, and a rugose surface is developed behind the horn bases. Within the Boselaphini, *Miotragocerus* was not close to *Boselaphus* or Bovini.

MIOTRAGOCERUS CYRENAICUS Thomas, 1979

Synonymy Miotragocerus cyrenaicus Thomas, 1979:268, plate 1, figure 5. *Sahabi. Late Miocene.

Remarks This is a large and late *Miotragocerus* species with diverging and backwardly curved horn cores. New skull remains from Member U-1 of the Sahabi Formation were more like the Pikermi *M. amalthea* than was the holotype thought to be from higher in the succession (Gentry, 2008; Boaz, 2008).

MIOTRAGOCERUS sp. aff. *M. CYRENAICUS*

Synonymy *Tragoportax* aff. *T. cyrenaicus* (Thomas, 1979); Harris, 2003:537, figures 11.5–11.6. Lothagam: Nawata Fm, Nachukui Fm, Apak Mb. Late Miocene–early Pliocene.

Remarks This boselaphine has less divergent horn cores than in *Miotragocerus cyrenaicus*. A similar form in the Adu-Asa Formation has a stronger anterior keel and more compressed horn cores (Haile-Selassie et al., 2004; more details in Haile-Selassie et al., 2009). Harris recorded and illustrated two smaller boselaphine species in the Nawata Formation, and Tsujikawa (2005: figure 11A) recorded several species in the Lower Member of the Namurungule Formation. It is possible that some late Miocene *Miotragocerus* have horned females (with smaller horn cores than the males), but it is equally possible that at least two species are present at Lothagam.

MIOTRAGOCERUS sp.

Synonymy Boselaphini gen. et sp. indet. Thomas and Petter, 1986:365, figure 4. Menacer. Late Miocene.

Remarks The lower tooth row in question does indeed appear to be boselaphine and to date from a period before the end of the late Miocene.

MIOTRAGOCERUS ACRAE (Gentry, 1974)

Synonymy *Mesembriportax acrae* Gentry, 1974:146, figures 1–23. *Langebaanweg. Early Pliocene.

Remarks This species has divergent and short horn cores, not curving backward. The frontals are much raised above the level of the dorsal orbital rims and have extensive internal sinuses. Skull, dental and postcranial material was described by Gentry (1974, 1980).

ASSESSMENT OF BOSELAPHINI

The Boselaphini of today are relicts, but they were prominent in Siwalik middle and late Miocene faunas. They were also present in Africa well before the end of the middle Miocene, but as yet we have no evidence of a dominant role for them before the latest Miocene. The existing allocation of species names among late Miocene *Miotragocerus* from various localities may yet have to be changed. The tribes of modern African antelopes were present by the end of the Miocene, and the African Boselaphini did not survive the end of that epoch or only narrowly managed to do so.

Tribe TRAGELAPHINI Blyth, 1863
Figure 38.5

Type Genus *Tragelaphus* Blainville, 1816.
Present-day Tragelaphini embrace *Tragelaphus* with seven species and *Taurotragus* with one or two. They are entirely African except for possible records of lesser kudu, *Tragelaphus imberbis*, in Arabia (Büttiker, 1982). *Tragelaphus* is known back to various late Miocene sites perhaps as early as ~6.5 Ma. The horn cores have keels and heteronymous torsion. Females are horned only in *Taurotragus* and *Tragelaphus eurycerus*. Postcornual fossae are weak or absent. Frontals are without internal sinuses and little elevated between the horn insertions. The front of the orbit lacks a rim along the lachrymal edge and usually has two lachrymal foramina, the latter being like Bovini. Nasals are long and narrow, ethmoidal fissures large, and preorbital fossae lacking. The infraorbital foramen is low and anterior. Cranial roofs are little inclined as in Boselaphini. The basioccipital is long with large anterior tuberosities passing in front of the foramina ovalia. Molar teeth are caprine-like in their simple occlusal pattern, but not hypsodont. Basal pillars are absent on upper molars and small or absent on lower molars, upper molars with only weak ribs between the styles, lower molars without goat folds, premolar rows quite long with large P2s and p2s, and p4s with paraconid-metaconid fusion to form a closed lingual wall. Figure 38.5 shows the temporal distribution of African fossil species.

Genus *TRAGELAPHUS* Blainville, 1816

Type Species *Tragelaphus scriptus* (Pallas, 1766).
Living *Tragelaphus* are tightly knit morphologically and live in habitats with bushes or trees.

TRAGELAPHUS sp. or spp.

Localities and Age Adu-Asa Fm, Lukeino, Langebaanweg. Late Miocene–early Pliocene.

Remarks Early *Tragelaphus* horn cores (Thomas, 1980; Gentry, 1980) are lyrate or only loosely spiraled and usually with a posterolateral keel stronger than the anterior one. They resemble modern nyala or sitatunga, but the fossils are smaller, less anteroposteriorly compressed, and less postorbitally inserted. Two species of moderate or larger size are present in the Adu-Asa Formation, and more details are available in Haile-Selassie et al. (2009). Possible tragelaphine teeth occur at Mpesida (Thomas, 1980), which would be a very early record.

TRAGELAPHUS KYALOAE Harris, 1991

Synonymy *Tragelaphus kyaloae* Harris, 1991:145, figures 5.7, 5.8. Koobi Fora Fm. *Lower Lokochot Mb., Lonyumun and Moiti Mbs; Kanapoi; Lothagam, Upper Nawata, Apak and Kaiyumung Mbs. Late Miocene–middle Pliocene.

Remarks This species was named on a frontlet with cranial pieces. It has the usual lyrate horn cores with a strong posterolateral keel and a weaker anterior one. It was a common species of the East African early to mid-Pliocene, as, for instance, at Kanapoi. It differs from the Lukeino and Langebaanweg *Tragelaphus* by horn cores more compressed anteroposteriorly and with an even weaker anterior keel arising from a more anterolateral basal insertion. The tips may be closer. The Mursi Formation "*Tragelaphus* aff. *gaudryi*" of Gentry (1985) may be *T. kyaloae*.

TRAGELAPHUS NKONDOENSIS Geraads and Thomas, 1994

Synonymy *Tragelaphus nkondoensis* Geraads and Thomas, 1994:387, plate 1, figures 5–6. Nkondo and *Warwire Fms. Early–middle Pliocene.

Remarks This small *Tragelaphus* has short, straight, and well-inclined horn cores showing a degree of torsion of the axis but no lyration or spiraling. They are about the size of extant bushbuck, *T. scriptus*, but less compressed anteroposteriorly. The posterolateral keel is stronger than the anterior one.

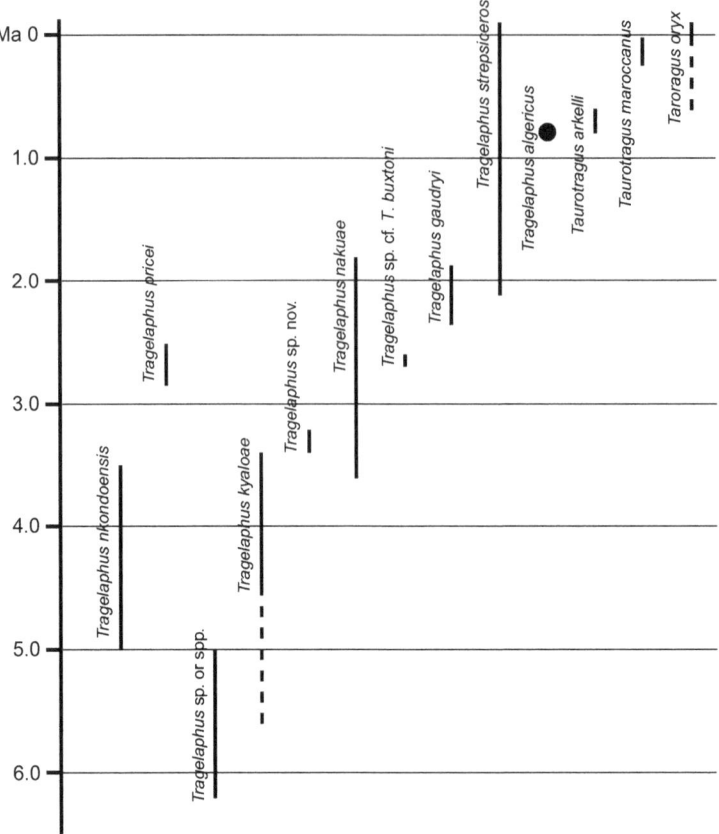

FIGURE 38.5 Temporal distribution of African fossil Tragelaphini. Vertical scale in millions of years.

TRAGELAPHUS sp. nov.

Synonymy Tragelaphus sp. nov. Gentry, 1981:6, plate 1. Hadar Fm, Mb SH. Middle Pliocene.

Remarks The horn cores have more mediolateral compression than *T. kyaloae* or the earlier and smaller Langebaanweg and Lukeino species. The horn cores also have a more definite curve outward (initially increasing divergence) from the base than at Langebaanweg or Lukeino. Both keels have disappeared proximally (as in modern *T. imberbis*) but survive distally where the posterolateral one is stronger.

TRAGELAPHUS NAKUAE Arambourg, 1941

Synonymy Tragelaphus nakuae Arambourg, 1941, *Bull. Mus. Natn. Hist. Nat.* 2, 13:343, figures 4a, 4b. *Shungura Fm. Shungura Fm Mbs B–H; Koobi Fora Fm Lokochot–KBS Mbs; Nachukui Fm Kataboi-Kalachoro Mbs. Middle–late Pliocene. *Tragelaphus* aff. *nakuae* Gentry, 1981:5. Hadar Fm DD Mb. Middle Pliocene.

Remarks The advanced form of this rather large species existed from around 3.0 Ma onward. By Shungura G, the horn cores are rather short, heavily built, little lyrated, with a strong posterolateral keel and anteroposterior flattening, much like modern *T. eurycerus*. The cranial roof is almost horizontal and strongly raised posteriorly into a transverse ridge astride the top of the occipital; an earlier stage of this character was present in *T. kyaloae* (Harris, 1991). The temporal ridges are strong. Older

forms of the species, as in the Hadar Formation, had longer horn cores in which the divergence first increased from the base then gave way to decreased divergence higher up, thereby giving a lyrated appearance. The cranial roof characters were also less advanced than they later became. Overall the horn cores were larger, longer, and more nearly parallel than in *T. kyaloae*.

TRAGELAPHUS sp. cf. T. BUXTONI. Dietrich, 1942

Synonymy Tragelaphus sp. cf. *buxtoni* Dietrich, 1942:118, figure 154. Laetoli. (?late) Pliocene.

Remarks Dietrich's specimen is a large frontlet. A second cranium from the Ndolanya Beds was described but not illustrated by Gentry (1987a) and has since been lost. Both specimens are larger than extant *T. buxtoni*. Their horn cores shared with *T. kyaloae* a strong posterolateral keel and wide lateral spread, and their divergence definitely increases in the basal sections, but they were larger and perhaps with less anterolateral compression. The second cranium was said to be approaching a more kudu-like appearance and was thought to resemble *T. buxtoni*.

TRAGELAPHUS PRICEI (Wells and Cooke, 1956)

Synonymy Cephalophus pricei Wells and Cooke, 1956:12, figures 5–6. *Makapansgat Limeworks (Mb 3; Vrba, 1995). Late Pliocene. *Tragelaphus ?pricei* Gentry, 1985:131, plate 4, figure 4. Shungura Fm Mb ca. Late Pliocene.

Remarks This species was founded for a holotype mandible, some other dentitions, and a neotragine paratype horn core. The Shungura C pair of horn cores is the size of a small extant bushbuck and not too small to belong to *T. nkondoensis*. They are less compressed anteroposteriorly and with less inclined insertions than in bushbuck. This last character is a surprising contrast with the older *T. nkondoensis*. Geraads and Thomas (1994) point out that conspecificity of Makapansgat Limeworks dentitions with Omo horn cores is unverifiable, but so is a species difference. A bushbuck cranial roof was described from Lothagam (Harris, 2003).

TRAGELAPHUS spp. aff. SPEKEI

Tragelaphus spekei is the living sitatunga, of moderate size. and with horns more lyrate than in bushbuck. *Tragelaphus angasi*, the nyala, is very similar although certainly a distinct species. Tragelaphine horn core pieces of similar species and teeth of an appropriate matching size are fairly often found as fossils and labeled "cf. *spekei*," but little else is known of them. They cannot be assumed to have belonged to swamp-living species like the sitatunga.

TRAGELAPHUS GAUDRYI (P. Thomas, 1884)

Synonymy Palaeoreas gaudryi P. Thomas, 1884:15, plate 1, figure 7. *Aïn Jourdel. Shungura E–G. Late Pliocene.
Remarks This species was founded on a kudu horn core base. Gentry (1985) extended the name to a kudu in the Shungura Formation, no larger than a lesser kudu but with the strong anterior keel of a greater kudu. The near-equal dimensions of the basal horn core diameters are like the Hadar SH new species described earlier, but *Tragelaphus gaudryi* exhibits the spiraled horn cores of a kudu and not just the lyration of earlier species. The cranium of an earlier small kudu with more divergent horn cores was found in Shungura C. North African tragelaphine premolars of late Pliocene age are also present at Ahl al Oughlam, but a kudu frontlet from Mansourah (Bayle, 1854; Gervais, 1867–69: plate 19, figure 4) may be of later age and evolving toward *T. algericus* (Chaîd-Saoudi et al., 2006). The lesser kudu, *T. imberbis*, has no fossil record.

TRAGELAPHUS STREPSICEROS (Pallas, 1766)

Synonymy Antilope strepsiceros Pallas 1766, *Misc. Zool.*:9. Extant greater kudu. Shungura G; Olduvai I–II; Koobi Fora Fm, Upper Burgi Mb upward; Nachukui Fm, Kalachoro Mb upward; Kabwe. Late Pliocene onward. *Strepsiceros maryanus* Leakey, 1965:40, plates 40–42. Olduvai I–lower II. Late Pliocene.
Remarks This species is found at Olduvai and other East African locations from about 2.0 Ma onward. Vrba (1987b) confined the Makapansgat Limeworks record to Member 5. An interesting middle–late Pleistocene smaller but probably conspecific form with tightly spiraled horn cores was found at Melkbos and Elandsfontein (Hendey, 1968; Klein and Cruz-Uribe, 1991), both in the Western Cape Province.

TRAGELAPHUS ALGERICUS Geraads, 1981

Synonymy Tragelaphus algericus Geraads, 1981:51, plate 1, figure 3. *Tighenif. Early–middle Pleistocene.

Remarks The holotype is a horn core from Tighenif. It is the size of a large *T. strepsiceros* and tightly spiraled, characters that would suit an ancestor of *Taurotragus*, especially given the hybridization between *Tragelaphus strepsiceros* and *Taurotragus oryx* (Boulineau, 1933; Van Gelder, 1977).

Genus TAUROTRAGUS Wagner, 1855

Type Species Taurotragus oryx (Pallas, 1766).
The type species is found in eastern and southern Africa and a possibly separate species, *Taurotragus derbianus*, in West Africa. They are the largest known antelopes, with horn cores much twisted along an otherwise straight axis and with a strong anterior keel and sometimes a posterolateral one. The horn cores are inserted a bit more widely apart than in *Tragelaphus*, and supraorbital pits are large. They browse and graze in herds and in more open and dry areas than used by *Tragelaphus*. They jump well despite their great size. Fossils occur only late in tragelaphine history.

TAUROTRAGUS ARKELLI Leakey, 1965

Synonymy Taurotragus arkelli Leakey, 1965: 43, plates 43–44. *Olduvai Bed IV surface. Middle Pleistocene.
Remarks The holotype cranium and only specimen differed from living eland in horn cores more uprightly inserted and the cranium longer and narrower and without a transverse ridge across the back of the cranial roof. These are primitive characters and suitable for a direct ancestor of *Taurotragus oryx*.

TAUROTRAGUS MAROCCANUS Arambourg, 1939

Synonymy Taurotragus maroccanus Arambourg, 1939: 42, plate 9, figure 3. *Casablanca. Late Pleistocene.
Remarks A fine frontlet is very like a modern eland, but also smaller than the earlier *Tragelaphus algericus*. The relatively open spiraling and slight divergence of its horn cores make it more like *Taurotragus derbianus* than *T. oryx* and thereby contrast with Olduvai *T. arkelli*. The more open spiraling of its horn cores in combination with the rather tight spiraling in *Tragelaphus algericus*, both of them north African species, also suggest a close relationship between kudu and eland. Additional dated fossils are needed to see whether and when the eland evolved from kudu, whether it happened only once, and whether the more open twisting of Lord Derby's eland is a character reversal.

*TAUROTRAGUS ORYX (Pallas, 1766)

Synonymy Antilope oryx Pallas, 1766, *Misc. Zool.*:9. Extant eland; *Taurotragus oryx pachyceros* Schwarz, 1937:33. *Olduvai. (?middle) Pleistocene.
Remarks The eland has been found at various late Pleistocene localities and also at some earlier ones but not Makapansgat (Vrba, 1987b). Elandsfontein eland are also advanced over *T. arkelli* but have slightly less shortened braincases and slightly more upright horn cores than in the present day. The *T. oryx* at Olduvai (Gentry et al., 1995: figure 1) is definitely advanced on *T. arkelli* but of unspecified stratigraphic origin.

ASSESSMENT OF TRAGELAPHINI

It seems that early tragelaphines were like modern *Tragelaphus angasi* or *spekei* in their lyrate and keeled horn cores. They were abundant in the late Miocene of the Adu-Asa Formation, where they co-existed with the last African boselaphines but were unknown in Chad (Vignaud et al., 2002; Haile-Selassie et al., 2004). The very rare fossils of the bushbuck line had less lyrate horn cores, probably because their small size allowed little scope for its expression, but torsion of their keels was present. Variation in keels and lyration will continue to aid or confuse the accurate diagnosis of tragelaphine fossil species. *Tragelaphus nakuae* was a specialized late Pliocene line, which went extinct. Kudus appeared in the mid- to late Pliocene with an intensified lyration of their horn cores, which amounted to spiraling and with more emphasis on the anterior keel of the horn cores. *Tragelaphus gaudryi* or a similar-sized species survived the incoming of greater kudu in Shungura G13. Gentry (1985) hypothesized that *T. imberbis* in its restricted eastern and northeastern African range, subsequently developed from *gaudryi*, and at some stage acquired its unusually long metacarpals.

Taurotragus may not have appeared until after the start of the Pleistocene. Its leaping ability might lead one to ponder a relationship with the Caprinae or with the Eurasian Pleistocene *Spirocerus*, the latter having eland-like horn cores but perhaps being an antilopine (Pilgrim, 1939). However, molecular evidence and the interbreeding of eland with the greater kudu militate against this speculation. Extra-African records of fossil Tragelaphini are unknown apart from tribal misalignments of spiral-horned antilopines in the early 20th century, but Kostopoulos (2006) made a new assertion of tragelaphine affinity for the late Miocene Greek *Pheraios*.

Tribe BOVINI Gray, 1821
Figure 38.6

Type Genus Bos Linnaeus, 1758.

Bovini descended from Boselaphini (Rütimeyer, 1877–78; Pilgrim, 1939), and if they did so more than once, the tribe would not be monophyletic. The main extant genera are the Eurasian *Bos* and *Bison*, South Asian *Bubalus*, and African *Syncerus*. They are the largest and heaviest of bovids, with low and wide skulls, horned females, horn cores emerging transversely from their postorbital insertions, internal sinuses in their frontals, a shortened braincase (analyzed in Hooijer, 1958a:44–50, 57–59, and significant in the ontogeny and phylogeny of *Bos* and *Bison* species), and occlusally complicated, hypsodont teeth. Living species of Bovini have internal sinuses extending far up their horn cores. Early Bovini were smaller and had horn cores with an anterior and possibly a posterolateral keel or traces thereof, a roughly triangular and uncompressed cross section, fairly strong divergence, low inclinations in side view, and slight backward curvature. They had a slightly sloping cranial roof, strong temporal ridges, and a rectangular platform at the back of the cranial roof. Teeth of Bovini go back to the late Miocene in Africa (Thomas, 1980; Vignaud et al., 2002 [perhaps as early as 7.0 Ma]; Harris, 2003), and in the Siwaliks (Bibi, 2007). It has yet to be seen whether skulls of such early species might have more of a boselaphine than a bovine appearance. The temporal distribution of African fossil bovines is shown in figure 38.6.

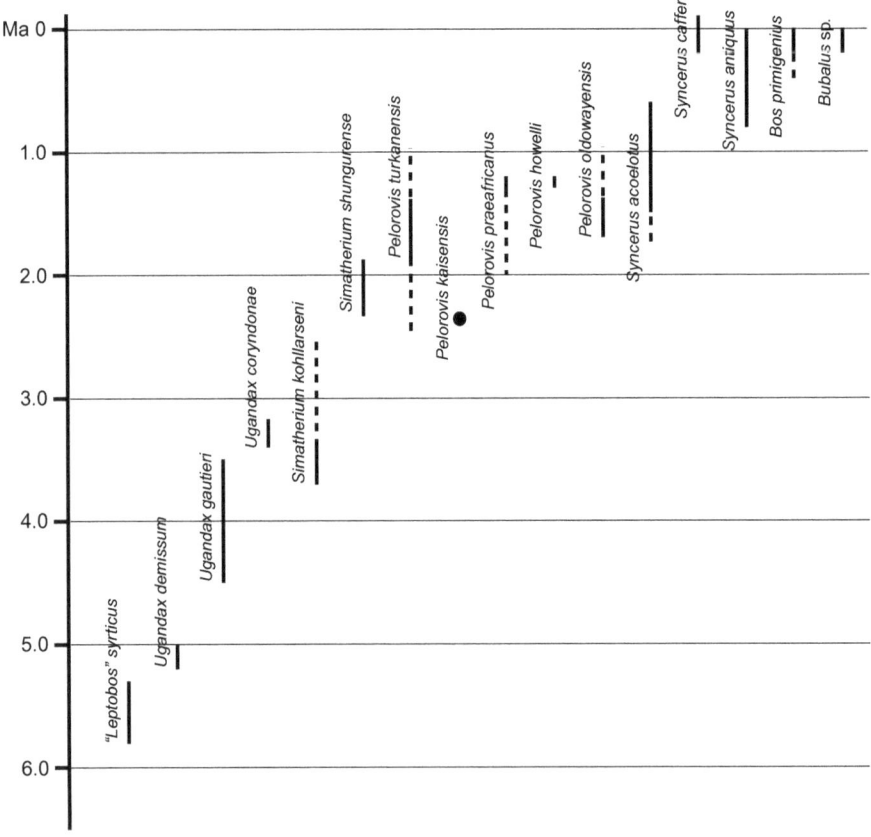

FIGURE 38.6 Temporal distribution of African fossil Bovini.

Gen. indet., *"LEPTOBOS" SYRTICUS* Petrocchi, 1956

Synonymy Leptobos syrticus Petrocchi, 1956, *Boll. Soc. Geol. Ital.* 75, 1:231, figures 1–7. *Sahabi. Late Miocene/?Pliocene.

Remarks This puzzling species, based on three crania, is not a *Leptobos* (Geraads, 1989) and was suspected by Geraads of being from a late stratigraphic level (also see Boaz, 2008). It is strikingly specialized in its large and almost transverse (= divergent and inclined) horn cores. Likely consequences of such a horn core morphology are that the cranial roof became horizontal, the frontals were not elevated between the horn bases, the dorsal orbital rims project because the insertions are so close behind, and the temporal ridges are strong. The closeness between the supraorbital pits is probably primitive.

Genus *UGANDAX* Cooke and Coryndon, 1970

Type Species Ugandax gautieri Cooke and Coryndon, 1970. This Pliocene genus may have appeared in the latest Miocene. Horn cores are short, of an approximately triangular cross section, with an anterior keel or remains of one, some signs of a posterolateral keel, and little compression. The horn cores have limited divergence and are quite strongly inclined backward and sometimes with backward curvature. The cranium is wide. The cranial roof is sloping and rather flat, with strong temporal ridges. Cheek teeth are somewhat hypsodont, basal pillars are present on the molars, upper molars develop large occlusal areas with labial ribs and widened lingual lobes, and central fossettes become more complicated than in boselaphines.

UGANDAX DEMISSUM (Gentry, 1980)

Synonymy Simatherium demissum Gentry, 1980:233, figures 8–13. *Langebaanweg. Early Pliocene; *Ugandax demissum* Gentry, 2006:44.

Remarks This species is primitive in having quite short horn cores with some tendency to a triangular cross-section, a strong anterior keel and some remaining signs of a posterolateral keel. Its cranium is less widened and shortened than in later bovines. It does not look too dissimilar to the European Pliocene *Alephis lyrix* or the Siwaliks *Proamphibos lachrymans*. The cheek teeth are lower crowned and less occlusally complex than in later Bovini. This or a similar species was already present in the latest Miocene of the Adu-Asa Formation (Haile-Selassie et al., 2004).

UGANDAX GAUTIERI Cooke and Coryndon, 1970

Synonymy Ugandax gautieri Cooke and Coryndon, 1970:206, plates 17–18. *Kaiso Fm, Kazinga Channel deposits of unknown age.

Remarks The description is based on a cranium and teeth. It is another primitive bovine, smaller than *Ugandax demissum* and with more irregular keels on its horn cores and less divergence. The braincase is not greatly shortened. The horn cores are still backwardly curved. They pass more upward and backward than outward, and the skull is less wide than in *Syncerus*.

UGANDAX CORYNDONAE Gentry, 2006

Synonymy Ugandax coryndonae Gentry, 2006, *Tr. Roy. Soc. S. Afr.*, 61:46, figures 1–5. Hadar Fm., Mbs SH–*DD. Middle Pliocene.

Remarks This *Ugandax* is more advanced than *U. gautieri* in its horn cores being more divergent, less backwardly curved in

side view, wider apart, and perhaps inserted more posteriorly. The braincase is shorter, lower and wider, the temporal ridges wider apart posteriorly, anterior tuberosities closer on the basioccipital, and the central longitudinal ridge of the basioccipital less pronounced. These characters may be representative of a mid-Pliocene African bovine on or close to the *Syncerus caffer* ancestry. Likely *Ugandax* teeth of an unknown species were reported from the middle–late Pliocene of the Chiwondo Beds (Sandrock et al., 2007), perhaps the same species as the "hippotragine" of Coryndon (1966) in the same beds.

Genus *SIMATHERIUM* Dietrich, 1941

Type Species Simatherium kohllarseni Dietrich, 1941. These are large Pliocene bovines with long horn cores, without obvious keels, of irregular or rounded rather than a neater, more or less triangular cross section, little compressed, divergent, well inclined backward in side view. The insertions are postorbital, the cranial roof slightly sloping, the frontals slightly raised between the horn bases, and the cranium wide and low.

SIMATHERIUM KOHLLARSENI Dietrich, 1941

Synonymy Simatherium kohllarseni Dietrich, 1941, Zntlbl. Miner. Geol. Paläont. B: 221. *Laetoli; also, as *S.* cf. *kohllarseni*, at Koobi Fora Fm. Lokochot and Tulu Bor Mbs; Makapansgat Limeworks Mbs 3 and 4. Middle–late Pliocene.

Remarks The type species is based on the large cranium described by Dietrich (1942: plate 20, figures 161, 163, 165). Another Laetolil Beds cranium attributed by Gentry (1987a: plate 10.3) to the same species has shorter and stockier horn cores. *Simatherium kohllarseni* differs from *Ugandax coryndonae* in longer horn cores with less sign of keels, greater divergence, and wider insertions of its horn cores, higher frontals between the horn insertions, and a wider and lower cranium. The Koobi Fora *Simatherium* cf. *S. kohllarseni* was quite small (Harris, 1991: figure 5.15) and may be the same species as at Kanapoi (Harris et al., 2003: figure 22—as "*S. demissum*"). The Makapansgat occurrence was published before the next species below had been described and may fit it better. Remains of unknown *Simatherium* species have been claimed from localities possibly going back to the early Pliocene—for example, Lothagam (Harris, 2003) and the Nkondo Formation (Geraads and Thomas, 1994).

SIMATHERIUM SHUNGURENSE Geraads, 1995

Synonymy Syncerus ?acoelotus Gentry, 1985: 139; *Simatherium shungurense* Geraads, 1995: 89, plate 1. *Shungura G. Late Pliocene.

Remarks Rediscovery of the horn cores of a fine bovine skull made *Simatherium* a better attribution than the "*Syncerus*" of Gentry. The horn cores were long, directed obliquely backward and slightly upward. They were also of relatively small diameter along their whole course, suggesting a female skull. In this case it would not be surprising if males had the more divergent and curved horn cores known in other Tertiary bovines. A massive Shungura G horn core (Gentry, 1985: plate 3, figure 1) may be a male of the species, its top left corner in the illustration being dorsoposterior in life. See also *Pelorovis kaisensis*.

Genus *PELOROVIS* RECK, 1925
Figure 38.6

Type Species Pelorovis oldowayensis Reck, 1925.

Pelorovis appeared in Africa in the late Pliocene c.2.4 Ma and survived into the early Pleistocene or later. The existence of four species around the time interval 2.0–1.5 Ma (figure 38.6) is unlikely. Later species were larger than *Syncerus*. Its horn cores are without keels, slightly compressed dorsoventrally, and forwardly curved from their close insertions. Such forwardly curving horn cores (concave edge anterior) must have evolved from primitive Bovini in which the horn cores originally had convex anterolateral edges. The change in curvature would have been effected by changes in distal divergence and course. The horn insertions are so far posterior as to overhang the occipital surface, so a temporal fossa exists on either side of the braincase between the orbits and the occipital surface. The face of *Pelorovis* was long. Geraads (1986) recorded a *Pelorovis* cranium with horn cores at Ubeidiya, Israel, and several other claims have been made for the genus outside Africa.

PELOROVIS TURKANENSIS Harris, 1991

Synonymy Pelorovis turkanensis Harris, 1991:149, figures 5.10–5.13. Koobi Fora Fm. *KBS and Okote Mbs; Nachukui Fm Kaitio Mb.; Melka Kunturé, Garba IV, Simbiro. Late Pliocene–middle Pleistocene.

Remarks This species is perhaps a junior synonym of *P. praeafricanus*. It differs from *P. oldowayensis* by its smaller size and narrower skull. Harris noted that it coexisted in its later time span with *P. oldowayensis*. The earliest *Pelorovis* horn cores from Shungura D and E were short and very flattened (and thereby unlike *Bos*), and slightly later Upper Burgi ones were similar. This morph has not been given a separate name. A right horn core, Shungura D L98-1, shows that the horn insertions were already positioned above the occipital. The short-horned *Pelorovis* at Garba IV was named *P. turkanensis brachyceras* (Geraads et al., 2004b: plate 12, figure 1) and apparently coexisted with *P. oldowayensis* at Simbiro (Geraads, 1979: plate 4, figures 1–2).

PELOROVIS KAISENSIS Geraads and Thomas, 1994

Synonymy Pelorovis kaisensis Geraads and Thomas, 1994:392, plate 2, figure 4. *Kaiso Village Beds, Loc. B. ?Late Pliocene.

Remarks If this large, long, straight, or slightly curved horn core should be a male of *Simatherium shungurense* Geraads, 1995, the Omo species would have to be known as *Simatherium kaisensis* (Geraads and Thomas, 1994). The premise of conspecificity may be disputed.

PELOROVIS PRAEAFRICANUS (Arambourg, 1979)

Synonymy Bos praeafricanus Arambourg, 1979:44, plate 35. *Aïn Hanech. Early Pleistocene. *Bos bubaloides* Arambourg, 1979:39, plates 33–34. Aïn Hanech. Early Pleistocene.

Remarks "*Bos*" *bubaloides* and *praeafricanus* are *Pelorovis*, the holotypes are from the same individual animal, and *Bos bubaloides* is a preoccupied name (Geraads, 1981; Geraads and Amani, 1998). The type cranium of *Bos palaethiopicus* Arambourg (1979: plate 31, figure 1; plate 32, figures 1, 1a) from Aïn Boucherit also looks like a *Pelorovis*. The likely age of Aïn Boucherit (up to 2.0 Ma) is too old for a bovine with a temporal fossa like that of *Bos primigenius* to belong to *Bos*. *Pelorovis* was also present at Mansourah and earlier at Ahl al Oughlam (Geraads and Amani, 1998; Chaïd-Saoudi et al., 2006).

PELOROVIS HOWELLI Hadjouis and Sahnouni, 2006

Synonymy Pelorovis howelli Hadjouis and Sahnouni, 2006, Geobios 39:674, figure 2. *Aïn Hanech (El-Kherba locus). Early Pleistocene.

Remarks The holotype cranium and horn cores are bigger than *Pelorovis praeafricanus* and about the size of *P. oldowayensis*. The span of the horn cores is not great, but they are robust basally.

*PELOROVIS OLDOWAYENSIS Reck, 1925

Synonymy Pelorovis oldowayensis Reck, 1925:451, figured. *Olduvai. Koobi Fora Fm: Upper Burgi, KBS and Okote Mbs. Melka Kunturé, Simbiro. Late Pliocene–early (?and middle) Pleistocene.

Remarks This, the type species, was first thought to be a giant sheep (Reck, 1925, 1928) and was probably the final species of *Pelorovis* to evolve. It is best known from Olduvai Bed II where a herd was discovered in 1952, but Dietrich (1933) gave Bed IV for the holotype. It was larger than *P. turkanensis*, and its horn cores had become longer. The teeth have less complex occlusal patterns than later African bovines. The p4 has close contact or fusion between paraconid and metaconid to form a closed wall on the lingual side, indicating a relationship or parallelism with *Syncerus*. The muzzle looks wider than in *P. turkanensis* as befits a specialized grazer. A very long-horned skull (Gentry, 1967: plate 5, figure 3) looked as if the individual would be in so much difficulty drinking or feeding at ground level that the species as a whole would be facing extinction or rapid evolution of a new horn shape. Gentry and Gentry (1978) opted for the latter and placed the African late Pleistocene "*Bubalus*" *antiquus* in *Pelorovis*. Many have disagreed (including an editor of this book), and *antiquus* is here listed under *Syncerus*.

Genus SYNCERUS Hodgson, 1847

Type Species Syncerus caffer (Sparrman, 1779).
This is a genus of African Bovini with wide skulls and short faces. The horn cores are of variable length and often short, dorsoventrally compressed, the supraorbital pits fairly close together, and the occipital surface low and wide. The extant eastern and South African *Syncerus caffer caffer* has short horn cores emerging transversely and then immediately downward from just behind the orbits. It also has massive basal bosses succeeded by reduced and upturned tips. African forest buffaloes, *S. caffer nanus*, are smaller and more primitive, and *S. c. brachyceras* is intermediate (Grubb, 1972). There is a problem with how antique such habitat-responsive differences might be. Have contemporaneous small and large morphs remained conspecific as *Ugandax* evolved into *Syncerus*? On this view. *Ugandax gautieri* and *coryndonae* might be conspecific but living in different habitats. If one denied an ancestral role for *Ugandax* and accepted on molecular (DNA) evidence that present-day *S. nanus* is a separate species from *S. caffer*, a problem remains with smaller and larger fossils at successive time periods. Do the fossils belong to long-lasting "small to small" and "large to large" lineages, or have small and large species repeatedly reevolved from a small (or large) immediately preceding ancestor?

SYNCERUS ACOELOTUS Gentry and Gentry, 1978

Synonymy Syncerus acoelotus Gentry and Gentry, 1978:313, plates 2–4 (figure 2), text figure 9. Olduvai, *upper Bed II, also middle Bed II and Beds III or IV. Late Pliocene–early (?and middle) Pleistocene.

Remarks Horn cores emerge transversely (more so than in *S. caffer nanus*) and then curve backward in a horizontal plane. They show anterior, upper, and lower surfaces with almost a keeled edge between the upper and anterior ones. Internal sinuses in the frontals pass only into the basal parts of the horn cores. The species exists possibly in Shungura Members B–G, with a smaller subspecies in Member C (Gentry, 1985). Bovine teeth of appropriate size in the Sterkfontein Valley cave localities belong to "one or more *Syncerus* species, probably evolving into *S. caffer*" (Vrba, 1976).

SYNCERUS CAFFER (Sparrman, 1779)

Synonymy Bos caffer Sparrman, 1779, *K. Svenska Vet.-Akad. Handl. Stockholm,* 40:79. Extant buffalo; *Bathyleptodon aberrans* Lönnberg, 1937:16, figures 6–10. Near Naivasha, Kenya. Late Pleistocene; *Homoioceras singae* Bate, 1949:397. Singa, Sudan. Late Pleistocene.

Remarks Numbers of taxonomically defective late Pleistocene fossils have been attributed to the extant buffalo, but it has never been validated north of the Sahara. Bate (1949; 1951, figures 4, 6, 8a) founded *Homoioceras singae* for a late Pleistocene skull from Singa which had been a door-stop in a government office for over 10 years. She intended *Homoioceras* to be the generic name for the late Pleistocene African long-horned buffalo, but J. W. Simons found that the holotype is probably a large race of *Syncerus caffer* (see Gentry and Gentry, 1978). Short metapodials give this species a more stocky appearance than a *Bos* or *Bubalus.*

SYNCERUS ANTIQUUS (Duvernoy, 1851)

Synonymy Bubalis antiquus Duvernoy, 1851: 597, no figure. *Setif. ?Late Pleistocene; *Bubalus bainii* Seeley, 1891: 201, figured. Modder River, Free State Province, South Africa. ?Late Pleistocene; *Bubalus nilssoni* Lönnberg, 1933: 28, plates 1–3. Near Naivasha, Kenya. ?Late Pleistocene; *Pelorovis antiquus* Gentry and Gentry, 1978:312, plate 4 figure 1. Olduvai, Bed IV. Middle Pleistocene; *Syncerus caffer antiquus* Gautier and Muzzolini, 1991:62. Late Pleistocene–Holocene.
NON: *Homoioceras singae* Bate, 1949:397.

Remarks This is a large mainly late Pleistocene bovine with long horn cores emerging transversely and slightly forward (concave edge to the rear unlike *Pelorovis*) from behind the orbits, insertions wide apart and without the large basal bosses seen in *Syncerus caffer caffer* today. It is portrayed from life in north African rock art with horn sheaths so curved that the tips can come close together. It survived until after 18,000 BP in East Africa, until 12,000–10,000 BP in South Africa, and even after 4,000 BP in north Africa (Klein, 1980; Gautier and Muzzolini, 1991; Marean, 1992).

The holotype cranium from Algeria was the first fossil bovid described from Africa. Subsequent finds have come from many other sites, for example Elandsfontein (Klein and Cruz-Uribe, 1991). Bate (1949, 1951) recognized that these buffaloes were not connected with the Asiatic *Bubalus,* as seen by their shorter faces, irregular or absent keels on the horn cores, and a tendency to paraconid-metaconid fusion on p4s. Gentry and Gentry (1978) used *Pelorovis* for *antiquus,* but fewer people have accepted than rejected this (Gautier and Muzzolini, 1991; Peters et al., 1994). Klein (1994)

emphasized that *antiquus* was a separate species from *S. caffer* irrespective of its generic affiliation, and that identifiable pieces of both species coexisted at Klasies River Mouth. He also found long-horned buffalo to be in drier more grassy settings and *Syncerus caffer* to be in bushier, better-watered ones. Possibly *S. caffer* moved into savannah after the extinction of *antiquus,* but it did not get as far as North Africa. Hadjouis (2002) founded *S. a. complexus* for a smaller and earlier form of the species in north Africa.

Genus *BOS* Linnaeus, 1758

Type Species Bos taurus Linaeus, 1758. Founded on domestic livestock and not applicable to wild species of *Bos* (ICZN Opinion 2027, *Bulletin of Zoological Nomenclature* 60: 81–84, 2003). *Bos primigenius* Bojanus, 1827 is correct for the wild ox, contrary to Wilson and Reeder (2005).

Bos is a Palaearctic and SE Asian genus with a good number of extant and fossil species, perhaps including those often assigned to *Bibos,* and it is also very close to *Bison.* The European *Bos primigenius* was a large and imposing species (Zeuner, 1963, figure 8:5) with moderately long, curving horn cores set widely apart and far back on the skull, and a long face.

BOS PRIMIGENIUS Bojanus, 1827

Synonymy Bos primigenius Bojanus, 1827, Nov. Acta Phys-Med. Acad. Caes. Leop. Car. Nat. Cur., 13:477, plate 24. Extant until the historical period; *Bubalus vignardi* Gaillard, 1934:37, plate 5, figures 1–2. Kom Ombo. Late Pleistocene; *Bos primigenius* Churcher, 1972:62, figures 21–27. Kom Ombo. Late Pleistocene.

Remarks The wild ox was present in north Africa (Pomel, 1894a) where it coexisted with *Syncerus antiquus.* Horn cores in north Africa are more dorsoventrally compressed than in Europe: examples in Gaillard (1934) and Churcher (1972) have compressions of about 75% or less, whereas a Danish sample (Degerbøl and Fredskild, 1970) gave means of 85% (male) and 82% (female). Geraads et al. (2004a: figure 10:1) illustrate a middle Pleistocene *Bos* in Ethiopia at about 0.8–0.6 Ma, based on a very large occipital top and forehead with bases of horn cores. They thought it was more like the South Asian *Bos namadicus* than like *B. primigenius.* At the end of the Pleistocene, *Bos* was still found as far south as the Atbara River at around 15°N in Sudan (Marks et al., 1987).

Genus *BUBALUS* H. Smith, 1827

Type Species Bubalus bubalis (Linnaeus, 1758). Founded as a species of *Bos* on domestic livestock. *Bubalus arnee* (Kerr, 1792) is to be used for the wild Asiatic water buffalo (ICZN Opinion 2027, *Bulletin of Zoological Nomenclature* 60: 81–84, 2003), contrary to Wilson and Reeder (2005).

Bubalus is an Asiatic genus with fossil relatives in the Siwaliks deposits of India. Several occurrences are known in the European Pleistocene (Koenigswald, 1986). The extant species *B. arnee* has long, wide-sweeping, dorsoventrally compressed horn cores.

BUBALUS sp.

Synonymy Buffelus palaeindicus Falconer; Solignac, 1924:176, plate 6, figure 1. Bizerte. Late Pleistocene.

Remarks Geraads (1992) noted that a bovine skull from Tunisia (Solignac, 1924), referred by Bate (1951) to *Homoioceras*, was probably truly a *Bubalus* as first thought by Solignac. He commented that the species could have come from either Europe or Asia and could account for a herd living near Tunis in the recent past and thought to have been introduced by the Carthaginians or Romans. European Pleistocene *Bubalus* have been put into different species from *Bubalus arnee*, so information on the DNA of the Tunisian herd might be interesting.

ASSESSMENT OF AFRICAN FOSSIL BOVINI

Many details of African bovine evolution remain misty. The Sahabi *"Leptobos" syrticus* is an early species in the Palaearctic fringe of Africa, about which nothing more can be said. The Langebaanweg *Ugandax demissum* is readily acceptable as a primitive bovine, but its relationship to bovines outside Africa is not known; Hernández Fernández and Vrba (2005) postulate a closer link of *Syncerus* with *Bubalus* than with *Bos,* so perhaps *U. demissum* is closest to the middle Pliocene Siwaliks *Proamphibos*. The later *U. coryndonae* could well be in a group evolving toward *Syncerus*, with *U. gautieri* as a variant or separate species in a more closed environment.

Simatherium is a difficult genus to deal with. *Simatherium kohllarseni* at Laetoli differs from *Ugandax*, but this is less certain for other *Simatherium* species and occurrences. Parts of large bovine horn cores of unknown orientation or original length often turn up in the second half of the Pliocene. We may note that bovid horns grow out from the skull and are not constrained in their variation by adjacent bones and organs. The only living wild bovines still found in large herds are the short-horned *Bison bison* and *Syncerus caffer*, neither of which reveals much about the scale of variation in a long-horned species.

The origin of *Pelorovis* in the late Pliocene is obscure, particularly whether it evolved in Africa, as the p4 character suggests, or immigrated from outside the continent. The relationship with *Bos* asserted on a cladogram of Geraads (1992) might seem plausible from resemblances like the curved horn cores with the concave edge anterior and insertions far back close to the plane of the occipital. The narrower skull is different from *Bos* but could be linked with close horn insertions, particularly since domesticated Ankole cattle of Uganda can likewise have close insertions and be narrow postorbitally. Martinez-Navarro et al. (2007) went so far as to postulate a direct transition from late *Pelorovis* to *Bos primigenius*. Their proposal distances *B. primigenius*, first known in the European earlier middle Pleistocene, from many earlier putative Eurasian relatives, such as *Bos gaurus, B. mutus, Bison palaeosinensis* "crâne A" (Teilhard de Chardin and Piveteau, 1930, plate 15), *Adjiderebos cantabilis* (Dubrovo and Burchak-Abramovich, 1986, plates 1,2), and a possible *Bison (Eobison)* recorded as *"Hemibos"* by Pilgrim (1941). None of these taxa are known to have such long faces as *Bos primigenius* or *Pelorovis*, and the horn insertions of many of them are certainly less posterior. Moreover, there are difficulties with *Pelorovis* as an ancestor. Its horn insertions may be too advanced in the extent to which they overhang the occipital, and the long face, so like *B. primigenius*, was present by ~1.6 Ma (Harris, 1991: figure 5.10B).

In cases where long-horned examples of *Pelorovis* appear to be present outside Africa (e.g., Thomas et al., 1998) or long-horned *Bos* in Africa, the Siwaliks *"Bos" acutifrons* (Lydekker, 1877, 1878; Pilgrim, 1939) should be taken into account. This long-horned species could be from an earlier time span than *Bos namadicus* or *B. primigenius* even if it postdates the European Villafranchian (Azzaroli and Napoleone, 1982:759). By the late Pleistocene, *Pelorovis oldowayensis* had been displaced by another long-horned bovine for which most people prefer the name *Syncerus antiquus*. Presumably after *Syncerus* had evolved transversely divergent horns seen in *S. acoelotus* and extant *S. caffer*, the way became open for it to give rise to a long-horned species with horn cores having a posteriorly concave edge of curvature unlike *Pelorovis*. Large bovine teeth of the (?early and) middle Pleistocene (e.g., De Ruiter, 2003) will be difficult to identify until the time span and relationships of the later long-horned buffaloes have been established.

?Subfamily BOVINAE or ANTILOPINAE
?Aff. Tribe BOVINI or CEPHALOPHINI
Genus *BRABOVUS* Gentry, 1987

Type Species Brabovus nanincisus Gentry, 1987.

***BRABOVUS NANINCISUS* Gentry, 1987**

Synonymy Brabovus nanincisus Gentry, 1987:382, plates 10.4–10.5, figure 10.1. *Laetolil Beds. Middle Pliocene.
Remarks Brabovus has only one species. The holotype is a medium-sized skull with short, little-compressed and little-divergent horn cores, internal sinuses in the frontals and horn pedicels, a little-inclined and not very long braincase roof, an extensive but shallow preorbital fossa without a clear dorsal border, low-crowned cheek teeth, long premolar rows, and small central incisors. I classified it as doubtfully Bovini, while Vrba (1987a) took it as a primitive hippotragine and later (Vrba and Gatesy, 1994) as not hippotragine. The small first incisors (Gentry, 1987a: figure 10.1) are a striking character, especially for an African bovid. Today the first incisors are small in *Bison, Pantholops*, and all Caprinae; intermediate in Boselaphini, Cephalophini, some Neotragini, *Saiga, Pelea*, and Hippotragini; and large in Tragelaphini, some Neotragini, Antilopini, Reduncini, and Alcelaphini. It looks as though primitively small central incisors could have enlarged to a variable extent in most bovids except Caprinae, but that this had not happened in *Brabovus*. Alternatively, early Pecora already had somewhat enlarged central incisors, and all subsequent enlargements and reductions are advanced.

Brabovus resembles the similarly large cephalophine *Cephalophus silvicultor* in several horn core and skull characters, perhaps because it lived in similar environments. Some modern duiker specializations are absent—for example, very short horn cores, strong inflation of the auditory bulla, and strong lingual outbowings on the lower molar walls. The frontal sinuses of *Brabovus* might be thought to rule it out as a cephalophine, yet sinuses do not prevent *Menelikia* being in the Reduncini or *Antidorcas* being in the Antilopini. Comparisons involving *Cephalophus* are difficult because it may be secondarily forest-dwelling and therefore the primitive-advanced polarities are uncertain. Within these limitations the best course is to classify *Brabovus nanincisus* as standing between Cephalophini and Bovinae.

Tribe CEPHALOPHINI Gray, 1871

Type Genus Cephalophus C. H. Smith, 1827

Cephalophini, the duikers, are small to medium-sized stocky antelopes feeding by frugivory and selective browsing. Females are hornless and slightly bigger than males. *Cephalophus* has many forest-living species across sub-Saharan Africa. Horn cores are short, not compressed, parallel, inclined backward, and inserted far postorbitally. The frontals are shallowly domed longitudinally in front of horn bases, and the supraorbital pits are in a longitudinally extended line. Cheek teeth are brachyodont, with basal pillars on the upper and lower molars, and with rounded lobes and weak or absent styles and stylids. The m3 has a small rear (third) lobe. Premolar rows are long and anterior premolars are relatively large. The first incisors are not much larger than the other incisors and canine. The less specialized *Sylvicapra* has only one species, and this has less shortened and more upright horn cores, less distinctive teeth, and longer legs and lives in areas with cover outside forest limits. Cephalophini are rarely found in fossil localities and several erroneous identifications have been made. Supposed *Cephalophus* at Menacer and in the Ngorora Formation (Arambourg, 1959: plate 17, figure 8; Gentry, 1978) were allocated to Neotragini (Thomas, 1981; Thomas and Petter, 1986). The Aïn Boucherit *C. leporinus* of Arambourg (1979) will be referred here to *Parantidorcas latifrons*.

Genus *CEPHALOPHUS* C. H. Smith, 1827

Type Species Cephalophus silvicultor (Afzelius, 1815). Often spelled *sylvicultor* by later writers.

CEPHALOPHUS sp. or spp.

Synonymy Cephalophus sp. Thomas, 1980:89, figure 1(2). Lukeino. Late Miocene. Cephalophini sp. indet. Gentry, 1987:386, plate 10.6. Laetolil Beds. Middle Pliocene. *Cephalophus* sp. Harris, 1991:228. Koobi Fora Fm, lower Burgi and KBS Mbs. Late Pliocene.
Remarks Cephalophus is uncommon in the Laetolil Beds but better represented than anywhere else. The teeth are intermediate in size between modern *C. spadix* and *C. silvicultor*, smaller than in *Brabovus*, and there is only one tooth of a smaller species. The KBS horn core is the size of a *C. silvicultor* or bigger, and the likely cephalophine upper molar from Lukeino would be from a smaller species unless it were an M1.

CEPHALOPHUS Small spp.

Synonymy Cf. *Cephalophus (Guevi) caerulus* Wells and Cooke, 1956: 15. Makapansgat Limeworks. Late Pliocene. *Cephalophus parvus* Broom, 1934: 477, figure7. *Taung. Late Pleistocene.
Remarks Rare dental fragments of a small *Cephalophus* in Makapansgat Limeworks Member 3 (Vrba, 1987b, 1995) are the size of the living blue duiker, *C. monticola*. The Taung record was thought to be of *C. monticola*, formerly called *C. caeruleus* (Wells, 1967).

Genus *SYLVICAPRA* Ogilby, 1837

Type Species Sylvicapra grimmia (Linnaeus, 1758).

SYLVICAPRA GRIMMIA (Linnaeus, 1758)

Synonymy Capra grimmia Linnaeus, 1758, *Syst. Nat.* 10th ed., 1:70. Extant; *Sylvicapra grimmia* Plug and Keyser, 1994a:141. Haasgat Cave. Early or Middle Pleistocene.

Remarks Wells (1967) wrote of provisional South African records at Cornelia, Vaal River deposits and Florisbad, but the Vaal River *S. grimmia* (Cooke, 1949) was later accepted as an *Antidorcas bondi* (Vrba, 1973). Cooke (1974) again referred to the Cornelia attribution, but Brink (1987) could not confirm the citation for Florisbad. Plug and Keyser (1994a) claimed the Haasgat Cave record as the oldest in South Africa. Late Pleistocene occurrences are known (Plug and Badenhorst, 2001).

ASSESSMENT OF CEPHALOPHINI

The possible or even probable cephalophine at Lukeino shows that this tribe could have emerged as long ago as the late Miocene. The only informative fossils are from Laetoli where they belong to quite a large *Cephalophus* and obviously have interesting ecological implications. Many zoologists have thought that Cephalophini may not be primitively forest dwelling (which would have implied an independence from other bovids since before *Eotragus*), and the debate continues (Kingdon, 1982; Heckner-Bisping, 2001). Any ancestor resembling the less specialized *Sylvicapra* could have arisen from a boselaphine like *Tetracerus* or from a neotragine.

Subfamily ANTILOPINAE Gray, 1821
Figure 38.7

Type Genus Antilope Pallas, 1766.
Extant Antilopinae are divided into the two tribes Neotragini and Antilopini. The former is a probably paraphyletic tribe of small African antelopes, from among whose early relatives sprang the Antilopini (Gentry, 1992). Neotragine horn cores take the form of small spikes, of small cross-sectional area and varying length. The temporal distribution of African fossil species (other than of *Gazella*) is shown in figure 38.7.

Genus *HOMOIODORCAS* Thomas, 1981

Type Species Homoiodorcas tugenium Thomas, 1981.

HOMOIODORCAS TUGENIUM Thomas, 1981

Synonymy Homoiodorcas tugenium Thomas, 1981:364, plate 4, text figures 10–13. *Ngorora Fm, Mb C. Also Mbs A–B and D, commonest in B–C. Middle (?and late) Miocene.
Remarks This was a smaller species than fossil or extant gazelles and was assigned to the Neotragini by Thomas. However, its horn core index (mean = 20.5 × 16.9 for eight specimens) is large, and it also shows slight backward curvature. Thomas (1981: plate 4, figures 5a–5b) shows a hornless female. One must suspect that it could be related to the ancestry of Antilopini rather than being closer to Neotragini than to any other tribe. Thomas noted that the major cross-sectional axis at the base is oblique, rather as if the lowest part of the anterolateral surface had been pushed to give a slight inward tilt to the horn core. A similar peculiarity exists in extant *Madoqua* and in the Langebaanweg *Gazella*.

HOMOIODORCAS spp.
Figure 38.8

Localities and Age Gebel Zelten, Maboko. Middle Miocene.

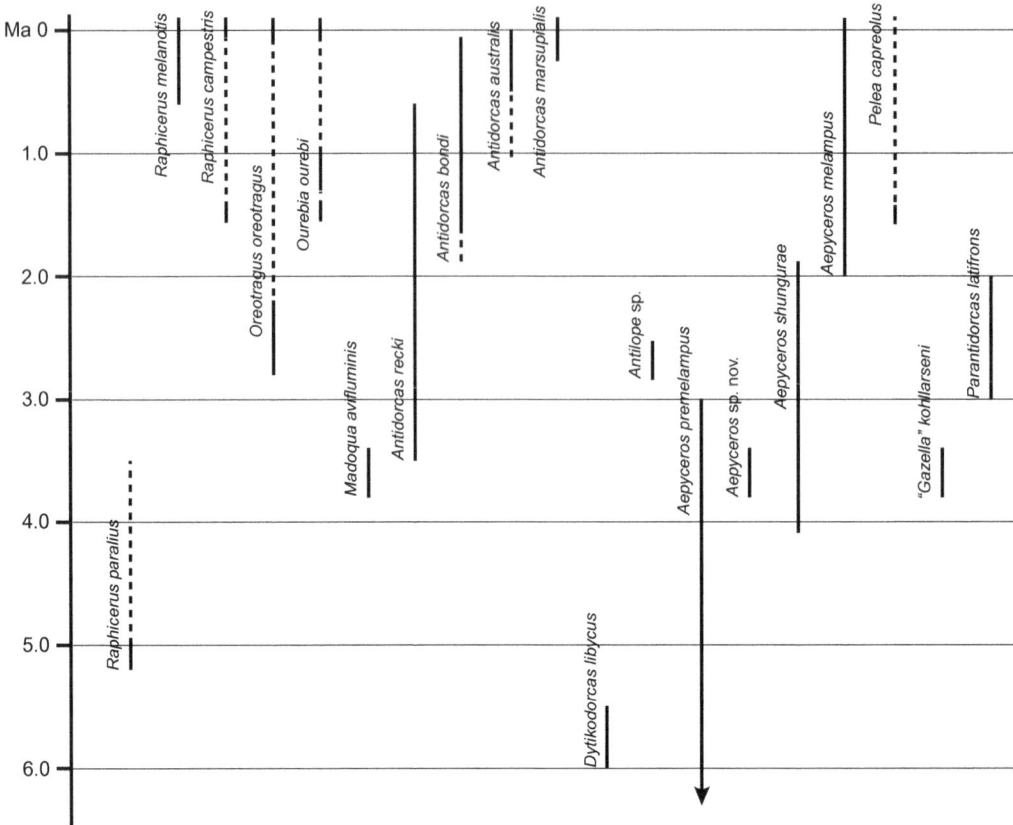

FIGURE 38.7   Temporal distribution of African fossil Antilopinae excluding *Gazella*. Earlier Antilopinae of uncertain tribal affiliation are *Homoiodorcas tugenium* at ~13.0–10.0 Ma and *Homoiodorcas* spp. back to the middle Miocene.

Remarks Earlier material has been assigned to *Homoiodorcas* (Gentry, 1994; Morales et al., 2003) e.g., Gebel Zelten horn core BMNH M26687 ("*Protragocerus* sp." of Hamilton, 1973: plate 13, figure 1 [left]) and an unpublished Maboko frontlet BMNH M15544 (figure 38.8). Thomas (1983) also described a middle Miocene horn core, mandible and other remains from Al Jadidah, Saudi Arabia, as *?Homoiodorcas* sp. The horn cores are more primitive than *H. tugenium* in having convex longitudinal profiles along their anterior edges rather than actual backward curvature. They are also appreciably bigger: basal DAP × DTs are 27.6 × 18.8 for Zelten, 23.0 × 22.3 for Maboko, and 23.4 × 19.7 for Al Jadidah. *Gazella negevensis* Tchernov et al. (1987) from Negev, Israel, is another probable *Homoiodorcas*. The teeth on the illustrated Negev mandible are as small as *Namacerus* or a large *Homoiodorcas*, and the premolar row appears to have been as long as in *Namacerus*, which makes it unlikely to be a hypsodontine (cf. Morales et al., 2003). Nothing as yet debars these earlier remains from being early Antilopini in which the horn cores were already larger than in extant Neotragini but not as long or backwardly curved as in Eurasian Turolian (late Miocene) *Gazella*. The Ngorora Formation species could be a smaller and late survivor of this stage of evolution.

Tribe NEOTRAGINI Sclater and Thomas, 1894

Type Genus Neotragus H. Smith, 1827.

Neotragini are African antelopes smaller than *Gazella*. Their horn cores are fairly upright and little divergent spikes, of small diameter, set widely apart, and plausibly primitive. The back part of the braincase roof often curves downward. Supraorbital pits are small and the preorbital fossa is large. There are no basal pillars on the teeth, the lingual walls of lower molars are straight, central fossettes disappear early in wear, and upper molars have small styles but rarely vertical ribs in between. *Neotragus* contains the royal antelope, *N. pygmaeus*, often cited as the smallest known bovid (see appendix note 2).

Genus RAPHICERUS C. H. Smith, 1827

Type Species Raphicerus campestris (Thunberg, 1811).

The short to moderately long horn cores have a slightly concave front edge. The back of the braincase roof is not very strongly curved downward, and temporal lines do not approach closely on the braincase roof. *Raphicerus campestris*, the steinbok, is in many ways the standard or most familiar of the Neotragini. The genus also includes the Cape and Sharpe's grysboks (*R. melanotis* and *sharpei*), the latter being the smallest and shortest horned and thereby more like a *Madoqua*.

RAPHICERUS PARALIUS Gentry, 1980

Synonymy Raphicerus paralius Gentry, 1980:300, figures 52–3, 55–7. *Langebaanweg. Early Pliocene.

Remarks This species is the largest known neotragine. It has short, thickened horn cores, often keeled posterolaterally and

FIGURE 38.8 Cranial roof of *Homoiodorcas* sp. from Maboko, BMNH M15544, in dorsal (A) and right lateral (B) views. Anterior to the right. Scale in millimeters. Photograph courtesy of The Natural History Museum, London.

with other irregular ridges. Insertions are more inclined than in living *Raphicerus*. *Raphicerus paralius* need not be close to living *Raphicerus* or any other neotragine genus. It cannot be a gazelle, so *Raphicerus* functions as a default generic name. A Makapansgat Limeworks neotragine horn core (*"Cephalophus" pricei* of Wells and Cooke, 1956: figure 6; from Member 3 according to Vrba, 1995) is as big as *R. paralius* but slightly compressed anteroposteriorly.

RAPHICERUS MELANOTIS (Thunberg, 1811)

Synonymy *Antilope melanotis* Thunberg, 1811, *Mem. Acad. Imp. Sci. St. Petersbourg* 3: 312. Extant.

Remarks Careful analysis by Klein (1976) demonstrated the presence of both *Raphicerus campestris* and *R. melanotis* early in the late Pleistocene of the southern Cape Province. An earlier and larger *Raphicerus* at Elandsfontein was also put into *R. melanotis* by Klein and Cruz-Uribe (1991), who attributed its large size to lack of competing species and not as a response to climate.

RAPHICERUS CAMPESTRIS (Thunberg, 1811)

Synonymy *Antilope campestris* Thunberg, 1811, *Mem. Acad. Imp. Sci. St. Petersbourg* 3:313. Extant; *Raphicerus campestris* Vrba, 1976: 29, plates 23–25. Swartkrans 2.

Remarks Vrba (1976) assigned a female skull from Swartkrans 2 to the living steenbok species. She assigned two larger horn cores from Swartkrans 2 to "cf. *Raphicerus* sp."

RAPHICERUS spp.

Localities and Age Many localities from late Miocene onward.

Remarks Neotragine horn cores and teeth larger than *Madoqua* have often been called "*Raphicerus* sp." Such identifications are quite likely to be correct for Pleistocene fossils, but less likely for earlier ones. Koobi Fora horn cores (Harris, 1991: table 5.75) are mostly large, about the size of the Elandsfontein *R. melanotis*, but backwardly curved unlike other *Raphicerus*. A Sahabi horn core (Lehmann and Thomas, 1987: figure 1E) is almost as big as at Langebaanweg. The Kanapoi record (Harris et al., 2003: figure 31) is not a *Raphicerus*; its horn cores are too large relative to their width apart, too inclined and with slight backward curvature. If it were a female gazelle, it would be a large species. Horn cores from Baard's Quarry, Langebaanweg (Gentry, 1980) are as small as in living *R. campestris* and *melanotis*. A Menacer m3 (Thomas and Petter, 1986: figure 5) is the size of *R. sharpei* or *Neotragus moschatus* (suni), but without the large third lobe of the latter.

Genus OREOTRAGUS A. Smith, 1834

Type Species *Oreotragus oreotragus* (Zimmermann, 1783).

OREOTRAGUS OREOTRAGUS (Zimmermann, 1783)

Synonymy *Antilope oreotragus* Zimmermann, 1783, , *Geogr. Gesch. Mensch. Vierf. Thiere* 3:269. Extant; *Palaeotragiscus longiceps* Broom, 1934:477, figure 6. Taung. Late Pliocene; *Oreotragus major* Wells, 1951, *S. Afr. J. Sci.* 47:167, figure 1. Makapan Valley. Plio-Pleistocene; *Oreotragus major* Wells and Cooke, 1956:35, figures 17–18. Makapansgat Limeworks. Late Pliocene; *Oreotragus oreotragus* Watson and Plug, 1995:183.

Remarks *Oreotragus* horn cores are more tapered from base to tip and more upright than in *Raphicerus*. The original maxilla fragment recorded by Broom (1934: figure 6) from Taung looks like an *Oreotragus*. *Oreotragus* from the South African Plio-Pleistocene caves was discussed by Vrba (1976) and Cooke (1990) and was considered to be indistinguishable from the living species by Watson and Plug (1995). This would make it the earliest record of an extant bovid in Africa.

Genus OUREBIA Laurillard, 1842

Type Species *Ourebia ourebi* (Zimmermann, 1783).
An extant monospecific genus with some reduncine similarities.

OUREBIA OUREBI (Zimmermann, 1783)

Synonymy *Antilope ourebi* Zimmermann, 1783, *Geogr. Gesch. Mensch. Vierf. Thiere* 3:268. Extant. Swartkrans Mb 3, ? Kanjera. Early Pleistocene.

Remarks Ourebia dentitions were found in the Swartkrans collections (Vrba, 1976, 1995). An upper molar from Kanjera is also probably an oribi (Gentry and Gentry, 1978).

Genus *MADOQUA* Ogilby, 1837

Type Species Madoqua saltiana (Desmarest, 1816).

Madoqua is made up of neotragines substantially smaller than *Raphicerus, Oreotragus* or *Ourebia*. Several very small species live in the Horn of Africa, and *M. kirki* and *damarensis* are more widespread.

MADOQUA AVIFLUMINIS (Dietrich, 1950)

Synonymy Praemadoqua avifluminis Dietrich, 1950:34, figures 3–4, 25–26. *Laetoli; *Madoqua avifluminis* Gentry and Gentry, 1978:425. Laetoli; *Madoqua avifluminis* Gentry, 1987:392. Laetolil Beds. Middle Pliocene.

Remarks Madoqua kirki still lives at Laetoli, and its remains are common among the fossils. The fossilized *Madoqua* of the Laetolil Beds is a different species. Horn cores are shorter and more thickened basally than in *M. kirki* and have a slight curvature; m3s have a reduced rear or third lobe, but it is not absent as in two of the living species; metatarsals are slightly shorter and relatively thicker. The syntypes of *M. avifluminis* are composite tooth rows, possibly of different dates. *Madoqua* in the Ndolanya Beds is less different from *M. kirki*. Gentry (1981) thought a Hadar KHT *Madoqua* horn core was closer to Ndolanya than to Laetolil Beds morphology. Harris (1991) noted a *Madoqua* maxilla in the KBS Member of the Koobi Fora Formation. *Madoqua* is listed for the late Miocene of Mpesida and the Adu-Asa Formation (Thomas, 1980: figure 1:10; Haile-Selassie et al., 2004).

ASSESSMENT OF NEOTRAGINI

Neotragini are not morphologically unified and could be a miscellany of small antelopes surviving in niches that are no longer within the grasp of larger bovid species. Fossil horn cores cited as Neotragini or *Raphicerus* are not always distinguishable from cephalophines, but one can be reassured by an absence of *Cephalophus* teeth at nearly all African localities. The less distinctive teeth of the bush duiker *Sylvicapra* might be confused with *Ourebia* but perhaps not with other Neotragini. The past existence of large species or variants of *Raphicerus* will need analysis, especially if they ever coexisted with *R. campestris* as seemed possible at Swartkrans 2 (Vrba, 1976). Small size in some Neotragini may be a secondary characteristic. A possible neotragine outside Africa is the late Miocene *Tyrrhenotragus gracillimus* at Baccinello, Italy (Weithofer, 1888; Thomas, 1984b).

Tribe ANTILOPINI Gray, 1821
Table 38.1

Type Genus Antilope Pallas, 1766.

This tribe contains the Indian blackbuck and its many relatives in the Eurasian late Miocene, as well as the widespread *Gazella* and some other genera. Molecular and genetic studies endorse the relationship of spiral-horned and non-spiral-horned species (Hernández Fernández and Vrba, 2005), and hybridization between *Antilope* and *Gazella* has been claimed (Ranjitsinh, 1989: sixth plate, figure 3).

Hitherto, a majority of recent cladistic studies have placed *Aepyceros* in an unresolved position close to the base of groups (B) and (C) in table 38.1. (See table 38.1 caption for references.) In this chapter I revert to placing it inside the Antilopini.

Genus *GAZELLA* Blainville, 1816
Figure 38.9

Type Species Gazella dorcas (Linnaeus, 1758).

Gazella is common from the late Miocene onward in Europe and China, and seemingly from the middle Miocene in the Siwaliks and Africa. Probably an unknown ancestor passed through a *Homoiodorcas*-like stage, acquired backward curvature of its horn cores, shortened its premolar rows, and became *Gazella*. In Europe some larger-sized descendants from a similar process became the genus *Tethytragus*.

Gazella species are small to moderate sized with moderate to long horn cores, the level of maximum mediolateral width lying at or slightly behind the anteroposterior midpoint, without keels or torsion, of subcircular or elliptical cross section, the lateral surface often flatter than the medial, moderately upright insertions, backwardly curved, and insertions moderately wide apart and placed above the back of the orbits. Some other characters are strong postcornual fossa, very limited sinuses within the frontals, horned females in most extant species, frontals between the horn bases at the same level as the dorsal orbital rims, large triangular supraorbital pits around the supraorbital foramina, braincase not shortened, its roof not greatly angled downward on the plane of the face, occipital surface with each half facing partly laterally as well as backward, moderate to large auditory bullae, m3s of living species often with an enlarged third lobe. The type species is one of a number found across drier habitats of the southern Palaearctic. Other modern species are sometimes split off as *Procapra* (East Asia), *Nanger* (three large African species), and *Eudorcas* (two sub-Saharan species larger than *dorcas* and living in slightly less dry conditions than *G. dorcas* or *Nanger*). In the following account, early and sub-Saharan *Gazella* will be listed before the later north African *Gazella* starting with *G. psolea*. The temporal distribution of African fossil *Gazella* species is shown in figure 38.9.

GAZELLA sp. (1)

Localities and Age Fort Ternan. Middle Miocene.

Remarks Several horn cores, a female partial skull and dental remains come from Fort Ternan. The complete horn core (Gentry, 1970: plate 15, figures 3–4 [appendix note 3]) is a member of the Antilopini lacking characters sufficient to remove it from *Gazella*. The female skull was hornless in life. Its temporal lines approach one another closely toward the back of the cranial roof, which is unusual in Antilopini and probably primitive. The teeth have occlusal lengths exceeding *Namacerus* by about 10% and shorter premolar rows than in that genus. Teeth are close in size and morphology to the European Turolian gazelles, but P2 is less shortened. Hypsodonty is little advanced, and upper and lower molars still have basal pillars. The p4 shows a metaconid growing forward toward the paraconid on the lingual side, which is not primitive.

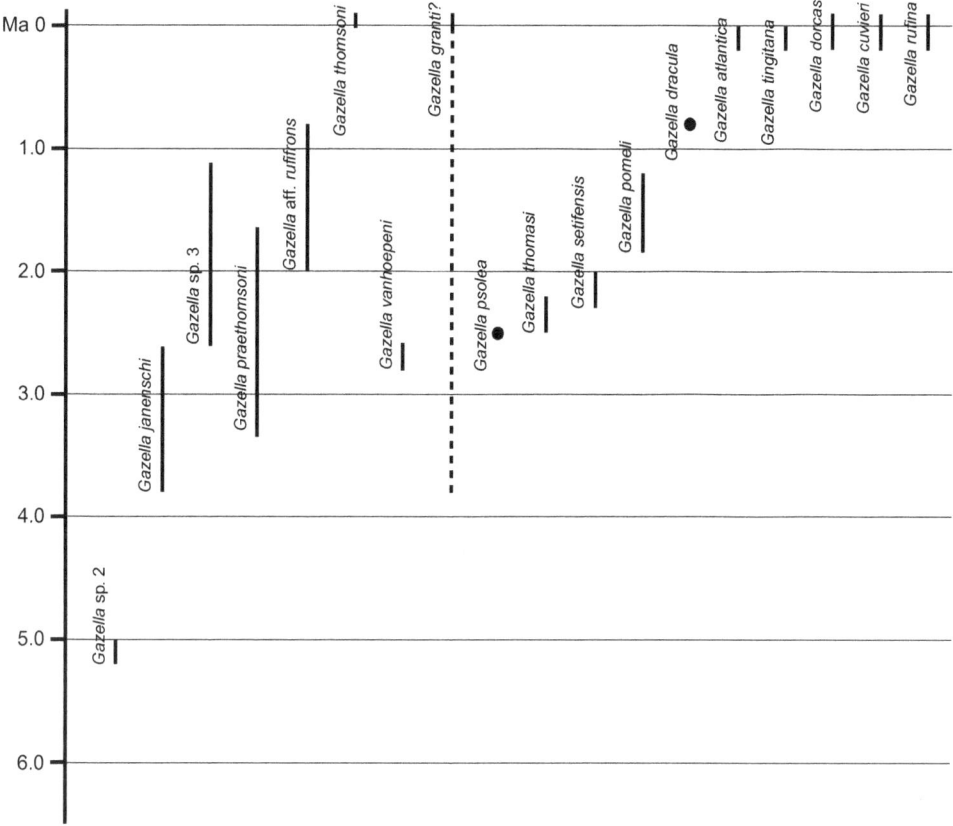

FIGURE 38.9 Temporal distribution of African fossil *Gazella*. Earlier records are *Gazella* sp. (1) at 14.0 Ma and *G. praegaudryi* at ~10.0–.0 Ma.

GAZELLA sp. (2)

Localities and Age Langebaanweg. Early Pliocene.

Remarks A larger gazelle at Langebaanweg (Gentry, 1980) has more robust and more compressed horn cores with flattened lateral surfaces, strong backward curvature, and a degree of divergence increasing distally. The lowest part of the horn cores often has a slight inward tilt as if deflected from an anterolateral direction (see *Homoiodorcas tugenium*, described earlier).

GAZELLA PRAEGAUDRYI Arambourg, 1959

Synonymy *Gazella praegaudryi* Arambourg, 1959:123, plate 17 figures 9–11. *Bou Hanifia. Late Miocene.

Remarks This little-known small species is much like that at Fort Ternan but seemingly without the p4 specialization. Similar remains have been found at Lothagam and in the Namurungule Formation.

GAZELLA JANENSCHI Dietrich, 1950

Synonymy *Gazella janenschi* Dietrich, 1950:25, figure 22. *Laetoli; *Gazella janenschi* Gentry, 1987:393, plate 10.9. Laetolil Beds, Ndolanya Beds. Middle–late Pliocene.

Remarks This is a small gazelle with horn cores little changed from the Fort Ternan species, but horn cores attributed to females have been found. The degree of divergence of the horn cores decreases distally. The species is common in the

Laetolil Beds and survived little changed in the Ndolanya Beds; it has also been reported from the Nachukui Formation (Harris et al., 1988).

GAZELLA sp. (3)
Figure 38.10

Synonymy Antilopini sp.1 Gentry and Gentry, 1978:444, plate 39, figure 2. Olduvai, Beds I–II. Early Pleistocene; Antilopini sp. indet. Gentry, 1985:180. Shungura Fm, Mbs K–L. Early Pleistocene; *Gazella praethomsoni* (in part, Omo 33 70.2680 + 70.2993) Gentry, 1985:179. Shungura Fm, Mb F. Late Pliocene; Antilopini "sp. 1" Gentry, 1987:401, plate 10.12. Laetoli, Ndolanya Beds. Late Pliocene.

Remarks Most of these fossils have not been previously referred to *Gazella*. The horn cores have some compression, divergence increasing from the base, backward curvature above fairly upright insertions, perhaps a faint anterior keel, and insertions probably close. Pedicels are low. The postcornual fossa is large, no sinuses are visible within the frontals, frontals are not raised between the horn bases, and the supraorbital pit is close below the anterior base of the horn core. The horn core in figure 38.10 is small and long and has a lyrated appearance.

Most of these characters would fit *Gazella*, but the more upright insertions, strong initial divergence, possible lyration distally, and perhaps the less pronounced flattening of the lateral surface differ from *G. janenschi* or the Olduvai *G.* aff. *rufifrons*. The horn cores are about the size of *G.* aff. *rufifrons* but with more compression.

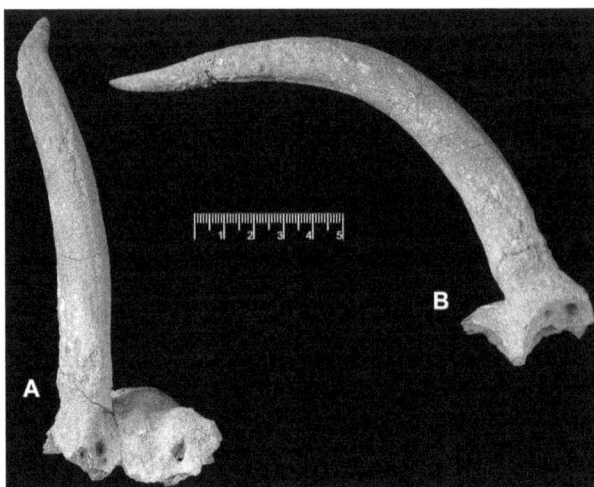

FIGURE 38.10 Frontlet of *Gazella* sp. (3) from Shungura Member F, Omo 33 70.2680 + 70.2993, in (A) anterior and (B) right lateral views. Scale in millimeters. Photograph courtesy of The Natural History Museum, London.

GAZELLA PRAETHOMSONI Arambourg, 1947

Synonymy *Gazella praethomsoni* Arambourg, 1947:387, plate 32 figures 4, 4a. *Shungura Fm.; *Gazella praethomsoni* Gentry, 1985:179. Shungura Fm, Mbs F–H. Late Pliocene; *Gazella praethomsoni* Harris, 1991:221, figure 5.71. Koobi Fora Fm, Lokochot Mb, Upper Burgi Mb to Okote Mb. Middle Pliocene–early Pleistocene.

Remarks This is a smaller species than extant *G. thomsoni*, and with horn cores only slightly less compressed. The holotype is a horn core base in poor condition and perhaps subadult, and the paratype mandible (Arambourg, 1947: plate 27, figure 1) may not be *Gazella*. *G. praethomsoni* is the most common gazelle in the upper part of the Koobi Fora Formation, and Harris (1991) comments that upper Burgi horn cores were shorter than in *G. thomsoni*.

GAZELLA aff. RUFIFRONS

Synonymy *Gazella* sp. Gentry and Gentry, 1978:438. Olduvai Beds I–II, Peninj, Elandsfontein. Late Pliocene–middle Pleistocene; ?(*Gazella* cf. *G. janenschi* Harris, 1991:223, figures 5.73–5.74. Koobi Fora Fm, Upper Burgi Mb to KBS Mbs. Late Pliocene–early Pleistocene); *Gazella* cf. *janenschi* Geraads, Eisenmann and Petter, 2004a:190, plate 12, figure 2. Melka Kunturé. Early Pleistocene.

Remarks The *Gazella* hitherto recognized at Olduvai is larger than most *G. janenschi* and like the living West African *G. rufifrons* in short horn cores with little divergence, well inclined backward and with little backward curvature. The Olduvai horn cores have less compression than in *rufifrons*, and a mandible SHK II 1957.793 probably had a premolar row as long as in *G. rufifrons*. This species may descend from or be connected with the more northern and partly earlier *G. praethomsoni* (in contrast to my own earlier opinion). Melka Kunturé gazelle horn cores may also belong here. The Elandsfontein gazelle (Klein and Cruz-Uribe, 1991: figure 18) could have been the latest in southern Africa, except that Klein (1984) had earlier mentioned late Pleistocene gazelles in

Zambia and Malawi. It would be interesting to know if these were more like *rufifrons* or *thomsoni*.

GAZELLA THOMSONI Günther, 1884

Synonymy *Gazella thomsoni* Günther, 1884, *Ann. Mag. Nat. Hist.* 5, 14:427. Extant; *Gazella praethomsoni* (in part) Harris, 1991:222, plate 5.72. Galana Boi Fm. Holocene.

Remarks Because of their late date, the Galana Boi horn cores may already be *G. thomsoni*. Since this species has or had a geographically separated subspecies *G. t. albonotata* in the southern Sudan and since it is a species of less arid country than *G. granti*, its past distribution is of much ecological interest. The descent of *G. thomsoni* is uncertain. Being related to *G. rufifrons* (Gentry, 1964), it could have descended from the Olduvai *G.* aff. *rufifrons* by acquiring more compressed horn cores and shorter premolar rows. Then again, the smaller *praethomsoni* in the Koobi Fora and Shungura Formation already had strongly compressed horn cores. Geraads et al. (2004a) appear to regard a gazelle at Asbole as intermediate between *praethomsoni* and *thomsoni*.

GAZELLA VANHOEPENI (Wells and Cooke, 1956)

Synonymy *Phenacotragus vanhoepeni* Wells and Cooke, 1956:43, figures 22–24. *Makapansgat Limeworks. Late Pliocene; *Gazella gracilior* Wells and Cooke, 1956:37, figures 20–21. Makapansgat Limeworks; *Gazella vanhoepeni* Wells, 1969, *S. Afr. J. Sci.* 65:162.

Remarks This large gazelle from Makapansgat Limeworks Member 3 (Vrba, 1995) had compressed horn cores curved backward fairly sharply in their mid course, and was a browser (Sponheimer et al., 1999). Thereafter it becomes encased in speculation. It may have been related to the three large living sub-Saharan gazelles sometimes placed in a separate genus *Nanger* and including *G. granti*. *Gazella gracilior* is probably its female. The Langebaanweg *Gazella* sp., with its rather large horn cores and teeth, could be linked with it. Gentry (1985) thought that an Usno Formation horn core, B377, could have belonged to the *Nanger* group.

GAZELLA GRANTI Brooke, 1872

Synonymy *Gazella granti* Brooke, 1872, *Proc. Zool. Soc. Lond.*:602. Extant.

Remarks This is the large East African gazelle of dry country. Fossil *Gazella granti* horn cores have been claimed at Laetoli from the Upper Ndolanya Beds and from both the upper and lower Laetolil Beds. I cannot believe that *G. granti* occurred in those faunas. but others with field experience remain adamant. Harris (1991: figure 5.75) similarly claims *Gazella* aff. *G. granti* from the Okote Member of the Koobi Fora Formation. A gazelle of the *Nanger* group at Asbole (Geraads et al., 2004a: figure 10:2) did have a much compressed horn core like *G. granti*, but it was a little smaller, and its backward curvature was like *G. vanhoepeni* or the modern *G. soemmerringi*.

GAZELLA PSOLEA Geraads and Amani, 1998

Synonymy *Gazella psolea* Geraads and Amani, 1998: 200, figure 5. *Ahl al Oughlam. Late Pliocene.

Remarks This gazelle has very large and little-compressed horn cores. The skull shows an inflated nasal opening flanked by the maxilla and a correspondingly reduced premaxilla, very like the condition of *Saiga tatarica* on the Eurasian steppes. Females are horned and the premolar rows are long for a gazelle. Subgenus *Deprezia* was founded for this highly specialized species.

GAZELLA THOMASI Pomel, 1895

Synonymy *Gazella atlantica* P. Thomas, 1884:17, plate 1 figure 9. Aïn Jourdel. Late Pliocene; *Gazella thomasi* Pomel, 1895:18. *Aïn Jourdel, Ahl al Oughlam. Late Pliocene.

Remarks Pomel gave a new name *thomasi* to a horn core misidentified by P. Thomas. This small gazelle has compressed horn cores and short premolar rows. Arambourg (1979) rightly regarded the type horn core as immature. Geraads and Amani (1998) put the smaller gazelle coexisting with *G. psolea* at Ahl al Oughlam into *G. thomasi*.

GAZELLA SETIFENSIS Pomel, 1895

Synonymy *Gazella setifensis* Pomel, 1895:15, plate 10, figures 14–15. *La route des Beni Fouda (= St. Arnaud, now El Eulma). Late Pliocene; *Gazella setifensis* Arambourg, 1979:56. Aïn Boucherit. Late Pliocene.

Remarks Arambourg (1979) used *setifensis* for the gazelle of Aïn Boucherit and took the *G. thomasi* type as a conspecific immature example. The Aïn Boucherit gazelles are larger than extant *G. dorcas* and have compressed and backwardly curved horn cores. Females are horned. Geraads (1981) restricted *setifensis* to Pomel's figured and slightly larger type horn core that he related to the recently exterminated *G. rufina*. He thought that several species could be present at Aïn Boucherit.

GAZELLA POMELI Arambourg, 1979

Synonymy *Gazella pomeli* Arambourg, 1979:62, plate 41 figures 8–9, plate 42. *Aïn Hanech. Early Pleistocene.

Remarks *Gazella pomeli* is another larger gazelle living somewhat later than *G. setifensis*. It has backwardly curved horn cores, but these are little compressed (ca. 82% for DT/DAP × 100 as against 67% at Aïn Boucherit) and more like the later and larger *Gazella atlantica*. The lateral surfaces of the horn cores are more obviously flat, and the degree of divergence increases toward the tips. It was probably at Mansourah (Chaîd-Saoudi et al., 2006).

GAZELLA DRACULA Geraads, 1981

Synonymy *Gazella dracula* Geraads, 1981:75, plate 5, figures 1–2. *Tighenif. Early–middle Pleistocene.

Remarks This species shows long, compressed and parallel horn cores with little backward curvature. Females have less compressed horn cores. Teeth are about the size of *G. atlantica*, although horn cores are somewhat smaller. The species has similarities to the geologically earlier European Villafranchian *G. borbonica*.

GAZELLA ATLANTICA Bourguignat, 1870

Synonymy *Gazella atlantica* Bourguignat, 1870:84, plate 10, figures 14–15. *Djebel-Thaya. Late Pleistocene.

Remarks This species is slightly larger than earlier North African gazelles, and has robust horn cores, little compressed and backwardly curved. For probable female horn cores, see Pomel (1895: plate 3, figures 1–5, and plate 10, figures 12–13) and appendix note 4. The geologically older "*Gazella atlantica*" of P. Thomas, 1884, is a homonym.

GAZELLA TINGITANA Arambourg, 1957

Synonymy *Gazella tingitana* Arambourg, 1957:68, plate 1, figures 1–4, and plate 2, figures 6, 8. *Mugharet el Aliya. Late Pleistocene.

Remarks The long slender horn cores are like the earlier and slightly smaller *G. dracula* but more backwardly curved and with degree of divergence increasing distally. The backward curvature is stronger than in extant *G. leptoceros*.

*GAZELLA DORCAS (Linnaeus, 1758)

Synonymy *Capra dorcas* Linnaeus, 1758, *Syst. Nat.*, 10th ed., 1: 69. Extant.

Remarks This once common living gazelle of north Africa has been reported from late Pleistocene localities (Arambourg, 1939, 1957). Many late fossils under various names might also belong here (Hopwood and Holleyfield, 1954:167).

GAZELLA CUVIERI (Ogilby, 1841)

Synonymy *Antilope cuvieri* Ogilby, 1841, *Proc. Zool. Soc. Lond.* 1840:35. Extant.

Remarks The larger upland gazelle of northwest Africa has been reported from late Pleistocene localities (Arambourg, 1957).

GAZELLA RUFINA O. Thomas, 1894

Synonymy *Gazella rufina* O. Thomas, 1894, *Proc. Zool. Soc. Lond.* 1894:467. Extant until 19th century AD; *Antilope pallaryi* Pomel, 1895:9, 27, plate 12, figures 1, 2. ?near Oran.; *Gazella rufina* Arambourg, 1957:63, plate 2, figures 3, 3a, 4. Mugharet el Aliya. Late Pleistocene.

Remarks Arambourg (1957) claimed a fossil record for this ghostly species founded on a skull and skin bought in Algiers in 1877. The holotype is in London and a few more specimens in Paris and Algeria (Lavauden, 1930), but no traveler ever wrote of sighting a living wild group. The holotype skull is like a large *G. rufifrons* (short and inclined horns, pointed central anterior tips on nasals, large preorbital fossa, and triangular supraorbital pits. Also the one surviving anterior tuberosity on the basioccipital is a localized structure as in *rufifrons* and *thomsoni* [Gentry, 1964]). *Gazella rufifrons* is a sub-Saharan West African species of more mesic habitats than any other gazelle, so *G. rufina* was probably a northern subspecies stranded in similar habitats amenable to human exploitation and threatened by desert advance from the south. Hernández Fernández and Vrba (2005) related it to *G. rufifrons* and *G. thomsoni*.

Genus ANTIDORCAS Sundevall, 1847

Type Species *Antidorcas marsupialis* (Zimmermann, 1780).

Antidorcas is like *Gazella* but with sinuses in the frontals, the frontals raised between the horn core bases, supraorbital pits small, braincase shorter, premolar rows shorter and p2s often absent. The horn cores often diverge distally, usually bend backward, and sometimes show a sign of homonymous torsion. Lower molars have noticeably flat lingual walls. All extinct species are smaller than the extant South African one. No reliable records predate the Ndolanya Beds at Laetoli.

ANTIDORCAS RECKI (Schwarz, 1932)

Synonymy *Adenota recki* Schwarz, 1932:1, plates 1–2. *Olduvai. Early–middle Pleistocene; *Phenacotragus recki* Schwarz, 1937:53, plate 1, figure 1; *Gazella wellsi* Cooke, 1949:38, figure 11. Vaal River Gravels. (?*A. recki* or *A. bondi*); *Gazella hennigi* Dietrich, 1950:25, plate 1, figures 1–2. Laetoli. Late Pliocene; *Antidorcas recki* Vrba, 1976:35, plates 32–35. Kromdraai A. Early (–middle?) Pleistocene; *Antidorcas recki* Gentry and Gentry, 1978:428, plate 39, figures 1 and 4, plate 40, figure 6. Olduvai Beds I–IV, Elandsfontein. Late Pliocene–middle Pleistocene; *Antidorcas recki* Gentry, 1985:177. Shungura Fm Mbs B–H. Late Pliocene; *Antidorcas recki* Harris, 1991:217, figures 5.67–5.70. Koobi Fora Fm, lower Tulu Bor–Okote Mbs. Nachukui Fm, Lomekwi, Kalachoro and Kaitio Mbs. Middle Pliocene–early Pleistocene; *Antidorcas recki* Geraads et al., 2001:342, figures 2I, 5F. Koro Toro. Middle–late Pliocene.

Remarks The holotype skull, figured by Schwarz (1932, 1937), was the first *Antidorcas* to be known outside southern Africa. Some populations may be separable as the *A. hennigi* of the Ndolanya Beds at Laetoli. Horn cores are often sharply bent backward in their distal parts, and upper molars have concave labial walls behind their mesostyles. Metatarsals are relatively less elongated than in living springbok. It was seemingly present in South Africa as far back as Sterkfontein Member 4, but its separation from the next two species can be problematic (Cooke, 1949; Vrba, 1976, 1995).

ANTIDORCAS BONDI (Cooke and Wells, 1951)

Synonymy *Gazella bondi* Cooke and Wells, 1951:207, figure 3. *Chelmer. Also Vlakkraal. Late Pleistocene; *Antidorcas bondi* Vrba, 1973:288, plates 16–24. Swartkrans. Pleistocene; *Antidorcas bondi* Vrba, 1976:13, 28. Swartkrans SKb (= Mb 2). Early Pleistocene; *Antidorcas bondi* Brown and Verhagen, 1985:102. Kruger Cave 35/83, Olifantsnek, Rustenburg. Holocene; *Antidorcas bondi* De Ruiter, 2003:34. Swartkrans Mbs 1–2. Early Pleistocene.

Remarks A small, markedly hypsodont *Antidorcas* was found at many Pleistocene sites of southern Africa north of the Cape zone (Klein, 1980) and was the most numerous bovid in Swartkrans 2 (Vrba, 1976). It was a specialized small grazer with teeth so hypsodont that the lower edge of the mandible in almost mature individuals is incompletely ossified. This condition does not quite come to pass in *A. recki*. De Ruiter (2003) claimed that *A. bondi* dates back to the early Pleistocene. It survived into the Holocene ca 7,500 BP (Brink and Lee-Thorp, 1992).

ANTIDORCAS AUSTRALIS Hendey and Hendey, 1968

Synonymy *Antidorcas marsupialis australis* Hendey and Hendey, 1968:56, plates 3–4. *Swartklip. Also Melkbos, Elandsfontein, Swartkrans. Early Pleistocene–early Holocene.

Remarks This form is smaller than living springbok, with horn cores more compressed and without a sharp bend backward. An *Antidorcas* sp. frontlet from Olduvai IV (Gentry and Gentry, 1978: plate 38) looked transitional from *A.recki* to *marsupialis* and may fall within the bounds of *A. australis*. The founders and most subsequent authors doubt that *A. australis* is separate from *A. marsupialis*. The analyses of Vrba (1976) and De Ruiter (2003) do not eliminate the scarcely believable possibility of coexistence of *A. recki, bondi* and *australis* at Swartkrans.

*ANTIDORCAS MARSUPIALIS (Zimmermann, 1780)

Synonymy *Antilope marsupialis* Zimmermann, 1780, *Geogr. Gesch. Mensch. Vierf. Thiere* 2:427. Extant.

Remarks This large *Antidorcas* with large m3s with noticeably enlarged third (hypoconulid) lobes, and with long metatarsals, was present at Florisbad (Brink, 1987) and other late sites in South Africa.

Genus DYTIKODORCAS Bouvrain and Bonis, 2007

Type Species *Dytikodorcas longicornis* Bouvrain and Bonis, 2007. *Dytiko 3, Greece, late Miocene (MN13).

DYTIKODORCAS LIBYCUS (Lehmann and Thomas, 1987)

Synonymy *Prostrepsiceros libycus* Lehmann and Thomas, 1987:330, figure 7. *Sahabi. Late Miocene.

Remarks This is a species with weakly to moderately lyrated horn cores showing strong longitudinal grooves posteriorly and sometimes a shallow longitudinal groove on the front surface. It differs from the gazelle-sized *D. longicornis* by its much larger size and shorter horn pedicels. The generic reattribution of Bouvrain and Bonis (2007) may be preferable to *Prostrepsiceros* in that the Greek type species of *Dytikodorcas* has very similar horn cores and is more nearly contemporaneous with the Sahabi species. The closest *Prostrepsiceros* was the Samos *P. fraasi* (Andree, 1926: plate 15, figure 1) which was closer in size to *D. libycus* but more different morphologically and probably from an earlier time in the Turolian. *Dytikodorcas* (as *Prostrepsiceros*) is also listed for the Adu-Asa Formation (Haile-Selassie et al., 2004).

Genus ANTILOPE Pallas, 1766

Type Species *Antilope cervicapra* (Linnaeus, 1758).
Antilope was the first genus of the Bovidae to be added to Linnaeus's original *Bos, Ovis,* and *Capra,* and was used for many bovids in the late 18th and early 19th centuries (Ogilby, 1841). Later it became restricted to the Indian blackbuck alone, *Antilope cervicapra,* but the corresponding English word *antelopes* continues to mean bovids beyond Europe and not in or closely related to *Bos, Ovis,* or *Capra. Antilope cervicapra* probably descends from a *Prostrepsiceros* species. Its horn cores are uncompressed, spirally horned, and absent in females.

ANTILOPE sp.

Synonymy *Antilope* aff. *subtorta* Gentry, 1985:180, plate 11, figure 3. Shungura Fm Mb. Ca. Late Pliocene.
Remarks A right and a left horn core probably from different individuals are the only record of *Antilope* in Africa.

They are less twisted than in living *A. cervicapra* and with traces of a posterolateral keel, but the same size as that species and without compression. They are not conspecific with the larger-sized *A. subtorta* Pilgrim, 1937 from the Pinjor Formation, India.

Genus *AEPYCEROS* Sundevall, 1847

Type Species Aepyceros melampus (Lichtenstein, 1812).

Horn cores are long, little compressed, often with transverse ridges, backwardly curved, and lyrated rather than spiraled. Females are hornless. The frontals contain sinuses and are slightly elevated between the horn pedicels. The postcornual fossa is large and supraorbital pits small. Cheek teeth are of antilopine aspect: quite high crowned, basal pillars absent on upper molars and tiny or absent on lowers, premolar rows short in comparison with molar rows. Upper molars have a fairly prominent mesostyle and only a weak rib on the labial wall of the paracone, the labial wall of the metacone is even flatter, and the M3 metastyle takes the form of a strong flange. Lower p2s are small, and p4s have paraconid-metaconid fusion to close the anterior part of the lingual wall and have a hypoconid tending to project labially. The teeth and limb bones of the living species have a number of unique or distinctive characters. *Aepyceros* shares lack of horn core compression with *Antilope,* some extinct spiral-horned species and some earlier *Gazella.* Temporal ridges on the cranial roof are less wide apart posteriorly than in *Antilope, Gazella* or *Antidorcas.*

AEPYCEROS PREMELAMPUS Harris, 2003

Synonymy Aepyceros premelampus Harris, 2003:551, figures 11.5G, 11.19–20. *Lothagam, upper Nawata Fm. Also lower Nawata and Nachukui Fms, Apak and Kaiyumung Mbs. Late Miocene–middle Pliocene.

Remarks This is the commonest bovid at Lothagam—an interesting contrast to its absence at Langebaanweg. The horn cores are reminiscent of an impala but seem to lack a flattened lateral surface or transverse ridges and are well inclined, the cranium is wide and rather long for an *Aepyceros,* and the supraorbital pits are large. The size of the species declines from the lower to the upper Nawata Formations. As with the Lothagam boselaphine, one wonders whether we have here an entrant to Africa from Eurasia, but this one would have had long-lasting success if its lineage really did turn into more fully evolved *Aepyceros.* The *Aepyceros* at Lukeino (Thomas, 1980) may be conspecific.

AEPYCEROS sp. nov.

Synonymy Aepycerotinae gen.et sp. indet. Dietrich, 1950:30, figure 45. Laetoli.; *Gazella kohllarseni (partim)* Dietrich, 1950:25, figures 16, 49. Laetoli; ?Hippotragini sp. *(partim)* Gentry and Gentry, 1978:351, 62. Laetolil Beds, Laetoli. Middle Pliocene; ?Hippotragini sp. nov. *(partim)* Gentry, 1987:388, plate 10.8. Laetolil Beds, Laetoli; sp. indet. aff. *Pelea* Gentry, 1987:394, plate 10.10. Laetolil Beds, Laetoli.

Remarks Confusion has long surrounded a Laetoli antelope, which Dietrich (1950: figure 45) first described. More details and a name will be given in a coming publication on Laetoli. The horn cores are substantially larger than in extant impala, and newly assigned teeth slightly larger. The species was outlasted in East Africa by smaller-horned relatives.

AEPYCEROS, sp. or spp. unknown

Localities and Age Karmosit, Mursi Fm, Hadar Fm, Nkondo Fm, Warwire Fm. Early–middle Pliocene.

Remarks Aepyceros at the first three localities (Gentry, 1978, 1981, 1985) show only modest lyration of their horn cores. The *Aepyceros* in the Nkondo and Warwire Formations (Geraads and Thomas, 1994) was thought to resemble the Mursi *Aepyceros.* All these occurrences probably coincide with the time span of the earlier and smaller *A. premelampus* at Lothagam. Lokochot horn cores attributed to *A. shungurae* (Harris, 1991: figure 5.65, table 5.64) also show the poor lyration and reportedly weak or absent transverse ridges appropriate for the Mursi species and are as small or smaller than the Karmosit and Hadar horn cores.

AEPYCEROS SHUNGURAE Gentry, 1985

Synonymy Aepyceros shungurae Gentry, 1985:171, plate 10, plate 11, figures 1–2, 7–8. *Shungura Fm, Mbs B–G. Usno Fm. 3.0–2.0 Ma. Middle–late Pliocene; *Aepyceros shungurae* Harris, 1991:306. Koobi Fora Fm, Moiti–Tulu Bor Mbs. Middle–late Pliocene; *Aepyceros shungurae* Geraads and Thomas, 1994:399, plate 3, figure 3. Kaiso Village Beds. ?Late Pliocene.

Remarks This species is smaller and more primitive than living *Aepyceros melampus.* The horn cores are becoming larger, more lyrated, and with more detectable transverse ridges than in the middle Pliocene *Aepyceros* of the preceding entry. The "cf. *Aepyceros* sp.nov." of Gentry (1985:175, plate 8, figures 1, 2) from Shungura F and lower G is not *A. shungurae;* it was slightly larger and with less lyrated horn cores and other minor differences. It occurred only in the more southerly area investigated by the French team of the 1967–74 Omo expedition. The "cf. *Aepyceros* sp." of Harris (1991:215, figure 5.66) from the Upper Burgi and KBS Members of the Koobi Fora Formation is strongly lyrate, but only as large as a smaller *A. shungurae.* It may be a small species living alongside the larger *Aepyceros* at the time of its transition to *A. melampus* (appendix note 5). The horn cores of Harris's "cf. *Aepyceros* sp." differ from the *Gazella* sp. 3 above by having transverse ridges and less compression.

**AEPYCEROS MELAMPUS* (Lichtenstein, 1812)

Synonymy Antilope melampus Lichtenstein, 1812, *Reisen Sudl. Africa* 2, plate 4, opp. p. 544. Extant.

Remarks It is possible that in the KBS Member of the Koobi Fora Formation and in levels above the top of Shungura Member G, impala attributable to *Aepyceros melampus* are found. In the East Turkana sequence, the Upper Burgi skull 1657 (Harris, 1991:211, figure 5.62) appears to be like *melampus* in having a straight parietofrontal suture and wider separation of supraorbital pits; it is less advanced in the distance of the supraorbital pits in front of the horn pedicels, and in facial length as judged by the relative anteroposterior positions of back of M3 to front of orbit. It continues to be like *shungurae* in smaller size, less widened back half of nasals, vestige of a preorbital fossa, and median indent at back of palate set further back. We should remember here that the holotype skull of *A. shungurae* is from Member B and that *shungurae* higher in the Shungura succession might have been like 1657. A pair of large and medially compressed horn cores from Shungura K ("cf. *Aepyceros* sp." of Gentry, 1985: 176) might be a temporal or geographic variant within *melampus.*

Gentry and Gentry (1978) noted *A. melampus* at Peninj. Occurrences of *Aepyceros* prior to the late Pleistocene are rare and uncertain in South Africa—for example, the horn core piece from Makapansgat Limeworks (Wells and Cooke, 1956: figure 19) thought by Vrba (1987b) to be conspecific with the species in the Laetolil Beds. The genus may never have occurred south of its present-day distribution.

Tribe PELEINI Gray, 1872

Type Genus Pelea Gray, 1851.

Genus *PELEA* Gray, 1851

Type Species Pelea capreolus (Forster, 1790). Extant. South Africa

Pelea is the only genus in its tribe and contains the single gazelle-sized species *Pelea capreolus*, the Vaal rhebok of southern Africa. It was long accepted by naturalists as a reduncine, but its skull is less robustly constructed than a reduncine, it has upright small-diameter horn cores, its teeth are like Antilopini, and the premolars are very small. Oboussier (1970) suggested membership in the Antilopini because of its large brain size, pattern of cerebral sulci, and neocortical development. Molecular and other studies have associated it with Reduncini (Vrba and Schaller, 2000) and sometimes with other tribes near to the Antilopini. It can best be taken as a survivor of early Antilopinae close to the origin of Reduncini.

PELEA CAPREOLUS (Forster, 1790)

Synonymy Antilopa capreolus Forster, 1790, In Levaillant, *Erste Reise Afrika*:71. Extant; *Pelea capreolus* Vrba, 1976:13, 27, 35, plates 17–20. Swartkrans a and b (= Mbs 1 and 2), Kromdraai A. Early (–middle?) Pleistocene.

Remarks Late Pleistocene *Pelea capreolus* is cited in Plug and Badenhorst (2001). In addition to Vrba's earlier records, there are more doubtful horn cores and teeth of "*?Pelea*" or "gen. indet. aff. *Pelea*" from Makapansgat 3 (Vrba, 1987b, 1995). Neither *Pelea* nor any other putative peleine comes from Laetoli, in contrast to the opinions of Gentry (1987).

Tribe Indet., ?aff. ANTILOPINI
Gen. indet., "*GAZELLA*" *KOHLLARSENI* Dietrich, 1950

Synonymy Gazella kohllarseni Dietrich, 1950:25, plate 1 figure 7. *Laetoli; ?Hippotragini sp. (partim) Gentry and Gentry, 1978:351, 62. Laetolil Beds, Laetoli. Middle Pliocene; ?Hippotragini sp.nov. (partim) Gentry, 1987:388. Laetolil Beds, Laetoli.

Remarks This is a hypothetical grouping of Laetoli antilopine-like teeth, larger than in *Gazella janenschi* but smaller than the large *Aepyceros*, and some more or less straight, little-compressed, and divergent horn cores. The pedicels of the horn cores contain sinuses.

PARANTIDORCAS Arambourg, 1979

Type Species Parantidorcas latifrons Arambourg, 1979.

PARANTIDORCAS LATIFRONS Arambourg, 1979

Synonymy Parantidorcas latifrons Arambourg, 1979:65, plates 46–47, plate 48, figures 1–11. *Aïn Boucherit, Aïn Brimba. Late Pliocene; *Cephalophus leporinus* (Pomel); Arambourg, 1979:78, plate 45, figure 4. *Aïn Boucherit.

Remarks The frontals of this gazelle-like species are not raised between the horn core bases, making it unlike *Antidorcas*. The horn cores are slightly spiraled, and with the torsion clockwise on the right or homonymous. This condition is present in *Menelikia* and the middle Miocene *Benicerus*, and sometimes detectable in *Antidorcas* and other genera. The species is like the Eurasian late Miocene *Oioceros* and its possible relatives, taken in table 38.1 as probably a separate tribe from Antilopini. Dentitions are slightly smaller than in the *Gazella* at Aïn Boucherit, the labial lobes of the lower molars more bluntly pointed and the mandibular ramus relatively deeper posteriorly. The "*Cephalophus leporinus*" partial lower dentition figured by Arambourg is a fairly well-worn *P. latifrons*.

ASSESSMENT OF ANTILOPINAE

Although a middle Miocene appearance of *Gazella* appears to be demonstrated at Fort Ternan, it is possible that the species there is in a phylogenetically earlier genus also ancestral to *Aepyceros* and other Antilopinae and perhaps even to Caprinae. (Any claims for Siwaliks middle Miocene *Gazella* will need similar questioning.) By the late Miocene, *Gazella* itself was in existence and by the middle Pliocene *G. janenschi* is probably a typical African gazelle. In the later Pliocene of East Africa, at least two species are known: (1) the *Gazella* sp. 3 with longer, backwardly curved horn cores and (2) the Olduvai and Koobi Fora *Gazella* aff. *rufifrons* with little compression or backward curvature. The second may descend from *G. praethomsoni* and be close to living *G. rufifrons* and *thomsoni*, but this is speculative.

In north Africa no gazelle is known between *G. praegaudryi* and the remarkable *G. psolea*, with the most specialized face structure known in a gazelle. The later late Pliocene produces *G. thomasi* (small), *setifensis* (larger and with compressed horn cores), and *pomeli* (larger and with less compressed horn cores). These are succeeded in the early–middle Pleistocene by *dracula* (long, compressed and straight horn cores), and in the late Pleistocene by *atlantica* (larger than *pomeli*, robust and little-compressed horn cores), and *tingitana* (larger than *dracula* and horn cores more curved backward). There are late Pleistocene occurrences of extant *G. cuvieri*, *rufina* (probably = *rufifrons*) and *dorcas*, but only *rufina* has been linked with an earlier fossil, namely the *setifensis* type horn core (Geraads, 1981). No likely ancestor of *G. leptoceros* has been found. It will be necessary to see if *G. cuvieri* and any of the late Pleistocene fossils are related to *G. gazella* of Israel and its fossil forms (Davis, 1980).

Antidorcas is not known much before the late Pliocene, and its last known occurrence in East Africa may be at Lainyamok at ~0.36 Ma (Potts and Deino, 1995). I could envisage *A. recki* of the middle Pleistocene evolving via *A. australis* into *A. marsupialis*, with *A. bondi* also appearing in the middle Pleistocene as a South African specialized side lineage, but the scrupulous analyses of the South African cave faunas do not support this story. No reliably identified fossils of *Litocranius* or *Ammodorcas* are known, although Gentry (1981: figure 2) plotted the dimensions of a notably gracile metacarpal from Hadar SH. The *Antilope* sp. in Shungura C is likely to be either temporally or geographically restricted in Africa, or both.

My former interpretation of *Aepyceros melampus* being descended from smaller ancestors with less lyrated horn cores and less specialized teeth or limb bones now looks like part of a bigger story involving earlier and other larger *Aepyceros* species. Both the *Dytikodorcas/Prostrepsiceros libycus* at Sahabi and the *Aepyceros premelampus* of the Nawata Formation are rather curious members of their genera, and my present expectation is that they are related to one another and more doubtfully to later *Aepyceros*. More solid information is required.

Subfamily REDUNCINAE Knottnerus-Meyer, 1907
Tribe REDUNCINI Knottnerus-Meyer, 1907
Figure 38.11

Type Genus Redunca H. Smith, 1827.

Reduncini contain two living genera, *Redunca*, reedbucks, and the larger-sized *Kobus*, kob, lechwes, and waterbuck. They live in habitats near water. Although the tribe is named Reduncini, *Kobus* species are more often and easily seen in African wildlife reserves and are commoner as fossils. A third main fossil lineage was *Menelikia*, which survived until the late Pliocene or just into the Pleistocene.

The tribe appeared in the late Miocene and African Reduncini have to be considered in relation to their long-known Siwaliks record. Asian Pliocene (–Pleistocene?) reduncines are reminiscent of lechwes or kobs, but with stronger temporal ridges on their cranial roofs. Earlier Siwaliks reduncines were found back to the Dhok Pathan and possibly earlier (Pilgrim, 1939). Gentry (1980, 1997) added to them the small *Dorcadoxa porrecticornis* (Lydekker, 1878). The occlusal complexity of present-day reduncine teeth agrees with the Bovini, Boselaphini and Hippotragini, and Pilgrim (1939) tentatively listed Reduncini close to those tribes. The temporal ridges of the Pliocene (–Pleistocene?) Siwaliks reduncines also prompted Gentry and Gentry (1978) to accept a reduncine-boselaphine relationship. This view did not last. The simpler and more antilopine-like characters of the reduncine teeth at Langebaanweg were a surprise, and eventually Gentry (1980, 1992) moved Reduncini away from Boselaphini and Bovini. Molecular and other studies led to the view that Reduncini were associated with Antilopinae and Cephalophini (Vrba and Schaller, 2000; Marcot, 2007), the next step up the cladistic ladder. This resurrected the possibility of *Pelea* being an interesting survival from close to the reduncine ancestry.

Extant forms show horn cores without keels or torsion, usually little compressed mediolaterally, and often with transverse ridges on the front. Some have a more upright insertion and backward curvature like most other bovids, while others are set at a low inclination in side view with upward and forward curvature. Postcornual fossae are present. Females are hornless. Frontals have no or limited development of internal sinuses and fail to become appreciably elevated between the horn insertions. A maxillary tuberosity is prominent in ventral view of the skull, the infraorbital foramen is placed low and anteriorly, and the palatal ridges on the maxillae in front of the tooth rows come close together. Temporal ridges on the cranial roof approach closely posteriorly, the basioccipital has large anterior tuberosities, and foramina ovalia are moderate to large. Teeth are moderately hypsodont, rather small in relation to skull and mandible size, upper and lower molars with basal pillars, P2s small, upper molars with small but marked ribs between the styles, lingual lobes of upper molars constricted, lower premolars with the appearance of anteroposterior compression, p2s small, p4s with strongly projecting hypoconid and often a deep and narrow labial valley in front of it, p4s without paraconid-metaconid fusion to form a closed lingual wall, labial lobes of lower molars constricted, and lower molars with goat folds.

The teeth of early reduncines differed from later ones in showing poorer ribs on labial walls of upper molars or lingual walls of lower molars, no basal pillars on upper molars and only tiny ones on lowers, P2 and P3 rather large relative to P4. These and other characteristics were similar to other early bovids, although a few features like small goat folds at the front of the lower molars foreshadowed later reduncines. The occipital bone and mastoids vary among fossil reduncines and may be significant for making identifications, but too few are known within any one species to be certain of the infraspecific variation. The temporal distribution of African fossil species is shown in figure 38.11.

?REDUNCINI spp.

Synonymy ?Reduncini gen. et sp. indet. Thomas, 1981:362, plate 3 figures 4–5, text figure 9. Ngorora Fm. Late Miocene; *?Antidorcas* sp. indet. Thomas, 1981:377, plate 3 figures 1–3, text figure 14. Ngorora Fm. Middle and late Miocene; *Pachytragus* aff. *solignaci*. Thomas, 1981:393, text figures 19–20. Ngorora Fm. Late Miocene.

Remarks These early remains may not belong to one species. Some of their characters suggest Reduncini. The cranium of "*Pachytragus* aff. *solignaci*" has horn cores too inclined backward and an insufficiently sloping cranial roof to fit that caprine species. No sinuses could be seen in the frontals or horn pedicel. Its horn cores are more compressed and perhaps larger than in *Kobus subdolus*. At least one of the "*?Antidorcas* sp." predates hipparionine horses and is therefore middle Miocene and all the more notable if it were reduncine.

Genus *ZEPHYREDUNCINUS* Vrba and Haile-Selassie, 2006

Type Species Zephyreduncinus oundagaisus Vrba and Haile-Selassie, 2006.

*ZEPHYREDUNCINUS OUNDAGAISUS Vrba
and Haile-Selassie, 2006

Synonymy Zephyreduncinus oundagaisus Vrba and Haile-Selassie, 2006:214, figure 2. *Middle Awash, Adu-Asa Fm, Alayla Vert. Paleont. Locality 2. Late Miocene.

Remarks The only known species was described on some small, strongly mediolaterally compressed horn cores including a short-horned holotype. Their compression is notable in a reduncine and at a pre-Pliocene date (5.8–5.5 Ma).

Genus *KOBUS* A. Smith, 1840

Type Species Kobus ellipsiprymnus (Ogilby, 1833).

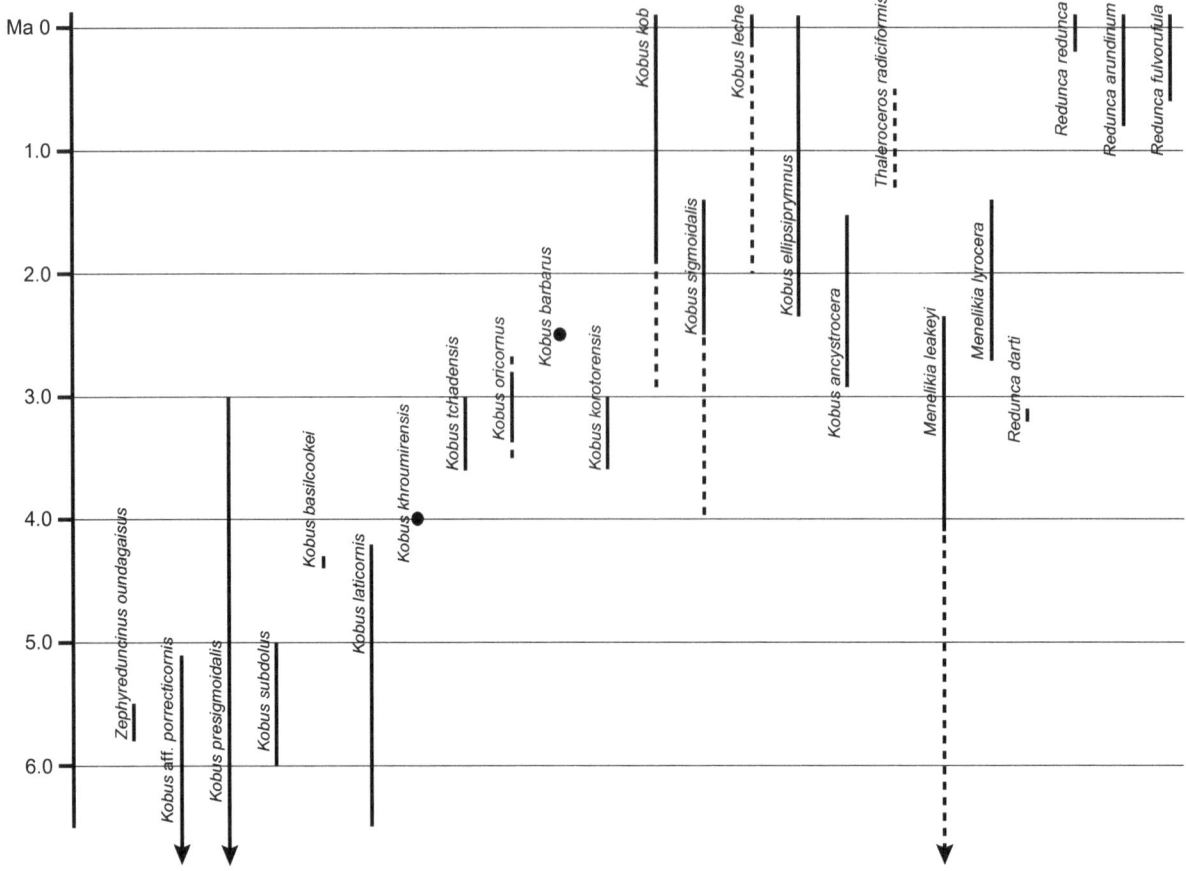

FIGURE 38.11 Temporal distribution of African fossil Reduncini. *Kobus* aff. *porrecticornis* and *K. presigmoidalis* go back to ~7.0Ma.

In extant faunas, it is difficult to pinpoint characters known in all *Kobus* and not in *Redunca*. Waterbuck and lechwes have long horns with moderate to much divergence and no or very little backward curvature. Kob and puku have backwardly curving horn cores with little divergence. The following account of *Kobus* lists ten named Pliocene species before beginning the Plio-Pleistocene and modern species with *Kobus kob*.

KOBUS aff. PORRECTICORNIS (Lydekker, 1878)

Synonymy Antilope porrecticornis Lydekker, 1878:158, plate 25, figure 4. *?Hasnot, Siwaliks. Late Miocene; *Gazella porrecticornis* Lydekker, 1886:11, figure 2; *Dorcadoxa porrecticornis* Pilgrim, 1939:44, plate 1, figure 9.

African Localities and Age Adu-Asa Fm, Mpesida, Lukeino, Manonga. Late Miocene.

Remarks The horn cores of this Siwaliks species look reduncine in their transverse ridges, wide divergence, inclined insertions, a deep localized postcornual fossa, and low pedicels (Gentry, 1970). Early reduncines or apparent reduncines exist in the Siwaliks well back into the late Miocene. Their appearance gradually becomes more suggestive of small kobs than of gazelles. Similar forms have been found in the latest Miocene of Africa (Thomas, 1980; Gentry, 1997; Haile-Selassie et al., 2004). Two of the Ngorora horn cores (Thomas, 1981: plate 3, figures 4–5), listed earlier under "?Reduncini spp.", might be an earlier record of *K.* aff.

porrecticornis. A claim for Baard's Quarry at Langebaanweg (Gentry, 1980) is temporally anomalous and presumably the species is a later small kob.

KOBUS PRESIGMOIDALIS Harris, 2003

Synonymy Kobus presigmoidalis Harris, 2003:541, figure 11.9. Lothagam: *Upper Nawata Mb, also Lower Nawata; Nachukui Fm Apak and Kaiyumung Mbs. Late Miocene–middle Pliocene.

Remarks This more completely known species is not very different from a large *K. porrecticornis* (Vrba and Haile-Selassie, 2006; I agree but have not seen the material). It is smaller and has stronger backward curvature than *K. subdolus*. The illustrated holotype cannot be an impala because the slight torsion is in the opposite direction. The species could span 7.0 to 4.5 Ma or younger, with the holotype ~6.0–5.0 Ma.

KOBUS SUBDOLUS Gentry, 1980

Synonymy Kobus subdolus Gentry, 1980:248, figures 18–19, 23–25. *Langebaanweg. Early Pliocene; *Redunca* aff. *darti* Lehmann and Thomas, 1987:327, figures 9B, 9C. Sahabi. Late Miocene.

Remarks Kobus subdolus is larger than *K. porrecticornis*, the horn cores short, their widest mediolateral diameter lying rather anteriorly, and the cross section narrowing behind to a posterolateral angle, often with deep grooves medially to this angle or edge, little compressed, with a flattened lateral surface, well-inclined backward and slightly curved backward, inserted close together and above the back of the orbits. The anterior edge of the horn pedicels is sometimes upright in side view and thus meets the anterior edge of the horn core at an angle. There is a large postcornual fossa and large supraorbital pits. Gentry (1980:261, figures 26–27) also postulated a smaller species, "*Kobus* sp. 2," at Langebaanweg. Reduncine teeth at Langebaanweg are more antilopine-like than are those of other early reduncines (Gentry, 2008), and this may necessitate the separation from *K. subdolus* of the reduncines at Sahabi, Manonga, and other localities.

KOBUS BASILCOOKEI Vrba, 2006

Synonymy *Kobus basilcookei* Vrba, 2006:64, figures 1–6. Middle Awash. *Sagantole Fm, Aramis Mb. Early Pliocene, ~4.4 Ma.

Remarks The horn cores are quite long, little compressed, and about the size of a larger *K. presigmoidalis*. They are longer than in *K. subdolus* and also have transverse ridges, very little divergence, and the supraorbital foramina are not in large or deep pits. The last character is also seen in the line or group of species, which may lead to *K. oricornus* but which have strongly divergent horn cores.

KOBUS LATICORNIS Harris, 2003

Synonymy *Kobus laticornis* Harris, 2003:542, figures 11.5E, 11.10. Lothagam: *Upper Nawata Mb., also Nachukui Fm Apak Mb. Late Miocene–early Pliocene.

Remarks The size probably overlaps and is larger than *K. presigmoidalis*, horn cores long, very divergent, widely inserted, with slight homonymous torsion, cranial roof inclined. The horn cores are said in the diagnosis to be anteroposteriorly compressed, although the measurements indicate slight lateromedial compression of ~75–80%. The species is most suggestive of *K. oricornus* among later Pliocene *Kobus*. Vrba (2006) questioned the tribal allocation of the holotype.

KOBUS KHROUMIRENSIS (Arambourg, 1979)

Synonymy *Redunca khroumirensis* Arambourg, 1979:72, plate 43, figures 3–4; plate 44, figures 1–3; plate 45, figures 1–3. *Lac Ichkeul. Early Pliocene.

Remarks This fairly small species has a wide cranium, horn cores strongly divergent, well inclined backward, and without backward curvature. The cranial roof slopes little but does curve downward posteriorly. *Kobus khroumirensis* is not likely to be related to *K. kob* but could be an early relative of either *K. oricornus* or *ancystrocera*.

KOBUS TCHADENSIS Geraads et al., 2001

Synonymy *Kobus tchadensis* Geraads et al., 2001:338, figures 2F–2H, 3A–3D. *Koro Toro. Middle Pliocene.

Remarks Like the earlier *Kobus laticornis*, this species is again and more decisively suggestive of *Kobus oricornus*. It has notably large and long horn cores, some of which reach almost to the size of the biggest *K. sigmoidalis* and *ellipsiprymnus*. The horn cores have anteroposterior compression, strong divergence, strong backward inclination in side view with some

upward and forward curvature distally, insertions wide apart (perhaps in consequence of the large horns being so divergent), a cranial roof that is not far off horizontal and slopes only very slightly downward posteriorly, the braincase widening posteriorly, a rounded occipital edge.

KOBUS ORICORNUS Gentry, 1985

Synonymy *Kobus* sp. A Gentry, 1981:9. Hadar Fm Mbs SH-3 –DD-3. Middle Pliocene; *Kobus oricornus* Gentry, 1985:155, plate 5, figures 1–4. *Shungura Fm, Mb B. Middle Pliocene; *Kobus oricornis* Harris, 1991:176, figure 5.37. Koobi Fora Fm, Lokochot and Tulu Bor Mbs. Nachukui Fm, Kataboi to Lokalalei Mbs. Middle–late Pliocene.

Remarks This species in Shungura B appears to represent the last of the *tchadensis* group at a date not younger than 2.65 Ma. It is very likely that the *Kobus* sp. A of Gentry (1981) is an earlier and overlapping subspecies. The p4s at Hadar are more advanced than are those of later reduncines in lingual approach and fusion of paraconid with metaconid.

KOBUS BARBARUS Geraads and Amani, 1998

Synonymy *Kobus barbarus* Geraads and Amani, 1998:194, figures 3C–3D. *Ahl al Oughlam. Late Pliocene.

Remarks The horn cores are large and long, not compressed in either direction, divergent, and inclined backward. They show slight backward curvature and a suggestion of a posterolateral keel. The species fails to have resemblances to *Kobus oricornus*, *K. ancystrocera*, or *K. kob* and the authors can only suggest a plausible survival of the Sahabi reduncine here included in *K. subdolus*.

KOBUS KOROTORENSIS Geraads et al., 2001

Synonymy *Kobus korotorensis* Geraads et al., 2001:336, figures 2B–2C, 3E–3I. *Koro Toro. Middle Pliocene.

Remarks This species is smaller but overlapping in size with *K. tchadensis* and with shorter and less diverging horn cores. It is little different from *K. subdolus* but lacks even the minimal mediolateral compression of that species. *Kobus korotorensis* might become a useful name for many koblike fossils in the middle or late Pliocene—for example, the *K.* aff. *patulicornis* in the Chiwondo Beds (Kaufulu et al., 1981).

KOBUS KOB (Erxleben, 1777)

Synonymy *Antilope kob* Erxleben, 1777: *Syst. Regni Anim.* 1:293. Extant kob; *Kobus kob* Gentry and Gentry, 1978:332, plate 7, figure 1, plates 8–9. Olduvai Beds II–III. Early Pleistocene; *Kobus kob* Gentry, 1985:150. Shungura Fm J–L, probably G, possibly B–E. (?Pliocene)–early Pleistocene; *Kobus kob* Harris, 1991:171, figure 5.34. Koobi Fora Fm, KBS and Okote Mbs. Late Pliocene–early Pleistocene; *Kobus* aff. *kob* Chaïd-Saoudi et al., 2006:969. Mansourah. Lower Pleistocene.

Remarks This extant species, taken here to include the puku, *K. vardoni*, is smaller than the waterbuck, *K. ellipsiprymnus*, and has horn cores with some mediolateral compression, less divergence, curving backward, and inserted close together. The skull is narrower than in waterbuck, the cranial roof more sloped, the occipital with a rounded edge, mastoids narrow, and with some limb bone characters more cursorial than in other reduncines. Gentry (1985) noted that

K. kob horn cores had little basal backward curvature while coexisting with *K. sigmoidalis*, but more of it after the advent of *K. ellipsiprymnus*. The kob is the most common antelope at Mansourah, an unusual situation for a fossil reduncine in North Africa.

KOBUS SIGMOIDALIS Arambourg, 1941

Synonymy *Kobus sigmoidalis* Arambourg, 1941: 346, figure 5. *Shungura Formation; *Kobus sigmoidalis* Arambourg, 1947: 411, plate 27, figure 4; plate 28, figure 3, and others. Shungura Formation; *Kobus sigmoidalis* Gentry and Gentry, 1978: 324, plates 5–6. Olduvai Bed I. Late Pliocene; *Kobus sigmoidalis* Gentry, 1985: 145, plate 5, figures 5–6. Shungura Fm D–G with a likely ancestral variant in Mb C. Late Pliocene; *Kobus sigmoidalis* Harris, 1991: 163, figures 5.21–5.25. Koobi Fora Fm, Upper Burgi and KBS Mbs. Also Moiti to Tulu Bur and ?Okote Mbs.

Remarks This is the size of a lechwe, but enlarging from Shungura G to Olduvai I (Gentry and Gentry, 1978: figure 11). The horn cores are long, with mediolateral compression, transverse ridges, and more divergence than in *K. ellipsiprymnus*. They curve weakly backward at the base in side view and then more upward, this being optimistically seen by Arambourg as a sigmoid course. Mastoid exposure is mostly within the occipital surface. Gentry (1985: figures 9–11) showed variation in size, divergence, and compression of horn cores through the Shungura sequence, either as a result of continuous change in the local norms or from movements of populations at intervals of 200,000 years or less. Gentry also thought that most reduncine teeth in Shungura D–G belonged to this species and noted differences from extant species in upper molars with simpler central fossettes, ribs between the styles not very localized or accentuated, and lower molars more frequently with less constricted labial lobes. Harris (1991) found Koobi Fora *K. sigmoidalis* to be mainly an Upper Burgi and KBS species but that a smaller and shorter-horned version was present as early as the Moiti Member.

KOBUS LECHE Gray, 1850

Synonymy *Kobus leche* Gray, 1850, Gleanings, Knowsley Menagerie 2, p. 23. Extant lechwe; *Cobus venterae* Broom, 1913:15, figure 3. Florisbad. Late Pleistocene; *Kobus* aff. *leche* Harris, 1991:170, figures 5.31–3. Koobi Fora Fm, Upper Burgi-Okote Mbs; Nachukui Fm Nariokotome Mb. Late Pliocene–early Pleistocene.

Remarks *Kobus leche* is probably descended from *K. sigmoidalis* with less change than was seen in the line giving rise to *K. ellipsiprymnus*. Gentry and Gentry (1978) redescribed *K. venterae* from Florisbad and referred it to *K. leche*, and Brink (1987) added further comments. The species is also present at other and earlier sites in South Africa (Lacruz et al., 2002; De Ruiter, 2003). Harris (1991) linked Koobi Fora horn cores with *K. leche*. The other extant lechwe, *K. megaceros* of the southern Sudan and Ethiopia, may not be closely related within the Reduncini to *K. leche* (appendix note 6).

*KOBUS ELLIPSIPRYMNUS (Ogilby, 1833)

Synonymy *Antilope ellipsiprymnus* Ogilby, 1833, *Proc. Zool. Soc. Lond.*:47. Extant waterbuck; *Kobus ellipsiprymnus* Gentry and Gentry, 1978:330, plate 7, figure 2. Olduvai Beds III–IV. Early–middle Pleistocene; *Kobus ellipsiprymnus* Gentry, 1985:149. Shungura Fm G, J–K. Late Pliocene–early Pleistocene;

Kobus ellipsiprymnus Harris, 1991:162, figure 5.20. Koobi Fora Fm, KBS Mb; Nachuikui Fm, Nariokotome Mb. Late Pliocene–early Pleistocene.

Remarks The modern waterbuck replaces *Kobus sigmoidalis* within Shungura G and appears in Olduvai Bed III and possibly earlier. It was very common at Buia (Martínez-Navarro et al., 2004). It is present but rare in the KBS Member at Koobi Fora, while *K. sigmoidalis* seems to continue into higher levels. *Kobus ellipsiprymnus* eventually differs in its larger size, less compressed horn cores, and loss of the basal backward curvature of the horn cores (which survived in *K. leche*), but the change appears to be gradual. Hadjouis (1986) illustrated a battered late Pleistocene frontlet from the Gisement des Phacochères near Algiers that looked like *K. ellipsiprymnus* but was too wide for the species.

KOBUS ANCYSTROCERA (Arambourg, 1947)

Synonymy *Redunca ancystrocera* Arambourg, 1947:416, plate 29, figure 4; plate 31, figures 2, 4, 4a; text figure 62. *Shungura Fm.; *Kobus ancystrocera* Gentry, 1985:152, plate 6. Shungura Fm B–C, E, G, J. Late Pliocene; *Kobus ancystrocera* Harris, 1991:166, figures 5.26–5.30. Koobi Fora Fm, Upper Burgi–KBS Mbs. Late Pliocene.

Remarks Horn cores are well inclined, divergent, with tips strongly recurved forward almost like a hook (the "crochet" of French texts), and an approach to a posterolateral keel. The species is smaller than *Kobus barbarus* and with more compressed horn cores. Harris (1991) also listed Tulu Bor horn cores that he considered closer to this than to other reduncine species, but with the following plausibly primitive differences: smaller size, shorter horn cores, less divergence, less backward curvature at their bases, less of a distal hook. The Tulu Bor skull *Kobus* aff. *K. kob* of Harris (1991: figure 5.36) could be related to the ancestry of *K. ancystrocera*. Gentry (1985) thought that the distribution of *K. ancystrocera* in Shungura G localities showed avoidance of *K. sigmoidalis*.

Genus THALEROCEROS Reck, 1925

Type Species *Thaleroceros radiciformis* Reck, 1925.

*THALEROCEROS RADICIFORMIS Reck, 1925

Synonymy *Thaleroceros radiciformis* Reck, 1925:451; Reck, 1935:218, figure 2; Reck, 1937, *Wiss. Ergebn. Oldoway-Exped.* 1913 (n.f.) 4:142, plate 8. *Olduvai Upper Beds II–IV. Early Pleistocene.

Remarks The only known specimen is a frontlet of a large and bizarre antelope with massive uncompressed horn cores, diverging little, and with an upward and anteriorly concave curvature from a large unitary pedicel. Gentry and Gentry (1978) took it as descended from *Kobus ancystrocera* via an intermediate pair of horn cores BMNH M15925 from Kanam.

Genus MENELIKIA Arambourg, 1941

Type Species *Menelikia lyrocera* Arambourg, 1941.
These medium-sized Pliocene–early Pleistocene reduncines have horn cores with transverse ridges, divergence that is sometimes considerable, backward inclinations, insertions

close together, and clockwise torsion on the right side. The homonymous torsion helps to distinguish isolated horn cores from *Aepyceros*. Sinuses within the frontals are more extensive than in any other reduncine. Postcornual fossae are small or absent.

MENELIKIA LEAKEYI Harris, 1991

Synonymy Menelikia sp. Gentry, 1985:162, plate 7, figures 2–3. Shungura Fm Mb C, perhaps D. Late Pliocene; *Menelikia leakeyi* Harris, 1991:183, figures 5.44–5.45; 2003:544, figure 11.11. *Koobi Fora Fm, Moiti Mb. Also Lothagam, Lower Nawata Fm; Koobi Fora Fm Lokochot to Tulu Bor Mbs; Nachukui Fm, Upper Lomekwi and Lokalalei Mbs. Late Miocene–late Pliocene.

Remarks This species has long horn cores, little compressed, divergent in their upper parts, long axis of basal cross section at a wide angle to the skull longitudinal midline, and inserted above the back of the orbits. The sinus system within the frontals is limited. Irregular deep longitudinal grooving on the horn cores has been noted but may be a postmortem change. The species may occur in the Warwire Formation (Geraads and Thomas, 1994: plate 1, figure 3). The Lothagam record (Harris, 2003: figure 11.11) is startlingly early.

**MENELIKIA LYROCERA* Arambourg, 1941

Synonymy Menelikia lyrocera Arambourg, 1941, *Bull. Mus. Natn. Hist. Nat.* 2, 13:341, figures 1–3. *Shungura Fm.; Menelikia lyrocera* Arambourg, 1947:392, plate 23; plate 24, figure 4; plate 29, figure 2; *Menelikia lyrocera* Gentry, 1985:157, plate 7, figure 1. Shungura Fm C, E–J, perhaps K. Late Pliocene–early Pleistocene; ?Caprini sp. A Harris et al., 1988:112, figure 58. Nachukui Formation, Lokalalei Mb. Late Pliocene (Harris, 1991:183); *Menelikia lyrocera* Harris, 1991:179, figures 5.39–5.42. Koobi Fora Fm, Upper Burgi–KBS Mbs (rare in Okote Mb). Late Pliocene–early Pleistocene.

Remarks Horn cores are moderately long (shortening later in the time span of the species), strongly curved backward and then diverging, postorbital insertions, frontals with extensive sinuses and raised high between the insertions, braincase short, basioccipital with strong longitudinal ridges behind the anterior tuberosities. Mediolateral compression of horn cores in Member F of the Shungura Formation almost disappears in Member G. From Member H onward the final stage of the species exhibits much shortened horn cores.

Gentry (1985) noted resemblances of *M. lyrocera* to the extant Nile lechwe, *Kobus megaceros*, but that species also has very long horns, and the existence of sinuses has not been decided. Gentry (1985), followed by Harris (1991), saw *M. lyrocera* as inhabiting drier habitats than *Kobus sigmoidalis*, but Spencer (1997) suggested much soft and seasonally unvarying grass in its diet, prompting the thought that it could have been feeding aquatically like *K. megaceros* in the Sudanese sudd (Kingdon, 1990:232).

Small horn cores in the Tulu Bor and Shungura C may be an early variant within or ancestral to *M. lyrocera* (Harris, 1991: figure 5.43; Gentry, 1985: plate 7, figure 2); they do not have a texture appropriate for immature examples of *M. leakeyi*. *Menelikia lyrocera* of Kaiso Village (Cooke and Coryndon, 1970; Geraads and Thomas, 1994:395, plate 3, figure 1) looks like specimens from the Shungura G time level in horn cores rising from their insertions, tapering rapidly, not having a good backward curvature, and the cranial roof angled and not

becoming more horizontal as it approaches the occipital top (Gentry, 1985:158–159).

Genus *REDUNCA* H. Smith, 1827

Type Species Redunca redunca (Pallas, 1767)
Redunca is frequently smaller and shorter horned than *Kobus*. Horn cores have little mediolateral compression, sometimes even becoming anteroposterior compression in later evolution, often with a medial or, again later, a posteromedial surface near the base, with an upward and anteriorly concave curvature in side view, frontals without internal sinuses, anterior tuberosities of the basioccipital very large and outwardly splayed. In South Africa, fossils of *Redunca* are more common than elsewhere on the continent, while *Kobus* is rare or absent between the early Pliocene and the appearance of *K. leche* and perhaps *ellipsiprymnus* in the late Pleistocene. Vrba and Haile-Selassie (2006) favored placing the Langebaanweg *Kobus subdolus* and the early Pliocene reduncines from Sahabi and Wadi Natrun in *Redunca*. Further discussion can be found in Haile-Selassie et al. (2009).

REDUNCA DARTI Wells and Cooke, 1956

Synonymy Redunca darti Wells and Cooke, 1956:17, figures 7–9. *Makapansgat Limeworks (Mb 3; Vrba, 1995). Late Pliocene.

Remarks It differs from extant *Redunca arundinum* and *R. redunca* by its horn cores being less laid back in side view and the posteromedial flattened surface lying more medially than posteriorly. Sparse dental remains possibly of this species are present at other Sterkfontein Valley localities (Vrba, 1976). The antecedents of this animal are unknown.

**REDUNCA REDUNCA* (Pallas, 1767)

Synonymy Antilope redunca Pallas, 1767, *Spicil. Zool.* 1:8. Extant reedbuck or behor reedbuck; *Antilope (Oegoceros) selenocera* Pomel, 1895, not mentioned in text, plate 6, figures 1–3; *Antilope (Dorcas) triquetricornis* Pomel, 1895:28, plate 11, figures 1–2; *Antilope (Nagor) maupasii* Pomel, 1895:38, plate 10, figures 1–11.

Remarks Arambourg (1939: plate 8, figures 1–2; 1957) documented this species from the late Pleistocene and Holocene of northwest Africa where it no longer occurs.

REDUNCA ARUNDINUM (Boddaert, 1785)

Synonymy Antilope arundinum Boddaert, 1785, *Elench. Anim.* 1:141. Extant southern reedbuck; *Redunca arundinum* Hendey, 1968:110. Melkbos. Late Pleistocene; *Redunca arundinum* Hendey and Hendey, 1968:51, plate 2. Swartklip. Early Holocene; *Redunca arundinum* Klein and Cruz-Uribe, 1991:36. Elandsfontein. Middle Pleistocene.

Remarks This species was found at South African localities back to the middle Pleistocene (Lacruz et al., 2002), but absent historically from the southwestern Cape. Late Pleistocene Western Cape Province fossils are larger (Klein and Cruz-Uribe, 1991: figure 7) and believed to differ slightly from present-day *R. arundinum* (Hendey, 1968; Hendey and Hendey, 1968). The earlier Elandsfontein representative had horn cores more anteroposteriorly compressed, less divergent, and perhaps shorter, the last two characters making it appreciably more like *R. redunca* in the north.

REDUNCA FULVORUFULA (Afzelius), 1815

Synonymy Antilope *fulvorufula* Afzelius, 1815, *Nova Acta Reg. Soc. Sci. Upsala*, 7:250. Extant mountain reedbuck.

Remarks This species shows less tendency to anteroposterior compression of the horn cores than in the other two living species. It is known by appropriately small teeth from some South African localities, such as the middle Pleistocene of Gladysvale (Lacruz et al., 2002).

ASSESSMENT OF REDUNCINI

It is probable that early reduncines were somewhat koblike, but *Zephyreduncinus* is significant in showing the extent of morphological diversity among early species. Vrba and Haile-Selassie (2006) believed there were at least four reduncines in the Adu-Asa Formation. By the middle of the Pliocene, there was one lineage centered on *Kobus tchadensis* (larger, long-horned, divergent horn cores, curving upward and forward distally, with minimal lateromedial compression becoming slightly anteroposterior, and advanced p4s) and another centered on *Kobus korotorensis*, continuing to look more or less like kobs, and perhaps eventually evolving into *K. kob*. *Menelikia* and *Kobus ancystrocera* constituted additional mid-Pliocene lineages. Later on, *Kobus sigmoidalis* and *ellipsiprymnus* superseded *K. tchadensis*, while both *Menelikia* and *Thaleroceros* (a likely descendant of *K. ancystrocera*) went extinct, unless the Nile lechwe should have a connection with *Menelikia* after all. Vrba and Haile-Selassie (2006) thought that *Redunca* could have existed among the reduncines present around the start of the Pliocene, but otherwise it is unknown until the late Pliocene.

In contrast to the picture just presented, a molecular phylogeny like that of Birungi and Arctander (2001) would imply that living kob and waterbuck differentiated from one another after the stem of both had diverged from lechwe, thereby disconnecting modern kob from any early koblike species. Other molecular phylogenies, however, may not agree (Hernández Fernández and Vrba, 2005: figure 6). Interestingly, Vrba et al. (1994) showed how allometrically corrected osteological characters suggested that *Redunca* originated within *Kobus*, but that adding nonosteological characters and allowing for paedomorphosis made *Kobus* as a whole more derived than *Redunca*. The timing and morphological changes involved in the differentiation of *Kobus* and *Redunca* remain problematical.

Subfamily indet., ?aff. HIPPOTRAGINAE and/or
ALCELAPHINAE
Genus *GENTRYTRAGUS* Azanza and Morales, 1994

Type Species Gentrytragus gentryi (Thomas, 1981).
Horn cores are long, without keels, transversely compressed, fairly uprightly inserted and backwardly curved. Supraorbital foramina are very small and situated close or adjacent to the bases of the horn pedicels. Primitive skull characters include: frontals not raised between the horn bases or pneumatized and the cranial roof very little inclined.

A winding but instructive history of nomenclature needs elucidating. Pilgrim (1939) named a Siwaliks frontlet, quite large for its period, as Hippotraginae gen. indet. (cf. *Tragoreas*) *potwaricus*. It may have come from the type section of the Nagri Formation (Barry, 1995), although a second horn core,

AMNH 96648 (Thomas, 1984a), was thought to be from the Chinji. Gentry (1970, 1978) referred similar material from Fort Ternan to the same species (using "?*Pseudotragus*" in place of "cf. *Tragoreas*"), included some notably hypsodont teeth, and later described material of a second species from the Ngorora Formation. He rejected Pilgrim's hippotragine affiliation and suggested congeneric status with the middle Miocene European "*Gazella*" *stehlini*, rescued from confusion with late Miocene gazelle species by Thenius (1951) but wrongly left within *Gazella*. Thenius (1979) founded *Caprotragoides* for these antelopes with *potwaricus* as type species (appendix note 7). Gentry (1990) was disconcerted when teeth of *Caprotragoides stehlini* at Paşalar, Turkey, turned out to be much smaller and less hypsodont than at Fort Ternan, gravely weakening the generic identity between *stehlini* and *potwaricus*. Azanza and Morales (1994) founded *Tethytragus* for European and Turkish middle Miocene species with teeth little different from *Eotragus*, and *Gentrytragus* for African and possibly Arabian (Thomas, 1983) species, some of which appear to have had hypsodont teeth and much longer M2s and M3s than in *Tethytragus*, and to have lasted into the late Miocene. They confined *Caprotragoides* to the Siwaliks fossils. This tortuous story arose because backward curvature of horn cores is probably an advanced character that appeared several times among early bovids. *Tethytragus* could be an antilopine, the earliest of the Caprinae, or something else, and *Gentrytragus* is probably unrelated to it; neither genus is boselaphine.

GENTRYTRAGUS THOMASI Azanza and Morales, 1994

Synonymy Gen. indet. (?*Pseudotragus*) *potwaricus* Gentry, 1970:284, plates 12–14; plate 16, figures 4–5. *Fort Ternan. Middle Miocene.

Remarks Assigned teeth are more hypsodont than in the sympatric *Hypsodontus tanyceras*. This species is also claimed for the Hofuf Formation in Saudi Arabia (Thomas, 1983: plate 1, figure 4).

GENTRYTRAGUS GENTRYI (Thomas, 1981)

Synonymy ?*Pseudotragus* sp. nov. Gentry, 1978:297. Ngorora Fm.; *Pseudotragus? gentryi* Thomas, 1981:381, plates 5–6, text figures 15–18. *Ngorora Fm Mb D3, loc. 2/11. Also Mbs B–C, E. Middle–late Miocene.

Remarks The holotype skull gives much of the morphological information about this species. It is slightly larger than *G. thomasi* and with less compression of the horn cores.

Assessment of Gentrytragus Without information on its teeth it would be unwise to make the Siwaliks *Caprotragoides potwaricus* again conspecific with either of the African *Gentrytragus* species. *Gentrytragus* existed over a long time period from Fort Ternan at 14.0 Ma until Ngorora levels postdating the appearance of hipparionine horses (which are not present at loc.2/11 itself, the type locality for *G. gentryi*). The long and straight cranial roof and rather upright horn insertions are less suggestive of *Hippotragus* than is the late Miocene *Tchadotragus* (discussed later), but the short premolar rows agree. Some advanced characters suggest an alcelaphine relationship, especially for *G. gentryi*: no ethmoidal fissure, hypsodont cheek teeth, semicircular upper tooth arcades with M3s less widely apart than M1s or M2s, very short premolar rows. However, it would be premature to extend

Alcelaphini back into the middle Miocene on the basis of *Gentrytragus*.

Subfamily HIPPOTRAGINAE Sundevall, 1845.
Tribe HIPPOTRAGINI Sundevall, 1845

Type Genus Hippotragus Sundevall, 1845.

Living Hippotragini contain *Hippotragus, Oryx,* and *Addax.* They are large and stocky antelopes with long horns and hypsodont teeth, feeding mostly by grazing. The two *Hippotragus* species live in or near the edges of woodland, *Oryx* in dry open, even semidesertic areas, and the little-studied *Addax* within the bounds of the Sahara Desert. Hippotragini are largely an African tribe but there is an *Oryx* in Arabia and Pliocene (–Pleistocene?) hippotragines in the Siwaliks. The long horn cores have no keels or transverse ridges, diverge little, have hollowed pedicels, and are present in both sexes. They are spiraled in *Addax.* Postcornual fossae are shallow when present. Despite their sinuses, frontals between the horn bases are substantially raised only in the extant *H. niger.* Braincase roofs are little sloped downward. Molar teeth have basal pillars and some sign of vertical ribs between the styles or stylids.

Genus *TCHADOTRAGUS* Geraads et al., 2008

Type Species Tchadotragus sudrei Geraads et al., 2008.

TCHADOTRAGUS SUDREI Geraads et al., 2008

Synonymy Tchadotragus sudrei Geraads et al., 2008:231, figures 1–2. *Toros-Menalla, Late Miocene.

Remarks The type species is the only one known and is abundant at the type locality and outstandingly informative as an early hippotragine. The close resemblance to modern *Hippotragus* is unmistakeable (perhaps arising from the inclination and proportional length of the cranial roof and the course of the horn core curvature), but it has primitive characters like smaller size, tooth row less anteriorly positioned, an extensive and shallow preorbital fossa, and the lower-crowned and occlusally simpler teeth. The premolar rows are short and thereby different from modern *Hippotragus* and from the *Miotragocerus* at that time still alive in Africa. Hippotragine remains from Djebel Krechem (Geraads, 1989) could turn out to be this species. So, too, could the later hippotragine at Sahabi (Lehmann and Thomas, 1987; Gentry, 2008) with an even shorter premolar row.

Genus *SAHELORYX* Geraads et al., 2008

Type Species Saheloryx solidus Geraads et al., 2008.

SAHELORYX SOLIDUS Geraads et al., 2008

Synonymy Saheloryx solidus Geraads et al., 2008:236, figures 4A–4E. *Toros-Menalla, Late Miocene.

Remarks Only the type species is known, and it was thought to be near the base of the Hippotragini by Geraads et al. (2008). It was less abundant than *Tchadotragus sudrei* in the same deposits. The horn cores were shorter than in *T. sudrei* and slightly less compressed; the braincase was also shorter. No pneumatization existed in the frontals or horn pedicels, and the supraorbital pits were large. There was no indication of a

relationship to *Oryx.* Some of the hippotragine teeth at Toros-Menalla are likely to belong to *Saheloryx.*

Genus *HIPPOTRAGUS* Sundevall, 1845

Type Species Hippotragus equinus (É. Geoffroy Saint-Hilaire, 1803). See ICZN Opinion 2030, *Bulletin of Zoological Nomenclature* 60:90–91, 2003.

The horn cores show lateromedial compression, varying between different species, and sometimes flattening of the lateral surface. They curve backward and are inserted fairly uprightly above the back of the orbits. Supraorbital pits are small and quite close together at the very base of the horn pedicels. Boödonty of the teeth becomes intensified late in the history of the genus.

HIPPOTRAGUS sp.

Locality Laetolil Beds. Middle Pliocene.

Remarks A *Hippotragus* is common in the Laetolil Beds in contrast to the more restricted representation of *Hippotragus* species in later faunas. It has smaller horn cores than in *Tchadotragus sudrei,* but the size of the teeth, the premolar/molar row length ratio, and the degree of hypsodonty are similar or only slightly more advanced. It is smaller than later *Hippotragus* species. The horn cores show some mediolateral compression, and the premolar rows were as short as in living *Oryx.* The teeth do not show the occlusal complexity of modern *Hippotragus* and have been variably identified (Dietrich, 1950: plate 1, figures 11–12; plate 3, figures 37–40, 42; Gentry and Gentry, 1978: plate 22, figure 3). Their primitive appearance could allow assignment to Tragelaphini, Hippotragini, a *Protoryx* (Caprinae of the Eurasian late Miocene), or even Boselaphini. They can be differentiated from Laetoli alcelaphine teeth by basal pillars, stronger styles on the upper molars, more rugose enamel, and later joining up of the crescentic crests in the center of the molars. More detailed information about this species will appear in Harrison (in press).

HIPPOTRAGUS COOKEI Vrba, 1987

Synonymy Hippotragus cookei Vrba, 1987a:49, figures 1–2. *Makapansgat Limeworks Mb 3, Sterkfontein Mb 4. Late Pliocene.

Remarks This, a larger species than the one in the Laetolil Beds, shows horn cores similar to *H. equinus* but more divergent and tending to have a flattened lateral surface. Teeth are higher crowned than in earlier *Hippotragus* and with goat folds on lower molars, but otherwise not evolving toward the occlusal complexity and lengthened premolar rows of modern *Hippotragus.* It was possibly descended from the species in the Laetolil Beds or perhaps related to the sparingly known *H. bohlini* (Pilgrim, 1939: plate 2, figures 3–6, text figure 6; Gentry, 2000, figure 5.3) of the Siwaliks.

HIPPOTRAGUS GIGAS Leakey, 1965

Synonymy Hippotragus gigas Leakey, 1965:49, plates 56, 58–61. *Olduvai Bed II. Also Olduvai Beds I and III, Elandsfontein. Late Pliocene–middle Pleistocene; *Hippotragus gigas* Vrba, 1987a:49. Makapansgat Limeworks Mb 5. Late Pliocene.

Remarks This was close to *Hippotragus cookei,* but its size became very large in Olduvai Bed II and declined thereafter.

The horn cores are less compressed than in living *Hippotragus*, especially *H. niger*, the braincase low and wide like roan and not as narrow as in *H. niger*, and the braincase and basioccipital both short. The teeth resemble those of *H. cookei*, and the premolar rows were as short as in the species of the Laetolil Beds. The final *H. gigas* at Elandsfontein (Klein and Cruz-Uribe, 1991: figure 8) had teeth on which no evolutionary tendencies toward living species could be detected. Vrba (1976, 1987a) accepted *H. gigas* in Makapansgat Limeworks Member 5, while De Ruiter (2003) considered that all *Hippotragus* from the Gauteng and Limpopo Province caves belonged to *H. gigas*. If both names *cookei* and *gigas* continue in use, then the Olduvai Bed I and Elandsfontein occurrences might be better placed in *cookei*, leaving *gigas* as a localized large variant in Olduvai Bed II. If only one name is required, then *gigas* has seniority. *Hippotragus* cf. *gigas* at Tighenif (Geraads, 1981: plate 2, figures 2–3) was a rare record of *Hippotragus* in northern Africa, but "*Hippotragus priscus*" is not hippotragine (appendix note 8).

HIPPOTRAGUS EQUINUS (É. Geoffroy Saint-Hilaire, 1803)

Synonymy Antilope equina É. Geoffroy Saint-Hilaire, 1803, *Cat. Mamm. Mus. Natn. Hist. Nat.*:259. Extant roan; *Hippotragoides broomi* Cooke, 1947:228, figure 2. Sterkfontein upper quarry (Cooke, 1938). ?Pleistocene.

Remarks The roan antelope is identified from some South African localities—for example, Gladysvale (Plug and Keyser, 1994b), from where Lacruz et al. (2002) also cite *H. niger*. The *Hippotragoides broomi* of Cooke (1947) was discussed by Vrba (1976). The lengthened premolar rows and boödont (oxlike) teeth of the two living *Hippotragus* and *H. leucophaeus* (Klein, 1974: figure 1 [3–4]), with enlarged basal pillars and much occlusal complexity, appear only to have become fully evolved in or after the middle Pleistocene.

HIPPOTRAGUS LEUCOPHAEUS (Pallas, 1766)

Synonymy Antilope leucophaea Pallas, 1766, *Misc. Zool.*:4. Extant until 1799. Cape Province, South Africa; *Hippotragus problematicus* Cooke, 1947:226, figure 1. Bloembos. ?Late Pleistocene.

Remarks Hippotragus leucophaeus, the bluebuck of the southern Cape Province, was smaller than *H. equinus* and was hunted to extinction more than two centuries ago. Its coat was said to look like blue velvet in life (Pennant, 1781). Klein (1974) detailed its possible late Pleistocene range in the southern Cape Province where it coexisted with the larger roan antelope. Wells (1967) suggested that *H. problematicus* might be *H. leucophaeus*.

Genus *ORYX* Blainville, 1816

Synonymy Praedamalis Dietrich, 1950:30.
Type Species Oryx gazella (Linnaeus, 1758). Extant gemsbok.

Horn cores show limited lateromedial compression and later acquiring slight anteroposterior compression. Compared with *Hippotragus*, the horn cores are set at a lower inclination and (a linked character) more postorbitally, nearly straight or with much less backward curvature, and inserted more widely apart. The occipital surface is lower and flatter. The longitudinal ridges behind the anterior tuberosities of the basioccipital are weaker than in modern *Hippotragus*. *Oryx* is a rarer fossil than *Hippotragus*. Pliocene (–Pleistocene?) Siwaliks fossils, referred by Pilgrim (1939) to two species of his genus *Sivoryx*, are likely to be oryxes and referable to *Oryx*.

ORYX sp. or spp.

Synonymy Praedamalis? sp. Harris, 2003:545, figures 11, 12. Nawata Fm, upper mb. Late Miocene.
Remarks A Lothagam horn core may be an early *Oryx*. The horn cores among fragmentary hippotragine remains in the later Miocene at Manonga (Gentry, 1997) also suggested *Oryx*.

ORYX DETURI (Dietrich, 1950)

Synonymy Praedamalis deturi Dietrich, 1950:30, plate 2, figure 23. *Laetoli; *Praedamalis deturi* Gentry, 1981:12, plate 5. Hadar Fm., Mb. DD. Middle Pliocene; *Praedamalis deturi* Gentry, 1987:387 (in part, not plate 10.7). Laetolil Beds. Middle Pliocene.
Remarks This, the type species of *Praedamalis*, was founded as a fossil alcelaphine. It has long, almost straight horn cores, with almost no compression of their cross section, and with upright insertions than in living *Oryx*. I had thought of *Praedamalis deturi* as a mid-Pliocene morphological intermediate and possible common ancestor of *Oryx* and *Hippotragus*. But realizing that the common hippotragine in the Laetolil Beds is already a *Hippotragus* and that this is probably also true of earlier horn cores from North and East Africa, I have to change this view, and it becomes reasonable to put *deturi* into *Oryx*. *Oryx deturi* is present but rare in the Laetolil Beds. The specimen of Gentry (1987a: plate 10.7) comes from the Upper Ndolanya Beds and was wrongly assigned at species level.

ORYX HOWELLI (Vrba and Gatesy, 1994)

Synonymy Praedamalis howelli Vrba and Gatesy, 1994:60, figures 1–3, 4a. *Maka. Middle Pliocene.
Remarks This is a smaller species than *Oryx deturi* with very large sinuses and thin covering bone in the horn pedicels. The second *Praedamalis?* sp. of Harris (2003: figure 11.13) from the Nachukui Fm Kaiyumung Mb would probably be assignable to either *Oryx deturi* or *howelli*.

ORYX sp. or spp.

Synonymy Oryx sp. Gentry, 1985:163, plate 7, figure 4. Shungura Fm, Mb G. Late Pliocene; *Praedamalis deturi* Dietrich; Gentry, 1987, plate 10.7. Laetoli, Upper Ndolanya Beds. Late Pliocene; *Oryx* sp. Harris, 1991:159, plates 5.18–5.19. Koobi Fora Fm: Upper Burgi and KBS Mbs. Late Pliocene.
Remarks These are later fossils than *Oryx deturi* or *howelli*. The inclination of the horn cores remains more upright than in modern *Oryx* species and the cranial roof less inclined. The Koobi Fora material shows probable increasing transverse basal diameter of its horn cores from the Upper Burgi to the KBS Members. Some horn cores, like the Ndolanya Beds example, show slight backward curvature like the living *O. leucoryx* and *dammah*. The puzzling Cornelia horn core of "*Gazella*" *helmoedi* (Van Hoepen, 1932: figure 2) may be an oryx.

ORYX ELEULMENSIS Arambourg, 1979

Synonymy Oryx eleulmensis Arambourg, 1979:82, plate 49, figures 1–4. *Aïn Hanech. Early Pleistocene; *Oryx* cf. *gazella* Geraads, 1981:58, plate 2, figures 1, 1a. Tighenif. Early–middle Pleistocene; *Oryx* cf. *gazella* Chaïd-Saoudi et al., 2006:969. Mansourah. Early Pleistocene.

Remarks The horn core and dental remains of Arambourg may be accepted as an oryx perhaps attaining the size of extant African species. The horn core had no evident curvature and thus resembled *O. gazella*. Two horn cores from Mansourah (Chaïd-Saoudi et al., 2006) could be oriented. They had cross sections closer to modern *Oryx* in their approach to wider transverse than anteroposterior diameters than is the case with oryx horn cores in the late Pliocene to early Pleistocene of the Turkana Basin. The slight curvature of one of them might suggest *O. dammah* (or *O. g. dammah* if the species separation is not accepted).

Genus *WELLSIANA* Vrba, 1987

Type Species Wellsiana torticornuta Vrba, 1987.

WELLSIANA TORTICORNUTA Vrba, 1987

Synonymy Damaliscus sp. (aff. *albifrons*) Wells and Cooke, 1956:23, figure 11. Makapansgat Limeworks. Late Pliocene; *Wellsiana torticornuta* Vrba, 1987a:53, figure 4. *Makapansgat Limeworks Mb 3. Late Pliocene.

Remarks This was founded as a hippotragine but excluded by Vrba and Gatesy (1994), who regarded it as only "possibly related" to the tribe. The holotype and only specimen is a medium-sized frontlet with short and compressed horns, abrupt transition to an even thinner cross section just above the base, long axis of basal cross section widely angled on median line of the skull, slight heteronymous torsion, and short pedicels.

ASSESSMENT OF HIPPOTRAGINI

An early hippotragine would be hard to characterize as different from an early gazelle, other than by larger size, or from an early caprine to which cladistic and molecular studies relate it. It looks as if the central arena for caprine evolution would have been Eurasia and for Hippotragini to the south in Africa. *Oryx* is more rarely fossilized than *Hippotragus* and would presumably always have come from sites sampling drier habitats than those with *Hippotragus*. *Addax* is closer to *Oryx* than to *Hippotragus*, for example, in its flatter occipital surface, but has no fossil history. Cranial lengths are variable in Hippotragini and await more precise analysis.

The mid-Pliocene *Hippotragus* in the Laetolil Beds is too small to fit *H. cookei*, which is a close relative of *H. gigas*. *Hippotragus gigas* reached a large size in Olduvai Bed II but may never have done so anywhere else. *Hippotragus gigas-cookei* survived into the middle Pleistocene at Elandsfontein. The important question regarding South African sites is to decide when *Hippotragus* teeth belong to the *cookei-gigas* lineage and when to the modern stock of *H. leucophaeus, equinus,* and *niger*. Did the modern species only evolve their boödont teeth and long premolars from the middle Pleistocene onward, while their once-larger *cookei-gigas* relative was finding it

impossible to continue? Ecologically, modern *Hippotragus* are nowhere such a numerous component of the faunas as they were at Laetoli. Perhaps they evolved their new dental characters in response to their displacement from a dominant role. The most likely agents to have displaced them would be alcelaphines.

Subfamily ALCELAPHINAE Brooke in Wallace, 1876:224
Figures 38.12 and 38.13

Type Genus Alcelaphus de Blainville, 1816:75.

Alcelaphines are medium to large grazing antelopes of more open country. They carry their heads vertically, and their withers are often high. Their horn core morphology is more fluid than is convenient for delimiting species, but it is often easy to say whether a fossil is alcelaphine or not. Overall, they are the commonest bovid tribe in numbers of fossils and species, but not invariably so at every locality. Modern species show several patterns of character differences: *Connochaetes taurinus* and *C. gnou* are congeneric species with strong differences in size and in skull and horn core morphology, whereas another congeneric pair, *Damaliscus lunatus* and *D. dorcas*, differ in little but size. Again, the single species *Alcelaphus buselaphus* shows much geographic variation of horns, but *Damaliscus lunatus* has only a difference in horn divergence between two subspecies.

The main features of alcelaphines are long skulls, horn cores often with transverse ridges but only rarely with keels, frontals raised between horn bases, frontals with extensive internal sinuses and one large sinus reaching up into the base of the horn core, frontals raised between horn bases, females horned. Supraorbital pits are small, ethmoidal fissures absent in adults, preorbital fossae are usually present and with an upper rim, and they are slightly deeper in males than females. The zygomatic arch deepens anteriorly under the orbits, and the jugal has two broad anterior lobes. Braincases are short and often strongly angled on the long face, mastoids are large, and the basioccipital has a central longitudinal groove. The upper tooth rows are set anteriorly and have curved arcades so that the P2s and M3s on opposite sides are closer to one another than are P4s or M1s. Teeth are hypsodont (sometimes very hypsodont), premolar rows short with p2s and sometimes P2s reduced or absent. Molars are without basal pillars. Upper molars have complicated central fossettes, lingual lobes of upper molars and labial lobes of lowers are rounded, ribs between the styles of upper molars are strong and rounded, lower molars have no goat folds, p4s have small hypoconids and paraconid-metaconid fusion to close the anterior part of the lingual wall. Mandibles are deep. Limb bones are cursorial and specialized to facilitate anteroposterior articulation.

An early alcelaphine (Gentry, 1980: figures 31–33, 38) might have little-compressed horn cores, slightly divergent, moderately inclined in side view, curved backward, inserted close together and over the back of the orbits, supraorbital pits close together and also close to the horn pedicels, and fairly wide dorsal orbital rims, all much the same as in an early gazelle, hippotragine or caprine. Two characters more suggestive of Alcelaphini would be raising of the frontals between the horn insertions and a sloping cranial roof. One frequent change from the primitive state, linked with the more vertical carriage of the head in alcelaphines, has been for the horn cores to become more backwardly inclined. This

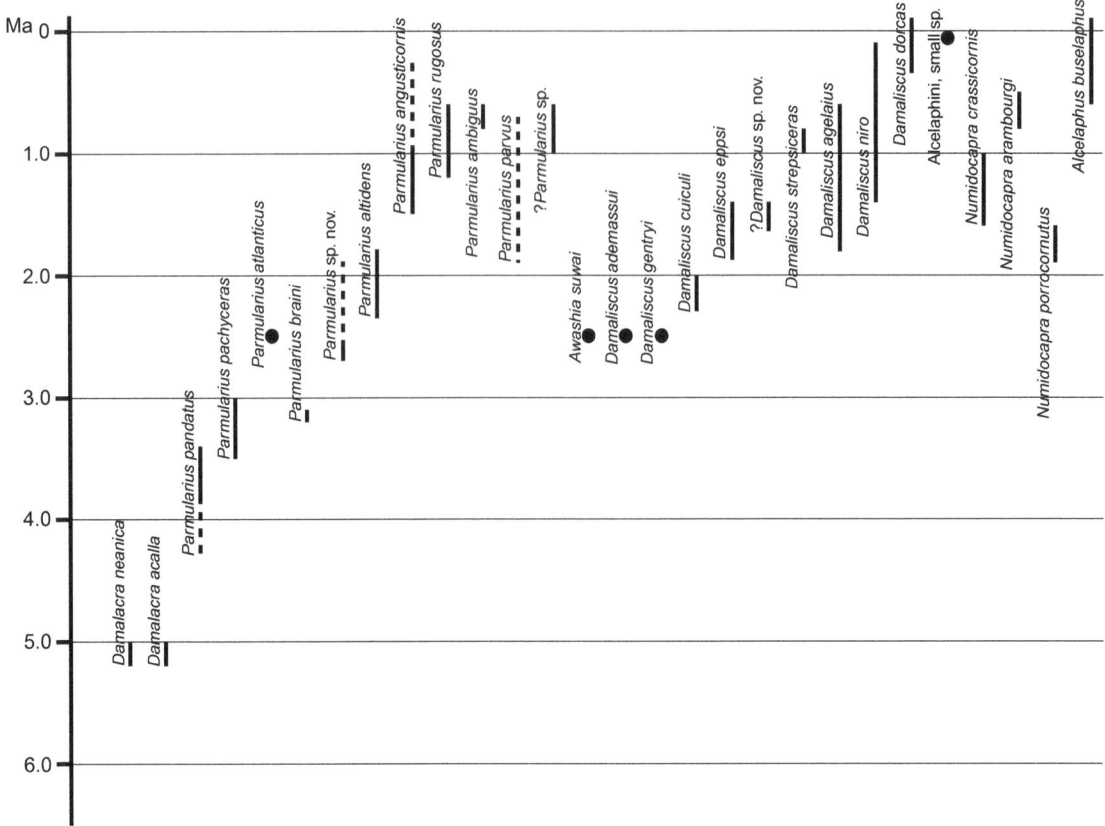

FIGURE 38.12 Temporal distribution of African fossil Alcelaphini. *Damalborea*, *Beatragus*, *Megalotragus*, and *Connochaetes* are excluded.

causes the insertion positions of the horn cores to look more posterior than they would otherwise have done, shortens the cranial roof and steepens its slope, increases the distance from the tops of the pedicels to the supraorbital pits and largely eliminates the possibility or need for projecting dorsal orbital rims. The bulkier horn cores of male animals have similar effects.

The earliest Alcelaphini are in the late Miocene ca. 7.5–7.0 Ma, as at Lothagam. This means that the late Miocene north Italian alcelaphines (Thomas, 1984b; tribal identity doubted by Vrba, 1997) are no longer a temporal as well as a geographic anomaly. Some of the Lothagam horn cores might be hippotragine, but others (Harris, 2003: figure 11.17) are reliably alcelaphine. Interestingly, alcelaphines are absent in the latest Miocene Adu-Asa Formation (Haile-Selassie et al., 2004). Fortunately, Langebaanweg has provided the well-preserved remains of two early alcelaphine species. In the following account of Alcelaphini, there are two sequences of genera, the second one starting with *Damalborea*. The temporal distribution of African fossil species of the *Parmularius-Damaliscus* group is shown in figure 38.12. Outline views of a number of crania of species in this group are shown on figure 38.13.

Genus *DAMALACRA* Gentry, 1980

Type Species Damalacra neanica Gentry, 1980:265.

The tribal identity of the genus is shown by the sinuses in the frontals and horn pedicels, both sexes horned, high

frontals between the horn bases, small supraorbital pits, basioccipital with a central longitudinal groove having its sides formed by ridges behind the anterior tuberosities, hypsodont teeth, and short premolar rows, among other characters. The teeth are primitive for alcelaphines. After Langebaanweg, alcelaphine teeth become increasingly unlikely to be confused with Tragelaphini or early Hippotragini. *Damalacra* may have had a wide distribution in Africa—for example, at Kanapoi (Harris et al., 2003, but surely not figure 26), Manonga (Gentry, 1997), Sahabi (Lehmann and Thomas, 1987: figure 4B), and Wadi Natrun (Stromer, 1907; Gentry, 1980).

DAMALACRA NEANICA Gentry, 1980

Synonymy Damalacra neanica Gentry, 1980:265, figures 28–30, 34, 37. *Langebaanweg. Early Pliocene.

Remarks This species has straight and divergent horn cores with slightly postorbital insertions. The postorbital insertion seems to have evolved without making the inclination lower in side view, but it must be linked with the steep cranial roof.

DAMALACRA ACALLA Gentry, 1980

Synonymy Damalacra acalla Gentry, 1980:272, figures 31–35, 37–38, 40. *Langebaanweg. Early Pliocene.

Remarks This *Damalacra* is more primitive and shows less divergence, backwardly curved horn cores, and insertions less postorbital. Pairs of species showing similar differences to the two *Damalacra* can be found at later time levels up to the living *Damaliscus lunatus* and *Alcelaphus buselaphus*, but routes of descent are not clear.

Genus *PARMULARIUS* Hopwood, 1934

Type Species Parmularius altidens Hopwood, 1934:550.

An extinct genus prominent in the later Pliocene and centered on Olduvai from where three species were described. *Parmularius* has high and narrow skulls. It shows the linked characters of much inclined horn cores, postorbital insertions, long pedicels, and short braincases with steep roofs. They often have posteromedial, posterior, or posterolateral swellings on the horn core bases, and often a median conical parietal boss on the cranial roof. The supraorbital pits remain relatively close. Preorbital fossae are small. Teeth are advanced on those of *Damalacra*, and the short premolar row is often without p2s. Many species have been named but, as with so many alcelaphines, a change in horn cores can indicate anything from an infraspecific variant up to an incoming pan-African novelty at genus level. *Damalops palaeindicus* (Falconer, 1859) of the late Pliocene Pinjor Formation, India (Lydekker, 1886: plate 4, figures 4–5) and Tadzhikistan (Dmitrieva, 1977) now seems likely to have been in or near *Parmularius*.

PARMULARIUS PANDATUS Gentry, 1987

Synonymy Reduncini gen. et sp. indet. Dietrich, 1950:364, figure 21. Laetoli; *?Parmularius* sp. Gentry and Gentry, 1978:382, 62, plates 21, 22, figure 2. Laetoli; *Parmularius pandatus* Gentry, 1987:389. *Laetolil Beds, Laetoli. Middle Pliocene.

Remarks This is a common species in the upper unit of the Laetolil Beds and probably present in the lower unit dating from before 4.0 Ma (and perhaps only about 1.0 myr after the Langebaanweg *Damalacra*). It differs from *P. altidens* by less inclined horn core insertions, posterolateral basal swellings on the horn cores, less shortened braincase, and a lower and less localized boss on the braincase roof. The fairly abrupt bending backward in the midcourse of the horn cores of the holotype is not found in all Laetoli horn cores. The occipital has a median vertical ridge so that the two flanking surfaces face partly laterally as well as backward, a character of earlier alcelaphines. The species may predate the origin of *Damaliscus*.

PARMULARIUS PACHYCERAS Geraads et al., 2001

Synonymy Parmularius pachyceras Geraads et al., 2001:339, figures 3O, 5D–5E. *Koro Toro. Middle Pliocene.

Remarks Another early species is larger than *Parmularius altidens* and shows specialized thick horn cores having a posterior surface toward their base. Such characters presumably prevent it being connected with later species elsewhere. Premolar rows are fairly short befitting a *Parmularius* of the period.

PARMULARIUS ATLANTICUS Geraads and Amani, 1998

Synonymy Parmularius atlanticus Geraads and Amani, 1998:198, figures 1D–1E. *Ahl al Oughlam. Late Pliocene.

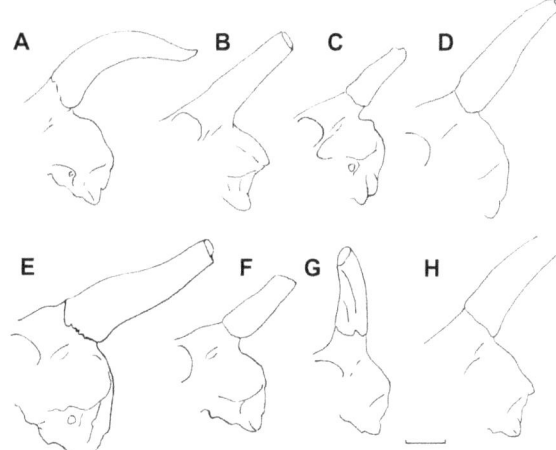

FIGURE 38.13 Left lateral views of alcelaphine crania. A) *Parmularius pandatus*. B) *Damalops palaeindicus*. C) *Parmularius altidens*. D) *Damaliscus eppsi*. E) *Parmularius angusticornis*. F) *Parmularius rugosus*. G) *Numidocapra arambourgi*. H) *Damaliscus niro*. Anterior to the left. Scale = 50 mm.

Remarks This *Parmularius* is larger than *P. pandatus*, with long horn cores curving backward and showing increasing divergence until shortly before the tips. The long axis of the cross section is oblique to the sagittal plane. The authors note difficulty in deciding a generic attribution, and I note similarity to the Siwaliks *Damalops palaeindicus*.

PARMULARIUS BRAINI Vrba, 1977

Synonymy Cf. *Gorgon taurinus* Wells and Cooke, 1956:24. Makapansgat Limeworks; cf. *Alcelaphus robustus* Wells and Cooke, 1956:25, figure 12. Makapansgat Limeworks; *Parmularius braini* Vrba, 1977:140, figures 3–5. *Makapansgat Limeworks, Mb 3. Late Pliocene.

Remarks Parmularius braini is a larger species about 1.0 Ma younger than *P. pandatus*. The horn cores of the holotype frontlet are well compressed, with posteromedial basal swellings, backwardly curved and closely inserted. The cranial roof slopes steeply and has a parietal boss. Vrba (1977, 1987b) assigned the *Connochaetes*-sized teeth in Wells and Cooke (1956) to this species and excluded *Connochaetes* from the bovid list for Makapansgat Limeworks. Vrba (1997:131) noted a smaller *Parmularius* or *Damaliscus* in Makapansgat Limeworks Member 3.

PARMULARIUS sp. nov. Gentry, in press

Synonymy Alcelaphini, small sp. Gentry, 1987:400; *?Pelea* sp. Gentry, 1987:402; *Parmularius* sp. nov. Gentry (in press). *Upper Ndolanya Beds, Laetoli. Late Pliocene.

Remarks This species is smaller than *Parmularius altidens* and with small-diameter, straight, and little-divergent horn cores. The braincase is primitive for *Parmularius* in being neither shortened nor well inclined. The species is perhaps an earlier relative of an alcelaphine with short and deep mandibles in the Shungura Formation and Olduvai Bed I ("*Antidorcas* sp." of Arambourg [1947] and "Alcelaphini species 4" of Gentry and Gentry [1978]). However, two assigned horn cores at Olduvai are not very similar.

PARMULARIUS ALTIDENS Hopwood, 1934

Synonymy Parmularius altidens Hopwood, 1934, *Ann. Mag. Nat. Hist.*, 10, 14:550. *Olduvai Bed I. Later recorded from Shungura G–H; Koobi Fora Fm, KBS Mb. Late Pliocene; *Redunca eulmensis* Arambourg, 1979:76, plate 44, figure 4. Aïn Boucherit. Late Pliocene.

Remarks The holotype skull of the type species (Leakey, 1965: plate 70) is almost certainly female and shows well the boss on the cranial roof after which the genus was named. Excavations in the 1960s showed that the slender, backwardly curved horn cores gradually acquired a straighter profile in the higher horizons of Olduvai Bed I. The common alcelaphine of the Ndolanya Beds (a different species from the *Parmularius* sp. nov. in the same beds) is most probably an ancestor of *P. altidens* continuing from *P. pandatus*. The North African *P. atlanticus*, larger and slightly more primitive than *altidens*, is another possible ancestor. I follow Geraads (1981) in attributing *Redunca eulmensis* to *P. altidens*.

PARMULARIUS ANGUSTICORNIS (Schwarz, 1937)

Synonymy Damaliscus angusticornis Schwarz, 1937:55, no figure. *Olduvai ?Bed II. *Parmularius angusticornis* Gentry and Gentry, 1978:382, plates 23, 24, 25, figure 3. Olduvai middle and later Bed II, Peninj, Isimila, Kanjera. Early (–middle?) Pleistocene; *Parmularius angusticornis* Vrba, 1997:168. Bouri-1. Early Pleistocene.

Remarks Crania of this species are larger and more heavily built than in *Parmularius altidens*, its temporal predecessor and presumed ancestor at Olduvai, and with a much shorter braincase. The horn core bases are far behind the orbits (or high above them if a vertical carriage of the head is assumed). Gentry and Gentry (1978) refer to further synonyms and illustrations in Leakey (1965). Harris (1991: figure 5.59) refers to a partial cranium with similarities to *P.angusticornis* from an unknown horizon in the Koobi Fora Formation, but the species is otherwise unknown or very uncommon there. Ditchfield et al. (1999) locate the Kanjera record at Kanjera North. The species looks like the end of an evolutionary line.

PARMULARIUS RUGOSUS Leakey, 1965

Synonymy Parmularius rugosus Leakey, 1965:59, plates 75–76. Olduvai Bed III and * IV. Early–middle Pleistocene.

Remarks The holotype skull is about the size of *Parmularius altidens*. The horn cores are short, have a posterolateral basal swelling, and curve outward. The cranial roof boss is small. The curled and short horn cores seem to be almost the only difference from *P. altidens*. Some problematic specimens perhaps linked with this species came from other levels at Olduvai (Gentry and Gentry, 1978: plate 26, figures 1–2; plate 32, figures 2–3; plate 35, figure 2).

PARMULARIUS AMBIGUUS (Pomel, 1894)

Synonymy Boselaphus ambiguus Pomel, 1894b:52, plate 6, figures 14–17. *Tighenif. Early–middle Pleistocene; *Connochaetes prognu* Pomel, 1894b, plate 3, figures 3–4; *Parmularius ambiguus* Geraads, 1981:64, plate 2, figure 6, plate 3.

Remarks I follow Geraads's (1981) analysis of the taxonomic confusion in Pomel (1894b) between this species and some *Connochaetes* specimens. As a result, the lectotype is a tooth rather than a previously misidentified and probably lost horn core. *Parmularius ambiguus* is the most common bovid at Tighenif and occurs also at Aïn Maarouf (Geraads and Amani, 1997), perhaps a slightly later site (Geraads, 2002). It is bigger than *P. altidens*. The strong angling of the planes of the frontals behind and in front of the horn bases indicate a steeply inclined cranial roof, and the supraorbital pits lie well anterior to the horn bases. Both characters suggest *Parmularius*. Moreover, the fairly localized backward bend of the horn cores can be found in earlier *Parmularius* and the marked thinning of their distal parts recalls *P. angusticornis*. The disturbing feature of this species is that the premolar rows are long for a *Parmularius*. The p2 is present in lower dentitions, but even examples of *P.altidens* and *P. pandatus* with p2s have p2–4/m1–3 ratios about 7% shorter than in *P. ambiguus*.

PARMULARIUS PARVUS Vrba, 1978

Synonymy Parmularius parvus Vrba, 1978:23, plates 2–6. *Kromdraai A. Early (–middle?) Pleistocene.

Remarks This interesting species is as small as *Damaliscus dorcas* and is represented by dental remains and the top of a face back to the top of the horn pedicels.

?PARMULARIUS sp.

Locality and Age Elandsfontein. Middle Pleistocene.

Remarks A frontlet and horn cores, EFT 20076, was possibly *Parmularius* for Gentry and Gentry (1978), possibly a new genus for Klein and Kruz-Uribe (1991: figures 12, 13), and possibly a caprine for Vrba (1997).

Genus *AWASHIA* Vrba, 1997

Type Species Awashia suwai Vrba, 1997:172.

AWASHIA SUWAI Vrba, 1997

Synonymy Awashia suwai Vrba, 1997:172, figures 14, 15. *Matabaietu 3. Late Pliocene.

Remarks This moderately large alcelaphine has quite a wide skull. Horn cores are not compressed, diverging at their bases but less so distally, and quite well inclined with a backward bend shortly above the base. The maximum diameters are set widely to the skull midline. The postcornual fossa is deeper and more localized than in most alcelaphines. A steep cranial roof. Nasals are broad and flat. Some characters fit *Parmularius*: supraorbital pits close together and rather far in front of the horn core bases, a steeply inclined braincase roof, and perhaps the well-inclined horn core insertions. However the preorbital fossae are very large with marked upper rims and any parietal boss is unclear. *Awashia* is certainly a puzzling alcelaphine and was thought by Vrba to descend from near the origin of *Damaliscus*. One can wonder whether it was native to northeast Africa and whether it outlasted the Pliocene.

Genus *DAMALISCUS* Sclater and Thomas, 1894

Type Species Damaliscus dorcas (Pallas, 1766). Extant.
Damaliscus shares high and narrow skulls and backwardly curved horn cores with early *Parmularius* and shows signs of

a parietal boss. It remains more primitive than *Parmularius* in the linked characters of horn cores inserted over the back of the orbits and more upright, pedicels shorter, frontals little raised between the horn insertions, dorsal orbital rims wider, supraorbital pits closer to the bases of the horn cores, and braincase longer. The preorbital fossa remains large, the premolar rows are not short for an alcelaphine, and p2s are often present. It is advanced in horn cores often compressed (rare in *Parmularius*) and, at the present day, in frontals being somewhat upbowed into a shallowly convex surface in front of the horn bases, and in the supraorbital pits becoming wider apart.

DAMALISCUS ADEMASSUI Vrba, 1997

Synonymy Damaliscus ademassui Vrba, 1997:170, figure 12a. *Gamedah. Late Pliocene.

Remarks This species is about the size of a large *Damaliscus dorcas*. The holotype cranium may be from a female. It is not greatly different from the earlier *Parmularius pandatus* but shows a plausible relationship with one or another *Damaliscus* in the compression of the horn cores (78%), their increased basal divergence, a flattened lateral surface, strong and widely spaced transverse ridges on the anterior surfaces, and their increased basal divergence. There are no localized basal swellings on the horn cores. The front of the horn pedicel looks quite long, suggesting that the short pedicels of later *Damaliscus* may be secondary.

DAMALISCUS GENTRYI Vrba, 1977

Synonymy Damaliscus gentryi Vrba, 1977:143, figures 6–7. *Makapansgat Limeworks Mb 5. Late Pliocene.

Remarks The holotype frontlet and only specimen was thought to resemble the informally designated type A variant of *Damaliscis niro* at Olduvai Gorge (Gentry and Gentry, 1978). This means that the horn cores appear to have had a more abrupt change in backward curvature not far above the base than is the case in "typical" *D. niro*. In this they are likely to resemble earlier *Parmularius*.

DAMALISCUS CUICULI Arambourg, 1979

Synonymy Damaliscus cuiculi Arambourg, 1979:84, plate 50, figure 1. *Aïn Boucherit. Late Pliocene.

Remarks A large sinus in the pedicels and some transverse ridges suggest an alcelaphine, and the insertions only just above the orbits suggest *Damaliscus*. Vrba (1995: appendix 27.2, fn. 30) regarded it as conspecific with *Parmularius braini* and *Damaliscus eppsi*. No other *Damaliscus* is known from north Africa, and *Parmularius* is present at Aïn Boucherit.

DAMALISCUS EPPSI Harris, 1991

Synonymy Damaliscus eppsi Harris, 1991:195, figures 5.51–2. *Koobi Fora Fm Okote Mb. Also KBS Mb. Late Pliocene–early Pleistocene.

Remarks Damaliscus eppsi predates and temporally overlaps *D. niro* further south in Africa. The holotype is very like some Olduvai horn cores ("type A") included in *niro* by Gentry and Gentry, 1978 (e.g. Leakey, 1965, plate 86, third from left) in having a localized backward bend in midcourse and a basal

posterolateral swelling. The two species are certainly very close, although the nondivergent horn cores of another Koobi Fora specimen (Harris, 1991: figure 5.52) are different. *Damaliscus eppsi* may be conspecific with *D. gentryi* or may be a related and more completely known northern form. I agree with Harris that it is not related to the earlier *Parmularius braini* of Makapansgat Limeworks Member 3.

?DAMALISCUS sp. nov. Harris, 1991

Synonymy ?Damaliscus sp.nov. Harris, 1991: 200, figures 5.55a, 5.55b. Koobi Fora Fm, Okote Mb. Early Pleistocene.

Remarks The figured skull looks hippotragine by the length of its braincase, but Harris thought that the teeth were undoubtedly alcelaphine and observed the loss of both P2 and p2. Could such losses happen in an aged *Hippotragus gigas*?

DAMALISCUS STREPSICERAS Geraads et al., 2004

Synonymy Damaliscus strepsiceras Geraads et al., 2004b:188, plate 13, figure 1. *Melka Kunture, Garba IV. Early Pleistocene.

Remarks A species with slightly larger horn cores than a male *Damaliscus agelaius* and with an unusual somewhat spiraled course. The premolar row is also longer relative to the molar row and p2 can be present. The authors wrote of it as close to *D. agelaius* but not part of an ancestor-descendant sequence.

DAMALISCUS AGELAIUS Gentry and Gentry, 1978

Synonymy Damaliscus agelaius Gentry and Gentry, 1978:402, plates 29, 30. *Olduvai Beds II–IV. Late Pliocene–middle Pleistocene. ("Late Pliocene" based on FLKW 1969.82a, Gentry and Gentry, 1978:404.)

Remarks The holotype is a female skull from a herd excavated in 1962. The species is about the size of *Damaliscus dorcas* with little compressed, rather divergent, and more upright horn cores. It is more primitive than *D. dorcas* in having no flattening of the side surfaces of the horn cores, frontals less upbowed in front of the horn bases, preorbital fossa deeper, and braincase longer. It is more advanced, however, in shorter premolar rows and no p2s. The supraorbital pits are wide apart as in *D. dorcas*. The preorbital fossa, although deep, is less extensive than in *D. lunatus*.

DAMALISCUS NIRO (Hopwood, 1936)

Synonymy Hippotragus niro Hopwood, 1936, *Ann. Mag. Nat. Hist.*, 10, 17:640, no figure. *Olduvai Bed III. Also Middle–Upper Bed II and IV and many other sites in East and South Africa. Early–late Pleistocene; *Hippotragus niro* Leakey, 1965: plate 54, showing the holotype; *Damaliscus niro* Gentry, 1965:335.

Remarks Damaliscus niro is large at Olduvai II and Peninj (perhaps larger than *Damaliscus lunatus*) and declines thereafter (Gentry and Gentry, 1978). This long-term decline opens the possibility of ancestry to *D. dorcas*. Horn cores often show flattened sides, strong and widely spaced transverse ridges on the front surface (Cooke, 1974: plates 2C, 2D), and backward curvature that is sometimes sharp in midcourse. Compared with *D. lunatus*, the horn cores are more compressed, the widest part of the cross section placed more anteriorly, transverse ridges stronger and wider apart, the insertions more upright, and the braincase more angled on the face.

*DAMALISCUS DORCAS (Pallas, 1766)

Synonymy Antilope dorcas Pallas, 1766, *Misc. Zool.*:6. Extant.

Remarks This species often appears in print as *D. pygargus* (see appendix note 9). The South African bontebok and blesbok (both of them *D. dorcas*) are present in late Pleistocene and later sites in South Africa. Descent from *Damaliscus niro* was denied by Gentry and Gentry (1978) but favored by Klein (1980). Brink (1987: figures 36. 37) claimed that both species were present in a Middle Stone Age assemblage at Florisbad. The interesting Lainyamok fauna in Kenya dates from ~0.36 Ma and contains a cranium unlike *D. agelaius* but like *Damaliscus dorcas* (Potts and Deino, 1995). Evidently a renewed wide survey of late smaller *Damaliscus* is needed, which would also have to take account of a short-horned *Damaliscus* at Cornelia and Elandsfontein (Cooke, 1974: plates 2A, 2B; Klein and Cruz-Uribe, 1991: figure 11), which L. H. Wells once intended to name as *hipkini*.

ALCELAPHINI, Small spp.

Localities and Age Semliki: Katwe Ashes; Lukenya Hill. Late Pleistocene and perhaps later.

Remarks Teeth and horn cores of small alcelaphines have been noted at several late East African localities by Gentry (1990) and Marean (1992). The records at Lukenya Hill and sites in northern Tanzania run from 40,000–14,000 or later BP. Such small species no longer live in East Africa.

Genus NUMIDOCAPRA Arambourg, 1949

Synonymy Rabaticeras Ennouchi, 1953:126.
Type Species Numidocapra crassicornis Arambourg, 1949:290.

These alcelaphines are moderate to large sized with skull proportions nearer to high and narrow than to low and wide. Horn cores curve upward and forward in side view. Frontals are little raised between horn bases, and the braincase roof is strongly sloping with a straight profile.

*NUMIDOCAPRA CRASSICORNIS Arambourg, 1949

Synonymy Numidocapra crassicornis Arambourg, 1949, *C. R. Somm. Séanc. Soc. Géol. Fr.* 13:290, figured. *Aïn Hanech. Early Pleistocene; (?) Gorgon mediterraneus Arambourg, 1979:87, plates 51–54. Aïn Hanech; *Numidocapra crassicornis* Bonis et al., 1988:329, plate 2, figure 3. Anabo Koma. Early Pleistocene; *Numidocapra crassicornis* Vrba, 1997:141, figures 6–7. Bouri 1 and 6. Early Pleistocene.

Remarks The type frontlet was refigured by Arambourg (1979, plate 38, figures 4–4b). It is large, much restored, and has long, upright, almost parallel horn cores with their concave edges anterior. It would have been a striking antelope in life. Geraads (1981) reallocated it from the Caprinae to the Alcelaphini.

NUMIDOCAPRA ARAMBOURGI (Ennouchi, 1953)

Synonymy Rabaticeras arambourgi Ennouchi, 1953:126, figures 1–2. *Quarry 8 on Rabat to Témara road. Middle Pleistocene.

Remarks Vrba (1997) postulated a trend of decreasing size within *Numidocapra*, and this second species is later and smaller than *N. crassicornis*. It has divergent horn cores showing homonymous torsion. It was the type species of *Rabaticeras* and occurred throughout Africa: Rabat, ?Tihoudaïne (Thomas, 1977), Aïn Maarouf (Geraads and Amani, 1997), Olduvai Beds III–IV (?and Bed II, Gentry and Gentry, 1978: plate 31, figure 3; plate 32, figure 1) and Elandsfontein. Hitherto unidentified Olduvai Bed II horn cores ("Alcelaphini spp. 2 and 3"; Gentry and Gentry, 1978: plate 31, figure 3; plate 32, figure 1; plate 40, figures 1–2) could be earlier East African *Numidocapra* species or subspecies, smaller than *N. crassicornis*. The "sp. 3" has straight, compressed, and nondivergent horn cores. Gentry and Gentry saw *N. arambourgi* as an ancestor for *Alcelaphus*, but any likely time gap between the Elandsfontein record (Klein and Cruz-Uribe, 1991: figures 14–15) and the *Alcelaphus* at Bodo is almost too tight. Vrba (1997) related *Alcelaphus* and its fossil relatives to *Megalotragus* and *Connochaetes*.

NUMIDOCAPRA PORROCORNUTUS (Vrba, 1971)

Synonymy Damaliscus porrocornutus Vrba, 1971, *Ann. Transv. Mus.* 27:59, pls.3–5. *Swartkrans Mb 1. Early Pleistocene.

Remarks This is very similar to the preceding species and was discussed in Vrba (1976).

Genus ALCELAPHUS de Blainville, 1816

Type Species Alcelaphus buselaphus (Pallas, 1766).

The horns in *A. buselaphus* are much inclined backward and appear from in front to be on united pedicels of moderate to extreme length, but the rear of the pedicel runs along the top of the cranium so that it is not freestanding. The laid-back horn cores have therefore shortened and steepened the cranial roof. The length, course, and curvature of the horn cores are subject to great geographic variability, on which many subspecific names have been founded. The nominate subspecies, the bubal hartebeest of northern Africa, is now extinct. The supposed second species, *Alcelaphus lichtensteini,* is probably a geographic variety of *A. buselaphus*, closest to *A. buselaphus caama* (Kingdon, 1982; Flagstad et al., 2001), in which the pedicel has widened instead of lengthened; Kingdon (1990) saw its pedicels as secondarily shortened. The preorbital fossae of *Alcelaphus* are slightly smaller than in *Damaliscus* but larger than in *Parmularius*.

*ALCELAPHUS BUSELAPHUS (Pallas, 1766)

Synonymy Antilope buselaphus Pallas, 1766, *Misc. Zool.*:7. Extant hartebeest.

Remarks Vrba (1997: figure 7) described and illustrated a complete skull from Bodo 1 at 0.64 Ma. The site is within the historic range of *Alcelaphus b. swaynei*, and the fossil skull differs from the latter by having larger and less divergent horns mounted on a wider pedicel. Possible *A. b.* aff. *lichtensteini* was recorded from late in the middle Pleistocene of Kabwe, and from a probably slightly earlier level in the Lower Terrace Complex, Semliki, both to the north of the historic range of that subspecies (Gentry and Gentry, 1978; Gentry, 1990). Late Pleistocene records of proven or likely *A. buselaphus* have come from many sites in north Africa and into the Near East (Clutton-Brock, 1970).

Genus *DAMALBOREA* gen. nov.

Type Species Damalborea elisabethae sp. nov.

The single species now placed in this genus needs a formal designation. It has already been extensively discussed by Vrba (1997) who intentionally used an unavailable informal name.

DAMALBOREA ELISABETHAE sp. nov.
Figure 38.14

Synonymy ?*Damalops* sp. (in part) Gentry, 1981:12; *(Damalops)* "*sidihakomae*" Vrba, 1997:132 (table 2), 135, figures 2(a–b), 3, 4.

Holotype A.L. 208–7, a skull with horn cores, right P4–M3, and left M2–3. See figure 38.14.

Horizon The holotype is from the Hadar Fm, Mb SH-3. Middle Pliocene, ca. 3.3 Ma. It is in the National Museum of Ethiopia, Addis Ababa.

Name The generic name indicates an alcelaphine of the north in contrast to the original *Damalacra* near the Cape of Good Hope. The species is named for Dr. E. S. Vrba who has contributed so much to the study of African fossil bovids and to the phylogeny of Alcelaphini.

Diagnosis A moderately large alcelaphine with high and narrow skull proportions. Horn cores moderately long, little compressed, without a flattened lateral surface, transverse ridges probably present distally. Horn cores taper markedly, diverge above the basal third of their length, curve slightly backward, and are inserted close together behind and above the orbits. No torsion. Frontals are raised between the horn bases. Supraorbital pits wide apart; nasals long, narrow and transversely domed; preorbital fossa large and shallow; zygomatic arch thick below the orbit; back of M3 forward of the vertical level of the front of the orbit; central rear palatal indentation forward of lateral ones. Braincase long for an

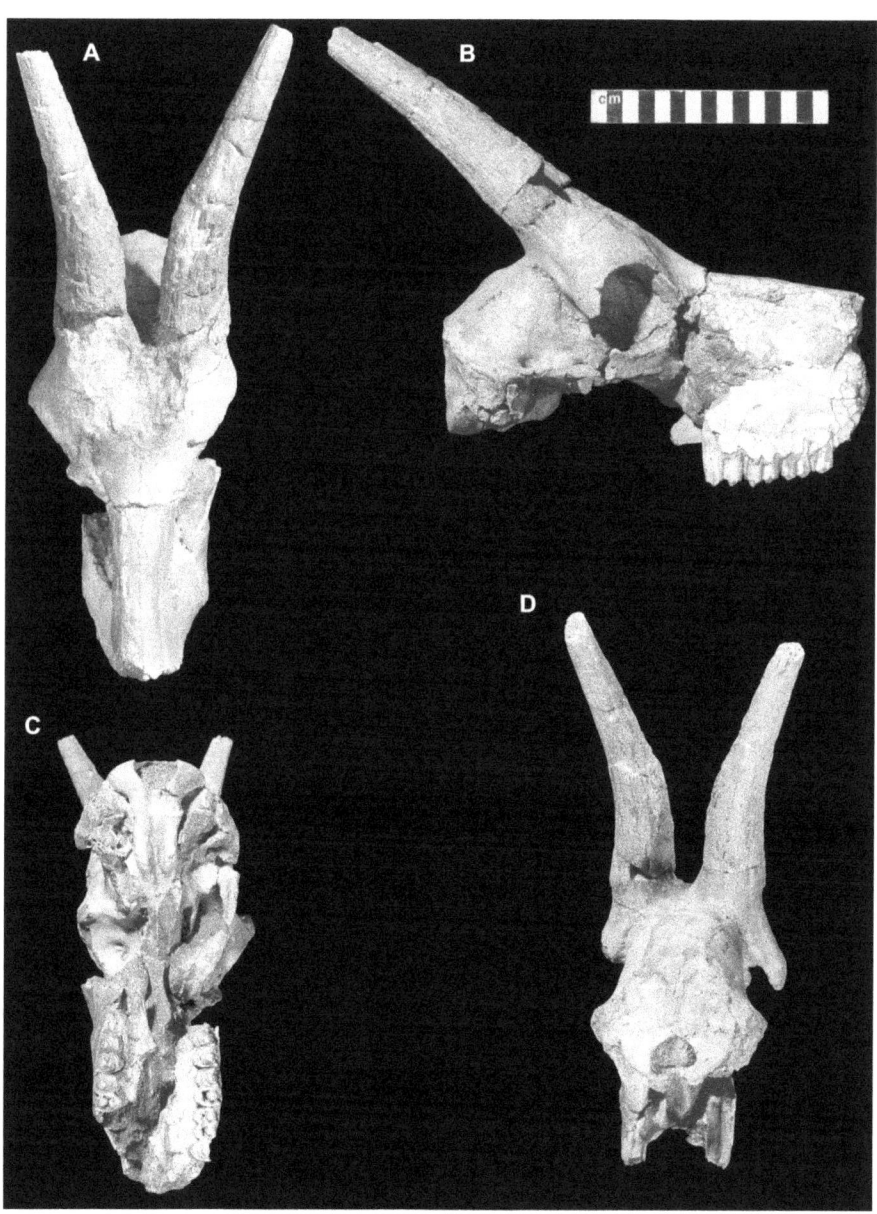

FIGURE 38.14   Holotype skull of *Damalborea elisabethae*, A.L. 208-7 from Hadar Formation Member SH-3. Shown in (A) dorsal, (B) right lateral, (C) ventral, and (D) posterior views. Scale in centimeters. Photograph courtesy of The Natural History Museum, London.

alcelaphine, widening posteriorly, the roof slightly sloping downward and with a slight parietal bump. A large mastoid exposure on the occipital. Teeth differ from *Damalacra* in greater hypsodonty, further decline of already-minimal entostyles and ectostylids, more complicated central fossettes of molars, labial ribs stronger in relation to styles on upper molars, more rounded lingual lobes of upper molars and labial lobes of lower molars.

Measurements of Holotype Anteroposterior diameter at base of horn core 48.5, lateromedial diameter at base of horn core 41.5, minimum width across lateral sides of horn pedicels 101.6, width across lateral edges of supraorbital foramina 63.7, length of frontals ca. 123.6, back of frontals to occipital top 67.2, skull width across mastoids behind external auditory meati 112.4, occipital height from dorsal edge of foramen magnum 48.9, width across anterior tuberosities of basioccipital 27.5, width across posterior tuberosities of basioccipital 36.7, occlusal length M1–3 67.9.

Remarks The holotype (figure 38.14) differs from *Damalacra* by larger size, longer face and nasals, supraorbital pits wider apart, and by tooth characters. Most of the changes are likely to be advances. However the braincase remains long and the horn cores show a slight backward curvature. It differs from earlier *Parmularius* by larger size, less backward curvature of its horn cores, and their distal attenuation and divergence. Gentry (1981) thought it resembled the Siwaliks *Damalops palaeindicus*—for example, in the large preorbital fossa—but it now seems better to associate *Damalops* with the *Parmularius-Damaliscus* group. Characters unlike *Awashia* include less basal divergence of horn cores, degree of divergence increasing distally, no sharp backward curve shortly above the base, maximum horn core diameters not set rather widely to the anteroposterior line of the skull, supraorbital pits perhaps wider apart, and cranial roof less steeply sloped.

Damalborea elisabethae or related species occur at other localities: Aramis, Wee-ee and Maka in the Middle Awash deposits (Vrba, 1997); lower and upper units of the Laetolil Beds (Kakesio skull 82/270; Gentry and Gentry, 1978: plate 22, figure 1); Tulu Bor Member and an unknown horizon of the Koobi Fora Formation (Harris, 1991: figures 5.61, 5.59). I am inclined to follow Vrba (1997: figure 22) in relating this species to later *Connochaetes* and *Megalotragus*.

DAMALBOREA sp.
Figure 38.15

Synonymy ?*Damalops* sp. (in part) Gentry, 1981:12; (*Damalops*) "*denendorae*" Vrba, 1997:132 (table 2), 135.

Remarks In higher Hadar levels, alcelaphine horn cores are more divergent and their curvature is stronger and upward and outward (figure 38.15). Both Gentry and Vrba implied that they evolved from the preceding species. Similar horn cores come from the Lokochot Member (Harris, 1991: figure 5.50) and Shungura Member B ("?*Damalops* sp." of Gentry, 1985). Similar curvature is seen in the horn cores of early *Connochaetes*, which are longer and may have appeared only slightly later.

Genus BEATRAGUS Heller, 1912

Type Species Beatragus hunteri (Sclater, 1889). Extant hirola, Hunter's or Tana River hartebeest.

Remarks These are large alcelaphines, later reducing to medium sized, with horn cores diverging near the base but with long straight distal parts directed upward and, in earlier times, outward. Any detectable torsion is heteronymous, so sufficiently complete horn core bases are distinguishable from *Connochaetes*. Supraorbital pits are wide apart, and p2 usually absent. The period of success for *Beatragus* may have preceded the rise of *Connochaetes*. Vrba (1997) postulated a relationship of *Beatragus* to *Damalacra neanica*. Two large fossil species have been described and the smaller living *B. hunteri* is a relic species.

BEATRAGUS WHITEI Vrba, 1997

Synonymy Beatragus whitei Vrba, 1997:160, figures 10b–10c, 11. *Matabaietu levels 3–5. Late Pliocene.

Remarks This is the largest and earliest *Beatragus* with the longest horn cores known in Alcelaphini. The distal straight parts of the horn cores are more divergent than in later *Beatragus* species. *Beatragus antiquus remotus* Geraads and Amani (1998: figures 2B–2E) from Ahl al Oughlam at about the same period looks very similar.

BEATRAGUS ANTIQUUS Leakey, 1965

Synonymy Beatragus antiquus Leakey, 1965:61. *Olduvai Bed I; Shungura G; Koobi Fora Fm, KBS and Okote Mbs. Late Pliocene.

Remarks This differs from *B. hunteri* by horn cores diverging immediately from their bases, a more gradual lessening of divergence distally, more upright insertions in side view, and frontals wider and more convex in front of the horn bases. Leakey (1965: plate 80) illustrated a complete left horn core, probably from the same individual as his holotype lower right horn core and frontal (Gentry and Gentry, 1978: plate 33). Other specimens are illustrated in Gentry (1985) and Harris (1991). Vrba (1997) thought that her "cf. *Connochaetes* sp." from Swartkrans 1 (Vrba, 1976: plate 6) could be a *Beatragus*, probably *B. antiquus*. A *Beatragus* species may have survived in the Swartkrans 2 and Elandsfontein faunas as the "cf. *Beatragus* sp."of Vrba (1976: plate 4B) and "*Damaliscus* aff. *lunatus*" of Klein and Cruz-Uribe (1991: figure 9). Vrba (1997) discussed the Elandsfontein record.

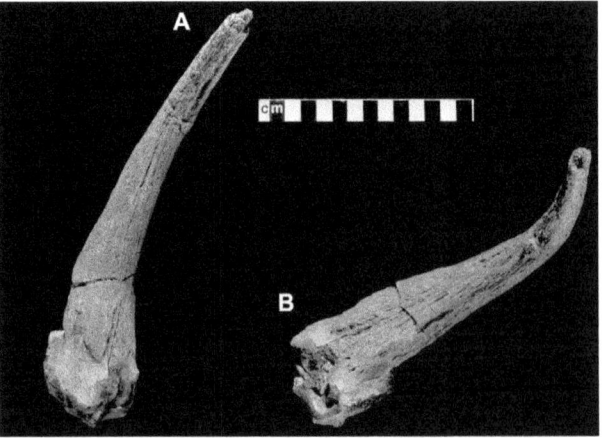

FIGURE 38.15 Anterodorsal (A) and lateral (B) views of left horn core of *Damalborea* sp., A.L. 120-2a from Hadar Formation, Member DD-3. Scale in centimeters. Photograph courtesy of The Natural History Museum, London.

Genus *CONNOCHAETES* Lichtenstein, 1814

Synonymy Oreonagor Pomel 1895:45.

Type Species Connochaetes gnou (Zimmermann, 1780).

This genus appears in the late Pliocene. It is characterized by its large size; skulls tending to be low and wide, especially postorbitally; horn cores diverging in earlier species and emerging transversely or forward in later species, and often with strong curvature. Horn insertions are postorbital.

Connochaetes gnou contrasts strongly with *C. taurinus* in its shorter face, tooth rows set less anteriorly, nasals widening anteriorly, and zygomatic arch not deepened toward its anterior end, nasals widening anteriorly. The horn cores have much expanded bases that greatly diminish their width apart. Above the bases they turn sharply anteriorly but less downward (the tooth rows being horizontal) than in *C. taurinus*. Their distal parts are more attenuated, as also happens in muskoxen with expanded horn bases. Front limbs are probably less elongated than in *C. taurinus*. In general, *C. gnou* looks the more primitive species, but it is advanced in its horn cores and in the more reduced preorbital fossae.

CONNOCHAETES TOURNOUERI (P. Thomas, 1884)

Synonymy Antilope tournoueri Thomas, 1884: 15, plate 7, figure 1. *Aïn Jourdel. Late Pliocene; *Oreonagor tournoueri* (Thomas); Pomel, 1895:45; *Oreonagor tournoueri* Arambourg, 1979:95, plates 55–57. Aïn Boucherit. Late Pliocene.

Remarks This species was founded on material of a primitive wildebeest including Thomas's illustrated type skull top (appendix note 10), which I saw in the 1970s. Pomel (1895) followed Thomas in thinking it was reduncine and founded *Oreonagor* for it. Divergence of the horn cores is strong but less than in extant *Connochaetes*. They are inserted widely apart and behind the orbits but in front of the occipital. The horn cores at Aïn Jourdel are straight in anterior view and curve upward and forward in side view, but at Aïn Boucherit they are more lyrated, bending backward then outward and then upward. The somewhat raised frontals in front of the horn insertions, very evident at Aïn Boucherit, look as if *C. tournoueri* was approaching *C. taurinus* rather than *C. gnou*. A later cranium from Bouri, judiciously discussed by Vrba (1997: figures 8b, 9b), is like those of Aïn Boucherit but looks from the fairly low level at which the horn cores diverge as if it might be more advanced.

CONNOCHAETES GENTRYI Harris, 1991

Synonymy Connochaetes sp. Gentry and Gentry, 1978:365, plate 15. Olduvai Beds I–middle II; *Connochaetes* new species. Harris et al., 1988:97, plates 51–56. Nachukui Fm, Kaitio Mb.; *Connochaetes gentryi* Harris, 1991:192, figure 5.49. *Koobi Fora Fm, upper Burgi Mb. Also KBS and Okote Mbs. Late Pliocene–early Pleistocene; *Connochaetes gentryi leptoceras* Geraads et al., 2004b:187, plate 13, figure 4. Melka Kunturé, Garba IV. Early Pleistocene.

Remarks The holotype skull shows much similarity to *C. taurinus*. Harris noted that its wide muzzle already indicated a grazing antelope. Horn cores turn outward very close to the base and certainly at a lower level than in *C. tournoueri*. Compared with extant *Connochaetes*, they bend less downward as they turn outward, have tips not recurved inward, and are inserted less posteriorly. The subspecies *C. g. leptoceras* has a slender horn core with a long and straight terminal portion

like an Olduvai horn core HWK EE II 2315 (Gentry and Gentry, 1978), here referred to the following species.

CONNOCHAETES AFRICANUS (Hopwood, 1934)
Figure 38.16

Synonymy Pultiphagonides africanus Hopwood, 1934, *Ann. Mag. Nat. Hist.* 10, 14:549, no figure. *Olduvai Bed II. Early Pleistocene; *Pultiphagonides africanus* Leakey, 1965: plates 93, 94. Holotype illustrated; *Connochaetes africanus* (Hopwood); Gentry and Gentry, 1978:364.

Remarks The holotype skull is a small *Connochaetes* with horn cores inserted less posteriorly than in either living species and then passing upward and backward more than outward. The face is short with nasals widening anteriorly and no preorbital fossa, and the cranial roof is steep. This is a puzzling skull with characters at variance with those of *C. tournoueri*, *gentryi*, and *taurinus*. The horn cores look subadult, but the permanent dentition is in place. Gentry and Gentry (1978) postulated ancestry to *C. gnou* of southern Africa, but the tooth row is relatively smaller than in *C. gnou*. An Olduvai horn core HWK EE II 2315 (figure 38.16) has a long and straight terminal portion like *C. gentryi leptoceras*, but a degree of lyration more befitting *C. tournoueri*. Perhaps it is a male of *C. africanus*.

CONNOCHAETES TAURINUS (Burchell, 1824)

Synonymy Antilope taurina (Burchell), 1824 Travels in the interior of southern Africa, 2:278 fn. (2:198 fn in the 1953 edition issued by Batchworth, London). Extant blue wildebeest.; *Damaliscus njarasensis* Lehmann, 1957:118, plate 9. figure 23; plate 10. figure 34. Mumba cave, Eyasi. Late Pleistocene; *Gorgon olduvaiensis* Leakey, 1965:45, plates 49–50, 52. Olduvai Beds III–IV

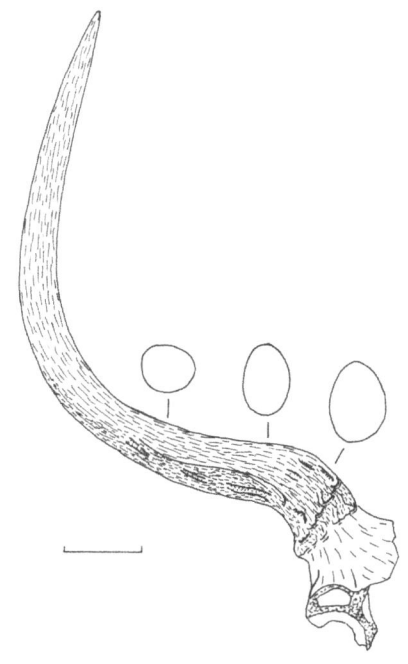

FIGURE 38.16 Anterior view of right horn core of *Connochaetes africanus* from Olduvai Gorge, HWK EE II 2315. Anterior to the right for cross sections of the horn core. Drawn from BMNH cast M35173. Scale = 50 mm.

junction; *Connochaetes taurinus olduvaiensis* Gentry and Gentry, 1978:368, plate 16, figure 3. Olduvai Beds II–III and possibly IV. Early–middle Pleistocene.

Remarks Horn cores of *C. t. olduvaiensis* are inserted slightly less posteriorly and pass less downward than in modern *C. taurinus*. The similar and senior subspecies *C. t. prognu* Pomel, 1894b from Tighenif has been carefully discussed by Geraads (1981:64, 71, 74) and also by Arambourg (1939), Gentry and Gentry (1978:371), and Harris (1991:194). *Connochaetes taurinus* is also present at Kabwe and in South Africa (Lacruz et al., 2002—unless their material could be of the next species). *Damaliscus njarasensis* was thought by Gentry (1990) to belong to *Connochaetes* and would be too far north to be *C. gnou*.

CONNOCHAETES LATICORNUTUS (Van Hoepen, 1932)

Synonymy Gorgon laticornutus van Hoepen, 1932:65, figure 3. *Cornelia. Middle Pleistocene.

Remarks Wildebeest crania and horn cores from Cornelia and Elandsfontein look more or less like *C. taurinus* but a little less advanced in the course of their horn cores. The species was founded on a Cornelia specimen and placed in *Gorgon*, at that time the customary generic attribution for *C. taurinus*. Gentry and Gentry (1978) noted the expanded horn bases with rugose bone beginning to spread across the frontals (Klein and Cruz-Uribe, 1991: figure 17), and they thought that this wildebeest might be an early form of *C. gnou* itself. Others might prefer the use of *laticornutus* to mark a reasonably distinct morphology and to avoid an awkward choice between the species names *taurinus* and *gnou*.

*CONNOCHAETES GNOU (Zimmermann, 1780)

Synonymy Antilope gnou Zimmermann, 1780, *Geogr. Gesch. Mensch. Vierf. Thiere* 2:102. Extant gnu or black wildebeest; *Connochaetes antiquus* Broom, 1913:14, figure 2. Florisbad. Late Pleistocene.

Remarks The living gnu or black wildebeest of South Africa south of 25°S is also known from late Pleistocene and later sites. Brink (1987) discussed the Florisbad subspecies *C. g. antiquus* which has less forwardly turned horn cores than exist today.

Genus MEGALOTRAGUS van Hoepen, 1932

Type Species Megalotragus priscus (Broom, 1909).
These are large extinct alcelaphines probably exceeding *Beatragus whitei* in size. The horn cores diverge and are inserted at a low inclination and postorbitally. Insertions are closer than in *Connochaetes*. Torsion is homonymous and variably expressed. Premolar rows are short and legs long.

MEGALOTRAGUS ISAACI Harris, 1991

Synonymy Megalotragus isaaci Harris, 1991:187, figures 5.46–5.48. *Koobi Fora Fm, KBS Mb. Also Upper Burgi and Okote Mbs. Late Pliocene–early Pleistocene.

Remarks Horn cores are moderately long. This species confirms that the nasals and anterior parts of the frontals are domed upward to form an inflated structure (Harris, 1991: figure 5.47), an unusual specialization of unknown use to the animal. *Megalotragus isaaci* was regarded as a subspecies of the next species by Vrba (1997). *Megalotragus* is also in Shungura G, K, and L and the Ndolanya Beds.

MEGALOTRAGUS KATTWINKELI (Schwarz, 1932)

Synonymy Rhynotragus semiticus Reck, 1925:451. Olduvai; *Alcelaphus kattwinkeli* Schwarz, 1932:4, no figure. *Olduvai Bed IV. Also Beds II and III. The name *kattwinkeli* has been preserved under Opinion 2029 of the ICZN and the rediscovered missing holotype has resumed the status of the name-bearing specimen (*Bulletin of Zoological Nomenclature* 60:88–89, 2003).

Remarks This was the first described *Megalotragus* in East Africa, and the holotype is a right horn core in Munich (Gentry et al., 1995: figure 2). The species differs from *M. isaaci* in shorter and more curved horn cores, presumably the result of a simple transition. It continued to possess the inflated nasal region. Olduvai material (Gentry and Gentry, 1978) demonstrates that the front legs were relatively longer than in *Connochaetes*. Either this or the preceding species was present in the Chiwondo Beds (Kaufulu et al., 1981; Sandrock et al., 2007).

*MEGALOTRAGUS PRISCUS (Broom, 1909)

Synonymy Bubalis priscus Broom, 1909, *Ann. S. Afr. Mus.* 7:280, figured. *Modder River, Free State Province, South Africa. Middle–late Pleistocene.

Remarks This South African species has many synonyms (Gentry and Gentry, 1978). The horn cores are inserted above the occipital surface. Many records are based on teeth larger than appropriate for *Connochaetes* and sometimes with a simpler occlusal pattern. The horn cores may become very long and acquire a curvature like that of *Pelorovis oldowayensis*. The state of the nasals is unknown. *Megalotragus priscus* survived until the end of the Pleistocene or perhaps 7,500 BP (Klein, 1980; Thackeray, 1983).

MEGALOTRAGUS ATOPOCRANION (Pickford and Thomas, 1984)

Synonymy Rusingoryx atopocranion Pickford and Thomas, 1984:446, figure 2. *Rusinga Island. Late Pleistocene or Holocene.

Remarks This is a *Megalotragus* of greatly reduced size. It shows the inflated nasals of other East African *Megalotragus*, although the authors did not use Reck (1935: figure 1) to help identify and orient their finds.

ASSESSMENT OF ALCELAPHINI

The nearly complete and often undistorted fossils at Langebaanweg reveal two species, *Damalacra acalla* and *D. neanica*, the latter being more specialized in its postorbitally inserted but still upright horn cores. By the middle Pliocene, alcelaphines acquired more advanced teeth. *Parmularius pandatus* is a good candidate for descent from a species like *Damalacra acalla* and for relationship and possible ancestry to later *Parmularius* and *Damaliscus* (a group that probably includes the Siwaliks *Damalops*). A slightly larger middle Pliocene species is *Damalborea elisabethae*. Early in the late Pliocene, the first *Beatragus* is known, very large for its period and perhaps descended from close to *Damalacra neanica*. *Megalotragus* and *Connochaetes* also appear. Their ancestry is not clear. If they are related, they could go back through a form like *Damalborea* to an early Pliocene or earlier basal alcelaphine, or they could have come from a Miocene form close to *Gentrytragus* thereby raising the spectre of alcelaphine diphyly. *Connochaetes* and *Megalotragus* fared better than *Beatragus* in the Pleistocene although *Beatragus* managed to outlast *Megalotragus* by a tiny margin. The relationships of *Connochaetes africanus* are an interesting problem.

The great central *Parmularius-Damaliscus* group is a source of uncertainty about how many times *Damaliscus*-like forms evolved. Characters of larger size, inclined horn core insertions, divergence, and inclined cranial roofs can appear in *Parmularius*, but if they fail to do so, the more chance the fossil has to be classified within *Damaliscus*. So was *D. ademassui* the first true *Damaliscus*, or was it a false start? And, turning to late Pleistocene events, we know that the type species *D. dorcas* has to belong to *Damaliscus,* but if it evolved from either *D. agelaius* or *D. niro,* would the rejected ancestor still be congeneric?

It still looks likely that *Alcelaphus* is linked with some smaller species of *Numidocapra* like *N. arambourgi* but the contemporaneous *Parmularius rugosus*, which appears to have supplanted *P. angusticornis* at Olduvai, had also acquired curly horn cores. Hence identifications of individual fossils may be difficult, and one can suspect that generic attributions for the post-*angusticornis* era may be problematic. The laid-back horn pedicels of *A. buselaphus* are visually striking and may have been a fast acquisition. Hybridization is known between species of *Alcelaphus* and *Damaliscus*, some cited instances (Gentry and Gentry, 1978:354) involving *Alcelaphus* with *D. dorcas* and not with the more appropriately sized *D. lunatus*. Vrba (1997), however, has continued to relate *Alcelaphus* to *Connochaetes* and *Megalotragus*. It is interesting that three of the seven alcelaphines at Elandsfontein (Klein and Kruz-Uribe, 1991: figures 9, 11, 12, 13) are not known elsewhere and may be new species in South Africa. They were cited above as ?*Parmularius* sp., a *Damaliscus* species (in the account of *D. dorcas*), and *Beatragus antiquus*.

Subfamily CAPRINAE Gray, 1821

Type Genus Capra Linnaeus, 1758.

Caprinae are one of the great divisions of the Bovidae and are largely Eurasian. They are a coherent subfamily zoogeographically and in molecular cladistics (Gatesy et al., 1997; Marcot, 2007), but it is hard to characterize the genera as a group or to arrange them in tribes. Teeth never show boödont tendencies, central incisors remain small, metapodials are often short, and the animals may be agile jumpers. They often live at higher altitudes. Many late Miocene bovids in Eurasia were acquiring characters of various later Caprinae (Gentry, 2000), and much parallel evolution seems to have occurred.

Genus *BENICERUS* Heintz, 1973

Type Species Benicerus theobaldi Heintz, 1973.

BENICERUS THEOBALDI Heintz, 1973

Synonymy Benicerus theobaldi Heintz, 1973, *Ann. Scient. Univ. Besançon Géol.* (3), 18:245, plate 1. *Beni Mellal. Middle Miocene.

Remarks The genus has only one species. A single left horn core is small, compressed, and has an anterior keel and heteronymous torsion. A few small teeth were also found

(Lavocat in Choubert and Faure-Muret, 1961). It could be an antilopine (Bouvrain and Bonis, 1985) or connected with *Tethytragus*, a European genus already mentioned under *Gentrytragus*. The torsion of the horn core and its early date match Hypsodontinae, but its backward curvature does not.

Genus *DAMALAVUS* Arambourg, 1959

Type Species Damalavus boroccoi Arambourg, 1959.

DAMALAVUS BOROCCOI Arambourg, 1959

Synonymy Damalavus boroccoi Arambourg, 1959:120, plate 18, figures 4, 4a. *Bou Hanifia. Late Miocene.

Remarks This species is known from a skull top and other horn cores. It is medium sized, and the horn cores show some compression and are divergent, well inclined, and slightly curved backward. Arambourg thought it was alcelaphine, while Gentry (1970) related it to the later and larger Eurasian Turolian *Palaeoryx*. I list it here as possibly Caprinae.

Genus *PROTORYX* Major, 1891

Type Species Protoryx carolinae Major, 1891. *Pikermi, Greece. Late Miocene.

Protoryx and related genera are nonboselaphine antelopes long known from the classical Eurasian Turolian faunas of Pikermi, Samos, and Maragheh. They are moderate sized (moderate to large for their period), and with higher frontals between the horn bases than in the smaller middle Miocene *Tethytragus*. They have horn cores without much divergence, inserted fairly uprightly and curving backward, no torsion, and rarely keeled. The braincase roof is inclined, and the teeth are often mesodont or hypsodont. Characters akin to those of Caprini only become well expressed in later species.

PROTORYX SOLIGNACI (Robinson, 1972)

Synonymy Pachytragus solignaci Robinson, 1972:75, figures 1–5, 7. *Beglia Formation, Bled Douarah. (?Middle)–late Miocene.

Remarks Much material came from the type locality in levels alongside and allegedly below hipparionine horses. This species predates many Eurasian *Protoryx* and is specialized in its very compressed horn cores. It was present at El Hamma du Djerid (Thomas, 1979) and perhaps also in the Namurungule Formation (Nakaya et al., 1984: plate 9, figure 4). A possible record in the Ngorora Formation (Thomas, 1981) may be a reduncine. A Vallesian *Protoryx* in Turkey has been identified with *P. solignaci* (Köhler, 1987; Gentry, 2000).

Tribe CAPRINI Gray, 1821

Type Genus Capra Linnaeus, 1758.

This tribe contains those Caprinae which are more fully like sheep or goats. They show frontals raised between horn insertions and with internal sinuses, cranial roofs often angled downward on the line of the facial axis, high-crowned cheek teeth with an uncomplicated occlusal pattern and no basal pillars, p4s often with paraconid-metaconid fusion to form a closed lingual wall anteriorly, and often shortened metapodials.

Genus *BOURIA* Vrba, 1997

Type Species Bouria anngettyae Vrba, 1997

BOURIA ANNGETTYAE Vrba, 1997

Synonymy Bouria anngettyae Vrba, 1997:182, figure 16. *Bouri. Early Pleistocene.

Remarks A cranium and other remains date from about 1.0 Ma. The horn cores are long, compressed, with an anterior keel, and backwardly curved, all as in *Capra*. However, the horn cores are more inclined backward than in *Capra* and more than would be expected for such long curving horns. Also the frontals between the horn bases are more elevated and the cranial roof shorter and more steeply sloped than in *Capra*. Such features are like the extant but short-horned *Hemitragus jemlahicus* and *H. hylocrius*, so comparisons will be needed with the early Pleistocene *Hemitragus* in Europe (Crégut-Bonnoure, 2007 and references therein). Alternatively, an independent evolution of a caprine or caprine-parallel in Ethiopia could be considered. The basioccipital is not expanded as in modern Caprini, and no teeth at Bouri could be recognized as caprine.

Genus *CAPRA* Linnaeus, 1758

Type Species Capra hircus Linnaeus, 1758. Founded on domestic livestock. *Capra aegagrus* Erxleben 1777 is to be used for the wild goat (ICZN Opinion 2027, *Bulletin of Zoological Nomenclature* 60: 81–84, 2003) contrary to Wilson and Reeder (2005).

The ibexes and goats are mainly Eurasian but have two living species in northeastern Africa, one being in Ethiopia.

CAPRA PRIMAEVA Arambourg, 1979

Synonymy Capra primaeva Arambourg, 1979:49, plate 36, figures 3–4; plate 37; plate 38, figures 1–3; plate 39. *Ain Brimba. Late Pliocene.

Remarks The horn cores are without keels, variably compressed, backwardly curved, and inadequate to sustain a tribal identity. They are smaller than horn cores of *Bouria anngettyae*. However, teeth, including those on the holotype palate, are hypsodont and caprine-like, and metapodials are rather short and somewhat flattened anteroposteriorly, again befitting Caprinae.

Genus *AMMOTRAGUS* Blyth, 1840

Type Species Ammotragus lervia (Pallas, 1777).

Only the type species is known in Africa. Its common name is "Barbary sheep," but it has often been seen as closer to or within *Capra* (Ansell, 1971).

AMMOTRAGUS LERVIA (Pallas, 1777)

Synonymy Antilope lervia Pallas, 1777, *Spicil. Zool.* 12:12. Extant Barbary sheep.

Remarks The species is known from many North African late Pleistocene localities. It includes a pair of crushed horn cores from Wadi Derna attributed by Bate (1955) to the bovine *Homoioceras*. Unfortunately, her posthumous paper muddled right and left sides. Moreover, the real right horn core (BMNH, no number) shows that the insertion is too close behind the orbital rim to fit *Bos* or *Pelorovis*. Moullé et al (2004) claimed an earlier *Ammotragus* from the early Pleistocene of France.

Tribe OVIBOVINI Gray, 1872

Type Genus Ovibos Blainville, 1816.

Hitherto this tribe has included the Nearctic (formerly Holarctic) muskox, *Ovibos moschatus*, and Chinese takin, *Budorcas taxicolor*. Bouvrain and Bonis (1984) doubted their closeness and were followed by Gentry (1992, but not 1996). Groves and Shields (1997) and some other molecular phylogenies now suggest a *Budorcas-Ovis* relationship. *Ovibos* has a number of fossil relatives (no longer including the Eurasian late Miocene *Urmiatherium* or *Criotherium*). Horn cores do not curve backward and are often strongly divergent.

Genus *MAKAPANIA* Wells and Cooke, 1956

Type Species Makapania broomi Wells and Cooke, 1956.

MAKAPANIA BROOMI Wells and Cooke, 1956

Synonymy Makapania broomi Wells and Cooke, 1956:26, figures 13–16. *Makapansgat Limeworks. Late Pliocene; *Makapania broomi* Vrba, 1976:48, plate 40, figures A–B, D, G. Sterkfontein Type Site (= Mb 4). Late Pliocene.

Remarks This Makapansgat Member 3 species is close to *Megalovis latifrons* and *Pliotragus ardeus* in the late Pliocene of Europe. It has long horn cores, diverging almost transversely from an elevated transverse frontal ridge. The basioccipital, however, is very similar to that of *Ovibos* itself. It seems that the modern muskox is a relict in an extreme habitat of a tribe with a former wider distribution in the Holarctic and Africa (Gentry, 2001). *Makapania* may occur in Swartkrans Member 1 and the Shungura Formation (Vrba, 1976; Gentry, 1985), and it is also claimed for the Middle Pleistocene of Gladysvale (Lacruz et al., 2002).

Tribe Indet.
Genus *BUDORCAS* Hodgson, 1850

Type Species Budorcas taxicolor Hodgson, 1850. Extant takin.

Unlike *Ovibos*, *Budorcas* has few fossil relatives. Three possible ones are in Africa.

BUDORCAS CHURCHERI Gentry, 1996

Synonymy Budorcas churcheri Gentry, 1996:575, figure 1. *Hadar Fm Mb DD. Middle Pliocene; Ovibovini sp. aff. *"Bos" makapaani* Broom; Gentry, 1981:17, plates 3, 4.

Remarks This is based on a single and distinctive skull dating from ~3.2 Ma. It shows some striking differences from living *B. taxicolor*: more massive and compressed horn cores mounted on very elevated frontals and emerging transversely at their bases. The height of the frontals necessitates a very steep slope for the cranial roof.

Genus Indet.
"BOS" MAKAPAANI Broom, 1937

Synonymy Bos makapaani Broom, 1937, *Ann. Mag. Nat. Hist.*, 10, 20:510, figured. *Buffalo Cave, Limpopo Province. Late Pliocene.

Remarks The sutures on Broom's frontlet suggest that the convex edge of the horn cores is anterior or anterodorsal and not posterior, so that a distant affinity with *Budorcas* is more likely than with *Bos*. Gentry and Gentry (1978: plate 41) thought a larger horn core from Olduvai Bed I was similar and also illustrated nonassociated caprine-like metapodials from Bed I (plate 19, figure 2). Kuykendall et al. (1995) regarded Buffalo Cave as Pleistocene, but this species might better fit the late Pliocene. Earlier enigmatic "ovibovine" remains occur at Langebaanweg (Gentry, 1980: figures 61–62).

Genus *NITIDARCUS* Vrba, 1997

Type Species Nitidarcus asfawi Vrba, 1997.

*NITIDARCUS ASFAWI Vrba, 1997

Synonymy Nitidarcus asfawi Vrba, 1997:177, figures 5b, 6b. *Bouri. Early Pleistocene.

Remarks This was founded on a cranium with highly distinctive horn cores inserted uprightly, diverging at their bases and then curving backward and becoming parallel, with homonymous torsion, compressed basally and with a posterolateral welt. There is a large sinus in the pedicel and horn core base, the braincase roof is strongly inclined (both characters being like Alcelaphini), and the basioccipital is triangular (unlike Alcelaphini). An alliance with *Budorcas* was rejected by Vrba.

ASSESSMENT OF CAPRINAE

Looking at distributions of living Caprinae, one would take them as a fringe group in Africa, remaining largely within the Palaearctic realm. The fossil record suggests several penetrations of Ovibovini and possible *Budorcas* relatives, perhaps longer lasting in some cases. The only abundant occurrence is the *"Bos" makapaani* at Buffalo Cave. More fossils are needed, but in the interim we could muse on the possibility of *Budorcas* having entered Asia from Africa. *Ovibos* and *Budorcas* are probably not related, although both differ from *Capra* and *Ovis* by never having backward curvature in the basal sectors of the horn cores, and neither do their possible fossil relatives. Within the Caprini, *Bouria* is a striking animal that will be relevant to future analysis of the differences between *Capra* and *Hemitragus*. Ropiquet and Hassanin (2005) have already produced a molecular phylogeny in which *Hemitragus* species are not a clade (although their new generic names need diagnoses before they can be used). No fossil *Ovis* have been found in Africa.

Discussion

EARLY DEVELOPMENT OF BOVIDAE IN AFRICA

The main problem in giving an account of bovid origins in Africa is that we do not know whether any pre-Miocene pecorans existed on the continent. Other drawbacks are (1) the lack of a comprehensive survey of the east Asian and European Oligocene and early Miocene Pecora, against which to compare the more limited African record; (2) the questionably bovid status of Hypsodontinae; and (3) the presence of brachyodont pecorans alongside the hypsodont ones in the Mongolian Oligocene. It may also be pointed out that genetic studies have revealed, first, that genes controlling the basic structural organization of animal phenotypes (hox genes) are shared across many phyla, and, second, that changes in adult organs can result from simple switches from one developmental pathway to another (Dawkins, 2004). Such a system, operating in early Neogene pecoran relatives of Bovidae and in early bovids themselves, would favor repeated and parallel evolutions from an underlying stable body plan in response to environmental changes. The role of unique new mutations is much diminished.

Pecorans are known in the early Miocene of Africa, and some have been thought to be bovids or their relatives (Cote, this volume, chapter 37). At the start of the middle Miocene, *Namacerus* is certainly a bovid and very informative. It does not fit into any later tribe within Bovidae. Hypsodontinae are present in Africa in the middle Miocene. Another appearance around that time (table 38.2) was the nonboselaphine *Homoiodorcas* with short and robust horn cores. It was probably a member of the Antilopinae, so this temporal order contrasts with most cladistic sequences in which Antilopinae follow Boselaphini. Fort Ternan produces the full representation of the four bovid groups A–D of table 38.1. It has a hypsodontine and the first African boselaphine, both species being common. In addition, there is the first member of the Antilopini, perhaps a *Gazella*, showing horn cores of a pattern frequently seen in postboselaphine bovid groups: a degree of mediolateral compression, no keels, backward curvature, and modest divergence. Finally it has the larger *Gentrytragus* of unknown subfamily or tribe but likely to be in group D of table 38.1. Premolar rows in all these bovids are shorter than in *Namacerus,* but teeth of Boselaphini and Antilopinae differ little at this time level.

Little is known of African bovids between Fort Ternan and the end of the Miocene (Cote, 2004). Geraads et al. (2002) reported on meagre unidentified bovids in the Chorora Formation and warned that late Miocene ones might be more diverse than so far known. Almost the only substantial information on this period comes from the Ngorora Formation (Thomas, 1981) within the 12.0–8.0 Ma time range, wherein even the bovids alongside *Hipparion* retain a mid-Miocene aspect. This is the level at which a larger boselaphine comes in, coexisting with or replacing an earlier boselaphine, and neither of them belonging to the late Miocene *Miotragocerus* dominant in Eurasia. *Homoiodorcas* and *Gentrytragus* continue. There also arrives *"Pachytragus aff. solignaci"* (BN1747), which I suspect of being a reduncine. None of the Ngorora species are unambiguous members of the later African tribes Tragelaphini, Hippotragini, or Alcelaphini. We have to wait until the Adu-Asa Formation, Toros-Menalla, and Lothagam for these tribes to appear in the last 2 myr of the Miocene. Tragelaphini then appeared from within group B of the table 38.1 bovid classification, and Boselaphini disappeared. New Antilopinae appeared in group C, and Hippotragini and Alcelaphini also appeared in group D, perhaps superseding *Gentrytragus.* Then the Pliocene opened.

COMPARISONS WITH EURASIA

This story offers some interesting comparisons with events on other continents. In the Miocene of Europe, *Andegameryx, Amphimoschus,* and *Hispanomeryx* have been seen as possible bovoids (Gentry, 1994). They are bigger or later than *Namibiomeryx* and perhaps less likely to be close to Bovidae. It will be more practicable in the near future to find out more about the relationship between *Namacerus* in Africa and the long-known *Eotragus* in Europe. With a keel sometimes visible on its horn cores, *Eotragus* may be boselaphine. So could *Namacerus* be the earliest nonboselaphine bovid? If so, how has it happened that it is on a different continent from *Eotragus*, and which continent was the bovid home? Moving on through the middle Miocene, we find Boselaphini and apparently a *Gazella* in the Siwaliks, a situation very like East Africa and contrasting with Europe wherein Boselaphini only appear near the end of the middle Miocene and *Gazella* in the early late Miocene. Before the end of the middle Miocene, Europe and perhaps the Siwaliks also have *Tethytragus*, a genus with the same basic nonboselaphine pattern of horn cores as *Gazella*, *Gentrytragus*, and *Hippotragus*. It may belong to group D of table 38.1 and be related to later Caprinae. In the Tethyan-Paratethyan zone or realm bordering Africa to the north, a big change and increase in numbers and diversity of species took place in the late Miocene. Many new forms of Antilopinae and Caprinae appeared alongside hipparionine horses, but the zenith of the new fauna was delayed until the Turolian (later late Miocene). Less is known of the end-Miocene and Pliocene bovids that came to share the ecological room for pecoran ruminants with cervids. This is the reverse of our understanding of African bovids, which is weak for most of the late Miocene and improved for the end of the Miocene onward.

FROM THE PLIOCENE TO THE PRESENT

From the start of the Pliocene, the history of bovids in Africa is that of the full range of modern tribes. One can see the broad outline, but it is usually difficult to pin down details. The Tragelaphini have an unknown origin, with their horn cores unlike the standard *Tethytragus*, *Gazella*, and *Hippotragus* model. The old idea of associating them with Boselaphini has been supported by molecular phylogenies, and keels on horn cores might be a significant character in common. The teeth became aegodont but without much tendency toward hypsodonty. One can only imagine the earliest tragelaphines coming from the boselaphines soon after Antilopinae had separated from the boselaphine ancestors. The main later innovations in the tribe were the appearance of kudus in the late Pliocene and, later again, of *Taurotragus*.

Bovini are most likely to have originated among the enlarged boselaphines that were appearing through the late Miocene in the Siwaliks and perhaps elsewhere. They evolved teeth of large occlusal area and with more hypsodonty, but the full boödont pattern didn't come until much later. Bovini present the conundrum of a succession of both smaller and primitive and larger and advanced morphologies that mask the lines of ancestry. An early bovine of closed habitats could well have looked like *Ugandax gautieri*, but how are we to relate it to the extant forest buffalo *Syncerus caffer nanus*? Were there separate long-lasting small forest and larger savanna bovine lineages or one that from time to time gave off more short-lived offshoots?

Cephalophini are likely to have developed from within the Antilopinae, although they have sometimes been placed close to Boselaphini by authors back to Rütimeyer (1877–78). They are rare as fossils. The difficulty with *Brabovus* is to know whether to exclude it from Cephalophini. Could its resemblances to a small bovine have arisen from having a similar way of life? One would not think to relate it to Boselaphini or Tragelaphini, the other tribes related to Bovini.

Neotragini are not well represented as fossils. They are rarer than larger bovids and have only received attention in

exceptional cases such as the large *Raphicerus* at Langebaanweg or the abundant *Madoqua* at Laetoli. *Gazella* has always been the most common of the Antilopini, and it was joined at the end of the Miocene in sub-Saharan Africa by a species that might have belonged to *Aepyceros*. *Antidorcas* is most likely to have appeared near the start of the late Pliocene, and more *Gazella* species differentiated in the late Pliocene.

Reduncini differentiated from Antilopini (or vice versa), which had scarcely advanced on primitive Boselaphini. They do not have keels on their horn cores like Boselaphini, but neither do most of them show horn cores like *Gazella*, *Tethytragus*, or *Hippotragus*. In the earlier part of the Pliocene, they were variably kob- or waterbuck-like, but the lineages of modern kob, lechwe, and waterbuck only become known in the late Pliocene. The teeth of reduncines gradually became boödont but remained small in relation to skull size.

The split in Hippotragini between *Hippotragus* and *Oryx* had occurred early, and presumably the latter has always been an inhabitant of drier habitats than the former. They both retained an unspecialized tooth pattern late into their history, and *Hippotragus* only advanced to full or fuller boödonty long after it had given up sharing faunal predominance with Alcelaphini in the middle Pliocene Laetolil Beds. *Addax* may have evolved from *Oryx* at a relatively late date.

The early alcelaphine *Damalacra acalla* at Langebaanweg has backwardly curved horn cores like many other bovids, but the accompanying species *D. neanica* has what must be a more advanced course of its horn cores. Once more the question concerns the descent of later members of the tribe. Would *Damaliscus* have come from an ancestor like *acalla* and other later genera from one like *neanica*? When are modern-looking characters in early fossils synapomorphies, and when are they parallels? This leads into the main classificatory question in Alcelaphini, which is to decide the limits of *Damaliscus* and *Parmularius* and how *Alcelaphus* is related to them. Alcelaphini gradually modified their teeth by increased hypsodonty, short premolars, and rounding of the external walls of the lobes of their molars. This is an obvious change from early bovid teeth but is not the same as either aegodonty or the boödonty of late *Hippotragus* and Bovini. Several new alcelaphines are notable in the later Pliocene, particularly the genera *Beatragus*, *Connochaetes*, and *Megalotragus*.

Caprinae are associated in modern cladistic classifications with either Hippotragini or a conjoined Hippotragini + Alcelaphini. The fossil record of Caprinae suggests many early lineages in the late Miocene of Eurasia. At various times they have managed to reach into Africa, especially north of the tropic of Cancer, and older relatives of *Budorcas* and *Ovibos* may have been or were present farther south.

Following this summary of tribal evolution, it may be repeated that recent findings in genetics show that it is simpler than formerly supposed to switch from one route to another in ontogeny. This reinforces suspicions that many of the characters used in this chapter could have evolved repeatedly and hence be unreliable indicators of relationship between species across different regions and time periods. The scope for future changes to current interpretations remains high.

Conclusions

Almost every paragraph in this chapter and every sentence in the discussion has a question built into it. Several large-scale questions will continue to engage attention. One is the timing

TABLE 38.4
Locations and approximate dates of some notable bovid
appearances in the late or later Pliocene

Taxon	Age
Kobus sigmoidalis	Moiti Mb 3.6 Ma, Shungura C 2.64 Ma
Antidorcas recki or sp.	Lower Tulu Bor Mb 3.13 Ma, Shungura B 3.11Ma, Koro Toro 3.00 Ma, Ndolanya Beds 2.6 Ma
Megalotragus isaaci or sp.	Upper Tulu Bor Mb 2.79 Ma, Shungura D 2.46 Ma
Gazella sp. 3	Ndolanya Beds 2.6 Ma
Tragelaphus gaudryi	Shungura C 2.64 Ma, Ahl al Oughlam 2.5 Ma
Beatragus whitei or sp.	Matabaietu 3–5 2.5 Ma, Ahl al Oughlam 2.5 Ma
Pelorovis spp.	Ahl al Oughlam 2.5 Ma, Shungura D 2.45 Ma
Connochaetes spp.	Aïn Boucherit 2.0 Ma, Upper Burgi 1.94 Ma
Gazella aff. *rufifrons*	Upper Burgi Mb 1.94 Ma, Olduvai Bed I 1.8 Ma

SOURCE: Gentry (1985, 1987); Gentry and Gentry (1978); Geraads and Amani (1998); Harris (1991), Vrba (1997).

NOTE: *Antidorcas* in Shungura B is a mandible; *Tragelaphus gaudryi* and *Connochaetes* are also known from Aïn Jourdel; *Tragelaphus strepsiceros* appears after 2.1 Ma, presumably from an ancestor like *T. gaudryi*.

and causes of the emergence of Bovidae and later of the bovid tribes. How can the cladistic association between Alcelaphini and Caprinae be fitted to a zoogeographic story of faunal movements between the continents? Are the Siwaliks part of the arena for the evolution of African bovids? A second set of questions is the faunal changes around the late Pliocene. At Laetoli the contrast is particularly obvious between the mid-Pliocene bovid fauna in the Laetolil Beds and the new lineages or genera in the Ndolanya Beds and later Plio-Pleistocene faunas. Table 38.4 shows the first dates for quite a lot of new bovids in this period comprising kudus, an advanced bovine, two gazelles, springbok, waterbuck and lechwe, two new large alcelaphines, and wildebeest. The table is defective to the extent that dates are only broadly known, that identifications of first appearances are often less certain than of later occurrences where the species have become commonplace, and that any new arrival alters the conditions of life for other sympatric species and could lead to further species changes. Nevertheless, it looks as though something happened at that time. The link of these and other mammal changes with environmental change and climatic cooling has been discussed (Hernández Fernández and Vrba, 2006; Bobe and Behrensmeyer, 2004; Bobe et al., 2007). From the late Pliocene onward, hypsodont and cursorial antilopines and alcelaphines often predominate in bovid faunas and can be taken as characterizing more open country faunas (Vrba, 1980; Geraads and Amani, 1998). A third question concerns the environmental or competitive pressures that called forth late intensifications of boödont teeth, and why alcelaphine teeth differ from the boödonty of Bovini, *Hippotragus*, and Reduncini. A final long-standing question is that of the timing of late Pleistocene extinctions.

ACKNOWLEDGMENTS

I thank my wife for support and advice on nomenclature and assorted other matters, also Bill Sanders and Lars Werdelin for their invitation to contribute to this book and for their editorial help, and Kaye Reed for her comments on the typescript. Susy Cote kindly supplied photographs and perceptive views on early bovids and Ngorora specimens. I thank Denis Geraads for information and guidance on North African locality and species names, Faysal Bibi for raising many significant points in conversations about fossil Bovidae, and Elisabeth Vrba, Maia Bukhsianidze, and Dimitris Kostopoulos for much help during a meeting in Addis Ababa when we all six came together. Chris Smeenk (Leiden) led me to some historical information. Photographs were supplied by Harry Taylor and Phil Hurst.

APPENDIX NOTES

1. The holotype of *Hypsodontus miocenicus* Sokolov, 1949, is a shallow mandibular ramus with m2 and m3. Gabunia (1973: figure 33) assigned to the species a horn core from the same locality. This horn core showed slight twisting of its axis but its overall course was little curved. Gabunia (1973: figure 34, plate 8; figure 6) referred a similar but smaller and slightly compressed horn core to another new genus and species *Kubanotragus sokolovi*. Gentry (1970) thought the *H. miocenicus* holotype mandible might be a boselaphine. Thomas (1984a) concurred and referred some Siwaliks hypsodontines to *Kubanotragus sokolovi*. Köhler (1987:155), however, used the name *Hypsodontus* for a new, large species, *pronaticornis*, in Turkey and founded *Turcocerus*, with type species *T. grangeri* (Pilgrim, 1934) from Mongolia, for hypsodontines with shorter and more grooved horn cores than *Hypsodontus* and more curved horn cores than *Kubanotragus*. The outcome is that *Hypsodontus* is used for large forms, *Kubanotragus* for smaller forms with slender and overall straighter horn cores, and *Turcocerus* for short-horned forms that, in western Eurasia, are also of small size. The assignment of *"Oioceros" tanyceras* from Fort Ternan remains undecided in such an arrangement while *Hypsodontus pickfordi* from Maboko resembles a *Kubanotragus*. I use only *Hypsodontus* in this chapter.

2. Among fossil bovids or possible bovids, at least one Oligocene *Palaeohypsodontus* from Mongolia (Dmitrieva, 2002) could be close to the size of *N. pygmaeus*; the tiny Siwaliks middle Miocene boselaphine *Elachistoceras khauristanensis* Thomas, 1977 (*Bull. Soc. Géol. Fr.* 7, 19:375–383—not the 1977 paper listed in this chapter)

is not as small; and an undescribed bovid from the late Miocene of Csákvár, Hungary (listed but not diagnosed as the cervid *Lagomeryx celer* by Kretzoi, 1954) could be close in size but perhaps with larger astragali. However it is very likely from a view of their skulls and some tooth lengths that another extant neotragine, *Madoqua piacentini* from the southeastern Somaliland coast (Yalden, 1978), is still smaller than *Neotragus pygmaeus*.

3. This gazelle horn core and its partner were noticed on the back wall of the dormant Fort Ternan excavation during a short stop there in 1967. Despite the professional hesitations of J. A. Van Couvering, A. C. Walker, and W. W. Bishop who were also present, my wife insisted on freeing the horn core and took it back to Nairobi. Many years later it seems to me that it is a significant specimen for tracing bovid evolution, and I appreciate her resolution in saving it.

4. Pomel's monographs of the 1890s provided a multitude of new names for Algerian Plio-Pleistocene mammals. Arambourg (1939, 1957) and others have made reassignments of the gazelles in Pomel (1895). Thus, *Gazella atlantica* probably includes *Antilope crassicornis* p. 19, plate 1, figures 2–7, plate 4, figures 1–8; plate 13, figures 3–6; *Antilope subgazella* p. 10, plate 3, figures 1–5, plate 10, figures 12–13; *Antilope massoessilia* Pomel, 1895, p. 21, plate 1, figure 1, plate 9; *Antilope nodicornis* p. 18, plate 5, figures 1–4. *Gazella cuvieri* probably includes *Antilope subkevella* p. 14, plate 5, figures 5–7; *Antilope kevella* p. 12, plate 12, figure 3, plate 13, figures 1–2; *Antilope oranensis* p. 25, plate 2, figures 1–2. When using Pomel (1895), it is essential to note discrepancies between plate captions and the text references to the plates.

5. Vrba (1995:421, n. 32) overcomes the problem of recognizing the *shungurae-melampus* transition by applying the one name *melampus* to the whole of the identified fossil lineage.

6. Some of the skull differences of *Kobus megaceros* from *K. leche* are horn cores less compressed; the proximal backward curve on the horn cores is more drawn out and the distal upward curve is correspondingly shorter; horn insertions closer and dorsal orbital rims wider; paired swellings on frontals in front of the supraorbital pits; larger ethmoidal fissures; lateral flanges on the front of the nasals; premaxillae contacting nasals; a more rounded anterior tip to the premaxillae; median indent at the back of the palate less narrowed anteriorly; stronger longitudinal ridges behind the anterior tuberosities of the basioccipital.

7. In founding *Caprotragoides*, Thenius nominated a Fort Ternan horned cranium (Gentry, 1970: plate 12 and plate 13, figures 1, 2) as holotype of *potwaricus*, but under the rules of nomenclature, the holotype must remain Pilgrim's frontlet.

8. Arambourg (1979:80, plate 50, figures 6–8) founded *Hippotragus priscus* for two upper molars from Bel Hacel and a third from Ichkeul. One from Bel Hacel (early Pleistocene) was designated "type" and the Ichkeul one (?middle Pliocene) "syntype," an impossible arrangement under the International Code of Zoological Nomenclature. I take the "type," Arambourg's figure 8, as a holotype. Geraads (1981) doubted that the

Bel Hacel teeth were Hippotragini, which was also my opinion when I saw them in Paris.

9. Linnaeus (1758) used *Capra dorcas* for the dorcas gazelle of Egypt. Pallas (1766) founded *Antilope* to include then-known antelopes, among them his new species *A. dorcas*, the bontebok of South Africa. Pallas (1767) substituted *Antilope pygargus* for the bontebok, which corrected the secondary homonymy of his own *A. dorcas* with the dorcas gazelle. Harper (1940) reinstated *Damaliscus dorcas* (Pallas, 1766) for the bontebok on the grounds that since 1894 gazelles had been in *Gazella* and the bontebok in *Damaliscus*, so homonymy no longer existed. The change was unnecessary and overturned 170 years of stable use of *pygargus*. Rookmaaker (1991) restored *pygargus*, because *A. dorcas* Pallas, 1766 was permanently invalid as a junior secondary homonym that had been replaced before 1961 (International Code for Zoological Nomenclature, 3rd ed., Article 59[b]). Even in the third edition, Article 59(b) was in awkward opposition to the requirement for stability of Article 23(b). Under the fourth edition of the Code (1999, Article 59.3), the 1991 change from *dorcas* would no longer have been required. I continue to use *dorcas*.

10. P. Thomas (1884) did not designate any of his material of *Antilope tournoueri* as a holotype. Arambourg (1979:95) wrote, "Le type d' "*Antilope*" *tournoueri* décrit par Ph. Thomas fait partie des collections du Muséum. Il s'agit d'une portion frontale de crâne munie de ses chevilles osseuses." This is a satisfactory lectotype designation, whereas his suggested neotype (Arambourg, 1979:101) is based on nonsyntypical material from Aïn Boucherit.

NOTE ADDED IN PROOF

Hispanomeryx was mentioned twice in this chapter as possibly having bovoid affinities, but its original 1981 attribution to the cervoid family Moschidae is being maintained Sánchez et al. (2009).

Haile-Selassie et al. (2009) recognise three new species of Middle Awash bovids from the Kuseralee Member of the Sagantole Formation:

Tragoportax abyssinicus has horn cores that are short, compressed, almost straight in side view, with a prominent demarcation and weak torsion, all these being potential differences from the *Miotragocerus* aff. *M. cyrenaicus* at Lothagam. Differences can also be seen from the Lothagam *Tragoportax* A and B of Harris (2003), including a more polygonal cross section of the horn cores than in form A. (In this chapter *Tragoportax* was included in *Miotragocerus*.)

Tragelaphus moroitu is a fairly small species also present in the Adu-Asa Member and not very different from the Lukeino *Tragelaphus*. The horn cores have a triangular cross section that is constant along their length, very slight anteroposterior compression, lyration, and some divergence basally. The anterior keel is stronger than in *T. kyaloae* and the latter may descend from *T. moroitu*. (In this chapter *T. moroitu* was mentioned under the heading *Tragelaphus* sp. or spp.)

Redunca ambae is about the size of *R. arundinum* or a small *Kobus kob* and the short horn cores have close insertions, weak divergence, and a tendency to an anterior swelling at their base. Compared with *K. porrecticornis* the insertions are more inclined in side view and the horn cores taper more rapidly above the base. They look as if they are losing any remaining basal backward curvature and thus approaching the condition of later *Redunca*. Once admitted to *Redunca*, *R. ambae* nearly doubles the known time span of that genus. As noted above, the authors believe that other early reduncines can be put into *Redunca*.

Geraads et al. (2009a) recognise two new species dating from about 3.75 Ma.

Tragelaphus saraitu has long and slender horn cores with an anterior keel distally and a posterolateral keel, slight antero-posterior compression, proximal divergence at about 90°, this divergence steady or slightly increasing distally, and a trapezoidally shaped occipital. It is bigger than *T. moroitu* and with a weaker anterior keel on the horn cores, and similar to *T. kyaloae* but without an incipient transverse ridge across the top of the occipital.

Aepyceros afarensis has horn cores as large as in the living *A. melampus* and with a similar morphology: long, with transverse ridges, seemingly upright insertions, and strong backwards curvature. The horn cores are so little compressed that some become wider than anteroposteriorly long in basal cross section. The sinuses in the frontal bones are less extensive and the premolar rows are less reduced than in *A. melampus*. This species corroborates the existence of large *Aepyceros* species predating *A. shungurae* and extends the range of *Aepyceros* further north in Africa than hitherto known.

Geraads et al. (2009b) recognise three new west African bovid species.

Tchadotragus fanonei, a hippotragine from the lowermost Pliocene of Kossom Bougoudi, is about 1.5 myr younger than *T. sudrei* at Toros Menalla. It differs in cranial proportions, longer horn pedicels, and stronger goat folds on lower molars. The authors believe that its evolutionary direction could be trending away from that towards modern *Hippotragus*.

Kobus ammolophi, also from Kossom Bougoudi, is medium sized, and the horn cores are quite short, little-compressed and with moderate divergence increasing distally. The teeth are exceedingly advanced, seemingly more so than in the Shungura Fomation at about 2.5-2.0 Ma (see Gentry 1997, figure 7), and certainly a strong contrast with supposed reduncine teeth at Langebaanweg.

Jamous kolleensis is a new bovine genus and species from the lower Pliocene of Kollé dating from 4.5-4.0 Ma and about a million years younger than Kossom Bougoudi. The horn cores are very straight and extend very nearly transversely from insertions above the back of the orbits. The authors look towards early Eurasian bovines for possible relationships and note that the north African "*Leptobos*" *syrticus* of unknown stratigraphic origin might be the most morphologically similar.

The new finds in Chad and elsewhere are continuing to reveal an ever-widening diversity of past bovid species across Africa. The animals themselves were intent only on staying alive and even in the Miocene to mid-Pliocene had evolved quite strongly varied morphologies to do so. This will add to their interest but also to our difficulties in tracing with any confidence the lines of phylogenetic descent from the constituent species of any one faunal level to those of the next.

Literature Cited

Alemseged, Z., J. G.Wynn, W. H. Kimbel, D. Reed, D. Geraads, and R. Bobe. 2005. A new hominin from the Basal Member of the Hadar Formation, Dikika, Ethiopia, and its geological context. *Journal of Human Evolution* 49:499–514.

Andree, J. 1926. Neue Cavicornier aus dem Pliocän von Samos. *Palaeontographica* 67:135–175.

Ansell, W. F. H. 1971. Part 15, Order Artiodactyla; pp. 1–84 in J. Meester and H. W. Setzer (eds.), *The Mammals of Africa: An Identification Manual*. Smithsonian Institution Press, Washington, D.C.

Arambourg, C. 1939. Mammifères fossiles du Maroc. *Mémoires de la Société des Sciences Naturelles, Maroc* 46:1–74.

———. 1947. Contribution à l'étude géologique et paléontologique du bassin du lac Rodolphe et de la basse vallée de l'Omo. 2, Paléontologie; pp. 232–562 in C. Arambourg (ed.), *Mission Scientifique de l'Omo, 1932–1933: Tome 1, Géologie-Anthropologie, fascicule 3*. Muséum National d'Histoire Naturelle, Paris.

———. 1957. Observations sur les Gazelles fossiles du Pléistocène supérieur de l'Afrique du Nord. *Bulletin de la Société d'Histoire Naturelle de l'Afrique du Nord* 48:49–77.

———. 1959. Vertébrés continentaux du Miocène supérieur de l'Afrique du Nord. *Publications du Service de la Carte Géologique de l'Algérie (n.s.) Paléontologie, Mémoires* 4:1–159.

———. 1979. *Vertébrés Villafranchiens d'Afrique du Nord*. Fondation Singer-Polignac, Paris, 141 pp.

Azanza, B., and J. Morales. 1994. *Tethytragus* nov.gen. et *Gentrytragus* nov. gen. Deux nouveaux Bovidés (Artiodactyla, Mammalia) du Miocène moyen. *Proceedings Koninklijke Nederlandse Akademie van Wetenschappen* 97:249–282.

Azzaroli, A., and G. Napoleone. 1982. Magnetostratigraphic investigation of the Upper Sivaliks near Pinjor, India. *Rivista Italiana di Paleontologia e Stratigrafia* 87:739–762.

Barry, J. C. 1995. Faunal turnover and diversity in the terrestrial Neogene of Pakistan; pp. 115–134 in E. S. Vrba, G. H. Denton, T. C. Partridge, and L. H. Burckle (eds.), *Paleoclimate and Evolution, with Emphasis on Human Origins*. Yale University Press, New Haven.

Bate, D. M. A. 1949. A new African fossil long-horned buffalo. *Annals and Magazine of Natural History, London* (12), 2:396–8.

———. 1951. The mammals from Singa and Abu Hugar: Fossil Mammals of Africa. *British Museum (Natural History), London* 2:1–28.

———. 1955. Appendix A: Vertebrate faunas of Quaternary deposits in Cyrenaica; pp. 274–291 in C. B. M. McBurney and R. W. Hey, *Prehistory and Pleistocene Geology in Cyrenaican Libya*. Cambridge University Press, Cambridge.

Bayle, E. 1854. Note sur des ossements fossiles trouvés près de Constantine. *Bulletin de la Société Géologique de France* (2), 11:343–345.

Bernor, R. L., and L. Rook. 2008. A current view of As Sahabi large mammal biogeographic relationships. *Garyounis Scientific Bulletin, Special Issue* 5: 283–290.

Bibi, F. 2007. Origin, paleoecology, and paleobiogeography of early Bovini. *Palaeogeography, Palaeoclimatology, Palaeoecology* 248:60–72.

Birungi, J., and P. Arctander. 2001. Molecular systematics and phylogeny of the Reduncini (Artiodactyla: Bovidae) inferred from the analysis of mitochondrial cytochrome *b* gene sequences. *Journal of Mammalian Evolution* 8:125–147.

Bishop, W. W., G. R. Chapman, A. Hill, and J. A. Miller. 1971. Succession of Cainozoic vertebrate assemblages from the northern Kenya Rift Valley. *Nature* 233:389–394.

Boaz, N. T. (ed.). 1990. *Evolution of Environments and Hominidae in the African Western Rift Valley*. Virginia Museum of Natural History, Martinsville, 356 pp.

———. 2008. A view to the south: Eo-Sahabi palaeoenvironments compared and implications for hominid origins in Neogene North Africa. *Garyounis Scientific Bulletin, Special Issue* 5: 291–308.

Bobe, R., and A. K. Behrensmeyer. 2004. The expansion of grassland ecosystems in Africa in relation to mammalian evolution and the origin of the genus *Homo*. *Palaeogeography, Palaeoclimatology, Palaeoecology* 207:399–420.

Bobe, R., A. K. Behrensmeyer, G. G. Eck, and J. M. Harris. 2007. Patterns of abundance and diversity in late Cenozoic bovids from the Turkana and Hadar Basins, Kenya and Ethiopia; pp.129–157 in R. Bobe, Z. Alemseged, and A. K. Behrensmeyer (eds.), *Hominin Environments in the East African Pliocene: An Assessment of the Faunal Evidence*. Springer, Dordrecht.

Bonis, L. de, D. Geraads, J.-J. Jaeger, and S. Sen. 1988. Vertébrés du Pléistocène de Djibouti. *Bulletin de la Société Géologique de France* (8), 4:323–334.

Bonis, L. de, G. D. Koufos, and S. Sen. 1998. Ruminants (Bovidae and Tragulidae) from the middle Miocene (MN 5) of the island of Chios, Aegean sea (Greece). *Neues Jahrbuch für Geologie und Paläontologie, Abhandlungen* 210:399–420.

Boulineau, P. 1933. Hybridations d'antilopides. *La Terre et la Vie* 3:690–691.

Bourguignat, J. R. 1870. *Histoire du Djebel-Thaya et des Ossements Fossiles Recueillis dans la Grande Caverne de la Mosquée*. Challamel, Paris, 108 pp.

Bouvrain, G., and L. de Bonis. 1984. Le genre *Mesembriacerus* (Bovidae, Artiodactyla, Mammalia): Un ovieboviné primitif du Vallésien (Miocène supérieur) de Macédoine (Grèce). *Palaeovertebrata* 14:201–223.

———. 1985. Le genre *Samotragus* (Artiodactyla, Bovidae): Une antilope du Miocène supérieur de Grèce. *Annales de Paléontologie* 71:257–299.

———. 2007. Ruminants (Mammalia, Artiodactyla: Tragulidae, Cervidae, Bovidae) des gisements du Miocène supérieur (Turolien) de Dytiko (Grèce). *Annales de Paléontologie* 93:121–147.

Brink, J. S. 1987. The archaeozoology of Florisbad, Orange Free State. *Memoirs van die Nasionale Museum Bloemfontein* 24:1–151.

Brink, J. S., and J. A. Lee-Thorp. 1992. The feeding niche of an extinct springbok, *Antidorcas bondi* (Antelopini, Bovidae), and its palaeoenvironmental meaning. *South African Journal of Science* 88:227–229.

Broom, R. 1934. On the fossil remains associated with *Australopithecus africanus*. *South African Journal of Science* 31:471–480.

Brown, F. H. 1994. Development of Pliocene and Pleistocene chronology of the Turkana Basin, East Africa, and its relation to other sites; pp. 285–312 in R. S. Corruccini and R. L. Ciochon, *Integrative Paths to the Past*. Prentice Hall, Englewood Cliffs, N.J.

Büttiker, W. 1982. Mammals of Saudi Arabia: The lesser kudu *(Tragelaphus imberbis)* (Blyth, 1869). *Fauna of Saudi Arabia* 4:483–487.

Chaîd-Saoudi, Y., D. Geraads, and J.-P. Raynal. 2006. The fauna and associated artefacts from the Lower Pleistocene site of Mansourah (Constantine, Algeria). *Comptes Rendus Palevol* 5:963–971.

Chen, G. 1988. Remarks on the *Oioceros* species (Bovidae, Artiodactyla, Mammalia) from the Neogene of China. *Vertebrata PalAsiatica* 26:157–172.

Choubert, G., and A. Faure-Muret. 1961. Le gisement de vertébrés miocènes de Beni Mellal. *Notes et Mémoires du Service Géologique du Maroc* 155:1–122.

Churcher, C. S. 1972. Late Pleistocene vertebrates from archaeological sites in the plain of Kom Ombo, Upper Egypt. *Life Sciences Contributions, Royal Ontario Museum, Toronto* 82:1–172.

Clarke, R. J. 2006. A deeper understanding of the stratigraphy of Sterkfontein fossil hominid site. *Transactions of the Royal Society of South Africa* 61:111–120.

Clutton-Brock, J. 1970. The fossil fauna from an Upper Pleistocene site in Jordan. *Journal of Zoology London* 162:19–29.

Cooke, H. B. S. 1938. The Sterkfontein bone breccia: A geological note. *South African Journal of Science* 35:204–208.

———. 1947. Some fossil hippotragine antelopes from South Africa. *South African Journal of Science* 43:226–231.

———. 1949. Fossil mammals of the Vaal River deposits. *Memoirs of the Geological Survey of South Africa* 35:1–109.

———. 1974. The geology, archaeology and fossil mammals of the Cornelia Beds, O.F.S. *Memoirs van die Nasionale Museum Bloemfontein* 9:63–84.

———. 1990. Taung fossils in the University of California collections; pp. 119–134 in G. H. Sperber (ed.), *From Apes to Angels: Essays in Anthropology in Honor of Phillip V. Tobias*. Wiley-Liss, New York.

———. 1991. *Dinofelis barlowi* (Mammalia, Carnivora, Felidae) cranial material from Bolt's Farm collected by the University of California African Expedition. *Palaeontologia Africana* 28:9–21.

Cooke, H. B. S., and S. C. Coryndon. 1970. Pleistocene mammals from the Kaiso Formation and other related deposits in Uganda; pp. 107–224 in L. S. B. Leakey and R. J. G. Savage (eds.), *Fossil Vertebrates of Africa*, vol. 2. Academic Press, London.

Cooke, H. B. S., and L. H. Wells. 1951. Fossil remains from Chelmer, near Bulawayo, Southern Rhodesia. *South African Journal of Science* 47:205–209.

Coryndon, S. C. 1966. Preliminary report on some fossils from the Chiwondo Beds of the Karonga District, Malawi. *American Anthropologist* 68:59–67.

Cote, S. M. 2004. Origins of the African hominoids: An assessment of the palaeobiogeographical evidence. *Comptes Rendus Palevol* 3:323–340.

Crégut-Bonnoure, E. 2007. Apport des Caprinae et Antilopinae (Mammalia, Bovidae) à la biostratigraphie du Pliocène terminal et du Pléistocène d'Europe. *Quaternaire* 18:73–97.

Davis, S. J. M. 1980. Late Pleistocene-Holocene gazelles of northern Israel. *Israel Journal of Zoology* 29:135–140.

Dawkins, R. 2004. *The Ancestor's Tale*. Weidenfeld and Nicolson, London, 528 pp.

Day, M. H. 1986. *Guide to Fossil Man*. 4th ed. Cassell, London, 432 pp.

Degerbøl, M., and B. Fredskild. 1970. The urus (*Bos primigenius* Bojanus) and Neolithic domesticated cattle (*Bos taurus domesticus* Linné) in Denmark. *Biologiske Skrifter København* 17:1–234.

Deino, A. L., L. Tauxe, M. Monaghan, and A. Hill. 2002. ^{40}Ar/^{39}Ar geochronology and paleomagnetic stratigraphy of the Lukeino and lower Chemeron Formations at Tabarin and Kapcheberek, Tugen Hills, Kenya. *Journal of Human Evolution* 42:117–140.

Dekeyser, P. L., and J. Derivot. 1956. Sur la présence de canines supérieures chez les Bovidés. *Bulletin de l'Institut Française d'Afrique Noire* (A) 18:1272–1281.

De Ruiter, D. J. 2003. Revised faunal lists for Members 1–3 of Swartkrans, South Africa. *Annals of the Transvaal Museum, Pretoria* 40:29–41.

Dietrich, W. O. 1933. Zur Altersfrage der Oldowaylagerstätte. *Zentralblatt für Mineralogie, Geologie und Paläontologie B* 1933:299–304.

———. 1942. Ältestquartäre Säugetiere aus der südlichen Serengeti, Deutsch Ostafrika. *Palaeontographica Abt. A* 94:43–133.

———. 1950. Fossile Antilopen und Rinder Äquatorialafrikas. *Palaeontographica Abt. A* 99:1–62.

Ditchfield, P., J. Hicks, T. Plummer, L. C. Bishop, and R. Potts. 1999. Current research on the Late Pliocene and Pleistocene deposits north of Homa mountain, southwestern Kenya. *Journal of Human Evolution* 36:123–150.

Dmitrieva, E. L. 1977. Tajikistan's and India's fossil Alcelaphinae. *Journal of the Palaeontological Society of India* 20:97–101.

———. 2002. On the early evolution of bovids. *Paleontological Journal* 36:204–206. (Translation from *Paleontologicheskii Zhurnal* 2 [2002]:86–88.)

Dubrovo, I. A., and N. I. Burchak-Abramovich 1986. New data on the evolution of Bovinae of the Tribe Bovini. *Quartärpaläontologie* 6:13–21.

Ennouchi, E. 1953. Un nouveau genre d'Ovicapriné dans un gisement Pléistocène de Rabat. *Compte Rendu Sommaire des Séances de la Société Géologique de France, Paris* 8:126–128.

Feibel, C. S., and F. H. Brown. 1991. Age of the primate-bearing deposits at Maboko Island, Kenya. *Journal of Human Evolution* 21:221–225.

Flagstad, O., P. O. Syvertsen, N. C. Stenseth, and K. S. Jakobsen. 2001. Environmental change and rates of evolution: The phylogeographic pattern within the hartebeest complex as related to climatic variation. *Proceedings of the Royal Society of London, B* 268:667–677.

Gabunia, L. K. 1973. *Fossil Vertebrates Fauna of Belometscheskaya*. Metsniereba, Tibilisi, 136 pp.

Gaillard, C. 1934. Contribution à l'étude de la faune préhistorique de l'Egypte. *Archives du Muséum d'Histoire Naturelle de Lyon* 14:1–126.

Gatesy, J., G. Amato, E. S. Vrba, G. B. Schaller, and R. DeSalle. 1997. A cladistic analysis of mitochondrial ribosomal DNA from the Bovidae. *Molecular Phylogenetics and Evolution* 7:303–319.

Gautier, A., and A. Muzzolini. 1991. The life and times of the giant buffalo alias *Bubalus/Homoioceras/Pelorovis antiquus* in North Africa. *ArchaeoZoologia* 4:39–92.

Gentry, A. W. 1964. Skull characters of African gazelles. *Annals and Magazine of Natural History* (13), 7:353–382.

———. 1967. *Pelorovis oldowayensis* Reck, an extinct bovid from East Africa. *Bulletin, British Museum (Natural History), Geology* 14:243–299.

———. 1970. The Bovidae (Mammalia) of the Fort Ternan fossil fauna; pp. 243–324 in L. S. B. Leakey and R. J. G. Savage (eds.), *Fossil Vertebrates of Africa*, vol. 2. Academic Press, London.

———. 1974. A new genus and species of Pliocene boselaphine (Bovidae, Mammalia) from South Africa. *Annals of the South African Museum* 65:145–188.

———. 1978. The fossil Bovidae of the Baringo area, Kenya; pp. 293–308 in W. W. Bishop (ed.), *Geological Background to Fossil Man*. Scottish Academic Press, Edinburgh.

———. 1980. Fossil Bovidae (Mammalia) from Langebaanweg, South Africa. *Annals of the South African Museum* 79:213–337.

———. 1981. Notes on Bovidae (Mammalia) from the Hadar Formation, and from Amado and Geraru, Ethiopia. *Kirtlandia* 33:1–30.

———. 1985. The Bovidae of the Omo Group deposits, Ethiopia; pp. 119–191 in *Les Faunes Plio-Pléistocènes de la Basse Vallée de l'Omo (Ethiopie): Tome 1. Périssodactyles-Artiodactyles (Bovidae)*. CNRS, Paris.

———. 1987a. Pliocene Bovidae from Laetoli; pp. 378–408 in M. D. Leakey and J. M. Harris (eds.), *Laetoli: A Pliocene Site in Northern Tanzania*. Clarendon Press, Oxford.

———. 1987b. Ruminants from the Miocene of Saudi Arabia. *Bulletin, British Museum (Natural History), Geology* 41:433–439.

———. 1990. The Semliki fossil bovids. *Memoirs, Virginia Museum of Natural History* 1:225–234.

———. 1992. The subfamilies and tribes of the family Bovidae. *Mammal Review* 22:1–32.

———. 1994. The Miocene differentiation of Old World Pecora (Mammalia). *Historical Biology* 7:115–158.

———. 1996. A fossil *Budorcas* (Mammalia, Bovidae) from Africa; pp. 571–87 in K. M. Stewart and K. L. Seymour (eds.), *Palaeoecology and Palaeoenvironments of Late Cenozoic Mammals.* University of Toronto Press, Toronto.

———. 1997. Fossil ruminants (Mammalia) from the Manonga Valley, Tanzania; pp. 107–135 in T. Harrison (ed.), *Neogene Paleontology of the Manonga Valley, Tanzania.* Plenum Press, New York.

———. 2000. Caprinae and Hippotragini (Bovidae, Mammalia) in the upper Miocene; pp. 65–83 in E. S. Vrba and G. B. Schaller (eds.), *Antelopes, Deer and Relatives: Fossil Record, Behavioral Ecology, Systematics and Conservation.* Yale University Press, New Haven.

———. 2001. An ovibovine (Mammalia, Bovidae) from the Neogene of Stratzing, Austria. *Annalen Naturhistorische Museum Wien A* 102:189–99.

———. 2008. New records of Bovidae from the Sahabi Formation. *Garyounis Scientific Bulletin*, Special Issue, 5:205–217.

Gentry, A. W., and A. Gentry. 1978. Fossil Bovidae (Mammalia) of Olduvai Gorge, Tanzania, parts I and II. *Bulletin British Museum (Natural History), Geology* 29:289–446; 30:1–83.

Gentry, A. W., A. Gentry, and H. Mayr. 1995. Rediscovery of fossil antelope holotypes (Mammalia, Bovidae) collected from Olduvai Gorge, Tanzania, in 1913. *Mitteilungen der Bayerische Staatssammlung für Paläontologie und historische Geologie, München* 35:125–135.

Geraads, D. 1979. La faune des gisements de Melka-Kunturé (Éthiopie): Artiodactyles, Primates. *Abbay* 10:21–49.

———. 1981. Bovidae et Giraffidae (Artiodactyla, Mammalia) du Pléistocène de Ternifine (Algérie). *Bulletin du Muséum National d'Histoire Naturelle* (4), 3, (C):47–86.

———. 1986. Les ruminants du Pléistocène d'Oubeidiyeh (Israel). *Mémoires et Travaux du Centre de Recherches Français de Jérusalem* 5:143–181.

———. 1989. Vertébrés fossiles du Miocène supérieur du Djebel Krechem el Artsouma (Tunisie Centrale). Comparaisons biostratigraphiques. *Geobios* 22:777–801.

———. 1992. Phylogenetic analysis of the tribe Bovini (Mammalia: Artiodactyla). *Zoological Journal of the Linnean Society* 104:193–207.

———. 2002. Plio-Pleistocene mammalian biostratigraphy of Atlantic Morocco. *Quaternaire* 13:43–53.

Geraads, D., Z. Alemseged, and H. Bellon. 2002. The late Miocene mammalian fauna of Chorora, Awash basin, Ethiopia: systematics, biochronology and the ⁴⁰K-⁴⁰Ar ages of the associated volcanics. *Tertiary Research* 21:113–122.

Geraads, D., Z. Alemseged, D. Reed, J. Wynn, and C. Roman. 2004a. The Pleistocene fauna (other than Primates) from Asbole, lower Awash Valley, Ethiopia, and its environmental and biochronological implications. *Geobios* 37:697–718.

Geraads, D., and F. Amani. 1997. La faune du gisement à *Homo erectus* de l'Aïn Maarouf, près de El Hajab (Maroc). *L'Anthropologie* 101:522–530.

———. 1998. Bovidae (Mammalia) du Pliocène final d'Ahl al Oughlam, Casablanca, Maroc. *Paläontologische Zeitschrift* 72:191–205.

Geraads, D., C. Blondel, A. Likius, H. T. Mackaye, P. Vignaud, and M. Brunet. 2008. New Hippotragini (Bovidae, Mammalia) from the Late Miocene of Toros-Menalla (Chad). *Journal of Vertebrate Paleontology* 28:231–242.

Geraads, D., C. Blondel, H. T. Mackaye, A. Likius, P.Vignaud, and M. Brunet. 2009b. Bovidae (Mammalia) from the lower Pliocene of Chad. *Journal of Vertebrate Paleontology* 29:923–933.

Geraads, D., M. Brunet, H. T. Mackaye, and P. Vignaud. 2001. Pliocene Bovidae (Mammalia) from the Koro Toro australopithecine sites, Chad. *Journal of Vertebrate Paleontology* 21:335–346.

Geraads, D., V. Eisenmann, and G. Petter. 2004b. The large mammal fauna of the Oldowan sites of Melka Kunturé pp. 169–192 in J. Chavaillon and M. Piperno (eds.), *Studies on the early Paleolithic site of Melka Kunturé Ethiopia.* Istituto Italiano di Preistoria e Protostoria, Florence.

Geraads, D., S. Melillo, and Y. Haile-Selassie. 2009a. Middle Pliocene Bovidae from hominid-bearing sites in the Woranso-Mille area, Afar Region, Ethiopia. *Palaeontologia Africana* 44:59–70.

Geraads, D., and H. Thomas. 1994. Bovidés du Plio-Pléistocène d'Ouganda: Geology and palaeobiology of the Albertine Rift Valley, Uganda-Zaire; pp. 383–407 in M. Pickford, B. Senut, and D. Hadoto (eds.), *Geology and Palaeobiology of the Albertine Rift Valley, Uganda-Zaire: Volume 2. Palaeobiology.* Publications Occasionnelles Centre International pour la Formation et les Echanges Géologiques, no. 29, Orléans.

Gervais, F. L. P. 1867–9. *Zoologie et Paléontologie générales*, Iᵉ série. Bertrand, Paris, 263 pp.

Groves, P., and G. F. Shields. 1997. Cytochrome *b* sequences suggest convergent evolution of the Asian takin and Arctic muskox. *Molecular Phylogenetics and Evolution* 8:363–374.

Grubb, P. 1972. Variation and incipient speciation in the African buffalo. *Zeitschrift für Säugetierkunde* 37:121–144.

———. 2001. Review of family-group names of living bovids. *Journal of Mammalogy* 82:374–388.

Hadjouis, D. 1986. Présence du genre *Kobus* (Bovidae, Artiodactyla) dans le Pléistocène supérieur d'Algérie. *L'Anthropologie* 90:317–320.

———. 2002. Un nouveau Bovini dans la faune du Pléistocène supérieur d'Algérie. *L'Anthropologie* 106:377–386.

Haile-Selassie, Y., E. S. Vrba, and F. Bibi. 2009. Bovidae; pp. 277–330 in Y. Haile-Selassie and G. WoldeGabriel (eds.), *Ardipithecus kaddaba: Late Miocene Evidence from the Middle Awash, Ethiopia.* University of California Press, Berkeley.

Haile-Selassie, Y., G. Woldegabriel, T. D. White, R. L. Bernor, D. Degusta, P. R. Renne, W. K. Hart, E. Vrba, A. Stanley, and F. C. Howell. 2004. Mio-Pliocene mammals from the Middle Awash, Ethiopia. *Geobios* 37:536–552.

Hamilton, W. R. 1973. The lower Miocene ruminants of Gebel Zelten, Libya. *Bulletin British Museum (Natural History), Geology* 21:73–150.

Harper, F. 1940. The nomenclature and type localities of certain Old World mammals. *Journal of Mammalogy* 21:322–332.

Harris, J. M. 1991. Family Bovidae; pp. 139–320 in J. M. Harris (ed.), *Koobi Fora Research Project*, vol. 3. Clarendon Press, Oxford.

———. 2003. Bovidae from the Lothagam succession; pp. 531–579 in M. G. Leakey and J. M. Harris (eds.), *Lothagam: The Dawn of Humanity in Eastern Africa.* Columbia University Press, New York.

Harris, J. M., F. H. Brown, and M. G. Leakey. 1988. Stratigraphy and paleontology of Pliocene and Pleistocene localities west of Lake Turkana, Kenya. *Contributions in Science, Natural History Museum of Los Angeles County* 399:1–128.

Harris, J. M., M. G. Leakey, and T. E. Cerling. 2003. Early Pliocene Tetrapod Remains. In J. M. Harris and M. G. Leakey (eds.), *Geology and Vertebrate Paleontology of the Early Pliocene site of Kanapoi, Northern Kenya. Contributions in Science, Natural History Museum of Los Angeles County* 498:39–113.

Harrison, T. (ed.). In press. *Paleontology and Geology of Laetoli, Tanzania: Human Evolution in Context.* Springer, New York.

Harrison, T., and E. Baker. 1997. Paleontology and biochronology of fossil localities in the Manonga Valley, Tanzania; pp. 361–393 in T. Harrison (ed.), *Neogene Paleontology of the Manonga Valley, Tanzania: A Window into the Evolutionary History of East Africa.* Plenum Press, New York.

Hassanin, A., and E. J. P. Douzery. 1999. The tribal radiation of the family Bovidae (Artiodactyla) and the evolution of the mitochondrial cytochrome *b* gene. *Molecular Phylogenetics and Evolution* 13:227–243.

Heckner-Bisping, U. 2001. Appendix E: The walking patterns of duikers in respect to their origin in evolution; pp. 743–752 in V. J. Wilson (ed.), *Duikers of Africa.* Chipangali Wildlife Trust, Bulawayo.

Hendey, Q. B. 1968. The Melkbos site: An upper Pleistocene fossil occurrence in the south-western Cape Province. *Annals of the South African Museum* 52:89–119.

Hendey, Q. B., and H. Hendey. 1968. New Quaternary fossil sites near Swartklip, Cape Province. *Annals of the South African Museum* 52:43–73.

Hernández Fernández, M., and E. S. Vrba. 2005. A complete estimate of the phylogenetic relationships in Ruminantia: A dated species-level supertree of the extant ruminants. *Biological Reviews* 80:269–302.

Herries, A. I. R., D. Curnoe, and J. W. Adams. 2009. Multi-disciplinary seriation of early *Homo* and *Paranthropus* bearing palaeocaves in southern Africa. *Quaternary International* 202:14–28.

———. 2006. Plio-Pleistocene climatic change in the Turkana Basin (East Africa). *Journal of Human Evolution* 50:595–626.

Hooijer, D. A. 1958. Fossil Bovidae from the Malay archipelago and the Punjab. *Zoologische Verhandelingen Leiden* 38:1–112.

Hopwood, A. T., and J. P. Holleyfield. 1954. An annotated bibliography of the fossil mammals of Africa (1742–1950). *Fossil Mammals of Africa* 8:1–194. British Museum (Natural History), London.

Howell, F. C., G. H. Cole, M. R. Kleindienst, B. J. Szabo, and K. P. Oakley. 1972. Uranium-series dating of bone from the Isimila prehistoric site, Tanzania. *Nature* 237:51–52.

Janis, C. M., and E. Manning. 1998. Antilocapridae; pp. 491–507 in C. M. Janis, K. M. Scott, and L. L. Jacobs (eds.), *Evolution of Tertiary Mammals of North America*, vol. 1. Cambridge University Press, Cambridge.

Kaufulu, Z., E. S. Vrba, and T. D. White. 1981. Age of the Chiwondo Beds, northern Malawi. *Annals of the Transvaal Museum* 33:1–8.

Kingdon, J. 1982a. *East African Mammals: Volume 3C. Bovidae*. Academic Press, London, New York, San Francisco, pp. 1–393.

———. 1982b. *East African Mammals: Volume 3D. Bovidae*. Academic Press, London, New York, San Francisco, pp. 394–746.

———. 1990. *Island Africa*, Collins, London, 287 pp.

Kingston, J. D., B. F. Jacobs, A. Hill, and A. Deino. 2002. Stratigraphy, age and environments of the late Miocene Mpesida Beds, Tugen Hills, Kenya. *Journal of Human Evolution* 42:95–116.

Klein, R. G. 1974. On the taxonomic status, distribution and ecology of the blue antelope, *Hippotragus leucophaeus* (Pallas, 1766). *Annals of the South African Museum* 65:99–143.

———. 1976. The fossil history of *Raphicerus* H. Smith, 1827 (Bovidae, Mammalia) in the Cape Biotic Zone. *Annals South African Museum* 71:169–191.

———. 1980. Environmental and ecological implications of large mammals from Upper Pleistocene and Holocene sites in southern Africa. *Annals of the South African Museum* 81:223–283.

———. 1984. Later stone age faunal samples from Heuningneskrans shelter (Transvaal) and Leopard's Hill Cave (Zambia). *South African Archaeological Bulletin* 39:109–116.

———. 1994. The long-horned African buffalo (*Pelorovis antiquus*) is an extinct species. *Journal of Archaeological Science* 21:725–733.

Klein, R. G., and K. Cruz-Uribe. 1991. The bovids from Elandsfontein, South Africa, and their implications for the age, palaeoenvironment, and origins of the site. *African Archaeological Review* 9:21–79.

Klein, R. G., G. Avery, K. Cruz-Uribe, and T. E. Steele. 2007. The mammalian fauna associated with an archaic hominin skullcap and later Acheulean artifacts at Elandsfontein, Western Cape Province, South Africa. *Journal of Human Evolution* 52:164–186.

Köhler, M. 1987. Boviden des türkischen Miozäns (Känozoikum und Braunkohlen der Türkei 28). *Paleontologia i Evolució* 21:133–246.

———. 1993. Skeleton and habitat of Recent and fossil ruminants. *Münchner Geowissenschaftlichen Abhandlungen A* 25:1–88.

Koenigswald, W. v. 1986. Beziehungen des pleistozänen Wasserbüffels (*Bubalus murrensis*) aus Europa zu den asiatischen Wasserbüffeln. *Zeitschrift für Säugetierkunde* 51:312–323.

Kostopoulos, D. S. 2005. The Bovidae (Mammalia, Artiodactyla) from the late Miocene of Akkaşdaǧi, Turkey. *Geodiversitas* 27:747–791.

Kostopoulos, D. S. and G. D. Koufos. 2006. *Pheraios chryssomallos*, gen. et sp. nov. (Mammalia, Bovidae, Tragelaphini), from the Late Miocene of Thessaly (Greece): Implications for tragelaphin biogeography. *Journal of Vertebrate Paleontology* 26:436–445.

Kovarovic, K., P. Andrews, and L. Aiello. 2002. The palaeoecology of the Upper Ndolanya Beds at Laetoli, Tanzania. *Journal of Human Evolution* 43:395–418.

Kretzoi, M. 1954. Rapport final des fouilles paléontologiques dans la grotte de Csákvár. *Evi Jelentése a Magyar Állami Földtani Intézet* (for 1952):37–69.

———. 1968. New generic names for homonyms. *Vertebrata Hungarica, Musei Historico-naturalis Hungarici* 10, (1–2):163–165.

Kuykendall, K. L., C. A. Toich and J. K. McKee. 1995. Preliminary analysis of the fauna from Buffalo Cave, northern Transvaal, South Africa. *Palaeontologia Africana* 32:27–31.

Lacruz, R. S., J. S. Brink, P. J. Hancox, A. R. Skinner, A. Herries, P. Schmid, and L. R. Berger. 2002. Palaeontology and geological context of a middle Pleistocene faunal assemblage from the Gladysvale Cave, South Africa. *Palaeontologia Africana* 38:99–114.

Lavauden, L. 1930. Notes de mammalogie Nord-Africaine. La gazelle rouge. *Bulletin de la Société Zoologique de France* 55:327–332.

Leakey, L. S. B. 1965. *Olduvai Gorge 1951–61: Fauna and Background*. Cambridge University Press, Cambridge, 118 pp.

Leakey, M. G., and J. M. Harris (eds.) 2003. *Lothagam: Dawn of Humanity in Eastern Africa*. Columbia University Press, New York. 688 pp.

Lehmann, U. 1957. Eine jungpleistozäne Wirbeltierfauna aus Ostafrika. *Mitteilungen der Geologischen Staatsinstitut Hamburg* 26:100–140.

Lehmann, U., and H. Thomas. 1987. Fossil Bovidae from the Mio-Pliocene of Sahabi, (Libya); pp. 323–335 in N. T. Boaz, A. El-Arnauti, A. W. Gaziry, J. de Heinzelin, and D. D. Boaz (eds.), *Neogene Paleontology and Geology of Sahabi*. Liss, New York.

Linnaeus, C. 1758. *Systema Naturae*. 10th ed. Laurentii Salvii, Stockholm, 824 pp.

Lydekker, R. 1877. Notices of new and other Vertebrata from Indian Tertiary and Secondary Rocks. *Records of the Geological Survey of India* 10:30–43.

———. 1878. Crania of ruminants from the Indian Tertiaries. *Memoirs of the Geological Survey of India, Palaeontologia Indica* (10), 1, 3:88–171.

Lydekker, R. 1886. Siwalik Mammalia, Supplement I. *Palaeontologia Indica* (10) 4:1–21.

Made, J. van der 1989. The bovid *Pseudoeotragus seegrabensis* nov. gen., nov. sp. from the Aragonian (Miocene) of Seegraben near Leoben (Austria). *Proceedings Koninklijke Nederlandse Akademie van Wetenschappen, B* 92:215–240.

Marcot, J.D. 2007. Molecular phylogeny of terrestrial artiodactyls; pp. 4–18 in D. R. Prothero and S. E. Foss (eds.), *The Evolution of Artiodactyls*. Johns Hopkins University Press, Baltimore.

Marean, C. W. 1992. Implications of Late Quaternary mammalian fauna from Lukenya Hill (South-Central Kenya) for paleoenvironmental change and faunal extinctions. *Quaternary Research* 37:239–255.

Marks, A. E., J. Peters, and W. Van Neer. 1987. Late Pleistocene and early Holocene occupations in the upper Atbara river valley, Sudan; pp. 137–161 in A. E. Close (ed.), *Prehistory of Arid North Africa: Essays in Honor of Fred Wendorf*. Southern Methodist University Press, Dallas.

Martínez-Navarro, B., J. A. Pérez-Claros, M. R. Palombo, L. Rook, and P. Palmqvist. 2007. The Olduvai buffalo *Pelorovis* and the origin of *Bos*. *Quaternary Research* 68:220–226.

Martínez-Navarro, B., L. Rook, A. Segid, D. Yosief, M. P. Ferretti, J. Shoshani, T. M. Tecle and Y. Libsekal. 2004. The large fossil mammals from Buia (Eritrea): systematics, biochronology and paleoenvironments. *Rivista Italiana di Paleontologia e Stratigrafia* 110 (suppl.):61–88.

Matthee, C. A., and S. K. Davis. 2001. Molecular insights into the evolution of the family Bovidae: a nuclear DNA perspective. *Molecular Biology and Evolution* 18:1220–1230.

Métais, G., P.-O. Antoine, L. Marivaux, J.-L. Welcomme, and S. Ducrocq. 2003. New artiodactyl ruminant mammal from the late Oligocene of Pakistan. *Acta Palaeontologia Polonica* 48:375–382.

Morales, J., D. Soria, and M. Pickford. 1995. Sur les origines de la famille des Bovidae (Artiodactyla, Mammalia). *Comptes Rendus de l'Académie des Sciences, Paris, Série II*, 321:1211–1217.

Morales, J., D. Soria, M. Pickford and M. Nieto. 2003. A new genus and species of Bovidae (Artiodactyla, Mammalia) from the early Middle Miocene of Arrisdrift, Namibia, and the origins of the family Bovidae. *Memoirs of the Geological Survey of Namibia* 19:371–384.

Moullé, P.-É., A. Echassoux, and B. Martinez-Navarro. 2004. *Ammotragus europaeus*: une nouvelle espèce de Caprini (Bovidae, Mammalia) du Pléistocène inférieur à la grotte du Vallonnet (France). *Comptes Rendus Palevol* 3:663–673.

Moyá Solá, S. 1983. Los Boselaphini (Bovidae Mammalia) del Neogeno de la península Ibérica. *Publicaciones de Geologia, Universitat Autonoma de Barcelona* 18:1–236.

———. 1986. El genero *Hispanomeryx* Morales et al. (1981): Posicion filogenética y systematica: Su contribucion al conocimiento de la evolución de los Pecora (Artiodactyla, Mammalia). *Paleontologia i Evolució* 20:267–287.

Nakaya, H. 1994. Faunal change of late Miocene Africa and Eurasia: Mammalian fauna from the Namurungule Formation, Samburu Hills, northern Kenya. *African Study Monographs* (suppl.) 20:1–112.

Nakaya, H., M. Pickford, Y. Nakano, and H. Ishida. 1984. The late Miocene large mammal fauna from the Namurungule Formation, Samburu Hills, northern Kenya. *African Study Monographs* (suppl.) 2:87–131.

Oboussier, H. 1970. Beiträge zur Kenntnis der *Pelea* (*Pelea capreolus*, Bovidae, Mammalia), ein Vergleich mit etwa gleichgrossen anderen Bovinae (*Redunca fulvorufula*, *Gazella thomsoni*, *Antidorcas marsupialis*). *Zeitschrift für Säugetierkunde* 35:342–353.

Ogilby, W. 1841. Monograph on the hollow-horned ruminants. *Proceedings Zoological Society of London* 1841:4–10.

Pallas, P. S. 1766. *Miscellanea Zoologica*. P. van Cleef, Hagae, Comitum, 224 pp.

———. 1767. *Spicilegia Zoologica*, fasc. 1. Gottl. August. Lange, Berlin, 44 pp.

Partridge, T. C., D. E. Granger, M. W. Caffee, and R. J. Clarke. 2003. Lower Pliocene hominid remains from Sterkfontein. *Science* 300:607–612.

Partridge, T. C., J. Shaw, and D. Heslop. 2000. Note on recent magnetostratigraphic analyses in Member 2 of the Sterkfontein Formation; pp. 129–130 in T. C. Partridge and R. R. Maud (eds.), *The Cenozoic of Southern Africa*. Oxford University Press, Oxford.

Pennant, T. 1781. *History of Quadrupeds*, vol. 1. B. White, London, 282 pp.

Peters, J., A. Gautier, J. S. Brink, and W. Haenen. 1994. Late Quaternary extinction of ungulates in sub-Saharan Africa: A reductionist's approach. *Journal of Archaeological Science* 21:17–28.

Pickford, M., and H. Thomas. 1984. An aberrant new bovid (Mammalia) in subrecent deposits from Rusinga Island, Kenya. *Proceedings Koninklijke Nederlandse Akademie van Wetenschappen B* 87:441–452.

Pilgrim, G. E. 1934. Two new species of sheep-like antelope from the Miocene of Mongolia. *American Museum Novitates* 716:1–29.

———. 1939. The fossil Bovidae of India. *Palaeontologia Indica* 26:1–356.

———. 1941. A fossil skull of *Hemibos* from Palestine. *Annals and Magazine of Natural History London* (11), 7:347–360.

Plug, I., and S. Badenhorst. 2001. The distribution of macromammals in southern Africa over the past 30,000 years. *Transvaal Museum Monographs* 12:1–234.

Plug, I., and A.W. Keyser. 1994a. Haasgat cave, a Pleistocene site in the central Transvaal: Geomorphological, faunal and taphonomic considerations. *Annals of the Transvaal Museum* 36:139–145.

———. 1994b. A preliminary report on the bovid species from recent excavations at Gladysvale, South Africa. *South African Journal of Science* 90:357–359.

Pomel, A. 1894a. Les boeufs taureaux. *Paléontologie Monographies, Carte Géologique de l'Algérie* 3:1–108.

———. 1894b. Les bosélaphes Ray. *Paléontologie Monographies, Carte Géologique de l'Algérie* 4:1–61.

———. 1895. Les antilopes Pallas. *Paléontologie Monographies, Carte Géologique de l'Algérie* 5:1–56.

Potts, R., and A. Deino. 1995. Mid-Pleistocene change in large mammal faunas of East Africa. *Quaternary Research* 43:106–113.

Ranjitsinh, M. K. 1989. *The Indian Blackbuck*. Natraj Publishers, Dehradun, 156 pp.

Reck, H. 1925. Aus der Vorzeit des innerafrikanischen Wildes. *Illustrierte Zeitung Leipzig* 164:451. (This article ends in the middle of a word, but no continuation of the text could be found in the same or the next two issues, or by reference to the index for the whole volume.)

———. 1928. *Pelorovis oldowayensis* n. g. n. sp. *Wissenschaftliche Ergebnisse der Oldoway-Expedition 1913, Berlin* (n.f.) 3:57–67.

Reck, H. 1935. Neue Genera aus der Oldoway-Fauna. *Zentralblatt für Mineralogie, Geologie und Paläontologie B* 1925:215–218.

Reed, K. E. 1998. Using large mammal communities to examine ecological and taxonomic structure and predict vegetation in extant and extinct assemblages. *Paleobiology* 24:384–408.

Robinson, P. 1972. *Pachytragus solignaci*, a new species of caprine bovid from the late Miocene Beglia Formation of Tunisia. *Notes du Service Géologique de Tunisie* 37:73–94.

———. 1986. Very hypsodont antelopes from the Beglia Formation (central Tunisia), with a discussion of the Rupicaprini. *Contributions in Geology, University of Wyoming, Special Papers* 3:305–315.

Rookmaaker, L. C. 1991. The scientific name of the bontebok. *Zeitschrift für Säugetierkunde* 56:190–191.

Ropiquet, A., and A. Hassanin. 2005. Molecular evidence for the polyphyly of the genus *Hemitragus* (Mammalia, Bovidae). *Molecular Phylogenetics and Evolution* 36:154–168.

Rütimeyer, L. 1877–78. Die Rinder der Tertiär-Epoche nebst Vorstudien zu einer natürlichen Geschichte der Antilopen. *Abhandlungen der Schweizerischen Paläontologischen Gesellschaft* 4–5:1–208.

Sánchez, I. M., M. Soledad Domingo, and J. Morales 2009. New data on the Moschidae (Mammalia, Ruminantia) from the Upper Miocene of Spain (MN 10–MN 11). *Journal of Vertebrate Paleontology* 29:567–575.

Sandrock, O., O. Kullmer, F. Schrenk, Y. M. Juwayeyi, and T. G. Bromage. 2007. Fauna, taphonomy, and ecology of the Plio-Pleistocene Chiwondo Beds, Northern Malawi; pp. 315–332 in R. Bobe, Z. Alemseged, and A. K. Behrensmeyer (eds.), *Hominin Environments in the East African Pliocene: An Assessment of the Faunal Evidence*. Springer, Dordrecht.

Schlosser, M. 1911. Mammalia; pp. 325–585 in K. A. Zittel (ed.), *Grundzüge der Paläontologie: Abteilung 2. Vertebrata*. Oldenbourg, Munich.

Schwarz, E. 1932. Neue diluviale Antilopen aus Ostafrika. *Zentralblatt für Mineralogie, Geologie und Paläontologie B* 1932:1–4.

———. 1937. Die fossilen Antilopen von Oldoway. *Wissenschaftliche Ergebnisse der Oldoway-Expedition 1913, Berlin* (n.f.) 4:8–104.

Solignac, M. 1924. Sur la présence de *Buffelus palaeindicus* Falc. dans le quaternaire ancien de la région de Bizerte (Tunisie). *Bulletin de la Société Géologique de France* (4), 24:176–192.

Solounias, N. 1990. A new hypothesis uniting *Boselaphus* and *Tetracerus* with the Miocene Boselaphini (Mammalia, Bovidae) based on horn morphology. *Annales Musei Goulandris* 8:425–439.

Spencer, L. M. 1997. Dietary adaptations of Plio-Pleistocene Bovidae: Implications for hominid habitat use. *Journal of Human Evolution* 32:201–228.

Sponheimer, M., K. E. Reed, and J. A. Lee-Thorp. 1999. Combining isotopic and ecomorphological data to refine bovid paleodietary reconstruction: A case study from the Makapansgat Limeworks hominin locality. *Journal of Human Evolution* 36:705–718.

Stromer, E. 1907. Fossile Wirbeltier-Reste aus dem Uadi Fâregh und Uadi Natrûn in Ägypten. *Abhandlungen der Senckenbergischen Naturforschenden Gesellschaft* 19:99–132.

Su, D. F., and T. Harrison. 2007. The paleoecology of the Upper Laetolil Beds at Laetoli; pp. 279–313 in R. Bobe, Z. Alemseged, and A. K. Behrensmeyer (eds.), *Hominin Environments in the East African Pliocene: An Assessment of the Faunal Evidence*. Springer, Dordrecht.

Tchernov, E., L. Gindburg, P. Tassy, and N. F. Goldsmith. 1987. Miocene mammals of the Negev (Israel). *Journal of Vertebrate Paleontology* 7:284–310.

Teilhard de Chardin, P., and J. Piveteau. 1930. Les mammifères fossiles de Nihowan (Chine). *Annales de Paléontologie* 19:1–134.

Thackeray, J. F. 1983. On Darwin, extinctions, and South African fauna. *Discovery, New Haven* 16, (2):2–11.

Thenius, E. 1951. *Gazella* cf. *deperdita* aus dem mitteleuropäischen Vindobonien und das Auftreten der Hipparionfauna. *Eclogae Geologicae Helvetiae* 44:381–394.

———. 1979. Zur systematischen Stellung und Verbreitung von "*Gazella*" *stehlini* (Bovidae, Mammalia) aus dem Miozän Europas. *Anzeiger der Oesterreichischen Akademie der Wissenschaften. Mathematisch-Naturwissenschaftliche Klasse, Wien* 116:9–13.

Thomas, H. 1977. Géologie et paléontologie du gisement acheuléen de l'erg Tihoudaïne. *Mémoires du Centre de Recherches Anthropologiques Préhistoriques et Ethnographiques, Algiers* 27:1–122.

———. 1979. *Miotragocerus cyrenaicus* sp.nov. (Bovidae, Artiodactyla, Mammalia) du Miocène supérieur de Sahabi (Libye) et ses rapports avec les autres *Miotragocerus*. *Geobios* 12:267–281.

———. 1980. Les bovidés du Miocène supérieur des couches de Mpesida et de la formation de Lukeino (district de Baringo, Kenya); pp. 82–91 in R. E. F. Leakey and B. A. Ogot (eds.), *Proceedings of the 8th Pan-African Congress of Prehistory, Nairobi 1977, Nairobi*.

———. 1981. Les Bovidés miocènes de la formation de Ngorora du Bassin de Baringo (Kenya). *Proceedings Koninklijke Nederlandse Akademie van Wetenschappen B* 84:335–409.

———. 1983. Les Bovidae (Artiodactyla, Mammalia) du Miocène moyen de la Formation Hofuf (province du Hasa, Arabie Saoudite). *Palaeovertebrata* 13:157–206.

———. 1984a. Les Bovidés anté-hipparions des Siwaliks inférieurs (plateau du Potwar, Pakistan). *Mémoires de la Société Géologique de France Paris* (n.s.) 145:1–68.

———. 1984b. Les origines africaines des Bovidae miocènes des lignites de Grosseto (Toscane, Italie). *Bulletin du Muséum National d'Histoire Naturelle Paris* (4), 6, (C): 81–101.

Thomas, H., and G. Petter. 1986. Révision de la faune de mammifères du miocène supérieur de Menacer (ex-Marceau), Algérie: discussion sur l'âge du gisement. *Geobios* 19:357–373.

Thomas, H., S. Sen, M. Khan, B. Battail, and G. Ligabue. 1982. The lower Miocene fauna of Al-Sarrar (Eastern province, Saudi Arabia). *ATLAL* 5:109–136.

Thomas, H., D. Geraads, D. Janjou, D. Vaslet, A. Memesh, D. Billiou, H. Bocherens, G. Dobigny, V. Eisenmann, M. Gayet, F. d. Lapparent de Broin, G. Petter, and M. Halawani. 1998. First Pleistocene faunas from the Arabian Peninsula: An Nafud desert, Saudi Arabia. *Comptes Rendus de l'Académie des Sciences, Paris* 326:145–152.

Thomas, P. 1884. Recherches stratigraphiques et paléontologiques sur quelques formations d'eau douce de l'Algérie. *Mémoires de la Société Géologique de France, Paris* (3), 3, 2:1–51.

Tsujikawa, H. 2005. The updated Late Miocene large mammal fauna from Samburu Hills, northern Kenya. *African Study Monographs* (suppl.) 32:1–50.

Van Gelder, R. G. 1977. An eland × kudu hybrid, and the contents of the genus *Tragelaphus. Lammergeyer* 23:1–6.

Van Hoepen, E. C. N. 1932. Voorlopige beskrywing van Vrystaatse soogdiere. *Paleontologiese Navorsing van die Nasionale Museum Bloemfontein* 2, 5:63–65.

Vignaud, P., P. Duringer, H. T. Mackaye, A. Likius, C. Blondel, J.-R. Boisserie, L. de Bonis, V. Eisenmann, M.-E. Etienne, D. Geraads, F. Guy, T. Lehmann, F. Lihoreau, N. Lopez-Martinez, C. Mourer-Chauviré, O. Otero, J.-C. Rage, M. Schuster, L. Viriot, A. Zazzo, and M. Brunet. 2002. Geology and paleontology of the Upper Miocene Toros-Menalla hominid locality, Chad. *Nature* 418:152–155.

Vrba, E. S. 1976. The fossil Bovidae of Sterkfontein, Swartkrans and Kromdraai. *Transvaal Museum Memoirs* 21:1–166.

———. 1977. New species of *Parmularius* Hopwood and *Damaliscus* Sclater and Thomas (Alcelaphini, Bovidae, Mammalia) from Makapansgat. *Palaeontologia Africana* 20:137–151.

———. 1978. Problematical alcelaphine fossils from the Kromdraai faunal site (Mammalia: Bovidae). *Annals of the Transvaal Museum* 31:21–28.

———. 1980. The significance of bovid remains as indicators of environment and predation patterns; pp. 247–271 in A. K. Behrensmeyer and A. Hill (eds.), *Fossils in the Making*. University of Chicago Press, Chicago.

———. 1987a. New species and a new genus of Hippotragini (Bovidae) from Makapansgat Limeworks. *Palaeontologia Africana* 26:47–58.

———. 1987b. A revision of the Bovini (Bovidae) and a preliminary revised checklist of Bovidae from Makapansgat. *Palaeontologia Africana* 26:33–46.

———. 1995. The fossil record of African antelopes (Mammalia, Bovidae) in relation to human evolution and paleoclimate; pp. 385–424 in E. S. Vrba, G. H. Denton, T. C. Partridge, and L. H. Burckle (eds.), *Paleoclimate and Evolution, with Emphasis on Human Origins*. Yale University Press, New Haven.

———. 1997. New fossils of Alcelaphini and Caprinae (Bovidae: Mammalia) from Awash, Ethiopia, and phylogenetic analysis of Alcelaphini. *Palaeontologia Africana* 34:127–198.

———. 2006. A possible ancestor of the living waterbuck and lechwes: *Kobus basilcookei* sp. nov. (Reduncini, Bovidae, Artiodactyla) from the early Pliocene of the Middle Awash, Ethiopia. *Transactions of the Royal Society of South Africa* 61:63–74.

Vrba, E. S., and J. Gatesy. 1994. New antelope fossils from Awash, Ethiopia, and phylogenetic analysis of Hippotragini (Bovidae, Mammalia). *Palaeontologia Africana* 31:55–72.

Vrba, E. S., and Y. Haile-Selassie. 2006. A new antelope fossil, *Zephyreduncinus oundagaisus* (Reduncini, Artiodactyla, Bovidae), from the Late Miocene of the Middle Awash, Afar Rift, Ethiopia. *Journal of Vertebrate Paleontology* 26:213–218.

Vrba, E. S., and G. B. Schaller. 2000. Phylogeny of Bovidae based on behavior, glands, skulls, and postcrania; pp. 203–222 in E. S. Vrba and G. B. Schaller (eds.), *Antelopes, Deer and Relatives: Fossil Record, Behavioral Ecology, Systematics and Conservation*. Yale University Press, New Haven.

Vrba, E. S., J. R. Vaisnys and J. E. Gatesy. 1994. Analysis of paedomorphosis using allometric characters: The example of Reduncini antelopes (Bovidae, Mammalia). *Systematic Biology* 43:92–116.

Wang, B. 1992. The Chinese Oligocene: A preliminary review of mammalian localities and local faunas; pp. 529–547 in D. R. Prothero and W. A. Berggren (eds.), *Eocene-Oligocene Climatic and Biotic Evolution*. Princeton University Press, Princeton, N.J.

Watson, W., and I. Plug. 1995. *Oreotragus major* Wells and *Oreotragus oreotragus* (Zimmermann) (Mammalia: Bovidae): Two species? *Annals of the Transvaal Museum* 36:183–191.

Weithofer, K. A. 1888. Alcune osservazioni sulla fauna delle ligniti di Casteani e di Montebamboli (Toscana). *Bollettino del R. Comitato Geologico d'Italia* 19:363–368.

Wells, L. H. 1967. Antelopes in the Pleistocene of southern Africa; pp. 99–107 in W. W. Bishop and J. D. Clark (eds.), *Background to Evolution in Africa*. University of Chicago Press, Chicago.

Wells, L. H., and H. B. S. Cooke. 1956. Fossil Bovidae from the Limeworks Quarry, Makapansgat, Potgietersrus. *Palaeontologia Africana* 4:1–55.

Whybrow, P. J., M. E. Collinson, R. Daams, A. W. Gentry, and H. A. McClure. 1982. Geology, fauna (Bovidae, Rodentia) and flora from the early Miocene of eastern Saudi Arabia. *Tertiary Research* 4:105–120.

Wilson, D. E., and D. M. Reeder. (eds.) 2005. *Mammal Species of the World: A Taxonomic and Geographic Reference*. 3rd ed., 2 vols. Johns Hopkins University Press, Baltimore, 2,142 pp.

Yalden, D. W. 1978. A revision of the dik-diks of the subgenus *Madoqua* (*Madoqua*). *Monitore Zoologico Italiano* NS (suppl.) 11, 10:245–264.

Zeuner, F. E. 1963. *A History of Domesticated Animals*. Hutchinson, London, 560 pp.

THIRTY-NINE

Giraffoidea

JOHN M. HARRIS, NIKOS SOLOUNIAS,
AND DENIS GERAADS

The extant giraffes are an iconic part of the African biota, their large size and elongate legs and neck providing an unmistakable silhouette against the African landscape. Their close relatives, the okapis, were among the latest of the large terrestrial mammals to be documented scientifically and are similarly iconic in terms of their rarity and cryptic nature. Giraffes are characterized by skin-covered ossicones attached to the frontals; only male okapis have ossicones, from which the skin may be worn off the distal portions in mature specimens. The nature of their relationship to each other and to the somewhat bewildering variety of African fossil pecorans is still a matter of debate.

The origins of the Giraffoidea remain uncertain although Janis and Scott (1987) suggested they could have originated from the Gelocidae before the early Miocene. The Giraffoidea have been variously allied with the Bovoidea and/or Cervoidea, but Hernández Fernández and Vrba (2005) construe them as a sister group of a clade containing both the Bovidae and Cervidae and suggest they are conceivably most closely related to the antilocaprids.

Systematic Paleontology

Superfamily GIRAFFOIDEA Gray, 1821
Figure 39.1 and Table 39.1

The giraffoids are ruminant artiodactyls with a bilobed lower canine (figure 39.1) but lacking first premolars. The stomach is four chambered in extant forms, but there is no gall bladder. Two families are recognized. The Climacoceratidae are known from the early and middle Miocene of eastern and southern Africa; early representatives of the Giraffidae first appear in the middle Miocene (table 39.1).

Family CLIMACOCERATIDAE Hamilton, 1978

Climacoceratids differ from the Palaeomerycidae by their bilobed lower canines and by their loss of the upper canine and occipital median ossicones. They differ from the Giraffidae by the presence of branched appendages (ossicones with tines) on the frontal.

Genus *CLIMACOCERAS* MacInnes, 1936

Diagnosis Giraffoids possessing hypsodont teeth and large frontal ossicones with many tines. The premolar row is shortened and the lower molars lack a palaeomeryx fold. In the upper molars there is external fusion of the lingual and buccal lobes.

Type Species *Climacoceras africanus* MacInnes, 1936.

Other Recognized Species *Climacoceras gentryi* Hamilton, 1978.

CLIMACOCERAS AFRICANUS MacInnes, 1936
Figure 39.2

Diagnosis Species of *Climacoceras* in which the thick, straight ossicones carry many short irregularly spaced tines (Hamilton, 1978).

Holotype KB. A.A.1, relatively complete ossicone with tines broken off, from Kiboko (now Maboko) Island, Kavirondo Gulf, Kenya; original conserved in the collections of the Natural History Museum, London.

Remarks *Climacoceras africanus* is abundantly represented at Maboko Island, Kenya. Its ossicone has a central thick straight beam with tines irregularly distributed throughout the surface (figure 39.2). The distal part of the ossicone has a fork with two thicker tines. The dentition of *Climacoceras* is simple and resembles that of very primitive ruminants such as *Propalaeoryx*. KNM-WK 18272 from Kalodirr, Kenya, is also identified as *C. africanus*.

CLIMACOCERAS GENTRYI Hamilton, 1978
Figures 39.3 and 39.4

Diagnosis Species of *Climacoceras* in which the ossicones have a slender curved beam bearing long tines (Hamilton 1978).

Holotype KNM-FT 2946, left mandible (i3, c1, p2–m3), from the middle Miocene site of Fort Ternan, Kenya; conserved in the National Museums of Kenya, Nairobi (figure 39.3).

Remarks The ossicones of *Climacoceras gentryi* have a slender curved beam (figure 39.4). At the base there is a long

TABLE 39.1
Sites in Africa and the Middle East from which fossil giraffoids have been reported

ALGERIA

1. Aïn Boucherit: Arambourg, 1979; late Pliocene
2. Aïn Hanech, near El Eulma, Constantine: Singer and Boné, 1960, Arambourg, 1979; early Pleistocene
3. Boulevard Bru, Mustapha Supérieur, Algiers: Romer, 1928; late Pleistocene/Holocene
4. Bou Hanifia, Oran: Arambourg, 1959; late Miocene
5. Menacer (Marceau): Thomas and Petter, 1986; late Miocene
6. Oued Mya 1: Sudre & Hartenberger, 1992; late Miocene
7. Smendou, Constantine: Singer and Boné, 1960, Churcher, 1978; late Miocene to early Pleistocene
8. St Charles: Pomel 1892. Singer and Boné, 1960; late Pliocene
9. Tighenif (Ternifine): Arambourg, 1952, Geraads, 1981; mid-Pleistocene

CHAD

10. Koro-Toro: Coppens, 1960; Brunet et al., 1995; early Pleistocene
11. Kollé, Djourab: Brunet et el., 1998; early Pliocene
12. Kossom Bougoudi: Brunet and MPFT, 2000; early Pliocene
13. Toros-Menalla: Vignaud et al., 2002; Likius et al., 2007; late Miocene

DJIBOUTI

14. Anabo Koma: Geraads, 1985; early Pleistocene

EGYPT

15. Garet el Moluk Hill, Wadi Natrun: Stromer, 1907; early Pliocene
16. Wadi Moghara: Miller, 1999, Pickford et al., 2001; early Miocene

ERITREA

17. Buia: Martinez-Navarro et al., 2004; early Pleistocene

ETHIOPIA

18. Wehaietu Fm, Middle Awash: Kalb et al., 1982; Pliocene
19. Sagantole Fm: Kalb et al., 1982, Woldegabriel et al., 1994, Haile-Selassie, 2009; Pliocene
20. As Duma, Gona: Semaw et al., 2005; early Pliocene
21. Adu-Asa Fm: Haile-Selassie, 2009
22. Asbole: Geraads et al., 2004; mid-Pleistocene
23. Bodo: Kalb et al., 1982; mid-Pleistocene
24. Bouri-Hata: de Heinzelin et al., 1999; late Pliocene
25. Bouri Daka: Asfaw et al., 2002; early Pleistocene
26. Chorora: Geraads et al., 2002; late Miocene
27. Dikika: Wynn et al., 2005; mid-Pliocene
28. Hadar Fm: Taieb et al., 1976; mid-Pliocene
29. Konso-Gardula: Suwa et al., 2003; early Pleistocene
30. Matabaeitu Formation, Middle Awash: Kalb et al., 1982; late Pliocene
31. Melka-Kunture: Geraads et al., 2004b; early Pleistocene
32. Omo, Shungura Fm: Arambourg, 1947; mid-Pliocene to early Pleistocene
33. Galili: Urbanek et al., 2005; early Pliocene

IRAQ

34. Gebel Hamrin: Heintz et al., 1981; late Miocene

ISRAEL

35. Bethlehem: Bar-Yosef and Tchernov, 1972.
36. Ubeidiya: Haas, 1966, Bar-Yosef and Tchernov, 1972, Geraads, 1986b; early Pleistocene

KENYA

37. Buluk: Leakey and Walker, 1985; early Miocene
38. Chepkesin: Aguirre and Leakey, 1974
39. Chiande Uyoma: Pickford, 1986; early Miocene
40. Fort Ternan: Hamilton, 1978; Churcher, 1970; mid-Miocene
41. Kagua: Kent, 1942b, Pickford, 1986; early Pleistocene
42. Kalodirr: Leakey and Leakey, 1986; early Miocene
43. Kanam, Homa Mountain: Pickford, 1986; early Pleistocene
44. Kanapoi: Harris et al., 2003; early Pliocene
45. Kanjera, Homa Mountain, Lake Victoria: Cooke, 1963; late Pliocene
46. Karungu: Pickford, 1986; early Miocene
47. Koobi Fora: Harris, 1976b; Harris, 1991; late Pliocene to early Pleistocene
48. Locherangan: Anyonge, 1991; early Miocene
49. Lothagam: Churcher, 1978, Harris, 2003; late Miocene
50. Lukeino: Pickford and Senut, 2001; late Miocene
51. Maboko: Thomas, 1984; MacInnes, 1936; Pickford, 1986; mid-Miocene
52. Majiwa: Andrews et al., 1981; early Miocene
53. Marsabit Road: Singer and Boné, 1960; late Pliocene
54. Mfwanganu: Pickford, 1986; early Miocene
55. Moruorot: Hamilton, 1978; Madden, 1972; early Miocene
56. Nachola Formation: Pickford et al., 1987; mid-Miocene
57. Nakali: Aguirre and Leakey, 1974; late Miocene
58. Ngorora: Hamilton, 1978, Bishop and Pickford, 1975 late Miocene
59. Nyakach Formation: Thomas, 1984
60. Olorgesaillie: Vaufrey, 1947; Cooke 1963; mid-Pleistocene
61. Ombo: Pickford, 1986; mid-Miocene
62. Rawe: Kent, 1942a; Leakey 1965, 1970; early Pleistocene
63. Rusinga: Whitworth, 1958; Hamilton, 1978; Pickford, 1986; early Miocene
64. Samburu Hills (Namurungule Formation): Nakaya et al., 1987; late Miocene
65. South Turkwell: Ward et al., 1999; late Pliocene
66. West Turkana: Harris et al., 1988; late Pliocene and early Pleistocene

LIBYA

67. Gebel Zelten: Hamilton, 1973, 1978; early Miocene
68. Sahabi: Harris, 1982; late Miocene

MALAWI

69. Chiwondo Beds: Mawby, 1970; Schrenk et al., 1993; Bromage et al., 1995; late Pliocene

MOROCCO

70. Ahl al Oughlam : Geraads, 1996; late Pliocene
71. Beni Mellal: Lavocat, 1961; Heintz, 1976; mid-Miocene

72. Lissasfa: Raynal et al., 1999; early Pliocene
73. Mugharet el 'Aliya, Cape Spartel, Tangier: Howe and
    Movius, 1947; late Pleistocene

NAMIBIA

74. Arrisdrift: Morales et al., 1999; Morales et al., 2003; basal
    middle Miocene
75. Elisabethfeld: Morales et al., 1999; early Miocene

NIGER

76. In Azaoua; Joleaud, 1936; Holocene

SAUDI ARABIA

77. Al Jadidah : Morales et al., 1987; mid-Miocene
78. Al-Sarrar: Thomas et al., 1985; mid-Miocene

SOUTH AFRICA

79. Barkley West, Cape Province: Cooke, 1963; mid-Pleistocene
80. Bloembosch, Darling District, Cape Province: Cooke, 1955;
    mid-Pleistocene?
81. Cornelia, OFS: Singer and Boné 1960, Cooke, 1963; late
    Pleistocene
82. Elandsfontein, Cape Province: Hendey, 1968, 1969; early
    Pleistocene
83. Florisbad: Singer and Boné, 1960; Cooke, 1963; mid- to late
    Pleistocene
84. Kalkbank, near Pietersburg, Transvaal: Mason, Dart and
    Kitching, 1958; late Pleistocene
85. Langebaanweg: Harris, 1976; Hendey, 1970, 1974, 1976,
    1981; early Pliocene
86. Makapansgat: Cooke, 1963; late Pliocene
87. Tierfontein, near Port Allen, Vet River: Cooke, 1974;
    mid-Pleistocene

SUDAN

88. Bahr el Ghazel. Arambourg, 1960.

TANZANIA

89. Laetoli: Dietrich, 1942; Harris, 1983; mid-Pliocene

90. Mkujuni Valley, Lake Manyara: Kent, 1942a
91. Manonga Valley: Gentry, 1997; late Pliocene
92. Mumba Hills: Cooke, 1963; late Pleistocene
93. Lake Natron: Geraads, 1987; early Pleistocene
94. Olduvai: Hopwood, 1934; Leakey, 1965; late Pliocene to
    early Pleistocene

TUNISIA

95. Aïn Brimba, near Chott el Djerid: Arambourg, 1979;
    Coppens, 1971; Pliocene
96. Bled Douarah, near Gafsa: Robinson and Black, 1974; mid to
    late Miocene
97. Djebel Krechem: Geraads, 1989; late Miocene
98. Djebel Sehib, south of Gafsa: Burrolet, 1956; late Miocene
99. Sud Tunisien, possibly near Djerid: Arambourg, 1948;
    Pleistocene
100. Douaria: Roman and Solignac, 1934, Churcher, 1978; late
    Miocene to early Pliocene
101. Garaet (Lake) Ichkeul, near Bizerte: Singer and Boné, 1960;
    Arambourg, 1979; Pliocene
102. Hamada Damous, near Grombalia: Coppens, 1971; Pliocene

UGANDA

103. Hohwa: Geraads, 1994; late Pliocene
104. Kaiso Fm: Cooke and Coryndon, 1970; late Pliocene
105. Napak: Bishop, 1962, 1967; early Miocene
106. Nkondo Fm: Geraads, 1994; early Pliocene
107. Oluka Fm: Geraads, 1994; late Miocene
108. Warwire Fm: Geraads, 1994; middle Pliocene

UNITED ARAB EMIRATES

109. Baynunah Fm, Abu Dhabi: Gentry, 1999; late Miocene

DEMOCRATIC REPUBLIC OF CONGO

110. Sinda Basin: Yasui et al., 1992; late Miocene?
111. Semliki Valley Russo Beds: Boaz, 1990; late Pliocene

ZAMBIA

112. Kabwe (Broken Hill): Cooke, 1963; late Pleistocene

anteriorly directed tine, above which are shorter, anteriorly positioned tines. The main shaft curves to be anteriorly concave. The overall shape resembles the antler of the mule deer (*Odocoileus hemionus*). The *C. gentryi* mandible KNM-FT 2946 has a downturned angle, an unusual feature in ruminants but which is also shared by *Prolibytherium*. Hamilton (1978) records *C. gentryi* from both Fort Ternan near the Kavirondo Gulf and from the Kabarsero Beds of the Ngorora Formation in the Lake Baringo Basin. Pickford et al. (1987) also document this species from the Nachola Formation in the Samburu Hills, northern Kenya.

Morales et al. (1999) interpreted *Sperrgebietomeryx wardi* Morales et al., 1999, *Propalaeoryx austroafricanus* Stromer, 1926, and *P. nyanzae* Whitworth, 1958 as climacoceratids that lacked frontal ossicones but had a palaeomeryx fold in the lower molars. However, as a bilobed canine has not yet been documented for any of these species, they do not qualify for inclusion in the Giraffoidea (see also Cote, this volume, chapter 37, regarding this taxon and the next).

Orangemeryx hendeyi from Arrisdrift is an early middle Miocene Namibian artiodactyl with cranial appendages that are short, slightly compressed, and conical in shape, have rounded tubercules at the base, and terminate distally in two or three points. Morales et al. (1999) interpreted *O. hendeyi* as a climacoceratid on the basis of its tined ossicones. The Arrisdrift sample of *Orangemeryx* includes about 50 incisiform teeth (Morales et al., 1999), but none of them is bilobed. Accordingly, until it can be demonstrated that *O. hendeyi* has bilobed canines, this species is excluded from the Giraffoidea.

Thomas (1984) erected *Nyanzameryx pickfordi* on the basis of a partial cranium from Maboko and additional material from the Nyakach Formation of Kenya and attributed this species to the Giraffidae. The "ossicones" of *Nyanzameryx* are cylindrical like those of *Climacoceras* rather than conical like those of *Orangemeryx*, but, as in *Orangemeryx* and unlike in *Climacoceras*, the "ossicones" lack tines. Geraads (1986a) thought the lower dentition that Thomas attributed to *N. pickfordi* actually

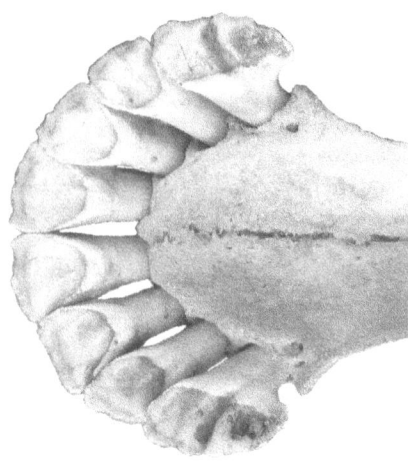

FIGURE 39.1 OM 2272; *Giraffa camelopardalis* mandibular symphysis showing bilobed lower canine. Scale = 5 cm.

belonged to *Climacoceras* and pointed out that there is no derived character that permits material currently assigned to *Nyanzameryx* to be included in the Giraffidae. McCrossin et al. (1998) suggested the cranium of *N. pickfordi* belonged to a bovid.

Pickford et al. (2001) included *Prolibytherium magnieri* from Wadi Moghara and Gebel Zelten in the Climacoceratidae, but *Prolibytherium* lacks tined ossicones and the bilobed (or other) nature of its canine has yet to be determined (see also Cote, this volume, chapter 37).

Family GIRAFFIDAE GRAY, 1821

Diagnosis Giraffoids in which the accessory lobe of the lower canine forms about one-third of the crown. Posterior region of the p4 separated from the central and anterior regions. Central lingual cuspid strongly developed on p4 and not usually joined to the central labial cuspid. Ossicones unbranched and attached to raised hollow bosses on the frontals (after Hamilton, 1978).

Remarks Constituent members of the Family Giraffidae are united by their bilobed lower canines and by their unbranched and tineless ossicones. Here we treat them in terms of seven subfamilies: Canthumerycinae, Bohlininae, Okapiinae, Giraffokerycinae, Sivatheriinae, Palaeotraginae, and Giraffinae, following the recent revision of Solounias (2007).

The ossicone bosses are less well developed in paleotragines and giraffokerycines. The ethmoidal fissure is reduced or closed in palaeotragines. The cervicals are short in okapiines, giraffokerycines, and sivatheres, moderately long in palaeotragines, and long in *Giraffa* and *Bohlinia*. Giraffids are for the most part larger than contemporaneous bovids. The deep thorax characteristic of extant giraffids is also documented (on the basis of the elongate scapula) in *Honanotherium* and *Samotherium*. Giraffid limbs are characteristically long but are proportionately shorter in the derived (and grazing) sivatheres and in the samothere-schansithere group of palaeotragines. They are also short in *Okapia* (Geraads, 1986a: figure 1; Solounias, 2007).

FIGURE 39.2 KNM-MB 36686; *Climacoceras africanus* right ossicone from Maboko Island, Kenya. Scale = 5 cm.

Subfamily CANTHUMERYCINAE Hamilton, 1978

This subfamily is represented by *Canthumeryx sirtensis* in the early Miocene of North and East Africa and by *Georgiomeryx georgalasi* (Paraskevaidis, 1940) in the middle Miocene of Greece. The ossicones take the form of small cones situated posterolaterally above the orbits. The occiput is narrow, suggesting a flexible neck. Lower p2 and p3 are shorter and more molariform than in the Climacoceratidae. The metapodials are of medium length; the posterior trough is moderately deep and extends down the proximal fourth of the shaft.

Genus *CANTHUMERYX* Hamilton, 1973

Type and Only Species Canthumeryx sirtensis Hamilton, 1973.

Diagnosis Canthumerycid in which the skull has a wide roof with ossicones in the supreme supraorbital position (Hamilton, 1978).

Holotype NHM M 26682, right mandibular fragment with d3, p2–4, m1–3 from Gebel Zelten, Libya; conserved in the Natural History Museum, London.

Remarks The species is represented by cranial material from Gebel Zelten, Libya (Hamilton, 1973) and by dental and postcranial material from Moruorot and Rusinga Island in Kenya (Hamilton, 1978). The skull is flat and broad in dorsal view, and the ossicones are short, conical, and inserted at the supraorbital margin. The ossicones are directed strongly laterally. There is at least one lacrimal canal in the orbit and this is open. A possible second lacrimal canal could be a break on the orbital surface. This is in contrast to more advanced giraffids in which all lacrimal canals are closed. The median base

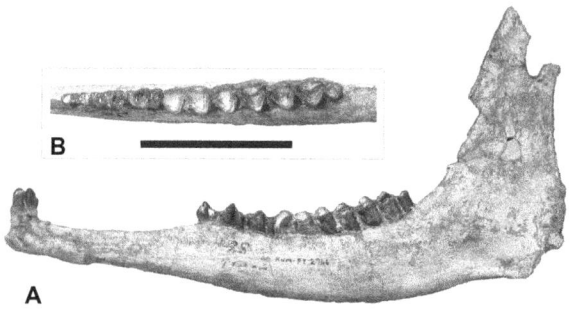

FIGURE 39.3   KNM-FT 2946; *Climacoceras gentryi*, holotype left mandible from Fort Ternan, Kenya: A) lateral and B) occlusal views. Scale = 5 cm.

of the occipital right above the foramen magnum is strongly developed and protrudes in lateral view.

Canthumeryx sirtensis has been reported from Egypt (16), Kenya (37, 42, 48, 51, 55, 61, 63), Libya (67), Saudi Arabia (78), and Uganda (105) and at the generic level from additional sites in Kenya (39, 46, 54)

Subfamily BOHLININAE Solounias, 2007

The subfamily includes *Injanatherium* species, *"Palaeotragus" tungurensis*, *Bohlinia* and *Birgerbohlinia*, none of which has yet been found in continental Africa although *Injanatherium* has been recovered from Iraq and Saudi Arabia. A giraffid from the late Miocene of Chad has been referred to *Bohlinia* (Likius et al., 2007), but its conical horns and long braincase are unlike this genus, and the identification is disputable. This subfamily is characterized by long slender metapodials with a deep posterior trough that extends for the proximal two-thirds of the diaphysis. In *Birgerbohlinia*, however, the metapodials are wide and lateral metacarpals II and V are well developed, and this genus could belong instead to the Sivatheriinae, as suggested by Crusafont (1952) and Montoya and Morales (1991).

Genus *INJANATHERIUM* Heintz et al., 1981

Diagnosis Giraffoids of small to medium size with two pairs of ossicones that differ from those of most other giraffoids by their orientation. The conical posterior ossicones are inserted above the orbit in the postorbital region, are directed laterally, and diverge at an angle of between 150° and 170°. The smaller anterior ossicones are triangular in transverse section, are inserted in front of the orbit and infraorbital canal level with M3, and diverge laterally and slightly upward.

Type Species Injanatherium hazimi Heintz et al., 1981.

Other Recognized Species Injanatherium arabicum Morales et al., 1987.

INJANATHERIUM HAZIMI Heintz et al., 1981

Diagnosis A species of *Injanatherium* a little larger than *I. arabicum*. Posterior ossicones conical, laterally projecting, and terminal knob only faintly developed. Anterior ossicones not known.

Holotype Calvaria with left ossicone from the late Miocene of Injana region, Gebel Hamrin, Iraq; conserved at the Muséum National d'Histoire Naturelle, Paris.

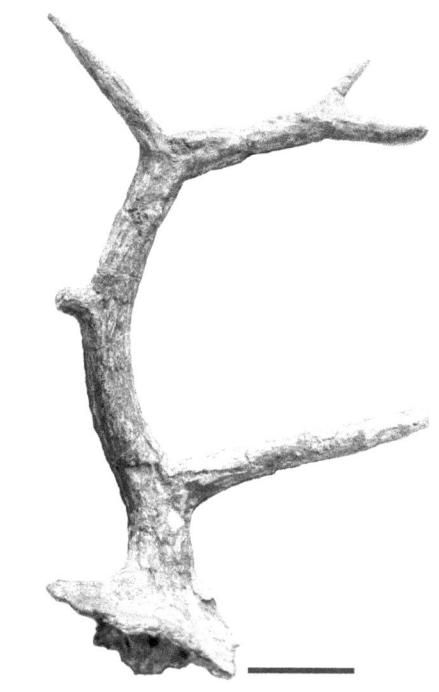

FIGURE 39.4   KNM-FT 3365; *Climacoceras gentryi*, left ossicone from Fort Ternan, Kenya. Scale = 5 cm.

INJANATHERIUM ARABICUM Morales et al., 1987.

Diagnosis A species of *Injanatherium* a little smaller than *I. hazimi* and with two pairs of subhorizontal ossicones. Posterior ossicones smaller than in *I. hazimi* and terminate in distal knob. Nuchal surface narrower, more elevated, and less concave than in *I. hazimi*, and temporal and parietal lines less well marked. Posterior tuberosities of basioccipital weaker and anterior tuberosities scarcely visible.

Holotype AJ 75, calvaria lacking left ossicone from the middle Miocene Hofuf Formation in Al Jadidah, Hasa Province, Saudi Arabia.

Paratype AJ 617 frontal region of cranium with proximal anterior ossicone and right M3, from the middle Miocene Hofuf Formation in Al Jadidah, Hasa Province, Saudi Arabia.

Remarks Although assigned to *Injanatherium* mainly on the basis of their ossicone orientation, it is possible that *I. hazimi* represents a species of *Samotherium* and that *I. arabicum* is close to *Giraffokeryx*.

Subfamily OKAPIINAE Bohlin, 1926

This subfamily contains the extant okapi and a specimen formerly identified as *Palaeotragus primaevus* from Ngorora (Hamilton, 1978). These two species have large tympanic bullae, ossicones with internal canals or distal pits, medium length metapodials and a metacarpal with a moderately deep trough.

Genus *OKAPIA* Lankester, 1901
Figure 39.5

Type and Only Species Okapia johnstoni (Sclater, 1901).

Diagnosis In part after Bodner and Rabb (1992). Medium-sized extant giraffid; one pair of supraorbital frontal ossicones present only in male; ossicones positioned medially to the

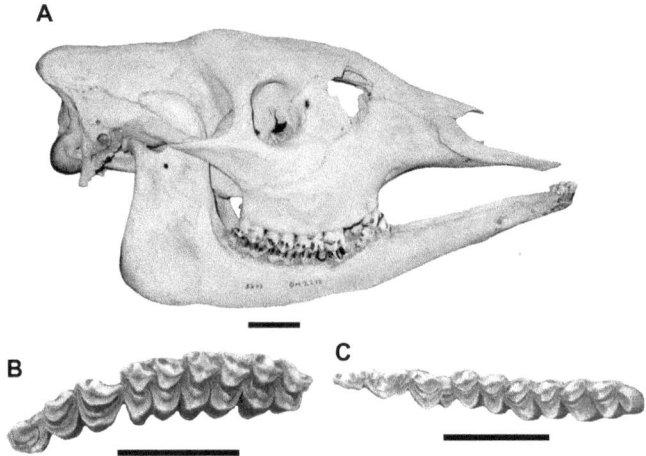

FIGURE 39.5 OM 2218; *Okapia johnstoni* skull: A) right lateral view; B) upper cheek teeth, occlusal view; and C) lower cheek teeth, occlusal view. Scale = 5 cm.

supraorbital margin; apices often bare of integument showing polished surfaces and with a constriction where the integument terminates. Apices often have pits and short canals, slightly smaller incisors and slightly larger permanent cheek teeth than in *Palaeotragus*; large palatine sinuses distinctive from other giraffids. Inferior margin of mandible uniformly convex. Cervical vertebrae not elongated (unlike in *Giraffa*) and five sacral vertebrae (vs. three or four in *Giraffa*). Naviculocuboid fused with cuneiform.

Remarks The rarity of *Okapia* fossils is readily understandable given the ecological adaptations of the extant species. A small giraffine from Laetoli represented by upper and lower cheek teeth was diagnosed as an okapi with palaeotragine characteristics and named *Okapia stillei* by Dietrich (1942). Leakey (1965) recognized the same species from Olduvai, and a lower third premolar from Kaiso was assigned to *Okapia* cf. *O. stillei* by Cooke and Coryndon (1970) on the basis of its similarity to *Okapia jacksoni* (*sic*). Harris (1976a) pointed out that the lower premolars of giraffes were quite different morphologically from those of okapis and transferred *Okapia stillei* to *Giraffa*. A small triangular "horn core" from the Wayland Collection from Kaiso (BMNH M 12582) was identified as the left ossicone of an okapi by Cooke and Coryndon (1970). It was said to be flattened laterally with a deep concavity at its base and distinct sinuses below and lateral to the concavity.

Genus *AFRIKANOKERYX* new genus
Figures 39.6 and 39.7

Type and Only Species *Afrikanokeryx leakeyi* sp. nov.

Diagnosis Ossicones flattened and oval in cross section with small bosses along the posterior margin, apex mediolaterally compressed and posterior margin sharper than anterior. Openings on the lateral and medial surface of the ossicone connect to the central canal. Auditory bullae large, auditory meatus long, retroarticular process of mandibular fossa is small, the mandibular fossa is convex, and the posterior basioccipital tuberosities are strongly developed. Mandibular symphysis very narrow and incisors tightly positioned. Lower fourth premolar with two buccolingually directed cuspids as in *Okapia* and *Samotherium* and with one central long lingual cuspid that is connected anteriorly; p4 cuspid crests thinner than those of *Okapia*. Curvature of the inferior jaw margin

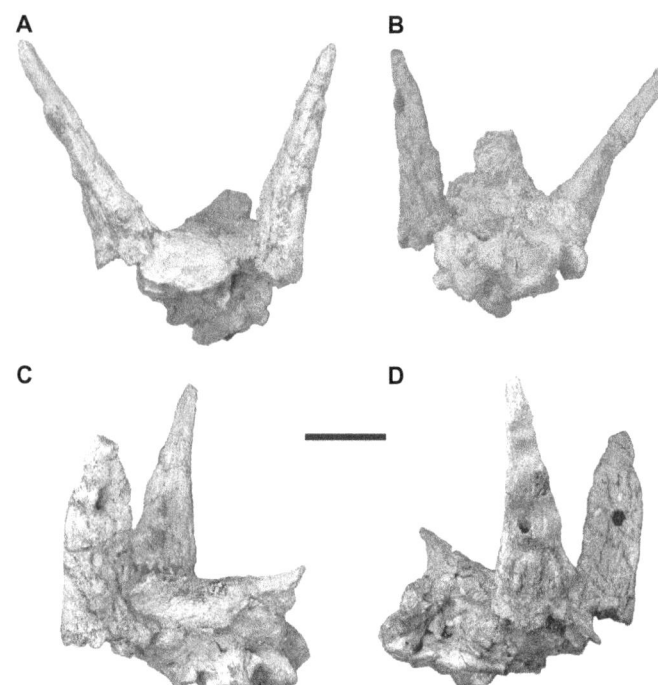

FIGURE 39.6 KNM-BN 1446; *Africanokeryx leakeyi*, holotype cranium from Ngorora, Kenya: A) anterior view, B) posterior view, C) left lateral view, and D) right lateral view. Scale = 5 cm.

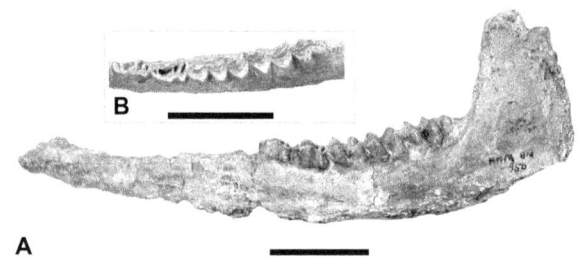

FIGURE 39.7 KNM-BN 950; *Afrikanokeryx leakeyi* paratype mandible from Ngorora, Kenya: A) left lateral view and B) occlusal view. Scale = 5 cm.

continues to the angle without a concave interruption as in *Okapia*.

Holotype KNM-BN 1446 a brain case with two posterior ossicones from Ngorora, Baringo Basin, Kenya, conserved in the collections of the National Museums of Kenya, Nairobi (figures 39.6A–39.6D).

Paratype NMK-BN 950 left mandible (figures 39.7A, 39.7B).

Etymology The species is named in honor of Louis Seymour Bassett Leakey (1903–1972).

Remarks KNM-BN 1446 was referred to *P. primaevus* by Hamilton (1978: figures 38–40). The ossicones are unique. The left ossicone shows a clear contact with the cranium, but the right ossicone has been wrongly restored (glued backward so that the internal surface is external and vice versa). The left ossicone is flat in cross section and maintains such a shape all the way to the apex. The inner side is flatter than the outer. It also has small bosses (bumps) along its posterior margin. The edge of the posterior margin is sharper than the anterior. The edge of the anterior margin is straighter than the posterior. The anterior edge gently curves back, making an arc.

Near the apex, a pronounced indentation occurs. The apex is a broad point. Surface grooving is present but not strongly developed. A canal runs through the middle of the ossicone that begins at a two-chambered supraorbital sinus and terminates a little higher than the middle of the ossicone's length. There are openings on both the lateral and medial surface of the ossicone that connect to the central canal; a connection formed by a Y-shaped canal junction within the ossicone. There are no other ruminants with such a canal. The right ossicone is similar to the left except that most features described are reversed. The braincase is small and similar in size to that of the white-tailed deer. The bullae are large as in *Okapia* and the posterior basioccipital tuberosities are strongly developed. Other Giraffidae have small bullae. The mandibular fossa is convex, the retromandibular process is small, and the auditory meatus is long.

Subfamily GIRAFFOKERYCINAE Solounias, 2007

Members of this subfamily are distinguished by crania with four ossicones, long metapodials, and a metacarpal with a moderately deep posterior trough that extends down the proximal two-thirds of the diaphysis. The subfamily is represented by *Giraffokeryx punjabiensis* in Asia, *G. primaevus* at Fort Ternan, and *G. anatoliensis* from Turkey.

Genus *GIRAFFOKERYX* Pilgrim, 1910

Diagnosis After Colbert (1933, 1935). Medium-sized giraffid with two pairs of ossicones, one at the anterior extremities of the frontal bones in front of the orbits and the others on the frontoparietal region overhanging the temporal fossae. Teeth brachyodont with rugose enamel.

Type Species *Giraffokeryx punjabiensis* Pilgrim, 1910.

Other Recognized Species *Giraffokeryx anatoliensis* Geraads and Aslan, 2003; *G. primaevus* (Churcher, 1970).

GIRAFFOKERYX PRIMAEVUS (Churcher, 1970)
Figures 39.8 and 39.9

Diagnosis Medium-sized giraffid with four ossicones. The anterior pair is rounded in cross section, notably long and slender with fine surficial braided grooving ending in a rounded apex that is separated by two constrictions from the shaft below. The posterior pair is longer, more curved and flatter at the base, and with a rounded apex. Their surface is covered with coarser surficial grooves and has two bosses (bumps) at the base. The location, either supraorbital or postorbital, of the posterior ossicone pair and their side is not certain. If the ossicone is a left, one boss is anterior and one medial. If the ossicone is a right, one boss is anterior and one posterior and in this manner the ossicone is more similar to *Giraffokeryx punjabiensis*. In either alternative, the posterior ossicones extend outward, curving either anteriorly or posteriorly. The choanae are situated posteriorly in relation to M3 and the orbit. The lower fourth premolar has a single separate central lingual cuspid—separate from the anterior transverse crests—the posterior cuspids are small and directed buccolingually. Upper second and third premolars are long.

Holotype KNM-FT 3065 and 3066, mandible (left and right dentaries) from Fort Ternan, Kenya; conserved in the collections of the National Museums of Kenya, Nairobi.

Remarks The Fort Ternan species that Churcher (1970) assigned to *Palaeotragus primaevus* is closely allied to *G. punjabiensis*, but, pending detailed comparison, we prefer to keep them as distinct because of the few differences mentioned below. Gentry (1994) suggested that the two giraffids that Churcher described from Fort Ternan, *P. primaevus* and *Samotherium africanum*, may be conspecific and later (Gentry, 1999) referred to the Fort Ternan species as *Giraffokeryx primaevus*.

Giraffokeryx primaevus has very long anterior ossicones; those of *G. punjabiensis* are shorter and those of *G. anatoliensis* are shorter still. *G. primaevus* differs from *Palaeotragus rouenii* in shape and position of the four ossicones, posteriorly positioned choanae in relation to the third molar, a lower fourth premolar with a small posterior region, and elongate upper second and third premolars.

The neck is short in *Giraffokeryx punjabiensis* but may be longer in the Fort Ternan *G. primaevus*. All known *Giraffokeryx* cervicals, however, are slightly longer than those of *Okapia* and the sivatheres. The metapodials resemble those of gazelline bovids. The anterior ossicones protrude laterally; the posterior pair lies behind the orbit, and each has a flange-like mass at its base. The surface of the ossicones is ornamented with irregular thick longitudinal ridges of secondary bone growth that are less prominent in *G. primaevus*. The ossicones

FIGURE 39.8 KNM-FT 3065; *Giraffokeryx primaevus* holotype left mandible from Fort Ternan, Kenya: A) left lateral view and B) occlusal view. Scale = 5 cm.

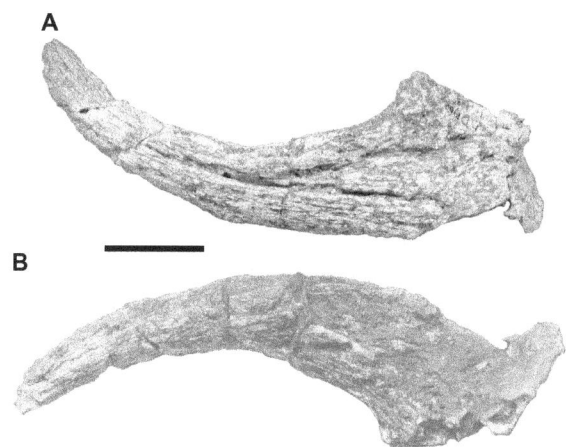

FIGURE 39.9 KNM-FT 3118; *Giraffokeryx primaevus* posterior ossicone from Fort Ternan, Kenya: A) anterior view and B) posterior view. Scale = 5 cm.

terminate in small knobs that differ from the characteristically enlarged knobs of *Giraffa* and from the pointed apices of *Palaeotragus* species.

Aguirre and Leakey (1974) recorded *Giraffokeryx* sp. nov. from the late Miocene Karbarsero Beds of Chepkesin, Kenya, based on three upper premolars, an upper molar, a mandible fragment with p3–4, and a poorly preserved metatarsal. However, Gentry (1997) reported the upper molar to be closer to *Giraffa* than to palaeotragines.

Palaeotragus primaevus has been reported from Moruorot (Hamilton, 1978) and Ngeringerowa (Benefit and Pickford, 1986), but further examination is needed to establish if these specimens properly belong to *Giraffokeryx primaevus*, *Afrikanokeryx leakeyi*, or some other species.

Subfamily SIVATHERIINAE Bonaparte, 1850

This subfamily comprises the large short-necked giraffids with medium and short metapodials and a metacarpal with posterior trough of medium depth. The posterior pair of the two pairs of ossicones is compressed in transverse section. The subfamily includes *Sivatherium*, *Bramatherium*, and *Helladotherium*. *Sivatherium* and *Bramatherium* occur in the Siwaliks of Pakistan and India. Only *Sivatherium* has been recovered from continental Africa, although Gentry (1999) reported ossicone and molar fragments from Abu Dhabi that he assigned to ?*Bramatherium*; otherwise, the closest record of *Bramatherium* is from Turkey (Geraads and Güleç, 1999) if *Bramatherium* does not also include *Helladotherium* and *Decennatherium*—a possibility proposed by Gentry (2003). The type of *Helladotherium* is from Pikermi in Greece, and this genus was reported from a number of late Miocene sites in the eastern Mediterranean and Iran. Reports of cf. *Helladotherium duvernoyi* Gaudry, 1860 from North Africa (Joleaud, 1937; Roman and Solignac, 1934) and Iran (Bosscha Erdbrink, 1977) probably constitute misidentified *Sivatherium* remains.

Genus *SIVATHERIUM* Falconer and Cautley, 1836

Type Species Sivatherium giganteum Falconer and Cautley, 1836.

Diagnosis Gigantic giraffid; skull brachycephalic, face short, nasals retracted. Ossicones in two pairs in males, absent in females; anterior pair conical and arising from the frontals above the orbits. The larger posterior pair is compressed and appears to be arising from the posterior part of the frontals near the parietals. In males the frontals are broad, flat, or slightly dished; nasals short, convex; sinuses in base of ossicones. The skull of females is longer and lower, not markedly broadened, frontals convex. Deep muscular pits in temporal and supraoccipital areas, Facial region relatively short, anterior margin of orbit above M2. Cranial region deeper than facial, especially in males; basicranial and palatal planes not parallel. Teeth large, enamel coarsely rugose; lower premolars molariform. Body, neck and limbs heavy, neck and limbs not elongated (in part after Churcher, 1978).

SIVATHERIUM MAURUSIUM (Pomel, 1892)

Diagnosis A species of *Sivatherium* with a shallower and narrower cranial region and a longer facial region than *S. giganteum*. Posterior ossicones less compressed and less palmate than those of *S. giganteum* and orientated in various directions. Anterior ossicone reduced to narrow flanges above and behind the orbits.

Holotype Left mandible with p3–m3 from the "Upper Villafranchian" of Aïn Hanech, Algeria; specimen housed in the Museum National d'Histoire Naturelle, Paris.

SIVATHERIUM HENDEYI Harris, 1976

Diagnosis A species of *Sivatherium* of similar size and dental morphology to the Asian *S. giganteum* Falconer and Cautley, 1836, or the later African species *S. maurusium* (Pomel, 1892), but the posterior ossicones are short, extending laterally and backward from the cranium, and are unornamented by knobs, flanges, or palmate digitations. The metacarpals are longer than those of *S. giganteum* or Pleistocene specimens of *S. maurusium*.

Holotype SAM-PQ-L12730, left posterior ossicone from the Quartzose Sand Member of the Varswater Formation in "E" Quarry, Langebaanweg; conserved in the collections of the South African Museum, Cape Town.

Remarks Sivatheres were widespread in Africa during the Pliocene and Pleistocene, though only in Langebaanweg were they abundant elements of the local assemblages. Most can be attributed to the single species *Sivatherium maurusium* (Pomel, 1892). Although the posterior ossicones of *S. maurusium* vary greatly in shape (Harris, 1974), they differ fundamentally in their shape and orientation from those of *S. giganteum*. Earlier African forms with simple ossicones and long metapodials have been referred to *S. hendeyi* (Harris, 1976b; Harris et al., 2003; Geraads, 1996; Vignaud et al., 2002), but Churcher (1978) thought this was just a variant of *S. maurusium*.

Sivatheres have been reported from Algeria (1, 2, 3, 6, 8), Chad (10–13), Djibouti (14), Ethiopia (19–21, 24, 26, 28–33), Kenya (41, 47, 49, 53, 60, 62, 66), Malawi (69), Morocco (70, 72), South Africa (79, 81–83, 85–87), Sudan (88), Tanzania (89, 91, 93, 94), Tunisia (96, 98–102), Uganda (106–108), the United Arab Emirates (109), Democratic Republic of Congo (110), and Zambia (112).

Subfamily PALAEOTRAGINAE Pilgrim, 1911

In palaeotragines, the ossicones are located medial to the orbital rim. The metapodials are long and the metacarpal has a shallow to moderately deep posterior trough that extends for the proximal two-thirds of the diaphysis. The subfamily includes *Palaeotragus*, *Samotherium*, and *Mitilanotherium* Samson and Radulesco, 1966 (= *Macedonitherium* Sickenberg, 1967 and *Sogdianotherium* Sharapov, 1974 [after Kostopoulos and Athanassiou, 2005]).

Genus *PALAEOTRAGUS* Gaudry, 1861

Type Species Palaeotragus rouenii Gaudry, 1861.

PALAEOTRAGUS GERMAINI Arambourg, 1959

Diagnosis Large giraffid with long neck and elongate legs of which the anterior are the longer. Cranium with long and pitted supraorbital ossicone that is triangular in transverse section. Dentition primitive, very brachyodont, with compressed upper milk teeth that are elongate and primitive like those of Protragulidae. Molars and premolars sloping strongly on their labial surface and with strong internal cingula. Metacone separated from the paracone. Lower milk teeth primitive on the basis of the wide separation of paraconid and metaconid. Radius longer than tibia; femur short;

metacarpals and metatarsals subequal, long, compressed laterally and subrectangular in transverse section. Tarsus with free ectocuneiform 1; cuneiforms II and III not fused with naviculocuboid. Naviculocuboid with lateral groove for the long peroneal tendon.

Holotype No. 282, adult palate from Ouad el Hammam, Algeria.

Paratypes No. 176, front leg (radius, carpals and metacarpal); no. 285 hind leg (tibia, tarsus and metatarsal).

Remarks Palaeotragus germaini has been reported from Algeria (4, 5, 7; and *Palaeotragus* cf. *P. germaini* from Algeria (6), Morocco (70), South Africa (85), and Tunisia (95, 98). The primitive teeth and large size of this species make it quite distinct from other late Miocene *Palaeotragus*, most of which are later in age; its metapodials are almost as elongated as those of *Bohlinia* and *Giraffa*, but too little is known of its cranial anatomy to demonstrate a close relationship with either of them. Churcher (1979) and Harris (2003) identified teeth from Lothagam in Kenya as *Palaeotragus germaini*, but Geraads (1986a) thought Churcher's tooth was closer to *Giraffa*.

PALAEOTRAGUS LAVOCATI Heintz, 1976

Diagnosis Small *Palaeotragus* about the size of a red deer (*Cervus elaphus*). Closer to *Palaeotragus tungurensis* than other species attributed to this genus. Distinguished from *P. tungurensis* by slightly smaller molars and slightly larger premolars, by the lesser development of anterior and posterior styles on the external face of the upper premolars, by the pinching of the lingual face of the protocone of M1, and by the less typically giraffoid form of the p4.

Holotype BML 128, incomplete right mandible (p4–m3) from middle Miocene lacustrine limestone at Beni Mellal, Morocco.

Remarks Heintz (1976) identified two giraffids from the middle Miocene locality of Beni Mellal; the larger he assigned to *Palaeotragus lavocati*, the smaller (and perhaps more primitive in terms of the external faces of its upper premolars) he identified only as *Palaeotragus* sp.

PALAEOTRAGUS ROBINSONI Crusafont-Pairó, 1979

Diagnosis Medium *Palaeotragus* species with very small ossicones; relatively short diastema, p4 very molarized; very aligned lower teeth; P3–P4 very wide, upper molars aligned but with very marked styles and very sinuous external walls; limbs dolichopodous.

Holotype T. 3371, left mandible (p2–m3) from section 17 of the late Miocene Beglia Formation, Bled Douarah, Tunisia.

Remarks Crusafont-Pairó (1979) described a small palaeotragine from Tunisia that he thought had affinities with *Giraffokeryx primaevus*, *Palaeotragus rouenii* Gaudry, 1861, and *P. microdon* Koken, 1885, but differed sufficiently to warrant a separate species.

PALAEOTRAGUS sp.

Haile-Selassie (2009) illustrated a few teeth that he referred to *Palaeotragus* sp. from three Middle Awash sites at the Mio-Pliocene boundary, but in the absence of an associated set of teeth, they could perhaps be better assigned to Giraffidae indet. cf. *Giraffa* sp. Robinson and Black (1974) recorded

?Palaeotragus sp. from the Beglia Formation at Bled Dourah in Tunisia, as did Andrews et al. (1981) from Majiwa Bluff, Bishop and Pickford (1975) from Ngorora, Nakaya et al. (1987) from the Namurungule Formation, and Gentry (1999) from Abu Dhabi. Bosscha Erdbrink (1977) and Campbell et al. (1980) recorded *Palaeotragus coelophrys* from Iran.

Genus SAMOTHERIUM Forsyth Major, 1888

Diagnosis Medium-sized giraffids with long facial region, especially between P2 and orbit. Ossicones paired, simple, pointed, and widely separated; borne on frontals above supraorbital bar; directed laterally, dorsally, or postero-dorsally; ossicones sexually dimorphic, better developed in males and poorly developed or absent in females. Rudimentary or small paired ossicones may lie anterior to the main ossicones. Forehead dished, nasals straight, orbit with superior margin at or above nasal-frontal plane and with anterior margin above or posterior to M3. Cheek teeth moderately hypsodont, with moderately rugose enamel. Neck and limbs elongate, metapodials broader than in the *Palaeotragus* species.

Type Species Samotherium boissieri Forsyth Major, 1888.

Remarks Churcher (1970) defined *Samotherium africanum* on the basis of two isolated, tapered, recurved ossicones from Fort Ternan, but it is now clear that he misidentified ossicones of *Giraffokeryx primaevus*—the common giraffid from that locality. Large teeth and/or postcranial material attributed to *Samotherium* have been reported from Egypt (15), Kenya (42, 56, 58), Libya (68), and Tunisia (95), but these identifications need reevaluation. Bosscha Erdbrink (1977) reported *S. boissieri* from Iran.

"?GIRAFFA" POMELI Arambourg, 1979

Diagnosis After Arambourg (1979). Giraffine characterized by its strongly brachyodont dentition, appearing to represent a species that is larger and had distinctly more robust limbs than the extant species.

Holotype 1938-10:36, left m2 from Aïn Hanech.

Syntypes 1938-10:33, left m3, and 1938-10:37; also from Aïn Hanech.

Remarks This very poorly known species was recorded from Aïn Hanech (1.2–.4 Ma.) by Arambourg (1979). He only described three lower teeth and a phalanx, but there are also two tarsal bones. The teeth are slightly less brachyodont than those of *Giraffa* (*contra* Arambourg, although there is some variation in the modern form), and the lingual lobes have weaker pillars, but they are similar to those of *Giraffa*. However, an upper molar from the early/middle Pleistocene of Tighenif (= Ternifine; Geraads, 1981) has flatter labial walls, while a metatarsal is slender but significantly shorter than that of *G. gracilis*. Two incomplete scapulae from Aïn Boucherit (1.8–2 Ma.), described by Arambourg (1979) as *Libytherium*, are probably too long and narrow for that genus and could also belong to "*G.*" *pomeli* (but the basioccipital illustrated under the same name is of a rhino). "*Giraffa*" *pomeli* is probably not a giraffine and could be a survivor of the Palaeotraginae, perhaps related to *Mitilanotherium* Samson and Radulesco, 1966, from the Plio-Pleistocene of the northern Mediterranean region. "*Giraffa*" *pomeli* has also been recorded from Aïn Brimba in Tunisia, and *Giraffa* cf. *G. pomeli* from Tighenif in Algeria (Geraads, 2002).

Subfamily GIRAFFINAE Gray, 1821

Members of this subfamily are characterized by paired lateral ossicones with blunt posterior keels and one or more median unpaired ossicones. The ossicones are more robust in males. The cervical vertebrae are elongated. The limbs are long, and the astragalus is squared. The metapodials are long and slender and the metacarpal has a very shallow posterior trough. The extant genus *Giraffa* first appears in Africa at the beginning of the Pliocene. It could have been derived from *Bohlinia* or *Palaeotragus rouenii*.

Genus *GIRAFFA* Brisson, 1762

Diagnosis Ossicones paired, short, on frontoparietal suture, sometimes with single frontal ossicone between or just behind level of orbits; all ossicones variable in form but usually straight and bluntly ended if paired and rounded or tumescent if single. Exostotic occipital "horns" may be developed from nuchal crest and azygous "horns" from the orbital boss. Lower canines and incisors robust, buccal enamel rugose; lower canine occasionally trifid. Cheek teeth variable in size and moderately brachyodont; premolars molariform and complex. Basicranial and basipalatal planes not parallel.

Remarks In Africa, representatives of this genus first appear in the early Pliocene. Thereafter they are persistent but uncommon elements of the local assemblages in which the giraffine teeth often fall into two or more discrete size groups. Extant giraffes are sexually dimorphic, males are larger with more robust limbs and have stouter ossicones with more secondary bone apposition in mature specimens, but the cheek teeth of males and females do not differ significantly in size. Several fossil *Giraffa* species have been proposed on the basis of tooth size and of the shape and orientation of ossicones. This may be taxonomically convenient, particularly as even moderately complete skeletons are rare, but the validity of the different species is difficult to assess.

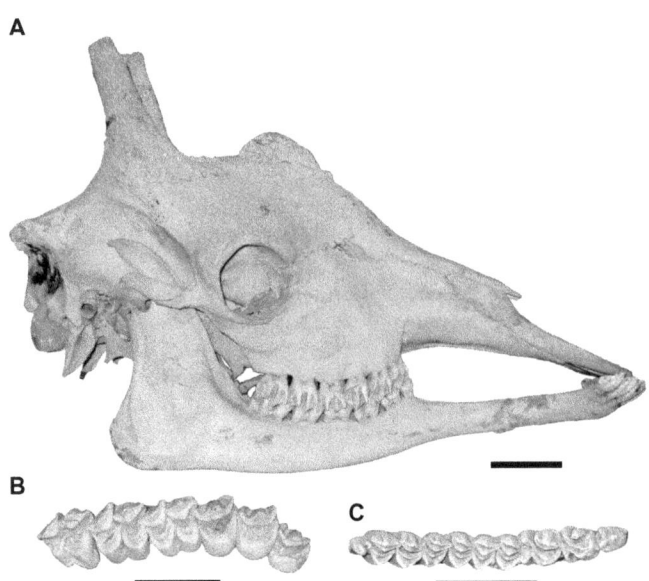

FIGURE 39.10 *Giraffa camelopardalis* skull: A) right lateral view; B) upper cheek teeth, occlusal view; and C) lower cheek teeth, occlusal view. Scale = 5 cm.

GIRAFFA CAMELOPARDALIS (Linnaeus, 1758)
Figure 39.10

Diagnosis After Dagg (1971). Height of single extant species up to 6 m, females smaller than males; legs and neck each exceed 1.5 m; both sexes have two unbranched ossicones fused to the skull above the frontoparietal suture, an anterior median ossicone better developed in males than females; facial region covered with other bony growths in males. Skull up to 73 cm long; molars brachyodont; upper molars lack inner accessory columns; canine teeth bilobed or trilobed.

Remarks Up to nine extant subspecies are recognized (Lydekker, 1904; Ansell, 1968; Dagg and Foster, 1981; Kingdon, 1997) and are separated on the basis of their coat pattern, ossicone morphology and geographic distribution:

- *G. c. camelopardalis* (Linnaeus, 1758), nubian giraffe; eastern Sudan, northeastern Congo
- *G. c. antiquorum* (Jardine, 1835), Kordofan giraffe; western and southwestern Sudan
- *G. c. peralta* Thomas, 1898, west African or Nigerian giraffe; Chad
- *G. c. reticulata* de Winton, 1899, reticulated or Somali giraffe; northeast Kenya, Ethiopia, Somalia
- *G. c. rothschildi* Lydekker, 1903, Rothschild's or Baringo or Ugandan giraffe; Uganda and north-central Kenya
- *G. c. tippelskirchi* Matschie, 1898, Masai or Kilimanjaro giraffe; central and southern Kenya, Tanzania
- *G. c. thornicrofti* Lydekker, 1911, Thorneycroft or Rhodesian giraffe; eastern Zambia
- *G. c. angolensis* Lydekker, 1903, Angolan giraffe; Angola, Zambia
- *G. c. giraffa* Schreber, 1784, southern giraffe; South Africa, Namibia, Botswana, Zimbabwe, Mozambique

Brown et al. (2007) analyzed mitochondrial DNA sequences and nuclear microsatellite loci and found at least six genealogically distinct giraffe lineages in Africa with little evidence of interbreeding. They postulated a mid to late Pleistocene radiation of extant giraffes and argued that the Angolan, West African, Rothschild's, reticulated, Masai, and South African giraffes should be recognized as separate species.

Specimens that are indistinguishable from equivalent parts of the extant species appear toward the end of the early Pleistocene, but the origin of the extant species is far from clear. Records of *Giraffa camelopardalis* older than Pleistocene are probably not valid. This species has been recorded from Algeria (2, 3, 9), Chad (12), Ethiopia (25), Israel (35, 36), Kenya (60), Malawi (69), Morocco (73), Niger (75), South Africa (79, 80, 82, 83), Tanzania (90, 92), and Zambia (112).

GIRAFFA JUMAE Leakey, 1965

Diagnosis A species of giraffe of similar size to the extant *G. camelopardalis*. The surface of the frontal bone between the external rims of the orbits is nearly flat. The width of the skull roof between the orbits is greater than in *G. camelopardalis*. The longitudinal median section from the posterior edge of the nasals to the lateral ossicone is flat or slightly concave. The lateral ossicones originate immediately above the orbit and project more posteriorly than in *G. camelopardalis*.

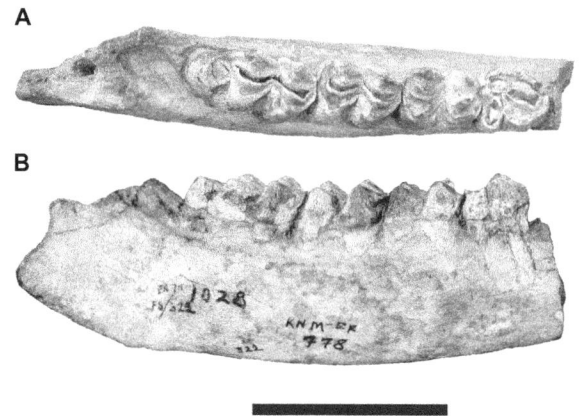

A

B

FIGURE 39.11   KNM-ER 778; *Giraffa pygmaea* holotype right mandible: A) occlusal view and B) lateral view. Scale = 5 cm.

No secondary bone apposition is known to occur on the lateral ossicones. The median ossicone is poorly developed. The basilar process of the occipital is longer than the external width of the palatal area at M2. The ascending ramus of the mandible is wide and stout. The corpus is deep and long, the anterior portion of the corpus being inclined upward from the premolars to the symphysis and then downward in the incisive region (after Leakey, 1970).

Holotype M. 14597, skull and partial skeleton from the middle Pleistocene of Rawi, Kenya; housed in the Natural History Museum, London.

Remarks *Giraffa jumae* has been recorded from Ethiopia (27–29, 32), Kenya (41, 43, 45, 47, 62, 66), South Africa (85, 86), Tanzania (89, 94) and with less certainty from Chad (11), Eritrea (17), Ethiopia (18, 19, 31), Kenya (50), Tanzania (91, 93), and Turkey (Geraads, 1998).

GIRAFFA PYGMAEA Harris, 1976
Figure 39.11

Emended Diagnosis A small species of *Giraffa* with teeth of similar morphology to those of *G. camelopardalis* but smaller than those of *G. stillei*. The lateral ossicones are similarly orientated to, but smaller than, those of the extant species and exhibit secondary bone apposition in male specimens (after Harris 1976a).

Holotype KNM-ER 778, left mandible fragment (p4–m3) from the KBS Member of Area 8A at Koobi Fora (figure 39.11).

Paratypes KNM-ER 656, frontlet with lateral ossicones from the Okote Member of Area 1 at Koobi Fora. Old 1960 F200, right mandible fragment (dp2–4, m1–2) from Bed I at site FLK in Olduvai Gorge, presently housed with the collections of the National Museums of Kenya.

Remarks *Giraffa pygmaea* has been documented from Ethiopia (28, 30, 32–33), Kenya (47), Malawi (69), and Tanzania (94), and something close to *G. pygmaea* is known from Asbole in Ethiopia (Geraads et al. 2004) and west of Lake Natron in Tanzania (Geraads 1987).

GIRAFFA STILLEI (Dietrich, 1942)

Emended Diagnosis A species of *Giraffa* with teeth that are often smaller than those of *G. camelopardalis* and *G. jumae* but always larger than those of *G. pygmaea*. The ossicones are uprightly inserted like those of *G. camelopardalis* but appreciably smaller than male specimens of the extant species and

often lack well developed terminal knobs; they are larger than specimens assigned to *G. pygmaea* but are not backwardly raked like those of *G. jumae*.

Holotype Unfortunately, Dietrich (1942) omitted to select a holotype when naming *"Okapia" stillei*. Harris (1987) suggested using the upper and lower tooth rows illustrated in Dietrich's description (Dietrich 1942:170, plate 21) as lectotypes without knowing that these were assembled from isolated teeth (D.G., pers. obs.). Perhaps it would be helpful to use Arambourg's (1947) holotype of *G. gracilis* for this purpose as *G. stillei* and *G. gracilis* are evidently conspecific.

Remarks *Giraffa stillei* is the senior synonym for *Giraffa gracilis* Arambourg, 1947, and is the most common of the three recognized African fossil species. It is known from Ethiopia (28, 32), Kenya (39, 47, 65, 66), Malawi (69), South Africa (82, 86), Tanzania (89, 94), and Uganda (103). Something close to *G. stillei* is known from Aramis in Ethiopia, and *Giraffa* cf. *G. gracilis* was reported by Boaz (1990) from the Pliocene Lusso Beds of Zaire.

Giraffoid Dietary Adaptations

Extant giraffes and okapis are browsers, and their long limbs and necks enable them to exploit taller vegetation than most other African ruminants. Although some had similarly elongate limbs and necks, the extinct giraffoids exploited a wider range of feeding adaptations. Dietary adaptations of fossil herbivores can be inferred from several different features that characterize their extant representatives. The siliceous phytoliths found in the cell walls of grasses are very abrasive, and consequently the crowns of grazing mammal teeth are usually taller than those of browsers. This also has an impact on the wear shape of the tooth cusps (mesowear), those of browsers being taller and more pointed whereas those of grazers quickly wear flat. The okapi is a characteristic browser in this respect but cusps on extant giraffe teeth show more abrasion than might be expected from their reported diet (Fortelius and Solounias, 2000).

The premaxillae of browsing mammals tend to be narrow and pointed for leaf selection, those of grazing mammals are square for grass feeding, and those of mixed feeders are intermediate in shape. Solounias et al. (1988) interpreted the broad premaxilla of *Samotherium boissieri* to indicate that this species may have been an intermediate feeder or a grazer, an interpretation that was supported by microwear analysis. A later study by Solounias and Moelleken (1993), also involving the dental arcade, determined that *Honanotherium, Bramatherium megacephalum, Sivatherium giganteum,* and *Samotherium* species were grazers, *Palaeotragus rouenii* was a mixed feeder, and *Palaeotragus coelophrys, Giraffokeryx primaevus* and *Afrikanomeryx leakeyi* were browsers.

The tooth enamel of herbivorous mammals becomes scored from abrasion by phytoliths and pitted by contact with hard seeds. Examination of tooth wear scars under the scanning electron microscope documents a variety of abrasion features in extant mammals that can be used to distinguish browsers, grazers, mixed feeders and frugivores. Solounias et al. (1988) compared microwear on teeth of *Samotherium boissieri* with that of a range of different bovids and concluded that *S. boissieri* was a grazer. A later detailed study (Solounias et al., 2000) indicated that in the late Turolian assemblages from Samos, *Samotherium* species changed from mixed feeding at 8.5 MA *(S. boissieri)* to grazing at 7.2 MA *(S. major).* Based on tooth microwear, Solounias et al. (2000) postulated that *Giraffokeryx primaevus* and *Helladotherium* were browsers; *Samotherium neumayri* was a mixed feeder; *Giraffokeryx punjabiensis, Paleotragus rouenii,* and *Samotherium boissieri* were seasonal mixed feeders; and *Samotherium major*, *Bramatherium*, and *Sivatherium* were grazers. Franz-Odendaal and Solounias (2004) subsequently established that *Sivatherium hendeyi* from Langebaanweg was a mixed feeder.

Additional information on giraffoid diets is provided by stable isotope analysis. Plants invoke two main methods of photosynthesis, and C_3 plants have a different chemical composition from C_4 plants. In Africa, C_4 plants have been present in the form of grasses since the late Miocene, whereas most trees, shrubs, and forbs are C_3 plants. Accordingly, it is possible to identify whether extant and Neogene African herbivores were grazers or browsers from the stable isotope geochemistry of their tooth enamel. Whereas the proportion of $\delta^{13}C$ indicates whether the diet was of C_3, C_4, or mixed vegetation, the proportion of $\delta^{18}O$ can be used to indicate if the animal was obtaining its water from a river or spring, or from a lake, or mostly from the vegetation that it ate (Cerling et al., 2005).

As documented by Cerling et al. (2005), the middle Miocene African giraffoids *Climacoceras africanus* and *Giraffokeryx primaevus* both seem to have had a pure or nearly pure C_3 diet. However, the $\delta^{18}O$ value for *Giraffokeryx* teeth indicated a more evaporated water source suggesting that, like extant giraffes, *G. primaevus* may have obtained much of the water it required from its food. Later specimens of *Paleotragus* from the Samburu Hills are slightly more positive than *G. primaevus* from Fort Ternan, although specimens from both localities plot as browsers. The difference may reflect a slight increase in the C_4 dietary component but could also suggest that the later and more northerly locality was drier and more water stressed.

Several species of *Giraffa* have been recognized in the East African Plio-Pleistocene. All plot out isotopically as dedicated browsers that obtain much of the water they need from their food. Although the samples are small, specimens of *Giraffa jumae* and *G. stillei* from Laetoli appear to be less negative in $\delta^{13}C$ and less positive in $\delta^{18}O$ than specimens from the Lake Turkana Basin; this may suggest that the Serengeti specimens drank more water and ate more grass but could also indicate that browse vegetation in the Laetoli region was stressed from lack of water at the time the giraffes were feeding on it. The consistency of their diets makes *Giraffa* species particularly valuable as evaporation-sensitive indicators when establishing stable isotope aridity indices for terrestrial environments (Levin et al., 2006).

Two species of *Sivatherium* have been recognized from sub-Saharan Africa—*Sivatherium hendeyi* from early Pliocene and the later, more widely distributed *Sivatherium maurusium*. Isotopically, *Sivatherium hendeyi* plots out as a browser, as do the samples of *Sivatherium* from Kanapoi and from the lower portions of the Koobi Fora Formation. However, a change in diet is indicated by specimens from the Upper Burgi Member, one of which plots as a browser but another plots as a mixed feeder. Specimens from higher in the Koobi Fora Formation plot as grazers. Corroboration of the dietary change is seen in the lengths of the sivathere metatarsals—those of browsing individuals from the early Pliocene being substantially longer than those from grazing individuals in the late Pliocene/early Pleistocene (Harris, 1976b; Geraads, 1996; Cerling et al., 2005). It is intriguing that African sivatheres adopted a grazing diet several millions of years after African proboscideans, perissodactyls, suids and bovids changed diets to exploit C_4 grasses in sub-Saharan Africa. It is also interesting that the Siwalik sivatheres became grazers somewhat earlier than their African relatives (Cerling et al., 1997).

Literature Cited

Aguirre, E., and P. Leakey. 1974. Nakali: Nueva fauna del *Hipparion* de Rift valley de Kenya. *Estudios Geológicos* 30:219–227.

Andrews, P., G. E. Meyer, D. R. Pilbeam, J. A. Van Couvering, and J. A. H. Van Couvering. 1981. The Miocene fossil beds of Maboko Island, Kenya: Geology, age, taphonomy, and palaeontology. *Journal of Human Evolution* 10:35–48.

Ansell, W. F. H. 1968. Artiodactyla (excluding the genus *Gazella*); in J. A. J. Meester (ed.), *Preliminary Identification for African Mammals*. Smithsonian Institution, Washington, D.C.

Anyonge, W. 1991. Fauna from a new lower Miocene locality west of Lake Turkana, Kenya. *Journal of Vertebrate Paleontology* 11:378–390.

Arambourg, C. 1947. Contribution a l'étude géologique et paléontologique du Bassin du Lac Rodolphe et de la Basse Vallée de L'Omo: Deuxième partie. Paléontologie; pp. 231–559 in C. Arambourg (ed.), *Mission Scientifique de L'Omo*. Muséum National d'Histoire Naturelle, Paris.

———. 1948. Les vertébrés fossiles des formations des plateaux Constantinois (Note préliminaire). *Bulletin de la Société d'Histoire Naturelle d'Afrique du Nord* 38:45–48.

———. 1952. *La Paléontologie des Vertébrés en Afrique du Nord Française.* 19ème Congrès Géologique Internationale, Monografies Régionales 92; 372 pp.

———. 1959. Vertébrés continentaux du Miocène supérieur de l'Afrique du Nord. *Publication du Service de la Carte Géologique de l'Algérie (n.s.), Paléontologie* 4:42–53.

———. 1960. Précisions nouvelles sur *Libytherium maurusium* Pomel, Giraffidé du Villafranchien d'Afrique. *Bulletin de la Société Géologique de France., Ser. 7*, 2:888–894.

———. 1979. *Vertébrés Villafranchiens d'Afrique du Nord (Artiodactyles, Carnivores, Primates, Reptiles, Oiseaux).* Fondation Singer Polignac, Paris, 141 pp.

Asfaw, B., W. H. Gilbert, Y. Beyene, W. K. Hart, P. R. Renne, G. WoldeGabriel, E. S. Vrba, and T. D. White. 2002. Remains of *Homo erectus* from Bouri, Middle Awash, Ethiopia. *Nature* 416:317–320.

Bar-Yosef, A., and E. Tchernov. 1972. *On the Paleo-ecological History of the Site of 'Ubeidiya*. Israel Academy of Sciences and Humanities, 35 pp.

Benefit, B. R., and M. Pickford. 1986. Miocene fossil cercopithecoids from Kenya. *American Journal of Physical Anthropology* 69:441–464.

Bishop, W. W. 1962. The mammalian fauna and geomorphological relations of the Napak volcanics, Karamoja. *Records of the Geological Survey of Uganda* 1957–58:1–18.

———. 1967. The later Tertiary in East Africa: Volcanics, sediments, and faunal inventory; pp. 31–54 in W. W. Bishop and J. D. Clark (eds.), *Background to Evolution in Africa*. University of Chicago Press, Chicago.

Bishop, W. W., and M. H. L. Pickford. 1975 Geology, fauna and palaeoenvironment of the Ngorora Formation, Kenya Rift Valley. *Nature* 254:185–192.

Boaz, N. T. 1990. The Semliki Research Expedition: History of Investigation, Results, and Background to Interpretation. *Virginia Museum of Natural History Memoirs* 1:3–14.

Bodner, R. E., and G. B. Rabb. 1992. *Okapia johnstoni*. *Mammalian Species* 422:1–8.

Bosscha Erdbrink, D. P. 1977. On the distribution in space and time of three giraffid genera with Turolian representatives at Maragheh in N.W. Iran. *Proceedings Koninklijke Nederlandse Akademie van Wetenschappen, Series B*, 80:337–355.

Bromage, T. G., F. Schrenk, and Y. M. Juwayeyi. 1995. Paleobiogeography of the Malawi Rift: Age and vertebrate paleontology of the Chiwondo Beds, northern Malawi. *Journal of Human Evolution* 28:37–57.

Brown, D. M., R. A. Brennerman, N. J. Georgiadis, K. Koepfli, J. P. Piollonger, B. Mila, E. E. Louis, G. F. Grether, D. K. Jacobs, and R. K. Wayne. 2007. Extensive population genetic structure in the giraffe. *BMC Biology* 5:57.

Burollet, P. F. 1956. Contribution a l'étude stratigraphique de la Tunisie Centrale. *Annales des Mines et da la Géologie, Tunisie* 18:1–350.

Brunet, M., A. Beauvilain, Y. Coppens, E. Heintz, A. Moutaye, and D. Pilbeam. 1995. The first australopithecine 2,500 kilometers west of the Rift Valley (Chad). *Nature* 378:273–275.

Brunet, M., A. Beauvilain, D. Geraads, F. Guy, M. Kasser, H. T. Mackaye, L. M. MacLatchy, G. Mouchelin, J. Sudre, and P. Vigneaud. 1998. Tchad: Découverte d'une faune de mammifères du Pliocène inférieur. *Comptes Rendus de l'Académie des Sciences, Paris, Sciences de la Terre et des Planètes* 326:153–158.

Brunet, M., and M.P.F.T. 2000. Chad: Discovery of a vertebrate fauna close to the Mio-Pliocene boundary. *Journal of Vertebrate Paleontology* 20:205–209.

Campbell, B. G., M. H. Amin, R. L. Bernor, W. Dickinson, R. Drake, R. Morris, J. A. Van Couvering, and J. A. H. Van Couvering. 1980. Maragheh: A classical late Miocene vertebrate locality in north-western Iran. *Nature* 287:837–841.

Cerling, T. E., J. M. Harris, and M. G. Leakey. 2005. Environmentally driven dietary adaptations in African mammals; pp. 258–272 in J. R. Ehleringer, M. D. Dearing, and T. E. Cerling (eds.), *History of Atmospheric CO₂ and Its Effects on Plants, Animals, and Ecosystems.* Springer, New York.

Cerling, T. E., J. M. Harris, B. J. MacFadden, M. G. Leakey, J. Quade, V. Eisenmann, and J. R. Ehleringer.1997. Global vegetation change through the Miocene/Pliocene boundary. *Nature* 389:153–158.

Churcher, C. S. 1970. Two new Upper Miocene giraffids from Fort Ternan, Kenya, East Africa; pp. 1–105 in L. S. B. Leakey and R. J. G. Savage (eds.), *Fossil Vertebrates of Africa*, vol. 2. Academic Press, London.

——. 1978. Giraffidae; pp. 509–535 in V. J. Maglio and H. B. S. Cooke (eds.), *Evolution of African Mammals.* Harvard University Press, Cambridge.

——. 1979 The large paleotragine giraffid *(Palaeotragus germaini)* from the Late Miocene of Lothagam Hill. *Breviora* 453:1–8.

Colbert, E. H. 1933. A skull and mandible of *Giraffokeryx punjabiensis* Pilgrim. *American Museum Novitates* 632:1–14.

——. 1935. Siwalik mammals in the American Museum of Natural History. *Transactions of the American Philosophical Society* (n.s.) 26:1–401.

Cooke, H. B. S. 1955. Some fossil mammals in the South African Museum collections. *Annals of the South African Museum* 42:161–168.

——. 1963. Pleistocene mammal faunas of Africa with particular reference to southern Africa; pp. 65–116 in F. C. Howell and F. Bourlière (eds.), *African Ecology and Human Evolution.* Aldine, Chicago.

——. 1974. The geology, archaeology and fossil mammals of the Cornelia Beds, O.F.S. *Memoirs van die Nasionale Museum Bloemfontein* 9:63–84.

Cooke, H. B. S., and S. C. Coryndon. 1970. Pleistocene mammals from the Kaisio Formation and other related deposits in Uganda; pp. 109–224 in L. S. B. Leakey and R. J. G. Savage (eds.), *Fossil Vertebrates of Africa*, vol. 2. Academic Press, London.

Coppens, Y. 1960. Le Quaternaire fossilifère de Koro-Toro (Tchad). Résultats d'une première mission. *Comptes Rendus de l'Académie des Sciences, Paris* 251:2385–2386.

——. 1971. Les vertébrés Villafranchiens de Tunisie: gisements nouveaux, signification. *Comptes Rendus de l'Académie des Sciences, Paris, Série D*, 273:51–54.

Crusafont-Pairó, M. 1952. Los Jiráfidos fósiles de España. *Memorias y Comunicaciones del Instituto Geológico* 8:1–239.

——. 1979. Les Girafidés des gisements du Bled Douarah (W. de Gafsa, Tunisie). *Notes du Service Géologique de Tunisie* 44:7–75.

Dagg, A. I. 1971. *Giraffa camelopardalis. Mammalian Species* 5:1–8.

Dagg, A. I., and J. B. Foster. 1981. *The Giraffe: Its Biology, Behavior and Ecology.* Krieger, Melbourne, Fla., 232 pp.

Dietrich, W. O. 1942. Ältestquartäre Säugetiere aus den südlichen Serengeti, Deutsch Ostafrika. *Palaeontographica Abt. A* 94:1–133.

Fortelius, M., and N. Solounias. 2000. Functional characterization of ungulate molars using the abrasion-attrition wear gradient: A new method for reconstructing paleodiets. *American Museum Novitates* 3301:1–36.

Franz-Odendaal, T. and N. Solounias. 2004. Comparative dietary evaluations of an extinct giraffid *(Sivatherium hendeyi)*(Mammalia, Giraffidae, Sivatheriinae) from Langebaanweg, South Africa (early Pliocene). *Geodiversitas* 26:1–11.

Gentry, A. 1994. The Miocene differentiation of Old World Pecora. *Historical Biology* 7:115–158.

——. 1997. Fossil ruminants (Mammalia) from the Manonga Valley, Tanzania; pp. 107–135 in T. Harrison (ed.), *Neogene Paleontology of the Manonga Valley, Tanzania: A Window into the Evolutionary History of East Africa.* Plenum Press, New York.

——. 1999. Fossil pecorans from the Baynunah Formation, Emirate of Abu Dhabi, United Arab Emirates; pp. 290–316 in P. J. Whybrow and A. Hill (eds.), *Fossil Vertebrates of Arabia.* Yale University Press, New Haven.

——. 2003. Ruminantia (Artiodactyla); pp. 332–379 in M. Fortelius, J. Kappelman, S. Sen, and R. L. Bernor (eds.), *Geology and Paleontology of the Miocene Sinap Formation, Turkey.* Columbia University Press, New York.

Geraads, D. 1981. Bovidae et Giraffidae (Artiodactyla, Mammalia) du Pléistocène de Ternifine (Algérie). *Bulletin du Muséum National d'Histoire Naturelle C*, 4ème sér., 3:47–86.

——. 1985. *Sivatherium maurusium* (Pomel)(Giraffidae, Mammalia) du Pléistocène de la République de Djibouti. *Paläontologische Zeitschrift* 59:311–321

——. 1986a. Remarques sur la systématique et la phylogénie des Giraffidae (Artiodactyla, Mammalia). *Geobios* 19:465–477.

——. 1986b. Les ruminants du Pléistocène d'Oubeidiyeh. *Mémoires et Travaux du Centre de Recherche Français de Jérusalem* 5:143–181.

——. 1987. La faune des dépôts Pléistocènes de l'ouest de Lac Natron (Tanzanie); interprétation biostratigraphique. *Bullétin des Sciences Géologiques* 40:167–184.

——. 1989. Vertébrés fossiles du Miocène supérieur du Djebel Krechem el Artsouma (Tunisie centrale): Comparaisons Biostratigraphiques. *Geobios* 22:777–801.

——. 1994. Girafes fossiles d'Ouganda; pp. 375–381 in M. Pickford and B. Senut (eds.), *Geology and Palaeontology of the Albertine Rift Valley, Uganda-Zaire.* Centre International pour la Formation et les Échanges Géologiques. CIFEG Publication Occasionnelle 29, Orléans, France.

——. 1996. Le *Sivatherium* (Giraffidae, Mammalia) du Pliocène final d'Ahl al Oughlam et l'évolution du genre en Afrique. *Paläontologische Zeitschrift* 70:623–629.

——. 1998. Le gisement de vertébrés pliocènes de Çalta, Ankara, Turquie: 9. Cervidae et Giraffidae. *Geodiversitas* 20:455–465.

——. 2002. Plio-Pleistocene mammalian stratigraphy of Atlantic Morocco. *Quaternaire* 13:43–53.

Geraads, D., Z. Alemseged, and H. Bellon. 2002. The late Miocene mammalian fauna of Chorora, Awash basin, Ethiopia: Systematics, biochronology and ⁴⁰K-⁴⁰Ar age of the associated volcanics. *Tertiary Research* 21:113–122.

Geraads, D., Z. Alemseged, D. Reed, J. Wynn, and D. C. Roman. 2004a. The Pleistocene fauna (other than Primates) from Asbole, lower Awash Valley, Ethiopia, and its environmental and biochronological implications. *Geobios* 37:697–718.

Geraads, D., V. Eisenmann, and G. Petter. 2004b. The large mammal fauna of the Oldowayan sites of Melka-Kunturé, Ethiopia; pp. 169–192 in J. Chavaillon and M. Piperno (eds.), *Studies on the Early Palaeolithic Site of Melka Kunturé, Ethiopia.* Istituto Italiano di Preistoria e Protostoria, Florence.

Geraads, D., and E. Güleç. 1999. A *Bramatherium* skull (Giraffidae, Mammalia) from the late Miocene of Kavakdere (Central Turkey): Biogeographic and phylogenetic implications. *Bulletin of the Mineral Research and Exploration* (foreign ed.) 121:51–56.

Haas, G. 1966. *On the Vertebrate Fauna of the Lower Pleistocene Site "Ubeidiya."* Publications of the Israeli Academy of Science and Humanities, 66 pp.

Haile-Selassie, Y. 2009. Giraffidae; pp. 389–395 in Y. Haile-Selassie and G. WoldeGabriel (eds.), *Ardipithecus kadabba. Late Miocene Evidence from the Middle Awash, Ethiopia.* University of California Press, Berkeley.

Hamilton, W. R. 1973. The Lower Miocene ruminants of Gebel Zelten, Libya. *Bulletin of the British Museum (Natural History), Geology* 21:76–150.

——. 1978. Fossil giraffes from the Miocene of Africa and a revision of the phylogeny of the Giraffoidea. *Philosophical Transactions of the Royal Society of London, B* 283:165–229.

Harris, J. M. 1974. Orientation and variability in the ossicones of African Sivatheriinae (Mammalia: Giraffidae). *Annals of the South African Museum* 65:189–198.

——. 1976a. Pleistocene Giraffidae (Mammalia, Artiodactyla) from East Rudolf, Kenya; pp. 283–32 in R. J. G. Savage and S. C. Coryndon (eds.), *Fossil Vertebrates of Africa*, vol. 4. Academic Press, London.

——. 1976b. Pliocene Giraffoidea (Mammalia, Artiodactyla) from the Cape Province. *Annals of the South African Museum* 69:325–353.

——. 1982. Fossil Giraffidae from Sahabi, Libya. *Garyounis Scientific Bulletin* 1982:95–100.

——. 1987. Fossil Giraffidae and Camelidae from Laetoli; pp. 358–377 in M. D. Leakey and J. M. Harris (eds.), *Laetoli: A Pliocene site in northern Tanzania.* Clarendon Press, Oxford.

———. 1991. Family Giraffidae; pp. 93–38 in J. M. Harris (ed.), *Koobi Fora Research Project:Volume 3. The Fossil Ungulates: Geology, Fossil Artiodactyls, and Palaeoenvironments.* Clarendon Press, Oxford.

———. 2003. Lothagam giraffids; pp. 523–30 in M. G. Leakey and J. M. Harris (eds.), *Lothagam: Dawn of Humanity in Eastern Africa.* Columbia University Press, New York.

Harris, J. M., F. H. Brown, and M. G. Leakey. 1988. Geology and paleontology of Plio-Pleistocene localities west of Lake Turkana, Kenya. *Contributions in Science* 399:1–128.

Harris, J. M., M. G. Leakey, and T. E. Cerling. 2003. Early Pliocene tetrapod remains from Kanapoi, Lake Turkana Basin, Kenya. *Contributions in Science* 498:39–113.

Heintz, E. 1976. Les Giraffidae (Artiodactyla, Mammalia) du Miocène de Beni Mellal. *Géologie Méditerranéenne* 3:91–104.

Heintz, E. M. Brunet, and S. Sen. 1981. Un nouveau Giraffidé du Miocène supérieur d'Irak: *Injanatherium hazimi* n. gen., n. sp. *Comptes rendus des Séances de l'Académie des Sciences, Paris,* 292:423–426.

Heinzelin, J. de, J. D. Clark, T. White, W. Hart, P. Renne, G. Woldegabriel, Y. Beyene, and E. Vrba. 1999. Environment and behavior of 2.5-million-year-old Bouri Hominids. *Science* 284:625–629.

Hendey, Q. B. 1968. New Quaternary fossil sites near Swartklip, Cape Province. *Annals of the South African Museum* 52:453–73.

———. 1969. Quaternary vertebrate fossil sites in the southwest Cape Province. *South African Archaeological Bulletin* 24:96–105.

———. 1970. A review of the geology and palaeontology of the Plio/Pleistocene deposits at Langebaanweg, Cape Province. *Annals of the South African Museum* 56:75–117.

———. 1974. Faunal dating of the late Cenozoic of southern Africa, with special reference to the Carnivora. *Quaternary Research* 4:49–161.

———. 1976. The Pliocene fossil occurrences in "E" Quarry, Langebaanweg, South Africa. *Annals of the South African Museum* 69:215–217.

———. 1981. Palaeoecology of the late Tertiary fossil occurrences in "E" Quarry, Langebaanweg, South Africa, and a reinterpetation of their geological context. *Annals of the South African Museum* 84:1–104.

Hernández Fernández, M., and E. S. Vrba. 2005. A complete estimate of the phylogenetic relationships in Ruminantia: A dated species-level supertree of extant ruminants. *Biological Reviews* 80:269–302

Hopwood, A. T. 1934. New fossil mammals from Olduvai, Tanganyika Territory. *Annals and Magazine of Natural History* (10) 14:546–550.

Howe, B., and H. L. Movius 1947. A stone age cave in Tangier. *Papers of the Peabody Museum* 28:1–23.

Janis, C. M., and K. M. Scott. 1987. The interrelationships of higher ruminant families with special emphasis in members of the Cervoidea. *American Museum Novitates* 2893:1–85.

Joleaud, L. 1936. Gisements de vertébrés quaternaires du Sahara. *Bulletin de la Société d'Histoire Naturelle d'Afrique du Nord* 26:23–29.

———. 1937. Remarques sur les giraffidés fossiles d'Afrique. *Mammalia* 1:85–96.

Kalb, J. E., C. J. Jolly, A. Mebrate, S. Tebedge, C. Smart, E. B.Oswald, D. Cramer, P. Whitehouse, C. B. Wood, G. C. Conroy, T. Adefris, L. Sperling, and B. Kana. 1982. Fossil mammals and artefacts from the Middle Awash Valley, Ethiopia. *Nature* 298:25–29.

Kent, P. E. 1942a. A note on Pleistocene deposits near Lake Manyara, Tanganyika. *Geological Magazine* 79:72–77.

———. 1942b. The Pleistocene Beds of Kanam and Kanjera, Kavirondo, Kenya. *Geological Magazine* 79:117–132.

Kingdon, J. 1997. *The Kingdon Field Guide to African Mammals.* Academic Press, San Diego, 464 pp.

Kostopoulos, D. M. and A. Athanassiou. 2005. In the shadow of bovids: suids, cervids, and giraffids from the Plio-Pleistocene of Greece. *Quaternaire* 2:179–190.

Lavocat, R. 1961. Le gisement de vertébrés de Beni Mellal (Maroc): Étude systématique de la faune de mammifères et conclusions générales. *Notes et Mémoires du Service Géologique du Maroc* 155:1–144.

Leakey, L. S. B. 1965. *Olduvai Gorge 1951–1961: Fauna and Background.* Cambridge University Press, Cambridge, 109 pp.

———. 1970. Additional information on the status of *Giraffa jumae* from East Africa; pp. 325–330 in L. S. B. Leakey and R. J. G. Savage (eds.), *Fossil Vertebrates of Africa,* vol. 2. Academic Press, London.

Leakey, R. E., and M. G. Leakey. 1986. A new Miocene hominoid from Kenya. *Nature* 324:143–146.

Leakey, R. E., and A. C. Walker. 1985. New higher primates from the early Miocene of Buluk, Kenya. *Nature* 318:173–175.

Levin, N. E., T. E. Cerling, B. H. Passey, J. M. Harris, and J. R. Ehleringer. 2006. A stable isotope aridity index for terrestrial environments. *Proceedings of the National Academy of Sciences, USA* 103:11201–11205.

Likius, A., P. Vignaud, and M. Brunet. 2007. Une nouvelle espèce du genre *Bohlinia* (Mammalia, Giraffidae) du Miocene supérieur de Toros-Menalla, Tchad. *Comptes Rendus Palevol* 6:211–220.

Lydekker, R. 1904. On the subspecies of *Giraffa camelopardalis. Proceedings of the Zoological Society of London* 1:202–207.

MacInnes, D. 1936. A new genus of fossil deer from the Miocene of Africa. *Journal of the Linnean Society* 39:521–530.

Madden, C. T. 1972. Miocene mammals, stratigraphy and environment of Moruorot Hill. *Paleobios* 14:1–12.

Martínez-Navarro, B., L. Rook, A. Segid, D. Yosieph, M. P. Ferretti, J. Shoshani, T. M. Tecle, and Y. Libsekal. 2004. The large fossil mammals from Buia (Eritrea): Systematics, biochronology and paleoenvironments. *Rivista Italiana di Paleontologia* 110 (suppl.):61–88.

Mason, R. J., R. A. Dart, and J. W. Kitching, 1958. Bone tools at the Kalkbank middle stone age tool site and the Makapansgat australopithecine locality, Central Transvaal, Pt 1. *South African Archaeological Bulletin* 13:85–93.

Mawby, J. E. 1970. Fossil vertebrates from northern Malawi: preliminary report. *Quaternaria* 13:319–323.

McCrossin, M. L., B. R. Benefit, S. N. Gitau, A. K. Palmer, and K. T. Blue. 1998. Fossil evidence for the origins of terrestriality among old world higher primates; pp. 353–396 in E. Strasser, J. Fleagle, A. Rosenberger, and H. McHenry (eds.), *Primate Locomotion: Recent Advances.* Plenum Press: New York.

Miller, E. R. 1999. Faunal correlation of Wadi Moghara, Egypt: Implications for the age of *Prohylobates tandyi. Journal of Human Evolution* 36:519–533.

Montoya, P., and J. Morales. 1991. *Birgerbohlinia schaubi* Crusafont, 1952 (Giraffidae, Mammalia) del Turoliense inferior de Crevillente-2 (Alicante, Espana). Filogenia e historia biogeographica de la subfamilia Sivatheriinae. *Bulletin du Muséum National d'Histoire Naturelle* 13, C:177–200.

Morales, J., D. Soria, M. Nieto, P. Pelaez-Campomanes, and Pickford, M. 2003. New data regarding *Orangemeryx hendeyi* Morales et al., 2000, from the type locality, Arrisdrift, Namibia. *Memoir of the Geological Survey of Namibia* 19:305–344.

Morales, J., D. Soria, and Pickford, M., 1999: New stem giraffoid ruminants from the early and middle Miocene of Namibia. *Geodiversitas* 21.229–253

Morales, J., D. Soria, and H. Thomas. 1987. Les Giraffidae (Artiodactyla, Mammalia) d'Al Jadidah du miocène moyen de la Formation Hofuf (Province du Hasa, Arabie Saoudite). *Geobios* 20:441–467

Nakaya, H., M. Pickford, K. Yasui, and Y. Nakano. 1987. Additional large mammalian fauna from the Namurungule Formation, Samburu Hills, Kenya. *African Study Monographs* (suppl.) 5:79–129.

Paraskevaidis, I. 1940. Eine obermiocäne Fauna von Chios. *Neues Jahrbuch für Mineralogie* 83:363–442.

Pickford, M. 1986. Cainozoic palaeontological sites of western Kenya. *Münchner Geowissenschaftliche Abhandlungen, A* 8:1–151.

Pickford, M., Y. S. Attia, and M. S. Abd El Ghany. 2001. Discovery of *Prolibytherium magnieri* (Climacoceratidae, Artiodactyla) in Egypt. *Geodiversitas* 23:647–652.

Pickford, M., H. Ishida, Y. Nakano, and K. Yasui. 1987. The middle Miocene fauna from the Nachola and Aka Aiteputh Formations, northern Kenya. *African Study Monographs* (suppl.) 5:141–154.

Pickford, M., and B. Senut. 2001. The geological and faunal context of Late Miocene hominid remains from Lukeino, Kenya. *Comptes Rendus de l'Académie des Sciences, Paris, Sciences de la Terre et des Planètes* 332:145–152.

Pomel, A. 1892. Sur le *Libytherium maurusium,* grand Ruminant du terrain pliocène plaisancien d'Algérie. *Comptes Rendus de l'Académie des Sciences, Paris* 115:100–102.

Raynal, J.-P., D. Lefèvre, D. Geraads, and M. El Graoui. 1999. Contribution du site paléontologique de Lissasfa (Casablanca, Maroc) à une nouvelle interprétation du Mio-Pliocène de la Meseta. *Comptes Rendus de l'Académie des Sciences, Paris, Sciences de la Terre et des Planètes* 329:617–622.

Robinson, P., and C. C. Black. 1974. Vertebrate faunas from the Neogene of Tunisia. *Annals of the Geological Survey of Egypt* 4:319–332.

Roman, F., and M. Solignac. 1934. Découverte d'un gisement de mammifères pontiens à Douaria (Tunisie septentrionale), *Comptes*

Rendus Hebdomadaires des Séances de l'Académie des Sciences, Paris, Série D, 199:1649–1650.

Romer, A. S. 1928. Pleistocene mammals of Algeria. Fauna of the Paleolithic station of Mechta-al-Arbi. *Bulletin of the Logan Museum* 1:80–163.

Schrenk, F., T. G. Bromage, C. G. Betzier, U. Ring, and Y. M. Jueayeyi. 1993. Oldest *Homo* and Pliocene biogeography of the Malawi Rift. *Nature* 365:833–835.

Semaw, S., S. W. Simpson, J. Quade, P. R. Renne, R. F. Butler, W. C. McIntosh, N. Levin, M. Dominguez-Rodrigo, and M. J. Rogers. 2005. Early Pliocene hominids from Gona, Ethiopia. *Nature* 433:301–305.

Singer, R., and E. L. Boné. 1960. Modern giraffes and the fossil giraffids of Africa. *Annals of the South African Museum* 45:375–603.

Solounias, N. 2007. Superfamily Giraffoidea; pp. 257–277 in D. R. Prothero and S. E. Foss (eds.), *The Evolution of Artiodactyls.* Johns Hopkins University Press, Baltimore.

Solounias, N., W. S. McGraw, L.-A. Hayek, and L. Werdelin. 2000. The paleodiet of the Giraffidae; pp. 84–95 in E. S. Vrba and G. B. Schaller (eds.), *Antelopes, Deer, and Relatives: Fossil Record, Behavioral Ecology, Systematics, and Conservation.* Yale University Press, New Haven.

Solounias, N., and S. M. C. Moelleken. 1993. Dietary adaptation of some extinct ruminants determined by premaxillary shape. *Journal of Mammalogy* 74:1059–1071.

Solounias, N., M. Teaford, and A. Walker. 1988. Interpreting the diet of extinct ruminants: The case of a non-browsing giraffid. *Paleobiology* 14:287–300.

Stromer, E. 1907. Fossile Wirbeltier-Reste aus dem Uadi Fâregh und Natrûn in Ägypten. *Abhandlungen der Senckenbergischen naturforschenden Gesellschaft* 247:1–97.

Sudre, J., and J.-L. Hartenberger. 1992. Oued Mya 1, nouveaux gisement de mammifères du Miocène supérieur dans le sud Algérien. *Geobios* 24:553–565.

Suwa, G., H. Nakaya, B. Asfaw, H. Saegusa, A. Amzaya, R. T. Kono, Y Beyene, and S. Katoh. 2003. Plio-Pleistocene terrestrial mammal assemblage from Konso, southern Ethiopia. *Journal of Vertebrate Paleontology* 23:901–916.

Taieb, M., D. C. Johanson, Y. Coppens, and J. L. Aronson. 1976. Geological and palaeontological background of Hadar hominid site, Afar, Ethiopia. *Nature* 260:288–293.

Thomas, H. 1984. Les Giraffoidea et les Bovidae miocènes de la Formation Nyakach (rift Nyanza, Kenya). *Palaeontographica, Abt. A* 183:64–89.

Thomas, H., and G. Petter. 1986. Révision de la faune mammifères du Miocène supérieur de Menacer (ex Marceau), Algérie; discussion sur l'âge du gisement. *Geobios* 19:357–373.

Thomas, H., S. Sen, M. Khan, B. Battail, and G. Ligabue. 1985. The Lower Miocene fauna of Al-Sarrar (Eastern Province, Saudi Arabia). *ATLAL* 5:109–136.

Urbanek, C., P. Faupl, W. Hujer, T. Ntaflos, W. Richter, G. Weber, K. Schaefer, B. Viola, P. Gunz, S. Neubauer, A. Stadlmayer, O. Kullmer, O. Sandrock, D. Nagel, G. Conroy, D. Falk, K. Woldearegay, H. Said, G. Assefa, and H. Seidler. 2005. Geology, paleontology and paleoanthropology of the Mount Galili Formation in the southern Afar Depression, Ethiopia—Preliminary results. *Joannea Geologie und Paläontologie* 6:29–43.

Vaufrey, R. 1947. Olorgesaillie. Un site acheuléen d'une exeptionnelle richesse. *Anthropologie* 51:367.

Vignaud, P., P. Duringer, H. T. Mackaye, A. Likius, C. Blondel, J.-R. Boisserie, L. de Bonis, V. Eisenmann, M.-E. Etienne, D. Geraads, F. Guy, T. Lehmann, F. Lihoreau, N. Lopez-Martinez, C. Mourer-Chauvire, O. Otero, J.-C. Rage, M. Schuster, L. Viriot, A. Zazzo, and M. Brunet. 2002. Geology and palaeontology of the Upper Miocene Toros-Menalla hominid locality, Chad. *Nature* 418:152–155.

Ward, C. V., M. G. Leakey, B. Brown, F. Brown, J. Harris, and A. Walker. 1999. South Turkwel: a new Pliocene hominid site in Kenya. *Journal of Human Evolution* 36:69–95.

Whitworth, T. 1958. Miocene ruminants of East Africa. *Fossil Mammals of Africa* 15:1–50.

Woldegabriel, G., T. D. White, G. Suwa, P. Renne, J. de Heinzelin, W. K. Hart, and G. Heiken. 1994. Ecological and temporal placement of early Pliocene hominids at Aramis, Ethiopia. *Nature* 371:330–333.

Wynn, J. G., Z. Alemseged, R. Bobe, D. Geraads, D. Reed, and D. C. Roman. 2006. Geological and palaeontological context of a Pliocene juvenile hominin at Dikika, Ethiopia. *Nature* 443:332–336.

Yasui, K., Y. Kunimatso, N. Kuga, B. Bajope, and H. Ishida. 1992. Fossil mammals from the Neogene strata in the Sinda Basin, eastern Zaire. *African Study Monographs* (suppl.) 17:87–107.

FORTY

Cervidae

ALAN W. GENTRY

Cervidae are pecoran ruminants in which most species have branched deciduous antlers inserted on the frontals. They are usually found in mesic and/or somewhat wooded habitats, and their teeth tend to be lower crowned than those of bovids. They evolved in Eurasia and are known from the early Miocene onward (but not from the Indian subcontinent until soon after 3.0 Ma; Barry, 1995). They entered North America at the start of the Pliocene and spread into South America around the start of the Pleistocene. They are known back to the beginning of the late Pleistocene in North Africa (Geraads, 1982). Two or possibly three species are represented in the African cohort of this group.

Systematic Paleontology

Genus *MEGALOCEROS* Brookes, 1828

Synonymy Megaceros Owen, 1844: 237; *Megaceroides* Joleaud, 1914:738.

Type Species Megaloceros giganteus, Blumenbach, 1799. ICZN Opinion 1566 (*Bulletin of Zoological Nomenclature* 46:219–220, 1989) validates the spelling of the generic name as given here and not as *Megalocerus*.

Remarks This genus contains large Pleistocene deer with large and often palmated antlers. The best known species is the "giant" deer or so-called Irish elk, *Megaloceros giganteus*, with an antler span of up to 3.5 m. It survived until about 8,000 years ago (Stuart et al., 2004).

MEGALOCEROS ALGERICUS (Lydekker, 1890)

Synonymy Cervus algericus, Lydekker, 1890:603, figured. Hammam Meskoutin (Joleaud, 1914a), near Guelma, Algeria. Late Pleistocene; *Cervus pachygenys* Pomel, 1892:213.

Remarks This species was founded on a left maxilla, BMNH M10647, with P4–M3 in early middle wear. Pomel (1892, 1893) added more material from other late Pleistocene North African sites and noted that some partial mandibles had thickened horizontal ramuses, a character of *Megaloceros* species. Other finds in Algeria and Morocco came to light in later years, including antler remains and a cranium (Arambourg, 1939: plate 2, figures 2, 2a; Hadjouis, 1990). Abbazzi (2004) judged that the totality of the material revealed a rather small megalocerine deer showing a degree of endemism in its northwestern African range. Its characters include a strongly

ossified skull, flattened and widely divergent antler beams without proximal tines, a short muzzle, well-marked cingula on molar teeth, and the strongly pachyostotic mandibles.

This might have been the end of the story, except that Joleaud (1914b and several later papers) had noted the megalocerine affinities of the fossils and placed *algericus* in a new subgenus *Megaceroides*. Ambrosetti (1967) then chose *Megaceroides* as a generic-level name for early–middle Pleistocene Eurasian species of megalocerine deer grouped around *Megaloceros verticornis*. Some subsequent authors, while agreeing with a generic-level separation from the predominantly late Pleistocene *Megaloceros*, preferred the name *Praemegaceros* Portis, 1920 in place of *Megaceroides*. The ensuing debate has expanded to include the number of lineages of megalocerine deer, whether any or all of them are related to the living fallow deer *Dama*, and whether their ancestry lies within or close to the Plio-Pleistocene *Eucladoceros* (Azzaroli and Mazza, 1993; Pfeiffer, 1999, 2002; Abbazzi, 2004; Lister et al., 2005). Some of the apomorphies of late *Megaloceros* are more upright pedicels in side view (their anterior edges descend more sharply to a more postorbital position), pedicels slightly closer together, a more localized transverse ridge between the two pedicels, and a more concave or scooped-out appearance of the frontals immediately anterior to the antler bases. Such characters must be functionally related to the mechanical support of the huge antlers. They are absent or less apparent in *M. algericus*. This situation could have arisen if this species were a late-surviving *Megaceroides/Praemegaceros* or secondarily if it were a *Megaloceros* of reduced size facing hard living conditions on the edge of its range. The first alternative would be strengthened if *M. algericus* were to be found in the earlier Pleistocene of North Africa. Hadjouis (1990) and Abbazzi (2004) discussed *M. algericus* in more detail and Hadjouis (1990) favored an East Asian origin for it.

Genus *CERVUS* Linnaeus, 1758

Type Species Cervus elaphus Linnaeus 1758.

CERVUS ELAPHUS Linnaeus, 1758

Remarks The red deer, *Cervus elaphus*, has survived into historical times in Algeria and Tunisia (Joleaud 1935:

figure 36) and until the present day in part of the border region between those two states. It is rather a small form, *C. elaphus barbarus* Bennett, 1833 (or 1837 or 1848; see Ellerman and Morrison-Scott, 1966), in which the bez tine (second one above the burr) is missing. *Cervus elaphus* and unidentified *Cervus* species are known from archaeological localities in Algeria and Tunisia back into the late Pleistocene. Laquay (1986) published a Moroccan record, but the illustrated teeth hint at belonging to *Megaloceros* instead—for example, in the relatively large p2 with well-formed transverse crests and in the basal pillars on the lower molars, said to be well marked.

Other Cervid Records and Claims in Africa

Roman (1931: plate 4, figure 5) illustrated a tooth and astragalus from the late Miocene of Djérid, Tunisia, as belonging to a contemporaneous European cervid *Pliocervus matheroni* (a species still placed in *Capreolus* in Roman's day). It is very likely that the tooth is a bovid dP4 (Dietrich, 1950:52, fn. 8) and perhaps conspecific with other bovid teeth at the locality. From the illustration of the tooth it appears to have an occlusal length of just over 16.0 mm, which is too large for upper molars or a dP4 of *Pliocervus matheroni* and *P.* aff. *matheroni* (Azanza, 1995: table 2; 2000: table 31).

The possibility of a fallow deer having occurred in Africa was discussed at length by Joleaud (1935). He relied on pictorial representations from Egyptian antiquity, but one would expect the ancient Egyptians to know of *Dama mesopotamica* in Mesopotamia and perhaps to have kept live examples. Zeuner (1963) discussed distribution of the fallow deer in classical times in southeast Europe and southwest Asia but mentioned no natural occurrence or introductions in Africa.

The Natural History Museum in London contains an unregistered partial upper molar from either Singa or Abu Hugar, Sudan, which might be cervid (figure 40.1). With an occlusal length of ca. 15.0 mm, it could fit a fallow deer in size and is certainly too small for *Cervus elaphus*. A piece of deer antler from Wadi Halfa, Sudan, mentioned by participants in a discussion of a short paper by Lydekker (1887), could not be traced later (Bate, 1951).

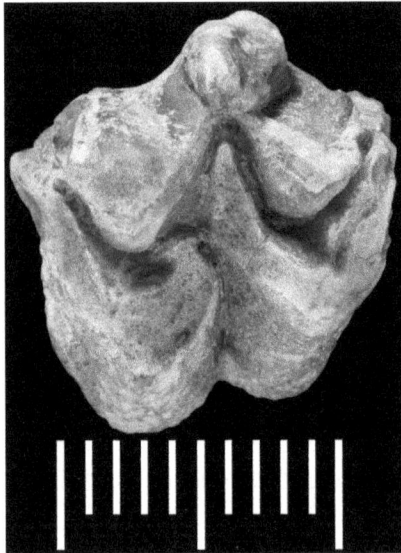

FIGURE 40.1 Occlusal view of partial right upper molar of a possible cervid from either Singa or Abu Hugar, Sudan. Scale in millimeters. Photograph courtesy of The Natural History Museum, London.

ACKNOWLEDGMENTS

Denis Geraads gave advice on this chapter.

Literature Cited

Abbazzi, L. 2004. Remarks on the validity of the generic name *Praemegaceros* Portis 1920, and an overview on *Praemegaceros* species in Italy. *Atti della Accademia Nazionale dei Lincei, Rendiconti Scienze Fisiche e Naturali* (9) 15:115–132.

Ambrosetti, P. 1967. Cromerian fauna of the Rome area. *Quaternaria* 9:267–284.

Arambourg, C. 1939. Mammifères fossiles du Maroc. *Mémoires de la Société des Sciences Naturelles Maroc* 46:1–74. (Publication date usually given as 1938, but a note on p.74 refers to the printing being completed on September 30, 1939. The introduction is dated December 15, 1938.)

Azanza, B. 1995. The vertebrate locality Maramena (Macedonia, Greece) at the Turolian-Ruscinian boundary (Neogene): 14. Cervidae (Artiodactyla, Mammalia). *Münchner Geowissenschaftliche Abhandlungen A* 28:157–166.

Azzaroli, A., and P. Mazza. 1993. Large early Pleistocene deer from Pietrafitta lignite mine, Central Italy. *Palaeontographia Italica* 80:1–24.

Barry, J. C. 1995. Faunal turnover and diversity in the terrestrial Neogene of Pakistan; pp. 115–134 in E. S. Vrba, G. H. Denton, T. C. Partridge, and L. H. Burckle (eds.), *Paleoclimate and Evolution, with Emphasis on Human Origins*. Yale University Press, New Haven.

Bate, D. M. A. 1951. The mammals from Singa and Abu Hugar. *Fossil Mammals of Africa: British Museum (Natural History), London* 2:1–28.

Dietrich, W. O. 1950. Fossile Antilopen und Rinder Äquatorialafrikas. *Palaeontographica Abt. A* 99:1–62.

Ellerman, J. R., and T. C. S. Morrison-Scott. 1966. *Checklist of Palaearctic and Indian Mammals 1758 to 1946*. 2nd ed. British Museum (Natural History), London, 810 pp.

Geraads, D. 1982. Paléobiogéographie de l'Afrique du nord depuis le Miocène terminal, d'après les grands mammifères. *Geobios, Mémoire Special* 6:473–481.

Hadjouis, D. 1990. *Megaceroides algericus* (Lydekker, 1890), du gisement des phacochères (Alger, Algérie): Étude critique de la position systématique de *Megaceroides*. *Quaternaire* 3–4:247–258.

Joleaud, L. 1914a. Notice géologique sur Hammam Meskoutin (Algérie). *Bulletin de la Société Géologique de France* (4) 14: 423–34.

———. 1914b. Sur le *Cervus (Megaceroides) algericus* Lydekker (1890). *Comptes Rendus de la Société de Biologie de Paris* 76:737–739.

———. 1935. Les ruminants cervicornes d'Afrique. *Mémoires de l'Institut d'Égypte* 27:1–85.

Laquay, G. 1986. *Cervus elaphus* (Mammalia, Artiodactyla) du Pléistocène supérieur de la carrière Doukkala II (Rabat—Maroc): Sa comparaison avec le cerf würmien de France. *Revue de Paléobiologie* 5:143–147.

Lister, A. M., C. J. Edwards, D. A. W. Nock, M. Bunce, I. A. van Pijlen, D. G. Bradley, M. G. Thomas, and I. Barnes. 2005. The phylogenetic position of the "giant deer" *Megaloceros giganteus*. *Nature* 438:850–853.

Lydekker, R. 1887. On a molar of a Pliocene type of *Equus* from Nubia. *Quarterly Journal of the Geological Society* 43:161–4.

Pfeiffer, T. 1999. Die Stellung von *Dama* (Cervidae, Mammalia) im System plesiometacarpaler Hirsche des Pleistozäns. *Courier Forschungsinstitut Senckenberg* 211:1–218.

———. 2002. The first complete skeleton of *Megaloceros verticornis* (Dawkins, 1868) Cervidae, Mammalia, from Bilshausen (Lower Saxony, Germany): Description and phylogenetic implications. *Mitteilungen aus dem Museum für Naturkunde in Berlin, Geowissenschaftliche Reihe* 5:289–308.

Pomel, A. 1892. Sur deux Ruminants de l'époque néolithique de l'Algérie. *Comptes Rendus hebdomadaires de Séances de l'Académie des Sciences, Paris* 115:213–216.

———. 1893. Caméliens et cervidés. *Carte Géologique de l'Algérie Paléontologie Monographies, Algiers*: 5–52.

Roman, F. 1931. Description de la faune pontique du Djerid (El Hamma et Nefta). *Annales de l'Université de Lyon* (n.s. 1) 48:30–42.

Stuart, A. J., P. Kosintsev, T. F. Higham, and A. M. Lister. 2004. Pleistocene to Holocene extinction dynamics in giant deer and woolly mammoth. *Nature* 431:684–9.

Zeuner, F. E. 1963. *A History of Domesticated Animals*. Hutchinson, London, 560 pp.

FORTY-ONE

Camelidae

JOHN M. HARRIS, DENIS GERAADS,
AND NIKOS SOLOUNIAS

Camels originated in North America in the middle Eocene (Uintan land mammal age) and the first 36 million years of their history is confined to that continent (Honey et al., 1998). They were a highly successful group in which some 95 species and 36 genera were distributed between five subfamilies: Stenomylinae, Floridatragulinae, Miolabinae, Protolabinae, and Camelinae (Honey et al., 1988), although McKenna and Bell (1997) also recognize Poebrodontinae, Poebrotheriinae, Pseudolabinae, and Aepycamelinae. The main radiation of this family took place in the Miocene, during which time at least 13 genera and 20 species were distributed over much of North America. Camelid generic diversity declined during the late Miocene, although camelids remained common elements of the North American biota until the late Pleistocene extinction event. Only the camelins (Camelinae) are known to have migrated into the Old World.

Domesticated dromedaries occur today throughout North Africa, the largest populations being in Somalia and Sudan. Their southernmost limit is determined by the degree of humidity and incidence of trypanosomiasis and is demarcated by the 400 mm isohyet that follows 15°N from Senegal to Niger and 13°N in Chad and Sudan (Köhler-Rollefson, 1991). Dromedaries are also found throughout Arabia and in Asia extend from Turkey to the western part of India, reaching as far north as Turkmenistan (40°N)(FAO, 1984). Their versatility as a riding animal, a pack animal, and a source of meat and milk in arid and semiarid conditions owes much to their unique physiology. The dromedary has the lowest water turnover of all mammals (MacFarlane, 1977); up to 30% of its body weight can be lost through desiccation but can be replenished in a matter of minutes (Schmidt-Nielsen, 1964; Yagil et al., 1974).

Systematic Paleontology

Family CAMELIDAE Gray, 1821

Early in their evolutionary history, camelids acquired long limbs and necks and lost the lateral metapodials. The proximal fibula fused to the tibia and the radius and ulna fused, but camelids differ from ruminants in the divergent pulleys of their fused central metapodials and in the unfused nature of the navicular and cuboid. Although some Miocene camelids retained complete dentitions, extant camels and llamas have lost their upper first and second incisors plus one or more premolars. Their selenodont cheek teeth are hypsodont.

Subfamily CAMELINAE Gray, 1821
Figures 41.1B and 41.1C

The subfamily Camelinae appeared in the late early Miocene (Hemingfordian) of North America and is represented by two tribes: Camelini (*Camelus* plus giant camels) and Lamini (llamas). The dental formula for camelins is 1/3, 1/1, 3/2. 3/3 (figures 41.1B, 41.C), and that for lamins is 1/3, 1/1, 2/1, 3/3. Camelins and lamins are united by the absence of I2 and p2 and by raised posterolateral edges on the proximal end of the first phalanx (Harrison, 1979). Camels migrated to Eurasia during the late Miocene. *Paracamelus*, which may have been derived from the North American *Procamelus* Leidy, 1858 (Schlosser, 1903; Zdansky, 1926), has been documented from Afghanistan (Raufi and Sickenberg, 1973), China (Zdansky, 1926; Flynn et al. 1991), Romania (Stefanescu, 1910), Hungary (Kretzoi, 1954), Russia (Khavesson, 1954), and Turkey (Kostopoulos and Sen, 1999). The earliest European occurrences appear to be those documented from Spain in MN 13 (Morales et al., 1980; Pickford et al., 1995).

With the recent discovery of fossil camels in Chad (Likius et al., 2003), their record in Africa now extends back to the latest Miocene. However, most of the documented records constitute fragmentary specimens and many of the Pleistocene or earlier specimens can only be identified to tribe or genus.

Tribe CAMELINI Gray, 1821

The Camelini are united by an enlarged and strongly inflected angular process on the mandible, large postglenoid foramen, long postglenoid process with matching large facet on mandibular condyle, enlarged canines that are rounded in cross section (especially in males), ventrally flattened auditory bullae, low and rounded diastemal crest on the mandible, and reduced maxillary fossa (Harrison 1979).

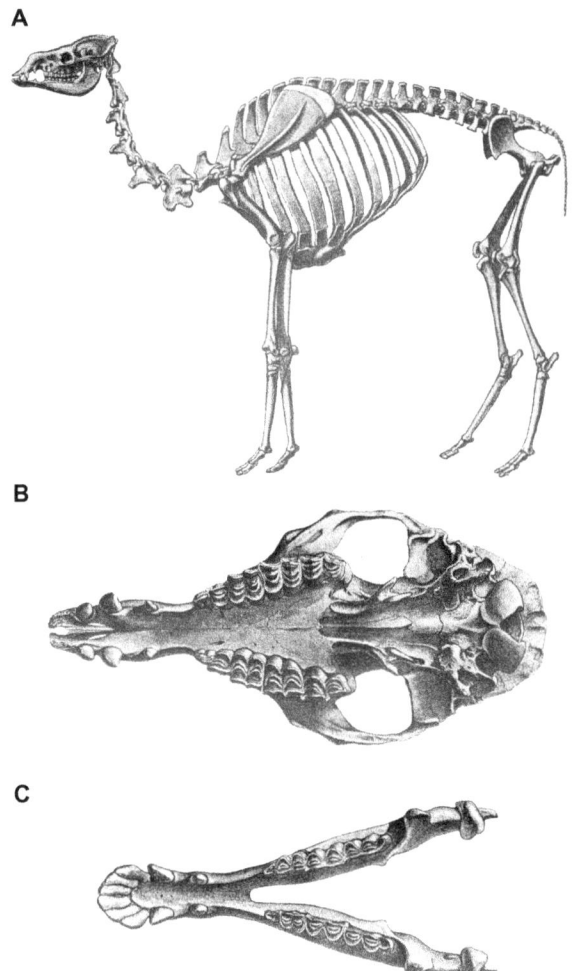

A

B

C

FIGURE 41.1 Extant *Camelus dromedarius* (after Smuts and Bezuidenhout, 1987). A) skeleton; B) cranium, palatal view; C) mandible, occlusal view. By permission of Oxford University Press.

Constituent genera include *Camelus* Linnaeus, 1758; *Procamelus* Leidy, 1858; *Megatylopus* Matthew and Cook, 1909; *Titanotylopus* Barbour and Schultz, 1934; and *Megacamelus* Frick, 1929 (McKenna and Bell, 1997). *Paracamelus* Schlosser, 1903, was considered a subgenus of *Camelus* by McKenna and Bell (1997) but has been treated as a genus in its own right by recent workers even though it has been diagnosed on mostly plesiomorphic characters.

Genus *PARACAMELUS* Schlosser, 1903

Diagnosis After Likius et al. (2003). Large camelin with three elongate upper and lower premolars that are less reduced than in the extant *Camelus* where p3 is vestigial or absent. The metapodials are massive and one fourth larger than those of *Camelus*.

Occurrence *Paracamelus* has been reported from the late Miocene of Chad and Egypt, and from the late Pliocene of Tunisia (table 41.1).

Type Species Paracamelus gigas Schlosser, 1903.

Other Recognized Species Paracamelus alexejevi Khavesson, 1954; *P. aguirrei* Morales, 1984; *P. alutensis* (Stefanescu, 1895).

PARACAMELUS GIGAS SCHLOSSER, 1903

Diagnosis After Likius et al. (2003). *Paracamelus* species larger than the extant *Camelus bactrianus* and *C. dromedarius* and characterized by the presence of p3, elongate metapodials with distinctive metacarpal proximal articular facets (subquadrangular in Mc IV and subtriangular in Mc III) and deep plantar gutters under the proximal epiphysis of the metatarsal.

Lectotype Isolated molar from Tientsin, China, published by Schlosser (1903) and selected as lectotype by Made and Morales (1999).

Remarks A fragmentary right mandible and two right metatarsals were recovered from latest Miocene sites at Kossom Bougoudi, Chad, by the Mission Paléoanthropologique Franco-Tchadienne (Likius et al., 2003). The mandible differs from those of *Camelus* species in depth, robusticity, and the presence of a well-developed p3. The metatarsals were much longer than those of extant camels. The Chad material is larger than that of *P. alexejevi* from the Pliocene (MN 15) of the Ukraine, and the mandible is deeper than that of *P. alutensis* from the early Pleistocene of Rumania (Stefanescu, 1910). The late Miocene *P. aguirrei* was described from elements that are not yet represented in the Chad sample, but the estimated length of the p3 in the Chad mandible is similar to that from Spain. The lengths of the tooth row and metatarsals fit into the range of variation documented for *P. gigas* from the late Miocene of China by Zdansky (1926) and Teilhard and Trassaert (1937), and therefore the Chad specimens were assigned to that species.

Arambourg (1979) identified a camelid calcaneum from Lake Ichkeul as *C. thomasi*, but Pickford et al. (1995) reinterpreted it to represent *Paracamelus*. The calcaneum is similar in length to that of a large modern camel, but the bone is more robust, and the nonarticular part is relatively slightly longer, though much shorter than in *P. aguirrei* (Pickford et al., 1995: plate 80), to which it was erroneously referred by Made and Morales (1999). Made and Morales (1999) also interpreted the camelid cuboid reported by Stromer (1902) from Wadi Natrun as *Paracamelus*. Although poorly preserved, the cuboid (housed in the Senckenberg Museum, Frankfurt, Germany) is definitely camelid rather than anthracotheriid by the horizontal orientation of the calcanear facet. It resembles *Paracamelus gigas* from China in its dorsoplantar elongation and L-shaped dorsal metatarsal facet, but it differs in its large, weakly concave talar facet, almost lacking the plantar process, which is more reminiscent of the cuboid of *P. aguirrei* from the Mio-Pliocene of Venta del Moro, Spain (Morales et al., 1980).

Genus *CAMELUS* Linnaeus, 1758

Diagnosis After Harrison (1979) and Orlov (1968). Camelin in which I1–2, P2, p2, and p3 are lost, I3 is present; P1, p1, and p4 reduced; internal crescent on P3 incomplete and molars hypsodont as in *Paracamelus*. Premaxilla moderate to heavy, lacrimal vacuity very reduced, maxillary fossa reduced or absent, nasals flattened, rostrum short, zygomatic arch straight, postglenoid foramen and process large. Postglenoid facet on mandibular condyle medially positioned and vertically elongated; diastemal crest on mandible reduced and rounded; angular process large and strongly inflected, dorsal surface of mandibular condyle convex, Metacarpal and metatarsal similar in length, metapodial elements III and IV fused as in *Paracamelus*, suspensory ligament scar on first phalanx extends to center of shaft, posterolateral edges

ALGERIA

1. Alger: Arambourg, 1932; *Camelus* sp.
2. Chaachas: Gautier, 1966; *Camelus* sp.; late Pleistocene
3. San Roch, near Oran: Gautier, 1966; *Camelus* sp.; late Pleistocene
4. Sintes, near Algiers: Gautier, 1966; *Camelus* sp.; late Pleistocene
5. Tighenif (=Ternifine): Thomas, 1884; Pomel, 1893; Vaufrey, 1955; *Camelus thomasi*; early middle Pleistocene

CHAD

6. Bochianga: Coppens, 1971; *Camelus* sp.; early Pliocene
7. Kossom Bougoudi: Likius et al., 2003. *Paracamelus gigas*; early Pliocene

EGYPT

8. Wadi Natrun: Stromer 1902; Made and Morales, 1999: *Paracamelus* sp.; late Miocene or early Pliocene
9. Dakhleh Oasis: Churcher et al., 1999; *Camelus* sp.; mid-Pleistocene

ETHIOPIA

10. Omo Shungura: Howell et al., 1969; *Camelus* sp.; late Pliocene

ISRAEL

11. Mugharet-el-Emireh Cave: Bate, 1927: *Camelus* sp.; 840 ± 80 BP
12. Tabun Cave: Payne and Garrard, 1983, *Camelus* sp.; 1060 ± 70 BP
13. Ubedeiyah: Haas, 1966; Geraads, 1986: Camelus sp.; early Pleistocene.

JORDAN

14. Azraq Oasis: Clutton-Brock, 1970: *Camelus* sp.; 3340 ± 200 BP

KENYA

15. Koobi Fora: Harris, 1991; *Camelus* sp.; late Pliocene
16. Marsabit Road: Gentry and Gentry, 1969; *Camelus* sp.; late Pliocene
17. West Turkana: Harris et al., 1988; *Camelus* sp.; late Pliocene

MALAWI

18. Chiwondo Beds: Schrenk et al., 1993; *Camelus* sp.; late Pliocene

MOROCCO

19. Ahl al Oughlam: Geraads et al., 1995; *Camelus* sp; late Pliocene
20. Oulad Hamida 1 (Rhinoceros Cave): Raynal et al., 1993; Geraads, 2002; *Camelus* cf. *C. thomasi*; mid-Pleistocene
21. Taza: Gautier, 1966; *Camelus* sp.; late Pleistocene

OMAN

22. Umm an-Nar: Hoch, 1979; *Camelus* sp.; Holocene

SAUDI ARABIA

23. An Nafud: Thomas et al., 1999; *Camelus* sp.; early Pleistocene

SUDAN

24. Site 1040, Nile terrace near Egypt-Sudan border: Gautier, 1966; *Camelus* sp.; late Pleistocene

SYRIA

25. Latamne: Hooijer, 1961; *Camelus* sp.; mid-Pleistocene
26. El Kowm: Reuters October 8, 2006; *Camelus* sp.; late Pleistocene

TANZANIA

26. Laetoli: Harris, 1987; *Camelus* sp.; mid-Pliocene
27. Olduvai: Gentry and Gentry, 1969; *Camelus* sp.; early Pleistocene

TUNISIA

28. El Guettar: Gautier, 1966; *Camelus* sp.; late Pleistocene
29. Lac Ichkeul: Arambourg, 1979; Pickford et al., 1995; *Paracamelus* sp.; late Pliocene

and center raised. Facial region of skull considerably shorter and relatively broader than in *Paracamelus*; m3 relatively long, anteroexternal folds of m2 and m3 rudimentary and may be absent.

Occurrence Camelus remains have been reported from the late Pliocene of Chad, Ethiopia, Kenya, Malawi, Morocco, and Tanzania, and the Pleistocene of Algeria, Egypt, Israel, Oman, Saudi Arabia, Sudan, and Tunisia (table 41.1).

Type Species Camelus bactrianus Linnaeus, 1758.

Other Recognized Species Camelus dromedarius Linnaeus, 1758; *C. thomasi* Pomel, 1893.

CAMELUS DROMEDARIUS Linnaeus, 1758
Figures 41.1A–41.1C

Diagnosis One-humped extant camelin differing from the two-humped *C. bactrianus* by a variety of characters including a U-shaped choana (vs. V shaped), several palatine foramina (vs. two pairs), more anteriorly extending palatine (to M2 vs. to M1), horizontal lower orbital border (vs. oblique), lingual crescent of P3 less complete, lingual wall of P4 absent, styles of upper molars stronger, p1 always present, and distal metacarpals and metatarsals less divergent. The two species are able to interbreed, although some first cross males may be sterile (Manefield and Tinson, 1996).

Remarks Although it is now widespread, the fossil history of the Arabian camel or dromedary is virtually unknown. Most of the late Pleistocene camels from Africa are too fragmentary to be identified to species, let alone to a domestic or wild form. A very large late Pleistocene camel from Sudan was referred to *C. thomasi* by Gautier (1966), who assumed that it was more closely related to the bactrian camel, but this relationship was rejected by Peters (1998), with whom we agree, although the differences between the calcanei of both modern species are far less clear-cut than he illustrated (Peters, 1998: figure 1). The very large size of the specimen, and its possible pre-Neolithic age, suggest that it was a wild form; if its age could be definitely established, this camel would be some of the best evidence for derivation of the dromedary from an autochthonous wild form. The specimen from Egypt that Churcher et al. (1999) questionably referred to *C. thomasi* might represent another individual. A large camel, 3 m tall and reputed to be 100,000 years old, was reported from El Kowm in Syria by J.-M. Le Tensorer of the University of Basel (Reuters, October 8, 2006).

As discussed by Köhler-Rollefson (1991), the earliest tentative evidence for domestication of dromedaries comes from an archaeological site on a small island off the Abu Dhabi coast, where camel bones and two stelae depicting dromedaries and dating to about 4,000 BP were unearthed (Hoch, 1979). By about 3,100 years ago, northern Arabian tribes had adopted the dromedary as a riding animal and made use of it on raids. By about 2,900 years ago, dromedary caravans transported incense from southern Arabia to the Mediterranean Sea. Trade along the Silk Route started about 2,000 years ago. Although at first both bactrians and dromedaries were used, the Parthians bred hybrids that proved to be superior beasts of burden (Bulliet, 1975). Rock drawings of dromedaries hunted by horsemen are known from the Arabian Peninsula; dated to about 3,000 years ago, they provide the most recent evidence for wild dromedaries (Köhler, 1981).

CAMELUS THOMASI Pomel, 1893
Figures 41.2A–41.2E

Diagnosis Species of *Camelus* slightly larger than the extant species but with relatively slender metapodials. Lower jaw pachyostotic. Teeth similar to those of the modern forms, but metastylid may be stronger (hence slightly convex lingual walls), and m3 hypoconulid more mesiodistally oriented.

Occurrence Camelus thomasi is known only from the early Pleistocene of Tighenif (Ternifine), Algeria, although comparable remains have been recovered from the mid-Pleistocene Oulad Hamida 1 in Morocco.

Holotype No. 7236001, maxilla from the early Pleistocene locality of Tighenif (Ternifine) figured by Pomel (1893: plate 3, figures 2, 3); housed in the Musée de Géologie, Algiers.

Remarks This species was originally described from maxilla, mandible, and metapodial fragments from Tighenif by Pomel (1893), who interpreted *C. thomasi* as larger than but close to, *C. dromedarius.* Arambourg and Hoffstetter reexcavated the type locality from 1954 through 1956, and their additions to the camelid sample are now housed in the Museum National d'Histoire Naturelle in Paris but are only partly available because of ongoing renovations of the collection facilities. This material has not yet been studied and the following notes are based upon preliminary observations by D. G.

Unfortunately, neither the type nor a second maxilla show the shape of the choanae, one of the best distinguishing

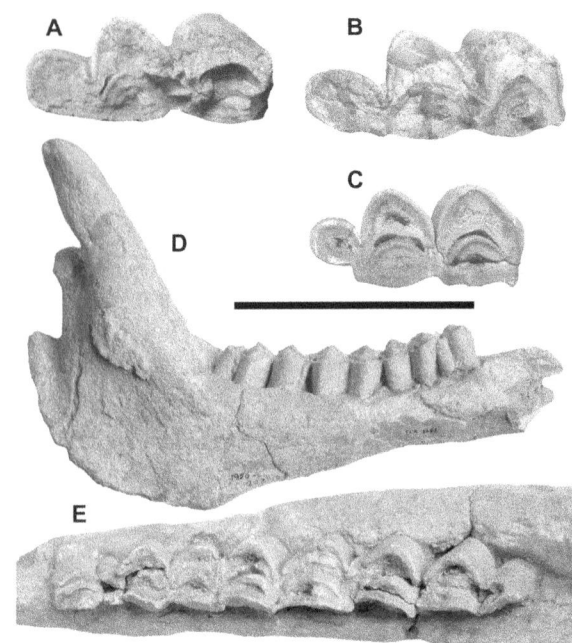

FIGURE 41.2 *Camelus thomasi*, Tighenif (=Ternifine). A) m3 TER-1688 (reversed from the right side); B) m3 TER-1900-27; C) m3 TER-1683; D) mandible TER-1685, lateral view; E) mandible TER-1685, occlusal view of the tooth row. Scale = 50 mm for A–C, 150 mm for D, 75 mm for E. All specimens in Muséum National d'Histoire Naturelle, Paris.

characters between *C. bactrianus* and *C. dromedarius*. On the type, the palatine extends back to the level of M1, but this is not a convincing resemblance with *C. bactrianus* (the palatine extends farther posteriorly in adult *C. dromedarius*) because the specimen is not fully adult.

All mandibular specimens are characterized by pachyostosis of both the corpus and ramus (figure 41.2E); modern camels also have robust lower jaws, but they are never so thick.

The teeth are very similar to those of modern *Camelus*, but on average stylids are better marked, so that the lingual walls of the metaconid and entoconid may appear slightly concave (figure 41.2E). The goat folds and styles of upper molars are also better indicated. On the m3s, the lingual wall of the hypoconulid (which varies in size) is usually in line with that of the entoconid (i.e., mesiodistally oriented), but on one m3, it is labially shifted in respect to the entoconid, with a marked step at the level of the entostylid (figures 41.2A–41.2C). In these features, the m3s definitely resemble the Turkana Basin camel teeth rather than the modern forms. The metapodials are slightly longer than those of the dromedary (themselves longer, on average, than those of the bactrian camel) and rather slender. Detailed study of the postcranials has yet to be performed. On the whole, using the criteria of Steiger (1990), they are more like those of the dromedary, but several features are unlike those of either extant species.

Other fragmentary specimens from northern Sudan were later assigned to *C. thomasi* by Gautier (1966) who, on the basis of the Sudanese but not the Algerian material, interpreted *C. thomasi* as close to, but larger than, *C. bactrianus*—as did Howell et al. (1969). Gautier cited Vaufrey (1955) and Zeuner (1963) as having documented this species at the Mousterian sites of Saint Roche and Sintes in Algeria, the Aterian site of El Guettar in Tunisia, and the Iberomaurusian site of Taza in Morocco. He also mentioned a fragmentary cannon bone collected by Yves Coppens from Bochianga, near Koro Toro, in

Chad (Coppens, 1971). Churcher et al. (1999) reported a camel questionably assigned to *Camelus ?thomasi* from a middle Pleistocene locality in Dakhleh Oasis, Egypt. We have reservations about all these identifications, as they were not supported by morphological comparisons with the Tighenif material. Only the fragment of maxilla from the slightly younger locality of the "Grotte des Rhinocéros" near Casablanca (Raynal et al., 1993) merits an identification as *Camelus* cf. *C. thomasi*.

CAMELUS sp.

Occurrence *Camelus* specimens that cannot be identified to species have been recovered from the early Pliocene (Chad), late Pliocene (Morocco, Ethiopia, Kenya, Tanzania and Malawi), early Pleistocene (Israel, Saudi Arabia and Tanzania), mid-Pleistocene (Egypt and Syria), late Pleistocene (Algeria, Morocco, Sudan, Syria and Tunisia), and Holocene (Israel, Jordan and Oman)(table 41.1).

The late Pliocene locality of Ahl al Oughlam, Morocco (Geraads et al., 1995), yielded a single *Camelus* phalanx that cannot be identified to species.

Only a few specimens of Plio-Pleistocene camelids have been recovered south of the Sahara at localities in Ethiopia, Kenya, Tanzania, and Malawi. These were readily identifiable to genus on the basis of cheek tooth morphology (Gentry and Gentry, 1969), divergent distal metapodials (Howell at al., 1969), incisor morphology (Harris, 1987), and reduction of the premolars (Harris, 1991), but none were sufficiently complete to permit attribution to species.

Gentry and Gentry (1969) reported a damaged upper right molar from the early Pleistocene locality of BKII at Olduvai Gorge I Tanzania and a left lower third molar from the late Pliocene locality of Marsabit Road in Kenya. They interpreted the two fossil teeth to be morphologically identical to the equivalent teeth of *C. dromedarius* but large enough to belong to *C. thomasi*.

Howell et al. (1969) reported a left lower *Camelus* molar from Shungura Member B in the lower Omo Valley, Ethiopia, and a distal right camelid metatarsal from Member F at the same locality that was similar in size to that of *C. bactrianus*. Additional specimens from Shungura Members B, D, F, and G were reported by Grattard et al. (1976) but could only be identified to genus. No camel has been reported from the Middle Awash, and there is no camel in the Hadar Formation of Ethiopia, but an unpublished incomplete m3, probably from Amado, could be of middle Pliocene age.

The southernmost camelid fossil yet described from sub-Saharan Africa is an isolated lower left incisor from the Laetolil Beds at Laetoli in Tanzania, and, at ca. 3.5 Ma, it is also one of the oldest. Although its root is of similar size to that of a *Sivatherium* incisor, its asymmetrical lanceolate shape with typical cameline torsion and the smoothness of its enamel preclude its assignation to anything other than a camel (Harris, 1987).

Harris (1991) reported a premaxilla, mandible and postcranial elements from the Tulu Bor Member of the Koobi Fora Formation (east of Lake Turkana, Kenya) and an isolated molar from the KBS Member. The mandible lacks a p3, indicating that it is *Camelus* rather than *Paracamelus*. The lower cheek teeth are superficially similar to, but appreciably larger than, those of *C. dromedarius*. Two somewhat younger and more fragmentary mandibles were recovered from the upper Lomekwi Member of the Nachukui Formation in the western part of the Lake Turkana basin, one of which was reported by Harris et al. (1988). All the Lake Turkana basin specimens seem to represent a single species.

Camelus sp. was reported in faunal lists from the Chiwondo Beds of Malawi (Schrenk et al., 1993; Bromage et al., 1995), but no published description has yet appeared. In addition to the localities already mentioned, table 41.1 lists camelid remains that have been recovered from Algeria (1–2), Israel (11–13), Jordan (14), Morocco (19), Oman (22), Saudi Arabia (23), and Syria (25).

Dietary Adaptations

Although the hypsodonty of camel teeth has been interpreted as denoting a grazing diet (Made and Morales, 1999), hypsodonty is also a reflection of open habitat (Stirton, 1947; Janis, 1988). Extant camelids are opportunistic and feed on a mixture of grass and other plants or plant parts. The premaxillary shape index of extant bactrian and dromedary camels is consistent with a browsing or intermediate diet, as is that of extinct camelins for which the premaxilla is known (Dompierre and Churcher, 1996). Mesowear analysis of camel teeth showed that, based on occlusal relief and the percentages of sharp versus blunt cusps, *C. dromedarius* clustered with other hypsodont and open-adapted mixed feeders (Fortelius and Solounias, 2000). Stable isotope analysis of extant dromedary teeth from the Lake Turkana basin is consistent with these being C_3 browsers that obtained the majority of their water requirements from their food. Dental enamel from Pliocene camels from Koobi Fora (Harris, 1991) and West Turkana (Harris et al., 1988) has yielded comparable results.

Summary

Camels migrated into Africa during the late Miocene and reached as far south as Tanzania, or perhaps Malawi, but remained rare and ephemeral elements of the East and North African assemblages throughout their known history. Today, the southernmost limit of their distribution is defined by annual precipitation and the incidence of insect-borne disease. Their rarity in the African fossil record may reflect their preference for (and adaptation to) arid habitats, rendering them less likely to be preserved than mesic-adapted species and, perhaps, less able to compete successfully in the kind of periaquatic assemblages that normally preserve as fossils. The earliest arrivals are attributed to *Paracamelus*. Specimens representing one or more large *Camelus* species are known from the late Pliocene onward but their relationship, if any, with either of the extant species remains elusive.

Literature Cited

Arambourg, C. 1932. Note préliminaire sur une nouvelle grotte à ossements des environs d'Alger. *Bulletin de la Société d'Histoire Naturelle d'Afrique du Nord, Alger* 23:154–162.
———. 1979. *Vertébrés Villafranchiens d'Afrique du Nord (Artiodactyles, Carnivores, Primates, Reptiles, Oiseaux)*. Fondation Singer Polignac, Paris, 141 pp.
Bate, D. M. A. 1927. On the animal remains obtained from the Mughare-el-Emireh in 1925; pp. 9–13 in F. Turville-Petre (ed.), *Research in Prehistoric Galilee 1925–26*. British School of Archaeology in Israel, London.
Bromage, T. G., F. Schrenk, and Y. M. Juwayeyi. 1995. Paleobiogeography of the Malawi Rift: Age and vertebrate paleontology of the Chiwondo Beds, northern Malawi. *Journal of Human Evolution* 28:37–57.
Bulliet, R. W. 1975. *The Camel and the Wheel*. Harvard University Press, Cambridge, 352 pp.
Churcher, C. S., M. R. Kleindeinst, and H. P. Schwartz. 1999. Faunal remains from a Middle Pleistocene lacustrine marl in Dakhleh Oasis, Egypt. *Palaeogeography, Palaeoclimatology, Palaeoecology* 124:301–312.

Clutton-Brock, J. 1970. The fossil fauna from an upper Pleistocene site in Jordan. *Journal of Zoology* 162:19–29.

Coppens, Y. 1971. Les vertébrés Villafranchiens de la Tunisie; gisements nouveaux, signification. *Comptes Rendus de l'Académie des Sciences, Paris, Série D*, 273:51–54.

Dompierre, H., and C. S. Churcher. 1996. Premaxillary shape as an indicator of the diet of seven extinct late Cenozoic New World camels. *Journal of Vertebrate Paleontology* 16:141–148.

Flynn, L. J., R. H. Tedford, and Z. Qiu. 1991. Enrichment and stability in the Pliocene mammalian faunas of North China. *Paleobiology* 17:246–280.

Fortelius, M., and N. Solounias. 2000. Functional characterization of ungulate molars using the abrasion-attrition wear gradient: A new method for reconstructing paleodiets. *American Museum Novitates* 3301:1–36.

Gautier, A. 1966. *Camelus thomasi* from the northern Sudan and its bearing on the relationship *C. thomasi–C. bactrianus*. *Journal of Paleontology* 40:1368–1372.

Gentry, A. W., and A. Gentry. 1969. Fossil camels in Kenya and Tanzania. *Nature* 222:898.

Geraads, D. 1986. Les ruminants du Pléistocène d'Oubeidiyeh. *Mémoires et Travaux du Centre de Recherche Français de Jérusalem* 5:143–181.

———. 2002. Plio-Pleistocene mammalian stratigraphy of Atlantic Morocco. *Quaternaire* 13:43–53.

Geraads, D., F. Amani, J.-P. Raynal, and F. Z. Sbihi Alaoui. 1995. La faune de mammifères du Pliocène terminal d'Ahl al Oughlam, Casablanca, Maroc. *Comptes Rendus de l'Académie des Sciences, Paris, Sciences de la Terre et des Planètes* 326:671–677.

Grattard, J. L., F. C. Howell, and Y. Coppens, 1976. Remains of *Camelus* from the Shungura Formation, lower Omo Valley, Ethiopia; pp. 268–274 in Y. Coppens, F. C. Howell, G. L. Isaac, and R. E. Leakey (eds.), *Earliest Man and Environments in the Lake Rudolf Basin*. University of Chicago Press, Chicago.

Haas, G. 1966. *On the Vertebrate Fauna of the Lower Pleistocene Site "Ubeidiya."* Publications of the Israeli Academy of Science and Humanities, 66 pp.

Harris, J. M. 1987. Fossil Giraffidae and Camelidae from Laetoli; pp. 358–377 in M. D. Leakey, and J. M. Harris (eds.), *Laetoli: A Pliocene Site in Northern Tanzania*. Clarendon Press, Oxford.

———. 1991. Family Camelidae; pp. 86–92 in J. M. Harris (ed.), *Koobi Fora Research Project: Volume 3. Geology, Fossil Artiodactyls and Paleoecology*. Clarendon Press, Oxford.

Harris, J. M., F. H. Brown, and M. G. Leakey. 1988. Geology and palaeontology of Plio-Pleistocene localities west of Lake Turkana, Kenya. *Contributions in Science* 399:1–128.

Harrison, J. A. 1979. Revision of the Camelinae (Artiodactyla, Tylopoda) and description of the new genus *Alforjas*. *University of Kansas Paleontological Contributions* 95:1–20.

Hoch, E. 1979. Reflections on prehistoric life at Umm an-Nar (Trucial Oman) based on faunal remains from the third millenium B.C. *South Asian Archaeology* 1977:589–638.

Honey, J. G., J. A. Harrison, D. R. Prothero, and M. S. Stevens. 1998. Camelidae; pp. 439–462 in C. M. Janis, K. M. Scott, and L. L. Jacobs (eds.), *Evolution of Tertiary Mammals of North America: Volume 1. Terrestrial Carnivores, Ungulates, and Ungulate-like Mammals*. Cambridge University Press, New York.

Hooijer, D. A. 1961. Middle Pleistocene mammals from Latamne, Orontes Valley, Syria. *Annales Archéologiques de Syrie* 11:117.

Howell, F. C., L. S. Fichter, and R. Wolff. 1969. Fossil camels in the Omo Beds, southern Ethiopia. *Nature* 223:150–152.

Janis, C. M. 1988. An estimation of tooth volume and hypsodonty indices in ungulate mammals and the correlation of these factors with dietary preference. *Mémoire du Muséum National d'Histoire Naturelle, Paris, Série C*, 53:367–387.

Khavesson, J. 1954. Tertiary camels from the Oriental Hemisphere. *Trudy Palaeontologicheskij Institut, Akademija Nauk SSSR* 47:100–162.

Köhler, I. E. 1981. Zur Domestikation der Kamels. Unpublished Ph.D. dissertation, Veterinary College, Hannover, Germany.

Köhler-Rollefson, I. E. 1991. *Camelus dromedarius*. *Mammalian Species* 375:1–8.

Kostopoulos, D. S. and S. Sen. 1999. Late Pliocene (Villafranchian) mammals from Sarikol Tepe, Ankara, Turkey. *Mitteilungen der Bayerische Staatssammlung für Paläontologie und historische Geologie* 39:165–202.

Kretzoi, M. 1954. Ostrich and camel remains from the central Danube basin. *Actes Géologiques de l'Academie des Sciences Hungary* 2:231–242.

Likius, A., M. Brunet, D. Geraads, and P. Vignaud. 2003. Le plus vieux Camelidae (Mammalia, Artiodactyla) d'Afrique: limite Mio-Pliocene, Tchad. *Bulletin de la Société Géologique de France* 174:187–193.

Macfarlane, W. V. 1977. Survival in an arid land. *Australian Natural History* 29:18–23,

Made, J. van der and J. Morales. 1999. Family Camelidae; pp. 221–224 in G. Rössner, and K. Heissig (eds.), *The Miocene Land Mammals of Europe*. Pfeil, Munich.

Manefield, G. W., and A. H. Tinson. 1996. *Camels: A compendium*. Vade Medicum, series C, No. 22. University of Sidney Post Graduate Foundation in Veterinary Science.

McKenna, M. C., and S. K. Bell. 1997. *Classification of Mammals above the Species Level*. Columbia University Press, New York, 640 pp.

Morales, J., D. Soria, and E. Aguirre 1980. Camélido Finimiocene en Venta del Moro. Primera cita para Europa occidental. *Estudios Geológicos* 26:139–142.

Orlov, J. A. 1968. *In the World of Ancient Animals* [in Russian]. 2nd ed. Nauka, Moscow.

Payne, S. and A. Garrard. 1983. *Camelus* from the upper Pleistocene of Mount Carmel. *Israeli Journal of Archaeological Science* 10:243–247.

Peters, J. 1998. *Camelus thomasi* Pomel, 1893, a possible ancestor of the one-humped camel? *Zeitschrift für Säugetierkunde* 63:372–376.

Pickford, M., J. Morales, and D. Soria. 1995. Fossil camels from the Upper Miocene of Europe: Implications for biogeography and faunal change. *Geobios* 28:641–650.

Pomel, A. 1893. Caméliens et cervidés. *Carte Géologie de l'Algérie, Monographies de Paléontologie* 1:1–52.

Raufi, F. and O. Sickenberg. 1973. Zür Geologie und Paläontologie der Becken von Lagman und Jalalabad. *Geologische Jahrbuch* 3:63–99.

Raynal, J.-P., D. Geraads, L. Magoga, A. El Hajraoui, J.-P. Texier, D. Lefevre, and F.-Z. Sbihi-Alaoui. 1993. La Grotte des Rhinocéros (Carrière Oulad Hamida 1, anciennement Thomas III, Casablanca), nouveau suite acheuléen du Maroc atlantique. *Comptes Rendus de l'Académie des Sciences, Paris, Série II*, 316:1477–1483.

Schlosser, M. 1903. Die fossilen Säugetiere Chinas nebst einer Odontographie der Rezenten Antilopen. *Abhandlungen der Königlichen bayerischen Akademie der Wissenschaften, Mathematisch-naturwissenschaftliche Klasse* 22:1–221.

Schmidt-Nielsen, K. 1964. *Desert Animals: Adaptation and Environment*. Oxford University Press, Oxford, 277pp.

Schrenk, F., T. G. Bromage, C. G. Betzier, U. Ring, and Y. M. Jueayeyi. 1993. Oldest *Homo* and Pliocene biogeography of the Malawi Rift. *Nature* 365:833–835.

Smuts, M. S., and A. J. Bezuidenhout. 1987. *Anatomy of the Dromedary*. Oxford University Press, New York, 244 pp.

Stefanescu, G. 1910. Le Chameau fossile de Roumanie est l'ancêtre des chameaux dromadaires et du chameau sauvage d'Afrique. *Annales du Muséum Géologique et Paléontologique de Bucarest* 4:46–70.

Steiger, C. 1990. Vergleichende morphologische Untersuchungen an Einzelknochen des postkranialen Skeletts der Altweltkamele. Unpublished veterinary medicine dissertation, University of Munich.

Stirton, R. A. 1947. Observations on evolutionary rates in hypsodonty. *Evolution* 1:32–41.

Stromer, E. 1902. Wirbeltierreste aus dem mittleren Pliozän des Natrontales und einige subfossile und rezenten Säugetierreste aus Ägypten. *Zeitschrift der Deutsche Geologische Gesellschaft* 54:108–115.

Teilhard de Chardin, P., and M. Trassaert. 1937. The Pliocene Camelidae, Giraffidae and Cervidae of southeastern Shansi. *Paleontologia Sinica* (n.s.) 1:1–54.

Thomas, H., D. Geraads, D. Janjou, D. Vaslet, A. Memensh, D. Billiou, H. Bocherens, G. Dobigny, V. Eisenmann, M. Gayet, F. de Lapparent de Broin, G. Petter, and M. Halawani. 1999. First Pleistocene faunas from the Arabian peninsula: An Nafud Desert, Saudi Arabia. *Comptes Rendus de l'Académie des Sciences, Paris* 326:145–152.

Thomas, P. 1884. Recherches stratigraphiques et paléontologiques sur quelques formations d'eau douce de l'Algérie. *Mémoires de la Société Géologique de France* (3) 3, 2:1–51.

Vaufrey, R. 1955. Préhistoire d'Afrique: I. Maghreb. *Institut de Haute Études Tunis (Paris) Publications* 4:1–458.

Yagil, R., U. A. Sod-Moriah, and N. Meyerstein. 1974. Dehydration and camel blood. I. The life span of the camel erythrocyte. *Journal of Physiology* 226:298–301.

Zdansky, O. 1926. *Paracamelus gigas* Schlosser. *Palaeontologia Sinica Ser. C*, 4 (4):1–44.

Zeuner, F. E. 1963. *A History of Domesticated Animals*. Harper & Row, New York, 560 pp.

FORTY-TWO

Suoidea

LAURA C. BISHOP

The superfamily Suoidea, to which pigs (Suidae), peccaries (Tayassuidae), and the extinct family Sanitheriidae belong, most likely originated in Europe and Asia during the Eocene. All suoids possess small incisors and have relatively enlarged canines (particularly the uppers) that are usually convex laterally. Their first premolars are triangular in shape, while the second, third, and fourth premolars are more molariform. The suoids are the sister group of all other Artiodactyls, including Hippopotamidae (Boisserie et al., 2005).

There were several movements of suoids from Eurasia; the earliest detectable one left identifiable pigs behind in Africa by the early Miocene. Two of the families of suoids will be discussed here, the Sanitheriidae (now extinct) and the Suidae, which were the dominant family of suoids in Africa during the Cenozoic. The Tayassuidae are an exclusively American family, although early research originally attributed some small African suid specimens to this family (Hendey, 1976; Made, 1997).

The true pigs, or Suidae, are far more common and are widespread through both time and space. The earliest known pigs are from Asia—the upper Eocene genera *Siamochoerus* from Thailand (Ducrocq et al., 1998) and *Oidochoerus* from China (Tong and Zhao, 1986). Pigs do not make their way to Africa until much later. The earliest true Suidae known from Africa are the Kubanochoerinae, which make their first appearance at the beginning of the Miocene in Namibia, in deposits that are approximately 21 Ma (Pickford, 1986, et seq.; Pickford, 2006). Subsequent appearances of the other suid subfamilies results from evolutionary change from this original founder sounder (group) as well as from a series of later migrations from Eurasia.

Pigs are medium-sized mammals. The largest living African suid weighs about 220 kg; skeletal evidence suggests that some extinct forms were much larger. Everyone has a platonic ideal of a pig—usually large, pinky white, with a corkscrew tail—but this domesticated Porky Pig look-alike has come a long way from the wild swine of the geological past. In general, pigs are ectomorphic animals with thick, torpedo-shaped bodies, relatively short legs and a short tail. The entire group has short, thick necks supporting more or less robust skulls, which are long, large, and flat on the dorsal surface. They have large snouts with a moist rhinarium,

supported by a terminal cartilaginous disk that is pierced by their nasal openings. There is a prenasal bone that supports this snout as well. The flat back of the skull has a pronounced occipital crest.

The eyes are framed by small, dorsally placed orbits; pigs in life have more use for their sense of smell than for their relatively poor eyesight. The external, fleshy ears of pigs are usually small and pointed; however, these ears are very mobile (and expressive), and pigs' sense of hearing is quite good. The primitive dental formula for pigs is (I3/3 C1/1 P4/4 M3/3), although there is considerable variation in this formula, particularly in more derived pigs. The upper central incisors are usually considerably larger than their lateral neighbors. All upper incisors are often curved toward the midline, with diagonally emplaced roots, giving isolated teeth a characteristically clubbed appearance. The lower incisors are usually emplaced in a strong V shape across a fused mandibular symphysis, oriented slightly procumbently. These teeth are usually long, straight, and narrow. Short incisors are probably primitive for pigs, with longer ones more derived in species that rootle for food (Made, 1996).

Pig canines are continuously growing through life, and pig canine size and to a lesser extent shape can be quite considerably sexually dimorphic, with males having larger and sometimes more robust and ornate canines. In fact, pig body size can also be sexually dimorphic. In some modern species the male can be 50% larger than females of the species (*Phacochoerus*, the warthog), while in the other two extant African genera there is sexual dimorphism in form if not obviously in body size. There are indications that sexual dimorphism was pronounced in some fossil species. Upper canines project laterally and posteriorly as curved "tusks"; they can be honed against the lower canines in occlusion, and sharp edges and points can result from this.

The first and second molars in suids are bunodont (rounded) or cuspidate, with pointed cusps. They are not in general high crowned, except in the third (and, very occasionally, the second) upper and lower molars of highly derived forms. The third molars, both upper and lower, have numerous cusps; it is the elaboration of this tooth in length, crown height, and number of cusps that forms the basis of biostratigraphic analyses for Pliocene and Pleistocene suids.

821

Unlike their fellow artiodactyls the Selenodontia, in pigs the metapodials are generally separate bones. There is also reduction of the digits and metapodials. Digit I is absent from both the forefoot (manus) and the hindfoot (pes). All remaining digits (II–V) terminate in ungules or hooves. Although pigs retain four digits, in modern pigs only the central two (digits III and IV) and their articulating metapodials (III and IV) truly function in locomotion; the other digits are not weight bearing.

Pig skin is thick and tough with relatively sparse, coarse hair. Their stomachs are relatively simple among artiodactyls, having only two chambers—they do not ruminate. In general, pigs have omnivorous diets; however, the modern warthog has a very high proportion of grass (not only leaves but seeds) in its diet, and there are indications from isotopic analyses of tooth enamel to suggest that some extinct species may have eaten grass almost exclusively (Harris and Cerling, 2002). With the exception of the warthog, pigs today are primarily nocturnal. Pigs and their close allies are unusual among the ungulates in giving birth to litters rather than to individual offspring. The individual offspring are altricial but develop very quickly.

Pig taxonomy is bewildering in its complexity. This is at least partially because the evolution of suids through time is characterized by rapid evolution and involved numerous parallel developments through time and space. Earlier suid evolution was rationalized, in the first instance, by Wilkinson (Wilkinson, 1976; Cooke and Wilkinson, 1978). Later efforts have also lead to simplified, if conflicting, perspectives on the Miocene suids, which largely retain species and reorganize on the higher taxonomic levels (Pickford, 1986, et seq.; Made, 1996; Liu, 2003). Made (1996) suggests that some of the taxa that Pickford (1995, et seq.) recognizes as subfamilially distinct (e.g., the Namachoerinae) are best considered as examples of parallel evolution within the Lopholistriodontinae, the subfamily erected by the former.

African suids from the latest Miocene into the Pleistocene (and those who study them) have benefited from a series of revisions and rationalizations, starting with the work of Cooke, who was principally responsible for the recognition of later suid evolution as a biostratigraphic and correlative indicator across the continent (Cooke, 1976, et seq.). Subsequent work by Harris and White (White and Harris, 1977; Harris and White, 1979) revised the taxa further and removed some of the chronological and geographic species that Cooke retained. However, further discoveries and the continuing field research of several groups has lead to resurrection of some of the species sunk by Harris and White, and also the naming of several new species in recent years.

Pigs, which have changed relatively rapidly in evolutionary time, are a valuable biostratigraphic indicator and are frequently used in the field to approximate the age of fossil deposits. In-depth analysis of the taxonomic identity and morphology of fossil pigs has proven more reliable than initial reports of radiometric "absolute" dates for fossil deposits in several cases. Perhaps the most famous of these was the KBS tuff controversy, centering around fossil deposits from the famous Koobi Fora sites in Kenya. This episode is particularly notorious in palaeoanthropology because the age of a new and different hominid skull, KNM-ER 1470, which had a particularly large brain size, hinged on the date of this tuff. Work by the eminent palaeontologist H. B. S. Cooke (1976) showed that the faunal context of the skull correlated well with fossils from 1.8-Ma deposits from the Omo, in Ethiopia,

contradicting an initial radiometric potassium-argon date from the overlying tuff of 2.6 Ma. Subsequent redating of the tuff showed that the faunal "age" determined by Cooke was completely accurate. This biostratigraphic correlation was probably facilitated by the fact that the Omo and Koobi Fora deposits were both part of the same basin during the late Pliocene and early Pleistocene; geographically the deposits are not that far away from one another.

Modern African Pigs

The classification of the living African suids is a topic of some debate in neontology. Although traditionally considered as monospecific, the genera of living pigs have been subdivided into several species, primarily for reasons of conservation (Oliver, 1993). Dividing the suid genera into several species (and subspecies) relies on population variations in characters (primarily pelage) invisible in the fossil record, which has interesting ramifications for the taxonomy of fossil suids. Worldwide, there are five genera of true pigs (or six, if the pygmy hog *Sus salvanius* is put in its own genus, *Porcula*). *Sus* is found throughout Asia, Europe, and North Africa and elsewhere when it is a domesticate or introduced species. It is the most speciose genus of living pigs. The one species of the genus *Babyrousa* is found only on Sulawesi and nearby islands. There are at present three indigenous suid genera in sub-Saharan Africa.

HYLOCHOERUS: THE (GIANT) FOREST HOG

Hylochoerus meinertzhageni, the giant forest hog, is the largest African suid. While there are differences of opinion as to whether the three genera of African suid are monospecific, all authorities seem to agree that there is only one species of *Hylochoerus*, *Hylochoerus meinertzhageni* Thomas, 1904 (Sjarmidi and Gerard, 1988). This taxon is distributed patchily throughout central Africa (e.g., Kingdon, 1979; d'Huart, 1978, 1993; Grubb, 1993). A recent revision of the taxonomy of living suids concludes that there are three valid subspecies of *Hylochoerus*: the nominate *H. m. meinertzhageni* or giant forest hog, *H. m. rimator*, the Congo forest hog and *H. m. ivoriensis*, the West African forest hog (Grubb, 1993). These subspecies are geographic, and the status of populations in Tanzania, Sudan, and Ethiopia is as yet unknown. It is acknowledged that there is an east-west cline in body size in this species (d'Huart, 1978, 1993) but, nonetheless, skull size is the primary differentiating character presented for separation of these subspecies. There are also differences in the distribution and density of hair accorded the different subspecies.

Exactly how giant this pig is has been a matter of some debate; some authors express wonderment at the animals' great size (Edmond-Blanc, 1960), while others merely accuse "the natives" and each other of exaggeration (Rothschild and Neuville, 1906). Any generalizations are confused by the east-west cline in body size; *H. m. ivoriensis*, the diminutive western subspecies, has a maximum recorded weight of 150 kg (Rode, 1944 in d'Huart, 1993). Only the easternmost races, with recorded body weights of up to 220 kg, can be truly considered "giant" (d'Huart, 1978, 1991, 1993). That an animal of this size remained undescribed until 1904 is testimony to the density of the tropical forest it inhabits in much of its range. Although it may become partly diurnal where it is unharassed, the forest hog is generally nocturnal and shy (d'Huart, 1978, 1993).

The patchy distribution of *H. meinertzhageni* is best attributed to the patchy distribution of its preferred habitats throughout Africa (Sjarmidi and Gerard, 1988). The forest hog ranges in a variety of forested ecotones, including bush and thickets, woodland savannas, gallery forests, lowland humid forests (including marshy areas), secondary forests, escarpment forests, lowland and montane dry forest, montane mosaics, and altitudinous montane forests (d'Huart, 1978). This testifies to a certain adaptability; however, in East Africa, the species is exclusively found in forests, which are generally confined to high altitudes there (Thomas, 1904; Rothschild and Neuville, 1906; Edmond-Blanc, 1960; Stewart and Stewart, 1963; Sale et al., 1976; Kingdon, 1979). Aspects of its eye morphology are well suited to a forest habitat, indicating a long-standing adaptation to this zone (Luck, 1965).

Throughout the range of potential habitats, it is most likely to be living where there is a variety of vegetation, a permanent source of water, and thick understory cover in a portion of its home range (d'Huart, 1978). d'Huart considers it to be an ecotonic species, which thrives in the gradations between habitat types, where there is maximum "edge effect" (d'Huart, 1993). *Hylochoerus* makes runs in vegetation throughout its range but is not known to use underground burrows (Copley, 1949).

Our understanding of this taxon is very incomplete: only one full-length study has been undertaken, and this on a population which spent considerable time on the savanna surrounding the Parc National des Virunga (d'Huart, 1978). Most information on the species behavior and ecology in the forest is anecdotal. Few individuals have survived in captivity; no examples are currently held in zoos (Leister, 1939; d'Huart, 1993).

There appears to be some measure of variety in the diets and habits of the different *H. meinertzhageni* subspecies. It browses nocturnally and diurnally on soft vegetation, grass, fallen fruits, berries, and roots (Ewer, 1970; Dorst and Dandelot, 1972; Cooke, 1976). Its forest foods can be considered to be far softer and less abrasive than the diet of the warthog (d'Huart, 1978). It is questionable whether *H. meinertzhageni* excavates vegetable foods. The soft rhinarium of the forest hog would seem to preclude using the snout to dig (Ewer, 1970). Some authorities say it does not dig (Copley, 1949), while others assert that it is a powerful digger (Leister, 1970). Excavations, which are characterized as "scrapes," are probably made by using the lower incisors, which show unusual wear (Ewer, 1970). At any rate, excavation of food is not a frequent occupation of *H. meinertzhageni*, which sets it apart from other living pigs (d'Huart, 1978).

Animal foods are also consumed by the forest hog; they have been observed eating carcasses of reptiles and mammals, and the eggs of ground-nesting birds (d'Huart, 1978). It has been suggested that the dentition of the forest hog is more suitable to eating insects than that of other pigs (Ewer, 1970; d'Huart, 1978). They have been observed turning over rotten logs, perhaps in search of insects, and eating maggots (Ewer, 1970). Insect remains have also been found in the feces of *H. meinertzhageni*, Coprophagy is uncommon in the forest hog, but they have been observed eating elephant feces (d'Huart, 1978). Salt licks are also important to some forest hogs, particularly in the Parc National des Virunga, and soil that has been found in fecal samples is attributed to eating at salt licks (d'Huart, 1978).

Although the fossil evidence for the origin of this species is very scant, some specimens from Behanga, Uganda, are found in a biostratigraphic context, suggesting an early Pleistocene first appearance for this taxon (Pickford, 1994). Similar material found by Leakey at Kanjera, Kenya, was initially assigned to a different species, *Hylochoerus antiquus*, subsequently sunk. However, this material may also support a relatively early origin for this taxon, which otherwise seems to arise from the *Kolpochoerus* lineage.

PHACOCHOERUS: THE WARTHOG

Phacochoerus aethiopicus, the warthog, is the most commonly seen Afrotropical suid, a fact that is attributable to its visibility in the open habitats it favors and its diurnal habits (Tookey, 1959). Warthogs are characterized by a large head, longish limbs, and a short neck. Their characteristic feeding posture is a sort of "kneel"; the animals rest on their carpometacarpal joint (Cumming, 1975; Kingdon, 1979). Animals are capable of walking along on their wrists from patch to patch. This stance characterizes most *Phacochoerus* feeding. Cumming (1975) considers this animal to be specialized for arid habitats, where its ability to rootle for underground resources allows it to compete successfully with ruminants. Warthog are sexually dimorphic and possess many specializations for grazing in the open savanna zones (Luck, 1965; Delaney and Happold, 1979). Warthog are rare among ungulates in having a clearly divided day/night activity pattern (Leuthold, 1977). Their 12-hour work day corresponds with the daylight hours (Clough, 1969), although they have been observed grazing on moonlit nights (Copley, 1949).

The number of species and subspecies comprising the genus *Phacochoerus* Cuvier, 1817 is the subject of a lively neontological debate. Some authors consider there to be only one species of living warthog, *Phacochoerus aethiopicus* (Gmelin, 1788)(e.g., Dorst and Dandelot, 1972; Kingdon, 1979; Sjarmidi and Gerard, 1988). Although this view holds that there are no species-level differences between the various geographical morphs of *Phacochoerus*, numerous regional subspecies are acknowledged (e.g., Sjarmidi and Gerard, 1988). This classification is disputed by some authorities (Ewer, 1957; Grubb, 1993) and others who believe there is evidence of two distinct warthog species—if not living, then recently extinct. This view is common in paleontology (e.g. Ewer, 1958; Cooke and Wilkinson, 1978). It has a complicated history linked to the recent extinction of the Cape warthog, for which the taxon *Ph. aethiopicus* was named by Pallas (1766). This form was never fully studied in life, nor was its geographic range determined, and no specimens have been obtained since the mid-19th century (Grubb, 1993). Following its local extinction, the designation was applied to all warthogs, and the species *Ph. africanus* (Gmelin, 1788), or common warthog, was collapsed into it. Evidence from mitochondrial DNA and extensive studies of crania from museum collections suggest that the original range of *Ph. aethiopicus sensu strictu* was discontinuous and that a relict population survives in the arid portions of eastern Africa (Lönnberg, 1908; Grubb, 1993; d'Huart and Grubb, 2001).

The characteristics that are said to distinguish the two species are dental and cranial (Lönnberg, 1908). The most obvious difference is the absence of upper incisors in specimens attributed to *Ph. aethiopicus*. *Ph. aethiopicus* is also considered to be smaller than the common warthog and to have a broader, more specialized cranium. The difference with most significance for paleontology is in the M3. The anterior roots of the M3 in *Ph. africanus* form early, preventing any further height increase along the anterior border of the tooth before the entire surface of the tooth is in occlusion. In the desert form, the third molar roots remain unfused and growing until the entire tooth is in occlusion (Ewer, 1958; Grubb, 1993).

These characteristics are tantalizing hints of diversity in the genus *Phacochoerus*. However, it might be premature to give Cape and common warthogs separate specific status. This is especially true given several potential stumbling blocks to the recognition of species from osteological collections made in the last century, including the lack of adequate geographical provenance. There is a tremendous amount of variability in the dental formulae of warthogs (Shaw, 1939; Child et al., 1965). Many warthogs lose their incisors early in life (Tookey, 1959). Additionally, there is evidence in the literature that the presence or absence of incisors in the common warthog is an unstable characteristic due not only to attrition but also to phenotypic variability. For example, Cumming (1975) reports that a tame warthog he studied had no upper incisors and that this had no detectable functional correlate. Furthermore, another member of the same litter had its full complement of incisors. That warthog dentition and cranial characteristics are notoriously variable is attested to repeatedly in the literature (see especially Shaw, 1939).

It is true that *Ph. aethiopicus* and *Ph. africanus* were considered separate species in the past, but many variants that are widely accepted as subspecies today were also accorded specific status by early authors (Lönnberg, 1908; Ansell, 1971). The significance of any populational differences in the context of overall suid variation must be considered.

The warthog is considered to be primarily a savanna animal, preferring the plains and thorn scrub and avoiding forest (Copley, 1949; Cumming, 1975; Kingdon, 1979). They inhabit woodland to arid areas (Sale et al., 1976). In southern Africa, a range of habitats is occupied in different densities, with seasonal differences (Cumming, 1975).

Warthog are unique among ungulates in their dependence on subterranean burrows (Kingdon, 1979). Holes are an important part of warthog home ranges, along with open water and adequate food (Cumming, 1975). These shelters are used as refuge from harassment, predators, weather, and also for sleeping and giving birth (Clough and Hassam, 1970; Bradley, 1971). Lions have been observed attempting to unearth warthogs from their burrows (Bradley, 1971). Underground burrows also maintain temperatures that are stable compared to the often violent daily fluctuations on the savanna; one study found that temperatures within burrows varied only 3°C as compared with 8°C externally (Bradley, 1971). Since warthog juveniles are extremely poor thermoregulators, the climatic stability of the holes may be necessary to their survival (Fradrich, 1965; Sowls and Phelps, 1966; Bradley, 1971). How *Phacochoerus* comes upon these holes is contested in the literature. Some authors consider the warthog to be a capable digger of its own holes (Geigy, 1955; Clough, 1969) or modifier of the holes of the aardvark (Bradley, 1971). Others consider the warthog to be dependent on hyena, aardvark, and porcupine burrows for shelter (Thomas and Kolbe, 1942; Copley, 1949). In eastern Africa, burrows are usually located near the top of watersheds or in the area of *Acacia drepanolobium* (Bradley, 1971). Bradley suggests three reasons for this. First, higher ground is less susceptible to flooding. Second, since the black cotton soil is more compressed in these areas, it is more difficult for predators to dig out a warthog. Third, it may be more difficult to excavate a burrow of sufficient size in the plains areas.

Warthog diets are markedly seasonal; they eat primarily grass leaves during the wet season and, in the dry season, dig for grass rhizomes while "kneeling" on their carpals (Cumming, 1975). The few comprehensive studies on diet in the warthog reveal that this animal specializes in the exploitation at all stages of the grass plant, while remaining open to the possibility of locally or seasonally available non-grass resources (Cumming, 1975). The proportion of different grass plant parts varies throughout the range: in eastern Africa, warthogs concentrate on leaves rather than underground rhizomes, as in southern Africa (Fradrich, 1965). In western Africa, however, *Ph. aethiopicus* is reported to be more omnivorous, eating roots of aquatic and land plants as well as grasses (Bigourdan, 1948; Ewer, 1958). In parts of Tanzania where it raids unripe rice crops, *Phacochoerus* has become a pest (Geigy, 1955).

The extent to which *Ph. aethiopicus* excavates vegetable foods is reported to differ seasonally (Cumming, 1975). The prevalence of rootling also differs geographically throughout the considerable geographic range of these animals. Rootling is nonetheless an important method of food procurement (Leister, 1939; Cumming, 1975; *contra* Ewer, 1958). Cumming (1975) describes the characteristic marking of the warthog rootle as a shallow and wide depression. Rhinarium digging by warthog in his study was primarily a cold, dry season activity, so it is not surprising that the animals would excavate hard, baked soil in addition to softer earth. Warthog rootling has only been observed to uncover grass rhizomes and the roots of other monocotyledons (Cumming, 1975; Leuthold, 1977). Warthogs also eat dicotyledonous plants, although apparently not as often. There is also anecdotal evidence that warthogs will opportunistically consume animal foods. Tame free-ranging warthog would sometimes eat raw meat offered to them (Cumming, 1975). A litter of rats found in a grain store was consumed by one female warthog, while another ignored them (Cumming, 1975). Small animals and large animal carcasses are also consumed; in one case a warthog group made forays on three successive days to feed on a carcass (Geigy, 1955; Cumming, 1975).

The antiquity of the warthog is unclear. There is fossil evidence for its presence in eastern Africa from at least the middle Pleistocene of Uganda (Pickford, 1994).

POTAMOCHOERUS: THE BUSHPIG OR RED RIVER HOG

The bushpig *Potamochoerus* is the smallest indigenous African pig. Current thinking on the genus *Potamochoerus* Gray, 1854, is leaning toward the presence of two forms, which are considered as separate species by some workers (Grubb, 1993) and as subspecies by others (Sjarmidi and Gerard, 1988). In the neontological literature, there is support for the view that *Potamochoerus* is a strictly monospecific genus (see Stewart and Stewart, 1963; Ansell 1971). The two species view holds that *Potamochoerus porcus* (Linnaeus, 1758), the red river hog, is restricted to the gallery forests and humid forests of western Africa. *Potamochoerus larvatus*, the bushpig, is found in eastern and southern Africa, where it occupies drier forest and savanna woodland. *Po. porcus* and *Po. larvatus* are also indicated to be parapatric throughout their ranges, although this might not be the case in the Aberdare National Forest in Kenya or in the Kibale forest of Uganda (Ghilgieri et al., 1982). Indeed, there is potential for contact between the two forms throughout much of their range (d'Huart, 1993; Grubb, 1993).

In one classification scheme that gives *Po. porcus* and *Po. larvatus* specific status, no subspecies of *Po. porcus* are recognized (Grubb, 1993). Instead, all formerly recognized subspecies are collapsed into one highly variable species (Grubb, 1993). In the same scheme, *Po. larvatus* is divided into five subspecies, two of which, *Po. l. larvatus* and *Po. l. hova*, are endemic to Madagascar. The other three proposed subspecies of *Po. larvatus* are geographic subspecies. The alternative to this is the sinking of all named subspecies into the species with priority, *Potamochoerus porcus* (Ansell, 1971).

The differences between these taxa are primarily based on details of coat color, which ranges from bright red with black "spectacles" and white dorsal stripe, to black or gray with white face (Forsyth Major, 1897; Grubb, 1993). There appears to be a geographic gradient in coloration, as well as substantial variation in coloration within individual populations. Ghiglieri et al. (1982) state that in the Kibale forest of Uganda, 49% have a *"porcus"* coat, 38% are *"larvatus,"* 7% are intermediate, and 6% are a blond phenotype that may be a hybrid. The authors hypothesize that *larvatus* and *porcus* may be separated by pre- rather than postzygotic mechanisms (Dobzhansky, 1937). However, they suggest that work with captive hybrids is necessary to determine whether hybrids are themselves capable of reproduction. "Morphological differences" have been cited as distinguishing *larvatus* and *porcus* on a specific level (Grubb, 1993). However, the only evidence presented for morphological difference other than coat color is overall skull size. Since available sample sizes are small, variability large, and there is potential for geographic size clines, differences in skull dimensions in the absence of discrete morphological differences will not be treated as significant here.

The bushpig is the most vividly colored of all pigs (Bigourdan, 1948). It occurs all over sub-Saharan Africa wherever the country is suitably dense and broken (Copley, 1949; Maberly, 1966; Cooke, 1976; Kingdon, 1979). In eastern Africa, these habitats include highland forest to 3,700 m above sea level, tree-grassland, riverine woodland, bush, and coastal bush (Stewart and Stewart, 1963; Sale et al., 1976). Favored habitats in southern Africa are similar, consisting of patches of forest or thicket, dense reedbeds, or rocky and well-wooded ravines (Maberly, 1966). A permanent source of water is necessary for the presence of bushpigs, and they are able to swim across even large rivers (Bigourdan, 1948; Maberly, 1966).

Potamochoerus is a nocturnal or crepuscular feeder, ranging into more open areas for its preferred foods (Copley, 1949; Maberly, 1950; Dorst and Dandelot, 1972; Cooke, 1976). Several morphological characteristics of the eye suggest that nocturnality is a recent habit for *Potamochoerus* (Luck, 1965). This is supported by its more diurnal habits in areas where it is undisturbed (Maberly, 1950; Scotcher, 1973; Kingdon, 1979). It is thus possible that nocturnality in this creature is a relatively recent adjustment to harassment by humans, which also enables it to raid crops more conveniently.

Potamochoerus occasionally uses underground burrows of hyenas and more often makes "tunnels" through dense undergrowth (Copley, 1949). These tunnels are used as refuge from their main predator, the leopard, for sleeping, and for nesting during farrowing (Copley, 1949). Bushpig juveniles are better thermoregulators than warthogs, which may explain their lesser reliance on shelter (Sowls and Phelps, 1966). Their denser body hair may also play a role in their thermal efficiency (Sowls and Phelps, 1966).

Although this pig is cosmopolitan in its distribution throughout Africa, surprisingly few studies have been made of it. Accounts of its diet and habits remain largely anecdotal despite its reviled and persecuted status as a devastating crop raider in sub-Saharan Africa (Maberly, 1950, 1966; Skinner et al., 1976). The principal vegetable foods of the bushpig in southern Africa are wild fruits, roots, tubers, rhizomes of forest and swamp ferns, mushrooms, grass, wild arum lilies, other bulbs, and the bark of some trees (Thomas and Kolbe, 1942; Maberly, 1966). Fallen figs are a particularly popular food (Maberly, 1950). In eastern Africa, roots, berries, and wild fruit form the main vegetable components of its diet (Copley, 1949). In western Africa, they favor fruits and leaves, only turning to crops when pickings are slim in the forest (Bigourdan, 1948). Bushpig are accused of destroying a staggering variety of domestic crops, including but not limited to sweet potato, maize, and papaya (Phillips, 1926; Thomas and Kolbe, 1942; Copley, 1949; Maberly, 1950; Sowls and Phelps, 1968; Milstein, 1971; Dorst and Dandelot, 1972; Skinner et al., 1976). During drought, bushpig eat succulents back to the point from which they cannot recover (Skinner et al., 1976).

Bushpig are inveterate rootlers, capable of plowing up and destroying the macadam bottom of cages in captivity (Leister, 1939). In the wild, they are ceaselessly excavating foods with their snouts and tusks, a task for which their rhinaria are ideally suited (Maberly, 1950; Ewer, 1958, 1970; Deane, 1962). They are capable of turning over the soil to a depth of 40 cm to obtain mainly buried seeds, roots, rhizomes, and insects (Phillips, 1926; Thomas and Kolbe, 1942; Skinner et al., 1976). Rootling has a positive outcome for the forest; soil is aerated, and seeds are passed in droppings and sown (Phillips, 1926; Maberly, 1950). As *Po. porcus* is characterized as the most omnivorous of an omnivorous bunch, it is no surprise that reports of animal foods in their diet are legion (Maberly, 1966; Cumming, 1975; Skinner et al., 1976). They will eat insects, reptiles, bird's eggs, young birds and mammals, disabled small antelope, carrion, domestic livestock, and poultry (Phillips, 1926; Thomas and Kolbe, 1942; Copley, 1949; Maberly, 1950; Milstein, 1971; Breytenbach, unpub. obs. in Skinner et al., 1971; Skinner et al., 1976).

There are also interesting polyspecific relations associated with bushpig feeding and habits. In southern Africa, chacma baboons and grey vervets follow rootling bushpigs in order to eat "choice tidbits" ignored by the swine (Phillips, 1926; Maberly, 1950). The association with *Cercopithecus aethiops* is mutually profitable; bushpig follow vervets to eat tree fruit, usually inaccessible to the pigs, which the vervets discard after eating only a few bites (Skinner et al., 1976; Durrell, 1993). *Potamochoerus* also follow the paths of the Knyssa elephant in southern Africa, eating their feces as well as plant and insect material dislodged by the elephants (Phillips, 1926).

It bears reiterating that the characters used to separate subspecies of the living Afrotropical suids are essentially invisible to paleontologists. The coat color of the possessors of isolated fossil teeth is not preserved. Even in the case of the proposed division between *Phacochoerus africanus* and *Ph. aethiopicus*, where there are postulated differences in anterior dentition (Grubb, 1993), it would be difficult to discern a difference between fossil specimens, since premaxillae are not particularly common in the fossil record. The temporal and geographic constraints imposed on paleontological research by limited fossil exposures usually preclude the consideration of even geographic variation. Moreover, within the constraints of the fossil record, it is difficult to separate temporal variation from geographic.

The geographic "species" discussed here do not usually exhibit ecological differences greater than those that are present within the habitat range of any one subspecies.

MODERN SUID ECOLOGY AND DISTRIBUTIONS

The distribution of suids throughout sub-Saharan Africa today is at best an approximation of what it was two or three centuries ago. All wild suids have been found to be susceptible to swine fever, a fact that has certainly affected their distributions and population sizes during historic times (Thomas, 1904; Tookey, 1959). *Hylochoerus* was perhaps more numerous before the European-imported bovine plague of 1890 (Thomas, 1904; Bigourdan, 1948). Warthogs have been systematically exterminated to eliminate them as a reservoir for tsetse-borne trypanosomiasis (Child et al., 1965, 1968). Bushpigs have been persecuted by farmers to prevent crop destruction by these persistent raiders (Thomas and Kolbe, 1942; Maberly, 1950). Habitat destruction and hunting for city "bush meat" markets are also threatening the survival of African suids (d'Huart, 1993).

The habitats and distributions of living suid taxa are largely constrained to particular habitats; they appear to range together only when these habitats are in close proximity (Behrensmeyer, 1975; Kingdon, 1979; Delaney and Happold, 1979). There appear to be few cases of habitat or resource overlap in modern suids. *Potamochoerus* is listed in some sources as sympatric with *Hylochoerus*; but there is never a question of their exploiting similar resources (Sjarmidi and Gerard, 1988). Apparently overlapping geographic ranges of suids are a function of scale; habitats favorable to different species are often interspersed on a level too small to be represented on a map (Thomas and Kolbe, 1942).

While specializing in a particular habitat or vegetational type, suids retain the ability to successfully exploit a wider range of habitats than those most favored. In some rare examples, one species of pig colonizes an area usually inhabited by another species. The invasive form assumes habits typically associated with the pigs more commonly found in that area (Bigourdan, 1949; d'Huart, 1978). The two cases mentioned in the literature, where forest hogs exploit savanna grasses and where warthogs eat soft roots and tubers, may be historical accidents. Population densities readily reveal which habitat is preferred by these animals; for *Hylochoerus*, the forest figure is 50 times that on the savanna (d'Huart, 1978). Another possibility is that population pressure in the primary habitat causes this and other cases of ecological expansionism.

Suids, especially juveniles, are intolerant of temperature extremes and high winds, and require standing water (Thomas and Kolbe, 1942; Sowls and Phelps, 1966). In historical times, the East African Plio-Pleistocene paleontological sites have fallen within the current range of a single taxon, *Phacochoerus*. In southern Africa, some palaeontological sites are currently in the range of both bushpigs and warthog. In the past these regions were host to many suid paleospecies at any one time (Cooke and Wilkinson, 1978; Harris and White, 1979). Changes in the environments of Africa since the Pliocene are reflected in the evolution and distribution changes of the Suidae.

Systematic Paleontology

What follows summarizes the available material on the African fossil suoid record. This chapter follows the higher taxonomy (to subfamily) of Harris and Liu (2008) but enumerates more species than in that work.

Order ARTIODACTYLA Owen, 1848
Superfamily SUOIDEA Gray, 1821
Family SUIDAE Gray, 1821
Subfamily LISTRIODONTINAE Simpson, 1945

Listriodontinae have incisors with low crowns and I1s that are oriented transversely and straight (as opposed to placement in a V shape; Made, 1996). Postcanine teeth in this subfamily are lophodont or nearly lophodont. Although Made (1996) includes some more bunodont taxa in his Listriodontinae, in this summary those taxa are considered Kubanochoerinae, where their cingula and pyramidal, bunodont molar cusps are a uniting feature. Listriodontinae replaced Kubanochoerinae in eastern Africa during the later part of the middle Miocene.

Genus LISTRIODON von Meyer, 1846
LISTRIODON AKATIKUBAS Wilkinson, 1976

Holotype KNM-MG 2, left m3 from Mbagathi, Kenya.
Distribution Middle Miocene of Kenya (including Maboko, Fort Ternan, Nyakach) and Democratic Republic of the Congo (Sinda).
Remarks *Listriodon akatikubas* is very similar to the slightly smaller Eurasian *Listriodon splendens* and distinguished chiefly by virtue of the labial cingula on the molars of the former taxon. Also, the loph structure is imperfect and the ridges connecting the cusps are relatively large and bunodont; even when the teeth are worn the separate cusps of the molars are visible (Cooke and Wilkinson, 1978). The m3 talonid is asymmetrical, and the dental enamel is smooth and without crenulations. The teeth have no accessory tubercles and weak or absent cingula. This species is slightly larger than *Listriodon akatidogus*.

LISTRIODON AKATIDOGUS Wilkinson, 1976

Holotype KNM-MG 9, left M3 from Mbagathi, Kenya.
Distribution Middle Miocene, Kenya.
Remarks Somewhat smaller than *Listriodon akatikubas*, this pig otherwise resembles it in morphology. P3 has a more strongly developed cingulum that encircles the tooth. The dental enamel is relatively thick. Upper molars can flare considerably, and the talonid of m3 is better developed. Pickford (2007) questions the assignment of this species to the genus *Listriodon*.

LISTRIODON BARTULENSIS Pickford, 2001

Holotype Mandible recovered from Ngorora Fm.
Distribution Middle Miocene of Kenya.

LISTRIODON JUBA Ginsburg, 1977

Holotype Bml 187, right p3 from Beni Mellal, Morocco.
Distribution Middle Miocene of Morocco.
Remarks There is little known about the two aforementioned taxa, and they are included here for completeness. Made (1992) places *Listriodon juba* within the genus *Lopholistriodon* and has it as ancestral to *Lo. kidogosana*, indicating trends in increased crown height and decreased body size.

Genus LOPHOLISTRIODON Pickford and Wilkinson, 1975

Pigs attributed to this genus are very small in size and have well-developed lophodont teeth, with strong, broad transverse crests in P4–M3. The cheek teeth do not generally have accessory cusps, which contributes to the lophodont appearance. The upper premolars have large and wide cingula. The talonid of m3 is well developed, and the small lower canines rest in depressions in the maxillae. The lower incisors are narrow. The taxon has a combination of relatively primitive cranial morphology combined with a specialized dentition (Cooke and Wilkinson, 1978).

LOPHOLISTRIODON PICKFORDI Made, 1995

Holotype KNM-WS 115, right and left mandible fragments, each with m2 and m3 from West Stephanie (Buluk), Kenya.

Distribution Early–middle Miocene of Kenya.

Remarks A larger species than *Lopholistriodon kidogosana*, which has less well-developed lophodonty in the molars.

LOPHOLISTRIODON KIDOGOSANA Pickford and Wilkinson, 1975
Figure 42.1

Holotype KNM-BN 992, male cranium with right and left C, M2–M3 from the Ngorora Fm, Kenya.

Distribution Middle Miocene of Kenya.

Remarks The cranium is gracile and slender, and smaller than Eurasian *Listriodon*. Weak development of the areas of

attachment for the muscles of the rhinarium suggests that this taxon was not a rooter. Instead, a developed *mm. levator lateralis* may have helped gather food along the cheektooth row when the lip was raised (Cooke and Wilkinson, 1978). This, combined with the lophodont dentition and elevated postion of the glenoid, indicates this suid was feeding on soft plant material that required cutting rather than grinding by its dentition (Cooke and Wilkinson, 1978). The upper canines are short and relatively straight (Made, 1995).

Subfamily KUBANOCHOERINAE Gabunia, 1958

Kubanochoerinae are the earliest known true pigs in Africa. Movements from Eurasia likely precipitated their appearance in the African fossil record around 21 Ma. The closest European relatives to these early African pigs are probably to be found in the genus *Aureliochoerus* from Europe (Pickford, 2006). While the subfamily was erected by Gabunia (1958) to contain the "horned" pig from the Caucasus, *Kubanochoerus robustus*, the African forms share with their Eurasian cousins bunodont molars with labial cingula on their upper teeth (Pickford, 2006). At least some of the African forms are thought to have the bony facial horns characteristic of the Caucasian *Kubanochoerus*, but relevant parts of the skull are not known for all taxa. It

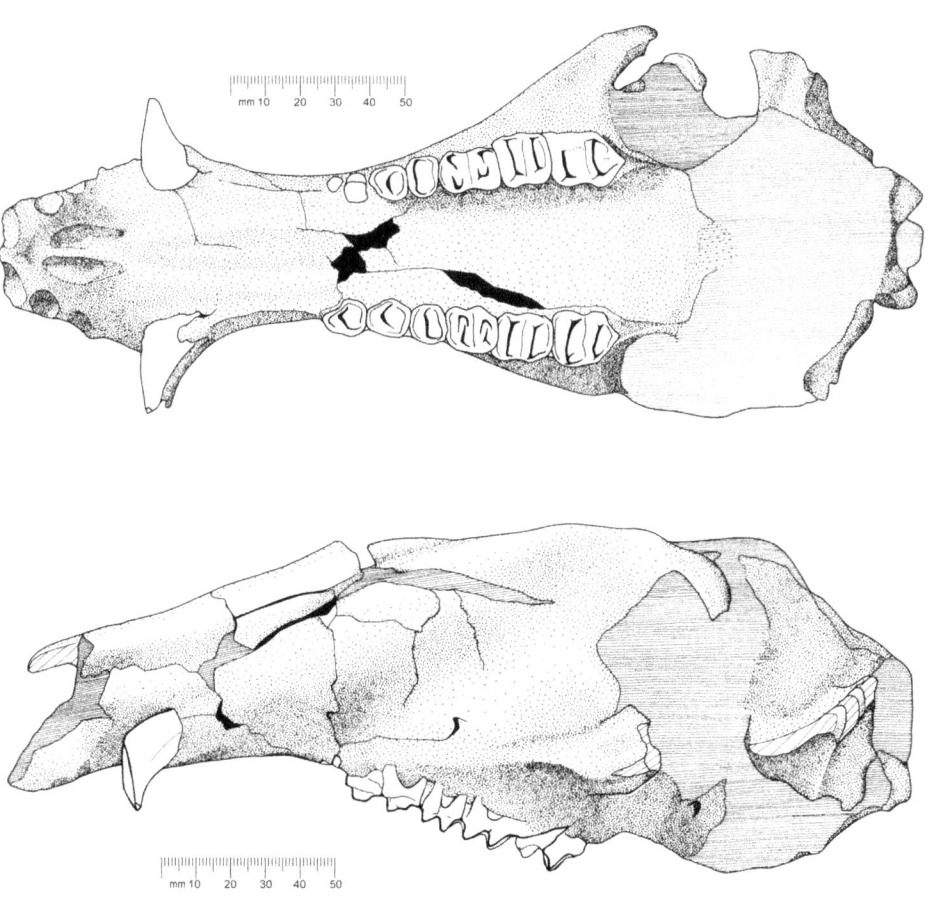

FIGURE 42.1 Inferior (A) and lateral (B) views of KNM-BN 992, *Lopholistriodon kidogosana*. Note extreme development of lophodont crests on postcanine teeth. After Pickford 1986:67–68, figures 63 and 64. Courtesy of M. Pickford and the Working Group of Tertiary to Quaternary Geology.

is also possible that the presence of these facial horns is a sexually dimorphic trait in the African taxa (Turner and Anton, 2004). Made (1996) includes the Kubanochoerinae in the Listriodontinae, but the view taken here is that they should remain distinct at the subfamilial level because of morphological discontinuities between two different subfamilies.

Kubanochoerinae increase in size with time, and this, along with some other morphological developments, may allow the group to be useful as biostratigraphic indicators in the Miocene of Africa (Made, 1996, Pickford, 2001). However, the material is far too scanty to fully test this idea at present. It appears possible that the kubanochoerines moved from Africa into Eurasia; this also fits with the temporal trend for increased body size (Pickford, 1986). Once they arrived in Eurasia, they appear to have spread both rapidly and widely. While kubanochoerines were the dominant suid during the middle Miocene, they were replaced comprehensively by the tetraconodontine suids in the later part of the Miocene. Harris and Leakey (2003) have described material from the late Miocene Nawata Formation at Lothagam that they attribute to the genus *Kubanochoerus*. This material may provide evidence of both late survivorship of this subfamily and co-occurrence of Kubanochoerinae and Tetraconodontinae. However, it has been considered elsewhere as *Nyanzachoerus* due to its massive premolars (Hill et al., 1992). It seems better accommodated in the Tetraconodontinae than the Kubanochoerinae, unless there is a subsequent migration of the latter from the Eurasia, undetected except for this example. The Kubanochoerinae otherwise disappear from the African fossil record in the middle Miocene.

Genus *NGURUWE* Pickford, 1986
NGURUWE NAMIBENSIS (Pickford, 1986)

Holotype SAM-PQ 20, cranium and mandible with dentition (left I2, C, P4–M3, right P3–M3, left i2, c, p3–m3, right p4–m3 from Langental (= Bogenfels), northern Sperrgebiet, Namibia.

Distribution Early–middle Miocene of Namibia.

Remarks The genus *Nguruwe* (from the Kiswahili for pig) refers to small species known only from the early–middle Miocene. These are the smallest of the kubanochoeres, weighing only 10–15 kg (Made, 1996) and having narrow incisors and bunodont molars. The known upper incisors are robust and triangular. The upper fourth premolar has a complete cingulum around the tooth. Enamel on the molars is very thick, and they have a labial cingulum and a weakly developed lingual cingulum. The third molars are unelaborated, having simple talon/ids and a complete labial cingulum. This suid is relatively poorly known and the species was originally put in the genus *Kenyasus*, but subsequent finds from further sites in the same region of Namibia allowed it to be characterized as *Nguruwe* (Pickford, 1997, 2001).

NGURUWE KIJIVIUM (Wilkinson, 1976)

Holotype Nap I'64, left maxilla fragment with M1–3 from Napak, Uganda.

Distribution Early Miocene of Egypt, Uganda and Kenya.

Remarks Material originally described as *Hyotherium kijivium* (Wilkinson, 1976) has subsequently been reassigned to the newer, African genus *Nguruwe* by Pickford (1986). Like its congener from southern Africa *Ng. namibensis*, *Ng. kijivium* is a small suid, known from middle Miocene localities of eastern Africa. At most of these sites, it is the only suid species, known from fragmentary dental remains. At Rusinga *Nguruwe* is joined by *Kenyasus rusingensis* and *Libycochoerus jeanelli* from which *Nguruwe kijivium* is separable on the basis of both size and morphology (Pickford, 1986).

Nguruwe kijivium differs from subsequent Kubanochoerinae in having a small and narrow I1 and a relatively high crowned i1. It also has a simple m3 with the four individual cusps closely arranged in transverse pairs, and a symmetrical taper to the talonid. The M3 is also simple, having a very small (or missing) talon (Cooke and Wilkinson, 1978). The canine enamel is thin but rugose, and the cingula of the premolars can have crenulations on their edges (Pickford, 1986). Once again, this suid is poorly known, with a limited distribution in time and space.

Genus *KENYASUS* Pickford, 1986
KENYASUS RUSINGENSIS Pickford, 1986

Holotype KNM-RU 2701 articulated subadult cranium and mandible (with associated skeleton) from Rusinga, Kenya.

Distribution Early Miocene of Kenya (Rusinga, Karungu and Arongo) and Namibia.

Remarks These pigs are small to medium sized, with the upper molar row approximately 50 mm in length (Pickford, 1986). Their I1 is not spatulate but peglike, differentiating them from other kubanochoerines. Their lower incisors are relatively small and thin. Like all kubanochoerines this pig has cingula on its molars and premolars; however, on the molars these can be relatively weak. The molars do not flare as much as some other species of kubanochoerines, and the enamel is thinner than in some examples. The molars also have cusps in pairs separated by a deep transverse valley and a centrally positioned hypoconulid (Pickford and Senut, 1997). The cingula have accessory cusplets. Pickford separated this taxon from *Hyotherium soemmeringi* because it lacks both a prezygomatic shelf and canine flange, and also the upper central incisors do not make contact mesially, all of which features are found in kubanochoerines but not in listriodontines. Made (1997) places this genus in Cainochoerinae, but Pickford (2006) lists it within the Kubanochoerinae, preferring to stress instead that the taxon may have given rise to the Namachoerinae, small suids with lophodont dentitions.

KENYASUS NAMAQUENSIS Pickford and Senut, 1997

Holotype SAM-PQ RK 1399, an upper left molar from Ryskop, Namaqualand, South Africa.

Distribution Early–middle Miocene Kenya, Uganda, and South Africa

Remarks This taxon is little known but has teeth larger than those of *Kenyasus rusingensis* from Kenya. The holotype molar has four main cusps and accessory cusplets, with a strong cingulum. The other referred specimen in the description, a right P4, is similarly larger than *Ke. rusingensis* with a single large lingual cusp surrounded by a cingulum; there are two labial cusps (Pickford and Senut, 1997). This species is thought to have a frugivorous/omnivorous diet (Pickford and Kunimatsu, 2005). It has very bunodont cheek teeth, separating it from the lophodont listriodontine pigs (Pickford, 2007).

Genus *LIBYCOCHOERUS* Arambourg, 1961

The genus *Libycochoerus* Arambourg, 1961 was first sunk into *Kubanochoerus* but subsequently revived by Pickford (1986) to accentuate the difference between Eurasian and African forms. Other authors consider this genus to be a synonym of *Kubanochoerus* but refer material assigned to species of the latter genus to *Bunolistriodon* and other taxa (Made, 1996). *Libycochoerus* are medium- to large-sized pigs with very robust teeth that are mesiodistally long. They also have stout canines and simple molars, with a small talon on the M3. The m3 has a large and continuous cingulum that is crenulated around its edge. The front of the face and zygomatic arches are large (Cooke and Wilkinson, 1978).

LIBYCOCHOERUS ANCHIDENS (Made, 1996)

Holotype KNM-RU 2785, mandible with left i1, partial c, p1–m3 and right i1–2, partial c and p1–p3 from Rusinga, Kenya.

Distribution Early Miocene of Kenya.

Remarks This taxon is thought to have derived from *Nguruwe* and is similar to it in morphology, although larger (Made, 1996). The incisors are narrow but higher crowned than in *Nguruwe*, and the postcanine teeth are bunodont. Material contained in it was formerly assigned to *Bunolistriodon jeanneli*, but it is much smaller than that species. It is unknown whether the skull of *Libycochoerus* would have possessed the "horn" between the frontals that characterizes other members of the Kubanochoerinae. The p4 of this taxon is more derived because it has a larger metaconid. *Libycochoerus anchidens* is known from a short period only, and occurrences are restricted to the sites of Rusinga and Karungu in Kenya.

LIBYCOCHOERUS JEANNELI (Arambourg, 1943)
Figure 42.2

Holotype MNHN (Paris) No 1933-9, Maxilla with left and right P3–M3 from Moruorot (Losodok), Turkana, Kenya.

Distribution Early–middle Miocene of Kenya.

Remarks While the original author thought this species was a *Listriodon* due to premolar morphology, Pickford (1986) ascribed it to Kubanochoerinae on the basis of its affinity to the more primitive members of that subfamily, *Nguruwe* and *Kenyasus*, which are smaller. The P4 has a strong, encircling cingulum. The lower molars are bunodont and the teeth do not tend toward lophodonty as is the case in Listriodontinae. The postcanine teeth possess cingula that are ridged with small cuspules. This species may have a distribution beyond Africa; a specimen from Fategad in India that has similar size and morphology to specimens from Moruorot may also be assigned to it (Made, 1996).

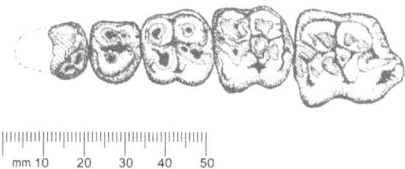

FIGURE 42.2 Dentition of *Libycochoerus jeanelli* (KNM-RU 2782 and KNM-RU 2780). After Pickford 1986:41, figure 37. Courtesy of M. Pickford and the Working Group of Tertiary to Quaternary Geology.

LIBYCOCHOERUS MASSAI Arambourg, 1961
Figure 42.3

Holotype MHNH (Paris) No. 1961–5–8, left mandible with p2–m3 from Jebel Zelten, Libya.

Distribution Early Miocene of Egypt and Libya and middle Miocene of Kenya and Libya.

Remarks This type species of the genus *Libycochoerus* is based on relatively complete material from Jebel Zelten, Libya. It has relatively robust P1 with strong roots, in contrast with *Listriodon*, to which it is sometimes referred. Like most kubanochoerines, this taxon also has bunodont teeth that are strongly cingulated. A male skull from Jebel Zelten has a bony "horn" on the frontal, and there appear to be bony bosses on the postorbital bars (figure 42.3); facial "horns" are a characteristic of this subfamily (Cooke and Wilkinson, 1978; Pickford, 1986). A specimen from the Tugen Hills, Kenya, has recently been attributed to this species, its first documented occurrence outside the type locality (Pickford, 2001). This species appears to be derived relative to the smaller *Libycochoerus jeanneli* in having larger third molar talon/ids.

Genus *MEGALOCHOERUS* Pickford, 1993

Megalochoerus is distinguished from *Libycochoerus* chiefly on the basis of its size and the flaring of its molars. The genus is also characterized by relatively large talonids on the lower third molars. *Megalochoerus* increases in size through time.

MEGALOCHOERUS MARYMUUNGUAE (Made, 1996)

Holotype KNM-WS 12595, right and left p2–m3 from Buluk (West Stephanie), Kenya.

Distribution Early–middle Miocene of Kenya.

Remarks This early representative of *Megalochoerus* is probably the smallest species of the genus in size; it was originally attributed to *Kubanochoerus* but is now considered as the early part of a small radiation of the genus *Megalochoerus*. The molars are large but relatively wide and flare out from the cervix, and the premolars are large. The lower premolars are also wide, and the incisors and premolars are relatively larger than in other species of this subfamily (Made, 1996).

MEGALOCHOERUS KHINZIKEBIRUS (Wilkinson, 1976)

Holotype BU 6416–82 a–e, right p2–p4 and m2–m3 from Jebel Zelten, Libya.

FIGURE 42.3 Reconstruction of *Libycochoerus massai* based on specimens from Gebel Zelten, Libya, by Mauricio Antón. Used with permission.

Distribution Middle Miocene, Jebel Zelten, Libya; Maboko, Kenya.

Remarks This extremely large suid, originally attributed to the genus *Kubanochoerus* by Wilkinson (1976) has now been referred to *Megalochoerus*, appropriately because of its enormous size. In other respects, this species is similar to others in the subfamily, with longer, simple premolars and fully bunodont molars, of which the upper molars are relatively wide. Once again this species is poorly known, although there appears to be a diastema between the robust P1 and P2 in the upper dentition. The P1 is relatively small, and the talons of the P3 and P4 are poorly developed (Cooke and Wilkinson, 1978). Based on material from Maboko, the lower incisors appear to be very large and the uppers both large and spatulate (Pickford, 1986). In addition to the material from Libya and Kenya, some fossils from Turkey have also been assigned to this species, making it one of the rare African forms to be found outside the continent in the Miocene (Pickford, 1986). This species is thought to be a very large-bodied omnivore (Pickford and Kanimatsu, 2005).

MEGALOCHOERUS HUMONGOUS Pickford, 1993

Holotype GSI B 450, right mandible with p4 through fragmentary m3 from Bugti, Baluchistan, Pakistan.

Distribution Middle Miocene of Kenya and Libya.

Remarks This truly huge member of the kubanochoerines was chiefly responsible both for the generic and trivial names of this species. Pickford (1993, 2001) interprets these animals as being very large, commensurate in size with some contemporaneous Proboscidea, with which they were initially confused. This is the terminal species in the genus found in Africa although there is evidence of its occurrence in Turkey and in Pakistan, from whence the type specimen derives. Made (1996) believes the type specimen to be an anthracothere. The species illustrates the trend in increased body size in kubanochoerines; it is both the latest and the largest of *Megalochoerus* species.

Subfamily TETRACONODONTINAE Lydekker, 1876
Genus *NYANZACHOERUS* Leakey, 1958

In Asia, tetraconodonts are widely represented by *Sivachoerus* (*S. prior*, not *S. giganteus*, which *contra* Pilgrim [1926] is a suine and not a tetraconodont; Pickford, 1986). There are early, unattributed specimens representing perhaps more primitive tetraconodont species found in the Ngorora Formation (and also at Maboko) that have very enlarged posterior premolars (e.g. KNM-BN 1491-2). Although the exact nature of the phylogenetic relationship of *Nyanzachoerus* to the Asian *Sivachoerus* is still debated, there are several craniodental characteristics that link the members of *Nyanzachoerus* to one another (Pickford, 1986; Bishop and Hill, 1999; Made, 1999). *Nyanzachoerus* has been identified in Arabia, which may provide a clue as to the movement of this subfamily between Africa and Eurasia (Bishop and Hill, 1999). In Africa, the Tetraconodontinae are represented by two genera: *Nyanzachoerus* and its descendant *Notochoerus*.

Like all tetraconodonts, nyanzachoeres are suids that possess enlarged posterior (fourth and third) premolars relative to their anterior (second and first, when preserved) premolars. All possess deep and robust mandibles, especially anteriorly, where the massive symphysis may extend as far posteriorly as the level of the premolars. Tooth enamel is relatively thick. Although there is a trend toward increasing elaboration of the third molars with time, through most of its temporal range *Nyanzachoerus*

is characterized by simple and bunodont M3s, with cusps which, when worn, show dentine lakes having a star shape. Evolutionary trends in the African tetraconodonts have made them useful for biochronology; several subspecies have been proposed to compartmentalize biogeographic and temporal variation (Made, 1999). *Nyanzachoerus* is a Mio-Pliocene genus.

NYANZACHOERUS DEVAUXI (Arambourg, 1968)

Holotype Right mandibular fragment from Oued el Hammam in Algeria's Bou Hanifia region. (originally *Propotamochoerus devauxi*).

Distribution Libya, Algeria, Tunisia, Chad, Kenya.

Remarks Ny. devauxi is the most primitive of the described species of Tetraconodontinae known from Africa. Recently recovered specimens from the lowest stratigraphic levels at Lothagam are very primitive representatives of *Nyanzachoerus*. They most closely resemble the type of *Ny. devauxi* (Harris and Leakey, 2003). The species may also be present in the Tugen Hills sequence, although remains are fragmentary. *Ny. devauxi* has also been recovered from Sahabi, Libya (Cooke, 1987). This taxon is the most likely ancestor for the later *Nyanzachoerus* species in East Africa.

The original description of the species presented a specimen with relatively larger posterior than anterior premolars, although the disparity was not huge, as with later *Ny. syrticus* (Arambourg, 1968). The molars are all relatively bunodont. The third molars possess an unelaborated, true talon/id, the posterior cusps of which are not in full occlusion. There appears to be slight cingulum formation on the cheek teeth. In contrast, *Sivachoerus prior* Pilgrim, 1926, illustrated in Pickford (1988), appears to have pronounced cingulum development. The specimen of *Propotamochoerus devauxi* illustrated in Arambourg (1968) appears to differ in size between its p3 and p2. Despite the apparent possession of this diagnostic tetraconodont feature of relatively large posterior premolars, Arambourg allied this specimen with a species, *Propotamochoerus hysudricus* Pilgrim, 1926, which is now considered to be Suinae (Pickford, 1988).

Leaving nomenclatural differences aside for a moment, the important thing to note is that *Ny. devauxi*, with its relatively generalized tetraconodont features, makes a very plausible ancestral taxon for the East African *Nyanzachoerus-Notochoerus* lineage. And, with its possible recognition at Lothagam, it is in the right place at the right time. Since there are few known occurrences of this taxon in East Africa, it would be difficult to deduce any time range for the taxon.

A study of the postcranial ecomorphology of *Ny. devauxi* suggests that this taxon preferred habitats intermediate between forest and open grasslands, such as bushland (Bishop, 1994; Bishop et al., 1999). These pigs were browsers to mixed feeders, as confirmed by carbon stable isotope analysis of their tooth enamel to reconstruct their diets (Harris and Cerling, 2002).

NYANZACHOERUS SYRTICUS (Leonardi, 1954)
Figures 42.4A and 42.5

Holotype Nearly complete mandible from Sahabi, Libya (Leonardi, 1952 Plate I, figure 2).

Distribution Libya, Tunisia, Algeria, Chad, Kenya, Ethiopia, Uganda.

Remarks Although also a tetraconodont, *Ny. syrticus* represents an extreme, with massive and broad P3/p3 and P4/p4 in the most derived specimens. Later *Ny. syrticus* specimens also

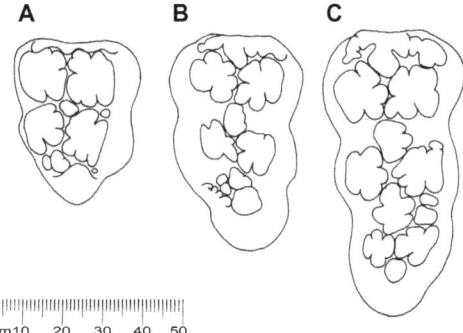

FIGURE 42.4 M3 in Tetraconodontinae: A) *Nyanzachoerus syrticus*; B) *Nyanzachoerus kanamensis*; C) *Notochoeurus jaegeri*. Note the increase in anterior cingulum area and elaboration in the talon through evolutionary time. After Made 1996:12, figure 7. Courtesy of the Working Group of Tertiary and Quaternary Geology.

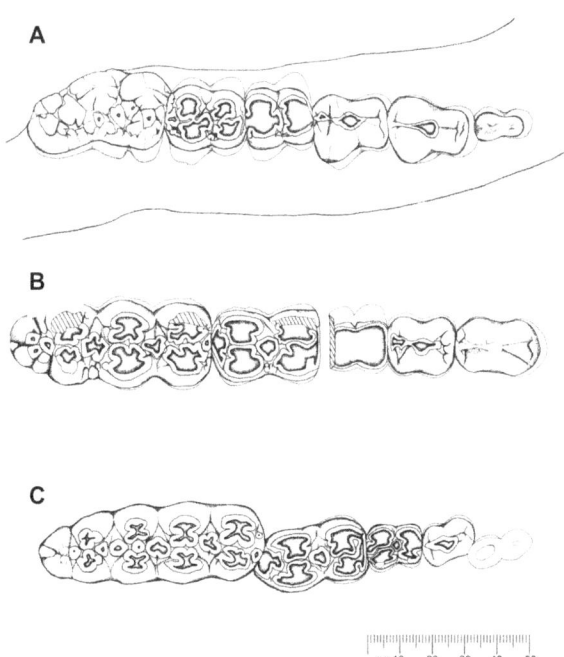

FIGURE 42.5 Lower cheek teeth in Tetraconodontinae: A) *Nyanzachoerus syrticus* (KNM-LT 295); B) *Notochoerus jaegeri* (Omo 1967[71]); C) *Notochoerus euilus* (KNM-ER 2773). Note the decrease in size in the posterior premolars, p3 and p4 co-occurring with the increased size, and elaboration in m3. After Harris and White 1979:18, 21, and 30, figures 13, 22, and 43. Courtesy of the American Philosophical Society.

possess well-developed molar cingula around the circumference of the tooth. This species occurs quite early in the temporal range of the *Nyanzachoerus* lineage (Harris and White, 1979), but it may be too derived to have been ancestral to later populations of *Ny. kanamensis* and *No. jaegeri* for reasons more fully explained below. At Lothagam, a large sample of *Nyanzachoerus syrticus* shows strong sexual dimorphism; males are larger and have more "ornamented" skulls with conspicuous bony bosses on their skull muzzle and widely flaring cheekbones that were probably enhanced by thick skin pads in life (Turner and Anton, 2004). Unlike later suine pigs, *Ny. syrticus* did not have very large canines.

Ny. syrticus has one named subspecies, *Ny. s. tulotus*, based on an incomplete cranium from Lothagam, Kenya (KNM-LT 316, the original holotype of *Ny. tulotus* Cooke and Ewer, 1972, subsequently demoted to subspecific status). With a much larger sample size of the species now available for study, the Lothagam form seems more properly attributed to a subspecies.

In North Africa, *Ny. syrticus* is known from the type locality in Sahabi, Libya, and has also been reported from Tunisia, Algeria, and Toros-Menalla, Chad (Vignaud et al., 2002). Harris and White (1979) report this taxon from the Afar region; further details are not available. In Kenya, it has been recognized in deposits from Lothagam, Kanam East, Kanam West, Ekora, and the Lukeino and Chemeron Formations of the Tugen Hills. *Ny. syrticus* has not been reported from southern Africa.

A study of diet in *Ny. syrticus* from the Tugen Hills suggests that they were mixed feeders to browsers (Bishop et al., 1999). Specimens from Lothagam show a similar range of dietary adaptations (Harris and Cerling, 2002). Although isotopic values suggest at least one specimen was consuming significant quantities of grass, they may also have focused on the more wooded end of the browsing spectrum, or perhaps ate fruit fallen from trees.

NYANZACHOERUS WAYLANDI (Cooke and Coryndon, 1970)

Holotype M26324, a left m3 from Nyaburogo Valley, Toro District, Uganda housed at the Natural History Museum, London.

Distribution Latest Miocene of Uganda.

Remarks The type specimen and small initial collection was originally described as *Sus* and then sunk into *Ko. limnetes* (Cooke, 1978b; Harris and White, 1979). More complete material was subsequently found at the type site in Uganda that suggested that *Sus waylandi* had tetraconodont features and was more properly attributed to *Nyanzachoerus* (Pickford, 1989). Like other tetraconodonts, it has large third and fourth premolars. *Ny. waylandi* is distinguished principally on the basis of its small size, which it shares with *Ny. devauxi*. However, compared with *Ny. devauxi*, *Ny. waylandi* has relatively smaller premolars and relatively longer m3s (Pickford, 1989).

NYANZACHOERUS KANAMENSIS Leakey, 1958
Figure 42.4B

Holotype BM M15882, a partial left mandibular corpus from Kanam, Kenya, housed at the Natural History Museum, London.

Distribution Egypt, Ethiopia, Chad, Kenya, Tanzania, Uganda, South Africa.

Remarks The genus *Nyanzachoerus* was founded on this taxon, a sexually dimorphic tetraconodont that has enlarged third and fourth premolars. Premolars of *Ny. kanamensis* are not as markedly large as in *Ny. syrticus*. There are also differences in emphasis in the development of the third molar. *Ny. syrticus* has low relief, compact, bunodont teeth with cingulum formation surrounding the tooth. There is no formation of an encircling cingulum on the posterior cheek teeth of *Ny. kanamensis*. Rather than being compact, the main cusps (or pillars) of *Ny. kanamensis* teeth are isolated and columnar. They are joined by small, cuspule-lined basins that rise approximately one-fourth of the height of the tooth above the cervix. *Ny. kanamensis* possesses an elaborated talon/id, which is trigonized—the talon/id pillars are at the same occlusal height as those of the trigon/id. There is also no waisting of the crown at the trigon/talon junction. The enamel is thick, but some derived specimens show increased crown height. Finally, enamel crenulation on the tooth cusps is more complex in *Ny. kanamensis* than in *Ny. syrticus* or *Ny. devauxi*. When worn down through occlusion, dentine lakes exposed on the cusps are a simple star shape.

A subspecies, *Ny. k. australis* Cooke and Hendey, 1992, which retains some primitive characteristics, has been recognized at Langebaanweg. Following Made (1999), this taxon was elevated to the species level based on material from Lothagam (Harris and Leakey, 2003). However, Bishop (1999) found that M3 metrics for known specimens of *Nyanzachoerus kanamensis sensu lato* were essentially continuous.

Some authors have confined the name *Ny. kanamensis* to specimens from the western Rift Valley, citing relatively narrower premolars in western versus eastern forms. In this case, *Ny. pattersoni* Cooke and Ewer, 1972 is used to describe the eastern form with wider premolars (Harris and Leakey, 2003; Harris et al., 2003). Specimens from Ethiopia have also been assigned to *Ny. pattersoni* (Kullmer et al., 2008). However, in light of the fact that the type of *Ny. kanamensis* is not from the western rift and following metric studies of third molar size through time, it may be more appropriate to use *pattersoni*, as now done for *australis*, as a subspecies of *Ny. kanamensis* (Bishop, 1999).

This suid is very well distributed in East African Pliocene localities. It has been recovered from sites with and without the potential for radiometric age determination. *Nyanzachoerus kanamensis* is known from the Omo and Afar, both east and west of Lake Turkana and from the Tugen Hills sequence. In addition, it is found at Kanam West, Kanapoi, Kanjera, Lothagam, and the Manonga Valley, Tanzania. It has been reported from Langebaanweg, South Africa as well. The earliest well-dated specimens of *Ny. kanamensis* derive from Tabarin, a Chemeron Formation site from the Tugen Hills of Kenya, in sediments that date to 4.3 Ma (Hill et al., 1985). Other specimens from the Tugen Hills may extend the radiometrically determined range further back in time. The youngest well-dated examples come from Ethiopia, from Upper Member B–Lower Member C of the Omo Shungura Formation, dated to approximately 2.85 Ma.

As inferred from their dentitions and from carbon stable isotope studies of their tooth enamel, *Ny. kanamensis* were probably browsers to mixed feeders, at least during the later part of their time range (Bishop et al., 1999; Harris and Cerling, 2002). Studies of fossil postcrania identified to this taxon suggest that they preferred intermediate habitat types such as woodlands or bushland (Bishop, 1994; Bishop et al., 1999).

Genus *NOTOCHOERUS* Broom, 1925

This genus is thought to have evolved from the genus *Nyanzachoerus* and its most primitive member is *Notochoerus jaegeri*. Originally named from South African material, *Notochoerus* are large suids with long crania. They exhibit progressive reduction in premolar size and most markedly a trend toward increasing hypsodonty and expansion of the third molars. Viewed laterally, m2s of *Notochoerus* flare markedly from the cervix to the crown. The lateral pillars of the elaborated third molars have greater mesial and distal folding of the enamel. On the later, more derived forms of *Notochoerus*, dentine lakes on worn cusps have an H-shaped appearance.

NOTOCHOERUS JAEGERI (Coppens, 1971)
Figures 42.4C and 42.5

Holotype Partial mandibular corpus with p2–m3 from Hamada Damous, Tunisia, housed at the Muséum National d'Histoire Naturelle, Paris.

Distribution Tunisia, Kenya, Ethiopia, Uganda, Malawi, South Africa.

Remarks *No. jaegeri* was originally thought to be the most advanced species of the genus *Nyanzachoerus* when it was first recognized in deposits from northern Africa. Similar specimens from East Africa were assigned to a new species *Ny. plicatus* by Cooke and Ewer (1972), but *jaegeri* was the taxon with nomenclatural precedence, as has been recognized thereafter (Cooke and Wilkinson, 1978; Harris and White, 1979).

Study of recently recovered material from Lothagam and Kanapoi in Kenya has led Harris and Leakey (2003, 2004) to conclude that *No. jaegeri* possesses several advanced features allying it more closely with the genus *Notochoerus* than to *Nyanzachoerus*. The posterior premolars are reduced in size, and there is increasing relative tooth length in the molar row. The third molars are more elaborate than in earlier nyanzachoeres. In particular, the tooth is relatively long and hypsodont, the latter being accomplished by the addition of a variable number of talon pillars, which rise to the occlusal level. Additionally, there is pronounced invagination of the enamel surface on the lateral pillars of the tooth; when worn, these cause the cusps to take on a complex star-shaped appearance. A broad mandibular symphysis is similar to the condition of *Notochoerus euilus*, to which *No. jaegeri* is doubtless closely related. The presence of zygomatic knobs in some crania has led to the conclusion that *No. jaegeri* was a sexually dimorphic species. It is often difficult to distinguish advanced specimens of *No. jaegeri* from early *No. euilus*.

No. jaegeri has been reported from a variety of localities, presumably those that have sediments formed during what may be a restricted time range. Although widely distributed in space, *No. jaegeri* has few known absolute dates associated with it. This is partially because many places where the species is found, such as Lothagam and Ekora, do not have firm radiometric dates, but it is also a relatively rare taxon (Kullmer, 2008). It has also been reported from Galili in Ethiopia (Kullmer et al., 2008). Specimens from dated sediments derive from Tabarin, a site in the Chemeron formation of the Tugen Hills sequence dating between 4.3 Ma and about 4.0 Ma, and from the Mursi Formation, dated to between 4.35 and 3.99 Ma. Numerous additional specimens derive from the Chemeron Formation and have yet to be fit into that chronostratigraphic framework. It has been recovered from the Chiwondo Beds of Malawi (Kullmer, 2008) and is known from Langebaanweg in South Africa (Cooke and Hendey, 1992). These specimens will doubtless increase our knowledge of the temporal range of *No. jaegeri*.

Based on carbon stable isotope studies of its diet, *No. jaegeri* is thought to have been a mixed feeder, with both grass and browse components consumed by the Tugen Hills specimens (Bishop et al., 1999). Specimens from Kanapoi appear to be more grass dependent, with isotopic values in the grazing end of the dietary spectrum (Harris and Cerling, 2002). This accords well with the trend seen in its molar morphology, with expansion of the third molar talon/ids linked to diets containing more grass.

NOTOCHOERUS EUILUS (Hopwood, 1926)
Figures 42.5 and 42.8

Holotype M12613A, a talonid of a right m3, from the Kaiso Formation of Uganda housed at the Natural History Museum, London.

Distribution Ethiopia, Kenya, Uganda, Kenya, Tanzania, Malawi, South Africa.

Remarks This species is more derived than its presumed ancestor *No. jaegeri* by virtue of expanded third molars and posterior premolars that are even further reduced relative to those of the genus *Nyanzachoerus*. There is greater expansion of the talon relative to *Nyanzachoerus*, with at least two of the lateral talon/id pillars equal in size and occlusal height with those of the trigon/id. The pillars are widely separated to the base and taper steeply. The species is variable in third molar size, but M3 tooth height and length have been said to increase with time. Crania are considered to be sexually dimorphic, with large zygomatic knobs on some (male) examples.

The species has a wide range in space and time, being found at most Pliocene localities. Particularly, *No. euilus* is present in Ethiopia in the Omo Shungura Formation, the Hadar Formation and Galili (Kullmer et al., 2008). In Kenya, the species is known from the Chemeron Formation, Lothagam, Karmosit Beds, the Nachukui and the Koobi Fora Formations. It is also known from the Laetolil Beds of Tanzania. Recently it was reported from the Chiwondo Beds of Malawi (Kullmer, 2008). It is well distributed temporally. The oldest well-dated specimen derives from Area 250 at Koobi Fora, within the Moiti Tuff dated at approximately 3.89 Ma. Specimens from the Tugen Hills may be older.

Notochoerus euilus is reconstructed to have stood approximately 120 cm at the shoulder, based on skeletons from Koobi Fora, Kenya (Turner and Anton, 2004). Based on postcranial locomotor ecomorphology, it appears that this taxon preferred closed habitats, such as forests (Bishop, 1994; Bishop et al., 1999). Carbon stable isotope analyses of several specimens from Kanapoi and Koobi Fora suggest that some specimens were consuming some nongrass plants, but the majority of them were grazing on tropical grasses (Harris and Cerling, 2002).

NOTOCHOERUS CAPENSIS Broom, 1925

Holotype Talon of a right M3 from an unrecorded locality in the Vaal River Gravels, South Africa. Located in the Bernard Price Institute for Palaeontological Research, Johannesburg, South Africa.

Distribution Likely restricted to Southern Africa.

Remarks *Notochoerus capensis* is advanced relative to *No. euilus*, with larger teeth and bigger crania. Particularly, there is an increased number of elaborated talon/id pillars in *No. capensis*, and enamel is both more crenelated and more hypsodont. Decreased emphasis on premolars continues with this taxon, which is considered intermediate in many ways between *No. euilus* and *No. scotti* (Harris and White, 1979).

This species is extremely rare in East Africa, and specimens that have been assigned to it from Koobi Fora might more properly belong to the recently recognized *No. clarkei* Suwa and White, 2004 (Harris and White, 1979; Harris, 1983). If *No. capensis* occurred only in South Africa, it would be the only Plio-Pleistocene suid species with such a restricted geographic range. If it does occur in eastern Africa, it is rare, offering limited opportunity to define its temporal range.

In South Africa, it is well-known from Makapansgat and the Vaal River Gravels (Harris and White, 1979).

NOTOCHOERUS SCOTTI (Leakey, 1943)
Figure 42.6

Holotype KNM-OS 5, left maxillary fragment with M2 and M3 from the Omo Shungura Formation, Ethiopia, housed in the National Museums of Kenya.

Distribution Ethiopia, Kenya, Tanzania, Malawi.

Remarks *Notochoerus scotti* is both the most advanced and one of the last species of its genus, and of the tetraconodont *Nyanzachoerus-Notochoerus* lineage. Crania are wider and more massive than is the case in the largest examples of *No. euilus* and *No. capensis*. However, the largest individuals of *No. scotti* are not as big as the largest individuals of the former species (Harris and White, 1979). Molar elaboration is distinctive, however. *Notochoerus scotti* possesses third molars that are extremely hypsodont and elongate, having between four and six pairs of major talon/id pillars. In advanced specimens there can be a thick cementum layer surrounding the crenelated enamel of the circumference of the tooth.

Notochoerus scotti is known from numerous Pliocene and Early Pleistocene sites in East Africa: the Omo Shungura Formation, the Chemeron Formation, the Koobi Fora, and Nachukui formations. It has also been reported from the Chiwondo Beds of Malawi (Kullmer, 2008) Specimens from the Tugen Hills may predate its currently known temporal range, which extends from just above the Tulu Bor tuff, dated at 3.36 Ma (West Turkana LO5) to the latest known specimen at 1.79 Ma (Koobi Fora Area 104). During all but the last 200,000 years of its long range, it coexisted with its presumed ancestor, *Notochoerus euilus*.

It appears that *No. scotti* was a grazer, as suggested by molar morphology, which is extremely long, hypsodont, and complex and confirmed by carbon stable isotope analysis of their dental enamel (Bishop et al., 1999; Harris and Cerling, 2002).

NOTOCHOERUS CLARKEI Suwa and White, 2004

Holotype GAM-VP-1/20 cranium with dentition from Gamedah Vertebrate Palaeontology Locality One, Middle Awash, Ethiopia.

Distribution Ethiopia.

Remarks This late member of the genus *Notochoerus* is in some ways very derived. Unlike other *Notochoerus*, its cheek teeth are small and gracile, although it shares with them the

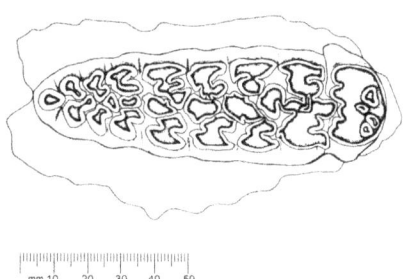

FIGURE 42.6 LM3 of *Notochoerus scotti* (KNM-ER 3438). This is the most extreme example of third molar elaboration in the tetraconodonts. Numerous, well-developed pillar pairs characterize the talon, and the anterior complex of the tooth is enlarged. The crown is also very high in this example. After Harris and White 1979:29, figure 41. Courtesy of the American Philosophical Society.

characteristic shape and arrangements of the cusps. The species has hypsodont third molars, which appear to be more organized and sharply folded in their cusp morphology. The third molars also have relatively thin enamel compared to other members of the genus. They are smaller than the most derived *No. scotti* and have less derived third molars (White and Suwa, 2004).

This recently described species has a known time range of 2.5–1.8 Ma in the localities from which it has currently been recognized, including the Hata Member of the Bouri Formation, the Middle Awash, Members D–H of the Omo Shungura Formation, and the Konso Formation, Ethiopia (White and Suwa, 2004). A thorough examination of the African *Notochoerus* hypodigm is required to determine the extent to which it is present in other sites of similar time periods.

Subfamily NAMACHOERINAE Pickford, 1995
Genus *NAMACHOERUS* Pickford, 1995
NAMACHOERUS MORUOROTI (Wilkinson, 1976)

Holotype KNM-MO 5 Mandible with left m1–m3 and roots of i1–i2, p3 and p4, and right i1–i2, c, p2–m3, and root of i3 from Moruorot, Kenya.

Distribution Early–middle Miocene of Kenya and Namibia.

Remarks This is a small pig with sublophodont cheek teeth and narrow central incisors (Made, 1996). The enamel on the teeth is smooth, and the m2s are simple in morphology (Cooke and Wilkinson, 1978). Originally ascribed to *Lopholistriodon* on the basis of its tooth morphology and tendency toward a loph structure in molar cusp arrangement, it has since been referred to a new genus *Namachoerus* on the basis of new material from Arrisdrift, Namibia. This new material established the extent of lophodonty on the postcanine teeth, which is not so developed in the holotype. Made (1996) considered it intermediate in morphology between *Nguruwe kijivium*, on the one hand, and *Lopholistriodon kidogosana*, on the other, and proposed that this species is more properly ascribed to *Listriodon*. *Namachoerus* demonstrates parallel evolution of lophodont dentition in the middle Miocene Suidae.

Subfamily CAINOCHOERINAE Pickford, 1995

The Cainochoerinae are a group of suids with extremely small body size. They persist into the latest Miocene/early Pliocene of Africa, providing a contrast to the very large tetraconodonts that otherwise dominate the suid fauna.

Genus *ALBANOHYUS* Ginsburg, 1974

Some specimens from Kenya with bunodont teeth have been attributed to *Albanohyus*, which is otherwise a Eurasian taxon (Pickford, 1986, 2006). This species is considered a more primitive member of the Cainochoerinae by Pickford (2006); however, the material is scanty, and no distinctly African taxon has been erected for it. Harris and Liu (2007) consider it to be a primitive genus of suid but do not assign it to a subfamily.

Genus *CAINOCHOERUS* Pickford, 1988
CAINOCHOERUS AFRICANUS (Hendey, 1976)

Holotype SAM PQ-L 31139, fragmented skull with complete upper and lower left dentition from Langebaanweg, South Africa.

Distribution Late Miocene–?early Pliocene of Kenya, Ethiopia, and South Africa.

Remarks Originally described as a tayassuid *?Pecarichoerus* Colbert, 1933, a new genus was erected to hold the specimens from Langebaanweg by Pickford (1988), who determined that this species was not a peccary but a true pig. Additional material from Lothagam, Kenya, has confirmed this observation; *Cainochoerus* is a small, cursorial pig as confirmed by its postcrania, which converge with peccaries and bovids (Pickford, 1988; Harris and Leakey, 2003; Turner and Anton, 2004). This species is very small, both cranially and postcranially, and the teeth are very simple and peccary-like, having single-cusped anterior premolars and four-cusped anterior molars. The m3, and to a lesser extent the other lower molars have slight development of an additional terminal cusp (Cooke and Wilkinson, 1978). It is unique in lacking postcanine diastemas in the the the upper and lower jaws (Hendey, 1976). Although most frequently compared with peccaries, a modern suid analogue might also be found in the diminutive *Sus (Porcula) salvanius*, the pygmy hog which is an endangered species of the Indian forest.

Subfamily SCHIZOCHOERINAE Thenius, 1979
Genus *MOROTOCHOERUS* Pickford, 1998
MOROTOCHOERUS UGANDENSIS Pickford, 1998

Holotype MOR 177-178 right mandible with p4–m3 from Moroto, Uganda.

Distribution Middle Miocene sites in Kenya (Muruyur, Maboko, Ngorora, Kirimun) and Uganda (Moroto).

Remarks The presence of *Morotochoerus* is documented from the earliest middle Miocene. Although when originally described, the species was thought to be a peccary, revision first placed the species within the Palaeochoeridae (Made 1997; Pickford, 1998, 2006). Subsequent analyses have found Palaeochoeridae to be a paraphyletic group that is therefore taxonomically invalid (Liu, 2003). Most recently it has been assigned to the Schizochoerinae within the family Suidae (Harris and Liu, 2007). The species is small with a number of peccary-like features, including relatively unspecialized third molars, canines that are oval in cross section and molars that are sublophodont. The maxillary cheek teeth abut the canine so that there is not a gap in the tooth row. The orbits are positioned forward on the skull.

The species is relatively poorly known because there are few fossils currently attributed to *Morotochoerus ugandensis*.

Subfamily SUINAE Gray, 1821
Genus *POTAMOCHOERUS* Gray, 1854
POTAMOCHOERUS AFARENSIS (Cooke, 1978)

Holotype AL 147-10, a partial cranium with P3–M3 from Hadar, Ethiopia, housed in the Ethiopian National Museum, Addis Ababa.

Distribution Kenya, Tanzania, Ethiopia.

Remarks Cooke (1978b) distinguishes *Po. afarensis* from both *Po. porcus* and *Ko. heseloni* on the basis of several cranial and dental characteristics. *Potamochoerus afarensis* possesses the characteristic heavy, downward-oriented sweeping zygomatics that ally the species with the genus *Kolpochoerus*. Mandibles are also inflated in the manner of *Kolpochoerus*. Premolars in *Po. afarensis* are not so reduced as is the case in the bushpig, and p1 (and sometimes P1) is retained. Third molars are larger in *Po. afarensis*, and the pillars tend to be columnar and distinct, giving the tooth a somewhat lophodont appearance. Unlike *Ko. heseloni*, there is little talon/id development in *Po. afarensis*; the only major pillars in the third molars are the two pairs from the trigon/id.

Originally attributed to the genus *Kolpochoerus*, this species was subsequently sunk into the modern bushpig genus *Potamochoerus* (Cooke, 1997). Pliocene specimens having small, simple, and bunodont third molars are attributed to the modern species by Cooke, but other workers attribute them solely to the genus *Potamochoerus*. With fragmentary material, there is the additional potential pitfall of attributing late *Ny. devauxi* specimens to *Ko. afarensis* (or indeed *Po. porcus*) in error. There are plesiomorphic similarities in molar morphology in these primitive examples of three genera.

Cranial and dental material from Laetoli and Hadar is attributed to *Po. afarensis*. Additionally, there is material from the Shungura, Nachukui, Koobi Fora, and Chemeron Formations that can be assigned to this taxon. The time range of this taxon is from approximately 4.3 Ma at Aterir (and other Chemeron sites) to 3 Ma (the Hadar material). More recent specimens from the Middle Awash have been reported (White, 1995). This species is primarily responsible for the long apparent history of the bushpigs in the African fossil record.

Genus *KOLPOCHOERUS* Van Hoepen and Van Hoepen, 1932

After years of confusion in the literature, Harris and White (1979) sank this genus into the later *"Mesochoerus"* Shaw and Cooke, 1941. According to the rules of zoological nomenclature, *Kolpochoerus* Van Hoepen and Van Hoepen, 1932, had priority, and this taxon was revived by Cooke (1974 et seq.). The genus is now uniformly known as *Kolpochoerus*. Cooke and Wilkinson (1978) consider *Ko. paiceae* Broom, 1931 the type species of the genus, but this taxon was sunk into *Ko. limnetes* by Harris and White (1979) which they considered to be the type species of the genus. However, Pickford (1994) reexamined the type material of *"Sus" limnetes* and concluded, on the basis of preservation and morphology, that this specimen was more likely a tetraconodont than a suine pig. Cooke (1997) agreed with this analysis and suggested the species *Kolpochoerus heseloni* was the more correct type species for this important genus.

Members of the genus *Kolpochoerus* are unified by their laterally and inferiorly expanded zygomatic arches. Their mandibles are narrow anteriorly, but they usually have a robust lateral prominence as far anteriorly as the p3. Anterior cheek teeth usually resemble in size and shape those of the modern bushpig *Potamochoerus*.

KOLPOCHOERUS DEHEINZELINI (Brunet and White, 2001)

Holotype ARA VP 1/986, maxillae with left C, P2–P3, M2–3, and right C, P2– P4 from Aramis, Ethiopia, housed at the National Museums of Ethiopia.

Distribution Ethiopia and Chad.

Remarks This is a small and primitive species of *Kolpochoerus* that is probably close in morphology to the earliest suines of this lineage to arrive in Africa, probably via Arabia (Bishop and Hill, 1999; Brunet and White, 2001). Similarly to *Po. afarensis* it possesses P1 and p1. It has a short diastema between the first and second lower premolars. It also has relatively large premolars and a short molar row. The molars are bunodont and low crowned. This species has been described from Mio-Pliocene sediments in Ethiopia and Chad that span an age from approximately 5.5 Ma to 3.8 Ma (Brunet and White, 2001; Kullmer et al., 2008).

KOLPOCHOERUS HESELONI (Leakey, 1943)
Figure 42.7

FIGURE 42.7 A reconstruction of *Kolpochoerus heseloni* based on a cranium from Koobi Fora (shown, inset) by Mauricio Antón. Used with permission.

Holotype (Syntypes) M17118a, left mandibular fragment with p4–m3 and M17118b, right mandibular fragment with p4–m3 from the Shungura Formation, Omo, Ethiopia, housed at the Natural History Museum, London.

Distribution Morocco, Sudan, Ethiopia, Kenya, Uganda, Tanzania, Malawi, South Africa.

Remarks This species encompasses a large amount of morphological variation during a long temporal and wide geographic range. It is plausible that there is more than one species represented in the variety of taxa and forms sunk into *Ko. limnetes* (and subsumed thereafter by *Kolpochoerus heseloni*) by Harris and White (1979; see, e.g., Cooke, 1985; White, 1995). It is the most common species in their monograph. According to their uniform definition, all attributed specimens have third molars in which the trigon/id–talon/id junction is formed by two triangular median pillars that, when the tooth is very worn, appear to be one. The upper third molar talon has between one and three pairs of major lateral pillars. The m3 talonid has between one and four pairs of major lateral pillars. Cheek teeth of advanced members of this species often have a cementum cover. The mandibles are narrow toward the symphysis but have robust lateral flaring that starts as far anteriorly as the level of the p3. The zygomatic arches are laterally and inferiorly expanded. Males, particularly in more advanced forms, have facial ornamentation, including enlarged zygomatics with projecting knobs.

This diverse species is known from every major late Pliocene and Pleistocene assemblage. Specimens have been recovered from Koobi Fora and West Turkana, the Chemeron Formation, the Omo, the Konso Formation, Olduvai Gorge,

FIGURE 42.8 Comparison (L-R) of *Phacochoerus africanus*, *Metridiochoeurus andrewsi*, and *Notochoerus euilus*. The size of *Notochoerus* is inferred from two excellent specimens from Koobi Fora, but *Metridiochoerus* size and proportions are estimated based on their cranial measurements and modern suid body plan. By Mauricio Antón, used with permission.

and Laetoli (Harris and White, 1979; Suwa et al., 2003). The species also occurs at the sites of Chesowanja, Marsabit Road, and Isimila. It has also been found at numerous South African fossil localities such as Cornelia, Elandsfontein and the Vaal River Gravels. It is possible that *Kolpochoerus* is the only Plio-Pleistocene genus to expand its range back into Eurasia; there is evidence of its presence in Israel (Geraads et al, 1986). The known time range spans from the earliest well-dated specimen from LO5 at West Turkana, 3.26 Ma, to the latest well-dated occurrences, in Bed IV at Olduvai Gorge, ca. 0.70 Ma. Specimens from the Tugen Hills may predate this range.

This taxon shows shifts in morphology through time that have been mirrored in studies of its palaeobiology. *Ko. heseloni* appears to exhibit change in both its diet (from more browse to more graze) and in its habitat preference, which shifts from more closed to more intermediate habitats during its long time range (Bishop, 1994; Bishop et al., 1999; Cerling and Harris, 2002; Bishop et al., 2006). Furthermore, the dental micowear of this taxon is significantly different from that of other pig species both living and dead, suggesting that niche partitioning among the Plio-Pleistocene species of pigs was at least partially dietary (Bishop et al., 2006).

KOLPOCHOERUS COOKEI (Brunet and White, 2001)

Holotype L116-14, a right m3 from Omo, Ethiopia site L-116, housed at the National Museums of Ethiopia.

Distribution Lower Omo Valley, Ethiopia.

Remarks This is another small and little known species of suid, with low-crowned molars and bunodont cheek teeth. Currently recognized from only two fossil sites in the Omo, Ethiopia, it is described as having higher crowns than *Ko. deheinzelini*, similar to those of *Po. afarensis*.

KOLPOCHOERUS MAJUS (Hopwood, 1934)

Holotype M.14682, a partial mandibular corpus with left p3–m1. From Olduvai Gorge, Tanzania, housed at the Natural History Museum, London.

Distribution Kenya, Tanzania, Uganda, Ethiopia.

Remarks This late species of *Kolpochoerus* possesses several derived characteristics which set it apart from the rest of the genus. *Kolpochoerus majus* is in many respects more primitive

than *Ko. heseloni*. There is no talon/id elaboration, but the cheek teeth are slightly more high crowned, while maintaining a basically bunodont aspect. Relative to smaller *Ko. heseloni* specimens, the enamel on the cheekteeth of *Ko. majus* is more rugose and thick, especially on the buccal aspect, where there is a protrusion of the crown elements.

Although long known from Olduvai, this species has more lately been recovered from numerous Pleistocene localities in East Africa, such as Olorgesailie, the Kapthurin Formation, the Nachukui Formation, the Konso Formation and Asbole in the Lower Awash, Bouri-Daka, and Garba IV (Suwa et al., 2003; Geraads et al., 2004). The known time range is from FLK I at Olduvai, 1.8 Ma, to Bed IV Olduvai, 0.70 Ma. One specimen from West Turkana, an M3 crown dated to 2.6 Ma, (800,000 years earlier than the next earliest example), has been attributed to this species by Harris et al. (1988) but may belong to *Ko. heseloni* instead. Another possible early example is from the Nyabusosi Fm of Uganda (Pickford, 1994; Geraads et al., 2004).

KOLPOCHOERUS PHACOCHOEROIDES (Thomas, 1884)

Holotype Mandible with m1–m3 and canine from Ain el Bey, Constantine, Algeria.

Distribution Morocco, Algeria.

Remarks This species is known primarily from North Africa, where a large sample from Ahl al Oughlam has been discovered and described by Geraads (1993). Problematically, this species overlaps in size with much of the material ascribed to *Ko. heseloni* but has some unique characteristics that some workers believe entitle the hypodigm to species status (Geraads, 1993; Cooke, 1997). Distinguishing it from the eastern and southern African forms are smaller premolars, a shortened muzzle and large canines in both males and females (Cooke, 1997). A thorough study of all the material would be important to understand the extent to which these characteristics are mirrored, if less common, in the hypodigm of *Ko. heseloni*, but at this point the species seems valid for, and restricted to, the North African examples.

Genus METRIDIOCHOERUS Hopwood, 1926

A number of genera were ultimately sunk into *Metridiochoerus* by Harris and White (1979). Confusion over the attribution of species and specimens was a result of the marked radiation undergone by this genus during the later part of the Pliocene and Pleistocene. In addition, many researchers failed to consider the apparent variations in cusp morphology that could be introduced by occlusal wear.

Metridiochoerus is the most advanced suid as related to a potential *Sus*-like suine ancestor. Species in the genus vary in size from small to the largest of East African extinct suids. Crania exhibit a variety of elaborations from none to an impressive array of cranial knobs and huge tusks. The length of the premolar row is reduced in all members of the genus, as is the size of the premolar teeth. In the most advanced specimens, the premolars are shed early in life, presumably as a result of wear and overcrowding. By the time the third molar has come into full occlusion, it is often the only postcanine tooth remaining in each quadrant. Increasing hypsodonty of the third molar is perhaps the most unifying trend in *Metridiochoerus*; fusion of their roots is usually delayed, often until the crown is in complete wear. Talon/id elaboration is often striking, with third molars having a highly variable number of main lateral pillars. Occlusal wear gives the dentine lakes on worn pillars a T- or mushroom-shaped appearance.

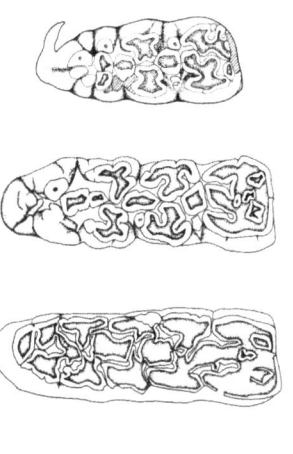

FIGURE 42.9 m3 elaboration in *Metridiochoerus andrewsi*. A) M.2013 from Makapansgat; B) KNM-ER 1169; C) KNM-ER 1089. The wear in the most recent specimen, KNM-ER 1089, has obliterated the borders of the pillars in the m3. This illustrates the difficulties in comparing cusp patterns between teeth of different wear stages. After Harris and White 1979:60 and 61, figures 102, 108, and 110. Courtesy of the American Philosophical Society.

METRIDIOCHOERUS ANDREWSI Hopwood, 1926
Figures 42.8 and 42.9

Holotype M14007, a right M3 from near Homa Mountain, Kenya, housed at the Natural History Museum, London.

Distribution Ethiopia, Kenya, Tanzania, Uganda, Malawi, South Africa.

Remarks This species encompasses a wide range of morphological variation. However, the continuum of morphological characters that unite this species and separate them from the other *Metridiochoerus* species appear to be robust. Many of these characteristics are primitive for the genus. Crania of *Met. andrewsi* are sexually dimorphic, with male crania having zygomatic knobs. The mandibular corpus of *Metridiochoerus* is neither as flared nor as constricted as in *Kolpochoerus*. Upper canines have enamel only in bands. The length and height of the premolar row are reduced in later forms. The molar row demonstrates an increase in hypsodonty coupled with a delay in fusion of the roots. Third molar length increases through the addition of talon/id lateral pillars and increased elaboration of the anterior cingulum trigon/id complex. The definition and separation of molar pillars does not persist throughout the crown height of the tooth, so that wear joins them together and gives the occlusal surface a complex and disorganized appearance.

Metridiochoerus andrewsi is a common suid at Pliocene and Pleistocene sites in eastern Africa and is known from the Shungura Formation, the Konso Formation, the Koobi Fora Formation, the Nachukui Formation, Chesowanja, western Kenya at Kanjera, Homa Mountain, and Olduvai Gorge, Tanzania (Harris and White, 1979; Suwa et al., 2003). Although some of these localities have not been radiometrically dated, numerous well-dated specimens from the others firmly establish a temporal range for this species. The earliest dated specimens come from the Usno Formation of Ethiopia, dated at approximately 3.4 Ma (White et al., 2006), and the youngest specimens are found at both West Turkana and Koobi Fora until 1.66 Ma. One surface specimen from Bed II Olduvai may postdate the latest Koobi Fora specimens. It has also been reported from the Chiwondo Beds of Malawi, Bolt's Farm, Swartkrans, Coopers, Makapansgat, and the Vaal River Gravels of South Africa (Harris and White, 1979; Kullmer, 2008)

Carbon stable isotope studies of *Met. andrewsi* diet have demonstrated that its diet might have changed during its time range, with earlier examples apparently a grass-dominated mixed feeder, while later examples were completely dependent on grazing tropical C_4 grasses (Bishop et al., 1999, Harris and Cerling, 2002).

METRIDIOCHOERUS MODESTUS (Van Hoepen and Van Hoepen, 1932)

Holotype C576, associated right and left M3s from Cornelia, RSA. Nasionale Museum, Bloemfontein.

Distribution Kenya, Ethiopia, Tanzania, South Africa.

Remarks Met. modestus, as the name implies, is the smallest of the *Metridiochoerus* species, resembling the modern warthog in size. The single known cranium is short and does not possess the zygomatic modifications of *Phacochoerus*. Third molars are very hypsodont, especially in relation to their crown area. In mature specimens, these are worn to a distinctly *Metridiochoerus* pattern, with obliteration of the internal pillar walls.

This suid is rare in numbers but well distributed geographically in the fossil record. It is found at Olduvai Gorge, West Turkana, Koobi Fora, the Konso Formation, the Omo, and Asbole in the Lower Awash (Harris and White, 1979; Suwa et al., 2003; Geraads et al., 2004). Its established temporal range is from Member G of the Shungura Formation, 2.21–0.70 Ma for the latest known specimens at Bed IV at Olduvai Gorge. It is known in South Africa from Cornelia and the Vaal River Gravels (Harris and White, 1979).

A study of the postcranial locomotor ecomorphology of *Met. modestus* has concluded that this diminutive member of the genus *Metridiochoerus* probably preferred more closed habitats (Bishop, 1999; Bishop et al., 1999). This contrasts with results of a study of its diet using carbon stable isotopes, which concludes that it was a tropical grass grazer (Harris and Cerling, 2002). This apparent discordance may be explained by more patchiness in past environments than is present today.

METRIDIOCHOERUS HOPWOODI (Leakey, 1958)

Holotype M14685, a partial left mandible with m2–m3 from the surface of Bed III, Olduvai Gorge. Housed at the Natural History Museum, London.

Distribution Eastern Africa: Kenya, Tanzania, Ethiopia.

Remarks Met. hopwoodi is another relatively rare species. The species is thought to be medium to large in size, although there is not enough cranial material to determine the extent of sexual dimorphism. The lower canine is verrucose. Molars are relatively narrow and hypsodont. Lateral pillars on the M3 display an unusual degree of symmetry about the central axis of the tooth. Third molar pillars are also separate and well defined to the base of the tooth root. When worn, the dentine lakes in third molar enamel pillars appear to be T shaped.

This taxon is found at Olduvai Gorge, Olorgesailie, Kanjera, Koobi Fora, West Turkana, the Konso Formation, and the Shungura Formation (Harris and White, 1979 [but not in list of specimens]; Suwa et al., 2003). The earliest specimen attributed to *Met. hopwoodi* is from site KL4 in the Nachukui Formation, dated to approximately 2.1 Ma. The youngest specimen is from Bed IV, Olduvai Gorge, and dates to 0.7 Ma.

Examples of *Met. hopwoodi* analysed for carbon stable isotope values have suggested that it was a C_4 grass grazer (Bishop et al., 1999; Cerling and Harris, 2002).

METRIDIOCHOERUS COMPACTUS (Van Hoepen and Van Hoepen, 1932)
Figure 42.10

Holotype C801, a right M3 from Cornelia, South Africa, Nasionale Museum, Bloemfontein.

Distribution Kenya, Tanzania, Ethiopia, South Africa.

Remarks Met. compactus is the largest and most advanced of the *Metridiochoerus* lineage. Its extremely derived and various nature—particularly the variation in tooth appearance caused by different stages of wear—was probably responsible for the number of generic and specific names given to this hypodigm. Cranial remains are very rare. The canines are laterally projecting, relatively straight, and frequently a foot (ca. 30 cm.) or more in length. In immature specimens, the premolar row is reduced; but in mature individuals, all postcanine teeth but the third molars are shed. The third molars are the largest of any *Metridiochoerus* species, and root fusion is very delayed, so the teeth are extremely hypsodont. The isolation of the talon/id pillars is not complete to the roots, so that worn teeth have a continuous outer margin of enamel. Worn, the dentine lakes exposed in the enamel pillars have a flattened Y shape. In the talon (but not the talonid), main lateral pillars pairs are often separated by a double row of minor medial (central) pillars.

This species is also well represented in space and time. It is known from the Shungura Formation, the Konso Formation, the Koobi Fora Formation, the Nachukui Formation, and Olduvai Gorge, as well as Kanjera, Chesowanja, and Olorgesailie (Harris and White, 1979; Suwa et al., 2003). The majority of

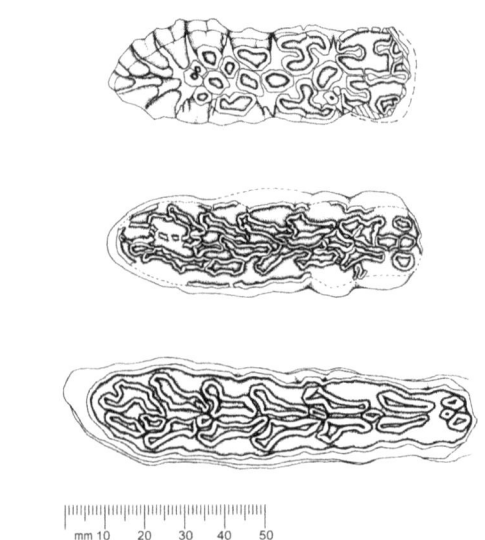

FIGURE 42.10 Examples of third molars in *Metridiochoerus compactus*. A) R. M3 KMN-ER 2659; B) L. M3 KNM-ER 2271; C) L. m3 KNM-ER 751. All examples from Koobi Fora, Kenya. After Harris and White 1979:57 and 62, figures 98, 99, and 120.

the earlier specimens come from Koobi Fora, while the most numerous sample is from Olduvai. Dated specimens range in age from the earliest at 1.92 Ma, from Bed I Olduvai, to the youngest 0.70 Ma ago from Bed IV Olduvai. There is only one Bed I *Met. compactus*, a M3 from DK I; the next oldest specimens derive from below the KBS tuff at Koobi Fora. *Metridiochoerus compactus* has been recovered from Cornelia, Elandsfontein, and the Vaal River Gravels (Harris and White, 1979).

Not surprisingly, given the extent of third molar elaboration in this taxon, carbon stable isotope analysis suggests that *Met. compactus* was an obligate grazer on tropical grasses (Harris and Cerling, 2002). Associated skeletons have not been attributed to this species, so to establish habitat preference has not so far been possible. However, the assumption since this species was identified has been that it preferred open grassland habitats.

Family SANITHERIIDAE Simpson, 1945

The sanitheres are a widely distributed but poorly understood family of small bodied suoids. Their patchy fossil record extends throughout the Miocene of Africa, Europe, and subcontinental Asia (Pickford, 1984, 1993, 2004). This group is first known from Africa in the early Miocene and survives there until ca. 14 Ma (Pickford, 2004). They are known from Eurasian contexts during the middle Miocene, persisting until the late Miocene in the Chinji Formation of Pakistan. The Sanitheriidae have unusual morphologies that separate them from other suoids. While the preserved basicrania have features similar to the Suidae, their facial and dental anatomy are unique, with a mosaic of features that recall various artiodactyl taxa (Harris and Liu, 2007). Enigmatic and poorly known, the taxonomic position of sanitheres has been disputed by recent research. They have been relegated to a subfamily of the Palaeochoeridae (Made, 1997); however, recent discoveries (Pickford, 2004) and a phylogenetic analysis of Liu (2003) suggest that sanitheres are a distinct family within the Suoidea. Recently described fossil remains from Kenya and Namibia have greatly enhanced the understanding of the African sanitheres (Pickford, 2004). Currently there is only one genus of sanithere, *Diamantohyus*, recognized from African deposits; no African fossils are considered sufficiently derived to be attributed to the later genus *Sanitherium* von Meyer, 1886.

Sanitheres have sexually dimorphic canines. The postcanine teeth of sanitheres are bunoselenodont. Upper molars have lingual cusps that are selenoid, an appearance enhanced by wear. Lingual cusps are separated only slightly by weak "furchen" or grooves. The enamel on the lingual side of the tooth has a wrinkled appearance. The buccal cusps are bunodont and can appear more conical, and they are aligned obliquely to the long axis of the molar. There is a buccal cingulum that has a distinctive, beaded appearance on its edge. Premolars are molarized and have extra cusps. This unique dentition in emplaced with an occlusal curvature which appears more like those of selenodont artiodactyls; the overall tooth row convex ventrally rather than dorsally as is the case for other suoids. This contradictory appearance is enhanced by various other features of sanithere dentition, such as the curvature of the inferior border of the mandible, wrinkled enamel, and a "*Palaeomeryx* fold" (a short enamel ridge running from the posterior of the protoconid) on the lower molars (Harris and Liu, 2007). However, despite the ruminant appearance of these gnathic features, the postcranial anatomy firmly allies sanitheres with the suoids (Pickford, 1984, 2004). Another significant feature is in the orientation of their crania as indicated by the position of their occipital condyles. Sanithere heads were perhaps held in a more horizontal orientation (Pickford and Tsujikawa, 2004).

Sanithere ecology might differ from many Suidae; Pickford (2004) suggests that they may have been more cursorial than pigs on the basis of their elongated and gracile limb anatomy. The presence of a "carnassial-like" shearing morphology on the buccal main cusps of P2 and P3 has lead to the suggestion that sanitheres may have been more carnivorous than their fellow suoids (Pickford, 2004).

Genus *DIAMANTOHYUS* Stromer, 1926

Diamantohyus is defined relatively; it has less molarization of the premolars than *Sanitherium* von Meyer, 1866, a more derived genus. The P4 has three main cusps and two subsidiary cusps. Postcanine teeth have relatively undeveloped lingual cusps. *Diamantohyus* incisors are low crowned and short rooted. Extensive new material attributed to this genus has been described recently (Pickford and Tsujikawa, 2005; Pickford et al., 2010). Short of repeating the descriptions of this material, some general observations to enable identification of these two rare Miocene taxa are made here.

DIAMANTOHYUS AFRICANUS Stromer, 1926

Holotype Right maxilla fragment with P3, M1, and M2 (Stromer 1922:332; after Pickford, 2004).
Distribution Namibia, Egypt, Libya, Kenya, Uganda.
Remarks Diamantohyus africanus has relatively undeveloped premolars and molars for this family, although the sanithere tendency toward polycuspy and polycristy is already apparent. The anterior origin of the zygomatic arch lies far anterior to those of other suoids.

DIAMANTOHYUS NADIRUS Wilkinson, 1976

Holotype KNM-OM 40 from Ombo.

Distribution Kenya.
Remarks Diamantohyus nadirus is more derived than *D. africanus* and has more complex molars with defined and separated crests and cusplets. Lower molars have a well-defined metastylid.

Suoid Evolutionary Relationships

For the most part, the evolution of African suoids during the Miocene is dominated by influxes of Eurasian taxa that subsequently evolve *in situ* in the African continent and then either die out there without issue or migrate back into Eurasia in slightly modified form. During the Miocene, Pickford (2006) estimates that there were at least five migrations of suoids into Africa from Eurasia and probably two back in the other direction; both the later Miocene Kubanochoerinae and the Pliocene Tetraconodontinae in Asia appear to have African forebears. In the case of the Listriodontinae, the picture is even more complicated, with similarities in the timing and degree of lophodonty development between African and Eurasian forms suggesting frequent, if not continuous, gene flow between populations on both continents (Pickford, 2006). The Sanitheriidae seem to originate in Africa in the early Miocene and then spread to Eurasia; this latter radiation considerably outlasts the known African examples, which disappear from the record ca. 14 Ma.

While the patchiness of the fossil record in the Miocene of Africa limits the extent to which biogeography and phylogenetic relationships can be reconstructed, the situation happily improves during the latest Miocene and Pliocene. Only two of the suid lineages, the Cainochoerinae and the Tetraconodontinae, survive across the Miocene-Pliocene boundary. For both of these unrelated lineages, this success is relatively short-lived. Cainochoerinae are rare in the fossil record; earlier bunodont forms are found in a few middle Miocene localities, and then the later, more lophodont but still tiny *Cainochoerus africanus* has only been described from a few very large faunal samples—primarily from Langebaanweg (the type site) and Lothagam. So while this subfamily was present for a long time in the fossil record, it has never been a common element. The Tetraconodontinae have a large Pliocene radiation and are extremely common in Pliocene African sites but go extinct without issue in the early part of the Pleistocene with the disappearance of the last and most specialized member of the subfamily, *Notochoerus scotti*.

The real success story of suoid evolution in African is the large flourishing of species of suine pig following their arrival in Africa, sometime before ca. 4.5 million years ago. There were most probably two migrations of these pigs around or after this time, one founding the *Potamochoerus/Kolpochoerus* lineage (first appearance ca. 4.3 Ma) and another founding the *Metridiochoerus* radiation, the earliest record for which is from the 3.4 Ma Usno Formation of Ethiopia (White et al., 2006). *Metridiochoerus* is relatively uncommon during the earlier part of its known range but after 2.5 Ma experiences both speciation events and also, in the most primitive species *Met. andrewsi*, increases in both third molar complexity and crown height through time. These two aspects of the *Metridiochoerus* radiation make it a very valuable biostratigraphic indicator. *Potamochoerus/Kolpochoerus* arrives in Africa from Eurasia perhaps slightly earlier than *Metridiochoerus* and also experiences a significant evolutionary radiation. *Kolpochoerus* is consistently present at late Pliocene and Pleistocene sites, and *Potamochoerus* makes only occasional tantalizing appearances.

While both the Pliocene immigrants experience radiations characterized by speciations and rapid increase in hypsodonty, *Potamochoerus/Kolpochoerus* always maintains a primitive streak; the modern relicts of this radiation (*Potamochoerus* and *Hylochoerus*) are closer to the primitive than to the craniodentally derived forms of this radiation. The same is true of the last surviving member of the *Metridiochoerus* lineage; the modern warthog *Phacochoerus* most closely resembles the diminutive *Met. modestus*, a slightly more specialized form than the basal *Metridiochoerus andrewsi*. The more derived (and much larger) extreme results of both *Kolpochoerus* and *Metridiochoerus* evolution went extinct some time in the later Pleistocene, and the smaller and more primitive forms survive to this day.

ACKNOWLEDGMENTS

This review relies heavily on work by suid taxonomists who have been working in this field for years: Basil Cooke, John Harris, Martin Pickford, and Jan van der Made. Without them and other pig palaeontologists, little of suoid taxonomy would make any sense to me. Thanks are due to the editors, both for asking me to contribute to this volume and for patiently awaiting this chapter. I would also like to thank the Leverhulme Trust for their continued support, and Andrew Hill for getting me interested in pigs in the first place.

Literature Cited

Ansell, W. F. H. 1971. Artiodactyla; 15:1–84. in J. Meester and H. Setzer (eds.), *The Mammals of Africa: An Identification Manual*. Smithsonian Institution Press, Washington, D.C.

Arambourg, C. 1968. Un suide fossile nouveau du Miocène superieur de l'Afrique du Nord. *Bulletin de la Société Géologique de France* 10:110–115.

Behrensmeyer, A. K. 1975. The taphonomy and paleoecology of Plio Pleistocene vertebrate assemblages east of Lake Rudolf, Kenya. *Bulletin of the Museum of Comparative Zoology.* 146:473–578.

Bigourdan, J. 1948. Le phacochère et les suides dans l'Ouest africain. *Bulletin de l'Institut Français d'Afrique Noire* 10:285–360.

Bishop, L. C. 1994. Pigs and the ancestors: hominids, suids and environments during the Plio-Pleistocene of East Africa. Unpublished Ph.D. dissertation, Yale University.

———. 1999. Suid palaeoecology and habitat preference at African Pliocene and Pleistocene hominid localities; pp. 216–225 in T. G. Bromage and F. Schrenk (eds.), *African Biogeography, Climate Change and Early Hominid Evolution*. Oxford University Press, Oxford.

Bishop, L. C., and A. Hill 1999. Fossil Suidae from the Baynunah Formation, Emirate of Abu Dhabi, UAE; pp. 252–270 in P. J. Whybrow and A. Hill (eds.), *Fossil Vertebrates of Arabia*. Yale University Press, New Haven.

Bishop, L. C., A. Hill, and J. Kingston 1999. Palaeoecology of Suidae from the Tugen Hills, Baringo, Kenya; pp. 99–111 in P. Andrews and P. Banham (eds.), *Late Cenozoic Environments and Hominid Evolution: A Tribute to Bill Bishop*. Special Publications of the Geological Society, London.

Bishop, L. C., T. King, A. Hill, and B. Wood. 2006. Palaeoecology of *Kolpochoerus heseloni* (= *K. limnetes*): A multiproxy approach. *Transactions of the Royal Society of South Africa* 61:81–88.

Boisserie, J.-R., F. Lihoreau, and M. Brunet 2005. The position of Hippopotamidae with Cetartiodactyla. *Proceedings of the National Academy of Sciences, USA* 102:1537–1541.

Bradley, R. M. 1971. Warthog (*Phacochoerus aethiopicus* Pallas) burrows in Nairobi National Park. *East African Wildlife Journal* 9:149–152.

Brunet, M., and T. D. White 2001. Deux nouvelles espèces de Suini (Mammalia, Suidae) du continent African (Éthiopie, Tchad). *Comptes Rendus de l'Académie des Sciences, Paris, Sciences de la Terrre et des Planètes* 332:51–57.

Child, G., H. H. Roth, and M. Kerr 1968. Reproduction and recruitment patterns in warthog (*Phacochoerus aethiopicus*) populations. *Mammalia* 32:6–29.

Child, G., L. Sowls, and B. L. Mitchell. 1965. Variations in the dentition, aging criteria and growth patterns in warthog. *Arnoldia* 1(38):1–23.

Clough, G. 1969. Some preliminary observations on reproduction in the warthog, *Phacochoerus aethiopicus* Pallas. *Journal of Reproduction and Fertility* (suppl.) 6:323–337.

Clough, G., and A. G. Hassam 1970. A quantitative study of the daily activity of the warthog in the Queen Elizabeth National Park, Uganda. *East African Wildlife Journal* 8:19–24.

Cooke, H. B. S. 1974. The fossil mammals of Cornelia, O.F.S., South Africa. *Memoirs of the National Museum of Bloemfontein* 9:63–84.

———. 1976. Suidae from the Plio-Pleistocene strata of the Rudolf Basin; pp. 251–263 in Y. Coppens, F. C. Howell, G. L. Isaac, and R. E. F. Leakey (eds.), *Early Man and Environments in the Lake Rudolph Basin*. University of Chicago Press, Chicago.

———. 1978. Suid evolution and correlation of African hominid localities: An alternate taxonomy. *Science* 201:460–463.

———. 1985. Plio-Pleistocene Suidae in relation to African hominid deposits; pp. 101–117 in *L'Environnement des Hominides ou Plio-Pleistocene*. Colloque International (Juin 1981) organisée par la Fondation Singer-Polignac. Masson, Paris.

———. 1987. Fossil Suidae from Sahabi, Libya; pp. 255–266 in N. T. Boaz, A. El-Arnauti, A. W. Gaziry, J. de Heinzelin, and D. D. Boaz (eds.), *Neogene Paleontology and Geology of Sahabi*. Liss, New York.

———. 1997. The status of the African fossil suids *Kolpochoerus limnetes* (Hopwood, 1926), *K. phacochoeroides* (Thomas, 1884) and "*K*" *afarensis* (Cooke, 1978) *Geobios* 30:121–126.

Cooke, H. B. S., and R. F. Ewer 1972. Fossil Suidae from Kanapoi and Lothagam, northwestern Kenya. *Bulletin of the Museum of Comparative Zoology* 143:149–296.

Cooke, H. B. S., and Q. B. Hendey 1992. *Nyanzachoerus* (Mammalia: Suidae: Tetraconodontinae) from Langebaanweg, South Africa. *Durban Museum Novitates* 17:1–20.

Cooke, H. B. S., and A. F. Wilkinson 1978. Suidae and Tayassuidae; pp. 435–482 in V. J. Maglio and H. B. S. Cooke (eds.), *Evolution of African Mammals*. Harvard University Press, Cambridge, Mass.

Copley, H. 1949. The pigs of Kenya. *Nature in East Africa* 2(2):6–10.

Cumming, D. H. M. 1975. *A Field Study of the Ecology and Behaviour of Warthog*. Museum Memoir 7. (Salisbury) Trustees of the National Museums and Monuments of Rhodesia, 175 pp.

Deane, N. N. 1962. Bushpig and warthog feeding. *The Lammergeyer* 2(1):67.

Delaney, M. J., and D. C. D. Happold 1979. *Ecology of African Mammals*. Longman, New York. 448 pp.

d'Huart, J.-P. 1978. Écologie de l'Hylochère (*Hylochoerus meinertzhageni* Thomas) au Parc National des Virunga: Exploration du Parc National des Virunga. *Fondation pour Favoriser les Recherches Scientifiques en Afrique, Brussels* 2(25):1–156.

———. 1991. Monographie des Reisenwaldschweines. *Bongo* 18:103–118.

———. 1993. The forest hog (*Hylochoerus meinertzhageni*); pp. 84–92 in W. L. R. Oliver (ed.), *Pigs, Peccaries and Hippos: Status Survey and Conservation Action Plan*. IUCN, Gland, Switzerland.

d'Huart, J.-P., and P. Grubb 2001 Distribution of the common warthog (*Phacochoerus africanus*) and the desert warthog (*Phacochoerus aethiopicus*) in the Horn of Africa. *African Journal of Ecology* 39:156–169.

Dobzhansky, T. D. 1937. *Genetics and the Origin of Species*. Columbia University Press, New York, 364 pp.

Dorst, J., and P. Dandelot, P. 1972. *A Field Guide to the Larger Mammals of Africa*. Collins, London, 287 pp.

Ducrocq, S., Y. Chaimanee, V. Suteethorn, and J.-J. Jaeger. 1998. The earliest known pig from the Upper Eocene of Thailand. *Palaeontology* 41:147–156

Durrell, G. 1993. Preface; in W. L. R. Oliver (ed.), *Pigs, Peccaries and Hippos:, Status Survey and Conservation Action Plan*. IUCN, Gland, Switzerland.

Edmond-Blanc, F. 1960. Contribution a l'étude du comportement et de la composition de la nourriture du bongo (*Boocercus eurycerus isaaci*) et de l'hylochère (*Hylochoerus meinerzhageni*) du versant sud du Mont Kenya. *Mammalia* 24:538–541.

Ewer, R. F. 1957. A collection of *Phacochoerus aethiopicus* teeth from the Kalkbank Middle Stone Age site, central Transvaal. *Palaeontologia Africana* 5:5–20.

———. 1958. Adaptive features in the skulls of African Suidae. *Proceedings of the Zoological Society, London* 131:135–155.

———. 1970. The head of the forest hog, *Hylochoerus meinertzhageni*. *East African Wildlife Journal* 8:43–52.

Forsyth Major, C. 1897. On the species of *Potamochoerus*, the bush-pigs of the Ethiopian region. *Proceedings of the Zoological Society, London* 1897:359–370.

Fradrich, H. 1965. Zur Biologie und Ethologie des Warzebschweines (*Phacochoerus aethiopicus* Pallas) unter Berücksichtigung des Verhaltens anderer Suiden. *Zeitschrift für Tierpsychologie* 22:328–393.

Gabunia, L. K. 1958. On a skull of a horned pig from the Middle Miocene of the Caucasus [in Russian]. *Doklady Akademia Nauk Azerbeidjan SSR* 118:1187–1190.

Geigy, R. 1955. Observations sur les Phacochères du Tanganyika. *Revue Suisse de Zoologie* 62:139–163.

Geraads, D. 1993. *Kolpochoerus phacochoeroides* (Thomas, 1884)(Suidae, Mammalia), du Pliocène supérieur de Ahl al Oughlam (Casablanca, Maroc). *Geobios* 26:731–743.

Geraads, D., Z. Alemseged, D. Reed, J. Wynn, and D. C. Roman 2004. The Pleistocene fauna (other than primates) from Asbole, lower Awash Valley, Ethiopia, and its environmental and biochronological implications. *Geobios* 37:697–718.

Geraads, D., C. Guérin, and M. Faure, 1986. Les suidés (Artiodactyla, Mammalia) du gisement Pléistocène Ancien d'Oubeidiyeh (Israel); pp. 93–105 in E. Tchernov (ed.), *Les Mammifères du Pléistocène Inférieur de la Vallée du Jourdain à Oubeidiyeh*. Association Paleorient, Paris.

Ghiglieri, M. P., T. M. Butynski, T. T. Struhsaker, L. Leland, S. J. Wallis, and P. Waser, 1982. Bush pig (*Potamochoerus porcus*) polychromatism and ecology in Kibale forest, Uganda. *African Journal of Ecology* 20:231–236.

Grubb, P. 1993. The Afrotropical suids: *Potamochoerus*, *Hylochoerus* and *Phacochoerus*; pp. 66–74 in W. L. R. Oliver (ed.), *Pigs, Peccaries and Hippos:, Status Survey and Conservation Action Plan*. IUCN, Gland, Switzerland.

Harris, J. M. 1983. Family Suidae; pp. 215–302 in J. M. Harris (ed.), *Koobi Fora Research Project: Volume 2. The Fossil Ungulates: Proboscidea, Perissodactyla, and Suidae*. Clarendon Press, Oxford.

Harris, J. M., F. H. Brown, and M. G. Leakey. 1988. Stratigraphy and paleontology of Pliocene and Pleistocene localities west of Lake Turkana, Kenya. *Contributions in Science* 399:1–128.

Harris, J. M., and T. E. Cerling 2002. Dietary adaptations of extant and Neogene African suids. *Journal of the Zoological Society of London* 256:45–56.

Harris, J. M., and M. G. Leakey 2003. Lothagam Suidae; pp. 485–519 in M. G. Leakey and J. M. Harris (eds.), *Lothagam: The Dawn of Humanity in Eastern Africa*. Columbia University Press, New York.

Harris, J. M., M. G. Leakey, T. E. Cerling, and A. J. Winkler 2003. Early Pliocene tetrapod remains from Kanapoi, Lake Turkana Basin, Kenya. *Contributions in Science, National History Museum of Los Angeles County* 498:39–113.

Harris, J. M., and L. Liu 2007. Superfamily Suoidea; pp. 130–150 in D. R. Prothero and S. E. Foss (eds.), *The Evolution of Artiodactyls* Johns Hopkins University Press, Baltimore.

Harris, J. M., and T. D. White 1979. Evolution of the Plio-Pleistocene African Suidae. *Transactions of the American Philosophical Society* 69(2):1–128.

Hendey, Q. B. 1976. Fossil peccary from the Pliocene of South Africa. *Science* 192:787–789.

Hill, A., R. Drake, L. Tauxe, M. Monaghan, J. C. Barry, A. K. Behrensmeyer, G. Curtis, B. F. Jacobs, L. Jacobs, N. Johnson, and D. Pilbeam. 1985. Neogene palaeontology and geochronology of the Baringo Basin, Kenya. *Journal of Human Evolution* 14:759–773.

Hill, A., S. Ward, and B. Brown 1992. Anatomy and age of the Lothagam mandible. *Journal of Human Evolution* 22:439–451.

Kingdon, J. 1979. *East African Mammals: Volume IIIB. Large Mammals*. University of Chicago Press, Chicago, 450 pp.

Kullmer, O. 2008. The fossil Suidae from the Plio-Pleistocene Chiwondo Beds of northern Malawi, Africa. *Journal of Vertebrate Paleontology* 28:208–216.

Kullmer, O., O. Sandrock, T. B. Viola, W. Hujer, H. Said, and H. Seidler 2008. Suids, elephantoids, palaeochronology, and palaeoecology of the Pliocene hominid site Galili, Somali Region, Ethiopia. *Palaios* 23:452–464.

Leister, C. W. 1939. The wild pigs of the world. *Bulletin New York Zoological Society* 42:131–139.

Leonardi, P. 1952. Resti fossili di "*Sivachoerus*" del giacimento di Sahabi in Cirenaica (Africa Settentrionale). *Atti della Accademia Nazionale dei Lincei* 13:166–169.

Leuthold, W. 1977. *African Ungulates: A Comparative Review of Their Ethology and Behavioral Ecology*. Springer, New York, 307 pp.

Liu, L. 2003. Chinese fossil Suoidea: Systematics, evolution, and paleoecology. Unpublished PhD dissertation, University of Helsinki, Department of Geology, Division of Geology and Paleontology, 40 pp.

Lönnberg, E. 1908. Remarks on some wart-hog skulls in the British Museum. *Proceedings of the Zoological Society, London* 1908:936–940.

Luck, C. P. 1965. The comparative morphology of the eyes of certain African Suiformes. *Vision Research* 5:283–297.

Maberly, C. T. A. 1950. The African bush-pig: sagacious and intelligent. *African Wildlife* 4:14–18.

———. 1966. African bushpigs. *Animals* 9:556–561.

Made, J. van der. 1992. African lower and middle Miocene Suoidea (pigs and peccaries): 3. Stratigraphy and correlations. *VIII Jornadas de Palaeontologia, Barcelona*, Ocotber 8–10.

———. 1996. Listriodontinae (Suidae, Mammalia), their evolution, systematics and distribution in time and space. *Contributions to Tertiary and Quaternary Geology* 33:1–254

———. 1997. Systematics and stratigraphy of the genera *Taucanamo* and *Schizochoerus* and a classification of the Palaeochoeridae (Suoidea, Mammalia). *Proceedings Koninklijke Nederlandse Akademie van Wetenschappen* 100:127–139.

Milstein, P. le S. 1971. The bushpig, *Potamochoerus porcus*, as a problem animal in South Africa; pp. 1–10 in *Proceedings of the Entomological Symposium, Pretoria. 27 September 1971*. (Mimeograph cited in Kingdon, 1979.)

Oliver, W. L. R. (ed.). *Pigs, Peccaries and Hippos: Status Survey and Conservation Action Plan*. IUCN, Gland, Switzerland, 294 pp.

Pallas, P. S. 1766. *Elenchus Zoophytorum Sistens Generum Adumbrationes Generaliores et Specierum Cognitarum Succinctas Descriptiones cum Selectis Auctorum Synonymis*. Hagae, 451 pp.

Phillips, J. F. V. 1926. "Wild pig" (*Potamochoerus choeropotamus*) at the Knysna: Notes by a naturalist. *South African Journal of Science* 23:655–660.

Pickford, M. 1984. A revision of the Sanitheriidae (Suiformes, Mammalia). *Geobios* 17:133–154.

———. 1986. A revision of the Miocene Suidae and Tayassuidae, (Artiodactyla, Mammalia) of Africa. *Tertiary Research Special Paper* 7:1–83.

———. 1988. Un étrange suidé nain du Néogène supérieur de Langebaanweg. *Annales de Paléontologie* 74:229–250.

———. 1989 New specimens of *Nyanzachoerus waylandi* (Mammalia, Tetraconodontinae) from the type area, Nyaburogo (Upper Miocene), Lake Albert Rift, Uganda. *Geobios* 22:641–651.

———. 1993. Old world suoid systematics, phylogeny, biogeography. and biostratigraphy. *Paleontologia i Evolució* 26–27:237–269

———. 1994. Fossil Suidae of the Albertine Rift, Uganda-Zaire; pp. 339–373 in M. Pickford and B. Senut (eds.), *Geology and Palaeobiology of the Albertine Rift Valley, Uganda-Zaire: Volume II. Palaeobiology*. Occasional Publications 29, CIFEG, Orléans.

———. 1995. Suidae (Mammalia, Artiodactyla) from the Early Middle Miocene of Arrisdrift, Namibia: *Namachoerus* (gen. nov.) *moruoroti*, and *Nguruwe kijivium*. *Comptes Rendus de l'Académie des Sciences, Paris, Série IIa*, 320:319–326.

———. 2001. New species of *Listriodon* (Suidae, Mammalia) from Bartule, Member A, Ngorora Formation (ca 13 Ma), Tugen Hills, Kenya. *Annales de Paléontologie* 87:209–223.

———. 2004. Miocene Sanitheriidae (Suiformes, Mammalia) from Namibia and Kenya: Systematic and phylogenetic implications. *Annales de Paléontologie* 90:223–278.

———. 2006. Synopsis of the biochronology of African Neogene and Quaternary Suiformes. *Transactions of the Royal Society of South Africa* 61:51–62.

———. 2007. A new suiform (Artiodactyla, Mammalia) from the Early Miocene of East Africa. *Comptes Rendus Palevol* 6:221–229.

Pickford, M., E. R. Miller, and A. N. El-Barkooky. 2010. Suidae and Sanitheriidae from Wadi Mghra, early Miocene, Egypt. *Acta Palaeontologica Polonica* 55:1–11.

Pickford, M., and B. Senut 1997. Cainozoic mammals from coastal Namaqualand, South Africa. *Paleontologia Africana* 34:199–217

Pickford, M., and Tsujikawa, H. 2005 A partial cranium of *Diamantohyus nadirus* (Sanitheriidae, Mammalia) from the Aka Aiteputh Formation (16–15 Ma), Kenya. *Palaeontological Research* 9:319–328.

Pickford, M., and Y. Kunimatsu 2005. Catarrhines from the Middle Miocene (ca. 14.5 Ma) of Kipsaraman, Tugen Hills, Kenya. *Anthropological Science* 113:189–224.

Pilgrim, G. E. 1926. The fossil Suidae of India. *Memoirs of the Geological Survey of India, Palaeontologica Indica* 4:1–68.

Rothschild, M., and H. Neuville 1906. L'*Hylochoerus meinertzhageni* O. Ths. *Bulletin de la Société Philomathique de Paris* 8:141–164.

Sale, J. B., et al. 1976. *Report of the Working Group on the Distribution and Status of East African Mammals: Phase I.* East African Wildlife Society, 110 pp.

Scotcher, J. S. B. 1973. Diurnal feeding by bushpig. *The Lammergeyer* 19:33–34.

Shaw, J. C. M. 1939. Growth changes and variations in wart hog third molars and their palaeontological importance. *Transactions of the Royal Society of South Africa* 28:259–299.

Sjarmidi, A., and J. F. Gerard 1988. Autour de la systématique et la distribution des suides. *Monitore Zoologico Italiano* (n.s.) 22:415–448.

Skinner, J. D., G. J. Breytenbach, and C. T. A. Maberly 1976. Observations on the ecology and biology of the bushpig *Potamochoerus porcus* Linn in the Northern Transvaal. *South African Journal of Wildlife Research* 6:123–128.

Sowls, L. K., and R. J. Phelps 1966. Body temperatures of juvenile warthogs *(Phacochoerus aethiopicus)* and bushpigs *(Potamochoerus porcus). Journal of Mammalogy* 47:134–137.

Stewart, D. R. M., and J. Stewart 1963. The distribution of some large mammals in Kenya. *Journal of the East Africa Natural History Society and Coryndon Museum* 24(3):1–52.

Suwa, G., H. Nakaya, B. Asfawe, H. Saegusa, A. Amzaye, R. T. Kono, Y. Beyene, and S. Katoh 2003. Plio-Pleistocene terrestrial mammal assemblage from Konso, southern Ethiopia. *Journal of Vertebrate Paleontology* 23:901–916.

Thomas, A. D., and F. F. Kolbe 1942. The wild pigs of South Africa: Their distribution and habits, and their significance as agricultural pests and carriers of disease. *Journal of the South African Veterinary Medical Association* 13:1–11.

Thomas, O. 1904. On *Hylochoerus,* the forest-pig of Central Africa. *Proceedings of the Zoological Society, London* 2(13):193–199.

Tong, Y. S. and Z. R. Zhao 1986. *Odoichoerus,* a new suoid (Artiodactyla, Mammalia) from the Early Tertiary of Guangxi. *Vertebrata PalAsiatica* 24:129–138.

Tookey, E. B. (1959). Pigs. *Wild Life* 1(2):17–20.

Turner, A., and M. Antón. 2004. *Evolving Eden.* Columbia University Press, New York, 269 pp.

Vignaud, P., P. Duringer, H. T. Mackaye, A. Likius, C. Blondel, J.-R. Boisserie, L. de Bonis, V. Eisenmann, M.-E. Etienne, D. Geraads, F. Guy, T. Lehmann, F. Lihoreau, N. Lopez-Martinez, C. Mourer-Chauviré, O. Otero, J.-C. Rage, M. Schuster, L. Viriot, A. Zazzo, and M. Brunet. 2002. Geology and paleontology of the Upper Miocene Toros-Menalla hominid locality, Chad. *Nature* 418:152–155.

White, T. D. 1995. African omnivores: global climatic change and Plio-Pleistocene hominids and suids; pp. 369–384 in E. S. Vrba, G. H. Denton, T. C. Partridge, and L. H. Burckle (eds.), *Paleoclimate and Evolution, with Emphasis on Human Origins.* Yale University Press, New Haven.

White, T. D., and Harris, J. M. 1977. Suid evolution and correlation of African hominid localities. *Science* 198:13–21.

White, T. D., and G. Suwa. 2004. A new species of *Notochoerus* (Artiodactyla, Suidae) from the Pliocene of Ethiopia. *Journal of Vertebrate Paleontology* 24:474–480.

White, T. D., G. WoldeGabriel, B. Asfaw, S. Ambrose, Y. Beyene, R. L. Bernor, J.-R. Boisserie, B. Currie, H. Gilbert, Y. Haile-Selassie, W. K. Hart, L. J. Hlusko, F. C. Howell, R. T. Kono, T. Lehmann, A. Louchart, C. O. Lovejoy, P. R. Renne, H. Saegusa, E. S. Vrba, H. Wesselman, and G. Suwa. 2006. Asa Issie, Aramis and the origin of *Ardipithecus. Nature* 440:883–889.

Wilkinson, A. 1976. The lower Miocene Suidae of Africa. *Fossil Vertebrates of Africa* 4:173–282.

FORTY-THREE

Anthracotheriidae

PATRICIA A. HOLROYD, FABRICE LIHOREAU,
GREGG F. GUNNELL, AND ELLEN R. MILLER

Anthracotheriidae Leidy, 1869, are a group of bunodont to selenodont artiodactyls distributed throughout the Old World and North America. The earliest anthracotheriids appear in the latest middle Eocene in Asia, and they survive into the late Miocene in Africa and Asia. Because members of the family are widespread, the group has often been important for interpretations of biogeography. Anthracotheres have also been pivotal in discussions of mammalian phylogeny, as some work suggests that the origin of Hippopotamidae may be rooted within Anthracotheriidae (Colbert, 1935a, 1935b; Black, 1978; Boisserie et al., 2005; Boisserie and Lihoreau, 2006), although this conclusion is not universally accepted (e.g., Pickford, 2007b, 2008).

In Africa, members of the family are first recorded from the late Eocene Qasr el Sagha Formation, Egypt (Holroyd et al., 1996), and the family persists through the late Miocene (Pickford, 1991b; Vignaud et al. 2002; Delmer et al., 2006). During the Miocene, anthracotheres had an extended range across eastern, central, southern, and northern Africa, although their diversity in Africa appears to have always been greatest in North Africa (figure 43.1).

Black (1978) provided the first review of the entire African record of this family. The Paleogene forms have since been considerably revised by Ducrocq (1997) and the Neogene ones by Pickford (1991b). Other important sources that have expanded knowledge of the systematics, paleoecology, and biogeographic relationships of African anthracotheres include Gaziry (1987a, 1987b), Ducrocq et al. (2001), Lihoreau and Ducrocq (2007), Lihoreau et al. (2006), Miller (1996), and Pickford (2006).

NOTE ON ANTHRACOTHERE SYSTEMATICS

Within Anthracotheriidae, species differ from each other primarily in size, specializations of the anterior dentition, and degree of brachydonty versus selenodonty. Anthracothere genera differ in molar morphology, but molar morphology can be conservative even when the size differences and specializations of the anterior dentition are considerable. Also, a large degree of sexual dimorphism has been documented for some anthracothere species, but the extent of sexual dimorphism in other species is not as clear.

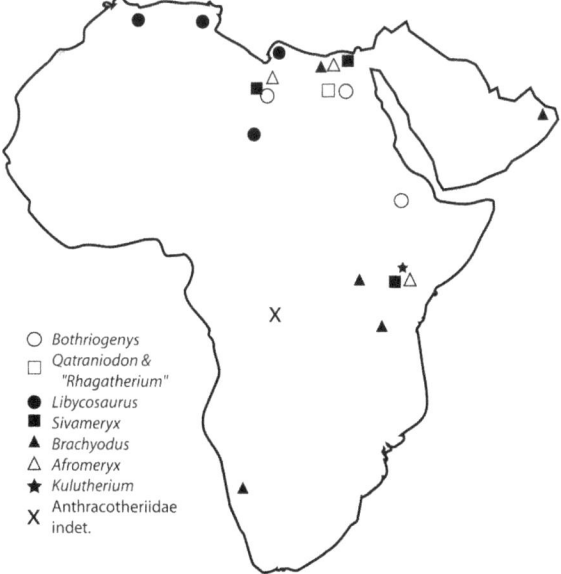

○ *Bothriogenys*
□ *Qatraniodon &*
   "Rhagatherium"
● *Libycosaurus*
■ *Sivameryx*
▲ *Brachyodus*
△ *Afromeryx*
★ *Kulutherium*
X Anthracotheriidae
   indet.

FIGURE 43.1 Map showing distribution of anthracotheriids in Afro-Arabia. For specific occurrences, see table 43.1.

Family ANTHRACOTHERIIDAE Leidy, 1869
Genus *BOTHRIOGENYS* Schmidt, 1913
Figures 43.2A, 43.2B, 43.2D–43.2G, and 43.2I

Selected Synonymy Brachyodus Schmidt, 1913, *Mixtotherium* Schmidt, 1913.

Included Species B. fraasi Schmidt, 1913 (*Brachyodus [Bothriogenys] fraasi* Schmidt, 1913); *B. andrewsi* Schmidt, 1913; *B. gorringei* (Andrews and Beadnell, 1902); *B. rugulosus* Schmidt, 1913.

Age and Occurrence Late Eocene to Oligocene of North Africa; see table 43.1.

Diagnosis After Lihoreau and Ducrocq (2007), with additional differential characters. Small to medium-sized anthracothere, no angular process on the shallow mandible, complete dental formula, no incisor enlargement, simple and elongated premolars, small canines, five-cusped upper molars with flattened parastyle and mesostyle, lower molars with short

TABLE 43.1
Major occurrences and ages of African anthracotheriids

Taxon	Occurrence (Site, Locality)	Stratigraphic Unit	Age
Bothriogenys sp.	Fayum, Egypt	Dir Abu Lifa Mbr., Qasr el Sagha Fm.	Late Eocene
B. fraasi	Fayum, Egypt	Upper sequence, Jebel Qatrani Fm.[1]	Early Oligocene
B. andrewsi	Fayum, Egypt	Jebel Qatrani Fm.	Early Oligocene
B. gorringei	Fayum, Egypt	Lower sequence, Jebel Qatrani Fm.[1]	Early Oligocene
	cf. Zellah, Libya		Early Oligocene
B. rugulosus	Fayum, Egypt	Jebel Qatrani Fm.	Early Oligocene
Qatraniodon parvus	Fayum, Egypt	lower sequence, Jebel Qatrani Fm.[1]	
Brachyodus depereti	Wadi Moghra and Siwa, Egypt		Early Miocene
	Sperrgebiet, Namibia		19 Ma
B. mogharensis	Wadi Moghra, Egypt		Early Miocene
B. aequatorialis	Kalodirr, Maboko, Rusinga, Loperot, Losodok, Baragoï, and Locherangan, Kenya Moroto, Uganda		Early Miocene
Sivameryx moneyi	Wadi Moghra, Egypt		Early Miocene
S. africanus	Karungu, Rusinga, and Kalodirr, Kenya Oued Bazina, Tunisia Jebel Zelten, Libya		Early Miocene
Afromeryx zelteni	Jebel Zelten, Libya Ombo, Nachola, Loperot, Buluk, and Wayondo, Kenya Ghaba, Sultanate of Oman		Miocene
A. africanus	Wadi Moghra, Egypt		Early Miocene
Libycosaurus petrocchii	As Sahabi, Libya Agranga and Toros-Ménalla, Chad	Anthracotheriid Unit	Late Miocene
L. anisae (Including L. Algeriensis)	Bir el Ater 2, Algeria Bled Douarah, Tunisia Uganda NY 32 and 33, Uganda	Nementcha Fm. Beglia Fm. Kisegi Fm. Kakara Fm.	Late middle Miocene Late middle Miocene 10 Ma
Anthracotheriidae indet.	Bir el Ater, Algeria Meswa Bridge, Kenya Moroto, Uganda	Muhoroni Agglomerates	Early late Eocene 23.5 Ma Early Miocene
Kulutherium kenyensis	Rusinga Island, Kenya Koru, Kenya	Kulu Fm. Chamtwara Fm.	20–18 Ma Early Miocene

SOURCES: Distribution based on Ducrocq et al. (2001), Pickford and Andrews (1981), Pickford et al. (1986), Pickford (1991a, 2003, 2007a), Jeddi et al. (1991), Roger et al. (1994), Miller (1996), Lihoreau (2003), Lihoreau et al. (2006), Delmer et al., (2006), and Holroyd and Gunnell (unpublished data). Ages for Wadi Moghra based on Miller (1996).

[1]. Precise type localities not known; stratigraphic distribution based on referred material.

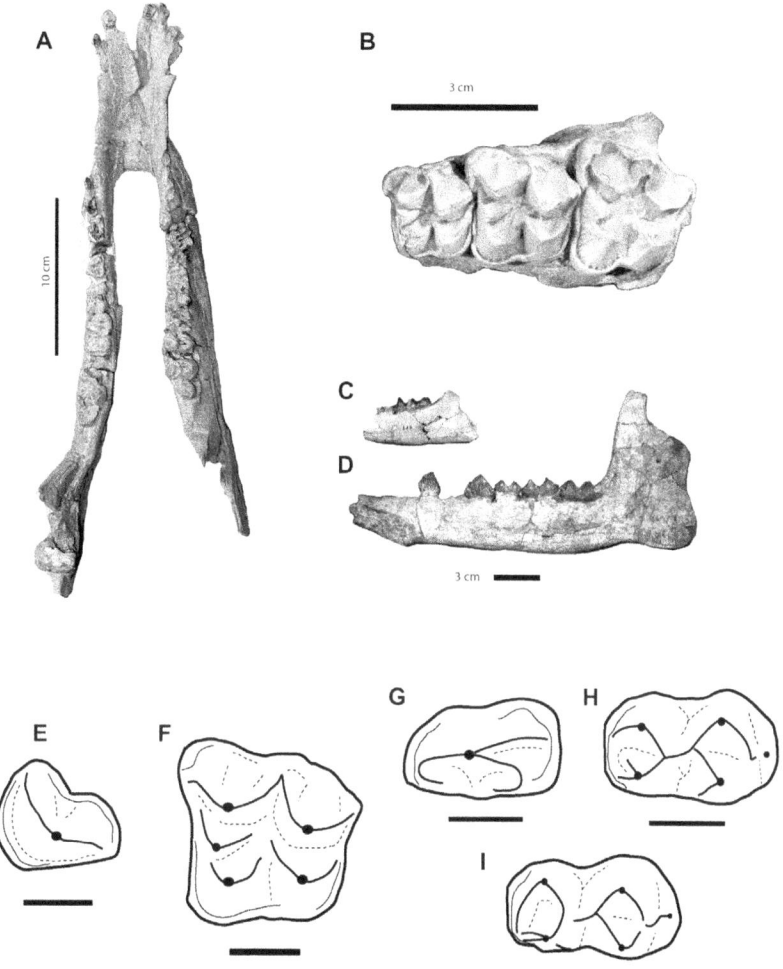

FIGURE 43.2 Examples of *Bothriogenys* and *Qatraniodon*. Dentary of *B. gorringei*, Yale Peabody Museum (YPM) 30548, in dorsal (A) view. Maxilla of *B. gorringei*, University of California Museum of Paleontology (UCMP) 41548, in occlusal view (B). Comparison of dentaries in lateral view of *Qatraniodon parvus*, YPM 18167 (C) and *B. gorringei*, YPM 18097 (D). Outline drawings of left P3 of *B. andrewsi*, YPM 18065 (E); right M2 of *B. gorringei*, UCMP 41548 (F); right p4 of *B. gorringei*, YPM 18097 (G); right m2 of *Q. parvus* (H); and right m2 of *B. gorringei*, UCMP 41492 (I).

preprotocristid and prehypocristid. Differs from *Qatraniodon* in larger size, more bulbous lower molars, deeper dentaries. Differs from *Libycosaurus* and *Afromeryx* by its five-cusped upper molars (i.e., retaining paraconule) and its reduced canine. Differs from *Sivameryx* and *Libycosaurus* by its short preprotocristid and prehypocristid. Differs from *Brachyodus* in generally smaller size and retaining noncaniniform incisors.

Description All described species are from the late Eocene Qasr el Sagha Formation and overlying late Eocene to early Oligocene Jebel Qatrani Formation in the Fayum Province of Egypt. Most species are known primarily from dental and some partial cranial remains. Abundant postcrania are also known (see, e.g., Schmidt, 1913), although since there are few associations in the Jebel Qatrani Formation, assigning any of these to a particular species can only be attempted based on size.

Species of *Bothriogenys* are primarily distinguished based on differences in size and cusp shape. The detailed differential diagnoses that follow are based on Ducrocq (1997).

Bothriogenys fraasi differs from other *Bothriogenys* spp. in its deeper lower jaw with lingual convexity under molars and lingual concavity under premolars, labial convexity of tooth row, strong increase in size from m1–m2, lower molars slightly bulbous and labially salient external wall, Y-shaped prehypocristid, strongly developed mesial and distal cingulids; upper molars as wide as long (as in *B. rugulosus*) but with more slanted cusps, interpremolar diastemata lacking, long diastema between C and P1; symphysis reaching p1.

Bothriogenys gorringei is a medium-sized species; differing from *B. fraasi* and *B. andrewsi* in smaller size and from *B. rugulosus* in larger size. Further differs from *B. fraasi* in having relatively smaller increase in size from m1–m3; more vertical distal wall of labial cusps, and poorly developed labial cingulid. Further differs from *B. rugulosus* in less bulbous labial walls of m1–m3 and more poorly developed premetacristid, m3 with longer and narrower hypoconulid lobe; upper molars wider than long and less wrinkled enamel.

B. andrewsi is the largest species of the genus, possessing quadratic upper molars with strong labial styles; labial wall of paracone steeper than that of metacone; P4 with oblique mesiolabial corner. Differs from *B. fraasi* and *B. rugulosus* in "more slender" lower molars and finely crenulated enamel.

Bothriogenys rugulosus differs from other *Bothriogenys* spp. by its smaller size, upper molars as wide as long (as in *B. fraasi*), no marked increase in size from M1 to M3, paracone steeper than metacone, deeper hypoflexid between trigonid and talonid on m1–m3; more rectilinear m3 hypoconulid lobe, almost complete labial cingulid, and strongly wrinkled enamel.

Remarks Originally described as a subgenus of *Brachyodus*, Black (1978) raised *Bothriogenys* to genus rank. Records of *Bothriogenys* sp. include postcrania from the Qasr el Sagha Formation, Egypt (Simons, 1968; Holroyd et al., 1996), and Schmidt's (1913) *"Mixtotherium mezi"* (Holroyd, 1994). Indeterminate anthracotheriid records from the Oligocene? of Gebel Bou Gobrine, Tunisia (Arambourg and Burollet, 1962), and an undescribed anthracothere from L-41 of the Jebel Qatrani Formation, Egypt (Liu et al., 2008), may also represent records of *Bothriogenys*.

Genus *QATRANIODON* Ducrocq, 1997
Figures 43.2C and 43.2H

Type and Only Species Qatraniodon parvus (Andrews, 1906).

Age and Occurrence Jebel Qatrani Formation, Egypt, type locality unknown but referred specimen from late Eocene lower sequence (Holroyd and Gunnell, unpub. data).

Diagnosis After Ducrocq (1997). Very small anthracotheriid; longer teeth relative to width than in other African anthracotheriids; cusps subselenodont.

Description The holotype is a right dentary with m1–m2 of unknown provenance. Additional specimens have been identified by two of us (P.A.H., G.F.G.) in collections at the American Museum of Natural History, Yale Peabody Museum, and University of California–Berkeley. All these specimens were found in the lower sequence of the Jebel Qatrani Formation, permitting us to establish an early Oligocene age for the taxon.

Remarks Qatraniodon is the smallest and one of the rarest of the Fayum anthracotheriids and is the most selenodont of Paleogene taxa.

Genus *BRACHYODUS* Depéret, 1895
Figure 43.3

Selected Synonymy Masritherium Fourtau, 1918, MacInnes, 1951, *Bothriogenys* (Black, 1978 in part). See Pickford (1991b) for full species level synonymies.

Included African Species B. depereti (Andrews, 1899), *B. aequatorialis* (MacInnes, 1951), *B. mogharensis* Pickford, 1991b.

Age and Occurrence Early Miocene, Libya, Egypt, Uganda, Kenya, and Namibia.

Diagnosis Modifiid after Pickford (1991b). Large size; tusklike upper central and lower lateral incisors, sexually dimorphic; lower and upper canine premolariform; short i–c diastema; canine-P1/p1 diastema long; fused mandibular symphysis; pentacuspidate upper molar with styles pinched rather than looplike; pentadactyl manus.

Differs from *Bothriogenys* by its anterior teeth reduction and morphology. Differs from *Afromeryx* and *Libycosaurus* in retaining a paraconule. Differs from the latter two genera

and from *Sivameryx* in having a more compressed ("pinched") rather than looplike mesostyle.

Description Brachyodus aequatorialis is a medium- to large-sized species, similar in size to the type species, European *B. onoideus*, but having small, spatulate lower incisors with wrinkled enamel rather than being caniniform. *B. depereti* is larger than *B. mogharensis* and differs in having a weaker labial cingulum on upper teeth and lower cusps on p3–4. *Brachyodus mogharensis* differs from other species of genus in having shorter upper premolars; smaller than *B. depereti*; differs from *B. aequatorialis* in having stronger labial cingulum on upper teeth, higher cusps on p3–4, m1–2 of more equal size (after Pickford, 1991b).

Discussion Brachyodus is best known from North African and East African sites. The single possible record of *Brachyodus* from southwest Africa is based on the presence of several large-sized artiodactyl postcranial elements from Namibia (Pickford, 2003). Most authors have suggested that *Brachyodus* stems from within the genus *Bothriogenys*, implying an African origin of the genus. This genus is considered to be hydrophilic based on taphonomic evidence. (Pickford 1983).

Genus *SIVAMERYX* Lydekker, 1877
Figures 43.4A–43.4F

Selected Synonymy Hyoboops Trouessart, 1904, Black, 1978; *Brachyodus* Fourtau, 1918, 1920; *Merycops* Pilgrim, 1910.

Included Species S. moneyi (Fourtau, 1918); *S. africanus* (Andrews, 1914; = cf. *S. palaeindicus* Lydekker, 1877; Lihoreau and Ducrocq, 2007).

Age and Occurrence Early Miocene, North and East Africa; see table 43.1.

Diagnosis From Pickford (1991b); Lihoreau (2003). Medium to small in size; quasi-pentacuspidate upper molars with looplike parastyles and mesostyles; two distal crests from protocone; paraconule almost fused with protocone and of similar height; canines sexually dimorphic; lower molars selenodont with anterior crests of labial cusps reaching lingual surface of crown, often ending in a small cuspule; four crests from the metaconid; postcanine diastema long, with a flange-like protuberance leaning laterally; p1 double rooted; symphysis reaches back to level of p1; no genial spine; lingual cusps of lower molars mediolaterally compressed; talonid of m3 looplike and strongly obliquely oriented (figure 43.4).

Description Sivameryx is restricted to the early Miocene, with distinct species in North and East Africa. *Sivameryx moneyi* (Egypt) appears to be a smaller version of *S. africanus* (East Africa, Libya)(Pickford, 1991b)

Discussion East African representatives of the genus *Sivameryx* have been considered members of *S. africanus* (Andrews, 1914), although all authors recognize strong morphological and metric similarities between *S. africanus* (Africa) and *S. palaeindicus* (Asia). The presence of *Sivameryx* in both Asian and African faunas is a result of early Miocene intercontinental dispersion events occurring around or before ca. 18 Ma. *Sivameryx moneyi*, which is a smaller species, may be related to possible isolation of the fauna from Moghra, Egypt (Pickford, 1991b; Miller, 1999).

Genus *AFROMERYX* Pickford, 1991b
Figures 43.4G–43.4I

Selected Synonymy Brachyodus Andrews, 1899 (in part); *Bothriogenys* (Black, 1978, in part).

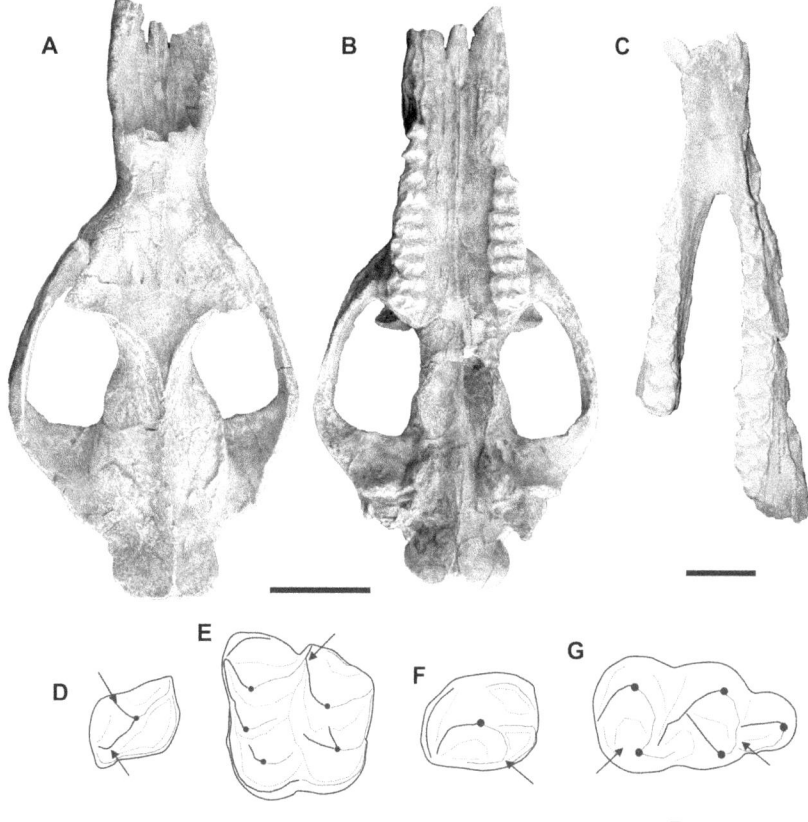

FIGURE 43.3 Main generic features of *Brachyodus*. Cast of the holotype of *Brachyodus aequatorialis* Kenyan National Museum KNM-RU 1009 from Kenya, skull in dorsal (A) and ventral (B) views; scale bar equals 10 cm. Cast of the mandible of *B. aequatorialis* KNM-RU 1014 in occlusal view (C); scale bar equals 5 cm. Occlusal outlines of P3 (D), M3 (E), p4 (F), and m3 (G) of *Brachyodus*; scale bars equal 1 cm. Small arrows indicate main diagnostic features of *Brachyodus*: two mesial crests on upper premolar (D), pinched mesostyle on upper molar (E), large lingually orientated distal crest on p4 (F), no anterior cristid from metaconid, and junction of posthypocristid and postentocristid (G).

Included African Species A. zelteni Pickford, 1991b (Type species); *A. africanus* (Andrews, 1899).

Age and Occurrence Early to middle Miocene, Libya, Egypt and Kenya; see table 43.1. Diagnosis (modified after Pickford 1991b)—Medium to small in size, upper molars with four cusps (lacking paraconule); two distal crests from the protocone; p1–4 low crowned with cuspate crests; anterior crest of labial cusps of m1–m3 do not reach lingual margin of tooth; p1 single rooted; genial spine on mandible present; short c–p1 diastema (figure 43.4).

Description The two described species differ primarily in size, with *Afromeryx zelteni* approximately one-half the size of *A. africanus*. Additionally, *A. zelteni* is characterized by I3–C diastema lacking; upper incisors separated by small gap; incisive foramen very large; sexually dimorphic canines; symphysis reaching to p1–p2; and m3 talonid centrally placed and only slightly obliquely oriented. By contrast, *A. africanus* possesses a c-p1 diastema of ca. 15mm; m3 talonid higher than proto- and metaconid; and the distal ridge of p4 is strong and reaches a centrally positioned central cusp.

Remarks A. africanus is known from only a few specimens recovered from Wadi Moghra, Egypt, while *A. zelteni* is more broadly distributed in Libya, Kenya, and Oman. Pickford (1991b) noted that *Afromeryx* (Africa) appears to share morphological and metric similarities with members of the *Gonotelma-Telmatodon* group (Asia), although this latter group is poorly known. Lihoreau and Ducrocq (2007) have suggested that *Afromeryx* may stem from an Eurasian group morphologically close to *Elomeryx*. All authors agree that *Afromeryx* represents an immigrant taxon that arrived in Africa from Eurasia, as part of the early Miocene faunal exchange.

Genus *LIBYCOSAURUS* Bonarelli, 1947
Figure 43.5

Selected Synonymy Merycopotamus Falconer & Cautley, in Owen, 1845 (in part), *Gelasmodon* Forster-Cooper in Hopwood and Hollyfield, 1954; *Merycopotamus* Black, 1978; Gaziry, 1978.

Included Species L. petrocchii Bonarelli, 1947; *L. anisae* (Black, 1972), and possibly *L. algeriensis* (Ducrocq et al., 2001).

Age and Occurrence Middle to late Miocene, Libya, Algeria, Tunisia, Chad, and Uganda; see table 43.1.

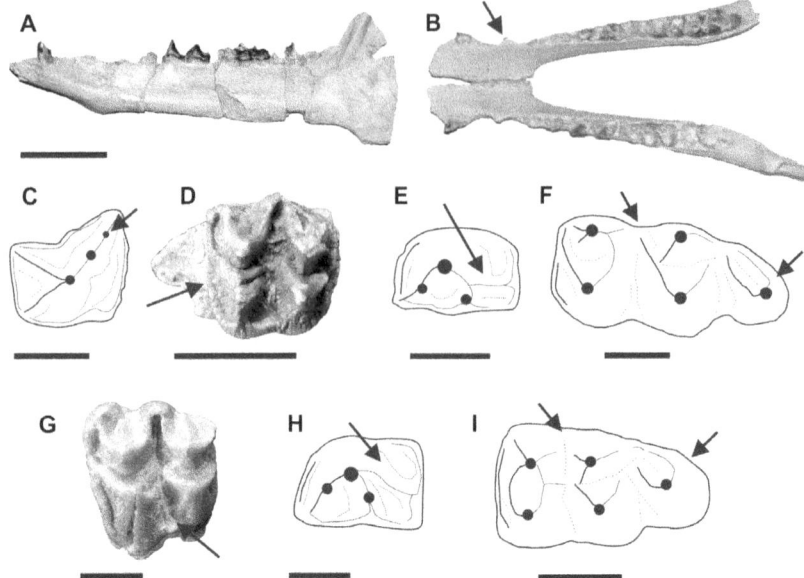

FIGURE 43.4 Main generic features of *Sivameryx*. Mandible of *Sivameryx africanus* from Gebel Zelten in lateral (A) and occlusal (B) views; scale bar equals 5 cm. Occlusal outlines of P3 (C), p4 (E), and m3 (F) and reversed photograph of M1, KNM WK 17109 (D), in occlusal view of *Sivameryx africanus*; scale bars equal 1 cm. Main generic features of *Afromeryx*. Reversed photograph of M3, KNM-OM 13287 (G) in occlusal view and, occlusal outlines of p4 (H) and m3 (I) of *Afromeryx zelteni*; scale bars equal 1 cm. Small arrows (B–F) indicate main diagnostic features of *Sivameryx*: dorsal protuberance on mandible (B), accessory cusps on upper premolar (C), reduced paraconule (D), lingually situated distal crest on p4 (E), anterior cristids of labial cusps that reach the lingual margin on m3, and hypoconulid labially positioned (F). Small arrows (G–I) indicate main diagnostic features of *Afromeryx*: second distal crest from protocone on tetracuspidate upper molar (G), centrally situated distal crest on p4 (H), short anterior cristid from labial cusps, and hypoconulid centrally positioned (I).

Diagnosis After Pickford (1991b, 2006) and Lihoreau (2003). Differs from all anthracothere genera by displaying a fifth upper premolar and the presence of an entoconid fold in lower molar (figure 43.5G). Differs from *Sivameryx*, *Brachyodus* and *Bothriogenys* by the lack of paraconule. Differs from *Afromeryx* by anterior aperture of the main palatine foramina, the lack of a second postprotocrista (also different than in *Sivameryx*), biradiculate p1, preprotocristid and prehypocristid that reach the lingual margin of the lower molar (also different than in *Brachyodus*, *Bothriogenys* and *Qatraniodon*; figure 43.5). Differs from *Sivameryx* in lacking premetacristid.

Description Large, sexually dimorphic anthracotheres with tetracuspidate upper molars in which the looplike mesostyles are undivided; sexually dimorphic canines; anterior palatal modifications include curved diastema folded inward over hard palate to form open-sided, tubelike structure; main palatine foramina open at canine level; smaller I3 than I1–2; upper and lower incisors with long roots; cuspate crests on premolars well developed; additional upper premolar anterior to P1 (Lihoreau et al., 2006); variable number of lower incisors; no anterior crests from the metaconid; developed entoconid fold (figure 43.5G); tetradactyl manus.

Libycosaurus species are known by cranial, dental, and postcranial remains from a number of sites in North, Central, and East Africa. *Libycosaurus petrocchii* differs from *L. anisae* in being approximately 30% larger and in having a single cusped hypoconulid (looplike in *L. anisae*).

Discussion Libycosaurus and its constituent species have had a complex taxonomic history. *Libycosaurus* was long synonymized with *Merycopotamus* (see, e.g., Black, 1978). However, Pickford (1991b), Vignaud et al. (2002), and Boisserie et al. (2005) recognized *Libycosaurus* as valid and restricted *Merycopotamus* to Asian species. Gaziry (1987b) erected the species *Merycopotamus ? maradensis* for a specimen from the Marada Fm., Jebel Zelten, Libya, but Pickford (1991b) synonymized this species with *Sivameryx moneyi*.

ANTHRACOTHERIIDAE indeterminate

In addition to the many records of *Afromeryx*, *Brachyodus*, and *Sivameryx* from Miocene localities, anthracothere generic diversity in the early Miocene is still likely to be underestimated (E.M. and G.F.G., pers. obs.). In addition, Hill (1995) has noted indeterminate anthracotheriids at Baringo, Ngorora Fm., and Pickford and Andrews (1981) have noted anthracotheres at Meswa Bridge locality 36 in the Muhoroni Agglomerate, which has been dated to approximately 23.5 Ma. An m3 from the early Miocene of Malembe, Angola, was reported as an anthracothere smaller than *Brachyodus* (Hooijer 1963), although it is more likely that this specimen represents *Bunohyrax* instead (F.L., pers. obs.).

Family ANTHRACOTHERIIDAE?
Genus *KULUTHERIUM* Pickford, 2007

Included Species K. kenyensis Pickford, 2007a.
Age and Occurrence Early Miocene, East Africa, Kenya; see table 43.1.

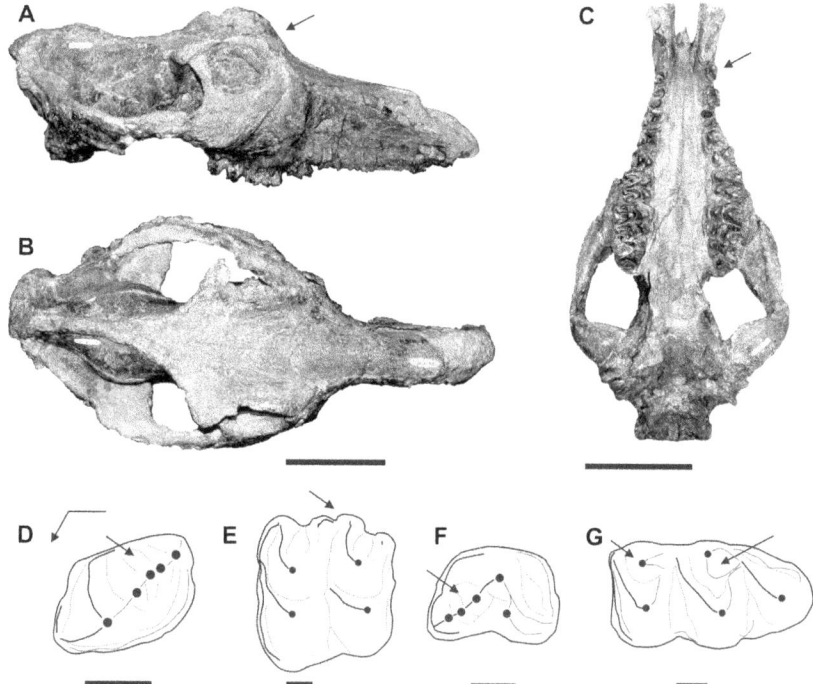

FIGURE 43.5 Main generic features of *Libycosaurus*. Skull of *Libycosaurus petrocchii* from Chad (TM90–00–68) in lateral (A) and dorsal (B) views; scale bar equals 10 cm. Skull of *Libycosaurus petrocchii* from Chad (TM134–01–06) in ventral view (C); scale bar equals 10 cm. Occlusal outlines of P1 (D), M3 (E), p4 (F), and m3 (G) of *Libycosaurus petrocchii*; scale bars equal 1 cm. Curved arrow is mesiolingually orientated. Small arrows indicate main diagnostic features of *Libycosaurus*: elevated orbits above the cranial roof (A), presence of a fifth premolar (C), several accessory cusps on distal crest of upper premolar (D), undivided looplike mesostyle (E), several accessory cusps on mesial crest of lower premolar (F), and absence of mesial crest on the metaconid and small mesiodistal crest between entoconid and hypoconid (entoconid fold) on lower molar (G).

Diagnosis From Pickford (2007a). Differs from *Brachyodus* in having thinner and smoother upper molar enamel, reduced paraconule, lacking para-, meta-, and mesostyles and labial flare; differs from Hippopotamidae in possessing an anterolingual ridge on metaconule, thinner enamel, and nontrefoliate cusp morphology.

Discussion Kulutherium is represented by only two specimens, both juveniles. We tentatively follow Pickford (2007a) in assigning *Kulutherium* to Anthracotheriidae. However, the lack of any stylar development and the molariform dP4 suggest that its affinities may lie elsewhere.

Discussion

Anthracotheriid artiodactyls first appear in the African fossil record as an immigrant group, arriving in Africa from Holarctic continents sometime during the Eocene. North Africa has been the primary area of anthracothere differentiation throughout most of their evolution on the continent. The oldest records of African anthracotheriids are species in the genus *Bothriogenys* from the late Eocene Qasr el Sagha Formation, Egypt. By the earliest Oligocene, anthracotheres had diversified into at least two genera, *Bothriogenys* and *Qatraniodon*, but *Bothriogenys* remained the most common and speciose. *Bothriogenys* persisted into the Oligocene and probably gave rise to *Brachyodus*, which flourished in the early Miocene. Around 20 Ma, *Brachyodus* invaded Europe and then reached Asia around 15 Ma (Ducrocq et al., 2003).

A second immigration of anthracotheres into Africa occurred during the early Miocene, as part of a larger, well-documented episode of faunal exchange permitted by the collision of Afro-Arabian and Turkish plates. This second wave included species with clear relatives on the Indian subcontinent, such as *Afromeryx* and *Sivameryx*. These forms persisted for about 3 Ma, making the early Miocene the time when African anthracotheres reached their greatest diversity, with a minimum of three genera and several species present. The third dispersal event is documented by the arrival of *Libycosaurus* in Africa, probably around 15–13 Ma. The last record of an anthracothere in Africa is from deposits in the 6–5 Ma range, after which anthracotheres appear to have gone extinct (figure 43.6).

Many authors have noted that the demise of anthracotheres in the late Miocene appears concomitant with the appearance and radiation of the Hippopotamidae, leading some researchers to suggest that anthracotheres may have been "hippo-like" in their adaptation and were outcompeted by hippopotamids (e.g., Coryndon, 1978; Pickford, 1991b). Based on phylogenetic analyses, other researchers have suggested that hippopotamids are derived from within anthracotheres, such that ancient anthracotheres live today in the form of the modern hippopotamus (Boisserie et al., 2005; Boisserie and Lihoreau, 2006). In contrast, Pickford (2007b, 2008) notes that Neogene anthracothere cranial and postcranial anatomy is divergent from that of hippopotamids, and that the functional morphology and behavior of Miocene anthracotheres was more

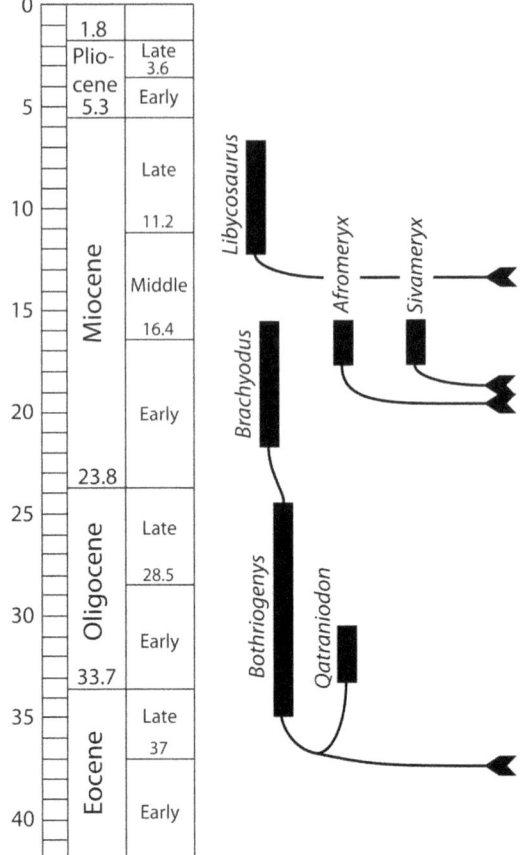

FIGURE 43.6 Possible relationships of anthracotheres in geographic and chronologic context (after Lihoreau and Ducrocq, 2007). Arrows indicate the main dispersal events of anthracotheres into Africa. © 2007 The Johns Hopkins University Press. Reprinted with permission of The Johns Hopkins University Press.

likely akin to that of swamp dwelling artiodactyls such as the Marshbuck *(Tragelaphus spekei),* Marsh Deer *(Blastocerus dichotomus),* or Lechwe *(Kobus leche).* However, no similar studies have examined Paleogene anthracothere postcrania.

Two recent papers have examined the hypothesis of semiaquatic habits in anthracotheres from examination of oxygen and carbon isotopes. Clementz et al. (2008) found that at least *Bothriogenys gorringei* from the early Oligocene part of the Jebel Qatrani Formation, Egypt, had low variation in oxygen isotopic values consistent with semiaquatic habits, while Liu et al. (2008) judged a slightly older undescribed anthracothere from the late Eocene portion of the Jebel Qatrani Formation to be terrestrial based on its carbon isotopic values. Multiple lines of evidence will need to be examined in all Old World anthracotheres to test these hypotheses more thoroughly.

Literature Cited

Arambourg, C., and P. F. Burollet. 1962. Reste de vertébrés oligocènes en Tunisie centrale. *Comptes Rendus Sommaire des Séances de la Société Géologique de France* 2:42–43.

Black, C. C. 1978. Anthracotheriidae; pp 423–434 in V. J. Maglio and H. B. S. Cooke (eds.), *Evolution of African Mammals.* Harvard University Press, Cambridge.

Boisserie, J.-R., and F. Lihoreau. 2006. Emergence of Hippopotamidae: New scenarios. *Comptes Rendus Palevol* 5:749–756.

Boisserie, J.-R., F. Lihoreau, and M. Brunet. 2005. The position of Hippopotamidae within Cetartiodactyla. *Proceedings of the National Academy of Sciences, USA* 102:1537–1541.

Clementz, M. A., P. A. Holroyd, and P. L. Koch. 2008. Identifying aquatic habits of herbivorous mammals through stable isotope analysis. *Palaios* 23:574–585.

Colbert, E. H. 1935a. Siwalik mammals in the American Museum of Natural History. *Transactions of the American Philosophical Society* 26:278–294.

———. 1935b. Distributional and phylogenetics studies on Indian fossil mammals: Part IV. Suidae and Hippopotamidae. *American Museum Novitates* 799:1–24.

Coryndon, S. C. 1978a. Hippopotamidae. pp. 284–395 in V. J. Maglio and H. B. S. Cooke (eds.), *Evolution of African Mammals.* Harvard University Press, Cambridge.

Delmer, C., M. Mahboubi, R. Tabuce, and P. Tassy. 2006. A new species of *Moeritherium* (Proboscidea, Mammalia) from the Eocene of Algeria: New perspectives on the ancestral morphotype of the genus. *Palaeontology* 49:421–434.

Ducrocq, S. 1997. The anthracotheriid genus *Bothriogenys* (Mammalia, Artiodactyla) in Africa and Asia during the Paleogene: Phylogenetical and paleobiogeographical relationships. *Stuttgarter Beiträge zur Naturkunde* 250:1–44.

Ducrocq, S., Y. Chaimanee, V. Suteethorn, and J.-J. Jaeger. 2003. Occurrence of the anthracotheriid *Brachyodus* (Artiodactyla, Mammalia) in the early Middle Miocene of Thailand. *Comptes Rendus Palevol* 2:261–268.

Ducrocq, S., B. Coiffait, P.-E. Coiffait, M. Mahboubi, and J.-J. Jaeger. 2001. The Miocene Anthracotheriidae (Artiodactyla, Mammalia) from the Nementcha, eastern Algeria. *Neues Jahrbuch für Geologie und Paläontologie, Monatshefte* 3:145–156.

Gaziry, A. W. 1987a. *Merycopotamus petrocchii* (Artiodactyla, Mammalia) from Sahabi, Libya; pp. 287–302 in N. T. Boaz, A. El-Arnauti, A. W. Gaziry, J. de Heinzelin, and D. D. Boaz (eds.), *Neogene Paleontology and Geology of Sahabi.* Liss, New York.

———. 1987b. New mammals from the Jabal Zaltan site, Libya. *Senckenbergiana Lethaea* 68:69–89.

Hill, A. 1995. Faunal and environmental change in the Neogene of East Africa: Evidence from the Tugen Hills sequence, Baringo District, Kenya; pp. 178–193 in E. S. Vrba, G. H. Denton, T. C. Partridge, and L. H. Burckle (eds.), *Paleoclimate and Evolution, with Emphasis on Human Origins.* Yale University Press, New Haven.

Holroyd, P. A. 1994. An examination of dispersal origins for Fayum Mammalia. Unpublished PhD dissertation, Duke University.

Holroyd, P. A., E. L. Simons, T. M. Bown, P. D. Polly, and M. J. Kraus. 1996. New records of terrestrial mammals from the upper Eocene Qasr El Sagha Formation, Fayum Depression, Egypt. *Paleobiologie et Evolution des Mammifères du Paléogène: Volume Jubilaire en Hommage à Donald E. Russell, Palaeovertebrata* 25:175–192.

Hooijer, D. A. 1963. Miocene Mammalia of the Congo. *Annales du Muséum Royal d'Afrique Centrale, Séries in 8°, Sciences Géologiques* 46:1–77.

Jeddi, R. S., M. Pickford, P. Tassy, and Y. Coppens. 1991. Discovery of mammals in the Tertiary of central Tunisia: Biostratigraphic and paleogeographic implications. *Comptes Rendus de l'Académie des Sciences, Paris* 312:543–548.

Lihoreau, F. 2003. Systématique et paléoécologie des Anthracotheriidae (Artiodactyla; Suiformes) du Mio-Pliocène de l'Ancien Monde: Implications paléobiogéographiques. Unpublished PhD dissertation, University of Poitiers.

Lihoreau, F., J.-R. Boisserie, L. Viriot, Y. Coppens, A. Likius, H. T. Mackaye, P. Tafforeau, P. Vignaud, and M. Brunet. 2006. Anthracothere dental anatomy reveals a late Miocene Chado-Libyan bioprovince. *Proceedings of the National Academy of Sciences, USA* 103:8763–8767.

Lihoreau, F., and S. Ducrocq. 2007. Family Anthracotheriidae: Systematics and Evolution; pp. 89–105 in D. Prothero and S. Foss (eds.), *The Evolution of Artiodactyls.* Johns Hopkins University Press, Baltimore.

Liu, A. G. S. C., E. R. Seiffert, and E. L. Simons. Stable isotope evidence for an amphibious phase in early proboscidean evolution. *Proceedings of the National Academy of Sciences, USA* 105:5786–5791.

Miller, E. R. 1996. Mammalian paleontology of an Old World monkey locality, Wadi Moghara, early Miocene, Egypt. Unpublished PhD dissertation, Washington University, St. Louis, Mo.

————. 1999. Faunal correlation of Wadi Moghara, Egypt: Implications for the age of *Prohylobates tandyi*. *Journal of Human Evolution* 36:519–533.

Pickford, M. 1983. On the origins of Hippopotamidae together with descriptions of two new species, a new genus and a new subfamily from the Miocene of Kenya. *Geobios* 16:193–217.

————. 1991a. Late Miocene anthracothere (Mammalia, Artiodactyla) from tropical Africa. *Comptes Rendus de l'Académie des Sciences, Série II*, 313:709–715.

————. 1991b. Revision of the Neogene Anthracotheriidae of Africa. *Geology of Libya* 4:1491–1525.

————. 2003. Early and middle Miocene Anthracotheriidae (Mammalia, Artiodactyla) from the Sperrgebiet, Namibia. *Memoir of the Geological Survey of Namibia* 19:283–289

————. 2006. Sexual and individual morphometric variation in *Libycosaurus* (Mammalia, Anthracotheriidae) from the Maghreb and Libya. *Geobios* 39:267–310.

————. 2007a. A new suiform (Artiodactyla, Mammalia) from the early Miocene of East Africa. *Comptes Rendus Palevol* 6:221–229.

————. 2007b. Suidae and Hippopotamidae from the middle Miocene of Kipsaraman, Kenya and other sites in East Africa. *Paleontological Research* 11:85–105.

————. 2008. The myth of the hippo-like anthracothere: The eternal problem of homology and convergence. [El mito de la similitud entre antracoterios e hipopótamos: El eterno problema entre homología y convergencia.] *Revista Española de Paleontología* 23:31–90.

Pickford, M., and P. Andrews. 1981. The Tinderet Miocene sequence in Kenya. *Journal of Human Evolution* 10:11–33.

Pickford, M., B. Senut, D. Hadoto, J. Musisi, and C. Kariira. 1986. Recent discoveries in the Miocene sites at Moroto, North-East Uganda: Biostratigraphical and palaeoecological implications. *Comptes Rendus de l'Académie des Sciences, Paris, Série II*, 9:681–686.

Roger, J., M. Pickford, H. Thomas, F. de Lapparent de Broin, P. Tassy, W. Van Neer, C. Bourdillon-de-Grissac, and S. Al-Busaidi. 1994. Découverte de vertébrés fossils dans le Miocène de la region du Huqf au Sultanat d'Oman. *Annales de Paléontologie* 80:253–273.

Schmidt, M. 1913. Über Paarhufer des fluviomarinen Schichten des Fajum, odontographisches und osteologisches Material. *Geologische und Paläontologische Abhandlungen* 11:153–264.

Simons, E. L. 1968. Early Cenozoic mammalian faunas, Fayum Province, Egypt: Part I. African Oligocene mammals: Introduction, history of study, and faunal succession. *Bulletin, Peabody Museum of Natural History* 28:1–22.

Vignaud, P., P. Duringer, H. T. Mackaye, A. Likius, C. Blondel, J.-R. Boisserie, L. de Bonis, V. Eisenmann, M.-E. Etienne, D. Geraads, F. Guy, T. Lehmann, F. Lihoreau, N. Lopez-Martinez, C. Mourer-Chauviré, O. Otero, J.-C. Rage, M. Schuster, L. Viriot, A. Zazzo, and M. Brunet. 2002. Geology and paleontology of the Upper Miocene Toros-Menalla hominid locality, Chad. *Nature* 418:152–155.

FORTY-FOUR

Hippopotamidae

ELEANOR WESTON AND JEAN-RENAUD BOISSERIE

Africa is both the place of emergence and the main center of hippopotamid evolution. Hippopotamidae is an exclusively Old World taxon that dispersed from Africa to Eurasia on several occasions (Coryndon, 1978b; Kahlke, 1990; Boisserie, 2005). The earliest hippopotamids have been found in Kenya dating back to the middle Miocene (Pickford, 1983; Behrensmeyer et al., 2002), and although today the family is only represented by two African species, *Hippopotamus amphibius* and *Choeropsis liberiensis*, it had a much greater past diversity with around 40 species known (Coryndon, 1978b; Boisserie, 2005). In Africa, hippos are often the most frequently preserved mammals in Neogene fossil assemblages (e.g., Coryndon, 1970b; Harris, 1991; Faure, 1994; Leakey et al., 1996; Harrison, 1997; Brunet et al., 1998). This abundance of fossil remains is in part linked to the animals' semiaquatic habit, but this unusual mode of life for a large mammal must also have contributed to the success of the group. There has recently been a resurgence of interest in hippopotamid evolution and palaeoecology, following the indication by molecular-based phylogenies that Cetacea form the extant sister group of the Hippopotamidae (Gatesy, 1997; Nikaido et al., 1999). This raises the intriguing possibility that the association with a semiaquatic habit is in fact ancient. The origin of hippos has been much debated (e.g., Pickford, 1983), with a number of contributions favoring the evolution of hippos from certain anthracotheres based on fossil evidence (Colbert, 1935; Gentry and Hooker, 1988; Boisserie et al., 2005a, 2005b). For many years, however, in spite of the wealth of fossil data available, the contribution of hippo research to broad scale palaeoecological and -biogeographic studies has been seriously hindered by their poorly resolved phylogeny. In the last decade the plethora of new fossil material that has been recovered from the Chad Basin (Boisserie et at., 2003, 2005c) and the Afar, Ethiopia (Boisserie, 2004; Boisserie and White, 2004), has radically broadened our knowledge of this group. Previous work in Africa had mainly focused on fossils from the Turkana Basin and the Western Rift of East Africa (e.g., Cooke and Coryndon, 1970; Gèze, 1980, 1985; Pavlakis, 1990; Harris, 1991; Faure 1994; Weston, 2003b).

The first cladistic revision of the group (Boisserie, 2005) has had a major impact on the classification of the African hippopotamids, with many taxa now assigned to different genera. This resulted partly from the discovery of new material and partly from the necessity of splitting the paraphyletic genus *Hexaprotodon* (*sensu* Coryndon, 1977). Recent work on the growth and development of hippos, addressing issues linked with body size adjustment, has also contributed to the reevaluation of some morphological characteristics (Weston, 2003a). In particular, trends in mandibular form corresponding to shifts in ontogeny have been identified as taxonomically informative (Weston, 2000; Boisserie, 2005). Our current understanding of these mammals' African evolutionary history is indebted to the contributions of many authors (e.g., Dietrich, 1928; Hopwood, 1939; Arambourg, 1947; Coryndon, 1977, 1978b; Pickford 1983; Gèze 1980, 1985; Stuenes 1989; Harris 1991; Harrison, 1997). Here we review the recent revisions to African hippopotamid taxonomy and phylogeny in the light of past contributions made to this field.

Systematic Paleontology

This overview of systematic palaeontology follows the revised taxonomy proposed by Boisserie (2005). The previous usage of the genus *Hexaprotodon* (e.g., Coryndon, 1977; Harris, 1991; Weston, 2003b) is not applied here. *Hippopotamus* is abbreviated *Hip.*, and *Hexaprotodon* is abbreviated *Hex.*

MAMMALIA Linnaeus, 1758
CETARTIODACTYLA Montgelard et al., 1997
CETANCODONTA Arnason et al., 2000
HIPPOPOTAMOIDEA Gray, 1821
(*sensu* Gentry and Hooker 1988)
Family HIPPOPOTAMIDAE Gray, 1821
Subfamily KENYAPOTAMINAE Pickford, 1983
Genus *KENYAPOTAMUS* Pickford, 1983

Type Species Kenyapotamus coryndonae Pickford, 1983, from the late Miocene (10–8 Ma) of Kenya (figure 44.1).
Other Species Kenyapotamus ternani Pickford, 1983.
Diagnosis Pickford (1983). Upper molars bunodont with strong lingual cingula; P4 with two main cusps with strong cingulum except labially; P3 with large distolingual cusp, outer enamel surface projects further rootward than lingual enamel, and with pustular enamel tubercles; P1 with two fused roots;

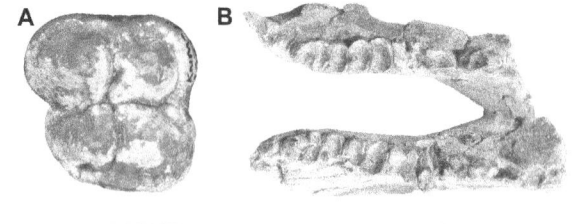

FIGURE 44.1 A) Occlusal view of a right M3 KNM-BN 1321 from Ngeringerowa, holotype of *Kenyapotamus coryndonae*, preserved at the National Museums of Kenya, Nairobi (NMK); scale bar is 2 cm. B) Dorsal view of mandible KNM-SH 15857 from Samburu Hills, *Kenyapotamus coryndonae*, preserved at the NMK; scale bar is 10 cm.

lower molars with distinct median accessory cusps strongly joined to hypoconids by a large crest; lower premolars triangular in side view; m3 with talonid; M3 without talon; astragalus with navicular facet smaller than cuboid facet.

Age Middle to late Miocene, between 15.6 and 8 Ma.

African Occurrence Kenya (Pickford, 1983; Nakaya et al., 1984, 1987; Behrensmeyer et al., 2002), possibly Ethiopia (Geraads et al., 2002) and Tunisia (Pickford, 1990).

Remarks Pickford (1983) created a separate subfamily Kenyapotaminae for the earliest hippopotamids that are now known from the middle and late Miocene of East and North Africa (Pickford, 1983; Nakaya et al., 1984, 1987; Pickford, 1990; Behrensmeyer et al., 2002). These fossils constitute the first record of the Hippopotamidae and should be pivotal in the debate over hippo origins. Unfortunately, they are fragmentary and include mostly isolated teeth or postcranial elements. The affinity of some of this material is not evident (Boisserie et al., 2005a, 2005b). Two species have been described. The older remains from Fort Ternan (ca. 14 Ma), possibly Maboko Island (ca. 15 Ma), and Kipsaraman (ca. 15.6 Ma) in Kenya were originally assigned to *Kenyapotamus ternani* (Pickford, 1983; Behrensmeyer et al., 2002). In a contribution published during the review process of this chapter, Pickford (2007) erected a new genus name *Palaeopotamus* for *K. ternani* on the basis of new material from Kipsaraman (Behrensmeyer et al., 2002). However, this species, the smaller of the two kenyapotamines, with a body weight estimated to be less than the extant pygmy hippopotamus (*Choeropsis*)(Pickford, 1983) remains poorly known (Pickford, 2007). Most of the kenyapotamine material dated between 8 and 10 Ma is attributed to *Kenyapotamus coryndonae* and is recorded from Kenya at Ngeringerowa, Ngorora, Nakali (Pickford, 1983), and the Namurungule Formation in the Samburu Hills (Nakaya et al., 1984, 1987). Isolated teeth of *K. coryndonae* have also been recorded from the Beglia Formation, Tunisia, dated at about 10–9 Ma (Pickford, 1990) and possibly at Chorora, dated at about 11–10 Ma, Ethiopia (Geraads et al., 2002).

The most complete fragment assigned to *K. coryndonae* is from the Samburu Hills, Kenya (Nakaya et al., 1987). This mandible (KNM-SH 15857, figure 44.1B), though very damaged, clearly has a narrow muzzle, and the lower jaw morphology is particularly reminiscent of the late Miocene hippopotamus, *Archaeopotamus lothagamensis*, from Lothagam, Kenya (Weston, 2000). However, all detail of the lower molar cusp morphology is not preserved, precluding the establishment of its precise taxonomic affinities. *Kenyapotamus* molars are extremely brachydont and exhibit a mixture of more derived hippopotamid and more general suiform characteristics. These suiform affinities are interpreted more specifically

by Pickford (1983) as tayassuid-like, but this interpretation is disputed by Boisserie et al. (2005a, 2005b).

Subfamily HIPPOPOTAMINAE Gray, 1821
Genus *ARCHAEOPOTAMUS* Boisserie, 2005
Figures 44.2A and 44.3

Synonymy "Ancestral hexaprotodont" Coryndon, 1976:248, figure 9; *Hexaprotodon* Coryndon, 1977:70, figure 6.

Type Species Archaeopotamus lothagamensis (Weston, 2000), from the upper Miocene (8.0–6.5 Ma) of Lothagam, Kenya (figure 44.2A).

Other Species Archaeopotamus harvardi (Coryndon, 1977); *Archaeopotamus* aff. *lothagamensis* (*Hex.* aff. *sahabiensis* in Gentry, 1999); *Archaeopotamus* aff. *harvardi* (Boisserie, 2005); *Hex. imaguncula* Hopwood, 1926 (in Kent, 1942).

Diagnosis After Boisserie (2005). Hexaprotodont with very elongate mandibular symphysis relative to its width; incisor alveolar process strongly projected frontally; very procumbent incisors; canine processes poorly extended laterally and not extended anteriorly; length of lower premolar row approaching length of molar row; horizontal ramus height is low compared to its length but tends to increase posteriorly; gonial angle of the ascending ramus is not laterally everted.

Age Late Miocene–late Pliocene.

African Occurrence Kenya (Coryndon, 1977; Weston, 2000, 2003b; Boisserie, 2005).

Remarks The genus *Archaeopotamus* Boisserie, 2005, includes the oldest hippopotamines known from East Africa and the Arabian Peninsula. These hexaprotodont species are all characterized by a sagittally long mandibular symphysis relative to their symphyseal breadth across the lower canines (Boisserie, 2005). In light of the considerable size variation that exists across fossil hippopotamid species, this difference in mandibular form is best illustrated when ontogenetic series of modern and fossil species are compared (figure 44.3). The ontogenetic trajectory of *A. harvardi* is displaced laterally relative to the ontogenetic trajectories of the extant hippopotamids that possess relatively short symphyses for their breadth (Weston, 2000, 2003b).

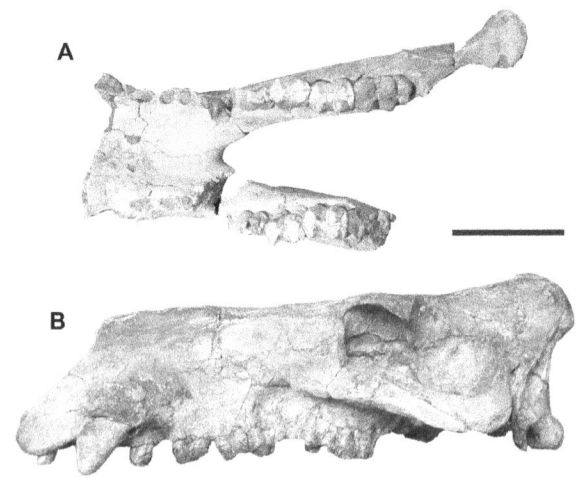

FIGURE 44.2 A) Dorsal view of mandible KNM-LT 23839 from Lothagam, holotype of *Archaoepotamus lothagamensis*, preserved at the NMK; B) lateral view of cranium KNM-LT 4 from Lothagam, holotype of *Archaoepotamus harvardi*, preserved at the NMK. Scale bars are 10 cm.

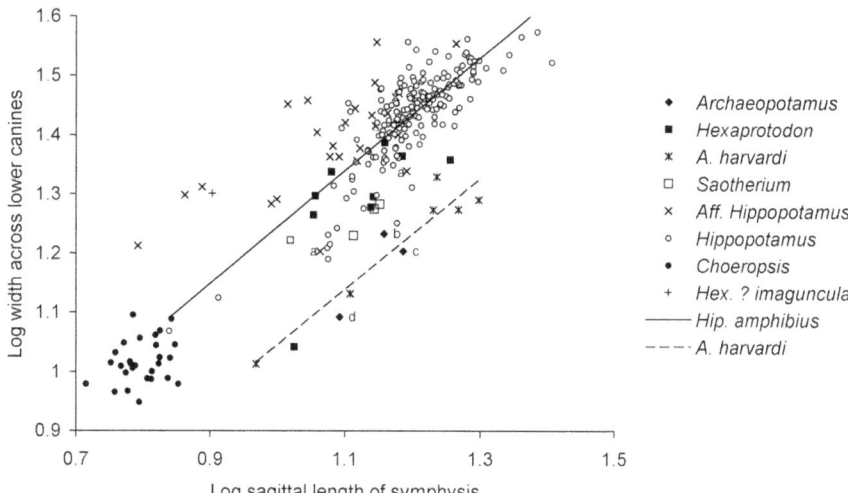

FIGURE 44.3 Relationship between the sagittal length of the mandibular symphysis and the width across the lower canines in hippopotamids. The slopes represent growth trajectories calculated from postnatal ontogenetic series of *Hip. amphibius* and *A. harvardi* mandibles, respectively. For comparison, specimens from all other hippopotamine genera have been included. Representatives of *Archaeopotamus* appear to plot on a separate ontogenetic trajectory relative to other hippopotamines, possessing relatively longer and narrower mandibular symphyses. *Hip. amphibius*: slope = 0.95, r = 0.82; *A. harvardi*: slope = 0.94, r = 0.96. For results of regression analyses that corroborate this ontogenetic trend see Weston (2003a: table 10.3). The included hippopotamids are *Choeropsis liberiensis, Hex. garyam, Hex. bruneti, Hex.* ? cf. *imaguncula* (UMP 6202 from Kazinga, Uganda), aff. *Hip. dulu* (a) aff. *Hip. afarensis*, aff. *Hip. coryndonae*, aff. *Hip. protamphibius*, aff. *Hip.* cf. *protamphibius*, aff. *Hip. aethiopicus*, aff. *Hip. karumensis, S. mingoz, S.* cf. *mingoz, Hip. amphibius, Hip. gorgops; A. harvardi, A.* aff. *harvardi* (b: M15939 from Rawi, Kenya), *A.* aff. *lothagamensis* (c: M496464 from Abu Dhabi, United Arab Emirates), and *A. lothagamensis* (d).

ARCHAEOPOTAMUS HARVARDI (Coryndon, 1977)
Figure 44.2B

A relatively large, gracile hippopotamus with unelevated, laterally facing orbits is by far the best-known species of *Archaeopotamus*, being represented at Lothagam, Kenya, by over 200 specimens from the late Miocene Nawata Formation (McDougall and Feibel, 2003; Weston, 2003b). Remains from other late Miocene and early Pliocene localities in East Africa, notably the Manonga Valley, Tanzania (Harrison, 1997), and the Middle Awash, Adu-Asa Formation, and the lower Sagantole Formation, Ethiopia (Kalb et al., 1982; Haile-Selassie et al., 2004), have also been attributed to *A. harvardi*. Boisserie (2004) considered the late Miocene hippopotamid material from the Adu-Asa Formation (5.2–5.8 Ma) as too fragmentary for accurate identification, whereas the material from the lower part of the Sagantole Formation (4.9–5.2 Ma) has now been assigned to a new species, aff. *Hip. dulu* (Boisserie, 2004). The Manonga Valley hippopotamids from Tanzania (Harrison, 1997) are also too incomplete for definite identification. A strong similarity between hippopotamid remains from the Lukeino Formation and the Mpesida Beds, Baringo, Kenya, and those from Lothagam has been reported (Coryndon, 1978a, 1978b; Harrison, 1997), but the presence of *A. harvardi* at Baringo needs to be clarified.

The *A. harvardi* hypodigm is morphologically variable and may include more than one species. A small partial adult cranium from the lower Nawata (KNM-LT 26236) differs markedly from the holotype (KNM-LT 4, figure 44.2B) from the upper Nawata; the zygoma is more rounded, the occipital is short, and the occipital condyles are small relative to the

breadth of mastoid (Weston, 2003b). Likewise, a mandibular symphysis (KNM-LT 23105) from the upper Nawata is very shallow in relation to its length and breadth, whereas KNM-LT 108 from the lower Nawata is much deeper and narrower (Weston, 2003b). The latter specimens are both subadult suggesting this distinction is not linked with ontogeny. In *A. harvardi* the upper and lower incisors are arranged in a shallow arc. The dorsal surface of the alveolar mandibular shelf is mediolaterally convex as opposed to flat or furrowed (concave; see Weston, 2003b: figure 10.15). This characteristic of the lower jaw is not shared by other *Archaeopotamus* species.

ARCHAEOPOTAMUS LOTHAGAMENSIS (Weston, 2000)
Figures 44.2A and 44.3

A narrow-muzzled small hippopotamus known only from the late Miocene lower Nawata member of Lothagam, northern Kenya (Weston, 2000, 2003b). The holotype (KNM-LT 23839, figure 44.2A) is a mandibular symphysis with c and p4–m3. The cranium is not known. The symphysis is shallow with equal sized procumbent incisors arranged horizontally. Figure 44.3 indicates that the length-to-breadth symphyseal proportions of the holotype scale ontogenetically relative to *A. harvardi*. However, the shallow depth of the symphysis is noteworthy as this feature distinguishes *A. lothagamensis* from the Arabian *Archaeopotamus* where even in an immature specimen (AUH 481: Gentry, 1999) the symphysis is relatively deep. The Arabian hippopotamid has a narrow, deep, and furrowed symphysis with a concave upward alveolar shelf with subequal incisors projecting obliquely from the front of the jaw (Gentry, 1999; Weston, 2000). The m3 of the holotype of *A. lothagamensis* has a central accessory cusp positioned between the mesial crest of the hypoconid and distal crest of the metaconid. Only one other hippopotamid m3 out of 21 examined from the Nawata Formation at Lothagam has a small but clearly demarcated central cusplet. Three other lower third molars have grooves cutting the prehypocristid (mesial crest of hypoconid), but no development of a centrally positioned accessory cusp on the m3 comparable to that of *A. lothagamensis* (KNM-LT 23839) is evident.

ARCHAEOPOTAMUS aff. HARVARDI
Figure 44.3

A mandible of a pygmy hippopotamus (M15939: NHM, London) collected in 1935 from the base of the late Pliocene Rawi beds, Homa Peninsula, Kenya (Ditchfield et al., 1999), has been referred to as *Archaeopotamus* aff. *harvardi* by Boisserie (2005). Figure 44.3 illustrates the similarity of the symphyseal proportions of M15939 with those of *A. harvardi*. The small Rawi hippo was initially identified by A. T. Hopwood as *Hip. imaguncula* (see *incertae sedis* later; Hopwood, 1926; Kent, 1942). The Rawi mandible (M15939) can be distinguished from *A. harvardi* by the differentiation of the incisors: the i2 and i3 are reduced in size in relation to i1, and the i2 is raised above the level of i1 and i3. The alveolar shelf is concave upward in anterior aspect reflecting the strongly furrowed symphysis. The m3 is small (mesiodistal length = 5.57 cm) falling below the range of *A. harvardi* (Weston, 2003b) and within that of *Hex. ?imaguncula*. The symphysis is deep and robust compared with some but not all of the material attributed to the *A. harvardi* hypodigm but differs from the Ugandan specimen (UMP 6202) from the Kazinga Channel (*Hex. ? imaguncula*; see figure 44.3), which has an extremely shallow symphysis (Cooke and Coryndon, 1970).

Genus CHOEROPSIS Leidy, 1853
Figure 44.4A

Synonymy *Hippopotamus* Morton, 1844, p. 15; *Choeropsis* Leidy, 1853:213; *Hexaprotodon* Coryndon, 1977:70, figure 6.

Type Species *Choeropsis liberiensis* (Morton, 1849), extant and only species known (figure 44.4A).

Diagnosis After Boisserie (2005). Small-sized genus; downwardly bent nasal anterior apex; orbits clearly below the cranial roof; strong posterior nasal spine of the palatine; large and elongated tympanic bulla that is apically rounded and without marked muscular process; presence of a lateral notch on the basilar part of the basioccipital; down-turned sagittal crest.

Age Genus exclusively known in the present.

African Occurrence West Africa: Liberia, Ivory Coast, Sierra Leone, Guinea, Nigeria (probably extinct), and possibly Guinea Bissau (Eltringham, 1993, 1999).

Remarks The extant pygmy hippopotamus from West Africa was originally described as a species of *Hippopotamus* (Morton, 1844) but was later assigned to its own genus *Choeropsis* Leidy, 1853. The specific name *minor* (Morton, 1844) was changed to *liberiensis* by Morton (1849) because of its earlier attribution to a fossil form (Desmarest, 1822) that was recognized much later by Major (1902) to be the extinct pygmy hippopotamus from Cyprus. Coryndon (1977) considered the extant pygmy hippo to have close affinities with Mio-Pliocene fossil taxa and transferred this modern species together with most of the extinct African hippopotamine species to the genus *Hexaprotodon* Falconer and Cautley, 1836. Subsequent workers have either followed Coryndon (1977), referring to the modern pygmy hippopotamus as *Hexaprotodon liberiensis* (e.g., Harris, 1991; Gentry, 1999; Weston, 2000, 2003b), or have chosen to maintain *Choeropsis*, differentiating it at the generic level from fossil representatives of the family Hippopotamidae (Pickford, 1983; Harrison, 1997, Boisserie, 2005). In the first cladistic revision of the Hippopotamidae, the genus *Choeropsis* is maintained and the generic diagnosis emended (Boisserie, 2005).

CHOEROPSIS LIBERIENSIS (Morton, 1849)
Figure 44.4A

Choeropsis liberiensis, the sole representative of the genus, with a body weight between 180 and 275 kg (Eltringham, 1993, 1999), is one of the smallest hippopotamids with only the dwarf Cypriot hippo, *Phanourios minor* (Boekschoten and Sondaar, 1972), and the earliest kenyapotamine estimated to be smaller in body size. *Choeropsis liberiensis* is restricted today to the dense rain forests of Liberia with small populations also occurring in neighboring Sierra Leone, Ivory Coast, and Guinea (Eltringham, 1993). Corbet (1969) identified a distinct subspecies from Nigeria, *C. liberiensis heslopi*, now believed to be extinct. The crania of the Nigeria subspecies are slightly smaller relative to those of *C. liberiensis liberiensis* (Weston, 1997).

The skull of *Choeropsis* has orbits positioned well below the cranial roof and a sagittal crest that is down-turned (figure 44.4A). These seemingly primitive traits occur only in *Choeropsis*, but it is noteworthy that such features do characterize the neonate *Hip. amphibius* skull and are potentially paedomorphic (Weston, 2003a). *Choeropsis* also exhibits some derived features (diprotodont mandible with short upright symphysis and laterally everted gonial angle, and single-cusped P4) that are likely convergences shared by some other hippopotamids (Boisserie, 2005). The postcranial elements (long bones, carpals, and tarsals)

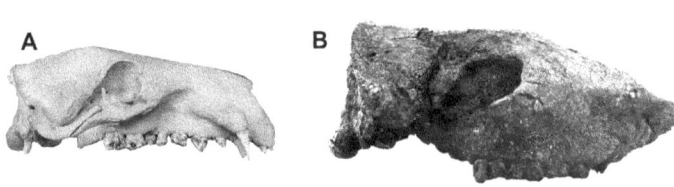

FIGURE 44.4   A) Lateral view of cranium M09.5.005.A of unknown provenience, *Choeropsis liberiensis*, preserved at the LGBPH/UMR CNRS 6041, University of Poitiers; B) lateral view of cranium KL09-98-049 from Kollé, holotype of *Saotherium mingoz*, preserved at the Centre National d'Appui à la Recherche, Ndjamena (CNAR). Scale bar is 10 cm.

are more gracile (slender for their length) than those of *Hip. amphibius*, although the metapodials compare more closely in relative length and robustness to the common hippopotamus (Weston, 1997, 2003b). A more gracile skeleton and extended curvature of the troclear facets of the metapodials and astragali typify *Choeropsis* and *Archaeopotamus* spp. (Weston, 1997, 2003a; Gentry, 1999). This type of postcranial variation can in terms of function be linked with varying hippopotamid lifestyles, but equally these postcranial characteristics appear to be plesiomorphic features of the hippopotamines that have yet to be analyzed in the context of phylogeny.

Genus *SAOTHERIUM* Boisserie, 2005
Figure 44.4B

Synonymy Hexaprotodon Boisserie, 2003:16.

Type Species Saotherium mingoz, from the early Pliocene of Kollé, Chad (Boisserie et al., 2003; see figure 44.4B).

Other Representative Saotherium cf. mingoz (Boisserie et al., 2003).

Diagnosis After Boisserie (2005). Hexaprotodont; antorbital angle of cranial roof present; high skull above the molars; slender mandibular symphysis in sagittal plane; orbits below cranial roof; slender zygomatic arches; laterally developed occipital plate; nasal and lacrimal separate; short extension of canine processes; lingual border of lower cheek tooth alveolar process lower than the labial border.

Age Early Pliocene (5–4 Ma).

African Occurrence Chad.

Remarks Saotherium is a new genus including two Pliocene hippopotamine forms from the Djurab Desert of Chad (Boisserie et al., 2003; Boisserie, 2005). *Saotherium mingoz* (Boisserie et al., 2003), a hexaprotodont hippopotamid with a unique antorbital angle characterizing the lateral profile of the cranial roof, was provisionally assigned to *Hexaprotodon* (sensu Coryndon, 1977) prior to the systematic revision of the family Hippopotamidae by Boisserie (2005). *S. mingoz* is known from abundant craniomandibular remains (and as yet undescribed postcranial elements) from the Kollé area of the Djurab Desert that is estimated to be between 5 and 4 Ma in age (Brunet et al., 1998). *Saotherium* cf. *mingoz* is slightly larger with a more developed premolar row and was recovered from the neighboring Kossom Bougoudi area of the Djurab Desert, a site with a vertebrate fauna estimated to be 5 Ma in age (Brunet and MPFT, 2000).

SAOTHERIUM MINGOZ Boisserie, 2003 and *SAOTHERIUM* cf. *MINGOZ*
Figures 44.3 and 44.4

These early Pliocene hippos from Chad exhibit a mixture of features considered to be derived (short premolar row, incisor differentiation, lateral expansion of the gonial angle), more primitive (low orbits, weak development of canine processes), and unique (antorbital angle, high skull above molars). As with *Choeropsis* the diagnostic aspects of *Saotherium* lie mainly in the "form" of the cranium (figure 44.4) and not in that of the mandible. Figure 44.3 indicates that the proportions of the symphysis of the *Saotherium* taxa lie between those of *A. harvardi* and the modern *Hip. amphibius*, relative to some hippopotamids from other genera. However, the cranium is more similar to that of *Choeropsis* (orbit position and size; cylindrical braincase) than to crania of other hippopotamine taxa (Boisserie, 2005).

Genus *HEXAPROTODON* Falconer and Cautley, 1836

Synonyms Hippopotamus (Hexaprotodon) Falconer and Cautley, 1836:51.

Type Species Hexaprotodon sivalensis (Falconer and Cautley, 1836), from Mio-Pliocene strata of the Siwalik Hills, India/Pakistan.

Other Species Africa: Hex. bruneti (Boisserie and White, 2004) and *Hex. garyam* (Boisserie et al., 2005c). Asia: several species known from India, Pakistan, Myanmar, and Indonesia; most were considered to be subspecies of *Hex. sivalensis* by Hooijer (1950), whereas other authors confer specific rank on them (Lydekker, 1884; Boisserie, 2005).

Diagnosis After Boisserie (2005). Hexaprotodont; very high, robust, relatively short mandibular symphysis with canine processes not particularly extended laterally; dorsal plane of symphysis angled obliquely; thick incisor alveolar process frontally projected; small differences in incisor diameter, i2 usually the smallest; laterally everted but not hooklike gonial angle; orbit with well-developed supraorbital process and deep narrow notch on anterior border; transversely thick zygomatic arches; elevated sagittal crest on a transversely compressed braincase.

Age Africa, late Miocene and late Pliocene (ca. 2.5 Ma); Asia, late Miocene–late Pleistocene.

African Occurrence Chad (Boisserie et al., 2005c), Ethiopia (Boisserie and White, 2004).

Remarks Hexaprotodon was first proposed as a subgenus (Falconer and Cautley, 1836) of *Hippopotamus* for the Siwalik material from Pakistan and India to signify the possession of six incisors (hexaprotodont) in the Asian hippopotamids as opposed to the four incisors (tetraprotodont) in the extant *Hippopotamus*. Owen (1845) elevated *Hexaprotodon* to full generic rank. Lydekker (1884) regarded incisor number alone to be an inadequate basis for generic distinction but Colbert (1935) recognized a list of other mainly cranial features that could distinguish *Hexaprotodon* from *Hippopotamus*. Coryndon (1977) showed that Colbert's (1935) traits characterized several East African hippopotamid taxa, and the genus *Hexaprotodon* was expanded further to include African (including the extant *Choeropsis*) as well as Asian representatives (Coryndon, 1977). This classification was adopted by subsequent workers (but see Pickford, 1983, and Stuenes, 1989) even though the integrity of genus *Hexaprotodon* (sensu Coryndon, 1977) was questioned given its probable paraphyletic nature (Harris, 1991; Harrison, 1997; Weston, 1997, 2000; Gentry, 1999). Boisserie's (2005) cladistic revision of the family Hippopotamidae confirmed that the genus *Hexaprotodon* sensu Coryndon (1977) was paraphyletic and Boisserie recognized a distinctive Asian hippopotamid clade (*Hexaprotodon* sensu Boisserie, 2005). This phylogenetic revision of the Hippopotamidae has resulted in the further emendation of the diagnosis of *Hexaprotodon* (Boisserie, 2005); the genus now excludes the taxa transferred to *Hexaprotodon* by Coryndon (1977). *Hexaprotodon* sensu Boisserie (2005) includes *Hex. sivalensis* Falconer and Cautley (1836), to which Hooijer (1950) attributed eight subspecies (specific rank is conserved by Boisserie, 2005) from the Indian subcontinent and the Indonesian archipelago (Bergh et al., 2001), *Hex. iravaticus* Falconer and Cautley, 1847, and related specimens from Myanmar (Colbert, 1938, 1943; Hooijer, 1950) and two new species from Africa, *Hex. bruneti* (Boisserie and White, 2004) and *Hex. garyam* (Boisserie et al., 2005c).

HEXAPROTODON GARYAM (Boisserie et al., 2005)
Figures 44.5A and 44.5B

A hexaprotodont hippopotamid known from the late Miocene levels of the Toros-Menalla region in Chad (Boisserie et al., 2005c). The holotype is an incomplete mandible that has a massive, high, furrowed symphysis with a projected incisor alveolar shelf bearing a smaller and more dorsally positioned i2 relative to the i1 and i3 (figures 44.5A, 44.5B). Uniquely, this species has laterally isolated lower canine processes that are separated from the incisor alveolar shelf by steep dorsal grooves while the mandibular corpus decreases in height posteriorly (Boisserie et al., 2005c). The symphyseal morphology of *Hex. garyam* is most similar to that of *Hex. sivalensis* from Asia. Boisserie et al. (2005c) suggested that *Hex. garyam* could be the first representative of a hippo lineage mostly known in Asia. The possible earliest appearance of *Hexaprotodon* in Asia was proposed to be ca. 6 Ma (Barry et al., 2002). However, a reappraisal of the Asian hippos, coupled with the discovery of more complete cranial fossils of *Hex. garyam*, as well as a comparison with late Miocene hippopotamid remains from southern Europe, are necessary to establish the exact affinities between *Hex. garyam* and the pre-Pleistocene Asian hippopotamids.

HEXAPROTODON BRUNETI Boisserie and White, 2004
Figures 44.5C and 44.5D

A hexaprotodont late Pliocene (ca. 2.5 Ma) hippo known from the Hata Member of the Bouri Formation and from upper Maka, Middle Awash, Ethiopia (Boisserie and White, 2004). The holotype consists of a partial cranium, a subcomplete mandible that has a high, massive, short symphysis with overhanging incisor alveolar process, and vertebrae of a single individual (figures 44.5C, 44.5D). Uniquely, *Hex. bruneti* has a greatly enlarged i3 relative to the first and second lower incisors. *Hex. bruneti* is more similar to the Asian *Hex. sivalensis* than any known African hippo, having identical symphyseal and anterior palatine proportions and many dental affinities (Boisserie and White, 2004). Even the tendency to develop an enlarged i3 relative to i1 has been noted in two Asian forms from the Pleistocene levels of central India (Hooijer, 1950), though the i3 in those hippopotamids never attained the exaggerated proportions observed in *Hex. bruneti* (Boisserie and White, 2004). The relatively small m3 and shallow distal groove on the upper canine further distinguish *Hex. bruneti* from *Hex. sivalensis* and its Asian relatives. *Hex. bruneti* is considered to be a likely Asian migrant (Boisserie and White, 2004).

Genus *HIPPOPOTAMUS* Linnaeus, 1758
Figure 44.6A

Type Species *Hip. amphibius* Linnaeus, 1758, extant (figure 44.6A).

Other Species Africa: *Hip. gorgops* Dietrich, 1926, *Hip. kaisensis* Hopwood, 1926, *Hip. lemerlei* Grandidier *in* Milne Edwards, 1868, *Hip. madagascariensis* Guldberg, 1883 and *Hip. laloumena* Faure and Guérin, 1990. Eurasia: several species known (Faure 1983, 1986; Vekua, 1986; Mazza, 1995) including most of those from the Mediterranean islands (Reese, 1975).

Diagnosis Tetraprotodont; elongated muzzle; upper canines with longitudinal and shallow posterior groove, narrow and enamel coated; lower canines with strong convergent enamel ridges; deep and widely open notch on the anterior orbital border; limbs short and robust with very large quadridigitigrade feet.

Age Ca. 5 Ma–present.

African Occurrence Africa including Madagascar.

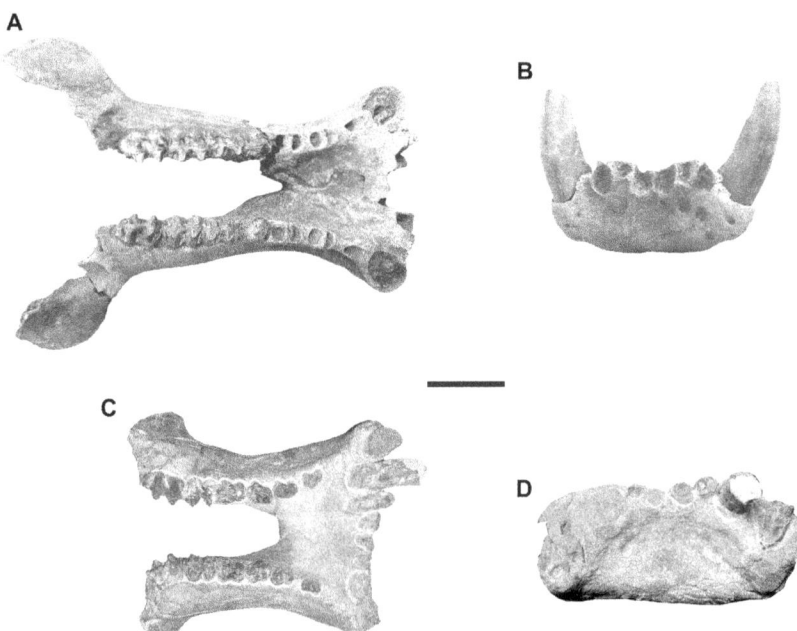

FIGURE 44.5 A) Dorsal view of mandible (canines have been removed to illustrate canine apophyses) TM337-01-001 from Toros-Menalla, holotype of *Hexaprotodon garyam*, preserved at the CNAR; B) rostral view with canines of TM337-01-001; C) dorsal view of mandible BOU-VP-8/79 from Bouri Hata, holotype of *Hexaprotodon bruneti*, preserved at the National Museum of Ethiopia, Addis Abeba (NME); D) rostral view of BOU-VP-8/79. Scale bar is 10 cm.

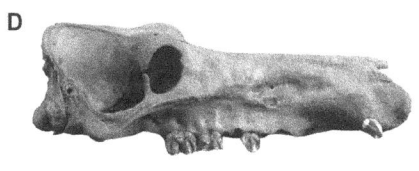

FIGURE 44.6   A) Lateral view of cranium MVZ 77804 of unknown provenience, *Hippopotamus amphibius*, preserved at the Museum of Vertebrate Zoology, University of California, Berkeley; B) left lateral view of cranium M15162, *Hippopotamus gorgops*, from Rawi, Kavirondo Gulf, Kenya, preserved at the Natural History Museum, London (image reversed); C) lateral view of cranium H11001, *Hippopotamus madagascariensis* from Antsirabé, Madagascar, preserved at the University Museum of Zoology, Cambridge, UK; D) lateral view of cranium VA3, *Hippopotamus lemerlei* from Ampoza, Madagascar, preserved at the Natural History Museum, London. Scale bar is 10 cm.

Remarks The tetraprotodont genus *Hippopotamus*, with a derived anterior dentition, can be distinguished relatively easily from other hippopotamines, and the monophyly of this taxon is well supported (Boisserie, 2005). Boisserie (2005) modified previous diagnoses of the genus (Gèze, 1980; Harris, 1991), listing, in particular, apomorphies linked with the canines (upper canine with shallow, narrow posterior groove; lower canine with strong convergent enamel ridges) and the presence of a deep and widely open notch on the anterior border of the orbit. However, the number of species and their relationships within the genus are not clearly established (Boisserie, 2005). The earliest record of *Hippopotamus*, *Hip. kaisensis* (Hopwood, 1926, 1939), is possibly from the lowest levels of the Nkondo Formation in the Western Rift, Uganda, at ~5 Ma (Faure, 1994). In the Turkana Basin, *Hip.* cf. *kaisensis* is first recorded from the Nachukui Formation West Turkana, Kenya, between 3.9 and 3.4 Ma (Harris et al., 1988; Leakey et al., 2001), and a *Hippopotamus* sp. similar to *Hip. kaisensis* is also reported from the Pliocene Chemeron Formation, Baringo, Kenya (Coryndon, 1978a; Gèze, 1985).

This hippopotamus, first described from two lower molars and postcranial elements from the Plio-Pleistocene Kaiso Formation, Uganda (Hopwood, 1926), is still a poorly known species with a partial tetraprotodont mandibular symphysis from Kaiso Village figured by Cooke and Coryndon (1970) being the most complete specimen known. Hippopotamid material very similar to *Hip. kaisensis* has also been recovered from the Lusso Beds (2.3–1.8 Ma; Pavlakis, 1990), in the Zaire portion of the Western rift. However, Pavlakis (1990) demonstrated that this material was indistinguishable from the modern *Hip. amphibius* and referred to it as *Hip.* aff. *Hip. amphibius*, possibly with more archaic molars, as Gentry (1999) noted their similarity to those of aff. *Hip. protamphibius* and *Hex. sivalensis*. However, material from Kaiso Village (Cooke and Coryndon, 1970) and West Turkana (Harris et al., 1988) is usually associated with strongly ribbed lower canines, confirming the occurrence of the genus *Hippopotamus*. More problematic is the distinction between *Hip. kaisensis* and the much more completely known *Hip. gorgops* (Dietrich, 1926, 1928), a hippopotamus recorded from many Pleistocene localities in East Africa and at Cornelia in South Africa (Hooijer, 1958; Coryndon, 1978b).

HIPPOPOTAMUS GORGOPS Dietrich, 1926
Figure 44.6B

In the Turkana Basin this species has been documented potentially as early as 2.5 Ma and up to around 0.7 Ma (Gèze 1985; Harris et al., 1988; Harris, 1991), in the Western Rift from 1.8 Ma (Faure, 1994), and from the middle Pleistocene at Cornelia in South Africa (Hooijer, 1958; Bender and Brink 1992). Hopwood (1939) first questioned the synonymy of *Hip. kaisensis* and *Hip. gorgops* and later formally stated that if these species proved to be synonymous, *Hip. gorgops* had priority (Hopwood and Hollyfield, 1954). Nevertheless, rather smaller specimens of *Hippopotamus* from the lower portion of the West Turkana sequence have been distinguished from larger material derived from higher levels of the same Formation and assigned to *Hip.* cf. *Hip. kaisensis* and *Hip. gorgops*, respectively (Harris et al., 1988; Harris, 1991). Faure (1994) also identified *Hip. kaisensis* from lower levels and *Hip. gorgops* from the highest level of the Albertine Rift Valley sediments in Uganda.

Hippopotamus gorgops was first described by Dietrich (1926, 1928) from Olduvai Gorge in Tanzania where it is recorded throughout the sequence, between 1.9 and 0.6 Ma (Bed I–Bed IV; Coryndon, 1970a). Coryndon (1970a, 1970b) described an evolutionary trend within *Hip. gorgops* at Olduvai, distinguishing a less specialized variety, more similar to *Hip. amphibius*, at the base of the sequence from a highly specialized more amphibious form, with very elevated orbits, high occipital crest, long diastema between p2 and p3, elongate and flattened rostrum and very shortened postorbital region, from the top of the sequence (figure 44.6B). In spite of this morphological variation typifying *Hip. gorgops*, early representatives of this taxon have been distinguished from *Hip. amphibius* in terms of their greater size, greater degree of orbit elevation, high-crowned molars with a less convoluted wear pattern and relatively low cingula, and upper molars with splayed multiple roots (Coryndon, 1970a; Harris, 1991). Dietrich (1928) and Harris (1991) describe the orbital roof as thickened or swollen, and this trait, though not unique to

Hip. gorgops, is not evident in *Hip. amphibius*. However, more comparative studies are necessary to resolve the relationship between *Hip. kaisensis*, *Hip. amphibius*, and *Hip. gorgops*, and it is probable that some material currently assigned to separate taxa is conspecific.

In addition to the relatively complete material known from the Turkana Basin (Shungura Fm. Omo, Nachukui Fm., Koobi Fora Fm.), other records of *Hip. gorgops* or similar forms in Africa include: Olorgesailie, Kenya, between 1.0 and 0.6 (Isaac, 1977), a highly specialized form of *Hip. gorgops* comparable with material from Bed IV, Olduvai (Coryndon, 1970a); Rawi and Kanjera, Plio-Pleistocene, southwestern Kenya (Ditchfield et al., 1999); Pleistocene deposits from Baringo, Kenya (Coryndon, 1978a; Gèze, 1985); Melka Kunturé, Ethiopia between 1.6 and 0.8 (Gèze, 1980, 1985); Buia, Eritrea ca. 1 Ma (Martínez-Navarro et al., 2004); Bouri (Daka Member), Ethiopia ca. 1 Ma (Heinzelin et al., 2000; Boisserie and Gilbert, 2008); ca. 0.6 Ma Bodo, Ethiopia (Gèze, 1980, 1985); and the middle Pleistocene material from Cornelia, South Africa (Hooijer, 1958).

HIPPOPOTAMUS AMPHIBIUS Linnaeus, 1758

This extant hippopotamus has been identified from the Nariokotome Member at the top of the West Turkana sequence (Harris et al., 1988) and in the Omo Shungura Formation from Member L, (1.38 Ma; Harris, 1991), but it has not been reported from East Turkana. *Hip. amphibius* has also been recorded from various other middle Pleistocene sites in East Africa, including Lainyamok and Isenya, Kenya and Asbole, Ethiopia (Geraads et al., 2004: table 3) but the earliest firm occurrence is from Gafalo, Djibouti dated at 1.6 Ma (Faure and Guérin, 1997). Pavlakis (1990) suggests that fossils from the Western Rift represent a single evolving lineage from *Hip. kaisensis* to *Hip. amphibius*, making the oldest record of *Hip.* aff. *Hip. amphibius* to be around 2.3 Ma (Pavlakis, 1990).

HIPPOPOTAMUS LEMERLEI Grandidier in Milne-Edwards, 1868, *HIPPOPOTAMUS MADAGASCARIENSIS* Guldberg, 1883, and *HIPPOPOTAMUS LALOUMENA* Faure and Guérin, 1990
Figures 44.6C and 44.6D

Three species of recently extinct (within the last 1,000 yrs) *Hippopotamus* have been described from Madagascar (Stuenes, 1989; Faure and Guérin, 1990). All hippopotamus material that has been dated is of Holocene age (Burney et al., 2004), and no earlier fossil record of hippopotamus exists on the island. *Hip. lemerlei* from the island's coastal lowlands, is more amphibious and expresses marked sexual dimorphism (Stuenes, 1989; figure 44.6D). A second more terrestrial species, *Hip. madagascariensis*, has been identified by Stuenes (1989) from the island's central highlands (figure 44.6C). A third species, *Hip. laloumena*, from Mananjary on the east coast of Madagascar, is close in size to the smallest *Hip. amphibius* (Faure and Guérin, 1990) and was initially described as a subspecies of the latter, *Hip. amphibius standini* (Monnier and Lamberton, 1922).

Hip. lemerlei and *Hip. madagascariensis* are dwarfed relative to the modern *Hip. amphibius*; their crania, though variable in size, are approximately a third the volume of the cranium of the extant common hippopotamus. The taxonomic interpretation of insular island forms is complicated by the correlated effects of size reduction, the tendency of certain adaptive traits to evolve in parallel, the timing of isolation of the species, and the potential number of invasions over time (Sondaar, 1977). Unlike the relatively small Mediterranean islands, Madagascar is a large

landmass that supports a diversity of habitats. It is, therefore, possible that hippo populations were exposed to different evolutionary pressures on the island, explaining the diversity of Malagasy species reported. The timing of arrival of hippos on Madagascar is not known (Stuenes, 1989). Judging from the similarity of *Hip. laloumena* to the modern *Hip. amphibius*, relative to the distinctiveness of the two dwarf species, there may have been more than one invasion. *Hip. lemerlei* and *Hip madagascariensis* both possess upper canines with a very shallow posterior groove and lower canines with strong and convergent ridges, features considered to be apomorphic traits of *Hippopotamus* (Boisserie, 2005). *Hip. lemerlei* has more elevated orbits with thickened supraorbital margins relative to those of *Hip. madagascariensis* (Stuenes, 1989). There are a suite of other cranial characteristics listed by Stuenes (1989) that distinguish the dwarf Malagasy species from each other and these features appear to be linked with adaptations to different lifestyles and possibly diet.

PLIO-PLEISTOCENE ENDEMIC FORMS FROM THE TURKANA AND AWASH BASINS WITH AFFINITY TO THE GENUS *HIPPOPOTAMUS*

In Boisserie's (2005) systematic revision of the family Hippopotamidae, hippopotamine taxa that had previously been assigned to *Hexaprotodon* (*sensu* Coryndon, 1977), showing some evolutionary trends that are well developed in *Hippopotamus*, and not included in the Asian hippopotamid clade, were provisionally reclassified as aff. *Hippopotamus*, in anticipation of thorough examination of their phylogenetic relationships. In Africa these include Plio-Pleistocene hippopotamid species from the Turkana Basin, Ethiopia, and Kenya: aff. *Hip. protamphibius* (Arambourg, 1944a), aff. *Hip.* cf. *protamphibius* (Weston, 2003a), aff. *Hip. karumensis* (Coryndon, 1977), and aff. *Hip. aethiopicus* (Coryndon and Coppens, 1975), and Pliocene species from the Afar depression, Ethiopia (aff. *Hip. afarensis*; Gèze, 1985), aff. *Hip. coryndonae* (Gèze, 1985) and aff. *Hip. dulu* (Boisserie, 2004).

aff. *HIPPOPOTAMUS PROTAMPHIBIUS* Arambourg, 1944
Figures 44.7A and 44.7B

This species was originally described from the Shungura Formation, Omo, Ethiopia, as a tetraprotodont hippopotamus of average size, with among other features, slightly elevated orbits and a lacrimal separate from the nasal (Arambourg, 1944a; figures 44.7A, 44.7B). Subsequently, much hippo material collected from the lower Omo Valley of southwestern Ethiopia, derived from the Mursi, Usno, and Shungura Formations (ca. 4–ca. 1.9 Ma) has been attributed to aff. *Hip. protamphibius* (Arambourg, 1947; Coryndon and Coppens, 1973; Gèze, 1985). Gèze (1985) recognized an early hexaprotodont variant from the lower part of the Shungura Formation and from the Mursi and Usno Formations (figured by Coryndon and Coppens, 1973; Gèze, 1980) as a separate subspecies, aff. *Hip. protamphibius turkanensis*. Gèze (1985) also recognized a second slightly smaller species, aff. *Hip. shungurensis* from the Shungura Formation (equivalent to *Hip.* sp. A; Coryndon and Coppens, 1973), but Harris (1991) and Harrison (1997) considered aff. *Hip. shungurensis* to be a probable female aff. *Hip. protamphibius* morphotype. aff. *Hip. protamphibius* has also been recovered from the Koobi Fora Formation, East Turkana, Kenya, and from the Nachukui Formation, West Turkana, Kenya (Harris et al., 1988; Harris, 1991). In East Turkana, analogous to the situation

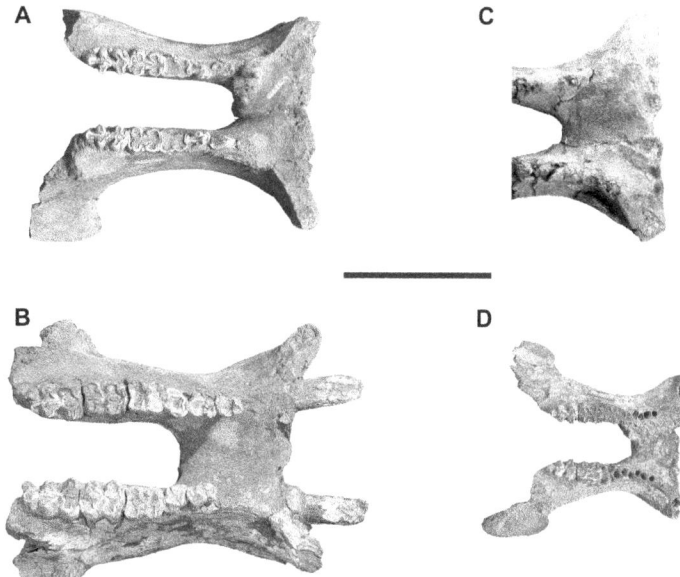

FIGURE 44.7   A) Dorsal view of mandible Omo 75-69-2798 from the Shungura Formation, aff. *Hippopotamus protamphibius*, preserved at the NME; B) dorsal view of
mandible KM-ER 798 from Koobi Fora, holotype of aff. *Hippopotamus karumensis*,
preserved at the NMK; C) dorsal view of the mandibular symphysis of KNM-ER
2738 from Koobi Fora, aff. *Hippopotamus protamphibius*, preserved at the NMK; D)
dorsal view of mandible P997-5 from the Shungura Formation, holotype of aff.
Hippopotamus aethiopicus, preserved at the NME. Scale bar is 10 cm.

in the Ethiopian Omo succession, both hexaprotodont and
tetraprotodont mandibular symphyses attributed to aff. *Hip.
protamphibius* have been recovered from the Tulu Bor Member
of the Koobi Fora Formation (Harris, 1991; figure 44.7B).

aff. *HIPPOPOTAMUS* cf. *PROTAMPHIBIUS* Weston, 2003
Figure 44.3

Kenyan hippopotamids from the Apak Member of the Nachukui Formation at Lothagam, from Kanapoi and from the base
of the Koobi Fora Formation, Allia Bay, have been referred to aff.
Hip. cf. *protamphibius* (Weston, 2003a). However, Harris et al.
(2003) distinguish the Kanapoi hippo as aff *Hip. protamphibius*. The Kanapoi material was initially identified by Coryndon (1977) as an "advanced" form of *A. harvardi*, but a suite of
features including the shorter, wider symphysis (figure 44.3),
differentiation of incisor size, laterally compressed lower
canine, depressions in frontal and maxilla/lacrimal region of
cranium linked with orbit elevation, simple P4 with a highly
reduced lingual cusp, and a less gracile skeleton distinguish
aff. *Hip.* cf. *protamphibius*. The degree of orbit elevation, extent
of incisor differentiation, and relative reduction in size of the
premolars are typically more advanced in aff. *Hip. protamphibius*. It is noteworthy that variation in the aff. *Hip. protamphibius* hypodigm from the Omo succession (distinguished as
aff. *Hip. shungurensis* and aff. *Hip. protamphibius turkanensis*;
Gèze, 1985) includes specimens with a proportionately longer
mandibular symphysis (Coryndon and Coppens, 1973; Gèze,
1985; Harris, 1991). In contrast, aff. *Hip.* cf. *protamphibius*,
although from slightly earlier Pliocene sediments, appears to
possess a relatively shorter mandibular symphysis compared
to the latter forms. A full revision and further description of
the Omo material is necessary to further resolve some of the
taxonomic issues evident in the Turkana Basin material.

aff. *HIPPOPOTAMUS KARUMENSIS* Coryndon, 1977
Figure 44.7C

A hippopotamus first described by Coryndon (1977) from the
upper members of the Koobi Fora Formation (~ 2–1.4 Ma) and
Ileret, East Turkana, Kenya. The holotype (KNM-ER 798: figure 44.7C) of this large hippopotamus had markedly elevated
orbits and a striking, two pronged "diprotodont" lower jaw
(Coryndon, 1977; Harris, 1991). Though originally diagnosed
with two lower incisors, Harris (1991) also recognized a less
progressive tetraprotodont form of aff. *Hip. karumensis* from
the upper Burgi and lower KBS Members of the Koobi Fora
Formation. aff. *Hip. karumensis* is also reported from the
Kaitio and Natoo Members of the Nachukui Formation of
West Turkana, Kenya (Harris et al., 1988), and Gèze (1980)
has identified tetraprodont material from the Omo Shungura
Formation as aff. *Hip.* cf. *Hip. karumensis*.

The mandible of aff. *Hip. karumensis* differs from that of aff.
Hip. protamphibius, the large incisors are widely separated by
a protruding shelf of bone, and the much smaller canines are
set in long slender alveolar processes that project anteriorly
beyond the midline of the symphysis. However, Harris (1991)
interprets aff. *Hip. karumensis* as a progressive form of aff.
Hip. protamphibius, and some of the early aff. *Hip. karumensis*
material is difficult to assign to one or the other species.

aff. *HIPPOPOTAMUS AETHIOPICUS*
Coryndon and Coppens, 1975
Figure 44.7D

A pygmy hippopotamus comparable in size to *Choeropsis liberiensis* (Coryndon and Coppens, 1975) that is exclusively
known from the Turkana basin (Ethiopia and Kenya) and
described originally from material from the upper part of the

Shungura Formation (~2.3–1.0 Ma), Ethiopia (Coryndon and Coppens, 1973, *Hip.* sp. B; Coryndon and Coppens, 1975). Although Coryndon and Coppens (1975) referred to pygmy hippopotamus material from East Turkana as aff. *Hip aethiopicus*, Gèze (1980, 1985) did not recognize it as the same species. Subsequently, Harris (1991) assigned further material from the Koobi Fora and Nachukui Formations to aff. *Hip. aethiopicus* (Harris et al., 1988; Harris, 1991).

Coryndon (1977) placed aff. *Hip aethiopicus* in the genus *Hippopotamus* based on the position of the lacrimal bone, but Harris (1991) recognized its close affinities with aff. *Hip. karumensis* and aff. *Hip. protamphibius*, a position now confirmed by Boisserie's (2005) systematic revision. The tetraprotodont short, broad symphysis is robust with the anterior face vertically oriented and relatively deep for its length (figure 44.7D). Although substantially scaled down, the mandibular morphology of this dwarf species resembles the "less progressive" aff. *Hip. karumensis* and the most progressive aff. *Hip. protamphibius* forms. Coryndon and Coppens (1975) describe the molars as lophodont, linking this development to the diet of a more forest adapted hippopotamus that potentially occupied a niche similar to *Choeropsis*, the living pygmy hippo.

aff. *HIPPOPOTAMUS AFARENSIS* Gèze, 1985, and aff. *HIPPOPOTAMUS CORYNDONAE* Gèze, 1985
Figures 44.8B and 44.8C

aff. *Hip. afarensis* (originally assigned to the monospecific genus *Trilobophorus* Gèze, 1985) and aff. *Hip. coryndonae* (Gèze, 1985) from the Afar, Ethiopia, are known from Hadar and Geraru between 3.4 and 2.33 Ma (Gèze, 1980; Boisserie, 2004). Gèze (1985) considered aff. *Hip. afarensis* to warrant placement in a separate genus based on the unique arrangement of the bony contacts of the facial bones. However, Boisserie (2005) on reexamination of the Hadar fossils established the presence of a contact between the lacrimal and the nasal, suggesting that Gèze's initial diagnosis was based on a misinterpretation of the anatomy. aff. *Hip. afarensis* had a large robust muzzle of comparable size to that of *Hip. amphibius* (figure 44.8B), whereas aff. *Hip. coryndonae* was smaller and more similar in size to aff. *Hip. protamphibius* (Gèze, 1985; figure 44.8C). The characters Gèze (1985) listed to distinguish aff. *Hip. coryndonae* from aff. *Hip. protamphibius* (variation in the lacrimal region, three incisors with the i2 more reduced than the i3, canine cross section, and mandibular form) do not adequately separate aff. *Hip. coryndonae* from aff. *Hip.* cf. *protamphibius* or the early hexaprotodont representatives of aff. *Hip. protamphibius* known from the Turkana basin (Harris, 1991; Weston, 2003a), and revision of the Afar and Turkana Basin Pliocene hippos is still required to establish their taxonomic affinities.

aff. *HIPPOPOTAMUS DULU* Boisserie, 2004
Figures 44.3 and 44.8A

Boisserie (2004) described another Afar hippo, aff. *Hip. dulu*, from the lower part of the Sagantole Formation (5.2–4.9 Ma). This species can be clearly distinguished from the later Pliocene Afar hippos and has closer affinities with *A. harvardi*. aff. *Hip. dulu* is slightly smaller than *A. harvardi*, the mandibular symphysis is shorter relative to its width (figure 44.3), the premolar rows are relatively shorter in comparison to the molar rows and the mandibular angular processes are laterally shifted (figure 44.8A). The occipital condyles of aff. *Hip.*

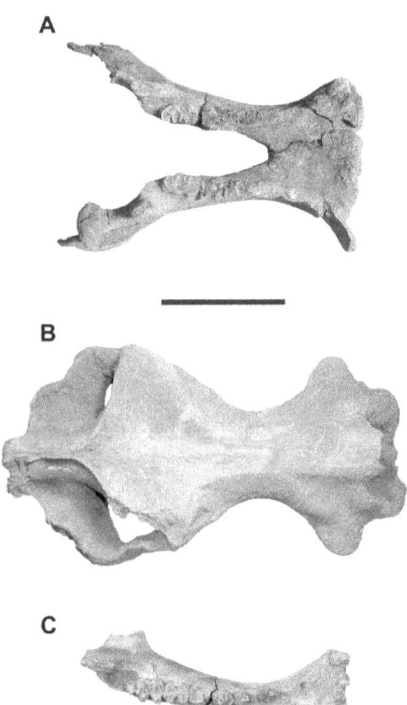

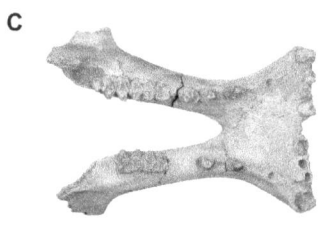

FIGURE 44.8 A) Dorsal view of mandible AME-VP-1/33 from Middle Awash, holotype of aff. *Hippopotamus dulu*, preserved at the NME; B) dorsal view of cranium AL109-3B from Hadar, paratype of aff. *Hippopotamus afarensis*, preserved at the NME; C) dorsal view of mandible AL170-1A from Hadar, holotype of aff. *Hippopotamus coryndonae*, preserved at the NME. Scale bar is 10 cm.

dulu are also larger than those of *A. harvardi*, but a relative increase in occipital condyle size relative to breadth of the posterior cranium is an evolutionary trend noted by Weston (1997, 2003a) in the *A. harvardi* sample from the late Miocene Nawata Formation, Kenya. The recovery of further material particularly indicating lower incisor size and arrangement is required to clarify the generic status of aff. *Hip. dulu*, which is provisionally left in open nomenclature.

Incertae Sedis
HEXAPROTODON ? *HIPPONENSIS* (Gaudry, 1876)

A hippopotamus that was first described by Gaudry (1876) from a selection of isolated lower teeth of probable association and likely upper Pliocene age, from Pont de Duvivier, south of Bône, Algeria (Joleaud, 1920; Arambourg, 1944b). This small hippopotamine with lower canines lacking ridged enamel was the first evidence of a hexaprotodont hippopotamus to be found in Africa, although Pomel (1890) did question whether the six incisors were derived from the same jaw. Further fragmentary hippopotamid remains recovered from the Pliocene (~4 Ma; Geraads, 1987) of Wadi Natrun (Gart el Muluk), Egypt have also been attributed to *Hex.* ? *hipponensis* (Andrews 1902; Stromer, 1914). Stromer (1914:5), however, considered the Egyptian hippopotamid to be tetraprotodont (possessing a first and second incisor of roughly equal size) based on the collection of associated incisors from a weathered lower jaw that was not illustrated. Subsequently, Arambourg (1947) disputed the allocation of the Egyptian material to *Hex. ? hipponensis* and referred to it as a new subspecies (*andrewsi*) of aff. *Hip. protamphibius*. Cooke and Coryndon (1970) considered the teeth described by Gaudry (1876) and Andrews (1902) to resemble closely those of *Hex. ? imaguncula* from the Kaiso deposits of Uganda, and Erdbrink and Krommenhoek (1975) considered *Hex. ? imaguncula* and *Hex. ? hipponensis* to be synonymous. In light of the incomplete nature of the North African Pliocene hippopotamine fossils, the identity of this relatively small hippopotamus remains indeterminate, although the dental remains can be clearly distinguished from those of *Hippopotamus*.

HEXAPROTODON ? IMAGUNCULA (Hopwood, 1926)
Figure 44.3

A hippopotamus that was first described by Hopwood (1926) as *Hip. imaguncula* from a sample of relatively small teeth and isolated postcranial elements from the Plio-Pleistocene Kaiso Formation, Kaiso Village, Uganda. Cooke and Coryndon (1970) emended the diagnosis of this species and the specific name to *imagunculus*. However, this emendation of the specific name, although adopted by nearly all subsequent workers with the exception of Erdbrink and Krommenhoek (1975) is not valid. "*Imaguncula*," meaning "little image," is a noun in apposition and not an adjective, so there is no requirement for the gender of the specifical name to agree with that of the generical name. In accordance with the international rules of zoological nomenclature, "*imaguncula*" should be preserved and is adopted by the current authors.

Fragmentary remains of a relatively small hippopotamine species referred to as *imaguncula* have been reported from the Western Rift of Uganda (Erdbrink and Krommenhoek, 1975; Faure, 1994), the Upper Semliki Valley of Zaire (Pavlakis, 1990), and from Rawi, Homa Peninsula, Kenya (Kent, 1942), a site located between the two branches of the East African Rift. The only relatively complete mandibular symphyses referred to *Hex. ? imaguncula* are specimens from the Kaiso Formation, Kazinga Channel, Uganda (UMP 6202; Erdbrink and Krommenhoek, 1975), and the Rawi hippo from Kenya, *A.* aff. *harvardi*. The symphyseal proportions of these two hexaprotodont mandibles indicate that two different species are represented (figure 44.3). Cooke and Coryndon (1970) also recognized a discrepancy between some of the smaller-sized hippopotamine craniomandibular remains recovered from the Kaiso Formation. A partial cranium (M14801) recovered from Behanga I retained primitive features such as low orbits, whereas the Kazinga Channel mandible (UMP 6202) possessed a relatively wide symphysis, considered a more derived condition (Cooke and Coryndon, 1970: figure 16, plate 14A). In light of this inconsistency, both specimens were referred by Cooke and Coryndon (1970) to *Hippopotamus* sp. as opposed to *Hippopotamus imaguncula*. The canines of *Hex. ? imaguncula* from the Kaiso Formation, with thin, finely rugose but nonridged enamel (Cooke and Coryndon, 1970), can be distinguished from *Hippopotamus*.

Boisserie (2005) questioned the validity of *Hex. ? imaguncula* as a taxon given that it is heterogeneous and is probably represented by several relatively small-sized species from the Western rift basins during the Pliocene or even extending back to the late Miocene (Faure, 1994). Some of the *Hex. ? imaguncula* material, such as the Rawi mandible from Kenya, has been classified as an *Archaeopotamus* and distinguished from other material (Boisserie, 2005). The exact status of *Hex. ? imaguncula* is still a taxonomic issue of some importance as it was one of the earliest named hippopotamid species from East Africa, and awaiting a revision of this material, Pavlakis (1990) and Boisserie (2005) favor the restriction of the nomen *imaguncula* to the material found at the type locality (Kaiso Village, according to Cooke and Coryndon, 1970) from the Ugandan Kaiso Formation.

HEXAPROTODON ? SAHABIENSIS Gaziry, 1987

A medium-sized hippopotamus known only from fragmentary remains from the late Miocene of Sahabi, Libya (Gaziry, 1987). The most complete specimen now known is a mandibular corpus fragment with p4–m3, although a report by Petrocchi (1952) recorded the discovery of a "hexaprotodont" hippopotamid skull from the Sahabi. The Sahabi hippopotamus was considered by Gaziry (1987) to have closest affinities with the late Miocene *A. harvardi*, but subsequently Gentry (1999) showed that the Libyan material more closely resembled the narrow-muzzled Abu Dhabi hippopotamus from the Baynunah Formation of the United Arab Emirates, initially referring the material to *Hex.* aff. *sahabiensis* (reclassified as *A.* aff. *lothagamensis* by Boisserie, 2005). Notably, the mandibular ramus is low, and the bulging profile of the lower margin of the corpus beneath the molar row is typical of *Archaeopotamus* but is found in some *Saotherium*, *Hexaprotodon* (*sivalensis* and *garyam*) and aff. *Hip. protamphibius* specimens. The dimensions of the cheek teeth and corpus height compare with the Abu Dhabi hippo but are smaller than those of *A. harvardi*. In contrast, the upper incisors are larger than those of *A. harvardi* though not of the magnitude reported by Gaziry (1987; see Harrison, 1997:183). Based on a plate-like, shallow premaxilla fragment bearing three relatively small incisor alveoli, the I3 is larger than the I1 and I2 (Gaziry, 1987). However, the small I1 diameter recorded from the premaxilla fragment contrasts strongly with the diameter taken from a much larger isolated I (1?). The premaxilla fragment may represent a young individual, but its size still exceeds that of the adult holotype of *A. harvardi*. Contrary to Coryndon (1977) the I1 and I2 of *A. harvardi* can be oval as opposed to cylindrical in cross section, and the premolar morphology is more variable than originally reported and comparable to that described for the Libyan hippo (Weston, 2003a). The isolated M1 described by Gaziry (1987), with a rudimentary protoconule and metaconule, is particularly uncommon among hippopotamines but has been reported in the molariform DP4 (Boisserie et al., 2005b). Overall, the fragmentary condition of Sahabi material makes it difficult to decipher its real affinities. It is hoped that additional material will be retrieved from Sahabi, given the interest in this hippopotamid's geographic and chronologic placement.

MISCELLANEOUS REMAINS FROM SOUTH OF THE EQUATOR

Late Miocene (ca. 6 Ma) hippopotamine remains were found at Lemudong'o, southern Kenya (Boisserie, 2007). These dental and postcranial isolated remains exhibit primitive features common to other late Miocene hippopotamine. Although some minor distinctive features were noted (Boisserie, 2007), the material is for now too fragmentary to allow more accurate identification.

A small and rare species of Pleistocene *Hippopotamus* of uncertain identity was found together with the large *Hip. gorgops* in Bed II at Olduvai Gorge, Tanzania (Coryndon, 1970a; Harris 1991).

Fragmentary dental and postcranial remains of potentially two species of hippopotamus have been reported from the Plio-Pleistocene Chiwondo Beds of Malawi (Coryndon, 1966, 1978b; Bromage et al., 1995). Postcranial elements including a complete tibia were initially identified as *Hippopotamus* by Hopwood (Dixie, 1927; Hopwood, 1931). Subsequently, Coryndon (1966) described some associated mandibular fragments with left and right m3's from Uraha as representing a hippopotamus that was slightly smaller than the modern *Hip. amphibius* but similar in size to *Hip. kaisensis* from the Western Rift of Uganda (Cooke and Coryndon, 1970). These two specimens of m3 from Malawi, accessioned in the NHM, London, possess a unique extension of the posterior cingulum that has not been recorded in any other *Hippopotamus*. In addition, Mawby (1970) distinguished a small hippopotamid maxillary fragment with molars found at Mwenirondo, from the more abundant remains of the larger hippopotamus, and assigned it to *Hex. ? imaguncula*. Like many other African Plio-Pleistocene fossil localities, two coeval species of large and small hippopotamus potentially occurred in Malawi, but prior to the full description and examination of the Malawi hippopotamid specimens, including new material (Bromage et al., 1995), their inferred closer affinity to the Western Rift hippopotamids cannot be verified.

Finally, an important collection of undescribed hippopotamid remains is known from the early Pliocene at Langebaanweg, South Africa (Franz-Odendaal et al., 2002). This assemblage represents the first significant pre-Pleistocene record of Hippopotamidae in southern Africa.

General Discussion

PHYLOGENETIC AND TAXONOMIC ISSUES

Previous Taxonomy

In the past, fossil and modern hippopotamus species have been classified into three genera. *Hippopotamus* was created first for the large extant *Hip. amphibius*, followed by *Hexaprotodon* (Falconer and Cautley, 1836; Owen, 1845) for the Siwalik fossil hippos from Asia, distinguished by having six as opposed to four incisors, and finally *Choeropsis* was created by Leidy (1853) for the extant Liberian pygmy hippopotamus, *C. liberiensis* (Morton, 1844, 1849). The original distinction between the Asian fossil taxa *(Hexaprotodon)* and African/European taxa *(Hippopotamus)* based on incisor number was contested by Lydekker (1884), but Colbert (1935), based on a suite of cranial distinctions, maintained this generic separation. Coryndon (1977) expanded the genus *Hexaprotodon* further to include African (including *Choeropsis*) as well as Asian representatives, the position of the lacrimal bone considered key in separating *Hexaprotodon* from *Hippopotamus*. Coryndon's (1977) classification was largely adopted by most subsequent authors, even though the integrity of the genus *Hexaprotodon* (*sensu* Coryndon 1977) was questioned given its probable paraphyletic nature (e.g., Harris, 1991; Weston, 1997, 2000; Gentry, 1999). Likewise, although some authors (e.g., Harris, 1991; Gentry, 1999; Weston, 2000, 2003a, 2003b) retained *Choeropsis* in a paraphyletic *Hexaprotodon* (*sensu* Coryndon, 1977), others chose to maintain *Choeropsis* as a distinct genus (Pickford, 1983; Harrison, 1997).

Current Taxonomy

The systematic palaeontology outlined here follows Boisserie's (2005) phylogenetic revision of the family Hippopotamidae. This cladistic analysis of cranial and dental features included 14 hippopotamine taxa (12 African and 2 Asian) and used as an outgroup a primitive anthracothere (*Anthracokeryx ulnifer* Pilgrim, 1928). Species known only from mandibles, such as the narrow-muzzled *A. lothagamensis*, the Arabian *A.* aff. *lothagamensis*, and the Chadian *Hex. garyam*, were not included. Also, due to difficulties in coding for continuous variables, the proportions of the mandibular symphysis (e.g., figure 44.3) did not form part of the character matrix (Boisserie, 2005). To circumvent some of these difficulties, Boisserie (2005) also performed a separate comparison of mandibular morphologies that to a large extent supported the clades identified in the parsimony analysis.

On the basis of this phylogenetic analysis, Boisserie (2005) proposed some taxonomic changes. This work clarified that the genus *Hexaprotodon* (*sensu* Coryndon, 1977) was paraphyletic and necessitated the splitting of the genus, with the name retained in the context of its initial usage for the Asian hippopotamids. Most remaining East African taxa previously assigned to *Hexaprotodon* form part of the same clade as the clearly defined genus *Hippopotamus* and are referred, prior to the further revision of these taxa, to aff. *Hippopotamus* (Boisserie, 2005). On the other hand, the genus *Choeropsis* was maintained, the modern pygmy hippo inferred to be part of an ancient independent lineage. Together with the Chadian hippos *(Saotherium)*, *Choeropsis* may have formed the sister group of all other hippopotamines (Boisserie, 2005). In addition to the Chadian *Saotherium*, another new genus, *Archaeopotamus* was created, mainly for the late Miocene narrow muzzled hippopotamuses but notably including *A. harvardi*, an anatomically well-known species not previously grouped with the narrow-muzzled species (Weston, 2000). It must be noted that there is still some discrepancy between evolutionary trends identified in the cranium (rostrum and neurocranium) and those that have been identified in the mandible that complicates phylogenetic analysis and interpretation. As a consequence, further improvement of hippopotamid phylogeny and taxonomy is still required.

Emergence of Hippopotamidae

The origin of the family has been vigorously debated since the 19th century. The principal hypotheses postulated that the first hippopotamids were derived from anthracotheriids (notably Falconer and Cautley, 1847; Colbert, 1935; Gentry and Hooker, 1988) or from suoids (e.g., Joleaud, 1920; Matthew, 1929; Pickford, 1983). This debate has acquired a new dimension in the last decades with numerous molecular-based phylogenies indicating that hippopotamids and cetaceans form a clade within Artiodactyla (e.g., Sarich, 1993; Gatesy et al., 1996; Gatesy, 1997; Arnason et al., 2004). This implies the paraphyly of Artiodactyla, and accordingly, a new name, Cetartiodactyla, was coined for the clade grouping artiodactyls and cetaceans (Montgelard et al., 1997). Part of the criticism of this molecular hypothesis came from the earliest cetaceans being known from the early Eocene (Bajpai and Gingerich, 1998), whereas the earliest hippopotamids are no older than the middle Miocene (Pickford, 1983; Behrensmeyer et al., 2002). The identification of the lineage that led to the Hippopotamidae is in this regard critical to resolve this issue and to better harmonize the molecular and paleontological data. On the basis of fossil data, Gingerich et al. (2001) and Geisler and Uhen (2003) have strengthened the idea that hippopotamids and cetaceans could have close affinities. Further morphological analyses favor the evolution of hippopotamids from certain anthracotheriids (Boisserie et al., 2005a, 2005b; Boisserie and Lihoreau, 2006) and have shown that this hypothesis represents a complementary step in closing the fossil gap between a still putative, early Cenozoic common ancestor to hippopotamids and cetaceans and the earliest hippopotamids (Boisserie et al., 2005b).

Affinities between Mio-Pliocene Hippopotaminae

The current phylogeny, here focusing on the African representatives of the Hippopotaminae, is outlined in figure 44.9. One of the main differences between this phylogeny and that proposed by Weston (2000) is the grouping of *A. harvardi* and *A.* aff. *harvardi* (Rawi, Kenya) with the extremely narrow-muzzled hippos *A. lothagamensis* and the Arabian hippopotamus. Weston's (2000) phylogeny considered the narrow-muzzled hippos as the sister group of all other hippopotamines, separating *A. lothagamensis*, the Arabian hippo, and potentially a poorly known Spanish species *Hex. crusafonti* (Lacomba et. al, 1986; Made, 1999) from *A. harvardi*. The position of *Archaeopotamus* in light of Boisserie's (2005) cladistic appraisal is determined solely on the affinities of *A. harvardi* as the other *Archaeopotamus* spp. are not known

from crania. Boisserie (2005) based *Archaeopotamus* on the symphyseal proportions (ratio of symphysis length vs. symphysis width) that differ from other hippopotamids and can be shown to scale ontogenetically (Weston, 2000; Boisserie 2005; figure 44.3). However, it is equally true that *C. liberiensis* has symphyseal proportions similar to those of *Hip. amphibius* that can also be shown to scale ontogenetically (Weston 2000: figure 5; figure 44.3). As the cranium of *C. liberiensis* is completely known and strikingly primitive for a number of traits, the more derived features attributable to the mandible are best interpreted as convergences (Boisserie, 2005). However, the crania of all *Archaeopotamus* spp. except *A. harvardi* are completely unknown and a comparable evaluation to that of *C. liberiensis* is not possible. Also of note is the single symphyseal specimen attributed to *Kenyapotamus*; though not preserved anteriorly, the overall form of this partial lower jaw suggests it was narrow (Nakaya et al., 1987; fig. 44.1B).

In Boisserie's (2005) phylogeny, *Choeropsis* is shown to represent a lineage distinct from all other hippopotamids, potentially diverging from its closest relatives the Chadian hippopotamines *(Saotherium)* before 5 Ma. Prior to the discovery of the Chadian hippos, the fossil record had appeared to lack any forms closely related to the extant *C. liberiensis*, although this affinity between *Saotherium* and *Choeropsis* is weakly supported by only one synapomorphy, the large size of the orbit (Boisserie, 2005). The inclusion of hippos from Chad has greatly enhanced our knowledge of the evolutionary

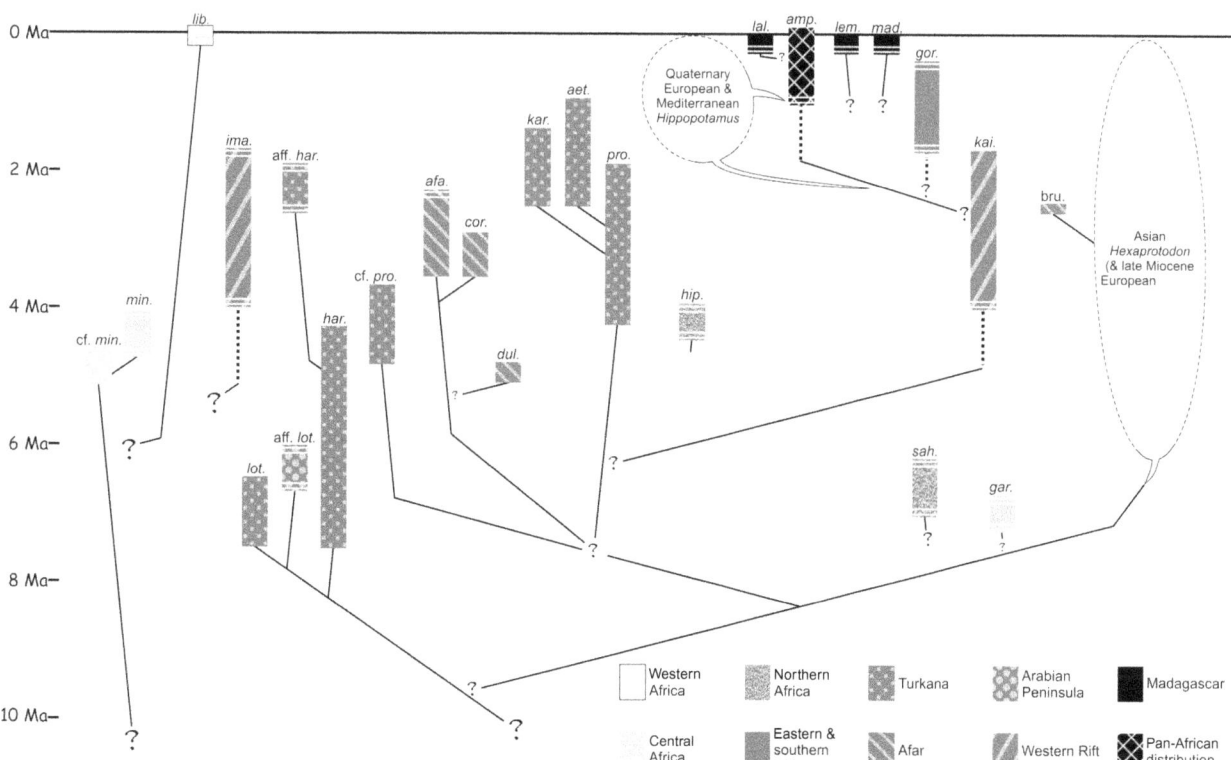

FIGURE 44.9 Phylogenetic relationships between African hippopotamines, with temporal and geographic placement. *aet.*: aff. *Hippopotamus aethiopicus*; *afa.*: aff. *Hip. afarensis*; aff. *har.*: *Archaeopotamus* aff. *harvardi*; aff. *lot.*: *A.* aff. *lothagamensis*; *amp.*: *Hip. amphibius*; *bru.*: *Hexaprotodon bruneti*; cf. *min.*: *Saotherium* cf. *mingoz*; cf. *pro.*: aff. *Hip.* cf. *protamphibius*; *cor.*: aff. *Hip. coryndonae*; *dul.*: aff. *Hip. dulu*; *gar.*: *Hex. garyam*; *gor.*: *Hip. gorgops*; *har.*: *A. harvardi*; *hip.*: *Hex.* ? *hipponensis*; *ima.*: *Hex.* ? *imaguncula*; *kai.*: *Hip. kaisensis*; *kar.*: aff. *Hip. karumensis*; *lal.*: *Hip. laloumena*; *lem.*: *Hip. lermelei*; *lib.*: *Choeropsis liberiensis*; *lot.*: *A. lothagamensis*; *mad.*: *Hip. madagascariensis*; *min.*: *S. mingoz*; *pro.*: aff. *Hip. protamphibius*; *sah*: *Hex.* ? *sahabiensis*.

history of the Hippopotamidae, but there are still some issues in relation to previous phylogenetic interpretations that deserve consideration. The early Pliocene Chadian hippos appear to represent a distinct lineage of hippopotamids reflecting African faunal provincialism at that time (Boisserie et al., 2003), but the exact relationship of these *Saotherium* taxa and the recently discovered late Miocene *Hex. garyam*, with closer affinities to Asian taxa than to African (Boisserie et al, 2005c) and known from the same Chad Basin, still needs to be carefully examined. The predominantly Asian distribution of *Hexaprotodon* (*sensu* Boisserie, 2005) could be challenged by a growing number of *Hexaprotodon* spp. (*sensu* Boisserie, 2005) being discovered in Africa. It remains uncertain whether poorly known taxa, based on material insufficient to allow accurate phylogenetic placement, such as that from Sahabi of Libya and those from the late Miocene of Europe (Made, 1999), have closer affinities with *Hexaprotodon* or *Archaeopotamus*.

A close relationship between the three Plio-Pleistocene Turkana hippos, aff. *Hip. protamphibius*, aff. *Hip. aethiopicus*, and aff. *Hip. karumensis*, as previously suggested by Harris (1991), was confirmed by Boisserie's (2005) systematic study, although the relationships between these taxa are still unresolved. Historically, a close relationship between aff. *Hip. protamphibius* and *Hippopotamus* was envisaged (Arambourg, 1944a, 1947), and this is still supported by cladistic analysis, with *Hippopotamus* forming the sister group of this *"protamphibius"* clade *(protamphibius-aethiopicus-karumensis)* (Boisserie, 2005). However, the inferred relationships of the other aff. *Hippopotamus* taxa from the Afar (aff. *Hip. coryndonae* and aff. *Hip. afarensis*) and from the Turkana Basin, Kenya (aff. *Hip.* cf. *protamphibius*), are more problematic. The position of the recently discovered aff. *Hip. dulu* from the Mio-Pliocene boundary of the Afar, Ethiopia (Boisserie, 2004), possessing affinities with *A. harvardi* is also uncertain, and extensive reexamination of the Plio-Pleistocene Turkana Basin taxa and those of the Afar depression, Ethiopia, is still required to better resolve the phylogenetic relationships between these East African Pliocene hippopotamids.

Affinities of and within *Hippopotamus*

The monophyletic genus *Hippopotamus* shares with related taxa from the Pliocene of eastern Africa a short and globular braincase, less than three lower incisors, and the possession of developed and anteriorly projecting canine processes (Boisserie, 2005). One incongruous character of note in relation to the position of *Hippopotamus* is the degree of lacrimal-nasal contact. A long contact between the medial border of the lacrimal and nasal bone was in the past considered diagnostic of *Hippopotamus* and important in separating *Hippopotamus* from *Hexaprotodon* (*sensu* Coryndon, 1977). However, Boisserie (2005) showed that such a long contact between these cranial bones was shared between *Hippopotamus* and aff. *Hip. afarensis*, a condition at odds with the inferred transition to minimal or variably developed contact in the *"protamphibius"* group. In addition, the Malagasy *Hippopotamus* spp. also exhibit variation in the development of the lacrimal-nasal contact, where in some specimens the contact is short or eliminated altogether by the presence of a supernumerary (intercalary) bone (Stuenes, 1989). Similar intercalary bones associated with the lacrimal have been noted in the Cypriot pygmy hippo and young *Hip. amphibius* (Reese, 1975), suggesting a possible link between their

development and the phenomenon of island dwarfing within *Hippopotamus*. The taxonomic value of bone contacts in the lacrimal area of the hippopotamus cranium is increasingly questionable and functionally poorly understood.

The origin of *Hippopotamus* is problematic mainly because *Hip. kaisensis*, the earliest record of the genus, is poorly known. The evolutionary relationships within *Hippopotamus* are not yet established. In Africa, by the basal Pleistocene, *Hip. gorgops* had become the most ubiquitous hippopotamus replacing all other large hippopotamids. By the middle Pleistocene, *Hip. amphibius*, apparently dentally more advanced than *gorgops* but cranially more archaic, appears to have supplanted the earlier species in Africa, where it is still common today. Several Pleistocene *Hippopotamus* spp. also colonized Europe and Western Asia prior to their final extinction by the Holocene.

EVOLUTIONARY TRENDS AND PALEOBIOLOGY

Ontogenetic Scaling and Dwarfism

Hippopotamids vary greatly in body size, as exemplified by the two living taxa, the large common hippopotamus being about six times heavier than the pygmy hippo (Eltringham, 1999). The range of body size encountered in fossil taxa is even greater. Growth-related variation is quite striking in hippos (Weston, 2003b), and interpreting the effects of ontogenetic scaling can be critical to establish correct taxonomic affinities. The assumption, however, that all small species of hippo are "dwarfed" in relation to a larger coeval taxon is not accurate (*contra* Gould, 1975). The extant pygmy hippo is not a dwarfed common hippopotamus and many cranial traits can be shown to deviate from an hypothesis of ontogenetic scaling (Weston 2003b). Nevertheless, *Choeropsis* may be an example of a secondarily adapted "dwarf" of an undetermined larger ancestral hippopotamus that was potentially less well adapted to a rain forest habitat.

Sondaar (1977, 1991) first recognized a suite of parallel adaptations common to the Pleistocene and Holocene insular dwarf hippos from the Mediterranean islands and Madagascar (Reese, 1975). Size decrease in island mammals is generally accompanied by postcranial adjustments, the hippos acquiring a more erect, shortened, stable foot and lengthened radius (Sondaar, 1977; Houtemaker and Sondaar, 1979). This has been interpreted as a cursorial adaptation to mountainous regions (Spaan, 1996; Caloi and Palombo, 1996) and not just a correlate of dwarfism peculiar to rapid phyletic evolution. Madagascar provides further support for hippos adopting different lifestyles, with two different dwarf species present, that are of similar cranial size but have different limb proportions (Stuenes, 1989; Faure and Guérin, 1990). Taxon-specific allometric changes do, however, characterize the growth of the skeleton in modern hippos (Weston, 1997; Weston 2003a,b). A better understanding of these developmental size adjustments and how they relate to dwarfism is still needed before these apparent affinities to a semiaquatic or terrestrial habit can be clarified.

Sexual Dimorphism

Hippopotamus amphibius is a sexually dimorphic species with male body weight and trunk length substantially exceeding those of the female (Kingdon, 1979). This type of size dimorphism results from bimaturism in which females attain physical majority earlier than males. It has been demonstrated that male and female skulls scale ontogenetically (Weston, 1997, 2003a). Allometric variation of skull form between the sexes is striking (Weston, 1997, 2003a). For example, the breadth of the lower jaw grows with positive allometry (Weston 2003b), the male's relatively broad, deep symphysis contrasting with the female's relatively narrow, shallow one. This jaw form corresponds to the sexually dimorphic anterior tusks (canines and incisors) that in males grow continuously throughout life (Weston, 1997). The cheek teeth (premolars and molars), in contrast, are not dimorphic in size and are similar in males and females (Laws, 1968; Weston, 1997, 2003a).

The extant pygmy hippo is not sexually dimorphic in terms of body weight, but shape changes in the skull are evident, and cranial breadth relative to skull length is greater in the male compared to that of the female (Weston, 2003a). In addition a reverse trend in dimorphism is apparent in the orbit, with females possessing larger eye sockets than males (Weston, 1997, 2003a). Neither of the latter sexually dimorphic features characterise *Hippopotamus amphibius*. The tusks of the pygmy hippo are also dimorphic in size (Weston, 1997). Weston (2003a) noted that the crania of the late Miocene *A. harvardi* exhibited similar cranial shape dimorphism to that of *Choeropsis*, but *A. harvardi* was not similar to the latter with regard to its orbit form. Stuenes (1989) described sexual variation in *Hip. lemerlei* from Madagascar that conforms to that of *Hip. amphibius*. Most fossil taxa are not complete enough to gauge sexual dimorphism but based on the evidence from modern species it is probable that a great deal of the variation noted in fossils is sex linked. Sexual dimorphism of the hippopotamid postcranium has not been fully studied, but provisional data indicate that trends in the pes, manus, and long bones differ (Weston, 1997, 2003a).

Semiaquatic Habitat

In general, environmental reconstructions based on fossil faunas uncritically classify hippopotamids among "aquatic" taxa. This assumption is largely based on the ecology of the living *Hippopotamus amphibius* (Elthringham, 1999), which spends most of its time in water except during periods of feeding, a way of life better depicted as "semiaquatic" or amphibious. Although its ecology is still imperfectly understood, *Choeropsis liberiensis* is reputedly more terrestrial than *Hip. amphibius*. Some of the morphological traits found in *Hippopotamus* and absent in *Choeropsis* have consequently been interpreted as adaptations to an aquatic environment: elevated orbits, robust limbs, large lacrimal in contact with nasal, elongated muzzle. On this basis, previous authors (Coryndon, 1972; Gèze, 1980; Harris, 1991) suggested a more terrestrial ecology for fossil hippopotamids not exhibiting these features (i.e., for most species classified here in *Hexaprotodon, Archaeopotamus, Saotherium*, and aff. *Hippopotamus*).

Several arguments can be made against this view. First is the inadequacy of the features selected to accurately characterize habitat preferences. There are no well-established functional interpretations of variation in lacrimal size and elongation of the muzzle in hippopotamids. Graviportalism, a condition typifying *Hip. amphibius*, is usually associated with terrestrial locomotion in other large mammals and hence the limb morphology of the specialist *Hip. amphibius* may not be representative of a generalized semiaquatic

habit. Of all osteological characters, the only one undoubtedly linked to life in water is the orbit elevation. This character clearly indicates living at the interface between water and air, but its absence does not necessarily indicate a lack of affinity for an aquatic habitat. Second, physiological features of *C. liberiensis* seem in contradiction with its more terrestrial habits. It exhibits a skin similar to that of *Hip. amphibius*, devoid of sweat glands and retaining few hairs (Olivier, 1975), features that are also found in aquatic mammals such as cetaceans. Like *Hip. amphibius*, it is also able to obstruct its external nasal and auditory conducts when diving (Robinson, in press). These adaptations, clearly linked to an aquatic habit, were probably inherited from a semiaquatic forerunner, and the occurrence of these physiological traits in phylogenetically distant taxa suggest their early evolution in the family. Third, the independent development of orbit elevation in various hippopotamid lineages (*Hexaprotodon*, Plio-Pleistocene hippopotamids from the Turkana, *Hippopotamus*), and the likely evolution of the family from anthracotheriids showing semiaquatic adaptations reinforces the idea of a primitive semiaquatic condition for the Hippopotamidae. Fourthly, fossil hippopotamines are generally abundant and well preserved in African late Neogene sites in close association with aquatic taxa such as crocodilians and osteichthyans.

With a likely predominant semiaquatic ecology, hippopotamids are particularly interesting tools for paleoenvironmental and paleogeographic reconstructions (Jablonski, 2003; Lihoreau et al., 2006). Such predominance does not exclude that some hippopotamids had, like *Choeropsis*, more terrestrial habits. This has clearly been the case for some insular Mediterranean hippopotamids, and this was also suggested for *Hippopotamus madagascariensis* (Stuenes, 1989) and *Archaeopotamus lothagamensis* (Weston 2003a).

Diet

Prior to the last decade, fossil hippopotamid diets were essentially interpreted on the basis of hypsodonty indices, incisor and canine morphology (Coryndon, 1977; Gèze, 1985), and general morphological comparisons with the modern species (Coryndon, 1972). However, their craniodental morphology does not favor dietary assessments. Hypsodonty indices are weakly variable in hippopotamids and quite low even in the extant *Hip. amphibius*, which is predominantly a grazer (Janis and Fortelius, 1988). Intraspecific competition considerably influences rostral dentition and cranial morphology, which is heavily specialized for biting/fighting (Herring, 1975; Kingdon, 1979). This is likely to obscure possible cranial adaptations to grazing or browsing as described for other ungulates (e.g., Janis, 1995). Until now, the most reliable data on fossil hippopotamid diet have been provided by carbon isotope analyses of tooth enamel (Morgan et al., 1994; Bocherens et al., 1996; Kingston, 1999; Zazzo et al., 2000; Franz-Odendaal et al., 2002; Cerling et al., 2003; Schoeninger et al., 2003; Boisserie et al., 2005d). Most of these studies indicate that fossil hippopotamids ate a large amount of C_4 plants, with the notable exception of those from Langebaanweg (early Pliocene, South Africa) that show a pure C_3 diet. The debate on ecological differences between hippopotamid species contemporaneously populating the same area (e.g., Plio-Pleistocene forms from Turkana) could greatly benefit from similar analyses, combined with the study of dental microwear.

The origin of the Hippopotamidae can be securely placed within Africa, with the earliest remains being found in eastern Africa. Around 10 Ma, *Kenyapotamus* is also found in Tunisia (Pickford, 1990), indicating a distribution that includes currently hyperarid Saharan areas (Boisserie and Lihoreau, 2006; Lihoreau et al., 2006). The origin of the Hippopotaminae is much more uncertain, as they appear abruptly in the fossil record already diversified and as a particularly abundant component of the terminal Miocene faunas of central and eastern Africa. The earliest certain evidence of the Hippopotaminae is from the Lower Member of the Nawata Formation and from Toros-Menalla in Chad—that is, younger than 7.5 Ma (Weston, 2000; Boisserie, et al. 2005c). It is tempting to link this "hippopotamine event" to the contemporaneous development of grass ecosystems boosted by the late Miocene expansion of C_4 plants in Africa (Boisserie et al., 2005d; Boisserie and Lihoreau, 2006).

After the first expansion into southern Eurasia at the end of the Miocene, the Pliocene distribution of hippopotamids is marked by basinal endemism (Boisserie et al., 2003; Boisserie and White, 2004; Boisserie, 2004; figure 44.9). This is probably related to the dependence of hippopotamids on water, limiting their ability to disperse. This situation makes Pliocene hippopotamids less useful for broad-range temporal correlation. However, they have locally experienced rapid evolution, such as in the Lake Turkana and Afar Basins, and could serve as good intrabasin chronological markers. In contrast, the Plio-Pleistocene transition recorded a dramatic expansion of *Hippopotamus* species (Kahlke, 1990), invading the whole continent and then western Eurasia for the second time. With the notable exception of *Choeropsis*, whose biogeographic history remains unknown, the last continental endemic forms apparently disappeared around 1 Ma, leaving the continent largely dominated by *Hippopotamus*.

Conclusions

In the later part of the Cenozoic, hippos are one of the best-represented mammal groups in the African fossil record, with abundant remains known from East, Central, and North Africa. Despite their abundance, hippopotamid evolution has received very little attention until recently. Although the taxonomic revisions outlined here are an essential first step in tackling this formidable task, much work remains to be done. Hippo remains continue to be unearthed and there are an increasing number of taxa awaiting description. The resurgence of interest in the origin of the family and growing recognition of its value as a tool for palaeoenvironmental reconstruction will hopefully encourage future studies and reveal more about the evolutionary history of this unusual group of predominantly semiaquatic large mammals.

ACKNOWLEDGMENTS

We are grateful to W. J. Sanders and L. Werdelin for the invitation to contribute to this volume. We would like to thank the following institutions for granting us access to their collections: Centre National d'Appui à la Recherche, Ndjamena, Chad; Institut International de Paléoprimatologie, Paléontologie Humaine: Evolution et Paléoenvironnements (IPHEP)/UMR CNRS 6046, University of Poitiers, France; Iziko Museums of Cape Town, South Africa; Museum of Vertebrate Zoology, University of California, Berkeley, USA; Natural History Museum, London; National Museum of Ethiopia, Addis Abeba; National Museums of Kenya, Nairobi; University of California Museum of Paleontology, Berkeley, USA; University Museum of Zoology, Cambridge, UK. We are deeply indebted to all contributors (field and lab work participants, assistance, and funding) to the following projects: Mission Paléoanthropologique Franco-Tchadienne (directors, M. Brunet, P. Vignaud); Middle Awash research project (directors, T. D. White, B. Asfaw, G. WoldeGabriel, Y. Beyene); Lothagam/Koobi Fora/Nachukui research projects (directors M. G. Leakey, J. M. Harris, L. N. Leakey). Support from BBSRC Research Grant and Balfour Fund, Department of Zoology, University of Cambridge (E.W.), and from Fondation Fyssen research grant (J.-R.B.). We would like to thank the reviewers for their helpful comments on this chapter. We would like to dedicate this chapter to the memory of S. C. Coryndon.

Literature Cited

Andrews, C. W. 1902. Note on a Pliocene vertebrate fauna from the Wadi-Natrun, Egypt. *The Geological Magazine* 9(10):432–439.

Arambourg, C. 1944a. Les Hippopotames fossiles d'Afrique. Comptes *Rendus de l'Académie des Sciences, Paris* 218:602–604.

———. 1944b. Au sujet de l'*Hippopotamus hipponensis* Gaudry. *Bulletin de la Société Géologique de France* 14:147–153.

———. 1947. Mission Scientifique de l'Omo 1932–1933; pp. 314–364 in C. Arambourg (ed.), *Mission Scientifique de l'Omo 1932–1933*. Editions du Muséum, Paris.

Arnason, U., A. Gullberg, and A. Janke. 2004. Mitogenomic analyses provide new insights into cetacean origin and evolution. *Gene* 333:27–34.

Bajpai, S., and P. D. Gingerich. 1998. A new Eocene archaeocete (Mammalia, Cetacea) from India and the time of origin of whales. *Proceedings of the National Academy of Sciences, USA* 95:15464–15468.

Barry, J. C., M. L. E. Morgan, L. J. Flynn, D. Pilbeam, A. K. Behrensmeyer, S. M. Raza, I. A. Khan, C. Badgley, J. Hicks, and J. Kelley. 2002. Faunal and environmental change in the late Miocene Siwaliks of northern Pakistan. *Paleobiology* 18(suppl. 2):1–71.

Behrensmeyer, A. K., A. L. Deino, A. Hill, J. D. Kingston, and J. J. Saunders. 2002. Geology and geochronology of the middle Miocene Kipsaraman site complex, Muruyur Beds, Tugen Hills, Kenya. *Journal of Human Evolution* 42:11–38.

Bender, P. A. and J. S. Brink. 1992. Preliminary report on new large-mammal fossil finds from the Cornelia-Uitzoek site, in the north-eastern Orange Free State. *South African Journal of Science* 88:512–515.

Bergh, G. D. van den, J. de Vos, and P. Y. Sondaar. 2001. The late Quaternary palaeogeography of mammal evolution in the Indonesian Archipelago. *Palaeogeography, Palaeoclimatology, Palaeoecology* 171:385–408.

Bocherens, H., P. L. Koch, A. Mariotti, D. Geraads, and J. J. Jaeger. 1996. Isotopic biogeochemistry (^{13}C, ^{18}O) of mammalian enamel from African Pleistocene hominid sites. *Palaios* 11:306–318.

Boekschoten, G. J., and P. Y. Sondaar. 1972. On the fossil Mammalia of Cyprus. *Proceedings Koninklijke Nederlandse Akademie van Wetenschappen* 75:306–338.

Boisserie, J.-R. 2004. A new species of Hippopotamidae (Mammalia, Artiodactyla) from the Sagantole Formation, Middle Awash, Ethiopia. *Bulletin de la Société Géologique de France* 175:525–533.

———. 2005. The phylogeny and taxonomy of Hippopotamidae (Mammalia: Artiodactyla): A review based on morphology and cladistic analysis. *Zoological Journal of the Linnean Society* 143:1–26.

———. 2007. Late Miocene Hippopotamidae from Lemudong'o, Kenya. *Kirtlandia* 56:158–162.

Boisserie, J.-R., M. Brunet, L. Andossa, and P. Vignaud. 2003. Hippopotamids from the Djurab Pliocene faunas, Chad, Central Africa. *Journal of African Earth Sciences* 36:15–27.

Boisserie, J.-R., and W. H. Gilbert. 2008. Hippopotamidae; pp. 179–191 in W. H. Gilbert and B. Asfaw (eds.), Homo erectus: *Pleistocene Evidence from the Middle Awash, Ethiopia*. University of California Press, Berkeley.

Boisserie, J.-R., and F. Lihoreau. 2006. Emergence of Hippopotamidae: New scenarios. *Comptes Rendus Palevol* 5:749–756.

Boisserie, J.-R., F. Lihoreau, and M. Brunet. 2005a. Origins of Hippopotamidae (Mammalia, Cetartiodactyla): Towards resolution. *Zoologica Scripta* 34:119–143.

———. 2005b. The position of Hippopotamidae within Cetartiodactyla. *Proceedings of the National Academy of Sciences, USA* 102:1537–1541.

Boisserie, J.-R., A. Likius, P. Vignaud, and M. Brunet. 2005c. A new late Miocene hippopotamid from Toros-Ménalla, Chad. *Journal of Vertebrate Paleontology* 25:665–673.

Boisserie, J.-R., and T. D. White. 2004. A new species of Pliocene Hippopotamidae from the Middle Awash, Ethiopia. *Journal of Vertebrate Paleontology* 24:464–473.

Boisserie, J.-R., A. Zazzo, G. Merceron, C. Blondel, P. Vignaud, A. Likius, H. T. Mackaye, and M. Brunet. 2005d. Diets of modern and late Miocene hippopotamids: Evidence from carbon isotope composition and micro-wear of tooth enamel. *Palaeogeography, Palaeoclimatology, Palaeoecology* 221:153–174.

Bromage, T. G., F. Schrenk and Y. M. Juwayeyi. 1995. Paleobiogeography of the Malawi Rift: Age and vertebrate paleontology of the Chiwondo Beds, northern Malawi. *Journal of Human Evolution* 28:37–57.

Brunet, M., A. Beauvilain, D. Geraads, F. Guy, M. Kasser, H. T. Mackaye, L. M. Maclatchy, G. S. J. Mouchelin, and P. Vignaud. 1998. Tchad: découverte d'une faune de mammifères du Pliocène inférieur. *Comptes Rendus de l'Académie des Sciences, Paris* 326:153–158.

Brunet, M., and MPFT. 2000. Chad: discovery of a Vertebrate fauna close to the Mio-Pliocene boundary. *Journal of Vertebrate Paleontology* 20:205–209.

Burney, D. A., B. L. Pigott, L. R. Godfrey, W. L. Jungers, S. M. Goodman, H. T. Wright, and A. J. T. Jull. 2004. A chronology for late prehistoric Madagascar. *Journal of Human Evolution* 45:25–63.

Caloi, L., and M. R. Palombo. 1996. Functional aspects and ecological implications in hippopotami and cervids of Crete; pp. 125–151 in D. S. Reese (ed.), *Pleistocene and Holocene fauna of Crete and Its First Settlers.* Monographs in World Archaeology 28. Prehistory Press, Madison.

Cerling, T. E., J. M. Harris, and M. G. Leakey. 2003. Isotope paleoecology of the Nawata and Nachukui Formations at Lothagam, Turkana Basin, Kenya; pp. 587–597 in J. M. Harris, and M. G. Leakey (eds.), *Lothagam: The Dawn of Humanity in Eastern Africa.* Columbia University Press, New York.

Colbert, E. H. 1935. Siwalik Mammals in the American Museum of Natural History. *Transactions of the American Philosophical Society* 26:278–294.

———. 1938. Fossil mammals from Burma. *Bulletin of the American Museum of Natural History* 74:419–424.

———. 1943. Pleistocene vertebrates collected in Burma by the American Southeast Asiatic expedition. *Transactions of the American Philosophical Society* 32:395–429.

Cooke, H. B. S., and S. C. Coryndon. 1970. Pleistocene Mammals from the Kaiso Formation and other related deposits in Uganda; pp. 107–224 in L. B. S. Leakey, and R. J. G. Savage (eds.), *Fossil Vertebrates of Africa,* vol. 2. Academic Press, London.

Corbet, G. B. 1969. The taxonomic status of the pygmy hippopotamus, *Choeropsis liberiensis,* from the Niger delta. *Journal of Zoology* 158:387–394.

Coryndon, S. C. 1966. Preliminary report on some fossils from the Chiwondo Beds of the Karonga district, Malawi. *American Anthropologist* 68:59–66.

———. 1970a. Evolutionary trends in East African Hippopotamidae. *Bulletin de Liaison de l'Association Sénégalaise pour l'Etude du Quaternaire de l'Ouest africain (ASEQUA), Dakar* 25:107–116.

———. 1970b. The extent of variation in fossil *Hippopotamus* from Africa. *Symposia of the Zoological Society of London* 26:135–147.

———. 1972. Hexaprotodont Hippopotamidae of East Africa and the phylogeny of the family; pp. 350–352 in *6 Congrès Panafricain de Préhistoire, Dakar 1967.* Les Imprimeries Réunies, Chambéry.

———. 1976. Fossil Hippopotamidae from Plio-Pleistocene successions of the Rudolf Basin; pp. 238–250 in Y. Coppens, F. C. Howell, G. L. Isaac, and R. E. F. Leakey (eds.), *Earliest Man and Environments in the Lake Rudolf Basin.* University of Chicago Press, Chicago.

———. 1977. The taxonomy and nomenclature of the Hippopotamidae (Mammalia, Artiodactyla) and a description of two new fossil species. *Proceedings Koninklijke Nederlandse Akademie van Wetenschappen* 80:61–88.

———. 1978a. Fossil Hippopotamidae from the Baringo Basin and relationships within the Gregory Rift, Kenya; pp. 279–292 in W. W. Bishop (ed.), *Geological Background to Fossil Man.* Scottish Academic Press, Edinburgh.

———. 1978b. Hippopotamidae; pp. 483–495 in V. J. Maglio and H. B. S. Cooke (eds.), *Evolution of African Mammals.* Harvard University Press, Cambridge.

Coryndon, S. C., and Y. Coppens. 1973. Preliminary report on Hippopotamidae (Mammalia, Artiodactyla) from the Plio/Pleistocene of the Lower Omo Basin, Ethiopia; pp. 139–157 in L. B. S. Leakey, and R. J. G. Savage (eds.), *Fossil Vertebrates of Africa,* vol. 3. Academic Press, London.

Coryndon, S. C., and Y. Coppens. 1975. Une espèce nouvelle d'hippopotame nain du Plio-Pléistocène du bassin du lac Rodolphe (Ethiopie, Kenya). *Comptes Rendus de l'Académie des Sciences, Paris* 280:1777–1780.

Desmarest, M. A. G. 1822. Mammalogie ou Description des Espèces de Mammifères. Mme Veuve Agasse, Imprimeur-Libraire, Paris, 388 pp.

Dietrich, W. O. 1926. Fortschritte der Säugetierpaläontologie Afrikas. *Forschungen und Fortschritte* 15:121–122.

———. 1928. Pleistocäne Deutsch-Ostafrikanische *Hippopotamus*-reste; pp. 2–41 in H. Reck (ed.), *Wissenschaftliche Ergebnisse des Oldoway Expedition herausgeben von Prof. Dr. Reck,* Neue Folge, Heft 3. Leipzig.

Ditchfield, P., J. Hicks, T. Plummer, L. C. Bishop, and R. Potts. 1999. Current research on the Late Pliocene and Pleistocene deposits north of Homa Mountain, southwestern Kenya. *Journal of Human Evolution* 36:123–150.

Dixie, T. 1927. The Tertiary and post-Tertiary lacustrine sediments of the Nyasan Rift Valley. Appendix I on Mammalian remains by A. T. Hopwood. Appendix II on Non-marine mollusca by M. Connelly. *Quarterly Journal of the Geological Society* 83: 432–447.

Eltringham, S. K. 1993. The pygmy hippopotamus (*Hexaprotodon liberiensis*); pp. 55–60 in W. L. R. Oliver (ed.), *Pigs, Peccaries, and Hippos.* IUCN, Gland.

———. 1999. *The Hippos.* Academic Press, London, 184 pp.

Erdbrink, D., and W. Krommenhoek. 1975. Contribution to the knowledge of the fossil Hippopotamidae from the Kazinga Channel area (Uganda). *Säugetierkundliche Mitteilungen* 23:258–294.

Falconer, H., and P. T. Cautley. 1836. Note on the fossil hippopotamus of the Siwalik Hills. *Asiatic Researches* 19:39–53.

———. 1847. *Fauna Antiqua Sivalensis.* Smith, Elder, London.

Faure, M. 1983. Les Hippopotamidae (Mammalia, Artiodactyla) d'Europe Occidentale. Unpublished PhD dissertation, Université Claude Bernard–Lyon I, Lyon, 223 pp.

———. 1986. Les hippopotamidés du Pléistocène ancien d'Oubeidiyeh (Israël). *Mémoires et Travaux du Centre de Recherche Français de Jérusalem* 5:107–142.

———. 1994. Les Hippopotamidae (Mammalia, Artiodactyla) du rift occidental (bassin du lac Albert, Ouganda). Etude Préliminaire; pp. 321–337 in B. Senut, and M. Pickford (eds.), *Geology and Paleobiology of the Albertine Rift Valley, Uganda-Zaïre: II. Paleobiology.* Cifeg, Orléans.

Faure, M., and C. Guérin. 1990. *Hippopotamus laloumena* nov. sp., la troisième espèce d'hippopotame holocène de Madagascar. *Comptes Rendus de l'Académie des Sciences, Paris* 310:1299–1305.

———. 1997. Gafalo, un nouveau site à *Palaeoloxodon* et *Hippopotamus amphibius* du Pléistocène ancien du Gobaad (République de Djibouti). *Comptes Rendus de l'Académie des Sciences, Paris* 324:1017–1021.

Franz-Odendaal, T. A., J. A. Lee-Thorp, and A. Chinsamy. 2002. New evidence for the lack of C4 grassland expansions during the early Pliocene at Langebaanweg, South Africa. *Paleobiology* 28:378–384.

Gatesy, J. 1997. More DNA support for a Cetacea/Hippopotamidae clade: the blood-clotting protein gene y-fibrinogen. *Molecular Biology and Evolution* 14:537–543.

Gatesy, J., C. Hayashi, M. A. Cronin, and P. Arctander. 1996. Evidence from milk casein genes that cetaceans are close relatives of hippopotamid artiodactyls. *Molecular Biology and Evolution* 13:954–963.

Gaudry, A. 1876. Sur un hippopotame fossile découvert à Bone (Algérie). *Bulletin de la Société Géologique de France* 4:147–154.

Gaziry, A. W. 1987. *Hexaprotodon sahabiensis* (Artiodactyla, Mammalia): A new hippopotamus from Libya; pp. 303–315 in N. T. Boaz, A. El-Arnauti, A. W. Gaziry, J. de Heinzelin, and D. D. Boaz (eds.), *Neogene Paleontology and Geology of Sahabi.* Liss, New York.

Geisler, J. H., and M. D. Uhen. 2003. Morphological support for a close relationship between hippos and whales. *Journal of Vertebrate Paleontology* 23:991–996.

Gentry, A. W. 1999. A fossil hippopotamus from the Emirate of Abu Dhabi, United Arab Emirates; pp. 271–289 in P. J. Whybrow and A. Hill (eds.), *Fossil Vertebrates of Arabia.* Yale University Press, New Haven.

Gentry, A. W., and J. J. Hooker. 1988. The phylogeny of the Artiodactyla; pp. 235–272 in M. J. Benton (ed.), *The Phylogeny and Classification of the Tetrapods: Volume 2. Mammals.* Clarendon Press, Oxford.

Geraads, D. 1987. Dating the northern African cercopithecid fossil record. *Human Evolution* 2:19–27.

Geraads, D., Z. Alemseged, and H. Bellon. 2002. The late Miocene mammalian fauna of Chorora, Awash Basin, Ethiopia: Systematics, biochronology, and the ⁴⁰K-⁴⁰Ar ages of the associated volcanics. *Tertiary Research* 21:113–122.

Geraads, D., Z. Alemseged, D. Reed, J. Wynn, and D. C. Roman. 2004. The Pleistocene fauna (other than Primates) from Asbole, lower Awash Valley, Ethiopia, and its environmental and biochronological implications. *Geobios* 37:697–718.

Gèze, R. 1980. Les Hippopotamidae (Mammalia, Artiodactyla) du Plio-Pléistocène de l'Ethiopie. Unpublished PhD dissertation, Université Pierre et Marie Curie—Paris VI, Paris, 116 p.

———. 1985. Répartition paléoécologique et relations phylogénétiques des Hippopotamidae (Mammalia, Artiodactyla) du néogène d'Afrique Orientale; pp. 81–100 in *L'environnement des Hominidés au Plio-Pléistocène.* Fondation Singer-Polignac Masson, Paris.

Gingerich, P. D., M. U. Haq, I. S. Zalmout, K. I. Hussain, and M. S. Malkani. 2001. Origin of whales from early artiodactyls: Hands and feet of Eocene Protocetidae from Pakistan. *Science* 293:2239–2242.

Gould S. J. 1975. On the scaling of tooth size in mammals. *American Zoologist* 15:351–362.

Guldberg, G. A. 1883. Undersøgelser over en subfossil flodhest fra Madagascar. *Videnskabs-Selskabets forhandlinger, Christiania* 6:1–24.

Haile-Selassie, Y., G. WoldeGabriel, T. D. White, R. L. Bernor, D. DeGusta, P. R. Renne, W. K. Hart, E. Vrba, A. Stanley, and F. C. Howell. 2004. Mio-Pliocene mammals from the Middle Awash, Ethiopia. *Geobios* 37:536–552.

Harris, J. M. 1991. Family Hippopotamidae; pp. 31–85 in J. M. Harris (ed.), *Koobi Fora Research Project,* vol. 3. Clarendon Press, Oxford.

Harris, J. M., F. H. Brown, and M. G. Leakey. 1988. Stratigraphy and paleontology of Pliocene and Pleistocene localities west of Lake Turkana, Kenya. *Contributions in Science, Natural History Museum of Los Angeles County* 399:1–128.

Harris, J. M., M. G. Leakey, T. E. Cerling, and A. J. Winkler, 2003. Early Pliocene tetrapod remains from Kanapoi, Lake Turkana Basin, Kenya; in J. M. Harris and M. G. Leakey (eds), *Geology and Vertebrate Paleontology of the Early Pliocene Site of Kanapoi, Northern Kenya. Contributions in Science, Natural History Museum of Los Angeles County* 498:39–113.

Harrison, T. 1997. The anatomy, paleobiology, and phylogenetic relationships of the Hippopotamidae (Mammalia, Artiodactyla) from the Manonga Valley, Tanzania; pp. 137–190 in T. Harrison (ed.), *Neogene Paleontology of the Manonga Valley, Tanzania.* Plenum Press, New York.

Heinzelin, J. de, J. D. Clark, K. D. Schick, and W. H. Gilbert. 2000. *The Acheulean and the Plio-Pleistocene Deposits of the Middle Awash Valley, Ethiopia.* Musée Royal de l'Afrique Centrale, Tervuren, 235 pp.

Herring, S. W. 1975. Adaptations for gape in the hippopotamus and its relatives. *Forma et Functio* 8:85–100.

Hooijer, D. A. 1950. The fossil Hippopotamidae of Asia, with notes on the recent species. *Zoologische Verhandelingen* 8:1–123.

———. 1958. Pleistocene remains of hippopotamus from the Orange Free State. *Navorsinge van die Nasionale Museum, Bloemfontein* 1:259–266.

Hopwood, A. T. 1926. The geology and palaeontology of the Kaiso Bone Beds, Uganda: Part II. Palaeontology, Fossil Mammalia. *Occasional Paper (Uganda Geological Survey)* 2:13–36.

———. 1931. Pleistocene Mammalia from Nyasaland and Tanganyika Territory. *Geological Magazine* 68:133–135.

———. 1939. The mammalian fossils; pp. 308–316 in T. P. O'Brien (ed.), *The Prehistory of Uganda Protectorate.* Cambridge University Press, Cambridge.

Hopwood A. T., and J. P. Hollyfield. 1954. An annotated bibliography of the fossil mammals of Africa (1742–1950). *British Museum of Natural History, Fossil Mammals of Africa* 8:1–194.

Houtemaker, J. L., and P. Y. Sondaar. 1979. Osteology of the fore limb of the Pleistocene dwarf hippopotamus from Cyprus with special reference to phylogeny and function. *Proceedings Koninklijke Nederlandse Akademie van Wetenschappen, Series B,* 82:411–448.

Isaac, G. L. 1977. *Olorgesailie: Archeological Studies of a Middle Pleistocene Lake Basin in Kenya.* University of Chicago Press, Chicago, 272 pp.

Jablonski, N. 2003. The hippo's tale: How the anatomy and physiology of late Neogene *Hexaprotodon* shed light on late Neogene environmental change. *Quaternary International* 117:119–123.

Janis, C. M. 1995. Correlations between craniodental morphology and feeding behavior in ungulates: Reciprocal illumination between living and fossil taxa; pp. 76–98 in J. Thomason (ed.), *Functional Morphology in Vertebrate Paleontology.* Cambridge University Press, Cambridge.

Janis, C. M., and M. Fortelius. 1988. On the means whereby mammals achieve increased functional durability of their dentitions, with special reference to limiting factors. *Biological Reviews* 63:197–230.

Joleaud, L. 1920. Contribution à l'étude des hippopotames fossiles. *Bulletin de la Société Géologique de France* 22:13–26.

Kahlke, R. D. 1990. Zum stand der Erforschung fossiler Hippopotamiden (Mammalia, Artiodactyla): Eine übersicht. *Quartärpaläontologie* 8:107–118.

Kalb, J. E., C. J. Jolly, S. Tebedge, A. Mebrate, C. Smart, E. B. Oswald, P. Whitehead, C. B. Wood, T. Adefris, and V. Rawn-Schatzinger. 1982. Vertebrate faunas from the Awash group, Middle Awash Valley, Afar, Ethiopia. *Journal of Vertebrate Paleontology* 2:237–258.

Kent, P. E. 1942. The Pleistocene Beds of Kanam and Kanjera, Kavirondo, Kenya. *Geological Magazine* 79:117–132.

Kingdon, J. 1979. *East African Mammals: An Atlas of Evolution in Africa: Volume III, Part B. Large Mammals.* Academic Press, London, 450 pp.

Kingston, J. D. 1999. Isotopes and environments of the Baynunah Formation, Emirate of Abu Dhabi, United Arab Emirates; pp. 354–372 in P. J. Whybrow and A. Hill (eds.), *Fossil Vertebrates of Arabia.* Yale University Press, New Haven.

Lacomba, J. I., J. Morales, F. Robles, C. Santisteban, and M. T. Alberdi. 1986. Sedimentologia y paleontologia del yacimiento finimioceno de La Portera (Valencia). *Estudios Geológicos* 42:167–180.

Laws, R. M. 1968. Dentition and ageing of the hippopotamus. *East African Wildlife Journal* 6:19–52.

Leakey, M. G., C. S. Feibel, R. L. Bernor, J. M. Harris, T. E. Cerling, K. M. Stewart, G. W. Storrs, A. Walker, L. Werdelin, and A. J. Winkler. 1996. Lothagam: A record of faunal change in the Late Miocene of East Africa. *Journal of Vertebrate Paleontology* 16:556–570.

Leakey, M. G., F. Spoor, F. H. Brown, P. N. Gathogo, C. Klarie, L. N. Leakey, and I. McDougall. 2001. New hominin genus from eastern Africa shows diverse middle Pliocene lineages. *Nature* 410:433–440.

Leidy, J. 1853. On the osteology of the head of *Hippopotamus. Journal of the Academy of Natural Sciences of Philadelphia* 2:207–224.

Lihoreau, F., J.-R. Boisserie, L. Viriot, Y. Coppens, A. Likius, H. T. Mackaye, P. Tafforeau, P. Vignaud, and M. Brunet. 2006. Anthracothere dental anatomy reveals a late Miocene Chado-Libyan bioprovince. *Proceedings of the National Academy of Sciences, USA* 103:8763–8767.

Lydekker, R. 1884. Siwalik and Narbada bunodont Suina. *Memoirs of the Geological Survey of India* 10(3):35–49.

Made, J. van der 1999. Superfamily Hippopotamoidea; pp. 203–208 in G. E. Rössner and K. Heissig (eds.), *The Miocene Land Mammals of Europe.* Pfeil, Munich.

Major, C. I. F. 1902. Some account of a nearly complete skeleton of *Hippopotamus madagascariensis,* Guldb., from Sirabé, Madagascar, obtained in 1895. *Geological Magazine* 9:193–199.

Martínez-Navarro, B., L. Rook, A. Segid, D. Yosieph, M. P. Ferretti, J. Shoshani, T. M. Tecle, and Y. Libsekal. 2004. The large fossil mammals from Buia (Eritrea). *Rivista Italiana di Paleontologia e Stratigrafia* 110 (suppl.):61–88.

Matthew, W. D. 1929. Critical observations upon Siwalik mammals. *Bulletin of the American Museum of Natural History* 56:437–560.

Mawby, J. E. 1970. Fossil vertebrates from northern Malawi: Preliminary Report. *Quaternaria* 13:319–323.

Mazza, P. 1995. New evidence on the Pleistocene hippopotamuses of western Europe. *Geologica Romana* 31:61–241.

McDougall, I., and C. S. Feibel. 2003. Numerical age control for the Miocene-Pliocene succession at Lothagam, a hominoid-bearing sequence in the Northern Kenya Rift; pp. 45–63 in J. M. Harris and M. G. Leakey (eds.), *Lothagam: The Dawn of Humanity in Eastern Africa.* Columbia University Press, New York.

Milne Edwards, A. 1868. Sur des découvertes zoologiques faites récemmment à Madagascar par M. Alfred Grandidier. *Comptes Rendus Hebdomadaires des Séances de l'Académie des Sciences Naturelles—Zoologie et Paléontologie* 10:375–378.

Monnier, L., and C. Lamberton. 1922. Note sur des ossements subfossiles de la region de Mananjary. *Bulletin de l'Académie Malagache* 3:211–212.

Montgelard, C., F. M. Catzeflis, and E. Douzery. 1997. Phylogenetic relationships of artiodactyls and cetaceans as deduced from the comparison of cytochrome b and 12s rRNA mitochondrial sequences. *Molecular Biology and Evolution* 14:550–559.

Morgan, M. E., J. D. Kingston, B. D. Marino. 1994. Carbon isotopic evidence for the emergence of C_4 plants in the Neogene from Pakistan and Kenya. *Nature* 367:162– 165.

Morton, S. G. 1844. On a supposed new species of hippopotamus. *Proceedings of the National Academy of Sciences, Philadelphia* 2:14–17.

———. 1849. Additional observations on a new living species of hippopotamus. *Journal of the Academy of Natural Sciences, Philadelphia* 2:231–239.

Nakaya, H., M. Pickford, Y. Nakano, and H. Ishida. 1984. The late Miocene large mammal fauna from the Namurungule Formation, Samburu Hills, northern Kenya. *African Study Monographs (suppl.)* 2:87–131.

Nakaya, H., M. Pickford, K. Yasui, and Y. Nakano. 1987. Additional large mammalian fauna from the Namurungule Formation, Samburu Hills, northern Kenya. *African Study Monographs (suppl.)* 5:47–98.

Nikaido M, A. P. Rooney, and N. Okada 1999. Phylogenetic relationships among cetartiodactyls based on insertions of short and long interspersed elements: Hippopotamuses are the closest extant relatives of whales. *Proceedings of the National Academy of Sciences, USA* 96:10261–10266.

Olivier, R. C. D. 1975. Aspects of skin physiology in the pygmy hippopotamus *Choeropsis liberiensis*. *Journal of Zoology* 176:211–213.

Owen R. 1845. *Odontography*. Bailliere Hyppolite, London, 655 p.

Pavlakis, P. P. 1990. Plio-Pleistocene Hippopotamidae from the Upper Semliki; pp. 203–223 in N. T. Boaz (ed.), *Results from the Semliki Research Expedition*. Virginia Museum of Natural History Memoir, Martinsville.

Petrocchi, C. 1952. Notizie generali sul giacimento fossilifero di Sahabi: Sotria de scavi-risultati. *Rendiconti della Accademia Nazionale Quaranta* 3:9–34.

Pickford, M. 1983. On the origins of Hippopotamidae together with descriptions of two species, a new genus and a new subfamily from the Miocene of Kenya. *Geobios* 16:193–217.

———. 1990. Découverte de *Kenyapotamus* en Tunisie. *Annales de Paléontologie* 76:277–283.

———. 2007. Suidae and Hippopotamidae from the middle Miocene of Kipsaraman, Kenya and other sites in East Africa. *Paleontological Research* 11:85–105.

Pomel, A. 1890. Sur les Hippopotames fossiles de l'Algérie. *Comptes Rendus de l'Académie des Sciences, Paris* 110:1112–1116.

Reese, D. S. 1975. Dwarfed hippos: Past and present. *Earth Science* 28:63–69.

Robinson, P. T. In press. *Choeropsis liberiensis* (Morton). In J. S. Kingdon and M. Hoffmann (eds), *The Mammals of Africa*. Academic Press, Amsterdam.

Sarich, V. M. 1993. Mammalian systematics: Twenty-five years among their albumins and transferrins; pp. 103–114 in F. S. Szalay, M. J. Novacek, and M. C. McKenna (eds.), *Mammal Phylogeny*. Springer, Berlin.

Schoeninger, M. J., H. Reeser, and K. Hallin. 2003. Paleoenvironment of *Australopithecus anamensis* at Allia Bay, East Turkana, Kenya: Evidence from mammalian herbivore enamel stable isotopes. *Journal of Anthropological Archeology* 22:200–207.

Sondaar, P. Y. 1991. Island mammals of the past. *Science Progress* 75:249–264.

———. 1977. Insularity and its effect on mammal evolution; pp. 671–707 in M. K. Hecht, P. C. Goody, and B. M. Hecht (eds.), *Major Patterns in Vertebrate Evolution*. Plenum, New York.

Spaan, A. 1996. *Hippopotamus creutzburgi*: The case of the Cretan hippopotamus; pp. 99–110 in D. S. Reese (ed.), *Pleistocene and Holocene fauna of Crete and its first settlers*. Monographs in World Archaeology 28. Prehistory Press, Madison, Wisc.

Stromer E. 1914. Mitteilungen über Wirbeltierreste aus dem Mittelpliocän des Natrontales (Ägypten). *Zeitschrift der Deutschen Geologischen Gesellschaft* 66:1–33.

Stuenes, S. 1989. Taxonomy, habits, and relationships of the subfossil Madagascan Hippopotami *Hippopotamus lemerlei* and *H. madagascariensis*. *Journal of Vertebrate Paleontology* 9:241–268.

Vekua, A. 1986. The lower Pleistocene mammalian fauna of Akhalkalaki (Southern Georgia, USSR). *Palaeontographia Italica* 74:63–96.

Weston, E. M. 1997. A biometrical analysis of evolutionary change within the Hippopotamidae. Unpublished PhD dissertation, Cambridge University, Cambridge, 141 pp.

———. 2000. A new species of hippopotamus *Hexaprotodon lothagamensis* (Mammalia: Hippopotamidae) from the late Miocene of Kenya. *Journal of Vertebrate Paleontology* 20(1):177–185.

———. 2003a. Evolution of ontogeny in the hippopotamus skull: Using allometry to dissect developmental change. *Biological Journal of the Linnean Society* 80:625–638.

———. 2003b. Fossil Hippopotamidae from Lothagam; pp. 380–410 in J. M. Harris, and M. G. Leakey (eds.), *Lothagam: The Dawn of Humanity in Eastern Africa*. Columbia University Press, New York.

Zazzo, A., H. Bocherens, M. Brunet, A. Beauvilain, D. Billiou, H. T. Mackaye, P. Vignaud, and A. Mariotti. 2000. Herbivore paleodiet and paleoenvironmental changes in Chad during the Pliocene using stable isotope ratios of tooth enamel carbonate. *Paleobiology* 26:294–309.

FORTY-FIVE

Cetacea

PHILIP D. GINGERICH

Cetacea, comprising the great whales and the smaller dolphins and porpoises, have special interest in mammalian evolution as one of the two orders of mammals that became fully aquatic. Much of the 200-million-year-long history of mammals is a history of life on land. Cetacea and Sirenia are exceptions and, of the two aquatic groups, Cetacea is the more diverse and broadly successful. Fifty years ago the transition linking cetaceans to a land-mammal ancestor was still largely hypothetical. Adaptation to life in water made cetacean morphology sufficiently different to preclude direct comparison to potential land-mammal ancestors. There were morphological and immunological suggestions that cetaceans might be related to Artiodactyla, but none of these claims was convincing by itself. In recent years, the fossil record has helped to clarify both the artiodactyl ancestry of cetaceans among land mammals, and also the nature of the transition (figure 45.1).

Several comprehensive reviews of cetacean evolution have been published in recent years by authors who are experts on Mysticeti and Odontoceti (Barnes and Mitchell, 1978; Fordyce and Barnes, 1994; Fordyce and Muizon, 2001; see also Gingerich, 2005). These are recommended for a general overview of the fossil record of cetacean evolution. Here I shall focus on Eocene Archaeoceti. Readers interested in scientific progress are encouraged to compare the review here with the Barnes and Mitchell (1978) review of African Cetacea. Much has been learned in the past 30 years, especially about the Eocene and about Archaeoceti.

Classification

Cetacea as an order is commonly divided into three suborders, Archaeoceti, Mysticeti, and Odontoceti. The first of these, Archaeoceti ('archaic cetaceans'), contains five families and about 30 genera, all of which are now extinct (or 'pseudoextinct' in the sense that one or more archaeocetes undoubtedly gave rise to later mysticetes and odontocetes). Archaeocetes are more primitive than other cetaceans, both in being older geologically and in retaining more generalized mammalian morphology. Archaeocetes made their first appearance and flourished during the Eocene epoch.

Mysticeti and Odontoceti, sometimes grouped as Neoceti, are younger than Archaeoceti, and each is more evolved in exhibiting greater specialization for life in water. Following Fordyce and Muizon (2001), the suborder Mysticeti ('mustache whales') contains seven families and about 50 genera, living and extinct, and the suborder Odontoceti ('toothed cetaceans') contains 15 families and about 130 genera (more have been added since 2001, but none of these is known from Africa). Mysticetes and odontocetes made their first appearances in the very latest Eocene or early Oligocene, then diversified in the Oligocene, and flourished through the subsequent Miocene, Pliocene, and Pleistocene epochs.

The three groups Archaeoceti, Mysticeti, and Odontoceti can be characterized in general terms as follows. Archaeoceti is the group that made the transition from life on land to life in the sea. Mysticeti is the group that abandoned teeth in favor of baleen and filter feeding. Odontoceti is the group that retains simplified teeth, but developed innovative echolocation enabling them to image or visualize their environment and virtually 'see' using high-frequency sound. The African fossil record is very important for understanding the evolution of Archaeoceti, but less so for understanding the evolution of Mysticeti and Odontoceti.

Stratigraphic ranges of African Archaeoceti, Mysticeti, and Odontoceti are summarized in figure 45.2. Solid black bars show the temporal and systematic representation of Cetacea in the African fossil record, which are first middle and late Eocene, representing Archaeoceti, and second late Miocene to Holocene Mysticeti and Odontoceti. Eocene archaeocetes are known in several instances from complete skeletons, while later cetaceans are generally known from skulls and more fragmentary bones and teeth.

Fossil Record of African Cetacea

The geological history of the African continent dictates where marine mammal fossils are preserved. The continent as a whole has been a stable craton through much of Cenozoic time. This means, generally, that marine mammal fossils are found on the periphery of the continent (figure 45.3), where marine mammals were preserved during times of flooding by rising oceans. The most important exception is on the northern margin of the continent from Tunisia to Egypt, where passive subsidence of the continental

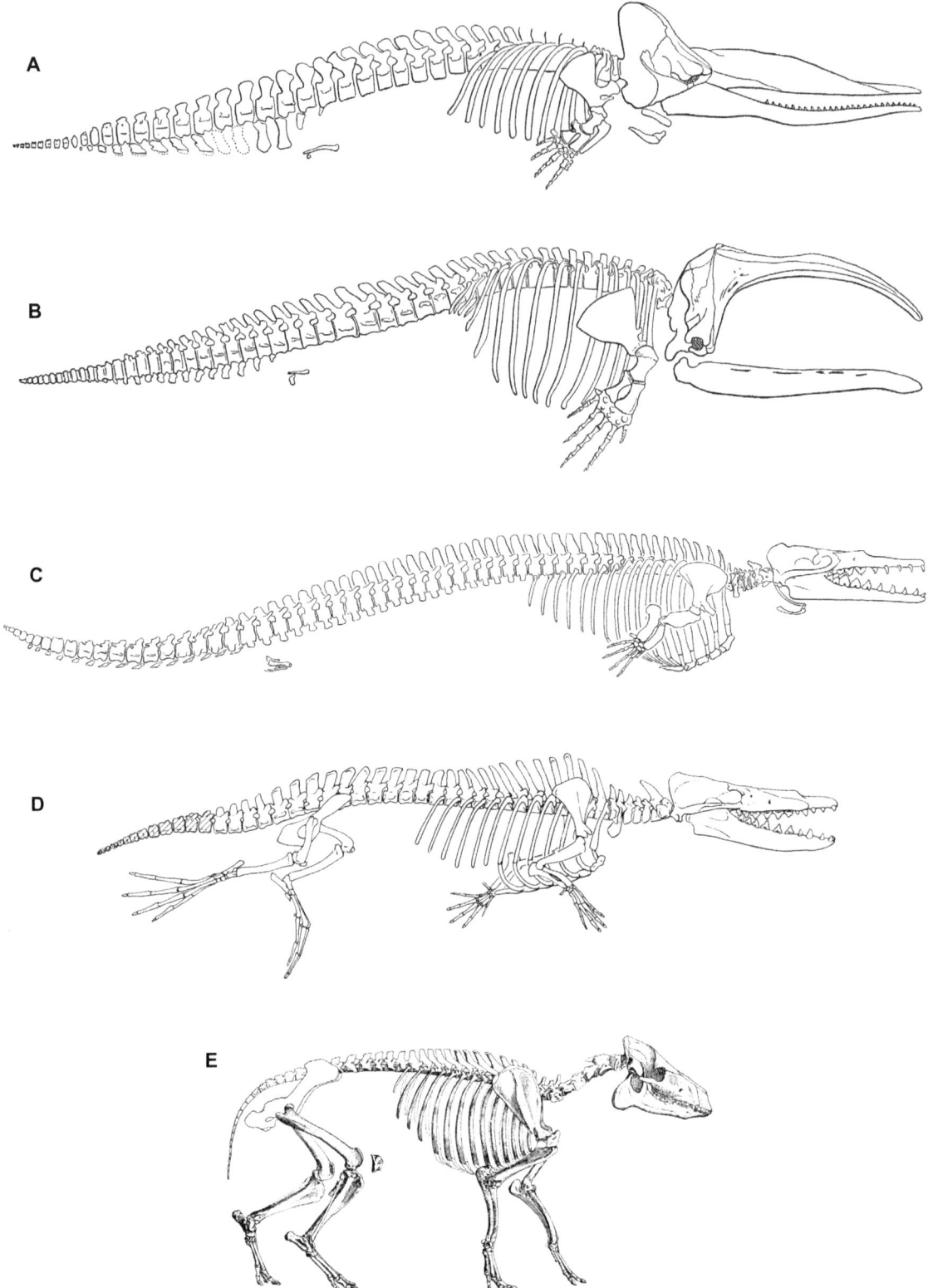

FIGURE 45.1 Skeletal drawings of a representative artiodactyl and archaeocete, mysticete, and odontocete cetaceans. A) Aquatic Recent physeterid odontocete *Physeter macrocephalus* (Gregory, 1951; ca. 12–19 m); B) aquatic Recent balaenid mysticete *Balaena mysticetus* (Eschricht and Reinhardt in Flower, 1866; ca. 14–18 m); C) aquatic late Eocene basilosaurid archaeocete *Dorudon atrox* (Gingerich and Uhen, 1996; ca. 5 m); D) semiaquatic early middle Eocene protocetid archaeocete *Rodhocetus balochistanensis* (Gingerich et al., 2001; ca. 3 m); E) terrestrial early Oligocene anthracotheriid *Elomeryx armatus* (from Scott, 1894; body length ca. 1.5 m). *Elomeryx* is representative of the primitive Artiodactyla from which Cetacea evolved. *Rodhocetus* and *Dorudon* are intermediate in geological age and morphology between early Artiodactyla and later Cetacea such as modern *Balaena* and *Physeter*.

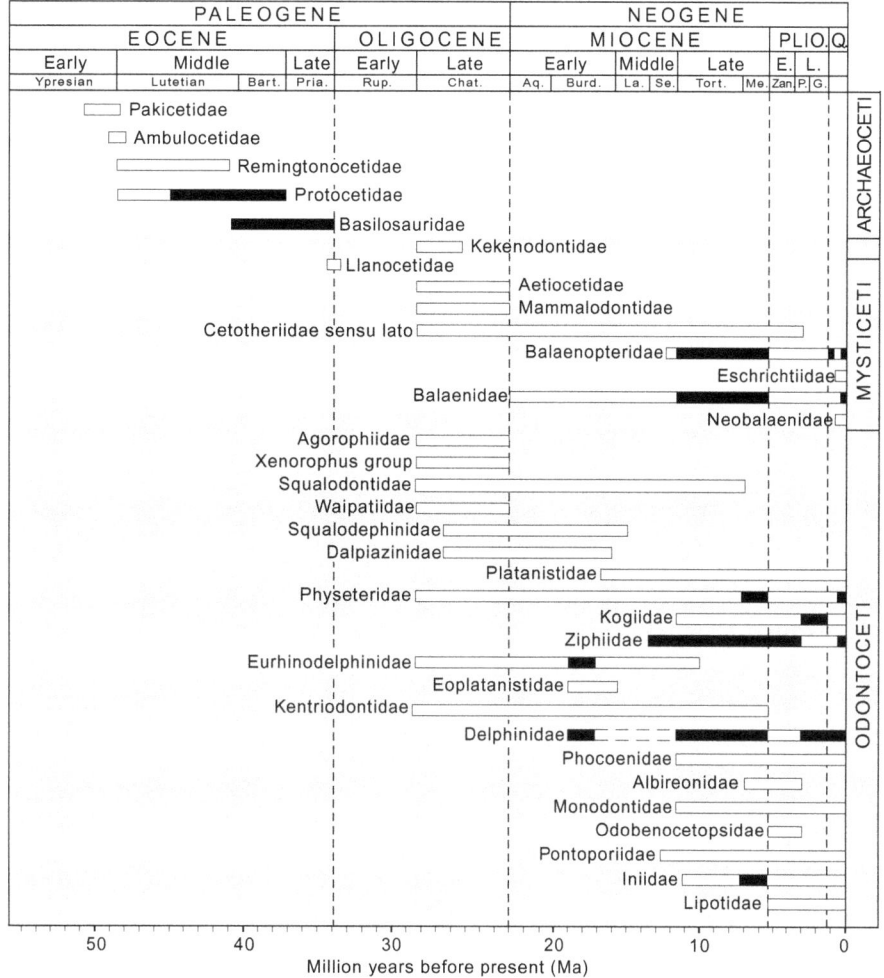

FIGURE 45.2 Stratigraphic range chart showing the distribution of most families of Archaeoceti, Mysticeti, and Odontoceti through geological time. Black bars are records from the African continent. Archaeoceti have a denser stratigraphic record in Africa than that of other cetaceans and are generally represented by better preserved and more complete remains. The geological time scale is from Gradstein et al. (2004). Stratigraphic ranges are modified from Fordyce and Muizon (2001; additional families of Cetacea have been described since 2001, but none involve African Cetacea).

margin during the middle and late Eocene allowed first extensive flooding and deposition of marine limestones, and later progradation and deposition of marginal-marine and fluvial sandstones and shales. Another exception is in the Miocene when a single fossil whale is known from freshwater sediments in the rift valley of Kenya (site 39 in figure 45.3). The rift itself provided a possible corridor into the interior of the continent, and caused the subsidence required for burial.

The greatest concentration of known cetacean-bearing fossil localities is in Egypt (figure 45.4). Most are Eocene in age, but two are Miocene. The oldest Egyptian archaeocete geologically is *Protocetus atavus* (discussed later) from Gebel Mokattam in Cairo (site 1 in figures 45.3 and 45.4), while the youngest is the small collection of fragmentary remains from Miocene localities in the vicinity of Wadi Moghara in the Western Desert (sites 36 and 37 in figure 45.4). The most important fossil cetaceans, from a morphological and evolutionary point of view, are those found in the northern and western parts of Fayum Province in Egypt (figure 45.5). All cetaceans found in Egypt lived and were buried on the shallow marine shelf of a passively subsiding continental margin.

Documentation for the maps in figures 45.3–45.5 is summarized in table 45.1. Entries are organized by geological epoch and subepoch (where known) and within each subepoch by the publication date of the site.

ABBREVIATIONS

Museum abbreviations: AMNH, American Museum of Natural History, New York, U.S.A.; BSPM, Paläontologische Museum, Bayerischen Staatssammlung für Paläontologie und Geologie, Munich, Germany; CGM, Egyptian Geological Museum, Cairo, Egypt; MNB, Museum für Naturkunde, Berlin, Germany; NHML, Natural History Museum, London; SFNF, Senckenberg Forschungsinstitut und Naturmuseum, Frankfurt; SMNS, Staatliches Museum für Naturkunde, Stuttgart, Germany; UCMP, University of California Museum of Paleontology, Berkeley, U.S.A.; UM, University of Michigan Museum of Paleontology, Ann Arbor, U.S.A.; YPM, Yale Peabody Museum, New Haven, U.S.A.

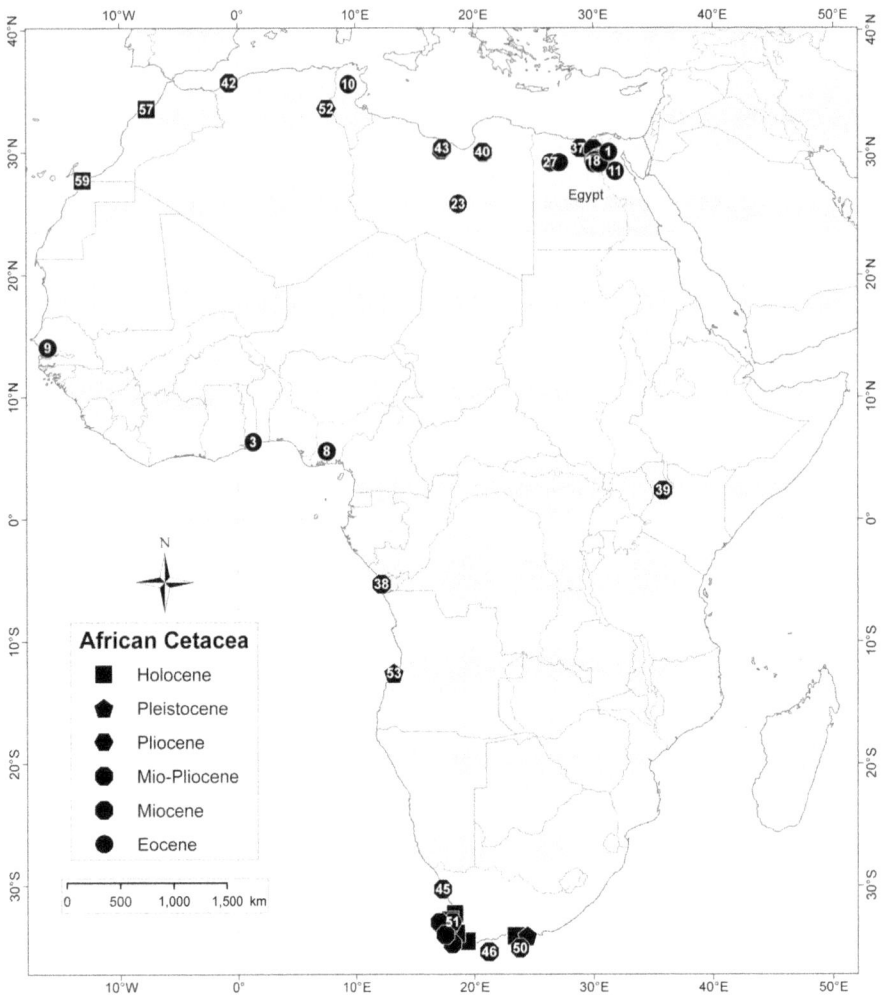

FIGURE 45.3 Geographic distribution of fossil cetaceans known from Africa. Site numbers correspond to those in table 45.1. See figures 45.4 and 45.5 for maps showing greater detail for Egypt and for Fayum Province in Egypt.

History of Study of African Cetacea

EOCENE ARCHAEOCETI

Most of the fossil cetaceans of Africa are Eocene Archaeoceti, and the history of their discovery is rich. Archaeoceti is the subordinal name proposed by Flower (1883) to include *Basilosaurus* and other archaic cetaceans first found in North America, while Basilosauridae is a family name coined by Cope (1868) for *Basilosaurus* and other archaic whales. Basilosaurids are fully aquatic, having lost any bony connection of the pelvic girdle with the vertebral column, and are found in the late middle Eocene (Bartonian) and late Eocene (Priabonian).

Protocetidae is a second family of Archaeoceti known from Africa. This was named by Stromer (1908) and includes geologically older early middle Eocene (Lutetian) and late middle Eocene (Bartonian) archaeocetes that retain hindlimbs and pelves connected to the vertebral column by a distinct sacrum. Protocetidae were semiaquatic and able to come out of the water to rest and reproduce (Gingerich et al., 2009).

Schweinfurth in Egypt

The first fossil cetaceans found in Africa are basilosaurid archaeocetes found by Georg Schweinfurth in 1879. Schweinfurth himself was a botanist by training. He is best known for the years he spent from 1863 to 1871 exploring the Nile from Egypt to Sudan, and eventually into Central Africa. Schweinfurth lived in Cairo for 14 years from 1875 through 1889. Here he continued to explore locally, and in 1879 he discovered the first vertebrate fossils in Fayum and the first archaeocete cetaceans from Africa (and indeed the first from the whole eastern hemisphere). These were found in what is now termed Birket Qarun Formation on the island of Geziret el-Qarn (site 13 in figure 45.4 and 45.5) in the middle of lake Birket Qarun. The first descriptions of these were published by Dames (1883a, 1883b).

Schweinfurth visited Qasr el-Sagha temple (sometimes called Schweinfurth's temple) and the Qasr el-Sagha escarpment on the north side of Birket Qarun, first in 1884 and again in 1886. The most important of Schweinfurth's fossils, the type dentary of *Zeuglodon osiris* (figure 45.6; now

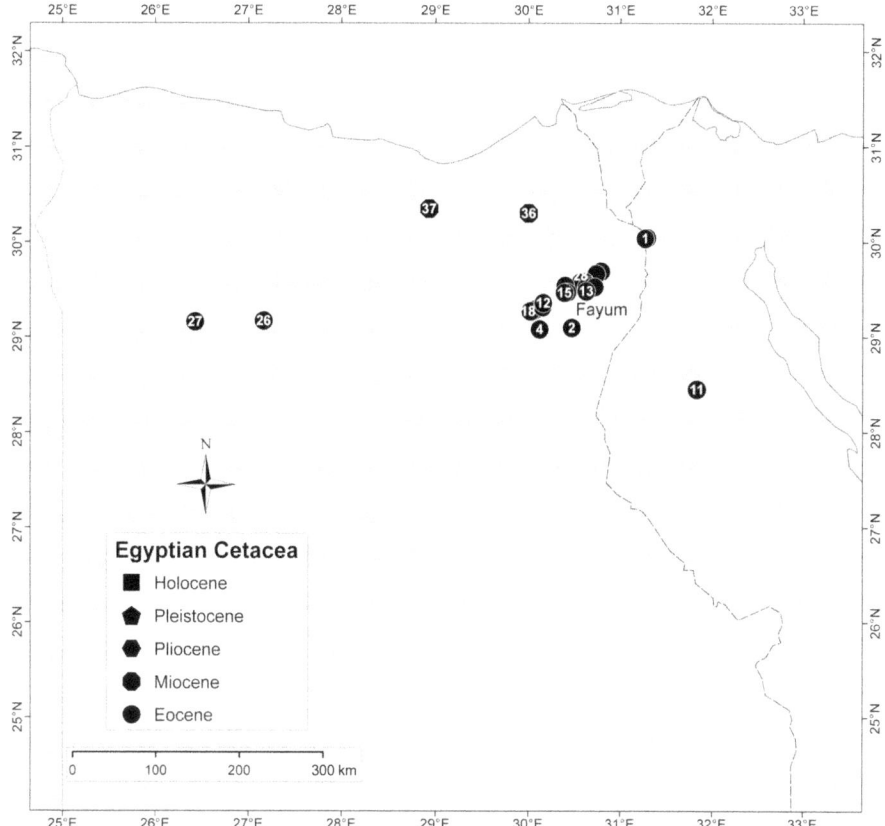

FIGURE 45.4 Geographic distribution of fossil cetaceans known from Egypt. Site numbers correspond to those in table 45.1. See figure 45.3 for a map of all African sites and figure 45.5 for greater detail in Fayum Province, Egypt.

Saghacetus osiris), came from what is now Qasr el-Sagha Formation at 'Zeuglodonberg' (site 28 in figure 45.5). *'Zeuglodon' osiris* was described and named by Dames (1894). The genus name *Zeuglodon* Owen, 1839, is a junior synonym of *Basilosaurus* Harlan, 1834. Owen proposed *Zeuglodon* as a replacement name when he reidentified *Basilosaurus* as a whale rather than a reptile (Owen, 1841), but *Basilosaurus* retains priority. *Zeuglodon* was so widely used as a name for so many different early-described archaeocetes that it retains no clear meaning today.

Schweinfurth published two studies important for Egyptian stratigraphy and archaeocete paleontology: "On the Geological Stratification of Mokattam Near Cairo" (1883), where he mentions cetacean remains, and "Travel in the Depression Circumscribing Fayum in January, 1886" (1886). On the expedition described in the latter, Schweinfurth came within a few kilometers of discovering Wadi Hitan, one of the most productive fossil whale sites in the world, but he was forced to turn back before entering the valley because of difficulties with his camels and staff.

Beadnell and Andrews in Egypt

Hugh J. L. Beadnell was British and employed by the Egyptian Geological Survey from 1896 to 1906. Beadnell started work in Fayum in October 1898. Much of Fayum is at or below sea level, and Beadnell's charge was to investigate the feasibility of storing Nile flood water in the Fayum Depression. He worked first in eastern Fayum, then extended exploration and mapping to

the escarpments north of Birket Qarun. Exploration extended west to Garet Gehannam during the spring of 1899. Charles W. Andrews of the British Museum (Natural History) published the first note on Beadnell's paleontological discoveries (Andrews, 1899), and Beadnell himself (1901) provided a summary of Fayum stratigraphy that has guided most subsequent work.

Andrews joined Beadnell in the field for the first time in April 1901, to investigate bone beds discovered in 1898. Many new specimens including archaeocetes were found during this expedition (Andrews, 1901). Collecting continued during the winters of 1901–1902, 1902–1903, and 1903–1904. Sometime in this interval Beadnell collected a large dentary for the Cairo Geological Museum that became the type of *Zeuglodon isis* Beadnell in Andrews, 1904 (now *Basilosaurus isis*; species name is from the Beadnell manuscript published in 1905). According to Beadnell (1905) the type of *'Zeuglodon' isis* came from the Birket Qarun Formation in the Birket Qarun escarpment near the west end of the lake (site 15 in figure 45.5).

Beadnell carried out a second phase of mapping in the winter of 1902–1903. This is when he made a traverse from Garet Gehannam west and southwest 12 kilometers into the desert to a valley where large skulls and other remains of fossil cetaceans were abundant (site 18 in figure 45.5). Beadnell (1905) coined the term 'Zeuglodon Valley' for the area west of Garet Gehannam, but this has been modernized as Wadi Hitan or Wadi Al Hitan: Arabic for 'Valley of Whales.'

Beadnell collected very little, but one skull of an immature whale was recovered for the Cairo Geological Museum,

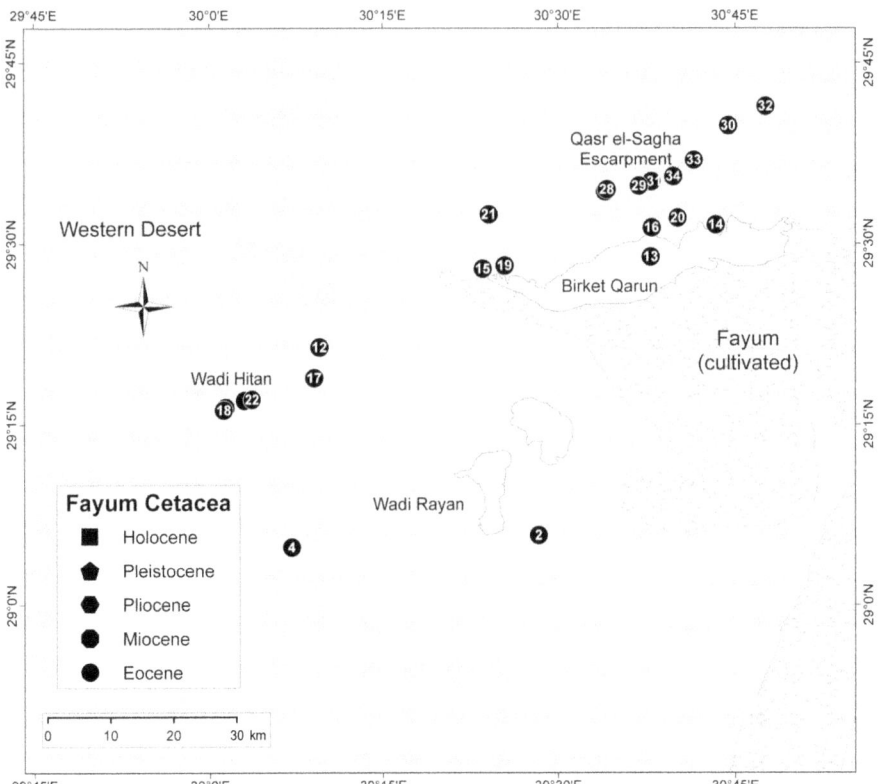

FIGURE 45.5 Geographic distribution of fossil cetaceans known from Fayum Province, Egypt. Wadi Hitan (site 18) is a UNESCO World Heritage site featuring hundreds of *Basilosaurus*, *Dorudon*, and other vertebrate skeletons, many of which are visible in natural stratigraphic position in the field. Site numbers correspond to those in table 45.1. See figure 45.3 for a map of all African sites and figure 45.4 for a map of all Egyptian sites.

and this was made the type of *Prozeuglodon atrox* by Andrews (1906). This taxon is now properly called *Dorudon atrox* (Uhen, 2004). Most fossils from Wadi Hitan come from the Birket Qarun Formation, and the type of *Dorudon atrox* almost certainly did as well.

The fossil cetaceans collected by Beadnell and Andrews were first studied in detail by Andrews (1906), in which new specimens of *'Zeuglodon' osiris* were described, description of *'Zeuglodon' isis* Beadnell was augmented, and *'Prozeuglodon' atrox* was named. Andrews diagnosed *'Prozeuglodon'* as "intermediate between *Protocetus* and *Zeuglodon* proper," but he seemingly did not recognize that the type is a juvenile with deciduous premolars (Andrews, 1908). This led to confusion when the type later was recognized to be juvenile, and for some time *Prozeuglodon atrox* was thought to be the juvenile form of *Zeuglodon isis*. Kellogg (1936), for example, combined these and referred to both as *Prozeuglodon isis* (see also Barnes and Mitchell, 1978). Surprisingly, following Beadnell's discovery of Zeuglodon Valley in 1902–1903, another 80 years would pass before anyone attempted systematic investigation of the fossil cetaceans that are so abundant there. Recovery of skeletons of mature *Dorudon atrox* (discussed later) showed that *Dorudon atrox* is clearly different from contemporary *Basilosaurus isis* (Gingerich et al., 1990).

G. Elliott Smith (1903) was the first to study a natural stone endocast of the brain of an archaeocete, based on a specimen of *Saghacetus ("Zeuglodon") osiris* collected in Fayum by Beadnell and Andrews. Later, Raymond Dart (1923) studied a series of natural stone endocasts collected in Fayum by Beadnell

and Andrews, together with plaster endocasts made from skulls in the Beadnell-Andrews collection. From these Dart described three new species: *'Zeuglodon' sensitivus*, *'Zeuglodon' elliotsmithii*, and *'Zeuglodon' intermedius*. He then focused on three endocasts forming what he termed a "phyletic series," with the provenance of the specimens corroborated by Andrews (1923). Dart (1923:635) concluded that the phyletic series showed a "dwindling in brain substance," supporting an idea popular at the time that "devolutional potentialities" might accompany specialization. Dart then argued that "devolution of the brain" precluded archaeocetes from giving rise to later cetaceans. This whole specious argument could have been avoided if Dart had been more cautious and worked with endocasts associated with more complete skeletons. Dart's "devolution" was a result of comparing brains of successively smaller cetaceans through time (Gingerich, 1998).

Remington Kellogg (1936) reviewed the Beadnell-Andrews collection, but his interpretation was handicapped somewhat by his lack of familiarity with the stratigraphic relationships of source localities.

Stromer and Fraas in Egypt

Ernst Stromer von Reichenbach of Munich organized a Royal Bavarian Academy of Science expedition to Egypt in 1901–1902. This, like the work of Beadnell mentioned earlier, was carried out in cooperation with the Geological Survey of Egypt, represented in this instance by geologist Max Blanckenhorn. Egypt was governed at the time by a British-French-German

TABLE 45.1
Localities of African fossil Cetacea
Localities numbered here are shown geographically on the maps in figures 45.3–45.5.

Location	Locality	Age and geological formation	North latitude (°N)	East longitude (°E)	Reference	Taxon (current identification)
		EARLY MIDDLE EOCENE (LUTETIAN)				
1	Schweinfurth XII, Cairo, Egypt	M. Eocene, Mokattam Fm.	30.036	31.272	Fraas (1904)	*Protocetus atavus* (Protocetidae)
2	Klein Rajan, Fayum, Egypt	M. Eocene, Midawara Fm.	29.097	30.472	Stromer (1908, 1914)	Protocetid(?)
3	Kpogamé, Togo	M. Eocene, Phosphate Bed	6.309	1.334	Gingerich et al., (1992)	Protocetid
4	Qusour el Arab, Egypt	M. Eocene, Midawara Fm.	29.080	30.119	Gingerich et al. (unpub. ms.)	Protocetid
		LATE MIDDLE EOCENE (BARTONIAN)				
5	Schweinfurth XXIII, Cairo, Egypt	M. Eocene, Giushi Fm.	30.043	31.295	Schweinfurth (1883)	Cetacea (Basilosauridae?)
6	Schweinfurth VII, Cairo, Egypt	M. Eocene, Giushi Fm.	30.030	31.275	Blanckenhorn (1900)	*Basilosaurus* sp. (Basilosauridae)
7	Schweinfurth IV, Cairo, Egypt	M. Eocene, Giushi Fm.	30.042	31.283	Fraas (1904)	*Eocetus schweinfurthi* (Protocetidae)
8	Ameke, Nigeria	M. Eocene (Bartonian?), Ameki Fm.	5.550	7.515	Andrews (1920)	*Pappocetus lugardi* (Protocetidae)
9	Tiavandou, Senegal	'Lutétien inf.' (Bartonian?), Eocene	14.050	−16.050	Elouard (1981)	Basilosaurid
10	Oued Kouki, Tunisia	Lutétien sup. (Bartonian), Eocene	35.578	9.403	Batik and Fejfar (1990)	Protocetid(?)
11	Khashm el-Raqaba, Egypt	M. Eocene, Gebel Hof Fm.	28.451	31.834	Bianucci et al. (2003)	Protocetid
12	Wadi Gehannam, Egypt	M. Eocene, Gehannam Fm.	29.357	30.158	Gingerich et al. (unpub. ms.)	Protocetid
		LATE EOCENE (EARLY PRIABONIAN)				
13	Geziret el Qarn, Fayum, Egypt	L. Eocene, Birket Qarun Fm.	29.482	30.631	Dames (1883)	*Dorudon atrox* (Basilosauridae)
14	El Kenissa, Fayum, Egypt	L. Eocene, Birket Qarun Fm.	29.526	30.723	Dames (1894)	*Dorudon atrox* (Basilosauridae)
15	Garet el Naqb, Fayum, Egypt	L. Eocene, Birket Qarun Fm.	29.466	30.394	Andrews (1904)	*Basilosaurus isis* (Basilosauridae)
16	Dimeh SW, Fayum, Egypt	L. Eocene, Birket Qarun Fm.	29.522	30.632	Beadnell (1905)	*Basilosaurus isis* (Basilosauridae)
17	Garet Gehannam, Fayum, Egypt	L. Eocene, Birket Qarun Fm.	29.314	30.151	Beadnell (1905)	*Basilosaurus isis* (Basilosauridae)
18	Wadi Hitan, Fayum, Egypt	L. Eocene, Birket Qarun Fm.	29.270	30.020	Beadnell (1905)	*Basilosaurus isis* (Basilosauridae)
	"	"	"	"	Andrews (1906)	*Dorudon atrox* (Basilosauridae)
19	W. corner B. Qarun, Fayum, Egypt	L. Eocene, Birket Qarun Fm.	29.470	30.424	Fraas (1906)	*Basilosaurus isis* (Basilosauridae)
20	Dimeh, Fayum, Egypt	L. Eocene, Birket Qarun Fm.	29.536	30.669	Kellogg (1936)	*Masracetus markgrafi* (Basilosauridae)
21	N. of Qasr Qarun, Fayum, Egypt	L. Eocene, Birket Qarun Fm.	29.540	30.402	Kellogg (1936)	*Dorudon atrox* (Basilosauridae)
22	UCMP 6788, Fayum, Egypt	L. Eocene, Birket Qarun Fm.	29.285	30.062	Phillips (1948)	*Basilosaurus isis* (Basilosauridae)
23	Dor el Talha, Libya	L. Eocene, Evaporite Unit	25.779	18.694	Savage (1969)	Cetacea indet.

TABLE 45.1 (CONTINUED)

Location	Locality	Age and geological formation	North latitude (°N)	East longitude (°E)	Reference	Taxon (current identification)
24	UCMP 6789, Fayum, Egypt	L. Eocene, Birket Qarun Fm.	29.283	30.050	Pilleri (1991)	*Dorudon atrox* (Basilosauridae)
25	WH-81, Fayum, Egypt	L. Eocene, Birket Qarun Fm.	29.274	30.023	Gingerich and Uhen (1996)	*Ancalecetus simonsi* (Basilosauridae)
26	Qattara Depression, Egypt	L. Eocene	29.168	27.163	Morrison (unpubl.)	*Masracetus markgrafi* (Basilosauridae)
27	Tabaghbagh, Egypt	L. Eocene (Priabonian)	29.158	26.428	Vliet and Paymans (unpub. ms.)	Basilosaurids

LATE EOCENE (MIDDLE PRIABONIAN)

Location	Locality	Age and geological formation	North latitude (°N)	East longitude (°E)	Reference	Taxon (current identification)
28	Zeuglodonberg, Fayum, Egypt	L. Eocene, Qasr el Sagha Fm.	29.575	30.569	Dames (1894)	*Saghacetus osiris* (Basilosauridae)
29	Gebel Hameier W, Fayum, Egypt	L. Eocene, Qasr el Sagha Fm.	29.580	30.614	Stromer (1902)	*Saghacetus osiris* (Basilosauridae)
30	Gebel Achdar, Fayum, Egypt	L. Eocene, Qasr el Sagha Fm.	29.664	30.740	Stromer (1903)	*Saghacetus osiris* (Basilosauridae)
31	Gebel Hameier, Fayum, Egypt	L. Eocene, Qasr el Sagha Fm.	29.586	30.632	Stromer (1903)	*Saghacetus osiris* (Basilosauridae)
32	Tamariskenbucht, Fayum, Egypt	L. Eocene, Qasr el Sagha Fm.	29.690	30.793	Stromer (1903)	*Stromerius nidensis* (Basilosauridae)
33	Beadnell Section, Fayum, Egypt	L. Eocene, Qasr el Sagha Fm.	29.616	30.692	Beadnell (1905)	*Saghacetus osiris* (Basilosauridae)
34	Qasr el-Sagha West, Fayum, Egypt	L. Eocene, Qasr el Sagha Fm.	29.593	30.662	Kellogg (1936)	*Saghacetus osiris* (Basilosauridae)
35	Garet el-Esh, Fayum, Egypt	L. Eocene, Qasr el-Sagha Fm.	29.572	30.566	Gingerich (2007)	*Stromerius nidensis* (Basilosauridae)

EARLY MIOCENE

Location	Locality	Age and geological formation	North latitude (°N)	East longitude (°E)	Reference	Taxon (current identification)
36	Wadi Faregh, Egypt	E. Miocene, Moghara Fm.	30.300	30.000	Stromer (1907)	*Schizodelphis* aff. *sulcatus* (Eurhinodel.)
37	Wadi Moghara	E. Miocene, Moghara Fm.	30.348	28.932	Fourtau (1918)	*Delphinus vanzelleri* (Delphinidae?)
	"	"	"	"	"	*Cyrtodelphis* aff. *sulcatus* (Eurhinodel.)

MIDDLE MIOCENE

Location	Locality	Age and geological formation	North latitude (°N)	East longitude (°E)	Reference	Taxon (current identification)
38	Malembe, Angola	Miocene, Malembe beds	-5.333	12.183	Dartevelle (1935)	Cetacea indet.
39	Loperot, Kenya	M. Miocene	2.333	35.833	Mead (1975)	Ziphiid (Ziphiidae)

LATE MIOCENE

Location	Locality	Age and geological formation	North latitude (°N)	East longitude (°E)	Reference	Taxon (current identification)
40	P8, Qasr as Sahabi, Libya	L. Miocene, Sahabi Fm.	30.028	20.788	Petrocchi (1941, 1951)	*Balaenoptera* sp. (Balaenopteridae)
	"	"	"	"	Whitmore (1987)	Cf. *Lagenorhynchus* sp. (Delphinidae)
41	P4A, Qasr as Sahabi, Libya	L. Miocene, Sahabi Fm.	29.952	20.812	White et al. (1983)	Odontoceti
	"	"	"	"	Whitmore (1987)	Iniidae
42	Raz-el-Ain, Algeria	L. Miocene	35.698	-0.709	Muizon (1981)	Odontocete (Physeteridae)
43	Langklip, S. Africa	L. Miocene	-30.359	17.328	Pickford and Senut (1997)	Cetacea indet.
44	Somnaas 2, S. Africa	L. Miocene	-30.168	17.235	Pickford and Senut (1997)	Cetacea indet.
45	Swartlintjies 2, S. Africa	L. Miocene	-30.285	17.290	Pickford and Senut (1997)	Cetacea indet.

Location	Locality	Age and geological formation	North latitude (°N)	East longitude (°E)	Reference	Taxon (current identification)

Location	Locality	Age and geological formation	North latitude (°N)	East longitude (°E)	Reference	Taxon (current identification)
46	Agulhas Bank, S. Africa	Mio-Pliocene	−35.400	21.200	Barnard (1954)	*Eubalaena australis* (Balaenidae)
	"	"	"	"	"	*Balaenoptera physalis* (Balaenopteridae)
	"	"	"	"	"	cf. *Orcinus orca* (Delphinidae)
	"	"	"	"	"	Cf. *Ziphius cavirostris* (Ziphiidae)
	"	"	"	"	"	*Mesoplodon* sp. (Ziphiidae)
	"	"	"	"	Bianucci et al. (2007)	*Khoikhoicetus agulhasis* (Ziphiidae)
47	Cape West Coast, S. Africa	Mio-Pliocene	−34.000	17.500	Haughton (1956)	*Balaena* sp. (Balaenidae)
	"	"	"	"	"	*Balaenoptera* sp. (Balaenopteridae)
	"	"	"	"	"	*Megaptera* sp. (Balaenopteridae)
	"	"	"	"	"	*Orca* sp. (Delphinidae)
	"	"	"	"	"	*Mesoplodon* spp. (Ziphiidae)
	"	"	"	"	"	*Ziphius?* sp. (Ziphiidae)
	"	"	"	"	Ross (1986)	*Mesoplodon* cf. *M. hectori* (Ziphiidae)
	"	"	"	"	"	*Tasmacetus* or *Berardius* (Ziphiidae)
	"	"	"	"	"	*Ziphius* sp. (Ziphiidae)
	"	"	"	"	Bianucci et al. (2007)	*Africanacetus ceratopsis* (Ziphiidae)
	"	"	"	"	"	*Ihlengesi saldanhae* (Ziphiidae)
	"	"	"	"	"	*Mesoplodon slangkopi* (Ziphiidae)
	"	"	"	"	"	*Nenga meganasalis* (Ziphiidae)
	"	"	"	"	"	*Pterocetus benguelae* (Ziphiidae)
48	Cape Columbine, S. Africa	Mio-Pliocene	−33.000	17.000	Bianucci et al. (2007)	*Ihlengesi saldanhae* (Ziphiidae)
49	Cape Town SW, S. Africa	Mio-Pliocene	−34.750	18.100	Bianucci et al. (2007)	*Izikoziphius angustus* (Ziphiidae)
50	South Coast, S. Africa	Mio-Pliocene	−35.100	23.750	Bianucci et al. (2007)	*Africanacetus ceratopsis* (Ziphiidae)
	"	"	"	"	"	*Xhosacetus hendeysi* (Ziphiidae)
	"	"	"	"	"	*Ziphius* sp. (Ziphiidae)

Location	Locality	Age and geological formation	North latitude (°N)	East longitude (°E)	Reference	Taxon (current identification)
51	Quarry E, Langebaanweg, S. Afr.	E. Pliocene	−32.963	18.111	Hendey (1976)	Mysticeti
	"	"	"	"	"	Odontoceti

Location	Locality	Age and geological formation	North latitude (°N)	East longitude (°E)	Reference	Taxon (current identification)
52	Ahl al Oughlam, Morocco	L. Pliocene, Messaoudian Fm.	33.579	−7.519	Geraads et al. (1998)	*Delphinus* or *Stenella* sp. (Delphinidae)
	"	"	"	"	"	*Kogia* sp. (Physeteridae)

Location	Locality	Age and geological formation	North latitude (°N)	East longitude (°E)	Reference	Taxon (current identification)
53	Baia Farta, Angola	Calabrian, e. Pleistocene	−12.600	13.205	Gutiérrez et al. (2001)	*Balaenoptera* sp. (Balaenopteridae)

TABLE 45.1 (CONTINUED)

Location	Locality	Age and geological formation	North latitude (°N)	East longitude (°E)	Reference	Taxon (current identification)
		LATE PLEISTOCENE				
54	Klasies River Mouth, S. Africa	L. Pleistocene, Middle Stone Age	–34.106	24.390	Klein (1976)	Delphinid
	"	"	"	"	"	Cetacea indet.
55	Sea Harvest Site, S. Africa	L. Pleistocene, Middle Stone Age	–33.023	17.950	Grine and Klein (1993)	Delphinid
		HOLOCENE				
56	Yzerplaats, South Africa	Holocene clay	–33.898	18.483	Gill (1928)	*Balaenoptera* sp. (Balaenopteridae)
57	Sidi Abderrahman, Morocco	Holocene	33.570	–7.692	Ennouchi (1961)	*Balaenoptera physalis* (Balaenopteridae)
58	Nelson Bay Cave, S. Africa	Holocene, Late Stone Age	–34.100	23.400	Klein (1972)	Delphinid
59	Site H, Tarfaya, Morocco	Holocene	27.730	–13.090	Saban (1974)	*Physeter macrocephalus* (Physeteridae)
60	Kasteelberg B, S. Africa	Holocene, Prehistoric	–32.804	17.930	Klein and Cruz-Uribe (1989)	Delphinid
61	Die Kelders Cave 1, S. Africa	Holocene, Late Stone Age	–34.533	19.375	Grine et al. (1991)	Delphinid
	"	"	"	"	"	Cetacea indet.
62	Elands Bay Cave, S. Africa	Holocene, Late Stone Age	–32.300	18.333	Grine and Kein (1993)	Delphinid

FIGURE 45.6 Left dentary of small basilosaurid *Saghacetus osiris* (holotype; Museum für Naturkunde, Berlin, MNB Ma.28388). Total length of dentary is 51 cm. Specimen is shown in lateral (A), medial (B), and occlusal (C) views. Note unfused mandibular symphysis; large mandibular canal making much of the dentary hollow (margin of medial foraminal opening not preserved); retention of a generalized placental-mammal dental formula of 3.1.4.3; simple anterior teeth (represented here by p2); and complex, multicusped cheek teeth (represented here by p2–4 and m1–3). Specimen was found in 1886 by Schweinfurth at 'Zeuglodonberg' (site 28 in figure 45.5; middle Priabonian late Eocene), and the species was named in 1894 by Dames (see Gingerich, 1992:73). Illustration is from Dames (1894).

consortium, and offices like the Geological Survey were staffed by representatives of all three countries. There is no indication that the work of Stromer and Blanckenhorn was coordinated in any way with that being carried out by Beadnell and Andrews.

Stromer and Blanckenhorn spent about 15 days of January 1902, engaged in paleontological field work north of Birket Qarun. Blanckenhorn (1902) provided a detailed description of many of the sites yielding the fossil cetaceans collected by Stromer. Important fossils included a new skull and lower jaw of *Saghacetus ('Zeuglodon') osiris* (Dames) from what is now the middle Qasr el-Sagha Formation (Stromer, 1902). Stromer (1903) named a new species *Zeuglodon zitteli* that was collected at or near the type locality of *Saghacetus osiris*, but this has proven indistinguishable from Dames' species. I described (Gingerich, 2007) a series of vertebrae collected by Stromer at 'Tamariskenbucht' (site 32 in figure 45.5) and a new specimen from Garet el-Esh (site 35 in figure 45.5) as *Stromerius nidensis*.

Another German paleontologist, Eberhard Fraas of Stuttgart, visited Egypt several times, starting in 1897. His greatest contribution may have been engaging Richard Markgraf as a private collector who worked first in the stone quarries of Gebel Mokattam and later in Fayum. Markgraf's contributions started in 1903 when he found the cranium and associated postcranial remains of a small archaeocete from the Mokattam Limestone of early middle Eocene age, near the base of the Gebel Mokattam section (site 1 in figures 45.3 and 45.4). This was described by Fraas (1904a) as a new genus and species *Protocetus atavus*. The skull is primitive, compared to later basilosaurids, in retaining the full complement of three upper molars and in having upper molars that retain medial roots and swellings of the crown in the position of a protocone (figures 45.7A, 45.7B). Associated vertebrae include a partial sacrum with auricular processes indicating retention of an articulation with left and right innominates of the pelvis (figure 45.7C, 45.7D). A second skull of a different whale from Gebel Mokattam was described as *Eocetus ('Mesocetus') schweinfurthi* (Fraas, 1904a, 1904b). *Protocetus*, the better preserved of the two genera, became the type of the new family Protocetidae (Stromer, 1908).

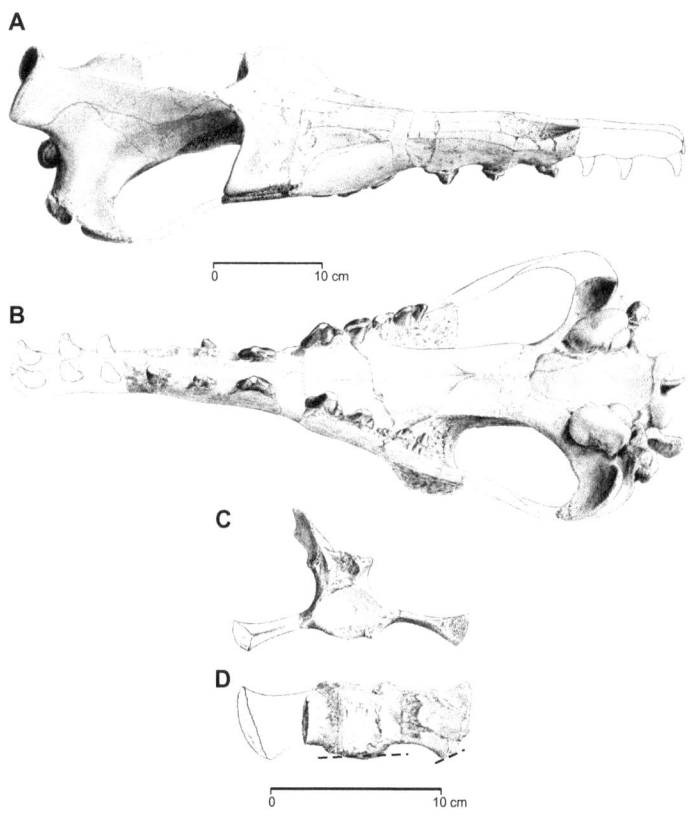

FIGURE 45.7 Cranium and anterior centrum of sacrum of *Protocetus atavus* (Staatliches Museum für Naturkunde, Stuttgart, SMNS 11084, holotype, and associated SMNS 11087). Cranium is shown in oblique dorsal (A) and palatal (B) views. Note external nares opening in the normal mammalian position above P1 (premaxillae extended far in front of this); broad frontal shield; retention of a generalized placental-mammal dental formula of 3.1.4.3 (incisors reconstructed); anteroposteriorly elongated and possibly double-rooted C1; retention of proto-cones on upper molars; and large and densely ossified tympanic bullae. Anterior centrum of sacrum is shown in anterior (C) and ventral (D) views. Note the well-developed auricular process for articulation with the ilium of a pelvis, and bro-ken surfaces (dashed lines) showing that the anterior centrum was originally part of a larger sacrum (contemporary protocetids generally have three or four centra solidly fused in the sacrum). Both were found together as a partial skele-ton, recovered by Markgraf in 1903 at Gebel Mokattam (site 1 in figures 45.3 and 45.4; Lutetian early middle Eocene). Illustrations are from Fraas (1904a).

Following publication of his 1903 monograph, Stromer made a second trip to Egypt to collect Eocene cetaceans and other vertebrates for the Senckenberg Museum of Frankfurt, starting in November 1903. This was a three-month expedition employing Markgraf and ranging widely. Two large archaeocete vertebrae were collected at Gebel Mokattam, and an archaeocete skull, jaws, and vertebrae were collected north of Birket Qarun (Stromer, 1904; the skull was presumably SFNF 4451). SFNF 4451 was studied by Pompeckj (1922) and Uhen (2004). Markgraf continued to collect fossil cetaceans in Fayum after the 1903–1904 expedition, and later in 1904 he sent another specimen from the Qasr el-Sagha escarpment to Munich. Stromer (1908) first identified this as *Zeuglodon osiris*, and it eventually became the type of *Dorudon stromeri* (Kellogg, 1928).

Eberhard Fraas made a final trip to Egypt to work with Markgraf in 1906. The two left Cairo on March 11, and they reached Qasr el-Sagha on March 13. Within days they moved to the west end of Birket Qarun (site 19 in figure 45.5), where they excavated much of the skeleton of a large '*Zeuglodon*' (*Basilosaurus isis*) with a 1.3-m skull and a 10-m sequence of vertebrae and ribs (Fraas, 1906). This specimen, described by Stromer (1908) and Slijper (1936), is in the Staatliches Museum für Naturkunde in Stuttgart (SMNS 11787). The skull was illustrated by Heizmann (1991).

Stromer (1908:126, 128, 136) described a dark brown humerus and radius as '*Zeuglodon*' *zitteli* (SMNS 11951b; 'St. 14'), and disarticulated pieces of a large immature cranium as *Prozeuglodon atrox* (SMNS 11951a; 'St. 10'), stating that these came from Wadi Rayan. Later, after discussing the finds again with the collector (Markgraf), Stromer (1914:8) corrected the locality as being in a place translated as Klein Rajan (site 2 in figure 45.5), rather than Wadi Rayan some 10 km farther to the northwest. This is important as cetaceans from the middle Eocene are rare in Egypt.

Osborn and Granger in Egypt

Henry Fairfield Osborn and Walter Granger of the American Museum of Natural History in New York organized a 1907 collecting expedition to Fayum to follow in the footsteps of Schweinfurth, Beadnell, Andrews, Stromer, and Fraas. Following publication of Andrews's 1906 'Descriptive Catalog of the Tertiary Vertebrata of the Fayum, Egypt,' Andrews himself encouraged Osborn to carry out further studies. The Egyptian Geological Survey assigned Hartley T. Ferrar (1879–1932), recently returned from Robert F. Scott's British National Antarctic Expedition of 1901–1903, to accompany and assist Osborn. The area that the American Museum party worked included Beadnell and Andrews's principal localities in continental beds above the Qasr el-Sagha escarpment. From here, Osborn and Ferrar made a 3-day camel march west to Garet Gehannam and Wadi Hitan on February 14–16, 1907. Osborn called this "the most famous fossil locality in the Fayum" and wrote, "We found [Wadi Hitan] strewn with the remains of monster zeuglodonts, including heads, ribs and long series of vertebrae, most tempting to the fossil hunter, yet too large and difficult of removal from this very remote and arid point" (Osborn, 1907).

On February 16, 1907, Granger had a chance meeting with Richard Markgraf, who was collecting fossils in the same area. Osborn met Markgraf on February 17 after his return from Wadi Hitan. At this time negotiations started for Markgraf to work the remainder of the season for the American Museum team. Two archaeocete specimens in the American Museum collection, a braincase and frontal of *Basilosaurus isis* (AMNH 14381) and a fine skull of *Saghacetus osiris* (AMNH 14382), were collected by Markgraf. Both are illustrated in Kellogg (1936). The Markgraf specimens came from the Birket Qarun and Qasr el-Sagha escarpments, respectively, but nothing more is known of their provenance. The only archaeocete that the American Museum party found themselves was a partial skull of *Saghacetus osiris* (AMNH 13720) collected on April 23 from 1–2 kms west of Qasr el-Sagha temple.

Lugard in Nigeria

The first African fossil cetaceans found outside Egypt were sent to the British Museum by Frederick Lugard, then governor-general of Nigeria. These comprised two left dentaries with teeth erupting, and an axis vertebra, all from middle Eocene strata near Ameke in Nigeria (site 8 in figure 45.3). Andrews (1920) described and illustrated all three specimens and named them *Pappocetus lugardi*. Kellogg (1936) classified *Pappocetus* in Protocetidae, where it has remained ever since. Halstead and Middleton (1974, 1976) revisited the type locality and described a number of additional vertebrae of *P. lugardi*.

Kellogg 1936

Remington Kellogg's (1936) 'Review of the Archaeoceti' was a milestone in understanding Eocene cetaceans. Kellogg reviewed virtually all of the specimens and taxa of Archaeoceti known at the time. Emphasis was placed on North American specimens that were the most complete skeletons known at the time, but Kellogg included the African specimens known from Egypt and Nigeria, many of which he was able to study during a 1930 trip to Europe, when he visited Berlin, Munich, Stuttgart, and London. The only substantial collection that he was unable to study firsthand was that in Cairo.

Denison and Deraniyagala in Egypt

A University of California African Expedition worked in the field in northern Fayum during the autumn of 1947 (Phillips, 1948). Most of the expedition's effort was concentrated on land mammal fossils, but on November 15–17, Robert H. Denison and Paules Deraniyagala worked in Wadi Hitan. Here they collected a partial skull of *Basilosaurus isis* (UCMP 93169; site 22 in figure 45.5) and an endocranial cast of *Dorudon atrox* (UCMP 41329). Deraniyagala (1948:3) reported finding 20 skeletons within a mile of the camp in Wadi Hitan and speculated that this represented a mass stranding (it is more likely to represent the passage of time in an interval with little sediment accumulation). The *Dorudon* endocranial cast was described by Pilleri (1991).

Simons and Meyer in Egypt

Elwyn Simons of Yale University sponsored Grant Meyer to work in Wadi Hitan on two occasions, once for several days during the winter of 1964–1965 and again for a week or so during the winter of 1966–1967. Meyer worked with Jeff Smith and Tom Walsh in 1964–1965, and with Lloyd Tanner and John Boyer in 1966–1967. The principal find collected in the first season was a skull of '*Prozeuglodon*' *isis* (presumably YPM 38454; YPM records indicating that this came from the Qasr el-Sagha Fm. are almost certainly wrong). A number of partial skulls were collected in 1966–1967, and these were divided equally between the Cairo Geological Museum and Yale Peabody Museum. None of these specimens has ever been described. Simons (1968:3) mentioned that "many additional archaeocete whales" were collected from the Qasr el-Sagha Formation but gave no details. Moustafa (1954) described the partial skull of a subadult *Prozeuglodon isis* (almost certainly *Dorudon atrox*) from the Birket Qarun Formation, and later (Moustafa, 1974) designated a '*Prozeuglodon*' zone for this and for other archaeocetes from the "lower Qasr el-Sagha Formation" (almost certainly Birket Qarun Formation).

Savage and Wight in Libya

R. J. G. Savage led two expeditions to the Dor el Talha escarpment in Libya, in 1968 and 1969 (Savage, 1969, 1971; Wight, 1980). Fragmentary cetacean remains were interpreted as late Eocene in age. Wight (1980) reported a cetacean from site 68-1 in his stratigraphic section 4, and there is another from site 69-51 in adjacent section 1 (J. Hooker, pers. comm.). These have not been studied in any detail, but they are almost certainly archaeocetes.

Barnes and Mitchell

Lawrence Barnes and Edward Mitchell did not collect any new African archaeocetes, but they did study collections in Cairo and London while preparing their review of African Cetacea (Barnes and Mitchell, 1978). They included *Protocetus atavus*, *Eocetus schweinfurthi*, and *Pappocetus lugardi* in Protocetidae, as I do here. However, they included '*Durodon*' *osiris* and five other species of *Dorudon*, together with '*Prozeuglodon*' *isis* and '*Zeuglodon*' *brachyspondylus*, in Basilosauridae. Reasons for revising the content of Basilosauridae developed as new specimens were collected in Fayum and compared to types and previously known specimens (see later discussion).

Elouard in Senegal

Pierre Elouard (1981) reported a partial skeleton of an archaeocete from limestone in a water well at Tiavandou near Kaolack in Senegal (site 9 in figure 45.3). The specimen included teeth, 18 vertebrae, and rib fragments published as 'Lutetian' in age (however, the teeth are clearly basilosaurid and, based on these, the age is more likely to be Bartonian or even Priabonian). The only sense of scale given for the Tiavandou whale is an unsubstantiated inference that it measured 8–10 m in length. Several years ago ornithologist Robert Payne attempted to locate and measure the specimen for me in the Laboratoire de Géologie, Faculté des Sciences, Dakar, but was unable to find it. Elouard identified the specimen as '*Zeuglodon*' cf. *osiris*, but it was clearly larger and seems to have had the proportions of *Dorudon atrox*.

Cappetta and Traverse in Togo

In 1985, Henri Cappetta and Michel Traverse made a large collection of Lutetian-age selachian teeth at Kpogamé-Hahotoé in Togo (site 3 in figure 45.3). This material included teeth and fragments of bone identified as *Pappocetus* (Cappetta and Traverse, 1988:362). However, it appears that several archaeocetes are present in this collection, and the most common is more the size and form of *Protocetus atavus*, with the distinction of retaining relatively long cervical centra (Gingerich et al., 1992).

Gingerich in Egypt

A new series of expeditions was initiated in 1983, with the support of Elwyn Simons, to investigate cetaceans of the Birket Qarun Formation in Zeuglodon Valley (Wadi Hitan), and of the Qasr el-Sagha Formation on the Qasr el-Sagha Escarpment (figure 45.5). The initial purpose was to collect one or two skulls and jaws for comparison with archaeocetes collected in Pakistan. However, it soon became clear that exceptionally well-preserved skeletons were abundant in both the Birket Qarun and Qasr el-Sagha formations. Thus the focus changed to a survey to map, identify, and count specimens. Collecting was targeted to document (1) form and ontogenetic development of archaeocete skulls and dentitions; (2) form, number of vertebrae, and total length of archaeocete skeletons; (3) morphology of the archaeocete hand; and (4) morphology of the archaeocete tail to determine whether a fluke was present. Morphology of archaeocete hindlimbs, feet, and toes became a fifth objective when it was discovered that these were retained (Gingerich et al., 1990).

Two archaeocetes are common in the Birket Qarun Formation of Wadi Hitan, and some 500 skeletons or partial skeletons have been mapped to date. The larger is a *Basilosaurus*, with a large cranium and dentary (figures 45.8A–45.8C), and posterior thoracic, lumbar, and anterior caudal vertebrae that are long relative to their diameter (figure 45.8D). Dentaries match that described by Andrews (1904) as the type of *Zeuglodon isis* Beadnell, and the correct name for this species is now *Basilosaurus isis* (Beadnell). All *B. isis* from Wadi Hitan are mature specimens with permanent teeth in place in the jaws. These were the first cetaceans to be found retaining hind legs, feet, and toes (figures 45.8D–45.8G).

The smaller archaeocete common in Wadi Hitan has posterior thoracic, lumbar, and anterior caudal vertebrae that are short relative to their diameter (figure 45.9B). For this the correct name was first thought to be *Prozeuglodon atrox* Andrews (1906); however, comparison with the type of *Dorudon serratus* (Gibbes, 1845), and with figures of vertebrae referred to this (Gibbes, 1847), indicated that *Prozeuglodon* is a junior synonym of *Dorudon*. Thus the correct name for the smaller archaeocete in Wadi Hitan is *Dorudon atrox* (Uhen, 2004). About one-half of the *Dorudon atrox* specimens from Wadi Hitan are mature adults, while the remainder, including the type (figures 45.9E, 45.9F), are immature. *Basilosaurus isis* and *D. atrox* are represented in Wadi Hitan by approximately equal numbers of specimens.

Because of their smaller size, it has been possible to collect whole skeletons of *Dorudon atrox*, and this is now one of the best known of all archaeocetes osteologically (Uhen, 2004). The hand skeleton of *Basilosaurus isis* (Gingerich and Smith, 1990) is larger but otherwise essentially the same as that of *D. atrox*. The hindlimb and foot skeleton of *D. atrox* is smaller, but otherwise essentially the same as that of *B. isis* (Gingerich et al., 1990). Morphology of the periotic and surrounding cranial bones has been described and analyzed by Luo and Gingerich (1999). Morphologically it is clear that *D. atrox* and *B. isis* were both fully aquatic, which is confirmed by analysis of stable isotopes (Clementz et al., 2006).

Several cetaceans are known in Wadi Hitan in addition to *B. isis* and *D. atrox*. These include *D. atrox*–sized *Ancalecetus simonsi* Gingerich and Uhen (1996), with fused elbows and unusual carpals (figures 45.9C, 45.9D), and *Masracetus markgrafi* Gingerich (2007). *Masracetus* is a larger whale, still incompletely known, that has *Dorudon*-like vertebral proportions but large vertebral centra approaching those of *Basilosaurus isis* in height and width.

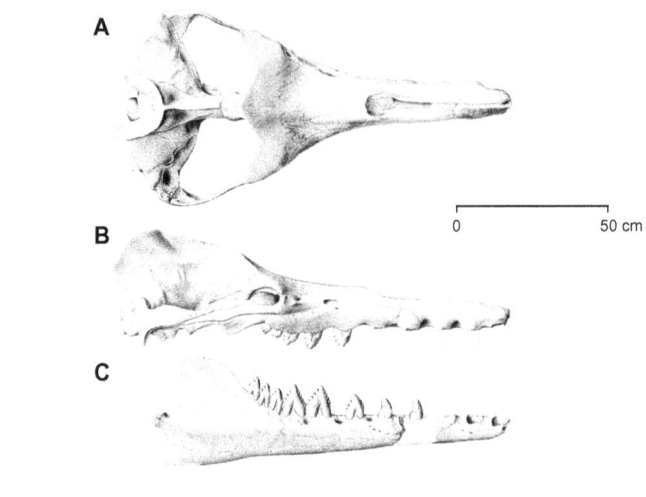

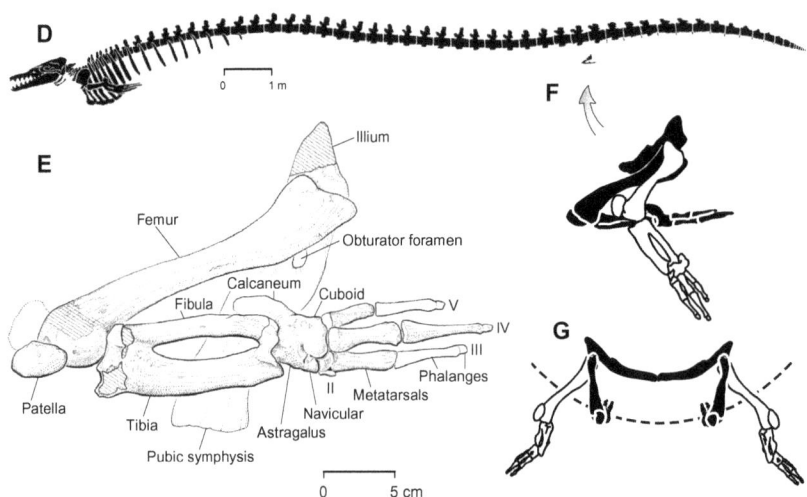

FIGURE 45.8   Skull of early Priabonian, late Eocene, *Basilosaurus isis*. Cranium in dorsal (A) and lateral (B) views (Staatliches Museum für Naturkunde, Stuttgart, 11787, St-9). C) Dentary in lateral view (Paläontologische Museum, Munich, BSPM, Mn-14, destroyed). These were first described and illustrated as *Zeuglodon isis* Beadnell by Stromer (1908; see also Heizmann, 1991). D) Graphical reconstruction of skeleton of *B. isis*, adapted from Gidley's (1913) reconstruction of *B. cetoides*, with the vertebral count taken from specimens in the field in Wadi Hitan (Egypt). E) Articulated hindlimb and foot of *B. isis* (based on CGM 42176 and UM 93231). Articulated hind limb and foot in left lateral (F) and anterior (G) views. Hindlimb has two positions, folded against the body wall (black) and extended (white). Hindlimbs and feet described in Gingerich et al. 1990 (©American Association for Advancement of Science; used by permission).

We were able to determine that one archaeocete is common in the middle part of the Qasr el-Sagha Formation on the Qasr el-Sagha escarpment. This is a small *'Zeuglodon'* for which the correct name is *Saghacetus osiris* (Dames). It is known from skulls, brain endocasts, and partial skeletons, and it differs postcranially from other small archaeocetes in having relatively longer posterior lumbar and anterior caudal vertebrae (Gingerich, 1992). Three Qasr el-Sagha species are synonyms of *S. osiris*: *Z. zitteli*, *Z. sensitivus*, and *Z. elliotsmithii*. In addition, there is at least one larger but still relatively small archaeocete from the middle Qasr el-Sagha escarpment: *Stromerius nidensis* Gingerich (2007).

Batik and Fejfar in Tunisia

Batik and Fejfar (1990) described an archaecete skull and partial vertebral column from Oued Kouki (site 10 in figure 45.3), some 65 km west-southwest of Kairouan in Tunisia. This was

found in a hard, compact shell bed and never extracted or collected. My own attempt to relocate the specimen in the field in 1997 failed. The age, based on ostracods, is interpreted as 'upper Lutetian,' which would be Bartonian, late middle Eocene on the current timescale. From the published description it could be a protocetid or a basilosaurid. It was intermediate in size between *Saghacetus osiris* and *Dorudon atrox*, but little more can be said about it.

Bianucci and an Archaeocete from Egypt

An interesting archaeocete was discovered recently when a block of marbleized limestone was imported to Italy from Egypt and cut into slabs for use as facing stone. The block contained an archaeocete skeleton, which was identified by Giovanni Bianucci (Bianucci et al., 2003). This was published as coming from Shaikh Fadl in Egypt, but further investigation showed that it came from the Gebel Hof Formation near Khashm

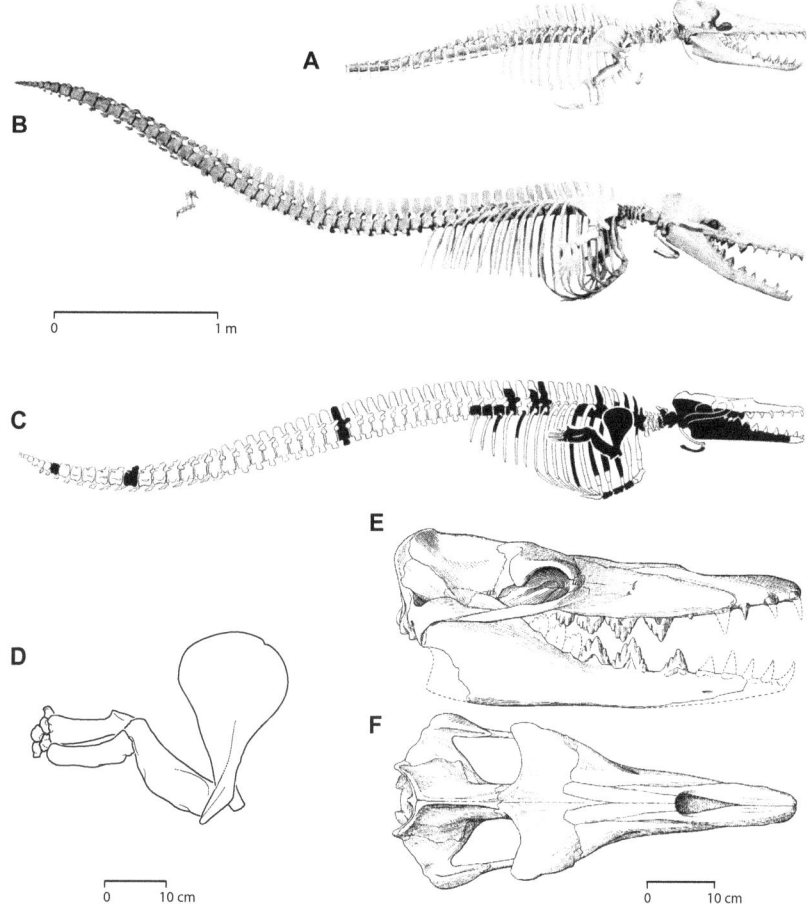

FIGURE 45.9 Comparison of skeletons of Egyptian late Eocene basilosaurids with short vertebral centra. A) Early attempt to reconstruct a composite skeleton of Priabonian *Zeuglodon osiris*
or *Dorudon stromeri* (Paläontologische Museum, Munich, BSPM, 1904 XII 134e, 'MN 9,' type of
D. stromeri, now destroyed; and Staatliches Museum für Naturkunde, Stuttgart, SMNS, 11237a,
'ST 11,' now placed in *Saghacetus osiris*). Skeleton was first described and illustrated as *Z. osiris*
by Stromer (1908), and then the skull and anterior skeleton were made the type of *D. stromeri*
by Kellogg (1928). B) Virtually complete skeleton of early Priabonian *Dorudon atrox* (Andrews,
1906) based on UM 101222 and 101225 described by Uhen (2004). Photograph shows the skeleton as mounted in the University of Michigan Exhibit Museum. C) Partial skeleton of early
Priabonian *Ancalecetus simonsi* (Egyptian Geological Museum, Cairo, CGM 42290). This was
described and named by Gingerich and Uhen (1996). Known elements are shown in black,
superimposed on an outline of the skeleton of *Dorudon atrox*. D) Right scapula and forelimb of
Ancalecetus simonsi (CGM 42290), showing immobile elbow characteristic of this genus and
species. Cranium of immature type specimen of *Dorudon atrox* (CGM 9319), in lateral (E) and
dorsal (F) views. Teeth shaded here are deciduous except for M1. Approximately one-half of
the specimens of *D. atrox* found in Wadi Hitan are immature. Mature specimens of *D. atrox*,
like UM 101222 and 101225, have vertebrae of distinctly smaller size and different proportion
than those of contemporary *Basilosaurus isis* (figure 45.8D).

el-Raqaba, north of Wadi Tarfa in the Eastern Desert of Egypt
(site 11 in figure 45.3 and 45.4). This is early Bartonian in
age, equivalent to the Sath el-Hadid and Guishi formations
elsewhere in Egypt (Gingerich et al., 2007).

New Sites in Egypt

In recent years several new areas have been found to yield
archaeocete skeletons in Egypt. These include (1) Qusour el
Arab in western Fayum Province (site 4 in figure 45.5) yielding
early middle Eocene protocetid archaeocete skulls and skeletons with limb bones from the Midawara Formation; (2) an
area north of Garet Gehannam (site 12 in figure 45.5) yielding
skeletons of late middle Eocene protocetids and basilosaurids

from the lower part of the Gehannam Formation; (3) Qattara
Depression (site 26 in figure 45.4) yielding vertebrae of late
Eocene *Masracetus markgrafi* (David Morrison, pers. comm.,
2006); and (4) Tabaghbagh near Siwa in the Western Desert of
Egypt (site 27 in figure 45.4) yielding numerous skeletons of
late Eocene *Basilosaurus isis* and other archaeocetes (H. J. van
Vliet and T. Paymans, pers. comm., 2008).

Neogene Mysticeti and Odontoceti

The African fossil record of Mysticeti and Odontoceti is not as
rich as that of Archaeoceti, but there are nevertheless important records of both modern suborders that are of interest.
These have generally been added to museum collections

incrementally and opportunistically rather than resulting from directed expeditions, and most represent taxa compared to living cetaceans rather than new taxa. Hence it makes sense to summarize the Neogene finds stratigraphically.

MIOCENE

Early Miocene cetaceans are known from two localities, Wadi Faregh and Wadi Moghara in Egypt (sites 36–37 in figure 45.4). The first is a fragmentary dentary described by Stromer (1907) and referred to *Schizodelphis* aff. *sulcatus*. Fourtau (1920) described two cetaceans from Wadi Moghara, both fragmentary dentaries. One specimen Fourtau, like Stromer, referred to '*Cyrtodelphis*' aff. *sulcatus*, and the other he named '*Delphinus*' *vanzelleri*.

Hamilton (1973) described fossils in the NHML collection purchased from a Lady Moon and said to have come from Siwa Oasis. This included a dentary fragment Hamilton identified as *Schizodelphis* aff. *S. sulcatus*. The 'Siwa' fossils almost certainly came from Wadi Moghara (E. Miller, pers. comm.). *Schizodelphis* is now included in Eurhinodelphinidae (Lambert, 2005).

Middle Miocene cetaceans are known from two specimens from two sites on the African continent. The first specimen, an uninformative vertebral fragment, was described by Dartevelle (1935) in a report on Miocene mammals of Malembe in the Cabinda Enclave, Angola (site 38 in figure 45.3). The second specimen, from Loperot in Kenya (site 39 in figure 45.3), is a weathered ziphiid rostrum described by Mead (1975). This was found on a 1964 Harvard University expedition to northern Kenya led by Bryan Patterson. The specimen represents a normally open-ocean form, here found far inland in freshwater deposits. It is consequently interpreted as having strayed up a large river (Mead, 1975).

Late Miocene cetaceans are known from many sites. The first to be discovered was found in 1937 by Carlo Petrocchi at Sahabi in Libya. Petrocchi (1941) reported a skeleton of a *"cetaceo quasi completo"* from Sahabi in Libya (site 40 in figure 45.3). This was subsequently illustrated in Petrocchi (1943, 1951), and identified as *"balenottera"* (rorqual *Balaenoptera*), but the specimen was never identified to species or described in detail. It came from locality P8 in Member U-2 of the Sahabi Formation (Heinzelin and Arnauti, 1987). This is presumably the skeleton on display in the museum in Tripoli.

Noel Boaz (1980) described a bone from locality P4A in Member T of the Sahabi Formation as a hominoid clavicle. This was reinterpreted by White et al. (1983) as the rib of an odontocete the size of *Tursiops* or *Lagenorhynchus*. Whitmore (1982, 1987) reported a periotic from P8A that he referred to the family Delphinidae as cf. *Lagenorhynchus*, and then skull fragments and vertebrae from locality P4A that he referred to the family Platanistidae, subfamily Iniinae (Whitmore, 1987).

Raz-el-Ain, near Oran in Algeria (site 42 in figure 45.3), is a late Miocene site that has yielded a whale. Muizon (1981) mentions that teeth and vertebrae of a physeterid were found here.

Additional late Miocene sites are known from the west coast of Namaqualand in northern Cape Province, South Africa (Pickford and Senut, 1997). These have yielded vertebrae, ribs, and teeth of unidentified cetaceans.

MIOCENE-PLIOCENE

The largest collection of Miocene-Pliocene African cetaceans is that dredged from the sea bottom at numerous sites off the western and southern shores of South Africa. The material was first mentioned by Keppel H. Barnard (1954), who listed a number of cetaceans found in phosphatic and glauconitic nodules on the floor of the sea of the Agulhas Bank off the south coast of South Africa (site 46 in figure 45.3). These included the balaenid *Eubalaena* sp., balaenopterid *Balaenoptera physalis*, delphinid cf. *Orcinus orca*, and ziphiids cf. *Ziphius cavirostris* and *Mesoplodon* sp.

Sidney H. Haughton (1956) listed additional specimens identified as *Balaena* sp., *Balaenoptera* sp., *Megaptera* sp., *Orca* sp., *Mesoplodon* spp., and *Ziphius*? sp. from the west coast of Cape Province, South Africa (see also Ross, 1986).

The first intensive study of this material, by Bianucci et al. (2007, 2008), involved strap-toothed cetaceans or Ziphiidae. Eight genera and 10 species were described as new: *Microberardius africanus*, *Izikoziphius rossi*, *Izikoziphius angustus*, *Khoikhoicetus agulhasis*, *Ihlengesi saldanhae*, *Africanacetus ceratopsis*, *Mesoplodon slangkopi*, *Nenga meganasalis*, *Xhosacetus hendeysi*, and *Pterocetus benguelae*. These came from five areas, some broadly defined, that border South Africa: (1) Agulhas Bank offshore from Cape Agulhas; (2) off the west coast of Cape Province north and west of Cape Agulhas; (3) off Cape Columbine on the west coast of South Africa; (4) southwest of Cape Town on the west coast of South Africa; and (5) off the south coast of South Africa east of Agulhas Bank.

Haughton (1956) considered material dredged offshore of South Africa to represent a composite geological history, with nodular fragments themselves being older, possibly early Miocene, and glauconitized specimens being younger. He included cetacean fossils in the later glauconitic greensand phase. From their preservation, Bianucci et al. (2007) considered that the ziphiid skulls they described to have undergone at least one phase of reworking on the seafloor. Bianucci et al. concluded that no precise age can be given for any of the skulls, and attributed all to a Miocene-Pliocene range of ages (followed here).

PLIOCENE

The principal site yielding cetaceans known to be Pliocene is Langebaanweg in South Africa (site 51 in figure 45.3). Here Hendey (1976, 1981, 1982) reported early Pliocene mysticetes and odontocetes, both being represented by largely undiagnostic postcranial bones. These specimens have never been studied.

Another Pliocene site yielding cetaceans is at the cave site of Ahl al Oughlam, in Morocco (site 52 in figure 45.3), where Geraads et al. (1998) indicated that small cetaceans, probably stranded, were scavenged and brought into the caves by carnivores. Two cetaceans are represented, one a delphinid, *Delphinus* or *Stenella* sp., represented by a dozen specimens including petrosals and fragments of jaws; and the other a pygmy sperm whale *Kogia* sp., represented by a petrosal.

PLEISTOCENE

The only early Pleistocene fossil cetacean known from Africa is the almost complete skeleton of a large rorqual (*Balaenoptera* sp.) found closely associated with lower Palaeolithic artefacts near Baia Farta near Benguela in Angola (site 53 in figure 45.3), 65 m above sea level and 3 km from the present shoreline (Gutiérrez et al., 2001). It is reported to represent the oldest evidence of human exploitation of a stranded whale.

Late Pleistocene fossil cetaceans are represented by fragmentary remains of delphinids and indeterminate whales in archaeological sites at Klasies River Mouth and Sea Harvest in South Africa (Klein, 1976; Grine and Klein, 1993).

Edwin L. Gill, director of the South African Museum in Cape-town, was the first to report an African Neogene whale (Gill, 1928). This was a skeleton of a rorqual the size of a full-grown sei whale *Balaenoptera borealis* that was discovered at Yzer-plaats near Capetown in South Africa in 1927. It was found in the middle of an extensive deposit of pure clay several meters thick, on a raised beach of Holocene age.

Charon et al. (1973) and Saban (1974) reported a single tooth of a sperm whale, *Physeter macrocephalis*, from a Neolithic archaeological site, site H, near Tarfaya in Morocco (site 59 in figure 45.3). Additional remains of Holocene delphinids and indeterminate cetaceans are included in reports on archaeological sites at Nelson Bay Cave, Kasteelberg B, Die Kelders Cave, and Elands Bay Cave in South Africa (Klein, 1972; Klein and Cruz-Uribe, 1989; Grine et al., 1991; and Grine and Klein, 1993).

Systematic Paleontology

More than half of all published records of Cetacea in the fossil record of Africa are archaeocetes classified in the family Basilosauridae. The greatest number of mysticetes known to date belong in Balaenopteridae, and the greatest number of odontocetes are Ziphiidae. The dominance of Basilosauridae is a result of favorable geological conditions in terms of a passive continental margin subsiding and accumulating sediment through the late Eocene. It is also a result of both favorable living conditions for cetaceans on the resulting shallow marine shelf, and excellent exposure of fossil beds today in relatively accessible desert in Fayum Province, Egypt. When other areas receive the same attention, it is likely that they will prove to be productive as well, and younger deposits are sure to yield a better representation of mysticetes and odontocetes than that known today.

Higher taxa are listed here from older and more primitive to younger and more derived. Where several within a family appear equally derived, genera and species are discussed in the order in which they were described.

Order CETACEA Brisson, 1762
Suborder ARCHAEOCETI Flower, 1883

Archaoceti are early, primitive cetaceans that retain generalized mammalian skulls with nares on the rostrum, cheek teeth with complex crowns and multiple roots, and periotics attached to surrounding bones. Skeletons range from semiaquatic to fully aquatic, with fore- and hindlimbs variably modified for swimming. The elbow joint is generally mobile. Early semiaquatic forms have skeletons with double-pulley astragali and other features indicating evolutionary derivation from early Artiodactyla (Gingerich et al., 2001).

Family PROTOCETIDAE Stromer, 1908

Protocetidae are early semiaquatic archaeocetes that retain the full complement of three molar teeth in both upper and lower jaws, but lack the anteriorly developed pterygoid sinuses seen in Basilosauridae. Protocetids known from articulated skeletons have 7 cervical vertebrae, 13 thoracics, 6 lumbars, 4 sacrals, and 21 caudals (Gingerich et al. 2009; the number of sacrals is sometimes reduced to 1). The sacrum retains auricular processes for articulation with innominates of more or less normal proportions for a generalized mammal.

Genus *PROTOCETUS* Fraas, 1904
PROTOCETUS ATAVUS Fraas, 1904
Figure 45.7

Age and Occurrence Lutetian (early middle Eocene) of Schweinfurth's locality XII in the Mokattam Formation of Gebel Mokattam near Cairo (site 1 in figures 45.3 and 45.4).

Diagnosis When *Protocetus atavus* was first described, two of the principal differences distinguishing it from later archaeocetes were (1) retention of three molars in left and right maxillae, and (2) retention of more primitive upper molars having three roots, distinct paracone and metacone cusps, and a lingual swelling of the crown reminiscent of a protocone. The postcranial skeleton was recognized to be primitive in retaining cervical centra longer than those of later basilosaurids, and in retaining a well-developed auricular process on a single centrum interpreted as the sacrum. Now the combination of relatively small size, narrow cranial rostrum, P1 apparently double rooted, and molars lacking a distinct protocone cusp distinguishes *Protocetus* from other protocetids, but interpretation of the sacrum as consisting of a single centrum is dubius (discussed later).

Description The holotype of *Protocetus atavus*, SMNS 11084-11086, includes an incomplete but well-preserved cranium (figures 45.7A, 45.7B) and a series of associated vertebrae including part of the sacrum (figures 45.7C, 45.7D). The cranium has an anteroposteriorly elongated rostrum; external nares opening above P1; a broad frontal shield; a relatively long, posteriorly extended nuchal crest; broadly flaring squamosals enclosing large temporal fossae; and somewhat enlarged and thickened tympanic bullae. The latter cover little-developed pterygoid sinuses. The anterior position of the external nares is sometimes cited as a distinctively primitive characteristic of *Protocetus* and other protocetids, but here the nares open over P1 as they do in basilosaurids (and many land mammals). *Protocetus* retains three molars in left and right maxillae, and retains more primitive upper molars with three roots, distinct paracone and metacone cusps, and a lingual swelling of the crown reminiscent of a protocone. Cervical centra are longer relative to their diameter than those of most later cetaceans, with centrum length being about two-thirds of centrum width and height. Anterior thoracics have long, posteriorly inclined neural spines, the 12th thoracic appears to be anticlinal, and anterior lumbars have relatively short anteriorly inclined neural spines. Lumbar vertebrae are on the order of 4.0–4.8 cm long. The sacrum was originally interpreted to comprise a single centrum with transversely expanded auricular processes, surmounted by a distinctively bifurcated neural spine (Fraas, 1904a). Stromer (1908) attributed the bifurcated neural spine to breakage and faulty reconstruction. My own comparisons indicate that the inferred reduction of the sacrum to comprise a single centrum is also an artifact of breakage (dashed lines in figure 45.7D).

Remarks Protocetus atavus was for many years the oldest and most primitive archaeocete known. This is no longer true, as older and more primitive cetaceans have been identified in Pakistan and India (Gingerich et al., 1983; Bajpai and Gingerich, 1998). *Protocetus* has also been supplanted as the best known protocetid by new specimens of *Rodhocetus* and related forms from Pakistan (Gingerich et al., 2001, 2009).

Genus *EOCETUS* Fraas, 1904
EOCETUS SCHWEINFURTHI (Fraas, 1904)

Age and Occurrence Bartonian (late middle Eocene) of Schweinfurth's locality IV in the Giushi Formation of Gebel Mokattam near Cairo (site 7 near site 1 in figures 45.3 and 45.4).

Diagnosis The type skull of *Eocetus schweinfurthi* cannot be identified with certainty to Protocetidae or Basilosauridae because it lacks diagnostic characteristics. As presently conceived (Uhen, 1999, 2001), *Eocetus* is a protocetid with distinctively elongated, pachyostotic vertebrae.

Description The type specimen is SMNS 10986, a poorly preserved cranium that has not been fully prepared. This has the rostrum better preserved than the rest of the skull. Crowns of I1 and I2 are present. These are conical and simple, with I1 being smaller and I2 larger in size. Alveoli for I3 indicate that it was small like I1, while the single-rooted C1 was large like I2. Part of the crown of P2 is preserved, and this has a well developed posterobasal cusp. The P4 in the type skull has a large posteromedial buttress, like a deciduous tooth, and it may even be three rooted. This is unlike the tooth referred to '*Mesocetus*' as a P4 by Fraas (1904a: 10, plate 2). External nares open above P1 in the type skull.

Remarks *Eocetus schweinfurthi* was first published as *Mesocetus schweinfurthi* by Fraas (1904a) and then moved to the new genus *Eocetus* by Fraas (1904b) when he realized that *Mesocetus* had been used previously.

As mentioned, the type skull of *E. schweinfurthi* cannot be identified with certainty as either protocetid or basilosaurid. *Eocetus* is listed here as a protocetid because this is where it is usually classified, but evidence for this is weak. Inference that *Eocetus* had three molars (and hence is protocetid) is based on identification of an isolated tooth as M3 (Fraas, 1904a:11, plate 2), but its very isolation means that identification as M3 is uncertain, as is its attribution to *E. schweinfurthi*.

Stromer (1903) described two large vertebrae (SMNS 10934) and then an additional two vertebrae (SFNF 4470; Stromer, 1908), all thought to come from the same stratigraphic interval as the type of *Eocetus schweinfurthi*. Uhen (1998) interpreted the former pair to represent *Basilosaurus drazindai*, and attributed the latter pair to *E. schweinfurthi* (Uhen, 1999). This is a reasonable inference, but it also assumes that the type skull of *Eocetus* is protocetid rather than basilosaurid. A clear determination will have to await discovery of skeletons with associated skulls and skeletons.

Genus *PAPPOCETUS* Andrews, 1920
PAPPOCETUS LUGARDI Andrews, 1920

Age and Occurrence Middle Eocene (Lutetian in the old sense, almost certainly Bartonian on a modern time scale). Type was found in the Ameki Formation 2 km southeast of Ameke (Ameki) in Nigeria (site 8 in figure 45.3).

Diagnosis The type and referred dentaries, both subadult, indicate a large archaeocete, much larger than *Protocetus* and most other protocetids. The type has portions of left and right dentaries joined so perfectly as to suggest fusion of the mandibular symphysis (Andrews calls these "united"; Andrews, 1920:309). However the referred specimen, ontogenetically older than the type, appears to indicate the opposite. *Pappocetus* is so poorly known that it cannot be clearly distinguished from other protocetids.

Description The holotype of *Pappocetus lugardi*, NHML 11414, includes much of a left dentary with crowns of deciduous premolars, the crown of a left m1 fully erupted, and m2 partially erupted. The right dentary, seemingly fused to the left at the mandibular symphysis, includes two deciduous premolars. The referred dentary, NHML 11086, is a left lower jaw with alveoli and/or broken roots for di1–3, dc1, a single-rooted dp1 or p1, double-rooted dp2–4, much of the crown of m1, a partial crown of m2 nearly fully erupted, and m3 partially erupted. This has the mandibular symphysis well preserved and open (unfused). Molars are typically protocetid, with a large trigonid cusp (protoconid) seemingly lacking accessory trigonid cusps, and a prominent laterally compressed talonid cusp (hypoconid). The anterior base of each lower molar crown is flattened and very slightly indented to accommodate the posterior base of the preceding tooth, while the posterior base of each crown is rounded where it fit into the indentation of the following tooth. A faint cingulid surrounds the entire crown. The molars are like those of protocetids rather than basilosaurids.

Andrews (1920) described an axis vertebra of *Pappocetus lugardi* (C2), and Halstead and Middleton (1974) described six vertebrae collected more recently. All are from subadult animals, and possibly part of the holotype. The most interesting are in a block consisting of three that are probably T13-L1-L2 or T12-T13-L1. All have centra that are short anteroposteriorly for their diameter (width and height). Anterior lumbar vertebrae are on the order of 4.8–5.6 cm long and have centra with length-to-width and length-to-height ratios of about 0.63 and 0.91, respectively.

Remarks African *Pappocetus* is much larger but, considering what is known of the dentaries and teeth, possibly most like South Asian *Babiacetus* (Gingerich et al., 1995).

PROTOCETIDAE indet.

Additional Protocetidae are known from several other African sites. A possible protocetid is known from a locality that Stromer (1908) first called Uadi Rojan and later changed to Klein Rajan (Stromer, 1914). Protocetids are known from the phosphate mine at Kpogamé in Togo (Gingerich et al., 1992), and a *Babiacetus*-like protocetid is known from Khashm el-Raqaba (Bianucci et al., 2003; Gingerich et al., 2007). Protocetids have been found in recent years in the Midawara Formation near Qusour el Arab, and in the Gehannam Formation near Garet Gehannam in western Fayum Province, Egypt.

Family BASILOSAURIDAE Stromer, 1908

Basilosauridae are generally later, more derived, and fully aquatic archaeocetes with only two molar teeth in the maxilla. Basilosaurids generally lack evidence of a distinct sacrum in the vertebral column and have innominates of reduced size and altered proportions. It is useful to separate *Basilosaurus* with extreme vertebral elongation from more typical basilosaurids. The former are sometimes classified in Basilosaurinae and the latter in Dorudontinae. However, the relative size of carpal bones and other differences cut across this simple dichotomy, and a better understanding of basilosaurid morphology in more taxa will be required before reliable subdivision will be possible.

Six genera and species of Basilosauridae are known from the African late Eocene, and all were described from Fayum Province, Egypt. It is useful to consider these in terms of overall size, and on a spectrum of relative length of lumbar vertebral centra relative to their width and height. *Saghacetus osiris* is the smallest of the Fayum basilosaurids, *Stromerius nidensis* is next, *Dorudon atrox* and *Ancalecetus simonsi* are medium sized, *Masracetus markgrafi* is larger, and *Basilosaurus isis* is largest. In terms of shape, *Masracetus markgrafi* has the anteroposteriorly shortest lumbar centra compared to their diameter, *Dorudon atrox* and *Ancalecetus simonsi* are next, *Stromerius nidensis* has more or less equidimensional centra, *Saghacetus osiris* has centra longer than their diameter, and *Basilosaurus isis* has the longest centra. Lumbar centrum size and shape vary independently, and they are thus useful for distinguishing taxa.

Genus *SAGHACETUS* Gingerich, 1992
SAGHACETUS OSIRIS (Dames, 1894)
Figure 45.6

Age and Occurrence Middle Priabonian (middle late Eocene) of the middle Qasr el-Sagha Formation, northern Fayum, Egypt. Type was found at site 28 in figure 45.5, but the species is common from this level the whole length of the Qasr el-Sagha escarpment.

Diagnosis Saghacetus osiris is distinctive as the smallest of the Fayum archaeocetes. Posterior thoracic and anterior lumbar vertebrae are relatively long in comparison to centrum height and width (but these vertebrae are not nearly so long as those of *Basilosaurus*). *Saghacetus* differs from *Dorudon* and *Basilosaurus* in having only 14 thoracic vertebrae.

Description Many specimens of *Saghacetus osiris* are known, but most are skulls and jaws, and few include good vertebral series. Much of the skeleton is not yet known, and this includes the full lumbus, cauda, forelimb, and hindlimb. The holotype, MNB Ma.28388, is the well-preserved dentary described by Dames (1894; figure 45.6). Three of the best-preserved skulls are BSPM 1902.XI.59 (Mn. 1) described by Stromer (1902, 1903); SMNS 11786 (St. 3; juv.) described by Stromer (1908), and AMNH 14382 described by Kellogg (1936). The best endocast is NHML 12123, type of *Zeuglodon sensitivus* Dart (1923). UM 83905 and 97550 are good axial skeletons, with crania, a complete series of cervical and thoracic vertebrae, ribs, and several associated lumbar and caudal vertebrae.

The cranium of *Saghacetus osiris* is like that of other basilosaurids in having a dental formula of 3.1.4.2/3.1.4.3 (Stromer, 1903). Well-developed pterygoid sinuses separate left and right middle ears in the basicranium (Pompeckj, 1922; Luo and Gingerich, 1999). There are 7 cervical vertebrae and 14 thoracics (UM 83905 and 97550), but the number of lumbars and caudals is not known.

Crania of *Saghacetus osiris* average about 67 cm in condylobasal length. If we take the natural logarithm of this length and make the common assumption of a 0.05-unit standard deviation on a natural-log scale, then, when exponentiated, the expected range of condylobasal length is 61–74 cm. Middle lumbar vertebrae are on the order of 5.2–5.3 cm long and have centra with length-to-width and length-to-height ratios of about 1.10 and 1.19, respectively. Body weight is estimated to have been 350 kg and brain weight 388 g, yielding an encephalization quotient of 0.49 by comparison to a baseline of terrestrial mammals as a class (Gingerich, 1998; meaning brain size is about one-half that expected for an extant terrestrial mammal of the same body weight).

Remarks Saghacetus osiris is the common archaeocete in the much-collected Temple Member in the middle part of the Qasr el-Sagha Formation, Fayum Province, Egypt. Several species described from this interval are synonyms: *Zeuglodon zitteli* Stromer (1903); *Zeuglodon sensitivus* Dart (1923), and *Zeuglodon elliotsmithii* Dart (1923).

The specimen described as *"Dorudon osiris"* and said to have come from Wadi Natrun in northern Egypt (Pilleri, 1985) is almost certainly a specimen of *Dorudon atrox* from an unknown locality in Fayum (Gingerich, 1991).

Genus *BASILOSAURUS* Harlan, 1834
BASILOSAURUS ISIS (Beadnell in Andrews, 1904)
Figure 45.8

Age and Occurrence Late Bartonian (late middle Eocene) and early Priabonian (early late Eocene) of the Gehannam and Birket Qarun formations, northern Fayum, Egypt. Type was found somewhere near site 15 in figure 45.5. The species is found at several sites north of Birket Qarun but is best known from Wadi Hitan (site 18 in figure 45.5).

Diagnosis Basilosaurus isis is distinctive as the largest of the Fayum archaeocetes. Posterior thoracic and anterior lumbar vertebrae are very long in comparison to centrum height and width (longer than those of any other Fayum archaeocete). *Basilosaurus* resembles *Dorudon* and differs from *Saghacetus* and *Stromerius* in having a larger number of thoracic vertebrae (18 in *Basilosaurus*).

Description Here again, many specimens of *Basilosaurus isis* are known, but those in collections are almost all skulls and jaws, and only two include reasonable vertebral series. Virtually the entire skeleton has been observed in the field, and the full vertebral series is known. Forelimbs including hands have been collected (Gingerich and Smith, 1990), and the pelvis and hindlimb are virtually completely known (figure 45.8; Gingerich et al., 1990). The holotype, CGM 10208, is the well-preserved dentary described by Andrews (1906). Three of the best-preserved skulls are SMNS 11787 (St. 9) described by Stromer (1908), CGM 42225 (not prepared), and UM 97507. The best endocast is UM 94800. SMNS 11787 and UM 97503 each include much of a vertebral column (the former being that described by Slijper, 1936).

The cranium of *Basilosaurus isis* is like that of other basilosaurids in having a dental formula of 3.1.4.2/3.1.4.3 (Stromer, 1908). Pterygoid sinuses are well developed in the basicranium (Luo and Gingerich, 1999). The number of vertebrae in a *Basilosaurus isis* skeleton is 7 cervicals, 16 thoracics, ca. 23 lumbars (including sacral homologues), and ca. 21 caudals, for a total of ca. 67 vertebrae; and the total length of the skeleton is approximately 18 m (Gingerich et al., 1990).

Crania of *Basilosaurus isis* average about 120 cm in condylobasal length, with an expected range of about 109–133 cm. Middle lumbar vertebrae are on the order of 33 cm long and have centra with length-to-width and length-to-height ratios of about 1.57 and 1.83, respectively. Body weight is estimated to have been 6,480 kg and brain weight 2,520 g, yielding an encephalization quotient of 0.37 by comparison to a baseline of terrestrial mammals as a class (Gingerich, 1998; meaning brain size is a little more than one-third that expected for an extant terrestrial mammal of the same body weight).

Remarks Basilosaurus isis is the common larger archaeocete in the Gehannam and Birket Qarun formations of northern and western Fayum Province, Egypt. Several authors (Kellogg, 1936; Barnes and Mitchell, 1978) recognized that the type specimen of *Prozeuglodon atrox* Andrews (1906) is a subadult and interpreted it as a young individual of '*Zeuglodon*' *isis*. However, new collections made in Egypt in recent years include adults of numerous individuals of both species, which are distinct, and it is clear that *Zeuglodon isis* Beadnell in Andrews (1904) belongs in *Basilosaurus*. *Prozeuglodon atrox* Andrews (1906) is a species of *Dorudon*.

Genus *DORUDON* Gibbs, 1845
DORUDON ATROX (Andrews, 1906)
Figures 45.5, 45.9B, 45.9E, and 45.9 F

Age and Occurrence Late Bartonian (late middle Eocene) and early Priabonian (early late Eocene) of the Gehannam and Birket Qarun formations, northern Fayum, Egypt. Type was found at Wadi Hitan (site 18 in figure 45.5), where it is common.

Diagnosis Dorudon atrox is distinctive as a medium-sized Fayum archaeocete with blocky posterior thoracic, lumbar, and anterior caudal vertebrae. Posterior thoracic and anterior lumbar vertebrae have centra that are approximately equal in length to their height and width. *Dorudon* resembles *Basilosaurus* and differs from *Saghacetus* and *Stromerius* in having a larger number of thoracic vertebrae (17 in *Dorudon*).

Description Dorudon atrox is the best-known species morphologically of all basilosaurids. It is represented by several exceptionally complete skeletons described by Uhen (2004; figure 45.9B). The holotype, CGM 9319, is the subadult skull described by Andrews (1906). Three of the best-preserved skulls are SFNF 4451, CGM 42183, and UM 101222. A well-preserved endocast is UM 94795. CGM 42183 and UM 101222 each include much of a vertebral column.

The cranium of *Dorudon atrox* is like that of other basilosaurids in having a dental formula of 3.1.4.2/3.1.4.3 (Uhen, 2004). Pterygoid sinuses are well developed in the basicranium (Pompeckj,1922; Luo and Gingerich, 1999). The number of vertebrae in a *Dorudon atrox* skeleton is 7 cervicals, 17 thoracics, 20 lumbars, and 21 caudals, for a total of ca. 65 vertebrae; the total length of the skeleton is approximately 5 m (Uhen, 2004).

Crania of *Dorudon atrox* average about 95 cm in condylobasal length, with an expected range of about 84–105 cm. Middle lumbar vertebrae are on the order of 7.6 cm long and have centra with length-to-width and length-to-height ratios of about 0.79 and 0.88, respectively (Uhen, 2004). Body weight is estimated to have been 1,140 kg and brain weight 960 g, yielding an encephalization quotient of 0.51 by comparison to a baseline of terrestrial mammals as a class (Gingerich, 1998; meaning brain size is close to one-half that expected for an extant terrestrial mammal of the same body weight).

Remarks Dorudon atrox is the common smaller archaeocete in the Gehannam and Birket Qarun formations of northern and western Fayum Province, Egypt. The type is a subadult with deciduous teeth (figures 45.9E, 45.9F), as are many specimens of this species referred to other taxa by Stromer (1908) and Kellogg (1936). Many described skulls with a condylobasal length in the range of 65–85 cm are subadult *D. atrox*. Two named species are synonyms of *D. atrox*: *Zeuglodon intermedius* Dart (1923) and *Prozeuglodon stromeri* Kellogg (1928; see Uhen, 2004:14).

Genus *ANCALECETUS* Gingerich and Uhen, 1996
ANCALECETUS SIMONSI Gingerich and Uhen, 1996
Figures 45.9C and 45.9D

Age and Occurrence Early Priabonian (early late Eocene) of the Birket Qarun Formation, northern Fayum, Egypt. Type was found at site 25 near site 18 in figure 45.5.

Diagnosis Ancalecetus simonsi is similar to *Dorudon atrox* in size and the form of its cranium and vertebrae. It differs conspicuously from *D. atrox* and all other basilosaurids in having anteroposteriorly narrow scapulae, very limited mobility of the humerus relative to the scapula at the shoulder joint, and fusion of the humerus, ulna, and radius at the elbow joint. Carpal bones of the wrist are small like those of *Zygorhiza*, but *Ancalecetus* differs from *Zygorhiza* in having the magnum and trapezoid conjoined as a single bone.

Description Ancalecetus simonsi is known from a single partial skeleton, the type CGM 42290 (Gingerich and Uhen, 1996; figure 45.9C). The basicranium is preserved, but the top of the braincase and anterior parts of the cranium were missing when the skull was collected. Pterygoid sinuses are well developed in the basicranium (Luo and Gingerich, 1999). Several vertebrae of *A. simonsi* are well preserved, but the number of vertebrae, conformation of the vertebral column as a whole, and total length are not known. The weight of the animal in life was probably about 1,140 kg, based on similarity in preserved parts to those of *D. atrox*, and we can imagine that relative brain size was similar as well.

The most distinctive characteristics of *Ancalecetus simonsi* are in its unusual forelimbs (figure 45.9D). The scapula is primitively narrow anteroposteriorly; the scapulohumeral articulation is mobile but greatly restricted in mobility compared to the shoulder of other archaeocetes; and the elbow joint has lost its mobility entirely, with the distal humerus, radius, and ulna forming a tight-fitting joint. The carpals are small compared to those of *Dorudon*, but the rest of the hand has not been recovered.

Remarks From a functional point of view, it is difficult to comprehend how a basilosaurid like *Ancalecetus simonsi* could have been an active swimmer without having had greater mobility of the forelimbs for stabilization and guidance. However, well-developed masticatory wear on teeth, and fusion of cranial bones and vertebral epiphyses, show that the one known individual of *A. simonsi* lived to adulthood. Hence, however poorly we understand its forelimb function, *Ancalecetus* appears to have represented a viable evolutionary experiment.

Genus *STROMERIUS* Gingerich, 2007
STROMERIUS NIDENSIS Gingerich, 2007

Age and Occurrence Middle Priabonian (middle late Eocene) of the middle Qasr el-Sagha Formation, northern Fayum, Egypt. Type was found at Garet el-Esh (site 35 near site 28 in figure 45.5).

Diagnosis The most distinctive feature of *Stromerius nidensis* is the presence of unusually long, anteriorly directed metapophyses on lumbar vertebrae. The penultimate thoracic vertebra has a vertical neural spine and appears to be anticlinal. The lumbus is short, comprising only 12 vertebrae, the last four of which are possibly identifiable as sacrals.

Description Stromerius nidensis is known from two specimens, one a group of associated vertebrae described by Stromer (1903), and the other, the type, UM 100140, a second group of associated vertebrae described by Gingerich (2007). The latter was found with a posteriormost thoracic and anterior lumbar neural spine showing on the surface. Tracing these into the outcrop yielded an articulated series of dorsal vertebrae comprising 11 additional lumbar vertebrae and 2 proximal caudals. The penultimate thoracic, described by Stromer (1903), has a vertical neural spine, while all of the more posterior neural spines, starting with the last thoracic, are anteriorly inclined. Many of the lumbars have distinctive, well-preserved, long, and anteriorly inclined metapophyses different from those described for other archaeocetes. Retention of transverse processes that enclose a pleurapophyseal space on lumbar vertebrae L9 and L10, and retention of posteroventral processes lacking chevron facets on L12 identify the last four lumbars as likely homologues of sacral vertebrae S1–S4 (morphologically the vertebrae are lumbars and they are not sacral in form).

Middle lumbar vertebrae of *Stromerius nidensis* are on the order of 6.1 cm long and have centra with length-to-width and length-to-height ratios of about 0.94 and 1.14, respectively

Remarks Stromer (1903) described the referred specimen, MNB 1902.XI.60a, as *Zeuglodon osiris*, and Kellogg (1936) identified this as *Dorudon zitteli*. *Stromerius nidensis* differs from both in having lumbar centra that are larger, longer, and at the same time both wider and less high relative to centrum length. *Stromerius nidensis* complements *Saghacetus osiris* in being the second archaeocete known from the Temple Member in the middle of the Qasr el-Sagha Formation.

Genus *MASRACETUS* Gingerich, 2007
MASRACETUS MARKGRAFI Gingerich, 2007

Age and Occurrence Early Priabonian (early late Eocene) of the Birket Qarun Formation, northern Fayum, Egypt. Type was found at or near Dimeh in northern Fayum (site 20 in figure 45.5).

Diagnosis *Masracetus markgrafi* is distinguished from other archaeocetes by the size and shape of its lumbar vertebrae. These are large, but relatively short compared to their width and height. Lumbar centra are nearly the diameter (width and height) of lumbar centra of *Basilosaurus isis*, but less than half the length of the latter. *Masracetus markgrafi* lumbar centra are similar to those of North American late Eocene '*Zeuglodon*' *brachyspondylus* Müller (1849: 26), but the centra are not as large and they also differ in being less wide relative to their height.

Description There are several good series of large but short lumbar vertebrae representing an archaeocete different from any described thus far in this chapter. The best of these is SMNS 11413 and 11414 (St. 8 of Stromer, 1908:129). Little can be said about the skull (SMNS 11413), which is substantially reconstructed in plaster. The vertebral column of the same specimen (numbered SMNS 11414) is described by Slijper (1936:319), who provides a good set of measurements of all of the vertebral centra.

Following Slijper, middle lumbar vertebrae of *Masracetus markgrafi* are on the order of 13–14 cm long and have centra with length-to-width and length-to-height ratios of about 0.77 and 0.94, respectively. These vertebrae also have centrum width to height ratios of about 1.20, making them relatively wider than centra of *Zeuglodon brachyspondylus* reported by Müller (1849:26).

Remarks Stromer (1908:129) identified the type specimen, SMNS 11413-11414, as *Zeuglodon isis*. Kellogg (1936:76) included SMNS 11413-11414 in *Prozeuglodon isis*. Slijper

(1936:319) considered SMNS 11413-11414 to represent *Zeuglodon brachyspondylus*, and Uhen (2005) included it in *Cynthiacetus maxwelli*. However, *Masracetus markgrafi* is different from all of these. It has vertebral proportions most similar to '*Zeuglodon*' *brachyspondylus*, and the latter may belong in this genus.

BASILOSAURIDAE indet.

Basilosauridae are also known from Dur et Talhah in Libya (Wight, 1980), Tiavandou in Senegal (Elouard, 1981), Oued Kouki in Tunisia (Batik and Fejfar, 1990), the Qattara Depression in Egypt (David Morrison, pers. comm. 2006), and Tabaghbagh near Siwa in Egypt (H. J. van Vliet and T. Paymans, pers. comm. 2008).

Suborder MYSTICETI Cope, 1891

Mysticetes known from African localities all come from late Miocene to Holocene deposits, and none has been referred to an extinct species.

Family BALAENOPTERIDAE Gray, 1864
Genus *BALAENOPTERA* Lacépède, 1804
BALAENOPTERA sp.

Age and Occurrence Late Miocene locality P8, U2 member of Sahabi Formation, at Sahabi in Libya (site 40 in figure 45.3; age is from Bernor and Scott, 2003); early Pleistocene of Baia Farta in Angola (site 53); and Miocene-Pliocene or Holocene of Agulhas Bank (site 46), and Holocene of coastal Yzerplaats, South Africa.

Description Each of the rorquals known from Libya (Petrocchi, 1941, 1951), Angola (Gutiérrez et al. (2001), and Yzerplaats in South Africa (Gill, 1928) is represented by a partial skeleton. The record from Agulhas Bank off the coast of South Africa is based on isolated periotic and/or tympanic ear bones, and this is the only rorqual identified to species (*Balaenoptera physalis*; Barnard, 1954).

Family BALAENIDAE Gray, 1821
Genus *BALAENA* Linnaeus, 1758
BALAENA sp.
Figure 45.3

Age and Occurrence Miocene-Pliocene of Agulhas Bank off the coast of South Africa (site 46 in figure 45.3).

Description Isolated periotic and/or tympanic ear bones of right whales (Barnard, 1954).

MYSTICETI indet.

Mysticete remains are reported from the early Pliocene of Langebaanweg in South Africa (Hendey, 1976), but these have not been studied.

Suborder ODONTOCETI Flower, 1869

Odontocetes known from African localities come from early Miocene through Holocene deposits. Eleven extinct species are represented, of which ten are based on African type specimens.

Family PHYSETERIDAE Gray, 1821
Genus *PHYSETER* Linnaeus, 1758
PHYSETER MACROCEPHALUS Linnaeus, 1758

Age and Occurrence Holocene of archaeological site H near Tarfaya in southwestern Morocco (site 59 in figure 45.3).

Description The only record of a sperm whale is a single tooth found in a coastal Paleolithic site (Charon et al., 1973; Saban, 1974).

PHYSETERIDAE indet.

Age and Occurrence Late Miocene of Raz-el-Ain in Algeria (site 42 in figure 45.3).

Description Muizon (1981) mentions teeth and vertebrae of a Physeterid from Raz-el-Ain in a report on a Miocene seal from Algeria but gives no further information about the cetacean.

Family KOGIIDAE Gill, 1871
Genus *KOGIA* Gray, 1846
KOGIA sp.

Age and Occurrence Late Pliocene of Ahl al Oughlam, near Casablanca in Morocco (site 52 in figure 45.3).

Description A pygmy sperm whale is represented by a periotic ear bone characteristic of *Kogia* (Geraads et al., 1998).

Family ZIPHIIDAE Gray, 1865
Genus *MICROBERARDIUS* Bianucci et al., 2007
MICROBERARDIUS AFRICANUS Bianucci et al., 2007

Age and Occurrence Miocene-Pliocene. No data with specimen: trawled off the South African coast.

Description The type and only specimen is a partial skull including a worn rostrum, much of the dorsal surface of the cranium, and a nearly complete vertex.

Genus *IZIKOZIPHIUS* Bianucci et al., 2007
IZIKOZIPHIUS ROSSI Bianucci et al., 2007

Age and Occurrence Miocene-Pliocene. No data with specimens: trawled off the South African coast.

Description The type specimen is a partial skull with rostrum, the anterior part of the cranium, and the vertex. A referred specimen is a second partial skull.

IZIKOZIPHIUS ANGUSTUS Bianucci et al., 2007

Age and Occurrence Miocene-Pliocene. Type specimen was trawled southwest of Cape Town, South Africa, at a depth of 450 m in the Atlantic Ocean.

Description The type specimen is a partial skull with the rostrum, much of the dorsal surface, and vertex.

Genus *ZIPHIUS* Cuvier, 1823
ZIPHIUS sp.

Age and Occurrence Miocene-Pliocene. Specimens described by Barnard (1954) were trawled off the coast of South Africa at Agulhas Bank and Cape West in the Atlantic Ocean. Specimen described by Bianucci et al. (2007) was trawled off the south coast of South Africa at a depth of 1,000 m in the Indian Ocean.

Description Partial skulls. Specimen described by Bianucci et al. (2007) is a fragment of a cranium with most of the vertex and parts of premaxillary sac fossae.

Genus *KHOIKHOICETUS* Bianucci et al., 2007
KHOIKHOICETUS AGULHASIS Bianucci et al., 2007

Age and Occurrence Miocene-Pliocene. Holotype was found offshore from Cape Agulhas.

Description Partial skull including much of the rostrum, anterior part of the cranium, and the vertex.

Genus *IHLENGESI* Bianucci et al., 2007
IHLENGESI SALDANHAE Bianucci et al., 2007

Age and Occurrence Miocene-Pliocene. Type specimen was trawled off Saldanha Bay on the west coast of South Africa in the Atlantic Ocean. Referred specimen was trawled off the coast of Cape Columbine.

Description Type is a partial skull including the base of the rostrum, anterior part of the cranium, and the vertex. Referred specimen is a rostrum with the anterior part of the cranium.

Discussion Haughton (1956) identified the referred specimen as cf. *Mesoplodon*.

Genus *AFRICANACETUS* Bianucci et al., 2007
AFRICANACETUS CERATOPSIS Bianucci et al., 2007

Age and Occurrence Miocene-Pliocene. Type specimen was trawled off the coast southwest of South Africa at a depth of less than 600 m in the Atlantic Ocean. Referred specimens are from various localities off the coast of South Africa.

Description Type is a partial skull including the rostrum and dorsal surface of the cranium with the vertex. Most referred specimens are rostra with the anterior cranium but no vertex.

Genus *MESOPLODON* Gervais, 1850
MESOPLODON SLANGKOPI Bianucci et al., 2007

Age and Occurrence Miocene-Pliocene. Type specimen was trawled off Slangkop on the west coast of Cape Province, South Africa.

Description Type specimen is a partial skull including the base of the rostrum, and the dorsal surface of the cranium with the vertex. Referred specimen is a deeply worn rostrum base with the maxillary sac fossae and the vertex.

Discussion It is not clear whether any of the specimens referred to *Mesoplodon* by Barnard (1954), Haughton (1956), and Ross (1986) belong to this species. One was included above in *Ihlengesi saldanhae*.

Genus *NENGA* Bianucci et al., 2007
NENGA MEGANASALIS Bianucci et al., 2007

Age and Occurrence Miocene-Pliocene. Type specimen was trawled west of Cape Town in the Atlantic Ocean. Referred specimens were all trawled off the coast of South Africa.

Description Type is a partial skull including the rostrum, premaxillary sac fossae, and vertex. Referred specimens are partial skulls and rostra.

Genus *XHOSACETUS* Bianucci et al., 2007
XHOSACETUS HENDEYSI Bianucci et al., 2007

Age and Occurrence Miocene-Pliocene. Type specimen has no data. This was trawled off the South African coast.

Description Type is a partial skull with most of the rostrum, anterior part of the cranium, and vertex.

Genus *PTEROCETUS* Bianucci et al., 2007
PTEROCETUS BENGUELAE Bianucci et al., 2007

Age and Occurrence Miocene-Pliocene. Type specimen was trawled of the west coast of South Africa, south of Saldanha Bay, at 700 m depth in the Atlantic Ocean. Referred specimens were trawled off the coast of South Africa.

Description Partial skull with most of the rostrum, anterior part of cranium, and vertex. Referred specimens are partial skulls with parts of the rostrum and anterior cranium.

ZIPHIIDAE indet.

Age and Occurrence Late middle Miocene fluvial sediments of Loperot in Kenya (site 39 in figure 45.3).

Description The oldest record of a beaked whale is the rostral part of a cranium found in inland riverine sediments in association with a freshwater and terrestrial fauna (Mead, 1975).

Family EURHINODELPHINIDAE Abel, 1901
Genus *SCHIZODELPHIS* Gervais, 1861
SCHIZODELPHIS aff. *SULCATUS* Gervais, 1853

Age and Occurrence Early Miocene of the Moghara Formation at Wadi Faregh (site 36 in figure 45.4) and Wadi Moghara (site 37) in northern Egypt.

Description The specimens from each site are pieces of mandibular symphysis broken where the fused dentaries separate behind the symphysis.

Remarks Eurhinodelphinids are an extinct group of late Oligocene to Miocene, small- to moderate-sized, long-necked, and long-snouted odontocetes. They are interesting in having a longer neck than is seen in most odontocetes, and edentulous premaxillae that extend well in front of the maxillae, making the rostrum longer than the dentaries. Such a specialized feeding apparatus suggests a coastal or even estuarine habitat (Lambert, 2005), which is consistent with recovery in estuarine deposits of the Moghara Formation.

Family DELPHINIDAE Gray, 1821
Genus *DELPHINUS* Linnaeus, 1758
DELPHINIS VANZELLERI Fourtau, 1920

Age and Occurrence Early Miocene of the Moghara Formation at Wadi Moghara in northern Egypt (site 37 in figure 45.4).

Description Delphinus vanzelleri is dolphin with relatively large and long teeth. Tooth crowns are flattened transversely, with the flattening being a little more prominent on the lingual surface. Each tooth has a pointed crown, but the apex of the crown is noticeably rounded. The type and only specimen is a mandible fragment with crowns or alveoli for nine teeth. This was illustrated by Fourtau (1920:36, figure 25).

Remarks It is difficult to say much about a species based on a specimen representing so little of the animal. Hence little has been written about this species. Making the species meaningful would require discovery of more complete specimens in the Moghara Formation.

Genus *DELPHINUS* or *STENELLA*

Age and Occurrence Late Pliocene fissure fillings in continental cave deposits overlying Messaoudian Pliocene marine beds at Ahl al Oughlam in Morocco (site 52 in figure 45.3).

Description Geraads et al. (1998) report periotics and jaw fragments of *Delphinus* sp. or *Stenella* sp. that they interpret as stranded cetaceans scavenged and brought into caves by carnivores.

Genus *LAGENORHYNCHUS* Gray, 1846

Age and Occurrence Late Miocene locality P8, U2 member of Sahabi Formation, at Sahabi in Libya (site 40 in figure 45.3). Age is from Bernor and Scott (2003).

Description An isolated periotic is known from the site that yielded the rorqual skeleton mentioned earlier. This was identified as cf. *Lagenorhynchus* sp. by Whitmore (1987).

Genus *ORCINUS* Fitzinger, 1860

Age and Occurrence Miocene-Pliocene of Agulhas Bank (site 46 in figure 45.3) and off the west coast of South Africa

Description Partial mandibles of killer whales are reported by both Barnard (1954) and Haughton (1956).

Family INIIDAE Gray, 1846

Age and Occurrence Late Miocene of Sahabi in Libya (site 41 near site 40 in figure 45.3).

Description Whitmore (1987) described a number of skeletal elements of a single individual odontocete specimen. These are fragmentary individually but informative when considered together. Whitmore interpreted them to represent a *Pontoporia*-like iniid with unfused cervical vertebrae and a relatively flexible neck.

Remarks Pontoporia and *Iniia* are 'river dolphins' living in South America today, but they evidently had a more cosmopolitan distribution in the late Miocene.

ODONTOCETI indet.

White et al. (1983) described a rib fragment identified only to suborder Odontoceti that came from locality P4A at Qasr as Sahabi in Libya (site 41 near site 40 in figure 45.3). This may represent the iniid from the same site described by Whitmore (1987). Odontocete remains are reported from the Miocene-Pliocene off the coast of South Africa (Bianucci et al., 2007) and from the early Pliocene of Langebaanweg in South Africa (Hendey, 1976), but these have not been studied in any detail.

CETACEA indet.

Cetacean remains of unknown affinities are reported from the Miocene site of Malembe in the Cabinda Enclave, Angola (site 38 in figure 45.3; Dartevelle, 1935; see also Antunes, 1964).

Discussion

Thirty-five of the 62 African localities listed in table 45.1 (56%) have yielded archaeocete cetaceans, as compared to only 27 (44%) yielding mysticetes and odontocetes. The African fossil record has always been important for understanding archaeocete diversity, morphology, and adaptations, but it has had somewhat less importance for understanding either mysticetes or odontocetes. Here I will focus on African Archaeoceti in summarizing what is important about the African record of Cetacea.

The last review of the fossil record of Cetacea in Africa (Barnes and Mitchell, 1978) focused on *Protocetus*, *Eocetus*, and *Pappocetus* within Protocetidae, and on *Dorudon* and *'Prozeuglodon'* within Basilosauridae. Protocetids remain poorly known, but there are promising new discoveries awaiting study. Our understanding of basilosaurids has changed substantially as a result of new discoveries in Egypt. Many specimens previously referred to *Dorudon* are now placed in *Saghacetus*. The differences became clear with recovery of new skulls and associated axial skeletons. Specimens formerly called *Prozeuglodon* are now divided among *Dorudon* and *Basilosaurus*, and here, too, recovery of new skulls with associated axial skeletons was required to enable adequate comparison of the genera. *Dorudon atrox* is now known from dozens of partial skeletons, of which about half are immature individuals with deciduous teeth virtually identical to those of the type of North American *Dorudon serratus* and to the type of Egyptian *Prozeuglodon atrox* (figures 45.9E, 45.9F), while the other half are fully adult (figure 45.9B). Adult *D. atrox* are medium in overall size, having a skeleton about 5 m long, with relatively short posterior thoracic and lumbar centra. (Uhen, 2004). *Basilosaurus isis* (figure 45.8D) is not represented by any immature individuals, but adult specimens are all large, having a skeleton about 18 m long, with relatively long posterior thoracic and lumbar centra.

The Barnes and Mitchell (1978) review of the fossil record of African cetaceans includes a table indicating that skeletal parts for each known species were represented by fewer specimens than could be counted on one hand. Forelimbs were known in a few specimens, but none preserved the hand or manus. The only hand bones known for archaeocetes were those described for North American *Zygorhiza* by Kellogg (1936). Hindlimbs were not represented by a single innominate, and archaeocetes were not known to have had a foot or pes. The only hindlimb elements known for archaeocetes at the time were two innominates and a partial femur of one North American *Basilosaurus* (Lucas, 1900). This changed in 1989 when fore- and hindlimb elements were found for *Dorudon* and *Basilosaurus*, and virtually complete hindlimbs were found for the latter (Gingerich et al., 1990).

Barnes and Mitchell (1978) traced the origin of Archaeoceti back to the archaic cursorial and hoofed mammals called Mesonychidae. Mesonychids had large but simple, laterally compressed teeth like those of early cetaceans, and they were thought at the time to be the land-mammals ancestral to cetaceans (Van Valen, 1966). Recovery of feet in Egyptian *Basilosaurus* in 1989 inspired a renewal of field work in Pakistan, which eventually led to discovery of skeletons of older and more primitive archaeocetes. Skeletons of the early protocetids *Rodhocetus* and *Artiocetus* were found to have had associated hindlimbs that retained characteristically artiodactyl ankle morphology with a double-pulley astragalus (Gingerich et al., 2001). Thus the ancestry of cetaceans is now generally understood to be from early artiodactyls rather than mesonychids (but see O'Leary and Gatesy, 2008).

The Pakistan specimens show that protocetids like *Rodhocetus* were semiaquatic, foot-powered swimmers, in contrast to basilosaurids like *Dorudon* that were fully aquatic, tail-powered swimmers (Gingerich, 2003). *Protocetus* of Lutetian age (early middle Eocene) from the Mokattam Formation of Egypt is similar to *Rodhocetus*, and was undoubtedly a semiaquatic, foot-powered swimmer as well.

The later middle Eocene (Bartonian) cetaceans *Eocetus schweinfurthi* and *Pappocetus lugardi* are larger and more advanced in some respects. It is possible that when they are better known postcranially they will prove to be transitional in some way from semiaquatic to fully aquatic.

The Birket Qarun Formation of early Priabonian (late Eocene) age is the only African geological formation that has been prospected extensively for cetaceans. Here four archaeocetes are known from partial to complete skeletons. These are, from smallest to largest: *Ancalecetus simonsi*, *Dorudon atrox*, *Masracetus markgrafi*, and *Basilosaurus isis*. One or two additional cetaceans are represented by vertebrae that seem to differ from all others, but these are not yet known from diagnostic specimens. Five to six species is the greatest archaeocete diversity known from any single formation in Egypt.

The overlying Qasr el-Sagha Formation of late Priabonian (late Eocene) age has two named archaeocetes. These are, from smaller to larger, *Saghacetus osiris* and *Stromerius nidensis*. Neither is well enough known postcranially to enable interpretation of their differences from *Dorudon* in terms of swimming locomotion, but they were presumably fully aquatic like *Dorudon*.

The northern part of the African continent is ringed by Eocene strata that have yielded a good fossil record of archaeocete cetaceans in the past. New localities in these beds promise to produce many cetaceans in the future as well. Egyptian localities are particularly important for having yielded the most completely known representatives of Basilosauridae, which include the earliest known fully aquatic cetaceans and early tail-powered swimming cetaceans. Some, like *Dorudon*, provide an important link between earlier semiaquatic Protocetidae and later Mysticeti and Odontoceti (figure 45.1). Others, like *Basilosaurus*, are so divergently specialized that they are not likely to be related to later whale evolution.

A 2005 excavation of the largest archaeocete from Egypt, *Basilosaurus isis*, is shown in figure 45.10. Even the largest fossil cetaceans are small in comparison to the scale of the North African desert, and many stratigraphic intervals offer great promise for recovery of new archaecete taxa represented by well preserved skeletons. Recovery of Neogene cetaceans in Africa appears less promising, but even here there is potential.

ACKNOWLEDGMENTS

I thank Elwyn Simons for help in initiating a long-term field project to study marine mammals of the Eocene of Egypt. This has been carried out in recent years under the auspices of the Egyptian Mineral Resources Authority and the Egyptian Environmental Affairs Agency. Many people helped with fieldwork in Egypt and Tunisia over the past 20 years, but sustained contributions by B. Holly Smith, William J. Sanders, M. Sameh Antar, and Iyad S. Zalmout stand out. Over the years I have been privileged to study virtually all of the known African archaeocete material, for which I am indebted to museums and curators too numerous to list.

FIGURE 45.10 Photograph of 2005 Egyptian Environmental Affairs Agency and University of Michigan excavation of *Basilosaurus isis* in the Birket Qarun Formation, late Eocene, of Wadi Hitan (Egypt). Skeleton was found by A. V. H. van Nievelt in 1987 when only a partial scapula was visible. Skull was collected in 1989 from lower right of photograph. Excavation shown here revealed an almost fully articulated skeleton, lying on a bed of sandstone overlain by shale. Workers are excavating the thorax. The lumbus and tail have been encased in white plaster jackets to protect them during transportation.

I thank colleagues Giovanni Bianucci, Ewan Fordyce, and Mark Uhen for reading the manuscript and providing many suggestions for improvement. Bonnie Miljour helped greatly in preparing the illustrations. Expert preparation of new archaeocete specimens by William J. Sanders at the University of Michigan underlies much of what we have learned since the last review of African Cetacea.

Fieldwork on African archaeocetes has been supported by the National Geographic Society (3424-86, 4154-89, 4624-91, 5072-93, and 7726-04), and in recent years by the U.S.-Egypt Joint Science and Technology Program and U.S. National Science Foundation (EAR-0517773 and OISE-0513544).

Literature Cited

Andrews, C. W. 1899. Fossil Mammalia from Egypt: Part I. *Geological Magazine* 6:481–484.

———. 1901. Preliminary note on some recently discovered extinct vertebrates from Egypt: Part II. *Geological Magazine* 8:436–444.

———. 1904. Further notes on the mammals of the Eocene of Egypt: Part III. *Geological Magazine* 1:211–215.

———. 1906. *A Descriptive Catalogue of the Tertiary Vertebrata of the Fayum, Egypt*. British Museum (Natural History), London, 324 pp.

———. 1908. Note on a model of the skull and mandible of *Prozeuglodon atrox*, Andrews. *Geological Magazine* 5:209–212.

———. 1920. A description of new species of zeuglodont and of leathery turtle from the Eocene of southern Nigeria. *Proceedings of the Zoological Society of London* 1919:309–319.

———. 1923. Note on the skulls from which the endocranial casts described by Dr. Dart were taken. *Proceedings of the Zoological Society of London* 1923:648–654.

Antunes, M. T. 1964. *O Neocretácico e i Cenozóico do littoral de Angola*. Junta de Investigações do Ultramar, Lisboa. 255 pp.

Bajpai, S., and P. D. Gingerich. 1998. A new Eocene archaeocete (Mammalia, Cetacea) from India and the time of origin of whales. *Proceedings of the National Academy of Sciences, USA* 95:15464–15468.

Barnard, K. H. 1954. *A Guide Book to South African Whales and Dolphins*. South African Museum, Cape Town, 33 pp.

Barnes, L. G., and E. D. Mitchell. 1978. Cetacea; pp. 582–602 in V. J. Maglio and H. B. S. Cooke (eds.), *Evolution of African Mammals*. Harvard University Press, Cambridge.

Batik, P., and O. Fejfar. 1990. Les vertébrés du Lutétien, du Miocène et du Pliocène de Tunisie centrale. *Service Géologique de Tunisie, Tunis* 56:69–82.

Beadnell, H. J. L. 1901. The Fayum depression: a preliminary notice of the geology of a district in Egypt containing a new Paleogene vertebrate fauna. *Geological Magazine* 8:540–546.

———. 1905. *The Topography and Geology of the Fayum Province of Egypt*. Survey Department of Egypt, Cairo, 101 pp.

Bernor, R. L., and R. S. Scott. 2003. New interpretations of the systematics, biogeography and paleoecology of the Sahabi hipparions (latest Miocene)(Libya). *Geodiversitas* 25:297–319.

Bianucci, G., O. Lambert, and K. Post. 2007. A high diversity in fossil beaked whales (Mammalia, Odontoceti, Ziphiidae) recovered by trawling from the sea floor off South Africa. *Geodiversitas* 29: 561–618.

Bianucci, G., C. Nocchi, C. Sorbini, and W. Landini. 2003. *L'Archeoceto nella roccia. Alle origini dei Cetacei*. Museo di Storia Naturale e del Territorio, Università degli Studi di Pisa, pp. 1–17.

Bianucci, G., K. Post, and O. Lambert. 2008. Beaked whale mysteries revealed by seafloor fossils trawled off South Africa. *South African Journal of Science* 104:140–142.

Blanckenhorn, M. 1902 (1903). Neue geologisch-stratigraphische Beobachtungen in Aegypten. *Sitzungsberichte der Mathematisch-physikalischen Classe der Königlichen Bayerischen Akademie der Wissenschaften, München* 32:353–433.

Boaz, N. T. 1980. A hominoid clavicle from the Mio-Pliocene of Sahabi, Libya. *American Journal of Physical Anthropology* 53:49–54.

Cappetta, H., and M. Traverse. 1988. Une riche faune de sélachiens dans le bassin à phosphate de Kpogamé-Hahotoé (Éocène moyen du Togo): Note préliminaire et précisions sur la structure et l'âge du gisement. *Geobios* 21:359–365.

Charon, M., L. Ortlieb, and N. Petit-Marie. 1973. Occupation humaine Holocène de la Région du Cap Juby. *Bulletin et Mémoires de la Société d'Anthropologie, Paris* 10:379–412.

Clementz, M. T., A. Goswami, P. D. Gingerich, and P. L. Koch. 2006. Isotopic records from early whales and sea cows: Contrasting patterns of ecological transition. *Journal of Vertebrate Paleontology* 26:355–370.

Cope, E. D. 1868. An addition to the vertebrate fauna of the Miocene period, with a synopsis of the extinct Cetacea of the United States. *Proceedings of the Academy of Natural Sciences, Philadelphia* 19:138–156.

Dames, W. B. 1883a. Ein Epistropheus von *Zeuglodon* sp. *Sitzungsberichte der Gesellschaft naturforschender Freunde, Berlin* 1883:3.

———. 1883b. Über eine tertiäre Wirbelthierfauna von der westlichen Insel der Birket-el-Qurun. *Sitzungsberichte der Königlich-preussischen Akademie der Wissenschaften, Berlin* 1883:129–153.

———. 1894. Über Zeuglodonten aus Aegypten und die Beziehungen der Archaeoceten zu den übrigen Cetaceen. *Geologische und Paläontologische Abhandlungen, Jena* 5:189–222.

Dart, R. A. 1923. The brain of the Zeuglodontidae (Cetacea). *Proceedings of the Zoological Society of London* 1923:615–648, 652–654.

Dartevelle, E. 1935. Les premiers restes de mammifères du Tertiaire du Congo: La faune Miocene de Malembe (première note sur les mammifères fossiles du Congo). *Comptes Rendus, 2e Congrés National des Sciences Belge, Bruxelles* 715–720.

Deraniyagala, P. E. P. 1948. Some scientific results of two visits to Africa. *Spolia Zeylanica* 25:1–42.

Elouard, P. 1981. Découverte d'un archéocète dans les environs de Kaolack. *Notes Africaines, Dakar* 109:8–10.

Ennouchi, E. 1961. A propos d'un arrière-crâne de cétacé. *Comptes Rendus des Séances Mensuelles, Societé des Sciences Naturelles et Physiques du Maroc, Tangiers* 27:103.

Flower, W. H. (ed.). 1866. *Recent Memoirs on the Cetacea by Professors Eschricht, Reinhardt and Lilljeborg.* Hardwicke, London, 312 pp.

———. 1883. On the arrangement of the orders and families of existing Mammalia. *Proceedings of the Zoological Society of London* 1883:178–186.

Fordyce, R. E., and L. G. Barnes. 1994. The evolutionary history of whales and dolphins. *Annual Review of Earth and Planetary Sciences* 22: 419–455.

Fordyce, R. E., and C. de Muizon. 2001. Evolutionary history of cetaceans: A review; pp. 169–233 in J.-M. Mazin and V. d. Buffrénil (eds.), *Secondary Adaptation of Tetrapods to Life in Water.* Pfeil, Munich.

Fourtau, R. 1920. *Contribution à l'Étude des Vertébrés Miocènes de l'Égypte.* Survey Department, Government Press, Cairo, 121 pp.

Fraas, E. 1904a. Neue Zeuglodonten aus dem unteren Mitteleocän vom Mokattam bei Cairo. *Geologische und Paläontologische Abhandlungen, Jena* 6:197–220.

———. 1904b. Neue Zeuglodonten aus dem unteren Mitteleocän vom Mokattam bei Cairo. *Geologisches Zentralblatt* 5:374.

———. 1906. Wüstenreise eines Geologen in Ägypten. *Kosmos, Handweiser für Naturfreunde, Stuttgart* 3:263–269.

Geraads, D., F. Amani, J.-P. Raynal, and F.-Z. Sbihi-Alaoui. 1998. La faune de mammiferes du Pliocene terminal d'Ahl al Oughlam, Casablanca, Maroc. *Comptes Rendus de l'Academie des Sciences, Paris, Serie II, Sciences de la Terre et des Planetes* 326:671–676.

Gibbes, R. W. 1845. Description of the teeth of a new fossil animal found in the green-sand of South Carolina. *Proceedings of the Academy of Natural Sciences, Philadelphia* 2:254–256.

———. 1847. On the fossil genus *Basilosaurus*, Harlan, (*Zeuglodon*, Owen) with a notice of specimens from the Eocene Green Sand of South Carolina. *Journal of the Academy of Natural Sciences, Philadelphia* 1:5–15.

Gidley, J. W. 1913. A recently mounted *Zeuglodon* skeleton in the United States National Museum. *Proceedings of the U.S. National Museum* 44:649–654.

Gill, E. L. 1928. Note on a whale buried on the Cape Flats. *Transactions of the Royal Society of South Africa* 16:53–54.

Gingerich, P. D. 1991. Provenance of Fourtau's Egyptian archaeocete. *Investigations on Cetacea, Paciano* 23:213–214.

———. 1992. Marine mammals (Cetacea and Sirenia) from the Eocene of Gebel Mokattam and Fayum, Egypt: Stratigraphy, age, and paleoenvironments. *University of Michigan Papers on Paleontology* 30:1–84.

———. 1998. Paleobiological perspectives on Mesonychia, Archaeoceti, and the origin of whales; pp. 423–449 in J. G. M. Thewissen (ed.), *Emergence of Whales: Evolutionary Patterns in the Origin of Cetacea.* Plenum Press, New York.

———. 2003. Land-to-sea transition of early whales: Evolution of Eocene Archaeoceti (Cetacea) in relation to skeletal proportions and locomotion of living semiaquatic mammals. *Paleobiology* 29:429–454.

———. 2005. Cetacea; pp. 234–252 in K. D. Rose and J. D. Archibald (eds.), *Placental Mammals: Origin, Timing, and Relationships of the Major Extant Clades.* Johns Hopkins University Press, Baltimore.

———. 2007. *Stromerius nidensis*, new archaeocete (Mammalia, Cetacea) from the upper Eocene Qasr el-Sagha Formation, Fayum, Egypt. *Contributions from the Museum of Paleontology, University of Michigan* 31:363–378.

Gingerich, P. D., M. Arif, M. A. Bhatti, H. A. Raza, and S. M. Raza. 1995. *Protosiren* and *Babiacetus* (Mammalia, Sirenia and Cetacea) from the middle Eocene Drazinda Formation, Sulaiman Range, Punjab (Pakistan). *Contributions from the Museum of Paleontology, University of Michigan* 29:331–357.

Gingerich, P. D., Y. Attia, F. A. Bedawi, and S. Sameeh. 2007. Khashm el-Raqaba: A new locality yielding middle Eocene whales and sea cows from Wadi Tarfa in the eastern desert of Egypt. *Journal of Vertebrate Paleontology*, 27 (suppl. 3):81.

Gingerich, P. D., H. Cappetta, and M. Traverse. 1992. Marine mammals (Cetacea and Sirenia) from the middle Eocene of Kpogamé-Hahotoé in Togo. *Journal of Vertebrate Paleontology* 12A:29–30.

Gingerich, P. D., and B. H. Smith. 1990. Forelimb and hand of *Basilosaurus isis* (Mammalia, Cetacea) from the middle Eocene of Egypt. *Journal of Vertebrate Paleontology* 10A:24.

Gingerich, P. D., B. H. Smith, and E. L. Simons. 1990. Hind limbs of Eocene *Basilosaurus isis*: Evidence of feet in whales. *Science* 249:154–157.

Gingerich, P. D., and M. D. Uhen. 1996. *Ancalecetus simonsi*, a new dorudontine archaeocete (Mammalia, Cetacea) from the early late Eocene of Wadi Hitan, Egypt. *Contributions from the Museum of Paleontology, University of Michigan* 29:359–401.

Gingerich, P. D., M. ul-Haq, W. v. Koenigswald, W. J. Sanders, B. H. Smith, and I. S. Zalmout. 2009. New protocetid whale from the middle Eocene of Pakistan: birth on land, precocial development, and sexual dimorphism. *PLoS ONE* 4 (e4366):1–20.

Gingerich, P. D., M. ul-Haq, I. S. Zalmout, I. H. Khan, and M. S. Malkani. 2001. Origin of whales from early artiodactyls: Hands and feet of Eocene Protocetidae from Pakistan. *Science* 293:2239–2242.

Gingerich, P. D., N. A. Wells, D. E. Russell, and S. M. I. Shah. 1983. Origin of whales in epicontinental remnant seas: New evidence from the early Eocene of Pakistan. *Science* 220:403–406.

Gradstein, F. M., J. G. Ogg, and A. G. Smith (eds.). 2004. *A Geological Time Scale 2004.* Cambridge University Press, Cambridge, 589 pp.

Gregory, W. K. 1951. *Evolution Emerging: A Survey of Changing Patterns from Primeval Life to Man*, vols. I–II. Macmillan, New York, 1,013 pp.

Grine, F. E., and R. G. Klein. 1993. Late Pleistocene human remains from the Sea Harvest site, Saldanha Bay, South Africa. *South African Journal of Science* 89:145–152.

Grine, F. E., R. G. Klein, and T. P. Volman. 1991. Dating, archaeology and human fossils from the Middle Stone Age levels of Die Kelders, South Africa. *Journal of Human Evolution* 21:363–395.

Gutiérrez, M., C. Guérin, M. Léna, and M. Piedade da Jesus. 2001. Exploitation d'un grand cétacé au Paléolithique ancien: Le site de Dungo V à Baia Farta (Benguela, Angola). *Comptes Rendus de l'Académie des Sciences, Paris, Sciences de la Terre et des Planètes* 332:357–362.

Halstead, L. B., and J. A. Middleton. 1974. New material of the archaeocete whale, *Pappocetus lugardi* Andrews, from the middle Eocene of Nigeria. *Journal of Mining and Geology* 8:81–85.

———. 1976. Fossil vertebrates of Nigeria. Part II, 3.4, Archaeocete whale: *Pappocetus lugardi* Andrews, 1920. *Nigerian Field* 41:131–133.

Hamilton, W. R. 1973. A lower Miocene mammalian fauna from Siwa, Egypt. *Palaeontology* 16:275–281.

Haughton, S. H. 1956. Phosphatic-glauconite deposits off the west coast of South Africa. *Annals of the South African Museum* 42:329–334.

Heinzelin, J. de, and A. El-Arnauti. 1987. The Sahabi Formation and related deposits (including a geological map of the Sahabi area); pp. 1–22 in N. T. Boaz, A. el-Arnauti, A. W. Gaziry, J. de Heinzelin, and D. D. Boaz (eds.), *Neogene Paleontology and Geology of Sahabi.* Liss, New York.

Heizmann, E. P. J. 1991. Durch die Wüste. Die Fayum-Expedition von Eberhard Fraas im Jahre 1906. *Stuttgarter Beiträge zur Naturkunde, Serie C (Populärwissenschaft)* 30:65–70.

Hendey, Q. B. 1976. The Pliocene fossil occurrence in 'E' Quarry Lange-baan, South Africa. *Annals of the South African Museum* 69:215–247.

———. 1981. Palaeoecology of the Late Tertiary fossil occurrences in "E" Quarry, Langebaanweg, South Africa, and a reinterpretation of their geological context. *Annals of the South African Museum* 84:1–104.

———. 1982. *Langebaanweg: A Record of Past Life*. South African Museum, Cape Town, 71 pp.

Kellogg, R. 1928. The history of whales: Their adaptation to life in the water. *Quarterly Review of Biology* 3:29–76, 174–208.

———. 1936. A review of the Archaeoceti. *Carnegie Institution of Washington Publications* 482:1–366.

Klein, R. G. 1972. The late Quaternary mammalian fauna of Nelson Bay Cave (Cape Province, South Africa): Its implications for megafaunal extinctions and environmental and cultural change. *Quaternary Research* 2:135–142.

———. 1976. The mammalian fauna of the Klasies River Mouth sites, southern Cape Province, South Africa. *South African Archaeological Bulletin* 31:75–98.

Klein, R. G., and K. Cruz-Uribe. 1989. Faunal evidence for prehistoric herder-forager activities at Kasteelberg, Western Cape Province, South Africa. *South African Archaeological Bulletin* 44:82–97.

Lambert, O. 2005. Review of the Miocene long-snouted dolphin *Priscodelphinus cristatus* du Bus, 1872 (Cetacea, Odontoceti) and phylogeny among eurhinodelphinids. *Bulletin de l'Institut Royal des Sciences naturelles de Belgique, Sciences de la Terre* 75:211–235.

Lucas, F. A. 1900. The pelvic girdle of *Zeuglodon*, *Basilosaurus cetoides* (Owen), with notes on the other portions of the skeleton. *Proceedings of the U.S. National Museum* 23:237–331.

Luo, Z., and P. D. Gingerich. 1999. Terrestrial Mesonychia to aquatic Cetacea: Transformation of the basicranium and evolution of hearing in whales. *University of Michigan Papers on Paleontology* 31:1–98.

Mead, J. G. 1975. A fossil beaked whale (Cetacea: Ziphiidae) from the Miocene of Kenya. *Journal of Paleontology* 49:745–751.

Moustafa, Y. S. 1954. Additional information on the skull of *Prozeuglodon isis* and the morphological history of the Archaeoceti. *Proceedings of the Egyptian Academy of Sciences* 9:80–88.

———. 1974. Critical observations on the occurrence of Fayum fossil vertebrates. *Annals of the Geological Survey of Egypt* 4:41–78.

Muizon, C. de. 1981. Premier signalement de Monachinae (Phocidae, Mammalia) dans le Sahelien (Miocene Superieur) d'Oran (Algerie). *Palaeovertebrata* 11:181–194.

Müller, J. 1849. *Über die Fossilen Reste der Zeuglodonten von Nordamerica, mit Rücksicht auf die Europäischen Reste aus Dieser Familie*. Reimer, Berlin, 38 pp.

O'Leary, M. A., and J. E. Gatesy. 2008. Impact of increased character sampling on the phylogeny of Cetartiodactyla (Mammalia): Combined analysis including fossils. *Cladistics* 24:397–442.

Osborn, H. F. 1907. The Fayum expedition of the American Museum. *Science* 25:513–516.

Owen, R. 1841. Observations on the *Basilosaurus* of Dr. Harlan (*Zeuglodon cetoides* Owen). *Transactions of the Geological Society of London* 2:69–79.

Petrocchi, C. 1941. Il giacimento fossilifero di Sahabi. *Bollettino della Società Geologica Italiana* 60:107–114.

———. 1943. Il giacimento fossilifero di Sahabi. *Collezione Scientifica e Documentaria dell'Africa Italiana* 12:1–167.

———. 1951. Paleontologia di Sahabi (Cirenaica): Notizie generali sul giacimento fossilifero di Sahabi: Storia degli scavi: Risultati; pp. 7–31 in R. Fabiani (ed.), *Paleontologia di Sahabi*. Rendiconti dell'Accademia Nazionale dei XL, Serie IV, 3, Rome.

Phillips, W. 1948. Recent discoveries in the Egyptian Fayum and Sinai. *Science* 107:666–670.

Pickford, M., and B. Senut. 1997. Cainozoic mammals from coastal Namaqualand, South Africa. *Palaeontologia Africana* 34:199–217.

Pilleri, G. E. 1985. Record of *Dorudon osiris* (Archaeoceti) from Wadi-el-Natrun, lower Nile Valley. *Investigations on Cetacea* 17:35–37.

———. 1991. Betrachtungen über das Gehirn der Archaeoceti (Mammalia, Cetacea) aus dem Fayûm Ägyptens. *Investigations on Cetacea, Paciano* 23:193–211.

Pompeckj, J. F. 1922. Das Ohrskelett von *Zeuglodon*. *Senckenbergiana* 4:43–102.

Ross, G. J. B. 1986. Fossil beaked whales (letter to editor). *National Geographic Research* 2:275.

Saban, R. 1974. Dent de cachalot du gisement Moghrebien de Tarfaya (Sud Marocain). *Mammalia* 38:315–323.

Savage, R. J. G. 1969. Early Tertiary mammal locality in southern Libya. *Proceedings of the Geological Society of London* 1657:167–171.

———. 1971. Review of the fossil mammals of Libya; pp. 215–225 in C. Gray (ed.), *Symposium on the Geology of Libya*. University of Libya, Tripoli.

Schweinfurth, G. A. 1883. Ueber die geologische Schichtengliederung des Mokattam bei Cairo. *Zeitschrift der Deutschen Geologischen Gesellschaft* 35:709–737.

———. 1886. Reise in das Depressionsgebiet im Umkreise des Fajum im Januar 1886. *Zeitschrift der Gesellschaft für Erdkunde zu Berlin* 21:96–149.

Scott, W. B. 1894. The structure and relationships of *Ancodus*. *Journal of the Academy of Natural Sciences, Philadelphia* 9:461–497.

Simons, E. L. 1968. Early Cenozoic mammalian faunas, Fayum Province, Egypt: Part 1. African Oligocene mammals: Introduction, history of study, and faunal succession. *Bulletin of the Peabody Museum of Natural History, Yale University* 28:1–21.

Slijper, E. J. 1936. Die Cetaceen, Vergleichend-Anatomisch und Systematisch. *Capita Zoologica* 6–7:1–590.

Smith, G. E. 1903. The brain of the Archaeoceti. *Proceedings of the Royal Society of London* 71:322–331.

Stromer von Reichenbach, E. 1902 (1903). Bericht über eine von den Privatdozenten Dr. Max Blanckenhorn und Dr. Ernst Stromer von Reichenbach ausgeführte Reise nach Aegypten. Einleitung: Ein Schädel und Unterkiefer von *Zeuglodon osiris* Dames. *Sitzungsberichte der Mathematisch-physikalischen Classe der Königlichen Bayerischen Akademie der Wissenschaften, München* 32:341–352.

———. 1903. *Zeuglodon*-Reste aus dem oberen Mitteleocän des Fajum. *Beiträge zur Paläontologie und Geologie Österreich-Ungarns und des Orients, Wien* 15:65–100.

———. 1904. Bericht über die Sammlungsergebnisse einer paläontologisch-geologischen Forschungsreise nach Ägypten. *Bericht über die Senckenbergische naturforschende Gesellschaft* 1904:111–113.

———. 1907. Fossile Wirbeltier-Reste aus dem Uadi Fâregh und Uadi Natrûn in Agypten. *Abhandlungen der Senckenbergischen Naturforschenden Gesellschaft* 29:97–132.

———. 1908. Die Archaeoceti des ägyptischen Eozäns. *Beiträge zur Paläontologie und Geologie Österreich-Ungarns und des Orients, Wien* 21:106–178.

———, E. 1914. Ergebnisse der Forschungsreisen Prof. E. Stromers in den Wüsten Ägyptens: I. Die Topographie und Geologie der Strecke Gharaq-Baharije nebst Ausführungen über de geologische Geschichte Ägyptens. *Abhandlungen der Königlichen Bayerischen Akademie der Wissenschaften, Mathematisch-physikalischen Classe, München* 26:1–78.

Uhen, M. D. 1998. Middle to late Eocene basilosaurines and dorudontines; pp. 29–61 in J. G. M. Thewissen (ed.), *The Emergence of Whales: Evolutionary Patterns in the Origin of Cetacea*. Plenum Press, New York.

———. 1999. New species of protocetid archaeocete whale, *Eocetus wardii* (Mammalia: Cetacea) from the middle Eocene of North Carolina. *Journal of Paleontology* 73:512–528.

———. 2001. New material of *Eocetus wardii* (Mammalia, Cetacea), from the middle Eocene of North Carolina. *Southeastern Geology* 40:135–148.

———. 2004. Form, function, and anatomy of *Dorudon atrox* (Mammalia, Cetacea): An archaeocete from the middle to late Eocene of Egypt. *University of Michigan Papers on Paleontology* 34:1–222.

———. 2005. A new genus and species of archaeocete whale from Mississippi. *Southeastern Geology* 43:157–172.

Van Valen, L. M. 1966. Deltatheridia, a new order of mammals. *Bulletin of the American Museum of Natural History* 132:1–126.

White, T. D., G. Suwa, G. Richards, J. P. Watters, and L. G. Barnes. 1983. "Hominoid clavicle" from Sahabi is actually a fragment of cetacean rib. *American Journal of Physical Anthropology* 61:239–244.

Whitmore, F. C. 1982. Remains of Delphinidae from the Sahabi Formation. *Garyounis Scientific Bulletin, Special Issue* 4:27–28.

———. 1987. Cetacea from the Sahabi Formation, Libya; pp. 145–151 in N. T. Boaz, A. El-Arnauti, A. W. Gaziry, J. de Heinzelin, and D. D. Boaz (eds.), *Neogene Paleontology and Geology of Sahabi*. Liss, New York.

Wight, A. W. R. 1980. Paleogene vertebrate fauna and regressive sediments of Dur at Talhah, southern Sirt Basin, Libya; pp. 309–325 in M. J. Salem and M. T. Busrewil (eds.), *The Geology of Libya*. Academic Press, New York.

BROADER PERSPECTIVES

FORTY-SIX

Systematics of Endemic African Mammals

ROBERT J. ASHER AND ERIK R. SEIFFERT

Shortly after the death of his coauthor, C. F. Sonntag, Sir Wilfrid Le Gros Clark published a collaborative monograph on the anatomy and relationships of the aardvark, *Orycteropus afer* (Le Gros Clark and Sonntag, 1926), completing a series of monographs on this animal initiated by his senior colleague (Sonntag, 1925; Sonntag and Wollard, 1925). Le Gros Clark was 30 years old at the time; and together with Sonntag, he reached some startlingly prescient conclusions about the affinities of an animal that in subsequent years received much less precise treatment. Despite the fact that during the 1920s many of their colleagues considered aardvarks to be closely related to other "edentates," Le Gros Clark and Sonntag stated clearly that *"Orycteropus* is quite unrelated to living Pholidota and Xenarthra" (1926:478). Based on the "anatomy of the unguiculate extremities, the axis, sacral vertebrae, carpus, tarsus, tongue, muscles and some features in the placenta . . . [aardvarks] should be placed beside the Hyracoidea and Proboscidea" (1926:483).

This conclusion differed considerably from those given in other treatments of *Orycteropus* during most of the 20th century (e.g., Broom, 1909; Gregory, 1910; Winge, 1941; Patterson, 1975; Thewissen, 1985). Only after the widespread application of molecular data to mammalian systematics (e.g., DeJong et al., 1981, 1993; Porter et al., 1996; Stanhope et al., 1998; Murphy et al., 2001a, 2001b) did a plurality of zoologists accept an expanded, "African" version of Le Gros Clark and Sonntag's evolutionary proposal regarding the aardvark.

To varying degrees, a similar history can be traced for two other groups of African mammals: sengis (elephant shrews) and golden moles. For both taxa, much disagreement and uncertainty characterized discussions of their interordinal relations throughout the 20th century; see, for example, Carlsson (1909), Evans (1942), Patterson (1965), and Sarich (1993) on sengis; Broom (1916), Butler (1988), and MacPhee and Novacek (1993) on golden moles; and Roux (1947) on both. While not all applications of molecular data have uniformly supported the African clade (cf. Sarich 1993; Corneli 2002), consensus regarding their membership in an "African clade" has now been reached (cf. Springer et al. 2004; Wildman et al. 2007).

Africa is home to a number of diverse, endemic radiations besides Afrotheria, including cetartiodactyls (e.g., bovids, hippopotamids, giraffids), primates (e.g., strepsirrhines, catarrhine anthropoids), rodents (e.g., bathyergids, thryonomyids, anom-

alurids, pedetids), and carnivorans (e.g., Malagasy euplerines). However, these groups do not share a common ancestry with one another to the exclusion of other mammals. Those African mammals that do share a unique bond of common descent, recently called the Afrotheria (Stanhope et al., 1998), form the focus of this chapter. This group consists of seven living radiations, all but two of which have been for many years accorded their own Linnean orders (Nowak, 1999): aardvarks (Tubulidentata), sengis (Macroscelidea), elephants (Proboscidea), hyraxes (Hyracoidea), sea cows (Sirenia), golden moles (Chrysochloridae), and tenrecs (Tenrecidae). In the following text, we explore the systematics and fossil history of those endemic African mammals that share a unique bond of phylogenetic history, fully appreciated only during the last 10 years, and that comprise one of the most novel hypotheses of animal classification since 1758.

Content and Distribution

EXTANT AFROTHERIA

Among the living members of the seven extant afrotherian clades, three are completely restricted to Africa (sengis, golden moles, and aardvarks), and two (hyraxes and tenrecs) have a distribution extending only slightly beyond the African mainland. Elephants and sea cows are widely distributed outside mainland Africa.

Living aardvarks, golden moles and tenrecs are unknown north of the Sahel, although aardvarks may have existed in ancient Egypt (Kingdon, 1974). Despite the relative abundance of *Chrysochloris asiatica* in South Africa's Western Cape Province, and of *Amblysomus hottentotus* in Western Cape, Eastern Cape, and Kwazulu-Natal provinces, the remaining estimated 16 species of golden moles have extremely limited and discontinuous ranges throughout sub-Saharan Africa (Meester et al., 1986; Bronner, 1995; Asher, this volume, chapter 9). Only two of these species, *Chrysochloris stuhlmanni* and *Huetia leucorhinus* (taxonomy follows Asher et al., in press), are known from the African tropics; the remainder are distributed primarily in South Africa, Mozambique, Namibia, and adjacent, subtropical countries. An isolated record from an owl pellet is known as far north as Somalia (Simonetta, 1968).

Tenrecs comprise the most morphologically diverse living afrotherian radiation due to the geographic isolation afforded to the eight tenrecid genera (and about 30 species) on the island of Madagascar. Within Madagascar, most taxa can be found in forest habitats along the east coast and in the central and northern highlands (Eisenberg and Gould, 1970; Goodman, 2003). Two species, the diminutive, shrew-like *Geogale aurita* (Stephenson, 2003) and the semiarboreal *Echinops telfairi*, are best known in the relatively arid regions of the Malagasy west and southwest.

The mainland African Tenrecidae consists of three species in two genera (*Micropotamogale* and *Potamogale*) that occupy the very specialized niche of semiaquatic carnivory (Vogel, 1983; Nicoll, 1985). Both have been reported from discontinuous localities in western and central Africa, including the Ivory Coast and Democratic Republic of Congo. *Potamogale* appears to have at least some populations farther east, near the Ugandan-Kenyan border. It may also be the only mammal, large or small, that shows a teleost-like style of locomotion, propelling itself through the water with lateral undulations of its tail (Kingdon, 1974).

Sengis are most diverse south of the Sahel, but *"Elephantulus" rozeti* (actually more closely related to *Petrodromus* than to other *Elephantulus* species) is present in the northwest, occupying much of Morocco, northern Algeria, and Tunisia (Douady et al., 2003). The remaining three genera and 14 species are sub-Saharan, with a few species of *Elephantulus* extending into southern Sudan, Ethiopia, and northern Somalia (Nowak, 1999).

Hyraxes are known throughout the African continent and *Procavia* extends into the Arabian Peninsula and Asia Minor. Following Nowak (1999), they consist of seven species in three genera: *Procavia*, *Heterohyrax*, and *Dendrohyrax*. *Heterohyrax* extends as far north as Algeria and Egypt; *Dendrohyrax* is sub-Saharan.

Elephants have historically occupied all but the most extreme habitats. The two extant genera, *Loxodonta* and *Elephas*, are traditionally accorded a single species each: *africana* and *maximus*, respectively. The former has, until recent times, been widely distributed throughout Africa, from the Mediterranean to South Africa. Presently, the range of the African elephant is much smaller due to human activity and includes Central and West African forests and more arid savannas and grasslands from Ethiopia to Namibia and northern South Africa. *Loxodonta* encompasses both forest- and savanna-adapted forms. Recently, Grubb et al. (2000), Roca et al. (2001), and Shoshani (2005) have recognized the forest elephant *L. cyclotis* as a full species, although controversy on this point persists (Sanders et al., this volume, chapter 15; Debruyne, 2005). *Elephas* has had historical records throughout Asia but is now restricted to India, Bangladesh, Bhutan, Nepal, Burma, Vietnam, Laos, Thailand, and the islands of Sri Lanka, Sumatra, and Borneo (Nowak, 1999).

Sea cows also have an extensive distribution outside of Africa and comprise four species in two extant genera (*Trichechus* and *Dugong*). *Trichechus manatus* is found in the Atlantic coastal regions of North and South America including the Caribbean; *T. inunguis* is known throughout the Amazon River basin. The remaining two species are known from riverine and coastal regions of West-Central Africa, the Indian Ocean and Oceania: *T. senegalensis* in the Congo River basin and *Dugong dugon* off the coasts of Africa, Madagascar, India, Taiwan, and Australia.

A third genus and species *(Hydrodamalis gigas)* appears to have existed into the 18th century in the northern Pacific (Nowak, 1999).

EXTINCT AFROTHERIA

Each of the seven extant lineages of afrotherians has a fossil record (figures 46.1, 46.2), though these records vary tremendously in quality. For more extensive discussion, see the respective chapters in this volume.

Paenungulates

By far the most geographically and taxonomically diverse fossil record is that of the Proboscidea, dating from the late Paleocene and including nearly 50 genera from all continents except Australia and Antarctica (McKenna and Bell, 1997; Shoshani and Tassy, 2005; Gheerbrant, 2009; Sanders et al., this volume, Ch. 15); but see Gheerbrant et al., 2009. Both of the other paenungulate clades (sirenians and hyracoids) also show a fossil record exceeding the geographic and taxonomic diversity of their living representatives (table 46.1). Fossil hyracoids are more geographically restricted than extinct proboscideans and sirenians, but later Neogene forms are nevertheless known outside of Africa in southern Europe, Pakistan, Afghanistan, and East Asia. In the past, they were also much more morphologically diverse. Their fossil record during the late Eocene–early Oligocene in northern Egypt alone includes cursorial taxa such as the springbok-like *Antilohyrax*, which resembles the extant bovid *Antidorcas* in (for example) its cranial shape, edentulous premaxilla, and cursorial hindlimb (Rasmussen and Simons, 2000). Also present was the massive *Titanohyrax ultimus* (Matsumoto, 1926), which rivaled the Sumatran rhinoceros in size, possibly exceeding 1,000 kg in body mass (Schwartz et al., 1995).

There are a number of condylarth-grade fossils that may nest within crown Afrotheria (Asher et al. 2003; Zack et al. 2005a; Tabuce et al. 2007). Some of these (i.e., *Phenacodus*, *Meniscotherium*, and *Hyopsodus*) have first appearances that predate any definitive record of fossil Afrotheria (figure 46.2) and raise the interesting possibility that this clade may not in fact be historically African. However, other data sets place these "condylarths" among northern ungulate radiations (e.g., Wible et al., 2007). Broadly speaking, a northern condylarth-afrotherian relationship must remain speculative until more is known about the continental African Cretaceous and Paleocene (see Robinson and Seiffert, 2004), including, for example, the affinities of Paleocene palaeoryctids and *Todralestes* from Morocco (Gheerbrant, 1991, 1992; Seiffert, 2007) and potential afrotherian relatives such as endemic South American litopterns and notoungulates. The latter groups are both known from the Paleocene and have yet to be examined in a modern cladistic framework that also includes the molecular data responsible for defining Afrotheria. Although missing for the fossils, molecular data shape major components of the mammalian tree and may indirectly influence the placement of these and/or other fossil taxa (Asher et al., 2005). Furthermore, extinct South American taxa are particularly important as they have implications for the hypothesis of placental mammal biogeography proposed by Murphy et al. (2001b).

Horovitz (2004) has made a good start to understanding the phylogeny of endemic South American ungulates by

TABLE 46.1
Summary of the fossil record for each living afrotherian radiation

Only undisputed crown-group fossils and/or immediate sister taxa are included. Based on Seiffert et al. (2007), the record of tenrecids and chrysochlorids may include two and one additional genera (respectively) from the Eocene-Oligocene boundary of Egypt.

Clade	Number of Extinct Genera	Age Range	Distribution
Proboscidea	48	Eocene–Pleistocene	Global
Sirenia	32	Eocene–Pleistocene	Global
Hyracoidea	18	Eocene–Pleistocene	Africa, Southwest and South Asia, Mediterranean islands
Macroscelidea	9	Eocene–Pleistocene	Africa
Tubulidentata	2	Miocene–Pleistocene	Africa, Southern Europe, South Asia
Tenrecidae	3	Miocene	Africa
Chrysochloridae	2	Miocene	Africa

SOURCES: McKenna and Bell (1997), the Paleobiology Database (http://paleodb.org), and references given in the text.

providing a large morphological matrix sampling several of these taxa, such as the early Eocene toxodont *Thomashuxleya*. Her study, which sampled morphological characters of the postcranium, did not support a sister group relation between "condylarths" and the two included paenungulates (*Numidotherium* and *Procavia*); but she did recover a clade with *Phenacodus, Meniscotherium* and *Hyopsodus* comprising successively distant sister taxa to notoungulates and litopterns. Requirements for further testing of this result include, for example, additional extant afrotherians and expansion of the data set to include DNA sequences (for living taxa) and craniodental characters. Importantly, a hyopsodontid origin for litopterns and other South American taxa, such as kollpaniines and didolodontids, has already been suggested by phylogenetic analysis of dental characters (Muizon and Cifelli, 2000).

Tenrecs and Golden Moles

A clade consisting of tenrecs (Tenrecidae) and golden moles (Chrysochloridae) was referred to as the "Afrosoricida" by Stanhope et al. (1998). This name, and not the arguably more suitable nomen "Tenrecoidea," has recently been used by Bronner and Jenkins (2005) in the third edition of *Mammal Species of the World* (Wilson and Reeder, 2005). In addition, some authors (Salton and Szalay, 2004; Seiffert et al., 2007) use the term "Tenrecoidea" to refer to mainland African otter shrews and Malagasy tenrecs, with both groups taxonomically elevated to family-level nomina: Potamogalidae and Tenrecidae. Debate on intra-afrotherian nomenclature has been detailed elsewhere (Malia et al., 2002; Bronner et al., 2003; Douady et al., 2004; Bronner and Jenkins, 2005; Asher, 2005). We will not repeat this discussion here beyond noting that, as used by McDowell (1958), "Tenrecoidea" indicates a clade consisting of the Recent families Tenrecidae and Chrysochloridae and is synonymous with "Afrosoricida" as used by Bronner and Jenkins (2005) or "African insectivorans" as an informal designation.

In contrast to the taxonomic and geographic diversity of fossil paenungulates, extinct tenrecs and golden moles have for many years been known from just a handful of fragmen-

tary specimens from Miocene to Pleistocene localities in Kenya, Namibia, and South Africa (cf. Butler, 1984; Avery, 2001; Mein and Pickford, 2003). Most recently, Seiffert et al. (2007) assigned several specimens from the Eo-Oligocene of Egypt to new taxa in both families. To the extent that their adaptive diversity can be understood from isolated craniodental material, none of these fossils indicates a departure from the niches occupied by their modern, insectivorangrade relatives.

Through 2007, there were three genera and four named species of fossil tenrecs: *Protenrec butleri, P. tricuspis, Erythrozootes chamerpes,* and *Parageogale aletris* (table 46.1; Butler, 1984; Mein and Pickford, 2003; Asher and Hofreiter, 2006). The latter three are restricted to the early Miocene of Kenya (Butler, 1984). Possibly the oldest tenrec fossils have been identified by Seiffert et al. (2007), who placed the North African Eo-Oligocene fossils *Widanelfarasia* and *Jawharia* closer to living tenrecs than to any other mammal.

Fragmentary dental remains of *Protenrec butleri*, as well as a golden mole (*Prochrysochloris* cf. *miocaenicus*), have been recovered from the Miocene of Namibia (Mein and Pickford, 2003). Poduschka and Poduschka (1985) proposed the generic name "*Butleriella*" for *Parageogale aletris*, a nomen rendered synonymous by Butler (1984). In addition, Poduschka and Poduschka (1985) disputed Butler's interpretation that *Parageogale* was a tenrecid, but they did not offer a taxonomic alternative. Furthermore, they did not add to the material described by Butler (1984), and the recent phylogenetic analysis of Asher and Hofreiter (2006) supports the original interpretation of Butler and Hopwood (1957) and Butler (1984, 1985) that *Parageogale* is in fact a tenrecid, closely related to the living genus *Geogale*. An anterior jaw fragment that had been called a "giant" fossil tenrec (*Ndamathaia kubwa*) by Jacobs et al. (1987) was regarded as a junior synonym of *Kelba quadeemae* by Morales et al. (2000), considered by the latter authors to be a viverrid carnivoran. More recently, Cote et al. (2007) suggested *Kelba* may belong to the enigmatic group Ptolemaiida, possibly also part of the afrotherian radiation. On this interpretation, *Ndamathaia* is unlikely to have anything to do with viverrid carnivorans or with *Kelba*.

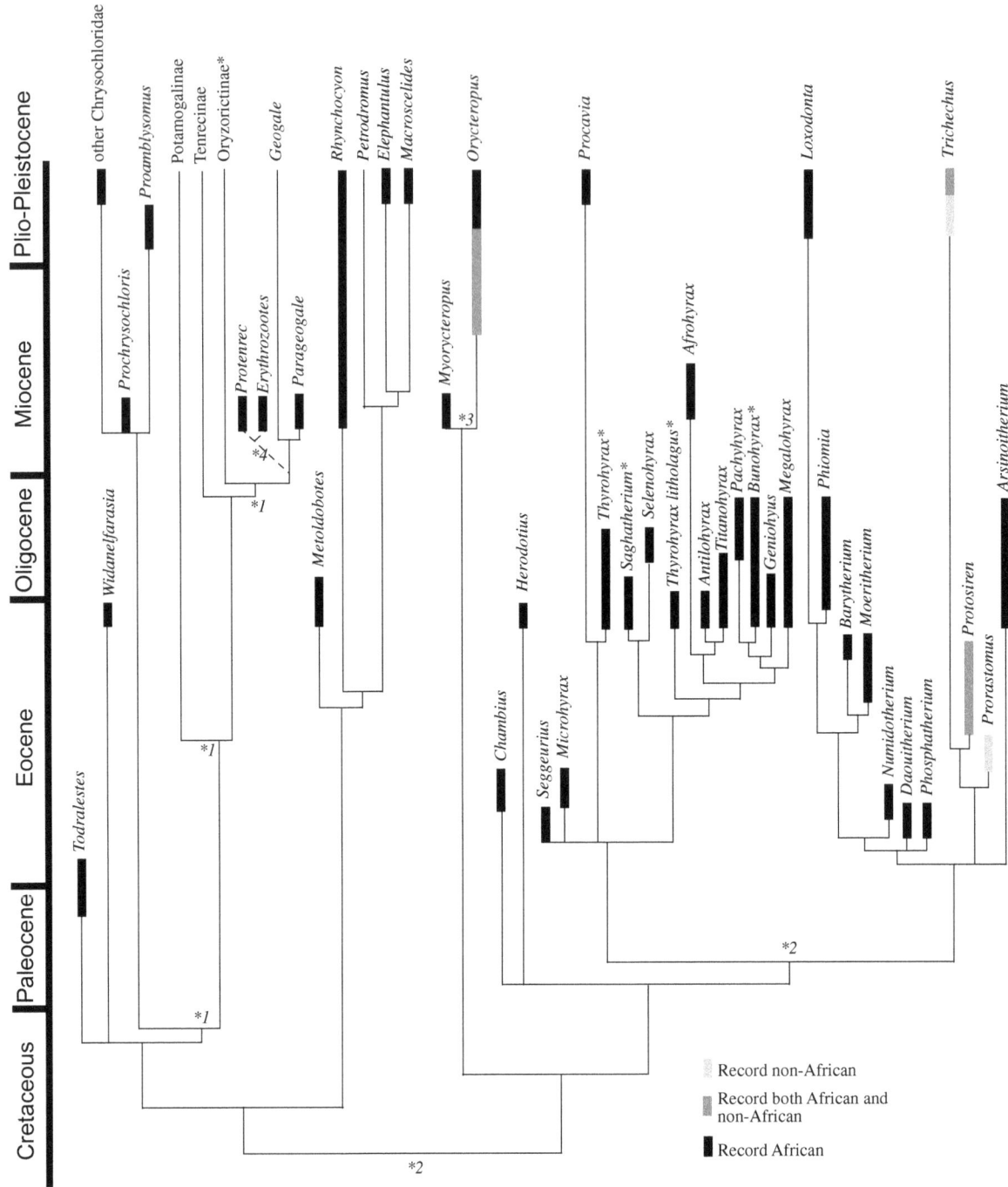

FIGURE 46.1 Afrotherian phylogeny adapted from Seiffert (2003: figure 3.8), summarizing afrotherian stratigraphic range and geographic distribution. Placement of fossil tenrecs is based on Asher and Hofreiter (2006); that of fossil golden moles on McKenna and Bell (1997).

FIGURE NOTES Asterisk, paraphyletic taxon; *1, divergence dates for crown tenrecids and tenrec + golden mole based on Poux et al. (2005). Note that following Seiffert et al. (2007), *Widanelfarasia* and *Jawharia* from the Eo-Oligocene boundary of Egypt are stem tenrecs and *Eochrysochloris* is a stem golden mole. *2, divergence dates for Paenungulata and Afrotheria from Springer et al. (2003); *3, not included are remains of *Archaeorycteropus* and *Palaeorycteropus* from Quercy, France, noted as questionable Oligocene tubulidentates by McKenna and Bell (1997); *4, depicted as dotted line to indicate uncertain result in phylogeny of Asher and Hofreiter (2006).

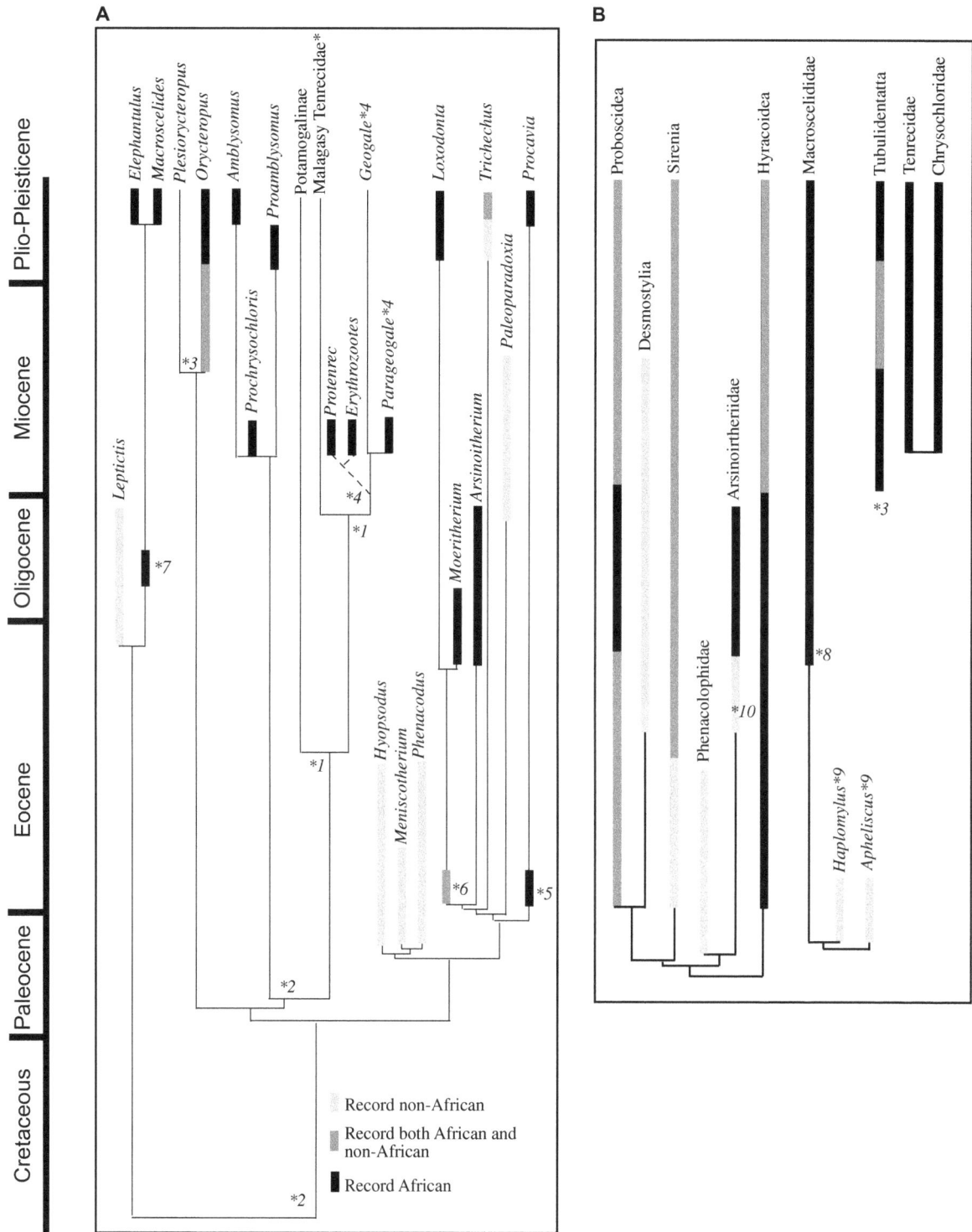

FIGURE 46.2   Afrotherian biostratigraphic distribution adapted from (A) Asher et al. (2003: figure 5) and (B) McKenna and Bell (1997), with additions as noted.

FIGURE NOTES   As in figure 46.1 plus *5, fossil hyracoids represented here by early Eocene *Seggeurius* from Morocco; *6, fossil proboscideans represented here by early Eocene numidotheriids from Morocco and anthracobunids from Pakistan; *7, fossil macroscelidids represented by *Metoldobotes*. Note that Eocene *Chambius* and *Herodotius* are not consistently recovered as macroscelidid relatives by Seiffert (2007). Also, *8, McKenna and Bell (1997) support inclusion of late Eocene *Herodotius* in Macroscelididae; *9, *Haplomylus* and *Apheliscus* association with macroscelidids based on Zack et al. (2005a) but questioned by Seiffert (2007); *10, mid-Eocene palaeoamasiine arsinoitheres from Turkey. Note also that following Seiffert et al. (2007) the Eo-Oligocene Fayum mammals *Widanelfarasia* and *Jawharia* comprise stem tenrecs, and *Eochrysochloris* is a stem golden mole.

Extinct species of golden moles include: *Prochrysochloris miocaenicus* (Butler, 1984), *Proamblysomus antiquus, Chlorotalpa spelea* (Broom, 1941, 1948), *Amblysomus ("Chrysotricha") hamiltoni* (DeGraaff, 1957), *Chrysochloris arenosa* (Asher and Avery, in press). The recently named *Eochrysochloris tribosphenus* from the early Oligocene of northern Egypt may comprise the oldest published record of the group (Seiffert et al., 2007). *Proamblysomus, C. spelea,* and *A. hamiltoni* are known from South African Plio-Pleistocene cave deposits at Sterkfontein and Makapansgat; *Prochrysochloris* is from the Miocene of Kenya and possibly Namibia (Mein and Pickford, 2003). Numerous, isolated remains of fossil *Chrysochloris* have been recovered from the Miocene/Pliocene of Langebaanweg in Western Cape Province, South Africa (Avery, 2000; Asher and Avery, in press). Furthermore, at least some extant golden mole species have a fossil record dating from the late Pliocene in South Africa (Pocock, 1987; Avery, 1998, 2001). Avery (2001) reports, for example, remains of *Amblysomus julianae* throughout all recorded Members of Sterkfontein and Swartkrans except one (Sterkfontein Member 4), indicating that this species has been present at least from ca. 2.6 to 0.1 Ma. Avery also notes the presence of *Chlorotalpa sclateri* from Sterkfontein Member 5E-O (2–1.7 Ma) and *Chrysospalax villosus* from Swartkrans Member SKX1, SKX2, and SKX3, spanning 1.7–1 Ma. Golden mole fossils are not common but are often found in faunal lists in Pleistocene localities throughout South Africa (e.g., Klein, 1977; Pickford and Senut, 1997).

Other references to fossil tenrecs and/or golden moles in Cretaceous through Miocene deposits of North America and Asia can be found sporadically in the literature. Trofimov (in Beliajeva et al., 1974:20), for example, referred the early Cretaceous eutherian *Prokennalestes* to the Tenrecidae. In fact, *Prokennalestes* is not a tenrecid but an endemic central Asian eutherian outside of the crown placental radiation (Wible et al., 2001, 2007). Hough (1956) referred to specimens of several insectivoran-grade genera (e.g., *Oligoryctes* and *Apternodus*) from North America as part of the "Oligocene Tenrecoidea". The phylogenetic analysis of Asher et al. (2002) indicates that these taxa, best known from the Eocene-Oligocene boundary in Wyoming and Montana, are more closely related to modern soricids. Matthew (1906) made reference to "fossil Chrysochloridae in North America" based on isolated humeri (see also Turnbull and Reed, 1967) and some cranial material of *Epoicotherium* (= "*Xenotherium*" Douglass), a taxon now regarded as a member of the extinct, fossorial Palaeanodonta and that appears to be related to extant pangolins (Rose and Emry, 1983).

The fauna and flora of Madagascar was even more extraordinary 1,000 years ago than it is today (Burney, 2003), yet specimens of "subfossil" tenrecs there are rare. Unlike other elements of the recently extinct Malagasy fauna (e.g., primates), they do not appear to be generically distinct from any modern tenrecid. To our knowledge, subfossil tenrecs are rarely encountered in museum collections and remain almost entirely undocumented in the literature (Goodman 2003; but see Muldoon, 2006).

In sum, the recent publication of the Eo-Oligocene tenrecs and golden moles from the Fayum (Seiffert et al., 2007) comprises the oldest, published record of the group, followed by the occurrence of undisputed craniodental material of both groups in the early Miocene of Kenya (Butler, 1984). As pointed out by Seiffert and Simons (2000) and Seiffert et al. (2007), older insectivoran-grade material from other North African localities (e.g., Gheerbrant, 1992) may comprise tenrecoid

relatives at some level. Importantly, afrotherian insectivorans from Egypt do not share the fully zalambdodont molar occlusal pattern (loss of upper molar metacones, reduction of lower molar talonids) that is seen in modern tenrecs and golden moles (discussed later). Zalambdodonty does not appear in the African fossil record until the early Miocene, implying either an extensive ghost lineage for tenrecs and golden moles that may extend back as far as the K-T boundary (figure 46.1), or convergent evolution of zalambdodonty in tenrecs and golden moles (a hypothesis favored by Seiffert, this volume, chapter 16). More evidence from the nonmolar dentition, cranium, and postcranium of Paleogene genera such as *Widanelfarasia, Todralestes,* and *Chambilestes* will be needed to test the latter hypothesis. In the meantime, bona fide fossil tenrecs and golden moles do exist by the early Miocene, but they remain very limited in their taxonomic and geographic diversity.

Tubulidentates

Two species of *Orycteropus* have been recorded from the middle/late Miocene Siwaliks of Pakistan: *O. pilgrimi* and *O. browni* (Moonen et al., 1978). Additional late Miocene remains of *O. gaudryi* are known from Turkey, Italy, Moldavia, and Iran (Kazanci, 1999; Lehmann, 2006) and the Mediterranean island of Samos (Lehmann, 1984). There are approximately nine other species of extinct *Orycteropus*, primarily in Africa (Avery, 2000; Lehmann, 2006). Two other genera of fossil aardvarks, both monotypic, are known from the early and late Miocene of Kenya (respectively): *Myorycteropus* and *Leptorycteropus* (Lehmann, 2006). Patterson (1975) listed the Malagasy subfossil *Plesiorycteropus* as a fourth genus of fossil aardvark; indeed, recent phylogenetic analyses including this taxon (e.g., Asher et al., 2003; Horovitz, 2004) have supported its position as the sister taxon to *Orycteropus*. However, the monographic study of MacPhee (1994) on *Plesiorycteropus* favored the hypothesis that it comprised an independent order, Bibymalagasia, *incertae sedis* relative to other placental mammals (see also Werdelin, this volume, chapter 11).

The oldest undisputed aardvark remains appear to be from the early Miocene of Kenya (Patterson, 1975; Pickford, 1975; Lehmann, 2006, 2009), although older material has been mentioned in the literature (e.g., Pickford and Andrews, 1981), including two Oligocene fossil "aardvarks," each of which is based on a single, isolated postcranial element from Quercy, France: *Archaeorycteropus* (based on a tibia; Ameghino, 1905) and *Palaeorycteropus* (based on a humerus; Filhol, 1894). Patterson (1975) argued that neither was a tubulidentate. Analysis of new and previously described remains of Miocene and later Paleogene ptolemaiids from Kenya and Egypt suggest the possibility of a distant relationship with Tubulidentata (Cote et al., 2007; Seiffert, 2007).

Macroscelideans

Fossil sengis are found with some frequency in Miocene or younger deposits throughout southern Africa (Avery, 2000). Late Eocene *Metoldobotes* is now universally considered to be a primitive macroscelidean (Patterson, 1965; Simons et al., 1991; Seiffert, 2007), but Seiffert (2003, 2007) questioned the alleged macroscelidean affinities of the Eocene African herodotiines *Chambius, Herodotius,* and *Nementchatherium* (e.g., Hartenberger, 1986; Simons et al., 1991) on the basis of available dental material, which is strikingly similar to that of primitive paenungulates. Tabuce et al. (2001) had earlier presented a dental character matrix for these taxa as well as a variety of other extinct placentals, including genera of "hyopsodontid"-grade, and found the African genera to be sister taxa of *Metoldobotes* and other macroscelidids. Interestingly, the European louisinine apheliscid (*sensu* Zack et al., 2005b) *Microhyus* was also placed as a basal macroscelidean, but other Laurasian hyopsodontids and apheliscids were placed far outside the afrotherian clade. McKenna and Bell (1997) place *Chambius* within the Louisininae.

Zack et al. (2005a) and Tabuce et al. (2007) have presented the most recent argument for a phylogenetic link between "hyopsodontids" and macroscelidids (but see Seiffert, 2007). Zack et al. (2005a) focused on postcranial similarities between the North American apheliscids *Haplomylus* and *Apheliscus*, both of which are known from the late Paleocene and early Eocene, and suggested that these "condylarths" share a common ancestry with macroscelidids to the exclusion of other afrotherians as well as ungulate-grade mammals, rodents and lagomorphs. They also highlighted the proposal that the so-called African clade may in fact not be African, as these Holarctic fossils have an older record than any undisputed afrothere (see Asher et al., 2003, and the earlier section on paenungulates).

However it is important to emphasize that there are no placental mammal localities in Africa that are older than ~60 Ma, whereas the Paleocene fossil records of Asia, Europe, and particularly North America are relatively well-known. Without older records from Africa, and if apheliscids are in fact afrotherians, it will be difficult to determine whether they represent out-migrants from Africa, as has been envisioned for adapisoriculids during the later Paleocene (Gheerbrant, 1992).

The Evidence

MORPHOLOGICAL SUPPORT

From an anatomical standpoint, characters that support Afrotheria in its entirety are rare, but several possibilities have recently been discussed (cf. Werdelin and Nilsonne, 1999; Asher, 2001; Mess and Carter, 2006; Seiffert, 2007; Sánchez-Villagra et al., 2007; Tabuce et al., 2007; Asher and Lehmann, 2008; Asher et al., 2009). Among morphologically based phylogenetic investigations of endemic African taxa, four character complexes are particularly notable: testicondy (Werdelin and Nilsonne, 1999), placentation (Mess and Carter, 2006), vertebral formula (Sánchez-Villagra et al., 2007; Asher et al., 2009), and dental eruption (Asher and Lehmann, 2008). One soft-tissue complex (the vomeronasal organ) deserves further investigation; and finally there are a number of potential osteological synapomorphies for Afrotheria (Seiffert, 2007; Tabuce et al., 2007).

Soft-Tissue Characters

Werdelin and Nilsonne (1999) were among the first to recognize the significance of the testicular descent character complex in the context of an explicit hypothesis of placental mammal phylogeny. Although the most recent consensus on the interrelationships of Placentalia (cf. Wildman et al., 2007) differs in several respects from the phylogeny figured in Werdelin and Nilsonne (1999: figures 2 and 3), their conclusion in terms of character support for Afrotheria remains timely. They noted variation in the position of the

male gonads and categorized this variation into three states: (1) "testicond" (i.e., testicles located intra-abdominally at or near the kidneys, (2) testicles descended but ascrotal, and (3) testicles descended and scrotal. Hyraxes, sea cows, elephants, sengis, and at least some tenrecs and golden moles are testicond (Werdelin and Nilsonne, 1999: figure 1). They also acknowledged variation: as pointed out by Sonntag (1925) and Whidden (2002), the aardvark does not have a scrotum but possesses "testes that protrude through the ventral abdominal wall" (Whidden, 2002:162). Whidden (2002) further noted the observations of Dobson (1883:86e, 125), which distinguished the testicond condition of the Malagasy hedgehog-tenrec *Setifer setosus* and the South African golden mole *Chrysospalax trevelyani* from the descended but intrapelvic position of the testes in another tenrecid, *Microgale dobsoni*. The extent to which golden moles show variation remains undocumented. Nevertheless, among all extant placental mammals, "true testicondy" with testicles and kidneys adjacent to one another occurs only among members of Afrotheria and remains an intriguing, likely morphological synapomorphy for the group.

Carter et al. (2006) and Mess and Carter (2006) have noted that extant afrotherian groups share several anatomical details regarding nutrition to the developing embryo. Specifically, they hypothesized that the placenta of the afrotherian common ancestor exhibited a large allantoic sac, a free, uninverted yolk sac, branching of allantoic vessels above the placental disk, and, perhaps of most interest, a partitioning of the allantois into four lobes by septa that carry umbilical blood vessels. According to these authors, the fourfold partitioning of the allantois would have most distinguished the afrotherian common ancestor from other placental mammals. Some diversity within Afrotheria does exist; for example, Mess and Carter (2006: figure 4A) note that *Echinops* and *Eremitalpa* show an undivided allantoic vesicle, not the four sacs present in other afrotherians. Nevertheless, *Echinops* still retains division of umbilical cord blood vessels (Mess and Carter, 2006:157); and furthermore, the fourfold division of the allantois still optimizes unambiguously as an afrotherian synapomorphy on virtually all recently proposed intra-afrotherian phylogenies.

Asher (2001) noted that *Chrysochloris, Procavia, Orycteropus, Setifer*, and potamogalines (but not *Echinops, Geogale, Tenrec*, or *Elephantulus*) possess a vomeronasal organ exhibiting a diffuse pattern of vascularization, without a single, large vessel traversing the organ anteroposteriorly (figure 46.3). However, because most afrotherian taxa are represented by only few or even single histological preparations (cf. table 2 of Asher, 2001), which to date represent the only means by which to code this character, it is not yet possible to rule out intraspecific variability in the distribution of the "diffuse" character state. Furthermore, the state identified in most afrotherian taxa is also evident in certain carnivorans, soricids, and *Solenodon*. In spite of this homoplasy, vomeronasal organ vasculature pattern (figure 46.3) was identified as the only unambiguously optimized afrotherian synapomorphy in the largest morphological data set yet assembled to investigate afrotherian systematics (Seiffert, 2007).

A number of other morphological changes regarding the dentition, petrosal, tarsal, and carpal anatomy and testicondy can also be reconstructed as afrotherian synapomorphies in Seiffert's (2007) study, but they are more dependent on how assumptions of character coding

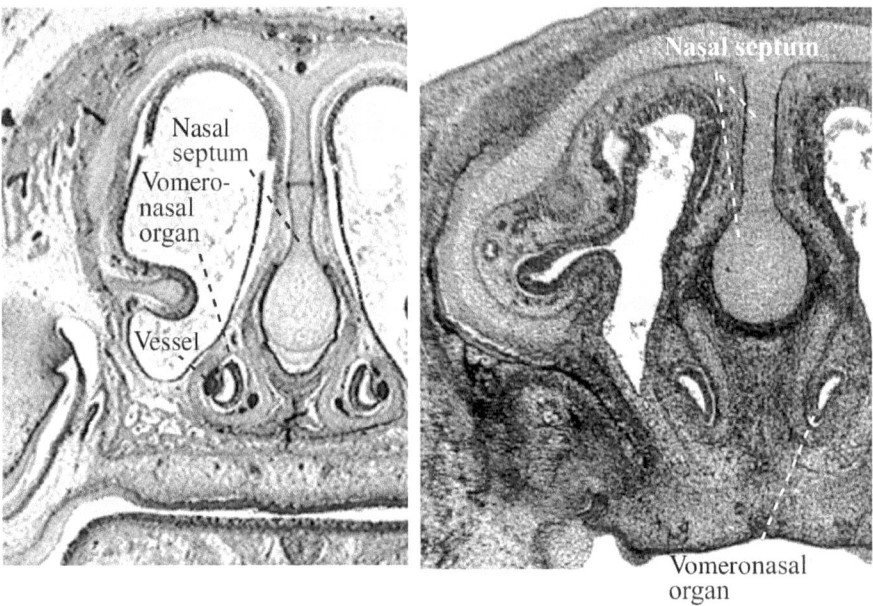

FIGURE 46.3 Coronal sections through the anterior nasal capsule of (A) *Echinops telfairi* (University of Tübingen W. Maier collection, slice 12.3.2) and (B) *Micropotamogale lamottei* (University of Göttingen H.-J. Kuhn collection 2273/3, slice 13.1.8) contrasting the large, anterior-posteriorly running blood vessel just lateral to the vomeronasal organ in *Echinops* with a more diffuse pattern, lacking a single, large anteroposteriorly directed blood vessel in *Micropotamogale*.

and optimization technique affect topology (tables 4.1, 4.2 in Seiffert, 2007). It is also worth noting resemblances between the stridulating organ of *Hemicentetes* and juvenile *Tenrec* (Eisenberg and Gould, 1970) with the skin and hairs surrounding the dorsal gland in hyracoids (Fischer, 1992). Investigation into the morphology and development of the dorsal integument throughout afrotherians may comprise a productive avenue by which the group may be better characterized morphologically.

Skeletal and Dental Characters

Seiffert (2003, 2007) and Robinson and Seiffert (2004) have highlighted the importance of resolving intra-afrotherian relationships for reconstructing character evolution at its base. Assuming, for example, that Murphy et al. (2001b) have correctly resolved the interrelationships of extant afrotherians, with the "Afroinsectiphillia" (i.e., tenrecs, golden moles, sengis, and aardvarks) comprising the sister group of paenungulates with elephant basal to hyrax-sea cow, several additional morphological characters (as coded by Seiffert, 2007) optimize unambiguously at the afrotherian base (table 46.2). These include a p4 talonid and trigonid of similar breadth, a prominent p4 hypoconid, presence of a P4 metacone, absence of parastyles on M1–2 and presence of a naviculocalcaneal facet. Two additional osteological characters used in the analyses of Asher et al. (2003, 2005) may be optimized as afrotherian synapomorphies: a carotid arterial sulcus on the petrosal and a fenestrated distal humerus. However, none of these characters is without homoplasy (table 46.2), and they result from a different intra-afrotherian phylogeny (figure 46.4) than those figured by Seiffert (2007) and Murphy et al. (2001b). Of particular note among the possible but ambiguously optimized, afrotherian synapomorphies is proximal fusion of the tibia and fibula, which is otherwise rare in placentals but present in hyracoids, aardvarks, sengis, golden moles, and potamogaline tenrecs, and the cuplike cotylar fossa of the astragalus, which is present in proboscideans, hyracoids, aardvarks, and sengis. Although Zack et al. (2005a) argued that a cotylar fossa is present in *Haplomylus*, this genus appears to bear only a very slight concavity of the medial aspect of the trochlea similar to that seen in numerous other nonafrotherians; the fossa observable in other louisinines, such as *Apheliscus*, *Paschatherium* (Godinot et al., 1996), and *Microhyus* (Tabuce et al., 2006) is somewhat more cuplike than that of *Haplomylus* but still does not approach the condition observable in the aforementioned afrotherians. Interestingly, *Haplomylus* also lacks proximal tibio-fibular fusion.

Sánchez-Villagra et al. (2007) have identified thoracolumbar vertebral count as a potential afrotherian synapomorphy. Remarkably, marsupial mammals apparently do not deviate from the single value of 19 thoracolumbar vertebrae. This number is a bit more variable among placental mammals, but with a few exceptions these, too, do not exceed 19. Most of these exceptions are found among afrotherians: Sánchez-Villagra et al. (2007) report at least 23 thoracolumbar vertebrae in elephants, 28–31 in hyraxes, 24 in the fossil sirenian *Pezosiren*, 21 in aardvarks, 21–24 in tenrecids, 22–24 in golden moles, and 20 in sengis. Carnivorans and erinaceids overlap the value of 20 present in sengis, but beyond Afrotheria, only certain primates, perissodactyls, and the bradypodid *Choloepus* exceed 21 or 22 thoracolumbar vertebrae. Seiffert (2007) also found an increased number of lumbar vertebrae (7 as opposed to 6 or fewer) to be a potential (ambiguously optimized) afrotherian synapomorphy.

The delayed eruption of the permanent dentition in elephants, hyraxes, and some sea cows is well documented;

TABLE 46.2
Potential morphological synapomorphies for Afrotheria

Publication	Character	Homoplasy
Werdelin and Nilsonne, 1999	Testicondy (abdominal testicles, lack of scrotum)	Descended but ascrotal in *Orycteropus* and some tenrecids (e.g., *Microgale*)
Asher, 2001: character #18	Diffuse vascularization of vomeronasal organ	Also present in several non afrotherians (e.g., *Canis*, *Crocidura*, *Solenodon*); central blood vessel in *Echinops*, *Geogale*, *Microgale* (Tenrecidae), and *Elephantulus*
Asher et al., 2003: character #11	Carotid sulcus on petrosal	*Elephantulus*, paenungulates, *Phenacodus*, *Meniscotherium*, numerous nonafrotherians
Asher et al., 2003: character #138	Fenestrated olecranon fossa of the humerus	Tethytheres including desmostylians, tenrecoids, numerous nonafrotherians
Seiffert, 2007	p4 talonid-trigonid similar in width	Tenrecs, golden moles, *Orycteropus*, primitive hyracoids, numerous non afrotherians
Seiffert, 2007	p4 hypoconid present, >50% height of protoconid	Tenrecs, golden moles, *Orycteropus*, numerous nonafrotherians
Seiffert, 2007	P4 metacone present, not well differentiated from paracone	Tenrecs, golden moles, *Orycteropus*, numerous nonafrotherians
Seiffert, 2007	M1–2 parastyles absent	Tenrecs, golden moles, numerous nonafrotherians
Seiffert, 2007	Naviculocalcaneal articulation	Potamogalines, procaviid hyracoids
Carter et al., 2006 and Mess and Carter, 2006	Four-lobed allantois, divided by septa carrying umbilical vessels	Single lobe in *Echinops* and *Eremitalpa*; not completely documented in all species of golden moles, tenrecs, sengis, and Asian elephants
Sánchez-Villagra et al., 2007	>21 thoracolumbar vertebrae	Sengis and some golden moles show 20, overlapping with carnivorans and erinaceids. Some primates, perissodactyls, and xenarthrans also exceed 21 thoracolumbar vertebrae.
Tabuce et al., 2007	Concave cotylar facet on astragalus	Absent in tenrecs and golden moles; present in some cercopithecoid primates, creodonts, and endemic South American "ungulates"
Asher and Lehmann, 2008	Complete eruption of permanent cheek teeth well after sexual maturity and attainment of adult body size	Some delay in eruption among terrestrial cetartiodactyls, perissodactyls, and anthropoid primates

these taxa may spend well over half of their life span without a completely erupted adult dentition (see Asher and Lehmann 2008). Reports of aberrant eruption patterns in small afrotherians are also known in the literature (Leche, 1907; MacPhee, 1987). However, the notion that delayed eruption of the permanent dentition comprises an afrotherian synapomorphy has only recently been made explicit (Asher and Lehmann, 2008). With the exception of dugongs and aardvarks, afrotherians tend to reach adult body size well in advance of the complete eruption of all permanent cheek teeth. While intriguing, this hypothesis requires further scrutiny of ontogenetic data from small afrotherians as well as an examination of the extent to which fossil clades (e.g., desmostylians, basal proboscideans) also exhibit delayed eruption of the permanent dentition.

In sum, although morphological phylogenies have not yet on their own recognized Afrotheria, there are several characters that, when mapped onto molecular phylogenies of placental mammals, can be viewed as morphological synapomorphies for the group (table 46.2).

MOLECULAR SUPPORT

As previously noted, support from molecular studies for the inclusion of African insectivorans in Afrotheria has been strong since 1996. Immunodiffusion and protein sequence studies of previous decades (e.g., DeJong et al., 1981, 1993; Rainey et al.; 1984; Kleinschmidt et al., 1986) supported a pared-down "African clade" including elephants, sea cows, hyraxes, sengis, and/or aardvarks; but no publication at that time argued that African insectivorans were also part of such a group.

Inclusion of tenrecs and/or golden moles in molecular analyses began with comparisons of alpha and beta hemoglobin chains (Piccinini et al., 1991) and 12S rRNA sequences (Allard and Miyamoto, 1992; Douzery and Catzeflis, 1995). However, these studies still did not foreshadow the Afrotheria due primarily to limited taxon sampling. The analysis by Lavergne et al. (1996) of 12S rRNA sequences recovered a golden mole–paenungulate clade with moderate support; however, they sampled only two other insectivoran-grade taxa: one of which (*Blarina*) was unresolved, and the other (*Atelerix*) was recovered with low support adjacent to murids at the base of Placentalia. Perhaps because of this, Lavergne et al. (1996) did not elaborate upon the golden mole–paeunungulate clade in the text of their paper.

The first molecular investigations broadly sampling ungulate- and insectivoran-grade taxa to test competing hypotheses of Afrotheria versus Insectivora (Springer et al., 1997;

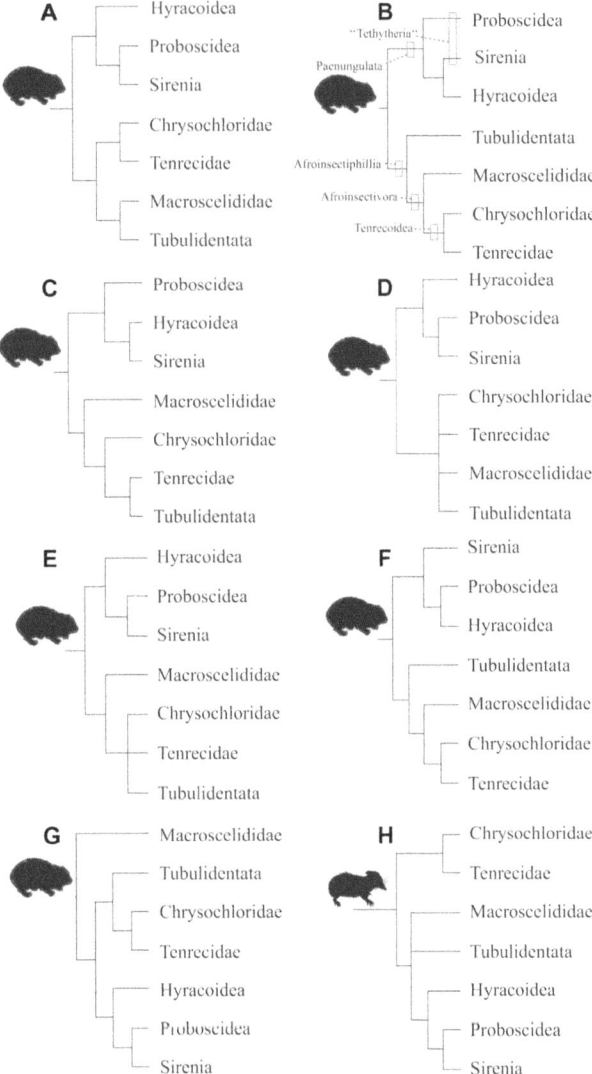

FIGURE 46.4 Recent phylogenetic hypotheses for internal relationships of Afrotheria based on the following: A) mtRNA, vWF, and A2AB (Stanhope et al., 1998); B) 19 nuclear genes and mtRNA (Murphy et al., 2001b); C) mtRNA, ND6, GHR, g-fibrinogin, RAG1 (Waddell and Shelley, 2003: figure 10); D) protein-coding mtDNA (Nikaido et al., 2003); E) retroposons (Nishihara et al., 2005; who favored Tethytheria in their text but recovered a SINE insertion supporting hyrax + sirenian); F) 20 nuclear genes and mtRNA (Amrine-Madsen et al., 2003); G) 19 nuclear genes, mtRNA, and 196 morphological characters (Asher et al. 2003: figure 5A); H) 378 morphological characters (Seiffert, 2003: figure 3.8), with afrotherian monophyly constrained. Image of hyrax (A–G) or tenrec (H) at base of each tree indicates reconstruction of "ungulate-like" or "insectivoran-like" anatomy of afrotherian common ancestor, respectively.

Stanhope et al., 1998) included approximately 2,200 nucleotides of the mitochondrial 12S and 16S ribosomal RNA genes and valine, a short transfer RNA (ca. 70 bases), in addition to slightly over 1 KB each of two nuclear genes: von Willebrand Factor (vWF) and alpha-2B adrenergic receptor (A2AB). For most of the 1990s, the genes with the best mammalian taxonomic representation were 12S and 16S rRNA, for which Stanhope et al. (1998) had just over 40 representatives, including seven afrotherians. By 2001, the sequence diversity of mammals was much better understood: the sample in

the landmark study of Murphy et al. (2001b) was over four times larger than that of Stanhope et al. (1998) and included approximately 20 KB from a similar set of 42 mammals, primarily from the nuclear genome. The mitogenomic studies of Mouchaty et al. (2000), Arnason et al. (2002), Murata et al. (2003), and Nikaido et al. (2003a) did not overlap at all with the data from Stanhope et al. (1998) or Murphy et al. (2001b), yet strongly supported their conclusions regarding afrotherian monophyly (but see Corneli, 2002). Nikaido et al. (2003a) included ca. 10 KB of mitochondrial, protein-coding genes for 69 mammals, including representatives of 7 distinct afrotherian genera. Waddell and Shelley (2003) analyzed sequences for ca. 90 mammalian taxa, of which 11 afrotherians were variably represented by sequences for the nuclear genes RAG1, gamma-fibrinogin, c-MYC, GHR, and epsilon-globin, plus mitochondrial sequences from ND-6, 12S and 16S rRNA.

Many of the studies listed in table 46.3 represent multiple applications of a single molecular data set to mammal phylogeny reconstruction. For example, the first GHR data set with a good sample of afrotherians was published by Malia et al. (2002). This data set was subsequently reanalyzed and expanded on by Douady et al. (2004) and Asher and Hofreiter (2006). In fact, the analysis of Douady et al. (2004) was highly critical of Malia et al. (2002); among other differences, Douady et al. emphasized a different tree-building methodology. Both Douady et al. (2004) and Asher and Hofreiter (2006) benefited greatly from the initial work of Malia et al. (2002) by incorporating their GHR sequences with a number of novel taxa, including all genera of the Tenrecidae in Asher and Hofreiter (2006). Importantly, as is the case in other instances where similar data are analyzed using different methods (e.g., the 19-gene nuclear data set first concatenated in Murphy et al., 2001b, and later reanalyzed in Asher et al., 2003, and Seiffert, 2007), each study emphatically supports the Afrotheria, despite substantial methodological differences.

Also impressive is the number and functional diversity of the genetic loci that support afrotherian monophyly. The data set of Stanhope et al. (1998) consisted of mitochondrial genes specifying the large (16S) and small (12S) ribosomal-RNA subunits and the transfer RNA valine, as well as two nuclear protein-coding genes: one involved in blood clotting (von Willebrand Factor or vWF) and the other linked to sympathetic nervous transmission (alpha 2 adrenergic receptor or A2AB). Adding to this diverse functional array, the studies of Madsen et al. (2001) and Murphy et al. (2001a) included sequences from a total of 17 additional nuclear genes, with functions (often poorly understood) that relate to DNA transcription regulation (BRCA1), sensitivity to marijuana (CNR1), cell differentiation (EDG1), vision (PLCB4), pigmentation (TYR), and cardiac tissue maintenance (ADORA3), to name a few. Most recently, genomic analyses including thousands of loci, with over a million nucleotides sampled across placental orders, support Afrotheria as well (e.g., Wildman et al. 2007).

In sum, the functional spectrum of genetic data that show support for Afrotheria is exceedingly broad. The Afrotheria hypothesis is supported not only from the aforementioned mitochondrial and nuclear genes, and the insertions and deletions (indels) present in those genes, but also from retroposons, chromosome morphology and Short Interspersed Nuclear Elements (SINEs; see table 46.3 and Robinson and Seiffert, 2004, and Springer et al., 2004).

TABLE 46.3
Partial list of phylogenetic analyses relevant to the Afrotheria hypothesis that incorporate molecular data and include
representatives of diverse placental orders, including at least one tenrec or golden mole
The "Data" and "Afrotherian DNA" columns are based on the most inclusive tree figured in the essay. The former lists
those genes/taxa identified as contributing to that tree; the latter includes recent taxa only.

Publication	Data	Afrotherian DNA
Piccinini et al., 1991	Alpha and beta haemoglobin	Tenrec(1)
Allard and Miyamoto, 1992	12S	Chryso(1)
Douzery and Catzeflis, 1995	12S	Chryso(1), Hyrac(1),
Lavergne et al., 1996	12S	Chryso(1), Hyrac(2), Probo(2), Siren(2)
Springer et al., 1997	12S-16S-valine, vWF, A2AB	Chryso(1), Hyrac(1), Macro(1), Probo(1), Siren(1), Tubul(1)
Stanhope et al., 1998	12S-16S-valine, vWF, A2AB	Chryso(1), Hyrac(1), Macro(1), Probo(1), Siren(1), Tenrec(1), Tubul(1)
Emerson et al., 1999	12S	Chryso(1), Hyrac(2), Macro(1), Probo(2), Siren(2), Tenrec(2), Tubul(1)
Mouchaty et al., 2000	12 mitochondrial genes	Probo(1), Tenrec(1), Tubul(1)
Madsen et al., 2001	BRCA1	Chryso(1), Hyrac(2), Macro(2), Probo(2), Siren(2), Tenrec(2), Tubul(1)
Murphy et al., 2001a	15 nuclear genes, 12S-16S-valine	Hyrac(1), Macro(2), Probo(1), Siren(1), Tenrec(1), Tubul(1)
Murphy et al., 2001b	19 nuclear genes, 12S-16S-valine	Chryso(1), Hyrac(1), Macro(2), Probo(1), Siren(1), Tenrec(1), Tubul(1)
Scally et al. 2001	BRCA1	Chryso(1), Hyrac(2), Macro(2), Probo(2), Siren(2), Tenrec(2), Tubul(1)
Malia et al., 2002	GHR	Chryso(1), Hyrac(1), Macro(1), Probo(1), Siren(1), Tenrec(4), Tubul(1)
Lin et al., 2002	12 mitochondrial genes	Probo(1), Tenrec(1), Tubul(1)
Corneli, 2002	12 mitochondrial genes	Probo(1), Tenrec(1), Tubul(1), Siren(1), Macro(1)
Douady et al., 2002a	A2AB, BRCA1, IRBP, vWF	Chryso(1), Hyrac(1), Macro(1), Probo(1), Siren(1), Tenrec(3), Tubul(1)
Douady et al., 2002b	vWF, 12S-16S	Chryso(1), Hyrac(1), Macro(1), Probo(2), Siren(1), Tenrec(3), Tubul(1)
Arnason et al., 2002	12 mitochondrial genes	Probo(1), Tenrec(1), Tubul(1), Siren(1), Macro(1)
Madsen et al., 2002	A2AB	Chryso(1), Hyrac(1), Macro(1), Probo(1), Siren(2), Tenrec(3), Tubul(1)
Amrine-Madsen et al., 2003	20 nuclear genes, 12S-16S-valine	Chryso(1), Hyrac(1), Macro(2), Probo(1), Siren(1), Tenrec(1), Tubul(1)
Asher et al., 2003	19 nuclear genes, 12S-16S-valine, 196 morphological characters	Chryso(1), Hyrac(1), Macro(2), Probo(1), Siren(1), Tenrec(2), Tubul(1)
DeJong et al., 2003	SCA1 sampled for both Chryso and Tenrec	Chryso(1), Hyrac(1), Macro(1), Probo(1), Siren(1), Tenrec(1), Tubul(1)
Douady et al., 2003	A2AB, BRCA1, vWF	Chryso(1), Hyrac(1), Macro(1), Probo(1), Siren(2), Tenrec(3), Tubul(1)
Murata et al., 2003	12 mitochondrial genes	Chryso(1), Hyrac(1), Macro(1), Probo(1), Siren(1), Tenrec(1), Tubul(1)
Nikaido et al., 2003a	12 mitochondrial genes	Chryso(1), Hyrac(1), Macro(1), Probo(1), Siren(1), Tenrec(1), Tubul(1)
Nikaido et al., 2003b	SINEs	Chryso(1), Hyrac(1), Macro(1), Probo(1), Siren(1), Tenrec(1), Tubul(1)
Seiffert, 2003	19 nuclear genes, 12S-16S-valine, 378 morphological characters	Chryso(1), Hyrac(1), Macro(2), Probo(1), Siren(1), Tenrec(1), Tubul(1)
Waddell and Shelley, 2003	RAG1, gam-fibrinogen, ND6, mtRNA, GHR	Chryso(2), Hyrac(1), Macro(2), Probo(1), Siren(2), Tenrec(1), Tubul(1)
Douady et al., 2004	GHR	Chryso(1), Hyrac(1), Macro(2), Probo(1), Siren(2), Tenrec(4), Tubul(1)
Roca et al., 2004	19 nuclear genes, 12S-16S-valine	Chryso(1), Hyrac(1), Macro(2), Probo(1), Siren(1), Tenrec(1), Tubul(1)
Robinson et al., 2004	Chromosome painting	Chryso(1), Macro(1), Probo(1), Tubul(1)
Nishihara et al., 2005	Retroposons	Chryso(1), Hyrac(1), Macro(1), Probo(2), Siren(1), Tenrec(1), Tubul(1)
Asher and Hofreiter, 2006	GHR	Chryso(1), Hyrac(1), Macro(1), Probo(2), Siren(1), Tenrec(10), Tubul(1)
Seiffert, 2007	19 nuclear genes, 12S-16S-valine, 400 morphological characters, 18 chromosomal and rare genomic events	Chryso(1), Hyrac(1), Macro(2), Probo(1), Siren(1), Tenrec(1), Tubul(1)

TABLE 46.3

(CONTINUED)

Publication	Data	Afrotherian DNA
Asher, 2007	19 nuclear genes, 12S-16S-valine, 196 morphological characters	Chryso(1), Hyrac(1), Macro(2), Probo(1), Siren(1), Tenrec(2), Tubul(1)
Wildman et al., 2007	1.4 million bases from ca. 1.7 thousand genes	Probo(1), Tenrec(1)
Kjer and Honeycutt, 2007	Coding and noncoding mitochondrial genes	Chryso(1), Hyrac(1), Macro(2), Probo(1), Siren(1), Tenrec(1), Tubul(1)
Poux et al., 2008	ADRA2B, AR, GHR, and vWF	Chryso(1), Hyrac(1), Macro(1), Probo(1), Siren(1), Tenrec(9), Tubul(1)

Phylogeny

In contrast to the near-unanimous support by molecular data of afrotherian monophyly, relations within the group remain ambiguous in several ways (figure 46.4). The best-supported intra-afrotherian taxon is the Paenungulata (Simpson, 1945), consisting of elephants, sea cows, and hyraxes. With the caveat that some anatomists favored a hyrax-perissodactyl clade into the 1990s (e.g., Fischer and Tassy, 1993), recognition of Paenungulata long predates support for Afrotheria itself (cf. Gill, 1870; Simpson, 1945; Novacek, 1986; Rasmussen et al., 1990). In addition, two other intra-afrotherian clades have enjoyed at least some level of support from several anatomical and/or sequence-based sources: Tethytheria (elephants and sea cows) and Tenrecoidea (tenrecs and golden moles).

Recent discussions of molecular data (Roca et al. 2004; Springer et al., 2004; Nishihara et al. 2005; Kjer and Honeycutt, 2007) have divided Afrotheria into paenungulate and nonpaenungulate clades, the latter named "Afroinsectiphillia" by Waddell et al. (2001) and consisting of (in order of increasing nestedness) aardvarks, sengis, tenrecs and golden moles (figure 46.4B). Waddell et al. (2001) named the two more-nested nodes "Afroinsectivora" and (following Stanhope et al., 1998) "Afrosoricida," the latter being equivalent to Tenrecoidea as used by McDowell (1958).

These intra-afrotherian hypotheses have major consequences for reconstructing the adaptations of the afrotherian common ancestor (Robinson and Seiffert, 2004). Was, for example, the ancestral afrotherian an "ungulate" or an "insectivoran"? Stated differently, did it possess cursorial and dental adaptations as in modern herbivores, or was it a small, terrestrial generalist with some faunivorous habits? Given, for example, a topology of modern taxa with aardvarks followed by sengis at the base of Afroinsectiphillia, comprising the sister taxon of paenungulates (figure 46.4B, following Murphy et al., 2001b), a parsimony algorithm would reconstruct an ungulate-grade afrotherian ancestor. Such an ancestor would lack the zalambdodont dentition, basisphenoid bulla, reduced pubic symphysis, small size and plantigrade posture that characterize African insectivorans (figures 46.4A–46.4F). The combined DNA-morphology analyses of Asher et al. (2003; but see Asher et al., 2005) and Seiffert (2007) do not support Afroinsectiphillia. When Seiffert (2007) constrained afrotherian monophyly, but not that of Afroinsectivora or Afroinsectiphillia, in a parsimony analysis of morphological data alone, tenrecs and golden moles were placed (figure 46.4H)

as the sister group of a clade containing aardvarks, sengis, and paenungulates, compatible with the hypothesis of an insectivoran-grade afrotherian common ancestor.

Macroscelidids combine some features of both "insectivorans" (e.g., small size and mobile proboscis) and "ungulates" (quadritubercular upper molars, cursorial skeletal adaptations and a caecum). If the paenungulate-afroinsectiphillian dichotomy is accurate, then macroscelideans comprise an interesting morphological intermediate between the two groups. This intermediate position is topologically supported in the phylogenies of Waddell and Shelley (2003: figure 10) and Nishihara et al. (2005). *Orycteropus* is hard to pigeonhole into this dichotomy but to some extent can also be viewed as intermediate, possessing, for example, an insectivorous diet but with large size and unguligrade posture (Lehmann et al., 2005). Aardvarks' mesiodistally elongate, bilobed upper and lower molars, which lack any occlusal features, potentially also indicate that these teeth have been modified from a quadritubercular plan. Interestingly, when occlusal dental features of *Orycteropus* and *Myorycteropus* are scored as missing (and thus optimized based on the scorings for its near relatives), the quadritubercular Eocene genus *Herodotius* is placed as the sister group of aardvarks in Seiffert's (2007) phylogenetic study.

The addition of fossil taxa to phylogenetic reconstructions of Afrotheria clearly has the potential to drastically alter morphological and ecological reconstructions of early members of this clade. The extreme diversity of fossil proboscideans (e.g., the dog-sized *Phosphatherium* and semiaquatic *Moeritherium*) and hyracoids (e.g., the springbok-like *Antilohyrax*) make it clear that living afrotherians are taxonomically impoverished compared to their extinct relatives. However, such material does not yet compel us to alter our perception of Paenungulata as historically consisting of primarily herbivorous, ungulate-grade mammals. Nor does the fossil record of African insectivorans provide evidence that this group previously occupied a noninsectivorous niche, although their fossil record is much worse than that of paenungulates, rendering any such statements of past diversity without much weight.

PAENUNGULATA

No one has seriously debated the validity of this clade for more than a decade. The most recent studies to do so include Fischer (1989), Fischer and Tassy (1993), and Prothero and Schoch (1989). At issue was the then-contentious hypothesis that hyracoids were more closely related to perissodactyls than to sea cows and elephants. Indeed, dating to Owen (1848) and briefly revived by Fischer (1989), hyracoids have even been classified within the Perissodactyla.

Anatomical support for this concept came in part from the dentition: fossil perissodactyls and Recent (but not Paleogene) procaviids both possess a distinctive "pi"-shaped pattern of crests on their cheek teeth (Radinsky, 1969). Fischer (1989) further identified a number of anatomical characters such as the eustachian sack, which comprises a similarity between equids, tapirids, and hyracoids. This diverticulum of the nasopharynx is exceedingly large in some taxa (Fischer, 1989: figure 4.3); however, it is notably absent in rhinocerotids. In addition, both hyracoids and perissodactyls are "mesaxonic"; that is, their distal hind- and forelimbs show an axis of support passing through a central digit, as opposed to the "paraxonic" condition in, for example, artiodactyls. There is even a biblical precedent for the emphasis on hyracoid distal limb morphology in its classification. Leviticus 11 identifies the hyrax (or "coney") as "unclean" for consumption by the Israelites because it is mesaxonic—it "cloveth not the hoof" (Gregory, 1910:7). Although the horse is not mentioned by name in Leviticus as nonkosher, it lacks a "cloven hoof" (not to mention rumination) and is thereby in clear violation of the Jewish law of kashrut—another, early intimation of similarity with the hyrax.

Nevertheless, although without biblical precedent and with its other members similarly nonkosher, Paenungulata has had a long history among vertebrate biologists. This history dates to Gill (1870) and much later Simpson (1945), who coined the superordinal name. Simpson envisioned Paenungulata as including a number of extinct taxa that are no longer considered to have a close relationship with hyraxes, sea cows, or elephants. A number of publications based on morphological data during the 1980s and 1990s strongly supported paenungulate monophyly (e.g., Novacek and Wyss, 1986; Rasmussen et al., 1990; Shoshani, 1993). The application of large molecular data sets in recent years has essentially solved this phylogenetic problem in their favor (cf. Murphy et al., 2001b; Waddell and Shelley, 2003).

TETHYTHERIA

While paenungulate monophyly was debated by morphologists into the 1990s, the monophyly of a sea cow–elephant clade had until recently never been seriously questioned. The designation Tethytheria dates to McKenna (1975). However, Linnaeus (1758) himself included both taxa in his "Bruta," along with a sloth, an anteater, and a pangolin. Although many 18th- and early 19th-century classifications placed sea cows with whales and/or pinnipeds, they were more frequently grouped with elephants, starting with Blainville (1834) and more commonly among post-Darwinian authors (see review in Gregory, 1910). In his monograph on Fayum mammals, Andrews (1906) detailed numerous similarities in the skeletons of *Moeritherium* (Proboscidea) and *Eosiren* (Sirenia), setting the tone for the general acceptance of an elephant–sea cow clade for the remainder of the 20th century.

Tethytheria has garnered some support in recent phylogenetic analyses, including several based on sequence data (e.g., Nikaido et al., 2003; Kjer and Honeycutt, 2007) and the combined morphology-DNA analyses of Seiffert (2007) and Asher (2007). However, many other sequence-based studies do not support this clade (Amrine and Springer, 1999; Murphy et al.,

2001b), with some recent studies favoring sirenian-hyracoid (e.g., Nishihara et al., 2006) over Tethytheria. The third permutation of this trichotomy, hyrax-elephant, has also been figured in a fairly recent DNA concatenation (Amrine-Madsen et al., 2003).

Morphological support for a close elephant–sea cow relationship has continued to surface in recent years. For example, Gaeth et al. (1999) noted the presence of nephrostomes in the kidneys of elephants throughout development. These structures are common throughout aquatic vertebrates, but not among placental mammals except for sirenians and, as Gaeth et al. (1999) noted, elephants. Hence, they argued not only for a close relationship between the two taxa but also that elephants have a semiaquatic ancestry. Such a hypothesis would also be consistent with functional interpretations of testicondy (see earlier discussion) and even of the proboscidean trunk, both of which have been related to a semiaquatic lifestyle (Shoshani and Tassy, 2005).

However, Gaeth et al. (1999) did not make explicit comparisons with hyraxes. Furthermore, at least some of the similarities of modern elephants and sea cows did not characterize some of the earliest members of each clade. For example, the persistent perilymphatic foramen and lack of a fenestra cochleae in living tethytheres do not characterize the fossil proboscideans *Numidotherium* (Court, 1994), *Phosphatherium* (Gheerbrant et al., 2005), or the fossil sirenian *Prorastomus* (Court, 1994), which instead resemble most other mammals in possessing a fenestra cochleae in adults. While the distribution of this character demonstrates the abundance of homoplasy, it does not invalidate the hypothesis that elephants and sea cows share a common ancestor, perhaps even a semiaquatic one, to the exclusion of other mammals. Indeed, stem proboscideans such as *Moeritherium* have been considered to show signs of an aquatic life for many years (Andrews, 1906; Matsumoto, 1926; Sanders et al., this volume, chapter 15), and the possible stem sirenian or tethythere group Desmostylia (Gingerich, 2005) also exhibits adaptations for some degree of existence in water (Clementz et al., 2003). Stable isotopes preserved in the tooth enamel of the primitive late Eocene genera *Moeritherium* and *Barytherium* have recently provided additional support for the hypothesis that these early proboscideans were semiaquatic (Liu et al., 2008). In sum, although recent molecular analyses yield surprisingly ambiguous results, the morphological case for tethythere monophyly remains strong.

TENRECOIDEA

Besides paenungulates and tethytheres, golden moles and tenrecs comprise the only other high-level clade of afrotheres that has had a relatively long taxonomic history. However, this has usually been in the context of their position within the "Lipotyphla" *sensu* Butler (1988) as members of a larger clade including hedgehogs, shrews, moles, and *Solenodon*.

One of the most conspicuous similarities between tenrecs and golden moles concerns the dentition. Both groups show "zalambdodont" cheek teeth (Asher and Sánchez-Villagra, 2005), in which the occlusal surface of the upper molars comprises a simple triangle, with a lingual cusp at its apex. Lower molars show a reduced trigonid, which lacks the mortar-pestle occlusion with the protocone seen in functionally tribosphenic mammals. In tenrecs and golden moles, the paracone comprises the main upper cusp and the protocone is reduced to a cingular structure or is absent altogether.

Since the mid-19th century, most zoologists have emphasized the dentition in classifying tenrecs and golden moles not only with each other but also with the zalambdodont Caribbean *Solenodon* and numerous North American fossils (see Asher et al., 2002). However, these taxa do not share unique nondental characters. Hence, long before anyone suspected that these "insectivores" might be related instead to elephants, the exceedingly autapomorphic anatomy of golden moles compelled Broom (1916) to recommend placing them in their own Linnean order, the Chrysochloridea, a recommendation also taken seriously by MacPhee and Novacek (1993).

In recent sequence-based and combined analyses of Afrotheria, tenrecs and golden moles are frequently supported in their own clade (e.g., Murphy et al., 2001b; Amrine-Madsen et al., 2003). However, in a few publications this has not been the case; for example, Murata et al.'s (2003) phylogenetic analysis of mtDNA sequences placed tenrecs as the sister group of a clade containing aardvarks, sengis, and golden moles (within which the latter two were sister taxa). Waddell and Shelley (2003: figure 10 based on their concatenated sequences) recovered an aardvark-tenrec clade to which their two golden mole genera comprised the sister taxon (figure 46.4C). Such an arrangement has no precedent in the literature and is highly dependent on the sampled genetic data and tree reconstruction methodology. Waddell and Shelley (2003) also figure a tenrec + sengi clade alongside golden moles + aardvarks (their figure 3 based on RAG1 plus G-fibrinogen) as well as a tree favoring golden moles + tenrecs (their figure 9 based on GHR). The latter arrangement remains the most common result across DNA-based studies and is the only one with some kind of taxonomic precedent. Although the weight of the morphological and molecular evidence favors a tenrec–golden mole clade, the instability of this grouping in some molecular phylogenetic analyses nevertheless suggests a relatively ancient divergence and brief period of common ancestry for the two lineages and leaves open the possibility that their shared zalambdodonty evolved independently (Seiffert, this volume, chapter 16).

Conclusions and Summary

High-level interrelationships of living mammals have never been so well understood as they are today. There is currently an unprecedented level of agreement among specialists on very specific hypotheses of mammalian phylogeny, including the placement of African insectivorans in a larger African clade, the Afrotheria. Intra-afrotherian relationships remain in comparison poorly understood. It is likely that additional sequence-based work will further improve the resolution of at least some nodes within Afrotheria; although the uncertainty regarding intrapaenungulate relations (for example) appears frustratingly impervious to molecular data sets of ever increasing size (cf. Waddell and Shelley, 2003 vs. Amrine-Madsen et al., 2003). There remains in any event a large body of data (anatomical, behavioral, genetic) with which competing hypotheses of living afrotherian phylogeny can be tested.

The same cannot be said for extinct afrotherians. The Paleogene fossil record of insectivoran-grade afrotherians is very poor and open to interpretation, and what does exist from the Miocene is not terribly informative. A decent case has been made for the recognition of new, insectivoran-grade fossils as stem tenrecs and golden moles from the Eo-Oligocene of Egypt (Seiffert et al., 2007); but even here decisive evidence for their placement will require more complete cranial and/or postcranial evidence. There is little indication as to how (or if) the few known extinct tenrecs and golden moles were adaptively different from their modern relatives; nor are there hints as to which of the other modern afrotherians might comprise their near relatives. Fortunately, the fossil record of paenungulates is better; and insights from the living radiations may also inform our understanding of afrotherian paleobiology (e.g., Poux et al., 2005, 2008).

One of the most drastic improvements to the "African" fossil record in recent years has been in the Late Cretaceous of Madagascar (Krause, 2003). However, as noted by Robinson and Seiffert (2004), Madagascar has had an independent history from the African mainland since the late Jurassic, and its Cretaceous fauna does not necessarily bear directly on biogeographic questions regarding the mainland. Fortunately, improved sampling of the continental African record is also underway, such as the Cretaceous (Krause et al., 2003), middle Eocene (Gunnell et al., 2003), and Oligocene (Stevens et al., 2006) of Tanzania; the Cretaceous of Mali (O'Leary et al., 2004); and the Oligocene of Ethiopia (Sanders et al., 2004). Ongoing work in the Paleocene/Eocene of North Africa (e.g., Gheerbrant et al., 2003, 2005; Tabuce et al., 2007) and Eocene/Oligocene of Egypt (e.g., Seiffert et al., 2007) continues to yield important discoveries that form the baseline of our growing understanding of Tertiary mammals of the African continent.

ACKNOWLEDGMENTS

For financial support R.J.A. thanks the Deutsche Forschungsgemeinschaft (grant AS 245/2-1), the European Commission's Research Infrastructure Action via the SYNTHESYS Project (GB-TAF 218), and the National Science Foundation USA (DEB 9800908). E.R.S.'s research has been funded by the Leakey Foundation and the U.S. National Science Foundation (BCS 0416164). For help with the manuscript and citations we thank Margaret Avery and Thomas Lehmann.

Literature Cited

Allard, M. W., and M. M. Miyamoto. 1992. Perspective: Testing phylogenetic approaches with empirical data, as illustrated with the parsimony method. *Molecular Biology and Evolution* 9:778–786.

Ameghino, F. 1905. Les édentés fossiles de France et d'Allemagne. *Annales du Muséum National de Buenos Aires* 13:175–250.

Amrine, H. M., and M. S. Springer. 1999. Maximum-likelihood analysis of the tethythere hypothesis based on a multigene data set and a comparison of different models of sequence evolution. *Journal of Mammalian Evolution* 6:161–176.

Amrine-Madsen, H., K.-P. Koepfli, R. K. Wayne, and M. S. Springer. 2003. A new phylogenetic marker, apolipoprotein B, provides compelling evidence for eutherian relationships. *Molecular Phylogenetics and Evolution* 28:225–240.

Andrews, C. W. 1906. *A Descriptive Catalogue of the Tertiary Vertebrata of the Fayum, Egypt.* British Museum (Natural History), London, 324 pp.

Arnason, U., J. A. Adegoke, K. Bodin, E. W. Born, Y. B. Esa, A. Gullberg, M. Nilsson, R. Short, X. Xu, and A. Janke. 2002. Mammalian mitogenomic relationships and the root of the eutherian tree. *Proceedings of the National Academy of Sciences, USA* 99:8151–8156.

Asher, R. J. 2001. Cranial anatomy in tenrecid insectivorans: Character evolution across competing phylogenies. *American Museum Novitates* 3352:1–54.

———. 2005. Insectivoran-grade placental mammals: Character evolution and fossil history; pp. 50–70 in K. D. Rose and J. D. Archibald (eds.), *The Rise of Placental Mammals: Origin and Relationships of the Major Clades.* Johns Hopkins University Press, Baltimore.

———. 2007. A web-database of mammalian morphology and a reanalysis of placental phylogeny. *BMC Evolutionary Biology* 7:108.

Asher, R. J., and D. M. Avery. 2010. New golden moles (Afrotheria, Chrysochloridae) from the Pliocene of South Africa. *Paleontologia Electronica* 13(1), 3A.

Asher, R. J., N. Bennett, and T. Lehmann. 2009. The new framework for understanding placental mammal evolution. *Bioessays* 31:853–864.

Asher, R. J., R. J. Emry, and M. C. McKenna. 2005. New material of *Centetodon* (Mammalia, Lipotyphla) and the importance of (missing) DNA sequences in systematic paleontology. *Journal of Vertebrate Paleontology* 25:911–923.

Asher, R. J., and M. Hofreiter. 2006. Tenrec phylogeny and the noninvasive extraction of nuclear DNA. *Systematic Biology* 55:181–194.

Asher, R. J., and T. Lehmann. 2008. Dental eruption in afrotherian mammals. *BMC Biology* 6:14.

Asher, R. J., S. Maree, G. Bronner, N. C. Bennett, P. Bloomer, P. Czechowski, M. Meyer, and M. Hofreiter. In press. A Phylogenetic Estimate for Golden Moles (Mammalia, Afrotheria, Chrysochloridae). *BMC Evolutionary Biology*.

Asher, R. J., M. C. McKenna, R. J. Emry, A. R. Tabrum, and D. G. Kron. 2002. Morphology and relationships of *Apternodus* and other extinct, zalambdodont, placental mammals. *Bulletin of the American Museum of Natural History* 243:1–117.

Asher, R. J., M. J. Novacek, and J. Geisler. 2003. Relationships of endemic African mammals and their fossil relatives based on morphological and molecular evidence. *Journal of Mammalian Evolution* 10:131–194.

Asher, R. J., and M. R. Sánchez-Villagra. 2005. Locking yourself out: Diversity among dentally zalambdodont therian mammals. *Journal of Mammalian Evolution* 12:265–282.

Avery, D. M. 1998. An assessment of the lower Pleistocene micromammalian fauna from Swartkrans Members 1–3, Gauteng, South Africa. *Geobios* 31:393–414.

———. 2000. Micromammals; pp. 305–338 in T. C. Partridge and R. R. Maud, (eds.), *The Cenozoic of Southern Africa*. Oxford University Press, New York.

———. 2001. The Plio-Pleistocene vegetation and climate of Sterkfontein and Swartkrans, South Africa, based on micromammals. *Journal of Human Evolution* 41:113–132.

Beliajeva, E. I., B. A. Trofimov, and V. J. Reshetov. 1974. General stages in evolution of late Mesozoic and early Tertiary mammalian fauna in central Asia; pp. 19–45 in N. N. Kramarenko et al. (eds.), *Mesozoic and Cenozoic Faunas and Biostratigraphy of Mongolia Joint Soviet-Mongolian Paleontological Expedition Transactions 1*. [In Russian]

Blainville, H. D. de. 1834. *Plan du Cours de Physiologie Générale et Comparée, Fait à la Faculté des Sciences de Paris, pendant les Années 1829, 1830, 1831 et 1832*. N.p.

Bronner, G. N. 1995. Cytogenetic properties of nine species of golden moles (Insectivora: Chrysochloridae). *Journal of Mammalogy* 76:957–971.

Bronner, G. N., M. Hoffman, P. J. Taylor, C. T. Chimimba, P. B. Best, C. A. Matthee, and T. J. Robinson. 2003. A revised systematic checklist of the extant mammals of the southern African subregion. *Durban Museum Novitates* 28: 56–95.

Bronner, G. N., and P. Jenkins. 2005. Afrosoricida; pp. 71–81 in D. E. Wilson and D. M. Reeder (eds.), *Mammal Species of the World*. Johns Hopkins University Press, Baltimore.

Broom, R. 1909. On the organ of Jacobson in *Orycteropus*. *Proceedings of the Zoological Society of London* 1909:680–683.

———. 1916. On the structure of the skull in *Chrysochloris*. *Proceedings of the Zoological Society of London* 1916:449–459.

———. 1941. On two Pleistocene golden moles. *Annals of the Transvaal Museum* 20:215–216.

———. 1948. Some South African Pliocene and Pleistocene mammals. *Annals of the Transvaal Museum* 21:1–38.

Burney, D. 2003. Madagascar's prehistoric ecosystems; pp. 47–51 in S. M. Goodman and J. P. Benstead (eds.), *The Natural History of Madagascar*. University of Chicago Press, Chicago.

Butler, P. M. 1984. Macroscelidea, Insectivora, and Chiroptera from the Miocene of East Africa. *Palaeovertebrata* 14:117–200.

———. 1985. The history of African insectivores. *Acta Zoologica Fennica* 173:215–217.

———. 1988. Phylogeny of the insectivores; pp. 117–141 in M. J. Benton (ed.), *The Phylogeny and Classification of the Tetrapods*, vol. 2. Clarendon Press, Oxford.

Butler, P. M., and A. T. Hopwood. 1957. Insectivora and Chiroptera from the Miocene rocks of Kenya colony. *Fossil Mammals of Africa* 13:1–35.

Carlsson, A. 1909. Die Macroscelididae und ihre Bezieungen zu den übrigen Insectivoren. *Zoologisches Jahrbuch* 28:249–400.

Carter, A. M., T. N. Blankenship, A. C. Enders, and P. Vogel. 2006. The fetal membranes of the otter shrews and a synapomorphy for Afrotheria. *Placenta* 27:258–268.

Clementz, M. T., K. A. Hoppe, and P. L. Koch. 2003. A paleoecological paradox: The habitat and dietary preferences of the extinct tethythere *Desmostylus*, inferred from stable isotope analysis. *Paleobiology* 29:506–519.

Corneli, P. S. 2002. Complete mitochondrial genomes and eutherian evolution. *Journal of Mammalian Evolution* 9:281–305.

Cote, S., L. Werdelin, E. R. Seiffert, and J. C. Barry. 2007. Additional material of the enigmatic early Miocene mammal *Kelba* and its relationship to the order Ptolemaiida. *Proceedings of the National Academy of Sciences, USA* 104:5510–5515.

Court, N. 1994. The periotic of *Moeritherium* (Mammalia, Proboscidea): Homology or homoplasy in the ear region of Tethytheria McKenna, 1975. *Zoological Journal of the Linnean Society* 112:13–28.

Debruyne, R. 2005. A case study of apparent conflict between molecular phylogenies: The interrelationships of African elephants. *Cladistics* 21:31–50.

DeGraaff, G. 1957. A new chrysochlorid from Makapansgat. *Palaeontologia Africana* 5:21–27.

DeJong, W. W., J. A. M. Leunissen, and G. J. Wistow. 1993. Eye lens crystallins and the phylogeny of placental orders: Evidence for a Macroscelid-Paenungulate clade? pp. 5–12 in F. S. Szalay, M. J. Novacek, and M. C. McKenna (eds.), *Mammal Phylogeny: Volume 1. Mesozoic Differentiation, Multituberculates, Monotremes, Early Therians, and Marsupials*. Springer, New York.

DeJong, W. W., M. A. M. Van Dijk, C. Poux, G. Kappé, T. Van Rheede, and O. Madsen. 2003. Indels in protein-coding sequences of Euarchontoglires constrain the rooting of the eutherian tree. *Molecular Phylogenetics and Evolution* 28:328–340.

DeJong, W. W., A. Zweers, and M. Goodman. 1981. Relationship of aardvark to elephants, hyraxes and sea cows from α-crystallin sequences. *Nature* 292:538–540.

Dobson, G. E. 1883. *A Monograph of the Insectivora: Part II. Potamogalidae, Chrysochloridae, Talpidae*. John van Voorst, London.

Douady, C. J., F. Catzeflis, D. J. Kao, M. S. Springer, and M. J. Stanhope. 2002a. Molecular evidence for the monophyly of Tenrecidae (Mammalia) and the timing of the colonization of Madagascar by Malagasy tenrecs. *Molecular Phylogenetics and Evolution* 22:357–363.

Douady, C. J., F. Catzeflis, J. Raman, M. S. Springer, and M. J. Stanhope. 2003. The Sahara as a vicariant agent, and the role of Miocene climatic events, in the diversification of the mammalian order Macroscelidea (elephant-shrews). *Proceedings of the National Academy of Sciences, USA* 100:8325–8330.

Douady, C. J. P. I. Chatelier, O. Madsen, W. W. DeJong, F. Catzeflis, M. S. Springer, and M. J. Stanhope. 2002b. Molecular phylogenetic evidence confirming the Eulipotyphla concept and in support of hedgehogs as the sister group to shrews. *Molecular Phylogenetics and Evolution* 25:200–209.

Douady, C. J., M. Scally, M. S. Springer, and M. J. Stanhope. 2004. "Lipotyphlan" phylogeny based on the growth hormone receptor gene: A reanalysis. *Molecular Phylogenetics and Evolution* 30:778–788.

Douzery, E., and F. M. Catzeflis. 1995. Molecular evolution of the mitochondrial 12S rRNA in Ungulata (Mammalia). *Journal of Molecular Evolution* 41:622–636.

Eisenberg, J. F., and E. Gould. 1970. The tenrecs: A study in mammalian behavior and evolution. *Smithsonian Contributions to Zoology* 27:1–137.

Emerson, G. L., C. W. Kilpatrick, B. E. McNiff, J. Ottenwalder, and M. W. Allard. 1999. Phylogenetic relationships of the order Insectivora based on complete 12S rRNA sequences from mitochondria. *Cladistics* 15:221–230.

Evans, F. G. 1942. The osteology and relationships of the elephant-shrews (Macroscelididae). *Bulletin of the American Museum of Natural History* 80:85–125.

Filhol, H. 1894. Observations concernant quelques mammifères fossiles nouveaux du Quercy. *Annales de Sciences Naturelles, Zoologie* 16(7):129–150.

Fischer, M. S. 1989. Hyracoids, the sister-group of perissodactyls; pp. 37–56 in D. R. Prothero and R. M. Schoch, R. M. (eds.), *The Evolution of Perissodactyls*. Oxford University Press, New York.

———. 1992. *Handbuch der Zoologie: Hyracoidea*. Walter de Gruyter, Berlin, 169 pp.

Fischer, M., and P. Tassy. 1993. The interrelation between Proboscidea, Sirenia, Hyracoidea, and Mesaxonia: The morphological evidence; pp. 217–234 in F. S. Szalay, M. J. Novacek, and McKenna, M. C. (eds.), *Mammal Phylogeny: Volume 2. Placentals.* Springer, New York.

Gaeth, A. P., R. V. Short, and M. B. Renfree. 1999. The developing renal, reproductive, and respiratory systems of the African elephant suggest an aquatic ancestry. *Proceedings of the National Academy of Sciences, USA* 96:5555–5558.

Gheerbrant, E. 1991. *Todralestes variabilis* n.g., n.sp., nouveau proteuthérien (Eutheria, Todralestidae fam. nov.) du paléocène du Maroc. *Comptes Rendus de l'Académie des Sciences, Série II,* 312:1249–1255.

———. 1992. Les mammifères Paléocènes du Bassin d'Ouarzazate (Maroc): I. Introduction générale et Palaeoryctidae. *Palaeontographica Abt. A* 224:67–132.

Gheerbrant, E. 2009. Paleocene emergence of elephant relatives and the rapid radiation of African ungulates. *Proceedings of the National Academy of Sciences, USA* 106:10717–10721.

Gheerbrant, E., J. Sudre, and H. Cappetta. 1996. A Paleocene proboscidean from Morocco. *Nature* 383:68–70.

Gheerbrant, E., J. Sudre, H. Cappetta, C. Mourer-Chauviré, E. Bourdon, M. Iarochène, M. Amaghzaz, and B. Bouya. 2003. The mammal localities of Grand Daoui Quarries, Ouled Abdoun Basin, Morocco, Ypresian: A first survey. *Bulletin de la Société Géologique de France* 174:279–293.

Gheerbrant, E., J. Sudre, M. Iarochène, and A. Moumni. 2001. First ascertained African "condylarth" mammals from the earliest Ypresian of the Ouled Abdoun Basin, Morocco. *Journal of Vertebrate Paleontology* 21:107–118.

Gheerbrant, E., J. Sudre, P. Tassy, M. Amaghzaz, B. Bouya, and M. Iarochène. 2005. New data on *Phosphatherium escuilliei* (Mammalia, Proboscidea) from the early Eocene of Morocco, and its impact on the phylogeny of Proboscidea and lophodont ungulates. *Geodiversitas* 27:239–333.

Gill, T. 1870. On the relations of the orders of mammals. *Science* 1870:267–270.

Gingerich, P. D. 2005. Aquatic adaptation and swimming mode inferred from skeletal proportions in the Miocene desmostylian *Desmostylus. Journal of Mammalian Evolution* 12:183–194.

Godinot, M., T. Smith, and R. Smith. 1996. Mode de vie et affinités de *Paschatherium* (Condylarthra, Hyopsodontidae) d'après ses os du tarse. *Palaeovertebrata* 25:225–242.

Goodman, S. 2003. Checklist to the extant land mammals of Madagascar; pp. 1187–1191 in S. M. Goodman and J. P. Benstead (eds.), *The Natural History of Madagascar.* University of Chicago Press, Chicago.

Gregory, W. K. 1910. The orders of mammals. *Bulletin of the American Museum of Natural History* 27:1–524.

Grubb, P., C. P. Groves, J. P. Dudley, and J. Shoshani. 2000. Living African elephants belong to two species: *Loxodonta africana* (Blumenbach, 1797) and *Loxodonta cyclotis* (Matschie, 1900). *Elephant* 2:1–4.

Gunnell G. F., B. F. Jacobs, P. S. Herendeen, J. J. Head, E. Kowalski, C. P. Msuya, F. A. Mizambwa, T. Harrison, J. Habersetzer, and G. Storch. 2003. Oldest placental mammal from sub-Saharan africa: Eocene microbat from Tanzania—Evidence for early evolution of sophisticated echolocation. *Palaeontologia Electronica* 5(2), 10 pp.

Hartenberger, J.-L. 1986. Hypothèse paléontologique sur l'origine des macroscelidea (Mammalia). *Comptes Rendus de l'Académie des Sciences, Paris, Série II,* 302:247–249.

Horovitz, I. 2004. Eutherian mammal systematics and the origins of South American ungulates as based on postcranial osteology. *Bulletin of the Carnegie Museum of Natural History* 36:63–79.

Hough, M. J. 1956. A new insectivore from the Oligocene of the Wind River Basin, Wyoming, with notes on the taxonomy of the Oligocene Tenrecoidea. *Journal of Paleontology* 30:531–541.

Jacobs, L., W. Anyonge, and C. Barry. 1987. A giant tenrecid from the Miocene of Kenya. *Journal of Mammalogy* 68:10–16.

Kazanci, N., S. Sen, G. Seyitoglu, L. de Bonis, G. Bouvrain, H. Araz, B. Varol, and L. Karadenizli. 1999. Geology of a new Late Miocene mammal locality in central Anatolia, Turkey. *Comptes Rendus de l'Académie des Sciences, Série II,* 329:503–510.

Kingdon, J. 1974. *East African Mammals: An Atlas of Evolution in Africa,* vol. IIA. University of Chicago Press, Chicago.

Kjer, K. M., and R. L. Honeycutt. 2007. Site specific rates of mitochondrial genomes and the phylogeny of Eutheria. *BMC Evolutionary Biology* 7:8.

Klein, R. G. 1977. The mammalian fauna from the Middle and Later Stone Age (Later Pleistocene) levels of Border Cave, Natal Province, South Africa. *South African Archaeological Bulletin* 32:14–27.

Kleinschmidt, T., J. Czelusniak, M. Goodman, and G. Braunitzer. 1986. Paenungulata: A comparison of the hemoglobin sequences from elephant, hyrax, and manatee. *Molecular Biology and Evolution* 3:427–435.

Krause, D. W. 2003. Late Cretaceous vertebrates of Madagascar: A window into Gondwanan biogeography at the end of the age of dinosaurs; pp. 40–47 in S. M. Goodman and J. P. Benstead (eds.), *The Natural History of Madagascar.* University of Chicago Press, Chicago.

Krause D. W., M. D. Gottfried, P. M. O'Connor, and E. M. Roberts. 2003. A Cretaceous mammal from Tanzania. *Acta Palaeontologica Polonica* 48:321–330.

Lavergne, A., E. Douzery, T. Stichler, F. M. Catzeflis, and M.S. Springer. 1996. Interordinal mammalian relationships: Evidence for paenungulate monophyly is provided by complete mitochondrial 12S rRNA sequences. *Molecular Phylogenetics and Evolution* 6:245–258.

Leche W. 1907. Zur Entwicklungsgeschichte des Zahnsystems der Säugetiere, zugleich ein Beitrag zur Stammengeschichte dieser Tiergruppe: Teil 2. *Zoologie (Stuttgart)* 49:1–157.

Le Gros Clark, W. E., and C. F. Sonntag. 1926. A monograph of *Orycteropus afer:* III. the skull, the skeleton of the trunk, and limbs. *Proceedings of the Zoological Society of London* 30:445–485.

Lehmann, T. 2006. Biodiversity of the Tubulidentata over geological time. *Afrotherian Conservation* 4:6–11.

Lehmann, T., P. Vignaud, A. Likius, and M. Brunet. 2005. A new species of Orycteropodidae (Mammalia, Tubulidentata) in the Mio-Pliocene of northern Chad. *Zoological Journal of the Linnean Society* 143:109–131.

Lehmann, U. 1984. Notiz über Säugetierreste von der Insel Samos in der Sammlung des Geologisch-Paläontologischen Instituts und Museums Hamburg. *Mitteilungen aus dem Geologisch-Paläontologischen Institut der Universität Hamburg* 57:147–156.

Lehmann, T. 2009. Phylogeny and systematics of the Orycteropodidae (Mammalia, Tubulidentata). *Zoological Journal of the Linnean Society* 155:649–702.

Lin, Y.-H., P. A. McLeanachan, A. R. Gore, M. J. Phillips, R. Ota, M. D. Hendy and M. Penny. 2002. Four new mitochondrial genomes and the increased stability of evolutionary trees of mammals from improved taxon sampling. *Molecular Biology and Evolution* 19:2060–2070.

Linnaeus, C. von. 1758 (1956). *Caroli Linnaei Systema Naturae: Regnum Animale. a Photographic Facsimile of the First Volume of the Tenth Edition (1758).* British Museum (Natural History), London.

Liu A. G. S. C., E. R. Seiffert, and E. L. Simons. 2008. Stable isotope evidence for an amphibious phase in early proboscidean evolution. *Proceedings of the National Academy of Sciences, USA* 105:5786–5791

MacPhee, R. D. E. 1987. The shrew tenrecs of Madagascar: Systematic revision and Holocene distribution of *Microgale* (Tenrecidae, Insectivora). *American Museum Novitates* 2889:1–45.

———. 1994. Morphology, adaptations, and relationships of *Plesiorycteropus,* and a diagnosis of a new order of eutherian mammals. *Bulletin of the American Museum of Natural History* 220:1–214.

MacPhee, R. D. E., and M. J. Novacek. 1993. Definition and relationships of Lipotyphla; pp. 13–31 in F. S. Szalay, M. J. Novacek, and M. C. McKenna (eds.), *Mammal Phylogeny: Volume 2. Placentals.* Springer, New York.

Madsen, O., M. Scally, C. J. Douady, D. J. Kao, R. W. DeBry, R. Adkins, H. M. Amrine, M. J. Stanhope, W. W. DeJong, and M. S. Springer. 2001. Parallel adaptive radiations in two major clades of placental mammals. *Nature* 409:610–614.

Madsen, O., D. Willemsen, B. M. Ursing, U. Arnason, and W. W. DeJong. 2002. Molecular evolution of the mammalian alpha 2B adrenergic receptor. *Molecular Biology and Evolution* 19:2150–2160.

Malia, M. J., R. M. Adkins, and M. W. Allard. 2002. Molecular support for Afrotheria and the polyphyly of Lipotyphla based on analyses of the growth hormone receptor gene. *Molecular Phylogenetics and Evolution* 24:91–101.

Matsumoto, H. 1926. Contribution to the knowledge of the fossil Hyracoidea of the Fayum, Egypt, with description of several new species. *Bulletin of the American Museum of Natural History* 56:253–350.

Matthew, W. D. 1906. Fossil Chrysochloridae in North America. *Science* 24:786–788.

McDowell, S. B. 1958. The Greater Antillean insectivores. *Bulletin of the American Museum of Natural History* 115:115–213.

McKenna, M. C. 1975. Toward a phylogeny and classification of the Mammalia; pp. 21–46 in W. P. Luckett and F. S. Szalay (eds.), *Phylogeny of the Primates: A Multidisciplinary Approach.* Plenum, New York.

McKenna, M. C., and S. K. Bell. 1997. *Classification of Mammals above the Species Level.* Columbia University Press, New York, 631 pp.

Meester, J. A. J., I. L. Rautenbach, N. J. Dippenaar, and L. M. Baker. 1986. Classification of southern African mammals. *Transvaal Museum Monograph* 5:1–359.

Mein, P., and M. Pickford. 2003. Insectivora from Arrisdrift, a basal Middle Miocene locality in southern Namibia. *Memoirs of the Geological Survey of Namibia* 19:143–146.

Mess, A., and A. M. Carter. 2006. Evolutionary transformations of fetal membrane characters in Eutheria with special reference to Afrotheria. *Journal of Experimental Zoology* 306B:140–163.

Moonen, J. J. M., P. Y. Sondaar, and S. T. Hussain. 1978. A comparison of larger fossil mammals in the stratotypes of the Chinji, Nagri and Dhok Pathan Formations (Punjab, Pakistan). *Proceedings of the Koninklijke Nederlandse Akademie van Wetenschappen, Series B: Palaeontology, Geology, Physics and Chemistry* 81:425–436.

Morales, J., M. Pickford, M. Salesa, and D. Soria. 2000. The systematic status of *Kelba* Savage, 1965, *Kenyalutra* Schmidt-Kittler, 1987, and *Ndamathaia* Jacobs et al., 1987 (Viverridae, Mammalia) and a review of early Miocene mongoose-like carnivores of Africa. *Annales de Paléontologie* 86:243–251.

Mouchaty, S. K, A. Gullberg, A. Janke, and U. Arnason. 2000. Phylogenetic position of the tenrecs (Mammalia: Tenrecidae) of Madagascar based on analysis of the complete mitochondrial genome sequence of *Echinops telfairi. Zoologica Scripta* 29:307–317.

Muizon C. de and R. L. Cifelli. 2000. The "condylarths" (archaic Ungulata, Mammalia) from the early Paleocene of Tiupampa (Bolivia): Implications on the origin of the South American ungulates. *Geodiversitas* 22:47–150.

Muldoon, K. M. 2006. Environmental change and human impact in southwestern Madagascar: Evidence from Ankilitelo cave. Unpublished PhD dissertation, Washington University, St. Louis, Mo.

Murata, Y., M. Nikaido, T. Sasaki, Y. Cao, Y. Fukumoto, M. Hasegawa, and N. Okada. 2003. Afrotherian phylogeny as inferred from complete mitochondrial genomes. *Molecular Phylogenetics and Evolution* 28:253–260.

Murphy, W. J., E. Eizirik, W. E. Johnson, Y. P. Zhang, O. A. Ryder, and S. J. O'Brien. 2001a. Molecular phylogenetics and the origin of placental mammals. *Nature* 409:614–618.

Murphy, W. J., E. Eizirik, S. J. O'Brien, O. Madsen, M. Scally, C. J. Douady, E. Teeling, O. A. Ryder, M. J. Stanhope, W. W. deJong, and M. S. Springer. 2001b. Resolution of the early placental mammal radiation using Bayesian phylogenetics. *Science* 294:2348–2351.

Nicoll, M. 1985. The biology of the giant otter shrew *Potamogale velox. National Geographic Society Research* 21:331–337.

Nikaido, M., Y. Cao, M. Harada, N. Okada, and M. Hasegawa. 2003a. Mitochondrial phylogeny of hedgehogs and monophyly of Eulipotyphla. *Molecular Phylogenetics and Evolution* 28:276–284.

Nikaido, M., and H. Nishihara, Y. Hukumoto and N. Okada. 2003b. Ancient SINEs from African endemic mammals. *Molecular Biology and Evolution* 20: 522–527.

Nishihara, H., M. Hasegawa, and N. Okada. 2006. Pegasoferae, an unexpected mammalian clade revealed by tracking ancient retroposon insertions. *Proceedings of the National Academy of Sciences, USA* 103:9929–9934.

Nishihara, H. Y. Satta, M. Nikaido, J. G. M. Thewissen, M. J. Stanhope, and N. Okada. 2005. A retroposon analysis of afrotherian phylogeny. *Molecular Biology and Evolution* 22:1823–1833.

Novacek, M. J. 1986. The skull of leptictid insectivorans and the higher-level classification of eutherian mammals. *Bulletin of the American Museum of Natural History* 183:1–111.

Novacek, M. J., and A. R. Wyss. 1986. Higher level relationships of the recent eutherian orders: the morphological evidence. *Cladistics* 2:257–287.

Nowak, R. M. 1999. *Walker's Mammals of the World.* 6th ed. Johns Hopkins University Press, Baltimore, 1,919 pp.

O'Leary M. A., E. M. Roberts, J. J. Head, F. Sissoko, and M. L. Bouaré. 2004. Titanosaurian (Dinosauria: Sauropoda) remains from the "continental intercalaire" of Mali. *Journal of Vertebrate Paleontology* 24:923–930.

Owen, R. 1848. Description of teeth and portions of jaws of two extinct anthracotheroid quadrupeds (*Hyopotamus vectianus* and *H. bovinus*) discovered by the Marchioness of Hastings in the Eocene deposits on the N.W. coast of the Isle of Wight. *Quarterly Journal of Geological Society of London* 4:104–141.

Patterson, B. 1965. The fossil elephant-shrews (Family Macroscelididae). *Bulletin of the Museum of Comparative Zoology* 135:295–335.

———. 1975. The fossil aardvarks (Mammalia: Tubulidentata). *Bulletin of the Museum of Comparative Zoology* 147:185–237.

Piccinini M, T. Kleinschmidt, T. Gorr, R. E. Weber, H. Kunzle, and G. Braunitzer. 1991. Primary structure and oxygen-binding properties of the hemoglobin from the lesser hedgehog tenrec (*Echinops telfairi*, Zalambdodonta): Evidence for phylogenetic isolation. *Biological Chemistry Hoppe-Seyler* 372:975–989.

Pickford, M. 1975. New fossil Orycteropodidae (Mammalia, Tubulidentata) from East Africa. *Netherlands Journal of Zoology* 25:57–88.

Pickford, M., and P. Andrews. 1981. The Tinderet Miocene sequence in Kenya. *Journal of Human Evolution* 10:11–33.

Pickford, M., and B. Senut. 1997. Cainozoic mammals from coastal Namaqualand, South Africa. *Paleontologia Africana* 34:199–217

Pocock, T. N. 1987. Plio-Pleistocene mammalian microfauna in southern Africa: A preliminary report including description of two new fossil muroid genera (Mammalia: Rodentia). *Paleontologia Africana* 26:69–91.

Poduschka, W., and C. Poduschka. 1985. Zur Frage des Gattungsnamens von *"Geogale" aletris* Butler und Hopwood, 1957 (Mammalia, Insectivora) aus dem Miozän Ostafrikas. *Zeitschrift für Säugetierkunde* 50:129–140.

Porter, C. A., M. Goodman, and M. J. Stanhope. 1996. Evidence on mammalian phylogeny from sequences of exon 28 of the von Willebrand factor gene. *Molecular Phylogenetics and Evolution* 5:89–101.

Poux, C., O. Madsen, J. Glos, W. W. DeJong, and M. Vences. 2008. Molecular phylogeny and divergence times of Malagasy tenrecs: Influence of data partitioning and taxon sampling on dating analyses. *BMC Evolutionary Biology* 8:102.

Poux, C., O. Madsen, E. Marquard, D. R. Vieites, W. W. DeJong, and M. Vences. 2005. Asynchronous colonization of Madagascar by the four endemic clades of primates, tenrecs, carnivores, and rodents as inferred from nuclear genes. *Systematic Biology* 54:719–730.

Prothero D. R., and R. M. Schoch (eds.). 1989. *The Evolution of Perissodactyls.* Oxford University Press, New York.

Radinsky, L. B. 1969. The early evolution of the Perissodactyla. *Evolution* 23:308–328.

Rainey W. E., J. M. Lowenstein, V. M. Sarich, and D. M. Magor. 1984. Sirenian molecular systematics-including the extinct Steller's sea cow (*Hydrodamalis gigas*). *Naturwissenschaften* 71:586–588.

Rasmussen, D. T., and E. L. Simons. 2000. Ecomorphological diversity among Paleogene hyracoids (Mammalia): A new cursorial browser from the Fayum, Egypt. *Journal of Vertebrate Paleontology* 20:167–176.

Rasmussen, D. T., M. Gagnon, and E. L. Simons. 1990. Taxeopody in the carpus and tarsus of Oligocene Pliohyracidae (Mammalia: Hyracoidea) and the phyletic position of hyraxes. *Proceedings of the National Academy of Sciences, USA* 87:4688–4691.

Robinson, T. J., B. Fu, M. A. Ferguson-Smith, and F. Yang. 2004. Cross-species chromosome painting in the golden mole and elephant-shrew: Support for the mammalian clades Afrotheria and Afroinsectiphillia but not Afroinsectivora. *Proceedings of the Royal Society of London, B* 271:1477–1484.

Robinson, T. J., and E. R. Seiffert. 2004. Afrotherian origins and interrelationships: New views and future prospects. *Current Topics in Developmental Biology* 63:37–60.

Roca, A. L., G. K. Bar-Gal, E. Eizirik, K. M. Helgen, R. Maria, M. S. Springer, S. J. O'Brien, and W. J. Murphy. 2004. Mesozoic origin for West Indian insectivores. *Nature* 429:649–651.

Roca, A. L., N. Georgiadis, J. Pecon-Slattery, and S. J. O'Brien. 2001. Genetic evidence for two species of elephant in Africa. *Science* 293:1473–1477.

Rose, K. D. and R. J. Emry. 1983. Extraordinary fossorial adaptations in the Oligocene palaeanodonts *Epoicotherium* and *Xenocranium. Journal of Morphology* 75:33–56.

Roux, G. H. 1947. The cranial development of certain Ethiopian "insectivores" and its bearing on the mutual affinities of the group. *Acta Zoologica* 28:165–397.

Salton, J., and F. S. Szalay. 2004. The tarsal complex of Afro-Malagasy Tenrecoidea: A search for phylogenetically meaningful characters. *Journal of Mammalian Evolution* 11:73–104.

Sánchez-Villagra, M. R., Y. Narita, and S. Kuratani. 2007. Thoracolumbar vertebral number in mammals and a potential morphological synapomorphy for Afrotheria. *Systematics and Biodiversity* 5:1–7.

Sanders W. J., J. Kappelman, and D. T. Rasmussen. 2004. New large-bodied mammals from the late Oligocene site of Chilga, Ethiopia. *Acta Palaeontologica Polonica*. 49:365–392.

Sarich, V. 1993. Mammalian systematics: Twenty-five years among their albumins and transferrins; pp. 103–113 in F. S. Szalay, M. J. Novacek, and M. C. McKenna (eds.), *Mammal Phylogeny: Volume 2. Placentals*. Springer, New York.

Scally, M., O. Madsen, C. J. Douady, W. W. DeJong, M. J. Stanhope, and M. S. Springer. 2001. Molecular evidence for the major clades of placental mammals. *Journal of Mammalian Evolution* 8:239–277.

Schwartz, G. T., D. T. Rasmussen, and R. J. Smith. 1995. Body size diversity and community structure of fossil hyracoids. *Journal of Mammalogy* 76:1088–1099.

Seiffert, E. R. 2003. A phylogenetic analysis of living and extinct afrotherian mammals. Unpublished PhD dissertation, Duke University.

———. 2007. A new estimate of afrotherian phylogeny based on simultaneous analysis of genomic, morphological, and fossil evidence. *BMC Evolutionary Biology* 7:224.

Seiffert E. R., and E. L. Simons. 2000. *Widanelfarasia*, a diminutive new placental from the late Eocene of Egypt. *Proceedings of the National Academy of Sciences, USA* 97:2646–2651.

Seiffert E. R., E. L. Simons, T. M. Ryan, T. M. Bown, and Y. Attia. 2007. New remains of Eocene and Oligocene Afrosoricida (Afrotheria) from Egypt, with implications for the origin(s) of afrosoricid zalambdodonty. *Journal of Vertebrate Paleontology* 27:963–972.

Shoshani, J. 1993. Hyracoidea-Tethytheria affinity based on myological data; pp. 235–256 in F. S. Szalay, M. J. Novacek, and M. C. McKenna (eds.), *Mammal Phylogeny: Volume 2. Placentals*. Springer, New York.

———. 2005. Proboscidea; pp. 90–91 in Wilson, D. E., and D. M. Reeder (eds.), 2005. *Mammal Species of the World*. Johns Hopkins University Press, Baltimore.

Shoshani, J., and P. Tassy. 2005. Advances in proboscidean taxonomy and classification, anatomy and physiology, and ecology and behavior. *Quaternary International* 126–128:5–20.

Simonetta, A. M. 1968. A new golden mole from Somalia with an appendix on the taxonomy of the family Chrysochloridae (Mammalia: Insectivora). *Monitore Zoologico Italiano* 2(suppl.):27–55.

Simons, E. L., P. A. Holroyd, and T. M. Bown. 1991. Early Tertiary elephant-shrews from Egypt and the origin of the Macroscelidea. *Proceedings of the National Academy of Sciences, USA* 88:9734–9737.

Simpson, G. G. 1945. The principles of classification and a classification of mammals. *Bulletin of the American Museum of Natural History* 85:1–350.

Sonntag, C. F. 1925. A monograph of *Orycteropus afer*: I. Anatomy except the nervous system, skin, and skeleton. *Proceedings of the Zoological Society of London* 23:331–437.

Sonntag, C. F., and H. H. Woolard. 1925. A monograph of *Orycteropus afer*: II. Nervous system, sense organs and hairs. *Proceedings of the Zoological Society of London* 1925:1185–1235.

Springer, M. S., G. C. Cleven, O. Madsen, W. W. DeJong, V. G. Waddell, H. M. Amrine, and M. J. Stanhope. 1997. Endemic African mammals shake the phylogenetic tree. *Nature* 388:61–64.

Springer, M. S., W. J. Murphy, E. Eizirik, and S. J. O'Brien. 2003. Placental mammal diversification and the Cretaceous-Tertiary boundary. *Proceedings of the National Academy of Sciences, USA* 100:1056–1061.

Springer, M. S., M. J. Stanhope, O. Madsen, and W.W. DeJong. 2004. Molecules consolidate the placental mammal tree. *Trends in Ecology and Evolution* 19:430–438.

Stanhope, M. J., V. G. Waddell, O. Madsen, W. W. DeJong, S. B. Hedges, G. C. Cleven, D. Kao, and M. S. Springer. 1998. Molecular evidence for multiple origins of the Insectivora and for a new order of endemic African mammals. *Proceedings of the National Academy of Sciences, USA* 95:9967–9972.

Stephenson, P. 2003. Lipotyphla (ex Insectivora): *Geogale aurita*, large eared tenrec; pp. 1265–1267 in S. M. Goodman and J. P. Benstead (eds.), *The Natural History of Madagascar*. University of Chicago Press, Chicago.

Stevens, N. J., P. M. O'Connor, M. D. Gottfried, E. M. Roberts, S. Ngasala, and M. R. Dawson. 2006. *Metaphiomys* (Rodentia: Phiomyidae) from the Paleogene of southwestern Tanzania. *Journal of Paleontology* 80:407–410.

Tabuce, R., M. T. Antunes, R. Smith, and T. Smith. 2006. Dental and tarsal morphology of the European Paleocene/Eocene "condylarth" mammal *Microhyus*. *Acta Palaeontologica Polonica* 51:37–52.

Tabuce, R., B. Coiffait, P. E. Coiffait, M. Mahboubi and J.-J. Jaeger. 2001. A new genus of Macroscelidea (Mammalia) from the Eocene of Algeria: A possible origin of elephant-shrews. *Journal of Vertebrate Paleontology* 21:535–546.

Tabuce, R., R. J. Asher, and T. Lehmann. 2008. Afrotherian mammals: a review of current data. *Mammalia* 72:2–14.

Tabuce, R., L. Marivaux, M. Adaci, M. Bensalah, J. L. Hartenberger, M. Mahboubi, F. Mebrouk, P. Tafforeau, and J.-J. Jaeger. 2007. Early Tertiary mammals from North Africa reinforce the molecular afrotheria clade. *Proceedings of the Royal Society of London, B* 274:1159–1166.

Thewissen, J. G. M. 1985. Cephalic evidence for the affinities of Tubulidentata. *Mammalia* 49:257–284.

Turnbull, W. D., and C. A. Reed. 1967. *Pseudochrysochloris*, a specialized burrowing mammal from the early Oligocene of Wyoming. *Journal of Paleontology* 41:623–631.

Vogel, P. 1983. Contribution a l'ecologie et a la zoogéographie de *Micropotamogale lamottei* (Mammalia, Tenrecidae). *Revue d'Écologie (La Terre et la Vie)* 38:37–48.

Waddell, P. J., H. Kishino, and R. Ota. 2001. A phylogenetic foundation for comparative mammalian genomics. *Workshop on Genome Informatics* 12:141–154.

Waddell, P. J., and S. Shelley. 2003. Evaluating placental inter-ordinal phylogenies with novel sequences including RAG1, γ-fibrinogen, ND6, and mt-tRNA, plus MCMC-driven nucleotide, amino acid, and codon models. *Molecular Phylogenetics and Evolution* 28:197–224.

Werdelin, L., and Å. Nilsonne. 1999. The evolution of the scrotum and testicular descent in mammals: A phylogenetic view. *Journal of Theoretical Biollogy* 196:61–72.

Whidden, H. P. 2002. Extrinsic snout musculature in Afrotheria and Lipotyphla. *Journal of Mammalian Evolution* 9:161 184.

Wible, J. R., G. W. Rougier, M. J. Novacek, and R. J. Asher. 2007. Cretaceous eutherians and Laurasian origin for placental mammals near the K/T boundary. *Nature* 447:1003–1006.

Wible, J. R., G. W. Rougier, M. J. Novacek, and M. C. McKenna. 2001. Earliest eutherian ear region: A petrosal referred to *Prokennalestes* from the Early Cretaceous of Mongolia. *American Museum Novitates* 3322:1–44.

Wildman, D. E., M. Uddin, J. C. Opazo, G. Liu, V. Lefort, S. Guindon, O. Gascuel, L. I. Grossman, R. Romero, and M. Goodman. 2007. Genomics, biogeography, and the diversification of placental mammals. *Proceedings of the National Academy of Sciences, USA* 104:14395–14400.

Wilson, D. E., and D. M. Reeder (eds). 2005. *Mammal Species of the World*. Johns Hopkins University Press, Baltimore, 2,142 pp.

Winge, H. 1941. *The Interrelationships of the Mammalian Genera*, vol. 1 [translated from Danish]. C. A. Reitzels Forlag, Copenhagen, 417 pp.

Zack, S. P., T. A. Penkrot, J. I. Bloch, and K. D. Rose. 2005a. Affinities of "hyopsodontids" to elephant-shrews and a Holarctic origin of Afrotheria. *Nature* 434:497–501.

Zack, S. P., T. A. Penkrot, D. W. Krause, and K. D. Rose. 2005b. A new apheliscine "condylarth" mammal from the late Paleocene of Montana and Alberta and the phylogeny of "hyopsodontids." *Acta Palaeontologica Polonica* 50:809–830.

FORTY-SEVEN

Mammal Species Richness in Africa

PETER ANDREWS AND EILEEN M. O'BRIEN

The distribution and diversity of mammals across Africa have long been attributed to differences in vegetation or climate (Pianka, 1966; Begon et al., 1990; Rosenzweig, 1997). Our earlier empirical study of southern African mammals strongly supports this interpretation (Andrews and O'Brien, 2000), even at the gross scale of spatial resolution needed to empirically measure differences in climate. In southern Africa, variability in woody plant species richness alone accounts for 75% of the variability in mammal species richness (Andrews and O'Brien, 2000). Of the climatic variables, only thermal seasonality approached this figure, accounting for 69% of mammal richness variability, while annual measures of temperature, precipitation and energy account for only 14 to 35% of variability. O'Brien (1993; O'Brien et al., 2000) has shown that 85% of the spatial variation in southern Africa's woody plant richness, and thus vegetation, is explained by the spatial variation in climate.

In our analyses, we investigated total mammal richness (termed *all mammals* here) and we also considered mammal distributions excluding bats. Bats are not commonly found in fossil deposits, so for comparison with fossil faunas they are normally excluded from analysis. We have further distinguished between different subsets of mammals based on size, spatial, and dietary guilds, as it is likely that some subsets are better indicators of change in vegetation and mammal diversity than others. The majority of mammal species are small (<10 kg) and have restricted foraging/distributional ranges, so that their diversity patterns should be affected differently by climate change compared with large mammals. Similarly, woody plants are a primary source of food and shelter for frugivorous and arboreal species, but less so for grazing mammals. Insect species richness increases as vegetation shifts from desert to rain forest (Mound and Waloff, 1978), and so the distribution of insectivorous mammals may co-vary with vegetation type. Overall we expect mammal richness to co-vary most strongly with differences in vegetation (woody plant richness) in a fashion similar to that observed for southern Africa, increasing as woody plant richness increases and as vegetation shifts from desert to lowland rain forest. Woody plant richness, however, is most strongly correlated with rainfall and energy (minimum potential evapotranspiration). Therefore, the mapped spatial variation in mammal richness should manifest a pattern broadly similar to that seen in maps of Africa's woody plant richness, floristic zones, and, to a lesser extent, energy and rainfall.

Given the significance of such findings for understanding and modeling the potential effects of environmental change on the distribution and diversity of mammals, both today and in the past, we have expanded our focus to the whole continent of Africa, following same methods and protocols of the southern African study. The aim is threefold: (1) to describe present-day patterns in mammal richness across Africa, in terms of both mammals in general and between ecologically meaningful subsets of mammals, and to compare these patterns with those for Africa's vegetation, woody plant richness, annual rainfall, and topography; (2) to evaluate data quality (e.g., whether it is representative); and (3) to test the implications of variations in mammal richness arising from the southern Africa study. Statistical analyses of how African mammal richness relates to geographic differences in (woody) plant species richness and climate will be reported elsewhere.

The Study Area

The study area encompasses the whole of continental Africa, extending from roughly 37°N to 35°S (figure 47.1). Its physiography is diverse (Moreau, 1966; O'Brien and Peters, 1999). Its most pronounced physical feature is the elevational division into High Africa (>1,000 m a.s.l.) and Low Africa (<200 m a.s.l.)(figure 47.1B). High Africa encompasses most of eastern, east-central, and the eastern half of South Africa. Low Africa encompasses the rest of Africa, with scattered uplands reaching elevations >500 m. In both cases, the terrain is further modified in three ways: first by tectonic features, including mountain ranges, volcanoes, plateaus, interior drainage basins (e.g., Lake Chad), tectonic rift zones (Ethiopia to Malawi), and so forth; second by drainage systems (e.g., Nile River), perennial (lakes/rivers, e.g., Lake Tanganyika, Congo River), seasonal (Sand River), or ephemeral (pans); and third by edaphic (soil) conditions that range from extremely fertile to extremely poor and even toxic (soils derived from serpentine rock), and from extremely wet to extremely dry soil (i.e., wetlands to bare rock and sand seas). This physiographic diversity, when combined with differences in climate, weather, and microclimate conditions,

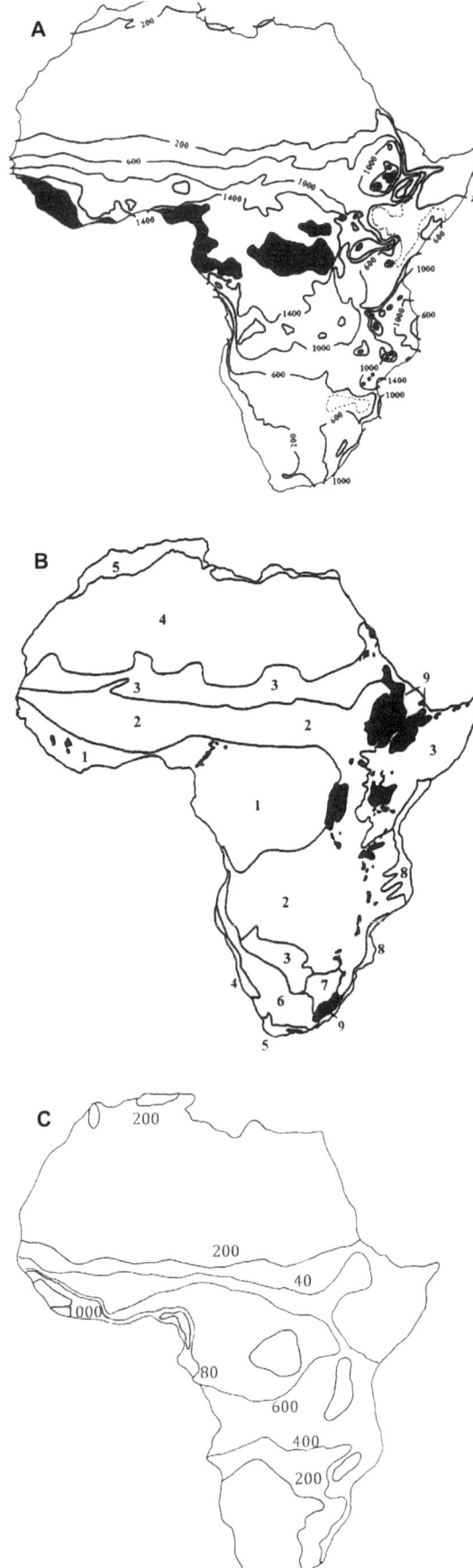

determines the abiotic environmental conditions limiting the distributional ranges of Africa's flora and fauna.

CLIMATE

Throughout Africa, the amount and timing of solar energy and rainfall (figure 47.1A) determine the timing and capacity for plant activity (photosynthesis). This in turn determines the availability of vegetation to support mammalian faunas and hence determines trophic structure. Year-round or bimodal rainfall occurs in low latitudes (e.g., west Central Africa and eastern Highland Africa, respectively). Summer rainfall (monsoon) conditions are present throughout the midlatitudes both north and south of the equator. Winter rainfall occurs along the northernmost and southwestern coastal regions of the continent (e.g., lands bordering the Mediterranean and Namibia, respectively). The range in annual rainfall is from close to zero to greater than 1,800 mm. (see figure 47.1A; Thornthwaite and Mather, 1955, 1962).

The geographic location of Africa means that virtually the whole continent (35°N and S) is located within the zone of surplus energy. Energy from the sun is always sufficient for plant growth within the tropics (23.5°N and S), and as a result the demand for water for transpiration/evaporation is high year-round. Poleward of the tropics, the intensity of insolation decreases while seasonal oscillations increase, contributing to near-freezing minimum (winter) temperatures at Africa's highest latitudes. In contrast, maximum insolation is almost the same throughout Africa. This is not the case for maximum temperatures, however, for temperature decreases as elevation increases above sea level; and evaporation cools air, with geographic differences in the amount and duration of rainfall causing differences in the capacity for atmospheric cooling. In High Africa (figure 47.1B), for example, elevation results in decreased temperatures year-round. For all of Africa, temperatures can vary dramatically at the same latitude. In equatorial latitudes, for example, annual temperature ranges from 25°C (equatorial wet) to greater than 45°C (equatorial dry), with variations depending on the interaction between Africa's topographic relief and atmospheric-oceanic thermodynamics. Precipitation may be either year-round or bimodal, or seasonal with most rain during the hot season (i.e., summer rainfall) or during the cold season (i.e. winter rainfall)(Thornthwaite and Mather 1955, 1962).

VEGETATION

Africa has more than 40,000 plant species, with more than 24,000 of them in southern Africa alone (White, 1983). As elsewhere in the world, woody plants (trees, shrubs, etc.) are

FIGURE 47.1 A) Mean annual precipitation in Africa. Areas shaded black have precipitation >1,800 mm. B) African physiognomic vegetation regions: 1, tropical lowland rainforest; 2, deciduous woodland; 3, Acacia bushland and woodland savanna; 4, desert; 5, Mediterranean sclerophyllous woodland; 6, Karoo desert; 7, highveldt grassland; 8, eastern coastal evergreen forest; 9, Afromontane vegetation. Based on White (1983). C) Species richness of woody plants based on predictions from IGM 2 (Field et al., 2005) based on climate station data from Thornthwaite and Mather (1962, 1965)(N = 980). These data are applied to the equal area grid as shown in figure 47.2. Modified from O'Brien and Peters, 1999.

the dominant and longest-living component of the vegetation (at least decades, usually hundreds of years), making this plant life form a good indicator of prevailing vegetation structure and environmental conditions. Based on empirical data for southern Africa (IGM-1 and 2; Field et al., 2005), woody plant richness (counted in 25,000 km² sampling units) varies across Africa following predictions from IGM-2 (figure 47.1C), increasing from near zero in absolute desert to greater than 1,100 species in equatorial lowland rain forest (O'Brien, 1998). These predictions were tested empirically in Kenya and in southern Africa, where woody plant distributions have been well recorded (15°–35°S), with species richness ranging from less than 10 to more than 450 species per unit area (Field et al., 2005, and O'Brien, 1993, respectively).

Ecophysiognomic vegetation zones from desert to tropical lowland rain forest are present in Africa (figure 47.1B). Excluding the Sahara, the dominant vegetation zone in Africa is deciduous forest and woodland (zone 2 in figure 47.1B: Sudano-Zambezian woodlands north and south of the equatorial zone), followed by lowland + coastal rain forest (zones 1 and 8), bushland, shrubland, and xeric scrubland (zones 3, 6, and 7). In mountainous areas and along escarpments, there is vertical zonation in vegetation, as well as leeward and windward differences, due to orographic rainfall and rainshadows. Throughout Africa, woody plant richness, structure, and complexity increase across Africa in a fashion similar to that of annual precipitation (compare figure 47.1A with 1C).

The same pattern can be seen in the distribution of floristic zones, or phytochoria (White, 1981, 1983). While vegetation physiognomy informs about vegetation structure (rain forest, bushland, etc.), phytochoria indicate their floristic complexity. For example, the Congo-Guinean phytochorion coincides with lowland rain forest (zone 1, figure 47.1B) and has the greatest number of floristic elements (species), many of which are endemics. The Congo-Guinean is surrounded by the Zambezian phytochorion in the Southern Hemisphere and its counterpart, the Sudanian, in the Northern Hemisphere. Both are characterized by deciduous forest and woodland (and coincide with zone 2, figure 47.1B). They are also floristically similar to each other, but the Sudanian has fewer species and fewer endemics than the Zambezian (White, 1983). All three phytochoria share some floristic elements with each other, but none of them have endemics in common. Similar relationships apply between other phytochoria and their associated vegetation zones. However, for reasons not yet established, there appear to be fewer plant species in the Northern Hemisphere of Africa than in the Southern Hemisphere, even when comparing analogous vegetation/floristic zones and climatic conditions Within each vegetation/floristic zone, heterogeneity of vegetation type (and thus habitat diversity) is indexed by differences in woody plant richness.

A striking feature of African plant physiognomy is the existence of a major ecotone complex in eastern High Africa (White, 1983). Roughly speaking, it stretches west to east from the Western Rift to Mt. Kenya, northward into Ethiopia and southward into Tanzania. In this region, several ecophysiognomic vegetation zones converge, as do five phytochoria: the Congo-Guinean, Zambezian, and Sudanian (mentioned earlier), plus Afromontane (zone 9, figure 47.1B, vertical vegetation zonation), and Somali-Masai Steppe (zone 3, low woodland and thicket). When coupled with this region's edaphic, hydrological, physiographic, and climatic complexity, the results are a complex and diverse mosaic of vegetation types that range from rain forest to semiarid scrub, and edaphic grasslands (e.g., Serengeti), with

a north-south corridor enabling dispersal of plant/animal species adapted to seasonal/arid climate conditions (Van Zinderen Bakker, 1969). In western Africa, the east-west stretch of lowland rain forest is a dispersal barrier to arid-adapted plants.

Methodology

This study follows the same methods and sampling strategy used to study mammal richness in southern Africa (detailed in Andrews and O'Brien, 2000). Our intention is also to provide data for future quantitative analyses of climate's relationship to the distribution and diversity of African mammals. At the present time, species range maps available for approximately half of Africa's mammal species are either incomplete or nonexistent. It is expected, however, that species richness values derived from these maps are sufficiently representative for describing general trends and relationships in mammal richness.

MAMMAL RICHNESS DATA

Species richness is the number of species per unit sampling area. It takes no account of relative abundance, which can only be measured at very discrete (local) scales of spatial resolution (i.e., ground sampling within a habitat). At the macroscale, it describes "geographic" richness, the richness resulting from differential overlap in the distributional ranges of extant species (cf. O'Brien, 1993). To allow systematic comparisons of climate data, plant richness, and mammal richness, the same methodology and sampling strategy employed by O'Brien (1993; Andrews and O'Brien 2000) has been used here to determine mammal richness patterns. A grid matrix of 1,236 equal-area grid cells (each cell equal to 25,000 km², cf. Griffiths, 1976) was laid out cartographically across continental Africa (figure 47.2), and this was overlain by mammal species range maps to determine the presence or absence of a mammal species per cell. Presence-absence data were then compiled to determine the total number of species, or species richness, per cell. Only 1,079 cells are truly equal-area, for 157 cells overlap the sea or large lakes. Our analyses of richness data are limited to these 1,079 equal-area cells. Richness data for higher taxonomic levels were compiled by aggregating species data per cell to generate genus richness values per cell, aggregating genus data per cell to obtain family richness values per cell, and so forth. The layout of the grid is shown in figure 47.2.

The sources of range maps for African mammals were Kingdon (1971–1982, 1997) and Smithers (1983), with additional data from Wilson and Reeder (1992), Frandsen (1992), and Stuart and Stuart (1988). These maps include only present-day ranges for each mammal species. No attempt has been made to take account of possible human impact beyond what has been done by the original sources. All exotic species and humans are excluded. Lastly, distribution maps for many of the smaller mammals are known to be incomplete, because some parts of Africa have been less intensively studied than others, the exceptions being Kingdon's (1971–1982, 1997) work in East Africa and Smithers's (1983) for southern Africa. Recognizing this problem, we recorded mammal species suspected or known to have incomplete distribution data by genus, with genus ranges being used in both species and genus-level analyses.

The Effects of Missing Distribution Data

Our results are based on the distributions of 562 taxa of mammals in Africa. The total number of species recorded by Kingdon (1997) is more than double that figure at 1,198,

FIGURE 47.2 Equal-area grid for Africa. Coordinates were denoted by letters, from A in the west to tt in the east, and by numbers from 1 in the south to 50 in the north. The total numbers of mammal species excluding bats are shown for each grid cell. The position of the equator is shown along the top of row 24.

Grid data (row number, Latitude labels on right; Column No. A B C D E F G H I J K L M N O P Q R S T U V W X Y Z aa bb cc dd ee ff gg hh ii jj kk ll mm nn oo pp qq rr ss tt uu; Longitude: 17 14 11 8 5 2 Degrees West, 1 4 7 10 13 16 19 22 25 28 31 34 37 40 43 46 49 52 Degrees East):

- Row 51 (Lat 37): 29 30 30 30 30 28 27
- Row 50: 30 29 29 32 32 33 34 34 34 31 28
- Row 49 (Lat 34): 28 29 31 31 32 33 35 35 33 33 32 30
- Row 48: 27 28 29 33 32 33 30 31 29 28 28 25 24 22 21   25 25   24
- Row 47 (Lat 31): 29 31 30 33 31 32 27 24 22 21 20 17 19 19 18 18 19 20 25 26 25 23 22 22 24 29 24
- Row 46: 31 30 30 30 26 25 22 18 18 19 19 19 18 19 18 20 20 21 21 18 17 15 18 19 21 26 24
- Row 45 (Lat 29): 24 27 25 25 24 23 21 18 16 15 14 14 15 16 17 18 15 14 14 14 12 10 9 8 7 8 12 20 22 17
- Row 44: 20 22 24 25 23 22 18 15 12 14 12 13 16 15 14 15 14 12 11 10 9 10 7 6 6 6 7 7 14 18 18
- Row 43 (Lat 26): 19 20 23 24 23 18 17 13 11 11 15 15 16 17 17 16 16 13 11 9 8 9 9 7 7 7 7 8 11 17 18 16
- Row 42: 18 19 20 22 21 18 15 17 12 10 11 15 17 17 18 19 18 16 16 13 8 9 11 11 11 10 10 8 9 9 10 14 17 19
- Row 41 (Lat 23): 18 17 20 21 15 14 13 12 13 14 17 18 17 19 20 19 15 14 13 12 16 17 16 15 13 13 9 11 11 10 14 19 19 18
- Row 40: 21 24 23 18 15 15 15 15 16 17 19 20 20 21 21 18 16 15 14 16 20 21 18 16 15 13 10 11 12 13 14 23 19 20
- Row 39 (Degrees North 20): 26 28 26 25 26 19 19 22 21 21 21 22 21 19 21 23 22 19 19 17 17 17 18 17 15 16 14 12 11 12 16 20 27 27 28 29
- Row 38: 39 35 35 33 32 30 30 30 33 35 28 26 23 22 28 29 28 21 22 21 22 18 20 18 20 23 23 23 19 19 22 26 28 30 34 39 38
- Row 37 (Lat 18): 44 45 46 44 42 40 39 40 40 44 45 43 35 33 32 29 34 33 32 30 30 29 30 30 29 32 29 28 26 26 28 30 30 30 35 39 43 46
- Row 36: 50 51 53 51 48 48 47 47 50 51 50 50 43 40 37 38 37 38 37 36 34 34 34 32 33 34 34 32 32 32 30 31 29 31 38 43 53 60 53 47
- Row 35 (Lat 15): 60 64 68 64 64 58 61 60 58 52 53 52 52 52 52 51 51 53 55 53 54 53 52 47 48 46 43 43 46 45 42 41 40 41 50 57 58 63 70 63 50
- Row 34: 68 74 78 74 69 66 66 66 63 62 62 63 63 63 63 61 62 62 63 62 62 62 59 57 55 53 54 54 51 50 48 49 53 61 58 69 77 74 54   43
- Row 33 (Lat 13): 71 76 78 77 71 67 65 69 71 69 70 70 69 70 72 72 71 71 75 74 69 69 66 66 64 64 66 66 63 63 62 66 69 64 81 67 54 51 50 53 51 47
- Row 32: 73 82 89 78 75 73 68 70 64 66 67 68 68 74 71 72 74 76 77 74 72 73 74 72 72 73 71 68 65 70 68 65 67 66 70 77 82 85 85 73 61 57 54 50
- Row 31 (Lat 10): 77 86 99 92 87 80 78 82 79 75 79 77 82 82 79 76 77 79 77 74 71 75 74 76 82 81 81 77 73 74 74 74 72 75 76 78 82 81 86 74 63 58 55 49
- Row 30: 83 89 91 95 91 88 93 90 90 84 82 83 83 94 101 99 95 84 82 85 81 81 84 86 86 81 82 84 81 77 81 81 88 86 85 86 87 82 72 62 57 52
- Row 29 (Lat 7): 80 83 84 79 84 82 77   63 74 82 95 106 102 96 92 91 90 88 93 93 94 94 95 96 92 88 92 91 95 95 90 85 88 79 77 68 65 60 53
- Row 28: 79 77   78 95 103 107 108 109 112 105 103 100 99 107 106 110 108 104 97 91 90 87 83 80 78 71 65 59 66 61
- Row 27 (Lat 4): 100 105 102 102 98 96 98 99 98 102 103 105 111 111 106 96 91 100 90 83 87 77 68 71 67
- Row 26: 94 94 95 95 96 96 91 87 84 87 96 105 111 114 108 103 103 106 107 102 91 83 78 72
- Row 25 (Equator, Lat 1): 90 94 95 100 99 100 97 85 81 80 81 85 104 111 120 108 95 94 106 113 112 103 91 78
- Row 24: 94 96 99 103 96 96 95 90 80 81 82 85 101 118 128 103 65 61 109 121 115 110 105
- Row 23 (Lat 1): 94 96 98 87 87 85 88 82 86 87 93 105 120 115 99 65 94 113 119 123 110
- Row 22: 97 91 90 91 93 92 94 96 98 98 110 124 104 104 90 105 114 112 120 114
- Row 21 (Lat 4): 97 91 88 94 89 92 100 101 102 109 111 100 102 100 105 96 106 113 99
- Row 20: 81 86 90 92 96 97 89 89 91 99 98 105 106 106 105 120 110 109 99
- Row 19 (Lat 7): 71 80 92 97 100 99 95 89 95 105 105 106 104 112 108 114 116 108 96
- Row 18: 75 83 92 102 100 99 94 94 101 102 106 106 101 103 111 105 102 96 93
- Row 17 (Lat 10): 81 92 94 94 96 98 96 95 92 91 95 98 101 105 104 97 91 87
- Row 16 (Degrees South): 79 89 87 94 93 95 97 97 102 97 98 98 101 103 103 104 100 92 87 83
- Row 15 (Lat 13): 73 88 92 94 92 92 94 98 100 98 100 98 96 98 101 103 97 90 84 80
- Row 14: 72 88 87 89 87 90 93 94 98 96 95 93 92 92 102 104 99 93 84
- Row 13 (Lat 16): 56 74 68 63 63 62 64 78 84 70 76 74 81 83 95 93
- Row 12: 38 76 70 70 08 77 84 84 84 85 82 76 82 91 102 93
- Row 11 (Lat 19): 57 70 71 62 64 73 81 78 72 77 83 86 93 99 78
- Row 10: 38 66 67 64 62 64 66 64 69 74 79 82 94 85 73
- Row 9 (Lat 22): 26 56 68 62 60 60 59 58 69 78 93 102 89 77 67
- Row 8: 18 40 61 58 61 60 59 63 75 83 83 96 85 76
- Row 7 (Lat 25): 29 49 58 60 56 58 58 75 81 81 86 93
- Row 6: 23 43 54 61 61 60 61 73 72 68 79 89
- Row 5 (Lat 28): 36 50 58 58 59 62 67 67 65 69 80 72
- Row 4: 56 59 50 49 61 69 69 65 73 65
- Row 3 (Lat 31): 63 59 55 61 71 73 67 67
- Row 2: 66 63 69 77 80 67
- Row 1 (Lat 34): 67 65 64 62 58

which includes 745 species in 163 genera of small-sized mammals of bats, rodents, and insectivores (table 47.1). Kingdon provides distributions for only 109 of these species, and for the remaining 636 species he provides distribution maps for the genus only (N = 73 genera). Some of the 636 species have recorded distributions for those parts of Africa that have been well investigated, but many studies do not record the entire distribution of the species concerned. In order to avoid bias in favor of these well-studied areas and against poorly studied areas, we have used the equivalent genus distributions for these species. In other words, the 636 "species distributions" are represented in our data by the 73 equivalent genus distributions (table 47.1). For example, Kingdon (1997) lists 103 species of the shrew genus *Crocidura*, known throughout Africa in all habitats and at all altitudes, but the distributions of most species are poorly known. Localities where distributions of *Crocidura* species are known almost certainly do not represent their full range, and to include such data in our records would bias results toward well-studied areas and reduce species richness in areas less well collected.

TABLE 47.1
Estimation of the effects of missing distribution data

	Genera	Species	Species with no distribution maps	Ranges of habitat types for the species with no distribution maps based on genus distributions
Megachiroptera	16	28	18 in 6 genera	44% woodland, 56% forest
Microchiroptera	33	174	162 in 21 genera	2% arid/desert, 11% forest, 87% wide range of habitats
Insectivora	25	179	180 in 14 genera	5% semiarid, 10% dry sandy soils, 75% wide range of habitats
Rodentia	89	364	276 in 32 genera	42% semi-arid, 5% upland grass, 9% swamp, 16% wood, 9% forest, 19% all
Totals	163	745	636	20.5% semiarid, 10.3% woodland, 13% forest, others 6.6%, wide range 49.6%

NOTE: Rodents and insectivores: 456 species represented by their 46 genera. These 456 species of rodents and genera out of a known total of 543 have no species distribution data—that is, 84%.

Extending the example of the genus *Crocidura*, Smithers (1983) records seven species in the southern African subregion. Two of these, *C. hirta* and *C. cyanea*, have distributions that cover most of the subregion, the former in the north and east (which also includes five of the other species), the latter in the south and west (which also includes two of the other species). These single species distributions were used in our earlier work on southern Africa (Andrews and O'Brien, 2000), which showed that some parts of southern Africa has 6 species present, other parts 3, and these contributed to the 285 species mammal fauna by 2.1% and 1.1%, respectively. This difference is lost in the present study, which combines all species of *Crocidura* so that the whole of the southern Africa subregion (except for the west coast littoral) is represented by one entry for the genus. By recording presence/absence of the genus *Crocidura* rather than individual species, we reduce this bias at the expense of loss of information (which in any case is not presently available).

In order to estimate the likely effect this missing information may have on different habitat types represented in Africa, we have summarized the habitat distributions of the 73 genera lacking species distributions (Kingdon, 1997). These are shown in the right-hand column of table 47.1, which shows the averaged values of habitat distributions for the four orders of small mammals. For example, there are 180 species of Insectivora in 14 genera that have no species ranges, and these 14 genera have habitat ranges of 1 genus (5%) exclusive to semiarid habitats, 2 genera (10%) requiring dry sandy soils, and 11 genera (75%) found in a wide range of habitats. Overall, half the genera with no comprehensive species distributions are found in wide ranges of habitats, and there is no clear preference for specific habitat types, whether forest, woodland, or semiarid, in the rest of the distributions. It may be concluded from this that there is no indication that any particular habitat is or is not favored by the substitution of genera for species. It will be seen later that the plotted distribution patterns of bats and insectivores are deficient in West and Central Africa, but rodents are better represented in these regions. There is every reason to believe, however, that distributions of both genera and species of small mammals will be extended and refined as more collections are made (Denys, pers. comm.), but at our present state of knowledge, the 73 genera must stand proxy both for distribution and habitat for the 636 species with no comprehensive distribution data.

These genus-for-species substitutions need to be kept in mind when interpreting results. First, in the following text, richness patterns will be designated species richness to avoid lengthy circumlocutions to the effect that they also include some genus distributions. Second, the substitutions introduce a bias toward lower than actual species richness values that is most likely to be a factor in poorly collected areas (e.g., Congo Basin) and for small mammals (those easiest to miss during field surveys). Unfortunately, it is at present impossible to distinguish between these two factors and genuinely low species richness. This bias should diminish at higher taxonomic levels, except where undercollection is a contributing factor to low richness. Future collections will increase the known numbers of species, so that current totals cannot be taken as final. Third, the substitutions should slightly inflate correlations between species and genus richness. However, given the large sample size and areal extent of study, general trends should swamp idiosyncratic ones between ecologically meaningful subsets of species richness, except where species richness is grossly underrepresented.

SCALE OF ANALYSIS

In accord with the southern African study, the scale of spatial analysis is macroscale, each sampling unit being 158×158 km or an area of ~25,000 km². This scale has been shown to be reasonable for measuring how climate and woody plant richness vary across Africa, as well as elsewhere in the world (O'Brien, 1993, 1998; Field et al., 2005).

ECOLOGICAL CATEGORIES OF MAMMAL RICHNESS

The ecological categories by which mammal richness has been analyzed follow Andrews and O'Brien (2000), with each species distinguished by three ecological parameters: size (body weight), dietary guild (primary diet), and space occupied (locomotor adaptations). The classes within each of these categories are depicted in table 47.2. Assignment to a size class was determined using the common size range (not extremes) for each species, and no species was assigned to more than one size class. Most mammals are small (<10 kg, $N = 434$), and large mammals (>90 kg) comprise only 33 species, with 95 species falling between these extremes.

TABLE 47.2
Numbers of mammal species for which distribution
data were available

	N
Total no. of species	562
Total species without bats	477

SPATIAL CATEGORIES

Terrestrial	213
Semiarboreal	112
Arboreal	92
Scansorial	45
Aerial	90
Aquatic	11
Fossorial	29
Arboreal/terrestrial	3
Scansorial/terrestrial	8
Semiarboreal/terrestrial	6

DIETARY CATEGORIES

Insectivore	221
Frugivore	155
Browser	182
Grazer	62
Carnivore	61
Omnivore	9
Frugivore/herbivore	22
Insectivore/herbivore	15
Carnivore/frugivore	4
Carnivore/insectivore	12

BODY WEIGHTS

0–100 g	205
100 g–1 kg	117
1–10 kg	112
10–45 kg	65
45–90 kg	30
90–180 kg	10
180–360 kg	15
>360 kg	8

NOTE: Some of the categories have double entry when some species were entered into more than one ecological category.

The seven space classes are based on the habitat space occupied and associated locomotor adaptations of each species. Terrestrial mammals are those that are restricted to the surface of the ground and lack digging, climbing, or flying adaptations. The class semiterrestrial applies to those mammals (usually small ones) that lack specialized digging, climbing, or flying adaptations, and they occupy both subsurface and above-surface spaces as if they were continuous with the surface of the ground. They use tunnels below the surface and rocks and trees above the surface to the same extent as the ground surface. Arboreal mammals are those having specialized anatomical or physiological adaptations for climbing trees, such as gripping feet. Scansorial mammals also climb trees but differ in having claws and reversible feet. Aquatic, fossorial, and aerial mammals are anatomically specialized for swimming, digging, and flying, respectively (Andrews et al., 1979). The majority of species were assigned to a single class based on their most common locomotor behavior, but in some cases, species were assigned to more than one spatial class. For example, a number of felid and viverrid carnivores (e.g., leopards and genets) were assigned to both scansorial and terrestrial categories. In the primates, baboons were assigned to arboreal and terrestrial categories, since both habitats fulfill important requirements of their lifestyles. In all cases, shared classes were on a 50-50 basis.

Six classes of dietary guild were distinguished (cf. Andrews et al., 1979; Janis, 1988; Shepherd, 1998). Although few mammals are exclusive to one dietary class, most fall primarily into one of them. Herbivores were divided into browsers eating mainly leaves and herbs and grazers eating mainly grass, and no attempt was made to distinguish mixed feeders. Species eating more than 50% fruit were designated frugivores, and where there was still a substantial element of fruit but mixed with herbage, such as baboons, they were entered as mixed herbivore-frugivores. Many rodent species are also mixed feeders of insects and plants, and they were similarly designated mixed insectivore-herbivore. Some species of Carnivora eat significant proportions of fruit or insects, and they were designated carnivore-frugivores or carnivore-insectivores. All such combinations were made on a 50-50 basis, and no species was assigned to more than two classes (see table 47.2). Only those mammal species with unequivocally mixed diets, such as *Potamochoerus* and civets, were designated as omnivores.

Numbers of species per equal-area grid cells were mapped for the 21 ecological subsets. Species richness isoclines were calculated to show their patterns of geographic variation, as far as possible to the same scale, but since some ecological categories have much greater numbers of species than others, the potential loss of information rendered this impractical. Correlations within and between ecological categories were calculated (Pearson's product moment), as well as with all-mammal richness. Some of the ecological subsets have few species (<10 species), in which case they were either combined with one or more similar classes (e.g., the three largest size classes were combined into one: >90 kg) or ignored in results/discussion of statistical analyses (e.g., omnivory, which has only 9 species). Other combinations include "small" species (e.g., <45 kg) and scansorial-arboreal (which are strongly correlated with each other), plus others that are self-explanatory.

Results

Mammal species richness per cell is presented in figure 47.2 and ranges from a minimum of 6 to a maximum of 179 species. Noteworthy is the north-south discrepancy in mammal richness. In effect, mammal richness differs markedly between the Northern and Southern hemispheres, being lower in the Northern Hemisphere and higher in the Southern Hemisphere, even in analogous vegetation and floristic zones (e.g., deciduous forest/woodland; Sudanian and Zambezian phytochoria, respectively; compare figures 47.1 and 47.2). The north-south discrepancy is greatest between arid areas (e.g., Sahara vs. Kalahari/Namib deserts).

GEOGRAPHIC PATTERNS IN MAMMAL RICHNESS IN RELATION TO VEGETATION

In order to illustrate the species richness patterns, we have mapped taxon richness by drawing isoclines connecting grid cells with similar species richness values. Figure 47.3

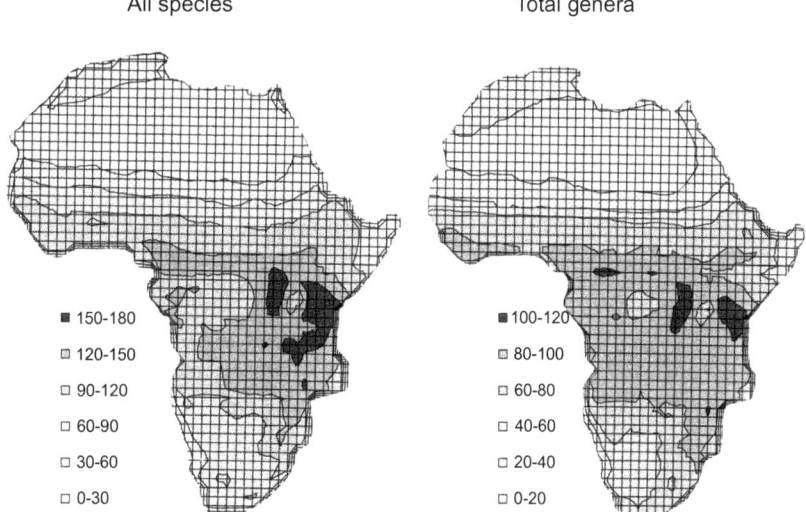

All species   Total genera

- 150-180   - 100-120
☐ 120-150   ☐ 80-100
☐ 90-120   ☐ 60-80
☐ 60-90   ☐ 40-60
☐ 30-60   ☐ 20-40
☐ 0-30   ☐ 0-20

FIGURE 47.3   Isoclines of species and genus richness per unit area for the equal-area grid shown in figure 2. Left: total numbers of mammal species; Right: total numbers of mammal genera.

illustrates the general pattern of mammal species richness across Africa. The general tendency is for mammal richness to increase across Africa as vegetation (physiognomically and floristically) shifts from desert to evergreen forest, as woody plant richness increases and as annual rainfall increases. This trend applies whether bats are excluded. The general pattern is consistent with findings for southern Africa with the exception of two differences in the equatorial zone. In the first place, maximum mammal species richness occurs in east High Africa, in two major peaks that are neither associated with areas having the highest annual rainfall or greatest woody plant richness, nor associated with lowland rain forest. Comparison with figure 47.1 indicates that the westernmost of these two peaks lies in an area that encompasses the eastern perimeter of the Congo Basin and the Western Rift zone where three major vegetation types and four phytochoria converge (Congo-Guinean, Sudanian, Zambezian, and Afromontane). In this area annual rainfall and predicted woody plant richness can be high, but not highest; topographic relief is high; the growing season is year-round; and vegetation ranges from lowland rain forest to deciduous forest/woodland and to alpine.

The easternmost peak is also associated with a macroecotone complex where four phytochoria converge (Sudanian, Zambezian, Somali-Masai, Afromontane). Vegetation zones range from scrubland to evergreen forest/woodland or alpine, annual rainfall and predicted woody plant richness are variable, topographic relief is highly variable, and plant growth is seasonal (bimodal/summer rainfall). The eastern peak forms a crescent that stretches southeastward from roughly Mt. Kenya to the Usambara Mountains in Tanzania; then southward, skirting the 600-mm rainfall isoline; and then westward to Lake Tanganyika, through Selous National Park. and Lake Ruaha to the Western Rift (compare figure 47.3 with figure 47.1). The southern portions of this crescent coincide with a narrow strip of annual rainfall greater than 1,000 mm (figure 47.1A), and with evergreen to deciduous forest/woodland belonging to the Zambezian floristic zone and disjunct patches of Afromontane vegetation. The entire crescent borders the central plateau and part of the Eastern Rift zone in Tanzania where mammal richness, woody plant richness, and annual rainfall are lower;

and where vegetation is mainly deciduous, low woodland, and thicket (bushland), with floristic elements belonging mainly to the Somali-Masai phytochorion. The disjunction between these two peaks of maximum richness is a sampling artifact since the cells within or bordering Lake Victoria fall mainly over water, making them unequal area samples.

The second difference from our southern African study is the abrupt decrease in mammal species richness west of High Africa that is sustained across equatorial Low Africa and the Congo Basin. Unlike equatorial High Africa, this part of equatorial Africa is confined to a single vegetation and floristic zone, tropical forest; and, moreover, parts of the Congo basin are floristically impoverished (Aubreville, 1938; Richards, 1952). This decrease in mammal richness is inconsistent with the trend in annual rainfall and predicted woody plant richness (compare figures 47.1 and 47.4), but it is consistent with recognized undercollection of animals and plants in this part of Africa. Further support for mammal richness being underrepresented in this region emerges when we examine ecological subsets (discussed later).

At the genus level of analysis, the pattern for mammal richness is similar to that of species richness (figure 47.3). There is still a decrease in richness west of High Africa, but it is less marked and is associated with a new peak in maximum richness. At the family level (figure 47.5), the pattern is very diffuse, with overall high richness in tropical Africa, again extending further to the south than in the north, but with no prominent peaks at any point. This is seen also in figure 47.4 where similar family richness values occur in cells along a longitudinal transect across the entire continent at 1°N (row 25 in figure 47.2). Consistent with findings for southern Africa, the general similarity in richness gradients between taxonomic levels is supported by a correlation of $r = 0.986$ between species and genus richness. Correlations between genus and family ($r = 0.913$), and between species and family ($r = 0.886$) are also high ($p < 0.0001$).

The longitudinal pattern of variation in species richness in African mammals in relation to plants is shown for west-east transects at 1°N in figure 47.4 (discussed later). The profile for projected numbers of woody plant species (O'Brien, 1998; Field et al., 2005) shows a west to east decline in plant species

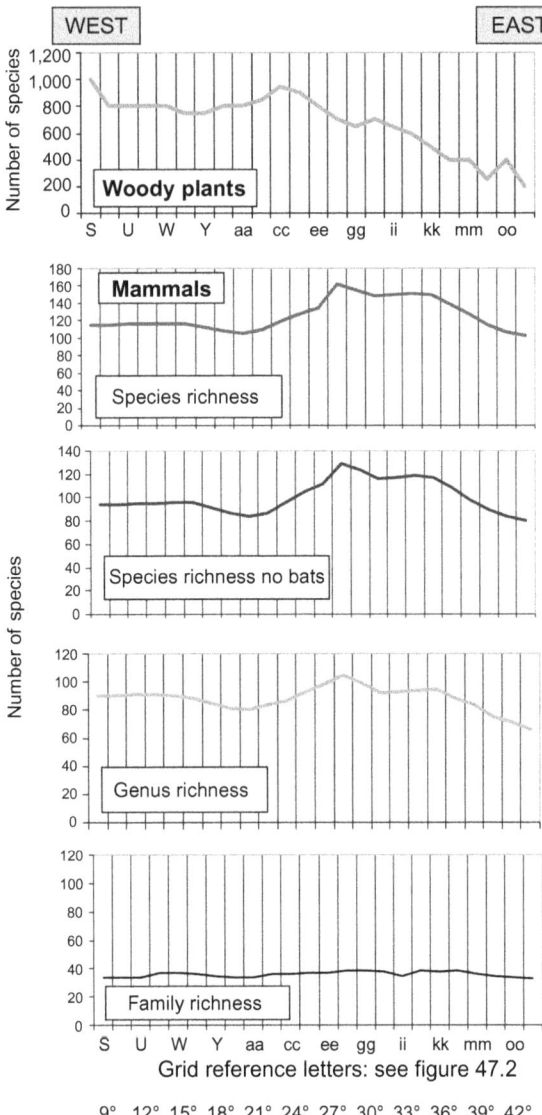

FIGURE 47.4 Profiles of woody plant and mammalian taxonomic richness from west to east at 1°N latitude across Africa. From top to bottom, species richness of woody plants, species richness of mammals, species richness of mammals excluding bats, genus richness of mammals, and family richness of mammals. The vertical scale gives numbers of taxonomic units; the horizontal scale shows the column letters indicated in figure 47.2 and degrees of longitude.

numbers, which reflects the high plant species richness in the lowland evergreen rain forests in western Africa and the lower projected numbers in the drier environments of eastern Africa. For the mammals, however, there is an opposite trend from east to west, with peaks as the profile crosses the western and eastern rift valleys as described above. This pattern is similar when bats are excluded and for numbers of genera (although the curve is smoother with less differentiation), and on the scale used here the family pattern varies little longitudinally.

ECOLOGICAL CORRELATES OF MAMMAL RICHNESS

All-mammal richness is positively correlated with richness of all of the ecological subsets (table 47.3). Most correlations are very strong (collinear; $r \geq 0.8$) and some indicate statistical synonymy ($r \geq 0.9$) in that trends of richness are

virtually identical to that of all mammal richness. The statistical synonyms for all-mammal richness, from strongest to weakest, are browsing herbivore, insectivory, aerial, body size 1–10 kg, terrestrial, body size <100 g, omnivory, and body size 45–90 kg. Essentially the same findings apply when bats are excluded (Nxbats). With the exception of the small size, aerial, and insectivore subsets, these results differ from those obtained for southern Africa. The weakest correlates ($r < 0.7$) are fossorial and body size 90–180 kg.

The ecological subsets that are statistically synonymous with mammal richness are also statistically synonymous with each other (table 47.4): the browsing herbivore subset is highly correlated with the insectivorous ($r = 0.921$) and terrestrial ($r = 0.913$) subsets; the aerial subset is highly correlated with the 1–10 kg ($r = 0.904$) and insectivorous ($r = 0.915$) subsets; and most strongly of all, the <100 g subset with the insectivorous subset ($r = 0.980$). Therefore, the ecological categories most representative of the mammalian fauna as a whole are the browsing, insectivorous, and aerial subsets.

Another major pattern in richness is indicated in table 47.4 by the high correlations between the frugivory subset and the arboreal ($r = 0.94$) and scansorial ($r = 0.93$) subsets. In addition, the arboreal subset is also highly correlated with the

TABLE 47.3
Correlations of ecological categories with total numbers of mammal species (N) and numbers excluding bats (Nxbats)

	All mammals	N excluding bats
N all mammals	1	
N excluding bats	**0.993**	1
LOCOMOTOR CATEGORIES		
Terrestrial	**0.925**	**0.929**
Semiterrestrial	0.697	0.722
Arboreal	0.827	0.823
Scansorial	0.712	0.715
Aerial	**0.950**	**0.908**
Aquatic	0.741	0.723
Fossorial	0.541	0.582
DIETARY CATEGORIES		
Insectivore	**0.954**	**0.936**
Frugivore	0.679	0.677
Browser	**0.967**	**0.976**
Grazer	0.835	0.832
Carnivore	0.791	0.782
Omnivore	**0.910**	**0.908**
BODY WEIGHTS		
0–100 g	**0.912**	**0.901**
100 g–1 kg	0.878	0.876
1–10 kg	**0.933**	**0.928**
10–45 kg	0.871	0.865
45–90 kg	0.899	0.889
90–180 kg	0.584	0.575
180–360 kg	0.693	0.698
>360 kg	0.773	0.780

All correlations are significant ($p = 0.001$) and values above 0.9 are highlighted in bold.

TABLE 47.4

Correlations of ecological categories

Ecological categories	Terr	Semiter	Arb	Scans	Aerial	Aquat	Foss	Insect	Frug	Brow	Graz	Carn	Omni	0–100 g	100 g–1 kg	1–10 kg	10–45 kg	45–90 kg	90–180 kg	180–360 kg	>360 kg
Terrestrial	1	0.597	0.642	0.615	0.836	0.662	0.623	0.865	0.501	0.913	0.852	0.902	0.85	0.817	0.713	0.855	0.863	0.92	0.717	0.765	0.844
Semiterrestrial		1	0.398	0.149	0.55	0.189	0.61	0.813	0.173	0.739	0.783	0.376	0.554	0.892	0.524	0.496	0.403	0.576	0.347	0.558	0.499
Arboreal			1	0.882	0.83	0.824	0.253	0.703	0.94	0.782	0.45	0.517	0.829	0.626	0.937	0.913	0.741	0.647	0.262	0.412	0.541
Scansorial				1	0.716	0.829	0.198	0.519	0.93	0.658	0.333	0.58	0.764	0.413	0.857	0.854	0.725	0.59	0.278	0.317	0.514
Aerial					1	0.722	0.373	0.915	0.711	0.877	0.745	0.737	0.866	0.849	0.864	0.904	0.833	0.849	0.532	0.599	0.685
Aquatic						1	0.263	0.586	0.795	0.67	0.477	0.672	0.711	0.489	0.775	0.829	0.77	0.646	0.394	0.375	0.621
Fossorial							1	0.553	0.136	0.583	0.547	0.482	0.537	0.604	0.323	0.4	0.375	0.46	0.303	0.623	0.58
Insectivory								1	0.5	0.921	0.874	0.698	0.824	0.980	0.781	0.828	0.752	0.851	0.559	0.690	0.705
Frugivory									1	0.627	0.269	0.41	0.73	0.410	0.886	0.820	0.651	0.494	0.134	0.259	0.420
Browsing herbivore										1	0.83	0.734	0.867	0.895	0.844	0.892	0.838	0.863	0.561	0.716	0.765
Grazing herbivore											1	0.694	0.713	0.878	0.603	0.696	0.642	0.838	0.761	0.742	0.769
Carnivore												1	0.703	0.636	0.582	0.745	0.856	0.852	0.730	0.564	0.712
Omnivore													1	0.775	0.843	0.901	0.773	0.828	0.460	0.638	0.763
0–100 g														1	0.707	0.742	0.674	0.801	0.531	0.683	0.678
100–1,000 g															1	0.923	0.747	0.711	0.316	0.457	0.566
1–10 kg																1	0.841	0.822	0.461	0.572	0.707
10–45 kg																	1	0.817	0.648	0.519	0.667
45–90 kg																		1	0.669	0.673	0.756
90–180 kg																			1	0.471	0.582
180–360 kg																				1	0.67
>360kg																					1

All correlations are significant ($p = 0.001$). Values above 0.9 are highlighted in bold.

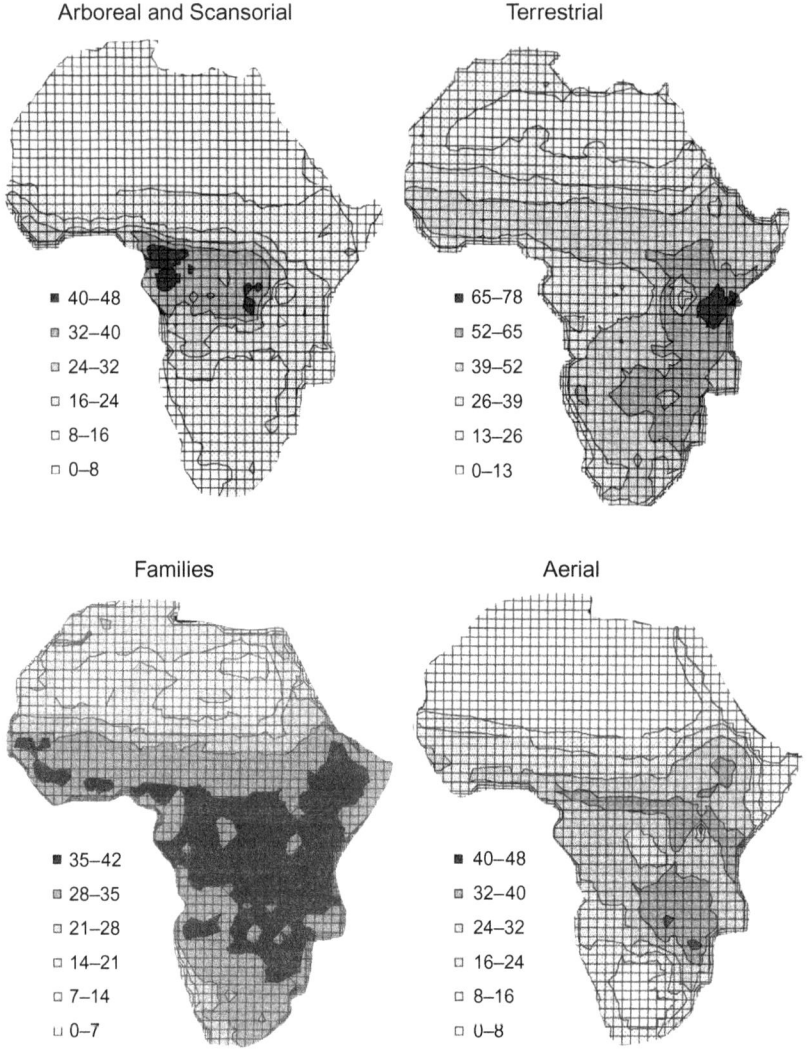

Arboreal and Scansorial

- ■ 40–48
- ▣ 32–40
- ▫ 24–32
- ▫ 16–24
- ▫ 8–16
- ▫ 0–8

Terrestrial

- ■ 65–78
- ▣ 52–65
- ▫ 39–52
- ▫ 26–39
- ▫ 13–26
- ▫ 0–13

Families

- ■ 35–42
- ▣ 28–35
- ▫ 21–28
- ▫ 14–21
- ▫ 7–14
- ▫ 0–7

Aerial

- ■ 40–48
- ▣ 32–40
- ▫ 24–32
- ▫ 16–24
- ▫ 8–16
- ▫ 0–8

FIGURE 47.5   Isoclines of species richness for the equal area grid shown in figure 47.2. Top left: arboreal/scansorial mammals; Top right: terrestrial mammals; Bottom left: family richness; Bottom right: aerial mammals.

100 g–1 kg (r = 0.94) and 1–10 kg (r = 0.91) subsets and less strongly correlated with the 10–45 kg (r = 0.741) subset. Such correlates are consistent with findings for southern Africa.

In order to illustrate the species richness patterns and make comparisons between ecological categories more apparent, we have mapped each category separately. Isoclines connect grid cells with similar species richness values, and the following sets of figures compare subsets of ecological variables. Figure 47.5 shows three spatial categories, and two distinct patterns are evident when we juxtapose the species richness map for the terrestrial subset with the map for a combined arboreal-scansorial subset. The latter are strongly correlated with each other (r = 0.882), but less so with the terrestrial subset (r = 0.642 and 0.615, respectively). In figure 47.6, we again see two distinct patterns when the browsing herbivore subset is compared with the map for the frugivore (excluding bats) subset, which are also less strongly correlated with each other (table 47.4). Two distinct equatorial maxima are evident in figures 47.5 and 47.6, one in west Low Africa, the other in High East Africa. Maximum species richness for the browsing herbivore and terrestrial subsets occurs in High Africa in the northern part of the peak described for all mammal richness (compare figures 47.5 and 47.6 with figure 47.3). Gradients of

decreasing richness extend north from this area in a gradual, latitudinal fashion that reaches further into northern Africa than do other subsets. This contrasts with the species richness pattern for carnivores in that high richness extends far to the south into South Africa (figure 47.6) while retaining a pattern similar overall to that of terrestrial species. South of Ethiopia/Somalia and the Congo Basin, gradients are longitudinal, tending to decrease from the east to the west and southwest, as described for southern Africa (Andrews and O'Brien, 2000). The north-south discrepancy in richness is evident in all ecological subsets.

For the arboreal-scansorial and frugivore subsets (figures 47.5 and 47.6), the greatest richness occurs in two disjunct peaks. One is in the western portion of the lowland rain forest/Congo-Guinean floristic zones. The other is in the eastern portion of this vegetation/floristic zone in the same area where the western peak in all-mammal richness occurs (compare with figures 47.1 and 47,3). Richness decreases east, north, and south of this area—steeply so to the north, more gradually so to the east and south. Again, the north-south discrepancy in richness is evident. This pattern is found only in arboreal frugivores, and it relates more clearly to the Congo-Guinean lowland forests than do any other of the ecological subsets. The more common

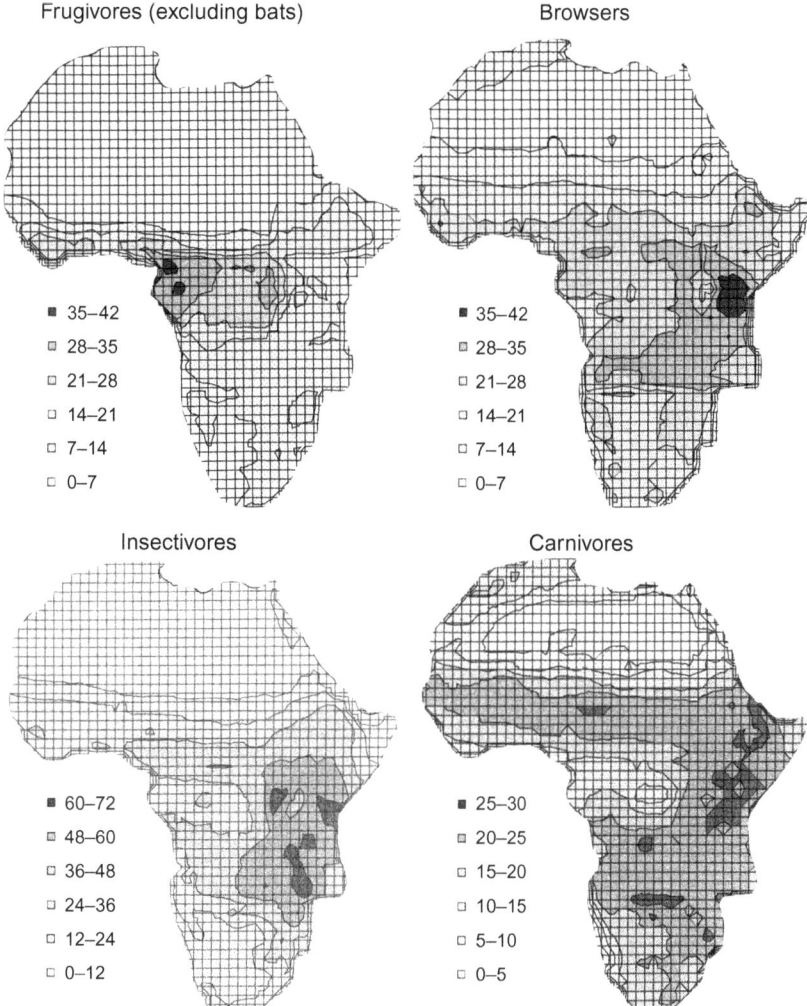

Frugivores (excluding bats)

■ 35–42
▫ 28–35
▫ 21–28
▫ 14–21
▫ 7–14
▫ 0–7

Browsers

■ 35–42
▫ 28–35
▫ 21–28
▫ 14–21
▫ 7–14
▫ 0–7

Insectivores

■ 60–72
▫ 48–60
▫ 36–48
▫ 24–36
▫ 12–24
▫ 0–12

Carnivores

■ 25–30
▫ 20–25
▫ 15–20
▫ 10–15
▫ 5–10
▫ 0–5

FIGURE 47.6   Isoclines of species richness for the equal-area grid shown in figure 47.2. Top left: frugivores (excluding frugivorous Bats; Top right: browsing herbivores; Bottom left: insectivores; Bottom right: carnivores.

pattern is that seen in terrestrial browsers (and grazers), the insectivorous subset (which nevertheless has a minor peak in the Congo basin) and small mammals (figures 47.5–47.7). Aerial mammals, mainly but not exclusively bats, show a richness belt running east-west just north of the equator (also present in all mammals and several other groups) and a richness peak in southeast Africa similar to the southerly peak in the 1–100 g subset. This distribution for aerial mammals is also the pattern for all-mammals (figure 47.3), hence the high correlations between them and these ecological subsets.

The greater contribution from small mammals rather than from large ones in generating the pattern for all-mammal richness is exemplified by figure 47.7, which compares the <100 g subset map with the >90 kg subset map (combination of three size subsets). The latter differs from all previous patterns, being more diffuse and with relatively high richness extending further northward than any other subset. The <100 g subset also shows a large area of maximum richness, but it extends southward into southern High Africa (Zimbabwe). Thereafter small mammal richness decreases to the south, north, and west. This decrease is consistent with decreasing annual rainfall and predicted woody plant richness and with associated changes in vegetation/floristic zones. The decrease to the west in equatorial Africa was unexpected, but it is a

pattern matched by aerial, insectivorous, carnivorous, and all-species. It would be expected that the diversity of insects should increase to maximum values in lowland rain forest and that the species richness of small mammals should correspondingly increase, and the fact that it does not indicates that small mammal, and thus all mammal richness, is underrepresented for the Congo-Guinean lowland rain forest zones. It is not clear at present whether this is a real feature or due to lack of collecting in these areas, and it may be that both are implicated.

The same patterns can be seen in figure 47.8, which presents east-west transects of species richness for various ecological subsets in equatorial regions (1°N, row 25 in figure 47.2). Two distinct trends in all mammal richness are apparent. First, the richness profiles for the insectivore and herbivore subsets (the latter combines browsers with grazers because they have similar distributions) are least across the lowland rain forest/Congo-Guinean floristic zone and then abruptly triples with the transition to High Africa. Thereafter, it slightly decreases eastward. A similar, but somewhat more gradual, richness profile can be seen for the terrestrial subset, and a hint of this trend can be seen in the richness profile for the >90 kg subset. None of these profiles matches the richness profiles for predicted woody plant species richness (shown at the top of figure 47.8) or for all-mammal richness across this transect

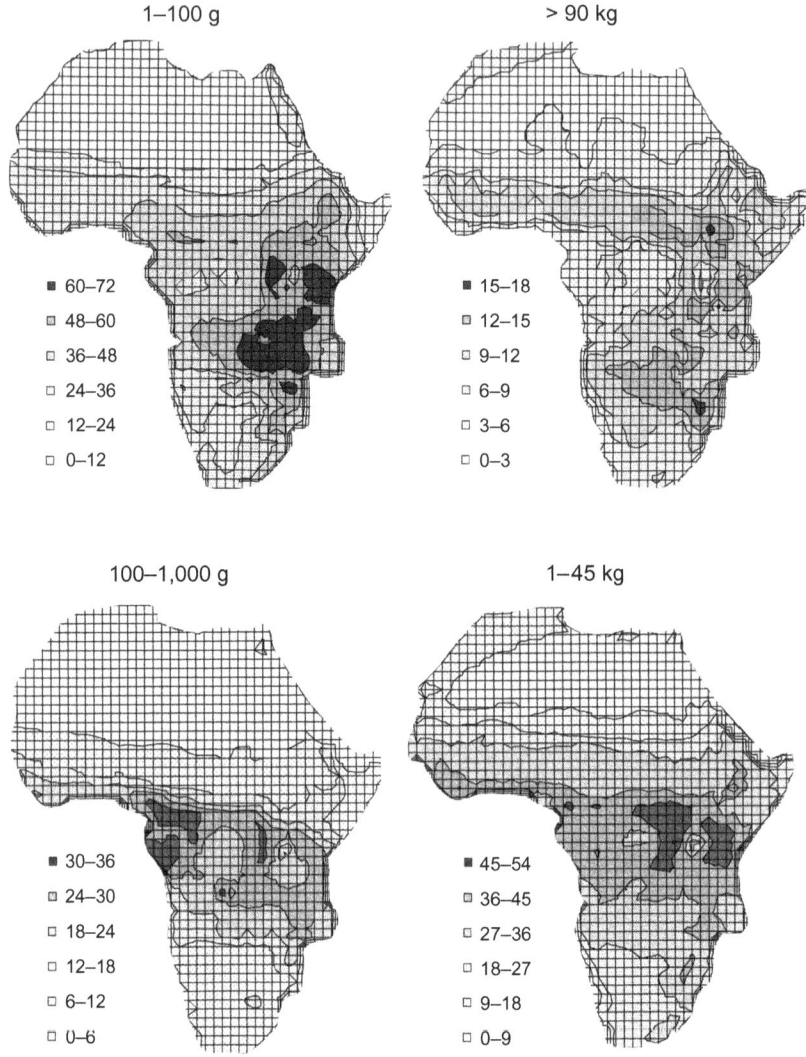

1–100 g

■ 60–72
▣ 48–60
▢ 36–48
▢ 24–36
▢ 12–24
▢ 0–12

> 90 kg

■ 15–18
▣ 12–15
▢ 9–12
▢ 6–9
▢ 3–6
▢ 0–3

100–1,000 g

■ 30–36
▣ 24–30
▢ 18–24
▢ 12–18
▢ 6–12
▢ 0–6

1–45 kg

■ 45–54
▣ 36–45
▢ 27–36
▢ 18–27
▢ 9–18
▢ 0–9

FIGURE 47.7 Isoclines of species richness for the equal-area grid shown in figure 47.2. Top left: small mammals 1–100 g; Top right: large mammals >90 kg; Bottom left: small mammals 100–1000 g; Bottom right: and mammals 1–45 kg.

(compare with figure 47.4). Second, richness profiles similar to that of predicted woody plant richness (and thus climate and vegetation) are seen for the frugivore, arboreal, and 100 g–1 kg subsets. The last of these is noteworthy because it contradicts the richness profile for the <100 g subset, which is instead similar to the richness profile for the insectivore and browsing herbivore subsets (not shown; see figure 47.7). This discrepancy suggests that the <100 g subset is likely to be underrepresented in our data set. Again, however, none of these profiles are similar to that described for all-mammal richness across this transect (compare with figure 47.4).

CORRELATIONS OF TAXONOMIC DIVERSITY WITH ECOLOGICAL PATTERNS OF SPECIES RICHNESS

Table 47.5 shows the correlations between the taxonomic groupings and the ecological ones described here. Highest correlations between taxonomic groupings and total numbers of species per grid cell are found for bats ($r = 0.916$), bovids ($r = 0.915$), and rodents ($r = 0.952$). These are the three largest groupings taxonomically, and the

taxonomic groups with few species have correspondingly low correlations with total species numbers. The highest correlations of Primates are with arboreal and frugivorous species ($r = 0.930$–0.962, table 47.5) and also with the scansorial species ($r = 0.900$), an interesting observation since none of the Old World primates are scansorial. Richness peaks in west central Africa (figure 47.9), and this differs dramatically from the all-species pattern (figure 47.3), as it does in arboreal (figure 47.5), scansorial, and frugivorous (figure 47.6) ecological subsets. Bovidae is most highly correlated with herbivorous and terrestrial species and with 45–90 kg weight classes, and when the distribution of bovid species richness is mapped (figure 47.9), the pattern is seen to be similar to these ecological subsets (figures 47.5–47.7) and to the all-species distribution shown in figure 47.3. Chiroptera is of course correlated with small body size, aerial locomotion, and insectivory (table 47.5), and their distribution pattern is again like that of all-mammals (figure 47.3). Bats are poorly known for the Congo Basin, and this is probably another reason why there is a marked drop in mammal species richness in that region (figure 47.3).

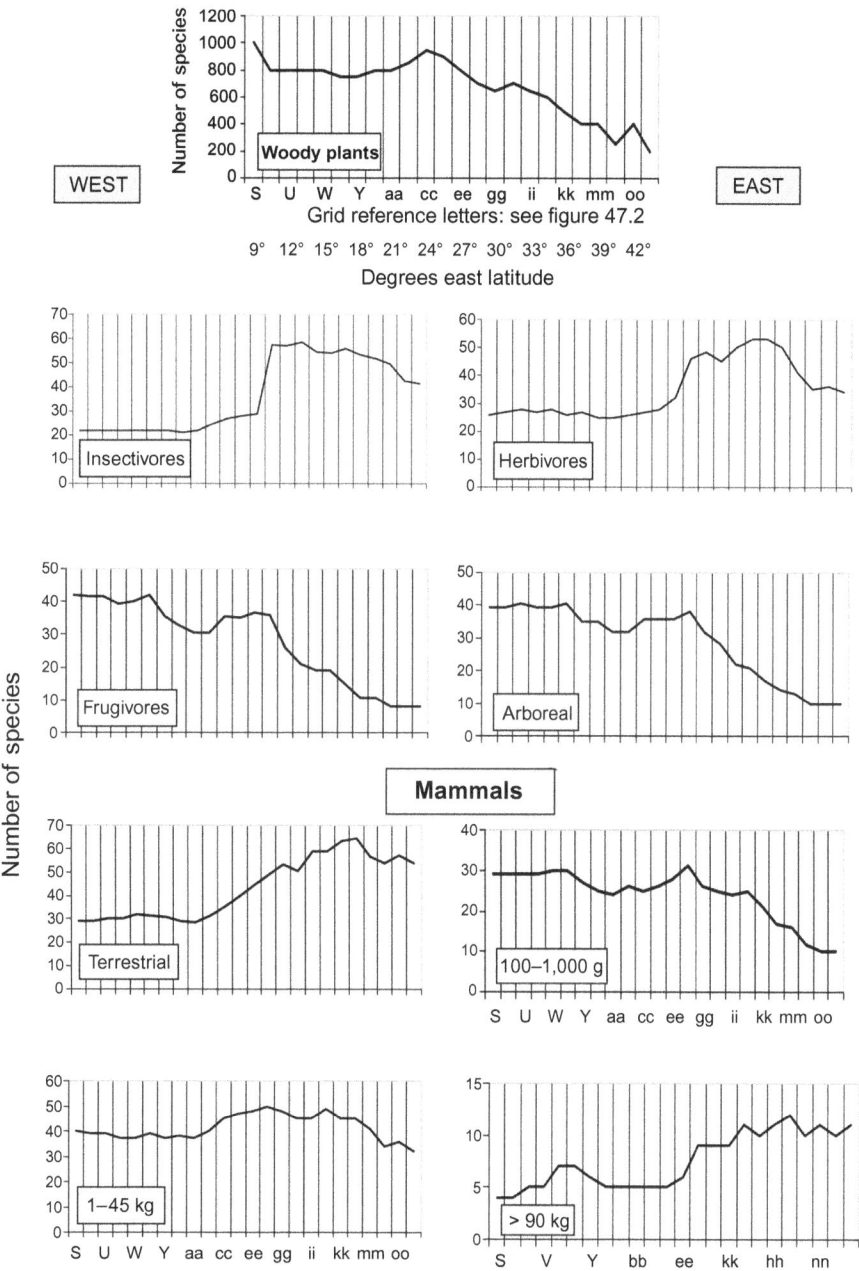

FIGURE 47.8   Profiles of woody plant and mammalian species richness from west to east at 1°N latitude across Africa. At the top is the profile of woody plant species richness from Figure 47.4 for reference with mammalian patterns; second row, profiles for insectivorous and herbivorous mammals; third row profiles for frugivorous and arboreal mammals; fourth row, profiles for terrestrial and small (100- to 1,000-g) mammals; and bottom row, profiles for medium-sized (1- to 45-kg) and large (>90-kg) mammals. The vertical scale gives numbers of taxonomic units; the horizontal scale shows the column letters indicated in Figure 47.2 and degrees of longitude.

The same pattern is seen in species of Insectivora (figure 47.9). Rodentia correlates most highly with small body size, semiterrestrial locomotion, and browsing species richness patterns (table 47.5). The mapped distribution of rodents is generally similar to that of total species, corresponding to their higher correlations with total species, but it differs in having a secondary peak in west Central Africa. Richness is not as high as in eastern Africa, but it is high enough to suggest that undercollecting is not the sole cause of low richness values of small mammals in western Africa and the Congo basin.

Discussion

The aims of this chapter have been to document the distribution of mammals in Africa today, to evaluate the quality of the data involved, and to test some of the implications for mammal richness arising from the southern Africa study (Andrews and O'Brien, 2000). Most animal and plant taxa show species richness gradients at the regional or continental scale (MacArthur, 1964, 1965; Pianka 1966; for North American mammals, see Simpson, 1964; Badgley and Fox 2000), and these have been interpreted either in terms of

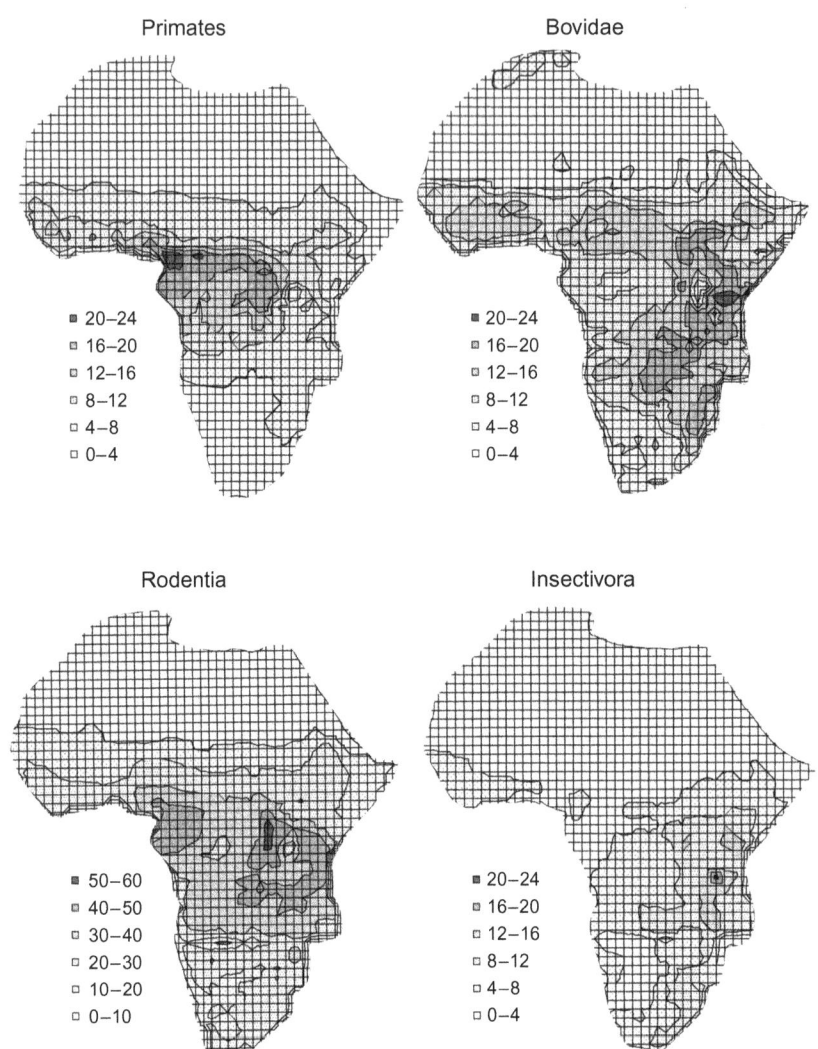

FIGURE 47.9   Isoclines of taxonomic species richness for the equal area grid shown in figure 47.2.
Top row: Primates and Bovidae; Bottom row: Rodentia and Insectivora.

physical factors, such as latitude or climate and their effects on habitat productivity, or as a result of biological interactions, such as competition or predation (MacArthur, 1965; Thiery, 1982; Begon et al., 1990). We have reviewed these factors in our earlier paper (Andrews and O'Brien, 2000; and see Badgley and Fox 2000), and few if any of the hypotheses explaining predictable richness gradients are independent of climate (Currie and Paquin, 1987; Currie, 1991; Kerr and Packer, 1997; Lawton et al., 1998). This will be discussed in a second paper (O'Brien and Andrews, unpub.), but it should be noted that variation in species richness also varies with such factors as area and time. For example, as unit area increases the amount of habitat variability increases, speciation rates may increase, and extinction rates decrease (Mayr, 1963; Van Valen, 1973, 1975; Kerr and Packer, 1997), leading to increase in species numbers. Limiting effects of area have had little effect on a large island continent such as Africa, and since our sampling has been based on an equal-area grid, area is not a major source of variation in our analyses. On the other hand, time is always a factor, for the accumulation of species as a result of the differential effects of speciation and extinction take time to affect species richness patterns. As a result, species numbers may be modified by historical factors

resulting from evolutionary processes and geological or topographic processes (Rosen, 1984). In these terms, the African continent has had a long and stable history at least since the early Miocene, with most if not all of the extant mammalian families established on the continent since that time. It has, however, undergone various perturbations of climate and tectonic movements that have had dramatic effects on faunal diversity, in particular the periodic loss of extensive areas of rain forest and the retreat of fauna and flora into limited refugia (Moreau, 1966).

In our earlier work on mammal richness patterns in southern Africa, we found that woody plant richness, and thus vegetation, alone accounts for 75% of the variation in all mammal richness (Andrews and O'Brien, 2000), and in a like fashion genus and family richness. Most in turn of the variation in woody plant species richness (85%; adj. r^2), and thus vegetation, is accounted for by climate (O'Brien et al., 1998, 2000). The same has been predicted for Africa in general (O'Brien, 1998); empirically so, for Kenya (Field et al., 2005). Following on from this, we have found or made predictions as follows.

Mammal richness patterns were expected to be broadly similar at all taxonomic levels: robustly so in the case of species and genus

TABLE 47.5
Correlations of taxonomic categories

	I	P	Ch	Ca	Eq	RH	H	Su	Tr	Gir	Bo	L	Rod
Total no. species	.761	.803	**.916**	.875	.600	.268	.748	.808	.332	.195	**.915**	.303	**.952**
No. exclud. bats	.760	.814	.877	.866	.608	.281	.752	.792	.363	.204	**.904**	.305	**.960**
Terrestrial	.738	.532	.827	**.941**	.656	.427	.678	.766	.026	.344	**.917**	.522	.797
Semiterrestrial	.890	.620	.885	.818	.674	.213	.653	.816	.175	.133	.810	.447	**.911**
Arboreal	.550	**.962**	.698	.546	.390	.103	.634	.623	.689	.023	.67	−.024	.885
Scansorial	.422	**.900**	.569	.516	.334	.133	.608	.503	.677	.041	.58	−.065	.820
Aerial	.699	.776	**.952**	.819	.518	.195	.696	.794	.312	.133	.874	.227	.889
Aquatic	.539	.818	.703	.639	.384	.198	.726	.604	.455	.137	.711	.121	.841
Fossorial	.585	.048	.337	.446	.523	.445	.438	.335	−.155	.117	.418	.425	.410
Insectivorous	.856	.577	**.950**	.876	.613	.246	.662	.856	.087	.157	.876	.464	.825
Frugivorous	.436	**.930**	.583	.426	.291	.085	.579	.516	.785	−.040	.565	−.146	.806
Browsing	.809	.686	.887	.877	.659	.301	.714	.833	.228	.243	**.916**	.422	.906
Grazing	.752	.374	.840	.854	.645	.320	.589	.784	−.151	.292	.869	.642	.716
Carnivorous	.505	.527	.691	**.900**	.494	.308	.570	.560	.042	.348	.753	.37	.704
Omnivorous	.709	.744	.800	.749	.515	.363	.732	.806	.335	.161	.829	.272	.885
0–100 g	.855	.463	.907	.833	.585	.263	.613	.820	.008	.147	.829	.506	.766
100–1,000 g	.699	.874	.830	.705	.502	.092	.655	.758	.546	.015	.754	.145	**.925**
1–10 kg	.722	.866	.864	.825	.557	.218	.717	.782	.446	.165	.853	.269	**.938**
10–45 kg	.539	.755	.732	.833	.448	.218	.660	.633	.319	.261	.829	.259	.805
45–90 kg	.674	.585	.846	.899	.561	.293	.639	.799	.075	.303	**.905**	.405	.786
90–180 kg	.308	.264	.524	.694	.378	.298	.371	.511	−.152	.388	.672	.463	.446
180–360 kg	.603	.322	.636	.661	.693	.432	.566	.599	−.060	.249	.786	.319	.585
>360 kg	.593	.450	.680	.728	.518	.563	.730	.621	.006	.493	.801	.429	.693

All correlations are significant ($p = 0.001$). Values above 0.9 are shown in bold. I, Insectivora, P, Primates, Ch, Chiroptera, Ca, Carnivora, Eq, Equidae, Rh, Rhinocerotidae, H, Hippopotamidae, Su, Suidae, Tr, Tragulidae, Gir, Giraffidae, Bo, Bovidae, L, Leporidae, Rod, Rodentia.

richness, diffusely so in the case of family richness. This has been found to be true for all-Africa, particularly in the comparison of species and genus distributions. In this case, however, it must be remembered that 636 small mammal species lacking recorded species distributions have been represented by their genus distributions (table 47.1).

Statistical correlations between taxonomic levels should be strong and ideally collinear ($r \geq 0.8$), especially between mammal species and genus levels, which should exhibit statistical singularity ($r \geq 0.9$). Again this is confirmed by our analyses but with the same caveat that some groups are represented by genus distributions (table 47.1).

Mammal richness was expected to be greatest in the Congo-Guinean lowland rain forest of Central and West Africa. Species richness of woody vegetation and insects is greatest in these habitats (Aubréville, 1938; Richards, 1952; Mound and Waloff, 1978), and they extend in a great swathe across the equatorial region of Africa west of the western rift valley. Our results, however, show species richness to be anomalously low, although this is less marked at the genus and family levels of analysis. Clearly there is underrepresentation of species known from this region, and while this might be expected for Bovidae, which is predominantly a family adapted to non-forest environments, it is strikingly evident for Insectivora (figure 47.9) and Chiroptera (not shown here). For example, the genus *Crocidura* has over 100 species so far described, but

most lack comprehensive distribution records, and so in our analyses we have represented them with a single entry at the genus level (table 47.1). While much information is lost in the process, it removes the likelihood of bias in favor of well-collected areas and against remoter parts of Africa that have yet to be investigated. The same is true for many species of bats and rodents, and it is accepted that small mammals in general are underrepresented, which may be a major factor in the richness lows in Central and West Africa. On the other hand, rodent species are well represented in west Central Africa, which is reflected in figure 47.9 by a diversity peak in this region. Since rodents are not usually collected to the exclusion of other small mammals, this indicates that that other issues may be important—for example, the fact that the entire area in Central and West Africa is represented by a single phytochorion, the Congo-Guinean rain forest. Moreover, parts of the rain forest in the central Congo basin are relatively poor in tree species (White, 1983), with extensive swamp forest that would not be a suitable environment for small mammals.

Mammal richness was expected to be lower in Africa north of the Congo-Guinean lowland rain forest than south of it, even in analogous vegetation/floristic zones. This prediction is based on the fact mentioned earlier that vegetation richness is higher south of the equator, even when comparing analogous vegetation zones (Werger, 1978; White, 1983; Cowling et al., 1997). Most ecological subsets of mammals exhibit a similar north-south discrepancy in richness, including mammals greater than 90 kg, and it is particularly marked in the family distribution (figure 47.5) and the carnivorous and terrestrial ecological subsets. It is surprising that this pattern is observed in large mammals, given their greater foraging, migratory, and/or distributional ranges, but such has been found to be the case (figure 47.7)

This difference in mammal richness is evident from the species-level distribution, is even greater in the genus distribution (figure 47.3) and is greatest in the family distribution, where the difference is quite marked (figure 47.5). Since the higher the taxonomic level, the longer in time the taxon has existed, this progression appears to show southern Africa as the center of dispersal of much of the present-day African mammalian fauna. This hypothesis could be tested by investigating the phylogenetic record of African mammals and their centers of endemism.

Mammal richness was expected to be greater in eastern Africa than would be predicted by woody plant richness (which is based on climate) or climate (annual rainfall) alone. Habitat heterogeneity in this part of Africa is high because of tectonic activity associated with the rift valleys, resulting in the convergence of several phytochoria within a limited geographic space, and species richness has been shown to increase with habitat heterogeneity (Kerr and Packer, 1997). This runs counter to the fact that predicted woody plant richness and precipitation are less than that of the rain forests in West and Central Africa, and therefore mammal richness should also be less as it is highly correlated with these factors. However, based on our southern Africa work, the habitat heterogeneity engendered by topographic relief results in an increase in mammal richness in that area (Andrews and O'Brien, 2000; O'Brien et al., 2000). We have found this to be the case for eastern Africa in our analyses for all-mammals and for many of the ecological subsets (figures 47.5–47.7).

All-mammal species richness was expected to be positively and significantly correlated with all ecological subsets, but most strongly correlated with small mammals and with insectivorous, aerial, arboreal, and frugivorous subsets. These were all shown to be the case for southern Africa (Andrews and O'Brien, 2000), with lowest correlations for fossorial (not significant), scansorial (not significant), and aquatic subsets. Our results here show high correlations between all-Africa and small ($\leq$100 g, r = 0.91, 1–10 kg r = 0.93), insectivorous (r = 0.95), and aerial (r = 0.95) subsets, but the arboreal (r = 0.83), and frugivore (r = 0.68) subsets have lower than expected correlations (table 47.3). On the contrary, the terrestrial subset, which had a slightly lower correlation with all-mammals in southern Africa, is highly correlated in the all-Africa sample (r = 0.93).

The ecological subsets of mammals that support the richness peaks in eastern High Africa are terrestrial, browsers, grazers, carnivores, insectivores, small mammals (<100 g) and small to medium-sized mammals (1–45 kg). These also show richness lows in Central Africa, expected in the case of terrestrial species but arising from other factors for the rest (as discussed earlier).

Increase in woody plant species richness within a limited geographic area is the product of convergence or diversity in vegetation zones. This leads to heterogeneity of the habitat (Kerr and Packer, 1997), and structural complexity of vegetation in turn is strongly related to mammalian faunas, in terms of both numbers of species and their adaptations (Fleming, 1973; Andrews et al., 1979). Habitat complexity in southern Africa was found

to be most closely matched in the arboreal (r = 0.88), insectivore (r = 0.85), frugivore (r = 0.82) and small mammals ($\leq$45 kg; r = 0.78) subsets. It was most poorly matched by large mammal subsets (>45 kg and >90 kg; r = 0.21 and 0.13) and by fossorial (r = −0.08), scansorial (r = −0.25), and aquatic (r = 0.35) subsets. Since the plant richness values used in the present study are based on predictions, we have not calculated correlations with mammal richness values.

SPECIES RICHNESS AND SPECIES ABUNDANCE

It has been observed that animal species richness may either increase or decrease with increased habitat productivity (Rosenzweig, 1997). It is necessary here to distinguish between species richness and species abundance (biomass), and distinction must also be made between different food chains in assessing the effects productivity may have on any one part of it. Ecosystems such as grasslands may be less productive than forests in terms of standing biomass, but they may have up to 50% of net primary production passing through the animal grazing food chain (Odum, 1983). As grasses and herbs grow on an annual cycle, with new resources available on a yearly basis, productivity *for mammals* may be high even where overall productivity is low. Forests carry a greater plant biomass and higher productivity, but most of it is locked up in long-lived trees, and 90% of net production may pass through the detritus food chain (Odum, 1983). The yearly production of leaves and fruit that is available for mammals to eat is only a small fraction of total biomass. Thus, animal biomass is greater in open, grassland habitats, because large amounts of plant food are available on an annual basis, but species richness is low because the lack of variety of plant types and habitat do not provide the niches for species with differing requirements. As a result, mammal species richness is not and should not be expected to be highly correlated with annual net primary productivity (Rosenzweig, 1997).

CORRELATIONS AT HIGHER TAXONOMIC LEVELS

Variation in woody plant species richness accounts for 77% of the variation in mammal species richness in southern Africa and nearly as much for mammal genus richness (70%), but it accounts for only 35%–50% of the variation in mammal family richness. It is usually the case that at higher taxonomic levels the distributional ranges of mammal taxa increase absolutely, including within them the distributional ranges of all subordinate taxa, and thus encompassing a wider range of climatic, vegetational (plant richness), and topographic conditions. The same applies with regard to the ecological and physiological characteristics of the taxa themselves. Such blending tends to homogenize the variations in climate, vegetation, and terrain and should result in lower correlations between mammal richness and these parameters at higher levels. Moreover, nearly 80% of family variability is accounted for by terrestrial species, with omnivores, insectivores, and frugivores only accounting for 63%–65%, a pattern distinct from that of species diversity. This blending only appears to be a significant factor at the family level of analysis for mammals, and it should caution against the use of family-level identifications for palaeoecological interpretation in the fossil record.

OMISSION OF BATS FROM ANALYSES

Most mammal faunas in the fossil record do not include bats for a variety of reasons. Their lifestyle is different, their manner of death and preservation are different, and their bones are more fragile than the majority of other mammal bones. It is only in cave deposits that bats are commonly found, and even there their mode of preservation is likely to be different (Andrews, 1990). Multiple regression of all mammal species on plant species richness in data from southern Africa reduced the R^2 value from 75% to 68% when bats were excluded (Andrews and O'Brien, 2000). In addition, multiple regression of different ecological guilds against the total mammalian fauna showed a distinct difference when bats were excluded. Over 93% of species richness variability is accounted for in the southern African mammalian faunas by insectivorous species, the great majority of which are bats, and 85% is accounted for by bats alone. Arboreal species account for 82% of variability in mammal species richness, and when bats are excluded, these remain as the best predictor of variability. In the models we put forward (Andrews and O'Brien, 2000), insectivorous and terrestrial species accounted for 97% of mammal species richness; but when bats were excluded, it was the combination of terrestrial and semiterrestrial species that accounted for the highest variability at 95%. In another model, we found that terrestrial and semiterrestrial species and bats account for nearly 99% of variability in mammal species richness; whereas when bats were excluded, arboreal species replaced bats with a similar R^2 value. Considering just the dietary guilds, frugivores accounted for the highest variability of mammal species richness when bats were excluded from the analysis, accounting for 75% compared with 81% when the small number of southern African frugivorous bats are included.

It is clear that inclusion of bats adds greatly to the discriminatory power of mammalian faunas, but their absence from most fossil faunas would be unlikely to alter their ecological interpretation. This is reinforced by the fact that multiple regression values between mammals and woody plants decrease by only about 7% when bats are excluded, by 10% between mammals and separate dietary and spatial guilds, and by little or no change with many climatic variables.

LIMITS OF DATA ON SMALL MAMMAL DISTRIBUTIONS

We have emphasized in our introductory remarks that the data available on African mammal distributions are far from comprehensive (see table 47.1). Large mammal distributions are well mapped throughout the continent, but distributions of mammals less than 5 kg are sporadic, being good for southern Africa and East Africa but poor for most of the rest of Africa (Kingdon 1971–82, Smithers, 1983). We gave the example of *Crocidura*, a genus of Insectivora with 103 known species in Africa, and few of these species have information on their geographic ranges. Similarly, 276 of 364 known species of rodent lack recorded species distributions. At generic and higher taxonomic levels, distributions are better documented for all small mammals, but rather than analysing richness at the genus level, or eliminating small mammals altogether from our analyses, we have chosen to use genus distributions for just those species whose distributions are not recorded. All other species have good data for their distributions, and accordingly their contributions to richness are mixed with generic data for some species. We recognize that this introduces biases into our analyses, in particular correlations between species and genus distributions. It also diminishes the species richness of areas having more than one species of any of the genera so treated, with the result that areas of

high potential richness will have lower than expected richness. These problems can be remedied over time as data are collected for other parts of Africa.

APPLICATIONS TO CONSERVATION

Species richness patterns in Africa have been investigated in the past mainly as a tool for conservation, seeking hot spots with high diversity in order to preserve them (Williams, 1996; Williams et al., 2000; Brooks et al., 2001). The patterns obtained in these investigations are extremely similar to the all-mammal map presented here in figure 47.3, with species richness peaks in eastern Africa along the rift valley highlands, greater diversity south of the equator than north, and relative richness lows in the central African forests in the Congo Basin. This similarity, obtained using different sampling techniques, corroborates our results, but in addition we can show that some areas are more important for different ecological categories of mammals. For example, the forests of Guinea and Cameroon have high species richness of arboreal (figure 47.5) and frugivorous (figure 47.6) mammals, while for aerial mammals, including bats, the diversity hot spot is further south in southern Tanzania, Mozambique, Malawi, and Zambia (figure 47.5). Other categories of mammal have distributions corresponding to that for all-mammals, such as medium sized and very small mammals (figure 47.7), while others have widely dispersed distributions, such as carnivores (figure 47.6) and large mammals (figure 47.7). There is no one part of Africa that preserves all types of mammal, and clearly conservation should be geared to the type of mammal to be preserved.

APPLICATIONS TO PALEOECOLOGY

There is increasing interest in reconstructing the habitats of fossil mammals and their ancestors, and methodology is being increasingly refined. There are issues of taphonomic biases introduced into all fossil faunas by taphonomic processes, but these will not be considered here (Behrensmeyer, 1975; Brain, 1981; Andrews, 1990; Lyman, 1994). Alternative approaches such as isotopes (Kingston and Harrison, 2007), pollen (Bonnefille et al., 2004), and vegetation reconstruction (Andrews and Bamford, 2008) provide independent evidence on environment, but these also are beyond the scope of this chapter. Our main concern here is that there is insufficient attention being paid to the range of habitats present today in Africa that could serve as a comparative base for reconstructions based on mammalian faunas, still one of the most common forms of environmental reconstruction (Andrews and Bamford, 2008). Different mammal species are used on an opportunistic basis to indicate past environments, often because they are the taxa that are most abundant or are most familiar to the workers concerned.

Two groups of mammal have been targeted in particular for environmental reconstruction, bovids and rodents (Kappelman, 1984, 1988; Avery, 1990, 1991; Plummer and Bishop, 1994; Gentry, 1996; Kappelman et al., 1997; Fernandez-Jalvo et al., 1998; DeGusta and Vrba, 2005; Kovarovic and Andrews, 2007). Both bovids and rodents are abundant in many fossil sites and have high species richness, and both have distributions of species richness very similar to that of all-mammals (compare figure 47.7 with figure 47.3, and see table 47.5). As such, they might appear to be good proxies for mammals as a whole, but they are less well suited as proxies for vegetation reconstruction.

The highest levels of species richness in Africa are not in the areas of greatest plant diversity, such as in the rain forests, but they are in areas where several vegetation types converge in one geographic area, this convergence being the consequence of topographic relief. Woody plant richness in southern Africa (data are not currently available for Africa as a whole) is most highly correlated with arboreal and aerial frugivores and insectivores ($r = 0.79$–0.88) but less highly correlated with terrestrial browsers and grazers ($r = 0.61$–0.69). If these results are corroborated for Africa as a whole, it indicates that while terrestrial bovids and semiterrestrial rodents may be good indicators for habitat complexity, they are not such good indicators for vegetation richness compared with primates, bats, and insectivores (i.e., arboreal and aerial frugivores and insectivores).

The application of species richness analysis to fossils has limitations, for the fossil record is too scattered for this to be possible at present. In addition, the assignment of ecological variables to fossils is problematic unless fossils are available that provide evidence on diet (from the dentition), locomotion, and size (both from postcranial remains). Some of these ecological variables can be determined through ecomorphology of fossil taxa, especially on bovid postcrania (Kappelman, 1988; Plummer and Bishop, 1994; DeGusta and Vrba, 2005; Kovarovic and Andrews, 2007), bovid teeth (Janis, 1988; Hunter and Fortelius, 1994), and carnivore postcrania (Van Valkenburgh, 1987; Turner 1990). In addition, microwear (Walker et al., 1978), mesowear (Fortelius and Solounias, 2000), and tooth structure (Jernvall and Fortelius, 2004) provide interpretations of diet on a wide range of taxa. Ecomorphology of small mammals has yet to be attempted with any success, but it is likely that useful results will be obtained in the future. When these measures of ecomorphology are combined with community analysis of fossil faunas, they will provide evidence of habitat structure in greater detail than is possible at present (Andrews et al., 1979; Andrews, 1996; Fernandez-Jalvo et al., 1998).

Conclusions

In the present study, patterns of species richness have shown marked differences between ecological categories of mammal, with some varying with vegetation but most varying with other factors. For example, arboreal frugivores and mammals in the small size class 100 g–1 kg have distribution patterns tracking woody plant distributions, with greatest richness in the West African lowland rain forests. Most mammals, however, have greatest richness where the habitat is most variable, as in East High Africa where five plant phytochoria converge as a result of high topographic relief. Variations in richness for the extant faunas provide a basis for interpreting fossil faunas to reconstruct past environments. This indeed is the basis for the community ecology approach, whereby relatively high proportions of one group or another is used as evidence of habitat. For example, arboreal frugivores have been shown here to be highly correlated with vegetation and to have greatest richness in tropical forest, and so their presence in a fossil fauna indicates an environment rich in trees, with fruit available for much of the year. This is making a readily understandable statement about the environment (Andrews et al., 1979). Similarly, high richness of terrestrial browsers in a fossil fauna provides less strong evidence of vegetation, for they are poorly correlated with distribution patterns of vegetation; but, on

the other hand, they are highly correlated with patterns of species richness as a whole, and they may therefore be representative of a fossil fauna, particularly with reference to the degree of complexity of the environment. Small mammals are also highly correlated with all-mammal species richness, and like terrestrial herbivores, they may be representative of environmental complexity; but they are less highly correlated with vegetation. The end conclusion of this work is that different ecological categories of mammals have different relationships with the environment, vegetation, and climate, and analyses of palaeoecology should take these differences into account.

ACKNOWLEDGMENTS

We are grateful to Sylvia Hixson, Brian Rosen, and Rob Whittaker for comments on this work. We are also glad to acknowledge the help of Jessica Pearson, Jennifer Scott, and Martin Strawbridge in the data analyses. We are also grateful to the editors of this volume for the invitation to submit a chapter to the volume, for their comments on the manuscript, and in particular to the anonymous referee who put much time and effort into improving the quality of this chapter.

Literature Cited

Andrews, P. 1990. *Owls, Caves and Fossils*. Natural History Museum, London, 231 pp.

———. 1996. Palaeoecology and hominoid palaeoenvironments. *Biological Reviews* 71:257–300.

Andrews, P., and M. Bamford 2008. Past and present vegetation ecology of Laetoli, Tanzania. *Journal of Human Evolution* 54:78–98.

Andrews, P., J. Lord, and E. M. N. Evans. 1979. Patterns of ecological diversity in fossil and modern mammalian faunas. *Biological Journal of the Linnean Society* 11:177–205.

Andrews P., and E. M. O'Brien. 2000. Climate, vegetation and predictable gradients in mammal species richness in southern Africa. *Journal of Zoology* 251:205–231.

Aubréville, A. 1938. La forêt coloniale: Les forêts de l'Afrique occidentale française. *Annales de l'Académie des Sciences Coloniales, Paris* 9:1–245.

Avery, D. M. 1991. Micromammals, owls and vegetation change in the Eastern Cape Midlands, South Africa, during the last millennium. *Journal of Arid Environments* 20:357–369.

Avery, M. 1990. Holocene climatic change in Southern Africa: The contribution of micromammals to its study. *South African Journal of Science* 86:407–412.

Badgley, C. and D. L. Fox. 2000. Ecological biogeography of North American mammals: Species density and ecological structure in relation to environmental gradients. *Journal of Biogeography* 27:1437–1467.

Begon, M., J. L. Harper, and C. R. Townsend. 1990. *Ecology: Individuals, Populations and Communities*. Sinauer, Sunderland, 1068 pp.

Behrensmeyer, A. K. 1975. Taphonomy and paleoecology of the Plio-Pleistocene vertebrate assemblages east of Lake Rudolf, Kenya. *Bulletin of the Museum of Comparative Zoology* 146:473–578.

Bonnefille, R., R. Potts, F. Chalié, D. Jolly, and O. Peyron. 2004. High resolution vegetation and climate change associated with Pliocene *Australopithecus afarensis*. *Proceedings of the National Academy of Sciences, USA* 101:12125–12129.

Brain, C. K. 1981. *The Hunters or the Hunted? An Introduction to African Cave Taphonomy*. University of Chicago Press, Chicago, 365 pp.

Brooks, T., A. Balmford, N. Burgess, J. Fjeldså, L. A. Hansen, J. Moore, C. Rahbek, and P. Williams. 2001. Toward a blueprint for conservation in Africa. *BioScience* 51:613–624.

Cowling, R. M., D. M. Richardson, and S. M. Pierce. 1997. *Vegetation of Southern Africa*. Cambridge University Press, Cambridge, 649 pp.

Currie, D. J. 1991. Energy and large-scale patterns of animal- and plant-species richness. *American Naturalist* 137:27–49.

Currie, D. J., and V. Paquin. 1987. Large scale biogeographic patterns of species richness in trees. *Nature* 329:326–327.

DeGusta, D., and E. S. Vrba. 2005. Methods for inferring paleohabitats from the functional morphology of bovid phalanges. *Journal of Archaeological Science* 32:1099–1113.

Gentry, A. W. 1996. A fossil *Budorcas* (Mammalia, Bovidae) from Africa; pp. 571–587 in K. M. Stewart and K. L. Seymour (eds.), *Palaeoecology and Palaeoenvironments of Late Cenozoic Mammals: Tributes to the Career of C. S. (Rufus) Churcher*. University of Toronto Press, Toronto.

Fernández-Jalvo, Y., C. Denys, P. Andrews, T. Williams, Y. Dauphin, and L. Humphrey. 1998. Taphonomy and palaeoecology of Olduvai Bed-I (Pleistocene, Tanzania). *Journal of Human Evolution* 34:137–172.

Field, R., E. M. O'Brien, and R. J. Whittaker. 2005. Global models for predicting the climatic potential for (tree-shrub) plant richness: Development and evaluation. *Ecology* 86:2263–2277.

Fleming, T. H. 1973. Numbers of mammal species in North and Central American forest communities. *Ecology* 54:555–563.

Fortelius, M., and N. Solounias. 2000. Functional characterization of ungulate molars using the abrasion-attrition wear gradient: A new method for reconstructing paleodiets. *American Museum Novitates* 3301:1–36.

Frandsen, R. 1992. *Southern Africa's Mammals: A Field Guide*. Frandsen Publishers, Sandton, 238 pp.

Griffiths, J. F. 1976. *Climate and the Environment*. Westview Press, Boulder, Colo., 152 pp.

Hunter, J. P., and M. Fortelius. 1994. Comparative dental occlusal morphology, facet development and microwear in two sympatric species of *Listriodon* (Mammalia, Suidae) from the middle Miocene of western Anatolia (Turkey). *Journal of Vertebrate Paleontology* 14:105–126.

Janis, C. 1988. An estimation of tooth volume and hypsodonty indices in ungulate mammals, and the correlation of these factors with dietary preference. *Mémoires du Muséum National d'Histoire Naturelle, Paris (Série C)* 53:367–387.

Jernvall, J., and M. Fortelius. 2004. Maintenance of trophic structure in fossil mammal communities: Site occupancy and taxon resilience. *American Naturalist* 164:614–624.

Kappelman, J. 1984. Plio-Pleistocene environments of Bed I and Lower Bed II, Olduvai Gorge, Tanzania. *Palaeogeography, Palaeoclimatology, Palaeoecology* 48:171–196.

———. 1988. Morphology and locomotor adaptations of the bovid femur in relation to habitat. *Journal of Morphology* 198:119–130.

Kappelman, J., T. Plummer, L. Bishop, A. Duncan, and S. Appleton. 1997. Bovids as indicators of Plio-Pleistocene paleoenvironments in East Africa. *Journal of Human Evolution* 32:229–256.

Kerr, J. T., and L. Packer. 1997. Habitat heterogeneity as a determinant of mammal species richness in high-energy regions. *Nature* 385:252–254.

Kingston, J. D., and T. Harrison. 2007. Isotopic dietary reconstructions of Pliocene herbivores at Laetoli: Implications for early hominin paleoecology. *Palaeogeography, Palaeoclimatology, Palaeoecology* 243:272–306.

Kingdon, J. 1971–1982. *East African Mammals: An Atlas of Evolution in Africa*. 7 parts. Academic Press, London.

———. 1997. *The Kingdon Field Guide to African Mammals*. Academic Press, London, 464 pp.

Kovarovic, K., and P. Andrews. 2007. Bovid postcranial ecomorphological survey of the Laetoli palaeoenvironments. *Journal of Human Evolution* 52:663–680.

Lawton, J. H., D. E. Bignell, B. Bolton, G. F. Bloemers, P. Eggleton, P. M. Hammond, M. Hodda, R. D. Holt, T. B. Larsen, N. A. Mawdsley, N. E. Stork, D. S. Srivastava, and A. D. Watt. 1998. Biodiversity inventories, indicator taxa and effects of habitat modification in tropical forest. *Nature* 391:72–76.

Lyman, L. 1994. *Vertebrate Taphonomy*. Cambridge University Press, Cambridge, 550 pp.

MacArthur, R. H. 1964. Environmental factors affecting bird species diversity. *American Naturalist* 98:387–397.

———. 1965. Patterns of species diversity. *Biological Reviews* 40:510–533.

Mayr, E. 1963. *Animal Species and Evolution*. Harvard University Press, Cambridge, 797 pp.

Moreau, R. E. 1966. *The Bird Faunas of Africa and Its Islands*. Academic Press, London, 424 pp.

Mound, L. A., and N. Waloff (eds.). 1978. *Diversity of Insect Faunas*. Symposia of the Royal Entomological Society number 9, Blackwell, Oxford, 204 pp.

O'Brien, E. M. 1993. Climatic gradients in woody plant species richness: Towards an explanation based on an analysis of southern Africa's woody flora. *Journal of Biogeography* 20:181–198.

———. 1998. Water-energy dynamics, climate and prediction of woody plant species richness: an interim general model. *Journal of Biogeography* 25:379–398.

O'Brien, E. M., and C. R. Peters. 1999. Landforms, climate, ecogeographic mosaics, and the potential for hominid diversity in Pliocene Africa; pp. 115–137 in T. G. Bromage and F. Schrenk (eds.), *African Biogeography, Climate Change and Human Evolution*. Oxford University Press, Oxford.

O'Brien, E. M., R. J. Whittaker, and R. Field. 1998. Climate and woody plant diversity in southern Africa: Relationships at species, genus and family levels. *Ecography* 21:495–509.

O'Brien, E. M., R. J. Whittaker, and R. Field. 2000. Climatic gradients in woody plant (tree and shrub) diversity: Water-energy dynamics, residual variation, and topography. *Oikos* 89:588–600.

Odum, E. P. 1983. *Basic Ecology*. Holt Saunders, New York, 320 pp.

Pianka, E. R. 1966. Latitudinal gradients in species diversity: a review of concepts. *American Naturalist* 100:33–46.

Plummer, T. W., and L. C. Bishop. 1994. Hominid paleoecology at Olduvai Gorge. Tanzania as indicated by antelope remains. *Journal of Human Evolution* 27:47–75.

Richards, P.W. 1952. *The Tropical Rain Forest: An Ecological Study*. Cambridge University Press, Cambridge, 468 pp.

Rosen, B. R. 1984. Reef coral biogeography and climate through the late Cainozoic: Just islands in the sun or a critical pattern of islands? *Geological Journal Special Issue* 11:201–262.

Rosenzweig, M. L. 1995. *Species Diversity in Space and Time*. Cambridge University Press, Cambridge, 436 pp.

Shepherd, U. L. 1998. A comparison of species diversity and morphological diversity across the North American latitudinal gradient. *Journal of Biogeography* 25:19–29.

Simpson, G. G. 1964. Species diversity of North American recent mammals. *Systematic Zoology* 13:57–73.

Smithers, R. H. N. 1983. *The Mammals of the Southern African Subregion*. University of Pretoria, Pretoria, 736 pp.

Stuart, C., and T. Stuart. 1988. *Field Guide to the Mammals of Southern Africa*. New Holland, London, 272 pp.

Thiery, R. G. 1982. Environmental instability and community diversity. *Biological Reviews* 57:671–710.

Thornthwaite, C. W., and J. R. Mather. 1955. *The Water Balance*. Publications in Climatology 8, Laboratory of Climatology, Centerton, N.J.

———. 1962. *Average Climatic Water Balance Data of the Continents: Part I. Africa*. Publications in Climatology 15, Laboratory of Climatology, Centerton, N.J.

Turner, A. 1990. The evolution of the guild of larger terrestrial carnivores during the Plio-Pleistocene in Africa. *Geobios* 23:349–368.

Van Valen, L. 1973. Body size and numbers of plants and mammals. *Evolution* 27:27–35.

———. 1975. Group selection, sex and fossils. *Evolutionary Theory* 1:1–30.

Van Valkenburgh, B. 1987. Skeletal indicators of locomotor behavior in living and extinct carnivores. *Journal of Vertebrate Paleontology* 7:162–182.

Van Zinderen Bakker, E. M. (ed.). 1969. *Palaeoecology of Africa and of the Surrounding Islands and Antarctica*, vol. 4. Balkema, Cape Town, 274 pp.

Walker, A. C., H. N. Hoeck, and L. Perez. 1978. Microwear of mammalian teeth as an indicator of diet. *Science* 201:908–910.

Werger, M. J., and A. C. Van Bruggen. 1978. *Biogeography and Ecology of Southern Africa*. Springer, Berlin, 1439 pp.

White, F. 1981. *Vegetation Map of Africa*. UNESCO/AETFAT/UNESCO, Paris.

———. 1983. *The Vegetation of Africa: A Descriptive Memoir to Accompany the UNESCO/AETFAT/UNSO Vegetation Map of Africa*. UNESCO, Paris, 356 pp.

Williams, P. H. 1996. Mapping variations in the strength and breadth of biogeographic transition zones using species turnover. *Proceedings of the Royal Society of London, B* 263:579–588.

Williams, P. H., N. D. Burgess, and C. Rahbek. 2000. Flagship species, ecological complementarity, and conserving the diversity of mammals and birds in sub-Saharan Africa. *Animal Conservation* 3:249–260.

Wilson, D. E., and D. M. Reeder. 1993. *Mammal Species of the World*. Smithsonian Institution Press, Washington, D.C., 1,206 pp.

FORTY-EIGHT

Stable Carbon and Oxygen Isotopes in East African Mammals: Modern and Fossil

THURE E. CERLING, JOHN M. HARRIS, MEAVE G. LEAKEY,
BENJAMIN H. PASSEY, AND NAOMI E. LEVIN

Stable isotopes have become an important tool to study diets and behavior of fossil mammals, but the path to acceptance has been long and arduous. In 1978, DeNiro and Epstein published an important paper showing that mammalian tissues recorded valuable dietary information in their $^{13}C/^{12}C$ ratios in bones, collagen, and other tissues (DeNiro and Epstein, 1978). Shortly thereafter, Sullivan and Krueger (1981) and Ericson et al. (1981) proposed that fossil bone could be used to reconstruct diet from their isotope ratios. In 1982, Schoeninger and DeNiro (1982) showed that bone apatite and bone collagen gave different estimates of diets; these results were interpreted to mean that bone apatite was susceptible to diagenesis and that no dietary information could be obtained from fossil bone apatite. The matter rested uneasily for a few years until Lee-Thorp and van der Merwe (1987) published observations that stable isotopes in tooth bioapatite preserved a signal through geological time; they interpreted these results to suggest that bioapatite recorded diet signals that were preserved through diagenesis. This study sparked additional studies, especially concentrating on tooth enamel (e.g., Lee-Thorp et al., 1989).

Several important developments followed in the early 1990s. Quade et al. (1992) gave conclusive evidence that initial $^{13}C/^{12}C$ ratios in tooth enamel were preserved during diagenesis by showing that the stable isotope composition of tooth enamel was different than diagenetic carbonates in the same sequences. Ambrose and Norr (1993) showed that collagen and bioapatite primarily recorded the protein and the carbohydrate sources of diet, respectively, and therefore collage and bioapatite may give different indications of diet. Ayliffe et al. (1994) used X-ray diffraction (XRD) to show that modern enamel was well crystallized but modern dentine and bone were poorly crystallized; in contrast, fossil enamel, bone, and dentine were all well crystallized, showing that bone and dentine underwent recrystallization during diagenesis. Wang and Cerling (1994) further showed that the isotope ratios for both $^{13}C/^{12}C$ and $^{18}O/^{16}O$ of fossil dentine and fossil bone could be compromised in diagenesis and that the process could be modeled using water/rock interaction. Some debate has ensued about the relative fidelity of the CO_3 versus PO_4 components in bioapatites, with some arguments being made that even during recrystallization the oxygen in the PO_4 component could preserve the original signal. In certain cases, the interior enamel exhibits different isotope ratios than the exterior (Schoeninger et al., 2003a), which may be due to diagenesis or to enamel maturation patterns (Passey and Cerling 2002). Studies of sample treatments (Koch et al., 1997), Fourier transform infrared spectrometry (FTIR; Sponheimer and Lee-Thorp, 1999), and trace element abundances (Sponheimer and Lee-Thorp, 2006) have also been used to study diagenesis processes with the conclusion that the isotope signal in enamel is much more robust than in dentine or bone.

There is general agreement now that the $^{13}C/^{12}C$ and $^{18}O/^{16}O$ ratios of fossil tooth enamel are preserved in the Neogene geological record, although there is as yet no diagnostic test to assure that no resetting of the stable carbon and oxygen isotope signal of enamel occurred during diagenesis; dentine and bone are almost always recrystallized and are likely to have been compromised during recrystallization. The confidence arises from several forms of evidence. One approach is to sample different taxa that might be expected to have a large range in $^{13}C/^{12}C$ or $^{18}O/^{16}O$ ratios. For example, deinotheres always have low $^{13}C/^{12}C$ ratios even when other mammals have high $^{13}C/^{12}C$ ratios (Cerling et al., 1999). Hippopotamids always have $^{18}O/^{16}O$ ratios lower than equids, even when sampled from the same sedimentary unit (Cerling et al., 2003a), and should, therefore, have experienced identical diagenetic conditions. These observations lend confidence that diagenesis has not significantly altered the original isotope ratios in enamel.

Many paleontological research projects in Africa now incorporate isotopic aspects to the work, and some isotope results are available from many study sites including Allia Bay (Schoeninger et al., 2003b), the Baringo Basin (Kingston et al., 1994), Chorora (Bernor et al., 2004), Gona (Levin et al., 2008), Fort Ternan (Cerling et al., 1996), Laetoli (Kingston and Harrison, 2007), Kanapoi (Harris et al., 2003), Langebaanweg (Franz-Odendaal et al., 2002), Lothagam (Leakey et al., 1996; Cerling et al., 2003a), Makapansgat (Sponheimer et al., 1999;

Hopley et al., 2006), Sterkfontein (van der Merwe et al., 2003), and Swartkrans (Lee-Thorp et al., 1994, 2003). Studies focused on mammalian lineages also often use isotope analysis as part of the study so that comparisons or histories are available for bovids (Cerling et al., 2003b; Sponheimer et al., 2003), hippos (Boisserie et al., 2005; Cerling et al., 2008; Harris et al., 2008), proboscideans (Cerling et al., 1999), suids (Harris and Cerling, 2002), and even including some hominids (Lee-Thorp et al., 1994; van der Merwe et al., 2003, 2008; Sponheimer et al., 2006). The increased application of stable isotopic studies is due to careful work to reduce sample size and working with museum curators to develop conservative approaches to sampling, including collecting samples that would otherwise not be collected specifically for isotope studies, profile work, and innovative analytical techniques (e.g., laser ablation).

In this review, we will discuss only the isotope ratios that are preserved in the carbonate component of tooth enamel. Phosphate studies apply only to oxygen isotopes and dietary information is lacking; likewise, collagen is rarely preserved in fossils older than some thousands of years.

We first provide a background by discussing general aspects of stable isotope ecology in Africa, principally those processes related to effects of $^{13}C/^{12}C$ in different ecosystems and then those processes affecting $^{18}O/^{16}O$ ratios in the water cycle. We review isotopic analysis of the diets of extant African mammals and consider the dietary history of different mammalian lineages. We then discuss the correspondence between the isotopic evidence for the transition from a C_3 to a C_4 world and the palaeontological evidence for faunal change at the end of the Miocene and beginning of the Pliocene as shown by the faunal assemblages from Lothagam, northern Kenya. This faunal turnover is one of the most marked in the Cenozoic. Lastly, we discuss future directions of stable isotope paleoecology.

Background

ISOTOPE TERMINOLOGY AND METHODS

Carbon occurs on Earth as two stable isotopes, ^{12}C and ^{13}C, with proportional abundances of 98.89% and 1.11%, respectively. Oxygen occurs as three stable isotopes, ^{16}O, ^{17}O, and ^{18}O, with proportional abundances of 99.759%, 0.037%, and 0.204%, respectively. That isotope ratios can be more accurately measured than absolute isotope abundances was recognized in the 1950s, and the now-traditional isotope definitions and terminology were developed as a result. Isotope ratios for carbon and oxygen are expressed as

$$\delta^{13}C = (R_{sample}/R_{standard} - 1)*1,000$$

and

$$\delta^{18}O = (R_{sample}/R_{standard} - 1)*1,000,$$

where R_{sample} and $R_{standard}$ are the $^{13}C/^{12}C$ or $^{18}O/^{16}O$ ratios of the sample and standard, and where the units are permil (‰). The $\delta^{13}C$ values of plants and bioapatite $\delta^{13}C$ and $\delta^{18}O$ values derived wholly from the CO_3 component in enamel are referenced relative to the international standard pee dee belemnite (PDB). The $\delta^{18}O$ of water values, on the other hand, are reported relative to the standard standard mean ocean water (SMOW). The two reference scales are related by

$$\delta^{18}O_{SMOW} = 1.03091 (d^{18}O_{PDB}) + 30.91.$$

Isotope fractionation describes the phenomenon that different isotope ratios are found in isotopic equilibrium with each other in different phases. For example, the heavy isotope ^{18}O is enriched in liquid water relative to water vapor; this enrichment is known as fractionation and is defined as

$$\alpha_{AB} = R_A/R_B = (1,000 + \delta_A)/(1,000 + \delta_B),$$

where α_{AB} is isotope fractionation, and A and B represent two different phases. Isotope enrichment describes fractionation in units of ‰:

$$\varepsilon_{AB} = (\alpha_{AB} - 1)*1,000.$$

Isotope discrimination is similar to isotope enrichment except that isotope equilibrium is not necessarily attained. This is a useful expression to describe nonequilibrium processes such as the isotope enrichment from diet to tissue. Isotope discrimination is here defined as

$$\varepsilon_{AB}^* = (R_A/R_B - 1)*1,000,$$

where the asterisk implies that the system is not necessarily at isotope equilibrium.

Carbon in bioapatite occurs as CO_3 substituted for PO_4 and OH in the bioapatite ($Ca_5(PO_4)_3(OH)$ structure; approximately 0.7 mmol/mg of CO_2 are evolved in the reaction of bioapatite with 100% H_3PO_4. Oxygen also exists in both the PO_4 and OH positions; the PO_4 can be chemically separated and the $\delta^{18}O$ of the PO_4 component can be determined. For tooth enamel, this is a redundant analysis because both the PO_4 and CO_3 components are in isotopic equilibrium with the same fluid: blood plasma. However, for dentine, bone, and cementum, the PO_4 component of bioapatite is more likely to be preserved through diagenesis. In this review we discuss only the results from tooth enamel, which, for most mammals, is sufficiently common that it can be analyzed for most purposes.

Climate parameters discussed in the text include mean annual precipitation (MAP) and mean annual temperature (MAT). The values discussed are taken from the East African Meteorological Department (1975).

CARBON ISOTOPES AND ECOLOGY

Plants use several photosynthetic pathways that have different $\delta^{13}C$ values. Plants using the C_3 photosynthetic pathway, which is used by most dicotyledonous plants, have an average $\delta^{13}C$ value of about −27.5‰; plants using the C_4 photosynthetic pathway, which is used by most tropical grasses and sedges, have an average $\delta^{13}C$ value of about −12.5‰ (Deines, 1980). In our discussion, we refer to C_4 grasses; in certain cases, C_4 sedges (and even more rarely, C_4 dicots or CAM plants) may be an important component in certain mammalian diets. Figure 48.1 shows the $\delta^{13}C$ of ca. 700 plants from East and Central Africa; the $\delta^{13}C$ of plants using the two different photosynthetic pathways is well separated, and thus $\delta^{13}C$ is a useful measure of the fraction of C_3 or C_4 biomass consumed by mammals. Fortunately, dicots make up browse, and C_4 plants are almost exclusively grasses in Africa so that the $\delta^{13}C$ value can be used to distinguish graze (grass) from browse (dicots). This distinction is blurred in higher latitudes and higher elevations because cool-season grasses use the C_3 photosynthetic pathway. Likewise, prior to the global

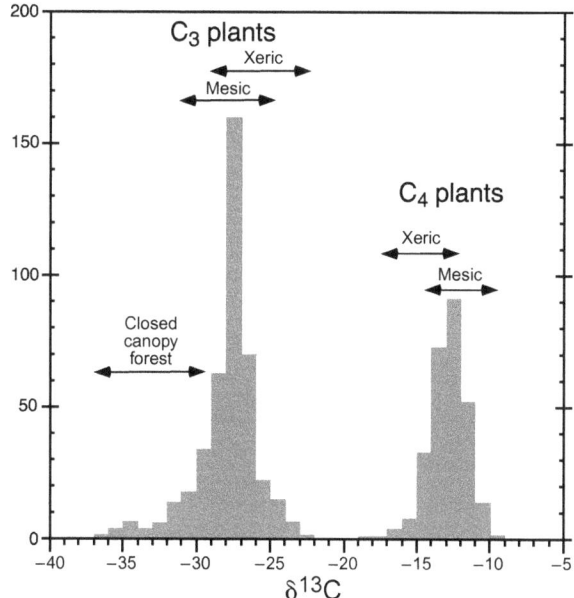

FIGURE 48.1 Histogram of $\delta^{13}C$ values of East and Central African plants ($n > 700$) collected between 1997 and 2006.

expansion of C_4 biomass in the late Miocene (Cerling et al., 1997), most grasses likely used the C_3 pathway even in tropical regions; thus, the distinction between grazing and browsing in the tropics can only be documented by stable isotopes in latest Miocene or younger fossils.

It is important to note that different ecosystems have differing $\delta^{13}C$ values for C_3 and C_4 plants. For example, previous authors have noted that C_4 grasses are more enriched in ^{13}C in mesic ecosystems than in arid ecosystems (e.g., Hattersley, 1982, 1992); the average $\delta^{13}C$ value of grasses in the Athi plains (MAP $\approx$ 800 mm, MAT $\approx$ 18°C) is about 2‰ enriched in ^{13}C compared grasses to the Turkana region (MAP $\approx$ 200 mm; MAT $\approx$ 29°C). The $\delta^{13}C$ of C_3 plants also is related to the aridity (Ehleringer and Cooper, 1988); however, more xeric sites tend to have more enriched $\delta^{13}C$ values than mesic sites. The xeric ecosystem in the Turkana region is enriched in ^{13}C compared to the relatively mesic ecosystem found in the Athi plains region. Closed canopy ecosystems can have very depleted ^{13}C ratios; the plants growing in the subcanopy of the Kakamega and Ituri forests have averaged $\delta^{13}C$ values of $-31‰$ and $-34‰$ (Cerling et al., 2003b, 2004), respectively.

The additional structure exhibited by these isotope differences in ecosystems provides both admonishment and reward. The admonishment is the lesson that there is no single mixing line between C_3 and C_4 ecosystems, but the reward is that the isotope method provides an independent check for a closed canopy ecosystem. Although sites of closed canopies are likely to be poorly preserved in the geological record, MacFadden and Higgins (2004) have interpreted closed canopy conditions based on very ^{13}C-depleted $\delta^{13}C$ values.

"You are what you eat" applies especially well to isotope ecology. The carbon isotopic composition of bioapatite is in isotopic equilibrium with blood bicarbonate. Several studies have documented the isotope enrichment from diet to bioapatite. Ambrose and Norr (1993) showed that the bioapatite $\delta^{13}C$ values are related to the bulk diet in

carbohydrate rich diets. Passey et al. (2005b) showed that the variation in isotope enrichment factors reported for different mammals (Tieszen and Fagre, 1993; Ambrose and Norr, 1993; Cerling and Harris, 1999) was related to the diet-breath enrichment: voles, rabbits, pigs, and cattle have isotopes enrichments of $11.5 \pm 0.3‰$, $12.8 \pm 0.7‰$, $13.3 \pm 0.3‰$, and $14.6 \pm 0.3‰$, respectively. Appropriate isotope enrichments are needed to interpret and compare fossil mammal assemblages.

Lastly, it is important to realize that the isotopic composition of the atmosphere, which is the source of carbon in plants, has not been constant over geological, or even human, time scales. For example, we know that humans have significantly changed the $\delta^{13}C$ of the atmosphere due to fossil fuel burning. Today, the $\delta^{13}C$ value of atmospheric CO_2 (-8 ‰) is shifted from the pre-1850 atmosphere, which had a $\delta^{13}C$ value of $-6.5‰$. Any interpretations of fossil material must take into account the isotope composition of the atmosphere at the time those fossils were preserved and its influence on the $\delta^{13}C$ of plants growing in that atmosphere. Figure 48.2 shows the change in $\delta^{13}C$ of deep ocean carbonates from the Atlantic Ocean through the past 20 million years based on the data of Zachos et al (2001); this represents a proxy for the isotopic composition of the atmosphere over geological time because Atlantic deep water has had little time for modification due to its recent downwelling. These data show that about the time of the global change from the "C_3 world" to the "C_4 world" at the end of the Miocene (Cerling and Ehleringer, 2000), the oceans underwent a significant shift in $\delta^{13}C$ perhaps related to that change in global terrestrial ecology (Cerling, 1997). Furthermore, the middle Miocene, equivalent to the Monterrey event in paleoceanography, was a period of very positive $\delta^{13}C$ values for the ocean and, presumably, the atmosphere. Therefore, in early and middle Miocene times, it is important to know the isotopic composition of the ecosystem making up the base of the food chain.

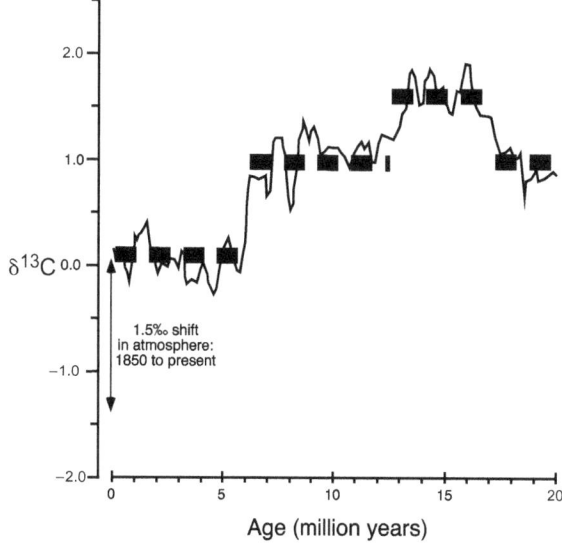

FIGURE 48.2 $\delta^{13}C$ of benthic foraminfera from the Atlantic Ocean (data from Zachos et al., 2001). Heavy lines show the long-term average $\delta^{13}C$ values for the periods from 20 to 17 Ma, 17 to 12.5 Ma, 12.5 to 6 Ma, and 6 Ma to pre-industrial times. The 1.5‰ shift in the atmosphere is shown as a solid line.

The meteoric water cycle determines the isotope composition of surface waters in East Africa. In present-day Africa, the $\delta^{18}O$ of rainfall varies from about 0‰ in the Ethiopian Highlands to about -6‰ in some high mountain regions (e.g., Mt. Kenya; Rietti-Shati et al., 2000). Globally, meteoric waters fall on or near to the relationship $\delta D = 8 \ \delta^{18}O + 10$. When water evaporates it is preferentially enriched in ^{18}O compared to D, so that the slope for evaporated waters is usually between ca. 3 and 5. Thus, Lake Turkana has a $\delta^{18}O$ value of about $+6\%$ although the inflowing Omo River, and local groundwater, has an average $\delta^{18}O$ value of about -2.3‰. Leaf water is highly enriched in D and ^{18}O compared to source waters, which usually are on, or near, the meteoric water line. Therefore, leaf water is highly enriched in ^{18}O and represents a source of water enriched in ^{18}O compared to unevaporated groundwater or surface water. It is important to consider the effects of evaporation if animals derive a significant portion of their water from lakes (e.g., topi or hippos from modern Lake Turkana) or from plants (e.g., giraffes). This has particular significance in the potential application to paleoaridity (discussed later).

Review of Isotopes in African Cenozoic Mammals

STABLE ISOTOPES IN EXTANT MAMMALS AND IMPLICATIONS FOR THE INTERPRETATION OF FOSSIL ASSEMBLAGES

Stable isotope analyses of extant mammals in Africa provide an important backdrop to the interpretation of fossil material. Modern analog studies show that stable isotopes are faithful recorders of animal diet and behavior today (e.g., Cerling et al., 2003b; Cerling et al., 2004; Sponheimer et al., 2003; Codron et al., 2006a, 2006b, 2007) and can be used as a template for interpreting isotopic data from the fossil record. In this section, we provide examples of studies done on extant mammals to illustrate the importance of understanding the stable isotope systematics of today's ecosystems.

Cerling et al. (2003b), Sponheimer et al. (2003), Codron et al. (2005), and Codron and Brink (2007) compiled extensive stable isotope data sets on extant African bovids. Similar compilations of stable isotope data have been produced for elephants (Cerling et al., 1999, 2007; Codron et al., 2006a), suids (Harris and Cerling, 2002; Cerling and Viehl, 2004), and hippos (Boisserie et al., 2005; and Cerling et al., 2008). Figure 48.3 shows box-and-whisker plots for a number of the large mammals of East and Central Africa.

The most important results from these studies have to do with the use of modern analogues for interpreting paleoenvironments and the relationship of mammals to environmental parameters. Isotope studies of bovids agree very well with the dietary data, which were usually obtained by observation and fecal analyses in the 1960s and 1970s. This confirms the utility of stable isotopes as indicators of dietary behavior.

However, there are several important differences between accepted dietary lore and stable isotope studies, and these have important implications for dietary reconstruction. Three large mammals stand out in this regard: elephants, forest hogs, and hippopotamuses. The diet of elephants has given rise to an enormous amount of both anecdotal data and observational data; stable isotope studies complement these

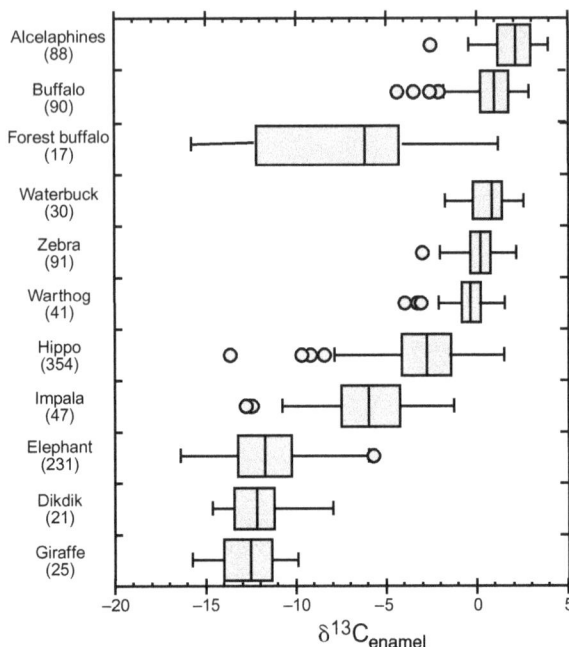

FIGURE 48.3 Box-and-whisker plots of $\delta^{13}C$ for representative large mammals from East and Central Africa. Forest and savanna buffalo are distinguished on the basis of where the samples were collected. Forest sites include the Aberdare, Mt. Kenya, Arobuke-Sokoke, and Ituri Forests. Figure includes data derived from elephant ivory (dentine) because diagenesis does not affect modern samples.

observational data by quantifying the amount of C_4 grass and C_3 browse in diets. Extensive surveys from all of Africa (van der Merwe et al., 1988, 1990; Cerling et al., 1999; Cerling et al., unpub. ms.) of ivory and tooth enamel, both of which give strongly long-term average dietary estimates, show that elephants are predominantly browsers with long-term average diets comprising 70%–100% browse. Isotope studies of hair and fecal material (Cerling et al., 2004, 2006; Codron et al. 2006), both of which give short-term diet estimates, show that grass is strongly favored in the wet season and browse is favored in the dry season; this agrees with observational data but gives quantitative estimates of the importance of grass in the diet on a seasonal basis. There has been significantly less observational data on forest hogs and hippos, in part because of the difficulty in observing their detailed behavior; both are reputed to be grazers (e.g., Kingdon, 1979). Stable isotope studies are in direct conflict with these conclusions: Boisserie et al (2005) and Cerling et al., (2008) show that C_3–plants make up an important (ca. 10%–20%) component of the diet in many hippos and that this value can vary seasonally or over long intervals. Likewise, stable isotope studies of forest hogs living in a savanna environment (Queen Elizabeth Park, Uganda; Cerling and Viehl, 2004) document a diet that contains up to 20% grass in the rainy season but otherwise is an essentially pure dicot-derived C_3 diet. In many environments, the forest hog has more negative $\delta^{13}C$ values than most other mammals, emphasizing its reliance on C_3 plants for its diet. Part of this may be due to the smaller fractionation factor for suids than for bovids (Passey et al., 2005b) and in part may be due to a canopy effect (Cerling et al., 2004).

Some of these aspects are illustrated in figure 48.4, which shows the relationship of grazers, browsers, mixed feeders, and carnivores in a single ecosystem. The average $\delta^{13}C$ value

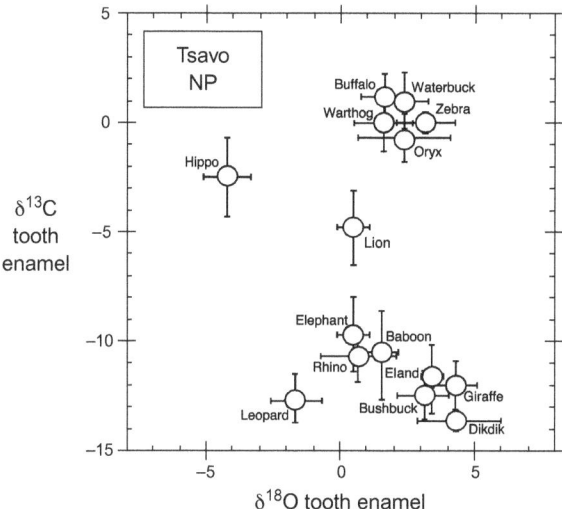

FIGURE 48.4 $\delta^{13}C$ and $\delta^{13}O_{PBD}$ of large mammals from Tsavo National Park, Kenya.

for hippos from Tsavo National Park is more negative than that for known grazers (oryx, buffalo, waterbuck, zebra), while elephants and rhinos are >2‰ enriched in ^{13}C compared to giraffes, dik-diks, and bushbuck indicating a measureable C_4 component to their diet.

Of course, there are always caveats. Figure 48.3 illustrates this point: warthogs from lower elevations (below 3-km elevation) have a diet that is nearly exclusively C_4 biomass. However, one individual from Nechisar Park in Ethiopia is very depleted in ^{13}C—not only one tooth, but two! We have reanalyzed the samples *ab initio*, and this specimen clearly had a C_3 diet. Thus, 98% of the warthogs from low elevations have a C_4-dominated diet and one specimen (of >50 individuals) has a C_3 diet. Is this an individual from high elevations that was translocated? Such a possibility cannot be ruled out but seems very unlikely. If not translocated, then what was the diet of this anomalous individual? Thus, in our interpretation of the modern and fossil records, we have to recognize that such anomalies exist and make our interpretations accordingly.

Oxygen isotopes have so far played a minor role in studies of modern African mammals compared to carbon isotopes; however, some studies show promise for developing oxygen isotopes to understand physiology such as obligate drinkers (e.g., Boisserie et al., 2005). However, different animals have different water strategies, and it is likely that oxygen isotopes will be used to compare animals using water resources in different ways. By example, figure 48.4 shows the $\delta^{13}C$ and $\delta^{18}O$ of mammals from Tsavo National Park, Kenya. The maximum $\delta^{18}O$ separation is between hippopotamus and giraffes, which differ by 8‰. Tsavo is semiarid bushland; in more mesic environments the $\delta^{18}O$ separation collapses to smaller values, suggesting a possible application to paleoaridity (Levin et al., 2006).

DIETARY HISTORIES OF MAMMALIAN LINEAGES FROM STABLE ISOTOPES

The Turkana Basin and the Suguta Basin to its south have a very long record of mammalian fossils, extending back to the mid-Miocene. The record is especially well represented from ca. 7 Ma to the present in the main Turkana Basin, but important older deposits occur in the Suguta Basin at Nakali

(10 Ma), and Namurungule (between 9.5 and 9 Ma). Here we briefly summarize those results and discuss the main dietary changes in some of the mammalian lineages through this time interval. Globally, the world changed in the late Miocene from a "C_3 world," where C_4 plants were not abundant enough to make an unequivocal contribution to mammalian diets, to a "C_4 world," where some mammals had diets comprised of >90% C_4 biomass. Recognition of an "unequivocal" contribution of C_4 biomass to a diet must be considered in the light of the known and estimated change in $\delta^{13}C$ of the atmosphere (figure 48.2) and the known shift of C_3 plants to more positive values in water-stressed conditions. In addition, there is an uncertainty in the isotope enrichment between diet and enamel, although lab and field studies indicate that ε^* is 14‰ for bovids and 12‰ for suids; the total range is somewhat greater than this with rodents having enrichment values of 11‰ (Passey et al., 2005b; Podlesak et al., 2008).

The first unequivocal indication of significant C_4 biomass in the diets is documented by equids (Cerling et al., 1997). Prior to ca. 9 Ma, there was little or no C_4 component in the diet, whereas by 6 Ma, equids from East Africa, Pakistan, and North America had diets dominated by C_4 plants. On the basis of large African mammals with $\delta^{13}C$ values between −8‰ and −10‰ and a single individual with a $\delta^{13}C$ value of −7.5‰, Morgan et al. (1994) argued that the radiation of C_4 plants occurred in the middle Miocene. Their results are consistent with C_4 plants being present in the mid-Miocene, but not in sufficient abundance to make up a substantial fraction of the diet of any mammalian group.

Figure 48.5 shows that that there was an important expansion of C_4 biomass in East Africa between the time of deposition of the Nakali and Namurungule formations. All mammal groups in the Nakali Formation, including equids, have $\delta^{13}C$ values that are consistent with an end-member C_3 diet. Kingston (data in Morgan et al., 1994) has also documented a C_3 diet for equids from Ngerngerwa in the Baringo Basin. Thus, it appears that when equids first appear in the East African fossil record they had a C_3 diet.

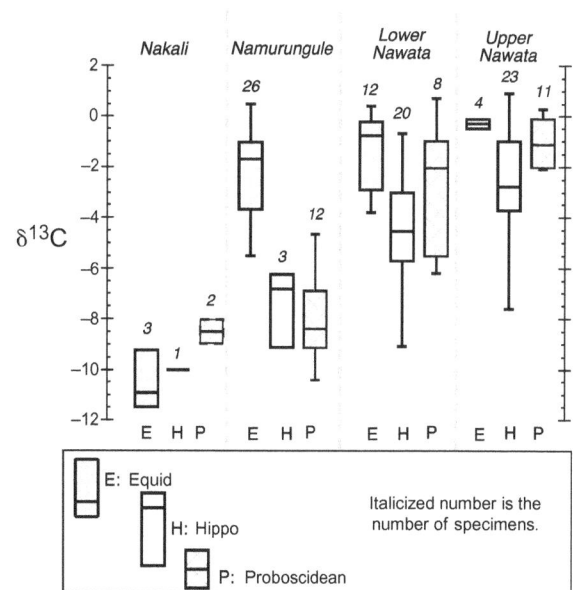

FIGURE 48.5 Box-and-whisker plots for late Miocene sites in the Turkana and Suguta Basins. Medians and quartiles are shown for the Nakali and Namurungule Formations and for the Upper and Lower Nawata Formations for equids, hippopotamids, and elephantids.

By Namurungule times, the diet of equids had a very strong C_4 component: 26 equids had an average $\delta^{13}C$ value of $-2.3 \pm 1.7\permil$, with a range from $+0.5\permil$ to $-5.5\permil$ (figure 48.5). For comparison, extant zebra in East Africa have an average $\delta^{13}C$ value of $+0.1 \pm 1.4\permil$, with a range from $+2.1\permil$ to $-6.6\permil$. In light of the $\delta^{13}C$ shift of the atmosphere due to human activities ($1.5\permil$ since 1750) and that observed in the late Miocene (figure 48.2), the Namurungule equids had attained a diet that was clearly dominated by C_4 biomass. Namurungule is now considered to be between 9.5 and 9.0 Ma (Saneyoshi et al., 2006) and may represent the oldest African assemblage to have a C_4-dominated diet. Chorora, in Ethiopia, is also a candidate for one of the earliest indications of a significant C_4 component in the diet of African equids (Bernor et al., 2004).

The emergent C_4 biomass was exploited sequentially by different mammalian groups. Figure 48.5 shows that equids began to change to a C_4 diet earlier than elephantids; hippopotamids show a steady increase in the fraction of C_4 biomass beginning in the Namurungule Formation but do not attain the high level (>90%) of C_4 biomass in their diet even by late Nawata time.

Thus, these three lineages show different responses to the possibility of C_4 grasses in their respective diets: equids change rapidly from a C_3 to a C_4 diet; elephantoids have a delayed response but attain a high level of C_4 biomass in their diet by Lower Nawata time; hippopotamids appear to have "discovered" C_4 grasses before the elephantids, but take a longer time for C_4 grasses to reach a maximum contribution to their diet.

Even longer-term trends in diets are summarized in figure 48.6, shows some striking features of diet histories of equids, suids, elephantids, and hippopotamids in East Africa. We summarize some of the salient features but note that with new data, these stories may change. Equids are the first large mammals to make the diet change from C_3 to C_4 biomass; this is in keeping with their hypsodont teeth. Elephantids make the transition to a C_4 diet after equids; this is a fruitful area of research: what anatomical changes are associated with such a diet change, and how rapid was it? Most striking, however, is that the modern elephant diet has C_4 grass only seasonally (Cerling et. al., 2004, 2006); grazing elephants, which were common for millions of years, only recently vanished from the African landscape. This has profound implications for the maintenance of savannas during the late Cenozoic: elephants must have played a very different role when they were predominantly grazers than they do today as browsers. Deinotheres, a more conservative proboscidean, consume C_3 biomass throughout their history. Suids make a gradual change to a C_4 diet, as noted by Harris and Cerling (2002), with teeth becoming more hypsodont as the fraction of C_4 biomass in the diet increases. Hippos have a wide range in diet; they show a tendency toward a C_4 component in their diet once C_4 plants become available (as indicated by equid diets) but have a mixed diet throughout their history (Harris et al., 2008). Sivatheres adopt a C_4 diet in the Siwaliks toward the end of the Miocene but in Africa remain browsers until the late Pliocene (Cerling et al., 2005).

These chronologies show that different mammalian groups responded at different times and in different ways to the new food source. Details about changes in seasonality associated with these changes remains to be studied. For example, elephants today are ca. 75% "browsers" on an annual basis, where they have short intense periods of grazing during and following the rainy season. In the past they were ca. 75% "grazers"—did they have a relatively constant diet throughout the year, or did they have intense periods of browsing, as the modern elephants do with respect to grazing? Likewise, in the tropics today, C_4 photosynthesis makes up 50% or more of the total primary productivity (NPP). How different were previous ecosystems where productivity was likely dominated by C_3 NPP: soils, erosion, and fire regimes all would be very different than today. We anticipate that future studies of changing diets with time will turn up more surprises and will lead to new understandings of mammalian behavior and competition.

LATE MIOCENE AND EARLY PLIOCENE FAUNAL TURNOVER AT LOTHAGAM

In this section we look at the evidence for marked faunal turnover from Lothagam, a richly fossiliferous late Miocene–early Pliocene site in the Turkana Basin, northern Kenya (Patterson et al., 1970; Leakey and Harris, 2003a). The Lothagam sequence provides evidence of major changes in East African landscapes and is one of the first in Africa where C_4 biomass contributes a significant dietary component for many large herbivorous mammals. Major faunal changes took place during the time interval represented by the Nawata Formation (~7.5–~5.0 Ma). This coincidentally corresponds to the time that the hominin and ape lineages diverged.

The Lothagam mammalian fossil record, comprising over 2,000 specimens, derives largely from three major stratigraphic units: the Nawata Formation and the Apak (~5–4.2 Ma), and Kaiyumung (~3.5 Ma) members of the Nachukui Formation (Feibel, 2003; McDougal and Feibel, 2003). The Lower Nawata is richly fossiliferous, but most of the fossils are from horizons just below or above the Lower Markers dated at ~7.44 Ma. Thus, the Lower Nawata fauna is slightly younger than that from the Namurungule Formation in the Suguta Valley (Nakaya et al., 1984, 1987; Itaya and Sawada, 1987) where the first evidence for C_4 diets has been documented.

The Lothagam succession documents many last appearances of lineages common earlier in the Miocene, together with the first appearances of lineages that dominate the later Plio-Pleistocene and the modern biota. Eleven Lower Nawata species do not persist into the Upper Nawata, and 10 new species are first recorded in the Upper Nawata. Excluding

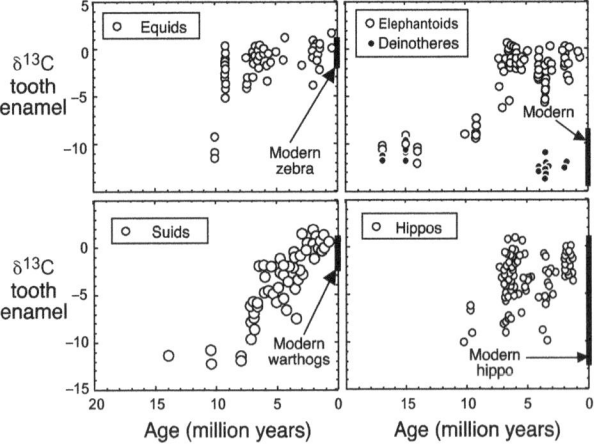

FIGURE 48.6 Long-term diet histories of equids, suids, proboscideans, and hippopotamids in the Turkana Basin and related deposits. Data from Fort Ternan and Maboko are also included.

microfauna, 27 mammalian species make their last appearance in the Upper Nawata, and 13 new species appear in the Apak Member. The geological marker units do not necessarily coincide with times of faunal change. For example, primitive taxa typical of the Nawata Formation such as *Stegotetrabelodon orbus, Brachypotherium lewisi, Palaeotragus germaini,* and *Nyanzachoerus syrticus* persist into the Apak Member but are extremely rare. And species such as *Nyanzachoerus* cf. *Ny. australis,* which is the dominant suid in the Apak Member, are rare in the Upper Nawata. A possible hiatus in sedimentation just above the Purple Marker may exaggerate these differences in occurrences and could in turn be due to climate change affecting precipitation (Leakey and Harris, 2003b).

The equids were the first mammalian herbivores to consistently exploit C_4 vegetation. The Lothagam Equidae include *Hippotherium* cf. *H. primigenium,* the last representative of the first wave of three-toed hipparionines to invade Africa and known only from the Lower Nawata, and *Eurygnathohippus* species that represent a second migratory phase (Bernor and Harris, 2003). Only the earliest Lothagam equids had some C_3 component in their diet. *Eurygnathohippus* species from the Upper Nawata and later horizons were all C_4 grazers.

The proboscideans were the second mammalian group to switch to a predominantly C_4 diet. Proboscideans were diverse in Lothagam times (Tassy, 2003), although today they are represented in Africa only by *Loxodonta africana.* Deinotheres, which remained C_3 browsers throughout their history, are represented by *Deinotherium bosazi,* which is present but rare throughout the Lothagam sequence (Harris, 2003a). Like deinotheres, the gomphotheres were also uncommon, but *Anancus kenyensis* is a rare component of the Nawata and Apak assemblages, and a trilophodont gomphothere is known from the Upper Nawata. The transition from a C_3 to a C_4 diet in gomphotheres appears to have been initiated during the time of the (earlier) Namurungule Formation and was completed by, or during, the Nawata Formation. Feeding adaptations in the Elephantidae include specialized modifications of their masticatory apparatus, dental eruption patterns, and dental morphology, the molars increasing in crown height and plate number and decreasing in enamel thickness through time. These adaptations increased the efficiency of processing a graminivorous diet, especially during dry intervals when food resources were short. In the Lower Nawata, the primitive *Stegotetrabelodon orbus* (with tusks in both upper and lower jaws) is the dominant elephantid and co-occurs with the rare *Primelephas gomphotheroides.* Only one specimen of *S. orbus* was recovered from the Upper Nawata and two in the Apak Member. The modern genus *Elephas* first appears in the Upper Nawata, and by the Apak Member, both *Elephas* and *Loxodonta* have largely replaced *Stegotetrabelodon* (Tassy, 2003). Isotopic analyses show that although the elephantids in the Lower Nawata were largely feeding on C_4 vegetation three specimens, one identifiable as *Stegotetrabelodon,* had a significant C_3 component. Later in the sequence, like the gomphotheres, elephantids fed almost exclusively on C_4 vegetation (Cerling et al., 2003a).

African suid assemblages underwent a major transformation at the end of the Miocene (Harris and Leakey, 2003b; Harris and Liu, 2007). At this time, immigrant representatives of the Eurasian Tetraconodontinae replaced the genera common earlier in the Miocene. *Nyanzachoerus syrticus* and *Ny. devauxi,* the two dominant species in the Nawata Formation are early examples of these immigrants. In the Apak Member, the early nyanzachoeres are largely replaced by the more progressive *Ny.* cf. *Ny. australis* (only two specimens of *Ny. syrticus* were recovered from the Apak Member). Later in the middle Pliocene, a second wave of immigration from Eurasia gave rise to the three lineages of extant suids (Harris and Leakey, 2003b). The isotopic data shows that suids were slower than equids and proboscideans to exploit C_4 vegetation. The Nawata Formation suids had diets dominated by C_3 grasses although with some C_4 component, and *Ny. syrticus* shows more positive $\delta^{13}C$ values in the Upper than the Lower Nawata. Suids with an undoubted C_4 diet do not appear until the Apak Member. At Lothagam the *Nyanzachoerus* third molars increased in size, hypsodonty, and complexity as more C_4 biomass was incorporated in their diet (Harris and Cerling, 2002; Cerling et al., 2005).

The Hippopotamidae also underwent marked faunal change at Lothagam, with the more primitive Miocene forms being replaced by hippos with closer affinities to the modern species and more developed aquatic adaptations (Weston, 2003). Isotopic evidence shows hippos to have been opportunistic mixed feeders throughout their evolutionary history. Interestingly, the earliest hippo *Kenyapotamus* had a mixed C_3/C_4 diet as early as the Namurungule Formation. In the Nawata Formation, the slender limbed cursorial *Archaeopotamus harvardi* is the dominant hippo but two less common species, the small *Archaeopotamus lothagamensis* and an unnamed larger species, co-occur. Both have disappeared by the Apak Member when *A. harvardi* coexisted with aff. *Hippopotamus* cf. *Hippopotamus protamphibius,* the common hippo in the early Pliocene of the Koobi Fora Formation. The isotopic data indicate that *A. harvardi* and the larger unnamed species had largely C_4 diets, and this together with the cranial evidence of slightly raised orbits and a wider mandibular symphysis indicates the onset of transitions to a more aquatic and grazing mode of life. The diet of the narrow-muzzled *A. lothagamensis* included a significant fraction of C_3 browse.

The Rhinocerotidae were common in the Miocene but are generally rare in the Pliocene and Pleistocene and represented today in Africa by the browsing *Diceros bicornis* and the grazing *Ceratotherium simum.* In the Nawata Formation, the primitive hornless *Brachypotherium lewisi* is relatively common and the isotopic analyses of three teeth indicate a C_3 diet. *B. lewisi* persists in the lower horizons of the Apak Member where two specimens represent the last record of this species. The analysis of one of these teeth indicates a mixed diet of C_3 and C_4 vegetation. The two modern rhino genera, *Ceratotherium* and *Diceros,* appear rarely in the Nawata Formation. In the Lower Nawata, all the rhinos had C_3 diets, but in the Apak Member, five of the six teeth sampled were grazers, suggesting that early *Ceratotherium* had switched to a C_4 diet by this time (Harris and Leakey, 2003a).

The Giraffidae, a predominantly browsing family that was common in the Miocene, is sparsely represented at Lothagam (Harris, 2003b). In the Nawata Formation two species of *Palaeotragus* occur, and a molar of *P. germaini* from the base of the Apak Member is the last known record of this primitive giraffid. The earliest recorded *Giraffa, G. stillei,* makes its first known appearance at Lothagam in the Apak Member. Giraffids persist as C_3 browsers throughout the Lothagam succession. Much later in time, in the early Pleistocene, *Sivatherium maurusium* evolved shorter metapodials, and isotopic analyses of specimens from upper part of the Koobi Fora Formation demonstrate the transition from C_3 browsing to C_4 grazing (Harris and Cerling, 1998).

Bovidae are common at Lothagam, where the transition from a boselaphine dominated assemblage to one in which extant tribes predominate took place, but isotopic analyses of Lothagam bovids are still at an early stage. Initial results suggest broad agreement between other measures of diet such a hypsodonty index, mesowear, and microwear.

The Carnivora show many first and last appearances in the Lothagam succession. The most notable of the last appearances are the Amphicyonidae, a family known primarily from the Northern Hemisphere from the Eocene to the late Miocene, and recorded sparsely in Africa. The Lothagam records are as young or possibly even younger than any Northern Hemisphere records. First appearances are more extensive and are mostly the earliest representatives of taxa with a significant subsequent presence in Africa. Several hold phylogenetically basal positions relative to later members of their respective genera (Werdelin, 2003). Isotopic studies of carnivores are difficult to interpret because they largely subsist on omnivorous or carnivorous diets. The turnover in the Carnivora likely mirrors changes in the herbivores on which they prey.

A family that shows less obvious faunal turnover in the late Miocene and early Pliocene is the Cercopithecidae. Monkeys are largely folivores or frugivores, so the increase in C_4 vegetation would not be expected to have had a significant influence on their diets. It is not until later in the Pliocene that changes in the cercopithecid fauna become more marked with the increasing dominance of Theropithecus and the radiation of the large bodied colobines (Jablonski and Leakey, 2008). One lineage of cercopithecids did switch to grazing, although detailed isotopic studies have yet to indicate exactly when and how this happened. At ~2.5 Ma, the highly successful and widespread, specialized graminivorous Theropithecus oswaldi becomes the most common monkey. It replaced the earlier Theropithecus brumpti, which appears to have been a more closed habitat species. Modern gelada baboons, Theropithecus gelada, are the only remaining modern graminivorous cercopithecids.

Some of the most intriguing questions pertaining to the late Miocene-Pliocene faunal turnover relate to the emergence of the hominins. The hominin and ape lineages are believed to have diverged somewhere between 5 and 7 Ma ago. Thus the Lothagam evidence of faunal and habitat change at this time is extremely pertinent. The hominin record at Lothagam is unfortunately very sparse; there are only three specimens from sediments older than 4.2 Ma, and all were just above or just below the Purple Marker and thus close to the Upper Nawata–Apak Member boundary (Leakey and Walker, 2003). The isotopic evidence for a marked radiation of C_4 grasses between 7 and 5 million years ago, and the evidence for changing taxa and diets in so many lineages of large herbivorous mammals coincides with the time of this crucial split. It is logical to conjecture that the radiation of C_4 grasses would have provided a wealth of new feeding niches for large mammals and also invertebrates, amphibians, reptiles, and birds—presenting new and plentiful dietary opportunities for early hominins venturing into these expanding and open C_4 grassy habitats. At the time of writing, early hominins dating between approximately 6 and 5 million years are frustratingly rare—being represented by three taxa from sites widely dispersed geographically in Chad, Kenya and Ethiopia. Little is known of the life histories of these species, but their presence shows that by ~6 million years ago the earliest hominins had appeared

and were occupying habitats that were probably dominated by C_4 grasses.

In summary, Lothagam documents major faunal turnover at the end of the Miocene and early Pliocene, when isotope analyses indicate significant ecological change. Lothagam records the last appearances of many taxa more abundant in the middle to late Miocene, when the large mammal herbivores were subsisting on C_3 vegetation, and numerous first appearances of the basal components of taxa typical of the Pliocene and Pleistocene, when many new lineages made the transition to specialized C_4 grazing. However, the true reasons for the timing of and the response of different mammalian lineages to the change in C_4 biomass remain an active area of research and speculation. The weight of the evidence on the history of atmospheric CO_2 levels indicates that CO_2 levels have been favorable to C_4 plants for at least 25 million years (Tipple and Pagani, 2007), with C_4 photosynthesis arising independently several times (Vicenti et al., 2008).

Future Directions for Isotope Studies

There are some exciting new developments in stable isotopes with potential to yield more information about mammalian behavior. We briefly mention some of those here and anticipate their future realization.

INDIVIDUAL HISTORIES

The study of sequential samples has already yielded detailed histories of individual mammals. So far, this has been applied principally to modern examples, but there is no reason why these methods cannot be applied to the fossil record. Recent developments in the understanding the tooth enamel maturation and the incorporation of stable isotopes into tooth enamel shows how the "input" signal is attenuated during enamel formation (Passey and Cerling, 2002; Passey et al., 2005a). The problem to be addressed before extensive applications can be made to either the modern or the fossil record is to understand the enamel maturation parameters and to develop inverse modeling methods to recover the original input signal. Figure 48.7 shows a short history of a fossil hippo canine on a scale approaching a single year: this individual had a diet ranging from almost pure C_4 to pure C_3 although the total isotope range in the tooth was less than 4‰.

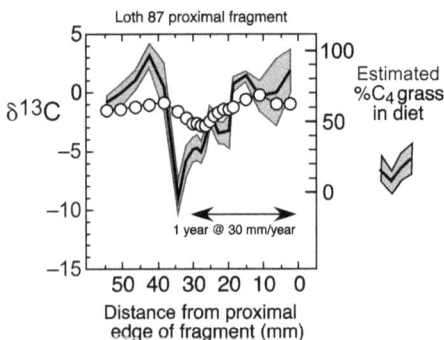

FIGURE 48.7 Diet history reconstructed from a fossil hippo canine (Lothagam). Diet estimate based on inverse modeling method of Passey et al. (2005a); shaded area shows 1-sigma uncertainty for 75 inversion with maturation, measurement, and sampling uncertainties as described in Passey et al. (2005a).

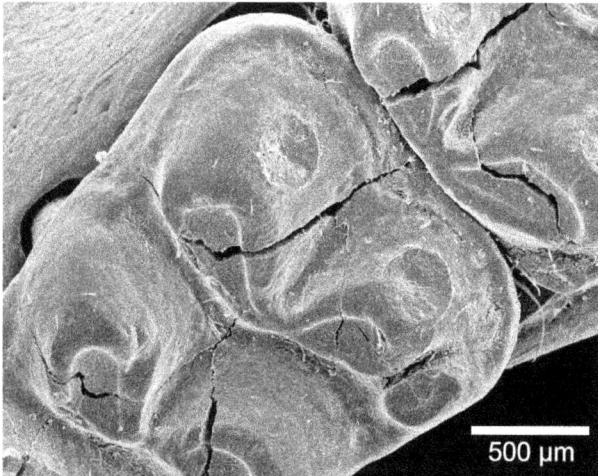

FIGURE 48.8 Rodent molar showing laser ablation pits used for stable isotope analysis.

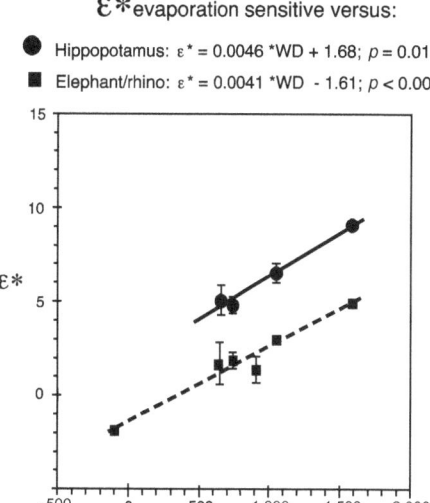

FIGURE 48.9 Modern data from East and Central Africa show increasing isotope enrichment from hippopotamus' and elephant/rhinoceros tooth enamel as water deficit increases. Modified from Levin et al. (2006).

MICROMAMMALS: SAMPLING RODENTS BY LASER ABLATION

Although the laser had been hailed as opening new possibilities for paleodiet studies a decade ago (Cerling and Sharp, 1996), it has found little use because microsampling using computer-controlled drilling has made it possible to sample small parts of teeth. Passey and Cerling (2006) improved on the earlier laser ablation methodology so that analyses could be made on teeth with very thin enamel. They showed that it is possible to sample individual rodent molars using laser ablation (figure 48.8) and that it is possible to obtain diet histories of rodent and lagomorph incisors. This opens up an entire new group of mammals, the "micromammals," which have hitherto been sampled only rarely. None of the previous laser experiments have sampled pure enamel because of geometrical considerations. This hurdle appears to be behind us now. Likewise, this method has been used on hominin teeth because damage to the specimens is less than by conventional methods (Sponheimer et al., 2006).

WATER RESOURCE PARTITIONING: IMPLICATIONS FOR PALEOARIDITY

Craig and Gordon (1963) found that evaporated waters are enriched in ^{18}O compared to their source waters. Roden and Ehleringer (1999) applied the Craig-Gordon model to plant systems and showed that ^{18}O is enriched in environments with low humidities. Carbohydrates produced during photosynthesis are enriched in ^{18}O compared to their source water (Roden and Ehleringer, 1999). Animals derive oxygen from four sources: drinking water, water in food, metabolic water, which itself is derived from water in the food (primarily carbohydrates for herbivores), and atmospheric oxygen. Obligate drinkers will be less enriched in ^{18}O than those animals that rely on leaf water for a significant fraction of their body water.

Environmental aridity is characterized in terms of water deficit (WD), the difference between potential evapotranspiration (PET), and mean annual precipitation (MAP): WD = PET − MAP (Levin et al., 2006). Certain extant African mammals can be shown to be sensitive to aridity (e.g., giraffids), while others do not show a significant $\delta^{18}O$ change with

water deficit (e.g., hippopotamids, elephantids, rhinocerotids). Thus, the difference between taxa that are "evaporation sensitive" and those that are not can be used as an indicator of aridity (figure 48.9). Levin et al. (2006) show that this is promising for species living in East Africa, where meteoric water shows little seasonal variation at low latitudes. We anticipate that stable isotope analysis of fossils may be used as a measure of aridity for paleoenvironments. As of the writing of this chapter, we are beginning those studies.

Summary

Stable isotopes are a relatively new tool in the paleontologists' toolbox. Tooth enamel (but not dentine or bone) locks in a record of diet and environment for millions of years, can be used to evaluate assumptions about diet or environment produced by other methodologies, and can give new insights into the past. We now know, through stable isotopes, that the late Miocene witnessed a new food source for mammals—that of plants using the C_4 photosynthetic pathway, which are, primarily, the tropical grasses.

Stable isotope studies of modern faunas show that the assumptions used for fossil reconstructions of diets is generally correct, but a few important discoveries have been made using stable isotopes. For example, hippopotamus' diet contains more C_3 biomass than is generally believed, and the forest hog is almost a pure browser rather than a grazer.

Studies of modern fauna show a few important differences from the fossil record. Elephants, for example, rely on C_3 biomass for the bulk of their nutrition, but fossil elephants in East Africa were predominantly C_4 grazers for millions of years. Why did *Loxodonta*, the only remaining modern elephant, switch to browsing in Africa in the late Pleistocene?

Studies of dietary lineages show that equids were the earliest large mammal group, apparently, to have a diet that was essentially fully C_4 in nature. Other mammalian groups (elephantoids, hippopotamids, suids) took longer to take advantage of this new resource on the landscape. Some giraffids even tried grazing: African sivatheres began grazing in the late Pliocene, and some became fully C₄ grazers. Clearly, the next step is to combine additional stable isotope, micro- and mesowear, and morphometric measurements to understand the details of how different mammalian taxa adapted to a changing diet.

The future of stable isotopes for refining our understanding of the Cenozoic history of East African mammalian paleontology is bright. Microsampling affords the opportunity to study rodents, mathematical modeling allows recovery of primary dietary input signals, and comparison of taxa shows promise as a paleoaridity indicator.

ACKNOWLEDGMENTS

This work was funded by the National Science Foundation and the Packard Foundation. Work in Kenya was facilitated by the National Museums of Kenya and Kenya Wildlife Service. We are grateful to numerous individuals for assistance in the collection of samples for analysis. We thank Jim Zachos for making his carbon isotope database available. T.E.C. and J.M.H. thank Jonathon Leakey and Dena Crain for hospitality.

Literature Cited

Ambrose S. H., and L. Norr. 1993. Experimental evidence for the relationship of the carbon isotope ratios of whole diet and dietary protein to those of bone collagen and carbonate; pp. 1–37 in J. B. Lambert and G. Grupe (eds.), *Molecular Archaeology of Prehistoric Human Bone.* Springer, Berlin.

Ayliffe, L. K., A. R. Chivas, and M. G. Leakey. 1994. The retention of primary oxygen isotope compositions of fossil elephant skeletal phosphate. *Geochimica et Cosmochimica Acta* 58:5291–5298.

Bernor, R. L., and J. M. Harris. 2003. Systematics and evolutionary biology of the Late Miocene and early Pliocene hipparionine equids from Lothagam; pp. 388–438 in M. G. Leakey and J. M. Harris (eds.), *Lothagam: The Dawn of Humanity in Eastern Africa.* Columbia University Press, New York.

Bernor, R.L., T. Kaiser, and S. Nelson. 2004. The oldest Ethiopian hipparion (Equinae, Perissodactyla) from Chorora: Systematics, paleodiet and paleoclimate. *Courier Forschungsinstitut Senckenberg* 246:213–226.

Boisserie, J. R, A. Zazzo, G. Merceron, C. Blondel, P. Vignaud, A. Likius, H. T. Mackaye, and M. Brunet. 2005. Diets of modern and late Miocene hippopotamids: Evidence from carbon isotope composition and microwear of tooth enamel. *Palaeogeography, Palaeoclimatology, Palaeoecology* 221:153–174.

Cerling, T. E. 1997. Late Cenozoic vegetation change, atmospheric CO_2, and tectonics; pp. 313–327 in W. F. Ruddiman (ed.), *Tectonic Uplift and Climate Change.* Plenum Press, New York.

Cerling, T. E., and J. R. Ehleringer. 2000. Welcome to the C₄ world; pp. 273–286 in R. A. Gastaldo and W. M. DiMichele (eds.), *Phanerozoic Terrestrial Ecosystems.* The Paleontology Society Papers 6.

Cerling, T. E., and J. M. Harris 1999. Carbon isotope fractionation between diet and bioapatite in ungulate mammals and implications for ecological and paleoecological studies. *Oecologia* 120:247–363.

Cerling, T. E., J. M. Harris, S. H. Ambrose, M. G. Leakey, and N. Solounias. 1997. Dietary and environmental reconstruction with stable isotope analyses of herbivore tooth enamel from the Miocene locality of Fort Ternan, Kenya. *Journal of Human Evolution* 33:635–650.

Cerling, T. E., J. M. Harris, and M. G. Leakey. 1999. Browsing and grazing in modern and fossil proboscideans. *Oecologia* 120:364–374.

———. 2003a. Isotope paleoecology of the Nawata and Apak Formations at Lothagam, Turkana Basin, Kenya; pp. 605–624 in M. G. Leakey and J. M. Harris (eds.), *Lothagam: The Dawn of Humanity in Eastern Africa.* Columbia University Press, New York.

———. 2005. Environmentally driven dietary adaptations in African mammals; pp. 258–272 in J. R. Ehleringer, T. E. Cerling, and M. D. Dearing (eds.), *A History of Atmospheric CO2 and Its Effects on Plants, Animals, and Ecosystems.* Springer, New York.

Cerling, T. E., J. M. Harris, B. J. MacFadden, M. G. Leakey, J. Quade, V. Eisenmann, and J. R. Ehleringer. 1997. Global change through the Miocene/Pliocene boundary. *Nature* 389;153–158.

Cerling, T. E., J. M. Harris, and B. H. Passey. 2003b. Dietary preferences of East African Bovidae based on stable isotope analysis. *Journal of Mammalogy* 84:456–471.

Cerling, T. E., J. A. Hart, and T. B. Hart. 2004. Stable isotope ecology in the Ituri Forest. *Oecologia* 138:5–12.

Cerling, T. E., J. A. Hart, P. Kaleme, H. Klingel, M. G. Leakey, N. E. Levin, R. L. Lewison, B. H. Passey, and J. M. Harris. 2008. Stable isotope ecology of modern *Hippopotamus amphibius* in East Africa. *Journal of Zoology* 276:204–212.

Cerling, T. E., P. Omondi, and A. N. Macharia. 2007. Diets of Kenyan elephants from stable isotopes and the origin of confiscated ivory in Kenya. *Journal of African Ecology* 45:614–623

Cerling, T. E., and Z. D. Sharp. 1996. Stable carbon and oxygen isotope analysis of fossil tooth enamel using laser ablation. *Palaeogeography, Palaeoclimatology, Palaeoecology* 125:173–186.

Cerling, T. E., and K. Viehl. 2004. Seasonal diet changes of the giant forest hog (*Hylochoerus meinertzhagani* Thomas) based on the carbon isotopic composition of hair. *African Journal of Ecology* 42: 88–92.

Cerling, T. E., G. Wittemyer, H. B. Rasmussen, F. Vollrath, C. E. Cerling, T. J. Robinson, and I. Douglas-Hamilton. 2006. Stable isotopes in elephant hair document migration patterns and diet changes. *Proceedings of the National Academy of Sciences, USA* 103:371–373.

Codron, D., and J. S. Brink. 2007. Trophic ecology of two savanna grazers, blue wildebeest *Connochaetes taurinus* and black wildebeest *Connochaetes gnou. European Journal of Wildlife Research* 53: 90–99.

Codron D., J. Codron, J. A. Lee-Thorp, M. Sponheimer, D. de Ruiter, and J. Brink. 2007. Stable isotope characterization of mammalian predator-prey relationships in a South African savanna. *European Journal of Wildlife Research* 53:161–170.

Codron, D., J. Codron, M. Sponheimer, R. C. Grant, and D. de Ruiter. 2005. Assessing diet in savanna herbivores using stable carbon isotope ratios of faeces. *Koedoe* 48:115–124.

Codron, J. J., A. Lee-Thorp, M. Sponheimer, D. Codron, R. C. Grant, and D. J. De Ruiter. 2006a. Elephant (*Loxodonta africana*) diets in Kruger National Park, South Africa: Spatial and landscape differences. *Journal of Mammalogy* 87:27–34.

Codron, D., J. A. Lee-Thorp, M. Sponheimer, D. de Ruiter, and J. Codron. 2006b. Inter- and intrahabitat dietary variability of Chacma baboons (*Papio ursinus*) in South African savannas based on fecal $\delta^{13}C$, $\delta^{15}N$, and %N. *American Journal of Physical Anthropology* 129:204–214.

Craig, H., and L. I. Gordon. 1965. Deuterium and oxygen-18 variations in the ocean and the marine atmosphere; pp. 9–130 in *Proceedings of the Congress on Stable Isotopes in Oceanographic Studies and Paleotemperatures, Spoleto, Italy.* N.p.

Deines, P. 1980. The isotopic composition of reduced organic carbon. pp. 329–406 in P. Fritz and J. C. Fontes (eds.), *Handbook of Environmental Isotope Geochemistry,* vol. 1. Elsevier, New York.

DeNiro M. J., and S. Epstein. 1978. Influence of diet on the distribution of carbon isotopes in animals. *Geochimica et Cosmochimica Acta* 42:495–506.

East African Meteorological Department. 1975. *Climatological Statistics for East Africa.* Nairobi, Kenya.

Ehleringer, J. R., and T. A. Cooper. 1988. Correlations between carbon isotope ratio and microhabitat in desert plants. *Oecologia* 76:562–566.

Ericson, J. E., C. H. Sullivan, and N. T. Boaz. 1981. Diets of Pliocene mammals from Omo deduced from carbon isotopic ratios in tooth apatite. *Palaeogeography, Palaeoclimatology, Palaeoecology* 36:69–73

Feibel, C. 2003. Stratigraphy and depositional history of the Lothagam sequence; pp. 17–29 in M. G. Leakey and J. M. Harris (eds.), *Lothagam: The Dawn of Humanity in Eastern Africa.* Columbia University Press, New York.

Franz-Odendaal T. A., J. A. Lee-Thorp, and A. Chinsamy. 2002, New evidence for the lack of C4 grassland expansions during the early Pliocene at Langebaanweg, South Africa. *Paleobiology* 28:378–388.

Harris, J. M. 2003a. Deinotheres from the Lothagam succession; pp. 359–361 in M. G. Leakey and J. M. Harris (eds.), *Lothagam: The Dawn of Humanity in Eastern Africa.* Columbia University Press, New York.

———. 2003b. Lothagam giraffids; pp. 523–531 in M. G. Leakey and J. M. Harris (eds.). *Lothagam: The Dawn of Humanity in Eastern Africa*. Columbia University Press, New York.

Harris, J. M., and T. E. Cerling. 1998. Isotopic changes in the diet of giraffids. *Journal of Vertebrate Paleontology* 18 (suppl. to no. 3):49A.

———. 2002, Dietary adaptations of extant and Neogene African suids. *Journal of Zoology*, 256:45–54.

Harris J. M., T. E. Cerling, M. G. Leakey, and B. H. Passey. 2008. Stable isotope ecology of fossil hippopotamids from the Lake Turkana Basin region of East Africa. *Journal of Zoology* 275:323–331.

Harris, J. M. and M. G. Leakey. 2003a. Lothagam Rhinocerotidae; pp. 371–385 in M. G. Leakey and J. M. Harris (eds.), *Lothagam: The Dawn of Humanity in Eastern Africa*. Columbia University Press, New York.

———. 2003b. Lothagam Suidae; pp. 485–519 in M. G. Leakey and J. M. Harris (eds.), *Lothagam: The Dawn of Humanity in Eastern Africa*. Columbia University Press, New York.

Harris, J. M., M. G. Leakey, T. E. Cerling, and A. J. Winkler. 2003. Early Pliocene tetrapod remains from Kanapoi, Lake Turkana Basin, Kenya; pp. 39–113 in J. M. Harris, M. G. Leakey, C. S. Feibel, K. Stewart, L. Werdelin, A. J. Winkler, and T. E. Cerling (eds.), *Geology and Vertebrate Paleontology of the Early Pliocene Site of Kanapoi, Northern Kenya*. Natural History Museum of Los Angeles County, Contributions in Science 498.

Harris, J. M., and L.-P. Liu. 2007. The Suoidea: Suidae, Tayassuidae and Sanitheriidae; pp. 130–150 in D. R. Prothero, and S. E. Foss (eds.), *The Evolution of Artiodactyls*. Johns Hopkins University Press, Baltimore.

Hattersley, P. W. 1982. ^{13}C values of C_4 types in grasses. *Australian Journal of Plant Physiology* 9:139–154.

———. 1992. C_4 photosynthetic pathway variation in grasses (Poaceae): Its significance for arid and semi-arid lands; pp. 181–212 in G. P. Chapman (ed.), *Desertified Grasslands: Their Biology and Management*. Linnean Society of London, London.

Hopley P. J., A. G. Latham, and J. D. Marshall. 2006. Palaeoenvironments and palaeodiets of mid-Pliocene micromammals from Makapansgat Limeworks, South Africa: A stable isotope and dental microwear approach. *Palaeogeography, Palaeoclimatology, Palaeoecology* 233:235–251.

Itaya, T., and Y. Sawada. 1987. K-Ar ages of volcanic rocks in the Samburu Hills area, northern Kenya. *African Study Monographs* (suppl.) 5:27–45.

Jablonski, N. G., and M. G. Leakey. 2008. *Koobi Fora Research Project: Volume 6. The Fossil Monkeys*. California Academy of Sciences, San Francisco. 469pp.

Kingdon, J. 1979. *Atlas of East African Mammals: Volume IIIB. Large Mammals*. University of Chicago Press, Chicago, 450 pp.

Kingston J. D., and T. Harrison. 2007. Isotopic dietary reconstructions of Pliocene herbivores at Laetoli: Implications for early hominin paleoecology. *Palaeogeography, Palaeoclimatology, Palaeoecology* 243:272–306

Kingston J. D., A. Hill, and B. D. Marino. 1994. Isotopic evidence for Neogene Hominid paleoenvironments in the Kenya Rift Valley. *Science* 264:955–959.

Koch P. L., N. Tuross, and M. L. Fogel. 1997. The effects of sample treatment and diagenesis on the isotopic integrity of carbonate in biogenic hydroxylapatite. *Journal of Archaeological Science* 24:417–429

Leakey, M. G., C. S. Feibel, R. L. Bernor, J. M. Harris, T. E. Cerling, K. M. Stewart, G. W. Stoors, A. Walker, L. Werdelin, and A. J. Winkler. 1996. Lothagam: A record of faunal change in the Late Miocene of East Africa. *Journal of Vertebrate Paleontology* 16: 556–570.

Leakey, M. G., and J. M. Harris. 2003a. *Lothagam: The Dawn of Humanity in Eastern Africa*. Columbia University Press, 678 pp.

———. 2003b. Lothagam: Its significance and contributions; pp. 625–660 in M. G. Leakey and J. M. Harris (eds.), *Lothagam: The Dawn of Humanity in Eastern Africa*. Columbia University Press, New York.

Leakey, M. G., and A. Walker. 2003. The Lothagam hominids; pp. 249–257 in M. G. Leakey and J. M. Harris (eds.), *Lothagam: The Dawn of Humanity in Eastern Africa*. Columbia University Press, New York.

Lee-Thorp J. A., M. Sponheimer, and N. J. van der Merwe. 2003. What do stable isotopes tell us about hominid dietary and ecological niches in the Pliocene? *International Journal of Osteoarchaeology* 13:104–113.

Lee-Thorp J., and N. J. van der Merwe. 1987. Carbon isotope analysis of fossil bone apatite. *South African Journal of Science* 83:712–715.

Lee-Thorp J. A., N. J. van der Merwe, and C. K. Brain. 1989. Isotopic evidence for dietary differences between two extinct baboon species from Swartkrans. *Journal of Human Evolution* 18:183–190.

———. 1994. Diet of *Australopithecus robustus* at Swartkrans from stable carbon isotopic analysis. *Journal of Human Evolution* 27: 361–372.

Levin, N. E., T. E. Cerling, B. H. Passey, J. M. Harris, J. R. Ehleringer. 2006. Stable isotopes as a proxy for paleoaridity. *Proceedings of the National Academy of Sciences, USA* 103:11201–11205.

Levin N. E., S. W. Simpson, J. Quade, T. E. Cerling, S. Semaw, and S. R. Frost. 2008. Herbivore enamel carbon isotopic composition and the environmental context of *Ardipithecus* at Gona, Ethiopia; pp. 215–234 in J. Quade and J. G. Wynn (eds.), *The Geology of Early Humans in the Horn of Africa*. Geological Society of America Special Paper 446.

MacFadden, B. J., and P. Higgins. 2004. Ancient ecology of 15-million-year-old browsing mammals within C_3 plant communities from Panama. *Oecologia* 140:169–182.

McDougal, I., and C. S. Feibel, C.S. 2003. Numerical age control for the Miocene-Pliocene succession at Lothagam, a hominoid-bearing sequence in the northern Kenya Rift; pp. 29–64 in M. G. Leakey and J. M. Harris (eds.), *Lothagam: The Dawn of Humanity in Eastern Africa*. Columbia University Press, New York.

Morgan, M. E., J. D. Kingston, and B. D. Marino. 1994. Carbon isotopic evidence for the emergence of C4 plants in the Neogene from Pakistan and Kenya. *Nature* 367:162–165.

Nakaya, H., M. Pickford, Y. Nakano, and H. Ishida. 1984. The late Miocene large mammal fauna from the Namurungule Formation, Samburu Hills, northern Kenya. *African Study Monographs* (suppl.) 2:87–131.

Nakaya H., M. Pickford, K. Yasu, and Y. Nakano. 1987. Additional large mammalian fauna from the Namurungule Formation, Samburu Hills, northern Kenya. *African Study Monographs* (suppl.) 5:79–129.

Passey, B. H., and T. E. Cerling. 2002. Tooth enamel mineralization in ungulates: Implications for recovering a primary isotopic time-series. *Geochimica et Cosmochimica Acta* 18:3225–3234.

———. 2006. In situ stable isotope analysis ($\delta^{13}C$ and $\delta^{18}O$) of very small teeth using laser ablation GC/IRMS. *Chemical Geology* 235:238–249.

Passey, B. H., T. E. Cerling, G. T. Schuster, T. F. Robinson, B. L. Roeder, and S. K. Krueger. 2005a. Inverse methods for estimating primary input signals from time-averaged intra-tooth profiles. *Geochimica et Cosmochimica Acta* 69:4101–4116.

Passey B. H., T. F. Robinson, L. K. Ayliffe, T. E. Cerling, M. Sponheimer, M. D. Dearing, B. L. Roeder, and J. R. Ehleringer. 2005b. Carbon isotopic fractionation between diet, breath, and bioapatite in different mammals. *Journal of Archaeological Science* 32:1459–1470.

Patterson, B., A. K. Behrensmeyer, and W. D. Sill. 1970. Geology of a new Pliocene locality in northwestern Kenya. *Nature* 256:279–284.

Podlesak, D. W., A. M. Torregrossa, J. R. Ehleringer, M. D. Dearing, B. H. Passey, and T. E. Cerling. 2008. Turnover of oxygen and hydrogen isotopes in the body water, CO_2, hair, and enamel of a small mammal. *Geochimica et Cosmochimica Acta* 72:19–35.

Quade, J., T. E. Cerling, M. M. Morgan, D. R. Pilbeam, J. Barry, A. R. Chivas, J. A. Lee-Thorp, and N. J. van der Merwe. 1992. A 16 million year record of paleodiet using carbon and oxygen isotopes in fossil teeth from Pakistan. *Chemical Geology (Isotope Geoscience Section)* 94:183–192.

Rietti-Shati, M., R. Yam, W. Karlén, A. Shemsh. 2000. Stable isotope composition of tropical high-altitude fresh-waters on Mt. Kenya, Equatorial East Africa. *Chemical Geology: Geology* 166:341–350.

Roden, J. S. and J. R. Ehleringer. 1999. Observations of hydrogen and oxygen isotopes in leaf water confirm the Craig-Gordon model under wide-ranging environmental conditions. *Plant Physiology* 120:1165–1173.

Saneyoshi, M., K. Nakayama, T. Sakai, Y. Sawada, and H. Ishida. 2006. Half graben filling processes in the early phase of continental rifting: The Miocene Namurungule Formation of the Kenya Rift. *Sedimentary Geology* 186:111–131.

Schoeninger M. J., and M. J. DeNiro. 1982. Carbon isotope ratios of apatite from fossil bone cannot be used to reconstruct diets of animals. *Nature* 297:577–578.

Schoeninger M. J., K. Hallin, H. Reeser, J. W. Valley, and J. Fournelle. 2003a. Isotopic alteration of mammalian tooth enamel. *International Journal of Osteoarchaeology* 13:11–19.

Schoeninger M. J., H. Reeser, and K. Hallin. 2003b. Paleoenvironment of *Australopithecus anamensis* at Allia Bay, East Turkana, Kenya: Evidence from mammalian herbivore enamel stable isotopes. *Journal of Anthropological Archaeology* 22:200–207.

Sponheimer M., and J. A. Lee-Thorp. 1999. Alteration of enamel carbonate environments during fossilization. *Journal of Archaeological Science* 26:143–150.

———. 2006. Enamel diagenesis at South African Australopith sites: Implications for paleoecological reconstruction with trace elements. *Geochimica et Cosmochimica Acta* 70:1644–1654.

Sponheimer, M., J. A. Lee-Thorp, D. J. DeRuiter, J. M. Smith, N. J. Van der Merwe, K. Reed, C. C. Grant, L. K. Ayliffe, T. F. Robinson, C. Heidelberger, and W. Marcus. 2003. Diets of southern African Bovidae: Stable isotope evidence. *Journal of Mammalogy* 84:471–479.

Sponheimer, M., B. H. Passey, D. J. de Ruiter, D. Guatelli-Steinberg, T. E. Cerling, and J. A. Lee-Thorp. 2006. Isotopic evidence for dietary variability in the early hominin *Paranthropus robustus*. *Science* 314:980–982.

Sponheimer M., K. E. Reed, and J. A. Lee-Thorp. 1999. Combining isotopic and ecomorphological data to refine bovid paleodietary reconstruction: A case study from the Makapansgat Limeworks hominin locality. *Journal of Human Evolution* 36:705–718.

Sullivan C. H., and H. W. Krueger. 1981. Carbon isotope analysis of separate chemical phases in modern and fossil bone. *Nature* 292:333–335.

Tassy, P. 2003. Elephantoidea from Lothagam; pp. 331–360 in M. G. Leakey and J. M. Harris (eds.), *Lothagam: The Dawn of Humanity in Eastern Africa*. Columbia University Press, New York.

Tieszen L. L., and T. Fagre. 1993. Effect of diet quality and composition on the isotopic composition of respiratory CO_2, bone collagen, bioapatite, and soft tissues; pp. 121–155 in J. B. Lambert and F. Grupe (eds.), *Molecular Archaeology of Prehistoric Human Bone*. Springer, Berlin.

Tipple B. J., and M. Pagani. 2007. The Early Origins of Terrestrial C4 Photosynthesis. *Annual Review of Earth and Planetary Sciences* 35:435–461/

van der Merwe, N. J., J. A. Lee-Thorp, and R. H. V. Bell. 1988. Carbon isotopes as indicators of elephant diets and African environments. *African Journal of Ecology* 26:163–172.

van der Merwe, N. J., F. T. Masao, and M. K. Bamford. 2008. Isotopic evidence for contrasting diets of early hominins *Homo habilis* and *Australopithecus boisei* of Tanzania. *South African Journal of Science* 104:153–155.

van der Merwe N. J., J. F. Thackeray, J.A. Lee-Thorp, and J. Luyt. 2003. The carbon isotope ecology and diet of *Australopithecus africanus* at Sterkfontein, South Africa. *Journal of Human Evolution* 44:581–597.

Vincenti A., J. C. Barber, S. Aliscioni, L. M. Giussani, E. A. Kellogg. 2008. The age of the grasses and clusters of origins of C4 photosynthesis. *Global Change Biology* 14:2963–2977.

Wang, Y., and T. E. Cerling. 1994, A model of fossil tooth enamel and bone diagenesis: Implications for stable isotope studies and paleoenvironment reconstruction. *Palaeogeography, Palaeoclimatology, Palaeoecology* 107:281–289.

Werdelin, L. 2003. Mio-Pliocene Carnivora from Lothagam, Kenya; pp. 261–328 in M. G. Leakey and J. M. Harris (eds.), *Lothagam: The Dawn of Humanity in Eastern Africa*. Columbia University Press, New York.

Weston, E. 2003. Fossil Hippopotamidae from Lothagam; pp. 441–482 in M. G. Leakey and J. M. Harris (eds.), *Lothagam: The Dawn of Humanity in Eastern Africa*. Columbia University Press, New York.

Zachos, J., M. Pagani, L. Sloan, E. Thomas, and K. Billups. 2001. Trends, rhythms, and aberrations in global climate 65 Ma to present. *Science* 292:686–693.

INDEX

Abbreviations following page numbers indicate the following:

f, figures

t, tables

Aardvarks, 107–111, 113, 903, 908, 909, 910, 911, 914, 916
Abdounodus, 563–566, 568, 569, 571
  hamdii, 563–566, 564f, 565t
Aboletylestes, 253–254
  hypselus, 253, 254, 254f, 255, 259
  robustus, 253–254, 254f, 259
Abounodus, 19
Abu Dhabi (United Arab Emirates)
  Giraffoidea in, 805
  Hippopotamidae in, 863
  Proboscidea in, 224
Abudhabia, 289
Abu Hugar (Sudan), Cervidae in, 814, 814f
Abundu Formation (Kenya), 34
Abuqatrania, 377, 388
  basiodontos, 377
Aceratheriina, 669, 671–676
Aceratherium, 29, 675, 676
  acutirostratum, 676
  campbelli, 670
  incisivum, 675
Acinonyx, 639, 652
  aicha, 639
  jubatus, 639
Acmeodon, 562
Acomys, 282, 288f, 289, 296
Acritophiomys, 293
Adapidae, 322–324, 325
Adapiformes, 319, 322–324
Adapis, 388
Adapisoriculidae, 255
Adapisoriculus minimus, 255
Adcrocuta, 632, 651, 652
  eximia, 632
Ad Dabtiyah (Saudi Arabia)
  catarrhine primates in, 444
  Proboscidea in, 197, 203
Addax, 773, 775
Adelphailurus, 646
Adi Ugri (Eritrea), 38f, 39
Adjiderebos cantabilis, 756
Adrar Mgorn 1 (Morocco)
  anthropoid primates in, 369, 370
  Creodonta in, 553, 556
  insectivores in, 253, 254, 254f, 255, 256
  primitive ungulates in, 570
Adrocuta eximia, 29
Adu-Asa Formation of Middle Awash (Ethiopia), 37–38, 38f
  Bovidae in, 787, 790
    Alcelaphinae, 776
    Antilopinae, 760, 764

Bovinae, 749, 752, 753
  Reduncinae, 767, 768, 772
Carnivora in, 649, 651
Cercopithecoidea in, 404
Hippopotamidae in, 855
Hominini in, 484
Lagomorpha in, 311, 312, 313
Rodentia in, 288, 292, 294
Aegodontia, 743
Aegyptopithecus, 371, 372, 373f, 378, 384–385, 388
  and *Catopithecus* compared, 383
  and *Parapithecus* compared, 381
  postcranial elements and skeletal reconstruction of, 379f
  and *Propliopithecus* compared, 386, 387
  zeuxis, 130, 373f, 384–385, 385f, 386, 387
Aeolopithecus chirobates, 386
Aepycamelinae, 815
Aepyceros, 760, 765–766, 767, 771, 788, 791
  afarensis, 791
  melampus, 741, 765–766, 767, 791
  premelampus, 765, 767
  shungurae, 765, 791
Aerial mammals, 926, 928, 930f, 931, 936, 937, 938
Aethomys, 289, 296
Afar (Ethiopia), 6, 10, 63
  Cercopithecoidea in, 399, 403, 404, 409, 421
  Hippopotamidae in, 853, 866, 868
  Hominini in, 491
  plume activity in, 62, 63
  Suoidea in, 831, 832
Afoud (Morocco), 28f, 29
Afradapis, 323, 324, 330
  longicristatus, 324
Aframonius, 322–324, 330
  dieides, 322f, 322–324
Africanacetus, 894
  ceratopsis, 888, 894
African Humid Period, 52
Africanictis, 621–622, 623
  hyaenoides, 622
  meini, 622
  schmidtkittleri, 622
Africanomys, 291
African Plate, 3–6
  collision with Eurasian plate, 4, 5, 7
  northward movement of, 11
  rotation of, 3, 5
African Superswell, 3, 8, 9f, 10, 16
African Surface, 13–14, 15, 16

Africanthropus njarasensis, 515
Afrikanokeryx, 802–803
  leakeyi, 802f, 802–803, 804, 807
Afrochoerodon, 197, 203–205
  chioticus, 205
  kisumuensis, 199f, 203–205, 204f, 205
  ngorora, 205
  zaltaniensis, 197, 205
Afrocricetodon, 287
Afrocricetodontinae, 287, 287f, 297
Afrocyon, 603, 605
  burolleti, 603, 603f
Afrodon, 253, 255, 258, 259
  chleuhi, 255, 256
  germanicus, 255
  tagourtensis, 255
Afrohyrax, 130, 137, 138
  championi, 136f, 138
  new species, 138
Afroinsectiphillia, 914
Afroinsectivora, 259, 914
Afromastodon, 199, 201
  coppensi, 197, 198, 199f, 200f, 201, 202, 203
  lybicus, 197
Afromeryx, 845, 846–847, 848, 848f, 849
  africanus, 27, 847
  zelteni, 847, 848f
Afromontane vegetation, 923, 927
Afropithecidae, 448
Afropithecinae, 429, 441–448, 463
Afropithecus, 384, 432, 441–444, 445, 447, 448, 454, 463
  turkanensis, 440, 441, 442f, 442–444, 443f, 444f
Afropterus gigas, 586, 589
Afrosmilus, 637, 638, 644
  africanus, 637
  hispanicus, 637
  turkanae, 637, 638, 638f
Afrosorex, 578
Afrosoricida, 99, 253, 256, 257, 259, 905, 914
Afrotarsius, 24, 319, 320–321, 330
  chatrathi, 320f, 320–321
Afrotheria, 81–260, 563, 903
  Bibymalagasia, 113–114
  distribution of, 906f, 907f
  Embrithopoda, 115–120
  extant, 903–904
  extinct, 904–909, 905t
  Hyracoidea, 123–144
  Macroscelidea, 89–97
  molecular studies, 911–912, 913t–914t

Stephanorhinus, 677, 681
  africanus, 677
  etruscus, 677
  hemitoechus, 677
  mercki, 677
Sterkfontein (South Africa), 30f, 31
  Bovidae in, 744, 764, 771, 773, 774, 786
  Carnivora in, 639, 644
  Cercopithecoidea in, 402, 413, 414, 415
  Chiroptera in, 583, 588, 590
  Equidae in, 708
  Hominini in, 496, 497, 499, 509, 514
  Hyracoidea in, 143
  Macroscelidea in, 95, 96
  Rodentia in, 296, 297
  Soricidae in, 577, 578
Strepsiceros maryanus, 751
Strepsirrhini, 319, 321–324, 330, 369
  classification of, 334t
  Lorisiformes, 333–347
Striped possum, 362
Strogulognathus sansaniensis, 736
Stromerius, 891, 892–893
  nidensis, 883, 886, 891, 892–893, 896
Stromer von Reichenbach, Ernst, 878, 883, 884
Stylohipparion, 697
  albertense, 697
Submarine shelves, 7
Sudan
  Bovidae in, 755
  Camelidae in, 817, 818, 819
  Cercopithecoidea in, 403
  Cervidae in, 814, 814f
  Giraffoidea in, 799t, 804
  Homo in, 521
  Lorisiformes in, 344
  Suoidea in, 822, 835
  vegetation in, 923
    historical, 62, 63, 64
Sudanian phytochoria, 923, 926, 927
Sugata Basin (Kenya), isotope studies of
  mammals in, 945, 945f, 946
Suidae, 821, 826–838, 839
  stable isotope studies of, 944, 946, 946f, 947
Suinae, 834–838
Suncus, 577, 578
  barbarus, 577
  dayi, 577
  etruscus, 577
  haessertsi, 577
  infinitesimus, 577
  leakeyi, 577
  lixus, 577
  remyi, 577
  shungurensis, 577
  varilla, 577
  varilla meesteri, 577, 578f
Suoidea, 821–840
  evolutionary relationships of, 839–840
Surdisorex, 577
Suricata, 631
  major, 631
  suricatta, 631
Sus, 822, 831
  limnetes, 835
  salvanius, 822, 834
  waylandi, 831
Swartklip (South Africa), Bovidae in, 764, 771
Swartkrans (South Africa), 30f, 31
  Bovidae in, 759–760, 764, 766, 780, 782, 786
  Carnivora in, 632f
  Cercopithecoidea in, 410
  Chiroptera in, 588, 590
  Hominini in
    Homo, 509, 513, 514
    Paranthropus, 505, 506–507
  Hyracoidea in, 143
  Lagomorpha in, 314
  Suoidea in, 837
  Tubulidentata in, 109
Swaziland, 8, 14, 15

Sylvicapra, 757, 760
  grimmia, 757
Sylvilagus, 305
Sylvisorex, 577, 578
  granti, 577
  johnsoni, 577
  megalura, 577
  olduvaiensis, 577
Syncerus, 752, 753, 754–755, 756
  acoelotus, 754–755, 756
  antiquus, 755, 756
  antiquus complexus, 755
  caffer, 753, 754, 755, 756
  caffer antiquus, 755
  caffer brachyceros, 754
  caffer caffer, 754, 755
  caffer nanus, 754, 755, 788
  nanus, 754
Syndesmotis, 585
Syria, Camelidae in, 818, 819
Syrian Arc, compression and formation of, 157
Syrtosmilus, 638
  syrtensis, 638, 638f

————————

Tabaghbagh (Egypt), Cetacea in, 887
Tabarin (Kenya)
  Hominini in, 487
  Suoidea in, 832
Tabelia, 22
  hammadae, 322
Tachyoryctes, 264, 279, 286–287, 484
  macrocephalus, 287
  makooka, 286
  pliocaenicus, 287
Tachyoryctinae, 279, 286
Tadelia, 22
  hammadae, 322
Tadarida, 587
  rusingae, 587
Tamaguilelt (Mali), 21t, 22
Tamariskenbucht (Egypt), Cetacea in, 883
Tana River hartebeest, 782
Tanzania, 916
  anthropoid primates in, 369
  Bovidae in, 784
  Camelidae in, 817, 819
  Carnivora in
    Canidae, 609, 612, 613
    Felidae, 639, 644, 645, 646
    Herpestidae, 624, 625, 626, 630
    Hyaenidae, 632, 636, 637
    Mustelidae, 613, 615, 616
    Viverridae, 622, 623, 624
  Cercopithecoidea in, 418
  Chalicotheriidae in, 661
  Chiroptera in, 590
  Equidae in, 705
  Giraffoidea in, 799t, 804, 806, 807
  Hominini in, 516, 519, 520
  Laetoli site. See Laetoli (Tanzania)
  Lorisiformes in, 335, 346
  Manonga site. See Manonga (Tanzania)
  Ndolanya Beds. See Ndolanya Beds
    (Tanzania)
  Neogene mammal localities in, 36, 37f
  Olduvai Gorge. See Olduvai Gorge
    (Tanzania)
  Paleogene mammal localities in, 20f, 21t,
    22, 24
  Peninj site. See Peninj (Tanzania)
  Proboscidea in, 208, 209
  Rodentia in, 295
  Suoidea in, 822, 831, 833, 834, 835, 836,
    837, 838
  vegetation history of, 62, 63, 68
Tanzanycteridae, 590
Tanzanycteris, 22, 581, 587, 590, 591
  mannardi, 590
Taphozous, 583–584
  abitus, 583
  incognita, 583
  nudiventris, 583

Taqah (Oman), 20f, 21t, 23
  anthropoid primates in, 378, 382f, 383, 386
  Chiroptera in, 581, 582, 584, 586, 587, 589
  Marsupialia in, 77, 79
  Oligocene vegetation in, 63
  prosimians in, 329
Taramsa Hill (Egypt), Homo in, 520
Tarfaya (Morocco), Cetacea in, 889, 894
Tarsiidae, 321, 369
Tarsius, 321, 371, 376, 380
  eocaenus, 321
  and Parapithecus compared, 381
Tatera, 282
  indica, 282
Taterillus, 296
Taung (South Africa), 30f, 31
  Bovidae in, 757, 759
  Cercopithecoidea in, 412, 413, 416
  Dart deposits in, 31
  Hominini in, 488, 495, 497
  Hrdlicka deposits in, 31
  Hyracoidea in, 143
  Macroscelidea in, 95
  Rodentia in, 294
  Soricidae in, 578
Taurotragus, 749, 751, 752, 788
  arkelli, 751
  derbianus, 751
  maroccanus, 751
  oryx, 741, 751
  oryx pachyceros, 751
Taxeopoda, 569
Tayassuidae, 821
Taza (Morocco), Camelidae in, 818
Tchadanthropus uxoris, 512
Tchadotragus, 772, 773
  fanonei, 791
  sudrei, 773, 791
Tébessa (Eritrea), 38f, 39
Tectonics, 3–10
  and climatic events, 45, 46f, 50
  major localities in Africa and Arabia, 5f
  and Proboscidea evolutionary phases, 236
  and vegetation history, 63, 65, 69
Tectonomys, 289
Teilhardina, 370
Telanthropus, 508
  capensis, 512
Teleoceratina, 669, 670–671
Tell range, 5
Telmatodon, 847
Temperature, global and regional, 45–53
  and mammal species richness in Africa, 922
Tenrec, 102, 104, 909, 910
  ecaudatus, 104
Tenrecidae, 99, 101–105, 258, 573, 904,
  905–908
Tenrecoidea, 99–105, 253, 258, 573, 905,
  914, 915–916
Tenrecoids, 99
  phylogeny of, 100f
Tenrecomorpha, 257
Tenrecs, 99, 101–105, 103f, 253, 903, 904,
  905–908
  distribution of, 101f
  molecular analysis of, 911, 913t–914t
  phylogeny, 914, 915–916
  skeletal and dental characteristics, 910
  soft, 104
  soft-tissue characteristics, 909
  spiny, 104
Tephrochronologic dating, 27
Teratodon, 549, 552, 553–554, 554f
  enigmae, 553
  spekei, 553
Teratodontidae, 554
Teratodontinae, 553–554, 557
Ternania, 287
Terrestrial mammals, 926, 928, 930, 930f,
  931, 932, 933f, 936, 938
Tertiary Period, Lorisiformes in, 333–347

Text   8/10.75 Stone Serif

Display   Stone Serif

Production Management   Michael Bass Associates